PHYSICAL CONSTANTS

Quantity	Symbol	Value
Universal gravitational constant	G	$6.674 \times 10^{-11}\ \mathrm{m^3/(kg \cdot s^2)}$
Speed of light in vacuum	c	$2.998 \times 10^{8}\ \mathrm{m/s}$
Elementary charge	e	$1.602 \times 10^{-19}\ \mathrm{C}$
Planck's constant	h	$6.626 \times 10^{-34}\ \mathrm{J \cdot s}$
		$4.136 \times 10^{-15}\ \mathrm{eV \cdot s}$
	$\hbar = h/(2\pi)$	$1.055 \times 10^{-34}\ \mathrm{J \cdot s}$
		$6.582 \times 10^{-16}\ \mathrm{eV \cdot s}$
Universal gas constant	R	$8.314\ \mathrm{J/(mol \cdot K)}$
Avogadro's number	N_{A}	$6.022 \times 10^{23}\ \mathrm{mol^{-1}}$
Boltzmann constant	k_{B}	$1.381 \times 10^{-23}\ \mathrm{J/K}$
		$8.617 \times 10^{-5}\ \mathrm{eV/K}$
Coulomb force constant	$k = 1/(4\pi\epsilon_0)$	$8.988 \times 10^{9}\ \mathrm{N \cdot m^2/C^2}$
Permittivity of free space (electric constant)	ϵ_0	$8.854 \times 10^{-12}\ \mathrm{C^2/(N \cdot m^2)}$
Permeability of free space (magnetic constant)	μ_0	$4\pi \times 10^{-7}\ \mathrm{T \cdot m/A}$
Electron mass	m_{e}	$9.109 \times 10^{-31}\ \mathrm{kg}$
		$0.000\ 548\ 580\ \mathrm{u}$
Electron rest energy	$m_{\mathrm{e}}c^2$	$0.5110\ \mathrm{MeV}$
Proton mass	m_{p}	$1.673 \times 10^{-27}\ \mathrm{kg}$
		$1.007\ 276\ 5\ \mathrm{u}$
Proton rest energy	$m_{\mathrm{p}}c^2$	$938.272\ \mathrm{MeV}$
Neutron mass	m_{n}	$1.675 \times 10^{-27}\ \mathrm{kg}$
		$1.008\ 664\ 9\ \mathrm{u}$
Neutron rest energy	$m_{\mathrm{n}}c^2$	$939.565\ \mathrm{MeV}$
Compton wavelength of electron	λ_{C}	$2.426 \times 10^{-12}\ \mathrm{m}$
Stefan-Boltzmann constant	σ	$5.670 \times 10^{-8}\ \mathrm{W/(m^2 \cdot K^4)}$
Rydberg constant	R	$1.097 \times 10^{7}\ \mathrm{m^{-1}}$
Bohr radius of hydrogen atom	a_0	$5.292 \times 10^{-11}\ \mathrm{m}$
Ionization energy of hydrogen atom	$-E_1$	$13.61\ \mathrm{eV}$

SECOND EDITION

Physics

Alan Giambattista

Cornell University

Betty McCarthy Richardson

Cornell University

Robert C. Richardson

Cornell University

McGraw Hill Higher Education

Boston Burr Ridge, IL Dubuque, IA New York San Francisco St. Louis
Bangkok Bogotá Caracas Kuala Lumpur Lisbon London Madrid Mexico City
Milan Montreal New Delhi Santiago Seoul Singapore Sydney Taipei Toronto

The McGraw-Hill Companies

Higher Education

PHYSICS, SECOND EDITION

This book is printed on acid-free paper.

3 4 5 6 7 8 9 0 VNH/VNH 0

ISBN 978–0–07–733866–4
MHID 0–07–733866–9

Publisher: *Thomas D. Timp*
Sponsoring Editor: *Debra B. Hash*
Vice-President New Product Launches: *Michael Lange*
Senior Developmental Editor: *Mary E. Hurley*
Senior Marketing Manager: *Lisa Nicks*
Senior Project Manager: *Gloria G. Schiesl*
Senior Production Supervisor: *Laura Fuller*
Senior Media Project Manager: *Tammy Juran*
Senior Designer: *David W. Hash*
Cover/Interior Designer: *Rokusek Design*
(USE) Cover Image: ©*Kazuya Shiota/Aflo/Jupiterimages*
Lead Photo Research Coordinator: *Carrie K. Burger*
Photo Research: *Danny Meldung/Photo Affairs, Inc*
Supplement Producer: *Mary Jane Lampe*
Compositor: *Laserwords Private Limited*
Typeface: *10/12 Times*
Printer: *R. R. Donnelley, Jefferson City, MO*

The credits section for this book begins on page C-1 and is considered an extension of the copyright page.

Library of Congress Cataloging-in-Publication Data

Giambattista, Alan.
Physics / Alan Giambattista, Betty McCarthy Richardson, Robert C. Richardson.—2nd ed.
p. cm.
Includes index.
ISBN 978–0–07–340453–0 — ISBN 0–07–340453–5 (hard copy : alk. paper)
1. Physics–Textbooks. I. Richardson, Betty McCarthy. II. Richardson, Robert C. III. Title.
QC21.3.G537 2010
530–dc22

2008034667

www.mhhe.com

About the Authors

Alan Giambattista grew up in Nutley, New Jersey. In his junior year at Brigham Young University he decided to pursue a physics major, after having explored math, music, and psychology. He did his graduate studies at Cornell University and has taught introductory college physics for over 20 years. When not found at the computer keyboard working on *Physics*, he can often be found at the keyboard of a harpsichord or piano. He is a member of the Cayuga Chamber Orchestra and has given performances of the Bach harpsichord concerti at several regional Bach festivals. He met his wife Marion in a singing group. They live in an 1824 parsonage built for an abolitionist minister, which is now surrounded by an organic dairy farm. Besides making music and taking care of the house, gardens, and fruit trees, they love to travel together.

Betty McCarthy Richardson was born and grew up in Marblehead, Massachusetts, and tried to avoid taking any science classes after eighth grade but managed to avoid only ninth grade science. After discovering that physics tells how things work, she decided to become a physicist. She attended Wellesley College and did graduate work at Duke University. While at Duke, Betty met and married fellow graduate student Bob Richardson and had two daughters, Jennifer and Pamela. Betty began teaching physics at Cornell in 1977. Many years later, she is still teaching the same course, Physics 101/102, an algebra-based course with all teaching done one-on-one in a Learning Center. From her own early experience of math and science avoidance, Betty has empathy with students who are apprehensive about learning physics. Betty's hobbies include collecting old children's books, reading, enjoying music, travel, and dining with royalty. A highlight for Betty during the Nobel Prize festivities in 1996 was being escorted to dinner on the arm of King Carl XVI Gustav of Sweden. Currently she is spending spare time enjoying grandsons Jasper (the 1-m child in Chapter 1), Dashiell and Oliver (the twins of Chapter 12), and Quintin, the newest arrival.

Robert C. Richardson was born in Washington, D.C., attended Virginia Polytechnic Institute, spent time in the United States Army, and then returned to graduate school in physics at Duke University where his thesis work involved NMR studies of solid helium-3. In the fall of 1966 Bob began work at Cornell University in the laboratory of David M. Lee. Their research goal was to observe the nuclear magnetic phase transition in solid ^{3}He that could be predicted from Richardson's thesis work with Professor Horst Meyer at Duke. In collaboration with graduate student Douglas D. Osheroff, they worked on cooling techniques and NMR instrumentation for studying low-temperature helium liquids and solids. In the fall of 1971, they made the accidental discovery that liquid ^{3}He undergoes a pairing transition similar to that of superconductors. The three were awarded the Nobel Prize for that work in 1996. Bob is currently the F. R. Newman Professor of Physics and the Senior Science Advisor at Cornell. In his spare time he enjoys gardening and photography.

In loving memory of Dad and of my niece, Natalie

Alan

In memory of our daughter Pamela,
and for Quintin, Oliver, Dashiell, Jasper,
Jennifer, and Jim Merlis

Bob and Betty

Brief Contents

Contents

PART FIVE

Quantum and Particle Physics and Relativity

List of Selected Applications

Note: Within the Problems, (#) = Chapter Number; CQ = Conceptual Question; MC = Multiple-Choice Question; P = Problem; R&S = Review & Synthesis.

Biology/Life Science

Number of cells in the body, Ex. 1.9, p. 14
Can the lion catch the buffalo? Sec. 2.3, p. 34
Gull dropping a clam Sec. 3.5, p. 73
Tensile forces in the body, Sec. 4.7, p. 111
Speed in centrifuge, Ex. 5.2, p. 150
The centrifuge, Sec. 5.7, p. 171
Energy conversion in animal jumping, Ex. 6.12, p. 210
Energy transformation in a jumping flea, Sec. 6.7, p. 210
Conditions for equilibrium in the human body, Sec. 8.5, p. 281
Flexor versus extensor muscles, Sec. 8.5, p. 281
Force to hold arm horizontal, Ex. 8.10, p. 281
Forces on the human spine during heavy lifting, Sec. 8.5, p. 283
Sphygmomanometer and blood pressure, Sec. 9.5, p. 327
Specific gravity measurements in medicine, Sec. 9.6, p. 329
Floating and sinking of fish and animals, Ex. 9.8, p. 331
Speed of blood flow, Ex. 9.9, p. 334
Plaque buildup and narrowed arteries, Sec. 9.8, p. 337
Narrowing arteries and high blood pressure, Sec. 9.9, p. 340
Arterial blockage, Ex. 9.12, p. 341
How insects can walk on the surface of a pond, Sec. 9.11, p. 343
Surface tension of alveoli in the lungs, Ex. 9.14, p. 344
Surfactant in the lungs, Sec. 9.11, p. 344
Tension and compression in bone, Ex. 10.2, p. 359
Osteoporosis, Sec. 10.3, p. 361
Size limitations on organisms, Sec. 10.3, p. 363
Comparison of walking speeds for various creatures, Ex. 10.10, p. 378
Sensitivity of the human ear, Sec. 11.1, p. 394
Sound waves from a songbird, Ex. 12.2, p. 425
Human ear, Sec. 12.6, p. 432
Echolocation of bats and dolphins, Sec. 12.10, p. 443
Ultrasound and ultrasonic imaging, Sec. 12.10, p. 444
Temperature conversion, Ex. 13.1, p. 460
Warm-blooded vs.cold-blooded animals, Ex. 13.8, Sec. 13.7, p. 476
Diffusion of O_2 into the bloodstream, Sec. 13.8, Ex. 13.9, pp. 478-479
Using ice to protect buds from freezing, Sec. 14.5, p. 500
Temperature regulation in the human body, Sec. 14.7, p. 510
Thermal radiation, Sec. 14.8, p. 515
Thermal radiation from the human body, Ex. 14.15, p. 515
Electrolocation in fish, Sec. 16.4, p. 581
Electrocardiogram and electroencephalogram, Sec. 17.2, p. 612
Neuron capacitance, Ex. 17.11, p. 624
Defibrillator, Ex. 17.12, Sec. 18.11, pp. 627, 672
RC circuits in neurons, Sec. 18.10, p. 671
Defibrillator, Sec. 18.11, p. 672
Magnetotactic bacteria, Sec. 19.1, p. 697
Medical uses of cyclotrons, Sec. 19.3, p. 705
Electromagnetic blood flowmeter, Sec. 19.5, p. 710
Magnetic resonance imaging, Sec. 19.7, p. 723
Magnetoencephalography, Sec. 20.3, p. 753
Fluorescence, Sec. 22.3, p. 818
Thermograms of the human body, Sec. 22.3, p. 818
X-rays in medicine and dentistry, CAT scans, Sec. 22.3, p. 820
Navigation of bees, Sec. 22.7, p. 837
Endoscope, Sec. 23.4, p. 863
Kingfisher looking for prey, Ex. 23.4, p. 866
Human eye, Sec. 24.3, p. 898
Correcting myopia, Sec. 24.3, Ex. 24.4, pp. 900-901
Correcting hyperopia, Ex. 24.5, p. 901
Iridescent colors in butterfly wings, Sec. 25.3, p. 934
Resolution of the human eye, Sec. 25.8, p. 950
Positron emission tomography, Sec. 27.9, p. 1021
Electron microscopes, Sec. 28.3, p. 1036
Lasers in medicine, Sec. 28.9, p. 1052
Radiocarbon dating, Sec. 29.4, p. 1082
Dating archeological sites, Ex. 29.9, p. 1083
Biological effect of radiation, Sec. 29.5, p. 1085
Radioactive tracers in medical diagnosis, Sec. 29.5, p. 1088
Gamma knife radio surgery, Sec. 29.5, p. 1089
Radiation therapy, Sec. 29.5, p. 1089
Problems (1) P: 49, 51-52, 59-60, 66, 69, 87. (2) P: 58, 73. (4) P: 81, 123. (5) P: 11, 53, 54, 77, 83; R&S: 4, 18. (6) P: 50, 57, 73, 103, 108. (7) CQ: 8; P: 87. (8) CQ: 9-11, 16; P: 39-46, 85, 99, 105, 107, 112-113; MCAT: 6-17. (9) CQ: 7, 12, 14; P: 8, 13, 18, 20, 22, 26, 36, 38, 42, 55, 59, 63, 70, 82, 85-88, 100, 101. (10) P: 2, 6, 8, 11, 15, 38, 76, 86-88, 91, 96. (11) CQ: 2; P: 16. (12) CQ: 4, 5, 8; P: 1, 2, 11, 19, 42, 49-52, 56-58, 60, 67, 68; R&S: 4, 15. (13) P: 31, 51, 57-58, 77, 89, 90, 102-103, 105. (14) CQ: 3; P: 14-15, 20, 41-42, 46, 53-55, 58, 63-64, 68, 70-71, 76-80, 90-91, 93, 97. (15) P: 55, 60, 67, 75. (16) P: 26, 53, 75. (17) P: 36, 60, 72, 84-85, 102, 108, 107, 111, 114. (18) CQ: 11, 17, 19; P: 24, 25, 86, 95-96, 103, 118. (19) P: 41-42, 97, 106. (20) CQ: 8. (21) P: 69. (22) P: 23. (23) CQ: 20. (24) CQ: 10, 13, 14, 17, 18; MC: 4-5; P: 22-28, 39, 72, 75, 76. (25) CQ: 16; P: 10, 54, 57, 61, 71. (26) CQ: 5. (27) CQ: 3, 21; P: 59, 62, 63. (28) P: 14-15. (29) CQ: 9, 11-12, 14; MC: 10; P: 35, 36, 41, 46-48, 69, 75, 79. (30) R&S: 10-13.

Chemistry

Collision between krypton atom and water molecule, Ex. 7.9, p. 242
Why reaction rates increase with temperature, Sec. 13.7, Ex. 13.8, p. 474
Polarization of charge in water, Sec. 16.1, p. 565
Current in neon signs and fluorescent lights, Sec. 18.1, p. 643
Spectroscopic analysis of elements, Sec. 27.6, p. 1010
Fluorescence, phosphorescence, and chemiluminescence, Sec. 27.7, p. 1018
Electronic configurations of atoms, Sec. 28.7, p. 1045
Understanding the periodic table, Sec. 28.7, p. 1046
Problems (7) P: 39. (13) CQ: 13-14; P: 27-30, 32-39, 76, 79, 86, 87. (14) P: 10. (16) P: 17, 92. (17) P: 2, 3, 49, 92. (18) MC: 1; P: 7; R&S: 11. (19) P: 28-32, 88, 96. (26) P: 43, 86. (27) P: 5, 6, 9-10, 31-44, 46-48, 51, 52, 64, 66, 68, 71-72, 74, 76, 78-79, 81-82, 85. (28) CQ: 12-18; MC: 4; P: 6, 11, 20, 30, 36, 40-41, 43-45, 47, 57, 64, 65. (29) P: 4-17, 19, 21-32, 37-40, 42, 51-57, 59-67. (30) R&S: 11-12, 16, 17, 27; MCAT: 1-2, 6-8.

Geology/Earth Science

Angular speed of Earth, Ex. 5.1, p. 149
Hidden depths of an iceberg, Ex. 9.7, p. 330
Why ocean waves approach shore nearly head on, Sec. 11.8, p. 406
Resonance and damage caused by earthquakes, Sec. 11.10, p. 411
Ocean currents and global warming, Sec. 14.7, p. 511
Global climate change, Sec. 14.8, p. 516
Second law and evolution, Sec. 15.8, p. 548
Second law and the "energy crisis", Sec. 15.8, p. 548
Electric potential energy in a thundercloud, Ex. 17.1, p. 604
Thunderclouds and lightning, Sec. 17.6, p. 624
Earth's magnetic field, Sec. 19.1, p. 696
Deflection of cosmic rays, Ex. 19.1, p. 700
Magnetic force on an ion in the air, Ex. 19.2, p. 701
Intensity of sunlight reaching the Earth, Ex. 22.6, p. 829
Colors of the sky during the day and at sunset, Sec. 22.7, p. 836
Rainbows, Sec. 23.3, p. 858
Cosmic rays, Ex. 26.2, 26.4, pp. 979-982
Radioactive dating of geologic formations, Sec. 29.4, p. 1084
Neutron activation analysis, Sec. 29.6, p. 1095
Angular momentum of hurricanes and pulsars, Sec. 8.8, p. 291
Problems (1) P: 55. (5) P: 70. (8) CQ 21. (9) CQ: 9; P: 46, 67, 80, 98-99. (11) CQ: 9; P: 61-62, 66, 68-70. (12) P: 4, 5, 46-48. (13) P: 49. (14) CQ: 4, 6; P: 89, 92, 94. (15) MCAT: 2-3. (16) P: 61, 75, 80. (17) CQ: 19; P: 66, 77, 86, 87. (18) P: 125, 126. (22) CQ: 6, 7, 15; P: 53, 54, 63. (29) CQ: 6; P: 70.

Astronomy/Space Science

Architecture

Technology/Machines

Transportation

Sports

Everyday Life

Preface

Physics is intended for a two-semester college course in introductory physics using algebra and trigonometry. Our main goals in writing this book are

- to present the basic concepts of physics that students need to know for later courses and future careers,
- to emphasize that physics is a tool for understanding the real world, and
- to teach transferable problem-solving skills that students can use throughout their lives.

We have kept these goals in mind while developing the main themes of the book.

NEW TO THIS EDITION

Although the fundamental philosophy of the book has not changed, detailed feedback from almost 60 reviewers (many of whom used the first edition in the classroom) has enabled us to fine-tune our approach to make the text even more user-friendly, conceptually based, and relevant for students. The second edition also has some added features to further facilitate student learning.

A greater emphasis has been placed on fundamental physics concepts:

- **Connections identify areas in each chapter where important concepts are revisited.** A marginal Connections heading and summary adjacent to the coverage in the main text help students easily recognize that a previously introduced concept is being applied to the current discussion. Knowledge is being revisited and further developed—not newly introduced.
- **Checkpoint** questions have been added to applicable sections of the text to **allow students to pause and test their understanding of the concept** explored within the current section. The answers to the Checkpoints are found at the end of the chapter so that students can confirm their knowledge without jumping too quickly to the provided answer.
- The exercises in the **Review & Synthesis** sections have been revised to concentrate even more heavily on **helping students to realize through practice problems how the concepts in the previously covered group of chapters are interrelated**. The number of problems in the Review & Synthesis sections has also been increased in the new edition. (The MCAT review problems have been retained to also help premed students focus on the concepts covered in the upcoming exam.)
- **Nonessential coverage and derivations have been moved to the text's website.** This will help students not only to focus further on the fundamental, core concepts in their reading of the text but also allow them to go online for additional information or explanation on topics of interest. identifiers in the text direct students to additional information online.

"G/R/R is as good as it gets as far as a college textbook in physics goes. One of the coauthors of this book has been teaching a course at this level for 30 years. This book is a direct result of her 30 years' worth of personal experience, and there is no better substitute for that. It is, without any doubt, one of the best of its kind."

Dr. Abu Fasihuddin, University of Connecticut

In addition, the following general revisions occur in chapters of the text:

- The topical question from the chapter-opening vignette now appears in the margin (along with a reduced version of the chapter-opening image) to help students identify where in the main text the answer to the chapter-opening question is addressed.
- Applications have been clearly identified as such in the text with a complete listing in the front matter.
- Many helpful subheadings have been added to the text to help students quickly identify new subtopics.
- Portions of the text now caption images to establish a visual connection between the text's concepts and terms and the art and photos.

- Great care was taken by both the authors and the contributors to the second edition to revise the end-of-chapter and Review & Synthesis problems. Approximately 150 problems are new, and an emphasis has been placed on progressing difficulty level to help students gain confidence and reinforce new skills before tackling more challenging problems.

The following lists major chapter-specific revisions to the text:

Chapter 2: Vector notation has been removed from Chapter 2. Discussion of vectors and components of vectors now begins in Chapter 3.
Chapter 3: A discussion of Unit Vectors has been added to Section 3.2. A new example for finding average velocity has been added.
Chapter 4: A more concise section on air resistance is provided with a more detailed discussion available online. A new Figure 4.20 emphasizes the normal and frictional forces as perpendicular components of a contact force.
Chapter 7: Section 7.6 Motion of the Center of Mass has been simplified.
Chapter 8: Example 8.1 has been replaced with a new problem on the rotational inertia of a barbell.
Chapter 10: Section 10.8 The Pendulum has been made much more concise with a more detailed discussion of the physical pendulum available online.
Chapter 11: A new "law box" highlights the physical properties that determine wave speed. The discussion on interference has been expanded for added clarity.
Chapter 12: In Section 12.9, the discussion of shock waves has been shortened. A more detailed discussion is available online.
Chapter 14: A detailed discussion of convection and Example 14.12 Roller Blading in Still Air have been moved online. Section 14.7 is now a brief, conceptual description of convection. Section 14.8 Thermal Radiation has been revised with a clearer description of solar radiation and global warming.
Chapter 15: Section 15.5 Heat Engines has been revised to include a more accurate description of the development of the steam engine. The process of the internal combustion engine is now illustrated in Figure 15.12. Details of the Carnot cycle and discussion of the statistical interpretation of entropy are available online.
Chapter 16: A new Example 16.7 Electric Field due to Three Point Charges has been added.
Chapter 22: Section 22.1 has been simplified and is now titled Maxwell's Equations and Electromagnetic Waves. A more detailed discussion appears online. The material on antennas has been made more concise.
Chapter 27: The derivation of the radii of the Bohr orbits has been moved online. The section on atomic energy levels has been revised and made more concise.
Chapter 28: Section 28.8 Electron Energy Levels in a Solid has been made much more concise with a more detailed discussion available online.
Chapter 30: The discussions of quarks and leptons have been expanded and clarified. The discussion of the standard model is significantly more concise. Twenty-first-century particle physics has been updated, and the most recent information will be provided online.

Please see your McGraw-Hill sales representative for a more detailed list of revisions.

ORGANIZATION OF CHAPTERS 2 THROUGH 4

In spite of the more traditional organization, Chapters 2–4 retain much of the flavor of the approach in *College Physics*. In particular, we use correct vector notation, diagrams, terminology, and methods from the very beginning. For example, we carefully distinguish components from magnitudes by writing "$v_x = -5$ m/s" and never "$v = -5$ m/s," even if the object moves only along the x-axis.

COMPREHENSIVE COVERAGE

Students should be able to get the whole story from the book. The text works well in our self-paced course, where students must rely on the textbook as their primary learning resource. Nonetheless, completeness and clarity are equally advantageous when the book is used in a more traditional classroom setting. *Physics* frees the instructor from having to try to "cover" everything. The instructor can then tailor class time to more important student needs—reinforcing difficult concepts, working through example problems, engaging the students in cooperative learning activities, describing applications, or presenting demonstrations.

INTEGRATING CONCEPTUAL PHYSICS INTO A QUANTITATIVE COURSE

Some students approach introductory physics with the idea that physics is just the memorization of a long list of equations and the ability to plug numbers into those equations. We want to help students see that a relatively small number of basic physics concepts are applied to a wide variety of situations. Physics education research has shown that students do not automatically acquire conceptual understanding; the concepts must be explained and the students given a chance to grapple with them. Our presentation, based on years of teaching this course, blends conceptual understanding with analytical skills. The **Conceptual Examples** and **Conceptual Practice Problems** in the text and a variety of Conceptual and Multiple-Choice Questions at the end of each chapter give students a chance to check and to enhance their conceptual understanding.

"Conceptual ideas are important, ideas must be motivated, physics should be integrated, a coherent problem-solving approach should be developed. I'm not sure other books are as explicit in these goals, or achieve them as well as Giambattista, Richardson, and Richardson."

Dr. Michael G. Strauss, University of Oklahoma

INTRODUCING CONCEPTS INTUITIVELY

We introduce key concepts and quantities in an informal way by establishing why the quantity is needed, why it is useful, and why it needs a precise definition. Then we make a transition from the informal, intuitive idea to a formal definition and name. Concepts motivated in this way are easier for students to grasp and remember than are concepts introduced by seemingly arbitrary, formal definitions.

For example, in Chapter 8, the idea of rotational inertia emerges in a natural way from the concept of rotational kinetic energy. Students can understand that a rotating rigid body has kinetic energy due to the motion of its particles. We discuss why it is useful to be able to write this kinetic energy in terms of a single quantity common to all the particles (the angular speed), rather than as a sum involving particles with many different speeds. When students understand why rotational inertia is defined the way it is, they are better prepared to move on to the concepts of torque and angular momentum.

We avoid presenting definitions or formulas without any motivation. When an equation is not derived in the text, we at least describe where the equation comes from or give a plausibility argument. For example, Section 9.9 introduces Poiseuille's law with two identical pipes in series to show why the volume flow rate must be proportional to the pressure drop per unit length. Then we discuss why $\Delta V/\Delta t$ is proportional to the fourth power of the radius (rather than to r^2, as it would be for an ideal fluid).

"The authors are clearly very able to communicate in written English. The text is well written, not concise to the point of density, but not discursive to the point of long-windedness. A real pleasure to read."

Dr. Galen T. Pickett, California State University, Long Beach

WRITTEN IN CLEAR AND FRIENDLY STYLE

We have kept the writing down-to-earth and conversational in tone—the kind of language an experienced teacher uses when sitting at a table working one-on-one with a student. We hope students will find the book pleasant to read, informative, and accurate without seeming threatening, and filled with analogies that make abstract concepts easier to grasp. We want students to feel confident that they can learn by studying the textbook.

While learning correct physics terminology is essential, we avoid all *unnecessary* jargon—terminology that just gets in the way of the student's understanding. For example, we never use the term *centripetal force,* since its use sometimes leads students to add a spurious "centripetal force" to their free-body diagrams. Likewise, we use *radial component of acceleration* because it is less likely to introduce or reinforce misconceptions than *centripetal acceleration.*

ACCURACY ASSURANCE

The authors and the publisher acknowledge the fact that inaccuracies can be a source of frustration for both the instructor and students. Therefore, throughout the writing and production of this edition, we have worked diligently to eliminate errors and inaccuracies. Bill Fellers of Fellers Math & Science conducted an independent accuracy check and worked all end-of-chapter questions and problems in the final draft of the manuscript. He then coordinated the resolution of discrepancies between accuracy checks, ensuring the accuracy of the text, the end-of-book answers, and the solutions manuals. Corrections were then made to the manuscript before it was typeset.

The page proofs of the text were double-proofread against the manuscript to ensure the correction of any errors introduced when the manuscript was typeset. The textual examples, practice problems and solutions, end-of-chapter questions and problems, and problem answers were accuracy checked by Fellers Math & Science again at the page proof stage after the manuscript was typeset. This last round of corrections was then cross-checked against the solutions manuals.

PROVIDING STUDENTS WITH THE TOOLS THEY NEED

Problem-Solving Approach

Problem-solving skills are central to an introductory physics course. We illustrate these skills in the example problems. Lists of problem-solving strategies are sometimes useful; we provide such strategies when appropriate. However, the most elusive skills—perhaps the most important ones—are subtle points that defy being put into a neat list. To develop real problem-solving expertise, students must learn how to think critically and analytically. Problem solving is a multidimensional, complex process; an algorithmic approach is not adequate to instill real problem-solving skills.

"The major strength of this text is its approach, which makes students think out the problems, rather than always relying on a formula to get an answer. The way the authors encourage students to investigate whether the answer makes sense, and compare the magnitude of the answer with common sense is good also."

Dr. Jose D'Arruda,
University of North Carolina,
Pembroke

Strategy We begin each example with a discussion—in language that the students can understand—of the *strategy* to be used in solving the problem. The strategy illustrates the kind of analytical thinking students must do when attacking a problem: How do I decide what approach to use? What laws of physics apply to the problem and which of them are *useful* in this solution? What clues are given in the statement of the question? What information is implied rather than stated outright? If there are several valid approaches, how do I determine which is the most efficient? What assumptions can I make? What kind of sketch or graph might help me solve the problem? Is a simplification or approximation called for? If so, how can I tell if the simplification is valid? Can I make a preliminary estimate of the answer? Only after considering these questions can the student effectively solve the problem.

Solution Next comes the detailed *solution* to the problem. Explanations are intermingled with equations and step-by-step calculations to help the student understand the approach used to solve the problem. We want the student to be able to follow the mathematics without wondering, "Where did that come from?"

Discussion The numerical or algebraic answer is not the end of the problem; our examples end with a *discussion.* Students must learn how to determine whether their answer is consistent and reasonable by checking the order of magnitude of the answer, comparing the answer to a preliminary estimate, verifying the units, and doing an independent calculation when more than one approach is feasible. When there are several different approaches, the discussion looks at the advantages and disadvantages of each approach. We also discuss the implications of the answer—what can we learn from it? We look at special cases and look at "what if" scenarios. The discussion sometimes generalizes the problem-solving techniques used in the solution.

"I understood the math, mostly because it was worked out step-by-step, which I like."
Student, Bradley University

Practice Problem After each Example, a Practice Problem gives students a chance to gain experience using the same physics principles and problem-solving tools. By comparing their answers to those provided at the end of each chapter, they can gauge their understanding and decide whether to move on to the next section.

Our many years of experience in teaching the college physics course in a one-on-one setting has enabled us to anticipate where we can expect students to have difficulty. In addition to the consistent problem-solving approach, we offer several other means of assistance to the student throughout the text. A boxed problem-solving strategy gives detailed information on solving a particular type of problem, while an icon for problem-solving tips draws attention to techniques that can be used in a variety of contexts. A hint in a worked example or end-of-chapter problem provides a clue on what approach to use or what simplification to make. A warning icon emphasizes an explanation that clarifies a possible point of confusion or a common student misconception.

An important problem-solving skill that many students lack is the ability to extract information from a graph or to sketch a graph without plotting individual data points. Graphs often help students visualize physical relationships more clearly than they can do with algebra alone. We emphasize the use of graphs and sketches in the text, in worked examples, and in the problems.

"The warning signs about many of the misconceptions, traps, and common mistakes is a very helpful and novel idea. Those of us who have taught undergraduate students in service courses have spent considerable time on these. It is good to see them in a book."
Dr. H.R. Chandrasekhar,
University of Missouri, Columbia

Review & Synthesis with MCAT Review®

Eight **Review & Synthesis** sections appear throughout the text, following groups of related chapters. The ***MCAT® Review*** includes actual reading passages and questions written for the **Medical College Admission Test (MCAT)**. The ***Review Exercises*** are intended to serve as a bridge between textbook problems that are linked to a particular chapter and exam problems that are not. These exercises give students practice in formulating a problem-solving strategy without an external clue (section or chapter number) that indicates which concepts are involved. Many of the problems draw on material from more than one chapter to help the student integrate new concepts and skills with what has been learned previously.

Using Approximation, Estimation, and Proportional Reasoning

Physics is forthright about the constant use of simplified models and approximations in solving physics problems. One of the most difficult aspects of problem solving that students need to learn is that some kind of simplified model or approximation is usually required. We discuss how to know when it is reasonable to ignore friction, treat g as constant, ignore viscosity, treat a charged object as a point charge, or ignore diffraction.

Some Examples and Problems require the student to make an estimate—a useful skill both in physics problem solving and in many other fields. Similarly, we teach proportional reasoning as not only an elegant shortcut but also as a means to understanding patterns. We frequently use percentages and ratios to give students practice in using and understanding them.

"I have tried a number of texts in this course over the past 30 years that I have taught Physics 116–117, and I can assure you that G/R/R is the one I (and the students . . .) like the best. The explanations are clear, and the graphics are excellent—the best I have seen anywhere. And the structure of the question and problem sets is very good. G/R/R is the best standard algebra-based text I have ever seen."
Dr. Carey E. Stronach,
Virginia State University

Showcasing an Innovative Art Program

To help show that physics is more than a collection of principles that explain a set of contrived problems, in every chapter we have developed a system of illustration's, ranging from simpler diagrams to ellaborate and beautiful illustrations, that brings to life the connections between physics concepts and the complex ways in which they are applied. We believe these illustrations, with subjects ranging from three-dimensional views of electric field lines to the biomechanics of the human body and from representations of waves to the distribution of electricity in the home, will help students see the power and beauty of physics.

Helping Students See the Relevance of Physics in Their Lives

Students in an introductory college physics course have a wide range of backgrounds and interests. We stimulate interest in physics by relating the principles to applications relevant to students' lives and in line with their interests. The text, examples, and end-of-chapter problems draw from the everyday world; from familiar technological applications; and from other fields such as biology, medicine, archaeology, astronomy, sports, environmental science, and geophysics. (Applications in the text are identified with a text heading or marginal note. An icon (⚕) identifies applications in the biological or medical sciences.)

The **Physics at Home** experiments give students an opportunity to explore and see physics principles operate in their everyday lives. These activities are chosen for their simplicity and for the effective demonstration of physics principles.

Each **Chapter Opener** includes a photo and vignette, designed to capture student interest and maintain it throughout the chapter. The vignette describes the situation shown in the photo and asks the student to consider the relevant physics. A reduced version of the chapter opener photo and question marks where the topic from the vignette is addressed within the chapter.

Focusing on the Concepts

To focus on the basic, core concepts of physics and reinforce for students that all of physics is based on a few, fundamental ideas, within chapters we have developed **Connections** to identify areas where important concepts are revisited. A marginal Connections heading and summary adjacent to the coverage in the main text help students easily recognize that a previously introduced concept is being applied to the current discussion. Knowledge is being built-up—not newly introduced.

The exercises in the **Review & Synthesis sections** have been revised to increase the number of available exercises and to also concentrate even more heavily on helping students to realize through practice problems how the concepts in the previously covered group of chapters are interrelated.

Checkpoint questions have been added to applicable sections of the text to allow students to pause and test their understanding of the concept explored within the current section. The answers to the Checkpoints are found at the end of the chapter so that students can confirm their knowledge without jumping too quickly to the provided answer.

Applications are clearly identified as such in the text with a complete listing in the front matter. With Applications, students have the opportunity to see how physics concepts are experienced through their everyday lives.

icons identify opportunities for students to access additional information or explanation of topics of interest online. This will help students to focus even further on just the very fundamental, core concepts in their reading of the text.

ADDITIONAL RESOURCES FOR INSTRUCTORS AND STUDENTS

Online Homework and Resources

McGraw-Hill's *Physics* website offers online electronic homework along with a myriad of resources for both instructors and students. Instructors can create homework with easy-to-assign algorithmically generated problems from the text and the simplicity of automatic grading and reporting:

- The end-of-chapter problems and Review & Synthesis exercises appear in the online homework system in diverse formats and with various tools.
- The online homework system incorporates new and exciting interactive tools and problem types: ranking problems, a graphing tool, a free-body diagram drawing tool, symbolic entry, a math palette, and multi-part problems.

Instructors also have access to PowerPoint lecture outlines, an Instructor's Resource Guide with solutions, suggested demonstrations, electronic images from the text, clicker questions, quizzes, tutorials, interactive simulations, and many other resources directly tied to text-specific materials in *Physics*. Students have access to self-quizzing, interactive simulations, tutorials, selected solutions for the text's problems, and more.

See www.mhhe.com/grr to learn more and to register.

Electronic Media Integrated with the Text

McGraw-Hill is proud to bring you an assortment of outstanding interactives and tutorials like no other. These activities offer a fresh and dynamic method to teach the physics basics by providing students with activities that work with real data. icons identify areas in the text where additional understanding can be gained through work with an interactive or tutorial on the text website.

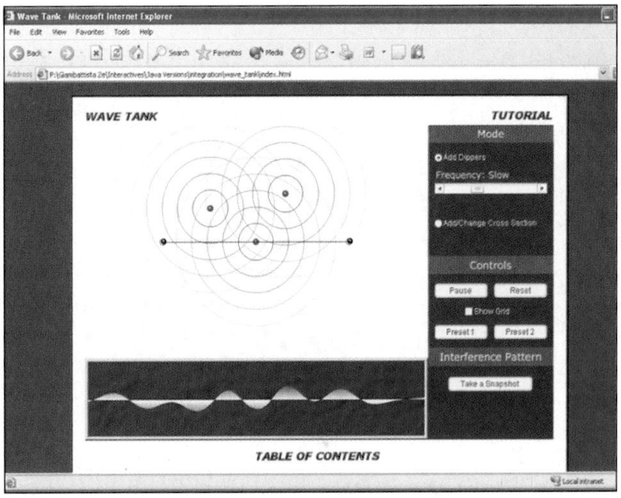

The interactives allow students to manipulate parameters and gain a better understanding of the more difficult physics concepts by watching the effect of these manipulations. Each interactive includes:

- Analysis tool (interactive model)
- Tutorial describing its function
- Content describing its principle themes

The text website contains accompanying interactive quizzes. An instructor's guide for each interactive with a complete overview of the content and navigational tools, a quick demonstration description, further study with the textbook, and suggested end-of-chapter follow-up questions is also provided as an online instructor's resource.

The tutorials, developed and integrated by Raphael Littauer of Cornell University, provide the opportunity for students to approach a concept in steps. Detailed feedback is provided when students enter an incorrect response, which encourages students to further evaluate their responses and helps them progress through the problem.

Electronic Book Images and Assets for Instructors

Build instructional materials wherever, whenever, and however you want!

Accessed from the *Physics* website, an online digital library containing photos, artwork, interactives, and other media types can be used to create customized lectures, visually enhanced tests and quizzes, compelling course websites, or attractive printed support materials. Assets are copyrighted by McGraw-Hill Higher Education, but can be used by instructors for classroom purposes. The visual resources in this collection include

- **Art** Full-color digital files of all illustrations in the book can be readily incorporated into lecture presentations, exams, or custom-made classroom materials. In addition, all files are preinserted into PowerPoint slides for ease of lecture preparation.

- **Active Art Library** These key art pieces—formatted as PowerPoint slides—allow you to illustrate difficult concepts in a step-by-step manner. The artwork is broken into small, incremental pieces, so you can incorporate the illustrations into your lecture in whatever sequence or format you desire.
- **Photos** The photos collection contains digital files of photographs from the text, which can be reproduced for multiple classroom uses.
- **Worked Example Library, Table Library, and Numbered Equations Library** Access the worked examples, tables, and equations from the text in electronic format for inclusion in your classroom resources.
- **Interactives** Flash files of the physics interactives described earlier are included so that you can easily make use of the interactives in a lecture or classroom setting.

Also residing on your textbook's website are

- **PowerPoint Lecture Outlines** Ready-made presentations that combine art and lecture notes are provided for each chapter of the text.
- **PowerPoint Slides** For instructors who prefer to create their lectures from scratch, all illustrations and photos are preinserted by chapter into blank PowerPoint slides.

Computerized Test Bank Online

A comprehensive bank of over 2000 test questions in multiple-choice format at a variety of difficulty levels is provided within a computerized test bank powered by McGraw-Hill's flexible electronic testing program—EZ Test Online (www.eztestonline.com). EZ Test Online allows you to create paper and online tests or quizzes in this easy-to-use program!

Imagine being able to create and access your test or quiz anywhere, at any time without installing the testing software. Now, with EZ Test Online, instructors can select questions from multiple McGraw-Hill test banks or create their own, and then either print the test for paper distribution or give it online. See www.mhhe.com/grr for more information.

Electronic Books

If you or your students are ready for an alternative version of the traditional textbook, McGraw-Hill brings you innovative and inexpensive electronic textbooks. By purchasing E-books from McGraw-Hill, students can save as much as 50% on selected titles delivered on the most advanced E-book platforms available.

E-books from McGraw-Hill are smart, interactive, searchable, and portable, with such powerful built-in tools as detailed searching, highlighting, note taking, and student-to-student or instructor-to-student note sharing. E-books from McGraw-Hill will help students to study smarter and quickly find the information they need. E-books also saves students money. Contact your McGraw-Hill sales representative to discuss E-book packaging options.

Personal Response Systems

Personal response systems, or "clickers," bring interactivity into the classroom or lecture hall. Wireless response systems give the instructor and students immediate feedback from the entire class. The wireless response pads are essentially remotes that are easy to use and engage students, allowing instructors to motivate student preparation, interactivity, and active learning. Instructors receive immediate feedback to gauge which concepts students understand. Questions covering the content of the *Physics* text (formatted in PowerPoint) are available on the website for *Physics*.

Instructor's Resource Guide

The *Instructor's Resource Guide* includes many unique assets for instructors, such as demonstrations, suggested reform ideas from physics education research, and ideas for incorporating just-in-time teaching techniques. It also includes answers to the end-of-chapter conceptual questions and complete, worked-out solutions for all the end-of-chapter problems from the text. The Instructors Resource Guide is available in the Instructor Resources on the text's website.

ALEKS®

Help students master the math skills needed to understand difficult physics problems. ALEKS® [Assessment and LEarning in Knowledge Spaces] is an artificial intelligence–based system for individualized math learning available via the World Wide Web.

ALEKS® is

- A robust course management system. It tells you exactly what your students know and don't know.
- Focused and efficient. It enables students to quickly master the math needed for college physics.
- Artificial intelligence. It totally individualizes assessment and learning.
- Customizable. Click on or off each course topic.
- Web based. Use a standard browser for easy Internet access.
- Inexpensive. There are no setup fees or site license fees.

ALEKS® is a registered trademark of ALEKS Corporation.

Student Solutions Manual

The *Student Solutions Manual* contains complete worked-out solutions to selected end-of-chapter problems and questions, selected Review & Synthesis problems, and the MCAT Review Exercises from the text. The solutions in this manual follow the problem-solving strategy outlined in the text's examples and also guide students in creating diagrams for their own solutions.

For more information, contact a McGraw-Hill customer service representative at (800) 338–3987, or by email at www.mhhe.com. To locate your sales representative, go to www.mhhe.com for Find My Sales Rep.

To the Student

HOW TO SUCCEED IN YOUR PHYSICS CLASS

It's true—how much you get out of your studies depends on how much you put in. Success in a physics class requires:

- Commitment of time and perseverance
- Knowing and motivating yourself
- Getting organized
- Managing your time

This section will help you learn how to be effective in these areas, as well as offer guidance in:

- Getting the most out of your lecture
- Finding extra help when you need it
- Getting the most out of your textbook
- How to study for an exam

Commitment of Time and Perseverance

A good rule of thumb is to allow 2 hours of study time for every hour you spend in lecture. For instance, a 3-hour lecture deserves 6 hours of study time per week. If you commit to studying for this course daily, you're investing a little less than one hour per day, including the weekend.

Learning and mastering takes time and patience. Nothing worthwhile comes easily. Be committed to your studies and you will reap the benefits in the long run. A regular, sustained effort is much more effective than sporadic bouts of cramming.

Knowing and Motivating Yourself

What kind of learner are you? When are you most productive? Know yourself and your limits, and work within them. Know how to motivate yourself to give your all to your studies and achieve your goals.

There are many types of learners, and no right or wrong way of learning. Which category do you fall into?

Begin each of the tasks assigned in your course with the goal of understanding the material. Simply completing the assignment does not mean that learning has taken place. Your fellow students, your instructor, and this textbook can all be important resources in broadening your knowledge.

- **Visual learner** You respond best to "seeing" processes and information. Focus on text illustrations and graphs. Use course handouts and the animations on the course and text websites to help you. Draw diagrams in your notes to illustrate concepts.
- **Auditory learner** You work best by listening to—and possibly recording—the lecture and by talking information through with a study partner.
- **Tactile/Kinesthetic Learner** You learn best by being "hands on." You'll benefit by applying what you've learned during lab time. Writing and drawing are physical activities, so don't neglect taking notes on your reading and the lecture to explain the content in your own words. Try pacing while you read the text. Stand up and write on a chalkboard during discussions in your study group.

Identify your own personal preferences for learning and seek out the resources that will best help you with your studies. Also remember, even though you have a preferred style of learning, most learners benefit when they engage in all styles of learning.

Getting Organized

It's simple, yet it's fundamental. It seems the more organized you are, the easier things come. Take the time before your course begins to analyze your life and your study habits. Get organized now and you'll find you have a little more time—and a lot less stress.

- **Find a calendar system that works for you.** The best kind is one that you can take with you everywhere. To be truly organized, you should integrate all aspects of your life into this one calendar—school, work, and leisure. Some people also find it helpful to have an additional monthly calendar posted by their desk for "at a glance" dates and to have a visual planner. If you do this, be sure you are consistently synchronizing both calendars so as not to miss anything. *More tips for organizing your calendar can be found in the time management discussion below.*

- By the same token, **keep everything for your course or courses in one place**—and at your fingertips. A three-ring binder works well because it allows you to add or organize handouts and notes from class in any order you prefer. Incorporating your own custom tabs helps you flip to exactly what you need at a moment's notice.
- **Find your space.** Find a place that helps you be organized and focused. If it's your desk in your dorm room or in your home, keep it clean. Clutter adds confusion and stress and wastes time. Perhaps your "space" is at the library. If that's the case, keep a backpack or bag that's fully stocked with what you might need—your text, binder or notes, pens, highlighters, Post-its, phone numbers of study partners. [*Hint:* A good place to keep phone numbers is in your "one place for everything calendar."]

Add extra "padding" into your personal deadlines. If you have a report due on Friday, set a goal for yourself to have it done on Wednesday. Then, take time on Thursday to look over your project with a fresh eye. Make any corrections or enhancements and have it ready to turn in on Friday.

Managing Your Time

Managing your time is the single most important thing you can do to help yourself, but it's probably one of the most difficult tasks to successfully master.

In college, you are expected to work much harder and to learn much more than you ever have before. To be successful you need to invest in your education with a commitment of time. We all lead busy lives, but we all make choices as to how we spend our time. Choose wisely.

- **Know yourself and when you'll be able to study most efficiently.** When are you most productive? Are you a night owl? Or an early bird? Plan to study when you are most alert and can have uninterrupted segments. This could include a quick 5-minute review before class or a one-hour problem-solving study session with a friend.
- **Create a set daily study time for yourself.** Having a set schedule helps you commit to studying and helps you plan instead of cram. Find—and use—a planner that is small enough that you can take it with you everywhere. This may be a simple paper calendar or an electronic version. They all work on the same premise: **organize *all* of your activities in one place**.
- **Schedule study time using shorter, focused blocks with small breaks.** Doing this offers two benefits: (1) You will be less fatigued and gain more from your effort and (2) Studying will seem less overwhelming, and you will be less likely to procrastinate.
- **Plan time for leisure, friends, exercise, and sleep.** Studying should be your main focus, but you need to balance your time—and your life.
- **Log your homework deadlines and exam dates** in your personal calendar.
- Try to **complete tasks ahead of schedule**. This will give you a chance to carefully review your work before it is due. You'll feel less stressed in the end.
- **Know where help can be found.** At the beginning of the semester, find your instructor's office hours, your lab partner's contact information, and the "Help Desk" or Learning Resource Center if your course offers one. Make use of all of the support systems that your college or university has to offer. Ask questions both in class and during your instructor's office hours. Don't be shy—your instructor is there to help you learn.
- **Prioritize!** In your calendar or planner, highlight or number key projects; do them first, and then cross them off when you've completed them. Give yourself a pat on the back for getting them done!
- **Review your calendar and reprioritize *daily*.**
- **Resist distractions by setting and sticking to a designated study time.**
- **Multitask when possible.** You may find a lot of extra time you didn't think you had. Review material in your head or think about how to tackle a tough problem while walking to class or doing laundry.

Plan to study and plan for leisure. Being well balanced will help you focus when it is time to study.

Try combining social time with studying in a group, or social time with mealtime or exercise. Being a good student doesn't mean you have to be a hermit. It does mean you need to know how to smartly budget your time.

Getting the Most Out of Lectures

Your instructors want you to succeed. They put a lot of effort into preparing their lectures and other materials designed to help you learn. Attending class is one of the simplest, most valuable things you can do to help yourself. But it doesn't end there—getting the most out of your lectures means being organized. Here's how:

Prepare Before You Go to Class Study the text on the lecture topic *before* attending class. Familiarizing yourself with the material gives you the ability to take notes selectively rather than scrambling to write everything down. You'll be able to absorb more of the subtleties and difficult points from the lecture. You may also develop some good questions to ask your instructor.

Don't feel overwhelmed by this task. Spend time the night before class gaining a general overview of the topics for the next lecture using your syllabus. If your schedule does not allow this, plan to arrive at class 5–15 minutes before lecture. Bring your text with you and skim the chapter before lecture begins.

Don't try to read an entire chapter in one sitting; study one or two sections at a time. It's difficult to maintain your concentration in a long session with so many new concepts and skills to learn.

Be a Good Listener Most people think they are good listeners, but few really are. Are you?

Important points to remember:

- You can't listen if you are talking.
- You aren't listening if you are daydreaming or constantly distracted by other concerns.
- Listening and comprehending are two different things. Listen carefully in class. The language of science is precise; be sure you understand your instructor. If you don't understand something your instructor is saying, ask a question or jot a note and visit the instructor during office hours. You are likely doing others a favor when you ask questions because there are probably others in the class who have the same questions.

Take Good Notes

- Use a standard size notebook, or better yet, a three-ring binder with loose leaf notepaper. The binder will allow you to organize and integrate your notes and handouts, integrate easy-to-reference tabs, and the like.
- Color-code your notes. Use one color of ink pen to take your initial notes. You can annotate later using a pencil, which can be erased if need be.
- Start a new page with each lecture or note-taking session.
- Label each page with the date and a heading for each day.
- Focus on main points and try to use an outline format to take notes to capture key ideas and organize sub-points.
- Take your text to lecture, and keep it open to the topics being discussed. You can also take brief notes in your textbook margin or reference textbook pages in your notebook to help you study later.
- Review and edit your notes shortly after class—within 24 hours—to make sure they make sense and that you've recorded core thoughts. You may also want to compare your notes with a study partner later to make sure neither of you have missed anything.
- This is a very IMPORTANT point: *You can and should also add notes from your reading of the textbook.*

Get a Study Partner Find a few study partners and get together regularly. Four or five study partners to a group is a good number. Too many students make the group unwieldy, but you want enough students to ensure the group can meet even if one or two people can't make it. Having study partners has many benefits. First, they can help you keep your commitment to this class. By having set study dates, you can combine study and social time, and maybe even make it fun! In addition, you now have several minds to help digest the information from the lecture and the text:

- Talk through concepts and go over the difficulties you may be having. Take turns explaining things to each other. You learn a tremendous amount when you teach someone else.
- Compare your notes and solutions with the Practice Problems.
- Try a new approach to a problem or look at the problem from the perspective of your partner. There are often many ways to do the same problem. You can benefit from the insights of others—and they from you—but resist the temptation to simply copy solutions. You need to learn how to solve the problem yourself.
- Quiz each other and discuss some of the Conceptual Questions from the end of the chapter.
- Don't take advantage of your study partner by skipping class or skipping study dates. You obviously won't have a study partner—or a friend—much longer if it's not a mutually beneficial arrangement!

Getting the Most Out of Your Textbook

We hope that you enjoy your physics course using this text. While studying physics does require hard work, we have tried to remove the obstacles that sometimes make introductory physics unnecessarily difficult. We have also tried to reveal the beauty inherent in the principles of physics and how these principles are manifest all around you.

In our years of teaching experience, we have found that studying physics is a skill that must be learned. It's much more effective to *study* a physics textbook, which involves active participation on your part, than to read through passively. Even though active study takes more time initially, in the long run it will save you time; you learn more in one active study session than in three or four superficial readings.

As you study, take particular note of the following elements:

Consider the **chapter opener**. It will help you make the connection between the physics you are about to study and how it affects the world around you. Each chapter opener includes a photo and vignette designed to pique your interest in the chapter. The vignette describes the situation shown in the photo and asks you to consider the relevant physics. The question is then answered within the chapter. Look for the reduced opener photo and question on the referenced page.

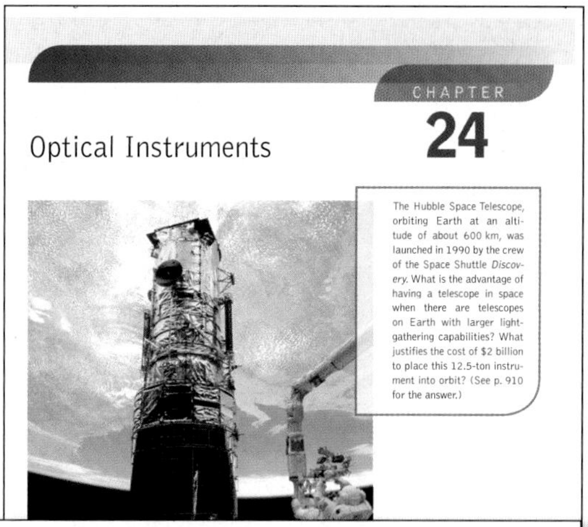

Evaluate the **Concepts & Skills to Review** on the first page of each chapter. It lists important material from previous chapters that you should understand before you start reading. If you have problems recalling any of the concepts, you can revisit the sections referenced in the list.

Concepts & Skills to Review

- distinction between real and virtual images (Section 23.6)
- magnification (Section 23.8)
- refraction (Section 23.3)
- thin lenses (Section 23.9)
- finding images with ray diagrams (Section 23.6)
- small-angle approximations (Appendix A.7)

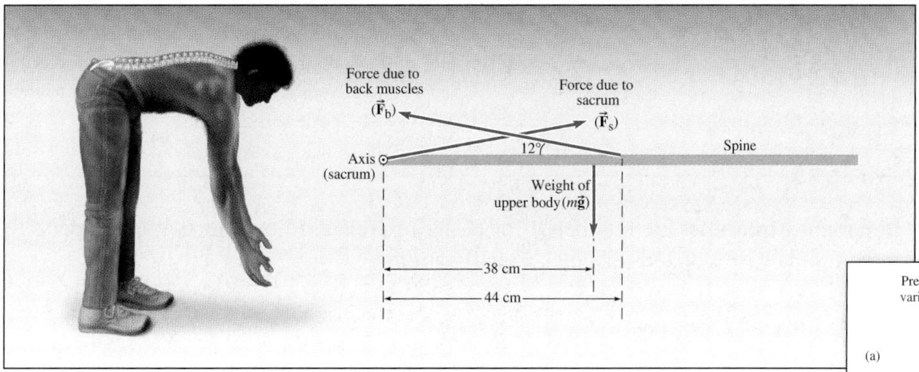

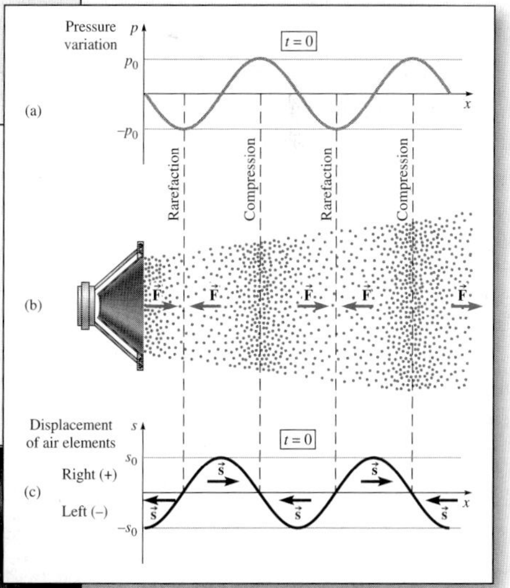

Study the figures and graphs carefully. **Some elaborate illustrations** and more straightforward **diagrammatic illustrations** are used in combination throughout the text to help you grasp concepts. Complex illustrations help you visualize the most difficult concepts. When looking at graphs, try to see the wealth of information displayed. Ask yourself about the physical meaning of the slope, the area under the curve, the overall shape of the graph, the vertical and horizontal intercepts, and any maxima and minima.

CONNECTION:

Rotational and translational kinetic energies have the same form: $\frac{1}{2}$inertia × speed^2.

Marginal **Connections** headings and summaries adjacent to the coverage in the main text identify areas where important concepts are revisited. Consider the notes carefully to help you recognize how a previously introduced concept is being applied to the current discussion.

Checkpoint questions appear in applicable sections of the text to allow you to test your understanding of the concept explored within the current section. The answers to the Checkpoints are found at the end of the chapter so that you can confirm your knowledge without jumping too quickly to the provided answer.

CHECKPOINT 8.2

You are trying to loosen a nut, without success. Why might it help to switch to a wrench with a longer handle?

WWW icons identify opportunities for you to access additional information or explanation of topics of interest online.

Various **Reinforcement Notes** appear in the margin to emphasize the important points in the text.

$$\vec{\mathbf{v}} = \lim_{\Delta t \to 0} \frac{\Delta \vec{\mathbf{r}}}{\Delta t} \qquad (3\text{-}12)$$

($\Delta\vec{\mathbf{r}}$ is the displacement during a *very short* time interval Δt)

If an object moves along a curved path, the direction of the velocity vector at any point is tangent to the path at that point.

Important **Equations** are numbered for easier reference. Equations that correspond to important laws are boxed for quick identification.

Statements of important physics **Rules and Laws** are boxed to highlight the most important and central concepts.

The Law of Conservation of Energy

The total energy in the universe is unchanged by any physical process:

total energy before = total energy after.

Boxed **Problem-Solving Strategies** give detailed information on solving a particular type of problem. These are supplied for the most fundamental physical rules and laws.

Problem-Solving Strategy for Newton's Second Law

- Decide what objects will have Newton's second law applied to them.
- Identify all the *external* forces acting on that object.
- Draw an FBD to show all the forces acting on the object.
- Choose a coordinate system. If the direction of the net force is known, choose axes so that the net force (and the acceleration) are along one of the axes.
- Find the net force by adding the forces as vectors.
- Use Newton's second law to relate the net force to the acceleration.
- Relate the acceleration to the change in the velocity vector during a time interval of interest.

A **warning note** describes possible points of confusion or any common misconceptions that may apply to a particular concept.

tor; the length of the arrow is proportional to the magnitude of the vector. By contrast, a **scalar** quantity can have magnitude, algebraic sign, and units, but not a direction in space. It wouldn't make sense to draw an arrow to represent a scalar such as mass!

In this book, an arrow over a boldface symbol indicates a vector quantity ($\vec{\mathbf{r}}$). (Some books use boldface without the arrow or the arrow without boldface.) When writing by hand, always draw an arrow over a vector symbol to distinguish it from a scalar. When the symbol for a vector is written without the arrow and in italics rather than boldface (r), it stands for the *magnitude* of the vector (which is a scalar). Absolute value bars are also used to stand for the magnitude of a vector, so $r = |\vec{\mathbf{r}}|$. The magnitude of a vector may have units and is never negative; it can be positive or zero.

A **problem-solving tip** will guide you in applying problem-solving techniques.

When scalars are added or subtracted, they do so in the usual way: 3 kg of water plus 2 kg of water is equal to 5 kg of water. Adding or subtracting vectors is different. Vectors follow rules of addition and subtraction that take into account the *directions* of the vectors as well as their magnitudes. Whenever you need to add or subtract quantities, check whether they are vectors. If so, be sure to add or subtract them correctly *as vectors. Do not just add or subtract their magnitudes.*

Example 6.4

Bungee Jumping

A bungee jumper makes a jump in the Gorge du Verdon in southern France. The jumping platform is 182 m above the bottom of the gorge. The jumper weighs 780 N. If the jumper falls to within 68 m of the bottom of the gorge, how much work is done by the bungee cord on the jumper during his descent? Ignore air resistance.

Strategy Ignoring air resistance, only two forces act on the jumper during the descent: gravity and the tension in the cord. Since the jumper has zero kinetic energy at both the highest and lowest points of the jump, the change in kinetic energy for the descent is zero. Therefore, the total work done by the two forces on the jumper must equal zero.

Solution Let W_g and W_c represent the work done on the jumper by gravity and by the cord. Then

$$W_{total} = W_g + W_c = \Delta K = 0$$

The work done by gravity is

$$W_g = F_y \Delta y = -mg \Delta y$$

where the weight of the jumper is $mg = 780$ N. With $y = 0$ at the bottom of the gorge, the vertical component of the displacement is

$$\Delta y = y_f - y_i = 68 \text{ m} - 182 \text{ m} = -114 \text{ m}$$

Then the work done by gravity is

$$W_g = -(780 \text{ N}) \times (-114 \text{ m}) = +89 \text{ kJ}$$

The work done by the cord is $W_c = W_{total} - W_g = -89$ kJ.

Discussion The work done by gravity is positive, since the force and the displacement are in the same direction (downward). If not for the negative work done by the cord, the jumper would have a kinetic energy of 89 kJ after falling 114 m.

The length of the bungee cord is not given, but it does not affect the answer. At first the jumper is in free fall as the cord plays out to its full length; only then does the cord begin to stretch and exert a force on the jumper, ultimately bringing him to rest again. Regardless of the length of the cord, the total work done by gravity and by the cord must be zero since the change in the jumper's kinetic energy is zero.

Practice Problem 6.4 The Bungee Jumper's Speed

Suppose that during the jumper's descent, at a height of 111 m above the bottom of the gorge, the cord has done −21.7 kJ of work on the jumper. What is the jumper's speed at that point?

CHECKPOINT 6.3

Kinetic energy and work are related. Can kinetic energy ever be negative? Can work ever be negative?

6.4 GRAVITATIONAL POTENTIAL ENERGY (1)

Gravitational Potential Energy When Gravitational Force Is Constant

Toss a stone up with initial speed v_i. Ignoring air resistance, how high does the stone go? We can solve this problem with Newton's second law, but let's use work and energy instead. The stone's initial kinetic energy is $K_i = \frac{1}{2}mv_i^2$. For an upward displacement Δy, gravity does negative work $W_{grav} = -mg \Delta y$. No other forces act, so this is the total work done on the stone. The stone is momentarily at rest at the top, so $K_f = 0$. Then

When you come to an **Example**, pause after you've read the problem. Think about the strategy you would use to solve the problem. See if you can work through the problem on your own. Now study the ***Strategy***, ***Solution***, and ***Discussion*** in the textbook. Sometimes you will find that your own solution is right on the mark; if not, you can focus your attention on the areas of misunderstanding or any mistakes you may have made.

Work the ***Practice Problem*** after each Example to practice applying the physics concepts and problem-solving skills you've just learned. Check your answer with the one given at the end of the chapter. If your answer isn't correct, review the previous section in the textbook to try to find your mistake.

Application headings identify places in the text where physics can be applied to other areas of your life. Familiar topics and interests are discussed in the accompanying text, including examples from biology, archaeology, astronomy, sports, and the everyday world. The biology/life science examples have a special icon.

Banked Curves To help prevent cars from going into a skid or losing control, the roadway is often banked (tilted at a slight angle) around curves so that the outer portion of the road—the part farthest from the center of curvature—is higher than the inner portion. Banking changes the angle and magnitude of the normal force, $\vec{\mathbf{N}}$, so that it has a horizontal component N_x directed toward the center of curvature (in the

Application of radial acceleration and contact forces: banked roadways

Application of the manometer: measuring blood pressure

Try the ***Physics at Home*** experiments in your dorm room or at home. They reinforce key physics concepts and help you see how these concepts operate in the world around you.

PHYSICS AT HOME

Drop a very tiny speck of dust or lint into a container of water and push the speck below the surface. The motion of the speck—called *Brownian motion*—is easily observed as it is pushed and bumped about randomly by collisions with water molecules. The water molecules themselves move about randomly, but at much higher speeds than the speck of dust due to their much smaller mass.

Master the Concepts

- Fluids are materials that flow and include both liquids and gases. A liquid is nearly incompressible, whereas a gas expands to fill its container.
- Pressure is the perpendicular force per unit area that a fluid exerts on any surface with which it comes in contact ($P = F/A$). The SI unit of pressure is the pascal (1 Pa = 1 N/m^2).
- The average air pressure at sea level is 1 atm = 101.3 kPa.
- Pascal's principle: A change in pressure at any point in a confined fluid is transmitted everywhere throughout the fluid.
- The average density of a substance is the ratio of its mass to its volume

$$\rho = \frac{m}{V} \quad (9\text{-}2)$$

- The specific gravity of a material is the ratio of its density to that of water at 4°C.
- Pressure variation with depth in a static fluid:

$$P_2 = P_1 + \rho g d \quad (9\text{-}3)$$

where point 2 is a depth ...
- Instruments to measure p... and the barometer. The ... sure of the atmosphere. ... pressure difference.

equal in magnitude to the weight of the volume of fluid displaced by the object:

$$F_B = \rho g V \quad (9\text{-}7)$$

where V is the volume of the part of the object that is submerged and ρ is the density of the fluid.

- In steady flow, the velocity of the fluid *at any point* is constant in time. In laminar flow, the fluid flows in neat layers so that each small portion of fluid that passes a particular point follows the same path as every other portion of fluid that passes the same point. The path that the fluid follows, starting from any point, is called a streamline. Laminar flow is steady. Turbulent flow is chaotic and unsteady. The viscous force opposes the flow of the fluid; it is the counterpart to the frictional force for solids.
- An ideal fluid exhibits laminar flow, has no viscosity, and is incompressible. The flow of an ideal fluid is governed by two principles: the continuity equation and Bernoulli's equation.

5.1 Description of Uniform Circular Motion

1. A carnival swing is fixed on the end of an 8.0-m-long beam. If the swing and beam sweep through an angle of 120°, what is the distance through which the riders move?
2. A soccer ball of diameter 31 cm rolls without slipping at a linear speed of 2.8 m/s. Through how many revolutions has the soccer ball turned as it moves a linear distance of 18 m?
3. Find the average angular speed of the second hand of a

Problems

- Ⓒ Combination conceptual/quantitative problem
- Biological or medical application
- ✦ Challenging problem
- Blue # Detailed solution in the Student Solutions Manual
- (1 2) Problems paired by concept
- Text website interactive or tutorial

5.1 Description of Uniform Circular Motion

1. A carnival swing is fixed on the end of an 8.0-m-l... beam. If the swing and beam sweep through an angl... 120... mo...

✦114. A student's head is bent over her physics book. The head weighs 50.0 N and is supported by the muscle force $\vec{\mathbf{F}}_m$ exerted by the neck extensor muscles and by the contact force $\vec{\mathbf{F}}_c$ exerted at the atlantooccipital joint. Given that the magnitude of $\vec{\mathbf{F}}_m$ is 60.0 N and is directed 35° below the horizontal, find (a) the magnitude and (b) the direction of $\vec{\mathbf{F}}_c$.

Write your *own* chapter summary or outline, adding notes from class where appropriate, and then compare it with the ***Master the Concepts*** provided at the end of the chapter. This will help you identify the most important and fundamental concepts in each chapter.

Along with working the problems assigned by your instructor, try quizzing yourself on the **Multiple-Choice Questions.** Check your answers against the answers at the end of the book. Consider the **Conceptual Questions** to check your qualitative understanding of the key ideas from the chapter. Try writing some responses to practice your writing skills and to help prepare for any essay problems on the exam.

When working the **Problems** and **Comprehensive Problems** assigned by your instructor, pay special attention to the explanatory paragraph below the Problem heading and the keys accompanying each problem.

- ***Paired Problems*** are connected with a bracket. Your instructor may assign the even-numbered problem, which has no answer at the end of the book. However, working the connected odd-numbered problem will allow you to check your answer at the back of the book and apply what you have learned to working the even-numbered problem.
- Problem numbers highlighted in blue have a solution available in the ***Student Solutions Manual*** if you need additional help or would like to double-check your work.
- The ***difficulty level*** for each problem is indicated. The least difficult problems and problems of intermediate difficulty have no diamond. The more challenging problems have one diamond ✦.

Read through all of the assigned problems and budget your time accordingly.

- Ⓒ indicates a combination **Conceptual and Quantitative** problem.
- indicates a problem with a biological or medical application.
- indicates a problem that has an accompanying interactive or tutorial online.

While working your solutions to problems, try to **keep your work in symbolic form** until the very end. Symbolic solutions will allow you to view which factors affect the results and how the answer would change should any one of the variables in the problem change their value. In this fashion, your solution to any one problem becomes a solution to a whole series of similar problems. Substituting values into your final symbolic solution will then enable you to judge if your answer is reasonable and provide greater ease in troubleshooting your error if it is not. Always perform a "reality check" at the end of each problem. Did you obtain a reasonable answer given the question being asked?

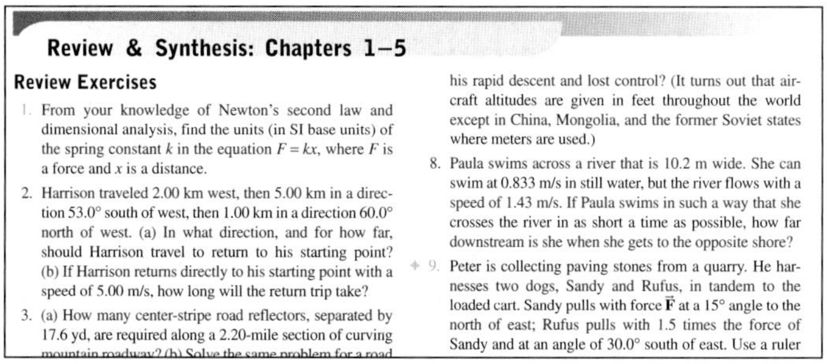

Review & Synthesis: Chapters 1–5

Review Exercises

1. From your knowledge of Newton's second law and dimensional analysis, find the units (in SI base units) of the spring constant k in the equation $F = kx$, where F is a force and x is a distance.
2. Harrison traveled 2.00 km west, then 5.00 km in a direction 53.0° south of west, then 1.00 km in a direction 60.0° north of west. (a) In what direction, and for how far, should Harrison travel to return to his starting point? (b) If Harrison returns directly to his starting point with a speed of 5.00 m/s, how long will the return trip take?
3. (a) How many center-stripe road reflectors, separated by 17.6 yd, are required along a 2.20-mile section of curving

his rapid descent and lost control? (It turns out that aircraft altitudes are given in feet throughout the world except in China, Mongolia, and the former Soviet states where meters are used.)

8. Paula swims across a river that is 10.2 m wide. She can swim at 0.833 m/s in still water, but the river flows with a speed of 1.43 m/s. If Paula swims in such a way that she crosses the river in as short a time as possible, how far downstream is she when she gets to the opposite shore?
9. Peter is collecting paving stones from a quarry. He harnesses two dogs, Sandy and Rufus, in tandem to the loaded cart. Sandy pulls with force $\vec{\mathbf{F}}$ at a 15° angle to the north of east; Rufus pulls with 1.5 times the force of Sandy and at an angle of 30.0° south of east. Use a ruler

After a group of related chapters, you will find a **Review & Synthesis** section. This section will provide ***Review Exercises*** that require you to combine two or more concepts learned in the previous chapters. Working these problems will help you to prepare for cumulative exams. This section also contains ***MCAT Review*** exercises. These problems were written for the actual MCAT exam and will provide additional practice if this exam is part of your future plans.

How to Study for an Exam

- Be an active learner:
  - read
  - be an active participant in class; ask questions
  - apply what you've learned; think through scenarios rather than memorizing your notes
- Finish reading all material—text, notes, handouts—at least three days prior to the exam.
- Three days prior to the exam, set aside time each day to do self-testing, work practice problems, and review your notes. Useful tools to help:
  - end-of-chapter summaries
  - questions and practice problems
  - text website
  - your professor's course website
  - the Student Solutions Manual
  - your study partner
- Analyze your weaknesses, and create an "I don't know this yet" list. Focus on strengthening these areas and narrow your list as you study.
- If you find that you were unable to allow the full three days to study for the exam, the most important thing you can do is try some practice problems that are similar to those your instructor assigned for homework. Choose odd-numbered problems so that you can check your answer. The Review & Synthesis problems are designed to help you prepare for exams. Try to solve each problem under exam conditions—use a formula sheet, if your instructor provides one with the exam, but don't look at the book or your notes. If you can't solve the problem, then you have found an area of weakness. Study the material needed to solve that problem and closely related material. Then try another similar problem.
- VERY IMPORTANT—Be sure to sleep and eat well before the exam. Staying up late and memorizing the night before an exam doesn't help much in physics. On a physics exam, you will be asked to demonstrate reasoning and analytical skills acquired by much practice. If you are fatigued or hungry, you won't perform at your highest level.

We hope that these suggestions will help you get the most out of your physics course. After many years working with students, both in the classroom and one-on-one in a self-paced course, we wrote this book so you could benefit from our experience. In *Physics*, we have tried to address the points that have caused difficulties for our students in the past. We also wish to share with you some of the pleasure and excitement we have found in learning about the physical laws that govern our world.

Alan Giambattista
Betty Richardson
Bob Richardson

Acknowledgments

We are grateful to the faculty, staff, and students at Cornell University, who helped us in a myriad of ways. We especially thank our friend and colleague Bob Lieberman who shepherded us through the process as our literary agent and who inspired us as an exemplary physics teacher. Donald F. Holcomb, Persis Drell, Peter Lepage, and Phil Krasicky read portions of the manuscript and provided us with many helpful suggestions. Raphael Littauer contributed many innovative ideas and served as a model of a highly creative, energetic teacher.

We are indebted to Jeevak Parpia, David G. Cassel, Edith Cassel, Richard Galik, Lou Hand, Chris Henley, and Tomás Arias for many helpful discussions while they taught Physics 101–102 using the second edition. We also appreciate the assistance of Leonard J. Freelove and Rosemary French. We thank our enthusiastic and capable graduate teaching assistants and, above all, the students in Physics 101–102, who patiently taught us how to teach physics.

We are grateful for the guidance and enthusiasm of Debra Hash and Mary Hurley at McGraw-Hill, whose tireless efforts were invaluable in bringing this project to fruition. We would like to thank the entire team of talented professionals assembled by McGraw-Hill to publish this book, including Traci Andre, Tammy Ben, Carrie Burger, Linda Davoli, Laura Fuller, David Hash, Tammy Juran, Mary Jane Lampe, Lisa Nicks, Mary Reeg, Gloria Schiesl, Thomas Timp, Dan Wallace, and many others whose hard work has contributed to making the book a reality.

We are grateful to Bill Fellers for accuracy-checking the manuscript and for many helpful suggestions.

Our thanks to Janet Scheel, Warren Zipfel, Rebecca Williams, and Mike Nichols for contributing some of the medical and biological applications; to Nick Taylor and Mike Strauss for contributing to the end-of-chapter and Review & Synthesis problems; and to Nick Taylor for writing answers to the Conceptual Questions.

From Alan: Above all, I am deeply grateful to my family. Marion, Katie, Charlotte, Julia, and Denisha, without your love, support, encouragement, and patience, this book could never have been written.

From Bob and Betty: We thank our daughter Pamela's classmates and friends at Cornell and in the Vanderbilt Master's in Nursing program who were an early inspiration for the book, and we thank Dr. Philip Massey who was very special to Pamela and is dear to us. We thank our friends at *blur,* Alex, Damon, Dave, and Graham, who love physics and are inspiring young people of Europe to explore the wonders of physics through their work with the European Space Agency's Mars mission. Finally we thank our daughter Jennifer, our grandsons Jasper, Dashiell, Oliver, and Quintin, and son-in-law Jim who endured our protracted hours of distraction while this book was being written.

REVIEWERS, CLASS TESTERS, AND ADVISORS

This text reflects an extensive effort to evaluate the needs of college physics instructors and students, to learn how well we met those needs, and to make improvements where we fell short. We gathered information from numerous reviews, class tests, and focus groups.

The primary stage of our research began with commissioning reviews from instructors across the United States and Canada. We asked them to submit suggestions for improvement on areas such as content, organization, illustrations, and ancillaries. The detailed comments of these reviewers constituted the basis for the revision plan.

We organized focus groups across the United States from 2006 through 2008. Participants reviewed our text in comparison to other books and suggested improvements to *Physics* and ways in which we as publishers could help to improve the content of the college physics course.

Finally, we received extremely useful advice on the instructional design, quality, and content of the print and media ancillary packages from Pete Anderson, Gerry Feldman, Ajawad Haija, Hong Luo, David Mast, John Prineas, Michael Pravica, and Craig Wiegert.

Considering the sum of these opinions, the Giambattista/Richardson/Richardson texts now embody the collective knowledge, insight, and experience of hundreds of college physics instructors. Their influence can be seen in everything from the content, accuracy, and organization of the text to the quality of the illustrations.

We are grateful to the following instructors for their thoughtful comments and advice:

REVIEWERS AND FOCUS GROUP ATTENDEES

David Aaron *South Dakota State University*
Rhett Allain *Southeastern Louisiana University*
Peter Anderson *Oakland Community College*
Natalie Batalha *San Jose State*
Thomas K. Bolland *The Ohio State University*
Juan Burciaga *Whitman College*
Peng Chen Dai *University of Tennessee—Knoxville*
Carl Covatto *Arizona State University*
Michael Crescimanno *Youngstown State*
Steven Ellis *University of Kentucky—Lexington*
Abbas Faridi *Orange Coast College*
Gerald Feldman *George Washington University*
David Gerdes *University of Michigan*
Robert Hagood *Washtenaw Community College*
Ajawad Haija *Indiana University of Pennsylvania*
Grady Hendricks *Blinn Community College*
Klaus Honschied *The Ohio State University*
John Hopkins *The Pennsylvania State University*
Brad Johnson *Western Washington University*
Kyungseon Joo *University of Connecticut*
Linda Jones *College of Charleston*
Arya Karamjeet *San Jose State*
Daniel Kennefick *University of Arkansas*
Yuri Kholodenko *Albany College of Pharmacy*
Dana Klinck *Hillsborough Community College, Tampa*
Allen Landers *Auburn University*
Paul Lee *California State University Northridge*
Hong Luo *University at Buffalo*
Stephanie Magelby *Brigham Young University*
George Marion *Texas State University, San Marcos*
David Mast *University of Cincinnati*
Dan Mazilu *Virginia Polytechnic Institute & State University*
Rahul Mehta *University of Central Arkansas*
Meredith Newby *Clemson University*
Miroslav Peric *California State University Northridge*
Amy Pope *Clemson University*
Michael Pravica *University of Nevada, Las Vegas*
Kent Price *Morehead State*
John Prineas *University of Iowa*
Oren Quist *South Dakota State University*
Larry Rowan *University of North Carolina—Chapel Hill*
Ajit Rupaal *Western Washington University*
Douglas Sherman *San Jose State University*
Bjoern Siepel *Portland State University*
Michael Sobel *Brooklyn College*
Xiang-Ning Song *Richland College*
Tim Stelzer *University of Illinois*
James Taylor *University of Central Missouri*
Marshall Thomsen *Eastern Michigan University*
Ralf Widenhorn *Portland State University*
Craig Wiegert *University of Georgia*
Karen Williams *East Central University*
Scott Wissink *Indiana University*
Pei Xiong-Skiba *Austin Peay State University*
Capp Yess *Morehead State*
David Young *Louisiana State University—Baton Rouge*
Michael Ziegler *The Ohio State University*

We are also grateful to our international reviewers for their comments and suggestions:

Goh Hock Leong *National Junior College—Singapore*

Mohammed Saber Musazay *King Fahd University of Petroleum and Minerals*

Contributors

We are deeply indebted to:

Professor Suzanne Willis of Northern Illinois University and Professor Susanne M. Lee, Visiting Scientist Rensselaer Polytechnic Institute, for creating the instructor resources and demonstrations in the *Physics Instructors' Resource Guide.*

Professor Jack Cuthbert of Holmes Community College, Ridgeland for the Test Bank to accompany *Physics.*

Professor Lorin Swint Matthews of Baylor University for the clicker questions to accompany *Physics.*

Professor Carl Covatto of Arizona State University for the PowerPoint Lectures to accompany *Physics.*

Professor Allen Landers of Auburn University for his work on the *Physics* collection of Active Art on the text's website.

CHAPTER

1

Introduction

In 2004, the exploration rovers *Spirit* and *Opportunity* landed on sites on opposite sides of Mars. The primary goal of the mission was to examine a wide variety of rocks and soils that might provide evidence of the past presence of water on Mars and clues to where the water went. The mission sent back tens of thousands of photographs and a wealth of geologic data. By contrast, in a previous mission to Mars, a simple mistake caused the loss of the Mars Climate Orbiter as it entered orbit around Mars. In this chapter, you will learn how to avoid making this same mistake. (See p. 9.)

The Mars Exploration Rover *Opportunity* looks back toward its lander in "Eagle Crater" on the surface of Mars.

Concepts & Skills to Review

- algebra, geometry, and trigonometry (Appendix A)
- To the Student: How to Succeed in Your Physics Class (p. xxii)

1.1 WHY STUDY PHYSICS?

Physics is the branch of science that describes matter, energy, space, and time at the most fundamental level. Whether you are planning to study biology, architecture, medicine, music, chemistry, or art, some principles of physics are relevant to your field.

Physicists look for patterns in the physical phenomena that occur in the universe. They try to explain what is happening, and they perform experiments to see if the proposed explanation is valid. The goal is to find the most basic laws that govern the universe and to formulate those laws in the most precise way possible.

The study of physics is valuable for several reasons:

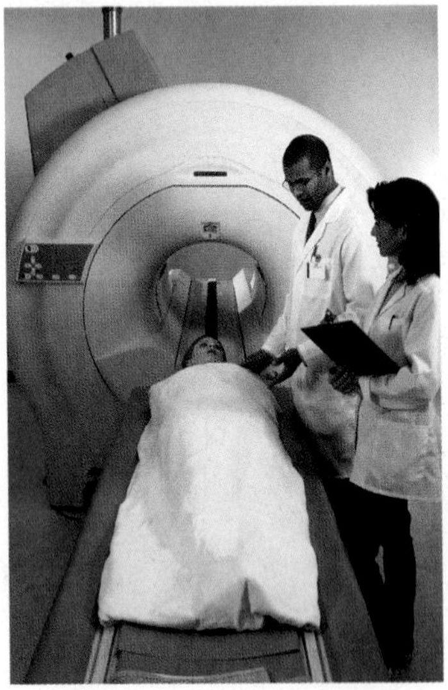

A patient being prepared for magnetic resonance imaging (MRI). MRI provides a detailed image of the internal structures of the patient's body.

- Since physics describes matter and its basic interactions, all natural sciences are built on a foundation of the laws of physics. A full understanding of chemistry requires a knowledge of the physics of atoms. A full understanding of biological processes in turn is based on the underlying principles of physics and chemistry. Centuries ago, the study of *natural philosophy* encompassed what later became the separate fields of biology, chemistry, geology, astronomy, and physics. Today there are scientists who call themselves biophysicists, chemical physicists, astrophysicists, and geophysicists, demonstrating how thoroughly the sciences are intertwined.
- In today's technological world, many important devices can be understood correctly only with a knowledge of the underlying physics. Just in the medical world, think of laser surgery, magnetic resonance imaging, instant-read thermometers, x-ray imaging, radioactive tracers, heart catheterizations, sonograms, pacemakers, microsurgery guided by optical fibers, ultrasonic dental drills, and radiation therapy.
- By studying physics, you acquire skills that are useful in other disciplines. These include thinking logically and analytically; solving problems; making simplifying assumptions; constructing mathematical models; using valid approximations; and making precise definitions.
- Society's resources are limited, so it is important to use them in beneficial ways and not squander them on scientifically impossible projects. Political leaders and the voting public are too often led astray by a lack of understanding of scientific principles. Can a nuclear power plant supply energy safely to a community? What is the truth about the greenhouse effect, the ozone hole, and the danger of radon in the home? By studying physics, you learn some of the basic scientific principles and acquire some of the intellectual skills necessary to ask probing questions and to formulate informed opinions on these important matters.
- Finally, by studying physics, we hope that you develop a sense of the beauty of the fundamental laws governing the universe.

1.2 TALKING PHYSICS

Some of the words used in physics are familiar from everyday speech. This familiarity can be misleading, since the scientific definition of a word may differ considerably from its common meaning. In physics, words must be precisely defined so that anyone reading a scientific paper or listening to a science lecture understands exactly what is meant. Some of the basic defined quantities, whose names are also words used in everyday speech, include time, length, force, velocity, acceleration, mass, energy, momentum, and temperature.

In everyday language, *speed* and *velocity* are synonyms. In physics, there is an important distinction between the two. In physics, *velocity* includes the *direction* of motion as well as the distance traveled per unit time. When a moving object changes direction, its velocity changes even though its speed may not have changed. Confusion of the scientific definition of *velocity* with its everyday meaning will prevent a correct understanding of some of the basic laws of physics and will lead to incorrect answers.

Mass, as used in everyday language, has several different meanings. Sometimes *mass* and *weight* are used interchangeably. In physics, mass and weight are *not* interchangeable. Mass is a measure of inertia—the tendency of an object at rest to remain at rest or, if moving, to continue moving with the same velocity. Weight, on the other hand, is a measure of the gravitational pull on an object. (Mass and weight are discussed in more detail in Chapter 4.)

There are two important reasons for the way in which we define physical quantities. First, physics is an experimental science. The results of an experiment must be stated unambiguously so that other scientists can perform similar experiments and compare their results. Quantities must be defined precisely to enable experimental measurements to be uniform no matter where they are made. Second, physics is a mathematical science. We use mathematics to quantify the relationships among physical quantities. These relationships can be expressed mathematically only if the quantities being investigated have precise definitions.

1.3 THE USE OF MATHEMATICS

A working knowledge of algebra, trigonometry, and geometry is essential to the study of introductory physics. Some of the more important mathematical tools are reviewed in Appendix A. If you know that your mathematics background is shaky, you might want to test your mastery by doing some problems from a math textbook. You may find it useful to visit www.mhhe.com to explore the Schaum's Outline series, especially the Schaum's Outlines of *Precalculus, College Physics,* or *Physics for Pre-Med, Biology, and Allied Health Students.*

Mathematical equations are shortcuts for expressing concisely in symbols relationships that are cumbersome to describe in words. Algebraic symbols in the equations stand for quantities that consist of numbers *and units.* The number represents a measurement and the measurement is made in terms of some standard; the unit indicates what standard is used. In physics, a number to specify a quantity is useless unless we know the unit attached to the number. When buying silk to make a sari, do we need a length of 5 millimeters, 5 meters, or 5 kilometers? Is the term paper due in 3 minutes, 3 days, or 3 weeks? Systems of units are discussed in Section 1.5.

There are not enough letters in the alphabet to assign a unique letter to each quantity. The same letter V can represent volume in one context and voltage in another. Avoid attempting to solve problems by picking equations that seem to have the correct letters. A skilled problem-solver understands *specifically* what quantity each symbol in a particular equation represents, can specify correct units for each quantity, and understands the situations to which the equation applies.

Ratios and Proportions In the language of physics, the word **factor** is used frequently, often in a rather idiosyncratic way. If the power emitted by a radio transmitter has doubled, we might say that the power has "increased by a factor of two." If the concentration of sodium ions in the bloodstream is half of what it was previously, we might say that the concentration has "decreased by a factor of two," or, in a blatantly inconsistent way, someone else might say that it has "decreased by a factor of one-half." The *factor* is the number by which a quantity is multiplied or divided when it is changed from one value to another. In other words, the factor is really a ratio. In the case of the radio transmitter, if P_0 represents the initial power and P represents the power after new equipment is installed, we write

$$\frac{P}{P_0} = 2$$

It is also common to talk about "increasing 5%" or "decreasing 20%." If a quantity increases $n\%$, that is the same as saying that it is multiplied by a factor of $1 + (n/100)$. If a quantity decreases $n\%$, then it is multiplied by a factor of $1 - (n/100)$. For example, an increase of 5% means something is 1.05 times its original value and a decrease of 4% means it is 0.96 times the original value.

$A \propto B$ means $A_1/A_2 = B_1/B_2$

Physicists talk about increasing "by some factor" because it often simplifies a problem to think in terms of **proportions**. When we say that A is proportional to B (written $A \propto B$), we mean that if B increases by some factor, then A must increase by the same factor. For instance, the circumference of a circle equals 2π times the radius: $C = 2\pi r$. Therefore $C \propto r$. If the radius doubles, the circumference also doubles. The area of a circle is proportional to the *square* of the radius ($A = \pi r^2$, so $A \propto r^2$). The area must increase by the same factor as the radius *squared*, so if the radius doubles, the area increases by a factor of $2^2 = 4$.

Example 1.1

Effect of Increasing Radius on the Volume of a Sphere

The volume of a sphere is given by the equation

$$V = \tfrac{4}{3}\pi r^3$$

where V is the volume and r is the radius of the sphere. If a basketball has a radius of 12.4 cm and a tennis ball has a radius of 3.20 cm, by what factor is the volume of the basketball larger than the volume of the tennis ball?

Strategy The problem gives the values of the radii for the two balls. To keep track of which ball's radius and volume we mean, we use subscripts "b" for basketball and "t" for tennis ball. The radius of the basketball is r_b and the radius of the tennis ball is r_t. Since $\tfrac{4}{3}$ and π are constants, we can work in terms of proportions.

Solution The ratio of the basketball radius to that of the tennis ball is

$$\frac{r_b}{r_t} = \frac{12.4\ [\text{cm}]}{3.20\ [\text{cm}]} = 3.875$$

The volume of a sphere is proportional to the cube of its radius:

$$V \propto r^3$$

Since the basketball radius is larger by a factor of 3.875, and volume is proportional to the cube of the radius, the new volume should be bigger by a factor of $3.875^3 \approx 58.2$.

Discussion A slight variation on the solution is to write out the proportionality in terms of ratios of the corresponding sides of the two equations:

$$\frac{V_b}{V_t} = \frac{\tfrac{4}{3}\pi r_b^3}{\tfrac{4}{3}\pi r_t^3} = \left(\frac{r_b}{r_t}\right)^3$$

Substituting the ratio of r_b to r_t yields

$$\frac{V_b}{V_b} = 3.875^3 \approx 58.2$$

which says that V_b is approximately 58.2 times V_t.

Practice Problem 1.1 Power Dissipated by a Lightbulb

The electric power P dissipated by a lightbulb of resistance R is $P = V^2/R$, where V represents the line voltage. During a brownout, the line voltage is 10.0% less than its normal value. How much power is drawn by a lightbulb during the brownout if it normally draws 100.0 W (watts)? Assume that the resistance does not change.

CHECKPOINT 1.3

If the radius of the sphere is increased by a factor of 3, by what factor does the volume of the sphere change?

1.4 SCIENTIFIC NOTATION AND SIGNIFICANT FIGURES

In physics, we deal with some numbers that are very small and others that are very large. It can get cumbersome to write numbers in conventional decimal notation. In **scientific notation**, any number is written as a number between 1 and 10 times an integer power of ten. Thus the radius of Earth, approximately 6 380 000 m at the equator, can be written 6.38×10^6 m; the radius of a hydrogen atom, 0.000 000 000 053 m, can be written 5.3×10^{-11} m. Scientific notation eliminates the need to write zeros to locate the decimal point correctly.

Learn how to use the button on your calculator (usually labeled EE) to enter a number in scientific notation. To enter 1.2×10^8, press 1.2, EE, 8.

In science, a measurement or the result of a calculation must indicate the **precision** to which the number is known. The precision of a device used to make a measurement is limited by the finest division on the scale. Using a meterstick with millimeter divisions as the smallest separations, we can measure a length to a precise number of millimeters and we can estimate a fraction of a millimeter between two divisions. If the meterstick has centimeter divisions as the smallest separations, we measure a precise number of centimeters and estimate the fraction of a centimeter that remains.

Significant Figures The most basic way to indicate the precision of a quantity is to write it with the correct number of **significant figures**. The significant figures are all the digits that are known accurately plus the one estimated digit. If we say that the distance from here to the state line is 12 km, that does not mean we know the distance to be *exactly* 12 kilometers. Rather, the distance is 12 km *to the nearest kilometer.* If instead we said that the distance is 12.0 km, that would indicate that we know the distance to the nearest *tenth* of a kilometer. More significant figures indicate a greater degree of precision.

Rules for Identifying Significant Figures

1. Nonzero digits are always significant.
2. Final or ending zeros written to the right of the decimal point are significant.
3. Zeros written to the right of the decimal point for the purpose of spacing the decimal point are not significant.
4. Zeros written to the left of the decimal point may be significant, or they may only be there to space the decimal point. For example, 200 cm could have one, two, or three significant figures; it's not clear whether the distance was measured to the nearest 1 cm, to the nearest 10 cm, or to the nearest 100 cm. On the other hand, 200.0 cm has four significant figures (see rule 5). Rewriting the number in scientific notation is one way to remove the ambiguity. In this book, when a number has zeros to the left of the decimal point, you may *assume a minimum of two significant figures.*
5. Zeros written between significant figures are significant.

Example 1.2

Identifying the Number of Significant Figures

For each of these values, identify the number of significant figures and rewrite it in standard scientific notation.

(a) 409.8 s
(b) 0.058700 cm
(c) 9500 g
(d) 950.0×10^1 mL

Strategy We follow the rules for identifying significant figures as given. To rewrite a number in scientific notation, we move the decimal point so that the number to the left of the decimal point is between 1 and 10 and compensate by multiplying by the appropriate power of ten.

continued on next page

Example 1.2 continued

Solution (a) All four digits in 409.8 s are significant. The zero is between two significant figures, so it is significant. To write the number in scientific notation, we move the decimal point two places to the left and compensate by multiplying by 10^2: 4.098×10^2 s.

(b) The first two zeros in 0.058700 cm are not significant; they are used to place the decimal point. The digits 5, 8, and 7 are significant, as are the two final zeros. The answer has five significant figures: 5.8700×10^{-2} cm.

(c) The 9 and 5 in 9500 g are significant, but the zeros are ambiguous. This number could have two, three, or four significant figures. If we take the most cautious approach and assume the zeros are not significant, then the number in scientific notation is 9.5×10^3 g.

(d) The final zero in 950.0×10^1 mL is significant since it comes after the decimal point. The zero to its left is also significant since it comes between two other significant digits. The result has four significant figures. The number is not in *standard* scientific notation since 950.0 is not between 1 and 10; in scientific notation we write 9.500×10^3 mL.

Discussion Scientific notation clearly indicates the number of significant figures since all zeros are significant; none are used only to place the decimal point. In (c), if we want to show that the zeros were significant, we would write 9.500×10^3 g.

Practice Problem 1.2 Identifying Significant Figures

State the number of significant figures in each of these measurements and rewrite them in standard scientific notation.

(a) 0.000 105 44 kg (b) 0.005 800 cm (c) 602 000 s

Significant Figures in Calculations

1. When two or more quantities are added or subtracted, the result is as precise as the *least precise* of the quantities (Example 1.3). If the quantities are written in scientific notation with different powers of ten, first rewrite them with the same power of ten. After adding or subtracting, round the result, keeping only as many decimal places as are significant in *all* of the quantities that were added or subtracted.
2. When quantities are multiplied or divided, the result has the same number of significant figures as the quantity with the *smallest number of significant figures* (Example 1.4).
3. In a series of calculations, rounding to the correct number of significant figures should be done only at the end, *not at each step.* Rounding at each step would increase the chance that roundoff error could snowball and have an adverse effect on the accuracy of the final answer. It's a good idea to keep *at least two* extra significant figures in calculations, then round at the end.

Example 1.3

Significant Figures in Addition

Calculate the sum 44.56005 s + 0.0698 s + 1103.2 s.

Strategy The sum cannot be more precise than the least precise of the three quantities. The quantity 44.56005 s is known to the nearest 0.00001 s, 0.0698 s is known to the nearest 0.0001 s, and 1103.2 s is known to the nearest 0.1 s. Therefore the least precise is 1103.2 s. The sum has the same precision; it is known to the nearest tenth of a second.

Solution According to the calculator,

$$44.56005 + 0.0698 + 1103.2 = 1147.82985$$

continued on next page

Example 1.3 continued

We do *not* want to write all of those digits in the answer. That would imply greater precision than we actually have. Rounding to the nearest tenth of a second, the sum is written

$$= 1147.8 \text{ s}$$

and there are five significant figures in the result.

Discussion Note that the least precise measurement is not necessarily the one with the fewest number of significant figures. The least precise is the one whose rightmost significant figure represents the largest unit: the "2" in 1103.2 s represents 2 tenths of a second. In addition or subtraction, we are concerned with the precision rather than the number of significant figures. The three quantities to be added have seven, three, and five significant figures, respectively, while the sum has five significant figures.

Practice Problem 1.3 Significant Figures in Subtraction

Calculate the difference 568.42 m − 3.924 m and write the result in scientific notation. How many significant figures are in the result?

Example 1.4

Significant Figures in Multiplication

Find the product of 45.26 m/s and 2.41 s. How many significant figures does the product have?

Strategy The product should have the same number of significant figures as the factor with the least number of significant figures.

Solution A calculator gives

$$45.26 \times 2.41 = 109.0766$$

Since the answer should have only three significant figures, we round the answer to

$$45.26 \text{ m/s} \times 2.41 \text{ s} = 109 \text{ m}$$

Discussion Writing the answer as 109.0766 m would give the false impression that we know the answer to a precision of about 0.0001 m, whereas we actually have a precision of only about 1 m.

Note that although both factors were known to two decimal places, our solution is properly given with no decimal places. It is the number of significant figures that matters in multiplication or division. In scientific notation, we write 1.09×10^2 m.

Practice Problem 1.4 Significant Figures in Division

Write the solution to 28.84 m divided by 6.2 s with the correct number of significant figures.

When an integer, or a fraction of integers, is used in an equation, the precision of the result is not affected by the integer or the fraction; the number of significant figures is limited only by the measured values in the problem. The fraction $\frac{1}{2}$ in an equation is *exact*; it does not reduce the number of significant figures to one. In an equation such as $C = 2\pi r$ for the circumference of a circle of radius r, the factors 2 and π are exact. We use as many digits for π as we need to maintain the precision of the other quantities.

Order-of-Magnitude Estimates Sometimes a problem may be too complicated to solve precisely, or information may be missing that would be necessary for a precise calculation. In such a case, an **order-of-magnitude** solution is the best we can do. By *order of magnitude*, we mean "roughly what power of ten?" An order of magnitude calculation is done to at most one significant figure. Even when a more precise solution is feasible, it is often a good idea to start with a quick, "back-of-the-envelope estimate." Why? Because we can often make a good guess about the correct order of magnitude of the answer to a problem, even before we start solving the problem. If the answer comes out with a different order of magnitude, we go back and search out an error. Suppose a problem concerns a vase that is knocked off a fourth-story window ledge. We can guess by experience the order of magnitude of the time it takes the vase to hit the ground. It might be 1 s, or 2 s, but we are certain that it is *not* 1000 s or 0.00001 s.

Back-of-the-envelope estimate: a calculation so short that it could easily fit on the back of an envelope

CHECK POINT 1.4

What are some of the reasons for making order-of-magnitude estimates?

1.5 UNITS

kg·m/s^2 can also be written kg·m·s^{-2}

A **metric system** of units has been used for many years in scientific work and in European countries. The metric system is based on powers of ten (Fig. 1.1). In 1960, the General Conference of Weights and Measures, an international authority on units, proposed a revised metric system called the *Système International d'Unités* in French (abbreviated **SI**), which uses the meter (m) for length, the kilogram (kg) for mass, the second (s) for time, and four more base units (Table 1.1). **Derived units** are constructed from combinations of the base units. For example, the SI unit of force is kg·m/s^2; the combination of kg·m/s^2 is given a special name, the newton (N), in honor of Isaac Newton. The newton is a derived unit because it is composed of a combination of base units. When units are named after famous scientists, the name of the unit is written with a lowercase letter, even though it is based on a proper name; the *abbreviation* for the unit is written with an uppercase letter. The inside front cover of the book has a complete listing of the derived SI units used in this book.

As an alternative to explicitly writing powers of ten, SI uses prefixes for units to indicate power of ten factors. Table 1.2 shows some of the powers of ten and the SI prefixes used for them. These are also listed on the inside front cover of the book. Note that when an SI unit with a prefix is raised to a power, the prefix is *also* raised to that power. For example, 8 cm^3 = 2 cm × 2 cm × 2 cm.

SI units are preferred in physics and are emphasized in this book. Since other units are sometimes used, we must know how to convert units. Various scientific fields, even in physics, do use units other than SI units, whether for historical or practical reasons.

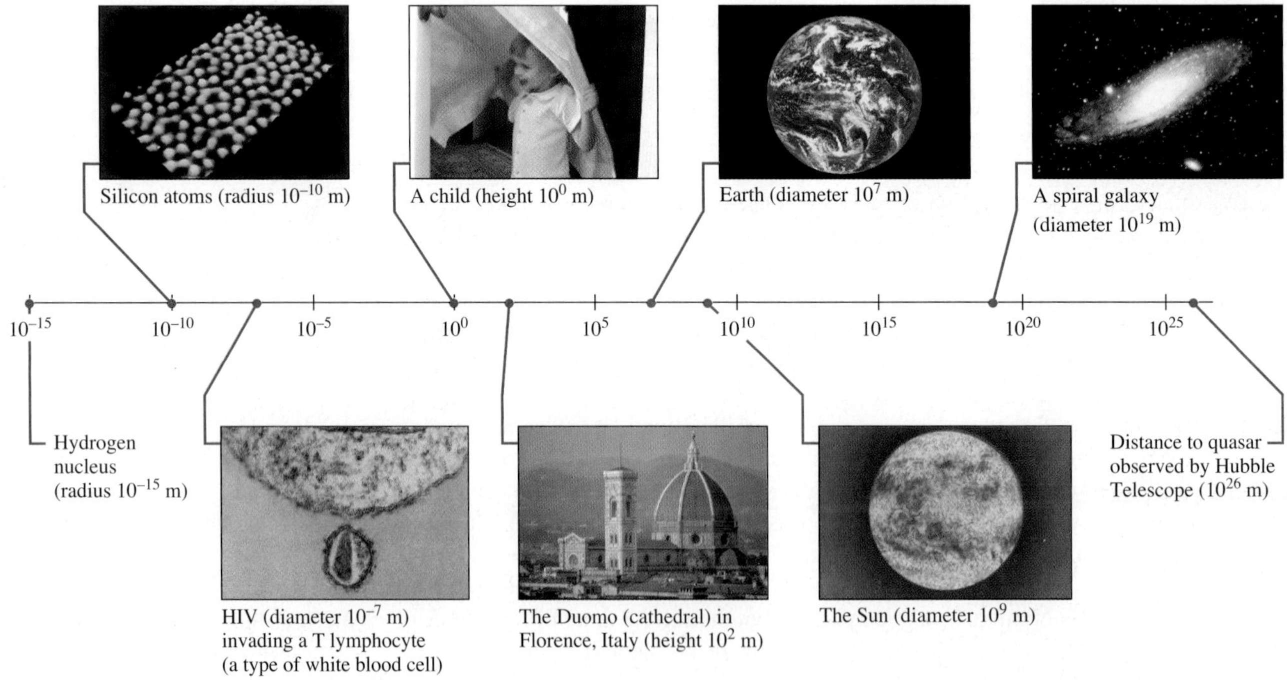

Figure 1.1 Scientific notation uses powers of ten to express quantities that have a wide range of values.

Table 1.1 SI Base Units

Quantity	Unit Name	Symbol	Definition
Length	meter	m	The distance traveled by light in vacuum during a time interval of 1/299 792 458 s.
Mass	kilogram	kg	The mass of the international prototype of the kilogram.
Time	second	s	The duration of 9 192 631 770 periods of the radiation corresponding to the transition between the two hyperfine levels of the ground state of the cesium-133 atom.
Electric current	ampere	A	The constant current in two long, thin, straight, parallel conductors placed 1 m apart in vacuum that would produce a force on the conductors of 2×10^{-7} N per meter of length.
Temperature	kelvin	K	The fraction 1/273.16 of the thermodynamic temperature of the triple point of water.
Amount of substance	mole	mol	The amount of substance that contains as many elementary entities as there are atoms in 0.012 kg of carbon-12.
Luminous intensity	candela*	cd	The luminous intensity, in a given direction, of a source that emits radiation of frequency 540×10^{12} Hz and that has a radiant intensity in that direction of 1/683 watts per steradian.

*Not used in this book

For example, in atomic and nuclear physics, the SI unit of energy (the joule, J) is rarely used; instead the energy unit used is usually the electron-volt (eV). Biologists and chemists use units that are not familiar to physicists. One reason that SI is preferred is that it provides a common denominator—all scientists are familiar with the SI units.

In most of the world, SI units are used in everyday life and in industry. In the United States, the U.S. customary units—sometimes called English units—are still used. The base units for this system are the foot, the second, and the pound. The pound is legally defined in the United States as a unit of mass, but it is also commonly used as a unit of force (in which case it is sometimes called *pound-force*). Since mass and force are entirely different concepts in physics, this inconsistency is one good reason to use SI units.

What happened to the Mars Climate Orbiter?

In the autumn of 1999, to the chagrin of NASA, a $125 million spacecraft was destroyed as it was being maneuvered into orbit around Mars. The company building the booster rocket provided information about the rocket's thrust in U.S. customary units, but the NASA scientists who were controlling the rocket thought the figures provided were in metric units. Arthur Stephenson, chairman of the Mars Climate Orbiter Mission Failure Investigation Board, stated that, "The 'root cause' of the loss of the spacecraft was the failed translation of English units into metric units in a segment of ground-based, navigation-related mission software." After a journey of 122 million miles, the Climate Orbiter dipped about 15 miles too deep into the Martian atmosphere, causing the propulsion system to overheat. The discrepancy in units unfortunately caused a dramatic failure of the mission.

Table 1.2 SI Prefixes

Prefix (abbreviation)	Power of Ten
peta- (P)	10^{15}
tera- (T)	10^{12}
giga- (G)	10^{9}
mega- (M)	10^{6}
kilo- (k)	10^{3}
deci- (d)	10^{-1}
centi- (c)	10^{-2}
milli- (m)	10^{-3}
micro- (μ)	10^{-6}
nano- (n)	10^{-9}
pico- (p)	10^{-12}
femto- (f)	10^{-15}

Converting Units If the statement of a problem includes a mixture of different units, the units must be converted to a single, consistent set before the problem is solved. Quantities to be added or subtracted *must be expressed in the same units.* Usually the best way is to convert everything to SI units. Common conversion factors are listed on the inside front cover of this book.

Examples 1.5 and 1.6 illustrate the technique for converting units. The quantity to be converted is multiplied by one or more conversion factors written as a fraction equal to 1. The units are multiplied or divided as algebraic quantities.

Some conversions are exact by definition. One meter is defined to be *exactly* equal to 100 cm; all SI prefixes are exactly a power of ten. The use of an exact conversion factor such as 1 m = 100 cm, or 1 foot = 12 inches, does not affect the precision of the result; the number of significant figures is limited only by the other quantities in the problem.

Example 1.5

Buying Clothes in a Foreign Country

Michel, an exchange student from France, is studying in the United States. He wishes to buy a new pair of jeans, but the sizes are all in *inches*. He does remember that 1 m = 3.28 ft and that 1 ft = 12 in. If his waist size is 82 cm, what is his waist size in inches?

Strategy Each conversion factor can be written as a fraction. If 1 m = 3.28 ft, then

$$\frac{3.28 \text{ ft}}{1 \text{ m}} = 1$$

We can multiply any quantity by 1 without changing its value. We arrange each conversion factor in a fraction and multiply one at a time to get from centimeters to inches.

Solution We first convert cm to meters.

$$82 \cancel{\text{cm}} \times \frac{1 \text{ m}}{100 \cancel{\text{cm}}}$$

Now, we convert meters to feet.

$$82 \cancel{\text{cm}} \times \frac{1 \cancel{\text{m}}}{100 \cancel{\text{cm}}} \times \frac{3.28 \text{ ft}}{1 \cancel{\text{m}}}$$

Finally, we convert feet to inches.

$$82 \cancel{\text{cm}} \times \frac{1 \cancel{\text{m}}}{100 \cancel{\text{cm}}} \times \frac{3.28 \cancel{\text{ft}}}{1 \cancel{\text{m}}} \times \frac{12 \text{ in}}{1 \cancel{\text{ft}}} = 32 \text{ in}$$

In each case, the fraction is written so that the unit we are converting *from* cancels out.

As a check:

$$\cancel{\text{cm}} \times \frac{\cancel{\text{m}}}{\cancel{\text{cm}}} \times \frac{\cancel{\text{ft}}}{\cancel{\text{m}}} \times \frac{\text{in}}{\cancel{\text{ft}}} = \text{in}$$

Discussion This problem could have been done in one step using a direct conversion factor from inches to cm (1 in = 2.54 cm). One of the great advantages of SI units is that all the conversion factors are powers of ten (see Table 1.2); there is no need to remember that there are 12 inches in a foot, 4 quarts in a gallon, 16 ounces in a pound, 5280 feet in a mile, and so on.

Practice Problem 1.5 Driving on the Autobahn

A BMW convertible travels on the German autobahn at a speed of 128 km/h. What is the speed of the car (a) in meters per second? (b) in miles per hour?

Example 1.6

Conversion of Volume

A beaker of water contains 255 mL of water. (1 mL = 1 milliliter; 1 L = 1000 cm^3.) What is the volume of the water in (a) cubic centimeters? (b) cubic meters?

Strategy First convert milliliters to liters; then convert liters to cubic centimeters. To convert cubic centimeters to cubic meters, use 100 cm = 1 m. Since there are *three* factors of centimeters to convert, we have to multiply by $\left(\frac{1 \text{ m}}{100 \text{ cm}}\right)$ *three times.*

Solution (a) The prefix milli- means 10^{-3}, so 1 mL = 10^{-3} L. Then

$$255 \cancel{\text{mL}} \times \frac{10^{-3} \cancel{\text{L}}}{1 \cancel{\text{mL}}} \times \frac{1000 \text{ cm}^3}{1 \cancel{\text{L}}} = 255 \text{ cm}^3$$

(b) 1 m = 100 cm. Since we need to convert *cubic* centimeters to *cubic* meters, we must raise the conversion factor to the third power:

$$255 \text{ cm}^3 \times \left(\frac{1 \text{ m}}{100 \text{ cm}}\right)^3 = 255 \text{ cm}^3 \times \frac{(1 \text{ m})^3}{(100 \text{ cm})^3}$$

$$255 \cancel{\text{cm}^3} \times \frac{1 \text{ m}^3}{100^3 \cancel{\text{cm}^3}} = 2.55 \times 10^{-4} \text{ m}^3$$

Discussion Be careful when a unit is raised to a power other than one; the conversion factor must be raised to the same power. Writing out the units to make sure they cancel prevents mistakes. When a quantity is raised to a power, both the number and the unit must be raised to the same power. (100 cm)^3 is equal to 100^3 cm^3 = 10^6 cm^3; it is *not* equal to 100 cm^3, nor is it equal to 10^6 cm.

Practice Problem 1.6 Surface Area of Earth

The radius of Earth is 6.4×10^3 km. Find the surface area of Earth in square meters and in square miles. (Surface area of a sphere = $4\pi r^2$.)

Whenever a calculation is performed, always write out the units with each quantity. Combine the units algebraically to find the units of the result. This small effort has three important benefits:

1. It shows what the units of the result are. A common mistake is to get the correct numerical result of a calculation but to write it with the wrong units, making the answer wrong.
2. It shows where unit conversions must be done. If units that should have canceled do not, we go back and perform the necessary conversion. When a distance is calculated and the result comes out with units of meter-seconds per hour (m·s/h), we should convert hours to seconds.
3. It helps locate mistakes. If a distance is calculated and the units come out as m/s, we know to look for an error.

CHECKPOINT 1.5

If 1 fluid ounce (fl oz) is approximately 30 mL, how many liters are in a half gallon (64 fl oz) of milk?

1.6 DIMENSIONAL ANALYSIS

Dimensions are basic *types* of units, such as time, length, and mass. (Warning: The word *dimension* has several other meanings, such as in "three-dimensional space" or "the dimensions of a soccer field.") Many different units of length exist: meters, inches, miles, nautical miles, fathoms, leagues, astronomical units, angstroms, and cubits, just to name a few. All have dimensions of length; each can be converted into any other.

We can add, subtract, or equate quantities only if they have the same dimensions (although they may not necessarily be given in the same units). It is possible to add 3 meters to 2 inches (after converting units), but it is not possible to add 3 meters to 2 kilograms. To analyze dimensions, treat them as algebraic quantities, just as we did

Example 1.7

Dimensional Analysis for a Distance Equation

Analyze the dimensions of the equation $d = vt$, where d is distance traveled, v is speed, and t is elapsed time.

Strategy Replace each quantity with its dimensions. Distance has dimensions [L]. Speed has dimensions of length per unit time [L/T]. The equation is dimensionally consistent if the dimensions are the same on both sides.

Solution The right side has dimensions

$$\frac{[\mathrm{L}]}{[\mathrm{T}]} \times [\mathrm{T}] = [\mathrm{L}]$$

Since both sides of the equation have dimensions of length, the equation is dimensionally consistent.

Discussion If, by mistake, we wrote $d = v/t$ for the relation between distance traveled and elapsed time, we could quickly catch the mistake by looking at the dimensions. On the right side, v/t would have dimensions $[\mathrm{L/T}^2]$, which is not the same as the dimensions of d on the left side.

A quick dimensional analysis of this sort is a good way to catch algebraic errors. Whenever we are unsure whether an equation is correct, we can check the dimensions.

Practice Problem 1.7 Testing Dimensions of Another Equation

Test the dimensions of the following equation:

$$d = \frac{1}{2}at$$

where d is distance traveled, a is acceleration (which has SI units $\mathrm{m/s^2}$), and t is the elapsed time. If incorrect, can you suggest what might have been omitted?

with units in Section 1.5. Usually [M], [L], and [T] are used to stand for mass, length, and time dimensions, respectively. Equivalently, we can use the SI base units: kg for mass, m for length, and s for time.

Applying Dimensional Analysis Dimensional analysis is good for more than just checking equations. In some cases, we can completely solve a problem—up to a dimensionless factor like $1/(2\pi)$ or $\sqrt{3}$—using dimensional analysis. To do this, first list all the relevant quantities on which the answer might depend. Then determine what combinations of them have the same dimensions as the answer for which we are looking. If only one such combination exists, then we have the answer, except for a possible dimensionless multiplicative constant.

Example 1.8

Violin String Frequency

While it is being played, a violin string produces a tone with frequency f in s^{-1}; the frequency is the number of vibrations *per second* of the string. The string has mass m, length L, and tension T. If the tension is increased 5.0%, how does the frequency change? Tension has SI unit kg·m/s^2.

Strategy We could make a study of violin strings, but let us see what we can find out by dimensional analysis. We want to find out how the frequency f can depend on m, L, and T. We won't know if there is a dimensionless constant involved, but we can work by proportions so any such constant will divide out.

Solution The unit of tension T is kg·m/s^2. The units of f do not contain kg or m; we can get rid of them from T by dividing the tension by the length and the mass:

$$\frac{T}{mL} \text{ has SI unit s}^{-2}$$

That is almost what we want; all we have to do is take the square root:

$$\sqrt{\frac{T}{mL}} \text{ has SI unit s}^{-1}$$

Therefore,

$$f = C\sqrt{\frac{T}{mL}}$$

where C is some dimensionless constant. To answer the question, let the original frequency and tension be f and T and the new frequency and tension be f' and T', where $T' = 1.050T$. Frequency is proportional to the square root of tension, so

$$\frac{f'}{f} = \sqrt{\frac{T'}{T}} = \sqrt{1.050} = 1.025$$

The frequency increases 2.5%.

Discussion We'll learn in Chapter 11 how to calculate the value of C, which is 1/2. That is the *only* thing we cannot get by dimensional analysis. There is *no* other way to combine T, m, and L to come up with a quantity that has the units of frequency.

Practice Problem 1.8 Increase in Kinetic Energy

When a body of mass m is moving with a speed v, it has kinetic energy associated with its motion. Energy is measured in kg·m^2·s^{-2}. If the speed of a moving body is increased by 25% while its mass remains constant, by what percentage does the kinetic energy increase?

CHECKPOINT 1.6

If two quantities have different dimensions, is it possible to (a) multiply; (b) divide; (c) add; (d) subtract them?

1.7 PROBLEM-SOLVING TECHNIQUES

No single method can be used to solve every physics problem. We demonstrate useful problem-solving techniques in the examples in every chapter of this text. Even for a particular problem, there may be more than one correct way to approach the solution. Problem-solving techniques are *skills* that must be *practiced* to be learned.

Think of the problem as a puzzle to be solved. Only in the easiest problems is the solution method immediately apparent. When you do not know the entire path to a solution, see where you can get by using the given information—find whatever you can. Exploration of this sort may lead to a solution by suggesting a path that had not been considered. Be willing to take chances. You may even find the challenge enjoyable!

When having some difficulty, it helps to work with a classmate or two. One way to clarify your thoughts is to put them into words. After you have solved a problem, try to explain it to a friend. If you can explain the problem's solution, you really do understand it. Both of you will benefit. But do not rely too much on help from others; the goal is for each of you to develop your own problem-solving skills.

General Guidelines for Problem Solving

1. Read the problem *carefully* and *all the way through.*
2. Reread the problem one sentence at a time and draw a sketch or diagram to help you visualize what is happening.
3. Write down and organize the given information. Some of the information can be written in labels on the diagram. Be sure that the labels are unambiguous. Identify in the diagram the object, the position, the instant of time, or the time interval to which the quantity applies. Sometimes information might be usefully written in a table beside the diagram. Look at the wording of the problem again for information that is implied or stated indirectly.
4. Identify the goal of the problem. What quantities need to be found?
5. If possible, make an estimate to determine the order of magnitude of the answer. This estimate is useful as a check on the final result to see if it is reasonable.
6. Think about how to get from the given information to the final desired information. Do not rush this step. Which principles of physics can be applied to the problem? Which will help get to the solution? How are the known and unknown quantities related? Are all of the known quantities relevant, or might some of them not affect the answer? Which equations are relevant and may lead to the solution to the problem? This step requires skills developed only with much practice in problem solving.
7. Frequently, the solution involves more than one step. Intermediate quantities might have to be found first and then used to find the final answer. Try to map out a path from the given information to the solution. Whenever possible, a good strategy is to divide a complex problem into several simpler subproblems.
8. Perform algebraic manipulations with algebraic symbols (letters) as far as possible. Substituting the numbers in too early has a way of hiding mistakes.
9. Finally, if the problem requires a numerical answer, substitute the known numerical quantities, *with their units*, into the appropriate equation. Leaving out the units is a common source of error. Writing the units shows when a unit conversion needs to be done—and also may help identify an algebra mistake.
10. Once the solution is found, don't be in a hurry to move on. Check the answer—is it reasonable? Try to think of other ways to solve the same problem. Many problems can be solved in several different ways. Besides providing a check on the answer, finding more than one method of solution deepens our understanding of the principles of physics and develops problem-solving skills that will help solve other problems.

1.8 APPROXIMATION

Physics is about building conceptual and mathematical models and comparing observations of the real world with the model. Simplified models help us to analyze complex situations. In various contexts we assume there is no friction, or no air resistance, no heat loss, or no wind blowing, and so forth. If we tried to take all these things into consideration with every problem, the problems would become vastly more complicated to solve. We never can take account of *every* possible influence. We freely make approximations whenever possible to turn a complex problem into an easier one, as long as the answer will be accurate enough for our purposes.

A valuable skill to develop is the ability to know when an assumption or approximation is reasonable. It might be permissible to ignore air resistance when dropping a stone, but not when dropping a beach ball. Why? We must always be prepared to justify any approximation we make by showing the answer is not changed very much by its use.

As well as making simplifying approximations in models, we also recognize that measurements are approximate. Every measured quantity has some uncertainty; it is impossible for a measurement to be exact to an arbitrarily large number of significant figures. Every measuring device has limits on the precision and accuracy of its measurements.

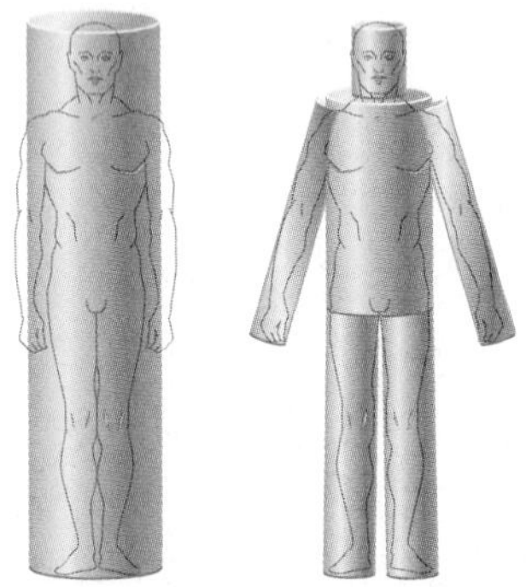

(a) (b)

Figure 1.2 Approximation of human body by one or more cylinders.

Approximating the Surface Area of the Human Body Sometimes it is difficult or impossible to measure precisely a quantity that is needed for a problem. Then we have to make a reasonable estimate. Suppose we need to know the surface area of a human being to determine the heat loss by radiation in a cold room. We can estimate the height of an average person. We can also estimate the average distance around the waist or hips. Approximating the shape of a human body as a cylinder, we can estimate the surface area by calculating the surface area of a cylinder with the same height and circumference (Fig. 1.2a).

If we need a better estimate, we use a slightly more refined model. For instance, we might approximate the arms, legs, trunk, and head and neck as cylinders of various sizes (Fig. 1.2b). How different is the sum of these areas from the original estimate? That gives an idea of how close the first estimate is.

Example 1.9

Number of Cells in the Human Body

Average-sized cells in the human body are about 10 μm in length (Fig. 1.3). How many cells are in the human body? Make an order-of-magnitude estimate.

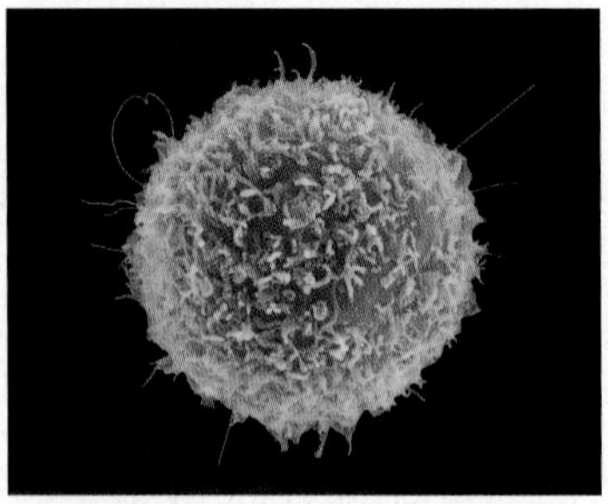

Figure 1.3 Scanning electron micrograph of a precursor T lymphocyte (a type of white blood cell in the human body). The cell is approximately 12 μm in diameter.

Strategy We divide this problem into three subproblems: estimating the volume of a human, estimating the volume of the average cell, and finally estimating the number of cells.

To find the volume of a human body, we approximate the body as a cylinder, as previously discussed. Next we assume the cells are cubical to find the volume of a cell. Third, the ratio of the two volumes (volume of the body to volume of the cell) shows how many cells are in the body.

Solution Model the body as a cylinder. A typical height is about 2 m. A typical *maximum* circumference (think hip size) is about 1 m. The corresponding radius is $1/(2\pi)$ m, or about 1/6 m. The *average* radius is somewhat smaller; say 0.1 m. The volume of a cylinder is the height times the cross-sectional area:

$$V = Ah = \pi r^2 h \approx 3 \times (0.1\text{ m})^2 \times (2\text{ m}) = 0.06\text{ m}^3$$

continued on next page

Example 1.9 continued

The volume of a cube is $V = s^3$. Then the volume of an average cell is about

$$V_{\text{cell}} \approx (1 \times 10^{-5}\ \text{m})^3 = 1 \times 10^{-15}\ \text{m}^3$$

The number of cells is the ratio of the two volumes:

$$N = \frac{\text{volume of body}}{\text{average volume of cell}} \approx \frac{6 \times 10^{-2}\ \text{m}^3}{1 \times 10^{-15}\ \text{m}^3} \approx 6 \times 10^{13}$$

Discussion Based on this rough estimate, we cannot rule out the possibility that a better estimate might be 3×10^{13}. On the other hand, we *can* rule out the possibility that the number of cells is, say, 100 million ($= 10^8$).

Practice Problem 1.9 Drinking Water Consumed in the United States

How many liters of water are swallowed by the people living in the United States in one year? This is a type of problem made famous by the physicist Enrico Fermi (1901–1954), who was a master at this sort of back-of-the-envelope calculation. Such problems are often called *Fermi problems* in his honor. (1 liter $= 10^{-3}\ \text{m}^3 \approx 1$ quart.)

1.9 GRAPHS

Graphs are used to help us see a pattern in the relationship between two quantities. It is much easier to see a pattern on a graph than to see it in a table of numerical values. When we do experiments in physics, we change one quantity (the **independent variable**) and see what happens to another (the **dependent variable**). We want to see how one variable *depends on* another. The value of the independent variable is usually plotted along the horizontal axis of the graph. In a plot of p versus q, which means p is plotted on the vertical axis and q on the horizontal axis, normally p is the dependent variable and q is the independent variable.

Some general guidelines for recording data and making graphs are given next.

Recording Data and Making Data Tables

1. Label columns with the names of the data being measured and be sure to include the units for the measurements. Do not erase any data, but just draw a line through data that you think are erroneous. Sometimes you may decide later that the data were correct after all.
2. Try to make a realistic estimate of the precision of the data being taken when recording numbers. For example, if the timer says 2.3673 s, but you know your reaction time can vary by as much as 0.1 s, the time should be recorded as 2.4 s. When doing calculations using measured values, remember to round the final answer to the correct number of significant figures.
3. Do not wait until you have collected all of your data to start a graph. It is much better to graph each data point as it is measured. By doing so, you can often identify equipment malfunction or measurement errors that make your data unreliable. You can also spot where something interesting happens and take data points closer together there. Graphing as you go means that you need to find out the range of values for both the independent and dependent variables.

Graphing Data

1. Make *large, neat* graphs. A tiny graph is not very illuminating. Use at least half a page. A graph made carelessly obscures the pattern between the two variables.
2. Label axes with the name of the quantities graphed and their units. Write a meaningful title.

The equation of a straight line on a graph of y versus x can be written $y = mx + b$, where m is the slope and b is the y-intercept (the value of y corresponding to $x = 0$).

The symbol Δ, the Greek uppercase letter delta, stands for the difference between two measurements. The notation Δy is read aloud as "delta y" and represents a change in the value of y.

3. When a linear relation is expected, use a ruler or straightedge to draw the best-fit straight line. Do not *assume* that the line must go through the origin—make a measurement to find out, if possible. Some of the data points will probably fall above the line and some will fall below the line.
4. Determine the slope of a best-fit line by measuring the ratio $\Delta y/\Delta x$ using as large a range of the graph as possible. Do not choose two data points to calculate the slope; instead, read values from two points on the best-fit line. Show the calculations. Do not forget to write the units; slopes of graphs in physics have units, since the quantities graphed have units.
5. When a nonlinear relationship is expected between the two variables, the best way to test that relationship is to manipulate the data algebraically so that a linear graph is expected. The human eye is a good judge of whether a straight line fits a set of data points. It is not so good at deciding whether a curve is parabolic, cubic, or exponential. To test the relationship $x = \frac{1}{2}at^2$, where x and t are the quantities measured, graph x versus t^2 instead of x versus t.
6. If one data point does not lie near the line or smooth curve connecting the other data points, that data point should be investigated to see whether an error was made in the measurement or whether some interesting event is occurring at that point. If something unusual is happening there, obtain additional data points in the vicinity.
7. When the slope of a graph is used to calculate some quantity, pay attention to the equation of the line and the units along the axes. The quantity to be found may be the inverse of the slope or twice the slope or one half the slope. For example, if you wish to find the value of a in the relationship $x = \frac{1}{2}at^2$, and you make a graph of x versus t^2, then the slope of the line is $\frac{1}{2}a$. The value of a you seek is twice the slope.

Example 1.10

Length of a Spring

In an introductory physics laboratory experiment, students are investigating how the length of a spring varies with the weight hanging from it. Various weights (accurately calibrated to 0.01 N) ranging up to 6.00 N can be hung from the spring; then the length of the spring is measured with a meterstick (Fig. 1.4). The goal is to see if the weight F and length L are related by

$$F = kx$$

where $x = (L - L_0)$, L_0 is the length of the spring when no weight is hanging from it, and k is called the *spring constant* of the spring. Graph the data in the table and calculate k for this spring.

F (N):	0	0.50	1.00	2.50	3.00	3.50	4.00	5.00	6.00
L (cm):	9.4	10.2	12.5	17.9	19.7	22.5	23.0	28.8	29.5

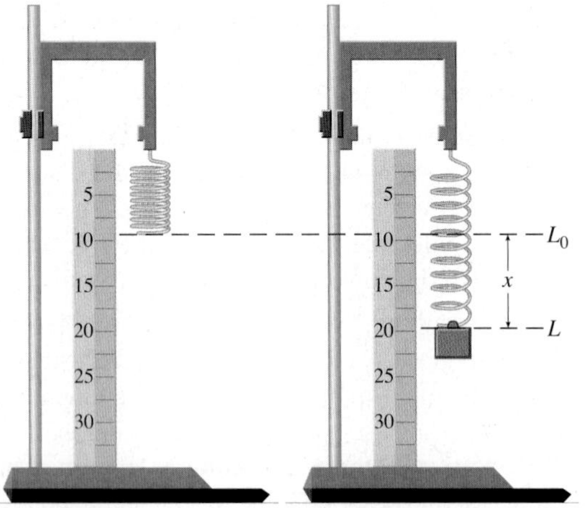

Figure 1.4

A weight causes an extension in the length of a spring.

Strategy Weight is the independent variable, so it is plotted on the horizontal axis. After plotting the data points, we draw the best-fit straight line. Then we calculate the slope of the line, using two points on the line that are widely separated and that cross gridlines of the graph (so the values are easy to read). The slope of the graph is not k; we must solve the equation for L, since length is plotted on the vertical axis.

continued on next page

Example 1.10 continued

Solution Figure 1.5 shows a graph with data points and a best-fit straight line. There is some scatter in the data, but a linear relationship is plausible.

Two points where the line crosses gridlines of the graph are (0.80 N, 12.0 cm) and (4.40 N, 25.0 cm). From these, we calculate the slope:

$$\text{slope} = \frac{\Delta L}{\Delta F} = \frac{25.0\text{ cm} - 12.0\text{ cm}}{4.40\text{ N} - 0.80\text{ N}} = 3.61\ \frac{\text{cm}}{\text{N}}$$

By analyzing the units of the equation $F = k(L - L_0)$, it is clear that the slope cannot be the spring constant; k has the same units as weight divided by length (N/cm). Is the slope equal to $1/k$? The units would be correct for that case. To be sure, we solve the equation of the line for L:

$$L = \frac{F}{k} + L_0$$

We recognize the equation of a line with a slope of $1/k$. Therefore,

$$k = \frac{1}{3.61\text{ cm/N}} = 0.277\text{ N/cm}$$

Discussion As discussed in the graphing guidelines, the slope of the straight-line graph is calculated from two widely spaced values *along the best-fit line.* We do not subtract values of actual data points. We are looking for an average value from the data; using two data points to find the slope would defeat the purpose of plotting a graph or of taking more than two data measurements. The values read from the graph, including the units, are indicated in Fig. 1.5. The units for the slope are cm/N, since we plotted centimeters versus newtons. For this particular problem the *inverse* of the slope is the quantity we seek, the spring constant in N/cm.

Practice Problem 1.10 Another Weight on Spring

What is the length of the spring of Example 1.10 when a weight of 8.00 N is suspended? Assume that the relationship found in Example 1.10 still holds for this weight.

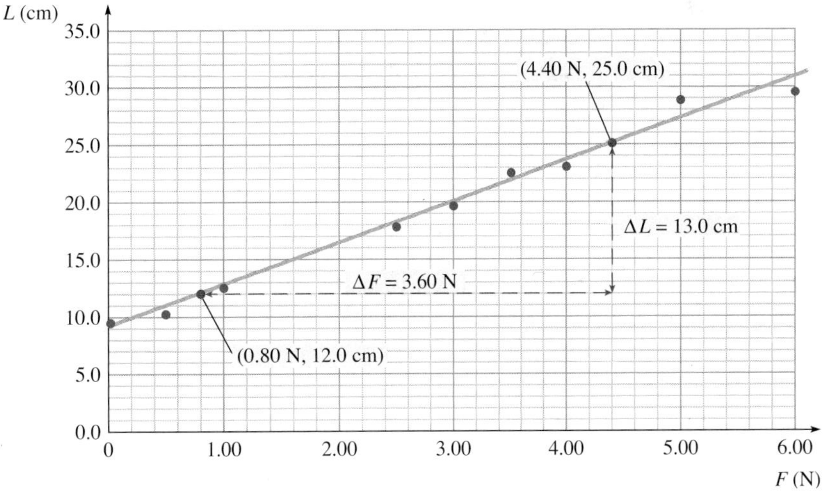

Figure 1.5
Spring length versus weight hanging.

Master the Concepts

- Terms used in physics must be precisely defined. A term may have a different meaning in physics from the meaning of the same word in other contexts.
- A working knowledge of algebra, geometry, and trigonometry is essential in the study of physics.
- The *factor* by which a quantity is increased or decreased is the ratio of the new value to the original value.
- When we say that A is *proportional* to B (written $A \propto B$), we mean that if B increases by some factor, then A must increase by the same factor.
- In *scientific notation*, a number is written as the product of a number between 1 and 10 and a whole-number power of ten.
- *Significant figures* are the basic *grammar* of precision. They enable us to communicate quantitative information and indicate the precision to which that information is known.
- When two or more quantities are added or subtracted, the result is as precise as the *least precise* of the quantities. When quantities are multiplied or divided, the result has

continued on next page

Master the Concepts continued

the same number of significant figures as the quantity with the *smallest number of significant figures.*

- Order-of-magnitude estimates and calculations are made to be sure that the more precise calculations are realistic.
- The units used for scientific work are those from the *Système International (SI)*. SI uses seven *base units*, which include the meter (m), the kilogram (kg), and the second (s) for length, mass, and time, respectively. Using combinations of the base units, we can construct other *derived units.*
- If the statement of a problem includes a mixture of different units, the units should be converted to a single, consistent set before the problem is solved. Usually the best way is to convert everything to SI units.
- Dimensional analysis is used as a quick check on the validity of equations. Whenever quantities are added, subtracted, or equated, they must have the same dimensions (although they may not necessarily be given in the same units).
- Mathematical approximations aid in simplifying complicated problems.
- Problem-solving techniques are *skills* that must be *practiced* to be learned.
- A graph is plotted to give a picture of the data and to show how one variable changes with respect to another. Graphs are used to help us see a pattern in the relationship between two variables.
- Whenever possible, make a careful choice of the variables plotted so that the graph displays a linear relationship.

Conceptual Questions

1. Give a few reasons for studying physics.
2. Why must words be carefully defined for scientific use?
3. Why are simplified models used in scientific study if they do not exactly match real conditions?
4. By what factor does tripling the radius of a circle increase (a) the circumference of the circle? (b) the area of the circle?
5. What are some of the advantages of scientific notation?
6. After which numeral is the decimal point usually placed in scientific notation? What determines the number of numerical digits written in scientific notation?
7. Are all the digits listed as "significant figures" precisely known? Might any of the significant digits be less precisely known than others? Explain.
8. Why is it important to write quantities with the correct number of significant figures?
9. List three of the base units used in SI.
10. What are some of the differences between the SI and the customary U.S. system of units? Why is SI preferred for scientific work?
11. Sort the following units into three groups of dimensions and identify the dimensions: fathoms, grams, years, kilometers, miles, months, kilograms, inches, seconds.
12. What are the first two steps to be followed in solving almost any physics problem?
13. Why do scientists plot graphs of their data instead of just listing values?
14. A student's lab report concludes, "The speed of sound in air is 327." What is wrong with that statement?
15. Once the solution of a problem has been found, what should be done before moving on to solve another problem?

Multiple-Choice Questions

1. One kilometer is approximately
   (a) 2 miles (b) 1/2 mile (c) 1/10 mile (d) 1/4 mile
2. 55 mi/h is approximately
   (a) 90 km/h (b) 30 km/h (c) 10 km/h (d) 2 km/h
3. By what factor does the volume of a cube increase if the length of the edges are doubled?
   (a) 16 (b) 8 (c) 4 (d) 2 (e) $\sqrt{2}$
4. If the length of a box is reduced to one third of its original value and the width and height are doubled, by what factor has the volume changed?
   (a) 2/3 (b) 1 (c) 4/3 (d) 3/2
   (e) depends on relative proportion of length to height and width
5. If the area of a circle is found to be half of its original value after the radius is multiplied by a certain factor, what was the factor used?
   (a) $1/(2\pi)$ (b) 1/2 (c) $\sqrt{2}$ (d) $1/\sqrt{2}$ (e) 1/4
6. In terms of the original diameter d, what new diameter will result in a new spherical volume that is a factor of eight times the original volume?
   (a) $8d$ (b) $2d$ (c) $d/2$ (d) $d \times \sqrt[3]{2}$ (e) $d/8$
7. An equation for potential energy states $U = mgh$. If U is in $kg \cdot m^2 \cdot s^{-2}$, m is in kg, and g is in $m \cdot s^{-2}$, what are the units of h?
   (a) s (b) s^2 (c) m^{-1} (d) m (e) g^{-1}

8. The equation for the speed of sound in a gas states that $v = \sqrt{\gamma k_B T/m}$. Speed v is measured in m/s, γ is a dimensionless constant, T is temperature in kelvins (K), and m is mass in kg. What are the units for the Boltzmann constant, k_B?
(a) $kg{\cdot}m^2{\cdot}s^2{\cdot}K$ (b) $kg{\cdot}m^2{\cdot}s^{-2}{\cdot}K^{-1}$ (c) $kg^{-1}{\cdot}m^{-2}{\cdot}s^2{\cdot}K$
(d) kg·m/s (e) $kg{\cdot}m^2{\cdot}s^{-2}$

9. How many significant figures should be written in the sum 4.56 g + 9.032 g + 580.0078 g + 540.439 g?
(a) 3 (b) 4 (c) 5 (d) 6 (e) 7

10. How many significant figures should be written in the product 0.007 840 6 m × 9.450 20 m?
(a) 3 (b) 4 (c) 5 (d) 6 (e) 7

Problems

- Combination conceptual/quantitative problem
- Biological or medical application
- Challenging problem
- Blue # Detailed solution in the Student Solutions Manual
- (1 2) Problems paired by concept
- Text website interactive or tutorial

1.3 The Use of Mathematics

1. The gardener is told that he must increase the height of his fences 37% if he wants to keep the deer from jumping in to eat the foliage and blossoms. If the current fence is 1.8 m high, how high will the new fence be?

2. What is the ratio of the number of seconds in a day to the number of hours in a day?

3. A spherical balloon expands when it is taken from the cold outdoors to the inside of a warm house. If its surface area increases 16.0%, by what percentage does the radius of the balloon change?

4. A spherical balloon is partially blown up and its surface area is measured. More air is then added, increasing the volume of the balloon. If the surface area of the balloon expands by a factor of 2.0 during this procedure, by what factor does the radius of the balloon change? (tutorial: car on curve)

5. For any cube with edges of length s, what is the ratio of the surface area to the volume?

6. Samantha is 1.50 m tall on her eleventh birthday and 1.65 m tall on her twelfth birthday. By what factor has her height increased? By what percentage?

7. The "scale" of a certain map is 1/10 000. This means the length of, say, a road as represented on the map is 1/10 000 the actual length of the road. What is the ratio of the *area* of a park as represented on the map to the actual area of the park? (tutorial: scaling)

8. On Monday, a stock market index goes up 5.00%. On Tuesday, the index goes down 5.00%. What is the net percentage change in the index for the two days?

9. According to Kepler's third law, the orbital period T of a planet is related to the radius R of its orbit by $T^2 \propto R^3$. Jupiter's orbit is larger than Earth's by a factor of 5.19. What is Jupiter's orbital period? (Earth's orbital period is 1 yr.)

10. If the radius of a circular garden plot is increased by 25%, by what percentage does the area of the garden increase?

11. A poster advertising a student election candidate is too large according to the election rules. The candidate is told she must reduce the length and width of the poster by 20.0%. By what percentage will the area of the poster be reduced?

12. An architect is redesigning a rectangular room on the blueprints of the house. He decides to double the width of the room, increase the length by 50%, and increase the height by 20%. By what factor has the volume of the room increased?

1.4 Scientific Notation and Significant Figures

13. Perform these operations with the appropriate number of significant figures.
(a) 3.783×10^6 kg + 1.25×10^8 kg
(b) (3.783×10^6 m) ÷ (3.0×10^{-2} s)

14. Write these numbers in scientific notation: (a) the U.S. population, 290 000 000; (b) the diameter of a helium nucleus, 0.000 000 000 000 003 8 m.

15. In the following calculations, be sure to use an appropriate number of significant figures.
(a) 3.68×10^7 g − 4.759×10^5 g
(b) $\dfrac{6.497 \times 10^4\ \text{m}^2}{5.1037 \times 10^2\ \text{m}}$

16. Write your answer to the following problems with the appropriate number of significant figures.
(a) 6.85×10^{-5} m + 2.7×10^{-7} m
(b) 702.35 km + 1897.648 km
(c) 5.0 m × 4.3 m
(d) $(0.04/\pi)$ cm
(e) $(0.040/\pi)$ m

17. Solve the following problem and express the answer in scientific notation with the appropriate number of significant figures: (3.2 m) × (4.0×10^{-3} m) × (1.3×10^{-8} m).

18. How many significant figures are in each of these measurements?
(a) 7.68 g (b) 0.420 kg
(c) 0.073 m (d) 7.68×10^5 g
(e) 4.20×10^3 kg (f) 7.3×10^{-2} m
(g) 2.300×10^4 s

19. Solve the following problem and express the answer in meters per second (m/s) with the appropriate number of significant figures. (3.21 m)/(7.00 ms) = ? [*Hint:* Note that ms stands for milliseconds.]

20. Solve the following problem and express the answer in meters with the appropriate number of significant figures and in scientific notation:

$$3.08 \times 10^{-1}\ \text{km} + 2.00 \times 10^{3}\ \text{cm}$$

1.5 Units

21. A cell membrane is 7.0 nm thick. How thick is it in inches?

22. The label on a small soda bottle lists the volume of the drink as 355 mL. (a) How many fluid ounces are in the bottle? A competitor's drink is labeled 16.0 fl oz. (b) How many milliliters are in that drink?

23. The length of the river span of the Brooklyn Bridge is 1595.5 ft. The total length of the bridge is 6016 ft. Find the length and the order of magnitude in meters of (a) the river span and (b) the total bridge length?

24. Convert 1.00 km/h to meters per second (m/s).

25. A sprinter can run at a top speed of 0.32 miles per minute. Express her speed in (a) m/s and (b) mi/h.

26. The first modern Olympics in 1896 had a marathon distance of 40 km. In 1908, for the Olympic marathon in London, the length was changed to 42.195 km to provide the British royal family with a better view of the race. This distance was adopted as the official marathon length in 1921 by the International Amateur Athletic Federation. What is the official length of the marathon in miles?

27. At the end of 2006 an expert economist from the Global Economic Institute in Kiel, Germany, predicted a drop in the value of the dollar against the euro of 10% over the next 5 years. If the exchange rate was $1.27 to 1 euro on November 5, 2006, and was $1.45 to 1 euro on November 5, 2007, what was the actual drop in the value of the dollar over the first year?

28. The intensity of the Sun's radiation that reaches Earth's atmosphere is 1.4 kW/m^2 (kW = kilowatt; W = watt). Convert this to W/cm^2.

29. Density is the ratio of mass to volume. Mercury has a density of 1.36×10^4 kg/m^3. What is the density of mercury in units of g/cm^3?

30. A molecule in air is moving at a speed of 459 m/s. How many meters would the molecule move during 7.00 ms (milliseconds) if it didn't collide with any other molecules?

31. Express this product in units of km^3 with the appropriate number of significant figures: (3.2 km) × (4.0 m) × (13×10^{-3} mm).

32. (a) How many square centimeters are in 1 square foot? (1 in. = 2.54 cm.) (b) How many square centimeters are in 1 square meter? (c) Using your answers to parts (a) and (b), but without using your calculator, roughly how many square feet are in one square meter?

33. A snail crawls at a pace of 5.0 cm/min. Express the snail's speed in (a) ft/s and (b) mi/h.

34. An average-sized capillary in the human body has a cross-sectional area of about 150 μm^2. What is this area in square millimeters (mm^2)?

1.6 Dimensional Analysis

35. An equation for potential energy states $U = mgh$. If U is in joules, with m in kg, h in m, and g in m/s^2, find the combination of SI base units that are equivalent to joules.

36. One equation involving force states that $F_{net} = ma$, where F_{net} is in newtons, m is in kg, and a is in m·s^{-2}. Another equation states that $F = -kx$, where F is in newtons, k is in kg·s^{-2}, and x is in m. (a) Analyze the dimensions of ma and kx to show they are equivalent. (b) What are the dimensions of the force unit newton?

37. An equation for the period T of a planet (the time to make one orbit about the Sun) is $4\pi^2 r^3/(GM)$, where T is in s, r is in m, G is in m^3/(kg·s^2), and M is in kg. Show that the equation is dimensionally correct.

38. The relationship between kinetic energy K (SI unit kg·m^2·s^{-2}) and momentum p is $K = p^2/(2m)$, where m stands for mass. What is the SI unit of momentum?

39. An expression for buoyant force is $F_B = \rho g V$, where F_B has dimensions [MLT^{-2}], ρ (density) has dimensions [ML^{-3}], and g (gravitational field strength) has dimensions [LT^{-2}]. (a) What must be the dimensions of V? (b) Which could be the correct interpretation of V: velocity or volume?

40. Use dimensional analysis to determine how the linear speed (v in m/s) of a particle traveling in a circle depends on some, or all, of the following properties: r is the radius of the circle; ω is an angular frequency in s^{-1} with which the particle orbits about the circle, and m is the mass of the particle. There is no dimensionless constant involved in the relation.

1.8 Approximation

41. What is the approximate distance from your eyes to a book you are reading?

42. What is the approximate volume of your physics textbook in cubic centimeters (cm^3)?

43. (a) Estimate the average mass of a person's leg. (b) Estimate the length of a full-size school bus.

44. Estimate the number of times a human heart beats during its lifetime.

45. Estimate the number of automobile repair shops in the city you live in by considering its population, how often an automobile needs repairs, and how many cars each shop can service per day. Then look in the yellow pages of your phone directory to see how accurate your estimate is. By what percentage was your estimate off?

46. What is the order of magnitude of the number of seconds in one year?

47. What is the order of magnitude of the height (in meters) of a 40-story building?

1.9 Graphs

48. You have just performed an experiment in which you measured many values of two quantities, A and B. According to theory, $A = cB^3 + A_0$. You want to verify that the values of c and A_0 are correct by making a graph of your data that enables you to determine their values from a slope and a vertical axis intercept. What quantities do you put on the vertical and horizontal axes of the plot?

49. A nurse recorded the values shown in the temperature chart for a patient's temperature. Plot a graph of temperature versus elapsed time and from the graph find (a) an estimate of the temperature at noon and (b) the slope of the graph. (c) Would you expect the graph to follow the same trend over the next 12 hours? Explain.

Time	Temp (°F)
10:00 A.M.	100.00
10:30 A.M.	100.45
11:00 A.M.	100.90
11:30 A.M.	101.35
12:45 P.M.	102.48

50. A graph of x versus t^4, with x on the vertical axis and t^4 on the horizontal axis, is linear. Its slope is 25 m/s^4 and its vertical axis intercept is 3 m. Write an equation for x as a function of t.

51. A patient's temperature was 97.0°F at 8:05 A.M. and 101.0°F at 12:05 P.M. If the temperature change with respect to elapsed time was linear throughout the day, what would the patient's temperature be at 3:35 P.M.?

52. The weight of a baby measured over an 11-mon period is given in the weight chart for this problem. (a) Plot the baby's weight versus age over the 11 mon. (b) What was the average monthly weight gain for this baby over the period from birth to 5 mon? How do you find this value from the graph? (c) What was the average monthly weight gain for the baby over the period from 5 mon to 10 mon? (d) If a baby continued to grow at the same rate as in the first five months of life, what would the child weigh at age 12 yr?

Weight (lb)	Age (mon)
6.6	0 (birth)
7.4	1.0
9.6	2.0
11.2	3.0
12.0	4.0
13.6	5.0
13.8	6.0
14.8	7.0
15.0	8.0
16.6	9.0
17.5	10.0
18.4	11.0

53. A physics student plots results of an experiment as v versus t. The equation that describes the line is given by $at = v - v_0$. (a) What is the slope of this line? (b) What is the vertical axis intercept of this line?

54. A linear plot of speed versus elapsed time has a slope of 6.0 m/s^2 and a vertical intercept of 3.0 m/s. (a) What is the change in speed in the time interval between 4.0 s and 6.0 s? (b) What is the speed when the elapsed time is equal to 5.0 s?

55. In a laboratory you measure the decay rate of a sample of radioactive carbon. You write down the following measurements:

Time (min)	0	15	30	45	60	75	90
Decays/s	405	237	140	90	55	32	19

(a) Plot the decays per second versus time. (b) Plot the natural logarithm of the decays per second versus the time. Why might the presentation of the data in this form be useful?

56. An object is moving in the x-direction. A graph of the distance it has moved as a function of time is shown. (a) What are the slope and vertical axis intercept? (Be sure to include units.) (b) What physical significance do the slope and intercept on the vertical axis have for this graph?

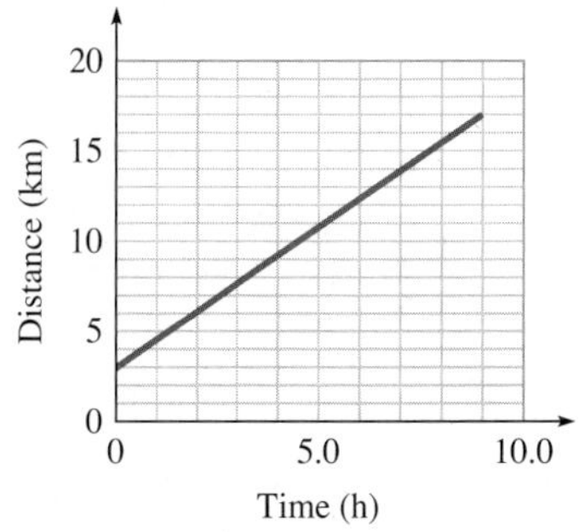

Comprehensive Problems

57. It is useful to know when a small number is negligible. Perform the following computations. (a) 186.300 + 0.0030 (b) 186.300 − 0.0030 (c) 186.300 × 0.0030 (d) 186.300/0.0030 (e) For cases (a) and (b), what percent error will result if you ignore the 0.0030? Explain why you can never ignore the smaller number, 0.0030, for case (c) and case (d)? (f) What rule can you make about ignoring small values?

58. The weight of an object at the surface of a planet is proportional to the planet's mass and inversely proportional to the square of the radius of the planet. Jupiter's radius is 11 times Earth's and its mass is 320 times Earth's. An apple weighs 1.0 N on Earth. How much would it weigh on Jupiter?

59. In cleaning out the artery of a patient, a doctor increases the radius of the opening by a factor of 2.0. By what factor does the cross-sectional area of the artery change?

60. A scanning electron micrograph of xylem vessels in a corn root shows the vessels magnified by a factor of 600. In the micrograph the xylem vessel is 3.0 cm in diameter. (a) What is the diameter of the vessel itself? (b) By what factor has the cross-sectional area of the vessel been increased in the micrograph?

61. The average speed of a nitrogen molecule in air is proportional to the square root of the temperature in kelvins (K). If the average speed is 475 m/s on a warm summer day (temperature = 300.0 K), what is the average speed on a cold winter day (250.0 K)?

62. A furlong is 220 yd; a fortnight is 14 d. How fast is 1 furlong per fortnight (a) in μm/s? (b) in km/day?

63. Given these measurements, identify the number of significant figures and rewrite in scientific notation.
(a) 0.00574 kg (b) 2 m (c) 0.450×10^{-2} m
(d) 45.0 kg (e) 10.09×10^{4} s (f) 0.09500×10^{5} mL

64. A car has a gas tank that holds 12.5 U.S. gal. Using the conversion factors from the inside front cover, (a) determine the size of the gas tank in cubic inches. (b) A cubit is an ancient measurement of length that was defined as the distance from the elbow to the tip of the finger, about 18 in. long. What is the size of the gas tank in cubic cubits?

65. You are given these approximate measurements: (a) the radius of Earth is 6×10^{6} m, (b) the length of a human body is 6 ft, (c) a cell's diameter is 10^{-6} m, (d) the width of the hemoglobin molecule is 3×10^{-9} m, and (e) the distance between two atoms (carbon and nitrogen) is 3×10^{-10} m. Write these measurements in the simplest possible metric prefix forms (in either nm, Mm, μm, or whatever works best).

66. A typical virus is a packet of protein and DNA (or RNA) and can be spherical in shape. The influenza A virus is a spherical virus that has a diameter of 85 nm. If the volume of saliva coughed onto you by your friend with the flu is 0.010 cm^3 and 10^{-9} of that volume consists of viral particles, how many influenza viruses have just landed on you?

67. The smallest "living" thing is probably a type of infectious agent known as a viroid. Viroids are plant pathogens that consist of a circular loop of single-stranded RNA, containing about 300 bases. (Think of the bases as beads strung on a circular RNA string.) The distance from one base to the next (measured along the circumference of the circular loop) is about 0.35 nm. What is the diameter of a viroid in (a) m, (b) μm, and (c) in.?

68. The largest living creature on Earth is the blue whale, which has an average length of 70 ft. The largest blue whale on record (and therefore the largest animal ever found) was 1.10×10^{2} ft long. (a) Convert this length to meters. (b) If a double-decker London bus is 8.0 m long, how many double-decker-bus lengths is the record whale?

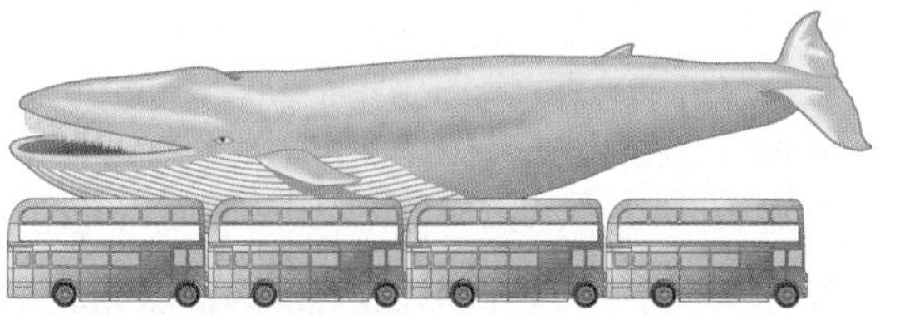

69. The record blue whale in Problem 68 had a mass of 1.9×10^{5} kg. Assuming that its average density was 0.85 g/cm^3, as has been measured for other blue whales, what was the volume of the whale in cubic meters (m^3)? (Average density is the ratio of mass to volume.)

70. A sheet of paper has length 27.95 cm, width 8.5 in., and thickness 0.10 mm. What is the volume of a sheet of paper in m^3? (Volume = length × width × thickness.)

71. An object moving at constant speed v around a circle of radius r has an acceleration a directed toward the center of the circle. The SI unit of acceleration is m/s^2. (a) Use dimensional analysis to find a as a function of v and r. (b) If the speed is increased 10.0%, by what percentage does the radial acceleration increase?

72. The speed of ocean waves depends on their wavelength λ (measured in meters) and the gravitational field strength g (measured in m/s^2) in this way:

$$v = K\lambda^{p}g^{q}$$

where K is a dimensionless constant. Find the values of the exponents p and q.

73. In the United States, we often use miles per hour (mi/h) when discussing speed, but the SI unit of speed is m/s. What is the conversion factor for changing m/s to mi/h? If you want to make a quick approximation of the speed in mi/h given the speed in m/s, what might be the easiest conversion factor to use?

74. How many cups of water are required to fill a bathtub?

75. Without looking up any data, make an order-of-magnitude estimate of the annual consumption of gasoline (in gallons) by passenger cars in the United States. Make reasonable order-of-magnitude estimates for any quantities you need. Think in terms of average quantities. (1 gal ≈ 4 L.)

76. Some thieves, escaping after a bank robbery, drop a sack of money on the sidewalk. (a) Estimate the mass of the sack if it contains \$5000 in half-dollar coins. (b) Estimate the mass if the sack contains \$1 000 000 in \$20 bills.

77. The weight W of an object is given by $W = mg$, where m is the object's mass and g is the gravitational field strength. The SI unit of field strength g, expressed in SI base units, is m/s^2. What is the SI unit for weight, expressed in base units?

78. Kepler's law of planetary motion says that the square of the period of a planet (T^2) is proportional to the cube of the distance of the planet from the Sun (r^3). Mars is about twice as far from the Sun as Venus. How does the period of Mars compare with the period of Venus?

79. One morning you read in the *New York Times* that the net worth of the richest man in the world, Carlos Slim Helu of Mexico, is \$59 000 000 000. Later that day you see him on the street, and he gives you a \$100 bill. What is his net worth now? (Think of significant figures.)

80. Estimate the number of hairs on the average human head. [*Hint:* Consider the number of hairs in an area of 1 in^2 and then consider the area covered by hair on the head.]

81. Suppose you have a pair of Seven League Boots. These are magic boots that enable you to stride along a distance of 7.0 leagues with each step. (a) If you march along at a military march pace of 120 paces/min, what will be your speed in km/h? (b) Assuming you could march on top of the oceans when you step off the continents, how long (in minutes) will it take you to march around the Earth at the equator? (1 league = 3 mi = 4.8 km.)

82. The electrical power P drawn from a generator by a lightbulb of resistance R is $P = V^2/R$, where V is the line voltage. The resistance of bulb B is 42% greater than the resistance of bulb A. What is the ratio P_B/P_A of the power drawn by bulb B to the power drawn by bulb A if the line voltages are the same?

83. Three of the fundamental constants of physics are the speed of light, $c = 3.0 \times 10^8$ m/s, the universal gravitational constant, $G = 6.7 \times 10^{-11}\ m^3 \cdot kg^{-1} \cdot s^{-2}$, and Planck's constant, $h = 6.6 \times 10^{-34}\ kg \cdot m^2 \cdot s^{-1}$.

(a) Find a combination of these three constants that has the dimensions of time. This time is called the *Planck time* and represents the age of the universe before which the laws of physics as presently understood cannot be applied. (b) Using the formula for the Planck time derived in part (a), what is the time in seconds?

84. Use dimensional analysis to determine how the period T of a swinging pendulum (the elapsed time for a complete cycle of motion) depends on some, or all, of these properties: the length L of the pendulum, the mass m of the pendulum bob, and the gravitational field strength g (in m/s^2). Assume that the amplitude of the swing (the maximum angle that the string makes with the vertical) has no effect on the period.

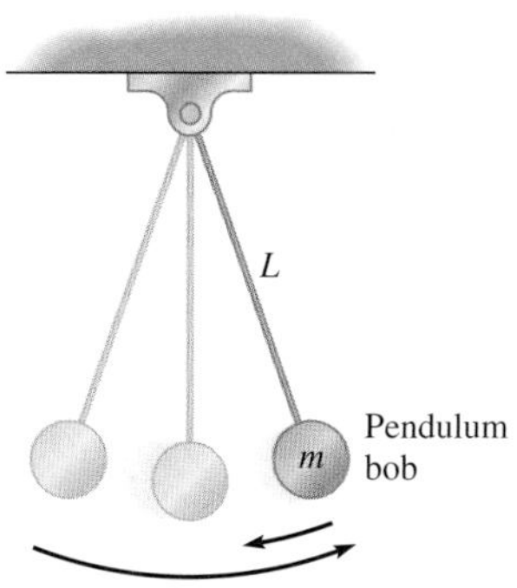

85. The Space Shuttle astronauts use a *massing chair* to measure their mass. The chair is attached to a spring and is free to oscillate back and forth. The frequency of the oscillation is measured and that is used to calculate the total mass m attached to the spring. If the spring constant of the spring k is measured in kg/s^2 and the chair's frequency f is 0.50 s^{-1} for a 62-kg astronaut, what is the chair's frequency for a 75-kg astronaut? The chair itself has a mass of 10.0 kg. [*Hint:* Use dimensional analysis to find out how f depends on m and k.]

86. The average depth of the oceans is about 4 km, and oceans cover about 70% of Earth's surface. Make an order-of-magnitude estimate of the volume of water in the oceans. Do not look up any data in books. (Use your ingenuity to estimate the radius or circumference of Earth.)

87. The population of a culture of yeast cells is studied in the laboratory to see the effects of limited resources (food, space) on population growth. At 2-h intervals, the size of the population (measured as total mass of yeast cells) is recorded (see table on p. 24). (a) Make a graph of the yeast population as a function of elapsed time. Draw a best-fit smooth curve. (b) Notice from the graph of part (a) that after a long time, the population asymptotically approaches a maximum known as the *carrying capacity*. From the graph, estimate the carrying capacity for this population. (c) When the population is much smaller than the carrying capacity, the growth is expected to be exponential: $m(t) = m_0 e^{rt}$, where m is the population at any time t, m_0 is the initial population, r is the *intrinsic growth rate* (i.e., the growth rate in the absence of limits), and e is the base of natural logarithms (see Appendix A.3). To obtain a straight line graph from this exponential relationship, we can plot the natural logarithm of m/m_0:

$$\ln \frac{m}{m_0} = \ln e^{rt} = rt$$

Make a graph of $\ln(m/m_0)$ versus t from $t = 0$ to $t = 6.0$ h, and use it to estimate the intrinsic growth rate r for the yeast population. (The term ln stands for the natural logarithm; see Appendix A.3 if you need help with natural logs.)

Time (h)	Mass (g)
0.0	3.2
2.0	5.9
4.0	10.8
6.0	19.1
8.0	31.2
10.0	46.5
12.0	62.0
14.0	74.9
16.0	83.7
18.0	89.3
20.0	92.5
22.0	94.0
24.0	95.1

Answers to Practice Problems

1.1 81.0 W

1.2 (a) five; 1.0544×10^{-4} kg; (b) four; 5.800×10^{-3} cm; (c) ambiguous, three to six; if three, 6.02×10^5 s

1.3 The least precise value is to the nearest hundredth of a meter, so we round the result to the nearest hundredth of a meter: 564.50 m or, in scientific notation, 5.6450×10^2 m; five significant figures.

1.4 4.7 m/s

1.5 (a) 35.6 m/s; (b) 79.5 mi/h

1.6 5.1×10^{14} m^2; 2.0×10^8 mi^2

1.7 The equation is dimensionally inconsistent; the right side has dimensions [L/T]. To have matching dimensions we must multiply the right side by [T]; the equation must involve time squared: $d = \frac{1}{2}at^2$.

1.8 kinetic energy = (constant) × mv^2; kinetic energy increases by 56%.

1.9 10^{11} L (Make a rough estimate of the population to be about 3×10^8 people, each drinking about 1.5 L/day.)

1.10 38.3 cm

Answers to Checkpoints

1.3 The volume increases by a factor of 27.

1.4 Order-of-magnitude estimates provide a quick method for obtaining limited precision solutions to problems. Even if greater accuracy is required, order-of-magnitude calculations are still useful as they provide a check as to the accuracy of the higher precision calculation.

1.5 1.9 L

1.6 (a) and (b) It is possible to multiply or divide quantities with different dimensions. (c) and (d) To be added or subtracted, quantities *must* have the same dimensions.

CHAPTER

2

Motion Along a Line

Despite its enormous mass (425 to 900 kg), the Cape buffalo is capable of running at a top speed of about 55 km/h (34 mi/h). Since the top speed of the African lion is about the same, how is it ever possible for a lion to catch the buffalo, especially since the lion typically makes its move from a distance of 20 to 30 m from the buffalo? (See p. 34 for the answer.)

Concepts & Skills to Review

- scientific notation and significant figures (Section 1.4)
- converting units (Section 1.5)
- problem-solving techniques (Section 1.7)
- meaning of *velocity* in physics (Section 1.2)

2.1 POSITION AND DISPLACEMENT

CONNECTION:

The topic of Chapters 2 and 3 is **kinematics**: the mathematical description of motion. Beginning in Chapter 4, we will learn the principles of physics that predict and explain *why* objects move the way they do.

Position

To describe motion unambiguously, we need a way to say *where* an object is located. Suppose that at 3:00 P.M. a train stops on an east-west track as a result of an engine problem. The engineer wants to call the railroad office to report the problem. How can he tell them where to find the train? He might say something like "three kilometers east of the old trestle bridge." Notice that he uses a point of reference: the old trestle bridge. Then he states how far the train is from that point and in what direction. If he omits any of the three pieces (the reference point, the distance, and the direction), then his description of the train's whereabouts is ambiguous.

The same thing is done in physics. First, we choose a reference point, called the **origin**. Then, to describe the location of something, we give its distance from the origin and the direction. For motion along a line, we can choose the line of motion to be the x-axis of a coordinate system. The origin is the point $x = 0$. The position of an object can be described by its x-coordinate, which tells us both how far the object is from the origin and on which side. For the train in Fig. 2.1, we choose the origin at the center of the bridge and the $+x$-direction to the east. Then $x = +3$ km means the train is 3 km east of the bridge and $x = -26$ km means the train is 26 km west of the bridge.

Displacement

Once the train's engine is repaired and it goes on its way, we might want to describe its motion. At 3:14 P.M., it leaves its initial position, 3 km east of the origin (see Fig. 2.1). At 3:56 P.M., the train is 26 km west of the origin, which is 29 km to the west of its initial position. **Displacement** is defined as the change of the position—the final position minus the initial position. The displacement is written Δx where the symbol Δ (the uppercase Greek letter delta) means *the change in* the quantity that follows.

Displacement: final position minus initial position

Displacement:

$$\Delta x = x_f - x_i \tag{2-1}$$

Figure 2.1 Initial (x_i) and final (x_f) positions of a train. (Train not to scale.)

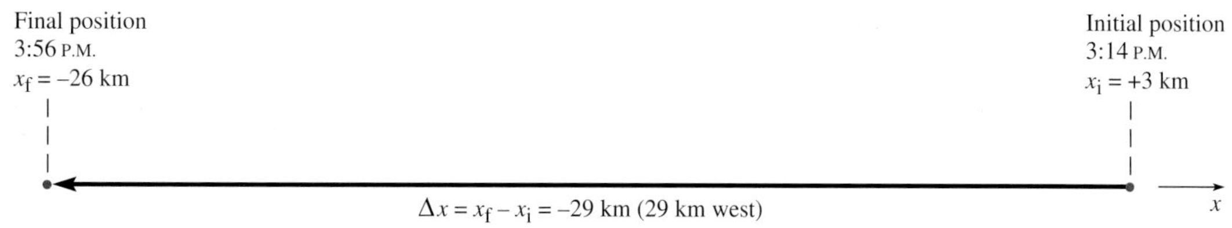

Figure 2.2 With the x-axis pointing east, $\Delta x = x_f - x_i = -26\text{ km} - (+3\text{ km}) = -29$ km. The train's displacement is 29 km west.

We can subtract x-coordinates to find the displacement of the train. If we choose the x-axis to the east, then $x_i = +3$ km and $x_f = -26$ km. The displacement is

$$\Delta x = x_f - x_i = (-26\text{ km}) - (+3\text{ km}) = -29\text{ km}$$

The displacement is 29 km in the $-x$-direction (west) (Fig. 2.2).

Displacement Versus Distance Notice that the magnitude of the displacement is not necessarily equal to the *distance traveled*. Suppose the train first travels 7 km to the east, putting it 10 km east of the origin, and then reverses direction and travels 36 km to the west. The total distance traveled in that case is (7 km + 36 km) = 43 km, but the magnitude of the displacement—which is the distance between the initial and final positions—is 29 km. The displacement depends only on the starting and ending positions, not on the path taken.

Example 2.1

A Mule Hauling Corn to Market

A mule hauls the farmer's wagon along a straight road for 4.3 km directly east to the neighboring farm where a few bushels of corn are loaded onto the wagon. Then the farmer drives the mule back along the same straight road, heading west for 7.2 km to the market. Find the displacement of the mule from the starting point to the market. (The train first travels 7 km to the east, then reverses direction and travels 36 km to the west.)

Strategy The problem gives us two successive displacements along a straight line. Let's choose the $+x$-axis to point east and an arbitrary point along the road to be the origin. Suppose the mule starts at position x_1 (Fig. 2.3). It goes east until it reaches the neighbor's farm at position x_2. The displacement to the neighbor's farm is $x_2 - x_1 = 4.3$ km east.

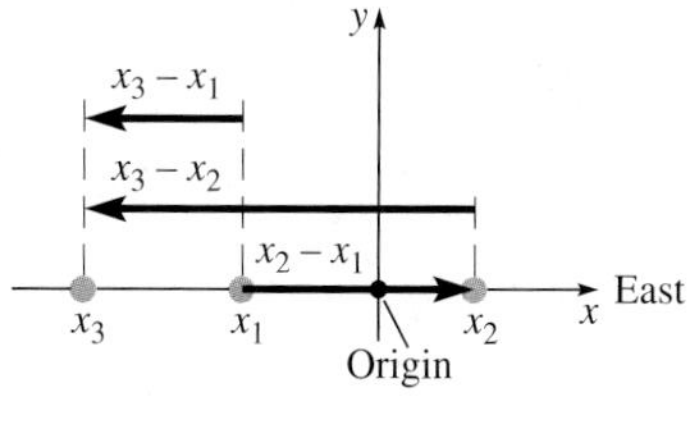

Figure 2.3
The total displacement is the sum of two successive displacements.

Then the mule goes 7.2 km west to reach the market at position x_3. The displacement from the neighbor's farm to the market is $x_3 - x_2 = -7.2$ km (negative because the displacement is in the $-x$-direction). The problem asks for the displacement of the mule from x_1 to x_3.

Solution We can eliminate x_2, the intermediate position, by adding the two displacements:

$$(x_3 - x_2) + (x_2 - x_1) = -7.2\text{ km} + 4.3\text{ km}$$
$$x_3 - x_1 = -2.9\text{ km}$$

The displacement is 2.9 km west.

Discussion When we added the two displacements, the intermediate position x_2 dropped out, as it must since the displacement is independent of the path taken from the initial position to the final position. The result does not depend on the choice of origin.

Practice Problem 2.1 A Nervous Squirrel

A nervous squirrel, trying to cross a road, first moves 3.0 m east, then 4.0 m west, then 1.2 m west, then 6.0 m east. What is the squirrel's total displacement?

Adding Displacements Generalizing the result of Example 2.1, the total displacement for a trip with several parts is the sum of the displacements for each part of the trip. Although x-coordinates depend on the choice of origin, displacements (*changes* in x-coordinates) do *not* depend on the choice of origin.

CHECKPOINT 2.1

In Example 2.1, is the magnitude of the displacement equal to the distance traveled? Explain.

2.2 VELOCITY: RATE OF CHANGE OF POSITION

We introduced *velocity* as a quantity with magnitude and direction in Section 1.2. The magnitude is the speed with which the object moves and the direction is the direction of motion. Now we develop a mathematical definition of velocity that fits that description. Note that displacement indicates by how much and in what direction the position has changed, but implies nothing about *how long* it took to move from one point to the other. Velocity depends on both the displacement and the time interval.

Average Velocity

Reminder: the symbol Δ stands for *the change in.* If the initial value of a quantity Q is Q_i and the final value is Q_f, then $\Delta Q = Q_f - Q_i$. ΔQ is read "delta Q."

When a displacement Δx occurs during a time interval Δt, the **average velocity** during that time interval is

Average velocity:

$$v_{av,x} = \frac{\Delta x}{\Delta t} \tag{2-2}$$

Since Δt is always positive, the direction of the average velocity is the same as the direction of the displacement.

The symbol Δ does not stand alone and cannot be canceled in equations because it *modifies* the quantity that follows it; $\dfrac{\Delta x}{\Delta t}$ means $\dfrac{x_f - x_i}{t_f - t_i}$, which is *not* the same as x/t.

Example 2.2

Average Velocity of a Train

Find the average velocity in kilometers per hour of the train shown in Fig. 2.1 during the time interval between 3:14 P.M., when the train is 3 km east of the origin, and 3:56 P.M., when it is 26 km west of the origin.

Strategy We choose the $+x$-axis to the east, as before. Then the displacement is $\Delta x = -29$ km, which means 29 km to the west. The average velocity is also to the west, so $v_{av,x}$ is negative. We convert Δt to hours to find the average velocity in kilometers per hour.

Solution The time interval is $\Delta t = 56\text{ min} - 14\text{ min} = 42\text{ min}$. Converting to hours,

$$\Delta t = 42\text{ min} \times \frac{1\text{ h}}{60\text{ min}} = 0.70\text{ h}$$

The average velocity is

$$v_{av,x} = \frac{\Delta x}{\Delta t} = \frac{-29\text{ km}}{0.70\text{ h}} = -41\text{ km/h}$$

The negative sign means that the average velocity is directed along the negative x-axis, or to the west.

continued on next page

Example 2.2 continued

Discussion If the train had started at the same instant of time, 3:14 P.M., and had traveled directly west at a constant 41 km/h, it would have ended up in exactly the same place—26 km west of the trestle bridge—at 3:56 P.M.

Had we started measuring time from when we first spotted the motionless train at 3:00 P.M., instead of 3:14 P.M., we would have found the average velocity over a different time interval, changing the average velocity. The average velocity depends on the time interval considered.

The magnitude of the train's average velocity is *not* equal to the total distance traveled divided by the time interval for the complete trip. The latter quantity is called the average *speed*:

$$\text{average speed} = \frac{\text{distance traveled}}{\text{total time}} = \frac{43 \text{ km}}{0.70 \text{ h}} = 61 \text{ km/h}$$

The distinction arises because the average velocity is the constant velocity that would result in the same *displacement* (during the given time interval), while the average speed is the constant speed that would result in the same *distance traveled* (during the same time interval).

Practice Problem 2.2 Average Velocity for a Different Time Interval

What is the average velocity of the same train during the time interval from 3:28 P.M., when it is at $x = 10$ km, to 3:56 P.M., when it is at $x = -26$ km?

Average Speed Versus Average Velocity The *average* velocity does not convey detailed information about the motion during the corresponding time interval Δt. The average velocity would be the same for any other motion that takes the object through the same displacement in the same amount of time. However, the average *speed*, defined as the total *distance* traveled divided by the time interval, depends on the path traveled.

CHECKPOINT 2.2

Can average speed ever be greater than the magnitude of the average velocity? Explain.

Instantaneous Velocity

The speedometer of a car does not indicate the average speed for an entire trip. When a speedometer reads 55 mi/h, it does *not* necessarily mean that the car travels 55 miles in the next hour; the car could change its speed or direction or stop during that hour. The speedometer reading can be used to calculate how far the car travels during a *very short time interval*—short enough that the speed does not change appreciably. For instance, at 55 mi/h (= 25 m/s), we can calculate that in 0.010 s the car moves 25 m/s × 0.010 s = 0.25 m—as long as the speed does not change significantly during that 0.010-s interval.

Similarly, the **instantaneous velocity** is a quantity whose magnitude is the speed and whose direction is the direction of motion. The instantaneous velocity can be used to calculate the *displacement* of the object *during a very short time interval*, as long as *neither the speed nor the direction of motion* change significantly during that time interval. Repeating the word *instantaneous* can get cumbersome. When we refer simply to *the velocity*, we always mean the *instantaneous* velocity.

CONNECTION:

Couldn't we omit "x" subscripts in average ($v_{av,x}$) and instantaneous (v_x) velocity? If we wanted to understand only motion along a line, then we certainly would. However, in Chapter 3 we generalize the definitions of position, displacement, velocity, and acceleration as vector quantities in three dimensions. Using the "x" subscripts now lets us carry forward everything in Chapter 2 *without requiring a change in notation*. Then, when you look back to review Chapter 2, you won't have to remember different definitions for the same symbol. For example, in Chapter 3 we'll learn that v (without the subscript) stands for the *magnitude* of the velocity (the speed), which can never be negative.

Thus, the velocity at some instant of time t is the average velocity during a *very short* time interval:

Instantaneous velocity:

$$v_x = \lim_{\Delta t \to 0} \frac{\Delta x}{\Delta t} \quad (2\text{-}3)$$

(Δx is the displacement during a *very short* time interval Δt)

The notation $\lim_{\Delta t \to 0}$ is read "the limit, as Δt approaches zero, of . . . ." In other words, let the time interval get smaller and smaller, *approaching*—but never reaching—zero. This notation in Eq. (2-3) reminds you that Δt must be a *very short* time interval. How short a time interval is short enough? If you use a shorter time interval and the calculation of v_x always gives the same value (to within the precision of your measurements), then Δt is short enough. In other words, Δt must be short enough that we can treat the velocity as constant during that time interval. When v_x is constant, cutting Δt in half also cuts the displacement in half, giving the same value for $\Delta x/\Delta t$.

Graphical Relationships Between Position and Velocity

For motion along the x-axis, the displacement is Δx. The average velocity can be represented on the graph of $x(t)$ as the slope of a line connecting two points (called a *chord*). In Fig. 2.4a, the displacement $\Delta x = x_3 - x_1$ is the *rise* of the graph (the change along the vertical axis) and the time interval $\Delta t = t_3 - t_1$ is the *run* of the graph (the change along the horizontal axis). The slope of the chord is the rise over the run:

$$\text{slope of chord} = \frac{\text{rise}}{\text{run}} = \frac{\Delta x}{\Delta t} = v_{av,x} \quad (2\text{-}4)$$

The slope of the chord is the average velocity for that time interval.

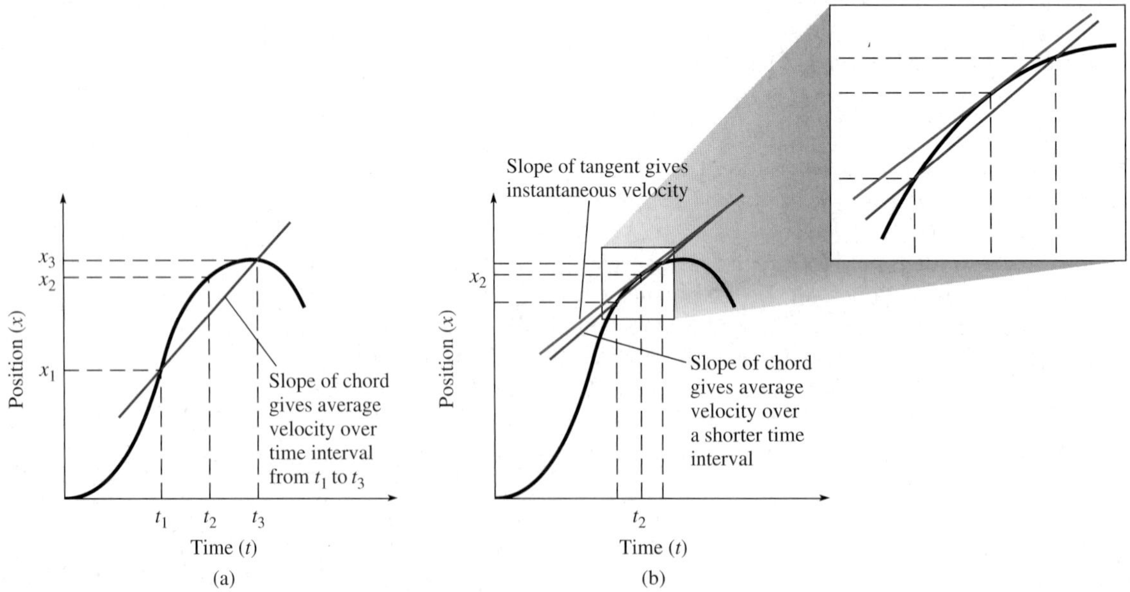

Figure 2.4 A graph of $x(t)$ for an object moving along the x-axis. (a) The average velocity $v_{x,av}$ for the time interval t_1 to t_3 is the slope of the chord connecting those two points on the graph. (b) The average velocity measured over a shorter time interval. As the time interval gets shorter and shorter, the average velocity approaches the *instantaneous* velocity v_x at the instant t_2. The slope of the *tangent line* to the graph is v_x at that instant.

Finding v_x on a Graph of $x(t)$ To find the *instantaneous* velocity at some time $t = t_2$, we draw lines showing the average velocity for shorter and shorter time intervals. As the time interval is reduced (Fig. 2.4b), the average velocity changes. As Δt gets shorter and shorter, the chord approaches a tangent line to the graph at t_2. Thus, v_x is the *slope of the line tangent to the graph of* $x(t)$ at the chosen time.

The slope of the tangent line on a graph of $x(t)$ is v_x.

In Fig. 2.5, the position of the train considered in Example 2.2 is graphed as a function of time, where 3:00 P.M. is chosen as $t = 0$.

The graph of position versus time shows a curving line, but that does not mean the train travels along a curved path. The motion of the train is along a straight line since the track runs in an east-west direction. The graph shows the train's position as a function of time.

A horizontal portion of the graph (as from $t = 0$ to $t = 14$ min and from $t = 23$ min to $t = 28$ min) indicates that the position is not changing during that time interval and, therefore, it is at rest (its velocity is zero). Sloping portions of the graph indicate that the train is moving. The steeper the graph, the larger the speed of the train. The sign of the slope indicates the direction of motion. A positive slope ($t = 14$ min to $t = 23$ min) indicates motion in the $+x$-direction, and a negative slope ($t = 28$ min to $t = 56$ min) indicates motion in the $-x$-direction.

Example 2.3

Velocity of the Train

Use Fig. 2.5 to estimate the velocity of the train in kilometers per hour at $t = 40$ min.

Strategy Figure 2.5 is a graph of $x(t)$. The slope of a line tangent to the graph at $t = 40$ min is v_x at that instant. After sketching a tangent line on the graph, we find its slope from the rise divided by the run.

Solution Figure 2.6 shows a tangent line drawn on the graph. Using the endpoints of the tangent line, the rise is (−25 km) − (15 km) = −40 km. The run is approximately (57 min) − (30 min) = 27 min = 0.45 h. Then

$$v_x \approx -40\text{ km}/(0.45\text{ h}) \approx -89\text{ km/h}$$

The velocity is approximately 89 km/h in the $-x$-direction (west).

Discussion Since the slope of a line is constant, any two points *on the tangent line* would give the same value for the slope. Using widely spaced points tends to give a more accurate estimate for the slope.

Practice Problem 2.3 Maximum Eastward Velocity

Estimate the maximum velocity of the train in kilometers per hour during the time it moves east ($t = 14$ min to $t = 23$ min).

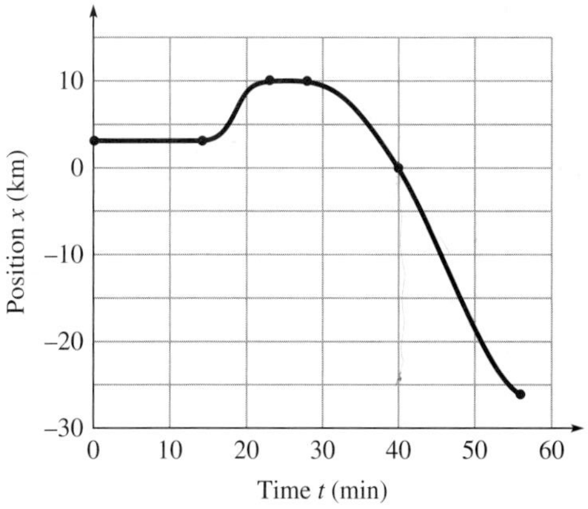

x (km)	t (min)
+3	0
+3	14
+10	23
+10	28
0	40
−26	56

Figure 2.5

Graph of position x versus time t for the train. The positions of the train at various times are marked with a dot. The position of the train would have to be measured at more frequent time intervals to accurately trace out the shape of the graph.

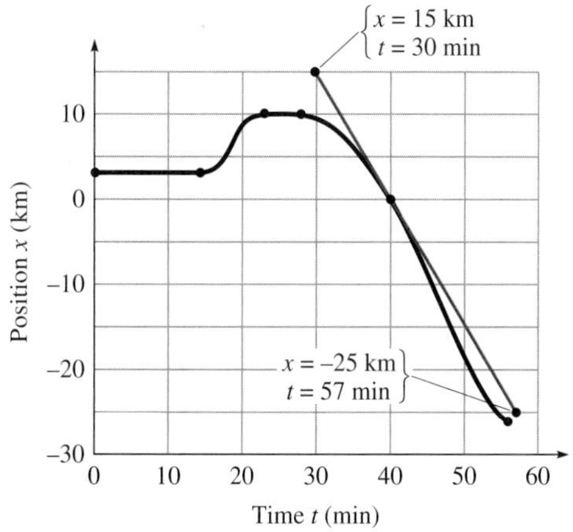

Figure 2.6

On the graph of $x(t)$, the slope of a line tangent to the graph at $t = 40$ min is v_x at $t = 40$ min.

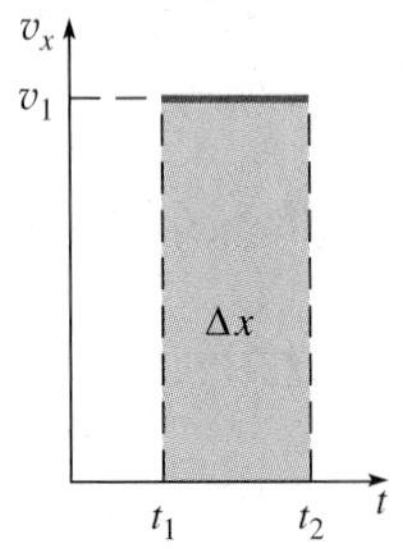

Figure 2.7 Displacement Δx between t_1 and t_2 is represented by the shaded area under the red $v_x(t)$ graph.

Finding Δx with Constant Velocity What about the other way around? Given a graph of $v_x(t)$, how can we determine the displacement (change in position)? If v_x is constant during a time interval, then the average velocity is equal to the instantaneous velocity:

$$v_x = v_{av,x} = \frac{\Delta x}{\Delta t} \quad \text{(for constant } v_x\text{)} \qquad (2\text{-}5)$$

and therefore

$$\Delta x = v_x \, \Delta t \quad \text{(for constant } v_x\text{)} \qquad (2\text{-}6)$$

The graph of Fig. 2.7 shows v_x versus t for an object moving along the x-axis with constant velocity v_1 from time t_1 to t_2. The displacement Δx during the time interval $\Delta t = t_2 - t_1$ is $v_1 \Delta t$. The shaded rectangle has "height" v_1 and "width" Δt. Since the area of a rectangle is the product of the height and width, the displacement Δx is represented by the area of the rectangle between the graph of $v_x(t)$ and the time axis for the time interval considered.

When we speak of the area under a graph, we are not talking about the literal number of square centimeters of paper or computer screen. The figurative area under a graph usually does not have dimensions of an ordinary area [L^2]. In a graph of $v_x(t)$, v_x has dimensions [L/T] and time has dimensions [T]; areas on such a graph have dimensions [L/T] × [T] = [L], which is correct for a displacement. The *units* of Δx are determined by the units used on the axes of the graph. If v_x is in meters per second and t is in seconds, then the displacement is in meters.

Δx is the area under the graph of $v_x(t)$. The area is negative when the graph is beneath the time axis ($v_x < 0$).

Finding Δx with Changing Velocity What if the velocity is not constant? The displacement Δx during a *very small* time interval Δt can be found in the same way as for constant velocity since, during a short enough time interval, the velocity does not change appreciably. Then v_x and Δt are the height and width of a narrow rectangle (Fig. 2.8a) and the displacement during that short time interval is the area of the rectangle. To find the total displacement during any time interval, the areas of all the narrow rectangles are added together (Fig. 2.8b). To improve the approximation, we let the time interval Δt approach zero and find that the displacement Δx during any time interval equals the area under the graph of $v_x(t)$ (Fig. 2.8c). When v_x is negative, x is decreasing and the displacement is in the $-x$-direction, so we must count the area as negative when it is below the time axis.

The magnitude of the train's displacement is represented as the shaded areas in Fig. 2.9. The train's displacement from $t = 14$ min to $t = 23$ min is +7 km (area *above* the t-axis means displacement in the $+x$-direction) and from $t = 28$ min to $t = 56$ min it is −36 km (area *below* the t-axis means displacement in the $-x$-direction). The total displacement from $t = 0$ to $t = 56$ min is $\Delta x = (+7 \text{ km}) + (-36 \text{ km}) = -29$ km.

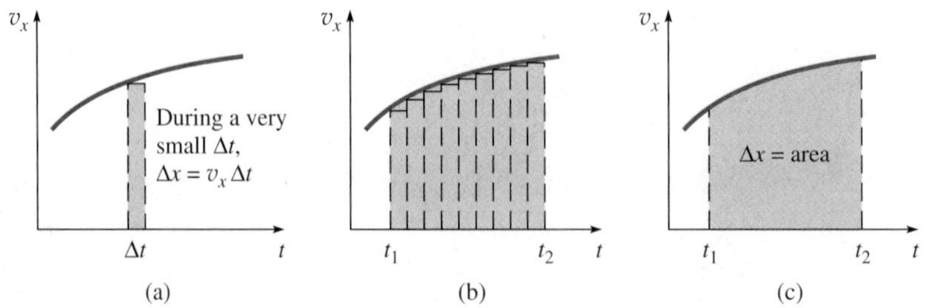

Figure 2.8 (a) Displacement Δx during a short time interval is approximately the area of a rectangle of height v_x and width Δt. (b) During a longer time interval, the displacement is approximately the sum of the areas of the rectangles. (c) The area under the v_x versus t graph for any time interval represents the displacement during that interval.

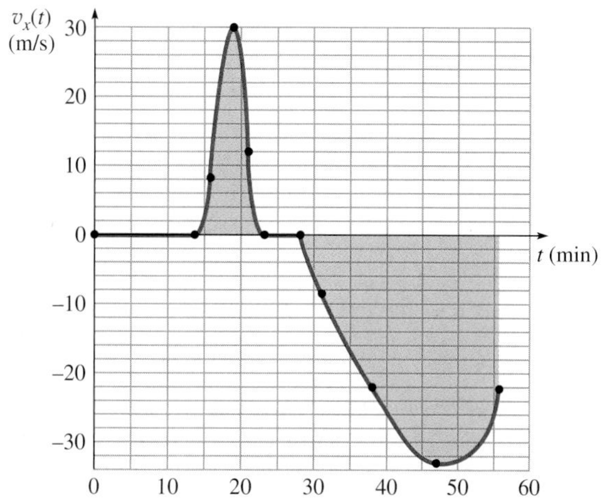

Figure 2.9 A graph of train velocity versus time. The train's displacement from $t = 14$ min to $t = 23$ min is the shaded area under the graph during that time interval. To estimate the area, count the number of grid boxes under the curve, estimating the fraction of the boxes that are only partly below the curve. Each box is 2 m/s in height and 5 min (= 300 s) in width, so each box represents an "area" (displacement) of 2 m/s × 300 s = 600 m = 0.60 km. The total number of shaded boxes for this time interval is about 12, so the displacement is about $\Delta x \approx 12 \times 0.60$ km = +7.2 km, which is close to the actual value of 7 km (during this time interval the train went from +3 km to +10 km). The shaded area for the time interval $t = 28$ min to $t = 56$ min is below the time axis; this negative area represents displacement in the $-x$-direction (west). The number of shaded grid boxes in this interval is about 60, so the displacement during this time interval is $\Delta x \approx -(60) \times 0.60$ km = −36 km.

2.3 ACCELERATION: RATE OF CHANGE OF VELOCITY

The rate of change of the velocity is called the **acceleration**. The use of the word *acceleration* in everyday language is often imprecise and not in accord with its scientific definition. In everyday language, it usually means "an increase in speed" but sometimes is used almost as a synonym for speed itself. In physics, acceleration does not necessarily indicate an increase in speed. Acceleration can indicate any kind of change in velocity.

The concept of acceleration is much less intuitive for most people than the concept of velocity. Keep reminding yourself that the acceleration tells you how the velocity *is changing*. The direction of the *change* in velocity is not necessarily the same as the direction of either the initial or final velocities.

Average Acceleration

The **average acceleration** during a time interval Δt is:

$$a_{\text{av},x} = \frac{\Delta v_x}{\Delta t} \qquad (2\text{-}7)$$

Since average acceleration is the change in velocity divided by the corresponding time interval, the SI units of acceleration are (m/s)/s = m/s^2, read as "meters per second squared." Thinking of m/s^2 as (m/s)/s can help you develop an understanding of what acceleration is. Suppose an object has a constant acceleration $a_x = +3.0$ m/s^2. Then v_x increases 3.0 m/s during every second of elapsed time (the change in v_x is +3.0 m/s per second). If $a_x = -2.0$ m/s^2, then v_x would decrease 2.0 m/s during every second (the change in v_x is −2.0 m/s per second).

CONNECTION:

Compare average acceleration [Eq. (2-7)] and average velocity [Eq. (2-2)]. Each is the change in a quantity divided by the time interval during which the change occurs. Each can have different values for different time intervals.

For example, suppose it takes 30 s for a truck to slow down from 25 m/s to 10 m/s while traveling east. With the x-axis pointing east, the truck's average acceleration during that time interval is

$$a_{\text{av},x} = \frac{\Delta v_x}{\Delta t} = \frac{-15 \text{ m/s}}{30 \text{ s}} = -0.50 \text{ m/s}^2$$

or 0.50 m/s^2 to the west.

Instantaneous Acceleration

To find the **instantaneous acceleration**, we calculate the average acceleration during a *very short time interval*:

Definition of instantaneous acceleration:

$$a_x = \lim_{\Delta t \to 0} \frac{\Delta v_x}{\Delta t} \tag{2-8}$$

(Δv_x is the change in velocity during a *very short* time interval Δt)

The time interval Δt must be short enough that we can treat the acceleration as constant during that time interval. Just as with instantaneous velocity, the word *instantaneous* is not always repeated. *Acceleration* without the adjective means *instantaneous* acceleration.

Can the lion catch the buffalo?

The chapter opener asked how an African lion can ever catch a Cape buffalo. Although Cape buffaloes and African lions have about the same top *speed*, lions are capable of much larger *accelerations* than are buffaloes. Starting from rest, it takes a buffalo much longer to get to its top speed. On the other hand, lions have much less stamina. Once the buffalo reaches its top speed, it can maintain that speed much longer than can the lion. Thus, a Cape buffalo is capable of outrunning a lion unless the stalking lion can get fairly close before charging.

Conceptual Example 2.4

Direction of Acceleration While Slowing Down

Damon moves in the $-x$-direction on his motor scooter. He "decelerates" as he approaches a stop sign. While slowing down, is the scooter's acceleration a_x positive or negative? What is the direction of the acceleration?

Strategy The acceleration has the same direction as the *change* in the velocity.

Solution and Discussion The term *decelerate* is not a scientific term. In common usage it means the scooter is slowing: the scooter's velocity is decreasing in magnitude. Damon is moving in the $-x$-direction, so v_x is negative. He is slowing down, so the *absolute value* of v_x, $|v_x|$, is getting smaller. To reduce the magnitude of a negative number, we have to add a positive number. Therefore, the change in v_x is positive ($\Delta v_x > 0$). In other words, v_x is increasing. Since Δv_x is positive, a_x is positive. The acceleration is in the $+x$-direction.

Conceptual Practice Problem 2.4 Continuing on His Way

As Damon pulls away from the stop sign, continuing in the $-x$-direction, his speed gradually increases. What is the sign of a_x? What is the direction of the acceleration?

The Direction of the Acceleration

Generalizing Example 2.4, suppose an object moves along the x-axis. When the acceleration is in the same direction as the velocity, the object is speeding up. If v_x and a_x are both positive, the object is moving in the $+x$-direction and is speeding up. If they are both negative, the object is moving in the $-x$-direction and is speeding up.

When the acceleration and velocity are in opposite directions, the object is slowing down. When v_x is positive and a_x is negative, the object is moving in the positive x-direction and is slowing down. When v_x is negative and a_x is positive, the object is moving in the negative x-direction and is slowing down.

In straight-line motion, the acceleration is always in the same direction as the velocity, in the direction opposite to the velocity, or zero.

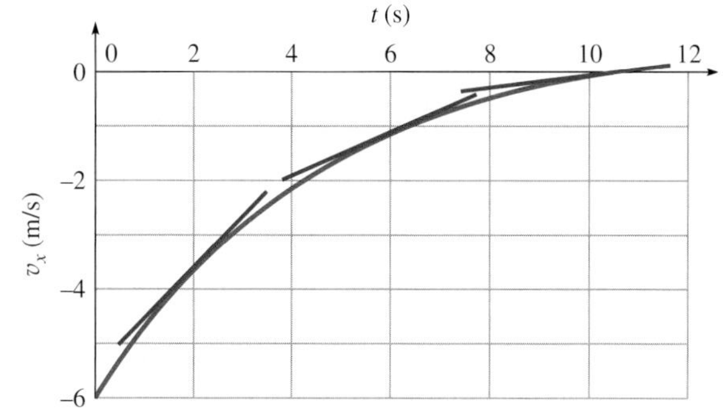

Figure 2.10 In this graph of v_x versus t, as Damon is stopping, v_x is negative, but a_x (the slope) is positive. The value of v_x is increasing, but—since it is less than zero to begin with and is getting closer to zero as time goes on—the speed is *decreasing*. The slopes of the three tangent lines shown represent the instantaneous accelerations (a_x) at three different times.

Graphical Relationships Between Velocity and Acceleration

Both velocity and acceleration measure rates of change: velocity is the rate of change of position and acceleration is the rate of change of velocity. Therefore, the graphical relationship of acceleration to velocity is the same as the graphical relationship of velocity to position: a_x is the slope on a graph of $v_x(t)$ and Δv_x is the area under a graph of $a_x(t)$.

Figure 2.10 shows a graph of v_x versus t for Damon slowing down on his scooter. He is moving in the $-x$-direction, so $v_x < 0$, and his speed is decreasing, so $|v_x|$ is decreasing. The slope of a tangent line to the graph is a_x at that instant. Three tangent lines are drawn, showing that a_x is positive (the slopes are positive) and is not constant (the slopes are not all the same).

CONNECTION:

On a graph of *any* quantity Q as a function of time, the slope of the graph represents the instantaneous rate of change of Q. On a graph of the *rate of change of Q* as a function of time, the area under the graph represents ΔQ.

Example 2.5

Acceleration of a Sports Car

A sports car starting at rest can achieve 30.0 m/s in 4.7 s according to the advertisements. Figure 2.11 shows data for v_x as a function of time as the sports car starts from rest and travels in a straight line in the $+x$-direction. (a) What is the average acceleration of the sports car from 0 to 30.0 m/s? (b) What is the maximum acceleration of the car? (c) What is the car's displacement from $t = 0$ to $t = 19.1$ s (when it reaches 60.0 m/s)? (d) What is the car's average velocity during the entire 19.1 s interval?

Strategy (a) To find the average acceleration, the change in velocity for the time interval is divided by the time interval. (b) The instantaneous acceleration is the slope of the velocity graph, so it is maximum where the graph is steepest. At that point, the velocity is changing at a high rate. We expect the maximum acceleration to take place early on; the magnitude of acceleration must decrease as the velocity gets higher and higher—there is a maximum velocity for the car, after all. (c) The displacement Δx is the area under the $v_x(t)$ graph. The graph is not a simple shape such as a triangle or rectangle, so an estimate of the area is made. (d) Once we have a value for the displacement, we can apply the definition of average velocity.

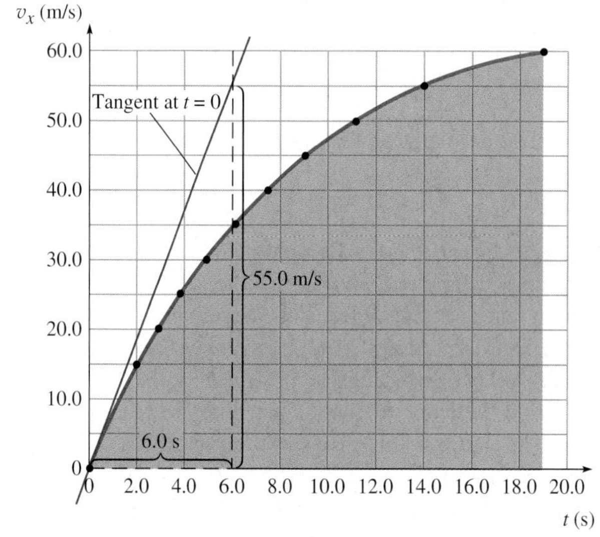

v_x (m/s)	0	15.0	20.0	25.0	30.0	35.0	40.0	45.0	50.0	55.0	60.0
t (s)	0	2.0	2.9	3.8	4.9	6.2	7.6	9.1	11.2	14.0	19.1

Figure 2.11

Data table and graph of $v_x(t)$ for a sports car.

continued on next page

Example 2.5 continued

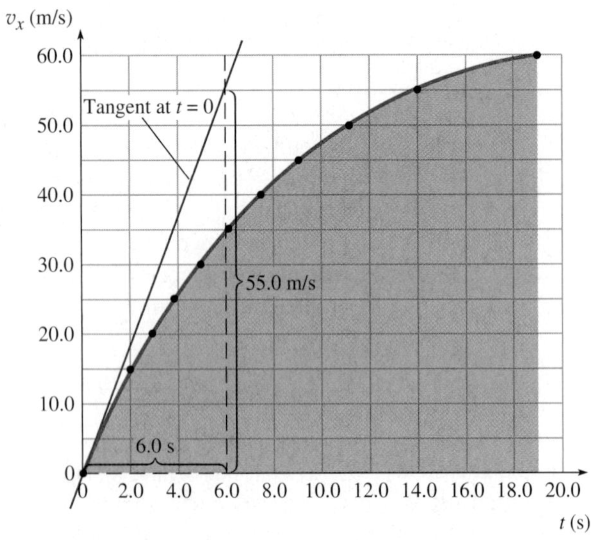

v_x (m/s)	0	15.0	20.0	25.0	30.0	35.0	40.0	45.0	50.0	55.0	60.0
t (s)	0	2.0	2.9	3.8	4.9	6.2	7.6	9.1	11.2	14.0	19.1

Figure 2.11

Data table and graph of $v_x(t)$ for a sports car.

Given: Graph of $v_x(t)$ in Fig. 2.11.

To find: (a) $a_{av,x}$ for $v_x = 0$ to 30.0 m/s; (b) maximum value of a_x; (c) Δx from $v_x = 0$ to 60.0 m/s; (d) $v_{av,x}$ from $t = 0$ to 19.1 s

Solution (a) The car starts from rest, so $v_{xi} = 0$. It reaches $v_x = 30.0$ m/s at $t = 4.9$ s, according to the data table. Then for this time interval,

$$a_{av,x} = \frac{\Delta v_x}{\Delta t} = \frac{30.0 \text{ m/s} - 0 \text{ m/s}}{4.9 \text{ s} - 0 \text{ s}} = 6.1 \text{ m/s}^2$$

The average acceleration for this time interval is 6.1 m/s^2 in the +x-direction.

(b) The acceleration a_x, at any instant of time, is the slope of the tangent line to the $v_x(t)$ graph at that time. To find the maximum acceleration, we look for the steepest part of the graph. In this case, the largest slope occurs near $t = 0$, just as the car is starting out. In Fig. 2.11, a tangent line to the $v_x(t)$ graph at $t = 0$ passes through $t = 0$. Values for the rise and run to calculate the slope of the tangent line are read from the graph. The tangent line passes through the two points ($t = 0$, $v_x = 0$) and ($t = 6.0$ s, $v_x = 55.0$ m/s) on the graph, so the rise is 55.0 m/s for a run of 6.0 s. The slope of this line is

$$a_x = \frac{\text{rise}}{\text{run}} = \frac{55.0 \text{ m/s} - 0 \text{ m/s}}{6.0 \text{ s} - 0 \text{ s}} = +9.2 \text{ m/s}^2$$

The maximum acceleration is 9.2 m/s^2 in the +x-direction.

(c) Δx is the area under the $v_x(t)$ graph shown shaded in Fig. 2.11. The area can be estimated by counting the number of grid boxes under the curve. Each box is 5.0 m/s in height and 2.0 s in width, so each represents an "area" (displacement) of 10 m. When counting the number of boxes under the curve, a best estimate is made for the fraction of the boxes that are only partly below the curve. Approximately 75 boxes lie below the curve, so the displacement is $\Delta x = 75 \times 10 \text{ m} = 750$ m. Since the car travels along a straight line and does not change direction, 750 m is also the distance traveled. (d) The average velocity during the 19.1-s interval is

$$v_{av,x} = \frac{\Delta x}{\Delta t} = \frac{750 \text{ m}}{19.1 \text{ s}} = 39 \text{ m/s}$$

Discussion The graph of velocity as a function of time is often the most helpful graph to have when solving a problem. If that graph is not given in the problem, it is useful to sketch one. The $v_x(t)$ graph shows displacement, velocity, and acceleration at once: the velocity v_x is given by the points or the curve graphed, the displacement Δx is the area under the curve, and the acceleration a_x is the slope of the curve.

Why is the average velocity 39 m/s? Why is it not halfway between the initial velocity (0 m/s) and the final velocity (60 m/s)? If the acceleration were constant, the average velocity would indeed be $\frac{1}{2}(0 + 60 \text{ m/s}) = 30$ m/s. The actual average velocity is somewhat higher than that—the acceleration is greater at the start, so less of the time interval is spent going (relatively) slow and more is spent going fast. The speed is less than 30 m/s for only 4.9 s, but is greater than 30 m/s for 14.2 s.

Practice Problem 2.5 Braking a Car

An automobile is traveling along a straight road heading to the southeast at 24 m/s when the driver sees a deer begin to cross the road ahead of her. She steps on the brake and brings the car to a complete stop in an elapsed time of 8.0 s. A data recording device, triggered by the sudden braking action, records the following velocities and times as the car slows. Let the positive x-axis be directed to the southeast. Plot a graph of v_x versus t and find (a) the average acceleration as the car comes to a stop and (b) the instantaneous acceleration at $t = 2.0$ s.

v_x (m/s)	24	17.3	12.0	8.7	6.0	3.5	2.0	0.75	0
t (s)	0	1.0	2.0	3.0	4.0	5.0	6.0	7.0	8.0

CHECKPOINT 2.3

What physical quantity does the slope of the tangent to a graph of v_x versus time represent?

2.4 MOTION ALONG A LINE WITH CONSTANT ACCELERATION

The graphical and mathematical relationships between position, velocity, and acceleration presented so far apply regardless of whether the acceleration is changing or is constant. In the important special case of an object whose acceleration is *constant* (both in magnitude and direction), we can write these relationships as algebraic equations. First, let us agree on a consistent notation:

- Choose an origin and a direction for the positive axis. (For vertical motion, it is conventional to use the y-axis instead of the x-axis, where the $+y$-direction is up.)
- At an initial time t_i, the initial position and velocity are x_i and v_{ix}.
- At a later time $t_f = t_i + \Delta t$, the final position and velocity are x_f and v_{fx}.

From the following two essential relationships the others can be derived:

1. Since the acceleration a_x is constant, the change in velocity over a given time interval $\Delta t = t_f - t_i$ is the acceleration—the rate of change of velocity—times the elapsed time:

$$\Delta v_x = v_{fx} - v_{ix} = a_x \Delta t \qquad (2\text{-}9)$$

(if a_x is constant during the entire time interval)

Equation (2-9) is the definition of a_x [Eq. (2-8)] with the assumption that a_x is constant.

2. Since the velocity changes linearly with time, the average velocity is given by:

$$v_{av,x} = \tfrac{1}{2}(v_{fx} + v_{ix}) \quad \text{(constant } a_x\text{)} \qquad (2\text{-}10)$$

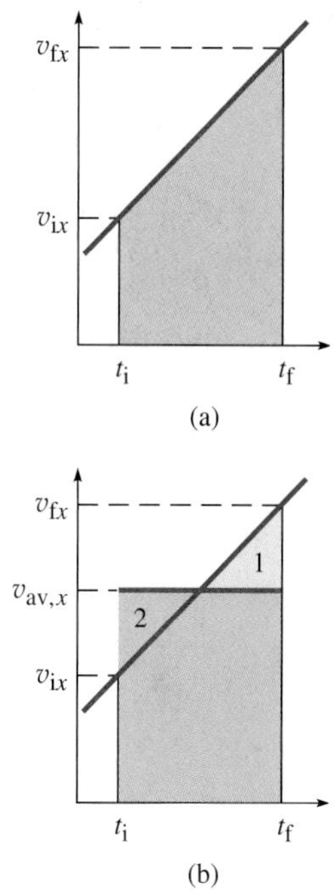

Figure 2.12 Finding the average velocity when the acceleration is constant.

Equation (2-10) is *not* true *in general*, but it is true for constant acceleration. To see why, refer to the $v_x(t)$ graph in Fig. 2.12a. The graph is linear because the acceleration—the slope of the graph—is constant. The displacement during any time interval is represented by the area under the graph. The average velocity is found by forming a rectangle with an area equal to the area under the curve in Fig. 2.12a, because the average velocity should give the same displacement in the same time interval. Figure 2.12b shows that, to make the excluded area above $v_{av,x}$ (triangle 1) equal to the extra area under $v_{av,x}$ (triangle 2), the average velocity must be exactly halfway between the initial and final velocities. Combining Eq. (2-10) with the definition of average velocity,

$$\Delta x = x_f - x_i = v_{av,x}\,\Delta t \qquad (2\text{-}2)$$

gives our second essential relationship for constant acceleration:

$$\Delta x = \tfrac{1}{2}(v_{fx} + v_{ix})\,\Delta t \qquad (2\text{-}11)$$

(if a_x is constant during the entire time interval)

If the acceleration is *not* constant, there is no reason why the average velocity has to be exactly halfway between the initial and the final velocity. As an illustration, imagine a trip where you drive along a straight highway at 80 km/h for 50 min and

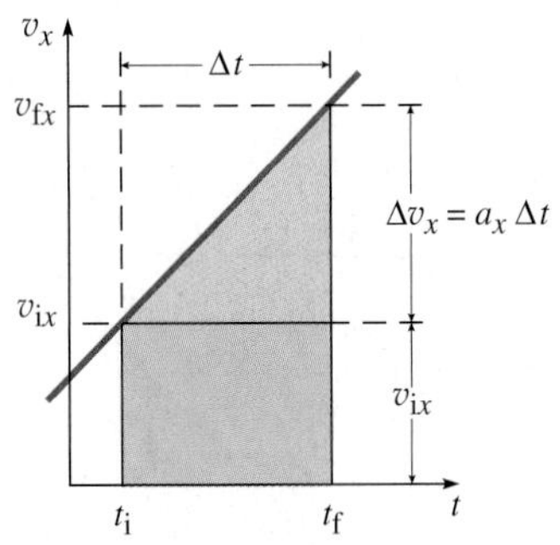

Figure 2.13 Graphical interpretation of Eq. (2-12).

then at 60 km/h for 30 min. Your acceleration is zero for the entire trip *except* during the few seconds while you slowed from 80 km/h to 60 km/h. The magnitude of your average velocity is *not* 70 km/h. You spent more time going 80 km/h than you did going 60 km/h, so the magnitude of your average velocity would be greater than 70 km/h.

Other Useful Relationships for Constant Acceleration Two more useful relationships can be formed between the various quantities (displacement, initial and final velocities, acceleration, and time interval) by eliminating some quantity from Eqs. (2-9) and (2-11). For example, suppose we don't know the final velocity v_{fx}. Then we can solve Eq. (2-9) for v_{fx}, substitute into Eq. (2-11), and simplify:

$$\Delta x = \tfrac{1}{2}(v_{fx} + v_{ix})\,\Delta t = \tfrac{1}{2}[(v_{ix} + a_x \Delta t) + v_{ix}]\,\Delta t$$

$$\Delta x = v_{ix}\Delta t + \tfrac{1}{2}a_x(\Delta t)^2 \quad \text{(constant } a_x\text{)} \qquad (2\text{-}12)$$

We can interpret Eq. (2-12) graphically. Figure 2.13 shows a $v_x(t)$ graph for motion with constant acceleration. The displacement that occurs between t_i and a later time t_f is the area under the graph for that time interval. Partition this area into a rectangle plus a triangle. The area of the rectangle is

$$\text{base} \times \text{height} = v_{ix}\,\Delta t$$

The height of the triangle is the change in velocity, which is equal to $a_x\,\Delta t$. The area of the triangle is

$$\tfrac{1}{2}\text{base} \times \text{height} = \tfrac{1}{2}\Delta t \times a_x\,\Delta t = \tfrac{1}{2}a_x(\Delta t)^2$$

Adding these areas gives Eq. (2-12).

Another useful relationship comes from eliminating the time interval Δt:

$$\Delta x = \tfrac{1}{2}(v_{fx} + v_{ix})\,\Delta t = \tfrac{1}{2}(v_{fx} + v_{ix})\left(\frac{v_{fx} - v_{ix}}{a_x}\right) = \frac{v_{fx}^2 - v_{ix}^2}{2a_x}$$

Rearranging terms,

$$v_{fx}^2 - v_{ix}^2 = 2a_x\,\Delta x \quad \text{(constant } a_x\text{)} \qquad (2\text{-}13)$$

CHECKPOINT 2.4

At 3:00 P.M., an airplane is moving due west at 460 km/h. At 3:05 P.M., it is moving due west at 480 km/h. Is its average velocity during the time interval necessarily 470 km/h west? Explain.

Example 2.6

A Sliding Brick

Starting from rest, a brick slides along a straight line down an icy roof with a constant acceleration of magnitude 4.9 m/s^2 (Fig. 2.14). How fast is the brick moving when it reaches the edge of the roof 0.90 s later?

Strategy What is the direction of the acceleration? It has to be downward along the roof, in the same direction as the brick's velocity. An acceleration opposite the velocity would make the brick slow down, but since it starts from rest, a

continued on next page

Example 2.6 continued

constant acceleration can only make it speed up. We choose the $+x$-axis in the direction of the acceleration. Then we use the acceleration to find how the velocity changes during the time interval.

Solution With the x-axis in the direction of the acceleration, $a_x = +4.9\ \text{m/s}^2$. The brick is initially at rest so $v_{ix} = 0$.

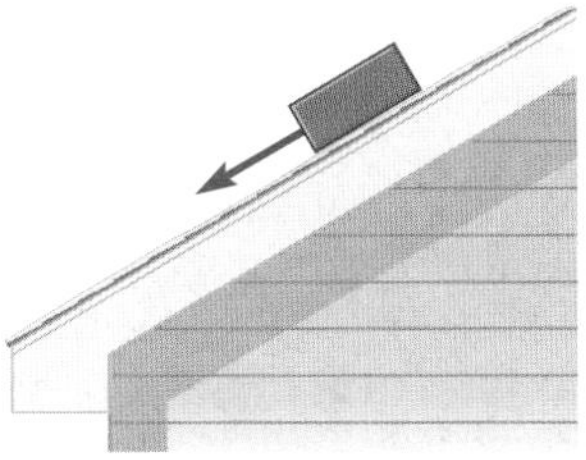

Figure 2.14
A brick sliding down an icy roof.

We want to know v_{fx} at the end of the time interval $\Delta t = 0.90$ s. Since a_x is constant, v_x changes at a constant rate:

$$\Delta v_x = v_{fx} - v_{ix} = a_x\,\Delta t = (+4.9\ \text{m/s}^2) \times (0.90\ \text{s}) = 4.4\ \text{m/s}$$

At the edge of the roof, the brick is moving at 4.4 m/s parallel to the roof.

Discussion Conceptual check: $a_x = +4.9\ \text{m/s}^2$ means that v_x increases 4.9 m/s every second. The brick slides for a bit less than 1 s, so the increase in v_x is a bit less than 4.9 m/s.

Practice Problem 2.6 Displacement of the Brick

How far from the edge of the roof was the brick when it started sliding?

Example 2.7

Displacement of a Motorboat

A motorboat starts from rest at a dock and heads due east with a constant acceleration of magnitude $2.8\ \text{m/s}^2$. After traveling for 140 m, the motor is throttled down to slow down the boat at $1.2\ \text{m/s}^2$ (while still moving east) until its speed is 16 m/s. Just as the boat attains the speed of 16 m/s, it passes a buoy due east of the dock. (a) Sketch a qualitative graph of $v_x(t)$ for the motorboat from the dock to the buoy. Let the $+x$-axis point east. (b) What is the distance between the dock and the buoy?

Strategy This problem involves two different values of acceleration, so it must be divided into two subproblems. The equations for constant acceleration cannot be applied to a time interval during which the acceleration changes. But for each of two time intervals, the acceleration of the boat is constant: from t_1 to t_2, $a_{1x} = +2.8\ \text{m/s}^2$; from t_2 to t_3, $a_{2x} = -1.2\ \text{m/s}^2$. The two subproblems are connected by the position and velocity of the boat at the instant the acceleration changes. This is reflected in the graph of $v_x(t)$: It consists of two different straightline segments with different slopes that connect with the same value of v_x at time t_2.

For subproblem 1, the boat speeds up with a constant acceleration of $2.8\ \text{m/s}^2$ to the east. We know the acceleration, the displacement (140 m east), and the initial velocity: the boat starts from rest, so the initial velocity v_{1x} is zero. We need to calculate the final velocity v_{2x}, which then becomes the initial velocity for the second subproblem. The boat is always headed to the east, so we choose east as the positive x-direction.

Subproblem 1:

Known: $v_{1x} = 0$; $a_{1x} = +2.8\ \text{m/s}^2$;
$\Delta x_{21} = x_2 - x_1 = 140$ m.

To find: v_{2x}.

For subproblem 2, we know acceleration, final velocity v_{3x}, and we have just found the initial velocity v_{2x} from subproblem 1. Because the boat is slowing down, its acceleration is in the direction opposite its velocity; therefore, $a_{2x} < 0$. From these three quantities we can find the displacement of the boat during the second time interval.

Subproblem 2:

Known: v_{2x} from subproblem 1;
$a_{2x} = -1.2\ \text{m/s}^2$; $v_{3x} = +16$ m/s.

To find: $\Delta x_{32} = x_3 - x_2$.

Adding the displacements for the two time intervals gives the total displacement. The magnitude of the total displacement is the distance between the dock and the buoy.

continued on next page

Example 2.7 continued

Solution (a) The graph starts with $v_x = 0$ at $t = t_1$. We choose $t_1 = 0$ for simplicity. The graph is a straight line with slope +2.8 m/s^2 until $t = t_2$. Then, starting from where the graph left off, the graph continues as a straight line with slope −1.2 m/s^2 until the graph reaches $v_x = 16$ m/s at $t = t_3$. Figure 2.15 shows the $v_x(t)$ graph. It is not quantitatively accurate because we have not calculated the values of t_2 and t_3.

(b1) To find v_{2x} without knowing the time interval, we eliminate Δt from Eqs. (2-9) and (2-11) for constant acceleration:

$$\Delta x_{21} = \frac{1}{2}(v_{2x} + v_{1x})\,\Delta t = \frac{1}{2}(v_{2x} + v_{1x})\left(\frac{v_{2x} - v_{1x}}{a_{1x}}\right) = \frac{v_{2x}^2 - v_{1x}^2}{2a_{1x}}$$

Solving for v_{2x},

$$v_{2x} = \pm\sqrt{v_{1x}^2 + 2a_{1x}\,\Delta x} = \pm\sqrt{0 + 2 \times 2.8\ \text{m/s}^2 \times 140\ \text{m}}$$
$$= \pm 28\ \text{m/s}$$

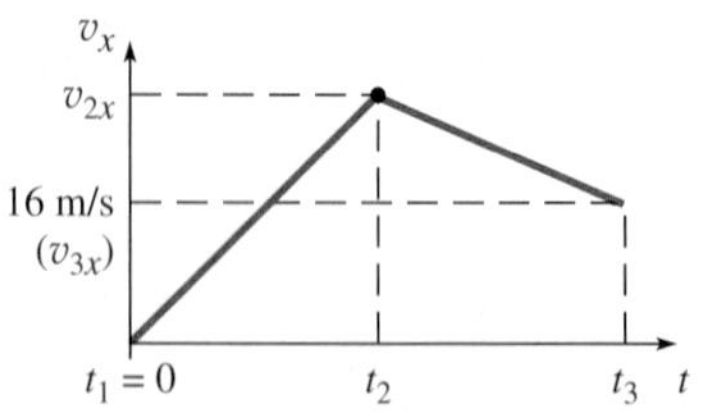

Figure 2.15 Graph of v_x versus t for the motorboat.

The boat is moving east, in the +x-direction, so the correct sign here is positive: $v_{2x} = +28$ m/s.

(b2) The final velocity for the first interval (v_{2x}) is the *initial* velocity for the second interval. The final velocity is v_{3x}. Using the same equation just derived for this time interval,

$$\Delta x_{32} = \frac{v_{3x}^2 - v_{2x}^2}{2a_{2x}} = \frac{(16\ \text{m/s})^2 - (28\ \text{m/s})^2}{2 \times (-1.2\ \text{m/s}^2)} = +220\ \text{m}$$

The total displacement is

$$x_3 - x_1 = (x_3 - x_2) + (x_2 - x_1) = 220\ \text{m} + 140\ \text{m} = +360\ \text{m}$$

The buoy is 360 m from the dock.

Discussion The natural division of the problem into two parts occurs because the boat has two different constant accelerations during two different time periods. In problems that can be subdivided in this way, the final velocity and position found in the first part becomes the initial velocity and position for the second part.

Practice Problem 2.7 Time to Reach the Buoy

What is the time required by the boat in Example 2.7 to reach the buoy?

2.5 VISUALIZING MOTION ALONG A LINE WITH CONSTANT ACCELERATION

Motion Diagrams In Fig. 2.16, three carts move in the same direction with three different values of constant acceleration. The position of each cart is depicted in a **motion diagram** as it would appear in a stroboscopic photograph with pictures taken at equal time intervals (here, the time interval is 1.0 s).

The yellow cart has zero acceleration and, therefore, constant velocity. During each 1.0-s time interval its displacement is the same: 1.0 m/s × 1.0 s = 1.0 m to the right.

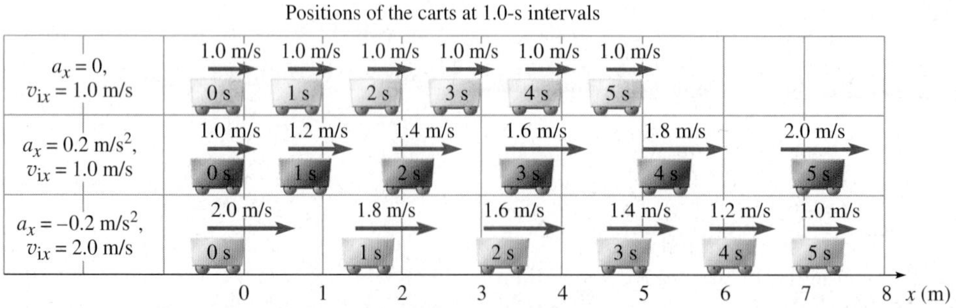

Figure 2.16 Each cart is shown as if photographs were taken at 1.0-s time intervals of 1.0 s. The arrows above each cart indicate the instantaneous velocities.

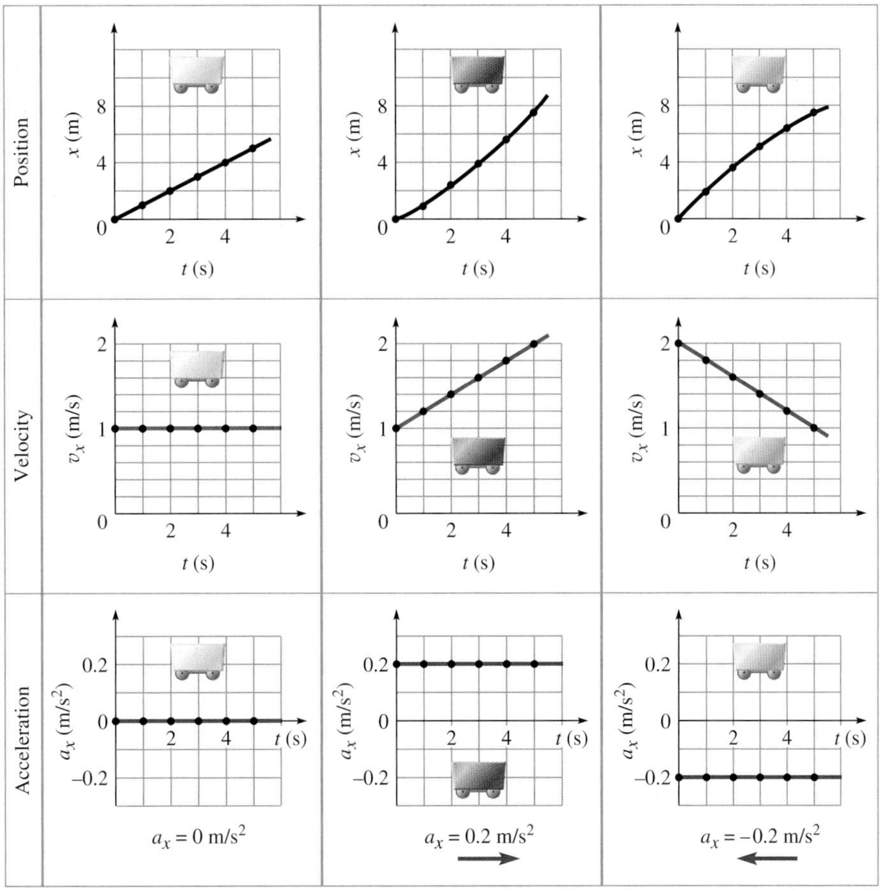

Figure 2.17 Graphs of position, velocity, and acceleration for the carts of Fig. 2.16.

The red cart has a constant acceleration of 0.2 m/s^2 to the right. Although m/s^2 is normally read "meters per second squared," it can be useful to think of it as "m/s per second": the cart's velocity changes by 0.2 m/s during each 1.0-s time interval. In this case, acceleration is in the same direction as the velocity, so the velocity increases. The displacement of the cart during successive 1.0-s time intervals gets larger and larger.

The blue cart experiences a constant acceleration of 0.2 m/s^2 in the $-x$-direction—the direction *opposite* to the velocity. The magnitude of the velocity then decreases; during each 1.0-s interval, the speed decreases by 0.2 m/s. Now the displacements during 1.0-s intervals get smaller and smaller.

Graphs Figure 2.17 shows graphs of $x(t)$, $v_x(t)$, and $a_x(t)$ for each of the carts. The acceleration graphs are horizontal since each of the carts has a constant acceleration. All three v_x graphs are straight lines. Since a_x is the rate of change of v_x, the slope of the v_x graph at any value of t is a_x at that value of t. With constant acceleration, the slope is the same everywhere and the graph is linear. Remember that a positive a_x does mean that v_x is increasing, but not necessarily that the *speed* is increasing. If v_x is negative, then a positive a_x indicates a *decreasing* speed. (See Conceptual Example 2.4.) Speed is increasing when the acceleration and velocity are in the same direction (a_x and v_x both positive *or* both negative). Speed is decreasing when acceleration and velocity are in opposite directions—when a_x and v_x have opposite signs.

The position graph is linear for the yellow cart because it has constant velocity. For the red cart, the $x(t)$ graph curves with increasing slope, showing that v_x is increasing. For the blue cart, the $x(t)$ graph curves with decreasing slope, showing that v_x is decreasing.

Example 2.8

Two Spaceships

Two spaceships are moving from the same starting point in the $+x$-direction with constant accelerations. The silver spaceship has an initial velocity of +2.00 km/s and an acceleration of +0.400 km/s^2. The black spaceship has an initial velocity of +6.00 km/s and an acceleration of −0.400 km/s^2. (a) Find the time at which the silver spaceship just overtakes the black spaceship. (b) Sketch graphs of $v_x(t)$ for the two spaceships. (c) Sketch a motion diagram (similar to Fig. 2.16) showing the positions of the two spaceships at 1.0-s intervals.

Strategy We can find the positions of the spaceships at later times from the initial velocities and the accelerations. At first, the black spaceship is moving faster, so it pulls out ahead. Later, the silver ship overtakes the black ship at the instant their *positions are equal.*

Solution (a) The position of either spaceship at a later time is given by Eq. (2-12):

$$x_f = x_i + \Delta x = x_i + v_{ix}\,\Delta t + \tfrac{1}{2}a_x(\Delta t)^2$$

We set the final position of the silver spaceship equal to that of the black spaceship ($x_{fs} = x_{fb}$):

$$x_{is} + v_{isx}\,\Delta t + \tfrac{1}{2}a_{sx}\,(\Delta t)^2 = x_{ib} + v_{ibx}\,\Delta t + \tfrac{1}{2}a_{bx}(\Delta t)^2$$

Subscripts are useful for preventing you from mixing up similar quantities. The subscripts s and b stand for silver and black, respectively. The subscripts i and f stand for initial and final, respectively. A skilled problem-solver must be able to come up with algebraic symbols that are explicit and unambiguous.

The initial positions are the same: $x_{is} = x_{ib}$. Subtracting the initial positions from each side, moving all terms to one side, and factoring out one power of Δt yields

$$\Delta t(v_{isx} + \tfrac{1}{2}a_{sx}\,\Delta t - v_{ibx} - \tfrac{1}{2}a_{bx}\,\Delta t) = 0$$

This equation has two solutions—there are two times at which the spaceships are at the same position. One solution is $\Delta t = 0$. We already knew that the two spaceships started at the same *initial* position. The other solution, which gives the time at which one spaceship overtakes the other, is found by setting the expression in parentheses equal to zero. Solving for Δt,

$$\Delta t = \frac{2(v_{isx} - v_{ibx})}{a_{bx} - a_{sx}} = \frac{2 \times (2.00\ \text{km/s} - 6.00\ \text{km/s})}{-0.400\ \text{km/s}^2 - 0.400\ \text{km/s}^2} = 10.0\ \text{s}$$

The silver spaceship overtakes the black spaceship 10.0 s after they leave the starting point.

(b) Figure 2.18 shows the $v_x(t)$ graphs with $t_i = 0$. Note that the area under the graphs from t_i to t_f is the same in the two graphs: the spaceships have the same displacement during that interval.

(c) Equation (2-12) can be used to find the position of each spaceship as a function of time. Choosing $x_i = 0$, $t_i = 0$, and $t = t_f$, the position at time t is

$$x(t) = 0 + v_{ix}t + \tfrac{1}{2}a_x t^2$$

Figure 2.19 shows the data table calculated this way and the corresponding motion diagram.

Discussion Quick check: the two ships must have the same displacement at $\Delta t = 10.0$ s.

$$\Delta x_s = v_{isx}\,\Delta t + \tfrac{1}{2}a_{sx}(\Delta t)^2$$
$$= 2.00\ \text{km/s} \times 10.0\ \text{s} + \tfrac{1}{2} \times 0.400\ \text{km/s}^2 \times (10.0\ \text{s})^2$$
$$= 40.0\ \text{km}$$

$$\Delta x_b = v_{ibx}\,\Delta t + \tfrac{1}{2}a_{bx}(\Delta t)^2$$
$$= 6.00\ \text{km/s} \times 10.0\ \text{s} + \tfrac{1}{2} \times (-0.400\ \text{km/s}^2) \times (10.0\ \text{s})^2$$
$$= 40.0\ \text{km}$$

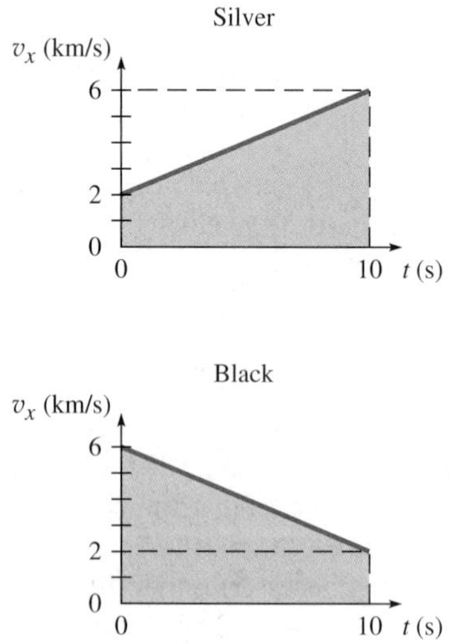

Figure 2.18
Graphs of v_x versus t for the silver and black spaceships. The shaded area under each graph represents the displacement Δx during the time interval.

continued on next page

Example 2.8 continued

Practice Problem 2.8 Time to Reach the Same Velocity

When do the two spaceships have the same *velocity*? What is the value of the velocity then?

t (s)	0	1.0	2.0	3.0	4.0	5.0	6.0	7.0	8.0	9.0	10.0
x_s (km)	0	2.2	4.8	7.8	11.2	15.0	19.2	23.8	28.8	34.2	40.0
x_b (km)	0	5.8	11.2	16.2	20.8	25.0	28.8	32.2	35.2	37.8	40.0

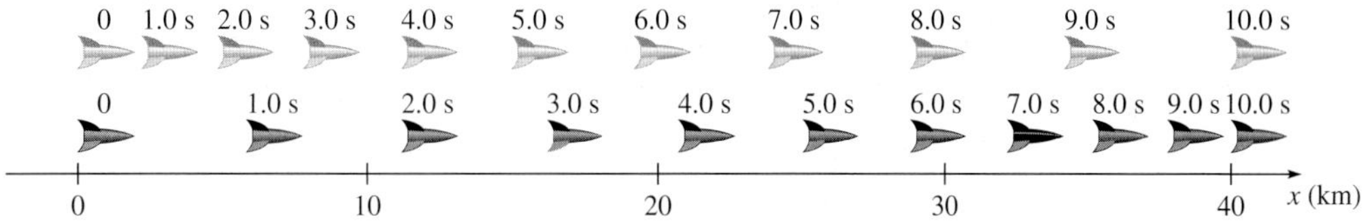

Figure 2.19
Calculated positions of the spaceships at 1.0-s time intervals and a motion diagram.

2.6 FREE FALL

Suppose you are standing on a bridge over a deep gorge. If you drop a stone into the gorge, how fast does it fall? You know from experience that it does not fall at a constant velocity; the longer it falls, the faster it goes. A better question is: What is the stone's acceleration?

CONNECTION:

Free fall is an example of motion with constant acceleration.

First, let us simplify the problem. If the stone were moving very fast, air resistance would oppose its motion. When it is not falling so fast, the effect of air resistance is negligibly small. In **free fall**, no forces act on an object other than the gravitational force that makes the object fall. On Earth, free fall is an idealization since there is always *some* air resistance. We also assume that the stone's change in altitude is small enough that Earth's gravitational pull on it is constant.

Free-fall Acceleration An object in free fall has a constant downward acceleration, called the *free-fall acceleration.* The magnitude of this acceleration varies a little from one place to another near Earth's surface, but at any given place, it has the same value for every object, regardless of the mass of the object. Unless another value is given in a particular problem, please assume that the magnitude of the free-fall acceleration near Earth's surface is

$$a_{\text{free fall}} = g = 9.80\ \text{m/s}^2 \qquad (2\text{-}14)$$

The symbol g represents the magnitude of the free-fall acceleration.

When dealing with vertical motion, the y-axis is usually chosen to be positive pointing upward. The direction of the free-fall acceleration is down, so $a_y = -g$. The same techniques and equations used for other constant acceleration situations are used with free fall.

In free fall, $a_y = -g$ (if the y-axis points up).

Earth's gravity always pulls downward, so the acceleration of an object in free fall is always downward and constant in magnitude, *regardless of whether the object is moving up, down, or is at rest, and independent of its speed.* If the object is moving downward, the downward acceleration makes it speed up; if it is moving upward, the downward acceleration makes it slow down.

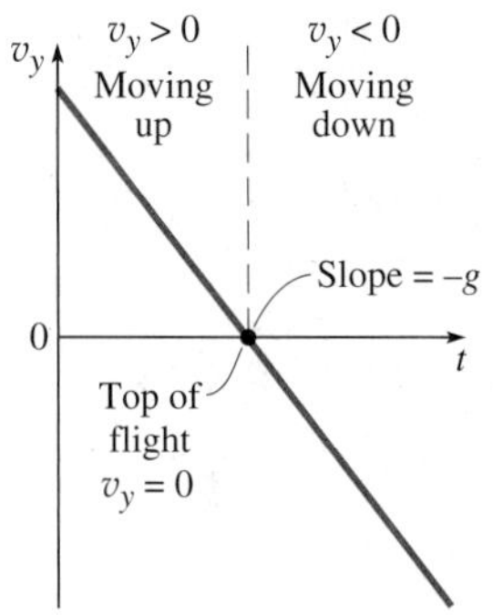

Figure 2.20 Graph of v_y versus t for an object thrown upward.

Acceleration at Highest Point If an object is thrown straight up, its velocity is zero at the highest point of its flight. Why? On the way up, its velocity v_y is positive (if the positive y-axis is pointing up). On the way down, v_y is negative. Since v_y changes continuously, it must pass through zero to change sign (Fig. 2.20). At the highest point, the velocity is zero but the *acceleration* is *not* zero. If the acceleration were to suddenly become zero at the top of flight, the velocity would no longer change; the object would get *stuck at the top* rather than fall back down. The velocity is zero at the top but it does not *stay* zero; it keeps changing at the same rate.

CHECKPOINT 2.6

Is it possible for an object in free fall to be moving upward? Explain.

Example 2.9

Throwing Stones

Standing on a bridge, you throw a stone straight upward. The stone hits a stream, 44.1 m below the point at which you release it, 4.00 s later. (a) What is the velocity of the stone just after it leaves your hand? (b) What is the velocity of the stone just before it hits the water? (c) Draw a motion diagram for the stone, showing its position at 0.1-s intervals during the first 0.9 s of its motion. (d) Sketch graphs of $y(t)$ and $v_y(t)$. The positive y-axis points up.

Strategy Ignoring air resistance, the stone is in free fall once your hand releases it and until it hits the water. For the time interval during which the stone is in free fall, the initial velocity is the velocity of the stone *just after* it leaves your hand and the final velocity is the velocity *just before* it hits the water. During free fall, the stone's acceleration is constant and assumed to be 9.80 m/s^2 downward. Known: $a_y = -9.80\ \text{m/s}^2$; $\Delta y = -44.1$ m at $\Delta t = 4.00$ s. To find: v_{iy} and v_{fy}.

Solution (a) Equation (2-12) can be used to solve for v_{iy} since all the other quantities in it (Δy, Δt, and a_y) are known and the acceleration is constant.

$$\Delta y = v_{iy}\,\Delta t + \tfrac{1}{2}a_y(\Delta t)^2$$

Solving for v_{iy},

$$v_{iy} = \frac{\Delta y}{\Delta t} - \frac{1}{2}a_y\Delta t \qquad (1)$$

$$= \frac{-44.1\ \text{m}}{4.00\ \text{s}} - \frac{1}{2}(-9.80\ \text{m/s}^2 \times 4.00\ \text{s})$$

$$= -11.0\ \text{m/s} + 19.6\ \text{m/s} = 8.6\ \text{m/s}$$

The initial velocity is 8.6 m/s upward.

(b) The change in v_y is $a_y\Delta t$ from Eq. (2-9):

$$v_{fy} = v_{iy} + a_y\Delta t$$

Substituting the expression for v_{iy} in the preceding equation,

$$v_{fy} = \left(\frac{\Delta y}{\Delta t} - \frac{1}{2}a_y\Delta t\right) + a_y\Delta t = \frac{\Delta y}{\Delta t} + \frac{1}{2}a_y\Delta t \qquad (2)$$

$$= \frac{-44.1\ \text{m}}{4.00\ \text{s}} + \frac{1}{2}(-9.80\ \text{m/s}^2 \times 4.00\ \text{s})$$

$$= -11.0\ \text{m/s} - 19.6\ \text{m/s} = -30.6\ \text{m/s}$$

The final velocity is 30.6 m/s downward.

continued on next page

Example 2.9 continued

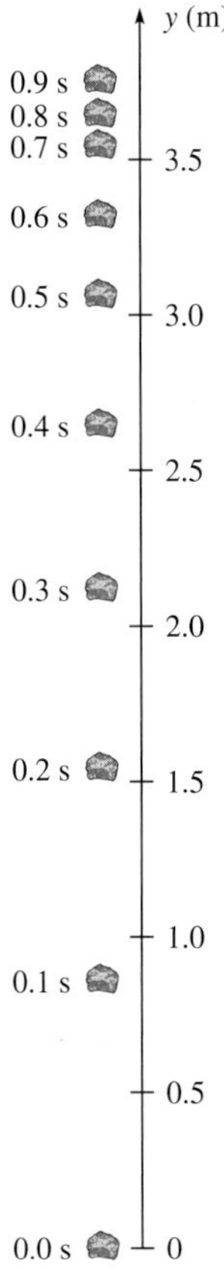

Figure 2.21

Motion diagram for a stone moving straight up.

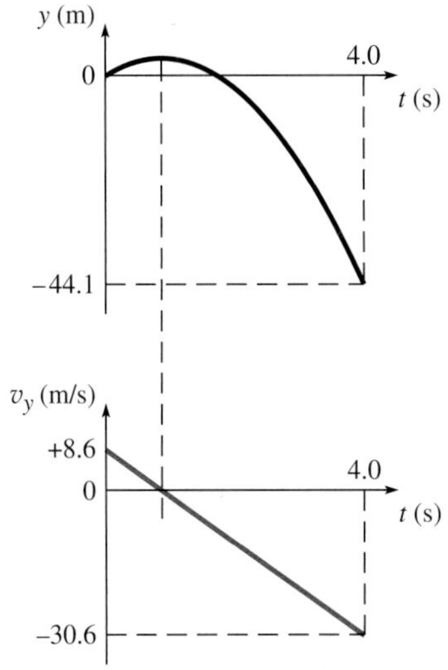

Figure 2.22

Graphs of $y(t)$ and $v_y(t)$ for the stone.

(c) Choosing $y_i = 0$ and $t_i = 0$, the position of the stone as a function of time is

$$y(t) = v_{iy}t + \tfrac{1}{2}a_y t^2$$

The motion diagram is shown in Fig. 2.21.

(d) The graphs are shown in Fig. 2.22.

Discussion The final speed is greater than the initial speed, as expected. Equations (1) and (2) have a direct interpretation, which is a good check on their validity. The first term, $\Delta y/\Delta t$, is the average velocity of the stone during the 4.00 s of free fall. The second term, $\tfrac{1}{2}a_y\Delta t$, is *half* the change in v_y since $\Delta v_y = a_y\,\Delta t$. Because the acceleration is constant, the average velocity is halfway between the initial and final velocities. Therefore, the initial velocity is the average velocity minus half of the change, while the final velocity is the average velocity plus half of the change.

Practice Problem 2.9 Height Attained by Stone

(a) How high above the bridge does the stone go? [*Hint:* What is v_y at the highest point?] (b) If you dropped the stone instead of throwing it, how long would it take to hit the water?

Master the Concepts

- Displacement is the change in position: $\Delta x = x_f - x_i$. The displacement depends only on the starting and ending positions, not on details of the motion. The magnitude of the displacement is not necessarily equal to the total distance traveled; it is the straight-line distance from the initial position to the final position.
- Average velocity is the constant velocity that would cause the same displacement in the same amount of time.

$$v_{av,x} = \frac{\Delta x}{\Delta t} \text{ (for any time interval } \Delta t) \qquad (2\text{-}2)$$

- Velocity is a measure of how fast and in what direction something moves. Its direction is the direction of the object's motion and its magnitude is the instantaneous speed. It is the instantaneous rate of change of the position.

$$v_x = \lim_{\Delta t \to 0} \frac{\Delta x}{\Delta t} \text{ (for a \textit{very short} time interval } \Delta t) \qquad (2\text{-}3)$$

- Average acceleration is the constant acceleration that would cause the same velocity change in the same amount of time.

$$a_{av,x} = \frac{\Delta v_x}{\Delta t} \text{ (for any time interval } \Delta t) \qquad (2\text{-}7)$$

- Acceleration is the instantaneous rate of change of the velocity.

$$a_x = \lim_{\Delta t \to 0} \frac{\Delta v_x}{\Delta t} \text{ (for a \textit{very short} time interval } \Delta t) \qquad (2\text{-}8)$$

  Acceleration does not necessarily mean speeding up. A velocity can change by decreasing speed or by changing direction.
- Interpreting graphs: On a graph of $x(t)$, the slope at any point is v_x. On a graph of $v_x(t)$, the slope at any point is a_x, and the area under the graph during any time interval is the displacement Δx during that time interval. If v_x is negative, the displacement is also

continued on next page

Master the Concepts continued

negative, so we must count the area as negative when it is below the time axis.

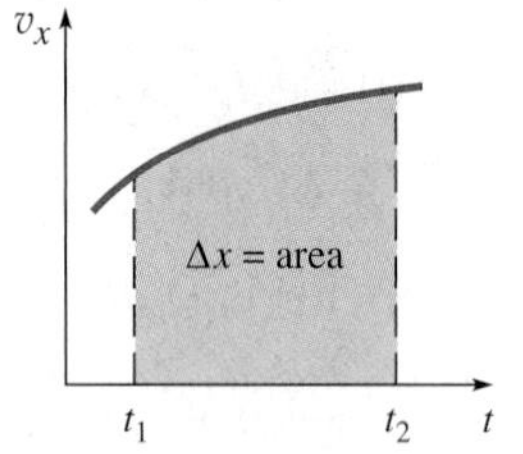

On a graph of $a_x(t)$, the area under the curve is Δv_x, the change in v_x during that time interval.

- Essential relationships for constant acceleration problems: if a_x is constant during the entire time interval Δt from t_i until a later time $t_f = t_i + \Delta t$,

$$\Delta v_x = v_{fx} - v_{ix} = a_x \Delta t \quad (2\text{-}9)$$

$$\Delta x = \tfrac{1}{2}(v_{fx} + v_{ix})\Delta t \quad (2\text{-}11)$$

$$\Delta x = v_{ix}\,\Delta t + \tfrac{1}{2}a_x(\Delta t)^2 \quad (2\text{-}12)$$

$$v_{fx}^2 - v_{ix}^2 = 2a_x\Delta x \quad (2\text{-}13)$$

These same relationships hold for position, velocity, and acceleration along the y-axis if a_y is constant.

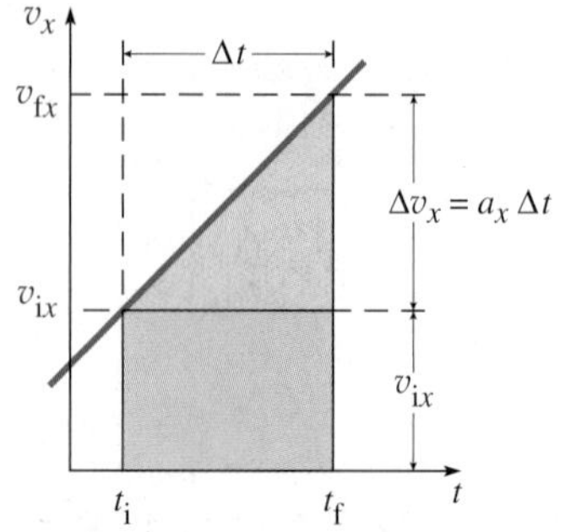

- An object in free fall has a constant downward acceleration. The magnitude of the acceleration g varies a little from place to place near Earth's surface. A typical value is $g = 9.80\ \text{m/s}^2$.

Conceptual Questions

1. Explain the difference between distance traveled, displacement, and displacement magnitude.
2. Explain the difference between speed and velocity.
3. On a graph of v_x versus time, what quantity does the area under the graph represent?
4. On a graph of v_x versus time, what quantity does the slope of the graph represent?
5. On a graph of a_x versus time, what quantity does the area under the graph represent?
6. On a graph of x versus time, what quantity does the slope of the graph represent?
7. What is the relationship between average velocity and instantaneous velocity? An object can have different instantaneous velocities at different times. Can the same object have different average velocities? Explain.
8. Can the velocity of an object be zero and the acceleration be nonzero at the same time? Explain.
9. You are bicycling along a straight north-south road. Let the x-axis point north. Describe your motion in each of the following cases. Example: $a_x > 0$ and $v_x > 0$ means you are moving north and speeding up. (a) $a_x > 0$ and $v_x < 0$. (b) $a_x = 0$ and $v_x < 0$. (c) $a_x < 0$ and $v_x = 0$. (d) $a_x < 0$ and $v_x < 0$. (e) Based on your answers, explain why it is not a good idea to use the expression "negative acceleration" to mean slowing down.
10. When a coin is tossed straight up, what can you say about its velocity and acceleration at the highest point of its motion?

Multiple-Choice Questions

1. A ball is thrown straight up into the air. Ignore air resistance. While the ball is in the air its acceleration
   (a) increases.
   (b) is zero.
   (c) remains constant.
   (d) decreases on the way up and increases on the way down.
   (e) changes direction.
2. Which car has a westward acceleration?
   (a) a car traveling westward at constant speed
   (b) a car traveling eastward and speeding up
   (c) a car traveling westward and slowing down
   (d) a car traveling eastward and slowing down
   (e) a car starting from rest and moving toward the east

Questions 3 and 4. A toy rocket is propelled straight upward from the ground and reaches a height Δy. After an elapsed time Δt, measured from the time the rocket was first fired off, the rocket has fallen back down to the ground, landing at the same spot from which it was launched.

Answer choices:

(a) zero (b) $2\dfrac{\Delta y}{\Delta t}$

(c) $\dfrac{\Delta y}{\Delta t}$ (d) $\dfrac{1}{2}\dfrac{\Delta y}{\Delta t}$

3. What is the magnitude of the average velocity of the rocket during this time?
4. What is the average speed of the rocket during this time?

5. A leopard starts from rest at $t = 0$ and runs in a straight line with a constant acceleration until $t = 3.0$ s. The distance covered by the leopard between $t = 1.0$ s and $t = 2.0$ s is
   (a) the same as the distance covered during the first second.
   (b) twice the distance covered during the first second.
   (c) three times the distance covered during the first second.
   (d) four times the distance covered during the first second.

Multiple-Choice Questions 6–15. A jogger is exercising along a long, straight road that runs north-south. She starts out heading north. A graph of $v_x(t)$ follows Question 10.

6. What is the displacement of the jogger from $t = 18.0$ min to $t = 24.0$ min?
   (a) 720 m, south
   (b) 720 m, north
   (c) 2160 m, south
   (d) 3600 m, north
7. What is the displacement of the jogger for the entire 30.0 min?
   (a) 3120 m, south
   (b) 2400 m, north
   (c) 2400 m, south
   (d) 3840 m, north
8. What is the total distance traveled by the jogger in 30.0 min?
   (a) 3840 m (b) 2340 m (c) 2400 m (d) 3600 m
9. What is the average velocity of the jogger during the 30.0 min?
   (a) 1.3 m/s, north
   (b) 1.7 m/s, north
   (c) 2.1 m/s, north
   (d) 2.9 m/s, north
10. What is the average speed of the jogger for the 30 min?
   (a) 1.4 m/s
   (b) 1.7 m/s
   (c) 2.1 m/s
   (d) 2.9 m/s

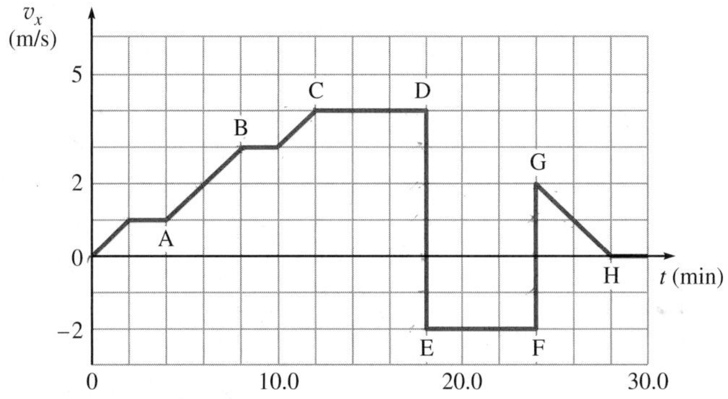

Multiple-Choice Questions 6–15

11. In what direction is she running at time $t = 20$ min?
   (a) south
   (b) north
   (c) not enough information
12. In which region of the graph is a_x positive?
   (a) A to B
   (b) C to D
   (c) E to F
   (d) G to H
13. In which region is a_x negative?
   (a) A to B
   (b) C to D
   (c) E to F
   (d) G to H
14. In which region is the velocity directed to the south?
   (a) A to B
   (b) C to D
   (c) E to F
   (d) G to H
15. What distance does the jogger travel during the first 10.0 min ($t = 0$ to 10.0 min)?
   (a) 8.5 m
   (b) 510 m
   (c) 900 m
   (d) 1020 m
16. The figure shown here has four graphs of x versus time. Which graph shows a constant, positive, nonzero velocity?

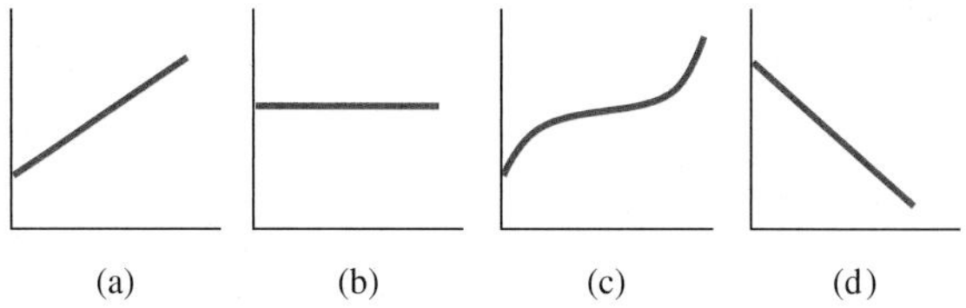

Multiple-Choice Questions 16 and 17

17. The four graphs show v_x versus time. (a) Which graph shows a constant velocity? (b) Which graph shows a_x constant and positive? (c) Which graph shows a_x constant and negative? (d) Which graph shows a changing a_x that is always positive?

Problems

- Ⓒ Combination conceptual/quantitative problem
- ⚕ Biological or medical application
- ✦ Challenging problem
- Blue # Detailed solution in the Student Solutions Manual
- ⎣1 2⎦ Problems paired by concept
- WWW Text website interactive or tutorial

2.1 Position and Displacement

1. A displacement of magnitude 32 cm toward the east is followed by displacements of magnitude 48 cm to the east and then 64 cm to the west. What is the total displacement?
2. A squirrel is trying to locate some nuts he buried for the winter. He moves 4.0 m to the right of a stone and digs unsuccessfully. Then he moves 1.0 m to the left of his hole, changes his mind, and moves 6.5 m to the right of that position and digs a second hole. No luck. Then he moves 8.3 m to the left and digs again. He finds a nut at last. What is the squirrel's total displacement from its starting point?
3. A runner, jogging along a straight line path, starts at a position 60 m east of a milestone marker and heads west. After a short time interval he is 20 m west of the mile marker. Choose east to be the positive x-direction. (a) What is the runner's displacement from his starting point? (b) What is his displacement from the milestone? (c) The runner then turns around and heads east. If at a later time the runner is 140 m east of the milestone, what is his displacement from the starting point at this time? (d) What is the total distance traveled from the starting point if the runner stops at the final position listed in part (c)?
4. Johannes bicycles from his dorm to the pizza shop that is 3.00 mi east. Darren's apartment is located 1.50 mi west of Johannes's dorm. If Darren is able to meet Johannes at the pizza shop by bicycling in a straight line, what is the distance and direction he must travel?
5. At 3 P.M. a car is located 20 km south of its starting point. One hour later it is 96 km farther south. After two more hours, it is 12 km south of the original starting point. (a) What is the displacement of the car between 3 P.M. and 6 P.M.? (b) What is the displacement of the car from the starting point to the location at 4 P.M.? (c) What is the displacement of the car between 4 P.M. and 6 P.M.?

2.2 Velocity: Rate of Change of Position

6. For the train of Example 2.2, find the average velocity between 3:14 P.M. when the train is at 3 km east of the origin and 3:28 P.M. when it is 10 km east of the origin.
7. A cyclist travels 10.0 km east in a time of 11 min 40 s. What is his average velocity in meters per second?
8. In a game against the White Sox, baseball pitcher Nolan Ryan threw a pitch measured at 45.1 m/s. If it was 18.4 m from Nolan's position on the pitcher's mound to home plate, how long did it take the ball to get to the batter waiting at home plate? Treat the ball's velocity as constant and ignore any gravitational effects.
9. Jason drives due west with a speed of 35.0 mi/h for 30.0 min, then continues in the same direction with a speed of 60.0 mi/h for 2.00 h, then drives farther west at 25.0 mi/h for 10.0 min. What is Jason's average velocity for the entire trip?
10. Two cars, a Toyota Yaris and a Jeep, are traveling in the same direction, although the Yaris is 186 m behind the Jeep. The speed of the Yaris is 24.4 m/s and the speed of the Jeep is 18.6 m/s. How much time does it take for the Yaris to catch the Jeep? [*Hint:* What must be true about the displacement of the two cars when they meet?] (tutorial: catchup)
11. Speedometer readings are obtained and graphed as a car comes to a stop along a straight-line path. How far does the car move between $t = 0$ and $t = 16$ s? (tutorial: start/stop traffic)

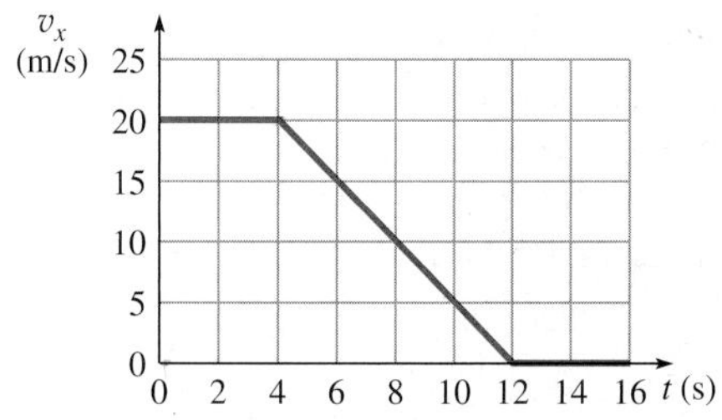

Problems 11 and 29

12. A graph is plotted of the vertical velocity v_y of an elevator versus time. The y-axis points up. (a) How high is the elevator above the starting point ($t = 0$) after 20 s has elapsed? (b) When is the elevator at its highest location above the starting point?

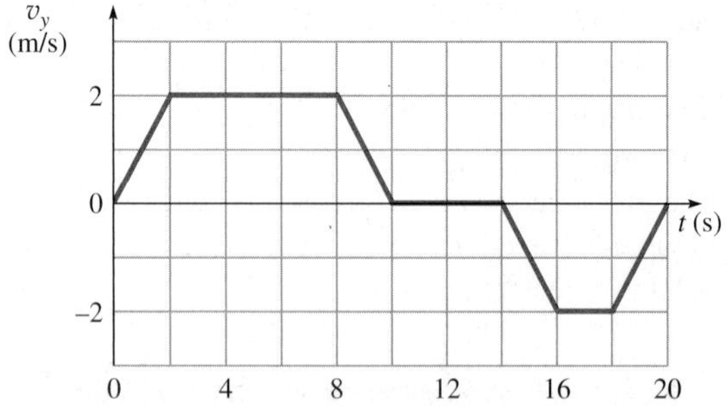

13. A bicycle is moving along a straight line. The graph in the figure shows its position from the starting point as a function of time. (a) In which section(s) of the graph does the object have the highest speed? (b) At which time(s) does the object reverse its direction of

motion? (c) How far does the object move from $t = 0$ to $t = 3$ s?

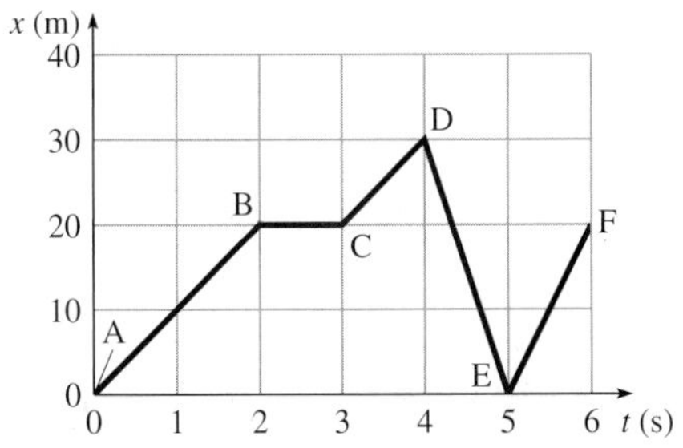

14. A ball thrown by a pitcher on a women's softball team is timed at 65.0 mph. The distance from the pitching rubber to home plate is 43.0 ft. In major league baseball the corresponding distance is 60.5 ft. If the batter in the softball game and the batter in the baseball game are to have equal times to react to the pitch, with what speed must the baseball be thrown? Assume the ball travels with a constant velocity. [*Hint:* There is no need to convert units; set up a ratio.]

15. A motor scooter travels east at a speed of 12 m/s. The driver then reverses direction and heads west at 15 m/s. What is the change in velocity of the scooter? Give magnitude and direction.

16. To pass a physical fitness test, Massimo must run 1000 m at an average rate of 4.0 m/s. He runs the first 900 m in 250 s. Is it possible for Massimo to pass the test? If so, how fast must he run the last 100 m to pass the test? Explain.

17. The graph shows speedometer readings, in meters per second (on the vertical axis), obtained as a skateboard travels along a straight-line path. How far does the board move between $t = 3.00$ s and $t = 8.00$ s?

18. The graph shows values of $x(t)$ in meters, on the vertical axis, for a skater traveling in a straight line. (a) What is $v_{av,x}$ for the interval from $t = 0$ to $t = 4.0$ s? (b) from $t = 0$ to $t = 5.0$ s?

19. The graph shows values of $x(t)$ in meters for a skater traveling in a straight line. What is v_x at $t = 2.0$ s?

20. The graph shows values of $x(t)$ in meters for an object traveling in a straight line. Plot v_x as a function of time for this object from $t = 0$ to $t = 8$ s.

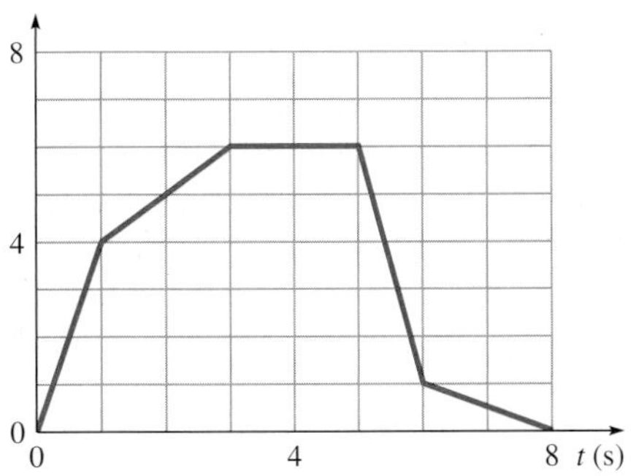

Problems 17, 18, 19, and 20

21. A chipmunk, trying to cross a road, first moves 80 cm to the right, then 30 cm to the left, then 90 cm to the right, and finally 310 cm to the left. (a) What is the chipmunk's total displacement? (b) If the elapsed time was 18 s, what was the chipmunk's average speed? (c) What was its average velocity?

22. Rita Jeptoo of Kenya was the first female finisher in the 110th Boston Marathon. She ran the first 10.0 km in a time of 0.5689 h. Assume the race course to be along a straight line. (a) What was her average speed during the first 10.0 km segment of the race? (b) She completed the entire race, a distance of 42.195 km, in a time of 2.3939 h. What was her average speed for the race?

23. A relay race is run along a straight-line track of length 300.0 m running south to north. The first runner starts at the south end of the track and passes the baton to a teammate at the north end of the track. The second runner races back to the start line and passes the baton to a third runner who races 100.0 m northward to the finish line. The magnitudes of the average velocities of the first, second, and third runners during their parts of the race are 7.30 m/s, 7.20 m/s, and 7.80 m/s, respectively. What is the average velocity of the baton for the entire race? [*Hint:* You will need to find the time spent by each runner in completing her portion of the race.]

2.3 Acceleration: Rate of Change of Velocity

24. If a pronghorn antelope accelerates from rest in a straight line with a constant acceleration of 1.7 m/s^2, how long does it take for the antelope to reach a speed of 22 m/s?

25. If a car traveling at 28 m/s is brought to a full stop in 4.0 s after the brakes are applied, find the average acceleration during braking.

26. An 1100-kg airplane starts from rest; 8.0 s later it reaches its takeoff speed of 35 m/s. What is the average acceleration of the airplane during this time?

27. A rubber ball is attached to a paddle by a rubber band. The ball is initially moving away from the paddle with a speed of 4.0 m/s. After 0.25 s, the ball is moving toward the paddle with a speed of 3.0 m/s. What is the average acceleration of the ball during that 0.25 s? Give magnitude and direction.

28. (a) In Fig. 2.11, what is the instantaneous acceleration of the sports car of Example 2.5 at the time of 14 s from the start? (b) What is the displacement of the car from $t = 12.0$ s to $t = 16.0$ s? (c) What is the average velocity of the car in the 4.0-s time interval from 12.0 s to 16.0 s?

29. The graph with Problem 11 shows speedometer readings as a car comes to a stop. What is the magnitude of the acceleration at $t = 7.0$ s?

30. The figure shows a plot of $v_x(t)$ for a car traveling in a straight line. (a) What is $a_{av,x}$ between $t = 6$ s and $t = 11$ s? (b) What is $v_{av,x}$ for the same time interval? (c) What is $v_{av,x}$ for the interval $t = 0$ to $t = 20$ s? (d) What is the increase in the car's speed between 10 s and 15 s? (e) How far does the car travel from time $t = 10$ s to time $t = 15$ s?

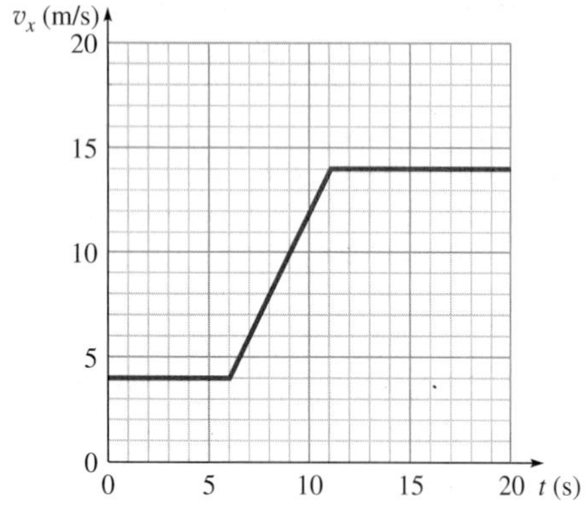

31. The graph shows v_x versus t for a body moving along a straight line. (a) What is a_x at $t = 11$ s? (b) What is a_x at $t = 3$ s? (c) How far does the body travel from $t = 12$ s to $t = 14$ s? (tutorial: x, v, a)

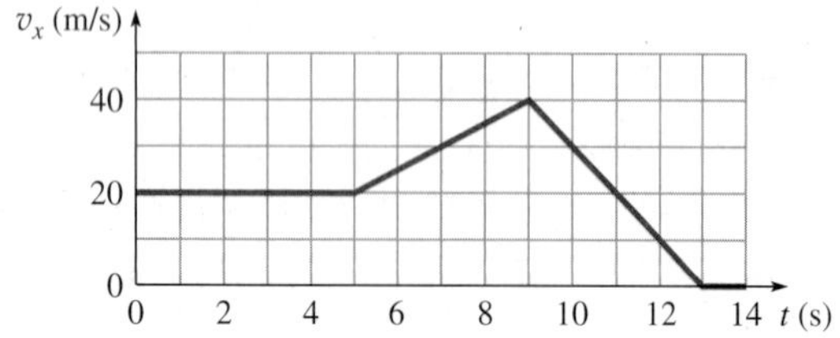

2.4 Motion Along a Line with Constant Acceleration; 2.5 Visualizing Motion Along a Line with Constant Acceleration

32. A toboggan is sliding in a straight line down a snowy slope. The table shows the speed of the toboggan at various times during its trip. (a) Make a graph of the speed as a function of time. (b) Judging by the graph, is it plausible that the toboggan's acceleration is constant? If so, what is the acceleration?

Time Elapsed, t (s)	Speed of Toboggan, v (m/s)
0	0
1.14	2.8
1.62	3.9
2.29	5.6
2.80	6.8

33. The St. Charles streetcar in New Orleans starts from rest and has a constant acceleration of 1.20 m/s^2 for 12.0 s. (a) Draw a graph of v_x versus t. (b) How far has the train traveled at the end of the 12.0 s? (c) What is the speed of the train at the end of the 12.0 s? (d) Draw a motion diagram, showing the streetcar's position at 2.0-s intervals.

34. An airplane lands and starts down the runway with a southwest velocity of 55 m/s. What constant acceleration allows it to come to a stop in 1.0 km?

35. A train is traveling south at 24.0 m/s when the brakes are applied. It slows down with a constant acceleration to a speed of 6.00 m/s in a time of 9.00 s. (a) Draw a graph of v_x versus t for a 12-s interval (starting 2 s before the brakes are applied and ending 1 s after the brakes are released). Let the x-axis point to the north. (b) What is the acceleration of the train during the 9.00-s interval? (c) How far does the train travel during the 9.00 s?

36. A 1200-kg airplane starts from rest and moves forward with a constant acceleration of magnitude 5.00 m/s^2 along a runway that is 250 m long. (a) How long does it take the plane to reach a speed of 46.0 m/s? (b) How far along the runway has the plane moved when it reaches 46.0 m/s?

37. A car is speeding up and has an instantaneous velocity of 1.0 m/s in the $+x$-direction when a stopwatch reads 10.0 s. It has a constant acceleration of 2.0 m/s^2 in the $+x$-direction. (a) What change in speed occurs between $t = 10.0$ s and $t = 12.0$ s? (b) What is the speed when the stopwatch reads 12.0 s?

38. You are driving your car along a country road at a speed of 27.0 m/s. As you come over the crest of a hill, you notice a farm tractor 25.0 m ahead of you on the road, moving in the same direction as you at a speed of 10.0 m/s. You immediately slam on your brakes and slow down with a constant acceleration of magnitude 7.00 m/s^2. Will you hit the tractor before you stop? How far will you travel before you stop or collide with the tractor? If you stop, how far is the tractor in front of you when you finally stop?

39. A train is traveling along a straight, level track at 26.8 m/s (60.0 mi/h). Suddenly the engineer sees a truck stalled on the tracks 184 m ahead. If the maximum possible braking acceleration has magnitude 1.52 m/s^2, can the train be stopped in time?

40. In a cathode ray tube in an old TV, electrons are accelerated from rest with a constant acceleration of magnitude 7.03×10^{13} m/s^2 during the first 2.0 cm of the tube's length; then they move at essentially constant velocity another 45 cm before hitting the screen. (a) Find the speed of the electrons when they hit the

screen. (b) How long does it take them to travel the length of the tube?

41. The graph is of v_x versus t for an object moving along the x-axis. How far does the object move between $t = 9.0$ s and $t = 13.0$ s? Solve using two methods: a graphical analysis and an algebraic solution.

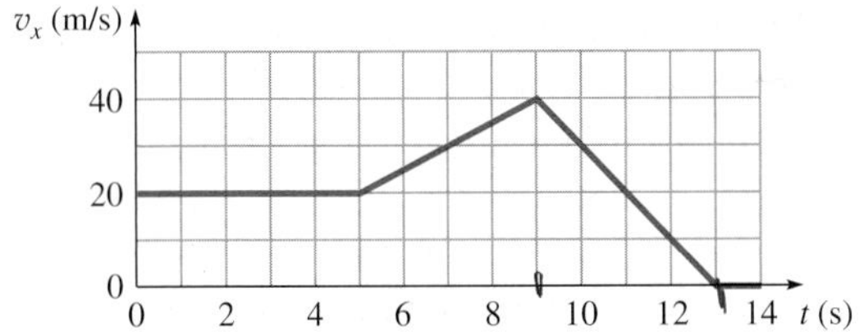

Problems 41–42

42. The graph is of v_x versus t for an object moving along the x-axis. What is the average acceleration between $t = 5.0$ s and $t = 9.0$ s? Solve using two methods: a graphical analysis and an algebraic solution.

43. A train, traveling at a constant speed of 22 m/s, comes to an incline with a constant slope. While going up the incline, the train slows down with a constant acceleration of magnitude 1.4 m/s^2. (a) Draw a graph of v_x versus t where the x-axis points up the incline. (b) What is the speed of the train after 8.0 s on the incline? (c) How far has the train traveled up the incline after 8.0 s? (d) Draw a motion diagram, showing the trains position at 2.0-s intervals.

2.6 Free Fall

In the problems, please assume the free-fall acceleration $g = 9.80$ m/s^2 unless a more precise value is given in the problem statement. Ignore air resistance.

44. A brick is thrown vertically upward with an initial speed of 3.00 m/s from the roof of a building. If the building is 78.4 m tall, how much time passes before the brick lands on the ground?

45. A penny is dropped from the observation deck of the Empire State building (369 m above ground). With what velocity does it strike the ground?

46. (a) How long does it take for a golf ball to fall from rest for a distance of 12.0 m? (b) How far would the ball fall in twice that time?

47. Grant Hill jumps 1.3 m straight up into the air to slam-dunk a basketball into the net. With what speed did he leave the floor?

48. During a walk on the Moon, an astronaut accidentally drops his camera over a 20.0-m cliff. It leaves his hands with zero speed, and after 2.0 s it has attained a velocity of 3.3 m/s downward. How far has the camera fallen after 4.0 s?

49. Glenda drops a coin from ear level down a wishing well. The coin falls a distance of 7.00 m before it strikes the water. If the speed of sound is 343 m/s, how long after Glenda releases the coin will she hear a splash?

50. A stone is launched straight up by a slingshot. Its initial speed is 19.6 m/s and the stone is 1.50 m above the ground when launched. (a) How high above the ground does the stone rise? (b) How much time elapses before the stone hits the ground?

51. A 55-kg lead ball is dropped from the leaning tower of Pisa. The tower is 55 m high. (a) How far does the ball fall in the first 3.0 s of flight? (b) What is the speed of the ball after it has traveled 2.5 m downward? (c) What is the speed of the ball 3.0 s after it is released? (d) If the ball is thrown vertically upward from the top of the tower with an initial speed of 4.80 m/s, where will it be after 2.42 s?

✦52. A balloonist, riding in the basket of a hot air balloon that is rising vertically with a constant velocity of 10.0 m/s, releases a sandbag when the balloon is 40.8 m above the ground. What is the bag's speed when it hits the ground?

✦53. Superman is standing 120 m horizontally away from Lois Lane. A villain throws a rock vertically downward with a speed of 2.8 m/s from 14.0 m directly above Lois. (a) If Superman is to intervene and catch the rock just before it hits Lois, what should be his minimum constant acceleration? (b) How fast will Superman be traveling when he reaches Lois?

✦54. A student, looking toward his fourth-floor dormitory window, sees a flowerpot with nasturtiums (originally on a window sill above) pass his 2.0-m high window in 0.093 s. The distance between floors in the dormitory is 4.0 m. From a window on which floor did the flowerpot fall?

✦55. You drop a stone into a deep well and hear it hit the bottom 3.20 s later. This is the time it takes for the stone to fall to the bottom of the well, plus the time it takes for the sound of the stone hitting the bottom to reach you. Sound travels about 343 m/s in air. How deep is the well?

Comprehensive Problems

In the problems, please assume the free-fall acceleration $g = 9.80$ m/s^2 unless a more precise value is given in the problem statement. Ignore air resistance.

56. (a) If a freestyle swimmer traveled 1500 m in a time of 14 min 53 s, how fast was his average speed? (b) If the pool was rectangular and 50 m in length, how does the

speed you found compare with his sustained swimming speed of 1.54 m/s during one length of the pool after he had been swimming for 10 min? What might account for the difference?

57. While passing a slower car on the highway, you accelerate uniformly from 17.4 m/s to 27.3 m/s in a time of 10.0 s. (a) How far do you travel during this time? (b) What is your acceleration magnitude?

58. A cheetah can accelerate from rest to 24 m/s in 2.0 s. Assuming the acceleration is constant over the time interval, (a) what is the magnitude of the acceleration of the cheetah? (b) What is the distance traveled by the cheetah in these 2.0 s? (c) A runner can accelerate from rest to 6.0 m/s in the same time, 2.0 s. What is the magnitude of the acceleration of the runner? By what factor is the cheetah's average acceleration magnitude greater than that of the runner?

59. A rocket is launched from rest. After 8.0 min, it is 160 km above the Earth's surface and is moving at a speed of 7.6 km/s. Assuming the rocket moves up in a straight line, what are its (a) average velocity and (b) average acceleration?

60. A streetcar named Desire travels between two stations 0.60 km apart. Leaving the first station, it accelerates for 10.0 s at 1.0 m/s^2 and then travels at a constant speed until it is near the second station, when it brakes at 2.0 m/s^2 in order to stop at the station. How long did this trip take? [*Hint:* What's the average velocity?]

61. An unmarked police car starts from rest just as a speeding car passes at a speed of v. If the police car speeds up with a constant acceleration of magnitude a, what is the speed of the police car when it catches up to the speeder, who does not realize she is being pursued and does not vary her speed?

62. A stone is thrown vertically downward from the roof of a building. It passes a window 16.0 m below the roof with a speed of 25.0 m/s. It lands on the ground 3.00 s after it was thrown. What was (a) the initial velocity of the stone and (b) how tall is the building?

63. A car traveling at 29 m/s (65 mi/h) runs into a bridge abutment after the driver falls asleep at the wheel. (a) If the driver is wearing a seat belt and comes to rest within a 1.0-m distance, what is his acceleration (assumed constant)? (b) A passenger who isn't wearing a seat belt is thrown into the windshield and comes to a stop in a distance of 10.0 cm. What is the acceleration of the passenger?

64. To pass a physical fitness test, Marcella must run 1000 m at an average speed of 4.00 m/s. She runs the first 500 m at an average of 4.20 m/s. (a) How much time does she have to run the last 500 m? (b) What should be her average speed over the last 500 m in order to finish with an overall average speed of 4.00 m/s?

65. At 3:00 P.M., a bank robber is spotted driving north on I-15 at milepost 126. His speed is 112.0 mi/h. At 3:37 P.M., he is spotted at milepost 185 doing 105.0 mi/h. During this time interval, what are the bank robber's displacement, average velocity, and average acceleration? (Assume a straight highway.)

66. Based on the information given in Problem 59, is it possible that the rocket moves with constant acceleration? Explain.

67. An elevator starts at rest on the ninth floor. At $t = 0$, a passenger pushes a button to go to another floor. The graph for this problem shows the acceleration a_y of the elevator as a function of time. Let the y-axis point upward. (a) Has the passenger gone to a higher or lower floor? (b) Sketch a graph of the velocity v_y of the elevator versus time. (c) Sketch a graph of the position y of the elevator versus time.

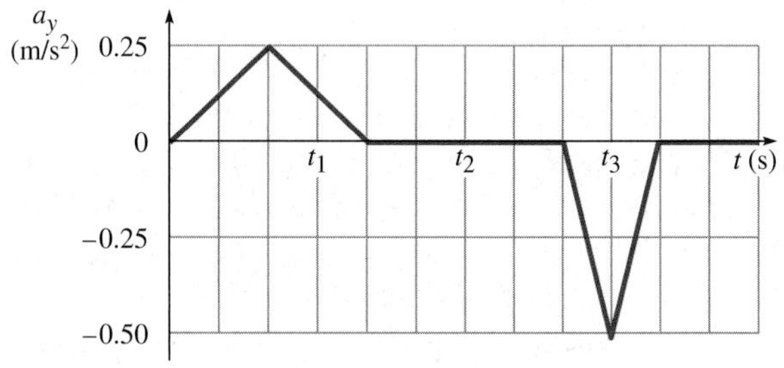

68. The graph for this problem shows the vertical velocity v_y of a bouncing ball as a function of time. The y-axis points up. Answer these questions based on the data in the graph. (a) At what time does the ball reach its maximum height? (b) For how long is the ball in contact with the floor? (c) What is the maximum height of the ball? (d) What is the acceleration of the ball while in the air? (e) What is the average acceleration of the ball while in contact with the floor?

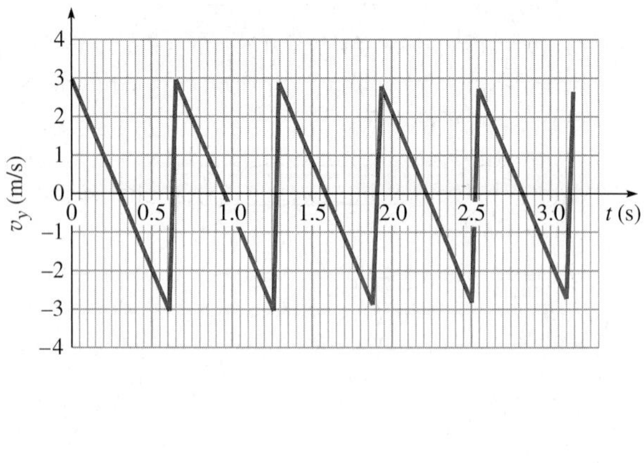

69. A rocket engine can accelerate a rocket launched from rest vertically up with an acceleration of 20.0 m/s^2. However, after 50.0 s of flight the engine fails. (a) What is the rocket's altitude when the engine fails? (b) When does it reach its maximum height? (c) What is the maximum height reached? [*Hint:* A graphical solution may be easiest.] (d) What is the velocity of the rocket just before it hits the ground?

70. The graph shows the position x of a switch engine in a rail yard as a function of time t. At which of the labeled times t_0 to t_7 is (a) $a_x < 0$, (b) $a_x = 0$, (c) $a_x > 0$, (d) $v_x = 0$, (e) the speed decreasing?

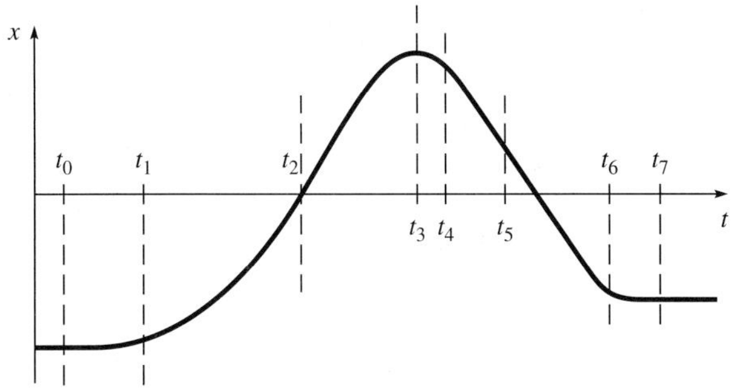

71. An airtrack glider, 8.0 cm long, blocks light as it goes through a photocell gate. The glider is released from rest on a frictionless inclined track and the gate is positioned so that the glider has traveled 96 cm when it is in the middle of the gate. The timer gives a reading of 333 ms for the glider to pass through this gate. Friction is negligible. What is the acceleration (assumed constant) of the glider along the track?

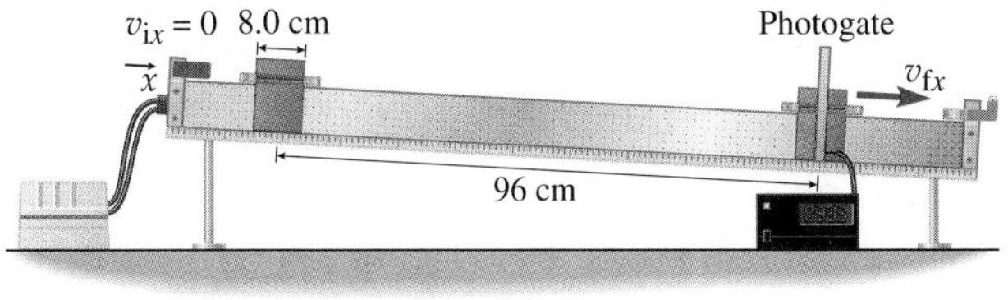

72. Find the point of no return for an airport runway of 1.50 mi in length if a jet plane can accelerate at 10.0 ft/s^2 and decelerate at 7.00 ft/s^2. The point of no return occurs when the pilot can no longer abort the takeoff without running out of runway. What length of time is available from the start of the motion in which to decide on a course of action?

73. In the human nervous system, signals are transmitted along neurons as *action potentials* that travel at speeds of up to 100 m/s. (An action potential is a traveling influx of sodium ions through the membrane of a neuron.) The signal is passed from one neuron to another by the release of neurotransmitters in the synapse. Suppose someone steps on your toe. The pain signal travels along a 1.0-m-long sensory neuron to the spinal column, across a synapse to a second 1.0-m-long neuron, and across a second synapse to the brain. Suppose that the synapses are each 100 nm wide, that it takes 0.10 ms for the signal to cross each synapse, and that the action potentials travel at 100 m/s. (a) At what average speed does the signal cross a synapse? (b) How long does it take the signal to reach the brain? (c) What is the average speed of propagation of the signal?

Answers to Practice Problems

2.1 3.8 m east

2.2 77 km/h in the $-x$-direction (west)

2.3 About 100 to 110 km/h in the $+x$-direction (east)

2.4 The velocity is increasing in magnitude, so the acceleration is in the same direction as the velocity (the $-x$-direction). Thus, a_x is negative; the acceleration is in the $-x$-direction.

2.5

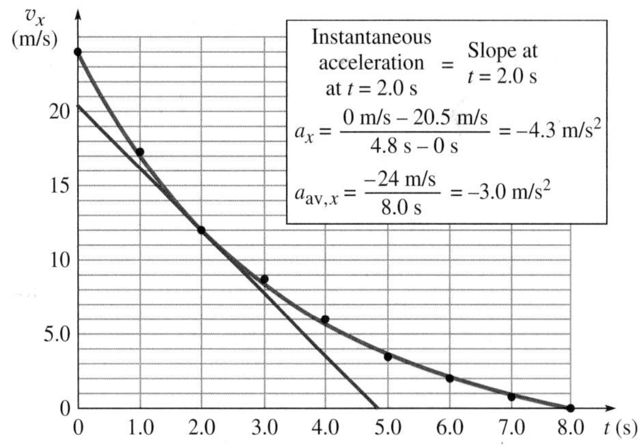

(a) $a_{av,x} = -3.0$ m/s^2 where the negative sign means the average acceleration is directed to the northwest;
(b) $a_x = -4.3$ m/s^2 (northwest)

2.6 2.0 m

2.7 20 s

2.8 5.00 s after they leave the starting point; 4.00 km/s in the $+x$-direction

2.9 (a) 3.8 m; (b) 3.00 s

Answers to Checkpoints

2.1 No. The magnitude of the displacement is the shortest distance between two points. The distance traveled can be greater than or equal to the displacement, depending on the path taken. In Example 2.1 the displacement is 2.9 km to the west, and the distance traveled is 11.5 km.

2.2 Yes. Average speed is the distance traveled divided by the time interval in moving from point A to point B. Average velocity is the displacement from point A to point B divided by the same time interval. The magnitude of the displacement is the shortest possible distance from A to B. Thus the average velocity magnitude is less than or equal to the average speed.

2.3 The slope of the tangent to a graph of v_x versus time is the instantaneous acceleration a_x at the time.

2.4 Only if the plane's acceleration is constant must its average velocity be 470 km/h west. If its acceleration is not constant, the average velocity is not necessarily 470 km/h west. To find the average velocity, we would divide the plane's displacement by the time interval.

2.6 Yes. If you throw a ball upward, it is in free fall as soon as it loses contact with your hand.

CHAPTER

3

Motion in a Plane

A gull scoops up a clam and takes it high above the ground. While flying parallel to the ground, the gull lets go of the clam. The clam lands on a rock below and cracks open. Then the gull alights and enjoys lunch. A beachcomber on the beach sees the clam fall along a parabolic path, just as a projectile would. Why does the clam not drop straight down? What does the path of the falling clam look like to the gull? (See pp. 73 and 76–77 for the answers.)

Concepts & Skills to Review

- trigonometric functions: sine, cosine, and tangent (Appendix A.7)
- Pythagorean theorem (Appendix A.6)
- position, displacement, velocity, and acceleration (Sections 2.1–2.3)
- average and instantaneous quantities (Sections 2.2–2.3)
- motion along a line with constant acceleration (Sections 2.4–2.6)

3.1 GRAPHICAL ADDITION AND SUBTRACTION OF VECTORS

CONNECTION:

Vector quantities must be added and subtracted according to special rules that take their directions into account. All vector quantities follow the *same* rules of addition and subtraction.

Chapter 2 introduced the quantities position, displacement, velocity, and acceleration to describe motion along a line—that is, motion in one dimension of space. To describe motion in more than one dimension, we need a full treatment of vector addition and subtraction because position, displacement, velocity, and acceleration are vectors. (Other vectors you will study in this book include force, momentum, angular momentum, torque, and the electric and magnetic fields.)

Vector quantities have both magnitude and direction.

Vectors and Scalars All **vectors** have a direction as well as a magnitude. The direction of any vector is always a *physical direction in space* such as up, down, north, or 35° south of west.

Vector quantities are usually drawn as arrows pointing in the direction of the vector; the length of the arrow is proportional to the magnitude of the vector. By contrast, a **scalar** quantity can have magnitude, algebraic sign, and units, but not a direction in space. It wouldn't make sense to draw an arrow to represent a scalar such as mass!

In this book, an arrow over a boldface symbol indicates a vector quantity ($\vec{\mathbf{r}}$). (Some books use boldface without the arrow or the arrow without boldface.) When writing by hand, always draw an arrow over a vector symbol to distinguish it from a scalar. When the symbol for a vector is written without the arrow and in italics rather than boldface (r), it stands for the *magnitude* of the vector (which is a scalar). Absolute value bars are also used to stand for the magnitude of a vector, so $r = |\vec{\mathbf{r}}|$. The magnitude of a vector may have units and is never negative; it can be positive or zero.

Conceptual Example 3.1

Vector or Scalar?

Is temperature a vector quantity?

Strategy If a quantity is a vector, it must have both a magnitude and a physical direction in space.

Solution and Discussion Does temperature have a direction? A temperature in Fahrenheit or Celsius can be above or below zero—is that a direction? No. A vector must have a *physical direction* in space. It does not make sense to say that the temperature of your coffee is "85 degrees Celsius in the southwest direction." "The temperature is up 5 degrees today," means that it has increased, not that it is pointing vertically upward. Temperature is a scalar, not a vector.

Conceptual Practice Problem 3.1 Bank Balance

When you deposit a paycheck, the balance of your checking account "goes up." When you pay a bill, it "goes down." Is the balance of your account a vector quantity?

When scalars are added or subtracted, they do so in the usual way: 3 kg of water plus 2 kg of water is equal to 5 kg of water. Adding or subtracting vectors is different. Vectors follow rules of addition and subtraction that take into account the *directions* of the vectors as well as their magnitudes. Whenever you need to add or subtract quantities, check whether they are vectors. If so, be sure to add or subtract them correctly *as vectors. Do not just add or subtract their magnitudes.*

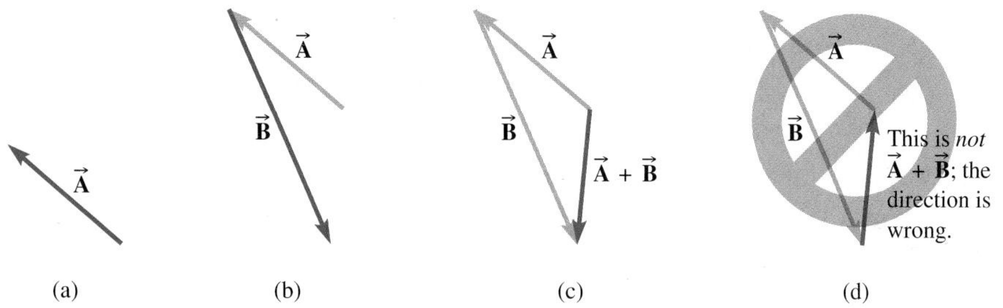

Figure 3.1 Adding two vectors graphically. (a) Draw one vector arrow. (b) Draw the second, starting where the first arrow ended. (c) The sum of the two. (d) A common mistake.

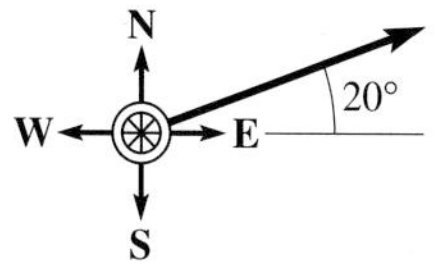

Figure 3.2 Measuring angles with respect to compass headings. The direction of this vector is 20° north of east (20° N of E).

Graphical Vector Addition We start with a graphical method to help develop your intuition. To add two vectors graphically, first draw an arrow to represent one of them (Fig. 3.1a). (It does not matter in what order vectors are added; $\vec{\mathbf{A}} + \vec{\mathbf{B}} = \vec{\mathbf{B}} + \vec{\mathbf{A}}$.) The arrow points in the direction of the vector and its length is proportional to the magnitude of the vector. It doesn't matter where you start drawing the arrow. The value of a vector is not changed by moving it as long as its direction and magnitude are not changed.

Now draw the second vector arrow starting where the first ends. In other words, place the "tail" of the second arrow at the "tip" of the first (Fig 3.1b). Finally, draw an arrow starting from the *tail* of the first and ending at the *tip* of the second. This arrow represents the sum of the two vectors (Fig. 3.1c). A common error is to draw the sum from the tip of the second to the tail of the first (Fig. 3.1d). If the lengths and directions of the vectors are drawn accurately to scale, using a ruler and a protractor, then the length and direction of the sum can be determined with the ruler and protractor. To add more than two vectors, continue drawing them tip to tail.

A plus sign (+) between vector quantities indicates *vector addition*, not ordinary addition. An equals sign (=) between vector quantities means that the vectors are identical in magnitude *and* direction, *not* simply that their magnitudes are equal.

Vector Subtraction *To subtract a vector is to add its opposite* (that is, a vector with the same magnitude but opposite direction): $\vec{\mathbf{r}}_f - \vec{\mathbf{r}}_i = \vec{\mathbf{r}}_f + (-\vec{\mathbf{r}}_i)$. Multiplying a vector by the scalar −1 reverses the vector's direction while leaving its magnitude unchanged, so $-\vec{\mathbf{r}}_i = -1 \times \vec{\mathbf{r}}_i$ is a vector equal in magnitude and opposite in direction to $\vec{\mathbf{r}}_i$.

Using Compass Headings It is common to use compass headings to specify vector directions in a horizontal plane. For example, the direction of the vector in Fig. 3.2 is "20° north of east," which means that the vector makes a 20° angle with the east direction and is on the north (rather than the south) side of east. The same direction could be described as "70° east of north," although it is customary to use the smaller angle. Northeast means "45° north of east" or, equivalently, "45° east of north."

Vector Subtraction: $\vec{\mathbf{A}} - \vec{\mathbf{B}} = \vec{\mathbf{A}} + (-\vec{\mathbf{B}})$, where $-\vec{\mathbf{B}}$ has the same magnitude as $\vec{\mathbf{B}}$ but is opposite in direction. Note that the order matters: $\vec{\mathbf{B}} - \vec{\mathbf{A}} = -(\vec{\mathbf{A}} - \vec{\mathbf{B}})$.

Position and Displacement

The position $\vec{\mathbf{r}}$ of an object can be represented as a vector arrow drawn from the origin to the location of the object (Fig. 3.3). Its magnitude is the distance from the origin. The displacement is literally the *change in position* (the final position vector minus the initial position vector):

$$\Delta\vec{\mathbf{r}} = \vec{\mathbf{r}}_f - \vec{\mathbf{r}}_i \qquad (3\text{-}1)$$

Figure 3.4 shows the graphical subtraction of two position vectors to illustrate the displacement for a trip from Killarney to Kenmare. This same procedure is used to subtract any kind of vector quantity (velocity, acceleration, etc.).

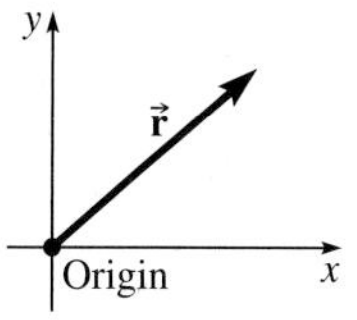

Figure 3.3 A position vector $\vec{\mathbf{r}}$.

Addition of Displacement Vectors As in Example 2.1, the total displacement for a trip with several parts is the vector sum of the displacements for each part of the trip because

$$\vec{\mathbf{r}}_3 - \vec{\mathbf{r}}_1 = (\vec{\mathbf{r}}_3 - \cancel{\vec{\mathbf{r}}_2}) + (\cancel{\vec{\mathbf{r}}_2} - \vec{\mathbf{r}}_1) \qquad (3\text{-}2)$$

Example 3.2 explores this idea further.

Figure 3.4 (a) Two position vectors, $\vec{\mathbf{r}}_i$ and $\vec{\mathbf{r}}_f$, drawn from an *arbitrary origin* to the starting point (Killarney) and to the ending point (Cork) of a trip. (b) The final position vector minus the initial position vector is the displacement $\Delta\vec{\mathbf{r}}$, found by adding $-\vec{\mathbf{r}}_i + \vec{\mathbf{r}}_f$.

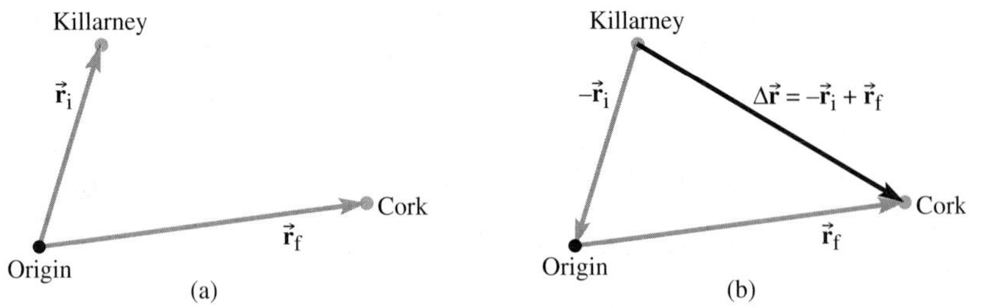

Example 3.2

An Irish Adventure (1)

In a trip from Killarney to Cork, Charlotte and Shona drive at a compass heading of 27° west of south for 18 km to Kenmare, then directly south for 17 km to Glengariff, then at a compass heading of 13° north of east for 48 km to Cork. Find the displacement vector for the entire trip by adding the three displacements graphically.

Blarney castle.

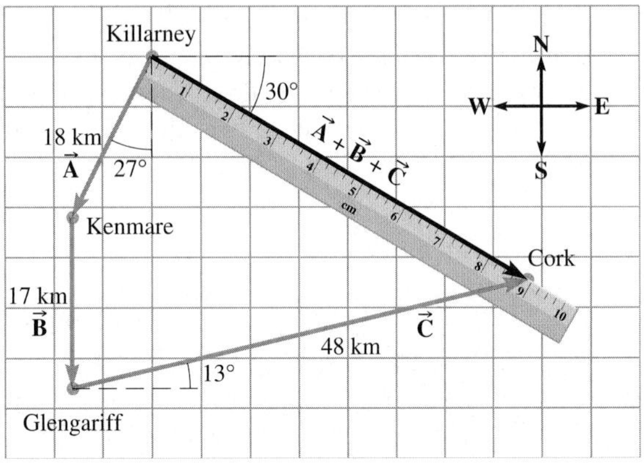

Figure 3.5

Graphical addition of the displacement vectors for the trip from Killarney to Cork via Kenmare and Glengariff.

Strategy To add the displacement vectors, place the tail of each successive vector at the tip of the preceding vector. The value of a vector is not changed by moving it as long as its direction and magnitude are not changed, so a vector can be drawn starting at any point. The sum of the three displacements is then drawn from the tail of the first vector to the tip of the last vector. To add vectors graphically and get an accurate result, we use a ruler and a protractor. The protractor is used to draw the vector arrows in the correct directions and the ruler is used to draw them with the correct lengths. Then the length and direction of the sum can be determined with the ruler and protractor.

Solution Let's call the four positions $\vec{\mathbf{r}}_1$ (Killarney), $\vec{\mathbf{r}}_2$ (Kenmare), $\vec{\mathbf{r}}_3$ (Glengariff), and $\vec{\mathbf{r}}_4$ (Cork). The displacement for the whole trip is $\vec{\mathbf{r}}_4 - \vec{\mathbf{r}}_1$. The problem gives the displacements for the three parts of the trip; let's call them $\vec{\mathbf{A}} = \vec{\mathbf{r}}_2 - \vec{\mathbf{r}}_1 =$ 18 km, 27° west of south; $\vec{\mathbf{B}} = \vec{\mathbf{r}}_3 - \vec{\mathbf{r}}_2 =$ 17 km, south; and $\vec{\mathbf{C}} = \vec{\mathbf{r}}_4 - \vec{\mathbf{r}}_3 =$ 48 km, 13° north of east. The sum of these three displacements is the total displacement because

$$\vec{\mathbf{A}} + \vec{\mathbf{B}} + \vec{\mathbf{C}} = (\cancel{\vec{\mathbf{r}}_2} - \vec{\mathbf{r}}_1) + (\cancel{\vec{\mathbf{r}}_3} - \cancel{\vec{\mathbf{r}}_2}) + (\vec{\mathbf{r}}_4 - \cancel{\vec{\mathbf{r}}_3}) = \vec{\mathbf{r}}_4 - \vec{\mathbf{r}}_1$$

Next we choose a convenient scale for the lengths of the vector arrows. Here we choose to represent 1 km as an arrow length of 0.2 cm, so the length of the vector arrow for $\vec{\mathbf{A}}$ should be

$$18 \text{ km} \times \frac{0.2 \text{ cm}}{1 \text{ km}} = 3.6 \text{ cm}$$

Similarly, the arrows for $\vec{\mathbf{B}}$ and $\vec{\mathbf{C}}$ should be 3.4 cm and 9.6 cm long, respectively.

After drawing the three vector arrows tip to tail, the arrow from the tail of the first vector to the tip of the last vector represents the sum (Fig. 3.5). This arrow is measured to have length 8.9 cm and its direction is 30° south of east. The total displacement has magnitude

$$8.9 \text{ cm} \times \frac{1 \text{ km}}{0.2 \text{ cm}} = 44.5 \text{ km}$$

Rounding to two significant figures, the total displacement $\vec{\mathbf{A}} + \vec{\mathbf{B}} + \vec{\mathbf{C}}$ has magnitude 45 km and is directed 30° south of east.

Discussion Note that the answer includes both the magnitude and direction of the displacement. If a homework or exam question has you calculate a vector quantity such as position or velocity, don't forget to specify the direction as well as the magnitude in your answer. One without the other is incomplete.

Although the magnitude and direction of a position vector depends on the choice of origin, the magnitude and

continued on next page

Example 3.2 continued

direction of a displacement (*change* of position) does *not* depend on the choice of origin.

The total *distance* traveled by Charlotte and Shona is 18 km + 17 km + 48 km = 83 km, which is *not* equal to the magnitude of the total displacement. Finding the total distance involves adding three *scalars*, while finding the total displacement involves adding three *vectors*. The magnitude of the total displacement is the *straight-line* distance from Killarney to Cork.

Practice Problem 3.2 A Traveling Executive

An executive flies from Kansas City to Chicago (displacement = 400 mi in the direction 30° north of east) and then from Chicago to Tulsa (600 mi, 45° south of west). Add the two displacements graphically to find the total displacement from Kansas City to Tulsa.

3.2 VECTOR ADDITION AND SUBTRACTION USING COMPONENTS

Components of a Vector

Any vector can be expressed as the sum of vectors parallel to the x-, y-, and (if needed) z-axes. The x-, y-, and z-components of a vector indicate the magnitude and direction of the three vectors along the three perpendicular axes. The sign of a component indicates the direction along that axis. The x-, y-, and z-components of vector $\vec{\mathbf{A}}$ are written with subscripts as follows: A_x, A_y, and A_z. One exception to this otherwise consistent notation is that the x-, y-, and z-components of a position vector $\vec{\mathbf{r}}$ are usually written x, y, and z (instead of r_x, r_y, and r_z). For now we will deal only with vectors in the xy-plane.

The x-component of a position vector $\vec{\mathbf{r}}$ is x, the x-coordinate. For all other vectors, the x-component is designated by a subscript x. For example, the x-component of a velocity vector $\vec{\mathbf{v}}$ is written v_x. Components of vectors have magnitude, units, and an algebraic sign. The sign indicates the direction: a positive x-component indicates the direction of the positive x-axis, while a negative x-component indicates the opposite direction (the negative x-axis).

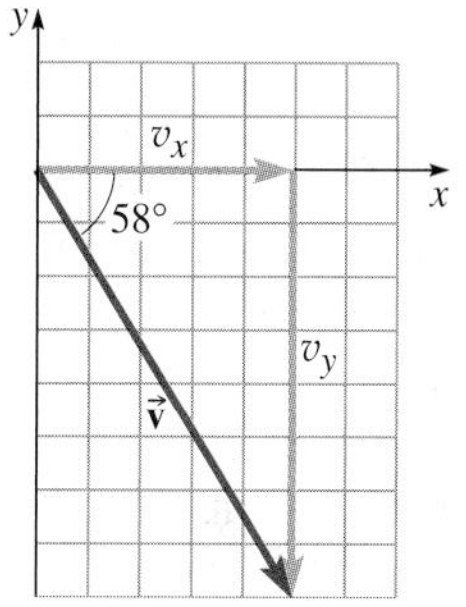

Figure 3.6 Resolving a velocity vector $\vec{\mathbf{v}}$ into x- and y-components.

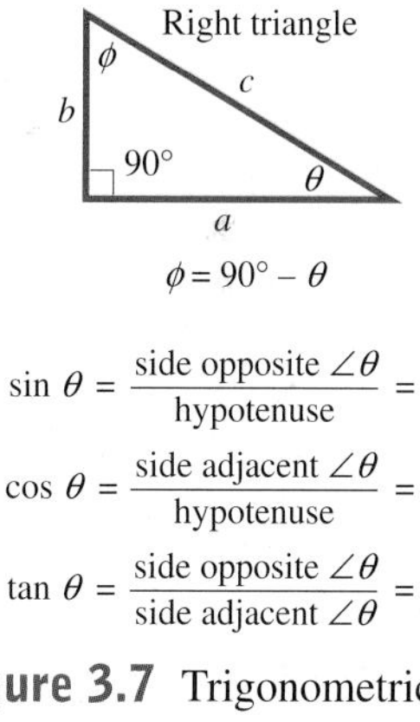

$\phi = 90° - \theta$

$$\sin\theta = \frac{\text{side opposite } \angle\theta}{\text{hypotenuse}} = \frac{b}{c}$$

$$\cos\theta = \frac{\text{side adjacent } \angle\theta}{\text{hypotenuse}} = \frac{a}{c}$$

$$\tan\theta = \frac{\text{side opposite } \angle\theta}{\text{side adjacent } \angle\theta} = \frac{b}{a}$$

Figure 3.7 Trigonometric functions (see Appendix A.7 for more information).

Finding Components The process of finding the components of a vector is called **resolving** the vector into its components. Consider the velocity vector $\vec{\mathbf{v}}$ in Fig. 3.6. We can think of $\vec{\mathbf{v}}$ as the sum of two vectors, one parallel to the x-axis and the other parallel to the y-axis. The magnitudes of these two vectors are the *magnitudes* (absolute values) of the x- and y-components of $\vec{\mathbf{v}}$. We can find the magnitudes of the components using the right triangle in Fig. 3.6 and the trigonometric functions in Fig. 3.7. The length of the arrow represents the magnitude of the vector ($v = 9.4$ m/s), so

$$\cos 58° = \frac{\text{adjacent}}{\text{hypotenuse}} = \frac{|v_x|}{v} \quad \text{and} \quad \sin 58° = \frac{\text{opposite}}{\text{hypotenuse}} = \frac{|v_y|}{v} \qquad (3\text{-}3)$$

Now we must determine the correct algebraic sign for each of the components. From Fig. 3.6, the vector along the x-axis points in the *positive* x-direction and the vector along the y-axis points in the *negative* y-direction, so in this case,

$$v_x = +v\cos 58° = 5.0 \text{ m/s} \quad \text{and} \quad v_y = -v\sin 58° = -8.0 \text{ m/s} \qquad (3\text{-}4)$$

Using the right triangle in Fig. 3.8 gives the same values for the x- and y-components of $\vec{\mathbf{v}}$ since $\cos 32° = \sin 58°$ and $\sin 32° = \cos 58°$.

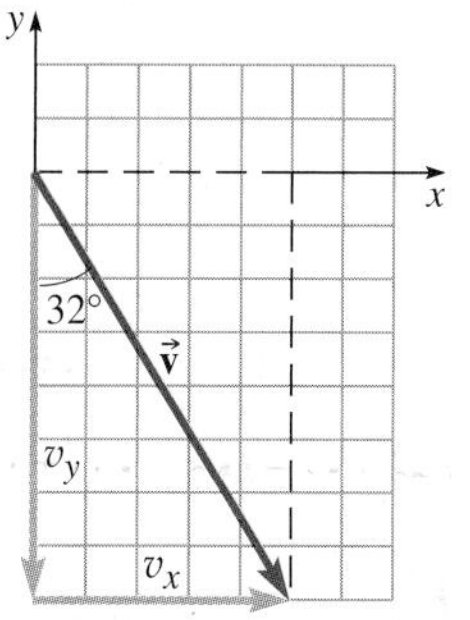

Figure 3.8 Resolving the velocity vector into components using a different right triangle.

Problem-Solving Strategy: Finding the *x*- and *y*-Components of a Vector from Its Magnitude and Direction

1. Draw a right triangle with the vector as the hypotenuse and the other two sides parallel to the *x*- and *y*-axes.
2. Determine one of the unknown angles in the triangle.
3. Use trigonometric functions to find the magnitudes of the components. Make sure your calculator is in "degree mode" to evaluate trigonometric functions of angles in degrees and "radian mode" for angles in radians.
4. Determine the correct algebraic sign for each component.

Finding Magnitude and Direction We must also know how to reverse the process to find a vector's magnitude and direction from its component.

Problem-Solving Strategy: Finding the Magnitude and Direction of a Vector $\vec{\mathbf{A}}$ from Its *x*- and *y*-Components

1. Sketch the vector on a set of *x*- and *y*-axes in the correct quadrant, according to the signs of the components.
2. Draw a right triangle with the vector as the hypotenuse and the other two sides parallel to the *x*- and *y*-axes.
3. In the right triangle, choose which of the unknown angles you want to determine.
4. Use the inverse tangent function to find the angle. The lengths of the sides of the triangle represent $|A_x|$ and $|A_y|$. If θ is opposite the side parallel to the *x*-axis, then $\tan\theta = \text{opposite/adjacent} = |A_x/A_y|$. If θ is opposite the side parallel to the *y*-axis, then $\tan\theta = \text{opposite/adjacent} = |A_y/A_x|$. If your calculator is in "degree mode," then the result of the inverse tangent operation will be in degrees. [In general, the inverse tangent has two possible values between 0 and 360° because $\tan\alpha = \tan(\alpha + 180°)$. However, when the inverse tangent is used to find one of the angles in a right triangle, the result can never be greater than 90°, so the value the calculator returns is the one you want.]
5. Interpret the angle: specify whether it is the angle below the horizontal, or the angle west of south, or the angle clockwise from the negative *y*-axis, etc.
6. Use the Pythagorean theorem to find the magnitude of the vector.

$$A = \sqrt{A_x^2 + A_y^2} \tag{3-5}$$

Suppose we knew the components of the velocity vector in Fig. 3.6, but not the magnitude and direction. Let us find the angle θ between $\vec{\mathbf{v}}$ and the $+x$-axis:

$$\theta = \tan^{-1}\frac{\text{opposite}}{\text{adjacent}} = \tan^{-1}\frac{|v_y|}{|v_x|} = \tan^{-1}\frac{8.0\text{ m/s}}{5.0\text{ m/s}} = 58° \tag{3-6}$$

The magnitude of $\vec{\mathbf{v}}$ is

$$v = \sqrt{v_x^2 + v_y^2} = \sqrt{(+5.0\text{ m/s})^2 + (-8.0\text{ m/s})^2} = 9.4\text{ m/s}$$

Adding Vectors Using Components

It is generally easier and more accurate to add vectors algebraically rather than graphically. The algebraic method relies on adding the components of the vectors. Remember that each vector is thought of as the sum of vectors parallel to the axes (Fig. 3.9a). When

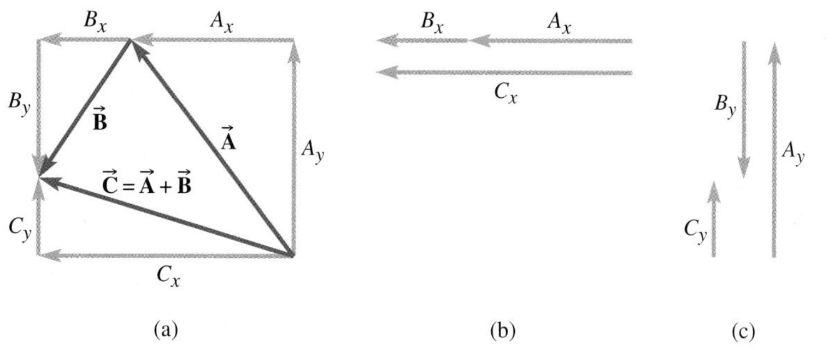

Figure 3.9 (a) $\vec{C} = \vec{A} + \vec{B}$, shown graphically with the x- and y-components of each vector illustrated. (b) $C_x = A_x + B_x$; (c) $C_y = A_y + B_y$.

adding vectors, we can add them in any order and group them as we please. So we can sum the x-components to find the x-component of the sum (Fig. 3.9b) and then do the same with the y-components (Fig. 3.9c):

$$\vec{\mathbf{C}} = \vec{\mathbf{A}} + \vec{\mathbf{B}} \quad \text{if and only if} \quad C_x = A_x + B_x \quad \text{and} \quad C_y = A_y + B_y \qquad (3\text{-}7)$$

In Eq. (3-7), remember that $A_x + B_x$ represents ordinary addition since the signs of the components carry the direction information.

Problem-Solving Strategy: Adding Vectors Using Components

1. Find the x- and y-components of each vector to be added.
2. Add the x-components (*with their algebraic signs*) of the vectors to find the x-component of the sum. (If the signs are not correct, the sum will not be correct.)
3. Add the y-components (with their algebraic signs) of the vectors to find the y-component of the sum.
4. If necessary, use the x- and y-components of the sum to find the magnitude and direction of the sum.

Estimation Using Graphical Addition Even when using the component method to add vectors, the graphical method is an important first step. A rough sketch of vector addition, even one made without carefully measuring the lengths or the angles, has important benefits. Sketching the vectors makes it much easier to get the signs of the components correct. The graphical addition also serves as a check on the answer—it provides an estimate of the magnitude and direction of the sum, which can be used to check the algebraic answer. Graphical addition gives you a mental picture of what is going on and an intuitive feel for the algebraic calculations.

CHECKPOINT 3.2

Two displacements $\vec{\mathbf{A}}$ and $\vec{\mathbf{B}}$ have x- and y-components as follows: $A_x = +3.0$ km, $A_y = -6.0$ km, $B_x = -8.5$ km, $B_y = -1.2$ km. The total displacement is $\vec{\mathbf{C}} = \vec{\mathbf{A}} + \vec{\mathbf{B}}$. What are the x- and y-components of $\vec{\mathbf{C}}$?

Choosing x- and y-Axes

A problem can be made easier to solve with a good choice of axes. We can choose any direction we want for the x- and y-axes, as long as they are perpendicular to one another. Three common choices are

- x-axis horizontal and y-axis vertical, when the vectors all lie in a vertical plane;
- x-axis east and y-axis north, when the vectors all lie in a horizontal plane; and
- x-axis parallel to an inclined surface and y-axis perpendicular to it.

Example 3.3

An Irish Adventure (2)

In the trip of Example 3.2, Charlotte and Shona drive at a compass heading of 27° west of south for 18 km to Kenmare, then directly south for 17 km to Glengariff, then at a compass heading of 13° north of east for 48 km to Cork. Use the component method to find the magnitude and direction of the displacement vector for the entire trip.

Strategy As before, let's call the three successive displacements $\vec{\mathbf{A}}$, $\vec{\mathbf{B}}$, and $\vec{\mathbf{C}}$, respectively. To add the vectors using components, we first choose directions for the *x*- and *y*-axes. Then we find the *x*- and *y*-components of the three displacements. Adding the *x*- or *y*-components of the three displacements gives the *x*- or *y*-component of the total displacement. Finally, from the components we find the magnitude and direction of the total displacement.

Solution A good choice is the conventional one: *x*-axis to the east and the *y*-axis to the north. The first displacement ($\vec{\mathbf{A}}$) is directed 27° west of south. Both of its components are negative since west is the $-x$-direction and south is the $-y$-direction. Using the right triangle in Fig. 3.10, the side of the triangle opposite the 27° angle is parallel to the *x*-axis. The sine function relates the opposite side to the hypotenuse:

$$A_x = -A \sin 27° = -18 \text{ km} \times 0.454 = -8.17 \text{ km}$$

where *A* is the magnitude of $\vec{\mathbf{A}}$. The cosine relates the adjacent side to the hypotenuse:

$$A_y = -A \cos 27° = -18 \text{ km} \times 0.891 = -16.0 \text{ km}$$

Displacement $\vec{\mathbf{B}}$ has no *x*-component since its direction is south. Therefore,

$$B_x = 0 \quad \text{and} \quad B_y = -17 \text{ km}$$

The direction of $\vec{\mathbf{C}}$ is 13° north of east. Both its components are positive. From Fig. 3.10, the side of the right triangle opposite the 13° angle is parallel to the *y*-axis, so

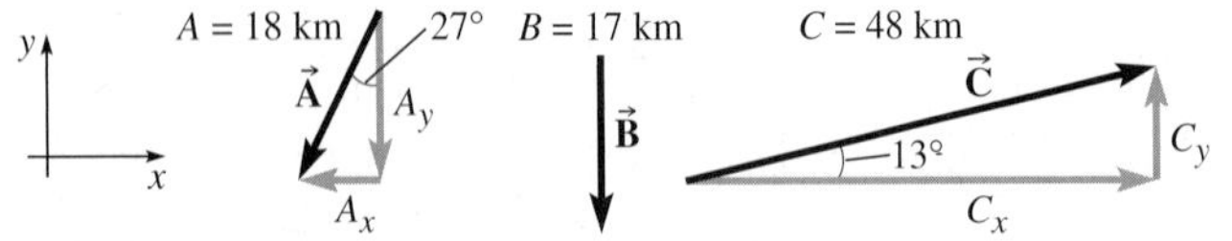

Figure 3.10
Resolving $\vec{\mathbf{A}}$, $\vec{\mathbf{B}}$, and $\vec{\mathbf{C}}$ into *x*- and *y*-components.

$$C_x = +C \cos 13° = +48 \text{ km} \times 0.974 = +46.8 \text{ km}$$
$$C_y = +C \sin 13° = +48 \text{ km} \times 0.225 = +10.8 \text{ km}$$

Now we sum the *x*- and *y*-components separately to find the *x*- and *y*-components of the total displacement:

$$\Delta x = A_x + B_x + C_x$$
$$= (-8.17 \text{ km}) + 0 + 46.8 \text{ km} = +38.63 \text{ km}$$
$$\Delta y = A_y + B_y + C_y$$
$$= (-16.0 \text{ km}) + (-17 \text{ km}) + 10.8 \text{ km} = -22.2 \text{ km}$$

The magnitude and direction of $\Delta\vec{\mathbf{r}}$ can be found from the right triangle in Fig. 3.11. The magnitude is represented by the hypotenuse:

$$\Delta r = \sqrt{(\Delta x)^2 + (\Delta y)^2} = \sqrt{(38.63 \text{ km})^2 + (-22.2 \text{ km})^2}$$
$$= 45 \text{ km}$$

The angle θ is

$$\theta = \tan^{-1}\frac{\text{opposite}}{\text{adjacent}} = \tan^{-1}\frac{22.2 \text{ km}}{38.63 \text{ km}} = 30°$$

Since $+x$ is east and $-y$ is south, the direction of the displacement is 30° south of east. The magnitude and direction of the displacement found using components agree with the displacement found graphically in Fig. 3.5.

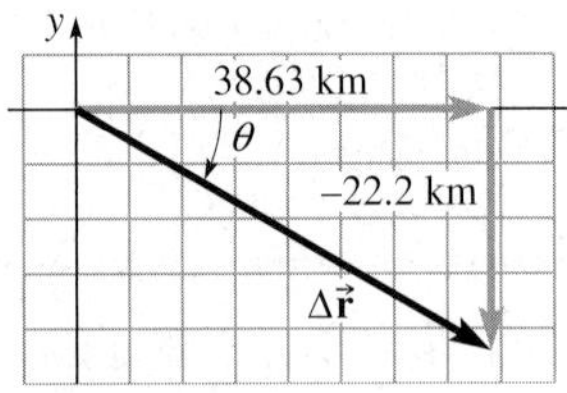

Figure 3.11
Finding the magnitude and direction of $\Delta\vec{\mathbf{r}}$.

Discussion Note that the *x*-component of one displacement was found using the sine function while another was found using the cosine. The *x*-component (or the *y*-component) of the vector can be related to *either* the sine or the cosine, depending on which angle in the right triangle is used.

Practice Problem 3.3 Changing the Coordinate Axes

Find the *x*- and *y*-components of the displacements for the three legs of the trip if the *x*-axis points south and the *y*-axis points east.

Unit Vectors

The connection between a vector and its components may be expressed using the **unit vectors** $\hat{\mathbf{x}}$ (read aloud as "x hat"), $\hat{\mathbf{y}}$, and $\hat{\mathbf{z}}$, which are defined as vectors of magnitude 1 that point in the $+x$-, $+y$-, and $+z$-directions, respectively. (In some books, you may see them written as $\hat{\mathbf{i}}$, $\hat{\mathbf{j}}$, and $\hat{\mathbf{k}}$.) They are called *unit* vectors because the magnitude of each is the pure number 1—they do *not* have physical units such as kilograms or meters. Any vector $\vec{\mathbf{A}}$ can be written as the sum of three vectors along the coordinate axes:

$$\vec{\mathbf{A}} = A_x\hat{\mathbf{x}} + A_y\hat{\mathbf{y}} + A_z\hat{\mathbf{z}} \qquad (3\text{-}8)$$

Here A_x is the x-component of $\vec{\mathbf{A}}$, which has physical units and can be positive or negative. $A_x\hat{\mathbf{x}}$ is a vector of magnitude $|A_x|$ directed in the $+x$-direction if $A_x > 0$ and in the $-x$-direction if $A_x < 0$. For example, consider the velocity vector $\vec{\mathbf{v}}$ of Fig. 3.8. $\vec{\mathbf{v}}$ has x-component $v_x = +5.0$ m/s and y-component $v_y = -8.0$ m/s, so $\vec{\mathbf{v}} = (+5.0 \text{ m/s})\hat{\mathbf{x}} + (-8.0 \text{ m/s})\hat{\mathbf{y}}$.

Using unit vector notation is one way to keep track of vector components in vector addition and subtraction without writing separate equations for each component. Adding two vectors in the xy-plane looks like this:

$$\vec{\mathbf{A}}_1 + \vec{\mathbf{A}}_2 = \left(A_{1x}\hat{\mathbf{x}} + A_{1y}\hat{\mathbf{y}}\right) + \left(A_{2x}\hat{\mathbf{x}} + A_{2y}\hat{\mathbf{y}}\right) \qquad (3\text{-}9)$$

Regrouping the terms shows that the x-component of the sum is the sum of the x-components and likewise for the y-components:

$$\vec{\mathbf{A}}_1 + \vec{\mathbf{A}}_2 = \left(A_{1x} + A_{2x}\right)\hat{\mathbf{x}} + \left(A_{1y} + A_{2y}\right)\hat{\mathbf{y}} \qquad (3\text{-}10)$$

3.3 VELOCITY

The definitions of average velocity, instantaneous velocity, average acceleration, and instantaneous acceleration from Chapter 2 still apply when the motion is not in a straight line as long as we add and subtract them as vectors. Suppose we want to know the instantaneous velocity of a race car at point P as it goes around a curved section of a racetrack (Fig. 3.12a). At a slightly later time the race car is at point Q. Let $\vec{\mathbf{r}}_i$ be the position of the car at P and $\vec{\mathbf{r}}_f$ be the position at point Q.

Average Velocity The displacement $\Delta\vec{\mathbf{r}} = \vec{\mathbf{r}}_f - \vec{\mathbf{r}}_i$ is represented as an arrow from P to Q. Alternatively, to subtract $\vec{\mathbf{r}}_i$ from $\vec{\mathbf{r}}_f$, the two vectors can be drawn with their tails at the same point. After reversing the direction of $\vec{\mathbf{r}}_i$ to represent $-\vec{\mathbf{r}}_i$, the arrows are tip to tail and ready to add $\vec{\mathbf{r}}_f + (-\vec{\mathbf{r}}_i)$—see Fig. 3.12b. The average velocity during this time interval is the displacement $\Delta\vec{\mathbf{r}}$ divided by the time interval:

$$\vec{\mathbf{v}}_{av} = \frac{\vec{\mathbf{r}}_f - \vec{\mathbf{r}}_i}{t_f - t_i} = \frac{\Delta\vec{\mathbf{r}}}{\Delta t} \qquad (3\text{-}11)$$

The direction of the average velocity is the direction of the displacement $\Delta\vec{\mathbf{r}}$.

Instantaneous Velocity The instantaneous velocity at P is the limit of the average velocity as Δt approaches zero. As we shorten the time interval between the initial and final positions by moving point Q closer and closer to P, the direction of the displacement

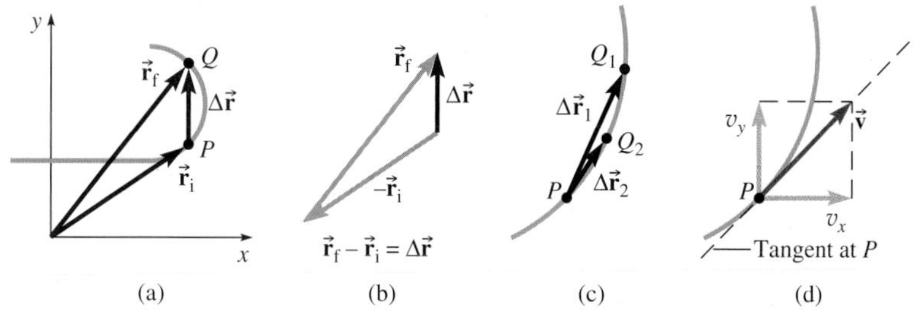

Figure 3.12 (a) Position vectors for two points on the curve. (b) The displacement $\Delta\vec{\mathbf{r}}$ from point P to point Q. (c) As the time interval is decreased, the final point moves closer and closer to P; the direction of the displacement $\Delta\vec{\mathbf{r}}$ approaches the tangent to the curve at P. (d) Instantaneous velocity can be resolved into components along perpendicular axes.

vector $\Delta\vec{\mathbf{r}}$ gradually changes, approaching the tangent to the curved path at P (Fig. 3.12c). Expressed in mathematical terminology, the instantaneous velocity is the limit of $\Delta\vec{\mathbf{r}}/\Delta t$ as the time interval approaches zero:

$$\vec{\mathbf{v}} = \lim_{\Delta t \to 0} \frac{\Delta\vec{\mathbf{r}}}{\Delta t} \quad (3\text{-}12)$$

($\Delta\vec{\mathbf{r}}$ is the displacement during a *very short* time interval Δt)

With this definition, the instantaneous velocity at P becomes tangent to the curve at P (Fig. 3.12d). Here we are talking about a tangent to the actual path through space, *not* a tangent line on a graph of position versus time. The magnitude of the velocity vector is the speed at which the object moves and the direction of the velocity vector is the direction of motion.

If an object moves along a curved path, the direction of the velocity vector at any point is tangent to the path at that point.

Component Equations A vector equation is always equivalent to a set of equations, one for each component. The x- and y-components of the average velocity are

$$v_{\text{av},x} = \frac{\Delta x}{\Delta t} \quad \text{and} \quad v_{\text{av},y} = \frac{\Delta y}{\Delta t} \quad (3\text{-}13)$$

The x- and y-components of the instantaneous velocity are

$$v_x = \lim_{\Delta t \to 0} \frac{\Delta x}{\Delta t} \quad \text{and} \quad v_y = \lim_{\Delta t \to 0} \frac{\Delta y}{\Delta t} \quad (3\text{-}14)$$

To put Eq. (3-14) into words, the x-component of an object's velocity is the rate of change of its x-coordinate and the y-component of its velocity is the rate of change of its y-coordinate.

Example 3.4

An Irish Adventure (3)

In their trip from Kenmare to Cork via Glengariff, Charlotte and Shona travel a total distance of 83 km in 1.4 h. The total displacement for the trip is 45 km, 30° south of east. What is their average velocity? Contrast it with their average speed, defined as the total distance divided by the time interval.

Strategy The average velocity is calculated from the displacement—not from the distance traveled.

Solution The magnitude of the average velocity is

$$|\vec{\mathbf{v}}_{\text{av}}| = \frac{|\Delta\vec{\mathbf{r}}|}{\Delta t} = \frac{45\text{ km}}{1.4\text{ h}} = 32\text{ km/h}$$

The average velocity has the same direction as the displacement, so $\vec{\mathbf{v}}_{\text{av}} = 32$ km/h, 30° south of east. The average speed is

$$\text{average speed} = \frac{83\text{ km}}{1.4\text{ h}} = 59\text{ km/h}$$

Therefore, $|\vec{\mathbf{v}}_{\text{av}}|$ is not equal to the average speed. Furthermore, average velocity is a vector quantity with a direction in space, and average speed is a scalar.

Practice Problem 3.4 Average Velocity Versus Average Speed

In Example 3.4, $|\vec{\mathbf{v}}_{\text{av}}|$ was less than the average speed. Can $|\vec{\mathbf{v}}_{\text{av}}|$ ever be greater than the average speed? Can $|\vec{\mathbf{v}}_{\text{av}}|$ ever be equal to the average speed? Explain.

3.4 ACCELERATION

The average acceleration $\vec{\mathbf{a}}_{\text{av}}$ is the change in velocity divided by the elapsed time:

$$\vec{\mathbf{a}}_{\text{av}} = \frac{\vec{\mathbf{v}}_{\text{f}} - \vec{\mathbf{v}}_{\text{i}}}{t_{\text{f}} - t_{\text{i}}} = \frac{\Delta\vec{\mathbf{v}}}{\Delta t} \quad (3\text{-}15)$$

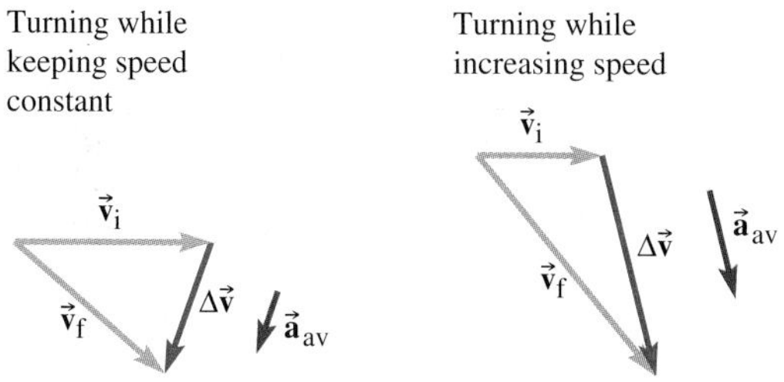

Figure 3.13 Two examples to illustrate that the average acceleration is always in the same direction as the change in velocity $\Delta\vec{\mathbf{v}}$ during the same time interval.

For motion in a plane, this vector equation is equivalent to two component equations:

$$a_{av,x} = \frac{\Delta v_x}{\Delta t} \quad \text{and} \quad a_{av,y} = \frac{\Delta v_y}{\Delta t} \qquad (3\text{-}16)$$

The direction of $\vec{\mathbf{a}}_{av}$ is the same as the direction of $\Delta\vec{\mathbf{v}}$ (Fig. 3.13).

Instantaneous acceleration is the limit of the average acceleration as the time interval approaches zero:

$$\vec{\mathbf{a}} = \lim_{\Delta t \to 0} \frac{\Delta\vec{\mathbf{v}}}{\Delta t} \qquad (3\text{-}17)$$

($\Delta\vec{\mathbf{v}}$ is the change in velocity during a *very short* time interval Δt)

In component form,

$$a_x = \lim_{\Delta t \to 0} \frac{\Delta v_x}{\Delta t} \quad \text{and} \quad a_y = \lim_{\Delta t \to 0} \frac{\Delta v_y}{\Delta t} \qquad (3\text{-}18)$$

In straight-line motion the acceleration is always along the same line as the velocity. For motion in two dimensions, the acceleration vector can make any angle with the velocity vector because the velocity vector can change in magnitude, in direction, or both. The direction of the acceleration is the direction of the *change* in velocity $\Delta\vec{\mathbf{v}}$ during a *very short* time interval.

CHECKPOINT 3.4

An airplane is initially moving due north at 400 km/h. After making a slight course correction, it is moving at the same speed but in a direction 2.0° east of north. Is the plane's average acceleration during this time interval zero? Explain.

Example 3.5

Skating Uphill

An inline skater is traveling on a level road with a speed of 8.94 m/s; 120.0 s later she is climbing a hill with a 15.0° angle of incline at a speed of 7.15 m/s. (a) What is the change in her velocity? (b) What is her average acceleration during the 120.0-s time interval?

Strategy The change in velocity is *not* 1.79 m/s (= 8.94 m/s −7.15 m/s). That is the change in *speed.* The change in velocity is found by subtracting the initial velocity *vector* from the final velocity *vector.* After first making a graphical sketch, we use the component method. The average acceleration is the change in velocity divided by the elapsed time.

continued on next page

Example 3.5 continued

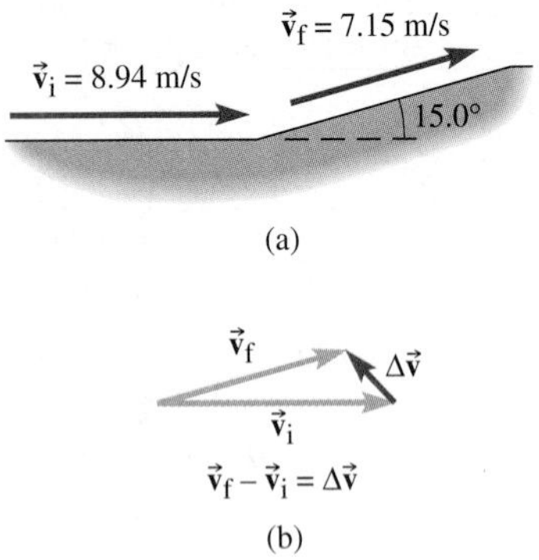

Figure 3.14
(a) Change in velocity as the skater slows going uphill and (b) graphical subtraction of velocity vectors.

Solution (a) Figure 3.14a shows the initial and final velocity vectors and the slope of the hill. The initial velocity is horizontal as the skater skates on level ground. The final velocity is 15.0° above the horizontal. To subtract the two velocity vectors graphically, we place the tails of the vectors together. The change in velocity $\Delta\vec{v}$ is found by drawing a vector arrow from the tip of $\vec{v}_i$ to the tip of $\vec{v}_f$. Judging by the graphical subtraction in Fig. 3.14b, the change in velocity is roughly at a 45° angle above the $-x$-axis. Its magnitude is smaller than the magnitudes of the initial and final velocity vectors—something like 2 to 3 m/s.

The components v_{fx} and v_{fy} can be found from a right triangle (Fig. 3.15):

$$v_{fx} = v_f \cos\theta = 7.15\ \text{m/s} \times 0.9659 = 6.91\ \text{m/s}$$

$$v_{fy} = v_f \sin\theta = 7.15\ \text{m/s} \times 0.2588 = 1.85\ \text{m/s}$$

Since v_i has only an x-component,

$$v_{iy} = 0 \quad \text{and} \quad v_{ix} = v_i = 8.94\ \text{m/s}$$

Now we subtract the components to find the components of $\Delta\vec{v}$:

$$\Delta v_x = v_{fx} - v_{ix} = (6.91 - 8.94)\ \text{m/s} = -2.03\ \text{m/s}$$

and

$$\Delta v_y = v_{fy} - v_{iy} = (1.85 - 0)\ \text{m/s} = +1.85\ \text{m/s}$$

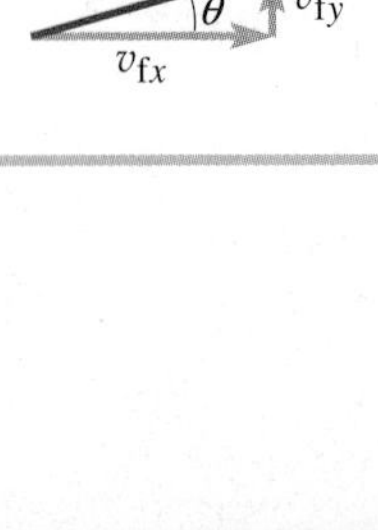

Figure 3.15
Initial and final velocity vectors resolved into components.

To find the magnitude of $\Delta\vec{v}$, we apply the Pythagorean theorem (Fig. 3.16):

$$|\Delta\vec{v}|^2 = (\Delta v_x)^2 + (\Delta v_y)^2 = (-2.03\ \text{m/s})^2 + (1.85\ \text{m/s})^2$$
$$= 7.54\ (\text{m/s})^2$$
$$|\Delta\vec{v}| = 2.75\ \text{m/s}$$

The angle is found from

$$\tan\phi = \frac{\text{opposite}}{\text{adjacent}} = \left|\frac{\Delta v_y}{\Delta v_x}\right| = \frac{1.85\ \text{m/s}}{2.03\ \text{m/s}} = 0.9113$$

$$\phi = \tan^{-1} 0.9113 = 42.3°$$

The direction of the change in velocity $\Delta\vec{v}$ is 42.3° above the negative x-axis.

(b) The magnitude of the average acceleration is

$$|\vec{a}_{av}| = \frac{|\Delta\vec{v}|}{\Delta t} = \frac{2.75\ \text{m/s}}{120.0\ \text{s}} = 0.0229\ \text{m/s}^2$$

The direction of the average acceleration is the same as the direction of $\Delta\vec{v}$: 42.3° above the negative x-axis.

Discussion Checking back with the graphical subtraction in Fig. 3.14b, the magnitude of $\Delta\vec{v}$ appears to be roughly $\frac{1}{4}$ to $\frac{1}{3}$ the magnitude of $\vec{v}_i$. Since $\frac{1}{4} \times 8.94$ m/s = 2.24 m/s and $\frac{1}{3} \times$ 8.94 m/s = 2.98 m/s, the answer of 2.75 m/s is reasonable.

Figure 3.14b also shows the direction of $\Delta\vec{v}$ to be roughly midway between the $+y$- and $-x$-axes. We found the direction of $\Delta\vec{v}$ to be 42.3° above the $-x$-axis and, therefore, 47.7° from the $+y$-axis. So the direction we calculated is also reasonable based on the graphical subtraction.

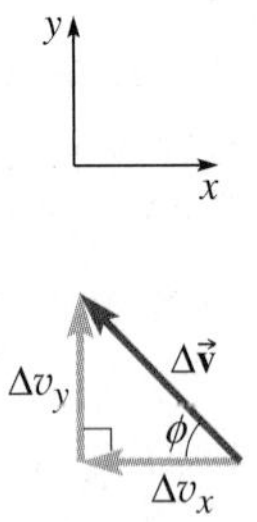

Figure 3.16
Reconstruction of $\Delta\vec{v}$ from its components (not to scale).

Practice Problem 3.5 Change in Sailboat Velocity

A C&C 30 sailboat is sailing at 12.0 knots (6.17 m/s) heading directly east across the harbor. When a gust of wind comes up, the boat changes its heading to 11.0° north of east and its speed increases to 14.0 knots (7.20 m/s). [A boat's speed is customarily expressed in knots, which means nautical miles per hour. A nautical mile (6076 ft) is a little longer than a statute mile (5280 ft).] (a) What is the magnitude and direction of the change in velocity of the sailboat in m/s? (b) If this velocity change occurs during a 2.0-s time interval, what is the average acceleration of the sailboat during that interval?

3.5 MOTION IN A PLANE WITH CONSTANT ACCELERATION

If an object moves in the *xy*-plane with constant acceleration, then both a_x and a_y are constant. By looking separately at the motion along two perpendicular axes, the *y*-direction and the *x*-direction, each component becomes a one-dimensional problem, which we studied in Chapter 2. We can apply any of the constant acceleration relationships from Section 2.4 separately to the *x*-components and to the *y*-components.

It is generally easiest to choose the axes so that the acceleration has only one nonzero component. Suppose we choose the axes so that the acceleration is in the positive or negative *y*-direction. Then $a_x = 0$ and v_x is constant. With this choice, the constant acceleration relationships [Eqs. (2-9) through (2-13)] become

x-axis: $a_x = 0$	**y-axis: constant a_y**	
$\Delta v_x = 0$ (v_x is constant)	$\Delta v_y = a_y \Delta t$	(3-19)
$\Delta x = v_x \Delta t$	$\Delta y = \frac{1}{2}(v_{fy} + v_{iy})\Delta t$	(3-20)
	$\Delta y = v_{iy}\Delta t + \frac{1}{2}a_y(\Delta t)^2$	(3-21)
	$v_{fy}^2 - v_{iy}^2 = 2a_y \Delta y$	(3-22)

Why are only two equations shown in the column for the *x*-axis? The other two are redundant when $a_x = 0$.

Note that there is no mixing of components in Eqs. (3-19) through (3-22). Each equation pertains either to the *x*-components or to the *y*-components; none contains the *x*-component of one vector quantity and the *y*-component of another. The only quantity that appears in both *x*- and *y*-component equations is the time interval—a scalar.

Motion of Projectiles

An object in free fall near the Earth's surface has a constant acceleration. As long as air resistance is negligible, the constant downward pull of gravity gives the object a constant downward acceleration with magnitude *g*. In Section 2.6 we considered objects in free fall, but only when they had no horizontal velocity component, so they moved straight up or straight down. Now we consider objects (called **projectiles**) in free fall that have a *nonzero* horizontal velocity component. The motion of a projectile takes place in a vertical plane.

CONNECTION:

Projectile motion is free fall for objects with a horizontal velocity component.

Suppose some medieval marauders are attacking a castle. They have a catapult that propels large stones into the air to bombard the walls of the castle (Fig. 3.17). Picture a stone leaving the catapult with initial velocity $\vec{\mathbf{v}}_i$. ($\vec{\mathbf{v}}_i$ is the *initial* velocity for the time interval *during which it moves as a projectile.* It is also the *final* velocity for the time interval during which it is in contact with the catapult.) The **angle of elevation** is the angle of the initial velocity above the horizontal. Once the stone is in the air, the only force acting on it is the downward gravitational force, provided that the air resistance has a negligible effect on the motion. The **trajectory** (path) of the stone is shown in Fig. 3.18. The positive *x*-axis is chosen in the horizontal direction (to the right) and the positive *y*-axis is upward.

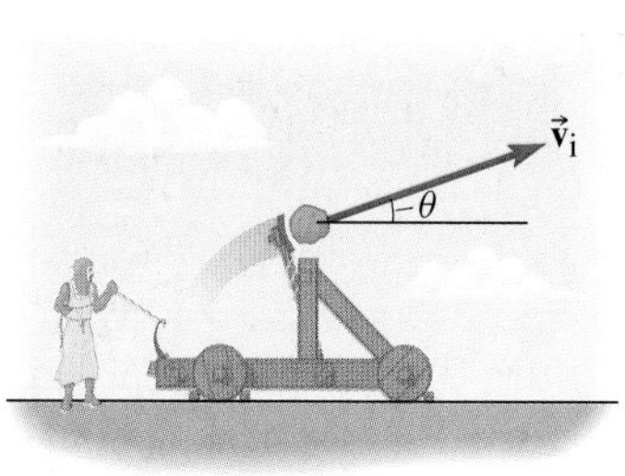

Figure 3.17 A medieval catapult.

If the initial velocity $\vec{\mathbf{v}}_i$ is at an angle θ above the horizontal, then resolving it into components gives

$$v_{ix} = v_i \cos\theta \quad \text{and} \quad v_{iy} = v_i \sin\theta \tag{3-23}$$

(+*y*-axis up, θ measured from the horizontal *x*-axis)

With the *y*-axis pointing up, $a_y = -g$ because the acceleration is downward (in the −*y*-direction). The acceleration has no *x*-component ($a_x = 0$), so the stone's horizontal

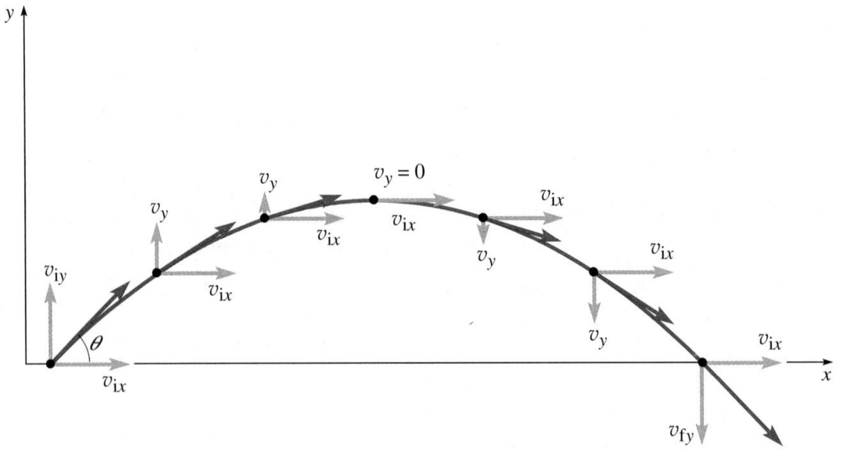

Figure 3.18 Motion diagram showing the trajectory of a projectile. The position is drawn at equal time intervals. Superimposed are the velocity vectors along with their x- and y-components.

velocity component v_x is *constant.* The vertical velocity component v_y changes at a constant rate, exactly as if the stone were propelled straight up with an initial speed of v_{iy}. The initially positive v_y decreases until, at the top of flight, $v_y = 0$. Then the pull of gravity makes the projectile fall back downward. During the downward trip, v_y is still changing at the same constant rate with which it changed on the way up and at the top of the path. The acceleration has the same constant value—magnitude and direction—for the entire path.

The motion of a projectile when air resistance is negligible is the superposition of horizontal motion with constant velocity and vertical motion with constant acceleration. The vertical and horizontal motions each proceed independently, as if the other motion were not present. In the experiment of Fig. 3.19, one ball was dropped and, at the same instant, another was projected horizontally. The strobe photo shows snapshots of the two balls at equally spaced time intervals. The *vertical* motion of the two is identical; at every instant, the two are at the same height. The fact that they have different horizontal motion does not affect their vertical motion. (This statement would *not* be true if air resistance were significant.)

The horizontal and vertical motions of a projectile can be treated separately; they are independent of each other.

PHYSICS AT HOME

Take a nickel and a penny to a room with a high table or countertop. Place the penny at the edge of the table and then slide the nickel so it collides with the penny. Listen for the sound of the two coins hitting the floor. The two coins will slide off the table with different horizontal velocities but will land at the same time.

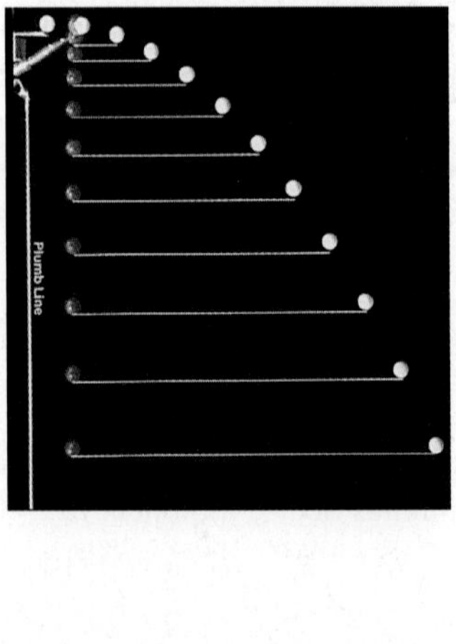

Figure 3.19 Independence of horizontal and vertical motion of a projectile in the absence of air resistance. The vertical motion of the projectile (white) is the same as that of an object (red) that falls straight down.

Conceptual Example 3.6

Trajectory of a Projectile

The graph of an equation of the form

$$y = kx^2,\ k = \text{a nonzero constant}$$

is a parabola. Show that the trajectory of a projectile is a parabola. [*Hint:* Choose the origin at the highest point of the trajectory and let $t_i = 0$ at that instant.]

Strategy and Solution We start at the high point of the path and look at displacements from there. The horizontal displacement is proportional to the elapsed time t since the horizontal velocity is constant. The vertical displacement is the average vertical velocity component times the elapsed time t. The average vertical velocity component is itself proportional to t since it changes at a constant rate. Therefore, the vertical displacement is proportional to t^2. Thus, the vertical displacement y is proportional to the square of the horizontal displacement x and $y = kx^2$, where k is a constant of proportionality. The path followed by a projectile in free fall is a parabola.

Discussion The same conclusion can be drawn algebraically. With the $+y$-axis upward and the origin and $t = 0$ at the top of flight, x_i, y_i, and v_{iy} are all zero. Then $x = v_{ix}t$ and

$$y = v_{iy}t + \frac{1}{2}a_y t^2 = -\frac{1}{2}gt^2 = -\frac{1}{2}g\left(\frac{x}{v_{ix}}\right)^2 = -\left(\frac{g}{2v_{ix}^2}\right)x^2$$

So y is proportional to x^2 and the constant of proportionality is $-g/(2v_{ix}^2)$.

Conceptual Practice Problem 3.6 Throwing Stones

You stand at the edge of a cliff and throw stones horizontally into the river below. To double the horizontal displacement of a stone from the cliff to where it lands, by what factor must you increase the stone's initial speed? Ignore air resistance.

Graphing Projectile Motion Figure 3.20 shows graphs of the x- and y-components of the velocity and position of a projectile as functions of time. In this case, the projectile is launched above flat ground at $t = 0$ and returns to the same elevation at a later time t_f. Note that the y-component graphs are *symmetrical* about the vertical line through the highest point in the trajectory. The y-component of velocity decreases linearly from its initial value; the slope of the line is $a_y = -g$. When $v_y = 0$, the projectile is at the apex of its trajectory. Then v_y continues to decrease at the same rate and is now negative with its magnitude getting larger and larger. At t_f, when the projectile has returned to its original altitude, the y-component of the velocity has the same magnitude as at $t = 0$ but with the opposite sign ($v_y = -v_{iy}$).

The graph of $y(t)$ indicates that the projectile moves upward, quickly at first and then gradually slowing, until it reaches the maximum height. The slope of the tangent to the $y(t)$ graph at any particular moment of time is v_y at that instant. At the highest point

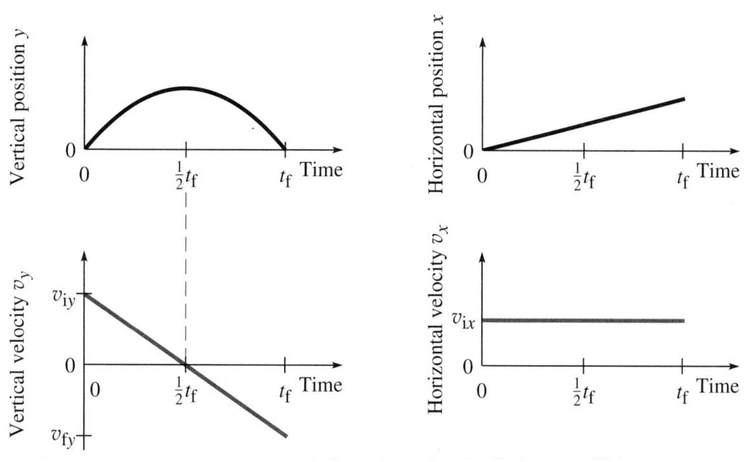

Figure 3.20 Projectile motion: separate vertical and horizontal quantities versus time.

of the $y(t)$ graph, the tangent is horizontal and $v_y = 0$. After that, gravity makes the projectile start to fall downward.

The horizontal velocity is constant, so the graph of $v_x(t)$ is a horizontal line. The horizontal position x increases uniformly in time because the object is moving with a constant v_x.

CHECKPOINT 3.5

When a basketball is thrown in an arc toward the net, what can you say about its velocity and acceleration at the highest point of the arc?

Example 3.7

Attacking the Castle Walls

The catapult used by the marauders hurls a stone with a velocity of 50.0 m/s at a 30.0° angle of elevation (Fig. 3.21). (a) What is the maximum height reached by the stone? (b) What is its *range* (defined as the horizontal distance traveled when the stone returns to its original height)? (c) How long has the stone been in the air when it returns to its original height?

Strategy The problem gives both the magnitude and direction of the initial velocity of the stone. Ignoring air resistance, the stone has a constant downward acceleration once it has been launched—until it hits the ground or some obstacle. We choose the positive y-axis upward and the positive x-axis in the direction of horizontal motion of the stone (toward the castle). When the stone reaches its maximum height, the velocity component in the y-direction is zero since the stone goes no higher. When the stone returns to its original height, $\Delta y = 0$ and $v_y = -v_{iy}$. The range can be found once the time of flight t_f is known—time is the quantity that connects the x-component equations to the y-component equations. Therefore, we solve (c) before (b). One way to find t_f is to find the time to reach maximum height and then double it (see Fig. 3.20). (Other methods include setting $\Delta y = 0$ or setting $v_y = -v_{iy}$.)

Solution (a) First we find the x- and y-components of the initial velocity for an angle of elevation $\theta = 30.0°$.

$$v_{iy} = v_i \sin\theta \quad \text{and} \quad v_{ix} = v_i \cos\theta$$

The maximum height is the vertical displacement Δy when $v_{fy} = 0$.

$$\Delta y = \tfrac{1}{2}(v_{fy} + v_{iy})\,\Delta t = \tfrac{1}{2}(0 + v_i \sin\theta)\,\Delta t$$

Eliminating the time interval using $v_{fy} - v_{iy} = a_y\Delta t$ yields

$$\Delta y = \frac{1}{2}(v_i \sin\theta)\left(\frac{0 - v_i \sin\theta}{a_y}\right) = -\frac{(v_i \sin\theta)^2}{2a_y}$$

$$= \frac{-(50.0\text{ m/s} \times \sin 30.0°)^2}{2 \times (-9.80\text{ m/s}^2)} = 31.9\text{ m}$$

The maximum height of the projectile is 31.9 m above its launch height.

(c) The initial and final heights are the same. Due to this symmetry, the time of flight (t_f) is *twice* the time it takes the projectile to reach its maximum height. The time to reach the maximum height can be found from

$$v_{fy} = 0 = v_{iy} + a_y\Delta t$$

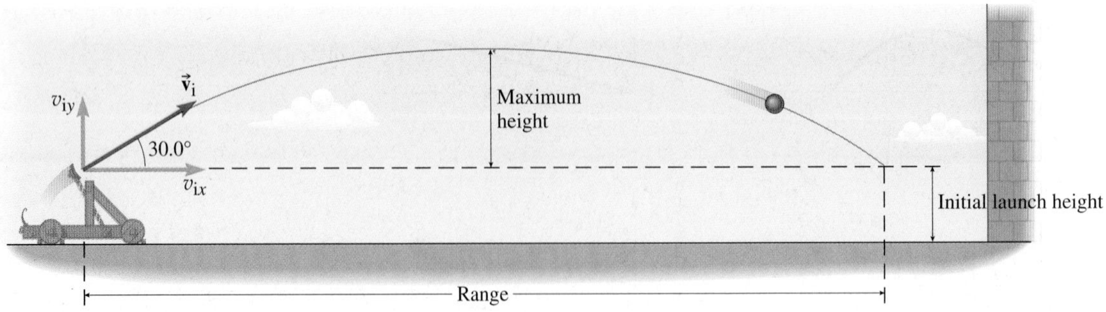

Figure 3.21

A catapult projects a stone into the air in an attack on a castle wall.

continued on next page

Example 3.7 continued

Solving for Δt,

$$\Delta t = \frac{-v_{iy}}{a_y}$$

The time of flight is

$$t_f = 2\,\Delta t = 2 \times \frac{-50.0\text{ m/s} \times \sin 30.0°}{-9.80\text{ m/s}^2} = 5.10\text{ s}$$

(b) The range is

$$\Delta x = v_{ix}\,t_f = (50.0\text{ m/s} \times \cos 30.0°) \times 5.10\text{ s} = 221\text{ m}$$

Discussion Quick check: using

$$y_f - y_i = v_{iy}\,\Delta t + \tfrac{1}{2}a_y\,(\Delta t)^2$$

we can check that $\Delta y = 31.9$ m when $\Delta t = \frac{1}{2} \times 5.10$ s and that $\Delta y = 0$ when $\Delta t = 5.10$ s. Here we check the first of these:

$$\Delta y = (50.0\text{ m/s} \times \sin 30.0°) \times 2.55\text{ s} + \tfrac{1}{2} \times (-9.80\text{ m/s}^2) \times (2.55\text{ s})^2$$
$$= 63.8\text{ m} + (-31.9\text{ m}) = 31.9\text{ m}$$

which is correct. This is not an *independent* check, since this equation can be derived from the others, but it can reveal algebra or calculation errors.

Since we analyze the horizontal motion independently from the vertical motion, we start by resolving the given initial velocity into x- and y-components. Time is what connects the horizontal and vertical motions.

Practice Problem 3.7 Maximum Height for Arrows

Archers have joined in the attack on the castle and are shooting arrows over the walls. If the angle of elevation for an arrow is 45°, find an expression for the maximum height of the arrow in terms of v_i and g. [*Hint:* Simplify the expression using $\sin 45° = \cos 45° = 1/\sqrt{2}$.]

PHYSICS AT HOME

On a warm day, take a garden hose and aim the nozzle so that the water streams upward at an angle above the horizontal. Set the nozzle for a fast, narrow stream for best effect. Once the water leaves the nozzle, it becomes a projectile with a constant downward acceleration (ignoring the small effect of air resistance). The continuous stream of water lets us see the parabolic path easily. Stand in one place and try aiming the nozzle at different angles of elevation to find an angle that gives the maximum range. Aim for a particular spot on the ground (at a distance less than the maximum range) and see if you can find two different angles of elevated nozzle position that allow the stream to hit the target spot (see Fig. 3.22).

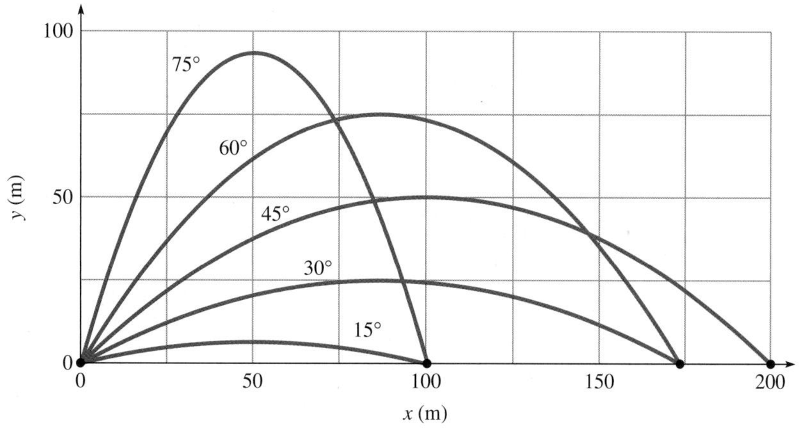

Figure 3.22 Parabolic trajectories of projectiles launched with the same initial speed ($v_i = 44.3$ m/s) at five different angles. The ranges of projectiles launched at angles θ and $90° - \theta$ are the same. The maximum range occurs for $\theta = 45°$.

Conceptual Example 3.8

Monkey and Hunter

An inexperienced hunter aims and shoots an arrow straight at a coconut that is being held by a monkey in a tree (Fig. 3.23). At the same instant that the arrow leaves the bow, the monkey drops the coconut. Ignoring air resistance, does the arrow hit the coconut, the monkey, or neither?

Strategy and Solution If there were no gravity, the arrow would fly straight to the coconut (along the dashed blue line in Fig. 3.23). Since gravity gives the dropped coconut and the released arrow the same constant acceleration downward, they each fall the same vertical distance below the positions they would have had with no gravity. The coconut falls along the dashed red line; the distance fallen at 0.25-s intervals is marked. The arrow falls below the blue dashed line by the same distances, marked along its trajectory at 0.25-s intervals.

The arrow ends up hitting the coconut no matter what the initial speed of the arrow (as long as the arrow's range is at least as large as the horizontal distance to the coconut). The higher the speed of the arrow, the sooner they meet and the shorter the vertical distance that the coconut falls before being hit.

Discussion An experienced hunter would have aimed *above* the initial position of the coconut to compensate for gravity; he would have missed the coconut but might have hit the monkey unless the monkey jumped down to retrieve the coconut.

Conceptual Practice Problem 3.8 Changes in Position and Velocity for Consecutive Arrows

An arrow is shot into the air. One second later, a second arrow is shot with the same initial velocity. While the two are both in the air, does the difference in their positions ($\vec{\mathbf{r}}_2 - \vec{\mathbf{r}}_1$) stay constant or does it change with time? Does the difference in their velocities ($\vec{\mathbf{v}}_2 - \vec{\mathbf{v}}_1$) stay constant or does it change with time?

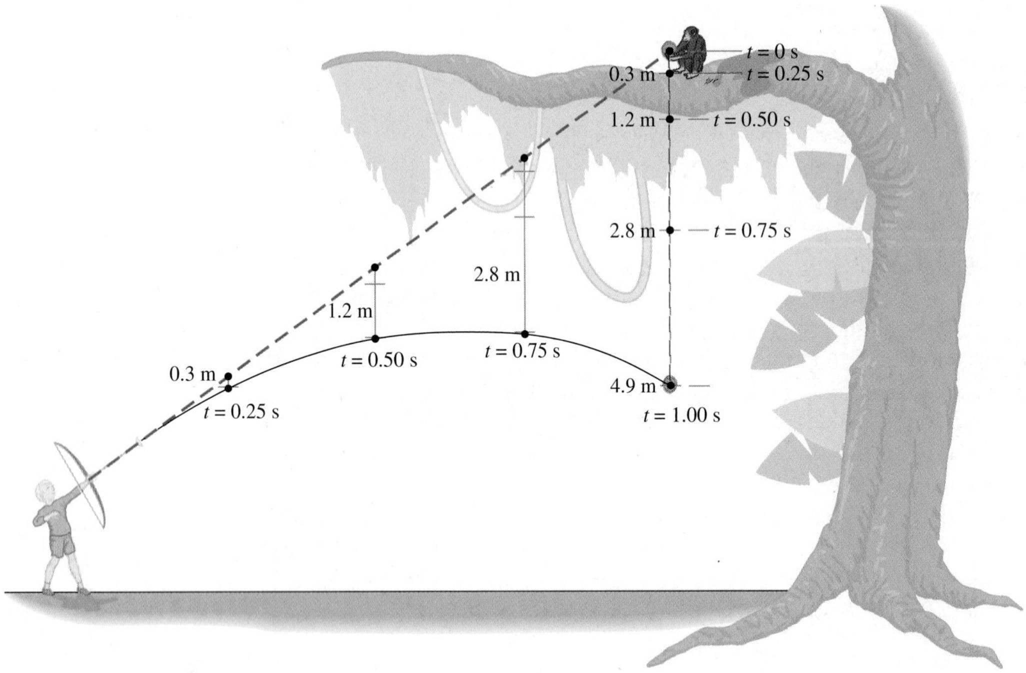

Figure 3.23
A monkey drops a coconut at the very instant an arrow is shot toward the coconut. In each quarter second, the coconut and arrow have fallen the same distance below where their positions would be if there were no gravity.

Example 3.9

A Bullet Fired Horizontally

A bullet is fired horizontally from the top of a cliff that is 20.0 m above a long lake. If the muzzle speed of the bullet is 500.0 m/s, how far from the bottom of the cliff does the bullet strike the surface of the lake? Ignore air resistance.

Strategy We need to find the total time of flight so that we can find the horizontal displacement. The bullet is starting from the high point of the parabolic path because $v_{iy} = 0$. As usual in projectile problems, we choose the y-axis to be the positive vertical direction.

Known: $\Delta y = -20.0$ m; $v_{iy} = 0$; $v_{ix} = 500.0$ m/s. To find: Δx.

Solution The vertical displacement through which the bullet falls is 20.0 m. The relationship between Δy and Δt is

$$\Delta y = \tfrac{1}{2}(v_{fy} + v_{iy})\,\Delta t$$

Substituting $v_{iy} = 0$ and $v_{fy} = v_{iy} + a_y\Delta t = a_y\Delta t$ yields

$$\Delta y = \frac{1}{2}a_y(\Delta t)^2 \Rightarrow \Delta t = \sqrt{\frac{2\,\Delta y}{a_y}}$$

The horizontal displacement of the bullet is

$$\Delta x = v_{ix}\Delta t = v_{ix}\sqrt{\frac{2\,\Delta y}{a_y}}$$

$$= 500.0\text{ m/s} \times \sqrt{\frac{2 \times (-20.0\text{ m})}{-9.8\text{ m/s}^2}} = 1.01\text{ km}$$

Discussion How did we know to start with the y-component equation when the question asks about the *horizontal* displacement? The question gives v_{ix} and asks for Δx. The missing information needed is the time during which the bullet is in the air; the time can be found from analysis of the *vertical* motion.

We ignored air resistance in this problem, which is not very realistic. The actual distance would be less than 1.01 km.

Practice Problem 3.9 Bullet Velocity

Find the horizontal and vertical components of the bullet's velocity just before it hits the surface of the lake. At what angle does it strike the surface?

At the beginning of the chapter, we asked why the clam does not fall straight down when the gull lets go. The gull is flying horizontally with the clam, so the clam has the same horizontal velocity as the gull. When the gull lets go, the clam falls toward Earth, but since $a_x = 0$ the clam retains the same horizontal component of velocity as the gull. Therefore, the clam is a projectile starting at the top of its parabolic trajectory.

Why does the clam not drop straight down?

3.6 VELOCITY IS RELATIVE; REFERENCE FRAMES

The idea of *relativity* arose in physics centuries before Einstein's theory. Nicole Oresme (1323–1382) wrote that motion of one object can only be perceived relative to some other object. Until now, we have tacitly assumed in most situations that displacements, velocities, and accelerations should be measured in a **reference frame** attached to Earth's surface—that is, by choosing an origin fixed in position relative to Earth's surface and a set of axes whose directions are fixed relative to Earth's surface. After learning about relative velocities, we will take another look at this assumption.

Relative Velocity

Suppose Wanda is walking down the aisle of a train moving along the track at a constant velocity (Fig. 3.24). Imagine asking, "How fast is Wanda moving?" This question is not well defined. Do we mean her speed as measured by Tim, a passenger on the train, or her speed as measured by Greg, who is standing on the ground and looking into the train as it passes by? The answer to the question "How fast?" depends on the observer.

Figure 3.25 shows Wanda walking from one end of the car to the other during a time interval Δt. The displacement of Wanda as measured by Tim—her displacement *relative*

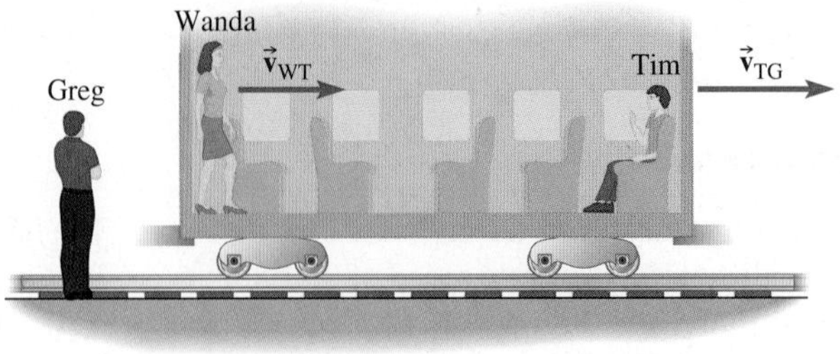

Figure 3.24 Tim and Greg watch Wanda walk down the aisle of a train. Wanda's velocity with respect to Tim (or with respect to the train) is $\vec{\mathbf{v}}_{WT}$; Tim's velocity with respect to Greg (or with respect to the ground) is $\vec{\mathbf{v}}_{TG}$.

to the train—is $\Delta\vec{\mathbf{r}}_{WT} = \vec{\mathbf{v}}_{WT}\Delta t$. During the same time interval, the *train's* displacement *relative to Greg* is $\Delta\vec{\mathbf{r}}_{TG} = \vec{\mathbf{v}}_{TG}\Delta t$. As measured by Greg, Wanda's displacement is partly due to her motion relative to the train and partly due to the motion of the train relative to the ground. Figure 3.25 shows that $\Delta\vec{\mathbf{r}}_{WT} + \Delta\vec{\mathbf{r}}_{TG} = \Delta\vec{\mathbf{r}}_{WG}$. Dividing by the time interval Δt gives the relationship between the three velocities:

$$\vec{\mathbf{v}}_{WT} + \vec{\mathbf{v}}_{TG} = \vec{\mathbf{v}}_{WG} \quad (3\text{-}24)$$

To be sure that you are adding the velocity vectors correctly, think of the subscripts as if they were fractions that get multiplied when the velocity vectors are added. In Eq. (3-24), $\frac{W}{\cancel{T}} \times \frac{\cancel{T}}{G} = \frac{W}{G}$ so the equation is correct.

Applications of Relative Velocities for Pilots and Sailors Relative velocities are of enormous practical interest to pilots of aircraft, sailors, and captains of ocean freighters. The pilot of an airplane is ultimately concerned with the motion of the plane with respect to the ground—the takeoff and landing points are fixed points on the ground. However, the controls of the plane (engines, rudder, ailerons, and spoilers) affect the motion of the plane *with respect to the air.* A sailor has to consider three different velocities of the boat: with respect to shore (for launching and landing), with respect to the air (for the behavior of the sails), and with respect to the water (for the behavior of the rudder).

CHECKPOINT 3.6

In Fig. 3.24, if the train is moving at 18.0 m/s with respect to the ground and Wanda walks at 1.5 m/s with respect to the train, how fast is Wanda moving (a) with respect to Greg and (b) with respect to Tim?

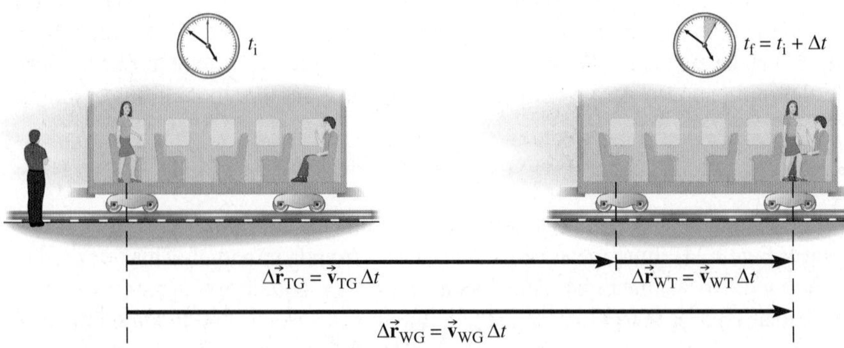

Figure 3.25 Wanda's displacement relative to the ground is the sum of her displacement relative to the train and the displacement of the train relative to the ground.

Example 3.10

Flight from Denver to Chicago

An airplane flies from Denver to Chicago (1770 km) in 4.4 h when no wind blows. On a day with a tailwind, the plane makes the trip in 4.0 h. (a) What is the wind speed? (b) If a headwind blows with the same speed, how long does the trip take?

Strategy We assume the plane has the same *airspeed*—the same speed relative to the air—in both cases. Once the plane is up in the air, the behavior of the wings, control surfaces, etc., depends on how fast the air is rushing by; the ground speed is irrelevant. But it is not irrelevant for the passengers, who are interested in a displacement relative to the ground.

Solution Let $\vec{\mathbf{v}}_{PG}$ and $\vec{\mathbf{v}}_{PA}$ represent the velocity of the plane relative to the ground and the velocity of the plane relative to the air, respectively. The wind velocity—the velocity of the air relative to the ground—can be written $\vec{\mathbf{v}}_{AG}$. Then $\vec{\mathbf{v}}_{PA} + \vec{\mathbf{v}}_{AG} = \vec{\mathbf{v}}_{PG}$. The equation is correct since $\frac{P}{\cancel{A}} \times \frac{\cancel{A}}{G} = \frac{P}{G}$. With no wind,

$$v_{PA} = v_{PG} = \frac{1770 \text{ km}}{4.4 \text{ h}} = 400 \text{ km/h}$$

(a) On the day with the tailwind,

$$v_{PG} = \frac{1770 \text{ km}}{4.0 \text{ h}} = 440 \text{ km/h}$$

We expect v_{PA} to be the same regardless of whether there is a wind or not. Since we are dealing with a tailwind, $\vec{\mathbf{v}}_{PA}$ and $\vec{\mathbf{v}}_{AG}$ are in the same direction, which we label as the $+x$-direction in Fig. 3.26. Then,

$$v_{PAx} + v_{AGx} = v_{PGx}$$

$$v_{AGx} = v_{PGx} - v_{PAx} = 440 \text{ km/h} - 400 \text{ km/h} = 40 \text{ km/h}$$

$v_{AGy} = 0$, so the wind speed is $v_{AG} = 40$ km/h.

(b) With a 40 km/h headwind, $\vec{\mathbf{v}}_{PA}$ and $\vec{\mathbf{v}}_{AG}$ are in opposite directions (Fig. 3.27). The velocity of the plane with respect to the ground is

$$v_{PGx} = v_{PAx} + v_{AGx} = 400 \text{ km/h} + (-40 \text{ km/h}) = 360 \text{ km/h}$$

The ground speed of the plane is 360 km/h and the trip takes

$$\frac{1770 \text{ km}}{360 \text{ km/h}} = 4.9 \text{ h}$$

Discussion Quick check: the trip takes longer with a headwind (4.9 h) than with no wind (4.4 h), as we expect.

Practice Problem 3.10 Rowing Across the Bay

Jamil, practicing to get on the crew team at school, rows a one-person racing shell to the north shore of the bay for a distance of 3.6 km to his friend's dock. On a day when the water is still (no current flowing), it takes him 20 min (1200 s) to reach his friend. On another day when a current flows southward, it takes him 30 min (1800 s) to row the same course. Ignore air resistance. (a) What is the speed of the current in m/s? (b) How long does it take Jamil to return home with that same current flowing?

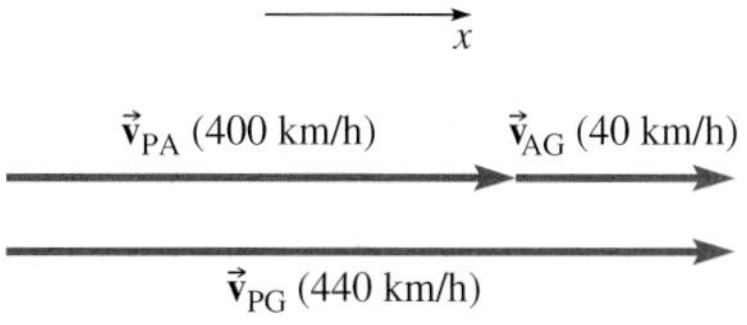

Figure 3.26 Addition of velocity vectors in the case of a tailwind. Lengths of vectors are not to scale.

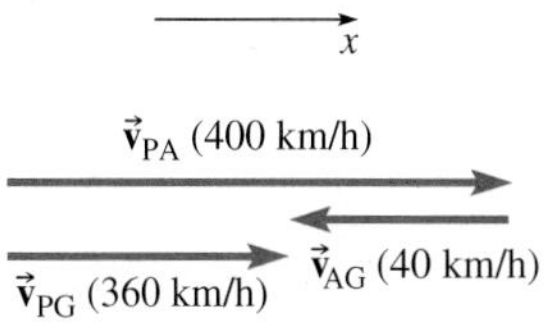

Figure 3.27 Addition of velocity vectors in the case of a headwind. Lengths of vectors are not to scale.

The vector equation (3-24) applies to situations where the velocities are not all along the same line, as illustrated in Example 3.11.

Example 3.11

Rowing Across a River

Jack wants to row directly across a river from the east shore to a point on the west shore. The width of the river is 250 m and the current flows from north to south at 0.61 m/s. The trip takes Jack 4.2 min. In what direction did he head his rowboat to follow a course due west across the river? At what speed with respect to still water is Jack able to row?

continued on next page

Example 3.11 continued

Strategy We start with a sketch of the situation (Fig. 3.28). To keep the various velocities straight, we choose subscripts as follows: R = rowboat; W = water; S = shore. The velocity of the current given is the velocity of the water relative to the shore: $\vec{\mathbf{v}}_{WS}$ = 0.61 m/s, south. The velocity of the rowboat relative to shore ($\vec{\mathbf{v}}_{RS}$) is due west. The magnitude of $\vec{\mathbf{v}}_{RS}$ can be found from the displacement relative to shore and the time interval, both of which are given. The question asks for the magnitude and direction of the velocity of the rowboat relative to the water ($\vec{\mathbf{v}}_{RW}$). The three velocities are related by

$$\vec{\mathbf{v}}_{RW} + \vec{\mathbf{v}}_{WS} = \vec{\mathbf{v}}_{RS}$$

To compensate for the current carrying the rowboat south with respect to shore, Jack heads (points) the rowboat upstream (against the current) at some angle to the north of west.

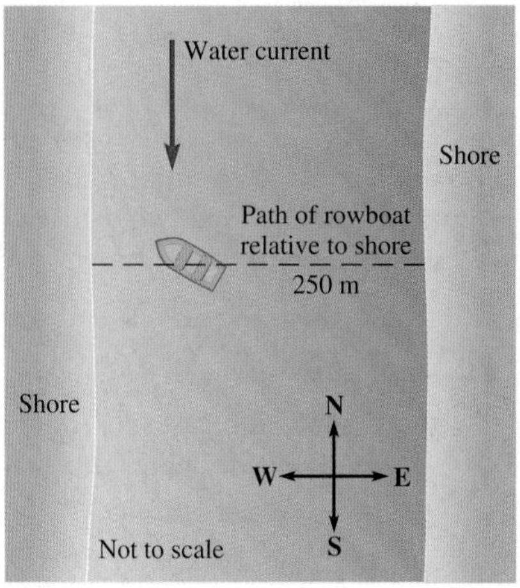

Figure 3.28
Rowing across a river.

Solution In a sketch of the vector addition (Fig. 3.29), the velocity of the rowboat with respect to the water is at an angle θ north of west. With respect to shore, Jack travels 250 m in 4.2 min, so his speed with respect to shore is

$$v_{RS} = \frac{250\ \text{m}}{4.2\ \text{min} \times 60\ \text{s/min}} = 0.992\ \text{m/s}$$

We can find the angle at which the rowboat should be headed by finding the tangent of the angle between $\vec{\mathbf{v}}_{RW}$ and $\vec{\mathbf{v}}_{RS}$:

$$\tan\theta = \frac{v_{WS}}{v_{RS}} = \frac{0.61\ \text{m/s}}{0.992\ \text{m/s}}$$

$$\theta = 32^\circ\ \text{N of W}$$

The speed at which Jack is able to row with respect to still water is the magnitude of $\vec{\mathbf{v}}_{RW}$. Since $\vec{\mathbf{v}}_{RS}$ and $\vec{\mathbf{v}}_{WS}$ are *perpendicular*, the Pythagorean theorem yields

$$|\vec{\mathbf{v}}_{RW}| = \sqrt{v_{WS}^2 + v_{RS}^2} = \sqrt{(0.61\ \text{m/s})^2 + (0.992\ \text{m/s})^2}$$
$$= 1.16\ \text{m/s}$$

Jack rows at a speed of 1.16 m/s with respect to the water.

Discussion If $\vec{\mathbf{v}}_{RS}$ and $\vec{\mathbf{v}}_{WS}$ had not been perpendicular, we could not have used the Pythagorean theorem in this way. Rather, we would use the component method to add the two vectors.

If Jack had headed the rowboat directly west, the current would have carried him south, so he would have traveled in a direction south of west relative to shore. He has to compensate by heading upstream at just such an angle that his velocity relative to shore is directed west.

Practice Problem 3.11 Heading Straight Across

If Jack were to head straight across the river, in what direction with respect to shore would he travel? How long would it take him to cross? How far downstream would he be carried? Assume that he rows at the same speed with respect to the water as in Example 3.11.

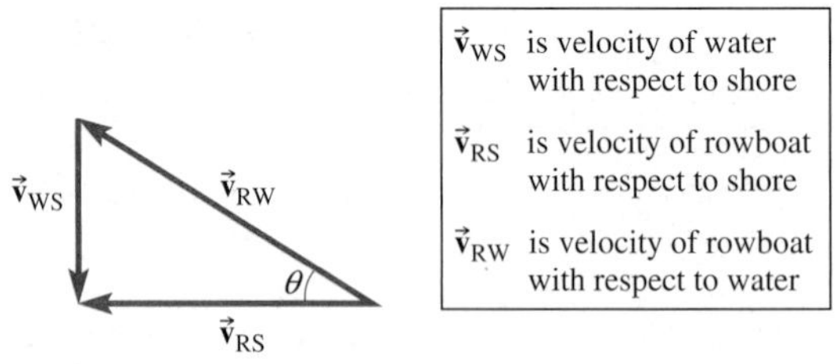

Figure 3.29
Graphical addition of the velocity vectors.

What does the path of the falling clam look like to the gull?

At the beginning of this chapter, we asked what the path followed by the falling clam looks like as seen by the gull flying through the air. With respect to a beachcomber on the ground and ignoring air resistance, the clam has a constant horizontal velocity component given to it by the gull and a changing vertical component of velocity due to gravity (Fig. 3.30a); the clam moves in a parabolic path. If the gull continues to fly at the same horizontal velocity after dropping the clam, it is directly overhead when the clam hits the rock because they both have the same constant horizontal component of velocity with respect to Earth.

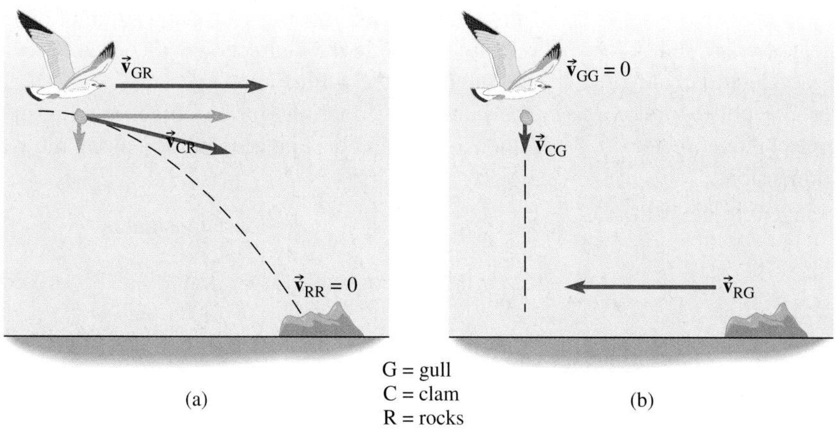

Figure 3.30 (a) Beachcomber view: The gull flies along a horizontal line while the clam follows a parabolic path. (b) Bird's eye view: The gull sees the rocks moving while the clam drops straight down, landing on the rocks just as the rocks move under the clam.

In its own reference frame—that is, using its own position as the origin of the coordinate axes—the gull sees the clam drop straight down toward the ground while rocks and other objects on the beach are moving horizontally (Fig. 3.30b). The bird sees a collision between the horizontally moving rocks and the vertically falling clam. At any instant, if the velocity of the clam with respect to the gull is $\vec{\mathbf{v}}_{CG}$, the velocity of the gull with respect to the rocks is $\vec{\mathbf{v}}_{GR}$, and the velocity of the clam with respect to the rocks is $\vec{\mathbf{v}}_{CR}$, then $\vec{\mathbf{v}}_{CG} + \vec{\mathbf{v}}_{GR} = \vec{\mathbf{v}}_{CR}$.

Master the Concepts

- Vectors are added graphically by drawing each vector so that its tail is placed at the tip of the previous vector. The sum is drawn as a vector arrow from the tail of the first vector to the tip of the last. Addition of vectors is commutative: $\vec{\mathbf{A}} + \vec{\mathbf{B}} = \vec{\mathbf{B}} + \vec{\mathbf{A}}$.

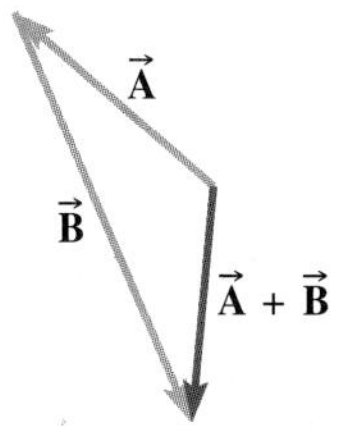

- Vectors are subtracted by adding the opposite of the second vector: $\vec{\mathbf{A}} - \vec{\mathbf{B}} = \vec{\mathbf{A}} + (-\vec{\mathbf{B}})$.
- Addition and subtraction of vectors algebraically using components is generally easier and more accurate than the graphical method. The graphical method is still a useful first step to get an approximate answer.
- To find the components of a vector, first draw a right triangle with the vector as the hypotenuse and the other two sides parallel to the x- and y-axes. Then use the trigonometric functions to find the magnitudes of the components. The correct algebraic sign must be determined for each component. The same triangle can be used to find the magnitude and direction of a vector if its components are known.

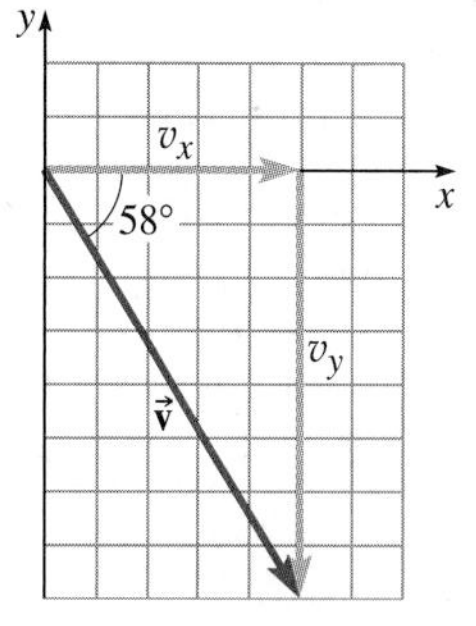

- To add vectors algebraically, add their components to find the components of the sum:

$$\vec{\mathbf{A}} + \vec{\mathbf{B}} = \vec{\mathbf{C}} \text{ if and only if}$$
$$A_x + B_x = C_x \text{ and } A_y + B_y = C_y$$

- The x- and y-axes are chosen to make the problem easiest to solve. Any choice is valid as long as the two are perpendicular. If the direction of the acceleration is known, choose x- and y-axes so that the acceleration vector is parallel to one of the axes.
- Position, displacement, velocity, and acceleration are vector quantities with both magnitude and direction. They must be added and subtracted as vectors.

continued on next page

Master the Concepts continued

- The equations for position, displacement, average velocity, instantaneous velocity, average acceleration, and instantaneous acceleration in Chapter 2 apply to *each perpendicular component* of the corresponding vector quantities for motion in two or three dimensions.
- The instantaneous velocity vector is tangent to the path of motion.

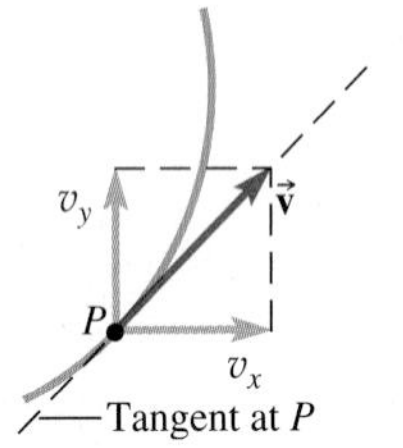

- The instantaneous acceleration vector does *not* have to be tangent to the path of motion, since velocities can change both in direction and in magnitude.
- For a projectile or any object moving with constant acceleration in the $\pm y$-direction, the motion in the x- and y-directions can be treated separately. Since $a_x = 0$, v_x is constant. Thus, the motion is a superposition of constant velocity motion in the x-direction and constant acceleration motion in the y-direction.
- The kinematic equations for an object moving in two dimensions with constant acceleration along the y-axis are

x-axis: $a_x = 0$	y-axis: constant a_y	
$\Delta v_x = 0$ (v_x is constant)	$\Delta v_y = a_y \Delta t$	(3-19)
$\Delta x = v_x \Delta t$	$\Delta y = \frac{1}{2}(v_{fy} + v_{iy})\Delta t$	(3-20)
	$\Delta y = v_{iy}\Delta t + \frac{1}{2}a_y(\Delta t)^2$	(3-21)
	$v_{fy}^2 - v_{iy}^2 = 2a_y \Delta y$	(3-22)

- To relate the velocities of objects measured in different reference frames, use the vector equation

$$\vec{\mathbf{v}}_{AC} = \vec{\mathbf{v}}_{AB} + \vec{\mathbf{v}}_{BC} \quad (3\text{-}24)$$

where $\vec{\mathbf{v}}_{AC}$ represents the velocity of A relative to C, and so forth.

Conceptual Questions

1. If two vectors have the same magnitude, are they necessarily equal? If not, why not? Can two vectors with different magnitudes ever be equal?
2. (a) Is it possible for the sum of two vectors to be smaller in magnitude than the magnitude of either vector? (b) Is it possible for the magnitude of the sum of two vectors to be larger than the sum of the magnitudes of the two vectors?
3. What is the distinction between a vector and a scalar quantity? Give two examples of each.
4. Is it possible for two identical projectiles with identical initial speeds, but with two different angles of elevation, to land in the same spot? Explain. Ignore air resistance and sketch the trajectories.
5. If the trajectory is parabolic in one reference frame, is it always, never, or sometimes parabolic in another reference frame that moves at constant velocity with respect to the first reference frame? If the trajectory can be other than parabolic, what else can it be?
6. You are standing on a balcony overlooking the beach. You throw a ball straight up into the air with speed v_i and throw an identical ball straight down with speed v_i. Ignoring air resistance, how do the speeds of the balls compare just before they hit the ground?
7. You throw a ball up with initial speed v_i and when it reaches its high point at height h, you throw another ball into the air with the same initial speed v_i. Will the two balls cross at half the height h, or more than half, or less than half? Explain.
8. If an object is traveling at a constant velocity, is it necessarily traveling in a straight line? Explain.
9. Can the average speed and the magnitude of the average velocity ever be equal? If so, under what circumstances?
10. Give an example of an object whose acceleration is (1) in the same direction as its velocity, (2) opposite its velocity, and (3) perpendicular to its velocity.
11. Name a situation where the speed of an object is constant while the velocity is not.
12. Tell whether or not each of the following objects has a constant velocity and explain your reasoning. (a) A car driving around a curve at constant speed on a flat road. (b) A car driving straight up a 6° incline at constant speed. (c) The Moon.
13. Explain how to add two displacement vectors of magnitudes $3L$ and $4L$ so that the vector sum has magnitude (a) L; (b) $7L$; (c) $5L$.
14. Compare the advantages and disadvantages of the two methods of vector addition (graphical and algebraic).
15. Can the x-component of a vector ever be greater than the magnitude of the vector? Explain.
16. Why is the muzzle of a rifle not aimed directly at the center of the target? Why is this more important at longer ranges?
17. Does the monkey, coconut, and hunter demonstration still work if the hunter is in a higher tree and the arrow is pointed *downward* at the monkey and coconut? Explain.

Multiple-Choice Questions

1. Vector $\vec{\mathbf{A}}$ in the drawing is equal to
   (a) $\vec{\mathbf{C}}+\vec{\mathbf{D}}$ (b) $\vec{\mathbf{C}}+\vec{\mathbf{D}}+\vec{\mathbf{E}}$ (c) $\vec{\mathbf{C}}+\vec{\mathbf{F}}$
   (d) $\vec{\mathbf{B}}+\vec{\mathbf{C}}$ (e) $\vec{\mathbf{B}}+\vec{\mathbf{F}}$

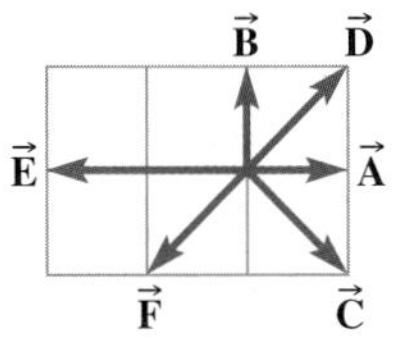

Multiple-Choice Questions 1 and 2

2. Which vector sum is *not* equal to zero?
   (a) $\vec{\mathbf{C}}+\vec{\mathbf{D}}+\vec{\mathbf{E}}$ (b) $\vec{\mathbf{B}}+\vec{\mathbf{C}}+\vec{\mathbf{F}}$
   (c) $\vec{\mathbf{D}}+\vec{\mathbf{F}}$ (d) $\vec{\mathbf{A}}+\vec{\mathbf{B}}+\vec{\mathbf{F}}$
3. A hunter spots a pheasant flying along horizontally. If he shoots the pheasant, the time interval between the bird being shot and the dead bird hitting the ground depends on
   (a) the speed with which the bird was flying.
   (b) the height of the bird above the ground.
   (c) the speed of the bird and its height above the ground.
4. A runner moves along a circular track at a constant speed.
   (a) Her acceleration is zero.
   (b) Her velocity is constant.
   (c) Both (a) and (b) are true.
   (d) Both her acceleration and her velocity are changing.
5. A boy plans to cross a river in a rubber raft. The current flows from north to south at 1 m/s. In what direction should he head to get across the river to the east bank in the least amount of time if he is able to paddle the raft at 1.5 m/s in still water?
   (a) directly to the east
   (b) south of east
   (c) north of east
   (d) The three directions require the same time to cross the river.
6. A boy plans to paddle a rubber raft across a river to the east bank while the current flows downriver from north to south at 1 m/s. He is able to paddle the raft at 1.5 m/s in still water. In what direction should he head the raft to go straight east across the river to the opposite bank?
   (a) directly to the east (b) south of east
   (c) north of east (d) north (e) south
7. A kicker kicks a football from the 5-yard line to the 45-yard line (both on the same half of the field). Ignoring air resistance, where along the trajectory is the speed of the football a minimum?
   (a) at the 5-yard line, just after the football leaves the kicker's foot
   (b) at the 45-yard line, just before the football hits the ground
   (c) at the 15-yard line, while the ball is still going higher
   (d) at the 35-yard line, while the ball is coming down
   (e) at the 25-yard line, when the ball is at the top of its trajectory
8. Two balls, identical except for color, are projected horizontally from the roof of a tall building at the same instant. The initial speed of the red ball is twice the initial speed of the blue ball. Ignoring air resistance,
   (a) the red ball reaches the ground first.
   (b) the blue ball reaches the ground first.
   (c) both balls land at the same instant with different speeds.
   (d) both balls land at the same instant with the same speed.
9. A person stands on the roof garden of a tall building with one ball in each hand. If the red ball is thrown horizontally off the roof and the blue ball is simultaneously dropped over the edge, which statement is true?
   (a) Both balls hit the ground at the same time, but the red ball has a higher speed just before it strikes the ground.
   (b) The blue ball strikes the ground first, but with a lower speed than the red ball.
   (c) The red ball strikes the ground first with a higher speed than the blue ball.
   (d) Both balls hit the ground at the same time with the same speed.
10. A ball is thrown into the air and follows a parabolic trajectory. At the highest point in the trajectory,
    (a) the velocity is zero, but the acceleration is not zero.
    (b) both the velocity and the acceleration are zero.
    (c) the acceleration is zero, but the velocity is not zero.
    (d) neither the acceleration nor the velocity are zero.
11. A ball is thrown into the air and follows a parabolic trajectory. Point A is the highest point in the trajectory and point B is a point as the ball is falling back to the ground. Choose the correct relationship between the speeds and the magnitudes of the acceleration at the two points.
    (a) $v_A > v_B$ and $a_A = a_B$ (b) $v_A < v_B$ and $a_A > a_B$
    (c) $v_A = v_B$ and $a_A \neq a_B$ (d) $v_A < v_B$ and $a_A = a_B$

Questions 12–14. Two projectiles launched with the same initial speed but at different launch angles 30° and 60° land at the same spot (see Fig. 3.22). Ignore air resistance. Answer choices:

(a) projectile launched at 30°
(b) projectile launched at 60°
(c) They are equal.

12. Which has the larger horizontal velocity component v_x?
13. Which has a longer time of flight Δt (time interval between launch and hitting the ground)?
14. For which is the product $v_x \Delta t$ larger?

Problems

- Combination conceptual/quantitative problem
- Biological or medical application
- Challenging problem
- Blue # Detailed solution in the Student Solutions Manual
- (1 2) Problems paired by concept
- Text website interactive or tutorial

3.1 Graphical Addition and Subtraction of Vectors

1. Displacement vector $\vec{\mathbf{A}}$ is directed to the west and has magnitude 2.56 km. A second displacement vector is also directed to the west and has magnitude 7.44 km. (a) What are the magnitude and direction of $\vec{\mathbf{A}} + \vec{\mathbf{B}}$? (b) What are the magnitude and direction of $\vec{\mathbf{A}} - \vec{\mathbf{B}}$? (c) What are the magnitude and direction of $\vec{\mathbf{B}} - \vec{\mathbf{A}}$?

2. Vector $\vec{\mathbf{A}}$ is directed along the positive x-axis and has magnitude 1.73 units. Vector $\vec{\mathbf{B}}$ is directed along the negative x-axis and has magnitude 1.00 unit. (a) What are the magnitude and direction of $\vec{\mathbf{A}} + \vec{\mathbf{B}}$? (b) What are the magnitude and direction of $\vec{\mathbf{A}} - \vec{\mathbf{B}}$? (c) What are the magnitude and direction of $\vec{\mathbf{B}} - \vec{\mathbf{A}}$?

3. Two vectors have magnitudes 3.0 and 4.0. How are the directions of the two vectors related if (a) the sum has magnitude 7.0, or (b) if the sum has magnitude 5.0? (c) What relationship between the directions gives the smallest magnitude sum and what is this magnitude?

4. A runner is practicing on a circular track that is 300 m in circumference. From the point farthest to the west on the track, he starts off running due north and follows the track as it curves around toward the east. (a) If he runs halfway around the track and stops at the farthest eastern point of the track, what is the distance he traveled? (b) What is his displacement?

5. Two displacement vectors each have magnitude 20 km. One is directed 60° above the $+x$-axis; the other is directed 60° below the $+x$-axis. What is the vector sum of these two displacements? Use graph paper to find your answer.

6. Orville walks 320 m due east. He then continues walking along a straight line, but in a different direction, and stops 200 m northeast of his starting point. How far did he walk during the second portion of the trip and in what direction?

7. Vectors $\vec{\mathbf{A}}$, $\vec{\mathbf{B}}$, and $\vec{\mathbf{C}}$ are shown in the figure. (a) Draw vectors $\vec{\mathbf{D}}$ and $\vec{\mathbf{E}}$, where $\vec{\mathbf{D}} = \vec{\mathbf{A}} + \vec{\mathbf{B}}$ and $\vec{\mathbf{E}} = \vec{\mathbf{A}} + \vec{\mathbf{C}}$. (b) Show that $\vec{\mathbf{A}} + \vec{\mathbf{B}} = \vec{\mathbf{B}} + \vec{\mathbf{A}}$ by graphical means.

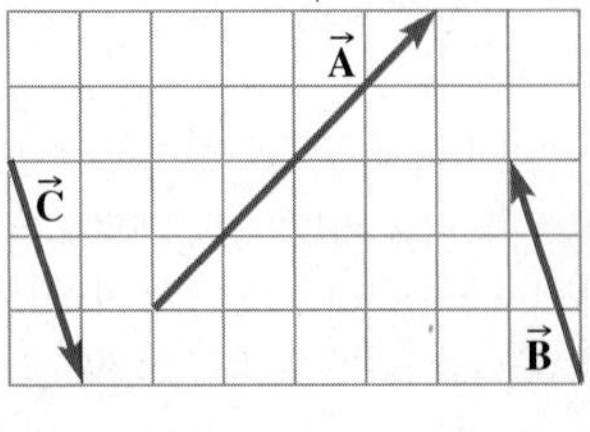

8. Two vectors, each of magnitude 4.0 cm, are directed at a small angle α below the horizontal as shown. (The grid is 1 cm on a side.) (a) Let $\vec{\mathbf{C}} = \vec{\mathbf{A}} + \vec{\mathbf{B}}$. Sketch the direction of $\vec{\mathbf{C}}$ and estimate its magnitude. (b) Let $\vec{\mathbf{D}} = \vec{\mathbf{A}} - \vec{\mathbf{B}}$. Sketch the direction of $\vec{\mathbf{D}}$ and estimate its magnitude. (WW tutorial: vectors)

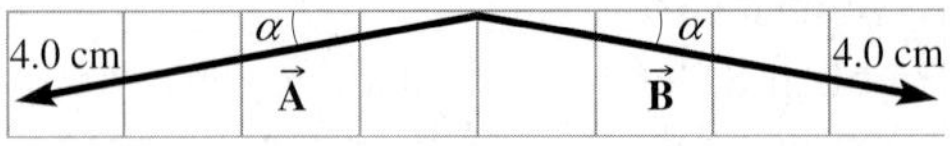

9. Michaela is planning a trip in Ireland from Killarney to Cork to visit Blarney Castle. (See Example 3.2.) She also wants to visit Mallow, which is located 39 km due east of Killarney and 22 km due north of Cork. Draw the displacement vectors for the trip when she travels from Killarney to Mallow to Cork. (a) What is the magnitude of her displacement once she reaches Cork? (b) How much additional distance does Michaela travel in going to Cork by way of Mallow instead of going directly from Killarney to Cork?

10. A scout troop is practicing its orienteering skills with map and compass. First they walk due east for 1.2 km. Next, they walk 45° west of north for 2.7 km. In what direction must they walk to go directly back to their starting point? How far will they have to walk? Use graph paper, ruler, and protractor to find a geometrical solution.

11. Prove that the displacement for a trip is equal to the vector sum of the displacements for each leg of the trip. [*Hint:* Imagine a trip that consists of n segments. The trip starts at position $\vec{\mathbf{r}}_1$, proceeds to $\vec{\mathbf{r}}_2$, then to $\vec{\mathbf{r}}_3, \ldots,$ then to $\vec{\mathbf{r}}_{n-1}$, then finally to $\vec{\mathbf{r}}_n$. Write an expression for each displacement as the difference of two position vectors and then add them.]

12. A sailboat sails from Marblehead Harbor directly east for 45 nautical miles, then 60° south of east for 20.0 nautical miles, returns to an easterly heading for 30.0 nautical miles, and sails 30° east of north for 10.0 nautical miles, then west for 62 nautical miles. At that time the boat becomes becalmed and the auxiliary engine fails to start. The crew decides to notify the Coast Guard of their position. Using graph paper, ruler, and protractor, sketch a graphical addition of the displacement vectors and estimate their position.

3.2 Vector Addition and Subtraction Using Components

13. A vector is 20.0 m long and makes an angle of 60.0° counterclockwise from the y-axis (on the side of the $-x$-axis). What are the x- and y-components of this vector?

14. Vector $\vec{\mathbf{A}}$ has magnitude 4.0 units; vector $\vec{\mathbf{B}}$ has magnitude 6.0 units. The angle between $\vec{\mathbf{A}}$ and $\vec{\mathbf{B}}$ is 60.0°. What is the magnitude of $\vec{\mathbf{A}} + \vec{\mathbf{B}}$?

15. Vector $\vec{\mathbf{A}}$ is directed along the positive y-axis and has magnitude $\sqrt{3.0}$ units. Vector $\vec{\mathbf{B}}$ is directed along the negative x-axis and has magnitude 1.0 unit. (a) What are

the magnitude and direction of $\vec{\mathbf{A}} + \vec{\mathbf{B}}$? (b) What are the magnitude and direction of $\vec{\mathbf{A}} - \vec{\mathbf{B}}$? (c) What are the x- and y-components of $\vec{\mathbf{B}} - \vec{\mathbf{A}}$?

16. Vector $\vec{\mathbf{a}}$ has components $a_x = -3.0$ m/s^2 and $a_y = +4.0$ m/s^2. (a) What is the magnitude of $\vec{\mathbf{a}}$? (b) What is the direction of $\vec{\mathbf{a}}$? Give an angle with respect to one of the coordinate axes.

17. In Problem 8, let $\alpha = 10°$ and find the magnitude of vector $\vec{\mathbf{C}}$ using the component method.

18. In Problem 8, let $\alpha = 10°$ and find the magnitude of vector $\vec{\mathbf{D}}$ using the component method.

19. Find the x- and y-components of the four vectors shown in the drawing.

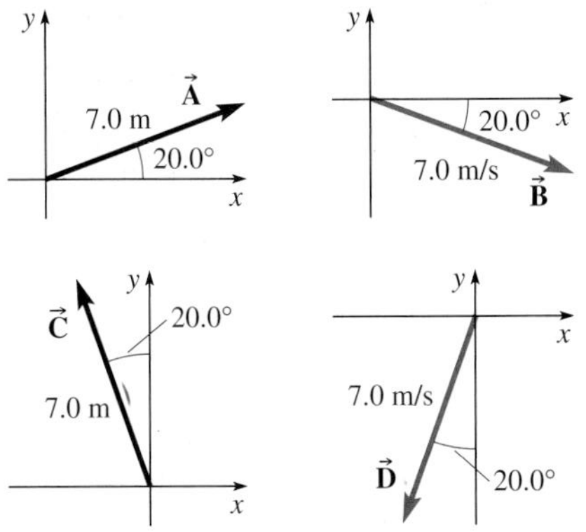

20. The velocity vector of a sprinting cheetah has x- and y-components $v_x = +16.4$ m/s and $v_y = -26.3$ m/s. (a) What is the magnitude of the velocity vector? (b) What angle does the velocity vector make with the $+x$- and $-y$-axes?

21. In each of these, the x- and y-components of a vector are given. Find the magnitude and direction of the vector. (a) $A_x = -5.0$ m/s, $A_y = +8.0$ m/s. (b) $B_x = +120$ m, $B_y = -60.0$ m. (c) $C_x = -13.7$ m/s, $C_y = -8.8$ m/s. (d) $D_x = 2.3$ m/s^2, $D_y = 6.5$ cm/s^2.

22. A vector $\vec{\mathbf{A}}$ has a magnitude of 22.2 cm and makes an angle of 130.0° with the positive x-axis. What are the x- and y-components of this vector?

23. Vector $\vec{\mathbf{B}}$ has magnitude 7.1 and direction 14° below the $+x$-axis. Vector $\vec{\mathbf{C}}$ has x-component $C_x = -1.8$ and y-component $C_y = -6.7$. Compute (a) the x- and y-components of $\vec{\mathbf{B}}$; (b) the magnitude and direction of $\vec{\mathbf{C}}$; (c) the magnitude and direction of $\vec{\mathbf{C}} + \vec{\mathbf{B}}$; (d) the magnitude and direction of $\vec{\mathbf{C}} - \vec{\mathbf{B}}$; (e) the x- and y-components of $\vec{\mathbf{C}} - \vec{\mathbf{B}}$.

24. Margaret walks to the store using the following path: 0.500 miles west, 0.200 miles north, 0.300 miles east. What is her total displacement? That is, what is the length and direction of the vector that points from her house directly to the store? Use vector components to find the answer.

25. Jerry bicycles from his dorm to the local fitness center: 3.00 miles east and 2.00 miles north. Cindy's apartment is located 1.50 miles west of Jerry's dorm. If Cindy is able to meet Jerry at the fitness center by bicycling in a straight line, what is the length and direction she must travel?

26. Repeat Problem 10 using the component (algebraic) method.

27. Use the component method to obtain a more accurate description of the sailboat's location in Problem 12.

28. You will be hiking to a lake with some of your friends by following the trails indicated on a map at the trailhead. The map says that you will travel 1.6 mi directly north, then 2.2 mi in a direction 35° east of north, then finally 1.1 mi in a direction 15° north of east. At the end of this hike, how far will you be from where you started, and what direction will you be from your starting point?

3.3 Velocity

29. A runner times his speed around a circular track with a circumference of 0.478 mi. At the start he is running toward the east and the track starts bending toward the north. If he goes halfway around, he will be running toward the west. He finds that he has run a distance of 0.750 mi in 4.00 min. What is his (a) average speed and (b) average velocity in m/s?

30. A runner times his speed around a track with a circumference of 0.50 mi. He finds that he has run a distance of 1.00 mi in 4.0 min. What is his (a) average speed and (b) average velocity magnitude in m/s?

31. Peggy drives from Cornwall to Atkins Glen in 45 min. Cornwall is 73.6 km from Illium in a direction 25° west of south. Atkins Glen is 27.2 km from Illium in a direction 15° south of west. Using Illium as your origin, (a) draw the initial and final position vectors, (b) find the displacement during the trip, and (c) find Peggy's average velocity for the trip.

32. To get to a concert in time, a harpsichordist has to drive 122 mi in 2.00 h. (a) If he drove at an average speed of 55.0 mi/h in a due west direction for the first 1.20 h, what must be his average speed if he is heading 30.0° south of west for the remaining 48.0 min? (b) What is his average velocity for the entire trip?

33. A bicycle travels 3.2 km due east in 0.10 h, then 4.8 km at 15.0° east of north in 0.15 h, and finally another 3.2 km due east in 0.10 h to reach its destination. The time lost in turning is negligible. What is the average velocity for the entire trip?

34. A car travels east at 96 km/h for 1.0 h. It then travels 30.0° east of north at 128 km/h for 1.0 h. (a) What is the average speed for the trip? (b) What is the average velocity for the trip?

35. A speedboat heads west at 108 km/h for 20.0 min. It then travels at 60.0° south of west at 90.0 km/h for 10.0 min. (a) What is the average speed for the trip? (b) What is the average velocity for the trip?

36. See Problem 9. During Michaela's travel from Killarney to Cork via Mallow, her actual travel time in the car is

48 min. (a) What is her average speed in m/s? (b) What is the magnitude of her average velocity in m/s?

37. Geoffrey drives from his home town due east at 90.0 km/h for 80.0 min. After visiting a friend for 15.0 min, he drives in a direction 30.0° south of west at 76.0 km/h for 45.0 min to visit another friend. (a) How far is it to his home from the second town? (b) If it takes him 45.0 min to drive directly home, what is his average velocity on the third leg of the trip? (c) What is his average velocity during the first two legs of his trip? (d) What is his average velocity over the entire trip? (e) What is his average speed during the entire trip if he spent 55.0 min visiting the second friend?

3.4 Acceleration

38. A hawk is flying north at 2.0 m/s with respect to the ground; 10.0 s later, it is flying south at 5.0 m/s. What is its average acceleration during this time interval?

39. A skydiver is falling straight down at 55 m/s when he opens his parachute and slows to 8.3 m/s in 3.5 s. What is the average acceleration of the skydiver during those 3.5 s?

40. A car travels three quarters of the way around a circle of radius 20.0 m in a time of 3.0 s at a constant speed. The initial velocity is west and the final velocity is south. (a) Find its average velocity for this trip. (b) What is the car's average acceleration during these 3.0 s? (c) Explain how a car moving at constant speed has a nonzero average acceleration.

41. At $t = 0$, an automobile traveling north begins to make a turn. It follows one-quarter of the arc of a circle with a radius of 10.0 m until, at $t = 1.60$ s, it is traveling east. The car does not alter its speed during the turn. Find (a) the car's speed, (b) the change in its velocity during the turn, and (c) its average acceleration during the turn.

42. At the beginning of a 3.0-h plane trip, you are traveling due north at 192 km/h. At the end, you are traveling 240 km/h in the northwest direction (45° west of north). (a) Draw your initial and final velocity vectors. (b) Find the change in your velocity. (c) What is your average acceleration during the trip?

43. John drives 16 km directly west from Orion to Chester at a speed of 90 km/h, then directly south for 8.0 km to Seiling at a speed of 80 km/h, then finally 34 km southeast to Oakwood at a speed of 100 km/h. Assume he travels at constant velocity during each of the three segments. (a) What was the change in velocity during this trip? [*Hint:* Do not assume he starts from rest and stops at the end.] (b) What was the average acceleration during this trip?

44. A particle's constant acceleration is south at 2.50 m/s^2. At $t = 0$, its velocity is 40.0 m/s east. What is its velocity at $t = 8.00$ s?

45. A particle's constant acceleration is north at 100 m/s^2. At $t = 0$, its velocity vector is 60 m/s east. At what time will the magnitude of the velocity be 100 m/s?

3.5 Motion in a Plane with Constant Acceleration

46. A baseball is thrown horizontally from a height of 9.60 m above the ground with a speed of 30.0 m/s. Where is the ball after 1.40 s has elapsed?

47. A clump of soft clay is thrown horizontally from 8.50 m above the ground with a speed of 20.0 m/s. Where is the clay after 1.50 s? Assume it sticks in place when it hits the ground.

48. A tennis ball is thrown horizontally from an elevation of 14.0 m above the ground with a speed of 20.0 m/s. (a) Where is the ball after 1.60 s? (b) If the ball is still in the air, how long before it hits the ground and where will it be with respect to the starting point once it lands?

49. A ball is thrown from a point 1.0 m above the ground. The initial velocity is 19.6 m/s at an angle of 30.0° above the horizontal. (a) Find the maximum height of the ball above the ground. (b) Calculate the speed of the ball at the highest point in the trajectory.

50. An arrow is shot into the air at an angle of 60.0° above the horizontal with a speed of 20.0 m/s. (a) What are the x- and y-components of the velocity of the arrow 3.0 s after it leaves the bowstring? (b) What are the x- and y-components of the displacement of the arrow during the 3.0-s interval?

51. You are working as a consultant on a video game designing a bomb site for a World War I airplane. In this game, the plane you are flying is traveling horizontally at 40.0 m/s at an altitude of 125 m when it drops a bomb. (a) Determine how far horizontally from the target you should release the bomb. (b) What direction is the bomb moving just before it hits the target?

52. You have been employed by the local circus to plan their human cannonball performance. For this act, a spring-loaded cannon will shoot a human projectile, the Great Flyinski, across the big top to a net below. The net is located 5.0 m lower than the muzzle of the cannon from which the Great Flyinski is launched. The cannon will shoot the Great Flyinski at an angle of 35.0° above the horizontal and at a speed of 18.0 m/s. The ringmaster has asked that you decide how far from the cannon to place the net so that the Great Flyinski will land in the net and not be splattered on the floor, which would greatly disturb the audience. What do you tell the ringmaster? (interactive: projectile motion)

53. A cannonball is catapulted toward a castle. The cannonball's velocity when it leaves the catapult is 40 m/s at an angle of 37° with respect to the horizontal and the cannonball is 7.0 m above the ground at this time. (a) What is the maximum height above the ground reached by the cannonball? (b) Assuming the cannonball makes it over the castle walls and lands back down on the ground, at what horizontal distance from its release point will it land? (c) What are the x- and y-components of the cannonball's velocity just before it lands? The y-axis points up.

54. After being assaulted by flying cannonballs, the knights on the castle walls (12 m above the ground) respond by propelling flaming pitch balls at their assailants. One ball lands on the ground at a distance of 50 m from the castle walls. If it was launched at an angle of 53° above the horizontal, what was its initial speed?

55. From the edge of the rooftop of a building, a boy throws a stone at an angle 25.0° above the horizontal. The stone hits the ground 4.20 s later, 105 m away from the base of the building. (Ignore air resistance.) (a) For the stone's path through the air, sketch graphs of x, y, v_x, and v_y as functions of time. These need to be only *qualitatively* correct—you need not put numbers on the axes. (b) Find the initial velocity of the stone. (c) Find the initial height h from which the stone was thrown. (d) Find the maximum height H reached by the stone.

56. Two angles are complementary when their sum is 90.0°. Find the ranges for two projectiles launched with identical initial speeds of 36.2 m/s at angles of elevation above the horizontal that are complementary pairs. (a) For one trial, the angles of elevation are 36.0° and 54.0°. (b) For the second trial, the angles of elevation are 23.0° and 67.0°. (c) Finally, the angles of elevation are both set to 45.0°. (d) What do you notice about the range values for each complementary pair of angles? At which of these angles was the range greatest?

57. The range R of a projectile is defined as the magnitude of the horizontal displacement of the projectile *when it returns to its original altitude.* (In other words, the range is the distance between the launch point and the impact point on flat ground.) A projectile is launched at $t = 0$ with initial speed v_i at an angle θ above the horizontal. (a) Find the time t at which the projectile returns to its original altitude. (b) Show that the range is

$$R = \frac{v_i^2 \sin 2\theta}{g}$$

[*Hint:* Use the trigonometric identity $\sin 2\theta = 2 \sin \theta \cos \theta$.] (c) What value of θ gives the maximum range? What is this maximum range?

58. Use the expression in Problem 57 to find (a) the maximum range of a projectile with launch speed v_i and (b) the launch angle θ at which the maximum range occurs.

59. A projectile is launched at $t = 0$ with initial speed v_i at an angle θ above the horizontal. (a) What are v_x and v_y at the projectile's highest point? (b) Find the time t at which the projectile reaches its maximum height. (c) Show that the maximum height H of the projectile is

$$H = \frac{(v_i \sin \theta)^2}{2g}$$

✦60. A ballplayer standing at home plate hits a baseball that is caught by another player at the same height above the ground from which it was hit. The ball is hit with an initial velocity of 22.0 m/s at an angle of 60.0° above the horizontal. (WW tutorial: projectile) (a) How high will the ball rise? (b) How much time will elapse from the time the ball leaves the bat until it reaches the fielder? (c) At what distance from home plate will the fielder be when he catches the ball?

✦61. You are planning a stunt to be used in an ice skating show. For this stunt a skater will skate down a frictionless ice ramp that is inclined at an angle of 15.0° above the horizontal. At the bottom of the ramp, there is a short horizontal section that ends in an abrupt drop off. The skater is supposed to start from rest somewhere on the ramp, then skate off the horizontal section and fly through the air a horizontal distance of 7.00 m while falling vertically for 3.00 m, before landing smoothly on the ice. How far up the ramp should the skater start this stunt?

✦62. A suspension bridge is 60.0 m above the level base of a gorge. A stone is thrown or dropped from the bridge. Ignore air resistance. At the location of the bridge g has been measured to be 9.83 m/s^2. (a) If you drop the stone, how long does it take for it to fall to the base of the gorge? (b) If you *throw* the stone straight down with a speed of 20.0 m/s, how long before it hits the ground? (c) If you throw the stone with a velocity of 20.0 m/s at 30.0° above the horizontal, how far from the point directly below the bridge will it hit the level ground?

✦63. A circus performer is shot out of a cannon and flies over a net that is placed horizontally 6.0 m from the cannon. When the cannon is aimed at an angle of 40° above the horizontal, the performer is moving in the horizontal direction and just barely clears the net as he passes over it. What is the muzzle speed of the cannon and how high is the net?

✦64. Show that for a projectile launched at an angle of 45° the maximum height of the projectile is one quarter of the range (the distance traveled on flat ground).

3.6 Velocity Is Relative; Reference Frames

65. Two cars are driving toward each other on a straight, flat Kansas road. The Jeep Wrangler is traveling at 82 km/h north and the Ford Taurus is traveling at 48 km/h south, both measured relative to the road. What is the velocity of the Jeep relative to an observer in the Ford?

66. Two cars are driving toward each other on a straight and level road in Alaska. The BMW is traveling at 100.0 km/h north and the VW is traveling at 42 km/h south, both velocities measured relative to the road. At a certain instant, the distance between the cars is 10.0 km. Approximately how long will it take from that instant for the two cars to meet? [*Hint:* Consider a reference frame in which one of the cars is at rest.]

67. A car is driving directly north on the freeway at a speed of 110 km/h and a truck is leaving the freeway driving 85 km/h in a direction that is 35° west of north. What is the velocity of the truck relative to the car?

68. A Nile cruise ship takes 20.8 h to go upstream from Luxor to Aswan, a distance of 208 km, and 19.2 h to

make the return trip downstream. Assuming the ship's speed relative to the water is the same in both cases, calculate the speed of the current in the Nile.

69. An airplane has a velocity relative to the ground of 210 m/s toward the east. The pilot measures his airspeed (the speed of the plane relative to the air) to be 160 m/s. What is the minimum wind velocity possible?

70. A small plane is flying directly west with an airspeed of 30.0 m/s. The plane flies into a region where the wind is blowing at 10.0 m/s at an angle of 30° to the south of west. (a) If the pilot does not change the heading of the plane, what will be the ground speed of the airplane? (b) What will be the new directional heading, relative to the ground, of the airplane? (tutorial: flight of crow)

71. A small plane is flying directly west with an airspeed of 30.0 m/s. The plane flies into a region where the wind is blowing at 10.0 m/s at an angle of 30° to the south of west. In that region, the pilot changes the directional heading to maintain her due west heading. (a) What is the change she makes in the directional heading to compensate for the wind? (b) After the heading change, what is the ground speed of the airplane?

72. A boat that can travel at 4.0 km/h in still water crosses a river with a current of 1.8 km/h. At what angle must the boat be pointed upstream to travel straight across the river? In other words, in what direction is the velocity of the boat relative to the water?

73. At an antique car rally, a Stanley Steamer automobile travels north at 40 km/h and a Pierce Arrow automobile travels east at 50 km/h. Relative to an observer riding in the Stanley Steamer, what are the x- and y-components of the velocity of the Pierce Arrow car? The x-axis is to the east and the y-axis is to the north.

74. Sheena can row a boat at 3.00 mi/h in still water. She needs to cross a river that is 1.20 mi wide with a current flowing at 1.60 mi/h. Not having her calculator ready, she guesses that to go straight across, she should head 60.0° upstream. (a) What is her speed with respect to the starting point on the bank? (b) How long does it take her to cross the river? (c) How far upstream or downstream from her starting point will she reach the opposite bank? (d) In order to go straight across, what angle upstream should she have headed?

75. A dolphin wants to swim directly back to its home bay, which is 0.80 km due west. It can swim at a speed of 4.00 m/s relative to the water, but a uniform water current flows with speed 2.83 m/s in the southeast direction. (a) What direction should the dolphin head? (b) How long does it take the dolphin to swim the 0.80-km distance home?

76. Demonstrate with a vector diagram that a displacement is the same when measured in two different reference frames that are at rest with respect to each other.

77. A boy is attempting to swim directly across a river; he is able to swim at a speed of 0.500 m/s relative to the water. The river is 25.0 m wide and the boy ends up at 50.0 m downstream from his starting point. (a) How fast is the current flowing in the river? (b) What is the speed of the boy relative to a friend standing on the riverbank?

78. An aircraft has to fly between two cities, one of which is 600.0 km north of the other. The pilot starts from the southern city and encounters a steady 100.0 km/h wind that blows from the northeast. The plane has a cruising speed of 300.0 km/h in still air. (a) In what direction (relative to east) must the pilot head her plane? (b) How long does the flight take?

Comprehensive Problems

79. Jason is practicing his tennis stroke by hitting balls against a wall. The ball leaves his racquet at a height of 60 cm above the ground at an angle of 80° with respect to the *vertical.* (a) The speed of the ball as it leaves the racquet is 20 m/s and it must travel a distance of 10 m before it reaches the wall. How far above the ground does the ball strike the wall? (b) Is the ball on its way up or down when it hits the wall?

80. Imagine a trip where you drive along an east-west highway at 80.0 km/h for 45.0 min and then you turn onto a highway that runs 38.0° north of east and travel at 60.0 km/h for 30.0 min. (a) What is your average velocity for the trip? (b) What is your average velocity on the return trip when you head the opposite way and drive 38.0° south of west at 60.0 km/h for the first 30.0 min and then west at 80.0 km/h for the last 45.0 min?

81. A jetliner flies east for 600.0 km, then turns 30.0° toward the south and flies another 300.0 km. (a) How far is the plane from its starting point? (b) In what direction could the jetliner have flown directly to the same destination (in a straight-line path)? (c) If the jetliner flew at a constant speed of 400.0 km/h, how long did the trip take? (d) Moving at the same speed, how long would the direct flight have taken?

82. An African swallow carrying a very small coconut is flying horizontally with a speed of 18 m/s. (a) If it drops the coconut from a height of 100 m above the Earth, how long will it take before the coconut strikes the ground? (b) At what horizontal distance from the release point will the coconut strike the ground?

83. A pilot starting from Athens, New York, wishes to fly to Sparta, New York, which is 320 km from Athens in the direction 20.0° N of E. The pilot heads directly for Sparta and flies at an airspeed of 160 km/h. After flying for 2.0 h, the pilot expects to be at Sparta, but instead he finds himself 20 km due west of Sparta. He has forgotten to correct for the wind. (a) What is the velocity of

the plane relative to the air? (b) Find the velocity (magnitude and direction) of the plane relative to the ground. (c) Find the wind speed and direction.

84. The citizens of Paris were terrified during World War I when they were suddenly bombarded with shells fired from a long-range gun known as Big Bertha. The barrel of the gun was 36.6 m long and it had a muzzle speed of 1.46 km/s. When the gun's angle of elevation was set to 55°, what would be the range? For the purposes of solving this problem, neglect air resistance. (The actual range at this elevation was 121 km; air resistance cannot be ignored for the high muzzle speed of the shells.)

85. You are serving as a consultant for the newest James Bond film. In one scene, Bond must fire a projectile from a cannon and hit the enemy headquarters located on the top of a cliff 75.0 m above and 350 m from the cannon. The cannon will shoot the projectile at an angle of 40.0° above the horizontal. The director wants to know what the speed of the projectile must be when it is fired from the cannon so that it will hit the enemy headquarters. What do you tell him? [*Hint:* Don't assume the projectile will hit the headquarters at the highest point of its flight.]

86. The pilot of a small plane finds that the airport where he intended to land is fogged in. He flies 55 mi west to another airport to find that conditions there are too icy for him to land. He flies 25 mi at 15° east of south and is finally able to land at the third airport. (a) How far and in what direction must he fly the next day to go directly to his original destination? (b) How many extra miles beyond his original flight plan has he flown?

87. A particle has a constant acceleration of 5.0 m/s^2 to the east. At time $t = 0$, it is 2.0 m east of the origin and its velocity is 20 m/s north. What are the components of its position vector at $t = 2.0$ s?

88. A baseball batter hits a long fly ball that rises to a height of 44 m. An outfielder on the opposing team can run at 7.6 m/s. What is the farthest the fielder can be from where the ball will land so that it is possible for him to catch the ball?

89. A locust jumps at an angle of 55.0° and lands 0.800 m from where it jumped. (a) What is the maximum height of the locust during its jump? Ignore air resistance. (b) If it jumps with the same initial speed at an angle of 45.0°, would the maximum height be larger or smaller? (c) What about the range? (d) Calculate the maximum height and range for this angle.

90. A helicopter is flying horizontally at 8.0 m/s and an altitude of 18 m when a package of emergency medical supplies is ejected horizontally backward with a speed of 12 m/s *relative to the helicopter.* Ignoring air resistance, what is the horizontal distance between the package and the helicopter when the package hits the ground?

91. An airplane is traveling from New York to Paris, a distance of 5.80×10^3 km. Ignore the curvature of the Earth. (a) If the cruising speed of the airplane is 350.0 km/h, how much time will it take for the airplane to make the round-trip on a calm day? (b) If a steady wind blows from New York to Paris at 60.0 km/h, how much time will the round-trip take? (c) How much time will it take if there is a crosswind of 60.0 km/h?

92. A gull is flying horizontally 8.00 m above the ground at 6.00 m/s. The bird is carrying a clam in its beak and plans to crack the clamshell by dropping it on some rocks below. Ignoring air resistance, (a) what is the horizontal distance to the rocks at the moment that the gull should let go of the clam? (b) With what speed relative to the rocks does the clam smash into the rocks? (c) With what speed relative to the gull does the clam smash into the rocks?

93. A beanbag is thrown horizontally from a dorm room window a height h above the ground. It hits the ground a horizontal distance h (the *same* distance h) from the dorm directly below the window from which it was thrown. Ignoring air resistance, find the direction of the beanbag's velocity just before impact.

✦94. In a plate glass factory, sheets of glass move along a conveyor belt at a speed of 15.0 cm/s. An automatic cutting tool descends at preset intervals to cut the glass to size. Since the assembly belt must keep moving at constant speed, the cutter is set to cut at an angle to compensate for the motion of the glass. If the glass is 72.0 cm wide and the cutter moves across the width at a speed of 24.0 cm/s, at what angle should the cutter be set?

✦95. A pilot wants to fly from Dallas to Oklahoma City, a distance of 330 km at an angle of 10.0° west of north. The pilot heads directly toward Oklahoma City with an air speed of 200 km/h. After flying for 1.0 h, the pilot finds that he is 15 km off course to the west of where he expected to be after one hour assuming there was no wind. (a) What is the velocity and direction of the wind? (b) In what direction should the pilot have headed his plane to fly directly to Oklahoma City without being blown off course?

96. A ball is thrown horizontally off the edge of a cliff with an initial speed of 20.0 m/s. (a) How long does it take for the ball to fall to the ground 20.0 m below? (b) How long would it take for the ball to reach the ground if it were dropped from rest off the cliff edge? (c) How long would it take the ball to fall to the ground if it were thrown at an initial velocity of 20.0 m/s but 18° below the horizontal?

✦97. A marble is rolled so that it is projected horizontally off the top landing of a staircase. The initial speed of the marble is 3.0 m/s. Each step is 0.18 m high and 0.30 m wide. Which step does the marble strike first?

✦98. A motor scooter rounds a curve on the highway at a constant speed of 20.0 m/s. The original direction of the scooter was due east; after rounding the curve the scooter is heading 36° north of east. The radius of curvature of the road at the location of the curve is 150 m. What is the average acceleration of the scooter as it rounds the curve?

✦ 99. You want to make a plot of the trajectory of a projectile. That is, you want to make a plot of the height y of the projectile as a function of horizontal distance x. The projectile is launched from the origin with an initial speed v_i at an angle θ above the horizontal. Show that the equation of the trajectory followed by the projectile is

$$y = \left(\frac{v_{iy}}{v_{ix}}\right)x + \left(\frac{-g}{2v_{ix}^2}\right)x^2$$

✦ 100. A person climbs from a Paris metro station to the street level by walking up a stalled escalator in 94 s. It takes 66 s to ride the same distance when standing on the escalator when it is operating normally. How long would it take for him to climb from the station to the street by walking up the moving escalator?

Answers to Practice Problems

3.1 No; the checkbook balance may increase or decrease, but there is no spatial direction associated with it. When we say it "goes down," we do not mean that it moves in a direction toward the center of Earth! Rather, we really mean that it decreases. The balance is a scalar.

3.2 240 mi 20° W of S

3.3 $A_x = +16$ km; $A_y = -8.2$ km; $B_x = +17$ km; $B_y = 0$ km; $C_x = -11$ km; $C_y = +47$ km

3.4 $|\vec{\mathbf{v}}_{av}|$ can never be greater than the average speed because the magnitude of the displacement cannot be greater than the distance traveled. $|\vec{\mathbf{v}}_{av}|$ can be equal to the average speed if the magnitude of the displacement is equal to the distance traveled, which is true when the motion is along a straight line with no change in direction.

3.5 (a) 1.64 m/s directed 33° east of north; (b) 0.82 m/s^2 directed 33° east of north

3.6 2

3.7 $v_i^2/(4g)$

3.8 Ignoring air resistance, the two arrows have the same constant horizontal velocity component: $v_{2x} - v_{1x} = 0$ (choosing the x-axis horizontal and the y-axis up). Their vertical velocity components are different, but they *change at the same rate*, so $v_{2y} - v_{1y}$ stays constant. The difference in their velocities $(\vec{\mathbf{v}}_2 - \vec{\mathbf{v}}_1)$ stays constant. This constant difference in their velocities makes the difference in their positions $(\vec{\mathbf{r}}_2 - \vec{\mathbf{r}}_1)$ change with time

3.9 $v_{fx} = 500.0$ m/s; $v_{fy} = -19.8$ m/s; bullet enters the water at an angle of 2.27° below the horizontal

3.10 (a) 1.0 m/s; (b) 15 min

3.11 28° south of west; 3.6 min; 130 m

Answers to Checkpoints

3.2 $C_x = -5.5$ km and $C_y = -7.2$ km

3.4 Velocity is a vector quantity. The plane's speed does not change, but its velocity does. Therefore, $\Delta\vec{\mathbf{v}} \neq 0$ and $\vec{\mathbf{a}}_{av} = \Delta\vec{\mathbf{v}}/\Delta t \neq 0$.

3.5 The horizontal velocity component does not change. The vertical component is zero at the highest point, so the velocity vector is directed horizontally. The acceleration is constant and directed vertically downward throughout the flight, including at the highest point.

3.6 (a) 19.5 m/s (b) 1.5 m/s

CHAPTER

4

Force and Newton's Laws of Motion

A sailplane (or "glider") is a small, unpowered, high-performance aircraft. A sailplane must be initially towed a few thousand feet into the air by a small airplane, after which it relies on regions of upward-moving air such as thermals and ridge currents to ascend further. Suppose a small plane requires about 120 m of runway to take off by itself. When it is towing a sailplane, how much runway does it need? (See p. 120 for the answer.)

Concepts & Skills to Review

- addition of vectors (Sections 3.1 and 3.2)
- vector components (Section 3.2)
- acceleration (Sections 2.3 and 3.4)
- motion with constant acceleration (Sections 2.4 and 3.5)
- motion diagrams (Section 2.5)

4.1 FORCE

Just as human life would be dull without social interactions, the physical universe would be dull without physical interactions. Social interactions with friends and family change our behavior; physical interactions change the "behavior" (motion, temperature, etc.) of matter.

Force: a push or pull that one object exerts on another

An interaction between two objects can be described and measured in terms of two *forces*, one exerted on each of the two interacting objects. A **force** is a push or a pull. When you play soccer, your foot exerts a force on the ball while the two are in contact, thereby changing the speed and direction of the ball's motion. At the same time, the ball exerts a force on your foot, the effect of which you can feel. To understand the motion of an object, whether it be a soccer ball or the International Space Station, we need to analyze the forces acting on the object.

The **weight** of an object near a planet or moon is the magnitude of the gravitational force exerted on it by that planet or moon.

Long-Range Forces Forces exerted on macroscopic objects—objects that are large enough for us to observe without instrumentation—can be either long-range forces or contact forces. **Long-range forces** do not require the two objects to be touching. These forces can exist even if the two objects are far apart and even if there are other objects between the two. For example, gravity is a long-range force. The gravitational force exerted on the Earth by the Sun keeps the Earth in orbit around the Sun, despite the great distance between them and despite other planets that occasionally come between them. The Earth also exerts a long-range gravitational force on objects on or near its surface. We call the magnitude of the gravitational force that a planet or moon exerts on a nearby object the object's **weight**.

PHYSICS AT HOME

Besides gravity, other long-range forces are electric or magnetic in nature. On a dry day, run a comb vigorously through your hair until you hear some crackling. Now hold the comb a few centimeters from small pieces of a torn paper napkin. Observe the long-range electrical interaction between the paper and the comb.

Now take a refrigerator magnet. Hold it near but not touching the refrigerator door. You can feel the effect of a long-range magnetic interaction.

Part 3 of this book treats electromagnetic forces in detail. Until then, you can safely assume that gravity is the only significant long-range interaction unless the statement of a problem indicates otherwise.

Contact forces exist only as long as the objects are touching one another.

Contact Forces All forces exerted on macroscopic objects, other than long-range gravitational and electromagnetic forces, involve contact. **Contact forces** exist only as long as the objects are touching one another. Your foot has no noticeable effect on a soccer ball's motion until the two come into contact, and the force lasts only as long as they are in contact. Once the ball moves away from your foot, your foot has no further influence over the ball's motion.

The idea of contact is a useful simplification for macroscopic objects. What we call a single contact force is really the net effect of enormous numbers of electromagnetic forces between atoms on the surfaces of the two objects. On an atomic scale, the idea of "contact" breaks down. There is no way to define "contact" between two atoms—in other words, there is no unique distance between the atoms at which the forces they exert on one another suddenly become zero.

Measuring Forces

If the concept of force is to be useful in physics, there must be a way to measure forces. Consider a simple spring scale (Fig. 4.1). As the scale's pan is pulled down, a spring is stretched. The harder you pull, the more the spring stretches. As the spring stretches, an attached pointer moves. Then all we have to do to measure the applied force is to calibrate the scale so the amount of stretch measures the magnitude of the force. For many springs, the extension is approximately proportional to the force, which makes calibration easy.

In the United States, supermarket scales are generally calibrated to measure forces in pounds (lb). In the SI system, the unit of force is the **newton** (N). To convert pounds to newtons, use the approximate conversion factors

$$1 \text{ lb} = 4.448 \text{ N} \qquad \text{or} \qquad 1 \text{ N} = 0.2248 \text{ lb} \tag{4-1}$$

There are more sophisticated means for measuring forces than a supermarket scale. Even so, many operate on the same principle as the supermarket scale: a force is measured by the deformation—change of size or shape—it produces in some object.

Force Is a Vector Quantity

The magnitude of a force is *not* a complete description of the force. The *direction* of the force is equally important. The direction of the brief contact force exerted by a soccer player's foot on the ball can make the difference between scoring a goal or not (Fig. 4.2). Force is a vector quantity that must be added (or subtracted) using the same methods used for other vector quantities such as position, velocity, and acceleration.

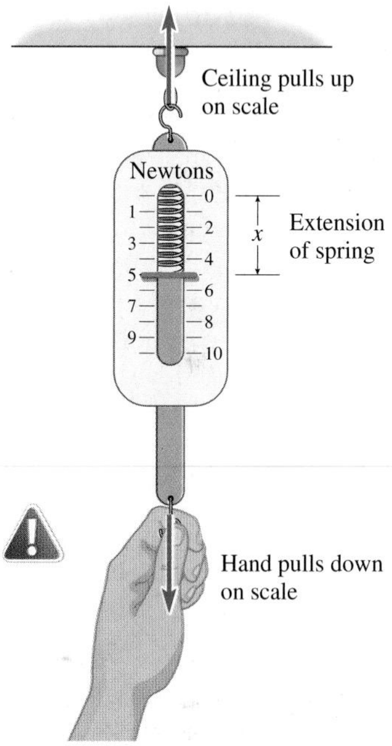

Figure 4.1 As the bottom of a spring scale is pulled downward, the spring stretches. We can measure the force by measuring the extension of the spring. For many springs, the extension is approximately proportional to the force, which makes calibration easy. Note that there is a pull on *both* ends of the scale. The ceiling pulls up on the scale and supports the scale from above.

Figure 4.2 A soccer player's foot exerts a force on the ball only when they are touching.

Example 4.1

Traction on a Foot

In a traction apparatus, three cords pull on the central pulley, each with magnitude 22.0 N, in the directions shown in Fig. 4.3. What is the sum of the forces exerted on the central pulley by the three cords? Give the magnitude and direction of the sum.

Strategy First, we sketch the graphical addition of the three forces to get an estimate of the magnitude and direction of the sum. Then, to get an accurate answer, we resolve the three forces into their x- and y-components, sum the components, and then calculate the magnitude and direction of the sum.

Solution Figure 4.4 shows the graphical addition of the three forces exerted on the central pulley by the cords. From this sketch, we can tell that the sum of the three forces is at a relatively small angle above the horizontal (roughly half of 45°) and has a magnitude a bit larger than 44 N.

To find an algebraic solution, we find the components along the x- and y-axes and add them (Fig. 4.5). The x-components of the forces are

$$F_{1x} = F_{2x} = (22.0\text{ N}) \cos 45.0°$$

$$F_{3x} = (22.0\text{ N}) \cos 30.0°$$

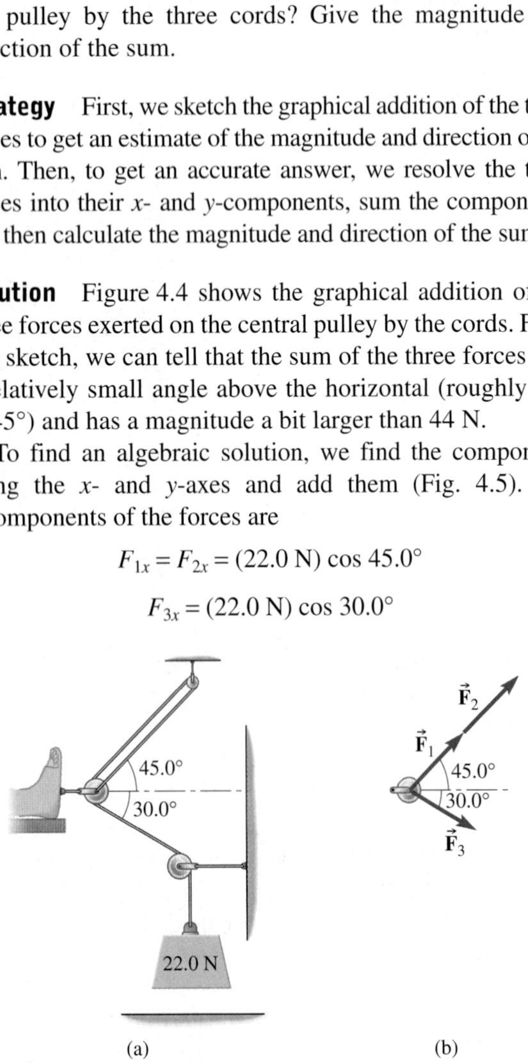

Figure 4.3
(a) A foot in traction; (b) the three forces exerted on the central pulley by the cords.

The y-components of the forces are

$$F_{1y} = F_{2y} = (22.0\text{ N}) \sin 45.0°$$

$$F_{3y} = (-22.0\text{ N}) \sin 30.0°$$

The sum of the x-components is

$$\begin{aligned} F_x &= F_{1x} + F_{2x} + F_{3x} \\ &= 2 \times (22.0\text{ N}) \cos 45.0° + (22.0\text{ N}) \cos 30.0° \\ &= 31.11\text{ N} + 19.05\text{ N} = 50.16\text{ N} \end{aligned}$$

We keep an extra decimal place for now to minimize round-off error. The sum of the y-components is

$$\begin{aligned} F_y &= F_{1y} + F_{2y} + F_{3y} \\ &= 2 \times (22.0\text{ N}) \sin 45.0° + (-22.0\text{ N}) \sin 30.0° \\ &= 31.11\text{ N} - 11.00\text{ N} = 20.11\text{ N} \end{aligned}$$

The magnitude of the sum is (Fig. 4.6):

$$F = \sqrt{F_x^2 + F_y^2} = \sqrt{(50.16\text{ N})^2 + (20.11\text{ N})^2} = 54.0\text{ N}$$

and the direction of the sum is

$$\theta = \tan^{-1} \frac{\text{opposite}}{\text{adjacent}} = \tan^{-1} \frac{20.11\text{ N}}{50.16\text{ N}} = 21.8°$$

The sum of the forces exerted on the pulley by the three cords is 54.0 N at an angle 21.8° above the $+x$-axis.

Discussion To check the answer, look back at the graphical estimate. The magnitude of the sum (54 N) is somewhat larger than 44 N and the direction is at an angle very nearly half of 45° above the horizontal.

Practice Problem 4.1 Changing the Pulley Angles

The pulleys are moved, after which $\vec{\mathbf{F}}_1$ and $\vec{\mathbf{F}}_2$ are at an angle of 30.0° above the x-axis and $\vec{\mathbf{F}}_3$ is 60.0° below the x-axis. (a) What is the sum of these three forces in component form? (b) What is the magnitude of the sum? (c) At what angle with the horizontal is the sum?

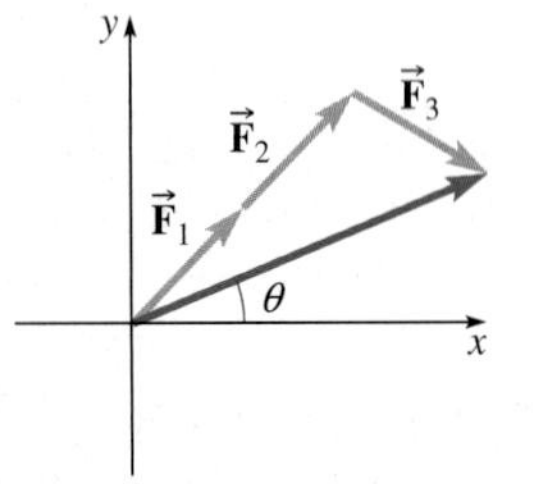

Figure 4.4
Graphical sum of the forces on the pulley due to the cords.

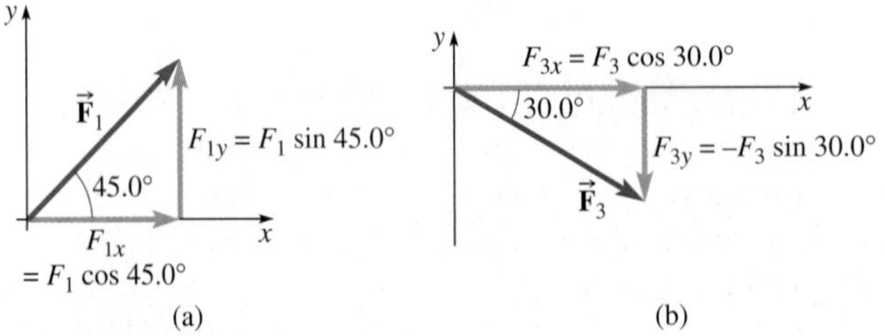

Figure 4.5
Finding the components of (a) $\vec{\mathbf{F}}_1$ and (b) $\vec{\mathbf{F}}_3$. For clarity, the vector arrows are drawn twice as long as they were in Fig. 4.4.

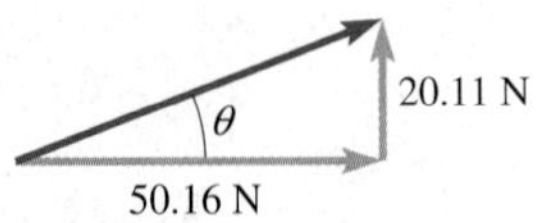

Figure 4.6
Finding the sum from its components.

Net Force

When more than one force acts on an object, the subsequent motion of the object is determined by the *net force* acting on the object. The **net force** is the vector sum of all the forces acting on an object.

Definition of net force:

If $\vec{\mathbf{F}}_1, \vec{\mathbf{F}}_2, \ldots, \vec{\mathbf{F}}_n$ are *all* the forces acting on an object, then the net force $\vec{\mathbf{F}}_{\text{net}}$ acting on that object is the vector sum of those forces:

$$\vec{\mathbf{F}}_{\text{net}} = \sum \vec{\mathbf{F}} = \vec{\mathbf{F}}_1 + \vec{\mathbf{F}}_2 + \cdots + \vec{\mathbf{F}}_n \quad (4\text{-}2)$$

The symbol Σ is a capital Greek letter sigma that stands for "sum."

Free-Body Diagrams

An essential tool used to find the net force acting on an object is a **free-body diagram** (FBD): a simplified sketch of a single object with force vectors drawn to represent *every* force *acting on that object.* (For example, the sum of three forces calculated in Example 4.1 is *not* the net force on the central pulley because the forces on the pulley due to the patient's leg and due to gravity are not included.) The net force must *not* include any forces that act on other objects. To draw an FBD:

- Draw the object in a simplified way—you don't have to be Michelangelo to solve physics problems! Almost any object can be represented as a box or a circle, or even a dot.
- Identify all the forces that are exerted on the object. Take care not to omit any forces that are exerted on the object. Consider that everything touching the object may exert one or more contact forces. Then identify long-range forces (for now, just gravity unless electric or magnetic forces are specified in the problem).
- Check your list of forces to make sure that each force is exerted *on* the object of interest *by* some other object. Make sure you have not included any forces that are exerted *on other objects.*
- Draw vector arrows representing all the forces acting on the object. We usually draw the vectors as arrows that start on the object and point away from it. Draw the arrows so they correctly illustrate the directions of the forces. If you have enough information to do so, draw the lengths of the arrows so they are proportional to the magnitudes of the forces.

Example 4.2

Net Force on an Airplane

The forces on an airplane in flight heading eastward are as follows: gravity = 16.0 kN (kilonewtons), downward; lift = 16.0 kN, upward; thrust = 1.8 kN, east; and drag = 0.8 kN, west. (Lift, thrust, and drag are three forces that the air exerts on the plane.) What is the net force on the plane?

Strategy All the forces acting on the plane are given in the statement of the problem. After drawing these forces in the FBD for the plane, we add the forces to find the net force. To resolve the force vectors into components, we choose x- and y-axes pointing east and north, respectively. All four forces are then lined up with the axes, so each will have only one nonzero component, with a sign that indicates the direction along that axis. For example, the drag force points in the $-x$-direction, so its x-component is negative and its y-component is zero.

continued on next page

Example 4.2 continued

Solution Figure 4.7a is the FBD for the plane, using $\vec{\mathbf{L}}$, $\vec{\mathbf{T}}$, and $\vec{\mathbf{D}}$ for the lift, thrust, and drag, respectively. $\vec{\mathbf{W}}$ stands for the gravitational force on the plane; its magnitude is the plane's weight *W*. The sum of the *x*-components of the forces is

$$\sum F_x = L_x + T_x + W_x + D_x$$
$$= 0 + (1.8\text{ kN}) + 0 + (-0.8\text{ kN}) = 1.0\text{ kN}$$

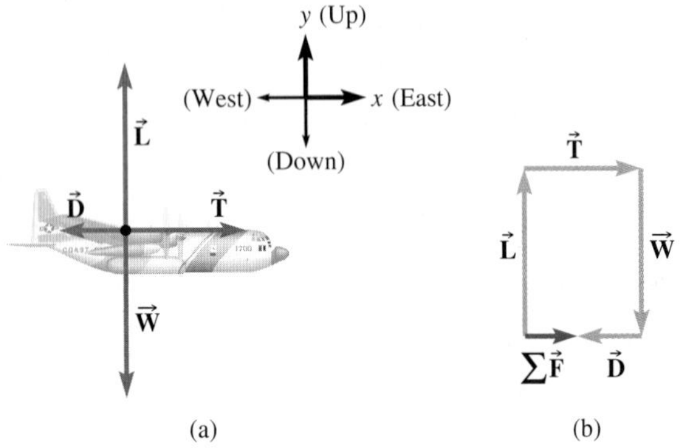

The sum of the *y*-components of the forces is

$$\sum F_y = L_y + T_y + W_y + D_y$$
$$= (16\text{ kN}) + 0 + (-16\text{ kN}) + 0 = 0$$

The net force is 1.0 kN east.

Discussion A graphical check of the vector addition is a good idea. Figure 4.7b shows that the sum of the four forces is indeed in the +*x*-direction (east).

Practice Problem 4.2 New Forces on the Airplane

Find the net force on the airplane if the forces are gravity = 16.0 kN, downward; lift = 15.5 kN, upward; thrust = 1.2 kN, north; drag = 1.2 kN, south.

Figure 4.7
(a) FBD for the airplane. (b) Graphical addition of the four force vectors.

4.2 INERTIA AND EQUILIBRIUM: NEWTON'S FIRST LAW OF MOTION

In 1687, Isaac Newton (1643–1727) published one of the greatest scientific works of all time, his *Philosophiae Naturalis Principia Mathematica* (or *Principia* for short). The Latin title translates as *The Mathematical Principles of Natural Philosophy.* In the *Principia*, Newton stated three laws of motion that form the basis of classical physics.

To pre-Newtonian thinkers, it seemed that there must be two different sets of physical laws: one set to describe the motion of the heavenly bodies, thought to be perfect and enduring, and another to describe the motion of earthly bodies that always come to rest. Together with his law of universal gravitation, Newton's laws of motion showed for the first time that the motion of the heavenly bodies (the Sun, the planets, and their satellites) and the motion of earthly bodies can be understood using the same physical principles.

Newton's First Law of Motion

Newton's first law says that an object acted on by zero net force moves in a straight line with constant speed, or, if it is at rest, remains at rest. Using the concept of the velocity vector, which is a measure of both the speed *and the direction of motion* of an object, we can state the first law:

Newton's First Law of Motion

An object's velocity vector $\vec{\mathbf{v}}$ remains constant if and only if the net force acting on the object is zero.

This concise statement of Newton's first law includes both the case of an object at rest (zero velocity) and a moving object (nonzero velocity). Certainly it makes sense that an object at rest remains at rest unless some force acts on it to make it start to move. On the

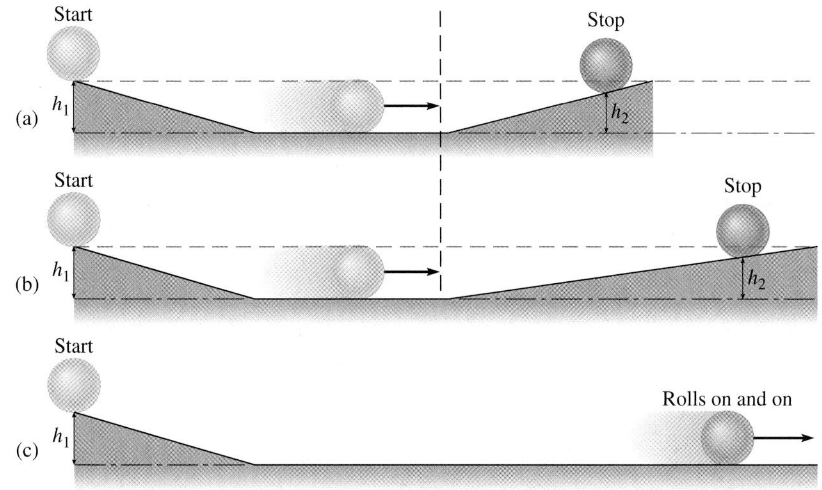

Figure 4.8 (a) Galileo found that a ball rolled down an incline stops when it reaches *almost* the same height on the second incline. He decided that it would reach the *same* height if resistive forces could be eliminated. (b) As the second incline is made less and less steep, the ball rolls farther and farther before stopping. (c) If the second incline is horizontal, and there are no resistive forces, the ball would never stop.

other hand, it may not be obvious that an object can continue to move with constant speed in a straight line without forces acting to keep it moving. In our experience, most moving objects come to rest because of forces that oppose motion, like friction and air resistance. A hockey puck can slide the entire length of a rink with very little change in speed or direction because the ice is slippery (frictional forces are small). If we could remove *all* the resistive forces, including friction and air resistance, the puck would slide without changing its speed or direction at all.

No force is required to keep an object in motion if there are no forces opposing its motion. When a hockey player strikes the puck with his stick, the brief contact force exerted on the puck by the stick changes the puck's velocity, but once the puck loses contact with the stick, it slides along the ice even though the stick no longer exerts a force on it.

Inertia Newton's first law is also called the **law of inertia.** In physics, **inertia** means resistance to *changes* in velocity. It does *not* mean resistance to the continuation of motion (or the tendency to come to rest). Newton based the law of inertia on the ideas of some of his predecessors, including Galileo Galilei (1564–1642) and René Descartes (1596–1650). In a series of clever experiments in which he rolled a ball up inclines of different angles, Galileo postulated that, if he could eliminate all resistive forces, a ball rolling on a horizontal surface would never stop (Fig. 4.8). Galileo made a brilliant conceptual leap from the real world with friction to an imagined, ideal world, free of friction. The law of inertia contradicted the view of the Greek philosopher Aristotle (384–322 B.C.E.). Almost 2000 years before Galileo, Aristotle had formulated his view that the natural state of an object is to be at rest; and, for an object to remain in motion, a force would have to act on it continuously. Galileo conjectured that, in the absence of friction and other resistive forces, no continued force is needed to keep an object moving.

However, Galileo thought that the sustained motion of an object would be in a great circle around the Earth. Shortly after Galileo's death, Descartes argued that the motion of an object free of any forces should be along a straight line rather than a circle. Newton acknowledged his debt to Galileo, Descartes, and others when he wrote: "If I have seen farther, it is because I was standing on the shoulders of giants."

Conceptual Example 4.3

Snow Shoveling

The task of shoveling newly fallen snow from the driveway can be thought of as a struggle against the inertia of the snow. Without the application of a net force, the snow remains at rest on the ground. However, there is an important way that the inertia of the snow makes it *easier* to shovel. Explain.

continued on next page

Conceptual Example 4.3 continued

Strategy Think about the physical motions used when shoveling snow. (If you live where there is no snow, think about shoveling gravel from a wheelbarrow to line a garden path.) In order for the shoveling to be facilitated by the snow's inertia, there must be a time when the snow is moving on its own, without the shovel pushing it.

Solution and Discussion Imagine scooping up a shovelful of snow and swinging the shovel forward toward the side of the driveway. The snow and the shovel are both in motion. Then suddenly the forward motion of the shovel stops, but the snow continues to move forward because of its inertia; it slides forward off the shovel, to be pulled down to the ground by gravity. The snow does not stop moving forward when the forward force due to the shovel is removed.

This procedure works best with fairly dry snow. Wet sticky snow tends to cling to the shovel. The frictional force on the snow due to the shovel keeps it from moving forward and makes the job far more difficult. In this case, it might help to give the shovel a thin coating of cooking oil to reduce the frictional force the shovel exerts on the snow.

Conceptual Practice Problem 4.3 Inertia on the Subway

Negar, a college student, stands on a subway car, holding on to an overhead strap. As the train starts to pull out of the station, she feels thrust toward the rear of the car; as the train comes to a stop at the next station, she feels thrust forward. Explain the role played by inertia in this situation.

PHYSICS AT HOME

For an easy demonstration of inertia, place a quarter on top of an index card, or a credit card, balanced on top of a drinking glass (Fig. 4.9a). With your thumb and forefinger, flick the card so it flies out horizontally from under the quarter. What happens to the quarter? The horizontal force on the coin due to friction is small. With a negligibly small horizontal force, the coin tends to remain motionless while the card slides out from under it (Fig. 4.9b). Once the card is gone, gravity pulls the coin down into the glass (Fig. 4.9c).

Equilibrium

An object in translational equilibrium has a net force of zero acting on it.

When the net force acting on an object is zero, the object is said to be in **translational equilibrium**. *Equilibrium* conveys the idea that the forces are in balance; there is as much force upward as there is downward, as much to the right as to the left, and so forth. Any object moving with a constant velocity, whether at rest or moving in a

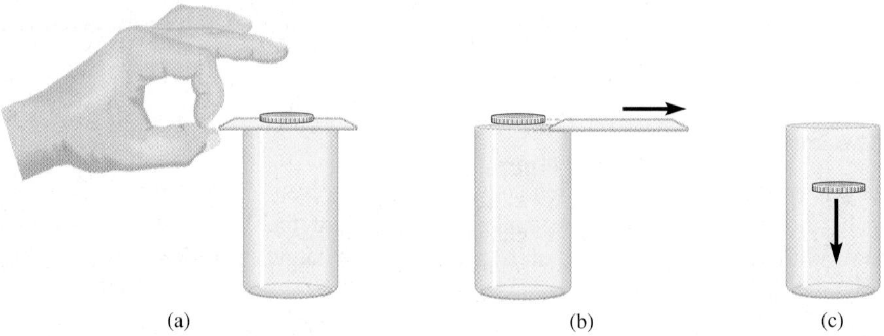

Figure 4.9 A demonstration of inertia.

straight line at constant speed, is in translational equilibrium. A vector can only have zero magnitude if all of its components are zero, so

For an object in equilibrium,

$$\sum F_x = 0 \quad \text{and} \quad \sum F_y = 0 \quad (\text{and} \quad \sum F_z = 0) \tag{4-3}$$

In an equilibrium problem, choose x- and y-axes so the fewest number of force vectors have both x- and y-components. It is always good practice to make a conscious *choice* of axes and then to draw them in the FBDs and any other sketches that you make in solving the problem.

Example 4.4

Sliding a Chest

In order to slide a chest that weighs 750 N across the floor at constant velocity, you must push it horizontally with a force of 450 N (Fig. 4.10). Find the contact force that the floor exerts on the chest.

Strategy The chest moves with constant velocity, so it is in equilibrium. The net force acting on it is zero. We will identify all the forces acting on the chest, draw an FBD, do a graphical addition of the forces, choose x- and y-axes, resolve the forces into their x- and y-components, and then set $\Sigma F_x = 0$ and $\Sigma F_y = 0$.

Solution There are three forces acting on the chest. The gravitational force $\vec{\mathbf{W}}$ has magnitude 750 N and is directed downward. Your push $\vec{\mathbf{F}}$ has magnitude 450 N and its direction is horizontal. The contact force due to the floor $\vec{\mathbf{C}}$ has unknown magnitude and direction. However, remembering that the chest is in equilibrium, upward and downward force components must balance, as must the horizontal force components. Therefore, $\vec{\mathbf{C}}$ must be roughly in the direction shown in the FBD (Fig. 4.11a), as is confirmed by adding the three forces graphically (Fig. 4.11b). The sum is zero because the tip of the last vector ends up at the tail of the first one.

Figure 4.10
Sliding a chest across the floor.

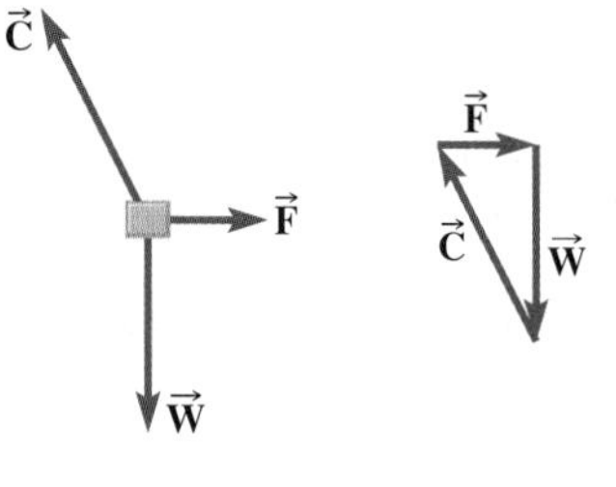

(a) (b)

Figure 4.11
(a) An FBD for the chest; (b) graphical addition of the three forces showing that the sum is zero.

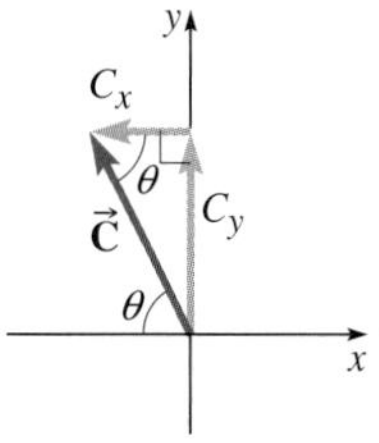

Figure 4.12
Finding the magnitude and direction of the contact force.

Choosing the x-axis to the right and the y-axis up means that two of the three force vectors, $\vec{\mathbf{W}}$ and $\vec{\mathbf{F}}$, have one component that is zero:

$$W_x = 0 \quad \text{and} \quad W_y = -750\text{ N}$$

$$F_x = 450\text{ N} \quad \text{and} \quad F_y = 0$$

Now we set the x- and y-components of the net force each equal to zero because the chest is in equilibrium.

$$\sum F_x = W_x + F_x + C_x = 0 + 450\text{ N} + C_x = 0$$

$$\sum F_y = W_y + F_y + C_y = -750\text{ N} + 0 + C_y = 0$$

These equations tell us the components of $\vec{\mathbf{C}}$: $C_x = -450$ N and $C_y = +750$ N. Then the magnitude of the contact force is (Fig. 4.12)

$$C = \sqrt{C_x^2 + C_y^2} = \sqrt{(-450\text{ N})^2 + (750\text{ N})^2} = 870\text{ N}$$

$$\theta = \tan^{-1}\frac{\text{opposite}}{\text{adjacent}} = \tan^{-1}\frac{750\text{ N}}{450\text{ N}} = 59°$$

The contact force due to the floor is 870 N, directed 59° above the leftward horizontal ($-x$-axis).

Discussion The x- and y-components of the contact force and its magnitude and direction are all reasonable based on the graphical addition, so we can be confident that we did not make an error such as a sign error with one of the components.

Practice Problem 4.4 The Chest at Rest

Suppose the same chest is at rest. You push it horizontally with a force of 110 N but it does not budge. What is the contact force on the chest due to the floor during the time you are pushing?

Using Newton's first law, we can understand how a spring scale can be used to measure weight (the magnitude of the gravitational force exerted on an object). If a melon remains at rest in the pan of the scale, the net force on the melon must be zero. There are only two forces acting on the melon: gravity pulls down and the scale pulls up. Then these two forces must be equal in magnitude and opposite in direction. The scale measures the magnitude of the force it exerts on the melon, which is equal to the weight of the melon.

4.3 NET FORCE, MASS, AND ACCELERATION: NEWTON'S SECOND LAW OF MOTION

When a *nonzero* net force acts on an object, the object's velocity changes. Newton's second law says that the *rate of change of the object's velocity*—that is, the object's acceleration—is proportional to the net force acting on it and inversely proportional to its mass:

Newton's Second Law

$$\vec{\mathbf{a}} = \frac{1}{m}\sum\vec{\mathbf{F}} \quad \text{or} \quad \sum\vec{\mathbf{F}} = m\vec{\mathbf{a}} \tag{4-4}$$

If the net force is zero, then the acceleration is zero, in accordance with Newton's first law. If the net force is not zero, then the acceleration has the same direction as the net force. When the net force is constant, the acceleration is also constant. In component form, Newton's second law is

$$\sum F_x = ma_x \quad \text{and} \quad \sum F_y = ma_y \tag{4-5}$$

If all the forces acting on an object are known, then Eq. (4-4) can be used to calculate its acceleration. Alternatively, sometimes we know the object's acceleration but we have incomplete information about the forces acting on it; then Eq. (4-4) provides information about the unknown forces.

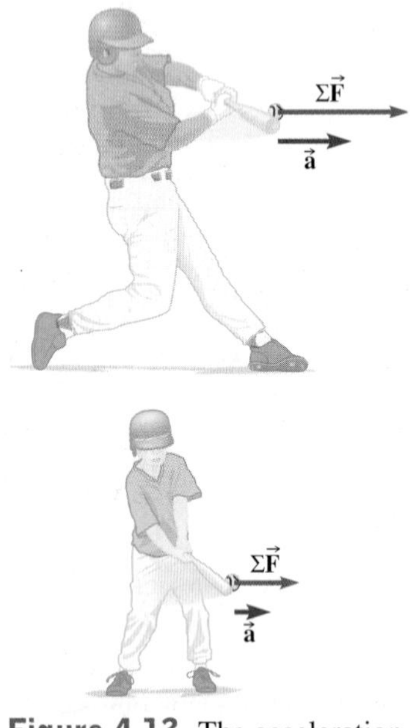

Figure 4.13 The acceleration of a baseball is proportional to the net force acting on it.

SI Unit of Force

The SI unit of force, the newton, is *defined* so that a net force of 1 N gives a 1-kg mass an acceleration of 1 m/s^2:

$$1\ \text{N} = 1\ \text{kg}\cdot\text{m/s}^2 \tag{4-6}$$

Defining the unit of force in this way makes it possible to write Eqs. (4-4) and (4-5) without needing a constant of proportionality to convert between the force unit and kg·m/s^2.

What Is Mass?

The acceleration of an object is proportional to the net force on it and is in the same direction (Fig. 4.13). A larger net force causes a more rapid change in the velocity vector. Newton's second law also says that the acceleration is inversely proportional to the object's mass. The same net force acting on two different objects causes a smaller acceleration on the object with greater mass (Fig. 4.14). Mass is a measure of an object's inertia—the amount of resistance to *changes in velocity*. Newton's second law serves as our *definition* of mass.

In everyday language mass and weight are sometimes used as synonyms, but in physics, mass and weight are different physical properties. The mass of an object is a measure of its inertia, while weight is the magnitude of the gravitational force acting on it. Imagine taking a shuffleboard puck to the Moon. Since the Moon's surface gravity is

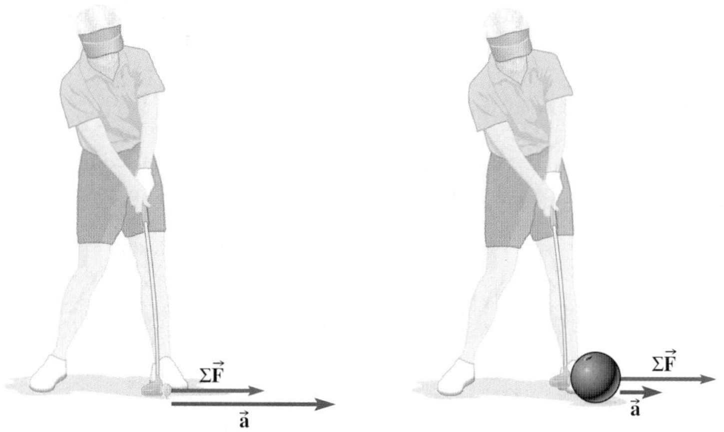

Figure 4.14 The same net force acting on two different objects produces accelerations in inverse proportion to the masses.

weaker than the Earth's, the puck's weight would be smaller on the Moon, but the puck's *mass* would be the same as on Earth. Ignoring the effects of friction, an astronaut playing shuffleboard on the Moon would have to exert the same horizontal force on the puck as on Earth to give it the same acceleration (Fig. 4.15).

4.4 INTERACTION PAIRS: NEWTON'S THIRD LAW OF MOTION

Forces always exist in pairs. Every force is part of an interaction between two objects and each of the interacting objects exerts a force on the other. We call the two forces an **interaction pair**; each force is the **interaction partner** of the other. When you push open a door, the door pushes you. When two cars collide, each exerts a force on the other. Note that interaction partners *act on different objects*—the two objects that are interacting.

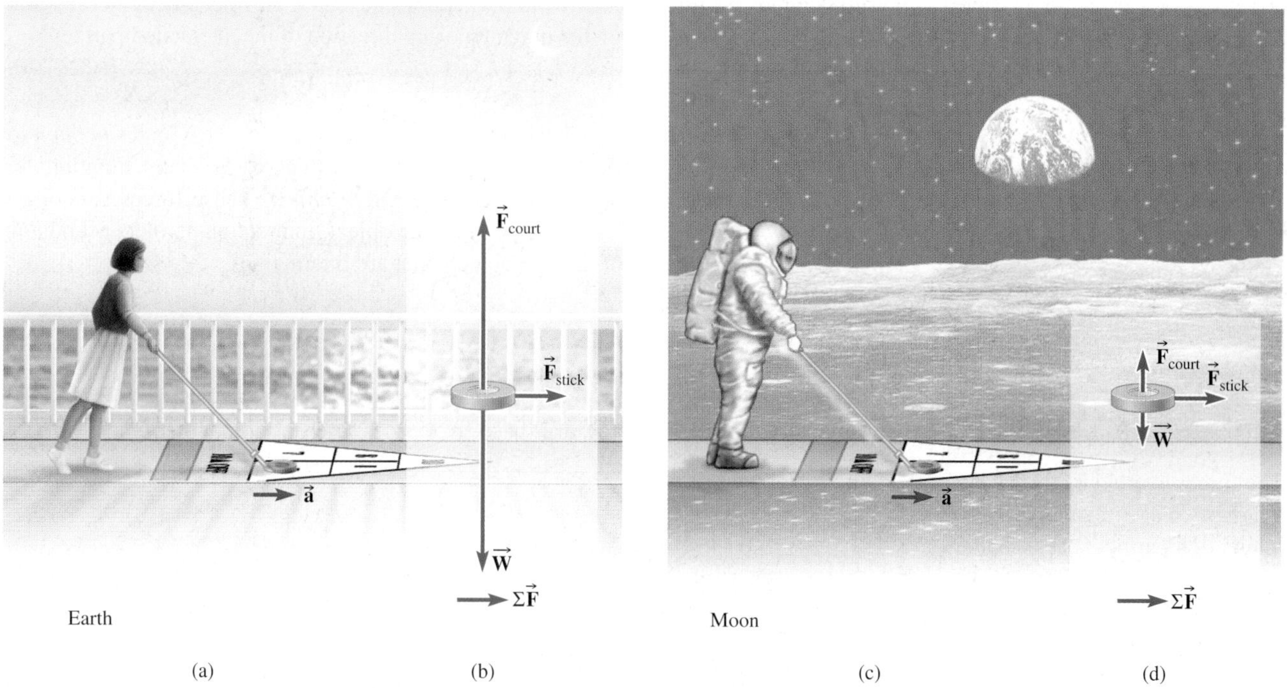

Figure 4.15 An astronaut playing shuffleboard (a) on Earth and (c) on the Moon. FBDs for a puck of mass m being given the same acceleration $\vec{\mathbf{a}}$ on a *frictionless* court on (b) Earth and (d) on the Moon. The contact force on the puck due to the *pushing stick* ($\vec{\mathbf{F}}_{stick}$) must be the same since the mass of the puck is the same: $\Sigma\vec{\mathbf{F}} = \vec{\mathbf{F}}_{stick} = m\vec{\mathbf{a}}$.

Newton's third law of motion says that interaction partners always have the *same magnitude* and are in *opposite directions*.

Newton's Third Law of Motion

In an interaction between two objects, each object exerts a force on the other. These two forces are equal in magnitude and opposite in direction.

Conceptual Example 4.5

An Orbiting Satellite

Earth exerts a gravitational force on an orbiting communications satellite. What is the interaction partner of this force?

Strategy The question concerns a gravitational interaction between two objects: Earth and the satellite. In this interaction, each object exerts a gravitational force on the other.

Solution The interaction partner is the gravitational force exerted on the Earth by the satellite.

Discussion Does the satellite really exert a force on the Earth with the same magnitude as the force Earth exerts on the satellite? If so, why does the satellite orbit Earth rather than Earth orbiting the satellite? Newton's third law says that the interaction partners are equal in magnitude, but does not say that these two forces have equal *effects*. The effect of a net force on an object's motion depends on the object's mass. These two forces of equal magnitude have vastly different effects due to the great discrepancy between the masses of the Earth and the satellite.

On the other hand, if a massive planet orbits a star in a relatively small orbit, the gravitational force that the planet exerts on the star can make the star wobble enough to be observed. The wobble enables astronomers to discover planets orbiting stars other than the Sun. The planets do not reflect enough light toward Earth to be seen, but their presence can be inferred from the effect they have on the star's motion.

Conceptual Practice Problem 4.5 Interaction Partner of a Surface Contact Force

In Example 4.4, the contact force exerted on the chest by the floor was 870 N, directed 59° above the leftward horizontal ($-x$-axis). Describe the interaction partner of this force—in other words, what object exerts it on what other object? What are the magnitude and direction of the interaction partner?

⚠ Do not assume that Newton's third law is involved *every* time two forces *happen* to be equal and opposite—*it ain't necessarily so!* You will encounter many situations in which two equal and opposite forces act *on a single object*. These forces cannot be *interaction partners* because they act on the same object. Interaction partners act on *different objects*, one on each of the two objects that are interacting.

The forces exerted by these two children on a toy cannot be interaction partners because they act on the same object (the toy). The interaction of the force exerted by a child on the toy is the force that the *toy* exerts *on that child.*

✓ CHECKPOINT 4.4

In the photo, two children are pulling on a toy. If they are exerting equal and opposite forces on the toy, are these two forces interaction partners?

PHYSICS AT HOME

The next time you go swimming, notice that you use Newton's third law to get the water to push you forward. When you push down and backward on the water with your arms and legs, the water pushes up and forward on you. The various swimming strokes are devised so that you exert as large a force as possible backward on the water during the power part of the stroke, and then as small a force as possible forward on the water during the return part of the stroke.

Internal and External Forces

When we say that a baseball has interactions with the Earth (gravity), with a baseball bat, and with the air, we are treating the baseball as a single entity. But the ball really consists of an enormous number of protons, neutrons, and electrons, all interacting with each other. The protons and neutrons interact with each other to form atomic nuclei; the nuclei interact with electrons to form atoms; interactions between atoms form molecules; and the molecules interact to form the structure of the thing we call a baseball. It would be difficult to have to deal with all of these interactions to predict the motion of a baseball.

Defining a System Let us call the set of particles comprising the baseball a **system**. Once we have defined a system, we can classify all the interactions that affect the system as either **internal** or **external** to the system. For an internal interaction, *both* interacting objects are part of the system. When we add up all the forces acting on the system to find the net force, every internal interaction contributes two forces—an interaction pair—that always add to zero. For an external interaction, *only one of the two interaction partners is exerted on the system.* The other partner is exerted on an object outside the system and does not contribute to the net force on the system. Therefore, to find the net force on the system, we can ignore all the internal forces and just add the external forces. The insight that internal forces always add to zero is particularly powerful because the choice of what constitutes a system is completely arbitrary. We can choose *any* set of objects and define it to be a system. In one problem, it may be convenient to think of the baseball as a system; in another, we may choose a system consisting of both the baseball and the bat. The second choice might be useful if we do not have detailed information about the interaction between the bat and the ball.

4.5 GRAVITATIONAL FORCES

Newton's Law of Universal Gravitation

Now we turn our attention to learning about some forces in more detail, beginning with gravity. According to **Newton's law of universal gravitation**, any two objects exert gravitational forces on each other that are proportional to the masses (m_1 and m_2) of the two objects and inversely proportional to the square of the distance (r) between their centers. Strictly speaking, the law of gravitation as presented here only applies to point particles and symmetrical spheres. (The *point particle* is a common model in physics used when the size of an object is negligibly small and the internal structure is irrelevant.) Nevertheless, the law of gravitation is *approximately* true for any two objects if the distance between their centers is large compared with their sizes.

In mathematical language, the magnitude of the gravitational force is written:

$$F = \frac{Gm_1m_2}{r^2} \tag{4-7}$$

where the constant of proportionality ($G = 6.674 \times 10^{-11}$ N·m^2/kg^2) is called the **universal gravitational constant**. Equation (4-7) is only part of the law of universal gravitation because it gives only the magnitudes of the gravitational forces that each object exerts on the other. The directions are equally important: each object is pulled toward the other's center (Fig. 4.16). In other words, gravity is an attractive force. The forces on the two objects are equal in magnitude and the directions are opposite, as they must be since they form an interaction pair.

Gravitational forces exerted *by* ordinary objects on each other are so small as to be negligible in most cases (see Practice Problem 4.6). Gravitational forces exerted by Earth, on the other hand, are much larger due to Earth's large mass.

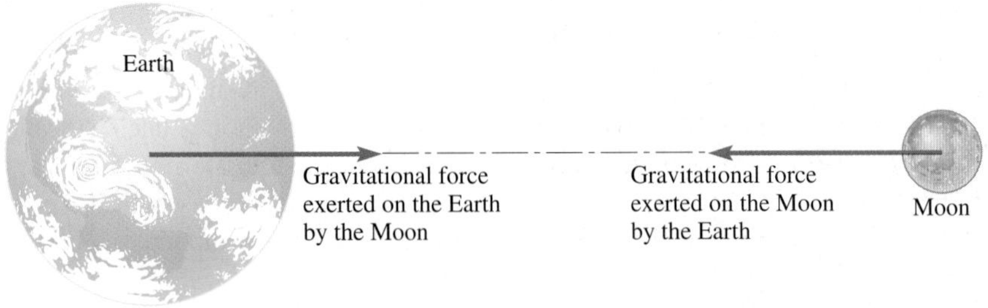

Figure 4.16 Gravity is always an attractive force. The force that each body exerts on the other is equal in magnitude, even though the masses may be very different. The force exerted *on the Moon by the Earth* is of the same magnitude as the force exerted on the Earth by the Moon. The directions are opposite.

Example 4.6

Weight at High Altitude

When you are in a commercial airliner cruising at an altitude of 6.4 km, by what percentage has your weight (as well as the weight of the airplane) changed compared with your weight on the ground?

Strategy Your weight is the magnitude of Earth's gravitational force exerted on you. Newton's law of universal gravitation gives the magnitude of the gravitational force at a distance r from the center of the Earth. For your weight on the ground W_1, we can use the mean radius of the Earth R_E as the distance between the Earth's center and you: $r_1 = R_E = 6.37 \times 10^6$ m (Fig. 4.17). At an altitude of $h = 6.4 \times 10^3$ m above the surface, your weight is W_2 and your distance from Earth's center is $r_2 = R_E + h$. Your mass m, the mass of the Earth M_E ($= 5.97 \times 10^{24}$ kg), and G are the same in the two cases, so it is efficient to write a ratio of the weights and let those factors cancel out.

Figure 4.17
The gravitational force depends on the distance r to the *center* of the Earth.

Solution The ratio of your weight in the airplane to your weight on the ground is

$$\frac{W_2}{W_1} = \frac{\dfrac{GM_E m}{r_2^2}}{\dfrac{GM_E m}{r_1^2}} = \frac{r_1^2}{r_2^2} = \frac{R_E^{\,2}}{(R_E + h)^2}$$

$$= \left(\frac{6.37 \times 10^6 \text{ m}}{6.37 \times 10^6 \text{ m} + 6.4 \times 10^3 \text{ m}}\right)^2 = 0.998$$

Since $0.998 = 1 - 0.002$ and $0.002 = 0.2/100$, your weight decreases by 0.2%.

Discussion Although 6400 m may seem like a significant altitude to us, it's a small fraction of the Earth's radius (0.10%), so the weight change is a small percentage. When judging whether a quantity is small or large, always ask: "Small (or large) compared to what?"

Practice Problem 4.6 A Creative Defense

After an automobile collision, one driver claims that the gravitational force between the two cars caused the collision. Estimate the magnitude of the gravitational force exerted by one car on another when they are driving side-by-side in parallel lanes and comment on the driver's claim.

Gravitational Field Strength

For an object near Earth's surface, the distance between the object and the Earth's center is very nearly equal to the Earth's mean radius, $R_E = 6.37 \times 10^6$ m. The mass of the Earth is $M_E = 5.97 \times 10^{24}$ kg, so the weight of an object of mass m near Earth's surface is

$$W = \frac{GM_E m}{R_E^2} = m\left(\frac{GM_E}{R_E^2}\right) \tag{4-8}$$

Notice that for objects near Earth's surface, the constants in the parentheses are always the same and the weight of the object is proportional to its mass. Rather than recalculate that combination of constants over and over, we call the combination the **gravitational field strength** g near Earth's surface:

$$g = \frac{GM_E}{R_E^2} = \frac{6.674 \times 10^{-11}\ \text{N·m}^2\text{·kg}^{-2} \times (5.97 \times 10^{24}\ \text{kg})}{(6.37 \times 10^6\ \text{m})^2} \approx 9.8\ \text{N/kg} \quad (4\text{-}9)$$

The units *newtons per kilogram* reinforce the conclusion that weight is proportional to mass: g tells us how many newtons of gravitational force are exerted on an object for every kilogram of the object's mass. The weight of a 1.0-kg object near Earth's surface is 9.8 N (2.2 lb). Using g, the weight of an object of mass m near Earth's surface is usually written

Relationship between mass and weight:

$$W = mg \quad (4\text{-}10)$$

Variations in Earth's Gravitational Field The Earth is not a perfect sphere; it is slightly flattened at the poles. Since the distance from the surface to the center of the Earth is smaller there, the field strength at sea level is greatest at the poles (9.832 N/kg) and smallest at the equator (9.814 N/kg). Altitude also matters; as you climb above sea level, your distance from Earth's center increases and the field strength decreases. Tiny local variations in the field strength are also caused by geologic formations. On top of dense bedrock, g is a little greater than above less dense rock. Geologists and geophysicists measure these variations to study Earth's structure and also to locate deposits of various minerals, water, and oil. The device they use, a *gravimeter*, is essentially a mass hanging on a spring. As the gravimeter is carried from place to place, the extension of the spring increases where g is larger and decreases where g is smaller. The mass hanging from the spring does not change, but its weight does ($W = mg$).

Furthermore, due to Earth's rotation, the *effective* value of g that we measure in a coordinate system attached to Earth's surface is slightly less than the true value of the field strength. This effect is greatest at the equator, where the effective value of g is 9.784 N/kg, about 0.3% smaller than the true value of g. The effect gradually decreases with latitude to zero at the poles. We learn more about this effect in Chapter 5.

The most important thing to remember from this discussion is that, unlike G, g is *not* a universal constant. The value of g is a function of position. Near Earth's surface, the variations are small, so we can adopt an average value $g = 9.80$ N/kg as a default.

Gravitational Field and Free-Fall Acceleration

An object in free fall is assumed to have only one force acting on it: gravity. Other forces, such as air resistance, must be negligibly small for this approximation to be valid. We can write the gravitational force on the object as $\vec{\mathbf{W}} = m\vec{\mathbf{g}}$, where the gravitational field vector $\vec{\mathbf{g}}$ has magnitude g and is directed downward (in the direction of the gravitational force). From Newton's second law,

$$\vec{\mathbf{F}}_{\text{net}} = m\vec{\mathbf{g}} = m\vec{\mathbf{a}}$$

Dividing by the mass yields

$$\vec{\mathbf{a}} = \vec{\mathbf{g}} \quad (4\text{-}11)$$

Therefore, the acceleration of an object in free fall is $\vec{\mathbf{g}}$, regardless of the object's mass. Since 1 N = 1 kg·m/s^2, 9.80 N/kg = 9.80 m/s^2—the magnitude of the free-fall acceleration near Earth's surface has average value 9.80 m/s^2.

More massive objects have the same free-fall acceleration as less massive objects. True, a more massive object is harder to accelerate: the acceleration of an object subjected to a given force is inversely proportional to its mass. However, the stronger gravitational force on a more massive object compensates for its greater inertia, giving it the same free-fall acceleration as a less massive object.

Gravitational Field Strength on Other Planets

Equation (4-10) can be used to find the weight of an object at or above the surface of *any* planet or moon, but the value of g will be different due to the different mass M of the planet or moon and the different distance r from the planet's center:

$$g = \frac{GM}{r^2} \tag{4-12}$$

CHECKPOINT 4.5

If you climb Mt. McKinley, what happens to the weight of your gear? What happens to its mass?

Example 4.7

"Weighing" Figs in Kilograms

In most countries other than the United States, produce is sold in mass units (grams or kilograms) rather than in force units (pounds or newtons). The scale still measures a force, but the scale is calibrated to show the mass of the produce instead of its weight. What is the weight of 350 g of fresh figs, in newtons and in pounds?

Strategy Weight is mass times the gravitational field strength. We will assume $g = 9.80$ N/kg. The weight in newtons can be converted to pounds using the conversion factor 1 N = 0.2248 lb.

Solution The weight of the figs in newtons is

$$W = mg = 0.35 \text{ kg} \times 9.80 \text{ N/kg} = 3.43 \text{ N}$$

Converting to pounds,

$$\text{W} = 3.43 \text{ N} \times 0.2248 \text{ lb/N} = 0.771 \text{ lb}$$

The figs weigh 3.4 N or 0.77 lb.

Discussion This is the weight of the figs at a location where g has its average value of 9.80 N/kg. The figs would weigh a little more in the northern city of St. Petersburg, Russia, and a little less in Quito, Ecuador, which is near the equator.

Practice Problem 4.7 Figs on the Moon

What would those figs weigh on the surface of the Moon, where $g = 1.62$ N/kg?

CONNECTION:

In Example 4.4, we resolved the contact force on a sliding chest into components perpendicular to and parallel to the contact surface. It is often convenient to think of these components as two separate but related contact forces: the *normal force* and the *frictional force.*

Normal force: a contact force between two solid objects that is perpendicular to the contact surfaces. Each object pushes the other one away.

4.6 CONTACT FORCES

We have already solved some problems involving forces exerted between two solid objects in contact. Now we look at contact forces in more detail.

Normal Force

A contact force perpendicular to the contact surface that prevents two objects from passing through one another is called the **normal force**. (In geometry, the word *normal* means *perpendicular.*) Consider a book resting on a horizontal table surface. The normal force due to the table must have just the right magnitude to keep the book from falling through the table. If no other vertical forces act, the normal force on the book is equal in magnitude to the book's weight because the book is in equilibrium (Fig. 4.18a).

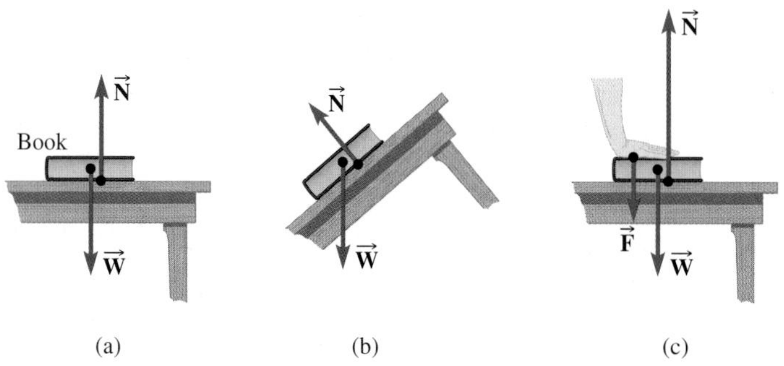

Figure 4.18 (a) The normal force is equal in magnitude to the weight of the book; the two forces sum to zero. (b) On an incline, the normal force is smaller than the weight of the book and is not vertical. (c) If you push down on the book ($\vec{F}$), the normal force on the book due to the table is larger than the book's weight.

According to Newton's third law, there is also a normal force exerted on the table by the book; this normal force acts downward and is of equal magnitude. In *everyday* language, we might say that the table "feels the book's weight." That is not an accurate statement in the language of physics. The table cannot "feel" the gravitational force on the book; the table can only feel forces exerted *on the table.* What the table does "feel" is the normal force—a *contact* force—exerted on the table by the book.

If the table's surface is horizontal, the normal force on the book will be vertical and equal in magnitude to the book's weight. If the surface of the table is *not* horizontal, the normal force is not vertical and is not equal in magnitude to the weight of the book. Remember that the normal force is *perpendicular to the contact surface* (Fig. 4.18b). Even on a horizontal surface, if there are other vertical forces acting on the book, then the normal force is *not* equal in magnitude to the book's weight (Fig. 4.18c). Never *assume* anything about the magnitude of the normal force. In general, we can figure out what the magnitude of the normal force must be in various situations if we have enough information about other forces.

What Causes Normal Forces? How does the table "know" how hard to push on the book? First imagine putting the book on a bathroom scale instead of the table. A spring inside the scale provides the upward force. The spring "knows" how hard to push because, as it is compressed, the force it exerts increases. When the book reaches equilibrium, the spring is exerting just the right amount of force, so there is no tendency to compress it further. The spring is compressed until it pushes up with a force equal to the book's weight. If the spring were stiffer, it would exert the same upward force but with less compression.

The forces that bind atoms together in a rigid solid, like the table, act like extremely stiff springs that can provide large forces with little compression—so little that it's usually not noticed. The book makes a tiny indentation in the surface of the table (Fig. 4.19); a heavier book would make a slightly larger indentation. If the book were to be placed on a soft foam surface, the indentation would be much more noticeable.

Figure 4.19 The book compresses the "atomic springs" in the table until they push up on the book to hold it up. The slight decrease in the distance between atoms is greatly exaggerated here.

CHECKPOINT 4.6

Your laptop is resting on the surface of your desk, which stands on four legs on the floor. Identify the normal forces acting on the desk and give their directions.

Friction

A contact force *parallel* to the contact surface is called **friction**. We distinguish two types: **static friction** and **kinetic** (or **sliding**) **friction**. When the two objects are slipping or sliding across one another, as when a loose shingle slides down a roof, the friction is kinetic. When no slipping or sliding occurs, such as between the tires of a car parked on a hill and the road surface, the friction is called static. Static friction acts to prevent objects from *starting* to slide; kinetic friction acts to try to make sliding objects stop sliding. Note that two objects in contact with one another that move with the same velocity exert *static* frictional forces on one another, because there is no *relative* motion between the two. For example, if a conveyor belt carries an air freight package up an incline and the package is not sliding, the two move with the same velocity and the friction is *static*.

Static Friction Frictional forces are complicated on the microscopic level and are an active field of current research. Despite the complexities, we can make some approximate statements about the frictional forces between dry, solid surfaces. In a simplified model, the maximum magnitude of the force of static friction $f_{s,max}$ that can occur in a particular situation is proportional to the magnitude of the normal force N acting between the two surfaces.

$$f_{s,max} \propto N$$

If you want better traction between the tires of a rear-wheel-drive car and the road, it helps to put something heavy in the trunk to increase the normal force between the tires and the road.

The constant of proportionality is called the **coefficient of static friction** (symbol μ_s):

Maximum force of static friction:

$$f_{s,max} = \mu_s N \qquad (4\text{-}13)$$

Since $f_{s,max}$ and N are both magnitudes of forces, μ_s is a dimensionless number. Its value depends on the condition and nature of the surfaces. Equation (4-13) provides only an *upper limit* on the force of static friction in a particular situation. The actual force of friction in a given situation is not necessarily the maximum possible. It tells us only that, if sliding does not occur, the magnitude of the static frictional force is less than or equal to this upper limit:

$$f_s \leq \mu_s N \qquad (4\text{-}14)$$

Kinetic (Sliding) Friction For sliding or kinetic friction, the force of friction is only weakly dependent on the speed and is roughly proportional to the normal force. In the simplified model we will use, the force of kinetic friction is assumed to be proportional to the normal force and independent of speed:

Force of kinetic (sliding) friction:

$$f_k = \mu_k N \qquad (4\text{-}15)$$

where f_k is the magnitude of the force of kinetic friction and μ_k is called the **coefficient of kinetic friction**. The coefficient of static friction is always larger than the coefficient of kinetic friction for an object on a given surface. On a horizontal surface, a larger force is required to start the object moving than is required to keep it moving at a constant velocity.

Direction of Frictional Forces Equations (4-13) through (4-15) relate only the *magnitudes* of the frictional and normal forces on an object. Remember that the frictional force is perpendicular to the normal force between the same two surfaces. Friction is always parallel to the contact surface, but there are many directions parallel to a given contact surface. Here are some rules of thumb for determining the direction of a frictional force.

- The static frictional force acts in whatever direction necessary to prevent the objects from beginning to slide or slip.

- Kinetic friction acts in a direction that tends to make the sliding stop. If a book slides to the left along a table, the table exerts a kinetic frictional force on the book to the right, in the direction opposite to the motion of the book.
- From Newton's third law, frictional forces come in interaction pairs. If the table exerts a frictional force on the sliding book to the right, the book exerts a frictional force on the table to the *left* with the same magnitude.

Example 4.8

Coefficient of Kinetic Friction for the Sliding Chest

Example 4.4 involved sliding a 750-N chest to the right at constant velocity by pushing it with a horizontal force of 450 N. We found that the contact force on the chest due to the floor had components $C_x = -450$ N and $C_y = +750$ N, where the x-axis points to the right and the y-axis points up (see Fig. 4.20). What is the coefficient of kinetic friction for the chest-floor surface?

Strategy To find the coefficient of friction, we need to know what the normal and frictional forces are. They are the components of the contact force that are perpendicular and parallel to the contact surface. Since the surface is horizontal (in the x-direction), the x-component of the contact force is friction and the y-component is the normal force.

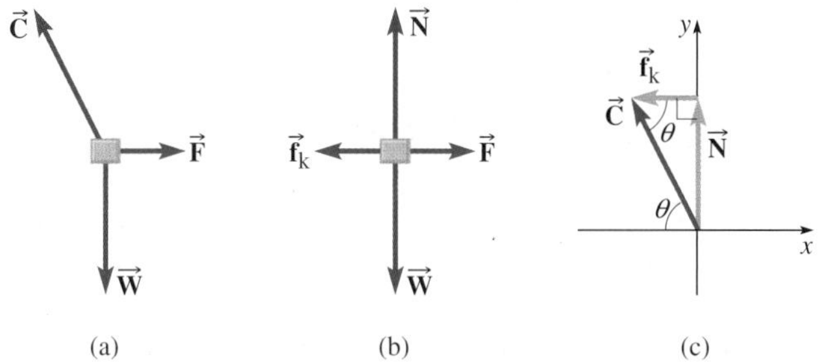

Figure 4.20 (a) FBD for the chest. $\vec{\mathbf{C}}$ is the contact force due to the floor. (b) FBD in which the contact force is replaced by two perpendicular forces, the normal force $\vec{\mathbf{N}}$ and the kinetic frictional force $\vec{\mathbf{f}}_k$. (c) Resolving $\vec{\mathbf{C}}$ into normal and frictional components.

Solution The magnitude of the force due to sliding friction is $f_k = |C_x| = 450$ N. The magnitude of the normal force is $N = |C_y| = 750$ N. Now we can calculate the coefficient of kinetic friction from $f_k = \mu_k N$:

$$\mu_k = \frac{f_k}{N} = \frac{450\text{ N}}{750\text{ N}} = 0.60$$

Discussion If we had written $f_k = C_x = -450$ N, we would have ended up with a negative coefficient of friction. The coefficient of friction is a relationship between the *magnitudes* of two forces, so it cannot be negative.

Practice Problem 4.8 Chest at Rest

Suppose the same chest is at rest. You push to the right with a force of 110 N but the chest does not budge. What are the normal and frictional forces on the chest due to the floor while you are pushing? Explain why you do not need to know the coefficient of static friction to answer this question.

Conceptual Example 4.9

Horse and Sleigh

A horse pulls a sleigh to the right at constant velocity on level ground. The horse exerts a horizontal force $\vec{\mathbf{F}}_{sh}$ on the sleigh. (The subscripts indicate the force on the sleigh due to the horse.) (a) Draw three FBDs, one for the horse, one for the sleigh, and one for the system horse + sleigh. (b) To make the sleigh increase its velocity, there must be a nonzero net force to the right acting on the sleigh. Suppose the horse pulls harder (F_{sh} increases in magnitude). According to Newton's third law, the sleigh always pulls back on the horse with a force of *the same* magnitude as the force with which the horse pulls the sleigh. Does this mean that no matter how hard it pulls, the horse can never make the net force on the sleigh nonzero? Explain. (c) Identify the interaction partner of each force acting on the sleigh.

continued on next page

Conceptual Example 4.9 continued

Strategy (a) In each FBD, we include only the *external* forces acting on that system. All three systems move with constant velocity, so the net force on each is zero. (b) Looking at the FBD for the sleigh, we can determine the conditions under which the net force on the sleigh can be nonzero. (c) For a force exerted on the sleigh by *X*, its interaction partner must be the force exerted on *X* by the sleigh.

Solution and Discussion (a) If we think of the normal and frictional forces as separate forces, then there are four forces acting on the sleigh: the force exerted by the horse $\vec{\mathbf{F}}_{sh}$, the gravitational force due to Earth $\vec{\mathbf{F}}_{sE}$, the normal force on the sleigh due to the ground $\vec{\mathbf{N}}_{sg}$, and kinetic (sliding) friction due to the ground $\vec{\mathbf{f}}_{sg}$. Figure 4.21 shows the FBD for the sleigh. The net force is zero, so its horizontal and vertical components must each be zero: $\vec{\mathbf{F}}_{sh} + \vec{\mathbf{f}}_{sg} = 0$ and $\vec{\mathbf{N}}_{sg} + \vec{\mathbf{F}}_{sE} = 0$.

Similarly, four forces are acting on the horse: the force exerted by the sleigh $\vec{\mathbf{F}}_{hs}$, the gravitational force $\vec{\mathbf{F}}_{hE}$, the normal force due to the ground $\vec{\mathbf{N}}_{hg}$, and friction due to the ground $\vec{\mathbf{f}}_{hg}$. Newton's third law says that $\vec{\mathbf{F}}_{hs} = -\vec{\mathbf{F}}_{sh}$; the sleigh pulls back on the horse with a force equal in magnitude to the forward pull of the horse on the sleigh. Therefore, $\vec{\mathbf{F}}_{hs}$ is to the left and has the same magnitude as $\vec{\mathbf{F}}_{sh}$. The horse is in equilibrium, so $\vec{\mathbf{F}}_{hs} + \vec{\mathbf{f}}_{hg} = 0$ and $\vec{\mathbf{N}}_{hg} + \vec{\mathbf{F}}_{hE} = 0$. The first of these equations means that the frictional force has to be to the *right*. How does the horse get friction to push it *forward*? By pushing *backward* on the ground with its feet. We all do the same thing when taking a step; by pushing backward on the ground, we get the ground to push forward on us. This is *static* friction because the horse's hoof is not sliding along the ground. If there were no friction (imagine the ground to be icy), the hoof might slide backward. Static friction acts to prevent sliding, so the frictional force on the hoof is forward. Figure 4.22 shows the FBD for the horse.

Of the eight forces acting either on the horse or on the sleigh, two are internal forces for the horse + sleigh system: $\vec{\mathbf{F}}_{sh}$ and $\vec{\mathbf{F}}_{hs}$. They add to zero since they are interaction partners, so we can omit them from the FBD for the system (Fig. 4.23). The two frictional forces on the system horse + sleigh are *not* interaction partners, but they are equal in magnitude and opposite in direction. From the FBDs, $\vec{\mathbf{f}}_{hg} = -\vec{\mathbf{F}}_{hs}$ and $\vec{\mathbf{f}}_{sg} = -\vec{\mathbf{F}}_{sh}$. Because $\vec{\mathbf{F}}_{hs}$ and $\vec{\mathbf{F}}_{sh}$ are interaction partners, they are equal and opposite. Therefore, $\vec{\mathbf{f}}_{hg}$ and $\vec{\mathbf{f}}_{sg}$ are equal and opposite. The system is in equilibrium.

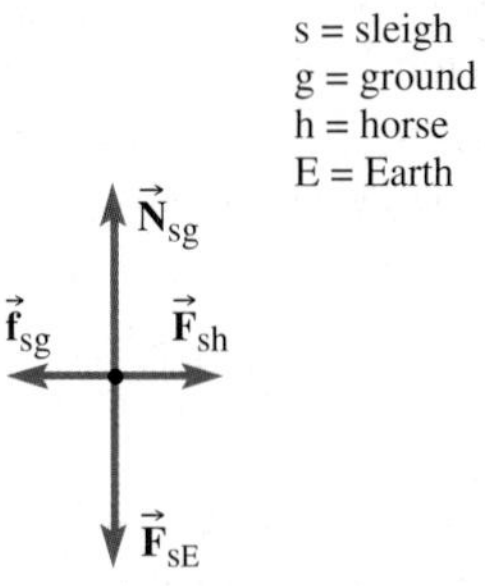

Figure 4.21
FBD for the sleigh.

(b) The FBD for the sleigh (see Fig. 4.21) shows that if the horse pulls the sleigh with a force greater in magnitude than the force of friction on the sleigh ($F_{sh} > f_{sg}$), then the net force on the sleigh is nonzero and to the right. From Fig. 4.22, we need $f_{hg} > F_{hs}$ to have a nonzero net force to the right on the horse. So the frictional force on the horse would have to increase to enable it to pull the sleigh with a greater force. Then in Fig. 4.23, the two frictional forces are no longer equal in magnitude. The forward frictional force on the horse is greater than the backward frictional force on the sleigh, so the net force on the system horse + sleigh is to the right.

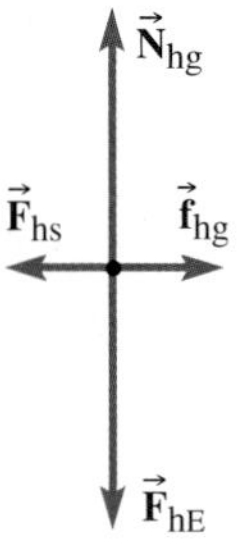

Figure 4.22
FBD for the horse.

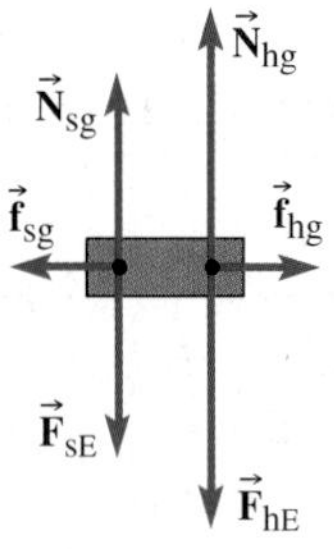

Figure 4.23
FBD for the system, horse and sleigh. The internal forces $\vec{\mathbf{F}}_{sh}$ and $\vec{\mathbf{F}}_{hs}$ are omitted—they form an interaction pair, so they add to zero.

(c)

Force Exerted on Sleigh	Interaction Partner
Force on the sleigh due to the horse $\vec{\mathbf{F}}_{sh}$	Force on the horse due to the sleigh $\vec{\mathbf{F}}_{hs}$
Gravitational force on the sleigh due to Earth $\vec{\mathbf{F}}_{sE}$	Gravitational force on Earth due to the sleigh $\vec{\mathbf{F}}_{Es}$
Normal force on the sleigh due to the ground $\vec{\mathbf{N}}_{sg}$	Normal force on the ground due to the sleigh $\vec{\mathbf{N}}_{gs}$
Friction on the sleigh due to the ground $\vec{\mathbf{f}}_{sg}$	Friction on the ground due to the sleigh $\vec{\mathbf{f}}_{gs}$

Practice Problem 4.9 Passing a Truck

A car is moving north and speeding up to pass a truck on a level road. The combined contact force exerted *on the road by all four tires* has vertical component 11.0 kN downward and horizontal component 3.3 kN southward. The drag force exerted on the car by the air is 1.2 kN southward. (a) Draw the FBD for the car. (b) What is the weight of the car? (c) What is the net force acting on the car?

Microscopic Origin of Friction What looks like the smooth surface of a solid to the unaided eye is generally quite rough on a microscopic scale (Fig. 4.24). Friction is caused by atomic or molecular bonds between the "high points" on the surfaces of the two objects. These bonds are formed by microscopic electromagnetic forces that hold the atoms or molecules together. If the two objects are pushed together harder, the surfaces deform a little more, enabling more "high points" to bond. That is why the force of kinetic friction and the maximum force of static friction are proportional to the normal force. A bit of lubricant drastically decreases the frictional forces, because the two surfaces can float past one another without many of the "high points" coming into contact.

In static friction, when these molecular bonds are stretched, they pull back harder. The bonds have to be broken before sliding can begin. Once sliding begins, molecular bonds are continually made and broken as "high points" come together in a hit-or-miss fashion. These bonds are generally not as strong as those formed in the absence of sliding, which is why $\mu_s > \mu_k$.

For dry, solid surfaces, the amount of friction depends on how smooth the surfaces are and how many contaminants are present on the surface. Does polishing two steel surfaces decrease the frictional forces when they slide across each other? Not necessarily. In an extreme case, if the surfaces are extremely smooth and all surface contaminants are removed, the steel surfaces form a "cold weld"—essentially, they become one piece of steel. The atoms bond as strongly with their new neighbors as they do with the old.

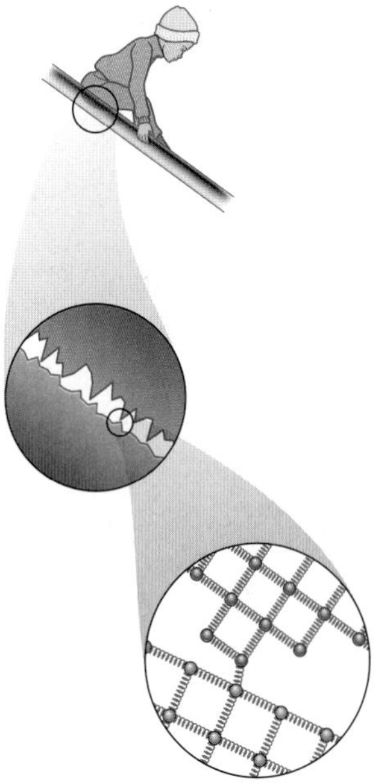

Figure 4.24 Friction is caused by bonds between atoms that form between the "high points" of the two surfaces that come into contact.

Equilibrium on an Inclined Plane

Suppose we wish to pull a large box up a *frictionless* incline to a loading dock platform. Figure 4.25 shows the three forces acting on the box. $\vec{\mathbf{F}}_a$ represents the applied force with which we pull. The force is parallel to the incline. If we choose the x- and y-axes to be horizontal and vertical, respectively, then two of the three forces have both x- and y-components. On the other hand, if we choose the x-axis parallel to the incline and the y-axis perpendicular to it, then only one of the three forces has both x- and y-components (the gravitational force).

With axes chosen, the weight of the box is then resolved into two perpendicular components (Fig. 4.26a). To find the x- and y-components of the gravitational force $\vec{\mathbf{W}}$, we must determine the angle that $\vec{\mathbf{W}}$ makes with one of the axes. The right triangle of Fig. 4.26b shows that $\alpha + \phi = 90°$, since the interior angles of a triangle add up to 180°. The x- and y-axes are perpendicular, so $\alpha + \beta = 90°$. Therefore, $\beta = \phi$.

The y-component of $\vec{\mathbf{W}}$ is perpendicular to the surface of the incline. From Fig. 4.26a, the side parallel to the y-axis is adjacent to angle β, so

$$\cos\beta = \frac{\text{adjacent}}{\text{hypotenuse}} = \frac{|W_y|}{|W|}$$

Since W_y is negative and $W = mg$,

$$W_y = -mg\cos\beta = -mg\cos\phi$$

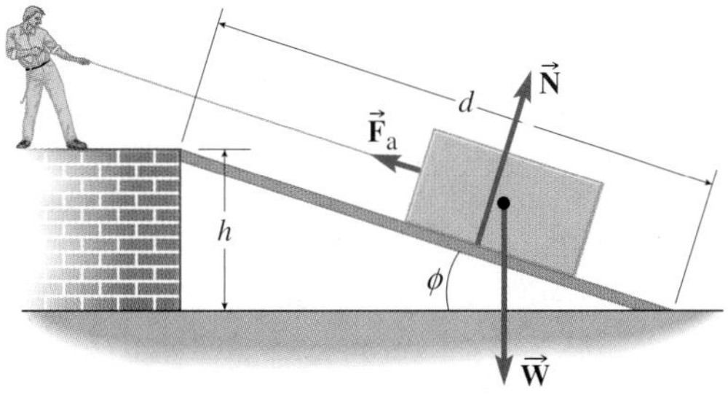

Figure 4.25 A box of mass m pulled up an incline.

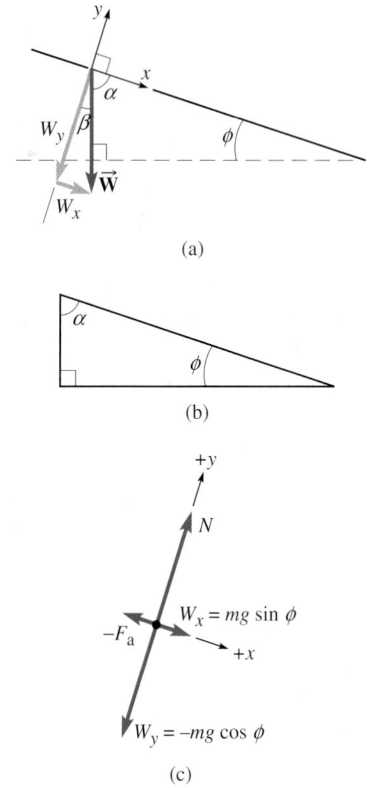

Figure 4.26 (a) Resolving the weight into components parallel to and perpendicular to the incline. (b) A right triangle shows that $\alpha + \phi = 90°$. (c) FBD for the box on the incline.

The x-component of the weight tends to make the box slide down the incline (in the positive x-direction). Using the same triangle,

$$W_x = +mg \sin \phi$$

When the box is pulled with a force equal in magnitude to W_x up the incline (in the negative x-direction), it will slide up with constant velocity. The component of the box's weight perpendicular to the incline is supported by the normal force $\vec{\mathbf{N}}$ that pushes the box away from the incline. Figure 4.26c is an FBD in which the gravitational force is separated into its x- and y-components.

If the box is in equilibrium, whether at rest or moving along the incline at constant velocity, the force components along each axis sum to zero:

$$\sum F_x = (-F_a) + mg \sin \phi = 0$$

and

$$\sum F_y = N + (-mg \cos \phi) = 0$$

On an incline, the normal force is *not* equal in magnitude to the weight and it does not point straight up. If the applied force has magnitude $mg \sin \phi$, we can pull the box up the incline at constant velocity. If friction acts on the box, we must pull with a force greater than $mg \sin \phi$ to slide the box up the incline at constant velocity.

Example 4.10

Pushing a Safe up an Incline

A new safe is being delivered to the Corner Book Store. It is to be placed in the wall at a height of 1.5 m above the floor. The delivery people have a portable ramp, which they plan to use to help them push the safe up and into position. The mass of the safe is 510 kg, the coefficient of static friction along the incline is $\mu_s = 0.42$, and the coefficient of kinetic friction along the incline is $\mu_k = 0.33$. The ramp forms an angle $\theta = 15°$ above the horizontal. (a) How hard do the movers have to push to start the safe moving up the incline? Assume that they push in a direction parallel to the incline. (b) To slide the safe up at a constant speed, with what magnitude force must the movers push?

Strategy (a) When the safe *starts* to move, its velocity is changing, so the safe is *not* in equilibrium. Nevertheless, to find the minimum applied force to start the safe moving, we can find the *maximum* applied force for which the safe *remains at rest*—an equilibrium situation. (b) The safe is in equilibrium as it slides with a constant velocity. Both parts of the problem can be solved by drawing the FBD, choosing axes, and setting the x- and y-components of the net force equal to zero.

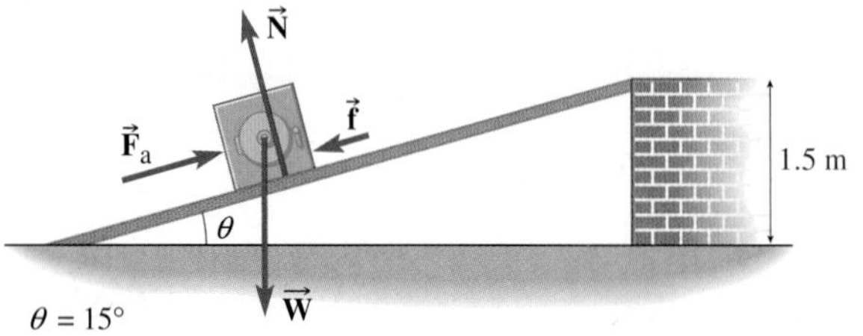

Figure 4.27

Forces acting on the safe as it is moved up the incline.

Solution First we draw a diagram to show forces acting (Fig. 4.27). Before resolving the forces into components, we must choose x- and y-axes. To use the coefficient of friction, we have to resolve the contact force on the safe due to the incline into components *parallel and perpendicular to the incline*—friction and the normal force, respectively—rather than into horizontal and vertical components. Therefore, we choose x- and y-axes parallel and perpendicular to the incline so friction is along the x-axis and the normal force is along the y-axis.

The gravitational force $\vec{\mathbf{W}}$ can be resolved into its components: $W_x = -mg \sin \theta$ and $W_y = -mg \cos \theta$ (Fig. 4.28a).

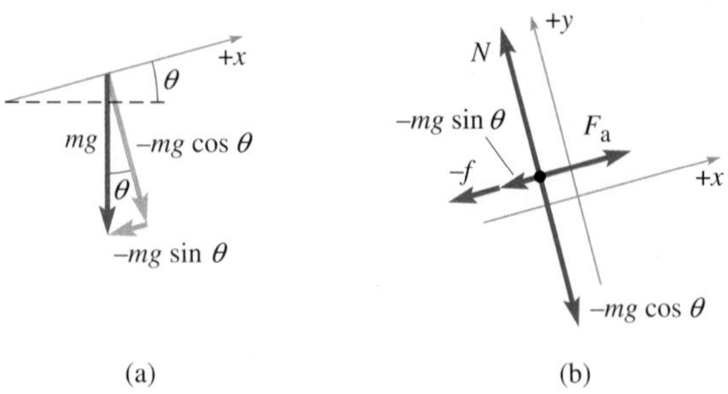

Figure 4.28

(a) Resolving the weight into x- and y-components, and (b) an FBD in which the weight is replaced with its x- and y-components.

continued on next page

Example 4.10 continued

Now we draw the FBD with $\vec{\mathbf{W}}$ replaced by its components (Fig. 4.28b).

(a) Suppose that the safe is initially at rest. As the movers start to push, F_a gets larger and the force of static friction gets larger to "try" to keep the safe from sliding. Eventually, at some value of F_a, static friction reaches its maximum possible value $\mu_s N$. If the movers continue to push harder, increasing F_a further, the force of static friction cannot increase past its maximum value $\mu_s N$, so the safe starts to slide. The direction of the frictional force is along the incline and downward since friction is "trying" to keep the safe from sliding *up* the incline.

The normal force is *not* equal in magnitude to the weight of the safe. To find the normal force, sum the y-components of the forces:

$$\sum F_y = N + (-mg \cos \theta) = 0$$

Then $N = mg \cos \theta$. The normal force is *less than the weight* since $\cos \theta < 1$.

When the movers push with the largest force for which the safe does *not* slide,

$$\sum F_x = F_{ax} + f_x + W_x = 0$$

The applied force is in the $+x$-direction, so $F_{ax} = +F_a$. The frictional force has its maximum magnitude and is in the $-x$-direction, so $f_x = -f_{s,max} = -\mu_s N = -\mu_s mg \cos \theta$. From the FBD, $W_x = -mg \sin \theta$. Then,

$$\sum F_x = F_a - \mu_s mg \cos \theta - mg \sin \theta = 0$$

Solving for F_a,

$$\begin{aligned} F_a &= mg\,(\mu_s \cos \theta + \sin \theta) \\ &= 510 \text{ kg} \times 9.80 \text{ m/s}^2 \times (0.42 \times \cos 15° + \sin 15°) \\ &= 3300 \text{ N} \end{aligned}$$

An applied force that *exceeds* 3300 N starts the box moving up the incline.

(b) Once the safe is sliding, the movers need only push hard enough to make the net force on the safe equal to zero if they want the safe to slide at constant velocity. We are now dealing with sliding friction, so the frictional force is now $f_x = -\mu_k N = -\mu_k mg \cos \theta$.

$$\begin{aligned} \sum F_x &= F_{ax} + f_x + W_x \\ &= F_a - \mu_k mg \cos \theta - mg \sin \theta \\ &= 0 \end{aligned}$$

$$\begin{aligned} F_a &= mg\,(\mu_k \cos \theta + \sin \theta) \\ &= 510 \text{ kg} \times 9.80 \text{ m/s}^2 \times (0.33 \times \cos 15° + \sin 15°) \\ &= 2900 \text{ N} \end{aligned}$$

The movers push with a force $\vec{\mathbf{F}}_a$ of magnitude 2900 N directed up the incline.

Discussion In (b), the expression $F_a = mg\,(\mu_k \cos \theta + \sin \theta)$ shows that the applied force up the incline has to balance the sum of two forces down the incline: the frictional force ($\mu_k mg \cos \theta$) and the component of the gravitational force down the incline ($mg \sin \theta$). This balance of forces is shown graphically in the FBD (Fig. 4.28b).

Practice Problem 4.10 Smoothing the Infield Dirt

During the seventh-inning stretch of a baseball game, groundskeepers drag mats across the infield dirt to smooth it. A groundskeeper is pulling a mat at a constant velocity by applying a force of 120 N at an angle of 22° above the horizontal. The coefficient of kinetic friction between the mat and the ground is 0.60. Find (a) the magnitude of the frictional force between the dirt and the mat and (b) the weight of the mat.

PHYSICS AT HOME

To estimate the coefficient of static friction between a penny and the cover of your physics book, place the penny on the book and slowly lift the cover. Note the angle of the cover when the penny starts to slide. Explain how you can use this angle to find the coefficient of static friction. Can you devise an experiment to find the coefficient of kinetic friction?

4.7 TENSION

Consider a heavy chandelier hanging by a chain from the ceiling (Fig. 4.29a). The chandelier is in equilibrium, so the upward force on it due to the chain is equal in magnitude to the chandelier's weight. With what force does the chain pull downward on the ceiling? The ceiling has to pull up with a force equal to the total weight of the chain and the chandelier. The interaction partner of this force—the force the chain exerts on the ceiling—is

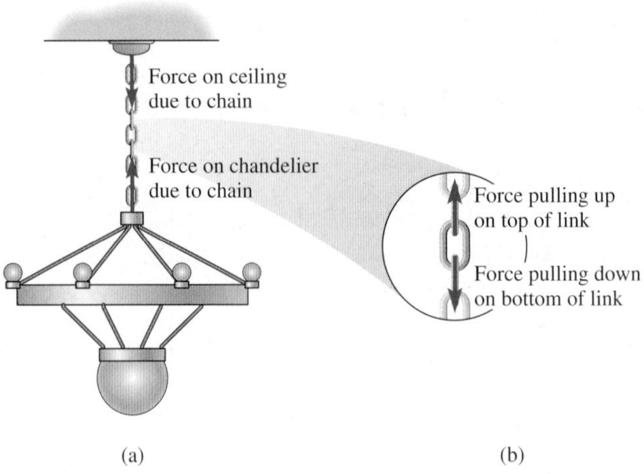

Figure 4.29 (a) The chain pulls up on the chandelier and pulls down on the ceiling. (b) The chain is under tension. Each link is pulled in opposite directions by its neighbors.

equal in magnitude and opposite in direction. Therefore, if the weight of the chain is negligibly small compared with the weight of the chandelier, then the chain exerts forces of equal magnitude at its two ends. The forces at the ends would *not* be equal, however, if you grabbed the chain in the middle and pulled it up or down or if we could not neglect the weight of the chain. We can generalize this observation:

An *ideal* cord (or rope, string, tendon, cable, or chain) pulls in the direction of the cord with forces of equal magnitude on the objects attached to its ends as long as no external force is exerted on it anywhere between the ends. An ideal cord has zero mass and zero weight.

A single link of the chain (Fig. 4.29b) is pulled at both ends by the neighboring links. The magnitude of these forces is called the **tension** in the chain. Similarly, a little segment of a cord is pulled at both its ends by the tension in the neighboring pieces of the cord. If the segment is in equilibrium, then the net force acting on it is zero. As long as there are no other forces exerted on the segment, the forces exerted by its neighbors must be equal in magnitude and opposite in direction. Therefore, the tension has the same value everywhere and is equal to the force that the cord exerts on the objects attached to its ends.

Example 4.11

Archery Practice

Figure 4.30 shows the bowstring of a bow and arrow just before it is released. The archer is pulling back on the midpoint of the bowstring with a horizontal force of 162 N. What is the tension in the bowstring?

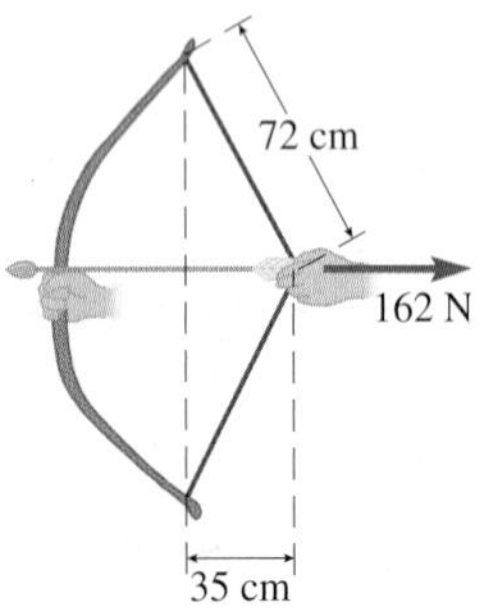

Figure 4.30
The force applied to the bowstring by an archer.

Strategy Consider a small segment of the bowstring that touches the archer's finger. That piece of the string is in equilibrium, so the net force acting on it is zero. We draw the FBD, choose coordinate axes, and apply the equilibrium condition: $\Sigma F_x = 0$ and $\Sigma F_y = 0$. We know the force exerted on the segment of string by the archer's fingers. That segment is also pulled on each end by the tension in the string. Can we assume the tension in the string is the same everywhere? The weight of the string is small compared with the other forces acting on it. The archer pulls sideways on the bowstring, exerting little or no *tangential* force, so we can assume the tension is the same everywhere.

Solution Figure 4.31a is an FBD for the segment of bowstring being considered. The forces are labeled with their magnitudes: F_a for the force applied by the archer's finger and T for each of the tension forces. Figure 4.31b shows these three forces adding to zero. From this sketch, we expect the tension T to be roughly the same as F_a. We choose the x-axis to the right and the y-axis upward. To find the

continued on next page

Example 4.11 continued

components of the forces due to tension in the string, we draw a triangle (Fig. 4.31c). From the measurements given, we can find the angle θ.

$$\sin\theta = \frac{\text{opposite}}{\text{hypotenuse}} = \frac{35\text{ cm}}{72\text{ cm}} = 0.486$$

$$\theta = \sin^{-1} 0.486 = 29.1°$$

The x-component of the tension force exerted on the upper end of the segment is

$$T_x = -T\sin\theta$$

The x-component of the force exerted on the lower end of the string is the same. Therefore,

$$\sum F_x = -2T\sin\theta + F_a = 0$$

Solving for T,

$$T = \frac{F_a}{2\sin\theta} = \frac{162\text{ N}}{2\times 0.486} = 170\text{ N}$$

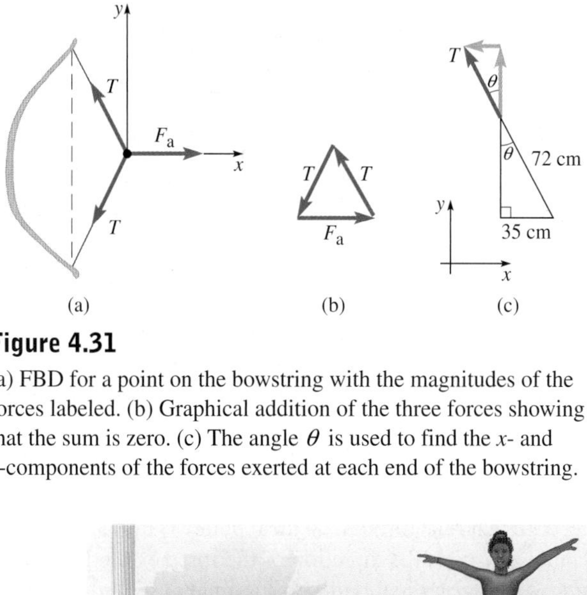

Figure 4.31

(a) FBD for a point on the bowstring with the magnitudes of the forces labeled. (b) Graphical addition of the three forces showing that the sum is zero. (c) The angle θ is used to find the x- and y-components of the forces exerted at each end of the bowstring.

Discussion The tension is only slightly larger than F_a, a reasonable result given the picture of graphical vector addition in Fig. 4.31b.

In this problem, only the x-components of the forces had to be used. The y-components must also add to zero. At the upper end of the string, the y-component of the force exerted by the bow is $+T\cos\theta$, while at the lower end it is $-T\cos\theta$. Therefore, $\Sigma F_y = 0$.

The expression $T = F_a/(2\sin\theta)$ can be evaluated for limiting values of θ to make sure that the expression is correct. As θ approaches 90°, the tension approaches

$$\frac{F_a}{2\sin 90°} = \frac{1}{2}F_a$$

That is correct because the archer would be pulling to the right with a force F_a, while each side of the bowstring would pull to the left with a force of magnitude T. For equilibrium, $F_a = 2T$ or $T = \frac{1}{2}F_a$.

As θ gets smaller, $\sin\theta$ decreases and the tension increases (for a fixed value of F_a). That agrees with our intuition. The larger the tension, the smaller the angle the string needs to make in order to supply the necessary horizontal force.

Practice Problem 4.11 Tightrope Practice

Jorge decides to rig up a tightrope in the backyard so his children can develop a good sense of balance (Fig. 4.32). For safety reasons, he positions a horizontal cable only 0.60 m above the ground. If the 6.00-m-long cable sags by 0.12 m from its taut horizontal position when Denisha (weight 250 N) is standing on the middle of it, what is the tension in the cable? Ignore the weight of the cable.

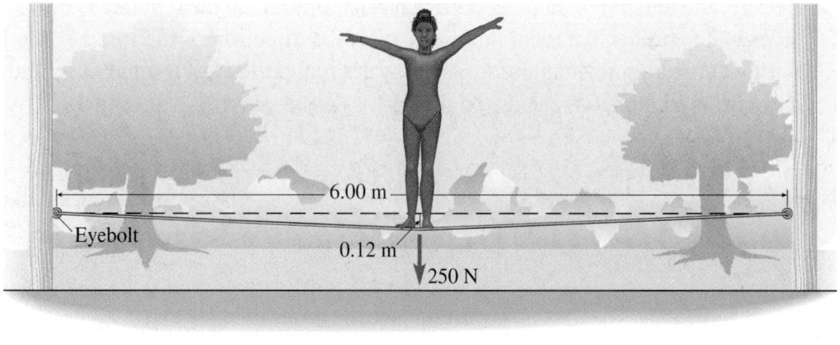

Figure 4.32

Tightrope for balancing practice.

Application: Tensile Forces in the Body Tensile forces are central in the study of animal motion, or biomechanics. Muscles are usually connected by tendons, one at each end of the muscle, to two different bones, which in turn are linked at a joint (Fig. 4.33). Usually one of the bones is more easily moved than the other. When the muscle contracts, the tension in the tendons increases, pulling on both of the bones.

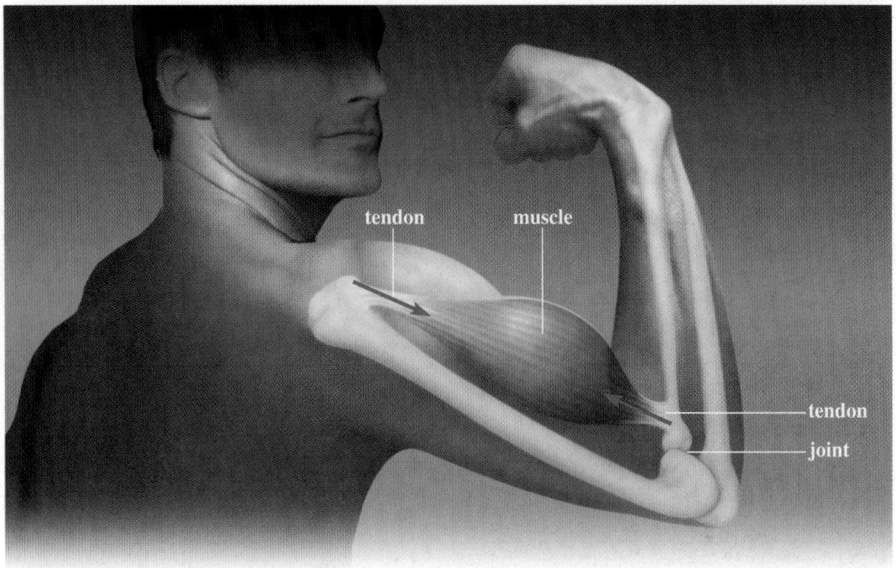

Figure 4.33 A muscle contracts, increasing the tension in the attached tendons. The tendons exert forces on two different bones.

PHYSICS AT HOME

Sit with your arm bent at the elbow with a heavy object on the palm of your hand. You can feel the contraction of the biceps muscle. With your other hand, feel the tendon that connects the biceps muscle to your forearm.

Now place your hand palm down on the desktop and push down. Now it is the triceps muscle that contracts, pulling up on the bone on the other side of the elbow joint. Muscles and tendons cannot push; they can only pull. The biceps muscle cannot push the forearm downward, but the triceps muscle can pull on the other side of the joint. In both cases, the arm acts as a lever.

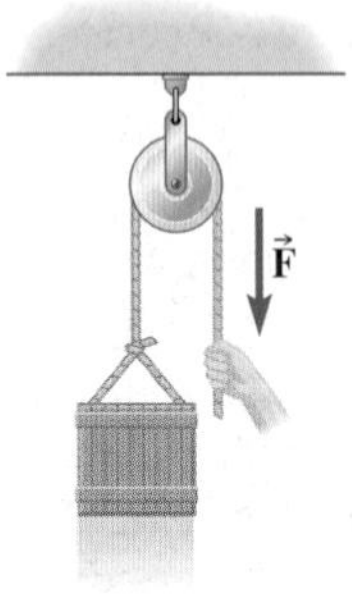

Figure 4.34 Using a pulley to lift an object by pulling *downward* on a rope with force $\vec{\mathbf{F}}$.

Application: Ideal Pulleys A pulley can change the direction of the force exerted by a cord under tension. To lift something heavy, it is easier to stand on the ground and pull *down* on the rope than to get above the weight on a platform and pull up on the rope (Fig. 4.34).

An *ideal* pulley has no mass and no friction. An ideal pulley exerts no forces on the cord that are *tangent* to the cord—it is not pulling in either direction along the cord. As a result, the tension of an ideal cord that runs through an ideal pulley is the same on both sides of the pulley. An ideal pulley changes the direction of the force exerted by a cord without changing its magnitude. As long as a real pulley has a small mass and negligible amount of friction, we can approximate it as an ideal pulley.

Example 4.12

A Two-Pulley System

A 1804-N engine is hauled upward at constant speed (Fig. 4.35). What are the tensions in the three ropes labeled A, B, and C? Assume the ropes and the pulleys labeled L and R are ideal.

Strategy The engine and pulley L move up at constant speed, so the net force on each of them is zero. Pulley R is at rest, so the net force on it is also zero. We can draw the FBD for any or all of these objects and then apply the equilibrium condition. If the pulleys are ideal, the tension in the rope is the same on both sides of the pulley. Therefore, rope C—which is attached to the ceiling, passes around both pulleys, and is pulled downward at the other end—has the same tension

continued on next page

Example 4.12 continued

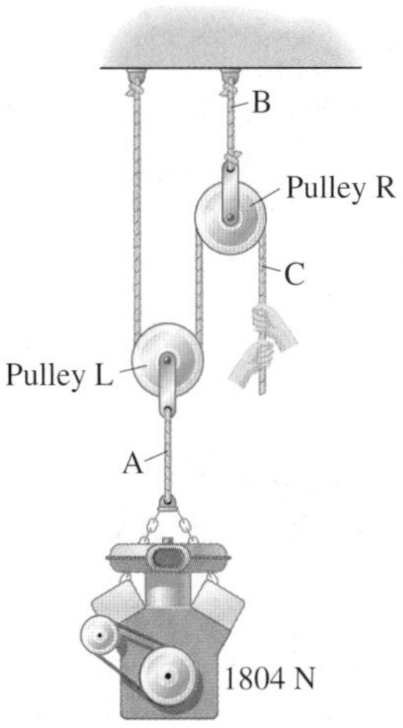

Figure 4.35
A system of pulleys used to raise a heavy weight.

throughout. Call the tensions in the three ropes T_A, T_B, and T_C. To analyze the forces exerted on a pulley, we define our system so the part of the rope wrapped around the pulley is considered part of the pulley. Then there are two cords pulling on the pulley, each with the same tension.

Solution There are two forces acting on the engine: the gravitational force (1804 N, downward) and the upward pull of rope A. These must be equal and opposite (Fig. 4.36a), since the net force is zero. Therefore $T_A = 1804$ N.

The FBD for pulley L (Fig. 4.36b) shows rope A pulling down with a force of magnitude T_A and rope C pulling upward on *each side.* The rope has the same tension throughout, so all forces labeled T_C in Fig. 4.36b,c have the same magnitude. For the net force to equal zero,

$$2T_C = T_A$$
$$T_C = \tfrac{1}{2}T_A = 902.0 \text{ N}$$

Figure 4.36c is the FBD for pulley R. Rope B pulls upward on it with a force of magnitude T_B. On *each side* of the pulley, rope C pulls downward. For the net force to equal zero,

$$T_B = 2T_C = 1804 \text{ N}$$

Discussion The engine is raised by pulling *down* on a rope—the pulleys change the direction of the applied force needed to lift the engine. In this case they also change the *magnitude* of the required force. They do that by making the rope pull up on the engine twice, so the person pulling the rope only needs to exert a force equal to half the engine's weight.

Practice Problem 4.12 System of Ropes, Pulleys, and Engine

Consider the entire collection of ropes, pulleys, and the engine to be a single system. Draw the FBD for this system and show that the net force on the system is zero. [*Hint:* Remember that only forces exerted by objects *external* to the system are included in the FBD.]

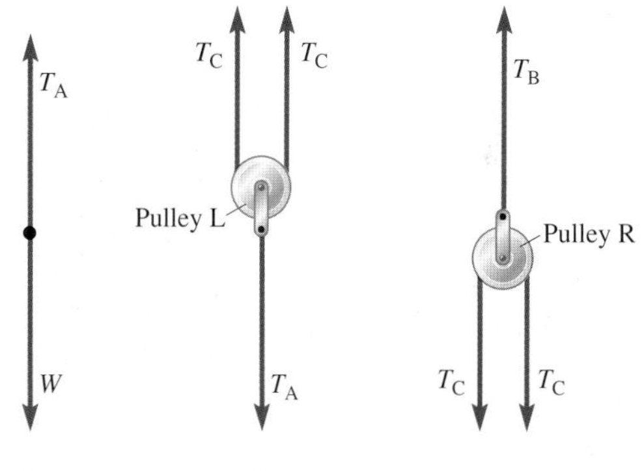

Figure 4.36
(a) FBD for the engine. (b) FBD for pulley L and (c) FBD for pulley R.

4.8 APPLYING NEWTON'S SECOND LAW

We can now apply Newton's second law to a great variety of situations involving the forces we have encountered so far—gravity, contact forces, and tension. The following steps are helpful in most problems that involve Newton's second law.

Problem-Solving Strategy for Newton's Second Law

- Decide what object will have Newton's second law applied to it.
- Identify all the *external* forces acting on that object.
- Draw an FBD to show all the forces acting on the object.
- Choose a coordinate system. If the direction of the net force is known, choose axes so that the net force (and the acceleration) are along one of the axes.
- Find the net force by adding the forces as vectors.
- Use Newton's second law to relate the net force to the acceleration.
- Relate the acceleration to the change in the velocity vector during a time interval of interest.

Example 4.13

The Broken Suitcase

The wheels fall off Beatrice's suitcase, so she ties a rope to it and drags it along the floor of the airport terminal (Fig. 4.37). The rope makes a 40.0° angle with the horizontal. The suitcase has a mass of 36.0 kg and Beatrice pulls on the rope with a force of 65.0 N. (a) What is the magnitude of the normal force acting on the suitcase due to the floor? (b) If the coefficient of kinetic friction between the suitcase and the marble floor is $\mu_k = 0.13$, find the frictional force acting on the suitcase. (c) What is the acceleration of the suitcase while Beatrice pulls with a 65.0 N force at 40.0°? (d) Starting from rest, for how long a time must she pull with this force until the suitcase reaches a comfortable walking speed of 0.5 m/s?

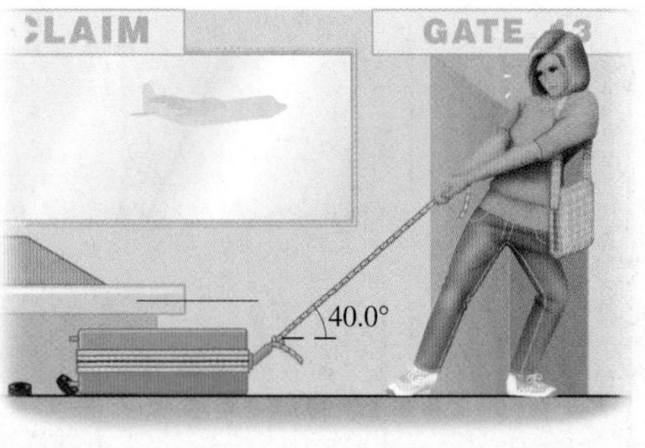

Figure 4.37
Beatrice dragging her suitcase.

Strategy Since the suitcase is dragged horizontally along the floor, the vertical component of its velocity is always zero. The vertical acceleration component of the suitcase is zero because the vertical velocity component does not change. (If it did have a vertical acceleration component, the suitcase would begin to move either down through the floor or up into the air.) If we choose the $+y$-axis up and the $+x$-axis to be horizontal, then $a_y = 0$. We resolve the forces acting on the suitcase into their components, draw a free-body diagram for the suitcase, and apply Newton's second law.

Solution (a) Figure 4.38 shows the forces acting on the suitcase, where $\vec{\mathbf{F}}$ is the force exerted by Beatrice. All the other forces are either parallel or perpendicular to the floor, so only $\vec{\mathbf{F}}$ needs to be resolved into x- and y-components.

$$F_x = F \cos 40.0° = 65.0 \text{ N} \times 0.766 = 49.8 \text{ N}$$

$$F_y = F \sin 40.0° = 65.0 \text{ N} \times 0.643 = 41.8 \text{ N}$$

Figure 4.39 is an FBD in which $\vec{\mathbf{F}}$ is replaced by its components. The vertical force components add to zero since $a_y = 0$.

$$\sum F_y = ma_y = 0$$

$$N + F \sin 40.0° - W = 0$$

We can solve this equation for the magnitude of the normal force. The magnitude of the gravitational force is $W = mg$, so

$$\begin{aligned} N &= mg - F \sin 40.0° \\ &= (36.0 \text{ kg} \times 9.80 \text{ N/kg}) - (65.0 \text{ N} \times \sin 40.0°) \\ &= 352.8 \text{ N} - 41.8 \text{ N} = 311 \text{ N} \end{aligned}$$

(b) The magnitude of the kinetic frictional force is

$$f_k = \mu_k N = 0.13 \times 311 \text{ N} = 40.43 \text{ N}$$

Rounding to two significant figures, the frictional force is 40 N in the $-x$-direction (opposite the motion of the suitcase).

(c) The y-component of the acceleration is zero. To find the x-component, we apply Newton's second law to the x-components of the forces acting on the suitcase:

$$\begin{aligned} \sum F_x &= +F \cos 40.0° + (-f_k) \\ &= 49.79 \text{ N} - 40.43 \text{ N} = 9.36 \text{ N} \end{aligned}$$

$$a_x = \frac{\sum F_x}{m} = \frac{9.36 \text{ N}}{36.0 \text{ kg}} = 0.260 \text{ m/s}^2$$

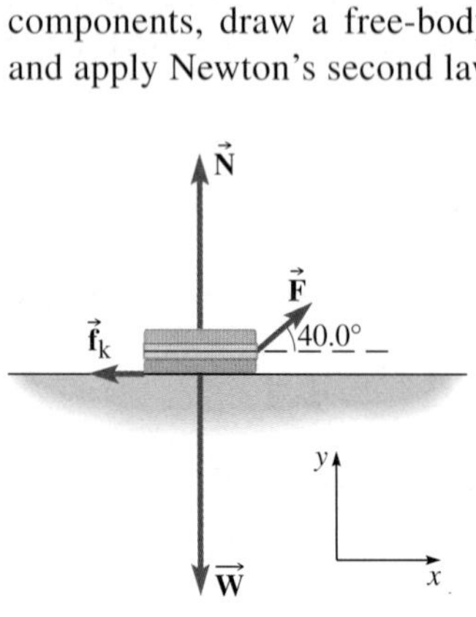

Figure 4.38
Forces acting on a suitcase dragged along the floor. The lengths of the vector arrows are not to scale.

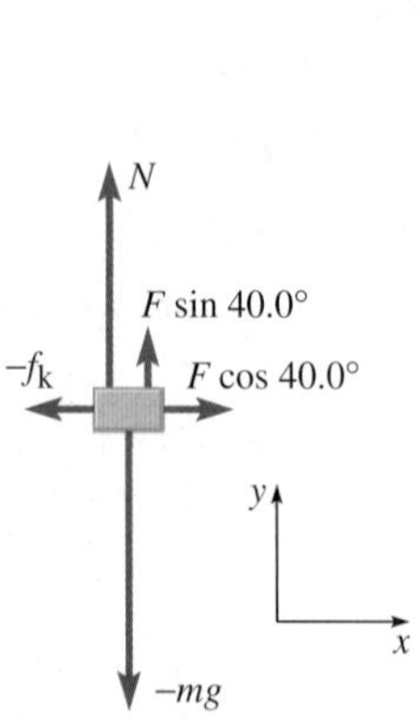

Figure 4.39
FBD for the suitcase, with the forces represented by their x- and y-components.

continued on next page

Example 4.13 continued

Here we have replaced newtons per kilogram with the equivalent meters per second squared, the usual way to write the SI units of acceleration. The acceleration is 0.3 m/s^2 in the $+x$-direction.

(d) With constant a_x,

$$\Delta v_x = a_x \Delta t$$

The suitcase starts from rest so $v_{ix} = 0$ and $\Delta v_x = v_{fx} - v_{ix} = v_{fx}$. Then,

$$\Delta t = \frac{v_{fx}}{a_x} = \frac{0.5 \text{ m/s}}{0.260 \text{ m/s}^2} = 2 \text{ s}$$

Discussion What Beatrice probably wants to do is to drag the suitcase along at constant velocity. To do that, she must first accelerate the suitcase from rest. Once the suitcase is moving at the desired velocity, she pulls a little less hard, so the net force is zero and the suitcase slides at constant speed. She would do so without thinking much about it, of course!

Practice Problem 4.13 The Continuing Story . . .

(a) How hard does Beatrice pull at a 40.0° angle while the suitcase slides along the floor at constant velocity? [*Hint:* Do *not* assume that the normal force is the same as in the previous discussion.] (b) The suitcase is moving at 0.50 m/s. Beatrice changes the force to 42 N at 40.0°. How long does it take the suitcase to come to rest?

Sometimes two or more objects are constrained to have the same acceleration by the way they are connected. In Example 4.14, we look at a train engine pulling five freight cars. The couplings maintain a fixed distance between the cars, so at any instant the cars move with the same velocity; if they didn't, the distance between them would change. The velocities don't have to be constant, they just have to change in exactly the same way, which implies that the accelerations must also be the same at any instant.

Example 4.14

Coupling Force on First and Last Freight Cars

A train engine pulls out of a station along a straight horizontal track with five identical freight cars behind it, each of which weighs 90.0 kN. The train reaches a speed of 15.0 m/s within 5.00 min of starting out. Assuming the engine pulls with a constant force during this interval, with what magnitude of force does the coupling between cars pull forward on the first and last of the freight cars? Ignore air resistance and friction on the freight cars.

Strategy A sketch of the situation is shown in Fig. 4.40. To find the force exerted by the first coupling, we consider all five cars to be one system so we do not have to worry about the force exerted on the first car by the second car. The only *external* forces on the group of five cars are the normal force, gravity, and the pull of the first coupling. To find the force exerted by the fifth coupling, we consider car five by itself to be a system. In each case, once we identify a system, we draw a free-body diagram, choose a coordinate system, and then apply Newton's second law.

As discussed previously, the engine and the cars must all have the same acceleration at any instant. We expect the acceleration to be *constant* because the engine pulls with a constant force. We can calculate the acceleration of the train from the initial and final velocities and the elapsed time.

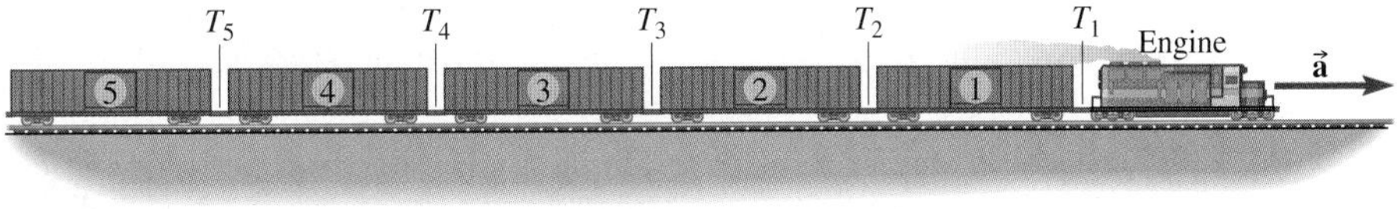

Figure 4.40

An engine pulling five identical freight cars. The entire train has a constant acceleration $\vec{\mathbf{a}}$ to the right.

continued on next page

Example 4.14 continued

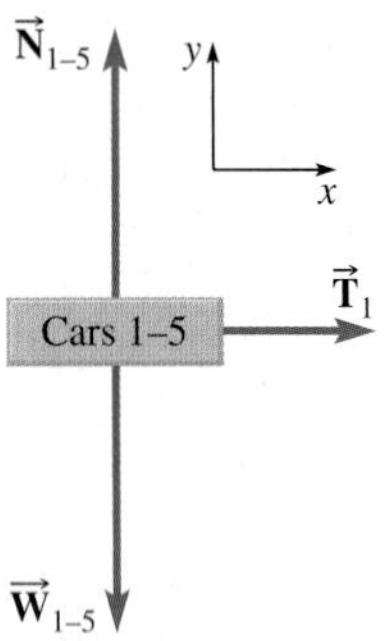

Figure 4.41
FBD for the system consisting of cars 1–5 (but not the engine).

Solution For the tension T_1 in the first coupling, we consider the five cars as *one system* of mass *M*. Figure 4.41 shows the FBD in which cars 1 to 5 are treated as a single object. We choose the *x*-axis in the direction of motion of the train and the *y*-axis up. Since the train moves along the *x*-axis, the acceleration vector is along the *x*-axis. Therefore, $a_y = 0$. Using the *y*-component of Newton's second law, the vertical forces add to zero:

$$\sum F_y = Ma_y = N_{1-5} - W_{1-5} = 0$$

The only external horizontal force is the force $\vec{\mathbf{T}}_1$ due to the tension in the first coupling. This force is constant according to the problem statement, so we know that the acceleration a_x is constant:

$$\sum F_x = T_1 = Ma_x$$

The mass of the system *M* is five times the mass of one car *m*. We are given the *weight* of one car ($W = 90.0$ kN $= 9.00 \times 10^4$ N). From the relation between mass and weight, $W = mg$, the mass of one car is $m = W/g$ and the mass of five cars is $M = 5W/g$.

The constant acceleration of the train is

$$a_x = \frac{\Delta v_x}{\Delta t} = \frac{v_{fx} - v_{ix}}{t_f - t_i} = \frac{15.0 \text{ m/s} - 0}{300 \text{ s} - 0} = 0.0500 \text{ m/s}^2$$

Therefore,

$$T_1 = Ma_x = \frac{5W}{g} \times \frac{\Delta v_x}{\Delta t} = \frac{5 \times 9.00 \times 10^4 \text{ N}}{9.80 \text{ m/s}^2} \times \frac{15.0 \text{ m/s}}{300 \text{ s}}$$

$$= 2.30 \text{ kN}$$

Now consider the last freight car (car 5). If we ignore friction and air resistance, the only external forces acting are the force $\vec{\mathbf{T}}_5$ due to the tension in the fifth coupling, the normal force $\vec{\mathbf{N}}_5$, and the gravitational force $\vec{\mathbf{W}}_5$; the FBD is shown in Fig. 4.42. Since $\vec{\mathbf{N}}_5 + \vec{\mathbf{W}}_5 = 0$, the net force is equal to $\vec{\mathbf{T}}_5$. From Newton's second law,

$$\sum F_x = T_5 = ma_x = \frac{W}{g} a_x$$

$$T_5 = \frac{W}{g} \times \frac{\Delta v_x}{\Delta t} = \frac{9.00 \times 10^4 \text{ N}}{9.80 \text{ m/s}^2} \times \frac{15.0 \text{ m/s}}{300 \text{ s}} = 459 \text{ N}$$

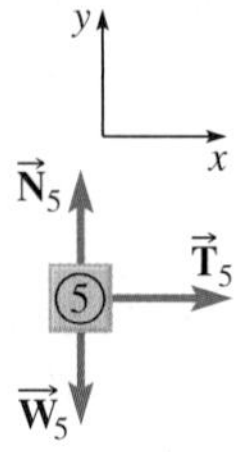

Figure 4.42
FBD for car 5. (Vector lengths are not to the same scale as those in Fig. 4.41.)

Discussion We considered two systems (cars 1 to 5 and car 5) that have the same acceleration and different masses. As expected, the net force is proportional to the mass: the net force on five cars is five times the net force on one car.

The solution to this problem is much simpler when Newton's second law is applied to a system comprised of all five cars, rather than to each car individually. Although the problem can be solved by looking at individual cars, to find the tension in the first coupling you would have to draw five FBDs (one for each car) and apply Newton's second law five times. That's because each car, except the fifth, is acted on by the unequal tensions in the couplings on either side. You'd have to first find the tension in the fifth coupling, then the fourth, then the third, and so on.

Practice Problem 4.14 Coupling Force Between First and Second Freight Cars

With what force does the coupling between the first and second cars pull forward on the second car? [*Hint:* Try two methods. One of them is to draw the FBD for the first car and apply Newton's *third* law as well as the second.]

Example 4.15 deals with two objects connected by an ideal cord. Although it may have a nonzero acceleration, the net force on an *ideal* cord is still zero because it has *zero mass*: if $m = 0$, then $\sum \vec{\mathbf{F}} = m\vec{\mathbf{a}} = 0$. As a result, the tension is the same at the two ends as long as no external force acts on the cord between the ends (Fig. 4.43a). An ideal cord that passes over an ideal pulley has the same tension at its ends. The pulley exerts an external force on part of the cord, but this force is everywhere *perpendicular to the cord.* As Fig. 4.43b shows, an external force that has no component tangent to the cord does not affect the tension in the cord.

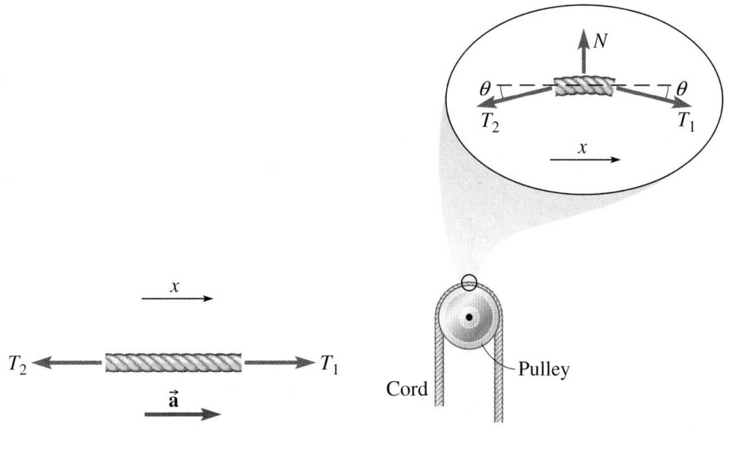

Figure 4.43 (a) FBD for an ideal cord with acceleration $\vec{\mathbf{a}}$. Applying Newton's second law along the x-axis: $\Sigma F_x = T_1 - T_2 = ma_x$. The ideal cord has mass $m = 0$, so $T_1 = T_2$: the tensions at the ends are equal. (b) An ideal cord passing around an ideal pulley and the FBD for a short segment of the cord at the top of the pulley. Choosing the x-axis to be horizontal, the normal force has no x-component. Applying Newton's second law along the x-axis: $\Sigma F_x = T_1 \cos\theta - T_2 \cos\theta = ma_x$. With $m = 0$, $T_1 = T_2$. The same reasoning can be applied to any segment of cord in contact with the pulley to show that the tensions are the same on either side of the pulley.

Example 4.15

Two Blocks Hanging on a Pulley

In Fig. 4.44, two blocks are connected by an ideal cord that does not stretch; the cord passes over an ideal pulley. If the masses are $m_1 = 26.0$ kg and $m_2 = 42.0$ kg, what are the accelerations of each block and the tension in the cord?

Strategy Since m_2 is greater than m_1, the downward force of gravity is stronger on the right side than on the left. We expect block 2's acceleration to be downward and block 1's to be upward.

The cord does not stretch, so blocks 1 and 2 move at the same speed at any instant (in opposite directions). Therefore, the accelerations of the two blocks are equal in magnitude and opposite in direction. If the accelerations had different magnitudes, then soon the two blocks would be moving with different speeds. That could only happen if the cord either stretches or contracts.

The tension in the cord must be the same everywhere along the cord since the masses of the cord and pulley are negligible and the pulley turns without friction.

We treat each block as a separate system, draw FBDs for each, and then apply Newton's second law to each. It is convenient to choose the positive y-direction differently for the two blocks since we know their accelerations are in opposite directions. For each block, we choose the $+y$-axis in the direction of the acceleration of that block: upward for block m_1 and downward for m_2. Doing so means that a_y has the same magnitude *and sign* (both positive) for the two blocks.

Solution Figure 4.45 shows FBDs for the two blocks. Two forces act on each: gravity and the pull of the cord. The

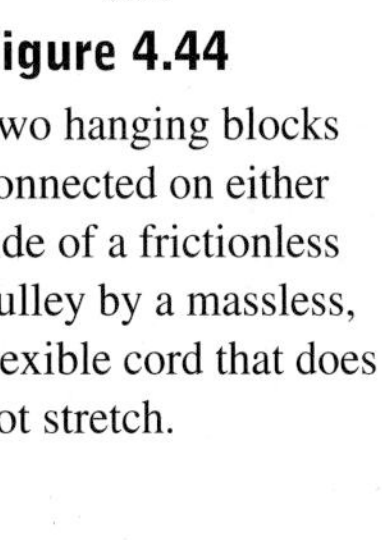

Figure 4.44
Two hanging blocks connected on either side of a frictionless pulley by a massless, flexible cord that does not stretch.

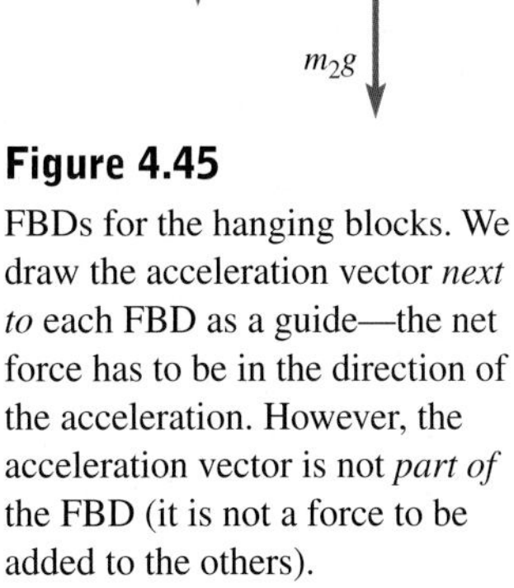

Figure 4.45
FBDs for the hanging blocks. We draw the acceleration vector *next to* each FBD as a guide—the net force has to be in the direction of the acceleration. However, the acceleration vector is not *part of* the FBD (it is not a force to be added to the others).

continued on next page

Example 4.15 continued

acceleration vectors are drawn *next to* the FBDs. Thus, we know the direction of the net force: it is always the same as the direction of the acceleration. Then we know that the tension must be greater than m_1g to give block 1 an upward acceleration and less than m_2g to give block 2 a downward acceleration. The $+y$-axes are drawn for each block to be in the direction of the acceleration.

From the FBD of block 1, the pull of the cord is in the $+y$-direction and the gravitational force is in the $-y$-direction. Then Newton's second law for block 1 is

$$\sum F_{1y} = T - m_1 g = m_1 a_{1y}$$

For block 2, the pull of the cord is in the $-y$-direction and the gravitational force is in the $+y$-direction. Newton's second law for block 2 is

$$\sum F_{2y} = m_2 g - T = m_2 a_{2y}$$

The tension T in the cord is the same in the two equations. Also a_{1y} and a_{2y} are identical, so we write them simply as a_y. We then have a system of two equations with two unknowns. We can add the equations to obtain

$$m_2 g - m_1 g = m_2 a_y + m_1 a_y$$

Solving for a_y, we find

$$a_y = \frac{(m_2 - m_1)g}{m_2 + m_1}$$

Substituting numerical values,

$$a_y = \frac{(42.0\text{ kg} - 26.0\text{ kg}) \times 9.80\text{ N/kg}}{42.0\text{ kg} + 26.0\text{ kg}}$$

$$= 2.31\text{ m/s}^2$$

since

$$1\,\frac{\text{N}}{\text{kg}} = 1\,\frac{\text{kg}\cdot\text{m/s}^2}{\text{kg}} = 1\text{ m/s}^2$$

The blocks have the same magnitude acceleration. For block 1 the acceleration points upward and for block 2 it points downward.

To find T we can substitute the expression for a_y into either of the two original equations. Using the first equation,

$$T - m_1 g = m_1 \frac{(m_2 - m_1)g}{m_2 + m_1}$$

Solving for T yields

$$T = \frac{2m_1 m_2}{m_1 + m_2} g$$

Substituting,

$$T = \frac{2 \times 26.0\text{ kg} \times 42.0\text{ kg}}{68.0\text{ kg}} \times 9.80\text{ N/kg} = 315\text{ N}$$

Discussion A few quick checks:

- a_y is positive, which means that the accelerations are in the directions we expect.
- The tension (315 N) is between m_1g (255 N) and m_2g (412 N), as it must be for the accelerations to be in opposite directions.
- The units and dimensions are correct for all equations.
- We can check algebraic expressions in special cases for which we have some intuition. For example, if the masses had been *equal*, we expect the blocks to hang in equilibrium (either at rest or moving at constant velocity) due to the equal pull of gravity on the two blocks. Substituting $m_1 = m_2$ into the expressions for a_y and T gives $a_y = 0$ and $T = m_1g = m_2g$, which is just what we expect.

Note that we did *not* find out which way the blocks move. We found the directions of their *accelerations*. If the blocks start out at rest, then the block of mass m_2 moves downward and the block of mass m_1 moves upward. However, if initially m_2 is moving up and m_1 down, they continue to move in those directions, slowing down since their accelerations are opposite to their velocities. Eventually they come to rest and then reverse directions.

Practice Problem 4.15 Another Check

Using the numerical values of the tension and the acceleration calculated in Example 4.15, verify Newton's second law directly for each of the two blocks.

Examples 4.16, 4.17, and 4.18 illustrate how different concepts and problem-solving techniques from Chapters 2–4 can be brought together to find the solution to a physics problem.

Example 4.16

Hauling a Crate up to a Third-Floor Window

A student is moving into a dorm room on the third floor and he decides to use a block and tackle arrangement (Fig. 4.46) to move a crate of mass 91 kg from the ground up to his window. If the breaking strength of the available rope is 550 N, what is the minimum time required to haul the crate to the level of the window, 30.0 m above the ground, without breaking the rope?

Strategy The tension in the rope is T and is the same at both ends or anywhere along the rope, assuming the rope and pulleys are ideal. Two pieces of rope support the lower pulley, each pulling upward with a force of magnitude T. The gravitational force acts downward. We draw an FBD for the system consisting of the crate and the lower pulley and set the tension equal to the breaking force of the rope to find the maximum possible acceleration of the crate. Then we use the maximum acceleration to find the minimum time to move the required distance to the third-floor window. We choose the y-axis to be upward. Known: $m = 91$ kg; $\Delta y = 30.0$ m; $T_{max} = 550$ N; $v_{iy} = 0$. To find: Δt, the time to raise the crate 30.0 m with the maximum tension in the cable.

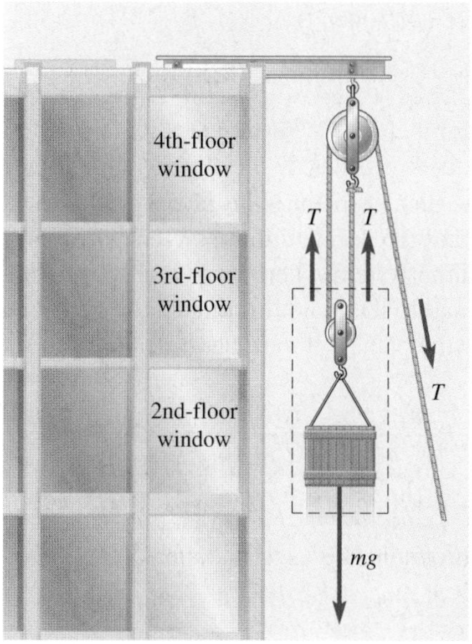

Figure 4.46
Block and tackle setup.

Solution From the FBD (Fig. 4.47), if the forces acting up are greater than the force acting down, the net force is upward and the crate's acceleration is upward. In terms of components, with the $+y$-direction chosen to be upward,

$$\sum F_y = T + T - mg = ma_y$$

Solving for the acceleration,

$$a_y = \frac{T + T - mg}{m}$$

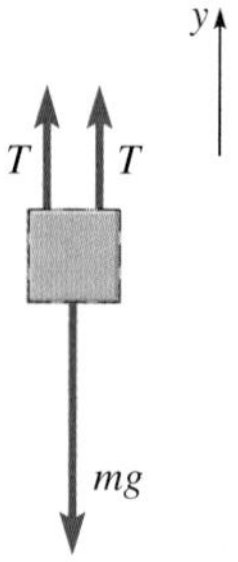

Figure 4.47
FBD for the crate and lower pulley. (This system is outlined by dashed lines in Fig. 4.46.)

Setting $T = 550$ N, the maximum possible value before the cable breaks, and substituting the other known values:

$$a_y = \frac{550\ \text{N} + 550\ \text{N} - 91\ \text{kg} \times 9.80\ \text{m/s}^2}{91\ \text{kg}} = 2.288\ \text{m/s}^2$$

The time to move the crate up a distance Δy starting from rest can be found from

$$\Delta y = v_{iy}\,\Delta t + \tfrac{1}{2}a_y(\Delta t)^2 \qquad (3\text{-}21)$$

Setting $v_{iy} = 0$ and solving for Δt, we find

$$\Delta t = \pm\sqrt{\frac{2\,\Delta y}{a_y}}$$

Our equation applies only for $\Delta t \geq 0$ (the crate reaches the window *after* it leaves the ground). Taking the positive root and substituting numerical values,

$$\Delta t = \sqrt{\frac{2 \times 30.0\ \text{m}}{2.288\ \text{m/s}^2}} = 5.1\ \text{s}$$

This is the minimum possible to haul the crate up without breaking the rope.

Discussion In reality, the student is not likely to achieve this *minimum possible* time. To do so would mean pulling the rope at an unrealistic speed. At the end of the 5.1-s interval, $v_{fy} = 2.288\ \text{m/s}^2 \times 5.1\ \text{s} = 12$ m/s! More likely, the student would hoist the crate at a roughly constant velocity (except at the beginning, to get it moving, and at the end, to let it come to rest). For motion with a constant velocity, the tension in the rope would be equal to half the weight of the crate (450 N).

Practice Problem 4.16 Hauling the Crate with a Single Pulley

If only a single pulley, attached to the beam above the fourth floor, were available and if the student had a few friends to help him pull on the cable, could they haul the crate up to the third-floor window using the same rope? If so, what is the minimum time required to do so?

Example 4.17

Towing a Glider

What length runway does the plane need?

A small plane of mass 760 kg requires 120 m of runway to take off by itself. (120 m is the horizontal displacement of the plane just before it lifts off the runway, not the entire length of the runway.) As a simplified model, ignore friction and drag forces and assume the plane's engine exerts a constant forward force on the plane. (a) When the plane is towing a 330-kg glider, how much runway does it need? (b) If the final speed of the plane just before it lifts off the runway is 28 m/s, what is the tension in the tow cable while the plane and glider are moving along the runway?

Strategy We draw FBDs for the two cases: plane alone, then plane + glider. The motion in both cases is horizontal (along the runway), because we are told the displacement *before it lifts off the runway.* Until the plane begins to lift off the runway, its vertical acceleration component is zero. We need not be concerned with the vertical forces (gravity, the normal force, and lift—the upward force on the plane's wings due to the air) since they cancel one another to produce zero vertical acceleration. We use Newton's second law to compare the accelerations in the two cases and then use the accelerations to compare the displacements.

Solution (a) When the plane takes off by itself, four forces act on it (see Fig. 4.48). Three are vertical and the third—the thrust due to the engine—is horizontal. Choosing the x-axis to be horizontal, Newton's second law says

$$\sum F_{1x} = F = m_1 a_{1x}$$

where F is the thrust, m_1 is the plane's mass, and a_{1x} is its horizontal acceleration component.

When the glider is towed, we can consider the plane, glider, and cable to be a single system (see Fig. 4.49). There is still only one horizontal external force and it is the same thrust as before. The tension in the cable is an *internal* force. Therefore,

$$\sum F_{2x} = F = (m_1 + m_2)a_x$$

where $m_1 + m_2$ is the total mass of the system (plane mass m_1 plus glider mass m_2) and a_x is the horizontal acceleration component of plane and glider. We ignore the mass of the cable.

The problem statement gives neither the thrust nor either of the accelerations. We can continue by setting the thrusts equal and finding the ratio of the accelerations:

$$m_1 a_{1x} = (m_1 + m_2)a_x \Rightarrow \frac{a_x}{a_{1x}} = \frac{m_1}{m_1 + m_2}$$

The magnitude of the acceleration is inversely proportional to the mass of the system for the same net force.

How is the acceleration related to the runway distance? The plane must get to the same final speed in order to lift off the runway. From our two basic constant acceleration equations

$$\Delta v_x = v_{fx} - v_{ix} = a_x \Delta t \qquad (2\text{-}9)$$

$$\Delta x = \tfrac{1}{2}(v_{fx} + v_{ix})\,\Delta t \qquad (2\text{-}11)$$

we can substitute $v_{ix} = 0$ and eliminate Δt to find

$$\Delta x = \frac{1}{2}(v_{fx} + 0)\left(\frac{v_{fx}}{a_x}\right) = \frac{v_{fx}^2}{2a_x}$$

In both cases, the displacement is inversely proportional to the acceleration and the acceleration is inversely proportional to the mass of the system. Therefore, the displacement is *directly* proportional to the mass. Letting $\Delta x_1 = 120$ m be the displacement of the plane without the glider, we can set up a proportion:

$$\frac{\Delta x}{\Delta x_1} = \frac{a_{1x}}{a_x} = \frac{m_1 + m_2}{m_1} = \frac{1090 \text{ kg}}{760 \text{ kg}} = 1.434$$

$$\Delta x = 1.434 \times 120 \text{ m} = 172.08 \text{ m} \rightarrow 170 \text{ m}$$

(b) The final speed given enables us to find the acceleration:

$$\Delta x = \frac{v_{fx}^2}{2a_x} \quad \text{or} \quad a_x = \frac{v_{fx}^2}{2\Delta x}$$

With $v_{fx} = 28$ m/s, $v_{ix} = 0$, and $\Delta x = 172.08$ m,

$$a_x = \frac{(28 \text{ m/s})^2}{2 \times 172.08 \text{ m}} = 2.278 \text{ m/s}^2$$

The tension in the cable is the only horizontal force acting on the glider. Therefore,

$$\sum F_x = T = m_2 a_x = 330 \text{ kg} \times 2.278 \text{ m/s}^2 = 751.7 \text{ N} \rightarrow 750 \text{ N}$$

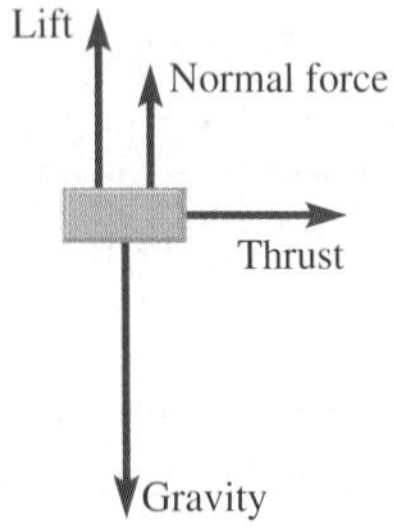

Figure 4.48
FBD for the plane.

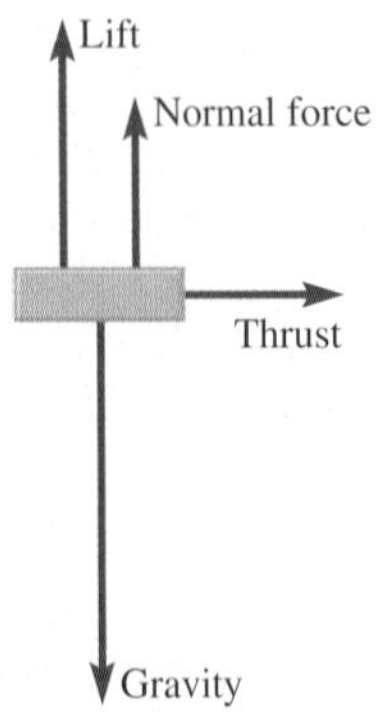

Figure 4.49
FBD for the system plane + glider.

continued on next page

Example 4.17 continued

Discussion This solution is based on a simplified model, so we can only regard the answers as approximate. Nevertheless, it illustrates Newton's second law. The same net force produces an acceleration inversely proportional to the mass of the object upon which it acts. Here we have the same net force acting on two different objects: first the plane alone, then the plane and glider together.

Alternatively, we can look at forces acting only on the plane. When towing the glider, the cable pulls backward on the plane. The net force *on the plane* is smaller, so its acceleration is smaller. The smaller acceleration means that it takes more time to reach takeoff speed and travels a longer distance before lifting off the runway.

Practice Problem 4.17 Engine Thrust

What is the thrust provided by the airplane's engines in Example 4.17?

Example 4.18

A Pulley, an Incline, and Two Blocks

A block of mass $m_1 = 2.60$ kg rests on an incline that is angled at 30.0° above the horizontal (Fig. 4.50). An ideal cord is connected from block 1 over an ideal, frictionless pulley to another block of mass $m_2 = 2.20$ kg that is hanging 2.00 m above the ground. The coefficient of kinetic friction between the incline and block 1 is 0.180. The blocks are initially at rest. (a) How long does it take for block 2 to reach the ground? (b) Sketch a motion diagram for block 2 with a time interval of 0.5 s.

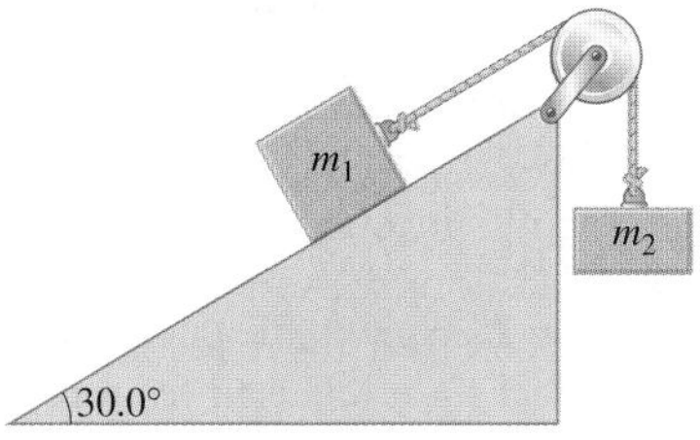

Figure 4.50
Block on an incline connected to a hanging block by a cord passing over a pulley.

Strategy The problem says that the blocks start from rest and that block 2 hits the floor, so block 2's acceleration is downward and block 1's is up the incline. For block 1, we choose axes parallel and perpendicular to the incline so that its acceleration has only one nonzero component. The magnitudes of the accelerations of the two blocks are equal since they are connected by an ideal cord that does not stretch. Since the cord and pulley are ideal, the tension is the same at the two ends.

Solution (a) We start by drawing separate FBDs for each block (Figs. 4.51 and 4.52). Since block 1 slides up the incline, the frictional force $\vec{\mathbf{f}}_k$ acts down the incline to oppose the sliding. The gravitational force on block 1 is resolved into two components, one along the incline and one perpendicular to the incline.

Using the FBDs, we write Newton's second law in component form for each block. Block 1 has no acceleration

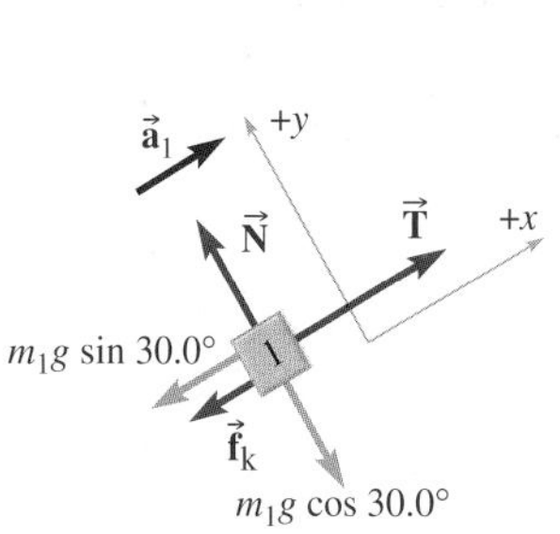

Figure 4.51
FBD for block 1.

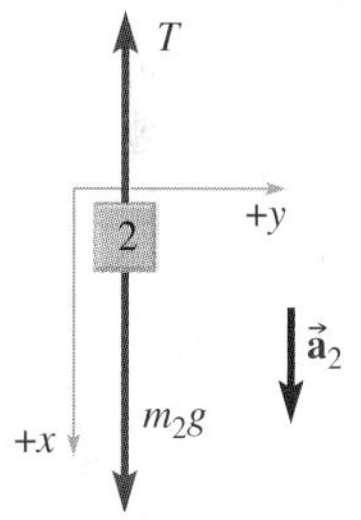

Figure 4.52
FBD for block 2 with the downward direction chosen as $+x$.

component perpendicular to the incline. It does not sink into the incline or rise above it; it can only slide along the incline. Thus, the net force on block 1 in the direction perpendicular to the incline—the direction we have chosen as the y-axis for block 1—is zero.

$$\sum F_y = N - m_1 g \cos\theta = 0$$

or

$$N = m_1 g \cos\theta$$

Here $\theta = 30.0°$. Along the incline, in the x-direction for block 1, the acceleration is nonzero:

$$\sum F_x = T - m_1 g \sin\theta - f_k = m_1 a_x$$

The kinetic frictional force is related to the normal force:

$$f_k = \mu_k N = \mu_k m_1 g \cos\theta$$

By substitution,

$$T - m_1 g \sin\theta - \mu_k m_1 g \cos\theta = m_1 a_x \qquad (1)$$

continued on next page

Example 4.18 continued

For block 2, we choose an x-axis pointing downward. Doing so simplifies the solution, since then the two blocks have the same a_x. Applying Newton's second law,

$$\sum F_x = m_2 g - T = m_2 a_x \qquad (2)$$

The tension in the cord T and the x-component of acceleration a_x are both unknown in Eqs. (1) and (2). We solve for T in Eq. (2) and substitute into Eq. (1):

$$T = m_2 g - m_2 a_x = m_2(g - a_x)$$

$$m_2(g - a_x) - m_1 g \sin\theta - \mu_k m_1 g \cos\theta = m_1 a_x$$

Rearranging and solving for a_x yields

$$a_x = \frac{m_2 - m_1(\sin\theta + \mu_k \cos\theta)}{m_1 + m_2} g \qquad (3)$$

Substituting the known and given values,

$$a_x = \frac{2.20\text{ kg} - 2.60\text{ kg} \times (0.50 + 0.180 \times 0.866)}{2.60\text{ kg} + 2.20\text{ kg}} \times 9.80\text{ m/s}^2$$

$$= 1.01\text{ m/s}^2$$

Block 2 has a distance of 2.00 m to travel starting from rest with a constant downward acceleration of 1.01 m/s^2. From Eq. (2-12) with $v_{ix} = 0$,

$$\Delta x = \tfrac{1}{2} a_x (\Delta t)^2$$

The time to travel that distance is

$$\Delta t = \sqrt{\frac{2\Delta x}{a_x}} = \sqrt{\frac{2 \times 2.00\text{ m}}{1.01\text{ m/s}^2}} = 2.0\text{ s}$$

(b) Figure 4.53 shows the motion diagram for block 2. Choosing $x_i = 0$ and $t_i = 0$, the position as a function of time is $x = \frac{1}{2} a_x t^2$.

Discussion One advantage to solving for a_x algebraically in Eq. (3) before substituting numerical values is that dimensional analysis can easily be used to check for errors. In Eq. (3), the quantity in parentheses is dimensionless—the values of trigonometric functions are pure numbers as are coefficients of friction. Therefore, the numerator is the sum of two quantities with dimensions of force, the denominator is the sum of two masses, and force divided by mass gives an acceleration.

What if the problem did not tell us the directions of the blocks' accelerations? We could figure it out by comparing the force with which gravity pulls down on block 2 ($m_2 g$) with the component of the gravitational force pulling block 1 down the incline ($m_1 g \sin\theta$). Whichever is greater "wins the tug-of-war," assuming that static friction doesn't prevent the blocks from starting to slide. Once we know the direction of block 1's acceleration, we can determine the direction of the kinetic frictional force. If block 1 is not initially at rest, the kinetic frictional force opposes the direction of sliding, even though that may be opposite to the direction of the acceleration.

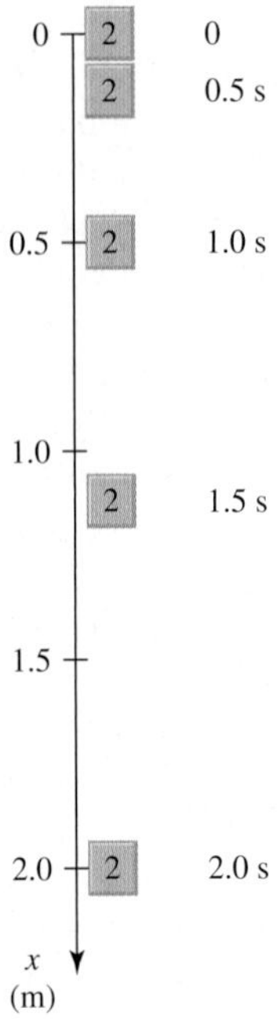

Figure 4.53
Motion diagram for block 2.

t (s)	x (m)
0	0
0.5	0.125
1.0	0.50
1.5	1.125
2.0	2.0

Practice Problem 4.18 More Fun with a Pulley and an Incline

Suppose that $m_1 = 3.8$ kg and $m_2 = 1.2$ kg and the coefficient of kinetic friction is 0.18. The blocks are released from rest and block 1 starts to slide. (a) Does block 1 slide up or down the incline? (b) In which direction does the kinetic frictional force act? (c) Find the acceleration of block 1.

CHECKPOINT 4.8

Is it ever useful to choose the x- and y-axes so the x-axis is not horizontal? If yes, give an example.

4.9 REFERENCE FRAMES

Imagine a train moving at constant velocity with respect to the ground (Fig. 4.54). Suppose Tim does some experiments using the train's reference frame for his measurements. Greg does similar experiments using the reference frame of the ground. Tim and Greg disagree about the numerical value of an object's velocity, but since their velocity

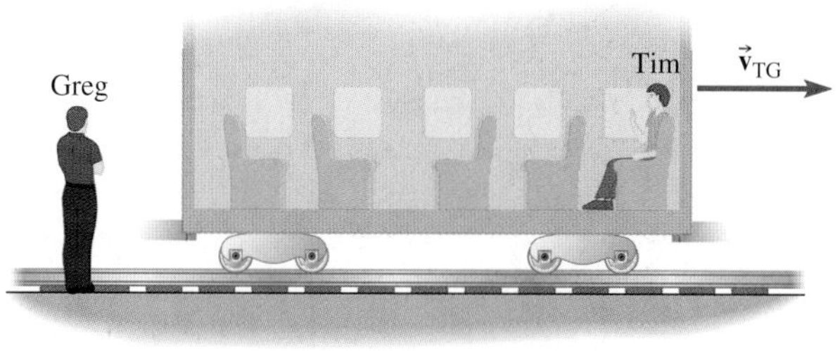

Figure 4.54 Greg's frame of reference is that of the ground; Tim's is that of the train, which moves at constant velocity $\vec{\mathbf{v}}_{TG}$ with respect to the ground.

measurements *differ by a constant*, they will always agree about *changes* in velocity and about accelerations. Both observers can use Newton's second law to relate the net force to the acceleration. The basic laws of physics, such as Newton's laws of motion, work equally well in any two reference frames if they move with a constant relative velocity.

Newton's First Law Defines an Inertial Reference Frame You might wonder why we need Newton's first law—isn't it just a special case of the second law when $\sum \vec{\mathbf{F}} = 0$? No, the first law *defines* what kind of reference frame we can use when applying the second law. For the second law to be valid, we must use an *inertial reference frame*—a reference frame in which the law of inertia holds—to observe the motion of objects. The law of inertia is a *postulate* of classical mechanics—an assumption that is used as a starting point. It is not something we can prove experimentally.

Is a reference frame attached to Earth's surface truly inertial? No, but it is close enough in many circumstances. When analyzing the motion of a soccer ball, the fact that Earth rotates about its axis does not have much effect. But if we want to analyze the motion of a meteor falling from a great distance toward Earth, Earth's rotation must be considered. We will take a closer look at the effect of Earth's rotation in Chapter 5.

4.10 APPARENT WEIGHT

Imagine being in an elevator when the cable snaps. Assume that some safety mechanism brings you to rest after you have been in free fall for a while. While you are in free fall, you *seem* to be "weightless," but your weight has not changed; the Earth still pulls downward with the same gravitational force. In free fall, gravity gives the elevator and everything in it a downward acceleration equal to $\vec{\mathbf{g}}$. If you jump up from the elevator floor, you seem to "float" up to the ceiling of the elevator. Your *weight* hasn't changed, but your *apparent* weight is zero while you are in free fall.

Similarly, astronauts in a space station in orbit around the Earth are in free fall (their acceleration is equal to the local value of $\vec{\mathbf{g}}$). Earth exerts a gravitational force on them so they are not weightless; their *apparent* weight is zero.

Imagine an object that appears to be resting on a bathroom scale. The scale measures the object's *apparent* weight W', which is equal to the true weight only if the object and the scale have zero acceleration. Newton's second law requires that

$$\sum \vec{\mathbf{F}} = \vec{\mathbf{N}} + m\vec{\mathbf{g}} = m\vec{\mathbf{a}}$$

where $\vec{\mathbf{N}}$ is the normal force of the scale pushing up. The apparent weight W' is the reading of the scale—that is, the magnitude of $\vec{\mathbf{N}}$:

$$W' = |\vec{\mathbf{N}}| = N$$

In Fig. 4.55a, the acceleration of the elevator is upward. The normal force must be larger than the weight for the net force to be upward (Fig. 4.55b). Writing the forces in component form where the $+y$-direction is upward

$$\sum F_y = N - mg = ma_y$$

or

$$N = mg + ma_y$$

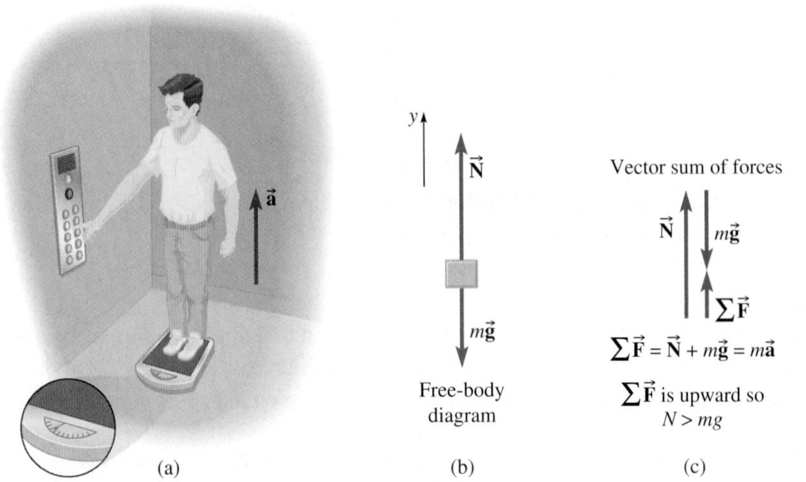

Figure 4.55 (a) Apparent weight in an elevator with acceleration upward. (b) FBD for the passenger. (c) The normal force must be greater than the weight to have an upward net force.

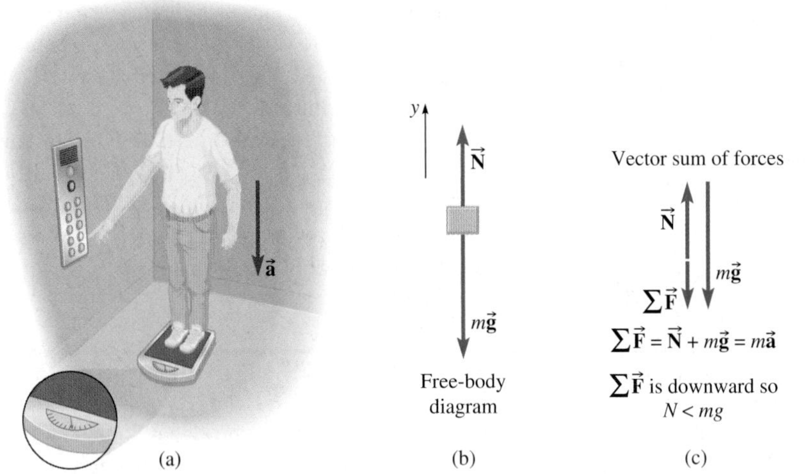

Figure 4.56 (a) Apparent weight in an elevator with acceleration downward. (b) FBD for the passenger. (c) The normal force must be less than the weight to have a downward net force.

Therefore,

$$W' = N = m(g + a_y) \tag{4-16}$$

Since the elevator's acceleration is upward, $a_y > 0$; the apparent weight is greater than the true weight (Fig. 4.55c).

In Fig. 4.56a, the acceleration is downward. Then the net force must also point downward. The normal force is still upward, but it must be smaller than the weight in order to produce a downward net force (Fig. 4.56b). It is still true that $W' = m(g + a_y)$, but now the acceleration is downward ($a_y < 0$). The apparent weight is less than the true weight (Fig. 4.56c). If the elevator is in free fall, then $a_y = -g$ and the apparent weight of the unfortunate passenger is zero.

Example 4.19

Apparent Weight in an Elevator

A passenger weighing 598 N rides in an elevator. What is the apparent weight of the passenger in each of the following situations? In each case, the magnitude of the elevator's acceleration is 0.500 m/s^2. (a) The passenger is on the first floor and has pushed the button for the fifteenth floor; the elevator is beginning to move upward. (b) The elevator is slowing down as it nears the fifteenth floor.

continued on next page

Example 4.19 continued

Strategy In each case, we sketch the FBD for the passenger. The apparent weight is equal to the magnitude of the normal force exerted by the floor on the passenger. The only other force acting is gravity. Newton's second law lets us find the normal force from the weight and the acceleration. Known: $W = 598$ N; magnitude of the acceleration is $a = 0.500\ \text{m/s}^2$. To find: W'.

Solution (a) Let the $+y$-axis be upward. When the elevator starts up from the first floor it has acceleration in the upward direction as its speed increases. Since the elevator's acceleration is upward, $a_y > 0$ (as in Fig. 4.55). We expect the apparent weight $W' = N$ to be greater than the true weight—the floor must push up with a force greater than W to cause an upward acceleration. Figure 4.57 is the FBD. Newton's second law says

$$\sum F_y = N - W = ma_y$$

Since $W = mg$, we can substitute $m = W/g$.

$$W' = N = W + ma_y = W + \frac{W}{g}a_y = W\left(1 + \frac{a_y}{g}\right)$$

$$= 598\ \text{N} \times \left(1 + \frac{0.500\ \text{m/s}^2}{9.80\ \text{m/s}^2}\right) = 629\ \text{N}$$

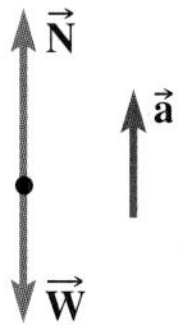

Figure 4.57

FBD for the passenger in an elevator with upward acceleration.

(b) When the elevator approaches the fifteenth floor, it is slowing down while still moving upward; its acceleration is downward ($a_y < 0$) as in Fig. 4.56. The apparent weight is less than the true weight. Figure 4.58 is the FBD. Again, $\sum F_y = N - W = ma_y$, but this time $a_y = -0.500\ \text{m/s}^2$.

$$N = W\left(1 + \frac{a_y}{g}\right)$$

$$= 598\ \text{N} \times \left(1 + \frac{-0.500\ \text{m/s}^2}{9.80\ \text{m/s}^2}\right) = 567\ \text{N}$$

Discussion The apparent weight is greater when the direction of the elevator's acceleration is upward. That can happen in two cases: either the elevator is moving up with increasing speed, or it is moving down with decreasing speed.

Practice Problem 4.19 Elevator Descending

What is the apparent weight of a passenger of mass 42.0 kg traveling in an elevator in each of the following situations? In each case, the magnitude of the elevator's acceleration is $0.460\ \text{m/s}^2$. (a) The passenger is on the fifteenth floor and has pushed the button for the first floor; the elevator is beginning to move downward. (b) The elevator is slowing down as it nears the first floor.

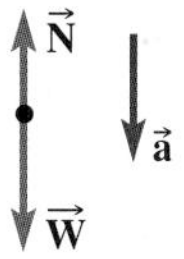

Figure 4.58

FBD for the passenger in an elevator with downward acceleration.

PHYSICS AT HOME

Take a bathroom scale to an elevator. Stand on the scale inside the elevator and push a button for a higher floor. When the elevator's acceleration is upward, you can feel the increase in your apparent weight and can see the increase by the reading on the scale. When the elevator slows down to stop, the elevator's acceleration is downward and your apparent weight is less than your true weight.

What is happening in your body while the elevator accelerates? The inertia principle means that your blood and internal organs cannot have the same acceleration as the elevator until the correct net force acts on them. Blood tends to collect in the lower extremities during acceleration upward and in the upper body during acceleration downward until the forces exerted on the blood by the body readjust to give the blood the same acceleration as the elevator. Likewise, the internal organs shift position within the body cavity, resulting in a funny feeling in the gut as the elevator starts and stops. To avoid this problem, high-speed express elevators in skyscrapers keep the acceleration relatively small, but maintain that acceleration long enough to reach high speeds. That way, the elevator can travel quickly to the upper floors without making the passengers feel too uncomfortable.

CHECKPOINT 4.10

You are standing on a bathroom scale in an elevator that is moving downward. Nearing your stop, the elevator's speed is decreasing. Is the scale reading greater or less than your weight?

4.11 AIR RESISTANCE

So far we have ignored the effect of air resistance on falling objects and projectiles. A skydiver relies on a parachute to provide a large force of air resistance (also called **drag**). Even with the parachute closed, drag is not negligible when the skydiver is falling rapidly. The drag force is similar to friction between two solid surfaces in that the direction of the force *opposes the motion* of the object through the air. However, in contrast to the force of friction, the magnitude of the drag force is strongly dependent on the speed of the object. In many cases, air drag is proportional to the square of the speed. Drag also depends on the size and shape of the object.

Since the drag force increases as the speed increases, a falling object approaches an equilibrium situation in which the drag force is equal in magnitude to the weight but opposite in direction. The velocity at which this equilibrium occurs is called the object's *terminal velocity.* (See text website for a more detailed treatment of drag.)

PHYSICS AT HOME

Drop a basket-style paper coffee filter (or a cupcake paper) and a penny simultaneously from as close to the ceiling as you can safely do so. Air resistance on the penny is negligible unless it is dropped from a very high balcony. At the other extreme, the effect of air resistance on the coffee filter is very noticeable; it reaches its terminal speed almost immediately. Stack several (two to four) coffee filters together and drop them simultaneously with a single coffee filter. Why is the terminal speed higher for the stack? Crumple a coffee filter into a ball and drop it simultaneously with the penny. Air resistance on the coffee filter is now reduced, but still noticeable.

4.12 FUNDAMENTAL FORCES

One of the main goals of physics has been to understand the immense variety of forces in the universe in terms of the fewest number of fundamental laws. Physics has made great progress in this quest for *unification*; today all forces are understood in terms of just four fundamental interactions (Fig. 4.59). At the high temperatures present in the early universe, two of these interactions—the electromagnetic and weak forces—are now understood as the effects of a single electroweak interaction. The ultimate goal is to describe all forces in terms of a single interaction.

Gravity You may be surprised to learn that gravity is by far the *weakest* of the fundamental forces. Any two objects exert gravitational forces on one another, but the force is tiny unless at least one of the masses is large. We tend to notice the relatively large gravitational forces exerted by planets and stars, but not the feeble gravitational

Figure 4.59 All forces result from just four fundamental forces: gravity, electromagnetism, and the weak and strong forces.

forces exerted by smaller objects, such as the gravitational force this book exerts on your body.

Gravity has an unlimited range. The force gets weaker as the distance between two objects increases, but it never drops exactly to zero, no matter how far apart the objects get.

Newton's law of gravity is an early example of unification. Before Newton, people did not understand that the same kind of force that makes an apple fall from a tree also keeps the planets in their orbits around the Sun. A single law—Newton's law of universal gravitation—describes both.

Electromagnetism The electromagnetic force is unlimited in range, like gravity. It acts on particles with electric charge. The electric and magnetic forces were unified into a single theoretical framework in the nineteenth century. We study electromagnetic forces in detail in Part 3 of this book.

Electromagnetism is the fundamental interaction that binds electrons to nuclei to form atoms and binds atoms together in molecules and solids. It is responsible for the properties of solids, liquids, and gases and forms the basis of the sciences of chemistry and biology. It is the fundamental interaction behind all macroscopic contact forces such as the frictional and normal forces between surfaces and forces exerted by springs, muscles, and the wind.

The electromagnetic force is *much* stronger than gravity. For example, the electrical repulsion of two electrons at rest is about 10^{43} times as strong as the gravitational attraction between them. Macroscopic objects have a nearly perfect balance of positive and negative electric charge, resulting in a nearly perfect balance of attractive and repulsive electromagnetic forces between the objects. Therefore, despite the fundamental strength of the electromagnetic forces, the net electromagnetic force between two macroscopic objects is often negligibly small except when atoms on the two surfaces come very close to each other—what we think of as *in contact.* On a microscopic level, there is no fundamental difference between contact forces and other electromagnetic forces.

The Strong Force The strong force holds protons and neutrons together in the atomic nucleus. The same force binds quarks (a family of elementary particles) in combinations so they can form protons and neutrons and many more exotic subatomic particles. The strong force is the strongest of the four fundamental forces—hence its name—but its range is short: its effect is negligible at distances much larger than the size of an atomic nucleus (about 10^{-15} m).

The Weak Force The range of the weak force is even shorter than that of the strong force (about 10^{-17} m). It is manifest in many radioactive decay processes.

Master the Concepts

- A *force* is a push or a pull. Gravity and electromagnetic forces have unlimited range. All other forces exerted on macroscopic objects involve contact. Force is a vector quantity.
- The SI unit of force is the newton: 1 N = 1 kg·m/s^2.
- The *net force* on a system is the vector sum of all the forces acting on it:

$$\vec{\mathbf{F}}_{\text{net}} = \sum \vec{\mathbf{F}} = \vec{\mathbf{F}}_1 + \vec{\mathbf{F}}_2 + \cdots + \vec{\mathbf{F}}_n \quad (4\text{-}2)$$

  Since all the internal forces form interaction pairs, we need only sum the external forces.
- *Newton's first law of motion*: If zero net force acts on an object, then the object's velocity does not change. Velocity is a vector whose magnitude is the speed at which the object moves and whose direction is the direction of motion.
- *Newton's second law of motion* relates the net force acting on an object to the object's acceleration and its mass:

$$\vec{\mathbf{a}} = \frac{\sum \vec{\mathbf{F}}}{m} \quad \text{or} \quad \sum \vec{\mathbf{F}} = m\vec{\mathbf{a}} \quad (4\text{-}4)$$

  The acceleration is always in the same direction as the net force. Many problems involving Newton's second law—whether equilibrium or nonequilibrium—can be solved by treating the *x*- and *y*-components of the forces and the acceleration separately:

$$\sum F_x = ma_x \quad \text{and} \quad \sum F_y = ma_y \quad (4\text{-}5)$$

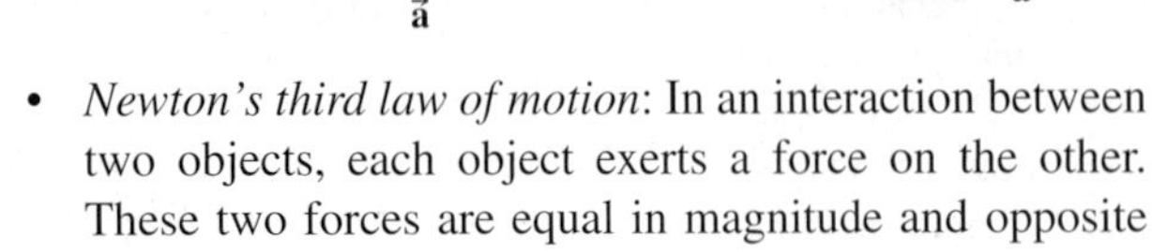

- *Newton's third law of motion*: In an interaction between two objects, each object exerts a force on the other. These two forces are equal in magnitude and opposite in direction.
- A *free-body diagram* (FBD) includes vector arrows representing every force acting on the chosen object due to some other object, but no forces acting on other objects.

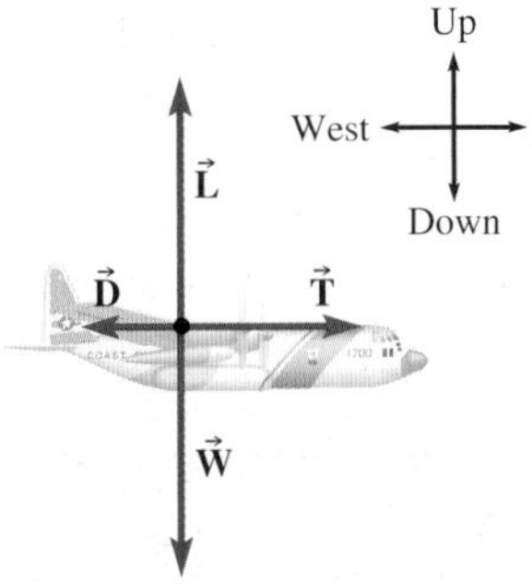

- The magnitude of the *gravitational force* between two objects is

$$F = \frac{Gm_1 m_2}{r^2} \quad (4\text{-}7)$$

  where r is the distance between their centers. Each object is pulled toward the other's center.
- The *weight* of an object is the magnitude of the gravitational force acting on it. An object's weight is proportional to its mass: $W = mg$ [Eq. (4-10)], where g is the gravitational field strength. Near Earth's surface, $g \approx 9.80$ N/kg.
- The *normal force* is a contact force perpendicular to the contact surfaces that pushes each object away from the other.

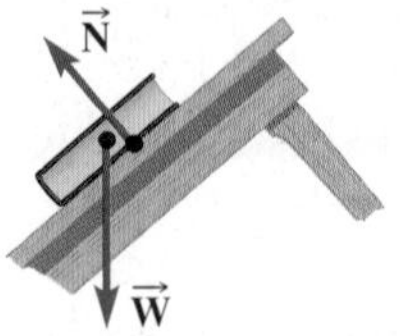

- *Friction* is a contact force parallel to the contact surfaces. In a simplified model, the kinetic frictional force and the maximum static frictional force are proportional

continued on next page

Master the Concepts continued

to the normal force acting between the same contact surfaces.

$$f_s \le \mu_s N \quad (4\text{-}14)$$

$$f_k = \mu_k N \quad (4\text{-}15)$$

The static frictional force acts in the direction that tends to keep the surfaces from beginning to slide. The direction of the kinetic frictional force is in the direction that would tend to make the sliding stop.

- An ideal cord pulls in the direction of the cord with forces of equal magnitude on the objects attached to its ends as long as no external force tangent to the cord is exerted on it anywhere between the ends. The tension of an ideal cord that runs through an ideal pulley is the same on both sides of the pulley.
- An object that is accelerating has an apparent weight that differs from its true weight. The apparent weight is equal to the normal force exerted by a supporting surface with the same acceleration. A helpful trick is to think of the apparent weight as the reading of a bathroom scale that supports the object.
- The drag force exerted on an object moving through air opposes the motion of the object but, unlike kinetic friction, is strongly dependent on the object's speed. When an object falls at its terminal velocity, the drag force is equal and opposite to the gravitational force, so the acceleration is zero.
- At the fundamental level, there are four interactions: gravity, the strong and weak interactions, and the electromagnetic interaction. Contact forces are large-scale manifestations of many microscopic electromagnetic interactions.

Conceptual Questions

1. Explain the need for automobile seat belts in terms of Newton's first law.
2. An American visitor to Finland is surprised to see heavy metal frames outside of all the apartment buildings. On Saturday morning the purpose of the frames becomes evident when several apartment dwellers appear, carrying rugs and carpet beaters to each frame. What role does the principle of inertia play in the rug beating process? Do you see a similarity to the role the principle of inertia plays when you throw a baseball?
3. You are lying on the beach after a dip in the ocean where the waves were buffeting you around. Is it true that there are now no forces acting on you? Explain.
4. A dog goes swimming at the beach and then shakes himself all over to get dry. What principle of physics aids in the drying process? Explain.
5. In an attempt to tighten the loosened steel head of a hammer, a carpenter holds the hammer vertically, raises it up, and then brings it down rapidly, hitting the bottom end of the wood handle on a two-by-four board. Explain how this tightens the head back onto the handle.

6. When a car begins to move forward, what force makes it do so? Remember that it has to be an *external* force; the internal forces all add to zero. How does the engine facilitate the propelling force?
7. Two cars are headed toward each other in opposite directions along a narrow country road. The cars collide head-on, crumpling up the hoods of both. Describe what happens to the car bodies in terms of the principle of inertia. Does the rear end of the car stop at the same time as the front end?
8. Can a body in free fall be in equilibrium? Explain.
9. (a) What assumptions do you make when you call the reading of a bathroom scale your "weight"? What does the scale really tell you? (b) Under what circumstances might the reading of the scale *not* be equal to your weight?
10. A freight train consists of an engine and several identical cars on level ground. Determine whether each of these statements is correct or incorrect and explain why. (a) If the train is moving at constant speed, the engine must be pulling with a force greater than the train's weight. (b) If the train is moving at constant speed, the engine's pull on the first car must exceed that car's backward pull on the engine. (c) If the train is coasting, its inertia makes it slow down and eventually stop.
11. (a) Does a man weigh more at the North Pole or at the equator? (b) Does he weigh more at the top of Mt. Everest or at the base of the mountain?
12. What is the acceleration of an object thrown straight up into the air at the highest point of its motion? Does the answer depend on whether air resistance is negligible or not? Explain.
13. If a wagon starts at rest and pulls back on you with a force equal to the force you pull on it, as required by Newton's third law, how is it possible for you to make the wagon start to move? Explain.

14. You are standing on a bathroom scale in an elevator. In which of these situations must the scale read the same as when the elevator is at rest? Explain. (a) Moving up at constant speed. (b) Moving up with increasing speed. (c) In free fall (after the elevator cable has snapped).
15. A heavy ball hangs from a string attached to a sturdy wooden frame. A second string is attached to a hook on the bottom of the lead ball. You pull slowly and steadily on the lower string. Which string do you think will break first? Explain.
16. An SUV collides with a Mini Cooper convertible. Is the force exerted on the Mini by the SUV greater than, equal to, or less than the force exerted on the SUV by the Mini? Explain.
17. You are standing on one end of a light wooden raft that has floated 3 m away from the pier. If the raft is 6 m long by 2.5 m wide and you are standing on the raft end nearest to the pier, can you propel the raft back toward the pier where a friend is standing with a pole and hook trying to reach you? You have no oars. Make suggestions of what to do without getting yourself wet.

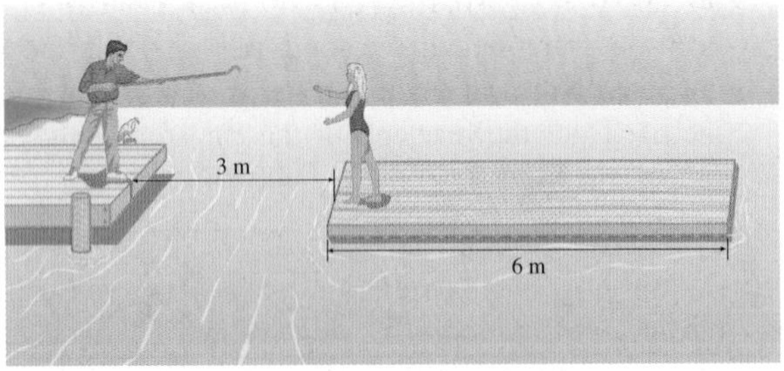

18. What does it mean when we refer to a cord as an "ideal cord" and a pulley as an "ideal pulley"?
19. If a feather and a lead brick are dropped simultaneously from the top of a ladder, the lead brick hits the ground first. What would happen if the experiment is repeated on the surface of the Moon?
20. A baseball is tossed straight up. Taking into consideration the force of air resistance, is the magnitude of the baseball's acceleration zero, less than g, equal to g, or greater than g on the way up? At the top of the flight? On the way down? Explain. [*Hint:* The force of air resistance is directed opposite to the velocity. Assume in this case that its magnitude is less than the weight.]
21. Why might an elevator cable break during acceleration when lifting a lighter load than it normally supports at rest or at constant velocity?
22. If air resistance is ignored, what force(s) act on an object in free fall?
23. The net force acting on an object is constant. Under what circumstances does the object move along a straight line? Under what circumstances does the object move along a curved path?
24. Pulleys and inclined planes are examples of *simple machines*. Explain what these machines do in Examples 4.10, 4.12, and 4.16 to make a task easier to perform.
25. For a problem about a crate sliding along an inclined plane, is it possible to choose the x-axis so that it is parallel to the incline?
26. A bird sits on a stretched clothesline, causing it to sag slightly. Is the tension in the line greatest where the bird sits, greater at either end of the line where it is attached to poles, or the same everywhere along the line? Treat the line as an ideal cord with negligible weight.
27. You decide to test your physics knowledge while going over a waterfall in a barrel. You take a baseball into the barrel with you and as you are falling vertically downward, you let go of the ball. What do you expect to see for the motion of the ball relative to the barrel? Will the ball fall faster than you and move toward the bottom of the barrel? Will it move slower than you and approach the top of the barrel, or will it hover apparently motionless within the falling barrel? Explain. [*Warning:* Do not try this.]

Multiple-Choice Questions

1. Interaction partners
   (a) are equal in magnitude and opposite in direction and act on the same object.
   (b) are equal in magnitude and opposite in direction and act on different objects.
   (c) appear in an FBD for a given object.
   (d) always involve gravitational force as one partner.
   (e) act in the same direction on the same object.
2. Within a given system, the internal forces
   (a) are always balanced by the external forces.
   (b) all add to zero.
   (c) are determined only by subtracting the external forces from the net force on the system.
   (d) determine the motion of the system.
   (e) can never add to zero.
3. A friction force is
   (a) a contact force that acts parallel to the contact surfaces.
   (b) a contact force that acts perpendicular to the contact surfaces.
   (c) a scalar quantity since it can act in any direction along a surface.
   (d) always proportional to the weight of an object.
   (e) always equal to the normal force between the objects.

4. When a force is called a "normal" force, it is
   (a) the usual force expected given the arrangement of a system.
   (b) a force that is perpendicular to the surface of the Earth at any given location.
   (c) a force that is always vertical.
   (d) a contact force perpendicular to the contact surfaces between two solid objects.
   (e) the net force acting on a system.

5. Your car won't start, so you are pushing it. You apply a horizontal force of 300 N to the car, but it doesn't budge. What force is the interaction partner of the 300 N force you exert?
   (a) the frictional force exerted on the car by the road
   (b) the force exerted on you by the car
   (c) the frictional force exerted on you by the road
   (d) the normal force on you by the road
   (e) the normal force on the car by the road

6. Which of these is *not* a long-range force?
   (a) the force that makes raindrops fall to the ground
   (b) the force that makes a compass point north
   (c) the force that a person exerts on a chair while sitting
   (d) the force that keeps the Moon in its orbital path around the Earth

7. When an object is in translational equilibrium, which of these statements is *not* true?
   (a) The vector sum of the forces acting on the object is zero.
   (b) The object must be stationary.
   (c) The object has a constant velocity.
   (d) The speed of the object is constant.

8. To make an object start moving on a surface with friction requires
   (a) less force than to keep it moving on the surface.
   (b) the same force as to keep it moving on the surface.
   (c) more force than to keep it moving on the surface.
   (d) a force equal to the weight of the object.

9. A thin string that can support a weight of 35.0 N, but breaks under any larger weight, is attached to the ceiling of an elevator. How large a mass can be attached to the string if the initial acceleration as the elevator starts to ascend is 3.20 m/s^2?
   (a) 3.57 kg
   (b) 2.69 kg
   (c) 4.26 kg
   (d) 2.96 kg
   (e) 5.30 kg

10. A woman stands on a bathroom scale in an elevator that is not moving. The scale reads 500 N. The elevator then moves downward at a constant velocity of 4.5 m/s. What does the scale read while the elevator descends with constant velocity?
    (a) 100 N
    (b) 250 N
    (c) 450 N
    (d) 500 N
    (e) 750 N

11. A 70.0-kg man stands on a bathroom scale in an elevator. What does the scale read if the elevator is slowing down at a rate of 3.00 m/s^2 while descending?
    (a) 70 kg (b) 476 N (c) 686 N
    (d) 700 N (e) 896 N

12. A space probe leaves the solar system to explore interstellar space. Once it is far from any stars, when must it fire its rocket engines?
    (a) All the time, in order to keep moving.
    (b) Only when it wants to speed up.
    (c) When it wants to speed up or slow down.
    (d) Only when it wants to turn.
    (e) When it wants to speed up, slow down, or turn.

13. A small plane climbs with a constant velocity of 250 m/s at an angle of 28° with respect to the horizontal. Which statement is true concerning the magnitude of the net force on the plane?
    (a) It is equal to zero.
    (b) It is equal to the weight of the plane.
    (c) It is equal to the magnitude of the force of air resistance.
    (d) It is less than the weight of the plane but greater than zero.
    (e) It is equal to the component of the weight of the plane in the direction of motion.

14. Two blocks are connected by a light string passing over a pulley (see the figure and tutorial: pulley). The block with mass m_1 slides on the frictionless horizontal surface, while the block with mass m_2 hangs vertically. ($m_1 > m_2$.) The tension in the string is
    (a) zero.
    (b) less than m_2g.
    (c) equal to m_2g.
    (d) greater than m_2g, but less than m_1g.
    (e) equal to m_1g.
    (f) greater than m_1g.

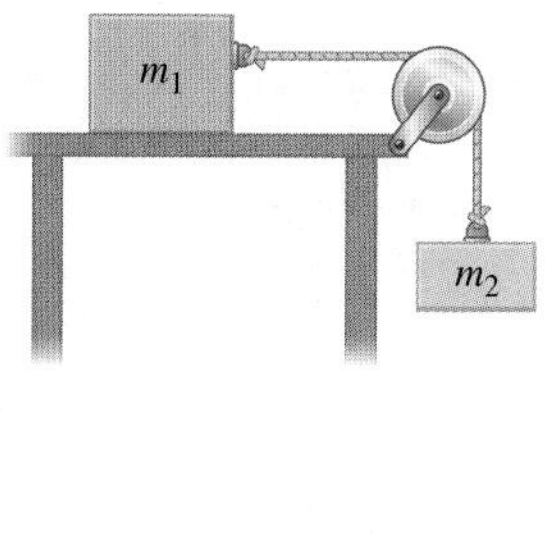

Problems

- Ⓒ Combination conceptual/quantitative problem
- Biological or medical application
- Challenging problem
- Blue # Detailed solution in the Student Solutions Manual
- ⓵ ⓶ Problems paired by concept
- Text website interactive or tutorial

4.1 Force

1. A person is standing on a bathroom scale. Which of the following is *not* a force exerted *on the scale*: a contact force due to the floor, a contact force due to the person's feet, the weight of the person, the weight of the scale?
2. A sack of flour has a weight of 19.8 N. What is its weight in pounds?
3. An astronaut weighs 175 lb. What is his weight in newtons?
4. Does the concept of a contact force apply to both a macroscopic scale and an atomic scale? Explain.
5. A force of 20 N is directed at an angle of 60° above the x-axis. A second force of 20 N is directed at an angle of 60° below the x-axis. What is the vector sum of these two forces?
6. Juan is helping his mother rearrange the living room furniture. Juan pushes on the armchair with a force of 30 N directed at an angle of 15° above a horizontal line while his mother pushes with a force of 40 N directed at an angle of 20° below the same horizontal. What is the vector sum of these two forces?
7. In the drawing, what is the vector sum of forces $\vec{\mathbf{A}} + \vec{\mathbf{B}} + \vec{\mathbf{C}}$ if each grid square is 2 N on a side?

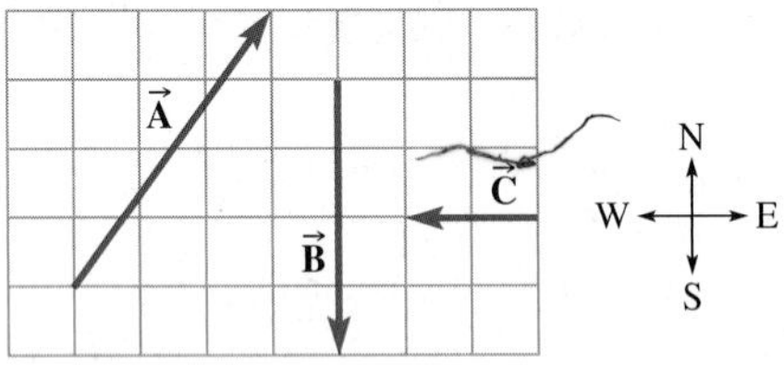

8. In the drawing, what is the vector sum of forces $\vec{\mathbf{D}} + \vec{\mathbf{E}} + \vec{\mathbf{F}}$ if each grid square is 2 N on a side?

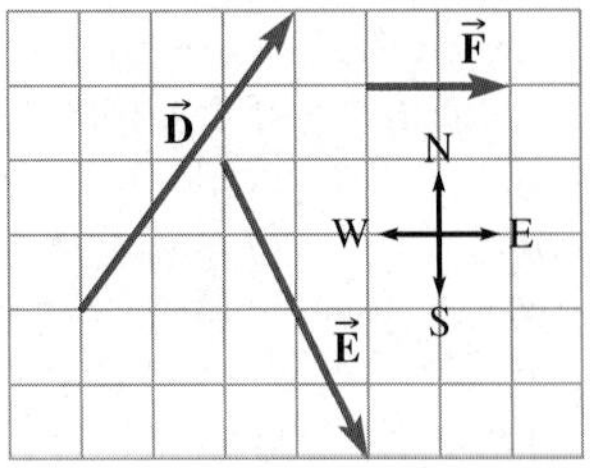

9. Two of Robin Hood's men are pulling a sledge loaded with some gold along a path that runs due north to their hideout. One man pulls his rope with a force of 62 N at an angle of 12° east of north and the other pulls with the same force at an angle of 12° west of north. Assume the ropes are parallel to the ground. What is the sum of these two forces on the sledge?
10. A barge is hauled along a straight-line section of canal by two horses harnessed to tow ropes and walking along the tow paths on either side of the canal. Each horse pulls with a force of 560 N at an angle of 15° with the centerline of the canal. Find the sum of the two forces exerted by the horses on the barge.
11. On her way to visit Grandmother, Red Riding Hood sat down to rest and placed her 1.2-kg basket of goodies beside her. A wolf came along, spotted the basket, and began to pull on the handle with a force of 6.4 N at an angle of 25° with respect to vertical. Red was not going to let go easily, so she pulled on the handle with a force of 12 N. If the net force on the basket is straight up, at what angle was Red Riding Hood pulling?
12. A parked automobile slips out of gear, rolls unattended down a slight incline, and then along a level road until it hits a stone wall. Draw an FBD to show the forces acting on the car while it is in contact with the wall.
13. Two objects, A and B, are acted on by the forces shown in the FBDs. Is the magnitude of the net force acting on object B greater than, less than, or equal to the magnitude of the net force acting on object A? Make a scale drawing on graph paper and explain the result.

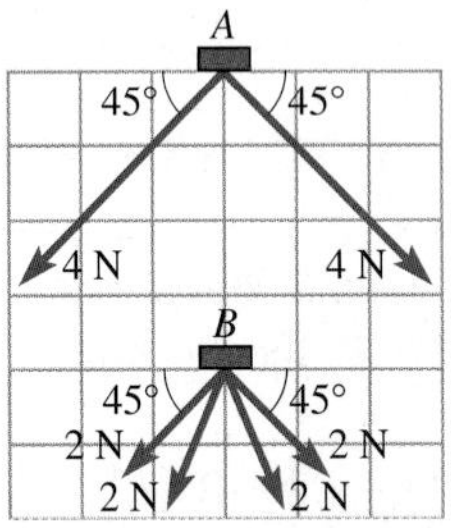

14. Find the magnitude and direction of the net force on the object in each of the FBDs for this problem.

(a)

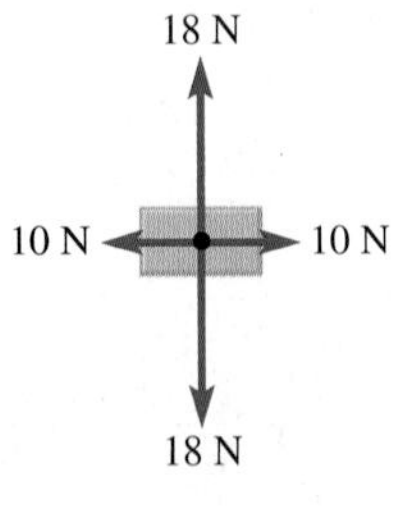

(b)

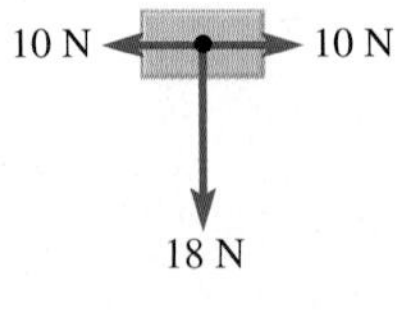

(c)

15. A truck driving on a level highway is acted on by the following forces: a downward gravitational force of 52 kN (kilonewtons); an upward contact force due to the road of 52 kN; another contact force due to the road of 7 kN, directed east; and a drag force due to air resistance of 5 kN, directed west. What is the net force acting on the truck?

4.2 Inertia and Equilibrium: Newton's First Law of Motion; 4.3 Net Force, Mass, and Acceleration: Newton's Second Law of Motion

16. A sailboat, tied to a mooring with a line, weighs 820 N. The mooring line pulls horizontally toward the west on the sailboat with a force of 110 N. The sails are stowed away and the wind blows from the west. The boat is moored on a still lake—no water currents push on it. Draw an FBD for the sailboat and indicate the magnitude of each force.

17. A hummingbird is hovering motionless beside a flower. The blur of its wings shows that they are rapidly beating up and down. If the air pushes upward on the bird with a force of 0.30 N, what is the weight of the hummingbird?

18. You are pulling a suitcase through the airport at a constant speed. The handle of the suitcase makes an angle of 60° with respect to the horizontal direction. If you pull with a force of 5.0 N parallel to the handle, what is the contact force due to the floor acting on the suitcase?

19. A model sailboat is slowly sailing west across a pond at 0.33 m/s. A gust of wind blowing at 28° south of west gives the sailboat a constant acceleration of magnitude 0.30 m/s^2 during a time interval of 2.0 s. (a) If the net force on the sailboat during the 2.0-s interval has magnitude 0.375 N, what is the sailboat's mass? (b) What is the new velocity of the boat after the 2.0-s gust of wind?

20. A man is lazily floating on an air mattress in a swimming pool. If the weight of the man and air mattress together is 806 N, what is the upward force of the water acting on the mattress?

21. A bag of potatoes with weight 39.2 N is suspended from a string that exerts a force of 46.8 N. If the bag's acceleration is upward at 1.90 m/s^2, what is the mass of the potatoes?

22. A 2010-kg elevator moves with an upward acceleration of 1.50 m/s^2. What is the force exerted by the cable on the elevator?

23. While an elevator of mass 2530 kg moves upward, the force exerted by the cable is 33.6 kN. (a) What is the acceleration of the elevator? (b) If at some point in the motion the velocity of the elevator is 1.20 m/s upward, what is the elevator's velocity 4.00 s later?

24. The vertical component of the acceleration of a sailplane is zero when the air pushes up against its wings with a force of 3.0 kN. (a) Assuming that the only forces on the sailplane are that due to gravity and that due to the air pushing against its wings, what is the gravitational force on the Earth due to the sailplane? (b) If the wing stalls and the upward force decreases to 2.0 kN, what is the acceleration of the sailplane?

25. A man lifts a 2.0-kg stone vertically with his hand at a constant upward velocity of 1.5 m/s. What is the magnitude of the total force of the man's hand on the stone?

26. A man lifts a 2.0-kg stone vertically with his hand at a constant upward *acceleration* of 1.5 m/s^2. What is the magnitude of the total force of the man's hand on the stone?

27. What is the acceleration of an automobile of mass 1.40×10^3 kg when it is subjected to a forward force of 3.36×10^3 N?

28. A large wooden crate is pushed along a smooth, frictionless surface by a force of 100 N. The acceleration of the crate is measured to be 2.5 m/s^2. What is the mass of the crate?

29. The forces on a small airplane (mass 1160 kg) in horizontal flight heading eastward are as follows: gravity = 16.000 kN downward, lift = 16.000 kN upward, thrust = 1.800 kN eastward, and drag = 1.400 kN westward. At $t = 0$, the plane's speed is 60.0 m/s. If the forces remain constant, how far does the plane travel in the next 60.0 s?

30. While an elevator of mass 832 kg moves downward, the tension in the supporting cable is a constant 7730 N. Between $t = 0$ and $t = 4.00$ s, the elevator's displacement is 5.00 m downward. What is the elevator's speed at $t = 4.00$ s?

4.4 Interaction Pairs: Newton's Third Law of Motion

31. A hanging potted plant is suspended by a cord from a hook in the ceiling. Draw an FBD for each of these: (a) the system consisting of plant, soil, and pot; (b) the cord; (c) the hook; (d) the system consisting of plant, soil, pot, cord, and hook. Label each force arrow using subscripts (for example, $\vec{\mathbf{F}}_{ch}$ would represent the force exerted on the cord by the hook).

32. A bike is hanging from a hook in a garage. Consider the following forces: (a) the force of the Earth pulling down on the bike, (b) the force of the bike pulling up on the Earth, (c) the force of the hook pulling up on the bike, and (d) the force of the hook pulling down on the ceiling. Which two forces are equal and opposite because of Newton's third law? Which two forces are equal and opposite because of Newton's first law?

33. A woman who weighs 600 N sits on a chair with her feet on the floor and her arms resting on the chair's armrests. The chair weighs 100 N. Each armrest exerts an upward force of 25 N on her arms. The seat of the chair exerts an upward force of 500 N. (a) What force does the floor exert on her feet? (b) What force does the floor exert on the chair? (c) Consider the woman and the chair to be a single system. Draw an FBD for this system that includes all of the *external* forces acting on it.

34. A fisherman is holding a fishing rod with a large fish suspended from the line of the rod. Identify the forces acting on the rod and their interaction partners.

35. A fish is suspended by a line from a fishing rod. Choose two forces acting on the fish and describe the interaction partner of each.

Problems 34 and 35

36. A skydiver, who weighs 650 N, is falling at a constant speed with his parachute open. Consider the apparatus that connects the parachute to the skydiver to be part of the parachute. The parachute pulls upward with a force of 620 N. (a) What is the force of the air resistance acting on the skydiver? (b) Identify the forces and the interaction partners of each force exerted on the skydiver. (c) Identify the forces and interaction partners of each force exerted on the parachute.

37. Margie, who weighs 543 N, is standing on a bathroom scale that weighs 45 N. (a) With what force does the scale push up on Margie? (b) What is the interaction partner of that force? (c) With what force does the Earth push up on the scale? (d) Identify the interaction partner of that force.

38. Refer to Problem 36. Consider the skydiver and parachute to be a single system. What are the external forces acting on this system?

4.5 Gravitational Forces

39. (a) Calculate your weight in newtons. (b) What is the weight in newtons of 250 g of cheese? (c) Name a common object whose weight is about 1 N.

40. A young South African girl has a mass of 40.0 kg. (a) What is her weight in newtons? (b) If she came to the United States, what would her weight be in pounds as measured on an American scale? Assume $g = 9.80$ N/kg in both locations.

41. A man weighs 0.80 kN on Earth. What is his mass in kilograms?

42. An astronaut stands at a position on the Moon such that Earth is directly over head and releases a Moon rock that was in her hand. (a) Which way will it fall? (b) What is the gravitational force exerted by the Moon on a 1.0-kg rock resting on the Moon's surface? (c) What is the gravitational force exerted by the Earth on the same 1.0-kg rock resting on the surface of the Moon? (d) What is the net gravitational force on the rock?

43. Alex is on stage playing his bass guitar. Estimate the magnitude of the *gravitational* attraction between Alex and Pat, a fan who is standing 8 m from Alex. Alex has a mass of 55 kg and Pat has a mass of 40 kg.

44. The Space Shuttle carries a satellite in its cargo bay and places it into orbit around the Earth. Find the ratio of the Earth's gravitational force on the satellite when it is on a launch pad at the Kennedy Space Center to the gravitational force exerted when the satellite is orbiting 6.00×10^3 km above the launch pad.

45. How far above the surface of the Earth does an object have to be in order for it to have the same weight as it would have on the surface of the Moon? (Ignore any effects from the Earth's gravity for the object on the Moon's surface or from the Moon's gravity for the object above the Earth.)

46. Find and compare the weight of a 65-kg man on Earth with the weight of the same man on (a) Mars, where $g = 3.7$ N/kg; (b) Venus, where $g = 8.9$ N/kg; and (c) Earth's Moon, where $g = 1.6$ N/kg.

47. Find the altitudes above the Earth's surface where Earth's gravitational field strength would be (a) two thirds and (b) one third of its value at the surface. [*Hint:* First find the radius for each situation; then recall that the altitude is the distance from the *surface* to a point above the surface. Use proportional reasoning.]

48. During a balloon ascension, wearing an oxygen mask, you measure the weight of a calibrated 5.00-kg mass and find that the value of the gravitational field strength at your location is 9.792 N/kg. How high above sea level, where the gravitational field strength was measured to be 9.803 N/kg, are you located?

49. At what altitude above the Earth's surface would your weight be half of what it is at the Earth's surface?

50. (a) What is the magnitude of the gravitational force that the Earth exerts on the Moon? (b) What is the magnitude of the gravitational force that the Moon exerts on the Earth? See the inside front and back covers for necessary information.

51. What is the approximate magnitude of the gravitational force between the Earth and the Voyager spacecraft when they are separated by 15 billion km? Each spacecraft has a mass of approximately 825 kg during the mission, although the mass at launch was 2100 kg because of expendable Titan-Centaur rockets.

✦52. In free fall, we assume the acceleration to be constant. Not only is air resistance ignored, but the gravitational field strength is assumed to be constant. From what height can an object fall to the Earth's surface such that the gravitational field strength changes less than 1.000% during the fall?

4.6 Contact Forces

53. A book rests on the surface of the table. Consider the following four forces that arise in this situation: (a) the force of the Earth pulling on the book, (b) the force of the table pushing on the book, (c) the force of the book pushing on the table, and (d) the force of the book pulling on the Earth. The book is not moving. Which pair of forces must be equal in magnitude and opposite in direction even though they are *not* an interaction pair?

54. A crate full of artichokes rests on a ramp that is inclined 10.0° above the horizontal. Give the direction of the normal force and the friction force acting on the crate in each of these situations. (a) The crate is at rest. (b) The crate is being pushed and is sliding up the ramp. (c) The crate is being pushed and is sliding down the ramp.

55. Mechanical advantage is the ratio of the force required without the use of a simple machine to that needed when using the simple machine. Compare the force to lift an object with that needed to slide the same object up a frictionless incline and show that the mechanical advantage of the inclined plane is the length of the incline divided by the height of the incline (d/h in Fig. 4.25).

56. An 80.0-N crate of apples sits at rest on a ramp that runs from the ground to the bed of a truck. The ramp is inclined at 20.0° to the ground. (a) What is the normal force exerted on the crate by the ramp? (b) The interaction partner of this normal force has what magnitude and direction? It is exerted *by* what object *on* what object? Is it a contact or a long-range force? (c) What is the static frictional force exerted on the crate by the ramp? (d) What is the minimum possible value of the coefficient of static friction? (e) The normal and frictional forces are perpendicular components of the contact force exerted on the crate by the ramp. Find the magnitude and direction of the contact force.

57. An 85-kg skier is sliding down a ski slope at a constant velocity. The slope makes an angle of 11° above the horizontal direction. (a) Ignoring any air resistance, what is the force of kinetic friction acting on the skier? (b) What is the coefficient of kinetic friction between the skis and the snow?

Problems 58–60. A crate of potatoes of mass 18.0 kg is on a ramp with angle of incline 30° to the horizontal. The coefficients of friction are $\mu_s = 0.75$ and $\mu_k = 0.40$. Find the frictional force (magnitude and direction) on the crate if

58. the crate is at rest.

59. the crate is sliding down the ramp.

60. the crate is sliding *up* the ramp.

61. You grab a book and give it a quick push across the top of a horizontal table. After a short push, the book slides across the table, and because of friction, comes to a stop. (a) Draw an FBD of the book while you are pushing it. (b) Draw an FBD of the book after you have stopped pushing it, while it is sliding across the table. (c) Draw an FBD of the book after it has stopped sliding. (d) In which of the preceding cases is the net force on the book not equal to zero? (e) If the book has a mass of 0.50 kg and the coefficient of friction between the book and the table is 0.40, what is the net force acting on the book in part (b)? (f) If there were no friction between the table and the book, what would the free-body diagram for part (b) look like? Would the book slow down in this case? Why or why not?

62. (a) In Example 4.10, if the movers stop pushing on the safe, can static friction hold the safe in place without having it slide back down? (b) If not, what minimum force needs to be applied to hold the safe in place?

✦63. A 3.0-kg block is at rest on a horizontal floor. If you push horizontally on the 3.0-kg block with a force of 12.0 N, it just starts to move. (a) What is the coefficient of static friction? (b) A 7.0-kg block is stacked on top of the 3.0-kg block. What is the magnitude F of the force, acting horizontally on the 3.0-kg block as before, that is required to make the two blocks start to move?

64. A horse is trotting along pulling a sleigh through the snow. To move the sleigh, of mass m, straight ahead at a constant speed, the horse must pull with a force of magnitude T. (a) What is the net force acting on the sleigh? (b) What is the coefficient of kinetic friction between the sleigh and the snow?

✦65. Before hanging new William Morris wallpaper in her bedroom, Brenda sanded the walls lightly to smooth out some irregularities on the surface. The sanding block weighs 2.0 N and Brenda pushes on it with a force of 3.0 N at an angle of 30.0° with respect to the vertical, and angled toward the wall. Draw an FBD for the sanding block as it moves straight up the wall at a constant speed. What is the coefficient of kinetic friction between the wall and the block?

66. Four separate blocks are placed side by side in a left-to-right row on a table. A horizontal force, acting toward the right, is applied to the block on the far left end of the row. Draw FBDs for (a) the second block on the left and for (b) the system of four blocks.

67. A box sits on a horizontal wooden ramp. The coefficient of static friction between the box and the ramp is 0.30. You grab one end of the ramp and lift it up, keeping the other end of the ramp on the ground. What is the angle between the ramp and the horizontal direction when the box begins to slide down the ramp? (tutorial: crate on ramp)

68. In a playground, two slides have different angles of incline θ_1 and θ_2 ($\theta_2 > \theta_1$). A child slides down the first at constant speed; on the second, his acceleration down the slide is a. Assume the coefficient of kinetic friction is the same for both slides. (a) Find a in terms of θ_1, θ_2, and g. (b) Find the numerical value of a for $\theta_1 = 45°$ and $\theta_2 = 61°$.

4.7 Tension

69. A sailboat is tied to a mooring with a horizontal line. The wind is from the southwest. Draw an FBD and identify all the forces acting on the sailboat.

70. A towline is attached between a car and a glider. As the car speeds due east along the runway, the towline exerts a horizontal force of 850 N on the glider. What is the magnitude and direction of the force exerted by the glider on the towline?

71. In Example 4.14, find the tension in the coupling between cars 2 and 3. (tutorial: towing a train)

72. A 200.0-N sign is suspended from a horizontal strut of negligible weight. The force exerted on the strut by the wall is horizontal. Draw an FBD to show the forces acting on the strut. Find the tension T in the diagonal cable supporting the strut.

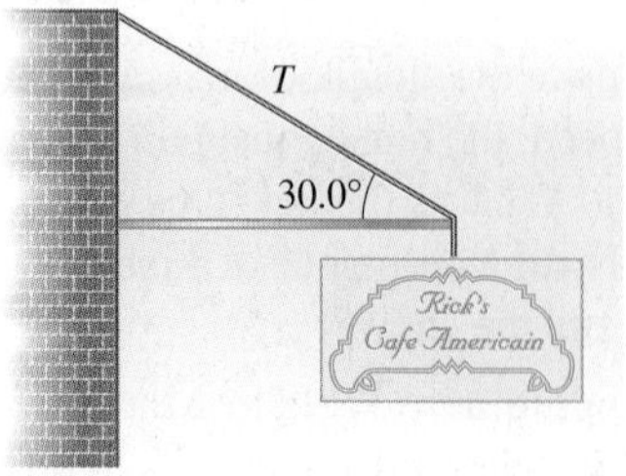

73. Two boxes with different masses are tied together on a frictionless ramp surface. What is the tension in each of the cords?

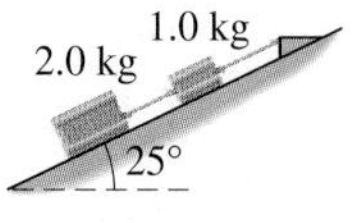

74. A pulley is attached to the ceiling. Spring scale A is attached to the wall and a rope runs horizontally from it and over the pulley. The same rope is then attached to spring scale B. On the other side of scale B hangs a 120-N weight. What are the readings of the two scales A and B? The weights of the scales are negligible.

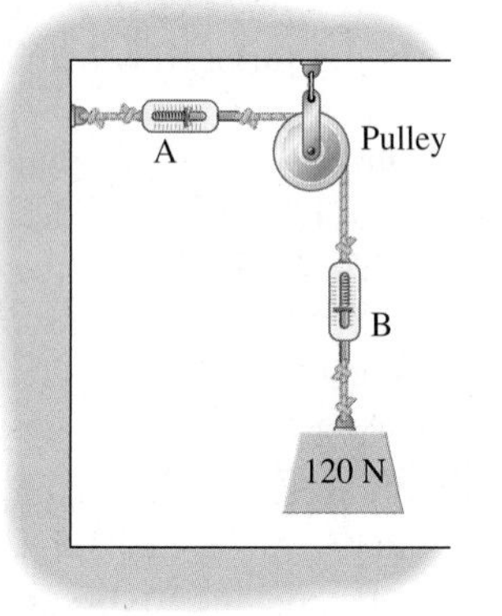

75. Spring scale A is attached to the floor and a rope runs vertically upward, loops over the pulley, and runs down on the other side to a 120-N weight. Scale B is attached to the ceiling and the pulley is hung below it. What are the readings of the two spring scales, A and B? Neglect the weights of the pulley and scales.

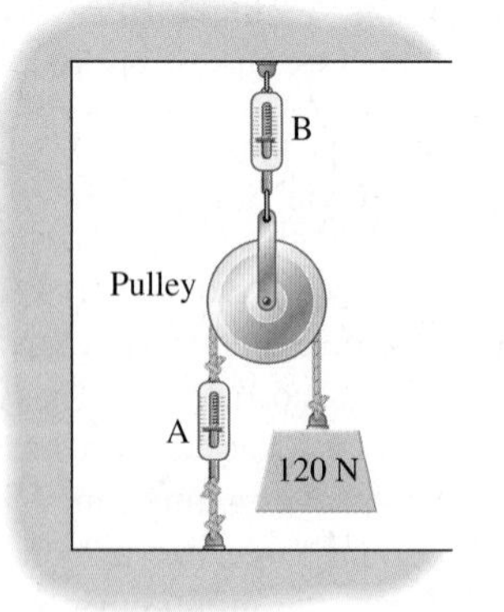

76. Two springs are connected in series so that spring scale A hangs from a hook on the ceiling and a second spring scale, B, hangs from the hook at the bottom of scale A.

Apples weighing 120 N hang from the hook at the bottom of scale B. What are the readings on the upper scale A and the lower scale B? Ignore the weights of the scales.

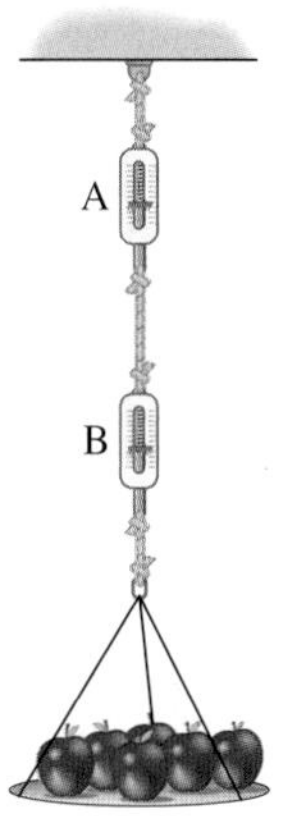

77. A pulley is hung from the ceiling by a rope. A block of mass M is suspended by another rope that passes over the pulley and is attached to the wall. The rope fastened to the wall makes a right angle with the wall. Ignore the masses of the rope and the pulley. Find (a) the tension in the rope from which the pulley hangs and (b) the angle θ that the rope makes with the ceiling.

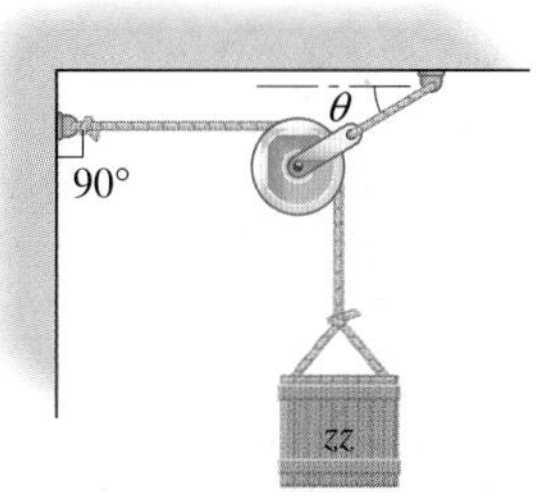

78. A 2.0-kg ball tied to a string fixed to the ceiling is pulled to one side by a force $\vec{\mathbf{F}}$. Just before the ball is released and allowed to swing back and forth, (a) how large is the force $\vec{\mathbf{F}}$ that is holding the ball in position and (b) what is the tension in the string?

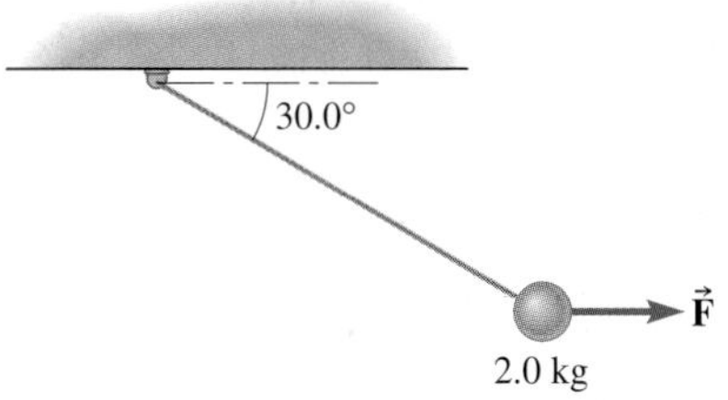

79. A 45-N lithograph is supported by two wires. One wire makes a 25° angle with the vertical and the other makes a 15° angle with the vertical. Find the tension in each wire. (tutorial: hanging picture)

80. A crow perches on a clothesline midway between two poles. Each end of the rope makes an angle of θ below the horizontal where it connects to the pole. If the weight of the crow is W, what is the tension in the rope? Ignore the weight of the rope.

81. The drawing shows an elastic cord attached to two back teeth and stretched across a front tooth. The purpose of this arrangement is to apply a force $\vec{\mathbf{F}}$ to the front tooth. (The figure has been simplified by running the cord straight from the front tooth to the back teeth.) If the tension in the cord is 1.2 N, what are the magnitude and direction of the force $\vec{\mathbf{F}}$ applied to the front tooth?

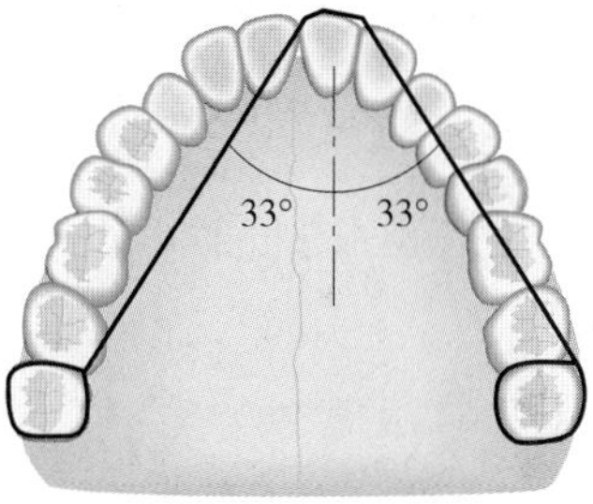

82. A cord, with a spring balance to measure forces attached midway along, is hanging from a hook attached to the ceiling. A mass of 10 kg is hanging from the lower end of the cord. The spring balance indicates a reading of 98 N for the force. Then two people hold the opposite ends of the same cord and pull against each other horizontally until the balance in the middle again reads 98 N. With what force must each person pull to attain this result?

83. Two blocks, masses m_1 and m_2, are connected by a massless cord. If the two blocks are pulled with a constant tension on a frictionless surface by applying a force of magnitude T_2 to a second cord connected to m_2, what is the ratio of the tensions in the two cords T_1/T_2 in terms of the masses?

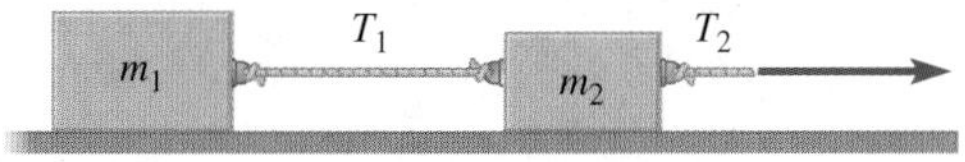

4.8 Applying Newton's Second Law

84. A 6.0-kg block, starting from rest, slides down a frictionless incline of length 2.0 m. When it arrives at the bottom of the incline, its speed is v_f. At what distance from the top of the incline is the speed of the block 0.50 v_f?

85. The coefficient of static friction between a block and a horizontal floor is 0.40, while the coefficient of kinetic friction is 0.15. The mass of the block is 5.0 kg. A horizontal force is applied to the block and slowly increased. (a) What is the value of the applied horizontal force at the instant that the block starts to slide? (b) What is the net force on the block after it starts to slide?

86. A 2.0-kg toy locomotive is pulling a 1.0-kg caboose. The frictional force of the track on the caboose is 0.50 N backward along the track. If the train's acceleration forward is 3.0 m/s^2, what is the magnitude of the force exerted by the locomotive on the caboose?

87. A block of mass $m_1 = 3.0$ kg rests on a frictionless horizontal surface. A second block of mass $m_2 = 2.0$ kg hangs from an ideal cord of negligible mass that runs over an ideal pulley and then is connected to the first block. The blocks are released from rest. (a) Find the acceleration of the two blocks after they are released. (b) What is the velocity of the first block 1.2 s after the release of the blocks, assuming the first block does not run out of room on the table and the second block does not land on the floor? (c) How far has block 1 moved during the 1.2-s interval? (d) What is the displacement of the blocks from their initial positions 0.40 s after they are released?

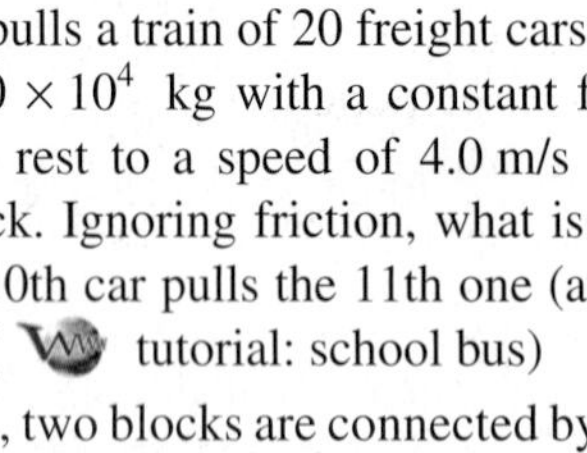

Problems 87 and 153

88. An engine pulls a train of 20 freight cars, each having a mass of 5.0×10^4 kg with a constant force. The cars move from rest to a speed of 4.0 m/s in 20.0 s on a straight track. Ignoring friction, what is the force with which the 10th car pulls the 11th one (at the middle of the train)? (tutorial: school bus)

89. In Fig. 4.44, two blocks are connected by a lightweight, flexible cord that passes over a frictionless pulley. (a) If $m_1 = 3.0$ kg and $m_2 = 5.0$ kg, what are the accelerations of each block? (b) What is the tension in the cord?

90. A rope is attached from a truck to a 1400-kg car. The rope will break if the tension is greater than 2500 N. Ignoring friction, what is the maximum possible acceleration of the truck if the rope does not break? Should the driver of the truck be concerned that the rope might break?

91. Two blocks are connected by a lightweight, flexible cord that passes over a frictionless pulley. If $m_1 = 3.6$ kg and $m_2 = 9.2$ kg, and block 2 is initially at rest 140 cm above the floor, how long does it take block 2 to reach the floor?

92. A 10.0-kg watermelon and a 7.00-kg pumpkin are attached to each other via a cord that wraps over a pulley, as shown. Friction is negligible everywhere in this system. (tutorial: pulley) (a) Find the accelerations of the pumpkin and the watermelon. Specify magnitude and direction. (b) If the system is released from rest, how far along the incline will the pumpkin travel in 0.30 s? (c) What is the speed of the watermelon after 0.20 s?

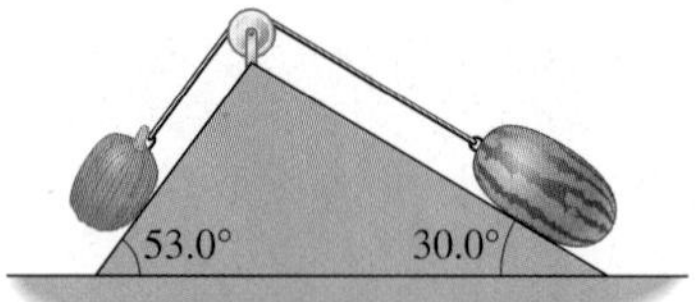

93. In the physics laboratory, a glider is released from rest on a frictionless air track inclined at an angle. If the glider has gained a speed of 25.0 cm/s in traveling 50.0 cm from the starting point, what was the angle of inclination of the track? Draw a graph of $v_x(t)$ when the positive x-axis points down the track.

94. A 10.0-kg block is released from rest on a frictionless track inclined at an angle of 55°. (a) What is the net force on the block after it is released? (b) What is the acceleration of the block? (c) If the block is released from rest, how long will it take for the block to attain a speed of 10.0 m/s? (d) Draw a motion diagram for the block. (e) Draw a graph of $v_x(t)$ for values of velocity between 0 and 10 m/s. Let the positive x-axis point down the track.

95. A box full of books rests on a wooden floor. The normal force the floor exerts on the box is 250 N. (a) You push horizontally on the box with a force of 120 N, but it refuses to budge. What can you say about the coefficient of static friction between the box and the floor?

(b) If you must push horizontally on the box with a force of at least 150 N to start it sliding, what is the coefficient of static friction? (c) Once the box is sliding, you only have to push with a force of 120 N to keep it sliding. What is the coefficient of kinetic friction?

96. A helicopter is lifting two crates simultaneously. One crate with a mass of 200 kg is attached to the helicopter by a cable. The second crate with a mass of 100 kg is hanging below the first crate and attached to the first crate by a cable. As the helicopter accelerates upward at a rate of 1.0 m/s^2, what is the tension in each of the two cables?

4.10 Apparent Weight

97. Oliver has a mass of 76.2 kg. He is riding in an elevator that has a downward acceleration of 1.37 m/s^2. With what magnitude force does the elevator floor push upward on Oliver?

98. While on an elevator, Jaden's apparent weight is 550 N. When he is on the ground, the scale reading is 600 N. What is Jaden's acceleration?

99. When on the ground, Ian's weight is measured to be 640 N. When Ian is on an elevator, his apparent weight is 700 N. What is the net force on the system (Ian and the elevator) if their combined mass is 1050 kg?

100. Refer to Example 4.19. What is the apparent weight of the same passenger (weighing 598 N) in the following situations? In each case, the magnitude of the elevator's acceleration is 0.50 m/s^2. (a) After having stopped at the 15th floor, the passenger pushes the 8th floor button; the elevator is beginning to move downward. (b) The elevator is moving downward and is slowing down as it nears the 8th floor.

101. You are standing on a bathroom scale inside an elevator. Your weight is 140 lb, but the reading of the scale is 120 lb. (a) What is the magnitude and direction of the acceleration of the elevator? (b) Can you tell whether the elevator is speeding up or slowing down?

102. Yolanda, whose mass is 64.2 kg, is riding in an elevator that has an upward acceleration of 2.13 m/s^2. What force does she exert on the floor of the elevator?

103. Felipe is going for a physical before joining the swim team. He is concerned about his weight, so he carries his scale into the elevator to check his weight while heading to the doctor's office on the 21st floor of the building. If his scale reads 750 N while the elevator has an upward acceleration of 2.0 m/s^2, what does the nurse measure his weight to be?

104. Luke stands on a scale in an elevator that has a constant acceleration upward. The scale reads 0.960 kN. When Luke picks up a box of mass 20.0 kg, the scale reads 1.200 kN. (The acceleration remains the same.) (a) Find the acceleration of the elevator. (b) Find Luke's weight.

4.12 Fundamental Forces

105. Which of the fundamental forces has the shortest range, yet is responsible for producing the sunlight that reaches Earth?

106. Which of the fundamental forces governs the motion of planets in the solar system? Is this the strongest or the weakest of the fundamental forces? Explain.

107. Which of the following forces have an unlimited range: strong force, contact force, electromagnetic force, gravitational force?

108. Which of the following forces bind electrons to nuclei to form atoms: strong force, contact force, electromagnetic force, gravitational force?

109. Which of the fundamental forces binds quarks together to form protons, neutrons, and many exotic subatomic particles?

Comprehensive Problems

110. A car is driving on a straight, level road at constant speed. Draw an FBD for the car, showing the significant forces that act upon it.

111. A skier with a mass of 63 kg starts from rest and skis down an icy (frictionless) slope that has a length of 50 m at an angle of 32° with respect to the horizontal. At the bottom of the slope, the path levels out and becomes horizontal, the snow becomes less icy, and the skier begins to slow down, coming to rest in a distance of 140 m along the horizontal path. (a) What is the speed of the skier at the bottom of the slope? (b) What is the coefficient of kinetic friction between the skier and the horizontal surface?

112. You want to push a 65-kg box up a 25° ramp. The coefficient of kinetic friction between the ramp and the box is 0.30. With what magnitude force parallel to the ramp should you push on the box so that it moves up the ramp at a constant speed?

113. An airplane is cruising along in a horizontal level flight at a constant velocity, heading due west. (a) If the weight of the plane is 2.6×10^4 N, what is the net force on the plane? (b) With what force does the air push upward on the plane?

114. A young boy with a broken leg is undergoing traction. (a) Find the magnitude of the total force of the traction apparatus applied to the leg, assuming the weight of the leg is 22 N and the weight hanging from the traction apparatus is also 22 N. (b) What is the horizontal

component of the traction force acting on the leg? (c) What is the magnitude of the force exerted on the femur by the lower leg?

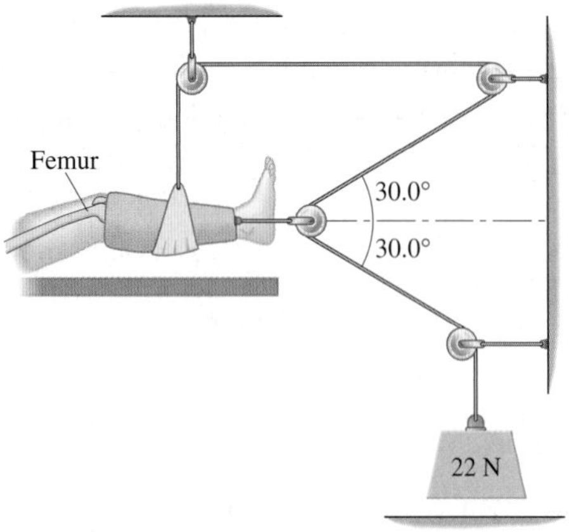

115. When you hold up a 100-N weight in your hand, with your forearm horizontal and your palm up, the force exerted by your biceps is much larger than 100 N—perhaps as much as 1000 N. How can that be? What other forces are acting on your arm? Draw an FBD for the forearm, showing all of the forces. Assume that all the forces exerted on the forearm are purely vertical—either up or down.

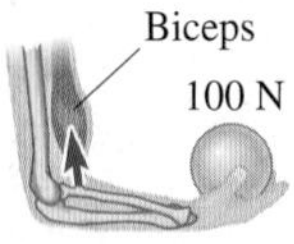

116. In the sport of curling, popular in Canada and Ireland, a player slides a 20.0-kg granite stone down a 38-m-long ice rink. Draw FBDs for the stone (a) while it sits at rest on the ice; (b) while it slides down the rink; (c) during a head-on collision with an opponent's stone that was at rest on the ice.

117. A truck is towing a 1000-kg car at a constant speed up a hill that makes an angle of $\alpha = 5.0°$ with respect to the horizontal. A rope is attached from the truck to the car at an angle of $\beta = 10.0°$ with respect to horizontal. Ignore any friction in this problem. (a) Draw an FBD showing all the forces on the car. Indicate the angle that each force makes with either the vertical or horizontal direction. (b) What is the tension in the rope?

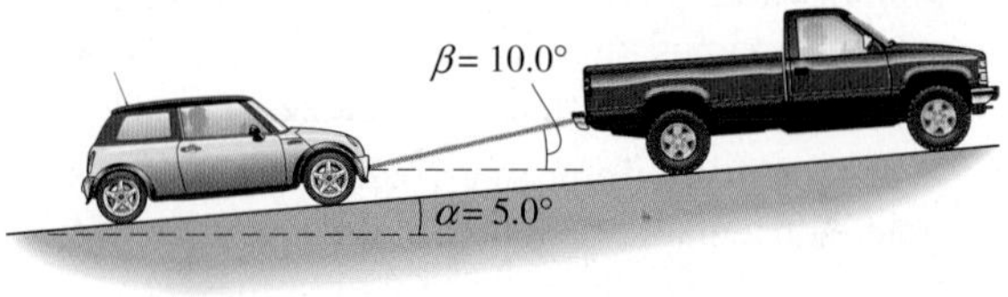

118. The readings of the two spring scales shown in the drawing are the same. (a) Explain why they are the same. [*Hint:* Draw free-body diagrams.] (b) What is the reading?

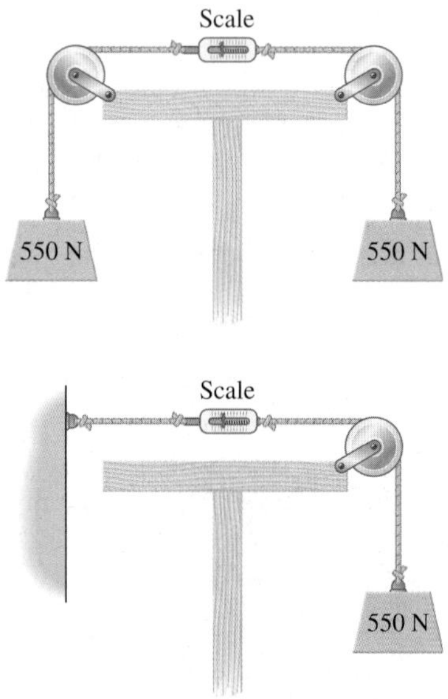

119. The tallest spot on Earth is Mt. Everest, which is 8850 m above sea level. If the radius of the Earth to sea level is 6370 km, how much does the gravitational field strength change between the sea level value at that location (9.826 N/kg) and the top of Mt. Everest?

120. By what percentage does the weight of an object change when it is moved from the equator at sea level, where the effective value of g is 9.784 N/kg, to the North Pole where $g = 9.832$ N/kg?

121. Two canal workers pull a barge along the narrow waterway at a constant speed. One worker pulls with a force of 105 N at an angle of 28° with respect to the forward motion of the barge and the other worker, on the opposite tow path, pulls at an angle of 38° relative to the barge motion. Both ropes are parallel to the ground. (a) With what magnitude force should the second worker pull to make the sum of the two forces be in the forward direction? (b) What is the magnitude of the force on the barge from the two tow ropes?

122. A large wrecking ball of mass m is resting against a wall. It hangs from the end of a cable that is attached at its upper end to a crane that is just touching the wall.

The cable makes an angle of θ with the wall. Ignoring friction between the ball and the wall, find the tension in the cable.

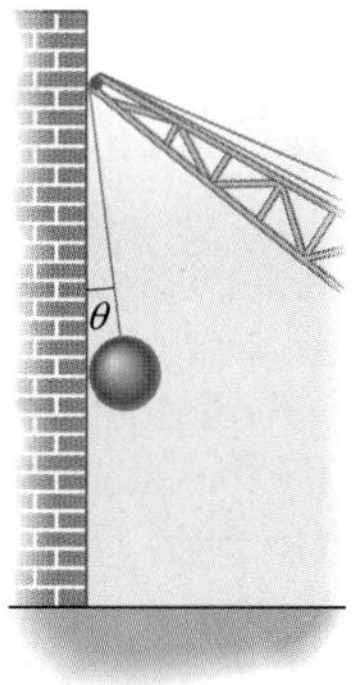

123. The figure shows the quadriceps and the patellar tendons attached to the patella (the kneecap). If the tension T in each tendon is 1.30 kN, what is (a) the magnitude and (b) the direction of the contact force $\vec{\mathbf{F}}$ exerted on the patella by the femur?

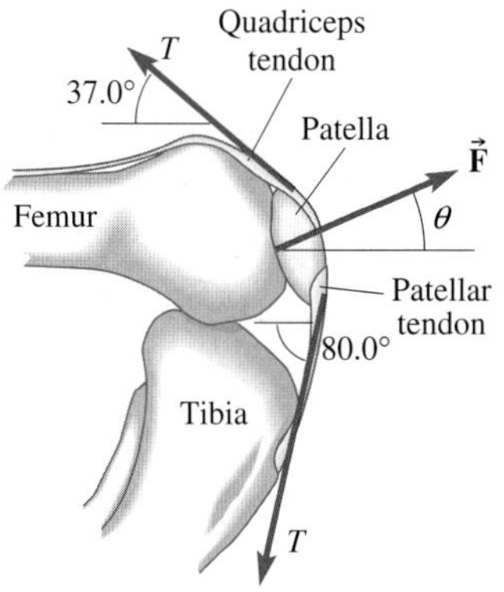

124. The coefficient of static friction between a block and a horizontal floor is 0.35, while the coefficient of kinetic friction is 0.22. The mass of the block is 4.6 kg and it is initially at rest. (a) What is the minimum horizontal applied force required to make the block start to slide? (b) Once the block is sliding, if you keep pushing on it with the same minimum starting force as in part (a), does the block move with constant velocity or does it accelerate? (c) If it moves with constant velocity, what is its velocity? If it accelerates, what is its acceleration?

125. Two blocks lie side by side on a frictionless table. The block on the left is of mass m; the one on the right is of mass $2m$. The block on the right is pushed to the left with a force of magnitude F, pushing the other block in turn. What force does the block on the left exert on the block to its right?

126. A locomotive pulls a train of 10 identical cars, on a track that runs east-west, with a force of 2.0×10^6 N directed east. What is the force with which the *last* car to the west pulls on the rest of the train?

127. The coefficient of static friction between a brick and a wooden board is 0.40 and the coefficient of kinetic friction between the brick and board is 0.30. You place the brick on the board and slowly lift one end of the board off the ground until the brick starts to slide down the board. (a) What angle does the board make with the ground when the brick starts to slide? (b) What is the acceleration of the brick as it slides down the board?

128. A woman of mass 51 kg is standing in an elevator. (a) If the elevator floor pushes up on her feet with a force of 408 N, what is the acceleration of the elevator? (b) If the elevator is moving at 1.5 m/s as it passes the fourth floor on its way down, what is its speed 4.0 s later?

129. In Fig. 4.15 an astronaut is playing shuffleboard on Earth. The puck has a mass of 2.0 kg. Between the board and puck the coefficient of static friction is 0.35 and of kinetic friction is 0.25. (a) If she pushes the puck with a force of 5.0 N in the forward direction, does the puck move? (b) As she is pushing, she trips and the force in the forward direction suddenly becomes 7.5 N. Does the puck move? (c) If so, what is the acceleration of the puck along the board if she maintains contact between puck and stick as she regains her footing while pushing steadily with a force of 6.0 N on the puck? (d) She carries her game to the Moon and again pushes a moving puck with a force of 6.0 N forward. Is the acceleration of the puck during contact more, the same, or less than on Earth? Explain. (tutorial: rough table)

130. You want to hang a 15-N picture as in part (a) using some very fine twine that will break with more than 12 N of tension. Can you do this? What if you have it as illustrated in part (b) of the figure?

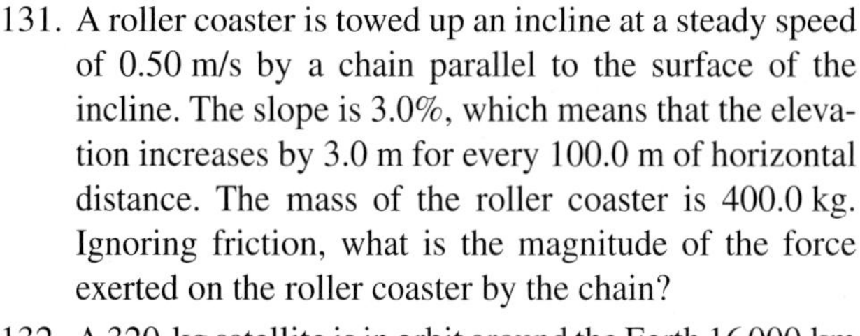

131. A roller coaster is towed up an incline at a steady speed of 0.50 m/s by a chain parallel to the surface of the incline. The slope is 3.0%, which means that the elevation increases by 3.0 m for every 100.0 m of horizontal distance. The mass of the roller coaster is 400.0 kg. Ignoring friction, what is the magnitude of the force exerted on the roller coaster by the chain?

132. A 320-kg satellite is in orbit around the Earth 16000 km above the Earth's surface. (a) What is the weight of the satellite when in orbit? (b) What was its weight when it was on the Earth's surface, before being launched? (c) While it orbits the Earth, what force does the satellite exert on the Earth?

133. The mass of the Moon is 0.0123 times that of the Earth. A spaceship is traveling along a line connecting the centers of the Earth and the Moon. At what distance from the Earth does the spaceship find the gravitational pull of the Earth equal in magnitude to that of the Moon? Express your answer as a percentage of the distance between the centers of the two bodies.

134. A model rocket is fired vertically from rest. It has a net acceleration of 17.5 m/s^2. After 1.5 s, its fuel is exhausted and its only acceleration is that due to gravity. (a) Ignoring air resistance, how high does the rocket travel? (b) How long after liftoff does the rocket return to the ground?

135. The model rocket in Problem 134 has a mass of 87 g and you may assume the mass of the fuel is much less than 87 g. (a) What was the net force on the rocket during the first 1.5 s after liftoff? (b) What force was exerted on the rocket by the burning fuel? (c) What was the net force on the rocket after its fuel was spent? (d) The rocket's vertical velocity was zero instantaneously when it was at the top of its trajectory. What were the net force and acceleration on the rocket at this instant?

136. A toy freight train consists of an engine and three identical cars. The train is moving to the right at constant speed along a straight, level track. Three spring scales are used to connect the cars as follows: spring scale A is located between the engine and the first car; scale B is between the first and second cars; scale C is between the second and third cars. (a) If air resistance and friction are negligible, what are the relative readings on the three spring scales A, B, and C? (b) Repeat part (a), taking air resistance and friction into consideration this time. [*Hint:* Draw an FBD for the car in the middle.] (c) If air resistance and friction together cause a force of magnitude 5.5 N on each car, directed toward the left, find the readings of scales A, B, and C.

137. Four *identical* spring scales, A, B, C, and D are used to hang a 220.0-N sack of potatoes. (a) Assume the scales have negligible weights and all four scales show the same reading. What is the reading of each scale? (b) Suppose that each scale has a weight of 5.0 N. If scales B and D show the same reading, what is the reading of each scale?

138. A computer weighing 87 N rests on the horizontal surface of your desk. The coefficient of friction between the computer and the desk is 0.60. (a) Draw an FBD for the computer. (b) What is the magnitude of the frictional force acting on the computer? (c) How hard would you have to push on it to get it to start to slide across the desk?

139. A refrigerator magnet weighing 0.14 N is used to hold up a photograph weighing 0.030 N. The magnet attracts the refrigerator door with a magnetic force of 2.10 N. (a) Identify the interactions between the magnet and other objects. (b) Draw an FBD for the magnet, showing all the forces that act on it. (c) Which of these forces are long-range and which are contact forces? (d) Find the magnitudes of all the forces acting on the magnet.

140. A 50.0-kg crate is suspended between the floor and the ceiling using two spring scales, one attached to the ceiling and one to the floor. If the lower scale reads 120 N, what is the reading of the upper scale? Ignore the weight of the scales.

141. Spring scale A is attached to the ceiling. A 10.0-kg mass is suspended from the scale. A second spring scale, B, is hanging from a hook at the bottom of the 10.0-kg mass and a 4.0-kg mass hangs from the second spring scale. (a) What are the readings of the two scales if the masses of the scales are negligible? (b) What are the readings if each scale has a mass of 1.0 kg?

142. A crate of oranges weighing 180 N rests on a flatbed truck 2.0 m from the back of the truck. The coefficients of friction between the crate and the bed are $\mu_s = 0.30$ and $\mu_k = 0.20$. The truck drives on a straight, level highway at a constant 8.0 m/s. (a) What is the force of friction acting on the crate? (b) If the truck speeds up with an acceleration of 1.0 m/s^2, what is the force of the friction on the crate? (c) What is the maximum acceleration the truck can have without the crate starting to slide?

143. A crate of books is to be put on a truck by rolling it up an incline of angle θ using a dolly. The total mass of the crate and the dolly is m. Assume that rolling the dolly up the incline is the same as sliding it up a frictionless surface. (a) What is the magnitude of the *horizontal* force that must be applied just to hold the crate in place on the incline? (b) What horizontal force must be applied to roll the crate up at constant speed? (c) In order to start the dolly moving, it must be accelerated from rest. What horizontal force must be applied to give the crate an acceleration up the incline of magnitude a? (tutorial: cart on ramp)

144. A toy cart of mass m_1 moves on frictionless wheels as it is pulled by a string under tension T. A block of mass m_2 rests on top of the cart. The coefficient of static friction between the cart and the block is μ. Find the maximum tension T that will not cause the block to slide on the

cart if the cart rolls on (a) a horizontal surface; (b) up a ramp of angle θ above the horizontal. In both cases, the string is parallel to the surface on which the cart rolls.

145. A helicopter of mass M is lowering a truck of mass m onto the deck of a ship. (a) At first, the helicopter and the truck move downward together (the length of the cable doesn't change). If their downward speed is decreasing at a rate of $0.10g$, what is the tension in the cable? (b) As the truck gets close to the deck, the helicopter stops moving downward. While it hovers, it lets out the cable so that the truck is still moving downward. If the truck's downward speed is decreasing at a rate of $0.10g$, while the helicopter is at rest, what is the tension in the cable?

146. The coefficient of static friction between block A and a horizontal floor is 0.45 and the coefficient of static friction between block B and the floor is 0.30. The mass of each block is 2.0 kg and they are connected together by a cord. (a) If a horizontal force $\vec{\mathbf{F}}$ pulling on block B is slowly increased, in a direction parallel to the connecting cord, until it is barely enough to make the two blocks start moving, what is the magnitude of $\vec{\mathbf{F}}$ at the instant that they start to slide? (b) What is the tension in the cord connecting blocks A and B at that same instant?

147. Tamar wants to cut down a large, dead poplar tree with her chain saw, but she does not want it to fall onto the nearby gazebo. Yoojin comes to help with a long rope. Yoojin, a physicist, suggests they tie the rope taut from the poplar to the oak tree and then pull *sideways* on the rope as shown in the figure. If the rope is 40.0 m long and Yoojin pulls sideways at the midpoint of the rope with a force of 360.0 N, causing a 2.00-m sideways displacement of the rope at its midpoint, what force will the rope exert on the poplar tree? Compare this with pulling the rope directly away from the poplar with a force of 360.0 N and explain why the values are different. [*Hint:* Until the poplar is cut through enough to start falling, the rope is in equilibrium.]

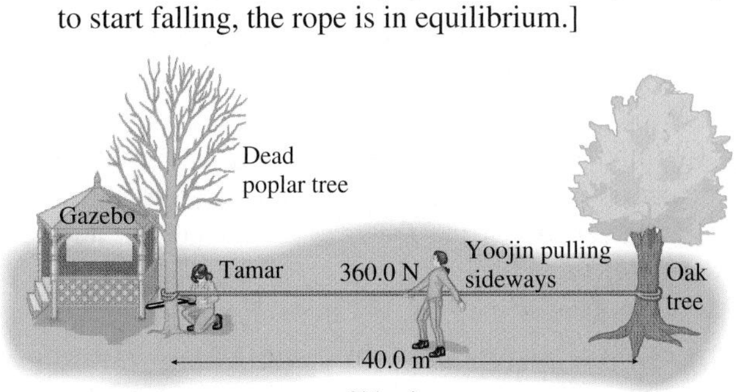

Side view

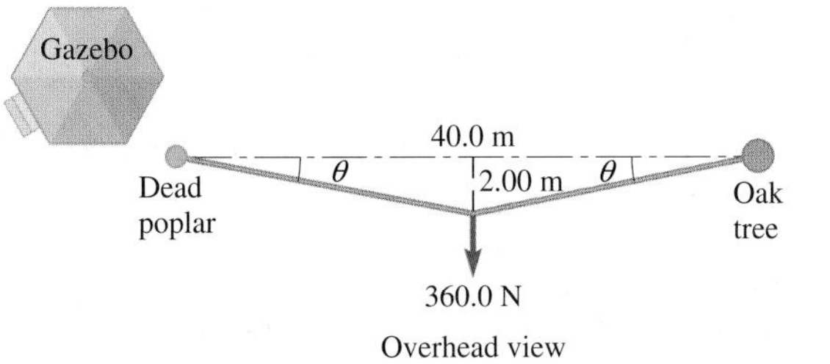

Overhead view

148. A student's head is bent over her physics book. The head weighs 50.0 N and is supported by the muscle force $\vec{\mathbf{F}}_m$ exerted by the neck extensor muscles and by the contact force $\vec{\mathbf{F}}_c$ exerted at the atlantooccipital joint. Given that the magnitude of $\vec{\mathbf{F}}_m$ is 60.0 N and is directed 35° below the horizontal, find (a) the magnitude and (b) the direction of $\vec{\mathbf{F}}_c$.

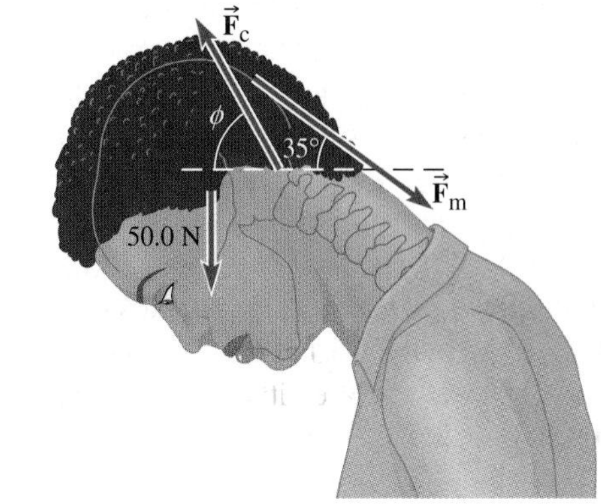

149. (a) If a spacecraft moves in a straight line between the Earth and the Sun, at what point would the force of gravity on the spacecraft due to the Sun be as large as that due to the Earth? (b) If the spacecraft is close to, but not at, this equilibrium point, does the net force on the spacecraft tend to push it toward or away from the equilibrium point? [*Hint:* Imagine the spacecraft a small distance d closer to the Earth and find out which gravitational force is stronger.]

150. While trying to decide where to hang a framed picture, you press it against the wall to keep it from falling. The picture weighs 5.0 N and you press against the frame with a force of 6.0 N at an angle of 40° from the vertical. (a) What is the direction of the normal force exerted on the picture by your hand? (b) What is the direction of the normal force exerted on the picture by the wall? (c) What is the coefficient of static friction between the wall and the picture? The frictional force exerted on the picture by the wall can have two possible directions. Explain why.

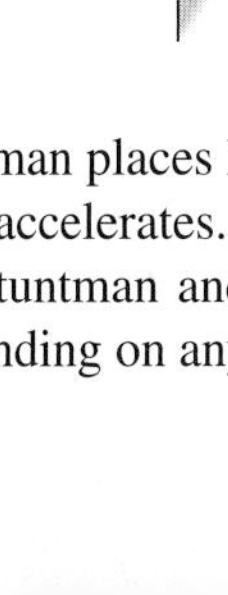

151. In a movie, a stuntman places himself on the front of a truck as the truck accelerates. The coefficient of friction between the stuntman and the truck is 0.65. The stuntman is not standing on anything but can "stick" to

the front of the truck as long as the truck continues to accelerate. What minimum forward acceleration will keep the stuntman on the front of the truck?

✦152. An airplane of mass 2800 kg has just lifted off the runway. It is gaining altitude at a constant 2.3 m/s while the horizontal component of its velocity is increasing at a rate of 0.86 m/s^2. Assume $g = 9.81$ m/s^2. (a) Find the direction of the force exerted on the airplane by the air. (b) Find the horizontal and vertical components of the plane's acceleration if the force due to the air has the same magnitude but has a direction 2.0° closer to the vertical than its direction in part (a).

✦153. In the figure with Problem 87, the block of mass m_1 slides to the right with coefficient of kinetic friction μ_k on a horizontal surface. The block is connected to a hanging block of mass m_2 by a light cord that passes over a light, frictionless pulley. (a) Find the acceleration of each of the blocks and the tension in the cord. (b) Check your answers in the special cases $m_1 \ll m_2$, $m_1 \gg m_2$, and $m_1 = m_2$. (c) For what value of m_2 (if any) do the two blocks slide at constant velocity? What is the tension in the cord in that case?

Answers to Practice Problems

4.1 (a) $F_x = 49.1$ N, $F_y = 2.9$ N; (b) $F = 49.2$ N; (c) 3.4° above the horizontal

4.2 0.5 kN downward

4.3 In the first case, the principle of inertia says that Negar tends to stay at rest with respect to the ground as the subway car begins to move forward, until forces acting on her (exerted by the strap and the floor) make her move forward. In the second case, Negar keeps moving forward with respect to the ground with constant speed as the subway car slows down, until forces acting on her make her slow down as well.

4.4 760 N, 81.7° above the $-x$-axis or 8.3° to the left of the $+y$-axis

4.5 The contact force exerted on the floor by the chest; 870 N, 59° below the rightward horizontal ($+x$-axis)

4.6 For $m_1 = m_2 = 1000$ kg and $r = 4$ m, $F \approx 4$ μN, which is about the same magnitude as the weight of a mosquito. The claim that this tiny force caused the collision is ridiculous.

4.7 0.57 N or 0.13 lb

4.8 The chest is in equilibrium, so the net force on it is zero. Setting the net force equal to zero separately for the horizontal and vertical components gives the answer: the normal force is 750 N, up, and the frictional force is 110 N, to the left. The quantity $\mu_s N$ is the *maximum* possible magnitude of the force of static friction for a surface. In this problem, the frictional force does not necessarily have the maximum possible magnitude.

4.9 (a)

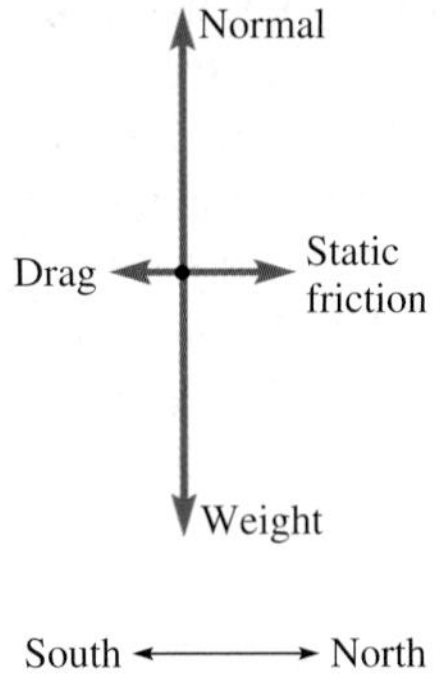

(b) Weight of the car = 11.0 kN; (c) 2.1 kN northward

4.10 (a) 110 N; (b) 230 N

4.11 3100 N

4.12

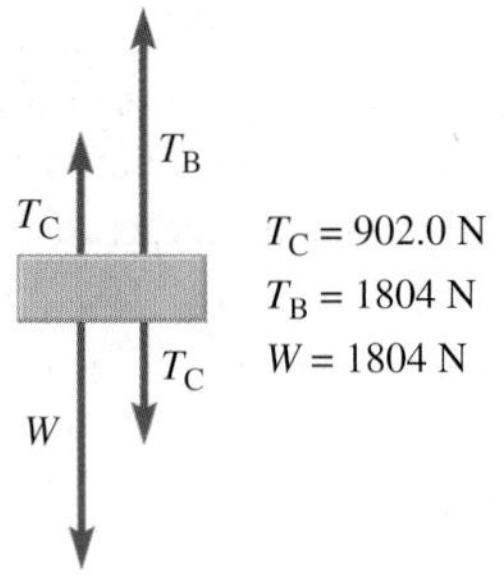

4.13 (a) 54 N; (b) 1.8 s

4.14 1.84 kN

4.15 Block 1: $\Sigma F_{1y} = T - m_1 g = 315\text{ N} - 255\text{ N} = 60\text{ N}$; $m_1 a_{1y} = 60$ N. Block 2: $\Sigma F_{2y} = m_2 g - T = 412\text{ N} - 315\text{ N} = 97\text{ N}$; $m_2 a_{2y} = 97$ N.

4.16 Impossible to pull the crate up with a single pulley. The entire weight of the crate would be supported by a single strand of cable and that weight exceeds the breaking strength of the cable.

4.17 2500 N

4.18 (a) down the incline; (b) up the incline; (c) 0.2 m/s^2 down the incline

4.19 (a) 392 N; (b) 431 N

Answers to Checkpoints

4.4 The two forces exerted by the two children on a toy cannot be interaction partners because they act on the *same* object (the toy), not on two different objects. Interaction partners act on different objects, one on each of the two objects that are interacting. The interaction partner of the force exerted by one child on the toy is the force that the toy exerts on that child.

4.5 The weight of the gear decreases as the value of g decreases. The mass of the gear does not change.

4.6 One upward normal force on each leg due to the floor and one downward normal force on the desktop due to the laptop.

4.8 Yes. For motion along an incline, it simplifies the problem to choose one axis parallel to the incline and the other perpendicular to the incline.

4.10 Your velocity is downward and decreasing in magnitude, so your acceleration is upward. Then the upward normal force exerted on you by the scale must be greater than your weight. The scale reading is greater than your weight.

CHAPTER

5 Circular Motion

German athlete Susanne Keil throws the hammer during the German Athletics championships. Keil qualified for the 2004 Olympics in Athens with a 67.77-m throw.

In the track and field event called the *hammer throw,* the "hammer" is actually a metal ball (mass 4.00 kg for women or 7.26 kg for men) attached by a cable to a grip. The athlete whirls the hammer several times around while not leaving a circle of radius 2.1 m and then releases it. The winner is the athlete whose hammer lands the greatest distance away. How large a force does an athlete have to exert on the grip to whirl the massive hammer around in a circle? What kind of path does the hammer follow once it is released? (See pp. 155–156 for the answer.)

Concepts & Skills to Review

- gravitational forces (Section 4.5)
- Newton's second law: force and acceleration (Sections 4.3 and 4.8)
- velocity and acceleration (Sections 2.2 and 2.3)
- apparent weight (Section 4.10)
- normal and frictional forces (Section 4.6)

5.1 DESCRIPTION OF UNIFORM CIRCULAR MOTION

Ask someone to name the most important machine ever invented by humans and you are likely to get the *wheel* as a response. Rotating objects are so important to modern—and even not-so-modern—technology that we barely notice them. Examples include wheels on cars, bicycles, trains, and lawnmowers; propellers on airplanes and helicopters; CDs and DVDs; computer hard drives; the gears and hands of an analog clock; amusement park rides and centrifuges—the list is endless.

Rotation of a Rigid Body To describe circular motion, we could use the familiar definitions of displacement, velocity, and acceleration. But much of the circular motion around us occurs in the rotation of a rigid object. A **rigid body** is one for which the distance between any two points of the body remains the same when the body is translated or rotated. When such an object rotates, every point on the object moves in a circular path. The radius of the path for any point is the distance between that point and the axis of rotation. When a compact disk spins inside a CD player, different points on the CD have different velocities and accelerations. The velocity and acceleration of a given point keep changing direction as the CD spins. It would be clumsy to describe the rotation of the CD by talking about the motion of arbitrary points on it. However, some quantities are the *same* for every point on the CD. It is much simpler, for instance, to say "the CD spins at 210 rpm" instead of saying "a point 6.0 cm from the rotation axis of the CD is moving at 1.3 m/s."

In a **rigid body,** the distance between any two points is constant.

The abbreviation rpm means *revolutions per minute.*

Angular Displacement and Angular Velocity To simplify the description of circular motion, we concentrate on *angles* instead of distances. If a CD spins through $\frac{1}{4}$ of a turn, every point moves through the same angle (90°), but points at different radii move different linear distances. On the CD shown in Fig. 5.1, point 1 near the axis of rotation moves through a smaller distance than point 4 on the circumference. For this reason we define a set of variables that are analogous to displacement, velocity, and acceleration, but use angular measure instead of linear distance. Instead of displacement, we speak of **angular displacement** $\Delta\theta$, the angle through which the CD turns. A point on the CD moves along the circumference of a circle. As the point moves from the angular position θ_i to the angular position θ_f, a radial line drawn between the center of the circle and that point sweeps out an angle $\Delta\theta = \theta_f - \theta_i$, which is the angular displacement of the CD during that time interval (Fig. 5.2).

CONNECTION:

Equations (5-1) through (5-3) have a familiar form because ω is the *rate of change* of θ, just as velocity is the rate of change of position.

Definition of angular displacement:

$$\Delta\theta = \theta_f - \theta_i \tag{5-1}$$

The sign of the angular displacement indicates the sense of the rotation. The usual convention is that a positive angular displacement represents counterclockwise rotation and a negative angular displacement represents clockwise rotation. Counterclockwise and clockwise are only well defined for a particular viewing direction; counterclockwise rotation viewed from above is clockwise when viewed from below.

+ Counterclockwise
– Clockwise

Remember that the notation $\lim_{\Delta t \to 0}$ indicates that $\Delta\theta$ is the angular displacement during a *very short* time interval Δt (short enough that the ratio $\Delta\theta/\Delta t$ doesn't change significantly if we make the time interval even shorter).

The **average angular velocity** ω_{av} is the average rate of change of the angular displacement.

Definition of average angular velocity:

$$\omega_{av} = \frac{\Delta\theta}{\Delta t} \qquad (5\text{-}2)$$

If we let the time interval Δt become shorter and shorter, we are averaging over smaller and smaller time intervals. In the limit $\Delta t \to 0$, ω_{av} becomes the **instantaneous angular velocity** ω.

Definition of instantaneous angular velocity:

$$\omega = \lim_{\Delta t \to 0} \frac{\Delta\theta}{\Delta t} \qquad (5\text{-}3)$$

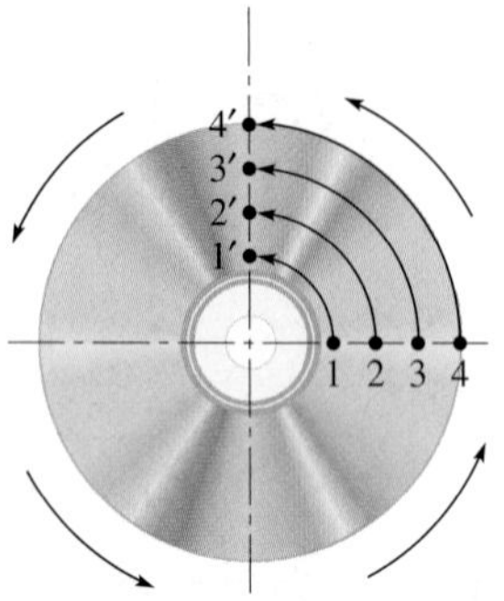

Figure 5.1 A CD rotates through $\frac{1}{4}$ turn; points 1, 2, 3, and 4 travel through the same angle but different distances to reach their new positions, marked 1′, 2′, 3′, and 4′, respectively.

The angular velocity also indicates—through its algebraic sign—in what direction the CD is spinning. Since angular displacements can be measured in degrees or radians, angular velocities have units such as degrees/second, radians/second, degrees/day, and the like.

Radian Measure You may be most familiar with measuring angles in degrees, but in many situations the most convenient measure is the **radian**. One such situation is when we relate the angular displacement or angular velocity of a rotating object with the distance traveled by, or the speed of, some point on the object.

In Fig. 5.3, an angle θ between two radii of a circle define an arc of length s. We say that θ is the angle *subtended* by the arc. The arc length is proportional to both the radius of the circle and to the angle subtended. The angle θ in radians is *defined* as

$$\theta \text{ (in radians)} = \frac{s}{r} \qquad (5\text{-}4)$$

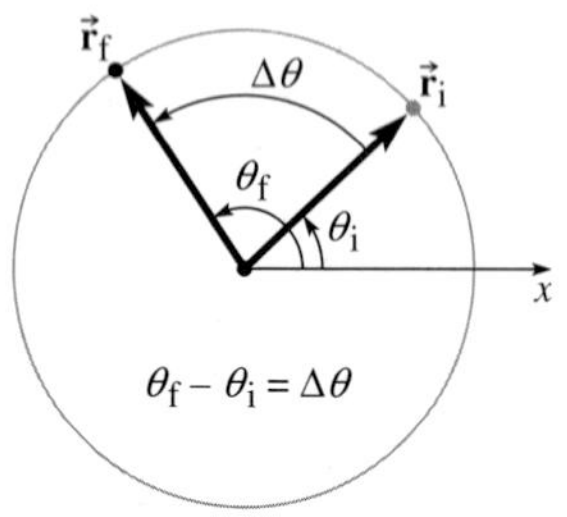

Figure 5.2 Angular positions such as θ_i and θ_f are measured counterclockwise from a reference axis (usually the x-axis).

where r is the radius of the circle. Since an angle in radians is defined by the ratio of two lengths, it is dimensionless (a pure number). We use the term radians, abbreviated "rad," to keep track of the angular measure used. Since "rad" is not a physical unit like meters or kilograms, it does not have to balance in Eq. (5-4). For the same reason, we can drop "rad" whenever there is no chance of being misunderstood. We can write $\omega = 23\ \text{s}^{-1}$ as long as context makes it clear that we mean 23 radians per second.

In equations that relate linear variables to angular variables [such as Eq. (5-4)], think of r as the number of meters of arc length per radian of angle subtended. In other words, think of r as having units of meters per radian. Doing so, the radians cancel out in these equations. For example, if $\theta = 2.0$ rad and $r = 1.2$ m, then the arc length is

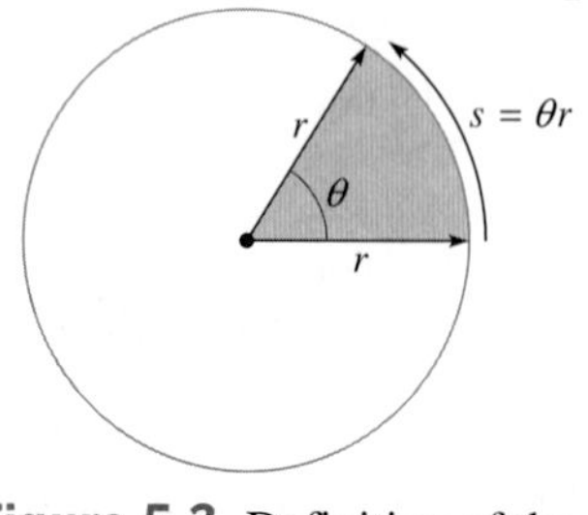

Figure 5.3 Definition of the radian: angle θ in radians is the arc length s divided by the radius r. The angle shown is 1 rad ≈ 57.3°.

$$s = \theta r = 2.0\ \cancel{\text{rad}} \times 1.2\ \frac{\text{m}}{\cancel{\text{rad}}} = 2.4\ \text{m}$$

Since the arc length for an angle of 360° is the circumference of the circle, the radian measure of an angle of 360° is

$$\theta = \frac{s}{r} = \frac{2\pi r}{r} = 2\pi \text{ rad}$$

Therefore, the conversion factor between degrees and radians is

$$360° = 2\pi \text{ rad} \qquad (5\text{-}5)$$

Example 5.1

Angular Speed of Earth

Earth is rotating about its axis. What is its angular speed in rad/s? (The question asks for angular *speed*, so we do not have to worry about the direction of rotation.)

Strategy The Earth's angular velocity is constant, or nearly so. Therefore, we can calculate the average angular velocity for any convenient time interval and, in turn, the Earth's instantaneous angular speed $|\omega|$.

Solution It takes the Earth 1 day to complete one rotation, during which the angular displacement is 2π rad. More formally, during a time interval $\Delta t = 1$ day, the angular displacement of the Earth is $\Delta\theta = 2\pi$ rad. So the angular speed of the Earth is 2π rad/day, and then convert days to seconds.

$$1 \text{ day} = 24 \text{ h} = 24 \text{ h} \times 3600 \text{ s/h} = 86\,400 \text{ s}$$

$$|\omega| = \frac{2\pi \text{ rad}}{86\,400 \text{ s}} = 7.3 \times 10^{-5} \text{ rad/s}$$

Discussion Notice that this problem is analogous to a problem in linear motion such as: "A car travels in a straight line at constant speed. In 3 h, it has traveled 192 mi. What is its velocity in m/s?" Just about everything in circular motion and rotation has this kind of analog—which means we can draw heavily on what we have already learned.

Earth actually completes one rotation in 23.9345 h (see inside back cover) rather than in 24 h due to Earth's motion around the Sun. This distinction would be important only if we needed a more precise value of $|\omega|$ (more than two significant figures).

Practice Problem 5.1 Angular Speed of Venus

Venus completes one rotation about its axis every 5816 h. What is the angular speed of the rotation of Venus in rad/s?

Relation Between Linear and Angular Speed

For a point moving in a circular path of radius r, the linear distance traveled along the circular path during an angular displacement of $\Delta\theta$ (in radians) is the arc length s where

$$s = r|\Delta\theta| = r|\theta_f - \theta_i| \quad \text{(angles in radians)} \qquad (5\text{-}6)$$

The point in question could be a point particle moving in a circular path, or it could be any point on a rotating rigid object. Since Eq. (5-6) comes directly from the definition of the radian, any equation derived from Eq. (5-6) is valid only when the angles are measured in radians.

What is the linear speed at which the point moves? The average linear speed is the distance traveled divided by the time interval,

$$v_{av} = \frac{s}{\Delta t} = \frac{r|\Delta\theta|}{\Delta t} \quad (\Delta\theta \text{ in radians})$$

We recognize $\Delta\theta/\Delta t$ as the average angular velocity ω_{av}. If we take the limit as Δt approaches zero, both average quantities (v_{av} and ω_{av}) become instantaneous quantities. Therefore, the relationship between linear speed and angular speed is

$$v = r|\omega| \quad (\omega \text{ in radians per unit time}) \qquad (5\text{-}7)$$

Equation (5-7) relates only the *magnitudes* of the linear and angular speeds. The direction of the velocity vector $\vec{\mathbf{v}}$ is tangent to the circular path. For a rotating object, points farther from the axis move at higher linear speeds; they have a circle of bigger radius to travel and, therefore, cover more distance in the same time interval. For example, a person standing at the equator has a much higher linear speed due to Earth's rotation than does a person standing at the Arctic Circle (see Fig. 5.4).

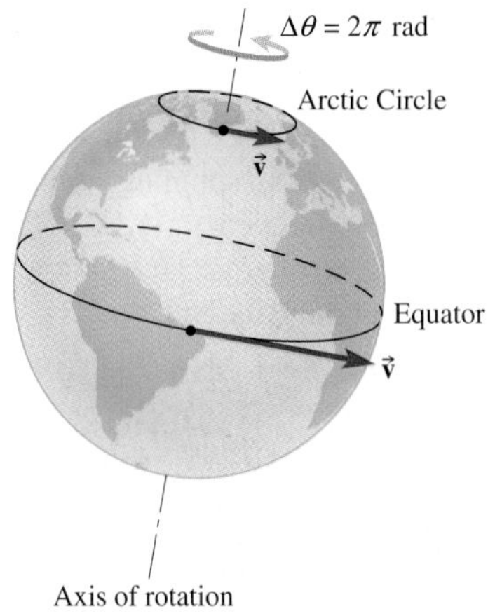

Figure 5.4 A person standing at the Equator is moving much faster than another person standing at the Arctic Circle, but their *angular* speeds are the same.

In uniform circular motion, speed is constant but velocity is *not* constant because the *direction* of the velocity is changing.

Period and Frequency

When the speed of a point moving in a circle is constant, its motion is called **uniform circular motion**. Even though the speed of the point is constant, the velocity is not: the direction of the velocity is changing. This distinction is important when we find the acceleration of an object in uniform circular motion (Section 5.2). The time for the point to travel completely around the circle is called the **period** of the motion, T. The **frequency** of the motion, which is the number of revolutions per unit time, is defined as

$$f = \frac{1}{T} \tag{5-8}$$

since

$$\frac{\text{revolutions}}{\text{second}} = \frac{1}{\text{second/revolution}}$$

CHECKPOINT 5.1

If it takes $\frac{1}{7200}$ of a second for a computer hard drive to spin around once, what is its frequency?

The speed is the total distance traveled divided by the time taken,

$$v = \frac{2\pi r}{T} = 2\pi r f$$

Then, for uniform circular motion

$$|\omega| = \frac{v}{r} = 2\pi f \quad (\omega \text{ in radians per unit time}) \tag{5-9}$$

SI unit of frequency: 1 Hz = 1 rev/s

where, in SI units, angular velocity ω is measured in rad/s and frequency f is measured in hertz (Hz). The hertz is a derived unit equal to 1 rev/s. The dimensions of Eq. (5-9) are correct since both revolutions and radians are pure numbers. The physical dimensions on both sides are a number per second (s^{-1}).

Example 5.2

Speed in a Centrifuge

A centrifuge is spinning at 5400 rpm. (a) Find the period (in s) and frequency (in Hz) of the motion. (b) If the radius of the centrifuge is 14 cm, how fast (in m/s) is an object at the outer edge moving?

Strategy Remember that rpm means *revolutions per minute.* 5400 rpm *is* the frequency, but in a unit other than Hz. After a unit conversion, the other quantities can be found using the relations already discussed.

Solution (a) First convert rpm to Hz:

$$f = 5400\ \frac{\text{rev}}{\text{min}} \times \frac{1\ \text{min}}{60\ \text{s}} = 90\ \text{rev/s}$$

so the frequency is $f = 90\ \text{Hz} = 90\ \text{s}^{-1}$. The period is

$$T = 1/f = 0.011\ \text{s}$$

(b) To find the linear speed, we first find the angular speed in rad/s:

$$|\omega| = 90\ \frac{\text{rev}}{\text{s}} \times 2\pi\ \frac{\text{rad}}{\text{rev}} = 180\pi\ \text{rad/s}$$

continued on next page

Example 5.2 continued

So $|\omega| = 2\pi f = 180\pi$ rad/s. The linear speed is

$$v = |\omega|r = 180\pi\ \text{s}^{-1} \times 0.14\ \text{m} = 79\ \text{m/s}$$

Discussion Notice that much of this problem was done with unit conversions. Instead of memorizing a formula such as $|\omega| = 2\pi f$, an understanding of where the formula came from (in this case, that 2π radians correspond to one revolution) is more useful and less prone to error.

Practice Problem 5.2 Clothing in the Drier

An automatic clothing drier spins at 51.6 rpm. If the radius of the drier drum is 30.5 cm, how fast is the outer edge of the drum moving?

Rolling Without Slipping: Rotation and Translation Combined

When an object is rolling, it is both rotating and translating. The wheel rotates about an axle, but the axle is not at rest; it moves forward or backward. What is the relationship between the angular speed of the wheel and the linear speed of the axle? You might guess that $v = |\omega|r$ is the answer. You would be right, as long as the object rolls without slipping or skidding.

There is no fixed relationship between the linear and angular speeds of a wheel if it is allowed to skid or slip. When an impatient driver guns the engine the instant a traffic light turns green, the automobile wheels are likely to slip. The rubber sliding against the road surface makes the squealing sound and leaves tracks on the road. The driver could actually make the acceleration of the car greater by giving the engine *less* gas. When the wheels are skidding or slipping, *kinetic* friction propels the car forward instead of the potentially larger force of *static* friction.

For a wheel that rolls *without* slipping, as the wheel turns through one complete rotation, the axle moves a distance equal to the circumference of the wheel (Fig. 5.5). Think of a paint roller leaving a line of paint as it rolls along a wall. After one complete

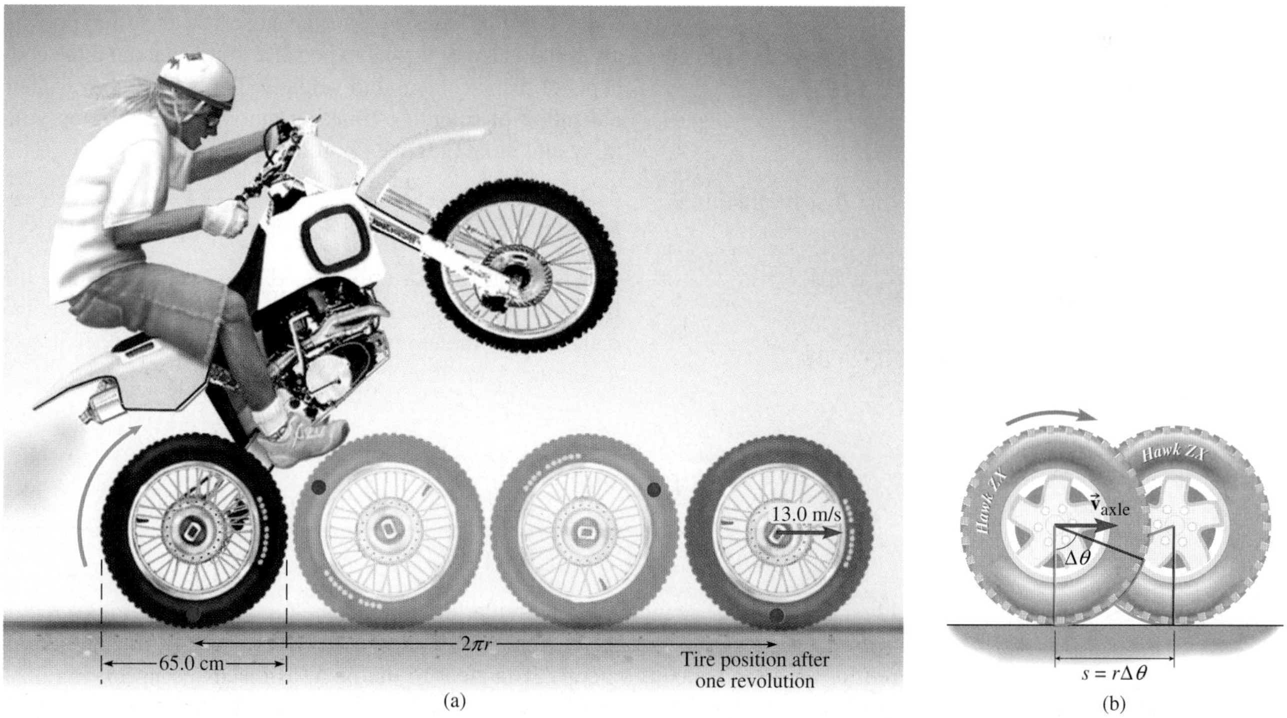

Figure 5.5 (a) As a wheel of radius r that rolls without slipping turns through one complete revolution, the distance its axle moves is equal to the circumference of the wheel ($2\pi r$). (b) As a wheel rolls without slipping through an angle $\Delta\theta$, the distance the axle moves is equal to the arc length s.

rotation, the same point on the roller wheel is touching the wall as was initially touching it. The length of the line of paint is $2\pi r$. The elapsed time is T, so the axle's speed is

$$v_{\text{axle}} = \frac{2\pi r}{T}$$

while the angular speed of the roller is

$$|\omega| = \frac{2\pi}{T}$$

Thus,

$$v_{\text{axle}} = |\omega| r \quad (\omega \text{ in radians per unit time}) \tag{5-10}$$

Example 5.3

Angular Speed of a Rolling Wheel

Kevin is riding his motorcycle at a speed of 13.0 m/s. If the diameter of the rear tire is 65.0 cm, what is the angular speed of the rear wheel? Assume that it rolls without slipping.

Strategy The given diameter of the tire enables us to find the circumference and, thus, the distance traveled in one revolution of the wheel. From the speed of the motorcycle we can find how many revolutions the tire must make per second.

Solution During one revolution of the wheel, the motorcycle travels a distance equal to the tire's circumference $2\pi r$ (Fig. 5.5). Then the time to make one revolution is T and the speed v is

$$v = \frac{\text{distance}}{\text{time}} = \frac{2\pi r}{T}$$

Therefore, $T = 2\pi r/v$. For each revolution there is an angular displacement of $\Delta\theta = 2\pi$ radians, so

$$|\omega| = \frac{|\Delta\theta|}{\Delta t} = \frac{2\pi}{T}$$

Substituting $T = 2\pi r/v$ and remembering that the radius is half the diameter,

$$|\omega| = \frac{2\pi}{2\pi r/v} = \frac{v}{r} = \frac{13.0 \text{ m/s}}{(0.650 \text{ m})/2} = 40.0 \frac{\text{rad}}{\text{s}}$$

Discussion Check: Time for one revolution is

$$\frac{2\pi \text{ rad}}{40.0 \text{ rad/s}} = 0.157 \text{ s.}$$

Time to travel a distance $2\pi r = 2.04$ m is

$$\frac{2.04 \text{ m}}{13.0 \text{ m/s}} = 0.157 \text{ s.}$$

Looks good.

You could have obtained this answer immediately by looking back through the text for the equation $|\omega| = v/r$ and plugging in numbers, but the solution here shows that you can re-create that equation. Here, and in many cases, there is no need to memorize a formula if you understand the concepts behind the formula. You are then less apt to make a mistake by forgetting a factor or constant in the equation, or by using an inappropriate formula. For another example, if an object moves along a straight line at a constant velocity, you know instantly that the displacement is the velocity times the time interval—not because you have memorized an equation ($\Delta\vec{\mathbf{r}} = \vec{\mathbf{v}}\,\Delta t$), but because you understand the concepts of displacement and velocity. This is the sort of internalization of scientific thinking that you will develop with more and more practice in problem solving.

Practice Problem 5.3 Rolling Drum

A cylindrical steel drum is tipped over and rolled along the floor of a warehouse. If the drum has a radius of 0.40 m and makes one complete turn every 8.0 s, how long does it take to roll the drum 36 m?

5.2 RADIAL ACCELERATION

For a particle undergoing uniform circular motion, the *magnitude* of the velocity vector is constant, but its direction is continuously changing. At any instant of time, the direction of the instantaneous velocity is tangent to the path, as discussed in Section 3.2. Since the *direction* of the velocity continually changes, the particle has a nonzero acceleration.

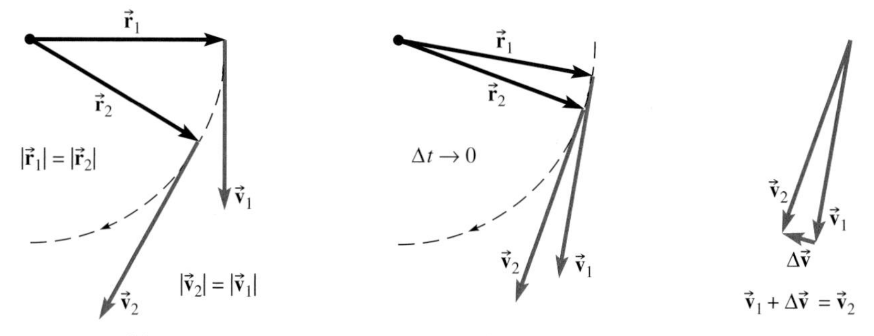

Figure 5.6 Uniform circular motion at constant speed. (a) The velocity vector is always tangent to the circular path and perpendicular to the radius at that point. (b) As the time interval between two velocity measurements decreases, the angle between the velocity vectors decreases. (c) The change in velocity ($\Delta\vec{v}$) is found by placing the tails of the two velocity vectors together. Then $\Delta\vec{v}$ is drawn from the tip of the initial velocity ($\vec{v}_1$) to the tip of the final velocity ($\vec{v}_2$) so that $\vec{v}_1 + \Delta\vec{v} = \vec{v}_2$.

In Fig. 5.6a, two velocity vectors of equal magnitude are drawn tangent to a circular path of radius r, representing the velocity at two different times of an object moving around a circular path with constant speed. At any instant, the velocity vector is perpendicular to a radius drawn from the center of the circle to the position of the object. As the time between velocity measurements approaches zero, the radii become closer together (Fig. 5.6b). To find the acceleration, $\vec{a} = \lim_{\Delta t \to 0} \frac{\Delta\vec{v}}{\Delta t}$, we must first find the change in velocity $\Delta\vec{v}$ for a very short time interval. Figure 5.6c shows that as the time interval Δt approaches zero, the angle between the two velocities also approaches zero and $\Delta\vec{v}$ becomes perpendicular to the velocity.

Since $\Delta\vec{v}$ is perpendicular to the velocity, it is directed along a radius of the circle. Inspection of Figs. 5.6b and 5.6c shows that $\Delta\vec{v}$ is radially *inward* (toward the center of the circle). Since the acceleration $\vec{a}$ has the same direction as $\Delta\vec{v}$ (in the limit $\Delta t \to 0$), the acceleration is also directed radially inward (Fig. 5.7)—that is, along a radius of the circular path toward the center of the circle. The acceleration of an object undergoing *uniform* circular motion is often called the **radial acceleration $\vec{a}_r$**. The word *radial* here just reminds us of the direction of the acceleration. (A synonym for radial acceleration is *centripetal acceleration. Centripetal* means "toward the center.")

CONNECTION:

Radial acceleration is not a new kind of acceleration. The acceleration vector for an object moving in uniform circular motion is directed radially inward toward the center of the circle.

In uniform circular motion, the direction of the acceleration is radially inward (toward the center of the circular path).

CHECKPOINT 5.2

Does a radial acceleration mean that the speed of the object is changing?

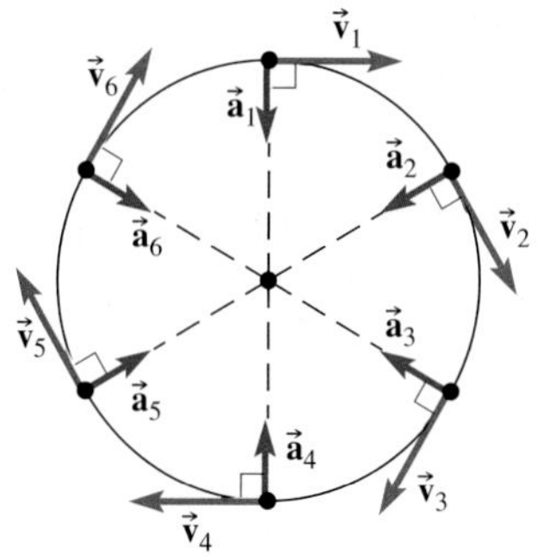

Figure 5.7 In uniform circular motion, the acceleration is always directed toward the center of the circle, perpendicular to the velocity (see interactive: circular motion).

Magnitude of the Radial Acceleration

To find the magnitude of the radial acceleration for uniform circular motion, we must find the change in velocity $\Delta\vec{v}$ for a time interval Δt in the limit $\Delta t \to 0$. The velocity keeps the same magnitude but changes direction at a steady rate, equal to the angular velocity ω. In a time interval Δt, the velocity $\vec{v}$ rotates through an angle equal to the angular displacement $\Delta\theta = \omega\Delta t$. During this time interval, the velocity vector sweeps out an arc of a circle of "radius" v (Fig. 5.8). In the limit $\Delta t \to 0$, the

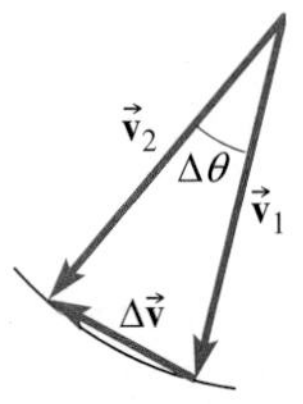

Figure 5.8 The velocity vector sweeps out an arc of a circle whose "length" is nearly equal to that of the chord $\Delta\vec{v}$.

magnitude of $\Delta\vec{\mathbf{v}}$ becomes equal to the arc length, since a very short arc approaches a straight line. Then

$$|\Delta\vec{\mathbf{v}}| = \text{arc length} = \text{radius of circle} \times \text{angle subtended}$$
$$= v\,|\Delta\theta| = v\,|\omega|\,\Delta t$$

Acceleration is the rate of change of velocity, so the magnitude of the radial acceleration is

$$a_r = |\vec{\mathbf{a}}| = \frac{|\Delta\vec{\mathbf{v}}|}{\Delta t} = v|\omega| \quad (\omega \text{ in radians per unit time}) \qquad (5\text{-}11)$$

where absolute value symbols are used with the vector quantities to indicate their magnitudes. Velocity and angular velocity are not independent; $v = |\omega|r$. It is usually most convenient to write the magnitude of the radial acceleration in terms of one or the other of these two quantities. So we write the radial acceleration in two other equivalent ways using $v = |\omega|r$:

$$a_r = \frac{v^2}{r} \quad \text{or} \quad a_r = \omega^2 r \quad (\omega \text{ in radians per unit time}) \qquad (5\text{-}12)$$

Note that Eqs. (5-11) and (5-12) assume that ω is expressed in *radians* per unit time (normally rad/s, but rad/min or rad/h would be correct).

Example 5.4

A Spinning CD

If a CD spins at 210 rpm, what is the radial acceleration of a point on the outer rim of the CD? The CD is 12 cm in diameter.

Strategy From the number of revolutions per minute, we can find the frequency and the angular velocity. The angular velocity and the radius of the CD enable us to calculate the radial acceleration.

Solution We convert 210 rpm into a frequency in revolutions per second (Hz).

$$f = 210\,\frac{\text{rev}}{\text{min}} \times \frac{1}{60}\,\frac{\text{min}}{\text{s}} = 3.5\,\frac{\text{rev}}{\text{s}} = 3.5\text{ Hz}$$

For each revolution, the CD rotates through an angle of 2π radians. The angular velocity is

$$|\omega| = 2\pi f = 2\pi\,\frac{\text{radians}}{\text{rev}} \times 3.5\,\frac{\text{rev}}{\text{s}} = 7.0\pi\text{ rad/s}$$

Then using Eq. (5-12), the radial acceleration is

$$a_r = \omega^2 r = (7.0\pi\text{ rad/s})^2 \times 0.060\text{ m} = 29\text{ m/s}^2$$

Discussion When finding the radial acceleration, use whichever form of Eq. (5-12) is more convenient. For rotating objects such as the spinning CD, it's usually easiest to think in terms of the angular velocity. For an object moving around a circle, such as a satellite in orbit whose speed is known, it might be easier to use v^2/r. Since the two equations are equivalent, either can be used in any situation.

Practice Problem 5.4 Radial Acceleration of a Point on an Old Record

What is the radial acceleration of a point 25.4 cm from the center of a record that is rotating at 78 rpm on a turntable?

Applying Newton's Second Law to Uniform Circular Motion

Now that we know the magnitude and direction of the acceleration of any object in uniform circular motion, we can use Newton's second law to relate the net force acting on the object to the speed and radius of its motion. The net force is found in the usual way: each of the individual forces acting on the object is identified and then the forces are added as vectors. Every force acting must be exerted *by some other object.* Resist the temptation to add in a new, separate force just because something moves in a circle. For an object to move in a circle at constant speed, real, physical forces such as gravity, tension, normal forces, and friction must act on it; these forces combine to produce a net force that has the correct magnitude and is always perpendicular to the velocity of the object.

Problem-Solving Strategy for an Object in Uniform Circular Motion

1. Begin as for any Newton's second law problem: identify all the forces acting on the object and draw an FBD.
2. Choose perpendicular axes at the point of interest so that one is radial and the other is tangent to the circular path.
3. Find the radial component of each force.
4. Apply Newton's second law as follows:

$$\sum F_r = ma_r$$

where ΣF_r is the radial component of the net force and the radial component of the acceleration is

$$a_r = \frac{v^2}{r} = \omega^2 r$$

(For uniform circular motion, neither the net force nor the acceleration has a tangential component.)

Example 5.5

The Hammer Throw

What force does the athlete exert on the grip? What path does the hammer follow after release?

An athlete whirls a 4.00-kg hammer six or seven times around and then releases it. Although the purpose of whirling it around several times is to increase the hammer's speed, assume that *just before* the hammer is released, it moves at constant speed along a circular arc of radius 1.7 m. At the instant she releases the hammer, it is 1.0 m above the ground and its velocity is directed 40° above the horizontal. The hammer lands a horizontal distance of 74.0 m away. What force does the athlete apply to the grip just before she releases it? Ignore air resistance.

Strategy After release, the only force acting on the hammer is gravity. The hammer moves in a parabolic trajectory like any other projectile. By analyzing the projectile motion of the hammer, we can find the speed of the hammer just after its release. Just *before* release, the forces acting on the hammer are the tension in the cable and gravity. We can relate the net force on the hammer to its radial acceleration, calculated from the speed and radius of its path. The problem becomes two subproblems, one dealing with circular motion and the other with projectile motion. The final velocity for the circular motion is the initial velocity for the projectile motion.

Solution During its projectile motion, the initial velocity has magnitude v_i (to be determined) and direction $\theta = 40°$ above the horizontal. Choosing the $+y$-axis pointing up, the displacement of the hammer (in component form) is $\Delta x = 74.0$ m and $\Delta y = -1.0$ m (Fig. 5.9), the acceleration of the hammer is $a_x = 0$ and $a_y = -g$, and the initial velocity is $v_{ix} = v_i \cos\theta$ and $v_{iy} = v_i \sin\theta$. Then, from Eqs. (4-8) and (4-9),

$$\Delta x = (v_i \cos\theta)\,\Delta t \quad \text{and} \quad \Delta y = (v_i \sin\theta)\,\Delta t - \tfrac{1}{2}g(\Delta t)^2$$

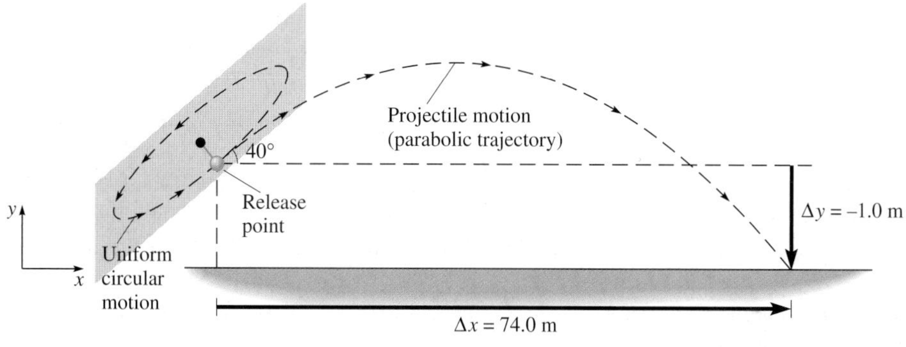

Figure 5.9
Path of the hammer from just before its release until it hits the ground. (Distances are *not* to scale.)

continued on next page

Example 5.5 continued

Solving the left equation for Δt and substituting into the right equation gives

$$\Delta y = \cancel{v_i} \sin\theta \, \frac{\Delta x}{\cancel{v_i}\cos\theta} - \frac{1}{2} g\left(\frac{\Delta x}{v_i \cos\theta}\right)^2$$

After a bit of algebra, we can solve for v_i. First we multiply through by $2v_i^2 \cos^2\theta$:

$$2v_i^2 \cos^2\theta\,\Delta y = 2v_i^2 \cos^2\theta\,\frac{\Delta x \sin\theta}{\cos\theta} - \frac{\cancel{2}\,v_i^2 \cancel{\cos^2\theta}}{\cancel{2}}\, g\left(\frac{\Delta x}{\cancel{v_i \cos\theta}}\right)^2$$

Subtracting the first term on the right side from both sides and factoring out v_i^2,

$$v_i^2(2\,\Delta y \cos^2\theta - 2\,\Delta x \cos\theta \sin\theta) = -g(\Delta x)^2$$

Now we solve for v_i:

$$v_i = \sqrt{\frac{g(\Delta x)^2}{2\Delta x \cos\theta \sin\theta - 2\Delta y \cos^2\theta}}$$

$$= \sqrt{\frac{9.80\ \text{m/s}^2 \times (74.0\ \text{m})^2}{2(74.0\ \text{m})\cos 40° \sin 40° - 2(-1.0\ \text{m})\cos^2 40°}}$$

$$= 26.9\ \text{m/s}$$

The net force on the hammer can be found from Newton's second law. The two forces acting on the hammer are due to the tension in the cable and to gravity (Fig. 5.10). We ignore the gravitational force, assuming that the hammer's weight is small compared with the tension in the cable. Then the tension in the cable is the only significant force acting on the hammer. Assuming uniform circular motion, the cable pulls radially inward and causes a radial acceleration of magnitude v^2/r. Newton's second law in the radial direction is

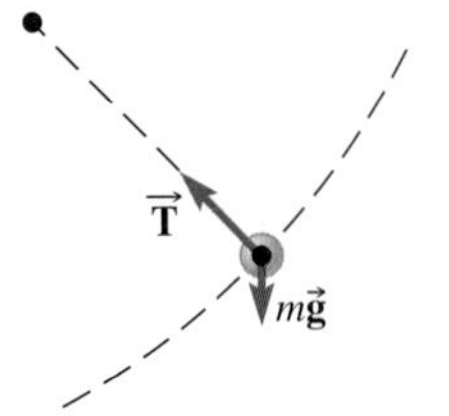

Figure 5.10
FBD for the hammer just before its release. (Not to scale.)

$$\sum F_r = T = ma_r = \frac{mv^2}{r}$$

Substituting numerical values,

$$T = \frac{4.00\ \text{kg} \times (26.9\ \text{m/s})^2}{1.7\ \text{m}} = 1700\ \text{N}$$

The tension is much larger than the weight of the hammer ($\approx$40 N), so the assumption that we could ignore the weight is justified. The athlete must apply a force of magnitude 1700 N—almost 400 lb—to the grip.

Discussion This example demonstrates the cumulative nature of physics concepts. The basic concepts keep reappearing, to be used over and over and to be extended for use in new contexts. Part of the problem involves new concepts (radial acceleration); the rest of the problem involves old material (Newton's second law, projectile motion, and tension in a cord).

Practice Problem 5.5 Rotating Carousel

A horse located 8.0 m from the central axis of a rotating carousel moves at a speed of 6.0 m/s. The horse is at a fixed height (it does not move up and down). What is the net force acting on a child seated on this horse? The child's weight is 130 N.

Example 5.6

Conical Pendulum

Suppose you whirl a stone in a horizontal circle at a slow speed so that the weight of the stone is *not* negligible compared with the tension in the cord. Then the cord cannot be horizontal—the tension must have a vertical component to cancel the weight and leave a horizontal net force (Fig. 5.11). If the cord has length L, the stone has mass m, and the cord makes an angle ϕ with the vertical direction, what is the constant angular speed of the stone?

Strategy The net force must point toward the center of the circle, since the stone is in uniform circular motion. With the stone in the position depicted in Fig. 5.11a, the direction of the net force is along the $+x$-axis. This time the tension in the cord does not pull toward the center, but the *net* force does.

Solution Start by drawing an FBD (Fig. 5.11b). Now apply Newton's second law in component form. The acceleration has components $a_x = \omega^2 r$ and $a_y = 0$. For the x-components,

$$\sum F_x = T \sin\phi = ma_x = m\omega^2 r$$

continued on next page

Example 5.6 continued

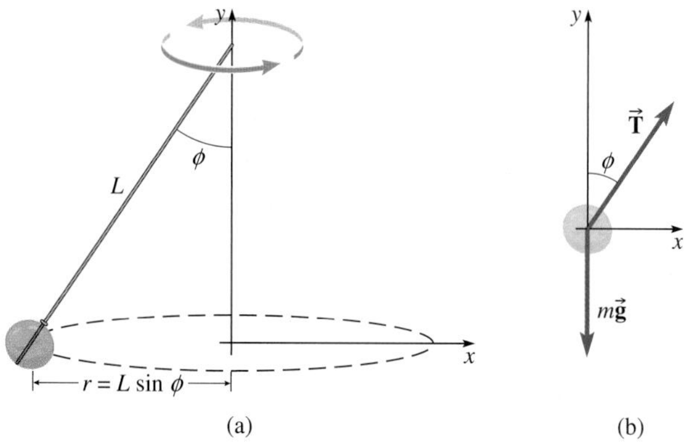

Figure 5.11

(a) A stone is whirled in a horizontal circle of radius $r = L \sin \phi$. (b) An FBD for the stone.

Since the problem does not specify r, we must express r in terms of L and ϕ. In Fig. 5.11a, the radius forms a right triangle with the cord and the y-axis. Then

$$r = L \sin \phi$$

and

$$\sum F_x = T \cancel{\sin \phi} = m\omega^2 L \cancel{\sin \phi}$$

Therefore, $T = m\omega^2 L$. For the y-components,

$$\sum F_y = T \cos \phi - mg = ma_y = 0 \quad \Rightarrow \quad T \cos \phi = mg$$

Now we eliminate the tension:

$$(m\omega^2 L) \cos \phi = mg$$

Solving for $|\omega|$,

$$|\omega| = \sqrt{\frac{g}{L \cos \phi}}$$

Discussion We should check the dimensions of the final expression. Since $\cos \phi$ is dimensionless,

$$\sqrt{\frac{[\mathrm{L/T^2}]}{[\mathrm{L}]}} = \frac{1}{[\mathrm{T}]}$$

which is correct for ω (SI unit rad/s).

Another check is to ask how ω and ϕ are related for a given length cord. As ϕ increases toward 90°, the cord gets closer to horizontal and the radius increases. In our expression, as ϕ increases, $\cos \phi$ decreases and, therefore, ω increases, in accordance with experience: the stone would have to be whirled faster and faster to make the cord more nearly horizontal.

Conceptual Practice Problem 5.6 Conical Pendulum on the Moon

Examine the result of Example 5.6 to see how ω depends on g, all other things being equal. Where the gravitational field is weaker, do you have to whirl the stone faster or more slowly to keep the cord at the same angle ϕ? Is that in accord with your intuition?

5.3 UNBANKED AND BANKED CURVES

Unbanked Curves When you drive an automobile in a circular path along an unbanked roadway, friction acting on the tires due to the pavement acts to keep the automobile moving in a curved path. This frictional force acts *sideways*, toward the center of the car's circular path (Fig. 5.12). The frictional force might also have a tangential component; for example, if the car is braking, a component of the frictional force makes the car slow down by acting backward (opposite to the car's velocity). For now we assume that the car's speed is constant and that the forward or backward component of the frictional force is negligibly small.

As long as the tires roll without slipping, there is no relative motion between the bottom of the tires and the road, so it is the force of *static* friction that acts (see Section 4.6). If the car is in a skid, then it is the smaller force of kinetic friction that acts as the bottom portion of the tire slides along the pavement. As the speed of the car increases, or for slippery surfaces with low coefficients of friction, the static frictional force may not be enough to hold the car in its curved path.

Application of radial acceleration and contact forces: banked roadways

Banked Curves To help prevent cars from going into a skid or losing control, the roadway is often banked (tilted at a slight angle) around curves so that the outer portion of the road—the part farthest from the center of curvature—is higher than the inner portion. Banking changes the angle and magnitude of the normal force, $\vec{\mathbf{N}}$, so that it has a horizontal component N_x directed toward the center of curvature (in the

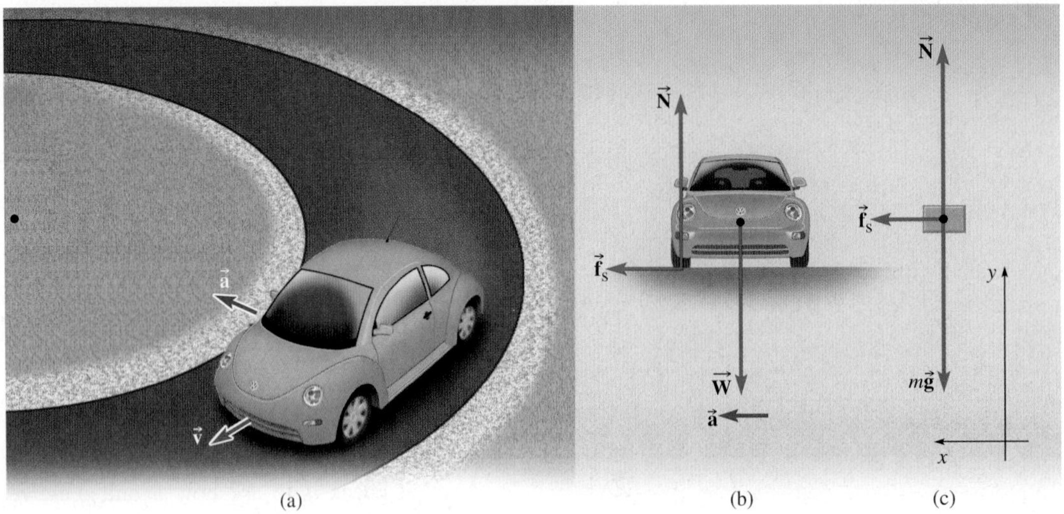

Figure 5.12 (a) A car negotiating a curve at constant speed on an *unbanked* roadway. The car's acceleration is toward the center of the circular path. (b) A head-on view of the same car. The center of the circular path is to the left as viewed here. The force vectors $\vec{\mathbf{N}}$ and $\vec{\mathbf{f}}_s$ are shown acting on one tire, but they represent the *total* normal and frictional forces acting on all four tires. (c) FBD for the car.

radial direction—see Fig. 5.13). Then we need no longer rely solely on friction to keep the car moving in a circular path as it negotiates the curve; this component of the normal force acts to help the car remain on the curved path. Figure 5.13 shows a banked road with the normal force, the gravitational force, and, in parts (b) and (c), the radial component of the normal force N_x. We choose the axes so that the x-axis is in the direction of the acceleration, which is to the left; the axes are *not* parallel and perpendicular to the incline.

Figure 5.13 (a) Head-on view of a car negotiating a curve at constant speed on a *banked* roadway. The car's acceleration is toward the center of the circular path (to the left as viewed here). $\vec{\mathbf{N}}$ represents the *total* normal force acting on all four tires. The car moves at just the right speed so that the frictional force is zero. (b) Resolving the normal force into x- and y-components. (c) FBD for the car with the normal force represented by its components.

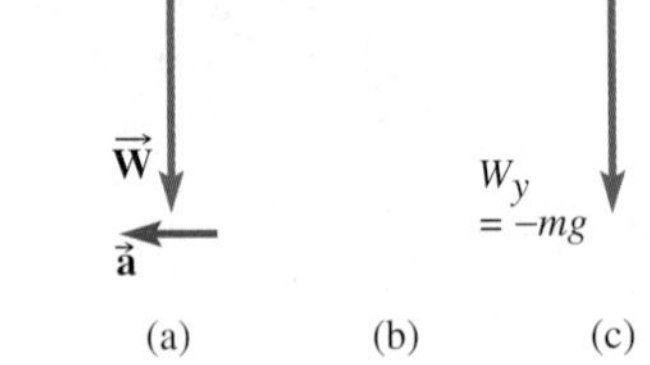

Example 5.7

A Possible Skid: Unbanked and Banked Curves

A car is going around an unbanked curve at the recommended speed of 11 m/s (see Fig. 5.12). (a) If the radius of curvature of the path is 25 m and the coefficient of static friction between the rubber and the road is $\mu_s = 0.70$, does the car skid as it goes around the curve? (b) What happens if the driver ignores the highway speed limit sign and travels at 18 m/s? (c) What speed is safe for traveling around the curve if the road surface is wet from a recent rainstorm and the

continued on next page

Example 5.7 continued

coefficient of static friction between the wet road and the rubber tires is $\mu_s = 0.50$? (d) For a car to safely negotiate the curve in icy conditions at a speed of 13 m/s, what banking angle would be required (see Fig. 5.13)?

Strategy The force of static friction is the only horizontal force acting on the car when the curve is not banked. The maximum force of static friction, which depends on road conditions, determines the maximum possible radial acceleration of the car. Therefore, we can compare the radial acceleration necessary to go around the curve at the specified speeds with the maximum possible radial acceleration determined by the coefficient of static friction. For part (d), in icy conditions we cannot rely much on friction, but the normal force has a horizontal component when the road is banked.

Solution (a) We find the radial acceleration required for a speed of 11 m/s:

$$a_r = \frac{v^2}{r} = \frac{(11\ \text{m/s})^2}{25\ \text{m}} = 4.8\ \text{m/s}^2$$

In order to have that acceleration, the component of the net force acting toward the center of curvature must be

$$\sum F_r = ma_r = m\frac{v^2}{r}$$

The only force with a horizontal component is the static frictional force acting on the tires due to the road (see the FBD in Fig. 5.12c). Therefore,

$$\sum F_r = f_s = m\frac{v^2}{r}$$

We must check to make sure that the maximum frictional force is not exceeded:

$$f_s \le \mu_s N$$

Since $N = mg$, the car can go around the curve without skidding as long as

$$\cancel{m}\frac{v^2}{r} \le \mu_s \cancel{m}g$$

Thus, the radial acceleration cannot exceed $\mu_s g$. That limits the car to speeds satisfying

$$v \le \sqrt{\mu_s g r}$$

Substituting numerical values,

$$v \le \sqrt{0.70 \times 9.80\ \text{m/s}^2 \times 25\ \text{m}} = 13\ \text{m/s}$$

Since 11 m/s is less than the maximum safe speed of 13 m/s, the car safely negotiates the curve.

(b) At 18 m/s, the car moves at a speed higher than the maximum safe speed of 13 m/s. The frictional force cannot supply the radial acceleration needed for the car to go around the curve—the car goes into a skid.

(c) In part (a), we found that the car is limited to speeds satisfying

$$v \le \sqrt{\mu_s g r}$$

With $\mu_s = 0.50$, the maximum safe speed is

$$v_{\text{max}} = \sqrt{\mu_s g r} = \sqrt{0.50 \times 9.80\ \text{m/s}^2 \times 25\ \text{m}} = 11\ \text{m/s}$$

which is the same maximum speed recommended by the road sign. The highway engineer knew what she was doing when she had the sign placed along the road.

(d) Finally, we find the banking angle that would enable cars to travel around the curve at 13 m/s in icy conditions. Assuming that friction is negligible, the horizontal component of the normal force is the only horizontal force. With the x-axis pointing toward the center of curvature and the y-axis vertical (Fig. 5.13),

$$\sum F_x = N \sin\theta = mv^2/r \qquad (1)$$

and

$$\sum F_y = N \cos\theta - mg = 0 \qquad (2)$$

Dividing Eq. (1) by Eq. (2) gives

$$\frac{N \sin\theta}{N \cos\theta} = \tan\theta = \frac{mv^2/r}{mg} = \frac{v^2}{rg}$$

$$\theta = \tan^{-1}\frac{v^2}{rg} = \tan^{-1}\frac{(13\ \text{m/s})^2}{25\ \text{m} \times 9.80\ \text{m/s}^2} = 35° \qquad (3)$$

Discussion Notice that the mass of the car does not appear in Eq. (3); the same banking angle holds for a scooter, motorcycle, car, or tractor-trailer. Notice also that the banking angle depends on the square of the speed. Automobile racetracks and bicycle racetracks have highly banked road surfaces at hairpin curves to minimize skidding of the high-speed vehicles. However, a banking angle of 35° is far greater than those used in practice along public roadways. Careful drivers would not try to drive around this curve in icy conditions at 13 m/s. What do you think might happen in icy conditions to a car that is traveling *very slowly* along a road banked at such a steep angle?

Highway curves are banked at slight angles to help drivers who are driving at reasonable speeds for the road conditions. They are not banked to save speed demons from their folly.

Practice Problem 5.7 A Bobsled Race

A bobsled races down an icy hill and then comes on a horizontal curve, located 60.0 m from the bottom of the hill. The sled is traveling at 22.4 m/s (50 mph) as it approaches the curve that has a radius of curvature of 50.0 m. The curve is banked at an angle of 45° and the frictional force on the sled runners is negligible. Does the sled make it safely around the curve?

If there is *no* friction between the road and the tires, then there is only one speed at which it is safe to drive around a given curve. *With* friction, there is a *range* of safe speeds. The static frictional force can have any magnitude from 0 to $\mu_s N$ and it can be directed either up or down the bank of the road.

Application of radial acceleration: banking angle of an airplane

When an airplane pilot makes a turn in the air, the pilot makes use of a banking angle. The airplane itself is tilted as if it were traveling over an inclined surface. Because of the shape of the wings, an aerodynamic force called *lift* acts upward when the plane is in level flight. To go around a turn, the wings are tilted; the lift force stays perpendicular to the wings and, therefore, now has a horizontal component (Fig. 5.14), just as the normal force has a horizontal component for a car on a banked curve. This component supplies the necessary radial acceleration, while the vertical component of the lift holds the plane up. Therefore,

Figure 5.14 The lift force $\vec{\mathbf{L}}$ is perpendicular to the wings of the plane. To turn, the pilot tilts the wings so a component of the lift force is directed toward the center of the circular path of the plane.

$$L_x = ma_r = \frac{mv^2}{r} \quad \text{and} \quad L_y = mg$$

where the x-axis is horizontal and the y-axis is vertical. The lift force is different in its physical origin from the normal force, but its components split up the same way, so a plane in a turn banks its wings at the same angle that a road would be banked for the same speed and radius of curvature. Of course, planes usually move much faster than cars and use large radii of curvature when they turn.

CHECKPOINT 5.3

A plane can't make a turn without tilting its wings. Why can a car turn on a flat road?

5.4 CIRCULAR ORBITS OF SATELLITES AND PLANETS

Application of radial acceleration: circular orbits

A satellite can orbit Earth in a circular path because of the long-range gravitational force on the satellite due to the Earth. The magnitude of the gravitational force on the satellite is

$$F = \frac{Gm_1m_2}{r^2} \qquad \text{(2-6)}$$

where the universal gravitational constant is $G = 6.67 \times 10^{-11}\ \text{N}\cdot\text{m}^2/\text{kg}^2$. We can use Newton's second law to find the speed of a satellite in circular orbit at constant speed. Let m be the mass of the satellite and M_E be the mass of the Earth. The direction of the gravitational force on the satellite is always toward the center of the Earth, which is the center of the orbit (Fig. 5.15). Since gravity is the only force acting on the satellite,

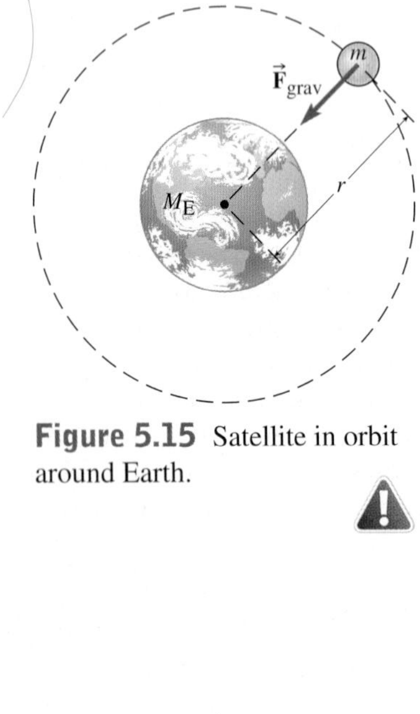

Figure 5.15 Satellite in orbit around Earth.

$$\sum F_r = G\frac{mM_E}{r^2}$$

where r is the distance from the *center* of the Earth to the satellite. Then, from Newton's second law,

$$\sum F_r = ma_r = \frac{mv^2}{r}$$

Setting these equal,

$$G\frac{mM_E}{r^2} = \frac{mv^2}{r}$$

Solving for the speed yields

$$v = \sqrt{\frac{GM_E}{r}} \qquad \text{(5-13)}$$

Notice that the mass of the satellite does not appear in the equation for speed; it has been algebraically canceled. The greater inertia of a more massive satellite is overcome by a proportionally greater gravitational force acting on it. Thus, the speed of a satellite in a circular orbit does not depend on the mass of the satellite. Equation (5-13) also shows that satellites in lower orbits (smaller radii) have greater speeds.

We have been discussing satellites orbiting Earth, but the same principles apply to the circular orbits of satellites around other planets and to the orbits of the planets around the Sun. For planetary orbits, the mass of the Sun would appear in Eq. (5-13) instead of the Earth's mass, because the *Sun's* gravitational pull keeps the planets in their orbits. The planetary orbits are actually ellipses (Fig. 5.16) instead of circles, although for most of the planets in the solar system the ellipses are nearly circular. Mercury is the exception; its orbit is markedly different from a circle.

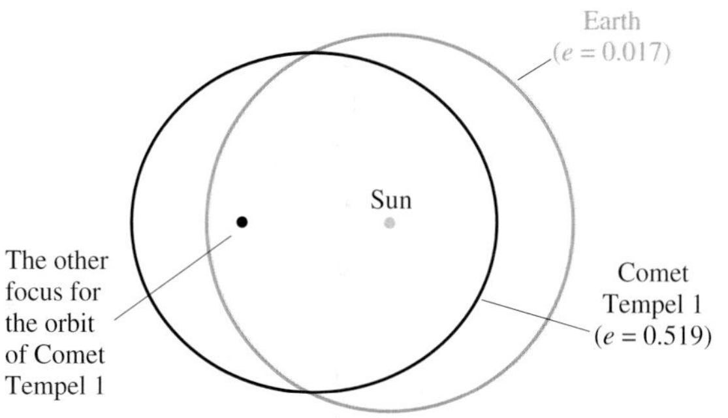

Figure 5.16 The shapes of two elliptical orbits around the Sun. (The *sizes* of the orbits are not to scale.) An ellipse looks like an elongated circle. The degree of elongation is measured by a quantity called the eccentricity e. A circle is a special case of an ellipse with $e = 0$. Most of the planetary orbits are nearly circular, with the exception of Mercury. The sum of the distances from any point on an ellipse to each of two fixed points (called the *foci*) is constant. The Sun is at one focus of each orbit. Since Earth's orbit is nearly circular, the second focus is very near the Sun.

Example 5.8

Speed of a Satellite

The Hubble Space Telescope is in a circular orbit 613 km above Earth's surface. The average radius of the Earth is 6.37×10^3 km and the mass of Earth is 5.97×10^{24} kg. What is the speed of the telescope in its orbit?

Strategy We first need to find the orbital radius of the telescope. It is not 613 km; that is the distance from the *surface* of Earth to the telescope. We must add the radius of the Earth to 613 km to find the orbital radius, which is measured from the center of the Earth to the telescope. Then we use Newton's second law, along with what we know about radial acceleration.

Solution The radius of the telescope's orbit is

$$r = 6.13 \times 10^2 \text{ km} + 6.37 \times 10^3 \text{ km} = (0.613 + 6.37) \times 10^3 \text{ km}$$

$$= 6.98 \times 10^3 \text{ km}$$

The net force on the telescope is equal to the gravitational force, given by Newton's law of gravity. Newton's second law relates the net force to the acceleration. Both are directed radially inward.

$$\sum F_r = \frac{GmM_E}{r^2} = \frac{mv^2}{r}$$

where m is the mass of the telescope. Solving for the speed, we find

$$v = \sqrt{\frac{GM_E}{r}}$$

$$v = \sqrt{\frac{6.67 \times 10^{-11} \text{ N}\cdot\text{m}^2/\text{kg}^2 \times 5.97 \times 10^{24} \text{ kg}}{6.98 \times 10^6 \text{ m}}}$$

$$v = 7550 \text{ m/s} = 27\,200 \text{ km/h}$$

Discussion *Any* satellite orbiting Earth at an altitude of 613 km has this same speed, regardless of its mass.

Practice Problem 5.8 Speed of Earth in Its Orbit

What is the speed of Earth in its approximately circular orbit about the Sun? The average Earth–Sun distance is 1.50×10^{11} m and the mass of the Sun is 1.987×10^{30} kg. Once you find the speed, use it along with the distance traveled by the Earth during one revolution about the Sun to calculate the time in seconds for one orbit.

Kepler's Laws of Planetary Motion

At the beginning of the seventeenth century, Johannes Kepler (1571–1630) proposed three laws to describe the motion of the planets. These laws predated Newton's laws of motion and his law of gravity. They offered a far simpler description of planetary motion than anything that had been proposed previously. We turn history on its head and look at one of Kepler's laws as a consequence of Newton's laws. The fact that Newton could derive Kepler's laws from his own work on gravity was seen as a confirmation of Newtonian mechanics.

Kepler's laws of planetary motion are

- The planets travel in elliptical orbits (Fig. 5.16) with the Sun at one focus of the ellipse.
- A line drawn from a planet to the Sun sweeps out equal areas in equal time intervals.
- The square of the orbital period is proportional to the cube of the average distance from the planet to the Sun.

Kepler's first law can be derived from the inverse square law of gravitational attraction. The derivation is a bit complicated, but for any two objects that have such an attraction, the orbit of one about the other is an ellipse, with the stationary object located at one focus. (Planetary orbits are also affected by gravitational interactions with other planets; Kepler's laws ignore these small effects.) The circle is a special case of an ellipse where the two foci coincide. We discuss Kepler's second law in Chapter 8.

Application of radial acceleration: Kepler's third law for a circular orbit

We can derive Kepler's third law from Newton's law of universal gravitation for the special case of a circular orbit. The gravitational force gives rise to the radial acceleration:

$$\sum F_{\mathrm{r}} = \frac{GmM_{\mathrm{Sun}}}{r^2} = \frac{mv^2}{r}$$

Solving for v yields

$$v = \sqrt{\frac{GM_{\mathrm{Sun}}}{r}}$$

The distance traveled during one revolution is the circumference of the circle, which is equal to $2\pi r$. The speed is the distance traveled during one orbit divided by the period:

$$v = \sqrt{\frac{GM_{\mathrm{Sun}}}{r}} = \frac{2\pi r}{T}$$

Now we solve for T:

$$T = 2\pi\sqrt{\frac{r^3}{GM_{\mathrm{Sun}}}}$$

Squaring both sides yields

$$T^2 = \frac{4\pi^2}{GM_{\mathrm{Sun}}} r^3 = \text{constant} \times r^3 \qquad (5\text{-}14)$$

Equation (5-14) is Kepler's third law: the square of the period of a planet is directly proportional to the cube of the average orbital radius.

Application of radial acceleration: geostationary orbits

Although Kepler's laws were derived for the motion of planets, they apply to satellites orbiting the Earth as well. Many satellites, such as those used for communications, are placed in a *geostationary* (or *geosynchronous*) orbit—a circular orbit in Earth's equatorial plane whose period is equal to Earth's rotational period (Fig. 5.17). A satellite in geostationary orbit remains directly above a particular point on the equator; to observers on the ground, it seems to hover above that point without moving. Due to their fixed positions with respect to Earth's surface, geostationary satellites are used as relay stations for communication signals. In Example 5.9, we find the speed of a geostationary satellite.

✓ CHECKPOINT 5.4

Do all geostationary satellites, no matter their masses, have to be the same height above Earth? Explain.

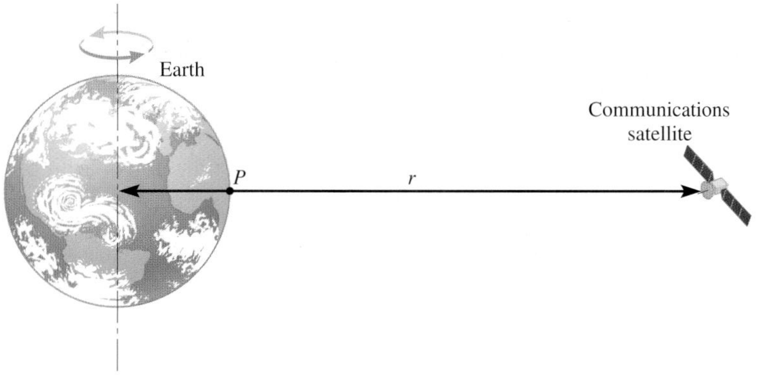

Figure 5.17 Geostationary satellite orbiting the Earth. The satellite has the same angular velocity as Earth, so it is always directly above point *P*.

Example 5.9

Geostationary Satellite

A 300.0-kg communications satellite is placed in a geostationary orbit 35,800 km above a relay station located in Kenya. What is the speed of the satellite in orbit?

Strategy The period of the satellite is 1 d or approximately 24 h. To find the speed of the satellite in orbit we use Newton's law of gravity and his second law of motion along with what we know about radial acceleration.

Solution Let m be the mass of the satellite and let M_E be the mass of the Earth. Gravity is the only force acting on the satellite in its orbit. From Newton's law of universal gravitation, Newton's second law, and the expression for radial acceleration,

$$\sum F_r = \frac{GmM_E}{r^2} = \frac{mv^2}{r}$$

Solving for the speed yields

$$v = \sqrt{\frac{GM_E}{r}}$$

We must add the mean radius of the Earth, $R_E = 6.37 \times 10^6$ m, to the height of the satellite above the Earth's surface to find the orbital radius.

$$r = h + R_E = 3.58 \times 10^7 \text{ m} + 0.637 \times 10^7 \text{ m}$$

$$= 4.217 \times 10^7 \text{ m}$$

Substituting numerical values into the speed equation,

$$v = \sqrt{\frac{6.67 \times 10^{-11} \text{ N·m}^2/\text{kg}^2 \times 5.97 \times 10^{24} \text{ kg}}{4.217 \times 10^7 \text{ m}}}$$

$$= \sqrt{9.443 \times 10^6 \text{ m}^2/\text{s}^2}$$

$$v = 3.07 \times 10^3 \text{ m/s}$$

Discussion This result, an orbital speed of 3.07 km/s and a distance above Earth's surface of 35,800 km, applies to *all* geostationary satellites. The mass of the satellite does not matter; it cancels out of the equations for orbital radius and for speed.

If we were actually putting a satellite into orbit, we would use a more accurate value for the period. We should use a time of 23 h and 56 min, which is the length of a *sidereal day*—the time for Earth to complete one rotation about its axis relative to the fixed stars. The solar day, 24 h, is the period of time between the daily appearances of the Sun at its highest point in the sky. The fact that Earth moves around the Sun is what causes the difference between these two ways of measuring the length of a day. The error introduced by using the longer time is negligible in this problem.

We can use Kepler's third law to check the result. Examples 5.8 and 5.9 both concern circular orbits around the Earth. Is the square of the period proportional to the cube of the orbital radius? From Example 5.8, $r_1 = 6.98 \times 10^3$ km and

$$T_1 = \frac{2\pi r_1}{v} = \frac{2\pi \times 6.98 \times 10^3 \text{ km}}{7.55 \text{ km/s}} = 5810 \text{ s}$$

From the present example, $r_2 = 4.22 \times 10^7$ m and

$$T_2 = 24 \text{ h} \times \frac{3600 \text{ s}}{1 \text{ h}} = 86400 \text{ s}$$

The ratio of the squares of the periods is

$$\left(\frac{T_2}{T_1}\right)^2 = \left(\frac{86400 \text{ s}}{5810 \text{ s}}\right)^2 = 221$$

The ratio of the cubes of the radii is

$$\left(\frac{r_2}{r_1}\right)^3 = \left(\frac{4.22 \times 10^7 \text{ m}}{6.98 \times 10^6 \text{ m}}\right)^3 = 221$$

Practice Problem 5.9 Orbital Radius of Venus

The period of the orbit of Venus around the Sun is 0.615 Earth years. Using this information, find the radius of its orbit in terms of R, the radius of Earth's orbit around the Sun.

Example 5.10

Orbiting Satellites

A satellite revolves about Earth with an orbital radius of r_1 and speed v_1. If an identical satellite were set into circular orbit with the same speed about a planet of mass three times that of Earth, what would its orbital radius be?

Strategy We can apply Newton's law of universal gravitation and set up a ratio to solve for the new orbital radius.

Solution From Newton's second law, the magnitude of the gravitational force on the satellite is equal to the satellite's mass times the magnitude of its radial acceleration:

$$\frac{G\cancel{m}M_E}{r_1^{\cancel{2}}} = \frac{\cancel{m}v_1^2}{\cancel{r_1}}$$

where M_E and m are the masses of Earth and of the satellite, respectively. Solving for r_1 yields

$$r_1 = \frac{GM_E}{v_1^2}$$

Now we apply Newton's second law to the orbit of the second satellite about the planet of mass $3M_E$:

$$\frac{G\cancel{m} \times 3M_E}{r_2^{\cancel{2}}} = \frac{\cancel{m}v_1^2}{\cancel{r_2}}$$

$$r_2 = \frac{G \times 3M_E}{v_1^2}$$

The ratio of r_2 to r_1 is

$$\frac{r_2}{r_1} = \frac{G \times 3M_E/v_1^2}{GM_E/v_1^2} = 3$$

Thus, $r_2 = 3r_1$.

Discussion Notice that we did not rush to substitute numerical values for the constants G and M_E into the equations. We took the ratio r_2/r_1 so that these constants cancel.

Practice Problem 5.10 Period of Lunar Lander

A lunar lander is orbiting about the Moon. If the radius of its orbit is $\frac{1}{3}$ the radius of Earth, what is the period of its orbit?

5.5 NONUNIFORM CIRCULAR MOTION

So far we have focused on *uniform* circular motion. Now we can extend the discussion to nonuniform circular motion, where the angular velocity changes with time.

Figure 5.18a shows the velocity vectors $\vec{\mathbf{v}}_1$ and $\vec{\mathbf{v}}_2$ at two different times for an object moving in a circle with changing speed. In this case, the speed is increasing ($v_2 > v_1$). In Fig. 5.18b, we subtract $\vec{\mathbf{v}}_1$ from $\vec{\mathbf{v}}_2$ to find the change in velocity. In the limit $\Delta t \to 0$, $\Delta\vec{\mathbf{v}}$ does *not* become perpendicular to the velocity, as it did for uniform circular motion. Thus, the direction of the acceleration is *not* radial if the speed is changing. However, we can resolve the acceleration into tangential and radial components

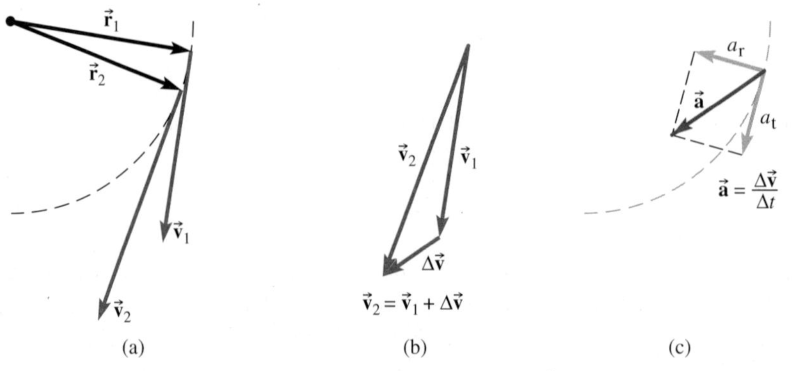

Figure 5.18 Motion along a circular path with a changing speed: (a) the magnitude of velocity $\vec{\mathbf{v}}_2$ is greater than the magnitude of velocity $\vec{\mathbf{v}}_1$, (b) the direction of $\Delta\vec{\mathbf{v}}$ is not radial when the speed is changing, and (c) components of $\vec{\mathbf{a}}$ can be taken along a tangent to the curved path (a_t) and along a radius (a_r).

(Fig. 5.18c). The radial component a_r changes the *direction* of the velocity, and the tangential component a_t changes the *magnitude* of the velocity. Since these are perpendicular components of the acceleration, the magnitude of the acceleration is

$$a = \sqrt{a_r^2 + a_t^2}$$

Using the same method as in Section 5.2 to find the radial acceleration, but working here with only the radial *component* of the acceleration, we find that

$$a_r = \frac{v^2}{r} = \omega^2 r \quad (\omega \text{ in radians per unit time}) \qquad (5\text{-}12)$$

For circular motion, *whether uniform or nonuniform*, the radial component of the acceleration is given by Eq. (5-12). However, in *uniform* circular motion the radial component of the acceleration a_r is constant in magnitude, but for nonuniform circular motion a_r changes as the speed changes.

Also still true for nonuniform circular motion is the relationship between speed and angular speed:

$$v = r\,|\omega| \qquad (5\text{-}7)$$

Many problems involving nonuniform circular motion are solved in the same way as for uniform circular motion. We find the *radial component of the net force* and then apply Newton's second law along the radial direction:

$$\sum F_r = ma_r$$

Problem-Solving Strategy for an Object in Nonuniform Circular Motion

1. Begin as for any Newton's second law problem: Identify all the forces acting on the object and draw an FBD.
2. Choose perpendicular axes at the point of interest so that one axis is radial and the other is tangent to the circular path.
3. Find the radial component of each force.
4. Apply Newton's second law along the radial direction:

   $$\sum F_r = ma_r$$

   where

   $$a_r = \frac{v^2}{r} = \omega^2 r$$

5. If necessary, apply Newton's second law to the tangential force components:

   $$\sum F_t = ma_t$$

   The tangential acceleration component a_t determines how the speed of the object changes.

CHECKPOINT 5.5

For an object in circular motion, what is it about the radial acceleration that distinguishes between uniform and nonuniform circular motion?

Example 5.11

Vertical Loop-the-Loop

A roller coaster includes a vertical circular loop of radius 20.0 m (Fig. 5.19a). What is the minimum speed at which the car must move at the top of the loop so that it doesn't lose contact with the track?

Strategy A roller coaster car moving around a vertical loop is in nonuniform circular motion; its speed decreases on the way up and increases on the way back down. Nevertheless, it is moving in a circle and has a radial acceleration component as given in Eq. (5-12) as long as it moves in a circle. The only forces acting on the car are gravity and the normal force of the track pushing the car. Even if frictional or drag forces are present, at the top of the loop they act in the tangential direction and, thus, do not contribute to the radial component of the net force. At the top of the loop, the track exerts a normal force on the car as long as the car moves with a speed great enough to stay on the track. If the car moves too slowly, it loses contact with the track and the normal force is then zero.

Solution The normal force exerted by the track on the car at the top pushes the car *away* from the track (downward); the normal force cannot pull up on the car. Then, at the top of the loop, the gravitational force and the normal force both point straight down toward the center of the loop. Figure 5.19b is an FBD for the car. From Newton's second law,

$$\sum F_r = N + mg = ma_r = \frac{mv_{top}^2}{r}$$

or

$$N = \frac{mv_{top}^2}{r} - mg$$

continued on next page

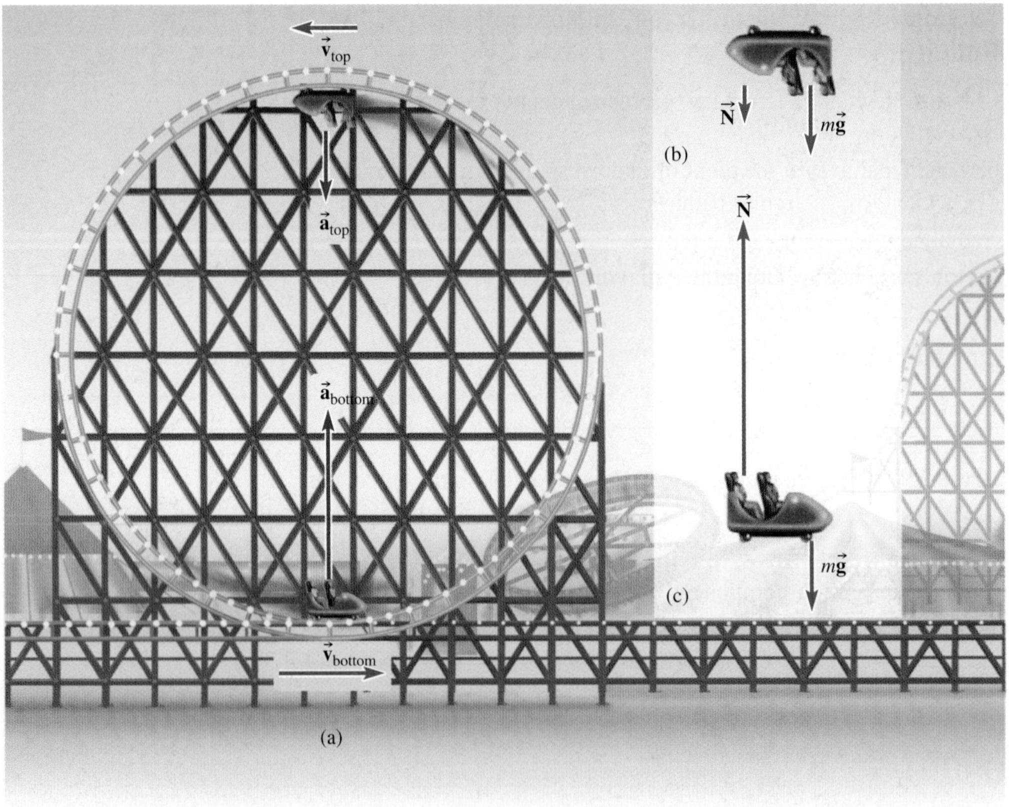

Figure 5.19 (a) A roller coaster car on a vertical circular loop. At the bottom of the loop, the car's acceleration $\vec{a}_{bottom}$ points upward toward the center of the circle. At the top of the loop, the car's acceleration $\vec{a}_{top}$ points downward. The magnitude of $\vec{a}_{top}$ is smaller than that of $\vec{a}_{bottom}$ because the speed is smaller at the top than at the bottom. (b) FBD for the car at the top of the loop. The track is above the car, so the normal force on the car due to the track is *downward.* (c) FBD for the car at the bottom of the loop.

Example 5.11 continued

where v_{top} stands for the speed at the top. In this expression, N stands for the magnitude of the normal force. Since $N \geq 0$,

$$m\left(\frac{v_{top}^2}{r} - g\right) \geq 0$$

or

$$v_{top} \geq \sqrt{gr}$$

Imagine sending a roller coaster car around the loop many times with a slightly smaller speed at the top each time. As v_{top} approaches $\sqrt{gr}$, the normal force at the top gets smaller and smaller. When $v_{top} = \sqrt{gr}$, the normal force just becomes zero at the top of the loop. Any slower and the car loses contact with the track *before* getting to the highest point and would fall off the track unless prevented from falling by a backup safety mechanism. Therefore, the minimum speed at the top is

$$v_{top} = \sqrt{gr} = \sqrt{9.80 \text{ m/s}^2 \times 20.0 \text{ m}} = 14.0 \text{ m/s}$$

Discussion If the car is going faster than 14 m/s at the top, its radial acceleration is larger. The track pushing on the car provides the additional net force component that results in a larger radial acceleration. The minimum speed occurs when gravity alone provides the radial acceleration at the top of the loop. In other words, $a_r = g$ at the top of the loop for minimum speed.

Practice Problem 5.11 Normal Force at the Bottom of the Track

If the speed of the roller coaster at the *bottom* of the loop is 25 m/s, what is the normal force exerted on the car by the track in terms of the car's weight mg? (See Fig. 5.19c.)

PHYSICS AT HOME

Go outside on a warm day and fill a bucket with water. Swing the bucket around in a vertical circle over your head. What, if anything, keeps the water in the bucket when the bucket is upside down over your head? Why doesn't the water spill out? Do any upward forces act on the water at that point? [*Hint:* The FBD for the water when it is directly overhead is similar to the FBD for a roller coaster car at the top of a loop.]

Conceptual Example 5.12

Acceleration of a Pendulum Bob

A pendulum is released from rest at point A (Fig. 5.20). Sketch qualitatively an FBD and the acceleration vector for the pendulum bob at points B and C.

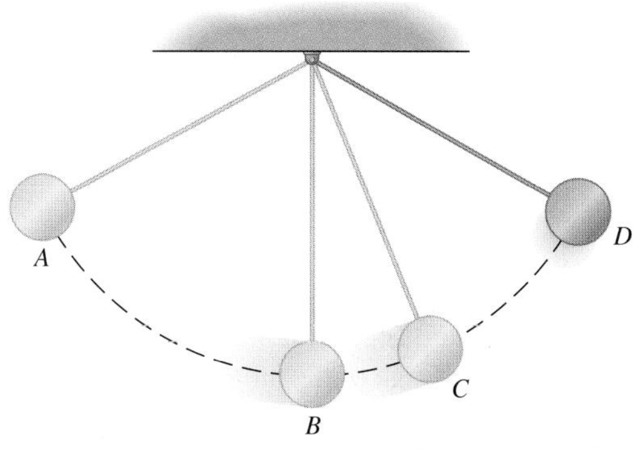

Figure 5.20

A pendulum swings to the right, starting from rest at point A.

Strategy Two forces appear on each FBD: gravity and the force due to the cord. The gravitational force is the same at both points (magnitude mg, direction down), but the force due to the cord varies in magnitude and in direction. Its direction is always along the cord. The net force on the bob is the sum of these two forces and its direction is the same as the direction of the acceleration. We can use what we know about the acceleration to guide us in drawing the forces.

The pendulum bob moves along the arc of a circle, but not at constant speed. At any point, the radial component of the acceleration is $a_r = v^2/r$. Unless $v = 0$, the radial acceleration component is nonzero. As the pendulum bob swings toward the bottom (from A to B), its speed is increasing; as it rises on the other side, its speed is decreasing. When the speed is increasing, the tangential component of the acceleration a_t is in the same direction as the velocity. From B to D, the speed is decreasing and a_t is in the direction *opposite* to

continued on next page

Conceptual Example 5.12 continued

the velocity. At point B, the speed is neither increasing nor decreasing and $a_t = 0$.

Solution and Discussion At point B, the tangential acceleration is zero, so the acceleration points in the radial direction: straight up (Fig. 5.21). The tension in the cord pulls straight up and gravity pulls down, so the tension must be larger than the weight of the bob to give an upward net force.

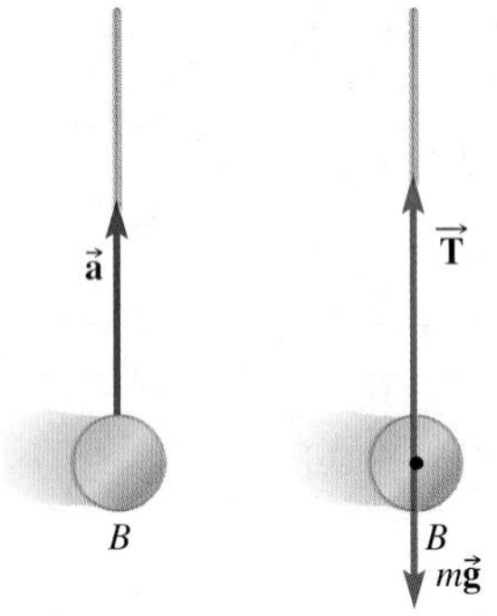

Figure 5.21
(a) Acceleration of the bob at point B. (b) FBD for the bob at B.

The acceleration at point C has both tangential and radial components. The tangential acceleration is opposite to the velocity because the bob is slowing down. Figure 5.22 shows the tangential and radial acceleration components added to form the acceleration vector $\vec{\mathbf{a}}$ and the FBD for the bob.

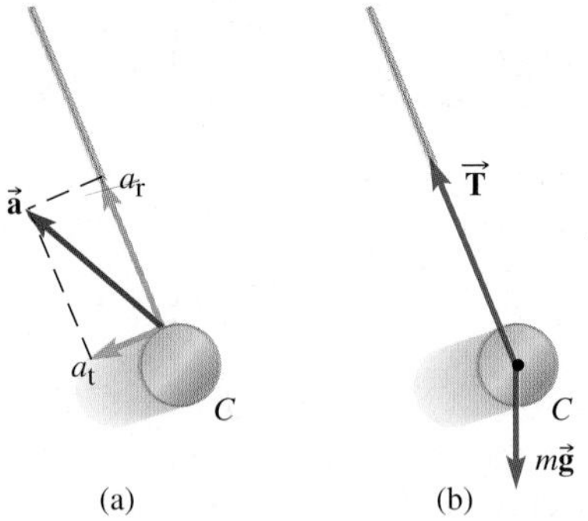

Figure 5.22
(a) At point C, the bob has both tangential and radial acceleration components. (b) FBD for the bob at C.

When the two forces are added, they give a net force in the same direction as the acceleration vector.

Conceptual Practice Problem 5.12 Analysis of the Bob at Point D

Sketch the FBD and the acceleration vector for the pendulum bob at point D, the highest point in its swing to the right.

5.6 TANGENTIAL AND ANGULAR ACCELERATION

An object in nonuniform circular motion has a changing speed and a changing angular velocity. To describe how the angular velocity changes, we define an angular acceleration. If the angular velocity is ω_1 at time t_1 and is ω_2 at time t_2, the change in angular velocity is

$$\Delta\omega = \omega_2 - \omega_1$$

The time interval during which the angular velocity changes is $\Delta t = t_2 - t_1$. The average rate at which the angular velocity changes is called the **average angular acceleration**, α_{av}.

$$\alpha_{av} = \frac{\omega_2 - \omega_1}{t_2 - t_1} = \frac{\Delta\omega}{\Delta t} \quad (5\text{-}15)$$

As we let the time interval become shorter and shorter, α_{av} approaches the **instantaneous angular acceleration**, α.

$$\alpha = \lim_{\Delta t \to 0} \frac{\Delta\omega}{\Delta t} \quad (5\text{-}16)$$

If ω is in units of rad/s, α is in units of rad/s^2.

The angular acceleration is closely related to the tangential component of the acceleration. The tangential component of velocity is

$$v_t = r\,|\omega| \quad (5\text{-}7)$$

Equation (5-7) gives us a way to relate tangential acceleration to the angular acceleration. The tangential acceleration is the rate of change of the tangential velocity, so

$$a_t = \frac{\Delta v_t}{\Delta t} = r\left|\frac{\Delta\omega}{\Delta t}\right| \quad \text{(in the limit } \Delta t \to 0\text{)}$$

Therefore,

$$a_t = r\,|\alpha| \quad (5\text{-}17)$$

Table 5.1 Relationships Between θ, ω, and α for Constant Angular Acceleration

Constant Acceleration Along x-Axis		Constant Angular Acceleration	
$\Delta v_x = v_{fx} - v_{ix} = a_x \Delta t$	(2-9)	$\Delta\omega = \omega_f - \omega_i = \alpha \Delta t$	(5-18)
$\Delta x = \frac{1}{2}(v_{fx} + v_{ix}) \Delta t$	(2-11)	$\Delta\theta = \frac{1}{2}(\omega_f + \omega_i) \Delta t$	(5-19)
$\Delta x = v_{ix} \Delta t + \frac{1}{2}a_x(\Delta t)^2$	(2-12)	$\Delta\theta = \omega_i \Delta t + \frac{1}{2}\alpha(\Delta t)^2$	(5-20)
$v_{fx}^2 - v_{ix}^2 = 2a_x \Delta x$	(2-13)	$\omega_f^2 - \omega_i^2 = 2\alpha\Delta\theta$	(5-21)

CONNECTION:

Because α is the rate of change of ω, and ω is the rate of change of θ, the equations for constant α have the same form as those for constant a_x.

Constant Angular Acceleration

The mathematical relationships between θ, ω, and α are the same as the mathematical relationships between x, v_x, and a_x that we developed in Chapter 2. Each quantity is the instantaneous rate of change of the preceding quantity. For example, a_x is the rate of change of v_x and α is the rate of change of ω. Because the mathematical relationships are the same, we can draw upon the skills and equations we developed to solve problems with constant acceleration a_x. All we have to do is take the equations for constant acceleration and replace x with θ, v_x with ω, and a_x with α (see Table 5.1).

Equation (5-18) is the definition of average angular acceleration, with α_{av} replaced by α since the angular acceleration is constant. Constant α means that ω changes linearly with time; therefore, the average angular velocity is halfway between the initial and final angular velocities for any time interval $\omega_{av} = \frac{1}{2}(\omega_i + \omega_f)$. Using this form for ω_{av} along with the definition of ω_{av} ($\omega_{av} = \Delta\theta/\Delta t$) yields Eq. (5-19). Equations (5-20) and (5-21) can be derived from the preceding two relations in a manner analogous to the derivations of Eqs. (2-12) and (2-13) in Section 2.4.

CHECKPOINT 5.6

A centrifuge is "spinning up" with a constant angular acceleration. Can the radial acceleration of a sample in the centrifuge be constant? Explain.

Example 5.13

A Rotating Potter's Wheel

A potter's wheel rotates from rest to 210 rpm in a time of 0.75 s. (a) What is the angular acceleration of the wheel during this time, assuming constant angular acceleration? (b) How many revolutions does the wheel make during this time interval? (c) Find the tangential and radial components of the acceleration of a point 12 cm from the rotation axis when the wheel is spinning at 180 rpm.

Strategy We know the initial and final frequencies, so we can find the initial and final angular velocities. We also know the time it takes for the wheel to get to the final angular velocity. That is all we need to find the average angular acceleration that, for constant angular acceleration, is equal to the instantaneous angular acceleration. To find the number of revolutions, we can find the angular displacement $\Delta\theta$ in radians and then divide by 2π rad/rev. We can find the angular velocity at $t = 0.75$ s and use it to find the radial acceleration component. The tangential acceleration is calculated from α.

continued on next page

Example 5.13 continued

Solution (a) Initially the wheel is at rest, so the initial angular velocity is zero.

$$\omega_i = 0 \text{ rad/s}$$

Converting 210 rpm to rad/s gives the final angular velocity:

$$\omega_f = 210 \frac{\text{rev}}{\text{min}} \times \frac{1}{60} \frac{\text{min}}{\text{s}} \times 2\pi \frac{\text{rad}}{\text{rev}} = 7.0\pi \text{ rad/s}$$

The angular acceleration is the rate of change of the angular velocity. Since α is constant, we can calculate it by finding the *average* angular acceleration for the time interval:

$$\alpha = \frac{\omega_f - \omega_i}{t_f - t_i} = \frac{7.0\pi \text{ rad/s} - 0}{0.75 \text{ s} - 0} = \frac{7.0\pi \text{ rad/s}}{0.75 \text{ s}} = 29 \text{ rad/s}^2$$

(b) The angular displacement is

$$\Delta\theta = \tfrac{1}{2}(\omega_f + \omega_i)\,\Delta t = \tfrac{1}{2}(7.0\pi \text{ rad/s} + 0)(0.75 \text{ s}) = 8.25 \text{ rad}$$

Since 2π rad = one revolution, the number of revolutions is

$$\frac{8.25 \text{ rad}}{2\pi \text{ rad/rev}} = 1.3 \text{ rev}$$

(c) At 180 rpm, the angular velocity is

$$\omega = 180 \frac{\text{rev}}{\text{min}} \times \frac{1}{60} \frac{\text{min}}{\text{s}} \times 2\pi \frac{\text{rad}}{\text{rev}} = 6.0\pi \text{ rad/s}$$

The radial acceleration component is

$$a_r = \omega^2 r = (6.0\pi \text{ rad/s})^2 \times 0.12 \text{ m} = 43 \text{ m/s}^2$$

and the tangential acceleration component is

$$a_t = \alpha r = 29 \text{ rad/s}^2 \times 0.12 \text{ m} = 3.5 \text{ m/s}^2$$

Discussion A quick check involves another of the equations for constant acceleration:

$$\omega_f^2 - \omega_i^2 = 2\alpha\Delta\theta$$

Since $\omega_i = 0$, we can check

$$\omega_f = \sqrt{2\alpha\Delta\theta}$$

From the answers to (a) and (b),

$$\sqrt{2\alpha\Delta\theta} = \sqrt{2 \times 29 \text{ rad/s}^2 \times 8.25 \text{ rad}} = 22 \text{ rad/s}$$

The original value for ω_f in rad/s was 7.0π rad/s. Since $\pi \approx 22/7$, the check is successful.

Practice Problem 5.13 The London Eye

The London Eye, a Ferris wheel on the banks of the Thames, has radius 67.5 m. At its cruising angular speed, it takes 30.0 min to make one complete revolution. Suppose that it takes 20.0 s to bring the wheel from rest to its cruising speed and that the angular acceleration is constant during startup. (a) What is the angular acceleration during startup? (b) What is the angular displacement of the wheel during startup?

The London Eye

5.7 APPARENT WEIGHT AND ARTIFICIAL GRAVITY

Application of apparent weight and circular motion: apparent weightlessness of orbiting astronauts

You are no doubt familiar with pictures of astronauts "floating" while in orbit around the Earth. It seems as if the astronauts are weightless. To be truly weightless, the force of gravity acting on the astronauts due to Earth would have to be zero, or at least close to zero. Is it? We can calculate the weight of an astronaut in orbit. The orbital altitude for the space shuttle is typically about 600 km above the Earth. Then the orbital radius is 600 km + 6400 km = 7000 km. Comparing the astronaut's weight in orbit to his or her weight on Earth's surface,

$$\frac{W_{\text{orbit}}}{W_{\text{surface}}} = \frac{\dfrac{GM_E m}{(R_E + h)^2}}{\dfrac{GM_E m}{R_E^2}} = \frac{R_E^2}{(R_E + h)^2} = \frac{(6400 \text{ km})^2}{(7000 \text{ km})^2} = 0.84$$

The weight in orbit is 0.84 times the weight on the surface. The astronaut weighs less but certainly isn't *weightless*! Then why does the astronaut *seem* to be weightless?

Recall Section 4.10 on the apparent weightlessness of someone unfortunate enough to be in an elevator when the cable snaps. In that situation, the elevator and the passenger both have the same acceleration ($\vec{\mathbf{a}} = \vec{\mathbf{g}}$). Similarly, the astronaut has the same acceleration as the space shuttle, which is equal to the *local* gravitational field $\vec{\mathbf{g}}$. Apparent weightlessness occurs when $\vec{\mathbf{a}} = \vec{\mathbf{g}}$, where $\vec{\mathbf{g}}$ is the *local* gravitational field.

Figure 5.23 A rotating space station from the movie *2001: A Space Odyssey.* Note jogger in the upper half running on the floor.

Application: Artificial Gravity In order for astronauts to spend long periods of time living in a space station without the deleterious effects of apparent weightlessness, *artificial gravity* would have to be created on the station. Many science fiction novels and movies feature ring-shaped space stations that rotate in order to create artificial gravity for the occupants. In a rotating space station, the acceleration of an astronaut is inward (toward the rotation axis), but the apparent gravitational field is outward. Therefore, the ceiling of rooms on the station are closest to the rotation axis and the floor is farthest away (Fig. 5.23).

The centrifuge is a device that creates artificial gravity on a smaller scale. Centrifuges are common not only in scientific and medical laboratories but also in everyday life. The first successful centrifuge was used to separate cream from milk in the 1880s. Water drips out of sopping wet clothes due to the pull of gravity when the clothes are hung on a clothesline, but the water is removed much faster by the artificial gravity created in the spin cycle of a washing machine.

The human body can be adversely affected not only by too little artificial gravity, but also by too much. Stunt pilots have to be careful about the accelerations to which they subject their bodies. An acceleration of about $3g$ can cause temporary blindness due to an inadequate supply of oxygen to the retina; the heart has difficulty pumping blood up to the head due to the blood's increased apparent weight. Larger accelerations can cause unconsciousness. Pressurized flight suits enable pilots to sustain accelerations up to about $5g$.

Example 5.14

Stunt Pilot

Dave wants to practice vertical circles for a flying show exhibition. (a) What must the minimum radius of the circle be to ensure that his acceleration at the bottom does not exceed $3.0g$? The speed of the plane is 78 m/s at the bottom of the circle. (b) What is Dave's apparent weight at the bottom of the circular path? Express your answer in terms of his true weight.

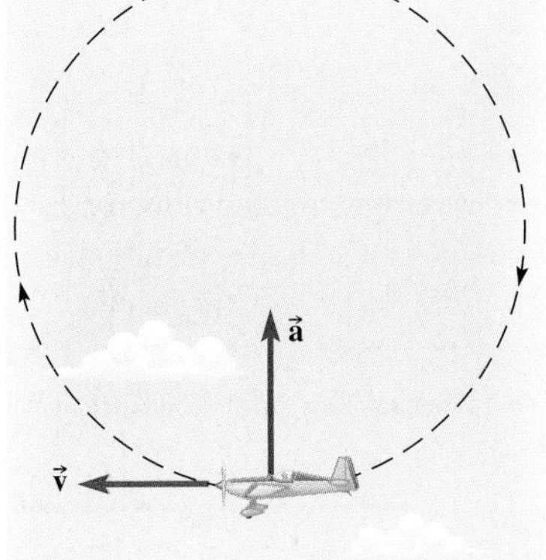

Figure 5.24

Velocity and acceleration vectors for the plane at the bottom of the circle.

Strategy For the *minimum* radius, we use the maximum possible radial acceleration since $a_r = v^2/r$. For the maximum radial acceleration, the *tangential* acceleration must be zero (Fig. 5.24)—the magnitude of the acceleration is $a = \sqrt{a_r^2 + a_t^2}$. Therefore, the radial acceleration component has magnitude $3.0g$ at the bottom. To find Dave's apparent weight, we do not need to use the numerical value of the radius found in part (a); we already know that his acceleration is upward and has magnitude $3.0g$.

Solution (a) The magnitude of the radial acceleration is

$$a_r = v^2/r$$

Solving for the radius,

$$r = \frac{v^2}{a_r} = \frac{v^2}{3.0g}$$

$$= \frac{(78\ \text{m/s})^2}{3.0 \times 9.8\ \text{m/s}^2} = 210\ \text{m}$$

continued on next page

Example 5.14 continued

(b) Dave's apparent weight is the magnitude of the normal force of the plane pushing up on him. Let the y-axis point upward. The normal force is up and the gravitational force is down (Fig. 5.25). Then

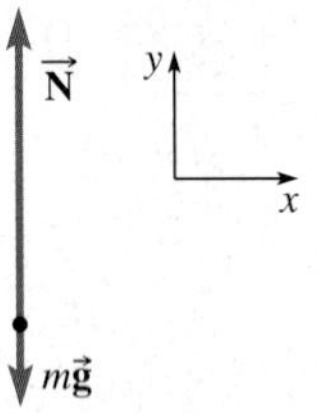

Figure 5.25
FBD for Dave.

$$\sum F_y = N - mg = ma_y$$

where $a_y = +3.0g$. Therefore,

$$W' = N = m(g + a_y) = 4.0mg$$

His apparent weight is 4.0 times his true weight.

Discussion It might have been tempting to jump to the conclusion that an acceleration of $3.0g$ means that his apparent weight is $3.0mg$. But is his apparent weight zero when his acceleration is zero? No.

Practice Problem 5.14 Astronaut's Apparent Weight

What is the apparent weight of a 730-N astronaut when her spaceship has an acceleration of magnitude $2.0g$ in the following two situations: (a) just above the surface of Earth, acceleration straight up; (b) far from any stars or planets?

Application of Apparent Weight to Objects at Rest with Respect to Earth's Surface Due to Earth's rotation, the *effective* value of g measured in a coordinate system attached to Earth's surface is slightly less than the true value of the gravitational field strength (see Section 4.5). The net force of an object placed on a scale is *not* zero because the object has a radial acceleration $a_r = \omega^2 r$ directed toward Earth's axis of rotation (Fig. 5.26). This relatively small effect is greatest where r is greatest—at the equator, where the effective value of g is about 0.3% smaller than the true value of g.

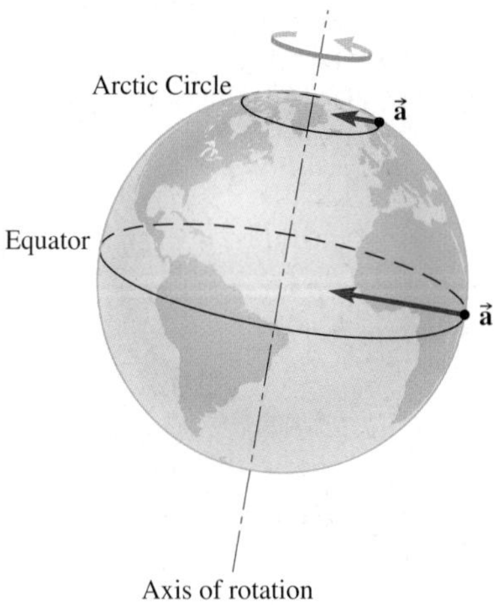

Figure 5.26 An object at rest with respect to Earth's surface has a radial acceleration due to Earth's rotation. The angular frequency ω is the same everywhere, so the radial acceleration $a_r = \omega^2 r$ is proportional to the distance from the axis of rotation.

Master the Concepts

- The angular displacement $\Delta\theta$ is the angle through which an object has turned. Positive and negative angular displacements indicate rotation in different directions. Conventionally, positive represents counterclockwise motion.

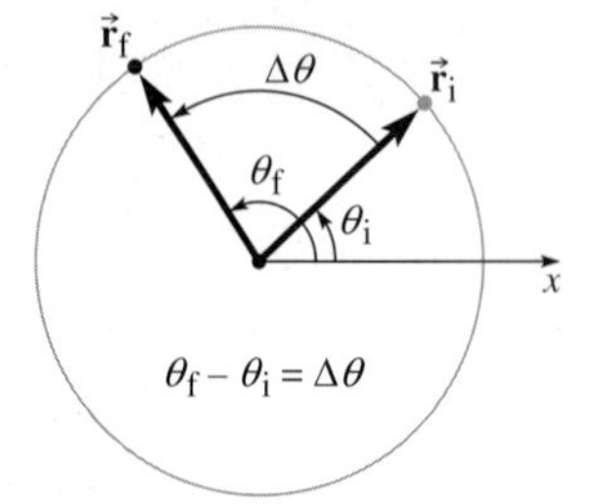

- Average angular velocity:

$$\omega_{av} = \frac{\theta_2 - \theta_1}{t_2 - t_1} = \frac{\Delta\theta}{\Delta t} \quad (5\text{-}2)$$

- Average angular acceleration:

$$\alpha_{av} = \frac{\omega_2 - \omega_1}{t_2 - t_1} = \frac{\Delta\omega}{\Delta t} \quad (5\text{-}15)$$

continued on next page

Master the Concepts continued

- The instantaneous angular velocity and acceleration are the limits of the average quantities as $\Delta t \to 0$.
- A useful measure of angle is the radian:

$$2\pi \text{ radians} = 360°$$

Using radian measure for θ, the arc length s of a circle of radius r subtended by an angle θ is

$$s = \theta r \quad (\theta \text{ in radian measure}) \qquad (5\text{-}4)$$

- Using radian measure for ω, the speed of an object in circular motion (including a point on a rotating object) is

$$v = r\,|\omega| \quad (\omega \text{ in radians per unit time}) \qquad (5\text{-}7)$$

- Using radian measure for α, the tangential acceleration component is related to the angular acceleration by

$$a_t = r\,|\alpha| \quad (\alpha \text{ in radians per time}^2) \qquad (5\text{-}17)$$

- An object moving in a circle has a radial acceleration component given by

$$a_r = \frac{v^2}{r} = \omega^2 r \quad (\omega \text{ in radians per unit time}) \qquad (5\text{-}12)$$

- The tangential and radial acceleration components are two perpendicular components of the acceleration vector. The radial acceleration component changes the direction of the velocity and the tangential acceleration component changes the speed.

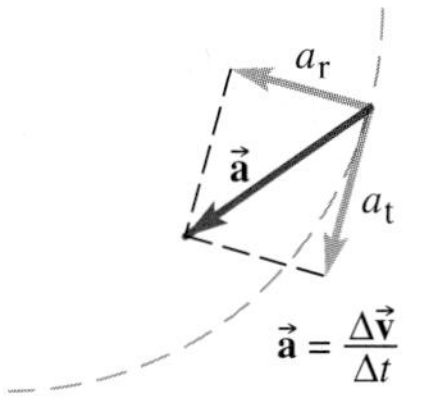

- Uniform circular motion means that v and ω are constant. In uniform circular motion, the time to complete one revolution is constant and is called the period T. The frequency f is the number of revolutions completed per second.

$$f = 1/T \qquad (5\text{-}8)$$

$$|\omega| = v/r = 2\pi f \qquad (5\text{-}9)$$

where the SI unit of angular velocity is rad/s and that of frequency is rev/s = Hz.

- A rolling object is both rotating and translating. If the object rolls without skidding or slipping, then

$$v_{\text{axle}} = r\,|\omega| \qquad (5\text{-}10)$$

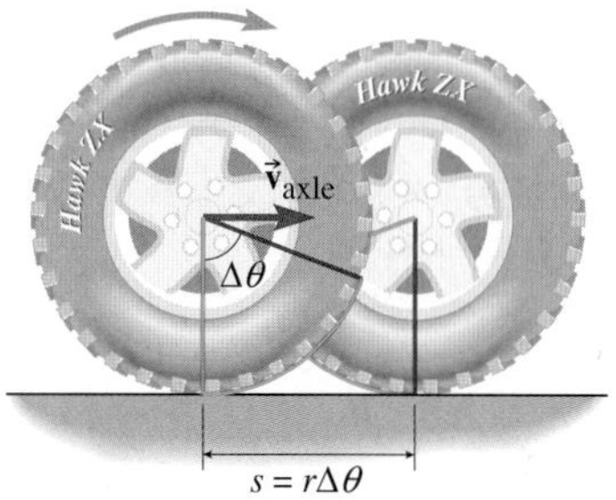

- Kepler's third law says that the square of the period of a planetary orbit is proportional to the cube of the orbital radius:

$$T^2 = \text{constant} \times r^3 \qquad (5\text{-}14)$$

- For constant angular acceleration, we can use equations analogous to those we developed for constant acceleration a_x:

$$\Delta\omega = \omega_f - \omega_i = \alpha\,\Delta t \qquad (5\text{-}18)$$

$$\Delta\theta = \tfrac{1}{2}(\omega_f + \omega_i)\,\Delta t \qquad (5\text{-}19)$$

$$\Delta\theta = \omega_i\,\Delta t + \tfrac{1}{2}\alpha(\Delta t)^2 \qquad (5\text{-}20)$$

$$\omega_f^2 - \omega_i^2 = 2\alpha\,\Delta\theta \qquad (5\text{-}21)$$

Conceptual Questions

1. Is depressing the "accelerator" (gas pedal) of a car the only way that the driver can make the car accelerate (in the physics sense of the word)? If not, what else can the driver do to give the car an acceleration?
2. Two children ride on a merry-go-round. One is 2 m from the axis of rotation and the other is 4 m from it. Which child has the larger (a) linear speed, (b) acceleration, (c) angular speed, and (d) angular displacement?
3. Explain why the orbital radius and the speed of a satellite in circular orbit are not independent.
4. In uniform circular motion, is the velocity constant? Is the acceleration constant? Explain.
5. In uniform circular motion, the net force is perpendicular to the velocity and changes the direction of the velocity but not the speed. If a projectile is launched horizontally, the net force (ignoring air resistance) is perpendicular to the initial velocity, and yet the projectile gains speed as it falls. What is the difference between the two situations?
6. The speed of a satellite in circular orbit around a planet does not depend on the mass of the satellite. Does it depend on the mass of the planet? Explain.
7. A flywheel (a massive disk) rotates with constant angular acceleration. For a point on the rim of the flywheel, is the tangential acceleration component constant? Is the radial acceleration component constant?
8. Explain why the force of gravity due to the Earth does not pull the Moon in closer and closer on an inward spiral until it hits Earth's surface.
9. When a roller coaster takes a sharp turn to the right, it feels as if you are pushed toward the left. Does a force push you to the left? If so, what is it? If not, why does there *seem* to be such a force?

10. Is there anywhere on Earth where a bathroom scale reads your true weight? If so, where? Where does your apparent weight due to Earth's rotation differ most from your true weight?

11. A physics teacher draws a cutaway view of a car rounding a banked curve as a rectangle atop a right triangle. A student draws a coordinate system based on the drawing. Is there another choice of axes that would make the problem easier to solve?

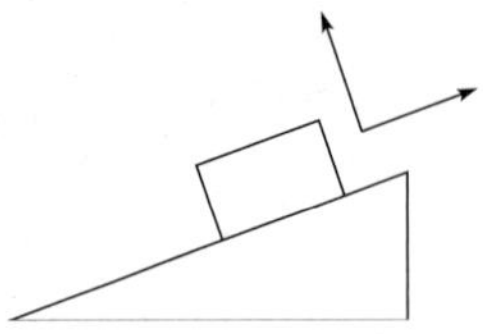

12. A bridal party is at a rehearsal dinner. The best man challenges the bridegroom to pick up an olive using only a brandy snifter. How does the groom accomplish this task?

Multiple-Choice Questions

1. A spider sits on a turntable that is rotating at a constant 33 rpm. The acceleration $\vec{a}$ of the spider is
   (a) greater the closer the spider is to the central axis.
   (b) greater the farther the spider is from the central axis.
   (c) nonzero and independent of the location of the spider on the turntable.
   (d) zero.

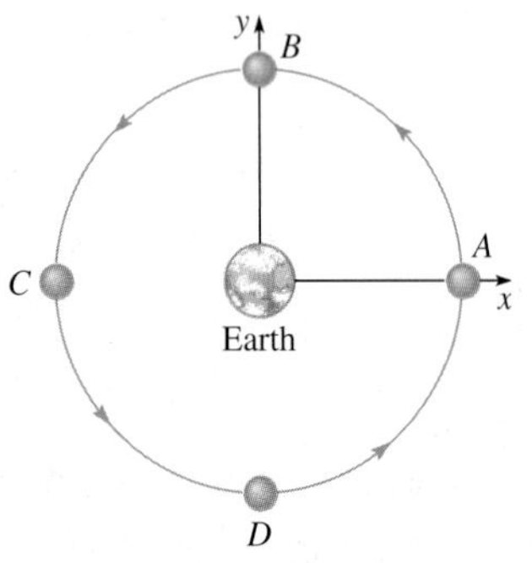

Multiple-Choice Questions 2–5 and Problem 36

Questions 2–5: A satellite in orbit travels around the Earth in uniform circular motion. In the figure, the satellite moves counterclockwise (*ABCDA*). Answer choices:

(a) $+x$ (b) $+y$ (c) $-x$ (d) $-y$
(e) 45° above $+x$ (toward $+y$)
(f) 45° below $+x$ (toward $-y$)
(g) 45° above $-x$ (toward $+y$)
(h) 45° below $-x$ (toward $-y$)

2. What is the direction of the satellite's instantaneous velocity at point *D*?
3. What is the direction of the satellite's average velocity for one quarter of an orbit, starting at *C* and ending at *D*?
4. What is the direction of the satellite's average acceleration for one half of an orbit, starting at *C* and ending at *A*?
5. What is the direction of the satellite's instantaneous acceleration at point *C*?
6. Two satellites are in orbit around Mars with the same orbital radius. Satellite 2 has twice the mass of satellite 1. The radial acceleration of satellite 2 has
   (a) twice the magnitude of the radial acceleration of satellite 1.
   (b) the same magnitude as the radial acceleration of satellite 1.
   (c) half the magnitude of the radial acceleration of satellite 1.
   (d) four times the magnitude of the radial acceleration of satellite 1.

Questions 7–8: A boy swings in a tire swing. Answer choices:
   (a) At the highest point of the motion
   (b) At the lowest point of the motion
   (c) At a point neither highest nor lowest
   (d) It is constant.

7. When is the tension in the rope the greatest?
8. When is the tangential acceleration the greatest?

Questions 9–10 concern these three statements:
   (1) Its acceleration is constant.
   (2) Its radial acceleration component is constant in magnitude.
   (3) Its tangential acceleration component is constant in magnitude.

9. An object is in uniform circular motion. Identify the correct statement(s).
   (a) 1 only (b) 2 only (c) 3 only
   (d) 1, 2, and 3 (e) 2 and 3 (f) 1 and 2
   (g) 1 and 3 (h) None of them
10. An object is in nonuniform circular motion with constant angular acceleration. Identify the correct statement(s). (Use the same answer choices as Question 9.)
11. An astronaut is out in space far from any large bodies. He uses his jets to start spinning, then releases a baseball he has been holding in his hand. Ignoring the gravitational force between the astronaut and the baseball, how would you describe the path of the baseball after it leaves the astronaut's hand?
   (a) It continues to circle the astronaut in a circle with the same radius it had before leaving the astronaut's hand.
   (b) It moves off in a straight line.
   (c) It moves off in an ever-widening arc.

12. An object moving in a circle at a constant speed has an acceleration that is
    (a) in the direction of motion
    (b) toward the center of the circle
    (c) away from the center of the circle
    (d) zero

Problems

- Combination conceptual/quantitative problem
- Biological or medical application
- ✦ Challenging problem
- Blue # Detailed solution in the Student Solutions Manual
- (1 2) Problems paired by concept
- Text website interactive or tutorial

5.1 Description of Uniform Circular Motion

1. A carnival swing is fixed on the end of an 8.0-m-long beam. If the swing and beam sweep through an angle of 120°, what is the distance through which the riders move?
2. A soccer ball of diameter 31 cm rolls without slipping at a linear speed of 2.8 m/s. Through how many revolutions has the soccer ball turned as it moves a linear distance of 18 m?
3. Find the average angular speed of the second hand of a clock.
4. Convert these to radian measure: (a) 30.0°, (b) 135°, (c) $\frac{1}{4}$ revolution, (d) 33.3 revolutions.
5. A bicycle is moving at 9.0 m/s. What is the angular speed of its tires if their radius is 35 cm? (tutorial: car tire)
6. An elevator cable winds on a drum of radius 90.0 cm that is connected to a motor. (a) If the elevator is moving down at 0.50 m/s, what is the angular speed of the drum? (b) If the elevator moves down 6.0 m, how many revolutions has the drum made?
7. Grace is playing with her dolls and decides to give them a ride on a merry-go-round. She places one of them on an old record player turntable and sets the angular speed at 33.3 rpm. (a) What is their angular speed in rad/s? (b) If the doll is 13 cm from the center of the spinning turntable platform, how fast (in m/s) is the doll moving?
8. A wheel is rotating at a rate of 2.0 revolutions every 3.0 s. Through what angle, in radians, does the wheel rotate in 1.0 s?
9. ✦ In the construction of railroads, it is important that curves be gentle, so as not to damage passengers or freight. Curvature is not measured by the radius of curvature, but in the following way. First a 100.0-ft-long chord is measured. Then the curvature is reported as the angle subtended by two radii at the endpoints of the chord. (The angle is measured by determining the angle between two tangents 100 ft apart; since each tangent is perpendicular to a radius, the angles are the same.) In modern railroad construction, track curvature is kept below 1.5°. What is the radius of curvature of a "1.5° curve"? [*Hint:* Since the angle is small, the length of the chord is approximately equal to the arc length along the curve.]

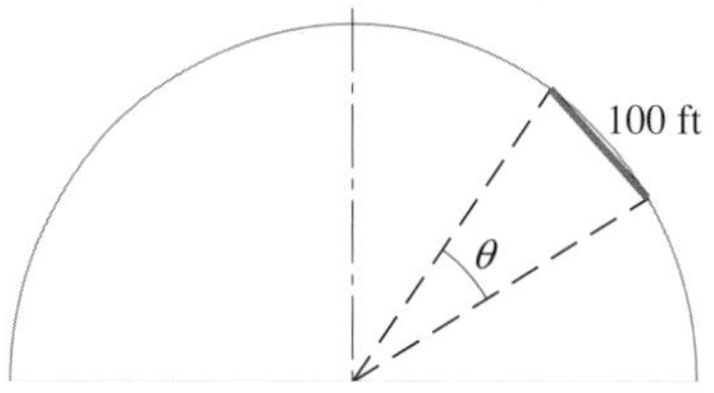

5.2 Radial Acceleration

10. Verify that all three expressions for radial acceleration ($v\omega$, v^2/r, and $\omega^2 r$) have the correct dimensions for an acceleration.
11. An apparatus is designed to study insects at an acceleration of magnitude 980 m/s^2 (= 100g). The apparatus consists of a 2.0-m rod with insect containers at either end. The rod rotates about an axis perpendicular to the rod and at its center. (a) How fast does an insect move when it experiences a radial acceleration of 980 m/s^2? (b) What is the angular speed of the insect? (tutorial: centrifuge)

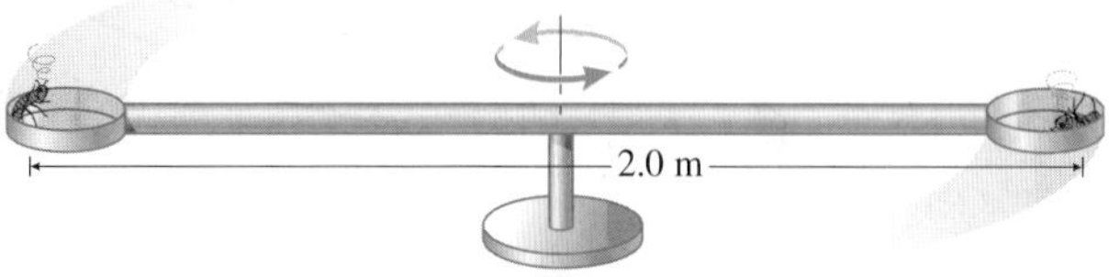

12. The rotor is an amusement park ride where people stand against the inside of a cylinder. Once the cylinder is spinning fast enough, the floor drops out. (a) What force keeps the people from falling out the bottom of the cylinder? (b) If the coefficient of friction is 0.40 and the cylinder has a radius of 2.5 m, what is the minimum angular speed of the cylinder so that the people don't fall out? (Normally the operator runs it considerably faster as a safety measure.)

13. Objects that are at rest relative to Earth's surface are in circular motion due to Earth's rotation. What is the radial acceleration of an African baobab tree located at the equator?

14. Earth's orbit around the Sun is nearly circular. The period is 1 yr = 365.25 d. (a) In an elapsed time of 1 d, what is Earth's angular displacement? (b) What is the change in Earth's velocity, $\Delta\vec{v}$? (c) What is Earth's average acceleration during 1 d? (d) Compare your answer for (c) to the magnitude of Earth's instantaneous radial acceleration. Explain.

15. A 0.700-kg ball is on the end of a rope that is 1.30 m in length. The ball and rope are attached to a pole and the entire apparatus, including the pole, rotates about the pole's symmetry axis. The rope makes an angle of 70.0° with respect to the vertical. What is the tangential speed of the ball?

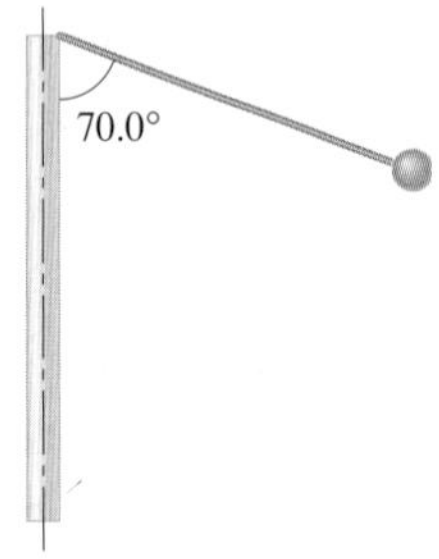

16. A child's toy has a 0.100-kg ball attached to two strings, *A* and *B*. The strings are also attached to a stick and the ball swings around the stick along a circular path in a horizontal plane. Both strings are 15.0 cm long and make an angle of 30.0° with respect to the horizontal. (a) Draw an FBD for the ball showing the tension forces and the gravitational force. (b) Find the magnitude of the tension in each string when the ball's angular speed is 6.00π rad/s.

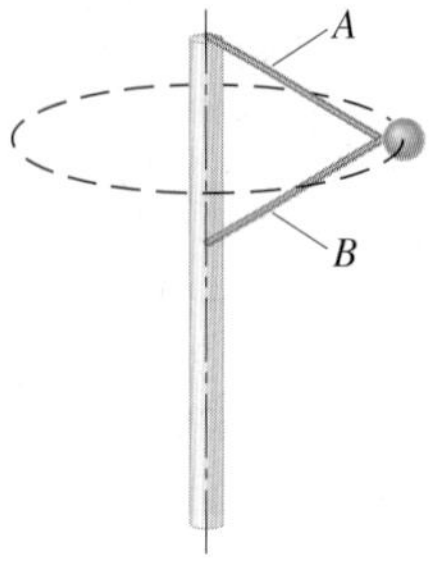

17. A child swings a rock of mass m in a horizontal circle using a rope of length L. The rock moves at constant speed v. (a) Ignoring gravity, find the tension in the rope. (b) Now include gravity (the weight of the rock is no longer negligible, although the weight of the rope still is negligible). What is the tension in the rope? Express the tension in terms of m, g, v, L, and the angle θ that the rope makes with the horizontal. (tutorial: skip rope)

18. A *conical pendulum* consists of a bob (mass m) attached to a string (length L) swinging in a horizontal circle (Fig. 5.11). As the string moves, it sweeps out the area of a cone. The angle that the string makes with the vertical is ϕ. (a) What is the tension in the string? (b) What is the period of the pendulum?

5.3 Unbanked and Banked Curves

19. A curve in a stretch of highway has radius R. The road is unbanked. The coefficient of static friction between the tires and road is μ_s. (a) What is the fastest speed that a car can safely travel around the curve? (b) Explain what happens when a car enters the curve at a speed greater than the maximum safe speed. Illustrate with an FBD. (interactive: banked curve)

20. A highway curve has a radius of 825 m. At what angle should the road be banked so that a car traveling at 26.8 m/s (60 mph) has no tendency to skid sideways on the road? [*Hint:* No tendency to skid means the frictional force is zero.]

21. A curve in a highway has radius of curvature 120 m and is banked at 3.0°. On a day when the road is icy, what is the safest speed to go around the curve?

22. A roller coaster car of mass 320 kg (including passengers) travels around a horizontal curve of radius 35 m. Its speed is 16 m/s. What is the magnitude and direction of the total force exerted on the car by the track?

23. A velodrome is built for use in the Olympics. The radius of curvature of the surface is 20.0 m. At what angle should the surface be banked for cyclists moving at 18 m/s? (Choose an angle so that no frictional force is needed to keep the cyclists in their circular path. Large banking angles *are* used in velodromes.)

24. A car drives around a curve with radius 410 m at a speed of 32 m/s. The road is not banked. The mass of the car is 1400 kg. (a) What is the frictional force on the car? (b) Does the frictional force necessarily have magnitude $\mu_s N$? Explain.

25. A car drives around a curve with radius 410 m at a speed of 32 m/s. The road is banked at 5.0°. The mass of the car is 1400 kg. (a) What is the frictional force on the car? (b) At what speed could you drive around this curve so that the force of friction is zero?

26. A curve in a stretch of highway has radius R. The road is banked at angle θ to the horizontal. The coefficient of static friction between the tires and road is μ_s. What is the fastest speed that a car can travel through the curve?

27. An airplane is flying at constant speed v in a horizontal circle of radius r. The lift force on the wings due to the air is perpendicular to the wings. At what angle to the vertical must the wings be banked to fly in this circle? (tutorial: plane in turn)

28. A road with a radius of 75.0 m is banked so that a car can navigate the curve at a speed of 15.0 m/s without any friction. When a car is going 20.0 m/s on this curve, what minimum coefficient of static friction is needed if the car is to navigate the curve without slipping?

5.4 Circular Orbits of Satellites and Planets

29. What is the average linear speed of the Earth about the Sun?

30. The orbital speed of Earth about the Sun is 3.0×10^4 m/s and its distance from the Sun is 1.5×10^{11} m. The mass of Earth is approximately 6.0×10^{24} kg and that of the Sun is 2.0×10^{30} kg. What is the magnitude of the force exerted by the Sun on Earth? [*Hint:* Two different methods are possible. Try both.]

31. Two satellites are in circular orbits around Jupiter. One, with orbital radius r, makes one revolution every 16 h. The other satellite has orbital radius $4.0r$. How long does the second satellite take to make one revolution around Jupiter?

32. The Hubble Space Telescope orbits Earth 613 km above Earth's surface. What is the period of the telescope's orbit?

33. Io, one of Jupiter's satellites, has an orbital period of 1.77 d. Europa, another of Jupiter's satellites, has an orbital period of about 3.54 d. Both moons have nearly circular orbits. Use Kepler's third law to find the distance of each satellite from Jupiter's center. Jupiter's mass is 1.9×10^{27} kg.

34. A spy satellite is in circular orbit around Earth. It makes one revolution in 6.00 h. (a) How high above Earth's surface is the satellite? (b) What is the satellite's acceleration?

35. Mars has a mass of about 6.42×10^{23} kg. The length of a day on Mars is 24 h and 37 min, a little longer than the length of a day on Earth. Your task is to put a satellite into a circular orbit around Mars so that it stays above one spot on the surface, orbiting Mars once each Mars day. At what distance from the center of the planet should you place the satellite?

36. A satellite travels around Earth in uniform circular motion at an altitude of 35 800 km above Earth's surface. The satellite is in geosynchronous orbit (that is, the time for it to complete one orbit is exactly 1 d). In the figure with Multiple-Choice Questions 2–5, the satellite moves counterclockwise (*ABCDA*). State directions in terms of the x- and y-axes. (a) What is the satellite's instantaneous velocity at point C? (b) What is the satellite's average velocity for one quarter of an orbit, starting at A and ending at B? (c) What is the satellite's average acceleration for one quarter of an orbit, starting at A and ending at B? (d) What is the satellite's instantaneous acceleration at point D?

37. A spacecraft is in orbit around Jupiter. The radius of the orbit is 3.0 times the radius of Jupiter (which is $R_J = 71\,500$ km). The gravitational field at the surface of Jupiter is 23 N/kg. What is the period of the spacecraft's orbit? [*Hint:* You don't need to look up any more data about Jupiter to solve the problem.]

5.5 Nonuniform Circular Motion

38. A roller coaster has a vertical loop with radius 29.5 m. With what minimum speed should the roller coaster car be moving at the top of the loop so that the passengers do not lose contact with the seats?

39. A pendulum is 0.80 m long and the bob has a mass of 1.0 kg. At the bottom of its swing, the bob's speed is 1.6 m/s. (a) What is the tension in the string at the bottom of the swing? (b) Explain why the tension is greater than the weight of the bob.

40. A 35.0-kg child swings on a rope with a length of 6.50 m that is hanging from a tree. At the bottom of the swing, the child is moving at a speed of 4.20 m/s. What is the tension in the rope?

41. A car approaches the top of a hill that is shaped like a vertical circle with a radius of 55.0 m. What is the fastest speed that the car can go over the hill without losing contact with the ground?

5.6 Tangential and Angular Acceleration

42. A child pushes a merry-go-round from rest to a final angular speed of 0.50 rev/s with constant angular acceleration. In doing so, the child pushes the merry-go-round 2.0 revolutions. What is the angular acceleration of the merry-go-round?

43. A cyclist starts from rest and pedals so that the wheels make 8.0 revolutions in the first 5.0 s. What is the angular acceleration of the wheels (assumed constant)?

44. During normal operation, a computer's hard disk spins at 7200 rpm. If it takes the hard disk 4.0 s to reach this angular velocity starting from rest, what is the average angular acceleration of the hard disk in rad/s^2?

45. Derive Eq. 5-20 from Eqs. 5-18 and 5-19. [*Hint:* See the derivation of Eq. (2-12) in Section 2.4.]

46. Derive Eq. 5-21 from Eqs. 5-18 and 5-19.

47. A pendulum is 0.800 m long and the bob has a mass of 1.00 kg. When the string makes an angle of $\theta = 15.0°$ with the vertical, the bob is moving at 1.40 m/s. Find the tangential and radial acceleration components and the tension in the string. [*Hint:* Draw an FBD for the bob. Choose the x-axis to be tangential to the motion of the bob and the y-axis to be radial. Apply Newton's second law.]

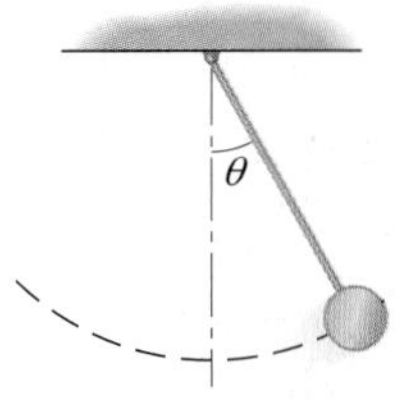

Problems 47 and 48

48. Find the tangential acceleration of a freely swinging pendulum when it makes an angle θ with the vertical.

49. A turntable reaches an angular speed of 33.3 rpm in 2.0 s, starting from rest. (a) Assuming the angular acceleration is constant, what is its magnitude? (b) How many revolutions does the turntable make during this time interval?

50. A wheel's angular acceleration is constant. Initially its angular velocity is zero. During the first 1.0-s time interval, it rotates through an angle of 90.0°. (a) Through what angle does it rotate during the next 1.0-s time interval? (b) Through what angle during the third 1.0-s time interval?

51. A car that is initially at rest moves along a circular path with a constant tangential acceleration component of 2.00 m/s^2. The circular path has a radius of 50.0 m. The initial position of the car is at the far west location on the circle and the initial velocity is to the north. (a) After the car has traveled $\frac{1}{4}$ of the circumference, what is the speed of the car? (b) At this point, what is the radial acceleration component of the car? (c) At this same point, what is the total acceleration of the car?

52. A disk rotates with constant angular acceleration. The initial angular speed of the disk is 2π rad/s. After the disk rotates through 10π radians, the angular speed is 7π rad/s. (a) What is the magnitude of the angular acceleration? (b) How much time did it take for the disk to rotate through 10π radians? (c) What is the tangential acceleration of a point located at a distance of 5.0 cm from the center of the disk?

53. In a Beams ultracentrifuge, the rotor is suspended magnetically in a vacuum. Since there is no mechanical connection to the rotor, the only friction is the air resistance due to the few air molecules in the vacuum. If the rotor is spinning with an angular speed of 5.0×10^5 rad/s and the driving force is turned off, its spinning slows down at an angular rate of 0.40 rad/s^2. (a) How long does the rotor spin before coming to rest? (b) During this time, through how many revolutions does the rotor spin?

54. The rotor of the Beams ultracentrifuge (see Problem 53) is 20.0 cm long. For a point at the end of the rotor, find the (a) initial speed, (b) tangential acceleration component, and (c) maximum radial acceleration component.

5.7 Apparent Weight and Artificial Gravity

55. If a washing machine's drum has a radius of 25 cm and spins at 4.0 rev/s, what is the strength of the artificial gravity to which the clothes are subjected? Express your answer as a multiple of g.

56. A space station is shaped like a ring and rotates to simulate gravity. If the radius of the space station is 120 m, at what frequency must it rotate so that it simulates Earth's gravity? [*Hint:* The apparent weight of the astronauts must be the same as their weight on Earth.] (WWW tutorial: space station)

57. A biologist is studying growth in space. He wants to simulate Earth's gravitational field, so he positions the plants on a rotating platform in the spaceship. The distance of each plant from the central axis of rotation is $r = 0.20$ m. What angular speed is required?

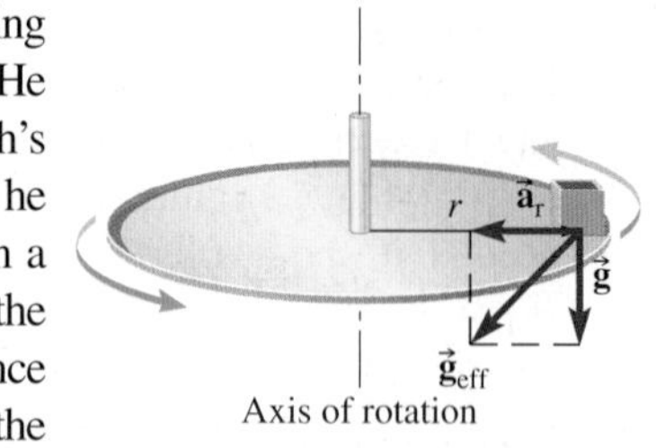

58. A biologist is studying plant growth and wants to simulate a gravitational field twice as strong as Earth's. She places the plants on a horizontal rotating table in her laboratory on Earth at a distance of 12.5 cm from the axis of rotation. What angular speed will give the plants an effective gravitational field $\vec{g}_{eff}$, whose magnitude is 2.0g? [*Hint:* Remember to account for Earth's gravitational field as well as the artificial gravity when finding the apparent weight.]

59. Objects that are at rest relative to the Earth's surface are in circular motion due to Earth's rotation. (a) What is the radial acceleration of an object at the equator? (b) Is the object's apparent weight greater or less than its weight? Explain. (c) By what percentage does the apparent weight differ from the weight at the equator? (d) Is there any place on Earth where a bathroom scale reading is equal to your true weight? Explain.

60. A person of mass M stands on a bathroom scale inside a Ferris wheel compartment. The Ferris wheel has radius R and angular velocity ω. What is the apparent weight of the person (a) at the top and (b) at the bottom?

61. A person rides a Ferris wheel that turns with constant angular velocity. Her weight is 520.0 N. At the top of the ride her apparent weight is 1.5 N different from her true weight. (a) Is her apparent weight at the top 521.5 N or 518.5 N? Why? (b) What is her apparent weight at the bottom of the ride? (c) If the angular speed of the Ferris wheel is 0.025 rad/s, what is its radius?

62. Objects that are at rest relative to Earth's surface are in circular motion due to Earth's rotation. What is the radial acceleration of a painting hanging in the Prado Museum in Madrid, Spain, at a latitude of 40.2° North? (Note that the object's radial acceleration is not directed toward the center of the Earth.)

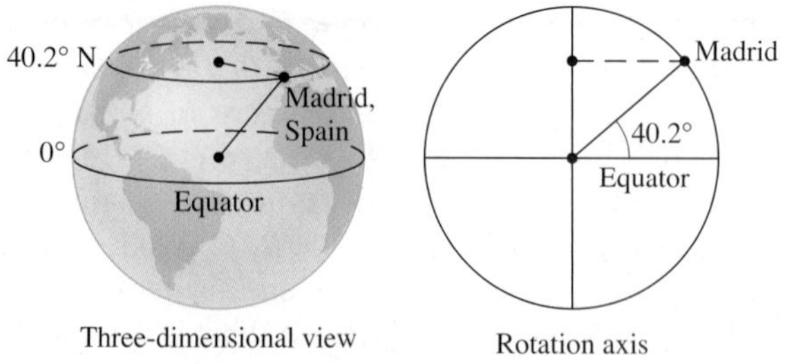

Three-dimensional view

Cross-sectional view

63. A rotating flywheel slows down at a constant rate due to friction in its bearings. After 1 min, its angular velocity has diminished to 0.80 of its initial value ω. At the end of the third minute, what is the angular velocity in terms of the initial value?

Comprehensive Problems

64. The Earth rotates on its own axis once per day (24.0 h). What is the tangential speed of the summit of Mt. Kilimanjaro (elevation 5895 m above sea level), which is located approximately on the equator, due to the rotation of the Earth? The equatorial radius of Earth is 6378 km.

65. A trimmer for cutting weeds and grass near trees and borders has a nylon cord of 0.23-m length that whirls about an axle at 660 rad/s. What is the linear speed of the tip of the nylon cord?

66. A high-speed dental drill is rotating at 3.14×10^4 rad/s. Through how many degrees does the drill rotate in 1.00 s?

67. A jogger runs counterclockwise around a path of radius 90.0 m at constant speed. He makes 1.00 revolution in 188.4 s. At $t = 0$, he is heading due east. (a) What is the jogger's instantaneous velocity at $t = 376.8$ s? (b) What is his instantaneous velocity at $t = 94.2$ s?

68. Two gears A and B are in contact. The radius of gear A is twice that of gear B. (a) When A's angular velocity is 6.00 Hz counterclockwise, what is B's angular velocity? (b) If A's radius to the tip of the teeth is 10.0 cm, what is the linear speed of a point on the tip of a gear tooth? What is the linear speed of a point on the tip of B's gear tooth?

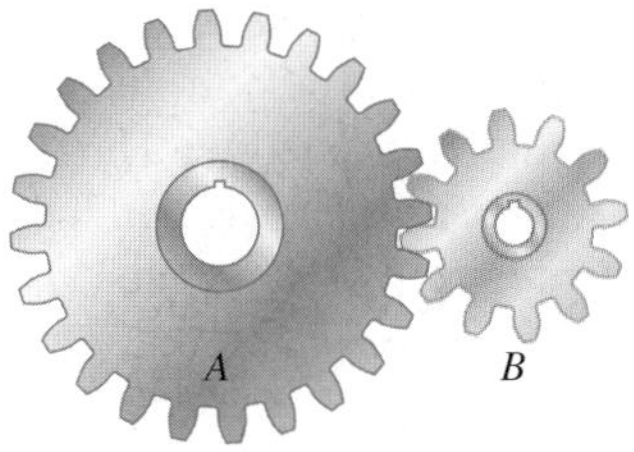

Problems 68 and 69

69. If gear A in Problem 68 has an initial frequency of 0.955 Hz and an angular acceleration of 3.0 rad/s^2, how many rotations does each gear go through in 2.0 s?

70. The time to sunset can be estimated by holding out your arm, holding your fingers horizontally in front of your eyes, and counting the number of fingers that fit between the horizon and the setting Sun. (a) What is the angular speed, in radians per second, of the Sun's apparent circular motion around the Earth? (b) Estimate the angle subtended by one finger held at arm's length. (c) How long in minutes does it take the Sun to "move" through this same angle?

71. In the professional videotape recording system known as quadriplex, four tape heads are mounted on the circumference of a drum of radius 2.5 cm that spins at 1500 rad/s. (a) At what speed are the tape heads moving? (b) Why are moving tape heads used instead of stationary ones, as in audiotape recorders? [*Hint:* How fast would the tape have to move if the heads were stationary?]

72. The Milky Way galaxy rotates about its center with a period of about 200 million yr. The Sun is 2×10^{20} m from the center of the galaxy. How fast is the Sun moving with respect to the center of the galaxy?

73. A small body of mass 0.50 kg is attached by a 0.50-m-long cord to a pin set into the surface of a frictionless table top. The body moves in a circle on the horizontal surface with a speed of 2.0π m/s. (a) What is the magnitude of the radial acceleration of the body? (b) What is the tension in the cord?

74. Two blocks, one with mass $m_1 = 0.050$ kg and one with mass $m_2 = 0.030$ kg, are connected to one another by a string. The inner block is connected to a central pole by another string as shown in the figure with $r_1 = 0.40$ m and $r_2 = 0.75$ m. When the blocks are spun around on a horizontal frictionless surface at an angular speed of 1.5 rev/s, what is the tension in each of the two strings?

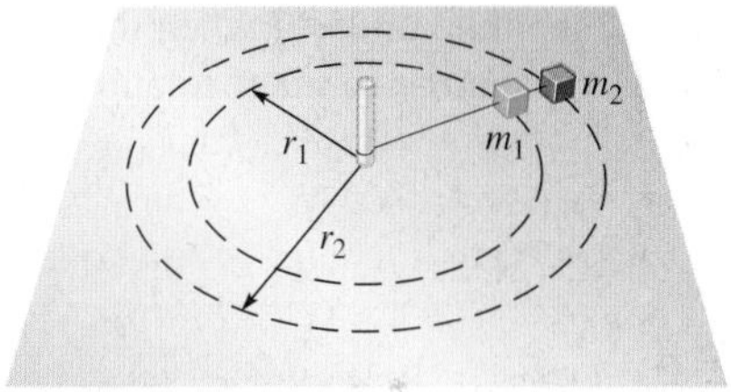

75. What's the fastest way to make a U-turn at constant speed? Suppose that you need to make a 180° turn on a circular path. The minimum radius (due to the car's steering system) is 5.0 m, while the maximum (due to the width of the road) is 20.0 m. Your acceleration must never exceed 3.0 m/s^2 or else you will skid. Should you use the smallest possible radius, so the distance is small, or the largest, so you can go faster without skidding, or something in between? What is the minimum possible time for this U-turn?

76. The Milky Way galaxy rotates about its center with a period of about 200 million yr. The Sun is 2×10^{20} m from the center of the galaxy. (a) What is the Sun's radial acceleration? (b) What is the net gravitational force on the Sun due to the other stars in the Milky Way?

77. Bacteria swim using a corkscrew-like helical flagellum that rotates. For a bacterium with a flagellum that has a pitch of 1.0 μm that rotates at 110 rev/s, how fast could it swim if there were no "slippage" in the medium in which it is swimming? The pitch of a helix is the distance between "threads."

78. You place a penny on a turntable at a distance of 10.0 cm from the center. The coefficient of static friction between the penny and the turntable is 0.350. The turntable's

angular acceleration is 2.00 rad/s^2. How long after you turn on the turntable will the penny begin to slide off of the turntable?

79. A coin is placed on a turntable that is rotating at 33.3 rpm. If the coefficient of static friction between the coin and the turntable is 0.1, how far from the center of the turntable can the coin be placed without having it slip off?

80. Grace, playing with her dolls, pretends the turntable of an old phonograph is a merry-go-round. The dolls are 12.7 cm from the central axis. She changes the setting from 33.3 rpm to 45.0 rpm. (a) For this new setting, what is the linear speed of a point on the turntable at the location of the dolls? (b) If the coefficient of static friction between the dolls and the turntable is 0.13, do the dolls stay on the turntable?

81. Your car's wheels are 65 cm in diameter and the wheels are spinning at an angular velocity of 101 rad/s. How fast is your car moving in kilometers per hour (assume no slippage)?

82. In an amusement park rocket ride, cars are suspended from 4.25-m cables attached to rotating arms at a distance of 6.00 m from the axis of rotation. The cables swing out at an angle of 45.0° when the ride is operating. What is the angular speed of rotation?

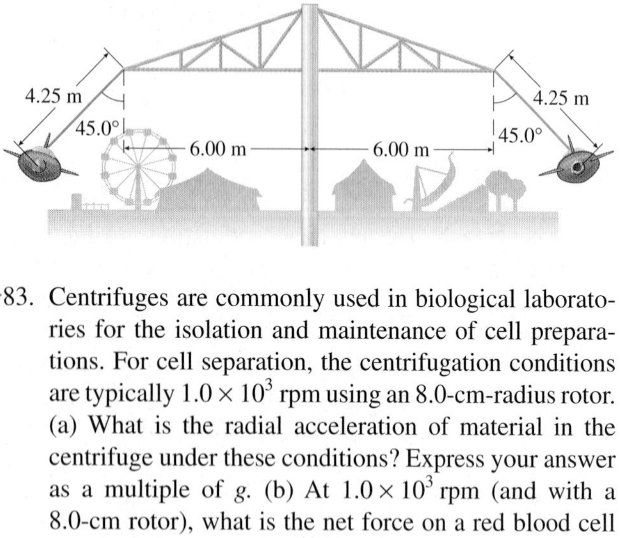

83. Centrifuges are commonly used in biological laboratories for the isolation and maintenance of cell preparations. For cell separation, the centrifugation conditions are typically 1.0×10^3 rpm using an 8.0-cm-radius rotor. (a) What is the radial acceleration of material in the centrifuge under these conditions? Express your answer as a multiple of g. (b) At 1.0×10^3 rpm (and with a 8.0-cm rotor), what is the net force on a red blood cell whose mass is 9.0×10^{-14} kg? (c) What is the net force on a virus particle of mass 5.0×10^{-21} kg under the same conditions? (d) To pellet out virus particles and even to separate large molecules such as proteins, superhigh-speed centrifuges called ultracentrifuges are used in which the rotor spins in a vacuum to reduce heating due to friction. What is the radial acceleration inside an ultracentrifuge at 75000 rpm with an 8.0-cm rotor? Express your answer as a multiple of g.

84. You take a homemade "accelerometer" to an amusement park. This accelerometer consists of a metal nut attached to a string and connected to a protractor, as shown in the figure. While riding a roller coaster that is moving at a uniform speed around a circular path, you hold up the accelerometer and notice that the string is making an angle of 55° with respect to the vertical with the nut pointing away from the center of the circle, as shown. (a) What is the radial acceleration of the roller coaster? (b) What is your radial acceleration expressed as a multiple of g? (c) If the roller coaster track is turning in a radius of 80.0 m, how fast are you moving?

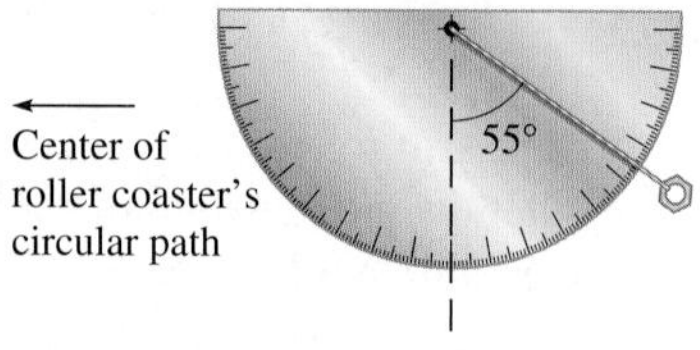

85. Massimo, a machinist, is cutting threads for a bolt on a lathe. He wants the bolt to have 18 threads per inch. If the cutting tool moves parallel to the axis of the would-be bolt at a linear velocity of 0.080 in./s, what must the rotational speed of the lathe chuck be to ensure the correct thread density? [*Hint:* One thread is formed for each complete revolution of the chuck.]

86. In Chapter 19 we will see that a charged particle can undergo uniform circular motion when acted on by a magnetic force and no other forces. (a) For that to be true, what must be the angle between the magnetic force and the particle's velocity? (b) The magnitude of the magnetic force on a charged particle is proportional to the particle's speed, $F = kv$. Show that two identical charged particles moving in circles at different speeds in the same magnetic field must have the same period. (c) Show that the radius of the particle's circular path is proportional to the speed.

87. Find the orbital radius of a geosynchronous satellite. Do not assume the speed found in Example 5.9. Start by writing an equation that relates the period, radius, and speed of the orbiting satellite. Then apply Newton's second law to the satellite. You will have two equations with two unknowns (the speed and radius). Eliminate the speed algebraically and solve for the radius.

Answers to Practice Problems

5.1 3.001×10^{-7} rad/s
5.2 1.65 m/s
5.3 1.9 min
5.4 17 m/s^2
5.5 60 N toward the center of the circular path
5.6 More slowly
5.7 No
5.8 29.7 km/s; 3.17×10^7 s
5.9 0.723R
5.10 2.44 h
5.11 4.2mg

5.12 Acceleration is purely tangential:

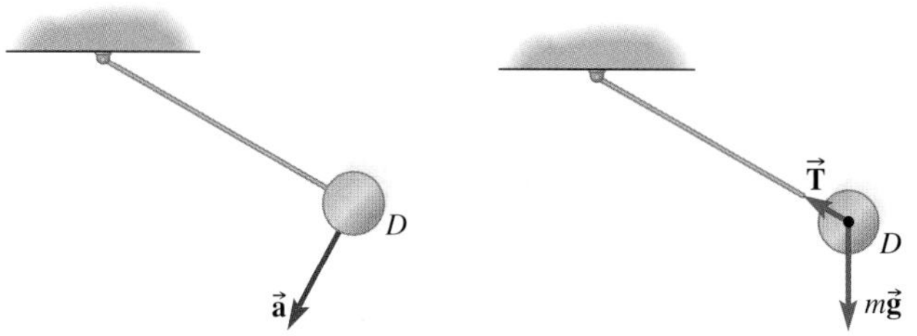

5.13 (a) 1.75×10^{-4} rad/s^2; (b) 0.0349 rad (2.00°)

5.14 (a) 2200 N; (b) 1500 N

Answers to Checkpoints

5.1 7200 Hz

5.2 No, for uniform circular motion the *direction* of the velocity vector is continuously changing but the magnitude of the velocity (the speed) is unchanged.

5.3 The car has friction between the road and the tires to exert a horizontal force that causes the radial acceleration.

5.4 To be geosynchronous the satellites must have an orbital period of 1 d. The only quantities that affect the period are the mass of Earth and the radial distance from Earth's center. These quantities are the same for all satellites no matter the mass.

5.5 For nonuniform circular motion, the direction and the magnitude of the velocity are both changing. There are tangential and radial components to the acceleration. The magnitude of the radial component changes as the speed changes. For uniform circular motion, the magnitude of the velocity is constant but the direction changes. The radial acceleration is constant in magnitude (and the tangential acceleration is zero).

5.6 The radial acceleration cannot be constant because the radius r is constant but the angular velocity ω is changing $a_r = \omega^2 r$.

Review & Synthesis: Chapters 1–5

Review Exercises

1. From your knowledge of Newton's second law and dimensional analysis, find the units (in SI base units) of the spring constant k in the equation $F = kx$, where F is a force and x is a distance.

2. Harrison traveled 2.00 km west, then 5.00 km in a direction 53.0° south of west, then 1.00 km in a direction 60.0° north of west. (a) In what direction, and for how far, should Harrison travel to return to his starting point? (b) If Harrison returns directly to his starting point with a speed of 5.00 m/s, how long will the return trip take?

3. (a) How many center-stripe road reflectors, separated by 17.6 yd, are required along a 2.20-mile section of curving mountain roadway? (b) Solve the same problem for a road length of 3.54 km with the markers placed every 16.0 m. Would you prefer to be the highway engineer in a country with a metric system or U.S. customary units?

4. A baby was spitting up after nursing and the pediatrician prescribed Zantac syrup to reduce the baby's stomach acid. The prescription called for 0.75 mL to be taken twice a day for a month. The pharmacist printed a label for the bottle of syrup that said "3/4 tsp. twice a day." By what factor was the baby overmedicated before the error was discovered at the baby's next office visit two weeks later? [*Hint:* 1 tsp = 4.9 mL.]

5. Mike swims 50.0 m with a speed of 1.84 m/s, then turns around and swims 34.0 m in the opposite direction with a speed of 1.62 m/s. (a) What is his average speed? (b) What is his average velocity?

6. You are watching a television show about Navy pilots. The narrator says that when a Navy jet takes off, it accelerates because the engines are at full throttle and because there is a catapult that propels the jet forward. You begin to wonder how much force is supplied by the catapult. You look on the Web and find that the flight deck of an aircraft carrier is about 90 m long, that an F-14 has a mass of 33000 kg, that each of the two engines supplies 27000 lb of force, and that the takeoff speed of such a plane is about 160 mi/h. Estimate the average force on the jet due to the catapult.

7. On April 15, 1999, a Korean cargo plane crashed due to a confusion over units. The plane was to fly from Shanghai, China, to Seoul, Korea. After take-off the plane climbed to 900 m. Then the first officer was instructed by the Shanghai tower to climb to 1500 m and maintain that altitude. The captain, after reaching 1450 m, twice asked the first officer at what altitude they should fly. He was twice told incorrectly they were to be at 1500 ft. The captain pushed the control column quickly forward and started a steep descent. The plane could not recover from the dive and crashed. How much above the correct altitude did the captain think they were when he started his rapid descent and lost control? (It turns out that aircraft altitudes are given in feet throughout the world except in China, Mongolia, and the former Soviet states where meters are used.)

8. Paula swims across a river that is 10.2 m wide. She can swim at 0.833 m/s in still water, but the river flows with a speed of 1.43 m/s. If Paula swims in such a way that she crosses the river in as short a time as possible, how far downstream is she when she gets to the opposite shore?

9. ✦ Peter is collecting paving stones from a quarry. He harnesses two dogs, Sandy and Rufus, in tandem to the loaded cart. Sandy pulls with force $\vec{\mathbf{F}}$ at a 15° angle to the north of east; Rufus pulls with 1.5 times the force of Sandy and at an angle of 30.0° south of east. Use a ruler and protractor to draw the force vectors to scale (choose a simple scale, such as 2.0 cm $\leftrightarrow F$). Find the sum of the two force vectors graphically. Measure its length and find the magnitude of the sum from the scale used and the direction with the protractor. Will the cart stay on the road that runs directly west to east?

10. A tire swing hangs at a 12° angle to the vertical when a stiff breeze is blowing. In terms of the tire's weight W, (a) what is the magnitude of the horizontal force exerted on the tire by the wind? (b) What is the tension in the rope supporting the tire? Ignore the weight of the rope.

11. An astronaut of mass 60.0 kg and a small asteroid of mass 40.0 kg are initially at rest with respect to the space station. The astronaut pushes the asteroid with a constant force of magnitude 250 N for 0.35 s. Gravitational forces are negligible. (a) How far apart are the astronaut and the asteroid 5.00 s after the astronaut stops pushing? (b) What is their relative speed at this time?

12. In the fairy tale, Rapunzel, the beautiful maiden let her long golden hair hang down from the tower in which she was held prisoner so that her prince could use her hair as a climbing rope to climb the tower and rescue her. (a) Estimate how much force is required to pull a strand of hair out of your head. (b) There are about 10^5 hairs growing out of Rapunzel's head. If the prince has a mass of 60 kg, estimate the average force pulling on each strand of hair. Will Rapunzel be bald by the time the prince reaches the top of the 30-m tower?

13. Marie slides a paper plate with a slice of pizza across a horizontal table to her friend Jaden. The coefficient of friction between the table and plate is 0.32. If the pizza must travel 44 cm to get from Marie to Jaden, what initial speed should Marie give the plate of pizza so that it just stops when it gets to Jaden?

14. Two wooden crates with masses as shown are tied together by a horizontal cord. Another cord is tied to the first crate and it is pulled with a force of 195 N at an angle of 20.0°, as shown. Each crate has a coefficient of

kinetic friction of 0.550. Find the tension in the rope between the crates and the magnitude of the acceleration of the system.

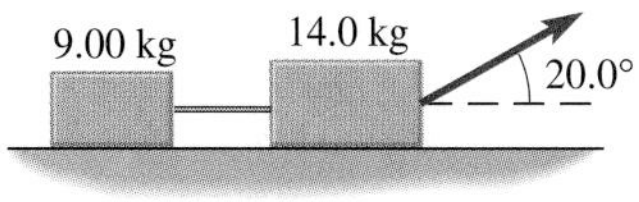

15. A boy has stacked two blocks on the floor so that a 5.00-kg block is on top of a 2.00-kg block. (a) If the coefficient of static friction between the two blocks is 0.400 and the coefficient of static friction between the bottom block and the floor is 0.220, with what minimum force should the boy push horizontally on the upper block to make both blocks start to slide together along the floor? (b) If he pushes too hard, the top block starts to slide off the lower block. What is the maximum force with which he can push without that happening if the coefficient of kinetic friction between the bottom block and the floor is 0.200?
16. A binary star consists of two stars of masses M_1 and $4.0M_1$ a distance d apart. Is there any point where the gravitational field due to the two stars is zero? If so, where is that point?
17. Two boys are trying to break a cord. Gerardo says they should each pull in opposite directions on the two ends; Stefan says they should tie the cord to a pole and both pull together on the opposite end. Which plan is more likely to work?
18. Fish don't move as fast as you might think. A small trout has a top swimming speed of only about 2 m/s, which is about the speed of a brisk walk (for a human, not a fish!). It may seem to move faster because it is capable of large *accelerations*—it can dart about, changing its speed or direction very quickly. (a) If a trout starts from rest and accelerates to 2 m/s in 0.05 s, what is the trout's average acceleration? (b) During this acceleration, what is the average net force on the trout? Express your answer as a multiple of the trout's weight. (c) Explain how the trout gets the water to push it forward.
19. A spotter plane sees a school of tuna swimming at a steady 5.00 km/h northwest. The pilot informs a fishing trawler, which is just then 100.0 km due south of the fish. The trawler sails along a straight-line course and intercepts the tuna after 4.0 h. How fast did the trawler move? [*Hint:* First find the velocity of the trawler relative to the tuna.]
20. Julia is delivering newspapers. Suppose she is driving at 15 m/s along a straight road and wants to drop a paper out the window from a height of 1.00 m so it slides along the shoulder and comes to rest in the customer's driveway. At what horizontal distance before the driveway should she drop the paper? The coefficient of kinetic friction between the newspaper and the ground is 0.40. Ignore air resistance and assume no bouncing or rolling.
21. Three rocks are thrown from a cliff with the same initial speeds but in different directions: one straight down, one straight up, and one horizontally. Ignore air resistance. (a) Compare the speeds of the three rocks just before they hit the flat ground at the bottom of the cliff. (b) Illustrate your answer by calculating the final speeds for three rocks thrown in the specified directions with initial speeds of 10.0 m/s from a cliff that is 15.00 m high. [*Hint:* Remember that the speed is the magnitude of the velocity vector.]
22. You are watching the Super Bowl where your favorite team is leading by a score of 21 to 20. The other team is lining up to try to kick the winning field goal. You watched their kicker warm up and you saw that he could kick the football with a velocity of 21 m/s. He lines up for a 45-yd kick. You watch as he kicks the ball at an angle of 35° above the horizontal. Assuming he kicks the ball straight and with the same speed as during the warmup, will the ball clear the 10-ft-high goal post, or will your favorite team win the Super Bowl?
23. A coin is placed on a turntable 13.0 cm from the center. The coefficient of static friction between the coin and the turntable is 0.110. Once the turntable is turned on, its angular acceleration is 1.20 rad/s^2. How long will it take until the coin begins to slide?
24. Carlos and Shannon are sledding down a snow-covered slope that is angled at 12° below the horizontal. When sliding on snow, Carlos's sled has a coefficient of friction $\mu_k = 0.10$; Shannon has a "supersled" with $\mu_k = 0.010$. Carlos takes off down the slope starting from rest. When Carlos is 5.0 m from the starting point, Shannon starts down the slope from rest. (a) How far have they traveled when Shannon catches up to Carlos? (b) How fast is Shannon moving with respect to Carlos as she passes by?
25. A proposed "space elevator" consists of a cable going all the way from the ground to a space station in geosynchronous orbit (always above the same point on Earth's surface). Elevator "cars" would climb the cable to transport cargo to outer space. Consider a cable connected between the equator and a space station at height H above the surface. Ignore the mass of the cable*. (a) Find the height H. (b) Suppose there is an elevator car of mass 100 kg sitting halfway up at height $H/2$. What tension T would be required in the cable to hold the car in place? Which part of the cable would be under tension (above the car or below it)?
26. Anthony is going to drive a flat-bed truck up a hill that makes an angle of 10° with respect to the horizontal direction. A 36.0-kg package sits in the back of the truck. The coefficient of static friction between the package

*More realistically, the mass of the cable is one of the primary engineering challenges of a space elevator. The cable is so long that it would have a very large mass and would have to withstand an enormous tension to support its own weight. The cable would need to be supported by a counterweight positioned beyond the geosynchronous orbit. Some believe *carbon nanotubes* hold the key to producing a cable with the required properties.

and the truck bed is 0.380. What is the maximum acceleration the truck can have without the package falling off the back?

27. A road with a radius of 75.0 m is banked so that a car can navigate the curve at a speed of 15.0 m/s without any friction. On a cold day when the street is icy, the coefficient of static friction between the tires and the road is 0.120. What is the *slowest* speed the car can go around this curve without sliding *down* the bank?

28. You want to lift a heavy box with a mass of 98.0 kg using the two-pulley system as shown. With what minimum force do you have to pull down on the rope in order to lift the box at a constant velocity? One pulley is attached to the ceiling and one to the box.

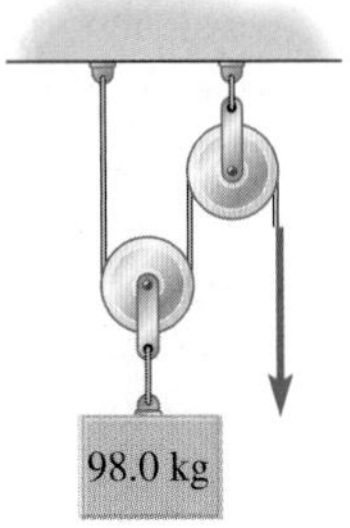

29. At time $t = 0$, block A of mass 0.225 kg and block B of mass 0.600 kg rest on a horizontal frictionless surface a distance 3.40 m apart, with block A located to the left of block B. A horizontal force of 2.00 N directed to the right is applied to block A for a time interval $\Delta t = 0.100$ s. During the same time interval, a 5.00-N horizontal force directed to the left is applied to block B. How far from B's initial position do the two blocks meet? How much time has elapsed from $t = 0$ until the blocks meet?

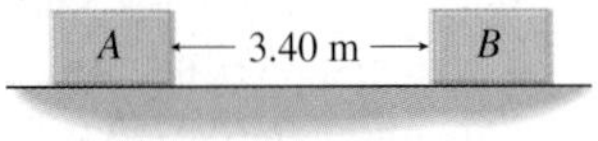

30. A hamster of mass 0.100 kg gets onto his 20.0-cm-diameter exercise wheel and runs along inside the wheel for 0.800 s until its frequency of rotation is 1.00 Hz. (a) What is the tangential acceleration of the wheel, assuming it is constant? (b) What is the normal force on the hamster just before he stops? The hamster is at the bottom of the wheel during the entire 0.800 s.

31. A pellet is fired from a toy cannon with a velocity of 12 m/s directed 60° above the horizontal. After 0.10 s, a second identical pellet is fired with the same initial velocity. After an additional 0.15 s have passed, what is the velocity of the first pellet with respect to the second? Ignore air resistance.

32. A crate is sliding down a frictionless ramp that is inclined at 35.0°. (a) If the crate is released from rest, how far does it travel down the incline in 2.50 s if it does not get to the bottom of the ramp before the time has elapsed? (b) How fast is the crate moving after 2.50 s of travel?

33. The invention of the cannon in the fourteenth century made the catapult unnecessary and ended the safety of castle walls. Stone walls were no match for balls shot from cannons. Suppose a cannonball of mass 5.00 kg is launched from a height of 1.10 m, at an angle of elevation of 30.0° with an initial velocity of 50.0 m/s, toward a castle wall of height 30 m and located 215 m away from the cannon. (a) The range of a projectile is defined as the horizontal distance traveled when the projectile returns to its original height. Derive an equation for the range in terms of v_i, g, and angle of elevation θ. (b) What will be the range reached by the projectile, if it is not intercepted by the wall? (c) If the cannonball travels far enough to hit the wall, find the height at which it strikes.

34. Two blocks are connected by a lightweight, flexible cord that passes over a single frictionless pulley. If $m_1 >> m_2$, find (a) the acceleration of each block and (b) the tension in the cord.

35. A runner runs three-quarters of the way around a circular track of radius 60.0 m, when she collides with another runner and trips. (a) How far had the runner traveled on the track before the collision? (b) What was the magnitude of the displacement of the runner from her starting position when the accident occurred?

36. A solar sailplane is going from Earth to Mars. Its sail is oriented to give a solar radiation force of 8.00×10^2 N. The gravitational force due to the Sun is 173 N and the gravitational force due to the Earth is 1.00×10^2 N. All forces are in the plane formed by Earth, Sun, and sailplane. The mass of the sailplane is 14 500 kg. (a) What is the net force (magnitude and direction) acting on the sailplane? (b) What is the acceleration of the sailplane?

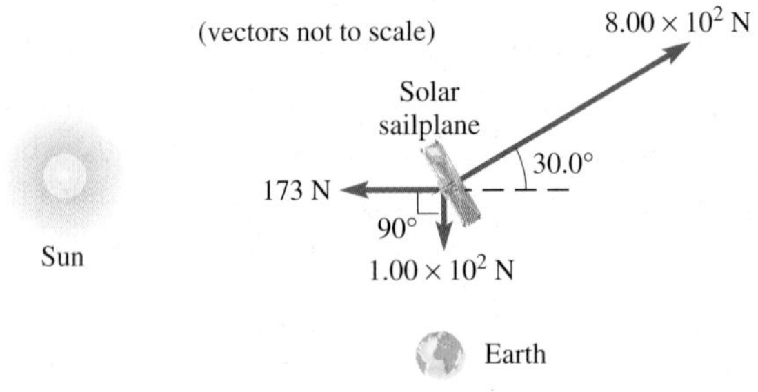

37. A star near the visible edge of a galaxy travels in a uniform circular orbit. It is 40 000 ly (light-years) from the galactic center and has a speed of 275 km/s. (a) Estimate the total mass of the galaxy based on the motion of the star? [*Hint:* For this estimate, assume the total mass to be concentrated at the galactic center and relate it to the gravitational force on the star.] (b) The total *visible* mass (i.e., matter we can detect via electromagnetic radiation) of the galaxy is 10^{11} solar masses. What fraction of the total mass of the galaxy is visible*, according to this estimate?

38. One of the tricky things about learning to sail is distinguishing the true wind from the apparent wind. When you are on a sailboat and you feel the wind on your face, you are experiencing the *apparent wind*—the motion of

*In many galaxies the stars appear to have roughly the *same orbital speed* over a large range of distances from the center. A popular hypothesis to explain such galaxy rotation velocities is the existence of *dark matter*—matter that we cannot detect via electromagnetic radiation. Dark matter is thought to account for the majority of the mass of some galaxies and nearly a fourth of the total mass of the universe.

the air relative to you. The true wind is the speed and direction of the air relative to the water while the apparent wind is the speed and direction of the air relative to the *sailboat.* The figure shows three different directions for the true wind along with one possible sail orientation as indicated by the position of the boom attached to the mast. (a) In each case, draw a vector diagram to establish the magnitude and direction of the apparent wind. (b) In which of the three cases is the apparent wind speed greater than the true wind speed? (Assume that the speed of the boat relative to the water is less than the true wind speed.) (c) In which of the three cases is the direction of the apparent wind direction forward of the true wind? ["Forward" means coming from a direction more nearly straight ahead. For example, (1) is forward of (2), which is forward of (3).]

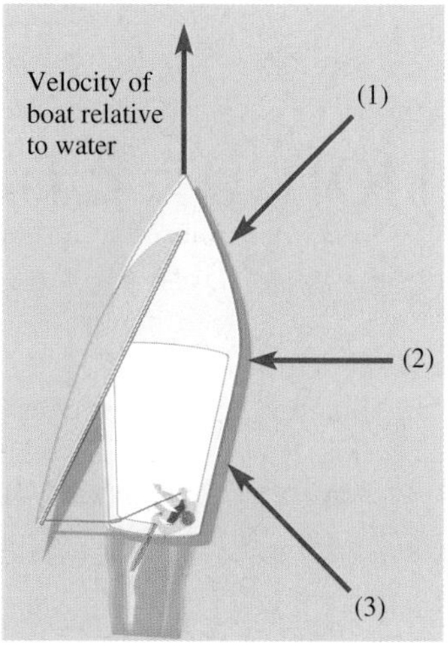

MCAT Review

The section that follows includes MCAT exam material and is reprinted with permission of the Association of American Medical Colleges (AAMC).

Read the paragraph and then answer the following four questions:

The study of the flight of projectiles has many practical applications. The main forces acting on a projectile are air resistance and gravity. The path of a projectile is often approximated by ignoring the effects of air resistance. Gravity is then the only force acting on the projectile. When air resistance is included in the analysis, another force, $\vec{\mathbf{F}}_R$, is introduced. F_R is proportional to the square of the velocity, v. The direction of the air resistance is exactly opposite the direction of motion. The equation for air resistance is $F_R = bv^2$, where b is a proportionality constant that depends on such factors as the density of the air and the shape of the projectile.

Air resistance was studied by launching a 0.5-kg projectile from a level surface. The projectile was launched with a speed of 30 m/s at a 40° angle to the surface. (Note: Assume air resistance is present unless otherwise specified.)

1. What is the magnitude of the vertical acceleration of the projectile immediately after it is launched? (Note: v_y = vertical velocity component.)
   A. $-g + (bvv_y)$
   B. $-g - (bvv_y)$
   C. $-g + (bvv_y)/(0.5\text{ kg})$
   D. $-g - (bvv_y)/(0.5\text{ kg})$

2. Approximately what horizontal distance does the projectile travel before returning to the elevation from which it was launched? (Note: Assume that the effects of air resistance are negligible.)
   A. 45 m B. 60 m C. 90 m D. 120 m

3. What is the magnitude of the *horizontal component of air resistance* on the projectile at any point during flight? (Note: v_x = horizontal velocity component.)
   A. $(bvv_x)\cos 40°$
   B. $(bvv_x)/2$
   C. $(bvv_x)\sin 40°$
   D. bvv_x

4. How does the amount of time it takes a projectile to reach its maximum height compare to the time it takes to fall from its maximum height back to the ground? (Note: b is greater than zero.)
   A. The times are the same.
   B. The time to reach its maximum height is greater.
   C. The time to fall back to the ground is greater.
   D. Either can be greater depending on the magnitude of b.

Read the paragraph and then answer the following questions:

A raft is constructed from wood and used in a river that varies in depth, width, and current at several points along its length. The river at point A has a current of 2 m/s, a width of 200 m, and an average depth of 3 m.

5. Near point A, the raft is rowed at a constant velocity of 2 m/s relative to the river current and perpendicular to it. How far does the raft travel before it reaches the other side?
   A. 224 m
   B. 250 m
   C. 283 m
   D. 400 m

6. A rower at point A rows the raft at 3 m/s relative to the river current and wants to end up directly across the river from the point of origin. At what angle to the shore should the rower direct the raft?
   A. $\cos^{-1}\frac{5}{3}$
   B. $\cos^{-1}\frac{2}{5}$
   C. $\cos^{-1}\frac{3}{2}$
   D. $\cos^{-1}\frac{2}{3}$

7. A rock is dropped from a cliff that is 100 m above ground level. How long does it take the rock to reach the ground? (Note: Use $g = 10\text{ m/s}^2$.)
   A. 4.5 s
   B. 10 s
   C. 14 s
   D. 20 s

CHAPTER

6 Conservation of Energy

As a kangaroo hops along, the maximum height of each hop might be around 2.8 m. This height is only slightly higher than that achieved by an Olympic high jumper, but the kangaroo is able to achieve this height hop after hop as it travels with a horizontal velocity of 15 m/s or more. What features of kangaroo anatomy make this feat possible? It cannot simply be a matter of having more powerful leg muscles. If it were, the kangaroo would have to consume large amounts of energy-rich food to supply the muscles with enough chemical energy for each jump, but in reality a kangaroo's diet consists largely of grasses that are poor in energy content. (See p. 210 for the answer.)

Concepts & Skills to Review

- gravitational forces (Section 4.5)
- Newton's second law: force and acceleration (Sections 4.3–4.8)
- components of vectors (Section 3.2)
- circular orbits (Section 5.4)
- area under a graph (Sections 2.2 and 2.3)

6.1 THE LAW OF CONSERVATION OF ENERGY

Until now, we have relied on Newton's laws of motion to be the fundamental physical laws used to analyze the forces that act on objects and to predict the motion of objects. Now we introduce another physical principle: the conservation of energy. A **conservation law** is a physical principle that identifies some quantity that does not change with time. Conservation of energy means that every physical process leaves the total energy in the universe unchanged. Energy can be converted from one form to another, or transferred from one place to another. If we are careful to account for all the energy transformations, we find that the total energy remains the same.

Conservation law: a physical law that identifies a quantity that does not change with time.

The Law of Conservation of Energy

The total energy in the universe is unchanged by any physical process:

total energy before = total energy after.

"Turn down the thermostat—we're trying to conserve energy!" In ordinary language, *conserving energy* means trying not to waste useful energy resources. In the scientific meaning of *conservation*, energy is *always* conserved no matter what happens. When we "produce" or "generate" electric energy, for instance, we aren't creating any new energy; we're just converting energy from one form into another that's more useful to us.

Conservation of energy is one of the few universal principles of physics. No exceptions to the law of conservation of energy have been found. Conservation of energy is a powerful tool in the search to understand nature. It applies equally well to radioactive decay, the gravitational collapse of a star, a chemical reaction, a biological process such as respiration, and to the generation of electricity by a wind turbine (Fig. 6.1). Think about the energy conversions that make life possible. Green plants use photosynthesis to convert the energy they receive from the Sun into stored chemical energy. When animals eat the plants, that stored energy enables motion, growth, and maintenance of body temperature. Energy conservation governs every one of these processes.

Figure 6.1 At a California "wind farm," these wind turbines convert the energy of motion of the air into electric energy.

Choosing Between Alternative Solution Methods Some problems can be solved using either energy conservation *or* Newton's second law. Usually the energy method is easier. We often don't know the details of all the forces acting on an object, making a direct application of Newton's second law difficult. Using conservation of energy enables us to solve some of these problems more easily. When deciding which of these two approaches to use to solve a problem, try using energy conservation first. If the energy method does not lead to the solution, then try Newton's second law.

Historical Development of the Principle of Energy Conservation While many scientists contributed to the development of the law of conservation of energy, the law's first clear statement was made in 1842 by the German surgeon Julius Robert von Mayer (1814–1878). As a ship's physician on a voyage to what is now Indonesia, Mayer had noticed that the sailors' venous blood was a much deeper red in the tropics than it was in Europe. He concluded that less oxygen was being used because they didn't need to "burn" as much fuel to keep the body warm in the warmer climate.

Figure 6.2 The stored chemical energy in food enables a weightlifter to lift the barbell over her head.

Table 6.1 Some Common Forms of Energy

Form of Energy	Brief Description
Translational kinetic	Energy of translational motion (Chapter 6)
Elastic	Energy stored in a "springy" object or material when it is deformed (Chapter 6)*
Gravitational	Energy of gravitational interactions (Chapter 6)
Rotational kinetic	Energy of rotational motion (Chapter 8)*
Vibrational, acoustic, seismic	Energy of the oscillatory motions of atoms and molecules in a substance caused by a mechanical wave passing through it (Chapters 11 and 12)*
Internal	Energies of motion and interaction of atoms and molecules in solids, liquids, and gases, related to our sensation of temperature (Chapters 14 and 15)*
Electromagnetic	Energy of interaction of electric charges and currents; energy of electromagnetic fields, including electromagnetic waves such as light (Chapters 14, 17–22)
Rest	The total energy of a particle of mass m when it is at rest, given by Einstein's famous equation $E = mc^2$ (Chapters 26, 29, and 30)
Chemical	Energies of motion and interaction of electrons in atoms and molecules (Chapter 28)*
Nuclear	Energies of motion and interaction of protons and neutrons in atomic nuclei (Chapters 29 and 30)

*Not a *fundamental* form of energy; made up of microscopic kinetic or electromagnetic energy.

In 1843, the English physicist James Prescott Joule (1818–1889), whose "day job" was running the family brewery, performed precise experiments to show that gravitational potential energy could be converted into a previously unrecognized form of energy (internal energy). It had previously been thought that forces like friction "use up" energy. Thanks to Mayer, Joule, and others, we now know that friction converts mechanical forms of energy into internal energy and that total energy is always conserved.

Forms of Energy

Kinetic energy: energy of motion.

Potential energy: stored energy due to interaction.

Energy comes in many different forms (Fig. 6.2). Table 6.1 summarizes the main forms of energy discussed in this text and indicates the principal chapters that discuss each one. At the most fundamental level, there are only three kinds of energy: energy due to motion (**kinetic energy**), stored energy due to interaction (**potential energy**), and rest energy.

Translation: motion of an object in which any point of the object moves with the same velocity as any other point. (That is, the object does not rotate or change shape.)

To apply the energy conservation principle, we need to learn how to calculate the amount of each form of energy. There isn't one formula that applies to all. Fortunately, we don't have to learn about all of them at once. This chapter focuses on three forms of macroscopic mechanical energy (kinetic energy, gravitational potential energy, and elastic potential energy). For now, we use energy conservation as a tool to understand the **translational** motion of objects, but we do not consider rotational motion or changes in the *internal* energy of an object. We assume that these moving objects are perfectly rigid, so every point on the object moves through the same displacement.

6.2 WORK DONE BY A CONSTANT FORCE

To apply the principle of energy conservation, we need to learn how energy can be converted from one form to another. We begin with an example. Suppose the trunk in Fig. 6.3a weighs 220 N and must be lifted a height $h = 4.0$ m. To lift it at constant speed, Rosie must exert a force of 220 N on the rope, assuming an ideal pulley and rope. (We

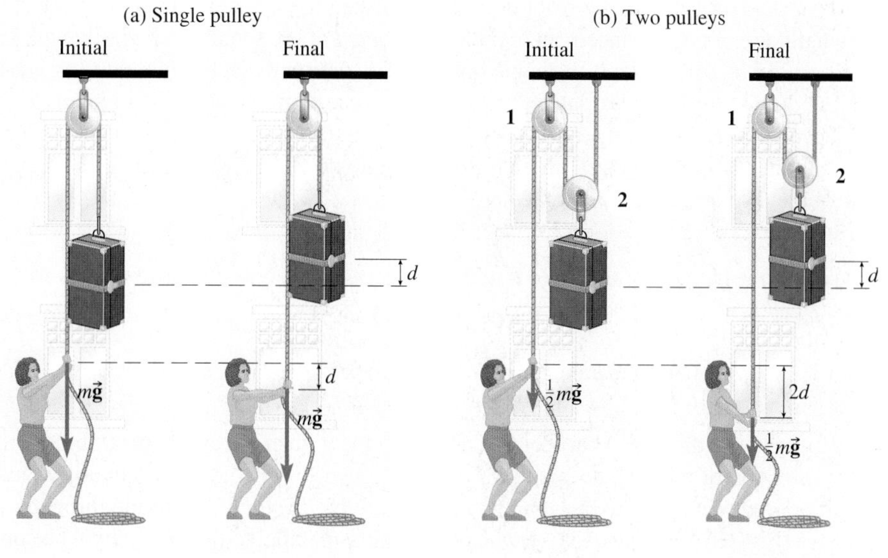

Figure 6.3 (a) Rosie moves a trunk into her dorm room through the window. (b) The two-pulley system makes it easier for Rosie to lift the trunk: the *force* she must exert is halved. Is she getting something for nothing, or does she still have to do the same amount of *work* to lift the trunk?

ignore for now the brief initial time when she pulls with more than 220 N to accelerate the trunk from rest to its constant speed and the brief time she pulls with less than 220 N to let it come to rest.)

As discussed in Example 4.12, she would only have to exert half the force (110 N) if she were to use the two-pulley system of Fig. 6.3b. She doesn't get something for nothing, though. To lift the trunk 4.0 m, the sections of rope on *both* sides of pulley 2 must be shortened by 4.0 m, so Rosie must pull an 8.0-m length of rope. The two-pulley system enables her to pull with half the force, but now she must pull the rope through *twice the distance.*

Notice that the *product* of the magnitude of the force and the distance is the same in both cases:

$$220 \text{ N} \times 4.0 \text{ m} = 110 \text{ N} \times 8.0 \text{ m} = 880 \text{ N·m} = W$$

Figure 6.4 While the trunk is held in place by tying the rope, no work is done and no energy transfers occur.

This product is called the **work** (W) done by Rosie on the rope. Work is a scalar quantity; it does not have a direction. The same symbol W is often used for the weight of an object. To avoid confusion, we write mg for weight and let W stand for work.

Don't be misled by the many different meanings the word *work* has in ordinary conversation. We talk about doing homework, or going to work, or having too much work to do. Not everything we call "work" in conversation is *work* as defined in physics.

The SI unit of work and energy is the newton-meter (N·m), which is given the name joule (symbol: J). Using either method, Rosie must do 880 J of work on the rope to lift the trunk. When we say that Rosie does 880 J of work, we mean that Rosie supplies 880 J of energy—the amount of energy required to lift the trunk 4.0 m. *Work is an energy transfer that occurs when a force acts on an object that moves.*

SI unit of work and energy is the joule: 1 J = 1 N·m.

Rosie does no work on the rope while she holds it in one place because the displacement is zero. She can just as well fasten it and walk away (Fig. 6.4). *If there is no displacement, no work is done and no energy is transferred.* Why then does she get tired if she holds the rope in place for a long time? Although Rosie does no work *on the rope* when holding it in place, work *is* done inside her body by muscle fibers, which have to contract and expand continually to maintain tension in the muscle. This internal work converts chemical energy into internal energy—the muscle warms up—but no energy is transferred *to the trunk.*

Work: an energy transfer that occurs when a force acts on an object that moves.

Work Done by a Force not Parallel to the Displacement The force that Rosie exerts on the rope is in the same direction as the displacement of that end of the rope. More generally, how much work is done by a constant force that is at some angle to the displacement? It turns out that only the *component* of the force *in the direction of the displacement*

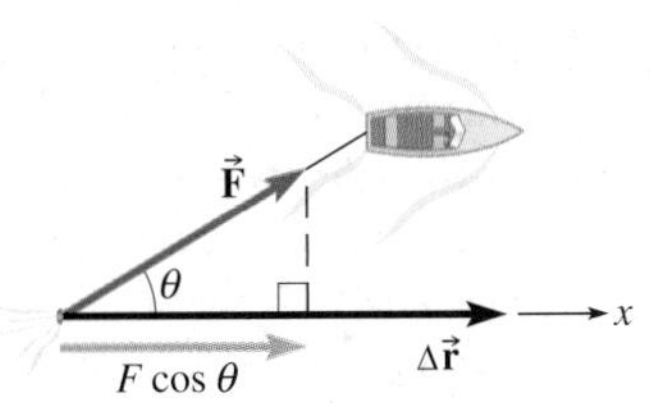

Figure 6.5 The work done by the force of the towrope on the water-skier during a displacement $\Delta\vec{\mathbf{r}}$ is $(F \cos \theta)\, \Delta r$, where $(F \cos \theta)$ is the component of $\vec{\mathbf{F}}$ in the direction of $\Delta\vec{\mathbf{r}}$.

does work. So, in general, the work done by a constant force is defined as the product of the magnitude of the displacement and the *component* of the force *in the direction of the displacement*. If θ represents the angle between the force and displacement vectors when they are drawn starting at the same point, then the force component in the direction of the displacement is $F \cos \theta$ (Fig. 6.5). Therefore, work done by a constant force on an object can be written $W = F\, \Delta r \cos \theta$, where F is the magnitude of the force and Δr is the magnitude of the displacement of the object.

Work done by a constant force $\vec{\mathbf{F}}$ acting on an object whose displacement is $\Delta\vec{\mathbf{r}}$:

$$W = F\, \Delta r \cos \theta \qquad (6\text{-}1)$$

(θ is the angle between $\vec{\mathbf{F}}$ and $\Delta\vec{\mathbf{r}}$)

If we choose the x-axis parallel to the displacement, then the component of the force in the direction of the displacement is $F_x = F \cos \theta$, so $W = F_x \Delta x$. Alternatively, we can identify $\Delta r \cos \theta$ in Eq. (6-1) as the component of the *displacement* in the direction of the *force* (Fig. 6.6). Therefore, if we choose the x-axis parallel to the *force*, then the component of the displacement in the direction of the force is Δx and $W = F_x \Delta x$, as before:

The **scalar product** (or **dot product**) of two vectors is defined by the equation $\vec{\mathbf{A}} \cdot \vec{\mathbf{B}} = AB \cos \theta$, where θ is the angle between $\vec{\mathbf{A}}$ and $\vec{\mathbf{B}}$ when they are drawn starting at the same point. The special name and notation are used because this pattern occurs often in physics and mathematics. Work can be expressed using the scalar product: $W = \vec{\mathbf{F}} \cdot \Delta\vec{\mathbf{r}}$. See Appendix A.8 for more information on the scalar product.

Work done by a constant force $\vec{\mathbf{F}}$ acting on an object whose displacement is $\Delta\vec{\mathbf{r}}$:

$$W = F_x\, \Delta x \qquad (6\text{-}2)$$

($\vec{\mathbf{F}}$ and/or $\Delta\vec{\mathbf{r}}$ parallel to the x-axis)

Work Can Be Positive, Negative, or Zero When the angle between $\vec{\mathbf{F}}$ and $\Delta\vec{\mathbf{r}}$ is less than 90°, $\cos \theta$ in Eq. (6-1) is positive, so the work done by the force is positive ($W > 0$). If the angle between $\vec{\mathbf{F}}$ and $\Delta\vec{\mathbf{r}}$ is greater than 90°, $\cos \theta$ is negative and the work done by the force is negative ($W < 0$). Pay careful attention to the algebraic sign when calculating work. For example, the rope pulls Rosie's trunk in the direction of its displacement, so $\theta = 0$ and $\cos \theta = 1$; the rope does positive work on the trunk. At the same time, gravity pulls downward in the direction opposite to the displacement, so $\theta = 180°$ and $\cos \theta = -1$; gravity does *negative* work on the trunk.

If the force is perpendicular to the displacement, $\theta = 90°$ and $\cos 90° = 0$, so the work done is zero. For example, the normal force exerted by a stationary surface on a sliding object does no work because it is perpendicular to the displacement of the object (Fig. 6.7a). Even if the surface is curved, at any instant the normal force is perpendicular to the velocity of the object. During a short time interval, then, the normal force is perpendicular to the displacement $\Delta\vec{\mathbf{r}} = \vec{\mathbf{v}}\, \Delta t$ (Fig. 6.7b), so the normal force still does zero work.

On the other hand, if the surface exerting the normal force is moving, then the normal force can do work. In Fig. 6.7c, the normal force exerted by the forklift on the pallet does positive work as it lifts the pallet.

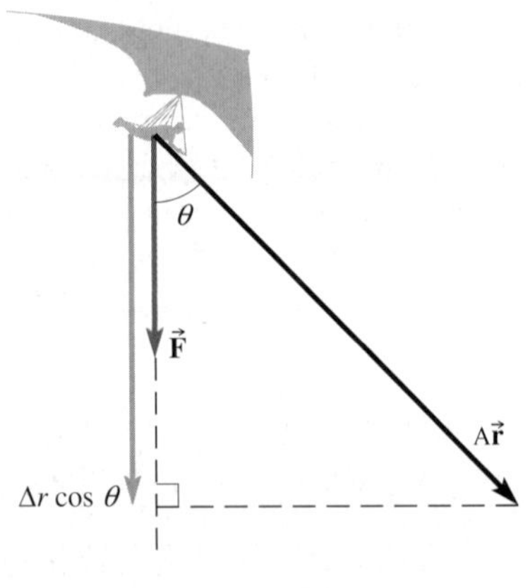

Figure 6.6 The work done by the force of gravity on the hang glider during a displacement $\Delta\vec{\mathbf{r}}$ is $F(\Delta r \cos \theta)$, which is F times the component of $\Delta\vec{\mathbf{r}}$ in the direction of $\vec{\mathbf{F}}$.

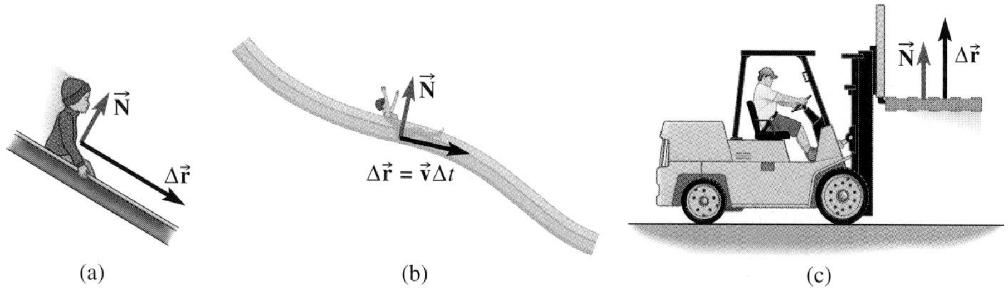

Figure 6.7 (a) The normal force does no work because it is perpendicular to the displacement. (b) Even while sliding on a curved surface, the direction of the normal force is always perpendicular to the displacement during a short Δt, so it does no work. (c) The normal force that the forklift exerts on the pallet does work; it is not perpendicular to the displacement.

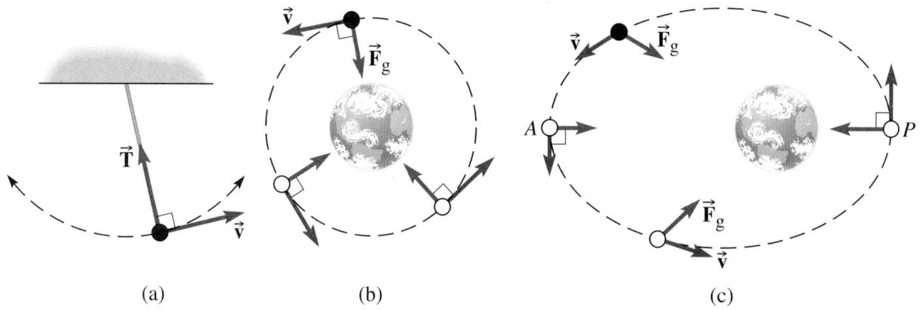

Figure 6.8 (a) The tension in the string of a pendulum is always perpendicular to the velocity of the pendulum bob, so the string does no work on the bob. (b) No matter where the satellite is in its circular orbit, it experiences a gravitational force directed toward the center of the Earth. This force is always perpendicular to the satellite's velocity; thus, gravity does no work on the satellite. (c) In an elliptical orbit, the gravitational force is *not* always perpendicular to the velocity. As the satellite moves counterclockwise in its orbit from point P to point A, gravity does negative work; from A to P, gravity does positive work.

No work is done by the tension in the string on a swinging pendulum bob because the tension is always perpendicular to the velocity of the bob (Fig. 6.8a). Similarly, no work is done by the Earth's gravitational force on a satellite in circular orbit (Fig. 6.8b). In a circular orbit, the gravitational force is always directed along a radius from the satellite to the center of the Earth. At every point in the orbit, the gravitational force is perpendicular to the velocity of the satellite (which is tangent to the circular orbit).

By contrast, gravity does work on a satellite in a noncircular orbit (Fig. 6.8c). Only at points A and P are the gravitational force and the satellite's velocity perpendicular. Wherever the angle between the gravitational force and the velocity is less than 90°, gravity is doing positive work, increasing the satellite's kinetic energy by making it move faster. Wherever the angle between the gravitational force and the velocity is greater than 90°, gravity is doing negative work, decreasing the satellite's kinetic energy by slowing it down.

Application of work: elliptical orbits

CHECKPOINT 6.2

A force is applied to a moving object, but no work is done. How is that possible?

Example 6.1

Antique Chest Delivery

A valuable antique chest, made in 1907 by Gustav Stickley, is to be moved into a truck. The weight of the chest is 1400 N. To get the chest from the ground onto the truck bed, which is 1.0 m higher, the movers must decide what to do. Should they lift it straight up, or should they push it up their 4.0-m-long ramp? Assume they push the chest on a wheeled dolly, which in a simplified model is equivalent to sliding it up a *frictionless* ramp.

(a) Find the work done by the movers on the chest if they lift it straight up 1.0 m at constant speed.
(b) Find the work done by the movers on the chest if they slide the chest up the 4.0-m-long *frictionless* ramp at constant speed by pushing parallel to the ramp.
(c) Find the work done by gravity on the chest in each case.
(d) Find the work done by the normal force of the ramp on the chest. Assume that all forces are constant.

Strategy To calculate work, we use either Eq. (6-1) or Eq. (6-2), whichever is easier. For (a) and (b), we must calculate the force exerted by the movers. Drawing the FBD helps us calculate the forces. The ramp is a simple machine—just as for Rosie's pulleys, the ramp cannot reduce the amount of work that must be done, so we expect the work done by the movers to be the same in both cases (ignoring friction). We expect the work done by gravity to be negative in both cases, since the chest is moving up while gravity pulls down. The normal force due to the ramp is perpendicular to the displacement, so it does zero work on the chest. Since more than one force does work on the chest, we use subscripts to clarify which work is being calculated.

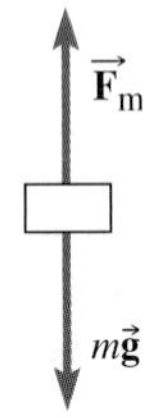

Figure 6.9
FBD for the chest as the movers lift it straight up at constant speed.

Given: Weight of chest $mg = 1400$ N; length of ramp $d = 4.0$ m; height of ramp $h = 1.0$ m
To find: Work done by movers on the chest W_m and work done by gravity on the chest W_g in the two cases; work done by the normal force on the chest W_N.

Solution (a) The displacement is 1.0 m straight up. The movers must exert an upward force $\vec{F}_m$ equal in magnitude to the weight of the chest to move it at constant speed (Fig. 6.9). The work done to lift it 1.0 m is

$$W_m = F_m \Delta r \cos \theta = 1400 \text{ N} \times 1.0 \text{ m} \times \cos 0 = +1400 \text{ J}$$

where $\theta = 0$ because $\vec{F}_m$ and $\Delta\vec{r}$ are in the same direction (upward).

(b) Figure 6.10 shows a sketch of the situation. We take the x-axis along the inclined ramp and the y-axis perpendicular to the ramp and resolve the gravitational force into its x- and y-components (Fig. 6.11a). Figure 6.11b is the FBD for the chest. Sliding along at constant speed, the chest's acceleration is zero, so the x-components of the forces add to zero.

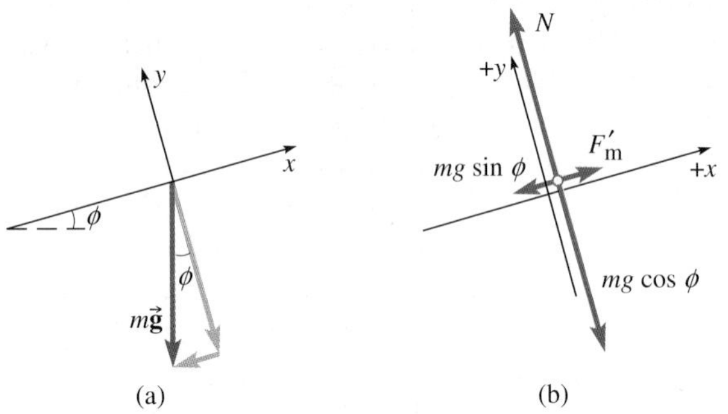

Figure 6.11
(a) Resolving $m\vec{g}$ into x- and y-components; (b) FBD for the chest.

continued on next page

Figure 6.10 An antique chest is pushed up a ramp into a truck.

Example 6.1 continued

The x-component of the gravitational force acts in the $-x$-direction and the force exerted by the movers $\vec{\mathbf{F}}'_m$ acts in the $+x$-direction. [The *prime* symbol indicates that the force exerted by the movers is different from what it was in part (a).]

$$\sum F_x = F'_m - mg \sin \phi = 0$$

From the right triangle formed by the ramp, the ground, and the truck bed in Fig. 6.12:

$$\sin \phi = \frac{\text{height of truck bed}}{\text{distance along ramp}} = \frac{h}{d}$$

We can now solve for F'_m:

$$F'_m = mg \sin \phi = \frac{mgh}{d}$$

The force and displacement are in the same direction, so $\theta = 0$:

$$W_m = F'_m\, d \cos 0 = \frac{mgh}{d} \times d \times 1 = mgh = +1400 \text{ J}$$

The work done by the movers is the same as in (a).

(c) In both cases, the force of gravity has magnitude mg and acts downward. Choosing the y-axis so it now points upward, $F_{gy} = -mg$. In both cases, the component of the displacement along the y-axis is $\Delta y = h = 1.0$ m. The work done by gravity is the same for the two cases. Using Eq. (6-2),

$$W_g = F_{gy}\, \Delta y = -mg\, \Delta y$$
$$= -1400 \text{ N} \times 1.0 \text{ m} = -1400 \text{ J}$$

(d) The normal force of the ramp on the chest does no work because it acts in a direction perpendicular to the displacement of the chest.

$$W_N = N\, \Delta r \cos 90° = 0$$

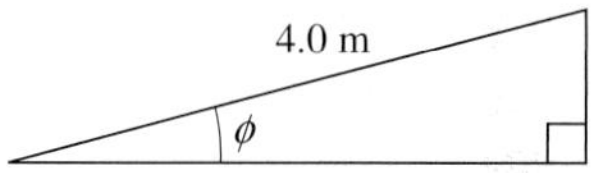

Figure 6.12 Finding the angle of the incline.

Discussion Since d, the length of the ramp, cancels when multiplying the force times the distance, the work done by the movers is the same for *any* length ramp (as long as the height is the same). Using the ramp, the movers apply one quarter the force over a displacement that is four times larger. With a *real* ramp, friction acts to oppose the motion of the chest, so the movers would have to do *more* than 1400 J of work to slide the chest up the ramp. There's no getting around it; if the movers want to get that chest into the truck, they're going to have to do *at least* 1400 J of work.

Practice Problem 6.1 Bicycling Uphill

A bicyclist climbs a 2.0-km-long hill that makes an angle of 7.0° with the horizontal. The total weight of the bike and the rider is 750 N. How much work is done on the bike and rider by gravity?

Total Work

When several forces act on an object, the total work is the sum of the work done by each force individually:

$$W_{total} = W_1 + W_2 + \cdots + W_N \qquad (6\text{-}3)$$

Total work is sometimes called *net* work because the work done by each force can be positive, negative, or zero, so the total work is often smaller than the work done by any one of the forces. Because we assume a rigid object with no rotational or internal motion, another way to calculate the total work is to find the work done by the *net* force as if there were a single force acting:

$$W_{total} = F_{net}\, \Delta r \cos \theta \qquad (6\text{-}4)$$

To show that these two methods give the same result, let's choose the x-axis in the direction of the displacement. Then the work done by each individual force is the x-component of the force times Δx. From Eq. (6-3),

$$W_{total} = F_{1x}\, \Delta x + F_{2x}\, \Delta x + \cdots + F_{Nx}\, \Delta x$$

Factoring out the Δx from each term,

$$W_{total} = (F_{1x} + F_{2x} + \cdots + F_{Nx})\, \Delta x = \left(\sum F_x\right) \Delta x$$

$\sum F_x$ is the x-component of the net force. In Eq. (6-4), $F_{net} \cos \theta$ is the component of the net force in the direction of the displacement, which is the x-component of the net force. The two methods give the same total work.

Example 6.2

Fun on a Sled

Diane pulls a sled along a snowy path on level ground with her little brother Jasper riding on the sled (Fig. 6.13). The total mass of Jasper and the sled is 26 kg. The cord makes a 20.0° angle with the ground. As a simplified model, assume that the force of friction on the sled is determined by $\mu_k = 0.16$, even though the surfaces are not dry (some snow melts as the runners slide along it). Find (a) the work done by Diane and (b) the work done by the ground on the sled while the sled moves 120 m along the path at a constant 3 km/h. (c) What is the total work done on the sled?

Strategy (a,b) To find the work done by a force on an object, we need to know the magnitudes and directions of the force and of the displacement of the object. The sled's acceleration is zero, so the vector sum of all the external forces (gravity, friction, rope tension, and the normal force) is zero. We draw the FBD and use Newton's second law to find the tension in the rope and the force of kinetic friction on the sled. Then we apply Eq. (6-1) or Eq. (6-2) to find the work done by each. (c) We have two methods to find the total work. We'll use Eq. (6-3) to calculate the total work and Eq. (6-4) as a check.

Solution (a) The FBD is shown in Fig. 6.14. The x- and y-axes are parallel and perpendicular to the ground, respectively. After resolving the tension into its components (Fig. 6.15), Newton's second law with zero acceleration yields

$$\sum F_x = +T\cos\theta - f_k = 0 \qquad (1)$$

$$\sum F_y = +T\sin\theta - mg + N = 0 \qquad (2)$$

where T is the tension and $\theta = 20.0°$. The force of kinetic friction is

$$f_k = \mu_k N$$

Substituting this into Eq. (1)

$$T\cos\theta - \mu_k N = 0 \qquad (3)$$

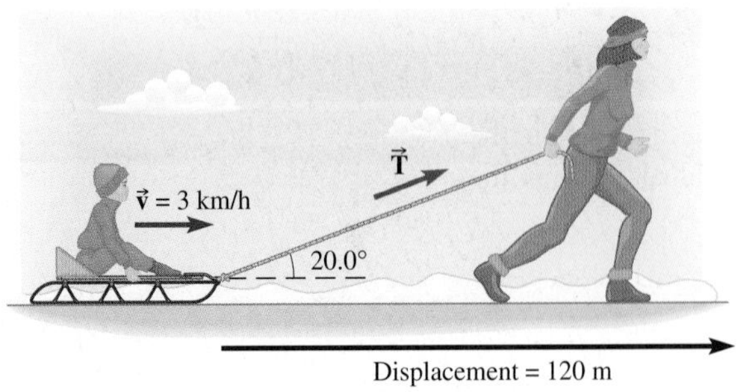

Figure 6.13
Jasper being pulled on a sled.

To find the tension, we need to eliminate the unknown normal force N. Equation (2) also involves the normal force N. We multiply Eq. (2) by μ_k,

$$\mu_k T\sin\theta - \mu_k mg + \mu_k N = 0 \qquad (4)$$

Adding Eqs. (3) and (4) eliminates N. Then we solve for T.

$$T\cos\theta + \mu_k T\sin\theta - \mu_k mg = 0$$

$$T = \frac{\mu_k mg}{\mu_k \sin\theta + \cos\theta}$$

$$= \frac{0.16 \times 26\text{ kg} \times 9.80\text{ m/s}^2}{0.16 \times \sin 20.0° + \cos 20.0°} = 41\text{ N}$$

Now that we know the tension, we find the work done by Diane. The component of the tension T acting parallel to the displacement is $T_x = T\cos\theta$ and the displacement is $\Delta x = 120$ m. The work done by Diane is

$$W_T = (T\cos\theta)\Delta x$$

$$= 41\text{ N} \times \cos 20.0° \times 120\text{ m} = +4600\text{ J}$$

(b) The force on the sled due to the ground has two components: N and f_k. The normal force does no work since it is perpendicular to the displacement of the sled. Friction acts in a direction opposite to the displacement, so the angle between the force and displacement is 180°. The work done by friction is

$$W_f = f_k\,\Delta x\cos 180° = -f_k\,\Delta x$$

From Eq. (1),

$$f_k = T\cos\theta$$

Therefore, the work done by the ground—the work done by the frictional force—is

$$W_f = -f_k\,\Delta x = -(T\cos\theta)\,\Delta x$$

Except for the negative sign, W_f is the same as W_T: $W_f = -4600$ J.

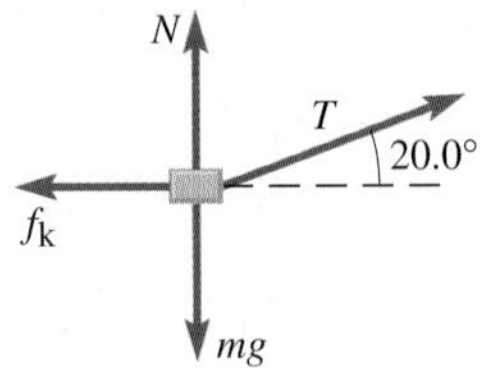

Figure 6.14
FBD.

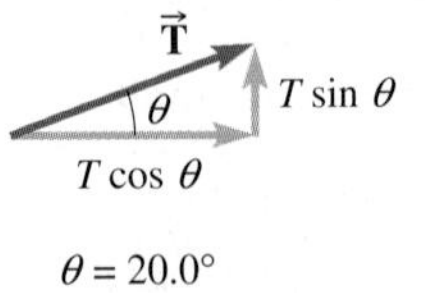

Figure 6.15
Resolving the tension into x- and y-components.

continued on next page

Example 6.2 continued

(c) The tension and friction are the only forces that do work on the sled. The normal force and gravity are both perpendicular to the displacement, so they do zero work.

$$W_{\text{total}} = W_{\text{T}} + W_{\text{f}} = 4600\text{ J} + (-4600\text{ J}) = 0$$

Discussion To check (c), note that the sled travels with constant velocity, so the net force acting on it is zero. $W_{\text{total}} = F_{\text{net}}\Delta r \cos\theta = 0$.

The speed (3 km/h) was not used in the solution. Assuming that the frictional force on the sled is independent of speed, Diane would exert the same force to pull the sled at *any* constant speed. Then the work she does is the same for a 120-m displacement. At a higher speed, though, she would have to do that amount of work in a shorter time interval.

Practice Problem 6.2 A Different Angle

Find the tension if Diane pulls at an angle $\theta = 30.0°$ instead of 20.0°, assuming the same coefficient of friction. What is the work done by Diane on the sled in this case for a 120-m displacement? Explain how the tension can be greater but the work done by Diane smaller.

Work Done by Dissipative Forces

The work done by kinetic friction was calculated in Example 6.2 according to a simplified model of friction. In this model, when friction truly does −4600 J of work on the sled, it transfers 4600 J of energy from the sled to the ground's internal energy—the ground warms up a bit. In reality, 4600 J of energy is converted into internal energy *shared* between the ground and the sled—both the ground and the sled warm up a little. So the 4600 J is not all transferred to the ground; some stays in the sled but is converted to a different form of energy.

Rather than saying friction does −4600 J of work, a more accurate statement is that friction *dissipates* 4600 J of energy. **Dissipation** is the conversion of energy from an organized form to a disorganized form such as the kinetic energy associated with the random motions of the atoms and molecules within an object, which is part of the object's internal energy. As a practical matter, we usually are not concerned with *where* the internal energy appears. When we can calculate the work done by friction using Eq. (6-1), we get the correct amount of energy dissipated; we just don't know how much of it is transferred to the stationary surface and how much remains in the sliding object. This is how we apply the term *work* to kinetic friction or to other dissipative forces such as air resistance. (In Chapters 14 and 15, when we study internal energy in detail, we will look at situations in which we *do* care where the internal energy appears.)

6.3 KINETIC ENERGY

Suppose a constant net force $\vec{\mathbf{F}}_{\text{net}}$ acts on a rigid object of mass m during a displacement $\Delta\vec{\mathbf{r}}$. Choosing the x-axis in the direction of the net force, the total work done on the object is

$$W_{\text{total}} = F_{\text{net}}\,\Delta x$$

where Δx is the x-component of the displacement. Newton's second law tells us that $\vec{\mathbf{F}}_{\text{net}} = m\vec{\mathbf{a}}$, so

$$W_{\text{total}} = ma_x\,\Delta x \tag{6-5}$$

Since the acceleration is constant, we can use any of the equations for constant acceleration from Chapter 2. From Eq. (2-13), $v_{\text{fx}}^2 - v_{\text{ix}}^2 = 2a_x\Delta x$ or

$$a_x\,\Delta x = \tfrac{1}{2}(v_{\text{fx}}^2 - v_{\text{ix}}^2)$$

Substituting into Eq. (6-5) yields

$$W_{\text{total}} = \tfrac{1}{2}m(v_{\text{fx}}^2 - v_{\text{ix}}^2)$$

Since the net force is in the x-direction, a_y and a_z are both zero. Only the x-component of the velocity changes; v_y and v_z are constant. As a result,

$$v_f^2 - v_i^2 = (v_{fx}^2 + \cancel{v_{fy}^2} + \cancel{v_{fz}^2}) - (v_{ix}^2 + \cancel{v_{iy}^2} + \cancel{v_{iz}^2}) = v_{fx}^2 - v_{ix}^2$$

Therefore, the total work done is

$$W_{total} = \tfrac{1}{2}m(v_f^2 - v_i^2) = \tfrac{1}{2}mv_f^2 - \tfrac{1}{2}mv_i^2$$

The total work done is equal to the change in the quantity $\frac{1}{2}mv^2$, which is called the object's **translational kinetic energy** (symbol K). (Often we just say *kinetic energy* if it is understood that we mean translational kinetic energy.) Translational kinetic energy is the energy associated with motion of the object as a whole; it does not include the energy of rotational or internal motion.

Translational kinetic energy:

$$K = \tfrac{1}{2}mv^2 \quad (6\text{-}6)$$

Relation between total work and kinetic energy

Work-kinetic energy theorem:

$$W_{total} = \Delta K \quad (6\text{-}7)$$

Kinetic energy is a scalar quantity and is always positive if the object is moving or zero if it is at rest. Kinetic energy is never negative, although a *change* in kinetic energy can be negative. The kinetic energy of an object moving with speed v is equal to the work that must be done on the object to accelerate it to that speed starting from rest. When the total work done is positive, the object's speed increases, increasing the kinetic energy. When the total work done is negative, the object's speed decreases, decreasing the kinetic energy.

Conceptual Example 6.3

Collision Damage

Why is the damage caused by an automobile collision so much worse when the vehicles involved are moving at high speeds?

Strategy When a collision occurs, the kinetic energy of the automobiles gets converted into other forms of energy. We can use the kinetic energy as a rough measure of how much damage can be done in a collision.

Solution and Discussion Suppose we compare the kinetic energy of a car at two different speeds: 60.0 mi/h and 72.0 mi/h (which is 20.0% greater than 60.0 mi/h). If kinetic energy were proportional to speed, then a 20.0% increase in speed would mean a 20.0% increase in kinetic energy. However, since kinetic energy is proportional to the *square* of the speed, a 20.0% speed increase causes an increase in kinetic energy greater than 20.0%. Working by proportions, we can find the percent increase in kinetic energy:

$$\frac{K_2}{K_1} = \frac{\frac{1}{2}\cancel{m}v_2^2}{\frac{1}{2}\cancel{m}v_1^2} = \left(\frac{72.0 \text{ mi/h}}{60.0 \text{ mi/h}}\right)^2 = 1.44$$

Therefore, a 20.0% increase in speed causes a 44% increase in kinetic energy. What seems like a relatively modest difference in speed makes a lot of difference when a collision occurs.

Practice Problem 6.3 Two Different Cars Collide with a Stone Wall

Suppose a sports utility vehicle and a small electric car both collide with a stone wall and come to a dead stop. If the SUV mass is 2.5 times that of the small car and the speed of the SUV is 60.0 mph while that of the other car is 40.0 mph, what is the ratio of the kinetic energy changes for the two cars (SUV to small car)?

Example 6.4

Bungee Jumping

A bungee jumper makes a jump in the Gorge du Verdon in southern France. The jumping platform is 182 m above the bottom of the gorge. The jumper weighs 780 N. If the jumper falls to within 68 m of the bottom of the gorge, how much work is done by the bungee cord on the jumper during his descent? Ignore air resistance.

Strategy Ignoring air resistance, only two forces act on the jumper during the descent: gravity and the tension in the cord. Since the jumper has zero kinetic energy at both the highest and lowest points of the jump, the change in kinetic energy for the descent is zero. Therefore, the total work done by the two forces on the jumper must equal zero.

Solution Let W_g and W_c represent the work done on the jumper by gravity and by the cord. Then

$$W_{total} = W_g + W_c = \Delta K = 0$$

The work done by gravity is

$$W_g = F_y \, \Delta y = -mg \, \Delta y$$

where the weight of the jumper is $mg = 780$ N. With $y = 0$ at the bottom of the gorge, the vertical component of the displacement is

$$\Delta y = y_f - y_i = 68 \text{ m} - 182 \text{ m} = -114 \text{ m}$$

Then the work done by gravity is

$$W_g = -(780 \text{ N}) \times (-114 \text{ m}) = +89 \text{ kJ}$$

The work done by the cord is $W_c = W_{total} - W_g = -89$ kJ.

Discussion The work done by gravity is positive, since the force and the displacement are in the same direction (downward). If not for the negative work done by the cord, the jumper would have a kinetic energy of 89 kJ after falling 114 m.

The length of the bungee cord is not given, but it does not affect the answer. At first the jumper is in free fall as the cord plays out to its full length; only then does the cord begin to stretch and exert a force on the jumper, ultimately bringing him to rest again. Regardless of the length of the cord, the total work done by gravity and by the cord must be zero since the change in the jumper's kinetic energy is zero.

Practice Problem 6.4 The Bungee Jumper's Speed

Suppose that during the jumper's descent, at a height of 111 m above the bottom of the gorge, the cord has done −21.7 kJ of work on the jumper. What is the jumper's speed at that point?

CHECKPOINT 6.3

Kinetic energy and work are related. Can kinetic energy ever be negative? Can work ever be negative?

6.4 GRAVITATIONAL POTENTIAL ENERGY (1)

Gravitational Potential Energy When Gravitational Force Is Constant

Toss a stone up with initial speed v_i. Ignoring air resistance, how high does the stone go? We can solve this problem with Newton's second law, but let's use work and energy instead. The stone's initial kinetic energy is $K_i = \frac{1}{2}mv_i^2$. For an upward displacement Δy, gravity does negative work $W_{grav} = -mg\,\Delta y$. No other forces act, so this is the total work done on the stone. The stone is momentarily at rest at the top, so $K_f = 0$. Then

$$W_{grav} = K_f - K_i$$

$$-mg\,\Delta y = -\frac{1}{2}mv_i^2 \quad \Rightarrow \quad \Delta y = \frac{v_i^2}{2g}$$

From the standpoint of energy conservation, where did the stone's initial kinetic energy go? If total energy cannot change, it must be "stored" somewhere. Furthermore,

The symbol for potential energy is U.

the stone gets its kinetic energy back as it falls from its highest point to its initial position, so the energy is stored in a way that is easily recovered as kinetic energy. Stored energy due to the interaction of an object with something else (here, Earth's gravitational field) that can easily be recovered as kinetic energy is called **potential energy** (symbol U).

The change in gravitational potential energy when an object moves up or down is the *negative* of the work done by gravity:

Change in gravitational potential energy:

$$\Delta U_{\text{grav}} = -W_{\text{grav}} \tag{6-8}$$

If the gravitational field is uniform, the work done by gravity is

$$W_{\text{grav}} = F_y\,\Delta y = -mg\,\Delta y$$

where the y-axis points up. Therefore,

Change in gravitational potential energy:

$$\Delta U_{\text{grav}} = mg\,\Delta y \tag{6-9}$$

(uniform $\vec{\mathbf{g}}$, y-axis up)

Equation (6-9) holds even if the object does not move in a straight-line path.

Significance of the Negative Sign in Eq. (6-8) When the stone moves up, Δy is positive. The gravitational force and the displacement of the stone are in opposite directions, so the work done by gravity is negative, gravity is taking away kinetic energy and adding it to its stored potential energy, so the potential energy increases (Fig. 6.16a). If the stone moves down, Δy is negative. The work done by gravity is positive; gravity is giving back kinetic energy by depleting its storage of potential energy, so the potential energy decreases (Fig. 6.16b).

CHECKPOINT 6.4

A stone is tossed straight up in the air and is moving upward. (a) Does the gravitational potential energy increase, decrease, or stay the same? (b) What about the kinetic energy? (c) What force, if any, does work on the stone once it leaves the hand of the one who threw it?

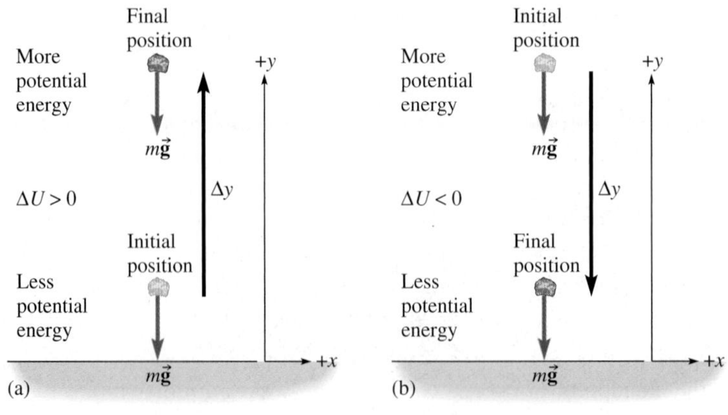

Figure 6.16 (a) When the stone moves up, the gravitational potential energy increases. (b) When the stone moves down, the gravitational potential energy decreases.

Other Forms of Potential Energy In addition to gravitational potential energy, other kinds of potential energy include elastic potential energy (Section 6.7) and electric potential energy (Chapter 17). Forces that have potential energies associated with them are called **conservative forces**, for reasons we explain shortly. Not every force has an associated potential energy. For instance, there is no such thing as "frictional potential energy." When kinetic friction does work, it converts energy into a disorganized form that is not easily recoverable as kinetic energy.

Mechanical Energy The total work done on an object can always be written as the sum of the work done by conservative forces (W_{cons}) plus the work done by nonconservative forces (W_{nc}). Since the total work is equal to the change in the object's kinetic energy [Eq. (6-7)],

$$W_{total} = W_{cons} + W_{nc} = \Delta K \quad \Rightarrow \quad W_{nc} = \Delta K - W_{cons} \tag{6-10}$$

Following the same reasoning we used for gravity [see Eq. (6-8)], the change in the total potential energy is equal to the negative of the work done by the conservative forces:

$$\Delta U = -W_{cons} \tag{6-11}$$

Combining Eqs. (6-10) and (6-11) yields

$$W_{nc} = \Delta K + \Delta U = \Delta E_{mech} \tag{6-12}$$

or

$$(K_i + U_i) + W_{nc} = (K_f + U_f)$$

The sum of the kinetic and potential energies ($K + U$) is called the **mechanical energy** (E_{mech}). W_{nc} is equal to the change in mechanical energy. When finding the change in mechanical energy, do not include the work done by conservative forces. Conservative forces such as gravity do not change the mechanical energy; they just change one form of mechanical energy into another. Work done by conservative forces is already accounted for by the change in potential energy.

Mechanical energy: the sum of the kinetic and potential energies

The term *conservative force* comes from a time before the general law of conservation of energy was understood and when no forms of energy other than mechanical energy were recognized. Back then, it was thought that certain forces conserved energy and others did not. Now we believe that *total* energy is *always* conserved. Nonconservative forces do not conserve *mechanical* energy, but they do conserve *total* energy.

Conservation of Mechanical Energy

When nonconservative forces do no work, mechanical energy is conserved:

$$E_i = E_f$$

Example 6.5

Rock Climbing in Yosemite

A team of climbers is rappelling down steep terrain in the Yosemite valley (Fig. 6.17). Mei-Ling (mass 60.0 kg) slides down a line starting from rest 12.0 m above a horizontal shelf. If she lands on the shelf below with a speed of 2.0 m/s, calculate the energy dissipated by the kinetic frictional forces acting between her and the line. The local value of g is 9.78 N/kg. Ignore air resistance.

continued on next page

Example 6.5 continued

Figure 6.17
Mei-Ling rappelling downward from a position 12.0 m above a shelf.

Strategy The forces acting on Mei-Ling are gravity and kinetic friction (Fig. 6.18). The only force whose work is not included in the change in potential energy is the work done by kinetic friction. Therefore, the change in the mechanical energy, $\Delta K + \Delta U$, is equal to the work done by friction. Since we know Mei-Ling's initial and final speeds as well as her mass, we can calculate the change in her kinetic energy. From the change in height, we can calculate the change in potential energy.

Figure 6.18
FBD for Mei-Ling.

Given: mass of climber, $m = 60.0$ kg; $\Delta y = -12.0$ m; $v_i = 0$ m/s and $v_f = 2.0$ m/s, just before stopping; local field strength $g = 9.78$ N/kg.

To find: change in mechanical energy ΔE.

Solution $W_{nc} = \Delta E_{mech} = \Delta K + \Delta U$, so we need to calculate the changes in kinetic and potential energy. Mei-Ling's kinetic energy is initially zero since she starts at rest. The change in her kinetic energy is

$$\Delta K = \tfrac{1}{2}mv_f^2 - \tfrac{1}{2}mv_i^2 = \tfrac{1}{2}mv_f^2 - 0 = \tfrac{1}{2}(60.0\text{ kg}) \times (2.0\text{ m/s})^2$$
$$= +120\text{ J}$$

The change in her potential energy is

$$\Delta U = mg\,\Delta y = 60.0\text{ kg} \times 9.78\text{ m/s}^2 \times (0 - 12.0\text{ m}) = -7040\text{ J}$$

The work done by friction is

$$\Delta E_{mech} = \Delta K + \Delta U = 120\text{ J} + (-7040\text{ J}) = -6920\text{ J}$$

The amount of energy dissipated by friction (converted from mechanical energy into internal energy) is 6920 J. Fortunately, Mei-Ling is wearing gloves, so her hands don't get burned.

Discussion If the line had broken when Mei-Ling was at the top, her final kinetic energy would have been +7040 J—disastrously large since it corresponds to a final speed of

$$v = \sqrt{\frac{K}{\frac{1}{2}m}} = \sqrt{\frac{7040\text{ J}}{30.0\text{ kg}}} = 15.3\text{ m/s}$$

Instead, kinetic friction reduces her final kinetic energy to a manageable +120 J (which corresponds to a final speed of 2.0 m/s). Mei-Ling can absorb this much kinetic energy safely by landing on the shelf while bending her knees.

Practice Problem 6.5 Energy Dissipated by Air Resistance

A ball thrown straight up at an initial speed of 14.0 m/s reaches a maximum height of 7.6 m. What fraction of the ball's initial kinetic energy is dissipated by air resistance as the ball moves upward?

Choosing Where the Potential Energy Is Zero

Notice that when we apply Eq. (6-12), only the *change* in potential energy enters the calculation. Therefore, we can always assign the value of the potential energy for any *one* position. Most often, we choose some convenient position and assign it to have zero potential energy. Once that choice is made, the potential energy of every other configuration is determined by Eq. (6-11).

For gravitational potential energy in a uniform gravitational field, we usually choose the potential energy to be zero at some convenient place: on the floor, on a table, or at the top of a ladder. After assigning $y = 0$ to that place, the potential energy at any other place is $U = mgy$.

Gravitational potential energy:

$$U_{\text{grav}} = mgy \tag{6-13}$$

(uniform $\vec{\mathbf{g}}$, y-axis up, assign $U = 0$ to $y = 0$)

Potential energy is then positive above $y = 0$ and negative below it. There is no special significance to the sign of the potential energy. What matters is the sign of the potential energy *change.*

Example 6.6

A Quick Descent

A ski trail makes a vertical descent of 78 m. A novice skier, unable to control his speed, skis down this trail and is lucky enough not to hit any trees. What is his speed at the bottom of the trail, ignoring friction and air resistance?

Strategy When nonconservative forces do no work, $W_{\text{nc}} = \Delta E_{\text{mech}} = 0$ and mechanical energy does not change. A skilled skier can control his speed by, in effect, controlling how much work the frictional force does on the skis. Here we assume *no* friction or air resistance. Then the only forces acting on the skier are the normal force and gravity (Fig. 6.19). The normal force does no work, since it is always perpendicular to the skier's velocity, so $W_{\text{nc}} = 0$.

continued on next page

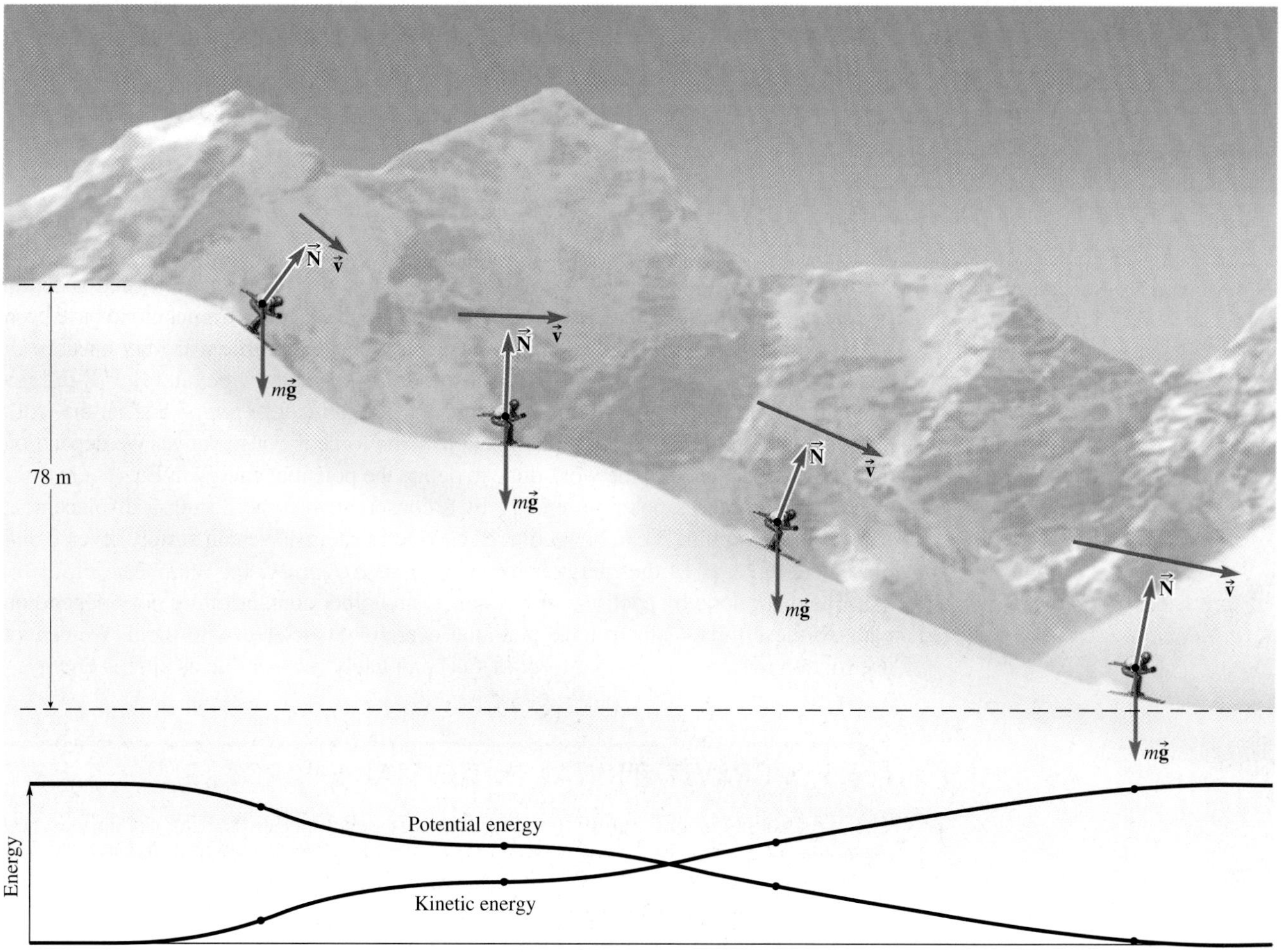

Figure 6.19 The final speed of the skier depends only on the initial and final altitudes if no friction acts.

Example 6.6 continued

Solution Because $W_{nc} = 0$, the mechanical energy does not change:

$$K_i + U_i = K_f + U_f$$

If we choose the y-axis up and $y = 0$ at the bottom of the hill, $y_i = 78$ m and $y_f = 0$. Then

$$U_i = mgy_i \quad \text{and} \quad U_f = 0$$

If the skier starts with zero kinetic energy, then $K_i = 0$ and $K_f = \frac{1}{2}mv_f^2$. Setting the mechanical energies equal,

$$0 + mgy_i = \tfrac{1}{2}mv_f^2 + 0$$

Solving for the final speed v_f,

$$v_f = \sqrt{2gy_i} = \sqrt{2 \times 9.80 \text{ m/s}^2 \times 78 \text{ m}} = 39 \text{ m/s}$$

Discussion Notice that the solution did not depend on the detailed shape of the path. If the slope were constant (Fig. 6.20), we could use Newton's second law to find the skier's acceleration and then the change in velocity:

$$\sum F_x = mg \sin\theta = ma_x \quad \Rightarrow \quad a_x = g \sin\theta$$

From Eq. (2-13),

$$\Delta x = \frac{v_{fx}^2 - \cancel{v_{ix}^2}}{2a_x} = \frac{v_{fx}^2}{2g\sin\theta} = \frac{h}{\sin\theta} \quad \Rightarrow \quad v_{fx} = \sqrt{2gh}$$

where $h = 78$ m.

This method shows that the final speed does not depend on the angle of the slope, but the energy method shows that the final speed is the same for *any* shape path, not just for constant slopes. On the other hand, the *time* that it takes the skier to reach the bottom *does* depend on the length and contour of the trail.

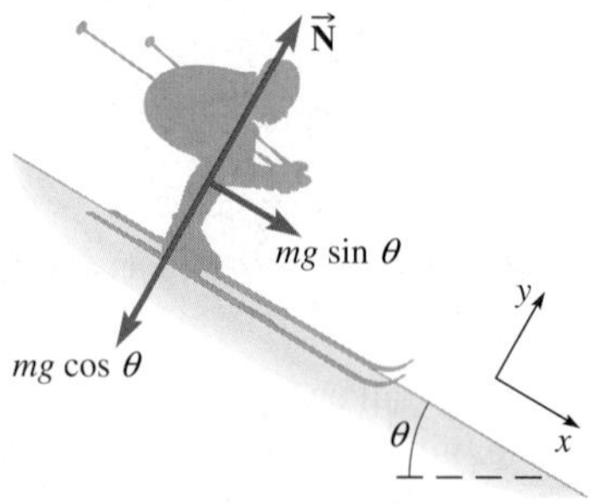

Figure 6.20
FBD for the skier on a constant slope.

A final speed of 39 m/s (87 mi/h) is dangerously fast. In reality, friction and air resistance would do negative work on the skier, so the final speed would be smaller.

Practice Problem 6.6 Speeding Roller Coaster

A roller coaster is hauled to the top of the first hill of the ride by a motorized chain drive. After that, the train of cars is released and no more energy is supplied by an external motor. The cars are moving at 4.0 m/s at the top of the first hill, 35.0 m above the ground. How fast are they moving at the top of the second hill, 22.0 m above the ground? Ignore friction and air resistance.

Recognizing a Conservative Force

In Example 6.6, the final speed doesn't depend on the shape of the trail: it could have been a steep descent, or a long gradual one, or have a complicated profile with varying slope. It could even be a vertical descent—the final speed is the same for free fall off a 78-m-high building. Any time the work done by a force is *independent of path*—that is, the work depends only on the initial and final positions—the force is conservative. We depend on the path-independence of the work done to define the potential energy in Eq. (6-11).

Energy stored as potential energy by a conservative force during a displacement from point A to point B can be recovered as kinetic energy. We can simply reverse displacement to get all of the energy back: $\Delta U_{B\to A} = -\Delta U_{A\to B}$.

The work done by friction, air resistance, and other contact forces *does* depend on path, so these forces cannot have potential energies associated with them. We cannot use friction to store energy in a form that is completely recoverable as kinetic energy.

6.5 GRAVITATIONAL POTENTIAL ENERGY (2)

The expressions for gravitational potential energy developed in Section 6.4 apply when the gravitational force is *constant* (or nearly constant). If the gravitational force is not constant, such as when a satellite is placed into orbit around the Earth, Eqs. (6-9) and (6-13) cannot be used. Instead, we need to use an expression for gravitational potential energy that corresponds to Newton's law of universal gravitation. Recall that the magnitude of the gravitational force that one body exerts on another is

$$F = \frac{Gm_1m_2}{r^2} \qquad (2\text{-}6)$$

where r is the distance between the centers of the bodies. The corresponding expression for gravitational potential energy in terms of the distance between two bodies is

Gravitational potential energy:

$$U = -\frac{Gm_1m_2}{r} \qquad (6\text{-}14)$$

(assign $U = 0$ when $r = \infty$)

A graph showing the gravitational potential energy as a function of r is shown in Fig. 6.21. Note that we have assigned the potential energy to be zero at infinite separation ($U = 0$ when $r = \infty$). Why this choice? Simply put, any other choice would mean adding a constant term to the expression for U. This constant term would *always subtract out* of our equations, which involve only *changes* in potential energy.

This choice ($U = 0$ when $r = \infty$) means that the gravitational potential energy is *negative* for any finite value of r, because potential energy decreases as the bodies get closer together and increases as they get farther apart.

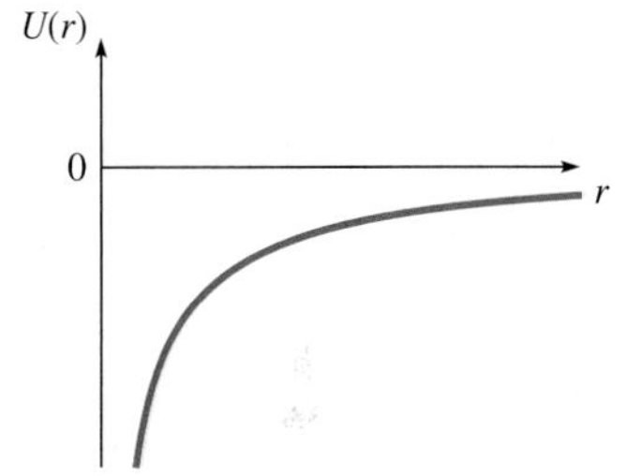

Figure 6.21 Gravitational potential energy as a function of r, the distance between the centers of the two bodies. The potential energy increases as the distance increases.

Does Eq. (6-14) Contradict Eq. (6-9)? Calculus is used to derive Eq. (6-14), but we can *verify* that it is consistent with Eq. (6-9) without using calculus. For a *very small* displacement from r_i to $r_f = r_i + \Delta y$ (Fig. 6.22), the potential energy change given by Eq. (6-14) must reduce to the constant-force case:

$$\Delta U = U_f - U_i = \left(-\frac{GM_Em}{r_i + \Delta y}\right) - \left(-\frac{GM_Em}{r_i}\right)$$

Rearranging and factoring out the common factors GM_Em and then rewriting with a common denominator,

$$\Delta U = GM_Em\left(\frac{1}{r_i} - \frac{1}{r_i + \Delta y}\right) = GM_Em\frac{\cancel{r_i} + \Delta y - \cancel{r_i}}{r_i(r_i + \Delta y)} \qquad (6\text{-}15)$$

For values of Δy that are small compared with r_i, $r_i + \Delta y \approx r_i$. Making that approximation in the denominator of Eq. (6-15),

$$\Delta U = m\left(\frac{GM_E}{r_i^2}\right)\Delta y \quad (\Delta y \ll r_i) \qquad (6\text{-}16)$$

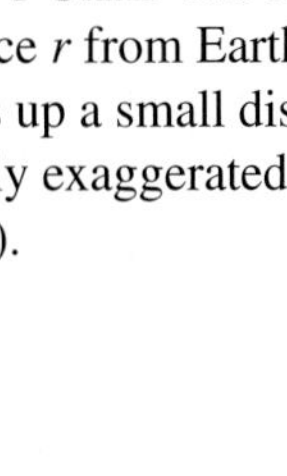

Figure 6.22 An object at a distance r from Earth's center moves up a small distance Δy (greatly exaggerated in the figure).

The quantity in the parentheses in Eq. (6-16) is the gravitational field strength g, the gravitational force on the object divided by its mass m. Then, $\Delta U = mg\,\Delta y$, in agreement with Eq. (6-9).

CHECKPOINT 6.5

As Mercury travels in its elliptical orbit about the Sun, how does its mechanical energy at its nearest point (*perihelion*) to the Sun compare with that at its farthest point (*aphelion*) from the Sun? How does its potential energy compare at the same two points?

Example 6.7

Orbital Speed of Mercury

The orbit of the planet Mercury around the Sun is an ellipse. At its perihelion (4.60×10^7 km), its orbital speed is 59 km/s. What is its orbital speed at aphelion (6.98×10^7 km)?

Strategy Ignoring the small gravitational forces exerted by other planets, the only force acting on Mercury is the gravitational force due to the Sun. Gravity is a conservative force, so the mechanical energy is constant. Figure 6.23 is a sketch of the orbit. At aphelion, Mercury is farther from the Sun than at perihelion, so the potential energy is greater. Then the kinetic energy must be smaller, so the answer must be less than 59 km/s.

Given: $v_p = 5.9 \times 10^4$ m/s, $r_p = 4.60 \times 10^{10}$ m, $r_a = 6.98 \times 10^{10}$ m.

To find: v_a.

Solution Mechanical energy is constant:

$$K_p + U_p = K_a + U_a$$

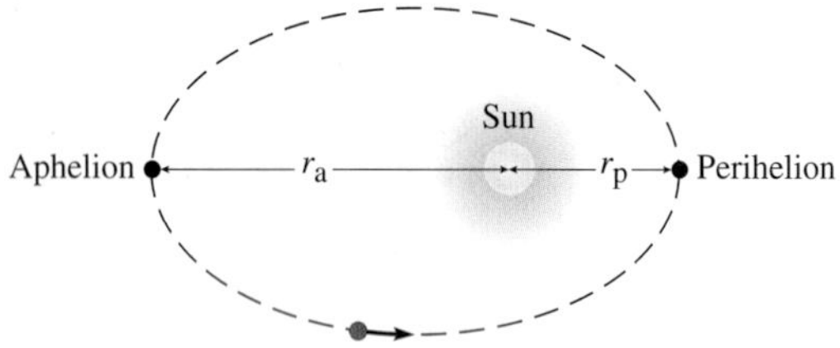

Figure 6.23
Sketch of Mercury's orbit.

The kinetic energy of Mercury at perihelion is $K_p = \frac{1}{2}mv_p^2$, where m is the mass of Mercury; the kinetic energy at aphelion is $K_a = \frac{1}{2}mv_a^2$. The potential energies at perihelion and at aphelion are

$$U_p = -\frac{GM_S m}{r_p} \quad \text{and} \quad U_a = -\frac{GM_S m}{r_a}$$

respectively, where $M_S = 1.99 \times 10^{30}$ kg is the mass of the Sun. From conservation of energy:

$$\frac{1}{2}\cancel{m}v_p^2 + \left(-\frac{GM_S\cancel{m}}{r_p}\right) = \frac{1}{2}\cancel{m}v_a^2 + \left(-\frac{GM_S\cancel{m}}{r_a}\right)$$

The mass of Mercury cancels out. Solving for v_a,

$$\frac{1}{2}v_a^2 = \frac{1}{2}v_p^2 + \left(-\frac{GM_S}{r_p}\right) - \left(-\frac{GM_S}{r_a}\right)$$

$$v_a = \sqrt{v_p^2 + 2GM_S\left(\frac{1}{r_a} - \frac{1}{r_p}\right)}$$

Substituting numerical values yields $v_a = 39$ km/s.

Discussion The speed at aphelion is less than the speed at perihelion, as expected.

Practice Problem 6.7 Speed at a Different Distance

What is Mercury's orbital speed when its distance from the Sun is 5.80×10^7 km?

Example 6.8

Escape Speed

Ignoring air resistance, what is the minimum initial speed a projectile must have at Earth's surface if the projectile is to escape Earth's gravitational pull?

Strategy What does "escape Earth's gravitational pull" mean? The gravitational force on the projectile due to Earth *approaches* zero at large distances, but never *reaches* zero. We are looking for the initial speed so that, even though Earth's gravity keeps pulling the projectile back, the projectile can keep moving away from Earth. The gravitational force is not constant, and the trajectory of the projectile may be complicated, so using $\Sigma\vec{\mathbf{F}} = m\vec{\mathbf{a}}$ is impractical. We try an energy approach.

The only force acting on the projectile is gravity, so the mechanical energy is constant. To escape, the projectile must have enough initial kinetic energy so that it can reach an unlimited distance from Earth.

Solution The mechanical energy is constant:

$$K_i + U_i = K_f + U_f$$

Initially the projectile is at a distance R_E from Earth's center and is moving at initial speed v_i. At some later time, the projectile has speed v_f at distance r_f from Earth. Then

$$\frac{1}{2}mv_i^2 + \left(-\frac{GM_E m}{R_E}\right) = K_f + U_f$$

continued on next page

Example 6.8 continued

To escape, the projectile must be able to reach any value of r_f, no matter how large. As r_f gets larger and larger, the potential energy approaches its maximum value, which is zero. (Mathematically, as $r_f \to \infty$, $U_f \to 0$.) The *minimum* value of v_i gives the projectile *just enough* energy. So we assume that the projectile can reach its maximum potential energy without any kinetic energy left over ($K_f = 0$):

$$\frac{1}{2}mv_i^2 + \left(-\frac{GM_E m}{R_E}\right) = 0 + 0$$

Solving for v_i,

$$\frac{1}{2}\not{m}v_i^2 = \frac{GM_E\not{m}}{R_E} \quad \Rightarrow \quad v_i = \sqrt{\frac{2GM_E}{R_E}} = 11.2 \text{ km/s}$$

Discussion This speed is called the **escape speed** of Earth. Note that the escape speed is independent of the mass of the projectile because both the kinetic energy and the potential energy are proportional to the projectile's mass.

The concept of escape speed helps explain why there is little hydrogen gas (H_2) or helium gas (He) in Earth's atmosphere. We will see in Chapter 13 that the molecules in a gas have an average kinetic energy determined by the temperature of the gas. In a mixture of gases, the molecules with the smallest mass have the highest average speeds. The average speeds of hydrogen and helium in our atmosphere are large enough that they can escape the atmosphere. A negligible fraction of the nitrogen, oxygen, or water molecules have speeds greater than the escape speed, so they persist in the atmosphere.

Practice Problem 6.8 Protons Streaming Away from the Sun

Particles such as protons and electrons are continually streaming away from the Sun in all directions. They carry off some of the energy released in the thermonuclear reactions occurring in the Sun. How fast must a proton be moving at a distance of 7.00×10^9 m from the center of the Sun for it to escape the Sun's gravitational pull and leave the solar system?

6.6 WORK DONE BY VARIABLE FORCES: HOOKE'S LAW

So far we have considered only constant forces when calculating work. The advantage of using energy methods really shines in problems dealing with variable forces, where it's difficult to use Newton's second law. How can we calculate the work done by a variable force? Consider an archer drawing back a compound bow (Fig. 6.24). The compound bow is designed to make it easier to draw the string back and hold it back because, at a certain point, the force required to draw the string farther stops increasing. A convenient way to describe how the force varies with string position is to plot a graph. Figure 6.25 shows the force that must be applied to hold the string back as a function of distance. How can we calculate the work done by the archer as he draws the string back 40 cm?

Figure 6.24 Application of work done by a variable force: drawing a compound bow.

We've asked analogous questions in previous chapters. Recall how we find the displacement Δx when the velocity v_x is not constant (Section 2.2). We divide the time interval into a series of *short* time intervals and sum up the displacements that occur during each one.

CONNECTION

See Sections 2.2 and 2.3 to review how we found that the area under a graph of $v_x(t)$ is Δx and that the area under a graph of $a_x(t)$ is Δv_x.

To approximate the work done by a variable force F_x, we divide the overall displacement into a series of small displacements Δx. During each small displacement, the work done is

$$\Delta W = F_x \, \Delta x \qquad (6\text{-}17)$$

On a graph of $F_x(x)$, each ΔW is the area of a rectangle of height F_x and width Δx (Fig. 6.26). The total work done is the sum of the areas of these rectangles. This approximation gets better as we make the rectangles thinner and thinner, so *the total work done is the area under the graph of $F_x(x)$ from x_i to x_f*.

W = the area under a graph of $F_x(x)$

In Fig. 6.25, the "area" of each rectangle represents (0.050 m × 20.0 N) = 1.0 J of work. There are approximately 36 rectangles under the graph between $x = 0$ and $x = 40$ cm, so the work done by the archer is +36 J.

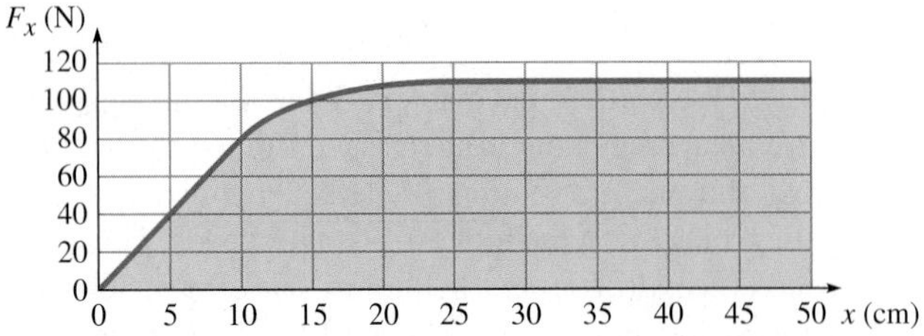

Figure 6.25 The force to draw back the compound bow depends on how far it is drawn. In this graph, the "area" represented by each rectangle is 0.050 m × 20.0 N = 1.0 J.

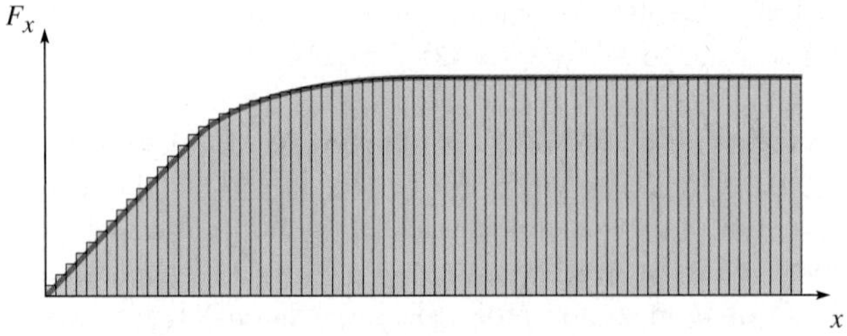

Figure 6.26 Each rectangle's area approximates the work done during a small displacement. The total area of the rectangles approximates the total work done.

Example 6.9

Archery Practice

To draw back a *simple* bow, the force the archer exerts on the string continues to increase as the displacement of the string increases and the bow bends slightly. The force-versus-position graph of Fig. 6.27 describes such a bow. Calculate the work done by the archer on the string as he draws the string back 40.0 cm.

Strategy The work done by the archer is the area under the force-versus-position graph. This time, instead of counting rectangles, we can calculate the triangular area formed by the force-versus-position graph.

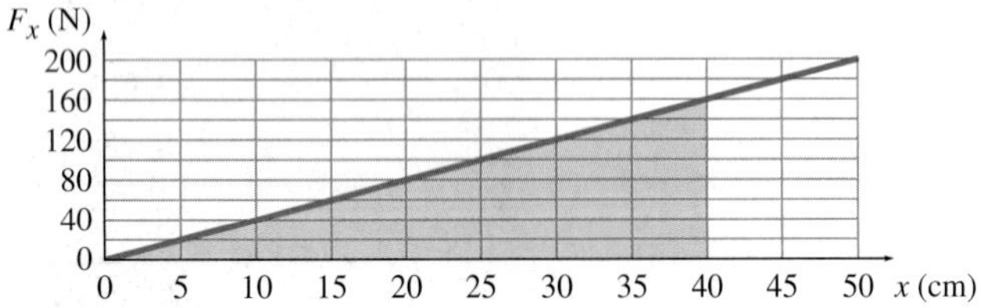

Figure 6.27
A simple bow requires a force proportional to the displacement of the string.

Solution We want to find the work done by the archer to draw the string back 40.0 cm, so the base of the triangle is 40.0 cm. The altitude of the triangle is the force at 40.0 cm: 160 N. The area of a triangle is $\frac{1}{2}$(base × altitude), so

$$W = \tfrac{1}{2}(0.400 \text{ m} \times 160 \text{ N}) = +32 \text{ J}$$

Discussion To check, we can count the number of rectangles (including the half rectangles) that lie under the graph. There are 32 rectangles and each represents 20 N × 0.05 m = 1 J of work, so the answer is correct.

By doing 32 J of work on the bowstring, the archer stores this much energy in the bow. When the arrow is released, the bowstring does 32 J of work on the arrow, giving the arrow a kinetic energy of 32 J.

Practice Problem 6.9 A Gentle Pull

How much work would you do to draw the string of the compound bow (Fig. 6.25) back 10.0 cm instead of 40.0 cm?

Hooke's Law and Ideal Springs

Hooke's law: the deformation is proportional to the deforming force.

In Example 6.9, the displacement of the bowstring is proportional to the force exerted by the archer. Robert Hooke (1635–1703) observed that, for many objects, the deformation—change in size or shape—of the object is proportional to the magnitude of the force that causes the deformation. This observation, called **Hooke's law**, is an approximation and is valid only within limits. For example, the compound bow of Fig. 6.25 is described by Hooke's law only for an applied force less than 80 N.

Many springs are described by Hooke's law as long as they are not stretched or compressed too far. That is, the extension or compression—the increase or decrease in length from the relaxed length—is proportional to the force applied to the ends of the spring. When we refer to an **ideal spring**, we mean a spring that is described by Hooke's law and is also massless.

Hooke's law for an ideal spring:

$$F = k\,\Delta L \qquad (6\text{-}18)$$

In Eq. (6-18), F is the *magnitude* of the force exerted *on each end* of the spring and ΔL is the distance that the spring is stretched or compressed from its relaxed length.

The constant k is called the **spring constant** for a particular spring. The SI unit of force is the newton and the SI unit of length is the meter, so the SI units of a spring constant are N/m. The spring constant is a measure of how hard it is to stretch or compress a spring. A stiffer spring has a larger spring constant because larger forces must be exerted on the ends of the spring to stretch or compress it. Example 1.10 describes an experiment to measure the spring constant of a real spring and shows a graph of length of the spring as a function of the forces applied to its ends (Fig. 1.5).

In many situations, we are more interested in the forces exerted *by* the spring than in the forces exerted *on* it. From Newton's third law, the forces exerted *by* the spring on whatever is attached to its ends are equal in magnitude and opposite in direction to the forces exerted *by* those objects *on* the ends of the spring. Suppose that an ideal spring is aligned with the x-axis. One end is fixed in place and the other end can move along the x-axis (Fig. 6.28). Choose the origin so the moveable end is at $x = 0$ when the spring is relaxed. Then the force exerted by the moveable end of the spring on whatever is attached to it is

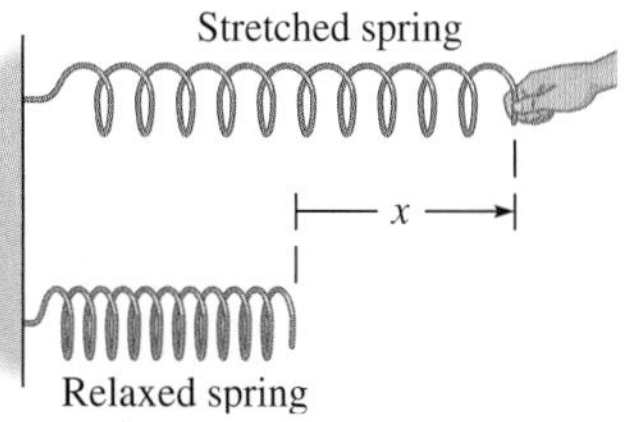

Figure 6.28 An ideal spring is stretched a distance x beyond its relaxed length.

Force exerted *by* an ideal spring (Hooke's law):

$$F_x = -kx \qquad (6\text{-}19)$$

(F_x is the force exerted *by* the moveable end when its position is x; the spring is relaxed at $x = 0$.)

The negative sign in Eq. (6-19) indicates the direction of the force. The moveable end of the spring always pushes or pulls toward its relaxed position. If it is displaced in the $+x$-direction, the force it exerts is in the $-x$-direction (back toward $x = 0$). If it is displaced in the $-x$-direction, the force it exerts is in the $+x$-direction (again, back toward $x = 0$).

Example 6.10

Getting Down to Nuts and Bolts

In many hardware stores, bulk nuts and bolts are sold by weight. A spring scale in the store stretches 4.8 cm when 24.0 N of bolts are weighed. On the scale, what is the distance in centimeters between calibration marks that are marked in increments of 1 N? Assume an ideal spring.

Strategy The bolts are in equilibrium, so the spring scale is pulling upward on them with a force of 24.0 N (see Fig. 6.29). Using Hooke's law and the data given, we can find the spring constant k. Then we can use Hooke's law again to find out how much the spring stretches when the applied force is increased by 1 N.

Solution Let the x-axis point up. When the pan of the scale is at $x = -4.8$ cm, it exerts a force $F_x = +24.0$ N on the bolts. From Hooke's law, $F_x = -kx$ and the spring constant is

$$k = -\frac{F_x}{x} = \frac{-24.0 \text{ N}}{-4.8 \text{ cm}} = 5.0 \text{ N/cm}$$

Now let $F_x = 1.00$ N and solve for x:

$$x = -\frac{F_x}{k} = -\frac{1.00 \text{ N}}{5.0 \text{ N/cm}} = -0.20 \text{ cm}$$

Since the relation between F and x is linear, the spring stretches an additional 0.20 cm for each additional newton of force. Therefore, the 1-N marks should be 0.20 cm apart.

Discussion A variation on the solution is to look back at the question and notice that we are asked how many centimeters the spring stretches for each newton of force, which is the *reciprocal* of the spring constant. The reciprocal of the spring constant is

$$\frac{1}{k} = -\frac{x}{F} = -\frac{-4.8 \text{ cm}}{24.0 \text{ N}} = 0.20 \text{ cm/N}$$

The answer is reasonable: since it takes 5 N to make the spring stretch 1 cm, 1 N makes the spring stretch $\frac{1}{5}$ cm.

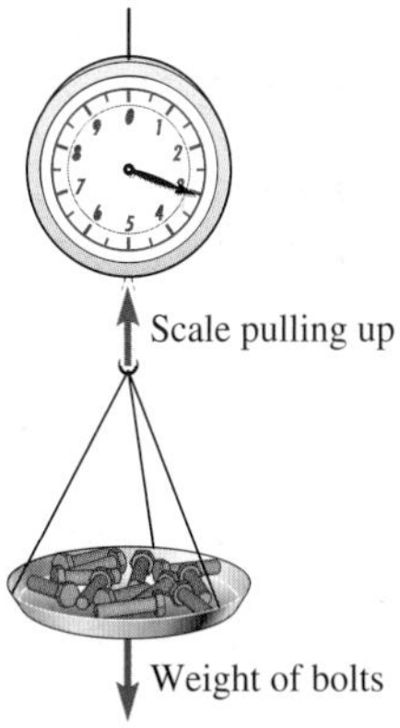

Figure 6.29 FBD for the pan of the scale.

Practice Problem 6.10 Stretching a Spring

16.0 N of nuts are placed in the pan of the scale of Example 6.10. How far does the spring stretch?

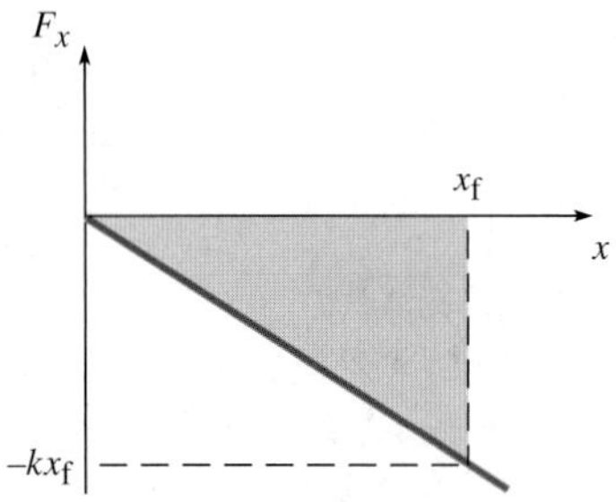

Figure 6.30 The work done by the spring is the (negative) area under the $F_x(x)$ graph.

Work Done by an Ideal Spring

To find the work done by an ideal spring, first we draw the $F_x(x)$ graph (Fig. 6.30). The unstretched position of the moveable end is $x = 0$. The work done by the spring as its moveable end moves from equilibrium ($x_i = 0$) to the final position x_f is the area of the shaded right triangle whose base is x and altitude is $-kx$:

$$W = \tfrac{1}{2}(\text{base} \times \text{altitude}) = -\tfrac{1}{2}kx^2 \quad (6\text{-}20)$$

The area is negative because the graph is underneath the x-axis. Think of $-\frac{1}{2}kx^2$ as the average force $(-\frac{1}{2}kx)$ times the displacement (x).

More generally, if the moveable end starts at position x_i, not necessarily at the equilibrium point, the work done by the spring is

$$W_{\text{spring}} = (-\tfrac{1}{2}kx_f^2) - (-\tfrac{1}{2}kx_i^2) = -\tfrac{1}{2}kx_f^2 + \tfrac{1}{2}kx_i^2 \quad (6\text{-}21)$$

Imagine the spring starting at equilibrium and ultimately ending up at a displacement x_f after passing through x_i. The total work done by the spring is $-\frac{1}{2}kx_f^2$; then we subtract the work that was done to get the spring to position x_i from equilibrium $(-\frac{1}{2}kx_i^2)$ to get the work done from x_i to x_f.

6.7 ELASTIC POTENTIAL ENERGY

The work done by an ideal spring [Eq. (6-21)] depends on the initial and final positions of the moveable end, but *not* on the path that was taken. Therefore, the force exerted by an ideal spring is *conservative* and we can associate a potential energy with it. The kind of potential energy stored in a spring is called **elastic potential energy**.

Just as for gravity [see Eqs. (6-8) and (6-11)], the change in elastic potential energy is the *negative* of the work done by the spring:

$$\Delta U_{\text{elastic}} = -W_{\text{spring}} \quad (6\text{-}22)$$

For example, if you increase the elastic energy stored in a spring by compressing it, the spring does *negative* work because the force its end exerts on your hand is in the direction opposite to its displacement. This stored elastic energy can be recovered as kinetic energy by, say, using the spring to shoot a stone. As the spring expands back to its original length, it does positive work on the stone to increase the stone's kinetic energy and the stored elastic energy decreases.

CONNECTION:

The change in potential energy is always equal to the negative of the work done by the associated force. See Eq. (6-11).

From Eqs. (6-21) and (6-22),

$$\Delta U_{\text{elastic}} = \tfrac{1}{2}kx_f^2 - \tfrac{1}{2}kx_i^2 \quad (6\text{-}23)$$

Remember that only changes in potential energy enter our calculations, so we can assign $U = 0$ to any convenient position. The most convenient choice is to assign $U = 0$ when the spring is relaxed ($x = 0$):

Elastic potential energy stored in an ideal spring:

$$U_{\text{elastic}} = \tfrac{1}{2}kx^2 \quad (6\text{-}24)$$

$U = 0$ when $x = 0$ (relaxed spring)

Conservation of Energy with More than One Form of Potential Energy When applying conservation of energy using $W_{nc} = \Delta K + \Delta U$ [Eq. (6-12)], ΔU must include the change in all forms of potential energy. For now, with two forms of potential energy,

$$\Delta U = \Delta U_{\text{grav}} + \Delta U_{\text{elastic}} \quad (6\text{-}25)$$

W_{nc} is the work done by all forces *other than* those included in the potential energy. When $W_{nc} = 0$, the mechanical energy $K + U$ is constant.

CHECKPOINT 6.7

If a spring is compressed horizontally on a table and then released so it expands to its original relaxed position, where does the spring have the greatest elastic potential energy?

Example 6.11

The Dart Gun

In a dart gun (Fig. 6.31), a spring with $k = 400.0$ N/m is compressed 8.0 cm when the dart (mass $m = 20.0$ g) is loaded (Fig. 6.31a). What is the muzzle speed of the dart when the spring is released (Fig. 6.31b)? Ignore friction.

Strategy The elastic energy initially stored in the spring is converted into the kinetic energy of the dart as the spring expands. There is no change in gravitational potential energy since the motion of the dart is horizontal. The vertical normal forces do no work because they are perpendicular to the displacement of the dart. The spring pushes the dart to the right until it reaches its relaxed length. Assuming the spring can't pull the dart to the left (as it would if they stick together), the dart loses contact with the spring when the spring is at its relaxed length. We choose the origin at the relaxed position of the spring; therefore, $x_f = 0$. Using the x-axis in Fig. 6.31, $x_i = -8.0$ cm. The dart starts from rest, so $v_i = 0$. To find: v_f.

Solution Since we ignore friction, no work is done by nonconservative forces. Therefore, the mechanical energy is constant:

$$K_i + U_i = K_f + U_f$$

We can ignore the gravitational potential energy because it does not change. Using Eq. (6-24) for the elastic potential energy in the spring,

$$\tfrac{1}{2}mv_i^2 + \tfrac{1}{2}kx_i^2 = \tfrac{1}{2}mv_f^2 + \tfrac{1}{2}kx_f^2$$

After setting $x_f = 0$ and $v_i = 0$,

$$0 + \tfrac{1}{2}kx_i^2 = \tfrac{1}{2}mv_f^2 + 0$$

Solving for v_f,

$$v_f = \sqrt{\frac{k}{m}}\,x_i = \sqrt{\frac{400.0 \text{ N/m}}{0.0200 \text{ kg}}} \times 0.080 \text{ m} = 11 \text{ m/s}$$

Discussion Checking the units,

$$\sqrt{\frac{\text{N/m}}{\text{kg}}} \times \text{m} = \sqrt{\frac{(\cancel{\text{kg}}\cdot\cancel{\text{m}}/\text{s}^2)/\cancel{\text{m}}}{\cancel{\text{kg}}}} \times \text{m} = \frac{\text{m}}{\text{s}}$$

Notice that the muzzle speed is proportional to the distance the spring is compressed when the gun is cocked. If the spring is compressed halfway, it stores only *one quarter* as much elastic energy. The dart then acquires one quarter the kinetic energy, which means its speed is half as much. A more massive dart fired from the same gun would have a smaller muzzle speed, but the *same* kinetic energy.

Practice Problem 6.11 A Misfire

The same dart gun is cocked by compressing the spring the same distance (8.0 cm). This time the spring gets caught inside the gun, stopping at the point where it is still compressed by 4.0 cm. The dart is not caught inside the gun, but is released. Find the muzzle speed of the dart. [*Hint:* What is x_f in this case?]

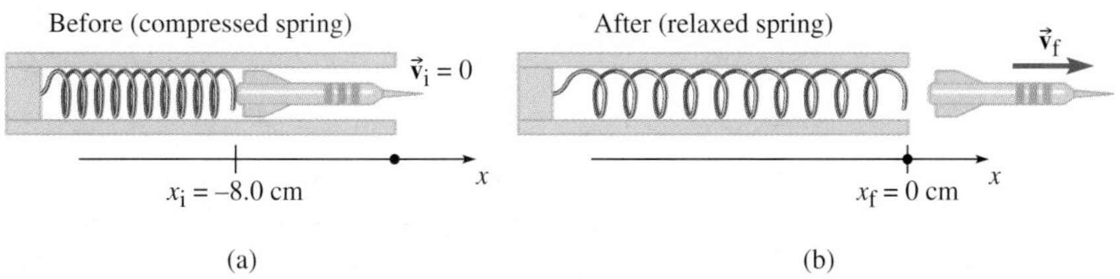

Figure 6.31 Dart gun (a) before and (b) after firing. The spring was compressed by 8.0 cm when the gun was cocked.

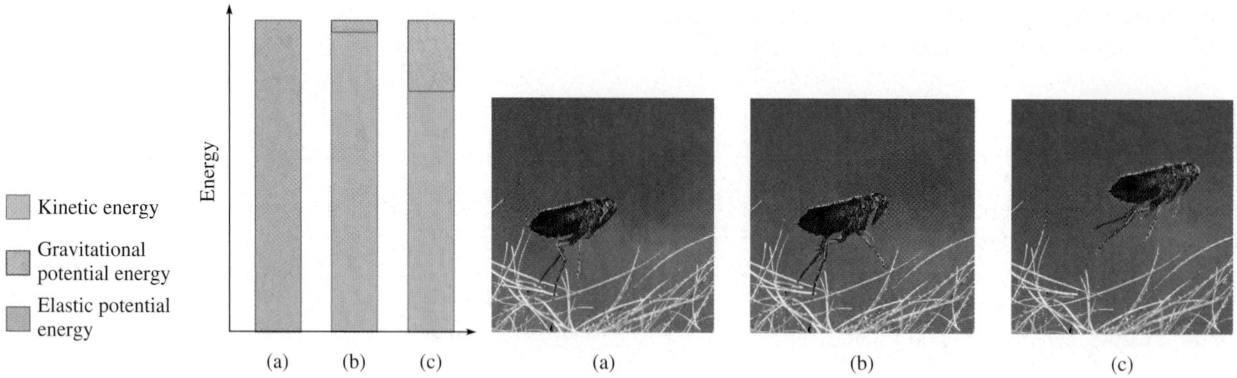

Figure 6.32 Energy transformations in the jump of a flea.

Application of Energy Conversion: Jumping

When a human jumps, the muscles supply the energy to propel the body upward. Try jumping as high as you can from a standing start. You no doubt start by crouching down. Then you accelerate upward, straightening your legs and your body; your muscles convert chemical energy into the mechanical energy of your jump. If you are very athletic, you might be able to jump about 1 m above the floor.

How does the kangaroo keep jumping?

The kangaroo uses a different mechanism. It has long, elastic tendons and small muscles in its hind legs, in contrast to the relatively large muscles and short, stiffer tendons found in humans. The kangaroo folds its legs before a jump, using its muscles to stretch the tendons and converting chemical energy into elastic potential energy. The kangaroo then quickly extends its legs, relaxing the tendons like a released spring. The elastic energy stored in the tendons supplies much of the energy needed for the jump; the rest is supplied by the kangaroo's leg muscles, which convert some more chemical energy into mechanical energy.

When the kangaroo lands on the ground, the tendons are stretched again as its legs bend. Thus, rather than dissipating all of the energy from the previous jump, a large fraction of it is recaptured as elastic energy in the tendons and then released to assist the next jump. This process reduces the amount of energy the muscles must supply for subsequent jumps and makes the kangaroo one of the most energy-efficient travelers among animals. The human body also stores some elastic energy in stretched tendons and in flexed foot bones when we run or jump, but not to the extent that its specialized anatomy enables the kangaroo to do.

Some insects jump using a catapult technique. The knee joint of a flea contains an elastic material called resilin (a rubber-like protein). The flea slowly bends its knee, stretching out the resilin and storing elastic energy, and then locks its knee in place (Fig. 6.32a). When the flea is ready to jump, the knee is unlocked and the resilin quickly contracts with a sudden conversion of the stored elastic energy into kinetic energy (Fig. 6.32b). Some of this kinetic energy is then converted into gravitational potential energy as the flea moves higher and higher (Fig. 6.32c). Ignoring air resistance and other dissipative forces, the total mechanical energy (kinetic energy + gravitational potential energy + elastic potential energy) does not change during the jump.

Example 6.12

The Hopping Kangaroo

Suppose the height h of a kangaroo's hop (Fig. 6.33) after it stretches its tendons a distance x_1 (beyond their unstretched length) is 2.0 m. How high would the hop be after it stretched the tendons 10% more than before (that is, a distance $1.10x_1$ beyond their unstretched length)? In a simplified model, we assume that all the energy for a kangaroo's hop comes from the elastic energy stored in the tendons, which behave as ideal springs. Ignore air resistance and other energy dissipation.

continued on next page

Example 6.12 continued

Strategy Ignoring dissipation, the mechanical energy does not change. We have to include both gravitational and elastic potential energies in the mechanical energy. At first we consider a kangaroo jumping straight up. Then we try to generalize to more typical hopping with forward motion as well as upward motion.

Solution The mechanical energy does not change:

$$K_i + U_{i,grav} + U_{i,elastic} = K_f + U_{f,grav} + U_{f,elastic}$$

Initially, when the kangaroo is crouched before the jump, it has zero kinetic energy. For convenience, we choose the initial gravitational potential energy to be zero. Thinking of the elastic potential energy as being stored in a single ideal spring with spring constant k, the initial mechanical energy is

$$K_i + U_{i,grav} + U_{i,elastic} = 0 + 0 + \tfrac{1}{2}kx_i^2$$

where x_i represents the initial stretch of the tendons. With the kangaroo at the high point of the jump, the kinetic energy is again zero if it jumped straight up. The tendons are no longer stretched, so the elastic potential energy is zero. But now there is gravitational potential energy. At a height h above the initial point, the final mechanical energy is

$$K_f + U_{f,grav} + U_{f,elastic} = 0 + mgh + 0$$

where m is the kangaroo's mass. Setting the mechanical energies equal,

$$\frac{1}{2}kx_i^2 = mgh \quad \Rightarrow \quad h = \frac{kx_i^2}{2mg}$$

We don't know all of the constants (mass, spring constant, initial amount of stretch), so we set up a ratio:

$$\frac{h_2}{h_1} = \frac{\cancel{k}x_2^2/(\cancel{2mg})}{\cancel{k}x_1^2/(\cancel{2mg})} = \frac{x_2^2}{x_1^2}$$

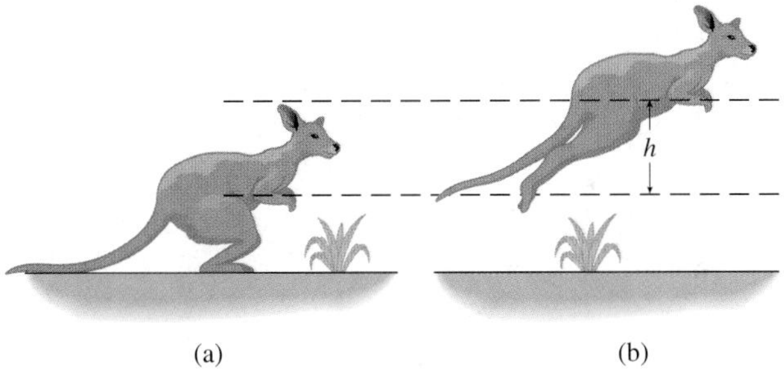

Figure 6.33
(a) Kangaroo crouched and ready to hop. (b) Kangaroo at the highest point in its hop.

For a 10% increase in stretch, $x_2 = 1.10x_1$ and

$$h_2 = \left(\frac{x_2}{x_1}\right)^2 h_1 = (1.10)^2 h_1 = 1.21 \times 2.0 \text{ m} = 2.4 \text{ m}$$

Using a 10% increase in the stretch of the tendon, the kangaroo jumps about 21% higher.

When the kangaroo is hopping along, it does not jump straight up. Will the kangaroo's jump still be 21% higher when jumping at another angle? Imagine the kangaroo hopping along so that it leaves the ground at a 45° angle, which gives the maximum horizontal range per hop in the absence of air resistance. The elastic energy in the tendon is first converted to kinetic energy. This time, not all of the kinetic energy is converted to gravitational potential energy. The kinetic energy at the highest point of the jump is *not* zero because the kangaroo is still moving forward. The initial velocity can be resolved into components:

$$v^2 = v_x^2 + v_y^2 = 2v_x^2 \quad \text{(since } v_x = v_y \text{ for a 45° angle)}$$

At the highest point of the jump, the kinetic energy is $\frac{1}{2}mv_x^2$, which is half of the initial kinetic energy. Overall, *half* of the elastic energy of the tendon is converted to gravitational potential energy:

$$\tfrac{1}{2} \times \left(\tfrac{1}{2}kx_i^2\right) = mgh$$

Since h is still proportional to x_i^2, the height of the jump still increases by 21% if the stretch of the tendon is increased by 10%.

Discussion The storage of elastic energy in the tendon is a clever way for the kangaroo to get more "miles per gallon." Without such an energy storage system, most of the kangaroo's mechanical energy would be converted to an unrecoverable form of energy at the end of each hop. The tendons store some of the energy that would otherwise be lost and then release it to help the next jump. Since less mechanical energy is "lost" on each landing, the energy supplied by the kangaroo's muscles is less than it would otherwise be. Humans use a similar energy-saving mechanism when running (see Problem 103).

Practice Problem 6.12 Jumping with Joey

Suppose the kangaroo has a baby kangaroo (a *joey*) riding in her pouch. If the joey has grown to be one sixth the mass of its mother, how high can the kangaroo jump with the additional load? Assume that, without the joey, she can jump 2.8 m.

6.8 POWER

Power: the *rate* of energy conversion

Sometimes the *rate* of energy conversion is important. When shopping for a sports car, you wouldn't ask the salesman how much work the engine can do. A tiny economy car like the Toyota Prius does more work than a Ferrari if the Prius is used for daily commuting while the Ferrari sits in the garage most of the time. But the Ferrari can do work *at a much faster rate* than the Prius can. In other words, it can change chemical energy in the gasoline into mechanical energy of the car at a much faster rate—it has a larger maximum power output. The higher power output enables the Ferrari to accelerate to high speeds much faster than the Prius. We give the name **power** (symbol P) to the rate of energy transfer. The average power is the amount of energy converted (ΔE) divided by the time the transfer takes (Δt):

Average power:

$$P_{av} = \frac{\Delta E}{\Delta t} \tag{6-26}$$

The SI unit of power, the joule per second, is given the name watt (1 W = 1 J/s), after James Watt (1736–1819), a Scottish inventor who greatly improved the efficiency of steam engines. Remember that the unit symbol W stands for *watt*, not *work*.

In the United States, the maximum power output of an electric motor or automobile engine is usually specified in horsepower, which is a non-SI unit of power (1 hp = 746 W).

The *kilowatt-hour* (kW·h) is a unit of energy, *not* a unit of power. One kilowatt-hour is the amount of energy transferred at a constant rate of 1 kW during a time interval of 1 h. The kilowatt-hour is commonly used by utility companies to measure the amount of electric energy used by consumers.

The work done by a force during a small time interval Δt is

$$W = F\,\Delta r \cos\theta \tag{6-1}$$

The magnitude of the displacement is

$$\Delta r = v\,\Delta t$$

Hence, the power—the rate at which the force does work—can be found from the force and the velocity.

$$P = \frac{W}{\Delta t} = \frac{F\Delta r\cos\theta}{\Delta t} = F\frac{\Delta r}{\Delta t}\cos\theta = Fv\cos\theta$$

Instantaneous power (rate at which work is done):

$$P = Fv\cos\theta \tag{6-27}$$

(θ is the angle between $\vec{\mathbf{F}}$ and $\vec{\mathbf{v}}$)

Example 6.13

Air Resistance on a Hill-Climbing Car

A 1000.0-kg car climbs a hill with a 4.0° incline at a constant 12.0 m/s (Fig. 6.34). (a) At what rate is the gravitational potential energy increasing? (b) If the mechanical power output of the engine is 20.0 kW, find the force of air resistance on the car. (Assume that air resistance is responsible for all of the energy dissipation.)

Strategy (a) We can find the rate of gravitational potential energy increase in two ways. One is to find the potential energy change during a time interval Δt and divide it by the time interval, which is equivalent to using the definition of average power [Eq. (6-26)]. The other possibility is to use Eq. (6-27) to find the rate at which the gravitational force does work.

continued on next page

Example 6.13 continued

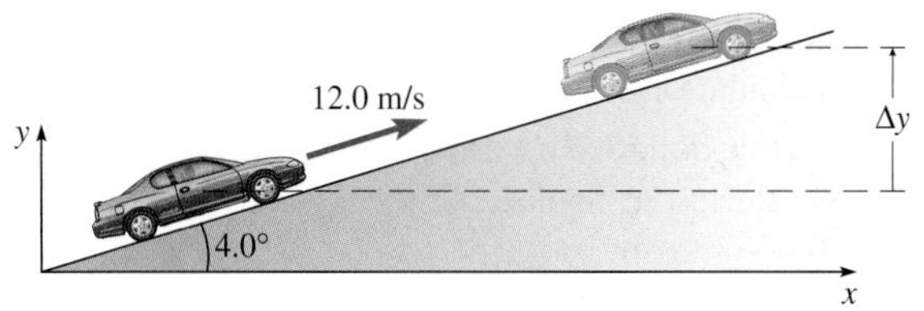

Figure 6.34
Car climbing a hill at constant speed.

(b) The car moves at constant speed, so its kinetic energy is not changing. Therefore, during any time interval, the work done by the engine (W_e) plus the (negative) work done by air resistance (W_a) is equal to the increase in the gravitational potential energy.

Given: car mass = 1000.0 kg; v = 12.0 m/s; 4.0° incline.

To find: (a) rate of potential energy change, $\Delta U/\Delta t$;
(b) force due to air resistance, $\vec{\mathbf{F}}_a$.

Solution (a) For a small change in elevation Δy, the change in potential energy is

$$\Delta U = mg\,\Delta y$$

The *rate* of potential energy change is

$$\frac{\Delta U}{\Delta t} = \frac{mg\,\Delta y}{\Delta t} = mg\frac{\Delta y}{\Delta t} = mgv_y$$

where $v_y = \Delta y/\Delta t$ is the y-component of the velocity. From Fig. 6.35, $v_y = v \sin \phi$, where $\phi = 4.0°$. Then,

$$\frac{\Delta U}{\Delta t} = mgv \sin \phi = 1000.0 \text{ kg} \times 9.80 \text{ m/s}^2 \times 12.0 \text{ m/s} \times \sin 4.0°$$
$$= 8200 \text{ W}$$

(b) During any time interval Δt, the (positive) work done by the engine plus the (negative) work done by air resistance must equal the increase in the gravitational potential energy:

$$W_{total} = W_e + W_a = \Delta U$$

Dividing each term by Δt, we find

$$\frac{W_e}{\Delta t} + \frac{W_a}{\Delta t} = \frac{\Delta U}{\Delta t} \quad \Rightarrow \quad P_e + P_a = \frac{\Delta U}{\Delta t}$$

where P_e and P_a represent the power output of the engine and the rate at which air resistance does (negative) work on the car, respectively. Then,

$$P_a = \frac{\Delta U}{\Delta t} - P_e = 8.2 \text{ kW} - 20.0 \text{ kW} = -11.8 \text{ kW}$$

So, of the 20.0 kJ of mechanical work that the engine does each second, 8.2 kJ goes into gravitational potential energy and 11.8 kJ goes into pushing air out of the way and stirring it up in the process.

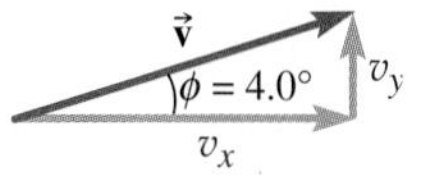

Figure 6.35
Resolving the velocity into x- and y-components.

The direction of the force of air resistance $\vec{\mathbf{F}}_a$ on the car is opposite to the car's velocity, so

$$P_a = F_a v \cos 180° = -F_a v$$

Solving for F_a,

$$F_a = -\frac{P_a}{v} = -\frac{-11\,800 \text{ W}}{12.0 \text{ m/s}} = 983 \text{ N}$$

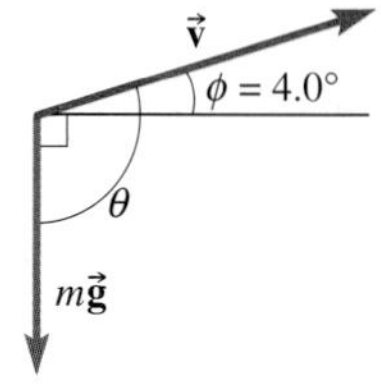

Figure 6.36
The angle between the force and the velocity is $\theta = 94.0°$. (The angle is exaggerated for clarity.)

Discussion We can check (a) by using Eq. (6-27) to find the rate at which the gravitational force does work: $P = Fv \cos \theta$, where $F = mg$. The angle θ is *not* the same as ϕ. In Eq. (6-27), θ is the angle between the force and velocity vectors, which is 94.0° (Fig. 6.36). Then,

$$P = mgv \cos 94.0°$$
$$= 1000.0 \text{ kg} \times 9.80 \text{ m/s}^2 \times 12.0 \text{ m/s} \times \cos 94.0°$$
$$= -8200 \text{ W}$$

Gravity does work on the car at a rate of −8200 W, which means the potential energy is *increasing* at a rate of +8200 W.

We can also figure out what mechanical power the engine must supply to go 12.0 m/s on level ground. With no change in potential energy, all of the mechanical power output of the engine goes into stirring up the air, so $P_e + P_a = 0$. The magnitude of the force of air resistance is the same (983 N) since the speed is the same. Then air resistance dissipates energy at the same rate as before:

$$P_a = -F_a v = -983 \text{ N} \times 12.0 \text{ m/s} = -11.8 \text{ kW}$$

Therefore, P_e = 11.8 kW. On level ground, the gravitational potential energy isn't increasing, so the engine only needs to do enough work to counteract the tendency of air resistance to slow down the car.

In this example, we have assumed that all of the mechanical power output of the engine is delivered to the wheels to propel the car forward. In reality, some of the engine's power output is used to run auxiliary devices such as headlights, radios, and windshield wipers. Friction (in the moving parts of the engine, transmission, and drivetrain) also reduces the amount of power that is actually delivered to the wheels.

Practice Problem 6.13 Mechanical Power Output on Flat Ground or Going Downhill

What mechanical power must the engine supply (a) to drive on level ground at 12.0 m/s and (b) to go down a 4.0° incline at 12.0 m/s? (Since this is the same speed as in Example 6.13, the force of air resistance is the same.)

Master the Concepts

- Conservation law: a physical law phrased in terms of a quantity that does not change with time.
- The law of conservation of energy: the total energy of the universe is unchanged by any physical process.
- Work is an energy transfer due to the application of a force. The work done by a constant force $\vec{\mathbf{F}}$ acting on an object during a displacement $\Delta\vec{\mathbf{r}}$ is

$$W = F\,\Delta r \cos\theta \qquad (6\text{-}1)$$

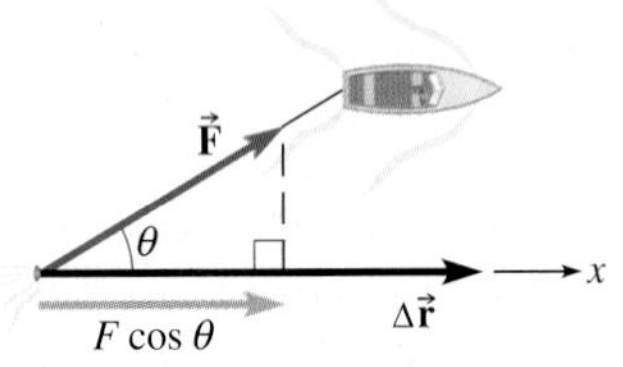

where θ is the angle between $\vec{\mathbf{F}}$ and $\Delta\vec{\mathbf{r}}$. If $\vec{\mathbf{F}}$ or $\Delta\vec{\mathbf{r}}$ is parallel to the x-axis,

$$W = F_x\,\Delta x \qquad (6\text{-}2)$$

- When several forces act on an object, the total work is the sum of the work done by each force individually.
- Translational kinetic energy is the energy associated with motion of the object as a whole. The translational kinetic energy of an object of mass m moving with speed v is

$$K = \tfrac{1}{2}mv^2 \qquad (6\text{-}6)$$

- The gravitational potential energy for an object of mass m in a *uniform* gravitational field is

$$U_{\text{grav}} = mgy \qquad (6\text{-}13)$$

where the $+y$-axis points up and we assign $U = 0$ to $y = 0$.
- The gravitational potential energy for two bodies of masses m_1 and m_2 whose centers are separated by a distance r is

$$U = -\frac{Gm_1m_2}{r} \qquad (6\text{-}14)$$

where we assign $U = 0$ to infinite separation ($r = \infty$).

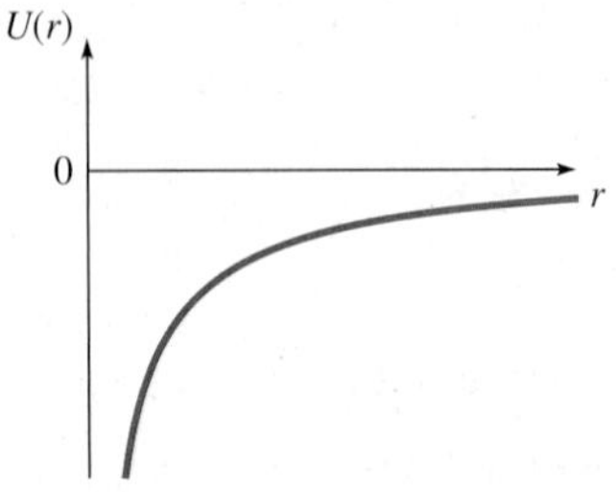

- There is no special significance to the sign of the potential energy. What matters is the sign of the potential energy *change*. Only *changes* in potential energy enter our calculations.
- The work done by a variable force directed along the x-axis during a displacement Δx is the area under the $F_x(x)$ graph from x_i to x_f.

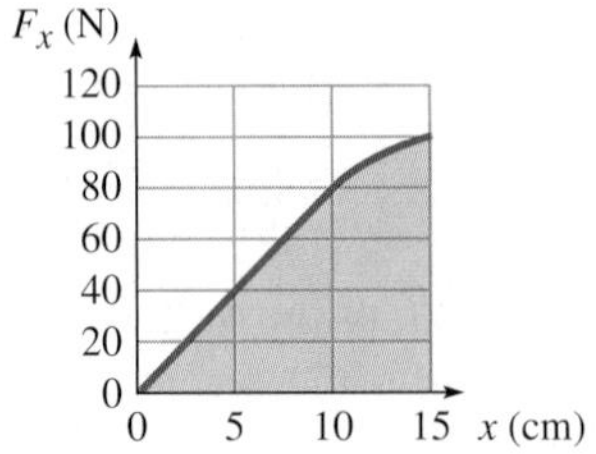

- Hooke's law: for many objects, the deformation is proportional to the magnitude of the force that causes the deformation. An ideal spring is massless and follows Hooke's law. The force exerted *by* the moveable end of an ideal spring when it is at position x is

$$F_x = -kx \qquad (6\text{-}19)$$

where the origin is chosen so the spring is relaxed at $x = 0$ and k is called the spring constant.
- If we assign $U = 0$ to the relaxed spring ($x = 0$), the elastic potential energy stored in an ideal spring of spring constant k is

$$U_{\text{elastic}} = \tfrac{1}{2}kx^2 \qquad (6\text{-}24)$$

- Mechanical energy is the sum of the kinetic and potential energies. The change in potential energy accounts for the work done by the forces associated with the potential energy. The work done by nonconservative forces is equal to the change in the mechanical energy:

$$W_{\text{nc}} = \Delta K + \Delta U = \Delta E_{\text{mech}} \qquad (6\text{-}12)$$

When nonconservative forces do no net work, the mechanical energy does not change.

$$\text{If } W_{\text{nc}} = 0,\ \Delta K + \Delta U = 0$$

- Average power is the average rate of energy conversion.

$$P_{\text{av}} = \frac{\Delta E}{\Delta t} \qquad (6\text{-}26)$$

The instantaneous rate at which a force $\vec{\mathbf{F}}$ does work when the object it acts on moves with velocity $\vec{\mathbf{v}}$ is

$$P = Fv\cos\theta \qquad (6\text{-}27)$$

where θ is the angle between $\vec{\mathbf{F}}$ and $\vec{\mathbf{v}}$.
- The SI unit of work and energy is the joule. 1 J = 1 N·m. The SI unit of power is the watt. 1 W = 1 J/s.

Conceptual Questions

1. An object moves in a circle. Is the total work done on the object by external forces necessarily zero? Explain.
2. You are walking to class with a backpack full of books. As you walk at constant speed on flat ground, does the force exerted on the backpack by your back and shoulders do any work? If so, is it positive or negative? Answer the same questions in two other situations: (1) you are walking down some steps at constant speed; (2) you start to run faster and faster on a level sidewalk to catch a bus.
3. Why do roads leading to the top of a mountain wind back and forth? [*Hint:* Think of the road as an inclined plane.]
4. A mango falls to the ground. During the fall, does the Earth's gravitational field do positive or negative work W_m on the mango? Does the mango's gravitational field do positive or negative work W_E on the Earth? Compare the signs and the magnitudes of W_m and W_E.
5. Can static friction do work? If so, give an example. [*Hint:* Static friction acts to prevent *relative* motion along the contact surface.]
6. In the design of a roller coaster, is it possible for any hill of the ride to be higher than the first one? If so, how?
7. When a ball is dropped to the floor from a height h, it strikes the ground and briefly undergoes a change of shape before rebounding to a maximum height less than h. Explain why it does not return to the same height h.
8. A gymnast is swinging in a vertical circle about a crossbar. In terms of energy conservation, explain why the speed of the gymnast's body is slowest at the top of the circle and fastest at the bottom.
9. A bicycle rider notices that he is approaching a steep hill. Explain, in terms of energy, why the bicyclist pedals hard to gain as much speed as possible on level road before reaching the hill.
10. You need to move a heavy crate by sliding it across a smooth floor. The coefficient of sliding friction is 0.2. You can either push the crate horizontally or pull the crate using an attached rope. When you pull on the rope, it makes a 30° angle with the floor. Which way should you choose to move the crate so that you do the least amount of work? How can you answer this question without knowing the weight of the crate or the displacement of the crate?
11. The main effort of running is the work done by the muscles to accelerate and decelerate the legs. When a foot strikes the ground, it is momentarily brought to rest while the remainder of the animal's body continues to move forward. When the foot is picked up, it is accelerated forward by one set of muscles in order to move ahead of the rest of the body. Then the foot is slowed down by a second set of muscles until it is brought to rest on the ground again. The muscles expend energy both when accelerating and when decelerating the leg. How are thoroughbred horses, deer, and greyhounds adapted so that they can run at great speed?

12. Explain why an ideal spring *must* exert forces of equal magnitude on the objects attached to each end, even if the spring itself has a nonzero acceleration. [*Hint:* Use one of Newton's laws of motion and remember that an ideal spring has zero mass.] Is the amount of work done by the spring on the two objects necessarily the same? Explain. If the answer is no, give an example to illustrate.
13. Zorba and Boris are at a water park. There are two water slides with straight slopes that start at the same height and end at the same height. Slide A has a more gradual slope than slide B. Boris says he likes slide B better because you reach a faster speed and he notes that he got to the bottom level in less time on slide B as measured with his stop watch. His brother Zorba says you reach the same speed with either slide. Who is correct and why? Both slides have negligible friction.

Multiple-Choice Questions

1. After getting on the Santa Monica Freeway, a sports car accelerates from 30 mi/h to 90 mi/h. Its kinetic energy
   (a) increases by a factor of $\sqrt{3}$.
   (b) increases by a factor of 3.
   (c) increases by a factor of 9.
   (d) increases by a factor that depends on the car's mass.
2. If a kangaroo on Earth can jump from a standing start so that its feet reach a height h above the surface, approximately how high can the same kangaroo jump from a standing start on the Moon's surface? $g_{\text{Moon}} \approx \frac{1}{6} g_{\text{Earth}}$. (Assume the kangaroo has an oxygen tank and pressure suit with negligible mass.)
   (a) h (b) $6h$ (c) $\frac{1}{6}h$
   (d) $36h$ (e) $\frac{1}{36}h$ (f) $\sqrt{6}h$

Questions 3–5. The orbit of Pluto is much more eccentric than the orbits of the planets. That is, instead of being nearly circular, the orbit is noticeably elliptical. The point in the orbit nearest the Sun is called the *perihelion* and the point farthest from the Sun is called the *aphelion.*

Answer choices for Questions 3–5:
(a) its maximum value (b) its minimum value
(c) the same value as at every other point in the orbit

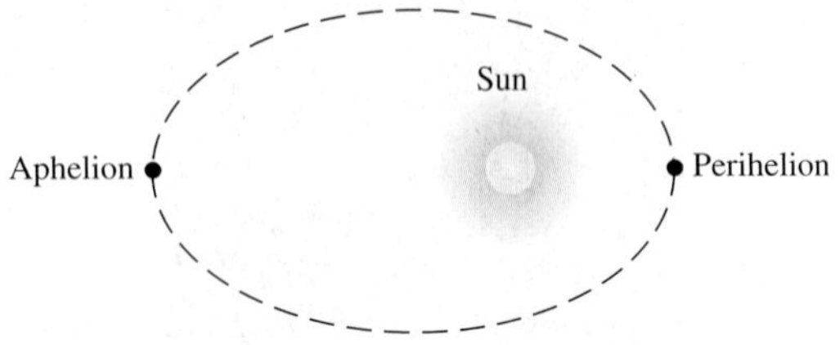

Multiple-Choice Questions 3–5

3. At perihelion, the gravitational potential energy of Pluto's orbit has
4. At perihelion, the kinetic energy of Pluto has
5. At perihelion, the mechanical energy of Pluto's orbit has
6. As Pluto moves from the perihelion to the aphelion, the work done by gravity on Pluto is
(a) zero. (b) positive. (c) negative.
7. Two balls are thrown from the roof of a building with the same initial speed. One is thrown horizontally while the other is thrown at an angle of 20° above the horizontal. Which hits the ground with the greatest speed? Ignore air resistance.
(a) The one thrown horizontally
(b) The one thrown at 20°
(c) They hit the ground with the same speed.
(d) The answer cannot be determined with the given information.
8. A hiker descends from the South Rim of the Grand Canyon to the Colorado River. During this hike, the work done by gravity on the hiker is
(a) positive and depends on the path taken.
(b) negative and depends on the path taken.
(c) positive and independent of the path taken.
(d) negative and independent of the path taken.
(e) zero.

Questions 9 and 10. A simple catapult, consisting of a leather pouch attached to rubber bands tied to two forks of a wooden Y, has a spring constant k and is used to shoot a pebble horizontally. When the catapult is stretched by a distance d, it gives a pebble of mass m a launch speed v. Answer choices for Questions 9 and 10:
(a) $\sqrt{3}v$ (b) $3v$ (c) $3\sqrt{3}v$ (d) $9v$ (e) $27v$

9. What speed does the catapult give a pebble of mass m when stretched to a distance $3d$?
10. What speed does the catapult give a pebble of mass $m/3$ when stretched to a distance d?
11. A projectile is launched at an angle θ above the horizontal. Ignoring air resistance, what fraction of its initial kinetic energy does the projectile have at the top of its trajectory?
(a) $\cos\theta$ (b) $\sin\theta$ (c) $\tan\theta$ (d) $\dfrac{1}{\tan\theta}$ (e) $\dfrac{1}{2}$
(f) $\cos^2\theta$ (g) $\sin^2\theta$ (h) 0 (i) 1

Problems

Combination conceptual/quantitative problem
Biological or medical application
Challenging problem
Blue # Detailed solution in the Student Solutions Manual
1 2 Problems paired by concept
Text website interactive or tutorial

Section 6.2 Work Done by a Constant Force

1. How much work must Denise do to drag her basket of laundry of mass 5.0 kg a distance of 5.0 m along a floor, if the force she exerts is a constant 30.0 N at an angle of 60.0° with the horizontal?
2. A sled is dragged along a horizontal path at a constant speed of 1.5 m/s by a rope that is inclined at an angle of 30.0° with respect to the horizontal. The total weight of the sled is 470 N. The tension in the rope is 240 N. How much work is done by the rope on the sled in a time interval of 10.0 s?

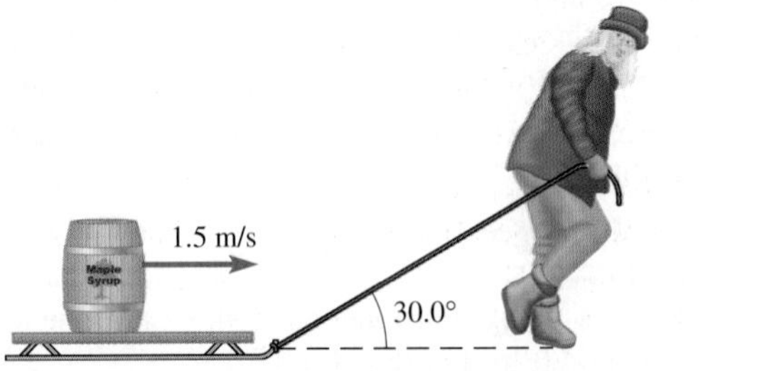

3. Hilda holds a gardening book of weight 10 N at a height of 1.0 m above her patio for 50 s. How much work does she do *on the book* during that 50 s?
4. The tension in the horizontal towrope pulling a water-skier is 240 N while the skier moves due west a distance of 54 m. How much work does the towrope do on the water-skier?
5. A barge of mass 5.0×10^4 kg is pulled along the Erie Canal by two mules, walking along towpaths parallel to the canal on either side of it. The ropes harnessed to the mules make angles of 45° to the canal. Each mule is pulling on its rope with a force of 1.0 kN. How much work is done on the barge by both of these mules together as they pull the barge 150 m along the canal?

6. A 402-kg pile driver is raised 12 m above ground. (a) How much work must be done to raise the pile driver? (b) How much work

does gravity do on the driver as it is raised? (c) The driver is now dropped. How much work does gravity do on the driver as it falls?

7. Jennifer lifts a 2.5-kg carton of cat litter from the floor to a height of 0.75 m. (a) How much *total* work is done on the carton during this operation? Jennifer then pours 1.2 kg of the litter into the cat's litter box on the floor. (b) How much work is done by gravity on the 1.2 kg of litter as it falls into the litter box?

8. Dirk pushes on a packing box with a horizontal force of 66.0 N as he slides it along the floor. The average friction force acting on the box is 4.80 N. How much *total* work is done on the box in moving it 2.50 m along the floor?

9. Juana slides a crate along the floor of the moving van. The coefficient of kinetic friction between the crate and the van floor is 0.120. The crate has a mass of 56.8 kg and Juana pushes with a horizontal force of 124 N. If 74.4 J of total work are done on the crate, how far along the van floor does it move?

Section 6.3 Kinetic Energy

10. An automobile with a mass of 1600 kg has a speed of 30.0 m/s. What is its kinetic energy?

11. A record company executive is on his way to a TV interview and is carrying a promotional CD in his briefcase. The mass of the briefcase and its contents is 5.00 kg. The executive realizes that he is going to be late. Starting from rest, he starts to run, reaching a speed of 2.50 m/s. What is the work done by the executive on the briefcase during this time? Ignore air resistance.

12. In 1899, Charles M. "Mile a Minute" Murphy set a record for speed on a bicycle by pedaling for a mile at an average of 62.3 mph (27.8 m/s) on a 3-mi track of plywood planks set over railroad ties in the draft of a Long Island Railroad train. In 1985, a record was set for this type of "motor pacing" by Olympic cyclist John Howard who raced at 152.2 mph (68.04 m/s) in the wake of a race car at Bonneville Salt Flats. The race car had a modified tail assembly designed to reduce the air drag on the cyclist. What was the kinetic energy of the bicycle plus rider in each of these feats? Assume that the mass of bicycle plus rider is 70.5 kg in each case.

13. Sam pushes a 10.0-kg sack of bread flour on a frictionless horizontal surface with a constant horizontal force of 2.0 N starting from rest. (a) What is the kinetic energy of the sack after Sam has pushed it a distance of 35 cm? (b) What is the speed of the sack after Sam has pushed it a distance of 35 cm?

14. Josie and Charlotte push a 12-kg bag of playground sand for a sandbox on a frictionless, horizontal, wet polyvinyl surface with a constant, horizontal force for a distance of 8.0 m, starting from rest. If the final speed of the sand bag is 0.40 m/s, what is the magnitude of the force with which they pushed?

15. A ball of mass 0.10 kg moving with speed of 2.0 m/s hits a wall and bounces back with the same speed in the opposite direction. What is the change in the ball's kinetic energy?

16. Jim rides his skateboard down a ramp that is in the shape of a quarter circle with a radius of 5.00 m. At the bottom of the ramp, Jim is moving at 9.00 m/s. Jim and his skateboard have a mass of 65.0 kg. How much work is done by friction as the skateboard goes down the ramp? (tutorial: energy, parts (a) and (b))

17. A 69.0-kg short-track ice skater is racing at a speed of 11.0 m/s when he falls down and slides across the ice into a padded wall that brings him to rest. Assuming that he doesn't lose any speed during the fall or while sliding across the ice, how much work is done by the wall while stopping the ice skater?

18. A plane weighing 220 kN (25 tons) lands on an aircraft carrier. The plane is moving horizontally at 67 m/s (150 mi/h) when its tailhook grabs hold of the arresting cables. The cables bring the plane to a stop in a distance of 84 m. (a) How much work is done on the plane by the arresting cables? (b) What is the force (assumed constant) exerted on the plane by the cables? (Both answers will be *underestimates*, since the plane lands with the engines full throttle forward; in case the tailhook fails to grab hold of the cables, the pilot must be ready for immediate takeoff.)

19. A shooting star is a meteoroid that burns up when it reaches Earth's atmosphere. Many of these meteoroids are quite small. Calculate the kinetic energy of a meteoroid of mass 5.0 g moving at a speed of 48 km/s and compare it to the kinetic energy of a 1100-kg car moving at 29 m/s (65 mi/h).

Section 6.4 Gravitational Potential Energy (1)

20. Sean climbs a tower that is 82.3 m high to make a jump with a parachute. The mass of Sean plus the parachute is 68.0 kg. If $U = 0$ at ground level, what is the potential energy of Sean and the parachute at the top of the tower? (tutorial: energy, parts (c) and (d))

21. Justin moves a desk 5.0 m across a level floor by pushing on it with a constant horizontal force of 340 N. (It slides for a negligibly small distance before coming to a stop when the force is removed.) Then, changing his mind, he moves it back to its starting point, again by pushing with a constant force of 340 N. (a) What is the change in the desk's gravitational potential energy during the round-trip? (b) How much work has Justin done on the desk? (c) If the work done by Justin is not equal to the change in gravitational potential energy of the desk, then where has the energy gone?

22. An airline executive decides to economize by reducing the amount of fuel required for long-distance flights. He orders the ground crew to remove the paint from the outer surface of each plane. The paint removed from a single plane has a mass of approximately 100 kg. (a) If the airplane cruises at an altitude of 12 000 m, how much energy is saved in not having to lift the paint to that altitude? (b) How much energy is saved by not having to move that amount of paint from rest to a cruising speed of 250 m/s?

23. Emil is tossing an orange of mass 0.30 kg into the air. (a) Emil throws the orange straight up and then catches it, throwing and catching it at the same point in space. What is the change in the potential energy of the orange during its trajectory? Ignore air resistance. (b) Emil throws the orange straight up, starting 1.0 m above the ground. He fails to catch it. What is the change in the potential energy of the orange during this flight?

24. A brick of mass 1.0 kg slides down an icy roof inclined at 30.0° with respect to the horizontal. (a) If the brick starts from rest, how fast is it moving when it reaches the edge of the roof 2.00 m away? Ignore friction. (b) Redo part (a) if the coefficient of kinetic friction is 0.10. (interactive: sliding brick)

25. An arrangement of two pulleys, as shown in the figure, is used to lift a 48.0-kg mass a distance of 4.00 m above the starting point. Assume the pulleys and rope are ideal and that all rope sections are essentially vertical. (a) What is the mechanical advantage of this system? (In other words, by what factor is the force you exert to lift the weight multiplied by the pulley system?) (b) What is the change in the potential energy of the weight when it is lifted a distance of 4.00 m? (c) How much work must be done to lift the 48.0-kg mass a distance of 4.00 m? (d) What length of rope must be pulled by the person lifting the weight 4.00 m higher in the air? (tutorial: block-and-tackle)

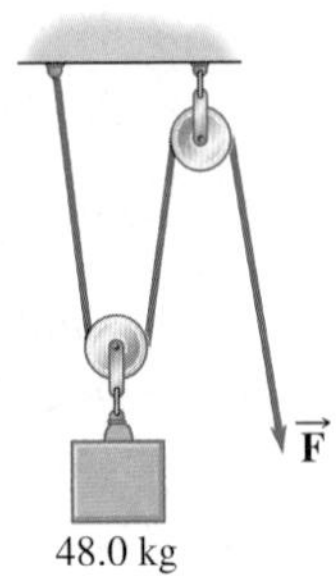

26. In Example 6.1, find the work done by the movers as they slide the chest up the ramp if the coefficient of friction between the chest and the ramp is 0.20. (tutorial: ramp)

27. A cart moving to the *right* passes point 1 at a speed of 20.0 m/s. Let $g = 9.81$ m/s^2. (a) What is the speed of the cart as it passes point 3? (b) Will the cart reach position 4? Ignore friction.

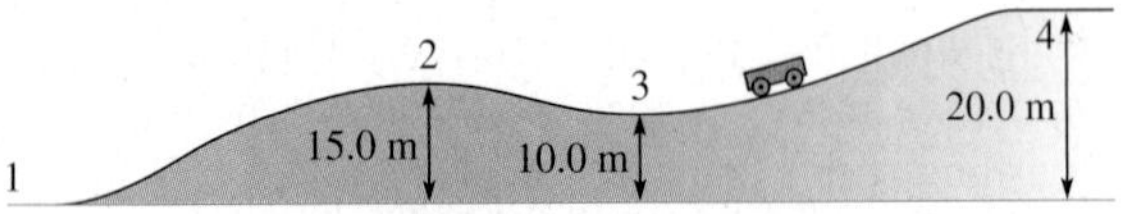

Problems 27 and 28

28. A cart starts from position 4 with a velocity of 15 m/s to the left. Find the speed with which the cart reaches positions 1, 2, and 3. Ignore friction.

29. Bruce stands on a bank beside a pond, grasps the end of a 20.0-m-long rope attached to a nearby tree and swings out to drop into the water. If the rope starts at an angle of 35.0° with the vertical, what is Bruce's speed at the bottom of the swing?

30. The maximum speed of a child on a swing is 4.9 m/s. The child's height above the ground is 0.70 m at the lowest point in his motion. How high above the ground is he at his highest point?

31. If the skier of Example 6.6 is moving at 12 m/s at the bottom of the trail, calculate the total work done by friction and air resistance during the run. The skier's mass is 75 kg.

32. A 750-kg automobile is moving at 20.0 m/s at a height of 5.0 m above the bottom of a hill when it runs out of gasoline. The car coasts down the hill and then continues coasting up the other side until it comes to rest. Ignoring frictional forces and air resistance, what is the value of h, the highest position the car reaches above the bottom of the hill?

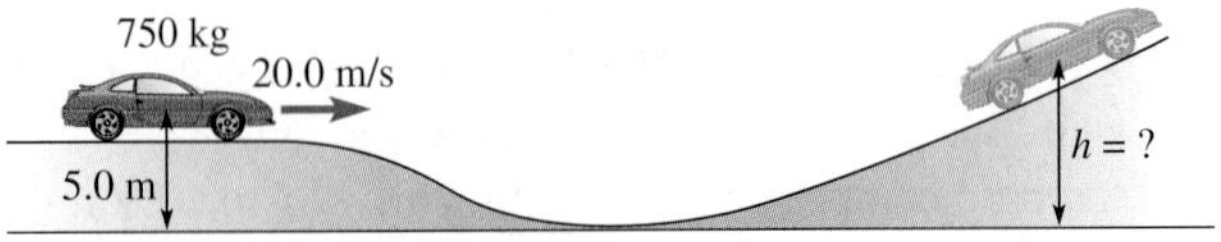

33. Rachel is on the roof of a building, h meters above ground. She throws a heavy ball into the air with a speed v, at an angle θ with respect to the horizontal. Ignore air resistance. (a) Find the speed of the ball when it hits the ground in terms of h, v, θ, and g. (b) For what value(s) of θ is the speed of the ball greatest when it hits the ground?

34. A crate of mass m_1 on a frictionless inclined plane is attached to another crate of mass m_2 by a massless rope. The rope passes over an ideal pulley so the mass m_2 is suspended in air. The plane is inclined at an angle $\theta = 36.9°$. Use conservation of energy to find how fast crate m_2 is moving after m_1 has traveled a distance of 1.4 m along the incline, starting from rest. The mass of m_1 is 12.4 kg and the mass of m_2 is 16.3 kg.

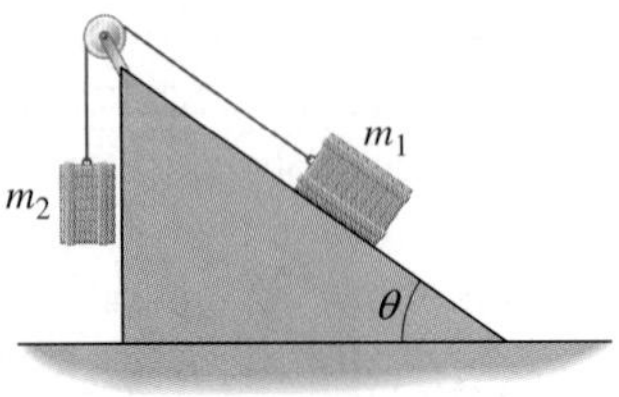

35. The forces required to extend a spring to various lengths are measured. The results are shown in the following table. Using the data in the table, plot a graph that helps you to answer the following two questions: (a) What is the spring constant? (b) What is the relaxed length of the spring?

Force (N)	1.00	2.00	3.00	4.00	5.00
Spring length (cm)	14.5	18.0	21.5	25.0	28.5

Section 6.5 Gravitational Potential Energy (2)

36. A 75.0-kg skier starts from rest and slides down a 32.0-m frictionless slope that is inclined at an angle of 15.0°

with the horizontal. Ignore air resistance. (a) Calculate the work done by gravity on the skier and the work done by the normal force on the skier. (b) If the slope is not frictionless so that the skier has a final velocity of 10.0 m/s, calculate the work done by gravity, the work done by the normal force, the work done by friction, the force of friction (assuming it is constant), and the coefficient of kinetic friction. (tutorial: water slide)

37. You are on the Moon and would like to send a probe into space so that it does not fall back to the surface of the Moon. What launch speed do you need?

38. A planet with a radius of 6.00×10^7 m has a gravitational field of magnitude 30.0 m/s^2 at the surface. What is the escape speed from the planet?

39. The escape speed from the surface of Planet Zoroaster is 12.0 km/s. The planet has no atmosphere. A meteor far away from the planet moves at speed 5.0 km/s on a collision course with Zoroaster. How fast is the meteor going when it hits the surface of the planet?

40. The escape speed from the surface of the Earth is 11.2 km/s. What would be the escape speed from another planet of the same density (mass per unit volume) as Earth but with a radius twice that of Earth?

41. A satellite is placed in a noncircular orbit about the Earth. The farthest point of its orbit (*apogee*) is 4 Earth radii from the center of the Earth, while its nearest point (*perigee*) is 2 Earth radii from the Earth's center. If we define the gravitational potential energy U to be zero for an infinite separation of Earth and satellite, find the ratio $U_{\text{perigee}}/U_{\text{apogee}}$.

42. What is the minimum speed with which a meteor strikes the top of the Earth's stratosphere (about 40 km above Earth's surface), assuming that the meteor begins as a bit of interplanetary debris far from Earth? Assume the drag force is negligible until the meteor reaches the stratosphere.

43. A projectile with mass of 500 kg is launched straight up from the Earth's surface with an initial speed v_i. What magnitude of v_i enables the projectile to just reach a maximum height of $5R_E$, measured from the *center* of the Earth? Ignore air friction as the projectile goes through the Earth's atmosphere.

44. The orbit of Halley's comet around the Sun is a long thin ellipse. At its aphelion (point farthest from the Sun), the comet is 5.3×10^{12} m from the Sun and moves with a speed of 10.0 km/s. What is the comet's speed at its perihelion (closest approach to the Sun) where its distance from the Sun is 8.9×10^{10} m?

45. Suppose a satellite is in a circular orbit 3.0 Earth radii above the surface of the Earth (4.0 Earth radii from the center of the Earth). By how much does it have to increase its speed in order to be able to escape Earth? [*Hint:* You need to calculate the orbital speed and the escape speed.]

46. An asteroid hits the Moon and ejects a large rock from its surface. The rock has enough speed to travel to a point between the Earth and the Moon where the gravitational forces on it from the Earth and the Moon are equal in magnitude and opposite in direction. At that point the rock has a very small velocity toward Earth. What is the speed of the rock when it encounters Earth's atmosphere at an altitude of 700 km above the surface?

Section 6.6 Work Done by Variable Forces: Hooke's Law

47. How much work is done on the bowstring of Example 6.9 to draw it back by 20.0 cm? [*Hint:* Rather than recalculate from scratch, use proportional reasoning.]

48. An ideal spring has a spring constant $k = 20.0$ N/m. What is the amount of work that must be done to stretch the spring 0.40 m from its relaxed length?

49. The force that must be exerted to drive a nail into a wall is roughly as shown in the graph. The first 1.2 cm are through soft drywall; then the nail enters the solid wooden stud. How much work must be done to hammer the nail a horizontal distance of 5.0 cm into the wall?

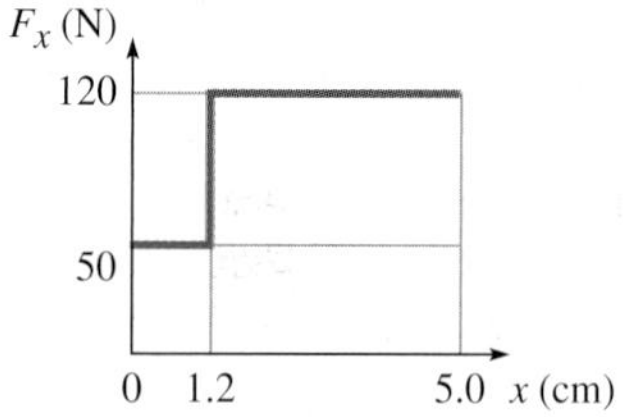

50. (a) If the length of the Achilles tendon increases 0.50 cm when the force exerted on it by the muscle increases from 3200 N to 4800 N, what is the "spring constant" of the tendon? (b) How much work is done by the muscle in stretching the tendon 0.50 cm as the force increases from 3200 N to 4800 N?

51. (a) If forces of magnitude 5.0 N applied to each end of a spring cause the spring to stretch 3.5 cm from its relaxed length, how far do forces of magnitude 7.0 N cause the same spring to stretch? (b) What is the spring constant of this spring? (c) How much work is done by the applied forces in stretching the spring 3.5 cm from its relaxed length? (tutorial: spring)

52. A block of wood is compressed 2.0 nm when inward forces of magnitude 120 N are applied to it on two opposite sides. (a) Assuming Hooke's law holds, what is the effective spring constant of the block? (b) Assuming Hooke's law still holds, how much is the same block compressed by inward forces of magnitude 480 N? (c) How much work is done by the applied forces during the compression of part (b)?

53. The length of a spring increases by 7.2 cm from its relaxed length when a mass of 1.4 kg is hanging in equilibrium from the spring. (a) What is the spring constant? (b) How much elastic potential energy is stored in the spring? (c) A different mass is suspended and the spring length increases by 12.2 cm from its relaxed length to its new equilibrium position. What is the second mass?

54. A spring fixed at one end is compressed from its relaxed position by a distance of 0.20 m. See the graph of the applied external force, F_x, versus the position, x, of the spring. (a) Find the work done by the external force in compressing the spring 0.20 m starting from its relaxed position. (b) Find the work done by the external force to compress the spring from 0.10 m to 0.20 m.

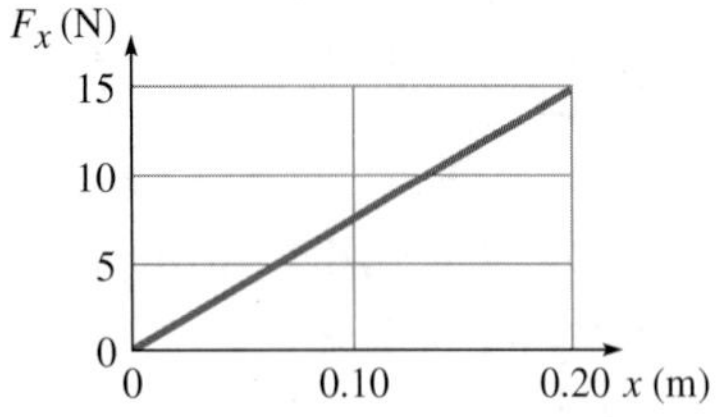

55. Rhonda keeps a 2.0-kg model airplane moving at constant speed in a horizontal circle at the end of a string of length 1.0 m. The tension in the string is 18 N. How much work does the string do on the plane during each revolution? (tutorial: circular motion)

56. The graph shows the force exerted on an object versus the position of that object along the x-axis. The force has no components other than along the x-axis. What is the work done by the force on the object as the object is displaced from 0 to 3.0 m?

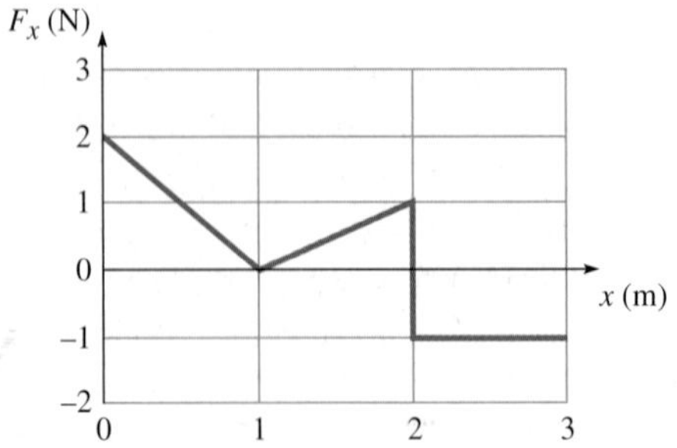

Section 6.7 Elastic Potential Energy

57. A kangaroo decides to see how high it can hop on *one leg*. Assuming the elastic energy stored in the tendon is the same as for Example 6.12, how high can it jump using a single leg?

58. When the spring on a toy gun is compressed by a distance x, it will shoot a rubber ball straight up to a height of h. Ignoring air resistance, how high will the gun shoot the same rubber ball if the spring is compressed by an amount $2x$? Assume $x << h$.

59. You shoot a 51-g pebble straight up with a catapult whose spring constant is 320 N/m. The catapult is initially stretched by 0.20 m. How high above the starting point does the pebble fly? Ignore air resistance.

60. A block (mass m) hangs from a spring (spring constant k). The block is released from rest a distance d above its *equilibrium* position. (a) What is the speed of the block as it passes through the equilibrium point? (b) What is the maximum distance below the equilibrium point that the block will reach?

61. A gymnast of mass 52 kg is jumping on a trampoline. She jumps so that her feet reach a maximum height of 2.5 m above the trampoline and, when she lands, her feet stretch the trampoline down 75 cm. How far does the trampoline stretch when she stands on it at rest? [*Hint:* Assume the trampoline obeys Hooke's law when it is stretched.]

62. Jorge is going to bungee jump from a bridge that is 55.0 m over the river below. The bungee cord has an unstretched length of 27.0 m. To be safe, the bungee cord should stop Jorge's fall when he is at least 2.00 m above the river. If Jorge has a mass of 75.0 kg, what is the minimum spring constant of the bungee cord? (tutorial: spring scale)

63. A 2.0-kg block is released from rest and allowed to slide down a frictionless surface and into a spring. The far end of the spring is attached to a wall, as shown. The initial height of the block is 0.50 m above the lowest part of the slide and the spring constant is 450 N/m. (a) What is the block's speed when it is at a height of 0.25 m above the base of the slide? (b) How far is the spring compressed? (c) The spring sends the block back to the left. How high does the block rise?

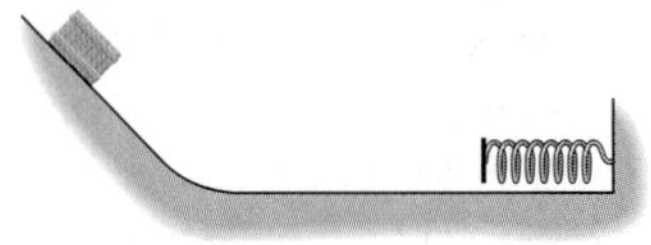

Problems 63 and 105

Section 6.8 Power

64. Lars, of mass 82.4 kg, has been working out and can do work for about 2.0 min at the rate of 1.0 hp (746 W). How long will it take him to climb three flights of stairs, a vertical height of 12.0 m?

65. Show that 1 kilowatt-hour (kW·h) is equal to 3.6 MJ.

66. If a man has an average useful power output of 40.0 W, what minimum time would it take him to lift fifty 10.0-kg boxes to a height of 2.00 m?

67. In Section 6.2, Rosie lifts a trunk weighing 220 N up 4.0 m. If it take her 40 s to lift the trunk, at what average rate does she do work?

68. A bicycle and its rider together has a mass of 75 kg. What power output of the rider is required to maintain a constant speed of 4.0 m/s (about 9 mph) up a 5.0% grade (a road that rises 5.0 m for every 100 m along the pavement)? Assume that frictional losses of energy are negligible.

69. The power output of a cyclist moving at a constant speed of 6.0 m/s on a level road is 120 W. (a) What is the force exerted on the cyclist and the bicycle by the air? (b) By bending low over the handlebars, the cyclist reduces the air resistance to 18 N. If she maintains a power output of 120 W, what will her speed be?

70. A car with mass of 1000.0 kg accelerates from 0 m/s to 40.0 m/s in 10.0 s. Ignore air resistance. The engine has a 22% efficiency, which means that 22% of the energy released by the burning gasoline is converted into mechanical energy. (a) What is the average mechanical power output of the engine? (b) What volume of gasoline is consumed? Assume that the burning of 1.0 L of gasoline releases 46 MJ of energy.

71. A motorist driving a 1200-kg car on level ground accelerates from 20.0 m/s to 30.0 m/s in a time of 5.0 s. Neglecting friction and air resistance, determine the *average* mechanical power in watts the engine must supply during this time interval.

72. A 62-kg woman takes 6.0 s to run up a flight of stairs. The landing at the top of the stairs is 5.0 m above her starting place. (a) What is the woman's average power output while she is running? (b) Would that be equal to her average power *input*—the rate at which chemical energy in food or stored fat is used? Why or why not?

73. How many grams of carbohydrate does a person of mass 74 kg need to metabolize to climb five flights of stairs (15 m height increase)? Each gram of carbohydrate provides 17.6 kJ of energy. Assume 10.0% efficiency—that is, 10.0% of the available chemical energy in the carbohydrate is converted to mechanical energy. What happens to the other 90% of the energy?

74. An object moves in the positive x-direction under the influence of a force F_x. A graph of F_x versus v_x is shown. (a) What is the instantaneous power on the object when its velocity is 11 m/s? (b) What is the instantaneous power on the object when its velocity is 16 m/s?

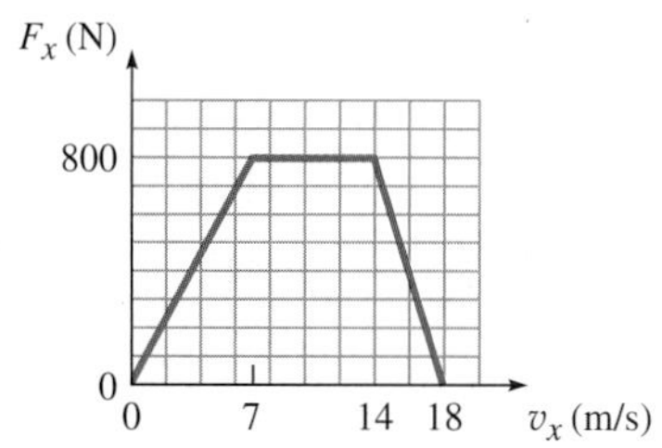

75. A top fuel drag racer with a mass of 500.0 kg completes a quarter-mile (402 m) drag race in a time of 4.2 s starting from rest. The car's final speed is 125 m/s. What is the engine's average power output? Ignore friction and air resistance.

76. (a) Calculate the change in potential energy of 1 kg of water as it passes over Niagara Falls (a vertical descent of 50 m). (b) At what rate is gravitational potential energy lost by the water of the Niagara River? The rate of flow is 5.5×10^6 kg/s. (c) If 10% of this energy can be converted into electric energy, how many households would the electricity supply? (An average household uses an average electrical power of about 1 kW.)

Comprehensive Problems

77. If a high jumper needs to make his center of gravity rise 1.2 m, how fast must he be able to sprint? Assume all of his kinetic energy can be transformed into potential energy. For an extended object, the gravitational potential energy is $U = mgh$, where h is the height of the center of gravity.

78. A pole-vaulter converts the kinetic energy of running to elastic potential energy in the pole, which is then converted to gravitational potential energy. If a pole-vaulter's center of gravity is 1.0 m above the ground while he sprints at 10.0 m/s, what is the maximum height of his center of gravity during the vault? For an extended object, the gravitational potential energy is $U = mgh$, where h is the height of the center of gravity. (In 1988, Sergei Bubka was the first pole-vaulter ever to clear 6 m.)

79. A hang glider moving at speed 9.5 m/s dives to an altitude 8.2 m lower. Ignoring drag, how fast is it then moving?

80. A car moving at 30 mi/h is stopped by jamming on the brakes and locking the wheels. The car skids 50 ft before coming to rest. How far would the car skid if it were initially moving at 60 mi/h? [*Hint:* You will not have to do any unit conversions if you set up the problem as a proportion.]

81. Prove that $U = -2K$ for any gravitational circular orbit. [*Hint:* Use Newton's second law to relate the gravitational force to the acceleration required to maintain uniform circular motion.]

82. A spring gun ($k = 28$ N/m) is used to shoot a 56-g ball horizontally. Initially the spring is compressed by 18 cm. The ball loses contact with the spring and leaves the gun when the spring is still compressed by 12 cm. What is the speed of the ball when it hits the ground, 1.4 m below the spring gun?

83. Two springs with equal spring constants k are connected first in series (one after the other) and then in parallel (side by side) with a weight hanging from the bottom of the combination. What is the effective spring constant of the two different arrangements? In other words, what would be the spring constant of a single spring that would behave exactly as (a) the series combination and (b) the parallel combination? Ignore the weight of the springs. [*Hint* for (a): *each* spring stretches an amount $x = F/k$, but only one spring exerts a force on the hanging object. *Hint* for (b): *each* spring exerts a force $F = kx$.]

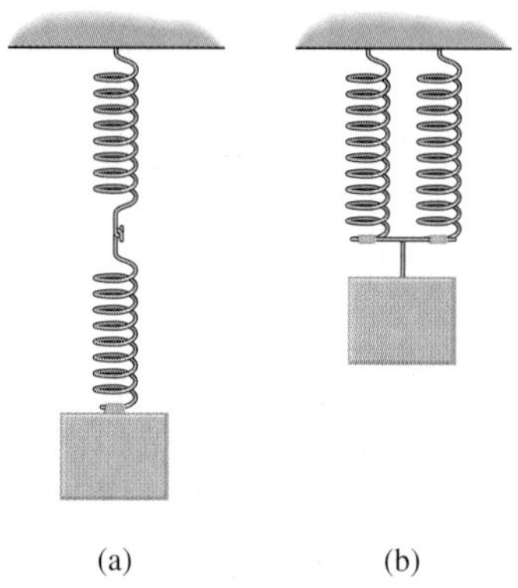

84. A roller coaster car (mass = 988 kg including passengers) is about to roll down a track. The diameter of the circular loop is 20.0 m and the car starts out from rest 40.0 m above the lowest point of the track. Ignore friction and air resistance. (a) At what speed does the car reach the top of the loop?

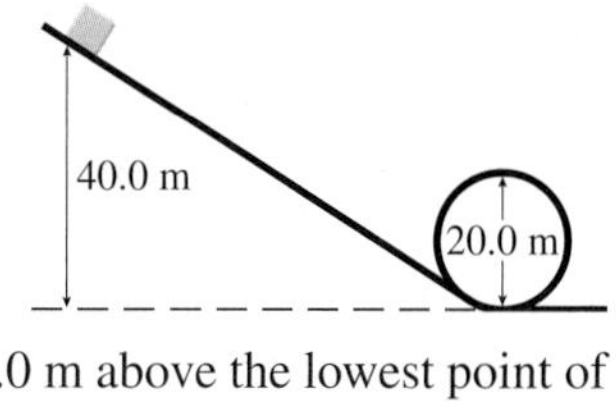

(b) What is the force exerted on the car by the track at the top of the loop? (c) From what minimum height above the bottom of the loop can the car be released so that it does not lose contact with the track at the top of the loop?

85. A 4.0-kg block is released from rest at the top of a frictionless plane of length 8.0 m that is inclined at an angle of 15° to the horizontal. A cord is attached to the block and trails along behind it. When the block reaches a point 5.0 m along the incline from the top, someone grasps the cord and exerts a constant tension parallel to the incline. The tension is such that the block just comes to rest when it reaches the bottom of the incline. (The person's force is a nonconservative force.) What is this constant tension? Solve the problem twice, once using work and energy and again using Newton's laws and the equations for constant acceleration. Which method do you prefer?

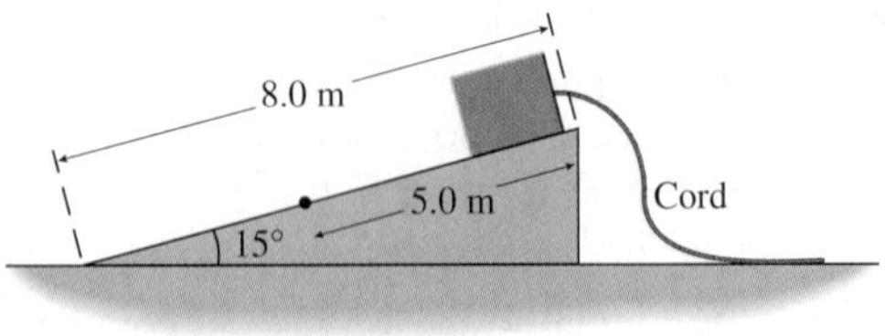

86. The bungee jumper of Example 6.4 made a jump into the Gorge du Verdon in southern France from a platform 182 m above the bottom of the gorge. The jumper weighed 780 N and came within 68 m of the bottom of the gorge. The cord's unstretched length is 30.0 m. (a) Assuming that the bungee cord follows Hooke's law when it stretches, find its spring constant. [*Hint:* The cord does not begin to stretch until the jumper has fallen 30.0 m.] (b) At what speed is the jumper falling when he reaches a height of 92 m above the bottom of the gorge?

87. A spring with $k = 40.0$ N/m is at the base of a frictionless 30.0° inclined plane. A 0.50-kg object is pressed against the spring, compressing it 0.20 m from its equilibrium position. The object is then released. If the object is not attached to the spring, how far up the incline does it travel before coming to rest and then sliding back down?

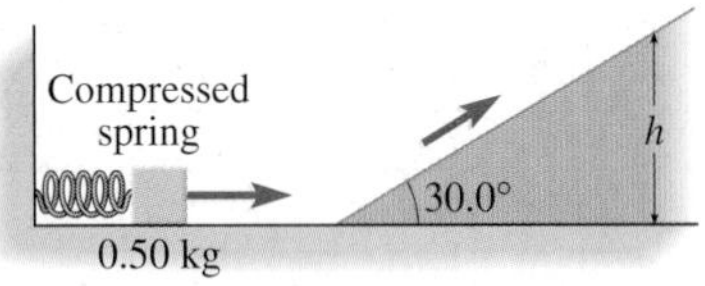

88. In an adventure movie, a 62.5-kg stunt woman falls 8.10 m and lands in a huge air bag. Her speed just before she hit the air bag was 10.5 m/s. (a) What is the total work done on the stunt woman during the fall? (b) How much work is done by gravity on the stunt woman? (c) How much work is done by air resistance on the stunt woman? (d) Estimate the magnitude of the average force of air resistance by assuming it is constant throughout the fall.

89. When a 0.20-kg mass is suspended from a vertically hanging spring, it stretches the spring from its original length of 5.0 cm to a total length of 6.0 cm. The spring with the same mass attached is then placed on a horizontal frictionless surface. The mass is pulled so that the spring stretches to a *total* length of 10.0 cm; then the mass is released and it oscillates back and forth. What is the maximum speed of the mass as it oscillates?

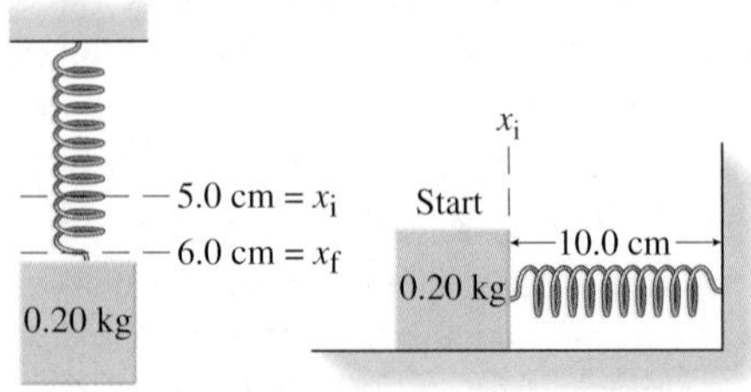

90. Yosemite Falls in California is about 740 m high. (a) What average power would it take for a 70-kg person to hike up to the top of Yosemite Falls in 1.5 h? (b) The human body is about 25% efficient at converting chemical energy to mechanical energy. How much chemical energy is used in this hike? (c) One food Calorie is equal to 4.186×10^3 J. How many Calories of food energy would a person use in this hike?

91. A 1500-kg car coasts in neutral down a 2.0° hill. The car attains a terminal speed of 20.0 m/s. (a) How much power must the engine deliver to drive the car on a *level* road at 20.0 m/s? (b) If the maximum useful power that can be delivered by the engine is 40.0 kW, what is the steepest hill the car can climb at 20.0 m/s?

92. A spring used in an introductory physics laboratory stores 10.0 J of elastic potential energy when it is compressed 0.20 m. Suppose the spring is cut in half. When one of the halves is compressed by 0.20 m, how much potential energy is stored in it? [*Hint:* Does the half spring have the same k as the original uncut spring?]

93. An elevator can carry a maximum load of 1202 kg (including the mass of the elevator car). The elevator has an 801-kg counterweight that always moves with the same speed but in the *opposite direction* to the car. (a) What is the average power that must be delivered by the motor to carry the maximum load up 40.0 m in 60.0 s? (b) How would your answer be different if there were no counterweight?

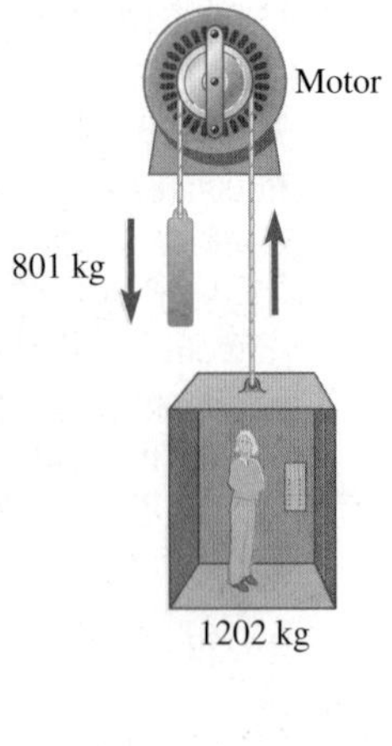

94. (a) How much work does a major-league pitcher do on the baseball when he throws a 90.0 mi/h (40.2 m/s) fastball? The mass of a baseball is 153 g. (b) How many fastballs would a pitcher have to throw to "burn off" a 1520-Calorie meal? (1 Calorie = 1000 cal = 1 kcal.)

Assume that 80.0% of the chemical energy in the food is converted to thermal energy and only 20.0% becomes the kinetic energy of the fastballs.

95. The number of kilocalories per day required by a person resting under standard conditions is called the basal metabolic rate (BMR). (a) To generate 1 kcal, Jermaine's body needs approximately 0.010 mol of oxygen. If Jermaine's net intake of oxygen through breathing is 0.015 mol/min while he is resting, what is his BMR in kcal/day? (b) If Jermaine fasts for 24 h, how many pounds of fat does he lose? Assume that only fat is consumed. Each gram of fat consumed generates 9.3 kcal.

96. Tarzan is running along the ground and approaching a deep gully. A tree branch with a vine hangs over the gully. Tarzan must grab the vine and swing across the gully to the other side, where the ground surface is 1.7 m higher than the ground surface from which Tarzan starts. How fast does Tarzan have to be running to accomplish this feat?

97. Jane is running from the ivory hunters in the jungle. Cheetah throws a 7.0-m-long vine toward her. Jane leaps onto the vine with a speed of 4.0 m/s. When she catches the vine, it makes an angle of 20° with respect to the vertical. (a) When Jane is at her lowest point, she has moved downward a distance h from the height where she originally caught the vine. Show that h is given by $h = L - L\cos 20°$, where L is the length of the vine. (b) How fast is Jane moving when she is at the lowest point in her swing? (c) How high can Jane swing above the lowest point in her swing?

98. The escape speed from Earth is 11.2 km/s, but that is only the minimum speed needed to escape *Earth's* gravitational pull; it does not give the object enough energy to leave the solar system. What is the minimum speed for an object near the Earth's surface so that the object escapes both the Earth's and the Sun's gravitational pulls? Ignore drag due to the atmosphere and the gravitational forces due to the Moon and the other planets. Also ignore the rotation and the orbital motion of the Earth.

99. A skier starts from rest at the top of a frictionless slope of ice in the shape of a hemispherical dome with radius R and slides down the slope. At a certain height h, the normal force becomes zero and the skier leaves the surface of the ice. What is h in terms of R?

100. Two springs with spring constants k_1 and k_2 are connected in series. (a) What is the effective spring constant of the combination? (b) If a hanging object attached to the combination is displaced by 4.0 cm from the relaxed position, what is the potential energy stored in the spring for $k_1 = 5.0$ N/cm and $k_2 = 3.0$ N/cm? [See Problem 83(a).]

101. Two springs with spring constants k_1 and k_2 are connected in parallel. (a) What is the effective spring constant of the combination? (b) If a hanging object attached to the combination is displaced by 2.0 cm from the relaxed position, what is the potential energy stored in the spring for $k_1 = 5.0$ N/cm and $k_2 = 3.0$ N/cm? [See Problem 83(b).]

102. A pendulum, consisting of a bob of mass M on a cord of length L, is interrupted in its swing by a peg a distance d below its point of suspension. (a) If the bob is to travel in a full circle of radius $(L - d)$ around the peg, what is the minimum possible speed it can have at the lowest point in its motion, just before it starts to go around? Ignore any decrease in the length of the string due to the peg's circumference. (b) From what minimum angle θ must the pendulum be released so that the bob attains the speed calculated in (a)?

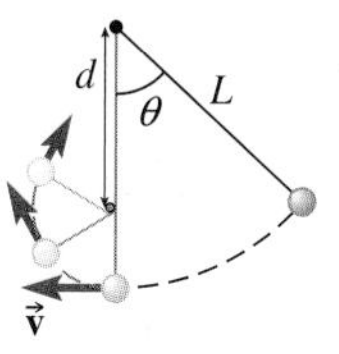

103. Human feet and legs store elastic energy when walking or running. They are not nearly as efficient at doing so as kangaroo legs, but the effect is significant nonetheless. If not for the storage of elastic energy, a 70-kg man running at 4 m/s would lose about 100 J of mechanical energy each time he sets down a foot. Some of this energy is stored as elastic energy in the Achilles tendon and in the arch of the foot; the elastic energy is then converted back into the kinetic and gravitational potential energy of the leg, reducing the expenditure of metabolic energy. If the maximum tension in the Achilles tendon when the foot is set down is 4.7 kN and the tendon's spring constant is 350 kN/m, calculate how far the tendon stretches and how much elastic energy is stored in it.

104. The graph shows the tension in a rubber band as it is first stretched and then allowed to contract. As you stretch a rubber band, the tension force at a particular length (on the way to a maximum stretch) is larger than the force at that same length as you let the rubber band contract. That is why the graph shows two separate lines, one for stretching and one for contracting; the lines are not superimposed as you might have thought they would be. (a) Make a rough estimate of the total work done by the external force applied to the rubber band for the entire process. (b) If the rubber band obeyed Hooke's law, what would the answer to (a) have to be? (c) While the rubber band is stretched, is all of the work done on it accounted for by the increase in elastic potential energy? If not, what happens to the rest of it? [*Hint:* Take a rubber band and stretch it rapidly several times. Then hold it against your wrist or your lip.]

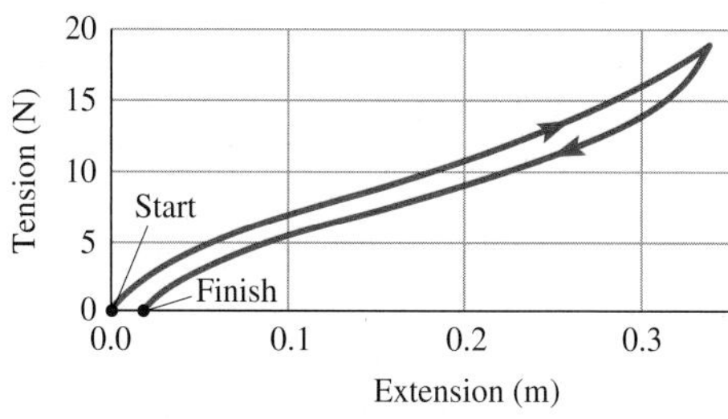

105. A 0.50-kg block, starting at rest, slides down a 30.0° incline with kinetic friction coefficient 0.25 (see the

figure with Problem 63). After sliding 85 cm down the incline, it slides across a frictionless horizontal surface and encounters a spring ($k = 35$ N/m). (a) What is the maximum compression of the spring? (b) After the compression of part (a), the spring rebounds and shoots the block back up the incline. How far along the incline does the block travel before coming to rest?

106. A wind turbine converts some of the kinetic energy of the wind into electric energy. Suppose that the blades of a small wind turbine have length $L = 4.0$ m. (a) When a 10 m/s (22 mi/h) wind blows head-on, what volume of air (in m^3) passes through the circular area swept out by the blades in 1.0 s? (b) What is the mass of this much air? Each cubic meter of air has a mass of 1.2 kg. (c) What is the translational kinetic energy of this mass of air? (d) If the turbine can convert 40% of this kinetic energy into electric energy, what is its electric power output? (e) What happens to the power output if the wind speed decreases to $\frac{1}{2}$ of its initial value? What can you conclude about electric power production by wind turbines?

107. Use dimensional analysis to show that the electric power output of a wind turbine is proportional to the *cube* of the wind speed. The relevant quantities on which the power can depend are the length L of the rotor blades, the density ρ of air (SI units kg/m^3), and the wind speed v.

108. Use this method to find how the speed with which animals of similar shape can run up a hill depends on the size of the animal. Let L represent some characteristic length, such as the height or diameter of the animal. Assume that the maximum rate at which the animal can do work is proportional to the animal's surface area: $P_{max} \propto L^2$. Set the maximum power output equal to the rate of increase of gravitational potential energy and determine how the speed v depends on L.

109. The potential energy of a particle constrained to move along the x-axis is shown in the graph. At $x = 0$, the particle is moving in the $+x$-direction with a kinetic energy of 200 J. Can this particle get into the region 3 cm $< x <$ 8 cm? Explain. If it can, what is its kinetic energy in that region? If it can't, what happens to it?

110. The potential energy of a particle constrained to move along the x-axis is shown in the graph. At $x = 0$, the particle is moving in the $+x$-direction with a kinetic energy of 400 J. Can this particle get into the region 3 cm $< x <$ 8 cm? Explain. If it can, what is its kinetic energy in that region? If it can't, what happens to it?

Answers to Practice Problems

6.1 −180 kJ

6.2 43 N; 4500 J; she pulls with a greater force but its component in the direction of the displacement is smaller.

6.3 $(2.5\ m)(1.50v)^2/(mv^2) = 5.6$

6.4 29 m/s

6.5 0.24

6.6 16.5 m/s

6.7 48 km/s

6.8 195 km/s

6.9 4.0 J

6.10 3.2 cm

6.11 9.8 m/s

6.12 2.4 m

6.13 (a) 11.8 kW (b) 3.6 kW

Answers to Checkpoints

6.2 The force is perpendicular to the displacement.

6.3 Kinetic energy is never negative. Work can be positive, negative, or zero, because kinetic energy can increase, decrease, or stay the same.

6.4 (a) The gravitational potential energy increases until it reaches its maximum value when the stone reaches its highest point above the ground. (b) The kinetic energy decreases as the potential energy increases. It is zero at the highest point. (c) The force of gravity does work on the stone throughout its motion.

6.5 The mechanical energy is the same throughout Mercury's orbit. The kinetic energy is greatest at the perihelion because the potential energy is smallest there.

6.7 The greatest elastic potential energy is at the maximum compression.

CHAPTER

7

Linear Momentum

After a collision, an accident investigator measures the lengths of skid marks on the road. How can the investigator use this information to figure out the velocities of the vehicles immediately *before* the collision? (See p. 246 for the answer.)

Concepts & Skills to Review

- conservation laws (Section 6.1)
- Newton's third law of motion (Section 4.4)
- Newton's second law of motion (Section 4.3)
- velocity (Section 2.2)
- components of vectors (Section 3.2)
- vector subtraction (Section 3.1)
- kinetic energy (Section 6.3)

7.1 A CONSERVATION LAW FOR A VECTOR QUANTITY

In Chapter 4 we learned how to determine the acceleration of an object by finding the net force acting on it and applying Newton's second law of motion. If the forces happen to be constant, then the resulting constant acceleration enables us to calculate changes in velocity and position. Calculating velocity and position changes when the forces are not constant is much more difficult. In many cases, the forces cannot even be easily determined. Conservation of energy is one tool that enables us to draw conclusions about motion without knowing all the details of the forces acting. Recall, for example, how easily we can calculate the escape speed of a projectile using conservation of energy, without even knowing the path the object takes. Now imagine how difficult the same calculation would be using Newton's second law, with a gravitational force that changes magnitude and direction depending on the path taken.

In this chapter we develop another conservation law. Conservation laws are powerful tools. If a quantity is conserved, then no matter how complicated the situation, we can set the value of the conserved quantity at one time equal to its value at a later time. The "before-and-after" aspect of a conservation law enables us to draw conclusions about the results of a complicated set of interactions without knowing all of the details.

CONNECTION:

Conservation laws can involve scalars, such as energy, or vectors, such as momentum.

The new conserved quantity, *momentum*, is a vector quantity, in contrast to energy, which is a scalar. When momentum is conserved, both the magnitude and the direction of the momentum must be constant. Equivalently, the *x*- and *y*-components of momentum are constant. When we find the total momentum of more than one object, we must add the momentum vectors according to the procedure by which vectors are always added.

7.2 MOMENTUM

The word *momentum* is often heard in broadcasts of sporting events. A sports broadcaster might say, "The home team has won five consecutive games; they have the momentum in their favor." The team with "momentum" is hard to stop; they are moving forward on a winning streak. A football player, running for the goal line with a football tucked under his arm, has momentum; he is hard to stop. This use of the word *momentum* is closer to the physics usage. In physics we would agree that the runner has momentum, but we have a precise definition in mind.

In everyday use, momentum has something to do with mass as well as with velocity. Would you rather have a running child bump into you, or a football player running with the same velocity? The child has much less momentum than the football player, even though their velocities are the same.

$\vec{\mathbf{F}}_{21}$ is the force exerted *on* object 2 *by* object 1.

Could a quantity combining mass and velocity be useful in physics? Imagine a collision between two spaceships (Fig. 7.1). Let the spaceships be so far from planets and stars that we can ignore gravitational interactions with celestial bodies. The spaceships exert forces on one another while they are in contact. According to Newton's third law, these forces are equal and opposite. The force on ship 2 exerted by ship 1 is equal and opposite to the force exerted on ship 1 by ship 2:

$$\vec{\mathbf{F}}_{21} = -\vec{\mathbf{F}}_{12}$$

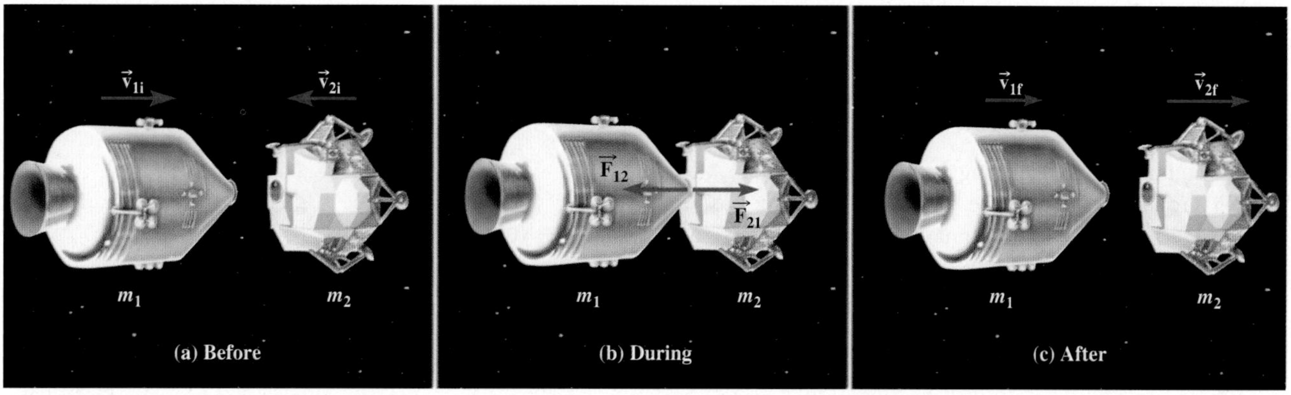

Figure 7.1 (a) Two spaceships about to collide. (b) During the collision, the spaceships exert forces on one another that are equal in magnitude and opposite in direction. (c) The velocities of the spaceships after the collision.

The changes in *velocities* of the two spaceships are *not* equal and opposite if the masses are different. Suppose a large spaceship (mass m_1) collides with a much smaller ship (mass $m_2 << m_1$). Assume for now that the forces are constant during the time interval Δt that the spaceships are in contact. Although the forces have the same magnitude, the magnitudes of the accelerations of the two ships are different because their masses are different. The ship with the larger mass has the smaller acceleration.

The acceleration of either spaceship causes its velocity to change by

$$\Delta\vec{\mathbf{v}} = \vec{\mathbf{a}}\,\Delta t = \frac{\vec{\mathbf{F}}}{m}\Delta t$$

The time interval Δt is the duration of the interaction between the two ships, so it must be the same for both ships.

Since the changes in velocity are inversely proportional to the masses, the changes in the *products* of mass and velocity are equal and opposite for the two bodies involved in the interaction:

$$m_1\Delta\vec{\mathbf{v}}_1 = \vec{\mathbf{F}}_{12}\Delta t$$

$$m_2\Delta\vec{\mathbf{v}}_2 = \vec{\mathbf{F}}_{21}\Delta t = (-\vec{\mathbf{F}}_{12})\Delta t = -(m_1\Delta\vec{\mathbf{v}}_1)$$

This is a useful insight, so we give the product of mass and velocity a name and symbol: **linear momentum** (symbol $\vec{\mathbf{p}}$, SI unit kg·m/s). Linear momentum (or just *momentum*) is a vector quantity having the same direction as the velocity.

Definition of linear momentum:

$$\vec{\mathbf{p}} = m\vec{\mathbf{v}} \tag{7-1}$$

The collision of the two spaceships causes changes in their momenta that are equal in magnitude and opposite in direction:

$$\Delta\vec{\mathbf{p}}_2 = -\Delta\vec{\mathbf{p}}_1$$

In any interaction between two objects, momentum can be transferred from one object to the other. The momentum changes of the two objects are always equal and opposite, so the total momentum of the two objects is unchanged by the interaction. (By *total momentum* we mean the vector sum of the individual momenta of the objects.)

Example 7.1 gives some practice in finding the change in momentum of an object whose velocity changes. Remember that momentum is a vector quantity, so changes in momentum must be found by subtracting momentum vectors, not by subtracting the magnitudes of the momenta.

CONNECTION:

Newton's third law implies that during an interaction momentum is transferred from one body to another.

Example 7.1

Change of Momentum of a Moving Car

A car weighing 12 kN is driving due north at 30.0 m/s. After driving around a sharp curve, the car is moving east at 13.6 m/s. What is the change in momentum of the car?

Strategy The definition of momentum is $\vec{\mathbf{p}} = m\vec{\mathbf{v}}$. We can start by finding the car's mass. There are two potential pitfalls:

1. momentum depends not on weight but on mass, and
2. momentum is a vector, so we must take its direction into consideration as well as its magnitude. To find the change in momentum, we need to do a *vector* subtraction.

Solution The car's mass is

$$m = \frac{W}{g} = \frac{1.2 \times 10^4\ \text{N}}{9.8\ \text{m/s}^2} = 1220\ \text{kg}$$

The car's initial velocity is

$$\vec{\mathbf{v}}_i = 30.0\ \text{m/s, north}$$

The car's initial momentum is then

$$\vec{\mathbf{p}}_i = m\vec{\mathbf{v}}_i = 1220\ \text{kg} \times 30.0\ \text{m/s north}$$
$$= 3.66 \times 10^4\ \text{kg·m/s north}$$

After the curve, the final velocity is

$$\vec{\mathbf{v}}_f = 13.6\ \text{m/s, east}$$

The final momentum is

$$\vec{\mathbf{p}}_f = m\vec{\mathbf{v}}_f = 1220\ \text{kg} \times 13.6\ \text{m/s east}$$
$$= 1.66 \times 10^4\ \text{kg·m/s east}$$

Momentum vectors are added and subtracted according to the same methods used for other vectors. To find the change in the momentum, we draw vector arrows representing the addition of $\vec{\mathbf{p}}_f$ and $-\vec{\mathbf{p}}_i$ (Fig. 7.2). Since *in this case* the three vectors in Fig. 7.2 form a right triangle, the magnitude of $\Delta\vec{\mathbf{p}}$ can be found from the Pythagorean theorem

$$|\Delta\vec{\mathbf{p}}| = \sqrt{p_i^2 + p_f^2}$$
$$= \sqrt{(3.66 \times 10^4\ \text{kg·m/s})^2 + (1.66 \times 10^4\ \text{kg·m/s})^2}$$
$$= 4.02 \times 10^4\ \text{kg·m/s}$$

From the vector diagram, $\Delta\vec{\mathbf{p}}$ is directed at an angle θ east of south. Using trigonometry,

$$\tan\theta = \frac{\text{opposite}}{\text{adjacent}} = \frac{p_f}{p_i} = \frac{1.66 \times 10^4\ \text{kg·m/s}}{3.66 \times 10^4\ \text{kg·m/s}} = 0.454$$

$$\theta = \tan^{-1} 0.454 = 24.4°$$

Since the weight is given with two significant figures, we report the change in momentum of the car as 4.0×10^4 kg·m/s directed 24° east of south.

Discussion As with displacements, velocities, accelerations, and forces, it is crucial to remember that momentum is a vector. When finding changes in momentum, we must find the difference between final and initial momentum *vectors*. If the initial and final momenta had not been perpendicular, we would have had to resolve the vectors into *x*- and *y*-components in order to subtract them.

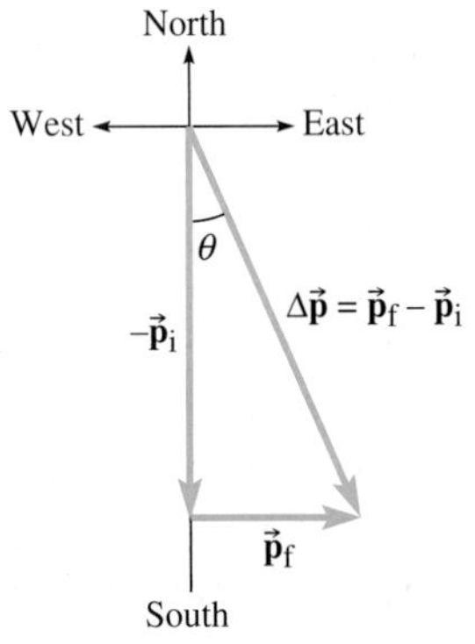

Figure 7.2 Vector subtraction to find the change in momentum.

Practice Problem 7.1 Falling Apple

(a) What is the momentum of an apple weighing 1.0 N just before it hits the ground, if it falls out of a tree from a height of 3.0 m? (b) The apple falls because of the gravitational interaction between the apple and the Earth. How much does this interaction change *Earth's* momentum? How much does it change Earth's velocity?

CHECKPOINT 7.2

In Example 7.1, if the speed of the car had remained constant, would $\Delta\vec{\mathbf{p}}$ have been zero?

7.3 THE IMPULSE-MOMENTUM THEOREM

We found that the change in momentum of an object when a single force acts on it is equal to the product of the force acting on the object and the time interval during which the force acts:

$$\Delta\vec{\mathbf{p}} = \vec{\mathbf{F}}\,\Delta t$$

The product $\vec{\mathbf{F}}\,\Delta t$ is given the name **impulse**. Since the impulse is the product of a vector (the force) and a positive scalar (the time), impulse is a vector quantity having the same direction as that of the force. In words, $\Delta\vec{\mathbf{p}} = \vec{\mathbf{F}}\,\Delta t$ can be read as "*the change in momentum equals the impulse.*" The SI units of impulse are newton-seconds (N·s) and those of momentum are kilogram-meters per second (kg·m/s). These are equivalent units, as can be demonstrated using the definition of the newton (Problem 3).

Impulse = $\vec{\mathbf{F}}\,\Delta t$

CONNECTION

Impulse is a momentum transfer due to a force; work is an energy transfer due to a force.

	Impulse	Work
Definition	$\vec{\mathbf{F}}\,\Delta t$	$\vec{\mathbf{F}}\cdot\Delta\vec{\mathbf{r}}$
Vector or Scalar?	Vector	Scalar*
Physical meaning	Momentum transfer	Energy transfer

*The scalar or dot product of two vectors is introduced in Section 6.2.

If an object is involved in more than one interaction, then its change in momentum during any time interval is equal to the *total* impulse during that time interval. The total impulse is the vector sum of the impulses due to each force. The total impulse is also equal to the net force times the time interval:

$$\text{total impulse} = \vec{\mathbf{F}}_1\,\Delta t + \vec{\mathbf{F}}_2\,\Delta t + \cdots$$
$$= (\vec{\mathbf{F}}_1 + \vec{\mathbf{F}}_2 + \cdots)\,\Delta t = \sum\vec{\mathbf{F}}\,\Delta t$$

The total impulse on an object is equal to the change in the object's momentum during the same time interval. This relationship between total impulse and momentum change is called the impulse-momentum theorem and is especially useful in solving problems that involve collisions and impacts.

Impulse-Momentum Theorem

$$\Delta\vec{\mathbf{p}} = \sum\vec{\mathbf{F}}\,\Delta t \qquad (7\text{-}2)$$

Impulse When Forces Are Not Constant Our discussion so far has assumed that the forces acting are constant or that Δt is very small so the change in $\vec{\mathbf{F}}$ is negligible. That is a rather unusual situation; the concept of momentum would be of limited use if it were applicable only when forces are constant. However, everything we have said still applies to situations where the forces are not constant, as long as we use the *average* force to calculate the impulse.

When the force is not constant, the impulse can be found using the average force.

$$\text{impulse} = \vec{\mathbf{F}}_{\text{av}}\,\Delta t \qquad (7\text{-}3)$$

Conceptual Example 7.2

Big Force–Short Time Versus Small Force–Long Time

Which causes the larger change in momentum of an object, an average force of 5 N acting for 4 s or an average force of 2 N acting for 10 s? How might this principle be used when designing products to protect the human body from injury? Give an example.

Figure 7.3

A stuntman lands safely in an air bag to break his fall. The air bag reduces the risk of injury in two ways. It changes the stuntman's momentum more gradually, so that forces of smaller magnitude act on his body. It also spreads these forces over a larger area so they are less likely to cause serious injury.

Solution and Discussion The change in momentum is equal to the impulse. The product of the force and the time interval gives the momentum change of the object. Over a period of 4 s, the 5-N force causes a momentum change of magnitude (5 N × 4 s) = 20 N·s, and the 2-N force acting for 10 s also causes a momentum change of magnitude (2 N × 10 s) = 20 N·s. The smaller force causes the same change in momentum because it acts for a longer time interval.

When designing products to protect the human body, one goal is to lengthen the time period during which a velocity change occurs. For example, when a movie stuntman falls from a great height, he lands on a large air bag (Fig. 7.3),

continued on next page

Example 7.2 continued

which changes his momentum much more gradually than if he were to fall onto concrete. The average force exerted by the air bag on the stuntman is much smaller than the average force exerted by concrete would be. Nets used under circus acrobats serve the same purpose. The net gives and dips downward when the acrobat falls into it, gradually reducing the speed of the fall over a longer time interval than if she fell directly onto the ground.

Practice Problem 7.2 Pole-Vaulter Landing on a Padded Surface

A pole-vaulter vaults over the bar and falls onto thick padding. He lands with a speed of 9.8 m/s; the padding then brings him to a stop in a time of 0.40 s. What is the average force on his body *due to the padding* during that time interval? Express your answer as a fraction or multiple of his weight. [*Hint:* The force due to the padding is not the only force acting on the vaulter during the 0.40-s interval.]

Application of momentum conservation to automotive design.

Designing a Safer Automobile One automotive design change implemented to minimize injury on collision is the foam padding built into automobile dashboards (Fig. 7.4). Automobile bumpers have shock absorbers built in to lessen damage to the car body in small collisions. The structure of the car itself is often a single piece of metal with reinforced supports (*unibody* construction) so that the entire body can crumple and absorb the change in momentum more slowly than it would if it were made of separate sections of metal that would slide into or over each other or fall off the car. The safety glass in a windshield has two advantages. One is that it does not shatter and send sharp shards of glass into human tissue, but the other is that it distorts when struck by solid objects like human bones or a human head. The glass doesn't give much, but in a crash every little bit helps.

The use of seat belts plus the air bag is better than either alone. Without a seat belt, the body continues moving with the same speed the car had before the crash. The rapidly inflating air bag moves toward the body and the effective velocity is then the sum of the two velocities (air bag velocity + body velocity) when the two collide. The body flying into the air bag can be injured more than a restrained body making more gradual contact with the air bag. An adult should sit at least 12 in. from the air bag container to avoid injury from the deploying air bag itself. Small children should always be placed in the back seat, in proper car seats for their size, to ensure their safety.

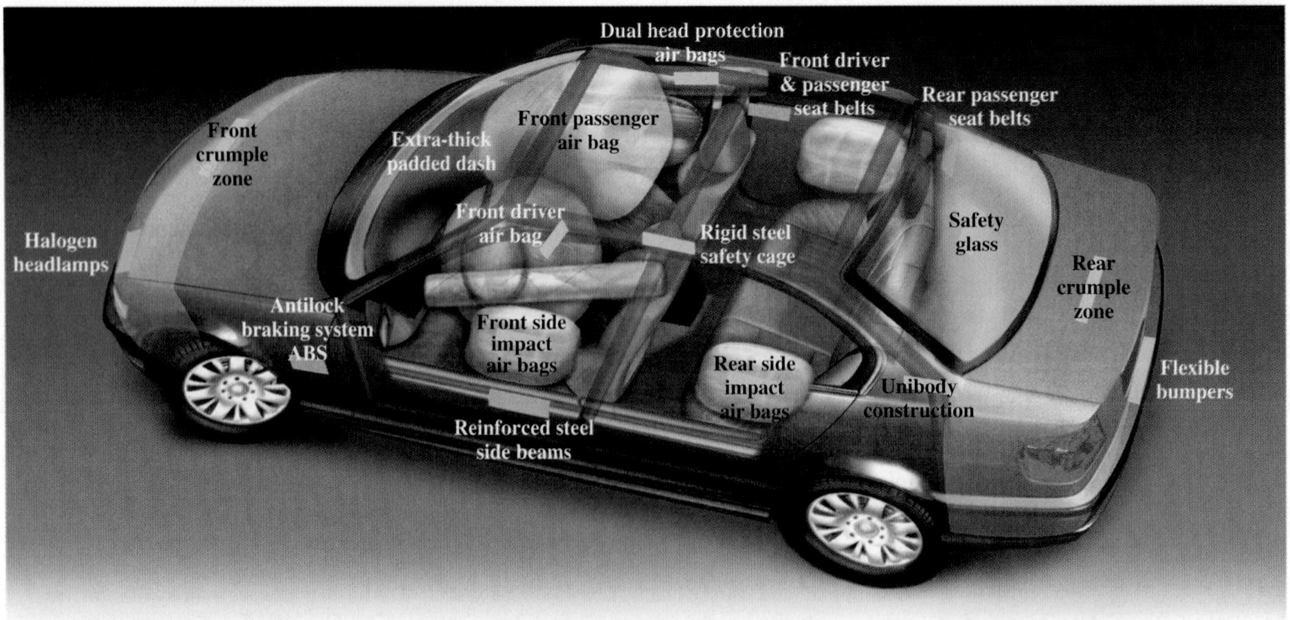

Figure 7.4 Some safety features of the modern automobile. Many of these features serve to lengthen the time interval during which a momentum change occurs in a crash, thereby lessening the forces acting on the passengers.

Example 7.3

Collision Between an Automobile and a Tree

A car moving at 20.0 m/s (44.7 mi/h) crashes into a tree. Find the magnitude of the average force acting on a passenger of mass 65 kg in each of the following cases. (a) The passenger is not wearing a seat belt. He is brought to rest by a collision with the windshield and dashboard that lasts 3.0 ms. (b) The car is equipped with a passenger-side air bag. The force due to the air bag acts for 30 ms, bringing the passenger to rest.

Strategy From the impulse-momentum theorem, $\Delta\vec{\mathbf{p}} = \vec{\mathbf{F}}_{av}\,\Delta t$, where $\vec{\mathbf{F}}_{av}$ is the average force acting on the passenger and Δt is the time interval during which the force acts. The change in the passenger's momentum is the same in the two cases. What differs is the time interval during which the change occurs. It takes a larger force to change the momentum in a shorter time interval.

Solution The magnitude of the passenger's initial momentum is

$$|\vec{\mathbf{p}}_i| = |m\vec{\mathbf{v}}_i| = 65\text{ kg} \times 20.0\text{ m/s} = 1300\text{ kg·m/s}$$

His final momentum is zero, so the magnitude of the momentum change is

$$|\Delta\vec{\mathbf{p}}| = 1300\text{ kg·m/s}$$

This momentum change divided by the time interval gives the magnitude of the average force in each case.

$$\text{(a) No seat belt: } |\vec{\mathbf{F}}_{av}| = \frac{|\Delta\vec{\mathbf{p}}|}{\Delta t} = \frac{1300\text{ kg·m/s}}{0.0030\text{ s}} = 4.3\times 10^5\text{ N}$$

$$\text{(b) Air bag: } |\vec{\mathbf{F}}_{av}| = \frac{|\Delta\vec{\mathbf{p}}|}{\Delta t} = \frac{1300\text{ kg·m/s}}{0.030\text{ s}} = 4.3\times 10^4\text{ N}$$

Discussion The average forces required to bring the passenger to rest are inversely proportional to the time interval over which those forces act. It is a far happier situation to have the momentum change over as long a period as possible to make the forces smaller. Automotive safety engineers design cars to minimize the average forces on the passengers during sudden stops and collisions.

The air bag also spreads the force over a much larger area than impact with a hard surface like the windshield, further reducing the risk of injury.

Practice Problem 7.3 Catching a Fastball

A baseball catcher is catching a fastball that is thrown at 43 m/s (96 mi/h) by the pitcher. If the mass of the ball is 0.15 kg and if the catcher moves his mitt backward toward his body by 8.0 cm as the ball lands in the glove, what is the magnitude of the average force acting on the catcher's mitt? Estimate the time interval required for the catcher to move his hands.

PHYSICS AT HOME

Try playing catch with a friend [on the lawn] while using a raw egg or a water balloon as a ball. How do you move your hands to minimize the chance of breaking the egg or balloon when you catch it? What is likely to happen if you forget that the "ball" is an egg or balloon and catch it as you would a ball?

Graphical Calculation of Impulse

When a force is changing, how can we find the impulse? We've asked similar questions in previous chapters. For simplicity we consider components along the x-axis. Recall:

- displacement $= \Delta x = v_{av,x}\,\Delta t =$ area under $v_x(t)$ graph
- change in velocity $= \Delta v_x = a_{av,x}\,\Delta t =$ area under $a_x(t)$ graph

In both cases, the mathematical relationship is that of a rate of change. Velocity is the rate of change of position with time and acceleration is the rate of change of velocity with time. Now we have force as the rate of change of momentum with time. By analogy:

- impulse $= F_{av,x}\,\Delta t =$ area under $F_x(t)$ graph

So to find the impulse for a variable force, we find the area under the $F_x(t)$ graph. Then, if we wish to know the average force, we can divide the impulse by the time interval during which the force is applied.

CONNECTION:

See Sections 2.2, 2.3, and 6.6 to review how we used the area under a graph to find displacement, change in velocity, and work done by a force.

A "graph of $v_x(t)$" means the quantity v_x is plotted as a function of the variable t with v_x on the vertical axis and t on the horizontal axis.

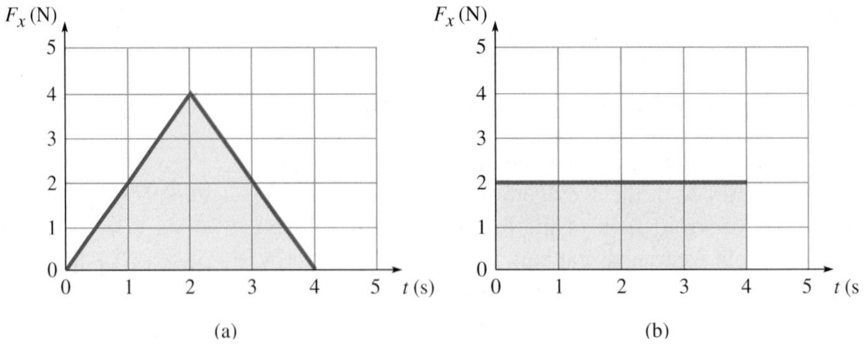

Figure 7.5 (a) The area under the $F_x(t)$ graph for a variable force is the impulse. (b) The average force for a given time interval is the constant force that would produce the same impulse.

The variable force of Fig. 7.5a increases linearly from 0 to 4 N in a time of 2 s; then it decreases from 4 N to 0 N in 2 s. The area under the $F_x(t)$ graph is found from the triangular area

$$\text{area} = \tfrac{1}{2}\,\text{base} \times \text{height} = 2\text{ s} \times 4\text{ N} = 8\text{ N·s} = \text{impulse}$$

The average force during the 4-s time interval is

$$\text{average force} = \frac{\text{impulse}}{\text{time interval}} = \frac{8\text{ N·s}}{4\text{ s}} = 2\text{ N}$$

Figure 7.5b shows the average force over the 4-s time interval; the area under the curve (the impulse) is the same as in Fig. 7.5a.

Example 7.4

Hitting the Wall

An experimental robotic car of mass 10.2 kg moving at 1.2 m/s in the $+x$-direction crashes into a brick wall and rebounds. A force sensor on the car's bumper records the force that the wall exerts on the car as a function of time. These data are shown in graphical form in Fig. 7.6. (a) What is the maximum magnitude of the force exerted on the car? (b) What is the average force on the car during the collision? (c) At what speed does the car rebound from the wall?

Strategy The maximum force can be read directly from the graph. To solve parts (b) and (c) of this problem, we must find the impulse exerted on the car. Since impulse is the area under the $F_x(t)$ curve, we'll make an estimate of the area. The impulse is then equal to the average force times the time interval and also to the car's change in momentum. Once we find the change in momentum, we use it to find the car's final speed.

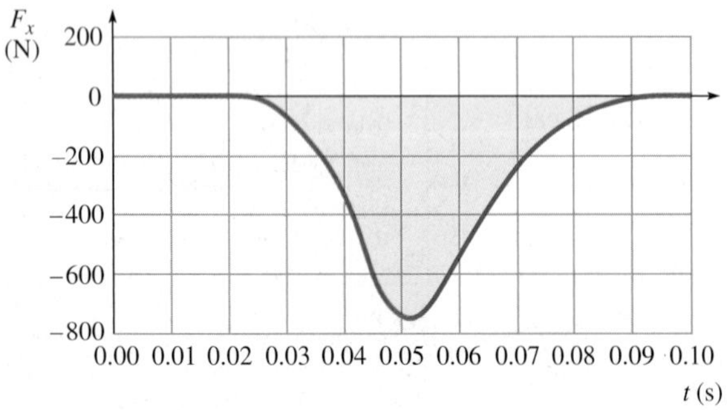

Figure 7.6

Force versus time for a car colliding with a wall.

Given: $m = 10.2$ kg; $v_{ix} = 1.2$ m/s; graph of $F_x(t)$

To find: (a) F_{max}; (b) $F_{av,x}$; (c) v_{fx}

Solution (a) From Fig. 7.6, the maximum force is approximately 750 N in magnitude.

(b) Each division on the horizontal axis represents 0.01 s, and each vertical division represents 200 N. Then the area of each grid box represents (200 N × 0.01 s) = 2 N·s. Counting the number of grid boxes between the $F_x(t)$ curve and the time axis, estimating fractions of boxes, yields about 10 boxes. Then the magnitude of the impulse is approximately

$$10\text{ boxes} \times 2\text{ N·s/box} = 20\text{ N·s}$$

The collision is underway when the force is nonzero. So the collision begins at about $t = 0.025$ s and ends at about $t = 0.095$ s. The duration of the collision is

$$\Delta t = 0.07\text{ s}$$

continued on next page

Example 7.4 continued

The magnitude of the average force is approximately

$$|F_{\text{av},x}| = \frac{|\text{impulse}|}{\Delta t} = \frac{20\ \text{N}\cdot\text{s}}{0.07\ \text{s}} = 300\ \text{N}$$

(c) The impulse gives us the momentum change. The force exerted by the wall is in the $-x$-direction. Thus, the x-component of the impulse is negative. In the graph of F_x versus t, the area lies under the time axis and so is counted as negative. So, working with x-components,

$$\Delta p_x = mv_{\text{f}x} - mv_{\text{i}x} = F_{\text{av},x}\,\Delta t = -20\ \text{N}\cdot\text{s}$$

Solving for $v_{\text{f}x}$,

$$v_{\text{f}x} = \frac{\Delta p_x + mv_{\text{i}x}}{m} = \frac{\Delta p_x}{m} + v_{\text{i}x}$$

Substituting numerical values in this expression yields

$$v_{\text{f}x} = \frac{-20\ \text{N}\cdot\text{s}}{10.2\ \text{kg}} + 1.2\ \text{m/s} = -0.8\ \text{m/s}$$

The car rebounds at a speed of 0.8 m/s.

Discussion As a check, we compare the average force with the maximum force. The average force is a bit less than half of the maximum force. If the force were a linear function of time, the average would be exactly half the maximum. Here, the average force is less than that because more time is spent at smaller values of force than at the larger values.

Practice Problem 7.4 Car-Van Collision

A car weighing 13.6 kN is moving at 10.0 m/s in the $+x$-direction when it collides head-on with a van weighing 33.0 kN. The horizontal force exerted on the car before, during, and after the collision is shown in Fig. 7.7. What is the car's velocity just after the collision?

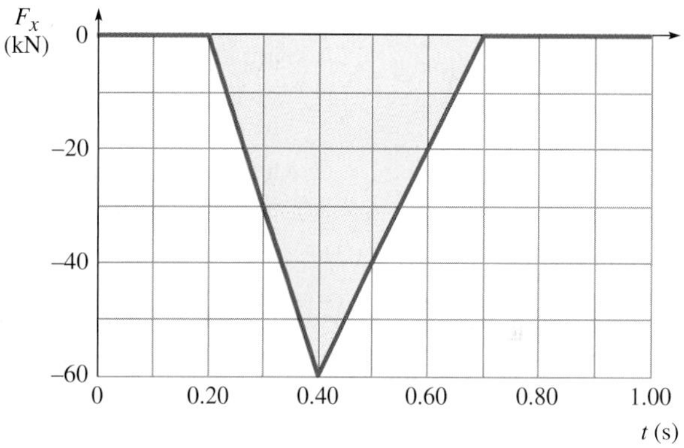

Figure 7.7
Varying force on a car during a car-van collision.

A Restatement of Newton's Second Law

We can use the relationship between impulse and momentum to find a new way to understand Newton's second law. Let's rewrite the impulse-momentum theorem this way:

$$\Sigma\vec{\mathbf{F}}_{\text{av}} = \frac{\Delta\vec{\mathbf{p}}}{\Delta t}$$

What happens if we let the time interval Δt get smaller and smaller, approaching zero? Then the average force is taken over a smaller and smaller time interval, approaching the instantaneous force:

$$\vec{\mathbf{F}} = \lim_{\Delta t\to 0}\frac{\Delta\vec{\mathbf{p}}}{\Delta t}$$

If more than one force acts, we must replace $\vec{\mathbf{F}}$ with $\Sigma\vec{\mathbf{F}}$. Then our restatement of Newton's second law becomes

Newton's Second Law

$$\Sigma\vec{\mathbf{F}} = \lim_{\Delta t\to 0}\frac{\Delta\vec{\mathbf{p}}}{\Delta t} \quad (7\text{-}4)$$

In words, *the net force is the rate of change of momentum.*

Equation (7-4) is *more general* than $\Sigma\vec{\mathbf{F}} = m\vec{\mathbf{a}}$, the form of Newton's second law used in Chapters 4 through 6, which holds only when mass is constant. One situation in which mass is not constant is the rocket engine. In a rocket engine, fuel combustion produces hot gases that are then expelled at high speeds (Fig. 7.8). The rocket's mass decreases as the exhaust gases are expelled.

CONNECTION:

Equation (7-4) is closer to Newton's original statement of his second law and is more general than $\Sigma\vec{\mathbf{F}} = m\vec{\mathbf{a}}$.

Figure 7.8 The Space Shuttle is propelled upward as hot gases are exhausted downward at high speeds.

When mass is constant, then it can be factored out:

$$\sum \vec{\mathbf{F}} = \lim_{\Delta t \to 0} \frac{\Delta \vec{\mathbf{p}}}{\Delta t} = \lim_{\Delta t \to 0} \frac{\Delta (m\vec{\mathbf{v}})}{\Delta t} = m \lim_{\Delta t \to 0} \frac{\Delta \vec{\mathbf{v}}}{\Delta t} = m\vec{\mathbf{a}}$$

Thus, Eq. (7-4) reduces to the familiar form of Newton's second law from Chapters 4–6 when mass is constant.

7.4 CONSERVATION OF MOMENTUM

Consider two pucks that bump into one another after sliding along a frictionless table. Figure 7.9 shows what happens to the two pucks before, during, and after their interaction. If we think of the two pucks as constituting a single system, then the gravitational interactions with the Earth and the contact interactions with the table are *external* interactions—interactions with objects external to the system. The force of gravity on each object is balanced by the normal force on the same object and, thus, there is no net impulse up or down. Together, these forces produce a net external force of zero, so they leave the system's momentum unchanged. Since these two always cancel, we can ignore these external interactions and just focus on the interaction between the pucks. Therefore, we omit the normal and gravitational forces in Fig. 7.9.

Until contact is made, there is no interaction between the pucks (ignoring the small gravitational interaction between the two). During the collision, the pucks exert forces on each other. Force $\vec{\mathbf{F}}_{12}$ is the contact force acting on mass m_1 and force $\vec{\mathbf{F}}_{21}$ is the contact force acting on mass m_2. If we continue to regard the two pucks as parts of a single interacting system, then those forces are *internal* forces of this system. When they collide, some momentum is transferred from one puck to the other. The changes in momentum of the two are equal and opposite:

$$\Delta \vec{\mathbf{p}}_1 = -\Delta \vec{\mathbf{p}}_2$$

Since the change in momentum is the final momentum minus the initial momentum, we write:

$$\vec{\mathbf{p}}_{1f} - \vec{\mathbf{p}}_{1i} = -(\vec{\mathbf{p}}_{2f} - \vec{\mathbf{p}}_{2i})$$

Moving the initial momenta to the right side and the final momenta to the left:

$$\vec{\mathbf{p}}_{1f} + \vec{\mathbf{p}}_{2f} = \vec{\mathbf{p}}_{1i} + \vec{\mathbf{p}}_{2i} \qquad (7\text{-}5)$$

Equation (7-5) says the sum of the momenta of the pucks after the interaction is equal to the sum of the momenta before the interaction; or, more simply, the total momentum of the objects is unchanged by the collision. This isn't surprising since, if some momentum is just transferred from one to the other, the total hasn't changed. We say that momentum is *conserved* for this collision. The interaction between the pucks changes the momentum of each puck, but the total momentum of the system is unchanged.

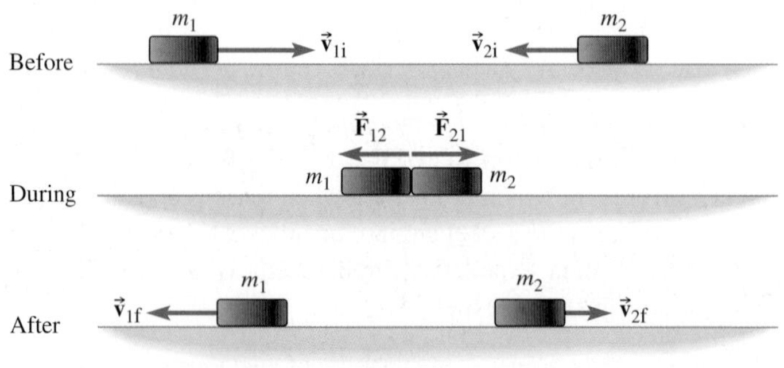

Figure 7.9 Two sliding pucks with different masses before, during, and after collision.

In a system composed of more than two objects, interactions between objects inside the system do not change the total momentum of the system—they just transfer some momentum from one part of the system to another. Only external interactions can change the total momentum of the system. To summarize:

- The total momentum of a system is the vector sum of the momenta of each object in the system.
- External interactions can change the total momentum of a system.
- Internal interactions do not change the total momentum of a system.

In the absence of external interactions, momentum is conserved:

Law of Conservation of Linear Momentum

If the net external force acting on a system is zero, then the momentum of the system is conserved.

$$\text{If } \sum \vec{\mathbf{F}}_{\text{ext}} = 0, \quad \vec{\mathbf{p}}_{\text{i}} = \vec{\mathbf{p}}_{\text{f}} \tag{7-6}$$

By definition, an *isolated*, or closed, system is subject to no external interactions; thus, *linear momentum is always conserved for an isolated system.* Remember that momentum is a vector quantity, so both the magnitude *and the direction* of the momentum at the beginning and end of the interaction must be the same. In component form, both p_x and p_y are unchanged by the interaction.

CHECKPOINT 7.4

When is the momentum of a system *not* conserved?

Example 7.5

Adrift on a Raft

Diana is standing on a raft of mass 100.0 kg that is floating on a still lake. She decides to walk the length of the raft (Fig. 7.10). If Diana's mass is 55 kg and she walks with a velocity of 0.91 m/s *with respect to the shore*, how fast and in what direction does the raft move while Diana is walking?

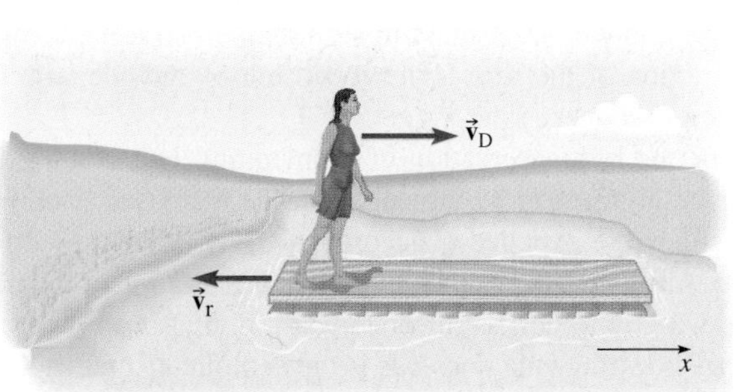

Figure 7.10

Diana walking along a raft. Velocities $\vec{\mathbf{v}}_D$ and $\vec{\mathbf{v}}_r$ are measured with respect to the shore.

Assume the raft is stationary with respect to the shore before Diana starts walking.

Strategy Diana and the raft can be considered to be a single *isolated system*: as long as frictional forces on the raft due to the water and air are small enough to ignore, the net external force on the system is zero. Then the momentum of this system (raft + Diana) is conserved. We let the subscripts D stand for Diana and r for the raft and set the change in momentum of the system equal to zero.

Solution To walk forward, Diana must exert a backward force on the raft: the static frictional force between her feet and the raft. This is an internal interaction within the isolated system, so it cannot change the total momentum of the system. Only something acting from outside the system could do that. As Diana walks in one direction, she acquires some momentum. The rest of the system (the raft) must acquire an equal and opposite momentum, because the momentum of the isolated system (Diana + raft) is conserved, which means that the change in momentum of the system is zero.

continued on next page

Example 7.5 continued

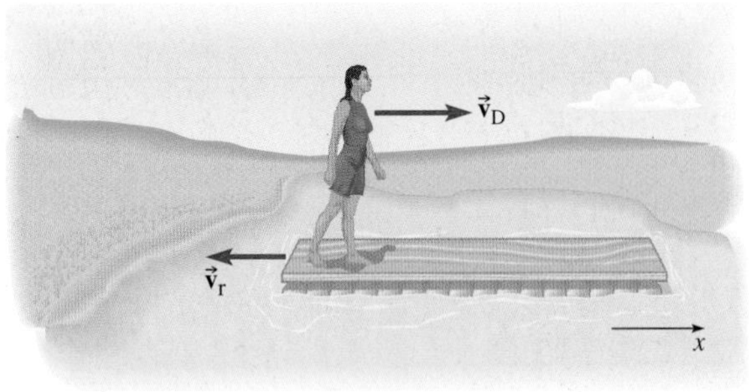

Figure 7.10
Diana walking along a raft. Velocities $\vec{\mathbf{v}}_D$ and $\vec{\mathbf{v}}_r$ are measured with respect to the shore.

First we set the change in momentum of the system equal to zero:

$$\Delta\vec{\mathbf{p}} = 0 = \Delta\vec{\mathbf{p}}_D + \Delta\vec{\mathbf{p}}_r$$

or

$$\Delta\vec{\mathbf{p}}_D = -\Delta\vec{\mathbf{p}}_r$$

This means that the momentum changes of Diana and of the raft are equal and opposite. Since momentum is the product of mass and velocity and the masses of the raft and Diana do not change,

$$m_D\,\Delta\vec{\mathbf{v}}_D = -m_r\,\Delta\vec{\mathbf{v}}_r$$

Solving for the change in velocity of the raft gives

$$\Delta\vec{\mathbf{v}}_r = -\frac{m_D}{m_r}\,\Delta\vec{\mathbf{v}}_D$$

Finally we substitute numerical values from the given information in the statement of the problem. Let Diana walk in the $+x$-direction.

$$\Delta\vec{\mathbf{v}}_r = \frac{-55 \text{ kg} \times 0.91 \text{ m/s (in the } +x\text{-direction)}}{100.0 \text{ kg}}$$

$$= 0.50 \text{ m/s in the } -x\text{-direction}$$

The negative sign reverses the direction: $\Delta\vec{\mathbf{v}}_r$ is in the $-x$-direction and has a magnitude of 0.50 m/s.

The raft moves in a direction opposite to Diana's motion to keep the momentum unchanged and thus conserved. Since the raft was originally stationary, this is the new velocity of the raft.

Discussion In any momentum conservation problem there are two equivalent ways to proceed. In this example we set the momentum change of the system equal to zero. We could just as well write an equation that sets the initial total momentum equal to the final momentum of the system. The raft and Diana are initially at rest, so the initial momentum is zero:

$$0 = m_D\vec{\mathbf{v}}_D + m_r\vec{\mathbf{v}}_r$$

where $\vec{\mathbf{v}}_D$ and $\vec{\mathbf{v}}_r$ are the final velocities of Diana and the raft.

Practice Problem 7.5 Skaters Pushing Apart

Two skaters on in-line skates, Lisa and Bart, are initially at rest. They push apart and start moving in opposite directions. If Lisa's speed just after they push apart is 2.0 m/s and her mass is 85% of Bart's mass, how fast is Bart moving at that time?

Application of momentum conservation: recoil of a rifle

When a bullet is fired from a rifle, the system of rifle plus bullet must conserve momentum. Suppose the rifle is at rest before the bullet is fired. The momentum of the system is zero. When the bullet is fired, part of the system's mass breaks away and travels in one direction with a certain momentum. The rifle, which is the remaining mass of the system, moves in the exact opposite direction such that the total momentum of the system is still zero. The rifle has a much larger mass than the bullet, so it has a much smaller speed. The backward motion of the rifle is the *recoil* felt by anyone who has held a rifle against her shoulder and squeezed the trigger.

Application of momentum conservation: jets, rockets, and airplane wings

Jet engines and rockets operate by conservation of momentum. Hot combustion gases are forced out of nozzles at high speed by the engines. The increased backward momentum of the hot gases as they are expelled is accompanied by an increased forward momentum of the engines. Airplane wings generate lift by conservation of momentum. The main purpose of the wing is to deflect air downward, giving it a downward momentum component. (Exactly how the wing does this is the complicated part.) Since the wing pushes air downward, air pushes the wing upward.

Conceptual Example 7.6

Escape on Slippery Ice

A pilot parachutes from his disabled aircraft and lands on the frozen surface of a lake. There is no breeze blowing and the lake surface is too slippery to walk on. What can the pilot do to reach the shore?

Strategy and Solution Since the person in jeopardy is a pilot, he begins to think about how hot gases forced backward from a jet engine cause the plane to move forward. That gives him an idea: he bundles the parachute into a package and pushes it as hard as possible in a direction away from the nearest point of the shore. If friction is negligible, the net external force on the system of pilot plus parachute is zero and the total momentum of the system cannot change. The momentum of the parachute plus the momentum of the pilot must still equal zero. By conservation of momentum, the pilot begins sliding in the opposite direction and glides toward the shore.

Discussion If friction brings the pilot to rest before he reaches the shore, he can search his pockets and belt loops for other items to throw away. Once he reaches shore, he can tie one end of a rope to a tree and, holding onto the other end, venture back out onto the ice to retrieve any essential items. The rope provides him with an external force so he can get back to shore.

Practice Problem 7.6 Recoil of a Rifle

During an afternoon of target practice, you fire a Winchester .308 rifle of mass 3.8 kg. The bullets have a mass of 9.72 g and leave the rifle at a muzzle velocity of 860 m/s. If you are sloppy and fire a round when the butt of the rifle is not firmly up against your shoulder, at what speed does the rifle butt smash into your shoulder? (Ouch!)

PHYSICS AT HOME

In case you and a friend ever end up stuck in the middle of the ice, practice the technique the pilot used to escape to the lakeshore. Bring a heavy medicine ball out to the middle of the ice rink and face each other with your skates aligned parallel. Toss the ball to your friend. What happens to you? What happens to your friend when he catches the ball? Can you both be "saved" by tossing the ball back and forth? (The same technique works using in-line skates.)

7.5 CENTER OF MASS

We have seen that the momentum of an isolated system is conserved even though parts of the system may interact with other parts; internal interactions transfer momentum between parts of the system but do not change the total momentum of the system. We can define a point called the **center of mass** (CM) that serves as an average location of the system. In Section 7.6, we prove that the center of mass of an isolated system must move with constant velocity, regardless of how complicated the motions of parts of the system may be. Then we can treat the mass of the system as if it were all concentrated at the CM, like a point particle. The CM of an object is not necessarily located within the object; for some objects, such as a boomerang, the center of mass is located outside of the object itself (Fig. 7.11a).

What if a system is not isolated, but has external interactions? Again imagine all of the mass of the system concentrated into a single point particle located at the CM. The motion of this fictitious point particle is determined by Newton's second law, where the net force is the sum of all of the external forces acting on *any part* of the system. In the case of a complex system composed of many parts interacting with each other, the motion of the CM is considerably simpler than the motion of an arbitrary particle of the system (Fig. 7.11b,c).

Figure 7.11 (a) The center of mass of a boomerang is a point outside of the boomerang. (b) The path followed by the center of mass when a hammer is tossed through the air. (c) British pole-vaulter Ben Challenger's center of mass actually passes *beneath* the bar as his body passes over the bar.

Location of Center of Mass For a system composed of two particles, the center of mass lies somewhere on a line between the two particles. In Fig. 7.12, particles of masses m_1 and m_2 are located at positions x_1 and x_2, respectively. We define the location of the CM for these two particles as

$$x_{CM} = \frac{m_1 x_1 + m_2 x_2}{m_1 + m_2} \tag{7-7}$$

The CM is a *weighted average* of the positions of the two particles. Here we use the word *weighted* in its statistical sense. The position of a particle with more mass counts more—carries more *statistical* weight—than does the position of a particle with a smaller mass. We can rewrite Eq. (7-7) as a weighted average:

$$x_{CM} = \frac{m_1}{M} x_1 + \frac{m_2}{M} x_2 \tag{7-8}$$

Here $M = m_1 + m_2$ represents the total mass of the system. The statistical weight used for the location of each particle is the mass of that particle as a fraction of the total mass of the system.

Suppose masses m_1 and m_2 are equal. Then we expect the CM to be located midway between the two particles (Fig. 7.12a). If $m_1 = 2m_2$, as in Fig. 7.12b, then the CM is closer to the particle of mass m_1. Figure 7.12b shows that, in this case, the CM is twice as far from m_2 as from m_1.

For a system of N particles, at arbitrary locations in three-dimensional space, the definition of the CM is a generalization of Eq. (7-7).

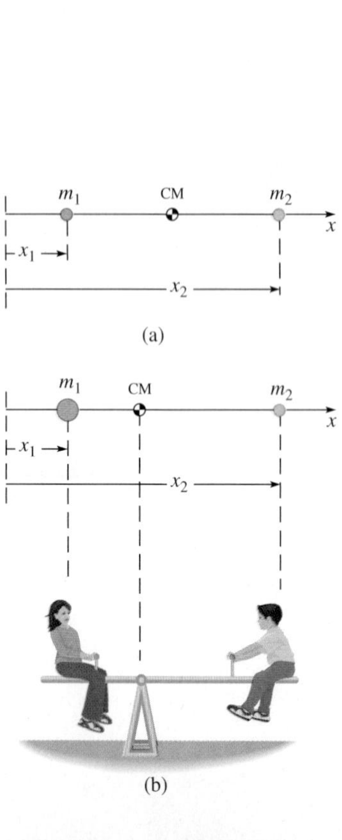

Figure 7.12 (a) Two particles of equal mass located at positions x_1 and x_2 from the origin. The CM is midway between the two. (b) Two particles of unequal mass. The CM is closer to the more massive particle. For two children balanced on a see-saw, the CM is at the fulcrum.

Definition of center of mass:

Vector form: $$\vec{\mathbf{r}}_{CM} = \frac{\sum m_i \vec{\mathbf{r}}_i}{M} \qquad (7\text{-}9)$$

Component form: $$x_{CM} = \frac{\sum m_i x_i}{M} \qquad y_{CM} = \frac{\sum m_i y_i}{M} \qquad z_{CM} = \frac{\sum m_i z_i}{M}$$

where $i = 1, 2, 3, \ldots, N$ and $M = \sum m_i$

Remember that the symbol $\sum$ stands for *sum*. The shorthand notation $\sum m_i x_i$ is interpreted as

$$\sum m_i x_i = m_1 x_1 + m_2 x_2 + \cdots + m_N x_N$$

For particles in two-dimensional space, we use only two of these equations for the *x*-*y* plane and find the *x*- and *y*-components of the CM.

Example 7.7

Center of Mass of a Binary Star System

Due to the gravitational interaction between the two stars in a binary star system, each moves in a circular orbit around their CM. One star has a mass of 15.0×10^{30} kg; its center is located at $x = 1.0$ AU and $y = 5.0$ AU. The other has a mass of 3.0×10^{30} kg; its center is at $x = 4.0$ AU and $y = 2.0$ AU. Find the CM of the system composed of the two stars. (AU stands for *astronomical unit.* 1 AU = the average distance between the Earth and the Sun = 1.5×10^{8} km.)

Strategy We treat the stars as point particles located at their centers. Since we are given *x*- and *y*-coordinates, the easiest way to proceed is to find the *x*- and *y*-coordinates of the CM. There is no particular advantage here in finding the position vector of the CM in terms of its length and direction.

Given: $m_1 = 15.0 \times 10^{30}$ kg $\quad x_1 = 1.0$ AU $\quad y_1 = 5.0$ AU
$m_2 = 3.0 \times 10^{30}$ kg $\quad x_2 = 4.0$ AU $\quad y_2 = 2.0$ AU

To find: x_{CM}; y_{CM}

Solution The total mass of the system is the sum of the individual masses:

$$M = m_1 + m_2 = 15.0 \times 10^{30}\text{ kg} + 3.0 \times 10^{30}\text{ kg} = 18.0 \times 10^{30}\text{ kg}$$

For the *x*-position, we find

$$x_{CM} = \frac{m_1}{M}x_1 + \frac{m_2}{M}x_2$$
$$= \frac{15.0 \times 10^{30}\text{ kg}}{18.0 \times 10^{30}\text{ kg}} \times 1.0\text{ AU} + \frac{3.0 \times 10^{30}\text{ kg}}{18.0 \times 10^{30}\text{ kg}} \times 4.0\text{ AU}$$
$$= 1.5\text{ AU}$$

and for the *y*-position, we find

$$y_{CM} = \frac{m_1}{M}y_1 + \frac{m_2}{M}y_2$$
$$= \frac{15.0}{18.0} \times 5.0\text{ AU} + \frac{3.0}{18.0} \times 2.0\text{ AU} = 4.5\text{ AU}$$

Discussion In Fig. 7.13, we mark the position of the CM. As we expect for the case of two particles, it is located closer to the larger mass and on a line connecting the two. Once the CM position is found in a problem, check to be sure its location is reasonable. Suppose we had made an error in this example and found the CM to be at $x = 1.5$ AU and $y = 1.7$ AU. This is not a reasonable location for the CM since it is not along the line connecting the two and is closer to the less massive star; we then would go back to look for the error.

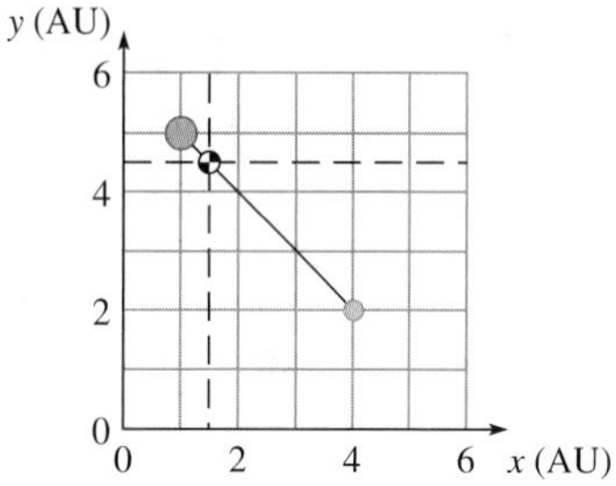

Figure 7.13
Finding the CM for the system of two stars.

Practice Problem 7.7 Three Balls with Unequal Masses

Three spherical objects are shown in Fig. 7.14. Their masses are $m_1 = m_3 = 1.0$ kg and $m_2 = 4.0$ kg. Find the location of the CM for the three objects.

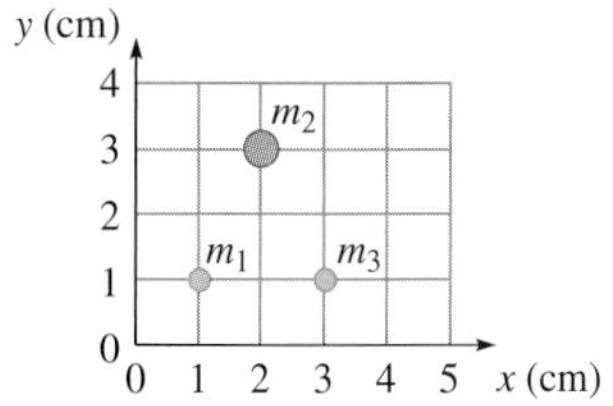

Figure 7.14
Three spheres located at *x*, *y* positions (1.0 cm, 1.0 cm), (2.0 cm, 3.0 cm), and (3.0 cm, 1.0 cm).

Using Symmetry to Locate the Center of Mass Most objects we deal with in real life are not composed of a small set of point particles or spherically symmetrical objects. In Example 7.7, we use the location of the center of each star to find the CM. Due to spherical symmetry, the CM of either star (by itself) is at its geometric center. The same technique can be applied to other shapes with symmetry. A standard 2 by 4, which is an 8-ft-long uniform piece of lumber used in building 1.5 in. deep × 3.5 in. high, has its center of mass at its geometric center. By contrast, a "loaded" die does *not* have its CM at its geometric center, since a small metal plug has been inserted near one face to make the distribution of mass in the die asymmetrical. The definition of the CM [Eq. (7-9)] still holds as long as (x_i, y_i, z_i) are the coordinates of the CM of a part of the system with mass m_i.

7.6 MOTION OF THE CENTER OF MASS

Now that we know how to find the position of the CM of a system, we turn our attention to the motion of the CM. How is the velocity of the CM related to the velocities of the various parts of the system?

During a short time interval Δt, the displacement of the i-th particle is $\Delta\vec{\mathbf{r}}_i = \vec{\mathbf{v}}_i\,\Delta t$ and the displacement of the center of mass is $\Delta\vec{\mathbf{r}}_{CM} = \vec{\mathbf{v}}_{CM}\,\Delta t$. From the definition of the CM [Eq. (7-9)], the displacements must be related as follows:

$$\Delta\vec{\mathbf{r}}_{CM} = \frac{\sum m_i \Delta\vec{\mathbf{r}}_i}{M} \quad \Rightarrow \quad \vec{\mathbf{v}}_{CM}\,\Delta t = \frac{\sum m_i \vec{\mathbf{v}}_i \Delta t}{M}$$

Dividing both sides by Δt and multiplying by M yields

$$M\vec{\mathbf{v}}_{CM} = \sum m_i \vec{\mathbf{v}}_i \tag{7-10}$$

The right side of Eq. (7-10) is the sum of the momenta of the particles that constitute the system—the total momentum of the system $\vec{\mathbf{p}}$. Therefore,

$$\vec{\mathbf{p}} = M\vec{\mathbf{v}}_{CM} \tag{7-11}$$

For two-dimensional motion, Eq. (7-11) is equivalent to two component equations

$$p_x = Mv_{CM,x} \quad \text{and} \quad p_y = Mv_{CM,y} \tag{7-12}$$

In Section 7.4, we showed that, for an isolated system, the total linear momentum is conserved. In such a system, Eq. (7-11) implies that the CM must move with constant velocity regardless of the motions of the individual particles. On the other hand, what if the system is not isolated? If a net external force acts on a system, the CM does not move with constant velocity. Instead, it moves as if all the mass were concentrated there into a fictitious point particle with all the external forces acting on that point. The motion of the CM obeys the following statement of Newton's second law:

$$\sum\vec{\mathbf{F}}_{ext} = M\vec{\mathbf{a}}_{CM} \tag{7-13}$$

where M is the total mass of the system, $\sum\vec{\mathbf{F}}_{ext}$ is the net external force, and $\vec{\mathbf{a}}_{CM}$ is the acceleration of the CM. [Eq. (7-13) is proved in Problem 38.]

CHECKPOINT 7.6

Turn back to Fig. 7.11b. Why does the CM of the hammer move along a parabolic path?

Example 7.8

An Exploding Rocket

A model rocket is fired from the ground in a parabolic trajectory. At the top of the trajectory, a horizontal distance of 260 m from the launch point, an explosion occurs within the rocket, breaking it into two fragments. One fragment, having one third of the mass of the rocket, falls straight down to Earth as if it had been dropped from rest at that point. At what horizontal distance from the launch point does the other fragment land? Ignore air resistance. [*Hint:* The two fragments land simultaneously.]

Strategy Two different strategies can be used to solve this problem.

Strategy 1: We apply conservation of momentum to the explosion. The momentum of the rocket *just before* the explosion is equal to the total momentum of the two fragments *just after* the explosion. Why can momentum conservation be assumed here? There is an external force—gravity—acting on the system. External forces change momentum. However, the explosion takes place in a *very short time interval.* From the impulse-momentum theorem [Eq. (7-2)], the momentum change of the system is the force of gravity multiplied by the time interval. As long as the time interval considered is sufficiently short, the momentum change of the system can be ignored.

Strategy 2: The explosion is caused by an *internal* interaction between two parts of the rocket. The motion of the CM of the system is unaffected by internal interactions, so it continues in the same parabolic path. Just before the explosion, the rocket is at the top of its trajectory, so it has $p_y = 0$ (with the y-axis pointing up). Just after the explosion, one fragment is at rest. Then the other fragment must have $p_y = 0$; otherwise, conservation of momentum would be violated. Then both fragments have $v_y = 0$ just after the explosion. Ignoring air resistance, they land simultaneously. At that same instant, the CM also reaches the ground.

Solution 1 First we make a sketch of the situation (Fig. 7.15). At the top of the trajectory, where the explosion occurs, $v_y = 0$; the rocket is moving in the x-direction. The initial momentum just before the explosion is entirely in the x-direction. If M is the mass of the rocket, then

$$p_{ix} = Mv_{ix}$$

Just after the explosion, one-third of the mass of the rocket is at rest; it then drops straight down under the influence of the gravitational force. This piece has zero momentum just after the explosion. To conserve momentum, the other two thirds of the rocket must have a momentum equal to the momentum just before the explosion.

$$p_{ix} = p_{1x} + p_{2x}$$
$$Mv_{ix} = 0 + (\tfrac{2}{3}M)\, v_{2x}$$

Solving for v_{2x}, we find

$$v_{2x} = \tfrac{3}{2} v_{ix}$$

The y-component of momentum must also be conserved:

$$p_{iy} = p_{1y} + p_{2y}$$

We know that both p_{iy} and p_{1y} are zero; therefore, p_{2y} is zero as well. Just after the explosion, both parts of the rocket have zero vertical components of velocity. Then both parts take the same time to fall to the ground as if the rocket had not exploded. With a horizontal velocity larger by a factor of $\frac{3}{2}$, the second piece of the rocket travels a horizontal distance from the explosion a factor of $\frac{3}{2}$ larger than 260 m (see Fig. 7.15). The distance from the launch point where this piece lands is

$$\Delta x = 260\text{ m} + \tfrac{3}{2} \times 260\text{ m} = 650\text{ m}$$

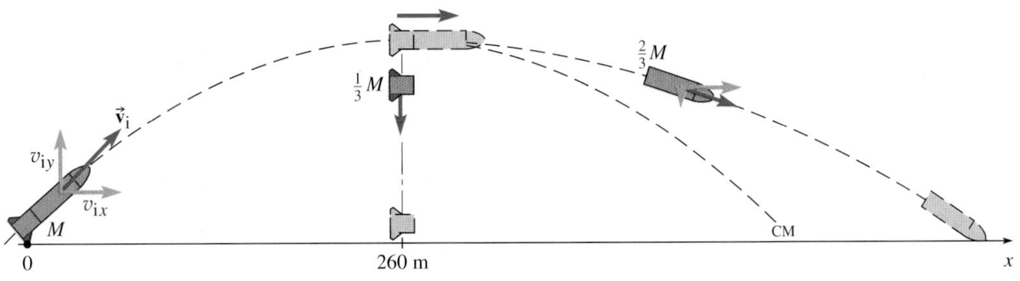

Figure 7.15
Rocket motion after explosion.

continued on next page

Example 7.8 continued

Solution 2 The piece with mass $\frac{1}{3}M$ falls straight down and lands 260 m from the launch point. After the explosion, the CM continues to travel just as the rocket itself would have done if it had not broken apart. From the symmetry of the parabola, the CM touches the ground at a distance of $2 \times 260 \text{ m} = 520 \text{ m}$ from the launch point. Since we know the location of the CM and that of one of the pieces, we can find where the second piece lands:

$$Mx_{CM} = \tfrac{1}{3}Mx_1 + \tfrac{2}{3}Mx_2$$

After canceling the common factor of M,

$$x_{CM} = \tfrac{1}{3}x_1 + \tfrac{2}{3}x_2$$

Solving for x_2 yields

$$x_2 = \frac{3x_{CM} - x_1}{2} = \frac{3 \times 520 \text{ m} - 260 \text{ m}}{2} = 650 \text{ m}$$

which is the same answer that we found in Solution 1.

Discussion The insight that the motion of the CM is unaffected by internal interactions can be of enormous help. Note, however, that solution 2 would not be so simple if the two fragments did not land simultaneously. As soon as one fragment (fragment 1) hits the ground, the external force on the system is no longer due exclusively to gravity, so the CM doesn't continue to follow the same parabolic path. The normal and frictional forces acting on fragment 1 affect its subsequent motion and the subsequent motion of the CM even though the motion of fragment 2 is unaffected.

Practice Problem 7.8 Diana and the Raft Revisited

In Example 7.5, Diana (mass 55 kg) walks at 0.91 m/s (relative to the water) on a raft of mass 100.0 kg. The raft moves in the opposite direction at 0.50 m/s. Suppose it takes her 3.0 s to walk from one end of the raft to the other. (a) How far does Diana walk (relative to the water)? (b) How far does the raft move while Diana is walking? (c) How far does the CM of Diana and the raft move during the 3.0 s?

7.7 COLLISIONS IN ONE DIMENSION

What Is a Collision? In the macroscopic world, a moving body bumps into another body that may be at rest or in motion. The two bodies exert forces on one another while they are in contact; as a result, their velocities change. In the microscopic and submicroscopic world, our picture of a collision is different. When atoms collide, they don't "touch" each other: the atom doesn't have a definite spatial boundary, so there are no surfaces to make "contact." However, the collision model is still useful for atoms and subatomic particles whenever there is an interaction in which the forces are strong over a short time interval, so that there is a clear "before collision" and a clear "after collision."

Analyzing Collisions Using Momentum Conservation We can often use conservation of momentum to analyze collisions even when external forces act on the colliding objects. If the net external force is small compared with the internal forces the colliding objects exert on each other during the collision, then the change in the total momentum of the two objects is small compared with the transfer of momentum from one object to the other. Then the total momentum after the collision is *approximately* the same as it was before the collision.

The same techniques that are used for collisions in the macroscopic world (car crashes, billiard ball collisions, baseball bats hitting balls) are also used in collisions in the microscopic world (gas molecules colliding with each other and with surfaces, radioactive decays of nuclei). First, we study collisions limited to motion along a line; later, we consider collisions limited to motion in a plane (in two dimensions).

Example 7.9

Collision in the Air

A krypton atom (mass 83.9 u) moving with a velocity of 0.80 km/s to the right and a water molecule (mass 18.0 u) moving with a velocity of 0.40 km/s to the left collide head-on. The water molecule has a velocity of 0.60 km/s to the right after

continued on next page

Example 7.9 continued

the collision. What is the velocity of the krypton atom after the collision? (The symbol "u" stands for the atomic mass unit.)

Strategy Since we know both initial velocities and one of the final velocities, we can find the second final velocity by applying momentum conservation. Let the subscript "1" refer to the krypton atom and let the subscript "2" refer to the water molecule. Let the *x*-axis point to the right. Figure 7.16 shows before and after pictures of the collision.

Solution Momentum conservation requires that the final momentum be equal to the initial momentum:

$$\vec{\mathbf{p}}_{1f} + \vec{\mathbf{p}}_{2f} = \vec{\mathbf{p}}_{1i} + \vec{\mathbf{p}}_{2i}$$

Now we substitute $\vec{\mathbf{p}} = m\vec{\mathbf{v}}$ for each momentum. It is easiest to work in terms of components. For simplicity we drop the "x" subscripts, remembering that all quantities refer to x-components:

$$m_1 v_{1f} + m_2 v_{2f} = m_1 v_{1i} + m_2 v_{2i}$$

Since $m_1/m_2 = 83.9/18.0 = 4.661$, we can substitute $m_1 = 4.661 m_2$:

$$4.661\, m_2 v_{1f} + m_2 v_{2f} = 4.661\, m_2 v_{1i} + m_2 v_{2i}$$

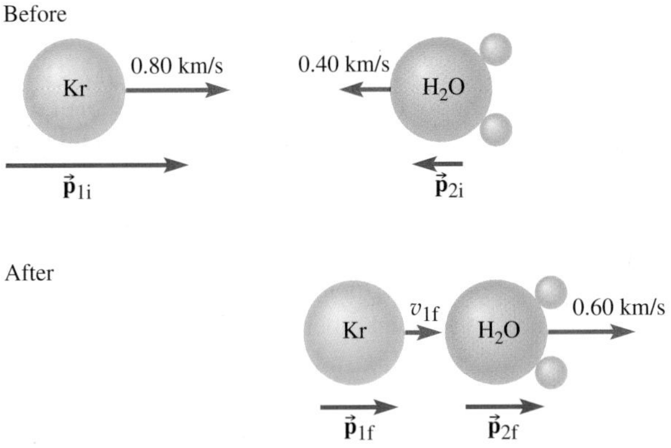

Figure 7.16
Before and after snapshots of a collision.

The common factor m_2 cancels out. Solving for v_{1f},

$$v_{1f} = \frac{4.661 v_{1i} + v_{2i} - v_{2f}}{4.661}$$

$$= \frac{4.661 \times 0.80\ \text{km/s} + (-0.40\ \text{km/s}) - 0.60\ \text{km/s}}{4.661}$$

$$= 0.59\ \text{km/s}$$

After the collision, the krypton atom moves to the right with a speed of 0.59 km/s.

Discussion To check this result, we calculate the total momentum (x-component) before and after the collision:

$$m_1 v_{1i} + m_2 v_{2i} = (83.9\ \text{u})(0.80\ \text{km/s}) + (18.0\ \text{u})(-0.40\ \text{km/s}) = 60\ \text{u·km/s}$$

$$m_1 v_{1f} + m_2 v_{2f} = (83.9\ \text{u})(0.59\ \text{km/s}) + (18.0\ \text{u})(0.60\ \text{km/s}) = 60\ \text{u·km/s}$$

Momentum is conserved. There is no need to convert u to kg since we only need to compare these two values.

If we made the mistake of thinking of momentum as a scalar, we would get the wrong answer. The sum of the *magnitudes* of the momenta before the collision is *not* equal to the sum of the *magnitudes* of the momenta after the collision. Conservation of energy is perhaps easier to understand intuitively since energy is a scalar quantity. Converting kinetic energy to potential energy is analogous to moving money from a checking account to a savings account; the total amount of money is the same before and after. This sort of analogy does *not* work with momentum!

Practice Problem 7.9 Head-On Collision

A 5.0-kg ball is at rest when it is struck head-on by a 2.0-kg ball moving along a track at 10.0 m/s. If the 2.0-kg ball is at rest after the collision, what is the speed of the 5.0-kg ball after the collision?

Suppose we observe a bumper car traveling at speed v_i toward a second car that is at rest. The masses of the two cars are equal. When the first car hits the second, what happens?

Based on momentum considerations *alone*, there are many possible outcomes. One possibility is that the first car stops moving and the second car moves off with the same velocity that the first one had to begin with (Fig. 7.17a). This possibility satisfies conservation of momentum because the total momentum is the same before and after.

Another possibility is that the two cars stick together, moving away together (Fig. 7.17b). With what speed do they move after the collision? If the momentum is to be the same with twice as much mass moving, the speed must be half the initial speed of the first car. There are many other possibilities. Conservation of momentum doesn't tell us which of these outcomes actually happens, but if we know one car's velocity after the collision, we can use momentum conservation to determine the other car's velocity.

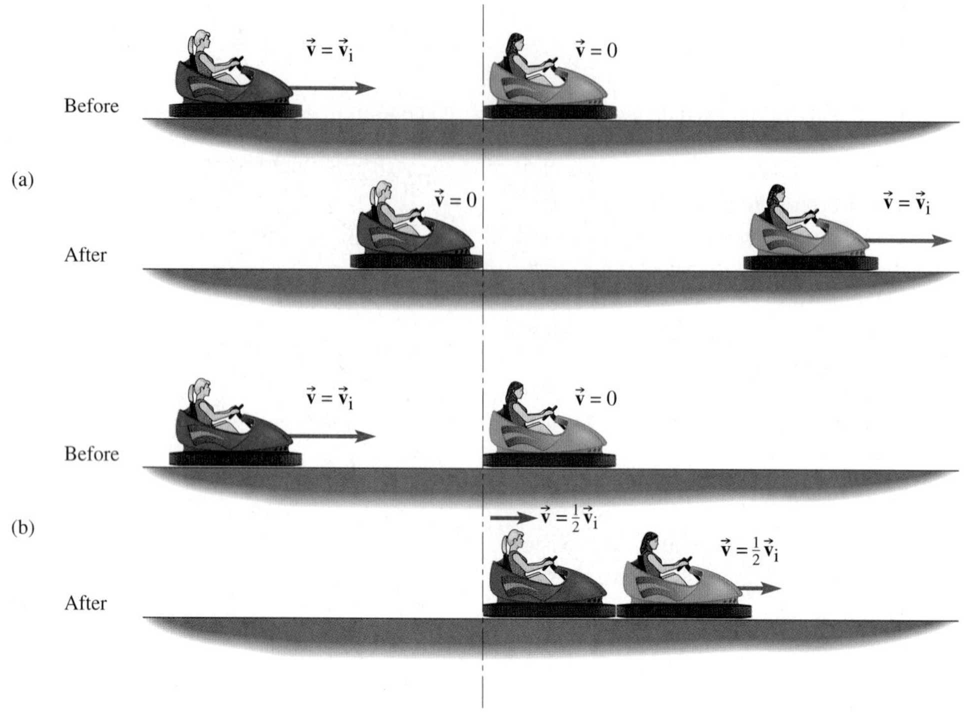

Figure 7.17 Two of the many possible outcomes of a collision between bumper cars of equal mass with one of them initially at rest.

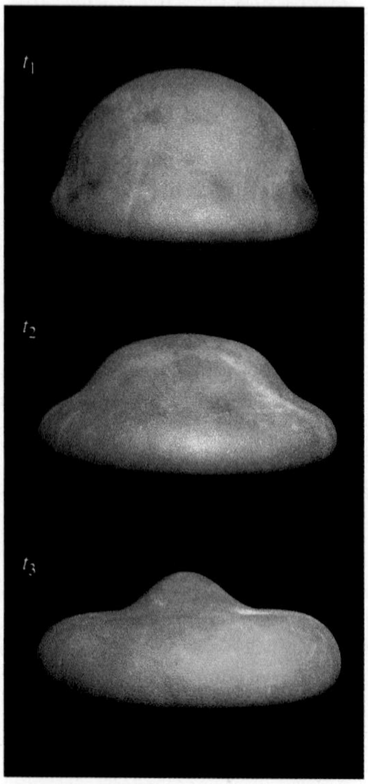

Figure 7.18 Three successive photos of a racquetball during its collision with the floor ($t_1 < t_2 < t_3$).

Elastic and Inelastic Collisions

Collisions are often classified based on what happens to the kinetic energy of the colliding objects. A ball dropped from a height *h* does not rebound to the same height. The kinetic energy of the ball just after the collision with the floor or ground is less than it was just before the collision; the amount of the kinetic energy decrease depends on the makeup of the ball and the ground. A racquetball dropped onto a hard wooden floor may rebound nearly to its original height, but a watermelon rebounds very little or not at all. Why do some objects rebound much better than others?

Imagine a racquetball colliding with the floor (Fig. 7.18). The bottom of the ball is flattened. What makes the ball rebound from the floor? The forces holding the ball together are like springs; the kinetic energy of the ball has been transformed largely into potential energy stored in these springs. When the ball bounces back up, this energy is transformed back into kinetic energy. Then why does the watermelon not rebound? The watermelon, too, is deformed when it collides with the floor, but this deformation is not reversible. The kinetic energy of the watermelon is changed mostly into thermal energy rather than into potential energy.

A collision in which the *total* kinetic energy is the same before and after is called **elastic**. When the final kinetic energy is less than the initial kinetic energy, the collision is said to be **inelastic**. Collisions between macroscopic objects are generally inelastic to some degree, but sometimes the change in kinetic energy is so small that we treat them as elastic. When a collision results in two objects sticking together, the collision is **perfectly inelastic**. The decrease of kinetic energy in a perfectly inelastic collision is as large as *possible* (consistent with the conservation of momentum). Now that we have defined elastic and inelastic collisions, we can put together a problem-solving strategy for collision problems.

Problem-Solving Strategy for Collisions Involving Two Objects

1. Draw before and after diagrams of the collision.
2. Collect and organize information on the masses and velocities of the two objects before and after the collision. Express the velocities in component form (with correct algebraic signs).
3. Set the sum of the momenta of the two before the collision equal to the sum of the momenta after the collision. Write one equation for each direction:

$$m_1 v_{1ix} + m_2 v_{2ix} = m_1 v_{1fx} + m_2 v_{2fx}$$

$$m_1 v_{1iy} + m_2 v_{2iy} = m_1 v_{1fy} + m_2 v_{2fy}$$

4. If the collision is known to be perfectly inelastic, set the final velocities equal:

$$v_{1fx} = v_{2fx} \quad \text{and} \quad v_{1fy} = v_{2fy}$$

5. If the collision is known to be perfectly elastic, then set the final kinetic energy equal to the initial kinetic energy:

$$\tfrac{1}{2}m_1 v_{1i}^2 + \tfrac{1}{2}m_2 v_{2i}^2 = \tfrac{1}{2}m_1 v_{1f}^2 + \tfrac{1}{2}m_2 v_{2f}^2$$

6. Solve for the unknown quantities.

There is no *conservation law* for kinetic energy by itself. Total energy is always conserved, but that does not preclude some kinetic energy being transformed into another type of energy. The elastic collision is just a special kind of collision in which no kinetic energy is changed into other forms of energy. Momentum is conserved regardless of whether a collision is elastic or inelastic.

It can be proved (see Problem 56) that for *any* elastic collision between two objects, the relative speed is the same before and after the collision. (This fact is most useful in one-dimensional collisions; in two-dimensional collisions the *direction* of the relative velocity changes due to the collision.) Since the relative velocity is in the opposite direction after a one-dimensional collision—first the objects move together, then they move apart—we can write:

$$v_{2ix} - v_{1ix} = -(v_{2fx} - v_{1fx}) \tag{7-14}$$

assuming the objects move along the x-axis. For a one-dimensional elastic collision, Eq. (7-14) is a useful alternative to setting the final kinetic energy equal to the initial kinetic energy.

CHECKPOINT 7.7

Is momentum conserved in a perfectly inelastic collision?

Example 7.10

Collision at the Highway Entry Ramp

At a Route 3 highway on-ramp, a car of mass 1.50×10^3 kg is stopped at a stop sign, waiting for a break in traffic before merging with the cars on the highway. A pickup of mass 2.00×10^3 kg comes up from behind and hits the stopped car. Assuming the collision is elastic, how fast was the pickup going just before the collision if the car is pushed straight ahead onto the highway at 20.0 m/s just after the collision?

continued on next page

Example 7.10 continued

Strategy Conservation of momentum provides one equation relating the initial and final velocities. That the collision is elastic provides another equation. With two unknown velocities, these two equations enable us to solve for both. Let "1" refer to the car stopped at the stop sign and "2" refer to the pickup. All motions are in one direction, which we call the x-axis. To simplify the notation, we drop the x subscripts and let all p's and v's refer to x-components. Figure 7.19 shows a before and after diagram for the collision.

Given: $m_1 = 1.50 \times 10^3$ kg; $m_2 = 2.00 \times 10^3$ kg; before the collision, $v_{1i} = 0$; after the collision, $v_{1f} = 20.0$ m/s

To find: v_{2i} (speed of the pickup just before the collision)

Solution From conservation of momentum,

$$\cancel{m_1 v_{1i}} + m_2 v_{2i} = m_1 v_{1f} + m_2 v_{2f} \qquad (1)$$

where we cross out the first term because $v_{1i} = 0$. The collision is elastic, so the relative velocity after the collision is equal and opposite to the relative velocity before the collision [Eq. (7-14)]:

$$v_{2i} - \cancel{v_{1i}} = -(v_{2f} - v_{1f}) \qquad (2)$$

We want to solve these two equations for v_{2i}, so we can eliminate v_{2f}. Multiplying Eq. (2) through by m_2 and rearranging yields

$$m_2 v_{2i} = m_2 v_{1f} - m_2 v_{2f} \qquad (3)$$

Adding Eqs. (1) and (3) gives

$$2m_2 v_{2i} = (m_1 + m_2) v_{1f} \qquad (4)$$

Finally, we solve Eq. (4) for v_{2i}:

$$v_{2i} = \frac{m_1 + m_2}{2m_2} v_{1f} = \frac{1500\text{ kg} + 2000\text{ kg}}{4000\text{ kg}} \times 20.0\text{ m/s} = 17.5\text{ m/s}$$

Discussion To check this answer, first solve for v_{2f}. Then you can verify that momentum is conserved [Eq. (1)] and that the relative velocity changes sign [Eq. (2)]. You can also calculate the total kinetic energy before and after the collision and show they are equal, as they must be for an elastic collision. We leave these checks to you for practice.

The road exerts frictional forces on the vehicles, so the net external force on the vehicles was *not* zero during the collision. We still use conservation of momentum because during the short time interval of the collision, friction doesn't have time to change the system's momentum significantly.

Practice Problem 7.10 Perfectly Inelastic Collision Between the Cars

Instead of colliding elastically, suppose the two vehicles lock bumpers when they collide. With the same initial conditions ($v_{1i} = 0$ and $v_{2i} = 17.5$ m/s), find the speed at which the car would be pushed out onto the highway.

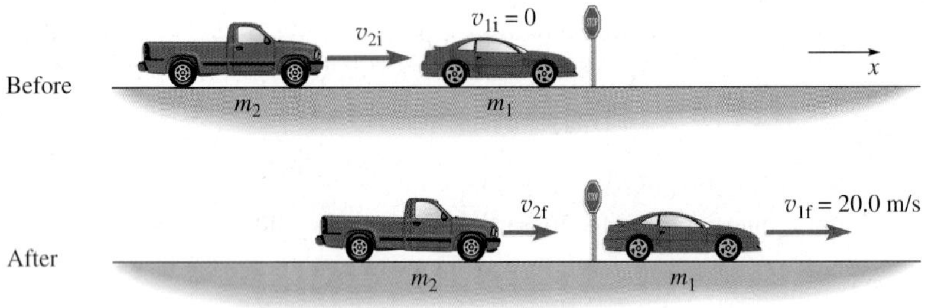

Figure 7.19
Before and after diagrams of the collision (side view).

How are skid marks used to find car velocities just before the collision?

Suppose in Example 7.10 that the entry ramp speed limit is 20 mi/h (8.94 m/s). By measuring the length of the skid marks from the stop sign and estimating the coefficient of friction, the accident investigator can determine that the car was pushed onto the highway at a speed of 20.0 m/s. Witnesses confirm that the car was stopped before the collision. Then the investigator calculates the speed of the pickup just before the collision using conservation of momentum. The duration Δt of the collision is so short that we can ignore momentum changes due to external forces and treat the two vehicles as an isolated system. Finding that the pickup exceeded the speed limit, the investigator adds speeding to the charges against the driver of the pickup.

7.8 COLLISIONS IN TWO DIMENSIONS

Most collisions are not limited to motion in one dimension in the absence of a track or other device to constrain motion to a single line. In a two-dimensional collision, we use the same techniques we used for one-dimensional collisions, as long as we remember that momentum is a vector. (WWW interactive: the virtual pool table.) To apply conservation of momentum, it is usually easiest to work with x- and y-components.

CONNECTION:

See the Problem-Solving Strategy in Section 7.7. The same strategy applies to collisions in two or three dimensions.

Example 7.11

Colliding Pucks on an Air Table

A small puck (mass $m_1 = 0.10$ kg) is sliding to the right with an initial speed of 8.0 m/s on an air table (Fig. 7.20a). An air table has many tiny holes through which air is blown; the resulting air cushion allows objects to slide with very little friction. The puck collides with a larger puck (mass $m_2 = 0.40$ kg), which is initially at rest. Figure 7.20b shows the outcome of the collision: the pucks move off at angles $\phi_1 = 60.0°$ above and $\phi_2 = 30.0°$ below the initial direction of motion of the small puck. (a) What are the final speeds of the pucks? (b) Is this an elastic collision or an inelastic collision? (c) If inelastic, what fraction of the initial kinetic energy is converted to other forms of energy in the collision?

Strategy The system of two pucks is an isolated system because the net external force is zero. Therefore, we can apply conservation of momentum. Since motions in two dimensions are involved, we treat the horizontal and vertical components of momentum separately.

Figure 7.20 shows the pucks before and after the collision. Now we collect information on the known quantities, writing velocities in component form.

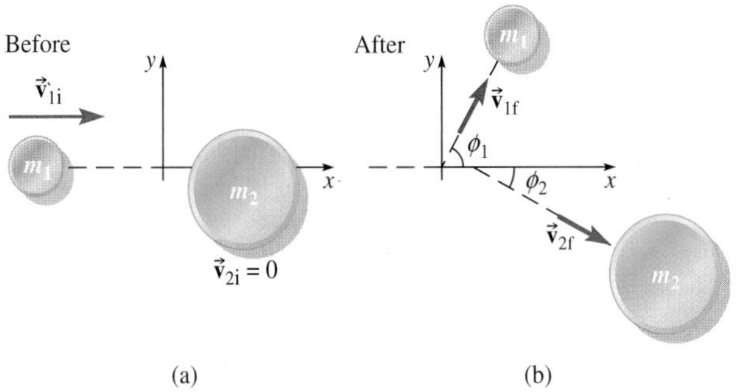

Figure 7.20

Snapshots in time, (a) before and (b) after a collision.

Masses:	$m_1 = 0.10$ kg; $m_2 = 0.40$ kg
Before collision:	$v_{1ix} = 8.0$ m/s; $v_{1iy} = v_{2ix} = v_{2iy} = 0$
After collision:	$v_{1fx} = v_{1f} \cos \phi_1$; $v_{1fy} = v_{1f} \sin \phi_1$; $v_{2fx} = v_{2f} \cos \phi_2$; $v_{2fy} = -v_{2f} \sin \phi_2$ ($\phi_1 = 60.0°$ and $\phi_2 = 30.0°$)
To find:	v_{1f} and v_{2f}; total kinetic energy before and after the collision

Solution (a) Working with components means that we set the total x-component of momentum before the collision equal to the total x-component of momentum after the collision. We treat the y-components in the same way. The initial momentum is in the x-direction only. Thus, the total momentum y-component after the collision must be zero.

First we set the x-component of the total momentum after the collision equal to the x-component of the total momentum before the collision:

$$p_{1fx} + p_{2fx} = p_{1ix} + p_{2ix}$$

Each momentum component is now rewritten using $p_x = mv_x$:

$$m_1 v_{1f} \cos \phi_1 + m_2 v_{2f} \cos \phi_2 = m_1 v_{1ix} + 0$$

Since $m_2 = 4m_1$,

$$m_1 v_{1f} \cos 60.0° + 4m_1 v_{2f} \cos 30.0° = m_1 v_{1ix}$$

After canceling the common factor m_1 and substituting numerical values for cos 60.0° and cos 30.0°, this reduces to

$$0.500v_{1f} + 3.46v_{2f} = 8.0 \text{ m/s} \quad (1)$$

For conservation of the y-component of the momentum:

$$p_{1fy} + p_{2fy} = p_{1iy} + p_{2iy} = 0$$

continued on next page

Example 7.11 continued

The y-component of $\vec{\mathbf{p}}_{2f}$ is negative because the y-component of $\vec{\mathbf{v}}_{2f}$ is negative.

$$m_1 v_{1f} \sin \phi_1 + (-4m_1 v_{2f} \sin \phi_2) = 0$$

$$v_{1f} \sin 60.0° - 4v_{2f} \sin 30.0° = 0$$

We solve for v_{2f} in terms of v_{1f}:

$$v_{2f} = \frac{\sin 60.0°}{4 \sin 30.0°} v_{1f} = 0.433 v_{1f} \qquad (2)$$

Equations (1) and (2) contain two unknowns. To eliminate one unknown, we substitute $0.433v_{1f}$ for v_{2f} in Eq. (1):

$$0.500 v_{1f} + 3.46(0.433 v_{1f}) = 2.00 v_{1f} = 8.0 \text{ m/s}$$

Solving this equation gives the value of v_{1f}:

$$v_{1f} = 4.0 \text{ m/s}$$

Then by substitution into Eq. (2), we find the value of v_{2f}:

$$v_{2f} = 0.433 v_{1f} = 1.73 \text{ m/s} \rightarrow 1.7 \text{ m/s}$$

(b) Now that we have the final speeds, we can compare the initial and final kinetic energies.

$$K_i = \tfrac{1}{2} m_1 v_{1i}^2$$

$$K_i = \tfrac{1}{2}(0.10 \text{ kg}) \times (8.0 \text{ m/s})^2 = 3.2 \text{ J}$$

and

$$K_f = \tfrac{1}{2} m_1 v_{1f}^2 + \tfrac{1}{2} m_2 v_{2f}^2$$

$$= \tfrac{1}{2}(0.10 \text{ kg}) \times (4.0 \text{ m/s})^2 + \tfrac{1}{2}(0.40 \text{ kg}) \times (1.73 \text{ m/s})^2$$

$$= 0.80 \text{ J} + 0.60 \text{ J} = 1.40 \text{ J}$$

The final kinetic energy is less than the initial kinetic energy, so the collision is inelastic.

(c) The amount of kinetic energy converted to other forms of energy (primarily internal energy of the pucks) is

$$3.2 \text{ J} - 1.40 \text{ J} = 1.8 \text{ J}$$

We divide by the initial kinetic energy to find the fraction of the initial kinetic energy converted to other forms:

$$\frac{1.8 \text{ J}}{3.2 \text{ J}} = 0.56$$

Less than half of the kinetic energy of the incident puck therefore survives the collision as the kinetic energies of the two pucks.

Discussion Although a two-dimensional collision problem tends to require more complicated algebra than a one-dimensional problem, the physical principles are the same. As long as the net external force on the system is zero (or negligibly small), the total vector momentum must be conserved.

Practice Problem 7.11 Colliding Balls

A ball of mass m_1 moves at speed v_i along the $+x$-axis toward a second ball of mass m_2, which is initially at rest. The second ball has five times the mass of the first ball. After the collision between these two objects, m_1 moves along the $+y$-axis at a speed v_1, and m_2 moves at a speed $v_2 = \frac{1}{4}v_i$ at an angle of 36.9° below the $+x$-axis. Find v_1 in terms of v_i.

Conceptual Example 7.12

Eric at the Pool Table

Playing a game of pool, Eric is trying to decide whether to attempt a shot to sink the 4-ball in the pocket at corner B without *scratching* (sinking the cue ball "C" in corner A). He notices that the lines from the 4-ball to the two corner pockets happen to make a right angle (Fig. 7.21). The collision of the balls is nearly elastic. Assume Eric is an amateur player and does not know how to do fancy things, like putting sidespin on a ball. Should he attempt the shot?

Strategy We assume a perfectly elastic collision between the balls. They have the same mass. The cue ball moves with an initial velocity $\vec{\mathbf{v}}_i$ and strikes the 4-ball, which is initially at rest. The 4-ball falls in pocket B if its velocity after the collision, $\vec{\mathbf{v}}_4$, points toward the pocket. Assuming it does, we use conservation of momentum and kinetic energy to find the direction of the cue ball velocity, $\vec{\mathbf{v}}_c$, after the collision.

Solution Conservation of momentum requires that

$$m\vec{\mathbf{v}}_i = m\vec{\mathbf{v}}_c + m\vec{\mathbf{v}}_4$$

or

$$\vec{\mathbf{v}}_i = \vec{\mathbf{v}}_c + \vec{\mathbf{v}}_4$$

This vector addition is shown graphically in Fig. 7.22a. Since the collision is elastic, the total kinetic energy doesn't change:

$$\tfrac{1}{2}mv_i^2 = \tfrac{1}{2}mv_c^2 + \tfrac{1}{2}mv_4^2$$

continued on next page

Conceptual Example 7.12 continued

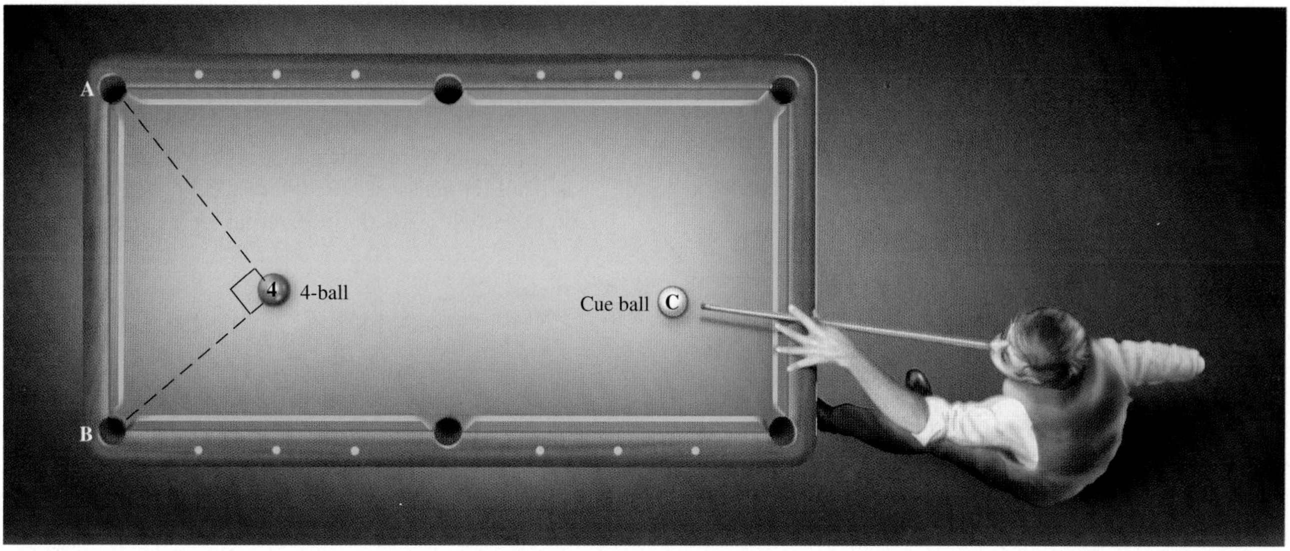

Figure 7.21 Should Eric try to sink the 4-ball?

or

$$v_i^2 = v_c^2 + v_4^2$$

Since v_i, v_c, and v_4 are the sides of a triangle, this is a statement of the Pythagorean theorem—the triangle must be a *right* triangle with v_i as the hypotenuse (Fig. 7.22b). Therefore, the velocities of the 4-ball and the cue ball after the collision are perpendicular to each other.

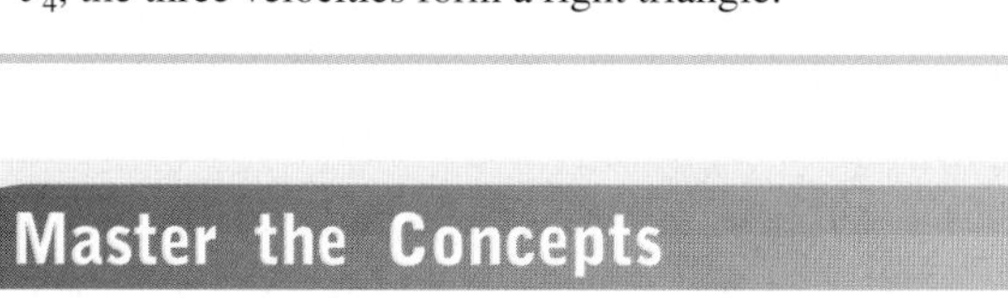

Figure 7.22
(a) Graphical addition of velocity vectors as required by the conservation of momentum. (b) Since $v_i^2 = v_c^2 + v_4^2$, the three velocities form a right triangle.

If Eric sinks the 4-ball, the cue ball falls into pocket A. He shouldn't attempt this shot until he learns how to put some spin on the ball.

Discussion Note that we did not resolve the velocities into *x*- and *y*-components. Doing so would have made the solution longer in this case.

We found that the two balls move at right angles after the collision. This result is true for any two-dimensional elastic collision between two objects of equal masses if one of them is initially at rest. In Example 7.11 the two pucks move at right angles after the collision, but the collision is inelastic—the masses are unequal.

Practice Problem 7.12 Finding the Speed Ratio

Suppose that the cue ball initially moves in the $-x$-direction. After the collision, the cue ball moves at 52.0° above the $-x$-axis and the 4-ball moves at 38.0° below the $-x$-axis. Find the ratio of the balls' speeds v_c/v_4 after the collision.

Master the Concepts

- Definition of linear momentum:

$$\vec{\mathbf{p}} = m\vec{\mathbf{v}} \quad (7\text{-}1)$$

- During an interaction, momentum is transferred from one body to another, but the total momentum of the two is unchanged.

$$\Delta\vec{\mathbf{p}}_2 = -\Delta\vec{\mathbf{p}}_1$$

- Impulse is the average force times the time interval.
- The total impulse equals the change in momentum:

$$\Delta\vec{\mathbf{p}} = \sum\vec{\mathbf{F}}\,\Delta t \quad (7\text{-}2)$$

continued on next page

Master the Concepts continued

- A conserved quantity is one that remains unchanged as time passes.
- Impulse is the area under a graph of force versus time.

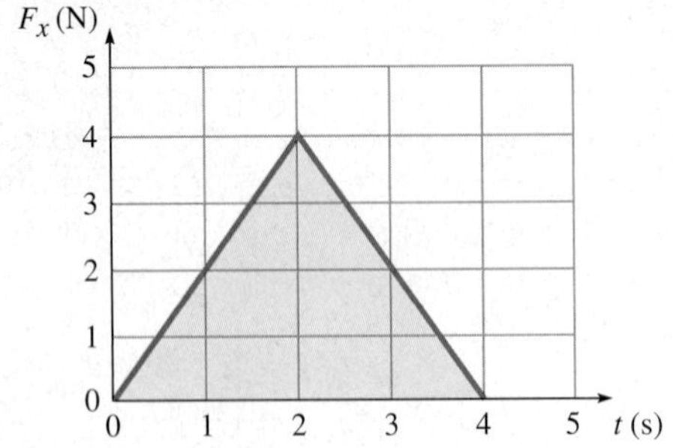

- The net force is the rate of change of momentum.

$$\sum \vec{\mathbf{F}} = \lim_{\Delta t \to 0} \frac{\Delta \vec{\mathbf{p}}}{\Delta t} \qquad (7\text{-}4)$$

- External interactions may change the total momentum of a system.
- Internal interactions do not change the total momentum of a system.
- Conservation of linear momentum: if the net external force acting on a system is zero, then the momentum of the system is conserved.
- The position of the CM of a system of N particles is

$$x_{CM} = \frac{m_1x_1 + m_2x_2 + \cdots + m_Nx_N}{M}$$

and

$$y_{CM} = \frac{m_1y_1 + m_2y_2 + \cdots + m_Ny_N}{M} \qquad (7\text{-}9)$$

where M is the total mass of the particles:

$$M = m_1 + m_2 + \cdots + m_N$$

- The total momentum of a system is equal to the total mass times the velocity of the center of mass:

$$\vec{\mathbf{p}} = \vec{\mathbf{p}}_1 + \vec{\mathbf{p}}_2 + \cdots + \vec{\mathbf{p}}_N = M\vec{\mathbf{v}}_{CM} \qquad (7\text{-}11)$$

- No matter how complicated a system is, the CM moves as if all the mass of the system were concentrated to a point particle with all the external forces acting on it:

$$\sum \vec{\mathbf{F}}_{ext} = M\vec{\mathbf{a}}_{CM} \qquad (7\text{-}13)$$

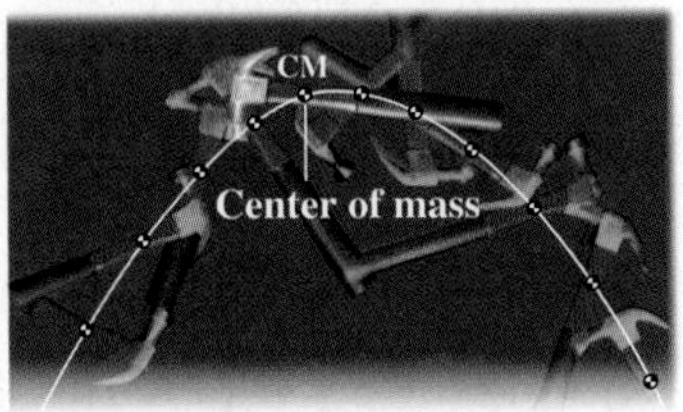

- The CM of an isolated system moves at constant velocity.
- Conservation of momentum is used to solve problems involving collisions, explosions, and the like. Even when external forces are acting, the momentum of the system just before a collision is nearly equal to the momentum just after if the collision interaction is brief. The impulse, and, therefore, the change in momentum of the system, is small since the time interval is small.

Conceptual Questions

1. You are trapped on the second floor of a burning building. The stairway is impassable, but there is a balcony outside your window. Describe what might happen in the following situations. (a) You jump from the second-story balcony to the pavement below, landing stiff-legged on your feet. (b) You jump into a privet hedge, landing on your back and rolling to your feet. (c) You jump into a firefighters' net, landing on your back. What happens to the net as you land in it? What do the firefighters do to cushion your fall even more?
2. A force of 30 N is applied for 5 s to each of two bodies of different masses. (a) Which one has the greatest momentum change? (b) The greatest velocity change? (c) The greatest acceleration?
3. If you take a rifle and saw off part of the barrel, the muzzle speed (the speed at which bullets emerge from the barrel) will be smaller. Why?
4. A firecracker at rest explodes, sending fragments off in all directions. Initially the firecracker has zero momentum, but after the explosion the fragments flying off each have quite a lot of momentum. Hasn't momentum been created? If not, explain why not.
5. An astronaut in deep space is taking a space walk when the tether connecting him to his spaceship breaks. How can he get back to the ship? He doesn't have a rocket propulsion backpack, unfortunately, but he is carrying a big wrench.
6. An astronaut hits a golf ball on the surface of the Moon. Is the momentum of the ball conserved while it is in flight? Is there a *component* of its momentum that is conserved?
7. Which would be more effective: a hammer that collides *elastically* with a nail, or one that collides perfectly *inelastically*? Assume that the mass of the hammer is much larger than that of the nail.

8. Squid are the fastest swimmers among invertebrates. A cavity within the squid is filled with water. The *mantle*, a powerful muscle, squeezes the cavity and expels the water through a narrow opening (the *siphon*) at high speed. Using momentum conservation, explain how this propels the squid forward. How is the squid's swimming mechanism like a rocket engine?
9. In your own words, phrase each of Newton's three laws of motion as a statement about momentum.
10. Two objects with different masses have the same kinetic energy. Which has the larger magnitude of momentum?
11. A woman is 1.60 m tall. When standing straight, is her CM necessarily 0.80 m above the floor? Explain.
12. The momentum of a system can only be changed by an external force. What is the external force that changes the momentum of a bicycle (with its rider) as it speeds up, slows down, or changes direction? Is it true that changes in the bicycle's kinetic energy must come from an external force? Explain.
13. In an egg toss, two people try to toss a raw egg back and forth without breaking it as they move farther and farther apart. Discuss a strategy in terms of impulse and momentum for catching the egg without breaking it.
14. In the "executive toy," two balls are pulled back and then released. After the collision, two balls move away on the opposite side. Why do we never see three balls move away following this action, although with a lower velocity so that linear momentum is still conserved?

15. A baseball batting coach emphasizes the importance of "follow-through" when a batter is trying for a home run. The coach explains that the follow-through keeps the bat in contact with the ball for a longer time so the ball will travel a greater distance. Explain the reasoning behind this statement in terms of the impulse-momentum theorem.
16. Micah is standing on his frictionless skateboard facing a concrete wall. He wants to project himself backward by throwing small balls at the wall. His friend Jeremy says that Micah need not throw the balls against the wall, he just needs to throw the balls away from himself, but Micah says the balls need something to push against if they are to propel him backward. Who is right and why?
17. Mary and Daryl are new to the sport of rock climbing. Mary says she wants a stiff rope because a stiff rope is a strong rope. Daryl insists that a good climbing rope must have some stretch. Who is correct, and why?

Multiple-Choice Questions

1. A ball of mass m with initial speed v collides with another ball of mass M, initially at rest. After the collision the two balls stick together, moving with speed V. The ratio of the final speed V to the initial speed v is $V/v =$

(a) $\dfrac{M}{M+m}$ (b) $\dfrac{M+m}{M}$

(c) $\dfrac{m}{M+m}$ (d) $\dfrac{M+m}{m}$

(e) $\sqrt{\dfrac{M}{M+m}}$ (f) $\sqrt{\dfrac{m}{M+m}}$

2. Two particles A and B of equal mass are located at some distance from each other. Particle A is at rest while B moves away from A at speed v. What happens to the center of mass of the system of two particles?
(a) It does not move.
(b) It moves with a speed v away from A.
(c) It moves with a speed v toward A.
(d) It moves with a speed $\frac{1}{2}v$ away from A.
(e) It moves with a speed $\frac{1}{2}v$ toward A.
3. Two uniform spheres with equal mass per unit volume are in contact with one another. The mass of sphere A is five times that of sphere B. The center of mass of the system is
(a) at the point where A and B touch.
(b) inside sphere B somewhere on the line joining the centers of A and B.
(c) inside sphere A somewhere on the line joining the centers.
(d) at the center of sphere A.
(e) outside of both spheres.
4. An object at rest suddenly explodes into three parts of equal mass. Two of the parts move away at right angles to each other and with equal speeds v. What is the velocity of the third part just after the explosion?
(a) Direction of vector 1 and magnitude $2v$
(b) Direction of vector 2 and magnitude $\sqrt{2}v$
(c) Direction of vector 3 and magnitude $\dfrac{1}{\sqrt{2}}v$
(d) Direction of vector 2 and magnitude $\dfrac{1}{\sqrt{2}}v$
(e) Direction of vector 1 and magnitude $\frac{1}{2}v$

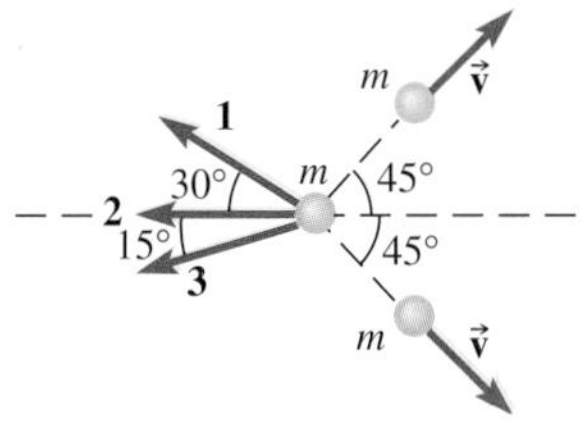

5. A 3.0-kg object is initially at rest. It then receives an impulse of magnitude 15 N·s. After the impulse, the object has
(a) a speed of 45 m/s.
(b) a momentum of magnitude 5.0 kg·m/s.
(c) a speed of 7.5 m/s.
(d) a momentum of magnitude 15 kg·m/s.

6. An object of mass m drops from rest a little above the Earth's surface for a time t. Ignore air resistance. After time t the magnitude of its momentum is
(a) mgt^2
(b) mgt
(c) $mg\sqrt{t}$
(d) $\sqrt{mgt}$
(e) $\frac{mgt^2}{2}$

Multiple-Choice Questions 7–12 refer to a situation in which a golf ball is projected straight upward in the $+y$-direction. Ignore air resistance. The answer choices are found in the figure.

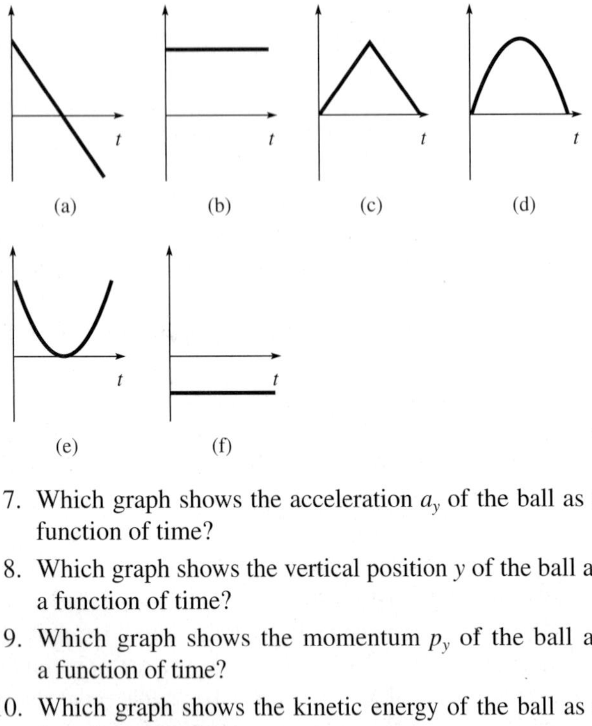

7. Which graph shows the acceleration a_y of the ball as a function of time?
8. Which graph shows the vertical position y of the ball as a function of time?
9. Which graph shows the momentum p_y of the ball as a function of time?
10. Which graph shows the kinetic energy of the ball as a function of time?
11. Which graph shows the potential energy of the ball as a function of time?
12. Which graph shows the total energy of the ball as a function of time?

Problems

Ⓒ Combination conceptual/quantitative problem
Biological or medical application
✦ Challenging problem
Blue # Detailed solution in the Student Solutions Manual
(1 2) Problems paired by concept
Text website interactive or tutorial

7.2 Momentum; 7.3 The Impulse-Momentum Theorem

1. Two cars, each of mass 1300 kg, are approaching each other on a head-on collision course. Each speedometer reads 19 m/s. What is the magnitude of the total momentum of the system?
2. What is the momentum of an automobile (weight = 9800 N) when it is moving at 35 m/s to the south?
3. Verify that the SI unit of impulse is the same as the SI unit of momentum.
4. A cue stick hits a cue ball with an average force of 24 N for a duration of 0.028 s. If the mass of the ball is 0.16 kg, how fast is it moving after being struck?
5. A system consists of three particles with these masses and velocities: mass 3.0 kg, moving north at 3.0 m/s; mass 4.0 kg, moving south at 5.0 m/s; and mass 7.0 kg, moving north at 2.0 m/s. What is the total momentum of the system?
6. A sports car traveling along a straight line increases its speed from 20.0 mi/h to 60.0 mi/h. (a) What is the ratio of the final to the initial magnitude of its momentum? (b) What is the ratio of the final to the initial kinetic energy?
7. A ball of mass 5.0 kg moving with a speed of 2.0 m/s in the $+x$-direction hits a wall and bounces back with the same speed in the $-x$-direction. What is the change of momentum of the ball?
8. An object of mass 3.0 kg is projected into the air at a 55° angle. It hits the ground 3.4 s later. What is its change in momentum while it is in the air? Ignore air resistance.
9. An object of mass 3.0 kg is allowed to fall from rest under the force of gravity for 3.4 s. What is the change in its momentum? Ignore air resistance.
10. What average force is necessary to bring a 50.0-kg sled from rest to a speed of 3.0 m/s in a period of 20.0 s? Assume frictionless ice.
11. For a safe re-entry into the Earth's atmosphere, the pilots of a space capsule must reduce their speed from 2.6×10^4 m/s to 1.1×10^4 m/s. The rocket engine produces a backward force on the capsule of 1.8×10^5 N. The mass of the capsule is 3800 kg. For how long must they fire their engine? [*Hint:* Ignore the change in mass of the capsule due to the expulsion of exhaust gases.]
12. A 0.15-kg baseball traveling in a horizontal direction with a speed of 20 m/s hits a bat and is popped straight up with a speed of 15 m/s. (a) What is the change in momentum (magnitude and direction) of the baseball? (b) If the bat was in contact with the ball for 50 ms, what was the average force of the bat on the ball?

13. An automobile traveling at a speed of 30.0 m/s applies its brakes and comes to a stop in 5.0 s. If the automobile has a mass of 1.0×10^3 kg, what is the average horizontal force exerted on it during braking? Assume the road is level.

14. A 3.0-kg body is initially moving northward at 15 m/s. Then a force of 15 N, toward the east, acts on it for 4.0 s. (a) At the end of the 4.0 s, what is the body's final velocity? (b) What is the change in momentum during the 4.0 s?

✦15. A boy of mass 60.0 kg is rescued from a hotel fire by leaping into a firefighters' net. The window from which he leapt was 8.0 m above the net. The firefighters lower their arms as he lands in the net so that he is brought to a complete stop in a time of 0.40 s. (a) What is his change in momentum during the 0.40-s interval? (b) What is the impulse on the net due to the boy during the interval? [*Hint:* Do not ignore gravity.] (c) What is the average force on the net due to the boy during the interval?

16. A 115-g ball is traveling to the left with a speed of 30 m/s when it is struck by a racket. The force on the ball, directed to the right and applied over 21 ms of contact time, is shown in the graph. What is the speed of the ball immediately after it leaves the racket? (WW tutorial: impulse)

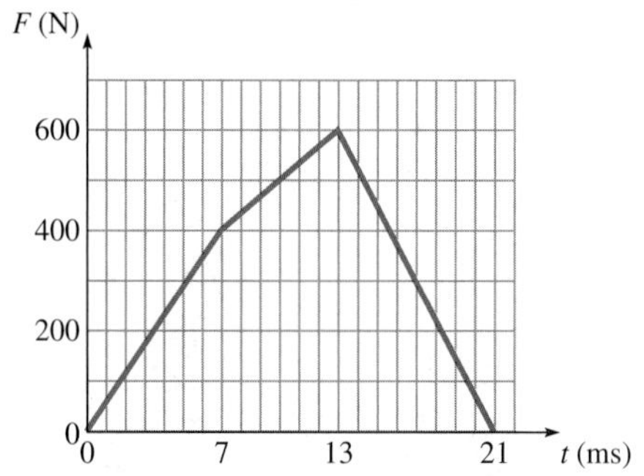

✦17. A pole-vaulter of mass 60.0 kg vaults to a height of 6.0 m before dropping to thick padding placed below to cushion her fall. (a) Find the speed with which she lands. (b) If the padding brings her to a stop in a time of 0.50 s, what is the average force on her body due to the padding during that time interval?

7.4 Conservation of Momentum

18. A rifle has a mass of 4.5 kg and it fires a bullet of mass 10.0 g at a muzzle speed of 820 m/s. What is the recoil speed of the rifle as the bullet leaves the gun barrel?

19. A 0.030-kg bullet is fired vertically at 200 m/s into a 0.15-kg baseball that is initially at rest. The bullet lodges in the baseball and, after the collision, the baseball/bullet rise to a height of 37 m. (a) What was the speed of the baseball/bullet right after the collision? (b) What was the average force of air resistance while the baseball/bullet was rising?

20. A submarine of mass 2.5×10^6 kg and initially at rest fires a torpedo of mass 250 kg. The torpedo has an initial speed of 100.0 m/s. What is the initial recoil speed of the submarine? Neglect the drag force of the water.

21. A uranium nucleus (mass 238 u), initially at rest, undergoes radioactive decay. After an alpha particle (mass 4.0 u) is emitted, the remaining nucleus is thorium (mass 234 u). If the alpha particle is moving at 0.050 times the speed of light, what is the recoil speed of the thorium nucleus? (Note: "u" is a unit of mass; it is *not* necessary to convert it to kg.)

22. Dash is standing on his frictionless skateboard with three balls, each with a mass of 100 g, in his hands. The combined mass of Dash and his skateboard is 60 kg. How fast should dash throw the balls forward if he wants to move backward with a speed of 0.50 m/s? Do you think Dash can succeed? Explain.

23. A 58-kg astronaut is in space, far from any objects that would exert a significant gravitational force on him. He would like to move toward his spaceship, but his jet pack is not functioning. He throws a 720-g socket wrench with a velocity of 5.0 m/s in a direction away from the ship. After 0.50 s, he throws a 800-g spanner in the same direction with a speed of 8.0 m/s. After another 9.90 s, he throws a mallet with a speed of 6.0 m/s in the same direction. The mallet has a mass of 1200 g. How fast is the astronaut moving after he throws the mallet?

24. A man with a mass of 65 kg skis down a frictionless hill that is 5.0 m high. At the bottom of the hill the terrain levels out. As the man reaches the horizontal section, he grabs a 20-kg backpack and skis off a 2.0-m-high ledge. At what horizontal distance from the edge of the ledge does the man land?

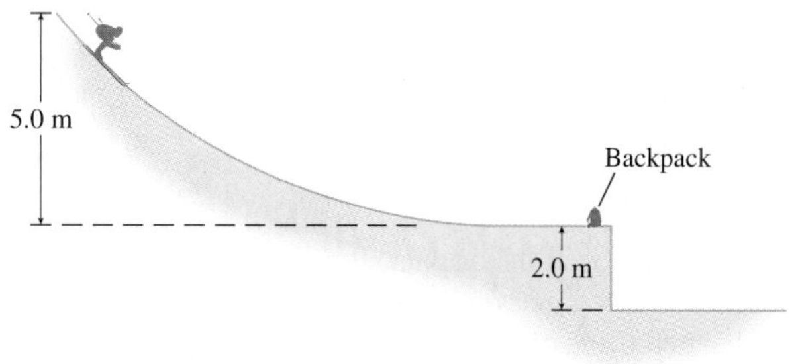

✦25. A cannon on a railroad car is facing in a direction parallel to the tracks. It fires a 98-kg shell at a speed of 105 m/s (relative to the ground) at an angle of 60.0° above the horizontal. If the cannon plus car have a mass of 5.0×10^4 kg, what is the recoil speed of the car if it was at rest before the cannon was fired? [*Hint:* A *component*

of a system's momentum along an axis is conserved if the net external force acting on the system has no component along that axis.]

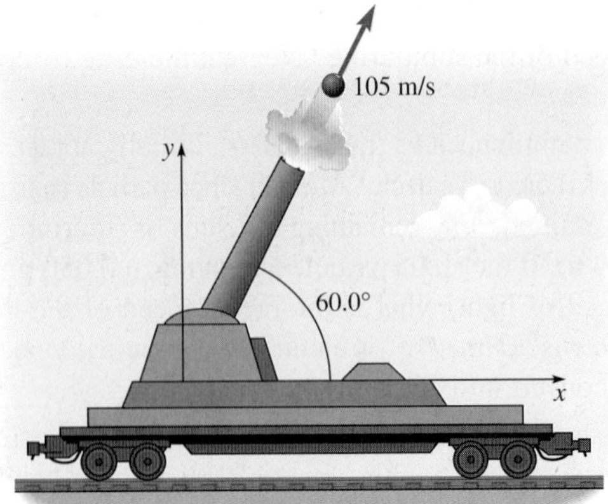

26. A marksman standing on a motionless railroad car fires a gun into the air at an angle of 30.0° from the horizontal. The bullet has a speed of 173 m/s (relative to the ground) and a mass of 0.010 kg. The man and car move to the left at a speed of 1.0×10^{-3} m/s after he shoots. What is the mass of the man and car? (See the hint in Problem 25.)

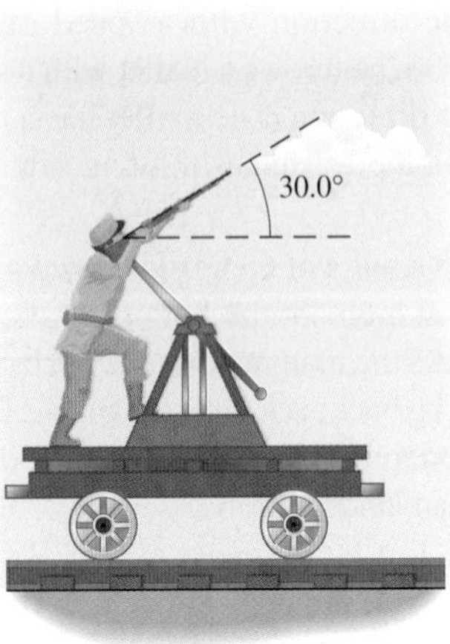

7.5 Center of Mass; 7.6 Motion of the Center of Mass

27. Particle A is at the origin and has a mass of 30.0 g. Particle B has a mass of 10.0 g. Where must particle B be located if the coordinates of the CM are $(x, y) = (2.0$ cm, 5.0 cm$)$?

28. Particle A has a mass of 5.0 g and particle B has a mass of 1.0 g. Particle A is located at the origin and particle B is at the point $(x, y) = (25$ cm, $0)$. What is the location of the CM? (tutorial: center of mass)

29. The three bodies in the figure each have the same mass. If one of the bodies is moved 12 cm in the positive x-direction, by how much does the CM move?

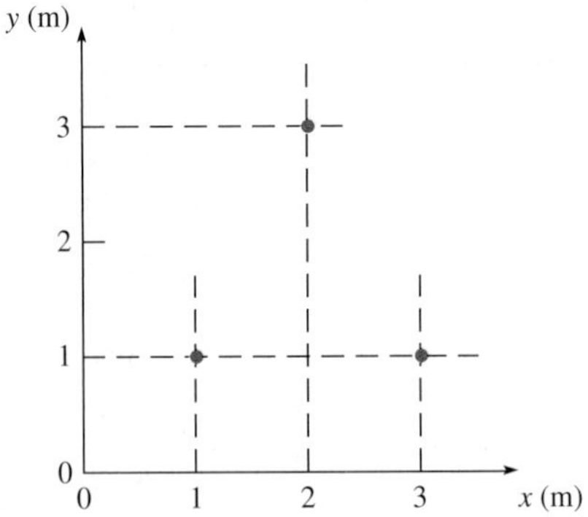

30. The positions of three particles, written as (x, y) coordinates, are: particle 1 (mass 4.0 kg) at (4.0 m, 0 m); particle 2 (mass 6.0 kg) at (2.0 m, 4.0 m); particle 3 (mass 3.0 kg) at (−1.0 m, −2.0 m). What is the location of the CM?

31. Belinda needs to find the CM of a sculpture she has made so that it will hang in a gallery correctly. The sculpture is all in one plane and consists of various shaped uniform objects with masses and sizes as shown. Where is the CM of this sculpture? Assume the thin rods connecting the larger pieces have no mass and place the reference frame origin at the top left corner of the sculpture.

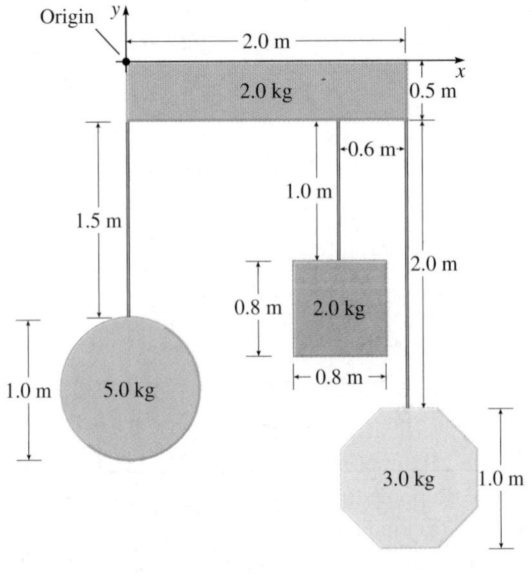

32. Jane is sitting on a chair with her lower leg at a 30.0° angle with respect to the vertical, as shown. You need to develop a computer model of her leg to assist in some medical research. If you assume that her leg can be modeled as two uniform cylinders, one with mass $M = 20$ kg and length $L = 35$ cm and one with mass $m = 10$ kg and length $l = 40$ cm, where is the CM of her leg?

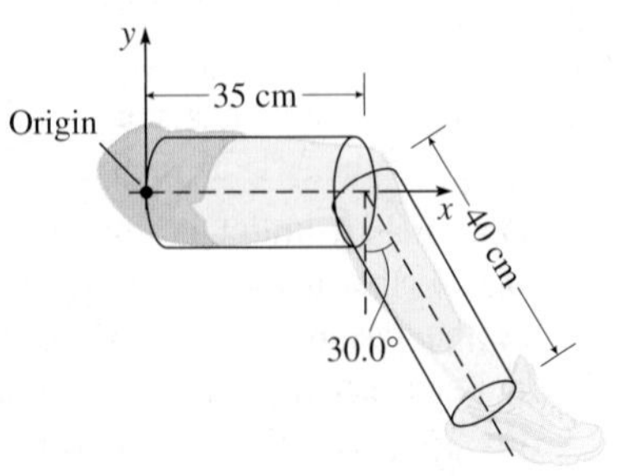

33. Find the x-coordinate of the CM of the composite object shown in the figure. The sphere, cylinder, and rectangular solid all have a uniform composition. Their masses and dimensions are: sphere: 200 g, diameter = 10 cm; cylinder: 450 g, length = 17 cm, radius = 5.0 cm; rectangular solid: 325 g, length in x-direction = 16 cm, height = 10 cm, depth = 12 cm.

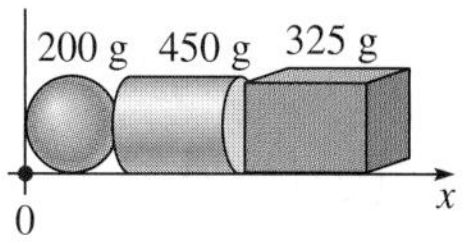

34. Consider two falling bodies. Their masses are 3.0 kg and 4.0 kg. At time $t = 0$, the two are released from rest. What is the velocity of their CM at $t = 10.0$ s? Ignore air resistance.

35. Body A of mass 3 kg is moving in the $+x$-direction with a speed of 14 m/s. Body B of mass 4 kg is moving in the $-y$-direction with a speed of 7 m/s. What are the x- and y-components of the velocity of the CM of the two bodies?

36. If a particle of mass 5.0 kg is moving east at 10 m/s and a particle of mass 15 kg is moving west at 10 m/s, what is the velocity of the CM of the pair?

37. An object located at the origin and having mass M explodes into three pieces having masses $M/4$, $M/3$, and $5M/12$. The pieces scatter on a horizontal frictionless xy-plane. The piece with mass $M/4$ flies away with velocity 5.0 m/s at 37° above the x-axis. The piece with mass $M/3$ has velocity 4.0 m/s directed at an angle of 45° above the $-x$-axis. (a) What are the velocity components of the third piece? (b) Describe the motion of the CM of the system after the explosion.

38. Prove Eq. (7-13) $\Sigma\vec{\mathbf{F}}_{ext} = M\vec{\mathbf{a}}_{CM}$. [*Hint:* Start with $\Sigma\vec{\mathbf{F}}_{ext} = \lim_{\Delta t \to 0} (\Delta\vec{\mathbf{p}}/\Delta t)$, where $\Sigma\vec{\mathbf{F}}_{ext}$ is the net external force acting on a system and $\vec{\mathbf{p}}$ is the total momentum of the system.]

7.7 Collisions in One Dimension

39. A helium atom (mass 4.00 u) moving at 618 m/s to the right collides with an oxygen molecule (mass 32.0 u) moving in the same direction at 412 m/s. After the collision, the oxygen molecule moves at 456 m/s to the right. What is the velocity of the helium atom after the collision?

40. A toy car with a mass of 120 g moves to the right with a speed of 0.75 m/s. A small child drops a 30.0-g piece of clay onto the car. The clay sticks to the car and the car continues to the right. What is the change in speed of the car? Consider the frictional force between the car and the ground to be negligible.

41. In the railroad freight yard, an empty freight car of mass m rolls along a straight level track at 1.0 m/s and collides with an initially stationary, fully loaded boxcar of mass $4.0m$. The two cars couple together on collision. (a) What is the speed of the two cars after the collision? (b) Suppose instead that the two cars are at rest after the collision. With what speed was the loaded boxcar moving before the collision if the empty one was moving at 1.0 m/s? (tutorial: sticking collision)

42. A 0.020-kg bullet traveling at 200.0 m/s east hits a motionless 2.0-kg block and bounces off it, retracing its original path with a velocity of 100.0 m/s west. What is the final velocity of the block? Assume the block rests on a perfectly frictionless horizontal surface.

43. A block of wood of mass 0.95 kg is initially at rest. A bullet of mass 0.050 kg traveling at 100.0 m/s strikes the block and becomes embedded in it. With what speed do the block of wood and the bullet move just after the collision?

44. A 0.020-kg bullet is shot horizontally and collides with a 2.00-kg block of wood. The bullet embeds in the block and the block slides along a horizontal surface for 1.50 m. If the coefficient of kinetic friction between the block and surface is 0.400, what was the original speed of the bullet?

45. A 2.0-kg block is moving to the right at 1.0 m/s just before it strikes and sticks to a 1.0-kg block initially at rest. What is the total momentum of the two blocks after the collision?

46. A 75-kg man is at rest on ice skates. A 0.20-kg ball is thrown to him. The ball is moving horizontally at 25 m/s just before the man catches it. How fast is the man moving just after he catches the ball?

47. A BMW of mass 2.0×10^3 kg is traveling at 42 m/s. It approaches a 1.0×10^3 kg Volkswagen going 25 m/s in the same direction and strikes it in the rear. Neither driver applies the brakes. Neglect the relatively small frictional forces on the cars due to the road and due to air resistance. (a) If the collision slows the BMW down to 33 m/s, what is the speed of the VW after the collision? (b) During the collision, which car exerts a larger force on the other, or are the forces equal in magnitude? Explain.

48. A 100-g ball collides elastically with a 300-g ball that is at rest. If the 100-g ball was traveling in the positive x-direction at 5.00 m/s before the collision, what are the velocities of the two balls after the collision? (tutorial: elastic collision)

49. A projectile of 1.0-kg mass approaches a stationary body of 5.0 kg at 10.0 m/s and, after colliding, rebounds in the reverse direction along the same line with a speed of 5.0 m/s. What is the speed of the 5.0-kg body after the collision?

50. A 2.0-kg object is at rest on a perfectly frictionless surface when it is hit by a 3.0-kg object moving at 8.0 m/s. If the two objects are stuck together after the collision, what is the speed of the combination?

51. A spring of negligible mass is compressed between two blocks, A and B, which are at rest on a frictionless horizontal surface at a distance of 1.0 m from a wall on the left and 3.0 m from a wall on the right. The sizes of the blocks and spring are small. When the spring is released, block A moves toward the left wall and strikes it at the same instant that block B strikes the right wall. The mass of A is 0.60 kg. What is the mass of B?

52. Two identical gliders on an air track are held together by a piece of string, compressing a spring between the gliders. While they are moving to the right at a common speed of 0.50 m/s, someone holds a match under the string and burns it, letting the spring force the gliders apart. One glider is then observed to be moving to the right at 1.30 m/s. (a) What velocity does the other glider have? (b) Is the total kinetic energy of the two gliders after the collision greater than, less than, or equal to the total kinetic energy before the collision? If greater, where did the extra energy come from? If less, where did the "lost" energy go?

53. A 0.010-kg bullet traveling horizontally at 400.0 m/s strikes a 4.0-kg block of wood sitting at the edge of a table. The bullet is lodged into the wood. If the table height is 1.2 m, how far from the table does the block hit the floor?

54. Two objects with masses m_1 and m_2 approach each other with equal and opposite momenta so that the total momentum is zero. Show that, if the collision is elastic, the final *speed* of each object must be the same as its initial speed. (The final *velocity* of each object is *not* the same as its initial velocity, however.)

55. A 6.0-kg object is at rest on a perfectly frictionless surface when it is struck head-on by a 2.0-kg object moving at 10 m/s. If the collision is perfectly elastic, what is the speed of the 6.0-kg object after the collision? [*Hint:* You will need two equations.]

56. Use the result of Problem 54 to show that in *any* elastic collision between two objects, the relative speed of the two is the same before and after the collision. [*Hints:* Look at the collision in its CM *frame*—the reference frame in which the CM is at rest. The *relative* speed of two objects is the same in any inertial reference frame.]

7.8 Collisions in Two Dimensions

57. A firecracker is tossed straight up into the air. It explodes into three pieces of equal mass just as it reaches the highest point. Two pieces move off at 120 m/s at right angles to each other. How fast is the third piece moving?

58. Body A of mass M has an original velocity of 6.0 m/s in the $+x$-direction toward a stationary body (body B) of the same mass. After the collision, body A has velocity components of 1.0 m/s in the $+x$-direction and 2.0 m/s in the $+y$-direction. What is the magnitude of body B's velocity after the collision?

59. (a) With reference to Practice Problem 7.11, find the momentum change of the ball of mass m_1 during the collision. Give your answer in x- and y-component form; express the components in terms of m_1 and v_i. (b) Repeat for the ball of mass m_2. How are the momentum changes related?

60. A hockey puck moving at 0.45 m/s collides elastically with another puck that was at rest. The pucks have equal mass. The first puck is deflected 37° to the right and moves off at 0.36 m/s. Find the speed and direction of the second puck after the collision.

61. Puck 1 sliding along the x-axis strikes stationary puck 2 of the same mass. After the elastic collision, puck 1 moves off at speed v_{1f} in the direction 60.0° above the x-axis; puck 2 moves off at speed v_{2f} in the direction 30.0° below the x-axis. Find v_{2f} in terms of v_{1f}.

62. Block A, with a mass of 220 g, is traveling north on a frictionless surface with a speed of 5.0 m/s. Block B, with a mass of 300 g travels west on the same surface until it collides with A. After the collision, the blocks move off together with a velocity of 3.13 m/s at an angle of 42.5° to the north of west. What was B's speed just before the collision?

63. A projectile of mass 2.0 kg approaches a stationary target body at 5.0 m/s. The projectile is deflected through an angle of 60.0° and its speed after the collision is 3.0 m/s. What is the magnitude of the momentum of the target body after the collision?

64. A 1500-kg car moving east at 17 m/s collides with a 1800-kg car moving south at 15 m/s and the two cars stick together. (a) What is the velocity of the cars right after the collision? (b) How much kinetic energy was converted to another form during the collision?

65. A car with a mass of 1700 kg is traveling directly northeast (45° between north and east) at a speed of 14 m/s (31 mph), and collides with a smaller car with a mass of 1300 kg that is traveling directly south at a speed of 18 m/s (40 mph). The two cars stick together during the collision. With what speed and direction does the tangled mess of metal move right after the collision?

66. In a nuclear reactor, a neutron moving at speed v_i in the positive x-direction strikes a deuteron, which is at rest. The neutron is deflected by 90.0° and moves off with speed $v_i/\sqrt{3}$ in the positive y-direction. Find the x- and y-components of the deuteron's velocity after the collision. (The mass of the deuteron is twice the mass of the neutron.)

67. Two identical pucks are on an air table. Puck A has an initial velocity of 2.0 m/s in the $+x$-direction. Puck B is at rest. Puck A collides elastically with puck B and A moves off at 1.0 m/s at an angle of 60° above the x-axis. What is the speed and direction of puck B after the collision?

68. In a circus trapeze act, two acrobats actually fly through the air and grab on to one another, then together grab a

swinging bar. One acrobat, with a mass of 60 kg, is moving at 3.0 m/s at an angle of 10° above the horizontal and the other, with a mass of 80 kg, is approaching her with a speed of 2.0 m/s at an angle of 20° above the horizontal. What is the direction and speed of the acrobats right after they grab on to each other? Let the positive x-axis be in the horizontal direction and assume the first acrobat has positive velocity components in the positive x- and y-directions.

69. Two African swallows fly toward one another, carrying coconuts. The first swallow is flying north horizontally with a speed of 20 m/s. The second swallow is flying at the same height as the first and in the opposite direction with a speed of 15 m/s. The mass of the first swallow is 0.270 kg and the mass of his coconut is 0.80 kg. The second swallow's mass is 0.220 kg and her coconut's mass is 0.70 kg. The swallows collide and lose their coconuts. Immediately after the collision, the 0.80-kg coconut travels 10° west of south with a speed of 13 m/s, and the 0.70-kg coconut moves 30° east of north with a speed of 14 m/s. The two birds are tangled up with one another and stop flapping their wings as they travel off together. What is the velocity of the birds immediately after the collision?

Comprehensive Problems

70. A sled of mass 5.0 kg is coasting along on a frictionless ice-covered lake at a constant speed of 1.0 m/s. A 1.0-kg book is dropped vertically onto the sled. At what speed does the sled move once the book is on it?

71. An automobile weighing 13.6 kN is moving at 17.0 m/s when it collides with a stopped car weighing 9.0 kN. If they lock bumpers and move off together, what is their speed just after the collision?

72. For a system of three particles moving along a line, an observer in a laboratory measures the following masses and velocities. What is the velocity of the CM of the system?

Mass (kg)	v_x (m/s)
3.0	+290
5.0	−120
2.0	+52

73. An intergalactic spaceship is traveling through space far from any planets or stars, where no human has gone before. The ship carries a crew of 30 people (of total mass 2.0×10^3 kg). If the speed of the spaceship is 1.0×10^5 m/s and its mass (excluding the crew) is 4.8×10^4 kg, what is the magnitude of the total momentum of the ship and the crew?

74. A baseball player pitches a fastball toward home plate at a speed of 41 m/s. The batter swings, connects with the ball of mass 145 g, and hits it so that the ball leaves the bat with a speed of 37 m/s. Assume that the ball is moving horizontally just before and just after the collision with the bat. (a) What is the magnitude of the change in momentum of the ball? (b) What is the impulse delivered to the ball by the bat? (c) If the bat and ball are in contact for 3.0 ms, what is the magnitude of the average force exerted on the ball by the bat?

75. A tennis ball of mass 0.060 kg is served. It strikes the ground with a velocity of 54 m/s (120 mi/h) at an angle of 22° below the horizontal. Just after the bounce it is moving at 53 m/s at an angle of 18° above the horizontal. If the interaction with the ground lasts 0.065 s, what average force did the ground exert on the ball?

76. A uniform rod of length 30.0 cm is bent into the shape of an inverted U. Each of the three sides is of length 10.0 cm. Find the location, in x- and y-coordinates, of the CM as measured from the origin.

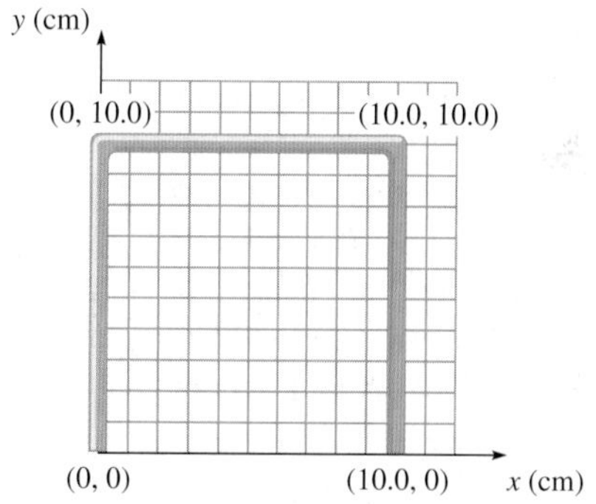

77. A child places 12 wooden blocks together, as shown in the figure. If each block has the same mass and density, where is the CM of these blocks? Each block is a cube with sides of 1.0 inch length. The origin of the coordinate system is at the center of the farthest block to the left.

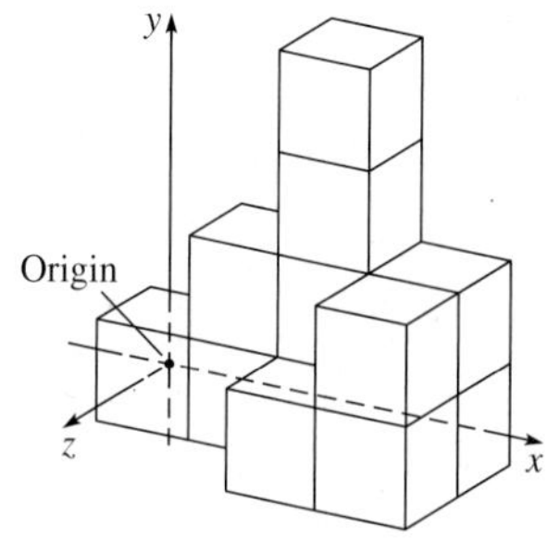

78. To contain some unruly demonstrators, the riot squad approaches with fire hoses. Suppose that the rate of flow of water through a fire hose is 24 kg/s and the stream of water from the hose moves at 17 m/s. What force is exerted by such a stream on a person in the crowd? Assume that the water comes to a dead stop against the demonstrator's chest.

79. An inexperienced catcher catches a 130 km/h fastball of mass 140 g within 1 ms, whereas an experienced catcher slightly retracts his hand during the catch, extending the

stopping time to 10 ms. What are the average forces imparted to the two gloved hands during the catches?

80. A stationary 0.1-g fly encounters the windshield of a 1000-kg automobile traveling at 100 km/h. (a) What is the change in momentum of the car due to the fly? (b) What is the change of momentum of the fly due to the car? (c) Approximately how many flies does it take to reduce the car's speed by 1 km/h?

81. A 0.15-kg baseball is pitched with a speed of 35 m/s (78 mph). When the ball hits the catcher's glove, the glove moves back by 5.0 cm (2 in.) as it stops the ball. (a) What was the change in momentum of the baseball? (b) What impulse was applied to the baseball? (c) Assuming a constant acceleration of the ball, what was the average force applied by the catcher's glove?

82. A projectile of mass 2.0 kg approaches a stationary target body at 8.0 m/s. The projectile is deflected through an angle of 90.0° and its speed after the collision is 6.0 m/s. What is the speed of the target body after the collision if the collision is perfectly elastic?

83. A radioactive nucleus is at rest when it spontaneously decays by emitting an electron and neutrino. The momentum of the electron is 8.20×10^{-19} kg·m/s and it is directed at right angles to that of the neutrino. The neutrino's momentum has magnitude 5.00×10^{-19} kg·m/s. (a) In what direction does the newly formed (daughter) nucleus recoil? (b) What is its momentum?

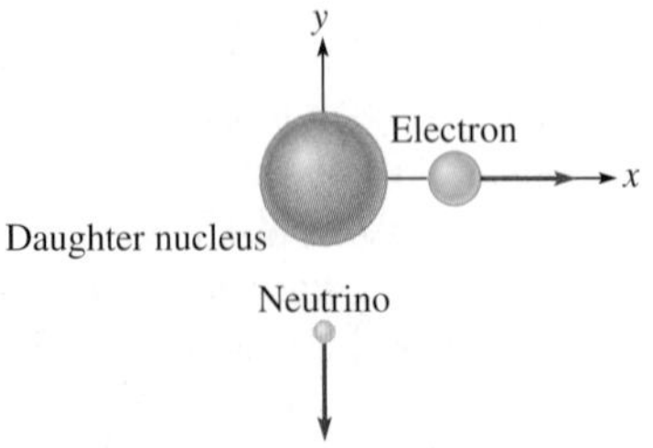

84. A 60.0-kg woman stands at one end of a 120-kg raft that is 6.0 m long. The other end of the raft is 0.50 m from a pier. (a) The woman walks toward the pier until she gets to the other end of the raft and stops there. Now what is the distance between the raft and the pier? (b) In (a), how far did the woman walk (relative to the pier)?

85. A police officer is investigating the scene of an accident where two cars collided at an intersection. One car with a mass of 1100 kg moving west had collided with a 1300-kg car moving north. The two cars, stuck together, skid at an angle of 30° north of west for a distance of 17 m. The coefficient of kinetic friction between the tires and the road is 0.80. The speed limit for each car was 70 km/h. Was either car speeding?

86. A jet plane is flying at 130 m/s relative to the ground. There is no wind. The engines take in 81 kg of air per second. Hot gas (burned fuel and air) is expelled from the engines at high speed. The engines provide a forward force on the plane of magnitude 6.0×10^4 N. At what speed relative to the ground is the gas being expelled? [*Hint:* Look at the momentum change of the air taken in by the engines during a time interval Δt.] This calculation is approximate since we are ignoring the 3.0 kg of fuel consumed and expelled with the air each second.

87. Within cells, small organelles containing newly synthesized proteins are transported along microtubules by tiny molecular motors called kinesins. What force does a kinesin molecule need to deliver in order to accelerate an organelle with mass 0.01 pg (10^{-17} kg) from 0 to 1 μm/s within a time of 10 μs?

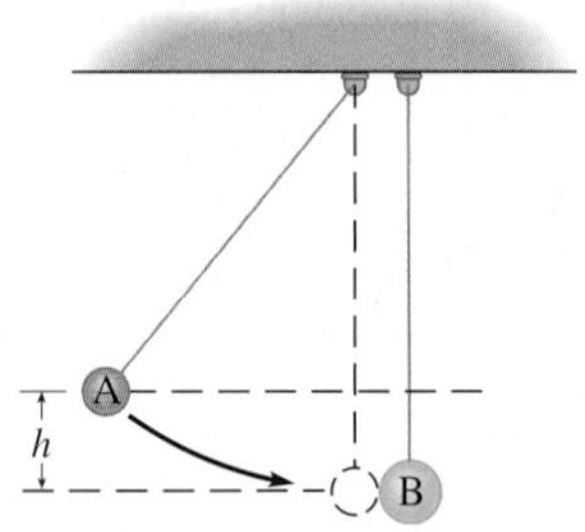

Problems 88 and 89.

88. The pendulum bobs in the figure are made of soft clay so that they stick together after impact. The mass of bob A is half that of bob B. Bob B is initially at rest. What is the ratio of the kinetic energy of the combined bobs, just after impact, to the kinetic energy of bob A just before impact?

89. The pendulum bobs in the figure are made of soft clay so that they stick together after impact. The mass of bob A is half that of bob B. Bob B is initially at rest. If bob A is released from a height h above its lowest point, what is the maximum height attained by bobs A and B after the collision?

90. A flat, circular metal disk of uniform thickness has a radius of 3.0 cm. A hole is drilled in the disk that is 1.5 cm in radius. The hole is tangent to one side of the disk. Where is the CM of the disk now that the hole has been drilled? [*Hint:* The original disk (before the hole is drilled) can be thought of as having two pieces—the disk with the hole plus the smaller disk of metal drilled out. Write an equation that expresses x_{CM} of the original disk in terms of the x_{CM}'s of the two pieces. Since the thickness is uniform, the mass of any piece is proportional to its area.]

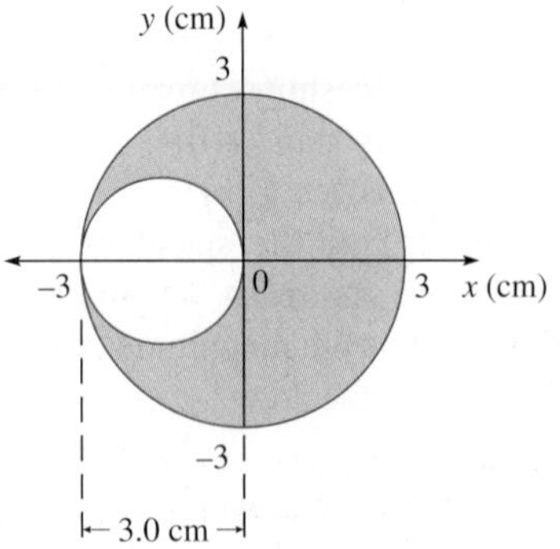

✦91. Two pendulum bobs have equal masses and lengths (5.1 m). Bob A is initially held horizontally while bob B hangs vertically at rest. Bob A is released and collides elastically with bob B. How fast is bob B moving immediately after the collision?

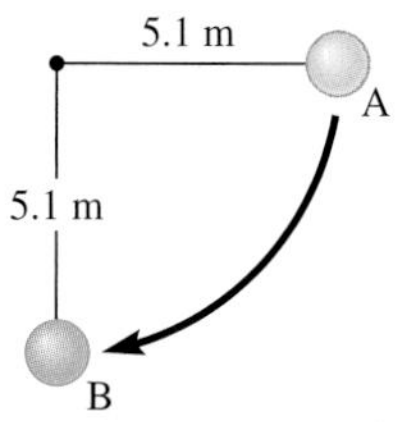

92. Two identical gliders, each with elastic bumpers and mass 0.10 kg, are on a horizontal air track. Friction is negligible. Glider 2 is stationary. Glider 1 moves toward glider 2 from the left with a speed of 0.20 m/s. They collide. After the collision, what are the velocities of glider 1 and glider 2?

✦93. A radium nucleus (mass 226 u) at rest decays into a radon nucleus (symbol Rn, mass 222 u) and an alpha particle (symbol α, mass 4 u). (a) Find the ratio of the speeds v_α/v_{Rn} after the decay. (b) Find the ratio of the magnitudes of the momenta p_α/p_{Rn}. (c) Find the ratio of the kinetic energies K_α/K_{Rn}. (Note: "u" is a unit of mass; it is *not* necessary to convert it to kg.)

Answers to Practice Problems

7.1 (a) 0.78 kg·m/s downward; (b) 0.78 kg·m/s toward the apple; 1.3×10^{-25} m/s

7.2 3.5 times his weight

7.3 1700 N; 0.0037 s

7.4 0.8 m/s in the $-x$-direction

7.5 1.7 m/s

7.6 2.2 m/s

7.7 (2.0 cm, 2.3 cm)

7.8 (a) 2.7 m; (b) 1.5 m in the other direction; (c) the CM does not move

7.9 4.0 m/s

7.10 10.0 m/s

7.11 0.751 v_i

7.12 0.781

Answers to Checkpoints

7.2 No, because the *direction* of the car's momentum would have changed.

7.4 When external forces act on a system, the momentum of the system is not conserved.

7.6 Despite the fact that the hammer is rotating, it is in free fall and its CM follows the same trajectory as a point particle in free fall.

7.7 Yes. Momentum is conserved in both elastic and inelastic collisions. In an inelastic collision, the initial and final *kinetic energies* are not equal.

CHAPTER

8 Torque and Angular Momentum

In gymnastics, the iron cross is a notoriously difficult feat requiring incredible strength. Why does it require such great strength? (See p. 282 for the answer.)

Concepts & Skills to Review

- translational equilibrium (Section 4.2)
- uniform circular motion and circular orbits (Sections 5.1 and 5.4)
- angular acceleration (Section 5.6)
- conservation of energy (Section 6.1)
- center of mass and its motion (Sections 7.5 and 7.6)
- rolling without slipping (Section 5.1)

8.1 ROTATIONAL KINETIC ENERGY AND ROTATIONAL INERTIA

When a rigid object is rotating about a fixed axis, it has kinetic energy because each particle other than those on the axis of rotation is moving in a circle around the axis. In principle, we can calculate the kinetic energy of rotation by summing the kinetic energy of each particle. To say the least, that sounds like a laborious task. We need a simpler way to express the rotational kinetic energy of such an object so that we don't have to calculate this sum over and over. Our simpler expression exploits the fact that the speed of each particle is proportional to the angular speed of rotation ω.

If a rigid object consists of N particles, the sum of the kinetic energies of the particles can be written mathematically using a subscript to label the mass and speed of each particle:

$$K_{rot} = \tfrac{1}{2}m_1v_1^2 + \tfrac{1}{2}m_2v_2^2 + \cdots + \tfrac{1}{2}m_Nv_N^2 = \sum_{i=1}^{N} \tfrac{1}{2}m_iv_i^2$$

The symbol Σ stands for a sum. $\sum_{i=1}^{N}$ means the sum for $i = 1, 2, \ldots, N$.

The speed of each particle is related to its distance from the axis of rotation. Particles that are farther from the axis move faster. In Section 5.1, we found that the speed of a particle moving in a circle is

$$v = r\omega \qquad (5\text{-}7)$$

where ω is the angular speed and r is the distance between the rotation axis and the particle (Fig. 8.1). By substitution, the rotational kinetic energy can be written

$$K_{rot} = \sum_{i=1}^{N} \tfrac{1}{2}m_ir_i^2\omega^2$$

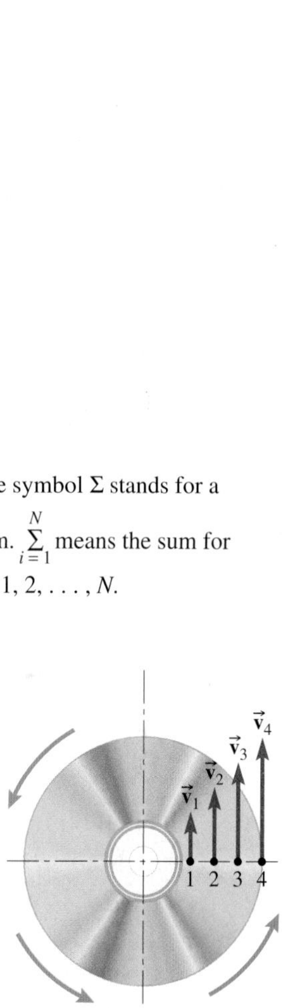

Figure 8.1 Four points on a spinning CD. Points at greater distances from the center are moving faster than points closer to the center.

The entire object rotates at the same angular velocity ω, so the constants $\frac{1}{2}$ and ω^2 can be factored out of each term of the sum:

$$K_{rot} = \tfrac{1}{2}\left(\sum_{i=1}^{N} m_ir_i^2\right)\omega^2$$

The quantity in the parentheses *cannot change* since the distance between each particle and the rotation axis stays the same if the object is rigid and doesn't change shape. However difficult it may be to compute the sum in the parentheses, we only need to do it *once* for any given mass distribution and axis of rotation.

Let's give the quantity in the brackets the symbol I. In Chapter 5, we found it useful to draw analogies between translational variables and their rotational equivalents. By using the symbol I, we can see that translational and rotational kinetic energy have similar forms: translational kinetic energy is

$$K_{tr} = \tfrac{1}{2}mv^2$$

and **rotational kinetic energy** is

Rotational kinetic energy:

$$K_{rot} = \tfrac{1}{2}I\omega^2 \qquad (8\text{-}1)$$

Since $v = r\omega$ was used to derive Eq. (8-1), ω must be expressed in radians per unit time (normally rad/s).

The quantity I is called the **rotational inertia**:

Rotational inertia:

$$I = \sum_{i=1}^{N} m_i r_i^2 \qquad \text{(8-2)}$$

(SI unit: kg·m^2)

CONNECTION:

Rotational and translational kinetic energies have the same form: $\frac{1}{2}$inertia × speed^2.

Comparing the expressions for translational and rotational kinetic energies, we see that angular speed ω takes the place of speed v and rotational inertia I takes the place of mass m. Mass is a measure of the inertia of an object, or, in other words, how difficult it is to change the object's velocity. Similarly, for a rigid rotating object, I is a measure of its rotational inertia—how hard it is to change its angular velocity. That is why the quantity I is called the rotational inertia; it is also sometimes called the **moment of inertia**.

When a problem requires you to find a rotational inertia, there are three principles to follow.

Finding the Rotational Inertia

1. If the object consists of a *small* number of particles, calculate the sum $I = \sum_{i=1}^{N} m_i r_i^2$ directly.
2. For symmetrical objects with simple geometric shapes, calculus can be used to perform the sum in Eq. (8-2). Table 8.1 lists the results of these calculations for the shapes most commonly encountered.
3. Since the rotational inertia is a sum, you can always mentally decompose the object into several parts, find the rotational inertia of each part, and then add them. This is an example of the *divide-and-conquer* problem-solving technique.

See the text website for more information on rotational inertia, including the parallel- and perpendicular-axis theorems.

Keep in mind that the rotational inertia of an object depends on the location of the rotation axis. For instance, imagine taking the hinges off the side of a door and putting them on the top so that the door swings about a horizontal axis like a cat flap door (Fig. 8.2b). The door now has a considerably larger rotational inertia than before the hinges were moved because the door's height is greater than its width. The door has the same mass as before, but its mass now lies on average much farther from the axis of rotation than that

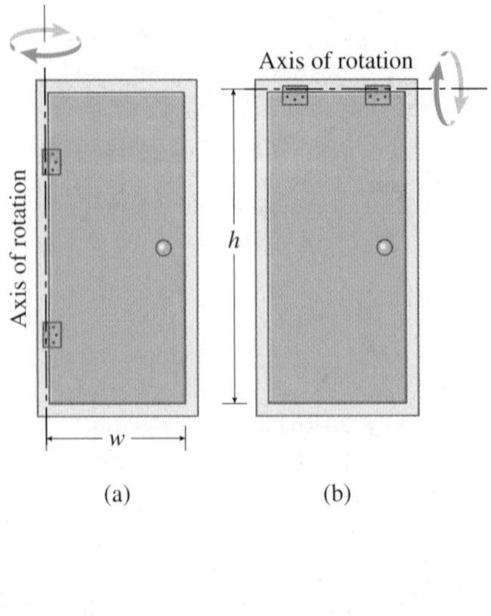

Figure 8.2 The rotational inertia of a door depends on the rotation axis. (a) The door with hinges at the side has a smaller rotational inertia, $I = \frac{1}{3}Mw^2$, than (b) the rotational inertia, $I = \frac{1}{3}Mh^2$, of the same door with hinges at the top, because the door is taller than it is wide.

Table 8.1 Rotational Inertia for Uniform Objects with Various Geometrical Shapes

Shape		Axis of Rotation	Rotational Inertia	Shape		Axis of Rotation	Rotational Inertia
Thin hollow cylindrical shell (or hoop)	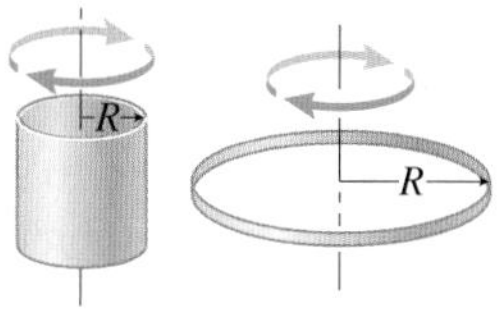	Central axis of cylinder	MR^2	Solid sphere	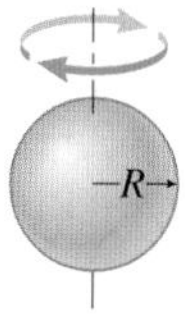	Through center	$\frac{2}{5}MR^2$
Solid cylinder (or disk)	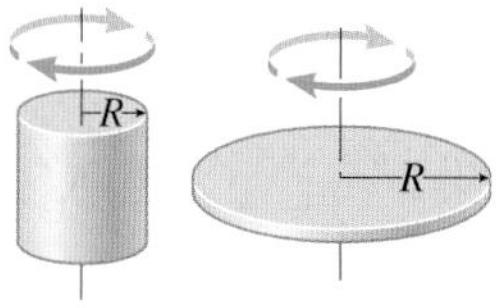	Central axis of cylinder	$\frac{1}{2}MR^2$	Thin hollow spherical shell	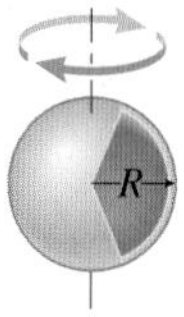	Through center	$\frac{2}{3}MR^2$
Hollow cylindrical shell or disk	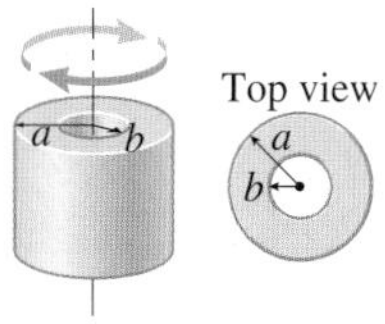	Central axis of cylinder	$\frac{1}{2}M(a^2 + b^2)$	Thin rod (or rectangular plate)	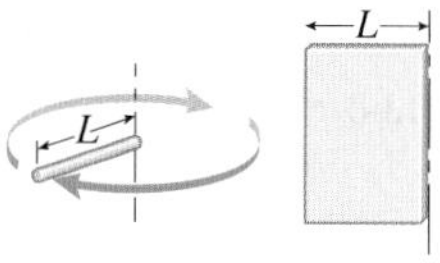	Perpendicular to rod through end (or along edge of plate)	$\frac{1}{3}ML^2$
Rectangular plate	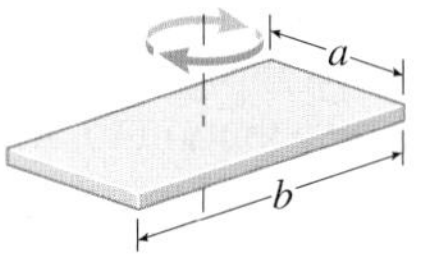	Perpendicular to plate through center	$\frac{1}{12}M(a^2 + b^2)$	Thin rod (or rectangular plate)	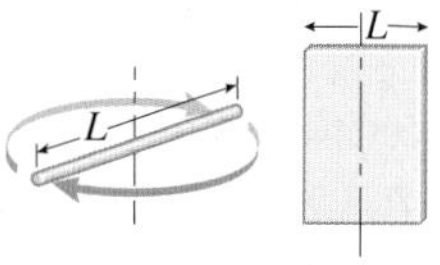	Perpendicular to rod through center (or parallel to edge of plate through center)	$\frac{1}{12}ML^2$

of the door in Fig. 8.2a. In applying Eq. (8-2) to find the rotational inertia of the door, the values of r_i range from 0 to the height of the door (h), whereas with the hinges in the normal position the values of r_i range from 0 only to the width of the door (w).

Figure 8.3 Hank Aaron choking up on the bat.

PHYSICS AT HOME

The change in rotational inertia of a rod as the rotation axis changes can be easily felt. Hold a baseball bat in the usual way, with your hands gripping the bottom of the bat. Swing the bat a few times. Now "choke up" on the bat—move your hands up the bat—and swing a few times. The bat is easier to swing because it now has a smaller rotational inertia. Children often choke up on a bat that is too massive for them. Even major league baseball players occasionally choke up on the bat when they want more control over their swing to place a hit in a certain spot (Fig. 8.3). On the other hand, choking up on the bat makes it impossible to hit a home run. To hit a long fly ball, you want the pitched baseball to encounter a bat that is swinging with a lot of rotational inertia.

CHECKPOINT 8.1

According to Table 8.1, the rotational inertia of a uniform cylinder or disk about its central axis depends only on the mass and radius. Why does it not depend on the height of the cylinder (or thickness of the disk)?

Example 8.1

Rotational Inertia of a Barbell

A barbell consists of two plates, each a uniform disk of mass 20 kg and radius 15 cm, attached 20 cm from each end of a uniform rod of mass 10 kg, radius 1.25 cm, and length 2.20 m (Fig. 8.4). Find the rotational inertia of the barbell about two different axes of rotation: (a) axis a, the central axis of the bar, and (b) axis b, perpendicular to the bar and through its midpoint. Ignore the thickness of the disks and the holes in the disks.

Strategy The rotational inertia of this composite object is the sum of the rotational inertias of the three parts (two disks and rod). Table 8.1 gives formulas for the rotational inertias of disks and rods, but *only for certain axes of rotation.* In particular, for axis b we have two disks rotating about an axis external to the disks, so none of the formulas in Table 8.1 apply; instead we'll return to the basic definition of rotational inertia [Eq. (8-2)] and make an approximation. Based on the distances between parts of the barbell and the two axes, we expect a smaller rotational inertia about axis a than about axis b. Let M and R be the mass and radius of each disk, and m, r, and L the mass, radius, and length of the rod, respectively.

Solution (a) Each of the three component parts, the two disks and the rod, are solid cylinders rotating about their central axes. (The two formulas in Table 8.1 for thin rods are for axes *perpendicular* to the rod, so they are not useful here.) From Table 8.1,

$$I = \tfrac{1}{2}MR^2 + \tfrac{1}{2}MR^2 + \tfrac{1}{2}mr^2$$
$$= 2 \times [\tfrac{1}{2} \times 20\text{ kg} \times (0.15\text{ m})^2] + \tfrac{1}{2} \times 10\text{ kg} \times (0.0125\text{ m})^2$$
$$= 2 \times 0.225\text{ kg·m}^2 + 0.00078\text{ kg·m}^2 = 0.45\text{ kg·m}^2$$

(b) Table 8.1 gives the rotational inertia of the rod about axis b as $\tfrac{1}{12}mL^2$. The center of each disk (assumed to have negligible thickness) is a distance $d = \tfrac{1}{2}(2.20\text{ m} - 0.40\text{ m}) = 0.90\text{ m}$ from the midpoint of the rod. If we think of breaking a disk into tiny pieces and applying Eq. (8-2), each of the distances r_i is at least $d = 0.90$ m (to the center) but no more than $\sqrt{d^2 + R^2} \approx 0.91$ m (to the edge). Therefore, to a good approximation, we can assume each disk to be a point mass at a distance d from the axis. Then

$$I = Md^2 + Md^2 + \tfrac{1}{12}mL^2$$
$$= 2 \times [20\text{ kg} \times (0.90\text{ m})^2] + \tfrac{1}{12} \times 10\text{ kg} \times (2.20\text{ m})^2$$
$$= 2 \times 16.2\text{ kg·m}^2 + 4.03\text{ kg·m}^2 = 36\text{ kg·m}^2$$

As expected, the rotational inertia is much smaller about axis a than about axis b.

Discussion The rod makes only a slight contribution to the rotational inertia about axis a because the radius of the rod is so much smaller than the radii of the disks, so its mass is on average much closer to the axis of rotation. The rod makes a more significant contribution to the rotational inertia about axis b because now the length, not radius, of the rod is relevant—its mass is distributed at distances from 0 to 1.10 m from the axis of rotation. Even if we account for the thickness of the disks, as long as their thicknesses are small relative to d, our estimate Md^2 of the contribution to I from each disk about axis b is still valid.

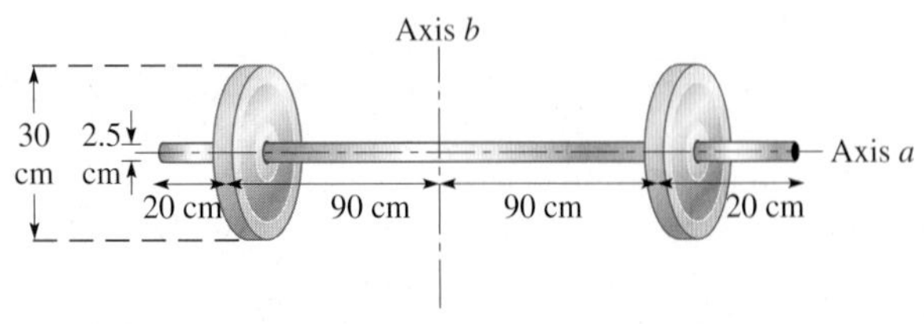

Figure 8.4
A barbell with two different rotation axes.

Practice Problem 8.1 Playground Merry-Go-Round

A playground merry-go-round is essentially a uniform disk that rotates about a vertical axis through its center (Fig. 8.5). Suppose the disk has a radius of 2.0 m and a mass of 160 kg; a child of mass 18.4 kg sits at the edge of the merry-go-round. What is the merry-go-round's rotational inertia, including the contribution due to the child? [*Hint:* Treat the child as a point mass at the edge of the disk.]

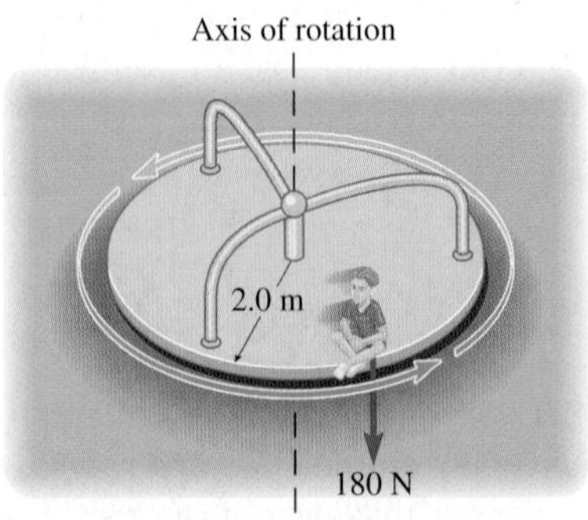

Figure 8.5
Child on a merry-go-round.

When applying conservation of energy to objects that rotate, the rotational kinetic energy is included in the mechanical energy. In Eq. (6-12),

$$W_{nc} = \Delta K + \Delta U = \Delta E_{mech} \qquad (6\text{-}12)$$

just as U stands for the sum of the elastic and gravitational potential energies, K stands for the sum of the translational and rotational kinetic energies:

$$K = K_{tr} + K_{rot}$$

Example 8.2

Atwood's Machine

Atwood's machine consists of a cord around a pulley of rotational inertia I, radius R, and mass M, with two blocks (masses m_1 and m_2) hanging from the ends of the cord as in Fig. 8.6. (Note that in Example 3.11 we analyzed Atwood's machine for the special case of a massless pulley; for a massless pulley $I = 0$.) Assume that the pulley is free to turn without friction and that the cord does not slip. Ignore air resistance. If the masses are released from rest, find how fast they are moving after they have moved a distance h (one up, the other down).

Strategy Ignoring both air resistance and friction means that no nonconservative forces act on the system; therefore, its mechanical energy is conserved:

$$\Delta U + \Delta K = 0$$

Gravitational potential energy is converted into the translational kinetic energies of the two blocks and the rotational kinetic energy of the pulley.

Solution For our convenience, we assume that $m_1 > m_2$. Mass m_1, therefore, moves down and m_2 moves up. After the masses have each moved a distance h, the changes in gravitational potential energy are

$$\Delta U_1 = -m_1 gh$$

$$\Delta U_2 = +m_2 gh$$

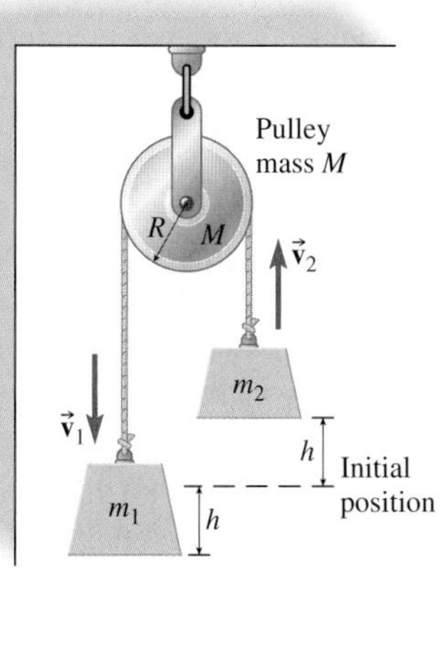

Figure 8.6
Atwood's machine.

The mechanical energy of the system includes the kinetic energies of three objects: the two masses and the pulley. All start with zero kinetic energy, so

$$\Delta K = \tfrac{1}{2}(m_1 + m_2)v^2 + \tfrac{1}{2}I\omega^2$$

The speed v of the masses is the same since the cord's length is fixed. The speed v and the angular speed of the pulley ω are related if the cord does not slip: the tangential speed of the pulley must equal the speed at which the cord moves. The tangential speed of the pulley is its angular speed times its radius:

$$v = \omega R$$

After substituting v/R for ω, the energy conservation equation becomes

$$\Delta U + \Delta K = [-m_1 gh + m_2 gh] + \left[\frac{1}{2}(m_1 + m_2)v^2 + \frac{1}{2}I\left(\frac{v}{R}\right)^2\right] = 0$$

or

$$\left[\frac{1}{2}(m_1 + m_2) + \frac{1}{2}\frac{I}{R^2}\right]v^2 = (m_1 - m_2)gh$$

Solving this equation for v yields

$$v = \sqrt{\frac{2(m_1 - m_2)gh}{m_1 + m_2 + I/R^2}}$$

Discussion This answer is rich in information, in the sense that we can ask many "What if?" questions. Not only do these questions provide checks as to whether the answer is reasonable, they also enable us to perform thought experiments, which could then be checked by constructing an Atwood's machine and comparing the results.

For instance: What if m_1 is only slightly greater than m_2? Then the final speed v is small—as m_2 approaches m_1, v approaches 0. This makes intuitive sense: a small imbalance in weights produces a small acceleration. You should practice this kind of reasoning by making other such checks.

It is also enlightening to look at terms in an algebraic solution and connect them with physical interpretations. The quantity $(m_1 - m_2)g$ is the imbalance in the gravitational

continued on next page

Example 8.2 continued

forces pulling on the two sides. The denominator $(m_1 + m_2 + I/R^2)$ is a measure of the total inertia of the system—the sum of the two masses plus an inertial contribution due to the pulley. The pulley's contribution is *not* simply equal to its mass. If, for example, the pulley is a uniform disk with $I = \frac{1}{2}MR^2$, the term I/R^2 would be equal to *half* the mass of the pulley.

The same principles used to analyze Atwood's machine have many applications in the real world. One such application is in elevators, where one of the hanging masses is the elevator and the other is the counterweight. Of course, the elevator and counterweight are not allowed to hang freely from a pulley—we must also consider the energy supplied by the motor.

Practice Problem 8.2 Modified Atwood's Machine

Figure 8.7 shows a modified form of Atwood's machine where one of the blocks slides on a table instead of hanging from the pulley. The blocks are released from rest. Find the speed of the blocks after they have moved a distance h in terms of m_1, m_2, I, R, and h. Ignore friction.

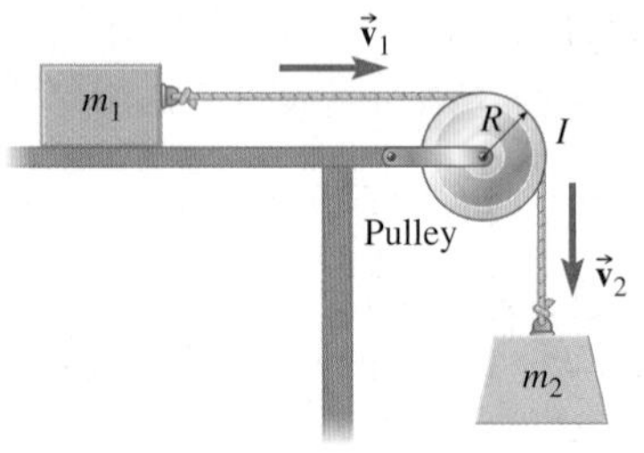

Figure 8.7 Modified Atwood's machine.

8.2 TORQUE

Suppose you place a bicycle upside down to repair it. First, you give one of the wheels a spin. If everything is working as it should, the wheel spins for quite a while; its angular acceleration is small. If the wheel doesn't spin for very long, then its angular velocity changes rapidly and the angular acceleration is large in magnitude; there must be excessive friction somewhere. Perhaps the brakes are rubbing on the rim or the bearings need to be repacked.

If we could eliminate *all* the frictional forces acting on the wheel, including air resistance, then we would expect the wheel to keep spinning without diminishing angular speed. In that case, its angular acceleration would be zero. The situation is reminiscent of Newton's first law: a body with no external interactions, or at least no net force acting on it, moves with constant velocity. We can state a "Newton's first law for rotation": a rotating body with no external interactions, and whose rotational inertia doesn't change, keeps rotating at constant angular velocity.

Of course, the hypothetical frictionless bicycle wheel does have external interactions. The Earth's gravitational field exerts a downward force and the axle exerts an upward force to keep the wheel from falling. Then is it true that, as long as there is no net external force, the angular acceleration is zero? No; it is easy to give the wheel an angular acceleration while keeping the net force zero. Imagine bringing the wheel to rest by pressing two hands against the tire on opposite sides. On one side, the motion of the rim of the tire is downward and the kinetic frictional force is upward (Fig. 8.8). On the other side, the tire moves upward and the frictional force is downward. In a similar way, we could apply equal and opposite forces to the opposite sides of a wheel at rest to make it start spinning. In either case, we exert equal magnitude forces, so that the net force is zero, and still give the wheel an angular acceleration.

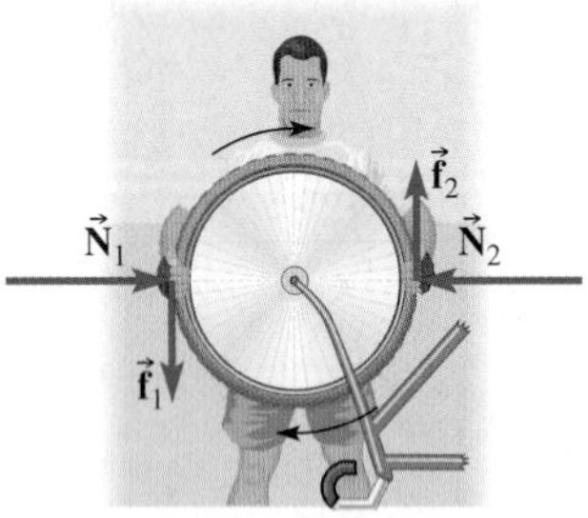

Figure 8.8 A spinning bicycle wheel slowed to a stop by friction. Each hand exerts a normal force and a frictional force on the tire. The two normal forces add to zero and the two frictional forces add to zero.

Torque A quantity related to force, called **torque**, plays the role in rotation that force itself plays in translation. A torque is not separate from a force; it is impossible to exert a torque without exerting a force. Torque is a measure of how effective a given force is at twisting or turning something. For something rotating about a fixed axis such as the bicycle wheel, a torque can *change* the rotational motion either by making it rotate faster or by slowing it down.

The *radial* direction is directly toward or away from the axis of rotation. The *perpendicular* or *tangential* direction is perpendicular to both the radial direction and the axis of rotation; it is tangent to the circular path followed by a point as the object rotates.

When stopping the bicycle wheel with two equal and opposite forces, as in Fig. 8.8, the net applied force is zero and, thus, the wheel is in translational equilibrium; but the net torque is not zero, so it is not in rotational equilibrium. Both forces tend to give the wheel the same sign of angular acceleration; they are both making the wheel slow down. The two torques are in fact equal, with the same sign.

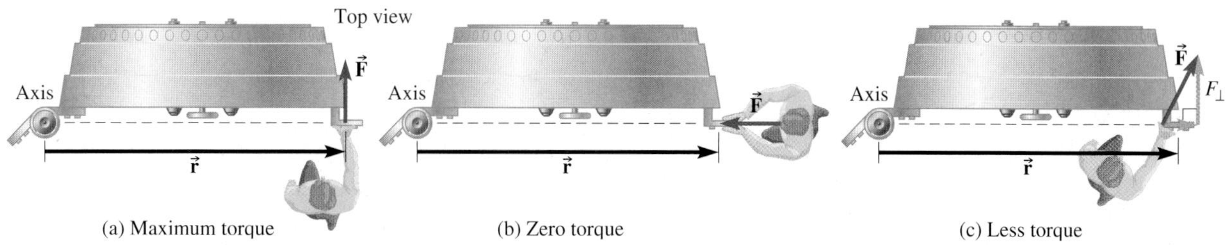

Figure 8.9 Torque on a bank vault door depends on the direction of the applied force. (a) Pushing perpendicularly gives the maximum torque. (b) Pushing radially inward with the same magnitude force gives zero torque. (c) The torque is proportional to the perpendicular component of the force ($F_\perp$).

Relationship Between Force and Torque What determines the torque produced by a particular force? Imagine trying to push open a massive bank vault door. Certainly you would push as hard as you can; the torque is proportional to the magnitude of the force. It also matters where and in what direction the force is applied. For maximum effectiveness, you would push perpendicularly to the door (Fig. 8.9a). If you pushed radially, straight in toward the axis of rotation that passes through the hinges, the door wouldn't rotate, no matter how hard you push (Fig. 8.9b). A force acting in any other direction could be decomposed into radial and perpendicular components, with the radial component contributing nothing to the torque (Fig. 8.9c). Only the perpendicular component of the force ($F_\perp$) produces a torque.

The symbol ⊥ stands for *perpendicular;* ∥ stands for *parallel.*

Furthermore, *where* you apply the force is critical (Fig. 8.10). Instinctively, you would push at the outer edge, as far from the rotation axis as possible. If you pushed close to the axis, it would be difficult to open the door. Torque is proportional to the distance between the rotation axis and the **point of application** of the force (the point at which the force is applied).

To satisfy the requirements of the previous paragraphs, we define the magnitude of the torque as the product of the distance between the rotation axis and the point of application of the force (r) with the perpendicular component of the force ($F_\perp$):

Definition of torque:

$$\tau = \pm r F_\perp \tag{8-3}$$

where r is the shortest distance between the rotation axis and the point of application of the force and $F_\perp$ is the perpendicular component of the force.

The symbol for torque is τ, the Greek letter tau. The SI unit of torque is the N·m. The SI unit of *energy*, the joule, is equivalent to N·m, but we do not write torque in joules. Even though both energy and torque can be written using the same SI base units, the two quantities have different meanings; torque is not a form of energy. To help maintain the distinction, the joule is used for energy but *not* for torque.

CHECKPOINT 8.2

You are trying to loosen a nut, without success. Why might it help to switch to a wrench with a longer handle?

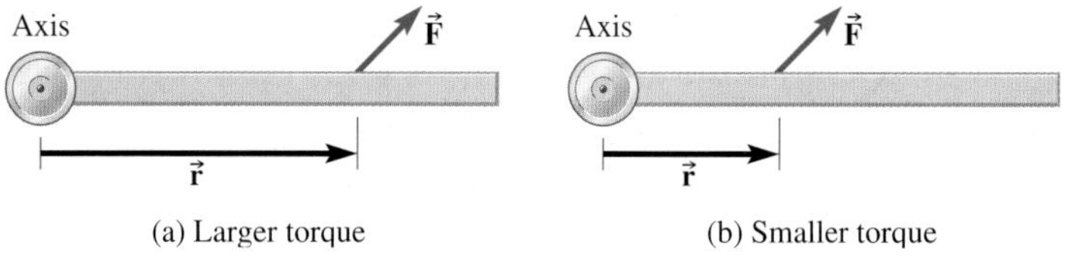

Figure 8.10 Torques; the same force at different distances from the axis.

Figure 8.11 (a) When the cyclist climbs a hill, the top half of the chain exerts a large force $\vec{\mathbf{F}}$ on the sprocket attached to the rear wheel. As viewed here, the torque about the axis of rotation (the axle) due to this force is clockwise. By convention, we call this a negative torque. (b) When the brakes are applied, the brake pads are pressed onto the rim, giving rise to frictional forces on the rim. As viewed here, the frictional force $\vec{\mathbf{f}}$ causes a counterclockwise (positive) torque on the wheel about the axle.

In a more general treatment of torque, torque is a vector quantity defined as the cross product $\vec{\boldsymbol{\tau}} = \vec{\mathbf{r}} \times \vec{\mathbf{F}}$. See Appendix A.8 for the definition of the cross product. For an object rotating about a fixed axis, Eq. (8-3) gives the component of $\vec{\boldsymbol{\tau}}$ along the axis of rotation.

Sign Convention for Torque The sign of the torque indicates the direction of the angular acceleration that torque would cause *by itself.* Recall from Section 5.1 that by convention a positive angular velocity ω means counterclockwise (CCW) rotation and a negative angular velocity ω means clockwise (CW) rotation. A positive angular acceleration α either increases the rate of CCW rotation (increases the magnitude of a positive ω) or decreases the rate of CW rotation (decreases the magnitude of a negative ω).

We use the same sign convention for torque. A force whose perpendicular component tends to cause rotation in the CCW direction gives rise to a positive torque; if it is the only torque acting, it would cause a positive angular acceleration α (Fig. 8.11). A force whose perpendicular component tends to cause rotation in the CW direction gives rise to a negative torque. The symbol ± in Eq. (8-3) reminds us to assign the appropriate algebraic sign each time we calculate a torque.

The sign of the torque is *not* determined by the sign of the angular velocity (in other words, whether the wheel is spinning CCW or CW); rather, it is determined by the sign of the angular *acceleration* the torque would cause if acting alone. To determine the sign of a torque, imagine which way the torque would make the object begin to spin if it is initially not rotating.

Example 8.3

A Spinning Bicycle Wheel

To stop a spinning bicycle wheel, suppose you push radially inward on opposite sides of the wheel, as shown in Fig. 8.8, with equal forces of magnitude 10.0 N. The radius of the wheel is 32 cm and the coefficient of kinetic friction between the tire and your hand is 0.75. The wheel is spinning in the CW sense. What is the net torque on the wheel?

continued on next page

Example 8.3 continued

Strategy The 10.0-N forces are directed radially toward the rotation axis, so they produce no torques themselves; only perpendicular components of forces give rise to torques. The forces of kinetic friction between the hands and the tire are tangent to the tire, so they do produce torques. The normal force applied to the tire is 10.0 N on each side; using the coefficient of friction, we can find the frictional forces.

Solution The frictional force exerted by each hand on the tire has magnitude

$$f = \mu_k N = 0.75 \times 10.0 \text{ N} = 7.5 \text{ N}$$

The frictional force is tangent to the wheel, so $f_\perp = f$. Then the magnitude of each torque is

$$|\tau| = rf_\perp = 0.32 \text{ m} \times 7.5 \text{ N} = 2.4 \text{ N·m}$$

The two torques have the same sign, since they are both tending to slow down the rotation of the wheel. Is the torque positive or negative? The angular velocity of the wheel is negative since it rotates CW. The angular acceleration has the opposite sign because the angular speed is decreasing. Since $\alpha > 0$, the net torque is also positive. Therefore,

$$\sum \tau = +4.8 \text{ N·m}$$

Discussion The trickiest part of calculating torques is determining the sign. To check, look at the frictional forces in Fig. 8.8. Imagine which way the forces would make the wheel begin to rotate if the wheel were not originally rotating. The frictional forces point in a direction that would tend to cause a CCW rotation, so the torques are positive.

Practice Problem 8.3 Disc Brakes

In the disc brakes that slow down a car, a pair of brake pads squeeze a spinning rotor; friction between the pads and the rotor provides the torque that slows down the car. If the normal force that each pad exerts on a rotor is 85 N and the coefficient of friction is 0.62, what is the frictional force on the rotor due to each of the pads? If this force acts 8.0 cm from the rotation axis, what is the magnitude of the torque on the rotor due to the pair of brake pads?

Lever Arms

There is another, completely equivalent, way to calculate torques that is often more convenient than finding the perpendicular component of the force. Figure 8.12 shows a force $\vec{\mathbf{F}}$ acting at a distance r from an axis. The distance r is the length of a line perpendicular to the axis that runs from the axis to the force's point of application. The force makes an angle θ with that line. The torque is then

$$\tau = \pm rF_\perp = \pm r(F \sin \theta)$$

The factor $\sin \theta$ could be grouped with r instead of with F. Then $\tau = \pm(r \sin \theta)F$, or

$$\tau = \pm r_\perp F \qquad (8\text{-}4)$$

The distance $r_\perp$ is called the **lever arm** (or **moment arm**). The magnitude of the torque is, therefore, the magnitude of the force times the lever arm.

Finding Torques Using the Lever Arm

1. Draw a line parallel to the force through the force's point of application; this line (dashed in Fig. 8.12) is called the force's **line of action**.
2. Draw a line from the rotation axis to the line of action. This line must be perpendicular to both the axis and the line of action. The distance from the axis to the line of action along this perpendicular line is the lever arm ($r_\perp$). If the line of action of the force goes through the rotation axis, the lever arm and the torque are both zero (see Fig. 8.9b).
3. The magnitude of the torque is the magnitude of the force times the lever arm:

$$\tau = \pm r_\perp F$$

4. Determine the algebraic sign of the torque as before.

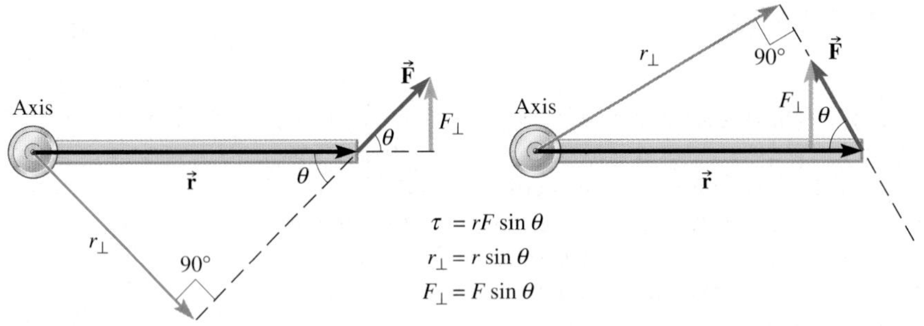

Figure 8.12 Finding the magnitude of a torque using the lever arm.

Example 8.4

Screen Door Closer

An automatic screen door closer attaches to a door 47 cm away from the hinges and pulls on the door with a force of 25 N, making an angle of 15° with the door (Fig. 8.13). Find the magnitude of the torque exerted on the door due to this force about the rotation axis through the hinges using (a) the perpendicular component of the force and (b) the lever arm. (c) What is the sign of this torque as viewed from above?

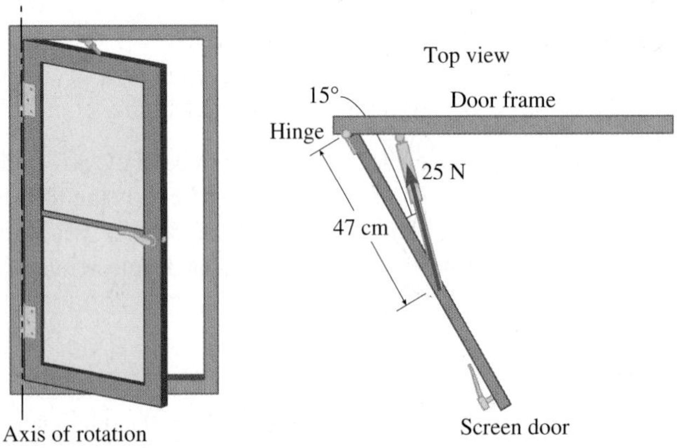

Figure 8.13
Screen door with automatic closing mechanism.

Strategy For method (a), we must find the component of the 25-N force perpendicular to the radial direction. Then this component is multiplied by the length of the radial line. For method (b), we draw in the line of action of the force. Then the lever arm is the perpendicular distance from the line of action to the rotation axis. The torque is the magnitude of the force times the lever arm. We must be careful not to combine the two methods: the torque is *not* equal to the perpendicular force component times the lever arm. For (c), we determine whether this torque would tend to make the door rotate CCW or CW.

Solution (a) As shown in Fig. 8.14a, the radial component of the force ($F_{\|}$) passes through the rotation axis. The angle labeled 15° would actually be a bit larger than 15°, but since the thickness of the door is much less than 47 cm, we approximate it as 15°. The perpendicular component is

$$F_\perp = F \sin 15°$$

The magnitude of the torque is

$$|\tau| = rF_\perp = 0.47 \text{ m} \times 25 \text{ N} \times \sin 15° = 3.0 \text{ N·m}$$

(b) Figure 8.14b shows the line of action of the force, drawn parallel to the force and passing through the point of application. The lever arm is the perpendicular distance between the rotation axis and the line of action. The distance r is approximately 47 cm (again neglecting the thickness of the door). Then the lever arm is

$$r_\perp = r \sin 15°$$

and the magnitude of the torque is

$$|\tau| = r_\perp F = 0.47 \text{ m} \times \sin 15° \times 25 \text{ N} = 3.0 \text{ N·m}$$

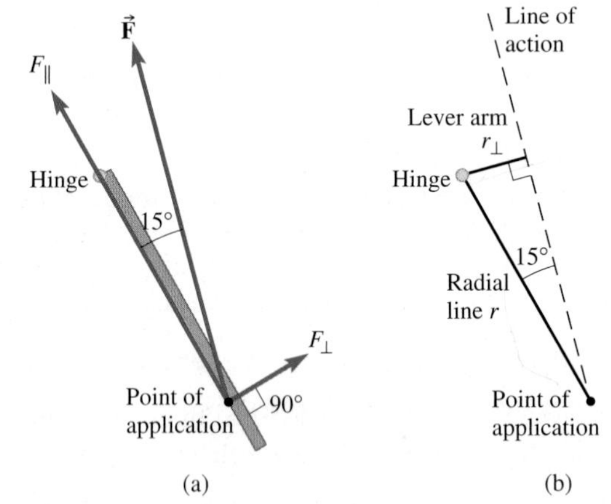

Figure 8.14
(a) Finding the perpendicular component of the force.
(b) Finding the lever arm.

(c) Using the top view of Fig. 8.13, the torque tends to close the door by making it rotate counterclockwise (assuming the door is initially at rest and no other torques act). The torque is therefore positive as viewed from above.

continued on next page

Example 8.4 continued

Discussion The most common mistake to make in either solution method would be to use cosine instead of sine (or, equivalently, to use the complementary angle 75° instead of 15°). A check is a good idea. If the automatic closer were more nearly parallel to the door, the angle would be less than 15°. The torque would be smaller because the force is more nearly pulling straight in toward the axis. Since the sine function gets smaller for angles closer to zero, the expression checks out correctly.

It might seem silly for a door closer to pull at such an angle that the perpendicular component is relatively small. The reason it's done that way is so the door closer does not get in the way. A closer that pulled in a perpendicular direction would stick straight out from the door. As discussed in Section 8.5, the situation is much the same in our bodies. In order to not inhibit the motion of our limbs, our tendons and muscles are nearly parallel to the bones. As a result, the forces they exert must be much larger than we might expect.

Practice Problem 8.4 Exercise Is Good for You

A person is lying on an exercise mat and lifts one leg at an angle of 30.0° from the horizontal with an 89-N (20-lb) weight attached to the ankle (Fig. 8.15). The distance between the ankle weight and the hip joint (which is the rotation axis for the leg) is 84 cm. What is the torque due to the ankle weight on the leg?

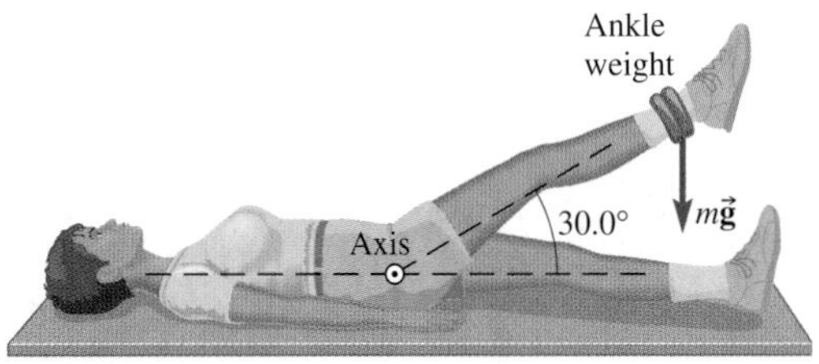

Figure 8.15
Exercise leg lifts.

Center of Gravity

We have seen that the torque produced by a force depends on the point of application of the force. What about gravity? The gravitational force on a body is not exerted at a single point, but is distributed throughout the volume of the body. When we talk of "the" force of gravity on something, we really mean the total force of gravity acting on each particle making up the system.

Fortunately, when we need to find the total torque due to the forces of gravity acting on an object, the total force of gravity can be considered to act at a single point. This point is called the **center of gravity**. The torque found this way is the same as finding all the torques due to the forces of gravity acting at every point in the body and adding them together. As you can verify in Problem 95, if the gravitational field is uniform in magnitude and direction, then the center of gravity of an object is located at the object's center of mass.

When calculating the torque due to gravity, consider the entire gravitational force to act at the center of gravity.

8.3 CALCULATING WORK DONE FROM THE TORQUE

Torques can do work, as anyone who has started a lawnmower with a pull cord can verify. Actually, it is the force that does the work, but in rotational problems it is often simpler to calculate the work done from the torque. Just as the work done by a constant force is the product of force and the parallel component of displacement, work done by a constant torque can also be calculated as the torque times the *angular* displacement.

CONNECTION:

We're not introducing a different kind of work, just a different way to calculate work.

Imagine a torque acting on a wheel that spins through an angular displacement $\Delta\theta$ while the torque is applied. The work done by the force that gives rise to the torque is the product of the perpendicular component of the force ($F_\perp$) with the arc length s through which the point of application of the force moves (Fig. 8.16). We use the perpendicular force component because that is the component parallel to the *displacement*, which is instantaneously tangent to the arc of the circle. Thus,

$$W = F_\perp s \quad (8\text{-}5)$$

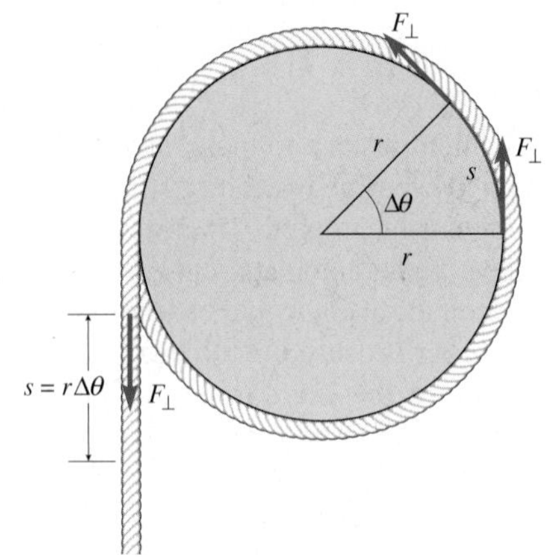

Figure 8.16 The work done by a torque is the product of the perpendicular force component $F_\perp$ and the arc length s.

To write the work in terms of torque, note that $\tau = rF_\perp$ and $s = r\Delta\theta$; then

$$W = F_\perp s = \frac{\tau}{r} \times r\Delta\theta = \tau\Delta\theta$$

$$W = \tau\Delta\theta \quad (\Delta\theta \text{ in radians}) \tag{8-6}$$

Work is indeed the product of torque and the angular displacement. If τ and $\Delta\theta$ have the same sign, the work done is positive; if they have opposite signs, the work done is negative. The *power* due to a constant torque—the rate at which work is done—is

$$P = \tau\omega \tag{8-7}$$

Example 8.5

Work Done on a Potter's Wheel

A potter's wheel is a heavy stone disk on which the pottery is shaped. Potter's wheels were once driven by the potter pushing on a foot treadle; today most potter's wheels are driven by electric motors. (a) If the potter's wheel is a uniform disk of mass 40.0 kg and diameter 0.50 m, how much work must be done by the motor to bring the wheel from rest to 80.0 rpm? (b) If the motor delivers a constant torque of 8.2 N·m during this time, through how many revolutions does the wheel turn in coming up to speed?

Strategy Work is an energy transfer. In this case, the motor is increasing the rotational kinetic energy of the potter's wheel. Thus, the work done by the motor is equal to the change in rotational kinetic energy of the wheel, ignoring frictional losses. In the expression for rotational kinetic energy, we must express ω in rad/s; we cannot substitute 80.0 rpm for ω. Once we know the work done, we use the torque to find the angular displacement.

Solution (a) The change in rotational kinetic energy of the wheel is

$$\Delta K = \tfrac{1}{2}I(\omega_f^2 - \omega_i^2) = \tfrac{1}{2}I\omega_f^2$$

Initially the wheel is at rest, so the initial angular velocity ω_i is zero. From Table 8.1, the rotational inertia of a uniform disk is

$$I = \tfrac{1}{2}MR^2$$

Substituting this for I,

$$\Delta K = \tfrac{1}{4}MR^2\omega_f^2$$

Before substituting numerical values, we convert 80.0 rpm to rad/s:

$$\omega_f = 80.0\,\frac{\text{rev}}{\text{min}} \times 2\pi\frac{\text{rad}}{\text{rev}} \times \frac{1}{60}\,\frac{\text{min}}{\text{s}} = 8.38\text{ rad/s}$$

Substituting the known values for mass and radius,

$$\Delta K = \tfrac{1}{4} \times 40.0\text{ kg} \times (\tfrac{0.50}{2}\text{ m})^2 \times (8.38\text{ rad/s})^2 = 43.9\text{ J}$$

Therefore, the work done by the motor, rounded to two significant figures, is 44 J.

(b) The work done by a constant torque is

$$W = \tau\Delta\theta$$

Solving for the angular displacement $\Delta\theta$ gives

$$\Delta\theta = \frac{W}{\tau} = \frac{43.9\text{ J}}{8.2\text{ N·m}} = 5.35\text{ rad}$$

Since 2π rad = 1 revolution,

$$\Delta\theta = 5.35\text{ rad} \times \frac{1\text{ rev}}{2\pi\text{ rad}} = 0.85\text{ rev}$$

continued on next page

Example 8.5 continued

Discussion As always, work is an energy transfer. In this problem, the work done by the motor is the means by which the potter's wheel acquires its rotational kinetic energy. But work done by a torque does not *always* appear as a change in rotational kinetic energy. For instance, when you wind up a mechanical clock or a windup toy, the work done by the torque you apply is stored as elastic potential energy in some sort of spring.

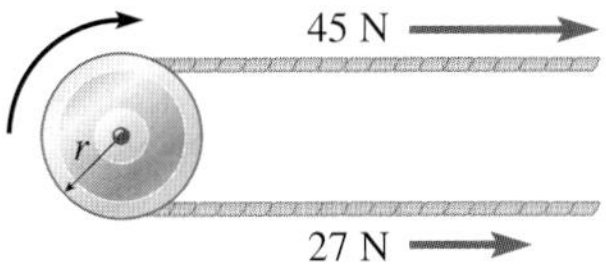

Figure 8.17
Air conditioner belt and pulley.

Practice Problem 8.5 Work Done on an Air Conditioner

A belt wraps around a pulley of radius 7.3 cm that drives the compressor of an automobile air conditioner. The tension in the belt on one side of the pulley is 45 N and on the other side of the pulley it is 27 N (Fig. 8.17). How much work is done by the belt on the compressor during one revolution of the pulley?

8.4 ROTATIONAL EQUILIBRIUM

In Chapter 4, we said that an object is in translational equilibrium when the net force acting on it is zero. It is quite possible for the net force acting to be zero, while the net torque is nonzero; the object would then have a nonzero angular acceleration. When designing a bridge or a new house, it would be unacceptable for any of the parts to have nonzero angular acceleration! Zero net force is sufficient to ensure *translational* equilibrium; if an object is also in *rotational* equilibrium, then the net torque acting on it must also be zero.

Conditions for equilibrium (both translational and rotational):

$$\sum \vec{\mathbf{F}} = 0 \quad \text{and} \quad \sum \tau = 0 \qquad (8\text{-}8)$$

Choosing an Axis of Rotation in Equilibrium Problems Before tackling equilibrium problems, we must resolve a conundrum: if something is not rotating, then where is the axis of rotation? How can we calculate torques without knowing where the axis of rotation is? In some cases, perhaps involving axles or hinges, there may be a clear axis about which the object would rotate if the balance of forces and torques is disturbed. In many cases, though, it is not clear what the rotation axis would be, and in general it depends on exactly how the equilibrium is upset. Fortunately, the axis can be chosen *arbitrarily* when calculating torques *in equilibrium problems.*

In equilibrium, the net torque about *any* rotation axis must be zero. Does that mean that we have to write down an infinite number of torque equations, one for each possible axis of rotation? Fortunately, no. Although the proof is complicated, it can be shown that if the net force acting on an object is zero and the net torque about one rotation axis is zero, then the net torque about every other axis parallel to that axis must also be zero. Therefore, one torque equation is all we need.

Since the torque can be calculated about any desired axis, a judicious choice can greatly simplify the solution of the problem. The best place to choose the axis is usually at the point of application of an unknown force so that the unknown force does not appear in the torque equation.

CHECKPOINT 8.4

Is it possible for the net torque on an object to be zero and the net force nonzero? Is it possible for the net force to be zero and the net torque nonzero?

Problem-Solving Steps in Equilibrium Problems

- Identify an object or system in equilibrium. Draw a diagram showing all the forces acting on that object, each drawn at its point of application. Use the center of gravity as the point of application of any gravitational forces.
- To apply the force condition $\Sigma\vec{\mathbf{F}} = 0$, choose a convenient coordinate system and resolve each force into its x- and y-components.
- To apply the torque condition $\Sigma\tau = 0$, choose a convenient rotation axis—generally one that passes through the point of application of an unknown force. Then find the torque due to each force. Use whichever method is easier: either the lever arm times the magnitude of the force or the distance times the perpendicular component of the force. Determine the direction of each torque; then either set the sum of all the torques (with their algebraic signs) equal to zero or set the magnitude of the CW torques equal to the magnitude of the CCW torques.
- Not all problems require all three equations (two force component equations and one torque equation). Sometimes it is easier to use more than one torque equation, with a different axis. Before diving in and writing down all the equations, think about which approach is the easiest and most direct.

Example 8.6

Carrying a 6 × 6 Beam

Two carpenters are carrying a uniform 6×6 beam. The beam is 8.00 ft (2.44 m) long and weighs 425 N (95.5 lb). One of the carpenters, being a bit stronger than the other, agrees to carry the beam 1.00 m in from the end; the other carries the beam at its opposite end. What is the upward force exerted on the beam by each carpenter?

Strategy The conditions for equilibrium are that the net external force equal zero and the net external torque equal zero:

$$\Sigma\vec{\mathbf{F}} = 0 \quad \text{and} \quad \Sigma\tau = 0$$

Should we start with forces or with torques? In this problem, it is easiest to start with torques. If we choose the axis of rotation where one of the unknown forces acts, then that force has a lever arm of zero and its torque is zero. The torque equation can be solved for the other unknown force. Then with only one force still unknown, we set the sum of the y-components of the forces equal to zero.

Solution The first step is to draw a force diagram (Fig. 8.18). Each force is drawn at the point where it acts. Known distances are labeled.

We choose a rotation axis perpendicular to the xy-plane and passing through the point of application of $\vec{\mathbf{F}}_2$. The simplest way to find the torques for this example is to multiply each force by its lever arm. The lever arm for $\vec{\mathbf{F}}_1$ is

$$2.44\ \text{m} - 1.00\ \text{m} = 1.44\ \text{m}$$

and the magnitude of the torque due to this force is

$$|\tau| = Fr_\perp = F_1 \times 1.44\ \text{m}$$

Since the beam is uniform, its center of gravity is at its midpoint. We imagine the entire gravitational force to act at this point. Then the lever arm for the gravitational force is

$$\tfrac{1}{2} \times 2.44\ \text{m} = 1.22\ \text{m}$$

and the torque due to gravity has magnitude

$$|\tau| = Fr_\perp = 425\ \text{N} \times 1.22\ \text{m} = 518.5\ \text{N·m}$$

The torque due to $\vec{\mathbf{F}}_1$ is negative since, if it were the only torque, it would make the beam start to rotate clockwise about our chosen axis of rotation. The torque due to gravity is positive since, if it were the only torque, it would make the beam start to rotate counterclockwise. Therefore,

$$\Sigma\tau = -F_1 \times 1.44\ \text{m} + 518.5\ \text{N·m} = 0$$

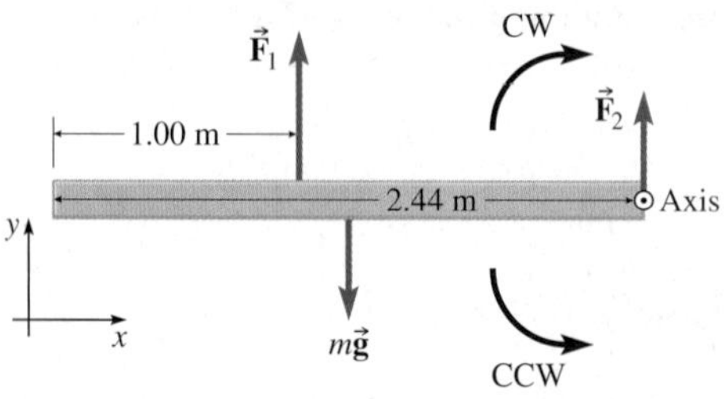

Figure 8.18 Diagram of the beam with rotation axis, forces, and distances shown.

continued on next page

Example 8.6 continued

Solving for F_1,

$$F_1 = \frac{518.5\ \text{N·m}}{1.44\ \text{m}} = 360\ \text{N}$$

Since another condition for equilibrium is that the net force be zero,

$$\Sigma F_y = F_1 + F_2 - mg = 0$$

Solving for F_2,

$$F_2 = 425\ \text{N} - 360\ \text{N} = 65\ \text{N}$$

Discussion A good way to check this result is to make sure that the net torque about a *different axis* is zero—for an object in equilibrium, the net torque about any axis must be zero. Suppose we choose an axis through the point of application of $\vec{\mathbf{F}}_1$. Then the lever arm for $m\vec{\mathbf{g}}$ is 1.22 m − 1.00 m = 0.22 m and the lever arm for $\vec{\mathbf{F}}_2$ is 2.44 m − 1.00 m = 1.44 m. Setting the net torque equal to zero:

$$\Sigma \tau = -425\ \text{N} \times 0.22\ \text{m} + F_2 \times 1.44\ \text{m} = 0$$

Solving for F_2 gives

$$F_2 = \frac{425\ \text{N} \times 0.22\ \text{m}}{1.44\ \text{m}} = 65\ \text{N}$$

which agrees with the value calculated before. We could have used this second torque equation to find F_2 instead of setting ΣF_y equal to zero.

Practice Problem 8.6 A Diving Board

A uniform diving board of length 5.0 m is supported at two points; one support is located 3.4 m from the end of the board and the second is at 4.6 m from the end (Fig. 8.19). The supports exert vertical forces on the diving board. A diver stands at the end of the board over the water. Determine the directions of the support forces. (tutorial: plank) [*Hint:* In this problem, consider torques about different rotation axes.]

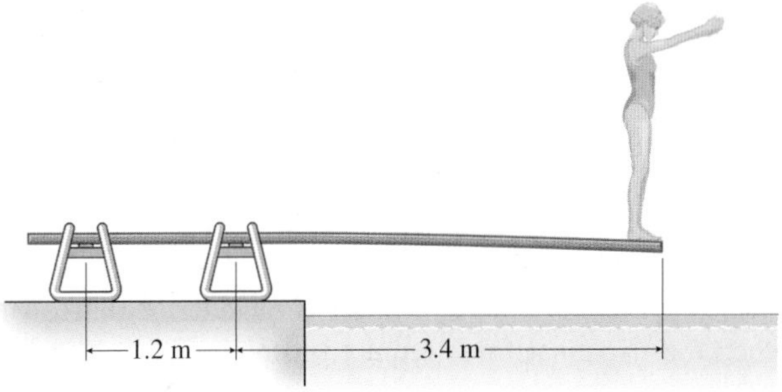

Figure 8.19
Diving board.

The Cantilever A diving board is an example of a cantilever—a beam or pole that extends beyond its support. The forces exerted by the supports on a diving board are considerably larger than if the same board were supported at both ends (see Problem 32). The advantage is that the far end of the board is left free to vibrate; as it does, the support forces adjust themselves to keep the board from tipping over. The architect Frank Lloyd Wright was fond of using cantilever construction to open up the sides and corners of a building, allowing corner windows that give buildings a lighter and more spacious feel (Fig. 8.20).

Application of Rotational Equilibrium: cantilever building construction

Figure 8.20 The cantilevered master bedroom in the north wing of Wingspread by Frank Lloyd Wright juts well out over its brick foundation. The cypress trellis extending even farther beyond the bedroom balcony filters the natural light and serves to emphasize the free-floating nature of the structure with views of the landscape below.

Example 8.7

The Slipping Ladder

A 15.0-kg uniform ladder leans against a wall in the atrium of a large hotel (Fig. 8.21a). The ladder is 8.00 m long; it makes an angle $\theta = 60.0°$ with the floor. The coefficient of static friction between the floor and the ladder is $\mu_s = 0.45$. How far along the ladder can a 60.0-kg person climb before the ladder starts to slip? Assume that the wall is frictionless. (interactive: ladder and tutorial: ladder)

Strategy Consider the ladder and the climber as a single system. Until the ladder starts to slip, this system is in equilibrium. Therefore, the net external force and the net external torque acting on the system are both equal to zero. Normal forces act on the ladder due to the wall ($\vec{\mathbf{N}}_w$) and the floor ($\vec{\mathbf{N}}_f$). A frictional force acts on the base of the ladder due to the floor ($\vec{\mathbf{f}}$), but no frictional force acts on the top of the ladder since the wall is frictionless. Gravitational forces act on the ladder and on the person climbing it. As the person ascends the ladder, the frictional force $\vec{\mathbf{f}}$ has to increase to keep the ladder in equilibrium. The ladder begins to slip when the frictional force required to maintain equilibrium is larger than its maximum possible value $\mu_s N_f$. The ladder is about to slip when $f = \mu_s N_f$.

$$\sum F_x = 0, \quad \sum F_y = 0, \quad \text{and} \sum \tau = 0$$

Solution The first step is to make a careful drawing of the ladder and label all distances and forces (Fig. 8.21b). Instead of cluttering the diagram with numerical values, we use L (= 8.00 m) for the length of the ladder, d for the unknown distance from the bottom of the ladder to the point where the person stands, and M (= 60.0 kg) and m (= 15.0 kg) for the masses of the person and ladder, respectively. The weight of the ladder acts at the ladder's center of gravity, which is the ladder's midpoint since it is uniform.

Now we apply the conditions for equilibrium. Starting with $\Sigma F_x = 0$, we find

$$N_w - f = 0$$

where, if the climber is at the highest point possible, the frictional force must have its maximum possible magnitude:

$$f = \mu_s N_f$$

Combining these two equations, we obtain a relationship between the magnitudes of the two normal forces:

$$N_w = \mu_s N_f$$

Next we use the condition $\Sigma F_y = 0$, which gives

$$N_f - Mg - mg = 0$$

The only unknown quantity in this equation is N_f, so we can solve for it:

$$N_f = Mg + mg = (M + m)g$$

Now we can find the other normal force, N_w:

$$N_w = \mu_s N_f = \mu_s (M + m)g$$

Figure 8.21
(a) A ladder and (b) forces acting on the ladder.

continued on next page

Example 8.7 continued

At this point, we have expressions for the magnitudes of all the forces. We do not know the distance d, which is the goal of the problem. To find d we must set the net torque equal to zero.

First we choose a rotation axis. The most convenient choice is an axis perpendicular to the plane of Fig. 8.21 and passing through the bottom of the ladder. Since two of the five forces ($\vec{\mathbf{N}}_f$ and $\vec{\mathbf{f}}$) act at the bottom of the ladder, these two forces have zero lever arms and, thus, produce zero torque. Another reason why this is a convenient choice of axis is that the distance d is measured from the bottom of the ladder.

In this situation, with the forces either vertical or horizontal, it is probably easiest to use lever arms to find the torques. In three diagrams (Fig. 8.22), we first draw the line of action for each force; then the lever arm is the perpendicular distance between the axis and the line of action.

Using the usual convention that CCW torques are positive, the torque due to $\vec{\mathbf{N}}_w$ is negative and the torques due to gravity are positive. The magnitude of each torque is the magnitude of the force times its lever arm:

$$\tau = Fr_\perp$$

Setting the net torque equal to zero yields

$$-N_w L \sin\theta + mg\left(\tfrac{1}{2}L\cos\theta\right) + Mgd\cos\theta = 0$$

Now we solve for d algebraically.

$$\frac{-N_w L\sin\theta}{Mg\cos\theta} + \frac{\frac{1}{2}mgL\cos\theta}{Mg\cos\theta} + d = 0$$

$$d = \frac{N_w L\tan\theta}{Mg} - \frac{mL}{2M}$$

Substituting for N_w, we have

$$d = L\left(\frac{\mu_s(M+m)\tan\theta}{M} - \frac{m}{2M}\right)$$

$$= 8.00\text{ m}\times\left(\frac{0.45\times 75.0\text{ kg}\times\tan 60.0°}{60.0\text{ kg}} - \frac{15.0\text{ kg}}{2\times 60.0\text{ kg}}\right)$$

$$= 6.8\text{ m}$$

The person can climb 6.8 m up the ladder without having it slip. This is the distance *along the ladder*, not the height above the ground, which is

$$h = 6.8\text{ m}\times\sin 60.0° = 5.9\text{ m}$$

Discussion If the person goes any higher, then his weight produces a larger CCW torque about our chosen rotation axis. To stay in equilibrium, the total CW torque would have to get larger. The only force providing a CW torque is the normal force due to the wall, which pushes to the right. However, if this force were to get larger, the frictional force would have to get larger to keep the net horizontal force equal to zero. Since friction already has its maximum magnitude, there is no way for the ladder to be in equilibrium if the person climbs any higher.

Practice Problem 8.7 Another Ladder Leaning on a Wall

A uniform ladder of mass 10.0 kg and length 3.2 m leans against a frictionless wall with its base located 1.5 m from the wall. If the ladder is not to slip, what must be the minimum coefficient of static friction between the bottom of the ladder and the ground? Assume the wall is frictionless.

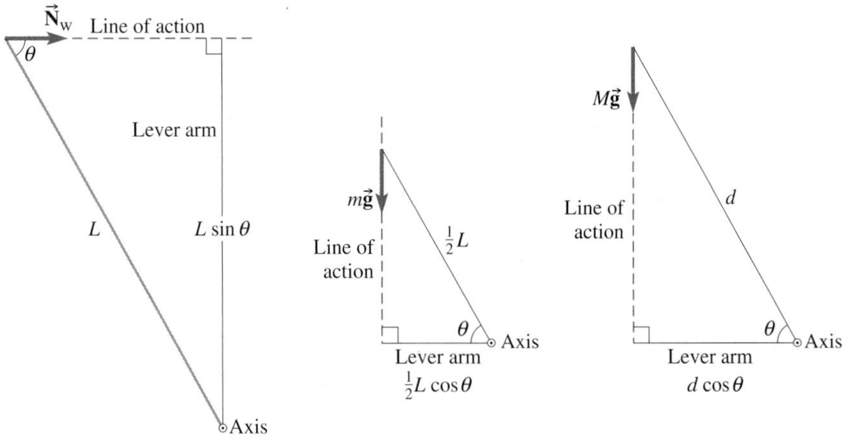

Figure 8.22
Finding the lever arm for each force.

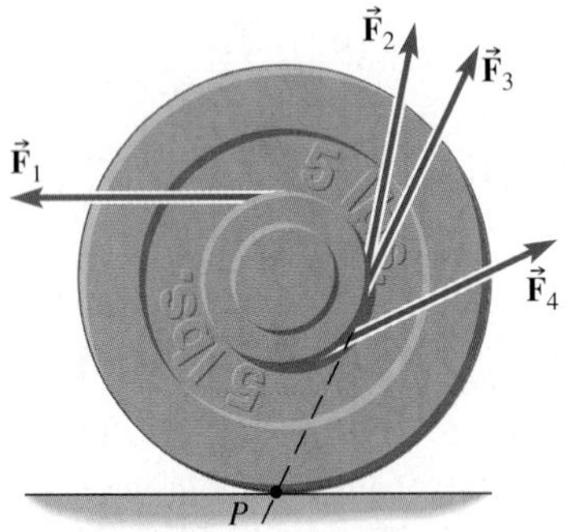

Figure 8.23 Forces $\vec{\mathbf{F}}_1$ and $\vec{\mathbf{F}}_2$ make the dumbbell roll to the left; $\vec{\mathbf{F}}_4$ makes it roll to the right; $\vec{\mathbf{F}}_3$ does not make it roll.

PHYSICS AT HOME

Take a dumbbell and wrap some string around the center of its axle. (An alternative: slide two spools of thread onto a pencil near its center with a small gap between the spools. Wrap some thread around the pencil between the two spools.) Place the dumbbell on a table (or on the floor). Unwind a short length of string and try pulling perpendicularly to the axle at different angles to the horizontal (Fig. 8.23). Depending on the direction of your pull, the dumbbell can roll in either direction. Try to find the angle at which the rolling changes direction; at this angle the dumbbell does not roll at all. (If using the pencil and spools of thread, pull gently and try to find the angle at which the whole thing *slides* along the table without any rotation.)

What is special about this angle? Since the dumbbell is in equilibrium when pulling at this angle, we can analyze the torques using any rotation axis we choose. A convenient choice is the axis that passes through point *P*, the point of contact with the table. Then the contact force between the table and the dumbbell acts at the rotation axis and its torque is zero. The torque due to gravity is also zero, since the line of action passes through point *P*. The dumbbell can only be in equilibrium if the torque due to the remaining force (the tension in the string) is zero. This torque is zero if the lever arm is zero, which means the line of action passes through point *P*.

Example 8.8

The Sign and the Breaking Cord

A uniform beam of weight 196 N and of length 1.00 m is attached to a hinge on the outside wall of a restaurant. A cord is attached at the center of the beam and is attached to the wall, making an angle of 30.0° with the beam (Fig. 8.24a). The cord keeps the beam perpendicular to the wall. If the breaking tension of the cord is 620 N, how large can the mass of the sign be without breaking the cord?

Strategy The beam is in equilibrium; both the net force and the net torque acting on it must be zero. To find the maximum weight of the sign, we let the tension in the cord have its maximum value of 620 N. We do not know the force exerted by the hinge on the beam, so we choose an axis of rotation through the hinge. Then the force exerted by the hinge on the beam has a zero lever arm and does not enter the torque equation.

Before doing anything else, we draw a diagram showing each force acting on the beam and the chosen rotation axis. The FBD in previous chapters often placed all the force vectors starting from a single point. Now we draw each force vector starting at its point of application so that we can find the torque—either by finding the lever arm or by finding the perpendicular force component and the distance from the axis to the point of application.

Solution Figure 8.24b shows the forces acting on the beam; three of these contribute to the torque. The gravitational

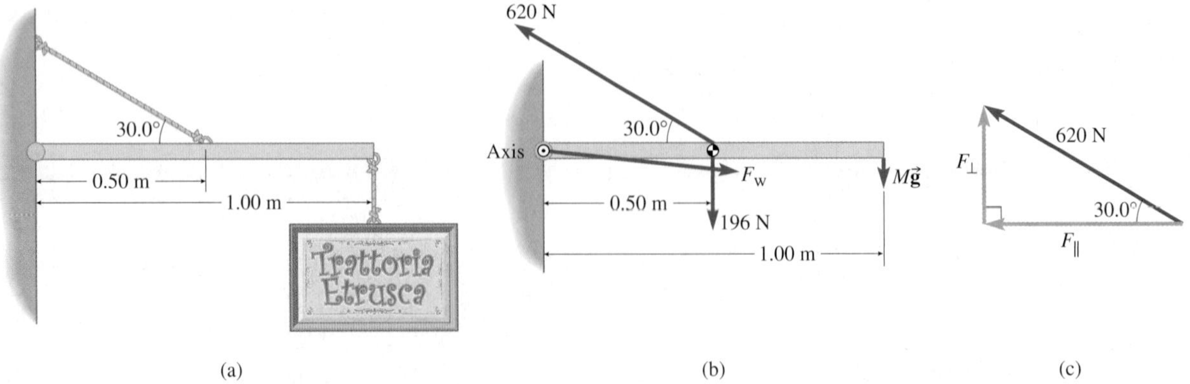

Figure 8.24

(a) A sign outside a restaurant. (b) Forces acting on the beam. (c) Finding the components of the tension in the cord.

continued on next page

Example 8.8 continued

force on the beam can be taken to act at the midpoint of the beam since it is uniform. The force due to the cord has a perpendicular component (Fig. 8.24c) of

$$F_\perp = 620 \text{ N} \times \sin 30.0° = 310 \text{ N}$$

The two gravitational forces tend to rotate the beam CW, while the tension in the cord tends to rotate it CCW. The net torque must be equal to zero:

$$-0.50 \text{ m} \times 196 \text{ N} - 1.00 \text{ m} \times Mg + 0.50 \text{ m} \times 310 \text{ N} = 0$$

or

$$1.00 \text{ m} \times Mg = 0.50 \text{ m} \times (310 \text{ N} - 196 \text{ N})$$

Now we solve for the unknown mass M:

$$M = \frac{0.50 \text{ m} \times (310 \text{ N} - 196 \text{ N})}{1.00 \text{ m} \times 9.80 \text{ N/kg}} = 5.8 \text{ kg}$$

Discussion In this problem, we did not have to set the net force equal to zero. By placing the axis of rotation at the hinge, we eliminated two of the three unknowns from the torque equation: the horizontal and vertical components of the hinge force (or, equivalently, its magnitude and direction). If we wanted to find the hinge force as well, setting the net force equal to zero would be necessary.

Practice Problem 8.8 Hinge Forces

Find the vertical component of the force exerted by the hinge in two different ways: (a) setting the net force equal to zero and (b) using a torque equation about a different axis.

Distributed Forces

Gravity is not the only force that is distributed rather than acting at a point. Contact forces, including both the normal component and friction, are spread over the contact surface. Just as for gravity, we can consider the contact force to act at a single point, but the location of that point is often not at all obvious. For a book sitting on a horizontal table, it seems reasonable that the normal force effectively acts at the geometric center of the book cover that touches the table. It is less clear where that effective point is if the book is on an incline or is sliding. As Example 8.9 shows, when something is about to topple over, contact is about to be lost everywhere except at the corner around which the toppling object is about to rotate. That corner then must be the location of the contact forces.

Example 8.9

The Toppling File Cabinet

A file cabinet of height a and width b is on a ramp at angle θ (Fig. 8.25a). The file cabinet is filled with papers in such a way that its center of gravity is at its geometric center. Find the largest θ for which the file cabinet does not tip over. Assume the coefficient of static friction is large enough to prevent sliding. (tutorial: file cabinet)

Strategy Until the file cabinet begins to tip over, it is in equilibrium; the net force acting on it must be zero and the total torque about any axis must also be zero. We first draw a force diagram showing the three forces (gravity, normal, friction) acting on the file cabinet. The point of application of the two contact forces (normal, friction) must be at the lower edge of the file cabinet if it is on the steepest possible incline, just about to tip over. In that case, contact has been lost over the rest of the bottom surface of the file cabinet so that only the lower edge makes good contact with the ramp.

As in all equilibrium problems, a good choice of rotation axis makes the problem easier to solve. We know that, at the maximum angle, the contact forces act at the bottom edge of the file cabinet. A good choice of rotation axis is along the bottom edge of the file cabinet, because then the normal and frictional forces have zero lever arm.

Solution Figure 8.25b shows the forces acting on the file cabinet at the maximum angle θ. The gravitational force is drawn at the center of gravity. Instead of drawing a single

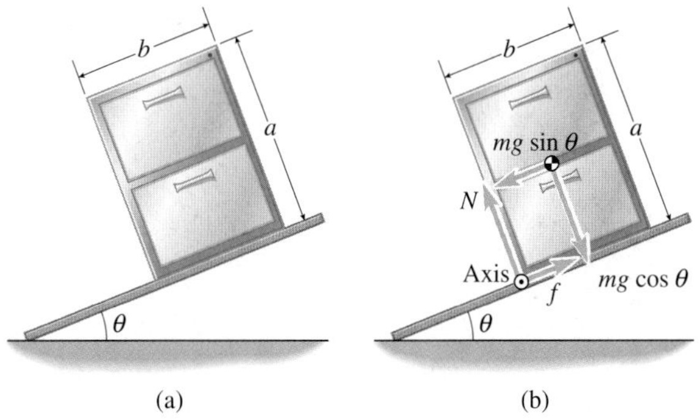

Figure 8.25
(a) File cabinet on an incline. (b) Forces acting on the file cabinet.

continued on next page

Example 8.9 continued

vector arrow for the gravitational force, we represent the gravitational force by its components parallel and perpendicular to the ramp. Then we find the lever arm for each of the components. The lever arm for the parallel component of the weight ($mg \sin \theta$) is $\frac{1}{2}a$ and the lever arm for the perpendicular component ($mg \cos \theta$) is $\frac{1}{2}b$. Setting the net torque equal to zero:

$$\sum \tau = -mg \cos \theta \times \tfrac{1}{2}b + mg \sin \theta \times \tfrac{1}{2}a = 0$$

After dividing out the common factors of $\frac{1}{2}mg$,

$$\cos \theta \times b = \sin \theta \times a$$

Solving for θ,

$$\theta = \tan^{-1} \frac{b}{a}$$

Discussion As a check, we can regard the normal and friction forces as two components of a single contact force. We can think of that contact force as acting at a single point—a "center of contact" analogous to the center of gravity. As the file cabinet is put on steeper and steeper surfaces, the effective point of application of the contact force moves toward the lower edge of the file cabinet (Fig. 8.26). If we take the rotation axis through the center of gravity so there is no gravitational torque, then the torque due to the contact force must be zero. The only way that can happen is if its lever arm is zero, which means that the contact force must point directly toward the center of gravity. If the angle θ has its maximum value, the contact force acts at the lower edge and $\tan \theta = b/a$. The file cabinet is about to tip when its center of gravity is directly above the lower edge. Any object supported only by contact forces can be in equilibrium only if the point of application of the total contact force is directly below the object's center of gravity.

Figure 8.27
Yuri Chechi of Italy holds the pike position on the rings at the World Gymnastic Championships in Sabae, Japan.

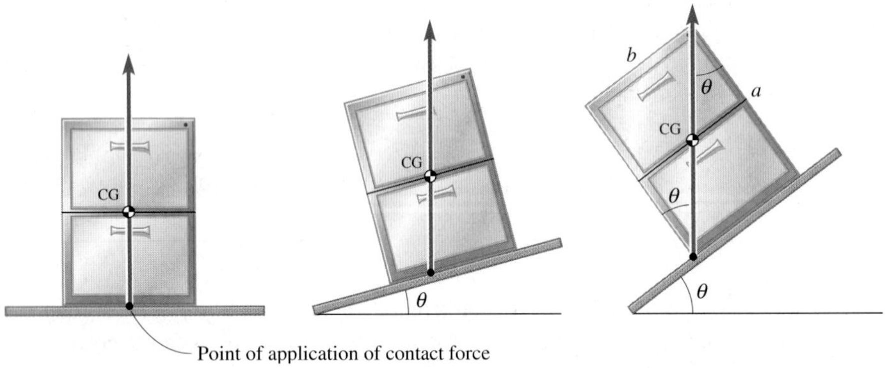

Conceptual Practice Problem 8.9 Gymnast Holding a Pike Position

Figure 8.27 shows a gymnast holding a pike position. What can you say about the location of the gymnast's center of gravity?

Figure 8.26
Contact force for various incline angles.

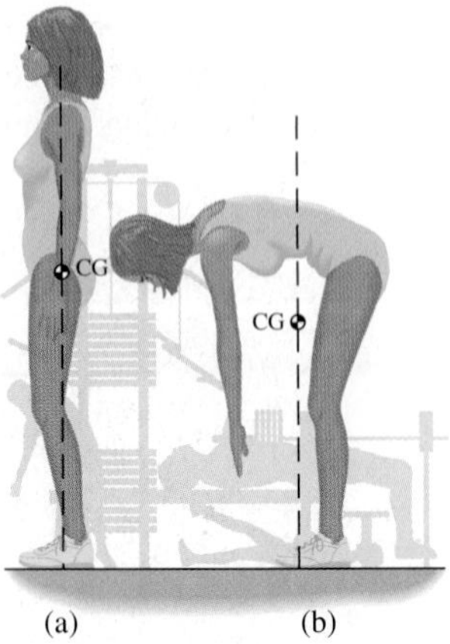

Figure 8.28 Location of the center of gravity when (a) standing and (b) reaching for the floor.

PHYSICS AT HOME

When a person stands up straight, the body's center of gravity lies directly above a point between the feet, about 3 cm in front of the ankle joint (see Fig. 8.28a). When a person bends over to touch her toes, the center of gravity lies outside the body (Fig. 8.28b). Note that the lower half of the body must move backward to keep the center of gravity from moving out in front of the toes, which would cause the person to fall over.

An interesting experiment can be done that illustrates what happens to your balance when you shift your center of gravity. Stand against a wall with the heels of your feet touching the wall and your back pressed against the wall. Then carefully try to bend over as if to touch your toes, without bending your knees. Can you do this without falling over? Explain.

8.5 EQUILIBRIUM IN THE HUMAN BODY

Application of Conditions for Equilibrium to the Human Body

We can use the concepts of torque and equilibrium to understand some of how the musculoskeletal system of the human body works. A muscle has tendons at each end that connect it to two different bones across a joint (the flexible connection between the bones). When the muscle contracts, it pulls the tendons, which in turn pull on the bones. Thus, the muscle produces a pair of forces of equal magnitude, one acting on each of the two bones. The biceps muscle (Fig. 8.29) in the upper arm attaches the scapula to the forearm (radius) across the inside of the elbow joint. When the biceps contracts, the forearm is pulled toward the upper arm. The biceps is a *flexor* muscle; it moves one bone closer to another.

A muscle can pull but not push, so a flexor muscle such as the biceps cannot reverse its action to push the forearm away from the upper arm. The *extensor* muscles make bones move apart from each other. In the upper arm (Fig. 8.29), an extensor muscle—the triceps—connects the scapula and humerus to the ulna (a bone in the forearm parallel to the radius) across the outside of the elbow. Since the biceps and triceps connect to the forearm on opposite sides of the elbow joint, they tend to cause rotation about the joint in opposite directions. When the triceps contracts it pulls the forearm away from the upper arm. Using flexor and extensor muscles on opposite sides of the joint, the body can produce both positive and negative torques, although both muscles pull in the same direction.

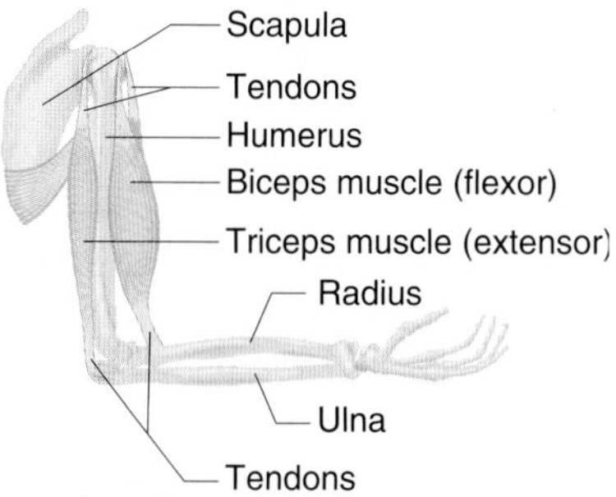

Figure 8.29 The biceps is a flexor muscle; the triceps is an extensor muscle.

Suppose the arm is held in a horizontal position. The deltoid muscle (the muscle shown in Fig. 8.30) exerts a force $\vec{\mathbf{F}}_m$ on the humerus at an angle of about 15° above the horizontal. This force has to do two things. The vertical component (magnitude $F_m \sin 15° \approx 0.26F_m$) supports the weight of the arm, while the horizontal component (magnitude $F_m \cos 15° \approx 0.97F_m$) stabilizes the joint by pulling the humerus in against the shoulder (scapula). In Example 8.10, we estimate the magnitude of $\vec{\mathbf{F}}_m$.

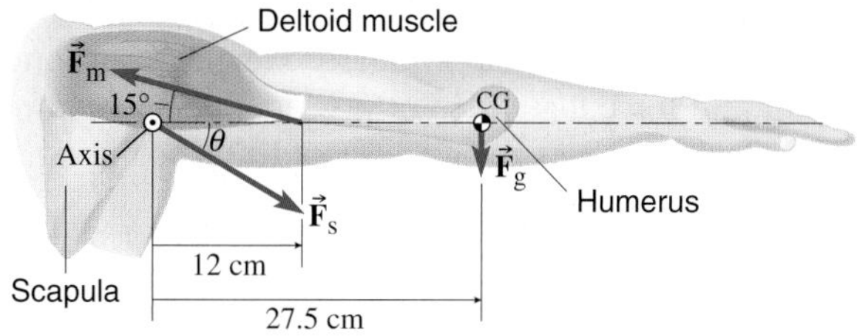

Figure 8.30 Forces exerted on an outstretched arm by the deltoid muscle ($\vec{\mathbf{F}}_m$), the scapula ($\vec{\mathbf{F}}_s$), and gravity ($\vec{\mathbf{F}}_g$).

Example 8.10

Force to Hold Arm Horizontal

A person is standing with his arm outstretched in a horizontal position. The weight of the arm is 30.0 N and its center of gravity is at the elbow joint, 27.5 cm from the shoulder joint (Fig. 8.30). The deltoid pulls on the upper arm at an angle of 15° above the horizontal and at a distance of 12 cm from the joint. What is the magnitude of the force exerted by the deltoid muscle on the arm?

Strategy The arm is in equilibrium, so we can apply the conditions for equilibrium: $\Sigma\vec{\mathbf{F}} = 0$ and $\Sigma\tau = 0$. When calculating torques, we choose the rotation axis at the shoulder joint because then the unknown force $\vec{\mathbf{F}}_s$, which acts on the arm at the joint, has a zero lever arm and produces zero torque. With only one unknown in the torque equation, we can solve immediately for F_m. We do not need to apply the condition $\Sigma\vec{\mathbf{F}} = 0$ unless we want to find $\vec{\mathbf{F}}_s$.

Solution The gravitational force is perpendicular to the line between its point of application and the rotation axis. Gravity produces a CW torque of magnitude

$$|\tau| = Fr = 30.0 \text{ N} \times 0.275 \text{ m} = 8.25 \text{ N·m}$$

For the torque due to $\vec{\mathbf{F}}_m$, we find the component of $\vec{\mathbf{F}}_m$ that is perpendicular to the line between its point of application and the rotation axis. Since this line is horizontal, we need the vertical component of $\vec{\mathbf{F}}_m$, which is $F_m \sin 15°$. Then the magnitude of the CCW torque due to $\vec{\mathbf{F}}_m$ is

$$|\tau| = F_\perp r = F_m \sin 15° \times 0.12 \text{ m}$$

The sum of these torques is zero. With the usual sign convention. (CCW is +),

$$F_m \sin 15° \times 0.12 \text{ m} - 8.25 \text{ N·m} = 0$$

continued on next page

Example 8.10 continued

Solving for F_m,

$$F_m = \frac{8.25\ \text{N·m}}{\sin 15° \times 0.12\ \text{m}} = 270\ \text{N}$$

Discussion The force exerted by the muscle is much larger than the 30.0-N weight of the arm. The muscle must exert a larger force because the lever arm is small; the point of application is less than half as far from the joint as the center of gravity [0.12 m/(0.275 m) ≈ 4/9]. Also, the muscle cannot pull straight up on the arm; the vertical component of the muscle force is only about $\frac{1}{4}$ of the magnitude of the force. These two factors together make the weight supported (30.0 N) only $\frac{4}{9} \times \frac{1}{4} = \frac{1}{9}$ as large as the force exerted by the muscle.

Practice Problem 8.10 Holding a Juice Carton

Find the force exerted by the same person's deltoid muscle when holding a 1-L juice carton (weight 9.9 N) with the arm outstretched and parallel to the floor (as in Fig. 8.30). Assume that the juice carton is 60.0 cm from the shoulder.

Why does the iron cross require great strength?

The Iron Cross When a gymnast does the iron cross (Fig. 8.31a), the primary muscles involved are the latissimus dorsi ("lats") and pectoralis major ("pecs"). Since the rings are supporting the gymnast's weight, they exert an upward force on the gymnast's arms. Thus, the task for the muscles is not to hold the arm up, but to pull it down. The lats pull on the humerus about 3.5 cm from the shoulder joint (Fig. 8.31b). The pecs pull on the humerus about 5.5 cm from the joint (Fig. 8.31c). The other ends of these two muscles connect to bone in many places, widely distributed over the back (lats) and chest (pecs). As a reasonable simplification, we can assume that these muscles pull at a 45° angle below the horizontal in the iron cross maneuver. We also assume that the two muscles exert equal forces, so we can replace the two with a single force acting at 4.5 cm from the joint.

To determine the force exerted, we look at the entire arm as a system in equilibrium. This time we can ignore the weight of the arm itself since the force exerted on the arm by the ring is much larger—half the gymnast's weight is supported by each ring. The ring exerts an upward force that acts on the hand about 60 cm from the shoulder joint (see Fig. 8.31d). Taking torques about the shoulder, in equilibrium we have

$$|\text{CW torque}| = |\text{CCW torque}|$$

$$F_m \times 0.045\ \text{m} \times \sin 45° = \tfrac{1}{2}W \times 0.60\ \text{m}$$

$$F_m = \frac{\frac{1}{2}W \times 0.60\ \text{m}}{0.045\ \text{m} \times \sin 45°} = 9.4W$$

Thus, the force exerted by the lats and pecs *on one side* of the gymnast's body is more than nine times his weight.

The structure of the human body makes large muscular forces necessary. Are there advantages to the structure? Due to the small lever arms, the muscle forces are much larger than they would otherwise be, but the human body has traded this for a wide range of movement of the bones. The biceps and triceps muscles can move the lower arms through almost 180° while they change their lengths by only a few centimeters. The muscles also remain nearly parallel to the bones. If the biceps and triceps muscles were attached to the lower arm much farther from the elbow, there would have to be a large flap of skin to allow them to move so far away from the bones. The arrangement of our bones and muscles favors a wide range of movement.

Another advantage of the body structure is that it tends to minimize the rotational inertia of our limbs. For example, the muscles that control the motion of the lower arm are contained mostly within the *upper* arm. This keeps the rotational inertia of the lower arms about the elbow smaller. It also keeps the rotational inertia of the entire arm about the shoulder smaller. Smaller rotational inertia means that the energy we have to expend to move our limbs around is smaller.

The biceps muscle with its tendons is almost parallel to the humerus. One interesting observation is that the tendon connects to the radius at different points in different people. In one person this point may be 5.0 cm from the elbow joint, while in another person whose arm is the same length it may be 5.5 cm from the elbow. Thus,

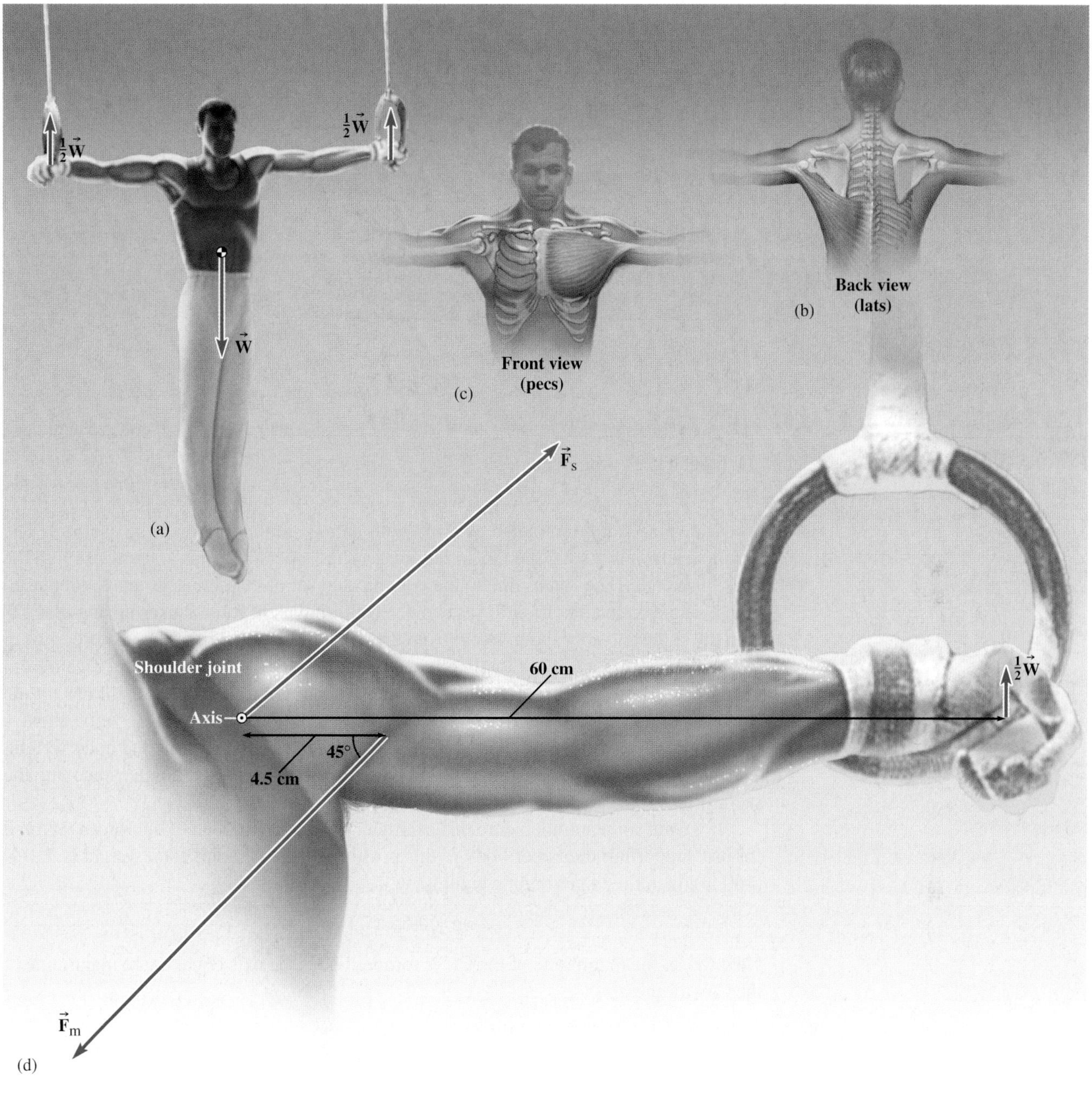

Figure 8.31 (a) Gymnast doing the iron cross. The principal muscles involved are (b) the "lats" and (c) the "pecs." (d) Simplified model of the forces acting on the arm of the gymnast.

some people are naturally stronger than others because of their internal structure. Chimpanzees have an advantage over humans because their biceps muscle has a longer lever arm. Do not make the mistake of arm wrestling with an adult chimp; challenge the chimp to a game of chess instead.

Application of Equilibrium Conditions: Heavy Lifting

When lifting an object from the floor, our first instinct is to bend over and pick it up. This is not a good way to lift something heavy. The spine is an ineffective lever and is susceptible to damage when a heavy object is lifted with bent waist. It is much better to squat down and use the powerful leg muscles to do the lifting instead of using our back muscles. Analyzing torques in a simplified model of the back can illustrate why.

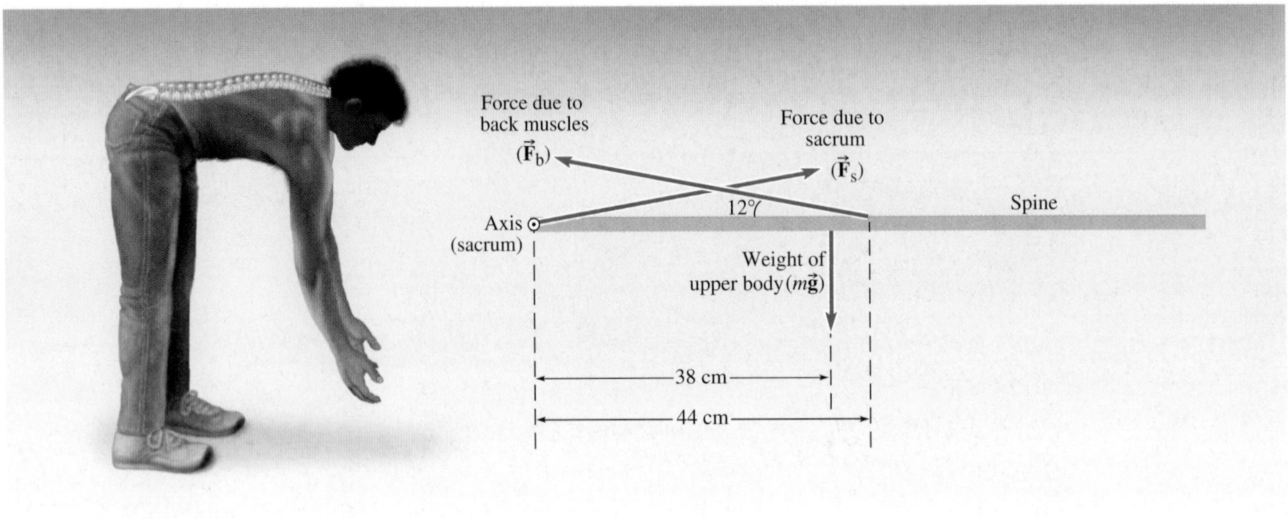

Figure 8.32 A simplified model of the human back when bent over.

The spine can be modeled as a rod with an axis at the tailbone (the sacrum). The sacrum exerts a force, marked $\vec{\mathbf{F}}_s$ in Fig. 8.32, when a person bends at the waist with the back horizontal. The forces due to the complicated set of back muscles can be replaced with a single equivalent force $\vec{\mathbf{F}}_b$ as shown. This equivalent force makes an angle of 12° with the spine and acts about 44 cm from the sacrum. The weight of the upper body, $m\vec{\mathbf{g}}$ in Fig. 8.32, is about 65% of total body weight; its center of gravity is about 38 cm from the sacrum. By placing an axis at the sacrum we can ignore the force $\vec{\mathbf{F}}_s$ in our torque equation. Since the vertical component of $\vec{\mathbf{F}}_b$ is $F_b \sin 12° \approx 0.21F_b$, only about $\frac{1}{5}$ the magnitude of the forces exerted by the back muscles is supporting the body weight. The much larger horizontal component is pressing the rod representing the spine into the sacrum.

If we put some numbers into this example, we can get an idea of the forces required for just supporting the upper body in this position. If the person's total weight is 710 N (160 lb), then the upper body weight is

$$mg = 0.65 \times 710 \text{ N}$$

Now we set the magnitude of the CCW torques about the axis equal to the magnitude of the CW torques:

$$F_b \times 0.44 \text{ m} \times \sin 12° = mg \times 0.38 \text{ m}$$

Substituting and solving,

$$F_b = \frac{0.65 \times 710 \text{ N} \times 0.38 \text{ m}}{0.44 \text{ m} \times \sin 12°} = 1920 \text{ N}$$

The muscular force that compresses the spine is the horizontal component of $\vec{\mathbf{F}}_b$:

$$F_b \cos 12° = 1900 \text{ N}$$

or about 430 lb. This is over four times the weight of the upper body.

Now if the person tries to lift something with his arms in this position, the lever arm for the weight of the load is even longer than for the weight of the upper body. The back muscles must supply a much larger force. The spine is now compressed with a dangerously large force. A cushioning disk called the lumbosacral disk, at the bottom of the spine, separates the last vertebra from the sacrum. This disk can be ruptured or deformed, causing great pain when the back is misused in such a fashion.

If, instead of bending over, we bend our knees and lower our body, keeping it vertically aligned as much as possible while lifting a load, the centers of gravity of the body and load are positioned more closely in a line above the sacrum, as in Fig. 8.33. Then

Figure 8.33 A safer way to lift a heavy object.

the lever arms of these forces with respect to an axis through the sacrum are relatively small and the force on the lumbosacral disk is roughly equal to the upper body weight plus the weight being lifted.

CONNECTION:

In Newton's second law for *rotation*, net torque takes the place of net force, and rotational inertia takes the place of mass, and α takes the place of $\vec{\mathbf{a}}$.

8.6 ROTATIONAL FORM OF NEWTON'S SECOND LAW

The concepts of torque and rotational inertia can be used to formulate a "Newton's second law for rotation"—a law that fills the role of $\Sigma\vec{\mathbf{F}} = m\vec{\mathbf{a}}$ for rotation about a fixed axis:

When calculating the net torque, remember to assign the correct algebraic sign to each torque before adding them.

Rotational form of Newton's second law:

$$\Sigma\tau = I\alpha \qquad (8\text{-}9)$$

The sum of the torques due to internal forces acting on a rigid object is always zero. Therefore, only *external* torques need be included in Eq. (8-9).

Thus, the angular acceleration of a rigid body is proportional to the net torque (more torque causes a larger α) and is inversely proportional to the rotational inertia (more inertia causes a smaller α). In rotational equilibrium, the angular acceleration must be zero; Eq. (8-9) then requires that the net torque be zero. We used $\Sigma\tau = 0$ as the condition for rotational equilibrium in Sections 8.4 and 8.5.

Equation (8-9) is proved in Problem 57. It is subject to an important restriction. Just as $\Sigma\vec{\mathbf{F}} = m\vec{\mathbf{a}}$ is valid only if the mass of the object is constant, $\Sigma\tau = I\alpha$ is valid only if the rotational inertia of the object is constant. For a *rigid* object rotating about a *fixed axis*, I cannot change, so Eq. (8-9) is always applicable.

Newton's second law for rotation explains why a tightrope walker carries a long pole to help maintain balance. Suppose the acrobat is about to topple over sideways. The pole has a large rotational inertia due to its length, so the angular acceleration of the system (acrobat plus pole) due to a small gravitational torque is much smaller than it would be without the pole. The smaller angular acceleration gives the acrobat more time to adjust his position and keep from falling.

Example 8.11

The Grinding Wheel

A grinding wheel is a solid, uniform disk of mass 2.50 kg and radius 9.00 cm. Starting from rest, what constant torque must a motor supply so that the wheel attains a rotational speed of 126 rev/s in a time of 6.00 s?

Strategy Since the grinding wheel is a uniform disk, we can find its rotational inertia using Table 8.1. After converting the revolutions per second to radians per second, we can find the angular acceleration from the change in angular velocity over the given time interval. Once we have I and α, we can find the net torque from Newton's second law for rotation.

Solution The grinding wheel is a uniform disk, so its rotational inertia is

$$I = \tfrac{1}{2}mr^2$$

$$\tfrac{1}{2} \times 2.50\ \text{kg} \times (0.0900\ \text{m})^2 = 0.010125\ \text{kg·m}^2$$

A single rotation of the wheel is equivalent to 2π radians, so

$$\omega = 126\ \frac{\text{rev}}{\text{s}} \times 2\pi\ \frac{\text{rad}}{\text{rev}}$$

The angular acceleration is

$$\alpha = \frac{\Delta\omega}{\Delta t}$$

continued on next page

Example 8.11 continued

Then the torque required is

$$\sum \tau = I\alpha = I\frac{\Delta\omega}{\Delta t}$$

$$= 0.010125 \text{ kg}\cdot\text{m}^2 \times \frac{126 \text{ rev/s} \times 2\pi \text{ rad/rev}}{6.00 \text{ s}}$$

$$= 1.34 \text{ N}\cdot\text{m}$$

If there are no other torques on the wheel, the motor must supply a constant torque of 1.34 N·m.

Discussion We assumed that no other torques are exerted on the wheel. There is certain to be at least a small frictional torque on the wheel with a sign opposite to the sign of the motor's torque. Then the motor would have to supply a torque larger than 1.34 N·m. The *net* torque would still be 1.34 N·m.

Practice Problem 8.11 Another Approach

Verify the answer to Example 8.11 by: (a) finding the angular displacement of the wheel using equations for constant α; (b) finding the change in rotational kinetic energy of the wheel; and (c) finding the torque from $W = \tau\Delta\theta$.

8.7 THE MOTION OF ROLLING OBJECTS

A rolling object combines translational motion of the center of mass with rotation about an axis that passes through the center of mass (see Section 5.1). For an object that is rolling without slipping, $v_{CM} = \omega R$. As a result, there is a specific relationship between the rolling object's translational and rotational kinetic energies. The total kinetic energy of a rolling object is the sum of its translational and rotational kinetic energies.

A wheel with mass M and radius R has a rotational inertia that is some pure number times MR^2; it couldn't be anything else and still have the right units. We can write the rotational inertia about an axis through the CM as $I_{CM} = \beta MR^2$ where β is a pure number that measures how far from the axis of rotation the mass is distributed. Larger β means the mass is, on average, farther from the axis. From Table 8.1, a hoop has $\beta = 1$; a disk, $\beta = \frac{1}{2}$; and a solid sphere, $\beta = \frac{2}{5}$.

Using $I_{CM} = \beta MR^2$ and $v_{CM} = \omega R$, the rotational kinetic energy for a rolling object can be written

$$K_{rot} = \frac{1}{2}I_{CM}\omega^2 = \frac{1}{2} \times \beta MR^2 \times \left(\frac{v_{CM}}{R}\right)^2 = \beta \times \frac{1}{2}Mv_{CM}^2$$

Since $\frac{1}{2}Mv_{CM}^2$ is the translational kinetic energy,

$$K_{rot} = \beta K_{tr} \tag{8-10}$$

This is convenient since β depends only on the shape, not on the mass or radius of the object. For a given shape rolling without slipping, the ratio of its rotational to translational kinetic energy is always the same (β).

The total kinetic energy can be written

$$K = K_{tr} + K_{rot}$$

$$K = \tfrac{1}{2}Mv_{CM}^2 + \tfrac{1}{2}I_{CM}\omega^2 \tag{8-11}$$

or in terms of β,

$$K = (1 + \beta)\, K_{tr}$$

$$K = (1 + \beta)\, \tfrac{1}{2}Mv_{CM}^2 \tag{8-12}$$

Thus, two objects of the same mass rolling at the same translational speed do *not* necessarily have the same kinetic energy. The object with the larger value of β has more rotational kinetic energy.

Conceptual Example 8.12

Hollow and Solid Rolling Balls

Starting from rest, two balls roll down a hill as in Fig. 8.34. One is solid, the other hollow. Which one is moving faster when it reaches the bottom of the hill? (WWW tutorial: rolling)

Strategy and Solution Energy conservation is the best way to approach this problem. As a ball rolls down the hill, its gravitational potential energy decreases as its kinetic energy increases by the same amount. The total kinetic energy is the sum of the translational and rotational contributions.

We do not know the mass or the radius of either ball and we cannot assume they are the same. Since both kinetic and potential energies are proportional to mass, mass does not affect the final speed. Also, the total kinetic energy does not depend on the radius of the ball [see Eq. (8-12)]. The final speeds of the two balls differ because different *fractions* of their total kinetic energies are translational.

One ball is a solid sphere and the other is approximately a spherical shell. The mass of a spherical shell is all concentrated on the surface of a sphere, while a solid sphere has its mass distributed throughout the sphere's volume. Therefore, the shell has a larger β than the solid sphere. When the shell rolls, it converts a bigger fraction of the lost potential energy into rotational kinetic energy; therefore, a smaller fraction becomes translational kinetic energy. The final speed of the solid sphere is larger since it puts a larger fraction of its kinetic energy into translational motion.

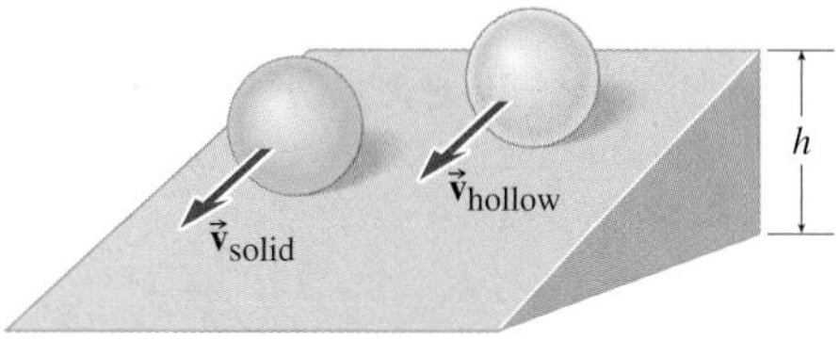

Figure 8.34
Rolling balls.

Discussion We can make this conceptual question into a quantitative one: what is the ratio of the speeds of the two balls at the bottom of the hill?

Let the height of the hill be h. Then for a ball of mass M, the loss of gravitational potential energy is Mgh. This amount of gravitational potential energy is converted into translational and rotational kinetic energy:

$$Mgh = K_{tr} + K_{rot} = (1 + \beta)K_{tr} = (1 + \beta)\frac{Mv_{CM}^2}{2}$$

Mass cancels out, as expected. We can solve for the final speed in terms of g, h, and β. The final speed is independent of the ball's mass and radius.

$$v_{CM} = \sqrt{\frac{2gh}{1+\beta}}$$

The ratio of the final speeds for two balls rolling down the same hill is, therefore,

$$\frac{v_1}{v_2} = \sqrt{\frac{1+\beta_2}{1+\beta_1}}$$

To evaluate the ratio, we look up the rotational inertias in Table 8.1. The solid sphere has $\beta = \frac{2}{5}$ and the spherical shell has $\beta = \frac{2}{3}$. Then

$$\frac{v_{solid}}{v_{hollow}} = \sqrt{\frac{1+\frac{2}{3}}{1+\frac{2}{5}}} \approx 1.091$$

The solid ball's final speed is, therefore, 9.1% faster than that of the hollow ball. This ratio depends neither on the masses of the balls, the radii of the balls, the height of the hill, nor the slope of the hill.

Practice Problem 8.12 Fraction of Kinetic Energy That Is Rotational Energy

What fraction of a rolling ball's kinetic energy is rotational kinetic energy? Answer both for a solid ball and a hollow one.

CHECKPOINT 8.7

Give an example of how a marble can move so that $K_{tr} > 0$ and $K_{rot} = 0$; (b) $K_{tr} = 0$ and $K_{rot} > 0$; (c) $K_{rot} = \frac{2}{5} K_{tr}$.

Acceleration of Rolling Objects What is the acceleration of a ball rolling down an incline? Figure 8.35 shows the forces acting on the ball. Static friction is the force that makes the ball rotate; if there were no friction, instead of rolling, the ball would just *slide* down the incline. This is true because friction is the only force acting that yields a

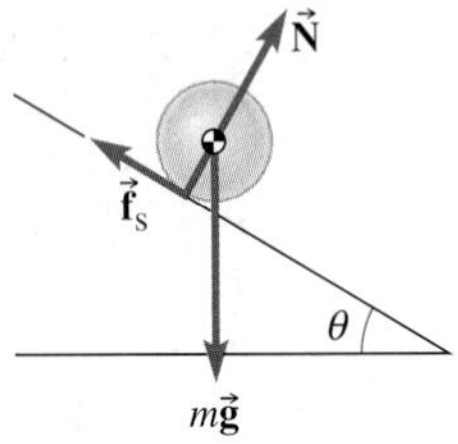

Figure 8.35 Forces acting on a ball rolling downhill.

nonzero torque about the rotation axis through the ball's center of mass. Gravity gives zero torque because it acts at the axis, so the lever arm is zero. The normal force points directly at the axis, so its lever arm is also zero.

The frictional force $\vec{\mathbf{f}}$ provides a torque

$$\tau = rf$$

where r is the ball's radius. An analysis of the forces and torques combined with Newton's second law in both forms enables us to calculate the acceleration of the ball in Example 8.13.

Example 8.13

Acceleration of a Rolling Ball

Calculate the acceleration of a solid ball rolling down a slope inclined at an angle θ to the horizontal (Fig. 8.36a).

Strategy The net torque is related to the angular acceleration by $\Sigma\tau = I\alpha$, Newton's second law for rotation. Similarly, the net force acting on the ball gives the acceleration of the center of mass: $\Sigma\vec{\mathbf{F}} = m\vec{\mathbf{a}}_{CM}$. The axis of rotation is through the ball's CM. As already discussed, neither gravity nor the normal force produce a torque about this axis; the net torque is $\Sigma\tau = rf$, where f is the magnitude of the frictional force. One problem is that the force of friction is unknown. We must resist the temptation to assume that $f = \mu_s N$; there is no reason to assume that static friction has its maximum possible magnitude. We do know that the two accelerations, translational and rotational, are related. We know that v_{CM} and ω are proportional since r is constant. To stay proportional they must change in lock step; their rates of change, a_{CM} and α, are proportional to each other by the same factor of r. Thus, $a_{CM} = \alpha r$. This connection should enable us to eliminate f and solve for the acceleration. Since the speed of a ball after rolling a certain distance was found to be independent of the mass and radius of the ball in Example 8.12, we expect the same to be true of the acceleration.

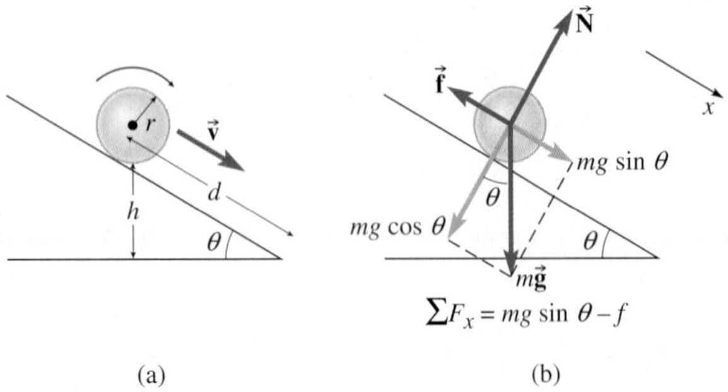

Figure 8.36
(a) A ball rolling downhill. (b) FBD for the ball, with the gravitational force resolved into components perpendicular and parallel to the incline.

Solution Since the net torque is

$$\Sigma\tau = rf$$

the angular acceleration is

$$\alpha = \frac{\Sigma\tau}{I} = \frac{rf}{I} \quad (1)$$

where I is the ball's rotational inertia about its CM.

Figure 8.36b shows the forces along the incline acting on the ball. The acceleration of the CM is found from Newton's second law. The component of the net force acting along the incline (in the direction of the acceleration) is

$$\Sigma F_x = mg \sin \theta - f = ma_{CM} \quad (2)$$

Because the ball is rolling without slipping, the acceleration of the CM and the angular acceleration are related by

$$a_{CM} = \alpha r$$

Now we try to eliminate the unknown frictional force f from the previous equations. Solving Eq. (1) for f gives

$$f = \frac{I\alpha}{r}$$

Substituting this into Eq. (2), we get

$$mg \sin \theta - \frac{I\alpha}{r} = ma_{CM}$$

Now to eliminate α, we can substitute $\alpha = a_{CM}/r$:

$$mg \sin \theta - \frac{Ia_{CM}}{r^2} = ma_{CM}$$

Solving for a_{CM},

$$a_{CM} = \frac{g \sin \theta}{1 + I/(mr^2)}$$

For a solid sphere, $I = \frac{2}{5}mr^2$, so

$$a_{CM} = \frac{g \sin \theta}{1 + \frac{2}{5}} = \frac{5}{7}g \sin \theta$$

continued on next page

Example 8.13 continued

Discussion The acceleration of an object *sliding* down an incline without friction is $a = g \sin \theta$. The acceleration of the rolling ball is smaller than $g \sin \theta$ due to the frictional force directed up the incline.

We can check the answer using the result of Example 8.12. The ball's acceleration is constant. If the ball starts from rest as in Fig. 8.36a, after it has rolled a distance d, its speed v is

$$v = \sqrt{2ad} = \sqrt{2\left(\frac{g \sin \theta}{1+\beta}\right)d}$$

where $\beta = \frac{2}{5}$. The vertical drop during this time is $h = d \sin \theta$, so

$$v = \sqrt{\frac{2gh}{1+\beta}}$$

Practice Problem 8.13 Acceleration of a Hollow Cylinder

Calculate the acceleration of a thin hollow cylindrical shell rolling down a slope inclined at an angle θ to the horizontal.

8.8 ANGULAR MOMENTUM

Newton's second law for translational motion can be written in two ways:

$$\sum \vec{\mathbf{F}} = \lim_{\Delta t \to 0} \frac{\Delta \vec{\mathbf{p}}}{\Delta t} \text{ (general form)} \quad \text{or} \quad \sum \vec{\mathbf{F}} = m\vec{\mathbf{a}} \text{ (constant mass)}$$

In Eq. (8-9) we wrote Newton's second law for rotation as $\sum \tau = I\alpha$, which applies only when I is constant—that is, for a rigid body rotating about a fixed axis. A more general form of Newton's second law for rotation uses the concept of **angular momentum** (symbol L).

The net external torque acting on a system is equal to the rate of change of the angular momentum of the system.

$$\sum \tau = \lim_{\Delta t \to 0} \frac{\Delta L}{\Delta t} \tag{8-13}$$

CONNECTION:

Note the analogy with

$$\sum \vec{\mathbf{F}} = \lim_{\Delta t \to 0} \frac{\Delta \vec{\mathbf{p}}}{\Delta t}$$

The angular momentum of a rigid body rotating about a fixed axis is the rotational inertia times the angular velocity, which is analogous to the definition of linear momentum (mass times velocity):

Angular momentum:

$$L = I\omega \tag{8-14}$$

(rigid body, fixed axis)

CONNECTION:

Note the analogy with $\vec{\mathbf{p}} = m\vec{\mathbf{v}}$. See the Master the Concepts section for a complete table of these analogies.

Either Eq. (8-13) or Eq. (8-14) can be used to show that the SI units of angular momentum are kg·m^2/s.

For a rigid body rotating around a fixed axis, angular momentum doesn't tell us anything new. The rotational inertia is constant for such a body since the distance r_i between every point on the object and the axis stays the same. Then any change in angular momentum must be due to a change in angular velocity ω:

$$\sum \tau = \lim_{\Delta t \to 0} \frac{\Delta L}{\Delta t} = \lim_{\Delta t \to 0} \frac{I \Delta \omega}{\Delta t} = I \lim_{\Delta t \to 0} \frac{\Delta \omega}{\Delta t} = I\alpha$$

Conservation of Angular Momentum However, Eq. (8-13) is *not* restricted to rigid objects or to fixed rotation axes. In particular, if the net external torque acting on a system is zero, then the angular momentum of the system cannot change. This is the **law of conservation of angular momentum**:

Conservation of angular momentum can be applied to any system if the net external torque on the system is zero (or negligibly small).

Conservation of angular momentum:

$$\text{If } \sum \tau = 0,\ L_i = L_f \quad (8\text{-}15)$$

Here L_i and L_f represent the angular momentum of the system at two different times. Conservation of angular momentum is one of the most basic and fundamental laws of physics, along with the two other conservation laws we have studied so far (energy and linear momentum). For an isolated system, the total energy, total linear momentum, and total angular momentum of the system are each conserved. None of these quantities can change unless some external agent causes the change.

CONNECTION:

Another conservation law

With conservation of energy, we add up the amounts of the different forms of energy (such as kinetic energy and gravitational potential energy) to find the *total* energy. The conservation law refers to the total energy. By contrast, linear momentum and angular momentum *cannot* be added to find the "total momentum." They are entirely different quantities, not two forms of the same quantity. They even have different dimensions, so it would be impossible to add them. Conservation of linear momentum and conservation of angular momentum are *separate* laws of physics.

Application of angular momentum: figure skater

Changing Rotational Inertia In this section, we restrict our consideration to cases where the axis of rotation is fixed but where the rotational inertia is not necessarily constant. One familiar example of a changing rotational inertia occurs when a figure skater spins (Fig. 8.37). To start the spin, the skater glides along with her arms outstretched and then begins to rotate her body about a vertical axis by pushing against the ice with a skate. The push of the ice against the skate provides the external torque that gives the skater her initial angular momentum. Initially the skater's arms and the leg not in contact with the ice are extended away from her body. The mass of the arms and leg when extended contribute more to her rotational inertia than they do when held close to the body. As the skater spins, she pulls her arms and leg close and straightens her body to decrease her rotational inertia. As she does, her angular velocity increases dramatically in such a way that her angular momentum stays the same.

(a) (b)

Figure 8.37 Figure skater Lucinda Ruh at the (a) beginning and (b) end of a spin. Her angular velocity is much higher in (b) than in (a).

CHECKPOINT 8.8

If the skater then extends her arms and leg back to their initial configuration, does her angular velocity decrease back to its initial value, ignoring friction?

Many natural phenomena can be understood in terms of angular momentum. In a hurricane, circulating air is sucked inward by a low-pressure region at the center of the storm (the *eye*). As the air moves closer and closer to the axis of rotation, it circulates faster and faster. An even more dramatic example is the formation of a pulsar. Under certain conditions, a star can implode under its own gravity, forming a neutron star (a collection of tightly packed neutrons). If the Sun were to collapse into a neutron star, its radius would be only about 13 km. If a star is rotating before its collapse, then as its rotational inertia decreases dramatically, its angular velocity must increase to keep its angular momentum constant. Such rapidly rotating neutron stars are called pulsars because they emit regular pulses of x-rays, at the same frequency as their rotation, that can be detected when they reach Earth. Some pulsars rotate in only a few thousandths of a second per revolution.

Applications of angular momentum: hurricanes and pulsars

Example 8.14

Mouse on a Wheel

A 0.10-kg mouse is perched at point *B* on the rim of a 2.00-kg wagon wheel that rotates freely in a horizontal plane at 1.00 rev/s (Fig. 8.38). The mouse crawls to point *A* at the center. Assume the mass of the wheel is concentrated at the rim. What is the frequency of rotation in rev/s when the mouse arrives at point *A*?

Strategy Assuming that frictional torques are negligibly small, there is no external torque acting on the mouse/wheel system. Then the angular momentum of the mouse/wheel system must be conserved; it takes an external torque to change angular momentum. The mouse and wheel exert torques on one another, but these *internal* torques only transfer some angular momentum between the wheel and the mouse without changing the total angular momentum. We can think of the system as initially being a rigid body with rotational inertia I_i. When the mouse reaches the center, we think of the system as a rigid body with a different rotational inertia I_f. The mouse changes the rotational inertia of the mouse/wheel system by moving from the outer rim, where its mass makes the maximum possible contribution to the rotational inertia, to the rotation axis, where its mass makes no contribution to the rotational inertia.

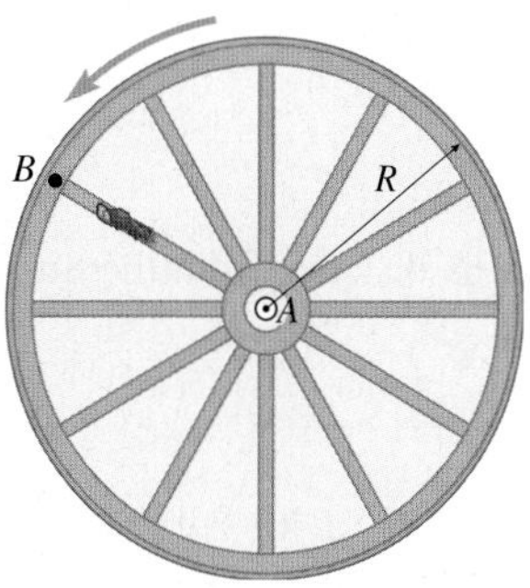

Figure 8.38
Mouse on a rotating wheel.

Solution Initially, all of the mass of the system is at a distance *R* from the rotation axis, where *R* is the radius of the wheel. Therefore,

$$I_i = (M + m)R^2$$

where *M* is the mass of the wheel and *m* is the mass of the mouse. After the mouse moves to the center of the wheel, its mass contributes nothing to the rotational inertia of the system:

$$I_f = MR^2$$

From conservation of angular momentum,

$$I_i\,\omega_i = I_f\,\omega_f$$

Substituting the rotational inertias and $\omega = 2\pi f$,

$$(M + m)R^2 \times 2\pi f_i = MR^2 \times 2\pi f_f$$

Factors of $2\pi R^2$ cancel from each side, leaving

$$(M + m)f_i = Mf_f$$

Solving for f_f,

$$f_f = \frac{M + m}{M} f_i = \frac{2.10\text{ kg}}{2.00\text{ kg}}(1.00\text{ rev/s}) = 1.05\text{ rev/s}$$

continued on next page

Example 8.14 continued

Discussion Conservation laws are powerful tools. We do not need to know the details of what happens as the mouse crawls along the spoke from the outer edge of the wheel; we need only look at the initial and final conditions.

A common mistake in this sort of problem is to assume that the initial rotational kinetic energy is equal to the final rotational kinetic energy. This is not true because the mouse crawling in toward the center expends energy to do so. In other words, the mouse converts some internal energy into rotational kinetic energy.

Practice Problem 8.14 Change in Rotational Kinetic Energy

What is the percentage change in the rotational kinetic energy of the mouse/wheel system?

Angular Momentum in Planetary Orbits

Application of angular momentum: planetary orbits

Conservation of angular momentum applies to planets orbiting the Sun in elliptical orbits. Kepler's second law says that the orbital speed varies in such a way that the planet sweeps out area at a constant rate (Fig. 8.39a). In Problem 104, you can show that Kepler's second law is a direct result of conservation of angular momentum, where the angular momentum of the planet is calculated using an axis of rotation perpendicular to the plane of the orbit and passing through the Sun. When the planet is closer to the Sun, it moves faster; when it is farther away, it moves more slowly. Conservation of angular momentum can be used to relate the orbital speeds and radii at two different points in the orbit. The same applies to satellites and moons orbiting planets.

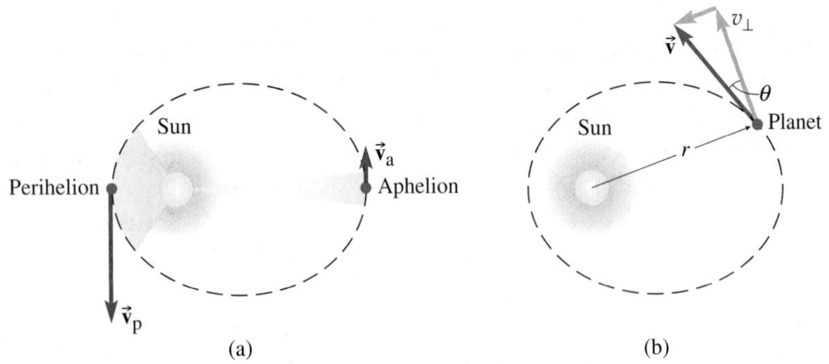

Figure 8.39 The planet's speed varies such that it sweeps out equal areas in equal time intervals. The eccentricity of the planetary orbit is exaggerated for clarity.

Example 8.15

Earth's Orbital Speed

At perihelion (closest approach to the Sun), Earth is 1.47×10^8 km from the Sun and its orbital speed is 30.3 km/s. What is Earth's orbital speed at aphelion (greatest distance from the Sun), when it is 1.52×10^8 km from the Sun? Note that at these two points Earth's velocity is perpendicular to a radial line from the Sun (see Fig. 8.39a).

Strategy We take the axis of rotation through the Sun. Then the gravitational force on Earth points directly toward the axis; with zero lever arm, the torque is zero. With no other external forces acting on the Earth, the net external torque is zero. Earth's angular momentum about the rotation axis through the Sun must therefore be conserved. To find Earth's rotational inertia, we treat it as a point particle since its radius is much less than its distance from the axis of rotation.

Solution The rotational inertia of the Earth is

$$I = mr^2$$

where m is Earth's mass and r is its distance from the Sun. The angular velocity is

$$\omega = \frac{v_\perp}{r}$$

where $v_\perp$ is the component of the velocity perpendicular to a radial line from the Sun. At the two points under consideration,

continued on next page

Example 8.15 continued

$v_\perp = v$. As the distance from the Sun r varies, its speed v must vary to conserve angular momentum:

$$I_i\omega_i = I_f\omega_f$$

By substitution,

$$mr_i^2 \times \frac{v_i}{r_i} = mr_f^2 \times \frac{v_f}{r_f}$$

or

$$r_i v_i = r_f v_f \qquad (1)$$

Solving for v_f,

$$v_f = r_i/r_f\, v_i = \frac{1.47 \times 10^8 \text{ km}}{1.52 \times 10^8 \text{ km}} \times 30.3 \text{ km/s} = 29.3 \text{ km/s}$$

Discussion Earth moves slower at a point farther from the Sun. This is what we expect from energy conservation. The potential energy is greater at aphelion than at perihelion. Since the mechanical energy of the orbit is constant, the kinetic energy must be smaller at aphelion.

Equation (1) implies that the orbital speed and orbital radius are inversely proportional, but strictly speaking this equation only applies to the perihelion and aphelion. At a general point in the orbit, the *perpendicular component* $v_\perp$ is inversely proportional to r (see Fig. 8.39b). The orbits of Earth and most of the other planets are nearly circular so that $\theta \approx 0°$ and $v_\perp \approx v$.

Practice Problem 8.15 Puck on a String

A puck on a frictionless, horizontal air table is attached to a string that passes down through a hole in the table. Initially the puck moves at 12 cm/s in a circle of radius 24 cm. If the string is pulled through the hole, reducing the radius of the puck's circular motion to 18 cm, what is the new speed of the puck?

8.9 THE VECTOR NATURE OF ANGULAR MOMENTUM

Until now we have treated torque and angular momentum as scalar quantities. Such a treatment is adequate in the cases we have considered so far. However, the law of conservation of angular momentum applies to *all* systems, including rotating objects whose axis of rotation changes direction. Torque and angular momentum are actually vector quantities. Angular momentum is conserved in *both magnitude and direction* in the absence of external torques.

An important special case is that of a symmetrical object rotating about an axis of symmetry, such as the spinning disk in Fig. 8.40. The magnitude of the angular momentum of such an object is $L = I\omega$. The direction of the angular momentum vector points along the axis of rotation. To choose between the two directions along the axis, a **right-hand rule** is used. Align your right hand so that, as you curl your fingers in toward your palm, your fingertips follow the object's rotation; then your thumb points in the direction of $\vec{\mathbf{L}}$.

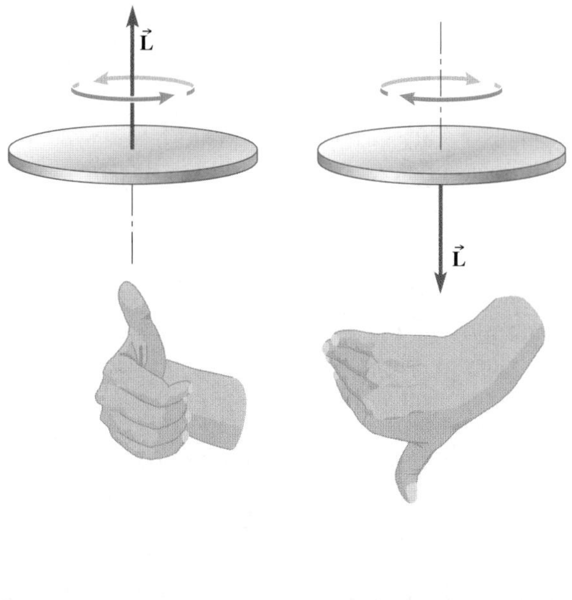

Figure 8.40 Right-hand rule for finding the direction of the angular momentum of a spinning disk.

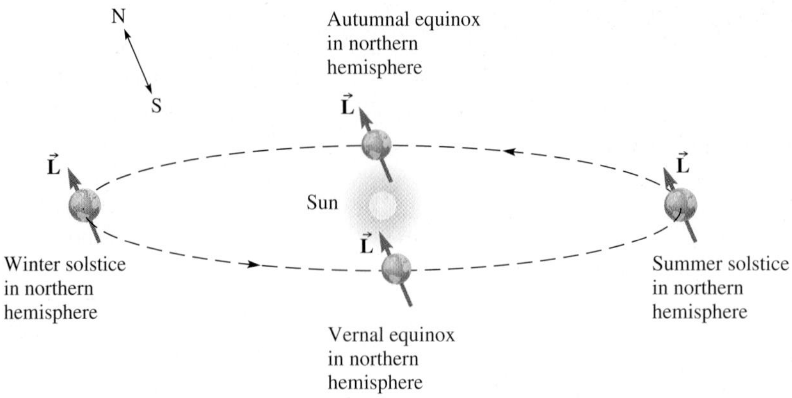

Figure 8.41 Spinning like a top, the Earth maintains the direction of its angular momentum due to rotation as it revolves around the Sun (not to scale).

For rotation about a *fixed* axis, the net torque is also along the axis of rotation, in the direction of the *change* in angular momentum it causes. The sign convention we have used up to now for angular momentum and torque gives the sign of the *z-component of the vector quantity*, where the *z*-axis points toward the viewer (out of the page).

Application of angular momentum: the gyroscope

A disk with a large rotational inertia can be used as a *gyroscope.* When the gyroscope spins at a large angular velocity, it has a large angular momentum. It is then difficult to change the orientation of the gyroscope's rotation axis, because to do so requires changing its angular momentum. To change the direction of a large angular momentum requires a correspondingly large torque. Thus, a gyroscope can be used to maintain stability. Gyroscopes are used in guidance systems in airplanes, submarines, and space vehicles to maintain a constant direction in space.

The same principle explains the great stability of rifle bullets and spinning tops. A rifle bullet is made to spin as it passes through the rifle's barrel. The spinning bullet then keeps its correct orientation—nose first—as it travels through the air. Otherwise, a small torque due to air resistance could make the bullet turn around randomly, greatly increasing air resistance and undermining accuracy. A properly thrown football is made to spin for the same reasons. A spinning top can stay balanced for a long time, while the same top soon falls over if it is not spinning.

The Earth's rotation gives it a large angular momentum. As the Earth orbits the Sun, the axis of rotation stays in a fixed direction in space. The axis points nearly at Polaris (the North Star), so even as the Earth rotates around its axis, Polaris maintains its position in the northern sky. The fixed direction of the rotation axis gives us the regular progression of the seasons (Fig. 8.41).

A Classic Demonstration

A demonstration often done in physics classes is for a student to hold a spinning bicycle wheel while standing on a platform that is free to rotate. The wheel's rotation axis is initially horizontal (Fig. 8.42a). Then the student repositions the wheel so that its axis of rotation is vertical (Fig. 8.42b). As he repositions the wheel, the platform begins to rotate opposite to the wheel's rotation. If we assume *no* friction acts to resist rotation of the platform, then the platform continues to rotate as long as the wheel is held with its axis vertical. If the student returns the wheel to its original orientation, the rotation of the platform stops.

The platform is free to rotate about a vertical axis. As a result, once the student steps onto the platform, *the vertical component* L_y of the angular momentum of the system (student + platform + wheel) is conserved. The horizontal components of $\vec{\mathbf{L}}$ are *not* conserved. The platform is not free to rotate about any horizontal axis since the floor can exert external torques to keep it from doing so. In vector language, we would say that only the vertical component of the external torque is zero, so only the vertical component of angular momentum is conserved.

Initially $L_y = 0$ since the student and the platform have zero angular momentum and the wheel's angular momentum is horizontal. When the wheel is repositioned so that it

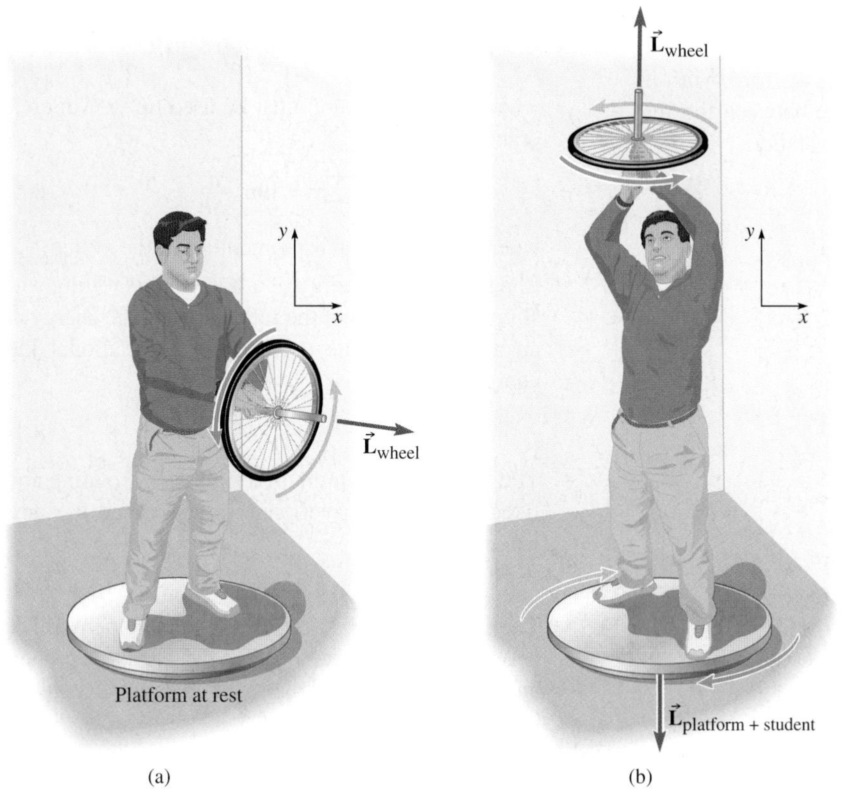

Figure 8.42 A demonstration of angular momentum conservation.

spins with an upward angular momentum ($L_y > 0$), the rest of the system (the student and the platform) must acquire an equal magnitude of downward angular momentum ($L_y < 0$) so that the vertical component of the total angular momentum is still zero. Thus, the platform and student rotate in the opposite sense from the rotation of the wheel. Since the platform and student have more rotational inertia than the wheel, they do not spin as fast as the wheel, but their vertical angular momentum is just as large.

The student and the wheel apply torques to each other to transfer angular momentum from one part of the system to the other. These torques are equal and opposite and they have both vertical and horizontal components. As the student lifts the wheel, he feels a strange twisting force that tends to rotate him about a horizontal axis. The platform prevents the horizontal rotation by exerting unequal normal forces on the student's feet. The horizontal component of the torque is so counterintuitive that, if the student is not expecting it, he can easily be thrown from the platform!

Master the Concepts

- The rotational kinetic energy of a rigid object with rotational inertia I and angular velocity ω is

$$K_{rot} = \tfrac{1}{2}I\omega^2 \qquad (8\text{-}1)$$

  In this expression, ω must be measured in *radians* per unit time.
- Rotational inertia is a measure of how difficult it is to change an object's angular velocity. It is defined as:

$$I = \sum_{i=1}^{N} m_i r_i^2 \qquad (8\text{-}2)$$

  where r_i is the perpendicular distance between a particle of mass m_i and the rotation axis. The rotational inertia depends on the location of the rotation axis.
- Torque measures the effectiveness of a force for twisting or turning an object. It can be calculated in two equivalent ways: either as the product of the perpendicular component of the force with the shortest distance between the rotation axis and the point of application of the force

$$\tau = \pm rF_\perp \qquad (8\text{-}3)$$

continued on next page

Master the Concepts continued

or as the product of the magnitude of the force with its lever arm (the perpendicular distance between the line of action of the force and the axis of rotation)

$$\tau = \pm r_{\perp} F \qquad (8\text{-}4)$$

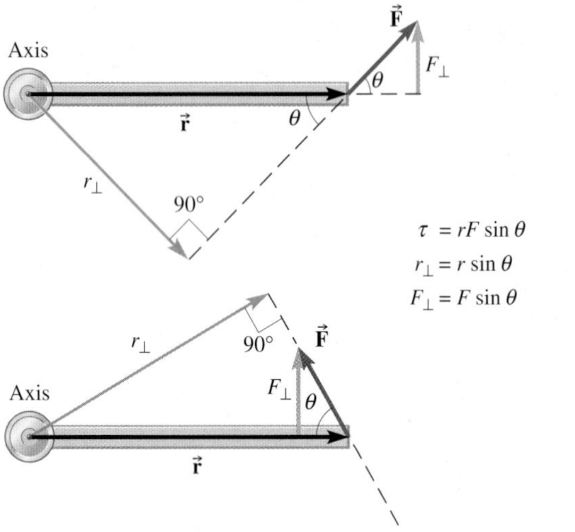

- A force whose perpendicular component tends to cause rotation in the CCW direction gives rise to a positive torque; a force whose perpendicular component tends to cause rotation in the CW direction gives rise to a negative torque.
- The work done by a constant torque is the product of the torque and the angular displacement:

$$W = \tau \Delta\theta \quad (\Delta\theta \text{ in radians}) \qquad (8\text{-}6)$$

- The conditions for translational and rotational equilibrium are

$$\Sigma \vec{\mathbf{F}} = 0 \text{ and } \Sigma \tau = 0 \qquad (8\text{-}8)$$

The rotation axis can be chosen *arbitrarily* when calculating torques in equilibrium problems. Generally, the best place to choose the axis is at the point of application of an unknown force so that the unknown force does not appear in the torque equation.

- Newton's second law for rotation is

$$\Sigma \tau = I\alpha \qquad (8\text{-}9)$$

where radian measure must be used for α. A more general form is

$$\Sigma \tau = \lim_{\Delta t \to 0} \frac{\Delta L}{\Delta t} \qquad (8\text{-}13)$$

where L is the angular momentum of the system.

- The total kinetic energy of a body that is rolling without slipping is the sum of the rotational kinetic energy about an axis through the CM and the translational kinetic energy:

$$K = \tfrac{1}{2} M v_{CM}^2 + \tfrac{1}{2} I_{CM} \omega^2 \qquad (8\text{-}11)$$

- The angular momentum of a rigid body rotating about a fixed axis is the rotational inertia times the angular velocity:

$$L = I\omega \qquad (8\text{-}14)$$

- The law of conservation of angular momentum: if the net external torque acting on a system is zero, then the angular momentum of the system cannot change.

$$\text{If } \Sigma \tau = 0, \; L_i = L_f \qquad (8\text{-}15)$$

- This table summarizes the analogous quantities and equations in translational and rotational motion.

Translation	Rotation
m	I
$\vec{\mathbf{F}}$	τ
$\vec{\mathbf{a}}$	α
$\Sigma \vec{\mathbf{F}} = m\vec{\mathbf{a}}$	$\Sigma \tau = I\alpha$
Δx	$\Delta\theta$
$W = F_x \Delta x$	$W = \tau \Delta\theta$
$\vec{\mathbf{v}}$	ω
$K = \frac{1}{2} m v^2$	$K = \frac{1}{2} I \omega^2$
$\vec{\mathbf{p}} = m\vec{\mathbf{v}}$	$L = I\omega$
$\Sigma \vec{\mathbf{F}} = \lim_{\Delta t \to 0} \frac{\Delta \vec{\mathbf{p}}}{\Delta t}$	$\Sigma \tau = \lim_{\Delta t \to 0} \frac{\Delta L}{\Delta t}$
If $\Sigma \vec{\mathbf{F}} = 0$, $\vec{\mathbf{p}}$ is conserved	If $\Sigma \tau = 0$, L is conserved

Conceptual Questions

1. In Fig. 8.2b, where should the doorknob be located to make the door easier to open?
2. Explain why it is easier to drive a wood screw using a screwdriver with a large diameter handle rather than one with a thin handle.
3. Why is it easier to push open a swinging door from near the edge away from the hinges rather than in the middle of the door?
4. A book measures 3 cm by 16 cm by 24 cm. About which of the axes shown in the figure is its rotational inertia smallest?

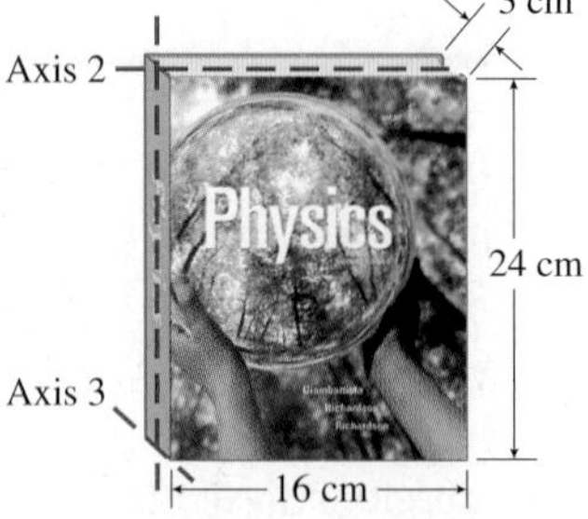

Conceptual Question 4

5. A body in equilibrium has only two forces acting on it. We found in Section 4.2 that the

forces must be equal in magnitude and opposite in direction in order to give a translational net force of zero. What else must be true of the two forces for the body to be in equilibrium? [*Hint:* Consider the lines of action of the forces.]

6. Why do many helicopters have a small propeller attached to the tail that rotates in a vertical plane? Why is this attached at the tail rather than somewhere else? [*Hint:* Most of the helicopter's mass is forward, in the cab.]

7. In the "Pinewood Derby," Cub Scouts construct cars and then race them down an incline. Some say that, everything else being equal (friction, drag coefficient, same wheels, etc.), a heavier car will win; others maintain that the weight of the car does not matter. Who is right? Explain. [*Hint:* Think about the fraction of the car's kinetic energy that is rotational.]

8. A large barrel lies on its side. In order to roll it across the floor, you apply a horizontal force, as shown in the figure. If the applied force points toward the axis of rotation, which runs down the center of the barrel through the center of mass, it produces zero torque about that axis. How then can this applied force make the barrel start to roll?

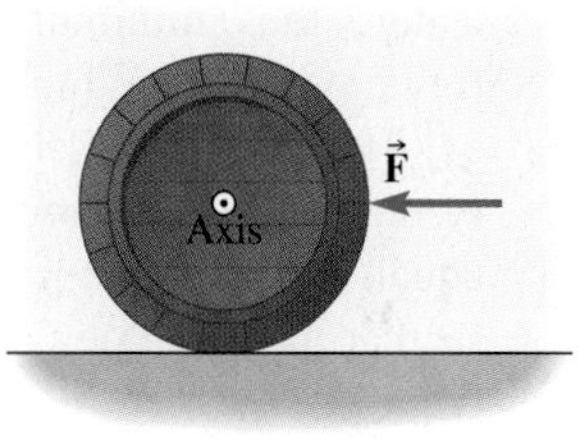

9. Animals that can run fast always have thin legs. Their leg muscles are concentrated close to the hip joint; only tendons extend into the lower leg. Using the concept of rotational inertia, explain how this helps them run fast.

10. Part (a) of the figure shows a simplified model of how the triceps muscle connects to the forearm. As the angle θ is changed, the tendon wraps around a nearly circular arc. Explain how this is much more effective than if the tendon is connected as in part (b) of the figure. [*Hint:* Look at the lever arm as θ changes.]

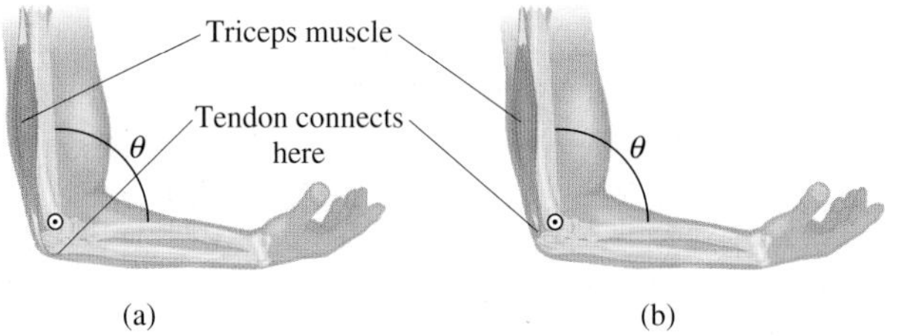

Question 10

11. Part (a) of the figure shows a simplified model of how the biceps muscle enables the forearm to support a load. What are the advantages of this arrangement as opposed to the alternative shown in part (b), where the flexor muscle is in the forearm instead of in the upper arm? Are the two equally effective when the forearm is horizontal? What about for other angles between the upper arm and the forearm? Consider also the rotational inertia of the forearm about the elbow and of the entire arm about the shoulder.

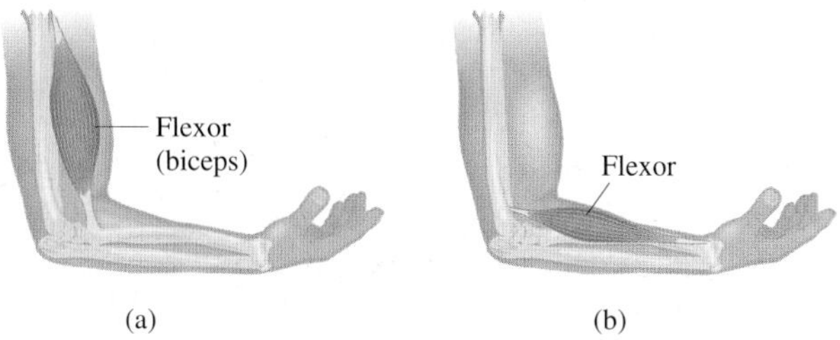

Question 11

12. In Section 8.6, it was asserted that the sum of all the internal torques (that is, the torques due to internal forces) acting on a rigid object is zero. The figure shows two particles in a rigid object. The particles exert forces $\vec{F}_{12}$ and $\vec{F}_{21}$ on each other. These forces are directed along a line that joins the two particles. Explain why the torques due to these two forces must be equal and opposite even though the forces are applied at different points (and, therefore, possibly different distances from the axis).

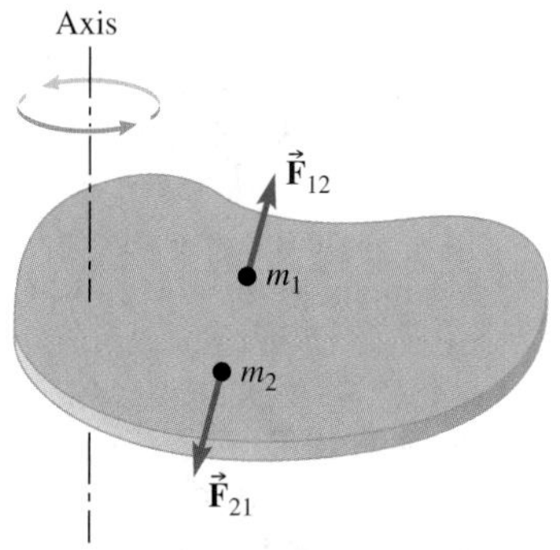

13. A playground merry-go-round (Fig. 8.5) spins with negligible friction. A child moves from the center out to the rim of the merry-go-round platform. Let the system be the merry-go-round plus the child. Which of these quantities change: angular velocity of the system, rotational kinetic energy of the system, angular momentum of the system? Explain your answer.

14. The figure shows a balancing toy with weights extending on either side. The toy is extremely stable. It can be pushed quite far off center one way or the other but it does not fall over. Explain why it is so stable.

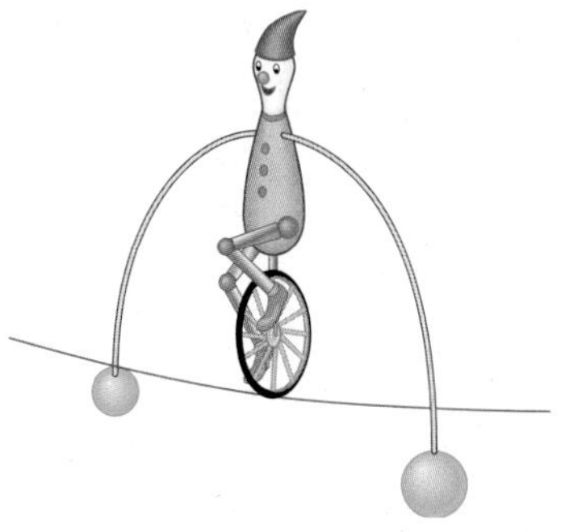

15. Explain why the posture taken by defensive football linemen makes them more difficult to push out of the way. Consider both the height of the center of gravity and the size of the support base (the area on the ground bounded

by the hands and feet touching the ground). In order to knock a person over, what has to happen to the center of gravity? Which do you think needs a more complex neurological system for maintaining balance: four legged animals or humans?

16. The center of gravity of the upper body of a bird is located below the hips; in a human, the center of gravity of the upper body is located well above the hips. Since the upper body is supported by the hips, are birds or humans more stable? Consider what happens if the upper body is displaced a little so that its center of gravity is not directly above or below the hips. In what direction does the torque due to gravity tend to make the upper body rotate about an axis through the hips?

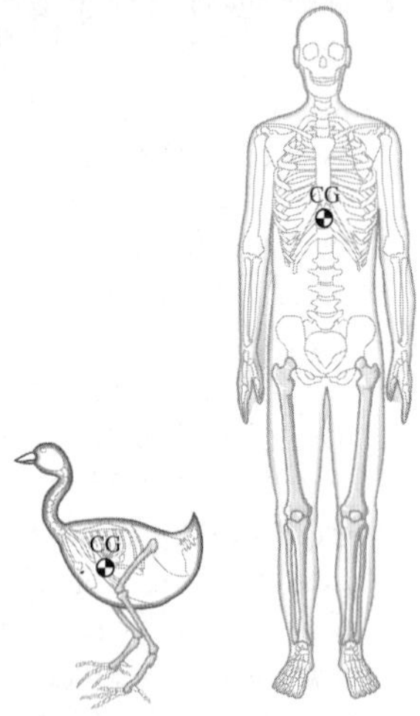

17. An astronaut wants to remove a bolt from a satellite in orbit. He positions himself so that he is at rest with respect to the satellite, then pulls out a wrench and attempts to remove the bolt. What is wrong with his method? How can he remove the bolt?

18. Your door is hinged to close automatically after being opened. Where is the best place to put a wedge-shaped door stopper on a slippery floor in order to hold the door open? Should it be placed close to the hinge or far from it?

19. You are riding your bicycle and approaching a rather steep hill. Which gear should you use to go uphill, a low gear or a high gear? With a low gear the wheel rotates less than with a high gear for one rotation of the pedals.

20. One way to find the center of gravity of an irregular flat object is to suspend it from various points so that it is free to rotate. When the object hangs in equilibrium, a vertical line is drawn downward from the support point. After drawing lines from several different support points, the center of gravity is the point where the lines all intersect. Explain how this works.

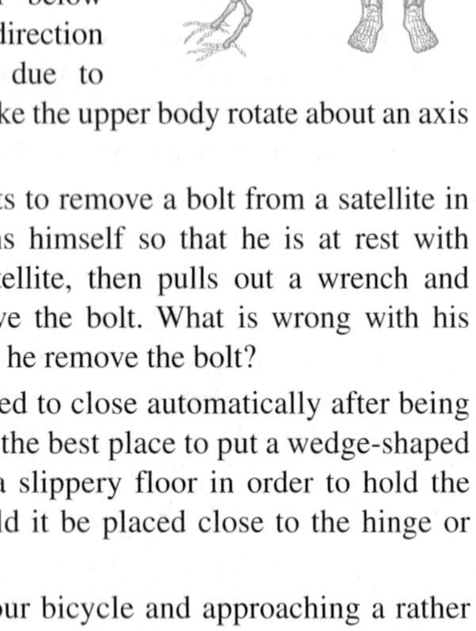

21. One of the effects of significant global warming would be the melting of part or all of the polar ice caps. This, in turn, would change the length of the day (the period of the Earth's rotation). Explain why. Would the day get longer or shorter?

Multiple-Choice Questions

1. A heavy box is resting on the floor. You would like to push the box to tip it over on its side, using the minimum force possible. Which of the force vectors in the diagram shows the correct location and direction of the force? The forces have equal horizontal components. Assume enough friction so that the box does not slide; instead it rotates about point P.

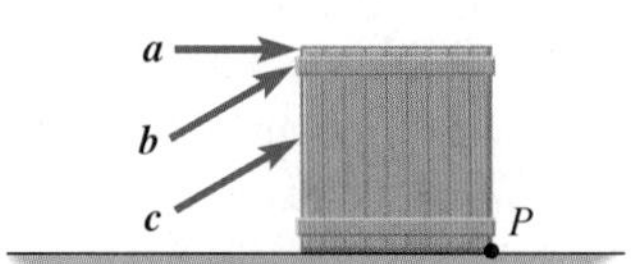

2. When both are expressed in terms of SI *base* units, torque has the same units as
   (a) angular acceleration (b) angular momentum
   (c) force (d) energy
   (e) rotational inertia (f) angular velocity

Questions 3–4: A uniform solid cylinder rolls without slipping down an incline. At the bottom of the incline, the speed, v, of the cylinder is measured and the translational and rotational kinetic energies (K_{tr}, K_{rot}) are calculated. A hole is drilled through the cylinder along its axis and the experiment is repeated; at the bottom of the incline the cylinder now has speed v' and translational and rotational kinetic energies K'_{tr} and K'_{rot}.

3. How does the speed of the cylinder compare with its original value?
   (a) $v' < v$ (b) $v' = v$ (c) $v' > v$
   (d) Answer depends on the radius of the hole drilled.

4. How does the ratio of rotational to translational kinetic energy of the cylinder compare to its original value?
   (a) $\dfrac{K'_{rot}}{K'_{tr}} < \dfrac{K_{rot}}{K_{tr}}$ (b) $\dfrac{K'_{rot}}{K'_{tr}} = \dfrac{K_{rot}}{K_{tr}}$ (c) $\dfrac{K'_{rot}}{K'_{tr}} > \dfrac{K_{rot}}{K_{tr}}$
   (d) Answer depends on the radius of the hole drilled.

5. The SI units of angular momentum are
   (a) $\dfrac{\text{rad}}{\text{s}}$ (b) $\dfrac{\text{rad}}{\text{s}^2}$ (c) $\dfrac{\text{kg}\cdot\text{m}}{\text{s}^2}$
   (d) $\dfrac{\text{kg}\cdot\text{m}^2}{\text{s}^2}$ (e) $\dfrac{\text{kg}\cdot\text{m}^2}{\text{s}}$ (f) $\dfrac{\text{kg}\cdot\text{m}}{\text{s}}$

6. Which of the forces in the figure produces the largest magnitude torque about the rotation axis indicated?
(a) 1 (b) 2 (c) 3 (d) 4

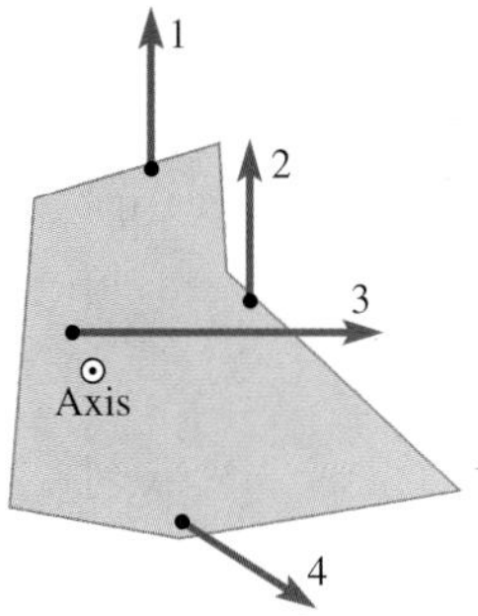

Multiple-Choice Questions 6–8

7. Which of the forces in the figure produces a CW torque about the rotation axis indicated?
(a) 3 only (b) 4 only (c) 1 and 2
(d) 1, 2, and 3 (e) 1, 2, and 4

8. Which pair of forces in the figure might produce equal magnitude torques with opposite signs?
(a) 2 and 3 (b) 2 and 4 (c) 1 and 2
(d) 1 and 3 (e) 1 and 4 (f) 3 and 4

9. A high diver in midair pulls her legs inward toward her chest in order to rotate faster. Doing so changes which of these quantities: her angular momentum L, her rotational inertia I, and her rotational kinetic energy K_{rot}?
(a) L only (b) I only (c) K_{rot} only
(d) L and I only (e) I and K_{rot} only (f) all three

10. A uniform bar of mass m is supported by a pivot at its top, about which the bar can swing like a pendulum. If a force F is applied perpendicularly to the lower end of the bar as in the diagram, how big must F be in order to hold the bar in equilibrium at an angle θ from the vertical?

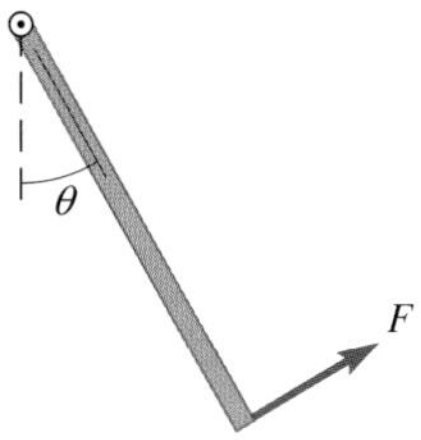

(a) $2mg$ (b) $2mg \sin \theta$
(c) $(mg/2) \sin \theta$ (d) $2mg \cos \theta$
(e) $(mg/2) \cos \theta$ (f) $mg \sin \theta$

Problems

- Combination conceptual/quantitative problem
- Biological or medical application
- Challenging problem
- Blue # Detailed solution in the Student Solutions Manual
- 1 2 Problems paired by concept
- Text website interactive or tutorial

8.1 Rotational Kinetic Energy and Rotational Inertia

1. Verify that $\frac{1}{2}I\omega^2$ has dimensions of energy.
2. What is the rotational inertia of a solid iron disk of mass 49 kg, with a thickness of 5.00 cm and radius of 20.0 cm, about an axis through its center and perpendicular to it?
3. A bowling ball made for a child has half the radius of an adult bowling ball. They are made of the same material (and therefore have the same mass *per unit volume*). By what factor is the (a) mass and (b) rotational inertia of the child's ball reduced compared with the adult ball?
4. Find the rotational inertia of the system of point particles shown in the figure assuming the system rotates about the (a) x-axis, (b) y-axis, (c) z-axis. The z-axis is perpendicular to the xy-plane and points out of the page. Point particle A has a mass of 200 g and is located at $(x, y, z) = (-3.0 \text{ cm}, 5.0 \text{ cm}, 0)$, point particle B has a mass of 300 g and is at (6.0 cm, 0, 0), and point particle C has a mass of 500 g and is at (−5.0 cm, −4.0 cm, 0). (d) What are the x- and y-coordinates of the center of mass of the system?

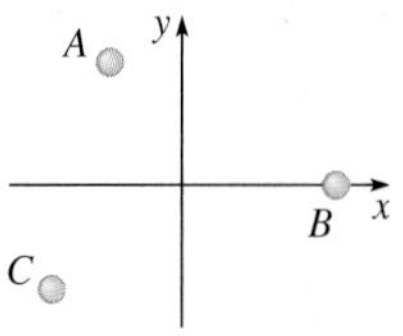

5. Four point masses of 3.0 kg each are arranged in a square on massless rods. The length of a side of the square is 0.50 m. What is the rotational inertia for rotation about an axis (a) passing through masses B and C? (b) passing through masses A and C? (c) passing through the center of the square and perpendicular to the plane of the square?

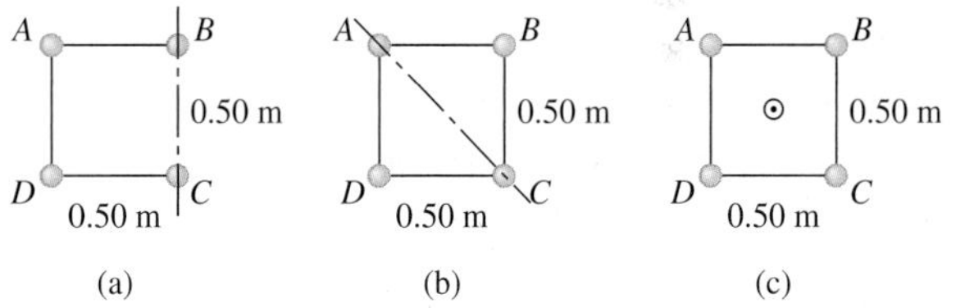

6. How much work is done by the motor in a CD player to make a CD spin, starting from rest? The CD has a diameter of 12.0 cm and a mass of 15.8 g. The laser scans at a constant tangential velocity of 1.20 m/s. Assume that the music is first detected at a radius of 20.0 mm from the center of the disk. Ignore the small circular hole at the CD's center.
7. Find the ratio of the rotational inertia of the Earth for rotation about its own axis to its rotational inertia for rotation about the Sun.
8. A bicycle has wheels of radius 0.32 m. Each wheel has a rotational inertia of 0.080 kg·m^2 about its axle. The total mass of the bicycle including the wheels and the rider is 79 kg. When coasting at constant speed, what fraction of the total kinetic energy of the bicycle (including rider) is the rotational kinetic energy of the wheels?

9. In many problems in previous chapters, cars and other objects that roll on wheels were considered to act as if they were sliding without friction. (a) Can the same assumption be made for a wheel rolling *by itself*? Explain your answer. (b) If a moving car of total mass 1300 kg has four wheels, each with rotational inertia of 0.705 kg·m^2 and radius of 35 cm, what fraction of the total kinetic energy is rotational?

10. A centrifuge has a rotational inertia of 6.5×10^{-3} kg·m^2. How much energy must be supplied to bring it from rest to 420 rad/s (4000 rpm)?

8.2 Torque

11. A mechanic turns a wrench using a force of 25 N at a distance of 16 cm from the rotation axis. The force is perpendicular to the wrench handle. What magnitude torque does she apply to the wrench?

12. The pull cord of a lawnmower engine is wound around a drum of radius 6.00 cm. While the cord is pulled with a force of 75 N to start the engine, what magnitude torque does the cord apply to the drum?

13. A child of mass 40.0 kg is sitting on a horizontal seesaw at a distance of 2.0 m from the supporting axis. What is the magnitude of the torque about the axis due to the weight of the child?

14. A 124-g mass is placed on one pan of a balance, at a point 25 cm from the support of the balance. What is the magnitude of the torque about the support exerted by the mass?

15. A uniform door weighs 50.0 N and is 1.0 m wide and 2.6 m high. What is the magnitude of the torque due to the door's own weight about a horizontal axis perpendicular to the door and passing through a corner?

16. A tower outside the Houses of Parliament in London has a famous clock commonly referred to as Big Ben, the name of its 13-ton chiming bell. The hour hand of each clock face is 2.7 m long and has a mass of 60.0 kg. Assume the hour hand to be a uniform rod attached at one end. (a) What is the torque on the clock mechanism due to the weight of one of the four hour hands when the clock strikes noon? The axis of rotation is perpendicular to a clock face and through the center of the clock. (b) What is the torque due to the weight of one hour hand about the same axis when the clock tolls 9:00 A.M.?

✦17. Any pair of equal and opposite forces acting on the same object is called a *couple*. Consider the couple in part (a) of the figure. The rotation axis is perpendicular to the page and passes through point *P*. (a) Show that the net torque due to this couple is equal to *Fd*, where *d* is the distance between the lines of action of the two forces. Because the distance *d* is independent of the location of the rotation axis, this shows that the torque is the same for any rotation axis. (b) Repeat for the couple in part (b) of the figure. Show that the torque is still *Fd* if *d* is the *perpendicular* distance between the lines of action of the forces.

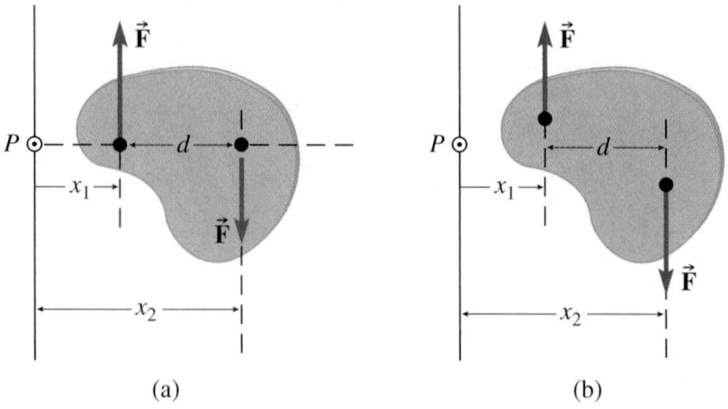

18. A 46.4-N force is applied to the outer edge of a door of width 1.26 m in such a way that it acts (a) perpendicular to the door, (b) at an angle of 43.0° with respect to the door surface, (c) so that the line of action of the force passes through the axis of the door hinges. Find the torque for these three cases.

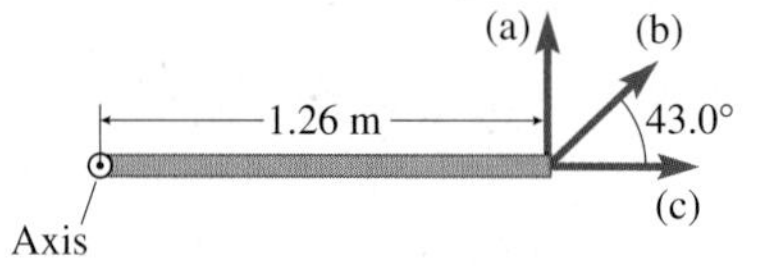

19. A trap door, of length and width 1.65 m, is held open at an angle of 65.0° with respect to the floor. A rope is attached to the raised edge of the door and fastened to the wall behind the door in such a position that the rope pulls perpendicularly to the trap door. If the mass of the trap door is 16.8 kg, what is the torque exerted on the trap door by the rope? (WW tutorial: deck hatch)

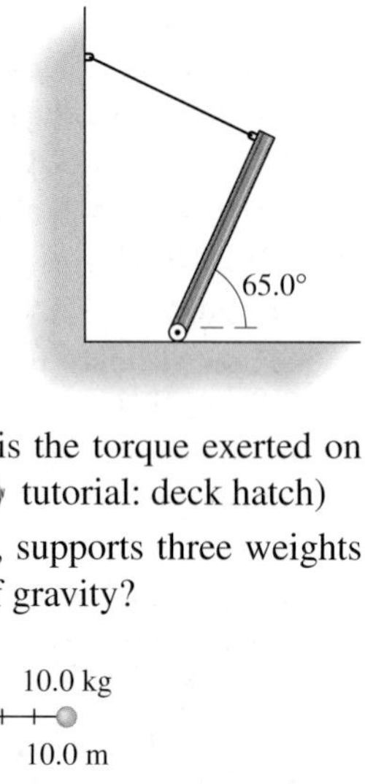

20. A weightless rod, 10.0 m long, supports three weights as shown. Where is its center of gravity?

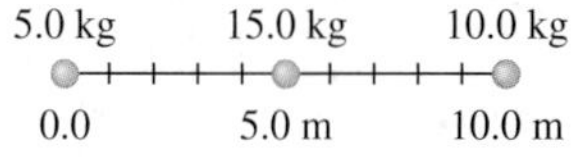

21. A door weighing 300.0 N measures 2.00 m × 3.00 m and is of uniform density; that is, the mass is uniformly distributed throughout the volume. A doorknob is attached to the door as shown. Where is the center of gravity if the doorknob weighs 5.0 N and is located 0.25 m from the edge?

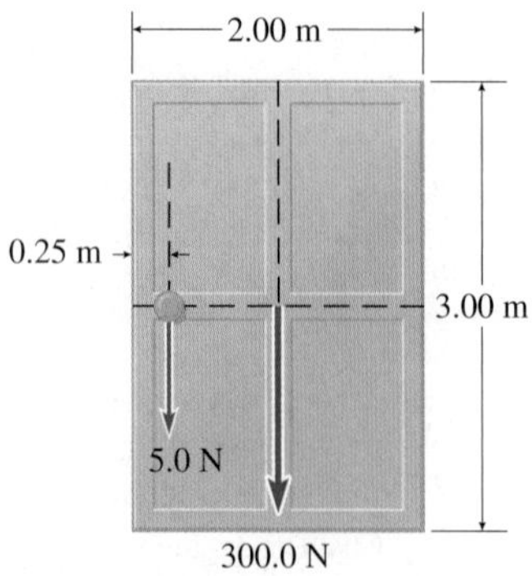

✦22. A plate of uniform thickness is shaped as shown. Where is the center of gravity? Assume the origin (0, 0) is located at the lower left corner of the plate; the upper left corner is at $(0, s)$ and upper right corner is at (s, s).

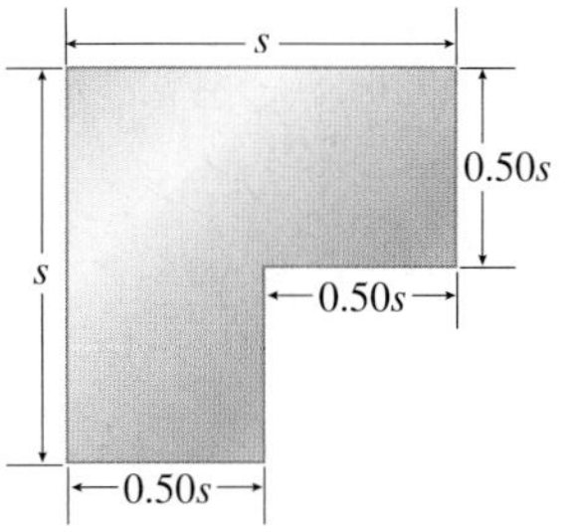

8.3 Calculating Work Done from the Torque

23. A stone used to grind wheat into flour is turned through 12 revolutions by a constant force of 20.0 N applied to the rim of a 10.0-cm-radius shaft connected to the wheel. How much work is done on the stone during the 12 revolutions?

24. The radius of a wheel is 0.500 m. A rope is wound around the outer rim of the wheel. The rope is pulled with a force of magnitude 5.00 N, unwinding the rope and making the wheel spin CCW about its central axis. Ignore the mass of the rope. (a) How much rope unwinds while the wheel makes 1.00 revolution? (b) How much work is done by the rope on the wheel during this time? (c) What is the torque on the wheel due to the rope? (d) What is the angular displacement $\Delta\theta$, in radians, of the wheel during 1.00 revolution? (e) Show that the numerical value of the work done is equal to the product $\tau\Delta\theta$.

✦25. A flywheel of mass 182 kg has an effective radius of 0.62 m (assume the mass is concentrated along a circumference located at the effective radius of the flywheel). (a) How much work is done to bring this wheel from rest to a speed of 120 rpm in a time interval of 30.0 s? (b) What is the applied torque on the flywheel (assumed constant)?

✦26. A Ferris wheel rotates because a motor exerts a torque on the wheel. The radius of the London Eye, a huge observation wheel on the banks of the Thames, is 67.5 m and its mass is 1.90×10^6 kg. The cruising angular speed of the wheel is 3.50×10^{-3} rad/s. (a) How much work does the motor need to do to bring the stationary wheel up to cruising speed? [*Hint:* Treat the wheel as a hoop.] (b) What is the torque (assumed constant) the motor needs to provide to the wheel if it takes 20.0 s to reach the cruising angular speed?

8.4 Rotational Equilibrium

27. A rod is being used as a lever as shown. The fulcrum is 1.2 m from the load and 2.4 m from the applied force. If the load has a mass of 20.0 kg, what force must be applied to lift the load?

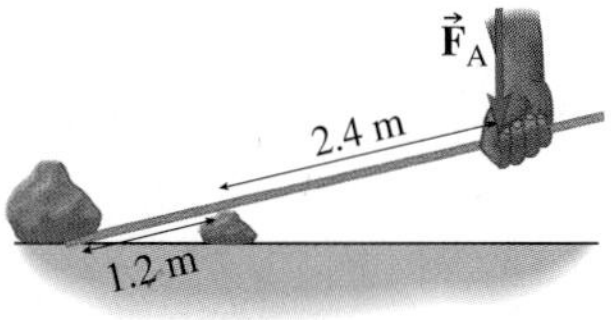

28. A weight of 1200 N rests on a lever at a point 0.50 m from a support. On the same side of the support, at a distance of 3.0 m from it, an upward force with magnitude F is applied. Ignore the weight of the board itself. If the system is in equilibrium, what is F?

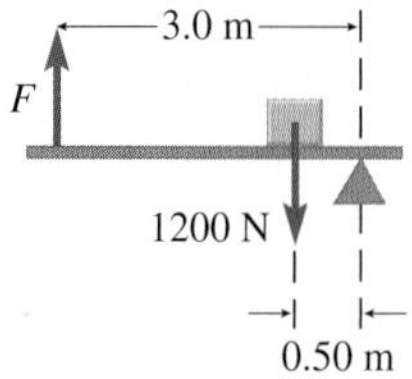

29. A sculpture is 4.00 m tall and has its center of gravity located 1.80 m above the center of its base. The base is a square with a side of 1.10 m. To what angle θ can the sculpture be tipped before it falls over? (W tutorial: filing cabinet)

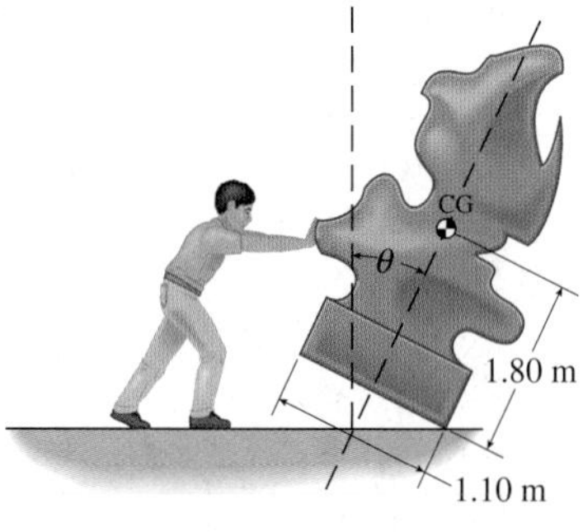

30. A house painter is standing on a uniform, horizontal platform that is held in equilibrium by two cables attached to supports on the roof. The painter has a mass of 75 kg and the mass of the platform is 20.0 kg. The distance from the left end of the platform to where the painter is standing is $d = 2.0$ m and the total length of the platform is 5.0 m. (a) How large is the force exerted by the left-hand cable on the platform? (b) How large is the force exerted by the right-hand cable?

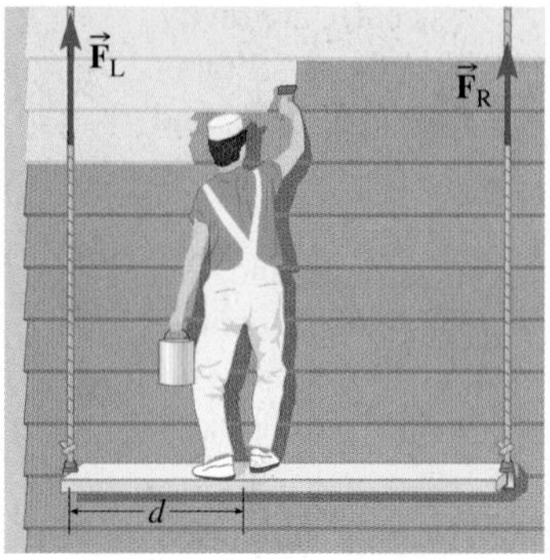

31. Four identical uniform metersticks are stacked on a table as shown. Where is the x-coordinate of the CM of the metersticks if the origin is chosen at the left end of the lowest stick? Why does the system balance?

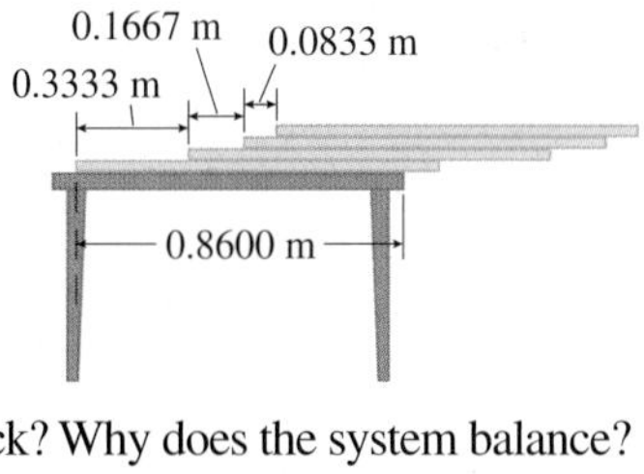

✦32. A uniform diving board, of length 5.0 m and mass 55 kg, is supported at two points; one support is located 3.4 m from the end of the board and the second is at 4.6 m from the end (see Fig. 8.19). What are the forces acting on the board due to the two supports when a diver of mass 65 kg stands at the end of the board over the water? Assume that these forces are vertical. (W tutorial: plank) [*Hint:* In this problem, consider using two different torque equations about different rotation axes. This may help you determine the directions of the two forces.]

✦33. A house painter stands 3.0 m above the ground on a 5.0-m-long ladder that leans against the wall at a point

4.7 m above the ground. The painter weighs 680 N and the ladder weighs 120 N. Assuming no friction between the house and the upper end of the ladder, find the force of friction that the driveway exerts on the bottom of the ladder. (interactive: ladder; tutorial: ladder)

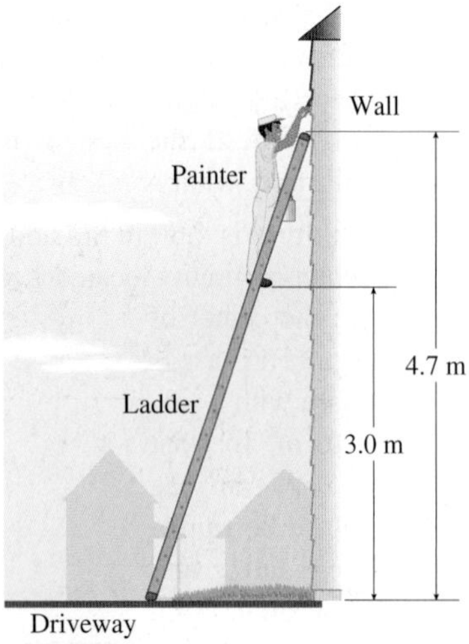

34. A mountain climber is rappelling down a vertical wall. The rope attaches to a buckle strapped to the climber's waist 15 cm to the right of his center of gravity. If the climber weighs 770 N, find (a) the tension in the rope and (b) the magnitude and direction of the contact force exerted by the wall on the climber's feet.

35. A sign is supported by a uniform horizontal boom of length 3.00 m and weight 80.0 N. A cable, inclined at an angle of 35° with the boom, is attached at a distance of 2.38 m from the hinge at the wall. The weight of the sign is 120.0 N. What is the tension in the cable and what are the horizontal and vertical forces F_x and F_y exerted on the boom by the hinge? Comment on the magnitude of F_y.

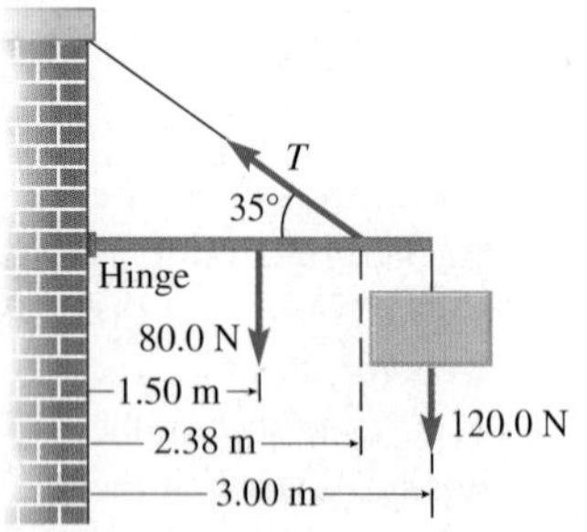

36. A boom of mass m supports a steel girder of weight W hanging from its end. One end of the boom is hinged at the floor; a cable attaches to the other end of the boom and pulls horizontally on it. The boom makes an angle θ with the horizontal. Find the tension in the cable as a function of m, W, θ, and g. Comment on the tension at $\theta = 0$ and $\theta = 90°$.

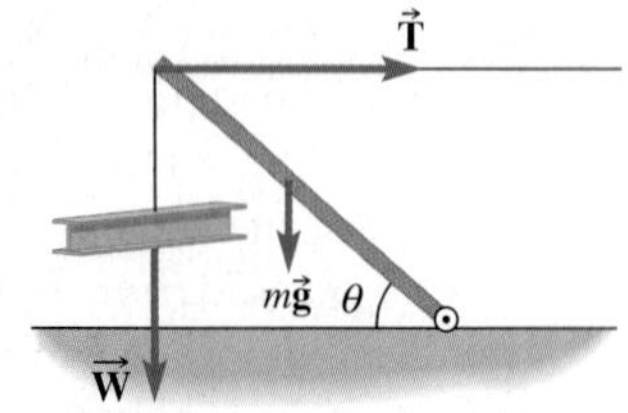

37. You are asked to hang a uniform beam and sign using a cable that has a breaking strength of 417 N. The store owner desires that it hang out over the sidewalk as shown. The sign has a weight of 200.0 N and the beam's weight is 50.0 N. The beam's length is 1.50 m and the sign's dimensions are 1.00 m horizontally × 0.80 m vertically. What is the minimum angle θ that you can have between the beam and cable?

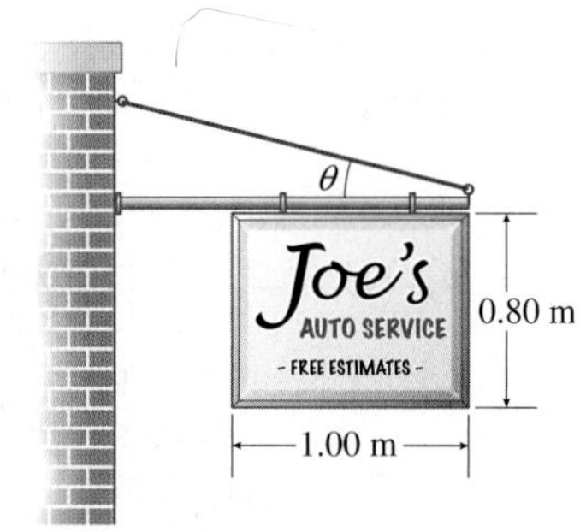

38. Refer to Problem 37. You chose an angle θ of 33.8°. An 8.7-kg cat has climbed onto the beam and is walking from the wall toward the point where the cable meets the beam. How far can the cat walk before the cable breaks?

39. A man is doing push-ups. He has a mass of 68 kg and his center of gravity is located at a horizontal distance of 0.70 m from his palms and 1.00 m from his feet. Find the forces exerted by the floor on his palms and feet.

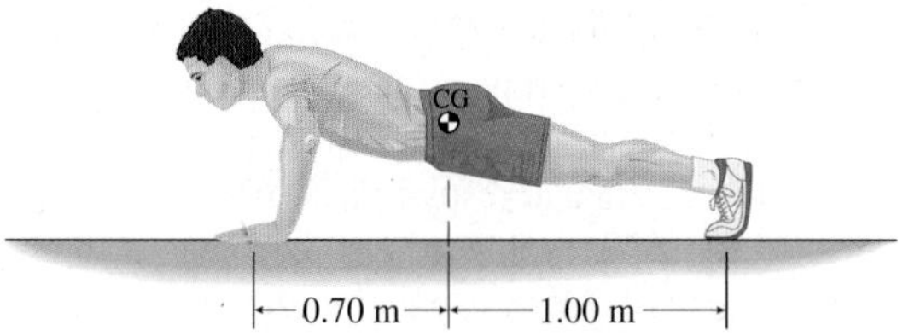

8.5 Equilibrium in the Human Body

40. Your friend balances a package with mass $m = 10$ kg on top of his head while standing. The mass of his upper body is $M = 55$ kg (about 65% of his total mass). Because the spine is vertical rather than horizontal, the force exerted by the sacrum on the spine ($\vec{\mathbf{F}}_s$ in Fig 8.32) is directed approximately straight up and the force exerted by the back muscles ($\vec{\mathbf{F}}_b$) is negligibly small. Find the magnitude of $\vec{\mathbf{F}}_s$.

41. Find the tension in the Achilles tendon and the force that the tibia exerts on the ankle joint when a person who weighs 750 N supports himself on the ball of one foot. The normal force $N = 750$ N pushes up on the ball of the foot on one side of the ankle joint, while the Achilles tendon pulls up on the foot on the other side of the joint.

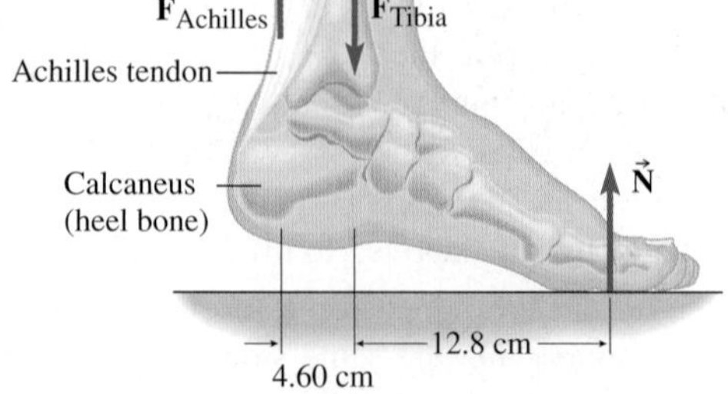

42. In the movie *Terminator*, Arnold Schwarzenegger lifts someone up by the neck and, with both arms fully extended and horizontal, holds the person off the ground. If the person being held weighs 700 N, is 60 cm from the shoulder joint, and Arnold has an anatomy analogous to that in Fig. 8.30, what force must *each* of the deltoid muscles exert to perform this task?

43. Find the force exerted by the biceps muscle in holding a 1-L milk carton (weight 9.9 N) with the forearm parallel to the floor. Assume that the hand is 35.0 cm from the elbow and that the upper arm is 30.0 cm long. The elbow is bent at a right angle and one tendon of the biceps is attached to the forearm at a position 5.00 cm from the elbow, while the other tendon is attached at 30.0 cm from the elbow. The weight of the forearm and empty hand is 18.0 N and the center of gravity of the forearm is at a distance of 16.5 cm from the elbow.

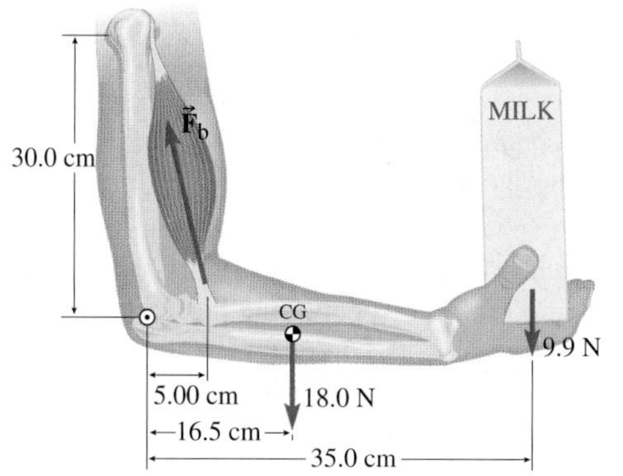

44. A person is doing leg lifts with 3.0-kg ankle weights. She is sitting in a chair with her legs bent at a right angle initially. The quadriceps muscles are attached to the patella via a tendon; the patella is connected to the tibia by the patellar tendon, which attaches to bone 10.0 cm below the knee joint. Assume that the tendon pulls at an angle of 20.0° *with respect to the lower leg*, regardless of the position of the lower leg. The lower leg has a mass of 5.0 kg and its center of gravity is 22 cm below the knee. The ankle weight is 41 cm from the knee. If the person lifts one leg, find the force exerted by the patellar tendon to hold the leg at an angle of (a) 30.0° and (b) 90.0° with respect to the vertical.

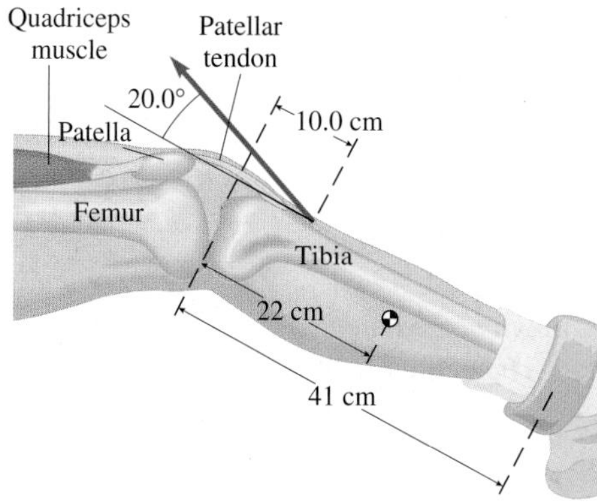

45. One day when your friend from Problem 40 is picking up a package, you notice that he bends at the waist to pick it up rather than keeping his back straight and bending his knees. You suspect that the lower back pain he complains about is caused by the large force on his lower vertebrae ($\vec{\mathbf{F}}_s$ in Fig. 8.32) when he lifts objects in this way. Suppose that when the spine is horizontal, the back muscles exert a force $\vec{\mathbf{F}}_b$ as in Fig. 8.32 (44 cm from the sacrum and at an angle of 12° to the horizontal). Assume that the CM of his upper body (including the arms) is at its geometric center, 38 cm from the sacrum. Find the horizontal component of $\vec{\mathbf{F}}_s$ when your friend is holding a 10-kg package at a distance of 76 cm from his sacrum. Compare this with the magnitude of $\vec{\mathbf{F}}_s$ found in Problem 40.

46. A man is trying to lift 60.0 kg off the floor by bending at the waist (see Fig. 8.32). Assume that the man's upper body weighs 455 N and the upper body's center of gravity is 38 cm from the sacrum (tailbone). (a) If, when bent over, the hands are a horizontal distance of 76 cm from the sacrum, what torque must be exerted by the erector spinae muscles to lift 60.0 kg off the floor? (The axis of rotation passes through the sacrum, as shown in Fig. 8.32.) (b) When bent over, the erector spinae muscles are a horizontal distance of 44 cm from the sacrum and act at a 12° angle above the horizontal. What force ($\vec{\mathbf{F}}_b$ in Fig. 8.32) do the erector spinae muscles need to exert to lift the weight? (c) What is the component of this force that compresses the spinal column?

8.6 Rotational Form of Newton's Second Law

47. Verify that the units of the rotational form of Newton's second law [Eq. (8-9)] are consistent. In other words, show that the product of a rotational inertia expressed in kg·m^2 and an angular acceleration expressed in rad/s^2 is a torque expressed in N·m.

48. A spinning flywheel has rotational inertia $I = 400.0$ kg·m^2. Its angular velocity decreases from 20.0 rad/s to zero in 300.0 s due to friction. What is the frictional torque acting?

49. A turntable must spin at 33.3 rpm (3.49 rad/s) to play an old-fashioned vinyl record. How much torque must the motor deliver if the turntable is to reach its final angular speed in 2.0 revolutions, starting from rest? The turntable is a uniform disk of diameter 30.5 cm and mass 0.22 kg.

50. A lawn sprinkler has three spouts that spray water, each 15.0 cm long. As the water is sprayed, the sprinkler turns around in a circle. The sprinkler has a total rotational inertia of

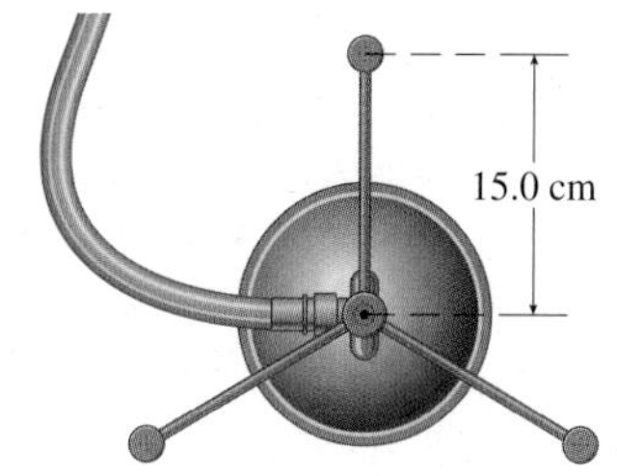

9.20×10^{-2} kg·m^2. If the sprinkler starts from rest and takes 3.20 s to reach its final speed of 2.2 rev/s, what force does each spout exert on the sprinkler?

51. A chain pulls tangentially on a 40.6-kg uniform cylindrical gear with a tension of 72.5 N. The chain is attached along the outside radius of the gear at 0.650 m from the axis of rotation. Starting from rest, the gear takes 1.70 s to reach its rotational speed of 1.35 rev/s. What is the total frictional torque opposing the rotation of the gear?

52. Four masses are arranged as shown. They are connected by rigid, massless rods of lengths 0.75 m and 0.50 m. What torque must be applied to cause an angular acceleration of 0.75 rad/s^2 about the axis shown?

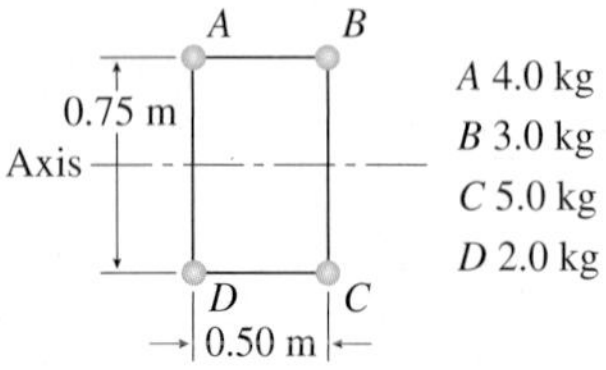

53. A bicycle wheel, of radius 0.30 m and mass 2 kg (concentrated on the rim), is rotating at 4.00 rev/s. After 50 s the wheel comes to a stop because of friction. What is the magnitude of the average torque due to frictional forces?

54. A playground merry-go-round (see Fig. 8.5), made in the shape of a solid disk, has a diameter of 2.50 m and a mass of 350.0 kg. Two children, each of mass 30.0 kg, sit on opposite sides at the edge of the platform. Approximate the children as point masses. (a) What torque is required to bring the merry-go-round from rest to 25 rpm in 20.0 s? (b) If two other bigger children are going to push on the merry-go-round rim to produce this acceleration, with what force magnitude must each child push? (tutorial: roundabout)

55. Two children standing on opposite sides of a merry-go-round (see Fig. 8.5) are trying to rotate it. They each push in opposite directions with forces of magnitude 10.0 N. (a) If the merry-go-round has a mass of 180 kg and a radius of 2.0 m, what is the angular acceleration of the merry-go-round? (Assume the merry-go-round is a uniform disk.) (b) How fast is the merry-go-round rotating after 4.0 s?

56. Refer to Atwood's machine (Example 8.2). (a) Assuming that the cord does not slip as it passes around the pulley, what is the relationship between the angular acceleration of the pulley (α) and the magnitude of the linear acceleration of the blocks (a)? (b) What is the net torque on the pulley about its axis of rotation in terms of the tensions T_1 and T_2 in the left and right sides of the cord? (c) Explain why the tensions cannot be equal if $m_1 \neq m_2$. (d) Apply Newton's second law to each of the blocks and Newton's second law for rotation to the pulley. Use these three equations to solve for a, T_1, and T_2. (e) Since the blocks move with constant acceleration, use the result of Example 8.2 along with the constant acceleration equation $v_{fy}^2 - v_{iy}^2 = 2a_y \Delta y$ to check your answer for a.

57. Derive the rotational form of Newton's second law as follows. Consider a rigid object that consists of a large number N of particles. Let F_i, m_i, and r_i represent the tangential component of the net force acting on the ith particle, the mass of that particle, and the particle's distance from the axis of rotation, respectively. (a) Use Newton's second law to find a_i, the particle's tangential acceleration. (b) Find the torque acting on this particle. (c) Replace a_i with an equivalent expression in terms of the angular acceleration α. (d) Sum the torques due to all the particles and show that

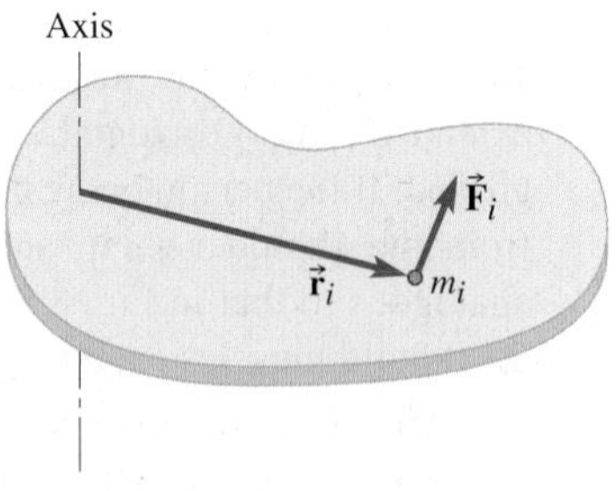

$$\sum_{i=1}^{N} \tau_i = I\alpha$$

8.7 The Motion of Rolling Objects

58. A solid sphere is rolling without slipping or sliding down a board that is tilted at an angle of 35° with respect to the horizontal. What is its acceleration?

59. A solid sphere is released from rest and allowed to roll down a board that has one end resting on the floor and is tilted at 30° with respect to the horizontal. If the sphere is released from a height of 60 cm above the floor, what is the sphere's speed when it reaches the lowest end of the board?

60. A hollow cylinder, a uniform solid sphere, and a uniform solid cylinder all have the same mass m. The three objects are rolling on a horizontal surface with identical translational speeds v. Find their total kinetic energies in terms of m and v and order them from smallest to largest.

61. A solid sphere of mass 0.600 kg rolls without slipping along a horizontal surface with a translational speed of 5.00 m/s. It comes to an incline that makes an angle of 30° with the horizontal surface. Ignoring energy losses due to friction, to what vertical height above the horizontal surface does the sphere rise on the incline?

62. A bucket of water with a mass of 2.0 kg is attached to a rope that is wound around a cylinder. The cylinder has a mass of 3.0 kg and is mounted horizontally on frictionless bearings. The bucket is released from rest. (a) Find its speed after it has fallen through a distance of 0.80 m. What are (b) the tension in the rope and (c) the acceleration of the bucket?

63. A 1.10-kg bucket is tied to a rope that is wrapped around a pole mounted horizontally on frictionless bearings. The cylindrical pole has a diameter of 0.340 m and a mass of 2.60 kg. When the

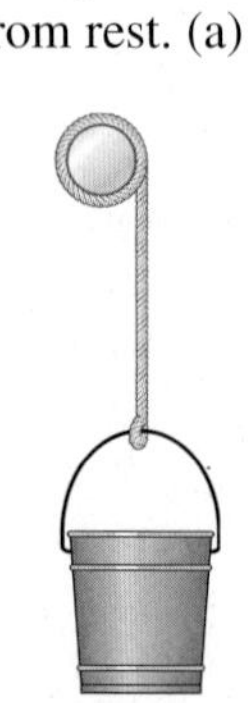

Problems 62 and 63

bucket is released from rest, how long will it take to fall to the bottom of the well, a distance of 17.0 m?

64. A uniform solid cylinder rolls without slipping down an incline. A hole is drilled through the cylinder along its axis. The radius of the hole is 0.50 times the (outer) radius of the cylinder. (a) Does the cylinder take more or less time to roll down the incline now that the hole has been drilled? Explain. (b) By what percentage does drilling the hole change the time for the cylinder to roll down the incline? (tutorial: rolling)

65. A solid sphere of radius R and mass M *slides* without friction down a loop-the-loop track. The sphere starts from rest at a height of h above the horizontal. Assume that the radius of the sphere is small compared to the radius r of the loop. (a) Find the minimum value of h in terms of r so that the sphere remains on the track all the way around the loop. (b) Find the minimum value of h if, instead, the sphere rolls without slipping on the track.

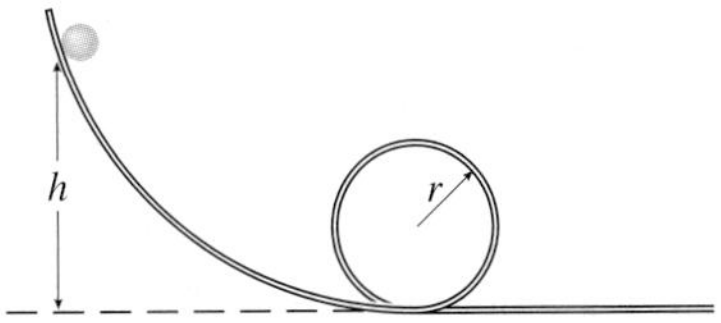

Problems 65 and 66

66. A hollow cylinder, of radius R and mass M, rolls without slipping down a loop-the-loop track of radius r. The cylinder starts from rest at a height h above the horizontal section of track. What is the minimum value of h so that the cylinder remains on the track all the way around the loop?

67. If the hollow cylinder of Problem 66 is replaced with a solid sphere, will the minimum value of h increase, decrease, or remain the same? Once you think you know the answer and can explain why, redo the calculation to find h.

68. The string in a yo-yo is wound around an axle of radius 0.500 cm. The yo-yo has both rotational and translational motion, like a rolling object, and has mass 0.200 kg and outer radius 2.00 cm. Starting from rest, it rotates and falls a distance of 1.00 m (the length of the string). Assume for simplicity that the yo-yo is a uniform circular disk and that the string is thin compared to the radius of the axle. (a) What is the speed of the yo-yo when it reaches the distance of 1.00 m? (b) How long does it take to fall? [*Hint:* The translational and rotational kinetic energies are related, but the yo-yo is *not* rolling on its outer radius.]

8.8 Angular Momentum

69. A turntable of mass 5.00 kg has a radius of 0.100 m and spins with a frequency of 0.550 rev/s. What is its angular momentum? Assume the turntable is a uniform disk.

70. Assume the Earth is a uniform solid sphere with radius of 6.37×10^6 m and mass of 5.97×10^{24} kg. Find the magnitude of the angular momentum of the Earth due to rotation about its axis.

71. The mass of a flywheel is 5.6×10^4 kg. This particular flywheel has its mass concentrated at the rim of the wheel. If the radius of the wheel is 2.6 m and it is rotating at 350 rpm, what is the magnitude of its angular momentum?

72. The angular momentum of a spinning wheel is 240 kg·m²/s. After application of a constant braking torque for 2.5 s, it slows and has a new angular momentum of 115 kg·m²/s. What is the torque applied?

73. How long would a braking torque of 4.00 N·m have to act to just stop a spinning wheel that has an initial angular momentum of 6.40 kg·m²/s?

74. A figure skater is spinning at a rate of 1.0 rev/s with her arms outstretched. She then draws her arms in to her chest, reducing her rotational inertia to 67% of its original value. What is her new rate of rotation?

75. A skater is initially spinning at a rate of 10.0 rad/s with a rotational inertia of 2.50 kg·m² when her arms are extended. What is her angular velocity after she pulls her arms in and reduces her rotational inertia to 1.60 kg·m²?

76. A uniform disk with a mass of 800 g and radius 17.0 cm is rotating on frictionless bearings with an angular speed of 18.0 Hz when Jill drops a 120-g clod of clay on a point 8.00 cm from the center of the disk, where it sticks. What is the new angular speed of the disk?

77. A spoked wheel with a radius of 40.0 cm and a mass of 2.00 kg is mounted horizontally on frictionless bearings. JiaJun puts his 0.500-kg guinea pig on the outer edge of the wheel. The guinea pig begins to run along the edge of the wheel with a speed of 20.0 cm/s with respect to the ground. What is the angular velocity of the wheel? Assume the spokes of the wheel have negligible mass.

78. A diver can change his rotational inertia by drawing his arms and legs close to his body in the tuck position. After he leaves the diving board (with some unknown angular velocity), he pulls himself into a ball as closely as possible and makes 2.00 complete rotations in 1.33 s. If his rotational inertia decreases by a factor of 3.00 when he goes from the straight to the tuck position, what was his angular velocity when he left the diving board?

79. The rotational inertia for a diver in a pike position is about 15.5 kg·m²; it is only 8.0 kg·m² in a tuck position. (a) If the diver gives himself an initial angular momentum of 106 kg·m²/s as he jumps off the board, how many turns can he make when jumping off a 10.0-m platform

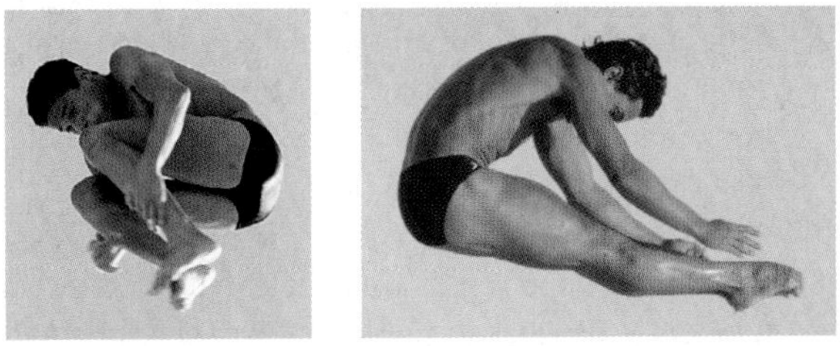
(a) (b)

Problem 79. (a) Mark Ruiz in the tuck position. (b) Gregory Louganis in the pike position.

in a tuck position? (b) How many in a pike position? [*Hint:* Gravity exerts no torque on the person as he falls; assume he is rotating throughout the 10.0-m dive.]

80. Consider the merry-go-round of Practice Problem 8.1. The child is initially standing on the ground when the merry-go-round is rotating at 0.75 rev/s. The child then steps on the merry-go-round. How fast is the merry-go-round rotating now? By how much did the rotational kinetic energy of the merry-go-round and child change?

8.9 The Vector Nature of Angular Momentum

Problems 81 and 82. A solid cylindrical disk is to be used as a stabilizer in a ship. By using a massive disk rotating in the hold of the ship, the captain knows that a large torque is required to tilt its angular momentum vector. The mass of the disk to be used is 1.00×10^5 kg and it has a radius of 2.00 m.

81. If the cylinder rotates at 300.0 rpm, what is the magnitude of the average torque required to tilt its axis by 60.0° in a time of 3.00 s? [*Hint:* Draw a vector diagram of the initial and final angular momenta.]

82. How should the disk be oriented to prevent rocking from side to side and from bow to stern? Does this orientation make it difficult to steer the ship? Explain.

Comprehensive Problems

83. The Moon's distance from Earth varies between 3.56×10^5 km at perigee and 4.07×10^5 km at apogee. What is the ratio of its orbital speed around Earth at perigee to that at apogee?

84. A ceiling fan has four blades, each with a mass of 0.35 kg and a length of 60 cm. Model each blade as a rod connected to the fan axle at one end. When the fan is turned on, it takes 4.35 s for the fan to reach its final angular speed of 1.8 rev/s. What torque was applied to the fan by the motor? Ignore torque due to the air.

85. The distance from the center of the breastbone to a man's hand, with the arm outstretched and horizontal to the floor, is 1.0 m. The man is holding a 10.0-kg dumbbell, oriented vertically, in his hand, with the arm horizontal. What is the torque due to this weight about a horizontal axis through the breastbone perpendicular to his chest?

86. A uniform rod of length L is free to pivot around an axis through its upper end. If it is released from rest when horizontal, at what speed is the lower end moving at its lowest point? [*Hint:* The gravitational potential energy change is determined by the change in height of the center of gravity.]

87. A gymnast is performing a giant swing on the high bar. In a simplified model of the giant swing, assume that the gymnast keeps his arms and body straight as he swings all the way around the upper bar. Assume also that the gymnast does no work during the swing. With what angular speed should he be moving at the bottom of the giant swing in order to make it all the way around? The distance from the bar to his feet is 2.0 m and his center of gravity is 1.0 m from his feet.

Problem 87. Notice that the angular speed is much greater at the bottom of the swing.

88. The 12.2-m crane weighs 18 kN and is lifting a 67-kN load. The hoisting cable (tension T_1) passes over a pulley at the top of the crane and attaches to an electric winch in the cab. The pendant cable (tension T_2), which supports the crane, is fixed to the top of the crane. Find the tensions in the two cables and the force $\vec{\mathbf{F}}_p$ at the pivot.

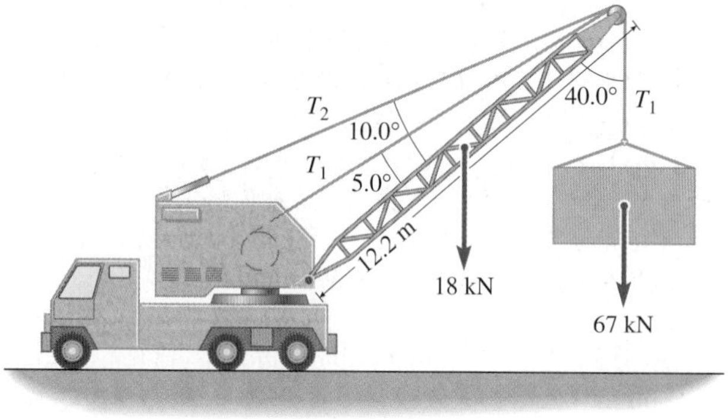

89. A collection of objects is set to rolling, without slipping, down a slope inclined at 30°. The objects are a solid sphere, a hollow sphere, a solid cylinder, and a hollow cylinder. A frictionless cube is also allowed to slide down the same incline. Which one gets to the bottom first? List the others in the order they arrive at the finish line.

90. A uniform cylinder with a radius of 15 cm has been attached to two cords and the cords are wound around it and hung from the ceiling. The cylinder is released from rest and the cords unwind as the cylinder descends. (a) What is the acceleration of the cylinder? (b) If the mass of the cylinder is 2.6 kg, what is the tension in each cord?

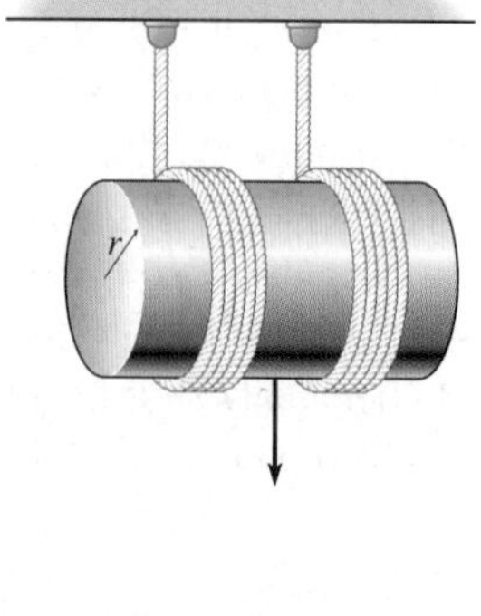

91. A modern sculpture has a large horizontal spring, with a spring constant of 275 N/m, that is attached to a 53.0-kg piece of uniform metal at its end and holds the metal at an angle of 50.0° above the horizontal direction. The other end of the metal is wedged into a corner as shown. By how much has the spring stretched?

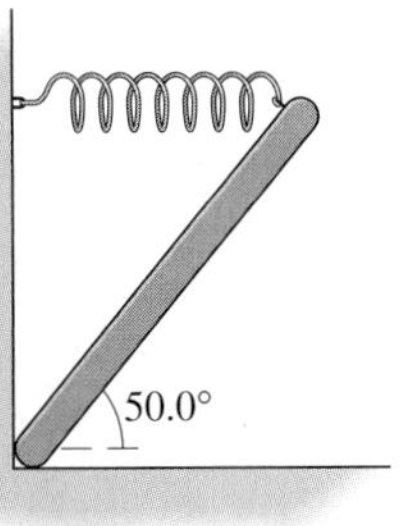

92. A painter (mass 61 kg) is walking along a trestle, consisting of a uniform plank (mass 20.0 kg, length 6.00 m) balanced on two sawhorses. Each sawhorse is placed 1.40 m from an end of the plank. A paint bucket (mass 4.0 kg, diameter 28 cm) is placed as close as possible to the right-hand edge of the plank while still having the whole bucket in contact with the plank. (a) How close to the right-hand edge of the plank can the painter walk before tipping the plank and spilling the paint? (b) How close to the left-hand edge can the same painter walk before causing the plank to tip? [*Hint:* As the painter walks toward the right-hand edge of the plank and the plank starts to tip clockwise, what is the force acting upward on the plank from the left-hand sawhorse support?]

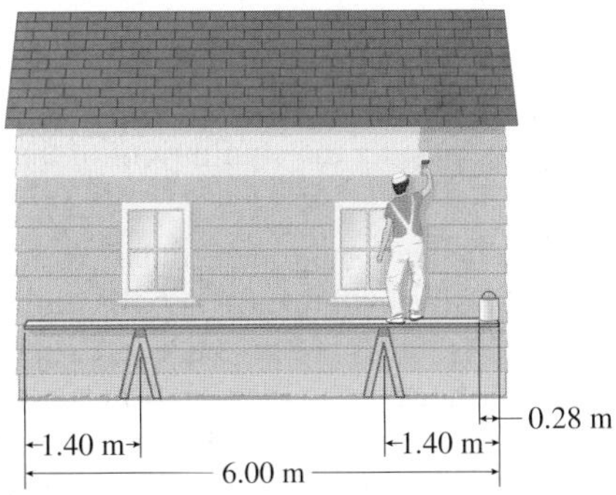

93. An experimental flywheel, used to store energy and replace an automobile engine, is a solid disk of mass 200.0 kg and radius 0.40 m. (a) What is its rotational inertia? (b) When driving at 22.4 m/s (50 mph), the fully energized flywheel is rotating at an angular speed of 3160 rad/s. What is the initial rotational kinetic energy of the flywheel? (c) If the total mass of the car is 1000.0 kg, find the ratio of the initial rotational kinetic energy of the flywheel to the translational kinetic energy of the car. (d) If the force of air resistance on the car is 670.0 N, how far can the car travel at a speed of 22.4 m/s (50 mph) with the initial stored energy? Ignore losses of mechanical energy due to means other than air resistance.

94. (a) Assume the Earth is a uniform solid sphere. Find the kinetic energy of the Earth due to its rotation about its axis. (b) Suppose we could somehow extract 1.0% of the Earth's rotational kinetic energy to use for other purposes. By how much would that change the length of the day? (c) For how many years would 1.0% of the Earth's rotational kinetic energy supply the world's energy usage (assume a constant 1.0×10^{21} J per year)?

95. A flat object in the *xy*-plane is free to rotate about the *z*-axis. The gravitational field is uniform in the $-y$-direction. Think of the object as a large number of particles with masses m_i located at coordinates (x_i, y_i), as in the figure. (a) Show that the torques on the particles about the *z*-axis can be written $\tau_i = -x_i m_i g$. (b) Show that if the center of gravity is located at (x_{CG}, y_{CG}), the total torque due to gravity on the object must be $\Sigma \tau_i = -x_{CG} M g$, where M is the total mass of the object. (c) Show that $x_{CG} = x_{CM}$. (This same line of reasoning can be applied to objects that are not flat and to other axes of rotation to show that $y_{CG} = y_{CM}$ and $z_{CG} = z_{CM}$.)

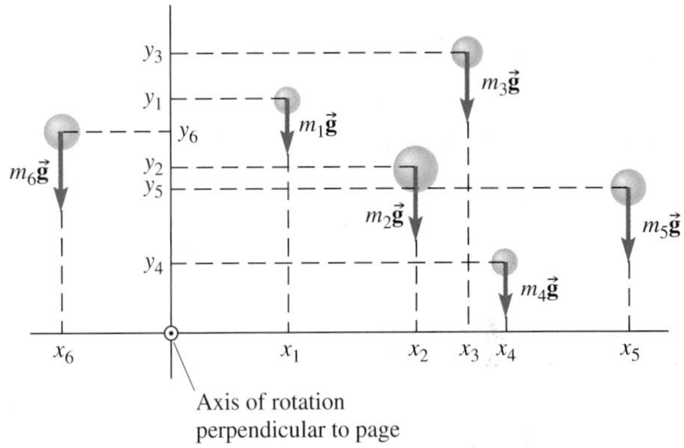

96. The operation of the Princeton Tokomak Fusion Test Reactor requires large bursts of energy. The power needed exceeds the amount that can be supplied by the utility company. Prior to pulsing the reactor, energy is stored in a giant flywheel of mass 7.27×10^5 kg and rotational inertia 4.55×10^6 kg·m^2. The flywheel rotates at a maximum angular speed of 386 rpm. When the stored energy is needed to operate the reactor, the flywheel is connected to an electrical generator, which converts some of the rotational kinetic energy into electric energy. (a) If the flywheel is a uniform disk, what is its radius? (b) If the flywheel is a hollow cylinder with its mass concentrated at the rim, what is its radius? (c) If the flywheel slows to 252 rpm in 5.00 s, what is the average power supplied by the flywheel during that time?

97. A box of mass 42 kg sits on top of a ladder. Ignoring the weight of the ladder, find the tension in the rope. Assume that the rope exerts horizontal forces on the ladder at each end. [*Hint:* Use a symmetry argument; then analyze the forces and torques on one side of the ladder.]

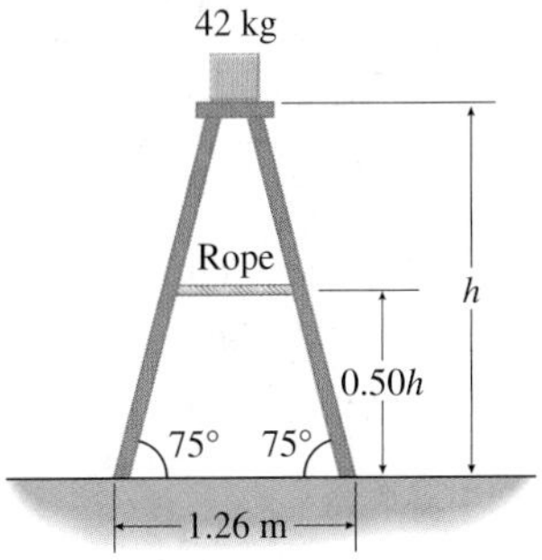

98. A person is trying to lift a ladder of mass 15 kg and length 8.0 m. The person is exerting a vertical force on the ladder at a point of contact 2.0 m from the center of

gravity. The opposite end of the ladder rests on the floor. (a) When the ladder makes an angle of 60.0° with the floor, what is this vertical force? (b) A person tries to help by lifting the ladder at the point of contact with the floor. Does this help the person trying to lift the ladder? Explain.

99. A crustacean (*Hemisquilla ensigera*) rotates its anterior limb to strike a mollusk, intending to break it open. The limb reaches an angular velocity of 175 rad/s in 1.50 ms. We can approximate the limb as a thin rod rotating about an axis perpendicular to one end (the joint where the limb attaches to the crustacean). (a) If the mass of the limb is 28.0 g and the length is 3.80 cm, what is the rotational inertia of the limb about that axis? (b) If the extensor muscle is 3.00 mm from the joint and acts perpendicular to the limb, what is the muscular force required to achieve the blow?

100. A block of mass m_2 hangs from a rope. The rope wraps around a pulley of rotational inertia I and then attaches to a second block of mass m_1, which sits on a frictionless table. What is the acceleration of the blocks when they are released?

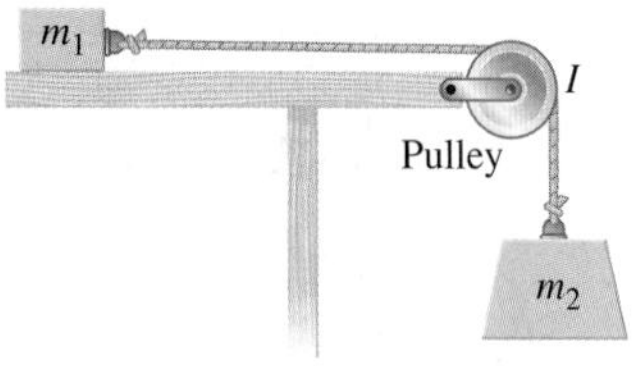

101. A 2.0-kg uniform flat disk is thrown into the air with a linear speed of 10.0 m/s. As it travels, the disk spins at 3.0 rev/s. If the radius of the disk is 10.0 cm, what is the magnitude of its angular momentum?

102. A hoop of 2.00-m circumference is rolling down an inclined plane of length 10.0 m in a time of 10.0 s. It started out from rest. (a) What is its angular velocity when it arrives at the bottom? (b) If the mass of the hoop, concentrated at the rim, is 1.50 kg, what is the angular momentum of the hoop when it reaches the bottom of the incline? (c) What force(s) supplied the net torque to change the hoop's angular momentum? Explain. [*Hint:* Use a rotation axis through the hoop's center.] (d) What is the magnitude of this force?

103. A large clock has a second hand with a mass of 0.10 kg concentrated at the tip of the pointer. (a) If the length of the second hand is 30.0 cm, what is its angular momentum? (b) The same clock has an hour hand with a mass of 0.20 kg concentrated at the tip of the pointer. If the hour hand has a length of 20.0 cm, what is its angular momentum?

104. A planet moves around the Sun in an elliptical orbit (see Fig. 8.39). (a) Show that the external torque acting on the planet about an axis through the Sun is zero. (b) Since the torque is zero, the planet's angular momentum is constant. Write an expression for the planet's angular momentum in terms of its mass m, its distance r from the Sun, and its angular velocity ω. (c) Given r and ω, how much area is swept out during a short time Δt? [*Hint:* Think of the area as a fraction of the area of a *circle*, like a slice of pie; if Δt is short enough, the radius of the orbit during that time is nearly constant.] (d) Show that the area swept out per unit time is constant. You have just proved Kepler's second law!

105. A 68-kg woman stands straight with both feet flat on the floor. Her center of gravity is a horizontal distance of 3.0 cm in front of a line that connects her two ankle joints. The Achilles tendon attaches the calf muscle to the foot a distance of 4.4 cm behind the ankle joint. If the Achilles tendon is inclined at an angle of 81° with respect to the horizontal, find the force that each calf muscle needs to exert while she is standing. [*Hint:* Consider the equilibrium of the part of the body *above* the ankle joint.]

106. A merry-go-round (radius R, rotational inertia I_i) spins with negligible friction. Its initial angular velocity is ω_i. A child (mass m) on the merry-go-round moves from the center out to the rim. (a) Calculate the angular velocity after the child moves out to the rim. (b) Calculate the rotational kinetic energy and angular momentum of the system (merry-go-round + child) before and after.

107. Since humans are generally not symmetrically shaped, the height of our center of gravity is generally not half of our height. One way to determine the location of the center of gravity is shown in the diagram. A 2.2-m-long uniform plank is supported by two bathroom scales, one at either end. Initially the scales each read 100.0 N. A 1.60-m-tall student then lies on top of the plank, with the soles of his feet directly above scale B. Now scale A reads 394.0 N and scale B reads 541.0 N. (a) What is the student's weight? (b) How far is his center of gravity from the soles of his feet? (c) When standing, how far above the floor is his center of gravity, expressed as a fraction of his height?

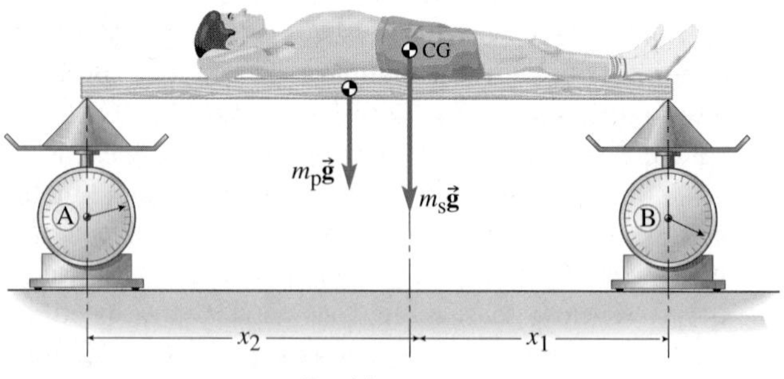

Problem 107

108. A spool of thread of mass m rests on a plane inclined at angle θ. The end of the thread is tied as shown. The outer radius of the spool is R and the inner radius (where the thread is wound) is r. The rotational

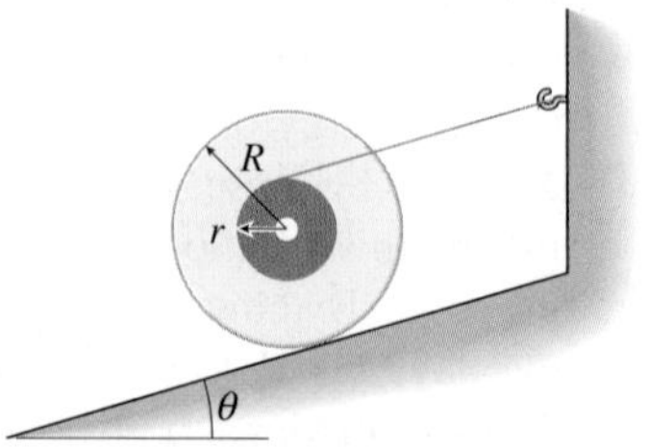

inertia of the spool is I. Give all answers in terms of m, θ, R, r, I, and g. (a) If there is no friction between the spool and the incline, describe the motion of the spool and calculate its acceleration. (b) If the coefficient of friction is large enough to keep the spool from slipping, calculate the magnitude and direction of the frictional force. (c) What is the minimum possible coefficient of friction to keep the spool from slipping in part (b)?

109. A bicycle travels up an incline at constant velocity. The magnitude of the frictional force due to the road on the rear wheel is $f = 3.8$ N. The upper section of chain pulls on the sprocket wheel, which is attached to the rear wheel, with a force $\vec{\mathbf{F}}_C$. The lower section of chain is slack. If the radius of the rear wheel is 6.0 times the radius of the sprocket wheel, what is the magnitude of the force $\vec{\mathbf{F}}_C$ with which the chain pulls?

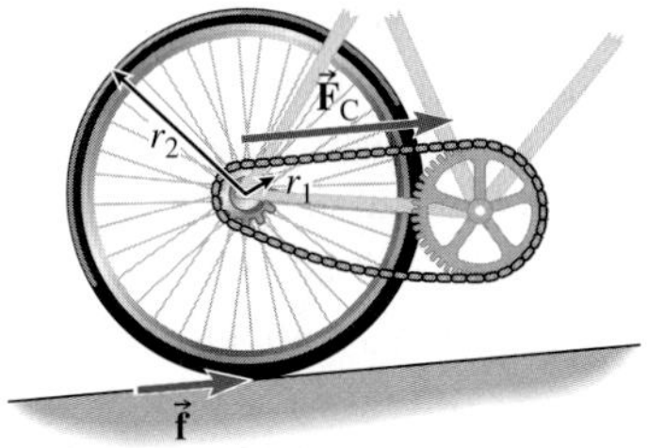

110. A circus roustabout is attaching the circus tent to the top of the main support post of length L when the post suddenly breaks at the base. The worker's weight is negligible relative to that of the uniform post. What is the speed with which the roustabout reaches the ground if (a) he jumps at the instant he hears the post crack or (b) if he clings to the post and rides to the ground with it? (c) Which is the safest course of action for the roustabout?

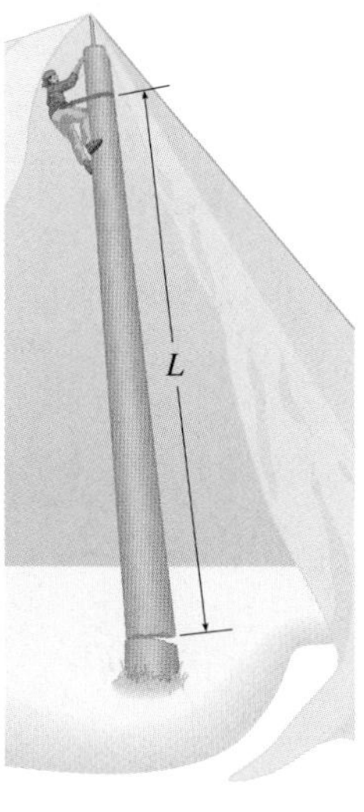

111. A student stands on a platform that is free to rotate and holds two dumbbells, each at a distance of 65 cm from his central axis. Another student gives him a push and starts the system of student, dumbbells, and platform rotating at 0.50 rev/s. The student on the platform then pulls the dumbbells in close to his chest so that they are each 22 cm from his central axis. Each dumbbell has a mass of 1.00 kg and the rotational inertia of the student, platform, and dumbbells is initially 2.40 kg·m². Model each arm as a uniform rod of mass 3.00 kg with one end at the central axis; the length of the arm is initially 65 cm and then is reduced to 22 cm. What is his new rate of rotation?

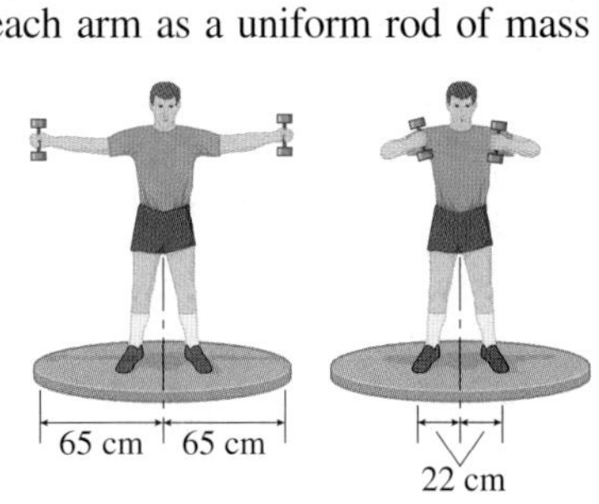

112. A person places his hand palm downward on a scale and pushes down on the scale until it reads 96 N. The triceps muscle is responsible for this arm extension force. Find the force exerted by the triceps muscle. The bottom of the triceps muscle is 2.5 cm to the left of the elbow joint and the palm is pushing at approximately 38 cm to the right of the elbow joint.

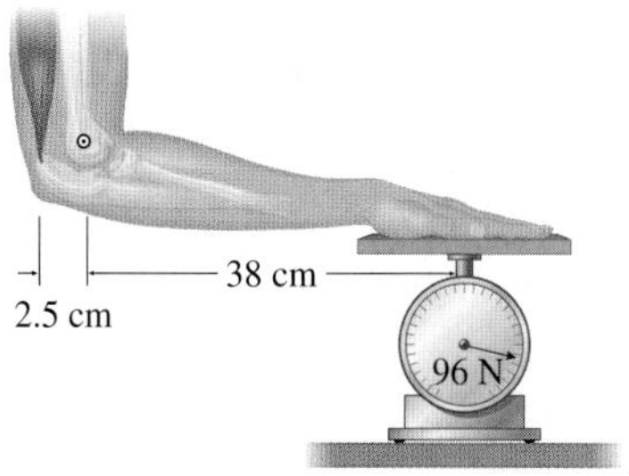

113. The posture of small animals may prevent them from being blown over by the wind. For example, with wind blowing from the side, a small insect stands with bent legs; the more bent the legs, the lower the body and the smaller the angle θ. The wind exerts a force on the insect, which causes a torque about the point where the downwind feet touch. The torque due to the weight of the insect must be equal and opposite to keep the insect from being blown over. For example, the drag force on a blowfly due to a sideways wind is $F_{wind} = cAv^2$, where v is the velocity of the wind, A is the cross-sectional area on which the wind is blowing, and $c \approx 1.3$ N·s²·m⁻⁴. (a) If the blowfly has a cross-sectional side area of 0.10 cm², a mass of 0.070 g, and crouches such that $\theta = 30.0°$, what is the maximum wind speed in which the blowfly can stand? (Assume that the drag force acts at the center of gravity.) (b) How about if it stands such that $\theta = 80.0°$? (c) Compare to the maximum wind velocity that a dog can withstand, if the dog stands such that $\theta = 80.0°$, has a cross-sectional area of 0.030 m², and weighs 10.0 kg. (Assume the same value of c.)

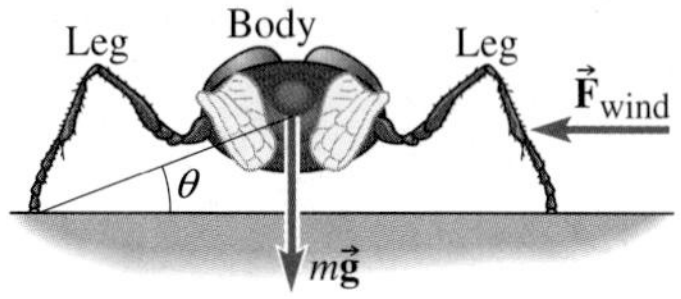

114. (a) Redo Example 8.7 to find an algebraic solution for d in terms of M, m, μ_s, L, and θ. (b) Use this expression to show that placing the ladder at a larger angle θ (that is, more nearly vertical) enables the person to climb farther up the ladder without having it slip, all other things being equal. (c) Using the numerical values from Example 8.7, find the minimum angle θ that enables the person to climb all the way to the top of the ladder.

Answers to Practice Problems

8.1 390 kg·m²

8.2 $v = \sqrt{\dfrac{2m_2gh}{m_1 + m_2 + I/R^2}}$

8.3 53 N; 8.4 N·m

8.4 −65 N·m

8.5 8.3 J

8.6 left support, downward; right support, upward

8.7 0.27

8.8 57 N, downward

8.9 It must lie in the same vertical plane as the two ropes holding up the rings. Otherwise, the gravitational force would have a nonzero lever arm with respect to a horizontal axis that passes through the contact points between his hands and the rings; thus, gravity would cause a net torque about that axis.

8.10 460 N

8.11 (a) 2380 rad; (b) 3.17 kJ; (c) 1.34 N·m

8.12 solid ball, $\frac{2}{7}$; hollow ball, $\frac{2}{5}$

8.13 $\frac{1}{2}g \sin \theta$

8.14 5% increase

8.15 16 cm/s

Answers to Checkpoints

8.1 Rotational inertia involves distances from masses to the rotation axis; distances *along* the rotation axis are irrelevant. Another way to see it: cut the cylinder or disk into a large number of thin disks with the same radius. Each thin disk has rotational inertia $I_i = \frac{1}{2}m_i R^2$. Now add up the rotational inertias of the thin disks: $I = \Sigma I_i = \Sigma \frac{1}{2} m_i R^2 = \frac{1}{2}R^2 \, \Sigma m_i = \frac{1}{2} MR^2$.

8.2 The longer handle lets you push at a greater distance from the rotation axis. Thus, you can exert a larger torque.

8.4 Yes in both cases. Torque depends not only on the magnitude and direction of the force but also on the point where the force is applied. Two forces that do not add to zero can produce torques that add to zero due to different lever arms. Then the net torque is zero and the net torque nonzero; the object is in rotational equilibrium but not in translational equilibrium. Similarly, two forces that add to zero can have different lever arms and produce torques that do not add to zero. In this case the net force is zero and the net torque is nonzero; the object is in translational equilibrium but not in rotational equilibrium.

8.7 (a) falling without spinning; (b) spinning about a fixed axis; (c) rolling without slipping along a surface

8.8 Yes. If friction is negligible, the external torque is zero so her angular momentum does not change. Extending her arms and leg makes her rotational inertia increase back to its initial value, so her angular velocity decreases to its initial value.

Review & Synthesis: Chapters 6–8

Review Exercises

1. A spring scale in a French market is calibrated to show the *mass* of vegetables in grams and kilograms. (a) If the marks on the scale are 1.0 mm apart for every 25 g, what maximum extension of the spring is required to measure up to 5.0 kg? (b) What is the spring constant of the spring? [*Hint:* Remember that the scale really measures *force.*]

2. Plot a graph of this data for a spring resting horizontally on a table. Use your graph to find (a) the spring constant and (b) the relaxed length of the spring.

Force (N)	0.200	0.450	0.800	1.500
Spring length (cm)	13.3	15.0	17.3	22.0

3. A pendulum consists of a bob of mass m attached to the end of a cord of length L. The pendulum is released from a point at a height of $L/2$ above the lowest point of the swing. What is the tension in the cord as the bob passes the lowest point?

4. How much energy is expended by an 80.0-kg person in climbing a vertical distance of 15 m? Assume that muscles have an efficiency of 22%; that is, the work done by the muscles to climb is 22% of the energy expended.

5. Ugonna stands at the top of an incline and pushes a 100-kg crate to get it started sliding down the incline. The crate slows to a halt after traveling 1.50 m along the incline. (a) If the initial speed of the crate was 2.00 m/s and the angle of inclination is 30.0°, how much energy was dissipated by friction? (b) What is the coefficient of sliding friction?

6. A packing carton slides down an inclined plane of angle 30.0° and of incline length 2.0 m. If the initial speed of the carton is 4.0 m/s directed down the incline, what is the speed at the bottom? Ignore friction.

7. A child's playground swing is supported by chains that are 4.0 m long. If the swing is 0.50 m above the ground and moving at 6.0 m/s when the chains are vertical, what is the maximum height of the swing?

8. A block slides down a plane that is inclined at an angle of 53° with respect to the horizontal. If the coefficient of kinetic friction is 0.70, what is the acceleration of the block?

9. Gerald wants to know how fast he can throw a ball, so he hangs a 2.30-kg target on a rope from a tree. He picks up a 0.50-kg ball of putty and throws it horizontally against the target. The putty sticks to the target and the putty and target swing up a vertical distance of 1.50 m from its original position. How fast did Gerald throw the ball of putty?

10. A hollow cylinder rolls without slipping or sliding along a horizontal surface toward an incline. If the cylinder's speed is 3.00 m/s at the base of the incline and the angle of inclination is 37.0°, how far along the incline does the cylinder travel before coming to a stop?

11. A grinding wheel, with a mass of 20.0 kg and a radius of 22.4 cm, is a uniform cylindrical disk. (a) Find the rotational inertia of the wheel about its central axis. (b) When the grinding wheel's motor is turned off, friction causes the wheel to slow from 1200 rpm to rest in 60.0 s. What torque must the motor provide to accelerate the wheel from rest to 1200 rpm in 4.00 s? Assume that the frictional torque is the same regardless of whether the motor is on or off.

12. An 11-kg bicycle is moving with a linear speed of 7.5 m/s. Each wheel can be modeled as a thin hoop with a mass of 1.3 kg and a diameter of 70 cm. The bicycle is stopped in 4.5 s by the action of brake pads that squeeze the wheels and slow them down. The coefficient of friction between the brake pads and a wheel is 0.90. There are four brake pads altogether; assume they apply equal magnitude normal forces on the wheels. What is the normal force applied to a wheel by one of the brake pads?

13. A 0.185-kg spherical steel ball is used in a pinball machine. The ramp is 2.05 m long and tilted at an angle of 5.00°. Just after a flipper hits the ball at the bottom of the ramp, the ball has an initial speed of 2.20 m/s. What is the speed of the ball when it reaches the top of the pinball machine?

14. A rotating star collapses under the influence of gravitational forces to form a pulsar. The radius of the star after collapse is 1.0×10^{-4} times the radius before collapse. There is no change in mass. In both cases, the mass of the star is uniformly distributed in a spherical shape. Find the ratios of the (a) angular momentum, (b) angular velocity, and (c) rotational kinetic energy of the star after collapse to the values before collapse. (d) If the period of the star's rotation before collapse is 1.0×10^{7} s, what is its period after collapse?

✦15. A 0.122-kg dart is fired from a gun with a speed of 132 m/s horizontally into a 5.00-kg wooden block. The block is attached to a spring with a spring constant of 8.56 N/m. The coefficient of kinetic friction between the block and the horizontal surface it is resting on is 0.630. After the dart embeds itself into the block, the block slides along the surface and compresses the spring. What is the maximum compression of the spring?

16. A 5.60-kg uniform door is 0.760 m wide by 2.030 m high, and is hung by two hinges, one at 0.280 m from the top and one at 0.280 m from the bottom of the door. If the vertical components of the forces on each of the two hinges are identical, find the vertical and horizontal force components acting on each hinge due to the door.

17. Consider the apparatus shown in the figure (not to scale). The pulley, which can be treated as a uniform disk, has a mass of 60.0 g and a radius of 3.00 cm. The spool also has a radius of 3.00 cm. The rotational inertia of the spool, axle, and paddles about their axis of rotation is $0.00140\ kg\cdot m^2$. The block has a mass of 0.870 kg and is released from rest. After the block has fallen a distance of 2.50 m, it has a speed of 3.00 m/s. How much energy has been delivered to the fluid in the beaker?

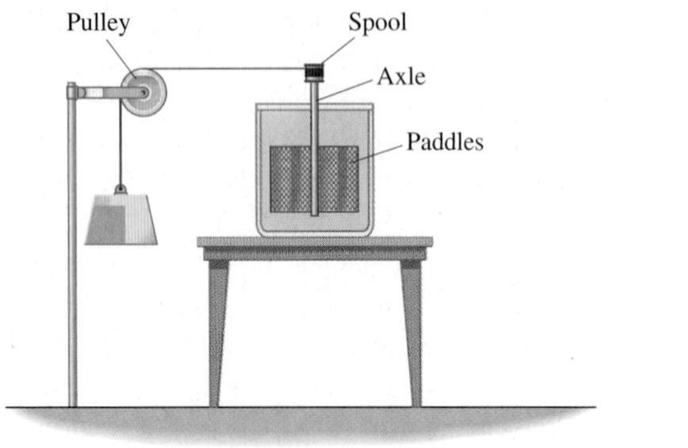

18. It is the bottom of the ninth inning at a baseball game. The score is tied and there is a runner on second base when the batter gets a hit. The 85-kg base runner rounds third base and is heading for home with a speed of 8.0 m/s. Just before he reaches home plate, he crashes into the opposing team's catcher, and the two players slide together along the base path toward home plate. The catcher has a mass of 95 kg and the coefficient of friction between the players and the dirt on the base path is 0.70. How far do the catcher and base runner slide?

19. Pendulum bob A has half the mass of pendulum bob B. Each bob is tied to a string that is 5.1 m long. When bob A is held with its string horizontal and then released, it swings down and, once bob A's string is vertical, it collides elastically with bob B. How high do the bobs rise after the collision?

✦20. During a game of marbles, the "shooter," a marble with three times the mass of the other marbles, has a speed of 3.2 m/s just before it hits one of the marbles. The other marble bounces off the shooter in an elastic collision at an angle of 40°, as shown, and the shooter moves off at an angle θ. Determine (a) the speed of the shooter after the collision, (b) the speed of the marble after the collision, and (c) the angle θ.

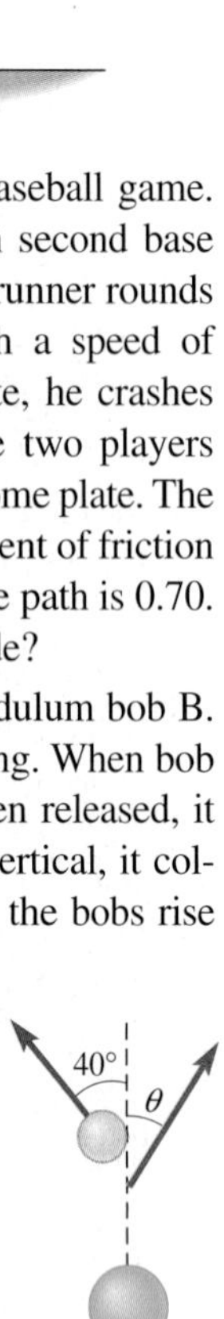

21. At the beginning of a scene in an action movie, the 78.0-kg star, Indianapolis Jones, will stand on a ledge 3.70 m above the ground and the 55.0-kg heroine, Georgia Smith, will stand on the ground. Jones will swing down on a rope, grab Smith around the waist, and continue swinging until they come to rest on another ledge on the other side of the set. At what height above the ground should the second ledge be placed? Assume that Jones and Smith remain nearly upright during the swing so that their CMs are always the same distance above their feet.

22. A uniform disk is rotated about its symmetry axis. The disk goes from rest to an angular speed of 11 rad/s in a time of 0.20 s with a constant angular acceleration. The rotational inertia and radius of the disk are $1.5\ kg\cdot m^2$ and 11.5 cm, respectively. (a) What is the angular acceleration during the 0.20-s interval? (b) What is the net torque on the disk during this time? (c) After the applied torque stops, a frictional torque remains. This torque has an associated angular acceleration of $9.8\ rad/s^2$. Through what total angle θ (starting from time $t = 0$) does the disk rotate before coming to rest? (d) What is the speed of a point halfway between the rim of the disk and its rotation axis 0.20 s after the applied torque is removed?

23. A block is released from rest and slides down an incline. The coefficient of sliding friction is 0.38 and the angle of inclination is 60.0°. Use energy considerations to find how fast the block is sliding after it has traveled a distance of 30.0 cm along the incline.

24. A uniform solid cylinder rolls without slipping or sliding down an incline. The angle of inclination is 60.0°. Use energy considerations to find the cylinder's speed after it has traveled a distance of 30.0 cm along the incline.

25. A block of mass 2.00 kg slides eastward along a frictionless surface with a speed of 2.70 m/s. A chunk of clay with a mass of 1.50 kg slides southward on the same surface with a speed of 3.20 m/s. The two objects collide and move off together. What is their velocity after the collision?

26. An ice-skater, with a mass of 60.0 kg, glides in a circle of radius 1.4 m with a tangential speed of 6.0 m/s. A second skater, with a mass of 30.0 kg, glides on the same circular path with a tangential speed of 2.0 m/s. At an instant of time, both skaters grab the ends of a lightweight, rigid set of rods, set at 90° to each other, that can freely rotate about a pole, fixed in place on the ice. (a) If each rod is 1.4 m long, what is the tangential speed of the skaters after they grab the rods? (b) What is the direction of the angular momentum before and after the skaters "collide" with the rods?

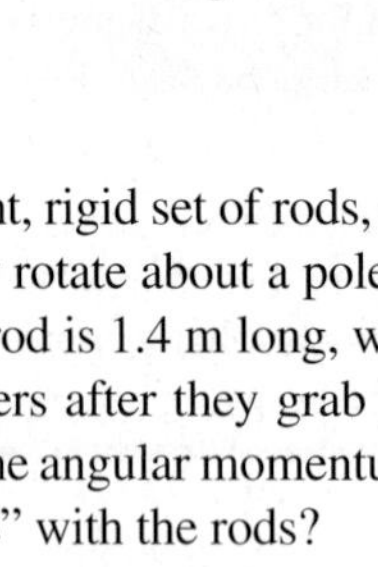

27. In a motor, a flywheel (solid disk of radius R and mass M) is rotating with angular velocity ω_i. When the clutch is released, a second disk (radius r and mass m) initially not rotating is brought into frictional contact with the flywheel. The two disks spin around the same axle with frictionless bearings. After a short time, friction between the two disks brings them to a common angular velocity. (a) Ignoring external influences, what is the final angular velocity?

(b) Does the total angular momentum of the two change? If so, account for the change. If not, explain why it does not. (c) Repeat (b) for the rotational kinetic energy.

28. A child's toy is made of a 12.0-cm-radius rotating wheel that picks up 1.00-g pieces of candy in a pocket at its lowest point, brings the candy to the top, then releases it. The frequency of rotation is 1.60 Hz. (a) How far from its starting point does the candy land? (b) What is the radial acceleration of the candy when it is on the wheel?

29. A Vulcan spaceship has a mass of 65 000 kg and a Romulan spaceship is twice as massive. Both have engines that generate the same total force of 9.5×10^6 N. (a) If each spaceship fires its engine for the same amount of time, starting from rest, which will have the greater kinetic energy? Which will have the greater momentum? (b) If each spaceship fires its engine for the same *distance*, which will have the greater kinetic energy? Which will have the greater momentum? (c) Calculate the energy and momentum of each spaceship in parts (a) and (b), ignoring any change in mass due to whatever is expelled by the engines. In part (a), assume that the engines are fired for 100 s. In part (b), assume that the engines are fired for 100 m.

30. Two blocks of masses m_1 and m_2, resting on frictionless inclined planes, are connected by a massless rope passing over an ideal pulley. Angle $\phi = 45.0°$ and angle $\theta = 36.9°$; mass m_1 is 6.00 kg and mass m_2 is 4.00 kg. (a) Using energy conservation, find how fast the blocks are moving after they travel 2.00 m along the inclines. (b) Now solve the same problem using Newton's second law. [*Hint:* First find the acceleration of each of the blocks. Then find how fast either block is moving after it travels 2.00 m along the incline with constant acceleration.]

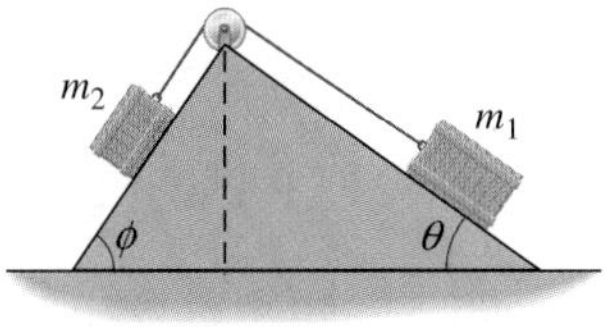

✦31. A particle, constrained to move along the x-axis, has a total mechanical energy of −100 J. The potential energy of the particle is shown in the graph. At time $t = 0$, the particle is located at $x = 5.5$ cm and is moving to the left. (a) What is the particle's potential energy at $t = 0$? What is its kinetic energy at this time? (b) What are the particle's total, potential, and kinetic energies when it is at $x = 1$ cm and moving to the right? (c) What is the particle's kinetic energy when it is at $x = 3$ cm and moving to the left? (d) Describe the motion of this particle starting at $t = 0$.

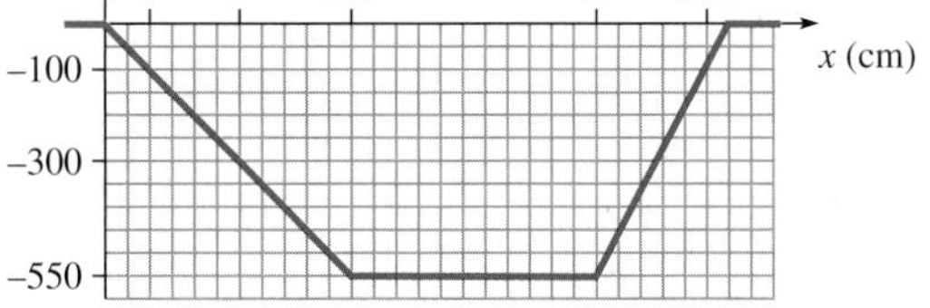

✦32. You are mowing the lawn on a hill near your house when the lawnmower blade strikes a stone of mass 100 g and sends it flying horizontally toward a window. The lawnmower blade can be modeled as a thin rod with a mass of 2.0 kg and a length of 50 cm rotating about its center. The stone impacts the blade near one end and is ejected with a velocity perpendicular to the rotation axis and the blade at the moment of collision. As a result of the impact, the blade slows from 60 rev/s to 55 rev/s. The window is 1.00 m in height, and its center is located 10.0 m away and at the same height as the lawnmower. (a) With what speed is the stone shot out by the mower? [*Hint:* The external force due to the lawnmower's drive shaft on the system (blade + stone) cannot be ignored during the collision, but the external *torque* about the shaft *can* be ignored. The angular momentum of the stone just after impact can be calculated from its tangential velocity and its distance from the rotation axis.] (b) Ignoring air resistance, will the stone hit the window?

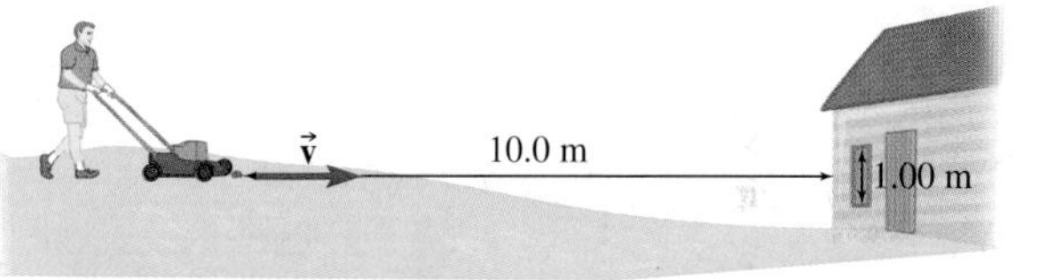

33. A person on a bicycle (combined total mass 80.0 kg) starts from rest and coasts down a hill to the bottom 20.0 m below. Each wheel can be treated as a hoop with mass 1.5 kg and radius 40 cm. Ignore friction and air resistance. (a) Find the speed of the bike at the bottom. (b) Would the speed at the bottom be the same for a less massive rider? Explain.

34. Tarzan wants to swing on a vine across a river. He is standing on a ledge 3.00 m above the water's edge, and the river is 5.00 m wide. The vine is attached to a tree branch that is 8.00 m directly above the opposite edge of the river. Initially the vine makes a 60.0° angle with the vertical as he is holding it. He swings across starting from rest, but unfortunately the vine breaks when the vine is 20.0° from the vertical. (a) Assuming Tarzan weighs 900.0 N, what was the tension in the vine just before it broke? (b) Does he land safely on the other side of the river?

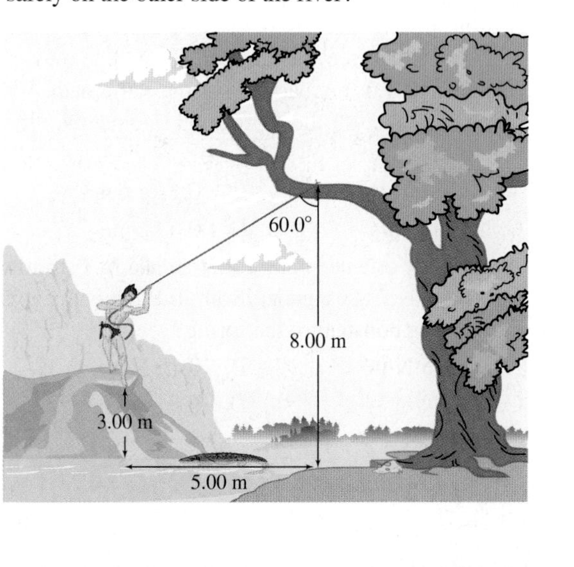

35. A boy of mass 60 kg is sledding down a 70-m slope starting from rest. The slope is angled at 15° below the horizontal. After going 20 m along the slope he passes his friend, who jumps on the sled. The friend has a mass of 50 kg and the coefficient of kinetic friction between the sled and the snow is 0.12. Ignoring the mass of the sled, find their speed at the bottom.

✦36. You want to throw a banana to a monkey hanging from a branch as shown in the figure. The banana has a mass of 200 g and the monkey has a mass of 3.00 kg. The monkey is startled and drops from the branch the moment you throw the banana. Ignore air resistance. (a) In what direction should you aim the banana so the monkey catches it in the air? (b) Explain why your answer to part (a) is the same for different values of the banana's launch speed. (c) If the monkey catches the banana at the point indicated in the figure, what was the banana's initial speed? (d) What is the horizontal distance d to the spot where the monkey lands?

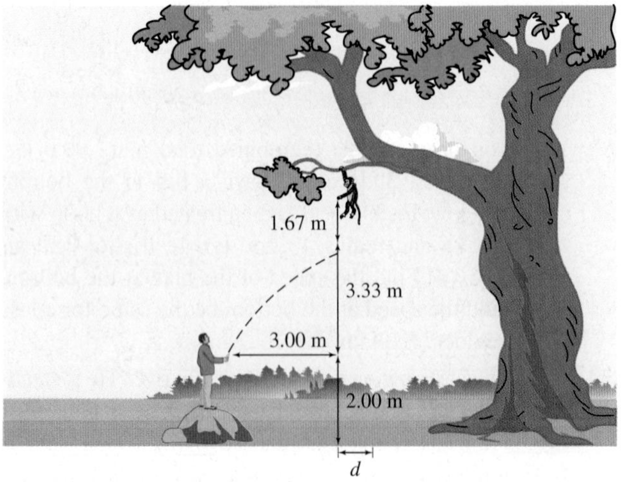

MCAT Review

The section that follows includes MCAT exam material and is reprinted with permission of the Association of American Medical Colleges (AAMC).

1. A projectile with a mass of 0.2 kg and a horizontal speed of 2.0 m/s hits a recycle bin (which is free to move), then rebounds at 1.0 m/s back along the same path. What is the magnitude of the horizontal momentum the bin receives?

   A. 0.2 kg·m/s B. 0.3 kg·m/s
   C. 0.5 kg·m/s D. 0.6 kg·m/s

2. A vertically oriented spring is stretched by 0.15 m when a 100-g mass is suspended from it. What is the approximate spring constant of the spring?

   A. 0.015 N/m B. 0.15 N/m
   C. 1.5 N/m D. 6.5 N/m

3. When a downward force is applied at a point 0.60 m to the left of a fulcrum, equilibrium is obtained by placing a mass of 1.0×10^{-7} kg at a point 0.40 m to the right of the fulcrum. What is the magnitude of the downward force?

   A. 1.5×10^{-7} N B. 6.5×10^{-7} N
   C. 9.8×10^{-7} N D. 1.5×10^{-6} N

4. A 0.50-kg ball accelerates from rest at 10 m/s^2 for 2.0 s. It then collides with and sticks to a 1.0-kg ball that is initially at rest. After the collision, approximately how fast are the balls going?

   A. 3.3 m/s B. 6.7 m/s
   C. 10.0 m/s D. 15.0 m/s

5. A 1000-kg car requires 10,000 W of power to travel at 15 m/s on a level highway. How much extra power in watts is required for the car to climb a 10° hill at the same speed? (Use $g = 10$ m/s^2.)

   A. $1.0 \times 10^4 \times \sin 10°$ B. $1.5 \times 10^4 \times \sin 10°$
   C. $1.0 \times 10^5 \times \sin 10°$ D. $1.5 \times 10^5 \times \sin 10°$

6. A 90-kg patient walks the treadmill at a speed of 2 m/s, and $\theta_{in} = 30°$ for 10 min (600 s). What is the total work done by the patient on the treadmill? (Use $g = 10$ m/s^2.)

   A. 1.80 kJ B. 18.0 kJ
   C. 0.54 MJ D. 1.08 MJ

7. A 100-kg patient walks the treadmill at a speed of 3 m/s, and $\theta_{in} = 30°$ for 5 min (300 s). What is the mechanical power output of the patient in watts? (Use $g = 10$ m/s^2.)

   A. 300 W B. 1500 W
   C. 3000 W D. 7500 W

Read the paragraphs and then answer the following questions:

An exercise bike has the basic construction of a bicycle with a single heavy disk wheel. In addition to friction in the bearings and the transmission system, resistance to pedaling is provided by two narrow friction pads that push with equal force on each side of the wheel. The coefficient of kinetic friction between the pads and the wheel is 0.4, and the pads provide a total retarding force of 20 N tangential to the wheel. The pads are located at a position 0.3 m from the center of the wheel. The distance, recorded on the odometer, is considered to be the distance that a point on the wheel 0.3 m from the center moves. The pedals move in a circle of 0.15 m in radius and complete one revolution, while a transmission system allows the wheel to rotate twice.

In human metabolic processes, the ratio of energy released to volume of oxygen consumed averages 20000 J/L. A cyclist with a basal metabolic rate of 85 W (rate of internal energy conversion while awake but inactive) pedals continuously for 20 min, registering 4800 m on the odometer. During this activity, the cyclist's average metabolic rate is 535 W. The cyclist's body converts the extra energy into mechanical work output with an efficiency of 20%.

8. What is the magnitude of the force pushing each friction pad onto the wheel?
   A. 10 N
   B. 25 N
   C. 40 N
   D. 50 N
9. Which of the following is closest to the radial acceleration of the part of the wheel that passes between the friction pads?
   A. 10 m/s^2
   B. 20 m/s^2
   C. 40 m/s^2
   D. 50 m/s^2
10. If the wheel has a kinetic energy of 30 J when the cyclist stops pedaling, how many rotations will it make before coming to rest?
   A. Less than 1
   B. Between 1 and 2
   C. Between 2 and 3
   D. Between 3 and 4
11. What is the difference between the average mechanical power output of the cyclist in the passage and the power dissipated by the wheel at the friction pads?
   A. 5 W
   B. 10 W
   C. 20 W
   D. 27 W
12. Which of the following actions would most likely increase the fraction of the cyclist's mechanical power output that is dissipated by the wheel at the friction pads?
   A. Reducing the force on the friction pads and pedaling at the same rate
   B. Maintaining the same force on the friction pads and pedaling at a slower rate
   C. Maintaining the same force on the friction pads and pedaling at a faster rate
   D. Increasing the force on the friction pads and pedaling at the same rate
13. Which of the following is the best estimate of the number of liters of oxygen the cyclist in the passage would consume in the 20 min of activity?
   A. 25 L
   B. 30 L
   C. 45 L
   D. 50 L
14. During a second workout, the cyclist reduces the force on the friction pads by 50%, then pedals for two times the previous distance in $\frac{1}{2}$ the previous time. How does the amount of energy dissipated by the pads in the second workout compare with energy dissipated in the first workout?
   A. One-eighth as much
   B. One-half as much
   C. Equal
   D. Two times as much
15. What is the ratio of the distance moved by a pedal to the distance moved by a point on the wheel located at a radius of 0.3 m in the same amount of time?
   A. 0.25
   B. 0.5
   C. 1
   D. 2
16. A cyclist's average metabolic rate during a workout is 500 W. If the cyclist wishes to expend at least 300 kcal (1 kcal = 4186 J) of energy, how long must the cyclist exercise at this rate?
   A. 0.6 min
   B. 3.6 min
   C. 36.0 min
   D. 41.9 min
17. If the friction pads are moved to a location 0.4 m from the center of the wheel, how does the amount of work done on the wheel, per revolution, change?
   A. It decreases by 25%
   B. It stays the same
   C. It increases by 33%
   D. It increases by 78%

CHAPTER

9 Fluids

A hippopotamus in Kruger National Park, South Africa, wants to feed on the vegetation growing on the bottom of a pond. When the hippo wades into the pond, it floats. How does a hippopotamus get its floating body to sink to the bottom of a pond? (See p. 329 for the answer.)

Concepts & Skills to Review

- conservation of energy (Chapter 6)
- force as rate of change of momentum (Section 7.3)
- conservation of momentum in collisions (Sections 7.7 and 7.8)
- equilibrium (Section 4.2)

9.1 STATES OF MATTER

Ordinary matter is usually classified into three familiar states or phases: solids, liquids, and gases. Solids tend to hold their shapes. Many solids are quite rigid; they are not easily deformed by external forces because forces due to neighboring atoms hold each atom in a particular position. Although the atoms vibrate around fixed equilibrium positions, they do not have enough energy to break the bonds with their neighbors. To bend an iron bar, for example, the arrangement of the atoms must be altered, which is not easy to do. A blacksmith heats iron in a forge to loosen the bonds between atoms so that he can bend the metal into the required shape.

In contrast to solids, liquids and gases do not hold their shapes. A liquid flows and takes the shape of its container and a gas expands to fill its container. **Fluids**—both liquids and gases—are easily deformed by external forces. This chapter deals mainly with properties that are common to both liquids and gases.

Fluids (liquids and gases) are materials that flow.

The atoms or molecules in a fluid do not have fixed positions, so a fluid does not have a definite shape. An applied force can easily make a fluid flow; for instance, the squeezing of the heart muscle exerts an applied force that pumps blood through the blood vessels of the body. However, this squeezing does not change the *volume* of the blood by much. In many situations we can think of liquids as **incompressible**—that is, as having a fixed volume that is impossible to change. The shape of the liquid can be changed by pouring it from a container of one shape into a container of a different shape, but the volume of the liquid remains the same.

In most liquids, the atoms or molecules are almost as closely packed as those in the solid phase of the same material. The intermolecular forces in a liquid are almost as strong as those in solids, but the molecules are not locked in fixed positions as they are in solids. That is why the volume of the liquid can remain nearly constant while the shape is easily changed. Water is one of the exceptions: in cold water, the molecules in the liquid phase are actually *more* closely packed than those in the solid phase (ice).

Gases, on the other hand, cannot be characterized by a definite volume nor by a definite shape. A gas expands to fill its container and can easily be compressed. The molecules in a gas are very far apart compared to the molecules in liquids and solids. The molecules are almost free of interactions with each other except when they collide.

Gases are much more compressible than liquids.

9.2 PRESSURE

Microscopic Origin of Pressure A **static** fluid does not flow; it is everywhere at rest. In the study of fluid statics (*hydrostatics*), we also assume that any solid object in contact with the fluid—whether a vessel containing the fluid or an object submerged in the fluid—is at rest. The atoms or molecules in a static fluid are not themselves static; they are continually moving. The motion of people bouncing up and down and bumping into each other in a mosh pit gives you a rough idea of the motion of the closely packed atoms or molecules in a liquid; in gases, the atoms or molecules are much farther apart than in liquids, so they travel greater distances between collisions.

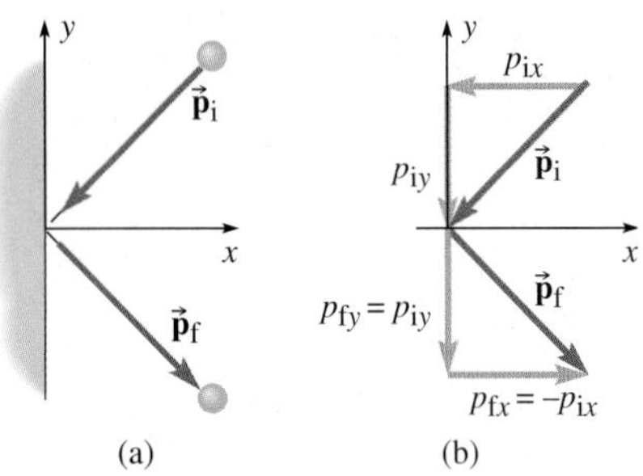

Figure 9.1 (a) A single fluid molecule bouncing off a container wall. (b) In this elastic collision, the *y*-component of the momentum is unchanged, while the *x*-component reverses direction.

Fluid pressure is caused by collisions of the fast-moving atoms or molecules of a fluid. When a single molecule hits a container wall and rebounds, its momentum changes due to the force exerted on it by the wall. Figure 9.1a shows a molecule of a fluid within a container making an elastic collision with one of the container walls. In this case, the *y*-component of momentum is unchanged, while the *x*-component reverses direction (Fig. 9.1b). The momentum change is in the +*x*-direction, which occurs because the wall exerts a force to the right on the molecule. By Newton's third law, the molecule exerts a force to the left on the wall during the collision. If we consider all the molecules colliding with this wall, *on average* they exert no force on the wall in the ± *y*-direction, but all exert a force in the −*x*-direction. The frequent collisions of fluid molecules with the walls of the container cause a net force pushing outward on the walls.

PHYSICS AT HOME

Drop a very tiny speck of dust or lint into a container of water and push the speck below the surface. The motion of the speck—called *Brownian motion*—is easily observed as it is pushed and bumped about randomly by collisions with water molecules. The water molecules themselves move about randomly, but at much higher speeds than the speck of dust due to their much smaller mass.

The force due to a static fluid on a surface is always perpendicular to the surface.

Definition of Pressure A static fluid exerts a force on any surface with which it comes in contact; the direction of the force is perpendicular to the surface (Fig. 9.2). A static fluid *cannot* exert a force *parallel* to the surface. If it did, the surface would exert a force on the fluid parallel to the surface, by Newton's third law. This force would make the fluid flow along the surface, contradicting the premise that the fluid is static.

The *average pressure* of a fluid at points on a planar surface is

Definition of average pressure:

$$P_{av} = \frac{F}{A} \qquad (9\text{-}1)$$

where F is the magnitude of the force acting perpendicularly to the surface and A is the area of the surface. By imagining a tiny surface at various points within the fluid and measuring the force that acts on it, we can define the pressure at any point within the fluid. In the limit of a small area A, $P = F/A$ is the **pressure** P of the fluid.

Pressure is a scalar quantity; it does not have a direction. The force acting on an object submerged in a fluid—or on some portion of the fluid itself—is a vector quantity; its direction is perpendicular to the contact surface. Pressure is defined as a scalar because, at a given location in the fluid, the magnitude of the force per unit area is the same for any orientation of the surface. The molecules in a static fluid are moving in random directions; there can be no preferred direction since that would constitute fluid flow. There is no reason that a surface would have a greater number of collisions, or collisions with more energetic molecules, for one particular surface orientation compared with any other orientation.

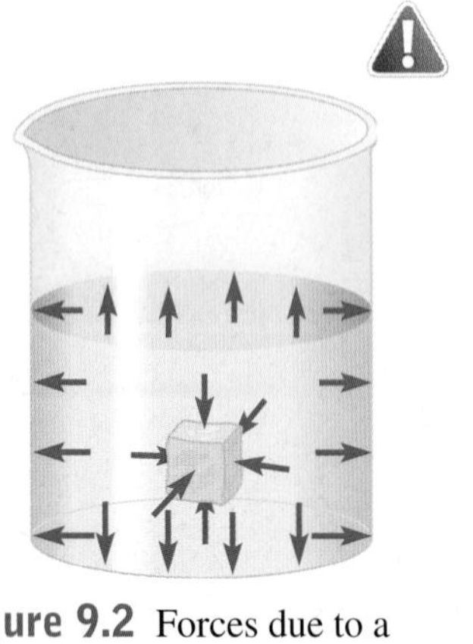

Figure 9.2 Forces due to a static fluid acting on the walls of the container and on a submerged object.

The SI unit for pressure is the newton per square meter (N/m^2), which is named the *pascal* (symbol Pa) after the French scientist Blaise Pascal (1623–1662). Another commonly used unit of pressure is the *atmosphere* (atm). One atmosphere is the *average* air pressure at sea level. The conversion factor between atmospheres and pascals is

$$1 \text{ atm} = 101.3 \text{ kPa}$$

Other units of pressure in common use are introduced in Section 9.5.

Example 9.1

Pressure due to Stiletto-Heeled Shoes

A young woman weighing 534 N (120 lb) walks to her bedroom while wearing tennis shoes. She then gets dressed for her evening date, putting on her new stiletto-heeled dress shoes. The area of the heel section of her tennis shoe is 60.0 cm^2 and the area of the heel of her dress shoe is 1.00 cm^2. For each pair of shoes, find the average pressure caused by the heel making contact with the floor when her entire weight is supported by one heel.

Strategy The average pressure is the force applied to the floor divided by the contact area. The force that the heel exerts on the floor is 534 N. To keep the units straight, we convert the areas from square centimeters to square meters.

Solution To convert the area of the tennis shoe heel and the dress shoe heel from cm^2 to m^2, we multiply by the conversion factor $\left(\frac{1\text{ m}}{10^2\text{ cm}}\right)^2$. For the tennis shoe heel:

$$A = 60.0\text{ cm}^2 \times \left(\frac{1\text{ m}}{10^2\text{ cm}}\right)^2 = 6.00 \times 10^{-3}\text{ m}^2$$

For the dress shoe heel:

$$A = 1.00\text{ cm}^2 \times \left(\frac{1\text{ m}}{10^2\text{ cm}}\right)^2 = 1.00 \times 10^{-4}\text{ m}^2$$

The average pressure is the woman's weight divided by the area of the heel. For the tennis shoe:

$$P = \frac{F}{A} = \frac{534\text{ N}}{6.00 \times 10^{-3}\text{ m}^2} = 8.90 \times 10^4\text{ N/m}^2 = 89.0\text{ kPa}$$

For the stilettos:

$$P = \frac{534\text{ N}}{1.00 \times 10^{-4}\text{ m}^2} = 5.34 \times 10^6\text{ N/m}^2 = 5.34\text{ MPa}$$

Discussion In atmospheres, these pressures are 0.879 atm and 52.7 atm, respectively. The pressure due to the dress shoe is 60 times the pressure due to the tennis shoe since the same force is spread over $\frac{1}{60}$ the area.

Practice Problem 9.1 Pressure from an Ordinary Dress Shoe Heel

Fortunately for floor manufacturers, and for women's feet, stiletto heels are out of fashion more often than they are in fashion. Suppose that a woman's dress shoes have heels that are each 4.0 cm^2 in area. Find the pressure on the floor, when the entire weight is on a single heel, for such a shoe worn by the same woman as in Example 9.1. Find the factor by which this pressure exceeds the pressure from the tennis shoe heel.

Atmospheric Pressure

On the surface of the Earth, we live at the bottom of an ocean of fluid called air. The forces exerted by air on our bodies and on surfaces of other objects may be surprisingly large: 1 atm is approximately 10 N/cm^2 of surface area, or nearly 15 lb/in^2. We are not crushed by this pressure because most of the fluids in our bodies are at approximately the same pressure as the air around us. As an analogy, consider a sealed bag of potato chips. Why is the bag not crushed by the air pushing in on all sides? Because the air inside the bag is at the same pressure and pushes out on the sides of the bag. The pressure of the fluids inside our cells matches the pressure of the surrounding fluids pushing in on the cell membranes, so the cells do not rupture.

By contrast, the blood pressure in the arteries is as much as 20 kPa higher than atmospheric pressure. The strong, elastic arterial walls are stretched by the pressure of the blood inside; the walls squeeze the arterial blood to keep its higher pressure from being transmitted to other fluids in the body.

Changing weather conditions cause variations of approximately 5% in the actual value of air pressure at sea level; 101.3 kPa (1 atm) is only the *average* value. Air pressure also decreases with increasing elevation. (In Section 9.4, we study the effect of gravity on fluid pressure in detail.) The average air pressure in Leadville, Colorado, the highest incorporated city in the United States (elevation 3100 m), is 70 kPa. Some Tibetans live at altitudes of over 5000 m, where the average air pressure is only half its value at sea level. In problems, please assume that the atmospheric pressure is 1 atm unless the problem states otherwise.

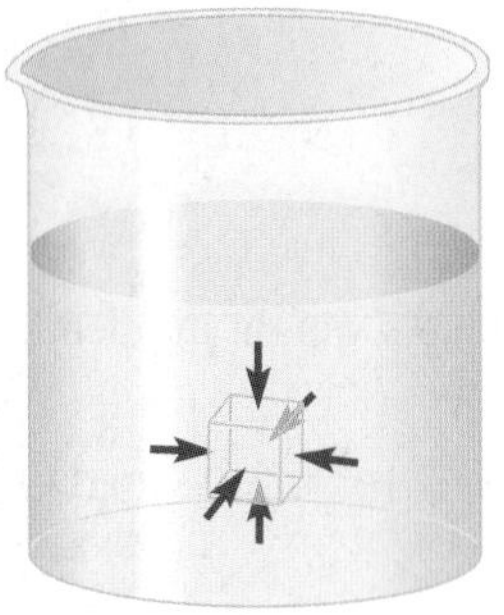

Figure 9.3 Forces acting on a cube of fluid.

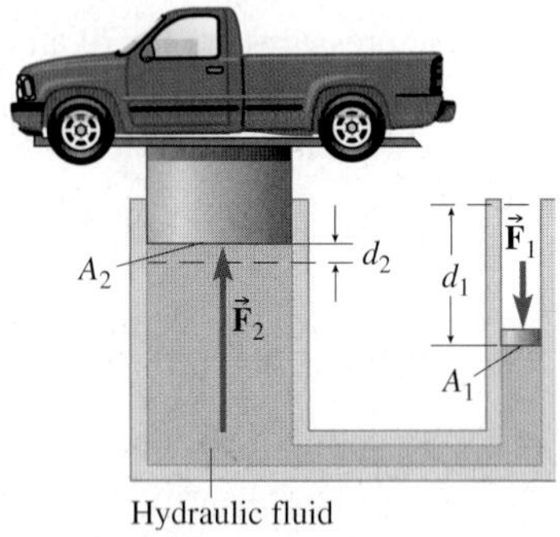

Figure 9.4 Simplified diagram of a hydraulic lift. Notice that piston 1 has to move a great distance (d_1) to lift the truck a much smaller distance (d_2). In a real hydraulic lift, piston 1 is usually replaced by a pump that draws fluid from a reservoir and pushes it into the hydraulic system.

9.3 PASCAL'S PRINCIPLE

If the weight of a static fluid is negligible (as, for example, in a hydraulic system under high pressure), then the pressure must be the same everywhere in the fluid. Why? In Fig. 9.3, imagine the submerged cube to be composed of the same fluid as its surroundings. Ignoring the fluid's weight, the only forces acting on the cubical piece of fluid are those due to the surrounding fluid pushing inward. The forces pushing on each pair of opposite sides of the cube must be equal in magnitude, since the fluid inside the cube is in equilibrium. Therefore, the pressure must be the same on both sides. Since we can extend this argument to any size and shape piece of fluid, *the fluid pressure must be the same everywhere in a weightless, static fluid.*

More generally, when the weight of the fluid is *not* negligible, the pressure is not the same everywhere. In this case, analysis of the forces acting on a piece of fluid (see Conceptual Question 15) leads to a more general result called **Pascal's principle**.

Pascal's Principle

A *change in pressure* at any point in a confined fluid is transmitted everywhere throughout the fluid.

Applications of Pascal's Principle: Hydraulic Lifts, Brakes, and Controls When a truck needs to have its muffler replaced, it is lifted into the air by a mechanism called a hydraulic lift (Fig. 9.4). A force is exerted on a liquid by a piston with a relatively small area; the resulting increase in pressure is transmitted everywhere throughout the liquid. Then the truck is lifted by the fluid pressure on a piston of much larger area. The upward force on the truck is much larger than the force applied to the small piston. Pascal's principle has many other applications, such as the hydraulic brakes in cars and trucks and the hydraulic controls in airplanes.

To analyze the forces in the hydraulic lift, let force F_1 be applied to the small piston of area A_1, causing a pressure increase:

$$\Delta P = \frac{F_1}{A_1}$$

A truck is supported by a piston of much larger area A_2 on the other side of the lift. The increase in pressure due to the small piston is transmitted everywhere in the liquid. Ignoring the weight of the fluid (or assuming the two pistons to be at the same height), the force F_2 exerted by the fluid on the large piston is related to F_1 by

$$\frac{F_1}{A_1} = \frac{F_2}{A_2}$$

Since A_2 is larger than A_1, the force exerted on the large piston (F_2) is larger than the force applied to the small piston (F_1). We are not getting something for nothing; just as for the two-pulley systems discussed in Section 6.2, the advantage of the smaller force applied to the small piston is balanced by a greater distance it must be moved. The small piston has to move a long distance d_1 while the large piston moves a short distance d_2. Assuming the liquid to be incompressible, the volume of fluid displaced by each piston is the same, so

$$A_1 d_1 = A_2 d_2$$

CONNECTION:

As for levers, systems of pulleys, and other simple machines, the hydraulic lift reduces the applied *force* needed to perform a task, but the *work* done is the same.

The displacements of the pistons are inversely proportional to their areas, while the forces are directly proportional to the areas; then the product of force and displacement is the same:

$$\frac{F_1}{A_1} \times A_1 d_1 = \frac{F_2}{A_2} \times A_2 d_2 \quad \Rightarrow \quad F_1 d_1 = F_2 d_2$$

The work (force times displacement) done by moving the small piston equals the work done by the large piston in raising the truck.

Example 9.2

The Hydraulic Lift

In a hydraulic lift, if the radius of the smaller piston is 2.0 cm and the radius of the larger piston is 20.0 cm, what weight can the larger piston support when a force of 250 N is applied to the smaller piston?

Strategy According to Pascal's principle, the pressure increases the same amount at every point in the fluid. A natural way to work is in terms of proportions since the forces are proportional to the areas of the pistons.

Solution Since the pressure on the two pistons increases by the same amount,

$$\frac{F_1}{A_1} = \frac{F_2}{A_2}$$

Equivalently, the forces are proportional to the areas:

$$\frac{F_2}{F_1} = \frac{A_2}{A_1}$$

The ratio of the radii is $r_2/r_1 = 10$, so the ratio of the areas is $A_2/A_1 = (r_2/r_1)^2 = 100$. Then the weight that can be supported is

$$F_2 = 100F_1 = 25\,000\text{ N} = 25\text{ kN}$$

Discussion One common error in this sort of problem is to think of the area and the force as a *trade-off*—in other words, that the piston with the *large* area has the *small* force and vice versa. Since the pressures are the same, the force exerted by the fluid on either piston is proportional to the piston's area. We make the piston that lifts the truck large because we know the force on it will be large, *in direct proportion to* its area.

Practice Problem 9.2 Application of Pascal's Principle

Consider the hydraulic lift of Example 9.2. (a) What is the increase in pressure caused by the 250-N force on the small piston? (b) If the larger piston moves 5.0 cm, how far does the smaller piston move?

9.4 THE EFFECT OF GRAVITY ON FLUID PRESSURE

On a drive through the mountains or on a trip in a small plane, the feeling of our ears popping is evidence that pressure is not the same everywhere in a static fluid. Gravity makes fluid pressure increase as you move down and decrease as you move up. To understand more about this variation, we must first define the density of a fluid.

Density The **density** of a substance is its mass per unit volume. The Greek letter ρ (rho) is used to represent density. The density of a uniform substance of mass m and volume V is

Density of a uniform substance: its mass divided by its volume.

$$\rho = \frac{m}{V} \tag{9-2}$$

The SI units of density are kilograms per cubic meter: kg/m^3. For a nonuniform substance, Eq. (9-2) defines the **average density**.

Table 9.1 lists the densities of some common substances. Note that temperatures and pressures are specified in the table. For solids and liquids, density is only weakly dependent on temperature and pressure. On the other hand, gases are highly compressible, so even a relatively small change in temperature or pressure can change the density of a gas significantly.

Pressure Variation with Depth due to Gravity Now, using the concept of density, we can find how pressure increases with depth due to gravity. Suppose we have a glass beaker containing a static liquid of uniform density ρ. Within this liquid, imagine a cylinder of liquid with cross-sectional area A and height d (Fig. 9.5a). The mass of the liquid in this cylinder is

$$m = \rho V$$

Table 9.1 Densities of Common Substances (at 0°C and 1 atm unless otherwise indicated)

Gases	Density (kg/m^3)	Liquids	Density (kg/m^3)	Solids	Density (kg/m^3)
Hydrogen	0.090	Gasoline	680	Polystyrene	100
Helium	0.18	Ethanol	790	Cork	240
Steam (100°C)	0.60	Oil	800–900	Wood (pine)	350–550
Nitrogen	1.25	Water (0°C)	999.87	Wood (oak)	600–900
Air (20°C)	1.20	Water (3.98°C)	1000.00	Ice	917
Air (0°C)	1.29	Water (20°C)	1001.80	Wood (ebony)	1000–1300
Oxygen	1.43	Seawater	1025	Bone	1500–2000
Carbon dioxide	1.98	Blood (37°C)	1060	Concrete	2000
		Mercury	13 600	Quartz, granite	2700
				Aluminum	2702
				Iron, steel	7860
				Copper	8920
				Lead	11 300
				Gold	19 300
				Platinum	21 500

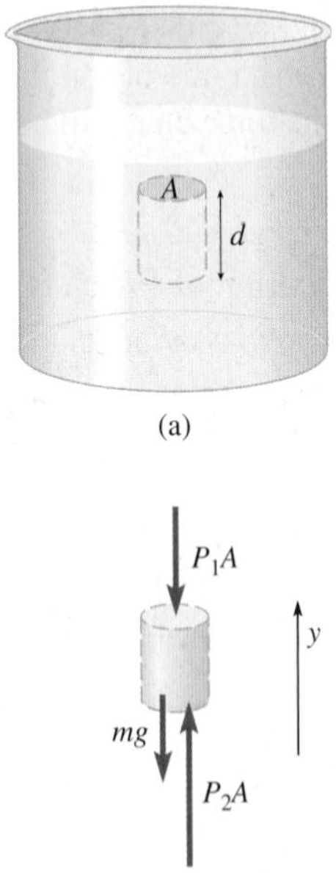

Figure 9.5 Applying Newton's second law to a cylinder of liquid tells us how pressure increases with increasing depth. (a) A cylinder of liquid of height d and area A. (b) Vertical forces on the cylinder of liquid.

where the volume of the cylinder is

$$V = Ad$$

The weight of the cylinder of liquid is therefore

$$mg = (\rho Ad)g$$

The vertical forces acting on this column of liquid are shown in Fig. 9.5b. The pressure at the top of the cylinder is P_1 and the pressure at the bottom is P_2. Since the liquid in the column is in equilibrium, the net vertical force acting on it must be zero by Newton's second law:

$$\sum F_y = P_2A - P_1A - \rho Adg = 0$$

Dividing by the common factor A and rearranging yields:

Pressure variation with depth in a static fluid with uniform density:

$$P_2 = P_1 + \rho gd \quad (9\text{-}3)$$

where point 2 is a depth d below point 1

Since we can imagine a cylinder anywhere we choose within the liquid, Eq. (9-3) relates the pressure at any two points in a static liquid where point 2 is a depth d below point 1.

For gases, Eq. (9-3) can be applied as long as the depth d is small enough that changes in the density due to gravity are negligible. Since liquids are nearly incompressible, Eq. (9-3) holds to great depths in liquids.

For a liquid that is open to the atmosphere, suppose we take point 1 at the surface and point 2 a depth d below. Then $P_1 = P_{atm}$, so the pressure at a depth d below the surface is

Pressure at a depth d below the surface of a liquid open to the atmosphere:

$$P = P_{atm} + \rho gd \quad (9\text{-}4)$$

CHECKPOINT 9.4

Pressure in a static fluid depends on vertical position. Can it also depend on horizontal position? Explain.

Example 9.3

A Diver

A diver swims to a depth of 3.2 m in a freshwater lake. What is the increase in the force pushing in on her eardrum, compared to what it was at the lake surface? The area of the eardrum is 0.60 cm^2.

Strategy We can find the increase in pressure at a depth of 3.2 m and then find the corresponding increase in force on the eardrum. If the force on the eardrum at the surface is P_1A and the force at a depth of 3.2 m is P_2A, then the increase in the force is $(P_2 - P_1)A$.

Solution The increase in pressure depends on the depth d and the density of water. From Table 9.1, the density of water is 1000 kg/m^3 to two significant figures for any reasonable temperature.

$$P_2 - P_1 = \rho g d$$

$$\Delta P = 1000\ \text{kg/m}^3 \times 9.8\ \text{m/s}^2 \times 3.2\ \text{m}$$

$$= 31.4\ \text{kPa}$$

The increase in force on the eardrum is

$$\Delta F = \Delta P \times A$$

where $A = 0.60\ \text{cm}^2 = 6.0 \times 10^{-5}\ \text{m}^2$. Then

$$\Delta F = (3.14 \times 10^4\ \text{Pa}) \times (6.0 \times 10^{-5}\ \text{m}^2)$$

$$= 1.9\ \text{N}$$

Discussion A force also pushes *outward* on the eardrum due to the pressure inside the ear canal. If the diver descends rapidly so that the pressure inside the ear canal does not change, then a 1.9-N net force due to fluid pressure pushes inward on the eardrum. When the diver's ear "pops," the pressure inside the ear canal increases to equal the fluid pressure outside the eardrum, so that the net force due to fluid pressure on the eardrum is zero.

Practice Problem 9.3 Limits on Submarine Depth

A submarine is constructed so that it can safely withstand a pressure of 1.6×10^7 Pa. How deep may this submarine descend in the ocean if the average density of seawater is 1025 kg/m^3?

Conceptual Example 9.4

The Hydrostatic Paradox

Three vessels have different shapes, but the same base area and the same weight when empty (Fig. 9.6). The vessels are filled with water to the same level and then the air is pumped out. The top surface of the water is then at a low pressure that, for simplicity, we assume to be zero. (a) Are the water pressures at the bottom of each vessel the same? If not, which is largest and which is smallest? (b) If the three vessels containing water are weighed on a scale, do they give the same reading? If not, which weighs the most and which weighs the least? (c) If the water exerts the same downward force on the bottom of each vessel, is that force equal to the weight of water in the vessel? Is there a paradox here?

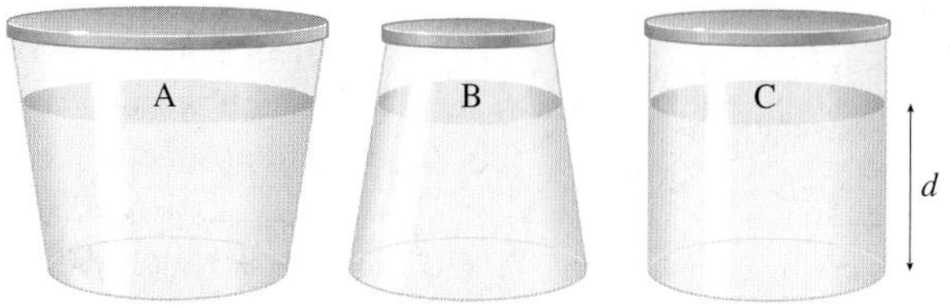

Figure 9.6

Three differently shaped vessels filled with water to same level.

[*Hint:* Think about the forces due to fluid pressure on the *sides* of the containers; do they have vertical components?]

continued on next page

Conceptual Example 9.4 continued

Solution and Discussion (a) The water at the bottom of each vessel is the same depth d below the surface. Water at the surface of each vessel is at a pressure $P_{surface} = 0$. Therefore, the pressures at the bottom must be equal:

$$P = P_{surface} + \rho g d = \rho g d$$

(b) The weight of each filled vessel is equal to the weight of the vessel itself plus the weight of the water inside. The vessels themselves have equal weights, but vessel A holds more water than C, whereas vessel B holds less water than C. Vessel A weighs the most and vessel B weighs the least.

(c) Each container supports the water inside by exerting an upward force equal in magnitude to the weight of the water. By Newton's third law, the water exerts a downward force on the container of the same magnitude. Figure 9.7 shows the forces acting on each container due to the water. In vessel C, the horizontal forces on any two diametrically opposite points on the walls of the container are equal and opposite; thus, the net force on the container walls is zero. The force on the bottom is

$$F = PA = (\rho g d)(\pi r^2)$$

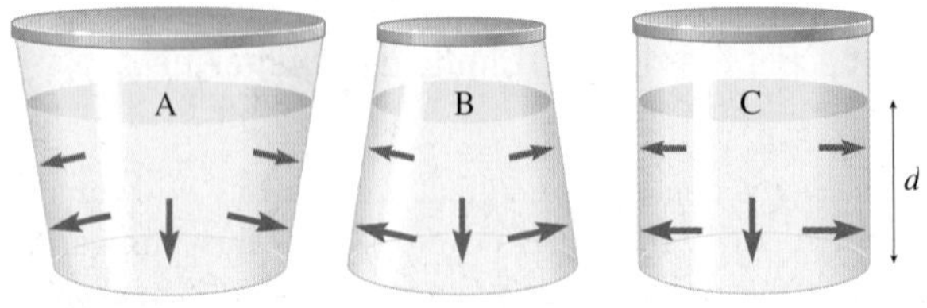

Figure 9.7
Forces exerted on the containers by the water.

The volume of water in the cylinder is $V = \pi r^2 d$, so

$$F = \rho g V = (\rho V)g = mg$$

The force on the bottom of vessel C is equal to the weight of the water, as expected. However, the force on the bottom of vessel A is less than the weight of the water in the container, while the force on the bottom of vessel B is greater than the weight of the water. Then how can the water be in equilibrium? In vessel A, the forces on the container walls have downward components as well as horizontal components. The horizontal components of the forces on any two diametrically opposite points are equal and opposite, so the horizontal components add to zero. The sum of the downward components of the forces on the walls and the downward force on the bottom of the container is equal to the weight of the water. Similarly, the forces on the walls of vessel B have upward components. In each case, the *total* force on the bottom *and sides* of the container due to the water is equal to the weight of the water.

Conceptual Practice Problem 9.4 Is Pressure Determined by Column Height?

Figure 9.8 shows a vessel with two points marked at the bottom of the water in the vessel. A narrow column of water is drawn above each point. (a) Is the pressure at point 2, P_2, the same as the pressure at point 1, P_1, even though the column of water above point 2 is not as tall? (b) Does $P = P_{atm} + \rho g d$ imply that $P_2 < P_1$? Explain.

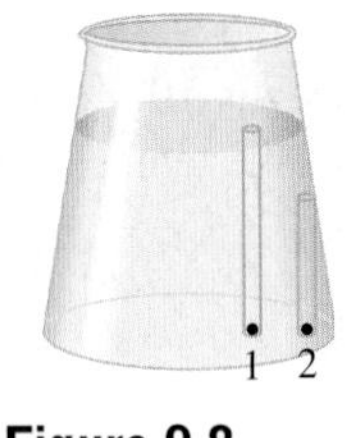

Figure 9.8
Two different points on the bottom of an open vessel.

9.5 MEASURING PRESSURE

Units of pressure: 1 atm = 101.3 kPa = 1.013 bar = 14.7 lb/in^2 = 760.0 mm Hg = 760.0 torr = 29.9 in Hg

Many other units are used for pressure besides atmospheres and pascals. In the United States, the pressure in an automobile tire is measured in pounds per square inch (lb/in^2). Weather bureaus record atmospheric pressure in bars or millibars. In the United States, television weather reports and home barometers measure pressure in inches of mercury. One atmosphere is equal to approximately 1 bar (1000 millibars), 76 cm of mercury, or 29.9 in. of mercury. Blood pressure, the *difference* between the pressure in the blood and atmospheric pressure, is measured in millimeters of mercury (mm Hg), also called the torr. Inches or millimeters of mercury may seem like strange units for pressure: how can a force per unit area be equal to a *distance* (so many mm Hg)? There is an assumption inherent in using these pressure units that we can understand by studying the mercury manometer.

Manometer

A mercury manometer consists of a vertical U-shaped tube, containing some mercury, with one side typically open to the atmosphere and the other connected to a vessel containing a gas whose pressure we want to measure. Figure 9.9 shows the manometer before it is connected to such a vessel. When both sides of the manometer are open to the atmosphere, the mercury levels are the same.

Now we connect an inflated balloon to the left side of the U-tube (Fig. 9.10). If the gas in the balloon is at a higher pressure than the atmosphere, the gas pushes the mercury down on the left side and forces it up on the right side. The density of a gas is small compared to the density of mercury, so every point within the gas is assumed to be at the same pressure no matter what the depth. At point B, the mercury pushes on the gas with the same magnitude force with which the gas pushes on the mercury, so point B is at the same pressure as the gas. Since point B' is at the same height within the mercury as point B, the pressure at B' is the same as at B. Point C is at atmospheric pressure.

The pressure at B is

$$P_B = P_B = P_C + \rho g d$$

where ρ is the density of mercury. The difference in the pressures on the two sides of the manometer is

$$\Delta P = P_B - P_C = \rho g d \qquad (9\text{-}5)$$

Thus, the difference in mercury levels d is a measure of the pressure *difference*—commonly reported in millimeters of mercury (mm Hg).

The pressure measured when one side of the manometer is open is the *difference* between atmospheric pressure and the gas pressure rather than the absolute pressure of the gas. This difference is called the **gauge pressure**, since it is what most gauges (not just manometers) measure:

Gauge pressure:

$$P_{gauge} = P_{abs} - P_{atm} \qquad (9\text{-}6)$$

Since the density of mercury is 13 600 kg/m^3, 1.00 mm Hg can be converted to pascals by substituting $d = 1.00$ mm in Eq. (9-5):

$$1.00 \text{ mm Hg} = \rho g d = (13\,600 \text{ kg/m}^3)(9.80 \text{ m/s}^2)(0.00100 \text{ m}) = 133 \text{ Pa}$$

The liquid in a manometer may be something other than mercury, such as water or oil. Equation (9-5) still applies, as long as we use the correct density ρ of the liquid in the manometer.

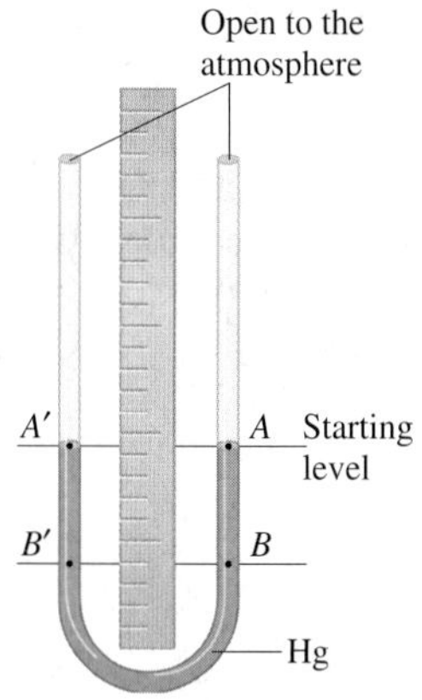

Figure 9.9 A mercury manometer open on both sides. Points A and A' are both at atmospheric pressure. Any two points (such as B and B') at the same height within the fluid are at the same pressure: $P_B = P_{B'}$.

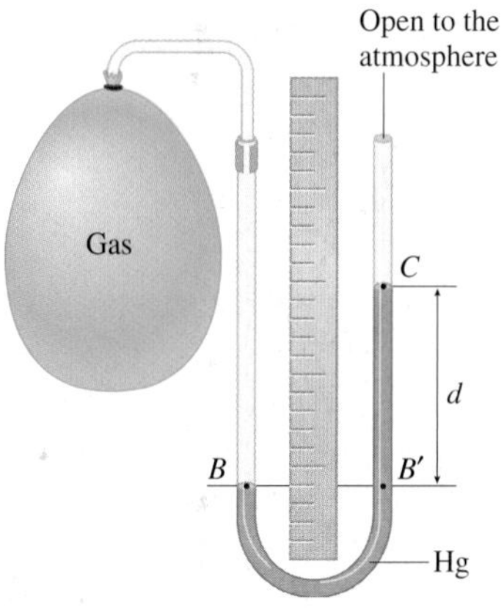

Figure 9.10 The manometer connected on one side to a container of gas at a pressure greater than atmospheric pressure.

Example 9.5

The Mercury Manometer

A manometer is attached to a container of gas to determine its pressure. Before the container is attached, both sides of the manometer are open to the atmosphere. After the container is attached, the mercury on the side attached to the gas container rises 12 cm above its previous level. (a) What is the gauge pressure of the gas in Pa? (b) What is the absolute pressure of the gas in Pa?

Strategy The mercury column is higher on the side connected to the container of gas, so we know that the pressure of the enclosed gas is lower than atmospheric pressure. We need to find the *difference* in levels of the mercury columns on the two sides. Careful: It is *not* 12 cm! If one side went up by 12 cm, then the other side has gone down by 12 cm, since the same volume of mercury is contained in the manometer.

Solution (a) The difference in the mercury levels is 24 cm (Fig. 9.11). Since the mercury on the gas side went *up*, the absolute pressure of the gas is *lower* than atmospheric pressure. Therefore, the gauge pressure of the gas is *less than zero*. The gauge pressure in Pa is

$$P_{gauge} = \rho g d$$

continued on next page

Example 9.5 continued

where the "depth" is $d = -24$ cm (the mercury is 24 cm higher on the gas side). Then

$$P_{gauge} = 13\,600 \text{ kg/m}^3 \times 9.8 \text{ m/s}^2 \times (-0.24 \text{ m}) = -32 \text{ kPa}$$

(b) The absolute pressure of the gas is

$$P = P_{gauge} + P_{atm}$$
$$= -32 \text{ kPa} + 101 \text{ kPa} = 69 \text{ kPa}$$

Discussion As a check, the manometer tells us directly that the gauge pressure of the gas is −240 mm Hg. Converting to pascals gives

$$-240 \text{ mm Hg} \times 133 \text{ Pa/mm Hg} = -32 \text{ kPa}$$

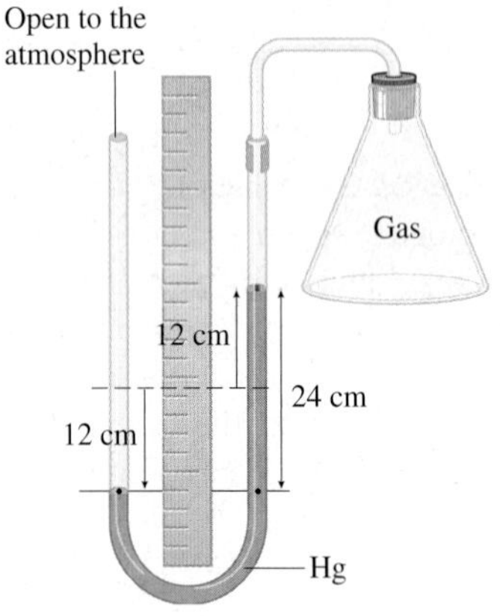

Figure 9.11
When a container of gas is attached to one side of the manometer, one side goes down 12 cm and the other side goes up 12 cm.

Practice Problem 9.5 Column Heights in Manometer

A mercury manometer is connected to a container of gas. (a) The height of the mercury column on the side connected to the gas is 22.0 cm (measured from the bottom of the manometer). What is the height of the mercury column on the open side if the gauge pressure is measured to be 13.3 kPa? (b) If the gauge pressure of the gas doubles, what are the new heights of the two columns?

Barometer

A manometer can act as a **barometer**—a device to measure atmospheric pressure. Instead of attaching a container with a gas to one end of the manometer, attach a container and a vacuum pump. Pump the air out of the container to get as close to a vacuum—zero pressure—as possible. Then the atmosphere pushes down on one side and pushes the fluid up on the other side toward the evacuated container.

Figure 9.12 shows a barometer in which the vacuum is not created by a vacuum pump. The barometer was invented by Evangelista Torricelli (1608–1647), an assistant to Galileo, in the 1600s; in his honor, one millimeter of mercury is called one torr.

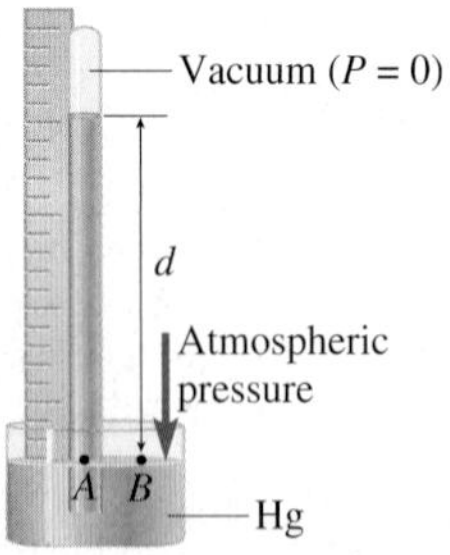

Figure 9.12 A simple barometer. A tube, of length greater than 76 cm and closed at one end, is filled with mercury. The tube is then inverted into an open container of mercury. Some mercury flows down from the tube into the bowl. The space left at the top of the tube is nearly a vacuum because nothing is left but a negligible amount of mercury vapor. Points A and B are at the same level in the mercury and, therefore, are both at atmospheric pressure since the bowl is open to the air. The distance d from A to the top of the mercury column in the closed tube is a measure of the atmospheric pressure (often called *barometric pressure* because it is measured with a barometer).

PHYSICS AT HOME

When you next have a drink with a straw, insert the straw into the drink and place your finger over the upper opening of the straw so that no more air can enter the straw. Raise the lower end of the straw up out of your drink. Does the liquid in the straw flow back down into your glass? What do you suppose is holding the liquid in place? Make an FBD on your paper napkin.

Some air is trapped between your finger and the top of the liquid in the straw; that air exerts a downward force on the liquid of magnitude P_1A (Fig. 9.13). A downward gravitational force mg also acts on the liquid. The air at the bottom of the straw exerts an upward force on the liquid of magnitude $P_{atm}A$; this upward force is what holds the liquid in place. Because the liquid does not pour out of the straw, but instead is in equilibrium,

$$P_{atm}A = P_1A + mg$$

Thus, the pressure P_1 of the air trapped above the liquid must be less than atmospheric pressure.

How did P_1 become less than atmospheric pressure? As you pulled the straw up and out, the liquid in the straw falls a bit, expanding the volume available to the air trapped above the liquid. When a gas expands under conditions like this, its pressure decreases.

When you remove your finger from the top of the straw, air can get in at the top of the straw. Then the pressures above and below the liquid are equal, so the gravitational force pulls the liquid down and out of the straw.

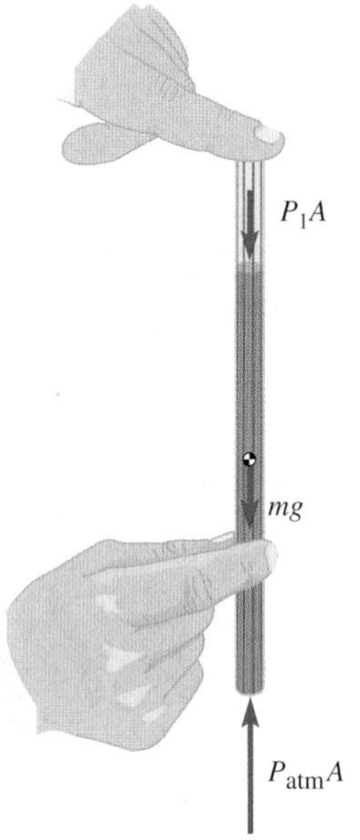

Figure 9.13 Force acting on the liquid inside a straw.

Sphygmomanometer

Blood pressure is measured with a sphygmomanometer (Fig. 9.14). The oldest kind of sphygmomanometer consists of a mercury manometer on one side attached to a closed bag—the cuff. The cuff is wrapped around the upper arm at the level of the heart and is then pumped up with air. The manometer measures the gauge pressure of the air in the cuff.

At first, the pressure in the cuff is higher than the *systolic* pressure—the maximum pressure in the brachial artery that occurs when the heart contracts. The cuff pressure squeezes the artery closed and no blood flows into the forearm. A valve on the cuff is then opened to allow air to escape slowly. When the cuff pressure decreases to just below the systolic pressure, a little squirt of blood flows past the constriction in the artery with each heartbeat. The sound of turbulent blood flow past the constriction can be heard through the stethoscope.

As air continues to escape from the cuff, the sound of blood flowing through the constriction in the artery continues to be heard. When the pressure in the cuff reaches the *diastolic* pressure in the artery—the minimum pressure that occurs when the heart muscle is relaxed—there is no longer a constriction in the artery, so the pulsing sounds cease. The *gauge* pressures for a healthy heart are nominally around 120 mm Hg (systolic) and 80 mm Hg (diastolic).

Application of the manometer: measuring blood pressure

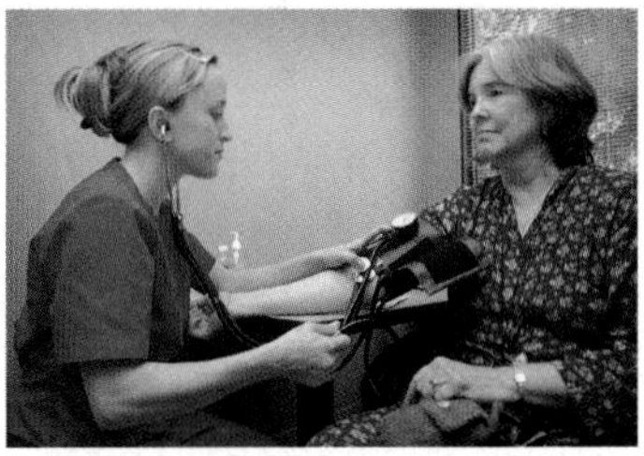

Figure 9.14 A sphygmomanometer being used to measure blood pressure.

9.6 THE BUOYANT FORCE

When an object is immersed in a fluid, the pressure on the lower surface of the object is higher than the pressure on the upper surface. The difference in pressures leads to an upward net force acting on the object due to the fluid pressure. If you try to push a beach ball underwater, you feel the effects of the buoyant force pushing the ball back up. It takes a rather large force to hold such an object completely underwater; the instant you let go, the object pops back up to the surface.

CONNECTION:

The buoyant force is not a new kind of force exerted by a fluid; it is the sum of forces due to fluid pressure.

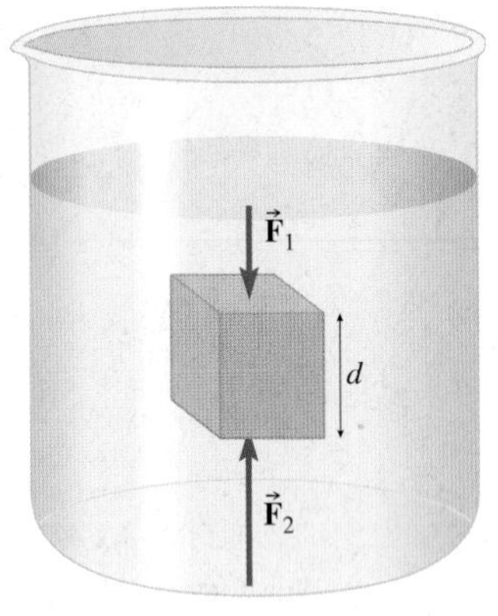

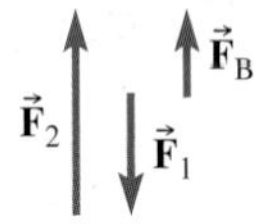

Figure 9.15 (a) Forces due to fluid pressure on the top and bottom of an immersed rectangular solid. (b) The buoyant force is the sum of $\vec{\mathbf{F}}_1$ and $\vec{\mathbf{F}}_2$. Since $|\vec{\mathbf{F}}_2| > |\vec{\mathbf{F}}_1|$, the net force due to fluid pressure is upward.

Consider a rectangular solid immersed in a fluid of uniform density ρ (Fig. 9.15a). For each vertical face (left, right, front, and back), there is a face of equal area opposite it. The forces on these two faces due to the fluid are equal in magnitude since the areas and the average pressures are the same. The directions are opposite, so the forces acting on the vertical faces cancel in pairs.

Let the top and bottom surfaces each have area A. The force on the lower face of the block is $F_2 = P_2A$; the force on the upper face is $F_1 = P_1A$. The total force on the block due to the fluid, called the **buoyant force** F_B, is upward since $F_2 > F_1$ (Fig. 9.15b).

$$\vec{\mathbf{F}}_B = \vec{\mathbf{F}}_1 + \vec{\mathbf{F}}_2$$

$$F_B = (P_2 - P_1)A$$

Since $P_2 - P_1 = \rho gd$, the magnitude of the buoyant force can be written

Buoyant force:

$$F_B = \rho gdA = \rho gV \qquad (9\text{-}7)$$

where $V = Ad$ is the volume of the block.

Note that ρV is the mass of the volume V of the fluid that the block displaces. Thus, the buoyant force on the submerged block is equal to the weight of an equal volume of fluid, a result called **Archimedes' principle**.

Archimedes' Principle

A fluid exerts an upward buoyant force on a submerged object equal in magnitude to the weight of the volume of fluid displaced by the object.

Archimedes' principle applies to a submerged object of *any shape* even though we derived it for a rectangular block. Why? Imagine replacing an irregular submerged object with enough fluid to fill the object's place. This "piece" of fluid is in equilibrium, so the buoyant force must be equal to its weight. The buoyant force is the net force exerted on the "piece" of fluid by the surrounding fluid, which is identical to the buoyant force on the irregular object since the two have the same shape and surface area.

The same argument can be used to show that if an object is only partly submerged, the buoyant force is still equal to the weight of fluid displaced. Equation (9-7) applies as long as V is the part of the object's volume below the fluid surface rather than the entire volume of the object.

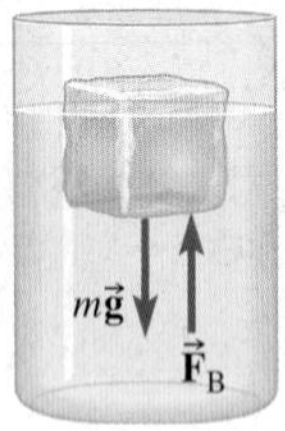

Figure 9.16 Forces acting on a floating ice cube. The ice cube is in equilibrium, so $\vec{\mathbf{F}}_B + m\vec{\mathbf{g}} = 0$.

Net Force due to Gravity and Buoyancy The net force due to gravity and buoyancy acting on an object totally or partially immersed in a fluid (Fig. 9.16) is

$$\vec{\mathbf{F}} = m\vec{\mathbf{g}} + \vec{\mathbf{F}}_B$$

The force of gravity on an object of volume V_o and average density ρ_o is

$$W = mg = \rho_o g V_o$$

and the buoyant force is

$$F_B = \rho_f g V_f$$

where V_f and ρ_f are the volume of fluid displaced and the fluid density, respectively. Choosing up to be the $+y$-direction, the net force due to gravity and buoyancy is

$$F_y = \rho_f g V_f - \rho_o g V_o \qquad (9\text{-}8)$$

Here F_y can be positive or negative, depending on which density is larger. Imagine releasing a pebble and an air bubble underwater. The pebble's average density is greater than the density of water, so the net force on it is downward; the pebble sinks. An air bubble's average density is less than the density of water, so the net force is upward, causing the bubble to rise toward the surface of the water.

If the object is completely submerged, the volumes of the object and the displaced fluid are the same and

$$F_y = (\rho_f - \rho_o)gV$$

If $\rho_o < \rho_f$, the object floats with only part of its volume submerged. In equilibrium, the object displaces a volume of fluid whose weight is equal to the object's weight. At that point the gravitational force equals the buoyant force and the object floats. Setting $F_y = 0$ in Eq. (9-8) yields

$$\rho_f g V_f = \rho_o g V_o$$

which can be rearranged as:

$$\frac{V_f}{V_o} = \frac{\rho_o}{\rho_f}$$

On the left side of this equation is the fraction of the object's volume that is submerged; it is equal to the ratio of the density of the object to the density of the fluid.

Specific Gravity This ratio of densities is called the **specific gravity** of the material when ρ_f is the density of water at 4°C. Specific gravity is without units because it is a ratio of two densities. Water at 4°C is chosen as the reference material because at that temperature, the density of water is a maximum (at atmospheric pressure). The gram was originally defined as the mass of one cubic centimeter of water at 4°C. Thus, water at 4°C has a density of 1 g/cm^3 (1000 kg/m^3). The specific gravity of seawater is 1.025, which means that seawater has a density of 1.025 g/cm^3 (1025 kg/m^3).

Specific gravity:

$$\text{S.G.} = \frac{\rho}{\rho_{\text{water}}} = \frac{\rho}{1000\ \text{kg/m}^3} \qquad (9\text{-}9)$$

Applications of specific gravity measurements in medicine

Blood tests often include determination of the specific gravity of the blood—normally around 1.040 to 1.065. A reading that is too low may indicate anemia, since the presence of red blood cells increases the average density of the blood. Before taking blood from a donor, a drop of the blood is placed in a solution of known density. If the drop does not sink, it is not safe for the donor to give blood because the concentration of red blood cells is too low. Urinalysis also includes a specific gravity measurement (normally 1.015 to 1.030); too high a value indicates an abnormally high concentration of dissolved salts, which can signal a medical problem.

Application of Archimedes' principle: how a ship can float

Freighters, aircraft carriers, and cruise ships float, although they are made from steel and other materials that are more dense than seawater. When a ship floats, the buoyant force acting on the ship is equal to the ship's weight. A ship is constructed so that it displaces a volume of seawater larger than the volume of the steel and other construction materials. The *average* density of the ship is its weight divided by its total volume. A large part of a ship's interior is filled with air. All of the "empty" space contributes to the total volume; the resulting average density is less than that of seawater, allowing the ship to float.

How can the hippo sink?

Now we can understand how a hippopotamus can sink to the bottom of a pond: it can expel some of the air in its body by exhaling. Exhalation increases the average density of the hippopotamus so that it is just slightly above the density of the water; thus, it sinks. (An armadillo does just the opposite: it swallows air, inflating its stomach and intestines, to increase the buoyant force for a swim across a large lake.) When the hippo needs to breathe, it swims back up to the surface.

Example 9.6

The Golden (?) Falcon

A small statue in the shape of a falcon has a weight of 24.1 N. The owner of the statue claims it is made of solid gold. When the statue is completely submerged in a container brimful of water, the weight of the water that spills over the top and into a bucket is 1.25 N. Find the density and specific gravity of the metal. Is the density consistent with the claim that the falcon is solid gold?

Strategy When the statue is completely submerged, it displaces a volume V of water equal to its own volume. The weight of the displaced water is equal to the buoyant force. Let $m_s g = 24.1$ N represent the weight of the statue (in terms of its mass m_s) and let $m_w g = 1.25$ N represent the weight of the water.

Solution The specific gravity of the statue is

$$\text{S.G.} = \frac{\rho_s}{\rho_w} = \frac{m_s/V}{m_w/V} = \frac{m_s}{m_w}$$

Rather than calculate the masses in kilograms, we recognize that a ratio of masses is equal to the ratio of the weights:

$$\text{S.G.} = \frac{m_s g}{m_w g} = \frac{24.1\text{ N}}{1.25\text{ N}} = 19.3$$

The density of the statue is

$$\rho_s = \text{S.G.} \times \rho_w = 19.3 \times 1000\text{ kg/m}^3 = 1.93 \times 10^4\text{ kg/m}^3$$

From Table 9.1, the statue has the correct density; it may *possibly* be gold.

Discussion According to legend, this method to determine the specific gravity of a solid was discovered by Archimedes in the third century B.C.E. King Hieron II asked Archimedes to find a way to check whether his crown was made of pure gold—without melting down the crown, of course! Archimedes came up with his method while he was taking a bath; he noticed the water level rising as he got in and connected the rising water level with the volume of water displaced by his body. In his excitement, he jumped out of the bath and ran naked through the streets of Siracusa (a city in Sicily) shouting "Eureka!"

Practice Problem 9.6 Identifying an Unknown Substance

An unknown solid substance has a weight of 142.0 N. The object is suspended from a scale and hung so that it is completely submerged in water (but not touching bottom). The scale reads 129.4 N. Find the specific gravity of the object and determine whether the substance could be anything listed in Table 9.1.

Example 9.7

Hidden Depths of an Iceberg

What percentage of a floating iceberg's volume is above water? The specific gravity of ice is 0.917 and the specific gravity of the surrounding seawater is 1.025.

Strategy The ratio of the density of ice to the density of seawater tells us the ratio of the volume of ice that is submerged in the seawater to the total volume of the iceberg. The rest of the ice is above the water.

Solution We could calculate the densities of seawater and of ice in SI units from their specific gravities, but that is unnecessary; the ratio of the specific gravities is equal to the ratio of the densities:

$$\frac{\text{S.G.}_{\text{ice}}}{\text{S.G.}_{\text{seawater}}} = \frac{\rho_{\text{ice}}/\rho_{\text{water}}}{\rho_{\text{seawater}}/\rho_{\text{water}}} = \frac{\rho_{\text{ice}}}{\rho_{\text{seawater}}}$$

The fraction of the iceberg's volume that is submerged is equal to the ratio of the densities of ice and seawater. Thus, the ratio of the volume submerged to the total volume of ice is

$$\frac{V_{\text{submerged}}}{V_{\text{ice}}} = \frac{\rho_{\text{ice}}}{\rho_{\text{seawater}}} = \frac{\text{S.G.}_{\text{ice}}}{\text{S.G.}_{\text{seawater}}}$$

$$= \frac{0.917}{1.025} = 0.895$$

89.5% of the ice is below the surface of the water, leaving only 10.5% above the surface.

Discussion An alternative solution does not depend on remembering that the ratio of the volumes is equal to the

continued on next page

Example 9.7 continued

ratio of the densities. The buoyant force is equal to the weight of a volume $V_{\text{submerged}}$ of water:

$$\text{buoyant force} = \rho_{\text{seawater}} V_{\text{submerged}} g$$

The weight of the iceberg is $mg = \rho_{\text{ice}} V_{\text{ice}}\, g$. From Newton's second law, the buoyant force must be equal in magnitude to the weight of the iceberg when it is floating in equilibrium:

$$\rho_{\text{seawater}} V_{\text{submerged}} g = \rho_{\text{ice}} V_{\text{ice}} g$$

or

$$\frac{V_{\text{submerged}}}{V_{\text{ice}}} = \frac{\rho_{\text{ice}}}{\rho_{\text{seawater}}}$$

The fact that ice floats is of great importance for the balance of nature. If ice were more dense than water, it would gradually fill up the ponds and lakes *from the bottom.* It would not form on top of lakes and remain there. The consequences for fish and other bottom dwellers of solidly frozen lakes would be catastrophic. The water below the surface layer of ice formed in winter remains just above freezing so that the fish are able to survive.

Practice Problem 9.7 Floating in Freshwater Versus Seawater

If the average density of a human being is 980 kg/m^3, what fraction of a human body floats above water in a freshwater pond and what fraction floats above seawater in the ocean? The specific gravity of seawater is 1.025.

Conceptual Example 9.8

A Hovering Fish

How is it that a fish is able to hover almost motionless in one spot—until some attractive food is spotted and, with a flip of the tail, off it swims after the food? Fish have a thin-walled bladder, called a swim bladder, located under the spinal column. The swim bladder contains a mixture of oxygen and nitrogen obtained from the blood of the fish. How does the swim bladder help the fish keep the buoyant and gravitational forces balanced so that it can hover?

Solution and Discussion If the fish's average density is greater than that of the surrounding water, it will sink; if its average density is smaller than that of the water, it will rise. By varying the volume of the swim bladder, the fish is able to vary its overall volume and, thus, its average density. By adjusting its average density to match the density of the surrounding water, the fish can remain suspended in position. The fish can also adjust the volume of the bladder when it wants to rise or sink.

Conceptual Practice Problem 9.8 The Diving Beetle

A diving beetle traps a bubble of air under its wings. While under the water, the beetle uses the air in the bubble to breathe, gradually exchanging the oxygen for carbon dioxide. (a) What does the beetle do to the air bubble so that it can dive under the water? (b) Once under water, what does the beetle do so that it can rise to the surface? [*Hint:* Treat the beetle and the air bubble as a single system. How can the beetle change the buoyant force acting on the system?]

Buoyant Forces on Objects Immersed in a Gas Gases such as air are fluids and exert buoyant forces just as liquids do. The buoyant force due to air is often negligible if an object's average density is much larger than the density of air. To see a significant buoyant force in air, we must use an object with a small average density. A hot air balloon has an opening at the bottom and a burner for heating the air within (Fig. 9.17). Many molecules of the heated air escape through the opening, decreasing the balloon's average density. When the balloon is less dense on average than the surrounding air, it rises; at higher altitudes, the surrounding air becomes less and less dense. At some particular altitude, the buoyant force is equal in magnitude to the weight of the balloon. Then, by Newton's second law, the net force on the balloon is zero. The balloon is in *stable* equilibrium at this altitude: if the balloon rises a bit, it experiences a net force downward, while if the balloon sinks down a bit, it is pushed back upward.

Application of buoyant forces: hot air balloons

Figure 9.17 The buoyant force due to the outside air keeps these balloons aloft.

9.7 FLUID FLOW

Types of Fluid Flow The study of *moving* fluids is a wonderfully complex subject. To illustrate some important ideas in less complex situations, we limit our study at first to fluids flowing under special conditions.

One difference between moving fluids and static fluids is that a moving fluid can exert a force *parallel* to any surface over or past which it flows; a static fluid cannot. Since the moving fluid exerts a force against a surface, the surface must also exert a force on the fluid. This **viscous force** opposes the flow of the fluid; it is the counterpart to the kinetic frictional force between solids. An external force must act on a viscous fluid (and thereby do work) to keep it flowing. Viscosity is considered in Section 9.9. Until then, we consider only nonviscous fluids—fluid flow where the viscous forces are negligibly small. We also ignore surface tension, which is considered in Section 9.11.

Fluid flow can be characterized as steady or unsteady. When the flow is **steady**, the velocity of the fluid *at any point* is constant in time. The velocity is not necessarily the same everywhere, but at any particular point, the velocity of the fluid passing that point remains constant in time. The density and pressure at any point in a steadily flowing fluid are also constant in time.

Steady flow is **laminar**. The fluid flows in neat layers so that each small portion of fluid that passes a particular point follows the same path as every other portion of fluid that passes the same point. The path that the fluid follows, starting from any point, is called a **streamline** (Fig. 9.18). The streamlines may curve and bend, but they cannot cross each other; if they did, the fluid would have to "decide" which way to go when it gets to such a point. The direction of the fluid velocity at any point must be tangent to the streamline passing through that point. Streamlines are a convenient way to depict fluid flow in a sketch.

Figure 9.18 A wind tunnel shows the streamlines in the flow of air past a car.

The Ideal Fluid The special case that we consider first is the flow of an **ideal fluid**. An ideal fluid is incompressible, undergoes laminar flow, and has no viscosity. Under some conditions, real fluids can be modeled as (nearly) ideal, but not under all conditions.

The flow of an ideal fluid is governed by two principles: the continuity equation and Bernoulli's equation. The continuity equation is an expression of conservation of mass for an incompressible fluid: since no fluid is created or destroyed, the total mass of the fluid must be constant. Bernoulli's equation, discussed in Section 9.8, is a form of the energy conservation law applied to fluid flow. Together, these two equations enable us to predict the flow of an ideal fluid.

The Continuity Equation

We start by deriving the continuity equation, which relates the speed of flow to the cross-sectional area of the fluid. Suppose an incompressible fluid flows into a pipe of nonuniform cross-sectional area under conditions of steady flow. In Fig. 9.19, the fluid on the left moves at speed v_1. During a time Δt, the fluid travels a distance

$$x_1 = v_1 \, \Delta t$$

If A_1 is the cross-sectional area of this section of pipe, then the mass of water moving past point 1 in time Δt is

$$\Delta m_1 = \rho V_1 = \rho A_1 x_1 = \rho A_1 v_1 \, \Delta t$$

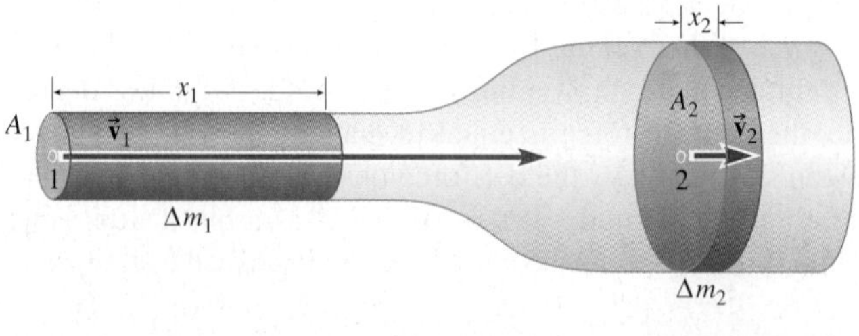

Figure 9.19 An incompressible fluid flowing horizontally through a nonuniform pipe.

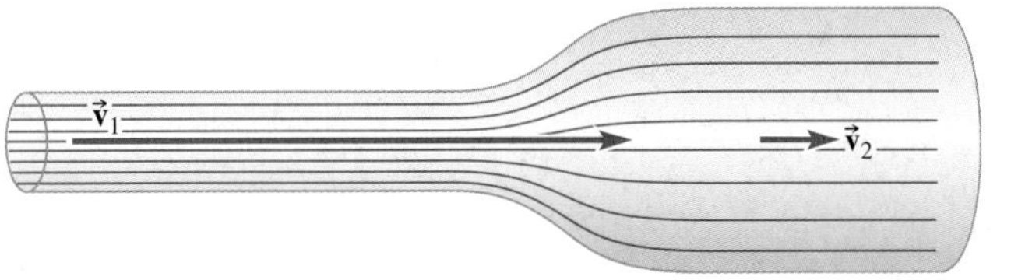

Figure 9.20 Streamlines in a pipe of varying cross-sectional area. Streamlines are closer together where the fluid velocity is larger and farther apart where the velocity is smaller.

During this same time interval, the mass of fluid moving past point 2 is

$$\Delta m_2 = \rho V_2 = \rho A_2 x_2 = \rho A_2 v_2 \, \Delta t$$

But, if the flow is steady, the mass passing through one section of pipe in time interval Δt must pass through any other section of the pipe in the same time interval. Therefore,

$$\Delta m_1 = \Delta m_2$$

or

$$\rho A_1 v_1 \, \Delta t = \rho A_2 v_2 \, \Delta t \qquad (9\text{-}10)$$

The quantity ρAv is the *mass flow rate* of the fluid:

Mass flow rate:

$$\frac{\Delta m}{\Delta t} = \rho Av \quad \text{(SI unit: kg/s)} \qquad (9\text{-}11)$$

Since the time intervals Δt are the same, Eq. (9-11) says that the mass flow rate past any two points is the same. Since the density of an incompressible fluid is constant, the volume flow rate past any two points must also be the same:

Volume flow rate:

$$\frac{\Delta V}{\Delta t} = Av \quad \text{(SI unit: m}^3\text{/s)} \qquad (9\text{-}12)$$

The **continuity equation** for an incompressible fluid equates the volume flow rates past two different points:

Continuity equation for incompressible fluid:

$$A_1 v_1 = A_2 v_2 \qquad (9\text{-}13)$$

The same volume of fluid that enters the pipe in a given time interval exits the pipe in the same time interval. Where the radius of the tube is large, the speed of the fluid is small; where the radius is small, the fluid speed is large. A familiar example is what happens when you use your thumb to partially block the end of a garden hose to make a jet of water. The water moves past your thumb, where the cross-sectional area is smaller, at a greater speed than it moves in the hose. Similarly, water traveling along a river speeds up, forming rapids, when the riverbed narrows or is partially blocked by rocks and boulders.

Streamlines are closer together where the fluid flows faster and farther apart where it flows more slowly (Fig. 9.20). Thus, streamlines help us visualize fluid flow. The fluid velocity at any point is tangent to a streamline through that point.

Figure 9.21 Demonstrating the continuity equation at a bathroom sink. Notice that the stream of water is narrower where the flow speed is faster.

PHYSICS AT HOME

The continuity equation applies to an ideal fluid even if it is not flowing through a pipe. Turn on a faucet so that the water flows out in a moderate stream (Fig. 9.21). The falling water is in free fall, accelerated by gravity until it hits the sink below. As the water falls, its speed increases. The stream of water gradually narrows as it falls so that the product of speed and cross-sectional area is constant, as predicted by the continuity equation.

Example 9.9

Speed of Blood Flow

The heart pumps blood into the aorta, which has an inner radius of 1.0 cm. The aorta feeds 32 major arteries. If blood in the aorta travels at a speed of 28 cm/s, at approximately what average speed does it travel in the arteries? Assume that blood can be treated as an ideal fluid and that the arteries each have an inner radius of 0.21 cm.

Strategy Since we have assumed blood to be an ideal fluid, we can apply the continuity equation. The main tube (the aorta) is connected to multiple tubes (the major arteries), so this problem seems to be more complicated than a single pipe with a constriction in it. What matters here is the total cross-sectional area into which the blood flows.

Solution We start by finding the cross-sectional area of the aorta

$$A_1 = \pi r^2_{\text{aorta}}$$

and then the total cross-sectional area of the arteries

$$A_2 = 32\pi r^2_{\text{artery}}$$

Now we apply the continuity equation and solve for the unknown speed.

$$A_1 v_1 = A_2 v_2$$

$$v_2 = v_1 \frac{A_1}{A_2} = 0.28 \text{ m/s} \times \frac{\pi \times (0.010 \text{ m})^2}{32\pi \times (0.0021 \text{ m})^2} = 0.20 \text{ m/s}$$

Discussion The blood flow slows in the arteries because the total cross-sectional area is greater than that of the aorta alone. From the arteries, the blood travels to the many capillaries of the body. Each capillary has a tiny cross-sectional area, but there are so many of them that the blood flow slows greatly once it reaches the capillaries—allowing time for the exchange of oxygen, carbon dioxide, and nutrients between the blood and the body tissues.

Practice Problem 9.9 Hosing Down a Wastebasket

A garden hose fills a 32-L wastebasket in 120 s. The opening at the end of the hose has a radius of 1.00 cm. (a) How fast is the water traveling as it leaves the hose? (b) How fast does the water travel if half the exit area is obstructed by placing a finger over the opening?

9.8 BERNOULLI'S EQUATION

The Bernoulli effect: Fluid flows faster where the pressure is lower.

The continuity equation relates the flow velocities of an ideal fluid at two different points, based on the change in cross-sectional area of the pipe. According to the continuity equation, the fluid must speed up as it enters a constriction (Fig. 9.22) and then slow down to its original speed when it leaves the constriction. Using energy ideas, we will show that the pressure of the fluid in the constriction (P_2) cannot be the same as the pressure before or after the constriction (P_1). For horizontal flow *the speed is higher where the pressure is lower.* This principle is often called the **Bernoulli effect.**

The Bernoulli effect can seem counterintuitive at first; isn't rapidly moving fluid at *high* pressure? For instance, if you were hit with the fast-moving water out of a firehose, you would be knocked over easily. The force that knocks you over is indeed due to fluid pressure; you would justifiably conclude that the pressure was high. However, the pressure is not high *until you slow down the water* by getting in its way. The rapidly moving water in the jet is, in fact, approximately at atmospheric pressure (zero gauge pressure), but when you *stop* the water, its pressure increases dramatically.

Let's find the quantitative relationship between pressure changes and flow speed changes for an ideal fluid. In Fig. 9.23, the shaded volume of fluid flows to the right.

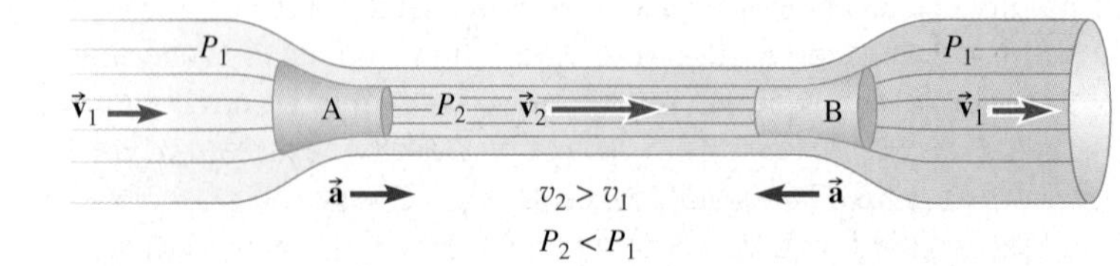

Figure 9.22 A small volume of fluid speeds up as it moves into a constriction (position A) and then slows down as it moves out of the constriction (position B).

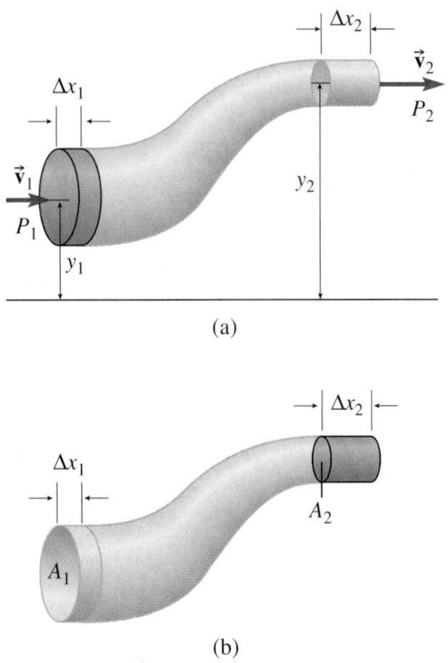

Figure 9.23 Applying conservation of energy to the flow of an ideal fluid. The shaded volume of fluid in (a) is flowing to the right; (b) shows the same volume of fluid a short time later.

If the left end moves a distance Δx_1, then the right end moves a distance Δx_2. Since the fluid is incompressible,

$$A_1\,\Delta x_1 = A_2\,\Delta x_2 = V$$

Work is done by the neighboring fluid during this flow. Fluid behind (to the left) pushes forward, doing positive work, while fluid ahead pushes backward, doing negative work. The total work done on the shaded volume by neighboring fluid is

$$W = P_1 A_1\,\Delta x_1 - P_2 A_2\,\Delta x_2 = (P_1 - P_2)V$$

Since no dissipative forces act on an ideal fluid, the work done is equal to the total change in kinetic and gravitational potential energy. The net effect of the displacement is to move a volume V of fluid from height y_1 to height y_2 and to change its speed from v_1 to v_2. The energy change is therefore

$$\Delta E = \Delta K + \Delta U = \tfrac{1}{2}m(v_2^2 - v_1^2) + mg(y_2 - y_1)$$

where the $+y$-direction is up. Substituting $m = \rho V$ and equating the work done on the fluid to the change in its energy yields

$$(P_1 - P_2)\,V = \tfrac{1}{2}\,\rho V(v_2^2 - v_1^2) + \rho V g(y_2 - y_1)$$

Dividing both sides by V and rearranging yields Bernoulli's equation, named after Swiss mathematician Daniel Bernoulli (1700–1782), but first derived by fellow Swiss mathematician Leonhard Euler (pronounced like *oiler*, 1707–1783).

CONNECTION:

Bernoulli's equation is a restatement of the principle of energy conservation applied to the flow of an ideal fluid.

Bernoulli's equation (for ideal fluid flow):

$$P_1 + \rho g y_1 + \tfrac{1}{2}\rho v_1^2 = P_2 + \rho g y_2 + \tfrac{1}{2}\rho v_2^2$$

$$(\text{or } P + \rho g y + \tfrac{1}{2}\rho v^2 = \text{constant}) \qquad (9\text{-}14)$$

Bernoulli's equation relates the pressure, flow speed, and height at two points in an ideal fluid. Although we derived Bernoulli's equation in a relatively simple situation, it applies to the flow of any ideal fluid as long as points 1 and 2 are on the same streamline.

Each term in Bernoulli's equation has units of pressure, which in the SI system is Pa or N/m^2. Since a joule is a newton-meter, the pascal is also equal to a joule per cubic meter (J/m^3). Each term represents work or energy per unit volume. The pressure is the work done by the fluid on the fluid ahead of it per unit volume of flow. The kinetic energy per unit volume is $\frac{1}{2}\rho v^2$ and the gravitational potential energy per unit volume is ρgy. (Text website tutorial: energies)

CHECKPOINT 9.8

Discuss Bernoulli's equation in two special cases: (a) horizontal flow ($y_1 = y_2$) and (b) a static fluid ($v_1 = v_2 = 0$).

Example 9.10

Torricelli's Theorem

A barrel full of rainwater has a spigot near the bottom, at a depth of 0.80 m beneath the water surface. (a) When the spigot is directed horizontally (Fig. 9.24a) and is opened, how fast does the water come out? (b) If the opening points upward (Fig. 9.24b), how high does the resulting "fountain" go? (tutorial: waterfall)

Strategy The water at the surface is at atmospheric pressure. The water emerging from the spigot is *also* at atmospheric pressure since it is in contact with the air. If the pressure of the emerging water were different than that of the air, the stream would expand or contract until the pressures were equal. We apply Bernoulli's equation to two points: point 1 at the water surface and point 2 in the emerging stream of water.

Solution (a) Since $P_1 = P_2$, Bernoulli's equation is

$$\rho g y_1 + \tfrac{1}{2}\rho v_1^2 = \rho g y_2 + \tfrac{1}{2}\rho v_2^2$$

Point 1 is 0.80 m above point 2, so

$$y_1 - y_2 = 0.80 \text{ m}$$

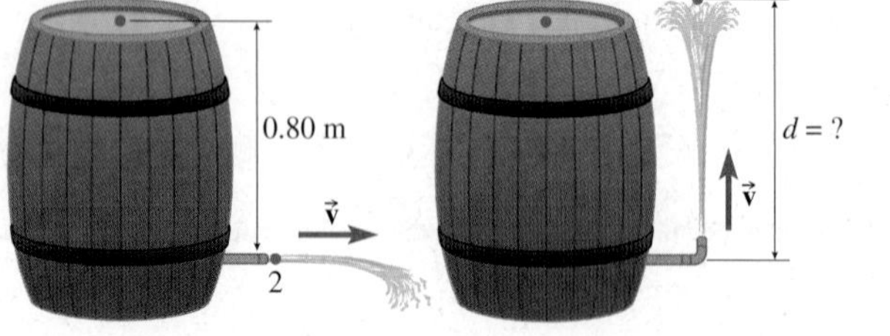

Figure 9.24

Full barrel of rainwater with open spigot (a) horizontal and (b) upward.

The speed of the emerging water is v_2. What is v_1, the speed of the water at the surface? The water at the surface is moving slowly, since the barrel is draining. The continuity equation requires that

$$v_1 A_1 = v_2 A_2$$

Since the cross-sectional area of the spigot A_2 is much smaller than the area of the top of the barrel A_1, the speed of the water at the surface v_1 is negligibly small compared with v_2. Setting $v_1 = 0$, Bernoulli's equation reduces to

$$\rho g y_1 = \rho g y_2 + \tfrac{1}{2}\rho v_2^2$$

After dividing through by ρ, we solve for v_2:

$$g(y_1 - y_2) = \tfrac{1}{2}v_2^2$$

$$v_2 = \sqrt{2g(y_1 - y_2)} = 4.0 \text{ m/s}$$

(b) Now take point 2 to be at the top of the fountain. Then $v_2 = 0$ and Bernoulli's equation reduces to

$$\rho g y_1 = \rho g y_2$$

The "fountain" goes right back up to the top of the water in the barrel!

Discussion The result of part (b) is called Torricelli's theorem. In reality, the fountain does not reach as high as the original water level; some energy is dissipated due to viscosity and air resistance.

Practice Problem 9.10 Fluid in Free Fall

Verify that the speed found in part (a) is the same as if the water just fell 0.80 m straight down. That shouldn't be too surprising since Bernoulli's equation is an expression of energy conservation.

Example 9.11

The Venturi Meter

A *Venturi meter* (Fig. 9.25) measures fluid speed in a pipe. A constriction (of cross-sectional area A_2) is put in a pipe of normal cross-sectional area A_1. Two vertical tubes, open to the atmosphere, rise from two points, one of which is in the constriction. The vertical tubes function like manometers, enabling the pressure to be determined. From this information the flow speed in the pipe can be determined.

Suppose that the pipe in question carries water, $A_1 = 2.0A_2$, and the fluid heights in the vertical tubes are $h_1 = 1.20$ m and $h_2 = 0.80$ m. (a) Find the ratio of the flow speeds v_2/v_1. (b) Find the gauge pressures P_1 and P_2. (c) Find the flow speed v_1 in the pipe.

Strategy Neither of the two flow speeds is given. We need more than Bernoulli's equation to solve this problem. Since we know the ratio of the areas, the continuity equation gives us the ratio of the speeds. The height of the water in the vertical tubes enables us to find the pressures at points 1 and 2. The fluid pressure at the bottom of each vertical tube is the same as the pressure of the moving fluid just beneath each tube—otherwise, water would flow into or out of the vertical tubes until the pressure equalized. The water in the vertical tubes is static, so the gauge pressure at the bottom is $P = \rho gd$. Once we have the ratio of the speeds and the pressures, we apply Bernoulli's equation.

Solution (a) From the continuity equation, the product of flow speed and area must be the same at points 1 and 2. Therefore,

$$\frac{v_2}{v_1} = \frac{A_1}{A_2} = 2.0$$

The water flows twice as fast in the constriction as in the rest of the pipe.

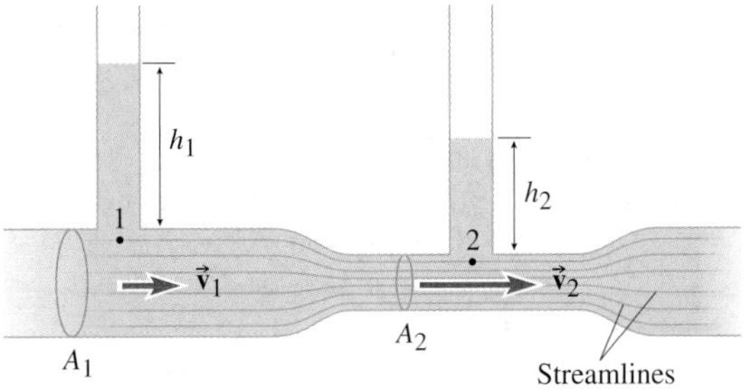

Figure 9.25
Venturi meter.

(b) The gauge pressures are:

$$P_1 = \rho gh_1 = 1000\ \text{kg/m}^3 \times 9.80\ \text{N/kg} \times 1.20\ \text{m} = 11.8\ \text{kPa}$$

$$P_2 = \rho gh_2 = 1000\ \text{kg/m}^3 \times 9.80\ \text{N/kg} \times 0.80\ \text{m} = 7.8\ \text{kPa}$$

(c) Now we apply Bernoulli's equation. We can use gauge pressures as long as we do so on both sides—in effect we are just subtracting atmospheric pressure from both sides of the equation:

$$P_1 + \rho gy_1 + \tfrac{1}{2}\rho v_1^2 = P_2 + \rho gy_2 + \tfrac{1}{2}\rho v_2^2$$

Since the tube is horizontal, $y_1 \approx y_2$ and we can ignore the small change in gravitational potential energy density ρgy. Then

$$P_1 + \tfrac{1}{2}\rho v_1^2 = P_2 + \tfrac{1}{2}\rho v_2^2$$

We are trying to find v_1, so we can eliminate v_2 by substituting $v_2 = 2.0v_1$:

$$P_1 + \tfrac{1}{2}\rho v_1^2 = P_2 + \tfrac{1}{2}\rho (2.0v_1)^2$$

Simplifying,

$$P_1 - P_2 = 1.5\rho v_1^2$$

$$v_1 = \sqrt{\frac{11\,800\ \text{Pa} - 7800\ \text{Pa}}{1.5 \times 1000\ \text{kg/m}^3}} = 1.6\ \text{m/s}$$

Discussion The assumption that $y_1 \approx y_2$ is fine as long as the pipe radius is small compared with the difference between the static water heights (40 cm). Otherwise, we would have to account for the different y values in Bernoulli's equation.

One subtle point: recall that we assumed that the fluid pressure at the bottom of the vertical tubes was the same as the pressure of the moving fluid just beneath. Does that contradict Bernoulli's equation? Since there is an abrupt change in fluid speed, shouldn't there be a significant difference in the pressures? No, because these points are *not on the same streamline.*

Practice Problem 9.11 Garden Hose

Water flows horizontally through a garden hose of radius 1.0 cm at a speed of 1.4 m/s. The water shoots horizontally out of a nozzle of radius 0.25 cm. What is the gauge pressure of the water inside the hose?

Application of Bernoulli's Principle: Arterial Flutter and Aneurisms Suppose an artery is narrowed due to buildup of plaque on its inner walls. The flow of blood through the constriction is similar to that shown in Fig. 9.22. Bernoulli's equation tells us that the pressure P_2 in the constriction is lower than the pressure elsewhere.

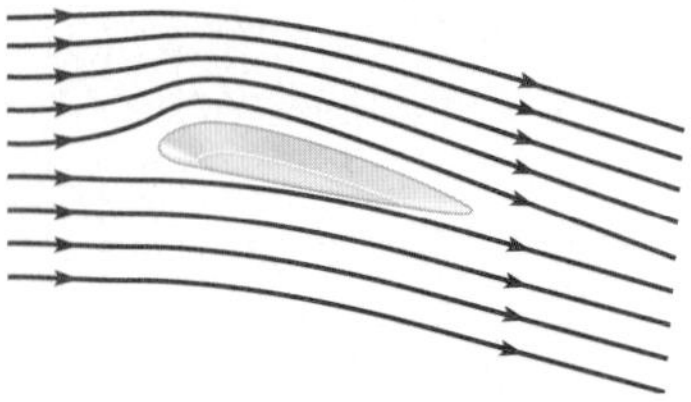

Figure 9.26 Streamlines showing the airflow past an airplane wing in a wind tunnel.

The arterial walls are elastic rather than rigid, so the lower pressure allows the arterial walls to contract a bit in the constriction. Now the flow velocity is even higher and the pressure even lower. Eventually the artery wall collapses, shutting off the flow of blood. Then the pressure builds up, reopens the artery, and allows blood to flow. The cycle of arterial flutter then begins again.

The opposite may happen where the arterial wall is weak. Blood pressure pushes the artery walls outward, forming a bulge called an aneurism. The lower flow speed in the bulge is accompanied by a higher blood pressure, which enlarges the aneurism even more (see Problem 88). Ultimately the artery may burst from the increased pressure.

Application of Bernoulli's Principle: Airplane Wings How does an airplane wing generate lift? Figure 9.26 is a sketch of some streamlines for air flowing past an airplane wing in a wind tunnel. The streamlines bend, showing that the wing deflects air downward. By Newton's third law (or conservation of momentum), if the wing pushes downward on the air, the air also pushes upward on the wing. This upward force on the wing is lift. However, the situation is not as simple as air "bouncing" off the bottom of the wing—note that air passing above the wing is also deflected downward.

We can use Bernoulli's equation to get more insight into the generation of lift. (Bernoulli's equation applies in an approximate way to moving air. Even though air is not incompressible, for subsonic flight the density changes are small enough to be ignored.) If the air exerts a net upward force on the wing, the air pressure must be lower above the wing than beneath the wing. In Fig. 9.26, the streamlines above the wing are closer together than beneath the wing, showing that the flow speed above the wing is faster than it is beneath. This observation confirms that the pressure is lower above the wing, because where the pressure is lower, the flow speed is faster.

9.9 VISCOSITY

CONNECTION:

Kinetic friction makes a sliding object slow down unless an applied force balances the force of friction. Similarly, viscous forces oppose the flow of a fluid. Steady flow of a viscous fluid requires an applied force to balance the viscous forces. The applied force is due to the pressure difference.

Bernoulli's equation ignores viscosity (fluid friction). According to Bernoulli's equation, an ideal fluid can continue to flow in a horizontal pipe at constant velocity on its own, just as a hockey puck would slide across frictionless ice at constant velocity without anything pushing it along. However, all real fluids have some viscosity; to maintain flow in a viscous fluid, we have to apply an external force since viscous forces oppose the flow of the fluid (Fig. 9.27). A *pressure difference* between the ends of the pipe must be maintained to keep a real liquid moving through a horizontal pipe. The pressure difference is important—in everything from blood flowing through arteries to oil pumped through a pipeline.

To visualize viscous flow in a tube of circular cross section, imagine the fluid to flow in cylindrical layers, or shells. If there were no viscosity, all the layers would move at the same speed (Fig. 9.28a). In viscous flow, the fluid speed depends on the distance from the tube walls (Fig. 9.28b). The fastest flow is at the center of the tube. Layers closer to the wall of the tube move more slowly. The outermost layer of fluid, which is in contact with the tube, does not move. Each layer of fluid exerts viscous forces on the

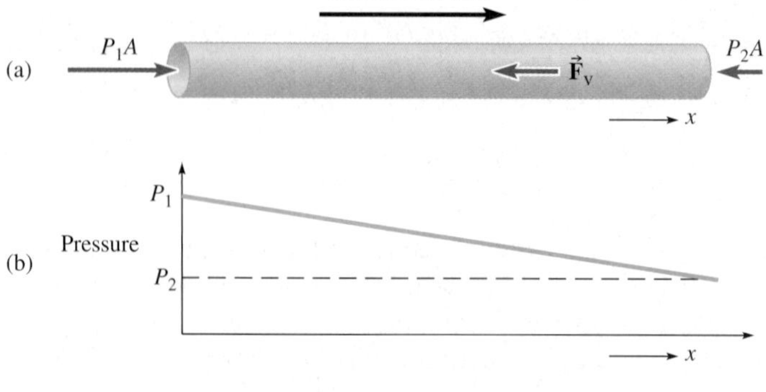

Figure 9.27 (a) To maintain viscous flow, a net force due to fluid pressure $(P_1 - P_2)A$ must be applied in the direction of flow to balance the viscous force F_v due to the pipe, which opposes flow. (b) The pressure in the fluid decreases from P_1 at the left end to P_2 at the right end.

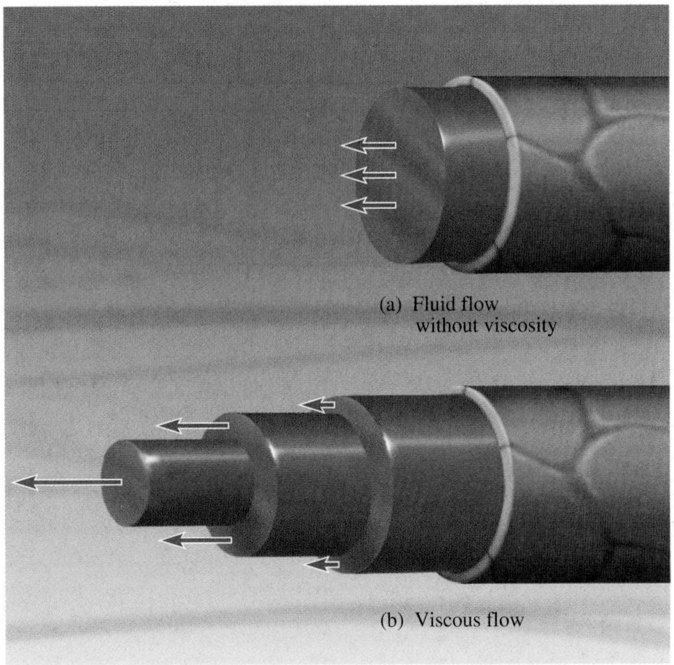

Figure 9.28 (a) In nonviscous flow through a tube, the flow speed is the same everywhere. (b) In viscous flow, the flow speed depends on distance from the tube wall. This simplified sketch shows layers of fluid each moving at a different speed, but in reality the flow speed increases continuously from zero for the outermost "layer" to a maximum speed at the center.

neighboring layers; these forces oppose the relative motion of the layers. The outermost layer exerts a viscous force on the tube.

A liquid is more viscous if the cohesive forces between molecules are stronger. The viscosity of a liquid decreases with increasing temperature because the molecules become less tightly bound. A decrease in the temperature of the human body is dangerous because the viscosity of the blood increases and the flow of blood through the body is hindered. Gases, on the other hand, have an increase in viscosity for an increase in temperature. At higher temperatures the gas molecules move faster and collide more often with each other.

The coefficient of viscosity (or simply the *viscosity*) of a fluid is written as the Greek letter eta (η) and has units of pascal-seconds (Pa·s) in SI. Other viscosity units in common use are the poise (pronounced *pwäz,* symbol P; 1 P = 0.1 Pa·s) and the centipoise (1 cP = 0.01 P = 0.001 Pa·s). Table 9.2 lists the viscosities of some common fluids.

Poiseuille's Law

The volume flow rate $\Delta V/\Delta t$ for laminar flow of a viscous fluid through a horizontal, cylindrical pipe depends on several factors. First of all, the volume flow rate is proportional to the *pressure drop per unit length* ($\Delta P/L$)—also called the pressure gradient. If a pressure drop ΔP maintains a certain flow rate in a pipe of length L, then a similar pipe of length $2L$ needs twice the pressure drop to maintain the same flow rate (ΔP across the first half and another ΔP across the second half). Thus, the flow rate ($\Delta V/\Delta t$) must be proportional to the pressure drop per unit length ($\Delta P/L$).

Next, the flow rate is inversely proportional to the viscosity of the fluid. The more viscous the fluid, the smaller the flow rate, if all other factors are equal.

The only other consideration is the radius of the pipe. In the nineteenth century, during a study of flow in blood vessels, French physician Jean-Léonard Marie Poiseuille (1799–1869) discovered that the flow rate is proportional to the *fourth power* of the pipe radius:

Poiseuille's Law (for Viscous Flow)

$$\frac{\Delta V}{\Delta t} = \frac{\pi}{8}\frac{\Delta P/L}{\eta}r^4 \qquad (9\text{-}15)$$

Table 9.2 Viscosities of Some Fluids

Substance	Temperature (°C)	Viscosity (Pa·s)
Gases		
Water vapor	100	1.3×10^{-5}
Air	0	1.7×10^{-5}
	20	1.8×10^{-5}
	30	1.9×10^{-5}
	100	2.2×10^{-5}
Liquids		
Acetone	30	0.30×10^{-3}
Methanol	30	0.51×10^{-3}
Ethanol	30	1.0×10^{-3}
Water	0	1.8×10^{-3}
	20	1.0×10^{-3}
	30	0.80×10^{-3}
	40	0.66×10^{-3}
	60	0.47×10^{-3}
	80	0.36×10^{-3}
	100	0.28×10^{-3}
Blood plasma	37	1.3×10^{-3}
Blood, whole	20	3.0×10^{-3}
	37	2.1×10^{-3}
Glycerin	20	0.83
	30	0.63
SAE 5W-30 motor oil	−30	≤ 6.6
	150	$\geq 2.9 \times 10^{-3}$

where $\Delta V/\Delta t$ is the volume flow rate, ΔP is the pressure difference between the ends of the pipe, r and L are the inner radius and length of the pipe, respectively, and η is the viscosity of the fluid. Poiseuille's name is pronounced *pwahzoy*, in a rough English approximation.

It isn't often that we encounter a *fourth-power* dependence. Why such a strong dependence on radius? First of all, if fluids are flowing through two different pipes at the *same speed*, the volume flow rates are proportional to radius squared (flow rate = speed multiplied by cross-sectional area). But, in viscous flow, the average flow speed is larger for wider pipes; fluid farther away from the walls can flow faster. It turns out that the average flow speed for a given pressure gradient is also proportional to radius squared, giving the overall fourth power dependence on the pipe radius of Poiseuille's law.

Application of viscous flow: high blood pressure

The strong dependence of flow rate on radius is important in blood flow. A person with cardiovascular disease has arteries narrowed by plaque deposits. To maintain the necessary blood flow to keep the body functioning, the blood pressure increases. If the diameter of an artery narrows to $\frac{1}{2}$ of its original value due to plaque deposits, the blood flow rate would decrease to $\frac{1}{16}$ of its original value if the pressure drop across it were to stay the same. To compensate for some of this decrease in blood flow, the heart pumps harder, increasing the blood pressure. High blood pressure is not good either; it introduces its own set of health problems, not least of which is the increased demands placed on the heart muscle.

Example 9.12

Arterial Blockage

A cardiologist reports to her patient that the radius of the left anterior descending artery of the heart has narrowed by 10.0%. What percent increase in the blood pressure drop across the artery is required to maintain the normal blood flow through this artery?

Strategy We assume that the viscosity of the blood has not changed, nor has the length of the artery. To maintain normal blood flow, the volume flow rate must stay the same:

$$\frac{\Delta V_1}{\Delta t} = \frac{\Delta V_2}{\Delta t}$$

Solution If r_1 is the normal radius and r_2 is the actual radius, a 10.0% reduction in radius means $r_2 = 0.900r_1$. Then, from Poiseuille's law,

$$\frac{\pi \Delta P_1 r_1^4}{8\eta L} = \frac{\pi \Delta P_2 r_2^4}{8\eta L}$$

$$r_1^4 \Delta P_1 = r_2^4 \Delta P_2$$

We solve for the ratio of the pressure drops:

$$\frac{\Delta P_2}{\Delta P_1} = \frac{r_1^4}{r_2^4} = \frac{1}{(0.900)^4} = 1.52$$

Discussion A factor of 1.52 means there is a 52% increase in the blood pressure difference across that artery. The increased pressure must be provided by the heart. If the normal pressure drop across the artery is 10 mm Hg, then it is now 15.2 mm Hg. The person's blood pressure either must increase by 5.2 mm Hg or there will be a reduction in blood flow through this artery. The heart is under greater strain as it works harder, attempting to maintain an adequate flow of blood.

Practice Problem 9.12 New Water Pipe

The town water supply is operating at nearly full capacity. The town board decides to replace the water main with a bigger one to increase capacity. If the maximum flow rate is to increase by a factor of 4.0, by what factor should they increase the radius of the water main?

Turbulence

When the fluid velocity at a given point changes, the flow is *unsteady.* **Turbulence** is an extreme example of unsteady flow (Fig. 9.29). In turbulent flow, swirling vortices—whirlpools of fluid—appear. The vortices are not stationary; they move with the fluid. The flow velocity at any point changes erratically; prediction of the direction or speed of fluid flow under turbulent conditions is difficult.

9.10 VISCOUS DRAG

When an object moves through a fluid, the fluid exerts a drag force on it. When the relative velocity between the object and the fluid is low enough for the flow around the object to be laminar, the drag force derives from viscosity and is called **viscous drag**.

Viscous drag: $F_D \propto v$
Turbulent drag: $F_D \propto v^2$

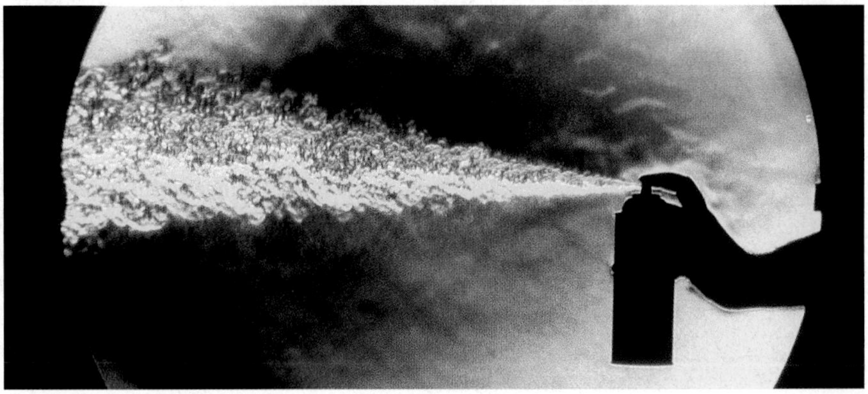

Figure 9.29 Turbulent flow of gas emerging from the nozzle of an aerosol can.

The viscous drag force is proportional to the speed of the object. For larger relative speeds, the flow becomes turbulent and the drag force is proportional to the square of the object's speed.

The viscous drag force depends also on the shape and size of the object. For a spherical object, the viscous drag force is given by Stokes's law:

Stokes's Law (viscous drag on a sphere)

$$F_D = 6\pi\eta r v \quad (9\text{-}16)$$

where r is the radius of the sphere, η is the viscosity of the fluid, and v is the speed of the object with respect to the fluid.

CHECKPOINT 9.10

Compare and contrast the viscous drag force with the kinetic frictional force.

An object's **terminal velocity** is the velocity that produces just the right drag force so that the net force is zero. An object falling at its terminal velocity has zero acceleration, so it continues moving at that constant velocity. Using Stokes's law, we can find the terminal velocity of a spherical object falling through a viscous fluid. When the object moves at terminal velocity, the net force acting on it is zero. If $\rho_o > \rho_f$, the object sinks; the terminal velocity is downward and the viscous drag force acts upward to oppose the motion. For an object, such as an air bubble in oil, that rises rather than sinks ($\rho_o < \rho_f$), the terminal velocity is *upward* and the drag force is *downward.*

Example 9.13

Falling Droplet

In an experiment to measure the electric charge of the electron, a fine mist of oil droplets is sprayed into the air and observed through a telescope as they fall. These droplets are so tiny that they soon reach their terminal velocity. If the radius of the droplets is 2.40 μm and the average density of the oil is 862 kg/m^3, find the terminal speed of the droplets. The density of air is 1.20 kg/m^3 and the viscosity of air is 1.8×10^{-5} Pa·s.

Strategy When the droplets fall at their terminal velocity, the net force on them is zero. We set the net force equal to zero and use Stokes's law for the drag force.

Solution We set the sum of the forces equal to zero when $v = v_t$.

$$\Sigma F_y = +F_D + F_B - W = 0$$

If m_{air} is the mass of displaced air, then

$$6\pi\eta r v_t + m_{air} g - m_{oil} g = 0$$

Solving for v_t,

$$v_t = \frac{g(m_{oil} - m_{air})}{6\pi\eta r}$$

$$= \frac{\frac{4}{3}\pi r^3 g(\rho_{oil} - \rho_{air})}{6\pi\eta r}$$

After dividing the numerator and denominator by πr, we substitute numerical values:

$$v_t = \frac{\frac{4}{3}(2.40 \times 10^{-6}\ \text{m})^2\,(9.80\ \text{N/kg})\,(862\ \text{kg/m}^3 - 1.20\ \text{kg/m}^3)}{6 \times 1.8 \times 10^{-5}\ \text{Pa·s}}$$

$$= 6.0 \times 10^{-4}\ \text{m/s} = 0.60\ \text{mm/s}$$

Discussion We should check the units in the final expression:

$$\frac{\text{m}^2\cdot(\text{N/kg})\cdot\text{kg/m}^3}{\text{Pa·s}} = \frac{\text{N/m}}{\text{N/m}^2 \times \text{s}} = \text{m/s}$$

Stokes's law was applied in this way by Robert Millikan (1868–1953) in his experiments in 1909–1913 to measure the charge of the electron. Using an atomizer, Millikan produced a fine spray of oil droplets. The droplets picked up electric charge as they were sprayed through the atomizer. Millikan kept a droplet suspended without falling by applying an upward electric force. After removing the electric force, he measured the terminal speed of the droplet as it fell through the air. He calculated the mass of the droplet from the terminal speed and the density of the oil using Stokes's law.

continued on next page

Example 9.13 continued

By setting the magnitude of the electric force equal to the weight of a suspended droplet, Millikan calculated the electric charge of the droplet. He measured the charges of hundreds of different droplets and found that they were all multiples of the same quantity—the charge of an electron.

Practice Problem 9.13 Rising Bubble

Find the terminal velocity of an air bubble of 0.500 mm radius in a cup of vegetable oil. The specific gravity of the oil is 0.840 and the viscosity is 0.160 Pa·s. Assume the diameter of the bubble does not change as it rises.

PHYSICS AT HOME

A demonstration of terminal velocity can be done at home. Climb up a small stepladder, or lean over an upstairs balcony, and drop two objects at the same time: a coin and two or three nested cone-shaped paper coffee filters. You will see the effects of viscous drag on the coffee filters as they fall with a constant terminal velocity. Enlist the help of a friend so you can get a side view of the two objects falling. Why do the coffee filters work so well?

For small particles falling in a liquid, the terminal velocity is also called the *sedimentation velocity.* The sedimentation velocity is often small for two reasons. First, if the particle isn't much more dense than the fluid, then the vector sum of the gravitational and buoyant forces is small. Second, notice that the terminal velocity is proportional to r^2; viscous drag is most important for small particles. Thus, it can take a long time for the particles to sediment out of solution. Because the sedimentation velocity is proportional to g, it can be increased by the use of a centrifuge, a rotating container that creates artificial gravity of magnitude $g_{eff} = \omega^2 r$ [see Eq. (5-12) and Section 5.7]. Ultracentrifuges are capable of rotating at 10^5 rev/min and produce artificial gravity approaching a million times g.

Application of viscous drag: sedimentation velocity and the centrifuge

9.11 SURFACE TENSION

The surface of a liquid has special properties not associated with the interior of the liquid. The surface acts like a stretched membrane under tension. The **surface tension** (symbol γ, the Greek letter gamma) of a liquid is the force per unit *length* with which the surface pulls *on its edge.* The direction of the force is tangent to the surface at its edge. Surface tension is caused by the cohesive forces that pull the molecules toward each other.

The high surface tension of water enables water striders and other small insects to walk on the surface of a pond. The foot of the insect makes a small indentation in the water surface (Fig. 9.30); the deformation of the surface enables it to push upward on the foot as if the water surface were a thin sheet of rubber. Visually it looks similar to a person walking across the mat of a trampoline. Other small water creatures, such as mosquito larvae and planaria, hang from the surface of water, using surface tension to hold themselves up. In plants, surface tension aids in the transport of water from the roots to the leaves.

Application of surface tension: how insects can walk on the surface of a pond

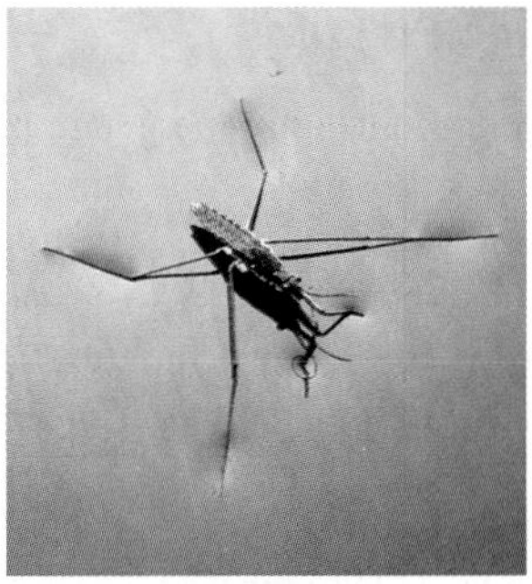

Figure 9.30 A water strider.

PHYSICS AT HOME

Place a needle (or a flat plastic-coated paper clip) gently on the surface of a glass of water. It may take some practice, but you should be able to get it to "float" on top of the water. Now add some detergent to the water and try again. The detergent reduces the surface tension of the water so it is unable to support the needle. Soaps and detergents are *surfactants*—substances that reduce the surface tension of a fluid. The reduced surface tension allows the water to spread out more, wetting more of a surface to be cleaned.

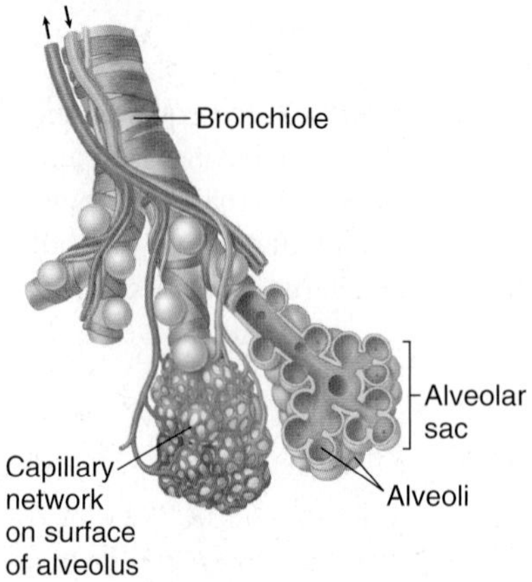

Figure 9.31 In the human lung, millions of tiny sacs called alveoli are inflated with each breath. Gas is exchanged between the air and the blood through the walls of the alveoli. The total surface area through which gas exchange takes place is about 80 m^2—about 40 times the surface area of the body.

Application of surface tension: surfactant in the lungs

The high surface tension of water is a hindrance in the lungs. The exchange of oxygen and carbon dioxide between inspired air and the blood takes place in tiny sacs called *alveoli*, 0.05 to 0.15 mm in radius, at the end of the bronchial tubes (Fig. 9.31). If the mucus coating the alveoli had the same surface tension as other body fluids, the pressure difference between the inside and outside of the alveoli would not be great enough for them to expand and fill with air. The alveoli secrete a surfactant that decreases the surface tension in their mucous coating so they can inflate during inhalation.

Bubbles

In an underwater air bubble, the surface tension of the water surface tries to contract the bubble while the pressure of the enclosed air pushes outward on the surface. In equilibrium, the air pressure inside the bubble must be larger than the water pressure outside so that the net outward force due to pressure balances the inward force due to surface tension. The excess pressure $\Delta P = P_{in} - P_{out}$ depends both on the surface tension and the size of the bubble. In Problem 72, you can show that the excess pressure is

$$\Delta P = \frac{2\gamma}{r} \tag{9-17}$$

Look closely at a glass of champagne and you can see strings of bubbles rising, originating from the same points in the liquid. Why don't bubbles spring up from random locations? A very small bubble would require an insupportably large excess pressure. The bubbles need some sort of nucleus—a small dust particle, for instance—on which to form so they can start out larger, with excess pressures that aren't so large. The strings of bubbles in the glass of champagne are showing where suitable nuclei have been "found."

Example 9.14

Lung Pressure

During inhalation the gauge pressure in the alveoli is about −400 Pa to allow air to flow in through the bronchial tubes. Suppose the mucous coating on an alveolus of initial radius 0.050 mm had the same surface tension as water (0.070 N/m). What lung pressure outside the alveoli would be required to begin to inflate the alveolus?

Strategy We model an alveolus as a sphere coated with mucus. Due to the surface tension of the mucus, the alveolus must have a lower pressure outside than inside, as for a bubble.

Solution The excess pressure is

$$\Delta P = \frac{2\gamma}{r} = \frac{2 \times 0.070\ \text{N/m}}{0.050 \times 10^{-3}\ \text{m}} = 2.8\ \text{kPa}$$

continued on next page

Example 9.14 continued

Thus, the pressure inside the alveolus would be 2.8 kPa higher than the pressure outside. The gauge pressure inside is –400 Pa, so the gauge pressure outside would be

$$P_{out} = -0.4 \text{ kPa} - 2.8 \text{ kPa} = -3.2 \text{ kPa}$$

Discussion The *actual* gauge pressure outside the alveoli is about –0.5 kPa rather than –3.2 kPa; then $\Delta P = P_{in} - P_{out} = -0.4 \text{ kPa} - (-0.5 \text{ kPa}) = 0.1$ kPa rather than 2.8 kPa. Here the surfactant comes to the rescue; by decreasing the surface tension in the mucus, it decreases ΔP to about 0.1 kPa and allows the expansion of the alveoli to take place. For a newborn baby, the alveoli are initially collapsed, making the required pressure difference about 4 kPa. That first breath is as difficult an event as it is significant.

Practice Problem 9.14 Champagne Bubbles

A bubble in a glass of champagne is filled with CO_2. When it is 2.0 cm below the surface of the champagne, its radius is 0.50 mm. What is the gauge pressure inside the bubble? Assume that champagne has the same average density as water and a surface tension of 0.070 N/m.

Master the Concepts

- Fluids are materials that flow and include both liquids and gases. A liquid is nearly incompressible, whereas a gas expands to fill its container.
- Pressure is the perpendicular force per unit area that a fluid exerts on any surface with which it comes in contact ($P = F/A$). The SI unit of pressure is the pascal ($1 \text{ Pa} = 1 \text{ N/m}^2$).
- The average air pressure at sea level is 1 atm = 101.3 kPa.
- Pascal's principle: A change in pressure at any point in a confined fluid is transmitted everywhere throughout the fluid.
- The average density of a substance is the ratio of its mass to its volume

$$\rho = \frac{m}{V} \quad \text{(9-2)}$$

- The specific gravity of a material is the ratio of its density to that of water at 4°C.
- Pressure variation with depth in a static fluid:

$$P_2 = P_1 + \rho g d \quad \text{(9-3)}$$

  where point 2 is a depth d below point 1.
- Instruments to measure pressure include the manometer and the barometer. The barometer measures the pressure of the atmosphere. The manometer measures a pressure difference.
- Gauge pressure is the amount by which the absolute pressure exceeds atmospheric pressure:

$$P_{gauge} = P_{abs} - P_{atm} \quad \text{(9-6)}$$

- Archimedes' principle: a fluid exerts an upward buoyant force on a completely or partially submerged object equal in magnitude to the weight of the volume of fluid displaced by the object:

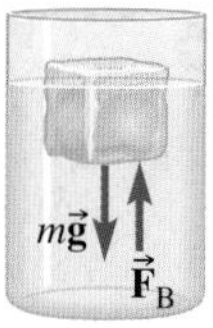

$$F_B = \rho g V \quad \text{(9-7)}$$

  where V is the volume of the part of the object that is submerged and ρ is the density of the fluid.
- In steady flow, the velocity of the fluid *at any point* is constant in time. In laminar flow, the fluid flows in neat layers so that each small portion of fluid that passes a particular point follows the same path as every other portion of fluid that passes the same point. The path that the fluid follows, starting from any point, is called a streamline. Laminar flow is steady. Turbulent flow is chaotic and unsteady. The viscous force opposes the flow of the fluid; it is the counterpart to the frictional force for solids.
- An ideal fluid exhibits laminar flow, has no viscosity, and is incompressible. The flow of an ideal fluid is governed by two principles: the continuity equation and Bernoulli's equation.
- The continuity equation states that the volume flow rate for an ideal fluid is constant:

$$\frac{\Delta V}{\Delta t} = A_1 v_1 = A_2 v_2 \quad \text{(9-12, 9-13)}$$

- Bernoulli's equation relates pressure changes to changes in flow speed and height:

$$P_1 + \rho g y_1 + \tfrac{1}{2}\rho v_1^2 = P_2 + \rho g y_2 + \tfrac{1}{2}\rho v_2^2 \quad \text{(9-14)}$$

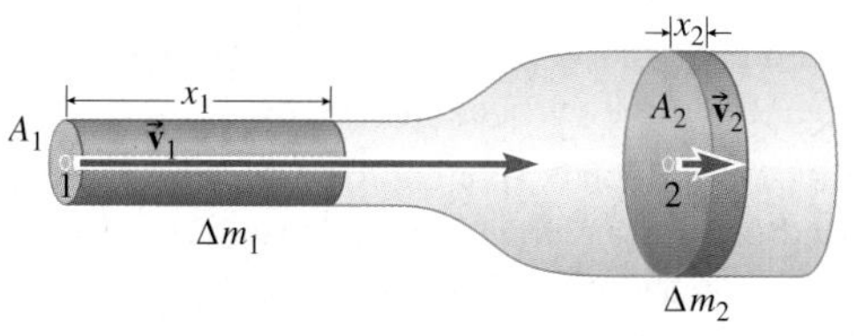

continued on next page

Master the Concepts continued

- Poiseuille's law gives the volume flow rate $\Delta V/\Delta t$ for viscous flow in a horizontal pipe:

$$\frac{\Delta V}{\Delta t} = \frac{\pi}{8}\frac{\Delta P/L}{\eta}r^4 \qquad (9\text{-}15)$$

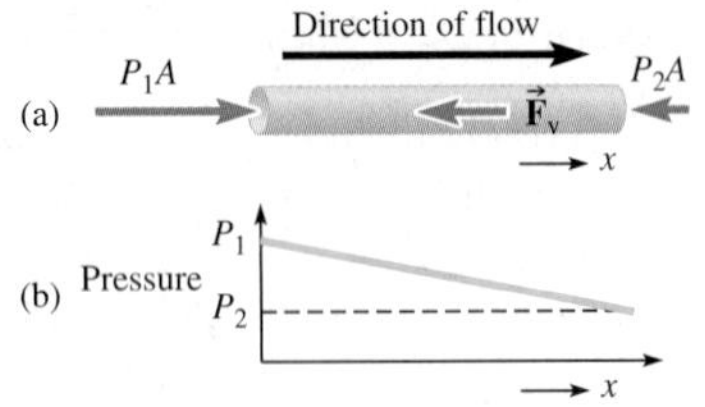

where ΔP is the pressure difference between the ends of the pipe, r and L are the inner radius and length of the pipe, respectively, and η is the viscosity of the fluid.

- Stokes's law gives the viscous drag force on a spherical object moving in a fluid:

$$F_D = 6\pi\eta r v \qquad (9\text{-}16)$$

- The surface tension γ (the Greek letter gamma) of a liquid is the force *per unit length* with which the surface pulls on its edge.

Conceptual Questions

1. Does a manometer (with one side open) measure absolute pressure or gauge pressure? How about a barometer? A tire pressure gauge? A sphygmomanometer?
2. A volunteer firefighter holds the end of a firehose as a strong jet of water emerges. (a) The hose exerts a large backward force on the firefighter. Explain the origin of this force. (b) If another firefighter steps on the hose, forming a constriction (a place where the area of the hose is smaller), the hose begins to pulsate wildly. Explain.
3. The weight of a boat is listed on specification sheets as its "displacement." Explain.
4. In tall buildings, the water supply system uses multiple pumping stations on different floors. At each station, water pumped up from below collects in a storage tank held at atmospheric pressure before it enters the pump. The storage tank supplies water to the floors below it. What are some of the reasons why these multiple pumping stations are used?
5. Can an astronaut on the Moon use a straw to drink from a normal drinking glass? How about if he pokes a straw through an otherwise sealed juice box? Explain.
6. It is commonly said that wood floats because it is "lighter than water" or that a stone sinks because it is "heavier than water." Are these accurate statements? If not, correct them.
7. Why must a blood pressure cuff be wrapped around the arm at the same vertical level as the heart?
8. A hot air balloon is floating in equilibrium with the surrounding air. (a) How does the pressure inside the balloon compare with the pressure outside? (b) How does the density of the air inside compare to the density outside?
9. When helium weather balloons are released, they are purposely underinflated. Why? [*Hint:* The balloons go to very high altitudes.]
10. Bernoulli's equation applies only to *steady flow.* Yet Bernoulli's equation allows the fluid velocity at one point to be different than the velocity at another point. For the fluid velocity to change, the fluid must be accelerated as it moves from one point to another. In what way is the flow *steady*, then?
11. Before getting an oil change, it is a good idea to drive a few miles to warm up the engine. Why?
12. Your ears "pop" when you change altitude quickly—such as during takeoff or landing in an airplane, or during a drive in the mountains. Curiously, if you are a passenger in a high-speed train, your ears sometimes pop as the speed of the train increases rapidly—even though there is little or no change in altitude. Explain.
13. It is easier to get a good draft in a chimney on a windy day than when the outside air is still, all other things being equal. Why?
14. Two soap bubbles of *different radii* are formed at the ends of a tube with a closed valve in the middle. What happens to the bubbles when the valve is opened? (If the alveoli in the lung did not have a surfactant that reduces surface tension in the smaller alveoli, the same thing would happen in the lung, with disastrous results!)
15. *Pascal's principle: proof by contradiction.* Points A and B are near each other at the same height in a fluid. Suppose $P_A > P_B$. (a) Can both v_A and v_B be zero? Explain. (b) Point C is just above point D in a static fluid. Suppose the pressure at C increases by an amount ΔP. What would happen if the pressure at D did not increase by the same amount?

16. What are the advantages of using hydraulic systems rather than mechanical systems to operate automobile brakes or the control surfaces of an airplane?
17. In any hydraulic system, it is important to "bleed" air out of the line. Why?
18. Is it possible for a skin diver to dive to any depth as long as his snorkel tube is sufficiently long? (A snorkel is a face mask with a breathing tube that sticks above the surface of the water.)
19. Is the buoyant force on a soap bubble greater than the weight of the bubble? If not, why do soap bubbles sometimes appear to float in air?
20. A plastic water bottle open at the top is three-fourths full of water and is placed on a scale. The bottle has an indentation for a label midway up the side and a strap has been placed around this indentation. If the strap is tightened, so the bottle is squeezed in at the middle and the water level is forced to rise, what happens to the reading on the scale? Is the water pressure at the bottom of the bottle the same?

Multiple-Choice Questions

1. A glass of ice water is filled to the brim with water; the ice cubes stick up above the water surface. After the ice melts, which is true?
   (a) The water level is below the top of the glass.
   (b) The water level is at the top of the glass but no water has spilled.
   (c) Some water has spilled over the sides of the glass.
   (d) Impossible to say without knowing the initial densities of the water and the ice.
2. A dam holding back the water in a reservoir exerts a horizontal force on the water. The magnitude of this force depends on
   (a) the maximum depth of the reservoir.
   (b) the depth of the water at the location of the dam.
   (c) the surface area of the reservoir.
   (d) both (a) and (b).
   (e) all three—(a), (b), and (c).
3. Bernoulli's equation applies to
   (a) any fluid.
   (b) an incompressible fluid, whether viscous or not.
   (c) an incompressible, nonviscous fluid, whether the flow is turbulent or not.
   (d) an incompressible, nonviscous, nonturbulent fluid.
   (e) a static fluid only.

Questions 4–5. Two spheres, A and B, fall through the same viscous fluid.

Answer choices for Questions 4 and 5:

(a) A has the larger terminal velocity.
(b) B has the larger terminal velocity.
(c) A and B have the same terminal velocity.
(d) Insufficient information is given to reach a conclusion.

4. A and B have the same radius; A has the larger mass. Which has the larger terminal velocity?
5. A and B have the same density; A has the larger radius. Which has the larger terminal velocity?
6. Bernoulli's equation is an expression of
   (a) conservation of mass.
   (b) conservation of energy.
   (c) conservation of momentum.
   (d) conservation of angular momentum.
7. The continuity equation is an expression of
   (a) conservation of mass.
   (b) conservation of energy.
   (c) conservation of momentum.
   (d) conservation of angular momentum.
8. What is the gauge pressure of the gas in the closed tube in the figure? (Take the atmospheric pressure to be 76 cm Hg.)
   (a) 20 cm Hg
   (b) −20 cm Hg
   (c) 96 cm Hg
   (d) 56 cm Hg
   (e) −96 cm Hg
   (f) −56 cm Hg

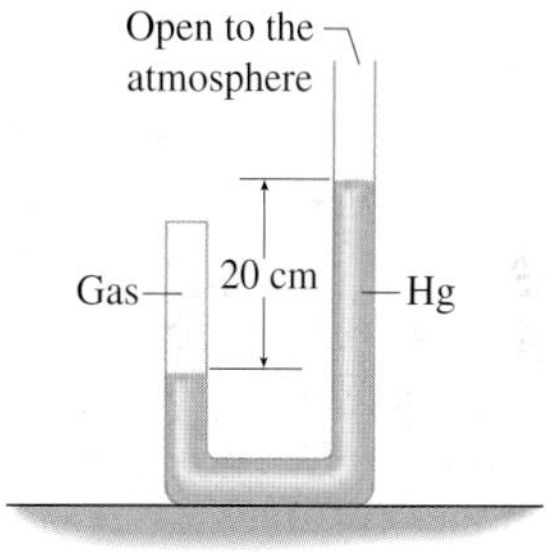

9. A manometer contains two different fluids of different densities. Both sides are open to the atmosphere. Which pair(s) of points in the figure have equal pressure?
   (a) $P_1 = P_5$
   (b) $P_2 = P_5$
   (c) $P_3 = P_4$
   (d) Both (a) and (c)
   (e) Both (b) and (c)

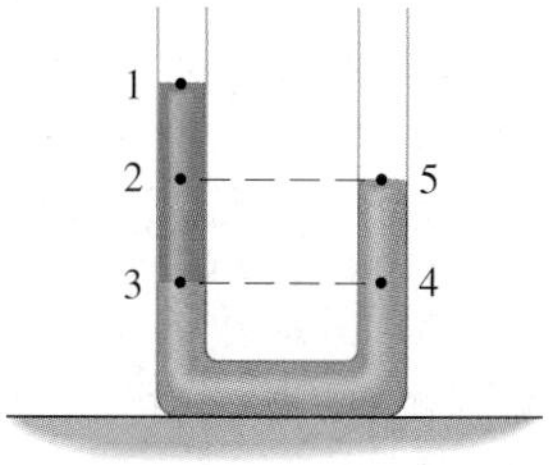

10. A Venturi meter is used to measure the flow speed of a *viscous* fluid. With reference to the figure, which is true?
(a) $h_3 = h_1$ (b) $h_3 > h_1$
(c) $h_3 < h_1$ (d) Insufficient information to determine

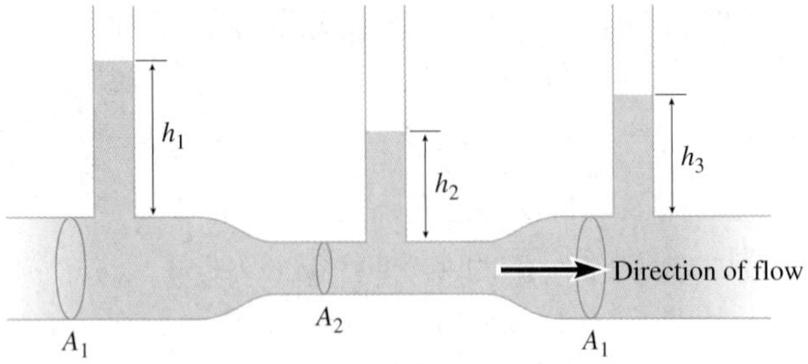

Problems

- Ⓒ Combination conceptual/quantitative problem
- Biological or medical application
- ✦ Challenging problem
- Blue # Detailed solution in the Student Solutions Manual
- ⓵ ⓶ Problems paired by concept
- Text website interactive or tutorial

9.2 Pressure

1. Someone steps on your toe, exerting a force of 500 N on an area of 1.0 cm^2. What is the average pressure on that area in atm?
2. The pressure inside a bottle of champagne is 4.5 atm higher than the air pressure outside. The neck of the bottle has an inner radius of 1.0 cm. What is the frictional force on the cork due to the neck of the bottle?
3. What is the average pressure on the soles of the feet of a standing 90.0-kg person due to the contact force with the floor? Each foot has a surface area of 0.020 m^2.
4. Atmospheric pressure is about 1.0×10^5 Pa on average. (a) What is the downward force of the air on a desktop with surface area 1.0 m^2? (b) Convert this force to pounds so you really understand how large it is. (c) Why does this huge force not crush the desk?
5. A 10-kg baby sits on a three-legged stool. The diameter of each of the stool's round feet is 2.0 cm. A 60-kg adult sits on a four-legged chair that has four circular feet, each with a diameter of 6.0 cm. Who applies the greater pressure to the floor and by how much?
6. A lid is put on a box that is 15 cm long, 13 cm wide, and 8.0 cm tall and the box is then evacuated until its inner pressure is 0.80×10^5 Pa. How much force is required to lift the lid (a) at sea level; (b) in Denver, on a day when the atmospheric pressure is 67.5 kPa ($\frac{2}{3}$ the value at sea level)?

9.3 Pascal's Principle

7. A container is filled with gas at a pressure of 4.0×10^5 Pa. The container is a cube, 0.10 m on a side, with one side facing south. What is the magnitude and direction of the force on the south side of the container due to the gas inside?
8. A nurse applies a force of 4.40 N to the piston of a syringe. The piston has an area of 5.00×10^{-5} m^2. What is the pressure increase in the fluid within the syringe?
9. A hydraulic lift is lifting a car that weighs 12 kN. The area of the piston supporting the car is A, the area of the other piston is a, and the ratio A/a is 100.0. How far must the small piston be pushed down to raise the car a distance of 1.0 cm? [*Hint:* Consider the work to be done.]
10. In a hydraulic lift, the radii of the pistons are 2.50 cm and 10.0 cm. A car weighing $W = 10.0$ kN is to be lifted by the force of the large piston. (a) What force F_a must be applied to the small piston? (b) When the small piston is pushed in by 10.0 cm, how far is the car lifted? (c) Find the mechanical advantage of the lift, which is the ratio W/F_a.
11. Depressing the brake pedal in a car pushes on a piston with cross-sectional area 3.0 cm^2. The piston applies pressure to the brake fluid, which is connected to two pistons, each with area 12.0 cm^2. Each of these pistons presses a brake pad against one side of a rotor attached to one of the rotating wheels. See the figure for this problem. (a) When the force applied by the brake pedal to the small piston is 7.5 N, what is the normal force applied to each side of the rotor? (b) If the coefficient of kinetic friction between a brake pad and the rotor is 0.80 and each pad is (on average) 12 cm from the rotation axis of the rotor, what is the torque on the rotor due to the two pads?

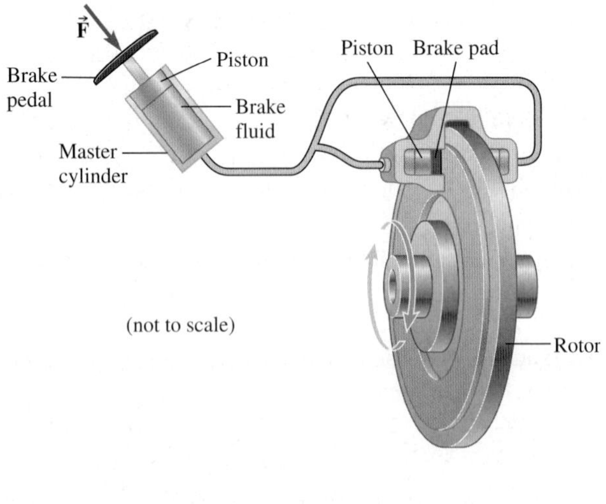

9.4 The Effect of Gravity on Fluid Pressure

12. At the surface of a freshwater lake the air pressure is 1.0 atm. At what depth under water in the lake is the water pressure 4.0 atm?

13. What is the pressure on a fish 10 m under the ocean surface?

14. How high can you suck water up a straw? The pressure in the lungs can be reduced to about 10 kPa below atmospheric pressure.

15. The density of platinum is 21 500 kg/m^3. Find the ratio of the volume of 1.00 kg of platinum to the volume of 1.00 kg of aluminum.

16. In the Netherlands, a dike holds back the sea from a town below sea level. The dike springs a leak 3.0 m below the water surface. If the area of the hole in the dike is 1.0 cm^2, what force must the Dutch boy exert to save the town?

17. A container has a large cylindrical lower part with a long thin cylindrical neck. The lower part of the container holds 12.5 m^3 of water and the surface area of the bottom of the container is 5.00 m^2. The height of the lower part of the container is 2.50 m and the neck contains a column of water 8.50 m high. The total volume of the column of water in the neck is 0.200 m^3. What is the magnitude of the force exerted by the water on the bottom of the container?

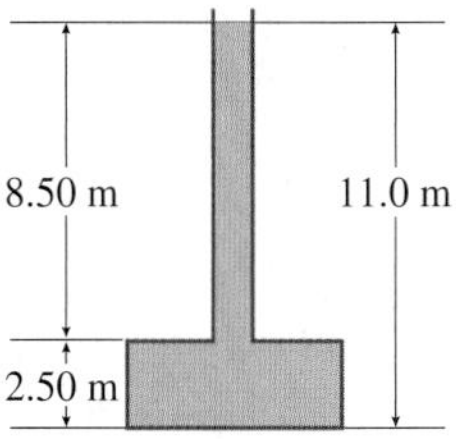

18. The maximum pressure most organisms can survive is about 1000 times atmospheric pressure. Only small, simple organisms such as tadpoles and bacteria can survive such high pressures. What then is the maximum depth at which these organisms can live under the sea (assuming that the density of seawater is 1025 kg/m^3)?

19. At the surface of a freshwater lake the pressure is 105 kPa. (a) What is the pressure increase in going 35.0 m below the surface? (b) What is the approximate pressure decrease in going 35 m above the surface? Air at 20°C has density of 1.20 kg/m^3.

9.5 Measuring Pressure

20. A woman's systolic blood pressure when resting is 160 mm Hg. What is this pressure in (a) Pa, (b) lb/in^2, (c) atm, (d) torr?

21. The gauge pressure of the air in an automobile tire is 32 lb/in^2. Convert this to (a) Pa, (b) torr, (c) atm.

22. An IV is connected to a patient's vein. The blood pressure in the vein has a gauge pressure of 12 mm Hg. At least how far above the vein must the IV bag be hung in order for fluid to flow into the vein? Assume the fluid in the IV has the same density as blood.

23. When a mercury manometer is connected to a gas main, the mercury stands 40.0 cm higher in the tube that is open to the air than in the tube connected to the gas main. A barometer at the same location reads 74.0 cm Hg. Determine the absolute pressure of the gas in cm Hg.

24. An experiment to determine the specific heat of a gas makes use of a water manometer attached to a flask. Initially the two columns of water are even. Atmospheric pressure is 1.0×10^5 Pa. After heating the gas, the water levels change to those shown. Find the change in pressure of the gas in Pa.

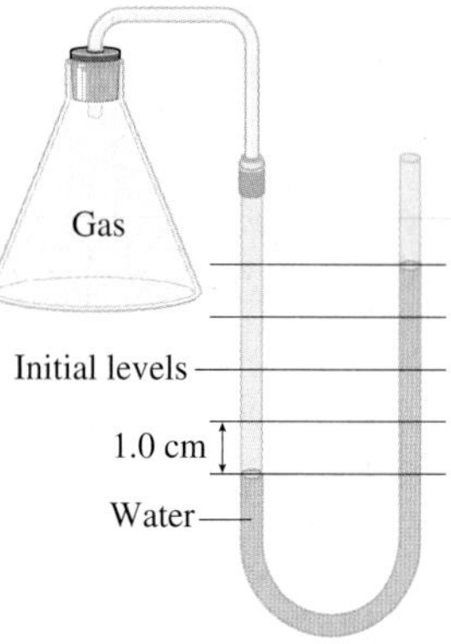

25. A manometer using oil (density 0.90 g/cm^3) as a fluid is connected to an air tank. Suddenly the pressure in the tank increases by 0.74 cm Hg. (a) By how much does the fluid level rise in the side of the manometer that is open to the atmosphere? (b) What would your answer be if the manometer used mercury instead?

26. Estimate the average blood pressure in a person's foot, if the foot is 1.37 m below the aorta, where the average blood pressure is 104 mm Hg. For the purposes of this estimate, assume the blood isn't flowing.

9.6 The Buoyant Force

27. A Canada goose floats with 25% of its volume below water. What is the average density of the goose?

28. A flat-bottomed barge, loaded with coal, has a mass of 3.0×10^5 kg. The barge is 20.0 m long and 10.0 m wide. It floats in fresh water. What is the depth of the barge below the waterline? (tutorial: boat)

29. (a) When ice floats in water at 0°C, what percent of its volume is submerged? (b) What is the specific gravity of ice?

30. (a) What is the density of an object that is 14% submerged when floating in water at 0°C? (b) What percentage of the object will be submerged if it is placed in ethanol at 0°C?

31. (a) What is the buoyant force on 0.90 kg of ice floating freely in liquid water? (b) What is the buoyant force on 0.90 kg of ice held completely submerged under water?

32. A block of birch wood floats in oil with 90.0% of its volume submerged. What is the density of the oil? The density of the birch is 0.67 g/cm^3.

33. When a block of ebony is placed in ethanol, what percentage of its volume is submerged?

34. A cylindrical disk has volume 8.97×10^{-3} m^3 and mass 8.16 kg. The disk is floating on the surface of some water with its flat surfaces horizontal. The area of each flat surface is 0.640 m^2. (a) What is the specific gravity of the disk? (b) How far below the water level is its bottom surface? (c) How far above the water level is its top surface?

35. An aluminum cylinder weighs 1.03 N. When this same cylinder is completely submerged in alcohol, the volume of the displaced alcohol is 3.90×10^{-5} m^3. If the cylinder is suspended from a scale while submerged in the alcohol, the scale reading is 0.730 N. What is the specific gravity of the alcohol? (tutorial: ball in beaker)

36. A fish uses a swim bladder to change its density so it is equal to that of water, enabling it to remain suspended under water. If a fish has an average density of 1080 kg/m^3 and mass 10.0 g with the bladder completely deflated, to what volume must the fish inflate the swim bladder in order to remain suspended in seawater of density 1060 kg/m^3?

37. While vacationing at the Outer Banks of North Carolina, you find an old coin that looks like it is made of gold. You know there were many shipwrecks here, so you take the coin home to check the possibility of it being gold. You suspend the coin from a spring scale and find that it has a weight in air of 1.75 oz (mass = 49.7 g). You then let the coin hang submerged in a glass of water and find that the scale reads 1.66 oz (mass = 47.1 g). Should you get excited about the possibility that this coin might really be gold?

38. The average density of a fish can be found by first weighing it in air and then finding the scale reading for the fish completely immersed in water and suspended from a scale. If a fish has weight 200.0 N in air and scale reading 15.0 N in water, what is the average density of the fish?

39. (a) A piece of balsa wood with density 0.50 g/cm^3 is released under water. What is its initial acceleration? (b) Repeat for a piece of maple with density 0.750 g/cm^3. (c) Repeat for a ping-pong ball with an average density of 0.125 g/cm^3.

40. A piece of metal is released under water. The volume of the metal is 50.0 cm^3 and its specific gravity is 5.0. What is its initial acceleration?

9.7 Fluid Flow; 9.8 Bernoulli's Equation

41. A garden hose of inner radius 1.0 cm carries water at 2.0 m/s. The nozzle at the end has radius 0.20 cm. How fast does the water move through the nozzle?

42. If the average volume flow of blood through the aorta is 8.5×10^{-5} m^3/s and the cross-sectional area of the aorta is 3.0×10^{-4} m^2, what is the average speed of blood in the aorta?

43. A nozzle of inner radius 1.00 mm is connected to a hose of inner radius 8.00 mm. The nozzle shoots out water moving at 25.0 m/s. (a) At what speed is the water in the hose moving? (b) What is the volume flow rate? (c) What is the mass flow rate?

44. Water entering a house flows with a speed of 0.20 m/s through a pipe of 1.0 cm inside radius. What is the speed of the water at a point where the pipe tapers to a radius of 2.5 mm?

45. A horizontal segment of pipe tapers from a cross-sectional area of 50.0 cm^2 to 0.500 cm^2. The pressure at the larger end of the pipe is 1.20×10^5 Pa and the speed is 0.040 m/s. What is the pressure at the narrow end of the segment?

46. In a tornado or hurricane, a roof may tear away from the house because of a difference in pressure between the air inside and the air outside. Suppose that air is blowing across the top of a 2000 ft^2 roof at 150 mph. What is the magnitude of the force on the roof?

47. Use Bernoulli's equation to estimate the upward force on an airplane's wing if the average flow speed of air is 190 m/s above the wing and 160 m/s below the wing. The density of the air is 1.3 kg/m^3 and the area of each wing surface is 28 m^2.

48. An airplane flies on a level path. There is a pressure difference of 500 Pa between the lower and upper surfaces of the wings. The area of each wing surface is about 100 m^2. The air moves below the wings at a speed of 80.5 m/s. Estimate (a) the weight of the plane and (b) the air speed above the wings.

49. A nozzle is connected to a horizontal hose. The nozzle shoots out water moving at 25 m/s. What is the gauge pressure of the water in the hose? Neglect viscosity and assume that the diameter of the nozzle is much smaller than the inner diameter of the hose.

50. Suppose air, with a density of 1.29 kg/m^3 is flowing into a Venturi meter. The narrow section of the pipe at point A has a diameter that is $\frac{1}{3}$ of the diameter of the larger section of the pipe at point B. The U-shaped tube is filled with *water* and the difference in height between the two sections of pipe is 1.75 cm. How fast is the air moving at point B?

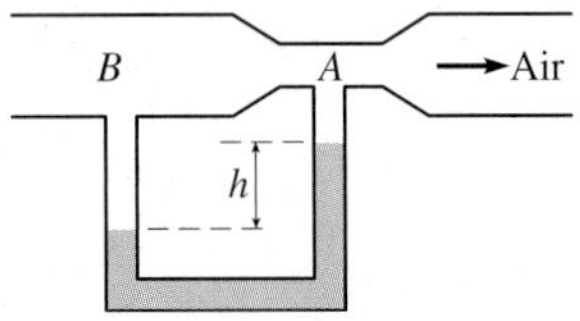

51. A water tower supplies water through the plumbing in a house. A 2.54-cm-diameter faucet in the house can fill a cylindrical container with a diameter of 44 cm and a height of 52 cm in 12 s. How high above the faucet is the top of the water in the tower? (Assume that the diameter of the tower is so large compared to that of the faucet that the water at the top of the tower does not move.)

52. The volume flow rate of the water supplied by a well is 2.0×10^{-4} m^3/s. The well is 40.0 m deep. (a) What is the power output of the pump—in other words, at what rate does the well do work on the water? (b) Find the pressure difference the pump must maintain. (c) Can the pump be at the top of the well or must it be at the bottom? Explain.

9.9 Viscosity

53. Using Poiseuille's law [Eq. (9-15)], show that viscosity has SI units of pascal-seconds.

54. A viscous liquid is flowing steadily through a pipe of diameter D. Suppose you replace it by two parallel pipes, each of diameter $D/2$, but the same length as the original pipe. If the pressure difference between the ends of these two pipes is the same as for the original pipe, what is the total rate of flow in the two pipes compared to the original flow rate?

55. A hypodermic syringe is attached to a needle that has an internal radius of 0.300 mm and a length of 3.00 cm. The needle is filled with a solution of viscosity 2.00×10^{-3} Pa·s; it is injected into a vein at a gauge pressure of 16.0 mm Hg. Ignore the extra pressure required to accelerate the fluid from the syringe into the entrance of the needle. (a) What must the pressure of the fluid in the syringe be in order to inject the solution at a rate of 0.250 mL/s? (b) What force must be applied to the plunger, which has an area of 1.00 cm^2?

Problems 56–58. Four identical sections of pipe are connected in various ways to pumps that supply water at the pressures indicated in the figure (in units of 10^5 Pa). The water exits at the right at atmospheric pressure. Assume viscous flow.

56. If the *total* volume flow rates in systems A and C are the same and the flow speed in each of the pipes in C is 3.0 m/s, what is the flow speed in system A?

57. If the total volume flow rate in system B is 0.020 m^3/s, what is the total volume flow rate in system C?

58. If the total volume flow rates in systems A and B are the same, at what pressure does the pump supply water in system A?

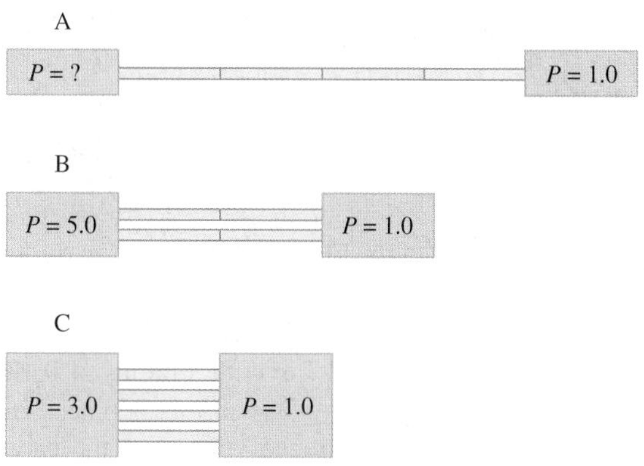

Problems 56–58

59. (a) What is the pressure difference required to make blood flow through an artery of inner radius 2.0 mm and length 0.20 m at a speed of 6.0 cm/s? (b) What is the pressure difference required to make blood flow at 0.60 mm/s through a capillary of radius 3.0 μm and length 1.0 mm? (c) Compare both answers to your average blood pressure, about 100 torr.

60. (a) Since the flow rate is proportional to the pressure difference, show that Poiseuille's law can be written in the form $\Delta P = IR$, where I is the volume flow rate and R is a constant of proportionality called the fluid flow *resistance.* (Written this way, Poiseuille's law is analogous to *Ohm's law* for electric current to be studied in Chapter 18: $\Delta V = IR$, where ΔV is the potential drop across a conductor, I is the electric current flowing through the conductor, and R is the electrical resistance of the conductor.) (b) Find R in terms of the viscosity of the fluid and the length and radius of the pipe.

9.10 Viscous Drag

61. Two identical spheres are dropped into two different columns: one column contains a liquid of viscosity 0.5 Pa·s, while the other contains a liquid of the same density but unknown viscosity. The sedimentation velocity in the second tube is 20% higher than the sedimentation velocity in the first tube. What is the viscosity of the second liquid?

62. A sphere of radius 1.0 cm is dropped into a glass cylinder filled with a viscous liquid. The mass of the sphere is 12.0 g and the density of the liquid is 1200 kg/m^3. The sphere reaches a terminal speed of 0.15 m/s. What is the viscosity of the liquid?

63. A dinoflagellate takes 5.0 s to travel 1.0 mm. Approximate a dinoflagellate as a sphere of radius 35.0 μm (ignoring the flagellum). (a) What is the drag force on the dinoflagellate in seawater of viscosity 0.0010 Pa·s? (b) What is the power output of the flagellate?

64. An air bubble of 1.0-mm radius is rising in a container with vegetable oil of specific gravity 0.85 and viscosity 0.12 Pa·s. The container of oil and the air bubble are at 20°C. What is its terminal velocity?

65. This table gives the terminal speeds of various spheres falling through the same fluid. The spheres all have the same radius.

$m =$	8	12	16	20	24	28	(g)
$v_t =$	1.0	1.5	2.0	2.5	3.0	3.5	(cm/s)

Is the drag force primarily viscous or turbulent? Explain your reasoning.

66. This table gives the terminal speeds of various spheres falling through the same fluid. The spheres all have the same radius.

$m =$	5.0	11.3	20.0	31.3	45.0	80.0	(g)
$v_t =$	1.0	1.5	2.0	2.5	3.0	4.0	(cm/s)

Is the drag force primarily viscous or turbulent? Explain your reasoning.

67. *What keeps a cloud from falling?* A cumulus (fair-weather) cloud consists of tiny water droplets of average radius 5.0 μm. Find the terminal velocity for these droplets at 20°C, assuming viscous drag. (Besides the viscous drag force, there are also upward air currents called *thermals* that push the droplets upward. tutorial: rain drop)

68. An aluminum sphere (specific gravity = 2.7) falling through water reaches a terminal speed of 5.0 cm/s. What is the terminal speed of an air bubble of the same radius rising through water? Assume viscous drag in both cases and ignore the possibility of changes in size or shape of the air bubble; the temperature is 20°C.

9.11 Surface Tension

69. An underwater air bubble has an excess inside pressure of 10 Pa. What is the excess pressure inside an air bubble with twice the radius?

70. Assume a water strider has a roughly circular foot of radius 0.02 mm. (a) What is the maximum possible upward force on the foot due to surface tension of the water? (b) What is the maximum mass of this water strider so that it can keep from breaking through the water surface? The strider has six legs.

71. The potential energy associated with surface tension is much like the elastic potential energy of a stretched spring or a balloon. Suppose we do work on a puddle of liquid, spreading it out through a distance of Δs along a line L perpendicular to the force. (a) What is the work done on the fluid surface in terms of γ, L, and Δs? (b) The work done is equal to the increase in surface energy of the fluid. Show that the increase in energy is proportional to the increase in area. (c) Show that we can think of γ as the surface energy per unit area. (d) Show that the SI units of surface tension can be expressed either as N/m (force per unit length) or J/m^2 (energy per unit area).

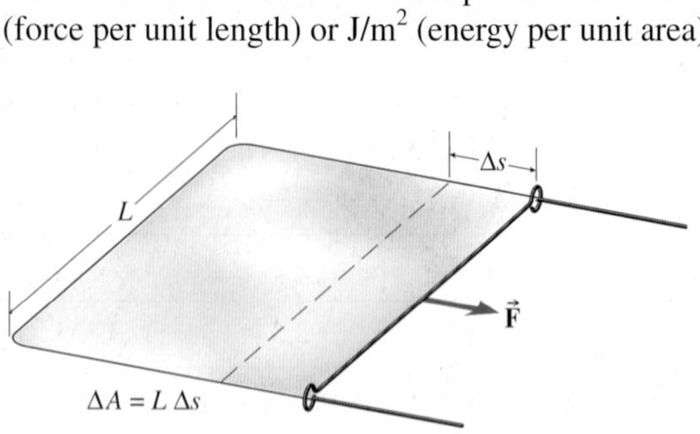

72. A hollow hemispherical object is filled with air as in part (a) of the figure. (a) Show that the magnitude of the force due to fluid pressure on the curved surface of the hemisphere has magnitude $F = \pi r^2 P$, where r is the radius of the hemisphere and P is the pressure of the air. Ignore the weight of the air. [*Hint:* First find the force on the *flat* surface. What is the net force on the hemisphere due to the air?] (b) Consider an underwater air bubble to be divided into two hemispheres along the circumference as in part (b) of the figure. The upper hemisphere of the water surface exerts a force of magnitude $2\pi r\gamma$ (circumference times force per unit length) on the lower hemisphere due to surface tension. Show that the air pressure inside the bubble must exceed the water pressure outside by $\Delta P = 2\gamma/r$.

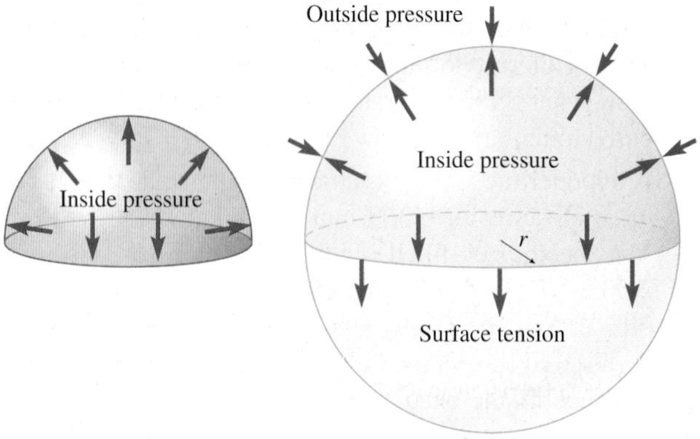

(a) (b)

Comprehensive Problems

73. A wooden barrel full of water has a flat circular top of radius 25.0 cm with a small hole in it. A tube of height 8.00 m and inner radius 0.250 cm is suspended above the barrel with its lower end inserted snugly in the hole. Water is poured into the upper end of the tube until it is full. (a) What is the weight of the water in the tube? (b) What is the force with which the water in the barrel pushes up on the top of the barrel? (c) How can adding such a small weight of water lead to such a large force on the top of the barrel? (As a demonstration of the principle now named for him, Pascal astonished spectators by showing that the addition of a small amount of water to the tube could make the barrel burst.)

74. A block of aluminum that has dimensions 2.00 cm by 3.00 cm by 5.00 cm is suspended from a spring scale. (a) What is the weight of the block? (b) What is the scale reading when the block is submerged in oil with a density of 850 kg/m^3?

75. A 85.0-kg canoe made of thin aluminum has the shape of half of a hollowed-out log with a radius of 0.475 m and a length of 3.23 m. (a) When this is placed in the water, what percentage of the volume of the canoe is below the waterline? (b) How much additional mass can be placed in this canoe before it begins to sink? (interactive: buoyancy)

76. Two identical beakers are filled to the brim and placed on balance scales. The base area of the beakers is large enough that any water that spills out of the beakers will fall onto the table the scales are resting on. A block of pine (density = 420 kg/m^3) is placed in one of

the beakers. The block has a volume of 8.00 cm^3. Another block of the same size, but made of steel, is placed in the other beaker. How does the scale reading change in each case?

77. A very large vat of water has a hole 1.00 cm in diameter located a distance 1.80 m below the water level. (a) How fast does water exit the hole? (b) How would your answer differ if the vat were filled with gasoline? (c) How would your answer differ if the vat contained water, but was on the Moon, where the gravitational field strength is 1.6 N/kg?

78. A cube that is 4.00 cm on a side and of density 8.00×10^2 kg/m^3 is attached to one end of a spring. The other end of the spring is attached to the base of a beaker. When the beaker is filled with water until the entire cube is submerged, the spring is stretched by 1.00 cm. What is the spring constant?

79. You are hiking through a lush forest with some of your friends when you come to a large river that seems impossible to cross. However, one of your friends notices an old metal barrel sitting on the shore. The barrel is shaped like a cylinder and is 1.20 m high and 0.76 m in diameter. One of the circular ends of the barrel is open and the barrel is empty. When you put the barrel in the water with the open end facing up, you find that the barrel floats with 33% of it under water. You decide that you can use the barrel as a boat to cross the river, as long as you leave about 30 cm sticking above the water. How much extra mass can you put in this barrel to use it as a boat?

80. The deepest place in the ocean is the Marianas Trench in the western Pacific Ocean, which is over 11.0 km deep. On January 23, 1960, the research sub *Trieste* went to a depth of 10.915 km, nearly to the bottom of the trench. This still is the deepest dive on record. The density of seawater is 1025 kg/m^3. (a) What is the water pressure at that depth? (b) What was the force due to water pressure on a flat section of area 1.0 m^2 on the top of the sub's hull?

81. The pressure in a water pipe in the basement of an apartment house is 4.10×10^5 Pa, but on the seventh floor it is only 1.85×10^5 Pa. What is the height between the basement and the seventh floor? Assume the water is not flowing; no faucets are opened.

82. The body of a 90.0-kg person contains 0.020 m^3 of body fat. If the density of fat is 890 kg/m^3, what percentage of the person's body weight is composed of fat?

83. Near sea level, how high a hill must you ascend for the reading of a barometer you are carrying to drop by 1.0 cm Hg? Assume the temperature remains at 20°C as you climb. The reading of a barometer on an average day at sea level is 76.0 cm Hg. (tutorial: gauge)

84. A stone of weight W has specific gravity 2.50. (a) When the stone is suspended from a scale and submerged in water, what is the scale reading in terms of its weight in air? (b) What is the scale reading for the stone when it is submerged in oil (specific gravity = 0.90)?

85. If you watch water falling from a faucet, you will notice that the flow decreases in radius as the water falls. This can be explained by the equation of continuity, since the cross-sectional area of the water decreases as the speed increases. If the water flows with an initial velocity of 0.62 m/s and a diameter of 2.2 cm at the faucet opening, what is the diameter of the water flow after the water has fallen 30 cm?

86. The average speed of blood in the aorta is 0.3 m/s and the radius of the aorta is 1 cm. There are about 2×10^9 capillaries with an average radius of 6 μm. What is the approximate average speed of the blood flow in the capillaries?

87. If the cardiac output of a small dog is 4.1×10^{-3} m^3/s, the radius of its aorta is 0.50 cm, and the aorta length is 40.0 cm, determine the pressure drop across the aorta of the dog. Assume the viscosity of blood is 4.0×10^{-3} Pa·s.

88. In an aortic aneurysm, a bulge forms where the walls of the aorta are weakened. If blood flowing through the aorta (radius 1.0 cm) enters an aneurysm with a radius of 3.0 cm, how much on average is the blood pressure higher inside the aneurysm than the pressure in the unenlarged part of the aorta? The average flow rate through the aorta is 120 cm^3/s. Assume the blood is nonviscous and the patient is lying down so there is no change in height.

89. Scuba divers are admonished not to rise faster than their air bubbles when rising to the surface. This rule helps them avoid the rapid pressure changes that cause the bends. Air bubbles of 1.0 mm radius are rising from a scuba diver to the surface of the sea. Assume a water temperature of 20°C. (a) If the viscosity of the water is 1.0×10^{-3} Pa·s, what is the terminal velocity of the bubbles? (b) What is the largest rate of pressure change tolerable for the diver according to this rule?

90. A shallow well usually has the pump at the top of the well. (a) What is the deepest possible well for which a surface pump will work? [*Hint:* A pump maintains a pressure difference, keeping the outflow pressure higher than the intake pressure.] (b) Why is there not the same depth restriction on wells with the pump at the bottom?

91. A plastic beach ball has radius 20.0 cm and mass 0.10 kg, not including the air inside. (a) What is the weight of the beach ball including the air inside? Assume the air density is 1.3 kg/m^3 both inside and outside. (b) What is the buoyant force on the beach ball in air? The thickness of the plastic is about 2 mm—negligible compared to the radius of the ball. (c) The ball is thrown straight up in the air. At the top of its trajectory, what is its acceleration? [*Hint:* When $v = 0$, there is no drag force.]

92. A block of wood, with density 780 kg/m^3, has a cubic shape with sides 0.330 m long. A rope of negligible

mass is used to tie a piece of lead to the bottom of the wood. The lead pulls the wood into the water until it is just completely covered with water. What is the mass of the lead? [*Hint:* Don't forget to consider the buoyant force on both the wood and the lead.]

93. Are evenly spaced specific gravity markings on the cylinder of a hydrometer equal distances apart? In other words, is the depth d to which the cylinder is submerged linearly related to the density ρ of the fluid? To answer this question, assume that the cylinder has radius r and mass m. Find an expression for d in terms of ρ, r, and m and see if d is a linear function of ρ.

94. A hydrometer is an instrument for measuring the specific gravity of a liquid. For example, vintners use a hydrometer to determine the density changes as wine is fermented, and producers of maple sugar and maple syrup use the hydrometer to find how much sugar is in the collected sap. Markings along a stem are calibrated to indicate the specific gravity for the level at which the hydrometer floats in a liquid. The weighted base ensures that the hydrometer floats vertically. Suppose the hydrometer has a cylindrical stem of cross-sectional area 0.400 cm^2. The total volume of the bulb and stem is 8.80 cm^3 and the mass of the hydrometer is 4.80 g. (a) How far from the top of the cylinder should a mark be placed to indicate a specific gravity of 1.00? (b) When the hydrometer is placed in alcohol, it floats with 7.25 cm of stem above the surface. What is the specific gravity of the alcohol? (c) What is the lowest specific gravity that can be measured with this hydrometer?

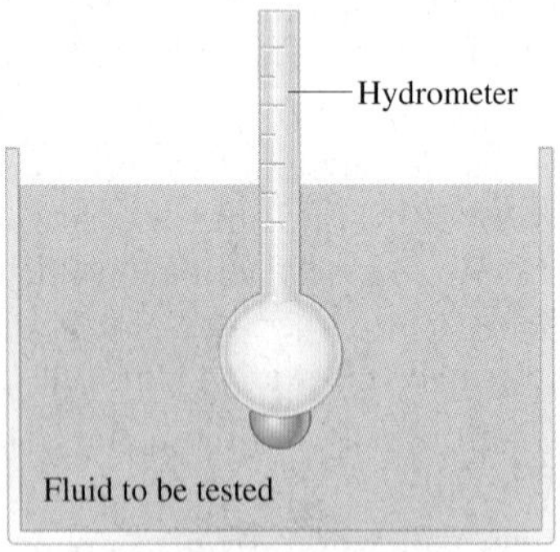

Simple hydrometer (Problems 93 and 94)

95. A house with its own well has a pump in the basement with an output pipe of inner radius 6.3 mm. Assume that the pump can maintain a gauge pressure of 410 kPa in the output pipe. A showerhead on the second floor (6.7 m above the pump's output pipe) has 36 holes, each of radius 0.33 mm. The shower is on "full blast" and no other faucet in the house is open. (a) Ignoring viscosity, with what speed does water leave the showerhead? (b) With what speed does water move through the output pipe of the pump?

96. To measure the airspeed of a plane, a device called a Pitot tube is used. A simplified model of the Pitot tube is a manometer with one side connected to a tube facing directly into the "wind" (stopping the air that hits it head-on) and the other side connected to a tube so that the "wind" blows across its openings. If the manometer uses mercury and the levels differ by 25 cm, what is the plane's airspeed? The density of air at the plane's altitude is 0.90 kg/m^3.

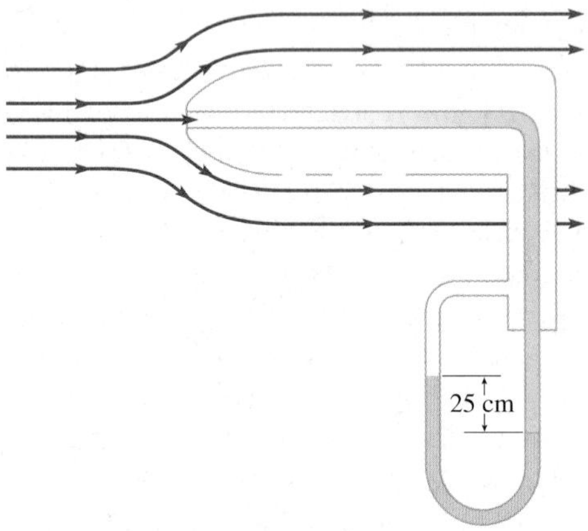

97. A U-shaped tube is partly filled with water and partly filled with a liquid that does not mix with water. Both sides of the tube are open to the atmosphere. What is the density of the liquid (in g/cm^3)?

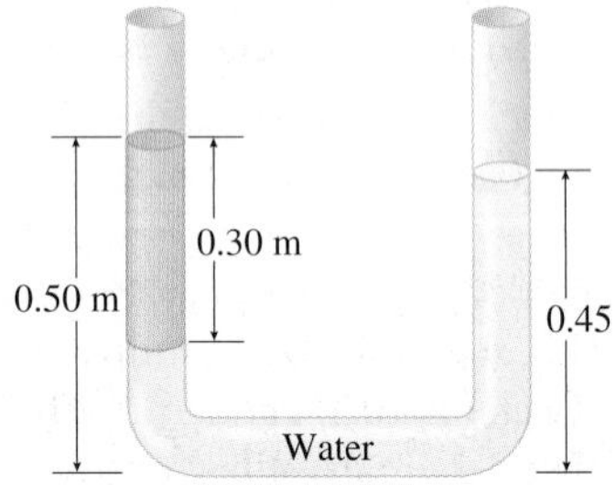

98. Atmospheric pressure is equal to the weight of a vertical column of air, extending all the way up through the atmosphere, divided by the cross-sectional area of the column. (a) Explain why that must be true. [*Hint:* Apply Newton's second law to the column of air.] (b) If the air all the way up had a uniform density of 1.29 kg/m^3 (the density at sea level at 0°C), how high would the column of air be? (c) In reality, the density of air decreases with increasing altitude. Does that mean that the height found in (b) is a lower limit or an upper limit on the height of the atmosphere?

99. On a nice day when the temperature outside is 20°C, you take the elevator to the top of the Sears Tower in Chicago, which is 440 m tall. (a) How much less is the air pressure at the top than the air pressure at the bottom? Express your answer both in pascals and atm.

(tutorial: gauge) [*Hint:* The altitude change is small enough to treat the density of air as constant.] (b) How many pascals does the pressure decrease for every meter of altitude? (c) If the pressure gradient—the pressure decrease per meter of altitude—were uniform, at what altitude would the atmospheric pressure reach zero? (d) Atmospheric pressure does *not* decrease with a uniform gradient since the density of air decreases as you go up. Which is true: the pressure reaches zero at a *lower* altitude than your answer to (c), or the pressure is nonzero at that altitude and the atmosphere extends to a higher altitude? Explain.

100. A bug from South America known as *Rhodnius prolixus* extracts the blood of animals. Suppose *Rhodnius prolixus* extracts 0.30 cm^3 of blood in 25 min from a human arm through its feeding tube of length 0.20 mm and radius 5.0 μm. What is the absolute pressure at the bug's end of the feeding tube if the absolute pressure at the other end (in the human arm) is 105 kPa? Assume the viscosity of blood is 0.0013 Pa·s. [*Note:* Negative absolute pressures are possible in liquids in very slender tubes.]

101. The diameter of a certain artery has decreased by 25% due to arteriosclerosis. (a) If the same amount of blood flows through it per unit time as when it was unobstructed, by what percentage has the blood pressure difference between its ends increased? (b) If, instead, the pressure drop across the artery stays the same, by what factor does the blood flow rate through it decrease? (In reality we are likely to see a combination of some pressure increase with some reduction in flow.)

Answers to Practice Problems

9.1 1.3×10^6 N/m^2 = 1.3 MPa; the pressure is a factor of 15 greater than the pressure from the tennis shoe heel.

9.2 (a) 2.0×10^5 Pa; (b) 5.0 m

9.3 1.6 km

9.4 (a) Yes, $P_2 = P_1$. The column above point 2 is not as tall, but the pressure at the top of that column is *greater than* atmospheric pressure. (b) No, $P = P_{atm} + \rho g d$ gives the pressure at a depth d below a point where the pressure is P_{atm}.

9.5 (a) 32.0 cm; (b) 17.0 cm and 37.0 cm

9.6 S.G. = 11.3; could be lead

9.7 2% and 4%

9.8 (a) The beetle can squeeze the air bubble with its wings, compressing the air to reduce the bubble volume and decreasing the buoyant force. (b) When it is time to rise to the surface, the beetle relaxes the pressure on the bubble, allowing it to expand again.

9.9 (a) 0.85 m/s; (b) 1.7 m/s

9.10 $\sqrt{2gh} = 4.0$ m/s

9.11 250 kPa

9.12 1.4

9.13 2.85 mm/s upward

9.14 480 Pa

Answers to Checkpoints

9.4 Pressure in a static fluid cannot depend on horizontal position. The net horizontal force on any part of the fluid must be zero—otherwise the horizontal acceleration would be nonzero and the fluid would begin to flow. The net vertical force *including the weight of the fluid* must also be zero, so pressure does depend on vertical position.

9.8 (a) For horizontal flow, Bernoulli's equation becomes $P_1 + \frac{1}{2}\rho v_1^2 = P_2 + \frac{1}{2}\rho v_2^2$; the pressure is lower where the flow speed is higher. (b) In a static fluid, Bernoulli's equation becomes $P_1 + \rho g y_1 = P_2 + \rho g y_2$. Letting $d = y_1 - y_2$, we have $P_2 - P_1 = \rho g y_1 - \rho g y_2 = \rho g d$, which is the pressure dependence with depth for a static fluid as discussed in Section 9.4.

9.10 Viscous drag and kinetic friction are both forces that oppose the motion of an object (relative to the surrounding fluid or relative to the surface on which the object slides, respectively). However, viscous drag depends strongly on the speed of the object ($F_D \propto v$), but kinetic friction does not.

CHAPTER

10 Elasticity and Oscillations

Near the top of the 241-m-tall Hancock Tower in Boston, two steel boxes filled with lead are part of a system designed to reduce the swaying and twisting of the building caused by the wind. The mass of each box is nearly 300 000 kg (weight 300 tons). It might seem that adding a large mass to the *top* of the building would make it more "top heavy" and might *increase* the amount of swaying. Why is such a large mass used and how does it reduce the swaying of the building? (See p. 381 for the answer.)

Concepts & Skills to Review

- Hooke's law (Section 6.6)
- graphical relationship of position, velocity, and acceleration (Sections 2.2 and 2.3)
- elastic potential energy (Section 6.7)

10.1 ELASTIC DEFORMATIONS OF SOLIDS

If the net force and the net torque on an object are zero, the object is in equilibrium—but that does not mean that the forces and torques have no effect. An object is deformed when contact forces are applied to it (Fig. 10.1). A **deformation** is a change in the size or shape of the object. Many solids are stiff enough that the deformation cannot be seen with the human eye; a microscope or other sensitive device is required to detect the change in size or shape.

When the contact forces are removed, an **elastic** object returns to its original shape and size. Many objects are elastic as long as the deforming forces are not too large. On the other hand, any object may be permanently deformed or even broken if the forces acting are too large. An automobile that collides with a tree at a low speed may not be damaged; but at a higher speed the car suffers a permanent deformation of the bodywork and the driver may suffer a broken bone.

CONNECTION:

The two topics of this chapter—elasticity and oscillations—may seem unrelated at first, but they are closely connected: many oscillations are caused by the kinds of elastic forces we study in Sections 10.1 through 10.4.

10.2 HOOKE'S LAW FOR TENSILE AND COMPRESSIVE FORCES

Suppose we stretch a wire by applying tensile forces of magnitude F to each end. The length of the wire increases from L to $L + \Delta L$. How does the elongation ΔL depend on the original length L? Conceptual Example 10.1 helps answer this question.

Figure 10.1 A tennis ball is flattened by the contact force exerted on it by the strings of the tennis racquet. Likewise, the strings of the racquet are deformed by the contact force exerted by the ball. The two forces are interaction partners.

Conceptual Example 10.1

Stretching Wires

If a given tensile force stretches a wire an amount ΔL, by how much would the same force stretch a wire twice as long but identical in thickness and composition?

Strategy and Solution Think of the wire of length $2L$ as two wires of length L placed end-to-end (Fig. 10.2). Under the same tension, each of the two imagined wires stretches by an amount ΔL, so the total deformation of the long wire is $2\,\Delta L$.

Practice Problem 10.1 Cutting a Spring in Half

If a spring (spring constant k) is cut in half, what is the spring constant of each of the two newly formed springs?

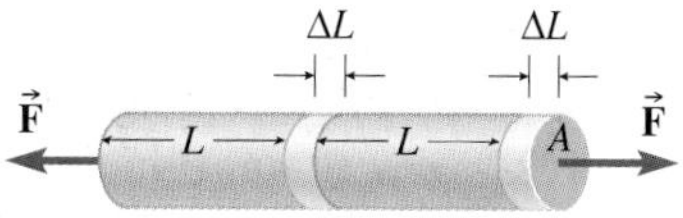

Figure 10.2

Two identical wires are joined end-to-end and stretched by tensile forces. Each wire stretches an amount ΔL.

Strain: fractional length change

Stress and Strain When stretched by the same tensile forces, the two wires in Conceptual Example 10.1 get longer by an amount proportional to their original lengths: $\Delta L \propto L$. In other words, the two wires have the same *fractional length change* $\Delta L/L$. The fractional length change is called the **strain**; it is a dimensionless measure of the degree of deformation.

$$\text{strain} = \frac{\Delta L}{L} \tag{10-1}$$

Suppose we had wires of the same composition and length but different thicknesses. It would require larger tensile forces to stretch the thicker wire the same amount as the thinner one; a thick steel cable is harder to stretch than the same length of a thin strand of steel. In Conceptual Question 13, we conclude that the tensile force required is proportional to the cross-sectional area of the wire ($F \propto A$). Thus, the same applied force *per unit area* produces the same deformation on wires of the same length and composition. The force per unit area is called the **stress**:

Stress: force per unit cross-sectional area

$$\text{stress} = \frac{F}{A} \tag{10-2}$$

The SI units of stress are the same as those of pressure: N/m^2 or Pa.

Hooke's Law Suppose that a solid object of initial length L is subjected to tensile or compressive forces of magnitude F. As a result of the forces, the length of the object is changed by magnitude ΔL. According to Hooke's law, the deformation is proportional to the deforming forces as long as they are not too large:

CONNECTION:

Hooke's law does not just apply to springs. The deformation of an object is often proportional to the forces applied to it.

$$F = k\Delta L \tag{10-3}$$

In Eq. (10-3), k is a measure of the object's stiffness; it is analogous to the spring constant of a spring. This constant k depends on the length and cross-sectional area of the object. A larger cross-sectional area A makes k larger; a greater length L makes k smaller.

We can rewrite Hooke's law in terms of stress (F/A) and strain ($\Delta L/L$):

Hooke's law: the strain is proportional to the stress

Hooke's Law

$$\text{stress} \propto \text{strain}$$

$$\frac{F}{A} = Y\frac{\Delta L}{L} \tag{10-4}$$

Equation (10-4) still says that the length change (ΔL) is proportional to the magnitude of the deforming forces (F). Stress and strain account for the effects of length and cross-sectional area; the proportionality constant Y depends only on the inherent stiffness of the material from which the object is composed; it is independent of the length and cross-sectional area. Comparing Eqs. (10-3) and (10-4), the "spring constant" k for the object is

$$k = \frac{YA}{L} \tag{10-5}$$

The constant of proportionality Y in Eqs. (10-4) and (10-5) is called the **elastic modulus** or **Young's modulus**; Y has the same units as those of stress (Pa or N/m^2), since strain is dimensionless. Young's modulus can be thought of as the inherent stiffness of a material; it measures the resistance of the material to elongation or compression. Material that is flexible and stretches easily (for example, rubber) has a *low* Young's modulus. A stiff material (such as steel) has a high Young's modulus; it takes a larger stress to produce the same strain. Table 10.1 gives Young's modulus for a variety of common materials.

Table 10.1 Approximate Values of Young's Modulus for Various Substances

Substance	Young's Modulus (10^9 Pa)	Substance	Young's Modulus (10^9 Pa)
Rubber	0.002–0.008	Wood, along the grain	10–15
Human cartilage	0.024	Brick	14–20
Human vertebra	0.088 (compression); 0.17 (tension)	Concrete	20–30 (compression)
		Marble	50–60
Collagen, in bone	0.6	Aluminum	70
Human tendon	0.6	Cast iron	100–120
Wood, across the grain	1	Copper	120
Nylon	2–6	Wrought iron	190
Spider silk	4	Steel	200
Human femur	9.4 (compression); 16 (tension)	Diamond	1200

CHECKPOINT 10.2

Which stretches more when put under the same tension: a steel wire 2.0 m long or a copper wire 1.0 m long with the same diameter? (See Table 10.1.)

Application of tensile and compressive forces: bone strength

Hooke's law holds up to a maximum stress called the *proportional limit.* For many materials, Young's modulus has the same value for tension and compression. Some composite materials, such as bone and concrete, have significantly different Young's moduli for tension and compression. The components of bone include fibers of collagen (a protein found in all connective tissue) that give it strength under tension and hydroxyapatite crystals (composed of calcium and phosphate) that give it strength under compression. The different properties of these two substances lead to different values of Young's modulus for tension and compression.

Example 10.2

Compression of the Femur

A man whose weight is 0.80 kN is standing upright. By approximately how much is his femur (thighbone) shortened compared with when he is lying down? Assume that the compressive force on each femur is about half his weight (Fig. 10.3). The average cross-sectional area of the femur is 8.0 cm^2 and the length of the femur when lying down is 43.0 cm.

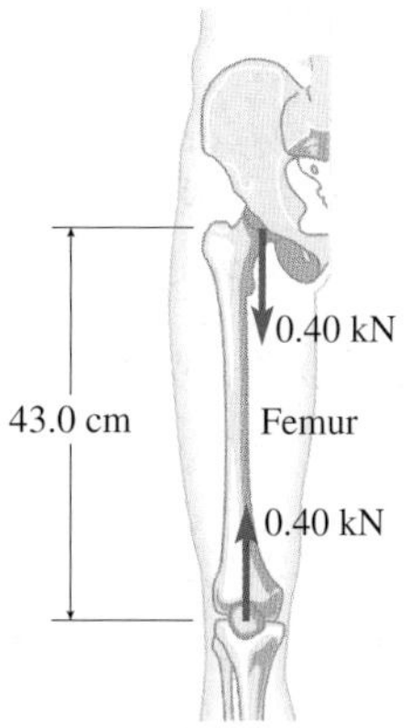

Figure 10.3 Compression of the femur.

Strategy A change in length of the femur involves a strain. After finding the stress and looking up the Young's modulus, we can find the strain using Hooke's law. We assume that each femur supports *half* the man's weight.

Solution The strain is proportional to the stress:

$$\frac{F}{A} = Y\frac{\Delta L}{L}$$

Solving this equation for ΔL gives

$$\Delta L = \frac{F/A}{Y} L$$

From Table 10.1, Young's modulus for a femur *in compression* is:

$$Y = 9.4 \times 10^9 \text{ Pa}$$

We need to convert the cross-sectional area to m^2 since 1 Pa = 1 N/m^2:

$$A = 8.0 \text{ cm}^2 \times \left(\frac{1 \text{ m}}{100 \text{ cm}}\right)^2 = 0.00080 \text{ m}^2$$

continued on next page

Example 10.2 continued

The force on each leg is 0.40 kN, or 4.0×10^2 N. The length change is then

$$\Delta L = \frac{F/A}{Y} L = \frac{(4.0 \times 10^2 \text{ N})/(0.00080 \text{ m}^2)}{9.4 \times 10^9 \text{ Pa}} \times 43.0 \text{ cm}$$

$$= 5.3 \times 10^{-5} \times 43.0 \text{ cm} = 0.0023 \text{ cm}$$

Discussion The strain—or fractional length change—is 5.3×10^{-5}. Since the strain is much smaller than 1, we are justified in not worrying about whether the length is 43.0 cm with or without the compressive load; we would calculate the same value of ΔL (to two significant figures) either way.

Practice Problem 10.2 Fractional Length Change of a Cable

A steel cable of diameter 3.0 cm supports a load of 2.0 kN. What is the fractional length increase of the cable compared to the length when there is no load if $Y = 2.0 \times 10^{11}$ Pa?

10.3 BEYOND HOOKE'S LAW

If the tensile or compressive stress exceeds the proportional limit, the strain is no longer proportional to the stress (Fig. 10.4). The solid still returns to its original length when the stress is removed as long as the stress does not exceed the *elastic limit.* If the stress exceeds the elastic limit, the material is permanently deformed. For still larger stresses, the solid fractures when the stress reaches the *breaking point.* The maximum stress that can be withstood without breaking is called the *ultimate strength.* The ultimate strength can be different for compression and tension; then we refer to the compressive strength or the tensile strength of the material.

A *ductile* material continues to stretch beyond its ultimate tensile strength without breaking; the stress then *decreases* from the ultimate strength (Fig. 10.4a). Examples of ductile solids are the relatively soft metals, such as gold, silver, copper, and lead. These metals can be pulled like taffy, becoming thinner and thinner until finally reaching the breaking point.

For a *brittle* substance, the ultimate strength and the breaking point are close together (Fig. 10.4b). Bone is an example of a brittle material; it fractures abruptly if the stress becomes too large (Fig. 10.4c). Under either tension or compression, its elastic limit, breaking point, and ultimate strength are approximately the same. Babies have

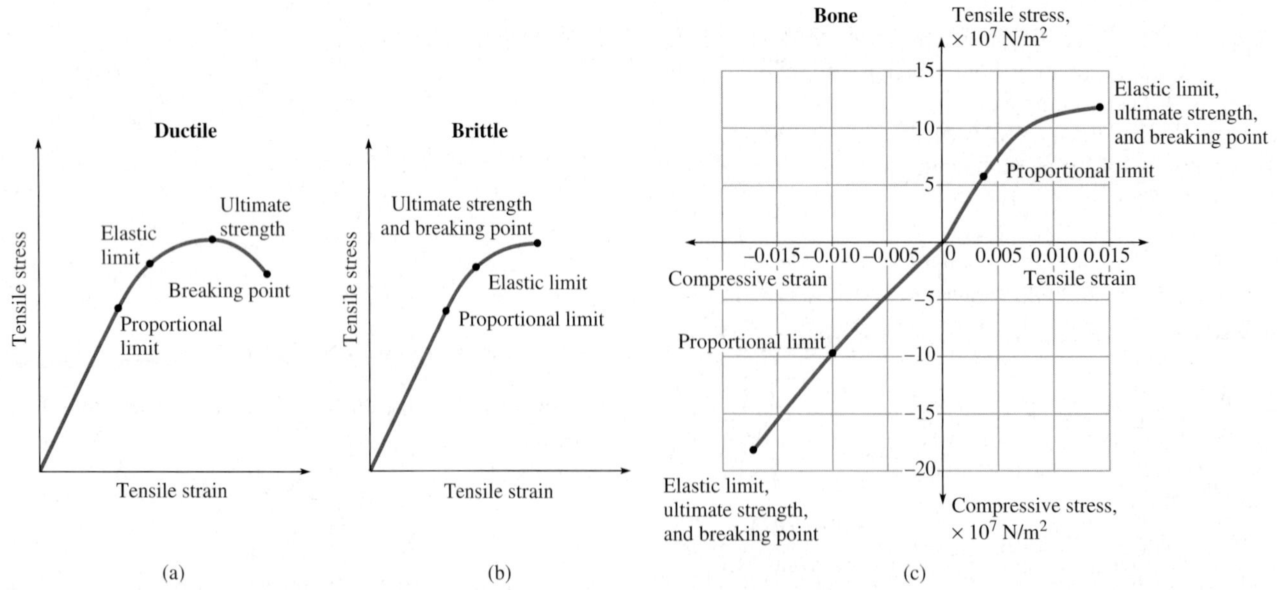

Figure 10.4 Stress-strain curves showing limits for (a) a ductile material, (b) a brittle material, and (c) compact bone. The elastic limit, ultimate strength, and breaking point are well separated for ductile materials, but close together for a brittle material.

Figure 10.5 In prestressed concrete, steel rods are stretched before the concrete is poured. After the concrete hardens, the frame holding the rods under tension is removed. The rods contract, compressing the concrete. Then, when the prestressed concrete is subject to a tensile force, the compression of the concrete is lessened but not eliminated so that the concrete itself is never subjected to a tensile stress.

more flexible bones than adults because they have built up less of the calcium compound hydroxyapatite. As people age, their bones become more brittle as the collagen fibers lose flexibility and their bones also become weaker as calcium gets reabsorbed (a condition called osteoporosis).

Applications of elastic properties of materials: osteoporosis and reinforced concrete

Like bone, reinforced concrete has one component for tensile strength and another for compressive strength. Reinforced concrete contains steel rods that provide tensile strength that concrete itself lacks (Fig. 10.5).

Human anatomy has special features for adapting to the compressive stress associated with standing upright. For example, the vertebrae in the spinal column gradually increase in size from the neck to the tailbone. Such an arrangement places the stronger vertebrae in the lower positions, where they must support more weight. The vertebrae are separated by fluid-filled disks, which have a cushioning effect by spreading out the compressive forces.

Application of compressive forces: the human vertebra

Example 10.3

Crane with Steel Cable

A crane is required to lift loads of up to 1.0×10^5 N (11 tons). (a) What is the minimum diameter of the steel cable that must be used? (b) If a cable of twice the minimum diameter is used and it is 8.0 m long when no load is present, how much longer is it when supporting a load of 1.0×10^5 N? (Data for steel: $Y = 2.0 \times 10^{11}$ Pa; proportional limit = 2.0×10^8 Pa; elastic limit = 3.0×10^8 Pa; tensile strength = 5.0×10^8 Pa.)

Strategy The data given for steel consists of four quantities that all have the same units. It would be easy to mix them up if we didn't understand what each one means. Young's modulus is the proportionality constant between stress and strain. That will be useful in part (b) where we find the elongation of the cable; the elongation is the strain times the original length. However, we should first check that the stress is less than the proportional limit before using Young's modulus to find the strain.

The elastic limit is the maximum stress so that no permanent deformation occurs; the tensile strength is the maximum stress so that the cable does not break. We certainly don't want the cable to break, but it would be prudent to keep the stress under the elastic limit to give the cable a long useful life. Therefore, we choose a minimum diameter in (a) to keep the stress below the elastic limit.

Solution (a) We choose the minimum diameter to keep the stress less than the elastic limit:

$$\frac{F}{A} < \text{elastic limit} = 3.0 \times 10^8 \text{ Pa}$$

for $F = 1.0 \times 10^5$ N. Then

$$A > \frac{F}{\text{elastic limit}} = \frac{1.0 \times 10^5 \text{ N}}{3.0 \times 10^8 \text{ Pa}} = 3.33 \times 10^{-4} \text{ m}^2$$

The minimum cross-sectional area corresponds to the minimum diameter. The cross-sectional area of the cable is πr^2 or $\pi d^2/4$, so

$$d = \sqrt{\frac{4A}{\pi}} = \sqrt{\frac{4 \times 3.33 \times 10^{-4} \text{ m}^2}{\pi}} = 2.1 \text{ cm}$$

The minimum *diameter* is therefore 2.1 cm.

continued on next page

Example 10.3 continued

(b) If we double the diameter and keep the same load, the stress is reduced by a factor of four since the cross-sectional area is proportional to the square of the diameter. Therefore, the stress is

$$\frac{F}{A} = \frac{3.0 \times 10^8 \text{ Pa}}{4} = 7.5 \times 10^7 \text{ Pa}$$

The strain is then

$$\frac{\Delta L}{L} = \frac{F/A}{Y} = \frac{7.5 \times 10^7 \text{ Pa}}{2.0 \times 10^{11} \text{ Pa}} = 0.000375$$

The strain is the fractional length change. Then the length change is

$$\Delta L = 0.000375L = 0.000375 \times 8.0 \text{ m} = 0.0030 \text{ m} = 3.0 \text{ mm}$$

Discussion By using a cable twice as thick as the minimum, we build in a safety factor. We don't want to be right at the edge of disaster! Since doubling the diameter of the cable increases the cross-sectional area of the cable by a factor of four, the maximum stress on the cable is one fourth of the elastic limit.

Practice Problem 10.3 Tuning a Harpsichord String

A harpsichord string is made of yellow brass (Young's modulus 9.0×10^{10} Pa, tensile strength 6.3×10^8 Pa). When tuned correctly, the tension in the string is 59.4 N, which is 93% of the maximum tension that the string can endure without breaking. What is the radius of the string?

The San Jacinto monument in Texas is the tallest stone column in the world.

Height Limits

What limits the height of a stone column? If the column is too tall, it could be crushed under its own weight. The maximum height of a column is limited since the compressive stress at the bottom cannot exceed the compressive strength of the material (see Problem 89). However, the maximum height at which a vertical column buckles is generally less than the height at which it would be crushed.

The bones of our limbs are hollow; the inside of the structural material is filled with marrow, which is structurally weak. A hollow bone is better able to resist fracture from bending and twisting forces than a solid bone with the same amount of structural material, although the hollow bone would buckle more easily under a compressive force along the central axis.

Figure 10.6 A column made from a rolled sheet of paper can support a book.

PHYSICS AT HOME

Challenge a friend to use a single sheet of 8.5 in. × 11 in. paper and two paper clips (or tape) to support a book at least 8 in. above a table. If your friend has no idea what to do, roll the sheet of paper into a narrow cylinder about 2.5 cm (an inch) in diameter; then fasten the cylinder at the top and bottom with paper clips (or with tape). Carefully place the book so that it is balanced on top of the cylinder (Fig. 10.6). If you have difficulty, try using thicker paper or a lighter book.

Use the same "apparatus" to get some insight into the buckling of columns. Try making the diameter of the paper cylinder twice as large. The walls of this column are thinner because there are fewer layers of the paper in the cylinder wall, although the same cross-sectional area *of paper* supports the book. If nothing happens, try again with a heavier book. You will likely see the walls crumple in on themselves as the cylinder buckles and the book falls to the table.

Why would the design of a giant's bones have to be different from a human's? If the giant's average density is the same as a human's, then his weight is larger by the same factor that his *volume* is larger. If the giant is five times as tall as a human, for instance, and has the same relative proportions, then his volume is $5^3 = 125$ times as large, since each of the three dimensions of any body part has increased by a factor of five. On the other hand, the cross-sectional area of a bone is proportional to the *square* of its radius. So while the leg bones must support 125 times as much weight, the maximum compressive force they can withstand has only increased by a factor of 25. The giant would need much thicker legs (in relation to their length) to support his increased weight. Similar analysis can be applied to the twisting and bending forces that are more likely to break bones than are compressive forces. The result is the same: the bones of a giant could not have human proportions.

Some science fiction or horror movies portray giant insects as greatly magnified versions of a normal insect. Such a giant insect's legs would collapse under the weight of the insect.

Application of compressive strength: size limitations on organisms

10.4 SHEAR AND VOLUME DEFORMATIONS

In this section we consider two other kinds of deformation. In each case we define a stress (force per unit area), a strain (dimensionless), and a modulus (the constant of proportionality between stress and strain).

Shear Deformation

Unlike tensile and compressive forces, which are perpendicular to two opposite surfaces of an object, a **shear deformation** is the result of a pair of equal and opposite forces that act *parallel* to two opposite surfaces (Fig. 10.7). The **shear stress** is the magnitude of the shear force divided by the area of the surface on which the force acts:

$$\text{shear stress} = \frac{\text{shear force}}{\text{area of surface}} = \frac{F}{A} \qquad (10\text{-}6)$$

Shear strain is the ratio of the relative displacement Δx to the separation L of the two surfaces:

$$\text{shear strain} = \frac{\text{displacement of surfaces}}{\text{separation of surfaces}} = \frac{\Delta x}{L} \qquad (10\text{-}7)$$

The shear strain is proportional to the shear stress as long as the stress is not too large. The constant of proportionality is the **shear modulus** S.

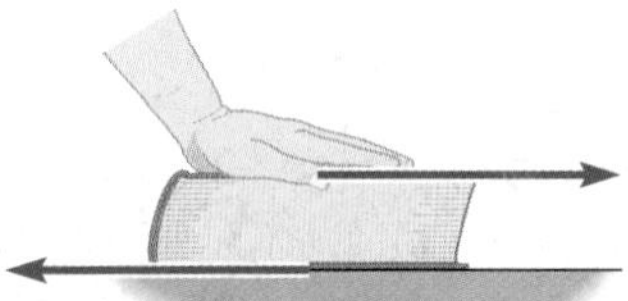

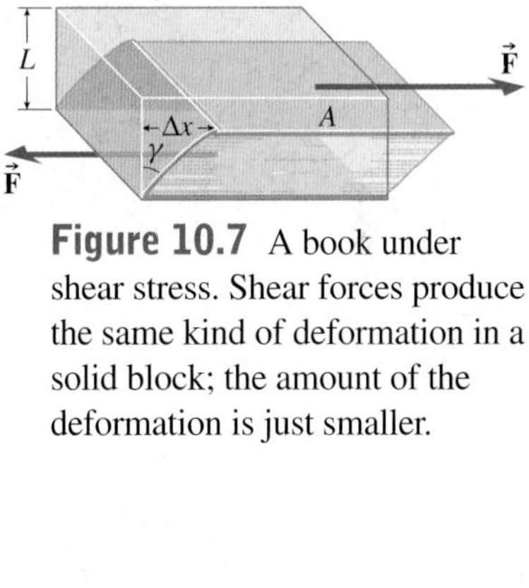

Figure 10.7 A book under shear stress. Shear forces produce the same kind of deformation in a solid block; the amount of the deformation is just smaller.

Hooke's Law for Shear Deformations

$$\text{shear stress} \propto \text{shear strain}$$

$$\frac{F}{A} = S\frac{\Delta x}{L} \qquad (10\text{-}8)$$

CONNECTION:

Hooke's law takes the same form for different kinds of stresses and strains. In each case, the strain is proportional to the stress.

The units of shear stress and the shear modulus are the same as for tensile or compressive stress and Young's modulus: Pa or N/m^2. The strain is once again dimensionless. Table 10.2 lists shear moduli for various materials.

An example of shear stress is the cutting action of a pair of scissors (or "shears") on a piece of paper. The forces acting on the paper from above and below are offset from each other and act parallel to the cross-sectional surfaces of the paper (Fig. 10.8).

Table 10.2 Shear and Bulk Moduli for Various Materials

Material	Shear Modulus S (10^9 Pa)	Bulk Modulus B (10^9 Pa)
Gases		
Air (1)		0.00010
Air (2)		0.00014
Liquids		
Ethanol		0.9
Water		2.2
Mercury		25
Solids		
Cast iron	40–50	60–90
Marble		70
Aluminum	25–30	70
Copper	40–50	120–140
Steel	80–90	140–160
Diamond		620

(1) At 0°C and 1 atm; constant temperature expansion or compression
(2) At 0°C and 1 atm; no heat flow during expansion or compression

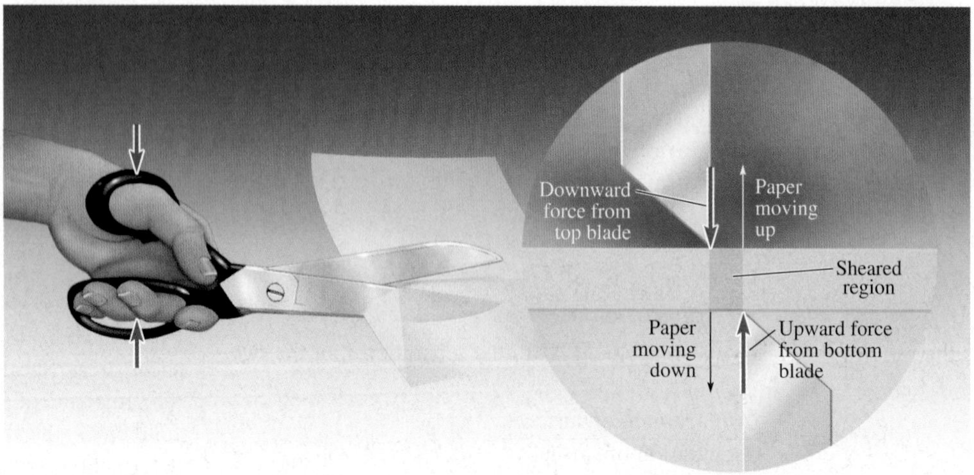

Figure 10.8 Scissors apply shear stress to a sheet of paper. The shear stress is the force exerted by a blade divided by the cross-sectional area of the paper—the thickness of the paper times the length of blade that is in contact with the paper.

Example 10.4

Cutting Paper

A sheet of paper of thickness 0.20 mm is cut with scissors that have blades of length 10.0 cm and width 0.20 cm. While cutting, the scissors blades each exert a force of 3.0 N on the paper; the length of each blade that makes contact with the paper is approximately 0.5 mm. What is the shear stress on the paper?

Strategy Shear stress is a force divided by an area. In this problem, identifying the correct area is tricky. The blades push two *cross-sectional* paper surfaces in opposite directions so they are displaced with respect to one another. The shear stress is the force exerted by each blade divided by this cross-sectional area—the thickness of the paper times the length of blade *in contact with the paper.* (Compare Figs. 10.7 and 10.8.) The total length and the width of the blades are irrelevant.

continued on next page

Example 10.4 continued

Solution The cross-sectional area is

$$A = \text{thickness} \times \text{contact length}$$
$$= 2.0 \times 10^{-4}\ \text{m} \times 5 \times 10^{-4}\ \text{m} = 1 \times 10^{-7}\ \text{m}^2$$

The shear stress is

$$\frac{F}{A} = \frac{3.0\ \text{N}}{1 \times 10^{-7}\ \text{m}^2} = 3 \times 10^{7}\ \text{N/m}^2$$

Discussion To identify the correct area, remember that shear forces act *in the plane of* the surfaces that are displaced with respect to each other. By contrast, tensile and compressive forces are perpendicular to the area used to find tensile and compressive stresses.

Practice Problem 10.4 Shear Stress due to a Hole Punch

A hole punch has a diameter of 8.0 mm and presses onto ten sheets of paper with a force of 6.7 kN. If each sheet of paper is of thickness 0.20 mm, find the shear stress. [*Hint:* Be careful in deciding what area to use. Remember that a shear force acts *parallel* to the surface whose area is relevant.]

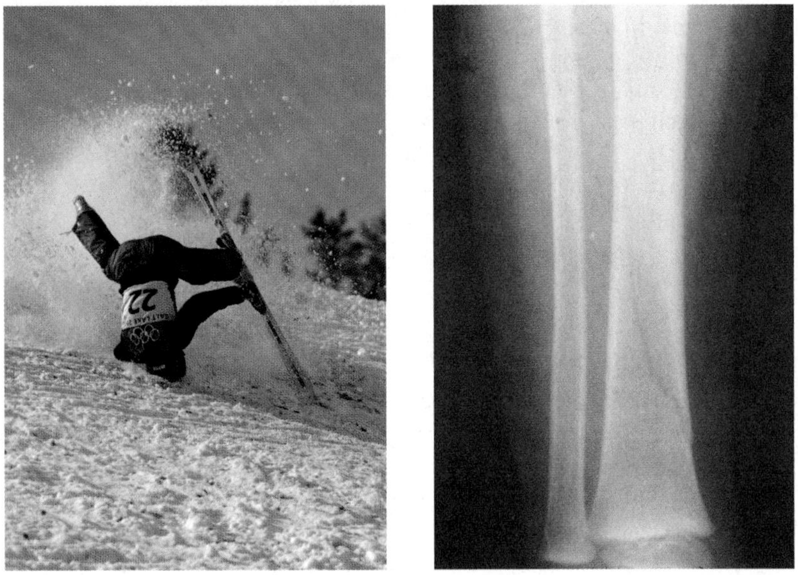

(a) (b)

Figure 10.9 (a) An Olympic skier falls and his leg is subjected to a shear stress. (b) X-ray of a spiral fracture of the tibia.

When a bone is twisted, it is subjected to a shear stress. Shear stress is a more common cause of fracture than a compressive or tensile stress along the length of the bone. The twisting of a bone can result in a spiral fracture (Fig. 10.9).

Volume Deformation

As discussed in Chapter 9, a fluid exerts inward forces on an immersed solid object. These forces are perpendicular to the surfaces of the object. Since the fluid presses inward on all sides of the object (Fig. 10.10), the solid is compressed—its volume is reduced. The fluid pressure P is the force per unit surface area; it can be thought of as the **volume stress** on the solid object. Pressure has the same units as the other kinds of stress: N/m^2 or Pa.

$$\text{volume stress} = \text{pressure} = \frac{F}{A} = P$$

The resulting deformation of the object is characterized by the **volume strain**, which is the fractional change in volume:

$$\text{volume strain} = \frac{\text{change in volume}}{\text{original volume}} = \frac{\Delta V}{V} \qquad (10\text{-}9)$$

Unless the stress is too large, the stress and strain are proportional within a constant of proportionality called the **bulk modulus** B. A substance with a large bulk modulus is more difficult to compress than a substance with a small bulk modulus.

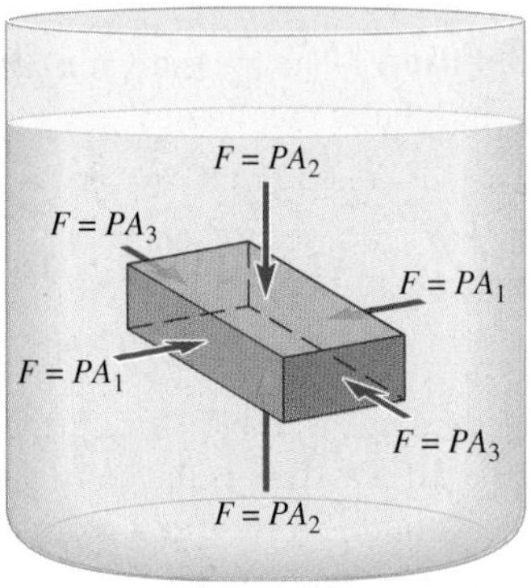

Figure 10.10 Forces on an object when submerged in a fluid.

An object at atmospheric pressure is already under volume stress: the air pressure already compresses the object slightly compared to what its volume would be in vacuum. For solids and liquids, the volume strain due to atmospheric pressure is, for most purposes, negligibly small (5×10^{-5} for water). Since we are usually concerned with the deformation due to a *change* in pressure ΔP from atmospheric pressure, we can write Hooke's law as:

Hooke's Law for Volume Deformation

$$\Delta P = -B\frac{\Delta V}{V} \qquad (10\text{-}10)$$

where V is the volume at atmospheric pressure. The negative sign in Eq. (10-10) allows the bulk modulus to be positive—an increase in the volume stress causes a *decrease* in volume, so ΔV is negative. Table 10.2 lists bulk moduli for various substances.

Unlike the stresses and strains discussed previously, volume stress can be applied to fluids (liquids and gases) as well as solids. The bulk moduli of liquids are generally not much less than those of solids, since the atoms in liquids are nearly as close together as those in solids. In Chapter 9 we assume that liquids are incompressible, which is often a good approximation since the bulk moduli of liquids are generally large. In gases, the atoms are much farther apart on average than in solids or liquids. Gases are much easier to compress than solids or liquids, so their bulk moduli are much smaller.

Example 10.5

Marble Statue Under Water

A marble statue of volume 1.5 m^3 is being transported by ship from Athens to Cyprus. The statue topples into the ocean when an earthquake-caused tidal wave sinks the ship; the statue ends up on the ocean floor, 1.0 km below the surface. Find the change in volume of the statue in cm^3 due to the pressure of the water. The density of seawater is 1025 kg/m^3.

Strategy The water pressure is the volume stress; it is the force per unit area pressing inward and perpendicular to all the surfaces of the statue. The water pressure at a depth d is greater than the pressure at the water surface; we can find the pressure using the given density of seawater. Then, using the bulk modulus of marble given in Table 10.2, we find the change in volume from Hooke's law.

Solution The pressure at a depth d = 1.0 km is larger than atmospheric pressure by

$$\Delta P = \rho g d$$

$$= 1025\ \text{kg/m}^3 \times 9.8\ \text{N/kg} \times 1000\ \text{m}$$

$$= 1.005 \times 10^7\ \text{Pa}$$

According to Table 10.2, the bulk modulus for marble is 70×10^9 Pa. This is the constant of proportionality between the volume stress (pressure increase) and the strain (fractional change in volume).

$$\Delta P = -B\frac{\Delta V}{V}$$

Solving for ΔV,

$$\Delta V = -\frac{\Delta P}{B}V = -\frac{1.005 \times 10^7\ \text{Pa}}{70 \times 10^9\ \text{Pa}} \times 1.5\ \text{m}^3$$

$$= -2.2 \times 10^{-4}\ \text{m}^3 \times \left(\frac{100\ \text{cm}}{1\ \text{m}}\right)^3 = -220\ \text{cm}^3$$

The statue's volume decreases approximately 220 cm^3.

Discussion The fractional decrease in volume is

$$\frac{1.005 \times 10^7\ \text{Pa}}{70 \times 10^9\ \text{Pa}} \approx \frac{1}{7000}$$

or a reduction of 0.014%.

In calculating the pressure increase, we assumed that the density of seawater is constant—the equation $\Delta P = \rho g d$ is derived for a constant fluid density ρ. Should we worry that our calculation of ΔP is wrong? The result of Practice Problem 10.5 shows that the density of seawater at a depth of 1.0 km is only about 0.43% greater than its density at the surface. The calculation of ΔP is inaccurate by less than 0.5%—negligible here since we only know the depth to two significant figures.

Practice Problem 10.5 Compression of Water

Show that a pressure increase of 1.0×10^7 Pa (100 atm) on 1 m^3 of seawater causes a 0.43% decrease in volume. The bulk modulus of seawater is 2.3×10^9 Pa.

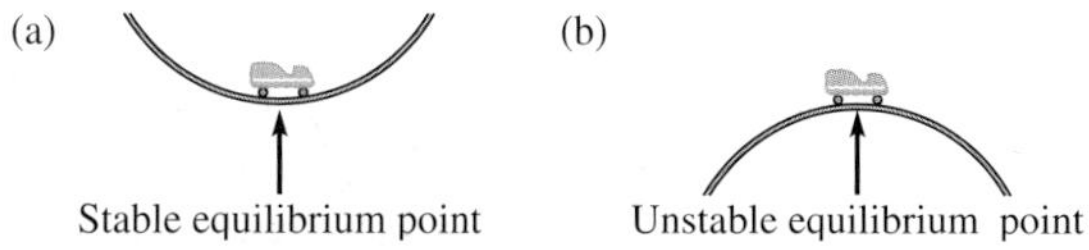

Figure 10.11 (a) A point of *stable* equilibrium for a roller-coaster car. If the car is displaced slightly from its position at the bottom of the track, the net force pulls the car back toward the equilibrium point. (b) A point of *unstable* equilibrium for a roller-coaster car. If the car is displaced slightly from the very top of the track, the net force pushes the car *away from* the equilibrium point.

10.5 SIMPLE HARMONIC MOTION

Vibration, one of the most common kinds of motion, is repeated motion back and forth along the same path. Vibrations occur in the vicinity of a point of **stable equilibrium**. An equilibrium point is *stable* if the net force on an object when it is displaced a small distance from equilibrium points back toward the equilibrium point (Fig. 10.11). Such a force is called a **restoring force** since it tends to restore equilibrium. A special kind of vibratory motion—called **simple harmonic motion** (or **SHM**)—occurs whenever the restoring force is proportional to the displacement from equilibrium.

Simple harmonic motion: vibratory motion that occurs when the restoring force is proportional to the displacement from equilibrium

CONNECTION:

As shown in Sections 10.2–10.4, Hooke's law applies to small deformations of many kinds of objects, not just springs. Thus, simple harmonic motion occurs in many situations as long as the vibrations are not too large.

Figure 10.12 shows a graph of F_x versus x for some restoring force. We choose $x = 0$ at the equilibrium position. Since the graph is not linear, the resulting oscillations are not SHM—unless the amplitude is small. For small amplitudes, we can approximate the graph near equilibrium by a straight line tangent to the curve at the equilibrium point. For small amplitude oscillations, the restoring force is approximately linear, so the resulting oscillations are (approximately) SHM. The ideal spring is a favorite model of physicists because the restoring force it provides is proportional to the displacement from equilibrium.

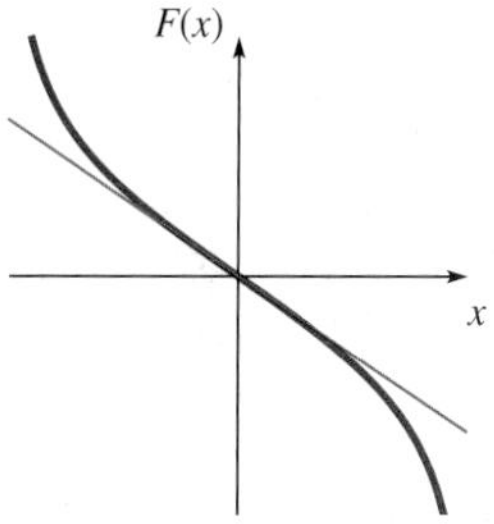

Figure 10.12 A nonlinear restoring force (red) can be approximated as a linear restoring force (blue) for small displacements.

Consider a relaxed ideal spring with spring constant k and zero mass. The spring is fixed at one end and attached at the other to an object of mass m (Fig. 10.13) that slides without friction. Since the normal force is equal and opposite to the weight of the object, the net force on the object is that due to the spring. When the spring is relaxed, the net force is zero; the object is in equilibrium.

If the object is now pulled to the right to the position $x = A$ and then released, the net force on the object is

$$F_x = -kx \qquad (10\text{-}11)$$

where the negative sign tells us that the spring force is opposite in direction to the displacement from equilibrium. At first the object is to the right of the equilibrium position and the spring pulls to the left. Notice that the force exerted by the spring is in the correct direction to restore the object to the equilibrium position; it always pushes or pulls back toward the equilibrium point.

Imagine taking a series of photos at equal time intervals as the object oscillates back and forth. In Fig. 10.14 the blue dots are the positions of the object at equal time intervals over one-half of a full cycle, from one endpoint to the other. (A full cycle would include the return trip.)

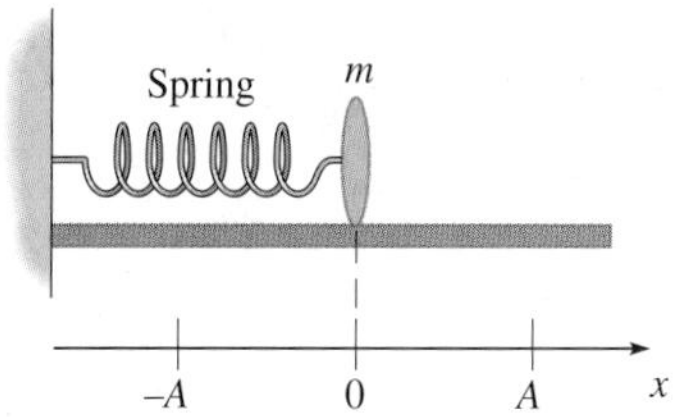

Figure 10.13 Spring in relaxed position. We choose the origin $x = 0$ at the object's equilibrium position.

Energy Analysis in SHM Figure 10.14 suggests that the speed is greatest as the object passes through the equilibrium position. The object slows as it approaches the endpoints and gains speed as it approaches the equilibrium point. At the endpoints ($x = \pm A$), the body is instantaneously at rest before heading back in the other direction. Conservation of energy supports these observations. The total mechanical energy of the mass and spring is constant.

$$E = K + U = \text{constant}$$

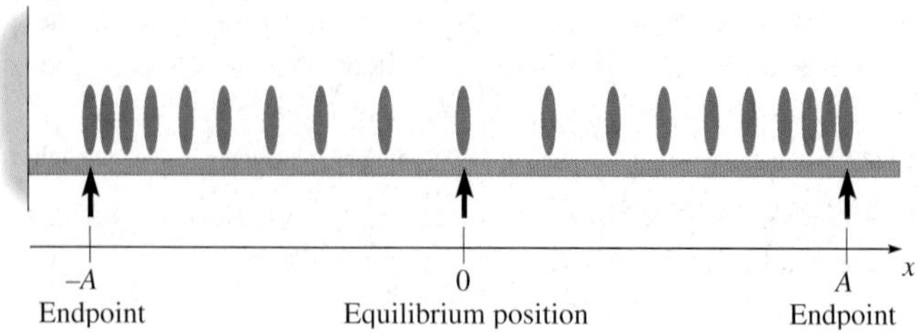

Figure 10.14 Positions of an oscillating body at equal time intervals over half a period. The spring is omitted for clarity.

CONNECTION:

Our study of SHM is based on familiar principles of energy conservation and Newton's second law, together with Hooke's law.

where K is the kinetic energy and U is the elastic potential energy stored in the spring. As the object oscillates back and forth, energy is converted from potential to kinetic and back to potential in the half-cycle shown in Fig. 10.14. From Section 6.7, the elastic potential energy of the spring is

$$U = \tfrac{1}{2}kx^2 \qquad (6\text{-}24)$$

The speed at any point x can be found from the energy equation

$$E = \tfrac{1}{2}mv_x^2 + \tfrac{1}{2}kx^2 \qquad (10\text{-}12)$$

Amplitude: maximum displacement from equilibrium

The maximum displacement of the body is the **amplitude** A. At the maximum displacement, where the motion changes direction, the velocity is zero. Since the kinetic energy is zero at $x = \pm A$, all the energy is elastic potential energy at the endpoints. Therefore, the total energy E at the endpoints is

$$E_{\text{total}} = \tfrac{1}{2}kA^2 \qquad (10\text{-}13)$$

and, since energy is conserved, this must be the total energy at any point in the object's motion. The maximum speed v_m occurs at $x = 0$ where all the energy is kinetic. Thus, at $x = 0$, the total energy equals the kinetic energy

$$E_{\text{total}} = \tfrac{1}{2}mv_m^2$$

and, from Eq. (10-13),

$$\tfrac{1}{2}mv_m^2 = \tfrac{1}{2}kA^2$$

Solving for v_m yields

$$v_m = \sqrt{\frac{k}{m}}A \qquad (10\text{-}14)$$

The maximum speed is proportional to the amplitude.

CHECKPOINT 10.5

What is the displacement of an object in SHM when the kinetic and potential energies are equal?

Acceleration in SHM The force on the object at any point x is given by Hooke's law; Newton's second law then gives the acceleration:

$$F_x = -kx = ma_x$$

Solving for the acceleration,

$$a_x(t) = -\frac{k}{m}x(t) \qquad (10\text{-}15)$$

Thus, the acceleration is a negative constant ($-k/m$) times the displacement; the acceleration and displacement are always in opposite directions. Whenever the acceleration is a negative constant times the displacement, the motion is SHM.

The acceleration has its maximum magnitude a_m, where the force is largest, which is at the maximum displacement $x = \pm A$:

$$a_m = \frac{k}{m} A \quad (10\text{-}16)$$

In SHM, the acceleration changes with time. Equations derived for constant acceleration do not apply.

Example 10.6

Oscillating Model Rocket

A model rocket of 1.0-kg mass is attached to a horizontal spring with a spring constant of 6.0 N/cm. The spring is compressed by 18.0 cm and then released. The intent is to shoot the rocket horizontally, but the release mechanism fails to disengage, so the rocket starts to oscillate horizontally. Ignore friction and assume the spring to be ideal. (a) What is the amplitude of the oscillation? (b) What is the maximum speed? (c) What are the rocket's speed and acceleration when it is 12.0 cm from the equilibrium point?

Strategy First, we sketch the situation (Fig. 10.15). Initially all of the energy is elastic potential energy and the kinetic energy is zero. The initial displacement must be the maximum displacement—or amplitude—of the oscillations since to get farther from equilibrium would require more elastic energy than the total energy available. The speed at any position can be found using energy conservation ($\frac{1}{2}kx^2 + \frac{1}{2}mv_x^2 = \frac{1}{2}kA^2$). The maximum speed occurs when all of the energy is kinetic. The acceleration can be found from Newton's second law.

Solution (a) The amplitude of the oscillation is the maximum displacement, so $A = 18.0$ cm.

(b) From energy conservation, the maximum kinetic energy is equal to the maximum elastic potential energy:

$$K_m = \tfrac{1}{2}mv_m^2 = E = \tfrac{1}{2}kA^2$$

Solving for v_m,

$$v_m = \sqrt{\frac{k}{m}}\,A = \sqrt{\frac{6.0 \times 10^2 \text{ N/m}}{1.0 \text{ kg}}} \times 0.180 \text{ m} = 4.4 \text{ m/s}$$

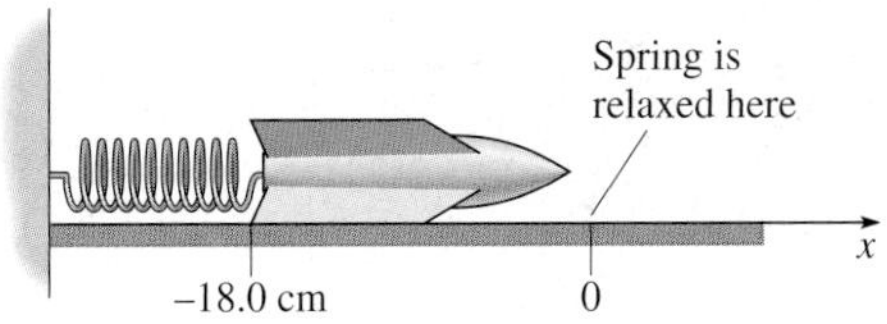

Figure 10.15 The model rocket before it is released.

(c) For the speed at a displacement of 0.120 m, we again use energy conservation.

$$\tfrac{1}{2}kx^2 + \tfrac{1}{2}mv^2 = \tfrac{1}{2}kA^2$$

Solving for v,

$$v = \sqrt{\frac{kA^2 - kx^2}{m}} = \sqrt{\frac{k}{m}(A^2 - x^2)}$$

$$= \sqrt{\frac{6.0 \times 10^2 \text{ N/m}}{1.0 \text{ kg}}[(0.180 \text{ m})^2 - (0.120 \text{ m})^2]} = 3.3 \text{ m/s}$$

From Newton's second law,

$$F_x = -kx = ma_x$$

At $x = \pm\, 0.120$ m,

$$a_x = -\frac{k}{m}x = \frac{6.0 \times 10^2 \text{ N/m}}{1.0 \text{ kg}} \times (\pm 0.120 \text{ m}) = \pm 72 \text{ m/s}^2$$

The magnitude of the acceleration is 72 m/s^2; the direction is toward the equilibrium point.

Discussion Note that at a given position (say $x = +0.120$ m), we can find the *speed* of the rocket, but the direction of the velocity can be either left or right; the rocket passes through each point (other than the endpoints) both on its way to the left and on its way to the right. By contrast, the *acceleration* at $x = +0.120$ m is always in the $-x$-direction, regardless of whether the rocket is moving to the left or to the right. If the rocket is moving to the right, then it is slowing down as it approaches $x = +A$; if it is moving to the left, then it is speeding up as it approaches $x = 0$.

Practice Problem 10.6 Maximum Acceleration of the Rocket

What is the maximum acceleration of the rocket in Example 10.6 and at what position(s) does it occur?

10.6 THE PERIOD AND FREQUENCY FOR SHM

CONNECTION:

The period and frequency are defined exactly as for uniform circular motion, which is another kind of periodic motion.

Definitions of Period and Frequency SHM is *periodic* motion because the same motion repeats over and over—a particle goes back and forth over the same path in exactly the same way. Each time the particle repeats its original motion, we say that it has completed another cycle. To complete one cycle of motion, the particle must be at the same point *and heading in the same direction* as it was at the start of the cycle. The **period** T is the time taken by one complete cycle. The **frequency** f is the number of cycles per unit time:

Period: time for one complete cycle

$$f = \frac{1}{T} \quad \text{(SI unit: Hz = cycles per second)} \qquad (5\text{-}8)$$

SHM is a special kind of periodic motion in which the restoring force is proportional to the displacement from equilibrium. Not all periodic vibrations are examples of simple harmonic motion since not all restoring forces are proportional to the displacement. Any restoring force can cause oscillatory motion. An electrocardiogram (Fig. 10.16) traces the periodic pattern of a beating heart, but the motion of the recorder needle is not simple harmonic motion. As we are about to show, in SHM the position is a *sinusoidal* function of time.

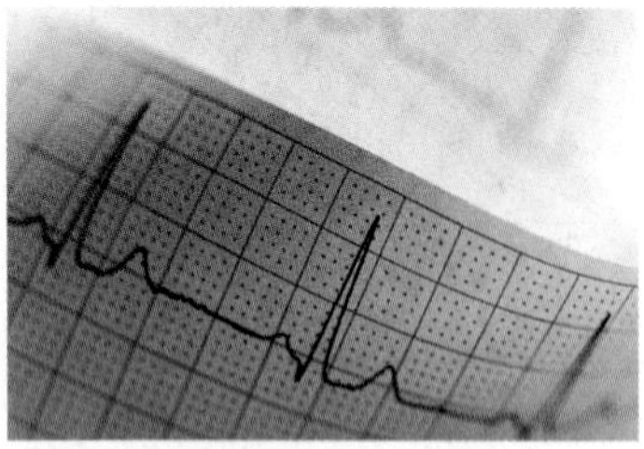

Figure 10.16 An electrocardiogram.

Most of the equations involving ω are correct only if ω is measured in *radians* per unit time (such as rad/s). Don't forget to put your calculator into radian mode.

Circular Motion and SHM To learn more about SHM, imagine setting up an experiment (Fig. 10.17). We attach an object to an ideal spring, move the object away from the equilibrium position, and then release it. The object vibrates back and forth in simple harmonic motion with amplitude A. At the same time a horizontal circular disk, of radius $r = A$ and with a pin projecting vertically up from its outer edge, is set into rotation with uniform circular motion. Both the pin and the object attached to the spring are illuminated so that shadows of the vibrating object and of the pin on the rotating disk are seen on a screen. The speed of the disk is adjusted until the shadows oscillate with the same period. We will show that the motion of the two shadows is identical, so the mathematical description of one can be used for the other.

To find the mathematical description of SHM, we analyze the uniform circular motion of the pin. Figure 10.17b shows the pin P moving counterclockwise around

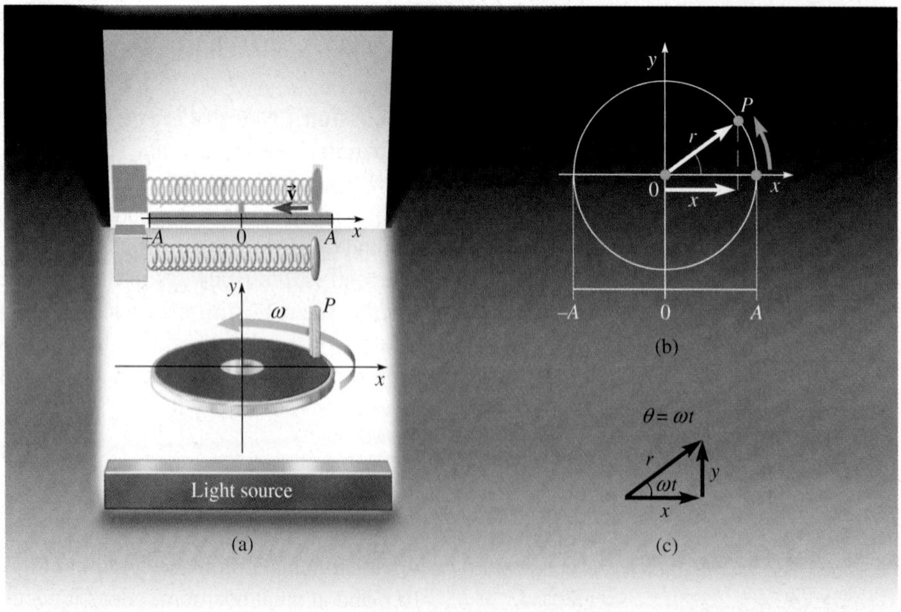

Figure 10.17 (a) An experiment to show the relation between uniform circular motion and SHM. (b) A pin P moving counterclockwise around a circle as a disk rotates with constant angular velocity ω. (c) Finding the x-component of the displacement.

a circle of radius A at a constant angular velocity ω in rad/s. For simplicity, let the pin start at $\theta = 0$ at time $t = 0$. The location of the pin at any time is then given by the angle θ:

$$\theta(t) = \omega t$$

The motion of the pin's shadow has the same x-component as the pin itself. Using a right triangle (Fig. 10.17c), we find that

$$x(t) = A \cos \theta = A \cos \omega t \qquad (10\text{-}17)$$

Since the pin moves in uniform circular motion, its acceleration is constant in *magnitude* but not in direction; the acceleration is toward the center of the circle. In Section 5.2, the magnitude of the radial acceleration is shown to be

$$a = \omega^2 r = \omega^2 A \qquad (5\text{-}12)$$

At any instant the direction of the acceleration vector is opposite to the direction of the displacement vector in Fig. 10.17b—that is, toward the center of the circle. Therefore,

$$a_x = -a \cos \theta = -\omega^2 A \cos \omega t \qquad (10\text{-}18)$$

Comparing Eqs. (10-17) and (10-18), we see that, at any time t,

$$a_x(t) = -\omega^2 x(t) \qquad (10\text{-}19)$$

In Eq. (10-15) we showed that in SHM the acceleration is proportional to the displacement:

$$a_x = -\frac{k}{m} x \qquad (10\text{-}15)$$

Comparing the right-hand sides of Eqs. (10-15) and (10-19), the motions of the two shadows are identical as long as

$$\omega = \sqrt{\frac{k}{m}} \qquad (10\text{-}20a)$$

The position and acceleration of an object in SHM are sinusoidal functions of time [Eqs. (10-17) and (10-18)]. The sinusoidal functions are sine and cosine. In Problem 54, you can show that v_x is also a sinusoidal function of time.

The term *harmonic* in *simple harmonic motion* refers to a sinusoidal vibration; this usage is related to similar usage in music and acoustics. The sinusoidal functions are also called harmonic functions. In Chapter 12, we show that a complex vibration can be formed by combining harmonic vibrations at different frequencies, which is why the study of SHM is the basis for understanding more complex vibrations. The term *simple* in SHM means that the amplitude of the vibration is constant; we assume there is no energy dissipation to cause the vibration to die out.

Period and Frequency for an Ideal Mass-Spring System Since the object in SHM and the pin in circular motion have the same frequency and period, the relationships between ω, f, and T still apply. Therefore, the frequency and period of a mass-spring system are

$$f = \frac{\omega}{2\pi} = \frac{1}{2\pi}\sqrt{\frac{k}{m}} \qquad (10\text{-}20b)$$

and

$$T = \frac{1}{f} = 2\pi\sqrt{\frac{m}{k}} \qquad (10\text{-}20c)$$

In the context of SHM, the quantity ω is called the **angular frequency**. Note that the angular frequency is determined by the mass and the spring constant but is independent of the amplitude.

With the identification of ω for a mass-spring system, we can write the maximum speed and acceleration from Eqs. (10-14) and (10-16):

$$v_m = \omega A \quad (10\text{-}21)$$

$$a_m = \omega^2 A \quad (10\text{-}22)$$

These expressions are more general than Eqs. (10-14) and (10-16)—they apply to any system in SHM, not just a mass-spring system.

To Find the Angular Frequency for Any Object in SHM

- Write down the restoring force as a function of the displacement from equilibrium. Since the restoring force is linear, it always takes the form $F = -kx$, where k is a constant.
- Use Newton's second law to relate the restoring force to the acceleration.
- Solve for ω using $a_x = -\omega^2 x$ [Eq. (10-19)].

A Vertical Mass and Spring

The mass and spring systems discussed so far oscillate horizontally. An oscillating mass on a vertical spring also exhibits SHM; the difference is that the equilibrium point is shifted downward by gravity. In our discussions, we assume ideal springs that obey Hooke's law and have a negligibly small mass of their own.

Suppose that an object of weight mg is hung from an ideal spring of spring constant k (Fig. 10.18). The object's equilibrium point is *not* the point at which the spring is relaxed. In equilibrium, the spring is stretched downward a distance d from its relaxed length so that the spring pulls up with a force equal to mg. Taking the $+y$-axis in the upward direction, the condition for equilibrium is

$$\Sigma F_y = +kd - mg = 0 \quad \text{(at equilibrium)} \quad (10\text{-}23)$$

Let us take the origin ($y = 0$) at the equilibrium point. If the object is displaced vertically from the equilibrium point to a position y, the spring force becomes

$$F_{\text{spring},y} = k(d - y)$$

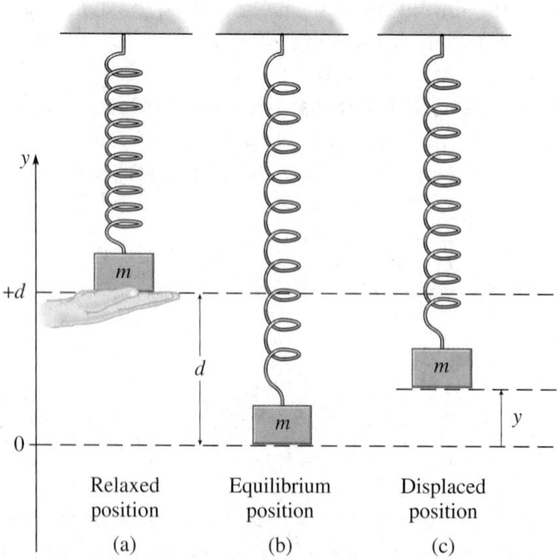

Figure 10.18 (a) A relaxed spring, of spring constant k, with mass m attached. (b) The same spring is extended to its equilibrium position, a distance d below the relaxed position, after mass m is allowed to hang freely. Note that we choose $y = 0$ at the equilibrium position, not at the relaxed position. (c) The spring is displaced from the equilibrium position.

If y is positive, the object is displaced upward and the spring force is less than kd. The y-component of the net force is then

$$\sum F_y = k(d - y) - mg = kd - ky - mg \quad \text{(at displacement } y \text{ from equilibrium)}$$

From Eq. (10-23), we know that $kd = mg$; therefore,

$$\sum F_y = -ky$$

The restoring force provided by the spring and gravity together is $-k$ times the displacement from equilibrium. Therefore, the vertical mass-spring exhibits SHM with the same period and frequency as if it were horizontal.

Example 10.7

A Vertical Spring

A spring with spring constant k is suspended vertically. A model goose of mass m is attached to the unstretched spring and then released so that the bird oscillates up and down. (Ignore friction and air resistance; assume an ideal massless spring.) Calculate the kinetic energy, the elastic potential energy, the gravitational potential energy, and the total mechanical energy at (a) the point of release and (b) the equilibrium point. Take the gravitational potential energy to be zero at the equilibrium point.

Strategy The bird oscillates in SHM about its equilibrium point $y = 0$ between two extreme positions $y = +A$ and $y = -A$ (Fig. 10.19). The amplitude A is equal to the distance the spring is stretched at the equilibrium point; it can be found by setting the net force on the bird equal to zero. The total mechanical energy is the sum of the kinetic energy, the elastic potential energy, and the gravitational potential energy. We expect the total energy to be the same at the two points; since no dissipative forces act, mechanical energy is conserved.

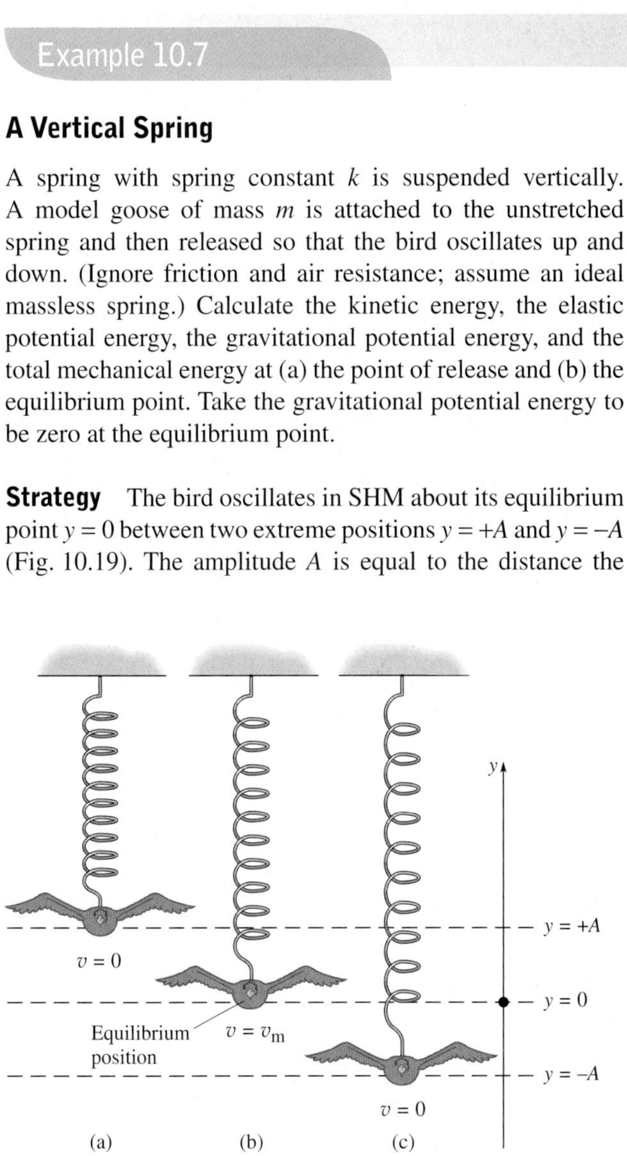

Figure 10.19

(a) The spring is unstretched before the model bird is released at position $y = +A$; (b) the model bird passes through the equilibrium position $y = 0$ with maximum speed; (c) the spring's maximum extension occurs when the bird is at $y = -A$.

Solution The equilibrium point is where the net force on the bird is zero:

$$\sum F_y = +kd - mg = 0 \qquad (10\text{-}23)$$

In this equation, d is the extension of the spring at equilibrium. Since the bird is released where the spring is relaxed, d is also the amplitude of the oscillations:

$$A = d = \frac{mg}{k}$$

(a) At the point of release, $v = 0$ and the kinetic energy is zero. The elastic energy is also zero—the spring is unstretched. The gravitational potential energy is

$$U_g = mgy = mgA = \frac{(mg)^2}{k}$$

The total mechanical energy is the sum of the kinetic and potential (elastic + gravitational) energies,

$$E = K + U_e + U_g = \frac{(mg)^2}{k}$$

(b) At the equilibrium point, the bird moves with its maximum speed $v_m = \omega A$. The angular frequency is the same as for a horizontal spring: $\omega = \sqrt{k/m}$. Then the kinetic energy is

$$K = \tfrac{1}{2}mv_m^2 = \tfrac{1}{2}m\omega^2 A^2$$

Substituting $A = mg/k$ and $\omega^2 = k/m$,

$$K = \frac{1}{2}m\frac{k}{m}\frac{(mg)^2}{k^2} = \frac{1}{2}\frac{(mg)^2}{k}$$

continued on next page

Example 10.7 continued

The spring is stretched a distance A, so the elastic energy is

$$U_e = \frac{1}{2}kA^2 = \frac{1}{2}k\frac{(mg)^2}{k^2} = \frac{1}{2}\frac{(mg)^2}{k}$$

The gravitational potential energy is zero at $y = 0$. Therefore, the total mechanical energy is

$$E = K + U_e + U_g = \frac{1}{2}\frac{(mg)^2}{k} + \frac{1}{2}\frac{(mg)^2}{k} + 0 = \frac{(mg)^2}{k}$$

which is the same as at $y = +A$.

Discussion As the bird moves down from the release point toward the equilibrium point, gravitational potential energy is converted into elastic energy and kinetic energy. After the bird passes the equilibrium point, both kinetic and gravitational energy are converted into elastic energy. At the lowest point in the motion, the gravitational potential energy has its lowest value, while the elastic potential energy has its greatest value. The *total* potential energy (gravitational plus elastic) has its minimum value at the equilibrium point since the kinetic energy is maximum there.

Practice Problem 10.7 Energy at Maximum Extension

Calculate the energies at the lowest point in the oscillations in Example 10.7.

10.7 GRAPHICAL ANALYSIS OF SHM

We have shown that the position of a particle moving in SHM along the x-axis is

$$x(t) = A\cos\omega t \qquad (10\text{-}17)$$

Since the cosine function goes from −1 to +1, multiplying it by A gives us a displacement from $-A$ to $+A$. Figure 10.20a is a graph of the position as a function of time.

The velocity at any time is the slope of the $x(t)$ graph. Note that the maximum slope in Fig. 10.20a occurs when $x = 0$, which confirms what we already know from energy

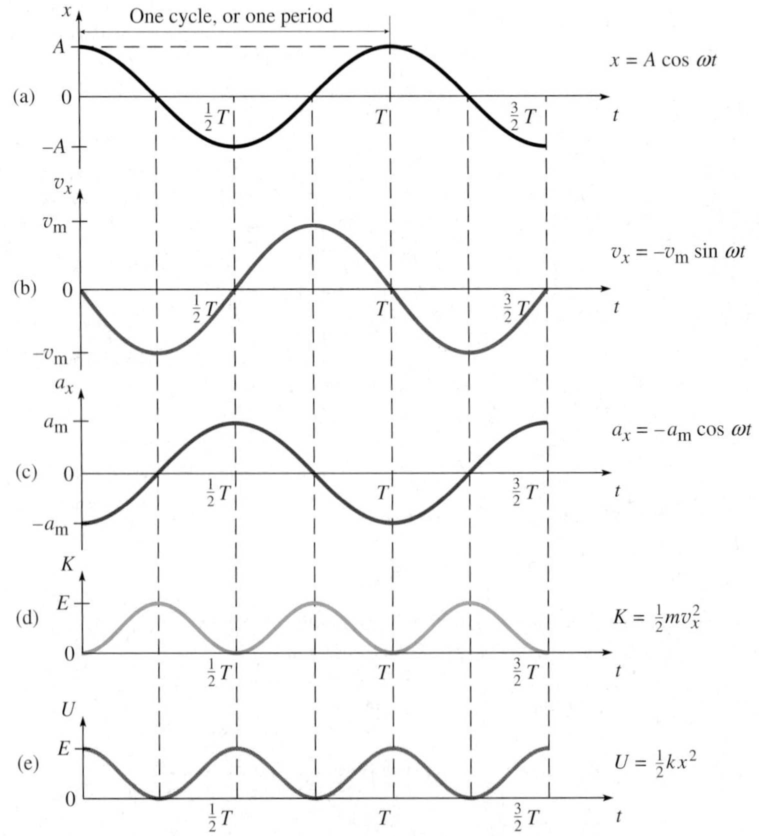

Figure 10.20 Graphs of (a) position, (b) velocity, and (c) acceleration as functions of time for a particle in simple harmonic motion. Observe the interrelationships between the three graphs. The velocity graph is one-quarter cycle ahead of the position graph; that is, $v_x(t)$ reaches its positive maximum one-quarter period before $x(t)$ reaches its positive maximum. Likewise, the acceleration is one-quarter cycle ahead of the velocity and one-half cycle ahead of the position. (d) Kinetic energy as a function of time. (e) Potential energy as a function of time.

conservation: the velocity is maximum at the equilibrium point. Note also that the velocity is zero when the displacement is a maximum ($+A$ or $-A$). Figure 10.20b shows a graph of $v_x(t)$. The equation describing this graph is (see Problem 54):

$$v_x(t) = -v_m \sin \omega t = -\omega A \sin \omega t \qquad (10\text{-}24)$$

The acceleration is the slope of the $v_x(t)$ graph. Figure 10.20c is a graph of $a_x(t)$, which is described by the equation

$$a_x(t) = -a_m \cos \omega t = -\omega^2 A \cos \omega t \qquad (10\text{-}18)$$

Figures 10.20d,e show the kinetic and potential energies as functions of time, respectively. The total energy $E = K + U = \frac{1}{2}kA^2$ is constant.

We have written the position as a function of time in terms of the cosine function, but we can just as correctly use the sine function. The difference between the two is the initial position at time $t = 0$. If the position is at a maximum ($x = A$) at $t = 0$, $x(t)$ is a cosine function. If the position is at the equilibrium point ($x = 0$) at $t = 0$, $x(t)$ is a sine function. By analyzing the slopes of the graphs, you can show (Problem 50) that if the position as a function of time is

$$x(t) = A \sin \omega t \qquad (10\text{-}25a)$$

then the velocity and acceleration are

$$v_x(t) = v_m \cos \omega t \qquad (10\text{-}25b)$$

$$a_x(t) = -a_m \sin \omega t \qquad (10\text{-}25c)$$

CHECKPOINT 10.7

(a) When the displacement of an object in SHM is zero, what is its speed?
(b) When the speed is zero, what is the displacement?

Example 10.8

A Vibrating Loudspeaker Cone

A loudspeaker has a movable diaphragm (the *cone*) that vibrates back and forth to produce sound waves. The displacement of a loudspeaker cone playing a sinusoidal test tone is graphed in Fig. 10.21. Find (a) the amplitude of the motion, (b) the period of the motion, and (c) the frequency of the motion. (d) Write equations for $x(t)$ and $v_x(t)$.

Strategy The amplitude and period can be read directly from the graph. The frequency is the inverse of the period.

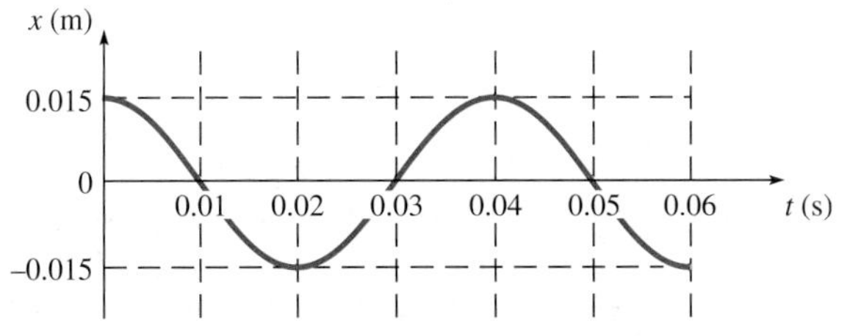

Figure 10.21
Horizontal displacement of a vibrating cone as a function of time.

Since $x(t)$ begins at the maximum displacement, it is described by a cosine function. By looking at the slope of $x(t)$, we can tell whether the velocity is a positive or negative sine function.

Solution (a) The amplitude is the maximum displacement shown on the graph: $A = 0.015$ m.

(b) The period is the time for one complete cycle. From the graph: $T = 0.040$ s.

(c) The frequency is the inverse of the period.

$$f = \frac{1}{T} = \frac{1}{0.040\text{ s}} = 25\text{ Hz}$$

(d) Since $x = +A$ at $t = 0$, we write $x(t)$ as a cosine function:

$$x(t) = A \cos \omega t$$

where $A = 0.015$ m and

$$\omega = 2\pi f = 160\text{ rad/s}$$

continued on next page

Example 10.8 continued

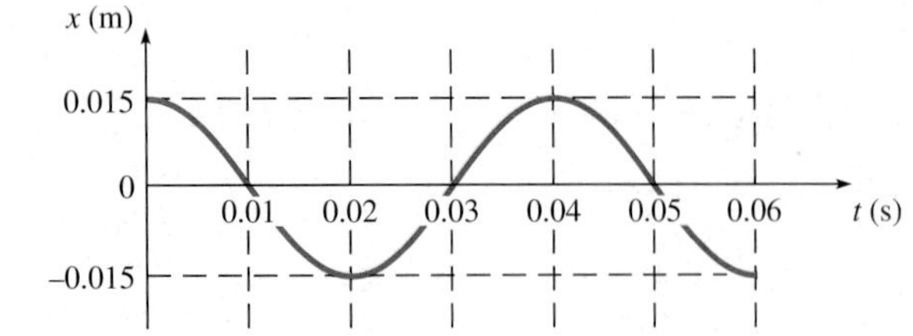

Figure 10.21
Horizontal displacement of a vibrating cone as a function of time.

The slope of $x(t)$ is initially zero and then goes negative. Therefore, $v_x(t)$ is a negative sine function:

$$v_x(t) = -v_m \sin \omega t$$

where $\omega = 2\pi f = 160$ rad/s and

$$v_m = \omega A = 160 \text{ rad/s} \times 0.015 \text{ m} = 2.4 \text{ m/s}$$

Discussion As a check, the velocity should be one-quarter cycle ahead of the position. If we imagine shifting the vertical axis to the right (ahead) by 0.01 s, the graph would have the shape of a negative sine function.

Practice Problem 10.8 Acceleration of the Speaker Cone

Sketch a graph and write an equation for $a_x(t)$.

10.8 THE PENDULUM

Simple Pendulum

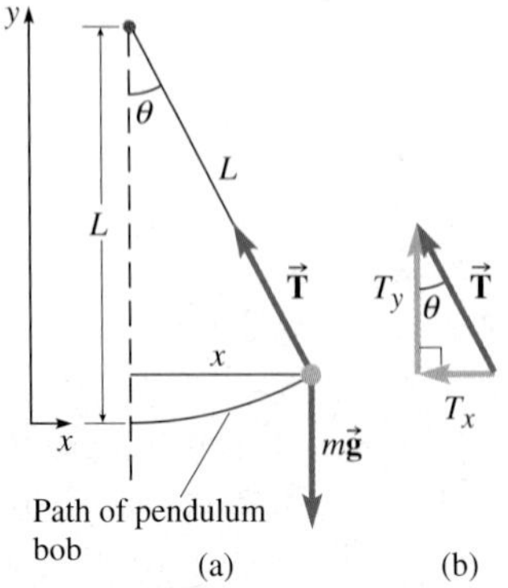

Figure 10.22 (a) Forces on a pendulum bob. (b) Finding the x-component of the force due to the string.

When a pendulum swings back and forth, a string or thin rod constrains the bob to move along a circular arc. However, *for oscillations with small amplitude*, we assume that the bob moves *back and forth along the x-axis*; the vertical motion of the bob is negligible.

Since the weight of the bob has no x-component, the restoring force is the x-component of the force due to the string. We expect the restoring force to be proportional to the displacement for small oscillations. From Fig. 10.22,

$$\sum F_x = -T \sin \theta = -\frac{Tx}{L}$$

where L is the length of the string and $\sin \theta = x/L$. The y-component of the acceleration is negligibly small, so

$$\sum F_y = T \cos \theta - mg = ma_y \approx 0$$

Since $\cos \theta \approx 1$ for small θ, $T \approx mg$. Then

$$\sum F_x \approx -\frac{mgx}{L} = ma_x$$

Solving for a_x:

$$a_x = -\frac{g}{L}x$$

To identify the angular frequency, we recall that $a_x = -\omega^2 x$ [Eq. (10-19)]. Therefore, the angular frequency is

$$\omega = \sqrt{\frac{g}{L}} \quad \text{(10-26a)}$$

and the period is

$$T = \frac{2\pi}{\omega} = 2\pi\sqrt{\frac{L}{g}} \quad \text{(10-26b)}$$

Note that the period depends on L and g but not on the mass of the pendulum. (Text website tutorial: change in period)

Be careful not to confuse the *angular frequency* of the pendulum with its *angular velocity*. Even though the two have the same units (rad/s in SI) and are written with the same symbol (ω), for a pendulum they are *not* the same. When dealing with the pendulum, we use the symbol ω to stand for the *angular frequency* only. The angular frequency $\omega = 2\pi f$ of a given pendulum is constant, while the angular velocity (the rate of change of θ) changes with time between zero (at the extremes) and its maximum magnitude (at the equilibrium point).

PHYSICS AT HOME

The relation between the period and the length of the pendulum is easily tested at home. Make a simple pendulum by tying a thin string to one end of a paper clip and sliding the clip over a coin. Some tape can be used to help hold the coin if it slips out of place. Holding the end of the string, let the coin swing through a small arc and note the time for the coin to make ten complete oscillations, starting from one extreme position and returning to the same position ten times. Divide the time by ten to get the period. (This gives a more accurate value than timing a single period.) Measure the length of the pendulum and test Eq. (10-26b).

Repeat the experiment by holding the string at a position closer to the coin, effectively shortening the length of the pendulum. What do you find? Is the period for the shorter pendulum longer, shorter, or the same as that measured for the longer pendulum?

The effect of a different mass on the period can also be tested by using two or three coins taped together, with the same length pendulum as used for the first measurement. Does a heavier coin affect the result?

Example 10.9

Grandfather Clock

A grandfather clock uses a pendulum with period 2.0 s to keep time. In one such clock, the pendulum bob has mass 150 g; the pendulum is set into oscillation by displacing it 33 mm to one side. (a) What is the length of the pendulum? (b) Does the initial displacement satisfy the small angle approximation?

Strategy The period depends on the length of the pendulum and on the gravitational field strength g. It does not depend on the mass of the bob. It also does not depend on the initial displacement, as long as it is small compared to the length.

Solution (a) Assuming small amplitudes, the period is

$$T = 2\pi\sqrt{\frac{L}{g}}$$

Solving for L,

$$L = \frac{T^2 g}{(2\pi)^2}$$

$$= \frac{(2.0\text{ s})^2 \times 9.80\text{ m/s}^2}{(2\pi)^2} = 0.99\text{ m}$$

(b) The small angle approximation is valid if the maximum displacement is small compared to the length of the pendulum.

$$\frac{x}{L} = \frac{33\text{ mm}}{990\text{ mm}} = 0.033$$

Is that small enough? If $\sin\theta = x/L = 0.033$, then

$$\theta = \sin^{-1} 0.033 = 0.033006$$

Sin θ and θ differ by less than 0.02%. Since we only know T to two significant figures, the approximation is good.

Discussion We should check that we didn't write the expression for the period "upside down," which is the most likely error we could make. Besides checking that the units work out, we know that a longer pendulum has a longer period, so L must go in the numerator. On the other hand, if g were larger, the restoring force would be larger and we would expect the period to shorten; thus, g belongs in the denominator.

Practice Problem 10.9 Pendulum on the Moon

A pendulum of length 0.99 m is taken to the Moon by an astronaut. The period of the pendulum is 4.9 s. What is the gravitational field strength on the surface of the Moon?

Large-Amplitude Motion of a Pendulum Is *Not* SHM The period of a pendulum as just determined is valid only for small amplitudes. For larger amplitudes, the pendulum's motion is still periodic (though not SHM). Why would the period be any different for large amplitudes? Remember that we assumed the bob was moving horizontally back and forth along the x-axis. This simplification breaks down for large amplitudes.

For instance, if we pull the pendulum out horizontally ($\theta = 90°$), the tangential component of the weight is mg, but $F_x = 0$! Since we have overestimated F_x, we have underestimated the time for the bob to return to $x = 0$; thus, the period for large amplitudes is greater than $2\pi\sqrt{L/g}$. Another way of looking at it is in terms of the tangential force. The expression for the tangential component of the weight is correct even for large amplitudes. However, the distance the bob must move to return to equilibrium is larger than x. For instance, starting at $\theta = 90°$, the bob must move one quarter of the circumference, a distance $\frac{1}{4}(2\pi L) \approx 1.6L$, to return to equilibrium. Assuming linear motion along the x-axis would make the distance only L. With a longer distance to travel, the time is longer.

Physical Pendulum

Imagine that you have a simple pendulum of length L. Beside it you have a uniform metal bar of the same length, which is free to swing about an axis at one end. Would the two have the same period if they are set into oscillation?

For the simple pendulum, the bob is assumed to be a point mass; all the mass of the pendulum is at a distance L from the rotation axis. For the metal bar, however, the mass is uniformly distributed from the axis to a *maximum* distance L away from the axis. The center of mass is located at the midpoint, a distance $d = \frac{1}{2}L$ from the axis (Fig. 10.23). Since the mass is on average closer to the axis, the period is shorter than that of the simple pendulum.

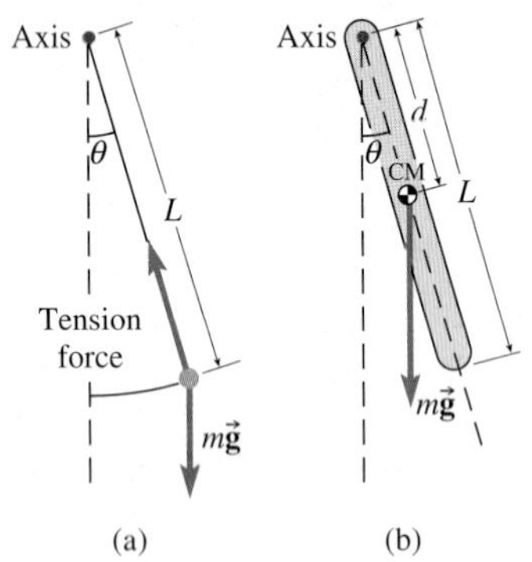

Figure 10.23 (a) A simple pendulum and (b) a physical pendulum.

Would this bar have a period equal to that of a simple pendulum of length $\frac{1}{2}L$? That is a good guess, since the center of mass of the bar is a distance $\frac{1}{2}L$ away from the rotation axis. Unfortunately, it isn't quite that easy. The gravitational force acts at the center of mass, but we *cannot* think of all the mass as being concentrated at that point—that would give the wrong rotational inertia. When set into oscillation, the bar, or any other rigid object free to rotate about a fixed axis, is called a **physical pendulum**. The period of a physical pendulum is

$$T = 2\pi\sqrt{\frac{I}{mgd}} \tag{10-27}$$

where d is the distance from the rotation axis to the CM of the object and I is the rotational inertia about that axis. [See text website for a derivation of Eq. (10-27).]

For a uniform bar of length L, the CM is halfway down the bar:

$$d = \tfrac{1}{2}L$$

From Table 8.1, the rotational inertia of a uniform bar rotating about an axis through an endpoint is $I = \frac{1}{3}mL^2$. The period of oscillation is

$$T = 2\pi\sqrt{\frac{I}{mgd}} = 2\pi\sqrt{\frac{\frac{1}{3}mL^2}{(mg)\frac{1}{2}L}} = 2\pi\sqrt{\frac{2L}{3g}}$$

The bar has the same period as a simple pendulum of length $\frac{2}{3}L$.

Example 10.10

Comparison of Walking Frequencies and Speeds for Various Creatures

During a relaxed walking pace, an animal's leg can be thought of as a physical pendulum of length L that pivots about the hip. (a) What is the relaxed walking frequency for a cat ($L = 30$ cm), dog (60 cm), human (1 m), giraffe (2 m), and a mythological titan (10 m)? (b) Derive an equation that gives the walking speed (amount of ground covered per unit time) for a given walking frequency f. [*Hint:* Start by drawing a picture of the leg position at the start of the swing (leg back) and the end of the swing (leg forward) and assume a comfortable angle of about 30° between these two positions. To how many steps does a complete period of the pendulum correspond?] (c) Find the walking speed for each of the animals listed in part (a).

continued on next page

Example 10.10 continued

Strategy We have to use an idealized model of the leg, since we don't know the exact location of the center of mass or the rotational inertia. The simple pendulum is not a good model, since it would assume all the mass of the leg at the foot! A much better model is to think of the leg as a uniform cylinder pivoting about one end.

Solution (a) For a uniform cylinder, the center of mass is a distance $d = \frac{1}{2}L$ from the pivot and the rotational inertia about an axis at one end is $I = \frac{1}{3}mL^2$. Then the period is

$$T = 2\pi\sqrt{\frac{I}{mgd}} = 2\pi\sqrt{\frac{\frac{1}{3}mL^2}{(mg)\frac{1}{2}L}} = 2\pi\sqrt{\frac{2L}{3g}}$$

and the frequency f is

$$f = \frac{1}{T} = \frac{1}{2\pi}\sqrt{\frac{3g}{2L}} \approx 0.2\sqrt{\frac{g}{L}}$$

Substituting the numerical values of L for each animal, we find the frequencies to be 1 Hz (cat), 0.8 Hz (dog), 0.6 Hz (human), 0.4 Hz (giraffe), and 0.2 Hz (titan).

(b) One period of the "pendulum" corresponds to two steps. In Fig. 10.24a, the right leg is about to step forward. The step occurs as the pendulum swings forward through half a cycle. In Fig. 10.24b, the right foot is about to touch the ground; in Fig. 10.24c, the right foot touches the ground and now the left leg is about to step forward. During this step, the right foot stays in place on the ground, but the right leg is swinging backward relative to the hip joint. During each step, the distance covered is approximately the length of a 30° arc of radius L, which is one twelfth the circumference of a circle of radius L. So during one period, the distance walked is

$$D = 2 \times \frac{1}{12} \times 2\pi L = \frac{\pi}{3}L \approx L$$

and the walking speed is

$$v = \frac{D}{T} = Lf = 0.2\sqrt{gL}$$

(c) The speeds are 0.3 m/s (cat), 0.5 m/s (dog), 0.6 m/s (human), 0.9 m/s (giraffe), and 2 m/s (titan).

Discussion You may be more familiar with walking speeds in mi/h. Converting the units, 0.6 m/s ≈ 1.3 mi/h, which is just about right for a leisurely walk. A brisk walk is about 3 mi/h for most people; to go much faster than that, you need to jog or run.

The solution says that longer legs walk faster, but the frequency of the steps is lower. You can verify that by walking beside a friend who is much taller or much shorter than you, or by taking your dog for a walk.

Practice Problem 10.10 Walking Speed for a Human

A more realistic model of a human leg of length 1.0 m has the center of mass 0.45 m from the hip and a rotational inertia of $\frac{1}{6}mL^2$. What is the walking speed predicted by this model?

(a) (b) (c)

Figure 10.24 The forward motion of a leg during walking is similar to the swing of a physical pendulum. From (a) to (b), the right leg swings forward like a pendulum. In (c), the right foot is on the ground and the left leg is about to swing forward.

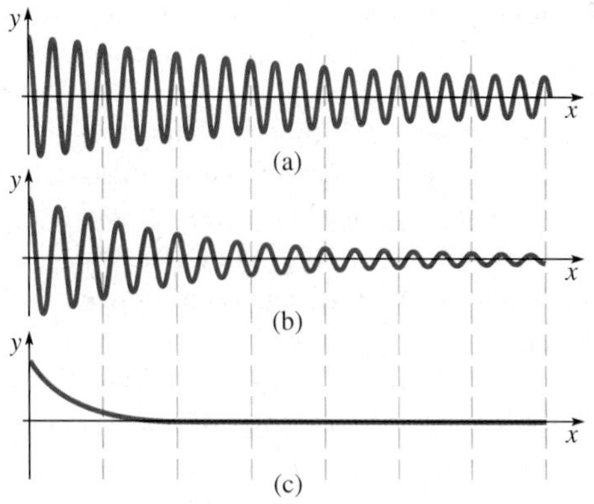

Figure 10.25 Graphs of $x(t)$ for a mass-spring system with increasing amounts of damping. In (c) the damping is sufficient to prevent oscillations from occurring.

10.9 DAMPED OSCILLATIONS

Application of damped oscillation: shock absorbers in a car

In SHM, we assume that no dissipative forces such as friction or viscous drag exist. Since the mechanical energy is constant, the oscillations continue forever with constant amplitude. SHM is a simplified model. The oscillations of a swinging pendulum or a vibrating tuning fork gradually die out as energy is dissipated. The amplitude of each cycle is a little smaller than that of the previous cycle (Fig. 10.25a). This kind of motion is called **damped oscillation**, where the word *damped* is used in the sense of *extinguished* or *restrained.* For a small amount of damping, oscillations occur at approximately the same frequency as if there were no damping. A greater degree of damping lowers the frequency slightly (Fig. 10.25b). Even more damping prevents oscillations from occurring at all (Fig. 10.25c).

Damping is not always a disadvantage. The suspension system of a car includes shock absorbers that cause the vibration of the body—a mass connected to the chassis by springs—to be quickly damped. The shock absorbers reduce the discomfort that passengers would otherwise experience due to the bouncing of an automobile as it travels along a bumpy road. Figure 10.26 shows how a shock absorber works. In order to compress or expand the shock absorber, a viscous oil must flow through the holes in the piston. The viscous force dissipates energy regardless of which direction the piston moves. The shock absorber enables the spring to smoothly return to its equilibrium length without oscillating up and down (Fig. 10.25c). When the oil leaks out of the shock absorber, the damping is insufficient to prevent oscillations. After hitting a bump, the body of the car oscillates up and down (Fig. 10.25b).

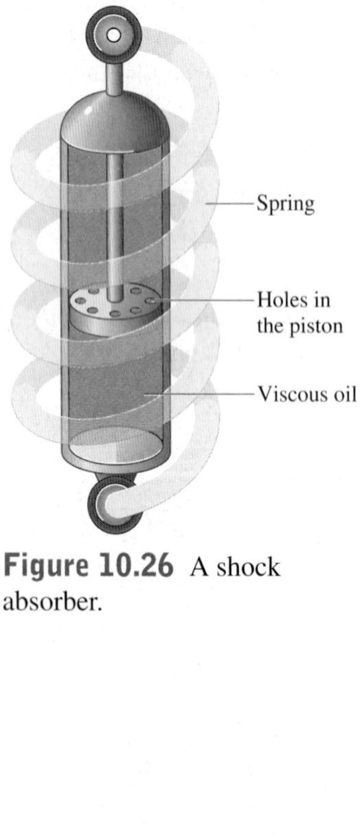

Figure 10.26 A shock absorber.

10.10 FORCED OSCILLATIONS AND RESONANCE

When damping forces are present, the only way to keep the amplitude of oscillations from diminishing is to replace the dissipated energy from some other source. When a child is being pushed on a swing, the parent replaces the energy dissipated with a small push. In order to keep the amplitude of the motion constant, the parent gives a little push once per cycle, adding just enough energy each time to compensate for the energy dissipated in one cycle. The frequency of the *driving force* (the parent's push) matches the *natural frequency* of the system (the frequency at which it would oscillate on its own).

Forced oscillations (or driven oscillations) occur when a periodic external driving force acts on a system that can oscillate. The frequency of the driving force does not have to match the natural frequency of the system. Ultimately, the system oscillates at the driving frequency, even if it is far from the natural frequency. However, the amplitude of the oscillations is generally quite small unless the driving frequency f is close to the natural frequency f_0 (Fig. 10.27). When the driving frequency is equal to the natural frequency of the system, the amplitude of the motion is a maximum. This condition is called **resonance**.

At resonance, the driving force is always in the same direction as the object's velocity. Since the driving force is always doing positive work, the energy of the oscillator builds up until the energy dissipated balances the energy added by the driving force. For an oscillator with little damping, this requires a large amplitude. When the driving and natural frequencies differ, the driving force and velocity are no longer synchronized; sometimes they are in the same direction and sometimes in opposite directions. The driving force is not at resonance, so it sometimes does negative work. The net work done by the driving force decreases as the driving frequency moves away from resonance. Therefore, the oscillator's energy and amplitude are smaller than at resonance.

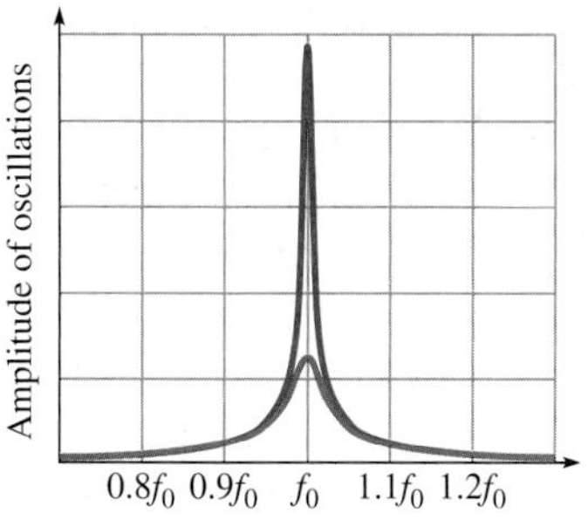

Figure 10.27 Two resonance curves for an oscillator with natural frequency f_0. The amplitude of the driving force is constant. In the red graph, the oscillator has one fourth as much damping as in the blue graph.

Applications of Resonance Large-amplitude vibrations due to resonance can be dangerous in some situations. Materials can be stressed past their elastic limits, causing permanent deformation or breaking. In 1940, the wind set the Tacoma Narrows Bridge in Washington state into vibration with increasing amplitude. Turbulence in the air as it flowed across the bridge caused the air pressure to fluctuate with a frequency matching one of the bridge's resonant frequencies. As the amplitude of the oscillations grew, the bridge was closed; soon after, the bridge collapsed (Fig. 10.28). Engineers now design bridges with much higher resonant frequencies so the wind cannot cause resonant vibrations.

Applications of resonance: vibration of bridges and buildings

In the nineteenth century, bridges were sometimes set into resonant vibration when the cadence of marching soldiers matched a resonant frequency of the bridge. After the collapse of several bridges due to resonance, soldiers were told to break step when crossing a bridge to eliminate the danger of their cadence setting the bridge into resonance.

Tall buildings sway back and forth at a particular resonant frequency determined by the structure. The vibration pattern is similar to what you see if you hold one end of a ruler to the edge of a desk and then pluck the other end. Engineers have many methods to reduce the amplitude of the swaying. One of the simplest and most widely used is the tuned mass damper (TMD). Building engineers attach a damped mass-spring system to the structure at a point where its vibration amplitude is largest—near the top. In the Hancock Tower, each of the 300 000-kg boxes is attached to the building frame with springs and shock absorbers and can slide back and forth, riding on a thin layer of oil that covers a 9-m-long steel plate. The resonant frequency of the TMD is matched to the resonant frequency of the swaying building. When the swaying of the building drives the TMD into oscillation, energy is dissipated in the shock absorbers. The TMD in the Hancock Tower reduces the amplitude of its swaying by about 50%.

How is the swaying of a tall building reduced?

(a) (b)

Figure 10.28 (a) The Tacoma Narrows Bridge begins to vibrate. (b) Ultimately the vibrations caused the bridge to collapse.

Master the Concepts

- A deformation is a change in the size or shape of an object.
- When deforming forces are removed, an *elastic* object returns to its original shape and size.
- Hooke's law, in a generalized form, says that the deformation of a material (measured by the strain) is proportional to the magnitude of the forces causing the deformation (measured by the stress). The definitions of stress and strain are as given in the following table.

Type of Deformation

	Tensile or Compressive	**Shear**	**Volume**
Stress	Force per unit cross-sectional area F/A	Shear force divided by the parallel area of the surface on which it acts F/A	Pressure P
Strain	Fractional length change $\Delta L/L$	Ratio of the relative displacement Δx to the separation L of the two parallel surfaces $\Delta x/L$	Fractional volume change $\Delta V/V$
Constant of proportionality	Young's modulus Y	Shear modulus S	Bulk modulus B

- If the tensile or compressive stress exceeds the *proportional limit*, the strain is no longer proportional to the stress. The solid still returns to its original length when the stress is removed as long as the stress does not exceed the *elastic limit*. If the stress exceeds the elastic limit, the material is permanently deformed. For larger stresses yet, the solid fractures when the stress reaches the *breaking point*. The maximum stress that can be withstood without breaking is called the *ultimate strength*.
- Vibrations occur in the vicinity of a point of stable equilibrium. An equilibrium point is *stable* if the net force on an object when it is displaced from equilibrium points back toward the equilibrium point. Such a force is called a restoring force since it tends to restore equilibrium.
- Simple harmonic motion is periodic motion that occurs whenever the restoring force is proportional to the displacement from equilibrium. In SHM, the position, velocity, and acceleration as functions of time are sinusoidal (i.e., sine or cosine functions). Any oscillatory motion is approximately SHM if the amplitude is small, because for small oscillations the restoring force is approximately linear.
- The period T is the time taken by one complete cycle of oscillation. The frequency f is the number of cycles per unit time:

$$f = \frac{1}{T} \quad (5\text{-}8)$$

The angular frequency is measured in radians per unit time:

$$\omega = 2\pi f \quad (5\text{-}9)$$

- The maximum velocity and acceleration in SHM are

$$v_m = \omega A \quad \text{and} \quad a_m = \omega^2 A \quad (10\text{-}21, 10\text{-}22)$$

where ω is the angular frequency. The acceleration is proportional to and in the opposite direction from the displacement:

$$a_x(t) = -\omega^2 x(t) \quad (10\text{-}19)$$

- The equations that describe SHM are

If $x = A$ at $t = 0$,	If $x = 0$ at $t = 0$,
$x = A \cos \omega t$	$x = A \sin \omega t$
$v_x = -v_m \sin \omega t$	$v_x = v_m \cos \omega t$
$a_x = -a_m \cos \omega t$	$a_x = -a_m \sin \omega t$

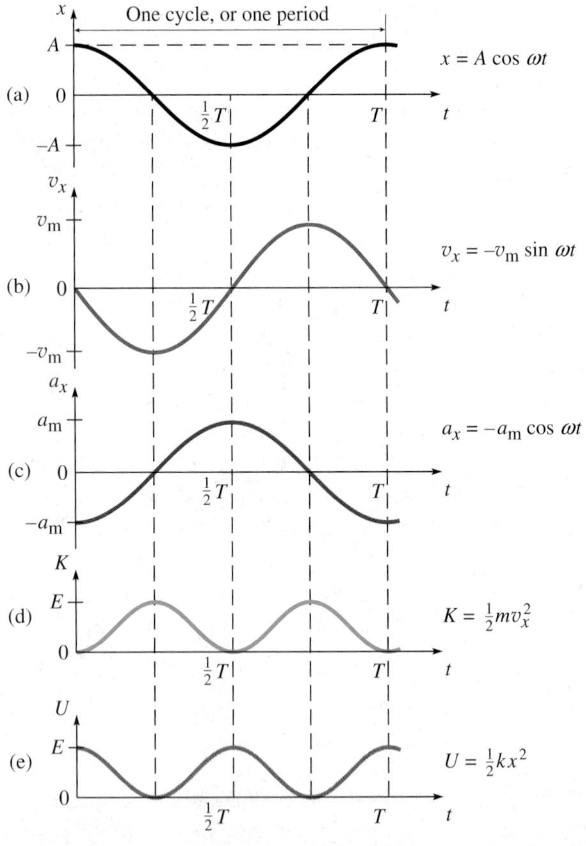

In either case, the velocity is one-quarter cycle ahead of the position and the acceleration is one-quarter cycle ahead of the velocity.

- The period of oscillation for a mass-spring system is

$$T = 2\pi\sqrt{\frac{m}{k}} \quad (10\text{-}20c)$$

continued on next page

Master the Concepts continued

For a simple pendulum it is

$$T = 2\pi\sqrt{\frac{L}{g}} \qquad (10\text{-}26b)$$

and for a physical pendulum it is

$$T = 2\pi\sqrt{\frac{I}{mgd}} \qquad (10\text{-}27)$$

- In the absence of dissipative forces, the total mechanical energy of a simple harmonic oscillator is constant and proportional to the square of the amplitude:

$$E = \tfrac{1}{2}kA^2 \qquad (10\text{-}13)$$

where the potential energy has been chosen to be zero at the equilibrium point. At any point, the sum of the kinetic and potential energies is constant:

$$E = \tfrac{1}{2}mv_x^2 + \tfrac{1}{2}kx^2 = \tfrac{1}{2}kA^2 \qquad (10\text{-}12)$$

Conceptual Questions

1. Young's modulus for diamond is about 20 times as large as that of glass. Does that tell you which is stronger? If not, what does it tell you?
2. A grandfather clock is running too fast. To fix it, should the pendulum be lengthened or shortened? Explain.
3. A karate student hits downward on a stack of concrete blocks supported at both ends. A block breaks. Explain where it starts to break first, at the bottom or at the top. (The block experiences shear, compressive, and tensile stresses. Recall that concrete has much less tensile strength than compressive strength. Which part of the block is stretched and which is compressed when the block bends in the middle?)

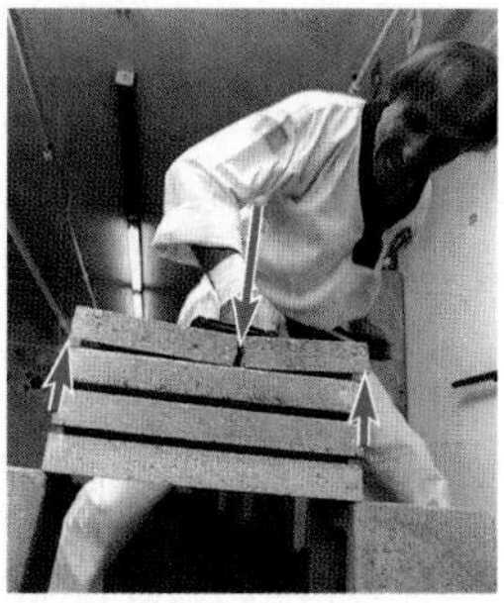

4. A cylindrical steel bar is compressed by the application of forces of magnitude F at each end. What magnitude forces would be required to compress by the same amount (a) a steel bar of the same cross-sectional area but one half the length? (b) a steel bar of the same length but one half the radius?
5. The columns built by the ancient Greeks and Romans to support temples and other structures are tapered; they are thicker at the bottom than at the top. This certainly has an aesthetic purpose, but is there an engineering purpose as well? What might it be?

6. Explain how the period of a mass-spring system can be independent of amplitude, even though the distance traveled during each cycle is proportional to the amplitude.
7. In a reciprocating saw, a *Scotch yoke* converts the rotation of the motor into the back-and-forth motion of the blade. The Scotch yoke is a mechanical device used to convert oscillatory motion to circular motion or *vice versa.* A wheel with a fixed knob rotates at constant angular velocity; the knob is constrained within a vertical slot causing the saw blade to move left and right without moving up and down. Is the motion of the saw blade SHM? Explain.

8. A mass hanging vertically from a spring and a simple pendulum both have a period of oscillation of 1 s on Earth. An astronaut takes the two devices to another planet where the gravitational field is stronger than that of Earth. For each of the two systems, state whether the period is now longer than 1 s, shorter than 1 s, or equal to 1 s. Explain your reasoning.
9. A bungee jumper leaps from a bridge and comes to a stop a few centimeters above the surface of the water below. At that lowest point, is the tension in the bungee cord equal to the jumper's weight? Explain why or why not.
10. Does it take more force to break a longer rope or a shorter rope? Assume the ropes are identical except for their lengths and are ideal—there are no weak points. Does it take more *energy* to break the long rope or the short rope? Explain.
11. A pilot is performing vertical loop-the-loops over the ocean at noon. The plane speeds up as it approaches the bottom of the circular loop and slows as it approaches the top of the loop. An observer in a helicopter is watching the shadow of the plane on the surface of the water. Does the shadow exhibit SHM? Explain.
12. Are you more likely to find steel rods in a horizontal concrete beam or in a vertical concrete column? Is concrete more in need of reinforcement under tensile or compressive stress?
13. Suppose that it takes tensile forces of magnitude F to produce a given strain $\Delta L/L$ in a steel wire of cross-sectional area A. If you had two such wires side by side and stretched them simultaneously, what magnitude tensile forces would be required to produce the same strain? By thinking of a thick wire as two (or more) thinner

wires side by side, explain why the force to produce a given strain must be proportional to the cross-sectional area. Thus, the strain depends on the stress—the force per unit area.

14. Think of a crystalline solid as a set of atoms connected by ideal springs. When a wire is stretched, how is the elongation of the wire related to the elongation of each of the interatomic springs? Use your answer to explain why a given tensile stress produces an elongation of the wire proportional to the wire's initial length—or, equivalently, that a given stress produces the same strain in wires of different lengths.

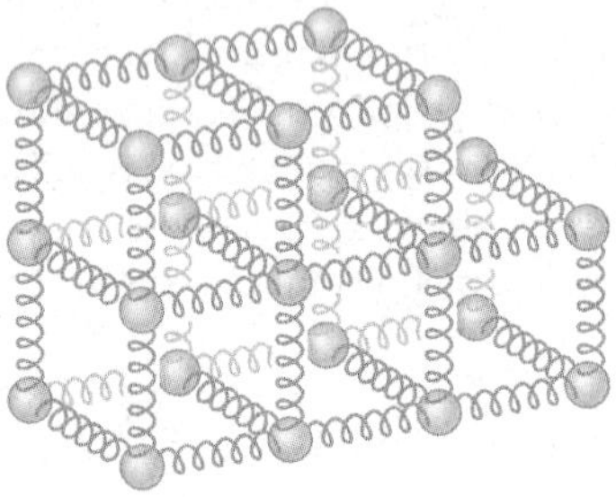

15. What are the advantages of using the concepts of stress and strain to describe deformations?
16. An old highway is built out of concrete blocks of equal length. A car traveling on this highway feels a little bump at the joint between blocks. The passengers in the car feel that the ride is uncomfortable at a speed of 45 mi/h, but much smoother at speeds either lower or higher than that. Explain.
17. The period of oscillation of a simple pendulum does not depend on the mass of the bob. By contrast, the period of a mass-spring system does depend on mass. Explain the apparent contradiction. [*Hint:* What provides the restoring force in each case? How does the restoring force depend on mass?]
18. A mass connected to an ideal spring is oscillating without friction on a horizontal surface. Sketch graphs of the kinetic energy, potential energy, and total energy as functions of time for one complete cycle.

Multiple-Choice Questions

Questions 1–4. A body is suspended vertically from an ideal spring. The spring is initially in its relaxed position. The body is then released and oscillates about the equilibrium position. Answer choices for Questions 1–4:

(a) The spring is relaxed.
(b) The body is at the equilibrium point.
(c) The spring is at its maximum extension.
(d) The spring is somewhere between the equilibrium point and maximum extension.

1. The acceleration is greatest in magnitude and is directed upward when:
2. The speed of the body is greatest when:
3. The acceleration of the body is zero when:
4. The acceleration is greatest in magnitude and is directed downward when:
5. Two simple pendulums, A and B, have the same length, but the mass of A is twice the mass of B. Their vibrational amplitudes are equal. Their periods are T_A and T_B, respectively, and their energies are E_A and E_B. Choose the correct statement.

   (a) $T_A = T_B$ and $E_A > E_B$ (b) $T_A > T_B$ and $E_A > E_B$
   (c) $T_A > T_B$ and $E_A < E_B$ (d) $T_A = T_B$ and $E_A < E_B$
6. A force F applied to each end of a steel wire (length L, diameter d) stretches it by 1.0 mm. How much does F stretch another steel wire, of length $2L$ and diameter $2d$?

   (a) 0.50 mm (b) 1.0 mm (c) 2.0 mm
   (d) 4.0 mm (e) 0.25 mm
7. A stiff material is characterized by
   (a) high ultimate strength.
   (b) high breaking strength.
   (c) high Young's modulus.
   (d) high proportional limit.
8. A brittle material is characterized by
   (a) high breaking strength and low Young's modulus.
   (b) low breaking strength and high Young's modulus.
   (c) high breaking strength and high Young's modulus.
   (d) low breaking strength and low Young's modulus.
9. Which pair of quantities can be expressed in the same units?

   (a) stress and strain
   (b) Young's modulus and strain
   (c) Young's modulus and stress
   (d) ultimate strength and strain
10. Two wires have the same diameter and length. One is made of copper, the other brass. The wires are connected together end to end. When the free ends are pulled in opposite directions, the two wires *must* have the same

    (a) stress. (b) strain. (c) ultimate strength.
    (d) elongation. (e) Young's modulus.

Questions 11–20. See the graph of $v_x(t)$ for an object in SHM. Answer choices for each question:

(a) 1 s, 2 s, 3 s (b) 5 s, 6 s, 7 s (c) 0 s, 1 s, 7 s, 8 s
(d) 3 s, 4 s, 5 s (e) 0 s, 4 s, 8 s (f) 2 s, 6 s
(g) 3 s, 5 s (h) 1 s, 3 s (i) 5 s, 7 s
(j) 3 s, 7 s (k) 1 s, 5 s

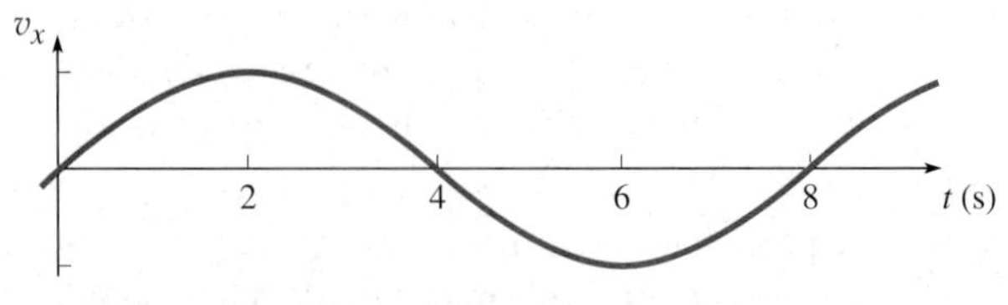

Multiple-Choice Questions 11–20

11. When is the kinetic energy maximum?
12. When is the kinetic energy zero?
13. When is the potential energy maximum?
14. When is the potential energy minimum?
15. When is the object at the equilibrium point?
16. When does the acceleration have its maximum magnitude?
17. Which answer specifies times when the net force is in the $+x$-direction?
18. Which answer specifies times when the object is on the $-x$-side of the equilibrium point ($x < 0$)?
19. Which answer specifies times when the object is moving away from the equilibrium point?
20. Which answer specifies times when the potential energy is decreasing?

Problems

- Combination conceptual/quantitative problem
- Biological or medical application
- Challenging problem
- Blue # Detailed solution in the Student Solutions Manual
- (1 2) Problems paired by concept
- Text website interactive or tutorial

10.2 Hooke's Law for Tensile and Compressive Forces

1. A steel beam is placed vertically in the basement of a building to keep the floor above from sagging. The load on the beam is 5.8×10^4 N, the length of the beam is 2.5 m, and the cross-sectional area of the beam is 7.5×10^{-3} m^2. Find the vertical compression of the beam.
2. A 91-kg man's thighbone has a relaxed length of 0.50 m, a cross-sectional area of 7.0×10^{-4} m^2, and a Young's modulus of 1.1×10^{10} N/m^2. By how much does the thighbone compress when the man is standing on both feet?
3. A brass wire with Young's modulus of 9.2×10^{10} Pa is 2.0 m long and has a cross-sectional area of 5.0 mm^2. If a weight of 5.0 kN is hung from the wire, by how much does it stretch?
4. A wire of length 5.00 m with a cross-sectional area of 0.100 cm^2 stretches by 6.50 mm when a load of 1.00 kN is hung from it. What is the Young's modulus for this wire?
5. Two steel wires (of the same length and different radii) are connected together, end to end, and tied to a wall. An applied force stretches the combination by 1.0 mm. How far does the *midpoint* move?

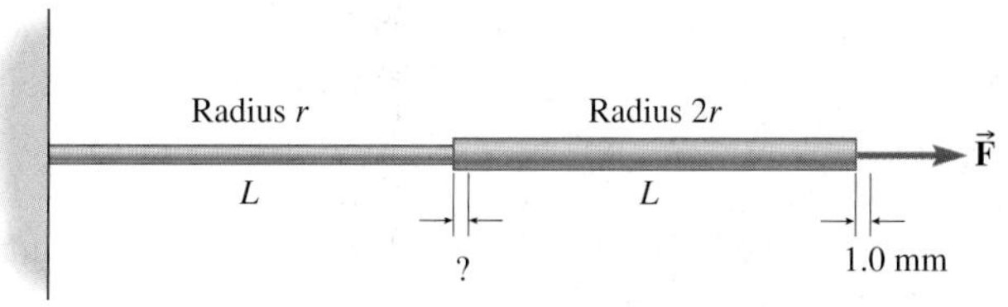

6. Abductin is an elastic protein found in scallops, with a Young's modulus of 4.0×10^6 N/m^2. It is used as an inner hinge ligament, with a cross-sectional area of 0.78 mm^2 and a relaxed length of 1.0 mm. When the muscles in the shell relax, the shell opens. This increases efficiency as the muscles do not need to exert any force to open the shell, only to close it. If the muscles must exert a force of 1.5 N to keep the shell closed, by how much is the abductin ligament compressed?
7. A 0.50-m-long guitar string, of cross-sectional area 1.0×10^{-6} m^2, has Young's modulus $Y = 2.0 \times 10^9$ N/m^2. By how much must you stretch the string to obtain a tension of 20 N?
8. It takes a flea 1.0×10^{-3} s to reach a peak speed of 0.74 m/s. (a) If the mass of the flea is 0.45×10^{-6} kg, what is the average power required? (b) Insect muscle has a maximum output of 60 W/kg. If 20% of the flea's weight is muscle, can the muscle provide the power needed? (c) The flea has a resilin pad at the base of the hind leg that compresses when the flea bends its leg to jump. If we assume the pad is a cube with a side of 6.0×10^{-5} m, and the pad compresses fully, what is the energy stored in the compression of the pads of the two hind legs? The Young's modulus for resilin is 1.7×10^6 N/m^2. (d) Does this provide enough power for the jump?

10.3 Beyond Hooke's Law

9. Using the stress-strain graph for bone (Fig. 10.4c), calculate Young's moduli for tension and for compression. Consider only small stresses.
10. An acrobat of mass 55 kg is going to hang by her teeth from a steel wire and she does not want the wire to stretch beyond its elastic limit. The elastic limit for the wire is 2.5×10^8 Pa. What is the minimum diameter the wire should have to support her?
11. A hair breaks under a tension of 1.2 N. What is the diameter of the hair? The tensile strength is 2.0×10^8 Pa.
12. The ratio of the tensile (or compressive) strength to the density of a material is a measure of how strong the material is "pound for pound." (a) Compare tendon (tensile strength 80.0 MPa, density 1100 kg/m^3) with steel (tensile strength 0.50 GPa, density 7700 kg/m^3): which is stronger "pound for pound" under tension? (b) Compare bone (compressive strength 160 MPa, density 1600 kg/m^3) with concrete (compressive strength 0.40 GPa, density 2700 kg/m^3): which is stronger "pound for pound" under compression?
13. What is the maximum load that could be suspended from a copper wire of length 1.0 m and radius 1.0 mm without permanently deforming the wire? Copper has an elastic limit of 2.0×10^8 Pa and a tensile strength of 4.0×10^8 Pa.
14. What is the maximum load that could be suspended from a copper wire of length 1.0 m and radius 1.0 mm

without breaking the wire? Copper has an elastic limit of 2.0×10^8 Pa and a tensile strength of 4.0×10^8 Pa.

15. The leg bone (femur) breaks under a compressive force of about 5×10^4 N for a human and 10×10^4 N for a horse. The human femur has a compressive strength of 1.6×10^8 Pa, while the horse femur has a compressive strength of 1.4×10^8 Pa. What is the effective cross-sectional area of the femur in a human and in a horse? (*Note:* Since the center of the femur contains bone marrow, which has essentially no compressive strength, the effective cross-sectional area is about 80% of the total cross-sectional area.)

16. The maximum strain of a steel wire with Young's modulus 2.0×10^{11} N/m^2, just before breaking, is 0.20%. What is the stress at its breaking point, assuming that strain is proportional to stress up to the breaking point?

17. A marble column with a cross-sectional area of 25 cm^2 supports a load of 7.0×10^4 N. The marble has a Young's modulus of 6.0×10^{10} Pa and a compressive strength of 2.0×10^8 Pa. (a) What is the stress in the column? (b) What is the strain in the column? (c) If the column is 2.0 m high, how much is its length changed by supporting the load? (d) What is the maximum weight the column can support?

18. A copper wire of length 3.0 m is observed to stretch by 2.1 mm when a weight of 120 N is hung from one end. (a) What is the diameter of the wire and what is the tensile stress in the wire? (b) If the tensile strength of copper is 4.0×10^8 N/m^2, what is the maximum weight that may be hung from this wire?

10.4 Shear and Volume Deformations

19. A sphere of copper is subjected to 100 MPa of pressure. The copper has a bulk modulus of 130 GPa. By what fraction does the volume of the sphere change? By what fraction does the radius of the sphere change?

20. By what percentage does the density of water increase at a depth of 1.0 km below the surface?

21. Atmospheric pressure on Venus is about 90 times that on Earth. A steel sphere with a bulk modulus of 160 GPa has a volume of 1.00 cm^3 on Earth. If it were put in a pressure chamber and the pressure were increased to that of Venus (9.12 MPa), how would its volume change?

22. How would the volume of 1.00 cm^3 of aluminum on Earth change if it were placed in a vacuum chamber and the pressure changed to that of the Moon (less than 10^{-9} Pa)?

23. Two steel plates are fastened together using four bolts. The bolts each have a shear modulus of 8.0×10^{10} Pa and a shear strength of 6.0×10^8 Pa. The radius of each bolt is 1.0 cm. Normally, the bolts clamp the two plates together and the frictional forces between the plates keep them from sliding. If the bolts are loose, then the frictional forces are small and the bolts themselves would be subject to a large shear stress. What is the maximum shearing force F on the plates that the four bolts can withstand?

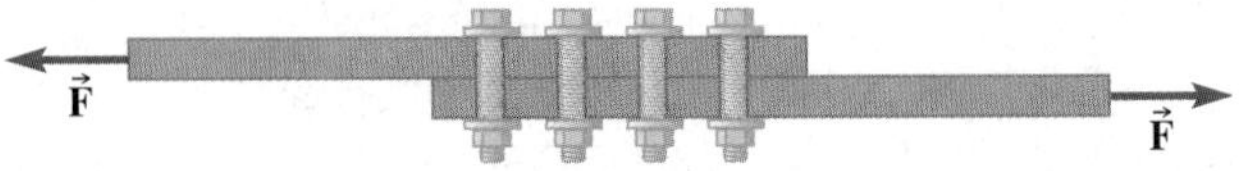

24. An anchor, made of cast iron of bulk modulus 60.0×10^9 Pa and of volume 0.230 m^3, is lowered over the side of the ship to the bottom of the harbor where the pressure is greater than sea level pressure by 1.75×10^6 Pa. Find the change in the volume of the anchor.

25. The upper surface of a cube of gelatin, 5.0 cm on a side, is displaced 0.64 cm by a tangential force. If the shear modulus of the gelatin is 940 Pa, what is the magnitude of the tangential force?

26. A large sponge has forces of magnitude 12 N applied in opposite directions to two opposite faces of area 42 cm^2 (see Fig. 10.7 for a similar situation). The thickness of the sponge (L) is 2.0 cm. The deformation angle (γ) is 8.0°. (a) What is Δx? (b) What is the shear modulus of the sponge?

10.5 Simple Harmonic Motion; 10.6 The Period and Frequency for SHM

27. The period of oscillation of a spring-and-mass system is 0.50 s and the amplitude is 5.0 cm. What is the magnitude of the acceleration at the point of maximum extension of the spring?

28. A sewing machine needle moves with a rapid vibratory motion, rather like SHM, as it sews a seam. Suppose the needle moves 8.4 mm from its highest to its lowest position and it makes 24 stitches in 9.0 s. What is the maximum needle speed?

29. The prong of a tuning fork moves back and forth when it is set into vibration. The distance the prong moves between its extreme positions is 2.24 mm. If the frequency of the tuning fork is 440.0 Hz, what are the maximum velocity and the maximum acceleration of the prong? Assume SHM.

30. The period of oscillation of an object in an ideal spring-and-mass system is 0.50 s and the amplitude is 5.0 cm. What is the speed at the equilibrium point?

31. Show that the equation $a = -\omega^2 x$ is consistent for units, and that $\sqrt{k/m}$ has the same units as ω.

32. A 170-g object on a spring oscillates left to right on a frictionless surface with a frequency of 3.00 Hz and an amplitude of 12.0 cm. (a) What is the spring constant? (b) If the object starts at $x = 12.0$ cm at $t = 0$ and the equilibrium point is at $x = 0$, what equation describes its position as a function of time?

33. The air pressure variations in a sound wave cause the eardrum to vibrate. (a) For a given vibration amplitude,

are the maximum velocity and acceleration of the eardrum greatest for high-frequency sounds or low-frequency sounds? (b) Find the maximum velocity and acceleration of the eardrum for vibrations of amplitude 1.0×10^{-8} m at a frequency of 20.0 Hz. (c) Repeat (b) for the same amplitude but a frequency of 20.0 kHz.

34. Show that, for SHM, the maximum displacement, velocity, and acceleration are related by $v_m^2 = a_m A$.

35. An empty cart, tied between two ideal springs, oscillates with $\omega = 10.0$ rad/s. A load is placed in the cart, making the total mass 4.0 times what it was before. What is the new value of ω?

36. A cart with mass m is attached between two ideal springs, each with the same spring constant k. Assume that the cart can oscillate without friction. (a) When the cart is displaced by a small distance x from its equilibrium position, what force magnitude acts on the cart? (b) What is the angular frequency, in terms of m, x, and k, for this cart? (tutorial: cart between springs)

Problems 35 and 36

37. In a playground, a wooden horse is attached to the ground by a stiff spring. When a 24-kg child sits on the horse, the spring compresses by 28 cm. With the child sitting on the horse, the spring oscillates up and down with a frequency of 0.88 Hz. What is the oscillation frequency of the spring when no one is sitting on the horse?

38. A small bird's wings can undergo a maximum displacement amplitude of 5.0 cm (distance from the tip of the wing to the horizontal). If the maximum acceleration of the wings is 12 m/s^2, and we assume the wings are undergoing simple harmonic motion when beating, what is the oscillation frequency of the wing tips?

39. Equipment to be used in airplanes or spacecraft is often subjected to a shake test to be sure it can withstand the vibrations that may be encountered during flight. A radio receiver of mass 5.24 kg is set on a platform that vibrates in SHM at 120 Hz and with a maximum acceleration of 98 m/s^2 (= 10g). Find the radio's (a) maximum displacement, (b) maximum speed, and (c) the maximum net force exerted on it.

40. In an aviation test lab, pilots are subjected to vertical oscillations on a shaking rig to see how well they can recognize objects in times of severe airplane vibration. The frequency can be varied from 0.02 to 40.0 Hz and the amplitude can be set as high as 2 m for low frequencies. What are the maximum velocity and acceleration to which the pilot is subjected if the frequency is set at 25.0 Hz and the amplitude at 1.00 mm?

41. The diaphragm of a speaker has a mass of 50.0 g and responds to a signal of frequency 2.0 kHz by moving back and forth with an amplitude of 1.8×10^{-4} m at that frequency. (a) What is the maximum force acting on the diaphragm? (b) What is the mechanical energy of the diaphragm?

42. An ideal spring has a spring constant $k = 25$ N/m. The spring is suspended vertically. A 1.0-kg body is attached to the unstretched spring and released. It then performs oscillations. (a) What is the magnitude of the acceleration of the body when the extension of the spring is a maximum? (b) What is the maximum extension of the spring?

43. An ideal spring with a spring constant of 15 N/m is suspended vertically. A body of mass 0.60 kg is attached to the unstretched spring and released. (a) What is the extension of the spring when the speed is a maximum? (b) What is the maximum speed?

44. A 0.50-kg object, suspended from an ideal spring of spring constant 25 N/m, is oscillating vertically. How much change of kinetic energy occurs while the object moves from the equilibrium position to a point 5.0 cm lower?

45. A small rowboat has a mass of 47 kg. When a 92-kg person gets into the boat, the boat floats 8.0 cm lower in the water. If the boat is then pushed slightly deeper in the water, it will bob up and down with simple harmonic motion (neglecting any friction). What will be the period of oscillation for the boat as it bobs around its equilibrium position?

46. A baby jumper consists of a cloth seat suspended by an elastic cord from the lintel of an open doorway. The unstretched length of the cord is 1.2 m and the cord stretches by 0.20 m when a baby of mass 6.8 kg is placed into the seat. The mother then pulls the seat down by 8.0 cm and releases it. (a) What is the period of the motion? (b) What is the maximum speed of the baby?

10.7 Graphical Analysis of SHM

47. The displacement of an object in SHM is given by $y(t) = (8.0 \text{ cm}) \sin [(1.57 \text{ rad/s})t]$. What is the frequency of the oscillations?

48. A body is suspended vertically from an ideal spring of spring constant 2.5 N/m. The spring is initially in its relaxed position. The body is then released and oscillates about its equilibrium position. The motion is described by

$$y = (4.0 \text{ cm}) \sin [(0.70 \text{ rad/s})t]$$

What is the maximum kinetic energy of the body?

49. An object of mass 306 g is attached to the base of a spring, with spring constant 25 N/m, that is hanging from the ceiling. A pen is attached to the back of the object, so that it can write on a paper placed behind the mass-spring system. Ignore friction. (a) Describe the pattern traced on the paper if the object is held at the

point where the spring is relaxed and then released at $t = 0$. (b) The experiment is repeated, but now the paper moves to the left at constant speed as the pen writes on it. Sketch the pattern traced on the paper. Imagine that the paper is long enough that it doesn't run out for several oscillations.

50. (a) Sketch a graph of $x(t) = A \sin \omega t$ (the position of an object in SHM that is at the equilibrium point at $t = 0$). (b) By analyzing the slope of the graph of $x(t)$, sketch a graph of $v_x(t)$. Is $v_x(t)$ a sine or cosine function? (c) By analyzing the slope of the graph of $v_x(t)$, sketch $a_x(t)$. (d) Verify that $v_x(t)$ is $\frac{1}{4}$ cycle ahead of $x(t)$ and that $a_x(t)$ is $\frac{1}{4}$ cycle ahead of $v_x(t)$. (tutorial: sinusoids)

51. A mass-and-spring system oscillates with amplitude A and angular frequency ω. (a) What is the *average* speed during one complete cycle of oscillation? (b) What is the maximum speed? (c) Find the ratio of the average speed to the maximum speed. (d) Sketch a graph of $v_x(t)$, and refer to it to explain why this ratio is greater than $\frac{1}{2}$.

52. A ball is dropped from a height h onto the floor and keeps bouncing. No energy is dissipated, so the ball regains the original height h after each bounce. Sketch the graph for $y(t)$ and list several features of the graph that indicate that this motion is *not* SHM.

53. A 230.0-g object on a spring oscillates left to right on a frictionless surface with a frequency of 2.00 Hz. Its position as a function of time is given by $x = (8.00 \text{ cm}) \sin \omega t$. (a) Sketch a graph of the elastic potential energy as a function of time. (b) The object's velocity is given by $v_x = \omega(8.00 \text{ cm}) \cos \omega t$. Graph the system's kinetic energy as a function of time. (c) Graph the sum of the kinetic energy and the potential energy as a function of time. (d) Describe qualitatively how your answers would change if the surface weren't frictionless.

54. (a) Given that the position of an object is $x(t) = A \cos \omega t$, show that $v_x(t) = -\omega A \sin \omega t$. [*Hint:* Draw the velocity vector for point P in Fig. 10.17b and then find its x-component.] (b) Verify that the expressions for $x(t)$ and $v_x(t)$ are consistent with energy conservation. [*Hint:* Use the trigonometric identity $\sin^2 \omega t + \cos^2 \omega t = 1$.]

10.8 The Pendulum

55. What is the period of a pendulum consisting of a 6.0-kg mass oscillating on a 4.0-m-long string?

56. A pendulum of length 75 cm and mass 2.5 kg swings with a mechanical energy of 0.015 J. What is the amplitude?

57. A 0.50-kg mass is suspended from a string, forming a pendulum. The period of this pendulum is 1.5 s when the amplitude is 1.0 cm. The mass of the pendulum is now reduced to 0.25 kg. What is the period of oscillation now, when the amplitude is 2.0 cm? (tutorial: change in period)

58. A bob of mass m is suspended from a string of length L, forming a pendulum. The period of this pendulum is 2.0 s. If the pendulum bob is replaced with one of mass $\frac{1}{3}m$ and the length of the pendulum is increased to $2L$, what is the period of oscillation?

59. A pendulum (mass m, unknown length) moves according to $x = A \sin \omega t$. (a) Write the equation for $v_x(t)$ and sketch one cycle of the $v_x(t)$ graph. (b) What is the maximum kinetic energy?

60. A clock has a pendulum that performs one full swing every 1.0 s (back *and* forth). The object at the end of the pendulum weighs 10.0 N. What is the length of the pendulum?

61. A pendulum of length L_1 has a period $T_1 = 0.950$ s. The length of the pendulum is adjusted to a new value L_2 such that $T_2 = 1.00$ s. What is the ratio L_2/L_1?

62. A pendulum clock has a period of 0.650 s on Earth. It is taken to another planet and found to have a period of 0.862 s. The change in the pendulum's length is negligible. (a) Is the gravitational field strength on the other planet greater than or less than that on Earth? (b) Find the gravitational field strength on the other planet.

63. A grandfather clock is constructed so that it has a simple pendulum that swings from one side to the other, a distance of 20.0 mm, in 1.00 s. What is the maximum speed of the pendulum bob? Use two different methods. First, assume SHM and use the relationship between amplitude and maximum speed. Second, use energy conservation.

64. Christy has a grandfather clock with a pendulum that is 1.000 m long. (a) If the pendulum is modeled as a simple pendulum, what would be the period? (b) Christy observes the actual period of the clock, and finds that it is 1.00% faster than that for a simple pendulum that is 1.000 m long. If Christy models the pendulum as two objects, a 1.000-m uniform thin rod and a point mass located 1.000 m from the axis of rotation, what percentage of the total mass of the pendulum is in the uniform thin rod?

65. A pendulum of length 120 cm swings with an amplitude of 2.0 cm. Its mechanical energy is 5.0 mJ. What is the mechanical energy of the same pendulum when it swings with an amplitude of 3.0 cm?

66. A thin circular hoop is suspended from a knife edge. Its rotational inertia about the rotation axis (along the knife) is $I = 2mr^2$. Show that it oscillates with the same frequency as a simple pendulum of length equal to the diameter of the hoop.

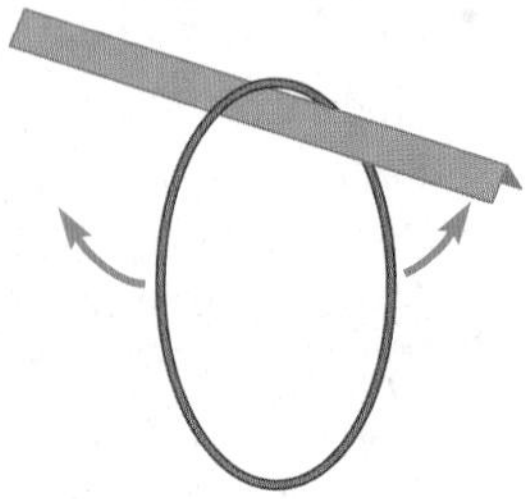

10.9 Damped Oscillations

67. (a) What is the energy of a pendulum ($L = 1.0$ m, $m = 0.50$ kg) oscillating with an amplitude of 5.0 cm? (b) The pendulum's energy loss (due to damping) is replaced in a

clock by allowing a 2.0-kg mass to drop 1.0 m in 1 week. What average percentage of the pendulum's energy is lost during one cycle?

68. The amplitude of oscillation of a pendulum decreases by a factor of 20.0 in 120 s. By what factor has its energy decreased in that time?

69. Because of dissipative forces, the amplitude of an oscillator decreases 5.00% in 10 cycles. By what percentage does its *energy* decrease in ten cycles?

Comprehensive Problems

70. Four people sit in a car. The masses of the people are 45 kg, 52 kg, 67 kg, and 61 kg. The car's mass is 1020 kg. When the car drives over a bump, its springs cause an oscillation with a frequency of 2.00 Hz. What would the frequency be if only the 45-kg person were present?

71. A pendulum passes $x = 0$ with a speed of 0.50 m/s; it swings out to $A = 0.20$ m. What is the period T of the pendulum? (Assume the amplitude is small.)

72. What is the length of a simple pendulum whose horizontal position is described by

$$x = (4.00 \text{ cm}) \cos [(3.14 \text{ rad/s})\, t]?$$

What assumption do you make when answering this question?

73. Martin caught a fish and wanted to know how much it weighed, but he didn't have a scale. He did, however, have a stopwatch, a spring, and a 4.90-N weight. He attached the weight to the spring and found that the spring would oscillate 20 times in 65 s. Next he hung the fish on the spring and found that it took 220 s for the spring to oscillate 20 times. (a) Before answering part (b), determine if the fish weighs more or less than 4.90 N. (b) What is the weight of the fish?

74. A naval aviator had to eject from her plane before it crashed at sea. She is rescued from the water by helicopter and dangles from a cable that is 45 m long while being carried back to the aircraft carrier. What is the period of her vibration as she swings back and forth while the helicopter hovers over her ship?

75. An object of mass m is hung from the base of an ideal spring that is suspended from the ceiling. The spring has a spring constant k. The object is pulled down a distance D from equilibrium and released. Later, the same system is set oscillating by pulling the object down a distance $2D$ from equilibrium and then releasing it. (a) How do the period and frequency of oscillation change when the initial displacement is increased from D to $2D$? (b) How does the total energy of oscillation change when the initial displacement is increased from D to $2D$? Give the answer as a numerical ratio. (c) The mass-spring system is set into oscillation a third time. This time the object is pulled down a distance of $2D$ and then given a push downward some more, so that it has an initial speed v_i downward. How do the period and frequency of oscillation compare to those you found in part (a)? (d) How does the total energy compare to when the object was released from rest at a displacement $2D$?

76. A spider's web can undergo SHM when a fly lands on it and displaces the web. For simplicity, assume that a web obeys Hooke's law (which it does not really as it deforms permanently when displaced). If the web is initially horizontal, and a fly landing on the web is in equilibrium when it displaces the web by 0.030 mm, what is the frequency of oscillation when the fly lands?

77. A mass-spring system oscillates so that the position of the mass is described by $x = -10 \cos (1.57t)$, where x is in cm when t is in seconds. Make a plot that has a dot for the position of the mass at $t = 0$, $t = 0.2$ s, $t = 0.4$ s, . . . , $t = 4$ s. The time interval between each dot should be 0.2 s. From your plot, tell where the mass is moving fastest and where slowest. How do you know?

78. A hedge trimmer has a blade that moves back and forth with a frequency of 28 Hz. The blade motion is converted from the rotation provided by the electric motor to an oscillatory motion by means of a Scotch yoke (see Conceptual Question 7). The blade moves 2.4 cm during each stroke. Assuming that the blade moves with SHM, what are the maximum speed and maximum acceleration of the blade?

79. The simple pendulum can be thought of as a special case of the physical pendulum where all of the mass is at a distance L from the rotation axis. For a simple pendulum of mass m and length L, show that the expression for the period of a physical pendulum (Eq. 10-27) reduces to the expression for the period of a simple pendulum (Eq. 10-26b).

80. Luke is trying to catch a pesky animal that keeps eating vegetables from his garden. He is building a trap and needs to use a spring to close the door to his trap. He has a spring in his garage and he wants to determine the spring constant of the spring. To do this, he hangs the spring from the ceiling and measures that it is 20.0 cm long. Then he hangs a 1.10-kg brick on the end of the spring and it stretches to 31.0 cm. (a) What is the spring constant of the spring? (b) Luke now pulls the brick 5.00 cm from the equilibrium position to watch it oscillate. What is the maximum speed of the brick? (c) When the displacement is 2.50 cm from the equilibrium position, what is the speed of the brick? (d) How long will it take for the brick to oscillate five times?

81. A 4.0-N body is suspended vertically from an ideal spring of spring constant 250 N/m. The spring is initially in its relaxed position. Write an equation to describe the motion of the body if it is released at $t = 0$. [*Hint:* Let $y = 0$ at the equilibrium point and take $+y$ = up.]

82. Show, using dimensional analysis, that the frequency f at which a mass-spring system oscillates radical is independent of the amplitude A and proportional to $\sqrt{k/m}$. [*Hint:* Start by assuming that f does depend on A (to some power).]

83. A horizontal spring with spring constant of 9.82 N/m is attached to a block with a mass of 1.24 kg that sits on a frictionless surface. When the block is 0.345 m from its equilibrium position, it has a speed of 0.543 m/s. (a) What is the maximum displacement of the block from the equilibrium position? (b) What is the maximum speed of the block? (c) When the block is 0.200 m from the equilibrium position, what is its speed?

84. A steel piano wire ($Y = 2.0 \times 10^{11}$ Pa) has a diameter of 0.80 mm. At one end it is wrapped around a tuning pin of diameter 8.0 mm. The length of the wire (not including the wire wrapped around the tuning pin) is 66 cm. Initially, the tension in the wire is 381 N. To tune the wire, the tension must be increased to 402 N. Through what angle must the tuning pin be turned?

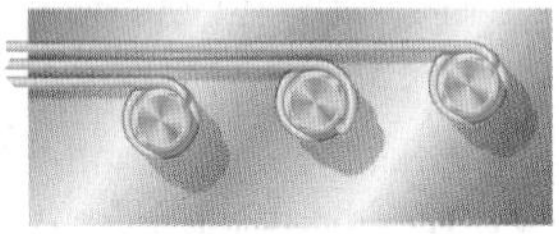

85. When the tension is 402 N, what is the tensile stress in the piano wire in Problem 84? How does that compare to the elastic limit of steel piano wire (8.26×10^8 Pa)?

86. A tightrope walker who weighs 640 N walks along a steel cable. When he is halfway across, the cable makes an angle of 0.040 rad below the horizontal. (a) What is the strain in the cable? Assume the cable is horizontal with a tension of 80 N before he steps onto it. Ignore the weight of the cable itself. (b) What is the tension in the cable when the tightrope walker is standing at the midpoint? (c) What is the cross-sectional area of the cable? (d) Has the cable been stretched beyond its elastic limit (2.5×10^8 Pa)?

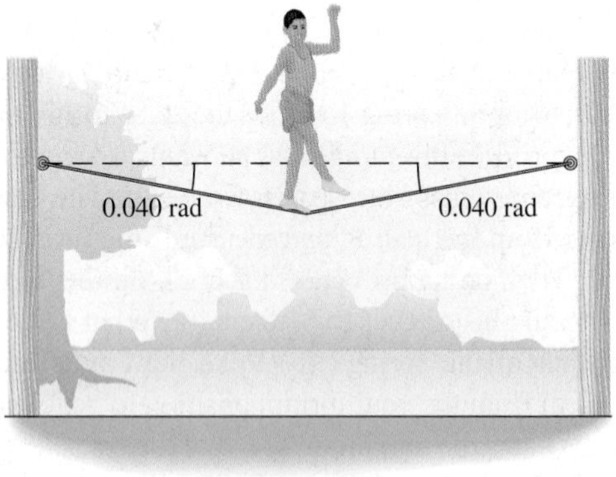

Problem 86 (the 0.040-rad angles are greatly exaggerated).

87. A gibbon, hanging onto a horizontal tree branch with one arm, swings with a small amplitude. The gibbon's CM is 0.40 m from the branch and its rotational inertia divided by its mass is $I/m = 0.25$ m^2. Estimate the frequency of oscillation.

88. In Problem 8.41, we found that the force of the tibia (shinbone) on the ankle joint for a person (of weight 750 N) standing on the ball of one foot was 2800 N. The ankle joint therefore pushes upward on the bottom of the tibia with a force of 2800 N, while the top end of the tibia must feel a net downward force of approximately 2800 N (ignoring the weight of the tibia itself). The tibia has a length of 0.40 m, an average inner diameter of 1.3 cm, and an average outer diameter of 2.5 cm. (The central core of the bone contains marrow that has negligible compressive strength.) (a) Find the average cross-sectional area of the tibia. (b) Find the compressive stress in the tibia. (c) Find the change in length for the tibia due to the compressive forces.

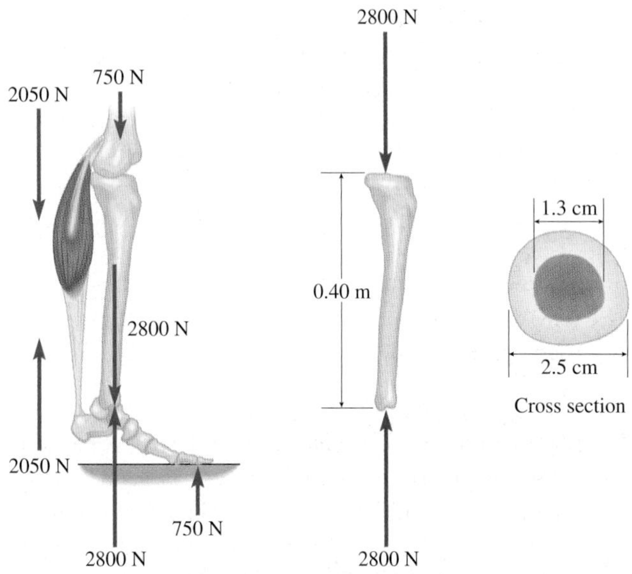

89. The maximum height of a cylindrical column is limited by the compressive strength of the material; if the compressive stress at the bottom were to exceed the compressive strength of the material, the column would be crushed under its own weight. (a) For a cylindrical column of height h and radius r, made of material of density ρ, calculate the compressive stress at the bottom of the column. (b) Since the answer to part (a) is independent of the radius r, there is an absolute limit to the height of a cylindrical column, regardless of how wide it is. For marble, which has a density of 2.7×10^3 kg/m^3 and a compressive strength of 2.0×10^8 Pa, find the maximum height of a cylindrical column. (c) Is this limit a practical concern in the construction of marble columns? Might it limit the height of a beanstalk?

90. A bungee jumper leaps from a bridge and undergoes a series of oscillations. Assume $g = 9.78$ m/s^2. (a) If a 60.0-kg jumper uses a bungee cord that has an unstretched length of 33.0 m and she jumps from a height of 50.0 m above a river, coming to rest just a few centimeters above the water surface on the first downward descent, what is the period of the oscillations? Assume the bungee cord follows Hooke's law. (b) The next jumper in line has a mass of 80.0 kg. Should he jump using the same cord? Explain.

91. Spider silk has a Young's modulus of 4.0×10^9 N/m^2 and can withstand stresses up to 1.4×10^9 N/m^2. A single web strand has a cross-sectional area of 1.0×10^{-11} m^2, and a web is made up of 50 radial strands. A bug lands in the center of a horizontal web so that the web stretches downward. (a) If the maximum stress is exerted on each strand, what angle θ does the web make with the horizontal? (b) What does the mass of a bug have to be in order to exert this maximum stress on the web? (c) If the web is 0.10 m in radius, how far down does the web extend?

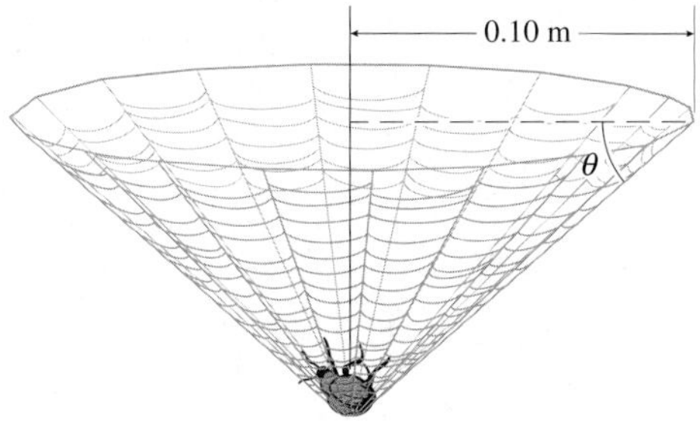

92. What is the period of a pendulum formed by placing a horizontal axis (a) through the end of a meterstick (100-cm mark)? (b) through the 75-cm mark? (c) through the 60-cm mark?

93. The motion of a simple pendulum is approximately SHM only if the amplitude is small. Consider a simple pendulum that is released from a horizontal position ($\theta_i = 90°$ in Fig. 10.22). (a) Using conservation of energy, find the speed of the pendulum bob at the bottom of its swing. Express your answer in terms of the mass m and the length L of the pendulum. Do *not* assume SHM. (b) Assuming (incorrectly, for such a large amplitude) that the motion *is* SHM, determine the maximum speed of the pendulum. Based on your answers, is the period of a pendulum for large amplitudes larger or smaller than that given by Eq. (10-26b)?

94. The gravitational potential energy of a pendulum is $U = mgy$. (a) Taking $y = 0$ at the lowest point, show that $y = L(1 - \cos\theta)$, where θ is the angle the string makes with the vertical. (b) If θ is small, $(1 - \cos\theta) \approx \frac{1}{2}\theta^2$ and $\theta \approx x/L$ (Appendix A.7). Show that the potential energy can be written $U \approx \frac{1}{2}kx^2$ and find the value of k (the equivalent of the spring constant for the pendulum).

95. A pendulum is made from a uniform rod of mass m_1 and a small block of mass m_2 attached at the lower end. (a) If the length of the pendulum is L and the oscillations are small, find the period of the oscillations in terms of m_1, m_2, L, and g. (b) Check your answer to part (a) in the two special cases $m_1 >> m_2$ and $m_1 << m_2$.

96. Resilin is a rubber-like protein that helps insects to fly more efficiently. The resilin, attached from the wing to the body, is relaxed when the wing is down and is extended when the wing is up. As the wing is brought up, some elastic energy is stored in the resilin. The wing is then brought back down with little muscular energy, since the potential energy in the resilin is converted back into kinetic energy. Resilin has a Young's modulus of 1.7×10^6 N/m^2. (a) If an insect wing has resilin with a relaxed length of 1.0 cm and with a cross-sectional area of 1.0 mm^2, how much force must the wings exert to extend the resilin to 4.0 cm? (b) How much energy is stored in the resilin?

Answers to Practice Problems

10.1 $2k$ (When the original spring is stretched an amount L, each of the half-springs stretches only $\frac{1}{2}L$. Each of the newly formed springs stretches half as far as the original spring for a given applied force.)

10.2 1.4×10^{-5}

10.3 0.18 mm

10.4 1.3×10^8 Pa

10.5 $-\frac{\Delta P}{B} = -\frac{1.0 \times 10^7 \text{ Pa}}{2.3 \times 10^9 \text{ Pa}} = -0.0043 = \frac{\Delta V}{V}$

and $\Delta V = -0.43\% \times V$

10.6 110 m/s^2 at $x = \pm A$

10.7 $K = 0$, $U_e = 2(mg)^2/k$, $U_g = -(mg)^2/k$, $E = (mg)^2/k$

10.8

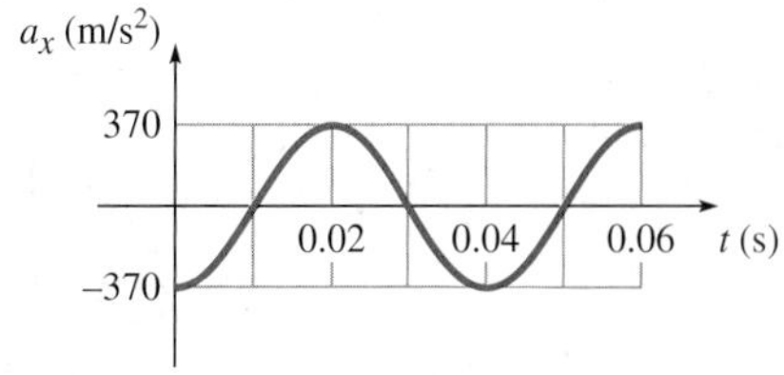

$a_x(t) = -a_m \cos\omega t$, where $\omega = 160$ rad/s and $a_m = 370$ m/s^2.

10.9 1.6 m/s^2 (about 1/6 that of the Earth)

10.10 0.82 m/s or 1.8 mi/h

Answers to Checkpoints

10.2 The two wires are under the same stress (same tensile force and same cross-sectional area). Young's modulus for steel is about $\frac{5}{3}$ times that for copper, so the *strain* for the steel wire is $\frac{3}{5}$ the strain of the copper wire. However, the strain is the *fractional* length change. The steel wire is twice as long, so its length change is $2 \times (3/5)$ times the length change of the copper wire. The steel wire stretches more.

10.5 When the kinctic and potential energies are equal, each is half of the total energy. When $U = \frac{1}{2}kx^2 = \frac{1}{2}E_{\text{total}} = \frac{1}{2}(\frac{1}{2}kA^2)$, $x = \pm A/\sqrt{2}$.

10.7 (a) When the displacement is zero, the potential energy has its minimum value. From conservation of energy, the kinetic energy then has its maximum value. Therefore, the speed has its maximum magnitude ($v = \pm v_m$), as shown in Fig. 10.20. (b) When the speed is zero, the kinetic energy is minimum and the potential energy is maximum. Therefore, the displacement has its maximum magnitude ($x = \pm A$).

CHAPTER

11 Waves

On January 17, 1995, a terrible earthquake struck the Hanshin region of Japan, killing over 6400 people and injuring about 40 000 others. Some 200 000 homes and buildings were damaged, causing the evacuation to shelters of 320 000 people. The heaviest damage occurred in the city of Kobe, including the buckling and collapse of an elevated highway. However, geologists found that the point of origin of the earthquake was 15–20 km below the northern tip of Awaji Island, about 20 km southwest of Kobe. How did the earthquake cause great devastation at locations many kilometers away? (See p. 393 for the answer.)

- period, frequency, angular frequency (Section 10.6)
- position, velocity, acceleration, and energy in simple harmonic motion (Section 10.5)
- resonance (Section 10.10)
- graphical analysis of SHM (Section 10.7)

Concepts & Skills to Review

11.1 WAVES AND ENERGY TRANSPORT

Basic Models: Particles and Waves Physicists use only a few basic models to describe the physical world. One such model is the particle: a pointlike object with no inner structure and with certain characteristics such as mass and electric charge. Another basic model is the **wave**. Water waves are familiar examples. When a pebble is dropped into a pond, it disturbs the surface of the water. Ripples on the surface of the pond travel away from the spot where the pebble landed.

Examples of Waves Any wave is characterized as some sort of "disturbance" that travels away from its source. In Chapters 11 and 12, we concentrate on mechanical waves traveling through a material medium, such as water waves, sound waves, and the seismic waves caused by earthquakes. Particles in the medium are disturbed from their equilibrium positions as the wave passes, returning to their equilibrium positions after the wave has passed. In Chapter 22, we discuss electromagnetic waves such as radio waves and light waves, in which the disturbance consists of oscillating electromagnetic fields. Two of our five human senses are wave detectors: the ear is sensitive to the tiny fluctuations in air pressure caused by compressional waves in air (sound) and the eye is sensitive to electromagnetic waves in a certain frequency range (light).

CONNECTION:

In wave motion, energy is transferred from one oscillating particle to another. Energy is conserved overall, but the energy of any one oscillating particle can change. Mechanical waves carry the same kinds of energy as a simple harmonic oscillator: kinetic energy and potential energy.

Energy Transport by a Wave

Suppose we drop a pebble into a still pond. The kinetic energy of the pebble just before it hits the pond is partly converted into the energy carried off by the water wave. That waves carry energy is clear to anyone who has been surfing or swimming in the ocean. Speaking of surfing, information on the Internet is carried by waves of various sorts: electrical waves in wires, microwaves between Earth and communications satellites, light waves in optical fibers. Microwaves in ovens carry energy from their source to the food; the electromagnetic energy of the microwaves is absorbed by water molecules in the food and appears as thermal energy. Electromagnetic waves from the Sun bring the energy that fuels the growth of green plants. Seismic waves carry energy released by an earthquake to other parts of the Earth, sometimes with devastating results.

How can an earthquake cause damage far from the epicenter?

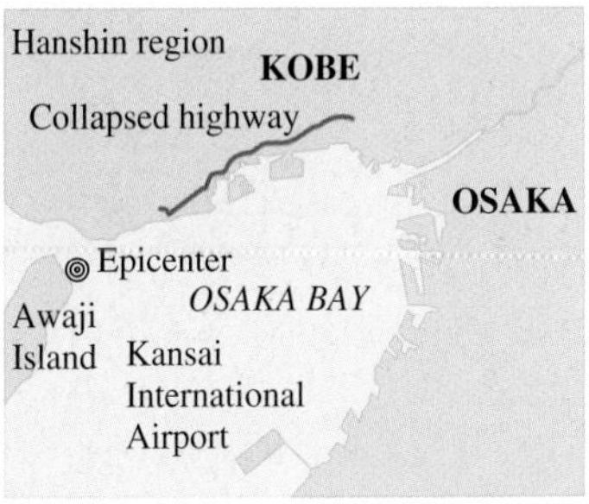

Seismic waves travel away from the *focus* of an earthquake (the point of origin) both through the Earth (*body waves*) and along the Earth's crust (*surface waves*), transporting vibrations and energy. However, the material through which the waves travel is *not* transported. Most earthquake damage is caused by seismic waves rather than caused by fault movement. In the Hanshin earthquake, damage to buildings was caused by seismic waves at distances over 100 km from the *epicenter* (the point on the surface directly above the focus), but the motion of the vibrating particles in the ground never moved more than about 1.5 m.

The sound of thunder travels for miles in all directions, but none of the air molecules zapped by lightning travels more than a meter or so during the few seconds that it takes the sound to reach our ears. A wave can transmit energy from one point to another without transporting any matter between the two points (Fig. 11.1).

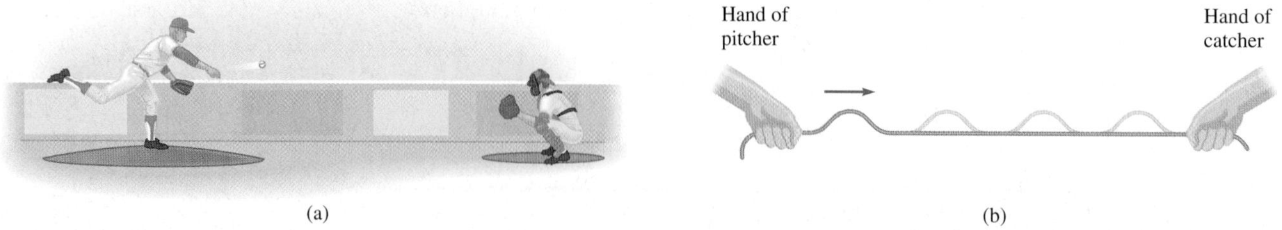

Figure 11.1 Two different ways to transfer energy. (a) When a baseball pitcher throws a ball to the catcher, the ball carries energy with it. The pitcher gives the ball kinetic energy; the catcher receives the energy when the ball hits his hand and his hand recoils. (b) Suppose instead that they hold a rope stretched between them. If the pitcher suddenly moves his hand up and down quickly, a wave pulse travels along the rope until it reaches the catcher's hand. Once again, the pitcher sends the energy and the catcher receives it when the rope makes his hand recoil. However, in this case the pitcher is still holding his end of the rope; it never leaves his hand. Energy is transferred *without any matter moving from the pitcher to the catcher.*

PHYSICS AT HOME

Stretch a heavy rope between yourself and a friend and test out the transfer of energy from one to the other by sending wave pulses down the rope. Can you feel the energy transfer when the pulse arrives?

PHYSICS AT HOME

Observe carefully what happens when you snap your fingers. You start by pressing your thumb against your fingers and then sideways, the thumb in one direction and the fingers in the opposite direction. Initially friction keeps them from moving sideways, but suddenly they slip, releasing the built-up energy.

Similarly, the rocks on two sides of a fault line are pressed together and sideways. Friction keeps them from moving sideways as elastic (or strain) energy builds up. Then suddenly they slip, releasing a tremendous amount of energy largely in the form of seismic waves that carry vibrations far from the focus of the earthquake.

Intensity

For a wave that travels in a three-dimensional medium (such as sound waves or seismic waves traveling through the Earth), the **intensity** (symbol I, SI unit W/m^2) is a measure of the *average power per unit area* carried by the wave past a surface perpendicular to the wave's direction of propagation.

$$I = P/A$$

For example, if a sound wave's intensity is a fairly loud $I = 10^{-5}$ W/m^2 when it reaches the eardrum and the area of the eardrum is $A = 10^{-4}$ m^2, then the power delivered to the eardrum is $P = IA = 10^{-9}$ W (assuming that all the energy incident on the eardrum is absorbed). The energy absorbed by the eardrum at this rate in *one hour* would be

$$10^{-9}\ \text{W} \times 3600\ \text{s} \approx 4\ \mu\text{J}$$

The human ear is a very sensitive detector indeed.

Application: sensitivity of the human ear

Intensity and Distance from the Wave Source For most waves, the intensity decreases as the distance from the source increases. Some of the energy can be absorbed (dissipated) by the wave medium. The amount of energy absorbed depends on the

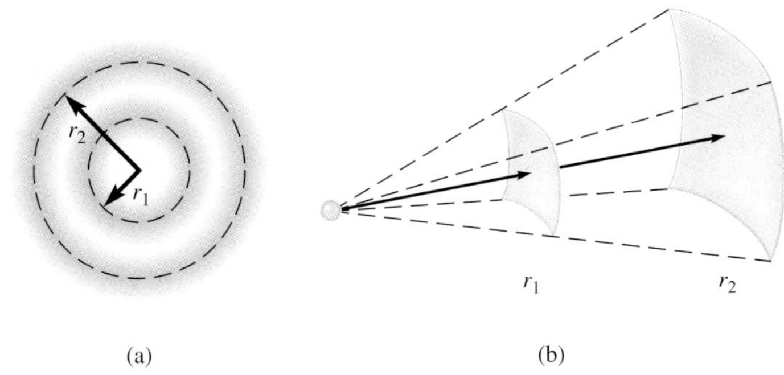

Figure 11.2 (a) A point source of sound radiating energy uniformly in all directions. (b) The intensity at a distance r_2 is smaller than the intensity at a distance r_1 since the same power is spread out over a greater area.

medium. Air absorbs relatively little sound energy, which is why we can hear sounds generated far away.

Another reason that intensity decreases with distance is that, as the wave spreads out, the energy gets spread over a larger and larger area. Consider a point source emitting a wave uniformly in all directions—an *isotropic* source (Fig. 11.2). The average power (energy per unit time) emitted is constant. Imagine a sphere surrounding the source; the rate at which energy passes through the surface of the sphere is the same no matter what the radius. The surface area of a sphere is $4\pi r^2$, so as the wave moves farther from the source, the energy spreads out over a larger and larger area. Thus, the power per unit area (intensity) decreases with distance. Assuming that no energy is absorbed by the medium and there are no obstacles to reflect or absorb sound,

$$I = \frac{\text{power}}{\text{area}} = \frac{P}{4\pi r^2} \qquad (11\text{-}1)$$

(point source emitting uniformly in all directions; no reflection or absorption)

Therefore, if energy absorption by the medium can be ignored, the intensity of the sound is inversely proportional to the square of the distance from the source. This "inverse square law" is the result of a conserved quantity (here, energy) radiating uniformly from a point source in three-dimensional space.

CHECKPOINT 11.1

A siren in a fire tower 20 m high generates a sound wave with intensity 0.090 W/m^2 at a point on the ground below the tower. What is the intensity of the sound wave 2.0 km from the tower? Assume the siren is an isotropic source.

11.2 TRANSVERSE AND LONGITUDINAL WAVES

A Slinky toy can be used to demonstrate two different kinds of wave. In a **transverse** wave, the motion of particles in the medium is perpendicular to the direction of propagation of the wave. To send a transverse wave down a Slinky, wiggle the end of the Slinky back and forth in a direction perpendicular to the length of the Slinky (Fig. 11.3a). In a **longitudinal** wave, the motion of particles in the medium is along the same line as the direction of propagation of the wave. To send a longitudinal wave down the Slinky, jiggle the end in and out along its length to alternately stretch and compress the coils (Fig. 11.3b). A red dot painted on one coil of the Slinky helps illustrate the difference. In a transverse wave, the dot moves back and forth about a fixed position with its motion perpendicular to the direction of propagation of the wave; in a longitudinal wave, the

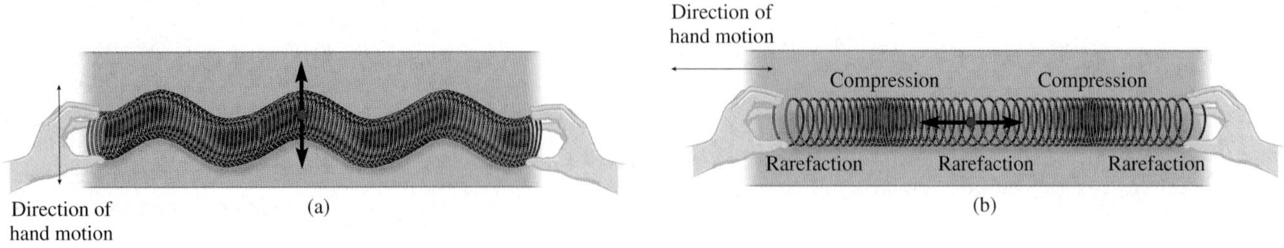

Figure 11.3 (a) Transverse and (b) longitudinal waves on a Slinky.

dot also moves back and forth about a fixed position but along the direction of propagation of the wave. In both cases, the wave itself moves from one end of the Slinky to the other while the dot is moving about its fixed position.

PHYSICS AT HOME

Ask a friend to sit at a table and hold one end of a long, loose spring (or a Slinky). With the spring supported by the table surface, grasp the other end and stretch the spring. Figure out how to move your hand to send transverse and longitudinal waves down the spring.

The Slinky—or any long spring—is a better approximation to solid materials than the stretched rope. In solids both types of waves can exist; a transverse wave results from a shear disturbance and a longitudinal wave from a compressional disturbance. Therefore, seismic body waves can be either longitudinal or transverse (Fig. 11.4).

Fluids can be compressed, but, because they flow, they do not sustain shear stresses. Therefore, longitudinal waves travel through fluids but transverse waves do not. However, gravity or surface tension can provide the transverse restoring force that allows a transverse wave to travel *along the surface* of a liquid.

A sound wave is longitudinal; each small volume of air vibrates back and forth along the direction of travel of the wave. Molecules are compressed together in some places and more thinly spaced (*rarefied*) in others; the air has regions of higher and lower density called **compressions** and **rarefactions** (see Fig. 11.3b).

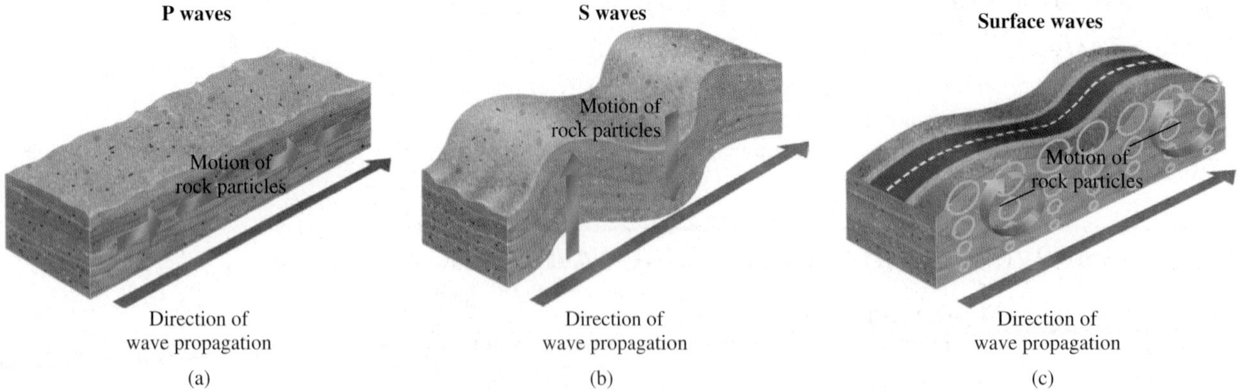

Figure 11.4 Three types of seismic waves. (a) Longitudinal body waves (*P waves*) are the fastest seismic waves (typically 4–8 km/s). They are similar to sound waves in air: particles in the Earth's interior are pushed together and pulled apart in the same direction that the wave propagates. (b) Transverse body waves (*S waves*) travel more slowly—typically 2–5 km/s. In an S wave, particles in the Earth's interior vibrate at right angles to the direction that the wave travels. By measuring the time between the first arrivals of these two types of waves at different detection stations, geologists are able to determine the point of origin of the earthquake. (c) In a surface wave, the motion of the ground combines longitudinal and transverse components.

Waves That Combine Transverse and Longitudinal Motion

Not all seismic waves are purely transverse or purely longitudinal. In a surface wave, the ground near the surface rolls approximately in a circle. Thus, the motion of the ground has components both parallel and perpendicular to the direction of propagation. The transverse component can either be up and down (as shown in Fig. 11.4c) or side to side. The motion of the ground is greatest at the surface.

Ocean waves are similar to the surface seismic wave shown in Fig. 11.4c. Deep underwater, the wave is mostly longitudinal (Fig. 11.5); as the wave passes, water moves back and forth along the direction of propagation of the wave. Higher up, the wave has both transverse and longitudinal components; water moves in an oval as the wave passes. Water near the surface moves approximately in a circle. The air above the surface presents little resistance, so water swells upward more easily there and then is pulled back downward by gravity (or, for small amplitudes, by surface tension). When the wave gets close to shore, the crest often collapses or *breaks*; the motion of the water particles is then much more complex.

When a guitar string is plucked *gently*, the wave on the string is almost purely transverse; stretching of the string is negligible. When it is plucked more forcefully, the resulting wave is a combination of transverse and longitudinal waves. At any instant, the string is stretched more in some places than in others; a point on the string has longitudinal motion as well as transverse motion.

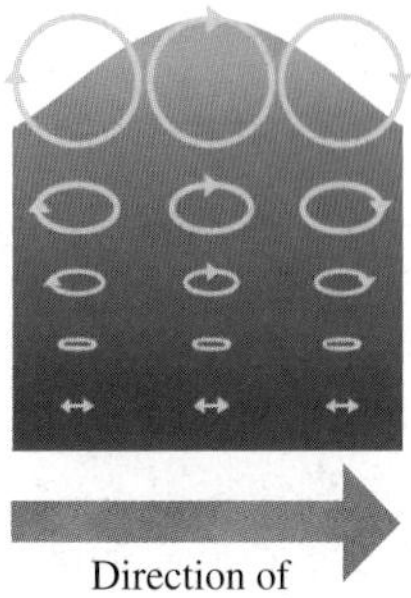

Figure 11.5 The motion of water in an ocean wave combines transverse and longitudinal motion.

11.3 SPEED OF TRANSVERSE WAVES ON A STRING

The speed of a mechanical wave depends on properties of the wave medium. What properties of a string determine the speed of a transverse wave moving along it? Suppose that a string of length L and mass m is under tension F. In Problem 56, you can show that $\sqrt{FL/m}$ is the only combination of those three quantities with the correct units for speed. There could be a dimensionless constant multiplier, but a derivation using more advanced mathematics shows that the constant is 1; the speed of a transverse wave on a string is

$$v = \sqrt{\frac{FL}{m}} \qquad (11\text{-}2)$$

We can rewrite Eq. (11-2) in another form. Length and mass are not independent; for a given string composition and diameter (say, a yellow brass string of 0.030 in. diameter), the mass of the string is proportional to its length. By defining the **linear mass density** (mass per unit length) of the string to be

$$\mu = \frac{m}{L} \qquad (11\text{-}3)$$

the speed of a transverse wave on a string can be written

$$v = \sqrt{\frac{F}{\mu}} \qquad (11\text{-}4)$$

An advantage of Eq. (11-4) over Eq. (11-2) is that it shows clearly that the wave speed depends on *local* properties of the medium; it does not depend on how much of the medium there is. The wave speed in the vicinity of some point P, for instance, does not depend on how long the string is; only properties of the string in the immediate vicinity of point P can determine how fast the wave travels past that point.

Note that as tension increases, wave speed increases; as mass density increases, wave speed decreases. A somewhat more general way to think about it, applicable to other waves as well, is:

> More restoring force makes faster waves; more inertia makes slower waves.

The wave speed is not the same as the speed of a particle in the wave medium.

The speed at which a wave propagates is not the same as the speed at which a particle in the medium moves. Suppose a horizontal string is stretched along the x-axis and a transverse pulse in the y-direction is sent down the string. The speed of propagation of the wave v is the speed at which the *pattern* or disturbance moves along the string (in the x-direction); for a uniform string, the wave speed is constant. A point on the string vibrates up and down in the $\pm y$-direction with a *different* speed that is *not* constant.

Example 11.1

A Piñata

A string of length 2.0 m has a mass of 125 mg. The string is attached to the ceiling and a piñata of mass 4.0 kg hangs from the other end. A child whacks the piñata sideways with a stick; as a result, a transverse pulse travels up the string toward the ceiling. At what speed does the pulse travel?

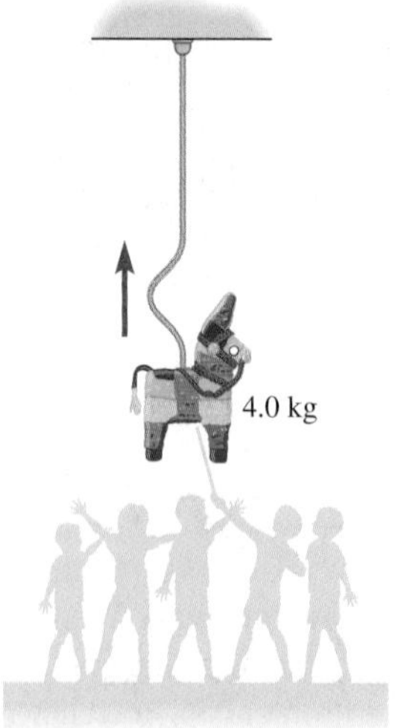

Strategy We start with a diagram of the situation (see the figure). The piñata puts the string under tension. The tension in the string is equal to the weight of the piñata because the weight of the string itself is negligible in comparison. The mass and length of the string are given, so the linear mass density can be found. Then we can find the wave speed.

Solution The speed of a transverse wave on a string is given by Eq. (11-4):

$$v = \sqrt{\frac{F}{\mu}}$$

where F is the tension in the string and μ is the linear mass density of the string. The tension is equal to the weight hanging on the string:

$$F = Mg$$

The linear mass density of the string is mass per unit length ($\mu = m/L$). Substituting the tension and mass density, we have

$$v = \sqrt{\frac{F}{m/L}} = \sqrt{\frac{(Mg)L}{m}}$$

$$= \sqrt{\frac{4.0\text{ kg} \times 9.8\text{ m/s}^2 \times 2.0\text{ m}}{125 \times 10^{-6}\text{ kg}}} = 790\text{ m/s}$$

Discussion The *weight* of the string (mg) is negligible in comparison with the weight hanging from the end of the string (Mg). That is not always the case, as can be seen in Practice Problem 11.1.

Practice Problem 11.1 Initial Velocity of Another Wave Pulse Traveling on a String

A string of length 10.0 m has a linear mass density of 25 g/m. The string is fixed at the top and has an object of mass 0.200 kg hanging from the bottom. (a) What is the *initial* wave speed of a pulse sent up the string from the bottom? (b) What is the speed of the pulse as it approaches the top of the string? [*Hint:* Does the weight of the string itself affect the tension in either case?]

11.4 PERIODIC WAVES

Application of periodic waves: the difference between musical sound and noise

A **periodic** wave repeats the same pattern over and over, each repeating section transporting the energy that was used to generate it. A periodic water wave can be produced by steadily dropping a series of pebbles into the water; a periodic wave on a cord can be produced by taking one end of the cord and moving it up and down, over and over, in a repeating pattern. As the wave propagates along the cord, every point on the cord oscillates with the same up and down pattern, though with a time delay that depends on the wave speed. Whereas musical sounds are often periodic waves, noise is *aperiodic*.

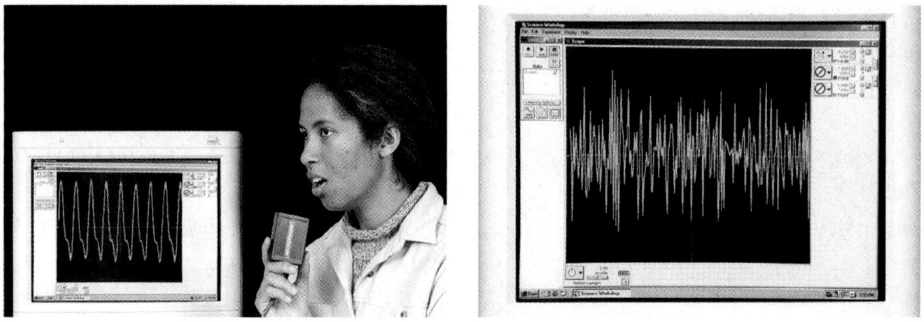

(a) (b)

Figure 11.6 (a) Periodic sound wave pattern produced by singing the vowel "ah." (b) Aperiodic sound wave pattern produced by hissing the consonant "s." (The microphone generates an electrical signal proportional to the pressure variations of the sound wave. This signal is displayed as a function of time on the screen.)

The human voice makes a periodic sound wave when a vowel is sung at a steady pitch (constant frequency); most of the consonant sounds are aperiodic (Fig. 11.6).

Period, Frequency, Wavelength, and Amplitude At any given point in space, a periodic wave repeats itself after a time T called the **period**. The inverse of the period is the **frequency** f.

$$f = \frac{1}{T} \quad (\text{SI unit Hz} = \text{s}^{-1}) \qquad (5\text{-}8)$$

The frequency tells how often the pattern of motion repeats itself at any single point. For instance, if the frequency is 20 Hz, then there are 20 repetitions, or cycles, per second. Each cycle takes a time $T = 1/f = 0.05$ s. The angular frequency is $\omega = 2\pi f$ and is measured in rad/s.

CONNECTION:

The terminology for periodic waves is similar to that used for uniform circular motion (Chapter 5) and for simple harmonic motion (Chapter 10).

During one period T, a periodic wave traveling at speed v moves a distance vT. In Fig. 11.7, note that, at any instant, points separated by a distance vT along the direction of propagation of a wave move "in sync" with each other. Thus, vT is the *repetition distance* of the wave, just as the period is the *repetition time*. This distance is called the **wavelength** (symbol λ, the Greek letter lambda).

$$\lambda = vT \qquad (11\text{-}5)$$

Combining this relation and the expression for frequency, we obtain

$$v = \frac{\lambda}{T} = f\lambda \qquad (11\text{-}6)$$

Equations (11-5) and (11-6) are true for all periodic waves, no matter how the wave is produced or what the shape of the wave.

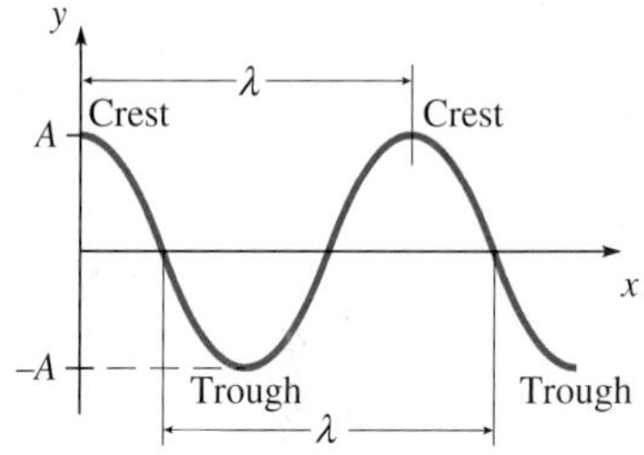

Figure 11.7 Snapshot graph of a sinusoidal wave moving with speed v in the x-direction. The graph shows the displacement y of particles in the wave medium as a function of x, their position along the direction of wave propagation, at one particular time t. The amplitude A and the wavelength λ are shown.

CHECKPOINT 11.4

A seismic wave travels at 4.0 km/s and has a wavelength of 20 km. How long does it take a rock particle to complete one cycle of oscillation?

The maximum displacement of any particle from its equilibrium position is the **amplitude** A of the wave. For a sinusoidal wave traveling along a stretched string in the x-direction, the amplitude A is the maximum displacement of a particle in the positive or negative y-direction. For surface water waves, the amplitude is the height of a crest (a high point) above or the depth of a trough (a low point) below the undisturbed water level.

Harmonic Waves **Harmonic waves** are a special kind of periodic wave in which the disturbance is sinusoidal (either a sine or cosine function). In a harmonic transverse wave on a string, for instance, every point on the string moves in SHM with the same amplitude and angular frequency, although different points reach their maximum displacements at different times. The maximum speed and maximum

In Eq. (10-21), v_m is the maximum speed at which a point on the string moves in the $\pm y$-direction. v_m is not the same as v, the speed of wave propagation in the $\pm x$-direction (see Section 11.3).

acceleration of a point on the string depend on both the angular frequency and the amplitude of the wave:

$$v_m = \omega A \quad (10\text{-}21)$$

$$a_m = \omega^2 A \quad (10\text{-}22)$$

Intensity and Amplitude Since the total energy of an object moving in SHM is proportional to the amplitude squared (Section 10.5), the total energy of a harmonic wave is proportional to the square of its amplitude. Intensity is the rate at which a wave transports energy per unit area perpendicular to the direction of propagation. The intensity of a harmonic wave is proportional to its total energy and, therefore, is proportional to the square of the amplitude. That turns out to be a general result not limited to harmonic waves:

> The intensity of a wave is proportional to the square of its amplitude.

11.5 MATHEMATICAL DESCRIPTION OF A WAVE

The notation $y(x, t)$ means that y is a *function* of x and t: the value of y depends on the values of x and t in such a way that only one value of y (the dependent variable) corresponds to a particular choice of x and t (the independent variables).

A wave is represented mathematically by a variation in some quantity (such as pressure or displacement) that is described as a function of both position and time. For a transverse wave on a guitar string, the function specifies the displacement of each point on the string from its equilibrium position. If the string is oriented along the x-axis and the displacement of any point on the string is in the $\pm y$-direction, then the wave is described by a function of two variables: $y(x, t)$.

$+x$-direction: $y(x, t) = h(t - x/v)$
$-x$-direction: $y(x, t) = h(t + x/v)$
To understand the notation $h(t - x/v)$, imagine that you have a computer program that calculates the function $h(t)$: you type in the value of t and the program returns the corresponding value of h. To find $h(t - x/v)$, you calculate $t - x/v$ and type *that* value in as input to the same computer program.

Traveling Waves Consider a long stretched string along the x-axis. One end of the string (at $x = 0$) is moved by an external agent according to some function $y = h(t)$; as a result, a transverse wave is produced that travels in the $+x$-direction with wave speed v. If the wave retains the same shape as it moves down the string, then the motion of any point x on the string copies the motion of the left end after a time delay x/v (the time it takes the wave to travel a distance x at speed v—see Fig. 11.8). Thus, $y(x, t) = h(t - x/v)$. Even though the function that describes the wave has two variables (x and t), these variables must occur in the particular combination $(t - x/v)$ in order to describe a wave that retains its form as it propagates in the $+x$-direction. For a wave moving in the $-x$-direction, the variables would occur in the combination $(t + x/v)$. A wave that retains its shape as it moves in a single direction is called a *traveling wave.*

Harmonic Traveling Waves Suppose the motion of the left end of the string is described by $y = A \cos \omega t$. By substituting $(t - x/v)$ for t, we obtain the function that describes the motion of *any* point $x > 0$:

$$y(x, t) = A \cos [\omega(t - x/v)]$$

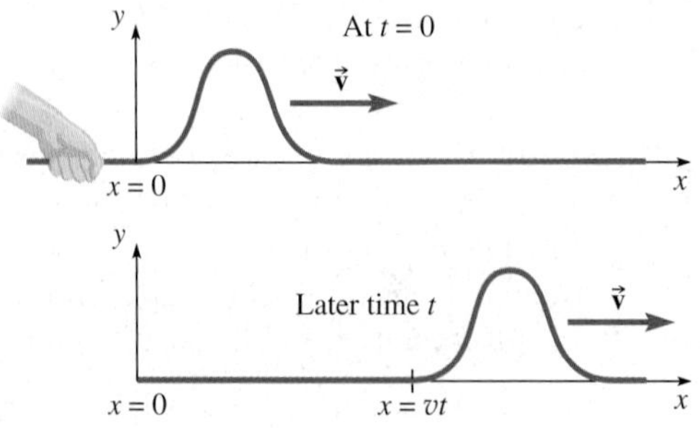

Figure 11.8 A wave pulse, with the same shape, at successive times. The motion of the point x repeats the motion of the point $x = 0$ with a time delay $\Delta t = x/v$.

To simplify the writing, we introduce a constant called the **wavenumber** (symbol k, SI unit rad/m):

$$k = \frac{\omega}{v} = \frac{2\pi f}{v} = \frac{2\pi}{\lambda} \quad (11\text{-}7)$$

Then the equation for the harmonic wave can be written

$$y(x, t) = A \cos (\omega t - kx) \quad (11\text{-}8)$$

The argument of the sine or cosine function, $(\omega t \pm kx)$, is called the **phase** of the wave at x and t. Phase is measured in units of angle (usually radians). The phase of a wave at a given point and instant of time tells us how far along that point is in the repeating pattern of its motion. Since a sine or cosine function repeats every 2π radians, the motions of two different points x_1 and x_2 that differ in phase by an integer times 2π are exactly the same; the points move "in sync" or *in phase* with each other. The distance between the two points is an integral number of wavelengths:

If

$$k(x_2 - x_1) = 2\pi n \quad \text{(where } n \text{ is any integer)}$$

then

$$x_2 - x_1 = \frac{2\pi n}{k} = \frac{2\pi n}{2\pi/\lambda} = n\lambda$$

CONNECTION:

Note the analogy between ω and k. $\omega = 2\pi/T$, where T is the repeat *time*; $k = 2\pi/\lambda$, where λ is the repeat *distance*. ω is measured in radians per *second*; k is measured in radians per *meter*. (WWW tutorial: sine wave)

Example 11.2

A Traveling Wave on a String

A wave on a string is described by $y(x, t) = a \sin (bt + cx)$, where a, b, and c are positive constants. (a) Does this wave retain its shape as it travels? (b) In what direction does the wave travel? (c) What is the wave speed?

Strategy We try to manipulate the function to see if it can be written as a function of either $(t - x/v)$ or $(t + x/v)$ as in the general harmonic wave equation $y(x, t) = A \cos \omega(t - x/v)$. The wave speed v does not appear explicitly in the function as written, but it may be some combination of the other constants in the function.

Solution The coefficient of t in our equation should be the constant that represents ω. Factoring out that constant, we have

$$y(x, t) = a \sin b\left(t + \frac{cx}{b}\right) = a \sin b\left(t + \frac{x}{b/c}\right)$$

Now we see that $y(x, t)$ is a function of $t + x/v$, where $v = b/c$:

$$y(x, t) = a \sin b\left(t + \frac{x}{v}\right) \text{ where } v = \frac{b}{c}$$

Therefore: (a) yes, the wave retains its shape since it is a function of $(t + x/v)$; (b) it travels in the $-x$-direction since the t and x/v terms have the same sign; and (c) the wave speed is b/c.

Discussion Before being completely satisfied with this solution, it is a good idea to check that b/c has the right units for wave speed. The two terms bt and cx that are added together must have the same units. In SI, the argument of a sine function is measured in radians. Then b is measured in rad/s and c is measured in rad/m. Then the units of b/c are (rad/s)/(rad/m) = m/s, which is correct for wave speed.

Practice Problem 11.2 Another Traveling Wave on a String

A wave on a string is described by

$$y(x, t) = (0.0050 \text{ m}) \sin [(4.0 \text{ rad/s})t - (0.50 \text{ rad/m})x]$$

(a) Does this wave retain its shape as it travels? (b) In what direction does the wave travel? (c) What is the wave speed?

11.6 GRAPHING WAVES

To graph a one-dimensional wave $y(x, t)$, only one of the two independent variables (x, t) can be plotted. The other must be "frozen"; it is treated as a constant. If x is held constant, then one particular point (determined by the value of x) is singled out; the graph shows the motion of *that point* as a function of time (Fig. 11.9a). If instead t is held constant and y is plotted as a function of x, then the graph is like a snapshot—an instantaneous picture of what the wave looks like *at that particular instant* (Fig. 11.9b).

Figure 11.9 Two graphs of a harmonic wave on a string described by the equation $y(x, t) = A \sin(\omega t - kx)$. (a) The vertical displacement y of a particular point on the string ($x = 0$) as a function of time t. (b) The vertical displacement y as a function of horizontal position x at a single instant of time ($t = 0$).

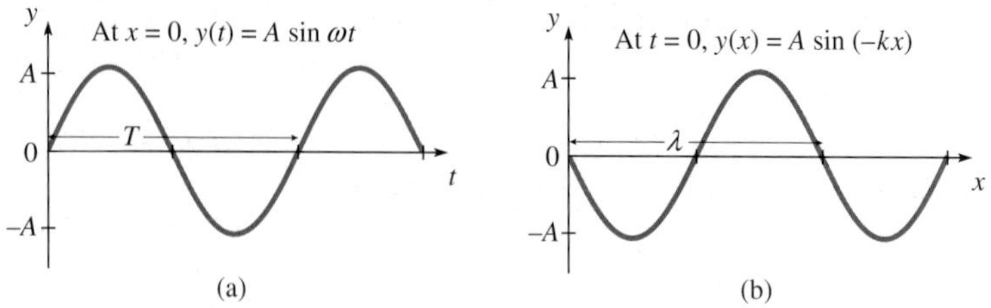

Example 11.3

A Transverse Harmonic Wave

A transverse harmonic wave travels in the $+x$-direction on a string at a speed of 5.0 m/s. Figure 11.10 shows a graph of $y(t)$ for the point $x = 0$. (a) What is the period of the wave? (b) What is the wavelength? (c) What is the amplitude? (d) Write the function $y(x, t)$ that describes the wave. (e) Sketch a graph of $y(x)$ at $t = 0$.

Strategy Since the graph uses time as the independent variable, the period can be read from the graph as the time for one cycle. The wavelength is the distance traveled by the wave during one period. The amplitude can be read from the graph as the maximum displacement. These are all the constants needed to write the function $y(x, t)$. We do have to think about the direction of travel and whether to write sine or cosine.

Solution (a) The period T is the time for one cycle. From the graph, $T = 2.0$ s.

(b) The wavelength λ is the distance traveled by the wave at speed $v = 5.0$ m/s during one period:

$$\lambda = vT = 5.0 \text{ m/s} \times 2.0 \text{ s} = 10 \text{ m}$$

(c) The amplitude A is the maximum displacement from equilibrium. From the graph, $A = 0.030$ m.

(d) Figure 11.10 is a sine function. The motion of the point $x = 0$ is

$$y(t) = A \sin\left(2\pi \frac{t}{T}\right)$$

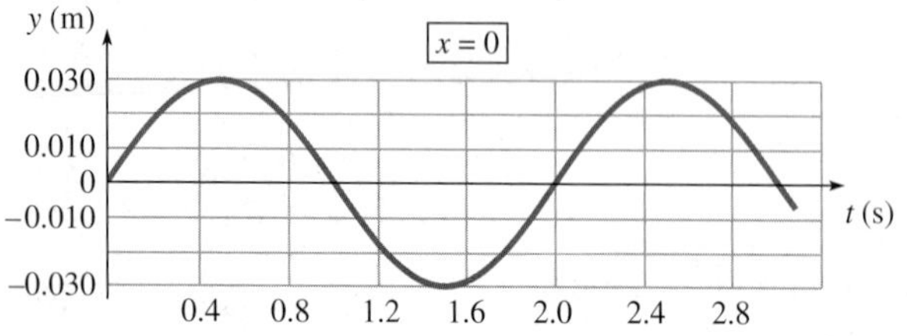

Figure 11.10
Graph of a transverse harmonic wave.

Since the wave moves in the $+x$-direction, a point at $x > 0$ duplicates the motion of $x = 0$ with a time delay of $\Delta t = x/v$ (the time for the wave to travel a distance x). Then

$$y(x, t) = A \sin\left(2\pi \frac{t - x/v}{T}\right)$$

where $v = 5.0$ m/s and $T = 2.0$ s.

(e) Substituting $t = 0$,

$$y(x) = A \sin\left(-2\pi \frac{x}{vT}\right)$$

Substituting $vT = \lambda$ and using the identity $\sin(-\theta) = -\sin\theta$ (Appendix A.7), we have

$$y(x, t = 0) = -A \sin\left(2\pi \frac{x}{\lambda}\right)$$

A graph of this function is an inverted sine function with amplitude $A = 0.030$ m and wavelength

$$\lambda = vT = 5.0 \text{ m/s} \times 2.0 \text{ s} = 10 \text{ m}$$

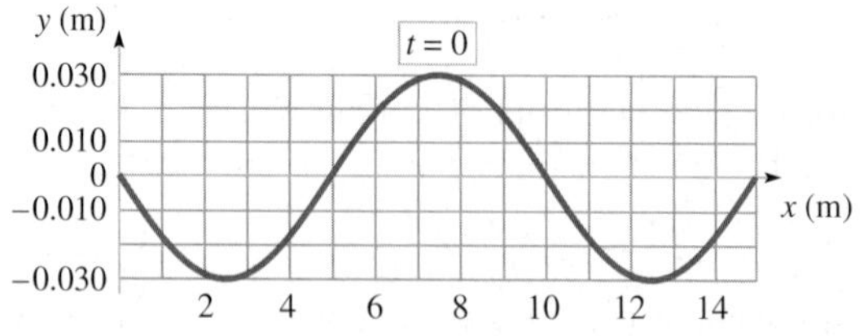

Discussion Figure 11.10 shows that the point $x = 0$ is initially at $y = 0$ and then moves up (in the $+y$-direction) until it reaches the crest (maximum y) at $t = 0.50$ s. Imagine the graph in (e) to represent the first frame (at $t = 0$) of a movie of the wave. Since the wave moves to the right, the sinusoidal pattern shifts a little to the right in each successive frame. The point $x = 0$ moves up until it reaches the crest when the wave has traveled 2.5 m to the right. Since the wave speed is 5.0 m/s, the point $x = 0$ reaches the crest at $t = (2.5 \text{ m})/(5.0 \text{ m/s}) = 0.50$ s.

continued on next page

Example 11.3 continued

Practice Problem 11.3 Another Harmonic Transverse Wave

A wave is described by $y(x, t) = (1.2\text{ cm}) \sin(10.0\pi t + 2.5\pi x)$, where x is in meters and t is in seconds. (a) Sketch a graph of $y(t)$ at $x = 0$. (b) Sketch a graph of $y(x)$ at $t = 0$. (c) What is the period of the wave? (d) What is the wavelength? (e) What is the amplitude? (f) What is the speed of the wave? (g) In what direction does the wave move?

11.7 PRINCIPLE OF SUPERPOSITION

Suppose two waves of the same type pass through the same region of space. Do the waves affect each other? If the amplitudes of the waves are large enough, then particles in the medium are displaced far enough from their equilibrium positions that Hooke's law (restoring force $\propto$ displacement) no longer holds; in that case, the waves *do* affect each other. However, for small amplitudes, the waves can pass through each other and emerge *unchanged.* More generally, when the amplitudes are not too large, the principle of superposition applies:

Principle of Superposition

When two or more waves overlap, the net disturbance at any point is the sum of the individual disturbances due to each wave.

Figure 11.11 illustrates the superposition principle for two wave pulses traveling toward one another on a string. The wave pulses pass right through one another without affecting one another; once they have separated, their shapes and heights are the same as before the overlap (Fig. 11.11a). The principle of superposition enables us to distinguish two voices speaking in the same room at the same time; the sound waves pass through each other unaffected.

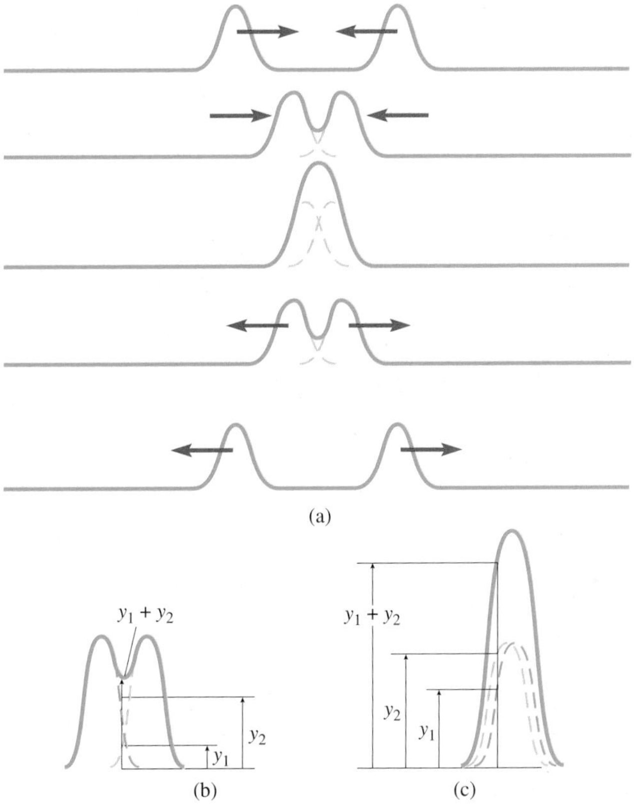

Figure 11.11 (a) Two identical wave pulses traveling toward and through each other. (b), (c) Applying the superposition principle at two different times; in each case, the dashed lines are the separate wave pulses and the solid line is the sum. If one of the pulses (acting alone) would produce a displacement y_1 at a certain point and the other would produce a displacement y_2 at the same point, the result when the two overlap is a displacement of $y_1 + y_2$.

Example 11.4

Superposition of Two Wave Pulses

Two identical wave pulses travel at 0.5 m/s toward each other on a long cord (Fig. 11.12). Sketch the shape of the cord at $t = 1.0$, 1.5, and 2.0 s.

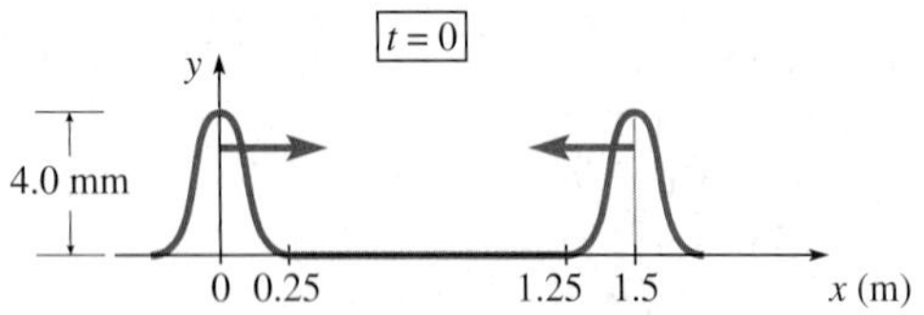

Figure 11.12
Two wave pulses at $t = 0$.

Strategy We start by sketching the two pulses in their new positions at each time given. Wherever they overlap, we apply superposition by adding the individual displacements at each point to find the net displacement of the cord at that point.

Solution Using graph paper, we draw the wave pulses at $t = 0$ (Fig. 11.13a). At $t = 1.0$ s, each pulse has moved 0.5 m toward the other. The leading edges of the pulses are just starting to overlap (Fig. 11.13b). At $t = 1.5$ s, each pulse has moved another 0.25 m; the crests overlap exactly. By adding the displacements point by point, we see that the string has the shape of a single pulse twice as high as either of the individual pulses (Fig. 11.13c). At $t = 2.0$ s, the pulses have each moved another 0.25 m (Fig. 11.13d).

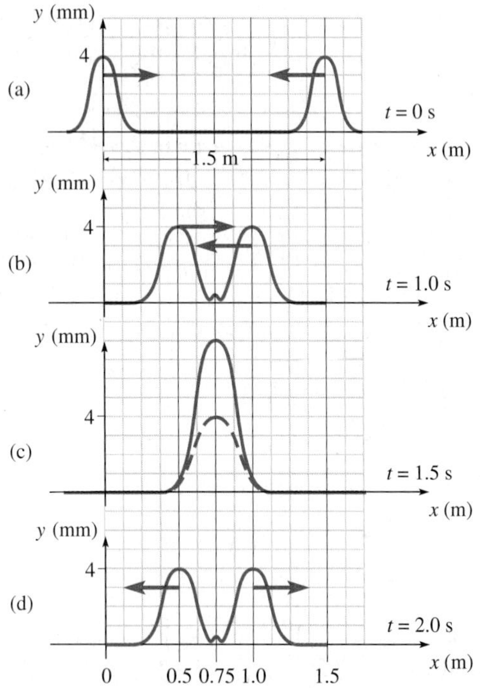

Figure 11.13
Wave positions at times $t = 0$, 1.0, 1.5, and 2.0 s.

Discussion When the two pulses exactly overlap, the displacement of points on the string is larger than for corresponding points on a single pulse because we add displacements *in the same direction* ($y > 0$ for both). However, superposition does not *always* produce larger displacements (see Practice Problem 11.4).

Practice Problem 11.4 Superposition of Two Opposite Wave Pulses

Repeat Example 11.4, except now let the pulse on the right be inverted (Fig. 11.14). [*Hint:* Points on the string below the x-axis have negative displacements ($y < 0$).]

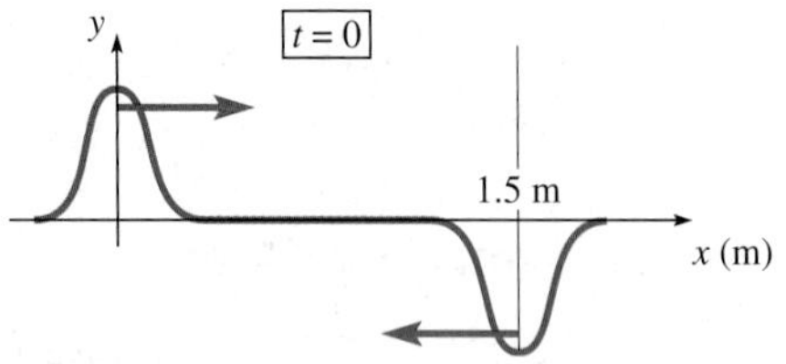

Figure 11.14
Wave pulses for Practice Problem 11.4.

11.8 REFLECTION AND REFRACTION

Reflection

At an abrupt boundary between one medium and another, **reflection** occurs; a reflected wave carrying some of the energy of the incident wave travels backward from the boundary. A sound wave in air, for instance, reflects when it reaches a wall.

A reflected wave can be inverted. Let's look at an extreme example: a string tied to a wall. If you send a wave pulse down the string, the reflected pulse is inverted (Fig. 11.15). By the principle of superposition, the shape of the string *at any point* is the sum of the incident and reflected waves, even at the fixed point at the end. The only way the end can stay in place is if the reflected wave is an upside down version of the incident wave. Another way to understand the inversion is by considering the force exerted on the string by the wall. When an upward pulse reaches the fixed end, the force exerted by the string on the wall has an upward component. By Newton's third law, the wall exerts a force on the string with a downward component. This downward force produces a downward reflected pulse.

Now, instead of tying the string to the wall, tie it to another string with an enormous linear mass density—so large that its motion is too small to measure. The original string doesn't know the difference; it just knows that one end is fixed in place. The second string with the huge density has a much slower wave speed than the first string. Now make the mass density of the second string not huge, but still greater than the first string. The greater inertia inhibits the motion of the boundary point and causes the reflected wave to be inverted. In general, when a transverse wave on a string reflects from a boundary with a region of slower wave speed, the reflected wave is inverted. On the other hand, when such a wave reflects from a boundary with a region of *faster* wave speed, the reflected wave is *not* inverted.

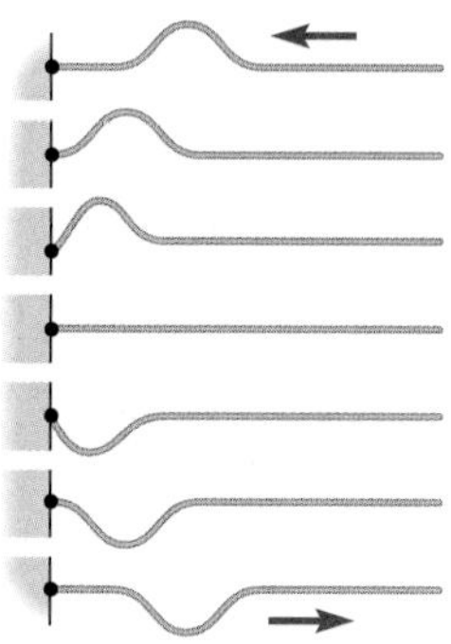

Figure 11.15 Snapshots of the reflection of a wave pulse from a fixed end. The reflected pulse is upside down.

Change in Wavelength at a Boundary

When there is an abrupt change in wave medium, an incident wave splits up at the boundary; part is reflected and part is transmitted past the boundary into the other medium. The frequencies of both the reflected and transmitted waves are the same as the frequency of the incident wave. To understand why, think of a wave incident on the knot between two different strings. Both the reflected and the transmitted waves are generated by the up-and-down motion of the knot; the knot vibrates at the frequency dictated by the incident wave. However, if the wave speed changes at the boundary, *the wavelength of the transmitted wave is not the same* as the wavelength of the incident and reflected waves. Since $v = \lambda f$ and the frequencies are the same,

$$f = \frac{v_1}{\lambda_1} = \frac{v_2}{\lambda_2} \qquad (11\text{-}9)$$

Equation (11-9) applies to any kind of wave and is of particular importance in the study of optics.

When a wave passes from one medium into another, the frequency of the transmitted wave is the same as that of the incident wave.

Example 11.5

Wavelength in Air and Under Water

A horn near the beach emits a 440-Hz sound wave. (a) What is the wavelength of the sound wave in air? The speed of sound in air is 340 m/s. (b) What is the wavelength of the sound wave in seawater? The speed of sound in seawater is 1520 m/s.

Strategy The *frequency* of the sound wave in water is the same as in air. The wavelengths depend on both the frequency and the speed of sound in the medium. Sound travels faster in solids and liquids than in gases; during one period, the wave travels farther in water than it does in air, so the wavelength is longer in water.

Solution (a) The wavelength in air is related to the speed of sound in air and the frequency:

$$\lambda_{air} = v_{air}T = \frac{v_{air}}{f}$$

Substituting numerical values,

$$\lambda_{air} = \frac{340\ \text{m/s}}{440\ \text{Hz}} = 0.77\ \text{m}$$

(b) The wave in the water has the *same frequency*, but the speed of sound is different:

$$\lambda_{water} = \frac{v_{water}}{f} = \frac{1520\ \text{m/s}}{440\ \text{Hz}} = 3.5\ \text{m}$$

Discussion The wavelength in water is longer, as expected. As a quick check, the ratio of the wavelengths should be equal to the ratio of the wave speeds:

$$\frac{0.77\ \text{m}}{3.5\ \text{m}} = 0.22; \quad \frac{340\ \text{m/s}}{1520\ \text{m/s}} = 0.22$$

continued on next page

Example 11.5 continued

Practice Problem 11.5 Working on the Railroad

A railroad worker, driving in spikes, misses the spike and hits the iron rail; a sound wave travels through the air and through the rail. (We ignore the *transverse* wave that also travels in the rail.) The wavelength of the sound in air is 0.548 m. The speed of sound in air is 340 m/s; the speed of sound in iron is 5300 m/s. (a) What is the frequency of the wave? (b) What is the wavelength of the sound wave in the rail?

Refraction

Application of refraction: why ocean waves approach shore nearly head-on

A transmitted wave not only has a different wavelength than the incident wave; it also travels in a different direction unless the incident wave's direction of propagation is along the *normal* (the direction perpendicular to the boundary). This change in propagation direction is called **refraction**. If the change in wave speed is gradual, then the change in direction is gradual as well. The speed of ocean waves depends on the depth of the water; the waves are slower in shallower water. As waves approach the shore, they gradually slow down; as a result, they gradually bend until they reach shore nearly head-on. A *sudden* change in wave speed, such as when a seismic wave is incident on a boundary between different kinds of rock, causes a sudden refraction (Fig. 11.16).

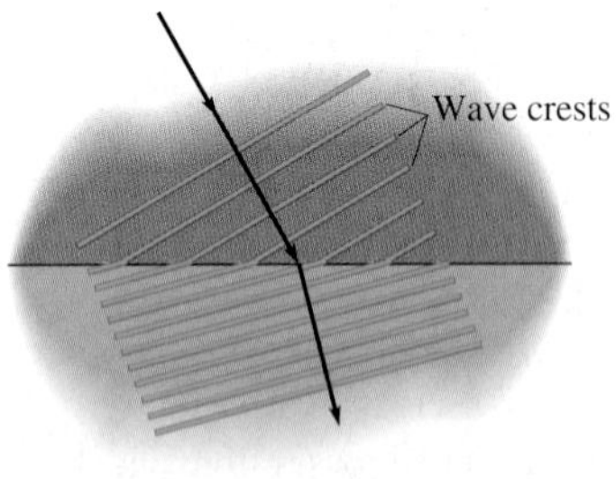

Figure 11.16 Wave crests for a seismic wave incident on a boundary between two different kinds of rock. Not only does the wavelength (distance between wave crests) change at the boundary, the wave also refracts (changes its direction of propagation). The reflected wave is omitted for clarity.

Application of Reflection and Refraction: Seismology Understanding the propagation of seismic waves, including reflection and refraction due to boundaries between geological features, is an essential part of the effort to reduce damage from future earthquakes. Scientists create small seismic waves with a large vibrator, then use seismographs to record ground vibrations at various locations. The goal is to produce a seismic hazard map so that preventative measures can be targeted to areas with the highest risk of earthquake damage.

11.9 INTERFERENCE AND DIFFRACTION

Interference

The principle of superposition leads to dramatic effects when applied to coherent waves. Two waves are **coherent** if they have the *same frequency* and they maintain a *fixed phase relationship* with one another. One way to obtain coherent waves is to get them from the same source. Such is the case, for example, if one *monophonic* amplifier sends the same signal to two speakers. Should some fluctuation occur in the amplifier driving the speakers, the same fluctuation occurs in both speakers at the same time and they maintain their coherence. Waves are **incoherent** if the phase relationship between them varies randomly. (As defined here, *coherent* and *incoherent* are idealized extremes. In reality, two waves do not have to have either perfect correlation between their phases or no correlation at all.)

Suppose coherent waves with amplitudes A_1 and A_2 pass through the same point in space. If the waves are *in phase* at that point—that is, the phase difference is any *even* integral multiple of π rad—then the two waves consistently reach their maxima at exactly the same instants of time (Fig. 11.17a). The superposition of the waves that are in phase with one another is called **constructive interference**; the amplitude of the combined waves is the sum of the amplitudes of the two individual waves ($A_1 + A_2$).

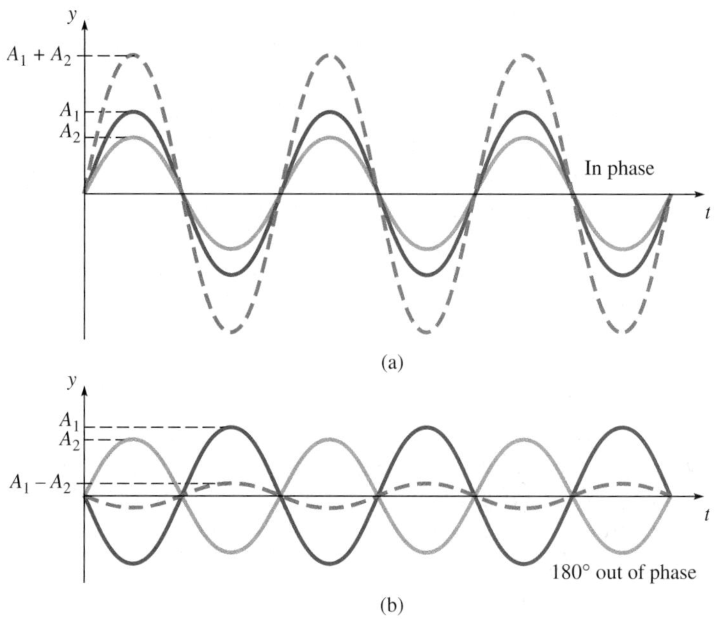

Figure 11.17 Coherent waves (a) in phase and (b) 180° out of phase. (One wave is drawn with a lighter line to distinguish it from the other.) The dashed curve is the superposition of the two waves.

Two waves that are *180° out of phase* at a given point have a phase difference of π rad, 3π rad, 5π rad, and so on. The waves are half a cycle apart; when one reaches its maximum, the other reaches its minimum (Fig. 11.17b). The superposition of waves that are 180° out of phase is called **destructive interference**—the amplitude of the combined waves is the *difference* of the amplitudes of the two individual waves ($|A_1 - A_2|$). For any other fixed phase relationship between the two waves, the superposition has an amplitude between $A_1 + A_2$ and $|A_1 - A_2|$.

Suppose two coherent waves start out in phase with one another. In Fig. 11.18, two rods vibrate up and down in step with one another to generate circular waves on the surface of the water. If the two waves travel the same distance to reach a point on the water surface, they arrive *in phase* and interfere constructively. At points where the distances are different, the phase difference is proportional to the path difference. One wavelength of path difference corresponds to a phase difference of 2π radians (one full cycle), so

$$\frac{d_1 - d_2}{\lambda} = \frac{\text{phase difference}}{2\pi\,\text{rad}} \quad (11\text{-}10)$$

Thus, the phase difference is

$$\text{phase difference} = \frac{2\pi\,\text{rad}}{\lambda} \times (d_1 - d_2) = k(d_1 - d_2) \quad (11\text{-}11)$$

If the path difference $d_1 - d_2$ is an integral number of wavelengths, then the phase difference is an even integral multiple of π rad and constructive interference occurs at point P. If the path difference is $\frac{1}{2}\lambda$, $\frac{3}{2}\lambda$, $\frac{5}{2}\lambda$, . . . , then the phase difference is an odd integral multiple of π rad and destructive interference occurs at point P. If the phase difference is not an integral multiple of π, the amplitude has a value between the maximum and minimum possible values.

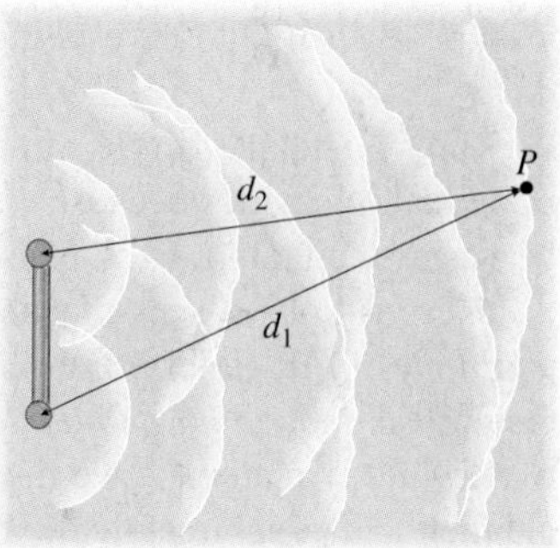

Figure 11.18 Overhead snapshot of two coherent surface water waves. The two waves travel different distances d_1 and d_2 to reach a point P. The phase difference between the waves at point P is $k(d_1 - d_2)$.

When two coherent waves are interfering and have a phase difference $= n\pi$, the interference is constructive for even n and the interference is destructive for odd n.

Intensity Effects for Interfering Waves

When coherent waves interfere, the *amplitudes* add (for constructive interference) or subtract (for destructive interference)—see Example 11.6. However, since intensity is proportional to amplitude *squared*, we cannot simply add or subtract the *intensities* of coherent waves when they interfere. *Incoherent* waves, on the other hand, have no fixed phase relationship; interference effects are averaged out due to the rapidly varying phase difference. In the superposition of incoherent waves, the total intensity is the sum of the individual intensities.

Why don't we see and hear interference effects all the time? Light from ordinary sources—incandescent bulbs, fluorescent bulbs, or the Sun—is incoherent because it is generated by large numbers of independent atomic sources. A single source of sound normally contains many different frequencies, so a point of constructive interference for one frequency is not a point of constructive interference for other frequencies. Furthermore, in most situations there are many different sound waves that reach our ears after traveling different paths due to the reflection of sound from walls, ceilings, chairs, and so forth.

Example 11.6

Intensity of Interfering Waves

Two coherent waves interfere. The intensity of one of them (alone) is 9.0 times the intensity of the other. What is the ratio of the maximum possible intensity to the minimum possible intensity of the resulting wave?

Strategy The intensity is *not* the sum or difference of the individual intensities because the waves are coherent. Since the waves maintain a fixed phase relationship, the principle of superposition tells us that the maximum and minimum *amplitudes* of the interfering waves are the sum and difference of the *individual amplitudes*. Intensity is proportional to amplitude squared, so we find the ratio of the amplitudes and then add or subtract them.

Solution The intensities of the two individual waves are related by $I_1 = 9.0I_2$ or $I_1/I_2 = 9.0$. Since intensity is proportional to amplitude squared,

$$\frac{A_1}{A_2} = \sqrt{\frac{I_1}{I_2}} = 3.0$$

Thus, $A_1 = 3.0A_2$. The maximum possible amplitude for the superposition occurs if the waves are in phase:

$$A_{\text{max}} = A_1 + A_2 = 4.0A_2$$

The minimum possible amplitude for the superposition occurs if the waves are 180° out of phase:

$$A_{\text{min}} = |A_1 - A_2| = 2.0A_2$$

The ratio of the maximum to minimum intensity is

$$\frac{I_{\text{max}}}{I_{\text{min}}} = \left(\frac{A_{\text{max}}}{A_{\text{min}}}\right)^2 = \left(\frac{4.0}{2.0}\right)^2 = 4.0$$

Discussion Had we added and subtracted the *intensities* instead of the amplitudes, we would have found a ratio of $10/8 = 1.25$ between the maximum and minimum intensities. We must be careful to add or subtract the *amplitudes* of the interfering waves instead of the intensities themselves when the waves are coherent.

Practice Problem 11.6 Two More Coherent Waves

Repeat Example 11.6, but change the ratio of the individual intensities to 4.0 (instead of 9.0).

Diffraction

Diffraction is the spreading of a wave around an obstacle in its path (Fig. 11.19). The amount of diffraction depends on the relative size of the obstacle and the wavelength of the waves. Diffraction enables you to hear around a corner but not to see around a corner. Sound waves, with typical wavelengths in air of around 1 m, diffract around the corner much more than do light waves with much smaller wavelengths (less than 1 μm). We will study interference and diffraction of electromagnetic waves (including light) in detail in Chapter 25.

Figure 11.19 Water waves incident from the right on a gap in a breakwater (Three Fathoms Cove, Hong Kong). Notice the shape of the wave crests to the left of the gap.

11.10 STANDING WAVES

Standing waves occur when a wave is reflected at a boundary and the reflected wave interferes with the incident wave so that the wave appears not to propagate. Suppose that a harmonic wave on a string, coming from the right, hits a boundary where the string is fixed. The equation of the incident wave is

$$y(x, t) = A \sin(\omega t + kx)$$

The + sign is chosen in the phase because the wave travels to the left.

The reflected wave travels to the right, so $+kx$ is replaced with $-kx$; and the reflected wave is inverted, so $+A$ is replaced with $-A$. Then the reflected wave is described by

$$y(x, t) = -A \sin(\omega t - kx)$$

Applying the principle of superposition, the motion of the string is described by

$$y(x, t) = A [\sin(\omega t + kx) - \sin(\omega t - kx)]$$

This can be rewritten in a form that shows the motion of the string more clearly. Using the trigonometric identity (Appendix A.7),

$$\sin \alpha - \sin \beta = 2 \cos [\tfrac{1}{2}(\alpha + \beta)] \sin [\tfrac{1}{2}(\alpha - \beta)]$$

where

$$\alpha = \omega t + kx \quad \text{and} \quad \beta = \omega t - kx$$

we have

$$y(x, t) = 2A \cos \omega t \sin kx$$

Notice that t and x are separated. Every point moves in SHM with the same frequency. However, in contrast to a *traveling* harmonic wave, every point reaches its maximum distance from equilibrium *simultaneously*. In addition, different points move with different amplitudes; the amplitude at any point x is $2A \sin kx$.

Figure 11.20 A standing wave at various times: $t = 0, \frac{1}{8}T, \frac{2}{8}T, \frac{3}{8}T$, and $\frac{4}{8}T$, where T is the period.

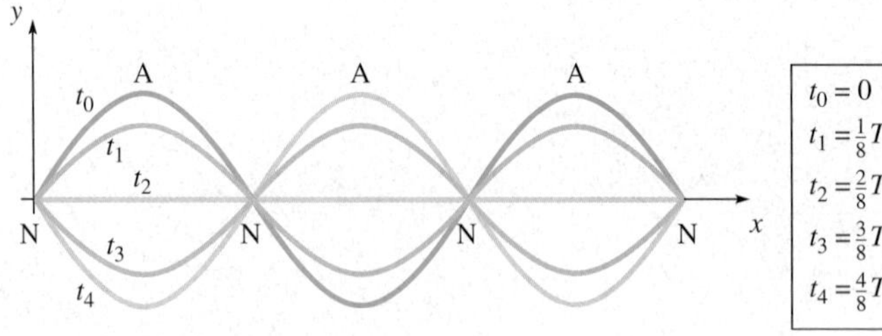

Node–node distance is $\frac{1}{2}\lambda$.
Node–antinode distance is $\frac{1}{4}\lambda$.

Figure 11.20 shows the string at time intervals of $\frac{1}{8}T$, where T is the period. What you actually see when looking at a standing wave is a blur of moving string, with points that never move (**nodes**, labeled "N") halfway between points of maximum amplitude (**antinodes**, labeled "A"). The nodes are the points where $\sin kx = 0$. Since $\sin n\pi = 0$ ($n = 0, 1, 2, \ldots$), the nodes are located at $x = n\pi/k = n\lambda/2$. Thus, the distance between two adjacent nodes is $\frac{1}{2}\lambda$. The antinodes occur where $\sin kx = \pm 1$, which is exactly halfway between a pair of nodes. So the nodes and antinodes alternate, with one quarter of a wavelength between a node and the neighboring antinode.

So far we have ignored what happens at the other end of the string. If the other end is fixed, then it is a node. The string thus has two or more nodes, with one at each end. The distance between each pair of nodes is $\frac{1}{2}\lambda$, so

$$n(\lambda/2) = L \qquad \text{(11-12a)}$$

where L is the length of the string and $n = 1, 2, 3, \ldots$. The possible wavelengths for standing waves on a string are

$$\lambda_n = \frac{2L}{n} \quad (n = 1, 2, 3, \ldots) \qquad \text{(11-12b)}$$

There is no need to memorize Eqs. (11-12) and (11-13). Start with a sketch like Fig. 11.21, find the wavelengths, and then use $v = f\lambda$ to find the frequencies.

The frequencies are

$$f_n = \frac{v}{\lambda_n} = \frac{nv}{2L} \quad (n = 1, 2, 3, \ldots) \qquad \text{(11-13)}$$

The lowest frequency standing wave ($n = 1$) is called the **fundamental**. Notice that the higher frequency standing waves are all integral multiples of the fundamental; the set of standing wave frequencies makes an evenly spaced set:

$$f_1, 2f_1, 3f_1, 4f_1, \ldots, nf_1, \ldots$$

CONNECTION:

An ideal mass-spring system has a single resonant frequency (Section 10.10), but extended objects generally have many different resonant frequencies.

These frequencies are called the *natural frequencies* or *resonant frequencies* of the string. *Resonance* occurs when a system is driven at one of its natural frequencies; the resulting vibrations are large in amplitude compared to when the driving frequency is not close to any of the natural frequencies.

✓CHECKPOINT 11.10

A standing wave on a string 1.0 m long has four nodes, not including the nodes at the two fixed ends. What is the wavelength?

Figure 11.21 shows the first four standing wave patterns on a string. The two ends are always nodes since they are fixed in place. Notice that each successive pattern has one more node and one more antinode than the previous one. The fundamental has the fewest possible number of nodes (2) and antinodes (1).

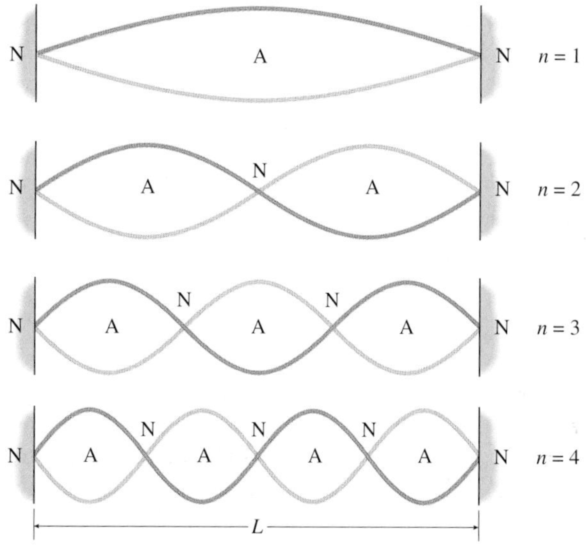

Figure 11.21 Four standing wave patterns for a string fixed at both ends. "N" marks the locations of the nodes and "A" marks the locations of the antinodes. In each case, the node-to-node distance is $\frac{1}{2}\lambda$ and n such "loops" fit into the length L of the string, so $n(\lambda/2) = L$.

Example 11.7

Wavelength of a Standing Wave

A string is attached to a vibrator driven at 1.20×10^2 Hz. A weight hangs from the other end of the string; the weight is adjusted until a standing wave is formed (Fig. 11.22). What is the wavelength of the standing wave on the string?

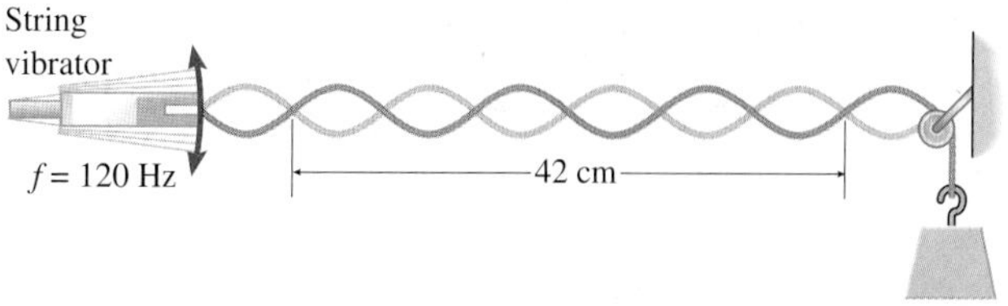

Figure 11.22
Measuring distance between nodes for a standing wave.

Strategy The measured distance of 42 cm encompasses six "loops"—that is, six segments of string between one node and the next. Each of the loops represents a length of $\frac{1}{2}\lambda$.

Solution The length of one loop is

$$42 \text{ cm} \times \tfrac{1}{6} = 7.0 \text{ cm}$$

Since the length of one loop is $\frac{1}{2}\lambda$, the wavelength is 14 cm.

Discussion This string is *not* fixed at both ends. The left end is connected to a moving vibrator, so it is not a node. The right end wraps around a pulley; it may not be easy to determine precisely where the "end" is. For this case, it is more accurate to measure the distance between two actual nodes rather than to assume that the ends are nodes.

Practice Problem 11.7 Standing Wave with Seven Loops

The vibrator frequency is increased until there are seven loops within the 42-cm length. What is the new standing wave frequency for this string (assuming the same tension)?

Resonance is responsible for much of the structural damage caused by seismic waves. If the frequency at which the ground vibrates is close to a resonant frequency of a structure, the vibration of the structure builds up to a large amplitude. Thus, to construct a building that can survive an earthquake, it is not enough to make it stronger. Either the building must be designed so it is isolated from ground vibrations, or a damping mechanism—something like a shock absorber—must be incorporated to dissipate energy and reduce the amplitude of the vibrations. Damping is b ecoming increasingly common in large buildings since it is just as effective and much less expensive than isolation.

Application of resonance: damage caused by earthquakes

Large sections of the Hanshin expressway vibrated in a twisting motion due to ground vibrations near a resonant frequency. The road now has rubber base isolators instead of steel bearings connecting the roadway to the concrete piers. Part of their function is to act like shock absorbers.

Master the Concepts

- An isotropic source radiates sound uniformly in all directions. Assuming that no energy is absorbed by the medium and there are no obstacles to reflect or absorb sound, the intensity I at a distance r from an isotropic source is

$$I = \frac{\text{power}}{\text{area}} = \frac{P}{4\pi r^2} \qquad (11\text{-}1)$$

- In a transverse wave, the motion of particles in the medium is perpendicular to the direction of propagation of the wave. In a longitudinal wave, the motion of particles in the medium is along the same line as the direction of propagation of the wave.

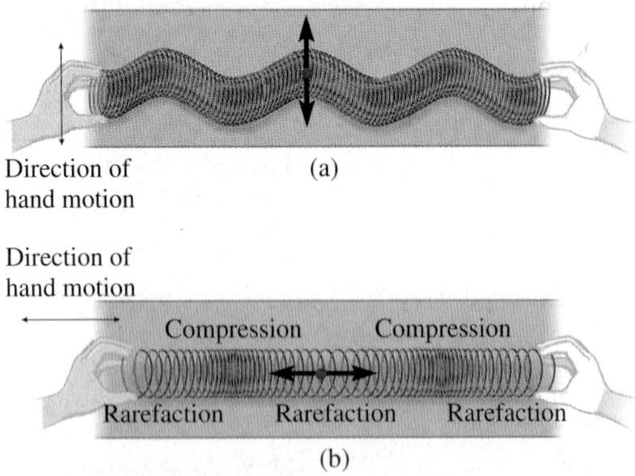

- The speed of a mechanical wave depends on properties of the wave medium. More restoring force makes faster waves; more inertia makes slower waves.
- The speed of a transverse wave on a string is

$$v = \sqrt{\frac{F}{\mu}} \qquad (11\text{-}4)$$

where

$$\mu = m/L \qquad (11\text{-}3)$$

- A periodic wave repeats the same pattern over and over. Harmonic waves are a special kind of periodic wave characterized by a sinusoidal function (either a sine or cosine function).
- If a periodic wave has period T and travels at speed v, the repetition distance of the wave is the wavelength:

$$\lambda = vT \qquad (11\text{-}5)$$

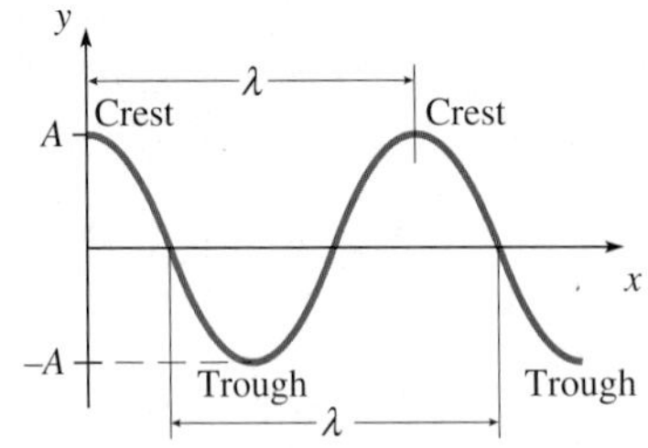

- The principle of superposition: When two or more waves overlap, the net disturbance at any point is the sum of the individual disturbances due to each wave.
- A harmonic traveling wave can be described by

$$y(x, t) = A \cos(\omega t - kx) \qquad (11\text{-}8)$$

The argument of the sinusoidal function, $(\omega t \pm kx)$, is called the phase of the wave at x and t. The constant k is the wave number

$$k = \frac{\omega}{v} = \frac{2\pi f}{v} = \frac{2\pi}{\lambda} \qquad (11\text{-}7)$$

- Reflection occurs at a boundary between different wave media. Some energy may be transmitted into the new medium and the rest is reflected. The wave transmitted past the boundary is refracted (propagates in a different direction).
- Coherent waves have the *same frequency* and maintain a *fixed phase relationship* with one another. Coherent waves that are in phase with one another interfere constructively; those that are 180° out of phase interfere destructively.
- Diffraction occurs when a wave bends around an obstacle in its path.
- In a standing wave on a string, every point moves in SHM with the same frequency. Nodes are points of zero amplitude; antinodes are points of maximum amplitude.

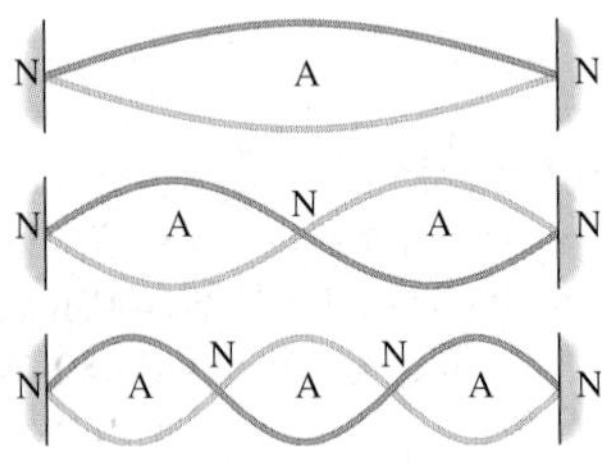

The distance between two adjacent nodes is $\frac{1}{2}\lambda$.

Conceptual Questions

1. The piano strings that vibrate with the lowest frequencies consist of a steel wire around which a thick coil of copper wire is wrapped. Only the inner steel wire is under tension. What is the purpose of the copper coil?
2. Is the vibration of a string in a piano, guitar, or violin a *sound* wave? Explain.
3. The wavelength of the fundamental standing wave on a cello string depends on which of these quantities: length of the string, mass per unit length of the string, or tension? The wavelength of the *sound wave* resulting from the string's vibration depends on which of the same three quantities?
4. If the length of a guitar string is decreased while the tension remains constant, what happens to each of these

quantities? (a) the wavelength of the fundamental, (b) the frequency of the fundamental, (c) the time for a pulse to travel the length of the string, (d) the maximum velocity of a point on the string (assuming the amplitude is the same both times), (e) the maximum acceleration of a point on the string (assuming the amplitude is the same both times).

5. Why is it possible to understand the words spoken by two people at the same time?
6. A cello player can change the frequency of the sound produced by her instrument by (a) increasing the tension in the string, (b) pressing her finger on the string at different places along the fingerboard, or (c) bowing a different string. Explain how each of these methods affects the frequency.
7. Why is a transverse wave sometimes called a shear wave?
8. The drawing shows a complex wave moving to the right along a cord. Draw the shape of the cord an instant later and determine which parts of the cord are moving upward and which are moving downward. Indicate the directions on your drawing with arrows.

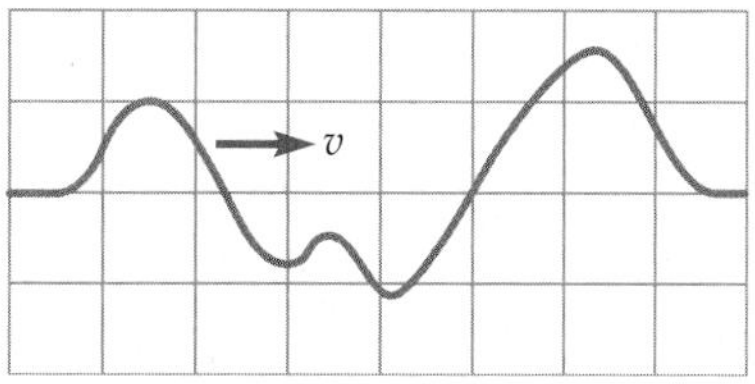

9. When an earthquake occurs, the S waves (transverse waves) are not detected on the opposite side of the Earth while the P waves (longitudinal waves) are. How does this provide evidence that the Earth's solid core is surrounded by liquid?
10. Simple ear-protection devices use materials that reflect or absorb sound before it reaches the ears. A newer technology, sometimes called *noise cancellation*, uses a microphone to produce an electrical signal that mimics the noise. The signal is modified electronically, then fed to the speakers in a pair of headphones. The speakers emit sound waves that *cancel* the noise. On what principle is this technology based? What kind of modification is made to the electrical signal?
11. When connecting speakers to a stereo, it is important to connect them with the correct polarity so that, if the same electrical signal is sent, they both move in the same direction. If the wires going to one speaker are reversed, the listener hears a noticeably weaker bass (low frequencies). Explain what causes this and why low frequencies are affected more than high frequencies.

Multiple-Choice Questions

1. Standing waves are produced by the superposition of two waves with
   (a) the same amplitude, frequency, and direction of propagation.
   (b) the same amplitude and frequency, and opposite propagation directions.
   (c) the same amplitude and direction of propagation, but different frequencies.
   (d) the same amplitude, different frequencies, and opposite directions of propagation.
2. A transverse wave travels on a string of mass m, length L, and tension F. Which statement here is correct?
   (a) The energy of the wave is proportional to the square root of the wave amplitude.
   (b) The speed of a moving point on the string is the same as the wave speed.
   (c) The wave speed is determined by the values of m, L, and F.
   (d) The wavelength of the wave is proportional to L.
3. A transverse wave on a string is described by $y(x, t) = A\cos(\omega t + kx)$. It arrives at the point $x = 0$ where the string is fixed in place. Which function describes the reflected wave?
   (a) $A\cos(\omega t + kx)$ (b) $A\cos(\omega t - kx)$
   (c) $-A\sin(\omega t + kx)$ (d) $-A\cos(\omega t - kx)$
   (e) $A\sin(\omega t + kx)$
4. A violin string of length L is fixed at both ends. Which one of these is *not* a wavelength of a standing wave on the string?
   (a) L (b) $2L$ (c) $L/2$ (d) $L/3$ (e) $2L/3$ (f) $3L/2$
5. The speed of waves in a stretched string depends on which one of the following?
   (a) The tension in the string
   (b) The amplitude of the waves
   (c) The wavelength of the waves
   (d) The gravitational field strength
6. The higher the frequency of a wave,
   (a) the smaller its speed.
   (b) the shorter its wavelength.
   (c) the greater its amplitude.
   (d) the longer its period.
7. In a transverse wave, the individual particles of the medium
   (a) move in circles. (b) move in ellipses.
   (c) move parallel to the direction of the wave's travel.
   (d) move perpendicularly to the direction of the wave's travel.
8. Which is the only one of these properties of a wave that could be changed without changing any of the others?
   (a) amplitude (b) wavelength
   (c) speed (d) frequency

9. Two successive transverse pulses, one caused by a brief displacement to the right and the other by a brief displacement to the left, are sent down a Slinky that is fastened at the far end. At the point where the first reflected pulse meets the second advancing pulse, the deflection (compared with that of a single pulse) is
(a) quadrupled. (b) doubled.
(c) canceled. (d) halved.

10. The intensity of an isotropic sound wave is
(a) directly proportional to the distance from the source.
(b) inversely proportional to the distance from the source.
(c) directly proportional to the square of the distance from the source.
(d) inversely proportional to the square of the distance from the source.
(e) none of the above.

Problems

- Ⓒ Combination conceptual/quantitative problem
- Biological or medical application
- ✦ Challenging problem
- Blue # Detailed solution in the Student Solutions Manual
- ⟨1 2⟩ Problems paired by concept
- Text website interactive or tutorial

11.1 Waves and Energy Transport

1. The intensity of sunlight that reaches Earth's atmosphere is 1400 W/m^2. What is the intensity of the sunlight that reaches Jupiter? Jupiter is 5.2 times as far from the Sun as Earth. [*Hint:* Treat the Sun as an isotropic source of light waves.]
2. Michelle is enjoying a picnic across the valley from a cliff. She is playing music on her radio (assume it to be an isotropic source) and notices an echo from the cliff. She claps her hands and the echo takes 1.5 s to return. (a) Given that the speed of sound in air is 343 m/s on that day, how far away is the cliff? (b) If the intensity of the music 1.0 m from the radio is 1.0×10^{-5} W/m^2, what is the intensity of the music arriving at the cliff?
3. The intensity of the sound wave from a jet airplane as it is taking off is 1.0×10^2 W/m^2 at a distance of 5.0 m. What is the intensity of the sound wave that reaches the ears of a person standing at a distance of 120 m from the runway? Assume that the sound wave radiates from the airplane equally in all directions.
4. At what rate in watts does the jet airplane in Problem 3 radiate energy in the form of sound waves?
5. The Sun emits electromagnetic waves (including light) equally in all directions. The intensity of the waves at Earth's upper atmosphere is 1.4 kW/m^2. At what rate does the Sun emit electromagnetic waves? (In other words, what is the power output?)

11.3 Speed of Transverse Waves on a String

6. (a) What is the speed of propagation of the pulse shown in the figure? (b) At what average speed does the point at $x = 2.0$ m move during this time interval?

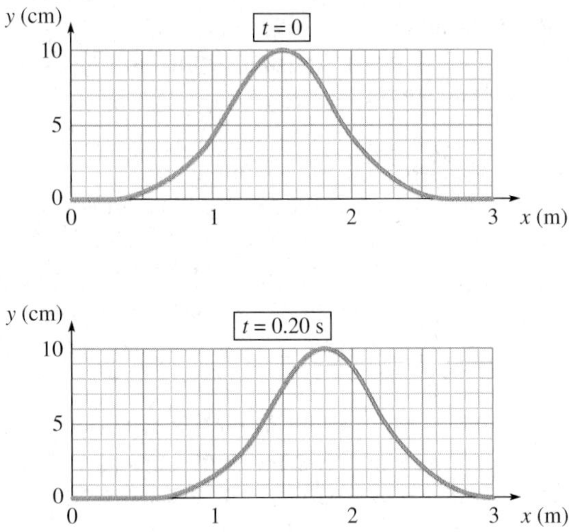

Problems 6 and 7

7. (a) What is the position of the peak of the pulse shown in the figure with Problem 6 at $t = 3.00$ s? (b) When does the peak of the pulse arrive at $x = 4.00$ m?
8. When the tension in a cord is 75 N, the wave speed is 140 m/s. What is the linear mass density of the cord?
9. A metal guitar string has a linear mass density of $\mu = 3.20$ g/m. What is the speed of transverse waves on this string when its tension is 90.0 N?
10. Two strings, each 15.0 m long, are stretched side by side. One string has a mass of 78.0 g and a tension of 180.0 N. The second string has a mass of 58.0 g and a tension of 160.0 N. A pulse is generated at one end of each string simultaneously. On which string will the pulse move faster? Once the faster pulse reaches the far end of its string, how much additional time will the slower pulse require to reach the end of its string?
11. A uniform string of length 10.0 m and weight 0.25 N is attached to the ceiling. A weight of 1.00 kN hangs from its lower end. The lower end of the string is suddenly displaced horizontally. How long does it take the resulting wave pulse to travel to the upper end? [*Hint:* Is the weight of the string negligible in comparison with that of the hanging mass?]

11.4 Periodic Waves

12. What is the speed of a wave whose frequency and wavelength are 500.0 Hz and 0.500 m, respectively?
13. What is the wavelength of a wave whose speed and period are 75.0 m/s and 5.00 ms, respectively?
14. What is the frequency of a wave whose speed and wavelength are 120 m/s and 30.0 cm, respectively?

15. The speed of sound in air at room temperature is 340 m/s. (a) What is the frequency of a sound wave in air with wavelength 1.0 m? (b) What is the frequency of a radio wave with the same wavelength? (Radio waves are electromagnetic waves that travel at 3.0×10^8 m/s in air or in vacuum.)

16. Light visible to humans consists of electromagnetic waves with wavelengths (in air) in the range 400–700 nm (4.0×10^{-7} m to 7.0×10^{-7} m). The speed of light in air is 3.0×10^8 m/s. What are the frequencies of electromagnetic waves that are visible?

17. A fisherman notices a buoy bobbing up and down in the water in ripples produced by waves from a passing speedboat. These waves travel at 2.5 m/s and have a wavelength of 7.5 m. At what frequency does the buoy bob up and down?

11.5 Mathematical Description of a Wave

18. You are swimming in the ocean as water waves with wavelength 9.6 m pass by. What is the closest distance that another swimmer could be so that his motion is exactly opposite yours (he goes up when you go down)?

19. What is the speed of the wave represented by $y(x, t) = A \sin (kx - \omega t)$, where $k = 6.0$ rad/cm and $\omega = 5.0$ rad/s?

20. The equation of a wave is

$$y(x, t) = (3.5 \text{ cm}) \sin \left\{\frac{\pi}{3.0 \text{ cm}} [x - (66 \text{ cm/s})t]\right\}$$

Find (a) the amplitude and (b) the wavelength of this wave.

✦21. A wave on a string has equation

$$y(x, t) = (4.0 \text{ mm}) \sin (\omega t - kx)$$

where $\omega = 6.0 \times 10^2$ rad/s and $k = 6.0$ rad/m. (a) What is the amplitude of the wave? (b) What is the wavelength? (c) What is the period? (d) What is the wave speed? (e) In which direction does the wave travel?

◎22. A transverse wave on a string is described by the equation $y(x, t) = (2.20 \text{ cm}) \sin [(130 \text{ rad/s})t + (15 \text{ rad/m})x]$. (a) What is the maximum transverse speed of a point on the string? (b) What is the maximum transverse acceleration of a point on the string? (c) How fast does the wave move along the string? (d) Why is your answer to (c) different from the answer to (a)?

23. Write an equation for a sine wave with amplitude 0.120 m, wavelength 0.300 m, and wave speed 6.40 m/s traveling in the $-x$-direction.

✦24. Write the equation for a transverse sinusoidal wave with a maximum amplitude of 2.50 cm and an angular frequency of 2.90 rad/s that is moving along the positive x-direction with a wave speed that is 5.00 times as fast as the maximum speed of a point on the string. Assume that at time $t = 0$, the point $x = 0$ is at $y = 0$ and then moves in the $-y$-direction in the next instant of time.

11.6 Graphing Waves

25. A sine wave is traveling to the right on a cord. The lighter line in the figure represents the shape of the cord at time $t = 0$; the darker line represents the shape of the cord at time $t = 0.10$ s. (Note that the horizontal and vertical scales are different.) What are (a) the amplitude and (b) the wavelength of the wave? (c) What is the speed of the wave? What are (d) the frequency and (e) the period of the wave?

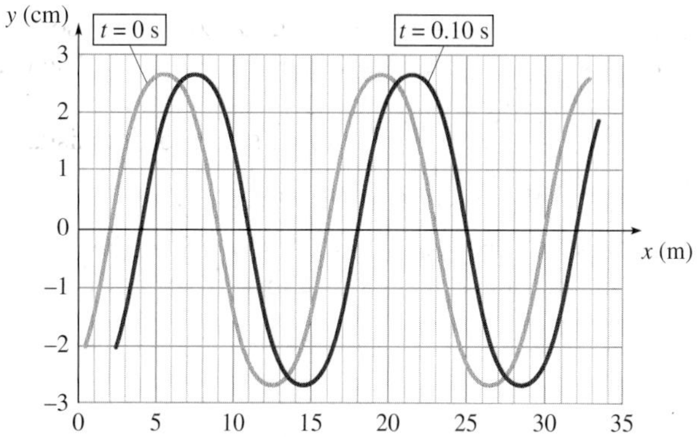

26. (a) Plot a graph for

$$y(x, t) = (4.0 \text{ cm}) \sin [(378 \text{ rad/s})t - (314 \text{ rad/cm})x]$$

versus x at $t = 0$ and at $t = \frac{1}{480}$ s. From the plots determine the amplitude, wavelength, and speed of the wave. (b) For the same function, plot a graph of $y(x, t)$ versus t at $x = 0$ and find the period of the vibration. Show that $\lambda = vT$.

27. For a transverse wave on a string described by

$$y(x, t) = (0.0050 \text{ m}) \cos [(4.0\pi \text{ rad/s})t - (1.0\pi \text{ rad/m})x]$$

find the maximum speed and the maximum acceleration of a point on the string. Plot graphs for one cycle of displacement y versus t, velocity v_y versus t, and acceleration a_y versus t at the point $x = 0$.

28. A transverse wave on a string is described by

$$y(x, t) = (1.2 \text{ mm}) \sin [(2.0\pi \text{ rad/s})t - (0.50\pi \text{ rad/m})x]$$

Plot the displacement y and the velocity v_y versus t for one complete cycle of the point $x = 0$ on the string.

29. (a) Sketch graphs of y versus x for the function

$$y(x, t) = (0.80 \text{ mm}) \sin (kx - \omega t)$$

for the times $t = 0$, 0.96 s, and 1.92 s. Make all three graphs of the same axes, using a solid line for the first, a dashed line for the second, and a dotted line for the third. Use the values $k = \pi/(5.0 \text{ cm})$ and $\omega = (\pi/6.0)$ rad/s. (b) Repeat part (a) for the function

$$y(x, t) = (0.50 \text{ mm}) \sin (kx + \omega t)$$

(c) Which function represents a wave traveling in the $-x$-direction and which represents a wave traveling in the $+x$-direction?

✦30. The drawing shows a snapshot of a transverse wave traveling along a string at 10.0 m/s. The equation for the wave is $y(x, t) = A \cos(\omega t + kx)$. (a) Is the wave moving to the right or to the left? (b) What are the numerical values of A, ω, and k? (c) At what times could this snapshot have been taken? (Give the three smallest nonnegative possibilities.)

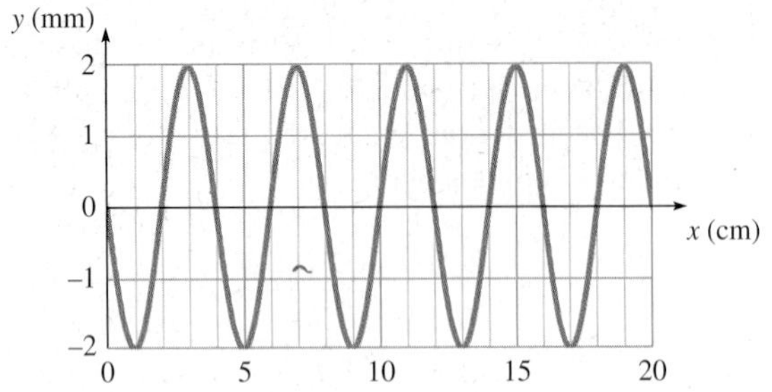

11.7 Principle of Superposition

31. Two pulses on a cord at time $t = 0$ are moving toward each other; the speed of each pulse is 40 cm/s. Sketch the shape of the cord at 0.15, 0.25, and 0.30 s.

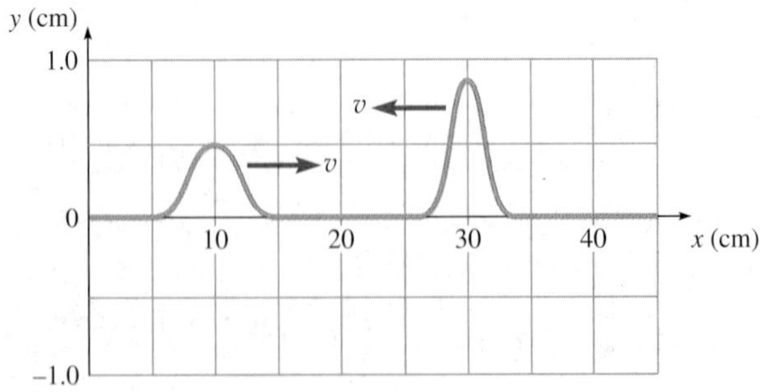

32. Two pulses on a cord at time $t = 0$ are moving toward one another; the speed of each pulse is 2.5 m/s. Sketch the shape of the cord at 0.60, 0.80, and 0.90 s.

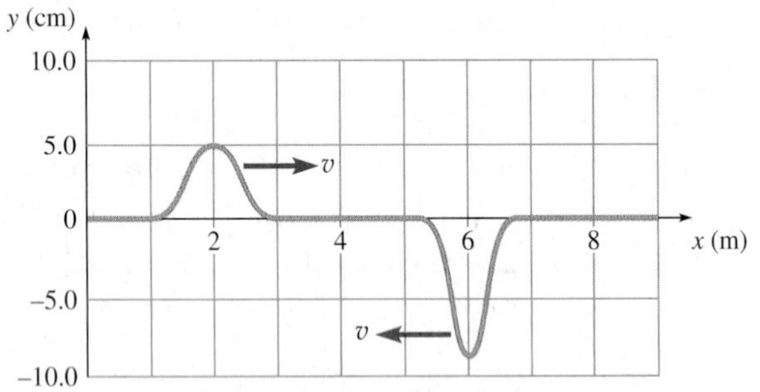

33. Using graph paper, sketch two identical sine waves of amplitude 4.0 cm that differ in phase by (a) $\pi/3$ rad (60°) and (b) $\pi/2$ rad (90°). Find the amplitude of the superposition of the two waves in each case.

34. Two traveling sine waves, identical except for a phase difference ϕ, add so that their superposition produces another traveling wave with the same amplitude as the two component waves. What is the phase difference between the two waves?

35. A traveling sine wave is the result of the superposition of two other sine waves with equal amplitudes, wavelengths, and frequencies. The two component waves each have amplitude 5.00 cm. If the superposition wave has amplitude 6.69 cm, what is the phase difference ϕ between the component waves? [*Hint:* Let $y_1 = A \sin(\omega t + kx)$ and $y_2 = A \sin(\omega t + kx - \phi)$. Make use of the trigonometric identity (Appendix A.7) for $\sin \alpha + \sin \beta$ when finding $y = y_1 + y_2$ and identify the new amplitude in terms of the original amplitude.]

11.8 Reflection and Refraction

36. Light of wavelength 0.500 μm (in air) enters the water in a swimming pool. The speed of light in water is 0.750 times the speed in air. What is the wavelength of the light in water?

Problems 37–38. The pulse of the figure travels to the right on a string whose ends at $x = 0$ and $x = 4.0$ m are both fixed in place. Imagine a reflected pulse that begins to move onto the string at an endpoint at the same time the incident pulse reaches that endpoint. The superposition of the incident and reflected pulses gives the shape of the string.

37. When does the string first look completely flat for $t > 0$?

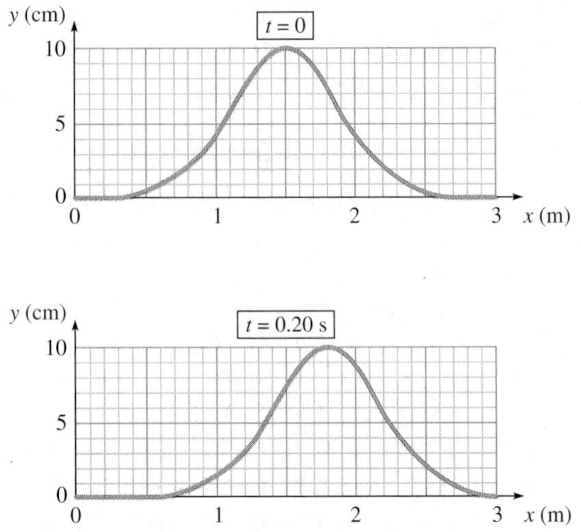

Problems 37, 38, 76, and 77

38. When is the first time for $t > 0$ that the string looks exactly as it does at $t = 0$?

11.9 Interference and Diffraction

39. Two waves with identical frequency but different amplitudes $A_1 = 5.0$ cm and $A_2 = 3.0$ cm, occupy the same region of space (are superimposed). (a) At what phase difference does the resulting wave have the largest amplitude? What is the amplitude of the resulting wave in that case? (b) At what phase difference does the resulting wave have the smallest amplitude and what is its amplitude? (c) What is the ratio of the largest and smallest amplitudes?

40. Two waves with identical frequency but different amplitudes $A_1 = 6.0$ cm and $A_2 = 3.0$ cm, occupy the same

region of space (i.e., are superimposed). (a) At what phase difference will the resulting wave have the highest intensity? What is the amplitude of the resulting wave in that case? (b) At what phase difference will the resulting wave have the lowest intensity and what will its amplitude be? (c) What is the ratio of the two intensities?

41. A sound wave with intensity 25 mW/m^2 interferes constructively with a sound wave that has an intensity of 15 mW/m^2. What is the intensity of the superposition of the two? (tutorial: superposition)

42. A sound wave with intensity 25 mW/m^2 interferes destructively with a sound wave that has an intensity of 28 mW/m^2. What is the intensity of the superposition of the two?

43. Two coherent sound waves have intensities of 0.040 W/m^2 and 0.090 W/m^2 where you are listening. (a) If the waves interfere constructively, what is the intensity that you hear? (b) What if they interfere destructively? (c) If they were incoherent, what would be the intensity? [*Hint:* If your answers are correct, then (c) is the average of (a) and (b).]

44. While testing speakers for a concert, Tomás sets up two speakers to produce sound waves at the same frequency, which is between 100 Hz and 150 Hz. The two speakers vibrate in phase with one another. He notices that when he listens at certain locations, the sound is very soft (a minimum intensity compared to nearby points). One such point is 25.8 m from one speaker and 37.1 m from the other. What are the possible frequencies of the sound waves coming from the speakers? (The speed of sound in air is 343 m/s.)

11.10 Standing Waves

45. In order to decrease the fundamental frequency of a guitar string by 4.0%, by what percentage should you reduce the tension?

46. The tension in a guitar string is increased by 15%. What happens to the fundamental frequency of the string?

47. A standing wave has wavenumber 2.0×10^2 rad/m. What is the distance between two adjacent nodes?

48. A harpsichord string of length 1.50 m and linear mass density 25.0 mg/m vibrates at a (fundamental) frequency of 450.0 Hz. (a) What is the speed of the transverse string waves? (b) What is the tension? (c) What are the wavelength and frequency of the sound wave in air produced by vibration of the string? (The speed of sound in air at room temperature is 340 m/s.)

49. A cord of length 1.5 m is fixed at both ends. Its mass per unit length is 1.2 g/m and the tension is 12 N. (a) What is the frequency of the fundamental oscillation? (b) What tension is required if the $n = 3$ mode has a frequency of 0.50 kHz?

50. Tension is maintained in a string by attaching one end to a wall and by hanging a 2.20-kg object from the other end of the string after it passes over a pulley that is 2.00 m from the wall. The string has a mass per unit length of 3.55 mg/m. What is the fundamental frequency of this string?

51. A guitar's E-string has length 65 cm and is stretched to a tension of 82 N. It vibrates at a fundamental frequency of 329.63 Hz. Determine the mass per unit length of the string.

52. A string 2.0 m long is held fixed at both ends. If a sharp blow is applied to the string at its center, it takes 0.050 s for the pulse to travel to the ends of the string and return to the middle. What is the fundamental frequency of oscillation for this string?

53. A 1.6-m-long string fixed at both ends vibrates at resonant frequencies of 780 Hz and 1040 Hz, with no other resonant frequency between these values. (a) What is the fundamental frequency of this string? (b) When the tension in the string is 1200 N, what is the total mass of the string?

54. A certain string has a mass per unit length of 0.120 g/m. It is attached to a vibrating device and weight similar to that shown in Figure 11.22. The vibrator oscillates at a constant frequency of 110 Hz. How heavy should the weight be in order to produce standing waves in a string of length 42 cm?

55. The longest "string" (a thick metal wire) on a particular piano is 2.0 m long and has a tension of 300.0 N. It vibrates with a fundamental frequency of 27.5 Hz. What is the total mass of the wire?

56. Suppose that a string of length L and mass m is under tension F. (a) Show that $\sqrt{FL/m}$ has units of speed. (b) Show that there is no other combination of L, m, and F with units of speed. [*Hint:* Of the dimensions of the three quantities L, m, and F, only F includes time.] Thus, the speed of transverse waves on the string can only be some dimensionless constant times $\sqrt{FL/m}$.

Comprehensive Problems

57. The speed of waves on a lake depends on frequency. For waves of frequency 1.0 Hz, the wave speed is 1.56 m/s; for 2.0-Hz waves, the speed is 0.78 m/s. The 2.0-Hz waves from a speedboat's wake reach you 120 s after the 1.0-Hz waves generated by the same boat. How far away is the boat?

58. A transverse wave on a string is described by

$$y(x, t) = (1.2\text{ cm}) \sin [(0.50\pi\text{ rad/s})t - (1.00\pi\text{ rad/m})x]$$

Find the maximum velocity and the maximum acceleration of a point on the string. Plot graphs for displacement y versus t, velocity v_y versus t, and acceleration a_y versus t at $x = 0$.

59. What is the wavelength of the radio waves transmitted by an FM station at 90 MHz? (Radio waves travel at 3.0×10^8 m/s.)

60. A longitudinal wave has a wavelength of 10 cm and an amplitude of 5.0 cm and travels in the y-direction. The wave speed in this medium is 80 cm/s. (a) Describe the

motion of a particle in the medium as the wave travels through the medium. (b) How would your answer differ if the wave were transverse instead?

61. An underground explosion sends out both transverse (S waves) and longitudinal (P waves) mechanical wave pulses (seismic waves) through the crust of the Earth. Suppose the speed of transverse waves is 8.0 km/s and that of longitudinal waves is 10.0 km/s. On one occasion, both waves follow the same path from a source to a detector (a seismograph); the longitudinal pulse arrives 2.0 s before the transverse pulse. What is the distance between the source and the detector?

62. The graph shows ground vibrations recorded by a seismograph 180 km from the focus of a small earthquake. It took the waves 30.0 s to travel from their source to the seismograph. Estimate the wavelength.

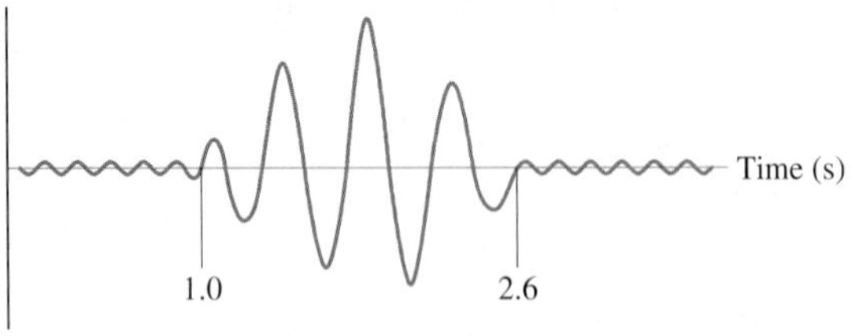

63. When the string of a guitar is pressed against a fret, the shortened string vibrates at a frequency 5.95% higher than when the previous fret is pressed. If the length of the part of the string that is free to vibrate is 64.8 cm, how far from one end of the string are the first three frets located?

64. A guitar string has a fundamental frequency of 300.0 Hz. (a) What are the next three lowest standing wave frequencies? (b) If you press a finger *lightly* against the string at its midpoint so that both sides of the string can still vibrate, you create a node at the midpoint. What are the lowest four standing wave frequencies now? (c) If you press *hard* at the same point, only one side of the string can vibrate. What are the lowest four standing wave frequencies?

65. A sign is hanging from a single metal wire, as shown in part (a) of the drawing. The shop owner notices that the wire vibrates at a fundamental resonance frequency of 660 Hz, which irritates his customers. In an attempt to fix the problem, the shop owner cuts the wire in half and hangs the sign from the two halves, as shown in part (b). Assuming the tension in the two wires to be the same, what is the new fundamental frequency of each wire?

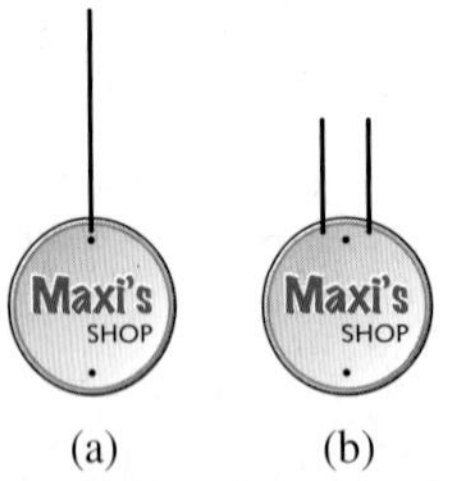

66. (a) Write an equation for a surface seismic wave moving along the $-x$-axis with amplitude 2.0 cm, period 4.0 s, and wavelength 4.0 km. Assume the wave is harmonic, x is measured in m, and t is measured in s. (b) What is the maximum speed of the ground as the wave moves by? (c) What is the wave speed?

67. The formula for the speed of transverse waves on a spring is the same as for a string. (a) A spring is stretched to a length much greater than its relaxed length. Explain why the tension in the spring is approximately proportional to the length. (b) A wave takes 4.00 s to travel from one end of such a spring to the other. Then the length is increased 10.0%. Now how long does a wave take to travel the length of the spring? [*Hint:* Is the mass per unit length constant?]

68. Deep-water waves are *dispersive* (their wave speed depends on the wavelength). The restoring force is provided by gravity. Using dimensional analysis, find out how the speed of deep-water waves depends on wavelength λ, assuming that λ and g are the only relevant quantities. (Mass density does not enter into the expression because the restoring force, arising from the weight of the water, is itself proportional to the mass density.)

69. In contrast to deep-water waves, shallow ripples on the surface of a pond are due to surface tension. The surface tension γ of water characterizes the restoring force; the mass density ρ of water characterizes the water's inertia. Use dimensional analysis to determine whether the surface waves are *dispersive* (the wave speed depends on the wavelength) or *nondispersive* (their wave speed is independent of wavelength). [*Hint:* Start by assuming that the wave speed is determined by γ, ρ, and the wavelength λ.]

70. A seismic wave is described by the equation

$$y(x, t) = (7.00 \text{ cm}) \cos [(6.00\pi \text{ rad/cm})x + (20.0\pi \text{ rad/s})t]$$

The wave travels through a uniform medium in the x-direction. (a) Is this wave moving right ($+x$-direction) or left ($-x$-direction)? (b) How far from their equilibrium positions do the particles in the medium move? (c) What is the frequency of this wave? (d) What is the wavelength of this wave? (e) What is the wave speed? (f) Describe the motion of a particle that is at $y = 7.00$ cm and $x = 0$ when $t = 0$. (g) Is this wave transverse or longitudinal?

71. The drawing shows a snapshot of a transverse wave moving to the left on a string. The wave speed is 10.0 m/s. At the instant the snapshot is taken, (a) in what direction is point A moving? (b) In what direction is point B moving? (c) At which of these points is the speed of the string segment (not the wave speed) larger? Explain.

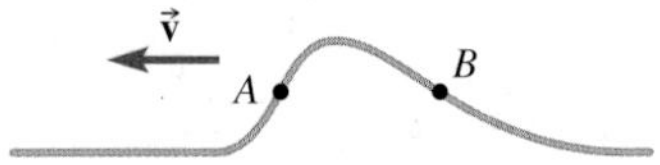

72. Consider a point just to the left of point A in the drawing with Problem 71. Plot the position of that point and the velocity of that point as a function of time as the wave passes the point.

73. Two speakers spaced a distance 1.5 m apart emit coherent sound waves at a frequency of 680 Hz in all directions. The waves start out in phase with each other. A listener walks in a circle of radius greater than one meter centered

on the midpoint of the two speakers. At how many points does the listener observe destructive interference? The listener and the speakers are all in the same horizontal plane and the speed of sound is 340 m/s. [*Hint:* Start with a diagram; then determine the *maximum* path difference between the two waves at points on the circle.] Experiments like this must be done in a special room so that reflections are negligible.

74. (a) Use a graphing calculator or computer graphing program to plot y versus x for the function

$$y(x, t) = (5.0 \text{ cm}) [\sin (kx - \omega t) + \sin (kx + \omega t)]$$

for the times $t = 0$, 1.0 s, and 2.0 s. Use the values $k = \pi/(5.0 \text{ cm})$ and $\omega = (\pi/6.0)$ rad/s. (b) Is this a traveling wave? If not, what kind of wave is it?

75 Show that the amplitudes of the graphs you made in Problem 74 satisfy the equation $A' = 2A \cos (\omega t)$, where A' is the amplitude of the wave you plotted and A is 5.0 cm, the amplitude of the waves that were added together.

Problems 76–77. The pulse of Problems 37–38 travels on a string that has fixed ends.

76. The pulse travels on a string whose ends at $x = 0$ and $x = 4.0$ m are both fixed in place. Sketch the shape of the string at $t = 2.2$ s.

77. The pulse travels on a string whose ends at $x = 0$ and $x = 4.0$ m are both fixed in place. Sketch the shape of the string at $t = 1.6$ s.

Answers to Practice Problems

11.1 (a) 8.9 m/s; (b) 13 m/s

11.2 (a) Yes, the traveling wave retains its shape; (b) it travels in the $+x$-direction because the t and x/v terms have *opposite* signs; (c) the wave speed is 8.0 m/s.

11.3

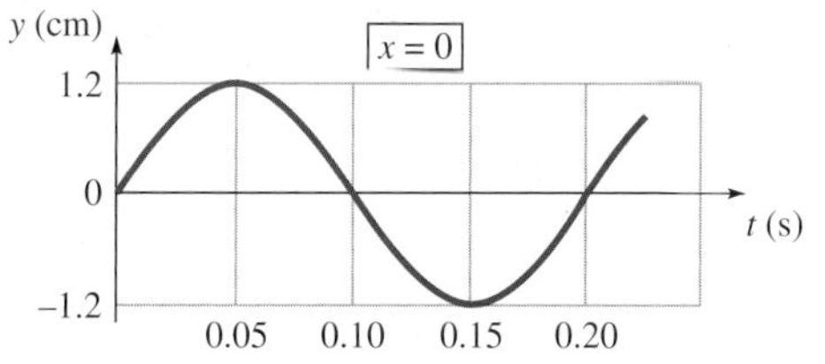

(a)

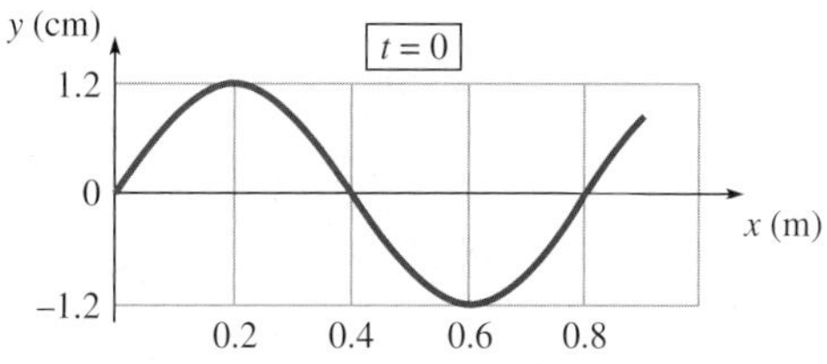

(b)

(c) $T = 0.200$ s; (d) $\lambda = 0.80$ m; (e) $A = 1.2$ cm; (f) $v = 4.0$ m/s; (g) the wave travels in the $-x$-direction because the signs of the terms containing x and t are the same.

11.4

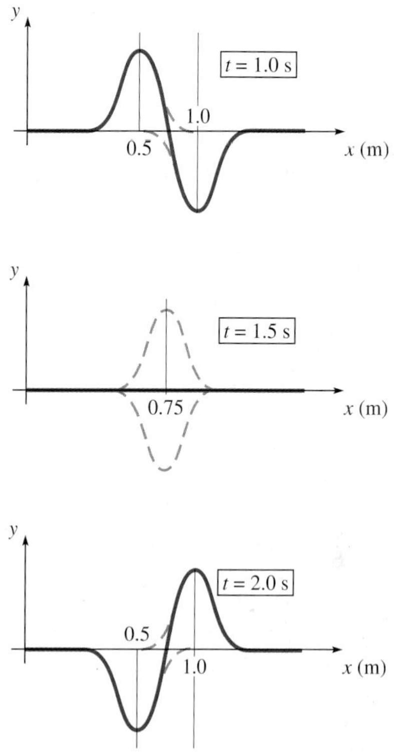

11.5 (a) 620 Hz; (b) 8.5 m

11.6 9.0

11.7 140 Hz

Answers to Checkpoints

11.1 For an isotropic source, $I \propto 1/r^2$. At a distance 10^2 times as far from the tower, the intensity is $10^{-4} \times 0.090 \text{ W/m}^2 = 9.0 \ \mu\text{W/m}^2$.

11.4 The period T is the time for one cycle. During one period, the wave travels 20 km at a speed of 4.0 km/s. Then the period is (20 km)/(4.0 km/s) = 5.0 s.

11.10 The nodes are evenly spaced, so the nodes are at $x = 0$, 20 cm, 40 cm, 60 cm, 80 cm, and 100 cm. The distance between nodes is half the wavelength, so the wavelength is 40 cm.

CHAPTER

12 Sound

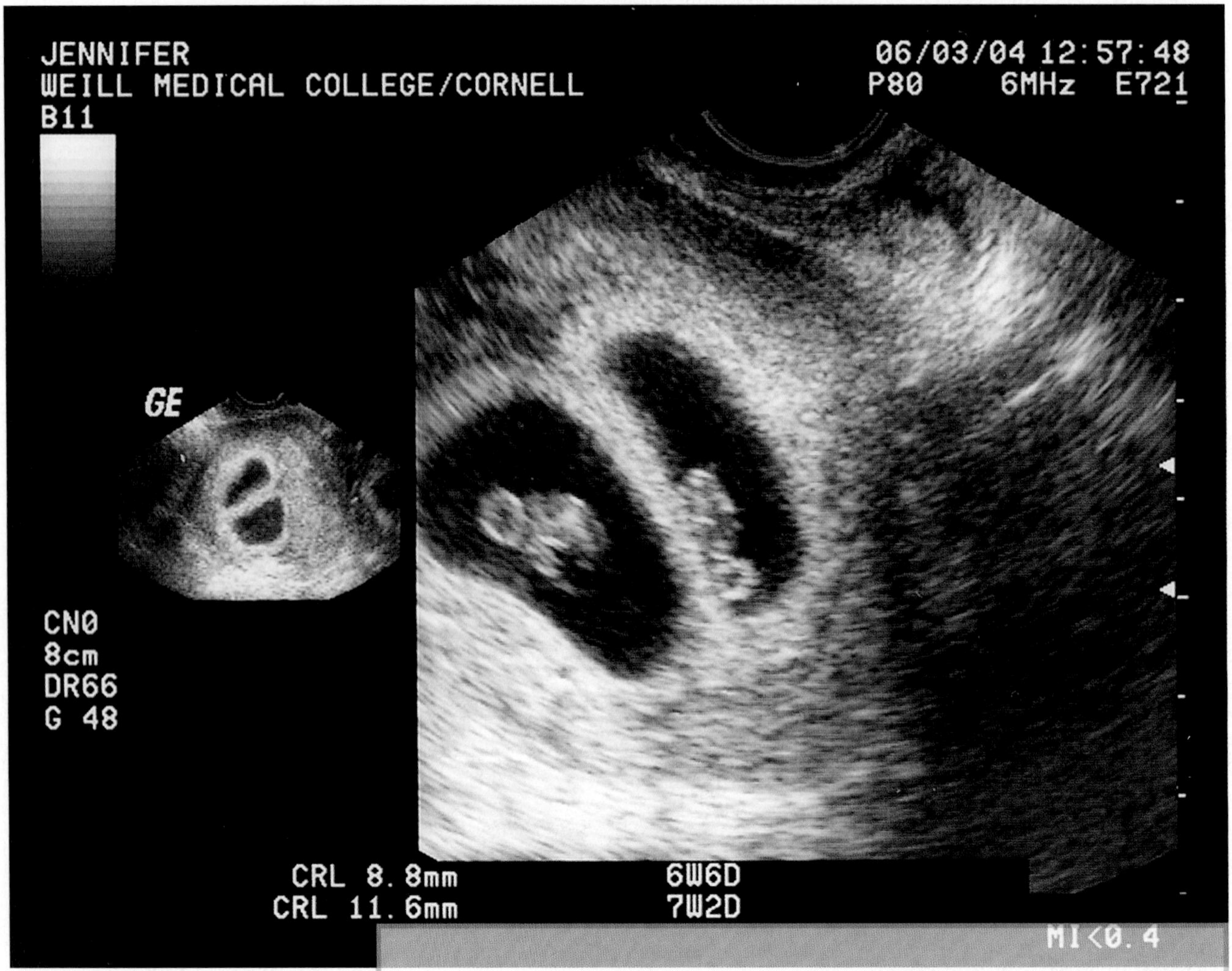

Congratulations! You're expecting twins!

Ultrasonic imaging of the fetus is an increasingly important part of prenatal care. Could an image of the fetus be produced just as well using sound in the audible range rather than ultrasound? Why is ultrasound used rather than some other imaging technology, such as x-rays? Are there other medical applications of ultrasound? (See p. 445 for the answer.)

Concepts & Skills to Review

- gauge pressure (Section 9.5)
- bulk modulus (Section 10.4)
- relation between energy and amplitude in SHM (Section 10.5)
- period and frequency in SHM (Section 10.6)
- longitudinal waves, intensity, standing waves, superposition principle (Chapter 11)
- logarithms (Appendix A.3)

12.1 SOUND WAVES

When a guitar string is plucked, a transverse wave travels along the string. The wave on the string is not what we hear, of course, since the string has no direct connection to our eardrums. The vibration of the string is transmitted through the bridge to the body of the guitar, which in turn transmits the vibration to the air—a sound wave. A transverse wave on a guitar string is not a sound wave, though it does *cause* a sound wave.

Application: how a guitar creates a sound wave

In the absence of a sound wave, molecules in the air dart around in random directions. On average, they are uniformly distributed and the pressure is the same everywhere (ignoring the insignificant variation of pressure due to small changes in altitude). In a sound wave, the uniform distribution of molecules is disturbed. A loudspeaker produces pressure fluctuations that travel through the air in all directions (Fig. 12.1). In some regions (*compressions*), the molecules are bunched together and the pressure is higher than the average pressure. In other regions (*rarefactions*), the molecules are spread out and the pressure is lower than average. The sound wave can be described mathematically by the gauge pressure p (the difference between the pressure at a given point and the average pressure in the surroundings) as a function of position and time (Fig. 12.2a).

The speaker cone produces these pressure variations by displacing molecules in the air from their uniform distribution (Fig. 12.2b). When the cone moves to the left of its equilibrium position, air spreads into a region of lower pressure (rarefaction).

Application: how a loudspeaker generates a sound wave

Figure 12.1 The vibrating speaker cones in this boombox create alternating regions of high and low pressure in the air. Air nearby is affected by a net force due to the nonuniform air pressure; as a result, variations in pressure travel in all directions away from the speakers. This traveling disturbance is a sound wave.

When the cone moves to the right, air is squeezed together into a region of higher pressure (compression).

Thus, the regions of higher and lower pressure are formed when molecules are displaced from a uniform distribution. A sound wave can be described equally well by the displacement s of an *element* of the air—a region of air that can be considered to move together as a unit (Fig. 12.2c). An element is much smaller than the wavelength of the wave, but still large enough to contain many molecules. For a sinusoidal wave, elements at points of maximum or minimum pressure have zero displacement, while the neighboring elements move in toward them (a compression) or away from them (a rarefaction). Conversely, where the gauge pressure is zero, the displacement of an element has its maximum magnitude.

If the pressure is higher on one side than on the other, the net force pushes air toward the side with lower pressure. The uneven distribution of pressure results in air molecules being pushed toward rarefactions and away from compressions, as shown by the force arrows in Fig. 12.2b. Note that the directions of these force arrows, pointing opposite to the displacement arrows in a corresponding region, are such that where there is a compression at a given instant, there will later be a rarefaction, and *vice versa*; the pressure at a given point fluctuates above and below the average pressure.

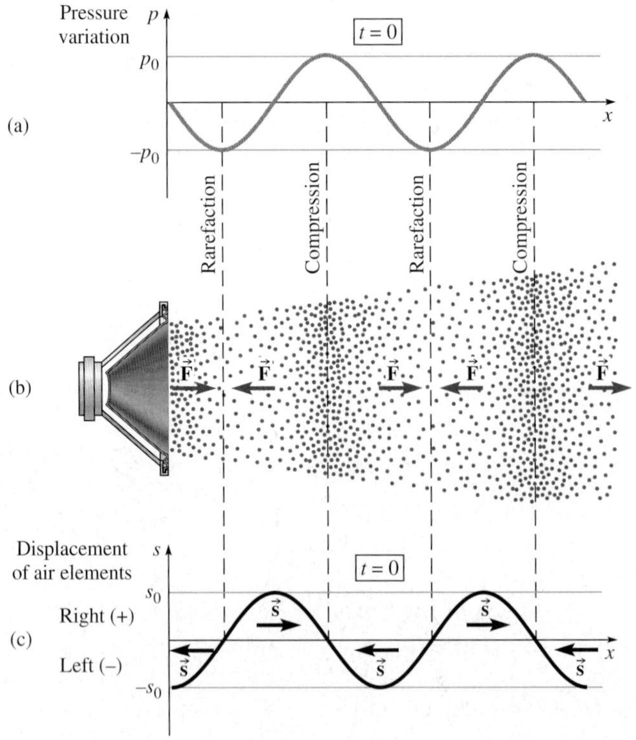

Figure 12.2 A sound wave generated by a loudspeaker. (a) Graph of the pressure variation p of the air as a function of position x. Pressure is high where air is squeezed together and low where it is more spread out. (b) Elements of the air are displaced from their equilibrium positions. Since the pressure is not uniform, air elements experience a net force due to air pressure; the force arrows indicate the direction of this net force. The force is always directed away from a compression (higher pressure) and toward a rarefaction. (c) Graph of the displacement s of an air element from its equilibrium position x as a function of x; the arrows indicate the directions of the displacements in each region. Air elements are displaced leftward or rightward toward compressions and away from rarefactions. Elements at the center of each compression or rarefaction are at their equilibrium positions ($s = 0$).

Frequencies of Sound Waves

The human ear responds to sound waves within a limited range of frequencies. We generally consider the **audible range** to extend from 20 Hz to 20 kHz. Very few people can actually hear sounds over that entire range. Even for a person with excellent hearing, the sensitivity of the ear declines rapidly below 100 Hz and above 10 kHz. The terms **infrasound** and **ultrasound** are used to describe sound waves with frequencies below 20 Hz and above 20 kHz, respectively.

The audible ranges for animals can be quite different. Dogs can hear frequencies as high as 50 kHz, which is why we can make a dog whistle that is inaudible to humans. Dolphins make use of frequencies as high as 250 kHz. Elephants communicate over long distances (up to 4 km) using sounds with fundamental frequencies as low as 14 Hz. A rhinoceros uses frequencies down to 10 Hz. Such low-frequency sounds cannot be heard by humans, but the vibrations can be felt and the sounds can be recorded using special equipment.

12.2 THE SPEED OF SOUND WAVES

For string waves, the restoring force is characterized by the tension in the string F and the inertia is characterized by the linear mass density μ (mass per unit length). The speed of transverse waves on a string is

$$v = \sqrt{\frac{F}{\mu}} \quad (11\text{-}4)$$

CONNECTION:

Just as for transverse waves on a string, the speed of sound waves is determined by a balance between two characteristics of the wave medium: the restoring force and the inertia.

For sound waves in a fluid, the restoring force is characterized by the bulk modulus B, defined in Section 10.4 as the constant of proportionality between an increase in pressure and the fractional volume change:

$$\Delta P = -B\frac{\Delta V}{V} \quad (10\text{-}10)$$

The inertia of the fluid is characterized by its mass density ρ. Following our dictum "more restoring force makes faster waves; more inertia makes slower waves," we expect the speed of sound to be faster in a medium with a larger bulk modulus (harder to compress means more restoring force) and slower in a medium with a larger density. By analogy with Eq. (11-4), we might *guess* that

More restoring force ⇒ faster waves; more inertia ⇒ slower waves.

$$v = \sqrt{\frac{\text{a measure of the restoring force}}{\text{a measure of the inertia}}} = \sqrt{\frac{B}{\rho}} \quad \text{(in fluids)} \quad (12\text{-}1)$$

This guess turns out to be correct; Eq. (12-1) is the correct expression for the speed of sound in fluids.

Temperature Dependence of the Speed of Sound in a Gas The bulk modulus B of an ideal gas turns out to be directly proportional to the density ρ and to T, the *absolute temperature* ($B \propto \rho T$). As a result, the speed of sound in an ideal gas is proportional to the square root of the absolute temperature, but is independent of pressure and density (at a fixed temperature):

$$v = \sqrt{\frac{B}{\rho}} \propto \sqrt{\frac{\rho T}{\rho}} \propto \sqrt{T} \quad \text{(ideal gas)}$$

The SI unit of absolute temperature is the kelvin (symbol K). To find absolute temperature in kelvins, add 273.15 to the temperature in degrees Celsius:

$$T\,(\text{in K}) = T_C\,(\text{in °C}) + 273.15 \quad (12\text{-}2)$$

Since $v \propto \sqrt{T}$, the speed of sound in an ideal gas at any absolute temperature T can be found if it is known at one temperature:

$$v = v_0\sqrt{\frac{T}{T_0}} \quad (12\text{-}3)$$

Table 12.1 Speed of Sound in Various Materials (at 0°C and 1 atm unless otherwise noted)

Medium	Speed (m/s)	Medium	Speed (m/s)
Carbon dioxide	259	Blood (37°C)	1570
Air	331	Muscle (37°C)	1580
Nitrogen	334	Lead	1322
Air (20°C)	343	Concrete	3100
Helium	972	Copper	3560
Hydrogen	1284	Bone (37°C)	4000
Mercury (25°C)	1450	Pyrex glass	5640
Fat (37°C)	1450	Aluminum	5100
Water (25°C)	1493	Steel	5790
Seawater (25°C)	1533	Granite	6500

where the speed of sound is v_0 at absolute temperature T_0. For example, the speed of sound in air at 0°C (or 273 K) is 331 m/s. At room temperature (20°C, or 293 K), the speed of sound in air is

$$v = 331 \text{ m/s} \times \sqrt{\frac{293 \text{ K}}{273 \text{ K}}} = 343 \text{ m/s}$$

An *approximate* formula that can be used for the speed of sound in air is

$$v = (331 + 0.606T_C) \text{ m/s} \tag{12-4}$$

where T_C is air temperature *in degrees Celsius* (see Problem 8). The speed of sound in air increases 0.606 m/s for each degree Celsius increase in temperature. Equation (12-4) gives speeds accurate to better than 1% all the way from –66°C to +89°C.

Speed of Sound in a Solid The speed of sound in a *solid* depends on the Young's modulus Y and the shear modulus S. For sound waves traveling along the length of a thin solid rod, the speed is approximately

$$v = \sqrt{\frac{Y}{\rho}} \quad \text{(thin solid rod)} \tag{12-5}$$

Table 12.1 gives the speed of sound in various materials.

Conceptual Example 12.1

Speed of Sound in Hydrogen and Mercury

From Table 12.1, the speed of sound in hydrogen gas at 0°C is almost as large as the speed of sound in mercury, even though the density of mercury is 150 000 times larger than the density of hydrogen. How is that possible? Shouldn't the speed in mercury be much smaller, since it has so much more inertia?

Solution and Discussion The speed of sound depends on *two* characteristics of the medium: the restoring force (measured by the bulk modulus) and the inertia (measured by the density). The bulk modulus of mercury is much larger than the bulk modulus of hydrogen. The bulk modulus is a measure of how hard it is to compress a material. Liquids (such as mercury) are much more difficult to compress than are gases. Thus, the restoring forces in mercury are much larger than those in hydrogen; this allows sound to travel a bit faster in mercury than it does in hydrogen gas.

Conceptual Practice Problem 12.1 Speed of Sound in Solids versus Liquids

Why does sound generally travel faster in a solid than in a liquid?

12.3 AMPLITUDE AND INTENSITY OF SOUND WAVES

Since there are two ways to describe a sound wave—pressure and displacement—the amplitude of a sound wave can take one of two forms: the pressure amplitude p_0 or the displacement amplitude s_0. The pressure amplitude p_0 is the maximum pressure fluctuation above or below the equilibrium pressure, the displacement amplitude s_0 is the maximum displacement of an element of the medium from its equilibrium position. The pressure amplitude is proportional to the displacement amplitude. For a harmonic sound wave at angular frequency ω, an advanced analysis shows that

$$p_0 = \omega v \rho s_0 \tag{12-6}$$

where v is the speed of sound and ρ is the mass density of the medium.

Is a larger amplitude sound wave perceived as *louder*? Yes, all other things being equal. However, the relationship between our perception of loudness and the amplitude of a sound wave is complex. Loudness is a subjective aspect of how sound is perceived; it has to do with how the ear responds to sound and how the brain interprets signals from the ear. Perceived loudness turns out to be *roughly* proportional to the logarithm of the amplitude. If the amplitude of a sound wave doubles repeatedly, the perceived loudness does not double; it increases by a series of roughly equal steps.

Discussions of loudness are more often phrased in terms of intensity rather than amplitude since we are interested in how much energy the sound wave carries. The intensity of a sinusoidal sound wave is

$$I = \frac{p_0^2}{2\rho v} \tag{12-7}$$

where ρ is the mass density of the medium and v is the speed of sound in that medium. The most important thing to remember is that *intensity is proportional to amplitude squared*, which is true for all waves, not just sound. It is closely related to the fact that energy in SHM is proportional to amplitude squared [see Eq. (10-13)].

Intensity $\propto$ (Amplitude)^2
Intensity is the average power per unit area carried by a wave (see Section 11.1).

Example 12.2

The Brown Creeper

The song of the Brown Creeper (*Certhia americana*) is very high in frequency—as high as 8 kHz. Many people who have lost some of their high-frequency hearing can't hear it at all. Suppose that you are out in the woods and hear the song. If the intensity of the song at your position is 1.4×10^{-8} W/m^2 and the frequency is 6.0 kHz, what are the pressure and displacement amplitudes? (Assume the temperature is 20°C.)

Strategy The displacement and pressure amplitudes are related through Eq. (12-6); the pressure amplitude is related to the intensity through Eq. (12-7). These relationships can be used to solve for both pressure amplitude, p_0, and displacement amplitude, s_0. The density of air at 20°C is $\rho = 1.20$ kg/m^3 (see Table 9.1). The speed of sound in air at 20°C is $v = 343$ m/s. We need to multiply the frequency by 2π to get the angular frequency ω.

Solution Intensity and pressure amplitude are related by

$$I = \frac{p_0^2}{2\rho v} \tag{12-7}$$

Solving for p_0,

$$p_0 = \sqrt{2I\rho v}$$
$$= \sqrt{2 \times 1.4 \times 10^{-8}\ \text{W/m}^2 \times 1.20\ \text{kg/m}^3 \times 343\ \text{m/s}}$$
$$= 3.4 \times 10^{-3}\ \text{Pa}$$

The pressure and displacement amplitudes are related by

$$p_0 = \omega v \rho s_0 \tag{12-6}$$

continued on next page

Example 12.2 continued

Substituting in Eq. (12-7) yields

$$I = \frac{(\omega v \rho s_0)^2}{2\rho v}$$

Solving for s_0,

$$s_0 = \sqrt{\frac{2I}{\rho \omega^2 v}} = \sqrt{\frac{2 \times 1.4 \times 10^{-8}\ \text{W/m}^2}{1.20\ \text{kg/m}^3 \times (2\pi \times 6000\ \text{Hz})^2 \times 343\ \text{m/s}}}$$

$$= 2.2 \times 10^{-10}\ \text{m}$$

Discussion This problem illustrates how sensitive the human ear is. The pressure amplitude is a fluctuation of one part in 30 million in the air pressure. Since the pressure amplitude is 3.4×10^{-3} Pa, the maximum force on the eardrum would be about

$$F_{max} = 3.4 \times 10^{-3}\ \text{N/m}^2 \times 10^{-4}\ \text{m}^2 \approx 3 \times 10^{-7}\ \text{N}$$

which is about the weight of a large amoeba. The displacement amplitude is about the size of an atom.

Practice Problem 12.2 Pressure and Intensity at an Outdoor Concert

At a distance of 5.0 m from the stage at an outdoor rock concert, the sound intensity is 1.0×10^{-4} W/m^2. Estimate the intensity and pressure amplitude at a distance of 25 m if there were no speakers other than those on stage. Explain the assumptions you make.

Decibels

Since the perception of loudness by the human ear is roughly proportional to the logarithm of the intensity, it is also roughly proportional to the logarithm of the amplitude (since $\log x^2 = 2 \log x$). An intensity of $I_0 = 10^{-12}$ W/m^2 is about the lowest intensity sound wave that can be heard under ideal conditions by a person with excellent hearing; it is therefore called the **threshold of hearing**. The threshold of hearing is used as a reference intensity in the definition of the intensity level.

A sound intensity I is compared to the reference level I_0 by taking the ratio of the two intensities. Suppose a sound has an intensity of 10^{-5} W/m^2; the ratio is

$$\frac{I}{I_0} = \frac{10^{-5}\ \text{W/m}^2}{10^{-12}\ \text{W/m}^2} = 10^7$$

so the intensity is 10^7 times that of the hearing threshold. The power to which 10 is raised is the **sound intensity level** β in units of bels (after Alexander Graham Bell). A ratio of 10^7 indicates a sound intensity of 7 bels or, as it is more commonly stated, 70 decibels (dB). Since $\log_{10} (10^x) = x$, the sound intensity level in decibels is

The notation $\log_{10}$ stands for the base-10 logarithm. See Appendix A.3 for a review of the properties of logarithms.

$$\beta = (10\ \text{dB}) \log_{10} \frac{I}{I_0} \qquad (12\text{-}8)$$

An intensity level of 0 dB corresponds to the threshold of hearing ($I = 10^{-12}$ W/m^2). Although the intensity level is really a pure number, the "units" (dB) remind us what the number means.

Table 12.2 gives the pressure amplitudes, intensities, and intensity levels for a wide range of sounds. Notice that, even for sounds that are quite loud, the pressure fluctuations due to sound waves are small compared to the "background" atmospheric pressure.

CHECKPOINT 12.3

Why doesn't Table 12.2 include a column listing the *displacement* amplitudes of the sound waves?

Table 12.2 Pressure Amplitudes, Intensities, and Intensity Levels of a Wide Range of Sounds in Air at 20°C

Sound	Pressure Amplitude (atm)	Pressure Amplitude (Pa)	Intensity (W/m^2)	Intensity Level (dB)
Threshold of hearing	3×10^{-10}	3×10^{-5}	10^{-12}	0
Leaves rustling	1×10^{-9}	1×10^{-4}	10^{-11}	10
Whisper (1 m away)	3×10^{-9}	3×10^{-4}	10^{-10}	20
Library background noise	1×10^{-8}	0.001	10^{-9}	30
Living room background noise	3×10^{-8}	0.003	10^{-8}	40
Office or classroom	1×10^{-7}	0.01	10^{-7}	50
Normal conversation at 1 m	3×10^{-7}	0.03	10^{-6}	60
Inside a moving car, light traffic	1×10^{-6}	0.1	10^{-5}	70
City street (heavy traffic)	3×10^{-6}	0.3	10^{-4}	80
Shout (at 1 m); or inside a subway train; risk of hearing damage if exposure lasts several hours	1×10^{-5}	1	10^{-3}	90
Car without muffler at 1 m	3×10^{-5}	3	10^{-2}	100
Construction site	1×10^{-4}	10	10^{-1}	110
Indoor rock concert; threshold of pain; hearing damage occurs rapidly	3×10^{-4}	30	1	120
Jet engine at 30 m	1×10^{-3}	100	10	130

Example 12.3

Decibels from a Jackhammer

The sound intensity 5 m from a jackhammer is 4.20×10^{-2} W/m². What is the sound intensity level in decibels? (Use the usual reference level of $I_0 = 1.00 \times 10^{-12}$ W/m².)

Strategy We are given the intensity in W/m² and asked for the intensity level in dB. First we find the ratio of the given intensity to the reference level. Then we take the logarithm of the result (to get the level in bels) and multiply by 10 (to convert from bels to dB).

Solution The ratio of the intensity to the reference level is

$$\frac{I}{I_0} = \frac{4.20 \times 10^{-2}\ \text{W/m}^2}{1.00 \times 10^{-12}\ \text{W/m}^2} = 4.20 \times 10^{10}$$

The intensity level in bels is

$$\log_{10} \frac{I}{I_0} = \log_{10} 4.20 \times 10^{10} = 10.6 \text{ bels}$$

The intensity level in decibels is

$$\beta = 10.6 \text{ bels} \times (10 \text{ dB/bel}) = 106 \text{ dB}$$

Discussion As a quick check, 100 dB corresponds to $I = 10^{-2}$ W/m² and 110 dB corresponds to $I = 10^{-1}$ W/m²; since the intensity is between 10^{-2} W/m² and 10^{-1} W/m², the intensity level must be between 100 dB and 110 dB.

Practice Problem 12.3 Consequences of a Hole in the Muffler

When rust creates a hole in the muffler of a car, the sound intensity level inside the car is 26 dB higher than when the muffler was intact. By what factor does the intensity increase?

As we saw in Section 11.9, when two sounds are coming from different sources, the waves are incoherent. If we know the intensity of each wave alone at a certain point, then the intensity due to the two waves together at that point is the sum of the two intensities:

$$I = I_1 + I_2 \quad \text{(incoherent waves)}$$

This is *not* true for two coherent waves, where the total intensity depends on the phase relationship between the waves. Since there is no fixed phase relationship between two incoherent waves, on average there is neither constructive nor destructive interference. The total power per unit area is the sum of the power per unit area of each wave.

Example 12.4

The Sound Intensity Level of Two Lathes

A metal lathe in a workshop produces a 90.0-dB sound intensity level at a distance of 1 m. What is the intensity level when a second identical lathe starts operating? Assume the listener is at the same distance from both lathes.

Strategy The noise is coming from two different machines and, thus, they are incoherent sources. We *cannot* add 90.0 dB to 90.0 dB to get 180.0 dB, which would be a senseless result—two lathes are not going to drown out a jet engine at close range (see Table 12.2). Instead, what doubles is the *intensity*. We must work in terms of intensity rather than intensity level.

Solution First find the intensity due to one lathe:

$$\beta = 90.0 \text{ dB} = (10 \text{ dB}) \log_{10} \frac{I}{I_0}$$

$$\log_{10} \frac{I}{I_0} = 9.00, \qquad \text{so} \quad \frac{I}{I_0} = 1.00 \times 10^9$$

We could solve for I numerically but it is not necessary. With two machines operating, the intensity doubles, so

$$\frac{I'}{I_0} = 2.00 \times 10^9$$

and the new intensity level is

$$\beta' = (10 \text{ dB}) \log_{10} \frac{I'}{I_0} = (10 \text{ dB}) \log_{10} (2.00 \times 10^9) = 93.0 \text{ dB}$$

Discussion The new intensity level is just 3 dB higher than the original one, even though the intensity is twice as big. This turns out to be a general result: a 3-dB increase represents a doubling of the intensity.

Practice Problem 12.4 Intensity Change for an Increment of 5 dB

The maximum recommended exposure time to a sound level of 90 dB is 8 h. For every increase of 5.0 dB in sound level up to 120 dB, the exposure time should be reduced by a factor of 2. (At 120 dB, damage occurs almost immediately; there is no safe exposure time.) What factor of intensity change does an intensity level increment of 5.0 dB represent?

Sound intensity level is useful because it roughly approximates the way we perceive loudness (since it is a logarithmic function of intensity). Equal increments in intensity level roughly correspond to equal increases in loudness. Two useful rules of thumb: every time the intensity increases by a *factor* of 10, the intensity level *adds* 10 dB; since $\log_{10} 2 = 0.30$, adding 3.0 dB to the intensity level *doubles* the intensity (see Problem 17). In Example 12.4, when both lathes are running at the same time, the intensity is twice as big as for one lathe, but the two do not sound twice as loud as one. Intensity *level* is a better guide to loudness; two lathes produce a level 3 dB higher than one lathe.

Decibels can also be used in a relative sense; instead of comparing an intensity to I_0, we can compare two intensities directly. Suppose we have two intensities I_1 and I_2 and two corresponding intensity levels β_1 and β_2. Then

$$\beta_2 - \beta_1 = 10 \text{ dB} \left(\log_{10} \frac{I_2}{I_0} - \log_{10} \frac{I_1}{I_0} \right)$$

Since $\log x - \log y = \log \frac{x}{y}$ [see Appendix A.3, Eq. (A-21)],

$$\beta_2 - \beta_1 = (10 \text{ dB}) \log_{10} \frac{I_2/I_0}{I_1/I_0} = (10 \text{ dB}) \log_{10} \frac{I_2}{I_1} \qquad (12\text{-}9)$$

Example 12.5

Variation of Intensity Level with Distance

At a distance of 30 m from a jet engine, the sound intensity level is 130 dB. Serious, permanent hearing damage occurs rapidly at intensity levels this high, which is why you see airport personnel using hearing protection out on the runway. Assume the engine is an isotropic source of sound and ignore reflections and absorption. At what distance is the intensity level 110 dB—still quite loud but below the threshold of pain?

Strategy The intensity level drops 20 dB. According to the rule of thumb, each 10-dB change represents a factor of 10 in intensity. Therefore, we must find the distance at which the intensity is 2 factors of 10 smaller—that is, $\frac{1}{100}$ the original intensity. The intensity is proportional to $1/r^2$ since we assume an isotropic source [see Eq. (11-1)].

Solution We set up a ratio between the intensities and the inverse square of the distances:

$$\frac{I_1}{I_2} = \left(\frac{r_2}{r_1}\right)^2$$

From the rule of thumb, we know that $I_2 = \frac{1}{100} I_1$. Then

$$\frac{r_2}{r_1} = \sqrt{\frac{I_1}{I_2}} = \sqrt{100} = 10$$

$$r_2 = 10r_1 = 300 \text{ m}$$

Discussion It is not necessary to use the rule of thumb. Let $\beta_1 = 130$ dB and $\beta_2 = 110$ dB. Then

$$\beta_2 - \beta_1 = -20 \text{ dB} = (10 \text{ dB}) \log_{10} \frac{I_2}{I_1}$$

From this, we find that

$$\log_{10} \frac{I_2}{I_1} = -2 \quad \text{or} \quad \frac{I_2}{I_1} = \frac{1}{100}$$

We can only consider 300 m an estimate. The jet engine may not radiate sound equally in all directions; it might be louder in front than on the side. Sound is partly absorbed and partly reflected by the runway, by the plane, and by any nearby objects. The air itself absorbs some of the sound energy—that is, some of the energy of the wave is dissipated.

Practice Problem 12.5 A Plane as Quiet as a Library

At what distance from the jet engine would the intensity level be comparable to the background noise level of a library (30 dB)? Is your answer realistic?

12.4 STANDING SOUND WAVES

Pipe Open at Both Ends

Recall (Section 11.8) that a transverse wave on a string is reflected from a fixed end. A string fixed at both ends reflects the wave at each end. A standing wave on a string is caused by the superposition of two waves traveling in opposite directions. Standing *sound* waves are also caused by reflections at boundaries. Standing wave patterns for sound waves can be more complex, since sound is a three-dimensional wave. However, the air inside a pipe open at both ends gives rise to standing waves closely analogous to those on a string, as long as the pipe's diameter is small compared with its length. Such a pipe is an excellent model of some organ pipes and flutes.

If the pipe is open at both ends, then the pipe has the same boundary condition at each end. At each open end, the column of air inside the pipe communicates with the outside air, so the pressure at the ends can't deviate much from atmospheric pressure. The open ends are therefore *pressure nodes* (Fig. 12.3). They are also *displacement*

Open ends are pressure nodes and displacement antinodes.

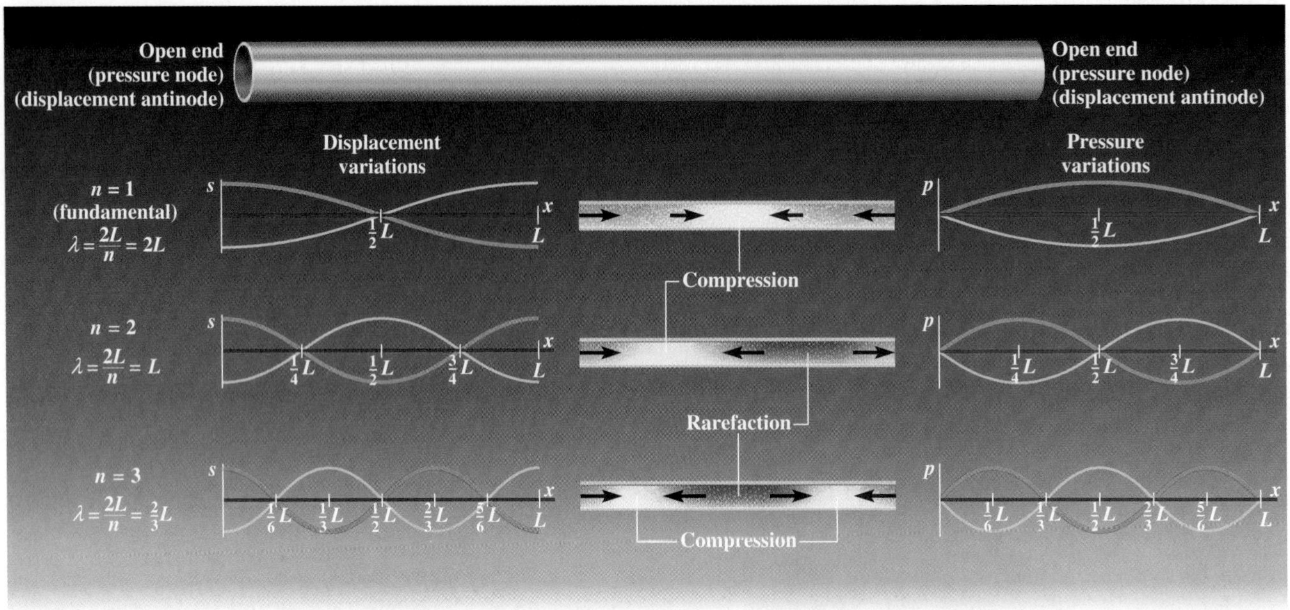

Figure 12.3 Standing waves in a pipe open at both ends. (Although the *graphs* show air displacement s on the vertical axis and x on the horizontal, remember that the displacements are in the $\pm x$-direction, as illustrated by the black vector arrows.)

antinodes—elements of air vibrate back and forth with maximum amplitude at the ends. Since nodes and antinodes alternate with equal spacing ($\lambda/4$), the wavelengths of standing sound waves in a pipe open at both ends are the same as for a string fixed at both ends (compare Fig. 12.3 with Fig. 11.21), regardless of whether you consider the pressure or the displacement description.

CONNECTION:

The same sketch used to find wavelengths of standing waves for a string fixed at both ends can be used to find the wavelengths for a pipe open at both ends. (The wave speeds are different, however, so a string and pipe of the same length do not have the same standing wave *frequencies.*)

Standing sound waves (thin pipe open at both ends):

$$\lambda_n = \frac{2L}{n} \qquad (11\text{-}12)$$

$$f_n = \frac{v}{\lambda_n} = n\frac{v}{2L} = nf_1 \qquad (11\text{-}13)$$

where $n = 1, 2, 3, \ldots$

Pipe Closed at One End

Some organ pipes are *closed at one end* and open at the other (Fig. 12.4). The closed end is a pressure *antinode*; the air at the closed end meets a rigid surface, so there is no restriction on how far the pressure can deviate from atmospheric pressure. The closed

Figure 12.4 Some organ pipes are open at the top; others are closed. A pipe closed at one end has a fundamental wavelength twice as large and therefore a fundamental frequency half as large as a pipe of the same length that is open at both ends, assuming the pipes are thin. (For musicians: the pitch of the pipe closed at one end sounds an octave lower than the other, since the interval of an octave corresponds to a factor of 2 in frequency.)

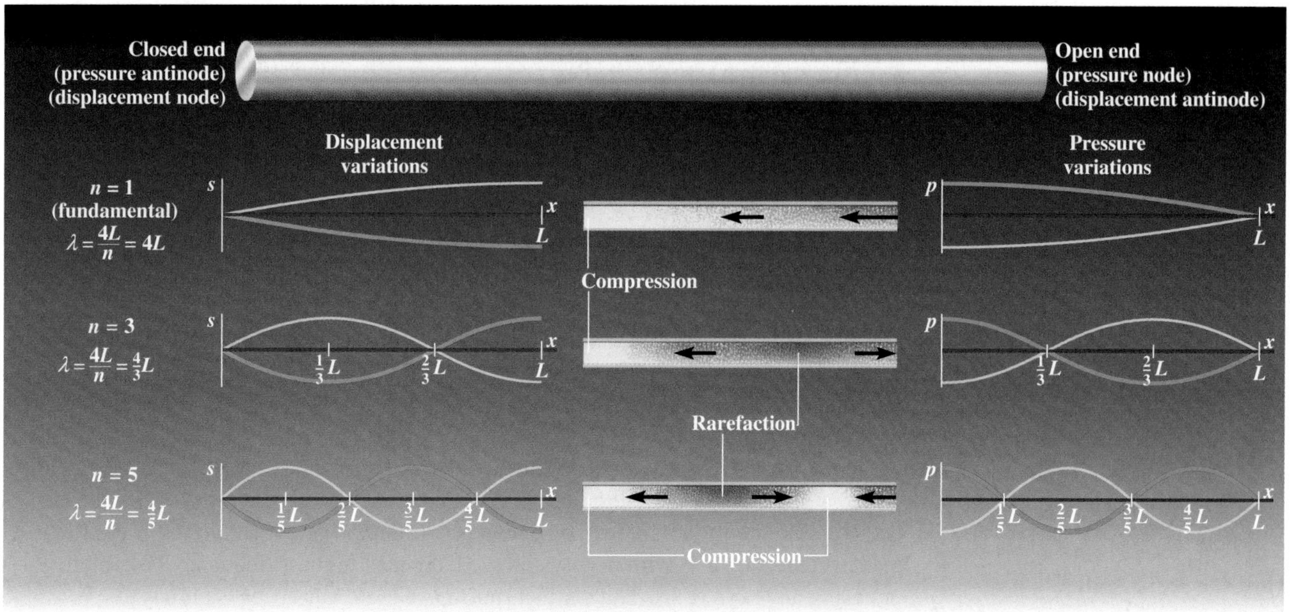

Figure 12.5 Standing waves in a pipe closed at one end.

end is also a *displacement node* since the air near it cannot move beyond that rigid surface. Some wind instruments are effectively pipes closed at one end. The reed of a clarinet admits only brief puffs of air into the instrument; the rest of the time the reed closes off that end of the pipe. The pressure at the reed end fluctuates above and below atmospheric pressure. The reed end is a pressure antinode and a displacement node.

Closed ends are pressure antinodes and displacement nodes.

The wavelengths and frequencies of the standing waves can be found using either the pressure or displacement descriptions of the wave. Using displacement, the fundamental has a node at the closed end, an antinode at the open end, and no other nodes or antinodes (Fig. 12.5). The distance from a node to the nearest antinode is always $\frac{1}{4}\lambda$, so for the fundamental

$$L = \tfrac{1}{4}\lambda \quad \text{or} \quad \lambda = 4L$$

which is twice as large as the wavelength ($2L$) of the fundamental in a pipe of the same length open at both ends. Two thin organ pipes of the same length, one open at both ends and one closed at one end, do not have the same fundamental wavelength (see Fig. 12.4).

What are the other standing wave frequencies? The next standing wave mode is found by adding one node and one antinode. Then the length of the pipe is 3 quarter cycles: $L = \frac{3}{4}\lambda$ or $\lambda = \frac{4}{3}L$. This is $\frac{1}{3}$ the wavelength of the fundamental and the frequency is 3 times that of the fundamental. Adding one more node and one more antinode, the wavelength is $\frac{4}{5}L$. Continuing the pattern, we find that the wavelengths and frequencies for standing waves are

Standing sound waves (thin pipe closed at one end):

$$\lambda_n = \frac{4L}{n} \qquad (12\text{-}10a)$$

$$f_n = \frac{v}{\lambda_n} = n\frac{v}{4L} = nf_1 \qquad (12\text{-}10b)$$

where $n = 1, 3, 5, 7, \ldots$

Note that the standing wave frequencies for a pipe closed at one end are only *odd* multiples of the fundamental. The "missing" standing wave patterns for even values of n require a clarinet to have many more keys and levers than a flute (Fig. 12.6). What the keys do is effectively shorten the length of the pipe, making the standing wave frequencies higher.

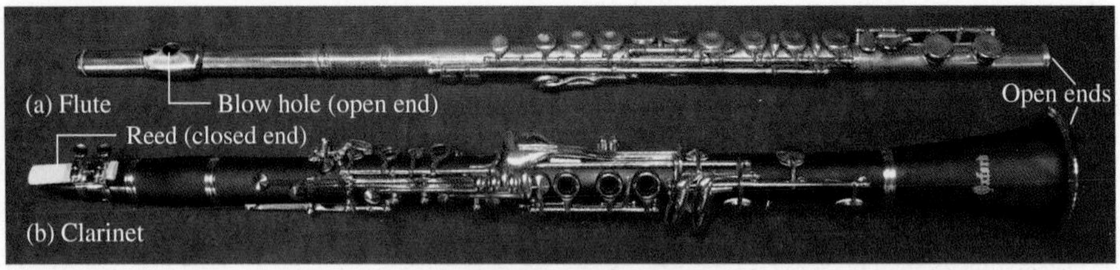

Figure 12.6 A flute can be modeled as a pipe open at both ends, while a clarinet can be modeled as a pipe closed at one end. Although the instruments are similar in length, the clarinet can play tones nearly an octave lower than the flute can. (a) The flute's open blow hole serves as one of its open "ends." If a flute's fundamental frequency is f_1 with no keys pressed, the next highest frequency possible without using any keys is $2f_1$—the flutist overblows, exciting the next highest standing wave frequency rather than the fundamental. The flute needs enough keys to fill in all the notes with frequencies between f_1 and $2f_1$. (b) The clarinet can be modeled as a pipe open at one end and closed at the other. The mouthpiece end with its vibrating reed is more like a closed end (pressure antinode) than an open end (pressure node). For a clarinet, if the fundamental frequency is f_1 with no keys pressed, the next highest frequency possible without using any keys is $3f_1$. The clarinet must have more keys because it has to accommodate all the notes with frequencies between f_1 and $3f_1$.

CHECKPOINT 12.4

Why can't a pipe of length L closed at one end support a standing wave with wavelength $2L$?

Example 12.6

A Demonstration of Resonance

A thin hollow tube of length 1.00 m is inserted vertically into a tall container of water (Fig. 12.7). A tuning fork ($f = 520.0$ Hz) is struck and held near the top of the tube as the tube is slowly pulled up and out of the water. At certain distances (L) between the top of the tube and the water surface, the otherwise faint sound of the tuning fork is greatly amplified. At what values of L does this occur? The temperature of the air in the tube is 18°C.

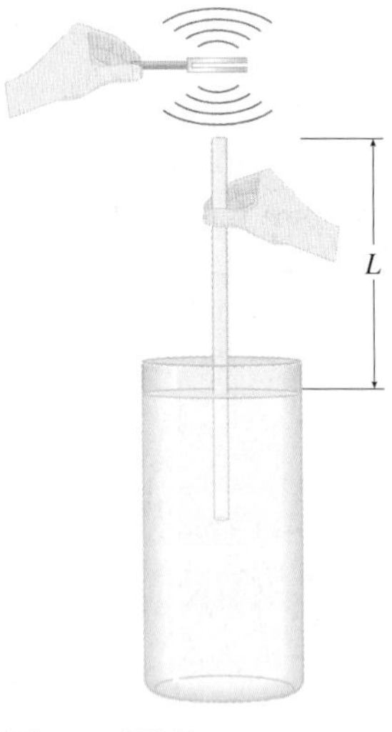

Figure 12.7 Experimental setup for Example 12.6.

Strategy Sound waves in the air inside the tube reflect from the water surface. Thus, we have an air column of variable length L, closed at one end by the water surface and open at the other end. The sound is amplified due to resonance; when the frequency of the tuning fork matches one of the natural frequencies of the air column, a large-amplitude standing wave builds up in the column. For standing waves in a column of air, the wavelength and frequency are related by the speed of sound in air. We start by finding the speed of sound in air from the temperature given. Then we can find the wavelength of the sound waves emanating from the tuning fork. Last, we find the column lengths that support standing waves of that wavelength.

Solution The speed of sound in air at 18°C is

$$v = (331 + 0.606 \times 18)\ \text{m/s} = 342\ \text{m/s}$$

With the speed of sound and the frequency known, we can find the wavelength. The wavelength is the distance traveled by a wave during one period:

$$\lambda = vT = \frac{v}{f}$$

$$\lambda = \frac{342\ \text{m/s}}{520.0\ \text{Hz}} = 0.6577\ \text{m} = 65.77\ \text{cm}$$

continued on next page

Example 12.6 continued

The first possible resonance for a tube closed at one end occurs when there is a pressure node at the open end, a pressure antinode at the closed end, and no other pressure nodes or antinodes. Therefore,

$$L_1 = \tfrac{1}{4}\lambda = \tfrac{1}{4} \times 65.77 \text{ cm} = 16.4 \text{ cm}$$

To reach other resonances, the tube must be pulled out to accommodate additional pressure nodes and antinodes. To add one node and one antinode, the additional distance is $\tfrac{1}{2}\lambda$ = 32.9 cm. The resonances occur at intervals of 32.9 cm:

$$L_2 = 16.4 \text{ cm} + 32.9 \text{ cm} = 49.3 \text{ cm}$$

$$L_3 = 49.3 \text{ cm} + 32.9 \text{ cm} = 82.2 \text{ cm}$$

The next one would require a tube longer than 1.00 m, so there are three values of L that produce resonance in this tube.

Discussion As a check, we can sketch the standing wave pattern for the third resonance (Figs. 12.8a,b). There are 5 quarter-wavelengths in the length of the column, so

$$L_3 = \tfrac{5}{4}\lambda = \tfrac{5}{4} \times 65.77 \text{ cm} = 82.2 \text{ cm}$$

At the open end of the tube, the node for pressure and the antinode for maximum displacement is actually a little *above* the opening. For this reason it is best to measure the distance between two successive resonances to find an accurate value for a half-wavelength rather than measuring the distance for the first possible resonance, the shortest distance between the opening and the water surface, and setting it equal to a quarter-wavelength.

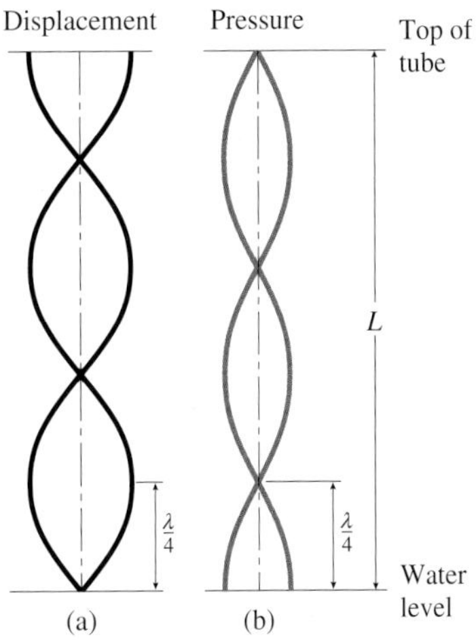

Figure 12.8

(a) Standing wave pattern, showing *displacement* nodes and antinodes, for the third resonance. (b) Standing wave pattern, showing *pressure* nodes and antinodes, for the third resonance.

Practice Problem 12.6 A Roundabout Way to Measure Temperature

A tuning fork of frequency 440.0 Hz is held above the hollow tube in Example 12.6. If the distance ΔL that the tube is moved between resonances is 39.3 cm, what is the temperature of the air inside the tube?

Problem-Solving Strategy for Standing Waves

There is no need to memorize equations for standing wave frequencies and wavelengths. Just sketch the standing wave patterns as in Figs. 12.3 and 12.5. Make sure that nodes and antinodes alternate and that the boundary conditions at the ends are correct. Then determine the wavelengths by setting the distance between a node and antinode equal to $\frac{1}{4}\lambda$. Once the wavelengths are known, the frequencies are found from $v = f\lambda$.

PHYSICS AT HOME

You can set up a resonance in an empty water bottle by blowing horizontally across the top of the bottle. Once you have heard one resonance, add varying amounts of water to raise the level within and listen for other resonances. The resonant sound is noticeably louder than the nonresonant sounds. Notice that the longer the air column within the bottle, the lower the pitch heard.

12.5 TIMBRE

The sound produced by the vibration of a tuning fork is nearly a pure sinusoid at a single frequency. In contrast, most musical instruments produce complex sounds that are the superposition of many different frequencies. The standing wave on a string or in a column of air is almost always the superposition of many standing wave patterns at

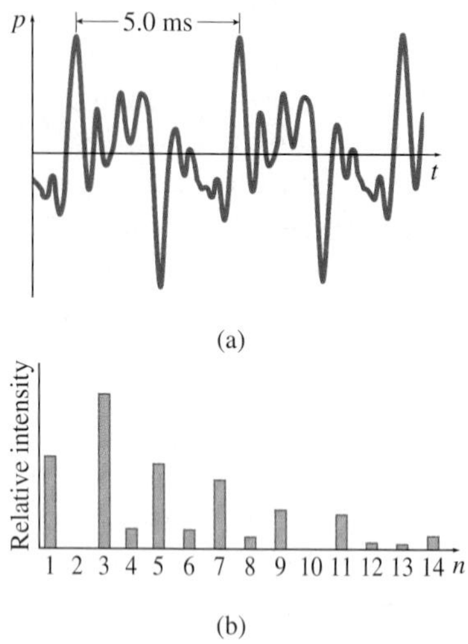

Figure 12.9 (a) A graph of the sound wave produced by a clarinet. (b) A bar graph showing the relative intensities of the harmonics, often called the *spectrum*. The frequency of each harmonic is nf_1, where $f_1 = 200$ Hz. Notice that *odd* multiples of the fundamental dominate the spectrum. A simple pipe closed at one end would show *only* odd multiples in its spectrum. (Data courtesy of P. D. Krasicky, Cornell University.)

different frequencies. The lowest frequency in a complex sound wave is called the fundamental; the rest of the frequencies are called **overtones**. All the overtones of a periodic sound wave have frequencies that are integral multiples of the fundamental; the fundamental and the overtones are then called **harmonics**.

Middle C played on an oboe does not sound the same as middle C played on a trumpet, even though the fundamental frequency is the same, largely because the two instruments produce overtones with different relative amplitudes. What is different about the two sounds is the **tone quality**, or **timbre** (pronounced "tamber").

Any periodic wave, no matter how complicated, can be decomposed into a set of harmonics, each of which is a simple sinusoid. The characteristic wave form for a note played on a clarinet, for example, can be decomposed into its harmonic series (Fig. 12.9). This process is called harmonic analysis, or Fourier analysis, in honor of the French mathematician, Jean Baptiste Joseph Fourier (1768–1830), who developed mathematical methods for analyzing periodic functions. Although the spectrum of a periodic wave consists only of members of a harmonic sequence, not all members of the sequence need be present, not even the fundamental (Fig. 12.10).

The opposite of harmonic analysis is harmonic synthesis: combining various harmonics to produce a complex wave. Electronic synthesizers can mimic the sounds of various instruments. Realistic-sounding synthesizers must also allow the adjustment of other parameters such as the attack and decay of the sound.

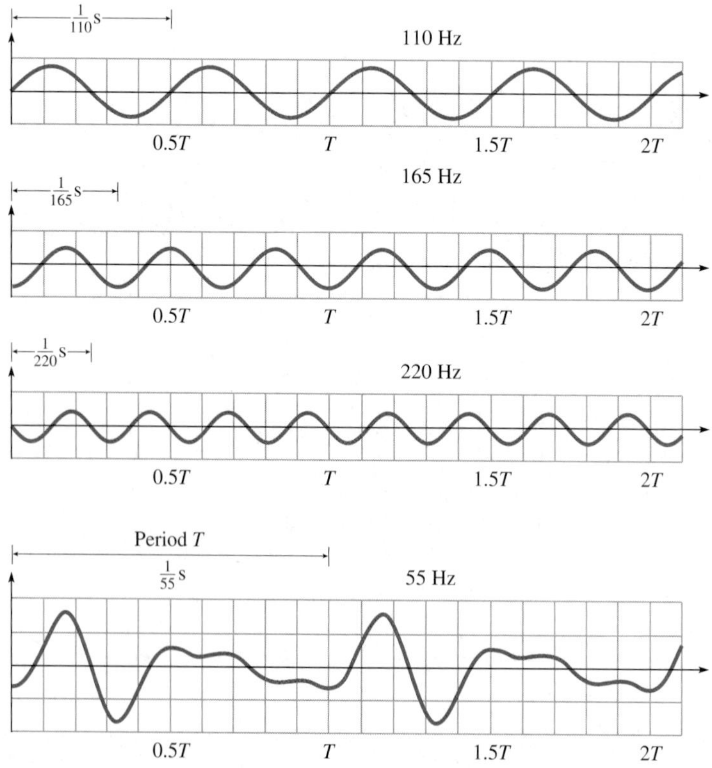

Figure 12.10 Complex wave form (bottom wave) composed by superposition of three sinusoidal waves (three upper waves). A wave with three harmonic components having frequencies of 110, 165, and 220 Hz repeats at a frequency of 55 Hz because each of these three frequencies is an integral multiple of 55 Hz. Even though the fundamental is missing—there is no harmonic component at 55 Hz—the ear is clever enough to "reconstruct" a 55-Hz tone. That's why you can listen to and recognize music on an inexpensive radio whose speaker may reproduce only a small range of frequencies.

12.6 THE HUMAN EAR

Figure 12.11 shows the structure of the human ear. The human ear has an external part or *pinna* that acts something like a funnel, collecting sound waves and concentrating them at the opening of the auditory canal. The pinna is better at collecting sound coming from in front than from behind, which helps with localization. Resonance in the *auditory canal* (see Problem 56) boosts the ear's sensitivity in the 2- to 5-kHz frequency range—a crucial range for understanding speech.

At the end of the auditory canal, the eardrum (*tympanum*) vibrates in response to the incident sound wave. The region just beyond the eardrum is called the middle ear. The vibrations of the eardrum are transmitted through three tiny bones of the middle ear (the *auditory ossicles*) to the *oval window* of the *cochlea*, a tapered spiral-shaped organ filled with fluid. The oval window is a membrane that is in contact with the fluid in the cochlea. The ossicles act as levers; the force exerted by the "stirrup" on the oval window is 1.5 to 2.0 times the force the eardrum exerts on the "hammer." The area of the oval window is one-twentieth that of the eardrum, so there is an overall amplification in pressure by a factor of 30 to 40. The ossicles protect the ear from damage: in response to a loud sound, a muscle pulls the stirrup away from the oval window. At the same time, another muscle increases the eardrum tension. These two changes make the ear temporarily less sensitive. It takes a few milliseconds for the muscles to respond in this way, so they provide no protection against *sudden* loud sounds.

The *cochlear partition* runs most of the length of the cochlea, separating it into two chambers (the *scala vestibuli* and the *scala tympani*). Vibration of the oval window sends a compressional wave down the fluid in the scala vestibuli, around the end of the partition, and back up the scala tympani to the *round window.* This wave sets the *basilar membrane*, located on the cochlear partition, into vibration. The basilar membrane is thinnest and under greatest tension near the oval and round windows; it gradually increases in thickness and decreases in tension toward its other end. High-frequency waves cause the membrane to vibrate with maximum amplitude near its thin, high-tension end; low-frequency

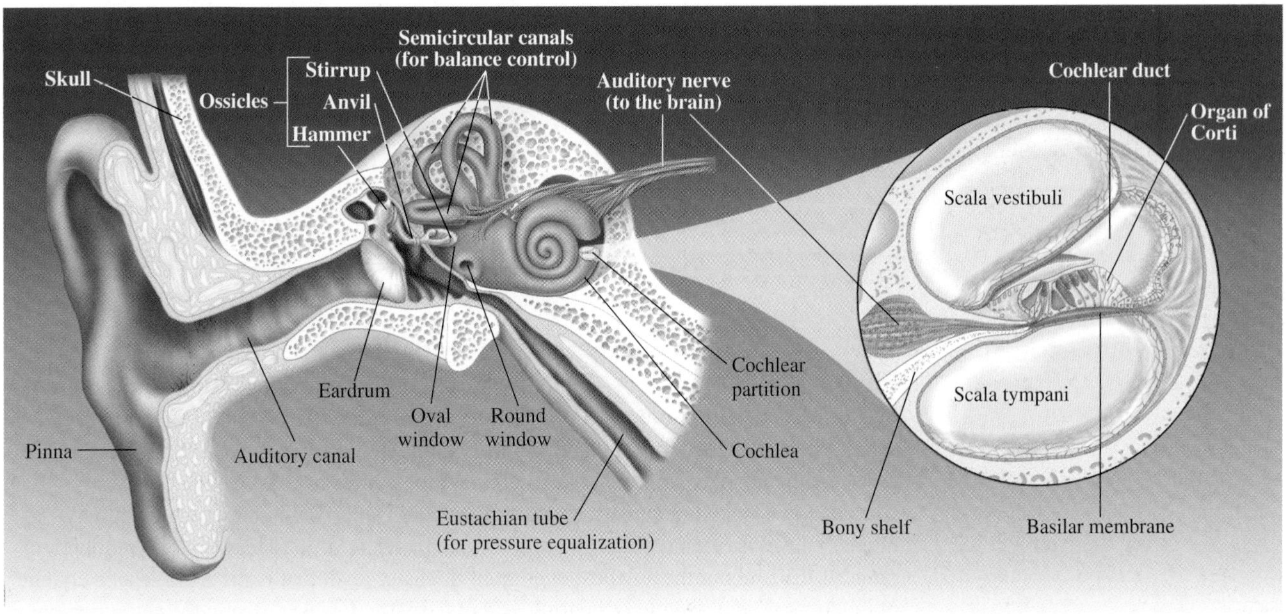

Figure 12.11 Structure of the human ear with a cross section of the cochlea.

waves cause maximum amplitude vibrations near its thicker, lower-tension end. The location of the maximum amplitude vibrations is one way the ear determines frequency; for low-frequency sounds (up to about 1 kHz), the ear sends periodic nerve signals to the brain at the frequency of the sound wave. For complex sounds, which consist of the superposition of many different frequencies (see Section 12.5), the ear performs a spectral analysis—it decomposes the complex sound into its constituent frequencies.

Located on the basilar membrane is the sensory organ (the *organ of Corti*). Rows of hair cells on the basilar membrane excite neurons when they bend in response to vibration. These neurons send electrical signals to the brain.

Loudness

Although loudness is most closely correlated to intensity level, it also depends on frequency (as well as other factors). In other words, the sensitivity of the ear is frequency-dependent. Figure 12.12 shows a set of *curves of equal loudness* for a typical person. Each curve shows the intensity levels at which sounds of different frequencies are perceived to be equally loud.

Pitch

Pitch is the perception of frequency. If you sing or play up and down a scale, it is the pitch that is rising and falling. Although pitch is the aspect of sound perception most closely tied to a single physical quantity, frequency, our sense of pitch is affected to a small extent by other factors such as intensity and timbre (Section 12.5).

Our sense of pitch is a *logarithmic* function of frequency, just as loudness is approximately a logarithmic function of intensity. If you start at the lowest note on the piano (which has a fundamental frequency of 27.5 Hz) and play a chromatic scale—every

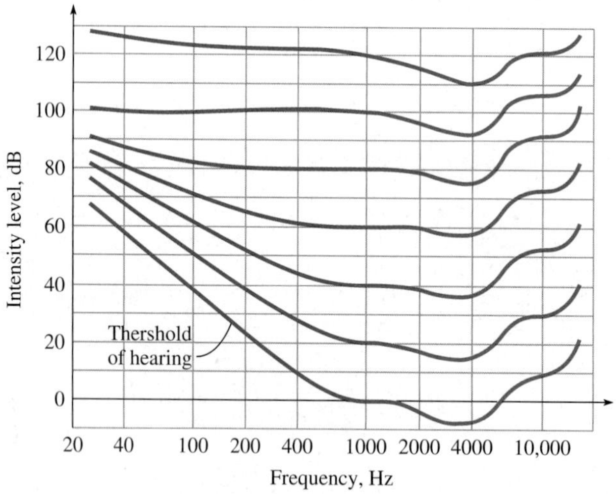

Figure 12.12 Curves of equal loudness. The curves show that the ear is most sensitive to frequencies between 3 kHz and 4 kHz, partly due to resonance in the auditory canal. The ear's sensitivity falls off rapidly below 800 Hz and above 10 kHz. At any given frequency between 800 Hz and 10 kHz, the curves are approximately evenly spaced: equal steps in intensity level produce equal steps in loudness, which is why intensity level is often used as an approximate measure of loudness. In this frequency range, 1 dB is about the smallest change in intensity level that is perceptible as a change in loudness. The threshold of hearing is shown by the lowest curve in the set; a person with excellent hearing cannot hear sounds with intensity levels below this curve. The threshold of hearing is at an intensity level of 0 dB only in the vicinity of 1 kHz.

white and black key in turn—all the way to the highest note (4190 Hz), you hear a series of equal steps in pitch. The frequencies do *not* increase in equal steps; the fundamental frequency of each note is 5.95% higher than the previous note. Under ideal conditions, most people can sense frequency changes as small as 0.3%. A trained musician can sense a frequency change of 0.1% or so.

Localization

How can you tell where a sound comes from? The ear has several different tools it uses to localize sounds:

- The principal method for high-frequency sounds (> 4 kHz) is the difference in intensity sensed by the two ears. The head casts a "sound shadow," so a sound coming from the right has a larger intensity at the right ear than at the left ear.
- The shape of the pinna makes it slightly preferential to sounds coming from the front. This helps with front-back localization for high-frequency sounds.
- For lower-frequency sounds, both the difference in arrival time and the phase difference between the waves arriving at the two ears are used for localization.

12.7 BEATS

When two sound waves are close in frequency (within about 15 Hz of one another), the superposition of the two produces a pulsation that we call **beats**. Beats can be produced by any kind of wave; they are a general result of the principle of superposition when applied to two waves of nearly the same frequency.

Beats are caused by the slow change in the phase difference between the two waves. Suppose that at one instant ($t = 0$ in Fig. 12.13), the two waves are in phase with one another and interfere constructively. The amplitude of the superposition is the sum of the amplitudes of the two waves shown in Fig. 12.13a. However, since the frequencies are different, the waves do not *stay* in phase. The higher-frequency wave has a shorter cycle, so it gets ahead of the other one. The phase difference between the two steadily increases; as it does, the amplitude of the superposition decreases. At a later time ($t = 5T_0$), the phase difference reaches 180°; now the waves are half a cycle out of phase and interfere destructively (Fig. 12.13b). Now the amplitude of the superposition is minimum—the difference

CONNECTION:

When two waves with different frequencies are superimposed, constructive interference alternates with destructive interference, causing *beats*.

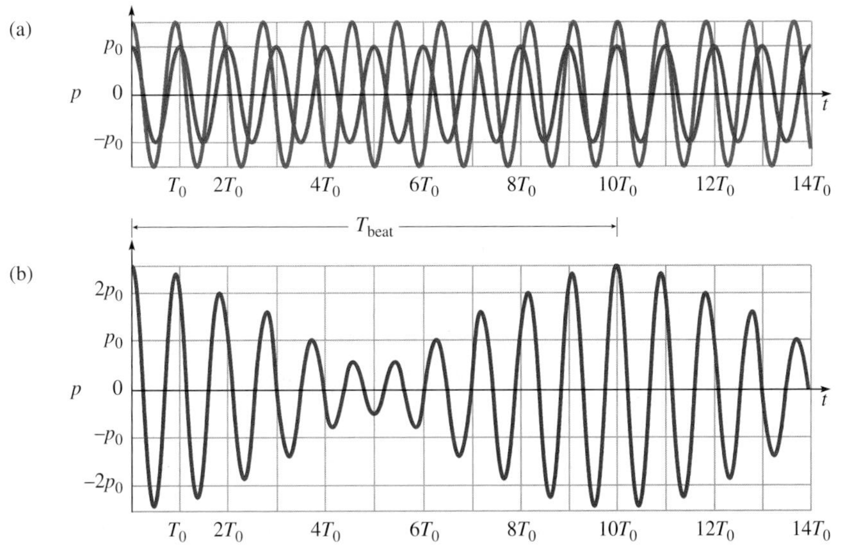

Figure 12.13 (a) Graph (red) of a sound wave with frequency $f_1 = 1/T_0$ and amplitude p_0. Graph (blue) of a second sound wave with frequency $f_2 = 1.1f_1$ and amplitude $1.5p_0$. (b) The superposition of the two has maximum amplitude $2.5p_0$ and minimum amplitude $0.5p_0$.

between the amplitudes of the two waves. As the phase difference continues to increase, the amplitude increases until constructive interference occurs again ($t = 10T_0$). The ear perceives the amplitude (and intensity) cycling from large to small to large to small as a pulsation—a repeating alternation of increasing and decreasing loudness.

At what frequency do the beats occur? It depends on how far apart the frequencies of the two waves are. We can measure the time between beats T_{beat} as the time to go from constructive interference to the next occurrence of constructive interference. During that time, each wave must go through a whole number of cycles, with one of them going through one more cycle than the other. Since frequency (f) is the number of cycles per second, the number of cycles a wave goes through during a time T_{beat} is fT_{beat}. (To illustrate: in Fig. 12.13, $T_{beat} = 10T_0$. During that time, wave 1 goes through $f_1T_{beat} = 10$ cycles, while wave 2 goes through $f_2T_{beat} = 1.1/T_0 \times 10T_0 = 11$ cycles.) If $f_2 > f_1$, then wave 2 goes through one cycle more than wave 1:

$$f_2T_{beat} - f_1T_{beat} = (\Delta f)T_{beat} = 1$$

The beat frequency f_{beat} is $1/T_{beat}$:

$$f_{beat} = 1/T_{beat} = \Delta f \qquad (12\text{-}11)$$

Thus, we obtain the remarkably simple result that the beat frequency is the difference between the frequencies of the two waves. If the difference in frequencies exceeds roughly 15 Hz, then the ear no longer perceives the beats; instead, we hear two tones at different pitches.

CHECKPOINT 12.7

(a) At what time(s) in Fig. 12.13 do the two waves interfere constructively? (b) At what time(s) do they interfere destructively?

Application of beats: tuning a piano

Piano tuners listen for beats as they tune. The tuner sounds two strings and listens for the beats. The beat frequency indicates whether the interval is correct or not. If the two strings are played by the same key, they are tuned to the same fundamental frequency, so the beat frequency should be (nearly) zero. If the two strings belong to two different notes, the beat frequency is nonzero. In this case the tuner listens to beats between two *overtones* that are close in frequency.

Example 12.7

The Piano Tuner

A piano tuner strikes his tuning fork ($f = 523.3$ Hz) and strikes a key on the piano at the same time. The two have nearly the same frequency; he hears 3.0 beats per second. As he tightens the piano string, he hears the beat frequency gradually decrease to 2.0 beats per second when the two sound together. (a) What was the frequency of the piano string before it was tightened? (b) By what percentage did the tension increase?

Strategy The beat frequency is the difference between the two frequencies; we only have to determine which is higher. The wavelength of the string is determined by its length, which does not change. The increase in tension increases the speed of waves on the string, which in turn increases the frequency.

Solution (a) Since the piano tuner heard 3.0 beats per second, the difference in the two frequencies was 3.0 Hz:

$$\Delta f = 3.0 \text{ Hz}$$

Is the piano string's frequency 3.0 Hz higher or 3.0 Hz lower than the tuning fork's frequency? As the tension increases gradually, the beat frequency decreases, which means that the frequency of the piano string is getting *closer* to the frequency

continued on next page

Example 12.7 continued

of the tuning fork. Therefore, the string frequency must be 3.0 Hz *lower* than the tuning fork frequency:

$$f_{\text{string}} = 523.3\text{ Hz} - 3.0\text{ Hz} = 520.3\text{ Hz}$$

(b) The tension (F) is related to the speed of the wave on the string (v) and the mass per unit length (μ) by

$$v = \sqrt{\frac{F}{\mu}} \qquad (11\text{-}4)$$

The mass per unit length does not change, so $v \propto \sqrt{F}$. The speed of the wave on the string is related to its wavelength and frequency by

$$v = \lambda f$$

The wavelength λ in this expression is the wavelength of the transverse wave on the string, *not* the wavelength of the sound wave in air. Since λ does not change, $v \propto f$. Therefore, $f \propto \sqrt{F}$ or

$$F \propto f^2$$

This means that the ratio of the tension F to the original tension F_0 is equal to the ratio of the frequencies squared:

$$\frac{F}{F_0} = \left(\frac{f}{f_0}\right)^2 = \left(\frac{521.3\text{ Hz}}{520.3\text{ Hz}}\right)^2 = 1.004$$

The tension was increased 0.4%.

Discussion We needed to find whether the original frequency was too high or too low. As the beat frequency decreases, the frequency of the string is getting closer to the frequency of the tuning fork. Tightening the string makes the string's frequency increase; since increasing the string's frequency brings it closer to the tuning fork's frequency, we know that the original frequency of the string was lower than the frequency of the tuning fork. Had an increase in tension *increased* the beat frequency instead, we would know that the original frequency was already too high; the tension would have to be relaxed to tune the string.

Practice Problem 12.7 Tuning a Violin

A tuning fork with a frequency of 440.0 Hz produces 4.0 beats per second when sounded together with a violin string of nearly the same frequency. What is the frequency of the string if a slight increase in tension increases the beat frequency?

12.8 THE DOPPLER EFFECT

A police car races by, its sirens screaming. As it passes, we hear the pitch change from higher to lower. The frequency change is called the **Doppler effect**, after the Austrian physicist Johann Christian Andreas Doppler (1803–1853). The observed frequency is different from the frequency transmitted by the source when the source or the observer are in motion relative to the wave medium.

We consider only the motion of the source and observer directly toward or away from one another in the reference frame in which the wave medium is at rest. Velocities of the source and observer are expressed as components along the direction of propagation of the sound wave (from source to observer). A positive component means the velocity is in the direction of propagation of the wave, but a negative component means the velocity is opposite the direction of propagation.

Moving Source

First we consider a moving source. A source emits a sound wave at frequency f_s, which means that wave crests (regions of maximum amplitude, indicated by circles in Fig. 12.14) leave the source spaced by a time interval $T_s = 1/f_s$. If the source is moving at velocity v_s toward a stationary observer on the right, Fig. 12.14a shows that the wavelength—the distance between crests—is smaller in front of the source and larger behind the source. In Fig. 12.14b, at the instant that crest 6 is emitted, crest 5 has traveled outward a distance vT_s from point 5, where v is the speed of sound. During the same time interval, the source has advanced a distance v_sT_s. The wavelength λ, measured by the observer on the right is the distance between crests 5 and 6:

$$\lambda = vT_s - v_sT_s$$

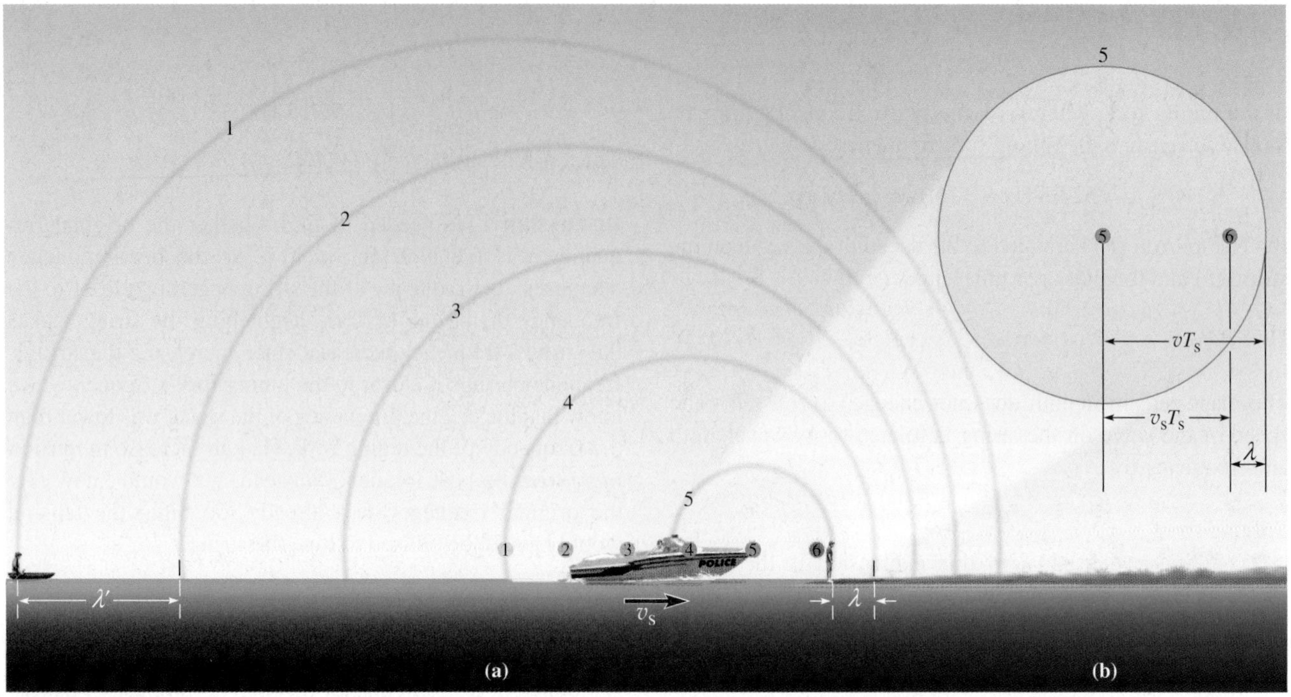

Figure 12.14 (a) A speedboat is moving to the right at speed v_s (exaggerated for clarity) while it blows its siren. The siren emits wave crests at positions 1, 2, 3, 4, 5, and 6; each wave crest moves outward in all directions, from the point at which it was emitted, at speed v. (b) Wave crest 6 is emitted a time T_s after wave crest 5 is emitted. During that time, wave crest 5 moves a distance vT_s and the boat moves a distance v_sT_s. The wavelength is the distance between wave crests: $\lambda = vT_s - v_sT_s$.

The frequency at which the crests arrive at the observer is the *observed* wave frequency f_o. The observed period T_o between the arrival of two crests is the time it takes sound to travel a distance $(v - v_s)T_s$:

$$T_o = \frac{\lambda}{v} = \frac{(v - v_s)T_s}{v}$$

The observed frequency is

$$f_o = \frac{1}{T_o} = \frac{v}{v - v_s} \times \frac{1}{T_s}$$

Dividing numerator and denominator by v and substituting $f_s = 1/T_s$ yields

Doppler effect (moving source):

$$f_o = \left(\frac{1}{1 - v_s/v}\right) f_s \qquad (12\text{-}12)$$

$v_s > 0$ for a source moving in the direction of the wave

Since the denominator $1 - v_s/v$ is less than 1, the observed frequency is higher than the source frequency when the source moves in the same direction as the wave (toward the observer). If the source instead moves *away* from the observer, the correct observed frequency is given by Eq. (12-12) as long as we make v_s negative (the source moves opposite the direction of the wave). With v_s negative, $1 - v_s/v$ is *greater* than 1, so the observed frequency is *less* than the source frequency.

Moving Observer

Now we consider motion of the observer. A stationary source emits a sound wave at frequency f_s and wavelength $\lambda = v/f_s$, where v is the speed of sound. A stationary observer would measure the arrival of wave crests spaced by a time interval $T_s = 1/f_s$. An observer

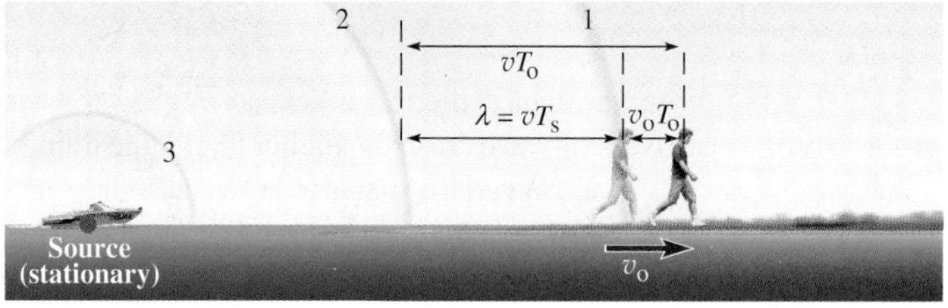

Figure 12.15 An observer moving at speed v_o (exaggerated for clarity) away from a stationary sound source. The observed frequency is lower than the source frequency.

moving away from the source at velocity v_o would observe a longer time interval between crests. Just as crest 1 reaches the observer, the next (crest 2) is a distance λ away. Crest 2 catches up with the observer at a time T_o later when the distance the wave crest travels toward the observer is equal to the distance the observer travels away from the wave crest plus the wavelength (Fig. 12.15):

$$vT_o = v_oT_o + \lambda \quad \text{or} \quad (v - v_o)T_o = \lambda = v/f_s$$

Solving for T_o,

$$T_o = \frac{v/f_s}{v - v_o}$$

The observed frequency is

$$f_o = \frac{1}{T_o} = \frac{v - v_o}{v} f_s$$

Dividing numerator and denominator by v yields

Doppler effect (moving observer):

$$f_o = (1 - v_o/v) f_s \qquad \text{(12-13)}$$

$v_o > 0$ for an observer moving in the direction of the wave

An observer moving away from the source measures a frequency lower than f_s. An observer moving *toward* the source moves opposite to the direction of the wave; in that case, v_o is negative and the observed frequency is *higher* than f_s.

✓ CHECKPOINT 12.8

(a) Does the motion of the source of a sound wave affect the wavelength?
(b) Does the motion of the observer affect the wavelength?

Example 12.8

Train Whistle and Doppler Shift

A monorail train approaches a platform at a speed of 10.0 m/s while it blows its whistle. A musician with perfect pitch standing on the platform hears the whistle as "middle C," a frequency of 261 Hz. There is no wind and the temperature is a chilly 0°C. What is the observed frequency of the whistle when the train is at rest?

Strategy In this case, the source—the whistle—is moving and the observer is stationary. The source is moving *toward* the observer, so v_s is *positive.* With the source approaching the observer, the observed frequency is higher than the source frequency. When the train is at rest, there is no Doppler shift; the observed frequency then is equal to the source frequency.

Solution For a moving source, the source (f_s) and observed (f_o) frequencies are related by

$$f_o = \left(\frac{1}{1 - v_s/v}\right) f_s$$

continued on next page

Example 12.8 continued

where $v = 331$ m/s (the speed of sound in air at 0°C), $v_s = +10.0$ m/s, and $f_o = 261$ Hz. Solving for f_s,

$$f_s = (1 - v_s/v)f_o$$
$$= \left(1 - \frac{10.0 \text{ m/s}}{331 \text{ m/s}}\right) \times 261 \text{ Hz}$$
$$= 253 \text{ Hz}$$

The source frequency is less than the observed frequency, as expected. The observed frequency when the train is at rest is equal to the source frequency: 253 Hz.

Discussion When the train is moving toward the platform, the distance between source and observer is decreasing. Wave crests emitted later take *less time* to reach the observer than if the train were at rest, so the time between arrivals of wave crests is smaller than if the train were stationary. When the distance between source and observer is decreasing, the observed frequency is higher than the source frequency; when the distance is increasing, the observed frequency is lower than the source frequency.

Practice Problem 12.8 A Sports Car Racing By

Justine is gardening in her front yard when a Mazda Miata races by at 32.0 m/s (71.6 mi/h). If she hears the sound of the Miata's engine at 220.0 Hz as it approaches her, what frequency does she hear after it passes? Assume the temperature is 20°C and there is no wind.

Motion of Both Source and Observer

If both source and observer are moving, we combine the two Doppler shifts (see Conceptual Question 10) to obtain

$$f_o = \left(\frac{1 - v_o/v}{1 - v_s/v}\right) f_s = \left(\frac{v - v_o}{v - v_s}\right) f_s \qquad (12\text{-}14)$$

Remember that the signs of v_o and v_s are positive for motion in the direction of propagation of the wave and negative for motion opposite the direction of propagation.

Example 12.9

Determining Speed from Horn Frequency

Two cars, with equal ground speeds, are moving in opposite directions away from each other on a straight highway. One driver blows a horn with a frequency of 111 Hz; the other measures the frequency as 105 Hz. If the speed of sound is 338 m/s and there is no wind, what is the ground speed of each car?

Strategy A sound wave travels from source to observer. The source moves opposite the direction of the wave, so v_s is negative. The observer moves in the direction of the wave, so v_o is positive. The speeds are the same, so $v_s = -v_o$.

Solution With both the source and observer moving, the frequencies are related by

$$f_o = \left(\frac{1 - v_o/v}{1 - v_s/v}\right) f_s$$

To simplify the algebra, we let $\alpha = v_o/v = -v_s/v$. Then

$$f_o = \left(\frac{1 - \alpha}{1 + \alpha}\right) f_s$$

Now we solve for α:

$$(1 + \alpha)\frac{f_o}{f_s} = 1 - \alpha$$

$$\alpha = \frac{1 - f_o/f_s}{1 + f_o/f_s} = \frac{1 - (105 \text{ Hz})/(111 \text{ Hz})}{1 + (105 \text{ Hz})/(111 \text{ Hz})} = 0.02778$$

Now we can find v_o:

$$v_o = \alpha v = 0.02778 \times 338 \text{ m/s} = 9.4 \text{ m/s}$$

The speed of each car is 9.4 m/s.

Discussion Quick check on the algebra: substituting $v = 338$ m/s, $f_s = 111$ Hz, $v_o = 9.4$ m/s, and $v_s = -9.4$ m/s directly into Eq. (12-14),

$$f_o = \frac{1 - (9.4 \text{ m/s})/(338 \text{ m/s})}{1 - (-9.4 \text{ m/s})/(338 \text{ m/s})} \times 111 \text{ Hz} = 105 \text{ Hz}$$

Practice Problem 12.9 Finding Speed from the Doppler Shift

A car is driving due west at 15 m/s and sounds its horn with a frequency of 260.0 Hz. A passenger in a car heading east away from the first car hears the horn at a frequency of 230.0 Hz. How fast is the second car traveling? The speed of sound is 350 m/s.

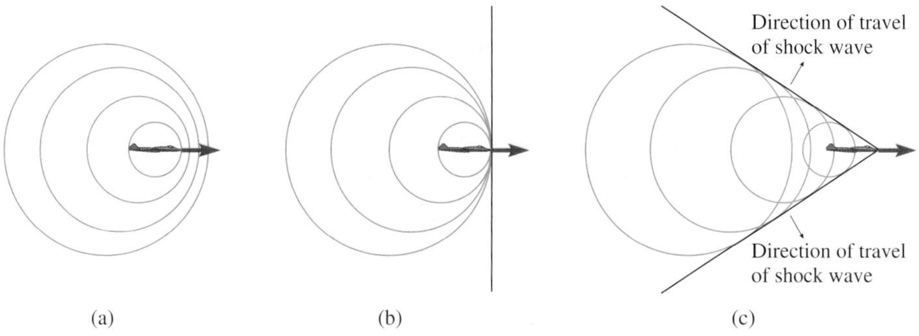

Figure 12.16 (a) Wave crests for a plane moving slower than sound. (b) A plane moving at the speed of sound; the wave crests pile up on each other since the plane moves to the right as fast as the wave crests. (c) Shock wave for a supersonic plane. The wave crests pile up along the cone indicated by the black lines.

Figure 12.17 A bullet moving through air faster than sound. Notice the two principal shock waves starting at either end of the bullet.

Shock Waves

Let's examine two interesting special cases of the Doppler formula [Eq. (12-14)]. First, what if the observer moves away from the source at the speed of sound ($v_o = v$)? The Doppler-shifted frequency would be zero according to Eq. (12-14). What does that mean? If the observer moves away from the source with a speed equal to (or greater than) the wave speed, the wave crests *never reach the observer.*

Second, what if the source moves toward the observer at a speed approaching the speed of sound ($v_s \rightarrow v$)? Then Eq. (12-14) gives an observed frequency that increases without bound ($f_o \rightarrow \infty$). Figure 12.16 helps us understand what that means. For a plane moving slower than sound, the wave crests in front of it are closer together due to the plane's motion (Fig. 12.16a). An observer to the right would measure a frequency higher than the source frequency. As the plane's speed increases, the wave crests in front of it get closer and closer together and the observed frequency increases. For a plane moving at the speed of sound (Fig. 12.16b), the wave crests pile up on top of each other; they move to the right at the same speed as the plane, so they can't get ahead of it. An observer to the right would measure a wavelength of zero—zero distance between wave crests—and therefore an infinite frequency.

What happens if the source moves at a speed *greater than* the speed of sound? Figure 12.16c shows that the wave crests pile up on top of one another to form cone-shaped *shock waves*, which travel outward in the direction indicated. There are two principal shock waves formed, one starting at the nose of the plane and one at the tail (Fig. 12.17). The sound of a shock wave is referred to as a *sonic boom.*

Application of shock waves: supersonic flight

See text website for more information about supersonic flight.

PHYSICS AT HOME

You can make a visible shock wave by trailing your finger along the surface of the water in a sink or tub. If your finger pushes the water faster than water waves travel, water piles up in front of your finger and forms a V-shaped shock wave. See if you can approximate the case of a plane moving at the speed of sound with rounded waves moving outward from your finger (Fig. 12.16b) instead of a V-shaped wave. The next time you are in a boat, notice the V-shaped bow wave that extends from the prow of the boat when you are moving faster than the speed of water waves.

12.9 ECHOLOCATION AND MEDICAL IMAGING

Bats, dolphins, whales, and some birds use *echolocation* to locate prey and to "see" their environment. To find their way around in the darkness of caves, oilbirds of northern South America and cave swiftlets of Borneo and East Asia emit sound waves and listen for the echoes. The time it takes for the echoes to return tells them how far they are from an obstacle or cave wall. Differences between the echoes that reach the two sides of the head provide information on the direction from which the echo comes.

Application of echolocation by bats and dolphins

The sounds used by oilbirds and cave swiftlets for echolocation are audible to humans, but dolphins, whales, and most bats use ultrasound (20 to 200 kHz) instead. Bats and dolphins can also determine an object's velocity by sensing the Doppler shift between the emitted and reflected waves—a clear advantage in locating prey that are darting around to avoid being eaten. Some horseshoe bats can detect frequency differences as small as 0.1 Hz.

Prey are not completely helpless. Moths, lacewings, and praying mantises have primitive ears containing a few nerve cells to detect the ultrasound emitted by a nearby bat. A group of moths fluttering about at some distance from a cave may, for no apparent reason, fold their wings and drop suddenly to the ground. Folding their wings both reduces the amount of reflected sound and helps them drop quickly to the ground to evade the swooping bat. The moths' bodies are furry rather than smooth to help absorb some of the sound waves and thus reduce the intensity of reflected sound.

When the tiger moth detects the ultrasound from a bat, it emits its own ultrasound by flexing a part of its exoskeleton. The extra sounds mixed in with the echoes tend to confuse the bat, perhaps encouraging it to hunt elsewhere.

Application of echolocation: sonar and radar

Echolocation is a useful navigational tool for seafarers. To find the depth of water below a ship, a *sonar* (**so**und **na**vigation and **r**anging) device sends out ultrasonic pulses (Fig. 12.18). The time delay Δt between an emitted ultrasonic pulse and the return of its reflection is used to determine the distance to the seafloor. Seismic P waves—sound waves traveling through the Earth—generated by explosions or air guns are used to study the interior structure of Earth and to find oil beneath the surface.

Radar is a form of echolocation that uses electromagnetic waves instead of sound waves, but otherwise the concept is similar. Weather forecasting relies on *Doppler radar* to show not only the location of a storm, but also the wind velocity.

Medical Applications of Ultrasound

Millions of expectant parents see their unborn child for the first time when the mother has an ultrasonic examination. Ultrasonic imaging uses a pulse-echo technique similar to that used by bats and in sonar. Pulses of ultrasound are reflected at boundaries between different types of tissue.

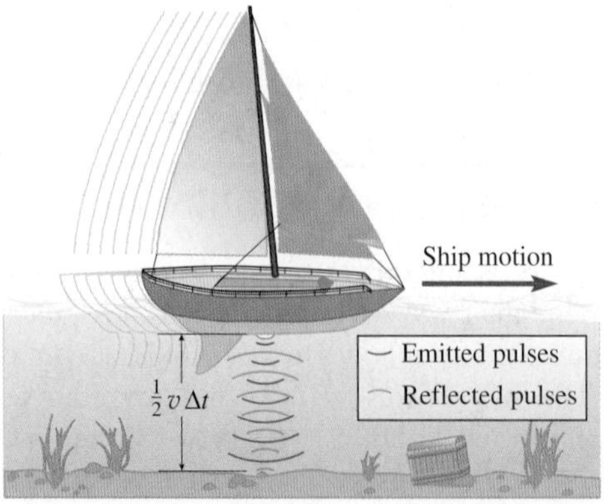

Figure 12.18 A ship with a sonar device to locate the depth of the seafloor; an ultrasound pulse, sent out from the ship by a transmitter, is reflected from the seafloor and detected by a receiver on the ship.

In the early stages of pregnancy (tenth to fourteenth weeks), the scan is used to verify that the fetus is alive and to check for twins. The length of the fetus is measured to help determine the due date more accurately. Some abnormalities can be discovered even at this early stage. For example, some chromosomal abnormalities can be detected by measuring the thickness of the skin at the back of the neck. After the eighteenth week, the fetus can be examined in even more detail. The major organs are examined to be sure they are developing normally. After the thirtieth week, the flow of blood in the umbilical cord is checked to ensure that oxygen and nutrients reach the fetus. The position of the placenta is also checked.

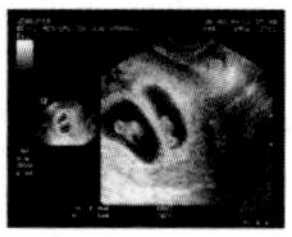

Why is ultrasound used to image the fetus?

Why are sound waves used rather than, say, electromagnetic waves such as x-rays? X-ray radiation is damaging to tissue—especially to rapidly growing fetal tissue. After decades of use, ultrasound has no known adverse effects. In addition, ultrasound images are captured in real time, so they are available immediately and can show movement. A third reason is that regular x-rays detect the amount of radiation that passes through tissue, but cannot resolve details at different depths, and so cannot produce an image of a "slice" of the abdomen; a more complicated and expensive diagnostic tool such as a CAT scan (computer-assisted tomography) would be required to resolve details at different depths. Fourth, some kinds of tissue are not detected well by x-rays but are clearly resolved in ultrasound.

Why is ultrasound used rather than sound waves of audible frequencies or lower? Sound waves with high frequencies have small wavelengths. Waves with small wavelengths diffract less around the same obstacle than do waves with larger wavelengths (see Section 11.9). Too much diffraction would obscure details in the image. As a rough rule of thumb, the wavelength is a lower limit on the smallest detail that can be resolved. The frequencies used in imaging are typically in the range 1 to 15 MHz, which means that the wavelengths in human tissue are in the range 0.1 to 1.5 mm. As a comparison, if sound waves at 15 kHz were used, the wavelength inside the body would be 10 cm. Higher frequencies give better resolution but at the expense of less penetration; sound waves are absorbed within a distance of about 500λ in tissue.

The medical applications of ultrasonic imaging are not limited to prenatal care. Ultrasound is also used to examine organs such as the heart, liver, gallbladder, kidneys, bladder, breasts, and eyes, and to locate tumors. It can be used to diagnose various heart conditions and to assess damage after a heart attack (Fig. 12.19). Ultrasound can show movement, so it is used to assess heart valve function and to monitor blood flow in large blood vessels. Because ultrasound provides real-time images, it is sometimes used to guide procedures such as biopsies, in which a needle is used to take a sample from an organ or tumor for testing.

Doppler ultrasound is a technique that is used to examine blood flow. It can help reveal blockages to blood flow, show the formation of plaque in arteries, and provide detailed information on the heartbeat of the fetus during labor and delivery. The Doppler-shifted reflections interfere with the emitted ultrasound, producing beats. The beat frequency is proportional to the speed of the reflecting object (see Problem 52).

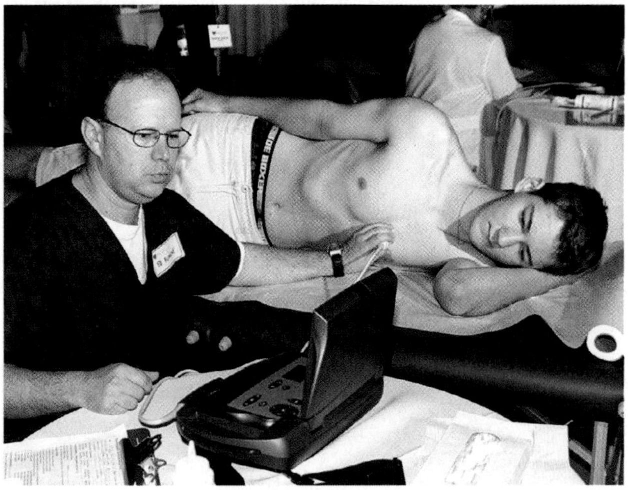

Figure 12.19 Ultrasonic imaging is used to diagnose heart disease.

Master the Concepts

- A sound wave can be described either by the gauge pressure p, which measures the pressure fluctuations above and below the ambient atmospheric pressure, or by the displacement s of each point in the medium from its undisturbed position.

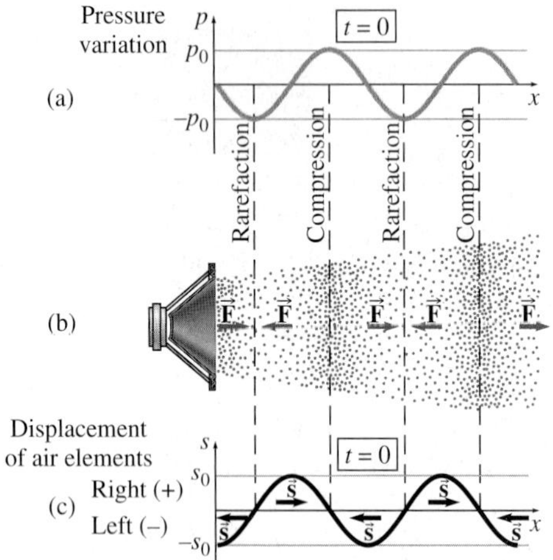

- Humans with excellent hearing can hear frequencies from 20 Hz to 20 kHz. The terms infrasound and ultrasound are used to describe sound waves with frequencies below 20 Hz and above 20 kHz, respectively.
- The speed of sound in a fluid is

$$v = \sqrt{\frac{B}{\rho}} \qquad (12\text{-}1)$$

- The speed of sound in an ideal gas at any absolute temperature T can be found if it is known at one temperature:

$$v = v_0\sqrt{\frac{T}{T_0}} \qquad (12\text{-}3)$$

where the speed of sound at absolute temperature T_0 is v_0.

- The speed of sound in air at 0°C (or 273 K) is 331 m/s.
- For sound waves traveling along the length of a thin solid rod, the speed is approximately

$$v = \sqrt{\frac{Y}{\rho}} \quad \text{(thin solid rod)} \qquad (12\text{-}5)$$

- The pressure amplitude of a sound wave is proportional to the displacement amplitude. For a harmonic sound wave at angular frequency ω,

$$p_0 = \omega v \rho s_0 \qquad (12\text{-}6)$$

where v is the speed of sound and ρ is the mass density of the medium.

- The intensity of a sound wave is related to the pressure amplitude as follows:

$$I = \frac{p_0^2}{2\rho v} \qquad (12\text{-}7)$$

where ρ is the mass density of the medium and v is the speed of sound in that medium. The most important thing to remember is that *intensity is proportional to amplitude squared*, which is true for all waves, not just sound.

- Sound intensity level in decibels is

$$\beta = (10\ \text{dB}) \log_{10} \frac{I}{I_0} \qquad (12\text{-}8)$$

where $I_0 = 10^{-12}\ \text{W/m}^2$. Sound intensity level is useful since it roughly corresponds to the way we perceive loudness. Equal increments in intensity level roughly correspond to equal increases in loudness.

- In a standing sound wave in a thin pipe, an open end is a pressure node and a displacement antinode; a closed end is a pressure antinode and a displacement node.

For a pipe open at both ends,

$$\lambda_n = \frac{2L}{n} \qquad (11\text{-}12)$$

$$f_n = n\frac{v}{2L} = nf_1 \qquad (11\text{-}13)$$

where $n = 1, 2, 3, \ldots$.

For a pipe closed at one end,

$$\lambda_n = \frac{4L}{n} \qquad (12\text{-}10a)$$

$$f_n = n\frac{v}{4L} = nf_1 \qquad (12\text{-}10b)$$

where $n = 1, 3, 5, 7, \ldots$.

- When two sound waves are close in frequency, the superposition of the two produces a pulsation called *beats*.

$$f_{beat} = \Delta f \qquad (12\text{-}11)$$

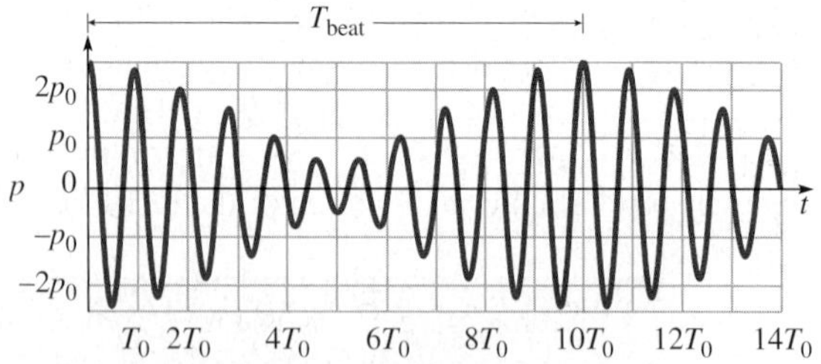

- Doppler effect: if v_s and v_o are the velocities of the source and observer, the observed frequency is

$$f_o = \left(\frac{1 - v_o/v}{1 - v_s/v}\right) f_s \qquad (12\text{-}14)$$

where v_s and v_o are positive in the direction of propagation of the wave and the wave medium is at rest.

Conceptual Questions

1. Explain why the pitch of a bassoon is more sensitive to a change in air temperature than the pitch of a cello. (That's why wind players keep blowing air through the instrument to keep it in tune.)
2. On a warm day, a piano is tuned to match an organ in an auditorium. Will the piano still be in tune with the organ the next morning, when the room is cold? If not, will the organ be higher or lower in pitch than the piano? (Assume that the piano's tuning doesn't change. Why is that a reasonable assumption?)
3. Many real estate agents have an ultrasonic rangefinder that enables them to quickly and easily measure the dimensions of a room. The device is held to one wall and reads the distance to the opposite wall. How does it work?
4. For high-frequency sounds, the ear's principal method of localization is the difference in intensity sensed by the two ears. Why can't the ear reliably use this method for low-frequency sounds? Doesn't the head cast a "sound shadow" regardless of the frequency? Explain. [*Hint:* Consider diffraction of sound waves around the head.]
5. For low-frequency sounds, the ear uses the phase difference between the sound waves arriving at the two ears to determine direction. Why can't the ear reliably use phase difference for high-frequency sounds? Explain.
6. A sign along the road in Tompkins County reads, "State Law: Noise Limit, 90 decibels." If you were subjected to such a noise level for an extended period of time, would you need to worry about your hearing being affected?
7. Why is it that your own voice sounds strange to you when you hear it played back on a tape recorder, but your friends all agree that it is just what your voice sounds like? [*Hint:* Consider the media through which the sound wave travels when you usually hear your own voice.]
8. What is the purpose of the gel that is spread over the skin before an ultrasonic imaging procedure? [*Hint:* The speed of sound in the gel is similar to the speed in the body, while the speed in air is much slower. What happens to a wave at an abrupt change in wave speed?]
9. A stereo system whose amplifier can produce 60 W per channel is replaced by one rated 120 W per channel. Would you expect the new stereo to be able to play twice as loudly as the old one? Explain.
10. A moving source emits a sound wave that is heard by a moving observer. Imagine a thin wall at rest between the source and observer. The wall completely absorbs the sound and instantaneously emits an *identical* sound wave. Use this scenario to explain why we can combine the Doppler shifts due to motion of the source and observer as in Eq. (12-14). [*Hint:* What is the net effect of this imaginary wall?]
11. Explain why the displacement of air elements at condensations and rarefactions is zero.
12. Why is the speed of sound in solids generally much faster than the speed of sound in air?
13. If the pressure amplitude of a sound wave is doubled, what happens to the displacement amplitude, the intensity, and the intensity level?
14. The source and observer of a sound wave are both at rest with respect to the ground. The wind blows in the direction from source to observer. Is the observed frequency Doppler-shifted? Explain.
15. Many brass instruments have valves that increase the total length of the pipe from mouthpiece to bell. When a valve is depressed, is the fundamental frequency raised or lowered? What happens to the pitch?

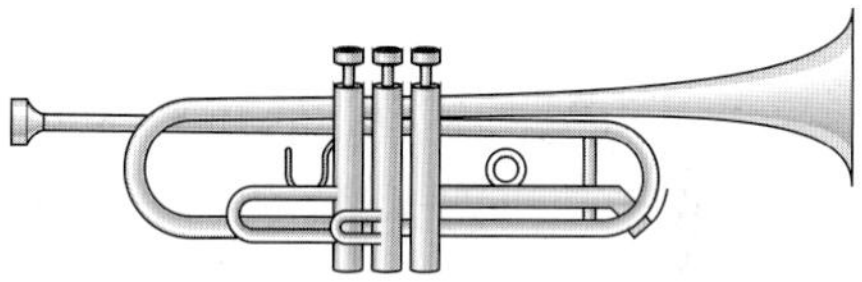

16. When the viola section of an orchestra with six members plays together, is the sound 6 times as loud as when a single viola plays? Explain. Is the intensity 6 times what it would be for a single viola? [*Hint:* The six sound waves are not coherent.]
17. The fundamental frequency of the highest note on the piano is 4.186 kHz. Most musical instruments do not go that high; only a few singers can produce sounds with fundamental frequencies higher than around 1 kHz. Yet a good-quality stereo system must reproduce frequencies up to at least 16 to 18 kHz. Explain.

Multiple-Choice Questions

1. An organ pipe is closed at one end. Several standing wave patterns are sketched in the drawing. Which one is not a possible standing wave pattern for this pipe?
2. Of the standing wave patterns sketched in the drawing, which shows the lowest frequency standing wave for an organ pipe closed at one end? (tutorial: standing waves)

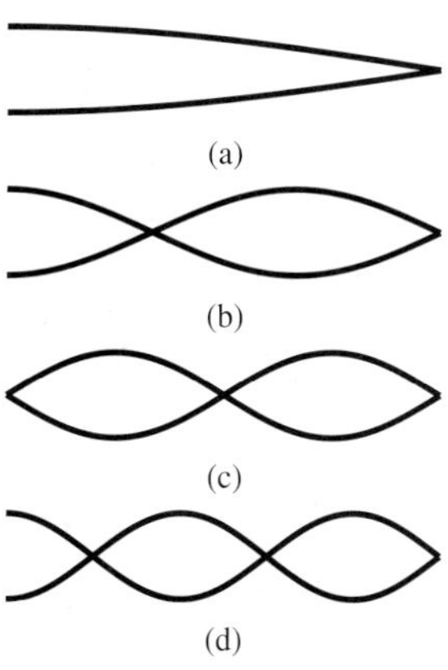

Multiple-Choice Questions 1 and 2

3. The speed of sound in water is 4.3 times the speed of sound in air. A whistle on land produces a sound wave with a frequency f_0. When this sound wave enters the water, its frequency becomes
   (a) $4.3f_0$
   (b) f_0
   (c) $\dfrac{f_0}{4.3}$
   (d) not enough information given
4. The intensity of a sound wave is directly proportional to
   (a) the frequency.
   (b) the amplitude.
   (c) the square of the amplitude.
   (d) the square of the speed of sound.
   (e) none of the above.
5. The fundamental frequency of a pipe closed at one end is f_1. How many nodes are present in a standing wave of frequency $9f_1$?
   (a) 4 (b) 5 (c) 6 (d) 8 (e) 9 (f) 10
6. The length of a pipe closed at one end is L. In the standing wave whose frequency is 7 times the fundamental frequency, what is the closest distance between nodes?
   (a) $\frac{1}{14}L$ (b) $\frac{1}{7}L$ (c) $\frac{2}{7}L$ (d) $\frac{4}{7}L$ (e) $\frac{8}{7}L$
   (f) none of the above
7. The three lowest resonant frequencies of a system are 50 Hz, 150 Hz, and 250 Hz. The system could be
   (a) a tube of air closed at both ends.
   (b) a tube of air open at one end.
   (c) a tube of air open at both ends.
   (d) a vibrating string with fixed ends.
8. A source of sound with frequency 620 Hz is placed on a moving platform that approaches a physics student at speed v; the student hears sound with a frequency f_1. Then the source of sound is held stationary while the student approaches it at the same speed v; the student hears sound with a frequency f_2. Choose the correct statement.
   (a) $f_1 = f_2$; both are greater than 620 Hz.
   (b) $f_1 = f_2$; both are less than 620 Hz.
   (c) $f_1 > f_2 > 620$ Hz.
   (d) $f_2 > f_1 > 620$ Hz.
9. A moving van and a small car are traveling in the same direction on a two-lane road. The van is moving at twice the speed of the car and overtakes the car. The driver of the car sounds his horn, frequency = 440 Hz, to signal the van that it is safe to return to the lane. Which is the correct statement?
   (a) The car driver and van driver both hear the horn frequency as 440 Hz.
   (b) The car driver hears 440 Hz, but the van driver hears a lower frequency.
   (c) The car driver hears 440 Hz, but the van driver hears a higher frequency.
   (d) Both drivers hear the same frequency and it is lower than 440 Hz.
10. A trombone and a bassoon play notes of equal loudness with the same fundamental frequency. The two sounds differ primarily in
    (a) pitch.
    (b) intensity level.
    (c) amplitude.
    (d) timbre.
    (e) wavelength.

Problems

- Combination conceptual/quantitative problem
- Biological or medical application
- Challenging problem
- Blue # Detailed solution in the Student Solutions Manual
- 1 2 Problems paired by concept
- Text website interactive or tutorial

Note: Assume a temperature of 20.0°C in all problems unless otherwise indicated.

12.2 The Speed of Sound Waves

1. Bats emit ultrasonic waves with a frequency as high as 1.0×10^5 Hz. What is the wavelength of such a wave in air of temperature 15°C?
2. Dolphins emit ultrasonic waves with a frequency as high as 2.5×10^5 Hz. What is the wavelength of such a wave in seawater at 25°C?
3. At a baseball game, a spectator is 60.0 m away from the batter. How long does it take the sound of the bat connecting with the ball to travel to the spectator's ears? The air temperature is 27.0°C.
4. A lightning flash is seen in the sky and 8.2 s later the boom of the thunder is heard. The temperature of the air is 12°C. (a) What is the speed of sound at that temperature? [*Hint:* Light is an electromagnetic wave that travels at a speed of 3.00×10^8 m/s.] (b) How far away is the lightning strike?
5. During a thunderstorm, you can easily estimate your distance from a lightning strike. Count the number of seconds that elapse from when you see the flash of lightning to when you hear the thunder. The rule of thumb is that 5 s elapse for each mile of distance. Verify that this rule of thumb is (approximately) correct. (One mile is 1.6 km and light travels at a speed of 3×10^8 m/s.)
6. A copper alloy has a Young's modulus of 1.1×10^{11} Pa and a density of 8.92×10^3 kg/m^3. What is the speed of sound in a thin rod made from this alloy? Compare your result with that given in Table 12.1.
7. Find the speed of sound in mercury, which has a bulk modulus of 2.8×10^{10} Pa and a density of 1.36×10^4 kg/m^3.
8. Derive Eq. (12-4) as: (a) Starting with Eq. (12-3), substitute $T = T_C + 273.15$. (b) Apply the binomial approximation to the square root (see Appendix A.5) and simplify.

9. (a) Show that since the bulk modulus has SI units N/m^2 and mass density has SI units kg/m^3, Eq. (12-1) gives the speed of sound in m/s. Thus, the equation is dimensionally consistent. (b) Show that no other combination of B and ρ can give dimensions of speed. Thus, Eq. (12-1) *must* be correct except for the possibility of a dimensionless constant.

10. Stan and Ollie are standing next to a train track. Stan puts his ear to the steel track to hear the train coming. He hears the sound of the train whistle through the track 2.1 s before Ollie hears it through the air. How far away is the train?

12.3 Amplitude and Intensity of Sound Waves

11. A sound wave with an intensity level of 80.0 dB is incident on an eardrum of area 0.600×10^{-4} m^2. How much energy is absorbed by the eardrum in 3.0 min?

12. The sound level 25 m from a loudspeaker is 71 dB. What is the rate at which sound energy is produced by the loudspeaker, assuming it to be an isotropic source?

13. In a factory, three machines produce noise with intensity levels of 85 dB, 90 dB, and 93 dB. When all three are running, what is the intensity level? How does this compare to running just the loudest machine?

14. At the race track, one race car starts its engine with a resulting intensity level of 98.0 dB at point P. Then seven more cars start their engines. If the other seven cars each produce the same intensity level at point P as the first car, what is the new intensity level with all eight cars running?

15. (a) What is the pressure amplitude of a sound wave with an intensity level of 120.0 dB in air? (b) What force does this exert on an eardrum of area 0.550×10^{-4} m^2?

16. An intensity level change of +1.00 dB corresponds to what percentage change in intensity?

17. (a) Show that if $I_2 = 10.0I_1$, then $\beta_2 = \beta_1 + 10.0$ dB. (A factor of 10 increase in intensity corresponds to a 10.0-dB increase in intensity level.) (b) Show that if $I_2 = 2.0I_1$, then $\beta_2 = \beta_1 + 3.0$ dB. (A factor of 2 increase in intensity corresponds to a 3.0-dB increase in intensity level. tutorial: decibels)

18. At a rock concert, the engineer decides that the music isn't loud enough. He turns up the amplifiers so that the amplitude of the sound, where you're sitting, increases by 50.0%. (a) By what percentage does the intensity increase? (b) How does the intensity level (in dB) change?

12.4 Standing Sound Waves

19. Humans can hear sounds with frequencies up to about 20.0 kHz, but dogs can hear frequencies up to about 40.0 kHz. Dog whistles are made to emit sounds that dogs can hear but humans cannot. If the part of a dog whistle that actually produces the high frequency is made of a tube open at both ends, what is the longest possible length for the tube?

20. (a) What should be the length of an organ pipe, closed at one end, if the fundamental frequency is to be 261.5 Hz? (b) What is the fundamental frequency of the organ pipe of part (a) if the temperature drops to 0.0°C?

21. Repeat Problem 20 for an organ pipe that is open at both ends.

22. An organ pipe that is open at both ends has a fundamental frequency of 382 Hz at 0.0°C. What is the fundamental frequency for this pipe at 20.0°C?

23. What is the length of the organ pipe in Problem 22?

24. A certain pipe has resonant frequencies of 234 Hz, 390 Hz, and 546 Hz, with no other resonant frequencies between these values. (a) Is this a pipe open at both ends or closed at one end? (b) What is the fundamental frequency of this pipe? (c) How long is this pipe?

25. In an experiment to measure the speed of sound in air, standing waves are set up in a narrow pipe open at both ends using a speaker driven at 702 Hz. The length of the pipe is 2.0 m. What is the air temperature inside the pipe (assumed reasonably near room temperature, 20°C to 35°C)? [*Hint:* The standing wave is not necessarily the fundamental.]

26. When a tuning fork is held over the open end of a very thin tube, as in Fig. 12.7, the smallest value of L that produces resonance is found to be 30.0 cm. (a) What is the wavelength of the sound? [*Hint:* Assume that the displacement antinode is at the open end of the tube.] (b) What is the next larger value of L that will produce resonance with the same tuning fork? (c) If the frequency of the tuning fork is 282 Hz, what is the speed of sound in the tube?

27. Two tuning forks, A and B, excite the next-to-lowest resonant frequency in two air columns of the same length, but A's column is closed at one end and B's column is open at both ends. What is the ratio of A's frequency to B's frequency?

28. How long a pipe is needed to make a tuba whose lowest note is low C (frequency 130.8 Hz)? Assume that a tuba is a long straight pipe open at both ends.

29. An aluminum rod, 1.0 m long, is held lightly in the middle. One end is struck head-on with a rubber mallet so that a longitudinal pulse—a sound wave—travels down the rod. The fundamental frequency of the longitudinal vibration is 2.55 kHz. (a) Describe the location of the node(s) and antinode(s) for the fundamental mode of vibration. Use either displacement or pressure nodes and antinodes. (b) Calculate the speed of sound in aluminum from the information given in the problem. (c) The vibration of the rod produces a sound wave in air that can be heard. What is the wavelength of the sound wave in the air? Take the speed of sound in air to be 334 m/s. (d) Do the two ends of the rod vibrate longitudinally in phase or out of phase with each other? That is, at any given instant, do they move in the same direction or in opposite directions?

12.7 Beats

30. A violin is tuned by adjusting the tension in the strings. Brian's A string is tuned to a slightly lower frequency than Jennifer's, which is correctly tuned to 440.0 Hz. (a) What is the frequency of Brian's string if beats of 2.0 Hz are heard when the two bow the strings together? (b) Does Brian need to tighten or loosen his A string to get in tune with Jennifer? Explain.

31. A piano tuner sounds two strings simultaneously. One has been previously tuned to vibrate at 293.0 Hz. The tuner hears 3.0 beats per second. The tuner increases the tension on the as-yet untuned string, and now when they are played together the beat frequency is 1.0 s^{-1}. (a) What was the original frequency of the untuned string? (b) By what percentage did the tuner increase the tension on that string?

32. An auditorium has organ pipes at the front and at the rear of the hall. Two identical pipes, one at the front and one at the back, have fundamental frequencies of 264.0 Hz at 20.0°C. During a performance, the organ pipes at the back of the hall are at 25.0°C, while those at the front are still at 20.0°C. What is the beat frequency when the two pipes sound simultaneously?

33. A musician plays a string on a guitar that has a fundamental frequency of 330.0 Hz. The string is 65.5 cm long and has a mass of 0.300 g. (a) What is the tension in the string? (b) At what speed do the waves travel on the string? (c) While the guitar string is still being plucked, another musician plays a slide whistle that is closed at one end and open at the other. He starts at a very high frequency and slowly lowers the frequency until beats, with a frequency of 5 Hz, are heard with the guitar. What is the fundamental frequency of the slide whistle with the slide in this position? (d) How long is the open tube in the slide whistle for this frequency?

34. A cello string has a fundamental frequency of 65.40 Hz. What beat frequency is heard when this cello string is bowed at the same time as a violin string with frequency of 196.0 Hz? [*Hint:* The beats occur between the third harmonic of the cello string and the fundamental of the violin.]

12.8 The Doppler Effect

35. An ambulance traveling at 44 m/s approaches a car heading in the same direction at a speed of 28 m/s. The ambulance driver has a siren sounding at 550 Hz. At what frequency does the driver of the car hear the siren?

36. At a factory, a noon whistle is sounding with a frequency of 500 Hz. As a car traveling at 85 km/h approaches the factory, the driver hears the whistle at frequency f_i. After driving past the factory, the driver hears frequency f_f. What is the change in frequency $f_f - f_i$ heard by the driver?

37. In parts of the midwestern United States, sirens sound when a severe storm that may produce a tornado is approaching. Mandy is walking at a speed of 1.56 m/s directly toward one siren and directly away from another siren when they both begin to sound with a frequency of 698 Hz. What beat frequency does Mandy hear? (WW tutorial: Doppler effect)

38. A source of sound waves of frequency 1.0 kHz is traveling through the air at 0.50 times the speed of sound. (a) Find the frequency of the sound received by a stationary observer if the source moves toward her. (b) Repeat if the source moves away from her instead.

39. A source of sound waves of frequency 1.0 kHz is stationary. An observer is traveling at 0.50 times the speed of sound. (a) What is the observed frequency if the observer moves toward the source? (b) Repeat if the observer moves away from the source instead.

40. A child swinging on a swing set hears the sound of a whistle that is being blown directly in front of her. At the bottom of her swing when she is moving toward the whistle, she hears a higher pitch, and at the bottom of her swing when she is moving away from the swing she hears a lower pitch. The higher pitch has a frequency that is 5.0% higher than the lower pitch. What is the speed of the child at the bottom of the swing?

41. A source and an observer are *each* traveling at 0.50 times the speed of sound. The source emits sound waves at 1.0 kHz. Find the observed frequency if (a) the source and observer are moving *toward* each other; (b) the source and observer are moving *away* from each other; (c) the source and observer are moving in the same direction.

42. Blood flow rates can be found by measuring the Doppler shift in frequency of ultrasound reflected by red blood cells (known as *angiodynography*). If the speed of the red blood cells is v, the speed of sound in blood is u, the ultrasound source emits waves of frequency f, and we assume that the blood cells are moving directly toward the ultrasound source, show that the frequency f_r of reflected waves detected by the apparatus is given by

$$f_r = f\frac{1 + v/u}{1 - v/u}$$

[*Hint:* There are *two* Doppler shifts. A red blood cell first acts as a moving observer; then it acts as a moving source when it reradiates the reflected sound at the same frequency that it received.]

43. Show that for a moving source, the fractional shift in observed frequency is equal to v_s/v, the source's speed as a fraction of the speed of sound. [*Hint:* Use the binomial approximation from Appendix A.5.]

44. The pitch of the sound from a race car engine drops the musical interval of a fourth when it passes the spectators. This means the frequency of the sound after passing is 0.75 times what it was before. How fast is the race car moving?

12.9 Echolocation and Medical Imaging

45. A ship is lost in a dense fog in a Norwegian fjord that is 1.80 km wide. The air temperature is 5.0°C. The captain fires a pistol and hears the first echo after 4.0 s. (a) How

far from one side of the fjord is the ship? (b) How long after the first echo does the captain hear the second echo?

46. A ship mapping the depth of the ocean emits a sound of 38 kHz. The sound travels to the ocean floor and returns 0.68 s later. (a) How deep is the water at that location? (b) What is the wavelength of the wave in water? (c) What is the wavelength of the reflected wave as it travels into the air, where the speed of sound is 350 m/s?

47. A boat is using sonar to detect the bottom of a freshwater lake. If the echo from a sonar signal is heard 0.540 s after it is emitted, how deep is the lake? Assume the temperature of the lake is uniform and at 25°C.

48. A geological survey ship mapping the floor of the ocean sends sound pulses down from the surface and measures the time taken for the echo to return. How deep is the ocean at a point where the echo time (down and back) is 7.07 s? The temperature of the seawater is 25°C.

49. A bat emits chirping sounds of frequency 82.0 kHz while hunting for moths to eat. If the bat is flying toward the moth at a speed of 4.40 m/s and the moth is flying away from the bat at 1.20 m/s, what is the frequency of the sound wave reflected from the moth as observed by the bat? Assume it is a cool night with a temperature of 10.0°C. [*Hint:* There are two Doppler shifts. Think of the moth as a receiver, which then becomes a source as it "retransmits" the reflected wave.]

50. The bat of Problem 49 emits a chirp that lasts for 2.0 ms and then is silent while it listens for the echo. If the beginning of the echo returns just after the outgoing chirp is finished, how close to the moth is the bat? [*Hint:* Is the change in distance between the two significant during a 2.0-ms time interval?]

51. Doppler ultrasound is used to measure the speed of blood flow (see Problem 42). The reflected sound interferes with the emitted sound, producing beats. If the speed of red blood cells is 0.10 m/s, the ultrasound frequency used is 5.0 MHz, and the speed of sound in blood is 1570 m/s, what is the beat frequency?

52. (a) In Problem 42, find the beat frequency between the outgoing and reflected sound waves. (b) Show that the beat frequency is proportional to the speed of the blood cell if $v \ll u$. [*Hint:* Use the binomial approximation from Appendix A.5.]

Comprehensive Problems

53. A 30.0-cm-long string has a mass of 0.230 g and is vibrating at its next-to-lowest natural frequency f_2. The tension in the string is 7.00 N. (a) What is f_2? (b) What are the frequency and wavelength of the sound in the surrounding air if the speed of sound is 350 m/s?

54. Kyle is climbing a sailboat mast and is 5.00 m above the surface of the ocean, while his friend Rob is scuba diving below the boat. Kyle shouts to someone on another boat and Rob hears him shout 0.0210 s later. The ocean temperature is 25°C and the air is at 20°C. How deep is Rob below the boat?

55. What are the four lowest standing wave frequencies for an organ pipe that is 4.80 m long and closed at one end?

56. The length of the auditory canal in humans averages about 2.5 cm. What are the lowest three standing wave frequencies for a pipe of this length open at one end? What effect might resonance have on the sensitivity of the ear at various frequencies? (Refer to Fig. 12.12. Note that frequencies critical to speech recognition are in the range 2 to 5 kHz.)

57. Some bats determine their distance to an object by detecting the difference in intensity between echoes.(a) If intensity falls off at a rate that is inversely proportional to the distance squared, show that the echo intensity is inversely proportional to the fourth power of distance. (b) The bat was originally 0.60 m from one object and 1.10 m from another. After flying closer, it is now 0.50 m from the first and at 1.00 m from the second object. What is the percentage increase in the intensity of the echo from each object?

58. The Vespertilionidae family of bats detect the distance to an object by timing how long it takes for an emitted signal to reflect off the object and return. Typically they emit sound pulses 3 ms long and 70 ms apart while cruising. (a) If an echo is heard 60 ms later ($v_{\text{sound}} = 331$ m/s), how far away is the object? (b) When an object is only 30 cm away, how long will it be before the echo is heard? (c) Will the bat be able to detect this echo?

59. At what frequency f does a sound wave in air have a wavelength of 15 cm, about half the diameter of the human head? Some methods of localization work well only for frequencies below f, while others work well only above f. (See Conceptual Questions 4 and 5.)

60. Horseshoe bats use the Doppler effect to determine their location. A Horseshoe bat flies toward a wall at a speed of 15 m/s while emitting a sound of frequency 35 kHz. What is the beat frequency between the emission frequency and the echo?

61. According to a treasure map, a treasure lies at a depth of 40.0 fathoms on the ocean floor due east from the lighthouse. The treasure hunters use sonar to find where the depth is 40.0 fathoms as they head east from the lighthouse. What is the elapsed time between an emitted pulse and the return of its echo at the correct depth if the water temperature is 25°C? [*Hint:* One fathom is 1.83 m.]

62. When playing *fortissimo* (very loudly), a trumpet emits sound energy at a rate of 0.800 W out of a bell (opening) of diameter 12.7 cm. (a) What is the sound intensity level right in front of the trumpet? (b) If the trumpet radiates sound waves uniformly in all directions, what is the sound intensity level at a distance of 10.0 m?

63. One cold and windy winter day, Zach notices a humming sound coming from his chimney when the chimney is open at the top and closed at the bottom. He opens the chimney at the bottom and notices that the sound

changes. He goes over to the piano to try to match the note that the chimney is producing with the bottom open. He finds that the "C" three octaves below middle "C" matches the chimney's fundamental frequency. Zach knows that the frequency of middle "C" is 261.6 Hz, and each lower octave is $\frac{1}{2}$ of the frequency of the octave above. From this information, Zach finds the height of the chimney and the fundamental frequency of the note that was produced when the chimney was *closed* at the bottom. Assuming that the speed of sound in the cold air is 330 m/s, reproduce Zach's calculations to find (a) the height of the chimney and (b) the fundamental frequency of the chimney when it is *closed* at the bottom.

64. A periodic wave is composed of the superposition of three sine waves whose frequencies are 36, 60, and 84 Hz. The speed of the wave is 180 m/s. What is the wavelength of the wave? [*Hint:* The 36 Hz is not necessarily the fundamental frequency.]

65. Analysis of the periodic sound wave produced by a violin's G string includes three frequencies: 392, 588, and 980 Hz. What is the fundamental frequency? [*Hint:* The wave on the string is the superposition of several different standing wave patterns.]

66. Your friend needs advice on her newest "acoustic sculpture." She attaches one end of a steel wire, of diameter 4.00 mm and density 7860 kg/m^3, to a wall. After passing over a pulley, located 1.00 m from the wall, the other end of the wire is attached to a hanging weight. Below the horizontal length of wire she places a 1.50-m-long hollow tube, open at one end and closed at the other. Once the sculpture is in place, air will blow through the tube, creating a sound. Your friend wants this sound to cause the steel wire to oscillate at the same resonant frequency as the tube. What weight (in newtons) should she hang from the wire if the temperature is 18.0°C?

67. A sound wave arriving at your ear is transferred to the fluid in the cochlea. If the intensity in the fluid is 0.80 times that in air and the frequency is the same as for the wave in air, what will be the ratio of the pressure amplitude of the wave in air to that in the fluid? Approximate the fluid as having the same values of density and speed of sound as water.

68. In this problem, you will estimate the smallest kinetic energy of vibration that the human ear can detect. Suppose that a harmonic sound wave at the threshold of hearing ($I = 1.0 \times 10^{-12}$ W/m^2) is incident on the eardrum. The speed of sound is 340 m/s and the density of air is 1.3 kg/m^3. (a) What is the maximum speed of an element of air in the sound wave? [*Hint:* See Eq. (10-21).] (b) Assume the eardrum vibrates with displacement s_0 at angular frequency ω; its maximum speed is then equal to the maximum speed of an air element. The mass of the eardrum is approximately 0.1 g. What is the *average* kinetic energy of the eardrum? (c) The average kinetic energy of the eardrum due to collisions with air molecules *in the absence of a sound wave* is about 10^{-20} J. Compare your answer with (b) and discuss.

69. During a rehearsal, all eight members of the first violin section of an orchestra play a very soft passage. The sound intensity level at a certain point in the concert hall is 38.0 dB. What is the sound intensity level at the same point if only one of the violinists plays the same passage? [*Hint:* When playing together, the violins are *incoherent* sources of sound.]

Answers to Practice Problems

12.1 Although solids usually have somewhat higher densities than liquids, they have *much* higher bulk moduli—they are much stiffer. The greater restoring forces in solids cause sound waves to travel faster.

12.2 Assumptions: Treat the stage as a point source; ignore reflection and absorption of waves. 4.0×10^{-6} W/m^2, 0.057 Pa.

12.3 400

12.4 a factor of 3.2

12.5 3000 km. No; it is not realistic to ignore absorption and reflection over such a great distance.

12.6 24°C

12.7 444.0 Hz

12.8 182.5 Hz

12.9 27 m/s

Answers to Checkpoints

12.3 The relationship between pressure and displacement amplitudes depends on the frequency and, therefore, does not have a unique value for a given pressure amplitude and intensity.

12.4 A pipe of length L closed at one end has a node at one end and an antinode at the other. The wavelength can be $2L$ only if both ends are nodes (or both are antinodes), because the distance between two successive nodes (or two successive antinodes) is $\frac{1}{2}\lambda$.

12.7 (a) Constructive interference means the two waves are *in phase*, which occurs at $t = 0$ and $t = 10T_0$. At those times, the superposition of the waves has its maximum amplitude. (b) Destructive interference means the two waves are *out of phase*, which occurs at $t = 5T_0$. At this time, the superposition has its minimum amplitude. Destructive interference would next occur at $t = 15T_0$ (not shown on the graph).

12.8 (a) The motion of the source does affect the wavelength: λ is shorter in front of the source and longer behind it (see Fig. 12.14). (b) The motion of the observer does not affect the wavelength, which is the instantaneous distance between two wave crests (see Fig. 12.15).

Review & Synthesis: Chapters 9–12

Review Exercises

1. (a) Which has more buoyant force acting on it in water, 1.0 kg of lead or 1.0 kg of aluminum? Explain. (b) Which has more buoyant force acting on it, 1.0 kg of steel that is sinking to the bottom of a lake or 1.0 kg of wood with a density of 500 kg/m^3 that is floating on the lake? Explain. (c) Once you have answered the qualitative questions, find the quantitative answers to parts (a) and (b).

2. A solid piece of plastic, with a density of 890 kg/m^3, is placed in oil with a density of 830 kg/m^3 and the plastic sinks (A). Then the plastic is placed in water and it floats (B). (a) What percentage of the plastic is submerged in the water? (b) Finally, the same oil in which the plastic sinks is poured over the plastic and the water. Will less (C) or more (D) of the plastic be submerged in the water compared to B? Explain. (c) Calculate the percentage of the plastic submerged in the water in figure C.

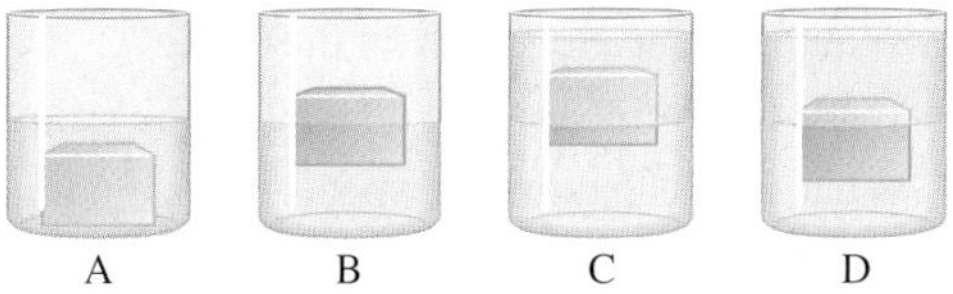

3. Water enters an apartment building 0.90 m below the street level with a gauge pressure of 52.0 kPa through the main pipe with a 5.00-cm radius. A second-story bathroom has a faucet with a 1.20-cm radius that is located 4.20 m above the street. How fast is the water moving through the main pipe?

4. To escape a burning building, Arnold has to jump from a third-story window that is about 10 m above the ground. Arnold is worried about breaking his leg. The largest bone in Arnold's leg is the femur, which has a minimum cross-sectional area of about 5×10^{-4} m and a maximum ultimate strength for compression of about 1.70×10^8 N/m^2. Arnold has a mass of 82 kg. (a) If Arnold lands on the ground with his legs stiff, then his femur can compress only about 5 mm. What will happen to Arnold's femur? (b) Suppose instead of landing on the ground, Arnold lands in deep snow so his legs can move about 30 cm between the time they first hit the snow and the time he comes to a complete stop. What will happen to Arnold's femur in this case?

5. A 5.0-kg block of wood is attached to a spring with a spring constant of 150 N/m. The block is free to slide on a horizontal frictionless surface once the spring is stretched and released. A 1.0-kg block of wood rests on top of the first block. The coefficient of static friction between the two blocks of wood is 0.45. What is the maximum speed that this set of blocks can have as it oscillates if the top block of wood is not to slip?

6. A child swinging on a swing set hears the sound of a whistle that is being blown directly in front of her. At the bottom of her swing when she is moving toward the whistle, she hears a higher pitch, and at the bottom of her swing when she is moving away from the whistle she hears a lower pitch. The higher pitch has a frequency that is 4.0% higher than the lower pitch. How high is the child swinging?

7. Consider the following equations for two different traveling waves:

   I. $y(x, t) = (1.50 \text{ cm}) \sin [(4.00 \text{ cm}^{-1})x + (6.00 \text{ s}^{-1})t]$

   II. $y(x, t) = (4.50 \text{ cm}) \sin [(3.00 \text{ cm}^{-1})x - (3.00 \text{ s}^{-1})t]$

   (a) Which wave has the fastest wave speed? What is that speed? (b) Which wave has the longest wavelength? What is that wavelength? (c) Which wave has the fastest maximum speed of a point in the medium? What is that speed? (d) Which wave is moving in the positive x-direction?

8. A Foucault pendulum has an object with a mass of 15.0 kg hung by a 14.0-m-long thin wire. (a) What is the oscillation frequency of this pendulum? (b) If the pendulum has a maximum oscillation angle of 6.10°, what is the maximum speed of this pendulum? (c) What is the maximum tension in the wire? (d) If the wire has a mass of 10.0 g, what is the fundamental frequency of the wire when it is at maximum tension?

9. The lowest frequency string on a guitar is 65.5 cm long and is tuned to 82 Hz. (a) If the string has a mass of 3.31 g, what is the tension in the string? (b) By fingering the guitar at the fifth fret, you shorten the vibrating length of the string, thereby changing the fundamental frequency of this string to match that of the next-highest-frequency string on the guitar, 110 Hz. How long is the lowest frequency string when it is fingered at the fifth fret?

10. Two children are playing with a tin-can telephone. The children are 12 m apart, the string connecting their tin cans has a linear mass density of 1.3 g/m, and it is stretched with a tension of 8.0 N. One child decides to pluck the string. How long does it take for the wave pulse to travel from one child to the other?

11. A sound wave with a frequency of 400.0 Hz is incident upon a set of stairs. The reflected waves from the vertical surfaces of adjacent steps cancel each other. What is the minimum tread depth of a step for this to occur?

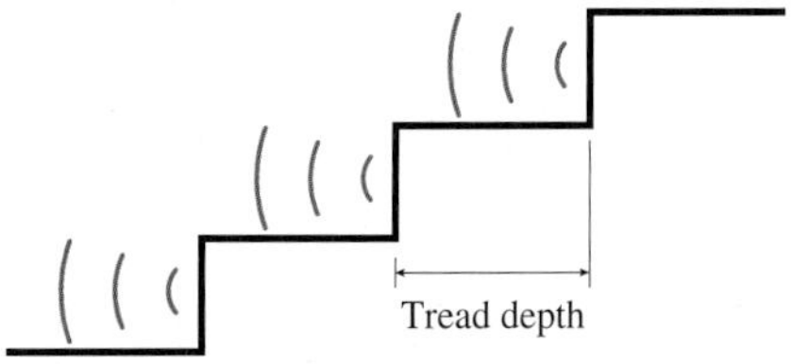

12. Akiko rides her bike toward a brick wall with a speed of 7.00 m/s while blowing a whistle that is emitting sound

with a frequency of 512.0 Hz. (a) What is the frequency of the sound that is reflected from the wall as heard by Haruki, who is standing still? (b) Junichi is walking away from the wall at a speed of 2.00 m/s. What is the frequency of the sound reflected from the wall that Junichi hears?

13. A siren has a circle of 25 equally sized, evenly spaced holes near the rim of a disk free to rotate about its center. Air is blown toward the plane of the disk as it rotates with a frequency of 60.0 Hz. What is the frequency and wavelength of the sound produced?

14. A stretched string has a fundamental frequency of 847 Hz. What is the fundamental frequency if the tension is tripled?

15. The average adult has about 5 L of blood and a healthy adult heart pumps blood at a rate of about 80 cm^3/s. Estimate how long it takes for medicine delivered intravenously to travel throughout a person's body.

16. A sound wave of frequency 1231 Hz travels through air directly toward a wall, then through the wall out into air again. If the initial speed of the sound wave is 341 m/s and its speed in the wall is 620 m/s, what are (a) the initial wavelength of the sound, (b) the wavelength of the sound in the wall, and (c) the wavelength of the sound when it exits the wall on the other side?

17. A speedboat is traveling at 20.1 m/s toward another boat moving in the opposite direction with a speed of 15.6 m/s. The speedboat pilot sounds his horn, which has a frequency of 312 Hz. What is the frequency heard by a passenger in the oncoming boat?

18. A glass tube is closed at one end and has a diaphragm covering the other end. The tube is filled with gas and some sawdust has been scattered along inside the tube. When the diaphragm is driven at a frequency of 1457 Hz, the sawdust forms small piles 20 cm apart. (a) What is the speed of the sound in the gas? (b) Do the piles of sawdust represent nodes or antinodes in the sound wave? Explain.

19. A section of pipe with an internal diameter of 10.0 cm tapers to an inner diameter of 6.00 cm as it rises through a height of 1.70 m at an angle of 60.0° with respect to the horizontal. The pipe carries water and its higher end is open to air. (a) If the speed of the water at the lower point is 15.0 cm/s, what are the pressure at the lower end and the speed of the water as it exits the pipe? (b) If the higher end of the pipe is 0.300 m above ground, at what horizontal distance from the pipe outlet does the water land?

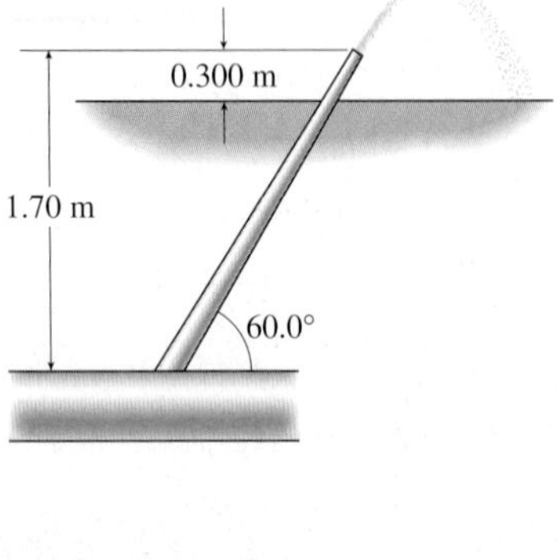

20. When a standing wave is produced in a string fixed at both ends, the string oscillates so fast that it looks like a blur. You want to photograph the string when it is at positions A, B, and C shown in the figure. The tension in the string is 2.00 N and its mass per unit length is 0.200 g/m. The string's length is 0.720 m. Assume that you take your first picture when the string is in position A and let that be time $t = 0$. What are the first two times after $t = 0$ at which you can photograph the string in each of the positions A, B, and C?

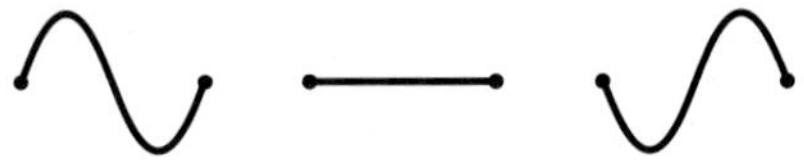

21. The A string on a guitar has length 64.0 cm and fundamental frequency 110.0 Hz. The string's tension is 133 N. It is vibrating in its fundamental standing wave mode with a maximum displacement from equilibrium of 2.30 mm. The air temperature is 20.0°C.
   (a) What is the wavelength of the fundamental mode of vibration?
   (b) What is the wave speed on the string?
   (c) What is the linear mass density of the string?
   (d) What is the maximum speed of any point on the oscillating string?
   (e) The string transmits vibrations through the bridge to the body of the instrument and then to the air. What is the frequency of the sound wave in air?
   (f) What is the wavelength of the sound wave in air?

22. At a grocery store, a spring scale (spring constant = 450 N/m) hangs near the produce section. The spring hangs vertically with a 0.250-kg pan suspended from its lower end. Jenna drops a 2.20-kg bag of oranges from a height of 30.0 cm above the pan. The pan and oranges start oscillating vertically in SHM.
   (a) What is the velocity of the pan immediately after the oranges land on the pan? Assume a perfectly inelastic collision.
   (b) How far is the new equilibrium point of the pan (with oranges) below its position before the oranges were dropped on it?
   (c) What is the amplitude of the oscillations?
   (d) What is the frequency of the oscillations?

23. A spherical balloon with a radius of 12.0 cm is filled with helium. The bottom of the balloon is attached to a 2.30 m length of ribbon that is anchored to the ground. The balloon alone has a mass of 2.80×10^{-3} kg. Ignore the mass of the ribbon.
   (a) What is tension in the ribbon?
   (b) After the balloon is displaced slightly to the side from its equilibrium position, it oscillates back and forth like an inverted pendulum. What is the period of oscillation? Ignore friction and air resistance.

24. An atomizer is a device that delivers a fine mist of some liquid such as perfume by blowing air horizontally over the top of a tube immersed in the liquid. Suppose a perfume with density 800 kg/m^3 has a 3.0-cm tube extending vertically from the top of the liquid. What minimum speed does air flow over the top of the tube when the liquid just reaches the top of the tube?

25. A tetherball set has a ball with mass 0.411 kg and a nylon string with diameter 2.50 mm, Young's modulus 4.00×10^9 Pa, and density 1150 kg/m^3. The nylon string has a length of 2.200 m when the ball is at rest (hanging straight down). While playing tetherball, Monty hits the ball around the pole so it moves in a horizontal circle with the string at an angle of 65.0° to the pole. (a) How much does the string stretch compared with when the ball is at rest? (b) What is the ball's kinetic energy? (c) How long would it take a transverse wave pulse to travel the length of the string from the ball to the top of the pole?

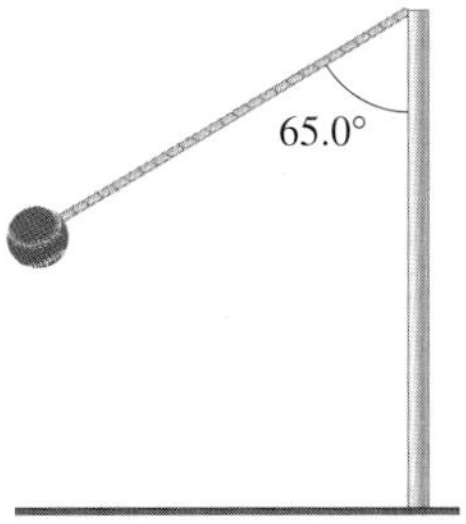

✦26. A harpsichord string is made of yellow brass (Young's modulus 9.0×10^{10} Pa, tensile strength 6.3×10^8 Pa, mass density 8500 kg/m^3). When tuned correctly, the tension in the string is 59.4 N, which is 93% of the maximum tension that the string can endure without breaking. The length of the string that is free to vibrate is 9.4 cm. What is the fundamental frequency?

MCAT Review

The section that follows includes MCAT exam material and is reprinted with permission of the Association of American Medical Colleges (AAMC).

1. What is the volume of a brick that weighs 30 N in air and 20 N when completely submerged in water? (Note: The density of water is 1000 kg/m^3 and let g be 10 m/s^2.)
   A. 1×10^{-3} m^3 B. 5×10^{-3} m^3
   C. 1×10^{-2} m^3 D. 5×10^{-2} m^3

2. The expansion of a particular cable when subjected to a tensile stress obeys $F = k\Delta L$, where F is the tension, $k = 5.0 \times 10^6$ N/m, and ΔL is the expansion. How far will a 100-m section of cable expand when placed under 5000 N of tension?
   A. 10^{-3} m B. 10^{-2} m
   C. 10^{-1} m D. 10 m

3. SL, the sound level in decibels, is defined as SL = $10 \log_{10}(I/I_0)$, where $I_0 = 1.0 \times 10^{-12}$ W/m^2 (the minimum sound intensity audible to humans). A fire siren has a sound level of about 100 dB. What is the intensity I of the fire siren?
   A. 1.0×10^{-22} W/m^2 B. 1.0×10^{-10} W/m^2
   C. 1.0×10^{-8} W/m^2 D. 1.0×10^{-2} W/m^2

4. Suppose that 2 cm of a liquid with a specific gravity of 0.5 is added to a 4-cm column of water. How does the new gauge pressure at the base of the column, P_n, compare with the original pressure, P_i?
   A. $P_n = \frac{3}{4}P_i$ B. $P_n = P_i$
   C. $P_n = \frac{5}{4}P_i$ D. $P_n = \frac{3}{2}P_i$

5. Consecutive resonances occur at wavelengths of 8 m and 4.8 m in an organ pipe closed at one end. What is the length of the organ pipe? (Note: Resonances occur at $L = n\lambda/4$, where L is the pipe length, λ is the wavelength, and $n = 1, 3, 5, \ldots$.)
   A. 3.2 m B. 4.8 m
   C. 6.0 m D. 8.0 m

6. Two mechanical waves of the same frequency pass through the same medium. The amplitude of wave A is 3 units, and the amplitude of wave B is 5 units. Which of the following describes the range of amplitudes possible when the two waves pass through the medium simultaneously?
   A. Always 4 units
   B. Between 2 and 8 units
   C. Between 3 and 5 units
   D. Between 5 and 8 units

7. A simple pendulum is swinging with an amplitude of 10°. As the bob of the pendulum swings through one oscillation, its linear acceleration
   A. remains constant in magnitude and direction.
   B. remains constant in magnitude but changes direction.
   C. changes in magnitude but remains constant in direction.
   D. changes in magnitude and direction.

8. In a simplified model of the blood flow, the velocity of blood flow through a coronary artery is inversely proportional to the fourth power of the radius of the artery. What is the ratio of kinetic energy of the blood in an artery of 2 cm radius to the kinetic energy of the same volume of blood in an artery 1 cm in radius?
   A. 1:2^4 B. 1:4^4
   C. 2^4:1 D. 2^4:1

Read the paragraph and then answer the following questions:

Three balls with the same volume of 1.0×10^{-6} m^3 are in an open tank of water that has a density (ρ) equal to 1.0×10^3 kg/m^3. The balls are in the water at different levels. Ball 1 floats in water with a part of it above the surface, ball 2 is completely submerged in the water, and ball 3 rests on the bottom of the tank. Any movement of the water obeys Bernoulli's equation:

$$P_1 + \tfrac{1}{2}\rho v_1^2 + \rho g y_1 = P_2 + \tfrac{1}{2}\rho v_2^2 + \rho g y_2$$

where P_1 and P_2 are the pressures at elevations y_1 and y_2, and v_1 and v_2 are the speeds of the water. (Note: Unless otherwise noted, the water and the balls are stationary.)

9. The buoyant forces B_1, B_2, and B_3 exerted by water on the balls are related by which of the following?

A. $B_1 < B_2 < B_3$

B. $B_1 < B_2 = B_3$

C. $B_1 = B_2 > B_3$

D. $B_1 > B_2 > B_3$

10. The densities of the balls ρ_1, ρ_2, and ρ_3 are related by which of the following?

A. $\rho_1 < \rho_2 < \rho_3$ B. $\rho_1 < \rho_2 = \rho_3$

C. $\rho_1 = \rho_2 < \rho_3$ D. $\rho_1 = \rho_2 > \rho_3$

11. Assume that the density of ball 3 is 7.8×10^3 kg/m^3. Ignoring atmospheric pressure, what is the supporting force exerted by the bottom of the tank on ball 3?

A. 1.0×10^{-2} N B. 6.7×10^{-2} N

C. 7.6×10^{-2} N D. 8.8×10^{-2} N

12. Assume that the density of ball 1 is 8.0×10^2 kg/m^3. Ignoring atmospheric pressure, what fraction of ball 1 is above the surface of the water?

A. $\frac{4}{5}$ B. $\frac{3}{4}$ C. $\frac{1}{4}$ D. $\frac{1}{5}$

13. Ball 2 is in the water 20 cm above ball 3. What is the approximate difference in pressure between the two balls?

A. 2×10^2 N/m^2 B. 5×10^2 N/m^2

C. 2×10^3 N/m^2 D. 5×10^3 N/m^2

14. If ball 3 is a hollow, iron ball and atmospheric pressure can be ignored, what should be the volume of the hollow portion of ball 3 such that the force exerted by it on the bottom of the tank is 0? (Note: Density of iron is 7.8×10^3 kg/m^3.)

A. 0.13×10^{-6} m^3 B. 0.78×10^{-6} m^3

C. 0.87×10^{-6} m^3 D. 1.15×10^{-6} m^3

CHAPTER

13

Temperature and the Ideal Gas

A crocodile basks on a rock in Lake Baringo (Kenya) to get warm.

In warm-blooded or homeothermic (constant temperature) animals, body temperature is carefully regulated. The hypothalamus, located in the brain, acts as the master thermostat to keep body temperature constant to within a fraction of a degree Celsius in a healthy animal. If the body temperature starts to deviate much from the desired constant level, the hypothalamus causes changes in blood flow and initiates other processes, such as shivering or perspiration, to bring the temperature back to normal. What evolutionary advantage does a constant body temperature give the warm-blooded animals (birds and mammals) over the cold-blooded (such as reptiles and insects)? What are the disadvantages? (See p. 476 for the answer.)

Concepts & Skills to Review

- energy conservation (Chapter 6)
- momentum conservation (Section 7.4)
- collisions (Sections 7.7 and 7.8)

13.1 TEMPERATURE AND THERMAL EQUILIBRIUM

The measurement of **temperature** is part of everyday life. We measure the temperature of the air outdoors to decide how to dress when going outside; a thermostat measures the air temperature indoors to control heating and cooling systems to keep our homes and offices comfortable. Regulation of oven temperature is important in baking. When we feel ill, we measure our body temperature to see if we have a fever. Despite how matter-of-fact it may seem, temperature is a subtle concept. Although our subjective sensations of hot and cold are related to temperature, they can easily mislead, as the next Physics at Home demonstrates.

Figure 13.1 It is easy to trick our sense of temperature.

PHYSICS AT HOME

Try an experiment described by the philosopher John Locke in 1690. Fill one container with water that is hot (but not too hot to touch); fill a second container with lukewarm water; and fill a third container with cold water. Put one hand in the hot water and one in the cold water (Fig. 13.1) for about 10 to 20 s. Then plunge both hands into the container of lukewarm water. Although both hands are now immersed in water that is at a single temperature, the hand that had been in the hot water feels cool while the hand that had been in the cold water feels warm. This demonstration shows that we cannot trust our subjective senses to measure temperature.

The definition of temperature is based on the concept of **thermal equilibrium**. Suppose two objects or systems are allowed to exchange energy. The net flow of energy is always from the object at the higher temperature to the object at the lower temperature. As energy flows, the temperatures of the two objects approach one another. When the temperatures are the same, there is no longer any net flow of energy; the objects are now said to be in thermal equilibrium. Thus, *temperature is a quantity that determines when objects are in thermal equilibrium.* (The objects do *not* necessarily have the same *energy* when in thermal equilibrium.) The energy that flows between two objects or systems due to a temperature difference between them is called **heat**. In Chapter 14 we discuss heat in detail. If heat can flow between two objects or systems, the objects or systems are said to be in **thermal contact**.

Heat: energy in transit due to a temperature difference. Heat flows spontaneously from the hotter object to the colder object.

To measure the temperature of an object, we put a thermometer into thermal contact with the object. Temperature measurement relies on the **zeroth law of thermodynamics**.

Zeroth Law of Thermodynamics

If two objects are each in thermal equilibrium with a third object, then the two are in thermal equilibrium with one another.

Without the zeroth law, it would be impossible to define temperature, since different thermometers could give different results. The rather odd name *zeroth* law of thermodynamics came about because this law was formulated historically *after* the first, second, and third laws of thermodynamics and yet it is so fundamental that it should come *before* the others. **Thermodynamics**, the subject of Chapters 13 to 15, concerns temperature, heat flow, and the internal energy of systems.

13.2 TEMPERATURE SCALES

Thermometers measure temperature by exploiting some property of matter that is temperature-dependent. The familiar liquid-in-glass thermometer relies on thermal expansion: the mercury or alcohol expands as its temperature rises (or contracts as its temperature drops) and we read the temperature on a calibrated scale. Since some materials expand more than others, these thermometers must be calibrated on a scale using some easily reproducible phenomenon, such as the melting point of ice or the boiling point of water. The assignment of temperatures to these phenomena is arbitrary.

The most commonly used temperature scale in the world is the Celsius scale. On the Celsius scale, 0°C is the freezing temperature of water at $P = 1$ atm (the *ice point*) and 100°C is the boiling temperature of water at $P = 1$ atm (the *steam point*).

The freezing and boiling temperatures of water depend on the pressure.

In the United States, the Fahrenheit scale is still commonly used (Fig. 13.2). At 1 atm, the ice point is 32°F and the steam point is 212°F, so the difference between the steam and ice points is 180°F. The size of the Fahrenheit degree interval is therefore smaller than the Celsius degree interval: a temperature difference of 1°C is equivalent to a difference of 1.8°F:

$$\Delta T_F = \Delta T_C \times 1.8 \frac{°F}{°C} \quad (13\text{-}1)$$

Since the two scales also have an offset (0°C is not the same temperature as 0°F), conversion between the two is:

$$T_F = (1.8°F/°C)\, T_C + 32°F \quad (13\text{-}2a)$$

$$T_C = \frac{T_F - 32°F}{1.8°F/°C} \quad (13\text{-}2b)$$

The SI unit of temperature is the **kelvin** (symbol K, *without* a degree sign). The kelvin has the same degree size as the Celsius scale; that is, a temperature *difference* of 1°C is the same as a difference of 1 K. However, 0 K represents *absolute zero*—there are no temperatures below 0 K. The ice point is 273.15 K, so temperature in °C (T_C) and temperature in kelvins (T) are related.

$$T_C = T - 273.15 \quad (13\text{-}3)$$

Equation (13-3) is the definition of the Celsius scale in terms of the kelvin. Table 13.1 shows some temperatures in kelvins, °C, and °F.

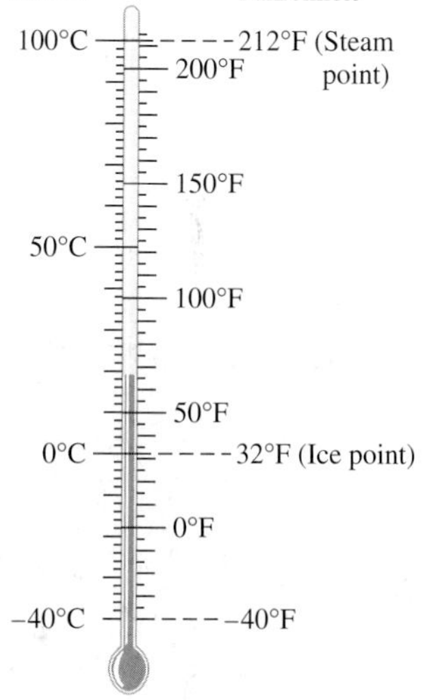

Figure 13.2 The Fahrenheit and Celsius temperature scales.

Table 13.1 Some Reference Temperatures in K, °C, and °F

	K	°C	°F
Absolute zero	0	−273.15	−459.67
Lowest transient temperature achieved (laser cooling)	10^{-9}		
Intergalactic space	3	−270	−454
Helium boils	4.2	−269	−452
Nitrogen boils	77	−196	−321
Carbon dioxide freezes ("dry ice")	195	−78	−108
Mercury freezes	234	−39	−38
Ice melts/water freezes	273.15	0	32.0
Human body temperature	310	37	98.6
Water boils	373.15	100.00	212.0
Campfire	1 000	700	1 300
Gold melts	1 337	1 064	1 947
Lightbulb filament	3 000	2 700	4 900
Surface of Sun; iron welding arc	6 300	6 000	11 000
Center of Earth	16 000	15 700	28 300
Lightning channel	30 000	30 000	50 000
Center of Sun	10^7	10^7	10^7
Interior of neutron star	10^9	10^9	10^9

Example 13.1

A Sick Friend

A friend suffering from the flu has a fever; her body temperature is 38.6°C. What is her temperature in (a) K and (b) °F?

Strategy (a) Kelvins and °C differ only by a shift of the zero point. Converting from °C to K requires only the addition of 273.15 K since 0°C (the ice point) corresponds to 273.15 K. (b) The °F is a different size than the °C, as well as having a different zero. In the Celsius scale, the zero is at the ice point. First multiply by 1.8°F/°C to find how many °F above the ice point. Then add 32°F (the Fahrenheit temperature of the ice point).

Solution (a) The temperature is 38.6 K *above* the ice point of 273.15 K. Therefore, the kelvin temperature is

$$T = 38.6\text{ K} + 273.15\text{ K} = 311.8\text{ K}$$

(b) First find how many °F above the ice point:

$$\Delta T_F = 38.6°\text{C} \times (1.8°\text{F}/°\text{C}) = 69.5°\text{F}$$

The ice point is 32°F, so

$$T_F = 32.0°\text{F} + 69.5°\text{F} = 101.5°\text{F}$$

Discussion The answer is reasonable since 98.6°F is normal body temperature.

Practice Problem 13.1 Normal Body Temperatures with Two Scales

Convert the normal human body temperature (98.6°F) to degrees Celsius and kelvins.

13.3 THERMAL EXPANSION OF SOLIDS AND LIQUIDS

Most objects expand as their temperature increases. Long before the cause of thermal expansion was understood, the phenomenon was put to practical use. For example, the cooper (barrel maker) heated iron hoops red hot to make them expand before fitting them around the wooden staves of a barrel. The iron hoops contracted as they cooled, pulling the staves tightly together to make a leak-tight barrel.

Linear Expansion

If the length of a wire, rod, or pipe is L_0 at temperature T_0 (Fig. 13.3), then

$$\frac{\Delta L}{L_0} = \alpha \Delta T \qquad (13\text{-}4)$$

where $\Delta L = L - L_0$ and $\Delta T = T - T_0$. The length at temperature T is

$$L = L_0 + \Delta L = (1 + \alpha \Delta T)\, L_0 \qquad (13\text{-}5)$$

The constant of proportionality α is called the **coefficient of linear expansion** of the substance. It plays a role in thermal expansion similar to that of the elastic modulus in tensile stress. If T is measured in kelvins or in degrees Celsius, then α has units of K^{-1} or $°C^{-1}$. Since only the *change* in temperature is involved in Eq. (13-4), either Celsius or Kelvin temperatures can be used to find ΔT; a temperature change of 1K is the same as a temperature change of 1°C.

As is true for the elastic modulus, the coefficient of linear expansion has different values for different solids and also depends to some extent on the starting temperature of the object. Table 13.2 lists the coefficients for various solids.

CONNECTION:

Recall that the fractional length change (strain) caused by a tensile or compressive stress is proportional to the stress that caused it [Hooke's law, Eq. (10-4)]. Similarly, the fractional length change caused by a temperature change is proportional to the temperature change, as long as the temperature change is not too great.

CHECKPOINT 13.3

A steel tower is 150.00 m tall at 40°C. How much shorter is it at –10°C?

Table 13.2 Coefficients of Linear Expansion α for Solids (at $T = 20°C$ unless otherwise indicated)

Material	$\alpha\,(10^{-6}\,K^{-1})$
Glass (Vycor)	0.75
Brick	1.0
Glass (Pyrex)	3.25
Granite	8
Glass, most types	9.4
Cement or concrete	12
Iron or steel	12
Copper	16
Silver	18
Brass	19
Aluminum	23
Lead	29
Ice (at 0°C)	51

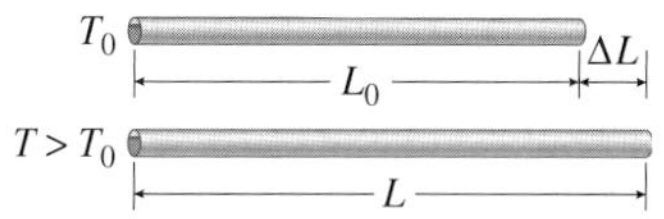

Figure 13.3 Expansion of a solid rod with increasing temperature.

Figure 13.4 is a graph of the relative length of a steel girder as a function of temperature over a *wide* range of temperatures. The curvature of this graph shows that the thermal expansion of the girder is in general *not* proportional to the temperature change. However, over a *limited* temperature range, the curve can be approximated by a straight line; the slope of the tangent line is the coefficient α at the temperature T_0. For small temperature changes near T_0, the change in length of the girder can be treated as being proportional to the temperature change with only a small error.

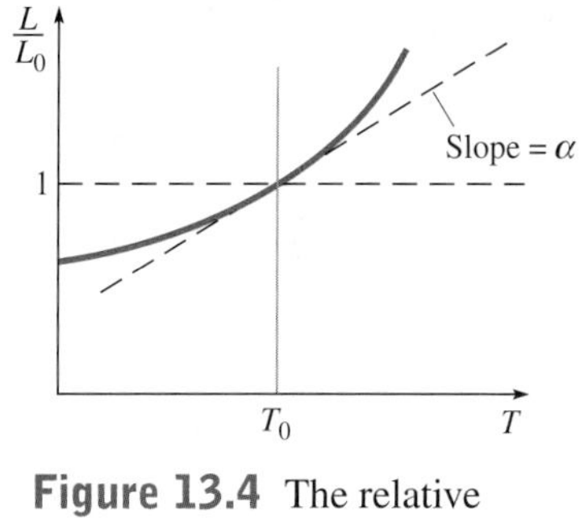

Figure 13.4 The relative length of a steel girder as a function of temperature. The dashed tangent line shows what Eq. (13-4) predicts for small temperature changes in the vicinity of T_0. The slope of this tangent line is the value of α at $T = T_0$.

Applications of Thermal Expansion: Expansion Joints in Bridges and Buildings Allowances must be made in building sidewalks, roads, bridges, and buildings to leave space for expansion in hot weather. Old subway tracks have small spaces left between rail sections to prevent the rails from pushing into each other and causing the track to bow. A train riding on such tracks is subject to a noticeable amount of "clickety-clack" as it goes over these small expansion breaks in the tracks. Expansion joints are easily observed in bridges (Fig. 13.5). Concrete roads and sidewalks have joints between sections. Homeowners sometimes build their own sidewalks without realizing the necessity for such joints; these sidewalks begin to crack almost immediately!

Allowances must also be made for contraction in cold weather. If an object is not free to expand or contract, then as the temperature changes it is subjected to *thermal stress* as its environment exerts forces on it to prevent the thermal expansion or contraction that would otherwise occur.

Figure 13.5 Expansion joints permit the roadbed of a bridge to expand and contract as the temperature changes.

Example 13.2

Expanding Rods

Two metal rods, one aluminum and one brass, are each clamped at one end (Fig. 13.6). At 0.0°C, the rods are each 50.0 cm long and are separated by 0.024 cm at their unfastened ends. At what temperature will the rods just come into contact? (Assume that the base to which the rods are clamped undergoes a negligibly small thermal expansion.)

Strategy Two rods of different materials expand by different amounts. The sum of the two expansions ($\Delta L_{br} + \Delta L_{Al}$) must equal the space between the rods. After finding ΔT, we add it to $T_0 = 0.0°C$ to obtain the temperature at which the two rods touch.

Known: $L_0 = 50.0$ cm, $T_0 = 0.0°C$ for both
Look up: $\alpha_{br} = 19 \times 10^{-6}\ K^{-1}$; $\alpha_{Al} = 23 \times 10^{-6}\ K^{-1}$
Requirement: $\Delta L_{br} + \Delta L_{Al} = 0.024$ cm
Find: $T_f = T_0 + \Delta T$

Solution The brass rod expands by

$$\Delta L_{br} = (\alpha_{br}\,\Delta T)L_0$$

and the aluminum rod by

$$\Delta L_{Al} = (\alpha_{Al}\,\Delta T)L_0$$

The sum of the two expansions is known:

$$\Delta L_{br} + \Delta L_{Al} = 0.024\ \text{cm}$$

Since both the initial lengths and the temperature changes are the same,

$$(\alpha_{br} + \alpha_{Al})\,\Delta T \times L_0 = 0.024\ \text{cm}$$

Solving for ΔT,

$$\Delta T = \frac{0.024\ \text{cm}}{(\alpha_{br} + \alpha_{Al})L_0}$$

$$= \frac{0.024\ \text{cm}}{(19 \times 10^{-6}\ K^{-1} + 23 \times 10^{-6}\ K^{-1}) \times 50.0\ \text{cm}}$$

$$= 11.4°C$$

The temperature at which the two touch is

$$T_f = T_0 + \Delta T = 0.0°C + 11.4°C \rightarrow 11°C$$

Discussion As a check on the solution, we can find how much each individual rod expands and then add the two amounts:

$$\Delta L_{Al} = \alpha_{Al}\,\Delta T\,L_0$$
$$= 23 \times 10^{-6}\ K^{-1} \times 11.4\ K \times 50.0\ \text{cm} = 0.013\ \text{cm}$$

$$\Delta L_{br} = \alpha_{br}\,\Delta T\,L_0$$
$$= 19 \times 10^{-6}\ K^{-1} \times 11.4\ K \times 50.0\ \text{cm} = 0.011\ \text{cm}$$

$$\text{total expansion} = 0.013\ \text{cm} + 0.011\ \text{cm} = 0.024\ \text{cm}$$

which is correct.

Practice Problem 13.2 Expansion of a Wall

The outer wall of a building is constructed from concrete blocks. If the wall is 5.00 m long at 20.0°C, how much longer is the wall on a hot day (30.0°C)? How much shorter is it on a cold day (−5.0°C)?

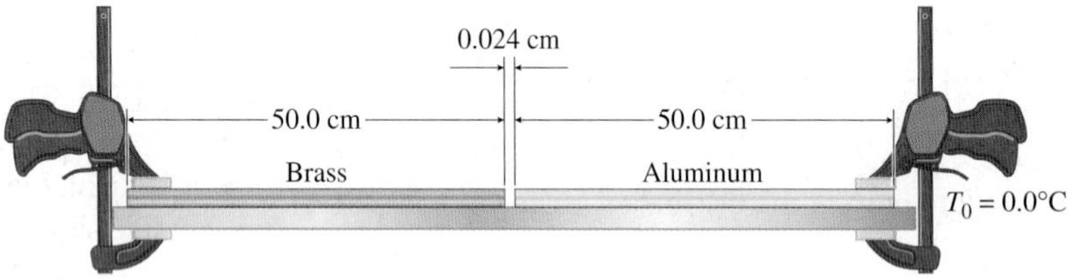

Figure 13.6
Two clamped rods.

Differential Expansion

When two strips made of different metals are joined together and then heated, one expands more than the other (unless they have the same coefficient of expansion). This differential expansion can be put to practical use: the joined strips bend into a curve, allowing one strip to expand more than the other.

Application of differential expansion: bimetallic strip in a thermostat

The bimetallic strip (Fig. 13.7) is made by joining a material with a lower coefficient of expansion, such as steel, and one of a higher coefficient of expansion, such as brass. Unequal expansions or contractions of the two materials force the bimetallic strip to bend. In Fig. 13.7, the brass expands more than the steel when the bimetallic strip is heated. As the strip is cooled, the brass contracts more than the steel.

The bimetallic strip is used in many wall thermostats. The bending of the bimetallic strip closes or opens an electrical switch in the thermostat that turns the furnace or air conditioner on or off. Inexpensive oven thermometers also use a bimetallic strip wound into a spiral coil; the coil winds tighter or unwinds as the temperature changes.

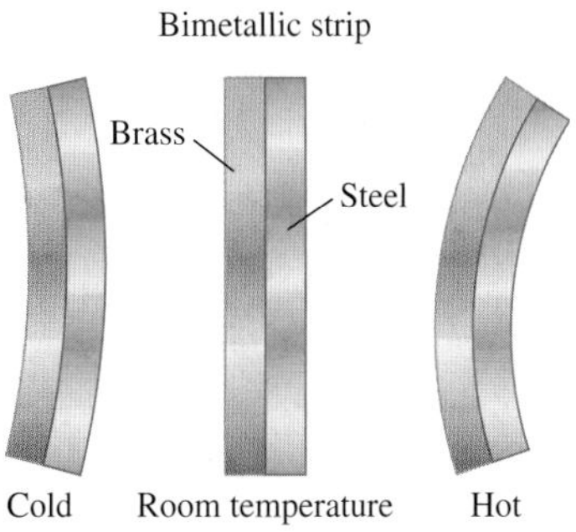

Figure 13.7 A bimetallic strip bends when its temperature changes; brass expands and contracts more than steel for the same temperature change.

Area Expansion

As you might suspect, *each dimension* of an object expands when the object's temperature increases. For instance, a pipe expands not only in length, but also in radius. An isotropic substance expands uniformly in all directions, causing changes in area and volume that leave the *shape* of the object unchanged. In Problem 25, you can show that, for small temperature changes, the area of any flat surface of a solid changes in proportion to the temperature change:

$$\frac{\Delta A}{A_0} = 2\alpha\Delta T \tag{13-6}$$

The factor of two in Eq. (13-6) arises because the surface expands in two perpendicular directions.

Volume Expansion

The fractional change in volume of a solid or liquid is also proportional to the temperature change as long as the temperature change is not too large:

$$\frac{\Delta V}{V_0} = \beta\Delta T \tag{13-7}$$

The coefficient of volume expansion, β, is the fractional change in volume per unit temperature change. For solids, the coefficient of volume expansion is three times the coefficient of linear expansion (as shown in Problem 26):

$$\beta = 3\alpha \tag{13-8}$$

The factor of three in Eq. (13-8) arises because the object expands in three-dimensional space. For liquids, the volume expansion coefficient is the only one given in tables. Since liquids do not necessarily retain the same shape as they expand, the quantity that is uniquely defined is the change in volume. Table 13.3 provides values of β for some common liquids and gases.

A hollow cavity in a solid expands exactly as if it were filled—the interior of a steel gasoline container expands when its temperature increases just as if it were a solid steel block. The steel wall of the can does *not* expand inward to make the cavity smaller.

CONNECTION:

Compare Eq. (10-10). There, the fractional volume change is proportional to the *pressure* change; here it is proportional to the *temperature* change.

Table 13.3 Coefficients of Volume Expansion β for Liquids and Gases (at $T = 20°C$ unless otherwise indicated)

Material	$\beta\,(10^{-6}\ K^{-1})$
Liquids	
Water (at 0°C)*	−68
Mercury	182
Water (at 20°C)	207
Gasoline	950
Ethyl alcohol	1120
Benzene	1240
Gases	
Air (and most other gases) at 1 atm	3340

*Below 3.98°C, water *contracts* with increasing temperature.

Application of volume expansion: thermometers

In an ordinary alcohol-in-glass or mercury-in-glass thermometer, it is not just the liquid that expands as temperature rises. The reading of the thermometer is determined by the difference in the volume expansion of the liquid and that of the interior of the glass. The calibration of an accurate thermometer must account for the expansion of the glass. Comparison of Tables 13.2 and 13.3 shows that, as is usually the case, the liquid expands much more than the glass for a given temperature change.

Example 13.3

Hollow Cylinder Full of Water

A hollow copper cylinder is filled to the brim with water at 20.0°C. If the water and the container are heated to a temperature of 91°C, what percentage of the water spills over the top of the container?

Strategy The volume expansion coefficient for water is greater than that for copper, so the water expands more than the interior of the cylinder. The cavity expands just as if it were solid copper. Since the problem does not specify the initial volume, we call it V_0. We need to find out how much a volume V_0 of water expands and how much a volume V_0 of copper expands; the difference is the water volume that spills over the top of the container.

Known: Initial copper cylinder interior volume = initial water volume = V_0

Initial temperature = $T_0 = 20.0$°C

Final temperature = 91°C; $\Delta T = 71$°C

Look up: $\alpha_{Cu} = 16 \times 10^{-6}\ °C^{-1}$; $\beta_{H_2O} = 207 \times 10^{-6}\ °C^{-1}$

Find: $\Delta V_{H_2O} - \Delta V_{Cu}$ as a percentage of V_0

Solution The volume expansion of the interior of the copper cylinder is

$$\Delta V_{Cu} = \beta_{Cu} \Delta T\, V_0$$

where $\beta_{Cu} = 3\alpha_{Cu}$. The volume expansion of the water is

$$\Delta V_{H_2O} = \beta_{H_2O} \Delta T\, V_0$$

The amount of water that spills is

$$\begin{aligned}\Delta V_{H_2O} - \Delta V_{Cu} &= \beta_{H_2O} \Delta T\, V_0 - \beta_{Cu} \Delta T\, V_0\\ &= (\beta_{H_2O} - \beta_{Cu}) \Delta T\, V_0\\ &= (207 \times 10^{-6}\ °C^{-1} - 3 \times 16 \times 10^{-6}\ °C^{-1})\\ &\quad \times 71°C \times V_0\\ &= 0.011 V_0\end{aligned}$$

The percentage of water that spills is therefore 1.1%.

Discussion As a check, we can find the change in volume of the copper container and of the water and find the difference.

$$\Delta V_{Cu} = \beta_{Cu} \Delta T\, V_0 = 3 \times 16 \times 10^{-6}\ °C^{-1} \times 71°C \times V_0 = 0.0034 V_0$$

$$\Delta V_{H_2O} = \beta_{H_2O} \Delta T\, V_0 = 207 \times 10^{-6}\ °C^{-1} \times 71°C \times V_0 = 0.0147 V_0$$

volume of water that spills $= 0.0147V_0 - 0.0034V_0 = 0.0113V_0$

which again shows that 1.1% spills.

Practice Problem 13.3 Overflowing Gas Can

A driver fills an 18.9-L steel gasoline can with gasoline at 15.0°C right up to the top. He forgets to replace the cap and leaves the can in the back of his truck. The temperature climbs to 30.0°C by 1 P.M. How much gasoline spills out of the can?

13.4 MOLECULAR PICTURE OF A GAS

Number Density As we saw in Chapter 9, the densities of liquids are generally not much less than the densities of solids. Gases are *much* less dense than liquids and solids because the molecules are, on average, much farther apart. The mass density—mass per unit volume—of a substance depends on the mass m of a single molecule and the number of molecules N packed into a given volume V of space (Fig. 13.8). The number of molecules per unit volume, N/V, is called the **number density** to distinguish it from mass density. In SI units, number density is written as the number of molecules per cubic meter, usually written simply as m^{-3} (read "per cubic meter"). If a gas has a total mass M, occupies a volume V, and each molecule has a mass m, then the number of gas molecules is

$$N = \frac{M}{m} \tag{13-9}$$

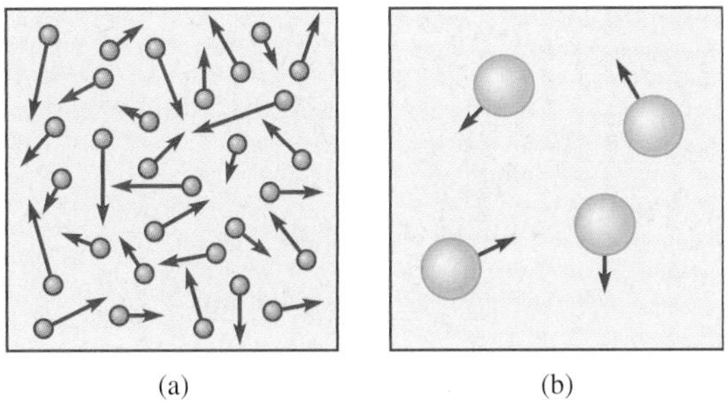

Figure 13.8 These two gases have the same mass per unit volume but different number densities. The red arrows represent the molecular velocities. In (a), there are a larger number of molecules in a given volume, but the mass of each molecule in (b) is greater.

and the average number density is

$$\frac{N}{V} = \frac{M}{mV} = \frac{\rho}{m} \quad (13\text{-}10)$$

where $\rho = M/V$ is the mass density.

Moles It is common to express the amount of a substance in units of **moles** (abbreviated mol). The mole is an SI base unit and is defined as follows: one mole of anything contains the same number of units as there are atoms in 12 *grams* (not kilograms) of carbon-12. This number is called **Avogadro's number** and has the value

The abbreviation "mol" stands for moles, *not* molecules.

$$N_A = 6.022 \times 10^{23}\ \text{mol}^{-1} \quad \text{(Avogadro's number)}$$

Avogadro's number is written with units, mol^{-1}, to show that this is the number *per mole.* The number of moles, n, is therefore given by

$$\text{number of moles} = \frac{\text{total number}}{\text{number per mole}}$$

$$n = \frac{N}{N_A} \quad (13\text{-}11)$$

Molecular Mass and Molar Mass The mass of a molecule is often expressed in units other than kg. The most common is the **atomic mass unit** (symbol u). By definition, one atom of carbon-12 has a mass of 12 u (exactly). Using Avogadro's number, the relationship between atomic mass units and kilograms can be calculated (see Problem 27):

$$1\ \text{u} = 1.66 \times 10^{-27}\ \text{kg} \quad (13\text{-}12)$$

The proton, neutron, and hydrogen atom all have masses within 1% of 1 u—which is why the atomic mass unit is so convenient. More precise values are 1.007 u for the proton, 1.009 u for the neutron, and 1.008 u for the hydrogen atom. The mass of an atom is *approximately* equal to the number of nucleons (neutrons plus protons)—the *atomic mass number*—times 1 u.

Instead of the mass of one molecule, tables commonly list the **molar mass**—the mass of the substance *per mole.* For an element with several isotopes (such as carbon-12, carbon-13, and carbon-14), the molar mass is averaged according to the naturally occurring abundance of each isotope. The atomic mass unit is chosen so that the mass of a molecule in "u" is numerically the same as the molar mass in g/mol. For example, the molar mass of O_2 is 32.0 g/mol and the mass of one molecule is 32.0 u.

The mass of a molecule is very nearly equal to the sum of the masses of its constituent atoms. The molar mass of a molecule is therefore equal to the sum of the molar masses of the atoms. For example, the molar mass of carbon is 12.0 g/mol and the molar mass of (atomic) oxygen is 16.0 g/mol; therefore, the molar mass of carbon dioxide (CO_2) is $(12.0 + 2 \times 16.0)$ g/mol = 44.0 g/mol.

CHECKPOINT 13.4

(a) What is the mass (in u) of a CO_2 molecule? (b) What is the mass (in g) of 3.00 mol of CO_2?

Example 13.4

A Helium Balloon

A helium balloon of volume 0.010 m^3 contains 0.40 mol of He gas. (a) Find the number of atoms, the number density, and the mass density. (b) Estimate the average distance between He atoms.

Strategy The number of moles tells us the number of atoms as a fraction of Avogadro's number. Once we have the number of atoms, N, the next quantity we are asked to find is N/V. To find the mass density, we can look up the atomic mass of helium in the periodic table. The mass per atom times the number density (atoms per m^3) equals the mass density (mass per m^3). To find the average distance between atoms, imagine a simplified picture in which each atom is at the center of a spherical volume equal to the total volume of the gas divided by the number of atoms. In this approximation, the average distance between atoms is equal to the diameter of each sphere.

Solution (a) The number of atoms is

$$N = nN_A$$
$$= 0.40 \text{ mol} \times 6.022 \times 10^{23} \text{ atoms/mol}$$
$$= 2.4 \times 10^{23} \text{ atoms}$$

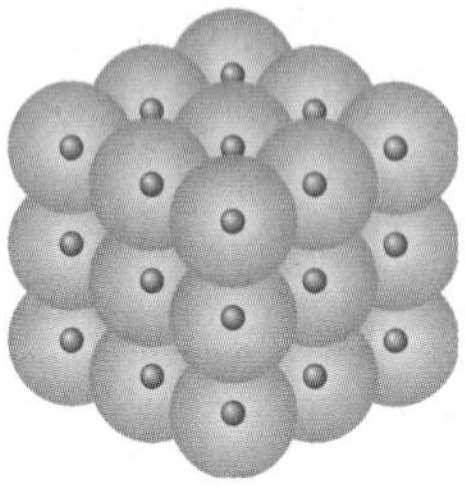

Figure 13.9
Simplified model in which equally spaced helium atoms sit at the centers of spherical volumes of space.

The number density is

$$\frac{N}{V} = \frac{2.4 \times 10^{23} \text{ atoms}}{0.010 \text{ m}^3} = 2.4 \times 10^{25} \text{ atoms/m}^3$$

The mass of a helium atom is 4.00 u. Then the mass in kg of a helium atom is

$$m = 4.00 \text{ u} \times 1.66 \times 10^{-27} \text{ kg/u} = 6.64 \times 10^{-27} \text{ kg}$$

and the mass density of the gas is

$$\rho = \frac{M}{V} = m \times \frac{N}{V}$$
$$= 6.64 \times 10^{-27} \text{ kg} \times 2.4 \times 10^{25} \text{ m}^{-3} = 0.16 \text{ kg/m}^3$$

(b) We assume that each atom is at the center of a sphere of radius r (Fig. 13.9). The volume of the sphere is

$$\frac{V}{N} = \frac{1}{N/V} = \frac{1}{2.4 \times 10^{25} \text{ atoms/m}^3} = 4.2 \times 10^{-26} \text{ m}^3 \text{ per atom}$$

Then

$$\frac{V}{N} = \frac{4}{3}\pi r^3 \approx 4r^3 \quad (\text{since } \pi \approx 3)$$

Solving for r,

$$r \approx \left(\frac{V}{4N}\right)^{1/3} = 2.2 \times 10^{-9} \text{ m} = 2.2 \text{ nm}$$

The average distance between atoms is $d = 2r \approx 4$ nm (since this is an estimate).

Discussion For comparison, in *liquid* helium the average distance between atoms is about 0.4 nm, so in the gas the average separation is about ten times larger.

Practice Problem 13.4 Number Density for Water

The mass density of liquid water is 1000.0 kg/m^3. Find the number density.

13.5 ABSOLUTE TEMPERATURE AND THE IDEAL GAS LAW

We have examined the thermal expansion of solids and liquids. What about gases? Is the volume expansion of a gas proportional to the temperature change? We must be careful; since gases are easily compressed, we must also specify what happens to the pressure. The French scientist Jacques Charles (1746–1823) found experimentally that, if the pressure of a gas is held constant, the change in temperature is indeed proportional to the change in volume (Fig. 13.10a).

$$\text{Charles's law: } \Delta V \propto \Delta T \quad (\text{for constant } P)$$

According to Charles's law, a graph of V versus T for a gas held at constant pressure is a straight line, but the line does not necessarily pass through the origin (Fig. 13.10b).

However, if we graph V versus T (at constant P) for various gases, something interesting happens. If we extrapolate the straight line to where it reaches $V = 0$, the

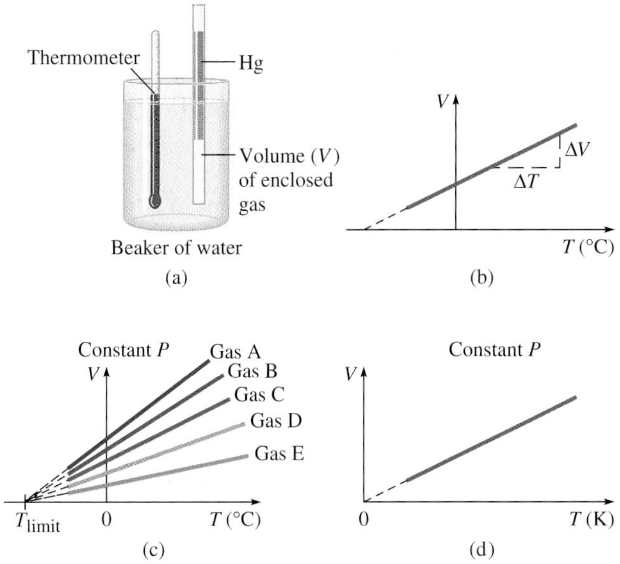

Figure 13.10 (a) Apparatus to verify Charles's law. The pressure of the enclosed gas is held constant by the fixed quantity of mercury resting on top of it and atmospheric pressure pushing down on the mercury. If the temperature of the gas is changed, it expands or contracts, moving the mercury column above it. (b) Charles's law: for a gas held at constant pressure, changes in temperature are proportional to changes in volume. (c) Volume versus temperature graphs for various gas samples, each at a constant pressure, are extrapolated to $V = 0$. The graphs intersect the temperature axis at the same temperature, T_{limit}, even though the gases may differ in composition and mass. (d) An absolute temperature scale sets $T_{limit} = 0$.

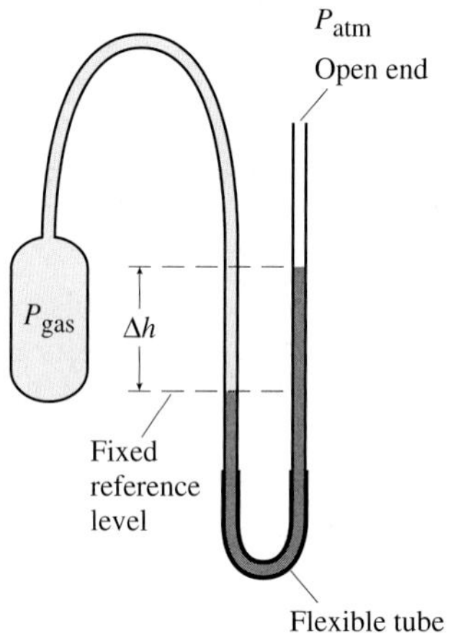

Figure 13.11 A constant volume gas thermometer. A dilute gas is contained in the vessel on the left, which is connected to a mercury manometer. The right side can be moved up or down to keep the mercury level on the left at a fixed level, so the volume of gas is kept constant. Then the manometer is used to measure the pressure of the gas: $P_{gas} = P_{atm} + \rho g \Delta h$.

temperature at that point is the *same* regardless of what gas we use, how many moles of gas are present, or what the pressure of the gas is (Fig. 13.10c). (One reason we have to extrapolate is that all gases become liquids or solids before they reach $V = 0$.) This temperature, −273.15°C or −459.67°F, is called **absolute zero**—the lower limit of attainable temperatures. In kelvins—an *absolute* temperature scale—absolute zero is defined as 0 K (Fig. 13.10d). As long as it is understood that an absolute temperature scale is to be used, then Charles's law can be written

Absolute zero: the lower limit of attainable temperatures.

$$V \propto T \quad \text{(for constant } P\text{)}$$

PHYSICS AT HOME

Take an empty 2-L soda bottle, cap it tightly, and put it in the freezer. Check it an hour later; what has happened? Estimate the percentage change in the volume of the air inside and compare with the percentage change in absolute temperature (if you don't have a thermometer handy, a typical freezer temperature is about −10°C).

Thermal expansion of a gas can be used to measure temperature. Gas thermometers are universal: it does not matter what gas is used or how many moles of gas are present, as long as the number density is sufficiently low. Gas thermometers give absolute temperature in a natural way and they are extremely accurate and reproducible. The main disadvantage of gas thermometers is that they are much less convenient to use than most other thermometers, so they are mainly used to calibrate other thermometers.

A thermometer based on Charles's law would be called a *constant pressure gas thermometer.* More common is the *constant volume gas thermometer* (Fig. 13.11), which is based on Gay-Lussac's law:

$$P \propto T \quad \text{(for constant } V\text{)}$$

Here we keep the volume of the gas constant, measure the pressure and use that to indicate the temperature. (It is much easier to keep the volume constant and measure the pressure than to do the opposite.)

Both Charles's law and Gay-Lussac's law are valid only for a **dilute** gas—a gas where the number density is low enough (and, therefore, the average distance between gas molecules is large enough) that interactions between the molecules are negligible except when they collide. Two other experimentally discovered laws that apply to dilute gases are Boyle's law and Avogadro's law. Boyle's law states that the pressure of a gas is inversely proportional to its volume at constant temperature:

$$P \propto \frac{1}{V} \quad \text{(for constant } T\text{)}$$

Avogadro's law states that the volume occupied by a gas at a given temperature and pressure is proportional to the number of gas molecules N:

$$V \propto N \quad \text{(constant } P, T\text{)}$$

(A constant number of gas molecules was assumed in the statements of Boyle's, Gay-Lussac's, and Charles's laws.)

One equation combines all four of these gas laws—the **ideal gas law**:

Ideal Gas Law (Microscopic Form)

$$PV = NkT \quad (N = \text{number of molecules}) \tag{13-13}$$

In the ideal gas law, T stands for *absolute* temperature (in K) and P stands for *absolute* (not gauge) pressure.

The constant of proportionality is a universal quantity known as **Boltzmann's constant** (symbol k); its value is

$$k = 1.38 \times 10^{-23} \text{ J/K} \tag{13-14}$$

The macroscopic form of the ideal gas law is written in terms of n, the number of *moles* of the gas, in place of N, the number of molecules. Substituting

$$N = nN_A$$

into the microscopic form yields

$$PV = nN_A kT$$

The product of N_A and k is called the **universal gas constant**:

$$R = N_A k = 8.31 \frac{\text{J/K}}{\text{mol}} \tag{13-15}$$

Then the ideal gas law in macroscopic form is written

Ideal Gas Law (Macroscopic Form)

$$PV = nRT \quad (n = \text{number of moles}) \tag{13-16}$$

Many problems deal with the changing pressure, volume, and temperature in a gas with a constant number of molecules (and a constant number of moles). In such problems, it is often easiest to write the ideal gas law as follows:

$$\frac{P_1V_1}{T_1} = \frac{P_2V_2}{T_2}$$

CHECKPOINT 13.5

Two containers with the same volume are filled with two different gases. The pressure of the two gases is the same. (a) Must their temperatures be the same? Explain. (b) If their temperatures are the same, must they have the same number density? The same mass density?

Example 13.5

Temperature of the Air in a Tire

Before starting out on a long drive, you check the air in your tires to make sure they are properly inflated. The pressure gauge reads 31.0 lb/in^2 (214 kPa), and the temperature is 15°C. After a few hours of highway driving, you stop and check the pressure again. Now the gauge reads 35.0 lb/in^2 (241 kPa). What is the temperature of the air in the tires now?

Strategy We treat the air in the tire as an ideal gas. We must work with absolute temperatures and absolute pressures when using the ideal gas law. The pressure gauge reads *gauge* pressure; to get absolute pressure we add 1 atm = 101 kPa. We don't know the number of molecules inside the tire or the volume, but we can reasonably assume that neither changes. The number is constant as long as the tire does not leak. The volume may actually change a bit as the tire warms up and expands, but this change is small. Since N and V are constant, we can rewrite the ideal gas law as a proportionality between P and T.

Solution First convert the initial and final gauge pressures to absolute pressures:

$$P_i = 214 \text{ kPa} + 101 \text{ kPa} = 315 \text{ kPa}$$

$$P_f = 241 \text{ kPa} + 101 \text{ kPa} = 342 \text{ kPa}$$

Now convert the initial temperature to an absolute temperature:

$$T_i = 15°\text{C} + 273 \text{ K} = 288 \text{ K}$$

According to the ideal gas law, pressure is proportional to temperature, so

$$\frac{T_f}{T_i} = \frac{P_f}{P_i} = \frac{342 \text{ kPa}}{315 \text{ kPa}}$$

Then

$$T_f = \frac{P_f}{P_i} T_i = \frac{342}{315} \times 288 \text{ K} = 313 \text{ K}$$

Now convert back to °C:

$$313 \text{ K} - 273 \text{ K} = 40°\text{C}$$

Discussion The final answer of 40°C seems reasonable since, after a long drive, the tires are noticeably warm, but not hot enough to burn your hand.

It is often most convenient to work with the ideal gas law by setting up a proportion. In this problem, we did not know the volume or the number of molecules, so we had no choice. In essence, what we used was Gay-Lussac's law. Starting with the ideal gas law, we can "rederive" Gay-Lussac's law or Charles's law or any other proportionality inherent in the ideal gas law.

Practice Problem 13.5 Air Pressure in the Tire After the Temperature Decreases

Suppose you now (unwisely) decide to bleed air from the tires, since the manufacturer suggests keeping the pressure between 28 lb/in^2 and 32 lb/in^2. (The manufacturer's specification refers to when the tires are "cold.") If you let out enough air so that the pressure returns to 31 lb/in^2, what percentage of the air molecules did you let out of the tires? What is the gauge pressure after the tires cool back down to 15°C?

PHYSICS AT HOME

The next time you take a car trip, check the tire pressure with a gauge just before the trip and then again after an hour or more of highway driving. Calculate the temperature of the air in the tires from the two pressure readings and the initial temperature. Feel the tire with your hand to see if your calculation is reasonable.

Example 13.6

Scuba Diver

A scuba diver needs air delivered at a pressure equal to the pressure of the surrounding water—the pressure in the lungs must match the water pressure on the diver's body to prevent the lungs from collapsing. Since the pressure in the air tank is much higher, a regulator delivers air to the diver at the appropriate pressure. The compressed air in a diver's tank lasts 80 min at the water's surface. About how long does the same tank last at a depth of 30 m under water? (Assume that the volume of air breathed per minute does not change and ignore the small quantity of air left in the tank when it is "empty.")

Strategy The compressed air in the *tank* is at a pressure much higher than the pressure at which the diver breathes, whether at the surface or at 30 m depth. The constant quantity is N, the number of gas molecules in the tank. We also assume that the temperature of the gas remains the same; it may change slightly, but much less than the pressure or volume.

Solution Since N and T are constant,

$$PV = \text{constant}$$

or

$$P \propto 1/V$$

The pressure at the surface is (approximately) 1 atm, while the pressure at 30 m under water is

$$P = 1 \text{ atm} + \rho g h$$

$$\rho g h = 1000 \text{ kg/m}^3 \times 9.8 \text{ m/s}^2 \times 30 \text{ m} = 294 \text{ kPa} \approx 3 \text{ atm}$$

Therefore, at a depth of 30 m,

$$P \approx 4 \text{ atm}$$

To match the pressure of the surrounding water, the pressure of the compressed air is four times larger at a depth of 30 m; then the volume of air is one fourth what it was at the surface. The diver breathes the same volume per minute, so the tank will last one fourth as long—20 min.

Discussion To do the same thing a bit more formally, we could write:

$$P_i V_i = P_f V_f$$

After setting $P_i = 1$ atm and $P_f = 4$ atm, we find that $V_f/V_i = \frac{1}{4}$.

In this problem, the only numerical values given (indirectly) were the initial and final pressures. Assuming that N and T remain constant, we then can find the ratio of the final and initial volumes. Whenever there *seems* to be insufficient numerical information given in a problem, think in terms of ratios and look for constants that cancel out.

Practice Problem 13.6 Pressure in the Air Tank After the Temperature Increases

A tank of compressed air is at an absolute pressure of 580 kPa at a temperature of 300.0 K. The temperature increases to 330.0 K. What is the pressure in the tank now?

Problem-Solving Tips for the Ideal Gas Law

- In most problems, some change occurs; decide which of the four quantities (P, V, N or n, and T) remain constant during the change.
- Use the microscopic form if the problem deals with the number of molecules and the macroscopic form if the problem deals with the number of moles.
- Use subscripts (i and f) to distinguish initial and final values.
- Work in terms of ratios so that constant factors cancel out.
- Write out the units when doing calculations.
- Remember that P stands for *absolute* pressure (not gauge pressure) and T stands for *absolute* temperature (in kelvins, not °C or °F).

13.6 KINETIC THEORY OF THE IDEAL GAS

In a gas, the interaction between two molecules weakens rapidly as the distance between the molecules increases. In a dilute gas, the average distance between gas molecules is large enough that we can ignore interactions between the molecules except when they collide. In addition, the volume of space occupied by the molecules themselves is a small fraction of the total volume of the gas—the gas is mostly "empty space." The **ideal gas** is a simplified model of a dilute gas in which we think of the molecules as pointlike particles that move *independently* in free space with no interactions except for elastic collisions.

This simplified model is a good approximation for many gases under ordinary conditions. Many properties of gases can be understood from this model; the microscopic theory based on it is called the **kinetic theory** of the ideal gas.

Microscopic Basis of Pressure

The force that a gas exerts on a surface is due to collisions that the gas molecules make with that surface. For instance, think of the air inside an automobile tire. Whenever an air molecule collides with the inner tire surface, the tire exerts an inward force to turn the air molecule around and return it to the bulk of the gas. By Newton's third law, the gas molecule exerts an outward force on the tire surface. The net force per unit area on the inside of the tire due to all the collisions of the many air molecules is equal to the air pressure in the tire. The pressure depends on three things: how many molecules there are, how often each one collides with the wall, and the momentum transfer due to each collision.

We want to find out how the pressure of an ideal gas is determined by the motions of the gas molecules. To simplify the discussion, consider a gas contained in a box of length L and side area A (Fig. 13.12a)—the result does not depend on the shape of the container. Figure 13.12b shows a gas molecule about to collide with the rightmost wall of the container. For simplicity, we assume that the collision is elastic; a more advanced analysis shows that the result is correct even though not all collisions are elastic.

For an elastic collision, the x-component of the molecule's momentum is reversed in direction since the wall is much more massive than the molecule. Since the gas exerts only an outward force on the wall (a static fluid exerts no tangential force on a boundary), the y- and z-components of the molecule's momentum are unchanged. Thus, the molecule's momentum change is $\Delta p_x = 2m|v_x|$.

CONNECTION:

We are using the principle that force is the rate of change of momentum (Newton's second law) to draw a conclusion about pressure in a gas.

When does this molecule next collide with the same wall? Ignoring for now collisions with other molecules, its x-component of velocity never changes magnitude—only the sign of v_x changes when it reverses direction (Fig. 13.12c). The time it takes the molecule to travel the length L of the container and hit the other wall is $L/|v_x|$. Then the round-trip time is

$$\Delta t = 2\frac{L}{|v_x|}$$

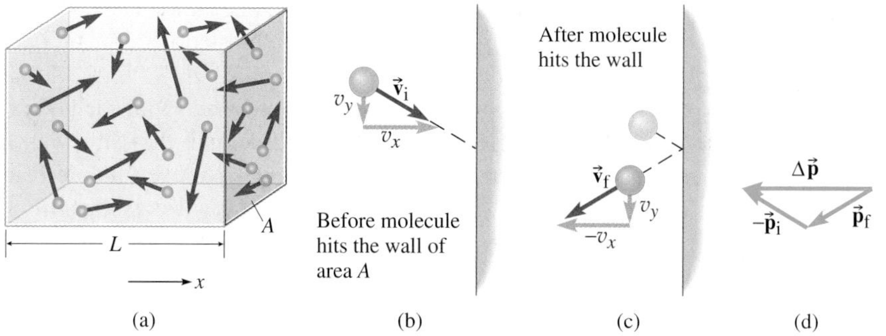

Figure 13.12 (a) Gas molecules confined to a container of length L and area A. (b) A molecule is about to collide with the wall of area A. (c) After an elastic collision, v_x has changed sign, while v_y and v_z are unchanged. (d) The change in momentum due to the collision has magnitude $2|p_x|$ and is perpendicular to the wall.

The *average* force exerted by the molecule on the wall is the change in momentum (Fig. 13.12d) divided by the time for one complete round-trip:

$$F_{\text{av},x} = \frac{\Delta p_x}{\Delta t} = \frac{2m|v_x|}{2L/|v_x|} = \frac{m|v_x|^2}{L} = \frac{mv_x^2}{L}$$

The total force on the wall is the sum of the forces due to each molecule in the gas. If there are N molecules in the gas, we can simply multiply N by the *average* force due to one molecule to get the total force on the wall. To represent such an average, we use angle brackets $\langle\ \rangle$; the quantity inside the brackets is averaged over all the molecules in the gas.

$$F = N\langle F_{\text{av}}\rangle = \frac{Nm}{L}\langle v_x^2\rangle$$

The pressure is then

$$P = \frac{F}{A} = \frac{Nm}{AL}\langle v_x^2\rangle$$

The volume of the box is $V = AL$, so

$$P = \frac{Nm}{V}\langle v_x^2\rangle \tag{13-17}$$

which is true regardless of the shape of the container enclosing the gas. Since we end up averaging over all the molecules in the gas, the simplifying assumption about no collisions with other molecules does not affect the result.

The product $m\langle v_x^2\rangle$ suggests kinetic energy. It certainly makes sense that if the average kinetic energy of the gas molecules is larger, the pressure is higher. The average translational kinetic energy of a molecule in the gas is $\langle K_{\text{tr}}\rangle = \frac{1}{2}m\langle v^2\rangle$. For any gas molecule, $v^2 = v_x^2 + v_y^2 + v_z^2$, since velocity is a vector quantity. The gas as a whole is at rest, so there is no preferred direction of motion. Then the average value of v_x^2 must be the same as the averages of v_y^2 and v_z^2, so

$$\langle v_x^2\rangle = \tfrac{1}{3}\langle v^2\rangle$$

Therefore,

$$m\langle v_x^2\rangle = \tfrac{1}{3}m\langle v^2\rangle = \tfrac{2}{3}\langle K_{\text{tr}}\rangle$$

The pressure of an ideal gas is proportional to the average translational kinetic energy of its molecules and to the number of molecules per unit volume.

Substituting this into Eq. (13-17), the pressure is

$$P = \frac{2}{3}\frac{N\langle K_{\text{tr}}\rangle}{V} = \frac{2}{3}\frac{N}{V}\langle K_{\text{tr}}\rangle \tag{13-18}$$

Equation (13-18) is written with the variables grouped in two different ways to give two different insights. The first grouping says that pressure is proportional to the kinetic energy density (the kinetic energy per unit volume). The second says that pressure is proportional to the product of the number density N/V and the average molecular kinetic energy. The pressure of a gas increases if either the gas molecules are packed closer together or if the molecules have more kinetic energy.

Note that $\langle K_{\text{tr}}\rangle$ is the average *translational* kinetic energy of a gas molecule and v is the CM speed of a molecule. A gas molecule with more than one atom (such as N_2), has vibrational and rotational kinetic energy *in addition to* its translational kinetic energy K_{tr}, but Eq. (13-18) still holds.

What about the assumption that the gas molecules never collide with each other? It certainly is *not* true that the same molecule returns to collide with the same wall at a fixed time interval and has the same v_x each time it returns! However, the derivation really only relies on average quantities. In a gas at equilibrium, an average quantity like $\langle v_x^2\rangle$ remains unchanged even though any one particular molecule changes its velocity components as a result of each collision.

Temperature and Translational Kinetic Energy

The temperature of an ideal gas has a direct physical interpretation that we can now bring to light. We found that in an ideal gas, the pressure, volume, and number of molecules are related to the average translational kinetic energy of the gas molecules:

$$P = \frac{2}{3}\frac{N}{V}\langle K_{tr}\rangle \tag{13-18}$$

Solving for the average kinetic energy,

$$\langle K_{tr}\rangle = \frac{3}{2}\frac{PV}{N} \tag{13-19}$$

The ideal gas law relates P, V, and N to the temperature:

$$PV = NkT \tag{13-13}$$

By rearranging the ideal gas law, we find that P, V, and N occur in the same combination as in Eq. (13-19):

$$\frac{PV}{N} = kT$$

Then by substituting kT for $(PV)/N$ in Eq. (13-19), we find that

$$\langle K_{tr}\rangle = \tfrac{3}{2}kT \tag{13-20}$$

Therefore, *the absolute temperature of an ideal gas is proportional to the average translational kinetic energy of the gas molecules.* Temperature then is a way to describe the average translational kinetic energy of the gas molecules. At higher temperatures, the gas molecules have (on average) greater kinetic energy.

CHECKPOINT 13.6

At what temperature in °C would molecules of O_2 have twice the average translational kinetic energy that molecules of H_2 have at 20°C?

RMS Speed The speed of a gas molecule that has the average kinetic energy is called the **rms** (root mean square) **speed**. The rms speed is *not* the same as the average speed. Instead, the rms speed is the square *root* of the *mean* (average) of the speed *squared.* Since

$$\langle K_{tr}\rangle = \tfrac{1}{2}m\langle v^2\rangle = \tfrac{1}{2}mv_{rms}^2 \tag{13-21}$$

the rms speed is

$$v_{rms} = \sqrt{\langle v^2\rangle}$$

Squaring before averaging emphasizes the effect of the faster-moving molecules, so the rms speed is a bit higher than the average speed—about 9% higher as it turns out.

Since the average kinetic energy of molecules in an ideal gas depends only on temperature, Eq. (13-21) implies that more massive molecules move more slowly on average than lighter ones at the same temperature. If two different gases are placed in a single chamber so that they reach equilibrium and are at the same temperature, their molecules must have the same average translational kinetic energies. If one gas has molecules of larger mass, its molecules must move with a slower average velocity than those of the gas with the lighter mass molecules. In Problem 74, you can show that

$$v_{rms} = \sqrt{\frac{3kT}{m}} \tag{13-22}$$

where k is Boltzmann's constant and m is the mass of a molecule. Therefore, at a given temperature, the rms speed is inversely proportional to the square root of the mass of the molecule.

Example 13.7

O_2 Molecules at Room Temperature

Find the average translational kinetic energy and the rms speed of the O_2 molecules in air at room temperature (20°C).

Strategy The average translational kinetic energy depends only on temperature. We must remember to use absolute temperature. The rms speed is the speed of a molecule that has the average kinetic energy.

Solution The absolute temperature is

$$20°\text{C} + 273\text{ K} = 293\text{ K}$$

Therefore, the average translational kinetic energy is

$$\langle K_{tr}\rangle = \tfrac{3}{2}kT$$
$$= 1.50 \times 1.38 \times 10^{-23}\text{ J/K} \times 293\text{ K}$$
$$= 6.07 \times 10^{-21}\text{ J}$$

From the periodic table, we find the atomic mass of oxygen to be 16.0 u; the molecular mass of O_2 is twice that (32.0 u). First we convert that to kg:

$$32.0\text{ u} \times 1.66 \times 10^{-27}\text{ kg/u} = 5.31 \times 10^{-26}\text{ kg}$$

The rms speed is the speed of a molecule with the average kinetic energy:

$$\langle K_{tr}\rangle = \tfrac{1}{2}mv_{rms}^2$$

$$v_{rms} = \sqrt{\frac{2\langle K_{tr}\rangle}{m}} = \sqrt{\frac{2 \times 6.07 \times 10^{-21}\text{ J}}{5.31 \times 10^{-26}\text{ kg}}} = 478\text{ m/s}$$

Discussion How can we decide if the result is reasonable, since we have no first-hand experience watching molecules bounce around? Recall from Chapter 12 that the speed of sound in air at room temperature is 343 m/s. Since sound waves in air propagate by the collisions that occur between air molecules, the speed of sound must be of the same order of magnitude as the average speeds of the molecules.

Practice Problem 13.7 CO_2 Molecules at Room Temperature

Find the average translational kinetic energy and the rms speed of the CO_2 molecules in air at room temperature (20°C).

Maxwell-Boltzmann Distribution

So far we have considered only the *average* kinetic energy and *rms* speed of a molecule. Sometimes we may want to know more: how many molecules have speeds in a certain range? The distribution of speeds is called the **Maxwell-Boltzmann distribution**. The distribution for oxygen at two different temperatures is shown in Fig. 13.13. The interpretation of the graphs is that the number of gas molecules having speeds between any two values v_1 and v_2 is proportional to the area under the curve between v_1 and v_2. In Fig. 13.13, the shaded areas represent the number of oxygen molecules having speeds

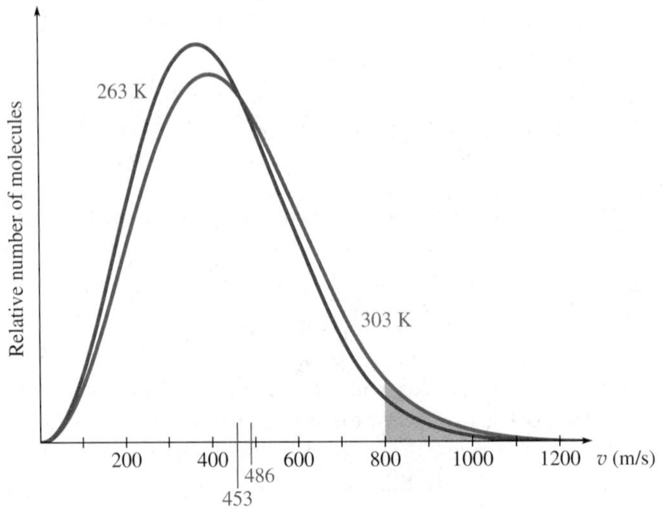

Figure 13.13 The probability distribution of kinetic energies in oxygen at two temperatures: −10°C (263 K) and +30°C (303 K). The area under either curve for any range of speeds is proportional to the number of molecules whose speeds lie in that range. Despite the relatively small difference in rms speeds (453 m/s at 263 K and 486 m/s at 303 K), the fraction of molecules in the high-speed tail is quite different.

above 800 m/s at the two selected temperatures. A relatively small temperature change has a significant effect on the number of gas molecules with high speeds.

Any given molecule changes its kinetic energy often—at each collision, which means billions of times per second. However, the total number of gas molecules in a given kinetic energy range in the gas stays the same, as long as the temperature is constant. In fact, it is the frequent collisions that maintain the stability of the Maxwell-Boltzmann distribution. The collisions keep the kinetic energy distributed among the gas molecules *in the most disordered way possible*, which is the Maxwell-Boltzmann distribution.

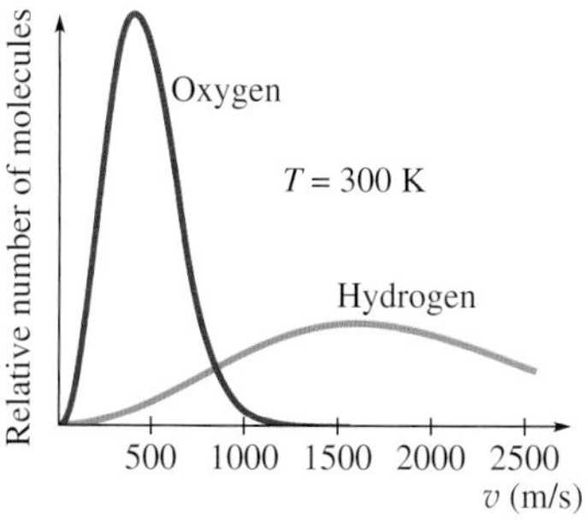

Figure 13.14 Maxwell-Boltzmann distributions for oxygen and hydrogen at $T = 300$ K. Escape speed from Earth is 11 200 m/s (not shown on the graph).

Application of the Maxwell-Boltzmann Distribution: Composition of Planetary Atmospheres The Maxwell-Boltzmann distribution helps us understand planetary atmospheres. Why does Earth's atmosphere contain nitrogen, oxygen, and water vapor, among other gases, but not hydrogen or helium, which are by far the most common elements in the universe? Molecules in the upper atmosphere that are moving faster than the *escape speed* (see Example 6.8) have enough kinetic energy to escape from the planetary atmosphere to outer space. Those that are heading away from the planet's surface will escape if they avoid colliding with another molecule. The high-energy tail of the Maxwell-Boltzmann distribution does not get depleted by molecules that escape. Other molecules will get boosted to those high kinetic energies as a result of collisions; these replacements will in turn also escape. Thus, the atmosphere gradually leaks away.

How fast the atmosphere leaks away depends on how far the rms speed is from the escape speed. If the rms speed is too small compared with the escape speed, the time for all the gas molecules to escape is so long that the gas is present in the atmosphere indefinitely. This is the case for nitrogen, oxygen, and water vapor in Earth's atmosphere. On the other hand, since hydrogen and helium are much less massive, their rms speeds are higher. Though only a tiny fraction of the molecules are above the escape speed, the fraction is sufficient for these gases to escape quickly from Earth's atmosphere (Fig. 13.14). The Moon is often said to lack an atmosphere. The Moon's low escape speed (2400 m/s) allows most gases to escape, but it does have an atmosphere about 1 cm tall composed of krypton (a gas with molecular mass 83.8 u, about 2.6 times that of oxygen).

CONNECTION:

The basic principle behind escape speed is conservation of energy (Sec. 6.5). At the escape speed, an atom or molecule has just enough kinetic energy to escape the planet's gravitational pull.

13.7 TEMPERATURE AND REACTION RATES

What we have learned about the distribution of kinetic energies and its relationship to temperature has a great relevance to the dependence of chemical reaction rates on temperature. Imagine a mixture of two gases, N_2 and O_2, which can react to form nitric oxide (NO):

$$N_2 + O_2 \rightarrow 2NO$$

In order for the reaction to occur, a molecule of nitrogen must collide with a molecule of oxygen. But the reaction does not occur every time such a collision takes place. The reactant molecules must possess enough kinetic energy to initiate the reaction, because the reaction involves the rearrangement of chemical bonds between atoms. Some chemical bonds must be broken before new ones form; the energy to break these bonds must come from the energy of the reactants. The minimum kinetic energy of the reactant molecules that allows the reaction to proceed is called the **activation energy** (E_a).

Note that energy must be *supplied* to break a bond. *Forming* a bond *releases* energy.

If a molecule of N_2 collides with one of O_2, but their total kinetic energy is less than the activation energy, then the two just bounce off one another. Some energy may be transferred from one molecule to the other, or converted between translational, rotational, and vibrational energy, but we are still left with one molecule of N_2 and one of O_2.

Now we begin to see why, with few exceptions, rates of reaction increase with temperature. At higher temperatures, the average kinetic energy of the reactants is higher and therefore a greater fraction of the collisions have total kinetic energies exceeding the activation energy. If the activation energy is much greater than the average translational kinetic energy of the reactants,

$$E_a \gg \tfrac{3}{2}kT \tag{13-23}$$

then the only candidates for reaction are molecules far off in the exponentially decaying, high-energy tail of the Maxwell-Boltzmann distribution. In this situation, a small increase in temperature can have a dramatic effect on the reaction rate: the reaction rate depends *exponentially* on temperature.

$$\text{reaction rate} \propto e^{-E_a/(kT)} \tag{13-24}$$

Although we have discussed reactions in terms of gases, the same general principles apply to reactions in liquid solutions. The temperature determines what fraction of the collisions have enough energy to react, so reaction rates are temperature-dependent whether the reaction occurs in a gas mixture or a liquid solution.

Example 13.8

Increase in Reaction Rate with Temperature Increase

The activation energy for the reaction $N_2O \rightarrow N_2 + O$ is 4.0×10^{-19} J. By what percentage does the reaction rate increase if the temperature is increased from 700.0 K to 707.0 K (a 1% increase in absolute temperature)?

Strategy We should first check that $E_a \gg \frac{3}{2}kT$; otherwise, Eq. (13-24) does not apply. Assuming that checks out, we can set up a ratio of the reaction rates at the two temperatures.

Solution Start by calculating $E_a/(kT_1)$, where $T_1 = 700.0$ K:

$$\frac{E_a}{kT_1} = \frac{4.0 \times 10^{-19}\ \text{J}}{1.38 \times 10^{-23}\ \text{J/K} \times 700.0\ \text{K}} = 41.41$$

So E_a is about 41 times kT, or about 28 times $\frac{3}{2}kT$. The activation energy is much greater than the average kinetic energy; thus, only a small fraction of the collisions might cause a reaction to occur.

At $T_2 = 707.0$ K,

$$\frac{E_a}{kT_2} = \frac{4.0 \times 10^{-19}\ \text{J}}{1.38 \times 10^{-23}\ \text{J/K} \times 707.0\ \text{K}} = \frac{41.41}{1.01} = 41.00$$

The ratio of the reaction rates is

$$\frac{\text{new rate}}{\text{old rate}} = \frac{e^{-41.00}}{e^{-41.41}} = e^{-(41.00-41.41)} = e^{0.41} = 1.5$$

The reaction rate at 707.0 K is 1.5 times the rate at 700.0 K—a 50% increase in reaction rate for a 1% increase in temperature!

Discussion Normally we might suspect an error when a 1% change in one quantity causes a 50% change in another! However, this problem illustrates the dramatic effect of an *exponential* dependence. Reaction rates can be *extremely* sensitive to small temperature changes.

Practice Problem 13.8 Decrease in Reaction Rate for Lower Temperature

What is the percentage decrease in the rate of the same reaction if the temperature is lowered from 700.0 K to 699.0 K?

What are the evolutionary advantages of warm-blooded versus cold-blooded animals?

At the beginning of this chapter, we asked about the necessity for temperature regulation in warm-blooded animals (Fig. 13.15). The temperature dependence of chemical reaction rates has a profound effect on biological functions. If our internal temperatures varied, we would have a varying metabolic rate, becoming sluggish in cold weather.

Application: Temperature and Metabolism By maintaining a constant body temperature higher than that of the environment, warm-blooded animals are able to tolerate a wider range of environmental temperatures than cold-blooded animals (such as

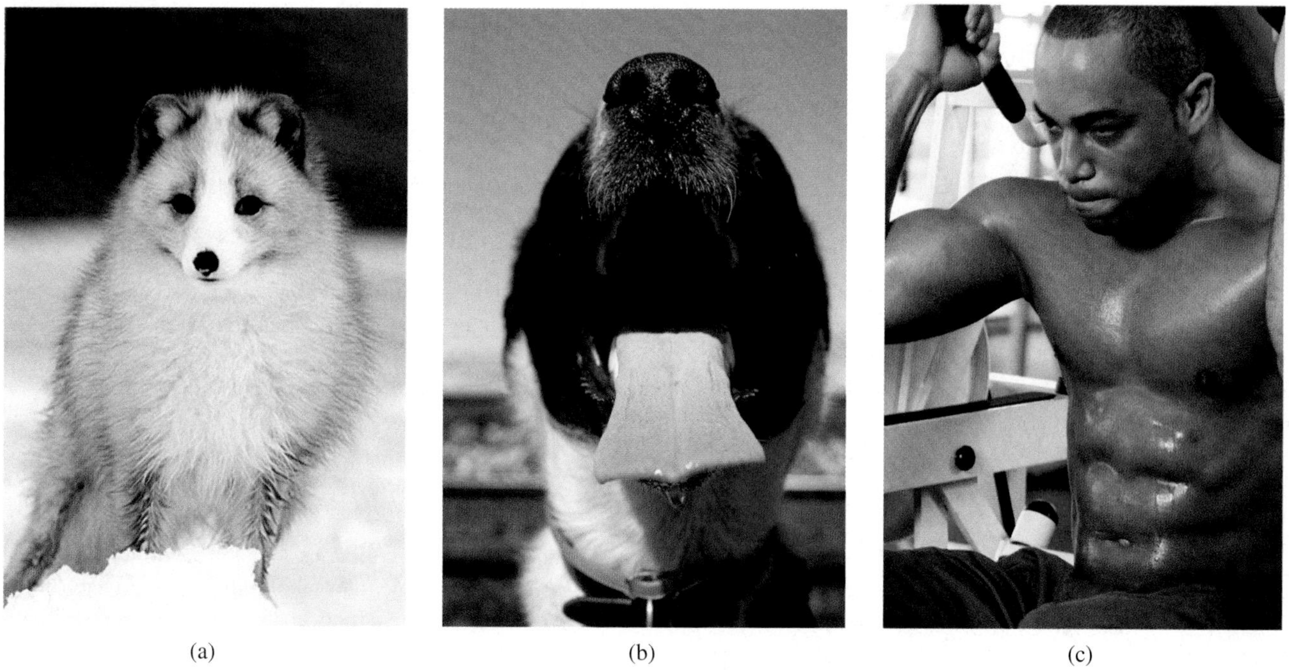

Figure 13.15 Warm-blooded animals use different strategies to maintain a constant body temperature. (a) The fur of an Arctic fox serves as a layer of insulation to help it stay warm. (b) Dogs pant and (c) people sweat when their bodies are in danger of overheating. In cases (b) and (c), the evaporation of water has a cooling effect on the body.

reptiles and insects). Temperature fluctuation in the environment is much more severe on land than in water; thus, land animals are more likely to be homeothermic than aquatic animals. Keeping muscles at their optimal temperatures contributes to the much larger effort required to move around on land or in the air as opposed to moving through water. Keeping the muscles and vital organs warm allows the high level of aerobic metabolism needed to sustain intense physical activity.

Cold-blooded animals depend on the environment for temperature regulation; thus, we see a snake lying on a rock heated by the Sun in an attempt to keep warm. As a snake's blood temperature goes down in cold weather, the snake becomes inactive and lethargic. Most insects are inactive below 10°C and many cannot survive the cold of winter.

However, if environmental conditions become too extreme, it may be difficult for homeotherms to maintain ideal body temperature. Hypothermia occurs when the central core of the body becomes too cold; bodily processes slow and eventually cease. People caught outside in blizzards are urged to stay awake and to keep moving; the energy produced by exercise may be up to 20 times that produced by the resting body and can compensate for heat loss in extreme cold.

Warm-blooded animals must consume much more food than cold-blooded animals of a similar size; metabolic processes in warm-blooded animals act like a furnace to keep the body warm. A human must consume about 1500 kcal of food energy per day just to keep warm when resting at 20°C; an alligator of similar body weight needs only 60 kcal/day at rest at 20°C.

13.8 DIFFUSION

Mean Free Path How far does a gas molecule move, on average, between collisions? The mean (average) length of the path traveled by a gas molecule as a free particle (no interactions with other particles) is called the **mean free path** (Λ, the Greek capital

Figure 13.16 Successive straight-line paths traveled by a molecule between collisions.

lambda). The mean free path depends on two things: how large the molecules are and how many of them occupy a given volume. A detailed calculation yields

Mean free path:

$$\Lambda = \frac{1}{\sqrt{2}\,\pi d^2\,(N/V)} \qquad (13\text{-}25)$$

Typically the mean free path is much larger than the average distance between neighboring molecules.Nitrogen molecules in air at room temperature have mean free paths of about 0.1 μm, which is about 25 times the average distance between molecules. Each molecule collides an average of 5×10^9 times per second. (For more information on mean free path, see text website.)

Diffusion

A gas molecule moves in a straight line between collisions—the effect of gravity on the velocity of the molecule is negligible during a time interval of only 0.2 ns. At each collision, both the speed and direction of the molecule's motion change. The mean free path tells us the *average* length of the molecule's straight line paths between collisions. The result is that a given molecule follows a *random walk* trajectory (Fig. 13.16).

After an elapsed time t, how far on average has a molecule moved from its initial position? The answer to this question is relevant when we consider **diffusion**. Someone across the room opens a bottle of perfume: how long until the scent reaches you? As gas molecules diffuse into the air, the frequent collisions are what determine how long it takes the scent to travel across the room (assuming, as we do here, that there are no air currents). When there is a difference in concentrations between different points in a gas, the random thermal motion of the molecules tends to even out the concentrations (other things being equal). The net flow from regions of high concentration (near the perfume bottle) to regions of lower concentration (across the room) is diffusion.

Consider a molecule of perfume in the air. It has a mean free path Λ. After a large number of collisions N, it has traveled a total *distance* $N\Lambda$. However, its displacement from its original position is much less than that, since at each collision it changes direction. It can be shown using statistical analysis of the random walk that the rms magnitude of its displacement after N collisions is proportional to $\sqrt{N}$. Since the number of collisions is proportional to the elapsed time, the rms displacement is proportional to $\sqrt{t}$.

The root mean squared displacement in one direction is

$$x_{\text{rms}} = \sqrt{2Dt} \qquad (13\text{-}26)$$

where D is a diffusion constant such as those given in Table 13.4. The diffusion constant D depends on the molecule or atom that is diffusing and the medium through which it is moving.

Application: diffusion of oxygen through cell membranes

Diffusion is crucial in biological processes such as the transport of oxygen. Oxygen molecules diffuse from the air in the lungs through the walls of the alveoli and then through the walls of the capillaries to oxygenate the blood. The oxygen is then carried by hemoglobin in the blood to various parts of the body, where it again diffuses through capillary walls into intercellular fluids and then through cell membranes into cells. Diffusion is a slow process over long distances but can be quite effective over short distances—which is why cell membranes must be thin and capillaries must have small diameters. Evolution has seen to it that the capillaries of animals of widely different sizes are all about the same size—as small as possible while still allowing blood cells to flow through them.

Table 13.4 Diffusion Constants at 1 atm and 20°C

Diffusing Molecule	Medium	D (m^2/s)
DNA	Water	1.3×10^{-12}
Oxygen	Tissue (cell membrane)	1.8×10^{-11}
Hemoglobin	Water	6.9×10^{-11}
Sucrose ($C_{12}H_{22}O_{11}$)	Water	5.0×10^{-10}
Glucose ($C_6H_{12}O_6$)	Water	6.7×10^{-10}
Oxygen	Water	1.0×10^{-9}
Oxygen	Air	1.8×10^{-5}
Hydrogen	Air	6.4×10^{-5}

Example 13.9

Diffusion Time for Oxygen into Capillaries

How long on average does it take an oxygen molecule in an alveolus to diffuse into the blood? Assume for simplicity that the diffusion constant for oxygen passing through the two membranes (alveolus and capillary walls) is the same: 1.8×10^{-11} m²/s. The total thickness of the two membranes is 1.2×10^{-8} m.

Strategy Take the x-direction to be through the membranes. Then we want to know how much time elapses until $x_{rms} = 1.2 \times 10^{-8}$ m.

Solution Solving Eq. (13-26) for t yields

$$t = \frac{x_{rms}^2}{2D}$$

Now substitute $x_{rms} = 1.2 \times 10^{-8}$ m and $D = 1.8 \times 10^{-11}$ m²/s:

$$t = \frac{(1.2 \times 10^{-8}\ \text{m})^2}{2 \times 1.8 \times 10^{-11}\ \text{m}^2/\text{s}} = 4.0 \times 10^{-6}\ \text{s}$$

Discussion The time is proportional to the *square* of the membrane thickness. It would take four times as long for an oxygen molecule to diffuse through a membrane twice as thick. The rapid increase of diffusion time with distance is a principal reason why evolution has favored thin membranes over thicker ones.

Practice Problem 13.9 Time for Oxygen to Get Halfway Through the Membrane

How long on average does it take an oxygen molecule to get *halfway* through the alveolus and capillary wall?

Master the Concepts

- Temperature is a quantity that determines when objects are in thermal equilibrium. The flow of energy that occurs between two objects or systems due to a temperature difference between them is called heat flow. If heat can flow between two objects or systems, the objects or systems are said to be in thermal contact. When two systems in thermal contact have the same temperature, there is no net flow of heat between them; the objects are said to be in thermal equilibrium.
- Zeroth law of thermodynamics: if two objects are each in thermal equilibrium with a third object, then the two are in thermal equilibrium with one another.
- The SI unit of temperature is the kelvin (symbol K, *without* a degree sign). The kelvin scale is an absolute temperature scale, which means that $T = 0$ is set to absolute zero.

continued on next page

Master the Concepts continued

- Temperature in °C (T_C) and temperature in kelvins (T) are related by

$$T_C = T - 273.15 \qquad (13\text{-}3)$$

- As long as the temperature change is not too great, the fractional length change of a solid is proportional to the temperature change:

$$\frac{\Delta L}{L_0} = \alpha \Delta T \qquad (13\text{-}4)$$

The constant of proportionality, α, is called the coefficient of linear expansion of the substance.

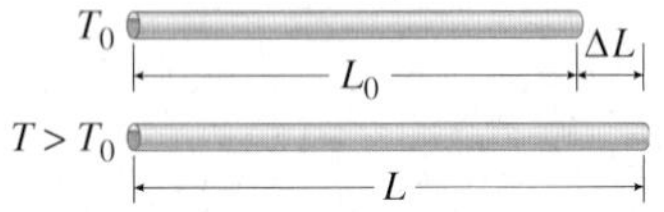

- The fractional change in volume of a solid or liquid is also proportional to the temperature change as long as the temperature change is not too large:

$$\frac{\Delta V}{V_0} = \beta \Delta T \qquad (13\text{-}7)$$

For solids, the coefficient of volume expansion is three times the coefficient of linear expansion: $\beta = 3\alpha$.

- The mole is an SI base unit and is defined as: one mole of anything contains the same number of units as there are atoms in 12 *grams* (not kilograms) of carbon-12. This number is called Avogadro's number and has the value

$$N_A = 6.022 \times 10^{23}\ \text{mol}^{-1}$$

- The mass of an atom or molecule is often expressed in the atomic mass unit (symbol u). By definition, one atom of carbon-12 has a mass of 12 u (exactly).

$$1\ \text{u} = 1.66 \times 10^{-27}\ \text{kg} \qquad (13\text{-}12)$$

The atomic mass unit is chosen so that the mass of an atom or molecule in "u" is numerically the same as the molar mass in g/mol.

- In an ideal gas, the molecules move independently in free space with no interactions except when two molecules collide. The ideal gas is a useful model for many real gases, provided that the gas is sufficiently dilute. The ideal gas law:

$$\text{microscopic form: } PV = NkT \qquad (13\text{-}13)$$

$$\text{macroscopic form: } PV = nRT \qquad (13\text{-}16)$$

where Boltzmann's constant and the universal gas constant are

$$k = 1.38 \times 10^{-23}\ \text{J/K} \qquad (13\text{-}14)$$

$$R = N_A k = 8.31\ \frac{\text{J/K}}{\text{mol}} \qquad (13\text{-}15)$$

- The pressure of an ideal gas is proportional to the average translational kinetic energy of the molecules:

$$P = \frac{2}{3}\frac{N}{V}\langle K_{tr}\rangle \qquad (13\text{-}18)$$

- The average translational kinetic energy of the molecules is proportional to the absolute temperature:

$$\langle K_{tr}\rangle = \tfrac{3}{2}kT \qquad (13\text{-}20)$$

- The speed of a gas molecule that has the average kinetic energy is called the rms speed:

$$\langle K_{tr}\rangle = \tfrac{1}{2}mv_{rms}^2 \qquad (13\text{-}21)$$

- The distribution of molecular speeds in an ideal gas is called the Maxwell-Boltzmann distribution.

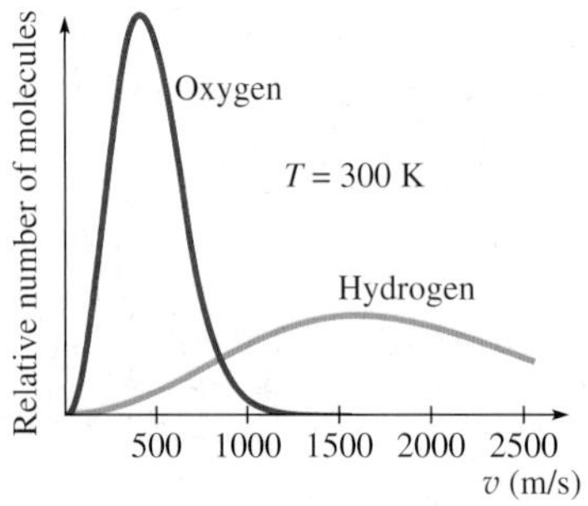

- If the activation energy for a chemical reaction is much greater than the average kinetic energy of the reactants, the reaction rate depends *exponentially* on temperature:

$$\text{reaction rate} \propto e^{-E_a/(kT)} \qquad (13\text{-}24)$$

- The mean free path (Λ) is the average length of the path traveled by a gas molecule as a free particle (no interactions with other particles) between collisions:

$$\Lambda = \frac{1}{\sqrt{2}\,\pi d^2 (N/V)} \qquad (13\text{-}25)$$

- The root mean square displacement of a diffusing molecule along the x-axis is

$$x_{rms} = \sqrt{2Dt} \qquad (13\text{-}26)$$

where D is a diffusion constant.

Conceptual Questions

1. Explain why it would be impossible to uniquely define the temperature of an object if the zeroth law of thermodynamics were violated?
2. Why do we call the temperature 0 K "absolute zero"? How is 0 K fundamentally different from 0°C or 0°F?
3. Under what special circumstances can kelvins or Celsius degrees be used interchangeably?
4. What happens to a hole in a flat metal plate when the plate expands on being heated? Does the hole get larger or smaller?
5. Why would silver and brass probably not be a good choice of metals for a bimetallic strip (leaving aside the question of the cost of silver)? (See Table 13.2.)
6. One way to loosen the lid on a glass jar is to run it under hot water. How does that work?
7. Why must we use absolute temperature (temperature in kelvins) in the ideal gas law ($PV = NkT$)? Explain how using the Celsius scale would give nonsensical results.
8. Natural gas is sold by volume. In the United States, the price charged is usually per cubic foot. Given the price per cubic foot, what other information would you need in order to calculate the price per mole?
9. What are the SI units of mass density and number density? If two different gases have the same number density, do they have the same mass density?
10. Suppose we have two tanks, one containing helium gas and the other nitrogen gas. The two gases are at the same temperature and pressure. Which has the higher number density (or are they equal)? Which has the higher mass density (or are they equal)?
11. The mass of an aluminum atom is 27.0 u. What is the mass of *one mole* of aluminum atoms? (No calculation required!)
12. A ping-pong ball that has been dented during hard play can often be restored by placing it in hot water. Explain why this works.
13. Why does a helium weather balloon expand as it rises into the air? Assume the temperature remains constant.
14. Explain why there is almost no hydrogen (H_2) or helium (He) in Earth's atmosphere, yet both are present in Jupiter's atmosphere. [*Hint:* Escape velocity from Earth is 11.2 km/s and escape velocity from Jupiter is 60 km/s.]
15. Explain how it is possible that more than half of the molecules in an ideal gas have kinetic energies less than the average kinetic energy. Shouldn't half have less and half have more?
16. In air under ordinary conditions (room temperature and atmospheric pressure), the average intermolecular distance is about 4 nm and the mean free path is about 0.1 μm. The diameter of a nitrogen molecule is about 0.3 nm. Explain how the mean free path can be so much larger than the average distance between molecules.
17. In air under ordinary conditions (room temperature and atmospheric pressure), the average intermolecular distance is about 4 nm and the mean free path is about 0.1 μm. The diameter of a nitrogen molecule is about 0.3 nm. Which two distances should we compare to decide that air is dilute and can be treated as an ideal gas? Explain.
18. In air under ordinary conditions (room temperature and atmospheric pressure), the average intermolecular distance is about 4 nm and the mean free path is about 0.1 μm. The diameter of a nitrogen molecule is about 0.3 nm. What would it mean if the intermolecular distance and the molecular diameter were about the same? In that case, would it make sense to speak of a mean free path? Explain.
19. Explain how an automobile airbag protects the passenger from injury. Why would the airbag be ineffective if the gas pressure inside is too low when the passenger comes into contact with it? What about if it is too high?
20. It takes longer to hard-boil an egg in Mexico City (2200 m above sea level) than it does in Amsterdam (parts of which are below sea level). Why? [*Hint:* At higher altitudes, water boils at less than 100°C.]

Multiple-Choice Questions

1. In a mixed gas such as air, the rms speeds of different molecules are
   (a) independent of molecular mass.
   (b) proportional to molecular mass.
   (c) inversely proportional to molecular mass.
   (d) proportional to $\sqrt{\text{molecular mass}}$.
   (e) inversely proportional to $\sqrt{\text{molecular mass}}$.
2. The average kinetic energy of the molecules in an ideal gas increases with the volume remaining constant. Which of these statements *must* be true?
   (a) The pressure increases and the temperature stays the same.
   (b) The number density decreases.
   (c) The temperature increases and the pressure stays the same.
   (d) Both the pressure and the temperature increase.
3. Which of these will increase the average kinetic energy of the molecules in an ideal gas?
   (a) reduce the volume, keeping P and N constant
   (b) increase the volume, keeping P and N constant
   (c) reduce the volume, keeping T and N constant
   (d) increase the pressure, keeping T and V constant
   (e) increase N, keeping V and T constant

4. The absolute temperature of an ideal gas is directly proportional to
   (a) the number of molecules in the sample.
   (b) the average momentum of a molecule of the gas.
   (c) the average translational kinetic energy of the gas.
   (d) the diffusion constant of the gas.
5. The rms speed is the
   (a) speed at which all the gas molecules move.
   (b) speed of a molecule with the average kinetic energy.
   (c) average speed of the gas molecules.
   (d) maximum speed of the gas molecules.
6. What are the most favorable conditions for real gases to approach ideal behavior?
   (a) high temperature and high pressure
   (b) low temperature and high pressure
   (c) low temperature and low pressure
   (d) high temperature and low pressure
7. An ideal gas has the volume V_0. If the temperature and the pressure are each tripled during a process, the new volume is
   (a) V_0.
   (b) $9V_0$.
   (c) $3V_0$.
   (d) $0.33V_0$.
8. The average kinetic energy of a gas molecule can be found from which of these quantities?
   (a) pressure only
   (b) number of molecules only
   (c) temperature only
   (d) pressure and temperature are both required
9. If the temperature of an ideal gas is doubled and the pressure is held constant, the rms speed of the molecules
   (a) remains unchanged.
   (b) is 2 times the original speed.
   (c) is $\sqrt{2}$ times the original speed.
   (d) is 4 times the original speed.
10. A metal box is heated until each of its sides has expanded by 0.1%. By what percent has the *volume* of the box changed?
    (a) −0.3% (b) −0.2% (c) +0.1%
    (d) +0.2% (e) +0.3%

Problems

- Combination conceptual/quantitative problem
- Biological or medical application
- Challenging problem
- Blue # Detailed solution in the Student Solutions Manual
- 1 2 Problems paired by concept
- Text website interactive or tutorial

13.2 Temperature Scales

1. On a warm summer day, the air temperature is 84°F. Express this temperature in (a) °C and (b) kelvins. (tutorial: sun's temperature)
2. The temperature at which liquid nitrogen boils (at atmospheric pressure) is 77 K. Express this temperature in (a) °C and (b) °F.
3. (a) At what temperature (if any) does the numerical value of Celsius degrees equal the numerical value of Fahrenheit degrees? (b) At what temperature (if any) does the numerical value of kelvins equal the numerical value of Fahrenheit degrees?
4. A room air conditioner causes a temperature change of −6.0°C. (a) What is the temperature change in kelvins? (b) What is the temperature change in °F?
5. Aliens from the planet Jeenkah have based their temperature scale on the boiling and freezing temperatures of ethyl alcohol. These temperatures are 78°C and −114°C, respectively. The people of Jeenkah have six digits on each hand, so they use a base-12 number system and have decided to have 144°J between the freezing and boiling temperatures of ethyl alcohol. They set the freezing point to 0°J. How would you convert from °J to °C?

13.3 Thermal Expansion of Solids and Liquids

6. A 2.4-m length of copper pipe extends directly from a hot-water heater in a basement to a faucet on the first floor of a house. If the faucet isn't fixed in place, how much will it rise when the pipe is heated from 20.0°C to 90.0°C. Ignore any increase in the size of the faucet itself or of the water heater.
7. Two 35.0-cm metal rods, one made of copper and one made of aluminum, are placed end to end, touching each other. One end is fixed, so that it cannot move. The rods are heated from 0.0°C to 150°C. How far does the other end of the system of rods move?
8. Steel railroad tracks of length 18.30 m are laid at 10.0°C. How much space should be left between the track sections if they are to just touch when the temperature is 50.0°C?

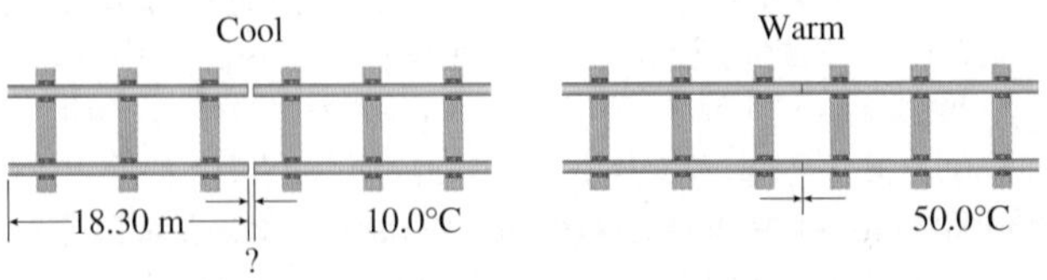

9. A highway is made of concrete slabs that are 15 m long at 20.0°C. (a) If the temperature range at the location of the highway is from −20.0°C to +40.0°C, what size expansion gap should be left (at 20.0°C) to prevent buckling of the highway? (b) How large are the gaps at −20.0°C?

10. A lead rod and a common glass rod both have the same length when at 20.0°C. The lead rod is heated to 50.0°C. To what temperature must the glass rod be heated so that they are again at the same length?

11. The coefficient of linear expansion of brass is 1.9×10^{-5} $°C^{-1}$. At 20.0°C, a hole in a sheet of brass has an area of 1.00 mm^2. How much larger is the area of the hole at 30.0°C? (tutorial: loop around the equator)

12. Aluminum rivets used in airplane construction are made slightly too large for the rivet holes to be sure of a tight fit. The rivets are cooled with dry ice (−78.5°C) before they are driven into the holes. If the holes have a diameter of 0.6350 cm at 20.5°C, what should be the diameter of the rivets at 20.5°C if they are to just fit when cooled to the temperature of dry ice?

13. A temperature change ΔT causes a volume change ΔV but has no effect on the mass of an object. (a) Show that the change in density $\Delta\rho$ is given by $\Delta\rho = -\beta\rho\Delta T$. (b) Find the fractional change in density ($\Delta\rho/\rho$) of a brass sphere when the temperature changes from 32°C to −10.0°C.

14. A cylindrical brass container with a base of 75.0 cm^2 and height of 20.0 cm is filled to the brim with water when the system is at 25.0°C. How much water overflows when the temperature of the water and the container is raised to 95.0°C?

15. An ordinary drinking glass is filled to the brim with water (268.4 mL) at 2.0°C and placed on the sunny pool deck for a swimmer to enjoy. If the temperature of the water rises to 32.0°C before the swimmer reaches for the glass, how much water will have spilled over the top of the glass? Assume the glass does not expand.

16. Consider the situation described in Problem 15. (a) Take into account the expansion of the glass and calculate how much water will spill out of the glass. Compare your answer with the case where the expansion of the glass was not considered. (b) By what percentage has the answer changed when the expansion of the glass is considered?

17. A steel sphere with radius 1.0010 cm at 22.0°C must slip through a brass ring that has an internal radius of 1.0000 cm at the same temperature. To what temperature must the brass ring be heated so that the sphere, still at 22.0°C, can just slip through?

18. A long, narrow steel rod of length 2.5000 m at 25°C is oscillating as a pendulum about a horizontal axis through one end. If the temperature changes to 0°C, what will be the fractional change in its period?

19. The George Washington Bridge crosses the Hudson River between New York and New Jersey. The span of the steel bridge is about 1.6 km. If the temperature can vary from a low of −15°F in winter to a high of 105°F in summer, by how much might the length of the span change over an entire year?

20. A square brass plate, 8.00 cm on a side, has a hole cut into its center of area 4.90874 cm^2 (at 20.0°C). The hole in the plate is to slide over a cylindrical steel shaft of cross-sectional area 4.91000 cm^2 (also at 20.0°C). To what temperature must the brass plate be heated so that it can just slide over the steel cylinder (which remains at 20.0°C)? [*Hint:* The steel cylinder is not heated so it does not expand; only the brass plate is heated.]

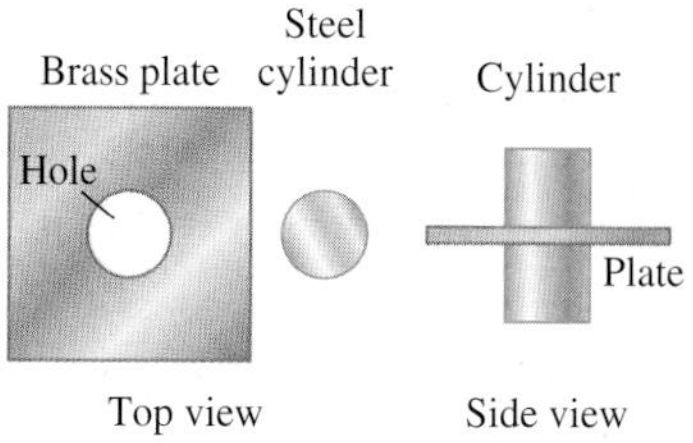

21. A copper washer is to be fit in place over a steel bolt. Both pieces of metal are at 20.0°C. If the diameter of the bolt is 1.0000 cm and the inner diameter of the washer is 0.9980 cm, to what temperature must the washer be raised so it will fit over the bolt? Only the copper washer is heated.

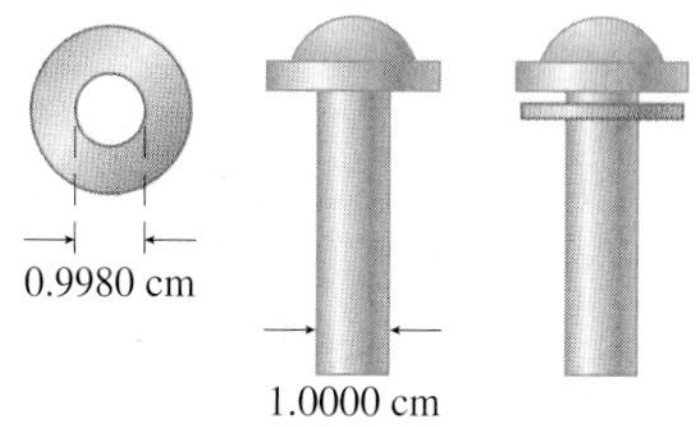

22. Repeat Problem 21, but now the copper washer and the steel bolt are both raised to the same temperature. At what temperature will the washer fit on the bolt?

23. A steel rule is calibrated for measuring lengths at 20.00°C. The rule is used to measure the length of a Vycor glass brick; when both are at 20.00°C, the brick is found to be 25.00 cm long. If the rule and the brick are both at 80.00°C, what would be the length of the brick as measured by the rule?

24. The fuselage of an Airbus A340 has a circumference of 17.72 m on the ground. The circumference increases by 26 cm when it is in flight. Part of this increase is due to the pressure difference between the inside and outside of the plane and part is due to the increase in the temperature due to air drag while it is flying along at 950 km/h. Suppose we wanted to heat a full-size model of the airbus made of aluminum to cause the same increase in circumference without changing the pressure. What would be the increase in temperature needed?

25. A flat square of side s_0 at temperature T_0 expands by Δs in both length and width when the temperature increases by ΔT. The original area is $s_0^2 = A_0$ and the final area is $(s_0 + \Delta s)^2 = A$. Show that if $\Delta s \ll s_0$,

$$\frac{\Delta A}{A_0} = 2\alpha\Delta T \qquad (13\text{-}6)$$

(Although we derive this relation for a square plate, it applies to a flat area of any shape.)

26. The volume of a solid cube with side s_0 at temperature T_0 is $V_0 = s_0^3$. Show that if $\Delta s \ll s_0$, the change in volume ΔV due to a change in temperature ΔT is given by

$$\frac{\Delta V}{V_0} = 3\alpha\Delta T \qquad (13\text{-}7, 8)$$

and therefore that $\beta = 3\alpha$. (Although we derive this relation for a cube, it applies to a solid of any shape.)

13.4 Molecular Picture of a Gas

27. Use the definition that 1 mol of ^{12}C (carbon-12) atoms has a mass of exactly 12 g, along with Avogadro's number, to derive the conversion between atomic mass units and kg.
28. Find the molar mass of ammonia (NH_3).
29. Find the mass (in kg) of one molecule of CO_2.
30. The mass of 1 mol of ^{13}C (carbon-13) is 13.003 g. (a) What is the mass in u of one ^{13}C atom? (b) What is the mass in kilograms of one ^{13}C atom?
31. Estimate the number of H_2O molecules in a human body of mass 80.2 kg. Assume that, on average, water makes up about 62% of the mass of a human body.
32. The mass density of diamond (a crystalline form of carbon) is 3500 kg/m^3. How many carbon atoms per cm^3 are there?
33. How many hydrogen atoms are present in 684.6 g of sucrose ($C_{12}H_{22}O_{11}$)?
34. How many moles of He are in 13 g of He?
35. The principal component of natural gas is methane (CH_4). How many moles of CH_4 are present in 144.36 g of methane?
36. What is the mass of one gold atom in kilograms?
37. Air at room temperature and atmospheric pressure has a mass density of 1.2 kg/m^3. The average molecular mass of air is 29.0 u. How many molecules are in 1.0 cm^3 of air?
38. At 0.0°C and 1.00 atm, 1.00 mol of a gas occupies a volume of 0.0224 m^3. (a) What is the number density? (b) Estimate the average distance between the molecules. (c) If the gas is nitrogen (N_2), the principal component of air, what is the total mass and mass density?
39. Sand is composed of SiO_2. Find the order of magnitude of the number of silicon (Si) atoms in a grain of sand. Approximate the sand grain as a sphere of diameter 0.5 mm and an SiO_2 molecule as a sphere of diameter 0.5 nm.

13.5 Absolute Temperature and the Ideal Gas Law

40. A flight attendant wants to change the temperature of the air in the cabin from 18°C to 24°C without changing the number of moles of air per m^3. What fractional change in pressure would be required?
41. A cylinder in a car engine takes $V_i = 4.50 \times 10^{-2}$ m^3 of air into the chamber at 30°C and at atmospheric pressure. The piston then compresses the air to one-ninth of the original volume (0.111 V_i) and to 20.0 times the original pressure (20.0 P_i). What is the new temperature of the air?
42. A tire with an inner volume of 0.0250 m^3 is filled with air at a gauge pressure of 36.0 psi. If the tire valve is opened to the atmosphere, what volume *outside of the tire* does the escaping air occupy? Some air remains within the tire occupying the original volume, but now that remaining air is at atmospheric pressure. Assume the temperature of the air does not change.
43. Verify, using the ideal gas law, the assertion in Problem 38 that 1.00 mol of a gas at 0.0°C and 1.00 atm occupies a volume of 0.0224 m^3.
44. Verify that the SI units of PV (pressure times volume) are joules.
45. Incandescent lightbulbs are filled with an inert gas to lengthen the filament life. With the current off (at $T =$ 20.0°C), the gas inside a lightbulb has a pressure of 115 kPa. When the bulb is burning, the temperature rises to 70.0°C. What is the pressure at the higher temperature?
46. What fraction of the air molecules in a house must be pushed outside while the furnace raises the inside temperature from 16.0°C to 20.0°C? The pressure does not change since the house is not 100% airtight.
47. What is the mass density of air at $P = 1.0$ atm and $T =$ (a) −10°C and (b) 30°C? The average molecular mass of air is approximately 29 u.
48. A constant volume gas thermometer containing helium is immersed in boiling ammonia (−33°C) and the pressure is read once equilibrium is reached. The thermometer is then moved to a bath of boiling water (100.0°C). After the manometer was adjusted to keep the volume of helium constant, by what factor was the pressure multiplied?
49. A hydrogen balloon at Earth's surface has a volume of 5.0 m^3 on a day when the temperature is 27°C and the pressure is 1.00×10^5 N/m^2. The balloon rises and expands as the pressure drops. What would the volume of the same number of moles of hydrogen be at an altitude of 40 km where the pressure is 0.33×10^3 N/m^2 and the temperature is −13°C?
50. An ideal gas that occupies 1.2 m^3 at a pressure of 1.0×10^5 Pa and a temperature of 27°C is compressed to a volume of 0.60 m^3 and heated to a temperature of 227°C. What is the new pressure?
51. A diver rises quickly to the surface from a 5.0-m depth. If she did not exhale the gas from her lungs before rising, by what factor would her lungs expand? Assume the temperature to be constant and the pressure in the lungs to match the pressure outside the diver's body. The density of seawater is 1.03×10^3 kg/m^3.

52. In intergalactic space, there is an average of about one hydrogen atom per cm^3 and the temperature is 3 K. What is the absolute pressure?

53. A tank of compressed air of volume 1.0 m^3 is pressurized to 20.0 atm at $T = 273$ K. A valve is opened and air is released until the pressure in the tank is 15.0 atm. How many air molecules were released?

54. A mass of 0.532 kg of molecular oxygen is contained in a cylinder at a pressure of 1.0×10^5 Pa and a temperature of 0.0°C. What volume does the gas occupy?

55. A bubble rises from the bottom of a lake of depth 80.0 m, where the temperature is 4°C. The water temperature at the surface is 18°C. If the bubble's initial diameter is 1.00 mm, what is its diameter when it reaches the surface? (Ignore the surface tension of water. Assume the bubble warms as it rises to the same temperature as the water and retains a spherical shape. Assume $P_{atm} = 1.0$ atm.)

56. A bubble with a volume of 1.00 cm^3 forms at the bottom of a lake that is 20.0 m deep. The temperature at the bottom of the lake is 10.0°C. The bubble rises to the surface where the water temperature is 25.0°C. Assume that the bubble is small enough that its temperature always matches that of its surroundings. What is the volume of the bubble just before it breaks the surface of the water? Ignore surface tension.

57. A scuba diver has an air tank with a volume of 0.010 m^3. The air in the tank is initially at a pressure of 1.0×10^7 Pa. Assuming that the diver breathes 0.500 L/s of air, find how long the tank will last at depths of (a) 2.0 m and (b) 20.0 m. (Make the same assumptions as in Example 13.6.)

58. An emphysema patient is breathing pure O_2 through a face mask. The cylinder of O_2 contains 0.60 ft^3 of O_2 gas at a pressure of 2200 lb/in^2. (a) What volume would the oxygen occupy at atmospheric pressure (and the same temperature)? (b) If the patient takes in 8 L/min of O_2 at atmospheric pressure, how long will the cylinder last?

59. Consider the expansion of an ideal gas at constant pressure. The initial temperature is T_0 and the initial volume is V_0. (a) Show that $\Delta V/V_0 = \beta \Delta T$, where $\beta = 1/T_0$. (b) Compare the coefficient of volume expansion β for an ideal gas at 20°C to the values for liquids and gases listed in Table 13.3.

13.6 Kinetic Theory of the Ideal Gas

60. What is the temperature of an ideal gas whose molecules have an average translational kinetic energy of 3.20×10^{-20} J?

61. What is the total translational kinetic energy of the gas molecules of 0.420 mol of air at atmospheric pressure that occupies a volume of 1.00 L (0.00100 m^3)?

62. What is the kinetic energy per unit volume in an ideal gas at (a) $P = 1.00$ atm and (b) $P = 300.0$ atm?

63. Show that, for an ideal gas,

$$P = \tfrac{1}{3}\rho v_{rms}^2$$

where P is the pressure, ρ is the mass density, and v_{rms} is the rms speed of the gas molecules.

64. Estimate the percentage of the O_2 molecules in air at 0.0°C and 1.00 atm that are moving faster than the speed of sound in air at that temperature (see Fig. 13.13).

65. What is the total internal kinetic energy of 1.0 mol of an ideal gas at 0.0°C and 1.00 atm?

66. If 2.0 mol of nitrogen gas (N_2) are placed in a cubic box, 25 cm on each side, at 1.6 atm of pressure, what is the rms speed of the nitrogen molecules?

67. There are two identical containers of gas at the same temperature and pressure, one containing argon and the other neon. What is the ratio of the rms speed of the argon atoms to that of the neon atoms? The atomic mass of argon is twice that of neon. (WW tutorial: RMS speed)

68. A smoke particle has a mass of 1.38×10^{-17} kg and it is randomly moving about in thermal equilibrium with room temperature air at 27°C. What is the rms speed of the particle?

69. Find the rms speed in air at 0.0°C and 1.00 atm of (a) the N_2 molecules, (b) the O_2 molecules, and (c) the CO_2 molecules.

70. What are the rms speeds of helium atoms, and nitrogen, hydrogen, and oxygen molecules at 25°C?

71. If the upper atmosphere of Jupiter has a temperature of 160 K and the escape speed is 60 km/s, would an astronaut expect to find much hydrogen there?

72. A sealed cylinder contains a sample of ideal gas at a pressure of 2.0 atm. The rms speed of the molecules is v_0. If the rms speed is then reduced to 0.90 v_0, what is the pressure of the gas?

73. What is the temperature of an ideal gas whose molecules in random motion have an average translational kinetic energy of 4.60×10^{-20} J?

74. Show that the rms speed of a molecule in an ideal gas at absolute temperature T is given by

$$v_{rms} = \sqrt{\frac{3kT}{m}} \qquad (13\text{-}22)$$

where k is Boltzmann's constant and m is the mass of a molecule.

75. Show that the rms speed of a molecule in an ideal gas at absolute temperature T is given by

$$v_{rms} = \sqrt{\frac{3RT}{M}}$$

where M is the *molar mass*—the mass of the gas per mole.

13.7 Temperature and Reaction Rates

76. The reaction rate for the hydrolysis of benzoyl-l-arginine amide by trypsin at 10.0°C is 1.878 times faster than that at 5.0°C. Assuming that the reaction rate is exponential as in Eq. (13-24), what is the activation energy?

77. The reaction rate for the prepupal development of male *Drosophila* is temperature-dependent. Assuming that the reaction rate is exponential as in Eq. (13-24), the activation energy for this development is then 2.81×10^{-19} J. A *Drosophila* is originally at 10.00°C and its temperature is increasing. If the rate of development has increased 3.5%, how much has its temperature increased?

78. At high altitudes, water boils at a temperature lower than 100.0°C due to the lower air pressure. A rule of thumb states that the time to hard-boil an egg doubles for every 10.0°C drop in temperature. What activation energy does this rule imply for the chemical reactions that occur when the egg is cooked?

13.8 Diffusion

79. Estimate the mean free path of a N_2 molecule in air at (a) sea level ($P \approx 100$ kPa and $T \approx 290$ K), (b) the top of Mt. Everest (altitude = 8.8 km, $P \approx 50$ kPa, and $T \approx 230$ K), and (c) an altitude of 30 km ($P \approx 1$ kPa and $T \approx 230$ K). For simplicity, assume that air is pure nitrogen gas. The diameter of a N_2 molecule is approximately 0.3 nm.

80. About how long will it take a perfume molecule to diffuse a distance of 5.00 m in one direction in a room if the diffusion constant is 1.00×10^{-5} m^2/s? Assume that the air is perfectly still—there are no air currents.

81. Estimate the time it takes a sucrose molecule to move 5.00 mm in one direction by diffusion in water. Assume there are no currents in the water.

82. Your friend is 3.0 m away from you in a room. There are no significant air currents. She opens a bottle of perfume and you first smell it 20 s later. How long would it have taken for you to smell it if she had been 6.0 m away instead? (tutorial: diffusion)

Comprehensive Problems

83. The driver from Practice Problem 13.3 fills his 18.9-L steel gasoline can in the morning when the temperature of the can and the gasoline is 15.0°C and the pressure is 1.0 atm, but this time he remembers to replace the tightly fitting cap after filling the can. Assume that the can is completely full of gasoline (no air space) and that the cap does not leak. The temperature climbs to 30.0°C. Ignoring the expansion of the steel can, what would be the pressure of the heated gasoline? The bulk modulus for gasoline is 1.00×10^9 N/m^2.

84. An iron bridge girder ($Y = 2.0 \times 10^{11}$ N/m^2) is constrained between two rock faces whose spacing doesn't change. At 20.0°C the girder is relaxed. How large a stress develops in the iron if the sun heats the girder to 40.0°C?

85. Consider the sphere and ring of Problem 17. What must the final temperature be if both the ring and the sphere are heated to the same final temperature?

86. Agnes Pockels (1862–1935) was able to determine Avogadro's number using only a few household chemicals, in particular oleic acid, whose formula is $C_{18}H_{34}O_2$. (a) What is the molar mass of this acid? (b) The mass of one drop of oleic acid is 2.3×10^{-5} g and the volume is 2.6×10^{-5} cm^3. How many moles of oleic acid are there in one drop? (c) Now all Pockels needed was to find the number of molecules of oleic acid. Luckily, when oleic acid is spread out on water, it lines up in a layer one molecule thick. If the base of the molecule of oleic acid is a square of side d, the height of the molecule is known to be $7d$. Pockels spread out one drop of oleic acid on some water, and measured the area to be 70.0 cm^2. Using the volume and the area of oleic acid, what is d? (d) If we assume that this film is one molecule thick, how many molecules of oleic acid are there in the drop? (e) What value does this give you for Avogadro's number?

87. A certain acid has a molecular mass of 63 u. By mass, it consists of 1.6% hydrogen, 22.2% nitrogen, and 76.2% oxygen. What is the chemical formula for this acid?

88. These data are from a constant-volume gas thermometer experiment. The volume of the gas was kept constant, while the temperature was changed. The resulting pressure was measured. Plot the data on a pressure versus temperature diagram. Based on these data, estimate the value of absolute zero in Celsius.

T (°C)	P (atm)
0	1.00
20	1.07
100	1.37
−33	0.88
−196	0.28

89. Given that our body temperature is 98.6°F, (a) what is the average kinetic energy of the molecules in the air in our lungs? (b) If our temperature has increased to 100.0°F, by what percentage has the kinetic energy of the molecules increased?

90. The volume of air taken in by a warm-blooded vertebrate in the Andes mountains is 210 L/day at standard temperature and pressure (i.e., 0°C and 1 atm). If the air in the lungs is at 39°C, under a pressure of 450 mm Hg, and we assume that the vertebrate takes in an average volume of 100 cm^3 per breath at the temperature and

pressure of its lungs, how many breaths does this vertebrate take per day?

91. As a Boeing 747 gains altitude, the passenger cabin is pressurized. However, the cabin is not pressurized fully to atmospheric (1.01×10^5 Pa), as it would be at sea level, but rather pressurized to 7.62×10^4 Pa. Suppose a 747 takes off from sea level when the temperature in the airplane is 25.0°C and the pressure is 1.01×10^5 Pa. (a) If the cabin temperature remains at 25.0°C, what is the percentage change in the number of moles of air in the cabin? (b) If instead, the number of moles of air in the cabin does not change, what would the temperature be?

92. An iron cannonball of radius 0.08 m has a cavity of radius 0.05 m that is to be filled with gunpowder. If the measurements were made at a temperature of 22°C, how much extra volume of gunpowder, if any, will be required to fill 500 cannonballs when the temperature is 30°C?

93. Ten students take a test and get the following scores: 83, 62, 81, 77, 68, 92, 88, 83, 72, and 75. What are the average value, the rms value, and the most probable value, respectively, of these test scores?

94. A hand pump is being used to inflate a bicycle tire that has a gauge pressure of 40.0 psi. If the pump is a cylinder of length 18.0 in. with a cross-sectional area of 3.00 in.^{2}, how far down must the piston be pushed before air will flow into the tire?

95. An ideal gas in a constant-volume gas thermometer (Fig. 13.11) is held at a volume of 0.500 L. As the temperature of the gas is increased by 20.0°C, the mercury level on the right side of the manometer must rise by 8.00 mm in order to keep the gas volume constant. (a) What is the slope of a graph of P versus T for this gas (in mm Hg/°C)? (b) How many moles of gas are present?

96. A cylinder with an interior cross-sectional area of 70.0 cm^2 has a moveable piston of mass 5.40 kg at the top that can move up and down without friction. The cylinder contains 2.25×10^{-3} mol of an ideal gas at 23.0°C. (a) What is the volume of the gas when the piston is in equilibrium? Assume the air pressure outside the cylinder is 1.00 atm. (b) By what factor does the volume change if the gas temperature is raised to 223.0°C and the piston moves until it is again in equilibrium?

97. Estimate the average distance between air molecules at 0.0°C and 1.00 atm.

98. If you wanted to make a scale model of air at 0.0°C and 1.00 atm, using ping-pong balls (diameter, 3.75 cm) to represent the N_2 molecules (diameter, 0.30 nm), (a) how far apart on average should the ping-pong balls be at any instant? (b) How far would a ping-pong ball travel on average before colliding with another?

99. For divers going to great depths, the composition of the air in the tank must be modified. The ideal composition is to have approximately the same number of O_2 molecules per unit volume as in surface air (to avoid oxygen poisoning), and to use helium instead of nitrogen for the remainder of the gas (to avoid nitrogen narcosis, which results from nitrogen dissolving in the bloodstream). Of the molecules in dry surface air, 78% are N_2, 21% are O_2, and 1% are Ar. (a) How many O_2 molecules per m^3 are there in surface air at 20.0°C and 1.00 atm? (b) For a diver going to a depth of 100.0 m, what percentage of the gas molecules in the tank should be O_2? (Assume that the density of seawater is 1025 kg/m^3 and the temperature is 20.0°C.)

100. Show that, in two gases at the same temperature, the rms speeds are inversely proportional to the square root of the molecular masses:

$$\frac{(v_{rms})_1}{(v_{rms})_2} = \sqrt{\frac{m_2}{m_1}}$$

101. The SR-71 Blackbird reconnaissance aircraft is primarily made of titanium and typically flies at speeds above Mach 3. In flight, the length of the SR-71 increases by about 0.20 m from its takeoff length of 32.70 m. The average coefficient of linear expansion for titanium over the temperature range experienced by the SR-71 is 10.1×10^{-6} K^{-1}. What is the approximate temperature of the SR-71 while it is in flight if it started at 20°C?

102. In plants, water diffuses out through small openings known as stomatal pores. If $D = 2.4 \times 10^{-5}$ m^2/s for water vapor in air, and the length of the pores is 2.5×10^{-5} m, how long does it take for a water molecule to diffuse out through the pore?

103. The alveoli (see Section 13.8) have an average radius of 0.125 mm and are approximately spherical. If the pressure in the sacs is 1.00×10^5 Pa, and the temperature is 310 K (average body temperature), how many air molecules are in an alveolus?

104. A 10.0-L vessel contains 12 g of N_2 gas at 20°C. (a) Estimate the nearest-neighbor distance. (b) Is the gas dilute? [*Hint:* Compare the nearest-neighbor distance to the diameter of an N_2 molecule, about 0.3 nm.]

105. During hibernation, an animal's metabolism slows down, and its body temperature lowers. For example, a California ground squirrel's body temperature lowers from 40.0°C to 10.0°C during hibernation. If we assume that the air in the squirrel's lungs is 75.0% N_2 and 25.0% O_2, by how much will the rms speed of the air molecules in the lungs have decreased during hibernation?

106. A steel ring of inner diameter 7.00000 cm at 20.0°C is to be heated and placed over a brass shaft of outer diameter 7.00200 cm at 20.0°C. (a) To what

temperature must the ring be heated to fit over the shaft? The shaft remains at 20.0°C. (b) Once the ring is on the shaft and has cooled to 20.0°C, to what temperature must the ring plus shaft combination be cooled to allow the ring to slide off the shaft again?

◆107. The inner tube of a Pyrex glass mercury thermometer has a diameter of 0.120 mm. The bulb at the bottom of the thermometer contains 0.200 cm^3 of mercury. How far will the thread of mercury move for a change of 1.00°C? Remember to take into account the expansion of the glass. (tutorial: thermometer)

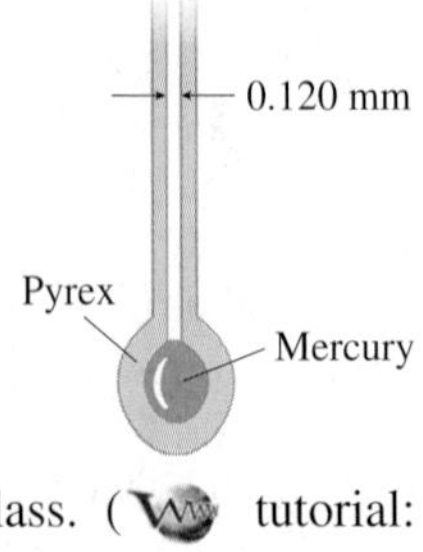

◆108. A wine barrel has a diameter at its widest point of 134.460 cm at a temperature of 20.0°C. A circular iron band, of diameter 134.448 cm, is to be placed around the barrel at the widest spot. The iron band is 5.00 cm wide and 0.500 cm thick. (a) To what temperature must the band be heated to be able to fit it over the barrel? (b) Once the band is in place and cools to 20.0°C, what will be the tension in the band?

◆109. A 12.0-cm cylindrical chamber has an 8.00-cm-diameter piston attached to one end. The piston is connected to an ideal spring as shown. Initially, the gas inside the chamber is at atmospheric pressure and 20.0°C and the spring is not compressed. When a total of 6.50×10^{-2} mol of gas is added to the chamber at 20.0°C, the spring compresses a distance of $\Delta x = 5.40$ cm. What is the spring constant of the spring?

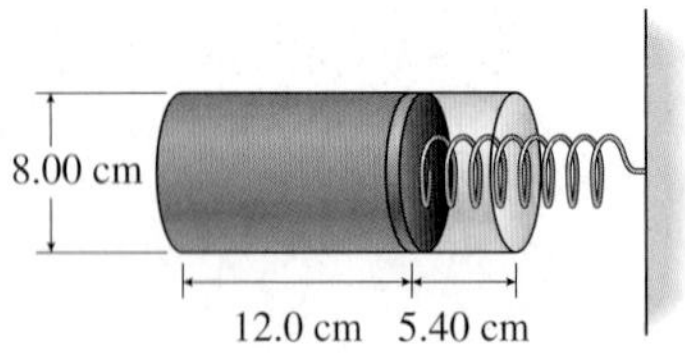

◆110. A bimetallic strip is made from metals with expansion coefficients α_1 and α_2 (with $\alpha_2 > \alpha_1$). The thickness of each layer is s. At some temperature T_0, the bimetallic strip is relaxed and straight. (a) Show that, at temperature $T_0 + \Delta T$, the radius of curvature of the strip is

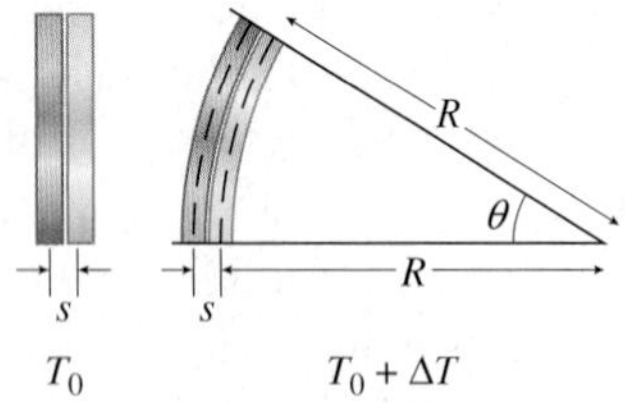

$$R \approx \frac{s}{(\alpha_2 - \alpha_1)\Delta T}$$

[*Hint:* At T_0, the lengths of the two layers are the same. At temperature $T_0 + \Delta T$, the layers form circular arcs of radii R and $R + s$, which subtend the same angle θ. Assume a small ΔT so that $\alpha \Delta T \ll 1$ (for either α).] (b) If the layers are made of iron and brass, with $s = 0.1$ mm, what is R for $\Delta T = 20.0$°C?

Answers to Practice Problems

13.1 37.0°C; 310.2 K

13.2 0.60 mm longer; 1.5 mm shorter

13.3 0.26 L

13.4 3.34×10^{28} molecules/m^3

13.5 7.9% of the air molecules; 189 kPa (27 lb/in^2)

13.6 640 kPa

13.7 $\langle K_{tr} \rangle = 6.07 \times 10^{-21}$ J (same as O_2) and $v_{rms} = 408$ m/s (lower than that of O_2 since the CO_2 molecule is more massive)

13.8 6% decrease

13.9 1.0×10^{-6} s

Answers to Checkpoints

13.3 From Table 13.2, $\alpha = 12 \times 10^{-6}$ K^{-1}. The temperature change is −50°C = −50 K and the fractional length change is $\Delta L/L_0 = \alpha \Delta T = -6.0 \times 10^{-4}$. Then $\Delta L = -6.0 \times 10^{-4} \times 150.00$ m = −0.090 m. The tower is 9.0 cm shorter.

13.4 (a) The molar mass is 44.0 g/mol, so one CO_2 molecule has a mass of 44.0 u. (b) 3.00 mol of CO_2 have a mass of (3.00 mol) × (44.0 g/mol) = 132 g.

13.5 (a) The temperatures do not have to be the same, because they could have different numbers of molecules (N) or moles (n). (b) If the temperatures are the same, then they have the same number of molecules, so they have the same number density N/V. They would have the same mass density only if their molar masses are the same.

13.6 The average translational kinetic energy of an ideal gas depends only on *absolute* temperature. The H_2 is at 20°C = 293 K, so to have twice the translational kinetic energy, the O_2 must be at 2 × 293 K = 586 K = 313°C.

CHAPTER

14

Heat

The weather forecast predicts a late spring hard freeze one night; the temperature is to fall several degrees below 0°C and the apple crop is in danger of being ruined. To protect the tender buds, farmers rush out and spray the trees with water. How does that protect the buds? (See p. 500 for the answer.)

Concepts & Skills to Review

- energy conservation (Chapter 6)
- thermal equilibrium (Section 13.1)
- absolute temperature and the ideal gas law (Section 13.5)
- kinetic theory of the ideal gas (Section 13.6)

14.1 INTERNAL ENERGY

From Section 13.6, the average translational kinetic energy $\langle K_{tr} \rangle$ of the molecules of an ideal gas is proportional to the absolute temperature of the gas:

$$\langle K_{tr} \rangle = \frac{3}{2}kT \qquad (13\text{-}20)$$

The molecules move about in random directions even though, on a macroscopic scale, the gas is neither moving nor rotating. Equation (13-20) also gives the average translational kinetic energy of the random motion of molecules in liquids, solids, and nonideal gases except at very low temperatures. This random microscopic kinetic energy is *part* of what we call the **internal energy** of the system:

Definition of Internal Energy

The internal energy of a system is the total energy of all of the molecules in the system *except* for the macroscopic kinetic energy (kinetic energy associated with macroscopic translation or rotation) and the external potential energy (energy due to external interactions).

CONNECTION:

We've used the idea of a *system* before, for instance when finding the net external force on a system or the momentum change of a system.

A **system** is whatever we define it to be: one object or a group of objects. Everything that is not part of the system is considered to be external to the system, or in other words, in the surroundings of the system.

Internal energy includes

- Translational and rotational kinetic energy of molecules *due to their individual random motions.*
- Vibrational energy—both kinetic and potential—of molecules and of atoms within molecules due to random vibrations about their equilibrium points.
- Potential energy due to interactions between the atoms and molecules of the system.
- Chemical and nuclear energy—the kinetic and potential energy associated with the binding of atoms to form molecules, the binding of electrons to nuclei to form atoms, and the binding of protons and neutrons to form nuclei.

CONNECTION:

Revisit Table 6.1 for an overview of the forms of energy discussed in this book.

Internal energy does *not* include

- The kinetic energy of the molecules due to translation, rotation, or vibration of the whole system or of a macroscopic part of the system.
- Potential energy due to interactions of the molecules of the system with something outside of the system (such as a gravitational field due to something outside of the system).

Example 14.1

Dissipation of Energy by Friction

A block of mass 10.0 kg starts at point A at a height of 2.0 m above the horizontal and slides down a frictionless incline (Fig. 14.1). It then continues sliding along the horizontal surface of a table that has friction. The block comes to rest at point C, a distance of 1.0 m along the table surface. How much has the internal energy of the system (block + table) increased?

continued on next page

Example 14.1 continued

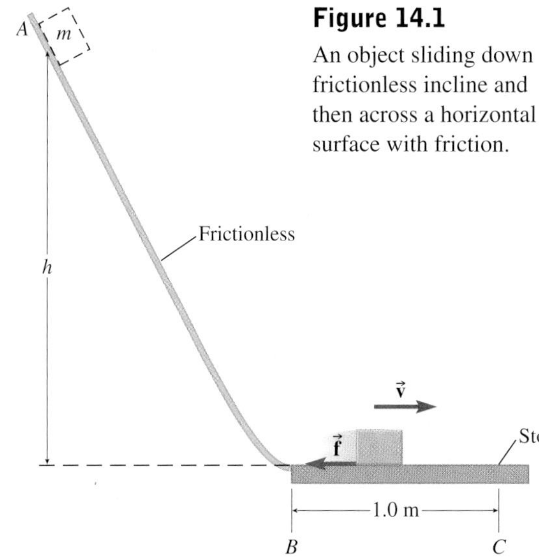

Figure 14.1

An object sliding down a frictionless incline and then across a horizontal surface with friction.

Strategy Gravitational potential energy is converted to macroscopic translational kinetic energy as the block's speed increases. Friction then converts this macroscopic kinetic energy into internal energy—some of it in the block and some in the table. Since total energy is conserved, the increase in internal energy is equal to the decrease in gravitational potential energy:

decrease in PE from A to B = increase in KE from A to B
= decrease in KE from B to C
= increase in internal energy from B to C

Solution The initial potential energy (taking $U_g = 0$ at the horizontal surface) is

$$U_g = mgh = 10.0 \text{ kg} \times 9.8 \text{ m/s}^2 \times 2.0 \text{ m} = 200 \text{ J}$$

The final potential energy is zero. The initial and final translational kinetic energies of the block are both zero. Neglecting the small transfer of energy to the air, the increase in the internal energy of the block and table is 200 J.

Discussion We do not know how much of this internal energy increase appears in the object and how much in the table; we can only find the total. We call friction a *nonconservative* force, but that only means that *macroscopic mechanical* energy is not conserved; total energy is always conserved. Friction merely converts some macroscopic mechanical energy into internal energy of the block and the table. This internal energy increase manifests itself as a slight temperature increase. We often say that mechanical energy is *dissipated* by friction or other nonconservative forces; in other words, energy in an ordered form (translational motion of the block) has been changed into disordered energy (random motion of molecules within the block and table).

Practice Problem 14.1 On the Rebound

If a rubber ball of mass 1.0 kg is dropped from a height of 2.0 m and rebounds on the first bounce to 0.75 of the height from which it was dropped, how much energy is dissipated during the collision with the floor?

A change in the internal energy of a system does not always cause a temperature change. As we explore further in Section 14.5, the internal energy of a system can change while the temperature of the system remains constant—for instance, when ice melts.

Conceptual Example 14.2

Internal Energy of a Bowling Ball

A bowling ball at rest has a temperature of 18°C. The ball is then rolled down a bowling alley. Ignoring the dissipation of energy by friction and drag forces, is the internal energy of the ball higher, lower, or the same as when the ball was at rest? Is the temperature of the ball higher, lower, or the same as when the ball was at rest?

Strategy, Solution, and Discussion The only change is that the ball is now rolling—the ball has macroscopic translational and rotational kinetic energy. However, the definition of *internal energy* does *not* include the kinetic energy of the molecules due to translation, rotation, or vibration *of the system as a whole.* Therefore, the *internal* energy of the ball is the same. Temperature is associated with the average translational kinetic energy due to the *individual random* motions of molecules; the temperature is still 18°C.

Conceptual Practice Problem 14.2
Total Translational KE

Is the *total* translational kinetic energy of the molecules in the ball higher, lower, or the same as when the ball was at rest?

14.2 HEAT

We defined heat in Section 13.1:

Definition of Heat

Heat is energy in transit between two objects or systems due to a temperature difference between them.

Many eighteenth-century scientists thought that heat was a fluid, which they called "caloric." The flow of heat into an object was thought to cause the object to expand in volume in order to accommodate the additional fluid; why no mass increase occurred was a mystery. Now we know that heat is not a substance but is a flow of energy. One experiment that led to this conclusion was carried out by Count Rumford (Benjamin Thompson, 1753–1814). While supervising the boring of cannon barrels, he noted that the drill doing the boring became quite hot. At the time it was thought that the grinding up of the cannon metal into little pieces caused caloric to be released because the tiny bits of metal could not hold as much caloric as the large piece from which they came. But Rumford noticed that the drill got hot even when it became so dull that metal was no longer being bored out of the cannon and that he could create a limitless amount of what we now call internal energy. He decided that "heat" must be a form of microscopic motion instead of a material substance.

It was not until later experiments were done by James Prescott Joule (1818–1889) that Rumford's ideas were finally accepted. In his most famous experiment (Fig. 14.2), Joule showed that a temperature increase can be caused by mechanical means. In a series of such experiments, Joule determined the "mechanical equivalent of heat," or the amount of mechanical work required to produce the same effect on a system as a given amount of heat. In those days heat was measured in calories, where one calorie was defined as the heat required to change the temperature of 1 g of water by 1°C (specifically from 14.5 to 15.5°C). Joule's experimental results were within 1% of the currently accepted value, which is

$$1 \text{ cal} = 4.186 \text{ J} \qquad (14\text{-}1)$$

Equation (14-1) is now the *definition* of the calorie. The Calorie (with an uppercase letter C) used by dietitians and nutritionists is actually a kilocalorie:

$$1 \text{ Calorie} = 1 \text{ kcal} = 10^3 \text{ cal} = 4186 \text{ J}$$

Although the calorie is still used, the SI unit for internal energy and for heat is the same as that used for all forms of energy and all forms of energy transfer: the joule.

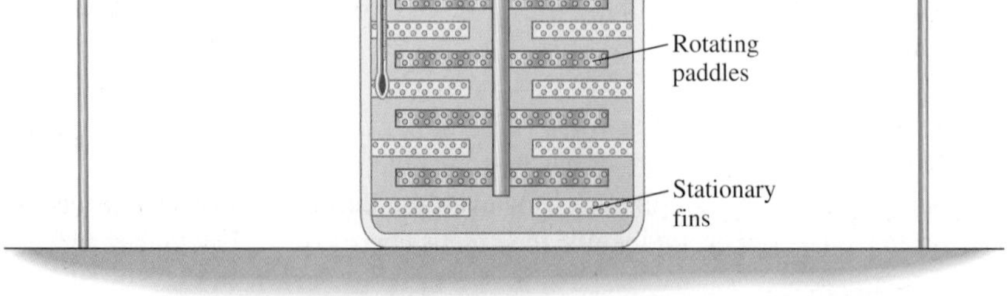

Figure 14.2 Joule's experiment. As the two masses fall, they cause paddles to rotate within an insulated container (not to scale) filled with water. The paddles agitate the water and cause its temperature to rise. By measuring the distance through which the masses fell and the temperature change of the known quantity of water, Joule determined the mechanical work done and the internal energy increase of the water.

Heat and Work Heat and work are similar in that both describe a particular kind of energy *transfer*. Work is an energy transfer due to a force acting through a displacement. Heat is a microscopic form of energy transfer involving large numbers of particles; the exchange of energy occurs due to the individual interactions of the particles. No macroscopic displacement occurs when heat flows and no macroscopic force is exerted by one object on the other.

CONNECTION:

Heat, like work, is a kind of energy transfer.

It does not make sense to say that a system *has* 15 kJ of heat, just as it does not make sense to say that a system *has* 15 kJ of work. Similarly, we cannot say that the heat of a system has changed (nor that the work of a system has changed). A system can possess *energy* in various forms (including internal energy), but it cannot possess heat or work. Heat and work are two ways of *transferring* energy from one system to another. Joule's experiments showed that a quantity of work done on a system or the same quantity of heat flowing into the system causes the same increase in the system's internal energy. If the internal energy increase comes from mechanical work, as from Joule's paddle wheel, no heat flow occurs.

CHECKPOINT 14.2

Take a rubber band and stretch it rapidly several times. Then hold it against your wrist or your lip. In everyday language, you might say the rubber band "heats up." Is the temperature increase caused by heat flow into the rubber band? If not, what has happened?

Direction of Heat Flow *Heat flows spontaneously from a system at higher temperature to one at lower temperature.* Temperature is associated with the microscopic translational kinetic energy of the molecules; thus, the flow of heat tends to equalize the average microscopic translational kinetic energy of the molecules. When two systems are in thermal contact and no net heat flow occurs, the systems are in thermal equilibrium and have the same temperature.

Example 14.3

A Joule Experiment

In an experiment similar to that done by Joule, an object of mass 12.0 kg descends a distance of 1.25 m at constant speed while causing the rotation of a paddle wheel in an insulated container of water. If the descent is repeated 20.0 times, what is the internal energy increase of the water in joules?

Strategy Each time the object descends, it converts gravitational potential energy into kinetic energy of the paddle wheel, which in turn agitates the water and converts kinetic energy into internal energy.

Solution The change in gravitational potential energy during 20.0 downward trips is

$$\Delta U_g = mg\,\Delta h$$
$$= 12.0\text{ kg} \times 9.80\text{ N/kg} \times 20.0\text{ descents} \times 1.25\text{ m/descent}$$
$$= 2.94\text{ kJ}$$

If all of this energy goes into the water, the internal energy increase of the water is 2.94 kJ.

Discussion To perform an experiment like Joule's, we can vary the amount of energy delivered to the water. One way is to change the number of times the object is allowed to descend. Other possibilities include varying the mass of the descending object or raising the apparatus so that the object can descend a greater distance. All of these variations allow a change in the amount of gravitational potential energy converted into internal energy without requiring any changes in the complicated mechanism involving the paddle wheel.

Practice Problem 14.3 Temperature Change of the Water

If the water temperature in the insulated container is found to have increased 2.0°C after 20.0 descents of the falling object, what mass of water is in the container? Assume all of the internal energy increase appears in the water (ignore any internal energy change of the paddle wheel itself). [*Hint:* Recall that the calorie was defined as the heat required to change the temperature of 1 g of water by 1°C.]

The Cause of Thermal Expansion

CONNECTION:

Section 13.3 discussed thermal expansion. Now we discuss *why* the expansion occurs.

If not to accommodate additional "caloric," then why do objects generally expand when their temperatures increase? (See Section 13.3.) An object expands when the *average* distance between the atoms (or molecules) increases. The atoms are not at rest; even in a solid, where each atom has a fixed equilibrium position, they *vibrate* to and fro about their equilibrium positions. The energy of vibration is part of the internal energy of the object. When heat flows into the object, raising its temperature, the internal energy increases. Some of the increase goes into vibration, so the average vibrational energy of an atom increases with increasing temperature.

The average distance between atoms usually increases with increasing vibrational energy because the forces between atoms are highly asymmetrical. Two atoms separated by *less* than their equilibrium distance repel one another *strongly*, while two atoms separated by *more* than their equilibrium distance attract one another much less strongly. Therefore, as vibrational energy increases, the maximum distance between the atoms increases more than the minimum distance decreases; the *average* distance between the atoms increases.

The coefficient of expansion varies from material to material because the strength of the interatomic (or intermolecular) bonds varies. As a general rule, the stronger the atomic bond, the smaller the coefficient of expansion. Liquids have much greater coefficients of volume expansion than do solids because the molecules are more loosely bound in a liquid than in a solid.

14.3 HEAT CAPACITY AND SPECIFIC HEAT

Heat Capacity

In Chapter 15, we consider cases where both work and heat change the internal energy of a system. We will see that work done on a system can change the system's temperature or cause a change of phase. For now, we assume that the work done is zero and consider changes due to heat flow.

Q is positive for heat flow *into* the system and negative for heat flow *out of* the system.

Suppose we have a system on which no mechanical work is done, but we allow heat to flow into the system by placing it in thermal contact with another system at higher temperature. As the internal energy of the system increases, its temperature increases (provided that no part of the system undergoes a change of phase, such as from solid to liquid). If heat flows *out* of the system rather than into the system, the internal energy of the system decreases. We account for that possibility by making Q negative if heat flows out of the system; since Q is defined as the heat *into* the system, a negative heat represents heat flow *out of* the system.

For a large number of substances, under normal conditions, the temperature change ΔT is approximately proportional to the heat Q. The constant of proportionality is called the **heat capacity** (symbol C):

$$C = \frac{Q}{\Delta T} \tag{14-2}$$

The heat capacity depends both on the substance and on how much of it is present: 1 cal of heat into 1 g of water causes a temperature increase of 1°C, but 1 cal of heat into 2 g of water causes a temperature increase of 0.5°C. The SI unit of heat capacity is J/K. We can write J/K or J/°C interchangeably since only temperature *changes* are involved; a temperature change of 1 K is equivalent to a temperature change of 1°C.

The term *heat capacity* is unfortunate since it has nothing to do with a capacity to *hold* heat, or a limited ability to absorb heat, as the name seems to imply. Instead, it relates the heat into a system to the temperature increase. Think of heat capacity as a measure of how much heat must flow into or out of the system to produce a given temperature change.

Specific Heat

The heat capacity of the water in a drinking glass is much smaller than the heat capacity of the water in Lake Superior. Since the heat capacity of a system is proportional to the

Table 14.1 Specific Heats of Common Substances at 1 atm and 20°C

Substance	Specific Heat $\left(\frac{\text{kJ}}{\text{kg·K}}\right)$	Substance	Specific Heat $\left(\frac{\text{kJ}}{\text{kg·K}}\right)$
Gold	0.128	Pyrex glass	0.75
Lead	0.13	Granite	0.80
Mercury	0.139	Marble	0.86
Silver	0.235	Aluminum	0.900
Brass	0.384	Air (50°C)	1.05
Copper	0.385	Wood (average)	1.68
Steel	0.45	Steam (110°C)	2.01
Iron	0.44	Ice (0°C)	2.1
Flint glass	0.50	Alcohol (ethyl)	2.4
Crown glass	0.67	Human tissue (average)	3.5
Vycor	0.74	Water (15°C)	4.186

mass of the system, the **specific heat capacity** (symbol c) of a substance is defined as the heat capacity per unit mass:

$$c = \frac{C}{m} = \frac{Q}{m\,\Delta T} \tag{14-3}$$

Specific heat capacity is often abbreviated to *specific heat.* The SI units of specific heat are J/(kg·K). In SI units, the specific heat is the number of joules of heat required to produce a 1 K temperature change in 1 kg of the substance. Again, since only temperature changes are involved, we can equivalently write J/(kg·°C).

Specific heat: heat capacity per unit mass

Table 14.1 lists specific heats for some common substances at 1 atm and 20°C (unless otherwise specified). For the range of temperatures in our examples and problems, assume these specific heat values to be valid. Note that water has a relatively large specific heat compared with most other substances. The relatively large specific heat of water causes the oceans to warm slowly in the spring and to cool slowly as winter approaches, moderating the temperature along the coast.

Equations (14-2) through (14-4) apply when no phase change occurs. The value of the specific heat is different for different phases of the same substance. That's why Table 14.1 lists different values for ice, liquid water, and steam.

Rearrangement of Eq. (14-3) leads to an expression for the heat required to produce a known temperature change in a system:

$$Q = mc\,\Delta T \tag{14-4}$$

Note that in Eqs. (14-3) and (14-4), the sign convention for Q is consistent: a temperature increase ($\Delta T > 0$) is caused by heat flowing *into* the system ($Q > 0$), while a temperature decrease ($\Delta T < 0$) is caused by heat flowing *out* of the system ($Q < 0$).

Example 14.4

Heating Water in a Saucepan

A saucepan containing 5.00 kg of water initially at 20.0°C is heated over a gas burner for 10.0 min. The final temperature of the water is 30.0°C. (a) What is the internal energy increase of the water? (b) What is the expected final temperature if the water were heated for an additional 5.0 min? (c) Is it possible to estimate the flow of heat from the burner during the first 10.0 min?

Strategy We are interested in the internal energy and the temperature of the *water*, so we define a system that consists

continued on next page

Example 14.4 continued

of the water in the saucepan. Although the pan is also heated, it is not part of this system. The pan, the burner, and the room are all outside the system.

Since no work is done on the water, the internal energy increase is equal to the heat flowing into the water. The heat can be found from the mass of the water, the specific heat of water, and the temperature change. As long as the burner delivers heat at a constant rate, we can find the additional heat delivered in the additional time. Since the temperature change is proportional to the heat delivered, the temperature changes at a constant rate (a constant number of °C per minute). So, in half the time, half as much energy is delivered and the temperature change is half as much.

Solution (a) First find the temperature change:

$$\Delta T = T_f - T_i = 30.0°C - 20.0°C = 10.0 \text{ K}$$

(A change of 10.0°C is equivalent to a change of 10.0 K.) The increase in the internal energy of the water is

$$\Delta U = Q = mc\,\Delta T$$

$$= 5.00 \text{ kg} \times 4.186 \text{ kJ/(kg·K)} \times 10.0 \text{ K} = 209 \text{ kJ}$$

(b) We assume that the heat delivered is proportional to the elapsed time. The temperature change is proportional to the energy delivered, so if the temperature changes 10.0°C in 10.0 min, it changes an additional 5.0°C in an additional 5.0 min. The final temperature is

$$T = 20.0°C + 15.0°C = 35.0°C$$

(c) Not all of the heat flows into the water. Heat also flows from the burner into the saucepan and into the room. All we can say is that *more than* 209 kJ of heat flows from the burner during the 10.0 min.

Discussion As a check, the heat capacity of the water is 5.00 kg × 4.186 kJ/(kg·K) = 20.9 kJ/K; 20.9 kJ of heat must flow for each 1.0 K change in temperature. Since the temperature change is 10.0 K, the heat required is

$$20.9 \text{ kJ/K} \times 10.0 \text{ K} = 209 \text{ kJ}$$

Practice Problem 14.4 Price of a Bubble Bath

If the cost of electricity is \$0.080 per kW·h, what does it cost to heat 160 L of water for a bubble bath from 10.0°C (the temperature of the well water entering the house) to 70.0°C? [*Hint:* 1 L of water has a mass of 1 kg. 1 kW·h = 1000 J/s × 3600 s.]

Heat Flow with More Than Two Objects Suppose some water is heated in a large iron pot by dropping a hot piece of copper into the pot. We can define the system to be the water, the copper, and the iron pot; the environment is the room containing the system. Heat continues to flow among the three substances (iron pot, water, copper) until thermal equilibrium is reached—that is, until all three substances are at the same temperature. If losses to the environment are negligible, all the heat that flows out of the copper flows into either the iron or the water:

CONNECTION:

Here we apply the principle of energy conservation.

$$Q_{Cu} + Q_{Fe} + Q_{H_2O} = 0$$

In this case, Q_{Cu} is negative since heat flows *out* of the copper; Q_{Fe} and Q_{H_2O} are positive since heat flows into both the iron and the water.

Calorimetry

A calorimeter is an insulated container that enables the careful measurement of heat (Fig. 14.3). The calorimeter is designed to minimize the heat flow to or from the surroundings. A typical constant volume calorimeter, called a *bomb calorimeter*, consists of a hollow aluminum cylinder of known mass containing a known quantity of water; the cylinder is inside a larger aluminum cylinder with insulated walls. An evacuated space separates the two cylinders. An insulated lid fits over the opening of the cylinders; often there are two small holes in the lid, one for a thermometer to be inserted into the contents of the inner cylinder and one for a stirring device to help the contents reach equilibrium faster.

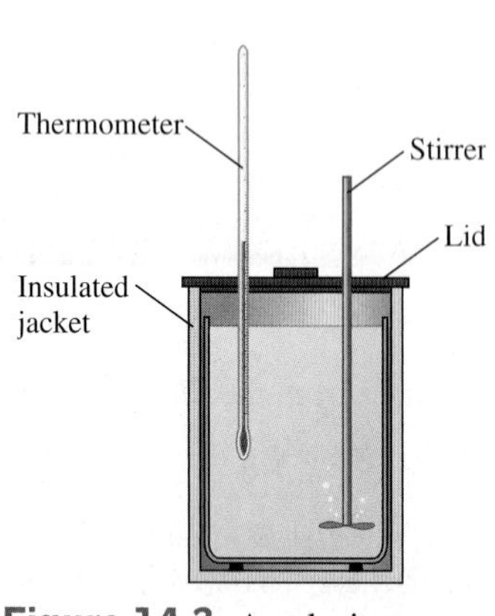

Figure 14.3 A calorimeter.

Suppose an object at one temperature is placed in a calorimeter with the water and aluminum cylinder at another temperature. By conservation of energy, all the heat that flows out of one substance ($Q < 0$) flows into some other substance ($Q > 0$). If no heat flows to or from the environment, the total heat into the object, water, and aluminum must equal zero:

$$Q_o + Q_w + Q_a = 0$$

Example 14.5 illustrates the use of a calorimeter to measure the specific heat of an unknown substance. The measured specific heat can be compared with a table of known values to help identify the substance.

Example 14.5

Specific Heat of an Unknown Metal

A sample of unknown metal of mass 0.550 kg is heated in a pan of hot water until it is in equilibrium with the water at a temperature of 75.0°C. The metal is then carefully removed from the heat bath and placed into the inner cylinder of an aluminum calorimeter that contains 0.500 kg of water at 15.5°C. The mass of the inner cylinder is 0.100 kg. When the contents of the calorimeter reach equilibrium, the temperature inside is 18.8°C. Find the specific heat of the metal sample and determine whether it could be any of the metals listed in Table 14.1.

Strategy Heat flows from the sample to the water and to the aluminum until thermal equilibrium is reached, at which time all three have the same temperature. We use subscripts to keep track of the three heat flows and three temperature changes. Let T_f be the final temperature of all three. Initially, the water and aluminum are both at 15.5°C while the sample is at 75.0°C. When thermal equilibrium is reached, all three are at 18.8°C. We assume negligible heat flow to the environment—in other words, that no heat flows into or out of the system of aluminum + water + sample.

Solution Heat flows out of the sample ($Q_s < 0$) and into the water and aluminum cylinder ($Q_w > 0$ and $Q_a > 0$). Assuming no heat into or out of the surroundings,

$$Q_s + Q_w + Q_a = 0$$

For each substance, the heat is related to the temperature change. Substituting $Q = mc\,\Delta T$ for each gives

$$m_s c_s\,\Delta T_s + m_w c_w\,\Delta T_w + m_a c_a\,\Delta T_a = 0 \qquad (1)$$

A table helps organize the given information:

	Sample	H_2O	Al
Mass (m)	0.550 kg	0.500 kg	0.100 kg
Specific Heat (c)	c_s (unknown)	4.186 kJ/(kg·°C)	0.900 kJ/(kg·°C)
Heat Capacity (mc)	0.550 kg × c_s	2.093 kJ/°C	0.0900 kJ/°C
T_i	75.0°C	15.5°C	15.5°C
T_f	18.8°C	18.8°C	18.8°C
ΔT	−56.2°C	3.3°C	3.3°C

Substituting known values into Eq. (1) yields

$$0.550\text{ kg} \times c_s \times (-56.2°\text{C}) + (2.093\text{ kJ/°C} + 0.0900\text{ kJ/°C}) \times 3.3°\text{C} = 0$$

Now we solve for c_s.

$$0.550\text{ kg} \times c_s \times 56.2°\text{C} = 7.204\text{ kJ}$$

$$c_s = \frac{7.204\text{ kJ}}{0.550\text{ kg} \times 56.2°\text{C}} = 0.233\ \frac{\text{kJ}}{\text{kg·°C}}$$

By comparing this result with the values in Table 14.1, it appears that the unknown sample could be silver.

Discussion As a quick check, the heat capacity of the sample is approximately $\frac{1}{17}$ that of the water since its temperature change is 56.2°C/3.3°C ≈ 17 times as much—ignoring the small heat capacity of the aluminum. Since the masses of the water and sample are about equal, the specific heat of the sample is roughly $\frac{1}{17}$ that of the water:

$$\frac{1}{17} \times 4.186\ \frac{\text{kJ}}{\text{kg·°C}} = 0.25\ \frac{\text{kJ}}{\text{kg·°C}}$$

That is quite close to our answer.

Practice Problem 14.5 Final Temperature

If 0.25 kg of water at 90.0°C is added to 0.35 kg of water at 20.0°C in an aluminum calorimeter with an inner cylinder of mass 0.100 kg, find the final temperature of the mixture.

14.4 SPECIFIC HEAT OF IDEAL GASES

Since the average translational kinetic energy of a molecule in an ideal gas is

$$\langle K_{tr} \rangle = \tfrac{3}{2}kT \qquad (13\text{-}20)$$

the *total* translational kinetic energy of a gas containing N molecules (n moles) is

$$K_{tr} = \tfrac{3}{2}NkT = \tfrac{3}{2}nRT$$

Suppose we allow heat to flow into a *monatomic* ideal gas—one where the gas molecules consist of single atoms—while keeping the volume of the gas constant. Since the volume is constant, no work is done on the gas, so the change in the internal energy is equal to the heat. If we think of the atoms as point particles, the only way for the internal energy to change when heat flows into the gas is for the translational kinetic energy of the atoms to change. The rest of the internal energy is "locked up" in the atoms and does not change unless something else happens, such as a phase transition or a chemical reaction—neither of which can happen in an ideal gas. Then

$$Q = \Delta K_{tr} = \tfrac{3}{2}nR\,\Delta T \qquad (14\text{-}5)$$

From Eq. (14-5), we can find the specific heat of the monatomic ideal gas. However, with gases it is more convenient to define the **molar specific heat** at constant volume (C_V) as

$$C_V = \frac{Q}{n\,\Delta T} \qquad (14\text{-}6)$$

The subscript "V" is a reminder that the volume of the gas is held constant during the heat flow. The molar specific heat is the heat capacity *per mole* rather than *per unit mass*. In one case, we measure the amount of substance by the number of moles; in the other case, by the mass.

From Eqs. (14-5) and (14-6), we can find the molar specific heat of a monatomic ideal gas:

$$Q = \tfrac{3}{2}nR\,\Delta T = nC_V\,\Delta T$$

$$C_V = \tfrac{3}{2}R = 12.5\,\frac{\text{J/K}}{\text{mol}} \qquad (\textit{monatomic}\ \text{ideal gas}) \qquad (14\text{-}7)$$

A glance at Table 14.2 shows that this calculation is remarkably accurate at room temperature for monatomic gases.

Diatomic gases have larger molar specific heats than monatomic gases. Why? We cannot model the diatomic molecule as a point mass; the two atoms in the molecule are separated, giving the molecule a much larger rotational inertia about two perpendicular axes (Fig. 14.4). The molar specific heat is *larger* because not all of the internal energy increase goes into the translational kinetic energy of the molecules; some goes into rotational kinetic energy.

Table 14.2 Molar Specific Heats at Constant Volume of Gases at 25°C

	Gas	C_V $\left(\frac{\text{J/K}}{\text{mol}}\right)$
Monatomic	He	12.5
	Ne	12.7
	Ar	12.5
Diatomic	H_2	20.4
	N_2	20.8
	O_2	21.0
Polyatomic	CO_2	28.2
	N_2O	28.4

CONNECTION:

Specific heat and molar specific heat can be thought of as the same quantity—heat capacity per amount of substance—expressed in different units.

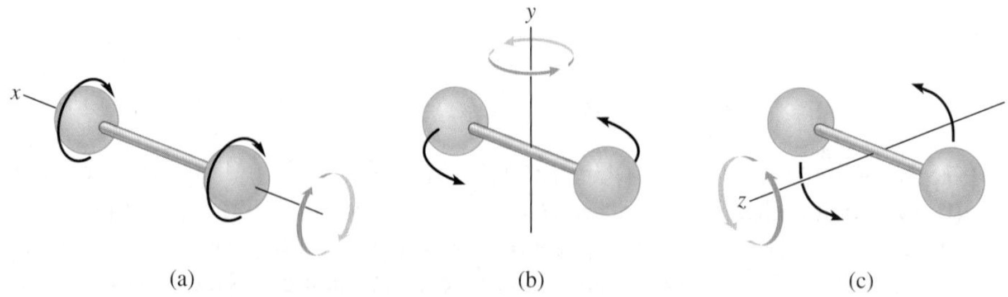

Figure 14.4 Rotation of a model diatomic molecule about three perpendicular axes. The rotational inertia about the x-axis (a) is negligible, so we can ignore rotation about this axis. The rotational inertias about the y- and z-axes (b) and (c) are much larger than for a single atom of the same mass because of the larger distance between the atoms and the axis of rotation.

It turns out that the molar specific heat of a diatomic ideal gas at room temperature is approximately

$$C_V = \tfrac{5}{2}R = 20.8\ \frac{\text{J/K}}{\text{mol}} \quad (\textit{diatomic}\ \text{ideal gas at room temperature}) \qquad (14\text{-}8)$$

Why $\frac{5}{2}R$ instead of $\frac{3}{2}R$? The diatomic molecule has rotational kinetic energy about two perpendicular axes (Fig. 14.4b and c) in addition to translational kinetic energy associated with motion in three independent directions. Thus, the diatomic molecule has five ways to "store" internal energy while the monatomic molecule has only three. The theorem of **equipartition of energy**—which we cannot prove here—says that internal energy is distributed equally among all the possible ways in which it can be stored (as long as the temperature is sufficiently high). Each independent form of energy has an average of $\frac{1}{2}kT$ of energy per molecule and contributes $\frac{1}{2}R$ to the molar specific heat at constant volume.

Example 14.6

Heating Some Xenon Gas

A cylinder contains 250 L of xenon gas (Xe) at 20.0°C and a pressure of 5.0 atm. How much heat is required to raise the temperature of this gas to 50.0°C, holding the volume constant? Treat the xenon as an ideal gas.

Strategy The molar heat capacity is the heat required per degree per mole. The number of moles of xenon (n) can be found from the ideal gas law, $PV = nRT$. Xenon is a monatomic gas, so we expect $C_V = \frac{3}{2}R$.

Solution First we convert the known quantities into SI units.

$$P = 5.0\text{ atm} = 5 \times 1.01 \times 10^5\text{ Pa} = 5.05 \times 10^5\text{ Pa}$$

$$V = 250\text{ L} = 250 \times 10^{-3}\text{ m}^3$$

$$T = 20.0°\text{C} = 293.15\text{ K}$$

From the ideal gas law, we find the number of moles,

$$n = \frac{PV}{RT} = \frac{5.05 \times 10^5\text{ Pa} \times 250 \times 10^{-3}\text{ m}^3}{8.31\text{ J/(mol·K)} \times 293.15\text{ K}} = 51.8\text{ mol}$$

We should check the units. Since Pa = N/m^2,

$$\frac{\text{Pa} \times \text{m}^3}{\text{J/(mol·K)} \times \text{K}} = \frac{\text{N/m}^2 \times \text{m}^3}{\text{J/mol}} = \frac{\text{N·m}}{\text{J}} \times \text{mol} = \text{mol}$$

For a monatomic gas at constant volume, the energy all goes into increasing the translational kinetic energy of the gas molecules. The molar specific heat is defined by $Q = nC_V\Delta T$, where, for a monatomic gas, $C_V = \frac{3}{2}R$. Then,

$$Q = \tfrac{3}{2}nR\,\Delta T$$

where

$$\Delta T = 50.0°\text{C} - 20.0°\text{C} = 30.0°\text{C}$$

Substituting,

$$Q = \tfrac{3}{2} \times 51.8\text{ mol} \times 8.31\text{ J/(mol·°C)} \times 30.0°\text{C} = 19\text{ kJ}$$

Discussion Constant volume implies that all the heat is used to increase the internal energy of the gas; if the gas were to expand it could transfer energy by doing work. When we find the number of moles from the ideal gas law, we must remember to convert the Celsius temperature to kelvins. Only when an equation involves a *change* in temperature can we use kelvin or Celsius temperatures interchangeably.

Practice Problem 14.6 Heating Some Helium Gas

A storage cylinder of 330 L of helium gas is at 21°C and is subjected to a pressure of 10.0 atm. How much energy must be added to raise the temperature of the helium in this container to 75°C?

You may wonder why we can ignore rotation for the monatomic molecule—which in reality is not a point particle—or why we can ignore rotation about one axis for the diatomic molecule. The answer comes from quantum mechanics. Energy cannot be added to a molecule in arbitrarily small amounts; energy can only be added in discrete amounts or "steps." At room temperature, there is not enough internal energy to excite the rotational modes with small rotational inertias, so they do not participate in the specific heat. We also ignored the possibility of vibration for the diatomic molecule. That is fine at room temperature, but at higher temperatures vibration becomes significant, adding two more energy modes (one kinetic and one potential). Thus, as temperature increases, the molar specific heat of a diatomic gas increases, approaching $\frac{7}{2}R$ at high temperatures.

Table 14.3 Heat to Turn 1 kg of Ice at –25°C to Steam at 125°C

Phase Transition or Temperature Change	Q (kJ)
Ice: –25°C to 0°C	52.3
Melting: ice at 0°C to water at 0°C	333.7
Water: 0°C to 100°C	419
Boiling: water at 100°C to steam at 100°C	2256
Steam: 100°C to 125°C	50

14.5 PHASE TRANSITIONS

If heat continually flows into the water in a pot, the water eventually begins to boil; liquid water becomes steam. If heat flows into ice cubes, they eventually melt and turn into liquid water. A **phase transition** occurs whenever a material is changed from one phase, such as the solid phase, to another, such as the liquid phase.

When some ice cubes at 0°C are placed into a glass in a room at 20°C, the ice gradually melts. A thermometer in the water that forms as the ice melts reads 0°C until all the ice is melted. At atmospheric pressure, ice and water can only coexist in equilibrium at 0°C. Once all the ice is melted, the water gradually warms up to room temperature. Similarly, water boiling on a stove remains at 100°C until all the water has boiled away. Suppose we change 1.0 kg of ice at –25°C into steam at 125°C. A graph of the temperature versus heat is shown in Fig. 14.5. During the two phase transitions, *heat flow continues, and the internal energy changes, but the temperature of the mixture of two phases does not change.* Table 14.3 shows the heat during each step of the process.

During a phase transition, the temperature of the mixture of two phases does not change.

Latent Heat The heat required *per unit mass* of substance to produce a phase change is called the **latent heat** (L). The word "latent" is related to the lack of temperature change during a phase transition.

Definition of latent heat:

$$|Q| = mL \tag{14-9}$$

The *sign* of Q in Eq. (14-9) depends on the direction of the phase transition. For melting or boiling, $Q > 0$ (heat flows *into* the system). For freezing or condensation, $Q < 0$ (heat flows *out of* the system).

The heat per unit mass for the solid-liquid phase transition (in either direction) is called the **latent heat of fusion** (L_f). From Table 14.3, it takes 333.7 kJ to change 1 kg of ice to water at 0°C, so for water $L_f = 333.7$ kJ/kg. For the liquid-gas phase transition (in either direction), the heat per unit mass is called the **latent heat of vaporization** (L_v). From Table 14.3, to change 1 kg of water to steam at 100°C takes 2256 kJ, so for water $L_v = 2256$ kJ/kg. Table 14.4 lists latent heats of fusion and vaporization for various materials.

Heat flowing into a substance can cause melting (solid to liquid) or boiling (liquid to gas). Heat flowing out of a substance can cause freezing (liquid to solid) or condensation (gas to liquid). If 2256 kJ must be supplied to turn 1 kg of water into steam, then 2256 kJ of heat is *released* from 1 kg of steam when it condenses to form water.

CHECKPOINT 14.5

Why is a burn caused by 100°C steam often much more severe than a burn caused by 100°C water?

How does spraying with water protect the buds?

The large latent heat of fusion of water is partly why spraying fruit trees with water can protect the buds from freezing. Before the buds can freeze, first the water must be cooled to 0°C and then it must freeze. In the process of freezing, the water gives up a large amount of heat and keeps the temperature of the buds from going below 0°C. Even

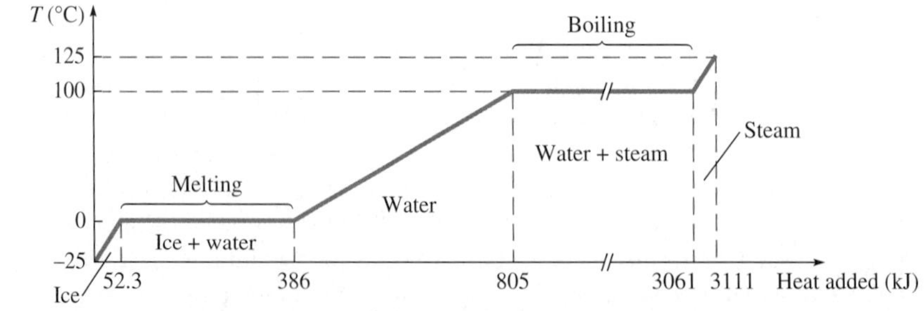

Figure 14.5 Temperature versus heat for 1 kg of ice that starts at a temperature below 0°C. (Horizontal axis not to scale.) During the two phase transitions—melting and boiling—the temperature does not change.

Table 14.4 Latent Heats of Some Common Substances

Substance	Melting Point (°C)	Heat of Fusion (kJ/kg)	Boiling Point (°C)	Heat of Vaporization (kJ/kg)
Alcohol (ethyl)	−114	104	78	854
Aluminum	660	397	2450	11 400
Copper	1083	205	2340	5070
Gold	1063	66.6	2660	1580
Lead	327	22.9	1620	871
Silver	960.8	88.3	1950	2340
Water	0.0	333.7	100	2256

if the water freezes, then the layer of ice over the buds acts like insulation since ice is not a particularly good conductor of heat.

Microscopic View of a Phase Change To understand what is happening during a phase change, we must consider the substance on the molecular level. When a substance is in solid form, bonds between the atoms or molecules hold them near fixed equilibrium positions. Energy must be supplied to break the bonds and change the solid into a liquid. When the substance is changed from liquid to gas, energy is used to separate the molecules from the loose bonds holding them together and to move the molecules apart. Temperature does not change during these phase transitions because the *kinetic energy* of the molecules is not changing. Instead, the *potential energy* of the molecules changes as work is done against the forces holding them together.

Note that energy must be *supplied* to break a bond. Forming a bond *releases* energy.

Example 14.7

Making Silver Charms

A jewelry designer plans to make some specially ordered silver charms for a commemorative bracelet. If the melting point of silver is 960.8°C, how much heat must the jeweler add to 0.500 kg of silver at 20.0°C to be able to pour silver into her charm molds?

Strategy The solid silver first needs to be heated to its melting point; then more heat has to be added to melt the silver.

Solution The total heat flow into the silver is the sum of the heat to raise the temperature of the solid and the heat that causes the phase transition:

$$Q = mc\,\Delta T + mL_\mathrm{f}$$

The temperature change of the solid is

$$\Delta T = 960.8°\mathrm{C} - 20.0°\mathrm{C} = 940.8°\mathrm{C}$$

We look up the specific heat of solid silver and the latent heat of fusion of silver. Substituting numerical values into the equation for Q yields

$$Q = 0.500\ \mathrm{kg} \times 0.235\ \mathrm{kJ/(kg{\cdot}°C)} \times 940.8°\mathrm{C} + 0.500\ \mathrm{kg} \times 88.3\ \mathrm{kJ/kg}$$

$$= 110.5\ \mathrm{kJ} + 44.15\ \mathrm{kJ} = 155\ \mathrm{kJ}$$

Discussion An easy mistake to make is to use the wrong latent heat. Here we were dealing with melting, so we needed the latent heat of fusion. Another possible error is to use the specific heat for the wrong phase: here we raised the temperature of *solid* silver, so we needed the specific heat of *solid* silver. With water, we must always be careful to use the specific heat of the correct phase; the specific heats of ice, water, and steam have three different values.

Practice Problem 14.7 Making Gold Medals

Some gold medals are to be made from 750 g of solid gold at 24°C (Fig. 14.6). How much heat is required to melt the gold so that it can be poured into the molds for the medals?

Figure 14.6
A gold medal: the Nobel Prize for physics.

Example 14.8

Turning Water into Ice

Ice cube trays are filled with 0.500 kg of water at 20.0°C and placed into the freezer compartment of a refrigerator. How much energy must be removed from the water to turn it into ice cubes at –5.0°C?

Strategy We can think of this process as three consecutive steps. First, the liquid water is cooled to 0°C. Then the phase change occurs at constant temperature. Now the water is frozen; the ice continues to cool to –5.0°C. The energy that must be removed for the whole process is the sum of the energy removed in each of the three steps.

Solution For liquid water going from 20.0°C to 0.0°C,

$$Q_1 = mc_w \Delta T_1$$

where

$$\Delta T_1 = 0.0°C - 20.0°C = -20.0°C$$

Since ΔT_1 is negative, Q_1 is negative: heat must flow *out of* the water in order for its temperature to decrease. Next the water freezes. The heat is found from the latent heat of fusion:

$$Q_2 = -m L_f$$

Again, heat flows *out* so Q_2 is negative. For phase transitions, we supply the correct sign of Q according to the direction of the phase transition (negative sign for freezing, positive sign for melting). Finally, the ice is cooled to –5.0°C:

$$Q_3 = mc_{ice}\, \Delta T_2$$

where

$$\Delta T_2 = -5.0°C - 0.0°C = -5.0°C$$

We use subscripts on the specific heats to distinguish the specific heat of ice from that of water. The total heat is

$$Q = m\,(c_w \Delta T_1 - L_f + c_{ice} \Delta T_2)$$

Now we look up c_w, L_f, and c_{ice} in Tables 14.1 and 14.4 and substitute:

$$Q = 0.500 \text{ kg} \times \left[4.186 \frac{\text{kJ}}{\text{kg·K}} \times (-20.0°C) - 333.7 \frac{\text{kJ}}{\text{kg}} + 2.1 \frac{\text{kJ}}{\text{kg·K}} \times (-5.0°C)\right]$$

$$= -0.500 \text{ kg} \times 427.9 \frac{\text{kJ}}{\text{kg}} = -214 \text{ kJ}$$

So 214 kJ of heat flows out of the water that becomes ice cubes.

Discussion We cannot consider the entire temperature change from +20°C to –5°C in one step. A phase change occurs, so we must include the flow of heat during the phase change. Also, the specific heat of ice is different from the specific heat of liquid water; we must find the heat to cool water 20°C and then the heat to cool ice 5°C.

Practice Problem 14.8 Frozen Popsicles

Nigel pulls a tray of frozen popsicles out of the freezer to share with his friends. If the popsicles are at –4°C and go directly into hungry mouths at 37°C, how much energy is used to bring a popsicle of mass 0.080 kg to body temperature? Assume the frozen popsicles have the same specific heat as ice and the melted popsicle has the specific heat of water.

Example 14.9

Cooling a Drink

Two 50-g ice cubes are placed into 0.200 kg of water in a Styrofoam cup. The water is initially at a temperature of 25.0°C and the ice is initially at a temperature of −15.0°C. What is the final temperature of the drink? The average specific heat for ice between −15°C and 0°C is 2.05 kJ/(kg·°C).

Strategy We need to raise the temperature of the ice from −15°C to 0°C before the ice can melt, so we first find how much heat this requires. Then we find how much heat is needed to melt all the ice. Once the ice is melted, the water from the melted ice can be raised to the final temperature of the drink. The heat for all three steps (raising temperature of ice, melting ice, raising temperature of water from melted ice) all comes from the water initially at 25°C. That water cools as heat flows out of it. Assuming no heat flow into or out of the room, the quantity of heat that flows out of the water initially at 25°C flows into the ice or melted ice (before, during, and after melting).

Given: $m_{ice} = 0.100$ kg at −15.0°C; $m_w = 0.200$ kg at 25.0°C; $c_{ice} = 2.05$ kJ/(kg·°C)
Look up: L_f for water = 333.7 kJ/kg; $c_w = 4.186$ kJ/(kg·°C)
Find: T_f

Solution Since heat flows out of the water and into ice, $Q_w < 0$ and $Q_{ice} > 0$. Their sum is zero:

$$Q_{ice} + Q_w = 0$$

The heat flow into the ice is the sum of three terms:

$$Q_{ice} = m_{ice}c_{ice}\Delta T_{ice} + m_{ice}L_f + m_{ice}c_w\,(T_f - 0.0°\text{C})$$

The heat flow out of the water is

$$Q_w = m_w c_w\,(T_f - 25.0°\text{C})$$

The heat required to bring the ice from −15.0°C to 0°C is

$$m_{ice}c_{ice}\Delta T_{ice} = 0.100\text{ kg} \times 2.05\,\frac{\text{kJ}}{\text{kg·°C}} \times 15.0°\text{C} = 3.075\text{ kJ}$$

The heat required to melt the ice at 0.0°C is

$$m_{ice}L_f = 0.100\text{ kg} \times 333.7\text{ kJ/kg} = 33.37\text{ kJ}$$

The heat to raise the temperature of the melted ice from 0.0°C to T_f is

$$m_{ice}c_w\,(T_f - 0.0°\text{C}) = 0.100\text{ kg} \times 4.186\,\frac{\text{kJ}}{\text{kg·°C}} \times T_f$$

$$= 0.4186\,\frac{\text{kJ}}{°\text{C}} \times T_f$$

The heat supplied by the water that was initially at 25.0°C is

$$m_w c_w\,(T_f - 25.0°\text{C}) = 0.200\text{ kg} \times 4.186\,\frac{\text{kJ}}{\text{kg·°C}} \times (T_f - 25.0°\text{C})$$

$$= 0.8372\,\frac{\text{kJ}}{°\text{C}} \times T_f - 20.93\text{ kJ}$$

Now we substitute these values back into the original equation, $Q_{ice} + Q_w = 0$.

$$3.075\text{ kJ} + 33.37\text{ kJ} + \left(0.4186\,\frac{\text{kJ}}{°\text{C}} \times T_f\right) + \left(0.8372\,\frac{\text{kJ}}{°\text{C}} \times T_f - 20.93\text{ kJ}\right) = 0$$

continued on next page

Example 14.9 continued

Simplifying yields

$$1.2558\ \frac{\text{kJ}}{°\text{C}} \times T_f + 15.515\ \text{kJ} = 0$$

Solving for T_f, we find

$$T_f = -\frac{15.515\ \text{kJ}}{1.2558\ \text{kJ/°C}} = -12.4°\text{C}$$

This result does not make sense: we assumed that all of the ice would melt and that the final mixture would be all liquid, but we cannot have liquid water at −12.4°C. Let's take another look at the solution.

What if the water initially at 25°C cools all the way to 0°C? From cooling the water, how much heat is available to warm the ice and melt it?

$$Q_w = m_w c_w\ (0°\text{C} - 25.0°\text{C})$$
$$= 0.200\ \text{kg} \times 4.186\ \frac{\text{kJ}}{\text{kg·°C}} \times (-25.0°\text{C}) = -20.93\ \text{kJ}$$

Thus, the water can supply 20.93 kJ when it cools to 0°C. But to warm the ice requires 3.075 kJ and to melt all of the ice requires another 33.37 kJ. The ice can be warmed to 0°C, but there is not enough heat available to melt all of the ice. Only some of the ice melts, so the drink ends up as a mixture of water and ice in equilibrium at 0°C.

Discussion This example shows the value of checking a result to make sure it is reasonable. We started by assuming incorrectly that all of the ice would melt. When we obtained an answer that was impossible, we went back to see if there was enough heat available to melt all of the ice. Since there was not, the final temperature of the drink can only be 0°C—the only temperature at which ice and water can be in thermal equilibrium at atmospheric pressure.

Practice Problem 14.9 Melting Ice

How much of the ice of Example 14.9 melts?

Evaporation

If you leave a cup of water out at room temperature, the water eventually evaporates. Recall that the temperature of the water reflects the average kinetic energy of the water molecules; some have higher than average energies and some have lower. The most energetic molecules have enough energy to break loose from the molecular bonds at the surface of the water. As these highest-energy molecules leave the water, the average energy of those left behind decreases—which is why evaporation is a cooling process. Approximately the same latent heat of vaporization applies to evaporation as to boiling, since the same molecular bonds are being broken. Perspiring basketball players cover up while sitting on the bench for a short time during a game to prevent getting a chill even though the air in the stadium may be warm.

Application of evaporation: chill caused by perspiration

When the humidity is high—meaning there is already a lot of water vapor in the air—evaporation proceeds more slowly. Water molecules in the air can also condense into water; the net evaporation rate is the difference in the rates of evaporation and condensation. A hot, humid day is uncomfortable because our bodies have trouble staying cool when perspiration evaporates slowly.

PHYSICS AT HOME

The effects of evaporation can easily be felt. Rub some water on the inside of your forearm and then blow on your arm. The motion of the air over your arm removes the newly evaporated molecules from the vicinity of your arm and allows other molecules to evaporate more quickly. You can feel the cooling effect. If you have some rubbing alcohol, repeat the experiment. Since the alcohol evaporates faster, the cooling effect is noticeably greater.

Phase Diagrams

A useful tool in the study of phase transitions is the **phase diagram**—a diagram on which pressure is plotted on the vertical axis and temperature on the horizontal axis. Figure 14.7a is a phase diagram for water. A point on the phase diagram represents water in a state determined by the pressure and the temperature at that point. The curves on the phase diagram are the demarcations between the solid, liquid, and gas phases. For most temperatures, there is one pressure at which two particular phases can coexist in equilibrium. Since point *P* lies on the fusion curve, water can exist as liquid, or as solid, or as a mixture of the two at that temperature and pressure. At point *Q*, water can only be a solid. Similarly, at *A* water is a liquid; at *B* it is a gas.

The one exception is at the **triple point**, where all three phases (solid, liquid, and gas) can coexist in equilibrium. Triple points are used in precise calibrations of thermometers. The triple point of water is precisely 0.01°C at 0.006 atm.

From the vapor pressure curve, we see that as the pressure is lowered, the temperature at which water boils decreases. It takes longer to cook a hard-boiled egg at high elevations because the temperature of the boiling water is less than 100°C; the chemical reactions proceed more slowly at a lower temperature. It might take as long as half an hour to hard-boil an egg on Pike's Peak, where the average pressure is 0.6 atm.

If either the temperature or the pressure or both are changed, the point representing the state of the water moves along some path on the phase diagram. If the path crosses one of the curves, a phase transition occurs and the latent heat for that phase transition is either absorbed or released (depending on direction). Crossing the fusion curve represents freezing or melting; crossing the vapor pressure curve represents condensation or vaporization.

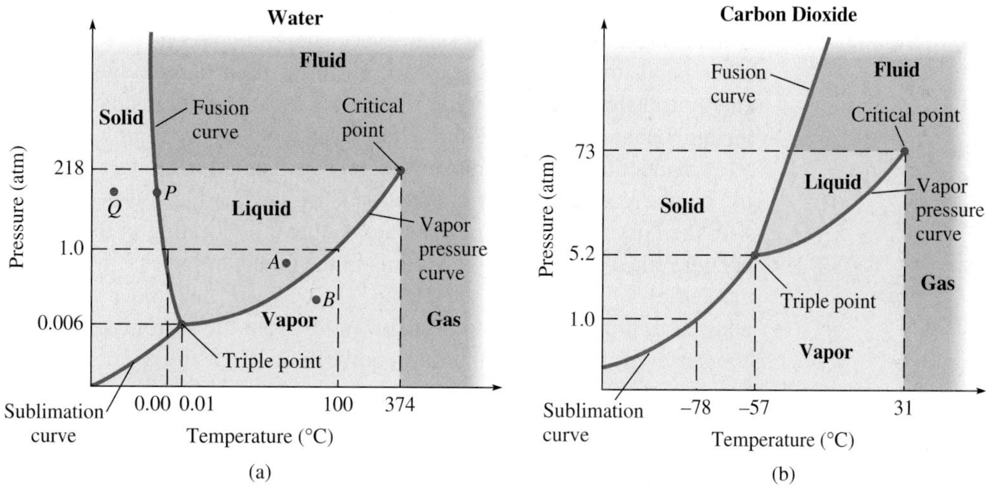

Figure 14.7 Phase diagrams for (a) water and (b) carbon dioxide. The term *vapor* is often used to indicate a substance in the gaseous state below its critical temperature; above the critical temperature it is called gas. (Note that the axes do not use a linear scale.)

Notice that the vapor pressure curve ends at the **critical point**. Thus, if the path for changing a liquid to a gas goes around the critical point without crossing the vapor pressure curve, no phase transition occurs. At temperatures above the critical temperature or pressures above the critical pressure, it is *impossible* to make a clear distinction between the liquid and gas phases.

Sublimation occurs when a solid becomes gas (or vice versa) without passing through the liquid phase. An example occurs when ice on a car windshield becomes water vapor on a cold dry day. Mothballs and dry ice (solid carbon dioxide) also pass directly from solid to gas. At atmospheric pressure, only the solid and gas phases of CO_2 exist (Fig. 14.7b). The liquid phase is not stable below 5.2 atm of pressure, so carbon dioxide does not melt at atmospheric pressure. Instead it sublimates; it goes from solid directly to gas. Solid CO_2 is called *dry ice* because it is cold and looks like ice, but does not melt. Sublimation has its own latent heat; the latent heat for sublimation is not the sum of the latent heats for fusion and vaporization.

The Unusual Phase Diagram of Water The phase diagram of water has an unusual feature: the slope of the fusion curve is negative. The fusion curve has a negative slope only for substances (such as water, gallium, and bismuth) that expand on freezing. In these substances the molecules are *closer together* in the liquid than they are in the solid! As liquid water starting at room temperature is cooled, it contracts until it reaches 3.98°C. At this temperature water has its highest density (at a pressure of 1 atm); further cooling makes the water *expand.* When water freezes, it expands even more; ice is less dense than water.

Figure 14.8 A Nunavut villager fishing for Arctic char.

One consequence of the expansion of water on freezing is that cell walls might rupture when foods are frozen and thawed. The taste of frozen food suffers as a result. Another consequence is that lakes, rivers, and ponds do not freeze solid in the winter. A layer of ice forms on *top* since ice is less dense than water; underneath the ice, liquid water remains, which permits fish, turtles, and other aquatic life to survive until spring (Fig. 14.8).

14.6 THERMAL CONDUCTION

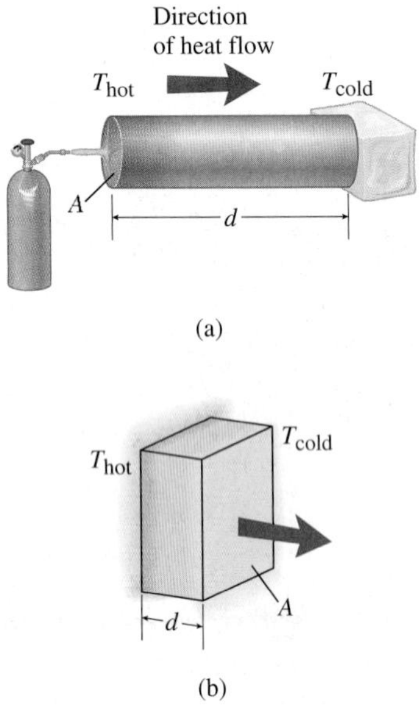

Figure 14.9 (a) Heat conduction along a cylindrical bar of length *d.* (b) Heat conduction through a slab of material of thickness *d.*

Until now we have considered the *effects* of heat flow, but not the mechanism of how heat flow occurs. We now turn our attention to three types of heat flow—*conduction, convection,* and *radiation.*

The **conduction** of heat can take place within solids, liquids, and gases. Conduction is due to collisions between atoms (or molecules) in which energy is exchanged. If the average energy is the same everywhere, there is no net flow of heat. If, on the other hand, the temperature is not uniform, then on average the atoms with more energy transfer some energy to those with less. The net result is that heat flows from the higher-temperature region to the lower-temperature region.

Conduction also occurs between objects that are in contact. A teakettle on an electric burner receives heat by conduction since the heating coil of the burner is in contact with the bottom of the kettle. The atoms that are vibrating in the object at higher temperature (the coil) collide with atoms in the object at lower temperature (the bottom of the kettle), resulting in a net transfer of energy to the colder object. If conduction is allowed to proceed, with no heat flow to or from the surroundings, then the objects in contact eventually reach thermal equilibrium when the average translational kinetic energies of the atoms are equal.

Fourier's Law of Heat Conduction Suppose we consider a simple geometry such as an object with uniform cross section in which heat flows in a single direction. Examples are a plate of glass, with different temperatures on the inside and outside surfaces, or a cylindrical bar, with its ends at different temperatures (Fig. 14.9). The rate of heat conduction depends on the temperature difference $\Delta T = T_{hot} - T_{cold}$, the length (or thickness) d, the cross-sectional area A through which heat flows, and the nature of the material itself. The greater the temperature difference, the greater the heat flow. The thicker the material, the longer it takes for the heat to travel through—since the energy transfer has to be passed along a longer "chain" of atomic collisions—making the rate of heat flow smaller. A larger cross-sectional area allows more heat to flow.

The nature of the material is the final thing that affects the rate of energy transfer. In metals the electrons associated with the atom are free to move about and they carry the heat. When a material has free electrons, the transfer rate is faster; if the electrons are all tightly bound, as in nonmetallic solids, the transfer is slower. Liquids, in turn, conduct heat less readily than solids, because the forces between atoms are weaker. Gases are even less efficient as conductors of heat than solids or liquids since the atoms of a gas are so much farther apart and have to travel a greater distance before collisions occur. The **thermal conductivity** (symbol κ, the Greek letter kappa) of a substance is directly proportional to the rate at which energy is transferred through the substance. Higher values of κ are associated with good conductors of heat, smaller values with *thermal insulators* that tend to prevent the flow of heat. Table 14.5 lists the thermal conductivities for several common substances.

Let $\mathcal{P} = Q/\Delta t$ represent the rate of heat flow (or *power*). (The script $\mathcal{P}$ is used to avoid confusing power with pressure.) The dependence of the *rate* of heat flow through a substance on all the factors mentioned is given by

Fourier's law of heat conduction:

$$\mathcal{P} = \kappa A \frac{\Delta T}{d} \qquad (14\text{-}10)$$

where κ is the thermal conductivity of the material, A is the cross-sectional area, d is the thickness (or length) of the material, and ΔT is the temperature difference between one side and the other. The quantity $\Delta T/d$ is called the *temperature gradient*; it tells how many °C or K the temperature changes per unit of distance moved along the path of heat flow. Inspection of Eq. (14-10) shows that the SI units of κ are W/(m·K).

In Fig. 14.9b, a slab of material is shown that conducts heat because of a temperature difference between the two sides. By rearranging Eq. (14-10) and solving for ΔT,

$$\Delta T = \mathcal{P}\frac{d}{\kappa A} = \mathcal{P}R \qquad (14\text{-}11)$$

The quantity $d/(\kappa A)$ is called the **thermal resistance** R.

$$R = \frac{d}{\kappa A} \qquad (14\text{-}12)$$

Thermal resistance has SI units of K/W (kelvins per watt). Notice that the thermal resistance depends on the nature of the material (through the thermal conductivity κ) and the geometry of the object (d/A). Equation (14-11) is useful for solving problems when heat flows through one material after another.

Conduction Through Two or More Materials in Series Suppose we have two layers of material between two temperature extremes as in Fig. 14.10. These layers are in *series* because the heat flows through one and then through the other. Looking at one layer at a time,

$$T_1 - T_2 = \mathcal{P}R_1 \quad \text{and} \quad T_2 - T_3 = \mathcal{P}R_2$$

Then, adding the two together

$$(T_1 - T_2) + (T_2 - T_3) = \mathcal{P}R_1 + \mathcal{P}R_2$$
$$\Delta T = T_1 - T_3 = \mathcal{P}(R_1 + R_2)$$

Table 14.5 Thermal Conductivities at 20°C

Material	$\kappa \left(\frac{\text{W}}{\text{m·K}}\right)$
Air	0.023
Rigid panel polyurethane insulation	0.023–0.026
Fiberglass insulation	0.029–0.072
Rock wool insulation	0.038
Cork	0.046
Wood	0.13
Soil (dry)	0.14
Asbestos	0.17
Snow	0.25
Sand	0.39
Water	0.6
Window glass (typical)	0.63
Pyrex glass	1.13
Vycor	1.34
Concrete	1.7
Ice	1.7
Stainless steel	14
Lead	35
Steel	46
Nickel	60
Tin	66.8
Platinum	71.6
Iron	80.2
Brass	122
Zinc	116
Tungsten	173
Aluminum	237
Gold	318
Copper	401
Silver	429

CONNECTION:

Fourier's law says that the rate of heat flow is proportional to the temperature gradient. Closely analogous is Poiseuille's law for viscous fluid flow [Eq. (9-15)], in which the volume flow rate is proportional to the pressure gradient.

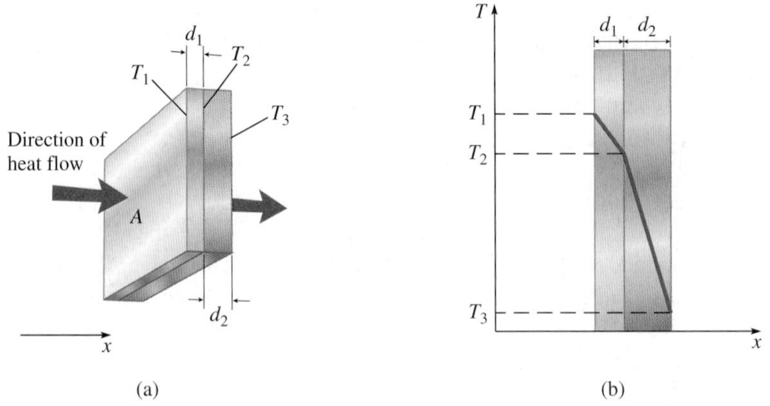

Figure 14.10 (a) Conduction of heat through two different layers ($T_1 > T_2 > T_3$). (b) Graph of temperature T as a function of position x. The slope of the graph in either material is the temperature gradient $\Delta T/d$ in that material. The temperature gradients are not the same because the materials have different thermal conductivities.

The rate of heat flow through the first layer is the same as the rate through the second layer because otherwise the temperatures would be changing. For n layers,

$$\Delta T = \mathcal{P}\sum R_n \qquad n = 1, 2, 3, \ldots \tag{14-13}$$

Equation (14-13) shows that the effective thermal resistance for layers in series is the sum of each layer's thermal resistance.

CHECKPOINT 14.6

In Fig. 14.10, which of the two materials has the larger thermal conductivity?

Example 14.10

The Rate of Heat Flow Through Window Glass

A windowpane that measures 20.0 cm by 15.0 cm is set into the front door of a house. The glass is 0.32 cm thick. The temperature outdoors is −15°C and inside is 22°C. At what rate does heat leave the house through that one small window?

Strategy We assume one side of the glass to be at the temperature of the air inside the house and the other to be at the outdoor temperature.

Given: $\Delta T = 22°\text{C} - (-15°\text{C}) = 37°\text{C}$; thickness of windowpane $d = 0.32 \times 10^{-2}$ m; area of windowpane $A = 0.200\ \text{m} \times 0.150\ \text{m} = 0.0300\ \text{m}^2$
Look up: thermal conductivity for glass = 0.63 W/(m·K)
Find: rate of heat flow, $\mathcal{P}$

Solution The temperature gradient is

$$\frac{\Delta T}{d} = \frac{37°\text{C}}{0.32 \times 10^{-2}\ \text{m}} = 1.16 \times 10^4\ \text{K/m}$$

Now we have all the information we need to find the rate of conductive heat flow:

$$\mathcal{P} = \kappa A \frac{\Delta T}{d}$$

$$= 0.63\ \text{W/(m·K)} \times 0.0300\ \text{m}^2 \times 1.16 \times 10^4\ \text{K/m}$$

$$= 220\ \text{W}$$

Discussion A loss of 220 W through one small window is significant. However, our assumption about the temperatures of the two glass surfaces exaggerates the temperature difference across the glass. In reality, the inside surface of the glass is colder than the air inside the house, while the outside surface is warmer than the air outside.

Practice Problem 14.10 An Igloo

A group of children build an igloo in their garden. The snow walls are 0.30 m thick. If the inside of the igloo is at 10.0°C and the outside is at −10.0°C, what is the rate of heat flow through the snow walls of area 14.0 m^2?

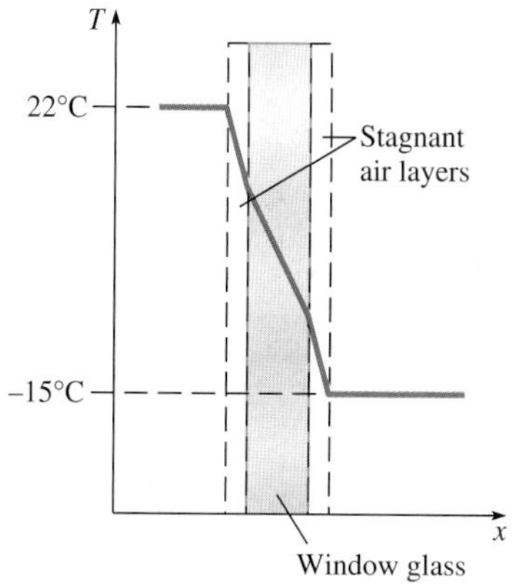

Figure 14.11 Temperature variation on either side of a windowpane. A plot of temperature versus position is superimposed on a cross section of the window glass and the air layers on either side.

Thermal Conductivity of Air Air has a low thermal conductivity; it is an excellent thermal insulator *when it is still.* An accurate calculation of the energy loss through a single-paned window *must* take into account the thin layer of stagnant air, due to viscosity, on each side of the glass. If the temperature is measured near a window, the temperature of the air just beside the window is intermediate in value between the temperatures of the room air and the outside air (see Fig. 14.11). Thus, the temperature gradient *across the glass* is considerably smaller than the difference between indoor and outdoor temperatures. In fact, much of the thermal resistance of a window is due to the stagnant air layers rather than to the glass.

Example 14.11

Heat Loss Through a Double-Paned Window

The single-paned window of Example 14.10 is replaced by a double-paned window with an air gap of 0.50 cm between the two panes. The inner surface of the inner pane is at 22°C and the outer surface of the outer pane is at −15°C. What is the new rate of heat loss through the double-paned window?

Strategy Now there are three layers to consider: two layers of glass and one layer of air. We find the thermal resistance of each layer and then add them together to find the total thermal resistance. Then we find the temperature difference between the inside of the house and the air outdoors and divide by the total thermal resistance to find the rate at which heat is lost through the replacement window.

Solution For the first layer of glass,

$$R_1 = \frac{d}{\kappa A} = \frac{0.32 \times 10^{-2}\ \text{m}}{0.63\ \text{W/(m·K)} \times 0.0300\ \text{m}^2} = 0.169\ \text{K/W}$$

For the air gap,

$$R_2 = \frac{d}{\kappa A} = \frac{0.50 \times 10^{-2}\ \text{m}}{0.023\ \text{W/(m·K)} \times 0.0300\ \text{m}^2} = 7.246\ \text{K/W}$$

The second layer of glass has the same thermal resistance as the first:

$$R_3 = R_1$$

The total thermal resistance is

$$\sum R_n = 0.169 + 7.246 + 0.169 = 7.584\ \text{K/W}$$

and the rate of conductive heat flow is

$$\mathcal{P} = \frac{Q}{\Delta t} = \frac{\Delta T}{\sum R_n} = \frac{37\ \text{K}}{7.584\ \text{K/W}} = 4.9\ \text{W}$$

Discussion The reduction in the rate of heat loss by replacing a single-paned window with a double pane is significant. This example, however, overestimates the reduction since we assume that heat can only be conducted through the air layer. In reality, heat can also flow through air by convection and radiation. A more accurate calculation would have to account for the other methods of heat flow.

Practice Problem 14.11 Two Panes of Glass Without the Air Gap

Repeat Example 14.11 if the two panes of glass are touching one another, without the intervening layer of air.

R-Factors The U.S. building industry rates materials used in construction with *R-factors.* The R-factor is not quite the same as the thermal resistance; thermal resistance cannot be specified without knowing the cross-sectional area. The R-factor is the thickness divided by the thermal conductivity:

$$\text{R-factor} = \frac{d}{\kappa} = RA$$

$$\frac{\mathcal{P}}{A} = \frac{\Delta T}{\text{R-factor}}$$

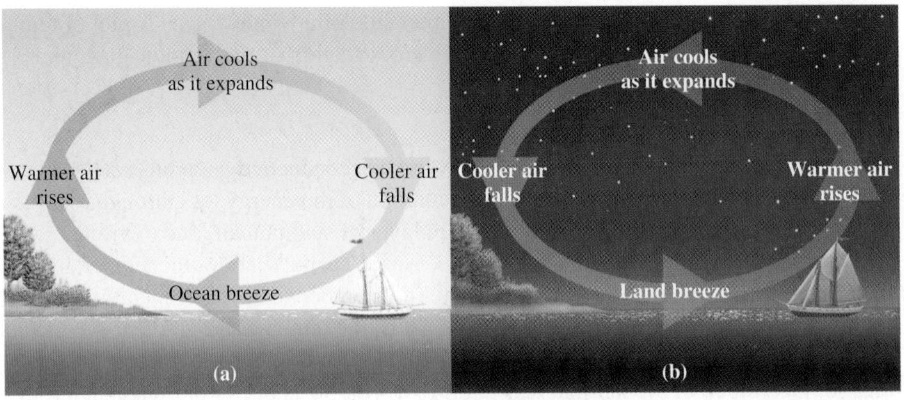

Figure 14.12 (a) During the day, air coming off the ocean is heated as it passes over the warm ground on shore; the heated air rises and expands. The expansion cools the air; it becomes more dense and falls back down. This cycle sets up a convection current that brings cool breezes from the sea to the shore. (b) The reverse circulation occurs at night when the land is cold and the sea is warmer, retaining heat absorbed during the day.

Unfortunately, SI units are not used. The R-factors quoted in the United States are in units of °F·ft^2/(Btu/h)! R-factors are added, just as thermal resistances are, when heat flows through several different layers.

14.7 THERMAL CONVECTION

Applications of convection: offshore and onshore breezes, double-paned windows, and down jackets

Convection involves *fluid currents* that carry heat from one place to another. In conduction, energy flows through a material but the material itself does not move. In convection, *the material itself moves* from one place to another. Thus, convection can occur only in fluids, not in solids. When a wood stove is burning, convection currents in the air carry heat upward to the ceiling. The heated air is less dense than cooler air, so the buoyant force causes it to rise, carrying heat with it. Meanwhile, cooler air that is more dense sinks toward the floor. An example of convection currents at the seashore is shown in Fig. 14.12. Air is a poor *conductor* of heat, but it can easily flow and carry heat by *convection.*

Applications of forced convection: building heating systems; temperature regulation in the human body.

The use of sealed, double-paned windows replaces the large air gap of about 6 or 7 cm between a storm window and regular window with a much smaller gap. The smaller air gap minimizes circulating convection currents between the two panes. Down jackets and quilts are good insulators because air is trapped in many little spaces among the feathers, minimizing heat flow due to convection. Materials such as rock wool, glass wool, or fiberglass are used to insulate walls; much of their insulating value is due to the air trapped around and between the fibers.

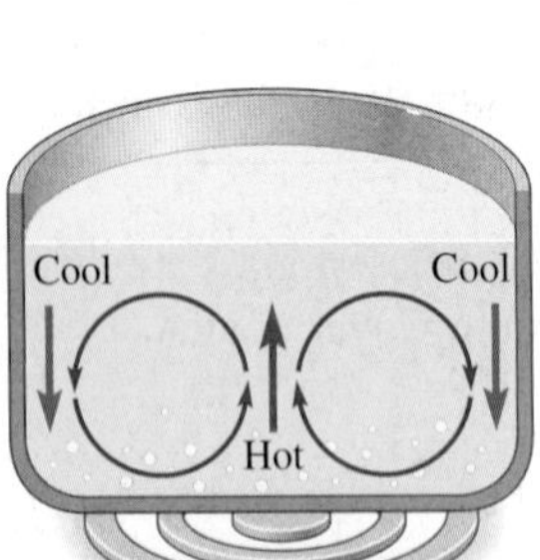

Figure 14.13 Convection currents in heated water. Heat flows through the bottom surface of the pot by conduction and then heats the layer of water in contact with the pot bottom. The heated water is less dense, so buoyant forces make it rise, setting up convection currents.

Natural and Forced Convection In *natural convection*, the currents are due to gravity. Fluid with a higher density sinks because the buoyant force is smaller than the weight; less dense fluid rises because the buoyant force exceeds the weight (Figs. 14.13 and 14.14). In *forced convection*, fluid is pushed around by mechanical means such as a fan or pump. In forced-hot-air heating, warm air is blown into rooms by a fan (Fig. 14.15); in hot water baseboard heating, hot water is pumped through baseboard radiators. Another example of forced convection is blood circulation in the body. The heart pumps blood around the body. When our body temperature is too high, the blood vessels near the skin dilate so that more blood can be pushed into them by the heart. The blood carries heat from the interior of the body to the skin; heat then flows from the skin into the cooler surroundings. If the surroundings are *hotter* than the skin, such as in a hot tub, this strategy backfires and can lead to dangerous overheating of the body. The hot water delivers heat to the dilated blood vessels; the blood carries the heat back to the central core of the body, *raising* the core temperature. (For more information on convection, see text website.)

Application of convection: ocean currents and global climate change

Convection and Global Climate Change

One worry for scientists studying global warming is that northern Europe might be plunged into a deep freeze—a seeming contradiction that results from an interruption of the natural convection cycle. Earth's climate is influenced by convection currents caused by temperature differences between the poles and the tropics (Fig. 14.16). Massive sea currents travel through the Pacific and Atlantic oceans, carrying about half of the heat from the tropics to the poles, where it is dissipated. Storms moving north from the tropics carry much of the rest of the heat. If the polar regions warm at a faster rate than the tropics, the smaller temperature difference between them changes the patterns of the prevailing winds, the tracks followed by storms, the speed of ocean currents, and the amount of precipitation.

For example, the melting of the ice shelves combined with increased precipitation could lead to a layer of fresh water lying on top of the more dense salt water in the North Atlantic. Normally, the cold ocean water at the surface sinks and starts the process of convection. With the buoyancy of the less dense freshwater layer keeping it from sinking, the convection currents slow down or are stopped. Without the pull of the convection current, the usual northward movement of water from the warm Gulf Stream would slow or cease, causing *colder* temperatures in northern Europe.

Such an effect on climate is not without precedent. At the end of the last Ice Age, freshwater from melting glaciers flowed out the St. Lawrence River and into the North Atlantic. A freshwater layer, buoyed up by the more dense salt water, disrupted the usual ocean currents. The Gulf Stream was effectively shut down and Europe experienced a thousand years of deep freeze.

Figure 14.14 Birds (and people flying sailplanes) take advantage of thermal updrafts.

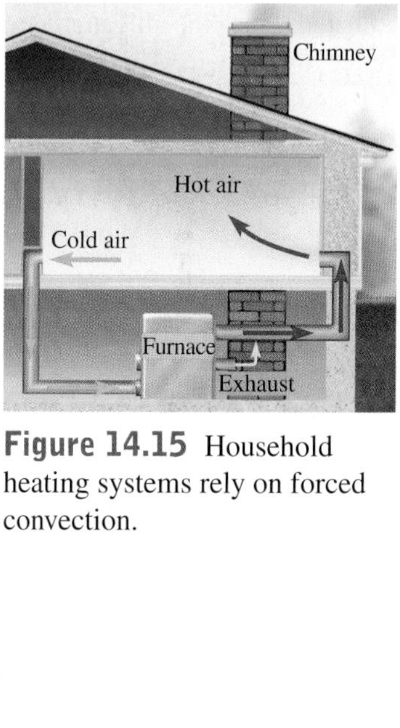

Figure 14.15 Household heating systems rely on forced convection.

14.8 THERMAL RADIATION

All bodies emit energy through electromagnetic radiation—due to the oscillation of electric charges in the atoms. Thermal radiation consists of electromagnetic waves that travel at the speed of light. Unlike conduction and convection, radiation does not require a material medium; the Sun radiates heat to Earth through the near vacuum of space.

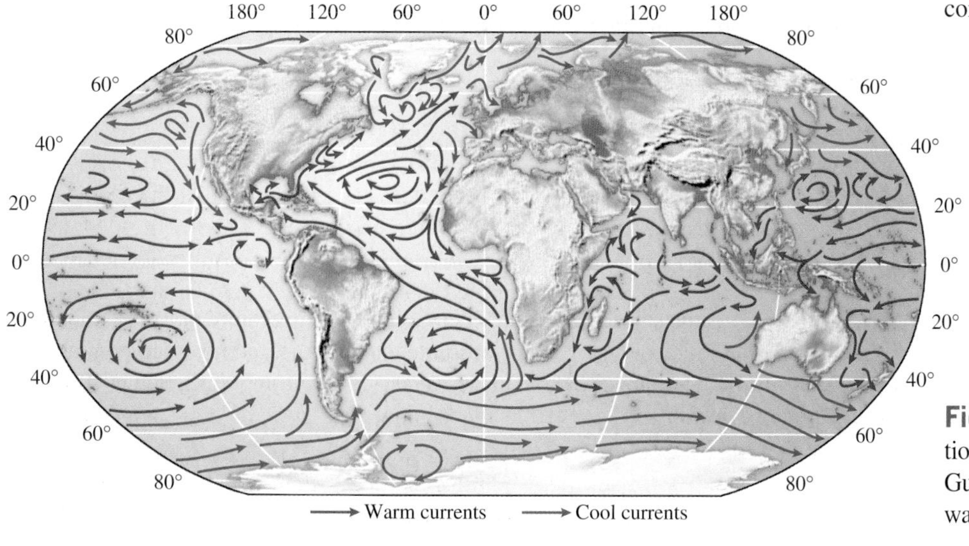

Figure 14.16 Surface convection currents in the oceans. The Gulf Stream is a current of warm water flowing across the Atlantic.

An object emits thermal radiation while absorbing some of the thermal radiation emitted by other objects. The rate of absorption may be less than, equal to, or greater than the rate of emission.

When solar radiation reaches Earth, it is partially absorbed and partially reflected. Earth also emits radiation at nearly the same average rate that it absorbs energy from the Sun. If there were an exact equilibrium between absorption and emission, Earth's average temperature would stay constant. However, increasing concentrations of CO_2 and other "greenhouse gases" in Earth's atmosphere cause energy to be emitted at a slightly lower rate than it is absorbed. As a result, Earth's average temperature is rising. Although the predicted temperature increase may seem small on an absolute scale, it will have dramatic consequences for life on Earth.

Conceptual Example 14.12

An Alligator Lying in the Sun

An alligator crawls out into the Sun to get warm. Solar radiation is incident on the alligator at the rate of 300 W; 70 W of it is reflected. (a) What happens to the other 230 W? (b) If the alligator emits 100 W, does its body temperature rise, fall, or stay the same? Ignore heat flow by conduction and convection.

Solution and Discussion (a) When radiation falls on an object, some can be absorbed, some can be reflected, and—for a transparent or translucent object—some can be transmitted through the object without being absorbed or reflected. Since the alligator is opaque, no radiation is transmitted through it. All the radiation is either absorbed or reflected, so the other 230 W is absorbed. (b) Since 230 W is absorbed while 100 W is emitted, the alligator absorbs more radiation than it emits. Absorption increases internal energy while emission decreases it, so the alligator's internal energy is increasing at a rate of 130 W. Thus, we expect the alligator's body temperature to rise. (The actual rate of increase of internal energy would be smaller since conduction and convection carry heat away as well.)

Conceptual Practice Problem 14.12 Maintaining Constant Temperature

After some time elapses, the alligator's body temperature reaches a constant level. The rate of absorption is still 230 W. If the alligator loses heat by conduction and convection at a rate of 90 W, at what rate does it emit radiation?

Stefan's Radiation Law

An idealized body that absorbs all the radiation incident upon it is called a **blackbody**. A blackbody absorbs not only all visible light, but infrared, ultraviolet, and all other wavelengths of electromagnetic radiation. It turns out (see Conceptual Question 23) that a good *absorber* is also a good *emitter* of radiation. A blackbody emits more radiant power per unit surface area than any real object at the same temperature. The rate at which a blackbody emits radiation per unit surface area is proportional to the fourth power of the *absolute* temperature:

Stefan's Law of Radiation (blackbody)

$$\mathcal{P} = \sigma A T^4 \tag{14-14}$$

In Eq. (14-14), A is the surface area and T is the surface temperature of the blackbody *in kelvins*. Since Stefan's law involves the absolute temperature and not a temperature difference, °C *cannot* be substituted. The universal constant σ (Greek letter sigma) is called *Stefan's constant*:

$$\sigma = 5.670 \times 10^{-8}\ \text{W/(m}^2\cdot\text{K}^4) \tag{14-15}$$

The fourth-power temperature dependence implies that the power emitted is extremely sensitive to temperature changes. If the absolute temperature of a body doubles, the energy emitted increases by a factor of $2^4 = 16$.

Emissivity Since real bodies are not perfect absorbers and therefore emit less than a blackbody, we define the **emissivity** (e) as the ratio of the emitted power of the body to that of a blackbody at the same temperature. Then Stefan's law becomes

Stefan's Law of Radiation

$$\mathcal{P} = e\sigma AT^4 \qquad (14\text{-}16)$$

The emissivity ranges from 0 to 1; $e = 1$ for a perfect radiator and absorber (a blackbody) and $e = 0$ for a perfect reflector. The emissivity for polished aluminum, an excellent reflector, is about 0.05; for soot (carbon black) it is about 0.95. Equation (14-16) is a refinement of Stefan's law, but it is still an approximation because it treats the emissivity as a constant. Emissivity is actually a function of the wavelength of the emitted radiation. Equation (14-16) is useful when the emissivity is approximately constant over the range of wavelengths in which most of the power is radiated.

Human skin, no matter what the pigmentation, has an emissivity of about 0.97 in the infrared part of the spectrum. Many objects have high emissivities in the infrared even though they may reflect much of the visible light incident on them and, therefore, have low emissivities in the visible range.

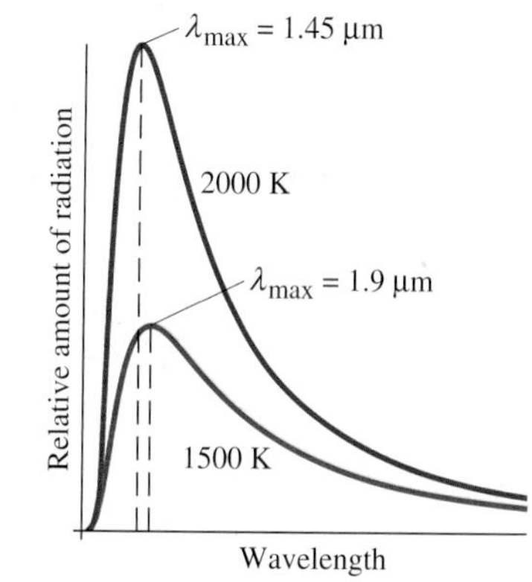

Figure 14.17 Graphs of blackbody radiation as a function of wavelength at two different temperatures. At the higher temperature, the wavelength of maximum radiation is shorter (Wien's law) and the total power radiated, represented by the area under the graph, increases (Stefan's law).

Radiation Spectrum

The electromagnetic radiation we are concerned with falls into three wavelength ranges. Infrared radiation includes wavelengths from about 100 μm down to 0.7 μm. The wavelengths of visible light range from about 0.7 μm to about 0.4 μm. Ultraviolet wavelengths are less than 0.4 μm.

The total power radiated is not the only thing that varies with temperature. Figure 14.17 shows the radiation spectrum—a graph of how much radiation occurs as a function of wavelength—for blackbodies at two different temperatures. The wavelength at which the maximum power is emitted decreases as temperature increases. Objects at ordinary temperatures emit primarily in the infrared—around 10 μm in wavelength for 300 K. The Sun, since it is much hotter, radiates primarily at shorter wavelengths. Its radiation peaks in the visible (no surprise there) but includes plenty of infrared and ultraviolet as well. The wavelength of maximum radiation is inversely proportional to the absolute temperature:

Wien's Law

$$\lambda_{max} T = 2.898 \times 10^{-3}\ \text{m·K} \qquad (14\text{-}17)$$

where the temperature T is the temperature in kelvins and λ_{max} is the wavelength of maximum radiation in meters.

As the temperature of the blackbody rises to 1000 K and above, the peak intensity shifts toward shorter wavelengths until some of the emitted radiation falls in the visible. Since the longest visible wavelengths are for red light, the heated body glows dull red. As the temperature of the blackbody continues to increase, the red glow becomes brighter red, then orange, then yellow-white, and eventually blue-white as the blackbody emits more and more visible light. When the body is emitting all the colors of the visible spectrum, the glow appears white to the eye. When something is red-hot, it is not as hot as something that is white-hot.

Example 14.13

Temperature of the Sun

The maximum rate of energy emission from the Sun occurs in the middle of the visible range—at about $\lambda = 0.5\ \mu\text{m}$. Estimate the temperature of the Sun's surface.

Strategy We assume the Sun to be a blackbody. Then the wavelength of maximum emission and the surface temperature are related by Wien's law.

Solution Given: $\lambda_{max} = 0.5\ \mu\text{m} = 5 \times 10^{-7}$ m. Then from Wien's law, we know that the product of the wavelength for maximum power emission and the corresponding temperature for the power emission is

$$\lambda_{max}T = 2.898 \times 10^{-3}\ \text{m·K}$$

We can solve for the temperature since we know λ_{max}:

$$T = \frac{2.898 \times 10^{-3}\ \text{m·K}}{5 \times 10^{-7}\ \text{m}}$$
$$= 6000\ \text{K}$$

Discussion Quick check: an object at 300 K has $\lambda_{max} \approx 10\ \mu\text{m}$, which is 20 times the λ_{max} in the radiation from the Sun (0.5 μm). Since λ_{max} and T are inversely proportional, the Sun's surface temperature is 20 times 300 K = 6000 K.

Practice Problem 14.13 Wavelengths of Maximum Power Emission for Skin

The temperature of skin varies from 30°C to 35°C depending on the blood flow near the skin surface. What is the range of wavelengths of maximum power emission from skin?

Simultaneous Emission and Absorption of Thermal Radiation

An object simultaneously emitting and absorbing thermal radiation has a *net* rate of heat flow due to thermal radiation given by $\mathcal{P}_{net} = \mathcal{P}_{emitted} - \mathcal{P}_{absorbed}$. Suppose an object with surface area A and temperature T is bathed in thermal radiation coming from its surroundings in all directions that are at a *uniform* temperature T_s. Then the *net* rate of heat flow due to thermal radiation is

Net rate of energy transfer due to emission and absorption of thermal radiation

$$\mathcal{P}_{net} = e\sigma AT^4 - e\sigma AT_s^4 = e\sigma A(T^4 - T_s^4) \quad (14\text{-}18)$$

A body emits energy even if it is at the same temperature as its surroundings; it just emits at the same rate that it absorbs, so $\mathcal{P}_{net} = 0$. If $T > T_s$, the object emits more thermal radiation than it absorbs. If $T < T_s$, the object absorbs more thermal radiation than it emits.

Why is the rate of *absorption* proportional to the *emissivity*? Because a good emitter is also a good absorber. The emissivity e measures not only how much the object emits compared to a blackbody; it also measures how much the object *absorbs* compared with a blackbody. A blackbody at the same temperature as its surroundings would have to absorb radiation at the rate $\mathcal{P}_{absorbed} = \sigma AT_s^4$ to exactly balance the rate of emission. However, emissivity does depend on temperature. Equation (14-18) assumes the emissivity at temperature T is the same as the emissivity at temperature T_s. If T and T_s are very different, we would have to modify Eq. (14-18) to use two different emissivities.

Do not substitute temperature in Celsius degrees into Eq. (14-18). The quantity inside the parentheses might look like a temperature difference, but it is not. The two kelvin temperatures are raised to the fourth power, *then* subtracted—which is not the same as the corresponding two Celsius temperatures subjected to the same mathematical operations. By the same token, do not subtract the temperatures in kelvins and then raise to the fourth power. The difference of the fourth powers is not equal to the difference raised to the fourth power, as can be readily demonstrated:

$$(2^4 - 1^4) = 15 \quad \text{but} \quad (2 - 1)^4 = 1$$

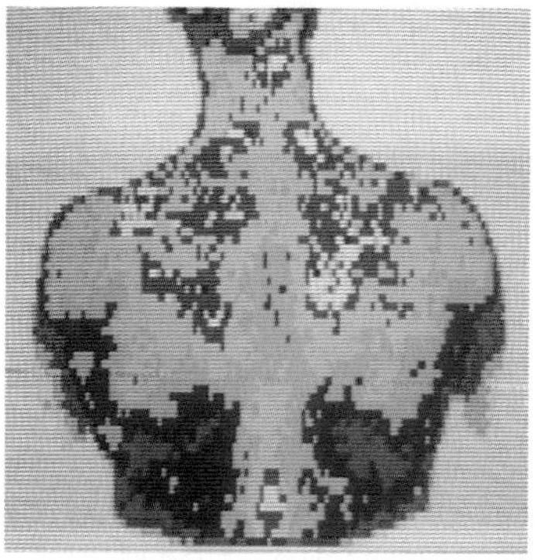

Figure 14.18 Thermography of a backache. The magenta areas are warmer than the surrounding tissue, revealing the location of the source of pain.

Medical Applications of Thermal Radiation

Thermal radiation from the body is used as a diagnostic tool in medicine. "Instant-read" thermometers work by measuring the intensity of thermal radiation in the patient's ear. A thermogram shows whether one area is radiating more heat than it should, indicating a higher temperature due to abnormal cellular activity. For example, when a broken bone is healing, heat can be detected at the location of the break just by placing a hand lightly on the area of skin over the break. Infrared detectors, originally developed for military uses (nightscopes, for example), can be used to detect radiation from the skin. The radiation is absorbed and an electrical signal is produced that is then used to produce a visual display (Fig. 14.18). Thermography has been used to screen travelers at airports in Asia for the high fever that accompanies infection with severe acute respiratory syndrome (SARS).

Example 14.14

Thermal Radiation from the Human Body

A person of body surface area 2.0 m^2 is sitting in a doctor's examining room with no clothing on. The temperature of the room is 22°C and the person's average skin temperature is 34°C. Skin emits about 97% as much as a blackbody at the same temperature for wavelengths in the infrared region, where most of the emission occurs. At what *net* rate is energy radiated away from the body?

Strategy Both radiation and absorption occur in the infrared—the absolute temperatures of the skin and the room are not very different. Therefore, we can assume that 97% of the incident radiation from the room is absorbed. Equation (14-18) therefore applies. We must convert the Celsius temperatures to kelvins.

Given: surface area, $A = 2.0$ m^2; $T_{\text{room}} = 22$°C; skin temperature, $T = 34$°C; fraction of energy emitted, $e = 0.97$
To find: net rate of energy transfer, $\mathcal{P}_{\text{net}}$

Solution The temperature of the skin surface is

$$T = 273 + 34 = 307 \text{ K}$$

and of the room is

$$T_s = 273 + 22 = 295 \text{ K}$$

The net rate of energy transfer between the room and the body is

$$\mathcal{P}_{\text{net}} = e\sigma A(T^4 - T_s^4)$$

Substituting,

$$\mathcal{P}_{\text{net}} = 0.97 \times 5.67 \times 10^{-8} \text{ W/(m}^2\cdot\text{K}^4) \times 2.0 \text{ m}^2 \times (307^4 - 295^4) \text{ K}^4$$

$$= 140 \text{ W}$$

Discussion 140 W is a significant heat loss because the body also loses about 10 W by convection and conduction. To stay at a constant body temperature, an inactive person must give off heat at a rate of 90 W to account for basal metabolic activity; if the rate of heat loss exceeds that, the body temperature starts to drop. The patient had better wrap a blanket around his body or start running in place.

We need only the fraction of energy emitted and absorbed by the body; the emissivity of the walls of the room is irrelevant. If the walls are poor emitters, then they also absorb poorly, so they reflect radiation. The amount of radiation incident on the body is the same.

Practice Problem 14.14 The Roller Blader Radiates

Find how much energy per unit time a roller blader loses by radiation from her body. Her skin temperature is 35°C and the air temperature is 30°C. Her surface area is 1.2 m^2, of which 75% is exposed to the air. Assume skin has $e = 0.97$.

Example 14.15

Radiative Equilibrium of Earth

Radiant energy from the Sun reaches Earth at a rate of 1.7×10^{17} W. An average of about 30% is reflected and the rest is absorbed. Energy is also radiated by the atmosphere. Assuming equal rates of absorption and emission, and that the atmosphere emits as a blackbody in the infrared ($e = 1$), calculate the temperature of the atmosphere. (The Sun's radiation peaks in the visible part of the spectrum, but Earth's radiation peaks in the infrared due to its much lower surface temperature.)

Strategy Earth must radiate the same power as it absorbs. We use Stefan's law to find the rate at which energy is radiated as a function of temperature and then equate that to the rate of energy absorption.

Solution Earth absorbs 70% of the incident solar radiation. To have a relatively constant temperature, it must emit radiation at the same rate:

$$\mathcal{P} = 0.70 \times 1.7 \times 10^{17}\ \text{W} = 1.2 \times 10^{17}\ \text{W}$$

From Stefan's law,

$$\mathcal{P} = e\sigma AT^4$$

where we take $e = 1$ since the atmosphere is assumed to emit as a blackbody. Earth's surface area is

$$A = 4\pi R_E^2$$

Solving Stefan's law for T yields

$$T = \left(\frac{\mathcal{P}}{e\sigma A}\right)^{1/4}$$

Now we substitute numerical values:

$$T = \left(\frac{\mathcal{P}}{e\sigma 4\pi R_E^2}\right)^{1/4}$$

$$= \left[\frac{1.2 \times 10^{17}\ \text{W}}{1 \times 5.67 \times 10^{-8}\ \text{W/(m}^2\cdot\text{K}^4) \times 4\pi (6.4 \times 10^6\ \text{m})^2}\right]^{1/4}$$

$$= 253\ \text{K} = -20°\text{C}$$

Discussion Remember that −20°C is supposed to be the average temperature of the *atmosphere*, not of Earth's surface. This relatively simple calculation gives impressively accurate results. To find the temperature of Earth's surface, we must take the greenhouse effect into account.

Practice Problem 14.15 Reflecting Less Incident Radiation

If Earth were to reflect 25% of the incident radiation instead of 30%, what would be the average temperature of the atmosphere?

Application of Thermal Radiation: Global Climate Change

Earth receives heat by radiation from the Sun. The atmosphere helps trap some of the radiation, acting rather like the glass in a greenhouse. When sunlight falls on the glass of a greenhouse, most of the visible radiation and short-wavelength infrared (*near-infrared*) travel right on through; the glass is transparent to those wavelengths. The glass absorbs much of the incoming ultraviolet radiation. The radiation that gets through the glass is mostly absorbed inside the greenhouse. Since the inside of the greenhouse is much cooler than the Sun, it emits primarily infrared radiation (IR). The glass is not transparent to this longer-wavelength IR; much of it is absorbed by the glass. The glass itself also emits IR, but in both directions: half of it is emitted back inside the greenhouse. The absorption of IR by the glass keeps the greenhouse warmer than it would otherwise be. (The glass in a greenhouse has a second function not mirrored in Earth's atmosphere—it prevents heat from being carried away by convection.)

Earth is something like a greenhouse, where the atmosphere fulfills the role of the glass. Like glass, the atmosphere is largely transparent to visible and near IR; the ozone layer in the upper atmosphere absorbs some of the ultraviolet. The atmosphere absorbs a great deal of the longer-wavelength IR emitted by Earth's surface. The atmosphere *radiates* IR in two directions: back toward the surface and out toward space (Fig. 14.19). "Greenhouse gases" such as CO_2 and water vapor are particularly good absorbers of IR. The higher the concentration of greenhouse gases in the atmosphere, the more IR is

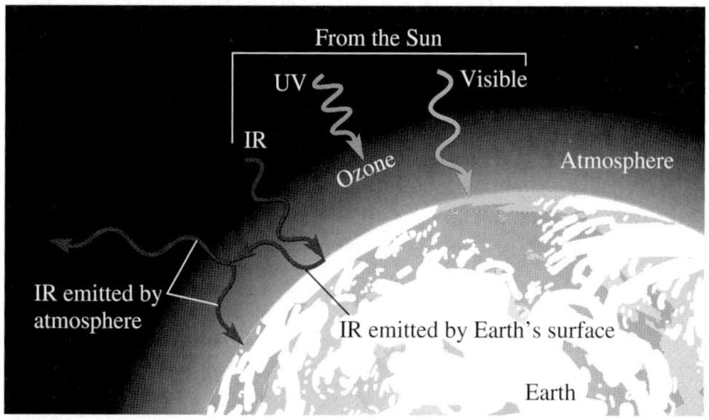

Figure 14.19 The global greenhouse effect. In this *simplified* diagram, all the UV from the Sun is absorbed by the atmosphere, while all the visible and IR from the Sun is transmitted. Earth absorbs the visible and IR and radiates longer-wavelength IR. The longer-wavelength IR is absorbed by the atmosphere, which itself radiates IR both back toward the surface and out toward space.

absorbed and the warmer Earth's surface becomes. Even small changes in the average surface temperature can have dramatic effects on climate.

In applying Stefan's radiation law to Earth, there are some complications. One is the effect of the cloud cover. Clouds are quite reflective, but they are sometimes there and sometimes not. The heating of the lakes and oceans causes water to evaporate and form clouds. The clouds then serve as a screen and reflect sunlight away from Earth, reducing the temperature again.

Master the Concepts

- The internal energy of a system is the total energy of all of the molecules in the system except for the macroscopic kinetic energy (kinetic energy associated with macroscopic translation or rotation) and the external potential energy (energy due to external interactions).
- Heat is a *flow* of energy that occurs due to a temperature difference.
- The joule is the SI unit for all forms of energy, for heat, and for work. An alternative unit sometimes used for heat and internal energy is the calorie:

$$1 \text{ cal} = 4.186 \text{ J} \qquad (14\text{-}1)$$

- The ratio of heat flow into a system to the temperature change of the system is the heat capacity of the system:

$$C = \frac{Q}{\Delta T} \qquad (14\text{-}2)$$

- The heat capacity per unit mass is the specific heat capacity (or specific heat) of a substance:

$$c = \frac{Q}{m\,\Delta T} \qquad (14\text{-}3)$$

- The *molar specific heat* is the heat capacity per mole:

$$C_V = \frac{Q}{n\,\Delta T} \qquad (14\text{-}6)$$

At room temperature, the molar heat capacity at constant volume for a monatomic ideal gas is approximately $C_V = \frac{3}{2}R$, and for a diatomic ideal gas it is approximately $C_V = \frac{5}{2}R$.

- Phase transitions occur at constant temperature. The heat *per unit mass* that must flow to melt a solid or to freeze a liquid is the latent heat of fusion L_f. The latent heat of vaporization L_v is the heat *per unit mass* that must flow to change the phase from liquid to gas or from gas to liquid.

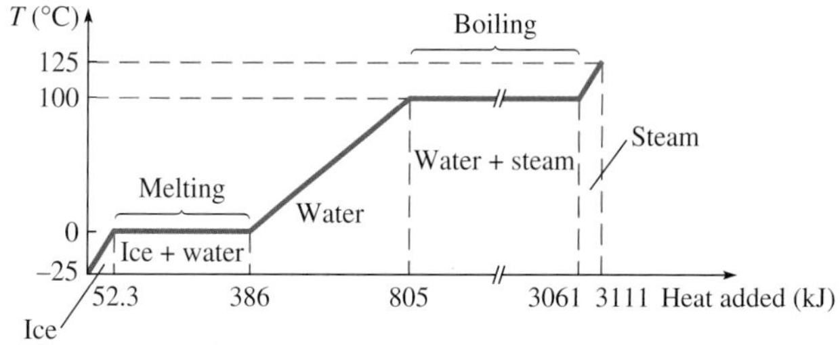

- *Sublimation* occurs when a solid changes directly to a gas without going into a liquid form.
- A phase diagram is a graph of pressure versus temperature that indicates solid, liquid, and gas regions for a

continued on next page

Master the Concepts continued

substance. The sublimation, fusion, and vapor pressure curves separate the three phases. Crossing one of these curves represents a phase transition.

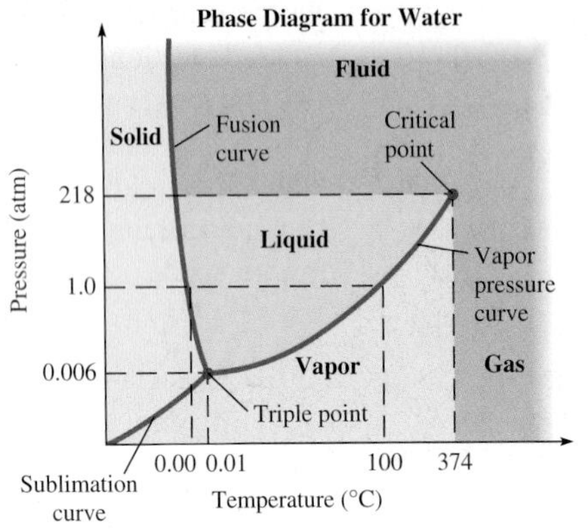

- Heat flows by three processes: conduction, convection, and radiation.
- Conduction is due to atomic (or molecular) collisions within a substance or from one object to another when they are in contact. The rate of heat flow within a substance is:

$$\mathcal{P} = \kappa A \frac{\Delta T}{d} \qquad (14\text{-}10)$$

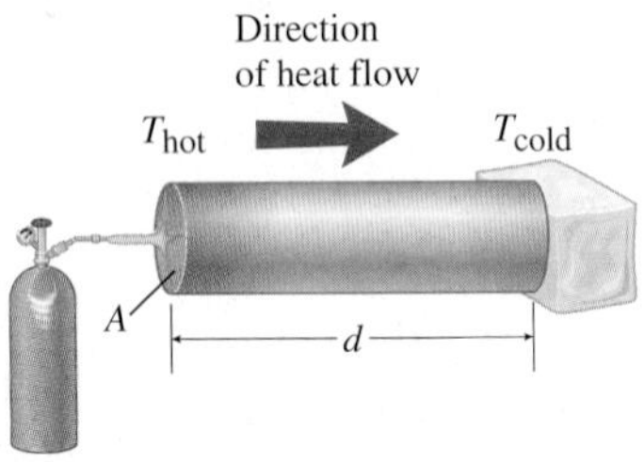

where $\mathcal{P}$ is the rate of heat flow (or power delivered), κ is the thermal conductivity of the material, A is the cross-sectional area, d is the thickness (or length) of the material, and ΔT is the temperature difference between one side and the other.

- Convection involves *fluid currents* that carry heat from one place to another. In convection, the material itself moves from one place to another.

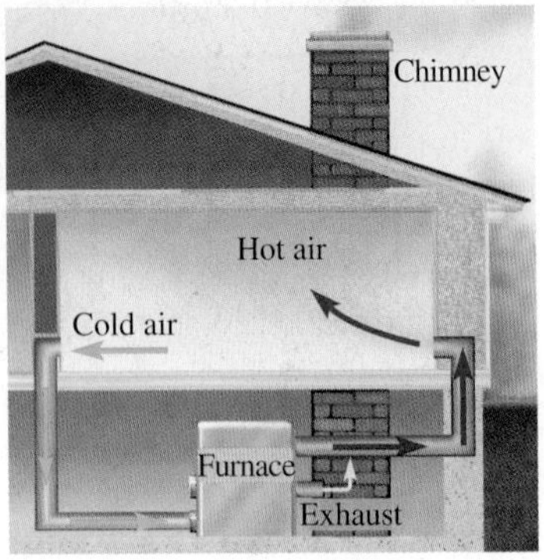

- Thermal radiation does not have to travel through a material medium. The energy is carried by electromagnetic waves that travel at the speed of light. All bodies emit energy through electromagnetic radiation. An idealized body that absorbs all the radiation incident on it is called a blackbody. A blackbody emits more radiant power per unit surface area than any real object at the same temperature. Stefan's law of thermal radiation is

$$\mathcal{P} = e\sigma A T^4 \qquad (14\text{-}16)$$

where the emissivity e ranges from 0 to 1, A is the surface area, T is the surface temperature of the blackbody *in kelvins*, and Stefan's constant is $\sigma = 5.670 \times 10^{-8}$ W/(m^2·K^4). The wavelength of maximum power emission is inversely proportional to the absolute temperature:

$$\lambda_{max} T = 2.898 \times 10^{-3} \text{ m·K} \qquad (14\text{-}17)$$

The difference between the power emitted by the body and that absorbed by the body from its surroundings is the net power emitted:

$$\mathcal{P}_{net} = e\sigma A(T^4 - T_s^4) \qquad (14\text{-}18)$$

Conceptual Questions

1. What determines the direction of heat flow when two objects at different temperatures are placed in thermal contact?
2. When an old movie has a scene of someone ironing, the person is often shown testing the heat of a hot flat iron with a moistened finger. Why is this safe to do?
3. Why do lakes and rivers freeze first at their surfaces?
4. Why is drinking water in a camp located near the equator often kept in porous jars?
5. Why are several layers of clothing warmer than one coat of equal weight?
6. Why are vineyards planted along lakeshores or riverbanks in cold climates?
7. A metal plant stand on a wooden deck feels colder than the wood around it. Is it necessarily colder? Explain.
8. Near a large lake, in what direction does a breeze passing over the land tend to blow at night?
9. What is the purpose of having fins on an automobile or motorcycle radiator?
10. Why do roadside signs warn that bridges ice before roadways? Explain.
11. Why do cooking directions on packages advise different timing to be followed for some locations?

12. Explain the theory behind the pressure cooker. How does it speed up cooking times?
13. When you eat a pizza that has just come from the oven, why is it that you are apt to burn the roof of your mouth with the first bite although the crust of the pizza feels only warm to your hand?
14. Explain why the molar specific heat of a diatomic gas such as O_2 is larger than that of a monatomic gas such as Ne.
15. At very low temperatures, the molar specific heat of hydrogen (H_2) is $C_V = 1.5R$. At room temperature, $C_V = 2.5R$. Explain.
16. When the temperature as measured in °C of a radiating body is doubled (such as a change from 20°C to 40°C), is the radiation rate necessarily increased by a factor of 16?
17. A cup of hot coffee has been poured, but the coffee drinker has a little more work to do at the computer before she picks up the cup. She intends to add some milk to the coffee. To keep the coffee hot as long as possible, should she add the milk at once, or wait until just before she takes her first sip?
18. Would heat loss be reduced or increased by increasing the usual air gap, 1 to 2 cm, between commercially made double-paned windows? Explain your reasoning. [*Hint:* Consider convection.]
19. A study of food preservation in Britain discovered that the temperature of meat that is kept in transparent plastic packages and stored in open and lighted freezers can be as much as 12°C above the temperature of the freezer. Why is this? How could this be prevented?
20. Which possesses more total internal energy, the water within a large, partially ice-covered lake in winter or a 6-cup teapot filled with hot tea? Explain.
21. A room in which the air temperature is held constant may feel warm in the summer but cool in the winter. Explain. [*Hint:* The walls are not necessarily at the same temperature as the air.]
22. Many homes are heated with "radiators," which are hollow metal devices filled with hot water or steam and located in each room of the house. They are sometimes painted with metallic, high-gloss silver paint so that they look well polished. Does this make them better radiators of heat? If not, what might be a more efficient finish to use?
23. Two objects with the same surface area are inside an evacuated container. The walls of the container are kept at a constant temperature. Suppose one object absorbs a larger fraction of incident radiation than the other. Explain why that object must emit a correspondingly greater amount of radiation than the other. Thus a good absorber must be a good emitter.
24. Even though heat is not a fluid, Eq. (14-11) has a close analogy in Poiseuille's law, which describes the viscous flow of a fluid through a pipe (see Problem 9.60). (a) Explain the analogy. (b) For two or more thermal conductors in series, the total thermal resistance is just the sum of the thermal resistances [Eq. (14-13)]. Is the total fluid flow resistance for two or more pipes in series equal to the sum of the resistances? Explain.

Multiple-Choice Questions

1. The main loss of heat from Earth is by
   (a) radiation.
   (b) convection.
   (c) conduction.
   (d) All three processes are significant modes of heat loss from Earth.
2. Assume the average temperature of Earth's atmosphere to be 253 K. What would be the eventual average temperature of Earth's atmosphere if the surface temperature of the Sun were to drop by a factor of 2?
   (a) 253 K
   (b) $\frac{253\text{ K}}{2} = 127\text{ K}$
   (c) $\frac{253\text{ K}}{4} = 63\text{ K}$
   (d) $\frac{253\text{ K}}{2^4} = 16\text{ K}$
3. In equilibrium, Mars emits as much radiation as it absorbs. If Mars orbits the Sun with an orbital radius that is 1.5 times the orbital radius of Earth about the Sun, what is the approximate atmospheric temperature of Mars? Assume the atmospheric temperature of Earth to be 253 K.
   (a) $\frac{253\text{ K}}{1.5} = 170\text{ K}$
   (b) $\frac{253\text{ K}}{1.5^2} = 112\text{ K}$
   (c) $\frac{253\text{ K}}{1.5^4} = 50\text{ K}$
   (d) $\frac{253\text{ K}}{\sqrt{1.5}} = 207\text{ K}$
4. Which term best represents the relation between a blackbody and radiant energy? A blackbody is an ideal _______ of radiant energy.
   (a) emitter (b) absorber
   (c) reflector (d) emitter and absorber
5. A window conducts power P from a house to the cold outdoors. What power is conducted through a window of *half* the area and *half* the thickness?
   (a) $4P$ (b) $2P$ (c) P (d) $P/2$ (e) $P/4$
6. Iron has a specific heat that is about 3.4 times that of gold. A cube of gold and a cube of iron, both of equal mass and at 20°C, are placed in two different Styrofoam cups, each filled with 100 g of water at 40°C. The Styrofoam cups have negligible heat capacities. After equilibrium has been attained,
   (a) the temperature of the gold is lower than that of the iron.
   (b) the temperature of the gold is higher than that of the iron.
   (c) the temperatures of the water in the two cups are the same.
   (d) Either (a) or (b), depending on the mass of the cubes.

7. Sublimation is involved in which of these phase changes?
   (a) liquid to gas (b) solid to liquid
   (c) solid to gas (d) gas to liquid
8. When a vapor condenses to a liquid,
   (a) its internal energy increases.
   (b) its temperature rises.
   (c) its temperature falls.
   (d) it gives off internal energy.
9. When a substance is at its triple point, it
   (a) is in its solid phase.
   (b) is in its liquid phase.
   (c) is in its vapor phase.
   (d) may be in any or all of these phases.
10. The phase diagram for water is shown in the figure. If the temperature of a certain amount of ice is increased by following the path represented by the dashed line from *A* to *B* in the phase diagram, which of the graphs of temperature as a function of heat added is correct?

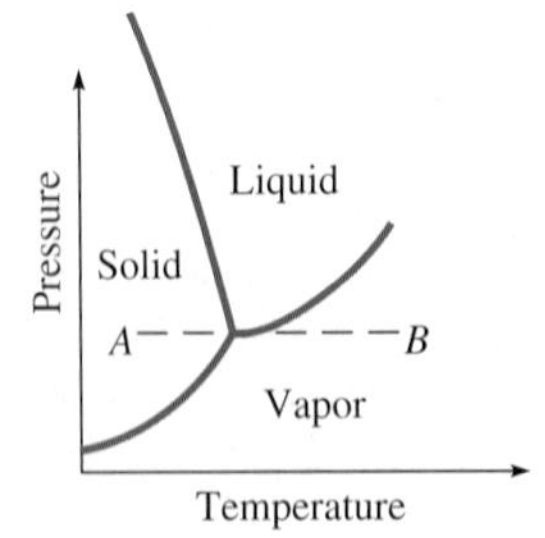

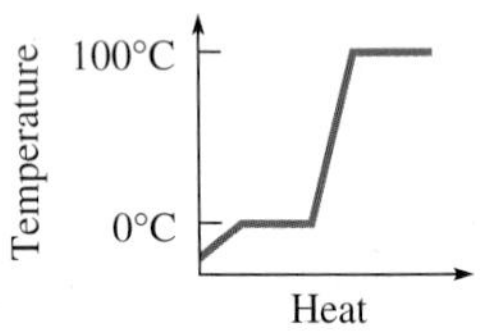

(a)

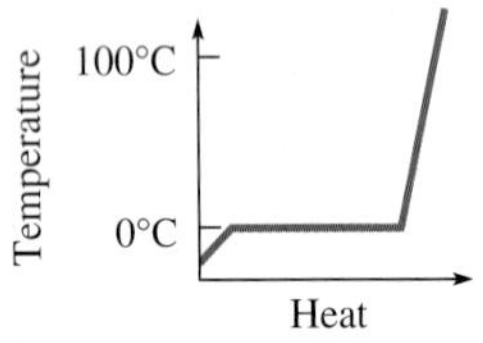

(b)

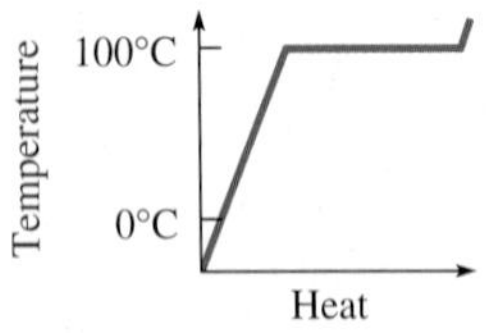

(c)

11. Two thin rods are made from the same material and are of lengths L_1 and L_2. The two ends of the rods have the same temperature difference. What should the relation be between their diameters and lengths so that they conduct equal amounts of heat energy in a given time?
   (a) $\frac{L_1}{L_2} = \frac{d_1}{d_2}$ (b) $\frac{L_1}{L_2} = \frac{d_2}{d_1}$
   (c) $\frac{L_1}{L_2} = \frac{d_1^2}{d_2^2}$ (d) $\frac{L_1}{L_2} = \frac{d_2^2}{d_1^2}$
12. If you place your hand underneath, but not touching, a kettle of hot water, you *mainly* feel the presence of heat from
   (a) conduction.
   (b) convection.
   (c) radiation.

Problems

- Ⓒ Combination conceptual/quantitative problem
- Biological or medical application
- ✦ Challenging problem
- Blue # Detailed solution in the Student Solutions Manual
- 1 2 Problems paired by concept
- Text website interactive or tutorial

14.1 Internal Energy

Ⓒ 1. A mass of 1.4 kg of water at 22°C is poured from a height of 2.5 m into a vessel containing 5.0 kg of water at 22°C. (a) How much does the internal energy of the 6.4 kg of water increase? (b) Is it likely that the water temperature increases? Explain.

2. The water passing over Victoria Falls, located along the Zambezi River on the border of Zimbabwe and Zambia, drops about 105 m. How much internal energy is produced per kilogram as a result of the fall?

3. How much internal energy is generated when a 20.0-g lead bullet, traveling at 7.00×10^2 m/s, comes to a stop as it strikes a metal plate?

4. Nolan threw a baseball, of mass 147.5 g, at a speed of 162 km/h to a catcher. How much internal energy was generated when the ball struck the catcher's mitt?

Ⓒ 5. A child of mass 15 kg climbs to the top of a slide that is 1.7 m above a horizontal run that extends for 0.50 m at the base of the slide. After sliding down, the child comes to rest just before reaching the very end of the horizontal portion of the slide. (a) How much internal energy was generated during this process? (b) Where did the generated energy go? (To the slide, to the child, to the air, or to all three?)

6. A 64-kg sky diver jumped out of an airplane at an altitude of 0.90 km. She opened her parachute after a while and eventually landed on the ground with a speed of 5.8 m/s. How much energy was dissipated by air resistance during the jump?

✦ 7. During basketball practice Shane made a jump shot, releasing a 0.60-kg basketball from his hands at a height of 2.0 m above the floor with a speed of 7.6 m/s. The ball swooshes through the net at a height of 3.0 m above the floor and with a speed of 4.5 m/s.

How much energy was dissipated by air drag from the time the ball left Shane's hands until it went through the net?

14.2 Heat; 14.3 Heat Capacity and Specific Heat

8. An experiment is conducted with a basic Joule apparatus, where a mass is allowed to descend by 1.25 m and rotate paddles within an insulated container of water. There are several different sizes of descending masses to choose among. If the investigator wishes to deliver 1.00 kJ to the water within the insulated container after 30.0 descents, what descending mass value should be used?
9. Convert 1.00 kJ to kilowatt-hours (kWh).
10. What is the heat capacity of 20.0 kg of silver?
11. What is the heat capacity of a gold ring that has a mass of 5.00 g?
12. If 125.6 kJ of heat are supplied to 5.00×10^2 g of water at 22°C, what is the final temperature of the water?
13. It is a damp, chilly day in a New England seacoast town suffering from a power failure. To warm up the cold, clammy sheets, Jen decides to fill hot water bottles to tuck between the sheets at the foot of the beds. If she wishes to heat 2.0 L of water on the wood stove from 20.0°C to 80.0°C, how much heat must flow into the water?
14. An 83-kg man eats a banana of energy content 1.00×10^2 kcal. If all of the energy from the banana is converted into kinetic energy of the man, how fast is he moving, assuming he starts from rest?
15. A high jumper of mass 60.0 kg consumes a meal of 3.00×10^3 kcal prior to a jump. If 3.3% of the energy from the food could be converted to gravitational potential energy in a single jump, how high could the athlete jump?
16. What is the heat capacity of a 30.0-kg block of ice?
17. What is the heat capacity of 1.00 m^3 of (a) aluminum? (b) iron? See Table 9.1 for density values.
18. What is the heat capacity of a system consisting of (a) a 0.450-kg brass cup filled with 0.050 kg of water? (b) 7.5 kg of water in a 0.75-kg aluminum bucket?
19. A 0.400-kg aluminum teakettle contains 2.00 kg of water at 15.0°C. How much heat is required to raise the temperature of the water (and kettle) to 100.0°C?
20. How much heat is required to raise the body temperature of a 50.0-kg woman from 37.0°C to 38.4°C?
21. It takes 880 J to raise the temperature of 350 g of lead from 0 to 20.0°C. What is the specific heat of lead?
22. A mass of 1.00 kg of water at temperature T is poured from a height of 0.100 km into a vessel containing water of the same temperature T, and a temperature change of 0.100°C is measured. What mass of water was in the vessel? Ignore heat flow into the vessel, the thermometer, etc.
23. A thermometer containing 0.10 g of mercury is cooled from 15.0°C to 8.5°C. How much energy left the mercury in this process?
24. A heating coil inside an electric kettle delivers 2.1 kW of electric power to the water in the kettle. How long will it take to raise the temperature of 0.50 kg of water from 20.0°C to 100.0°C? (WWW tutorial: heating)

14.4 Specific Heat of Ideal Gases

25. A cylinder contains 250 L of hydrogen gas (H_2) at 0.0°C and a pressure of 10.0 atm. How much energy is required to raise the temperature of this gas to 25.0°C?
26. A container of nitrogen gas (N_2) at 23°C contains 425 L at a pressure of 3.5 atm. If 26.6 kJ of heat are added to the container, what will be the new temperature of the gas?
27. Imagine that 501 people are present in a movie theater of volume 8.00×10^3 m^3 that is sealed shut so no air can escape. Each person gives off heat at an average rate of 110 W. By how much will the temperature of the air have increased during a 2.0-h movie? The initial pressure is 1.01×10^5 Pa and the initial temperature is 20.0°C. Assume that all the heat output of the people goes into heating the air (a diatomic gas).
28. A chamber with a fixed volume of 1.0 m^3 contains a monatomic gas at 3.00×10^2 K. The chamber is heated to a temperature of 4.00×10^2 K. This operation requires 10.0 J of heat. (Assume all the energy is transferred to the gas.) How many gas molecules are in the chamber?

14.5 Phase Transitions

29. As heat flows into a substance, its temperature changes according to the graph in the diagram. For what sections of the graph is the substance undergoing a phase change? For the sections you identified, what kind of phase change is occurring? (WWW tutorial: temperature graph)

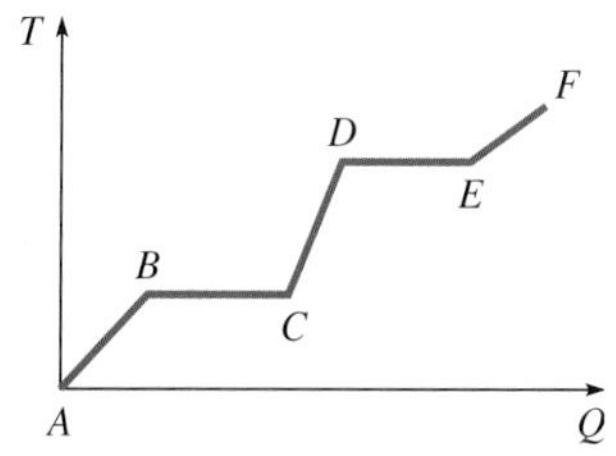

30. Given these data, compute the heat of vaporization of water. The specific heat capacity of water is 4.186 J/(g·K).

Mass of calorimeter = 3.00×10^2 g	Specific heat of calorimeter = 0.380 J/(g·K)
Mass of water = 2.00×10^2 g	Initial temperature of water and calorimeter = 15.0°C
Mass of condensed steam = 18.5 g	Initial temperature of steam = 100.0°C
	Final temperature of calorimeter = 62.0°C

31. Given these data, compute the heat of fusion of water. The specific heat capacity of water is 4.186 J/(g·K).

Mass of calorimeter = 3.00×10^2 g	Specific heat of calorimeter = 0.380 J/(g·K)
Mass of water = 2.00×10^2 g	Initial temperature of water and calorimeter = 20.0°C
Mass of ice = 30.0 g	Initial temperature of ice = 0°C
	Final temperature of calorimeter = 8.5°C

32. In a physics lab, a student accidentally drops a 25.0-g brass washer into an open dewar of liquid nitrogen at 77.2 K. How much liquid nitrogen boils away as the washer cools from 293 K to 77.2 K? The latent heat of vaporization for nitrogen is 199.1 kJ/kg.

33. What mass of water at 25.0°C added to a Styrofoam cup containing two 50.0-g ice cubes from a freezer at −15.0°C will result in a final temperature of 5.0°C for the drink?

34. How much heat is required to change 1.0 kg of ice, originally at −20.0°C, into steam at 110.0°C? Assume 1.0 atm of pressure.

35. Ice at 0.0°C is mixed with 5.00×10^2 mL of water at 25.0°C. How much ice must melt to lower the water temperature to 0.0°C?

36. Tina is going to make iced tea by first brewing hot tea, then adding ice until the tea cools. How much ice, at a temperature of −10.0°C, should be added to a 2.00×10^{-4} m^3 glass of tea at 95.0°C to cool the tea to 10.0°C? Ignore the temperature change of the glass. (tutorial: iced tea)

37. Repeat Problem 36 without neglecting the temperature change of the glass. The glass has a mass of 350 g and the specific heat of the glass is 0.837 kJ/(kg·K). By what percentage does the answer change from the answer for Problem 36?

38. The graph shows the change in temperature as heat is supplied to a certain mass of ice initially at −80.0°C. What is the mass of the ice?

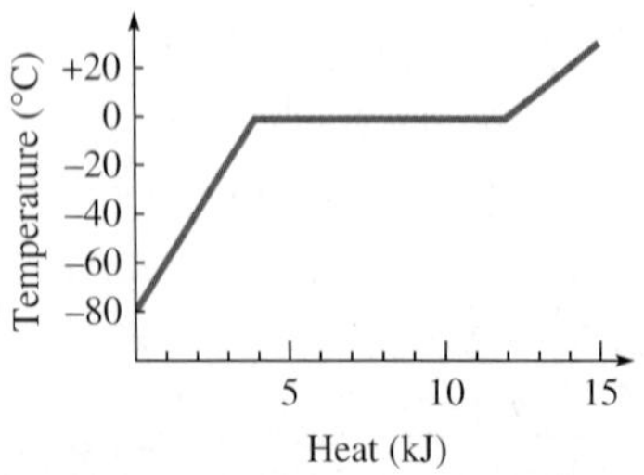

39. How many grams of aluminum at 80.0°C would have to be dropped into a hole in a block of ice at 0.0°C to melt 10.0 g of ice?

40. Is it possible to heat the aluminum of Problem 39 to a high enough temperature so that it melts an equal mass of ice? If so, what temperature must the aluminum have?

41. If a leaf is to maintain a temperature of 40°C (reasonable for a leaf), it must lose 250 W/m^2 by transpiration (evaporative heat loss). Note that the leaf also loses heat by radiation, but we will neglect this. How much water is lost after 1 h through transpiration only? The area of the leaf is 0.005 m^2.

42. A birch tree loses 618 mg of water per minute through transpiration (evaporation of water through stomatal pores). What is the rate of heat lost through transpiration?

43. You are given 250 g of coffee (same specific heat as water) at 80.0°C (too hot to drink). In order to cool this to 60.0°C, how much ice (at 0.0°C) must be added? Ignore heat content of the cup and heat exchanges with the surroundings.

44. A phase diagram is shown. Starting at point A, follow the dashed line to point E and consider what happens to the substance represented by this diagram as its pressure and temperature are changed. (a) Explain what happens for each line segment, AB, BC, CD, and DE. (b) What is the significance of point a and of point b?

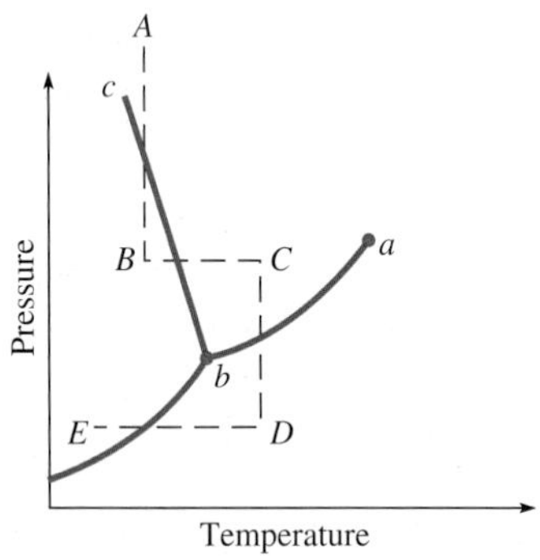

45. Compute the heat of fusion of a substance from these data: 31.15 kJ will change 0.500 kg of the solid at 21°C to liquid at 327°C, the melting point. The specific heat of the solid is 0.129 kJ/(kg·K).

46. A dog loses a lot of heat through panting. The air rushing over the upper respiratory tract causes evaporation and thus heat loss. A dog typically pants at a rate of 670 pants per minute. As a rough calculation, assume that one pant causes 0.010 g of water to be evaporated from the respiratory tract. What is the rate of heat loss for the dog through panting?

14.6 Thermal Conduction

47. (a) What thickness of cork would have the same R-factor as a 1.0-cm thick stagnant air pocket? (b) What thickness of tin would be required for the same R-factor?

48. A metal rod with a diameter of 2.30 cm and length of 1.10 m has one end immersed in ice at 32.0°F and the other end in boiling water at 212°F. If the ice melts at a rate of 1.32 g every 175 s, what is the thermal conductivity of this metal? Identify the metal. Assume there is no heat lost to the surrounding air.

49. Given a slab of material with area 1.0 m^2 and thickness 2.0×10^{-2} m, (a) what is the thermal resistance if the material is asbestos? (b) What is the thermal resistance if the material is iron? (c) What is the thermal resistance if the material is copper?

50. A copper rod of length 0.50 m and cross-sectional area 6.0×10^{-2} cm^2 is connected to an iron rod with the same

cross section and length 0.25 m. One end of the copper is immersed in boiling water and the other end is at the junction with the iron. If the far end of the iron rod is in an ice bath at 0°C, find the rate of heat transfer passing from the boiling water to the ice bath. Assume there is no heat loss to the surrounding air. (tutorial: composite rod)

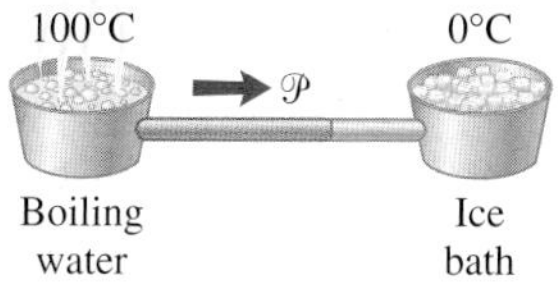

51. For a temperature difference $\Delta T = 20.0°C$, one slab of material conducts 10.0 W/m^2; another of the same shape conducts 20.0 W/m^2. What is the rate of heat flow per m^2 of surface area when the slabs are placed side by side with $\Delta T_{tot} = 20.0°C$?

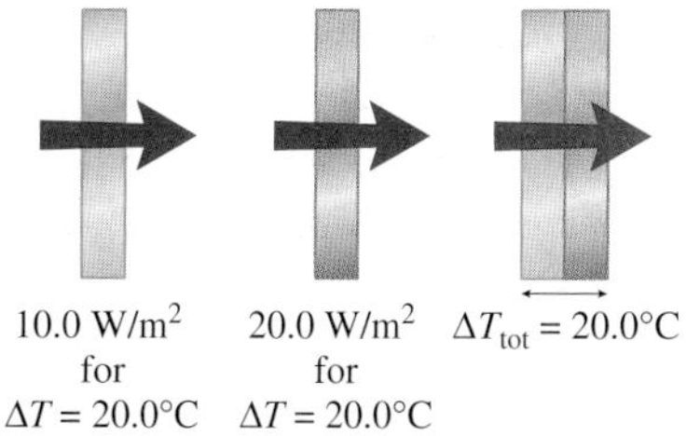

52. A wall consists of a layer of wood and a layer of cork insulation of the same thickness. The temperature inside is 20.0°C and the temperature outside is 0.0°C. (a) What is the temperature at the interface between the wood and cork if the cork is on the inside and the wood on the outside? (b) What is the temperature at the interface if the wood is inside and the cork is outside? (c) Does it matter whether the cork is placed on the inside or the outside of the wooden wall? Explain.

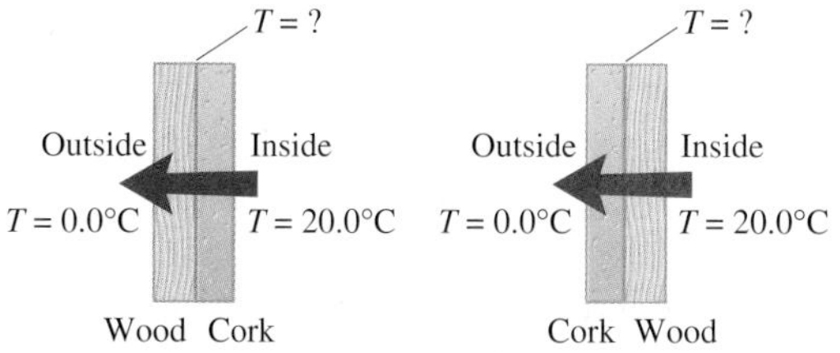

53. The thermal conductivity of the fur (including the skin) of a male Husky dog is 0.026 W/(m·K). The dog's heat output is measured to be 51 W, its internal temperature is 38°C, its surface area is 1.31 m^2, and the thickness of the fur is 5.0 cm. How cold can the outside temperature be before the dog must increase its heat output?

54. The thermal resistance of a seal's fur and blubber combined is 0.33 K/W. If the seal's internal temperature is 37°C and the temperature of the sea is about 0°C, what must be the heat output of the seal in order for it to maintain its internal temperature?

55. A hiker is wearing wool clothing of 0.50-cm thickness to keep warm. Her skin temperature is 35°C and the outside temperature is 4.0°C. Her body surface area is 1.2 m^2. (a) If the thermal conductivity of wool is 0.040 W/(m·K), what is the rate of heat conduction through her clothing? (b) If the hiker is caught in a rainstorm, the thermal conductivity of the soaked wool increases to 0.60 W/(m·K) (that of water). Now what is the rate of heat conduction?

56. A window whose glass has $\kappa = 1.0$ W/(m·K) is covered completely with a sheet of foam of the same thickness as the glass, but with $\kappa = 0.025$ W/(m·K). How is the rate at which heat is conducted through the window changed by the addition of the foam?

57. A copper bar of thermal conductivity 401 W/(m·K) has one end at 104°C and the other end at 24°C. The length of the bar is 0.10 m and the cross-sectional area is 1.0×10^{-6} m^2. (a) What is the rate of heat conduction, $\mathcal{P}$, along the bar? (b) What is the temperature gradient in the bar? (c) If two such bars were placed in series (end to end) between the same temperature baths, what would $\mathcal{P}$ be? (d) If two such bars were placed in parallel (side by side) with the ends in the same temperature baths, what would $\mathcal{P}$ be? (e) In the series case, what is the temperature at the junction where the bars meet?

58. One cross-country skier is wearing a down jacket that is 2.0 cm thick. The thermal conductivity of goose down is 0.025 W/(m·K). Her companion on the ski outing is wearing a wool jacket that is 0.50 cm thick. The thermal conductivity of wool is 0.040 W/(m·K). (a) If both jackets have the same surface area and the skiers both have the same body temperature, which one will stay warmer longer? (b) How much longer can the person with the warmer jacket stay outside for the same amount of heat loss?

14.8 Thermal Radiation

59. If a blackbody is radiating at $T = 1650$ K, at what wavelength is the maximum intensity?

60. Wien studied the spectral distribution of many radiating bodies to finally discover a simple relation between wavelength and intensity. Use the limited data shown in Fig. 14.17 to find the constant predicted by Wien for the product of wavelength of maximum emission and temperature.

61. An incandescent lightbulb has a tungsten filament that is heated to a temperature of 3.00×10^3 K when an electric current passes through it. If the surface area of the filament is approximately 1.00×10^{-4} m^2 and it has an emissivity of 0.32, what is the power radiated by the bulb?

62. A tungsten filament in a lamp is heated to a temperature of 2.6×10^3 K by an electric current. The tungsten has an emissivity of 0.32. What is the surface area of the filament if the lamp delivers 40.0 W of power?

63. A person of surface area 1.80 m^2 is lying out in the sunlight to get a tan. If the intensity of the incident sunlight is 7.00×10^2 W/m^2, at what rate must heat be lost by the person in order to maintain a constant body temperature? (Assume the effective area of skin exposed to the

Sun is 42% of the total surface area, 57% of the incident radiation is absorbed, and that internal metabolic processes contribute another 90 W for an inactive person.)

64. A student wants to lose some weight. He knows that rigorous aerobic activity uses about 700 kcal/h (2900 kJ/h) and that it takes about 2000 kcal per day (8400 kJ) just to support necessary biological functions, including keeping the body warm. He decides to burn calories faster simply by sitting naked in a 16°C room and letting his body radiate calories away. His body has a surface area of about 1.7 m^2 and his skin temperature is 35°C. Assuming an emissivity of 1.0, at what rate (in kcal/h) will this student "burn" calories?

65. An incandescent light bulb radiates at a rate of 60.0 W when the temperature of its filament is 2820 K. During a brownout (temporary drop in line voltage), the power radiated drops to 58.0 W. What is the temperature of the filament? Neglect changes in the filament's length and cross-sectional area due to the temperature change. (tutorial: light bulb)

66. If the maximum intensity of radiation for a blackbody is found at 2.65 μm, what is the temperature of the radiating body?

67. A black wood stove has a surface area of 1.20 m^2 and a surface temperature of 175°C. What is the net rate at which heat is radiated into the room? The room temperature is 20°C.

68. A lizard of mass 3.0 g is warming itself in the bright sunlight. It casts a shadow of 1.6 cm^2 on a piece of paper held perpendicularly to the Sun's rays. The intensity of sunlight at the Earth is 1.4×10^3 W/m^2, but only half of this energy penetrates the atmosphere and is absorbed by the lizard. (a) If the lizard has a specific heat of 4.2 J/(g·°C), what is the rate of increase of the lizard's temperature? (b) Assuming that there is no heat loss by the lizard (to simplify), how long must the lizard lie in the Sun in order to raise its temperature by 5.0°C?

69. At a tea party, a coffeepot and a teapot are placed on the serving table. The coffeepot is a shiny silver-plated pot with emissivity of 0.12; the teapot is ceramic and has an emissivity of 0.65. Both pots hold 1.00 L of liquid at 98°C when the party begins. If the room temperature is at 25°C, what is the rate of radiative heat loss from the two pots? [*Hint:* To find the surface area, approximate the pots with cubes of similar volume.]

70. If the total power per unit area from the Sun incident on a horizontal leaf is 9.00×10^2 W/m^2, and we assume that 70.0% of this energy goes into heating the leaf, what would be the rate of temperature rise of the leaf? The specific heat of the leaf is 3.70 kJ/(kg·°C), the leaf's area is 5.00×10^{-3} m^2, and its mass is 0.500 g.

71. Consider the leaf of Problem 70. Assume that the top surface of the leaf absorbs 70.0% of 9.00×10^2 W/m^2 of radiant energy, while the bottom surface absorbs all of the radiant energy incident on it due to its surroundings at 25.0°C. (a) If the only method of heat loss for the leaf were thermal radiation, what would be the temperature of the leaf? (Assume that the leaf radiates like a blackbody.) (b) If the leaf is to remain at a temperature of 25.0°C, how much power per unit area must be lost by other methods such as transpiration (evaporative heat loss)?

Comprehensive Problems

72. A hotel room is in thermal equilibrium with the rooms on either side and with the hallway on a third side. The room loses heat primarily through a 1.30-cm-thick glass window that has a height of 76.2 cm and a width of 156 cm. If the temperature inside the room is 75°F and the temperature outside is 32°F, what is the approximate rate (in kJ/s) at which heat must be added to the room to maintain a constant temperature of 75°F? Ignore the stagnant air layers on either side of the glass.

73. While camping, some students decide to make hot chocolate by heating water with a solar heater that focuses sunlight onto a small area. Sunlight falls on their solar heater, of area 1.5 m^2, with an intensity of 750 W/m^2. How long will it take 1.0 L of water at 15.0°C to rise to a boiling temperature of 100.0°C?

74. Five ice cubes, each with a mass of 22.0 g and at a temperature of −50.0°C, are placed in an insulating container. How much heat will it take to change the ice cubes completely into steam?

75. A 10.0-g iron bullet with a speed of 4.00×10^2 m/s and a temperature of 20.0°C is stopped in a 0.500-kg block of wood, also at 20.0°C. (a) At first all of the bullet's kinetic energy goes into the internal energy of the bullet. Calculate the temperature increase of the bullet. (b) After a short time the bullet and the block come to the same temperature *T*. Calculate *T*, assuming no heat is lost to the environment.

76. If the temperature surrounding the sunbather in Problem 63 is greater than the normal body temperature of 37°C and the air is still, so that radiation, conduction, and convection play no part in cooling the body, how much water (in liters per hour) from perspiration must be given off to maintain the body temperature? The heat of vaporization of water is 2430 J/g at normal skin temperature.

77. If 4.0 g of steam at 100.0°C condenses to water on a burn victim's skin and cools to 45.0°C, (a) how much heat is given up by the steam? (b) If the skin was originally at 37.0°C, how much tissue mass was involved in cooling the steam to water? See Table 14.1 for the specific heat of human tissue.

78. If 4.0 g of boiling water at 100.0°C was splashed onto a burn victim's skin, and if it cooled to 45.0°C on the 37.0°C skin, (a) how much heat is given up by the water? (b) How much tissue mass, originally at 37.0°C, was involved in cooling the water? See Table 14.1. Compare the result with that found in Problem 77.

79. The amount of heat generated during the contraction of muscle in an amphibian's leg is given by

$$Q = 0.544 \text{ mJ} + (1.46 \text{ mJ/cm})\Delta x$$

where Δx is the length shortened. If a muscle of length 3.0 cm and mass 0.10 g is shortened by 1.5 cm during a contraction, what is the temperature rise? Assume that the specific heat of muscle is 4.186 J/(g·°C).

80. Many species cool themselves by sweating, because as the sweat evaporates, heat is given up to the surroundings. A human exercising strenuously has an evaporative heat loss rate of about 650 W. If a person exercises strenuously for 30.0 min, how much water must he drink to replenish his fluid loss? The heat of vaporization of water is 2430 J/g at normal skin temperature.

81. A wall consists of a layer of wood outside and a layer of insulation inside. The temperatures inside and outside the wall are +22°C and −18°C; the temperature at the wood/insulation boundary is −8.0°C. By what factor would the heat loss through the wall increase if the insulation were not present?

82. Two 62-g ice cubes are dropped into 186 g of water in a glass. If the water is initially at a temperature of 24°C and the ice is at −15°C, what is the final temperature of the drink?

83. A 0.500-kg slab of granite is heated so that its temperature increases by 7.40°C. The amount of heat supplied to the granite is 2.93 kJ. Based on this information, what is the specific heat of granite?

84. A spring of force constant $k = 8.4 \times 10^3$ N/m is compressed by 0.10 m. It is placed into a vessel containing 1.0 kg of water and then released. Assuming all the energy from the spring goes into heating the water, find the change in temperature of the water.

85. One end of a cylindrical iron rod of length 1.00 m and of radius 1.30 cm is placed in the blacksmith's fire and reaches a temperature of 327°C. If the other end of the rod is being held in your hand (37°C), what is the rate of heat flow along the rod? The thermal conductivity of iron varies with temperature, but an average value between the two temperatures is 67.5 W/(m·K). (WW tutorial: conduction)

86. A blacksmith heats a 0.38-kg piece of iron to 498°C in his forge. After shaping it into a decorative design, he places it into a bucket of water to cool. If the available water is at 20.0°C, what minimum amount of water must be in the bucket to cool the iron to 23.0°C? The water in the bucket should remain in the liquid phase.

87. The student from Problem 64 realizes that standing naked in a cold room will not give him the desired weight loss results since it is much less efficient than simply exercising. So he decides to burn calories through conduction. He fills the bathtub with 16°C water and gets in. The water right next to his skin warms up to the same temperature as his skin, 35°C, but the water only 3.0 mm away remains at 16°C. At what rate (in kcal/h) would he "burn" calories? The thermal conductivity of water at this temperature is 0.58 W/(m·K). [*Warning:* Do not try this. Sitting in water this cold can lead to hypothermia and even death.]

88. A stainless steel saucepan, with a base that is made of 0.350-cm-thick steel [$\kappa = 46.0$ W/(m·K)] fused to a 0.150-cm thickness of copper [$\kappa = 401$ W/(m·K)], sits on a ceramic heating element at 104.00°C. The diameter of the pan is 18.0 cm and it contains boiling water at 100.00°C. (a) If the copper-clad bottom is touching the heat source, what is the temperature at the copper-steel interface? (b) At what rate will the water evaporate from the pan?

89. A 75-kg block of ice at 0.0°C breaks off from a glacier, slides along the frictionless ice to the ground from a height of 2.43 m, and then slides along a horizontal surface consisting of gravel and dirt. Find how much of the mass of the ice is melted by the friction with the rough surface, assuming 75% of the internal energy generated is used to heat the ice.

90. Small animals eat much more food per kg of body mass than do larger animals. The basal metabolic rate (BMR) is the minimal energy intake necessary to sustain life in a state of complete inactivity. The table lists the BMR, mass, and surface area for five animals. (a) Calculate the BMR/kg of body mass for each animal. Is it true that smaller animals must consume much more food per kg of body mass? (b) Calculate the BMR/m^2 of surface area. (c) Can you explain why the BMR/m^2 is approximately the same for animals of different sizes? Consider what happens to the food energy metabolized by an animal in a resting state.

Animal	BMR (kcal/day)	Mass (kg)	Surface Area (m^2)
Mouse	3.80	0.018	0.0032
Dog	770	15	0.74
Human	2050	64	2.0
Pig	2400	130	2.3
Horse	4900	440	5.1

91. Imagine a person standing naked in a room at 23.0°C. The walls are well insulated, so they also are at 23.0°C. The person's surface area is 2.20 m^2 and his basal metabolic rate is 2167 kcal/day. His emissivity is 0.97. (a) If the person's skin temperature were 37.0°C (the same as the internal body temperature), at what net rate would heat be lost through radiation? (Ignore losses by conduction and convection.) (b) Clearly the heat loss in (a) is not sustainable—but skin temperature is less than internal body temperature. Calculate the skin temperature such that the net heat loss due to radiation is equal to the basal metabolic rate. (c) Does wearing clothing slow the loss of heat by radiation, or does it only decrease losses by conduction and convection? Explain.

92. Bare, dark-colored basalt has a thermal conductivity of 3.1 W/(m·K), whereas light-colored sandstone's thermal

conductivity is only 2.4 W/(m·K). Even though the same amount of radiation is incident on both and their surface temperatures are the same, the temperature gradient within the two materials will differ. For the same patch of area, what is the ratio of the depth in basalt as compared with the depth in sandstone that gives the same temperature difference?

93. The power expended by a cheetah is 160 kW while running at 110 km/h, but its body temperature cannot exceed 41.0°C. If 70.0% of the energy expended is dissipated within its body, how far can it run before it overheats? Assume that the initial temperature of the cheetah is 38.0°C, its specific heat is 3.5 kJ/(kg·°C), and its mass is 50.0 kg.

94. A scientist working late at night in her low-temperature physics laboratory decides to have a cup of hot tea, but discovers the lab hot plate is broken. Not to be deterred, she puts about 8 oz of water, at 12°C, from the tap into a lab dewar (essentially a large thermos bottle) and begins shaking it up and down. With each shake the water is thrown up and falls back down a distance of 33.3 cm. If she can complete 30 shakes per minute, how long will it take to heat the water to 87°C? Would this really work? If not, why not?

95. A 2.0-kg block of copper at 100.0°C is placed into 1.0 kg of water in a 2.0-kg iron pot. The water and the iron pot are at 25.0°C just before the copper block is placed into the pot. What is the final temperature of the water, assuming negligible heat flow to the environment?

96. A piece of gold of mass 0.250 kg and at a temperature of 75.0°C is placed into a 1.500-kg copper pot containing 0.500 L of water. The pot and water are at 22.0°C before the gold is added. What is the final temperature of the water?

97. For a cheetah, 70.0% of the energy expended during exertion is internal work done on the cheetah's system and is dissipated within his body; for a dog only 5.00% of the energy expended is dissipated within the dog's body. Assume that both animals expend the same total amount of energy during exertion, both have the same heat capacity, and the cheetah is 2.00 times as heavy as the dog. (a) How much higher is the temperature change of the cheetah compared to the temperature change of the dog? (b) If they both start out at an initial temperature of 35.0°C, and the cheetah has a temperature of 40.0°C after the exertion, what is the final temperature of the dog? Which animal probably has more endurance? Explain.

98. A 20.0-g lead bullet leaves a rifle at a temperature of 87.0°C and hits a steel plate. If the bullet melts, what is the minimum speed it must have?

99. The inner vessel of a calorimeter contains 2.50×10^2 g of tetrachloromethane, CCl_4, at 40.00°C. The vessel is surrounded by 2.00 kg of water at 18.00°C. After a time, the CCl_4 and the water reach the equilibrium temperature of 18.54°C. What is the specific heat of CCl_4?

100. On a very hot summer day, Daphne is off to the park for a picnic. She puts 0.10 kg of ice at 0°C in a thermos and then adds a grape-flavored drink, which she has mixed from a powder using room temperature water (25°C). How much grape-flavored drink will just melt all the ice?

101. It requires 17.10 kJ to melt 1.00×10^2 g of urethane [$CO_2(NH_2)C_2H_5$] at 48.7°C. What is the latent heat of fusion of urethane in kJ/mol?

102. A 20.0-g lead bullet leaves a rifle at a temperature of 47.0°C and travels at a velocity of 5.00×10^2 m/s until it hits a large block of ice at 0°C and comes to rest within it. How much ice will melt?

Answers to Practice Problems

14.1 4.9 J

14.2 Higher. The molecules have the same amount of *random* translational kinetic energy plus the additional kinetic energy associated with the ball's translation and rotation.

14.3 350 g

14.4 at least \$0.89

14.5 48°C

14.6 92 kJ

14.7 150 kJ

14.8 40 kJ

14.9 53.5 g

14.10 230 W

14.11 110 W

14.12 To maintain constant temperature, the net heat must be zero. The rate at which energy is emitted is 140 W.

14.13 9.4 μm (at 35°C) to 9.6 μm (at 30°C)

14.14 28 W

14.15 −16°C

Answers to Checkpoints

14.2 No, the temperature increase is not caused by heat flow. When you stretch the rubber band, you do work on it. This increases its internal energy and its temperature. (If you now put the rubber band down, heat does flow *out* of the rubber band, decreasing its internal energy and its temperature until it is in thermal equilibrium with its surroundings.)

14.5 The steam releases a large quantity of heat as it condenses into water on the skin. Much more energy is transferred to the skin than would be the case for the same amount of water at 100°C.

14.6 The rate of heat flow through the two materials is the same, so the material with the larger thermal conductivity has the smaller temperature gradient. Figure 14.10b shows that the temperature gradient is smaller in the material on the left, so it has the larger thermal conductivity.

CHAPTER

15

Thermodynamics

The gasoline engines in cars are terribly inefficient. Of the chemical energy that is released in the burning of gasoline, typically only 20% to 25% is converted into useful mechanical work done on the car to move it forward. Yet scientists and engineers have been working for decades to make a more efficient gasoline engine. Is there some fundamental limit to the efficiency of a gasoline engine? Is it possible to make an engine that converts all—or nearly all—of the chemical energy in the fuel into useful work? (See p. 544 for the answer.)

Concepts & Skills to Review

- conservation of energy (Section 6.1)
- internal energy and heat (Sections 14.1–14.2)
- zeroth law of thermodynamics (Section 13.1)
- system and surroundings (Section 14.1)
- work done is the area under a graph of $F_x(x)$ (Section 6.6)
- heat capacity (Section 14.3)
- the ideal gas law (Section 13.5)
- specific heat of ideal gases at constant volume (Section 14.4)
- natural logarithm (Appendix A.3)

15.1 THE FIRST LAW OF THERMODYNAMICS

Both work and heat can change the internal energy of a system. Work can be done on a rubber ball by squeezing it, stretching it, or slamming it into a wall. Heat will flow into the ball if it is left out in the Sun or put into a hot oven. These two methods of changing the internal energy of a system lead to the **first law of thermodynamics**:

The choice of a *system* is made in any way convenient for a given problem.

First Law of Thermodynamics

The change in internal energy of a system is equal to the heat flow into the system plus the work done *on* the system.

CONNECTION:

The first law is not a new principle—just a specialized form of energy conservation.

The first law is a specialized statement of energy conservation applied to a thermodynamic system, such as a gas inside a cylinder that has a movable piston. The gas can exchange energy with its surroundings in two ways. Heat can flow between the gas and its surroundings when they are at different temperatures and work can be done on the gas when the piston is pushed in.

In equation form, we can write

The symbol U, previously used for potential energy, is used exclusively for *internal energy* in this chapter. Internal energy was defined in Section 14.1.

First Law of Thermodynamics

$$\Delta U = Q + W \tag{15-1}$$

CONNECTION:

Our sign conventions for Q and W are consistent with their definitions in previous chapters (Chapter 6 for work and Chapter 14 for heat).

In Eq. (15-1), ΔU is the change in internal energy of the system. The internal energy can increase or decrease, so ΔU can be positive or negative. The signs of Q and W have the same meaning we have used in previous chapters. If heat flows into the system, Q is positive, while if heat flows out of the system, Q is negative. W represents the work done *on* the system, which can be positive or negative, depending on the directions of the applied force and the displacement. Using the example of the gas in a cylinder, if the piston is pushed in, then the force on the gas due to the piston and the displacement of the gas are in the same direction (Fig. 15.1a) and W is positive. If the piston moves out, then the force and the displacement are in opposite directions, because the piston still pushes inward on the gas, and W is negative (Fig. 15.1b). Table 15.1 summarizes the meanings of the signs of ΔU, Q, and W.

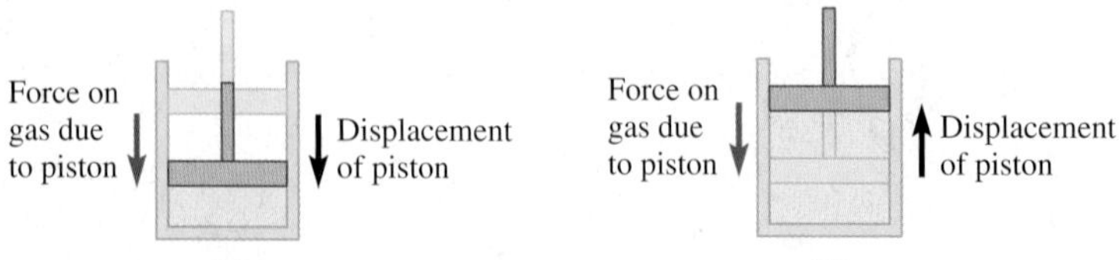

Figure 15.1 (a) When a gas is compressed, the work done *on* the gas is positive. (b) When a gas expands, the work done *on* the gas is *negative*.

Table 15.1 Sign Conventions for the First Law of Thermodynamics

Quantity	Definition	Meaning of + Sign	Meaning of − Sign
Q	Heat flow into the system	Heat flows *into* the system	Heat flows *out of* the system
W	Work done *on* the system	Surroundings do *positive* work on the system	Surroundings do *negative* work on the system (system does positive work on the surroundings)
ΔU	Internal energy change	Internal energy *increases*	Internal energy *decreases*

Example 15.1

Stirring a Can of Paint

A contractor uses a paddle stirrer to mix a can of paint (Fig. 15.2). The paddle turns at 28.0 rad/s and exerts a torque of 16.0 N·m on the paint, doing work at a rate

$$\text{power} = \tau\omega = 16.0\ \text{N}\cdot\text{m} \times 28.0\ \text{rad/s} = 448\ \text{W}$$

An internal energy increase of 12.5 kJ causes the temperature of the paint to increase by 1.00 K. (a) If there were no heat flow between the paint and the surroundings, what would be the temperature change of the paint as it is stirred for 5.00 min? (b) If the actual temperature change was 6.3 K, how much heat flowed from the paint to the surroundings?

Strategy From conservation of energy, the change in the internal energy of the paint is equal to the heat flow into the paint plus the work done on the paint.

Solution (a) In 5.00 min, the work done by the paddle on the paint is

$$W = 0.448\ \text{kJ/s} \times 5.00\ \cancel{\text{min}} \times 60\ \text{s}/\cancel{\text{min}} = 134.4\ \text{kJ}$$

Figure 15.2
An electric paint stirrer does work on the paint as it stirs.

Since we assume no heat flow ($Q = 0$), the internal energy of the paint changes by $\Delta U = Q + W = +134.4$ kJ. The temperature increases 1.00 K for every 12.5 kJ of increased internal energy, so

$$\Delta T = 134.4\ \text{kJ} \times \frac{1.00\ \text{K}}{12.5\ \text{kJ}} = 10.8\ \text{K}$$

(b) To apply the first law, we first find the internal energy change:

$$\Delta U = \frac{12.5\ \text{kJ}}{1.00\ \text{K}} \times 6.3\ \text{K} = 78.75\ \text{kJ}$$

Now we apply the first law:

$$\Delta U = Q + W$$

$$Q = \Delta U - W = 78.75\ \text{kJ} - 134.4\ \text{kJ} = -56\ \text{kJ}$$

Q is negative because 56 kJ of heat flow *out of* the paint.

Discussion How did we know the work done by the paddle on the paint was *positive*? Think of the force the paddle exerts on the paint as it pushes paint out of its way; the force and the displacement are in the same direction.

The quantity 12.5 kJ/K is the heat capacity of the paint—it tells us how many kJ the internal energy of the paint must increase for its temperature to increase 1 K, *regardless of whether the internal energy increase is caused by heat, work, or a combination of the two.*

Conceptual Practice Problem 15.1
Changing Internal Energy of a Gas

While 14 kJ of heat flows into the gas in a cylinder with a moveable piston, the internal energy of the gas increases by 42 kJ. Was the piston pulled out or pushed in? Explain. [*Hint:* Determine whether the piston does positive or negative work on the gas.]

15.2 THERMODYNAMIC PROCESSES

A thermodynamic process is the method by which a system is changed from one *state* to another. The state of a system is described by a set of **state variables** such as pressure, temperature, volume, number of moles, and internal energy. State variables describe the state of a system at some instant of time but not how the system got to that state. Heat and work are *not* state variables—they describe *how* a system gets from one state to another.

The *PV* Diagram

If a system is changed so that it is always very near equilibrium, the changes in state can be represented by a curve on a plot of pressure versus volume (called a ***PV* diagram**). Each point on the curve represents an equilibrium state of the system. The *PV* diagram is a useful tool for analyzing thermodynamic processes. One of the chief uses of a *PV* diagram is to find the work done on the system.

CONNECTION:

In Chapter 6, we saw that work is represented by the area under a graph of force versus displacement. Here we use the same concept; we just modify which variables are being graphed.

Work and Area Under a *PV* Curve Figure 15.3a shows the expansion of a gas, starting with volume V_i and pressure P_i; Fig. 15.3b is the *PV* diagram for the process. In Fig. 15.3, the force exerted by the piston on the gas is downward, while the displacement of the gas is upward, so the piston does negative work on the gas. This work represents a transfer of energy from the gas to its surroundings. (Equivalently, we can say the gas does positive work on the piston.) The piston pushes against the gas with a force of magnitude $F = PA$, where P is the pressure of the gas and A is the cross-sectional area of the piston. This force is not constant since the pressure decreases as the gas expands. As was shown in Section 6.6, the work done by a variable force is the area under a graph of $F_x(x)$.

To see how work is related to the area under the curve, first note that the units of $P \times V$ are those of work:

$$[\text{pressure} \times \text{volume}] = [\text{Pa}] \times [\text{m}^3] = \frac{[\text{N}]}{[\text{m}^2]} \times [\text{m}^3] = [\text{N}] \times [\text{m}] = [\text{J}]$$

So far, so good. Imagine that the piston moves out a *small* distance *d*—small enough that the pressure change is insignificant. The work done on the gas is

$$W = Fd \cos 180° = -PAd$$

The volume change of the gas is

$$\Delta V = Ad$$

So the work done on the gas is

$$W = -P\,\Delta V \qquad (15\text{-}2)$$

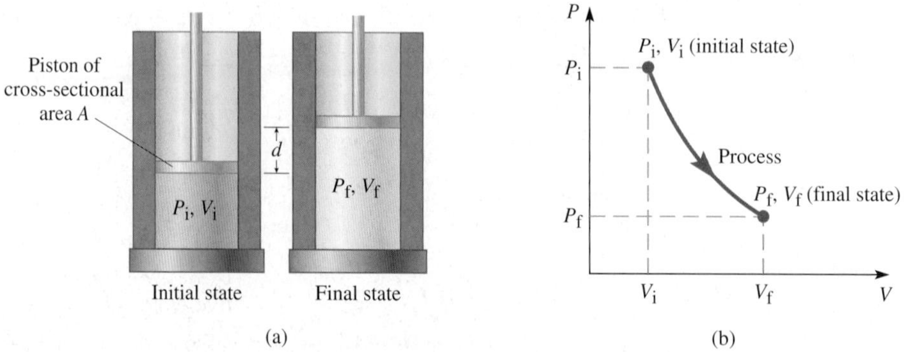

Figure 15.3 (a) Expansion of a gas from initial pressure P_i and volume V_i to final pressure P_f and volume V_f. During the expansion, *negative* work is done on the gas by the moving piston because the force exerted on the gas and the displacement are in opposite directions. (b) A *PV* diagram for the expansion shows the pressure and volume of the gas starting at the initial values, passing through intermediate values, and ending at the final values.

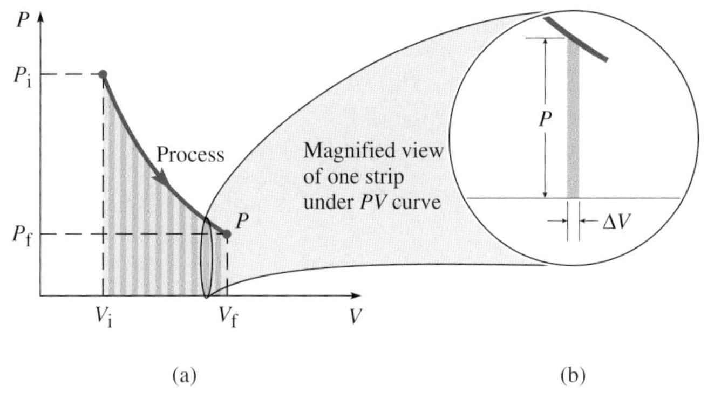

Figure 15.4 (a) The area under the *PV* curve is divided into many narrow strips of width ΔV and of varying heights *P*. The sum of the areas of the strips is the total area under the *PV* curve, which represents the magnitude of the work done on the gas. (b) An enlarged view of one strip under the curve. If the strip is very narrow, we can ignore the change in *P* and approximate its area as $P\Delta V$.

Magnitude of work done *on a system* = area under *PV* curve. $W > 0$ for compression and $W < 0$ for expansion.

To find the *total* work done on the gas, we add up the work done during each small volume change. During each small ΔV, the magnitude of the work done is the area of a thin strip of height *P* and width ΔV under the *PV* curve (Fig. 15.4). Therefore, the magnitude of the total work done on the gas is the area under the *PV* curve. During an increase in volume, ΔV is positive and the work done on the gas is negative. During a decrease in volume, ΔV is negative and the work done on the gas is positive.

The magnitude of the work done on a system depends on the *path* taken on the *PV* curve. Figure 15.5 shows two other possible paths between the same initial and final states as those of Fig. 15.4. In Fig. 15.5a, the pressure is kept constant at the initial value P_i while the volume is increased from V_i to V_f. Then the volume is kept constant while the pressure is reduced from P_i to P_f. The magnitude of the work done is represented by the shaded area under the *PV* curve; it is greater than the magnitude of the work done in Fig. 15.4a. Alternatively, in Fig. 15.5b the pressure is first reduced from P_i to P_f while the volume is held fixed; then the volume is allowed to increase from V_i to V_f while the pressure is kept at P_f. We see by the shaded area that the magnitude of the work done this way is less than the magnitude of the work done in Fig. 15.4a. The work done differs from one process to another, even though the initial and final states are the same in each case.

Work Done During a Closed Cycle Because the work done on a system depends on the path on the *PV* diagram, the net work done on a system during a **closed cycle**—a series of processes that leave the system in the same state it started in—can be nonzero. For example, during the cycle 1→2→3→4→1 in Fig. 15.5c, you can verify that the net work done on the gas is negative. Equivalently, the net work done *by* the gas is positive. A closed cycle during which the system does net work is the essential idea behind the heat engine (Section 15.5).

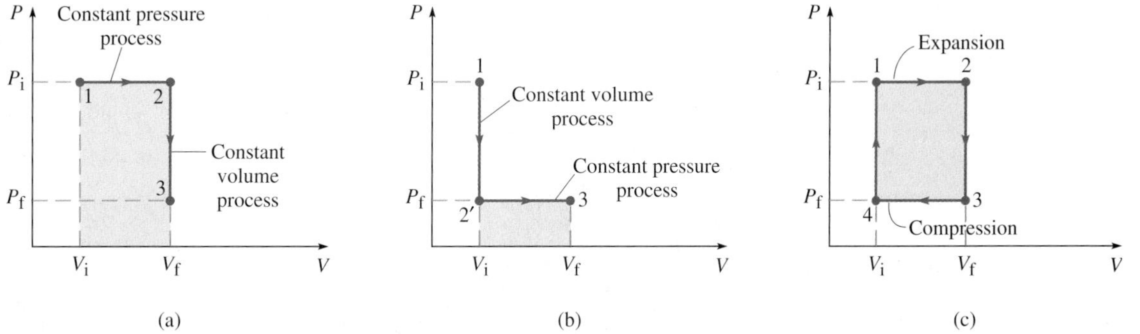

Figure 15.5 (a) and (b) Two different paths between the same initial and final states. (c) A closed cycle. The net work done on the gas during this cycle is the negative of the area inside the rectangle because the negative work done during expansion (1→2) is greater in magnitude than the positive work done during compression (3→4).

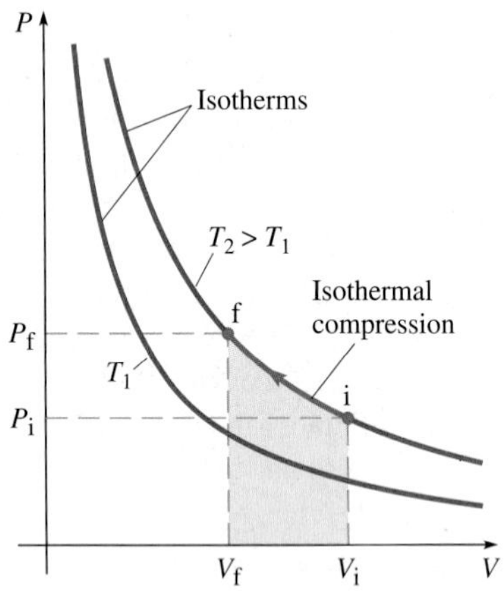

Figure 15.6 Isotherms for an ideal gas at two different temperatures. Each isotherm is a graph of $P = nRT/V$ for a constant temperature. The shaded area represents the work done by the gas during an isothermal compression at temperature T_2, which is positive. (The work done by the gas during an isothermal *expansion* would be *negative.*)

Constant Pressure Processes

A process by which the state of a system is changed while the pressure is held constant is called an *isobaric* process. The word *isobaric* comes from the same Greek root as the word "barometer." In Fig. 15.5a, the first change of state from V_i to V_f along the line from 1 to 2 occurs at the constant pressure P_i. A constant pressure process appears as a horizontal line on the *PV* diagram. The work done on the gas is

$$W = -P_i (V_f - V_i) = -P_i \Delta V \quad \text{(constant pressure)} \tag{15-3}$$

Constant Volume Processes

A process by which the state of a system is changed while the *volume* remains constant is called an *isochoric* process. Such a process is illustrated in Fig. 15.5a when the system moves along the line from 2 to 3 as the pressure changes from P_i to P_f at the constant volume V_f. No work is done during a constant volume process; without a displacement, work cannot be done. The area under the *PV* curve—a vertical line—is zero:

$$W = 0 \quad \text{(constant volume)} \tag{15-4}$$

If no work is done, then from the first law of thermodynamics, the change in internal energy is equal to the heat flow into the system:

$$\Delta U = Q \quad \text{(constant volume)} \tag{15-5}$$

Constant Temperature Processes

A process in which the temperature of the system remains constant is called an **isothermal** process. On a *PV* diagram, a path representing a constant temperature process is called an **isotherm** (Fig. 15.6). All the points on an isotherm represent states of the system with the same temperature.

Definition of heat reservoir

How can we keep the temperature of the system constant? One way is to put the system in thermal contact with a **heat reservoir**—something with a heat capacity so large that it can exchange heat in either direction without changing its temperature significantly. Then as long as the state of the system does not change too rapidly, the heat flow between the system and the reservoir keeps the system's temperature constant.

Adiabatic Processes

A process in which no heat is transferred into or out of the system is called an **adiabatic** process. An adiabatic process is *not* the same as a constant temperature (isothermal) process. In an isothermal process, heat flow into or out of a system is necessary to maintain a constant temperature. In an adiabatic process, *no* heat flow occurs, so if work is done, the temperature of the system may change. One way to perform an adiabatic process is to completely insulate the system so that no heat can flow in or out; another way is to perform the process so quickly that there is no time for heat to flow in or out.

For example, the compressions and rarefactions caused by a sound wave occur so fast that heat flow from one place to another is negligible. Hence, the compressions and rarefactions are adiabatic. Isaac Newton made a now-famous error when he assumed that these processes were isothermal and calculated a speed of sound that was about 20% lower than the measured value.

PHYSICS AT HOME

Hold an elastic band against your lip; it should feel cool. Now grasp the elastic and stretch it back and forth rapidly several times. Hold the stretched region to your lip. Does it feel warm? The elastic's temperature is higher because the work you did in stretching it increased its internal energy. The rapid stretching is approximately adiabatic—it occurs quickly so there is little time for heat to flow out of the elastic.

Table 15.2 Summary of Thermodynamic Processes

Process	Name	Condition	Consequences
Constant temperature	Isothermal	T = constant	(*For an ideal gas*, $\Delta U = 0$)
Constant pressure	Isobaric	P = constant	$W = -P\,\Delta V$
Constant volume	Isochoric	V = constant	$W = 0$; $\Delta U = Q$
No heat flow	Adiabatic	$Q = 0$	$\Delta U = W$

From the first law of thermodynamics

$$\Delta U = Q + W \quad (15\text{-}1)$$

With $Q = 0$,

$$\Delta U = W \text{ (adiabatic)}$$

Table 15.2 summarizes all of the thermodynamic processes discussed. (See also the interactive: thermodynamics.)

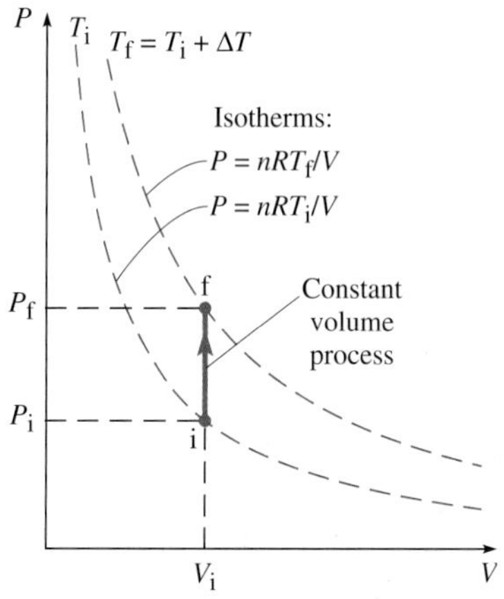

Figure 15.7 A *PV* diagram for a constant volume process for an ideal gas. Every point on an isotherm (red dashed lines) represents a state of the gas at the same temperature.

CHECKPOINT 15.2

(a) Can an adiabatic process cause a change in temperature? Explain. (b) Can heat flow during an isothermal process? (c) Can the internal energy of a system change during an isothermal process?

15.3 THERMODYNAMIC PROCESSES FOR AN IDEAL GAS

Constant Volume

Figure 15.7 is a *PV* diagram for heat flow into an ideal gas at constant volume. Since the temperature of the gas changes, the initial and final states are shown as points on two different isotherms. (Note that the higher-temperature isotherm is farther from the origin.) The area under the vertical line is zero; no work is done when the volume is constant. With $W = 0$, the heat flow increases the internal energy of the gas, so the temperature increases.

CONNECTION:

Section 15.2 described some general aspects of various thermodynamic processes. Now we find out what happens when the system undergoing the process is an ideal gas.

In Section 14.4, we discussed the molar specific heat of an ideal gas at constant volume. The first law of thermodynamics enables us to calculate the internal energy change ΔU. Since no work is done during a constant volume process, $\Delta U = Q$. For a constant volume process, $Q = nC_V\,\Delta T$ and therefore,

$$\Delta U = nC_V\,\Delta T \quad \text{(ideal gas)} \quad (15\text{-}6)$$

Internal energy is a state variable—its value depends only on the current state of the system, not on the path the system took to get there. Therefore, as long as the number of moles is constant, *the internal energy of an ideal gas changes only when the temperature changes.* Equation (15-6) therefore gives the internal energy change of an ideal gas for *any* thermodynamic process, not just for constant volume processes.

Constant Pressure

Another common situation is when the *pressure* of the gas is constant. In this case, work is done because the volume changes. The first law of thermodynamics enables us to calculate the molar specific heat at constant pressure (C_P), which is different from the molar specific heat at constant volume (C_V).

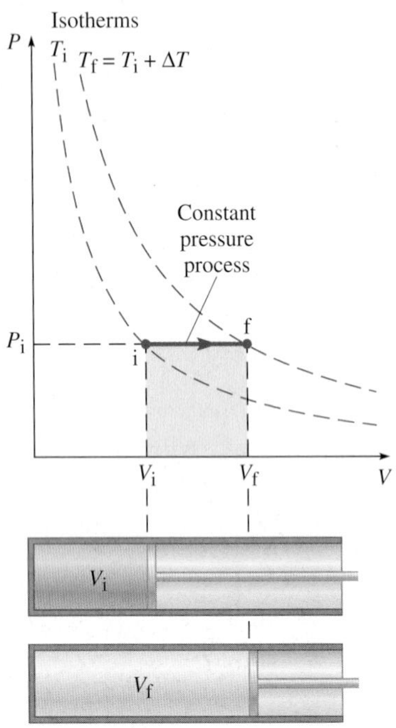

Figure 15.8 A PV diagram of a constant pressure expansion of an ideal gas. Heat flows into the ideal gas ($Q > 0$). The increase in the internal energy ΔU is less than Q because negative work is done on the expanding gas by the piston. The work done by the gas is the negative of the shaded area under the path.

Figure 15.8 shows a PV diagram for the constant pressure expansion of an ideal gas starting and ending at the same temperatures as for the constant volume process of Fig. 15.7. Applying the first law to the constant pressure process requires that

$$\Delta U = Q + W$$

where the work done on the gas is, from the ideal gas law,

$$W = -P\,\Delta V = -nR\,\Delta T$$

The definition of C_P is

$$Q = nC_P\,\Delta T \qquad (15\text{-}7)$$

Substituting Q and W into the first law, we obtain

$$\Delta U = nC_P\,\Delta T - nR\,\Delta T \qquad (15\text{-}8)$$

Since the internal energy of an ideal gas is determined by its temperature, ΔU for this constant pressure process is the same as ΔU for the constant volume process between the same two temperatures:

$$\Delta U = nC_V\,\Delta T \qquad (15\text{-}6)$$

Then

$$nC_V\,\Delta T = nC_P\,\Delta T - nR\,\Delta T$$

Canceling common factors of n and ΔT, this reduces to

$$C_P = C_V + R \qquad \text{(ideal gas)} \qquad (15\text{-}9)$$

Since R is a positive constant, the molar specific heat of an ideal gas at constant pressure is larger than the molar specific heat at constant volume.

Is this result reasonable? When heat flows into the gas at constant pressure, the gas expands, doing work on the surroundings. Thus, not all of the heat goes into increasing the internal energy of the gas. More heat has to flow into the gas at constant pressure for a given temperature increase than at constant volume.

Example 15.2

Warming a Balloon at Constant Pressure

A weather balloon is filled with helium gas at 20.0°C and 1.0 atm of pressure. The volume of the balloon after filling is measured to be 8.50 m^3. The helium is heated until its temperature is 55.0°C. During this process, the balloon expands at constant pressure (1.0 atm). What is the heat flow into the helium?

Strategy We can find how many moles of gas n are present in the balloon by using the ideal gas law. For this problem, we consider the helium to be a system. Helium is a monatomic gas, so its molar specific heat at constant *volume* is $C_V = \frac{3}{2}R$. The molar specific heat at constant *pressure* is then $C_P = C_V + R = \frac{5}{2}R$. Then the heat flow into the gas during its expansion is $Q = nC_P\Delta T$.

Solution The ideal gas law is

$$PV = nRT$$

We know the pressure, volume, and temperature: $P = 1.0$ atm $= 1.01 \times 10^5$ Pa, $V = 8.50$ m^3, and $T = 273$ K $+ 20.0$°C $= 293$ K. Solving for the number of moles yields

$$n = \frac{PV}{RT} = \frac{1.01 \times 10^5 \text{ Pa} \times 8.50 \text{ m}^3}{8.31 \text{ J/(mol·K)} \times 293 \text{ K}} = 352.6 \text{ mol}$$

For an ideal gas at constant pressure, the heat required to change the temperature is

$$Q = nC_P\,\Delta T$$

where $C_P = \frac{5}{2}R$. The temperature change is

$$\Delta T = 55.0°\text{C} - 20.0°\text{C} = 35.0 \text{ K}$$

Now we have everything we need to find Q:

$$Q = nC_P\,\Delta T = 352.6 \text{ mol} \times \tfrac{5}{2} \times 8.31 \text{ J/(mol·K)} \times 35.0 \text{ K}$$
$$= 260 \text{ kJ}$$

continued on next page

Example 15.2 continued

Discussion We do not have to find the work done on the gas separately and then subtract it from the change in internal energy to find Q. The work done is *already* accounted for by the molar specific heat at constant pressure. This simplifies the problem since we use the same method for constant pressure as we use for constant volume; the only change is the choice of C_V or C_P.

Practice Problem 15.2 Air Instead of Helium

Suppose the balloon were filled with dry air instead of helium. Find Q for the same temperature change. (Dry air is mostly N_2 and O_2, so assume an ideal diatomic gas.)

Constant Temperature

For an ideal gas, we can plot isotherms using the ideal gas law $PV = nRT$ (Fig. 15.6). Since the change in internal energy of an ideal gas is proportional to the temperature change,

$$\Delta U = 0 \quad \text{(ideal gas, isothermal process)} \qquad (15\text{-}10)$$

From the first law of thermodynamics, $\Delta U = 0$ means that $Q = -W$. Note that Eq. (15-10) is true for an *ideal gas* at constant temperature. Other systems can change internal energy without changing temperature; one example is when the system undergoes a phase change.

It can be shown (using calculus to find the area under the PV curve) that the work done on an ideal gas during a constant temperature expansion or contraction from volume V_i to volume V_f is

$$W = nRT \ln\left(\frac{V_i}{V_f}\right) \quad \text{(ideal gas, isothermal)} \qquad (15\text{-}11)$$

In Eq. (15-11), "ln" stands for the natural (or base-e) logarithm.

Example 15.3

Constant Temperature Compression of an Ideal Gas

An ideal gas is kept in thermal contact with a heat reservoir at 7°C (280 K) while it is compressed from a volume of 20.0 L to a volume of 10.0 L (Fig. 15.9). During the compression, an average force of 33.3 kN is used to move the piston a distance of 0.15 m. How much heat is exchanged between the gas and the reservoir? Does the heat flow into or out of the gas?

Strategy We can find the work done on the gas from the average force applied and the distance moved. For isothermal compression of an ideal gas, $\Delta U = 0$. Then $Q = -W$.

Solution The work done on the gas is

$$W = fd = 33.3\ \text{kN} \times 0.15\ \text{m} = 5.0\ \text{kJ}$$

This work adds 5.0 kJ to the internal energy of the gas. Then 5.0 kJ of heat must flow out of the gas if its internal energy

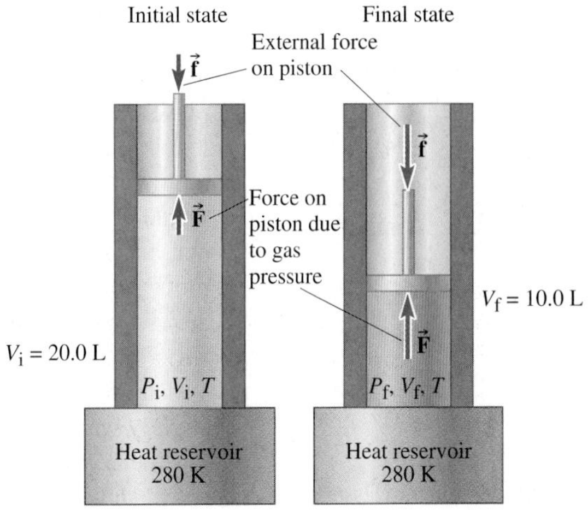

Figure 15.9
Isothermal compression of an ideal gas. Thermal contact with a heat reservoir keeps the gas at a constant temperature.

continued on next page

Example 15.3 continued

does not change. The work done on the gas is positive since the piston is pushed with an inward force as it moves inward.

$$Q = -W = -5.0 \text{ kJ}$$

Since positive Q represents heat flow *into* the gas, the negative sign tells us that heat flows out of the gas into the reservoir.

Discussion Although the temperature remains constant during the process, it does not mean that no heat flows. To maintain a constant temperature when work is done on the gas, some heat must flow out of the gas. If the gas were thermally isolated so no heat could flow, then the work done on the gas would increase the internal energy, resulting in an increase in the temperature of the gas.

Practice Problem 15.3 Work Done During Constant Temperature Expansion of a Gas

Suppose 2.0 mol of an ideal gas are kept in thermal contact with a heat reservoir at 57°C (330 K) while the gas expands from a volume of 20.0 L to a volume of 40.0 L. Does heat flow into or out of the gas? How much heat flows? [*Hint:* Use $W = nRT \ln (V_i/V_f)$, which applies to an ideal gas at constant temperature.]

15.4 REVERSIBLE AND IRREVERSIBLE PROCESSES

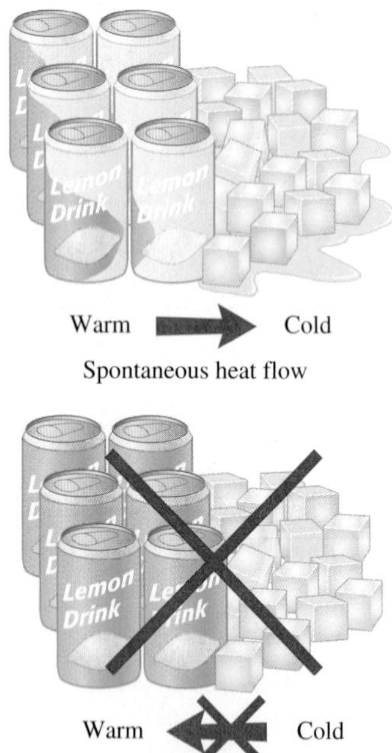

Figure 15.10 Spontaneous heat flow goes from warm to cool; the reverse does not happen spontaneously.

Have you ever wished you could make time go backward? Perhaps you accidentally broke an irreplaceable treasure in a friend's house, or missed a one-time opportunity to meet your favorite movie star, or said something unforgivable to someone close to you. Why can't the clock be turned around?

Imagine a perfectly elastic collision between two billiard balls. If you were to watch a movie of the collision, you would have a hard time telling whether the movie was being played forward or backward. The laws of physics for an elastic collision are valid even if the direction of time is reversed. Since the total momentum and the total kinetic energy are the same before and after the collision, the reversed collision is physically possible.

The perfectly elastic collision is one example of a **reversible** process. A reversible process is one that does not violate any laws of physics if "played in reverse." Most of the laws of physics do not distinguish forward in time from backward in time. A projectile moving in the absence of air resistance (on the Moon, say) is reversible: if we play the movie backward, the total mechanical energy is still conserved and Newton's second law ($\Sigma\vec{\mathbf{F}} = m\vec{\mathbf{a}}$) still holds at every instant in the projectile's trajectory.

Notice the caveats in the examples: "perfectly elastic" and "in the absence of air resistance." If friction or air resistance is present, then the process is **irreversible**. If you played *backward* a movie of a projectile with noticeable air resistance, it would be easy to tell that something is wrong. The force of air resistance on the projectile would act in the wrong direction—in the direction of the velocity, instead of opposite to it. The same would be true for sliding friction. Slide a book across the table; friction slows it down and brings it to rest. The macroscopic kinetic energy of the book—due to the orderly motion of the book in one direction—has been converted into disordered energy associated with the random motion of molecules; the table and book will be at slightly higher temperatures. The reversed process certainly would never occur, even though it does not violate the first law of thermodynamics (energy conservation). We would not expect a slightly warmed book placed on a slightly warmed table surface to spontaneously begin to slide across the table, gaining speed and cooling off as it goes, even if the total energy is the same before and after. It is easy to convert ordered energy into disordered energy, but not so easy to do the reverse. *The presence of energy dissipation (sliding friction, air resistance) always makes a process irreversible.*

Irreversible processes do *not* violate energy conservation.

As another example of an irreversible process, imagine placing a container of warm lemonade into a cooler with some ice (Fig. 15.10). Some of the ice melts and the lemonade gets cold as heat flows out of the lemonade and into the ice. The reverse would never happen: putting cold lemonade into a cooler with some partially melted ice, we would never find that the lemonade gets warmer as the liquid water freezes. *Spontaneous heat flow from a hotter body to a colder body is always irreversible.*

Conceptual Example 15.4

Irreversibility and Energy Conservation

Suppose heat *did* flow spontaneously from the cold ice to the warm lemonade, making the ice colder and the lemonade warmer. Would conservation of energy be violated by this process?

Solution and Discussion Heat flow from the ice to the lemonade would increase the internal energy of the lemonade by the same amount that the internal energy of the ice would decrease. The total internal energy of the ice and the lemonade would remain unchanged—energy would be conserved. The process would never occur, but *not* because energy conservation would be violated.

Conceptual Practice Problem 15.4 A Campfire

On a camping trip, you gather some twigs and logs and start a fire. Discuss the campfire in terms of irreversible processes.

As we will see later in this chapter, irreversible processes such as the frictional dissipation of energy and the spontaneous heat flow from a hotter to a colder body can be thought of in terms of a change in the amount of order in the system. A system never goes *spontaneously* from a disordered state to a more ordered state. Reversible processes are those that do not change the total amount of disorder in the universe; irreversible processes increase the amount of disorder.

Second Law of Thermodynamics According to the **second law of thermodynamics**, the total amount of disorder in the universe never decreases. Irreversible processes increase the disorder of the universe. We see in Section 15.8 that the second law is based on the statistics of systems with extremely large numbers of atoms or molecules. For now, we start with an equivalent statement of the second law, phrased in terms of heat flow:

Second Law of Thermodynamics (Clausius Statement)

Heat never flows spontaneously from a colder body to a hotter body.

Spontaneous heat flow from a colder body to a hotter body would decrease the total disorder in the universe.

The second law of thermodynamics determines what we sense as the direction of time—none of the other physical laws we have studied would be violated if the direction of time were reversed.

CHECKPOINT 15.4

A perfectly elastic collision is reversible. What about an inelastic collision? Explain.

15.5 HEAT ENGINES

We said in Section 15.4 that it is far *easier* to convert ordered energy into disordered energy than to do the reverse. Converting ordered into disordered energy occurs spontaneously, but the reverse does not. A **heat engine** is a device designed to convert disordered energy into ordered energy. We will see that there is a fundamental limitation on how much ordered energy (mechanical work) can be produced by a heat engine from a given amount of disordered energy (heat).

Application of thermodynamics: heat engines

The development of practical steam engines—heat engines that use steam as the working substance—around the beginning of the eighteenth century was one of the crucial elements in the industrial revolution. These steam engines were the first machines that produced a sustained work output using an energy source other than muscle, wind, or moving water. Steam engines are still used in many electric power plants.

The source of energy in a heat engine is most often the burning of some fuel such as gasoline, coal, oil, natural gas, and the like. A nuclear power plant is a heat engine using energy released by a nuclear reaction instead of a chemical reaction (as in burning). A geothermal engine uses the high temperature found beneath the Earth's crust (which comes to the surface in places such as volcanoes and hot springs).

Cyclical Engines The engines that we will study operate in cycles. Each cycle consists of several thermodynamic processes that are repeated the same way during each cycle. In order for these processes to repeat the same way, the engine must end the cycle in the same state in which it started. In particular, the internal energy of the engine must be the same at the end of a cycle as it was in the beginning. Then for one complete cycle,

$$\Delta U = 0$$

From the first law of thermodynamics (energy conservation),

$$Q_{\text{net}} + W_{\text{net}} = 0 \quad \text{or} \quad |W_{\text{net}}| = |Q_{\text{net}}|$$

Therefore, for a cyclical heat engine,

> The net work done by an engine during one cycle is equal to the net heat flow into the engine during the cycle.

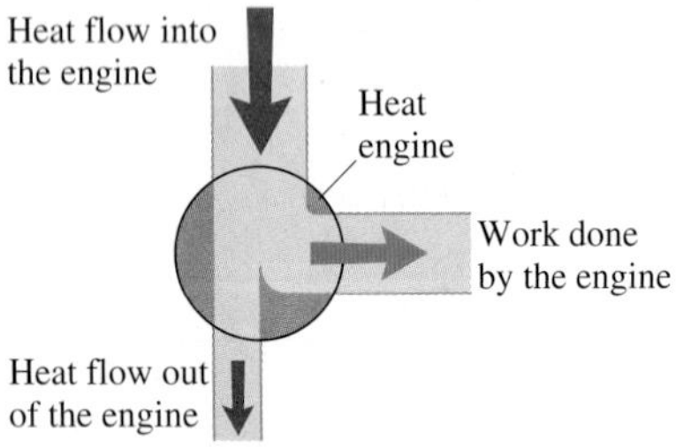

Figure 15.11 A heat engine. The engine is represented by a circle and the arrows indicate the direction of the energy flow. The total energy entering the engine during one cycle equals the total energy leaving the engine during the cycle.

We stress that it is the *net* heat flow since an engine not only takes in heat but exhausts some as well. Figure 15.11 shows the energy transfers during one cycle of a heat engine.

Application: the Internal Combustion Engine One familiar engine is the internal combustion engine found in automobiles. *Internal* combustion refers to the fact that gasoline is burned inside a cylinder; the resulting hot gases push against a piston and do work. A steam engine is an *external* combustion engine. The coal burned, for example, releases heat that is used to make steam; the steam is the working substance of the engine that drives the turbines.

Most automobile engines work in a cyclic thermodynamic process shown in Fig. 15.12. Of the energy released by burning gasoline, only about 20% to 25% is turned into mechanical work used to move the car forward and run other systems. The rest is discarded. The hot exhaust gases carry energy out of the engine, as does the liquid cooling system.

Efficiency of an Engine

To measure how effectively an engine converts heat into mechanical work, we define the engine's **efficiency** e as what you get (net useful work) divided by what you supply (heat input):

Efficiency of an engine:

$$e = \frac{\text{net work done by the engine}}{\text{heat input}} = \frac{W_{\text{net}}}{Q_{\text{in}}} \qquad (15\text{-}12)$$

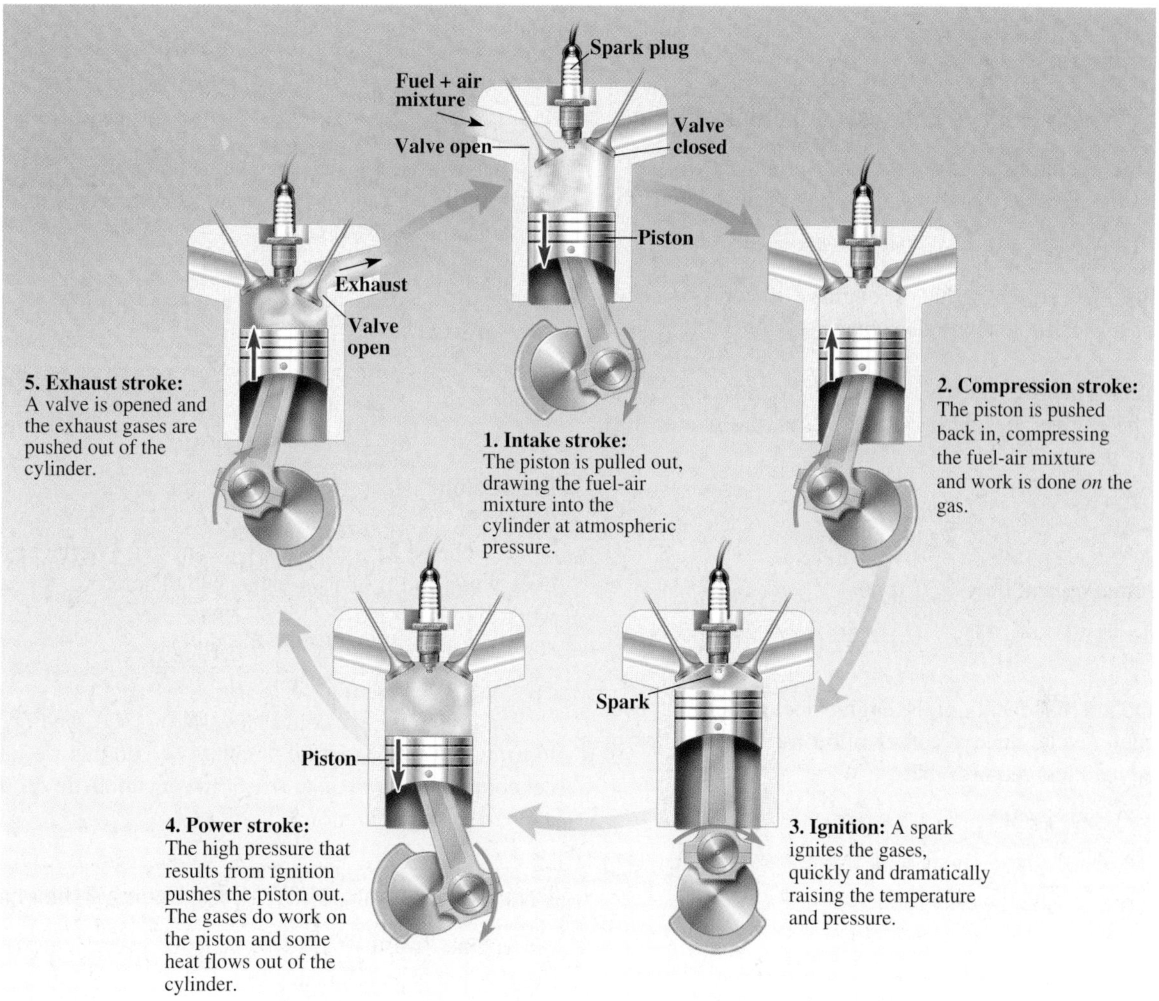

Figure 15.12 The four-stroke automobile engine. Each cycle has four strokes during which the piston moves (steps 1, 2, 4, and 5).

To avoid getting mixed up by algebraic signs, we let the symbols Q_{in}, Q_{out}, and W_{net} stand for the *magnitudes* of the heat flows into and out of the engine and the net work done by the engine *during one or more cycles*. Hence, Q_{in}, Q_{out}, and W_{net} are never negative. We supply minus signs in equations when necessary, based on the direction of energy flow. Doing so helps keep us focused on what is happening physically with the energy flows, rather than on a sign convention. (We will do the same when we discuss refrigerators and heat pumps later in this chapter.)

The efficiency is stated as either a fraction or a percentage. It gives the fraction of the heat input that is turned into useful work. Note that the heat input is *not* the same as the net heat flow into the engine; rather,

$$Q_{net} = Q_{in} - Q_{out} \qquad (15\text{-}13)$$

The efficiency of an engine is less than 100% because some of the heat input is exhausted, instead of being converted into useful work.

If an engine does work at a constant rate and its efficiency does not change, then it also takes in and exhausts heat at constant rates. The work done, heat input, and heat exhausted during any time interval are all proportional to the elapsed time. Therefore, all the same relationships that are true for the amounts of heat flow and work done apply to the *rates* at which heat flows and work is done. For example, the efficiency is

$$e = \frac{\text{net work done}}{\text{heat input}} = \frac{\text{net } rate \text{ of doing work}}{rate \text{ of taking in heat}} = \frac{W_{net}/\Delta t}{Q_{in}/\Delta t}$$

Example 15.5

Rate at Which Heat Is Exhausted from an Engine

An engine operating at 25% efficiency produces work at a rate of 0.10 MW. At what rate is heat exhausted into the surroundings?

Strategy We are given that the engine does work at a constant *rate*. The efficiency is also constant.

Solution The efficiency is the ratio of $W_{net}/\Delta t$, the rate at which the engine does net work, to $Q_{in}/\Delta t$, the rate of heat flow into the engine:

$$e = \frac{W_{net}}{Q_{in}} = \frac{W_{net}/\Delta t}{Q_{in}/\Delta t}$$

The *net* rate of heat flow $Q_{net}/\Delta t$ is

$$\frac{Q_{net}}{\Delta t} = \frac{Q_{in}}{\Delta t} - \frac{Q_{out}}{\Delta t}$$

Since the internal energy of the engine does not change over a complete cycle, energy conservation (or the first law of thermodynamics) requires that

$$Q_{net} = W_{net} \quad \text{or} \quad Q_{in} - Q_{out} = W_{net}$$

In terms of the rate at which heat is delivered or exhausted and the rate at which work is done,

$$\frac{Q_{in}}{\Delta t} - \frac{Q_{out}}{\Delta t} = \frac{W_{net}}{\Delta t}$$

In other words, the rate at which the engine does work is equal to the *net* rate of heat input. We are asked to find the rate of heat exhausted $Q_{out}/\Delta t$:

$$\frac{Q_{out}}{\Delta t} = \frac{Q_{in}}{\Delta t} - \frac{W_{net}}{\Delta t} = \frac{W_{net}/\Delta t}{e} - \frac{W_{net}}{\Delta t}$$

$$= \frac{W_{net}}{\Delta t}\left(\frac{1}{e} - 1\right) = 0.10 \text{ MW} \times \left(\frac{1}{0.25} - 1\right)$$

$$= 0.30 \text{ MW}$$

Discussion Heat flows *out* of the engine at a rate of 0.30 MW.

As a check: 25% efficiency means that $\frac{1}{4}$ of the heat input does work and $\frac{3}{4}$ of it is exhausted. Therefore, the ratio of work to exhaust is

$$\frac{1/4}{3/4} = \frac{1}{3} = \frac{0.10 \text{ MW}}{0.30 \text{ MW}}$$

For simplicity, we could have let W_{net}, Q_{in}, and Q_{out} refer to *rates* instead of to total amounts—to do this we just cancel common factors of Δt out of the equations for efficiency and energy conservation.

Practice Problem 15.5 Heat Engine Efficiency

An engine "wastes" 4.0 J of heat for every joule of work done. What is its efficiency?

Efficiency and the First Law According to the first law of thermodynamics, the efficiency of a heat engine cannot exceed 100%. An efficiency of 100% would mean that all of the heat input is turned into useful work and no "waste" heat is exhausted. It might seem theoretically possible to make a 100% efficient engine by eliminating all of the imperfections in design such as friction and lack of perfect insulation. However, it is not, as we see in Section 15.7.

15.6 REFRIGERATORS AND HEAT PUMPS

Application: refrigerators and heat pumps

The second law of thermodynamics says that heat cannot *spontaneously* flow from a colder body to a hotter body; but machines such as refrigerators and heat pumps can make that happen. In a refrigerator, heat is pumped out of the food compartment into the warmer room. That doesn't happen by itself; it requires the *input* of work. The electricity used by a refrigerator turns the compressor motor, which does the work required to make the refrigerator function (Fig. 15.13). An air conditioner is essentially the same thing: it pumps heat out of the house into the hotter outdoors.

CONNECTION:

A refrigerator or heat pump is like a heat engine with the directions of the energy transfers *reversed.*

The only difference between a refrigerator (or an air conditioner) and a heat pump is which end is performing the useful task. Refrigerators and air conditioners pump heat out of a compartment that they are designed to keep cool. Heat pumps pump heat from the colder outdoors into the warmer house. The idea is not to cool the outdoors; it is to warm the house.

Notice that the energy transfers in a heat pump are reversed in direction from those in a heat engine (Fig. 15.14). In the heat engine, heat flows from hot to cold, with work

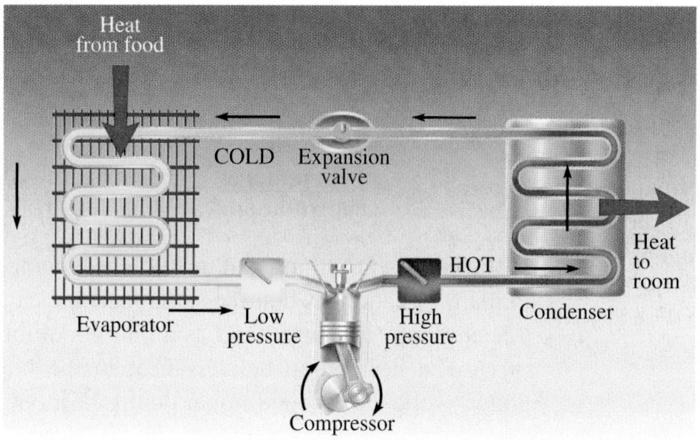

Figure 15.13 In a refrigerator, a fluid is compressed, increasing its temperature. Heat is exhausted as the fluid passes through the condenser. Now the fluid is allowed to expand; its temperature falls. Heat flows from the food compartment into the cold fluid. The fluid returns to the compressor to begin the same cycle again.

as the output. In a heat pump, heat flows from cold to hot, with work as the *input.* It will be most convenient to distinguish the heat transfers not by which is input and which output (since that switches in going from an engine to a heat pump), but rather by the temperature at which the exchange is made, using subscripts "H" and "C" for hot and cold. Q_H, Q_C, and W_{net} stand for the *magnitudes* of the energy transfers during one or more cycles and are never negative. We supply minus signs in equations when necessary, based on the directions of the energy transfers, as appropriate for the engine, heat pump, or refrigerator under consideration.

Thus, the efficiency of the heat engine can be rewritten

$$e = \frac{\text{net work output}}{\text{heat input}} = \frac{W_{net}}{Q_H} \qquad (15\text{-}12)$$

The efficiency can also be expressed in terms of the heat flows. Since $W_{net} = Q_H - Q_C$,

$$e = \frac{Q_H - Q_C}{Q_H} = 1 - \frac{Q_C}{Q_H} \qquad (15\text{-}14)$$

The efficiency of an engine is *less* than 1.

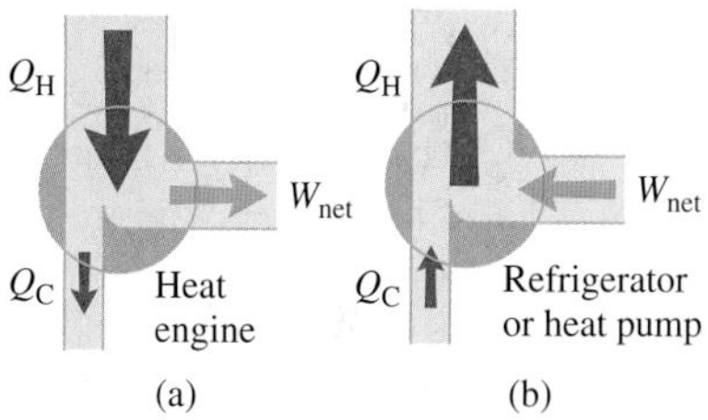

Figure 15.14 Energy transfers during one cycle for (a) a heat engine and (b) a refrigerator or heat pump. With our definition of Q_H, Q_C, and W_{net} as positive quantities, in either case conservation of energy requires that $Q_H = W_{net} + Q_C$.

Sign convention for engines, refrigerators, and heat pumps: Q_H, Q_C, and W_{net} are all positive.

Coefficient of Performance To measure the performance of a heat pump or refrigerator, we define a **coefficient of performance** *K*. Just as for the efficiency of an engine, the coefficients of performance are ratios of what you get divided by what you pay for:

- for a heat pump:

$$K_p = \frac{\text{heat delivered}}{\text{net work input}} = \frac{Q_H}{W_{net}} \qquad (15\text{-}15)$$

- for a refrigerator or air conditioner:

$$K_r = \frac{\text{heat removed}}{\text{net work input}} = \frac{Q_C}{W_{net}} \qquad (15\text{-}16)$$

A higher coefficient of performance means a better heat pump or refrigerator. Unlike the efficiency of an engine, coefficients of performance can be (and usually are) *greater than* 1.

The second law says that heat cannot flow spontaneously from cold to hot—we need to do some work to make that happen. That's equivalent to saying that the coefficient of performance can't be infinite.

Example 15.6

A Heat Pump

A heat pump has a performance coefficient of 2.5. (a) How much heat is delivered to the house for every joule of electrical energy consumed? (b) In an electric heater, for each joule of electric energy consumed, one joule of heat is delivered to the house. Where does the "extra" heat delivered by the heat pump come from? (c) What would its coefficient of performance be when used as an air conditioner instead?

continued on next page

Example 15.6 continued

Strategy There are two slightly different meanings of coefficient of performance. For a heat pump, whose object is to deliver heat to the house, the coefficient of performance is the heat delivered (Q_H) per unit of net work done to run the pump. With an air conditioner, the object is to *remove* heat from the house. The coefficient of performance is the heat *removed* (Q_C) per unit of work done.

Solution (a) As a heat pump,

$$K_p = \frac{\text{heat delivered}}{\text{net work input}} = \frac{Q_H}{W_{net}} = 2.5$$

$$Q_H = 2.5W_{net}$$

For every joule of electric energy (= work input), 2.5 J of heat are delivered to the house.

(b) The 2.5 J of heat delivered include the 1.0 J of work input plus 1.5 J of heat pumped in from the outside. The electric heater just transforms the joule of work into a joule of heat.

(c) From (b), the coefficient is 1.5:

$$K_r = \frac{\text{heat removed}}{\text{net work input}} = \frac{Q_C}{W_{net}} = \frac{1.5\ \text{J}}{1.0\ \text{J}} = 1.5$$

Discussion One thing that makes a heat pump economical in many situations is that the same machine can function as a heat pump (in winter) and as an air conditioner (in summer). The heat pump delivers heat to the interior of the house, while the air conditioner pumps heat out.

Practice Problem 15.6 Heat Exhausted by Air Conditioner

An air conditioner with a coefficient of performance $K_r = 3.0$ consumes electricity at an average rate of 1.0 kW. During 1.0 h of use, how much heat is exhausted to the outdoors?

15.7 REVERSIBLE ENGINES AND HEAT PUMPS

What limitation does the *second* law of thermodynamics place on the efficiencies of heat engines or on the coefficients of performance of heat pumps and refrigerators? To address that question, we first introduce a simplified model of engines and heat pumps. We assume the existence of two reservoirs, a hot reservoir at absolute temperature T_H and a cold reservoir at absolute temperature T_C (where $T_C < T_H$). In this model, an engine takes its heat input from the hot reservoir and exhausts heat into the cold reservoir (Fig. 15.15). A heat pump takes in heat from the cold reservoir and exhausts heat to the hot reservoir. The cold reservoir stays at temperature T_C and the hot reservoir stays at T_H.

Recall that a *reservoir* is a system with such a large heat capacity that it can exchange heat in either direction with a negligibly small temperature change.

Now imagine a hypothetical **reversible engine** exchanging heat with two reservoirs. In this engine, no irreversible processes occur: there is no friction or other dissipation of energy and heat only flows between systems that have the same temperature. In practice, there would have to be some small temperature difference to make heat flow from one system to another, but we can imagine making the temperature difference smaller and smaller. Hence, the reversible engine is an idealization, not something we can actually build. We can now show that

- the efficiency of this reversible engine depends only on the absolute temperatures of the two reservoirs; and
- the efficiency of a real engine that exchanges heat with two reservoirs cannot be greater than the efficiency of a reversible engine using the same two reservoirs.

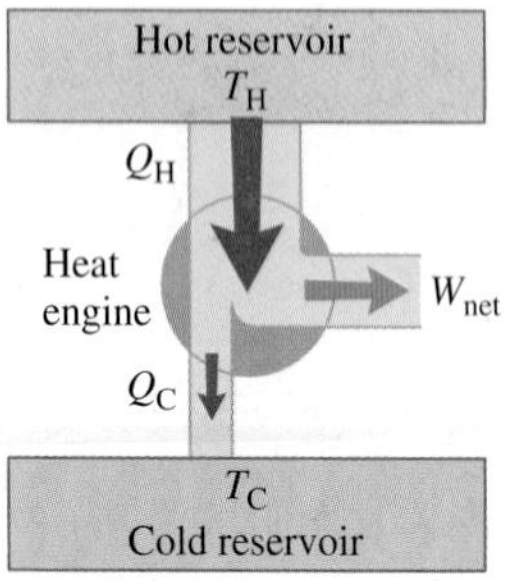

Figure 15.15 Simplified model of a heat engine. Heat flows into the engine from a reservoir at temperature T_H, and heat flows out of the engine into a reservoir at T_C.

A Reversible Engine Has the Maximum Possible Efficiency We can prove that no real engine can have a higher efficiency than a reversible engine using the same two reservoirs by the following thought experiment. Imagine two engines using the same hot and cold reservoirs that do the same amount of work per cycle (Fig. 15.16a). Suppose engine 1 is reversible and hypothetical engine 2 has a higher efficiency than engine 1 ($e_2 > e_1$). The more efficient engine does the same amount of work per cycle but takes in a smaller quantity of heat from the hot reservoir per cycle ($Q_{H2} < Q_{H1}$). Energy conservation for a cyclical engine requires that $Q_C = Q_H - W_{net}$, so the more efficient engine also exhausts a smaller quantity of heat to the cold reservoir ($Q_{C2} < Q_{C1}$).

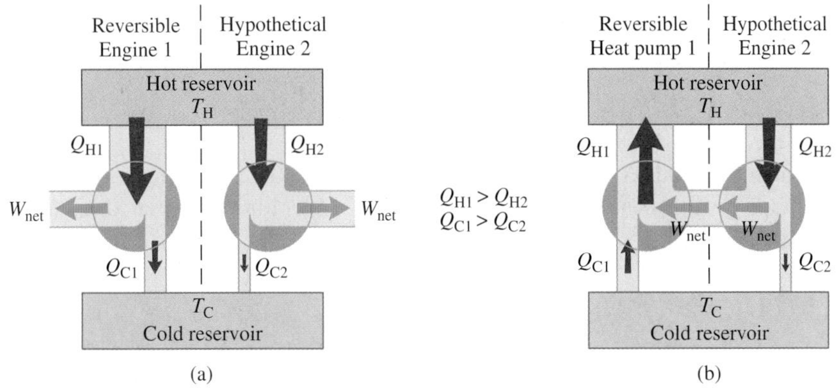

Figure 15.16 (a) Two engines that take in heat from the same hot reservoir and exhaust heat to the same cold reservoir. The two engines do the same amount of net work per cycle. Engine 1 is reversible, while hypothetical engine 2 is assumed to have an efficiency *higher* than that of engine 1, which we will show to be impossible. (b) Engine 1 is reversed, making it into a reversible heat pump. The work output of hypothetical engine 2 is used to run the heat pump. The net effect of the two connected devices is heat flow from the cold reservoir to the hot reservoir without any work input.

Now imagine reversing the energy flow directions for engine 1, turning it into a heat pump. Engine 1 is reversible, so the magnitudes of the energy transfers per cycle do not change. Connect this heat pump to engine 2, using the work output of the engine as the work input for the heat pump (Fig. 15.16b). Since $Q_{C1} > Q_{C2}$ and $Q_{H1} > Q_{H2}$, the net effect of the two devices is a flow of heat from the cold reservoir to the hot reservoir without the input of work, which is impossible—it violates the second law of thermodynamics. The conclusion is that according to the second law, no engine can have an efficiency greater than that of a reversible engine that uses the same two reservoirs.

Furthermore, every reversible engine exchanging heat with the same two reservoirs, no matter what the details of its construction, has the same efficiency. (To see why, use the same thought experiment with two reversible engines such that $e_2 > e_1$.) Therefore, the efficiency of such an engine can depend only on the temperatures of the hot and cold reservoirs. It turns out that e_r is given by the remarkably simple expression:

$$e_r = 1 - \frac{T_C}{T_H} \quad (15\text{-}17)$$

Efficiency of a Reversible Engine

Equation (15-17) was first derived by Sadi Carnot (see later in this section). The temperatures in Eq. (15-17) must be *absolute* temperatures. [Absolute temperature is also called *thermodynamic temperature* because you can use the efficiency of reversible engines to set a temperature scale. In fact, the definition of the kelvin is based on Eq. (15-17).]

Using Eq. (15-17), the ratio of the heat exhaust to the heat input *for a reversible engine* is

$$\frac{Q_C}{Q_H} = \frac{Q_H - W_{net}}{Q_H} = 1 - \frac{W_{net}}{Q_H} = 1 - e_r = \frac{T_C}{T_H} \quad (15\text{-}18)$$

For a reversible engine, the ratio of the heat magnitudes is equal to the temperature ratio.

The efficiency of a reversible engine is always less than 100%, assuming that the cold reservoir is not at absolute zero. Even an ideal, perfectly reversible engine must exhaust some heat, so the efficiency can never be 100%, *even in principle.* Efficiencies of real engines cannot be greater than those of reversible engines, so the second law of thermodynamics sets a limit on the theoretical maximum efficiency of an engine ($e < 1 - T_C/T_H$).

Reversible Refrigerators and Heat Pumps Equation (15-18) also applies to reversible heat pumps and refrigerators because they are just reversible engines with the directions of the energy transfers reversed. Using Eq. (15-18) and the first law, we can find the coefficients of performance for reversible heat pumps and refrigerators (see Problems 42 and 44):

$$K_{p,\,rev} = \frac{1}{1 - T_C/T_H} \quad \text{and} \quad K_{r,\,rev} = \frac{1}{T_H/T_C - 1} = K_{p,\,rev} - 1 \qquad (15\text{-}19)$$

Real heat pumps and refrigerators cannot have coefficients of performance greater than those of reversible heat pumps and refrigerators operating between the same two reservoirs.

Example 15.7

Efficiency of an Automobile Engine

In an automobile engine, the combustion of the fuel-air mixture can reach temperatures as high as 3000°C and the exhaust gases leave the cylinder at about 1000°C. (a) Find the efficiency of a *reversible* engine operating between reservoirs at those two temperatures. (b) Theoretically, we might be able to have the exhaust gases leave the engine at the temperature of the outside air (20°C). What would be the efficiency of the hypothetical reversible engine in this case?

Strategy First we identify the temperatures of the hot and cold reservoirs in each case. We must convert the reservoir temperatures to kelvins in order to find the efficiency of a reversible engine.

Solution (a) The reservoir temperatures in kelvins are found using

$$T = T_C + 273\text{ K}$$

Therefore,

$$T_H = 3000°\text{C} = 3273\text{ K}$$

$$T_C = 1000°\text{C} = 1273\text{ K}$$

The efficiency of a reversible engine operating between these temperatures is

$$e_r = 1 - \frac{T_C}{T_H} = 1 - \frac{1273\text{ K}}{3273\text{ K}} = 0.61 = 61\%$$

(b) The high-temperature reservoir is still at 3273 K, while the low-temperature reservoir is now

$$T_C = 293\text{ K}$$

This gives a higher efficiency:

$$e_r = 1 - \frac{T_C}{T_H} = 1 - \frac{293\text{ K}}{3273\text{ K}} = 0.910 = 91.0\%$$

Can an engine be 100% efficient?

Discussion As mentioned in the chapter opener, real gasoline engines achieve efficiencies of only about 20% to 25%. Although improvement is possible, the second law of thermodynamics limits the theoretical maximum efficiency to that of a reversible engine operating between the same temperatures. The *theoretical* maximum efficiency can only be increased by using a hotter hot reservoir or a colder cold reservoir. However, practical considerations may prevent us from using a hotter hot reservoir or colder cold reservoir. Hotter combustion gases might cause engine parts to wear out too fast, or there may be safety concerns. Letting the gases expand to a greater volume would make the exhaust gases colder, leading to an increase in efficiency, but might *reduce* the *power* the engine can deliver. (A reversible engine has the theoretical maximum efficiency, but the *rate* at which it does work is vanishingly small because it takes a long time for heat to flow across a small temperature difference.)

Practice Problem 15.7 Temperature of Hot Gases

If the efficiency of a reversible engine is 75% and the temperature of the outdoor world into which the engine sends its exhaust is 27°C, what is the combustion temperature in the engine cylinder? [*Hint:* Think of the combustion temperature as the temperature of the hot reservoir.]

Example 15.8

Coal-Burning Power Plant

A coal-burning electrical power plant burns coal at 706°C. Heat is exhausted into a river near the power plant; the average river temperature is 19°C. What is the minimum possible rate of thermal pollution (heat exhausted into the river) if the station generates 125 MW of electricity?

continued on next page

Example 15.8 continued

Strategy The minimum discharge of heat into the river would occur if the engine generating the electricity were *reversible.* As in Example 15.5, we can take all of the rates to be constant.

Solution First find the absolute temperatures of the reservoirs:

$$T_H = 706°C = 979\ K$$
$$T_C = 19°C = 292\ K$$

The efficiency of a reversible engine operating between these temperatures is

$$e_r = 1 - \frac{T_C}{T_H} = 1 - \frac{292\ K}{979\ K} = 0.702$$

We want to find the rate at which heat is exhausted, which is $Q_C/\Delta t$. The efficiency is equal to the ratio of the net work output to the heat input from the hot reservoir:

$$e = \frac{W_{net}}{Q_H}$$

Conservation of energy requires that

$$Q_H = Q_C + W_{net}$$

Solving for Q_C,

$$Q_C = Q_H - W_{net} = \frac{W_{net}}{e} - W_{net} = W_{net} \times \left(\frac{1}{e} - 1\right)$$

Assuming that all the rates are constant,

$$\frac{Q_C}{\Delta t} = 125\ MW \times \left(\frac{1}{0.702} - 1\right) = 53\ MW$$

The rate at which heat enters the river is 53 MW.

Discussion We expect the actual rate of thermal pollution to be higher. A real, irreversible engine would have a lower efficiency, so more heat would be dumped into the river.

Practice Problem 15.8 Generating Electricity from Coal

What is the minimum possible rate of heat *input* (from the burning of coal) needed to generate 125 MW of electricity in this same plant?

The Carnot Cycle

Sadi Carnot (1796–1832), a French engineer, published a treatise in 1824 that greatly expanded the understanding of how heat engines work. His treatment introduced a hypothetical, ideal engine that uses two heat reservoirs at different temperatures as the source and sink for heat and an ideal gas as the working substance of the engine. We now call this engine a **Carnot engine** and its cycle of operation the **Carnot cycle.** Remember that the Carnot engine is an *ideal* engine, not a real engine.

Carnot was able to calculate the efficiency of an engine operating in this cycle and obtained Eq. (15-17). Since *all* reversible engines operating between the same two reservoirs must have the same efficiency, deriving the efficiency for one particular kind of reversible engine is sufficient to derive it for all of them.

The Carnot engine is a particular kind of reversible engine. (Other reversible engines might use a working substance other than an ideal gas or might exchange heat with three or more reservoirs.) We must assume that all friction has somehow been eliminated—otherwise an irreversible process takes place. We also must avoid heat flow across a finite temperature difference, which would be irreversible. Therefore, whenever the ideal gas takes in or gives off heat, the gas must be at the same temperature as the reservoir with which it exchanges energy.

How can we get heat to flow without a temperature difference? Imagine putting the gas in good thermal contact with a reservoir at the same temperature. Now *slowly* pull a piston so that the gas expands. Since the gas does work, it must lose internal energy—and therefore its temperature drops, since in an ideal gas the internal energy is proportional to absolute temperature. As long as the expansion occurs slowly, heat flows into the gas fast enough to keep its temperature constant.

So, to keep every step reversible, we must exchange heat in *isothermal* processes. To take in heat, we expand the gas; to exhaust heat, we compress it. We also need reversible processes to change the gas temperature from T_H to T_C and back to T_H. These processes must be *adiabatic* (no heat flow) since otherwise an irreversible heat flow would occur. (For more detail on the Carnot cycle, see the text website.)

15.8 ENTROPY

When two systems of different temperatures are in thermal contact, heat flows out of the hotter system and into the colder system. There is no change in the total energy of the two systems; energy just flows out of one and into the other. Why then does heat flow in one direction but not in the other? As we will see, heat flow *into* a system not only increases the system's internal energy, it also increases the *disorder* of the system. Heat flow *out of* a system decreases not only its internal energy but also its disorder.

The **entropy** of a system (symbol S) is a quantitative measure of its disorder. Entropy is a state variable (like U, P, V, and T): a system in equilibrium has a unique entropy that does *not* depend on the past history of the system. (Recall that heat and work are *not* state variables. Heat and work describe *how* a system goes from one state to another.) The word *entropy* was coined by Rudolf Clausius (1822–1888) in 1865; its Greek root means *evolution* or *transformation.*

If an amount of heat Q flows into a system at constant absolute temperature T, the entropy change *of the system* is

$$\Delta S = \frac{Q}{T} \qquad (15\text{-}20)$$

The SI unit for entropy is J/K. Heat flowing into a system increases the system's entropy (both ΔS and Q are positive); heat leaving a system decreases the system's entropy (both ΔS and Q are negative). Equation (15-20) is valid as long as the temperature of the system is constant, which is true if the heat capacity of the system is large (as for a reservoir), so that the heat flow Q causes a negligibly small temperature change in the system.

Note that Eq. (15-20) gives only the *change* in entropy, not the initial and final values of the entropy. As with potential energy, the *change* in entropy is what's important in most situations.

If a small amount of heat Q flows from a hotter system to a colder system ($T_H > T_C$), the *total* entropy change of the systems is

$$\Delta S_{tot} = \Delta S_H + \Delta S_C = \frac{-Q}{T_H} + \frac{Q}{T_C}$$

Since $T_H > T_C$,

$$\frac{Q}{T_H} < \frac{Q}{T_C}$$

The increase in the colder system's entropy is larger than the decrease of the hotter system's entropy and the total entropy increases:

$$\Delta S_{tot} > 0 \quad \text{(irreversible process)} \qquad (15\text{-}21)$$

Thus, the flow of heat from a hotter system to a colder system causes an increase in the total entropy of the two systems. Every irreversible process increases the total entropy of the universe. A process that would decrease the total entropy of the universe is impossible. A reversible process causes no change in the total entropy of the universe. We can restate the second law of thermodynamics in terms of entropy:

Second Law of Thermodynamics (Entropy Statement)

The entropy of the universe never decreases.

For example, a reversible engine removes heat Q_H from a hot reservoir at temperature T_H and exhausts Q_C to a cold reservoir at T_C. The entropy of the engine itself is left unchanged since it operates in a cycle. The entropy of the hot reservoir decreases by an amount Q_H/T_H and that of the cold reservoir increases by Q_C/T_C. Since the entropy of the universe must be unchanged by a reversible engine, it must be true that

$$-\frac{Q_H}{T_H} + \frac{Q_C}{T_C} = 0 \quad \text{or} \quad \frac{Q_C}{Q_H} = \frac{T_C}{T_H}$$

The efficiency of the engine is therefore

$$e_r = \frac{W_{net}}{Q_H} = \frac{Q_H - Q_C}{Q_H} = 1 - \frac{Q_C}{Q_H} = 1 - \frac{T_C}{T_H}$$

as stated in Section 15.7.

Entropy is *not* a conserved quantity like energy. The entropy of the universe is always increasing. It is possible to decrease the entropy of a *system*, but only at the expense of increasing the entropy of the surroundings by at least as much (usually more).

CONNECTION:

Energy is a conserved quantity; entropy is not.

CHECKPOINT 15.8

The entropy of a system increases by 10 J/K. Does this mean the process is necessarily irreversible? Explain.

Example 15.9

Entropy Change of a Freely Expanding Gas

Suppose 1.0 mol of an ideal gas is allowed to freely expand into an evacuated container of equal volume so that the volume of the gas doubles (Fig. 15.17). No work is done on the gas as it expands, since there is nothing pushing against it. The containers are insulated so no heat flows into or out of the gas. What is the entropy change of the gas?

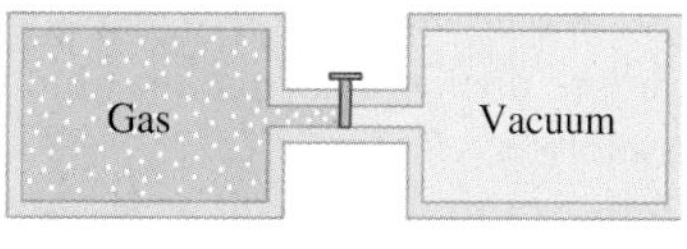

Figure 15.17
Two chambers connected by a valve. One chamber contains a gas and the other has been evacuated. When the valve is opened, the gas expands until it fills both chambers.

Strategy The only way to calculate entropy changes that we've learned so far is for heat flow at a constant temperature. In free expansion, there is no heat flow—but that does not necessarily mean there is no entropy change. Since entropy is a state variable, ΔS depends only on the initial and final states of the gas, not the intermediate states. We can therefore find the entropy change using *any* thermodynamic process with the same initial and final states. The initial and final temperatures of the gas are identical since the internal energy does not change; therefore we find the entropy change for an *isothermal* expansion.

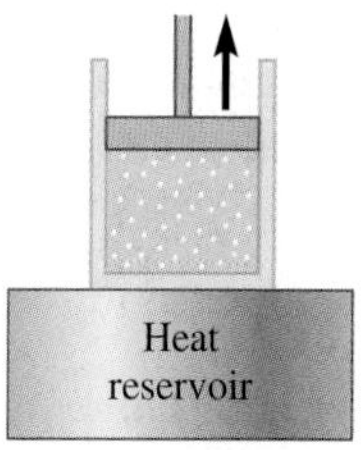

Figure 15.18
As the gas in the cylinder expands, heat flows into it from the reservoir and keeps its temperature constant.

Solution Imagine the gas confined to a cylinder with a moveable piston (Fig. 15.18). In an isothermal expansion, heat flows into the gas from a reservoir at a constant temperature T. As the gas expands, it does work on the piston. If the temperature is to stay constant, the work done must equal the heat flow into the gas:

$$\Delta U = 0 \quad \text{implies} \quad Q + W = 0$$

In Section 15.3, we found the work done by an ideal gas during an isothermal expansion:

$$W = nRT \ln\left(\frac{V_i}{V_f}\right)$$

The volume of the gas doubles, so $V_i/V_f = 0.50$:

$$W = nRT \ln 0.50$$

Since $Q = -W$, the entropy change is

$$\Delta S = \frac{Q}{T} = -nR \ln 0.50$$

$$= -(1.0 \text{ mol}) \times \left(8.31 \frac{\text{J}}{\text{mol·K}}\right) \times (-0.693) = +5.8 \text{ J/K}$$

Discussion The entropy change is positive, as expected. Free expansion is an irreversible process; the gas molecules do not spontaneously collect back in the original container. The reverse process would cause a decrease in entropy, without a larger increase elsewhere, and so violates the second law.

Practice Problem 15.9 Entropy Change of the Universe When a Lump of Clay Is Dropped

A room-temperature lump of clay of mass 400 g is dropped from a height of 2 m and makes a totally inelastic collision with the floor. Approximately what is the entropy change of the universe due to this collision? [*Hint:* The temperature of the clay rises, but only slightly.]

Application of the Second Law to Evolution

Some have argued that evolution cannot have occurred because it would violate the second law of thermodynamics. The argument views evolution as an increase in order: life spontaneously developed from simple life forms to more complex, more highly ordered organisms.

However, the second law says only that the *total entropy of the universe* cannot decrease. It does not say that the entropy of a particular system cannot decrease. When heat flows from a hot body to a cold body, the entropy of the hot body decreases, but the increase in the cold body's entropy is greater, so the entropy of the universe increases. A living organism is not a closed system and neither is the Earth. An adult human, for instance, requires roughly 10 MJ of chemical energy from food per day. What happens to this energy? Some is turned into useful work by the muscles, some more is used to repair body tissues, but most of it is dissipated and leaves the body as heat. The human body therefore is constantly increasing the entropy of its environment. As evolution progresses from simpler to more complicated organisms, the increase in order within the organisms must be accompanied by a larger increase in disorder in the environment.

Application of the Second Law to the "Energy Crisis"

When people speak of "conserving energy," they usually mean using fuel and electricity sparingly. In the physics sense of the word *conserve*, energy is *always* conserved. Burning natural gas to heat your house does not change the amount of energy around; it just changes it from one form to another.

What we need to be careful not to waste is *high-quality* energy. Our concern is not the total amount of energy, but rather whether the energy is in a form that is useful and convenient. The chemical energy stored in fuel is relatively high-quality (ordered) energy. When fuel is burned, the energy is degraded into lower-quality (disordered) energy.

Statistical Interpretation of Entropy

Thermodynamic systems are collections of huge numbers of atoms or molecules. How these atoms or molecules behave statistically determines the disorder in the system. In other words, the second law of thermodynamics is based on the statistics of systems with extremely large numbers of atoms or molecules.

The **microstate** of a thermodynamic system specifies the state of each constituent particle. For instance, in a monatomic ideal gas with N atoms, a microstate is specified by the position and velocity of each of the N atoms. As the atoms move about and collide, the system changes from one microstate to another. The **macrostate** of a thermodynamic system specifies only the values of the macroscopic state variables (such as pressure, volume, temperature, and internal energy).

Statistical analysis is the microscopic basis for the second law of thermodynamics. It turns out, remarkably, that the number of microstates corresponding to a given macrostate is related to the entropy of that macrostate in a simple way. Letting Ω (the Greek capital omega) stand for the number of microstates, the relationship is

$$S = k \ln \Omega \qquad (15\text{-}22)$$

where k is Boltzmann's constant. Equation (15-22) is inscribed on the tombstone of Ludwig Boltzmann (1844–1906), the Austrian physicist who made the connection between entropy and statistics in the late nineteenth century. The relationship between S and Ω has to be logarithmic because entropy is additive: if system 1 has entropy S_1 and system 2 has entropy S_2, then the total entropy is $S_1 + S_2$. However, the number of microstates is *multiplicative*. Think of dice: if die 1 has 6 microstates and die 2 also has 6, the total number of microstates when rolling two dice is not 12, but $6 \times 6 = 36$. The entropy is additive since $\ln 6 + \ln 6 = \ln 36$. (For more information on entropy and statistics see text website.)

15.9 THE THIRD LAW OF THERMODYNAMICS

Like the second law, the third law of thermodynamics can be stated in several equivalent ways. We will state just one of them:

Third Law of Thermodynamics

It is impossible to cool a system to absolute zero.

While it is impossible to *reach* absolute zero, there is no limit on how *close* we can get. Scientists who study low-temperature physics have attained equilibrium temperatures as low as 1 μK and have sustained temperatures of 2 mK; transient temperatures in the nano- and picokelvin range have been observed.

Master the Concepts

- The first law of thermodynamics is a statement of energy conservation:

$$\Delta U = Q + W \qquad (15\text{-}1)$$

where Q is the heat flow *into* the system and W is the work done *on* the system.
- Pressure, temperature, volume, number of moles, internal energy, and entropy are state variables; they describe the state of a system at some instant of time but *not* how the system got to that state. Heat and work are *not* state variables—they describe *how* a system gets from one state to another.
- The work done on a system when the pressure is constant—or for a volume change small enough that the pressure change is insignificant—is

$$W = -P\Delta V \qquad (15\text{-}2)$$

The magnitude of the work done is the total area under the PV curve.

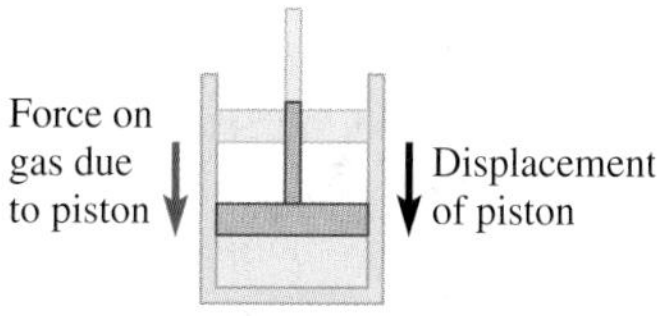

$W > 0$ for compression

- The change in internal energy of an ideal gas is determined solely by the temperature change. Therefore,

$$\Delta U = 0 \quad \text{(ideal gas, isothermal process)} \qquad (15\text{-}10)$$

- A process in which no heat is transferred into or out of the system is called an adiabatic process.
- The molar specific heats of an ideal gas at constant volume and constant pressure are related by

$$C_P = C_V + R \qquad (15\text{-}9)$$

- Spontaneous heat flow from a hotter body to a colder body is always irreversible.

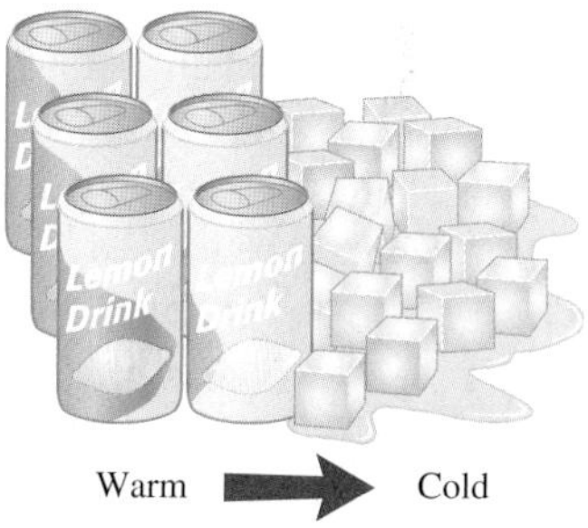

Spontaneous heat flow

- For one cycle of an engine, heat pump, or refrigerator, conservation of energy requires

$$Q_{net} = Q_H - Q_C = W_{net}$$

where Q_H, Q_C, and W_{net} are defined as positive quantities.

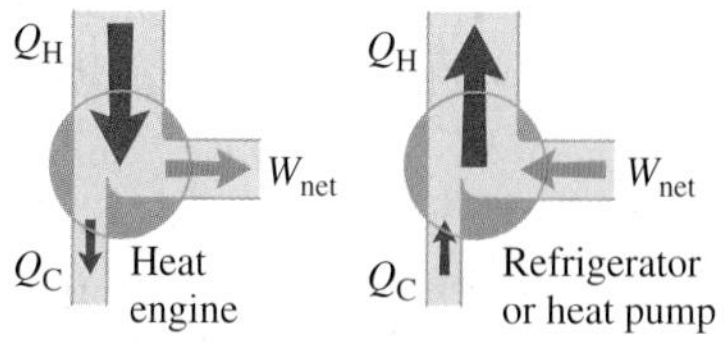

- The efficiency of an engine is defined as

$$e = \frac{W_{net}}{Q_H} \qquad (15\text{-}12)$$

- The coefficient of performance for a heat pump is

$$K_p = \frac{\text{heat delivered}}{\text{net work input}} = \frac{Q_H}{W_{net}} \qquad (15\text{-}15)$$

- The coefficient of performance for a refrigerator or air conditioner is

$$K_r = \frac{\text{heat removed}}{\text{net work input}} = \frac{Q_C}{W_{net}} \qquad (15\text{-}16)$$

continued on next page

Master the Concepts continued

- A reservoir is a system with such a large heat capacity that it can exchange heat in either direction with a negligibly small temperature change.

Reverse heat flow does not happen spontaneously

- The second law of thermodynamics can be stated in various equivalent ways. Two of them are: (1) heat never flows spontaneously from a colder body to a hotter body, and (2) the entropy of the universe never decreases.
- The efficiency of a *reversible* engine is determined only by the *absolute* temperatures of the hot and cold reservoirs:

$$e_r = 1 - \frac{T_C}{T_H} \quad (15\text{-}17)$$

- If an amount of heat Q flows into a system at constant absolute temperature T, the entropy change of the system is

$$\Delta S = \frac{Q}{T} \quad (15\text{-}20)$$

- The third law of thermodynamics: it is impossible to cool a system to absolute zero.

Conceptual Questions

1. Is it possible to make a heat pump with a coefficient of performance equal to 1? Explain.
2. An electric baseboard heater can convert 100% of the electric energy used into heat that flows into the house. Since a gas furnace might be located in a basement and sends exhaust gases up the chimney, the heat flow into the living space is less than 100% of the chemical energy released by burning. Does this mean that electric heating is better? Which heating method consumes less fuel? In your answer, consider how the electricity might have been generated and the efficiency of that process.
3. A whimsical statement of the laws of thermodynamics—probably not one favored by gamblers—goes like this:

   I. You can never win; you can only lose or break even.

   II. You can only break even at absolute zero.

   III. You can never get to absolute zero.

   What do we mean by "win," "lose," and "break even"? [*Hint:* Think about a heat engine.]
4. Why must all reversible engines (operating between the same reservoirs) have the same efficiency? Try an argument by contradiction: imagine that two reversible engines exist with $e_1 > e_2$. Reverse one of them (into a heat pump) and use the work output from the engine to run the heat pump. What happens? (If it seems fine at first, switch the two.)
5. When supplies of fossil fuels such as petroleum and coal dwindle, people might call the situation an "energy crisis." From the standpoint of physics, why is that not an accurate name? Can you think of a better one?
6. If you leave the refrigerator door open and the refrigerator runs continuously, does the kitchen get colder or warmer? Explain.
7. Most heat pumps incorporate an auxiliary electric heater. For relatively mild outdoor temperatures, the electric heater is not used. However, if the outdoor temperature gets very low, the auxiliary heater is used to supplement the heat pump. Why?
8. Why are heat pumps more often used in mild climates than in areas with severely cold winters?
9. Are entropy changes always caused by the flow of heat? If not, give some other examples of processes that increase entropy. (W tutorial: reversibility)
10. Can a heat engine be made to operate without creating any "thermal pollution," that is, without making its cold reservoir get warmer in the long run? The net work output must be greater than zero.
11. A warm pitcher of lemonade is put into an ice chest. Describe what happens to the entropies of lemonade and ice as heat flows from the lemonade to the ice within the chest.
12. A new dormitory is being built at a college in North Carolina. To save costs, it is proposed to not include air conditioning ducts and vents. A member of the board overseeing the construction says that stand-alone air conditioning units can be supplied to each room later. He has seen advertisements that claim these new units do not need to be vented to the outside. Can the claim be true? Explain.
13. After a day at the beach, a child brings home a bucket containing some salt water. Eventually the water evaporates, leaving behind a few salt crystals. The molecular order of the salt crystals is greater than the order of the dissolved salt sloshing around in the sea water. Is this a violation of the entropy principle? Explain.
14. Explain why the molar specific heat at constant volume is not the same as the molar specific heat at constant pressure for gases. Why is the distinction between constant volume and constant pressure usually insignificant for the specific heats of liquids and solids?

Multiple-Choice Questions

1. A heat engine runs between reservoirs at temperatures of 300°C and 30°C. What is its maximum possible efficiency?
   (a) 10% (b) 47% (c) 53%
   (d) 90% (e) 100%
2. If two different systems are put in thermal contact so that heat can flow from one to the other, then heat will flow until the systems have the same
   (a) energy.
   (b) heat capacity.
   (c) entropy.
   (d) temperature.
3. As moisture from the air condenses on the outside of a cold glass of water, the entropy of the condensing moisture
   (a) stays the same.
   (b) increases.
   (c) decreases.
   (d) not enough information
4. As a system undergoes a constant volume process
   (a) the pressure does not change.
   (b) the internal energy does not change.
   (c) the work done on or by the system is zero.
   (d) the entropy stays the same.
   (e) the temperature of the system does not change.
5. Which of these statements are implied by the *second* law of thermodynamics?
   (a) The entropy of an engine (including its fuel and/or heat reservoirs) operating in a cycle never decreases.
   (b) The increase in internal energy of a system in any process is the sum of heat absorbed plus work done on the system.
   (c) A heat engine, operating in a cycle, that rejects no heat to the low-temperature reservoir is impossible.
   (d) Both (a) and (c).
   (e) All three [(a), (b), and (c)].
6. On a summer day, you keep the air conditioner in your room running. From the list numbered 1 to 4, choose the hot reservoir and the cold reservoir.
   1. the air outside
   2. the compartment inside the air conditioner where the air is compressed
   3. the freon gas that is the working substance (expands and compresses in each cycle)
   4. the air in the room

   (a) 1 is the hot reservoir, 2 is the cold reservoir.
   (b) 1 is the hot reservoir, 3 is the cold reservoir.
   (c) 1 is the hot reservoir, 4 is the cold reservoir.
   (d) 2 is the hot reservoir, 3 is the cold reservoir.
   (e) 2 is the hot reservoir, 4 is the cold reservoir.
   (f) 3 is the hot reservoir, 4 is the cold reservoir.
7. The *PV* diagram illustrates several paths to get from an initial to a final state. For which path does the system do the most work?

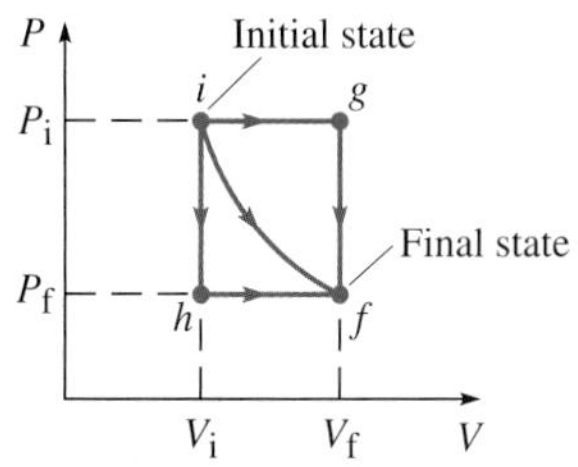

   (a) path *igf*
   (b) path *if*
   (c) path *ihf*
   (d) All paths represent equal work.
8. An ideal gas is confined to the left chamber of an insulated container. The right chamber is evacuated. A valve is opened between the chambers, allowing gas to flow into the right chamber. After equilibrium is established, the temperature of the gas ______ . [*Hint:* What happens to the internal energy?]
   (a) is lower than the initial temperature
   (b) is higher than the initial temperature
   (c) is the same as the initial temperature
   (d) could be higher than, the same as, or lower than the initial temperature
9. When the first law of thermodynamics ($\Delta U = Q + W$) is applied to a system *S*, the variables *Q* and *W* stand for
   (a) the heat flow *out of S* and the work done *on S*.
   (b) the heat flow *out of S* and the work done *by S*.
   (c) the heat flow *into S* and the work done *by S*.
   (d) the heat flow *into S* and the work done *on S*.
10. As an ideal gas is adiabatically expanding,
   (a) the temperature of the gas does not change.
   (b) the internal energy of the gas does not change.
   (c) work is not done on or by the gas.
   (d) no heat is given off or taken in by the gas.
   (e) both (a) and (d)
   (f) both (a) and (b)
11. As an ideal gas is compressed at constant temperature,
   (a) heat flows out of the gas.
   (b) the internal energy of the gas does not change.
   (c) the work done on the gas is zero.
   (d) none of the above
   (e) both (a) and (b)
   (f) both (a) and (c)
12. Given 1 mole of an ideal gas, in a state characterized by P_A, V_A, a change occurs so that the final pressure and volume are equal to P_B, V_B, where $V_B > V_A$. Which of these is true?
   (a) The heat supplied to the gas during the process is completely determined by the values P_A, V_A, P_B, and V_B.
   (b) The change in the internal energy of the gas during the process is completely determined by the values P_A, V_A, P_B, and V_B.
   (c) The work done by the gas during the process is completely determined by the values P_A, V_A, P_B, and V_B.
   (d) All three are true.
   (e) None of these is true.

13. Which choice correctly identifies the three processes shown in the diagrams?

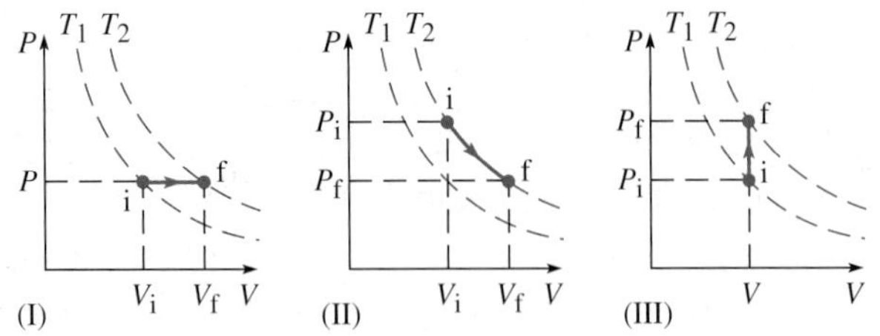

(a) I = isobaric; II = isochoric; III = adiabatic
(b) I = isothermal; II = isothermal; III = isobaric
(c) I = isochoric; II = adiabatic; III = isobaric
(d) I = isobaric; II = isothermal; III = isochoric

Problems

- Combination conceptual/quantitative problem
- Biological or medical application
- Challenging problem
- Blue # Detailed solution in the Student Solutions Manual
- (1 2) Problems paired by concept
- Text website interactive or tutorial

15.1 The First Law of Thermodynamics; 15.2 Thermodynamic Processes; 15.3 Thermodynamic Processes for an Ideal Gas

1. On a cold day, Ming rubs her hands together to warm them up. She presses her hands together with a force of 5.0 N. Each time she rubs them back and forth they move a distance of 16 cm with a coefficient of kinetic friction of 0.45. Assuming no heat flow to the surroundings, after she has rubbed her hands back and forth eight times, by how much has the internal energy of her hands increased?

2. A system takes in 550 J of heat while performing 840 J of work. What is the change in internal energy of the system?

3. The internal energy of a system increases by 400 J while 500 J of work are performed on it. What was the heat flow into or out of the system?

4. A model steam engine of 1.00-kg mass pulls eight cars of 1.00-kg mass each. The cars start at rest and reach a velocity of 3.00 m/s in a time of 3.00 s while moving a distance of 4.50 m. During that time, the engine takes in 135 J of heat. What is the change in the internal energy of the engine?

5. A monatomic ideal gas at 27°C undergoes a constant pressure process from *A* to *B* and a constant volume process from *B* to *C*. Find the total work done during these two processes.

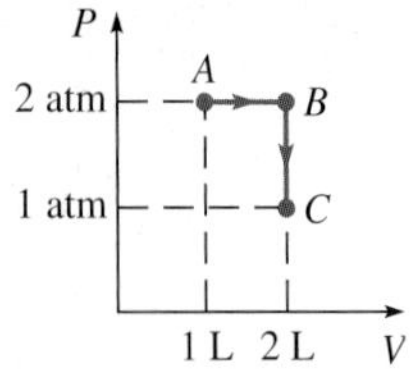

6. A monatomic ideal gas at 27°C undergoes a constant volume process from *A* to *B* and a constant pressure process from *B* to *C*. Find the total work done during these two processes.

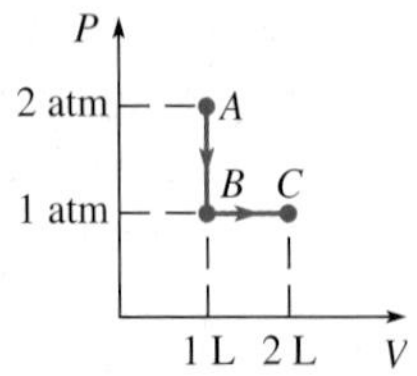

7. An ideal monatomic gas is taken through the cycle in the *PV* diagram. (a) If there are 0.0200 mol of this gas, what are the temperature and pressure at point *C*? (b) What is the change in internal energy of the gas as it is taken from *A* to *B*? (c) How much work is done by this gas per cycle? (d) What is the total change in internal energy of this gas in one cycle?

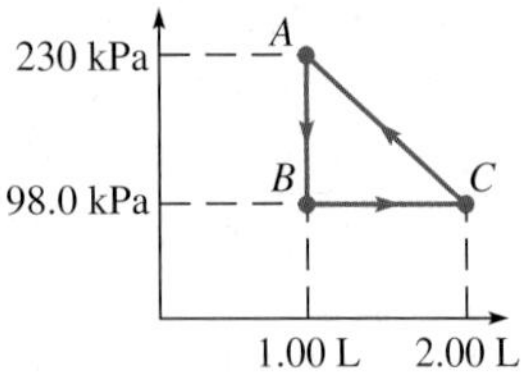

8. An ideal gas is in contact with a heat reservoir so that it remains at a constant temperature of 300.0 K. The gas is compressed from a volume of 24.0 L to a volume of 14.0 L. During the process, the mechanical device pushing the piston to compress the gas is found to expend 5.00 kJ of energy. How much heat flows between the heat reservoir and the gas and in what direction does the heat flow occur?

9. Suppose 1.00 mol of oxygen is heated at constant pressure of 1.00 atm from 10.0°C to 25.0°C. (a) How much heat is absorbed by the gas? (b) Using the ideal gas law, calculate the change of volume of the gas in this process. (c) What is the work done by the gas during this expansion? (d) From the first law, calculate the change of internal energy of the gas in this process.

10. Suppose a monatomic ideal gas is changed from state *A* to state *D* by one of the processes shown on the *PV* diagram. (a) Find the total work done on the gas if it follows the constant volume path *A–B* followed by the constant pressure path *B–C–D*. (b) Calculate the total change in internal energy of the gas during the entire process and the total heat flow into the gas.

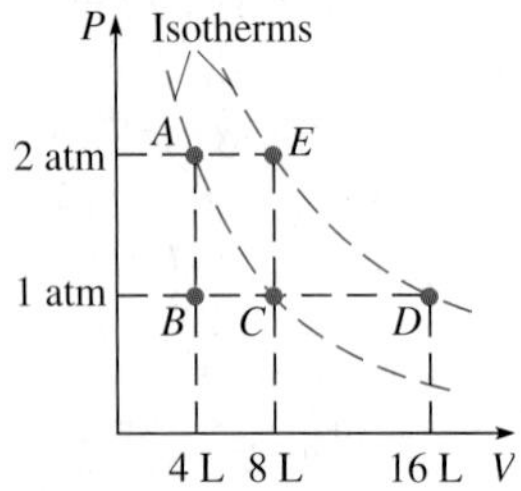

Problems 10–12

11. Repeat Problem 10 for the case when the gas follows the constant temperature path *A–C* followed by the constant pressure path *C–D*.

12. Repeat Problem 10 for the case when the gas follows the constant pressure path *A–E* followed by the constant temperature path *E–D*.

15.5 Heat Engines; 15.6 Refrigerators and Heat Pumps

13. A heat engine follows the cycle shown in the figure. (a) How much net work is done by the engine in one cycle? (b) What is the heat flow *into* the engine per cycle? (tutorial: closed cycle)

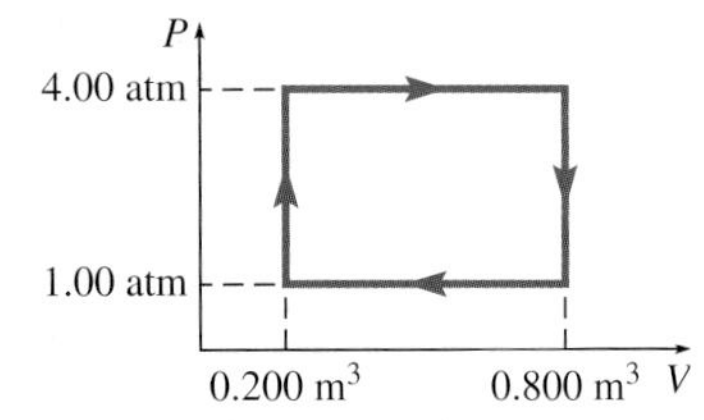

14. What is the efficiency of an electric generator that produces 1.17 kW·h per kg of coal burned? The heat of combustion of coal is 6.71×10^6 J/kg.
15. A heat pump delivers heat at a rate of 7.81 kW for 10.0 h. If its coefficient of performance is 6.85, how much heat is taken from the cold reservoir during that time?
16. (a) How much heat does an engine with an efficiency of 33.3% absorb in order to deliver 1.00 kJ of work? (b) How much heat is exhausted by the engine?
17. The efficiency of an engine is 0.21. For every 1.00 kJ of heat absorbed by the engine, how much (a) net work is done by it and (b) heat is released by it?
18. A certain engine can propel a 1800-kg car from rest to a speed of 27 m/s in 9.5 s with an efficiency of 27%. What are the rate of heat flow into the engine at the high temperature and the rate of heat flow out of the engine at the low temperature?
19. The United States generates about 5.0×10^{16} J of electric energy a day. This energy is equivalent to work, since it can be converted into work with almost 100% efficiency by an electric motor. (a) If this energy is generated by power plants with an average efficiency of 0.30, how much heat is dumped into the environment each day? (b) How much water would be required to absorb this heat if the water temperature is not to increase more than 2.0°C?
20. The intensity (power per unit area) of the sunlight incident on Earth's surface, averaged over a 24-h period, is about 0.20 kW/m^2. If a solar power plant is to be built with an output capacity of 1.0×10^9 W, how big must the area of the solar energy collectors be for photocells operating at 20.0% efficiency?
21. An engine releases 0.450 kJ of heat for every 0.100 kJ of work it does. What is the efficiency of the engine?
22. An engine works at 30.0% efficiency. The engine raises a 5.00-kg crate from rest to a vertical height of 10.0 m, at which point the crate has a speed of 4.00 m/s. How much heat input is required for this engine?
23. How much heat does a heat pump with a coefficient of performance of 3.0 deliver when supplied with 1.00 kJ of electricity?
24. An air conditioner whose coefficient of performance is 2.00 removes 1.73×10^8 J of heat from a room per day. How much does it cost to run the air conditioning unit per day if electricity costs \$0.10 per kilowatt-hour? (Note that 1 kilowatt-hour = 3.6×10^6 J.)

15.7 Reversible Engines and Heat Pumps

25. An ideal engine has an efficiency of 0.725 and uses gas from a hot reservoir at a temperature of 622 K. What is the temperature of the cold reservoir to which it exhausts heat?
26. A heat engine takes in 125 kJ of heat from a reservoir at 815 K and exhausts 82 kJ to a reservoir at 293 K. (a) What is the efficiency of the engine? (b) What is the efficiency of an ideal engine operating between the same two reservoirs?
27. In a certain steam engine, the boiler temperature is 127°C and the cold reservoir temperature is 27°C. While this engine does 8.34 kJ of work, what minimum amount of heat must be discharged into the cold reservoir?
28. Calculate the maximum possible efficiency of a heat engine that uses surface lake water at 18.0°C as a source of heat and rejects waste heat to the water 0.100 km below the surface where the temperature is 4.0°C.
29. An ideal refrigerator removes heat at a rate of 0.10 kW from its interior (+2.0°C) and exhausts heat at 40.0°C. How much electrical power is used?
30. A heat pump is used to heat a house with an interior temperature of 20.0°C. On a chilly day with an outdoor temperature of −10.0°C, what is the minimum work that the pump requires in order to deliver 1.0 kJ of heat to the house? (tutorial: heat pump)
31. A coal-fired electrical generating station can use a higher T_H than a nuclear plant; for safety reasons the core of a nuclear reactor is not allowed to get as hot as coal. Suppose that $T_H = 727$°C for a coal station but $T_H = 527$°C for a nuclear station. Both power plants exhaust waste heat into a lake at $T_C = 27$°C. How much waste heat does each plant exhaust into the lake to produce 1.00 MJ of electricity? Assume both operate as reversible engines. (tutorial: power stations)
32. Two engines operate between the same two temperatures of 750 K and 350 K, and have the same rate of heat input. One of the engines is a reversible engine with a power output of 2.3×10^4 W. The second engine has an efficiency of 42%. What is the power output of the second engine?
33. (a) Calculate the efficiency of a reversible engine that operates between the temperatures 600.0°C and 300.0°C. (b) If the engine absorbs 420.0 kJ of heat from the hot reservoir, how much does it exhaust to the cold reservoir?
34. A reversible engine with an efficiency of 30.0% has $T_C = 310.0$ K. (a) What is T_H? (b) How much heat is exhausted for every 0.100 kJ of work done?

35. An electric power station generates steam at 500.0°C and condenses it with river water at 27°C. By how much would its theoretical maximum efficiency decrease if it had to switch to cooling towers that condense the steam at 47°C?

36. An oil-burning electric power plant uses steam at 773 K to drive a turbine, after which the steam is expelled at 373 K. The engine has an efficiency of 0.40. What is the theoretical maximum efficiency possible at those temperatures?

37. An inventor proposes a heat engine to propel a ship, using the temperature difference between the water at the surface and the water 10 m below the surface as the two reservoirs. If these temperatures are 15.0°C and 10.0°C, respectively, what is the maximum possible efficiency of the engine?

38. A heat engine uses the warm air at the ground as the hot reservoir and the cooler air at an altitude of several thousand meters as the cold reservoir. If the warm air is at 37°C and the cold air is at 25°C, what is the maximum possible efficiency for the engine?

39. A reversible refrigerator has a coefficient of performance of 3.0. How much work must be done to freeze 1.0 kg of liquid water initially at 0°C?

40. An engine operates between temperatures of 650 K and 350 K at 65.0% of its maximum possible efficiency. (a) What is the efficiency of this engine? (b) If 6.3×10^3 J is exhausted to the low temperature reservoir, how much work does the engine do?

41. A town is planning on using the water flowing through a river at a rate of 5.0×10^6 kg/s to carry away the heat from a new power plant. Environmental studies indicate that the temperature of the river should only increase by 0.50°C. The maximum design efficiency for this plant is 30.0%. What is the maximum possible power this plant can produce?

42. Show that the coefficient of performance for a reversible heat pump is $1/(1 - T_C/T_H)$.

43. On a hot day, you are in a sealed, insulated room. The room contains a refrigerator, operated by an electric motor. The motor does work at the rate of 250 W when it is running. Assume the motor is ideal (no friction or electrical resistance) and that the refrigerator operates on a reversible cycle. In an effort to cool the room, you turn on the refrigerator and open its door. Let the temperature in the room be 320 K when this process starts, and the temperature in the cold compartment of the refrigerator be 256 K. At what *net* rate is heat added to (+) or subtracted from (−) the room and all of its contents?

44. Show that the coefficient of performance for a reversible refrigerator is $1/[(T_H/T_C) - 1]$.

45. Show that in a reversible engine the amount of heat Q_C exhausted to the cold reservoir is related to the net work done W_{net} by

$$Q_C = \frac{T_C}{T_H - T_C} W_{net}$$

15.8 Entropy

46. List these in order of increasing entropy: (a) 0.01 mol of N_2 gas in a 1-L container at 0°C; (b) 0.01 mol of N_2 gas in a 2-L container at 0°C; (c) 0.01 mol of liquid N_2.

47. List these in order of increasing entropy: (a) 0.5 kg of ice and 0.5 kg of (liquid) water at 0°C; (b) 1 kg of ice at 0°C; (c) 1 kg of (liquid) water at 0°C; (d) 1 kg of water at 20°C.

48. An ice cube at 0.0°C is slowly melting. What is the change in the ice cube's entropy for each 1.00 g of ice that melts?

49. From Table 14.4, we know that approximately 2256 kJ are needed to transform 1.00 kg of water at 100°C to steam at 100°C. What is the change in entropy of 1.00 kg of water evaporating at 100.0°C? (Specify whether the change in entropy is an increase, +, or a decrease, −.)

50. What is the change in entropy of 10 g of steam at 100°C as it condenses to water at 100°C? By how much does the entropy of the universe increase in this process?

51. A large block of copper initially at 20.0°C is placed in a vat of hot water (80.0°C). For the first 1.0 J of heat that flows from the water into the block, find (a) the entropy change of the block, (b) the entropy change of the water, and (c) the entropy change of the universe. Note that the temperatures of the block and water are essentially unchanged by the flow of only 1.0 J of heat.

52. A large, cold (0.0°C) block of iron is immersed in a tub of hot (100.0°C) water. In the first 10.0 s, 41.86 kJ of heat are transferred, although the temperatures of the water and the iron do not change much in this time. Ignoring heat flow between the system (iron + water) and its surroundings, calculate the change in entropy of the system (iron + water) during this time.

53. On a cold winter day, the outside temperature is −15.0°C. Inside the house the temperature is +20.0°C. Heat flows out of the house through a window at a rate of 220.0 W. At what rate is the entropy of the universe changing due to this heat conduction through the window?

54. Within an insulated system, 418.6 kJ of heat is conducted through a copper rod from a hot reservoir at +200.0°C to a cold reservoir at +100.0°C. (The reservoirs are so big that this heat exchange does not change their temperatures appreciably.) What is the net change in entropy of the system, in kJ/K?

55. A student eats 2000 kcal per day. (a) Assuming that all of the food energy is released as heat, what is the rate of heat released (in watts)? (b) What is the rate of change of entropy of the surroundings if all of the heat is released into air at room temperature (20°C)?

56. The motor that drives a reversible refrigerator produces 148 W of useful power. The hot and cold temperatures of the heat reservoirs are 20.0°C and −5.0°C. What is the maximum amount of ice it can produce in 2.0 h from water that is initially at 8.0°C?

57. An engineer designs a ship that gets its power in the following way: The engine draws in warm water from the ocean, and after extracting some of the water's internal energy, returns the water to the ocean at a temperature 14.5°C lower than the ocean temperature. If the ocean is at a uniform temperature of 17°C, is this an efficient engine? Will the engineer's design work?

58. A balloon contains 200.0 L of nitrogen gas at 20.0°C and at atmospheric pressure. How much energy must be added to raise the temperature of the nitrogen to 40.0°C while allowing the balloon to expand at atmospheric pressure?

59. An ideal gas is heated at a constant pressure of 2.0×10^5 Pa from a temperature of −73°C to a temperature of +27°C. The initial volume of the gas is 0.10 m^3. The heat energy supplied to the gas in this process is 25 kJ. What is the increase in internal energy of the gas?

60. If the pressure on a fish increases from 1.1 to 1.2 atm, its swim bladder decreases in volume from 8.16 mL to 7.48 mL while the temperature of the air inside remains constant. How much work is done on the air in the bladder?

Comprehensive Problems

61. A monatomic ideal gas follows the cyclic process shown in the figure. The temperature of the point at the bottom left of the triangle is 470.0 K. (a) How much net work does this engine do per cycle? (b) What is the maximum temperature of this engine? (c) How many moles of gas are used in this engine?

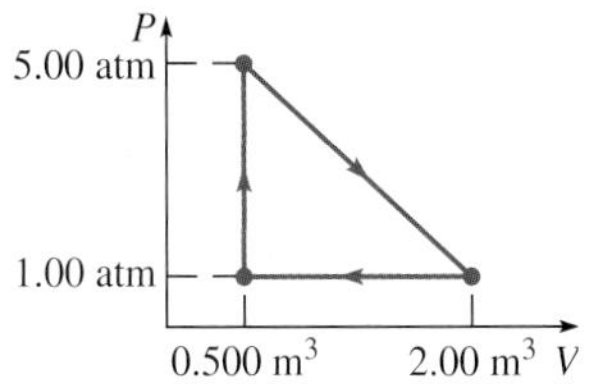

62. For a reversible engine, will you obtain a better efficiency by increasing the high-temperature reservoir by an amount ΔT or decreasing the low-temperature reservoir by the same amount ΔT?

63. A 0.50-kg block of iron [$c = 0.44$ kJ/(kg·K)] at 20.0°C is in contact with a 0.50-kg block of aluminum [$c =$ 0.900 kJ/(kg·K)] at a temperature of 20.0°C. The system is completely isolated from the rest of the universe. Suppose heat flows from the iron into the aluminum until the temperature of the aluminum is 22.0°C. (a) From the first law, calculate the final temperature of the iron. (b) Estimate the entropy change of the system. (c) Explain how the result of part (b) shows that this process is impossible. [*Hint:* Since the system is isolated, $\Delta S_{\text{System}} = \Delta S_{\text{Universe}}$.]

64. List these in order of increasing entropy: (a) 1 mol of water at 20°C and 1 mol of ethanol at 20°C in separate containers; (b) a mixture of 1 mol of water at 20°C and 1 mol of ethanol at 20°C; (c) 0.5 mol of water at 20°C and 0.5 mol of ethanol at 20°C in separate containers; (d) a mixture of 1 mol of water at 30°C and 1 mol of ethanol at 30°C.

65. Suppose you mix 4.0 mol of a monatomic gas at 20.0°C and 3.0 mol of another monatomic gas at 30.0°C. If the mixture is allowed to reach equilibrium, what is the final temperature of the mixture? [*Hint:* Use energy conservation.]

66. A balloon contains 160 L of nitrogen gas at 25°C and 1.0 atm. How much energy must be added to raise the temperature of the nitrogen to 45°C while allowing the balloon to expand at atmospheric pressure?

67. The efficiency of a muscle during weight lifting is equal to the work done in lifting the weight divided by the total energy output of the muscle (work done plus internal energy dissipated in the muscle). Determine the efficiency of a muscle that lifts a 161-N weight through a vertical displacement of 0.577 m and dissipates 139 J in the process.

68. (a) What is the entropy change of 1.00 mol of H_2O when it changes from ice to water at 0.0°C? (b) If the ice is in contact with an environment at a temperature of 10.0°C, what is the entropy change of the universe when the ice melts?

69. Estimate the entropy change of 850 g of water when it is heated from 20.0°C to 50.0°C. [*Hint:* Assume that the heat flows into the water at an average temperature.]

70. For a more realistic estimate of the maximum coefficient of performance of a heat pump, assume that a heat pump takes in heat from outdoors at *10°C below* the ambient outdoor temperature, to account for the temperature difference across its heat exchanger. Similarly, assume that the output must be *10°C hotter* than the house (which itself might be kept at 20°C) to make the heat flow into the house. Make a graph of the coefficient of performance of a reversible heat pump under these conditions as a function of outdoor temperature (from −15°C to +15°C in 5°C increments).

71. A 0.500-kg block of iron at 60.0°C is placed in contact with a 0.500-kg block of iron at 20.0°C. (a) The blocks soon come to a common temperature of 40.0°C. *Estimate* the entropy change of the universe when this occurs. [*Hint:* Assume that all the heat flow occurs at an average temperature for each block.] (b) Estimate the entropy change of the universe if, instead, the temperature of the hotter block increased to 80.0°C while the temperature of the colder block decreased to 0.0°C. [*Hint:* The answer is negative, indicating that the process is impossible.]

72. A container holding 1.20 kg of water at 20.0°C is placed in a freezer that is kept at −20.0°C. The water freezes and comes to thermal equilibrium with the interior of the freezer. What is the minimum amount of electrical energy required by the freezer to do this if it operates between reservoirs at temperatures of 20.0°C and −20.0°C?

73. A reversible heat engine has an efficiency of 33.3%, removing heat from a hot reservoir and rejecting heat to a cold reservoir at 0°C. If the engine now operates in reverse, how long would it take to freeze 1.0 kg of water at 0°C, if it operates on a power of 186 W?

74. Consider a heat engine that is *not* reversible. The engine uses 1.000 mol of a diatomic ideal gas. In the first step (A) there is a constant temperature expansion while in contact with a warm reservoir at 373 K from $P_1 = 1.55 \times 10^5$ Pa and $V_1 = 2.00 \times 10^{-2}$ m^3 to $P_2 = 1.24 \times 10^5$ Pa and $V_2 = 2.50 \times 10^{-2}$ m^3. Then (B) a heat reservoir at the cooler temperature of 273 K is used to cool the gas at constant volume to 273 K from P_2 to $P_3 = 0.91 \times 10^5$ Pa. This is followed by (C) a constant temperature compression while still in contact with the cold reservoir at 273 K from P_3, V_2 to $P_4 = 1.01 \times 10^5$ Pa, V_1. The final step (D) is heating the gas at constant volume from 273 K to 373 K by being in contact with the warm reservoir again, to return from P_4, V_1 to P_1, V_1. Find the change in entropy of the cold reservoir in step B. Remember that the gas is always in contact with the cold reservoir. (b) What is the change in entropy of the hot reservoir in step D? (c) Using this information, find the change in entropy of the total system of gas plus reservoirs during the whole cycle.

✦75. A fish at a pressure of 1.1 atm has its swim bladder inflated to an initial volume of 8.16 mL. If the fish starts swimming horizontally, its temperature increases from 20.0°C to 22.0°C as a result of the exertion. (a) Since the fish is still at the same pressure, how much work is done by the air in the swim bladder? [*Hint:* First find the new volume from the temperature change.] (b) How much heat is gained by the air in the swim bladder? Assume air to be a diatomic ideal gas. (c) If this quantity of heat is lost by the fish, by how much will its temperature decrease? The fish has a mass of 5.00 g and its specific heat is about 3.5 J/(g·°C).

✦76. Consider the heat engine described in Problem 74. (a) For each step in the cycle, find the work done by the gas, the heat flow into or out of the gas, and the change in internal energy of the gas. (b) Find the efficiency of this engine. (c) Compare to the efficiency of a reversible engine that uses the same two reservoirs.

✦77. A town is considering using its lake as a source of power. The average temperature difference from the top to the bottom is 15°C, and the average surface temperature is 22°C. (a) Assuming that the town can set up a reversible engine using the surface and bottom of the lake as heat reservoirs, what would be its efficiency? (b) If the town needs about 1.0×10^8 W of power to be supplied by the lake, how many m^3 of water does the heat engine use per second? (c) The surface area of the lake is 8.0×10^7 m^2 and the average incident intensity (over 24 h) of the sunlight is 200 W/m^2. Can the lake supply enough heat to meet the town's energy needs with this method?

✦78. In a heat engine, 3.00 mol of a monatomic ideal gas, initially at 4.00 atm of pressure, undergoes an isothermal expansion, increasing its volume by a factor of 9.50 at a constant temperature of 650.0 K. The gas is then compressed at a constant pressure to its original volume. Finally, the pressure is increased at constant volume back to the original pressure. (a) Draw a *PV* diagram of this three-step heat engine. (b) For each step of this process, calculate the work done on the gas, the change in internal energy, and the heat transferred into the gas. (c) What is the efficiency of this engine?

Answers to Practice Problems

15.1 The internal energy increase is greater than the heat flow into the gas, so positive work was done on the gas. Positive work is done by the piston when it moves inward.

15.2 360 kJ

15.3 Heat flows into the gas; $Q = 3.8$ kJ.

15.4 The fire is irreversible: smoke, carbon dioxide, and ash will not come together to form logs and twigs.

15.5 20%

15.6 4.0 kW·h = 14 MJ

15.7 1200 K

15.8 178 MW

15.9 0.03 J/K

Answers to Checkpoints

15.2 (a) Yes. The heat flow during an adiabatic process is zero ($Q = 0$), but work can be done. The work done on the system changes its internal energy, which can cause a temperature change. (b) Yes. If a system is in thermal contact with a heat reservoir, heat flows between the reservoir and the system to keep the temperature constant. (c) Yes. During a phase transition such as freezing or melting, the internal energy of the system changes but the temperature does not.

15.4 An inelastic collision involves the conversion of kinetic energy into internal energy, an irreversible process.

15.8 No, in an irreversible process the *total* entropy of the universe increases. If the entropy of one system increases by 10 J/K while the entropy of its surroundings decreases by 10 J/K, the process would be reversible ($\Delta S_{tot} = 0$).

Review & Synthesis: Chapters 13–15

Review Exercises

1. How much does the internal energy change for 1.00 m^3 of water after it has fallen from the top of a waterfall and landed in the river 11.0 m below? Assume no heat flow from the water to the air.
2. At what temperature will nitrogen gas (N_2) have the same rms speed as helium (He) when the helium is at 20.0°C?
3. A bit of space debris penetrates the hull of a spaceship traversing the asteroid belt and comes to rest in a container of water that was at 20.0°C before being hit. The mass of the space rock is 1.0 g and the mass of the water is 1.0 kg. If the space rock traveled at 8.4×10^3 m/s and if all of its kinetic energy is used to heat the water, what is the final temperature of the water?
4. A pot containing 2.00 kg of water is sitting on a hot stove and the water is stirred violently by a mixer that does 6.0 kJ of mechanical work on the water. The temperature of the water rises by 4.00°C. What quantity of heat flowed into the water from the stove during the process?
5. (a) How much ice at −10.0°C must be placed in 0.250 kg of water at 25.0°C to cool the water to 0°C and melt all of the ice? (b) If half that amount of ice is placed in the water, what is the final temperature of the water?
6. A Pyrex container is filled to the very top with 40.0 L of water. Both the container and the water are at a temperature of 90.0°C. When the temperature has cooled to 20.0°C, how much additional water can be added to the container?
7. A 75-g cube of ice at −10.0°C is placed in 0.500 kg of water at 50.0°C in an insulating container so that no heat is lost to the environment. Will the ice melt completely? What will be the final temperature of this system?
8. A hot air balloon with a volume of 12.0 m^3 is initially filled with air at a pressure of 1.00 atm and a temperature of 19.0°C. When the balloon air is heated, the volume and the pressure of the balloon remain constant because the balloon is open to the atmosphere at the bottom. How many moles are in the balloon when the air is heated to 40.0°C?
9. A star's spectrum emits more radiation with a wavelength of 700.0 nm than with any other wavelength. (a) What is the surface temperature of the star? (b) If the star's radius is 7.20×10^8 m, what power does it radiate? (c) If the star is 9.78 ly from Earth, what will an Earth-based observer measure for this star's intensity? Stars are nearly perfect blackbodies. [*Note:* ly stands for light-years.]
10. A wall that is 2.74 m high and 3.66 m long has a thickness composed of 1.00 cm of wood plus 3.00 cm of insulation (with the thermal conductivity approximately of wool). The inside of the wall is 23.0°C and the outside of the wall is at −5.00°C. (a) What is the rate of heat flow through the wall? (b) If half the area of the wall is replaced with a single pane of glass that is 0.500 cm thick, how much heat flows out of the wall now?
11. In a refrigerator, 2.00 mol of an ideal monatomic gas are taken through the cycle shown in the figure. The temperature at point A is 800.0 K. (a) What are the temperature and pressure at point D? (b) What is the net work done on the gas as it is taken through four cycles? (c) What is the internal energy of the gas when it is at point A? (d) What is the total change in internal energy of this gas during four complete cycles?

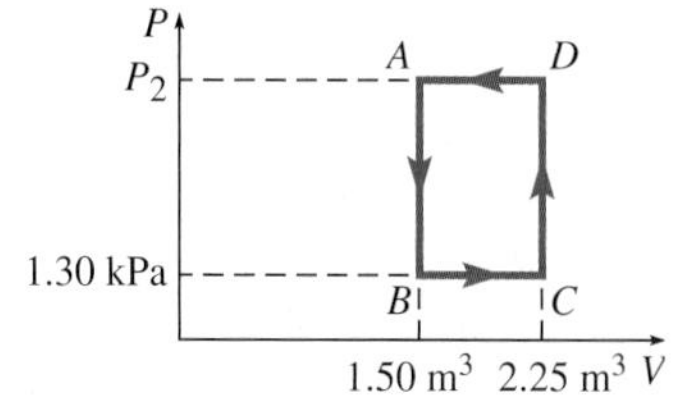

12. Boiling water in an aluminum pan is being converted to steam at a rate of 10.0 g/s. The flat bottom of the pan has an area of 325 cm^2 and the pan's thickness is 3.00 mm. If 27.0% of all heat that is transferred to the pan from the flame beneath it is lost from the sides of the pan and the remaining 73.0% goes into the water, what is the temperature of the base of the pan?
13. A 2.00-kg block of ice at 0.0°C melts. What is the change in entropy of the ice as a result of this process?
14. A sphere with a diameter of 80 cm is held at a temperature of 250°C and is radiating energy. If the intensity of the radiation detected at a distance of 2.0 m from the sphere's center is 102 W/m^2, what is the emissivity of the sphere?
15. A 7.30-kg steel ball at 15.2°C is dropped from a height of 10.0 m into an insulated container with 4.50 L of water at 10.1°C. If no water splashes, what is the final temperature of the water and steel?
16. Michael has set the gauge pressure of the tires on his car to 36.0 psi (lb/in^2). He draws chalk lines around the edges of the tires where they touch the driveway surface to measure the area of contact between the tires and the ground. Each front tire has a contact area of 24.0 in^2 and each rear tire has a contact area of 20.0 in^2. (a) What is the weight (in lb) of the car? (b) The center-to-center distance between front and rear tires is 7.00 ft. Taking the straight line between the centers of the tires on the left side (driver's side) to be the y-axis with the origin at the center of the front left tire (positive direction pointing forward), what is the y-coordinate of the car's CM?

17. Your hot water tank is insulated, but not very well. To reduce heat loss, you wrap some old blankets around it. With the water at 81°C and the room at 21°C, a thermometer inserted between the outside of the original tank and your blanket reads 36°C. By what factor did the blanket reduce the heat loss?

18. An ideal refrigerator keeps its contents at 0.0°C and exhausts heat into the kitchen at 40.0°C. For every 1.0 kJ of work done, (a) how much heat is exhausted? (b) How much heat is removed from the contents?

19. The outdoor temperature on a winter's day is –4°C. If you use 1.0 kJ of electric energy to run a heat pump, how much heat does that put into your house at 21°C? Assume that the heat pump is ideal.

20. A copper rod has one end in ice at a temperature of 0°C, the other in boiling water. The length and diameter of the rod are 1.00 m and 2.00 cm, respectively. At what rate in grams per hour does the ice melt? Assume no heat flows out the sides of the rod.

21. (a) Why is the coolant fluid in an automobile kept under high pressure? (b) Why do radiator caps have safety valves, allowing you to reduce the pressure before removing the cap? [*Hint:* See Fig. 14.7a, the phase diagram for water.]

22. A steam engine has a piston with a diameter of 15.0 cm and a stroke (the displacement of the piston) of 20.0 cm. The average pressure applied to this piston is 1.3×10^5 Pa. What operating frequency in cycles per second (Hz) would yield an average power output of 27.6 kW?

23. Two aluminum blocks in thermal contact have the same temperature. (a) Under what condition do they have the same internal energy? (b) Is there an energy transfer between the two blocks? (c) Are the blocks necessarily in physical contact?

24. A power plant burns coal to produce pressurized steam at 535 K. The steam then condenses back into water at a temperature of 323 K. (a) What is the maximum possible efficiency of this plant? (b) If the plant operates at 50.0% of its maximum efficiency and its power output is 1.23×10^8 W, at what rate must heat be removed by means of a cooling tower?

25. A heat engine consists of the following four step cyclic process. During step 1, 2.00 mol of a diatomic ideal gas at a temperature of 325 K are compressed isothermally to one-eighth of the original volume. In step 2, the temperature of the gas is increased to 985 K by an isochoric process. During step 3, the gas expands isothermally back to its original volume. Finally, in step 4, an isochoric process takes the gas back to its original temperature. (a) Sketch a qualitative *PV* diagram, showing the four steps in this cycle. (b) Make a table showing the values of W, Q, and ΔU for each of the four steps and the totals for one cycle of this process. (c) What is the efficiency of this engine? (d) What would be the efficiency of a Carnot engine operating at the same extreme temperatures?

26. On a day when the air temperature is 19°C, a 0.15-kg baseball is dropped from the top of a 24-m tower. After the ball hits the ground, bounces a few times, and comes to rest, by how much has the entropy of the universe increased?

27. In a certain bimetallic strip (see Fig. 13.7) the brass strip is 0.100% longer than the steel strip at a temperature of 275°C. At what temperature do the two strips have the same length?

28. A 0.360-kg piece of solid lead at 20°C is placed into an insulated container holding 0.980 kg of liquid lead at 420°C. The system comes to an equilibrium temperature with no loss of heat to the environment. Ignore the heat capacity of the container. (a) Is there any solid lead remaining in the system? (b) What is the final temperature of the system?

29. (a) Calculate Earth's escape speed—the minimum speed needed for an object near the surface to escape Earth's gravitational pull. [*Hint:* Use conservation of energy and ignore air resistance.] (b) Calculate the average speed of a hydrogen molecule (H_2) at 0°C. (c) Calculate the average speed of an oxygen molecule (O_2) at 0°C. (d) Use your answers from parts (a) through (c) along with what you know about the distribution of molecular speeds to explain why Earth's atmosphere contains plenty of oxygen but almost no hydrogen.

30. A 10.0-cm cylindrical chamber has a 5.00-cm-diameter piston attached to one end. The piston is connected to an ideal spring with a spring constant of 10.0 N/cm, as shown. Initially, the spring is not compressed but is latched in place so that it cannot move. The cylinder is filled with gas to a pressure of 5.00×10^5 Pa. Once the gas in the cylinder is at this pressure, the spring is unlatched. Because of the difference in pressure between the inside of the chamber and the outside, the spring moves a distance Δx. Heat is allowed to flow into the chamber as it expands so that the temperature of the gas remains constant; thus, you may assume T to be the same before and after the expansion. Find the compression of the spring, Δx.

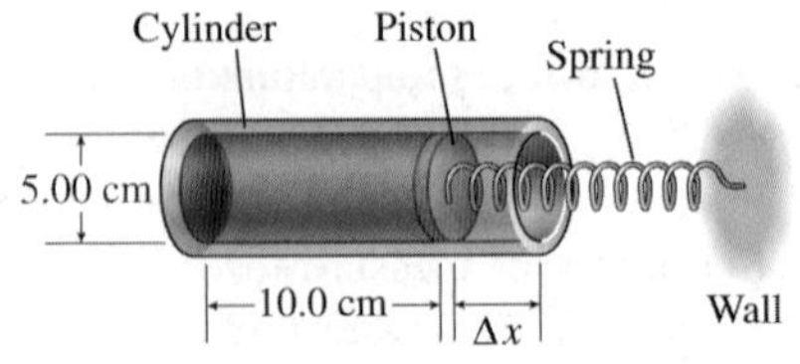

MCAT Review

The section that follows includes MCAT exam material and is reprinted with permission of the Association of American Medical Colleges (AAMC).

1. Suppose 2 identical copper bars, A and B, with initial temperatures of 25°C and 75°C, respectively, are placed in contact with each other. If the specific heat of copper is independent of temperature, and if A and B do *not* exchange heat or work with the surroundings, is it likely that A and B will reach 24°C and 76°C, respectively?
   A. Yes, because the bars are identical.
   B. Yes, because heat will flow from B to A.
   C. No, because heat will not flow from A to B.
   D. No, because energy will not be conserved.
2. Some ocean currents carry water from the polar regions to warmer seas. What is the approximate temperature of a solution resulting from mixing 1.00 kg of seawater at 0°C with 1.00 kg of seawater at 5°C?
   A. 1.25°C B. 2.50°C C. 3.25°C D. 4.00°C
3. How much energy is gained by 18.0 g of ice if it melts at the polar ice caps?
   A. 4.18 kJ
   B. 5.87 kJ
   C. 6.02 kJ
   D. 6.17 kJ

Read the paragraph and then answer the following questions:

The steam engine pictured here demonstrates principles of thermodynamics. Water boils, creating steam that pushes against a piston. The steam then changes back to water in the condenser, and the water circulates back to the boiler. The efficiency of this engine is

$$e = W/Q_H = 1 - Q_C/Q_H$$

where W is the output work, Q_H is the heat put in, and Q_C is the heat that flows out as exhaust. It is not possible to convert all of the input heat into output work.

A refrigerator works like a heat engine in reverse. Heat is absorbed from a refrigerator when the liquid that circulates through the refrigerator changes to gas. The gas is then changed back to liquid in a compressor, and the refrigerant is then recirculated.

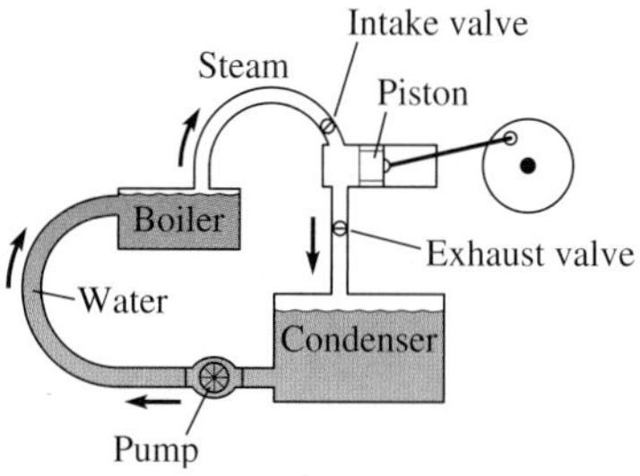

MCAT Review Questions 4–9

4. Making which of the following changes to the steam engine will increase its efficiency?
   A. Increasing the exhaust temperature
   B. Decreasing the exhaust temperature
   C. Increasing the amount of heat input
   D. Decreasing the amount of heat input
5. The amount of heat that a unit of mass of a refrigerant can remove from a refrigerator is primarily dependent on which of the following characteristics of the refrigerant?
   A. Heat of vaporization
   B. Heat of fusion
   C. Specific heat in liquid form
   D. Specific heat in gaseous form
6. Surrounding the condenser with which one of the following would be most effective for changing steam to water?
   A. High-pressure steam
   B. Low-pressure steam
   C. Stationary water
   D. Circulating water
7. The amount of useful work that can be generated from a source of heat can only be
   A. less than the amount of heat.
   B. less than or equal to the amount of heat.
   C. equal to the amount of heat.
   D. equal to or greater than the amount of heat.
8. What energy transformation causes the piston of the steam engine discussed in the passage to move to the right?
   A. Mechanical to internal
   B. Mechanical to chemical
   C. Internal to mechanical
   D. Internal to electrical
9. Which of the following accurately contrasts the boiling or freezing points of water and of a refrigerant used in a household refrigerator?
   A. The boiling point of the refrigerant should be higher than the boiling point of water.
   B. The boiling point of the refrigerant should be lower than the freezing point of water.
   C. The freezing point of the refrigerant should be higher than the boiling point of water.
   D. The freezing point of the refrigerant should be higher than the freezing point of water.

Read the paragraph and then answer the following questions:

An engineer was instructed to design a holding tank for synthetic lubricating oil. Two requirements were that the amount of time necessary to drain the tank and the force needed to lift the drain plug be minimized. In the initial design, the drain plug, which weighed 500 N, rested on the drain hole and was lifted by a thin rod that extended through the top of the tank. The tank was insulated and had 10 electric immersion heaters that each use 5 kW of power. The oil has a boiling point of 220°C, a specific gravity of 0.7, and a specific heat that is 60% that of water. The heat capacity of the tank was negligible compared to the fluids contained in it.

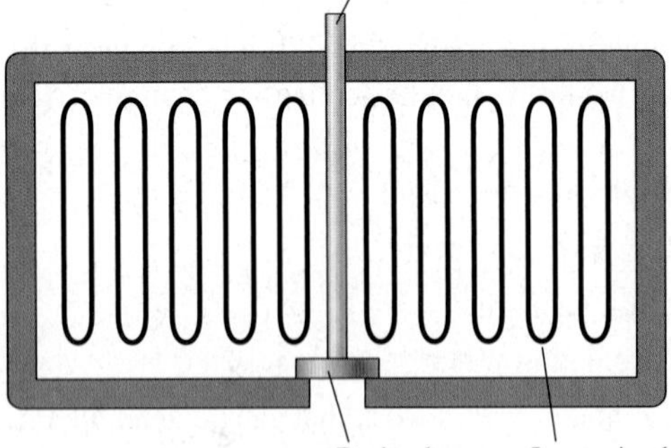

MCAT Review Questions 10 and 11

The tank was built and then tested by filling it with water. The air pressure inside and outside the tank was 1 atm. The force required to lift the plug was found to be 5310 N. In testing the heater capability, the tank was filled to the top with water at 20°C. With all 10 heaters operating, the water temperature reached 100°C 15 h later. The technician who conducted the evaluation reported that the full tank of water was completely discharged approximately 30 s after opening the drain.

10. With the heaters operating, how long would it take to raise the temperature of a full tank of oil from 20°C to 60°C?
    A. 3.2 h
    B. 6.3 h
    C. 7.5 h
    D. 9.0 h

11. It is suggested that the air in the tank above the oil be pressurized at 4 atm above normal air pressure. Which of the following is the *least* likely to occur along with this increase in pressure?
    A. The time required to heat the oil would be greatly extended.
    B. The drain plug would be more difficult to lift.
    C. Fluid velocity would be increased when the tank is drained.
    D. The time required to drain the tank would decrease.

CHAPTER 16

Electric Forces and Fields

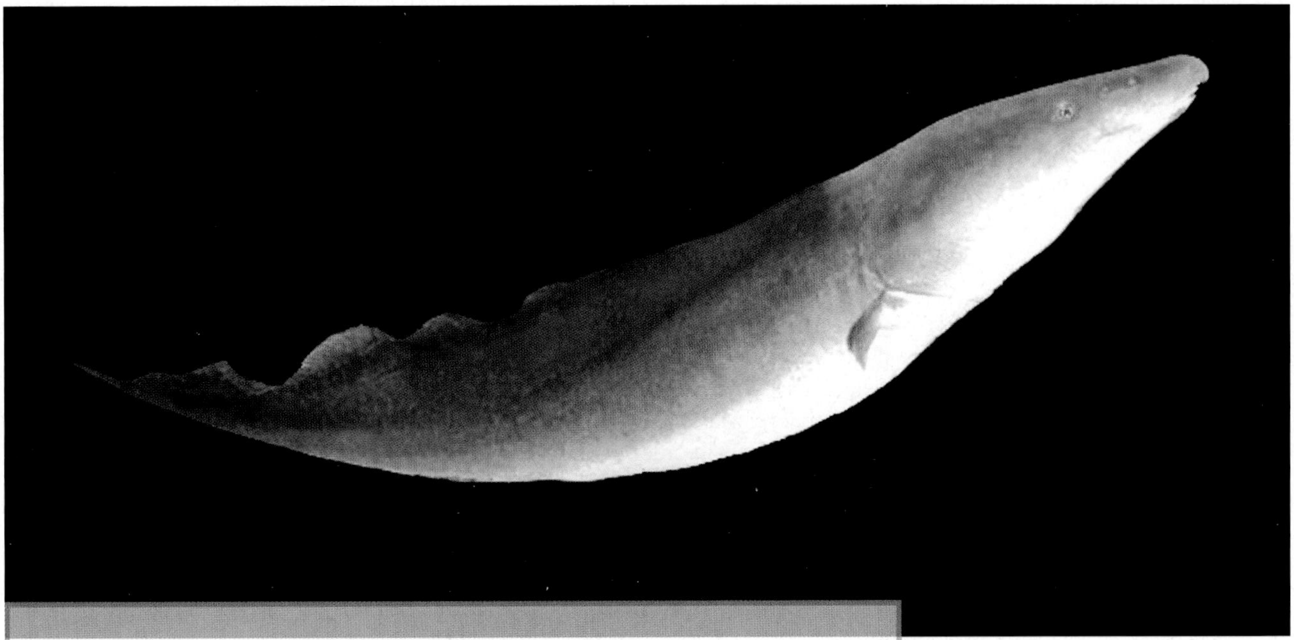

The elegant fish in the photograph is the *Gymnarchus niloticus,* a native of Africa found in the Nile River. *Gymnarchus* has some interesting traits. It swims gracefully with equal facility either forward or backward. Instead of propelling itself by lashing its tail sideways, as most fish do, it keeps its spine straight—not only when swimming straight ahead, but even when turning. Its propulsion is accomplished by means of the undulations of the fin along its back.

Gymnarchus navigates with great precision, darting after its prey and evading obstacles in its path. What is surprising is that it does so just as precisely when swimming backward. Furthermore, *Gymnarchus* is nearly blind; its eyes respond only to extremely bright light. How then, is it able to locate its prey in the dim light of a muddy river? (See p. 581 for the answer.)

Concepts & Skills to Review

- gravitational forces, fundamental forces (Sections 4.5 and 4.12)
- free-body diagrams (Section 4.1)
- Newton's second law: force and acceleration (Section 4.3)
- motion with constant acceleration (Sections 2.4 and 3.5)
- equilibrium (Section 4.2)
- adding vectors; resolving a vector into components (Sections 3.1–3.2)

16.1 ELECTRIC CHARGE

In Part Three of this book, we study electric and magnetic fields in detail. Recall from Chapter 4 that all interactions in the universe fall into one of four categories: gravitational, electromagnetic, strong, and weak. All of the familiar, everyday forces other than gravity—contact forces, tension in cables, and the like—are fundamentally electromagnetic. What we think of as a single interaction is really the net effect of huge numbers of microscopic interactions between electrons and atoms. Electromagnetic forces bind electrons to nuclei to form atoms and molecules. They hold atoms together to form liquids and solids, from skyscrapers to trees to human bodies. Technological applications of electromagnetism abound, especially once we realize that radio waves, microwaves, light, and other forms of electromagnetic radiation consist of oscillating electric and magnetic fields.

Many everyday manifestations of electromagnetism are complex; hence we study simpler situations in order to gain some insight into how electromagnetism works. The hybrid word *electromagnetism* itself shows that electricity and magnetism, which were once thought to be completely separate forces, are really aspects of the same fundamental interaction. This unification of the studies of electricity and magnetism occurred in the late nineteenth century. However, understanding comes more easily if we first tackle electricity (Chapters 16–18), then magnetism (Chapter 19), and finally see how they are closely related (Chapters 20–22).

The existence of electric forces has been familiar to humans for at least 3000 years. The ancient Greeks used pieces of amber—a hard, fossilized form of the sap from pine trees—to make jewelry. When a piece of amber was polished by rubbing it with a piece of fabric, it was observed that the amber would subsequently attract small objects, such as bits of string or hair. Using modern knowledge, we say that the amber is *charged* by rubbing: some electric charge is transferred between the amber and the cloth. Our word *electric* comes from the Greek word for amber (*elektron*).

A similar phenomenon occurs on a dry day when you walk across a carpeted room wearing socks. Charge is transferred between the carpet and your socks and between your socks and your body. Some of the charge you have accumulated may be unintentionally transferred from your fingertips to a doorknob or to a friend—accompanied by the sensation of a shock.

Types of Charge

Electric charge is not *created* by these processes; it is just transferred from one object to another. The law of **conservation of charge** is one of the fundamental laws of physics; no exceptions to it have ever been found.

CONNECTION:

Conservation of charge is a fundamental conservation law. Charge is a conserved *scalar* quantity, like energy. Momentum and angular momentum are conserved *vector* quantities.

Conservation of Charge

The net charge of a closed system never changes.

Experiments with amber and other materials that can be charged show that there are **two types of charge**; Benjamin Franklin (1706–1790) was the first to call them

Table 16.1 Masses and Electric Charges of the Proton, Electron, and Neutron

Particle	Mass	Electric Charge
Proton	$m_p = 1.673 \times 10^{-27}$ kg	$q_p = +e = +1.602 \times 10^{-19}$ C
Electron	$m_e = 9.109 \times 10^{-31}$ kg	$q_e = -e = -1.602 \times 10^{-19}$ C
Neutron	$m_n = 1.675 \times 10^{-27}$ kg	$q_n = 0$

positive (+) and *negative* (−). The **net charge** of a system is the algebraic sum—taking care to include the positive and negative signs—of the charges of the constituent particles in the system. When a piece of glass is rubbed by silk, the glass acquires a positive charge and the silk a negative charge; the net charge of the system of glass and silk does not change. An object that is **electrically neutral** has equal amounts of positive and negative charge and thus a net charge of zero. The symbols used for quantity of charge are q or Q.

Net charge: the algebraic sum of all the charges in a system

Ordinary matter consists of atoms, which in turn consist of electrons, protons, and neutrons. The protons and neutrons are called *nucleons* because they are found in the nucleus. The neutron is electrically neutral (thus the name *neutron*). The charges on the proton and the electron are of *equal magnitude* but of opposite sign. The charge on the proton is arbitrarily chosen to be positive; that on the electron is therefore negative. A neutral atom has equal numbers of protons and electrons, a balance of positive and negative charge. If the number of electrons and protons is not equal, then the atom is called an *ion* and has a nonzero net charge. If the ion has more electrons than protons, its net charge is negative; if the ion has fewer electrons than protons, its net charge is positive.

Elementary Charge

The *magnitude* of charge on the proton and electron is the same (Table 16.1). That amount of charge is called the **elementary charge** (symbol e). In terms of the SI unit of charge, the coulomb (C), the value of e is

$$e = 1.602 \times 10^{-19}\ \text{C} \qquad (16\text{-}1)$$

Since ordinary objects have only slight imbalances between positive and negative charge, the coulomb is often an inconveniently large unit. For this reason, charges are often given in millicoulombs (mC), microcoulombs (μC), nanocoulombs (nC), or picocoulombs (pC). The coulomb is named after the French physicist Charles Coulomb (1736–1806), who developed the expression for the electric force between two charged particles.

The net charge of any object is an integral multiple of the elementary charge. Even in the extraordinary matter found in exotic places like the interior of stars, the upper atmosphere, or in particle accelerators, the observable charge is always an integer times e.

CHECKPOINT 16.1

A glass rod and piece of silk are both electrically neutral. Then the rod is rubbed with the silk. If 4.0×10^9 electrons are transferred from the glass to the silk and no ions are transferred, what are the net charges of both objects?

Example 16.1

An Unintentional Shock

The magnitude of charge transferred when you walk across a carpet, reach out to shake hands, and unintentionally give a shock to a friend might be typically about 1 nC. (a) If the charge is transferred by electrons only, how many electrons are transferred? (b) If your body has a net charge of −1 nC, estimate the percentage of excess electrons. [*Hint:* See Table 16.1. The mass of the electron is only about 1/2000 that of a nucleon, so most of the mass of the body is in the nucleons. For an order-of-magnitude calculation, we can just assume that $\frac{1}{2}$ of the nucleons are protons and half are neutrons.]

Strategy Since the coulomb (C) is the SI unit of charge, the "n" must be the prefix "nano-" ($= 10^{-9}$). We know the value of the elementary charge in coulombs. For part (b), we first make an order-of-magnitude estimate of the number of electrons in the human body.

Solution (a) The number of electrons transferred is the quantity of charge transferred divided by the charge of each electron:

$$\frac{-1 \times 10^{-9}\ \text{C}}{-1.6 \times 10^{-19}\ \text{C per electron}} = 6 \times 10^{9}\ \text{electrons}$$

Notice that the *magnitude* of the charge transferred is 1 nC, but since it is transferred by electrons, the sign of the charge transferred is negative.

(b) We estimate a typical body mass of around 70 kg. Most of the mass of the body is in the nucleons, so

$$\text{number of nucleons} = \frac{\text{mass of body}}{\text{mass per nucleon}} = \frac{70\ \text{kg}}{1.7 \times 10^{-27}\ \text{kg}}$$
$$= 4 \times 10^{28}\ \text{nucleons}$$

Assuming that roughly $\frac{1}{2}$ of the nucleons are protons,

$$\text{number of protons} = \tfrac{1}{2} \times 4 \times 10^{28} = 2 \times 10^{28}\ \text{protons}$$

In an electrically neutral object, the number of electrons is equal to the number of protons. With a net charge of −1 nC, the body has 6×10^{9} extra electrons. The percentage of excess electrons is then

$$\frac{6 \times 10^{9}}{2 \times 10^{28}} \times 100\% = (3 \times 10^{-17})\%$$

Discussion As shown in this example, charged macroscopic objects have *tiny* differences between the magnitude of the positive charge and the magnitude of the negative charge. For this reason, electric forces between macroscopic bodies are often negligible.

Practice Problem 16.1 Excess Electrons on a Balloon

How many excess electrons are found on a balloon with a net charge of −12 nC?

One of the important differences between the gravitational force and the electric force comes about because charge has either a positive or a negative sign, but mass is always a positive quantity. The gravitational force between two massive bodies is always an attractive force, but the electric force between two charged particles can be attractive or repulsive depending on the signs of the charges. Two particles with charges of the same sign repel one another, but two particles with charges of opposite sign attract one another. More briefly,

> Like charges repel one another; unlike charges attract one another.

A common shorthand is to say "a charge" instead of saying "a particle with charge."

Polarization

Polarization: charge separation within an object

An electrically neutral object may have regions of positive and negative charge within it, separated from one another. Such an object is **polarized**. A polarized object can experience an electric force even though its net charge is zero. A rubber rod charged negatively after being rubbed with fur attracts small bits of paper. So does a glass rod that is *positively* charged after being rubbed with silk (Fig. 16.1a,b). The bits of paper are electrically neutral, but a charged rod polarizes the paper—it attracts the unlike charge in the paper a bit closer and pushes the like charge in the paper a bit farther away (Fig. 16.1c). The

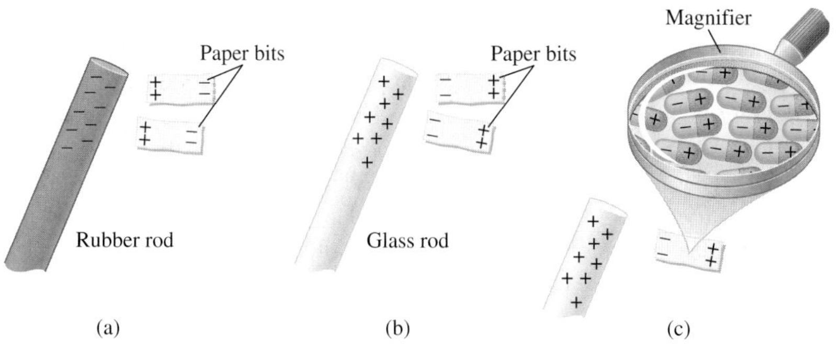

Figure 16.1 (a) Negatively charged rubber rod attracting bits of paper. (b) Positively charged glass rod attracting bits of paper. (c) Magnified view of polarized molecules within a bit of paper.

attraction between the rod and the unlike charge then becomes a little stronger than the repulsion between the rod and the like charge, since the electric force gets weaker as the separation increases and the like charge is farther away. Thus, the net force on the paper is always attractive, regardless of the sign of charge on the rod. In this case, we say that the paper is *polarized by induction*; the polarization of the paper is induced by the charge on the nearby rod. When the rod is moved away, the paper is no longer polarized.

Some molecules are *intrinsically* polarized. An electrically neutral water molecule, for example, has equal amounts of positive and negative charge (10 protons and 10 electrons), but the center of positive charge and the center of negative charge do not coincide. The electrons in the molecule are shared in such a way that the oxygen end of the molecule has a net negative charge, but the hydrogen atoms are positive.

Application: polarization of charge in water

PHYSICS AT HOME

On a dry day, run a comb through your hair (this works best if your hair is clean and dry and you have not used conditioner) or rub the comb on a wool sweater. When you are sure the comb is charged (by observing the behavior of your hair, listening for crackling sounds, etc.), go to a sink and turn the water on so that a *thin* stream of water comes out. It does not matter if the stream breaks up into droplets near the bottom. Hold the charged comb near the stream of water. You should see that the water experiences a force due to the charge on the comb (Fig. 16.2). Is the force attractive or repulsive? Does this mean that the water coming from the tap has a net charge? Explain why holding the comb near the top of the stream is more effective than holding it farther down (at the same horizontal distance from the stream).

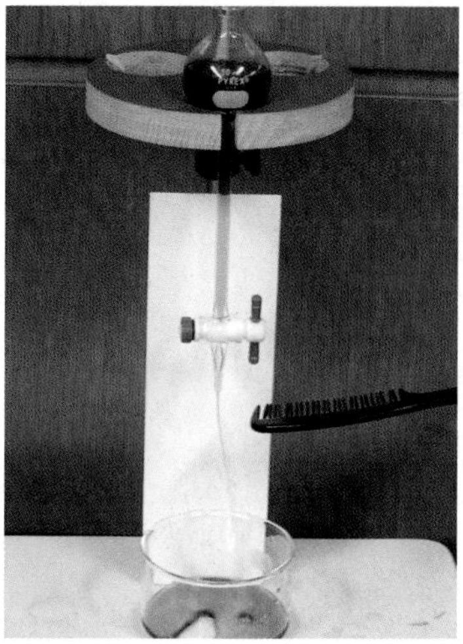

Figure 16.2 A stream of water is deflected by a charged comb.

16.2 ELECTRIC CONDUCTORS AND INSULATORS

Ordinary matter consists of atoms containing electrons and nuclei. The electrons differ greatly in how tightly they are bound to the nucleus. In atoms with many electrons, most of the electrons are tightly bound—under ordinary circumstances nothing can tear them away from the nucleus. Some of the electrons are much more weakly bound and can be removed from the nucleus in one way or another.

Materials vary dramatically in how easy or difficult it is for charge to move within them. Materials in which some charge can move easily are called electric **conductors**, whereas materials in which charge does not move easily are called electrical **insulators**.

Metals are materials in which *some* of the electrons are so weakly bound that they are not tied to any one particular nucleus; they are free to wander about within the metal. The *free electrons* in metals make them good conductors. Some metals are better conductors than others, with copper being one of the best. Glass, plastics, rubber, wood,

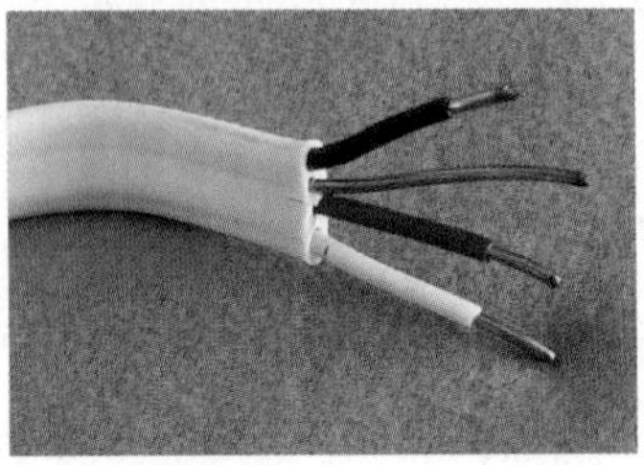

Figure 16.3 Some electric wires. The metallic conductors are surrounded by insulating material. The insulation must be stripped away where the wire makes an electric connection with something else.

paper, and many other familiar materials are insulators. Insulators do not have free electrons; each electron is bound to a particular nucleus.

The terms *conductor* and *insulator* are applied frequently to electric wires, which are omnipresent in today's society (Fig. 16.3). The copper wires allow free electrons to flow. The plastic or rubber insulator surrounding the wires keeps the electric current—the flow of charge—from leaving the wires (and entering your hand, for instance).

Water is usually thought of as an electric conductor. It is wise to assume so and take precautions such as not handling electric devices with wet hands. Actually, *pure* water is an electrical insulator. Pure water consists mostly of complete water molecules (H_2O), which carry no net charge as they move about; there is only a tiny concentration of ions (H^+ and OH^-). But tap water is by no means pure—it contains dissolved minerals. The mineral ions make tap water an electrical conductor. The human body contains many ions and therefore is a conductor.

Similarly, air is a good insulator, because most of the molecules in air are electrically neutral, carrying no charge as they move about. However, air does contain some ions; air molecules are ionized by radioactive decays or by cosmic rays.

Intermediate between conductors and insulators are the **semiconductors**. The part of the computer industry clustered in northern California is referred to as "Silicon Valley" because silicon is a common semiconductor used in making computer chips and other electronic devices. *Pure* semiconductors are good insulators, but by *doping* them—adding tiny amounts of impurities in a controlled way—their electrical properties can be fine-tuned.

Charging Insulators by Rubbing When different insulating objects are rubbed against one another, both electrons and ions (charged atoms) can be transferred from one object to the other. If both objects had zero net charge before they were rubbed together, they now have net charges of equal magnitudes and opposite signs, since charge is conserved. Charging by rubbing works best in dry air. When the humidity is high, a film of moisture condenses on the surfaces of objects; charge can then leak off more easily, so it is difficult to build up charge.

Notice that we rub two *insulators* together to separate charge. A piece of metal can be rubbed all day with fur or silk without charging the metal; it is too easy for charge in the metal to move around and avoid getting transferred. Once an insulator is charged, the charge remains where it is.

Charging a Conductor by Contact How can a conductor be charged? First rub two insulators together to separate charge; then touch one of the charged insulators to the conductor (Fig. 16.4). Since the charge transferred to the conductor spreads out, the process can be repeated to build up more and more charge on the conductor.

Grounding a Conductor How can a conductor be discharged? One way is to *ground* it. Earth is a conductor because of the presence of ions and moisture and is large enough

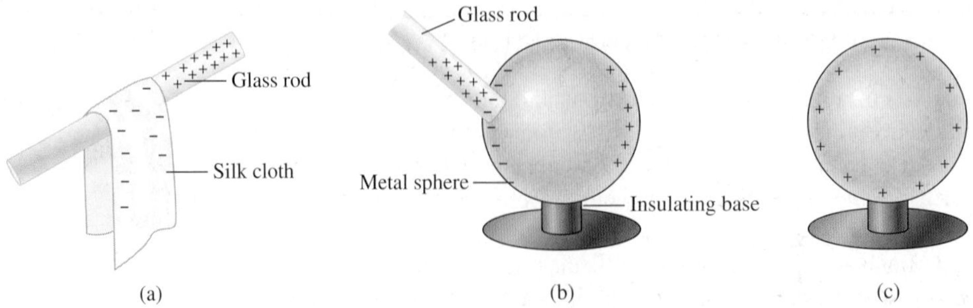

Figure 16.4 Charging a conductor. (a) After rubbing a glass rod with a silk cloth, the glass rod is left with a net positive charge and the silk is left with a net negative charge. (b) Touching the glass rod to a metal sphere. The positively charged glass attracts some of the free electrons from the metal onto the glass. (c) The glass rod is removed. The metal sphere now has fewer electrons than protons, so it has a net positive charge. Even though negative charge is actually transferred (electrons), it is often said that "positive charge is transferred to the metal" since the net effect is the same.

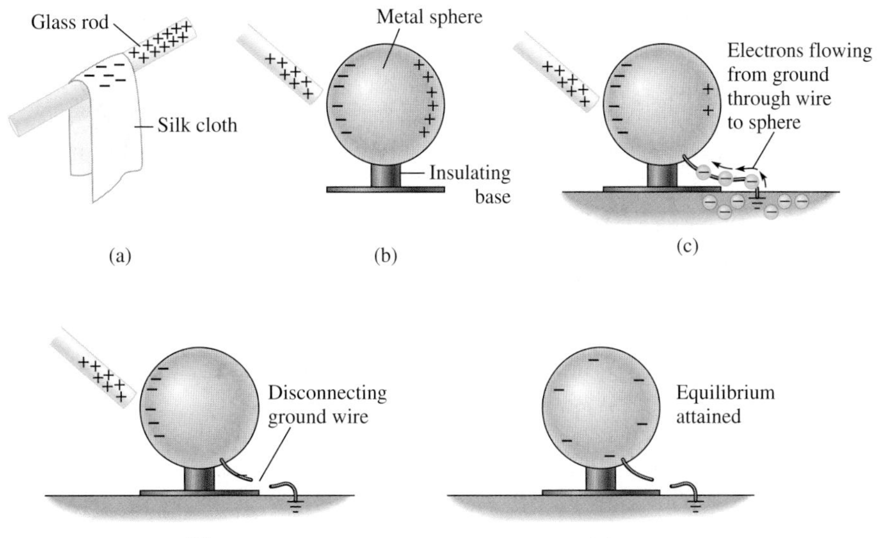

Figure 16.5 Charging by induction. (a) A glass rod is charged by rubbing it with silk. (b) The positively charged glass rod is held near a metal sphere, but does not touch it. The sphere is polarized as free electrons within the sphere are attracted toward the glass rod. (c) When the sphere is grounded, electrons from the ground move onto the sphere, attracted there by positive charges on the sphere. (d) The ground connection is broken without moving the glass rod. (e) Now the glass rod is removed with the ground wire still disconnected. Charge spreads over the metal surface as the like charges repel one another. The sphere is left with a net negative charge because of the excess electrons.

that for many purposes it can be thought of as a limitless reservoir of charge. To *ground* a conductor means to provide a conducting path between it and the Earth (or to another charge reservoir). A charged conductor that is grounded discharges because the charge spreads out by moving off the conductor and onto the Earth.

A buildup of even a relatively small amount of charge on a truck that delivers gasoline could be dangerous—a spark could trigger an explosion. To prevent such a charge buildup, the truck grounds its tank before starting to deliver gasoline to the service station.

The round opening of modern electric outlets is called *ground*. It is literally connected by a conducting wire to the ground, either through a metal rod driven into the Earth or through underground metal water pipes. The purpose of the ground connection is more fully discussed in Chapter 17, but you can understand one purpose already: it prevents static charges from building up on the conductor that is grounded.

Charging a Conductor by Induction A conductor is not necessarily discharged when it is grounded if there are other charges nearby. It is even possible to charge an initially neutral conductor by grounding it. In the process shown in Fig. 16.5, the charged insulator never touches the conducting sphere. The positively charged rod first polarizes the sphere, attracting the negative charges on the sphere while repelling the positive charges. Then the sphere is grounded. The resulting separation of charge on the conducting sphere causes negative charges from the Earth to be attracted along the grounding wire and onto the sphere by the nearby positive charges.

The symbol ⏚ represents a connection to ground.

Application of grounding: fuel trucks

CONNECTION:

The word *reservoir* may remind you of heat reservoirs. A heat reservoir has such a large heat capacity that it is possible to exchange heat with it without changing its temperature appreciably. Once we study electric potential in Chapter 17, we can describe a charge reservoir as something that can transfer charge of either sign without changing its potential.

Conceptual Example 16.2

The Electroscope

An electroscope is charged negatively and the gold foil leaves hang apart as in Fig. 16.6. What happens to the leaves as the following operations are carried out in the order listed? Explain what you see after each step. (a) You touch the metal bulb at the top of the electroscope with your hand. (b) You bring a glass rod that has been rubbed with silk *near* the bulb

continued on next page

Example 16.2 continued

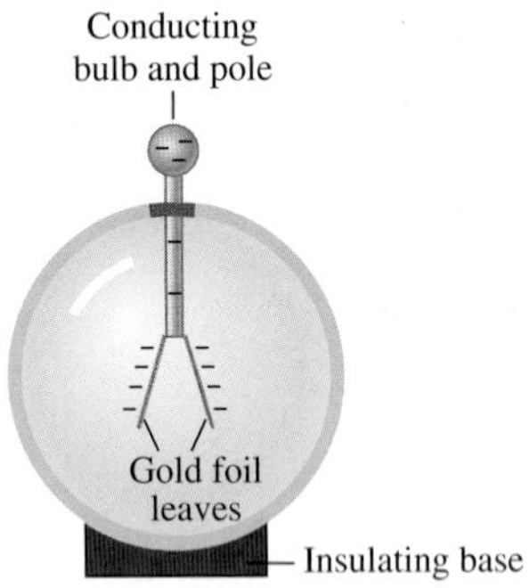

Figure 16.6

An electroscope is a device used to demonstrate the presence of charge. A conducting pole has a metallic bulb at the top and a pair of flexible leaves of gold foil at the bottom. The leaves are pushed apart due to the repulsion of the negative charges.

without touching it. [*Hint:* A glass rod rubbed with silk is positively charged.] (c) The glass rod touches the metal bulb.

Solution and Discussion (a) By touching the electroscope bulb with your hand, you ground it. Charge is transferred between your hand and the bulb until the bulb's net charge is zero. Since the electroscope is now discharged, the foil leaves hang down as in Fig. 16.7. (b) When the positively charged rod is held near the bulb, the electroscope becomes polarized by induction. Negatively charged free electrons are drawn toward the bulb, leaving the foil leaves with a positive net charge (Fig. 16.8). The leaves hang apart due to the mutual repulsion of the net positive charges on them. (c) When the positively charged rod touches the bulb, some negative charge is transferred from the bulb to the rod. The electroscope now has a positive net charge. The glass rod still has a positive net charge that repels the positive charge on the electroscope, pushing it as far away as possible—toward the foil leaves. The leaves hang farther apart, since they now have *more* positive charge on them than before.

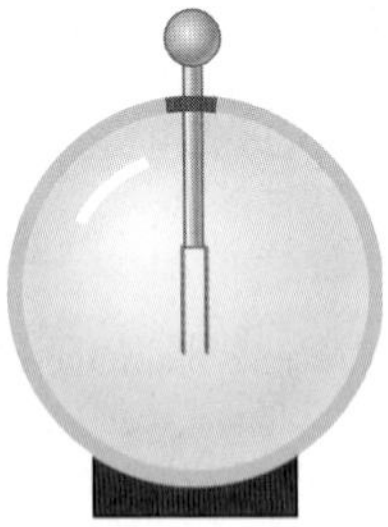

Figure 16.7

With no net charge, the leaves hang straight down.

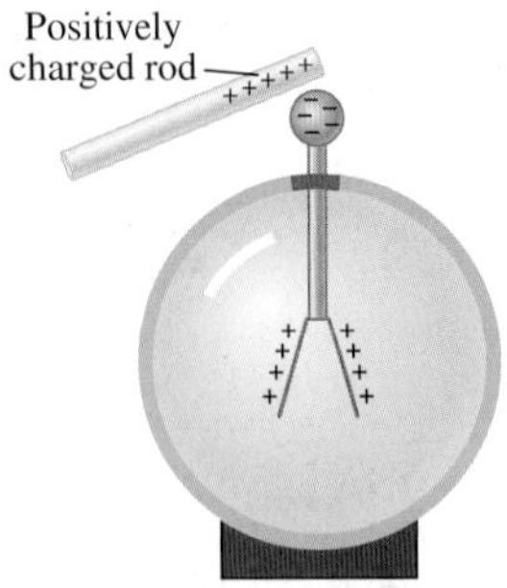

Figure 16.8

With a positively charged rod near the bulb, the electroscope has no net charge but it is polarized: the bulb is negative and the leaves are positive. Repulsion between the positive charges on the leaves pushes them apart.

Conceptual Practice Problem 16.2 Removing the Glass Rod

What happens to the leaves as the glass rod is moved away?

Application: electrostatic charge of adhesive tape

PHYSICS AT HOME

Ordinary transparent tape has an adhesive that allows it to stick to paper and many other materials. Since the sticking force is electric in nature, it is not too surprising that adhesive can be used to separate charge. If you have ever peeled a roll of tape too quickly and noticed that the strip of tape curls around and behaves strangely, you have seen effects of this charge separation—the strip of tape has a net charge (and so does the tape left behind, but of opposite sign). Tape pulled *slowly* off a surface does not tend to have a net charge. There are some instructive experiments you can perform:

- Pull a strip of tape quickly from the roll. How can you tell if the tape has a net charge?
- Take the roll of tape into a dark closet. What do you see when you pull a strip quickly from the roll?
- See if the strip is attracted or repelled when you hold it near a paper clip. Explain what you see.
- Rub the tape on both sides between your thumb and forefinger. Now try the paper clip again. What has happened? Explain.
- Pull a second strip of tape *slowly* from the roll. Is the force between the two strips attractive or repulsive? What does that tell you?
- Hold the second strip near the paper clip. Is there a net force? What can you conclude?
- Can you think of a way of reliably making two strips of tape with like charges? With unlike charges?
- Enough suggestions—have some fun and see what you can discover!

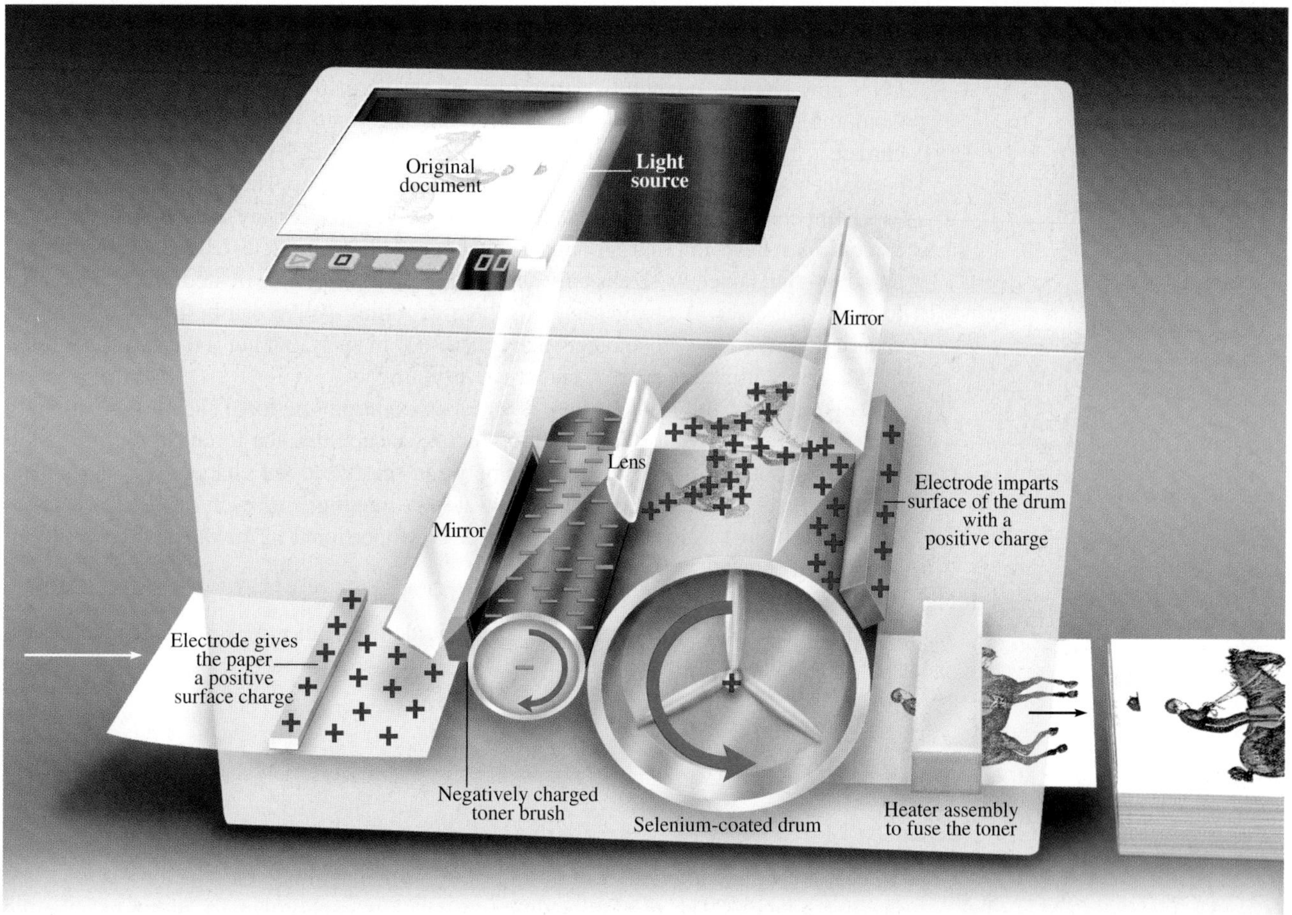

Figure 16.9 The operation of a photocopier is based on the attraction of negatively charged toner particles to regions on the drum that are positively charged.

Application: Photocopiers and Laser Printers

The operation of photocopiers (and laser printers) is based on the separation of charge and the attraction between unlike electric charges (Fig. 16.9). Positive charge is applied to a selenium-coated aluminum drum by rotating the drum under an electrode. The drum is then illuminated with a projected image of the document to be copied (or by a laser).

Selenium is a *photoconductor*—a light-sensitive semiconductor. When no light shines on the selenium, it is a good insulator; but when light shines on it, it becomes a good conductor. The selenium coating on the drum is initially in the dark. Behaving as an insulator, it can be electrically charged. When the selenium is illuminated, it becomes conducting wherever light falls on it. Electrons from the aluminum—a good conductor—pass into the illuminated regions of selenium and neutralize the positive charge. Regions of the selenium coating that remain dark do not allow electrons from the aluminum to flow in, so those regions remain positively charged.

Next, the drum is allowed to come into contact with a black powder called *toner.* The toner particles have been given a negative charge so they will be attracted to positively charged regions of the drum. Toner adheres to the drum where there is positive charge, but no toner adheres to the uncharged regions. A sheet of paper is now rotated onto the drum and positive charge is applied to the back surface of the paper. The charge on the paper is larger than that on the drum, so the paper attracts the negatively charged toner away from the drum, forming an image of the original document on the paper. The final step is to fuse the toner to the paper by passing the paper between hot rollers. With the ink sealed into the fibers of the paper, the copy is finished.

16.3 COULOMB'S LAW

Let's now begin a quantitative treatment of electrical forces among charged objects. Coulomb's law gives the electric force acting between two *point charges.* A **point charge** is a pointlike object with a nonzero electric charge. Recall that a pointlike object is small enough that its internal structure is of no importance. The electron can be treated as a point charge, since there is no experimental evidence for any internal structure. The proton *does* have internal structure—it contains three particles called *quarks* bound together—but, since its size is only about 10^{-15} m, it too can be treated as a point charge for most purposes. A charged metal sphere of radius 10 cm can be treated as a point charge if it interacts with another such sphere 100 m away, but not if the two spheres are only a few centimeters apart. Context is everything!

Like gravity, the electric force is an *inverse square law* force. That is, the strength of the force decreases as the separation increases such that the force is proportional to the inverse square of the separation r between the two point charges ($F \propto 1/r^2$). The strength of the force is also proportional to the *magnitude* of each of the two charges ($|q_1|$ and $|q_2|$) just as the gravitational force is proportional to the *mass* of each of two interacting objects.

Magnitude of Electric Force The magnitude of the electric force that each of two charges exerts on the other is given by

$$F = \frac{k|q_1|\,|q_2|}{r^2} \qquad (16\text{-}2)$$

Since we use the *magnitudes* of q_1 and q_2, F—the magnitude of a vector—is always a positive quantity. The proportionality constant k is experimentally found to have the value

$$k = 8.99 \times 10^9 \ \frac{\text{N}\cdot\text{m}^2}{\text{C}^2} \qquad (16\text{-}3\text{a})$$

The constant k, which we call the *Coulomb constant*, can be written in terms of another constant ϵ_0, the *permittivity of free space*:

$$\epsilon_0 = \frac{1}{4\pi k} = 8.85 \times 10^{-12} \ \frac{\text{C}^2}{\text{N}\cdot\text{m}^2} \qquad (16\text{-}3\text{b})$$

Using ϵ_0, the magnitude of the force is

$$F = \frac{|q_1|\,|q_2|}{4\pi\epsilon_0 r^2}$$

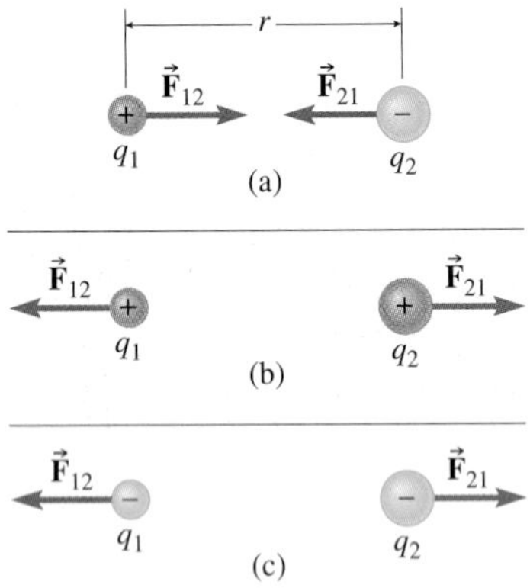

Figure 16.10 The electric force on (a) two opposite charges; (b) and (c) two like charges. Vectors are drawn showing the force on each of the two interacting charges. ($\vec{\mathbf{F}}_{12}$ is the force exerted on charge 1 due to charge 2. $\vec{\mathbf{F}}_{21}$ is the force exerted on charge 2 due to charge 1.)

Direction of Electric Force The direction of the electric force exerted on one point charge due to another point charge is always along the line that joins the two point charges. Remember that, unlike the gravitational force, the electric force can either be attractive or repulsive, depending on the signs of the charges. Coulomb's law is in agreement with Newton's third law: the forces on the two charges are equal in magnitude and opposite in direction (Fig. 16.10).

CHECKPOINT 16.3

(a) List some similarities between gravity and the electric force. (b) What is a major difference between them?

Problem-Solving Tips for Coulomb's Law

1. Use consistent units; since we know k in standard SI units ($N \cdot m^2/C^2$), distances should be in meters and charges in coulombs. When the charge is given in μC or nC, be sure to change the units to coulombs: $1\ \mu C = 10^{-6}$ C and $1\ nC = 10^{-9}$ C.
2. When finding the electric force on a single charge due to two or more other charges, find the force due to each of the other charges separately. The net force on a particular charge is the vector sum of the forces acting on that charge due to each of the other charges. Often it helps to separate the forces into x- and y-components, add the components separately, then find the magnitude and direction of the net force from its x- and y-components.
3. If several charges lie along the same line, do not worry about an intermediate charge "shielding" the charge located on one side from the charge on the other side. The electric force is long-range just as is gravity; the gravitational force on the Earth due to the Sun does not stop when the Moon passes between the two.

CONNECTION:

Electric forces are added the same way any other kind of forces are added—as *vector* quantities. When applying Newton's second law ($\Sigma \vec{\mathbf{F}} = m\vec{\mathbf{a}}$) to an object, *all* forces acting on the object are included in the FBD and all are added (as vectors) to find the net force.

Example 16.3

Electric Force on a Point Charge

Suppose three point charges are arranged as shown in Fig. 16.11. A charge $q_1 = +1.2$ μC is located at the origin of an (x, y) coordinate system; a second charge $q_2 = -0.60$ μC is located at (1.20 m, 0.50 m) and the third charge $q_3 = +0.20$ μC is located at (1.20 m, 0). What is the force on q_2 due to the other two charges?

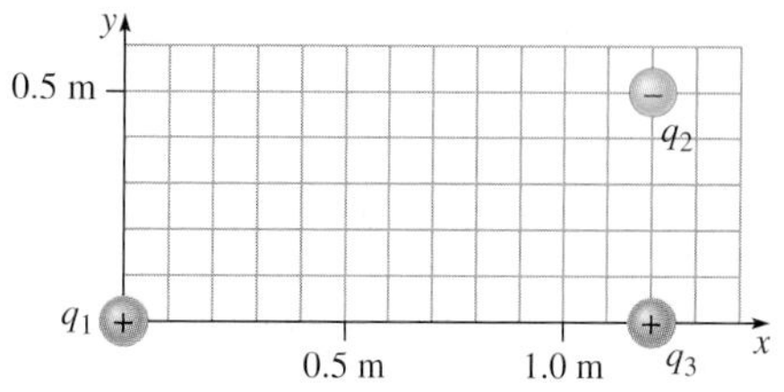

Figure 16.11
Location of point charges in Example 16.3.

Strategy The force on q_2 due to q_1 and the force on q_2 due to q_3 are determined separately. After sketching a free-body diagram, we add the two forces as vectors. Let the distance between charges 1 and 2 be r_{12} and the distance between charges 2 and 3 be r_{23}.

Solution Charges 1 and 3 are both positive, but charge 2 is negative. The forces acting on charge 2 due to charges 1 and 3 are both *attractive*. Figure 16.12a shows an FBD for charge 2 with force vectors pointing toward each of the other charges.

Now find the magnitude of force $\vec{\mathbf{F}}_{21}$ on q_2 due to q_1 from Coulomb's law and then repeat the same process to find the magnitude of force $\vec{\mathbf{F}}_{32}$ on q_2 due to q_3.

The distance between charges 1 and 2 is, from the Pythagorean theorem,

$$r_{12} = \sqrt{r_{13}^2 + r_{23}^2} = 1.30 \text{ m}$$

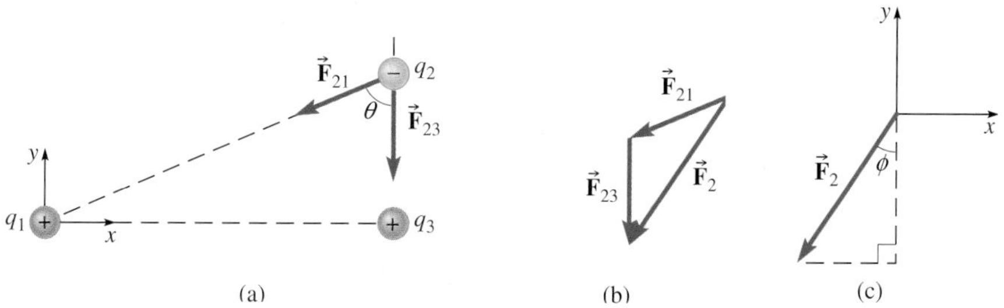

Figure 16.12
(a) Free-body diagram showing the directions of forces $\vec{\mathbf{F}}_{21}$ and $\vec{\mathbf{F}}_{23}$. (b) Vectors $\vec{\mathbf{F}}_{21}$ and $\vec{\mathbf{F}}_{23}$ and their sum $\vec{\mathbf{F}}_2$. (c) Finding the direction of $\vec{\mathbf{F}}_2$ from its x- and y-components.

continued on next page

Example 16.3 continued

From Coulomb's law,

$$F_{21} = \frac{k|q_1|\,|q_2|}{r_{12}^2}$$

$$= 8.99 \times 10^9 \frac{\text{N·m}^2}{\text{C}^2} \times \frac{(1.2 \times 10^{-6}\ \text{C}) \times (0.60 \times 10^{-6}\ \text{C})}{(1.30\ \text{m})^2}$$

$$= 3.83 \times 10^{-3}\ \text{N} = 3.83\ \text{mN}$$

Now for the force due to charge 3.

$$F_{23} = \frac{k\,|q_2|\,|q_3|}{r_{23}^2}$$

$$= 8.99 \times 10^9 \frac{\text{N·m}^2}{\text{C}^2} \times \frac{(0.20 \times 10^{-6}\ \text{C}) \times (0.60 \times 10^{-6}\ \text{C})}{(0.50\ \text{m})^2}$$

$$= 4.32 \times 10^{-3}\ \text{N} = 4.32\ \text{mN}$$

Adding the two force vectors gives the total force $\vec{\mathbf{F}}_2$. Finding x- and y-components:

$$F_{21x} = -F_{21} \sin\theta = -3.83\ \text{mN} \times \frac{1.20\ \text{m}}{1.30\ \text{m}} = -3.53\ \text{mN}$$

$$F_{21y} = -F_{21} \cos\theta = -3.83\ \text{mN} \times \frac{0.50\ \text{m}}{1.30\ \text{m}} = -1.47\ \text{mN}$$

$\vec{\mathbf{F}}_{23}$ is in the $-y$-direction, so $F_{23x} = 0$ and $F_{23y} = -4.32$ mN. Adding components, $F_{2x} = -3.53$ mN and $F_{2y} = (-1.47\ \text{mN}) + (-4.32\ \text{mN}) = -5.79$ mN. The magnitude of $\vec{\mathbf{F}}_2$ is

$$F_2 = \sqrt{F_{2x}^2 + F_{2y}^2} = 6.8\ \text{mN}$$

From Fig. 16.12c, $\vec{\mathbf{F}}_2$ is clockwise from the $-y$-axis by an angle

$$\phi = \tan^{-1} \frac{3.53\ \text{mN}}{5.79\ \text{mN}} = 31°$$

Discussion The net force has a direction compatible with the graphical addition in Fig. 16.12b—it has components in the $-x$- and $-y$-directions.

Practice Problem 16.3 Electric Force on Charge 3

Find the magnitude and direction of the electric force on charge 3 due to charges 1 and 2 in Fig. 16.11.

If we consider the forces acting on the microscopic building blocks of matter (such as atoms, molecules, ions, and electrons), we find that the electric forces are much stronger than the gravitational forces between them. Only when we put a large number of atoms and molecules together to make a massive object can the gravitational force dominate. This domination occurs only because there is an almost perfect balance between positive and negative charges in a large object, leading to nearly zero net charge.

Example 16.4

Two Charged Balls, Hanging in Equilibrium

Two Styrofoam balls of mass 10.0 g are suspended by threads of length 25 cm. The balls are charged, after which they hang apart, each at $\theta = 15.0°$ to the vertical (Fig. 16.13). (a) Are the signs of the charges the same or opposite? (b) Are the magnitudes of the charges necessarily the same? Explain. (c) Find the net charge on each ball, *assuming* that the charges are equal.

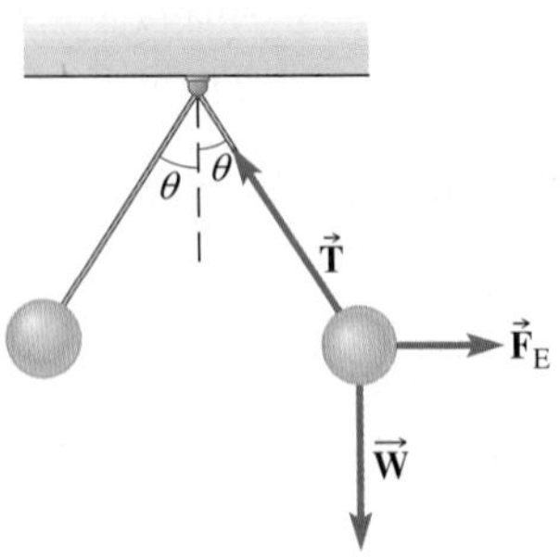

Figure 16.13
Sketch of the situation.

Strategy The situation is similar to the charged electroscope (see Fig. 16.6). Each ball exerts an electric force on the other since both are charged. The *gravitational* forces that the balls exert on one another are negligibly small, but the gravitational forces that Earth exerts on the balls are not negligible. The third force acting on each of the balls is due to the tension in a thread. We analyze the forces acting on the ball using an FBD. The sum of the three forces must be zero since the ball is in equilibrium.

Solution Each ball experiences three forces: the electric force, the gravitational force, and the pull of the thread, which is under tension. Figure 16.14 shows an FBD for one of the balls.

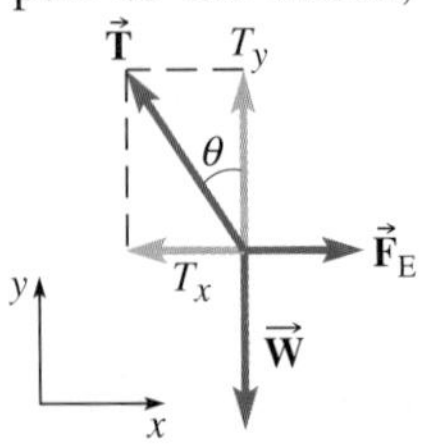

Figure 16.14
An FBD for the ball on the right.

(a) The electric force is clearly repulsive—the balls are pushed apart—so the charges must have the same sign. There is no way to tell whether they are both positive or both negative.

(b) At first glance it *might* appear that the charges must be the same; the

continued on next page

Example 16.4 continued

balls are hanging at the same angle, so there is no clue as to which charge is larger. But look again at Coulomb's law: the force on either of the balls is proportional to the product of the two charge magnitudes; $F \propto |q_1||q_2|$. In accordance with Newton's third law, Coulomb's law says that the two forces that make up the interaction are equal in magnitude and opposite in direction. The charges are not necessarily equal.

(c) Let us choose the x- and y-axes in the horizontal and vertical directions, respectively. Of the three forces acting on a ball, only one, that due to the tension in the thread, has both x- and y-components. From Fig. 16.14, the tension in the thread has a y-component equal in magnitude to the weight of the ball, and an x-component equal in magnitude to the electric force on the ball. The ball is in equilibrium, so the x- and y- components of the net force acting on it are both zero:

$$\sum F_x = F_E - T \sin \theta = 0$$

$$\sum F_y = T \cos \theta - mg = 0$$

Eliminating the unknown tension yields

$$F_E = T \sin \theta = \left(\frac{mg}{\cos \theta}\right) \sin \theta = mg \tan \theta \quad (1)$$

From Coulomb's law [Eq. (16-2)],

$$F_E = \frac{k|q|^2}{r^2}$$

where $|q|$ is the magnitude of the charge on each of the two balls (now assumed to be equal). The separation of the balls (Fig. 16.15) is

$$r = 2(d \sin \theta) \quad (2)$$

where $d = 25$ cm is the length of the thread.

Some algebra now enables us to solve for $|q|$. From Coulomb's law,

$$|q|^2 = \frac{F_E r^2}{k} \quad (3)$$

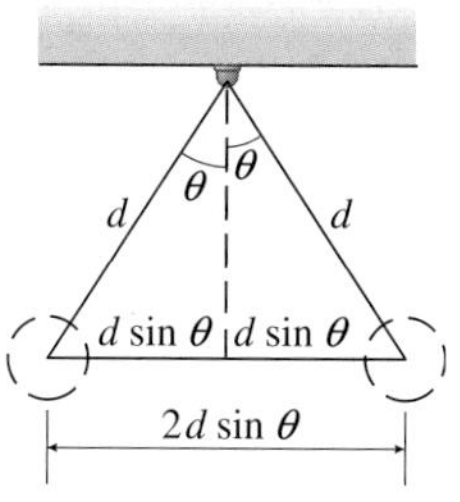

Figure 16.15
Finding the separation between the two balls.

We can substitute expressions (1) and (2) into Eq. (3) for F_E and r:

$$|q|^2 = \frac{(mg \tan \theta)(2d \sin \theta)^2}{k} = \frac{4d^2 mg \tan \theta \sin^2 \theta}{k}$$

$$|q| = \sqrt{\frac{4 \times (0.25\text{ m})^2 \times 0.0100\text{ kg} \times 9.8\text{ N/kg} \times \tan 15.0° \times \sin^2 15.0°}{8.99 \times 10^9\text{ N·m}^2/\text{C}^2}}$$

$$= 0.22\ \mu\text{C}$$

The charges can either be both positive or both negative, so the charges are either both +0.22 μC or both −0.22 μC.

Discussion We can check the units in the final expression for q:

$$\sqrt{\frac{\text{m}^2 \times \text{kg} \times \text{N/kg}}{\text{N·m}^2/\text{C}^2}} = \sqrt{\frac{\text{N·m}^2}{\text{N·m}^2/\text{C}^2}} = \sqrt{\text{C}^2} = \text{C} \quad \text{(OK!)}$$

Another check: if the balls were uncharged, they would hang straight down ($\theta = 0$). Substituting $\theta = 0$ into the final algebraic expression does give $q = 0$.

How large a charge would make the threads horizontal (assuming they don't break first)? As the charge on the balls is increased, the angle of the threads *approach* 90° but can never reach 90° because the tension in the thread must always have an upward component to balance gravity. In the algebraic answer, as $\theta \to 90°$, $\tan \theta \to \infty$ and $\sin \theta \to 1$, which would yield a charge q approaching ∞. The threads cannot be horizontal for any *finite* amount of charge.

Practice Problem 16.4 Three Point Charges

Three identical point charges $q = -2.0$ nC are at the vertices of an equilateral triangle with sides of length $L = 1.0$ cm (Fig. 16.16). What is the magnitude of the electric force acting on any one of them?

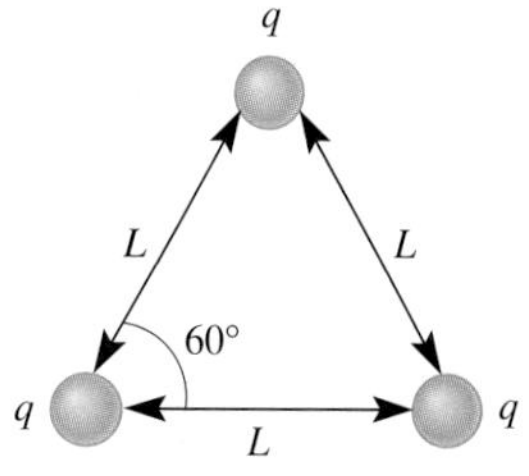

Figure 16.16
Practice Problem 16.4

16.4 THE ELECTRIC FIELD

Recall that the gravitational field at a point is defined to be the gravitational force per unit mass on an object placed at that point. If the gravitational force on an apple of mass m due to the Earth is $\vec{\mathbf{F}}_g$, then Earth's gravitational field $\vec{\mathbf{g}}$ at the location of the apple is given by

$$\vec{\mathbf{g}} = \frac{\vec{\mathbf{F}}_g}{m}$$

CONNECTION:

The definition of electric field is similar to the definition of gravitational field. Gravitational field is gravitational force per unit mass; electric field is electric force per unit charge.

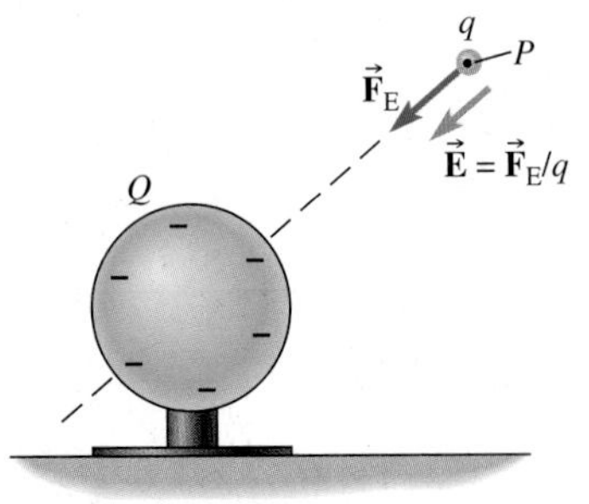

Figure 16.17 The electric field $\vec{\mathbf{E}}$ that exists at a point P due to a charged object with charge Q is equal to the electric force $\vec{\mathbf{F}}_E$ experienced by a small test charge q placed at that point divided by q.

The directions of $\vec{\mathbf{F}}_g$ and $\vec{\mathbf{g}}$ are the same since m is positive. The gravitational field we encounter most often is that due to the Earth, but the gravitational field could be due to any astronomical body, or to more than one body. For instance, an astronaut may be concerned with the gravitational field at the location of her spacecraft due to the Sun, the Earth, and the Moon combined. Since gravitational forces add as vectors—as do all forces—the gravitational field at the location of the spacecraft is the vector sum of the separate gravitational fields due to the Sun, the Earth, and the Moon.

Similarly, if a point charge q is in the vicinity of other charges, it experiences an electric force $\vec{\mathbf{F}}_E$. The **electric field** (symbol $\vec{\mathbf{E}}$) *at any point* is defined to be the electric force *per unit charge* at that point (Fig. 16.17):

$$\vec{\mathbf{E}} = \frac{\vec{\mathbf{F}}_E}{q} \quad (16\text{-}4a)$$

The SI units of the electric field are N/C.

In contrast to the gravitational force, which is always in the same direction as the gravitational field, the electric force can either be parallel or antiparallel to the electric field depending on the sign of the charge q that is sampling the field. *If q is positive*, the direction of the electric force $\vec{\mathbf{F}}_E$ is the same as the direction of the electric field $\vec{\mathbf{E}}$; if q is negative, the two vectors have opposite directions. To probe the electric field in some region, imagine placing a point charge q at various points. At each point you calculate the electric force on this *test charge* and divide the force by q to find the electric field at that point. It is usually easiest to imagine a *positive* test charge so that the field direction is the same as the force direction, but the field comes out the same regardless of the sign or magnitude of q, unless its magnitude is large enough to disturb the other charges and thereby change the electric field.

Why is $\vec{\mathbf{E}}$ defined as the force per unit *charge* instead of per unit mass as done for gravitational field? The gravitational force on an object is proportional to its mass, so it makes sense to talk about the force per unit mass (the SI units of $\vec{\mathbf{g}}$ are N/kg). In contrast to the gravitational force, the electrical force on a point charge is instead proportional to its *charge.*

Why is the electric field a useful concept? One reason is that once we know the electric field at some point, then it is easy to calculate the electric force $\vec{\mathbf{F}}_E$ on *any* point charge q placed there:

The electric force $\vec{\mathbf{F}}_E$ on a charge q at a point where the electric field is $\vec{\mathbf{E}}$ is $q\vec{\mathbf{E}}$.

$$\vec{\mathbf{F}}_E = q\vec{\mathbf{E}} \quad (16\text{-}4b)$$

Note that $\vec{\mathbf{E}}$ is the electric field at the location of point charge q due to all the *other* charges in the vicinity. Certainly the point charge produces a field of its own at nearby points; this field causes forces on *other* charges. In other words, a point charge exerts no force on itself.

Example 16.5

Charged Sphere Hanging in a Uniform $\vec{\mathbf{E}}$ Field

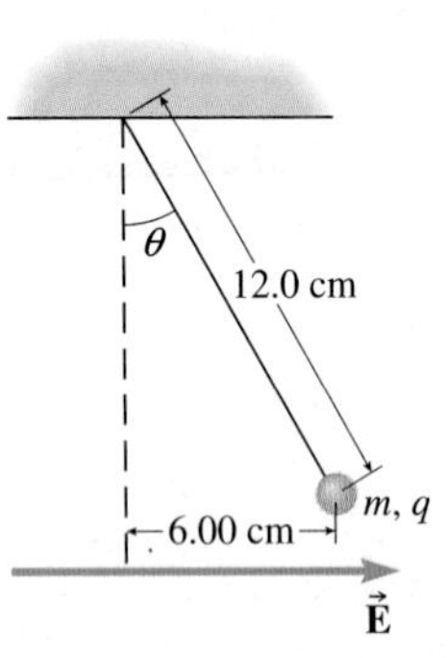

Figure 16.18 A charged sphere hanging in a uniform electric field $\vec{\mathbf{E}}$ (to the right) and a uniform gravitational field $\vec{\mathbf{g}}$ (downward).

A small sphere of mass 5.10 g is hanging vertically from an insulating thread that is 12.0 cm long. By charging some nearby flat metal plates, the sphere is subjected to a horizontal electric field of magnitude 7.20×10^5 N/C. As a result, the sphere is displaced 6.00 cm horizontally in the direction of the electric field (Fig. 16.18). (a) What is the angle θ that the thread makes with the vertical? (b) What is the tension in the thread? (c) What is the charge on the sphere?

Strategy We assume that the sphere is small enough to be treated as a point charge. Then the electric force on the sphere is given by $\vec{\mathbf{F}}_E = q\vec{\mathbf{E}}$. Figure 16.18 shows that the

continued on next page

Example 16.5 continued

sphere is pushed to the right by the field; therefore, $\vec{\mathbf{F}}_E$ is to the right. Since $\vec{\mathbf{F}}_E$ and $\vec{\mathbf{E}}$ have the same direction, the charge on the sphere is positive. After drawing an FBD showing all the forces acting on the sphere, we set the net force on the sphere equal to zero since it hangs in equilibrium.

Solution (a) The angle θ can be found from the geometry of Fig. 16.18. The thread's length (12.0 cm) is the hypotenuse of a right triangle. The side of the triangle opposite angle θ is the horizontal displacement (6.00 cm). Thus,

$$\sin\theta = \frac{6.00\text{ cm}}{12.0\text{ cm}} = 0.500 \quad \text{and} \quad \theta = 30.0°$$

(b) We start by drawing an FBD (Fig. 16.19a). The gravitational force must balance the vertical component of the thread's pull on the sphere ($\vec{\mathbf{F}}_T$). The electric force must balance the horizontal component of the same force. In Fig. 16.19b, we show the components of $\vec{\mathbf{F}}_T$. The magnitude of $\vec{\mathbf{F}}_T$ is the tension in the thread T.

The sphere is in equilibrium, so the x- and y- components of the net force acting on it are both zero. From the y-components, we can find the tension:

$$\sum F_y = T\cos\theta - mg = 0$$

$$T = \frac{mg}{\cos\theta} = \frac{5.10\times10^{-3}\text{ kg}\times9.80\text{ N/kg}}{\cos 30.0°} = 0.0577\text{ N}$$

This is the magnitude of $\vec{\mathbf{F}}_T$. The direction is along the thread toward the support point, at an angle of 30.0° from the vertical.

(c) The horizontal force components also add to zero. Because $F_E = |q|E$,

$$\sum F_x = |q|\,E - T\sin\theta = 0$$

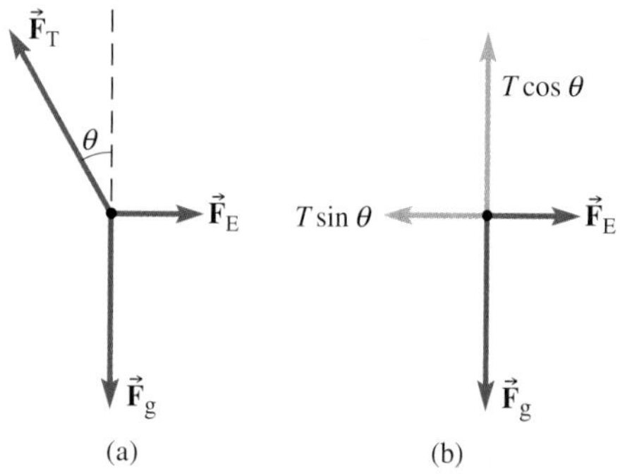

Figure 16.19 (a) An FBD showing forces acting on the sphere. (b) FBD in which the force due to the cord is replaced by its vertical and horizontal components.

We can now solve for $|q|$.

$$|q| = \frac{T\sin\theta}{E} = \frac{(5.77\times10^{-2}\text{ N})\sin 30.0°}{7.20\times10^{5}\text{ N/C}} = 40.1\text{ nC}$$

We have determined the magnitude of the charge. The sign of the charge is positive because the electric force on the sphere is in the direction of the electric field. Therefore,

$$q = 40.1\text{ nC}$$

Discussion This problem has many steps, but, taken one by one, each step helps to solve for one of the unknowns and leads the way to find the next unknown. At first glance, it may appear that not enough information is given, but after a figure is drawn to aid in the visualization of the forces and their components, the steps to follow are more easily determined.

Practice Problem 16.5 Effect of Doubling the Charge on the Hanging Mass

If the charge on the sphere were doubled in Example 16.5, what angle would the thread make with the vertical?

Electric Field due to a Point Charge

The electric field due to a single point charge Q can be found using Coulomb's law. Imagine a positive test charge q placed at various locations. Coulomb's law says that the force acting on the test charge is

$$F = \frac{k|q||Q|}{r^2} \tag{16-2}$$

The electric field strength is then

$$E = \frac{F}{|q|} = \frac{k|Q|}{r^2} \tag{16-5}$$

The field falls off as $1/r^2$, following the same inverse square law as the gravitational force (Fig. 16.20).

What is the direction of the field? If Q is positive, then a positive test charge would be repelled, so the field vector points away from Q (or *radially outward*). If Q is negative, then the field vector points toward Q (*radially inward*).

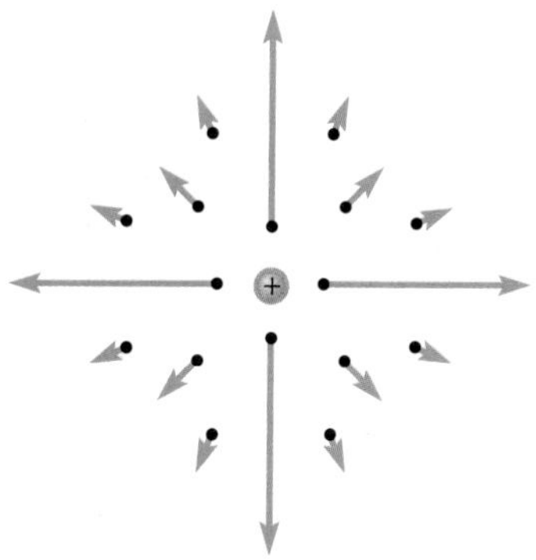

Figure 16.20 Vector arrows representing the electric field at a few points near a positive point charge. The length of the arrow is proportional to the magnitude of the field.

CONNECTION:

The principle of superposition for electric fields is a direct consequence of adding electric forces as vector quantities.

Principle of Superposition

The electric field due to more than one point charge can be found using the **principle of superposition**:

> The electric field at any point is the vector sum of the field vectors at that point caused by each charge separately.
> (Text website interactive: electric fields)

Example 16.6

Electric Field due to Two Point Charges

Two point charges are located on the x-axis (Fig. 16.21). Charge $q_1 = +0.60\ \mu\text{C}$ is located at $x = 0$; charge $q_2 = -0.50\ \mu\text{C}$ is located at $x = 0.40$ m. Point P is located at $x = 1.20$ m. What is the magnitude and direction of the electric field at point P due to the two charges?

Strategy We can determine the field at P due to q_1 and the field at P due to q_2 separately using Coulomb's law and the definition of the electric field. In each case, the electric field points in the direction of the electric force on a *positive* test charge at point P. The sum of these two fields is the electric field at P. We sketch a vector diagram to help add the fields correctly. Since there are two different distances in the problem, subscripts help to distinguish them. Let the distance between charge 1 and point P be $r_1 = 1.20$ m and the distance between charge 2 and point P be $r_2 = 0.80$ m.

Solution Charge 1 is positive. We imagine a tiny positive test charge, q_0, located at point P. Since charge 1 repels the positive test charge, the force $\vec{\mathbf{F}}_1$ on the test charge due to q_1 is in the positive x-direction (Fig. 16.22). The direction of the electric field due to charge 1 is also in the $+x$-direction since $\vec{\mathbf{E}}_1 = \vec{\mathbf{F}}_1/q_0$ and $q_0 > 0$. Charge q_2 is negative so it attracts the imaginary test charge along the line joining the two charges; the force $\vec{\mathbf{F}}_2$ on the test charge due to q_2 is in the negative x-direction. Therefore $\vec{\mathbf{E}}_2 = \vec{\mathbf{F}}_2/q_0$ is in the $-x$-direction.

We first find the magnitude of the field $\vec{\mathbf{E}}_1$ at P due to q_1 and then repeat the same process to find the magnitude of field $\vec{\mathbf{E}}_2$ at P due to q_2. From the given information,

$$E_1 = \frac{k|q_1|}{r_1^2}$$
$$= 8.99 \times 10^9\ \frac{\text{N}\cdot\text{m}^2}{\text{C}^2} \times \frac{0.60 \times 10^{-6}\ \text{C}}{(1.20\ \text{m})^2}$$
$$= 3.75 \times 10^3\ \text{N/C}$$

Now for the magnitude of field $\vec{\mathbf{E}}_2$ at P due to charge 2.

$$E_2 = \frac{k|q_2|}{r_2^2}$$
$$= 8.99 \times 10^9\ \frac{\text{N}\cdot\text{m}^2}{\text{C}^2} \times \frac{0.50 \times 10^{-6}\ \text{C}}{(0.80\ \text{m})^2}$$
$$= 7.02 \times 10^3\ \text{N/C}$$

Figure 16.23 shows the vector addition $\vec{\mathbf{E}}_1 + \vec{\mathbf{E}}_2 = \vec{\mathbf{E}}$, which points in the $-x$-direction since $E_2 > E_1$. The magnitude of E at point P is

$$E = 7.02 \times 10^3\ \text{N/C} - 3.75 \times 10^3\ \text{N/C} = 3.3 \times 10^3\ \text{N/C}$$

The electric field at P is 3.3×10^3 N/C in the $-x$-direction.

Discussion This same method is used to find the electric field at a point due to *any* number of point charges. The direction of the electric field due to each charge alone is the direction of the electric force on an imaginary positive test charge at that point. The magnitude of each electric field is found from Eq. (16-5). Then the electric field vectors are

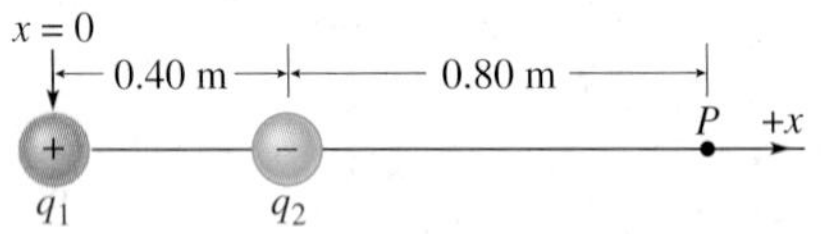

Figure 16.21
Two point charges on the x-axis, one at $x = 0$ and one at $x = +0.40$ m.

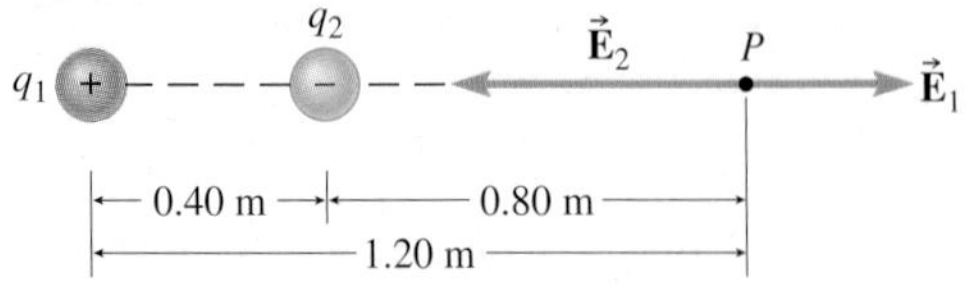

Figure 16.22
Directions of electric field vectors at point P due to charges q_1 and q_2.

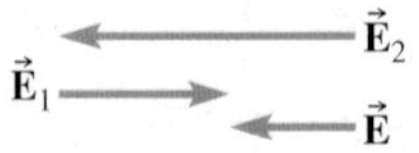

Figure 16.23
Vector addition of $\vec{\mathbf{E}}_1$ and $\vec{\mathbf{E}}_2$.

continued on next page

Example 16.6 continued

added. If the charges and the point do not all lie on the same line, then the fields can be added by resolving them into x- and y-components and summing the components.

Even when electric fields are not due to a small number of point charges, the principle of superposition still applies: the electric field at any point is the vector sum of the fields at that point caused by each charge or set of charges separately.

Practice Problem 16.6 Electric Field at Point P due to Two Charges

Find the magnitude and direction of the electric field at point P due to charges 1 and 2 located on the x-axis. The charges are $q_1 = +0.040$ μC and $q_2 = +0.010$ μC. Charge q_1 is at the origin, charge q_2 is at $x = 0.30$ m, and point P is at $x = 1.50$ m.

Example 16.7

Electric Field due to Three Point Charges

Three point charges are placed at the corners of a rectangle, as shown in Fig. 16.24. (a) What is the electric field due to these three charges at the fourth corner, point P? (b) What is the acceleration of an electron located at point P? Assume that no forces other than that due to the electric field act on it.

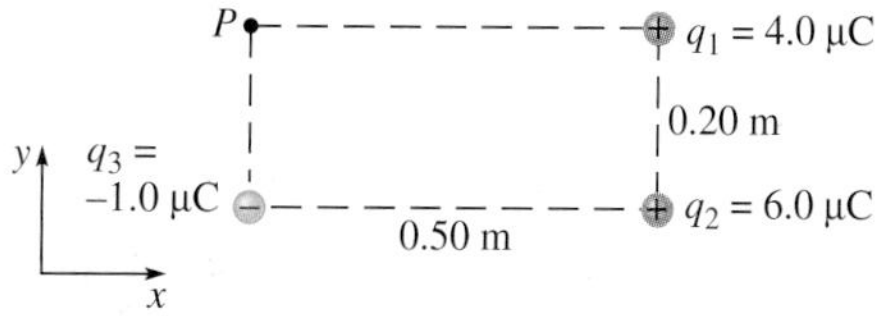

Figure 16.24 Three point charges at the corners of a rectangle.

Strategy (a) After determining the magnitude and direction of the electric field at point P due to each point charge individually, we use the principle of superposition to add them as vectors.

(b) Since we have already calculated $\vec{\mathbf{E}}$ at point P, the force on the electron is $\vec{\mathbf{F}} = q\vec{\mathbf{E}}$, where $q = -e$ is the charge of the electron.

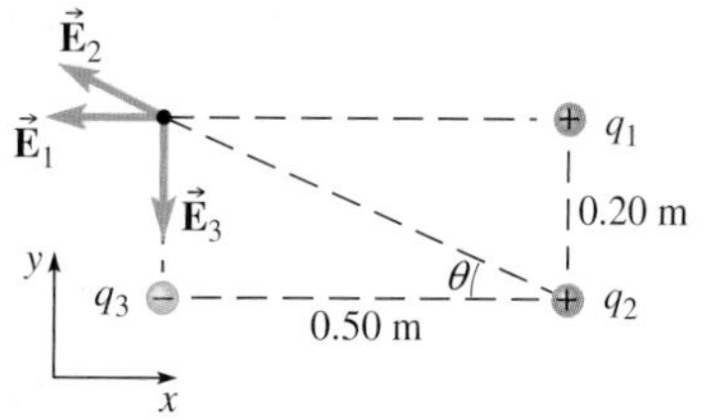

Figure 16.25 Directions of the electric field vectors at point P due to each of the point charges individually. (Lengths of vector arrows are *not* to scale.)

Solution (a) The electric field due to a single point charge is directed *away* from the point charge if it is positive and *toward* it if it is negative. The directions of the three electric fields are shown in Fig. 16.25. Equation (16-5) gives the magnitudes:

$$E_1 = \frac{k|q_1|}{r_1^2} = \frac{8.99\times10^9\ \text{N}\cdot\text{m}^2\cdot\text{C}^{-2}\times4.0\times10^{-6}\ \text{C}}{(0.50\ \text{m})^2} = 1.44\times10^5\ \text{N/C}$$

A similar calculation with $|q_3| = 1.0\times10^{-6}$ C and $r_3 = 0.20$ m yields $E_3 = 2.25\times10^5$ N/C. Using the Pythagorean theorem, to find $r_2 = \sqrt{(0.50\ \text{m})^2 + (0.20\ \text{m})^2}$, we have

$$E_2 = \frac{kq_2}{r_2^2} = \frac{8.99\times10^9\ \text{N}\cdot\text{m}^2\cdot\text{C}^{-2}\times6.0\times10^{-6}\ \text{C}}{(0.50\ \text{m})^2+(0.20\ \text{m})^2} = 1.86\times10^5\ \text{N/C}$$

Now we find the x- and y-components of $\vec{\mathbf{E}}$ due to all three. Using the angle θ in Fig. 16.25, we have $\cos\theta = r_1/r_2 = 0.928$ and $\sin\theta = 0.371$. Then

$$\sum E_x = E_{1x} + E_{2x} + E_{3x} = (-E_1) + (-E_2\cos\theta) + 0 = -3.17\times10^5\ \text{N/C}$$

$$\sum E_y = E_{1y} + E_{2y} + E_{3y} = 0 + E_2\sin\theta - E_3 = -1.56\times10^5\ \text{N/C}$$

The magnitude of the electric field is then $E = \sqrt{E_x^2 + E_y^2} = 3.5\times10^5$ N/C and the direction is at angle $\phi = \tan^{-1}|E_y/E_x| = 26°$ below the $-x$-axis (Fig. 16.26).

(b) The force on the electron is $\vec{\mathbf{F}} = q_e\vec{\mathbf{E}}$. Its acceleration is then $\vec{\mathbf{a}} = q_e\vec{\mathbf{E}}/m_e$. The electron charge $q_e = -e$ and mass m_e are given in Table 16.1. The acceleration has magnitude $a = eE/m_e = 6.2\times10^{16}$ m/s^2. The direction of the acceleration is the direction of the electric *force*, which is *opposite* the direction of $\vec{\mathbf{E}}$ since the electron's charge is negative.

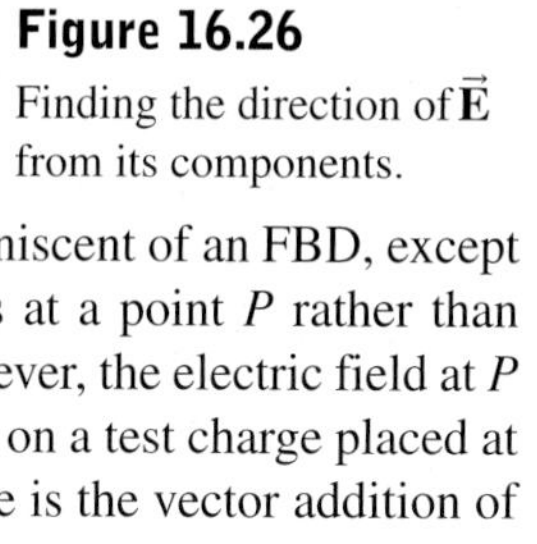

Figure 16.26 Finding the direction of $\vec{\mathbf{E}}$ from its components.

Discussion Figure 16.25 is reminiscent of an FBD, except that it shows electric field vectors at a point P rather than forces acting on some object. However, the electric field at P is the electric force per unit charge on a test charge placed at point P, so the underlying principle is the vector addition of forces.

Practice Problem 16.7 Electric Field due to Two Point Charges

If the point charge $q_1 = 4.0$ μC is removed, what is the electric field at point P due to the remaining two point charges?

Electric Field Lines

It is often difficult to make a visual representation of an electric field using vector arrows; the vectors drawn at different points may overlap and become impossible to distinguish. Another visual representation of the electric field is a sketch of the **electric field lines**, a set of continuous lines that represent both the magnitude and the direction of the electric field vector as follows:

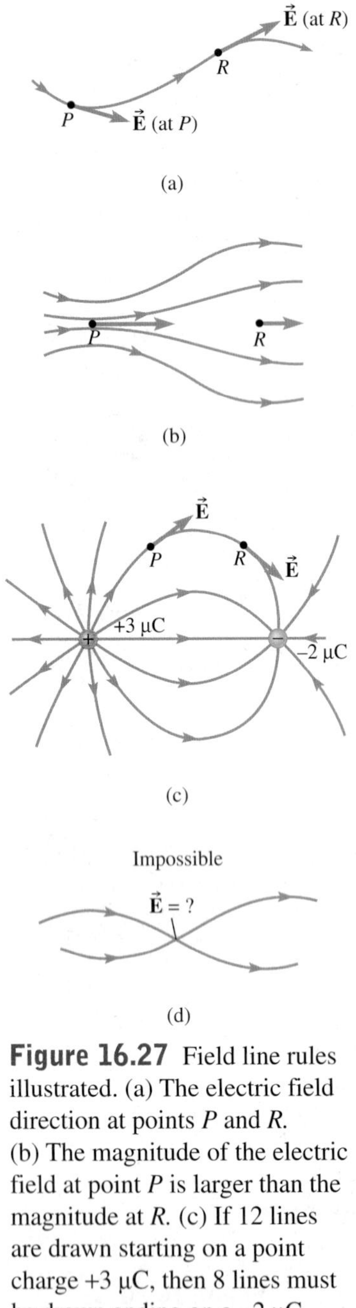

Figure 16.27 Field line rules illustrated. (a) The electric field direction at points P and R. (b) The magnitude of the electric field at point P is larger than the magnitude at R. (c) If 12 lines are drawn starting on a point charge +3 μC, then 8 lines must be drawn ending on a −2 μC point charge. (d) If field lines were to cross, the direction of $\vec{\mathbf{E}}$ at the intersection would be undetermined.

Interpretation of Electric Field Lines

- The direction of the electric field vector at any point is *tangent to the field line* passing through that point and in the direction indicated by arrows on the field line (Fig. 16.27a).
- The electric field is strong where field lines are close together and weak where they are far apart (Fig. 16.27b). (More specifically, if you imagine a small surface perpendicular to the field lines, the magnitude of the field is proportional to the number of lines that cross the surface divided by the area.)

To help sketch the field lines, these three additional rules are useful:

Rules for Sketching Field Lines

- For the electric fields we study in Chapters 16 through 19, the field lines always start on positive charges and always end on negative charges.
- The number of lines starting on a positive charge (or ending on a negative charge) is proportional to the magnitude of the charge (Fig. 16.27c). (The total number of lines you draw is arbitrary; the more lines you draw, the better the representation of the field.)
- Field lines never cross. The electric field at any point has a unique direction; if field lines crossed, the field would have two directions at the same point (Fig. 16.27d).

Field Lines for a Point Charge

Figure 16.28 shows sketches of the field lines due to single point charges. The field lines show that the direction of the field is radial (away from a positive charge or toward a negative charge). The lines are close together near the point charge, where the field is strong, and are more spread out farther from the point charge, showing that the field strength diminishes with distance. No other nearby charges are shown in these sketches, so the lines go out to infinity as if the point charge were the only thing in the universe. If the field of view is enlarged, so that other charges are shown, the lines starting on the positive point charge would end on some faraway negative charges, and those that end on the negative charge would start on some faraway positive charges.

Electric Field due to a Dipole

Dipole: two equal and opposite point charges

A pair of point charges with equal and opposite charges that are near one another is called a **dipole** (literally *two poles*). To find the electric field due to the dipole at various points by using Coulomb's law would be extremely tedious, but sketching some field lines immediately gives an approximate idea of the electric field (Fig. 16.29).

Because the charges in the dipole have equal charge magnitudes, the same number of lines that start on the positive charge end on the negative charge. Close to either of

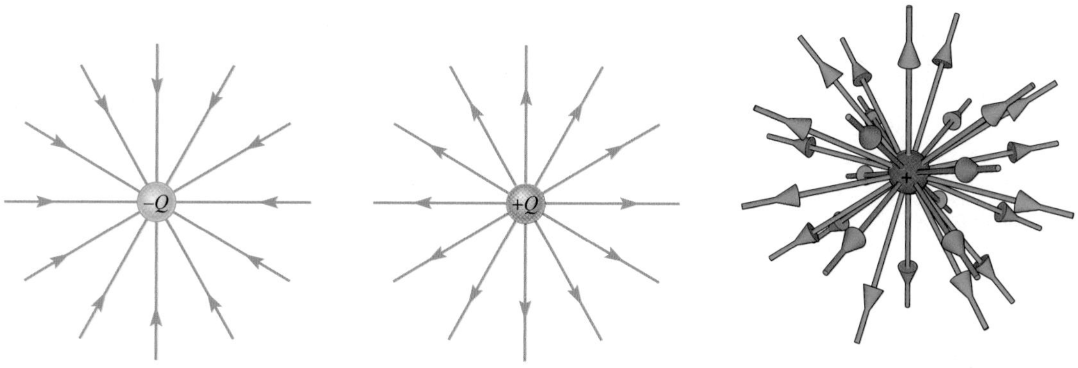

(a) (b) (c)

Figure 16.28 Electric field lines due to isolated point charges. (a) Field of negative point charge; (b) field of positive point charge. These sketches show only field lines that lie in a two-dimensional plane. (c) A three-dimensional illustration of electric field lines due to a positive charge. The electric field is strong where the field lines are close together and weak where they are far apart. Compare the lengths of the electric field vector arrows in Fig. 16.20.

the charges, the field lines are evenly spaced in all directions, just as if the other charge were not present. As we approach one of the charges, the field due to that charge gets so large ($F \propto 1/r^2$, $r \rightarrow 0$) that the field due to the other charge is negligible in comparison and we are left with the spherically symmetric field due to a single point charge.

The field at other points has contributions from both charges. Figure 16.29 shows, for one point P, how the field vectors ($\vec{\mathbf{E}}_-$ and $\vec{\mathbf{E}}_+$) due to the two separate charges add, following vector addition rules, to give the total field $\vec{\mathbf{E}}$ at point P. Note that the total field $\vec{\mathbf{E}}$ is tangent to the field line through point P.

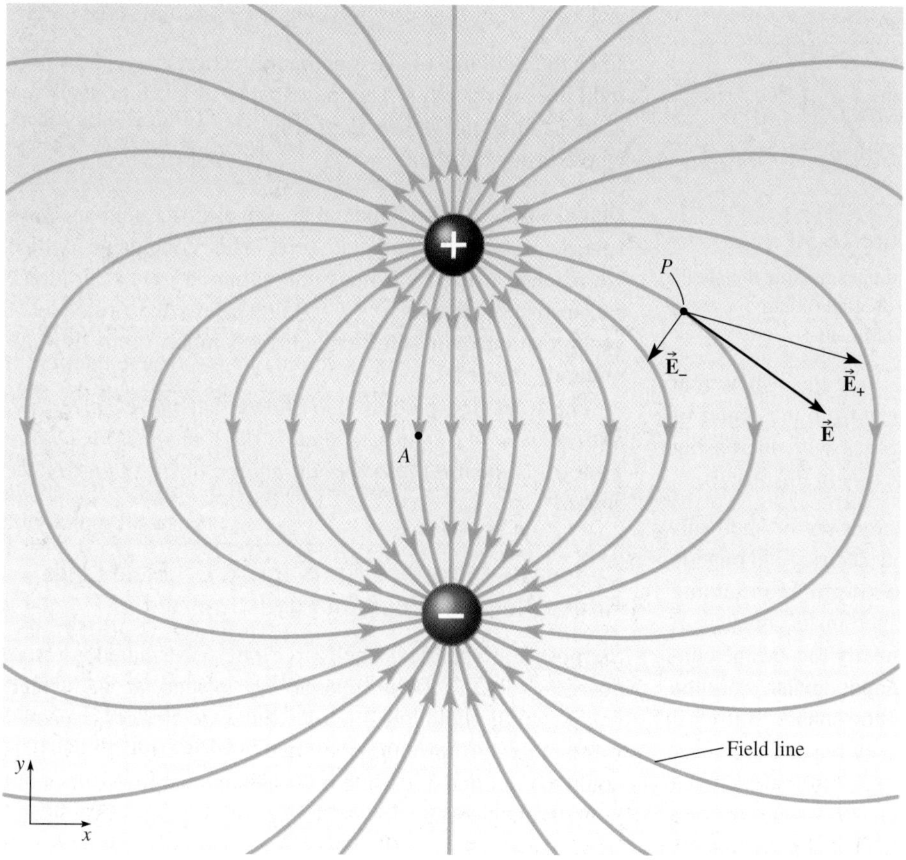

Figure 16.29 Electric field lines for a dipole. The electric field vector $\vec{\mathbf{E}}$ at a point P is tangent to the field line through that point and is the sum of the fields ($\vec{\mathbf{E}}_-$ and $\vec{\mathbf{E}}_+$) due to each of the two point charges. (Text website tutorial: $\vec{\mathbf{E}}$-field of dipole)

The principles of superposition and symmetry are two powerful tools for determining electric fields. The use of symmetry is illustrated in Conceptual Example 16.8.

CHECKPOINT 16.4

(a) What is the direction of the electric field at point *A* in Fig. 16.29? (b) At which point, *A* or *P*, is the magnitude of the field weaker?

Conceptual Example 16.8

Field Lines for a Thin Spherical Shell

A thin metallic spherical shell of radius *R* carries a total charge *Q*, which is positive. The charge is spread out evenly over the shell's outside surface. Sketch the electric field lines in two different views of the situation: (a) The spherical shell is tiny and you are looking at it from distant points; (b) you are looking at the field inside the shell's cavity. In (a), also sketch $\vec{\mathbf{E}}$ field vectors at two different points outside the shell.

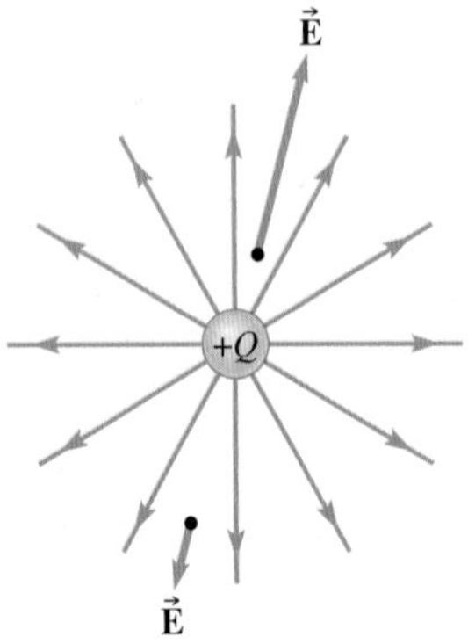

Figure 16.30
Field lines outside the shell are directed radially outward.

Strategy Since the charge on the shell is positive, field lines begin on the shell. A sphere is a highly symmetrical shape: standing at the center, it looks the same in any chosen direction. This symmetry helps in sketching the field lines.

Solution (a) A tiny spherical shell located far away cannot be distinguished from a point charge. The sphere looks like a point when seen from a great distance and the field lines look just like those emanating from a positive point charge (Fig. 16.30). The field lines show that the electric field is directed radially away from the center of the shell and that its magnitude decreases with increasing distance, as illustrated by the two $\vec{\mathbf{E}}$ vectors in Fig. 16.30.

(b) Field lines begin on the positive charges on the shell surface. Some go outward, representing the electric field outside the shell, while others may *perhaps* go inward, representing the field inside the shell. Any field lines inside must start evenly spaced on the shell and point directly toward the center of the shell (Fig. 16.31); the lines cannot deviate from the radial direction due to the symmetry of the sphere. But what would happen to the field lines when they reach the center? The lines can only end at the center if a negative point charge is found there—but there is no point charge. If the lines do not end, they would cross at the center. That cannot be right since the field must have a *unique* direction at every point—field lines never cross. The inescapable conclusion: *there are no field lines inside the shell* (Fig. 16.32), so $\vec{\mathbf{E}} = 0$ everywhere inside the shell.

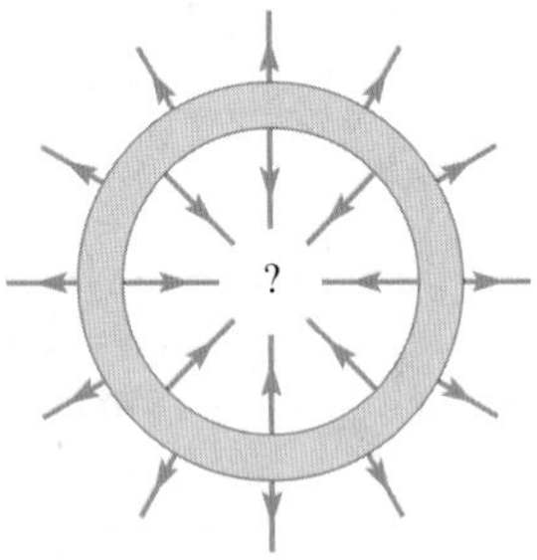

Figure 16.31
If there are field lines inside the shell, they must start on the shell and point radially inward. Then what?

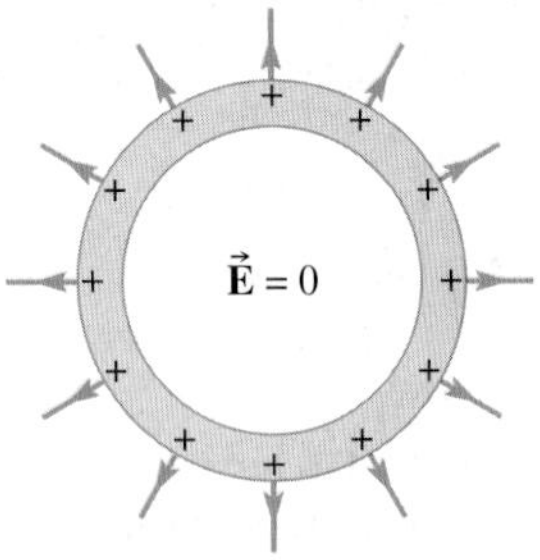

Figure 16.32
There can be no field lines—and therefore no electric field—inside the shell.

Discussion We conclude that the electric field *inside* a spherical shell of charge is zero. This conclusion, which we reached using field lines and symmetry considerations, can also be proved using Coulomb's law, the principle of superposition, and some calculus—a much more difficult method!

The field line picture also shows that *the electric field pattern outside a spherical shell is the same as if the charge were all condensed into a point charge at the center of the sphere.*

Conceptual Practice Problem 16.8 Field Lines After a Negative Point Charge Is Inserted

Suppose the spherical shell of evenly distributed positive charge *Q* has a point charge $-Q$ placed at its center. (a) Sketch the field lines. [*Hint:* Since the charges are equal in magnitude, the number of lines starting on the shell is equal to the number ending on the point charge.] (b) Defend your sketch using the principle of superposition (total field = field due to shell + field due to point charge).

Application of Electric Fields: Electrolocation

How does *Gymnarchus* navigate in muddy water?

Long before scientists learned how to detect and measure electric fields, certain animals and fish evolved organs to produce and detect electric fields. *Gymnarchus niloticus* (see the Chapter Opener) has electrical organs running along the length of its body; these organs set up an electric field around the fish (Fig. 16.33). When a nearby object distorts the field lines, *Gymnarchus* detects the change through sensory receptors, mostly near the head, and responds accordingly. This extra sense enables the fish to detect prey or predators in muddy streams where eyes are less useful.

Since *Gymnarchus* relies primarily on electrolocation, where slight changes in the electric field are interpreted as the presence of nearby objects, it is important that it be able to create the same electric field over and over. For this reason, *Gymnarchus* swims by undulating its long dorsal fin while holding its body rigid. Keeping the backbone straight keeps the negative and positive charge centers aligned and at a fixed distance apart. A swishing tail would cause variation in the electric field and that would make electrolocation much less accurate.

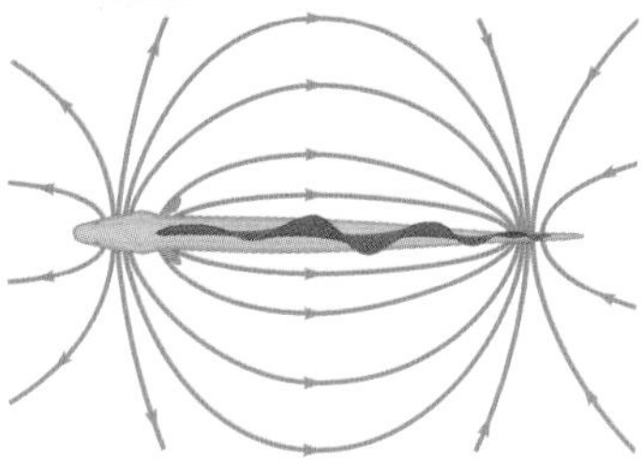

Figure 16.33 The electric field generated by *Gymnarchus.* The field is approximately that of a dipole. The head of the fish is positively charged and the tail is negatively charged.

16.5 MOTION OF A POINT CHARGE IN A UNIFORM ELECTRIC FIELD

The simplest example of how a charged object responds to an electric field is when the electric field (due to other charges) is **uniform**—that is, has the same magnitude and direction at every point. The field due to a single point charge is *not* uniform; it is radially directed and its magnitude follows the inverse square law. To create a uniform field requires a large number of charges. The most common way to create a (nearly) uniform electric field is to put equal and opposite charges on two parallel metal plates (Fig. 16.34). If the charges are $\pm Q$ and the plates have area A, the magnitude of the field between the plates is

Electric field between oppositely charged metal plates

$$E = \frac{Q}{\epsilon_0 A} \qquad (16\text{-}6)$$

(This expression can be derived using Gauss's law—see Section 16.7.) The direction of the field is perpendicular to the plates, from the positively charged plate toward the negatively charged plate.

Assuming the uniform field $\vec{\mathbf{E}}$ is known, a point charge q experiences an electric force

$$\vec{\mathbf{F}} = q\vec{\mathbf{E}} \qquad (16\text{-}4b)$$

If this is the only force acting on the point charge, then the net force is constant and therefore so is the acceleration:

$$\vec{\mathbf{a}} = \frac{\vec{\mathbf{F}}}{m} = \frac{q\vec{\mathbf{E}}}{m} \qquad (16\text{-}7)$$

With a constant acceleration, the motion can take one of two forms. If the initial velocity of the point charge is zero or is parallel or antiparallel to the field, then the motion is along a straight line. If the point charge has an initial velocity component perpendicular to the field, then the trajectory is parabolic (just like a projectile in a

CONNECTION:

If no forces act on a point charge other than the force due to a uniform electric field, then the acceleration is constant. All the principles we learned for motion with constant acceleration in a uniform gravitational field apply. However, the acceleration does not have the same magnitude and direction for all point charges in the same field—see Eq. (16-7).

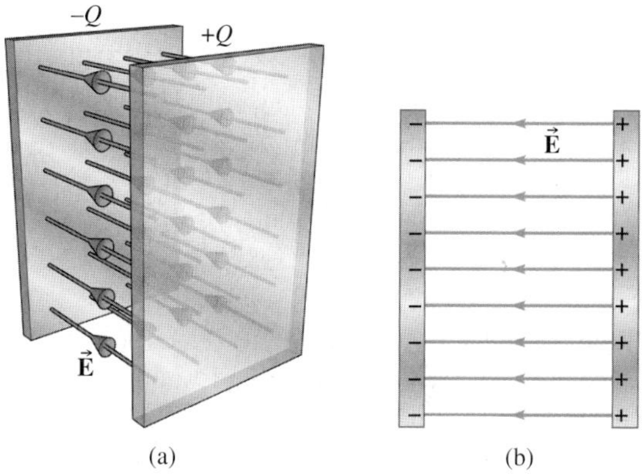

Figure 16.34 (a) Uniform electric field between two parallel metal plates with opposite charges $+Q$ and $-Q$. The field has magnitude $E = Q/(\epsilon_0 A)$ where A is the area of each plate. (b) Side view of the field lines.

uniform gravitational field if other forces are negligible). All the tools developed in Chapters 2 and 3 to analyze motion with constant acceleration can be used here. The direction of the acceleration is either parallel to $\vec{\mathbf{E}}$ (for a positive charge) or antiparallel to $\vec{\mathbf{E}}$ (for a negative charge).

✓ CHECKPOINT 16.5

An electron moves in a region of uniform electric field in the +*x*-direction. The electric field is also in the +*x*-direction. Describe the subsequent motion of the electron.

Example 16.9

Electron Beam

A cathode ray tube (CRT) is used to accelerate electrons in some televisions, computer monitors, oscilloscopes, and x-ray tubes. Electrons from a heated filament pass through a hole in the cathode; they are then accelerated by an electric field between the cathode and the anode (Fig. 16.35). Suppose an electron passes through the hole in the cathode at a velocity of 1.0×10^5 m/s toward the anode. The electric field is uniform between the anode and cathode and has a magnitude of 1.0×10^4 N/C. (a) What is the acceleration of the electron? (b) If the anode and cathode are separated by 2.0 cm, what is the final velocity of the electron?

Strategy Because the field is uniform, the acceleration of the electron is constant. Then we can apply Newton's second law and use any of the methods we previously developed for motion with constant acceleration.

Given: initial speed $v_i = 1.0 \times 10^5$ m/s;
separation between plates $d = 0.020$ m;
electric field magnitude $E = 1.0 \times 10^4$ N/C
Look up: electron mass $m_e = 9.109 \times 10^{-31}$ kg;
electron charge $q = -e = -1.602 \times 10^{-19}$ C
Find: (a) acceleration; (b) final velocity

Solution (a) First, check that gravity is negligible. The weight of the electron is

$$F_g = mg = 9.109 \times 10^{-31}\ \text{kg} \times 9.8\ \text{m/s}^2 = 8.9 \times 10^{-30}\ \text{N}$$

The magnitude of the electric force is

$$F_E = eE = 1.602 \times 10^{-19}\ \text{C} \times 1.0 \times 10^4\ \text{N/C} = 1.6 \times 10^{-15}\ \text{N}$$

which is about 14 orders of magnitude larger. Gravity is completely negligible. While between the plates, the electron's acceleration is therefore

$$a = \frac{F}{m_e} = \frac{eE}{m_e} = \frac{1.602 \times 10^{-19}\ \text{C} \times 1.0 \times 10^4\ \text{N/C}}{9.109 \times 10^{-31}\ \text{kg}}$$
$$= 1.76 \times 10^{15}\ \text{m/s}^2$$

To two significant figures, $a = 1.8 \times 10^{15}$ m/s^2. Since the charge on the electron is negative, the direction of the acceleration is opposite to the electric field, or to the right in the figure.

(b) The initial velocity of the electron is also to the right. We have a one-dimensional constant acceleration problem since the initial velocity and the acceleration are in the same direction. From Eq. (2-13), the final velocity is

$$v_f = \sqrt{v_i^2 + 2ad}$$
$$= \sqrt{(1.0 \times 10^5)^2 + 2 \times 1.76 \times 10^{15} \times 0.020}\ \text{m/s}$$
$$= 8.4 \times 10^6\ \text{m/s to the right}$$

Discussion The acceleration of the electrons seems large. This large value might cause some concern, but there is no law of physics against such large accelerations. Note that the final *speed* is less than the speed of light (3×10^8 m/s), the universe's ultimate speed limit.

You may suspect that this problem can also be solved using energy methods. We could indeed find the work done by the electric force and use the work done to find the change in kinetic energy. The energy approach for electric fields is developed in Chapter 17.

Practice Problem 16.9 Slowing Some Protons

If a beam of *protons* were projected horizontally to the right through the hole in the cathode (see Fig. 16.35) with an initial speed of $v_i = 3.0 \times 10^5$ m/s, with what speed would the protons reach the anode (if they do reach it)?

Application of electric field: oscilloscope

The electric field is used to speed up the electron beam in a CRT. In an oscilloscope—a device used to measure time-dependent quantities in circuits—it is also used to deflect the beam. An electric field is *not* used to *deflect* the electron beam in the CRT used in a TV or computer monitor; that function is performed by a magnetic field.

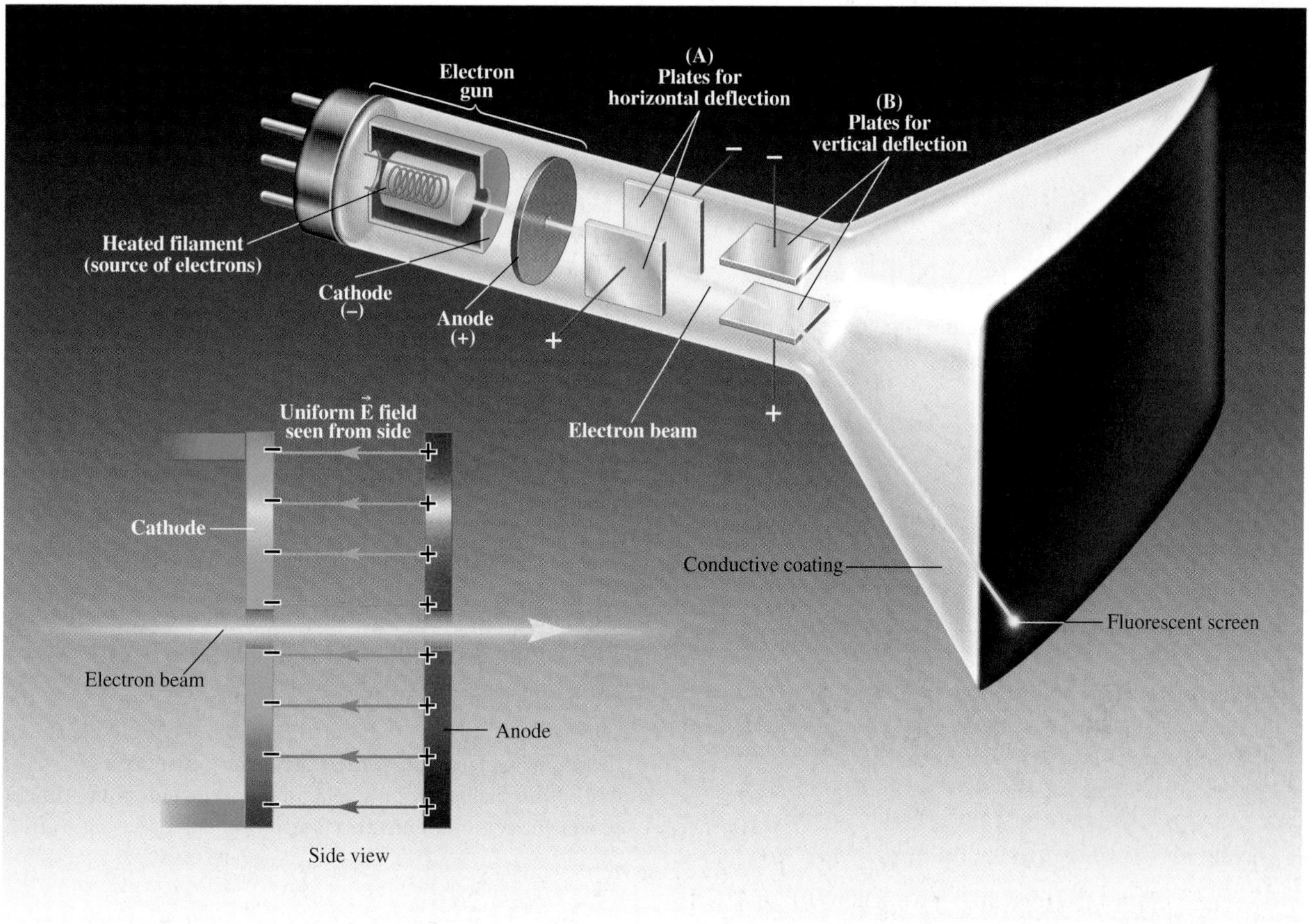

Figure 16.35 In a cathode ray tube (CRT), electrons are accelerated to high speeds by an electric field between the cathode and anode. This CRT, used in an oscilloscope, also has two pairs of parallel plates that are used to deflect the electron beam horizontally (A) and vertically (B).

Example 16.10

Deflection of an Electron Projected into a Uniform $\vec{E}$ Field

An electron is projected horizontally into the uniform electric field directed vertically downward between two parallel plates (Fig. 16.36). The plates are 2.00 cm apart and are of length 4.00 cm. The initial speed of the electron is $v_i = 8.00 \times 10^6$ m/s. As it enters the region between the plates, the electron is midway between the two plates; as it leaves, the electron just misses the upper plate. What is the magnitude of the electric field?

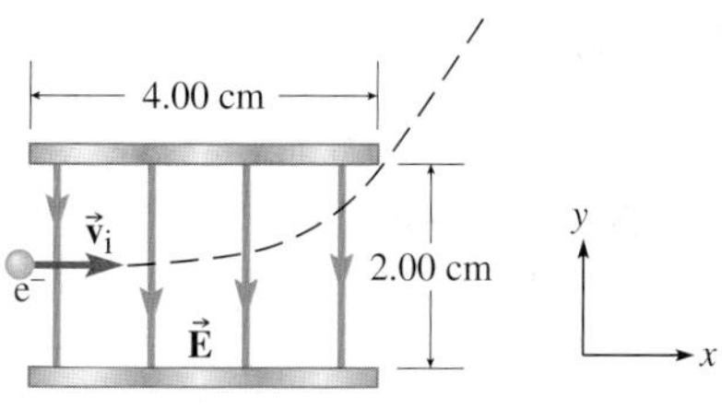

Figure 16.36 An electron deflected by an electric field.

Strategy Using the x- and y-axes in the figure, the electric field is in the $-y$-direction and the initial velocity of the electron is in the $+x$-direction. The electric force on the electron is *upward* (in the $+y$-direction) since it has a negative charge and is constant because the field is uniform. Thus, the acceleration of the electron is constant and directed upward. Since the acceleration is in the $+y$-direction, the x-component of the velocity is constant. The problem is similar to a projectile problem, but the constant acceleration is due to a uniform *electric* field instead of a uniform gravitational field. If the electron just misses the upper plate, its displacement is +1.00 cm in the y-direction and +4.00 cm in the x-direction. From v_x and Δx, we can find the time the electron spends between the plates. From Δy and the time, we can find a_y. From the acceleration we find the electric field using Newton's second law, $\Sigma\vec{F} = m\vec{a}$.

continued on next page

Example 16.10 continued

We ignore the gravitational force on the electron because we assume it to be negligible. We can test this assumption later.

Given: $\Delta x = 4.00$ cm; $\Delta y = 1.00$ cm; $v_x = 8.00 \times 10^6$ m/s

Find: electric field strength, E

Solution We start by finding the time the electron spends between the plates from Δx and v_x.

$$\Delta t = \frac{\Delta x}{v_x} = \frac{4.00 \times 10^{-2}\ \text{m}}{8.00 \times 10^6\ \text{m/s}} = 5.00 \times 10^{-9}\ \text{s}$$

From the time spent between the plates and Δy, we find the component of the acceleration in the y-direction.

$$\Delta y = \tfrac{1}{2} a_y (\Delta t)^2$$

$$a_y = \frac{2\,\Delta y}{(\Delta t)^2} = \frac{2 \times 1.00 \times 10^{-2}\ \text{m}}{(5.00 \times 10^{-9}\ \text{s})^2} = 8.00 \times 10^{14}\ \text{m/s}^2$$

This acceleration is produced by the electric force acting on the electron since we assume that no other forces act. From Newton's second law,

$$F_y = qE_y = m_e a_y$$

Solving for E_y, we have

$$E_y = \frac{m_e a_y}{q} = \frac{9.109 \times 10^{-31}\ \text{kg} \times 8.00 \times 10^{14}\ \text{m/s}^2}{-1.602 \times 10^{-19}\ \text{C}}$$

$$= -4.55 \times 10^3\ \text{N/C}$$

Since the field has no x-component, its magnitude is 4.55×10^3 N/C.

Discussion We have ignored the gravitational force on the electron because we suspect that it is negligible in comparison with the electric force. This should be checked to be sure it is a valid assumption.

$$\vec{\mathbf{F}} = m_e \vec{\mathbf{g}} = 9.109 \times 10^{-31}\ \text{kg} \times 9.80\ \text{N/kg downward}$$

$$= 8.93 \times 10^{-30}\ \text{N downward}$$

$$\vec{\mathbf{F}}_E = q\vec{\mathbf{E}} = -1.602 \times 10^{-19}\ \text{C} \times 4.55 \times 10^3\ \text{N/C downward}$$

$$= 7.29 \times 10^{-16}\ \text{N upward}$$

The electric force is stronger than the gravitational force by a factor of approximately 10^{14}, so the assumption is valid.

Practice Problem 16.10 Deflection of a Proton Projected into a Uniform $\vec{\mathbf{E}}$ Field

If the electron is replaced by a proton projected with the same initial velocity, will the proton exit the region between the plates or will it hit one of the plates? If it does not strike one of the plates, by how much is it deflected by the time it leaves the region of electric field?

16.6 CONDUCTORS IN ELECTROSTATIC EQUILIBRIUM

In Section 16.1, we described how a piece of paper can be polarized by nearby charges. The polarization is the paper's response to an applied electric field. By *applied* we mean a field due to charges *outside the paper*. The separation of charge in the paper produces an electric field of its own. The net electric field at any point—whether inside or outside the paper—is the sum of the applied field and the field due to the separated charges in the paper.

How much charge separation occurs depends on both the strength of the applied field and properties of the atoms and molecules that make up the paper. Some materials are more easily polarized than others. The *most* easily polarized materials are conductors because they contain highly mobile charges that can move freely through the entire volume of the material.

It is useful to examine the distribution of charge in a conductor, whether the conductor has a net charge or lies in an externally applied field, or both. We restrict our attention to a conductor in which the mobile charges are at rest in equilibrium, a situation called **electrostatic equilibrium**. If charge is put on a conductor, mobile charges move about until a stable distribution is attained. The same thing happens when an external field is applied or changed—charges move in response to the external field, but they soon reach an equilibrium distribution.

In **electrostatic equilibrium**, there is no net motion of charge.

If the electric field within a conducting material is nonzero, it exerts a force on each of the mobile charges (usually electrons) and makes them move preferentially in a

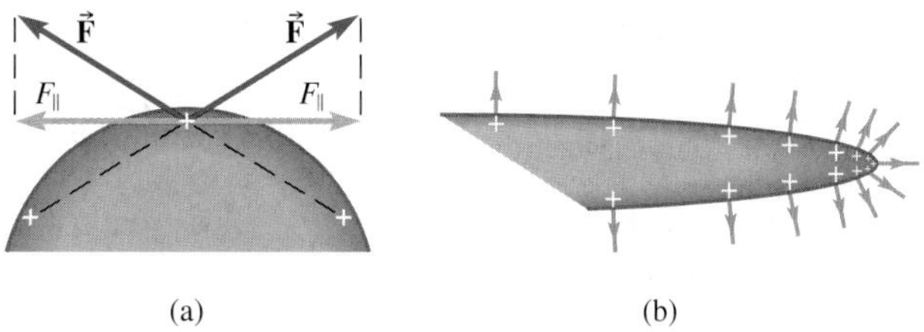

Figure 16.37 (a) Repulsive forces on a charge constrained to move along a curved surface due to two of its neighbors. The parallel components of the forces ($F_{\parallel}$) determine the spacing between the charges. (b) For a conductor in electrostatic equilibrium, the surface charge density is largest where the radius of curvature of the surface is smallest and the electric field just outside the surface is strongest there.

certain direction. With mobile charge in motion, the conductor cannot be in electrostatic equilibrium. Therefore, we can draw this conclusion:

> The electric field is zero at any point within a conducting material in electrostatic equilibrium.

Application: electrostatic shielding

Electronic circuits and cables are often shielded from stray electric fields produced by other devices by placing them inside metal enclosures (see Conceptual Question 6). Free charges in the metal enclosure rearrange themselves as the external electric field changes. As long as the charges in the enclosure can keep up with changes in the external field, the external field is canceled inside the enclosure.

The electric field is zero *within* the conducting material, but is not necessarily zero *outside.* If there are field lines outside but none inside, field lines must either start or end at charges on the surface of the conductor. Field lines start or end on charges, so

> When a conductor is in electrostatic equilibrium, only its surface(s) can have net charge.

At any point within the conductor, there are equal amounts of positive and negative charge. Imbalance between positive and negative charge can occur only on the surface(s) of the conductor.

It is also true that, in electrostatic equilibrium,

> The electric field at the surface of the conductor is perpendicular to the surface.

How do we know that? If the field had a component parallel to the surface, any free charges at the surface would feel a force parallel to the surface and would move in response. Thus, if there is a parallel component at the surface, the conductor cannot be in electrostatic equilibrium.

If a conductor has an irregular shape, the excess charge on its surface(s) is concentrated more at sharp points. Think of the charges as being constrained to move along the surfaces of the conductor. On flat surfaces, repulsive forces between neighboring charges push parallel to the surface, making the charges spread apart evenly. On a curved surface, only the components of the repulsive forces parallel to the surface, $F_{\parallel}$, are effective at making the charges spread apart (Fig. 16.37a). If charges were spread evenly over an irregular surface, the parallel components of the repulsive forces would be smaller for charges on the more sharply curved regions and charge would tend to move toward these regions. Therefore,

> The surface charge density (charge per unit area) on a conductor in electrostatic equilibrium is highest at sharp points. (Fig. 16.37b).

Application: lightning rods

Figure 16.38 A lightning rod protects a Victorian house in Mt. Horeb, Wisconsin.

The electric field lines just outside a conductor are densely packed at sharp points because each line starts or ends on a surface charge. Since the density of field lines reflects the magnitude of the electric field, the electric field outside the conductor is largest near the sharpest points of the conducting surface.

Lightning rods (invented by Franklin) are often found on the roofs of tall buildings and old farmhouses (Fig. 16.38). The rod comes to a sharp point at the top. When a passing thunderstorm attracts charge to the top of the rod, the strong electric field at the point ionizes nearby air molecules. Neutral air molecules do not transfer net charge when they move, but ionized molecules do, so ionization allows charge to leak gently off the building through the air instead of building up to a dangerously large value. If the rod did not come to a sharp point, the electric field might not be large enough to ionize the air.

The conclusions we have reached about conductors in electrostatic equilibrium can be restated in terms of field line rules:

For a conductor in electrostatic equilibrium,

- There are no field lines within the conducting material;
- Field lines that start or stop on the surface of a conductor are perpendicular to the surface where they intersect it; and
- The electric field just outside the surface of a conductor is strongest near sharp points.

Conceptual Example 16.11

Spherical Conductor in a Uniform Applied Field

Two oppositely charged parallel plates produce a uniform electric field between them (Fig. 16.39). An uncharged metal sphere is placed between the plates. Assume that the sphere is small enough that it does not affect the charge distribution on the plates. Sketch the electric field lines between the plates once electrostatic equilibrium is reached.

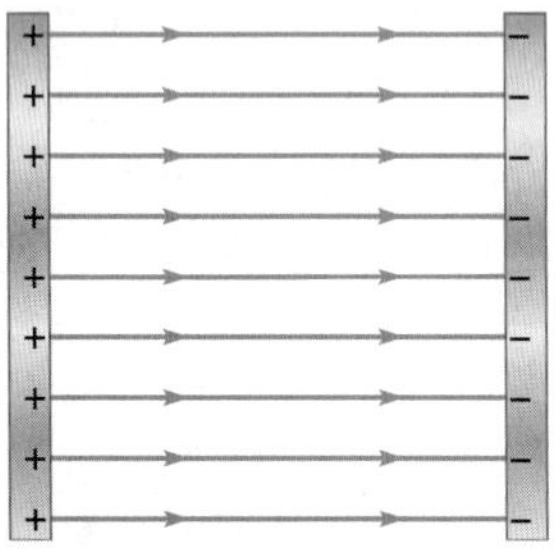

Figure 16.39 Uniform field between two plates.

Strategy We expect electrons in the metal sphere to be attracted to the positive plate, leaving the surface near the positive plate with a negative surface charge. The other side will have a positive surface charge. The electric field is changed by these surface charges, so that it is no longer uniform.

Solution and Discussion There are no field lines inside the metal sphere. The field lines cannot "go around" the sphere, since then there would be a field component parallel to the sphere's surface. Furthermore, since we already know that there is charge on the sphere's surface, some field lines must start on the positive side and others end on the negative side. The field lines must intersect the sphere perpendicular to the surface. Figure 16.40 shows a field line diagram for the sphere.

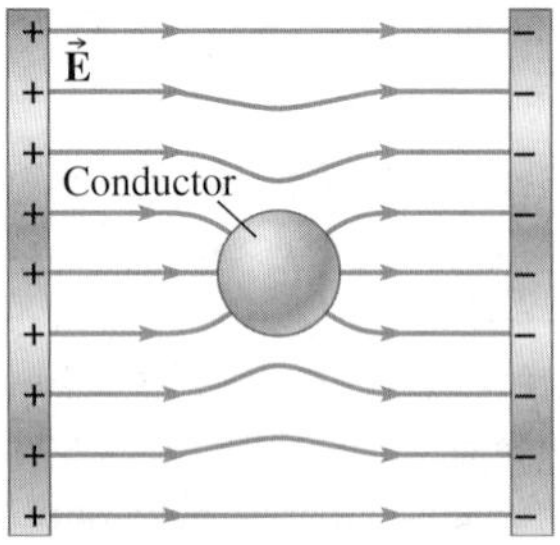

Figure 16.40 Field lines when a metal sphere is placed between the plates.

Conceptual Practice Problem 16.11
Oppositely Charged Spheres

Two metal spheres of the same radius R are given charges of equal magnitude and opposite sign. No other charges are nearby. Sketch the electric field lines when the center-to-center distance between the spheres is approximately $3R$.

Application: Electrostatic Precipitator

One direct application of electric fields is the *electrostatic precipitator*—a device that reduces the air pollution emitted from industrial smokestacks (Fig. 16.41). Many industrial processes, such as the burning of fossil fuels in electrical generating

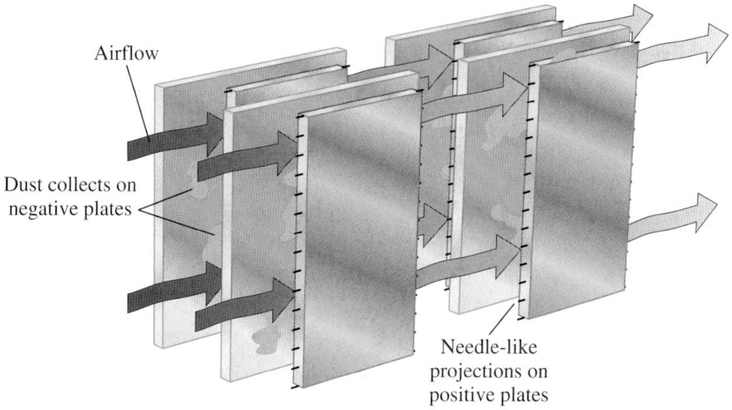

Figure 16.41 An electrostatic precipitator. Inside the precipitator chamber is a set of oppositely charged metal plates. The positively charged plates are fitted with needle-like wire projections that serve as discharge points. The electric field is strong enough at these points to ionize air molecules. The particulates are positively charged by contact with the ions. The electric field between the plates then attracts the particulates to the negatively charged collection plates. After enough particulate matter has built up on these plates, it falls to the bottom of the precipitator chamber from where it is easily removed.

plants, release flue gases containing particulates into the air. To reduce the quantity of particulates released, the gases are sent through a precipitator chamber before leaving the smokestack. Many air purifiers sold for use in the home are electrostatic precipitators.

16.7 GAUSS'S LAW FOR ELECTRIC FIELDS

Gauss's law, named after German mathematician Karl Friedrich Gauss (1777-1855), is a powerful statement of properties of the electric field. It relates the electric field on a closed surface—*any* closed surface—to the net charge inside the surface. A **closed surface** encloses a volume of space, so that there is an inside and an outside. The surface of a sphere, for instance, is a closed surface, whereas the interior of a circle is not. Gauss's law says: I can tell you how much charge you have inside that "box" without looking inside; I'll just look at the field lines that enter or exit the box.

If a box has no charge inside of it, then the same number of field lines that go into the box must come back out; there is nowhere for field lines to end or to begin. Even if there is charge inside, but the *net* charge is zero, the same number of field lines that start on the positive charge must end on the negative charge, so again the same number of field lines that go in must come out. If there is net positive charge inside, then there will be field lines starting on the positive charge that leave the box; then more field lines come out than go in. If there is net negative charge inside, some field lines that go in end on the negative charge; more field lines go in than come out.

Field lines are a useful device for visualization, but they are not quantifiable in any standard way. In order for Gauss's law to be useful, we formulate it mathematically so that numbers of field lines are not involved. To reformulate the law, there are two conditions to satisfy. First, a mathematical quantity must be found that is proportional to the number of field lines leaving a closed surface. Second, a proportionality must be turned into an equation by solving for the constant of proportionality.

Recall from Section 16.4 that the magnitude of the electric field is proportional to the number of field lines *per unit cross-sectional area*:

$$E \propto \frac{\text{number of lines}}{\text{area}}$$

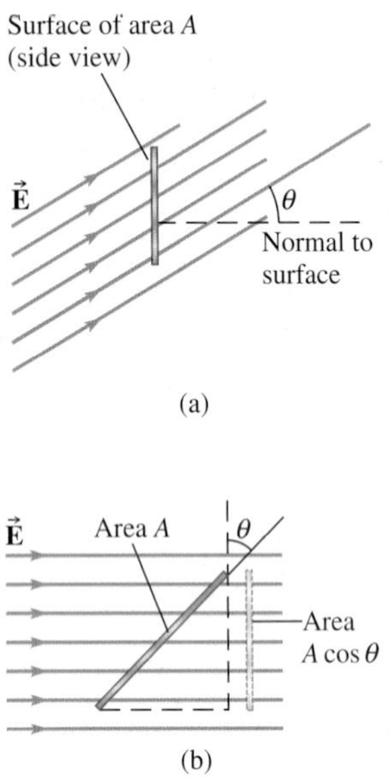

Figure 16.42 (a) Electric field lines crossing through a rectangular surface (side view). The angle between the field lines and the line *perpendicular* to the surface is θ. (b) The number of field lines that cross the surface of area A is the same as the number that cross the perpendicular surface of area $A \cos \theta$.

If a surface of area A is everywhere perpendicular to an electric field of uniform magnitude E, then the number of field lines that cross the surface is proportional to EA, since

$$\text{number of lines} = \frac{\text{number of lines}}{\text{area}} \times \text{area} \propto EA$$

This is only true if the surface is perpendicular to the electric field everywhere. As an analogy, think of rain falling straight down into a bucket. Less rainwater enters the bucket when it is tilted to one side than if the bucket rests with its opening perpendicular to the direction of rainfall. In general, the number of field lines crossing a surface is proportional to the *perpendicular component* of the field times the area:

$$\text{number of lines} \propto E_\perp A = EA \cos \theta$$

where θ is the angle that the field lines make with the normal (perpendicular) to the surface (Fig. 16.42a). Equivalently, Fig. 16.42b shows that the number of lines crossing the surface is the same as the number crossing a surface of area $A \cos \theta$, which is the area perpendicular to the field.

The mathematical quantity that is proportional to the number of field lines crossing a surface is called the **flux of the electric field** (symbol Φ_E; Φ is the Greek capital phi).

Definition of Flux

$$\Phi_E = E_\perp A = EA_\perp = EA \cos \theta \quad (16\text{-}8)$$

For a closed surface, flux is defined to be positive if more field lines leave the surface than enter, or negative if more lines enter than leave. Flux is then positive if the net enclosed charge is positive and it is negative if the net enclosed charge is negative.

Since the net number of field lines is proportional to the net charge inside a closed surface, Gauss's law takes the form

$$\Phi_E = \text{constant} \times q$$

where q stands for the *net charge enclosed by the surface.* In Example 16.12 (and Problem 65), you can show that the constant of proportionality is $4\pi k = 1/\epsilon_0$. Therefore,

Gauss's Law

$$\Phi_E = 4\pi kq = q/\epsilon_0 \quad (16\text{-}9)$$

Example 16.12

Flux Through a Sphere

What is the flux through a sphere of radius $r = 5.0$ cm that has a point charge $q = -2.0$ μC at its center?

Strategy In this case, there are two ways to find the flux. The electric field is known from Coulomb's law and it can be used to find the flux; or we can use Gauss's law.

Solution The electric field at a separation r from a point charge is

$$E = \frac{kq}{r^2}$$

For a negative point charge, the field is radially inward. The field has the same strength everywhere on the sphere, since the separation from the point charge is constant. Also, the field is always perpendicular to the surface of the sphere ($\theta = 0$ everywhere). Therefore,

$$\Phi_E = EA = \frac{kq}{r^2} \times 4\pi r^2 = 4\pi kq$$

This is exactly what Gauss's law tells us. The flux is independent of the radius of the sphere, since all the field lines cross the sphere regardless of its radius. A negative value

continued on next page

Example 16.12 continued

of q gives a negative flux, which is correct since the field lines go inward. Then

$$\Phi_E = 4\pi kq$$
$$= 4\pi \times 9.0 \times 10^9 \frac{\text{N}\cdot\text{m}^2}{\text{C}^2} \times (-2.0 \times 10^{-6}\ \text{C})$$
$$= -2.3 \times 10^5 \frac{\text{N}\cdot\text{m}^2}{\text{C}}$$

Discussion In this case, we can find the flux directly because the field at every point on the sphere is constant in magnitude and perpendicular to the sphere. However, Gauss's law tells us that the flux through *any* surface that encloses this charge, no matter what shape or size, must be the same.

Practice Problem 16.12 Flux Through a Side of a Cube

What is the flux through *one side* of a cube that has a point charge $-2.0\ \mu\text{C}$ at its center? [*Hint:* Of the total number of field lines, what fraction passes through one side of the cube?]

Using Gauss's Law to Find the Electric Field

As presented so far, Gauss's law is a way to determine how much charge is inside a closed surface given the electric field on the surface. Sometimes it can be turned around and used to *find the electric field* due to a distribution of charges. Why not just use Coulomb's law? In many cases there are such a large number of charges that the charge can be viewed as being continuously spread along a line, or over a surface, or throughout a volume. Microscopically, charge is still limited to multiples of the electronic charge, but when there are large numbers of charges, it is simpler to view the charge as a continuous distribution.

For a continuous distribution, the **charge density** is usually the most convenient way to describe how much charge is present. There are three kinds of charge densities:

- If the charge is spread throughout a volume, the relevant charge density is the charge per unit *volume* (symbol ρ).
- If the charge is spread over a two-dimensional surface, then the charge density is the charge per unit *area* (symbol σ).
- If the charge is spread over a one-dimensional line or curve, the appropriate charge density is the charge per unit *length* (symbol λ).

Gauss's law can be used to calculate the electric field in cases where there is enough *symmetry* to tell us something about the field lines. Example 16.13 illustrates this technique.

Example 16.13

Electric Field at a Distance from a Long Thin Wire

Charge is spread *uniformly* along a long thin wire. The charge per unit length on the wire is λ and is constant. Find the electric field at a distance r from the wire, far from either end of the wire.

Strategy The electric field at any point is the sum of the electric field contributions from the charge all along the wire. Coulomb's law tells us that the strongest contributions come from the charge on nearby parts of the wire, with contributions falling off as $1/r^2$ for faraway points. When concerned only with points near the wire, and far from either end, an approximately correct answer is obtained by assuming the wire is *infinitely long*.

How is it a simplification to *add* more charges? When using Gauss's law, a symmetrical situation is far simpler than a situation that lacks symmetry. An infinitely long wire with a uniform linear charge density has *axial symmetry.* Sketching the field lines first helps show what symmetry tells us about the electric field.

continued on next page

Example 16.13 continued

Solution We start by sketching field lines for an infinitely long wire. The field lines either start or stop on the wire (depending on whether the charge is positive or negative). Then what do the field lines do? The only possibility is that they move radially outward (or inward) from the wire. Figure 16.43a shows sketches of the field lines for positive and negative charges, respectively. The wire looks the same from all sides, so a field line could not start to curl around as in Fig. 16.43b: how would it determine which way to go? Also, the field lines cannot go along the wire as in Fig. 16.43c: again, how could the lines decide whether to go right or left? The wire looks exactly the same in both directions.

Once we recognize that the field lines are radial, the next step is to choose a surface. Gauss's law is easiest to handle if the electric field is constant in magnitude and either perpendicular or parallel to the surface. A cylinder with a radius r with the wire as its axis has the field perpendicular to the surface everywhere, since the lines are radial (Fig. 16.44). The magnitude of the field must also be constant on the surface of the cylinder because every point on the cylinder is located an equal distance from the wire. Since a *closed* surface is necessary, the two circular ends of the cylinder are included. The flux through the ends is zero since no field lines pass through; equivalently, the *perpendicular component* of the field is zero.

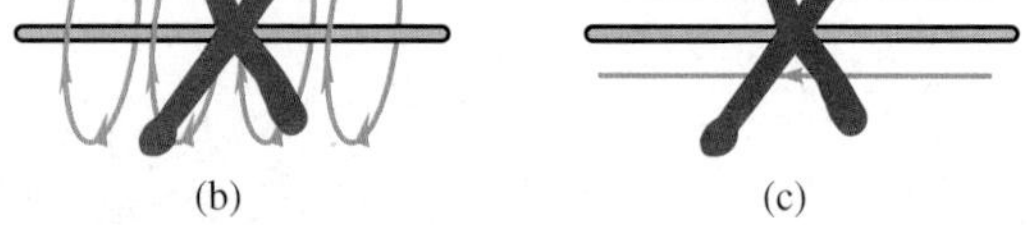

Figure 16.43
(a) Electric field lines emanating from a long wire, radially outward and radially inward; (b) hypothetical lines circling a wire; (c) hypothetical lines parallel to the wire.

Since the field is constant in magnitude and perpendicular to the surface, the flux is

$$\Phi_E = E_r A$$

where E_r is the radial component of the field. E_r is positive if the field is radially outward and negative if the field is radially inward. A is the area of a cylinder of radius r and . . . what length? Since the cylinder is imaginary, we can consider an arbitrary length denoted by L. The area of the cylinder is (Appendix A.6)

$$A = 2\pi rL$$

How much charge is enclosed by this cylinder? The charge per unit length is λ and a length L of the wire is inside the cylinder, so the enclosed charge is

$$q = \lambda L$$

which can be either positive or negative. Gauss's law and the definition of flux yield

$$4\pi kq = \Phi_E = E_r A$$

Substituting the expressions for A and q into Gauss's law yields

$$E_r \times (2\pi rL) = 4\pi k\lambda L$$

Solving for E_r,

$$E_r = \frac{2k\lambda}{r}$$

The field direction is radially outward for $\lambda > 0$ and radially inward for $\lambda < 0$.

Discussion The final expression for the electric field does not depend on the arbitrary length L of the cylinder. If L appeared in the answer, we would know to look for a mistake.

We should check the units of the answer: λ is the charge per unit length, so it has SI units

$$[\lambda] = \frac{\text{C}}{\text{m}}$$

The constant k has SI units

$$[k] = \frac{\text{N·m}^2}{\text{C}^2}$$

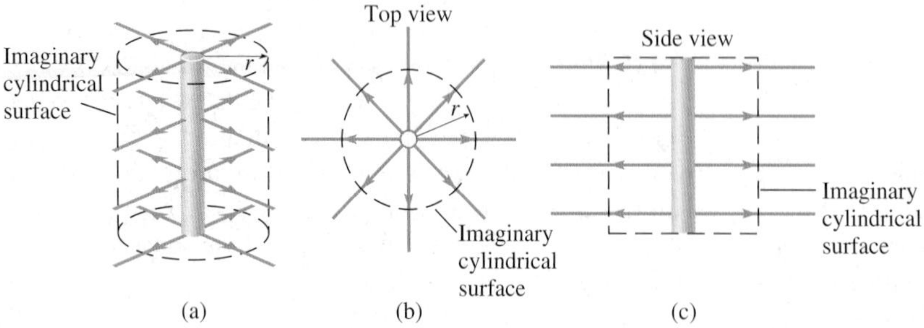

Figure 16.44
(a) Electric field lines from a wire located along the axis of a cylinder are perpendicular to the surrounding imaginary cylindrical surface. (b) Top and (c) side views of the cylinder and the field lines; the field lines are perpendicular to the cylindrical surface area but parallel to the planes of the top and bottom circular areas.

continued on next page

Example 16.13 continued

The factor of 2π is dimensionless and r is a distance. Then

$$\left[\frac{2k\lambda}{r}\right] = \frac{\text{C}}{\text{m}} \times \frac{\text{N}\cdot\text{m}^2}{\text{C}^2} \times \frac{1}{\text{m}} = \frac{\text{N}}{\text{C}}$$

which is the SI unit of electric field.

The electric field falls off as the inverse of the separation ($E \propto 1/r$). Wait a minute—does this violate Coulomb's law, which says $E \propto 1/r^2$? No, because that is the field at a separation r from a *point charge*. Here the charge is spread out in a line. The different geometry changes the field lines (they come radially outward from a line rather than from a point) and this changes how the field depends on distance.

Conceptual Practice Problem 16.13 Which Area to Use?

In Example 16.13, we wrote the area of a cylinder as $A = 2\pi rL$, which is only the area of the curved surface of the cylinder. The total area of a cylinder includes the area of the circles on each end (top and base): $A_{\text{total}} = 2\pi rL + 2\pi r^2$. Why did we not include the area of the ends of the cylinder when calculating flux?

Master the Concepts

- Coulomb's law gives the electric force exerted on one point charge due to another. The magnitude of the force is

$$F = \frac{k|q_1||q_2|}{r^2} \qquad (16\text{-}2)$$

where the Coulomb constant is

$$k = 8.99 \times 10^9 \ \frac{\text{N}\cdot\text{m}^2}{\text{C}^2} \qquad (16\text{-}3\text{a})$$

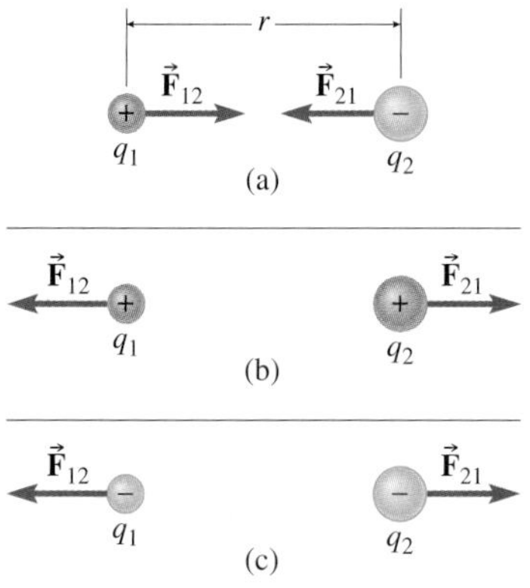

- The direction of the force on one point charge due to another is either directly toward the other charge (if the charges have opposite signs) or directly away (if the charges have the same sign).
- The electric field (symbol $\vec{\mathbf{E}}$) is the electric force per unit *charge*. It is a vector quantity.
- If a point charge q is located where the electric field due to all other charges is $\vec{\mathbf{E}}$, then the electric force on the point charge is

$$\vec{\mathbf{F}}_E = q\vec{\mathbf{E}} \qquad (16\text{-}4\text{b})$$

- The SI units of the electric field are N/C.
- Electric field lines are useful for representing an electric field.
- The direction of the electric field at any point is tangent to the field line passing through that point and in the direction indicated by the arrows on the field line.

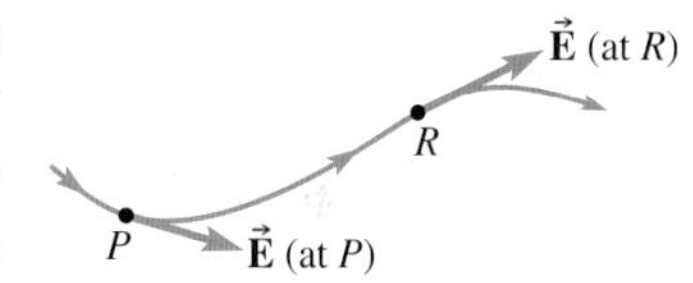

- The electric field is strong where field lines are close together and weak where they are far apart.
- Field lines never cross.
- Field lines start on positive charges and end on negative charges.
- The number of field lines starting on a positive charge (or ending on a negative charge) is proportional to the magnitude of the charge.
- The principle of superposition says that the electric field due to a collection of charges at any point is the vector sum of the electric fields caused by each charge separately.
- The uniform electric field between two parallel metal plates with charges $\pm Q$ and area A has magnitude

$$E = \frac{Q}{\epsilon_0 A} \qquad (16\text{-}6)$$

The direction of the field is perpendicular to the plates and away from the positively charged plate.

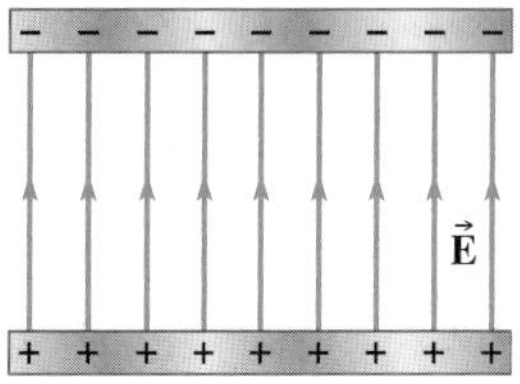

- Electric flux:

$$\Phi_E = E_\perp A = EA_\perp = EA\cos\theta \qquad (16\text{-}8)$$

- Gauss's law:

$$\Phi_E = 4\pi kq = q/\epsilon_0 \qquad (16\text{-}9)$$

Conceptual Questions

1. Due to the similarity between Newton's law of gravity and Coulomb's law, a friend proposes this hypothesis: perhaps there is no gravitational interaction at all. Instead, what we call gravity might be *electric* forces acting between objects that are almost, but not quite, electrically neutral. Think up as many counterarguments as you can.
2. What makes clothes cling together—or to your body—after they've been through the dryer? Why do they not cling as much if they are taken out of the dryer while slightly damp? In which case would you expect your clothes to cling more, all other things being equal: when the clothes in the dryer are all made of the same material, or when they are made of several different materials?
3. Explain why any net charge on a solid metal conductor in electrostatic equilibrium is found on the outside surface of the conductor instead of being distributed uniformly throughout the solid.
4. Explain why electric field lines begin on positive charges and end on negative charges. [*Hint:* What is the direction of the electric field near positive and negative charges?]
5. A metal sphere is initially uncharged. After being touched by a charged rod, the metal sphere is positively charged. (a) Is the mass of the sphere larger, smaller, or the same as before it was charged? Explain. (b) What sign of charge is on the rod?
6. Electronic devices are usually enclosed in metal boxes. One function of the box is to shield the inside components from external electric fields. (a) How does this shielding work? (b) Why is the degree of shielding better for constant or slowly varying fields than for rapidly varying fields? (c) Explain the reasons why it is not possible to shield something from gravitational fields in a similar way.
7. Your laboratory partner hands you a glass rod and asks if it has negative charge on it. There is an electroscope in the laboratory. How can you tell if the rod is charged? Can you determine the sign of the charge? If the rod is charged to begin with, will its charge be the same after you have made your determination? Explain.
8. A lightweight plastic rod is rubbed with a piece of fur. A second plastic rod, hanging from a string, is attracted to the first rod and swings toward it. When the second rod touches the first, it is suddenly repelled and swings away. Explain what has happened.
9. The following *hypothetical* reaction shows a neutron (n) decaying into a proton (p^+) and an electron (e^-):

$$n \rightarrow p^+ + e^-$$

At first there is no charge, but then charge seems to be "created." Does this reaction violate the law of charge conservation? Explain. (In Section 29.3, it is shown that the neutron does not decay into just a proton and an electron; the decay products include a third, electrically neutral particle.)

10. A fellow student says that there is *never* an electric field inside a conductor. Do you agree? Explain.
11. Explain why electric field lines never cross.
12. A truck carrying explosive gases either has chains or straps that drag along the ground, or else it has special tires that conduct electricity (ordinary tires are good insulators). Explain why the chains, straps, or conducting tires are necessary.
13. An electroscope consists of a conducting sphere, conducting pole, and two metal foils (Fig. 16.6). The electroscope is initially uncharged. (a) A positively charged rod is allowed to touch the conducting sphere and then is removed. What happens to the foils and what is their charge? (b) Next, another positively charged rod is brought near to the conducting sphere without touching it. What happens? (c) The positively charged rod is removed and a negatively charged rod is brought near the sphere. What happens?
14. A rod is negatively charged by rubbing it with fur. It is brought near another rod of unknown composition and charge. There is a repulsive force between the two. (a) Is the first rod an insulator or a conductor? Explain. (b) What can you tell about the charge of the second rod?
15. A negatively charged rod is brought near a grounded conductor. After the ground connection is broken, the rod is removed. Is the charge on the conductor positive, negative, or zero? Explain.
16. In some textbooks, the electric field is called the *flux density*. Explain the meaning of this term. Does flux density mean the flux per unit volume? If not, then what does it mean?
17. The word *flux* comes from the Latin "to flow." What does the quantity $\Phi_E = E_\perp A$ have to do with flow? The figure shows some streamlines for the flow of water in a pipe. The streamlines are actually field lines for the *velocity field.* What is the physical significance of the quantity $v_\perp A$? Sometimes physicists call positive charges *sources* of the electric field and negative charges *sinks.* Why?

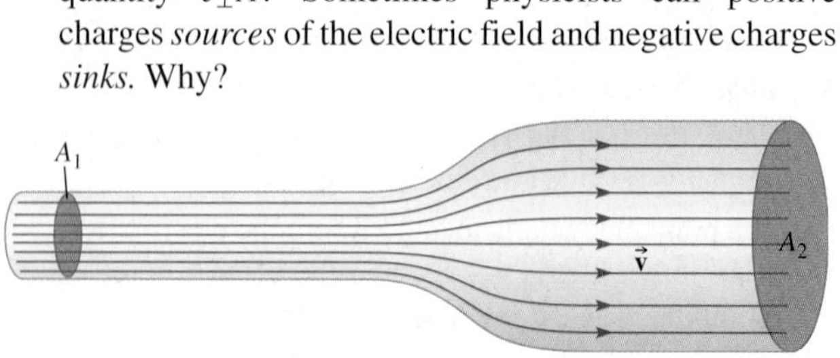

18. The flux through a closed surface is zero. Is the electric field necessarily zero? Is the net charge inside the surface necessarily zero? Explain your answers.
19. Consider a closed surface that surrounds Q_1 and Q_2 but not Q_3 or Q_4. (a) Which charges contribute to the electric field at point P? (b) Would the value obtained

for the flux through the surface, calculated using only the electric field due to Q_1 and Q_2, be greater than, less than, or equal to that obtained using the total field?

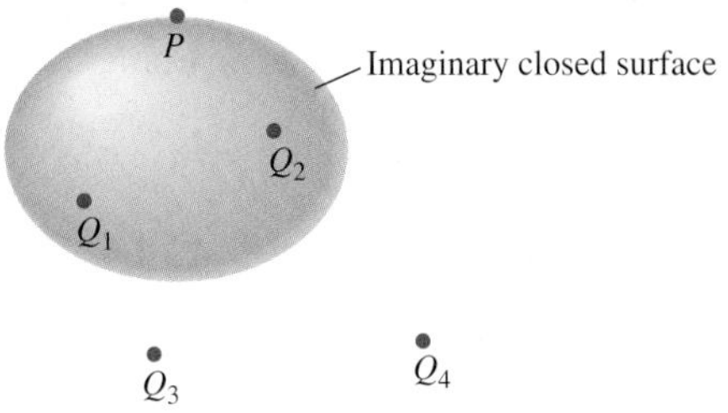

Multiple-Choice Questions

1. An α particle (charge $+2e$ and mass $4m_p$) is on a collision course with a proton (charge $+e$ and mass m_p). Assume that no forces act other than the electrical repulsion. Which one of these statements about the accelerations of the two particles is true?
(a) $\vec{\mathbf{a}}_\alpha = \vec{\mathbf{a}}_p$ (b) $\vec{\mathbf{a}}_\alpha = 2\vec{\mathbf{a}}_p$ (c) $\vec{\mathbf{a}}_\alpha = 4\vec{\mathbf{a}}_p$
(d) $2\vec{\mathbf{a}}_\alpha = \vec{\mathbf{a}}_p$ (e) $4\vec{\mathbf{a}}_\alpha = \vec{\mathbf{a}}_p$ (f) $\vec{\mathbf{a}}_\alpha = -\vec{\mathbf{a}}_p$
(g) $\vec{\mathbf{a}}_\alpha = -2\vec{\mathbf{a}}_p$ (h) $\vec{\mathbf{a}}_\alpha = -4\vec{\mathbf{a}}_p$ (i) $-2\vec{\mathbf{a}}_\alpha = \vec{\mathbf{a}}_p$
(j) $-4\vec{\mathbf{a}}_\alpha = \vec{\mathbf{a}}_p$
2. In electrostatic equilibrium, the excess electric charge on an irregularly shaped conductor is
(a) uniformly distributed throughout the volume.
(b) confined to the surfaces and is uniformly distributed.
(c) entirely on the surfaces, but is not uniformly distributed.
(d) dispersed throughout the volume of the object, but is not uniformly distributed.
3. The electric field at a point in space is a measure of
(a) the total charge on an object at that point.
(b) the electric force on any charged object at that point.
(c) the charge-to-mass ratio of an object at that point.
(d) the electric force per unit mass on a point charge at that point.
(e) the electric force per unit charge on a point charge at that point.
4. Two charged particles attract each other with a force of magnitude F acting on each. If the charge of one is doubled and the distance separating the particles is also doubled, the force acting on each of the two particles has magnitude
(a) $F/2$ (b) $F/4$ (c) F (d) $2F$ (e) $4F$
(f) None of the above.
5. A charged insulator and an uncharged metal object near one another
(a) exert no electric force on one another.
(b) repel one another electrically.
(c) attract one another electrically.
(d) attract or repel, depending on whether the charge is positive or negative.
6. A tiny charged pellet of mass m is suspended at rest by the electric field between two horizontal, charged metallic plates. The lower plate has a positive charge and the upper plate has a negative charge. Which statement in the answers here is *not* true?

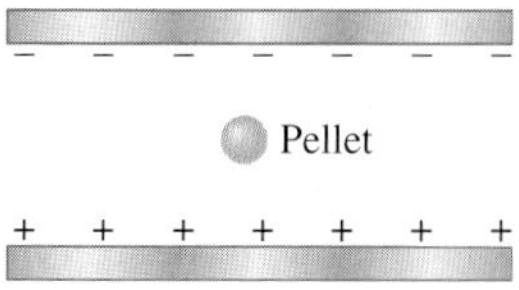

(a) The electric field between the plates points vertically upward.
(b) The pellet is negatively charged.
(c) The magnitude of the electric force on the pellet is equal to mg.
(d) If the magnitude of charge on the plates is increased, the pellet begins to move upward.
7. Which of these statements comparing electric and gravitational forces is correct?
(a) The direction of the electric force exerted by one point particle on another is always the same as the direction of the gravitational force exerted by that particle on the other.
(b) The electric and gravitational forces exerted by two particles on one another are inversely proportional to the separation of the particles.
(c) The electric force exerted by one planet on another is typically stronger than the gravitational force exerted by that same planet on the other.
(d) none of the above
8. In the figure, which best represents the field lines due to two point charges with opposite charges?

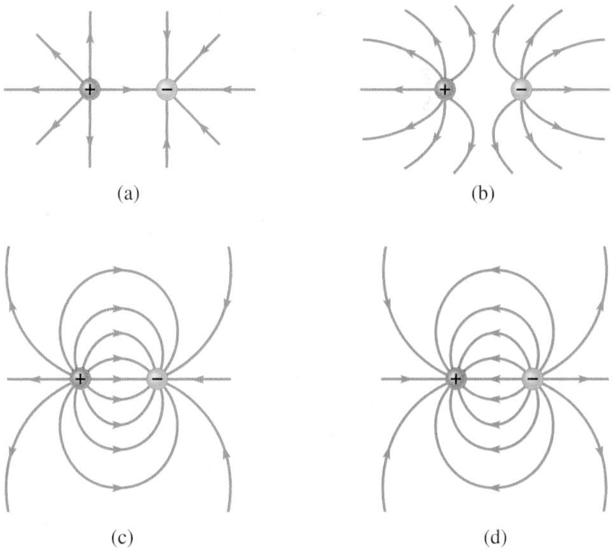

9. In the figure, put points 1–4 in order of increasing field strength.
(a) 2, 3, 4, 1
(b) 2, 1, 3, 4
(c) 1, 4, 3, 2
(d) 4, 3, 1, 2
(e) 2, 4, 1, 3

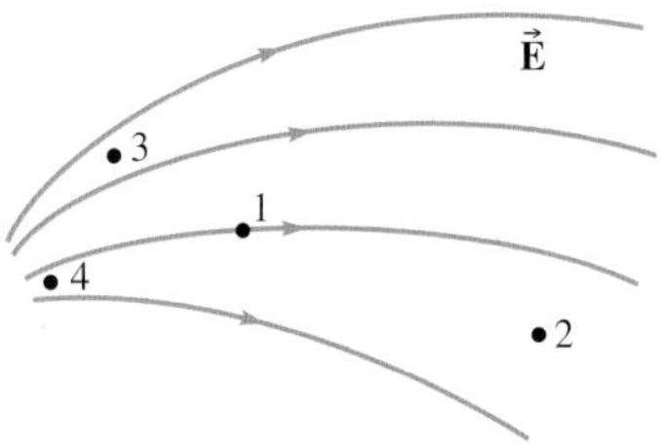

10. Two point charges q and $2q$ lie on the x-axis. Which region(s) on the x-axis include a point where the electric field due to the two point charges is zero?
(a) to the right of $2q$ (b) between $2q$ and point P
(c) between point P and q (d) to the left of q
(e) both (a) and (c) (f) both (b) and (d)

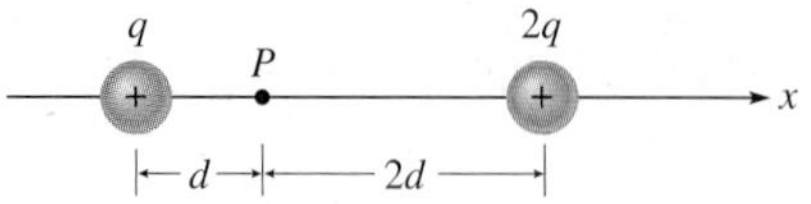

Problems

- Combination conceptual/quantitative problem
- Biological or medical application
- Challenging problem
- Blue # Detailed solution in the Student Solutions Manual
- 1 2 Problems paired by concept
- Text website interactive or tutorial

16.1 Electric Charge; 16.2 Electric Conductors and Insulators

1. Find the total positive charge of all the protons in 1.0 mol of water.

2. Suppose a 1.0-g nugget of pure gold has zero net charge. What would be its net charge after it has 1.0% of its electrons removed?

3. A balloon, initially neutral, is rubbed with fur until it acquires a net charge of –0.60 nC. (a) Assuming that only electrons are transferred, were electrons removed from the balloon or added to it? (b) How many electrons were transferred?

4. A metallic sphere has a charge of +4.0 nC. A negatively charged rod has a charge of –6.0 nC. When the rod touches the sphere, 8.2×10^9 electrons are transferred. What are the charges of the sphere and the rod now?

5. A positively charged rod is brought near two uncharged conducting spheres of the same size that are initially touching each other (diagram a). The spheres are moved apart and then the charged rod is removed (diagram b). (a) What is the sign of the net charge on sphere 1 in diagram b? (b) In comparison with the charge on sphere 1, how much and what sign of charge is on sphere 2?

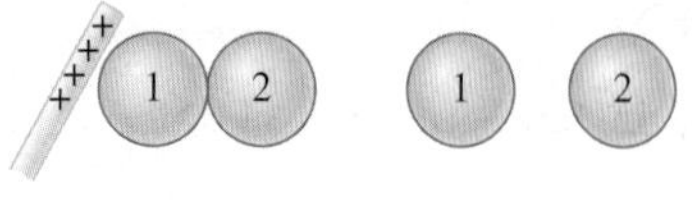

(a) (b)

6. A metal sphere A has charge Q. Two other spheres, B and C, are identical to A except they have zero net charge. A touches B, then the two spheres are separated. B touches C, then those spheres are separated. Finally, C touches A and those two spheres are separated. How much charge is on each sphere?

7. Repeat Problem 6 with a slight change. The difference this time is that sphere C is grounded when it is touching B, but C is not grounded at any other time. What is the final charge on each sphere?

8. Five conducting spheres are charged as shown. All have the same magnitude net charge except E, whose net charge is zero. Which pairs are attracted to each other and which are repelled by each other when they are brought near each other, but well away from the other spheres?

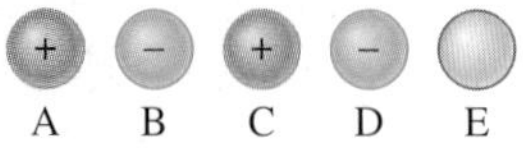

16.3 Coulomb's Law

9. If the electric force of repulsion between two 1-C charges is 10 N, how far apart are they?

10. Two small metal spheres are 25.0 cm apart. The spheres have equal amounts of negative charge and repel each other with a force of 0.036 N. What is the charge on each sphere?

11. What is the ratio of the electric force to the gravitational force between a proton and an electron separated by 5.3×10^{-11} m (the radius of a hydrogen atom)?

12. How many electrons must be removed from each of two 5.0-kg copper spheres to make the electric force of repulsion between them equal in magnitude to the gravitational attraction between them?

13. A +2.0-nC point charge is 3.0 cm away from a –3.0-nC point charge. (a) What are the magnitude and direction of the electric force acting on the +2.0-nC charge? (b) What are the magnitude and direction of the electric force acting on the –3.0-nC charge?

14. Two metal spheres separated by a distance much greater than either sphere's radius have equal mass m and equal electric charge q. What is the ratio of charge to mass q/m in C/kg if the electrical and gravitational forces balance?

15. In the figure, a third point charge $-q$ is placed at point P. What is the electric force on $-q$ due to the other two point charges?

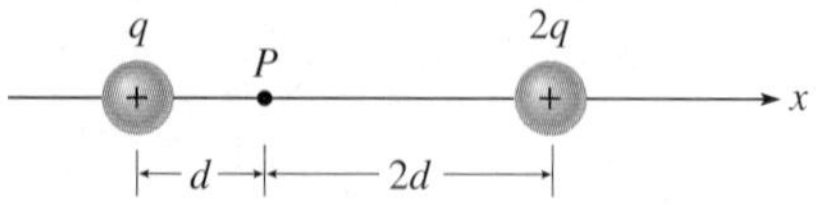

16. Two point charges are separated by a distance r and repel each other with a force F. If their separation is reduced to 0.25 times the original value, what is the magnitude of the force of repulsion between them?

17. A K^+ ion and a Cl^- ion are directly across from each other on opposite sides of a membrane 9.0 nm thick. What is the electric force on the K^+ ion due to the Cl^- ion? Ignore the presence of other charges.

18. Three point charges are fixed in place in a right triangle. What is the electric force on the -0.60-μC charge due to the other two charges?

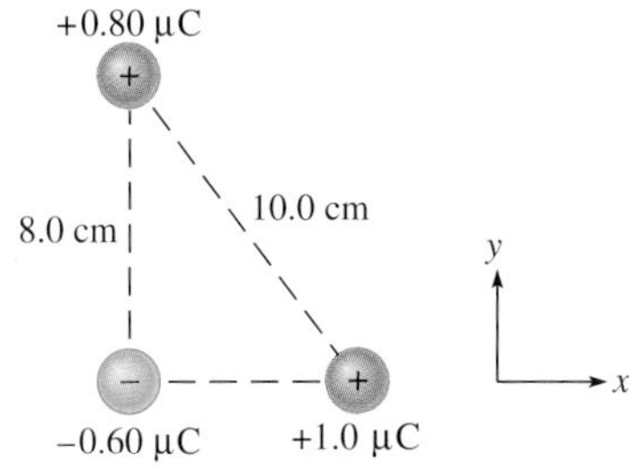

Problems 18 and 19

19. Three point charges are fixed in place in a right triangle. What is the electric force on the $+1.0$-μC charge due to the other two charges?

20. A tiny sphere with a charge of 7.0 μC is attached to a spring. Two other tiny charged spheres, each with a charge of -4.0 μC, are placed in the positions shown in the figure and the spring stretches 5.0 cm from its previous equilibrium position toward the two spheres. Calculate the spring constant.

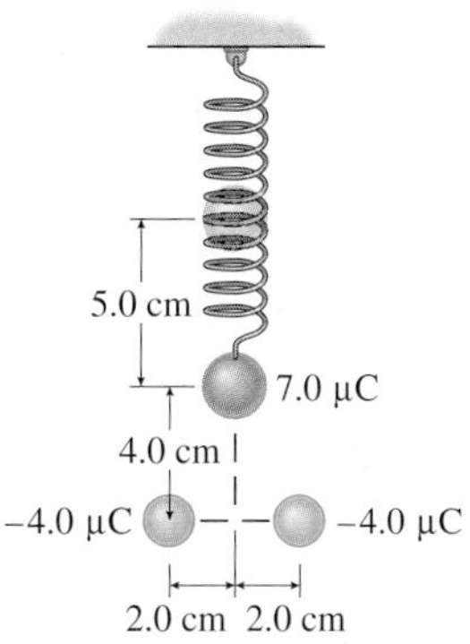

21. A total charge of 7.50×10^{-6} C is distributed on two different small metal spheres. When the spheres are 6.00 cm apart, they each feel a repulsive force of 20.0 N. How much charge is on each sphere?

22. Two Styrofoam balls with the same mass $m = 9.0 \times 10^{-8}$ kg and the same positive charge Q are suspended from the same point by insulating threads of length $L = 0.98$ m. The separation of the balls is $d = 0.020$ m. What is the charge Q?

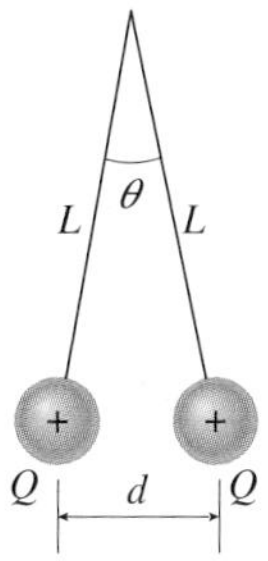

23. Using the three point charges of Example 16.3, find the magnitude of the force on q_2 due to the other two charges, q_1 and q_3. [*Hint:* After finding the force on q_2 due to q_1, separate that force into x- and y-components.]

24. An equilateral triangle has a point charge $+q$ at each of the three vertices (A, B, C). Another point charge Q is placed at D, the midpoint of the side BC. Solve for Q if the total electric force on the charge at A due to the charges at B, C, and D is zero.

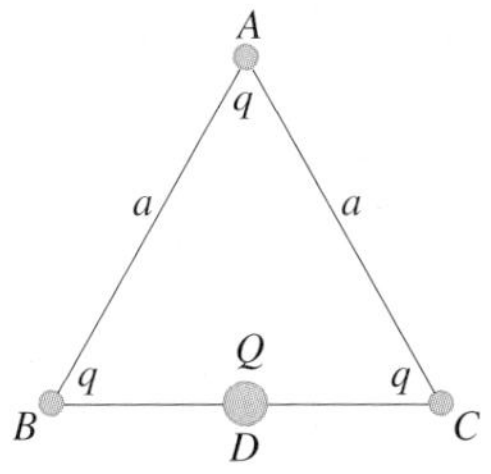

16.4 The Electric Field

25. A small sphere with a charge of -0.60 μC is placed in a uniform electric field of magnitude 1.2×10^6 N/C pointing to the west. What is the magnitude and direction of the force on the sphere due to the electric field?

26. The electric field across a cellular membrane is 1.0×10^7 N/C directed into the cell. (a) If a pore opens, which way do sodium ions (Na^+) flow—into the cell or out of the cell? (b) What is the magnitude of the electric force on the sodium ion? The charge on the sodium ion is $+e$.

27. What are the magnitude and direction of the acceleration of a proton at a point where the electric field has magnitude 33 kN/C and is directed straight up?

28. What are the magnitude and direction of the acceleration of an electron at a point where the electric field has magnitude 6100 N/C and is directed due north?

29. What are the magnitude and direction of the electric field midway between two point charges, -15 μC and $+12$ μC, that are 8.0 cm apart?

30. An electron traveling horizontally from west to east enters a region where a uniform electric field is directed upward. What is the direction of the electric force exerted on the electron once it has entered the field?

31. A negative point charge $-Q$ is situated near a large metal plate that has a total charge of $+Q$. Sketch the electric field lines.

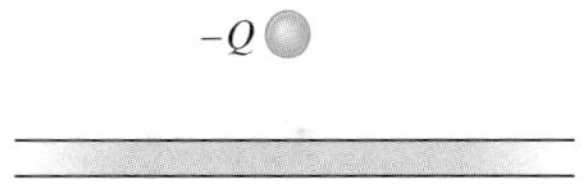

Problems 32–36. Positive point charges q and $2q$ are located at $x = 0$ and $x = 3d$, respectively.

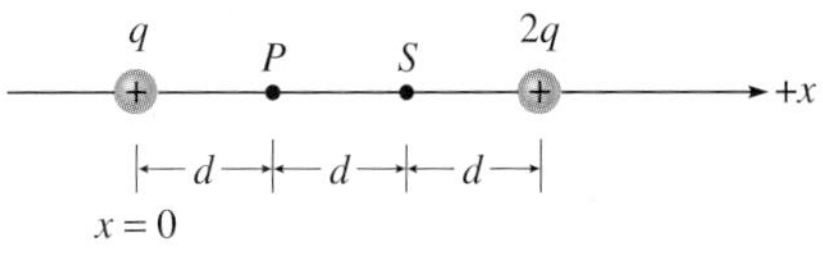

Problems 32–36

32. What is the electric field at $x = d$ (point P)?

33. What is the electric field at $x = 2d$ (point S)?

34. Are there any points *not* on the x-axis where $\vec{\mathbf{E}} = 0$? Explain.

35. On the x-axis, in which of the three regions $x < 0$, $0 < x < 3d$, and $x > 3d$ is there a point where $\vec{\mathbf{E}} = 0$? Explain.

36. Find the x-coordinates of the point(s) on the x-axis where $\vec{\mathbf{E}} = 0$.

37. Sketch the electric field lines in the plane of the page due to the charges shown in the diagram.

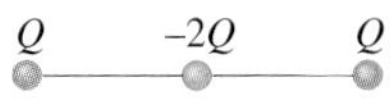

38. Sketch the electric field lines near two isolated and equal (a) positive point charges and (b) negative point charges. Include arrowheads to show the field directions.

Problems 39–42. Two tiny objects with equal charges of 7.00 μC are placed at the two lower corners of a square with sides of 0.300 m, as shown.

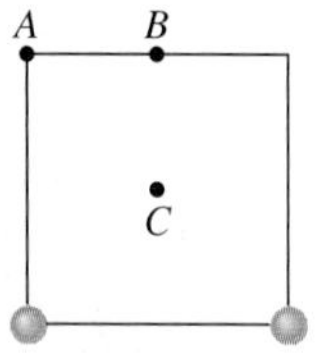

Problems 39–42

39. Find the electric field at point B, midway between the upper left and right corners.
40. Find the electric field at point C, the center of the square.
41. Find the electric field at point A, the upper left corner.
42. Where would you place a third small object with the same charge so that the electric field is zero at the corner of the square labeled A?
43. Three point charges are placed on the x-axis. A charge of 3.00 μC is at the origin. A charge of −5.00 μC is at 20.0 cm, and a charge of 8.00 μC is at 35.0 cm. What is the force on the charge at the origin?
44. Two equal charges (Q = +1.00 nC) are situated at the diagonal corners A and B of a square of side 1.0 m. What is the magnitude of the electric field at point D?

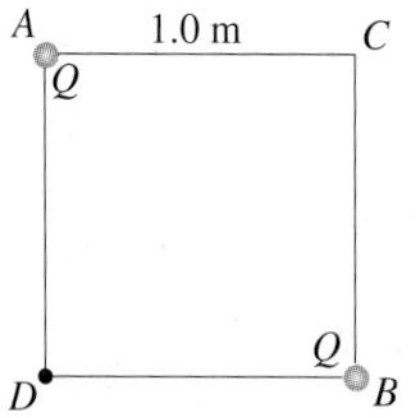

45. Suppose a charge q is placed at point $x = 0$, $y = 0$. A second charge q is placed at point $x = 8.0$ m, $y = 0$. What charge must be placed at the point $x = 4.0$ m, $y = 0$ in order that the field at the point $x = 4.0$ m, $y = 3.0$ m be zero?
46. Two point charges, $q_1 = +20.0$ nC and $q_2 = +10.0$ nC, are located on the x-axis at $x = 0$ and $x = 1.00$ m, respectively. Where on the x-axis is the electric field equal to zero?
47. Two electric charges, $q_1 = +20.0$ nC and $q_2 = +10.0$ nC, are located on the x-axis at $x = 0$ m and $x = 1.00$ m, respectively. What is the magnitude of the electric field at the point $x = 0.50$ m, $y = 0.50$ m?

16.5 Motion of a Point Charge in a Uniform Electric Field

48. An electron is placed in a uniform electric field of strength 232 N/C. If the electron is at rest at the origin of a coordinate system at $t = 0$ and the electric field is in the positive x-direction, what are the x- and y-coordinates of the electron at $t = 2.30$ ns?
49. An electron is projected horizontally into the space between two oppositely charged metal plates. The electric field between the plates is 500.0 N/C, directed up. (a) While in the field, what is the force on the electron? (b) If the vertical deflection of the electron as it leaves the plates is 3.00 mm, how much has its kinetic energy increased due to the electric field?
50. A horizontal beam of electrons initially moving at 4.0×10^7 m/s is deflected vertically by the vertical electric field between oppositely charged parallel plates. The magnitude of the field is 2.00×10^4 N/C. (a) What is the direction of the field between the plates? (b) What is the charge per unit area on the plates? (c) What is the vertical deflection d of the electrons as they leave the plates?

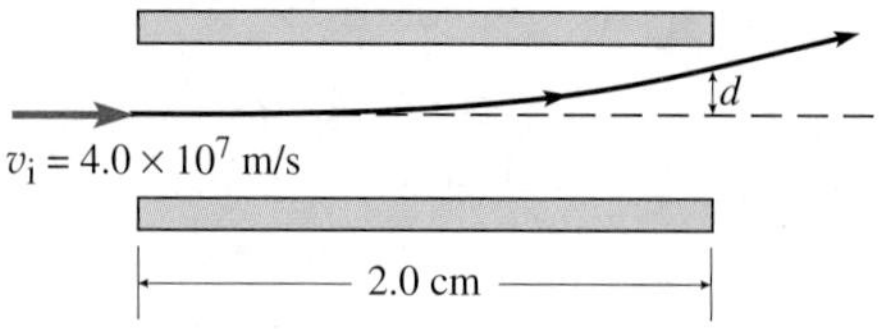

51. A particle with mass 2.30 g and charge +10.0 μC enters through a small hole in a metal plate with a speed of 8.50 m/s at an angle of 55.0°. The uniform $\vec{\mathbf{E}}$ field in the region above the plate has magnitude 6.50×10^3 N/C and is directed downward. The region above the metal plate is essentially a vacuum, so there is no air resistance. (a) Can you neglect the force of gravity when solving for the horizontal distance traveled by the particle? Why or why not? (b) How far will the particle travel, Δx, before it hits the metal plate?

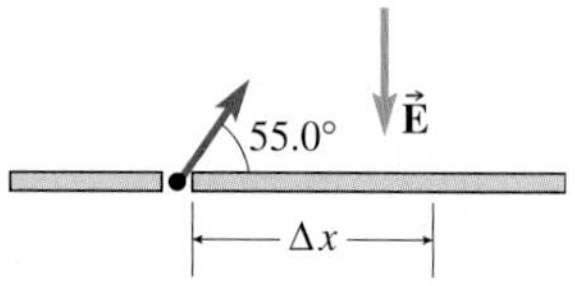

52. Consider the same situation as in Problem 51, but with a proton entering through the small hole at the same angle with a speed of $v = 8.50 \times 10^5$ m/s. (a) Can you ignore the force of gravity when solving this problem for the horizontal distance traveled by the proton? Why or why not? (b) How far will the proton travel, Δx, before it hits the metal plate?
53. Some forms of cancer can be treated using proton therapy in which proton beams are accelerated to high energies, then directed to collide into a tumor, killing the malignant cells. Suppose a proton accelerator is 4.0 m long and must accelerate protons from rest to a speed of 1.0×10^7 m/s. Ignore any relativistic effects (Chapter 26) and determine the magnitude of the average electric field that could accelerate these protons.
54. After the electrons in Example 16.9 pass through the anode, they are moving at a speed of 8.4×10^6 m/s. They next pass between a pair of parallel plates [(A) in Fig. 16.35]. The plates each have an area of 2.50 cm by 2.50 cm and they are separated by a distance of 0.50 cm. The uniform electric field between them is 1.0×10^3 N/C and the plates are charged as shown. (a) In what direction are the electrons deflected? (b) By how much are the electrons deflected after passing through these plates?
55. After the electrons pass through the parallel plates in Problem 54, they pass between another set of parallel plates [(B) in Fig. 16.35]. These plates also have an area of 2.50 cm by 2.50 cm and are separated by a distance of 0.50 cm. (a) In what direction must the field be oriented so that the electrons are deflected vertically upward? (b) If we neglect the gravitational force, how

strong must the field be between these plates in order for the electrons to be deflected by 2.0 mm? (c) How much less will the electrons be deflected if we *do* include the gravitational force?

16.6 Conductors in Electrostatic Equilibrium

56. A conducting sphere that carries a total charge of $-6\ \mu C$ is placed at the center of a conducting spherical shell that carries a total charge of $+1\ \mu C$. The conductors are in electrostatic equilibrium. Determine the charge on the *outer surface* of the shell. [*Hint:* Sketch a field line diagram.]

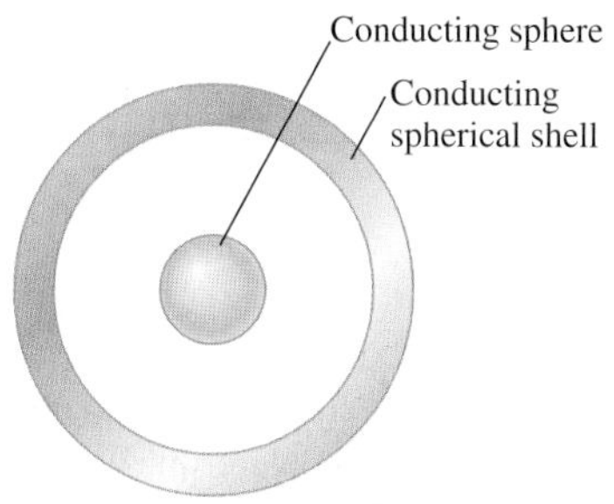

Problems 56 and 57

57. A conducting sphere that carries a total charge of $+6\ \mu C$ is placed at the center of a conducting spherical shell that also carries a total charge of $+6\ \mu C$. The conductors are in electrostatic equilibrium. (a) Determine the charge on the inner surface of the shell. (b) Determine the total charge on the outer surface of the shell.

58. A hollow conducting sphere of radius R carries a negative charge $-q$. (a) Write expressions for the electric field $\vec{\mathbf{E}}$ inside $(r < R)$ and outside $(r > R)$ the sphere. Also indicate the direction of the field. (b) Sketch a graph of the field strength as a function of r. [*Hint:* See Conceptual Example 16.8.]

59. A conducting sphere is placed within a conducting spherical shell. The conductors are in electrostatic equilibrium. The inner sphere has a radius of 1.50 cm, the inner radius of the spherical shell is 2.25 cm, and the outer radius of the shell is 2.75 cm. If the inner sphere has a charge of 230 nC, and the spherical shell has zero net charge, (a) what is the magnitude of the electric field at a point 1.75 cm from the center? (b) What is the electric field at a point 2.50 cm from the center? (c) What is the electric field at a point 3.00 cm from the center? [*Hint:* What must be true about the electric field inside a conductor in electrostatic equilibrium?]

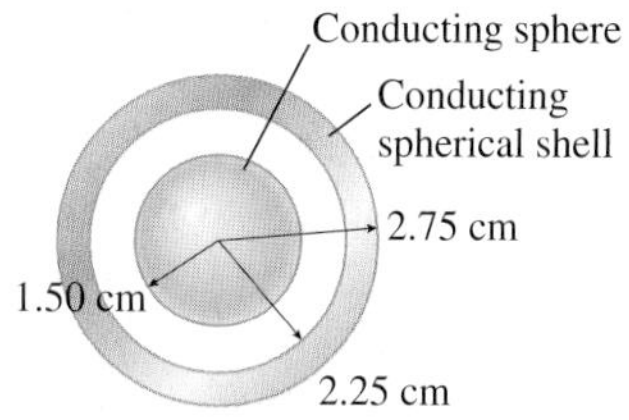

60. A conductor in electrostatic equilibrium contains a cavity in which there are two point charges: $q_1 = +5\ \mu C$ and $q_2 = -12\ \mu C$. The conductor itself carries a net charge $-4\ \mu C$. How much charge is on (a) the inner surface of the conductor? (b) the outer surface of the conductor?

61. In fair weather, over flat ground, there is a downward electric field of about 150 N/C. (a) Assume that the Earth is a conducting sphere with charge on its surface. If the electric field just outside is 150 N/C pointing radially inward, calculate the total charge on the Earth and the charge per unit area. (b) At an altitude of 250 m above Earth's surface, the field is only 120 N/C. Calculate the charge density (charge per unit volume) of the air (assumed constant). [*Hint:* See Conceptual Example 16.8.]

16.7 Gauss's Law for Electric Fields

62. (a) Find the electric flux through each side of a cube of edge length a in a uniform electric field of magnitude E. The field direction is perpendicular to two of the faces. (b) What is the total flux through the cube?

63. In a uniform electric field of magnitude E, the field lines cross through a rectangle of area A at an angle of $60.0°$ with respect to the plane of the rectangle. What is the flux through the rectangle?

64. An object with a charge of $0.890\ \mu C$ is placed at the center of a cube. What is the electric flux through one surface of the cube?

65. In this problem, you can show from Coulomb's law that the constant of proportionality in Gauss's law must be $1/\epsilon_0$. Imagine a sphere with its center at a point charge q. (a) Write an expression for the electric flux in terms of the field strength E and the radius r of the sphere. [*Hint:* The field strength E is the same everywhere on the sphere and the field lines cross the sphere perpendicular to its surface.] (b) Use Gauss's law in the form $\Phi_E = cq$ (where c is the constant of proportionality) and the electric field strength given by Coulomb's law to show that $c = 1/\epsilon_0$.

66. (a) Use Gauss's law to prove that the electric field *outside* any spherically symmetric charge distribution is the same as if all of the charge were concentrated into a point charge. (b) Now use Gauss's law to prove that the electric field *inside* a spherically symmetric charge distribution is zero if none of the charge is at a distance from the center less than that of the point where we determine the field.

67. Using the results of Problem 66, we can find the electric field at any radius for any spherically symmetrical charge distribution. A solid sphere of charge of radius R has a total charge of q uniformly spread throughout the sphere. (a) Find the magnitude of the electric field for $r \geq R$. (b) Find the magnitude of the electric field for $r \leq R$. (c) Sketch a graph of $E(r)$ for $0 \leq r \leq 3R$.

68. An electron is suspended at a distance of 1.20 cm above a uniform line of charge. What is the linear charge density of the line of charge? Ignore end effects.

69. A thin, flat sheet of charge has a uniform surface charge density σ ($\sigma/2$ on each side). (a) Sketch the field lines due to the sheet. (b) Sketch the field lines for an infinitely large sheet with the same charge density. (c) For the infinite sheet, how does the field strength depend on the distance from the sheet? [*Hint:* Refer to your field line sketch.] (d) For points close to the finite sheet and far from its edges, can the sheet be approximated by an infinitely large sheet? [*Hint:* Again, refer to the field line sketches.] (e) Use Gauss's law to show that the magnitude of the electric field near a sheet of uniform charge density σ is $E = \sigma/(2\epsilon_0)$.

70. A flat *conducting* sheet of area A has a charge q *on each surface.* (a) What is the electric field inside the sheet? (b) Use Gauss's law to show that the electric field just outside the sheet is $E = q/(\epsilon_0 A) = \sigma/\epsilon_0$. (c) Does this contradict the result of Problem 69? Compare the field line diagrams for the two situations.

71. A *parallel-plate capacitor* consists of two flat metal plates of area A separated by a small distance d. The plates are given equal and opposite net charges $\pm q$. (a) Sketch the field lines and use your sketch to explain why almost all of the charge is on the inner surfaces of the plates. (b) Use Gauss's law to show that the electric field between the plates and away from the edges is $E = q/(\epsilon_0 A) = \sigma/\epsilon_0$. (c) Does this agree with or contradict the result of Problem 70? Explain. (d) Use the principle of superposition and the result of Problem 69 to arrive at this same answer. [*Hint:* The inner surfaces of the two plates are thin, flat sheets of charge.]

72. A coaxial cable consists of a wire of radius a surrounded by a thin metal cylindrical shell of radius b. The wire has a uniform linear charge density $\lambda > 0$ and the outer shell has a uniform linear charge density $-\lambda$. (a) Sketch the field lines for this cable. (b) Find expressions for the magnitude of the electric field in the regions $r \le a$, $a < r < b$, and $b \le r$.

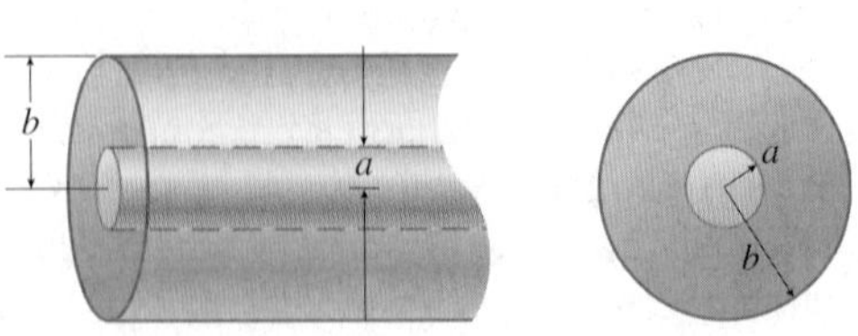

73. Use Gauss's law to derive an expression for the electric field outside the thin spherical shell of Conceptual Example 16.8.

Comprehensive Problems

74. Consider two protons (charge $+e$), separated by a distance of 2.0×10^{-15} m (as in a typical atomic nucleus). The electric force between these protons is equal in magnitude to the gravitational force on an object of what mass near Earth's surface?

75. In lab tests it was found that rats can detect electric fields of about 5.0 kN/C or more. If a point charge of 1.0 μC is sitting in a maze, how close must the rat come to the charge in order to detect it?

76. A raindrop inside a thundercloud has charge $-8e$. What is the electric force on the raindrop if the electric field at its location (due to other charges in the cloud) has magnitude 2.0×10^6 N/C and is directed upward?

77. An electron beam in an oscilloscope is deflected by the electric field produced by oppositely charged metal plates. If the electric field between the plates is 2.00×10^5 N/C directed downward, what is the force on each electron when it passes between the plates?

78. A point charge $q_1 = +5.0$ μC is fixed in place at $x = 0$ and a point charge $q_2 = -3.0$ μC is fixed at $x = -20.0$ cm. Where can we place a point charge $q_3 = -8.0$ μC so that the net electric force on q_1 due to q_2 and q_3 is zero?

79. The Bohr model of the hydrogen atom proposes that the electron orbits around the proton in a circle of radius 5.3×10^{-11} m. The electric force is responsible for the radial acceleration of the electron. What is the speed of the electron in this model?

80. In a thunderstorm, charge is separated through a complicated mechanism that is ultimately powered by the Sun. A simplified model of the charge in a thundercloud represents the positive charge accumulated at the top and the negative charge at the bottom as a pair of point charges. (a) What is the magnitude and direction of the electric field produced by the two point charges at point P, which is just above Earth's surface? (b) Thinking of Earth as a conductor, what sign of charge would accumulate on the surface near point P? (This accumulated charge increases the magnitude of the electric field near point P.)

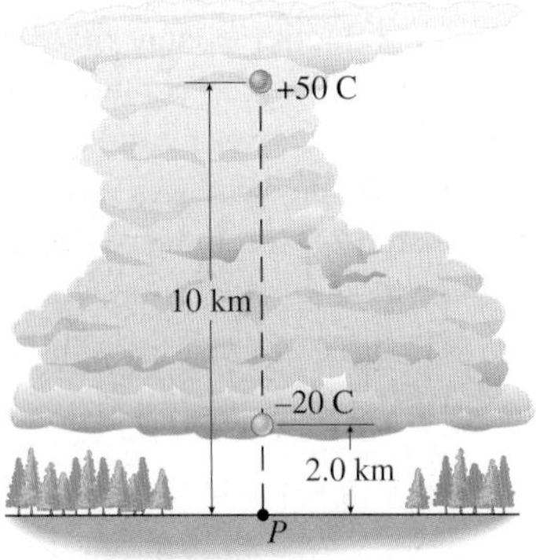

81. Two point charges are located on the x-axis: a charge of +6.0 nC at $x = 0$ and an unknown charge q at $x = 0.50$ m. No other charges are nearby. If the electric field is zero at the point $x = 1.0$ m, what is q?

82. Three equal charges are placed on three corners of a square. If the force that Q_a exerts on Q_b has magnitude F_{ba} and the force that Q_a exerts on Q_c has magnitude F_{ca}, what is the ratio of F_{ca} to F_{ba}?

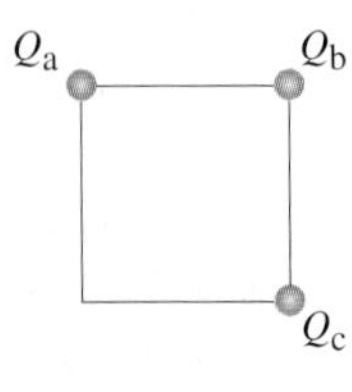

83. Two otherwise identical conducting spheres carry charges of +5.0 μC and −1.0 μC. They are initially a large distance L apart. The spheres are brought together, touched together, and then returned to their original

separation L. What is the ratio of the magnitude of the force on either sphere after they are touched to that before they were touched?

84. Two metal spheres of radius 5.0 cm carry net charges of +1.0 μC and +0.2 μC. (a) What (approximately) is the magnitude of the electrical repulsion on either sphere when their centers are 1.00 m apart? (b) Why cannot Coulomb's law be used to find the force of repulsion when their centers are 12 cm apart? (c) Would the actual force be larger or smaller than the result of using Coulomb's law with $r = 12$ cm? Explain.

85. A charge of 63.0 nC is located at a distance of 3.40 cm from a charge of −47.0 nC. What are the x- and y-components of the electric field at a point P that is directly above the 63.0-nC charge at a distance of 1.40 cm? Point P and the two charges are on the vertices of a right triangle.

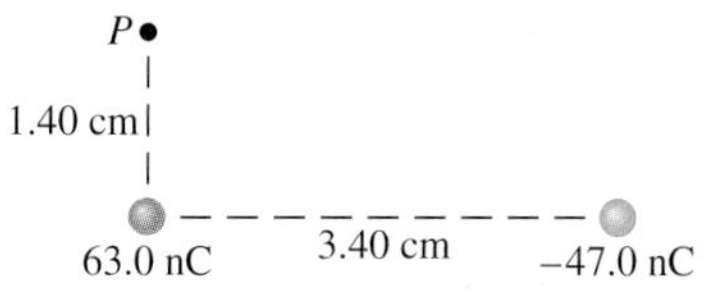

86. Point charges are arranged on the vertices of a square with sides of 2.50 cm. Starting at the upper left corner and going clockwise, we have charge A with a charge of 0.200 μC, B with a charge of −0.150 μC, C with a charge of 0.300 μC, and D with a mass of 2.00 g, but with an unknown charge. Charges A, B, and C are fixed in place, and D is free to move. Particle D's instantaneous acceleration at point D is 248 m/s^2 in a direction 30.8° below the negative x-axis. What is the charge on D?

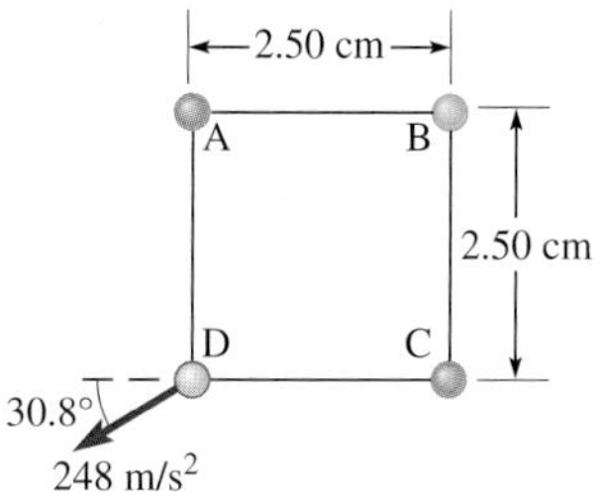

87. In a cathode ray tube, electrons initially at rest are accelerated by a uniform electric field of magnitude 4.0×10^5 N/C during the first 5.0 cm of the tube's length; then they move at essentially constant velocity another 45 cm before hitting the screen. (a) Find the speed of the electrons when they hit the screen. (b) How long does it take them to travel the length of the tube?

88. In the diagram, regions A and C extend far to the left and right, respectively. The electric field due to the two point charges is zero at some point in which region or regions? Explain.

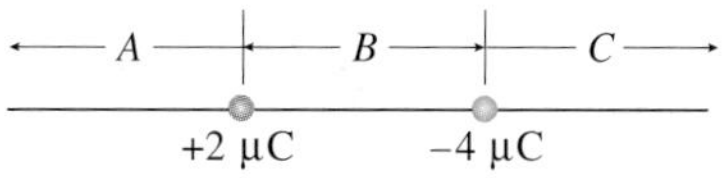

89. A thin wire with positive charge evenly spread along its length is shaped into a semicircle. What is the direction of the electric field at the center of curvature of the semicircle? Explain.

90. A very small charged block with a mass of 2.35 g is placed on an insulated, frictionless plane inclined at an angle of 17.0° with respect to the horizontal. The block does not slide down the plane because of a 465-N/C uniform electric field that points parallel to the surface downward along the plane. What is the sign and magnitude of the charge on the block?

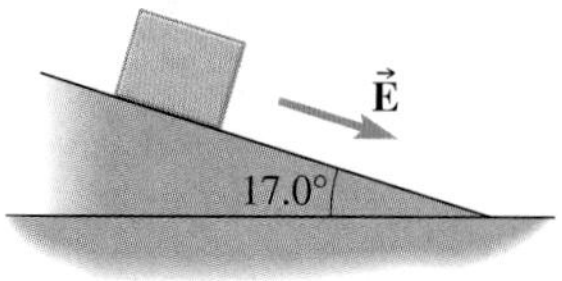

91. (a) What would the net charges on the Sun and Earth have to be if the electric force instead of the gravitational force were responsible for keeping Earth in its orbit? There are many possible answers, so restrict yourself to the case where the magnitude of the charges is proportional to the masses. (b) If the magnitude of the charges of the proton and electron were not exactly equal, astronomical bodies would have net charges that are approximately proportional to their masses. Could this possibly be an explanation for the Earth's orbit?

92. What is the electric force on the chloride ion in the lower right-hand corner in the diagram? Since the ions are in water, the "effective charge" on the chloride ions (Cl^-) is -2×10^{-21} C and that of the sodium ions (Na^+) is $+2 \times 10^{-21}$ C. The effective charge is a way to account for the partial shielding due to nearby water molecules. Assume that all four ions are coplanar.

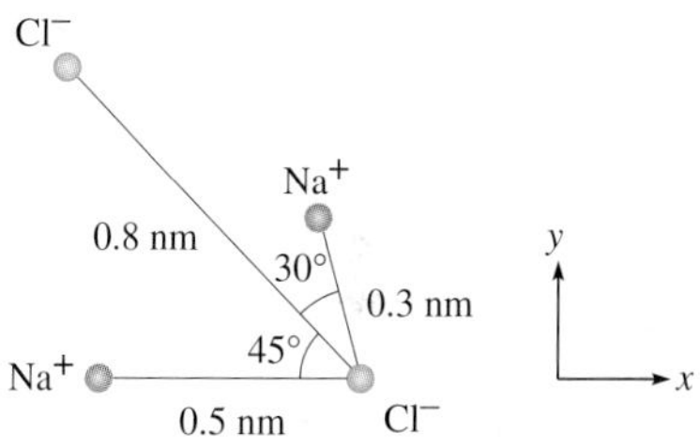

93. A dipole consists of two equal and opposite point charges ($\pm q$) separated by a distance d. (a) Write an expression for the magnitude of the electric field at a point $(x, 0)$ a large distance ($x \gg d$) from the midpoint of the charges on a line perpendicular to the dipole axis. [*Hint:* Use small angle approximations.] (b) Give the direction of the field for $x > 0$ and for $x < 0$.

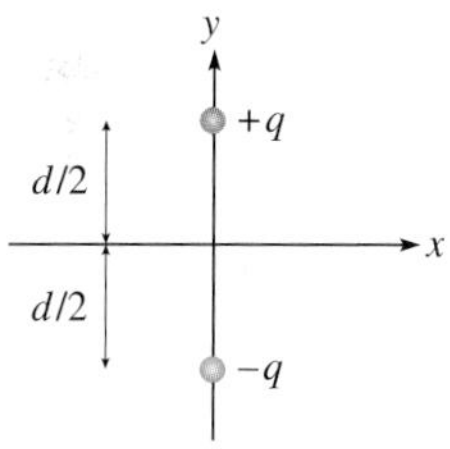

Problems 93 and 94

94. A dipole consists of two equal and opposite point charges ($\pm q$) separated by a distance d. (a) Write an expression for the electric field at a point $(0, y)$ on the dipole axis. Specify the direction of the field in all four regions: $y > \frac{1}{2}d$, $0 < y < \frac{1}{2}d$, $-\frac{1}{2}d < y < 0$, and $y < -\frac{1}{2}d$. (b) At distant points ($|y| \gg d$), write a simpler, approximate expression for the field. To what power of y is the field proportional? Does this conflict with Coulomb's law? [*Hint:* Use the binomial approximation $(1 \pm x)^n \approx 1 \pm nx$ for $x \ll 1$.]

95. A dipole consists of two opposite charges (q and $-q$) separated by a fixed distance d. The dipole is placed in an electric field in the +x-direction of magnitude E. The dipole axis makes an angle θ with the electric field as shown in the diagram. (a) Calculate the net electric force acting on the dipole. (b) Calculate the net torque

acting on the dipole due to the electric forces as a function of θ. Let counterclockwise torque be positive. (c) Evaluate the net torque for $\theta = 0°$, $\theta = 36.9°$, and $\theta = 90.0°$. Let $q = \pm 3.0\ \mu C$, $d = 7.0$ cm, and $E = 2.0 \times 10^4$ N/C.

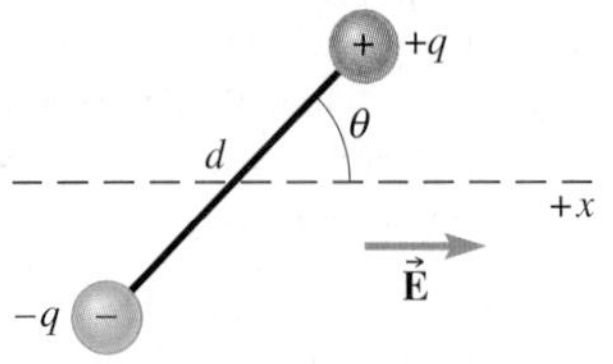

after spending 5.00×10^{-9} s between the plates, is 5.44×10^{-6} m, or 5.44×10^{-4} cm. The proton is barely deflected at all before leaving the region between the plates.

16.11

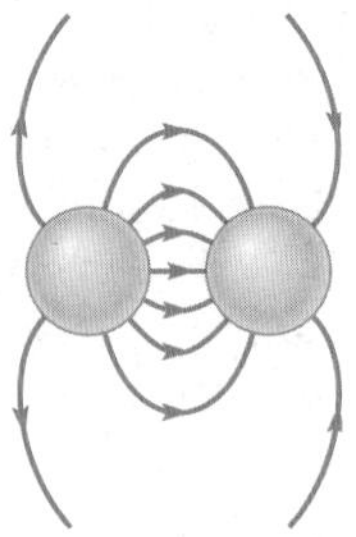

Answers to Practice Problems

16.1 7.5×10^{10} electrons

16.2 As the positively charged rod is moved away, the free electrons of the electroscope spread out more evenly. Since there is less net positive charge on the leaves, they do not hang as far apart.

16.3 4.6 mN, 71° CCW from the $+x$-axis

16.4 6.2×10^{-4} N

16.5 $\theta = 49.1°$

16.6 220 N/C to the right

16.7 2.3×10^5 N/C, 42° below the $-x$-axis

16.8

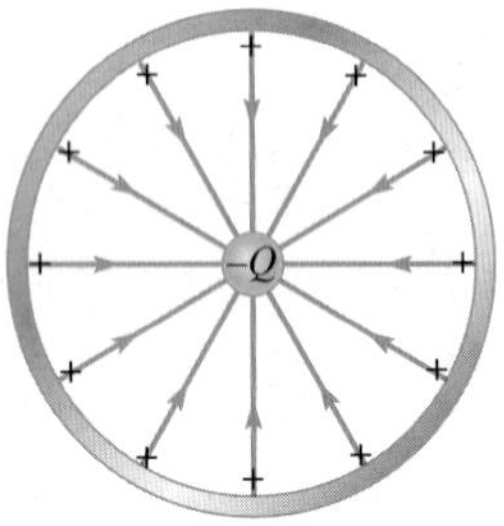

(a) Inside the shell, field lines run from the positive charge spread about the surface to the negative charge located at the center of the shell.

(b) Outside the shell, we can imagine the charge $+Q$ all concentrated at the center of the sphere where it cancels the $-Q$ of the point charge. Therefore, $E = 0$ outside. Inside, the shell produces no electric field (as we found in the Example), so the field is just that due to the point charge $-Q$.

16.9 2.3×10^5 m/s to the right

16.10 The proton is deflected downward, but it has a much smaller acceleration because it has a much larger mass than the electron ($m_p = 1.673 \times 10^{-27}$ kg). The proton's acceleration vertically downward is 4.36×10^{11} m/s^2. The y-displacement,

16.12 -3.8×10^4 N·m^2/C

16.13 On the ends, $\vec{\mathbf{E}}$ is *parallel* to the surface, so the component of $\vec{\mathbf{E}}$ perpendicular to the ends is zero and the flux through the ends is zero. No field lines *pass through* the ends of the cylinder.

Answers to Checkpoints

16.1 The glass and silk are left with opposite charges of equal magnitude because charge is conserved. Electrons have negative charge, so the silk's charge is negative and the rod's is positive. 4.0×10^9 electrons have a total charge of $4.0 \times 10^9 \times (-1.6 \times 10^{-19}\ C) = -0.64$ nC (nanocoulombs). Therefore, $Q_{silk} = -0.64$ nC and $Q_{rod} = +0.64$ nC.

16.3 (a) Gravity and the electric force are long-range forces. The magnitude of the force exerted on one point particle due to another has the same distance dependence in both cases ($F \propto 1/r^2$). Gravity and the electric force are proportional to the *product* of the masses or charges, respectively. (b) Gravity is always an attractive force, but the electric force can be attractive or repulsive. (In other words, mass cannot be negative, but electric charge can be positive or negative.)

16.4 (a) The electric field vector at any point is tangent to a field line through that point. At A, the field is downward ($-y$-direction). (b) The field is weaker where the field lines are spaced farther apart. The field is weaker at P.

16.5 The electron's charge is negative, so the electric force on it is in the direction opposite to the electric field ($-x$). The electron moves with constant acceleration in the $-x$-direction. While moving in the $+x$-direction, it slows down; then it turns around and moves in the $-x$-direction with increasing speed.

CHAPTER

17

Electric Potential

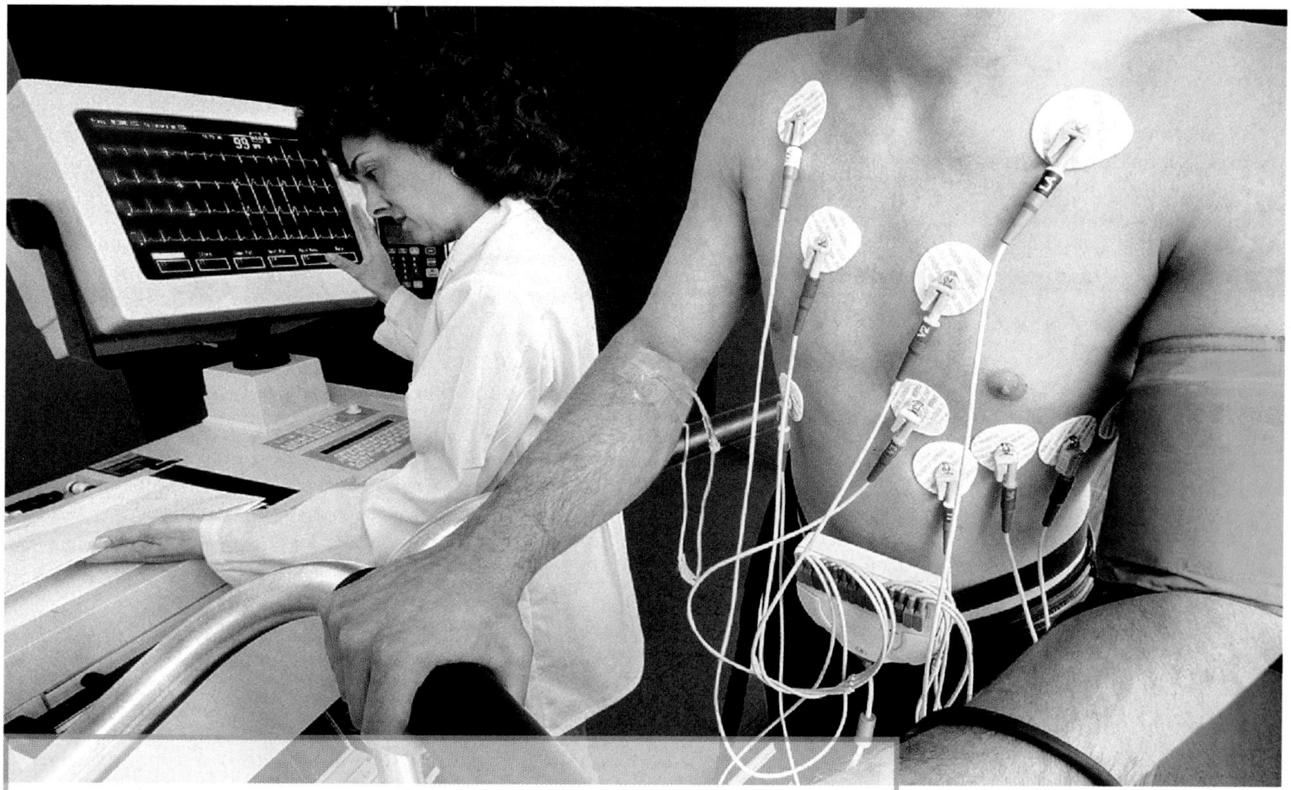

A tool widely used in medicine to diagnose the condition of the heart is the electrocardiogram (ECG). The ECG data is displayed on a graph that shows a pattern repeated with each beat of the heart. What physical quantity is measured in an ECG? (See p. 612 for the answer.)

Concepts & Skills to Review

- gravitational forces (Section 4.5)
- gravitational potential energy (Sections 6.4 and 6.5)
- Coulomb's law (Section 16.3)
- electric field inside a conductor (Section 16.6)
- polarization (Section 16.1)

17.1 ELECTRIC POTENTIAL ENERGY

CONNECTION:

Potential energy is energy stored in a field. Now, instead of energy stored in a gravitational field, we study energy stored in an *electric* field.

In Chapter 6, we learned about gravitational potential energy—energy stored in a gravitational field. **Electric potential energy** is the energy stored in an *electric* field (Fig. 17.1). For both gravitational and electric potential energy, the *change* in potential energy when objects move around is equal in magnitude but opposite in sign to the work done by the field:

$$\Delta U = -W_{\text{field}} \qquad (6\text{-}8)$$

The minus sign indicates that, when the field does positive work W_{field} on an object, the *object's* energy is increased by an amount W_{field}. That amount of energy is *taken from* stored potential energy. The field dips into its "potential energy bank account" and gives the energy to the object, so the potential energy decreases when the force does positive work.

CONNECTION:

Some of the many similarities between gravitational and electric potential energy include:

- In both cases, the potential energy depends on only the *positions* of various objects, not on the *path* they took to get to those positions.
- Only *changes* in potential energy are physically significant, so we are free to assign the potential energy to be zero at any *one* convenient point. The potential energy in a given situation depends on the choice of the point where $U = 0$, but *changes* in potential energy are *not* affected by this choice.
- For two point particles, we usually choose $U = 0$ when the particles are infinitely far apart.
- Both the gravitational and electrical forces exerted by one point particle on another are inversely proportional to the square of the distance between them ($F \propto 1/r^2$). As a result, the gravitational and electric potential energies have the *same distance dependence* ($U \propto 1/r$, with $U = 0$ at $r = \infty$).
- The gravitational force and the gravitational potential energy for a pair of point particles are proportional to the product of the masses of the particles:

$$F = \frac{Gm_1m_2}{r^2} \qquad (4\text{-}7)$$

$$U_g = -\frac{Gm_1m_2}{r} \quad (U_g = 0 \text{ at } r = \infty) \qquad (6\text{-}14)$$

The electric force and the electric potential energy for a pair of point particles are proportional to the product of the *charges* of the particles:

$$F = \frac{k|q_1||q_2|}{r^2} \qquad (16\text{-}2)$$

Electric potential energy for a pair of point charges a distance r apart.

$$U_E = \frac{kq_1q_2}{r} \quad (U_E = 0 \text{ at } r = \infty) \qquad (17\text{-}1)$$

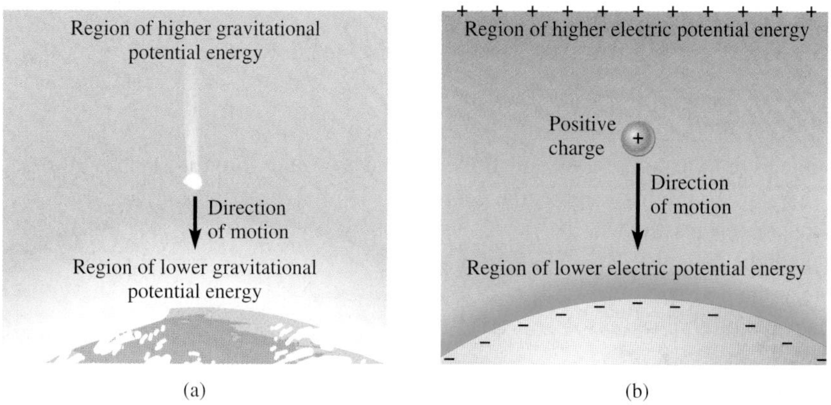

Figure 17.1 (a) An object moving through a gravitational field; the gravitational potential energy decreases when the object moves in the direction of the gravitational force. (b) A charged particle moving through an *electric* field; the *electric* potential energy decreases when the particle moves in the direction of the *electric* force.

Positive and Negative Potential Energy The negative sign in Eq. (6-14) indicates that gravity is always an attractive force: if two particles move closer together (r decreases), gravity does positive work and ΔU is negative—some gravitational potential energy is converted to other forms of energy. If the two particles move farther apart, the gravitational potential energy increases.

Why is there no negative sign in Eq. (17-1)? If the two charges have opposite signs, the force is an attractive one. The potential energy should be negative, as it is for the attractive force of gravity. With opposite signs, the product q_1q_2 is negative and the potential energy has the correct sign (Fig. 17.2). If the two charges instead have the same sign—both positive or both negative—the product q_1q_2 is positive. The electric force between two like charges is *repulsive*; the potential energy *increases* as they move closer together. Thus, Eq. (17-1) automatically gives the correct sign in every case.

If potential energy is negative, it *increases* when its absolute value gets smaller (just as −6 is *greater* than −8).

Coulomb's law is written in terms of the *magnitudes* of the charges ($|q_1||q_2|$) since it gives the *magnitude* of a vector quantity—the force. In the potential energy expression [Eq. (17-1)], we do *not* write the absolute value bars. The signs of the two charges determine the sign of the potential energy, a scalar quantity that can be positive, negative, or zero.

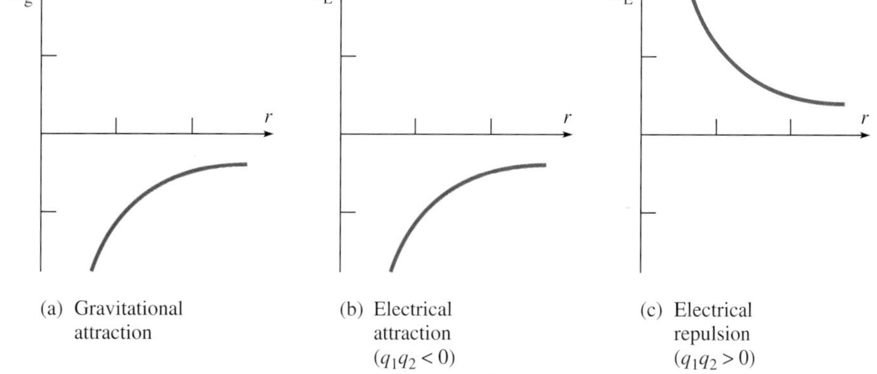

Figure 17.2 Potential energies for pairs of point particles as a function of separation distance r. In each case, we choose $U = 0$ at $r = \infty$. For an attractive force, (a) and (b), the potential energy is negative. If two particles start far apart where $U = 0$, they "fall" spontaneously toward one another as the potential energy *decreases*. For a repulsive force (c), the potential energy is positive. If two particles start far apart, they have to be pushed together by an external agent that does work to increase the potential energy.

Example 17.1

Electric Potential Energy in a Thundercloud

In a thunderstorm, charge is separated through a complicated mechanism that is ultimately powered by the Sun. A simplified model of the charge in a thundercloud represents the positive charge accumulated at the top and the negative charge at the bottom as a pair of point charges (Fig. 17.3). (a) What is the electric potential energy of the pair of point charges, assuming that $U = 0$ when the two charges are infinitely far apart?

(b) Explain the sign of the potential energy in light of the fact that *positive* work must be done by external forces in the thundercloud to *separate* the charges.

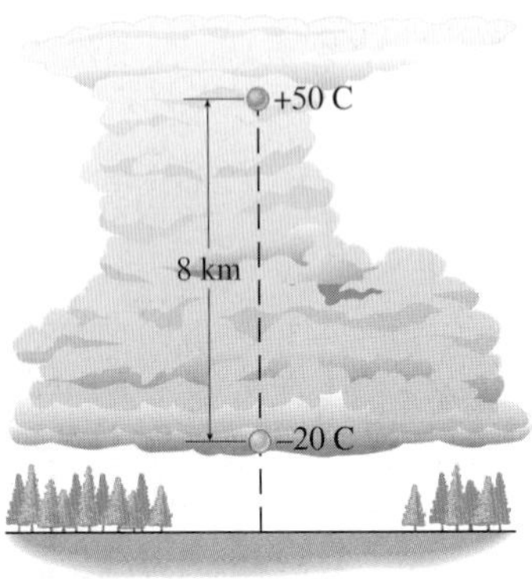

Figure 17.3
Charge separation in a thundercloud.

Strategy (a) The electric potential energy for a pair of point charges is given by Eq. (17-1), where $U = 0$ at infinite separation is assumed. The algebraic signs of the charges are included when finding the potential energy. (b) The work done by an external force to separate the charges is equal to the *change* in the electric potential energy as the charges are *moved apart* by forces acting within the thundercloud.

Solution and Discussion (a) The general expression for electric potential energy for two point charges is

$$U_E = \frac{kq_1q_2}{r}$$

We substitute the known values into the equation for electric potential energy.

$$U_E = 8.99 \times 10^9 \frac{\text{N}\cdot\text{m}^2}{\text{C}^2} \times \frac{(+50\text{ C}) \times (-20\text{ C})}{8000\text{ m}}$$

$$= -1 \times 10^9 \text{ J}$$

(b) Recall that we chose $U = 0$ at infinite separation. Negative potential energy therefore means that, *if the two point charges started infinitely far apart*, their electric potential energy would decrease as they are brought together—in the absence of other forces they would "fall" spontaneously toward one another. However, in the thundercloud, the unlike charges *start close together* and are moved *farther apart* by an external force; the external agent must do *positive* work to increase the potential energy and move the charges *apart*. Initially, when the charges are close together, the potential energy is *less than* -1×10^9 J; the *change* in potential energy as the charges are moved apart is *positive*.

Practice Problem 17.1 Two Point Charges with Like Signs

Two point charges, $Q = +6.0\ \mu\text{C}$ and $q = +5.0\ \mu\text{C}$, are separated by 15.0 m. (a) What is the electric potential energy? (b) Charge q is free to move—no other forces act on it—while Q is fixed in place. Both are initially at rest. Does q move toward or away from charge Q? (c) How does the motion of q affect the electric potential energy? Explain how energy is conserved.

Potential Energy due to Several Point Charges

To find the potential energy due to more than two point charges, we add the potential energies of each *pair* of charges. For three point charges, there are three pairs, so the potential energy is

Electric potential energy due to three point charges ($U_E = 0$ when all three are infinitely separated)

$$U_E = k\left(\frac{q_1q_2}{r_{12}} + \frac{q_1q_3}{r_{13}} + \frac{q_2q_3}{r_{23}}\right) \quad (17\text{-}2)$$

where, for instance, r_{12} is the distance between q_1 and q_2. The potential energy in Eq. (17-2) is the negative of the work done by the electric field as the three charges are put into their positions, *starting from infinite separation.* If there are more than three charges, the potential energy is a sum just like Eq. (17-2), which includes *one* term for each *pair* of charges. Be sure not to count the potential energy of the same pair twice. If the potential energy expression has a term $(q_1q_2)/r_{12}$, it must not also have a term $(q_2q_1)/r_{21}$.

CHECKPOINT 17.1

When finding the potential energy due to four point charges, how many *pairs* of charges are there? How many terms in the potential energy?

Example 17.2

Electric Potential Energy due to Three Point Charges

Find the electric potential energy for the array of charges shown in Fig. 17.4. Charge $q_1 = +4.0\ \mu C$ is located at (0.0, 0.0) m; charge $q_2 = +2.0\ \mu C$ is located at (3.0, 4.0) m; and charge $q_3 = -3.0\ \mu C$ is located at (3.0, 0.0) m.

Strategy With three charges, there are three pairs to include in the potential energy sum [Eq. (17-2)]. The charges are given; we need only find the distance between each pair. Subscripts are useful to identify the three distances; r_{12}, for example, means the distance between q_1 and q_2.

Solution From Fig. 17.4, $r_{13} = 3.0$ m and $r_{23} = 4.0$ m. The Pythagorean theorem enables us to find r_{12}:

$$r_{12} = \sqrt{3.0^2 + 4.0^2}\ \text{m} = \sqrt{25}\ \text{m} = 5.0\ \text{m}$$

The potential energy has one term for each pair:

$$U_E = k\left(\frac{q_1 q_2}{r_{12}} + \frac{q_1 q_3}{r_{13}} + \frac{q_2 q_3}{r_{23}}\right)$$

Substituting numerical values,

$$U_E = 8.99 \times 10^9\ \frac{\text{N·m}^2}{\text{C}^2} \times \left[\frac{(+4.0)(+2.0)}{5.0} + \frac{(+4.0)(-3.0)}{3.0} + \frac{(+2.0)(-3.0)}{4.0}\right] \times 10^{-12}\ \frac{\text{C}^2}{\text{m}}$$

$$= -0.035\ \text{J}$$

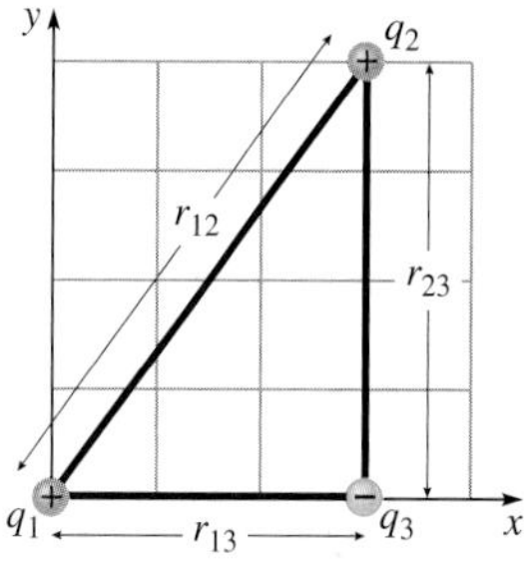

Figure 17.4 Three point charges.

Discussion To interpret the answer, assume that the three charges start far apart from each other. As the charges are brought together and put into place, the electric fields do a total work of +0.035 J. Once the charges are in place, an external agent would have to supply 0.035 J of energy to separate them again.

Conceptual Practice Problem 17.2 Three Positive Charges

What would the potential energy be if $q_3 = +3.0\ \mu C$ instead?

17.2 ELECTRIC POTENTIAL

Imagine that a collection of point charges is somehow fixed in place while another charge q can move. Moving q may involve changes in electric potential energy since the distances between it and the fixed charges may change. Just as the electric field is defined as the electric force per unit charge, the **electric potential** V is defined as the electric potential energy *per unit charge* (Fig. 17.5).

$$V = \frac{U_E}{q} \qquad (17\text{-}3)$$

Potential: electric potential energy per unit charge

In Eq. (17-3), U_E is the electric potential energy *as a function of the position of the moveable charge (q).* Then the electric potential V is also a function of the position of charge q.

The SI unit of electric potential is the joule per coulomb, which is named the volt (symbol V) after the Italian scientist Alessandro Volta (1745–1827). Volta invented the voltaic pile, an early form of battery. *Electric potential* is often shortened to *potential.* It is also

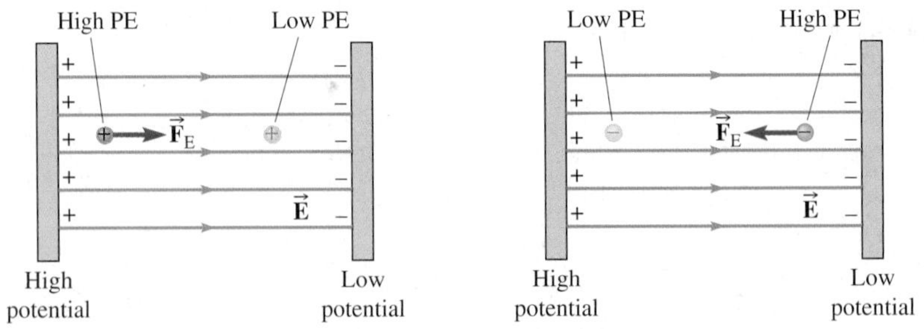

Figure 17.5 The electric *force* on a charge is always in the direction of lower electric potential energy. The electric *field* is always in the direction of lower *potential.*

informally called "voltage," especially in connection with electric circuits, just as weight is sometimes called "tonnage." Be careful to distinguish *electric potential* from *electric potential energy.* It is all too easy to confuse the two, but they are not interchangeable.

Volt: one joule per coulomb

$$1 \text{ V} = 1 \text{ J/C} \tag{17-4}$$

Since potential energy and charge are scalars, potential is also a scalar. The principle of superposition is easier to apply to potentials than to fields since fields must be added as vectors. Given the potential at various points, it is easy to calculate the potential energy change when a charge moves from one point to another. Potentials do not have direction in space; they are added just as any other scalar. Potentials can be either positive or negative and so must be added with their algebraic signs.

Since only changes in potential energy are significant, only changes in potential are significant. We are free to choose the potential arbitrarily at any one point. Equation (17-3) assumes that the potential is zero infinitely far away from the collection of fixed charges.

If the potential at a point due to a collection of fixed charges is V, then when a charge q is placed at that point, the electric potential energy is

$$U_E = qV \tag{17-5}$$

Potential Difference

When a point charge q moves from point A to point B, it moves through a *potential difference*

$$\Delta V = V_f - V_i = V_B - V_A \tag{17-6}$$

The potential difference is the change in electric potential energy per unit charge:

$$\Delta U_E = q\,\Delta V \tag{17-7}$$

Electric Field and Potential Difference The electric force on a charge is always directed toward regions of lower electric potential energy, just as the gravitational force on an object is directed toward regions of lower gravitational potential energy (that is, downward). For a positive charge, lower potential energy means lower potential (Fig. 17.5a), but for a negative charge, lower potential energy means *higher* potential (Fig. 17.5b). This shouldn't be surprising, since the force on a negative charge is opposite to the direction of $\vec{\mathbf{E}}$, while the force on a positive charge is in the direction of $\vec{\mathbf{E}}$. Since the electric field points toward lower potential energy for positive charges,

$\vec{\mathbf{E}}$ points in the direction of decreasing V.

In a region where the electric field is zero, the potential is constant.

CHECKPOINT 17.2

If the potential increases as you move from point P in the $+x$-direction, but the potential does not change as you move from P in the y- or z-directions, what is the direction of the electric field at P?

Example 17.3

A Battery-Powered Lantern

A battery-powered lantern is switched on for 5.0 min. During this time, electrons with total charge -8.0×10^2 C flow through the lamp; 9600 J of electric potential energy is converted to light and heat. Through what potential difference do the electrons move?

Strategy Equation (17-7) relates the change in electric potential energy to the potential difference. We could apply Eq. (17-7) to a single electron, but since all of the electrons move through the same potential difference, we can let q be the total charge of the electrons and ΔU_E be the total change in electric potential energy.

Solution The total charge moving through the lamp is $q = -800$ C. The change in electric potential energy is *negative* since it is converted into other forms of energy. Therefore,

$$\Delta V = \frac{\Delta U_E}{q} = \frac{-9600 \text{ J}}{-8.0 \times 10^2 \text{ C}} = +12 \text{ V}$$

Discussion The sign of the potential difference is positive: negative charges decrease the electric potential energy when they move through a potential increase.

Conceptual Practice Problem 17.3 An Electron Beam

A beam of electrons is deflected as it moves between oppositely charged parallel plates (Fig. 17.6). Which plate is at the higher potential?

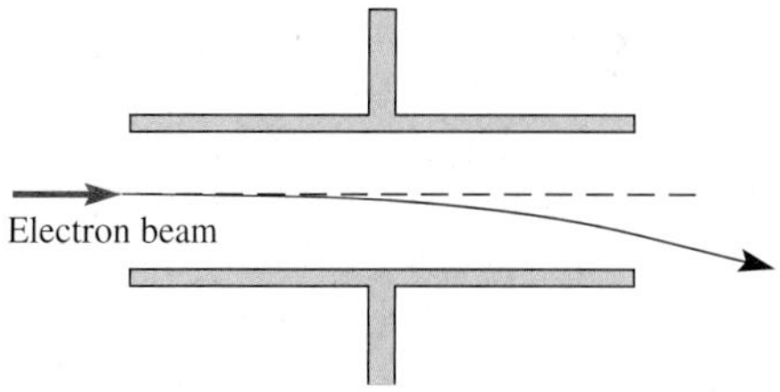

Figure 17.6
An electron beam deflected by a pair of oppositely charged plates.

Potential due to a Point Charge

If q is in the vicinity of one other point charge Q, the electric potential energy is

$$U = \frac{kQq}{r} \qquad (17\text{-}1)$$

when Q and q are separated by a distance r. Therefore, the electric potential at a distance r from a point charge Q is

$$V = \frac{kQ}{r} \quad (V = 0 \text{ at } r = \infty) \qquad (17\text{-}8)$$

Potential due to a point charge

Superposition of Potentials The potential at a point P due to N point charges is the sum of the potentials due to each charge:

$$V = \Sigma V_i = \Sigma \frac{kQ_i}{r_i} \quad \text{for } i = 1, 2, 3, \ldots, N \qquad (17\text{-}9)$$

where r_i is the distance from the ith point charge Q_i to point P.

Example 17.4

Potential due to Three Point Charges

Charge $Q_1 = +4.0\ \mu C$ is located at (0.0, 3.0) cm; charge $Q_2 = +2.0\ \mu C$ is located at (1.0, 0.0) cm; and charge $Q_3 = -3.0\ \mu C$ is located at (2.0, 2.0) cm (Fig. 17.7). (a) Find the electric potential at point A ($x = 0.0$, $y = 1.0$ cm) due to the three charges. (b) A point charge $q = -5.0$ nC moves from a great distance to point A. What is the change in electric potential energy?

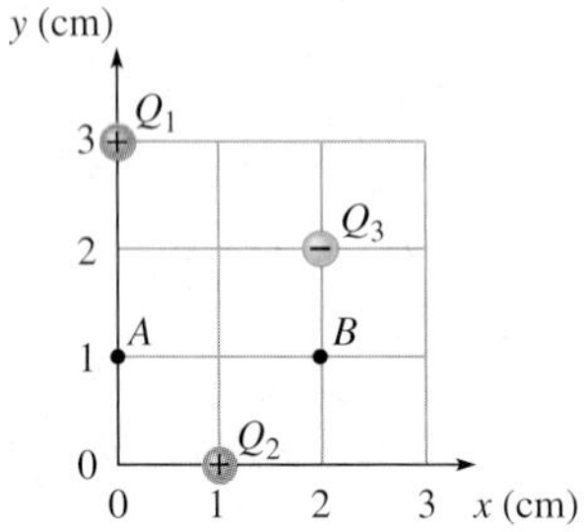

Figure 17.7
An array of three point charges.

Strategy The potential at A is the sum of the potentials due to each point charge. The first step is to find the distance from each charge to point A. We call these distances r_1, r_2, and r_3 to avoid using the wrong one by mistake. Then we add the potentials due to each of the three charges at A.

Solution (a) From the grid, $r_1 = 2.0$ cm. The distance from Q_2 to point A is the diagonal of a square that is 1.0 cm on a side. Thus, $r_2 = \sqrt{2.0}$ cm $= 1.414$ cm. The third charge is located at a distance equal to the hypotenuse of a right triangle with sides of 2.0 cm and 1.0 cm. From the Pythagorean theorem,

$$r_3 = \sqrt{1.0^2 + 2.0^2}\ \text{cm} = \sqrt{5.0}\ \text{cm} = 2.236\ \text{cm}$$

The potential at A is the sum of the potentials due to each point charge:

$$V = k\sum\frac{Q_i}{r_i}$$

Substituting numerical values:

$$V_A = 8.99 \times 10^9\ \frac{\text{N}\cdot\text{m}^2}{\text{C}^2} \times \left(\frac{+4.0 \times 10^{-6}\ \text{C}}{0.020\ \text{m}} + \frac{+2.0 \times 10^{-6}\ \text{C}}{0.01414\ \text{m}} + \frac{-3.0 \times 10^{-6}\ \text{C}}{0.02236\ \text{m}}\right)$$

$$= +1.863 \times 10^6\ \text{V}$$

To two significant figures, the potential at point A is $+1.9 \times 10^6$ V.

(b) The change in potential energy is

$$\Delta U_E = q\,\Delta V$$

Here ΔV is the potential difference through which charge q moves. If we assume that q starts from an infinite distance, then $V_i = 0$. Therefore,

$$\Delta U_E = q(V_A - 0) = (-5.0 \times 10^{-9}\ \text{C}) \times (+1.863 \times 10^6\ \text{J/C} - 0)$$
$$= -9.3 \times 10^{-3}\ \text{J}$$

Discussion The positive sign of the potential indicates that a positive charge at point A would have positive potential energy. To bring in a positive charge from far away, the potential energy must be increased and therefore positive work must be done by the agent bringing in the charge. A negative charge at that point, on the other hand, has negative potential energy. When q moves from a potential of zero to a positive potential, the potential increase causes a potential energy decrease ($q < 0$).

In Practice Problem 17.4, you are asked to find the work done by the field as q moves from A to B. The force is not constant in magnitude or direction, so we cannot just multiply force component times distance. In principle, the problem could be solved this way using calculus; but using the potential difference gives the same result without vector components or calculus.

Practice Problem 17.4 Potential at Point B

Find the potential due to the same array of charges at point B ($x = 2.0$ cm, $y = 1.0$ cm) and the work done by the electric field if $q = -5.0$ nC moves from A to B.

Conceptual Example 17.5

Field and Potential at the Center of a Square

Four equal positive point charges q are fixed at the corners of a square of side s (Fig. 17.8). (a) Is the electric field zero at the center of the square? (b) Is the potential zero at the center of the square?

Strategy and Solution (a) The electric field at the center is the *vector* sum of the fields due to each of the point charges. Figure 17.9 shows the field vectors at the center of the square due to each charge. Each of these vectors has the

continued on next page

Conceptual Example 17.5 continued

same magnitude since the center is equidistant from each corner and the four charges are the same. From symmetry, the vector sum of the electric fields is zero.

(b) Since potential is a scalar rather than a vector, the potential at the center of the square is the *scalar* sum of the potentials due to each charge. These potentials are all equal since the distances and charges are the same. Each is positive since $q > 0$. The total potential at the center of the square is

$$V = 4\frac{kq}{r}$$

where $r = s/\sqrt{2}$ is the distance from a corner of the square to the center.

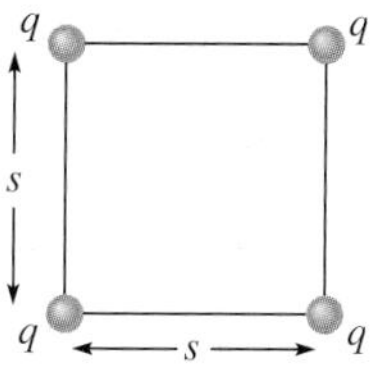

Figure 17.8
Four equal point charges at the corners of a square.

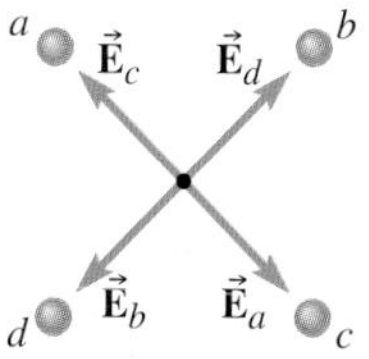

Figure 17.9
Electric field vectors due to each of the point charges at the center of the square.

Discussion In this example, the electric field is zero at a point where the potential is not zero. In other cases, there may be points where the potential is zero while the electric field at the same points is not zero. Never assume that the potential at a point is zero because the electric field is zero or vice versa. If the electric field is zero at a point, it means that a point charge placed at that point would feel no net electric force. If the potential is zero at a point, it means zero total work would be done by the electric field as a point charge moves from infinity to that point.

Practice Problem 17.5 Field and Potential for a Different Set of Charges

Find the electric field and the potential at the center of a square of side 2.0 cm with a charge of +9.0 μC at one corner and with charges of −3.0 μC at the other three corners (Fig. 17.10).

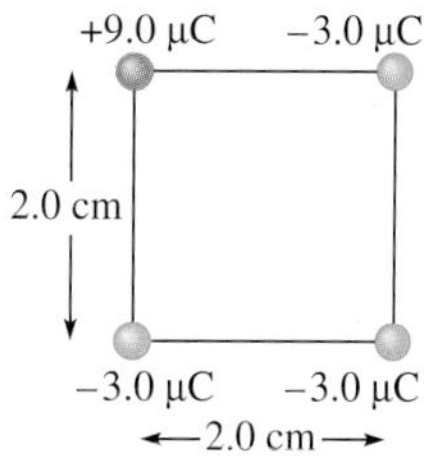

Figure 17.10
Charges for Practice Problem 17.5.

Potential due to a Spherical Conductor

In Section 16.4, we saw that the field outside a charged conducting sphere is the same as if all of the charge were concentrated into a point charge located at the center of the sphere. As a result, the electric potential due to a conducting sphere is similar to the potential due to a point charge.

Figure 17.11 shows graphs of the electric field strength and the potential as functions of the distance r from the center of a hollow conducting sphere of radius R and charge Q. The electric field inside the conducting sphere (from $r = 0$ to $r = R$) is zero. The magnitude of the electric field is greatest at the surface of the conductor and then drops off as $1/r^2$. Outside the sphere, the electric field is the same as for a charge Q located at $r = 0$.

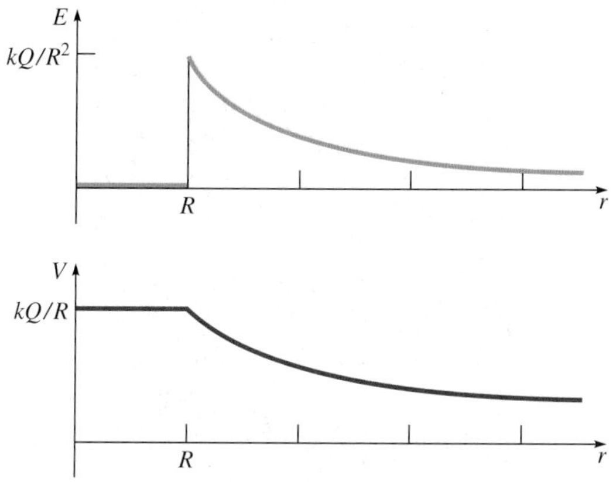

Figure 17.11 The electric field and the potential due to a hollow conducting sphere of radius R and charge Q as a function of r, the distance from the center. For $r \geq R$, the field and potential are the same as if there were a point charge Q at the origin instead. For $r < R$, the electric field is zero and the potential is *constant*.

The potential is chosen to be zero for $r=\infty$. The electric field outside the sphere ($r \geq R$) is the same as the field at a distance r from a *point charge* Q. Therefore, for any point at a distance $r \geq R$ from the center of the sphere, the potential is the same as the potential at a distance r from a point charge Q:

$$V=\frac{kQ}{r} \quad (r \geq R) \tag{17-8}$$

For a positive charge Q, the potential is positive; and it is negative for a negative charge. At the surface of the sphere, the potential is

$$V=\frac{kQ}{R}$$

Since the electric field inside the cavity is zero, no work would be done by the electric field if a test charge were moved around within the cavity. Therefore, the potential *anywhere inside* the sphere is the same as the potential at the *surface* of the sphere. Thus, for $r<R$, the potential is *not* the same as for a point charge. (The magnitude of the potential due to a *point charge* continues to increase as $r \rightarrow 0$.)

Application: van de Graaff Generator

An apparatus designed to charge a conductor to a high potential difference is the van de Graaff generator (Fig. 17.12). A large conducting sphere is supported on an insulating cylinder. In the cylinder, a motor-driven conveyor belt collects negative charge either by rubbing or from some other source of charge at the base of the cylinder. The charge is carried by the conveyor belt to the top of the cylinder, where it is collected by small metal rods and transferred to the conducting sphere. As more and more charge is deposited onto the conducting sphere, the charges repel each other and move as far away from each other as possible, ending up on the outer surface of the conducting sphere.

Inside the conducting sphere, the electric field is zero, so no repulsion from charges already on the sphere is felt by the charge on the conveyor belt. Thus, a large quantity of charge can build up on the conducting sphere so that an extremely high potential difference can be established. Potential differences of millions of volts can be attained with a large sphere (Fig. 17.13). Commercial van de Graaff generators supply the large potential differences required to produce intense beams of high-energy x-rays. The x-rays are used in medicine for cancer therapy; industrial uses include radiography (to detect tiny defects in machine parts) and the polymerization of plastics. Old science fiction movies often show sparks jumping from generators of this sort.

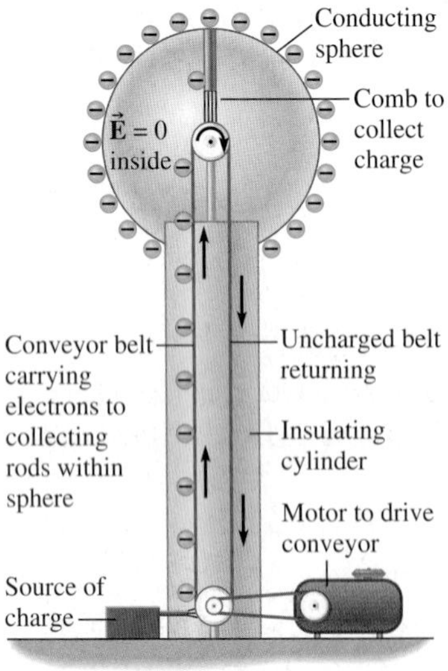

Figure 17.12 The van de Graaff generator.

Figure 17.13 A hair-raising experience. A person touching the dome of a van de Graaff while electrically isolated from ground reaches the same potential as the dome. Although the effects are quite noticeable, there is no danger to the person since the whole body is at the same potential. A large potential *difference* between two parts of the person's body would be dangerous or even lethal.

Example 17.6

Minimum Radius Required for a van de Graaff

You wish to charge a van de Graaff to a potential of 240 kV. On a day with average humidity, an electric field of 8.0×10^5 N/C or greater ionizes air molecules, allowing charge to leak off the van de Graaff. Find the minimum radius of the conducting sphere under these conditions.

Strategy We set the potential of a conducting sphere equal to $V_{max} = 240$ kV and require the electric field strength just outside the sphere to be less than $E_{max} = 8.0 \times 10^5$ N/C. Since both $\vec{\mathbf{E}}$ and V depend on the charge on the sphere and its radius, we should be able to eliminate the charge and solve for the radius.

Solution The potential of a conducting sphere with charge Q and radius R is

$$V = \frac{kQ}{R}$$

The electric field strength just outside the sphere is

$$E = \frac{kQ}{R^2}$$

Comparing the two expressions, we see that $E = V/R$ just outside the sphere. Now let $V = V_{max}$ and require $E < E_{max}$:

$$E = \frac{V_{max}}{R} < E_{max}$$

Solving for R,

$$R > \frac{V_{max}}{E_{max}} = \frac{2.4 \times 10^5 \text{ V}}{8.0 \times 10^5 \text{ N/C}}$$

$$R > 0.30 \text{ m}$$

The minimum radius is 30 cm.

Discussion To achieve a large potential difference, a large conducting sphere is required. A small sphere—or a conductor with a sharp point, which is like part of a sphere with a small radius of curvature—cannot be charged to a high potential. Even a relatively small potential on a conductor with a sharp point, such as a lightning rod, enables charge to leak off into the air since the strong electric field ionizes the nearby air.

The equation $E = V/R$ derived in this example is *not* a general relationship between field and potential. The general relationship is discussed in Section 17.3.

Practice Problem 17.6 **A Small Conducting Sphere**

What is the largest potential that can be achieved on a conducting sphere of radius 0.5 cm? Assume $E_{max} = 8.0 \times 10^5$ N/C.

Potential Differences in Biological Systems

Application: transmission of nerve impulses

In general, the inside and outside of a biological cell are *not* at the same potential. The potential difference across a cell membrane is due to different concentrations of ions in the fluids inside and outside the cell. These potential differences are particularly noteworthy in nerve and muscle cells.

A nerve cell or *neuron* consists of a cell body and a long extension, called an *axon* (Fig. 17.14a). Human axons are 10 to 20 μm in diameter. When the axon is in its

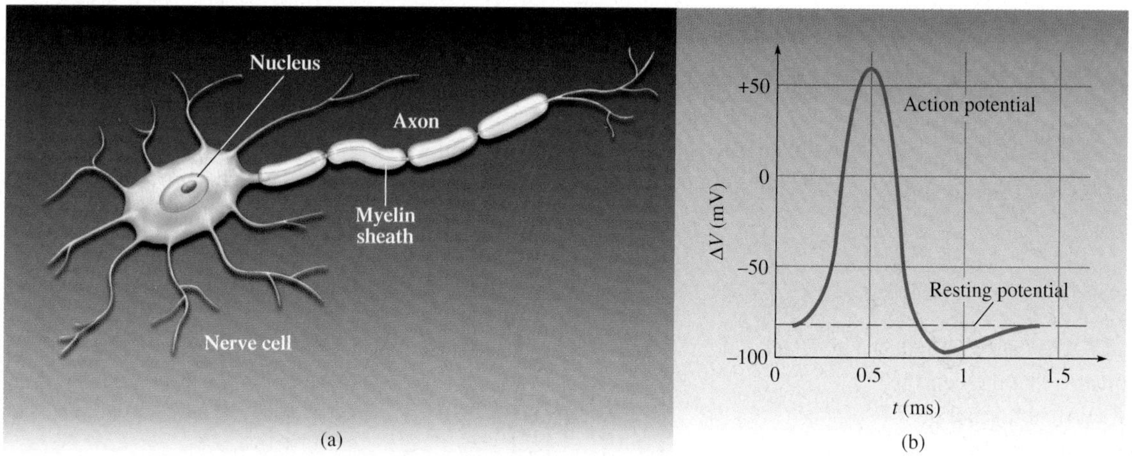

Figure 17.14 (a) The structure of a neuron. (b) The action potential. The graph shows the potential difference between the inside and outside of the cell membrane at a point along the axon as a function of time.

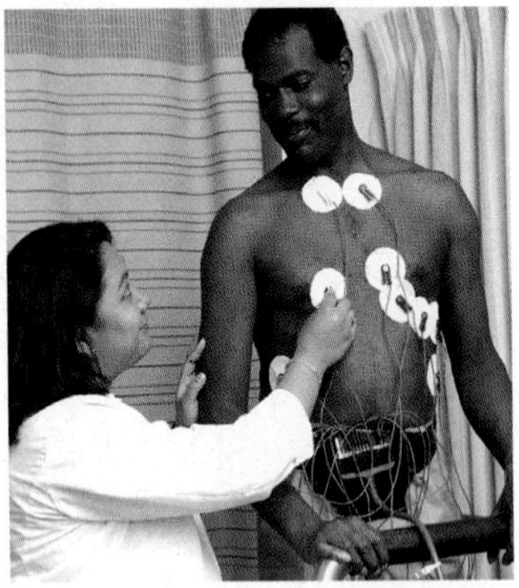

Figure 17.15 A stress test. The ECG is a graph of the potential difference measured between two electrodes as a function of time. These potential differences reveal whether the heart functions normally during exercise.

What physical quantity is measured in an ECG?

Application: electrocardiogram (ECG) and electroencephalogram (EEG)

resting state, negative ions on the inner surface of the membrane and positive ions on the outer surface cause the fluid inside to be at a potential of about −85 mV relative to the fluid outside.

A nerve impulse is a change in the potential difference across the membrane that gets propagated along the axon. The cell membrane at the end stimulated suddenly becomes permeable to positive sodium ions for about 0.2 ms. Sodium ions flow into the cell, changing the polarity of the charge on the inner surface of the membrane. The potential difference across the cell membrane changes from about −85 mV to +60 mV. The reversal of polarity of the potential difference across the membrane is called the *action potential* (Fig. 17.14b). The action potential propagates down the axon at a speed of about 30 m/s.

Restoration of the resting potential involves both the diffusion of potassium and the pumping of sodium ions out of the cell—called *active transport.* As much as 20% of the resting energy requirements of the body are used for the active transport of sodium ions.

Similar polarity changes occur across the membranes of muscle cells. When a nerve impulse reaches a muscle fiber, it causes a change in potential, which propagates along the muscle fiber and signals the muscle to contract.

Muscle cells, including those in the heart, have a layer of negative ions on the inside of the membrane and positive ions on the outside. Just before each heartbeat, positive ions are pumped into the cells, neutralizing the potential difference. Just as for the action potential in neurons, the *depolarization* of muscle cells begins at one end of the cell and proceeds toward the other end. Depolarization of various cells occurs at different times. When the heart relaxes, the cells are polarized again.

An electrocardiogram (ECG) measures the potential difference between points on the chest as a function of time (Fig. 17.15). The depolarization and polarization of the cells in the heart causes potential differences that can be measured using electrodes connected to the skin. The potential difference measured by the electrodes is amplified and recorded on a chart recorder or a computer (Fig. 17.16).

Potential differences other than those due to the heart are used for diagnostic purposes. In an electroencephalogram (EEG), the electrodes are placed on the head. The EEG measures potential differences caused by electrical activity in the brain. In an electroretinogram (ERG), the electrodes are placed near the eye to measure the potential differences due to electrical activity in the retina when stimulated by a flash of light.

17.3 THE RELATIONSHIP BETWEEN ELECTRIC FIELD AND POTENTIAL

In this section, we explore the relationship between electric field and electric potential in detail, starting with visual representations of each.

Equipotential Surfaces

A field line sketch is a useful visual representation of the electric field. To represent the electric potential, we can create something analogous to a contour map. An **equipotential surface** has the same potential at every point on the surface. The idea is similar to the lines of constant elevation on a topographic map, which show where the elevation is

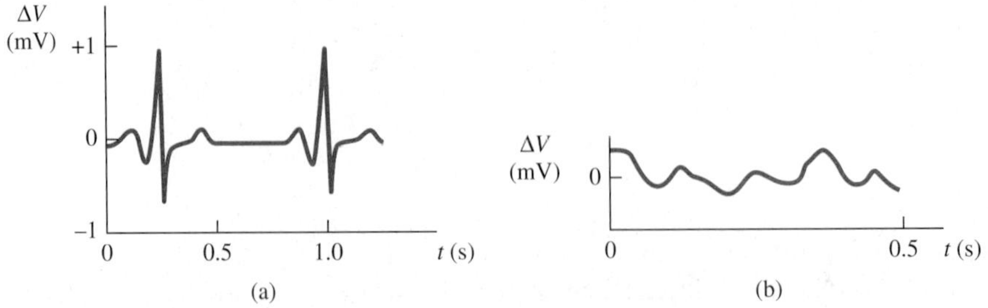

Figure 17.16 (a) A normal ECG indicates that the heart is healthy. (b) An abnormal or irregular ECG indicates a problem. This ECG indicates ventricular fibrillation, a potentially life threatening condition.

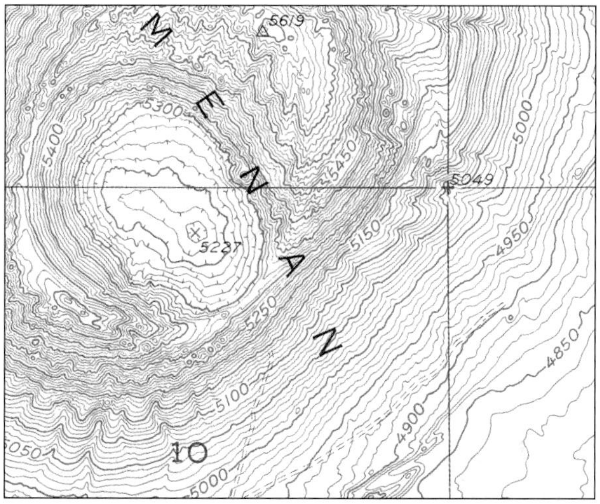

Figure 17.17 A topographic map showing lines of constant elevation in feet.

the same (Fig. 17.17). Since the potential difference between any two points on such an equipotential surface is zero, no work is done by the field when a charge moves from one point on the surface to another.

Equipotential surfaces and field lines are closely related. Suppose you want to move a charge in a direction so that the potential stays constant. In order for the field to do no work on the charge, the displacement must be perpendicular to the electric force (and therefore perpendicular to the field). As long as you always move the charge in a direction perpendicular to the field, the work done by the field is zero and the potential stays the same.

An equipotential surface is perpendicular to the electric field lines at all points.

Conversely, if you want to move a charge in a direction that *maximizes* the change in potential, you would move parallel or antiparallel to the electric field. Only the component of displacement perpendicular to an equipotential surface changes the potential. Think of a contour map: the steepest slope—the quickest change of elevation—is perpendicular to the lines of constant elevation. The electric field is the negative gradient of the potential (Fig. 17.18). The *gradient* points in the direction of maximum increase in potential, so the negative gradient—the electric field—points in the direction of maximum *decrease* in potential. On a contour map, a hill is steepest where the lines of constant elevation are close together; a diagram of equipotential surfaces is similar.

CONNECTION:

On a contour map, lines of constant elevation are lines of constant *gravitational* potential (gravitational P.E. per unit *mass*).

If equipotential surfaces are drawn such that the potential difference between adjacent surfaces is constant, then the surfaces are closer together where the field is stronger.

The electric field always points in the direction of maximum potential decrease.

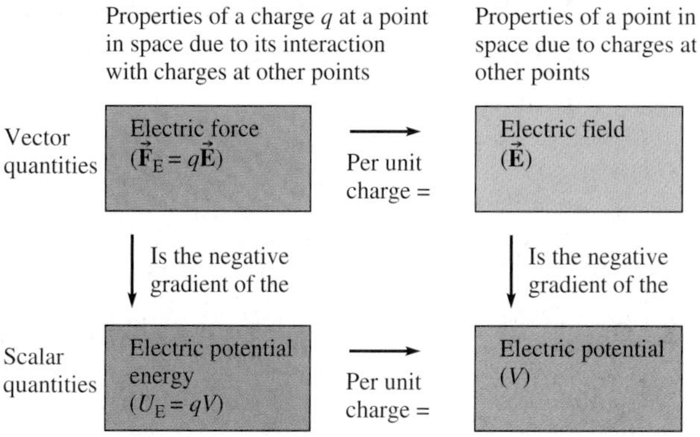

Figure 17.18 Relationships between force, field, potential energy, and potential.

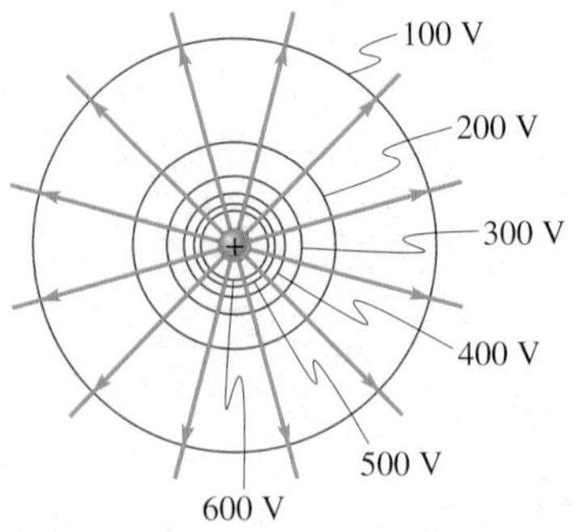

Figure 17.19 Equipotential surfaces near a positive point charge. The circles represent the intersection of the spherical surfaces with the plane of the page. The potential decreases as we move away from a positive charge. The electric field lines are perpendicular to the spherical surfaces and point toward lower potentials. The spacing between equipotential surfaces increases with increasing distance since the electric field gets weaker.

The simplest equipotential surfaces are those for a single point charge. The potential due to a point charge depends only on the distance from the charge, so the equipotential surfaces are spheres with the charge at the center (Fig. 17.19). There are an infinite number of equipotential surfaces, so we customarily draw a few surfaces equally spaced in potential—just like a contour map that shows places of equal elevation in 5-m increments.

Conceptual Example 17.7

Equipotential Surfaces for Two Point Charges

Sketch some equipotential surfaces for two point charges $+Q$ and $-Q$.

Strategy and Solution One way to draw a set of equipotential surfaces is to first draw the field lines. Then we construct the equipotential surfaces by sketching lines that are perpendicular to the field lines at all points. Close to either point charge, the field is primarily due to the nearby charge, so the surfaces are nearly spherical.

Figure 17.20 shows a sketch of the field lines and equipotential surfaces for the two charges.

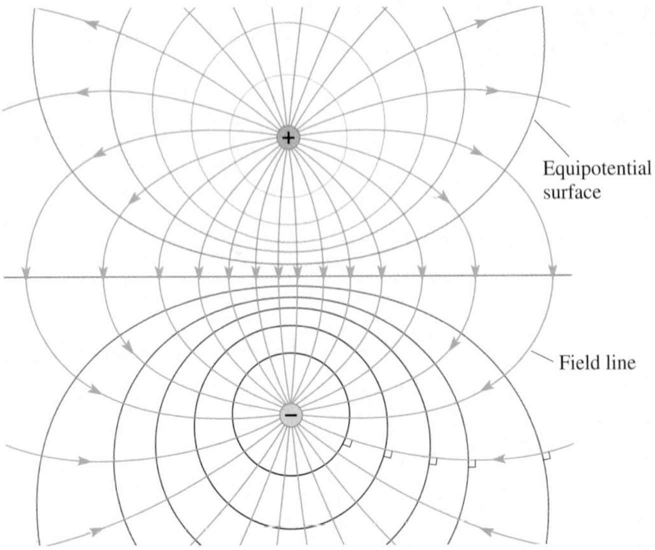

Discussion This two-dimensional sketch shows only the intersection of the equipotential surfaces with the plane of the page. Except for the plane midway between the two charges, the equipotentials are closed surfaces that enclose one of the charges. Equipotential surfaces very close to either charge are approximately spherical.

Conceptual Practice Problem 17.7 Equipotential Surfaces for Two Positive Charges

Sketch some equipotential surfaces for two equal positive point charges.

Figure 17.20

A sketch of some equipotential surfaces (purple) and electric field lines (green) for two point charges of the same magnitude but opposite in sign.

Potential in a Uniform Electric Field

In a uniform electric field, the field lines are equally spaced parallel lines. Since equipotential surfaces are perpendicular to field lines, the equipotential surfaces are a set of parallel planes (Fig. 17.21). The potential decreases from one plane to the next in the direction of $\vec{\mathbf{E}}$. Since the spacing of equipotential planes depends on the magnitude of $\vec{\mathbf{E}}$, in a uniform field planes at equal potential increments are equal distances apart.

To find a quantitative relationship between the field strength and the spacing of the equipotential planes, imagine moving a point charge $+q$ a distance d in the direction of an electric field of magnitude E. The work done by the electric field is

$$W_E = F_E d = qEd$$

The change in electric potential energy is

$$\Delta U_E = -W_E = -qEd$$

From the definition of potential, the potential change is

$$\Delta V = \frac{\Delta U_E}{q} = -Ed \quad (17\text{-}10)$$

Relationship between uniform electric field and potential difference

The negative sign in Eq. (17-10) is correct because potential *decreases* in the direction of the electric field.

Equation (17-10) implies that the SI unit of electric field (N/C) can also be written *volts per meter* (V/m):

$$1 \text{ N/C} = 1 \text{ V/m} \quad (17\text{-}11)$$

Where the field is strong, the equipotential surfaces are close together: with a large number of volts per meter, it doesn't take many meters to change the potential a given number of volts.

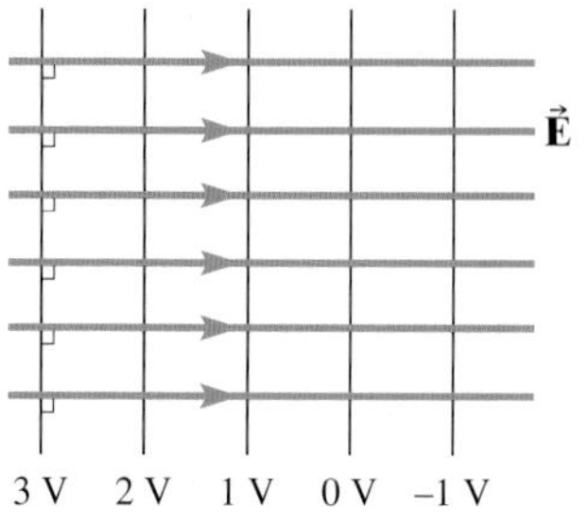

Figure 17.21 Field lines and equipotential surfaces (at 1-V intervals) in a uniform field. The equipotential surfaces are equally spaced *planes* perpendicular to the field lines.

CHECKPOINT 17.3

In Fig. 17.21, the equipotential planes differ in potential by 1.0 V. If the electric field magnitude is 25 N/C = 25 V/m, what is the distance between the planes?

Potential Inside a Conductor

In Section 16.6, we learned that $E = 0$ at every point inside a conductor in electrostatic equilibrium (when no charges are moving). If the field is zero at every point, then the potential does not change as we move from one point to another. If there were potential differences within the conductor, then charges would move in response. Positive charge would be accelerated by the field toward regions of lower potential and negative charge would be accelerated toward higher potential. If there are no moving charges, then the field is zero everywhere and no potential differences exist within the conductor. (WWW tutorial: E-field in conducting box) Therefore:

> In electrostatic equilibrium, every point within a conducting material must be at the same potential.

17.4 CONSERVATION OF ENERGY FOR MOVING CHARGES

CONNECTION:

This is the same principle of energy conservation; we're just applying it to another form of energy—electric potential energy.

When a charge moves from one position to another in an electric field, the change in electric potential energy must be accompanied by a change in other forms of energy so that the total energy is constant. Energy conservation simplifies problem solving just as it does with gravitational or elastic potential energy.

If no other forces act on a point charge, then as it moves in an electric field, the sum of the kinetic and electric potential energy is constant:

$$K_i + U_i = K_f + U_f = \text{constant}$$

Changes in gravitational potential energy are negligible compared with changes in electric potential energy when the gravitational force is much weaker than the electric force.

Example 17.8

Electron Gun in a CRT

In an electron gun, electrons are accelerated from the cathode toward the anode, which is at a potential higher than the cathode (see Fig. 16.35). If the potential difference between the cathode and anode is 12 kV, at what speed do the electrons move as they reach the anode? Assume that the initial kinetic energy of the electrons as they leave the cathode is negligible. (tutorial: electron gun)

Strategy Using energy conservation, we set the sum of the initial kinetic and potential energies equal to the sum of the final kinetic and potential energies. The initial kinetic energy is taken to be zero. Once we find the final kinetic energy, we can solve for the speed.

Known: $K_i = 0$; $\Delta V = +12$ kV
Find: v

Solution The change in electric potential energy is

$$\Delta U = U_f - U_i = q\,\Delta V$$

From conservation of energy,

$$K_i + U_i = K_f + U_f$$

Solving for the final kinetic energy,

$$K_f = K_i + (U_i - U_f) = K_i - \Delta U$$
$$= 0 - q\,\Delta V$$

To find the speed, we set $K_f = \frac{1}{2}mv^2$.

$$\tfrac{1}{2}mv^2 = -q\,\Delta V$$

Solving for the speed,

$$v = \sqrt{\frac{-2q\,\Delta V}{m}}$$

For an electron,

$$q = -e = -1.602 \times 10^{-19}\ \text{C}$$
$$m = 9.109 \times 10^{-31}\ \text{kg}$$

Substituting numerical values,

$$v = \sqrt{\frac{-2 \times (-1.602 \times 10^{-19}\ \text{C}) \times (12{,}000\ \text{V})}{9.109 \times 10^{-31}\ \text{kg}}}$$
$$= 6.5 \times 10^{7}\ \text{m/s}$$

Discussion The answer is more than 20% of the speed of light (3×10^8 m/s). A more accurate calculation of the speed, accounting for Einstein's theory of relativity, is 6.4×10^7 m/s.

Using conservation of energy to solve this problem makes it clear that the final speed depends only on the potential difference between the cathode and anode, not on the distance between them. To solve the problem using Newton's second law, even if the electric field is uniform, we have to assume some distance d between the cathode and anode. Using d, we can find the magnitude of the electric field

$$E = \frac{\Delta V}{d}$$

The acceleration of the electron is

$$a = \frac{F_E}{m} = \frac{eE}{m} = \frac{e\,\Delta V}{md}$$

Now we can find the final speed. Since the acceleration is constant,

$$v = \sqrt{v_i^2 + 2ad} = \sqrt{0 + 2 \times \frac{e\Delta V}{md} \times d}$$

The distance d cancels and gives the same result as the energy calculation.

Practice Problem 17.8 Proton Accelerated

A proton is accelerated from rest through a potential difference. Its final speed is 2.00×10^6 m/s. What is the potential difference? The mass of the proton is 1.673×10^{-27} kg.

Figure 17.22 The arrows indicate a few of the many capacitors on a circuit board from the inside of an amplifier.

17.5 CAPACITORS

Can a useful device be built to store electric potential energy? Yes. Many such devices, called *capacitors*, are found in every piece of electronic equipment (Fig. 17.22).

A **capacitor** is a device that stores electric potential energy by storing separated positive and negative charges. It consists of two conductors separated by either vacuum or an insulating material. Charge is separated, with positive charge put on one of the conductors and an equal amount of negative charge on the other conductor. Work must be done to separate positive charge from negative charge, since there is an attractive force between the two. The work done to separate the charge ends up as electric potential energy. An electric field arises between the two conductors, with field lines beginning on the conductor with positive charge and ending on the conductor with negative charge (Fig. 17.23). The stored potential energy is associated with this electric field. We can recover the stored energy—that is, convert it into some other form of energy—by letting the charges come together again.

The simplest form of capacitor is a **parallel plate capacitor**, consisting of two parallel metal plates, each of the same area A, separated by a distance d. A charge $+Q$ is put on one plate and a charge $-Q$ on the other. For now, assume there is air between the plates. One way to charge the plates is to connect the positive terminal of a battery to one and the negative terminal to the other. The battery removes electrons from one plate, leaving it positively charged, and puts them on the other plate, leaving it with an equal magnitude of negative charge. In order to do this, the battery has to do work—some of the battery's chemical energy is converted into electric potential energy.

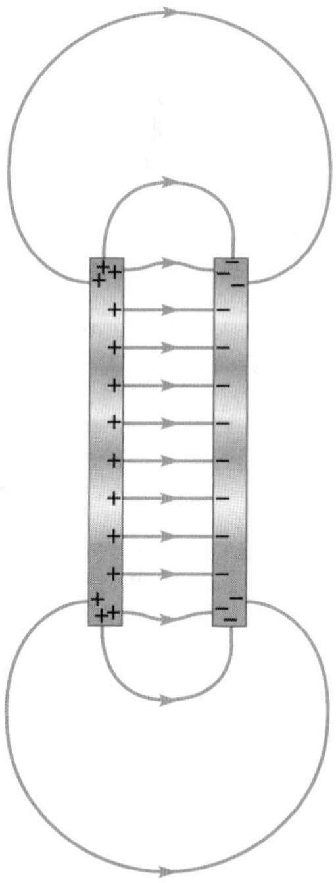

Figure 17.23 Side view of two parallel metal plates with charges of equal magnitude and opposite sign. There is a potential difference between the two plates; the positive plate is at the higher potential.

In general, the field between two such plates does not have to be uniform (Fig. 17.23). However, if the plates are close together, then a good approximation is to say that the charge is evenly spread on the inner surfaces of the plates and none is found on the outer surfaces. The plates in a real capacitor are almost always close enough that this approximation is valid.

With charge evenly spread on the inner surfaces, a uniform electric field exists between the two plates. We can neglect the nonuniformity of the field near the edges as long as the plates are close together. The electric field lines start on positive charges and end on negative charges. If charge of magnitude Q is evenly spread over each plate with surface of area A, then the *surface charge density* (the charge per unit area) is denoted by σ, the Greek letter sigma:

$$\sigma = Q/A \qquad (17\text{-}12)$$

In Problem 64, you can show that the magnitude of the electric field just outside a conductor is

Electric field just outside a conductor:

$$E = 4\pi k\sigma = \sigma/\epsilon_0 \qquad (17\text{-}13)$$

Recall that the constant $\epsilon_0 = 1/(4\pi k) = 8.85 \times 10^{-12}\ C^2/(N{\cdot}m^2)$ is called the *permittivity of free space* [Eq. (16-3b)]. Since the field between the plates of the capacitor is uniform, Eq. (17-13) gives the magnitude of the field *everywhere* between the plates.

What is the potential difference between the plates? Since the field is uniform, the *magnitude* of the potential difference is

In Eq. (17-10), ΔV stands for the potential difference between the positively and negatively charged plates: $\Delta V = V_+ - V_-$

$$\Delta V = Ed \qquad (17\text{-}10)$$

The field is proportional to the charge and the potential difference is proportional to the field; therefore, *the charge is proportional to the potential difference.* That turns out to be true for any capacitor, not just a parallel plate capacitor. The constant of proportionality between charge and potential difference depends only on geometric factors (sizes and shapes of the plates) and the material between the plates. Conventionally, this proportionality is written

Definition of capacitance:

$$Q = C\,\Delta V \qquad (17\text{-}14)$$

where Q is the magnitude of the charge on each plate and ΔV is the magnitude of the potential *difference* between the plates. The constant of proportionality C is called the **capacitance**. Think of capacitance as the capacity to hold charge for a given potential difference. The SI units of capacitance are coulombs per volt, which is called the *farad* (symbol F). Capacitances are commonly measured in μF (microfarads), nF (nanofarads), or pF (picofarads) because the farad is a rather large unit; a pair of plates with area 1 m^2 spaced 1 mm apart has a capacitance of only about 10^{-8} F = 10 nF.

SI unit of capacitance: 1 farad = 1 coulomb per volt

We can now find the capacitance of a parallel plate capacitor. The electric field is

$$E = \frac{\sigma}{\epsilon_0} = \frac{Q}{\epsilon_0 A}$$

where A is the inner surface area of each plate. If the plates are a distance d apart, then the *magnitude* of the potential difference is

$$\Delta V = Ed = \frac{Qd}{\epsilon_0 A}$$

By rearranging, this can be rewritten in the form $Q = \text{constant} \times \Delta V$:

$$Q = \frac{\epsilon_0 A}{d}\Delta V$$

Comparing with the definition of capacitance, the capacitance of a parallel plate capacitor is

Capacitance of parallel plate capacitor:

$$C = \frac{\epsilon_0 A}{d} = \frac{A}{4\pi k d} \qquad (17\text{-}15)$$

To produce a large capacitance, we make the plate area large and the plate spacing small. To get large areas while still keeping the physical size of the capacitor reasonable, the plates are often made of thin conducting foil that is rolled, with the insulating material sandwiched in between, into a cylinder (Fig. 17.24). The effect of using an insulator other than air or vacuum is discussed in Section 17.6.

CHECKPOINT 17.5

A capacitor is connected to a 6.0-V battery. When fully charged, the plates have net charges +0.48 C and −0.48 C. What are the net charges on the plates if the same capacitor is connected to a 1.5-V battery?

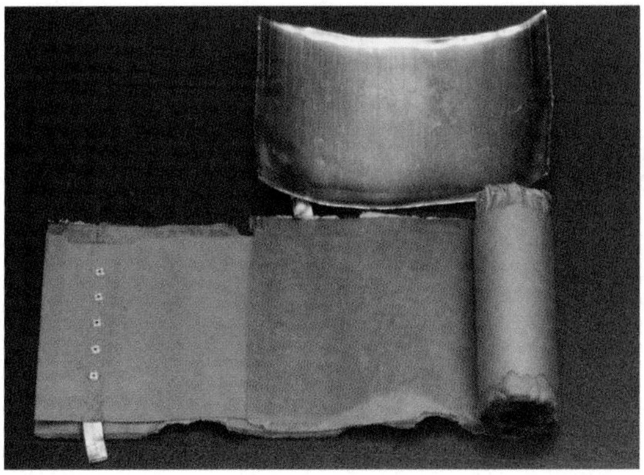

Figure 17.24 A disassembled capacitor, showing the foil conducting plates and the thin sheet of insulating material.

Example 17.9

Computer Keyboard

In one kind of computer keyboard, each key is attached to one plate of a parallel plate capacitor; the other plate is fixed in position (Fig. 17.25). The capacitor is maintained at a constant potential difference of 5.0 V by an external circuit. When the key is pressed down, the top plate moves closer to the bottom plate, changing the capacitance and causing charge to flow through the circuit. If each plate is a square of side 6.0 mm and the plate separation changes from 4.0 mm to 1.2 mm when a key is pressed, how much charge flows through the circuit? Does the charge on the capacitor increase or decrease? Assume that there is air between the plates instead of a flexible insulator.

Strategy Since we are given the area and separation of the plates, we can find the capacitance from Eq. (17-15). The charge is then found from the product of the capacitance and the potential difference across the plates: $Q = C\Delta V$.

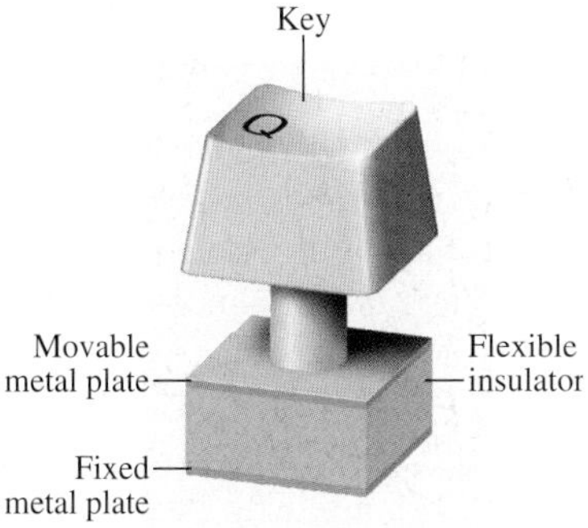

Figure 17.25

Basically, this kind of computer key is merely a capacitor with a variable plate spacing. A circuit detects the change in the plate spacing as charge flows from one plate through an external circuit to the other plate.

Solution The capacitance of a parallel plate capacitor is given by Eq. (17-15):

$$C = \frac{A}{4\pi kd}$$

The area is $A = (6.0\text{ mm})^2$. Since the potential difference ΔV is kept constant, the change in the magnitude of the charge on the plates is

$$Q_\text{f} - Q_\text{i} = C_\text{f}\Delta V - C_\text{i}\Delta V$$

$$= \left(\frac{A}{4\pi k d_\text{f}} - \frac{A}{4\pi k d_\text{i}}\right)\Delta V = \frac{A\,\Delta V}{4\pi k}\left(\frac{1}{d_\text{f}} - \frac{1}{d_\text{i}}\right)$$

Substituting numerical values,

$$Q_\text{f} - Q_\text{i} = \frac{(0.0060\text{ m})^2 \times 5.0\text{ V}}{4\pi \times 8.99 \times 10^9\text{ N·m}^2/\text{C}^2} \times \left(\frac{1}{0.0012\text{ m}} - \frac{1}{0.0040\text{ m}}\right)$$

$$= +9.3 \times 10^{-13}\text{ C} = +0.93\text{ pC}$$

Since ΔQ is positive, the magnitude of charge on the plates increases.

Discussion If the plates move closer together, the capacitance increases. A greater capacitance means that more charge can be stored for a given potential difference. Therefore, the magnitude of the charge increases.

Practice Problem 17.9 Capacitance and the Charge Stored

A parallel plate capacitor has plates of area 1.0 m^2 and a separation of 1.0 mm. The potential difference between the plates is 2.0 kV. Find the capacitance and the magnitude of the charge on each plate. Which of these quantities depends on the potential difference?

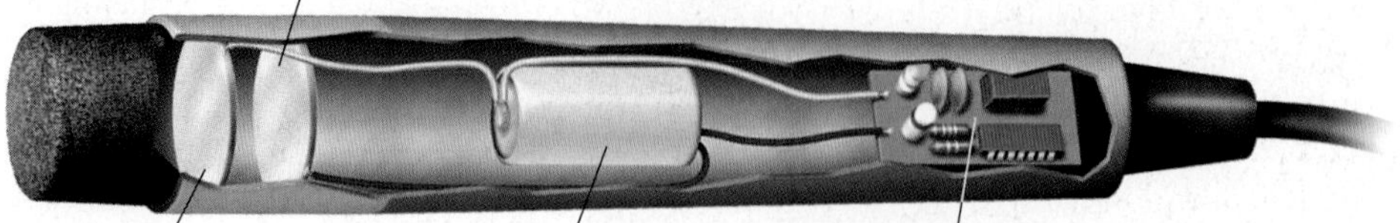

Figure 17.26 This microphone uses a capacitor with one moving plate to create an electrical signal.

Application: condenser microphone

Other devices are based on a capacitor with one moveable plate. In a *condenser microphone* (Fig. 17.26), one plate moves in and out in response to a sound wave. (*Condenser* is a synonym for capacitor.) The capacitor is maintained at a constant potential difference; as the plate spacing changes, charge flows onto and off the plates. The moving charge—an electric current—is amplified to generate an electrical signal. The design of many *tweeters* (speakers for high-frequency sounds) is just the reverse; in response to an electrical signal, one plate moves in and out, generating a sound wave.

Application: random-access memory (RAM) chips

Capacitors have many other uses. Each RAM (random-access memory) chip in a computer contains millions of microscopic capacitors. Each of the capacitors stores one bit (binary digit). To store a 1, the capacitor is charged; to store a 0, it is discharged. The insulation of the capacitors from their surroundings is not perfect, so charge would leak off if it were not periodically refreshed—which is why the contents of RAM are lost when the computer's power is turned off.

Application: oscilloscopes

Besides storing charge and electric energy, capacitors are also useful for the uniform electric field between the plates. This field can be used to accelerate or deflect charges in a controlled way. The oscilloscope—a device used to display time-dependent potential differences in electric circuits—is a cathode ray tube that sends electrons between the plates of two capacitors (see Fig. 16.35). One of the capacitors deflects the electrons vertically; the other deflects them horizontally.

Discharging a Capacitor

If we connect one plate of a charged capacitor to the other with a conducting wire, charge moves along the wire until there is no longer a difference in potential between the plates.

Application: camera flash

PHYSICS AT HOME

The next time you are taking flash pictures with a camera, try to take two pictures one right after the other. Unless you have a professional-quality camera, the flash does not work the second time. There is a minimum time interval of a few seconds between successive flashes. Many cameras have an indicator light to show when the flash is ready.

Did you ever wonder how the small battery in a camera produces such a bright flash? Compare the brightness of a flashlight with the same type of battery. By itself, a small battery cannot pump charge fast enough to produce the bright flash needed. During the time when the flash is inoperative, the battery charges a capacitor. Once the capacitor is fully charged, the flash is ready. When the picture is taken, the capacitor is discharged through the bulb, producing a bright flash of light.

17.6 DIELECTRICS

There is a problem inherent in trying to store a large charge in a capacitor. To store a large charge without making the potential difference excessively large, we need a large capacitance. Capacitance is inversely proportional to the spacing d between the plates. One problem with making the spacing small is that the air between the plates of the capacitor breaks down at an electric field of about 3000 V/mm with dry air (less for humid air). The breakdown allows a spark to jump across the gap so the stored charge is lost.

One way to overcome this difficulty is to put a better insulator than air between the plates. Some insulating materials, which are also called **dielectrics**, can withstand electric fields larger than those that cause air to break down and act as a conductor rather than as an insulator. Another advantage of placing a dielectric between the plates is that the capacitance itself is increased.

For a parallel plate capacitor in which a dielectric fills the entire space between the plates, the capacitance is

Capacitance of parallel plate capacitor with dielectric:

$$C = \kappa\frac{\epsilon_0 A}{d} = \kappa\frac{A}{4\pi k d} \qquad (17\text{-}16)$$

The effect of the dielectric is to increase the capacitance by a factor κ (Greek letter kappa), which is called the **dielectric constant**. The dielectric constant is a dimensionless number: the ratio of the capacitance with the dielectric to the capacitance without the dielectric. The value of κ varies from one dielectric material to another. Equation (17-16) is more general than Eq. (17-15), which applies only when $\kappa = 1$. When there is vacuum between the plates, $\kappa = 1$ by definition. Air has a dielectric constant that is only slightly larger than 1; for most practical purposes we can take $\kappa = 1$ for air also. The flexible insulator in a computer key (Example 17.9) increases the capacitance by a factor of κ. Thus, the amount of charge that flows when the key is pressed is larger than the value we calculated.

The dielectric constant depends on the insulating material used. Table 17.1 gives dielectric constants and the breakdown limit, or **dielectric strength**, for several materials. The dielectric strength is the electric field strength at which **dielectric breakdown** occurs and the material becomes a conductor. Since $\Delta V = Ed$ for a uniform field, the dielectric strength determines the maximum potential difference that can be applied across a capacitor per meter of plate spacing.

Do not confuse dielectric constant and dielectric strength; they are not related. The dielectric constant determines how much charge can be stored for a given potential difference, while dielectric strength determines how large a potential difference can be applied to a capacitor before dielectric breakdown occurs.

Polarization in a Dielectric

What is happening microscopically to a dielectric between the plates of a capacitor? Recall that polarization is a separation of the charge in an atom or molecule (Section 16.1). The atom or molecule remains neutral, but the center of positive charge no longer coincides with the center of negative charge.

Figure 17.27 is a simplified diagram to indicate polarization of an atom. The unpolarized atom with a central positive charge is encircled by a cloud of electrons, so that the center of the negative charge coincides with the center of the positive charge. When a positively charged rod is brought near the atom, it repels the positive charge in the atoms and attracts the negative. This separation of the charges means the centers of positive and negative charge no longer coincide; they are distorted by the influence of the charged rod.

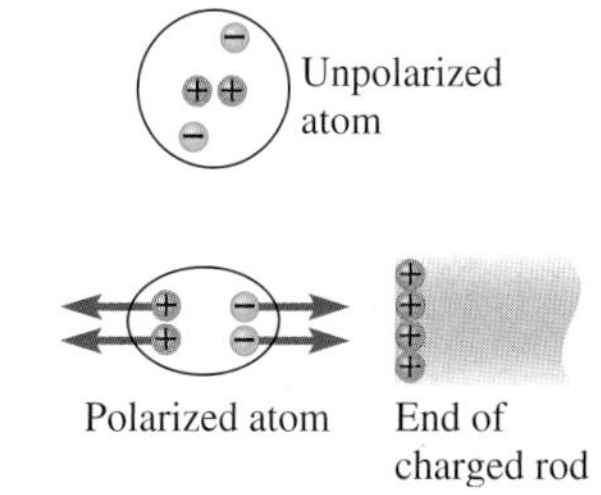

Figure 17.27 A positively charged rod induces polarization in a nearby atom.

Table 17.1 Dielectric Constants and Dielectric Strengths for Materials at 20°C (in order of increasing dielectric constant)

Material	Dielectric Constant κ	Dielectric Strength (kV/mm)
Vacuum	1 (exact)	—
Air (dry, 1 atm)	1.00054	3
Paraffined paper	2.0–3.5	40–60
Teflon	2.1	60
Rubber (vulcanized)	3.0–4.0	16–50
Paper (bond)	3.0	8
Mica	4.5–8.0	150–220
Bakelite	4.4–5.8	12
Glass	5–10	8–13
Diamond	5.7	100
Porcelain	5.1–7.5	10
Rubber (neoprene)	6.7	12
Titanium dioxide ceramic	70–90	4
Water	80	—
Strontium titanate	310	8
Nylon 11	410	27
Barium titanate	6000	—

In Fig. 17.28a, a slab of dielectric material has been placed between the plates of a capacitor. The charges on the capacitor plates induce a polarization of the dielectric. The polarization occurs throughout the material, so the positive charge is slightly displaced relative to the negative charge.

Throughout the bulk of the dielectric, there are still equal amounts of positive and negative charge. The net effect of the polarization of the dielectric is a layer of positive charge on one face and negative charge on the other (Fig. 17.28b). Each conducting plate faces a layer of opposing charge.

The layer of opposing charge induced on the surface of the dielectric helps attract more charge to the conducting plate, for the same potential difference, than would be there without the dielectric. Since capacitance is charge per unit potential difference, the capacitance must have increased. The dielectric constant of a material is a measure of the ease with which the insulating material can be polarized. A larger dielectric constant indicates a more easily polarized material. Thus, neoprene rubber ($\kappa = 6.7$) is more easily polarized than Teflon ($\kappa = 2.1$).

The induced charge on the faces of the dielectric reduces the strength of the electric field in the dielectric compared to the field outside. Some of the electric field lines end

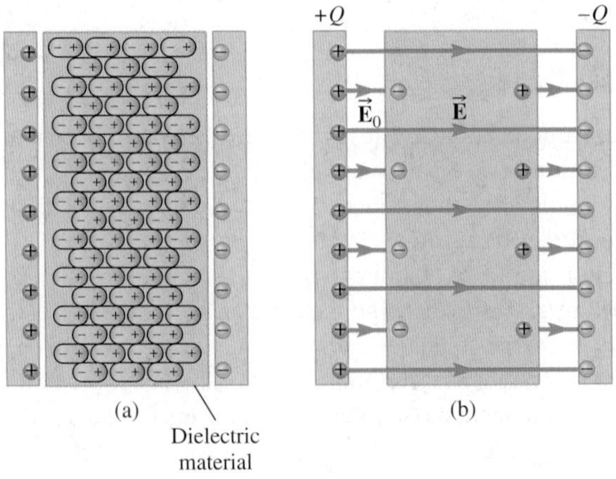

Figure 17.28 (a) Polarization of molecules in a dielectric material. (b) A dielectric with $\kappa = 2$ between the plates of a parallel plate capacitor. The electric field inside the dielectric ($\vec{\mathbf{E}}$) is smaller than the field outside ($\vec{\mathbf{E}}_0$).

on the surface of the insulating dielectric material; fewer lines penetrate the dielectric and thus the field is weaker. With a weaker field, there is a smaller potential difference between the plates (recall that for a uniform field, $\Delta V = Ed$). A smaller potential difference makes it easier to put more charge on the capacitor. We have succeeded in having the capacitor store more charge with a smaller potential difference. Since there is a limiting potential difference before breakdown occurs, this is an important factor for reaching maximum charge storage capability.

Dielectric Constant Suppose a dielectric is immersed in an external electric field E_0. The *definition* of the **dielectric constant** is the ratio of the electric field in vacuum E_0 to the electric field E inside the dielectric material:

Definition of dielectric constant:

$$\kappa = \frac{E_0}{E} \quad (17\text{-}17)$$

Polarization *weakens* the field, so $\kappa > 1$. The electric field inside the dielectric (E) is

$$E = E_0/\kappa$$

In a capacitor, the dielectric is immersed in an applied field E_0 due to the charges on the plates. By reducing the field between the plates to E_0/κ, the dielectric reduces the potential difference between the plates by the same factor $1/\kappa$. Since $Q = C\,\Delta V$, multiplying ΔV by $1/\kappa$ for a given charge Q means the capacitance is multiplied by a factor of κ due to the dielectric [see Eq. (17-16)].

Example 17.10

Parallel Plate Capacitor with Dielectric

A parallel plate capacitor has plates of area 1.00 m^2 and spacing of 0.500 mm. The insulator has dielectric constant 4.9 and dielectric strength 18 kV/mm. (a) What is the capacitance? (b) What is the maximum charge that can be stored on this capacitor?

Strategy Finding the capacitance is a straightforward application of Eq. (17-16). The dielectric strength and the plate spacing determine the maximum potential difference; using the capacitance we can find the maximum charge.

Solution (a) The capacitance is

$$C = \kappa \frac{A}{4\pi kd}$$
$$= 4.9 \times \frac{1.00\ \text{m}^2}{4\pi \times 8.99 \times 10^9\ \text{N·m}^2/\text{C}^2 \times 5.00 \times 10^{-4}\ \text{m}}$$
$$= 8.67 \times 10^{-8}\ \text{F} = 86.7\ \text{nF}$$

(b) The maximum potential difference is

$$\Delta V = 18\ \text{kV/mm} \times 0.500\ \text{mm} = 9.0\ \text{kV}$$

Using the definition of capacitance, the maximum charge is

$$Q = C\,\Delta V = 8.67 \times 10^{-8}\ \text{F} \times 9.0 \times 10^3\ \text{V} = 7.8 \times 10^{-4}\ \text{C}$$

Discussion Check: Each plate has a surface charge density of magnitude $\sigma = Q/A$ [Eq. (17-12)]. If the capacitor plates had this same charge density with no dielectric between them, the electric field between the plates would be [Eq. (17-13)]:

$$E_0 = 4\pi k\sigma = \frac{4\pi kQ}{A} = 8.8 \times 10^7\ \text{V/m}$$

From Eq. (17-17), the dielectric reduces the field strength by a factor of 4.9:

$$E = \frac{E_0}{\kappa} = \frac{8.8 \times 10^7\ \text{V/m}}{4.9} = 1.8 \times 10^7\ \text{V/m} = 18\ \text{kV/mm}$$

Thus, with the charge found in (b), the electric field has its maximum possible value.

Practice Problem 17.10 Changing the Dielectric

If the dielectric were replaced with one having twice the dielectric constant and half the dielectric strength, what would happen to the capacitance and the maximum charge?

Example 17.11

Neuron Capacitance

A neuron can be modeled as a parallel plate capacitor, where the membrane serves as the dielectric and the oppositely charged ions are the charges on the "plates" (Fig. 17.29). Find the capacitance of a neuron and the number of ions (assumed to be singly charged) required to establish a potential difference of 85 mV. Assume that the membrane has a dielectric constant of $\kappa = 3.0$, a thickness of 10.0 nm, and an area of 1.0×10^{-10} m^2.

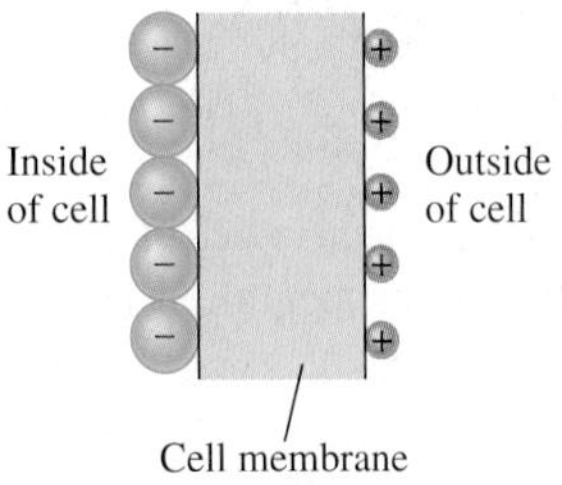

Figure 17.29
Cell membrane as a dielectric.

Strategy Since we know κ, A, and d, we can find the capacitance. Then, from the potential difference and the capacitance, we can find the magnitude of charge Q on each side of the membrane. A singly charged ion has a charge of magnitude e, so Q/e is the number of ions on each side.

Solution From Eq. (17-16),

$$C = \kappa \frac{A}{4\pi k d}$$

Substituting numerical values,

$$C = 3.0 \times \frac{1.0 \times 10^{-10}\ \text{m}^2}{4\pi \times 8.99 \times 10^9\ \text{N·m}^2/\text{C}^2 \times 10.0 \times 10^{-9}\ \text{m}}$$

$$= 2.66 \times 10^{-13}\ \text{F} = 0.27\ \text{pF}$$

From the definition of capacitance,

$$Q = C\,\Delta V = 2.66 \times 10^{-13}\ \text{F} \times 0.085\ \text{V}$$

$$= 2.26 \times 10^{-14}\ \text{C} = 0.023\ \text{pC}$$

Each ion has a charge of magnitude $e = +1.602 \times 10^{-19}$ C. The number of ions on each side is therefore,

$$\text{number of ions} = \frac{2.26 \times 10^{-14}\ \text{C}}{1.602 \times 10^{-19}\ \text{C/ion}} = 1.4 \times 10^5\ \text{ions}$$

Discussion To see if the answer is reasonable, we can estimate the average distance between the ions. If 10^5 ions are evenly spread over a surface of area 10^{-10} m^2, then the area per ion is 10^{-15} m^2. Assuming each ion to occupy a square of area 10^{-15} m^2, the distance from one ion to its nearest neighbor is the side of the square $s = \sqrt{10^{-15}\ \text{m}^2} = 30$ nm. The size of a typical atom or ion is 0.2 nm. Since the distance between ions is much larger than the size of an ion, the answer is plausible; if the distance between ions came out to be less than the size of an ion, the answer would not be plausible.

Practice Problem 17.11 Action Potential

How many ions must cross the membrane to change the potential difference from −0.085 V (with negative charge inside and positive outside) to +0.060 V (with negative charge outside and positive charge inside)?

Application: Thunderclouds and Lightning

Lightning (Fig. 17.30) involves the dielectric breakdown of air. Charge separation occurs within a thundercloud; the top of the cloud becomes positive and the lower part becomes negative (Fig. 17.31a). How this charge separation occurs is not completely understood, but one leading hypothesis is that collisions between ice particles or between an ice particle and a droplet of water tend to transfer electrons from the smaller particle to the larger. Updrafts in the thundercloud lift the smaller, positively charged particles to the top of the cloud, while the larger, negatively charged particles settle nearer the bottom of the cloud.

The negative charge at the bottom of the thundercloud induces positive charge on the Earth just underneath the cloud. When the electric field between the cloud and Earth reaches the breakdown limit for moist air (about 3.3×10^5 V/m), negative charge jumps from the cloud, moving in branching steps of about 50 m each. This stepwise progression of negative charges from the cloud is called a *stepped leader* (Fig. 17.31b).

Since the average electric field strength is $\Delta V/d$, the largest field occurs where d is the smallest—between tall objects and the stepped leader. *Positive streamers*—stepwise progressions of positive charge from the surface—reach up into the air from the tallest objects. If a positive streamer connects with one of the stepped leaders, a lightning

Figure 17.30 Lightning illuminates the sky near the West Virginia state capitol building.

channel is completed; electrons rush to the ground, lighting up the bottom of the channel. The rest of the channel then glows as more electrons rush down. The other stepped leaders also glow, but less brightly than the main channel because they contain fewer electrons. The flash of light starts at the ground and moves upward so it is called a *return stroke* (Fig. 17.31c). A total of about −20 to −25 C of charge is transferred from the thundercloud to the surface.

How can you protect yourself during a thunderstorm? Stay indoors or in an automobile if possible. If you are caught out in the open, keep low to prevent yourself from being the source of positive streamers. Do not stand under a tall tree; if lightning strikes the tree, charge traveling down the tree and then along the surface puts you in grave danger. Do not lie flat on the ground, or you risk the possibility of a large potential difference developing between your feet and head when a lightning strike travels through the ground. Go to a nearby ditch or low spot if there is one. Crouch with your head low and your feet as close together as possible to minimize the potential difference between your feet.

Figure 17.31 (a) Charge separation in a thundercloud. A thunderstorm acts as a giant heat engine; work is done by the engine to separate positive charge from negative charge. (b) A stepped leader extends from the bottom of the cloud toward the surface. (c) When a positive streamer from the surface connects to a stepped leader, a complete path—a column of ionized air—is formed for charge to move between the cloud and the surface.

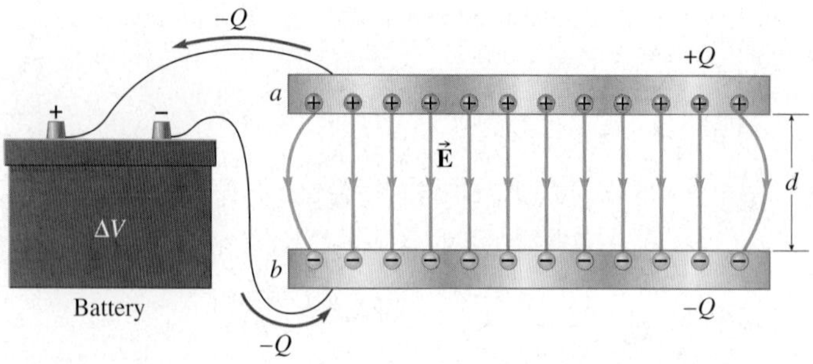

Figure 17.32 A parallel plate capacitor charged by a battery. Electrons with total charge $-Q$ are moved from the upper plate to the lower, leaving the plates with charges of equal magnitude and opposite sign.

17.7 ENERGY STORED IN A CAPACITOR

A capacitor not only stores charge; it also stores energy. Figure 17.32 shows what happens when a battery is connected to an initially uncharged capacitor. Electrons are pumped off the upper plate and onto the lower until the potential difference between the capacitor plates is equal to the potential difference ΔV maintained by the battery.

The energy stored in the capacitor can be found by summing the work done by the battery to separate the charge. As the amount of charge on the plates increases, the potential difference ΔV through which charge must be moved also increases. Suppose we look at this process at some instant of time when one plate has charge $+q_i$, the other has charge $-q_i$, and the potential difference between the plates is ΔV_i.

To avoid writing a collection of minus signs, we imagine transferring positive charge instead of the negative charge; the result is the same whether we move negative or positive charges. From the definition of capacitance,

$$\Delta V_i = \frac{q_i}{C}$$

Now the battery transfers a little more charge Δq_i from one plate to the other, increasing the electric potential energy. If Δq_i is small, the potential difference is approximately constant during the transfer. The increase in energy is

$$\Delta U_i = \Delta q_i \times \Delta V_i$$

The total energy U stored in the capacitor is the sum of all the electric potential energy increases, ΔU_i:

$$U = \sum \Delta U_i = \sum \Delta q_i \times \Delta V_i$$

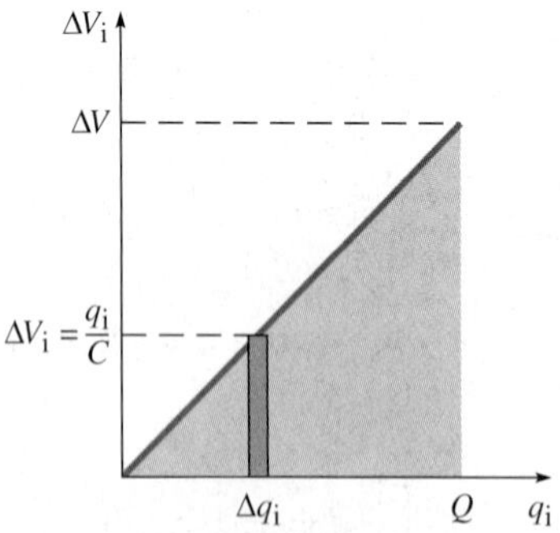

Figure 17.33 The total energy transferred is the area under the curve $\Delta V_i = q_i/C$.

We can find this sum using a graph of the potential difference ΔV_i as a function of the charge q_i (Fig. 17.33). The graph is a straight line since $\Delta V_i = q_i/C$. The energy increase $\Delta U_i = \Delta q_i \times \Delta V_i$ when a small amount of charge is transferred is represented on the graph by the area of a rectangle of height ΔV_i and width Δq_i.

Summing the energy increases means summing the areas of a series of rectangles of increasing height. Thus, the total energy stored in the capacitor is represented by the triangular area under the graph. If the final values of the charge and potential difference are Q and ΔV, then

Energy stored in a capacitor:

$$U = \text{area of triangle} = \tfrac{1}{2}(\text{base} \times \text{height})$$

$$U = \tfrac{1}{2} Q\,\Delta V \qquad \text{(17-18a)}$$

The factor of $\frac{1}{2}$ reflects the fact that the potential difference through which the charge was moved increases from zero to ΔV; the *average* potential difference through which the charge was moved is $\Delta V/2$. To move charge Q through an average potential difference of $\Delta V/2$ requires $Q\,\Delta V/2$ of work.

Equation (17-18a) can be written in other useful forms, using the definition of capacitance to eliminate either Q or ΔV.

$$U = \frac{1}{2}Q\Delta V = \frac{1}{2}(C\Delta V)\times\Delta V = \frac{1}{2}C(\Delta V)^2 \quad (17\text{-}18b)$$

$$U = \frac{1}{2}Q\Delta V = \frac{1}{2}Q\times\frac{Q}{C} = \frac{Q^2}{2C} \quad (17\text{-}18c)$$

CONNECTION:

We've used this kind of averaging before. For example, if an object starts from rest and reaches velocity v_x in a time Δt with constant acceleration, then $\Delta x = \frac{1}{2}v_x\,\Delta t$.

Example 17.12

A Defibrillator

Fibrillation is a chaotic pattern of heart activity that is ineffective at pumping blood and is therefore life-threatening. A device known as a *defibrillator* is used to shock the heart back to a normal beat pattern. The defibrillator discharges a capacitor through paddles on the skin, so that some of the charge flows through the heart (Fig. 17.34). (a) If an 11.0-μF capacitor is charged to 6.00 kV and then discharged through paddles into a patient's body, how much energy is stored in the capacitor? (b) How much charge flows through the patient's body if the capacitor discharges completely?

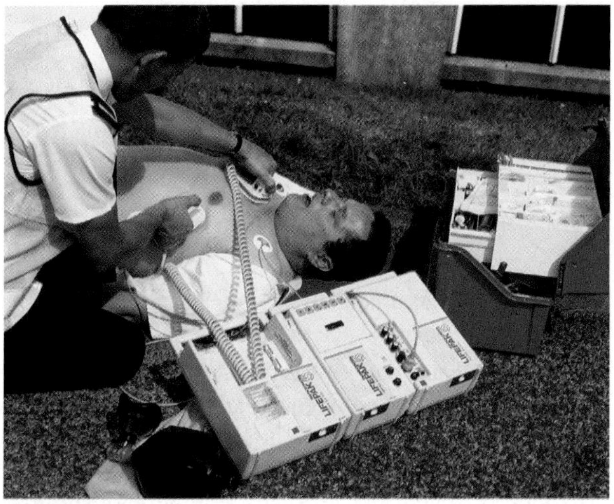

Figure 17.34

A paramedic uses a defibrillator to resuscitate a patient.

Strategy There are three equivalent expressions for energy stored in a capacitor. Since the capacitance and the potential difference are given, Eq. (17-18b) is the most direct. Since the capacitor is completely discharged, all of the charge initially on the capacitor flows through the patient's body.

Solution (a) The energy stored in the capacitor is

$$U = \tfrac{1}{2}C(\Delta V)^2 = \tfrac{1}{2}\times 11.0\times 10^{-6}\ \text{F}\times(6.00\times 10^3\ \text{V})^2 = 198\ \text{J}$$

(b) The charge initially on the capacitor is

$$Q = C\Delta V = 11.0\times 10^{-6}\ \text{F}\times 6.00\times 10^3\ \text{V} = 0.0660\ \text{C}$$

Discussion To test our result, we make a quick check:

$$U = \frac{Q^2}{2C} = \frac{(0.0660\ \text{C})^2}{2\times 11.0\times 10^{-6}\ \text{F}} = 198\ \text{J}$$

Practice Problem 17.12 Charge and Stored Energy for a Parallel Plate Capacitor

A parallel plate capacitor of area 0.24 m^2 has a plate separation, in air, of 8.00 mm. The potential difference between the plates is 0.800 kV. Find (a) the charge on the plates and (b) the stored energy.

Energy Stored in an Electric Field

Potential energy is energy of interaction or field energy. The energy stored in a capacitor is stored in the electric field between the plates. We can use the energy stored in a capacitor to calculate how much energy *per unit volume* is stored in an electric field E. Why energy per unit volume? Two capacitors can have the same electric field but store different amounts of energy. The larger capacitor stores more energy, proportional to the volume of space between the plates.

In a parallel plate capacitor, the energy stored is

$$U = \frac{1}{2}C(\Delta V)^2 = \frac{1}{2}\,\kappa\,\frac{A}{4\pi k d}\,(\Delta V)^2$$

Assuming the field is uniform between the plates, the potential difference is

$$\Delta V = Ed$$

Substituting Ed for ΔV,

$$U = \frac{1}{2}\kappa\frac{A}{4\pi kd}(Ed)^2 = \frac{1}{2}\kappa\frac{Ad}{4\pi k}E^2$$

We recognize Ad as the volume of space between the plates of the capacitor. This is the volume in which the energy is stored—$E = 0$ outside an ideal parallel plate capacitor. Then the **energy density** u—the electric potential energy per unit volume—is

$$u = \frac{U}{Ad} = \frac{1}{2}\kappa\frac{1}{4\pi k}E^2 = \frac{1}{2}\kappa\epsilon_0 E^2 \quad (17\text{-}19)$$

The energy density is proportional to the square of the field strength. This is true in general, not just for a capacitor; there is energy associated with any electric field.

Master the Concepts

- Electric potential energy can be stored in an electric field. The electric potential energy of two point charges separated by a distance r is

$$U_E = \frac{kq_1q_2}{r} \quad (U_E = 0 \text{ at } r = \infty) \quad (17\text{-}1)$$

- The signs of q_1 and q_2 determine whether the electric potential energy is positive or negative. For more than two charges, the electric potential energy is the scalar sum of the individual potential energies for each *pair* of charges.
- The electric potential V at a point is the electric potential energy per unit charge:

$$V = \frac{U_E}{q} \quad (17\text{-}3)$$

In Eq. (17-3), U_E is the electric potential energy due to the interaction of a moveable charge q with a collection of fixed charges and V is the electric potential due to that collection of fixed charges. Both U_E and V are functions of the position of the moveable charge q.

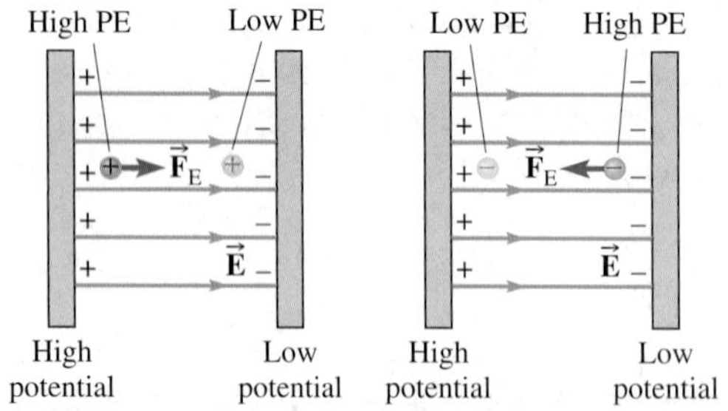

- Electric potential, like electric potential energy, is a scalar quantity. The SI unit for potential is the volt (1 V = 1 J/C).
- If a point charge q moves through a potential difference ΔV, then the change in electric potential energy is

$$\Delta U_E = q\,\Delta V \quad (17\text{-}7)$$

- The electric potential at a distance r from a point charge Q is

$$V = \frac{kQ}{r} \quad (V = 0 \text{ at } r = \infty) \quad (17\text{-}8)$$

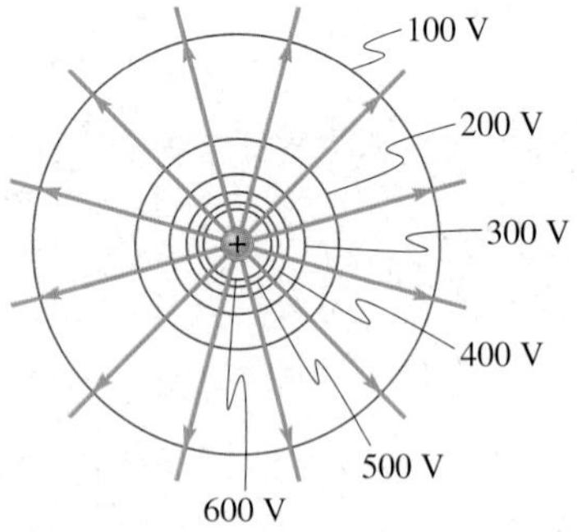

- The potential at a point P due to N point charges is the sum of the potentials due to each charge.
- An equipotential surface has the same potential at every point on the surface. An equipotential surface is perpendicular to the electric field at all points. No change in electric potential energy occurs when a charge moves from one position to another on an equipotential surface. If equipotential surfaces are drawn such that the potential difference between adjacent surfaces is constant, then the surfaces are closer together where the field is stronger.
- The electric field always points in the direction of maximum potential decrease.
- The potential difference that occurs when you move a distance d in the direction of a uniform electric field of magnitude E is

$$\Delta V = -Ed \quad (17\text{-}10)$$

- The electric field has units of

$$\text{N/C} = \text{V/m} \quad (17\text{-}11)$$

continued on next page

Master the Concepts continued

- In electrostatic equilibrium, every point in a conductor must be at the same potential.
- A capacitor consists of two conductors (the *plates*) that are given opposite charges. A capacitor stores charge and electric potential energy. Capacitance is the ratio of the magnitude of charge on each plate (Q) to the electric potential difference between the plates (ΔV). Capacitance is measured in farads (F).

$$Q = C\,\Delta V \qquad (17\text{-}14)$$

$$1\text{ F} = 1\text{ C/V}$$

- The capacitance of a parallel plate capacitor is

$$C = \kappa\frac{A}{4\pi kd} = \kappa\frac{\epsilon_0 A}{d} \qquad (17\text{-}16)$$

where A is the area of each plate, d is their separation, and ϵ_0 is the permittivity of free space [$\epsilon_0 = 1/(4\pi k) = 8.854\times10^{-12}\ \text{C}^2/(\text{N}\cdot\text{m}^2)$]. If vacuum separates the plates, $\kappa = 1$; otherwise, $\kappa > 1$ is the dielectric constant of the dielectric (the insulating material). If a dielectric is immersed in an external electric field, the dielectric constant is the ratio of the external electric field E_0 to the electric field E in the dielectric.

$$\kappa = \frac{E_0}{E} \qquad (17\text{-}17)$$

- The dielectric constant is a measure of the ease with which the insulating material can be polarized.

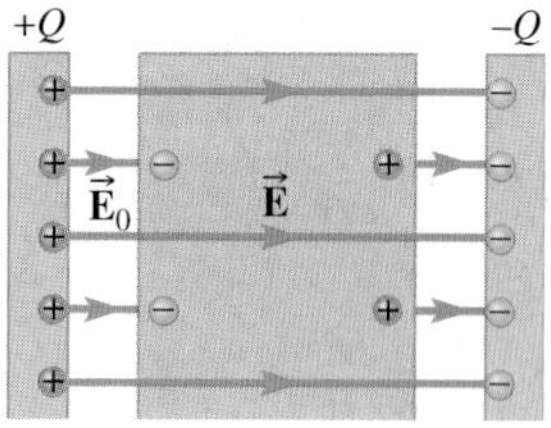

- The dielectric strength is the electric field strength at which dielectric breakdown occurs and the material becomes a conductor.
- The energy stored in a capacitor is

$$U = \tfrac{1}{2}Q\,\Delta V = \tfrac{1}{2}C(\Delta V)^2 = \frac{Q^2}{2C} \qquad (17\text{-}18)$$

- The energy density u—the electric potential energy per unit volume—associated with an electric field is

$$u = \frac{1}{2}\,\kappa\frac{1}{4\pi k}E^2 = \frac{1}{2}\,\kappa\epsilon_0 E^2 \qquad (17\text{-}19)$$

Conceptual Questions

1. A negatively charged particle with charge $-q$ is far away from a positive charge $+Q$ that is fixed in place. As $-q$ moves closer to $+Q$, (a) does the electric field do positive or negative work? (b) Does $-q$ move through a potential increase or a potential decrease? (c) Does the electric potential energy increase or decrease? (d) Repeat questions (a)–(c) if the fixed charge is instead negative ($-Q$).
2. Dry air breaks down for a voltage of about 3000 V/mm. Is it possible to build a parallel plate capacitor with a plate spacing of 1 mm that can be charged to a potential difference greater than 3000 V? If so, explain how.
3. A bird is perched on a high-voltage power line whose potential varies between −100 kV and +100 kV. Why is the bird not electrocuted?
4. A positive charge is initially at rest in an electric field and is free to move. Does the charge start to move toward a position of higher or lower potential? What happens to a negative charge in the same situation?
5. Points A and B are at the same potential. What is the total work that must be done by an external agent to move a charge from A to B? Does your answer mean that no external force need be applied? Explain.
6. A point charge moves to a region of higher potential and yet the electric potential energy *decreases*. How is this possible?
7. Why are all parts of a conductor at the same potential in electrostatic equilibrium?
8. If $E = 0$ at a single point, then a point charge placed at that point will feel no electric force. What does it mean if the *potential* is zero at a point? Are there any assumptions behind your answer?
9. If $E = 0$ everywhere throughout a region of space, what do we know is true about the potential at points in that region?
10. Explain why the woman's hair in Fig. 17.13 stands on end. Why are the hairs directed approximately radially away from her scalp? [*Hint:* Think of her head as a conducting sphere.]
11. If the potential is the same at every point throughout a region of space, is the electric field the same at every point in that region? What can you say about the magnitude of $\vec{\mathbf{E}}$ in the region? Explain.
12. If a uniform electric field exists in a region of space, is the potential the same at all points in the region? Explain.
13. When we talk about the potential difference between the plates of a capacitor, shouldn't we really specify two *points*, one on each plate, and talk about the potential difference between those points? Or doesn't it matter which points we choose? Explain.
14. A swimming pool is filled with water (total mass M) to a height h. Explain why the gravitational potential energy of the water (taking $U = 0$ at ground level) is

$\frac{1}{2}Mgh$. Where does the factor of $\frac{1}{2}$ come from? How much work must be done to fill the pool, if there is a ready supply of water at ground level? What does this have to do with capacitors? [*Hint:* Make an analogy between the capacitor and the pool. What is analogous to the water? What quantity is analogous to M? What quantity is analogous to gh?]

15. The charge on a capacitor doubles. What happens to its capacitance?

16. During a thunderstorm, some cows gather under a large tree. One cow stands facing directly toward the tree. Another cow stands at about the same distance from the tree, but it faces sideways (tangent to a circle centered on the tree). Which cow do you think is more likely to be killed if lightning strikes the tree? [*Hint:* Think about the potential difference between the cows' front and hind legs in the two positions.]

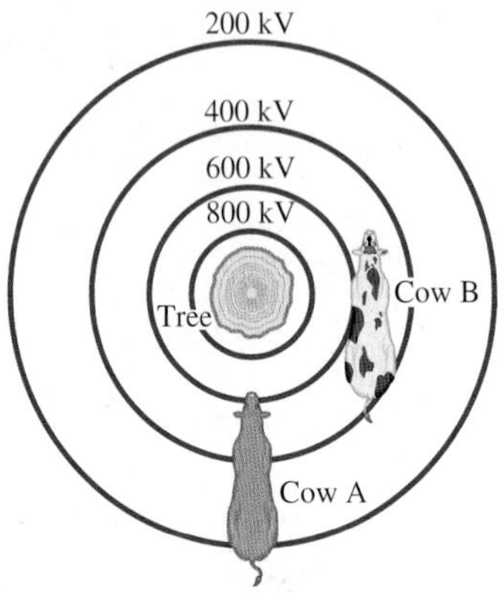

Conceptual Question 16 and Problem 67

17. If we know the potential at a single point, what (if anything) can we say about the magnitude of the electric field at that same point?

18. In Fig. 17.13, why is the person touching the dome of the van de Graaff generator not electrocuted even though there may be a potential difference of hundreds of thousands of volts between her and the ground?

19. The electric field just above Earth's surface on a clear day in an open field is about 150 V/m downward. Which is at a higher potential: the Earth or the upper atmosphere?

20. A parallel plate capacitor has the space between the plates filled with a slab of dielectric with $\kappa = 3$. While the capacitor is connected to a battery, the dielectric slab is removed. Describe *quantitatively* what happens to the capacitance, the potential difference, the charge on the plates, the electric field, and the energy stored in the capacitor as the slab is removed. [*Hint:* First figure out which quantities remain constant.]

21. Repeat Question 20 if the capacitor is charged and then disconnected from the battery before removing the dielectric slab.

22. A charged parallel plate capacitor has the space between the plates filled with air. The capacitor has been disconnected from the battery that charged it. Describe *quantitatively* what happens to the capacitance, the potential difference, the charge on the plates, the electric field, and the energy stored in the capacitor as the plates are moved closer together. [*Hint:* First figure out which quantities remain constant.]

23. A positive charge +2 μC and a negative charge −5 μC lie on a line. In which region or regions (*A*, *B*, *C*) is there a point on the line a finite distance away where the potential is zero? Explain your reasoning. Are there any points where both the electric field and the potential are zero?

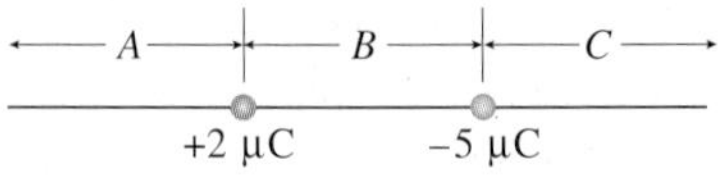

Multiple-Choice Questions

In all problems, we assign the potential due to a point charge to be zero at an infinite distance from the charge.

1. Two charges are located at opposite corners (*A* and *C*) of a square. We do not know the magnitude or sign of these charges. What can be said about the potential at corner *B* relative to the potential at corner *D*?

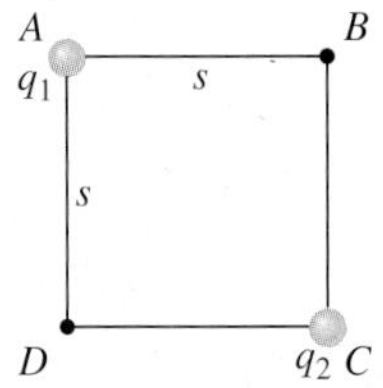

   (a) It is the same as that at *D*.
   (b) It is different from that at *D*.
   (c) It is the same as that at *D* only if the charges at *A* and *C* are equal.
   (d) It is the same as that at *D* only if the charges at *A* and *C* are equal in magnitude and opposite in sign.

2. Among these choices, which is/are correct units for electric field?
   (a) N/kg only (b) N/C only
   (c) N only (d) N·m/C only
   (e) V/m only (f) both N/C and V/m

3. In the diagram, the potential is zero at which of the points *A*–*E*?

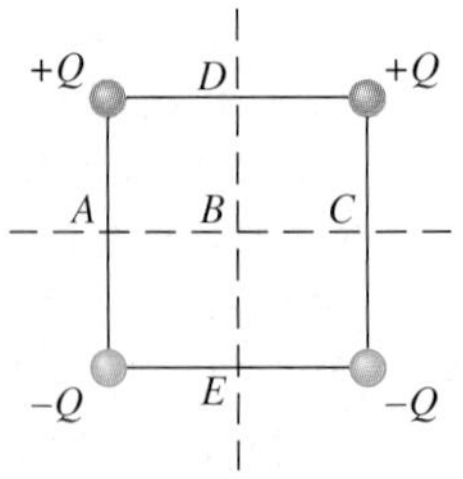

   (a) *B*, *D*, and *E*
   (b) *B* only
   (c) *A*, *B*, and *C*
   (d) all five points
   (e) all except *B*

4. Which of these units can be used to measure electric potential?
   (a) N/C (b) J (c) V·m (d) V/m (e) $\frac{\text{N·m}}{\text{C}}$

5. A parallel plate capacitor is attached to a battery that supplies a constant potential difference. While the battery is still attached, the parallel plates are separated a little more. Which statement describes what happens?
   (a) The electric field increases and the charge on the plates decreases.
   (b) The electric field remains constant and the charge on the plates increases.

(c) The electric field remains constant and the charge on the plates decreases.
(d) Both the electric field and the charge on the plates decrease.

6. A capacitor has been charged with $+Q$ on one plate and $-Q$ on the other plate. Which of these statements is true?
(a) The potential difference between the plates is QC.
(b) The energy stored is $\frac{1}{2}Q\,\Delta V$.
(c) The energy stored is $\frac{1}{2}Q^2C$.
(d) The potential difference across the plates is $Q^2/(2C)$.
(e) None of the previous statements is true.

7. Two solid metal spheres of different radii are far apart. The spheres are connected by a fine metal wire. Some charge is placed on one of the spheres. After electrostatic equilibrium is reached, the wire is removed. Which of these quantities will be the *same* for the two spheres?
(a) the charge on each sphere
(b) the electric field inside each sphere, at the same distance from the center of the spheres
(c) the electric field just outside the surface of each sphere
(d) the electric potential at the surface of each sphere
(e) both (b) and (c) (f) both (b) and (d)
(g) both (a) and (c)

8. A large negative charge $-Q$ is located in the vicinity of points A and B. Suppose a positive charge $+q$ is moved at constant speed from A to B by an external agent. Along which of the paths shown in the figure will the work done by the field be the greatest?

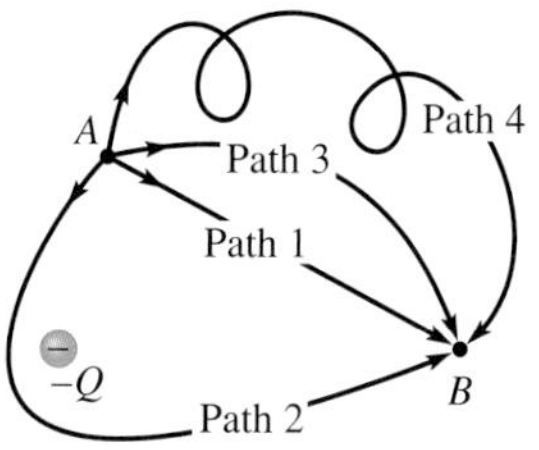

(a) path 1 (b) path 2 (c) path 3 (d) path 4
(e) Work is the same along all four paths.

9. A tiny charged pellet of mass m is suspended at rest between two horizontal, charged metallic plates. The lower plate has a positive charge and the upper plate has a negative charge. Which statement in the answers here is *not* true?

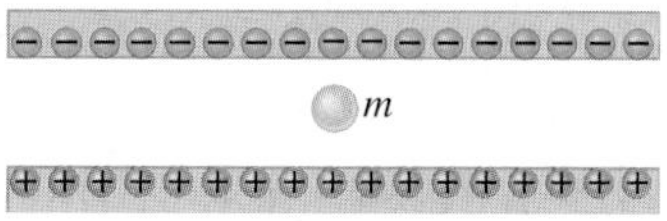

(a) The electric field between the plates points vertically upward.
(b) The pellet is negatively charged.
(c) The magnitude of the electric force on the pellet is equal to mg.
(d) The plates are at different potentials.

10. Two positive 2.0-μC point charges are placed as shown in part (a) of the figure. The distance from each charge to the point P is 0.040 m. Then the charges are rearranged as shown in part (b) of the figure. Which statement is now true concerning $\vec{\mathbf{E}}$ and V at point P?

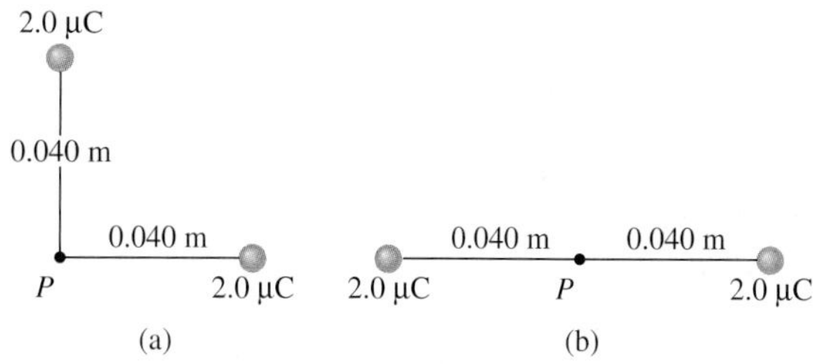

(a) The electric field and the electric potential are both zero.
(b) $\vec{\mathbf{E}} = 0$, but V is the same as before the charges were moved.
(c) $V = 0$, but $\vec{\mathbf{E}}$ is the same as before the charges were moved.
(d) $\vec{\mathbf{E}}$ is the same as before the charges were moved, but V is less than before.
(e) Both $\vec{\mathbf{E}}$ and V have changed and neither is zero.

11. In the diagram, which two points are closest to being at the same potential?
(a) A and D (b) B and C
(c) B and D (d) A and C

12. In the diagram, which point is at the lowest potential?
(a) A (b) B (c) C (d) D

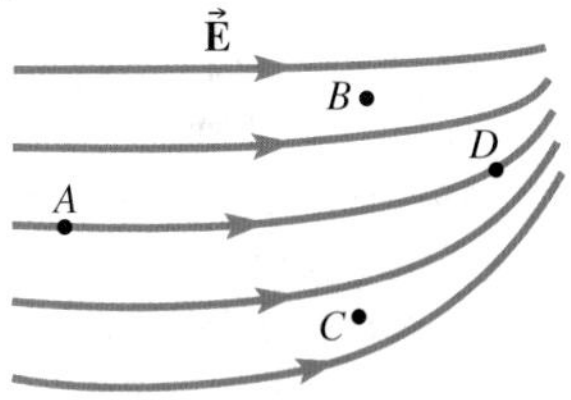

Multiple-Choice Questions 11 and 12

Problems

Combination conceptual/quantitative problem
Biological or medical application
Challenging problem
Blue # Detailed solution in the Student Solutions Manual
1 2 Problems paired by concept
Text website interactive or tutorial

17.1 Electric Potential Energy

1. Two point charges, +5.0 μC and −2.0 μC, are separated by 5.0 m. What is the electric potential energy?

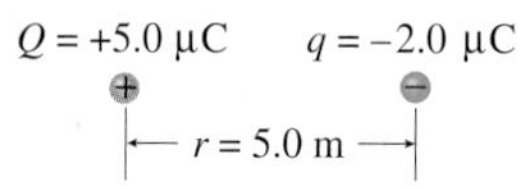

2. A hydrogen atom has a single proton at its center and a single electron at a distance of approximately 0.0529 nm from the proton. (a) What is the electric potential energy in joules? (b) What is the significance of the sign of the answer?

3. How much work is done by an applied force that moves two charges of 6.5 μC that are initially very far apart to a distance of 4.5 cm apart?

4. The nucleus of a helium atom contains two protons that are approximately 1 fm apart. How much work must be done by an external agent to bring the two protons from an infinite separation to a separation of 1.0 fm?

5. How much work does it take for an external force to set up the arrangement of charged objects in the diagram on the corners of a right triangle when the three objects are initially very far away from each other?

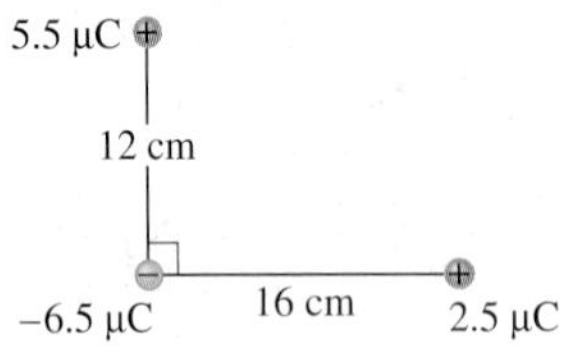

Problems 6–9. Two point charges (+10.0 nC and −10.0 nC) are located 8.00 cm apart. For each problem, let $U = 0$ when *all* of the charges are separated by infinite distances.

6. What is the potential energy for these two charges?

7. What is the potential energy if a third point charge $q = -4.2$ nC is placed at point a?

8. What is the potential energy if a third point charge $q = -4.2$ nC is placed at point b?

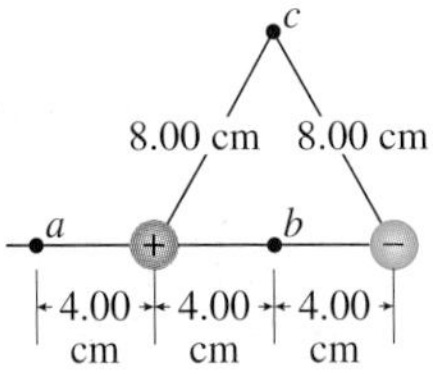

Problems 6–9

9. What is the potential energy if a third point charge $q = -4.2$ nC is placed at point c?

10. Find the electric potential energy for the following array of charges: charge $q_1 = +4.0$ μC is located at $(x, y) = (0.0, 0.0)$ m; charge $q_2 = +3.0$ μC is located at (4.0, 3.0) m; and charge $q_3 = -1.0$ μC is located at (0.0, 3.0) m.

11. In the diagram, how much work is done *by the electric field* as a third charge $q_3 = +2.00$ nC is moved from infinity to point a?

12. In the diagram, how much work is done *by the electric field* as a third charge $q_3 = +2.00$ nC is moved from infinity to point b?

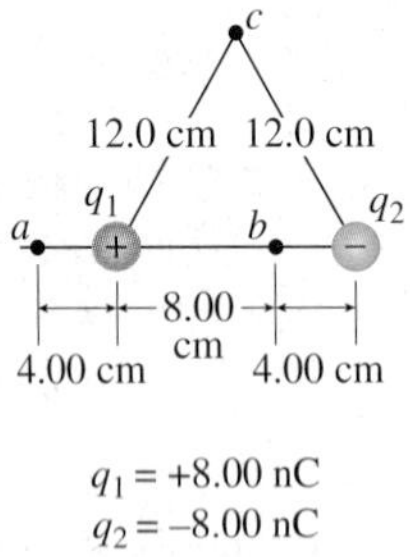

Problems 11–14

13. In the diagram, how much work is done *by the electric field* as a third charge $q_3 = +2.00$ nC is moved from point a to point b?

14. In the diagram, how much work is done *by the electric field* as a third charge $q_3 = +2.00$ nC is moved from point b to point c?

17.2 Electric Potential

15. A point charge $q = +3.0$ nC moves through a potential difference $\Delta V = V_f - V_i = +25$ V. What is the change in the electric potential energy?

16. An electron is moved from point A, where the electric potential is $V_A = -240$ V, to point B, where the electric potential is $V_B = -360$ V. What is the change in the electric potential energy?

17. Find the electric field and the potential at the center of a square of side 2.0 cm with charges of +9.0 μC at each corner.

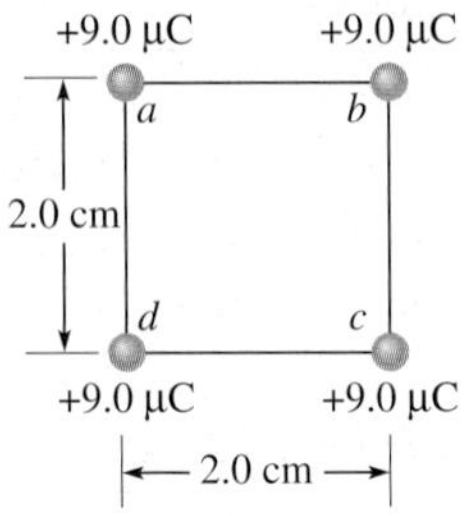

18. Find the electric field and the potential at the center of a square of side 2.0 cm with a charge of +9.0 μC at one corner of the square and with charges of −3.0 μC at the remaining three corners.

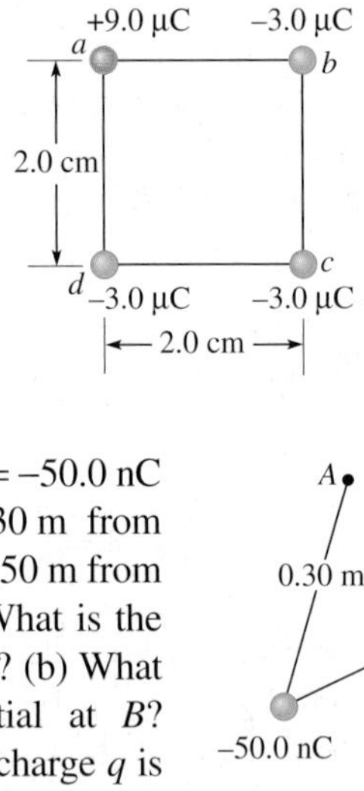

19. A charge $Q = -50.0$ nC is located 0.30 m from point A and 0.50 m from point B. (a) What is the potential at A? (b) What is the potential at B? (c) If a point charge q is moved from A to B while Q is fixed in place, through what potential difference does it move? Does its potential increase or decrease? (d) If $q = -1.0$ nC, what is the change in electric potential energy as it moves from A to B? Does the potential energy increase or decrease? (e) How much work is done by the electric field due to charge Q as q moves from A to B?

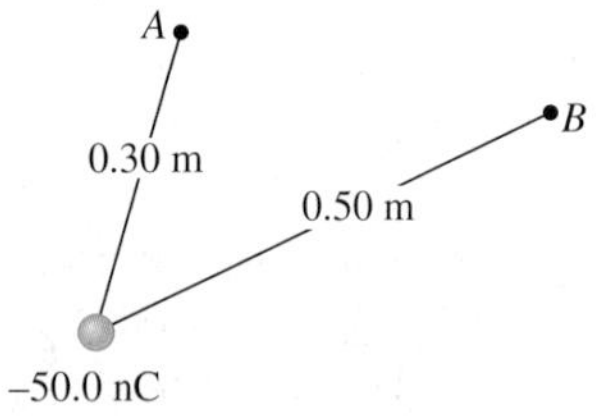

20. A charge of +2.0 mC is located at $x = 0$, $y = 0$ and a charge of −4.0 mC is located at $x = 0$, $y = 3.0$ m. What is the electric potential due to these charges at a point with coordinates $x = 4.0$ m, $y = 0$?

21. The electric potential at a distance of 20.0 cm from a point charge is +1.0 kV (assuming $V = 0$ at infinity). (a) Is the point charge positive or negative? (b) At what distance is the potential +2.0 kV? (WWW tutorial: field and potential of point charge)

22. A spherical conductor with a radius of 75.0 cm has an electric field of magnitude 8.40×10^5 V/m just outside its surface. What is the electric potential just outside the surface, assuming the potential is zero far away from the conductor?

23. An array of four charges is arranged along the x-axis at intervals of 1.0 m. (a) If two of the charges are $+1.0\ \mu$C and two are $-1.0\ \mu$C, draw a configuration of these charges that minimizes the potential at $x = 0$. (b) If three of the charges are the same, $q = +1.0\ \mu$C, and the charge at the far right is $-1.0\ \mu$C, what is the potential at the origin?

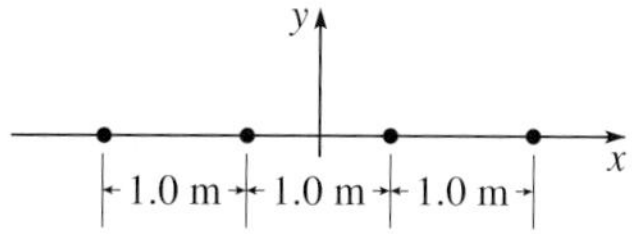

24. At a point P, a distance R_0 from a positive charge Q_0, the electric field has a magnitude $E_0 = 100$ N/C and the electric potential is $V_0 = 10$ V. The charge is now increased by a factor of three, becoming $3Q_0$. (a) At what distance, R_E, from the charge $3Q_0$ will the electric field have the same value, $E = E_0$; and (b) at what distance, R_V, from the charge $3Q_0$ will the electric potential have the same value, $V = V_0$?

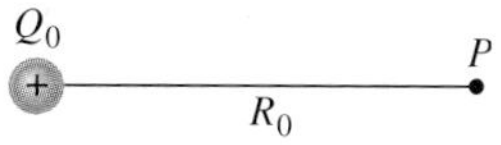

25. Charges of +2.0 nC and −1.0 nC are located at opposite corners, A and C, respectively, of a square which is 1.0 m on a side. What is the electric potential at a third corner, B, of the square (where there is no charge)?

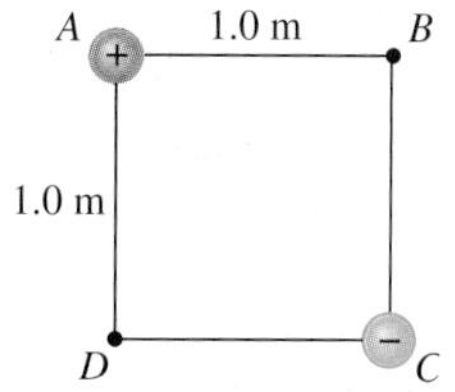

26. (a) Find the electric potential at points a and b for charges of +4.2 nC and −6.4 nC located as shown in the figure. (b) What is the potential difference ΔV for a trip from a to b? (c) How much work must be done by an external agent to move a point charge of +1.50 nC from a to b?

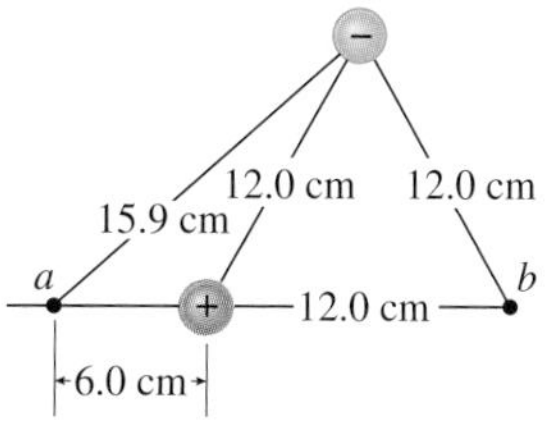

27. (a) Find the potential at points a and b in the diagram for charges $Q_1 = +2.50$ nC and $Q_2 = -2.50$ nC. (b) How much work must be done by an external agent to bring a point charge q from infinity to point b?

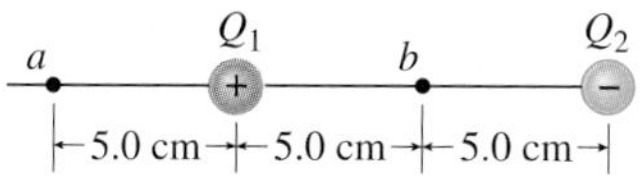

28. (a) In the diagram, what are the potentials at points a and b? Let $V = 0$ at infinity. (b) What is the change in electric potential energy if a third charge $q_3 = +2.00$ nC is moved from point a to point b? (If you have done Problem 13, compare your answers.)

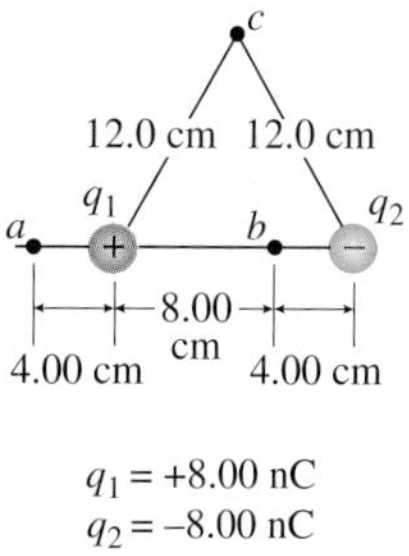

Problems 28 and 29

29. (a) In the diagram, what are the potentials at points b and c? Let $V = 0$ at infinity. (b) What is the change in electric potential energy if a third charge $q_3 = +2.00$ nC is moved from point b to point c? (If you have done Problem 14, compare your answers.)

17.3 The Relationship Between Electric Field and Potential

30. By rewriting each unit in terms of kilograms, meters, seconds, and coulombs, show that 1 N/C = 1 V/m.

31. A uniform electric field has magnitude 240 N/C and is directed to the right. A particle with charge +4.2 nC moves along the straight line from a to b. (a) What is the electric force that acts on the particle? (b) What is the work done on the particle by the electric field? (c) What is the potential difference $V_a - V_b$ between points a and b?

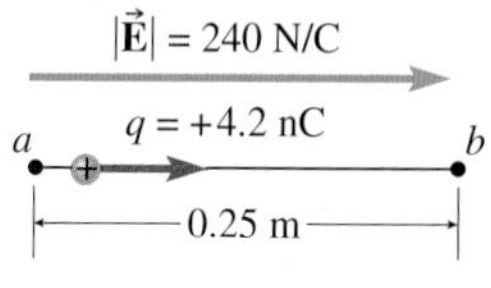

32. In a region where there is an electric field, the electric forces do $+8.0 \times 10^{-19}$ J of work on an electron as it moves from point X to point Y. (a) Which point, X or Y, is at a higher potential? (b) What is the potential difference, $V_Y - V_X$, between point Y and point X?

33. Suppose a uniform electric field of magnitude 100.0 N/C exists in a region of space. How far apart are a pair of equipotential surfaces whose potentials differ by 1.0 V?

34. Draw some electric field lines and a few equipotential surfaces outside a negatively charged hollow conducting sphere. What shape are the equipotential surfaces?

35. Draw some electric field lines and a few equipotential surfaces outside a positively charged conducting cylinder. What shape are the equipotential surfaces?

36. It is believed that a large electric fish known as *Torpedo occidentalis* uses electricity to shock its victims. A typical fish can deliver a potential difference of 0.20 kV for a duration of 1.5 ms. This pulse delivers charge at a rate of 18 C/s. (a) What is the rate at which work is done by the electric organs during a pulse? (b) What is the total amount of work done during one pulse?

37. A positive point charge is located at the center of a hollow spherical metal shell with zero net charge. (a) Draw some electric field lines and sketch some equipotential surfaces for this arrangement. (b) Sketch graphs of the electric field magnitude and the potential as functions of r.

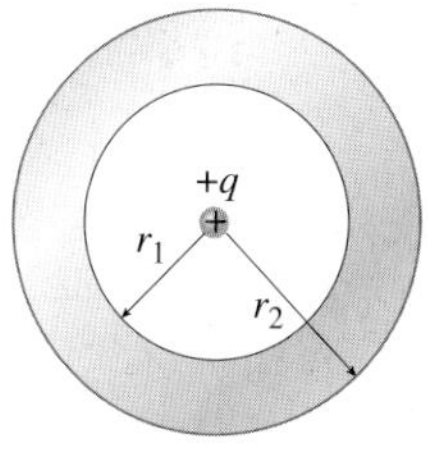

Problems 38 and 39. A positively charged oil drop is injected into a region of uniform electric field between two oppositely charged, horizontally oriented plates spaced 16 cm apart.

38. If the electric force on the drop is found to be 9.6×10^{-16} N and the potential difference between the plates is 480 V, what is the magnitude of the charge on the drop in terms of the elementary charge e? Ignore the small buoyant force on the drop. (tutorial: Millikan's experiment)

39. If the mass of the drop is 1.0×10^{-15} kg and it remains stationary when the potential difference between the plates is 9.76 kV, what is the magnitude of the charge on the drop? (Ignore the small buoyant force on the drop.)

17.4 Conservation of Energy for Moving Charges

40. Point P is at a potential of 500.0 kV and point S is at a potential of 200.0 kV. The space between these points is evacuated. When a charge of $+2e$ moves from P to S, by how much does its kinetic energy change?

41. An electron is accelerated from rest through a potential difference ΔV. If the electron reaches a speed of 7.26×10^6 m/s, what is the potential difference? Be sure to include the correct sign. (Does the electron move through an increase or a decrease in potential?)

42. As an electron moves through a region of space, its speed decreases from 8.50×10^6 m/s to 2.50×10^6 m/s. The electric force is the only force acting on the electron. (a) Did the electron move to a higher potential or a lower potential? (b) Across what potential difference did the electron travel?

43. In the electron gun of Example 17.8, if the potential difference between the cathode and anode is reduced to 6.0 kV, with what speed will the electrons reach the anode?

44. In the electron gun of Example 17.8, if the electrons reach the anode with a speed of 3.0×10^7 m/s, what is the potential difference between the cathode and the anode?

45. A beam of electrons of mass m_e is deflected vertically by the uniform electric field between two oppositely charged, parallel metal plates. The plates are a distance d apart and the potential difference between the plates is ΔV. (a) What is the direction of the electric field between the plates? (b) If the y-component of the electrons' velocity as they leave the region between the plates is v_y, derive an expression for the time it takes each electron to travel through the region between the plates in terms of ΔV, v_y, m_e, d, and e. (c) Does the electric potential energy of an electron increase, decrease, or stay constant while it moves between the plates? Explain.

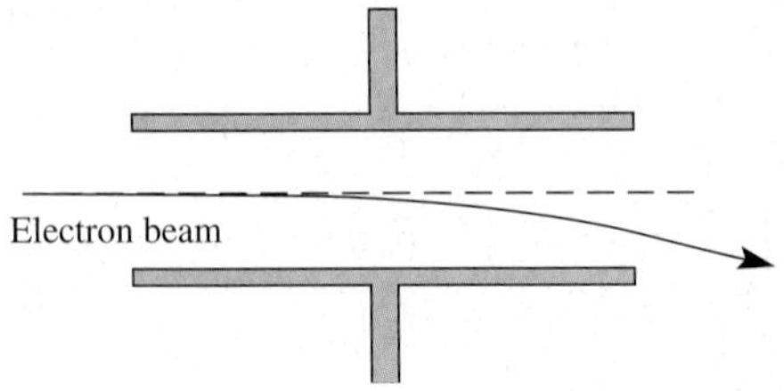

46. An electron (charge $-e$) is projected horizontally into the space between two oppositely charged parallel plates. The electric field between the plates is 500.0 N/C upward. If the vertical deflection of the electron as it leaves the plates has magnitude 3.0 mm, how much has its kinetic energy increased due to the electric field? [*Hint:* First find the potential difference through which the electron moves.]

47. An alpha particle (charge $+2e$) moves through a potential difference $\Delta V = -0.50$ kV. Its initial kinetic energy is 1.20×10^{-16} J. What is its final kinetic energy?

48. In 1911, Ernest Rutherford discovered the nucleus of the atom by observing the scattering of helium nuclei from gold nuclei. If a helium nucleus with a mass of 6.68×10^{-27} kg, a charge of $+2e$, and an initial velocity of 1.50×10^7 m/s is projected head-on toward a gold nucleus with a charge of $+79e$, how close will the helium atom come to the gold nucleus before it stops and turns around? (Assume the gold nucleus is held in place by other gold atoms and does not move.)

49. The figure shows a graph of electric potential versus position along the x-axis. A proton is originally at point A, moving in the positive x-direction. How much kinetic energy does it need to have at point A in order to be able to reach point E (with no forces acting on the electron other than those due to the indicated potential)? Points B, C, and D have to be passed on the way.

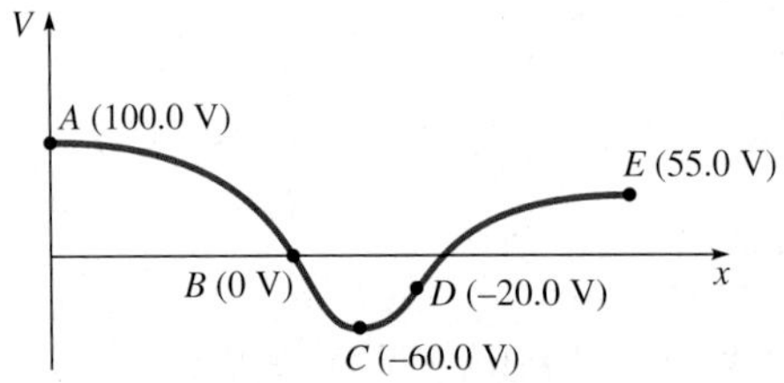

Problems 49 and 50

50. Repeat Problem 49 for an electron rather than a proton.

17.5 Capacitors

51. A 2.0-μF capacitor is connected to a 9.0-V battery. What is the magnitude of the charge on each plate?

52. The plates of a 15.0-μF capacitor have net charges of +0.75 μC and –0.75 μC, respectively. (a) What is the potential difference between the plates? (b) Which plate is at the higher potential?

53. If a capacitor has a capacitance of 10.2 μF and we wish to lower the potential difference across the plates by 60.0 V, what magnitude of charge will we have to remove from each plate?

54. A parallel plate capacitor has a capacitance of 2.0 μF and plate separation of 1.0 mm. (a) How much potential difference can be placed across the capacitor before dielectric breakdown of air occurs ($E_{max} = 3 \times 10^6$ V/m)? (b) What is the magnitude of the greatest charge the capacitor can store before breakdown?

55. A parallel plate capacitor is charged by connecting it to a 12-V battery. The battery is then disconnected from the capacitor. The plates are then pulled apart so the spacing between the plates is increased. What is the effect (a) on the electric field between the plates? (b) on the potential difference between the plates?

56. A parallel plate capacitor has a capacitance of 1.20 nF. There is a charge of magnitude 0.800 μC on each plate. (a) What is the potential difference between the plates? (b) If the plate separation is doubled, while the charge is kept constant, what will happen to the potential difference?

57. A parallel plate capacitor is connected to a 12-V battery. While the battery remains connected, the plates are pushed together so the spacing is decreased. What is the effect on (a) the potential difference between the plates? (b) the electric field between the plates? (c) the magnitude of charge on the plates? (tutorial: capacitor)

58. A parallel plate capacitor has a capacitance of 1.20 nF and is connected to a 12-V battery. (a) What is the magnitude of the charge on each plate? (b) If the plate separation is doubled while the plates remain connected to the battery, what happens to the charge on each plate and the electric field between the plates?

59. A variable capacitor is made of two parallel semicircular plates with air between them. One plate is fixed in place and the other can be rotated. The electric field is zero everywhere except in the region where the plates overlap. When the plates are directly across from one another, the capacitance is 0.694 pF. (a) What is the capacitance when the movable plate is rotated so that only one half its area is across from the stationary plate? (b) What is the capacitance when the movable plate is rotated so that two thirds of its area is across from the stationary plate?

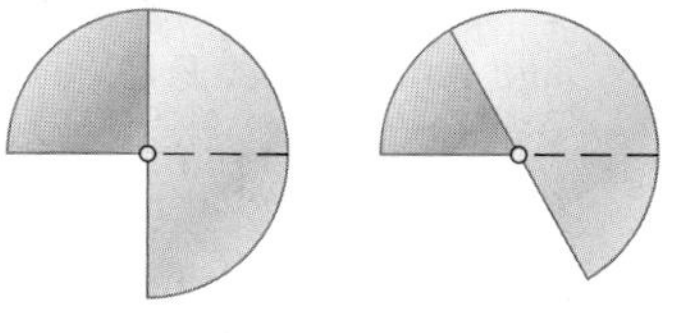

60. A shark is able to detect the presence of electric fields as small as 1.0 μV/m. To get an idea of the magnitude of this field, suppose you have a parallel plate capacitor connected to a 1.5-V battery. How far apart must the parallel plates be to have an electric field of 1.0 μV/m between the plates?

61. Two metal spheres have charges of equal magnitude, 3.2×10^{-14} C, but opposite sign. If the potential difference between the two spheres is 4.0 mV, what is the capacitance? [*Hint:* The "plates" are not parallel, but the definition of capacitance holds.]

62. Suppose you were to wrap the Moon in aluminum foil and place a charge Q on it. What is the capacitance of the Moon in this case? [*Hint:* It is not necessary to have two oppositely charged conductors to have a capacitor. Use the definition of potential for a spherical conductor and the definition of capacitance to get your answer.]

63. A tiny hole is made in the center of the negatively and positively charged plates of a capacitor, allowing a beam of electrons to pass through and emerge from the far side. If 40.0 V are applied across the capacitor plates and the electrons enter through the hole in the negatively charged plate with a speed of 2.50×10^6 m/s, what is the speed of the electrons as they emerge from the hole in the positive plate?

64. A spherical conductor of radius R carries a total charge Q. (a) Show that the magnitude of the electric field just outside the sphere is $E = \sigma/\epsilon_0$, where σ is the charge per unit area on the conductor's surface. (b) Construct an argument to show why the electric field at a point P just outside *any* conductor in electrostatic equilibrium has magnitude $E = \sigma/\epsilon_0$, where σ is the local surface charge density. [*Hint:* Consider a tiny area of an arbitrary conductor and compare it to an area of the same size on a spherical conductor with the same charge density. Think about the number of field lines starting or ending on the two areas.]

17.6 Dielectrics

65. A 6.2-cm by 2.2-cm parallel plate capacitor has the plates separated by a distance of 2.0 mm. (a) When 4.0×10^{-11} C of charge is placed on this capacitor, what is the electric field between the plates? (b) If a dielectric with dielectric constant of 5.5 is placed between the plates while the charge on the capacitor stays the same, what is the electric field in the dielectric?

66. Before a lightning strike can occur, the breakdown limit for *damp* air must be reached. If this occurs for an electric field of 3.33×10^5 V/m, what is the maximum possible height above the Earth for the bottom of a thundercloud, which is at a potential 1.00×10^8 V below Earth's surface potential, if there is to be a lightning strike?

67. Two cows, with approximately 1.8 m between their front and hind legs, are standing under a tree during a thunderstorm. See the diagram with Conceptual Question 16. (a) If the equipotential surfaces about the tree just after a lightning strike are as shown, what is the average electric field between Cow A's front and hind legs? (b) Which cow is more likely to be killed? Explain.

68. A parallel plate capacitor has a charge of 0.020 μC on each plate with a potential difference of 240 V. The parallel plates are separated by 0.40 mm of bakelite. What is the capacitance of this capacitor?

69. Two metal spheres are separated by a distance of 1.0 cm and a power supply maintains a constant potential difference of 900 V between them. The spheres are brought closer to one another until a spark flies between them. If the dielectric strength of dry air is 3.0×10^6 V/m, what is the distance between the spheres at this time?

70. To make a parallel plate capacitor, you have available two flat plates of aluminum (area 120 cm^2), a sheet of paper (thickness = 0.10 mm, $\kappa = 3.5$), a sheet of glass (thickness = 2.0 mm, $\kappa = 7.0$), and a slab of paraffin (thickness = 10.0 mm, $\kappa = 2.0$). (a) What is the largest capacitance possible using one of these dielectrics? (b) What is the smallest?

71. A capacitor can be made from two sheets of aluminum foil separated by a sheet of waxed paper. If the sheets of aluminum are 0.30 m by 0.40 m and the waxed paper, of slightly larger dimensions, is of thickness 0.030 mm and dielectric constant $\kappa = 2.5$, what is the capacitance of this capacitor?

72. In capacitive electrostimulation, electrodes are placed on opposite sides of a limb. A potential difference is applied to the electrodes, which is believed to be beneficial in treating bone defects and breaks. If the capacitance is measured to be 0.59 pF, the electrodes are 4.0 cm^2 in area, and the limb is 3.0 cm in diameter, what is the (average) dielectric constant of the tissue in the limb?

17.7 Energy Stored in a Capacitor

73. A certain capacitor stores 450 J of energy when it holds 8.0×10^{-2} C of charge. What is (a) the capacitance of this capacitor and (b) the potential difference across the plates?

74. What is the maximum electric energy density possible in dry air without dielectric breakdown occurring?

75. A parallel plate capacitor has a charge of 5.5×10^{-7} C on one plate and -5.5×10^{-7} C on the other. The distance between the plates is increased by 50% while the charge on each plate stays the same. What happens to the energy stored in the capacitor?

76. A large parallel plate capacitor has plate separation of 1.00 cm and plate area of 314 cm^2. The capacitor is connected across a voltage of 20.0 V and has air between the plates. How much work is done on the capacitor as the plate separation is increased to 2.00 cm?

77. Figure 17.31b shows a thundercloud before a lightning strike has occurred. The bottom of the thundercloud and the Earth's surface might be modeled as a charged parallel plate capacitor. The base of the cloud, which is roughly parallel to the Earth's surface, serves as the negative plate and the region of Earth's surface under the cloud serves as the positive plate. The separation between the cloud base and the Earth's surface is small compared to the length of the cloud. (a) Find the capacitance for a thundercloud of base dimensions 4.5 km by 2.5 km located 550 m above the Earth's surface. (b) Find the energy stored in this capacitor if the charge magnitude is 18 C.

78. A parallel plate capacitor of capacitance 6.0 μF has the space between the plates filled with a slab of glass with $\kappa = 3.0$. The capacitor is charged by attaching it to a 1.5-V battery. After the capacitor is disconnected from the battery, the dielectric slab is removed. Find (a) the capacitance, (b) the potential difference, (c) the charge on the plates, and (d) the energy stored in the capacitor after the glass is removed.

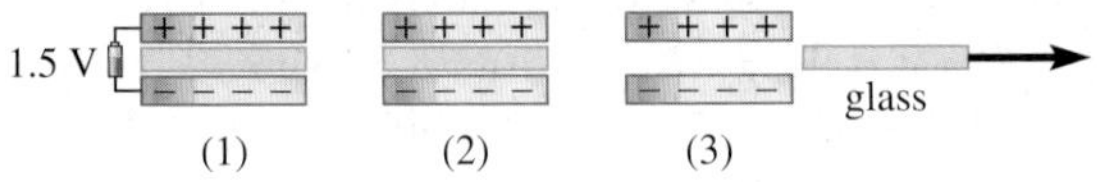

79. A parallel plate capacitor is composed of two square plates, 10.0 cm on a side, separated by an air gap of 0.75 mm. (a) What is the charge on this capacitor when there is a potential difference of 150 V between the plates? (b) What energy is stored in this capacitor?

80. The capacitor of Problem 79 is initially charged to a 150-V potential difference. The plates are then physically separated by another 0.750 mm in such a way that none of the charge can leak off the plates. Find (a) the new capacitance and (b) the new energy stored in the capacitor. Explain the result using conservation of energy.

81. Capacitors are used in many applications where you need to supply a short burst of energy. A 100.0-μF capacitor in an electronic flash lamp supplies an average power of 10.0 kW to the lamp for 2.0 ms. (a) To what potential difference must the capacitor initially be charged? (b) What is its initial charge?

82. A parallel plate capacitor has a charge of 0.020 μC on each plate with a potential difference of 240 V. The parallel plates are separated by 0.40 mm of air. What energy is stored in this capacitor?

83. A parallel plate capacitor has a capacitance of 1.20 nF. There is a charge of 0.80 μC on each plate. How much work must be done by an external agent to double the plate separation while keeping the charge constant?

84. A defibrillator is used to restart a person's heart after it stops beating. Energy is delivered to the heart by discharging a capacitor through the body tissues near the heart. If the capacitance of the defibrillator is 9 μF and the energy delivered is to be 300 J, to what potential difference must the capacitor be charged?

85. A defibrillator consists of a 15-μF capacitor that is charged to 9.0 kV. (a) If the capacitor is discharged in 2.0 ms, how much charge passes through the body tissues? (b) What is the average power delivered to the tissues?

86. The bottom of a thundercloud is at a potential of -1.00×10^8 V with respect to Earth's surface. If a charge of -25.0 C is transferred to the Earth during a lightning strike, find the electric potential energy released.

(Assume that the system acts like a capacitor—as charge flows, the potential difference decreases to zero.)

◆87. (a) If the bottom of a thundercloud has a potential of -1.00×10^9 V with respect to Earth and a charge of -20.0 C is discharged from the cloud to Earth during a lightning strike, how much electric potential energy is released? (Assume that the system acts like a capacitor—as charge flows, the potential difference decreases to zero.) (b) If a tree is struck by the lightning bolt and 10.0% of the energy released vaporizes sap in the tree, about how much sap is vaporized? (Assume the sap to have the same latent heat as water.) (c) If 10.0% of the energy released from the lightning strike could be stored and used by a homeowner who uses 400.0 kW·hr of electricity per month, for how long could the lightning bolt supply electricity to the home?

Comprehensive Problems

88. Charges of -12.0 nC and -22.0 nC are separated by 0.700 m. What is the potential midway between the two charges?

89. Two point charges (+10.0 nC and -10.0 nC) are located 8.00 cm apart. (a) What is the electric potential energy when a point charge of -4.2 nC is placed at points *a*, *b*, and *c* in turn? Let $U = 0$ when the -4.2 nC charge is far away (but the other two are still in place). (b) How much work would an external force have to do to move the point charge from *b* to *a*?

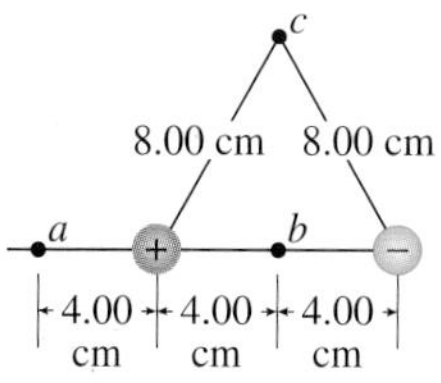

90. If an electron moves from one point at a potential of -100.0 V to another point at a potential of $+100.0$ V, how much work is done by the electric field?

91. A van de Graaff generator has a metal sphere of radius 15 cm. To what potential can it be charged before the electric field at its surface exceeds 3.0×10^6 N/C (which is sufficient to break down dry air and initiate a spark)?

92. Find the potential at the sodium ion, Na^+, which is surrounded by two chloride ions, Cl^-, and a calcium ion, Ca^{2+}, in water as shown in the diagram. The effective charge of the positive sodium ion in water is 2.0×10^{-21} C, of the negative chlorine ion is -2.0×10^{-21} C, and of the positive calcium ion is 4.0×10^{-21} C.

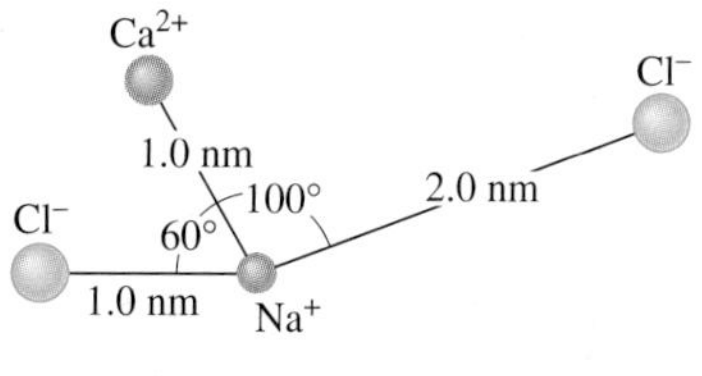

93. An infinitely long conducting cylinder sits near an infinite conducting sheet (side view in the diagram). The cylinder and sheet have equal and opposite charges; the cylinder is positive. (a) Sketch some electric field lines. (b) Sketch some equipotential surfaces.

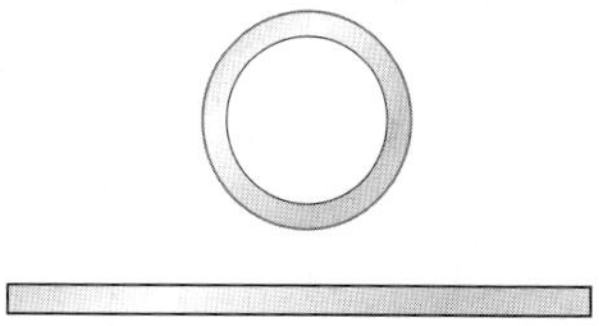

94. Two parallel plates are 4.0 cm apart. The bottom plate is charged positively and the top plate is charged negatively, producing a uniform electric field of 5.0×10^4 N/C in the region between the plates. What is the time required for an electron, which starts at rest at the upper plate, to reach the lower plate? (Assume a vacuum exists between the plates.)

95. The potential difference across a cell membrane is -90 mV. If the membrane's thickness is 10 nm, what is the magnitude of the electric field in the membrane? Assume the field is uniform.

96. A beam of electrons traveling with a speed of 3.0×10^7 m/s enters a uniform, downward electric field of magnitude 2.0×10^4 N/C between the deflection plates of an oscilloscope. The initial velocity of the electrons is perpendicular to the field. The plates are 6.0 cm long. (a) What is the direction and magnitude of the change in velocity of the electrons while they are between the plates? (b) How far are the electrons deflected in the $\pm y$-direction while between the plates?

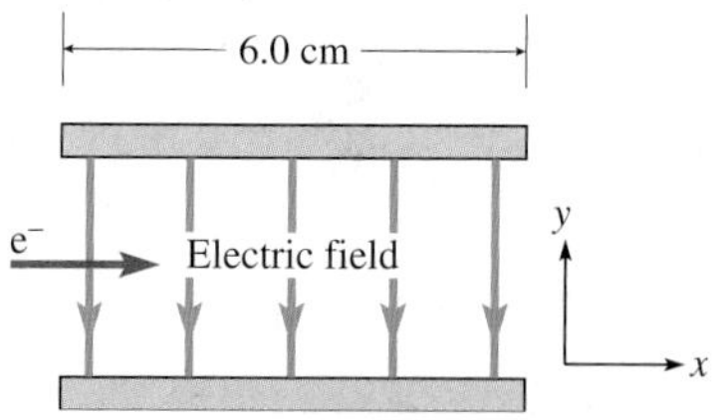

97. A negatively charged particle of mass 5.00×10^{-19} kg is moving with a speed of 35.0 m/s when it enters the region between two parallel capacitor plates. The initial velocity of the charge is parallel to the plate surfaces and in the positive *x*-direction. The plates are square with a side of 1.00 cm and the voltage across the plates is 3.00 V. If the particle is initially 1.00 mm from both plates, and it just barely clears the positive plate after traveling 1.00 cm through the region between the plates, how many excess electrons are on the particle? You may ignore gravitational and edge effects.

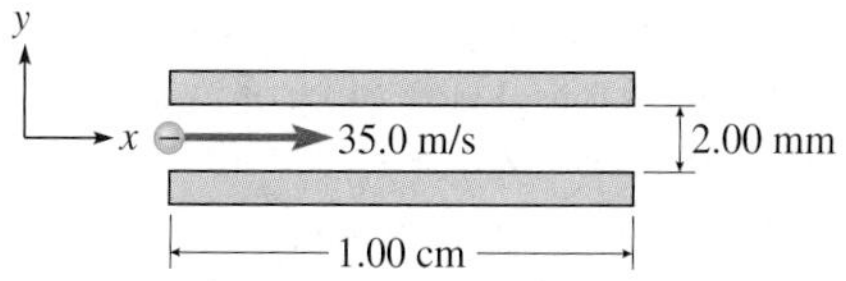

98. (a) Show that it was valid to ignore the gravitational force in Problem 97. (b) What are the components of velocity of the particle when it emerges from the plates?

99. Refer to Problem 97. One capacitor plate has an excess of electrons and the other has a matching deficit of electrons. What is the number of excess electrons?

100. A parallel plate capacitor has a charge of 0.020 μC on each plate with a potential difference of 240 V. The parallel plates are separated by 0.40 mm of air. (a) What is the capacitance for this capacitor? (b) What is the area of a single plate? (c) At what voltage will the air between the plates become ionized? Assume a dielectric strength of 3.0 kV/mm for air.

101. A 200.0-μF capacitor is placed across a 12.0-V battery. When a switch is thrown, the battery is removed from the capacitor and the capacitor is connected across a heater that is immersed in 1.00 cm^3 of water. Assuming that all the energy from the capacitor is delivered to the water, what is the temperature change of the water?

102. A cell membrane has a surface area of 1.0×10^{-7} m^2, a dielectric constant of 5.2, and a thickness of 7.5 nm. The membrane acts like the dielectric in a parallel plate capacitor; a layer of positive ions on the outer surface and a layer of negative ions on the inner surface act as the capacitor plates. The potential difference between the "plates" is 90.0 mV. (a) How much energy is stored in this capacitor? (b) How many positive ions are there on the outside of the membrane? Assume that all the ions are singly charged (charge $+e$).

103. An axon has the outer part of its membrane positively charged and the inner part negatively charged. The membrane has a thickness of 4.4 nm and a dielectric constant $\kappa = 5$. If we model the axon as a parallel plate capacitor whose area is 5 μm^2, what is its capacitance?

104. An electron beam is deflected upward through 3.0 mm while traveling in a vacuum between two deflection plates 12.0 mm apart. The potential difference between the deflecting plates is 100.0 kV and the kinetic energy of each electron as it enters the space between the plates is 2.0×10^{-15} J. What is the kinetic energy of each electron when it leaves the space between the plates?

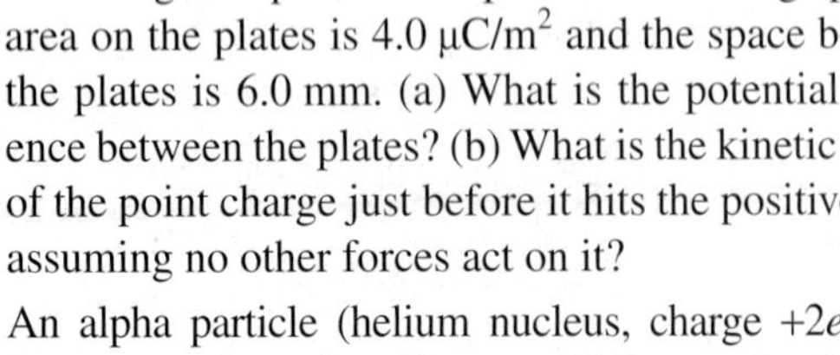

105. A point charge $q = -2.5$ nC is initially at rest adjacent to the negative plate of a capacitor. The charge per unit area on the plates is 4.0 $\mu C/m^2$ and the space between the plates is 6.0 mm. (a) What is the potential difference between the plates? (b) What is the kinetic energy of the point charge just before it hits the positive plate, assuming no other forces act on it?

106. An alpha particle (helium nucleus, charge $+2e$) starts from rest and travels a distance of 1.0 cm under the influence of a uniform electric field of magnitude 10.0 kV/m. What is the final kinetic energy of the alpha particle?

107. The inside of a cell membrane is at a potential of 90.0 mV lower than the outside. How much work does the electric field do when a sodium ion (Na^+) with a charge of $+e$ moves through the membrane from outside to inside?

108. Draw some electric field lines and a few equipotential surfaces outside a positively charged metal cube. [*Hint:* What shape are the equipotential surfaces close to the cube? What shape are they far away?]

109. A parallel plate capacitor is attached to a battery that supplies a constant voltage. While the battery remains attached to the capacitor, the distance between the parallel plates increases by 25%. What happens to the energy stored in the capacitor?

110. A parallel plate capacitor is attached to a battery that supplies a constant voltage. While the battery is still attached, a dielectric of dielectric constant $\kappa = 3.0$ is inserted so that it just fits between the plates. What is the energy stored in the capacitor after the dielectric is inserted in terms of the energy U_0 before the dielectric was inserted?

111. (a) Calculate the capacitance per unit length of an axon of radius 5.0 μm (see Fig. 17.14). The membrane acts as an insulator between the conducting fluids inside and outside the neuron. The membrane is 6.0 nm thick and has a dielectric constant of 7.0. (*Note:* The membrane is thin compared with the radius of the axon, so the axon can be treated as a parallel plate capacitor.) (b) In its resting state (no signal being transmitted), the potential of the fluid inside is about 85 mV lower than the outside. Therefore, there must be small net charges $\pm Q$ on either side of the membrane. Which side has positive charge? What is the magnitude of the charge density on the surfaces of the membrane?

112. A 4.00-μF air gap capacitor is connected to a 100.0-V battery until the capacitor is fully charged. The battery is removed and then a dielectric of dielectric constant 6.0 is inserted between the plates without allowing any charge to leak off the plates. (a) Find the energy stored in the capacitor before and after the dielectric is inserted. [*Hint:* First find the new capacitance and potential difference.] (b) Does an external agent have to do positive work to insert the dielectric or to remove the dielectric? Explain.

113. It has only been fairly recently that 1.0-F capacitors have been readily available. A typical 1.0-F capacitor can withstand up to 5.00 V. To get an idea why it isn't easy to make a 1.0-F capacitor, imagine making a 1.0-F parallel plate capacitor using titanium dioxide ($\kappa = 90.0$, breakdown strength 4.00 kV/mm) as the dielectric. (a) Find the minimum thickness of the titanium dioxide such that the capacitor can withstand 5.00 V. (b) Find the area of the plates so that the capacitance is 1.0 F.

114. The potential difference across a cell membrane from outside to inside is initially at −90 mV (when in its resting phase). When a stimulus is applied, Na^+ ions are allowed to move into the cell such that the potential changes to +20 mV for a short amount of time. (a) If the membrane capacitance per unit area is

1 μF/cm^2, how much charge moves through a membrane of area 0.05 cm^2? (b) The charge on Na^+ is $+e$. How many ions move through the membrane?

✦115. A parallel plate capacitor is connected to a battery. The space between the plates is filled with air. The electric field strength between the plates is 20.0 V/m. Then, *with the battery still connected*, a slab of dielectric ($\kappa = 4.0$) is inserted between the plates. The thickness of the dielectric is half the distance between the plates. Find the electric field inside the dielectric.

Answers to Practice Problems

17.1 (a) +0.018 J; (b) away from Q; (c) U decreases as the separation increases. The potential energy decrease accompanies an increase in kinetic energy as q moves faster and faster.

17.2 +0.064 J

17.3 the lower plate

17.4 $V_B = -1.5 \times 10^5$ V; work (done by $\vec{\mathbf{E}}$) = $-\Delta U_E$ = −0.010 J

17.5 $\vec{\mathbf{E}}$ = 5.4 × 10^8 N/C away from the +9.0-μC charge; $V = 0$

17.6 4 kV

17.7

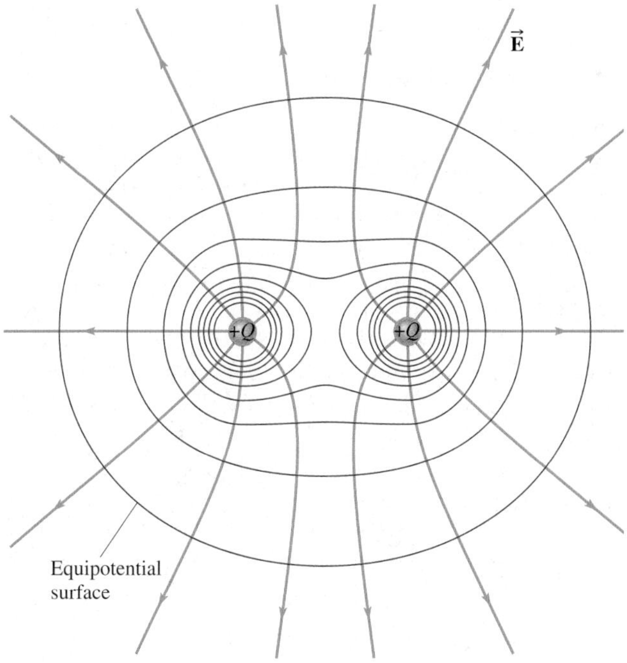

17.8 −20.9 kV (Note that a positive charge gains kinetic energy when it moves through a potential decrease; a negative charge gains kinetic energy when it moves through a potential increase.)

17.9 8.9 nF; 18 μC; charge (capacitance is independent of potential difference)

17.10 C doubles; maximum charge is unchanged

17.11 2.4 × 10^5 ions

17.12 (a) 0.21 μC; (b) 85 μJ

Answers to Checkpoints

17.1 Six pairs and therefore six terms in the potential energy (with subscripts 12, 13, 14, 23, 24, and 34).

17.2 $\vec{\mathbf{E}}$ points in the direction of *decreasing* potential, so the electric field is in the $-x$-direction.

17.3 The electric field magnitude is 25 V/m, so the potential decreases 25 V for each meter moved in the direction of the field. To move from one plane to another, the potential changes by 1.0 V and the distance must be

$$\frac{1.0\text{ V}}{25\text{ V/m}} = 0.040\text{ m}$$

17.5 The magnitude of the charge on each plate is proportional to the potential difference between them. With one quarter the potential difference, the plates have one quarter as much charge: +0.12 C and −0.12 C. (The capacitance of the capacitor is $C = Q/\Delta V = 0.080$ F.)

CHAPTER

18 Electric Current and Circuits

Graham's car won't start; the battery is dead after he left the headlights on overnight. In a kitchen drawer are several 1.5-V flashlight batteries. Graham decides to connect eight of them together, being careful to connect the positive terminal of one to the negative terminal of the next, just the way two 1.5-V batteries are connected inside a flashlight to provide 3.0 V. Eight 1.5-V batteries should provide 12 V, the same as a car battery, he reasons. Why won't this scheme work? (See p. 653 for the answer.)

Do *not* try this at home! If the car battery is not *completely* dead, it could send a dangerously large current through the flashlight batteries, causing one or more of them to explode.

Concepts & Skills to Review

- conductors and insulators (Section 16.2)
- electric potential (Section 17.2)
- capacitors (Section 17.5)
- solving simultaneous equations (Appendix A.2)
- power (Section 6.8)

18.1 ELECTRIC CURRENT

A net flow of charge is called an **electric current**. The *current* (symbol I) is defined as the *net* amount of charge passing per unit time through an area perpendicular to the flow direction (Fig. 18.1). The magnitude of the current tells us the rate of the net flow of charge. If Δq is the net charge that passes through the shaded surface in Fig. 18.1 during a time interval Δt, then the current in the wire is defined as

Definition of current:

$$I = \frac{\Delta q}{\Delta t} \quad (18\text{-}1)$$

Currents are not necessarily steady. In order for Eq. (18-1) to define the instantaneous current, we must use a sufficiently small time interval Δt.

The SI unit of current, equal to one coulomb per second, is the ampere (A), named for the French scientist André-Marie Ampère (1775–1836). The ampere is one of the SI base units; the coulomb is a derived unit defined as one ampere-second:

$$1\ \text{C} = 1\ \text{A·s}$$

Small currents are more conveniently measured in milliamperes (mA = 10^{-3} A) or in microamperes (μA = 10^{-6} A). The word *amperes* is often shortened to *amps*; for smaller currents, we speak of *milliamps* or *microamps*.

CONNECTION:

When a conductor is in electrostatic equilibrium, there are no currents; the electric field within the conducting material is zero and the entire conductor is at the same potential. If we can keep a conductor from reaching electrostatic equilibrium by maintaining a potential difference between two points of a conductor, then the electric field within the conducting material is not zero and a sustained current exists in the conductor.

Conventional Current According to convention, the direction of an electric current is defined as the direction in which *positive* charge is transported or would be transported to produce an equivalent movement of net charge. Benjamin Franklin established this convention (and decided which kind of charge would be called positive) long before scientists understood that the mobile charges (or *charge carriers*) in metals are electrons. If electrons move to the left in a metal wire, the direction of the current is to the *right*; negative charge moving to the left has the same effect on the net distribution of charge as positive charge moving to the right.

Direction of current = direction of flow of positive charge

In most situations, the motion of positive charge in one direction causes the same macroscopic effects as the motion of negative charge in the opposite direction. In circuit analysis, we always draw currents in the conventional direction regardless of the sign of the charge carriers.

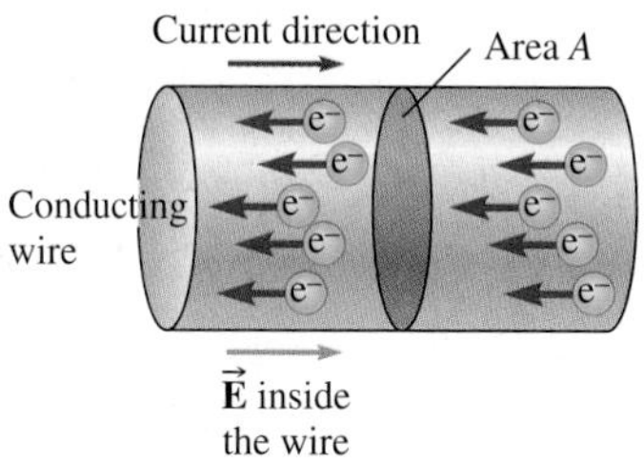

Figure 18.1 Close-up picture of a wire that carries an electric current. The current is the rate of flow of charge through an area perpendicular to the direction of flow.

CHECKPOINT 18.1

In a water pipe, there is an enormous amount of moving charge—the protons (charge $+e$) and electrons (charge $-e$) in the neutral water molecules all move with the same average velocity. Does the water carry an electric current? Explain.

Example 18.1

Current in a Clock

Two wires of cross-sectional area 1.6 mm^2 connect the terminals of a battery to the circuitry in a clock. During a time interval of 0.040 s, 5.0×10^{14} electrons move to the right through a cross section of one of the wires. (Actually, electrons pass through the cross section in both directions; the number that cross to the right is 5.0×10^{14} more than the number that cross to the left.) What is the magnitude and direction of the current in the wire?

Strategy Current is the rate of flow of charge. We are given the number N of electrons; multiplying by the elementary charge e gives the magnitude of moving charge Δq.

Solution The magnitude of the charge of 5.0×10^{14} electrons is

$$\Delta q = Ne = 5.0 \times 10^{14} \times 1.60 \times 10^{-19}\ \text{C} = 8.0 \times 10^{-5}\ \text{C}$$

The magnitude of the current is therefore,

$$I = \frac{\Delta q}{\Delta t} = \frac{8.0 \times 10^{-5}\ \text{C}}{0.040\ \text{s}} = 0.0020\ \text{A} = 2.0\ \text{mA}$$

Negatively charged electrons moving to the right means that the direction of conventional current—the direction in which positive charge is effectively being transported—is to the *left.*

Discussion To find the magnitude of the current, we use the *magnitude* of the charge on the electron. We *do* treat current as a signed quantity when analyzing circuits. We arbitrarily choose a direction for current when the actual direction is not known. If the calculations result in a negative current, the negative sign reveals that the actual direction of the current is opposite the chosen direction. The negative sign merely means the current flows in the direction opposite to the one we assumed.

In this problem, the cross-sectional area of the wire was extraneous information. To find the current, we need only the quantity of charge and the time for the charge to pass.

Practice Problem 18.1 Current in a Calculator

(a) If 0.320 mA of current flow through a calculator, how many electrons pass through per second? (b) How long does it take for 1.0 C of charge to pass through the calculator?

Electric Current in Liquids and Gases

Electric currents can exist in liquids and gases as well as in solid conductors. In an ionic solution, both positive and negative charges contribute to the current by moving in opposite directions (Fig. 18.2). The electric field is to the right, away from the positive electrode and toward the negative electrode. In response, positive ions move in the direction of the electric field (to the right) and negative ions move in the opposite direction (to the left). Since positive and negative charges are moving in opposite directions, they both contribute to current in the *same* direction. Thus, we can find the magnitudes of the currents separately due to the motion of the negative charges and the positive charges and *add* them to find the total current. The direction of the current in Fig. 18.2 is to the right. If positive and negative charges were moving in the *same* direction, they

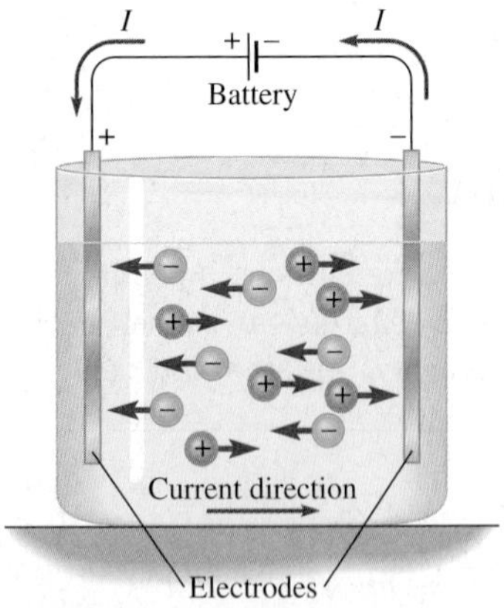

Figure 18.2 A current in a solution of potassium chloride consists of positive ions (K^+) and negative ions (Cl^-) moving in opposite directions. The direction of the current is the direction in which the positive ions move.

would represent currents in *opposite* directions and the individual currents would be *subtracted* to find the net current. (See Checkpoint 18.1.)

Currents also exist in gases. Figure 18.3 shows a neon sign. A large potential difference is applied to the metal electrodes inside a glass container of neon gas. There are always some positive ions present in a gas due to bombardment by cosmic rays and natural radioactivity. The positive ions are accelerated by the electric field toward the cathode; if they have sufficient energy, they can knock electrons loose when they collide with the cathode. These electrons are accelerated toward the anode; they ionize more gas molecules as they pass through the container. Collisions between electrons and ions produce the characteristic red light of a neon sign. Fluorescent lights are similar, but the collisions produce ultraviolet radiation; a coating on the inside of the glass absorbs the ultraviolet and emits visible light.

Application: current in neon signs and fluorescent lights

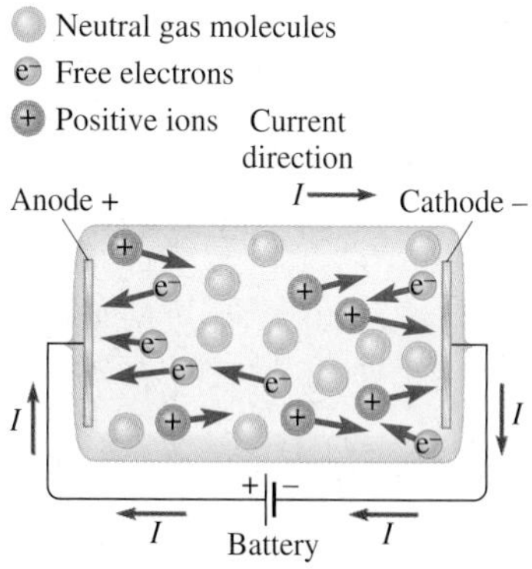

Figure 18.3 Simplified diagram of a neon sign. The neon gas inside the glass tube is ionized by a large potential difference between the electrodes.

18.2 EMF AND CIRCUITS

To maintain a current in a conducting wire, we need to maintain a potential difference between the ends of the wire. One way to do that is to connect the ends of the wire to the terminals of a battery (one end to each of the two terminals). An *ideal* battery maintains a constant potential difference between its terminals, regardless of how fast it must pump charge to do so. An ideal battery is analogous to an ideal water pump that maintains a constant pressure difference between intake and output regardless of the volume flow rate.

The potential difference maintained by an ideal battery is called the battery's **emf** (symbol $\mathscr{E}$). Emf originally stood for *electromotive force*, but emf is *not* a measure of the force applied to a charge or to a collection of charges; emf cannot be expressed in newtons. Rather, emf is measured in units of potential (volts) and is a measure of the work done by the battery per unit charge. To avoid this confusion, we just write "emf" (pronounced *ee-em-eff*). If the amount of charge pumped by an ideal battery of emf $\mathscr{E}$ is q, then the work done by the battery is

The circuit symbol for a battery is —|⊢ . Of the two vertical lines, the long line represents the terminal at higher potential and the short line represents the terminal at lower potential. Since many batteries consist of more than one chemical cell, an alternative form is —||⊢.

Work done by an ideal battery:

$$W = \mathscr{E}q \qquad (18\text{-}2)$$

Any device that pumps charge is called a *source of emf* (or just an *emf*). Generators, solar cells, and fuel cells are other sources of emf. Fuel cells, used in the Space Shuttle and perhaps someday in cars and homes, are similar to batteries, but their reactants are supplied externally. Many living organisms also contain sources of emf (Fig. 18.4). The signals transmitted by the human nervous system are electrical in nature, so our bodies

Figure 18.4 The South American electric eel (*Electrophorous electricus*) has hundreds of thousands of cells (called *electroplaques*) that supply emf. The current supplied by the electroplaques is used to stun its enemies and to kill its prey.

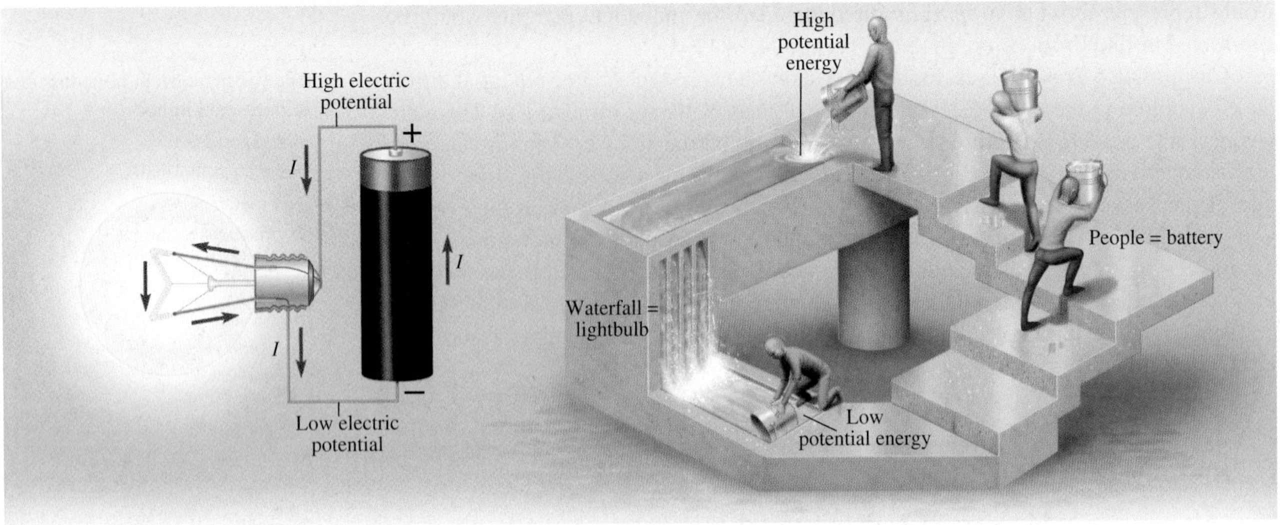

Figure 18.5 Using the flow of water as an analogy to what happens in an electric circuit.

contain sources of emf. The same circuit symbol is used for *any* source of constant emf (−|⊢). All emfs are energy conversion devices; they convert some other form of energy into electric energy. The energy sources used by emfs include chemical energy (batteries, fuel cells, biological sources of emf), sunlight (solar cells), and mechanical energy (generators).

Emf in an Electric Circuit In Fig. 18.5, imagine that the flow of water represents electric current (the flow of charge) in a circuit. The people act as a pump, taking water from the place where its potential energy is lowest and doing the work necessary to carry it uphill to the place where its potential energy is highest. The water then runs downhill, encountering resistance to its flow (the sluice gate) along the way. A battery (or other source of emf) plays a role something like that of the people who carry buckets of water. Thinking of current as the movement of positive charge, a battery takes positive charge from the place where its *electric potential* is lowest (the negative terminal of the battery) and does the work necessary to move it to the place where the electric potential is highest (the positive terminal). Then the charge flows through some device that offers resistance to the flow of current—perhaps a lightbulb or a CD player—before returning to the negative terminal of the battery.

Figure 18.6 Batteries come in many sizes and shapes. In the back is a lead-acid automobile battery. In front, from left to right are three types of rechargeable nickel-cadmium batteries, seven batteries commonly used in flashlights, cameras, and watches, and a zinc graphite dry cell.

Batteries A 9-V battery (such as the kind used in a smoke detector) maintains its positive terminal 9 V higher than its negative terminal—as long as conditions permit the battery to be treated as ideal. Since a volt is a joule per coulomb, the battery does 9 J of work for every coulomb of charge that it pumps. The battery does work by converting some of its stored chemical energy into electric energy. When a battery is dead, its supply of chemical energy has been depleted and it can no longer pump charge. Some batteries can be recharged by forcing charge to flow through them in the opposite direction, reversing the direction of the electrochemical reaction and converting electric energy into chemical energy.

Batteries come with various emfs (12 V, 9 V, 1.5 V, etc.) as well as in various sizes (Fig. 18.6). The size of a battery does *not* determine its emf. Common battery sizes AAA, AA, A, C, and D all provide the same emf (1.5 V). However, the larger batteries have a larger quantity of the chemicals and thus store more chemical energy. A larger battery can supply more energy by pumping more charge than a smaller one, even though the two do the same amount of work *per unit charge.* The amount of charge that a battery can pump is often measured in A·h (ampere-hours). Another difference is that larger batteries can generally pump charge *faster*—in other words, they can supply larger currents.

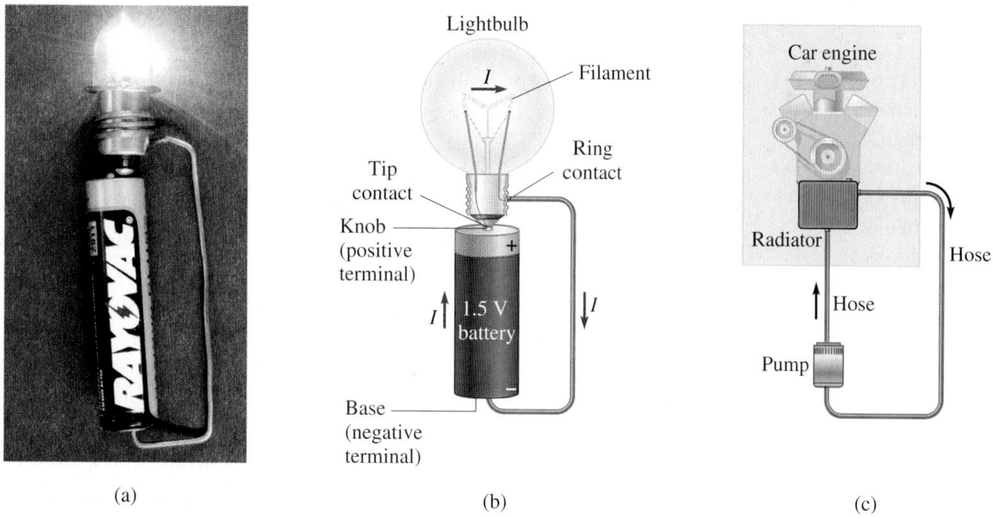

Figure 18.7 (a) Connecting a battery to a lightbulb. The bulb lights up only when current flows through its filament. (b) To maintain current flow, a complete circuit must exist. Note the use of the arrows to indicate the direction of current flow in the wires, lightbulb, and battery. (c) An analogous circuit dealing with the flow of water rather than of charge.

Circuits

For currents to continue to flow, a complete circuit is required. That is, there must be a continuous conducting path from one terminal of the emf to one or more devices and then back to the other terminal. In Fig. 18.7a,b there is one complete circuit for the current to travel from the positive terminal of the battery, through a wire, through the lightbulb filament, through another wire, into the battery at the negative terminal, and through the battery to return to the positive terminal. Since this circuit has only a single loop for current to flow, the current must be the same everywhere. Think of the battery as a water pump, the wires as hoses, and the lightbulb as the engine block and radiator of an automobile (Fig. 18.7c). Water must flow from the pump, through a hose, through the engine and radiator, through another hose, and back to the pump. The volume flow rate in this single-loop "water circuit" is the same everywhere. Current does not get "used up" in the lightbulb any more than water gets used up in the radiator.

In this chapter, we consider only circuits in which the current in any branch always moves in the same direction—a **direct current** (dc) circuit. In Chapter 21, we study **alternating current** (ac) circuits, in which the currents periodically reverse direction.

18.3 MICROSCOPIC VIEW OF CURRENT IN A METAL: THE FREE-ELECTRON MODEL

Figure 18.1 showed a simplified picture of the conduction electrons in a metal, all moving with the same constant velocity due to an electric field. Why do the electrons not move with a constant *acceleration* due to a constant electric force? To answer this question and to understand the relationship between electric field and current in a metal, we need a more accurate picture of the motion of the electrons.

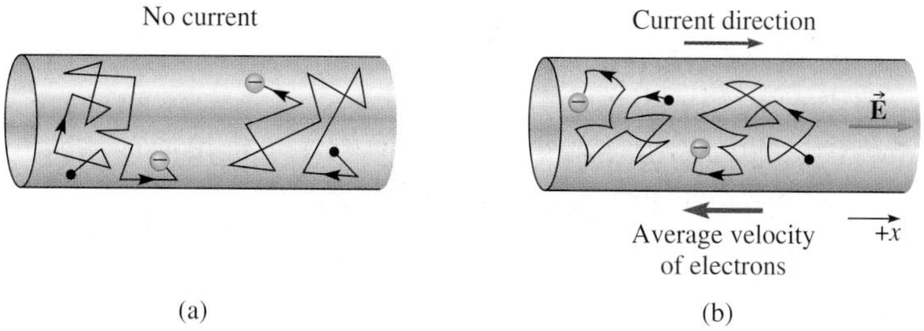

Figure 18.8 (a) Random paths followed by two conduction electrons in a metal wire in the absence of an electric field. (b) An electric field in the $+x$-direction gives the electrons a constant acceleration in the $-x$-direction between collisions. *On average*, the electrons drift in the $-x$-direction. The current in the wire is in the $+x$-direction.

CONNECTION:

The random motion of conduction electrons in a metal is reminiscent of the random motion of atoms or molecules in a gas. One difference is that the distribution of electron speeds is quite different from the Maxwell-Boltzmann distribution (Section 13.6).

In the absence of an applied electric field, the conduction electrons in a metal are in constant random motion at high speed—about 10^6 m/s in copper. The electrons suffer frequent collisions with each other and with ions (the atomic nuclei with their bound electrons). In copper, a given conduction electron collides 4×10^{13} times per second, traveling on average about 40 nm between collisions. A collision can change the direction of the electron's motion, so each electron moves in a random path similar to that of a gas molecule (Fig. 18.8a). The average *velocity* of the conduction electrons in a metal is zero in the absence of an electric field, so there is no net transport of charge.

If a uniform electric field exists within the metal, the electric force on the conduction electrons gives them a uniform acceleration between collisions (when the net force due to nearby ions and other conduction electrons is small). The electrons still move about in random directions like gas molecules, but the electric force makes them move on average a little faster in the direction of the force than in the opposite direction—much like air molecules in a gentle breeze. As a result, the electrons slowly drift in the direction of the electric force (Fig. 18.8b). The electrons now have a nonzero average velocity called the **drift velocity** $\vec{\mathbf{v}}_D$ (which corresponds to the wind velocity for air molecules). The magnitude of the drift velocity (the *drift speed*) is much smaller than the instantaneous speeds of the electrons—typically less than 1 mm/s—but since it is nonzero, there is a net transport of charge.

It might seem that a uniform acceleration should make the electrons move faster and faster. If there were no collisions, they would. An electron has a uniform acceleration *between collisions*, but every collision sends it off in some new direction with a different speed. Each collision between an electron and an ion is an opportunity for the electron to transfer some of its kinetic energy to the ion. The net result is that the drift velocity is constant and energy is transferred from the electrons to the ions at a constant rate.

CONNECTION:

Another situation where an applied force results in motion at constant *velocity* (rather than constant acceleration) is an object falling through a viscous fluid (Section 9.10). When falling at terminal velocity, the viscous drag force opposes the constant downward force of gravity so the *net* force is zero. To make an analogy, the electric field in a metal acts like gravity for the falling object (constant applied force) and collisions of electrons with ions acts like the drag force.

Relationship Between Current and Drift Velocity

To find out how current depends on drift velocity, we use a simplified model in which all the electrons move at a constant velocity $\vec{\mathbf{v}}_D$ (Fig. 18.9). The number of conduction electrons per unit volume (n) is a characteristic of a particular metal. Suppose we calculate the current by finding how much charge moves through the shaded area in a time Δt. During that time, every electron moves a distance $v_D \Delta t$ to the left. Thus, every conduction electron in a volume $Av_D \Delta t$ moves through the shaded area. The number of electrons in this volume is $N = nAv_D \Delta t$; the magnitude of the charge is

$$\Delta Q = Ne = neAv_D \Delta t$$

Therefore, the magnitude of the current in the wire is

$$I = \frac{\Delta Q}{\Delta t} = neAv_D \tag{18-3}$$

Remember that, since electrons carry negative charge, the direction of current flow is opposite the direction of motion of the electrons. The electric force on the electrons is opposite the electric field, so the current is in the direction of the electric field in the wire.

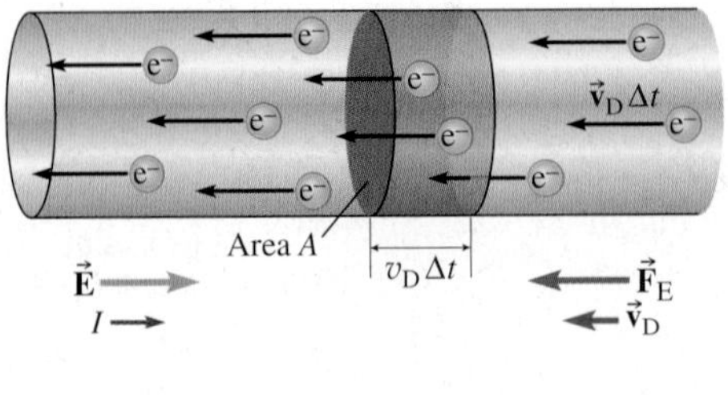

Figure 18.9 Simplified picture of the conduction electrons moving at a uniform velocity $\vec{\mathbf{v}}_D$. In a time Δt, each electron moves a distance $v_D \Delta t$. The black vector arrows show the displacement of each electron during Δt. All of the conduction electrons within a distance $v_D \Delta t$ pass through the shaded cross-sectional area in a time Δt.

Equation (18-3) can be generalized to systems in which the current carriers are not necessarily electrons, simply by replacing e with the charge of the carriers. In materials called semiconductors, there may be both positive and negative carriers. The negative carriers are electrons; the positive carriers are "missing" electrons (called *holes*) that act as particles with charge $+e$. The electrons and holes drift in opposite directions; both contribute to the current. Since the concentrations of electrons and holes may be different and they may have different drift speeds, the current is

$$I = n_+ eAv_+ + n_- eAv_- \quad (18\text{-}4)$$

In Eq. (18-4), v_+ and v_- are drift *speeds*—both are positive.

CHECKPOINT 18.3

Two copper wires with different diameters carry the same current. Compare the drift speeds of the conduction electrons in the two wires.

When we turn on a light by flipping a wall switch, current flows through the lightbulb almost instantaneously. We do *not* have to wait for electrons to move from the switch to the lightbulb—which is a good thing, since it would be a long wait (see Example 18.2). Conduction electrons are present all along the wires that form the circuit. When the switch is closed; the *electric field* extends into the entire circuit very quickly. The electrons start to drift as soon as the electric field is nonzero.

Example 18.2

Drift Speed in Household Wiring

A #12 gauge copper wire, commonly used in household wiring, has a diameter of 2.053 mm. There are 8.00×10^{28} conduction electrons per cubic meter in copper. If the wire carries a constant dc current of 5.00 A, what is the drift speed of the electrons?

Strategy From the diameter, we can find the cross-sectional area A of the wire. The number of conduction electrons per cubic meter is n in Eq. (18-3). Then Eq. (18-3) enables us to solve for the drift speed.

Solution The cross-sectional area of the wire is

$$A = \pi r^2 = \tfrac{1}{4}\pi d^2$$

The drift speed is given by

$$v_D = \frac{I}{neA} =$$

$$\frac{5.00\ \text{A}}{8.00 \times 10^{28}\ \text{m}^{-3} \times 1.602 \times 10^{-19}\ \text{C} \times \tfrac{1}{4}\pi \times (2.053 \times 10^{-3}\ \text{m})^2}$$

$$= 1.179 \times 10^{-4}\ \text{m·s}^{-1} \rightarrow 0.118\ \text{mm/s}$$

Discussion The drift speed may seem surprisingly small: at an average speed of 0.118 mm/s, it takes an electron over 2 h to move one meter along the wire! How can 5 C/s—a respectable amount of current—be carried by electrons with such small average velocities? Because there are so many of them. As a check: the number of conduction electrons per unit length of wire is

$$nA = 8.00 \times 10^{28}\ \text{m}^{-3} \times \tfrac{1}{4}\pi \times (2.053 \times 10^{-3}\text{m})^2$$

$$= 2.648 \times 10^{23}\ \text{electrons/m}$$

Then the number of conduction electrons in a 0.1179 mm length of wire is

$$2.648 \times 10^{23}\ \text{electrons/m} \times 0.1179 \times 10^{-3}\ \text{m} = 3.122 \times 10^{19}\ \text{electrons}$$

The magnitude of the total charge of these electrons is

$$3.122 \times 10^{19}\ \text{electrons} \times 1.602 \times 10^{-19}\ \text{C/electron} = 5.00\ \text{C}$$

Practice Problem 18.2 Current and Drift Speed in a Silver Wire

A silver wire has a diameter of 2.588 mm and contains 5.80×10^{28} conduction electrons per cubic meter. A battery of 1.50 V pushes 880 C through the wire in 45 min. Find (a) the current and (b) the drift speed in the wire.

18.4 RESISTANCE AND RESISTIVITY

Resistance

Suppose we maintain a potential difference across the ends of a conductor. How does the current I that flows through the conductor depend on the potential difference ΔV across the conductor? For many conductors, the I is proportional to ΔV. Georg Ohm (1789–1854) first observed this relationship, which is now called **Ohm's law**:

Ohm's Law

$$I \propto \Delta V \quad (18\text{-}5)$$

Ohm's law is not a universal law of physics like the conservation laws. It does not apply at all to some materials, whereas even materials that obey Ohm's law for a wide range of potential differences fail to do so when ΔV becomes too large. Hooke's law ($F \propto \Delta x$ or stress $\propto$ strain) is similar; it applies to many materials under many circumstances but is not a fundamental law of physics. Any *homogeneous* material follows Ohm's law for *some* range of potential differences; metals that are good conductors follow Ohm's law over a *wide* range of potential differences.

CONNECTION:

Ohm was inspired to look at the relationship between current and potential difference by Fourier's observation that the rate of heat flow through a conductor of heat is proportional to the temperature difference across it (Section 14.6). Another analogous situation is the flow of oil (or any viscous fluid) through a pipe. Poiseuille's law says that the rate of flow of the fluid is proportional to the pressure difference between the ends of the pipe (Section 9.9).

The electrical **resistance** R is *defined* to be the ratio of the potential difference (or *voltage*) ΔV across a conductor to the current I through the material:

Definition of resistance:

$$R = \frac{\Delta V}{I} \quad (18\text{-}6)$$

In SI units, electrical resistance is measured in ohms (symbol Ω, the Greek capital omega), defined as

$$1\ \Omega = 1\ \text{V/A} \quad (18\text{-}7)$$

For a given potential difference, a large current flows through a conductor with a small resistance, while a small current flows through a conductor with a large resistance.

An *ohmic* conductor—one that follows Ohm's law—has a resistance that is constant, regardless of the potential difference applied. Equation (18-6) is *not* a statement of Ohm's law, since it does not require that the resistance be constant; it is the *definition of resistance* for nonohmic conductors as well as for ohmic conductors. For an ohmic conductor, a graph of current versus potential difference is a straight line through the origin with slope $1/R$ (Fig. 18.10a). For some nonohmic systems, the graph of I versus ΔV is dramatically nonlinear (Fig. 18.10b,c).

Resistivity

Resistance depends on size and shape. We expect a long wire to have higher resistance than a short one (everything else being the same) and a thicker wire to have a lower resistance than a thin one. The electrical resistance of a conductor of length L and cross-sectional area A can be written:

CONNECTION:

Returning to the analogy with fluid flow: a longer pipe offers more resistance to fluid flow than does a short pipe and a wider pipe offers less resistance than a narrow one.

$$R = \rho\frac{L}{A} \quad (18\text{-}8)$$

Equation (18-8) assumes a uniform distribution of current across the cross section of the conductor.

Table 18.1 Resistivities and Temperature Coefficients at 20°C

	ρ (Ω·m)	α (°C^{-1})		ρ (Ω·m)	α (°C^{-1})
Conductors			**Semiconductors (pure)**		
Silver	1.59×10^{-8}	3.8×10^{-3}	Carbon	3.5×10^{-5}	-0.5×10^{-3}
Copper	1.67×10^{-8}	4.05×10^{-3}	Germanium	0.6	-50×10^{-3}
Gold	2.35×10^{-8}	3.4×10^{-3}	Silicon	2300	-70×10^{-3}
Aluminum	2.65×10^{-8}	3.9×10^{-3}			
Tungsten	5.40×10^{-8}	4.50×10^{-3}			
Iron	9.71×10^{-8}	5.0×10^{-3}	**Insulators**		
Lead	21×10^{-8}	3.9×10^{-3}	Glass	$10^{10} - 10^{14}$	
Platinum	10.6×10^{-8}	3.64×10^{-3}	Lucite	$> 10^{13}$	
Manganin	44×10^{-8}	0.002×10^{-3}	Quartz (fused)	$> 10^{16}$	
Constantan	49×10^{-8}	0.002×10^{-3}	Rubber (hard)	$10^{13} - 10^{16}$	
Mercury	96×10^{-8}	0.89×10^{-3}	Teflon	$> 10^{13}$	
Nichrome	108×10^{-8}	0.4×10^{-3}	Wood	$10^{8} - 10^{11}$	

The constant of proportionality ρ (Greek letter rho), which is an intrinsic characteristic of a particular material at a particular temperature, is called the **resistivity** of the material. The SI unit for resistivity is Ω·m. Table 18.1 lists resistivities for various substances at 20°C. The resistivities of good conductors are small. The resistivities of pure semiconductors are significantly larger. By doping semiconductors (introducing controlled amounts of impurities), their resistivities can be changed dramatically, which is one reason that semiconductors are used to make computer chips and other electronic devices (Fig. 18.11). Insulators have very large resistivities (about a factor of 10^{20} larger than for conductors). The inverse of resistivity is called *conductivity* [SI units (Ω·m)$^{-1}$].

Why is resistance proportional to length? Suppose we have two otherwise identical wires with different lengths. If the wires carry the same current, they must have the same drift speed; to have the same drift speed, the electric field must be the same. Since for a uniform field $\Delta V = EL$, the potential differences across the wires are proportional to their lengths. Therefore, $R = \Delta V/I$ is proportional to length.

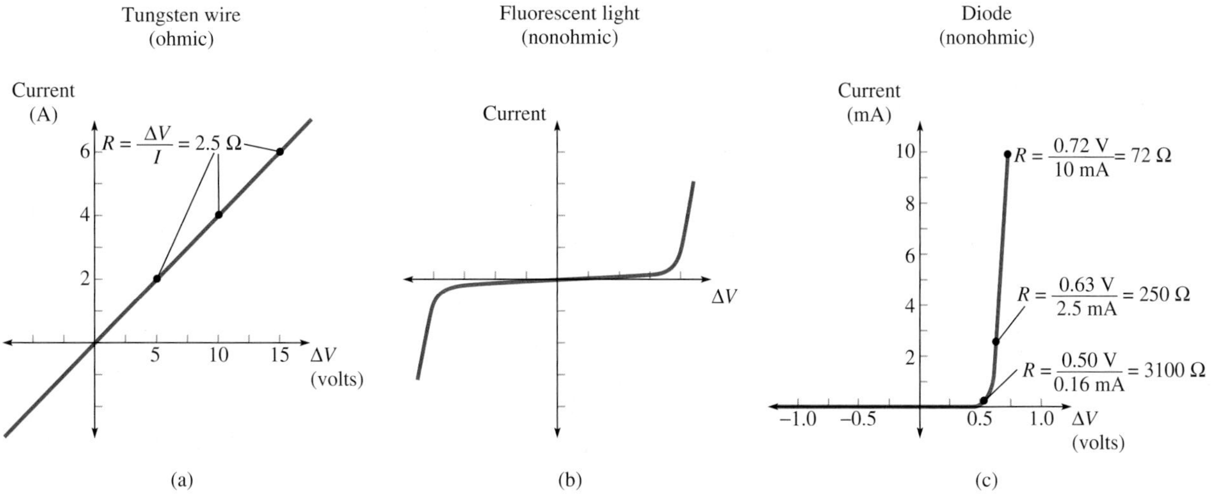

Figure 18.10 (a) Current as a function of potential difference for a tungsten wire at constant temperature. The resistance is the same for any value of ΔV on the graph, so the wire is an ohmic conductor. Similar graphs for (b) the gas in a fluorescent light and (c) a diode (a semiconductor device) are far from linear; these systems are nonohmic.

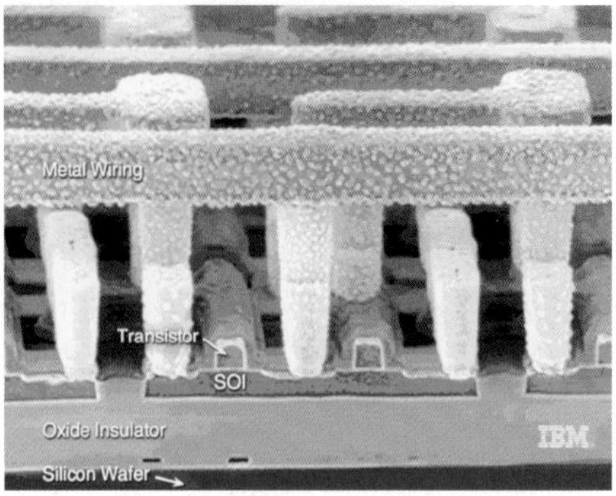

Figure 18.11 A scanning electron microscope view of a microprocessor chip. Much of the chip is made of silicon. By introducing impurities into the silicon in a controlled way, some regions act as insulating material, others as conducting wires, and others as the transistors—circuit elements that act as switches. SOI stands for silicon on insulator, a technology that reduces the heat generated within a chip.

Why is resistance inversely proportional to cross-sectional area? Suppose we have two otherwise identical wires with different areas. Applying the same potential difference produces the same drift speed, but the thicker wire has more conduction electrons per unit length. Since $I = neAv_D$ [Eq. (18-3)], the current is proportional to the area and $R = \Delta V/I$ is inversely proportional to area.

CHECKPOINT 18.4

Why can you look up in a table the resistivity of a substance (at a given temperature), but not the resistance?

Example 18.3

Resistance of an Extension Cord

(a) A 30.0-m-long extension cord is made from two #19 gauge copper wires. (The wires carry currents of equal magnitude in opposite directions.) What is the resistance of each wire at 20.0°C? The diameter of #19 gauge wire is 0.912 mm. (b) If the copper wire is to be replaced by an aluminum wire of the same length, what is the minimum diameter so that the new wire has a resistance no greater than the old?

Strategy After calculating the cross-sectional area of the copper wire from its diameter, we find the resistance of the copper wire from Eq. (18-8). The resistivities of copper and aluminum are found in Table 18.1.

Solution (a) From Table 18.1, the resistivity of copper is

$$\rho = 1.67 \times 10^{-8}\ \Omega\cdot\text{m}$$

The wire's cross-sectional area is

$$A = \tfrac{1}{4}\pi d^2 = \tfrac{1}{4}\pi(9.12 \times 10^{-4}\ \text{m})^2 = 6.533 \times 10^{-7}\ \text{m}^2$$

Resistance is resistivity times length over area:

$$R = \rho\frac{L}{A}$$

$$= \frac{1.67 \times 10^{-8}\ \Omega\cdot\text{m} \times 30.0\ \text{m}}{6.533 \times 10^{-7}\ \text{m}^2}$$

$$= 0.767\ \Omega$$

(b) We want the resistance of the aluminum wire to be less than or equal to the resistance of the copper wire ($R_a \le R_c$):

$$\frac{\rho_a L}{\frac{1}{4}\pi d_a^2} \le \frac{\rho_c L}{\frac{1}{4}\pi d_c^2}$$

which simplifies to $\rho_a d_c^2 \le \rho_c d_a^2$. Solving for d_a yields

$$d_a \ge d_c\sqrt{\frac{\rho_a}{\rho_c}} = 0.912\ \text{mm} \times \sqrt{\frac{2.65 \times 10^{-8}\ \Omega\cdot\text{m}}{1.67 \times 10^{-8}\ \Omega\cdot\text{m}}} = 1.15\ \text{mm}$$

continued on next page

Example 18.3 continued

Discussion Check: the resistance of an aluminum wire of diameter 1.149 mm is

$$R = \frac{\rho L}{A} = \frac{2.65 \times 10^{-8}\ \Omega\cdot\text{m} \times 30.0\ \text{m}}{\frac{1}{4}\pi(1.149 \times 10^{-3}\ \text{m})^2} = 0.767\ \Omega$$

Aluminum has a higher resistivity, so the wire must be thicker to have the same resistance.

Extension cords are rated according to the maximum safe current they can carry. For an appliance that draws a larger current, a thicker extension cord must be used; otherwise, the potential difference across the wires would be too large ($\Delta V = IR$).

Practice Problem 18.3 Resistance of a Lightbulb Filament

Find the resistance at 20°C of a tungsten lightbulb filament of length 4.0 cm and diameter 0.020 mm.

Resistivity Depends on Temperature

Resistivity does not depend on the size or shape of the material, but it does depend on temperature. Two factors primarily determine the resistivity of a metal: the number of conduction electrons per unit volume and the rate of collisions between an electron and an ion. The second of these factors is sensitive to changes in temperature. At a higher temperature, the internal energy is greater; the ions vibrate with larger amplitudes. As a result, the electrons collide more frequently with the ions. With less time to accelerate between collisions, they acquire a smaller drift speed; thus, the current is smaller for a given electric field. Therefore, as the temperature of a metal is raised, its resistivity increases. The metal filament in a glowing incandescent lightbulb reaches a temperature of about 3000 K; its resistance is significantly higher than at room temperature.

For many materials, the relation between resistivity and temperature is linear over a fairly wide range of temperatures (about 500°C):

$$\rho = \rho_0(1 + \alpha\Delta T) \qquad (18\text{-}9)$$

Temperature dependence of resistivity

Here ρ_0 is the resistivity at temperature T_0 and ρ is the resistivity at temperature $T = T_0 + \Delta T$. The quantity α is called the **temperature coefficient of resistivity** and has SI units $°C^{-1}$ or K^{-1}. Temperature coefficients for some materials are listed in Table 18.1.

Application: resistance thermometer

The relationship between resistivity and temperature is the basis of the *resistance thermometer.* The resistance of a conductor is measured at a reference temperature and at the temperature to be measured; the change in the resistance is then used to calculate the unknown temperature. For measurements over limited temperature ranges, the linear relationship of Eq. (18-9) can be used in the calculation; over larger temperature ranges, the resistance thermometer must be calibrated to account for the nonlinear variation of resistivity with temperature. Materials with high melting points (such as tungsten) can be used to measure high temperatures.

Semiconductors For semiconductors, $\alpha < 0$. A negative temperature coefficient means that the resistivity *decreases* with increasing temperature. It is still true, as for metals that are good conductors, that the collision rate increases with temperature. However, in semiconductors the number of carriers (conduction electrons and/or holes) per unit volume increases dramatically with increasing temperature; with more carriers, the resistivity is smaller.

Superconductors Some materials become *superconductors* ($\rho = 0$) at low temperatures. Once a current is started in a superconducting loop, it continues to flow indefinitely *without* a source of emf. Experiments with superconducting currents have lasted more than 2 years without any measurable change in the current. Mercury was the first superconductor discovered (by Dutch scientist Kammerlingh Onnes in 1911). As the temperature of mercury is decreased, its resistivity gradually decreases—as for any metal—but at mercury's critical temperature ($T_C = 4.15$ K) its resistivity suddenly becomes zero. Many other superconductors have since been discovered. In the past two decades, scientists have created ceramic materials with much higher critical temperatures than those previously known. Above their critical temperatures, the ceramics are insulators.

Example 18.4

Change in Resistance with Temperature

The nichrome heating element of a toaster has a resistance of 12.0 Ω when it is red-hot (1200°C). What is the resistance of the element at room temperature (20°C)? Ignore changes in the length or diameter of the element due to temperature.

Strategy Since we assume the length and cross-sectional area to be the same, the resistances at the two temperatures are proportional to the resistivities at those temperatures:

$$\frac{R}{R_0} = \frac{\rho L/A}{\rho_0 L/A} = \frac{\rho}{\rho_0}$$

Thus, we do not need the length or cross-sectional area of the heating element.

Given: $T_0 = 20$°C; $R = 12.0\ \Omega$ at $T = 1200$°C.
To find: R_0

Solution From Eq. (18-9),

$$\frac{R}{R_0} = \frac{\rho L/A}{\rho_0 L/A} = \frac{\rho}{\rho_0} = 1 + \alpha \Delta T$$

The change in temperature is

$$\Delta T = T - T_0 = 1200°\text{C} - 20°\text{C} = 1180°\text{C}$$

For nichrome, Table 18.1 gives

$$\alpha = 0.4 \times 10^{-3}\ °\text{C}^{-1}$$

Solving for R_0 yields

$$R_0 = \frac{R}{1 + \alpha \Delta T} = \frac{12.0\ \Omega}{1 + 0.4 \times 10^{-3}\ °\text{C}^{-1} \times 1180°\text{C}} = 8\ \Omega$$

Discussion Why do we write only one significant figure? Since the temperature change is so large (1180°C), the result must be considered an estimate. The relationship between resistivity and temperature may not be linear over such a large temperature range.

Practice Problem 18.4 Using a Resistance Thermometer

A platinum resistance thermometer has a resistance of 225 Ω at 20.0°C. When the thermometer is placed in a furnace, its resistance rises to 448 Ω. What is the temperature of the furnace? Assume the temperature coefficient of resistivity is constant over the temperature range in this problem.

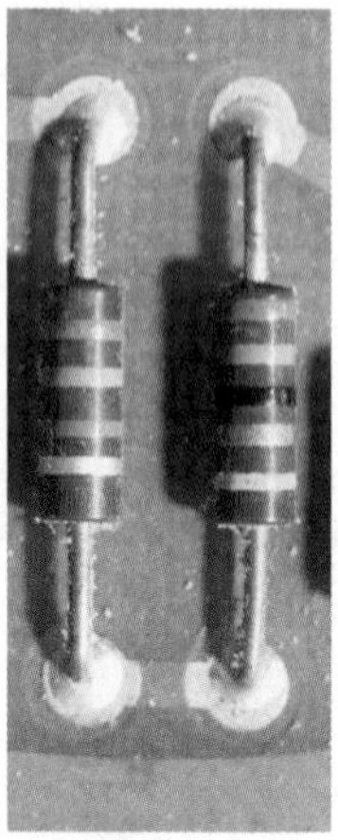

Figure 18.12 The little cylinders on this computer circuit board are resistors. The colored bands specify the resistance of the resistor.

Resistors

A **resistor** is a circuit element designed to have a known resistance. Resistors are found in virtually all electronic devices (Fig. 18.12). In circuit analysis, it is customary to write the relationship between voltage and current for a resistor as $V = IR$. Remember that V actually stands for the potential *difference* between the ends of the resistor even though the symbol Δ is omitted. Sometimes V is called the *voltage drop*. Current in a resistor flows in the direction of the electric field, which points from higher to lower potential. Therefore, if you move across a resistor in the direction of current flow, the voltage *drops* by an amount IR. Remember a useful analogy: water flows downhill (toward lower potential energy); electric current *in a resistor* flows toward lower potential.

Internal Resistance of a Battery

Figure 18.13a shows a circuit we've seen before. Figure 18.13b is a *circuit diagram* of the circuit. The lightbulb is represented by the symbol for a resistor (R). The battery is represented by two symbols surrounded by a dashed line. The battery symbol represents an *ideal* emf and the resistor (r) represents the *internal resistance* of the battery. If the internal resistance of a source of emf is negligible, then we just draw the symbol for an ideal emf.

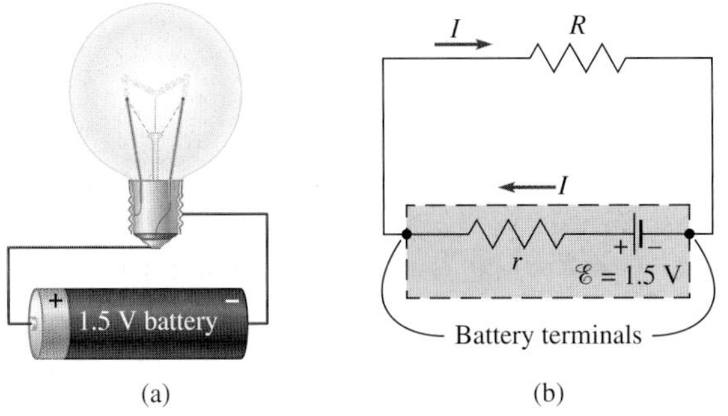

Figure 18.13 (a) A lightbulb connected to a battery by conducting wires. (b) A circuit diagram for the same circuit. The emf and the internal resistance of the battery are enclosed by a dashed line as a reminder that in reality the two are not separate; we can't make a connection to the "wire" between the two!

When the current through a source of emf is zero, the **terminal voltage**—the potential difference between its terminals—is equal to the emf. When the source supplies current to a *load* (a lightbulb, a toaster, or any other device that uses electric energy), its terminal voltage is less than the emf; there is a voltage drop due to the internal resistance of the source. If the current is I and the internal resistance is r, then the voltage drop across the internal resistance is Ir and the terminal voltage is

$$V = \mathscr{E} - Ir \qquad (18\text{-}10)$$

When the current is small enough, the voltage drop Ir due to the internal resistance is negligible compared with $\mathscr{E}$; then we can treat the emf as ideal ($V \approx \mathscr{E}$). A flashlight that is left on for a long time gradually dims because, as the chemicals in a battery are depleted, the internal resistance increases. As the internal resistance increases, the terminal voltage $V = \mathscr{E} - Ir$ decreases; thus, the voltage across the lightbulb decreases and the light gets dimmer.

In a circuit diagram, the symbol —/\/\/\— represents a resistor or any other device in a circuit that dissipates electric energy.

A straight line ——— represents a conducting wire with negligible resistance. (If a wire's resistance is appreciable, then we draw it as a resistor.)

Conceptual Example 18.5

Starting a Car Using Flashlight Batteries

Why won't Graham's scheme work?

Discuss the merits of Graham's scheme to start his car using eight D-cell flashlight batteries, each of which provides an emf of 1.50 V and has an internal resistance of 0.10 Ω. (A current of several hundred amps is required to turn the starter motor in a car, while the current through the bulb in a flashlight is typically less than 1 A.)

Strategy We consider not only the values of the emfs, but also whether the batteries can supply the required *current*.

Solution and Discussion Connecting eight 1.5-V batteries as in a flashlight—with the positive terminal of one connected to the negative terminal of the next—does provide an emf of 12 V. Each battery does 1.5 J of work per coulomb of charge; if the charge must pass through all eight batteries in turn, the total work done is 12 J per coulomb of charge.

When the batteries are used to power a device that draws a *small* current (because the resistance of the load R is large compared with the internal resistance r of each battery), the terminal voltage of each battery is nearly 1.5 V and the terminal voltage of the combination is nearly 12 V. For instance, in a flashlight that draws 0.50 A of current, the terminal voltage of a D-cell is

$$V = \mathscr{E} - Ir = 1.50\ \text{V} - 0.50\ \text{A} \times 0.10\ \Omega = 1.45\ \text{V}$$

However, the current required to start the car is large. As the current increases, the terminal voltage decreases. We can estimate the *maximum* current that a battery can supply by setting its terminal voltage to zero (the smallest possible value):

$$V = \mathscr{E} - I_{\text{max}} r = 0$$

$$I_{\text{max}} = \mathscr{E}/r = (1.5\ \text{V})/(0.10\ \Omega) = 15\ \text{A}$$

(This estimate is optimistic since the battery's chemical energy would be rapidly depleted and the internal resistance would increase dramatically.) The flashlight batteries cannot supply a current large enough to start the car.

Practice Problem 18.5 Terminal Voltage of a Battery in a Clock

The current supplied by an alkaline D-cell (1.500 V emf, 0.100 Ω internal resistance) in a clock is 50.0 mA. What is the terminal voltage of the battery?

PHYSICS AT HOME

Turn on the headlights of a car and then start the car. Notice that the headlights dim considerably. If the car battery were an *ideal* emf of 12 V, it would continue to supply 12 V to the headlights regardless of how much current is drawn from it. Due to the internal resistance of the battery, the terminal voltage of the battery is significantly less than 12 V when it supplies a few hundred amps of current to the starter.

18.5 KIRCHHOFF'S RULES

Two rules, developed by Gustav Kirchhoff (1824–1887), are essential in circuit analysis. **Kirchhoff's junction rule** states that the sum of the currents that flow into a junction—any electric connection—must equal the sum of the currents that flow out of the same junction. The junction rule is a consequence of the law of conservation of charge. Since charge does not continually build up at a junction, the *net* rate of flow of charge into the junction must be zero.

CONNECTION:

The junction rule is just the conservation of charge written in a convenient form for circuits.

Kirchhoff's Junction Rule

$$\sum I_{\text{in}} - \sum I_{\text{out}} = 0 \qquad (18\text{-}11)$$

Figure 18.14a shows two streams joining to form a larger stream. Figure 18.14b shows an analogous junction (point A) in an electric circuit. Applying the junction rule to point A results in the equation $I_1 + I_2 - I_3 = 0$.

Kirchhoff's loop rule is an expression of energy conservation applied to changes in potential in a circuit. Recall that the electric potential must have a unique value at any point; the potential at a point cannot depend on the path one takes to arrive at that point. Therefore, if a closed path is followed in a circuit, beginning and ending at the same point, the algebraic sum of the potential changes must be zero (Fig. 18.15). Think of taking a hike in the mountains, starting and returning at the same spot. No matter what path you take, the algebraic sum of all your elevation changes must equal zero.

CONNECTION:

The loop rule is just energy conservation written in a convenient form for circuits.

Kirchhoff's Loop Rule

$$\sum \Delta V = 0 \qquad (18\text{-}12)$$

for any path in a circuit that starts and ends at the same point. (Potential rises are positive; potential drops are negative.)

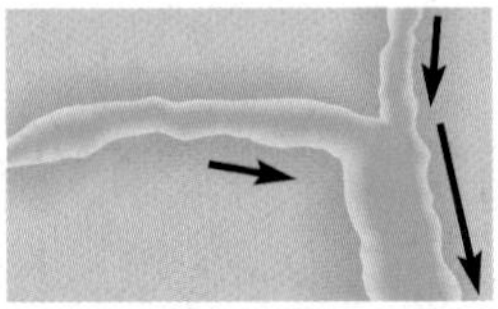
(a)

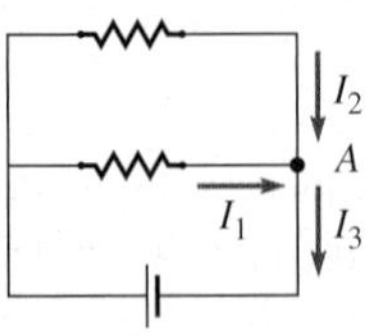

(b)

Figure 18.14 (a) The rate at which water flows into the junction from the two streams is equal to the rate at which water flows out of the junction into the larger stream. Equivalently, we can say that the net rate of flow of water into the junction is zero. (b) An analogous junction in an electric circuit.

Be careful to get the signs right when applying the loop rule. If you follow a path through a resistor going in the same direction as the current, the potential drops ($\Delta V = -IR$). If your path takes you through a resistor in a direction opposite to the current ("upstream"), the potential rises ($\Delta V = +IR$). For an emf, the potential drops if you move from the positive terminal to the negative ($\Delta V = -\mathscr{E}$); it rises if you move from the negative to the positive ($\Delta V = +\mathscr{E}$).

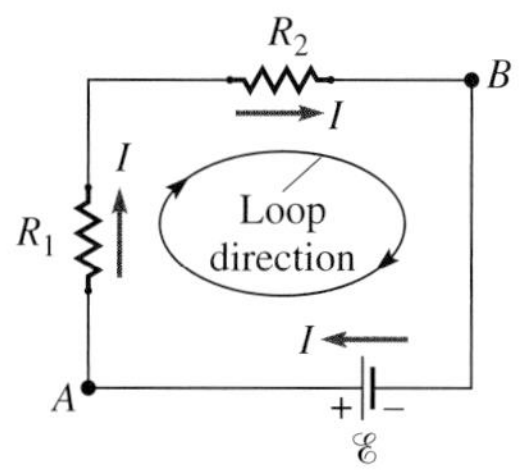

Figure 18.15 Applying the loop rule. If we start at point A and walk around the loop in the direction shown (clockwise), the loop rule gives $\Sigma \Delta V = -IR_1 - IR_2 + \mathscr{E} = 0$. (Starting at B and walking counterclockwise gives $\Sigma \Delta V = +IR_2 + IR_1 - \mathscr{E} = 0$, an equivalent equation.)

18.6 SERIES AND PARALLEL CIRCUITS

Resistors in Series

When one or more electric devices are wired so that the *same current* flows through each one, the devices are said to be wired in **series** (Figs. 18.16 and 18.17). The circuit of Fig. 18.17a shows two resistors in series. The straight lines represent wires, which we assume to have negligible resistance. Negligible resistance means negligible voltage drop ($V = IR$), so *points connected by wires of negligible resistance are at the same potential.* The junction rule, applied to any of the points *A–D*, tells us that the same current flows through the emf and the two resistors.

Series: same current through each device

Let's apply the loop rule to a clockwise loop *DABCD*. From *D* to *A* we move from the negative terminal to the positive terminal of the emf, so $\Delta V = +1.5$ V. Since we move around the loop *with* the current, the potential *drops* as we move across each resistor. Therefore,

$$1.5 \text{ V} - IR_1 - IR_2 = 0$$

The same current I flows through the two resistors in series. Factoring out the common current I,

$$I(R_1 + R_2) = 1.5 \text{ V}$$

The current I would be the same if a single equivalent resistor $R_{eq} = R_1 + R_2$ replaced the two resistors in series:

$$IR_{eq} = I(R_1 + R_2) = 1.5 \text{ V}$$

Figure 18.17b shows how the circuit diagram can be redrawn to indicate the simplified, equivalent circuit.

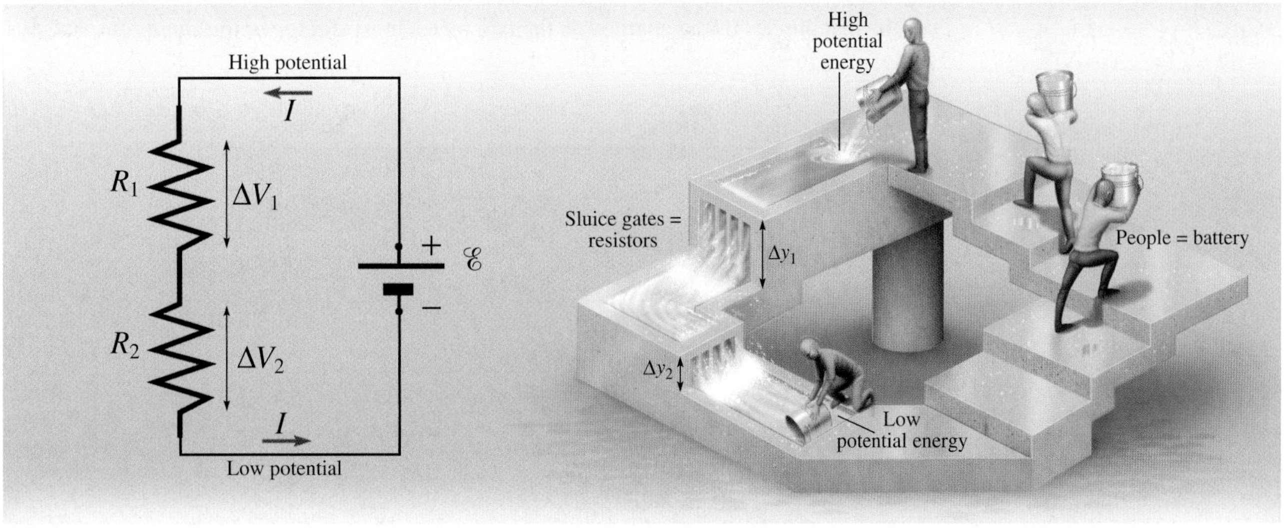

Figure 18.16 Just as water flows at the same mass flow rate through each of the two sluice gates, the same current flows through two resistors in series. Just as $\Delta y_1 + \Delta y_2 = \Delta y$, the potential difference ΔV across a series pair is the sum of the two potential differences. In this circuit, $\Delta V_1 + \Delta V_2 = \mathscr{E}$, the emf of the battery. If $R_1 \neq R_2$, the potential differences across the resistors (ΔV_1 and ΔV_2) are *not* equal, but the current through them (I) is still the same.

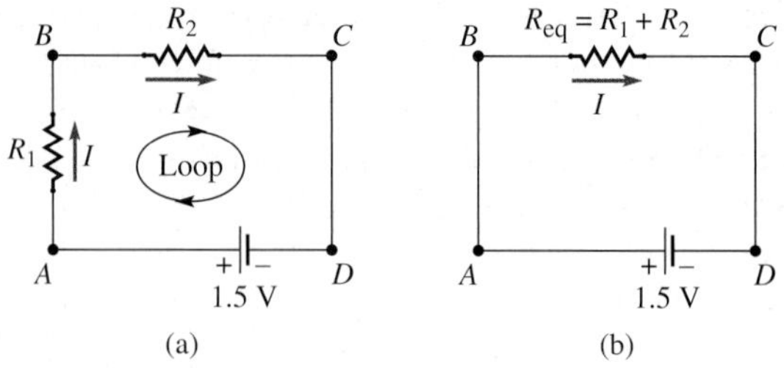

Figure 18.17 (a) A circuit with two resistors in series. (b) Replacing the two resistors with an equivalent resistor.

We can generalize this result to any number of resistors in series:

For any number N of resistors connected in series,

$$R_{eq} = \sum R_i = R_1 + R_2 + \cdots + R_N \quad (18\text{-}13)$$

Note that the equivalent resistance for two or more resistors in series is *larger* than *any* of the resistances.

Emfs in Series

Application: battery connection in a flashlight

In many devices, batteries are connected in series with the positive terminal of one connected to the negative terminal of the next. This provides a larger emf than a single battery can (Fig. 18.18). The emfs of batteries connected in this way are added just as series resistances are added. However, there is a disadvantage in connecting batteries in series: the internal resistance is larger because the internal resistances are in series as well.

Application: battery charger

Sources can be connected in series with the emfs in opposition. A common use for such a circuit is in a battery charger. In Fig. 18.19, as we move from point C to B to A, the potential decreases by $\mathscr{E}_2$ and then increases by $\mathscr{E}_1$, so the net emf is $\mathscr{E}_1 - \mathscr{E}_2$.

Capacitors in Series

The symbol —⁄•— represents an open switch (no electric connection). The symbol —•–•— represents a closed switch.

Figure 18.20a shows two capacitors connected in series. Although no charges can move *through* the dielectric of a capacitor from one plate to the other, the instantaneous currents I that flow onto one plate and from the other must be equal. Why? The two plates of a capacitor always have charges of equal magnitudes and opposite signs. Therefore, the magnitudes of the charges on the two plates must *change at the same rate.* The rate

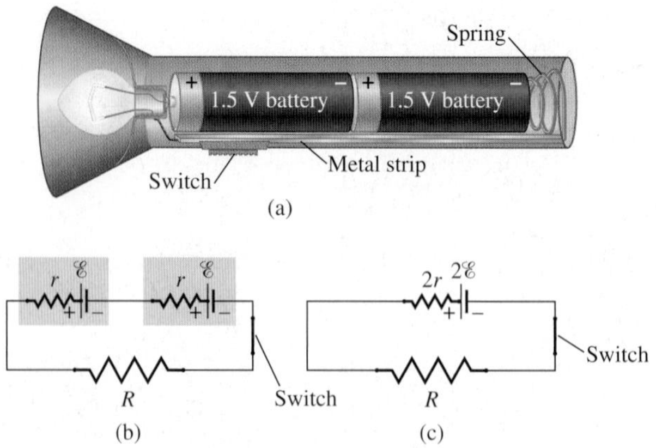

Figure 18.18 (a) Two 1.5-V batteries connected in series in a flashlight to supply 3.0 V. (b) Circuit diagram, including the internal resistances of the batteries. (c) Simplified circuit diagram, where the two batteries are combined into a single source of emf $2\mathscr{E}$ with internal resistance $2r$.

of change of the charge is equal to the current. Viewed from the outside, the capacitor behaves *as if* a current I flows through it.

The instantaneous currents "through" series capacitors C_1 and C_2 must be equal because no charge is created or destroyed and there is no junction between them to another branch of the circuit. Because their charges always change at the same rate, *the instantaneous charges on series capacitors are equal.*

We want to find the equivalent capacitance C_{eq} that would store the same amount of charge as each of the series capacitors for the same applied voltage. With the switch closed, the emf pumps charge so that the potential difference between points A and B is equal to the emf. The capacitors are fully charged and the current goes to zero. From Kirchhoff's loop rule,

$$\mathscr{E} - V_1 - V_2 = 0 \qquad (18\text{-}14)$$

The magnitude Q of the charges on series capacitors is the same, so

$$V_1 = \frac{Q}{C_1} \quad \text{and} \quad V_2 = \frac{Q}{C_2}$$

The equivalent capacitance (Fig. 18.20b) is defined by $\mathscr{E} = Q/C_{eq}$. Substituting into Eq. (18-14) yields

$$\frac{Q}{C_{eq}} - \frac{Q}{C_1} - \frac{Q}{C_2} = 0$$

The equivalent capacitance is given by

$$\frac{1}{C_{eq}} = \frac{1}{C_1} + \frac{1}{C_2}$$

This reasoning can be extended to the general case for any number of capacitors connected in series.

For N capacitors connected in series,

$$\frac{1}{C_{eq}} = \Sigma\frac{1}{C_i} = \frac{1}{C_1} + \frac{1}{C_2} + \cdots + \frac{1}{C_N} \qquad (18\text{-}15)$$

Note that the equivalent capacitor stores the same magnitude of charge as *each* of the capacitors it replaces.

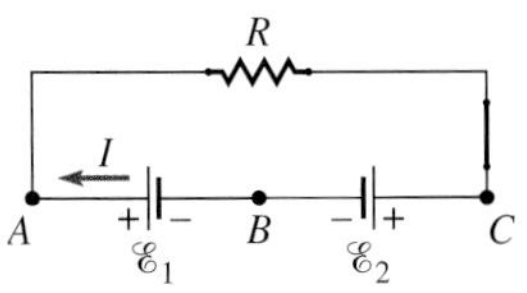

Figure 18.19 Circuit for charging a rechargeable battery (shown as emf $\mathscr{E}_2$). The source supplying the energy to charge the battery must have a larger emf ($\mathscr{E}_1 > \mathscr{E}_2$). The net emf in the circuit is $\mathscr{E}_1 - \mathscr{E}_2$; the current is $I = (\mathscr{E}_1 - \mathscr{E}_2)/R$ (where R includes the internal resistances of the sources).

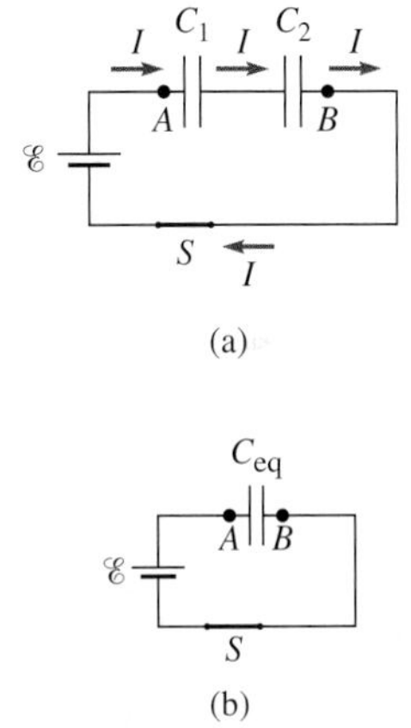

Figure 18.20 (a) Two capacitors connected in series. (b) Equivalent circuit.

Resistors in Parallel

Parallel: same potential difference across each device

When one or more electrical devices are wired so that the *potential difference across them is the same*, the devices are said to be wired in **parallel** (Fig. 18.21). In Fig. 18.22, an emf is connected to three resistors in parallel with each other. The left side of each resistor is at the same potential since they are all connected by wires of negligible resistance. Likewise, the right side of each resistor is at the same potential. Thus, there is a common potential difference across the three resistors. Applying the junction rule to point A yields

$$+I - I_1 - I_2 - I_3 = 0 \quad \text{or} \quad I = I_1 + I_2 + I_3 \qquad (18\text{-}16)$$

How much of the current I from the emf flows through each resistor? The current divides such that the potential difference $V_A - V_B$ must be the same along each of the three paths—and it must equal the emf $\mathscr{E}$. From the definition of resistance,

$$\mathscr{E} = I_1R_1 = I_2R_2 = I_3R_3$$

Therefore, the currents are

$$I_1 = \frac{\mathscr{E}}{R_1}, \quad I_2 = \frac{\mathscr{E}}{R_2}, \quad I_3 = \frac{\mathscr{E}}{R_3}$$

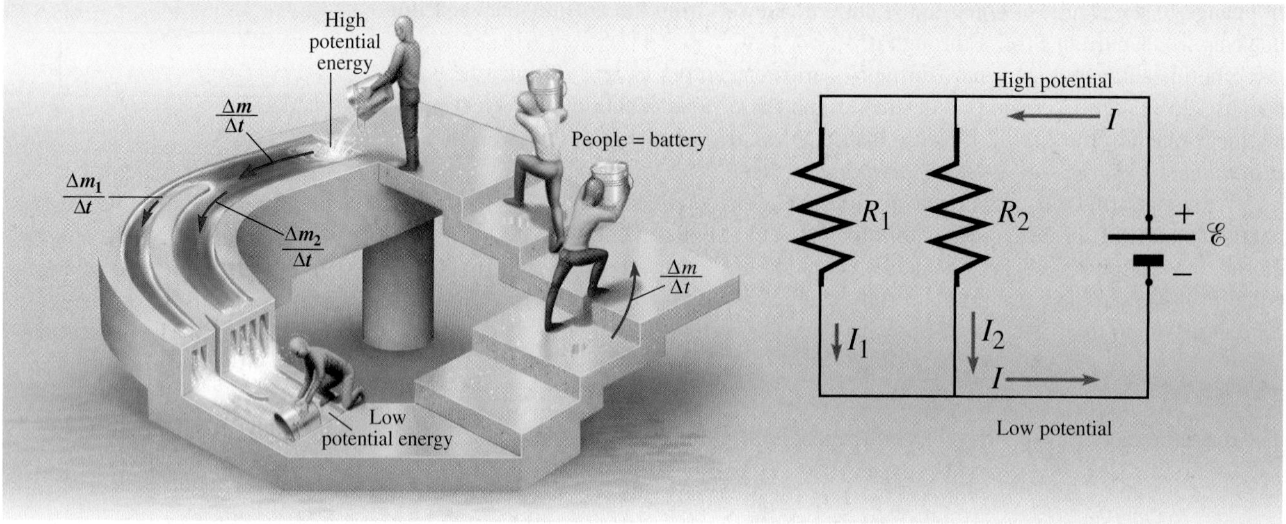

Figure 18.21 Some water flows through one branch and some through the other. The mass flow rate before the water channels divide and after they come back together is equal to the sum of the flow rates in the two branches. The elevation change Δy for the two branches is equal since they start and end at the same elevations. For two resistors in parallel, the currents *add* ($I = I_1 + I_2$); the potential differences are *equal* ($\Delta V_1 = \Delta V_2 = \mathscr{E}$). If $R_1 \neq R_2$, the currents I_1 and I_2 are *not* equal, but the potential differences are still equal.

Substituting the currents into Eq. (18-16) yields

$$I = \frac{\mathscr{E}}{R_1} + \frac{\mathscr{E}}{R_2} + \frac{\mathscr{E}}{R_3}$$

Dividing by $\mathscr{E}$ yields

$$\frac{I}{\mathscr{E}} = \frac{1}{R_1} + \frac{1}{R_2} + \frac{1}{R_3}$$

The three parallel resistors can be replaced by a single equivalent resistor R_{eq}. In order for the same current to flow, R_{eq} must be chosen so that $\mathscr{E} = IR_{eq}$. Then $I/\mathscr{E} = 1/R_{eq}$ and

$$\frac{1}{R_{eq}} = \frac{1}{R_1} + \frac{1}{R_2} + \frac{1}{R_3}$$

CONNECTION:

The same results for series and parallel resistors, Eqs. (18-13) and (18-17), are valid for thermal resistances (Section 14.6) and to the resistance of pipes to viscous fluid flow (Section 9.9).

Although we examined three resistors in parallel, the result applies to any number of resistors in parallel:

For N resistors connected in parallel,

$$\frac{1}{R_{eq}} = \sum \frac{1}{R_i} = \frac{1}{R_1} + \frac{1}{R_2} + \cdots + \frac{1}{R_N} \quad (18\text{-}17)$$

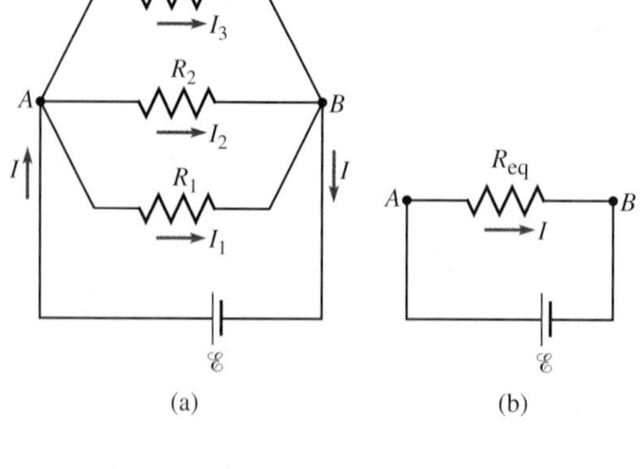

Figure 18.22 (a) Three resistors connected in parallel. (b) The equivalent circuit.

Note that the equivalent resistance for two or more resistors in parallel is *smaller* than *any* of the resistances ($1/R_{eq} > 1/R_i$, so $R_{eq} < R_i$). Note also that the equivalent resistance for resistors in *parallel* is found in the same way as the equivalent capacitance for capacitors in *series*. The reason is that resistance is defined as $R = \Delta V/I$ and capacitance as $C = Q/\Delta V$. One has ΔV in the numerator, the other in the denominator.

CHECKPOINT 18.6

What is the equivalent resistance for two equal resistors (R) in parallel?

Example 18.6

Current for Two Parallel Resistors

(a) Find the equivalent resistance for the two resistors in Fig. 18.23 if $R_1 = 20.0\ \Omega$ and $R_2 = 40.0\ \Omega$. (b) What is the ratio of the current through R_1 to the current through R_2?

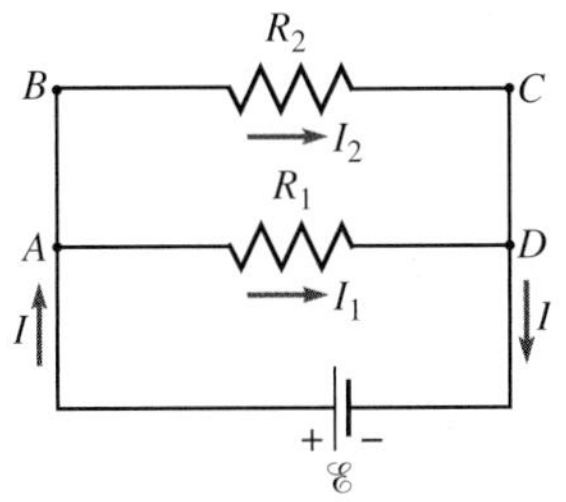

Figure 18.23
Circuit with parallel resistors for Example 18.6.

Strategy Points A and B are at the same potential; points C and D are at the same potential. Therefore, the voltage drops across the two resistors are equal; the two resistors are in parallel. The ratio of the currents can be found by equating the potential differences in the two branches in terms of the current and resistance.

Solution (a) The equivalent resistance for two parallel resistors is

$$\frac{1}{R_{eq}} = \frac{1}{R_1} + \frac{1}{R_2} = \frac{1}{20.0\ \Omega} + \frac{1}{40.0\ \Omega} = 0.0750\ \Omega^{-1}$$

$$R_{eq} = \frac{1}{0.0750\ \Omega^{-1}} = 13.3\ \Omega$$

(b) The potential differences across the resistors are equal

$$I_1 R_1 = I_2 R_2$$

Therefore,

$$\frac{I_1}{I_2} = \frac{R_2}{R_2} = \frac{40.0\ \Omega}{20.0\ \Omega} = 2.00$$

Discussion Note that the current in each branch of the circuit is inversely proportional to the resistance of that branch. Since R_2 is twice R_1, it has half as much current flowing through it. At the junction of two or more parallel branches, the current does not all flow through the "path of least resistance," but *more* current flows through the branch of least resistance than through the branches with larger resistances.

Practice Problem 18.6 Three Resistors in Parallel

Find the equivalent resistance from point A to point B for the three resistors in Fig. 18.24.

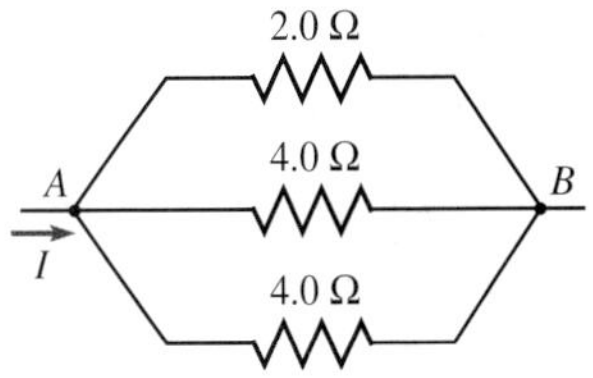

Figure 18.24
Three parallel resistors.

Example 18.7

Equivalent Resistance for Network in Series and Parallel

(a) Find the equivalent resistance for the network of resistors in Fig. 18.25. (b) Find the current through the resistor R_2 if $\mathscr{E} = 0.60$ V.

Strategy We simplify the network of resistors in a series of steps. At first, the only series or parallel combination is the two resistors (R_3 and R_4) in parallel between points B and C. No other pair of resistors has either the same current (for series) or the same voltage drop (for parallel). We replace those two with an equivalent resistor, redraw the circuit, and look for new series or parallel combinations, continuing until the entire network reduces to a single resistor.

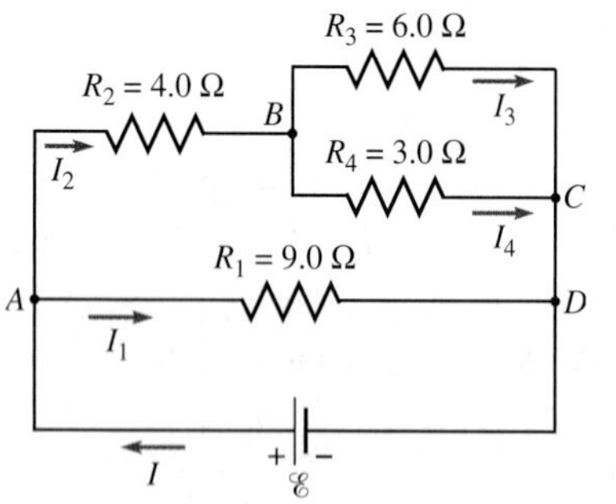

Figure 18.25
Network of resistors for Example 18.7.

Solution (a) For the two resistors in parallel between points B and C,

$$R_{eq} = \left(\frac{1}{R_3} + \frac{1}{R_4}\right)^{-1} = \left(\frac{1}{6.0\ \Omega} + \frac{1}{3.0\ \Omega}\right)^{-1} = 2.0\ \Omega$$

We redraw the circuit, replacing the two parallel resistors with an equivalent 2.0-Ω resistor.

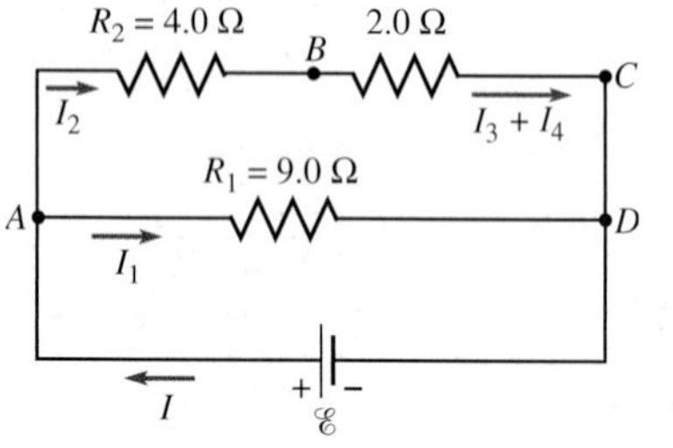

(1)

The 4.0-Ω and 2.0-Ω resistors are in series since the same current must flow through them. They can be replaced with a single resistor,

$$R_{eq} = 4.0\ \Omega + 2.0\ \Omega = 6.0\ \Omega$$

The network of resistors now becomes

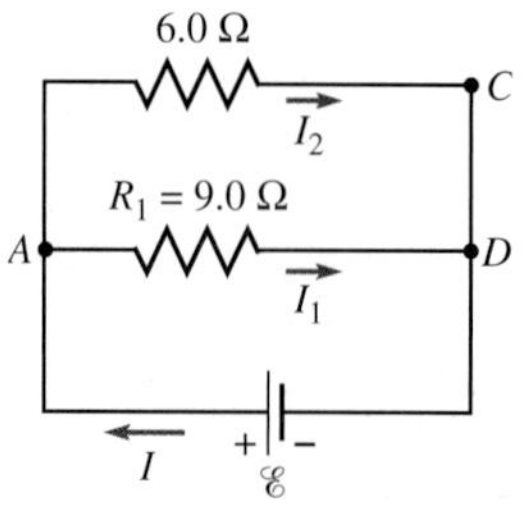

(2)

The two resistors in parallel have an equivalent resistance of

$$R_{eq} = \left(\frac{1}{6.0\ \Omega} + \frac{1}{9.0\ \Omega}\right)^{-1} = 3.6\ \Omega$$

The network of resistors reduces to a single equivalent 3.6-Ω resistor.

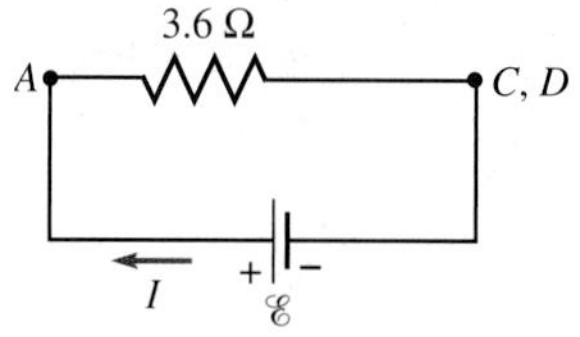

(3)

(b) The current through R_2 is I_2 (Fig. 18.25). From circuit diagram (2), when I_2 flows through an equivalent resistance of 6.0 Ω, the voltage drop is 0.60 V. Therefore,

$$I_2 = \frac{0.60\ \text{V}}{6.0\ \Omega} = 0.10\ \text{A}$$

Discussion To reduce complicated arrangements of resistors to an equivalent resistance, look for resistors in parallel (resistors connected so that they must have the same potential difference) and resistors in series (connected so that they must have the same current). Replace all parallel and series combinations of resistors with equivalents. Then look for new parallel and series combinations in the simplified circuit. Repeat until there is only one resistor remaining.

Practice Problem 18.7 Three Resistors Connected

Find the equivalent resistance that can be placed between points A and B to replace the three equal resistors shown in Fig. 18.26. First try to decide whether these resistors are in series or parallel. Label the black dots with A or B by tracing the straight lines from A or B to their connections at one side or another of the resistors. Redraw the diagram if that helps you decide.

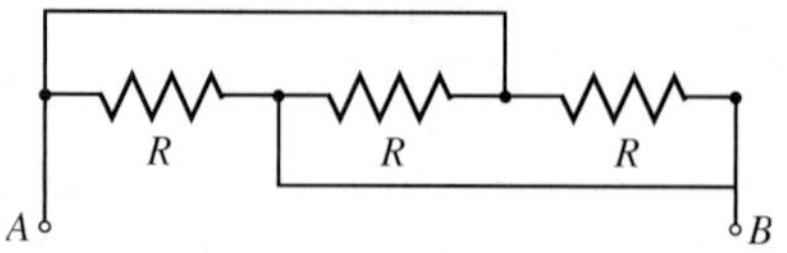

Figure 18.26
Three connected resistors.

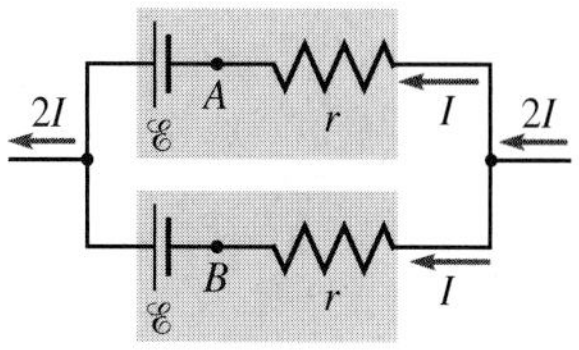

(a)

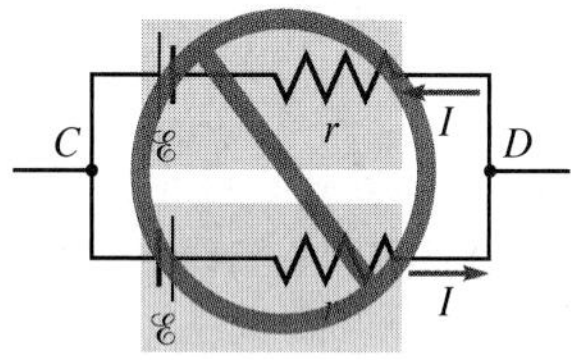

(b)

Figure 18.27 (a) Two identical batteries (with internal resistances r) in parallel. The combination provides an emf $\mathscr{E}$ and can supply twice as much current as one battery since the equivalent internal resistance is $\frac{1}{2}r$. (b) Never connect batteries in parallel with opposite polarities. In the case shown, the emfs are equal in magnitude, so points C and D are at the same potential. The batteries supply no emf to the rest of the circuit; they just drain one another. If two car batteries were connected in this way, a dangerously large current would flow through the batteries, possibly causing an explosion.

Emfs in Parallel

Two or more sources of *equal* emf are often connected in parallel with all the positive terminals connected together and all the negative terminals connected together (Fig. 18.27a). The equivalent emf for any number of equal sources in parallel is the same as the emf of each source. The advantage of connecting sources in this way is not to achieve a larger emf, but rather to lower the internal resistance and thus supply more current. In Fig. 18.27a, the two internal resistances (r) are equal. Since they are in parallel—note that points A and B are at the same potential—the equivalent internal resistance for the parallel combination is $\frac{1}{2}r$. To jump-start a car, one connects the two batteries in parallel, positive to positive and negative to negative.

Never connect unequal emfs in parallel or connect emfs in parallel with opposite polarities (Fig. 18.27b). In such cases the two batteries quickly drain one another and supply little or no current to the rest of the circuit.

Capacitors in Parallel

Capacitors in series have the same charge but may have different potential differences. Capacitors in parallel share a common potential difference but may have different charges. Suppose three capacitors are in parallel (Fig. 18.28). After the switch is closed, the source of emf pumps charge onto the plates of the capacitors until the potential difference across each capacitor is equal to the emf $\mathscr{E}$. Suppose that the total magnitude of charge pumped by the battery is Q. If the magnitude of charge on the three capacitors is q_1, q_2, and q_3, respectively, conservation of charge requires that

$$Q = q_1 + q_2 + q_3$$

The relation between the potential difference across a capacitor and the charge on either plate of the capacitor is $q = C\Delta V$. For each capacitor, $\Delta V = \mathscr{E}$. Therefore,

$$Q = q_1 + q_2 + q_3 = C_1\mathscr{E} + C_2\mathscr{E} + C_3\mathscr{E} = (C_1 + C_2 + C_3)\mathscr{E}$$

We can replace the three capacitors with a single equivalent capacitor. In order for it to store charge of magnitude Q for a potential difference $\mathscr{E}$, $Q = C_{eq}\mathscr{E}$. Therefore, $C_{eq} = C_1 + C_2 + C_3$. Once again, this result can be extended to the general case for any number of capacitors connected in parallel.

For N capacitors connected in parallel,

$$C_{eq} = \Sigma C_i = C_1 + C_2 + \cdots + C_N \quad (18\text{-}18)$$

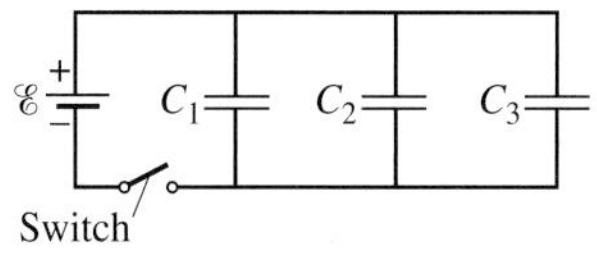

(a)

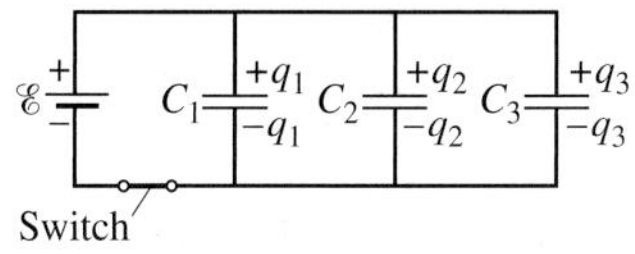

(b)

Figure 18.28 (a) Three capacitors in parallel. (b) When the switch is closed, each capacitor is charged until the potential difference between its plates is equal to $\mathscr{E}$. If the capacitances are unequal, the charges on the capacitors are unequal.

18.7 CIRCUIT ANALYSIS USING KIRCHHOFF'S RULES

Sometimes a circuit cannot be simplified by replacing parallel and series combinations alone. In such cases, we apply Kirchhoff's rules directly and solve the resulting equations simultaneously.

Problem-Solving Strategy: Using Kirchhoff's Rules to Analyze a Circuit

1. Replace any series or parallel combinations with their equivalents.
2. Assign variables to the currents in each branch of the circuit (I_1, I_2, . . .) and choose directions for each current. Draw the circuit with the current directions indicated by arrows. It does not matter whether or not you choose the correct direction.
3. Apply Kirchhoff's junction rule to *all but one* of the junctions in the circuit. (Applying it to every junction produces one redundant equation.) Remember that current into a junction is positive; current out of a junction is negative.
4. Apply Kirchhoff's loop rule to enough loops so that, together with the junction equations, you have the same number of equations as unknown quantities. For each loop, choose a starting point and a direction to go around the loop. Be careful with signs. For a resistor, if your path through a resistor goes *with* the current ("downstream"), there is a potential drop; if your path goes *against* the current ("upstream"), the potential rises. For an emf, the potential drops or rises depending on whether you move from the positive terminal to the negative or vice versa; the direction of the current is irrelevant. A helpful method is to write "+" and "–" signs on the ends of each resistor and emf to indicate which end is at the higher potential and which is at the lower potential.
5. Solve the loop and junction equations simultaneously. If a current comes out negative, the direction of the current is opposite to the direction you chose.
6. Check your result using one or more loops or junctions. A good choice is a loop that you did not use in the solution.

Example 18.8

A Two-Loop Circuit

Find the currents through each branch of the circuit of Fig. 18.29.

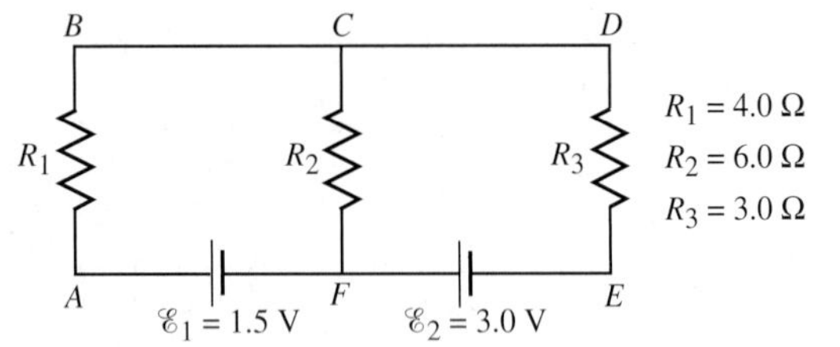

Figure 18.29
Circuit to be analyzed using Kirchhoff's rules.

Strategy First we look for series and parallel combinations. R_1 and $\mathscr{E}_1$ are in series, but since one is a resistor and one an emf we cannot replace them with a single equivalent circuit element. No pair of resistors is either in series or in parallel. R_1 and R_2 might look like they're in parallel, but the emf $\mathscr{E}_1$ keeps points *A* and *F* at different potentials, so they are not. The two emfs might look like they're in series, but the junction at point *F* means that the current through the two is not the same. Since there are no series or parallel combinations to simplify, we proceed to apply Kirchhoff's rules directly.

Solution First we assign the currents variable names and directions on the circuit diagram: Points *C* and *F* are junctions between the three branches of the circuit. We choose current I_1 for branch *FABC*, current I_3 for branch *FEDC*, and current I_2 for branch *CF*.

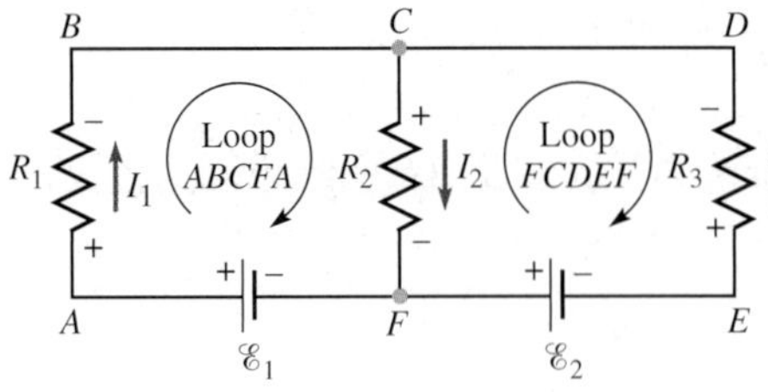

Now we can apply the junction rule. There are two junctions; we can choose either one. For point *C*, I_1 and I_3 flow

continued on next page

Example 18.8 continued

into the junction and I_2 flows out of the junction. The resulting equation is

$$I_1 + I_3 - I_2 = 0 \qquad (1)$$

Before applying the loop rule, we write "+" and "–" signs on each resistor and emf to show which side is at the higher potential and which at the lower, given the directions assumed for the currents. In a resistor, current flows from higher to lower potential. The emf symbol uses the longer line for the positive terminal and the shorter line for the negative terminal.

Now we choose a closed loop and add up the potential rises and drops as we travel around the loop. Suppose we start at point A and travel around loop $ABCFA$. The starting point and direction to go around the loop are arbitrary choices, but once made, we stick with it regardless of the directions of the currents. From A to B, we move in the same direction as the current I_1. The current through a resistor travels from higher to lower potential, so going from A to B is a potential drop: $\Delta V_{A\to B} = -I_1R_1$.

From B to C, since the wire is assumed to have negligible resistance, there is no potential rise or drop. From C to F, we move with current I_2, so there is another potential drop: $\Delta V_{C\to F} = -I_2R_2$.

Finally, from F to A, we move from the negative terminal of an emf to the positive terminal. The potential *rises*: $\Delta V_{F\to A} = +\mathscr{E}_1$. A was the starting point, so the loop is complete. The loop rule says that the sum of the potential changes is equal to zero:

$$-I_1R_1 - I_2R_2 + \mathscr{E}_1 = 0 \qquad (2)$$

We must choose another loop since we have not yet gone through resistor R_3 or emf $\mathscr{E}_2$. There are two choices possible: the right-hand loop (such as $FCDEF$) or the outer loop ($ABCDEFA$). Let's choose $FCDEF$.

From F to C, we move *against* the current I_2 ("upstream"). The potential rises: $\Delta V_{F\to C} = +I_2R_2$. From C to D, the potential does not change. From D to E, we again move upstream, so $\Delta V_{D\to E} = +I_3R_3$. From E to F, we move through a source of emf from the negative to the positive terminal. The potential increases: $\Delta V_{E\to F} = +\mathscr{E}_2$. Then the loop rule gives

$$+I_2R_2 + I_3R_3 + \mathscr{E}_2 = 0 \qquad (3)$$

Now we have three equations and three unknowns (the three currents). To solve them simultaneously, we first substitute known numerical values:

$$I_1 + I_3 - I_2 = 0 \qquad (1)$$
$$-(4.0\ \Omega)I_1 - (6.0\ \Omega)I_2 + 1.5\ \text{V} = 0 \qquad (2)$$
$$(6.0\ \Omega)I_2 + (3.0\ \Omega)I_3 + 3.0\ \text{V} = 0 \qquad (3)$$

To solve simultaneous equations, we can solve one equation for one variable and substitute into the other equations, thus eliminating one variable. Solving Eq. (1) for I_1 yields $I_1 = -I_3 + I_2$. Substituting in Eq. (2):

$$-(4.0\ \Omega)(-I_3 + I_2) - (6.0\ \Omega)I_2 + 1.5\ \text{V} = 0$$

Simplifying,

$$4.0I_3 - 10.0I_2 = -1.5\ \text{V}/\Omega = -1.5\ \text{A} \qquad (2a)$$

Eqs. (2a) and (3) now have only two unknowns. We can eliminate I_3 if we multiply Eq. (2a) by 3 and Eq. (3) by 4 so that I_3 has the same coefficient.

$$12.0I_3 - 30.0I_2 = -4.5\ \text{A} \quad 3 \times \text{Eq. (2a)}$$

$$12.0I_3 + 24.0I_2 = -12.0\ \text{A} \quad 4 \times \text{Eq. (3)}$$

Subtracting one from the other,

$$54.0I_2 = -7.5\ \text{A}$$

Now we can solve for I_2:

$$I_2 = -\frac{7.5}{54.0}\ \text{A} = -0.139\ \text{A}$$

Substituting the value of I_2 into Eq. (2a) enables us to solve for I_3:

$$4I_3 + 10 \times 0.139\ \text{A} = -1.5\ \text{A}$$
$$I_3 = \frac{-1.5 - 1.39}{4}\ \text{A} = -0.723\ \text{A}$$

Equation (1) now gives I_1:

$$I_1 = -I_3 + I_2 = +0.723\ \text{A} - 0.139\ \text{A} = +0.584\ \text{A}$$

Rounding to two significant figures, the currents are $I_1 = +0.58$ A, $I_3 = -0.72$ A, and $I_2 = -0.14$ A. Since I_3 and I_2 came out negative, the actual directions of the currents in those branches are opposite to the ones we arbitrarily chose.

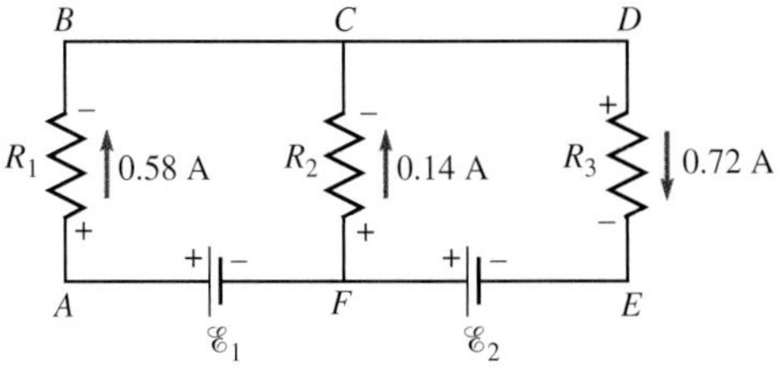

Discussion Note that it did not matter that we chose some of the current directions wrong. It also doesn't matter which loops we choose (as long as we cover every branch of the circuit), which starting point we use for a loop, or which direction we go around a loop.

The hardest thing about applying Kirchhoff's rules is getting the signs correct. It is also easy to make an algebraic

continued on next page

Example 18.8 continued

mistake when solving simultaneous equations. Therefore, it is a good idea to check the answer. A good way to check is to write down a loop equation for a loop that was not used in the solution (see Practice Problem 18.8).

Practice Problem 18.8 Verifying the Solution with the Loop Rule

Apply Kirchhoff's loop rule to loop *CBAFEDC* to verify the solution of Example 18.8.

18.8 POWER AND ENERGY IN CIRCUITS

From the definition of electric potential, if a charge q moves through a potential difference ΔV, the change in electric potential energy is

$$\Delta U_E = q\Delta V \qquad (17\text{-}7)$$

From energy conservation, a change in electric potential energy means that conversion between two forms of energy takes place. For example, a battery converts stored chemical energy into electric potential energy. A resistor converts electric potential energy into internal energy. The *rate* at which the energy conversion takes place is the *power P.* Since current is the rate of flow of charge, $I = q/\Delta t$ and

Power

$$P = \frac{\Delta U_E}{\Delta t} = \frac{q}{\Delta t}\Delta V = I\,\Delta V \qquad (18\text{-}19)$$

Thus, the power for *any circuit element* is the product of current and potential difference. We can verify that current times voltage comes out in the correct units for power by substituting coulombs per second for amperes and joules per coulomb for volts:

$$\text{A} \times \text{V} = \frac{\text{C}}{\text{s}} \times \frac{\text{J}}{\text{C}} = \frac{\text{J}}{\text{s}} = \text{W}$$

According to the definition of emf, if the amount of charge pumped by an ideal source of constant emf $\mathscr{E}$ is q, then the work done by the battery is

$$W = \mathscr{E}q \qquad (18\text{-}2)$$

The power supplied by the emf is the rate at which it does work:

$$P = \frac{\Delta W}{\Delta t} = \mathscr{E}\frac{q}{\Delta t} = \mathscr{E}I \qquad (18\text{-}20)$$

Since $\Delta V = \mathscr{E}$ for an ideal emf, Eqs. (18-20) and (18-19) are equivalent.

Power Dissipated by a Resistor

> The term *power dissipated* means *the rate at which energy is dissipated.* "Power" is not dissipated in a resistor; *energy* is.

If an emf causes current to flow through a resistor, what happens to the energy supplied by the emf? Why must the emf continue supplying energy to maintain the current?

Current flows in a metal wire when an emf gives rise to a potential difference between one end and the other. The electric field makes the conduction electrons drift in the direction of lower electric potential energy (higher potential). If there were no collisions between electrons and atoms in the metal, the average kinetic energy of the electrons would continually increase. However, the electrons frequently collide with atoms; each such collision is an opportunity for an electron to give away some of its kinetic energy. For a steady current, the average kinetic energy of the conduction electrons does not increase; the rate at which the electrons gain kinetic energy (due to the electric field) is equal to the rate at which they lose kinetic energy (due to collisions). The net effect is that the energy supplied by the emf increases the vibrational energy of the atoms. The vibrational energy of the atoms is part of the internal energy of the metal, so the temperature of the metal rises.

From the definition of resistance, the potential drop across a resistor is

$$V = IR$$

Then the rate at which energy is dissipated in a resistor can be written

Dissipation: the conversion of energy from an organized form to a disorganized form

$$P = I \times IR = I^2R \quad (18\text{-}21a)$$

or

$$P = \frac{V}{R} \times V = \frac{V^2}{R} \quad (18\text{-}21b)$$

Is the power dissipated in a resistor directly proportional to the resistance [Eq. (18-21a)] or inversely proportional to the resistance [Eq. (18-21b)]? It depends on the situation. For two resistors with the *same current* (such as two resistors in series), the power is directly proportional to resistance—the voltage drops are not the same. For two resistors with the same voltage drop (such as two resistors in parallel), the power is inversely proportional to resistance; this time the currents are not the same.

Dissipation in a resistor is not necessarily undesirable. In any kind of electric heater—in portable or baseboard heaters, electric stoves and ovens, toasters, hair dryers, and electric clothes dryers—and in incandescent lights, the dissipation of energy and the resulting temperature increase of a resistor are put to good use.

Application: resistive heating

Power Supplied by an Emf with Internal Resistance

If the source has internal resistance, then the net power supplied is *less* than $\mathscr{E}I$. Some of the energy supplied by the emf is dissipated by the internal resistance. The net power supplied to the rest of the circuit is

$$P = \mathscr{E}I - I^2r \quad (18\text{-}22)$$

where r is the internal resistance of the source. Equation (18-22) agrees with Eq. (18-19); remember that the potential difference is *not* equal to the emf when there is internal resistance (see Problem 71).

Example 18.9

Two Flashlights

A flashlight is powered by two batteries in series. Each has an emf of 1.50 V and an internal resistance of 0.10 Ω. The batteries are connected to the lightbulb by wires of total resistance 0.40 Ω. At normal operating temperature, the resistance of the filament is 9.70 Ω. (a) Calculate the power dissipated by the bulb—that is, the rate at which energy in the form of heat and light flows away from it. (b) Calculate the power dissipated by the wires and the net power supplied by the batteries. (c) A second flashlight uses *four* such batteries in series and the same resistance wires. A bulb of resistance 42.1 Ω (at operating temperature) dissipates approximately the same power as the bulb in the first flashlight. Verify that the power dissipated is nearly the same and calculate the power dissipated by the wires and the net power supplied by the batteries.

Strategy All the circuit elements are in series. We can simplify the circuit by replacing all the resistors (including the internal resistances of the batteries) with one series equivalent and the two emfs with one equivalent emf. Doing so enables us to find the current. Then we can use Eq. (18-21a) to find the power in the wires and in the filament. Equation (18-21b) could be used, but would require an extra step: finding the voltage drops across the resistors. Equation (18-22) gives the net power supplied by the batteries.

Solution (a) Figure 18.30 is a sketch of the circuit for the first flashlight. To find the power dissipated in the lightbulb, we need either the current through it or the voltage drop across it.

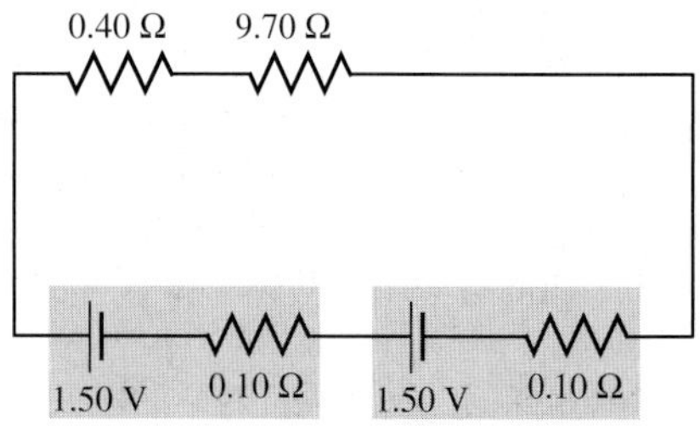

Figure 18.30 Circuit for the first flashlight.

continued on next page

Example 18.9 continued

We can find the current in this single-loop circuit by replacing the two ideal emfs with a series equivalent emf of $\mathscr{E}_{eq} = 3.00$ V and all the resistors by a series equivalent resistance of

$$R_{eq} = 9.70\ \Omega + 0.40\ \Omega + 2 \times 0.10\ \Omega = 10.30\ \Omega$$

Then the current is

$$I = \frac{\mathscr{E}_{eq}}{R_{eq}} = \frac{3.00\text{ V}}{10.30\ \Omega} = 0.2913\text{ A}$$

The power dissipated by the filament is

$$P_f = I^2R = (0.2913\text{ A})^2 \times 9.70\ \Omega = 0.823\text{ W}$$

(b) The power dissipated by the wires is

$$P_w = I^2R = (0.2913\text{ A})^2 \times 0.40\ \Omega = 0.034\text{ W}$$

The net power supplied by the batteries is

$$P_b = \mathscr{E}_{eq}I - I^2r_{eq}$$

where $r_{eq} = 0.20\ \Omega$ is the series equivalent for the two internal resistances. Then

$$P_b = 3.00\text{ V} \times 0.2913\text{ A} - (0.2913\text{ A})^2 \times 0.20\ \Omega = 0.857\text{ W}$$

(c) In the second circuit, $\mathscr{E}_{eq} = 6.00$ V and

$$R_{eq} = 42.1\ \Omega + 0.40\ \Omega + 4 \times 0.10\ \Omega = 42.90\ \Omega$$

The current is

$$I = \frac{\mathscr{E}_{eq}}{R_{eq}} = \frac{6.00\text{ V}}{42.90\ \Omega} = 0.13986\text{ A}$$

The power dissipated by the filament is

$$P_f = I^2R = (0.13986\text{ A})^2 \times 42.1\ \Omega = 0.824\text{ W}$$

which is only 0.1% more than the filament in the first flashlight. The power dissipated by the wires is

$$P_w = I^2R = (0.13986\text{ A})^2 \times 0.40\ \Omega = 0.0078\text{ W}$$

The series equivalent for the four internal resistances is $r_{eq} = 0.40\ \Omega$, so the net power supplied by the batteries is

$$P_b = \mathscr{E}_{eq}I - I^2r_{eq}$$

$$= 6.00\text{ V} \times 0.13986\text{ A} - 0.0078\text{ W} = 0.831\text{ W}$$

Discussion Note that in each case, the net power supplied by the batteries is equal to the total power dissipated in the wires and the filament. Since there is nowhere else for the energy to go, the wires and filament must dissipate energy—convert electric energy to light and heat—at the same rate that the battery supplies electric energy.

The power supplied to the two filaments is about the same in the two cases. However, the power dissipated by the wires in the second flashlight is a bit less than one-fourth as much as in the first. By using a larger emf, the current required to supply a given amount of power is smaller. The current is smaller because the load resistance (the resistance of the filament) is larger. A smaller current means the power dissipated in the wires is smaller. Utility companies distribute power over long distances using high-voltage wires for exactly this reason: the smaller the current, the smaller the power dissipated in the wires.

Practice Problem 18.9 A Simplified Flashlight Circuit

A flashlight takes two 1.5-V batteries connected in series. If the current that flows to the bulb in the flashlight is 0.35 A, find the power delivered to the lightbulb and the amount of energy dissipated after the light has been in the "on position" for 3 min. Treat the batteries as ideal and ignore the resistance of the wires. [*Hint:* It is not necessary to calculate the resistance of the filament since in this case the voltage drop across it is equal to the emf.]

18.9 MEASURING CURRENTS AND VOLTAGES

Figure 18.31 A digital multimeter being used to test a circuit board. A multimeter can function as an ammeter, as a voltmeter, or as an ohmmeter (to measure resistance). Most multimeters can measure both dc and ac currents and voltages.

Current and potential difference in a circuit can be measured with instruments called **ammeters** and **voltmeters**, respectively. A multimeter (Fig. 18.31) functions as an ammeter or a voltmeter, depending on the setting of a switch and which of its terminals are connected. Meters can be either digital or analog; the latter uses a rotating pointer to indicate the value of current or voltage on a calibrated scale. At the heart of an analog voltmeter or analog ammeter is a **galvanometer**, a sensitive detector of current whose operation is based on magnetic forces.

Suppose a particular galvanometer has a resistance of 100.0 Ω and deflects full scale for a current of 100 μA. We want to build an ammeter to measure currents from 0 to 10 A—when a current of 10 A passes through the meter, the needle should deflect full scale. Therefore, when a current of 10 A goes through the ammeter, 100 μA should go through the galvanometer; the other 9.9999 A must bypass the galvanometer. We put a resistor in parallel with the galvanometer so that the 10 A current branches, sending 100 μA to deflect the needle and 9.9999 A through the *shunt resistor* (Fig. 18.32a).

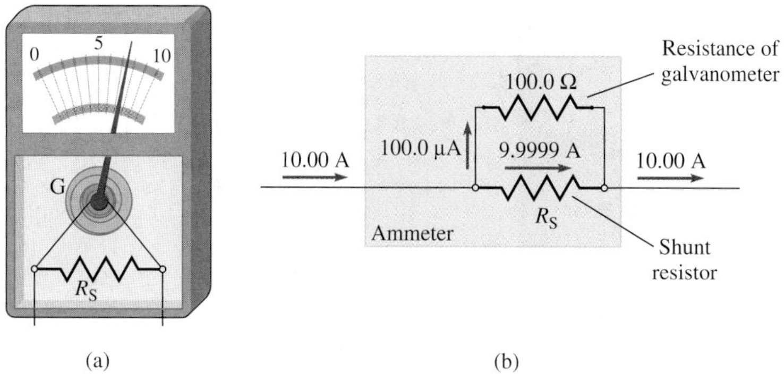

Figure 18.32 (a) An ammeter constructed from a galvanometer. (b) The circuit diagram for the ammeter. The galvanometer is represented as a 100.0-Ω resistor.

Example 18.10

Constructing an Ammeter from a Galvanometer

If the internal resistance of a galvanometer (Fig. 18.32a) is 100.0 Ω and it deflects full scale for a current of 100.0 μA, what resistance should the shunt resistor have to make an ammeter for measuring currents up to 10.00 A?

Strategy When 10.00 A flows into the ammeter, 100.0 μA should go through the galvanometer and the remaining 9.9999 A should go through the shunt resistor (Fig. 18.32b). Since the two are in parallel, the potential difference across the galvanometer is equal to the potential difference across the shunt resistor.

Solution The voltage drop across the galvanometer when it deflects full scale is

$$V = IR = 100.0\ \mu\text{A} \times 100.0\ \Omega$$

The voltage drop across the shunt resistor must be the same, so

$$V = 100.0\ \mu\text{A} \times 100.0\ \Omega = 9.9999\ \text{A} \times R_S$$

$$R_S = \frac{100.0\ \mu\text{A} \times 100.0\ \Omega}{9.9999\ \text{A}} = 0.001000\ \Omega = 1.000\ \text{m}\Omega$$

Discussion The resistance of the ammeter is

$$\left(\frac{1}{0.001000\ \Omega} + \frac{1}{100.0\ \Omega}\right)^{-1} = 1.000\ \text{m}\Omega$$

A good ammeter should have a small resistance. When an ammeter is used to measure the current in a branch of a circuit, it must be inserted *in series* in that branch—the ammeter can only measure whatever current passes through it. Adding a small resistance in series has only a slight effect on the circuit.

Practice Problem 18.10 Changing the Ammeter Scale

If the ammeter measures currents from 0 to 1.00 A, what shunt resistance should be used? What is the resistance of the ammeter? Use the same galvanometer as in Example 18.10.

In order to give accurate measurements, an ammeter must have a small resistance so its presence in the circuit does not change the current significantly from its value in the absence of the ammeter. An ideal ammeter has zero resistance.

An ammeter must have a small resistance.

It is also possible to construct a voltmeter by connecting a resistor (R_S) *in series* with the galvanometer (R_S, Fig. 18.33). The series resistor R_S is chosen so that the current through the galvanometer makes it deflect full scale when the desired full-scale voltage appears across the voltmeter. A voltmeter measures the potential difference between its terminals; to measure the potential difference across a resistor, for example, the voltmeter is connected in parallel with the resistor, with one terminal connected to each side of the resistor. So as not to affect the circuit too much, a good voltmeter must have a large resistance; then when measurements are taken, the current through the voltmeter (I_m) is small compared with I and the potential difference across the parallel combination is nearly the same as when the voltmeter is disconnected. An ideal voltmeter has infinite resistance.

A voltmeter must have a large resistance.

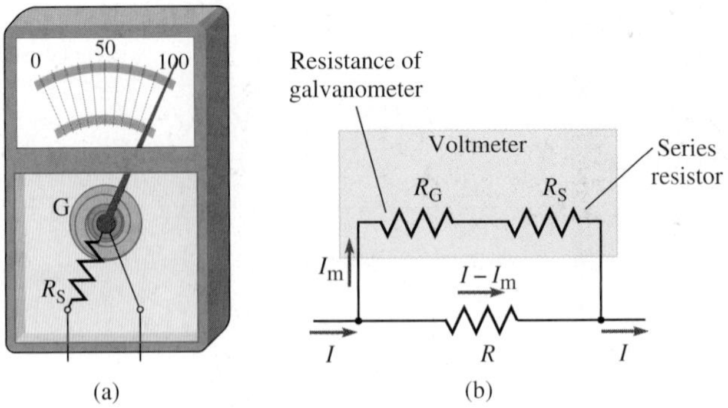

Figure 18.33 (a) A voltmeter constructed from a galvanometer. (b) Circuit diagram of the voltmeter measuring the voltage across the resistor *R*.

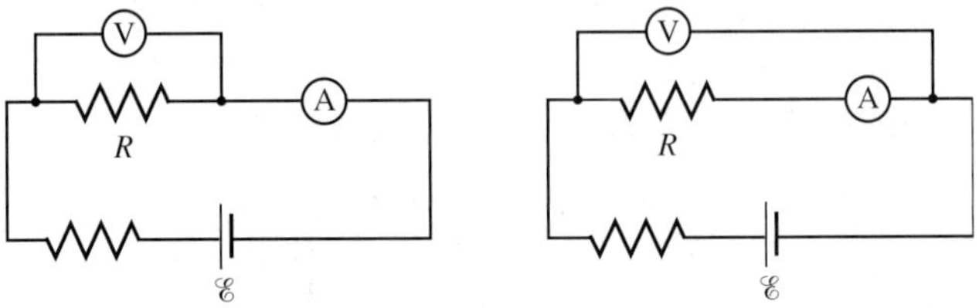

Figure 18.34 Two ways to arrange meters to measure a resistance *R*. If the meters were ideal (an ammeter with zero resistance and a voltmeter with infinite resistance), the two arrangements would give exactly the same measurement. Note the symbols used for the meters.

To measure a resistance in a circuit, we can use a voltmeter to measure the potential difference across the resistor and an ammeter to measure the current through the resistor (Fig. 18.34). By definition, the ratio of the voltage to the current is the resistance.

18.10 *RC* CIRCUITS

Circuits containing both resistors and capacitors have many important applications. *RC circuits* are commonly used to control timing. When windshield wipers are set to operate intermittently, the charging of a capacitor to a certain voltage is the trigger that turns them on. The time delay between wipes is determined by the resistance and capacitance in the circuit; adjusting a variable resistor changes the length of the time delay. Similarly, an *RC* circuit controls the time delay in strobe lights and in some pacemakers. We can also use the *RC* circuit as a simplified model of the transmission of nerve impulses.

Charging *RC* Circuit

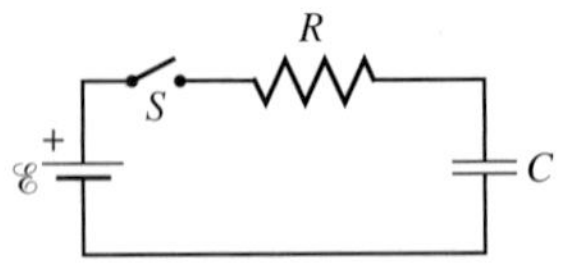

Figure 18.35 An *RC* circuit.

In Fig. 18.35, switch *S* is initially open and the capacitor is uncharged. When the switch is closed, current begins to flow and charge starts to build up on the plates of the capacitor. At any instant, Kirchhoff's loop law requires that

$$\mathscr{E} - V_R - V_C = 0$$

where V_R and V_C are the voltage drops across the resistor and capacitor, respectively. As charge accumulates on the capacitor plates, it becomes increasingly difficult to push more charge onto them.

Just after the switch is closed, the potential difference across the resistor is equal to the emf since the capacitor is uncharged. Initially, a relatively large current $I_0 = \mathscr{E}/R$ flows. As the voltage drop across the capacitor increases, the voltage drop across the resistor decreases, and thus the current decreases. Long after the switch is closed, the potential difference across the capacitor is nearly equal to the emf and the current is small.

Using calculus, it can be shown that the voltage across the capacitor involves an exponential function (see Fig. 18.36):

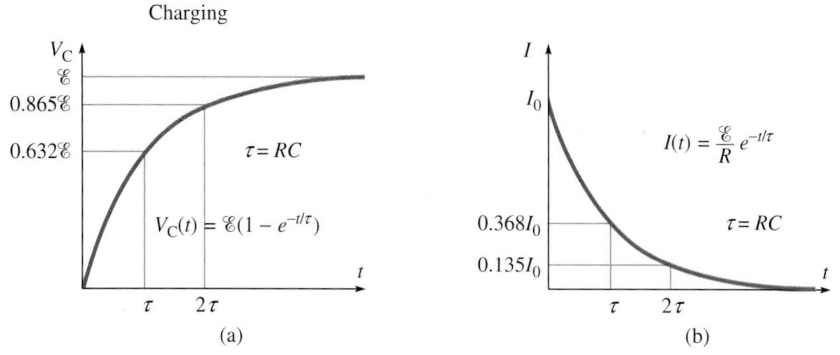

Figure 18.36 (a) The potential difference across the capacitor as a function of time as the capacitor is charged. (b) The current through the resistor as a function of time.

Charging capacitor

$$V_C(t) = \mathcal{E}(1 - e^{-t/\tau}) \quad (18\text{-}23)$$

where $e \approx 2.718$ is the base of the natural logarithm and the quantity $\tau = RC$ is called the **time constant** for the *RC* circuit.

Time constant

$$\tau = RC \quad (18\text{-}24)$$

The product *RC* has time units:

$$[R] = \frac{\text{volts}}{\text{amps}} \quad \text{and} \quad [C] = \frac{\text{coulombs}}{\text{volts}} \quad \text{so} \quad [RC] = \frac{\text{C}}{\text{A}} = \text{s}$$

The time constant is a measure of how fast the capacitor charges. At $t = \tau$, the voltage across the capacitor is

$$V_C(t = \tau) = \mathcal{E}(1 - e^{-1}) \approx 0.632\mathcal{E}$$

Since $Q = CV_C$, when one time constant has elapsed, the capacitor has 63.2% of its final charge.

From Eq. (18-23), we can use the loop rule to find the current.

$$\mathcal{E} - IR - \mathcal{E}(1 - e^{-t/\tau}) = 0$$

Solving for I,

Charging *or* discharging

$$I(t) = \frac{\mathcal{E}}{R}e^{-t/\tau} \quad (18\text{-}25)$$

At $t = \tau$, the current is

$$I(t = \tau) = \frac{\mathcal{E}}{R}e^{-1} \approx 0.368\frac{\mathcal{E}}{R}$$

When one time constant has elapsed, the current is reduced to 36.8% of its initial value. The voltage drop across the resistor as a function of time can be found from $V_R = IR$.

Example 18.11

An *RC* Circuit with Two Capacitors in Series

Two 0.500-μF capacitors in series are connected to a 50.0-V battery through a 4.00-MΩ resistor at $t = 0$ (Fig. 18.37). The capacitors are initially uncharged. (a) Find the charge on the capacitors at $t = 1.00$ s and $t = 3.00$ s. (b) Find the current in the circuit at the same two times.

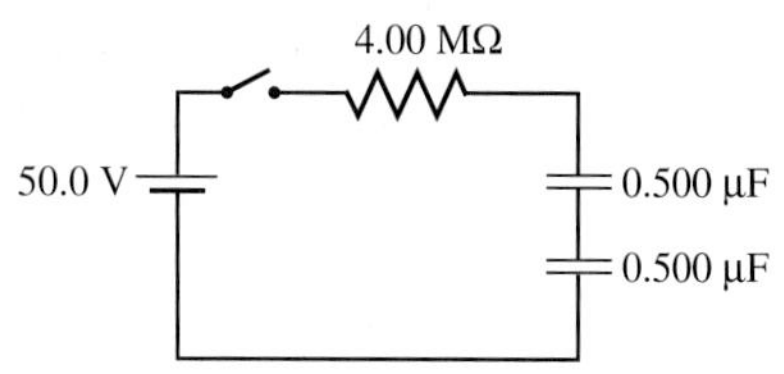

Figure 18.37 The circuit for Example 18.11.

continued on next page

Example 18.11 continued

Strategy First we find the equivalent capacitance of two 0.500-μF capacitors in series. Then we can find the time constant using the equivalent capacitance. Equation (18-23) gives the voltage across the equivalent capacitor at any time t; once we know the voltage, we can find the charge from $Q = CV_C$. The charge on each of the two capacitors is equal to the charge on the equivalent capacitor. The current decreases exponentially according to Eq. (18-25).

Solution (a) For two equal capacitors C in series,

$$\frac{1}{C_{eq}} = \frac{1}{C} + \frac{1}{C} = \frac{2}{C}$$

Then $C_{eq} = \frac{1}{2}C = 0.250\ \mu F$. The time constant is

$$\tau = RC_{eq} = 4.00 \times 10^6\ \Omega \times 0.250 \times 10^{-6}\ F = 1.00\ s$$

The final charge on the capacitor is

$$Q_f = C_{eq}\mathscr{E} = 0.250 \times 10^{-6}\ F \times 50.0\ V = 12.5 \times 10^{-6}\ C = 12.5\ \mu C$$

At any time t, the charge on each capacitor is

$$Q(t) = C_{eq}V_C(t) = C_{eq}\,\mathscr{E}(1 - e^{-t/\tau}) = Q_f(1 - e^{-t/\tau})$$

At $t = 1.00$ s, $t/\tau = 1.00$; the charge on each capacitor is

$$Q = Q_f(1 - e^{-1.00}) = 12.5\ \mu C \times (1 - e^{-1.00}) = 7.90\ \mu C$$

At $t = 3.00$ s, $t/\tau = 3.00$; the charge on each capacitor is

$$Q = Q_f(1 - e^{-3.00}) = 12.5\ \mu C \times (1 - e^{-3.00}) = 11.9\ \mu C$$

(b) The initial current is

$$I_0 = \frac{\mathscr{E}}{R} = \frac{50.0\ V}{4.00 \times 10^6\ \Omega} = 12.5\ \mu A$$

At a time t,

$$I = I_0 e^{-t/\tau}$$

At $t = 1.00$ s,

$$I = I_0 e^{-1.00} = 12.5\ \mu A \times e^{-1.00} = 4.60\ \mu A$$

At $t = 3.00$ s,

$$I = I_0 e^{-3.00} = 12.5\ \mu A \times e^{-3.00} = 0.622\ \mu A$$

Discussion The solution can be checked using the loop rule. At $t = \tau$, we found that $Q = 7.90\ \mu C$ and $I = 4.60\ \mu A$. Then at $t = \tau$,

$$V_C = \frac{Q}{C_{eq}} = \frac{7.90\ \mu C}{0.250\ \mu F} = 31.6\ V$$

and

$$V_R = IR = 4.60\ \mu A \times 4.00\ M\Omega = 18.4\ V$$

Since 31.6 V + 18.4 V = 50.0 V = $\mathscr{E}$, the loop rule is satisfied.

Notice the pattern: the current is multiplied by $1/e$ during a time interval equal to the time constant. Thus, for a current of 4.60 μA at $t = \tau$, we expect a current of 4.60 μA × $1/e$ = 1.69 μA at $t = 2\tau$ and a current of 1.69 μA × $1/e$ = 0.622 μA at $t = 3\tau$.

Practice Problem 18.11 Another *RC* Circuit

At $t = 0$ a capacitor of 0.050 μF is connected through a 5.0-MΩ resistor to a 12-V battery. Initially the capacitor is uncharged. Find the initial current, the charge on the capacitor at $t = 0.25$ s, the current at $t = 1.00$ s, and the final charge on the capacitor.

Discharging *RC* Circuit

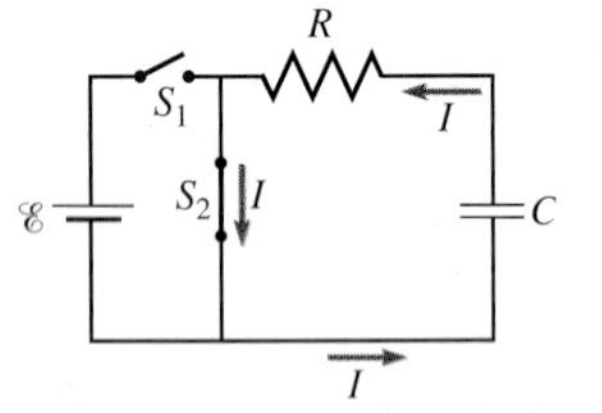

Figure 18.38 A capacitor is discharged through a resistor R.

In Fig. 18.38, the capacitor is first charged to a voltage $\mathscr{E}$ by closing switch S_1 with switch S_2 open. Once the capacitor is fully charged, S_1 is opened and then S_2 is closed at $t = 0$. Now the capacitor acts like a battery in the sense that it supplies energy in the circuit, though not at a constant potential difference. As the potential difference between the plates causes current to flow, the capacitor discharges.

The loop rule requires that the voltages across the capacitor and resistor be equal in magnitude. As the capacitor discharges, the voltage across it decreases. A decreasing voltage across the *resistor* means that the current must be decreasing. The current as a function of time is the same as for the charging circuit [Eq. (18-25)] with time constant $\tau = RC$. The voltage across the capacitor begins at its maximum value $\mathscr{E}$ and decreases exponentially (Fig. 18.39):

Discharging capacitor

$$V_C(t) = \mathscr{E}e^{-t/\tau} \quad (18\text{-}26)$$

Application: camera flash

The bulb in a camera flash needs a quick burst of current much larger than a small battery can supply (due to the battery's internal resistance). Therefore, the battery

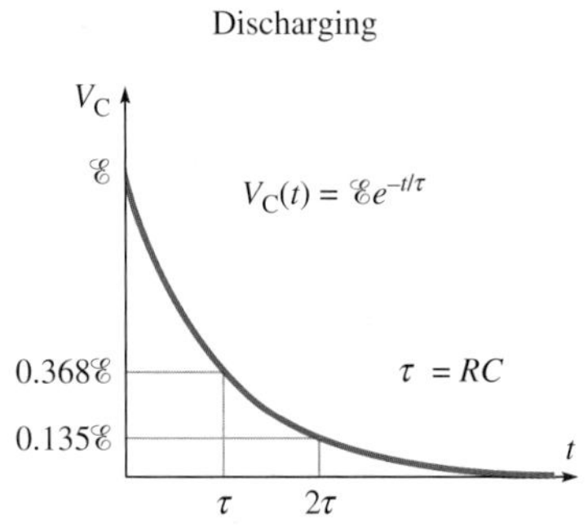

Figure 18.39 Decreasing voltage across a capacitor as it discharges through a resistor.

charges a capacitor (Fig. 18.40). When the capacitor is fully charged, the flash is ready; when the picture is taken, the capacitor is discharged quickly. After taking a picture, there is a delay of a second or two while the capacitor recharges. The time constant is longer for the charging circuit due to the internal resistance of the battery.

Application of *RC* Circuits in Neurons

An *RC* time constant also determines the speed at which nerve impulses travel. Figure 18.41a is a simplified model of a myelinated axon. Inside the axon is a fluid called the *axoplasm*, which is a conductor due to the presence of ions. Outside is the *interstitial fluid*, a conducting fluid with a much lower resistivity. Between the *nodes of Ranvier*, the cell membrane is covered with a *myelin sheath*—an insulator that reduces the capacitance of the section of axon (by increasing the distance between the conducting fluids) and reduces the leakage current that flows through the membrane.

A section of axon between nodes is modeled as an *RC* circuit in Fig. 18.41b. The interstitial fluid has little resistance, so it is modeled as a conducting wire. Current I travels inside the axon through the axoplasm (resistor R). The capacitor consists of the two conducting fluids as the plates, with the membrane and myelin sheath acting as insulator. For a section of axon 1 mm long with radius 5 μm, the resistance and capacitance are approximately $R = 13\ \text{M}\Omega$ and $C = 1.6$ pF. The time constant is therefore,

$$\tau = RC = 13\ \text{M}\Omega \times 1.6\ \text{pF} \approx 20\ \mu\text{s}$$

An estimate of how fast the electric impulse travels is

$$v \approx \frac{\text{length of section}}{\tau} = \frac{1\ \text{mm}}{20\ \mu\text{s}} = 50\ \text{m/s}$$

This simple estimate is remarkably accurate; nerve impulses in a human myelinated axon of radius 5 μm travel at speeds ranging from 60 to 90 m/s.

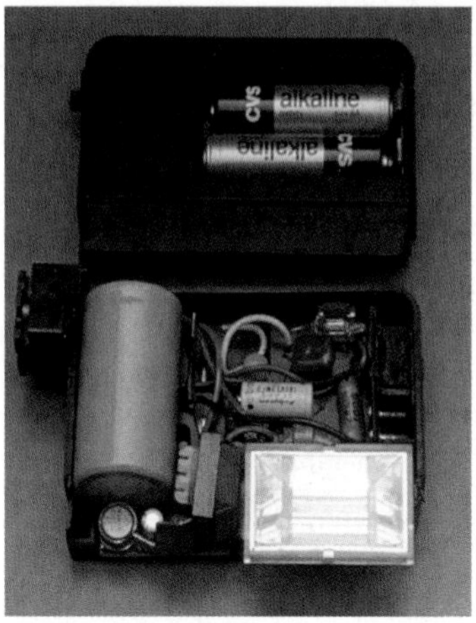

Figure 18.40 A flash attachment for a camera. The large gray cylinder is the capacitor.

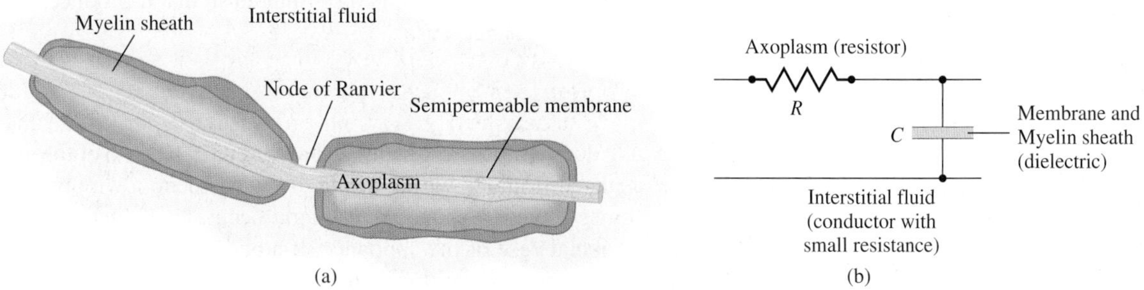

Figure 18.41 (a) A simplified picture of two sections of myelinated axon. (b) A simplified *RC* circuit model of a section of axon between nodes of Ranvier. The myelin sheath acts as a dielectric between two conductors, the axoplasm and the interstitial fluid.

Both R and C depend on the radius r of the axon. In humans, r ranges from under 2 μm to over 10 μm. The capacitance is proportional to r due to the larger plate area, but the resistance is inversely proportional to r^2 due to the larger cross-sectional area of the "wire." Thus, $RC \propto 1/r$ and $v \propto r$. The largest radius axons—those with the largest signal speeds—are those that must carry signals over relatively long distances.

18.11 ELECTRICAL SAFETY

Effects of Current on the Human Body

Electric currents passing through the body interfere with the operation of the muscles and the nervous system. Large currents also cause burns due to the energy dissipated in the tissues. A current of around 1 mA or less causes an unpleasant sensation but usually no other effect. The maximum current that can pass through the body without causing harm is about 5 mA. A current of 10 to 20 mA results in muscle contractions or paralysis; paralysis may prevent the person from letting go of the source of the current.

Application: defibrillator

Currents of 100 to 300 mA may cause ventricular fibrillation (uncontrolled, arrhythmic contractions of the heart) if they pass through or near the heart. In this condition, the person will die unless treated with a defibrillator to shock the heart back into a normal rhythm. Through the defibrillator paddles, a brief spurt of current of several amps is sent into the body near the heart (see Fig. 17.34). The shocked heart suffers a sudden muscular contraction, after which it may return to a normal state with regular contractions.

Most of the electrical resistance of the body is due to the skin. The fluids inside the body are good conductors due to the presence of ions. The total resistance of the body between distant points *when the skin is dry* ranges from around 10 kΩ to 1 MΩ. The resistance is much lower when the skin is wet—around 1 kΩ or even less.

A *short circuit* (a low-resistance path) may occur between the circuitry inside an appliance to metal on the outside of the appliance. A person touching the appliance would then have one hand at 120 V with respect to ground. (To simplify the discussion, we treat the emf as if it were dc rather than ac.) If his feet are in a wet tub, which makes good electric contact to the grounded water pipes, he might have a resistance as low as 500 Ω. Then a current of magnitude 120 V/500 Ω = 0.24 A = 240 mA flows through the body past the heart. Ventricular fibrillation is likely to occur. If the person were not standing in the tub, but had one hand on the hair dryer and another hand on a metal faucet, which is also grounded through the household plumbing, he is still in trouble. The electrical resistance of a person from one damp hand to the other might be around 1600 Ω, resulting in a current of 75 mA, which could still be lethal.

Remember that the symbol ⏚ represents a connection to ground.

Application: electrified fence

An electrified fence (Fig. 18.42) keeps farm animals in a pasture or wild animals out of a garden. One terminal of an emf is connected to the wire; the wire is insulated from the fence posts by ceramic insulators. The other terminal of the emf is connected to ground by a metal rod driven into the ground. When an animal or person touches the metal wire, the circuit is completed from the wire through the body and back to the ground. The current flowing through the body is limited so that it produces an unpleasant sensation without being dangerous.

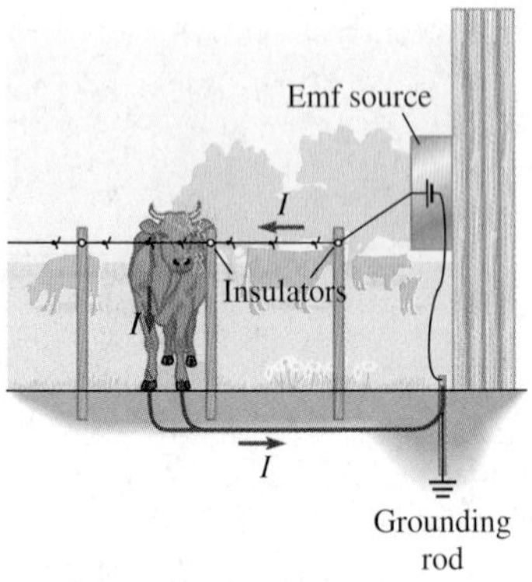

Figure 18.42 An electric fence. The circuit is completed when a person touches the wire.

Grounding of Appliances

A two-pronged plug does not protect against a short circuit. The case of the appliance is supposed to be insulated from the wiring inside. If, by accident, a wire breaks loose or its insulation becomes frayed, a short circuit might occur, providing a low-resistance path directly to the metal case of the appliance. If a person touches the case, which is now at a high potential, the current travels through the person and back to the ground (Fig. 18.43a).

With a three-pronged plug, the case of the appliance is connected directly to ground through the third prong (Fig. 18.43b). Then, if a short circuit occurs, the person touching the case does not complete the circuit to the ground. Instead the current travels from the

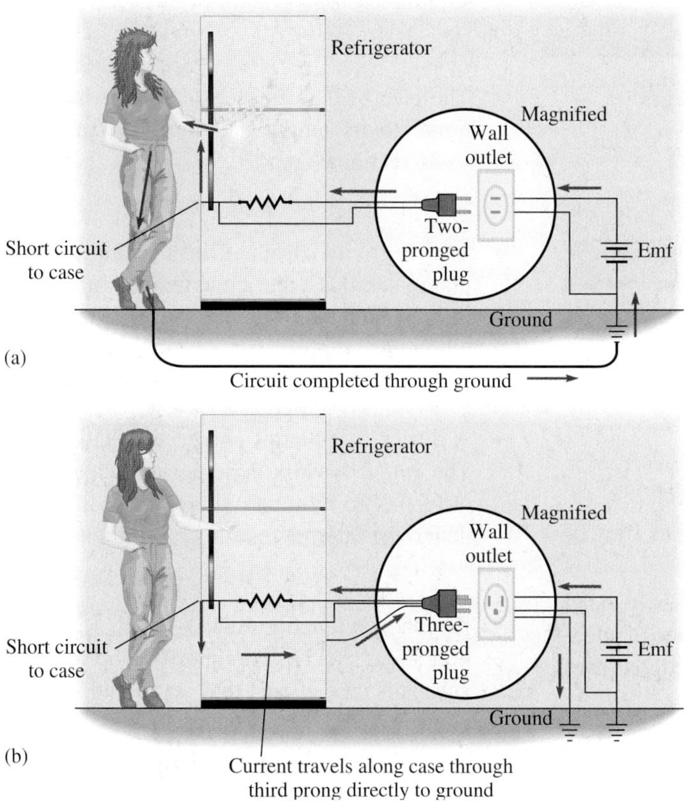

Figure 18.43 (a) If a refrigerator were connected with a two-pronged plug to a wall outlet, a short circuit to the case of the refrigerator allows the circuit to be completed through the body of a person touching the refrigerator. (b) If a short circuit occurs with a three-pronged plug, the person is safe.

case directly to the ground through low-resistance wiring via the third prong in the wall outlet. For safety reasons, the metal cases of many electric appliances are grounded.

Hospitals must take care that patients, connected to various monitors and IVs, are protected from a possible short circuit. For this reason the patient's bed, as well as anything else that the patient might touch, is insulated from the ground. Then if the patient touches something at a high potential, there is no ground connection to complete the circuit through the patient's body.

Fuses and Circuit Breakers

A simple fuse is made from an alloy of lead and tin that melts at a low temperature. The fuse is put in series with the circuit and is designed to melt—due to I^2R heating—if the current to the circuit exceeds a given value. The melted fuse is an open switch, interrupting the circuit and stopping the current. Many appliances are protected by fuses. Replacing a fuse with one of a higher current rating is dangerous because too much current may go through the appliance, damaging it or causing a fire.

Most household wiring is protected from overheating by circuit breakers instead of fuses. When too much current flows, perhaps because too many appliances are connected to the same circuit, a bimetallic strip or an electromagnet "trips" the circuit breaker, making it an open switch. After the problem causing the overload is corrected, the circuit breaker can be reset by flipping it back into the closed position.

Application: household wiring

Household wiring is arranged so that several appliances can be connected in parallel to a single circuit with one side of the circuit (the *neutral* side) grounded and the other side (the *hot* side) at a potential of 120 V with respect to ground (in our simplified dc model). Within one house or apartment, there are many such circuits; each one is protected by a circuit breaker (or fuse) placed in the hot side of the circuit. If a short circuit occurs, the large current that results trips the circuit breaker. If the breaker were placed on the grounded side, a blown circuit breaker would leave the hot side hot, possibly allowing a hazardous condition to continue. For the same reason, wall switches for overhead lights and for wall outlets are placed on the hot side.

Master the Concepts

- Electric current is the rate of net flow of charge:

$$I = \frac{\Delta q}{\Delta t} \qquad (18\text{-}1)$$

The SI unit of current is the ampere (1 A = 1 C/s), one of the base units of the SI. By convention, the direction of current is the direction of flow of positive charge. If the carriers are negative, the direction of the current is opposite the direction of motion of the carriers.

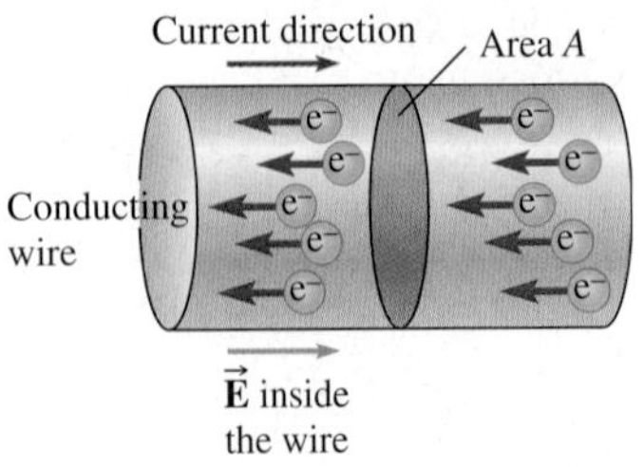

- A complete circuit is required for a continuous flow of charge.
- The current in a metal is proportional to the drift speed (v_D) of the conduction electrons, the number of electrons per unit volume (n), and the cross-sectional area of the metal (A):

$$I = \frac{\Delta Q}{\Delta t} = neAv_D \qquad (18\text{-}3)$$

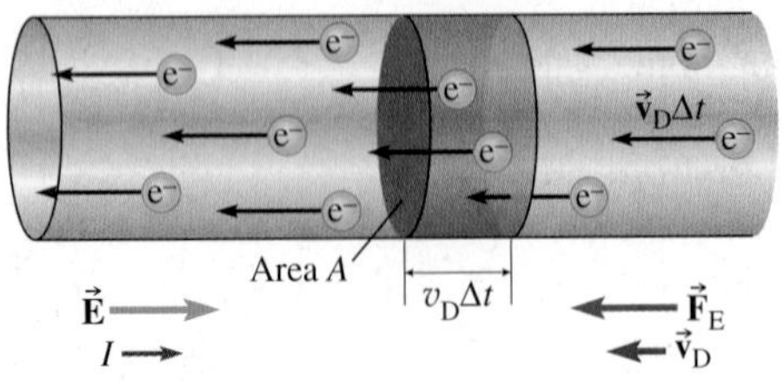

- Electrical resistance is the ratio of the potential difference across a conducting material to the current through the material. It is measured in ohms: 1 Ω = 1 V/A.

$$R = \frac{\Delta V}{I} \qquad (18\text{-}6)$$

For an ohmic conductor, R is independent of ΔV and I; then ΔV is proportional to I.

- The electrical resistance of a wire is directly proportional to its length and inversely proportional to its cross-sectional area:

$$R = \rho\frac{L}{A} \qquad (18\text{-}8)$$

- The resistivity ρ is an intrinsic characteristic of a particular material at a particular temperature and is measured in Ω·m. For many materials, resistivity varies linearly with temperature:

$$\rho = \rho_0(1 + \alpha\Delta T) \qquad (18\text{-}9)$$

- A device that pumps charge is called a source of emf. The emf $\mathscr{E}$ is work done per unit charge [$W = \mathscr{E}q$, Eq. (18-2)]. The terminal voltage may differ from the emf due to the internal resistance r of the source:

$$V = \mathscr{E} - Ir \qquad (18\text{-}10)$$

- Kirchhoff's junction rule: $\Sigma I_{in} - \Sigma I_{out} = 0$ at any junction [Eq. (18-11)]. Kirchhoff's loop rule: $\Sigma \Delta V = 0$ for any path in a circuit that starts and ends at the same point [Eq. (18-12)]. Potential rises are positive; potential drops are negative.
- Circuit elements wired in series have the same current through them. Circuit elements wired in parallel have the same potential difference across them.
- The power—the rate of conversion between electric energy and another form of energy—for any circuit element is

$$P = I\Delta V \qquad (18\text{-}19)$$

The SI unit for power is the watt (W). Electric energy is dissipated (transformed into internal energy) in a resistor.

- The quantity $\tau = RC$ is called the time constant for an RC circuit. The currents and voltages are

$$V_C(t) = \mathscr{E}(1 - e^{-t/\tau}) \quad \text{(charging)} \qquad (18\text{-}23)$$

$$V_C(t) = \mathscr{E}e^{-t/\tau} \quad \text{(discharging)} \qquad (18\text{-}26)$$

$$I(t) = \frac{\mathscr{E}}{R}e^{-t/\tau} \quad \text{(both)} \qquad (18\text{-}25)$$

Charging

Discharging

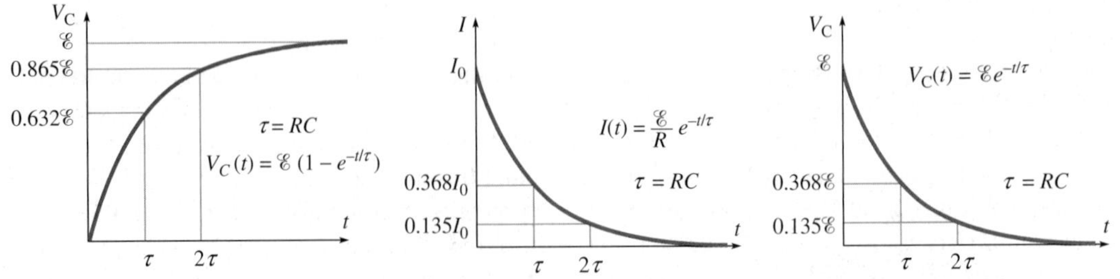

Conceptual Questions

1. Draw a circuit diagram for automobile headlights, connecting two separate bulbs and a switch to a single battery so that: (1) one switch turns both bulbs on and off and (2) one bulb still lights up even if the other bulb burns out.
2. Ammeters often contain fuses that protect them from large currents, whereas voltmeters seldom do. Explain.
3. Why do lightbulbs usually burn out just after they are switched on and not when they have been on for a while?
4. Jeff needs a 100-Ω resistor for a circuit, but he only has a box of 300-Ω resistors. What can he do?
5. A friend says that electric current "follows the path of least resistance." Is that true? Explain.
6. Compare the resistance of an ideal ammeter with that of an ideal voltmeter. Which has the larger resistance? Why?
7. Why does the resistivity of a metallic conductor increase with increasing temperature?
8. Suppose a battery is connected to a network of resistors and capacitors. What happens to the energy supplied by the battery?
9. Why are electric stoves and clothes dryers supplied with 240 V, but lights, radios, and clocks are supplied with 120 V?
10. Why are ammeters connected in series with a circuit element in which the current is to be measured and voltmeters connected in parallel across the element for which the potential difference is to be measured?
11. Is it more dangerous to touch a "live" electric wire when your hands are dry or wet, everything else being equal? Explain.
12. Is the electric field inside a conductor always zero? If not, when is it not zero? Explain.
13. Some batteries can be "recharged." Does that mean that the battery has a supply of charge that is depleted as the battery is used? If "recharging" does not literally mean to put charge back into the battery, what *does* it mean?
14. A battery is connected to a clock by copper wires as shown. What is the direction of current through the clock (B to C or C to B)? What is the direction of current through the battery (D to A or A to D)? Which terminal of the battery is at the higher potential (A or D)? Which side of the clock is at the higher potential (B or C)? Does current *always* flow from higher to lower potential? Explain.

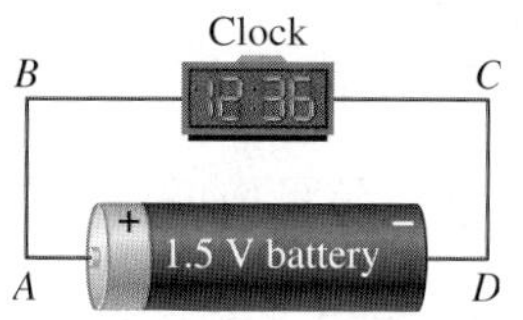

15. Think of a wire of length L as two wires of length $L/2$ in series. Construct an argument for why the resistance of a wire must be proportional to its length.
16. Think of a wire of cross-sectional area A as two wires of area $A/2$ in parallel. Construct an argument for why the resistance of a wire must be inversely proportional to its cross-sectional area.
17. An electrician working on "live" circuits wears insulated shoes and keeps one hand behind his or her back. Why?
18. A 15-A circuit breaker trips repeatedly. Explain why it would be dangerous to replace it with a 20-A circuit breaker.
19. A bird perched on a power line is not harmed, but if you are pruning a tree and your metal pole saw comes in contact with the same wire, you risk being electrocuted. Explain.
20. When batteries are connected in parallel, they should have the same emf. However, batteries connected in series need not have the same emf. Explain.
21. (a) If the resistance R_1 decreases, what happens to the voltage drop across R_3? The switch S is still open, as in the figure. (b) If the resistance R_1 decreases, what happens to the voltage drop across R_2? The switch S is still open, as in the figure. (c) In the circuit shown, if the switch S is closed, what happens to the current through R_1?

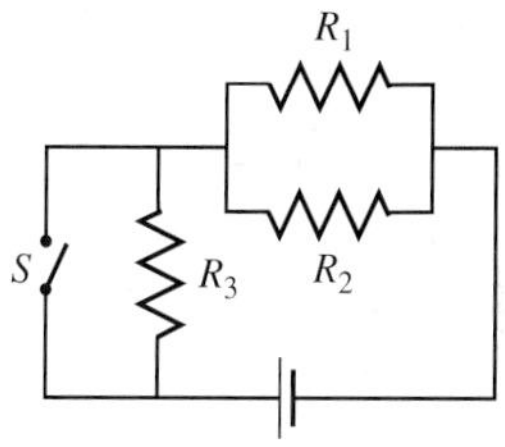

22. Four identical lightbulbs are placed in two different circuits with identical batteries. Bulbs A and B are connected in series with the battery. Bulbs C and D are connected in parallel across the battery. (a) Rank the brightness of the bulbs. (b) What happens to the brightness of bulb B if bulb A is replaced by a wire? (c) What happens to the brightness of bulb C if bulb D is removed from the circuit?

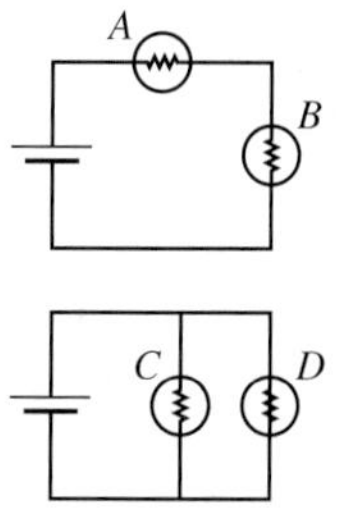

23. Three identical lightbulbs are connected in a circuit as shown in the diagram. (a) What happens to the brightness of the remaining bulbs if bulb A is removed from the circuit and replaced by a wire? (b) What happens to the brightness of the remaining bulbs if bulb B is removed from the circuit? (c) What happens to the brightness of the remaining bulbs if bulb B is replaced by a wire?

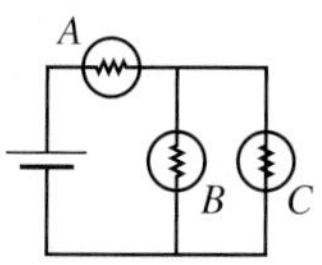

Multiple-Choice Questions

1. In an ionic solution, sodium ions (Na^+) are moving to the right and chlorine ions (Cl^-) are moving to the left. In which direction is the current due to the motion of (1) the sodium ions and (2) the chlorine ions?
   (a) Both are to the right.
   (b) Current due to Na^+ is to the left; current due to Cl^- is to the right.
   (c) Current due to Na^+ is to the right; current due to Cl^- is to the left.
   (d) Both are to the left.

2. A capacitor and a resistor are connected through a switch to an emf. At the instant just after the switch is closed,

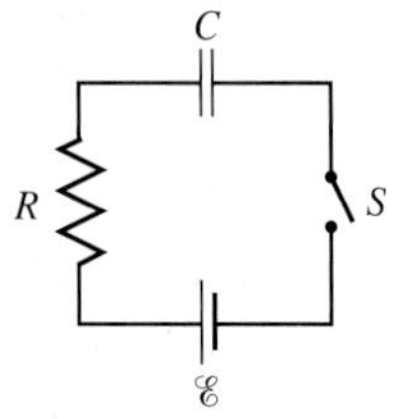

   (a) the current in the circuit is zero.
   (b) the voltage across the capacitor is $\mathscr{E}$.
   (c) the voltage across the resistor is zero.
   (d) the voltage across the resistor is $\mathscr{E}$.
   (e) Both (a) and (c) are true.

3. Which is a unit of energy?
   (a) $A^2{\cdot}\Omega$ (b) V·A (c) Ω·m
   (d) $\frac{N{\cdot}m}{V}$ (e) $\frac{A}{C}$ (f) V·C

4. How does the resistance of a piece of conducting wire change if both its length and diameter are doubled?
   (a) Remains the same
   (b) 2 times as much
   (c) 4 times as much
   (d) 1/2 as much
   (e) 1/4 as much

Questions 5 and 6. Each of the graphs shows a relation between the potential drop across (V) and the current through (I) a circuit element.

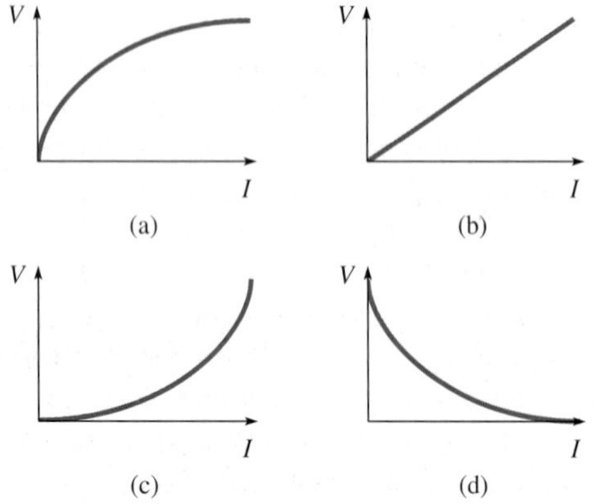

Multiple-Choice Questions 5 and 6

5. Which depicts a circuit element whose resistance increases with increasing current?

6. Which depicts an ohmic circuit element?

7. The electrical properties of copper and rubber are different because
   (a) the positive charges are free to move in copper and stationary in rubber.
   (b) many electrons are free to move in copper but nearly all are bound to molecules in rubber.
   (c) the positive charges are free to move in rubber but are stationary in copper.
   (d) many electrons are free to move in rubber but nearly all are bound to molecules in copper.

8. Consider these four statements. Choose true or false for each one in turn and then find the answer that matches your choices for all four together.
   (1) An ammeter should draw very little current compared with that in the rest of the circuit.
   (2) An ammeter should have a high resistance compared with the resistances of the other elements in the circuit.
   (3) To measure the current in a circuit element, the ammeter should be connected in series with that element.
   (4) Connecting the ammeter in series with a circuit element causes at least a small reduction of the current in that element.
   (a) (1) true, (2) true, (3) false, (4) false
   (b) (1) true, (2) false, (3) true, (4) true
   (c) (1) false, (2) false, (3) true, (4) false
   (d) (1) false, (2) false, (3) true, (4) true
   (e) (1) false, (2) true, (3) true, (4) true
   (f) (1) false, (2) false, (3) false, (4) true

9. A 12-V battery with internal resistance 0.5 Ω has initially no load connected across its terminals. Then the switches S_1 and S_2 are closed successively. The voltmeter (assumed ideal) has which set of successive readings?

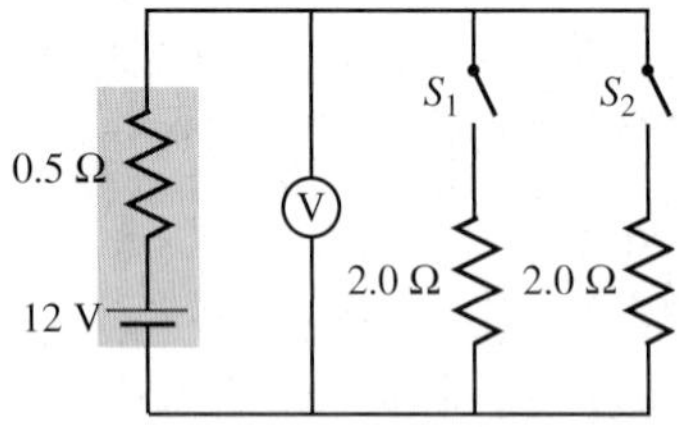

   (a) 12 V, 11 V, 10 V (b) 12 V, 12 V, 12 V
   (c) 12 V, 9.6 V, 7.2 V (d) 12 V, 9.6 V, 8 V
   (e) 12 V, 8 V, 4 V (f) 12 V, zero, zero

10. Which of these is equal to the emf of a battery?
    (a) the chemical energy stored in the battery
    (b) the terminal voltage of the battery when no current flows
    (c) the maximum current that the battery can supply
    (d) the amount of charge the battery can pump
    (e) the chemical energy stored in the battery divided by the net charge of the battery

Problems

- Combination conceptual/quantitative problem
- Biological or medical application
- Challenging problem
- Blue # Detailed solution in the Student Solutions Manual
- 1 2 Problems paired by concept
- Text website interactive or tutorial

18.1 Electric Current

1. A battery charger delivers a current of 3.0 A for 4.0 h to a 12-V storage battery. What is the total charge that passes through the battery in that time?

2. The current in a wire is 0.500 A. (a) How much charge flows through a cross section of the wire in 10.0 s? (b) How many electrons move through the same cross section in 10.0 s?

3. (a) What is the direction of the current in the vacuum tube shown in the figure? (b) Electrons hit the anode at a rate of 6.0×10^{12} per second. What is the current in the tube?

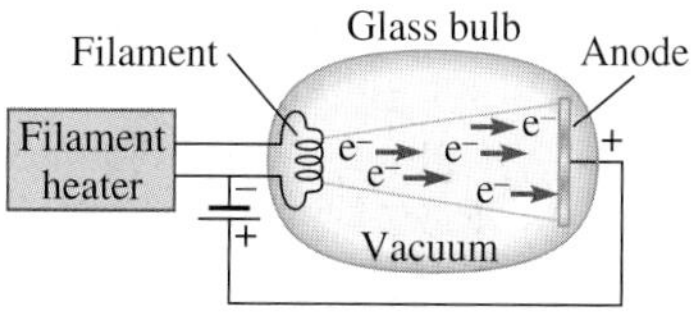

4. In an ion accelerator, 3.0×10^{13} helium-4 nuclei (charge $+2e$) per second strike a target. What is the beam current?

5. The current in the electron beam of a computer monitor is 320 μA. How many electrons per second hit the screen?

6. A potential difference is applied between the electrodes in a gas discharge tube. In 1.0 s, 3.8×10^{16} electrons and 1.2×10^{16} singly charged positive ions move in opposite directions through a surface perpendicular to the length of the tube. What is the current in the tube?

7. Two electrodes are placed in a calcium chloride solution and a potential difference is maintained between them. If 3.8×10^{16} Ca^{2+} ions and 6.2×10^{16} Cl^- ions per second move in opposite directions through an imaginary area between the electrodes, what is the current in the solution?

18.2 Emf and Circuits

8. A Vespa scooter and a Toyota automobile might both use a 12-V battery, but the two batteries are of different sizes and can pump different amounts of charge. Suppose the scooter battery can pump 4.0 kC of charge and the automobile battery can pump 30.0 kC of charge. How much energy can each battery deliver, assuming the batteries are ideal?

9. What is the energy stored in a small battery if it can move 675 C through a potential difference of 1.20 V?

10. The label on a 12.0-V truck battery states that it is rated at 180.0 A·h (ampere-hours). Treat the battery as ideal. (a) How much charge in coulombs can be pumped by the battery? [*Hint:* Convert A·h to A·s.] (b) How much electric energy can the battery supply? (c) Suppose the radio in the truck is left on when the engine is not running. The radio draws a current of 3.30 A. How long does it take to drain the battery if it starts out fully charged?

11. The starter motor in a car draws 220.0 A of current from the 12.0-V battery for 1.20 s. (a) How much charge is pumped by the battery? (b) How much electric energy is supplied by the battery?

12. A solar cell provides an emf of 0.45 V. (a) If the cell supplies a constant current of 18.0 mA for 9.0 h, how much electric energy does it supply? (b) What is the power—the rate at which it supplies electric energy?

18.3 Microscopic View of Current in a Metal: The Free-Electron Model

13. Two copper wires, one double the diameter of the other, have the same current flowing through them. If the thinner wire has a drift speed v_1, and the thicker wire has a drift speed v_2, how do the drift speeds of the charge carriers compare?

14. A current of 2.50 A is carried by a copper wire of radius 1.00 mm. If the density of the conduction electrons is $8.47 \times 10^{28}\ m^{-3}$, what is the drift speed of the conduction electrons?

15. A current of 10.0 A is carried by a copper wire of diameter 1.00 mm. If the density of the conduction electrons is $8.47 \times 10^{28}\ m^{-3}$, how long does it take for a conduction electron to move 1.00 m along the wire?

16. A silver wire of diameter 1.0 mm carries a current of 150 mA. The density of conduction electrons in silver is $5.8 \times 10^{28}\ m^{-3}$. How long (on average) does it take for a conduction electron to move 1.0 cm along the wire?

17. A strip of doped silicon 260 μm wide contains 8.8×10^{22} conduction electrons per cubic meter and an insignificant number of holes. When the strip carries a current of 130 μA, the drift speed of the electrons is 44 cm/s. What is the thickness of the strip?

18. A gold wire of 0.50 mm diameter has 5.90×10^{28} conduction electrons/m³. If the drift speed is 6.5 μm/s, what is the current in the wire?

19. A copper wire of cross-sectional area 1.00 mm² has a current of 2.0 A flowing along its length. What is the drift speed of the conduction electrons? Assume 1.3 conduction electrons per copper atom. The mass density of copper is 9.0 g/cm³ and its atomic mass is 64 g/mol.

20. An aluminum wire of diameter 2.6 mm carries a current of 12 A. How long on average does it take an electron to move 12 m along the wire? Assume 3.5 conduction electrons per aluminum atom. The mass density of aluminum is 2.7 g/cm³ and its atomic mass is 27 g/mol.

18.4 Resistance and Resistivity

21. A 12-Ω resistor has a potential difference of 16 V across it. What current flows through the resistor?

22. Current of 83 mA flows through the resistor in the diagram. (a) What is the resistance of the resistor? (b) In what direction does the current flow through the resistor?

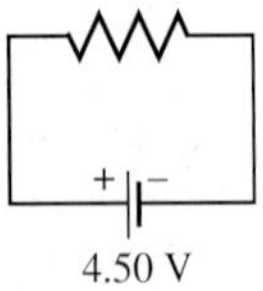

23. A copper wire and an aluminum wire of the same length have the same resistance. What is the ratio of the diameter of the copper wire to that of the aluminum wire?

24. A bird sits on a high-voltage power line with its feet 2.0 cm apart. The wire is made from aluminum, is 2.0 cm in diameter, and carries a current of 150 A. What is the potential difference between the bird's feet?

25. A person can be killed if a current as small as 50 mA passes near the heart. An electrician is working on a humid day with hands damp from perspiration. Suppose his resistance from one hand to the other is 1 kΩ and he is touching two wires, one with each hand. (a) What potential difference between the two wires would cause a 50-mA current from one hand to the other? (b) An electrician working on a "live" circuit keeps one hand behind his or her back. Why?

26. An electric device has the current-voltage (I-V) graph shown. What is its resistance at (a) point 1 and (b) point 2? [*Hint:* Use the definition of resistance.]

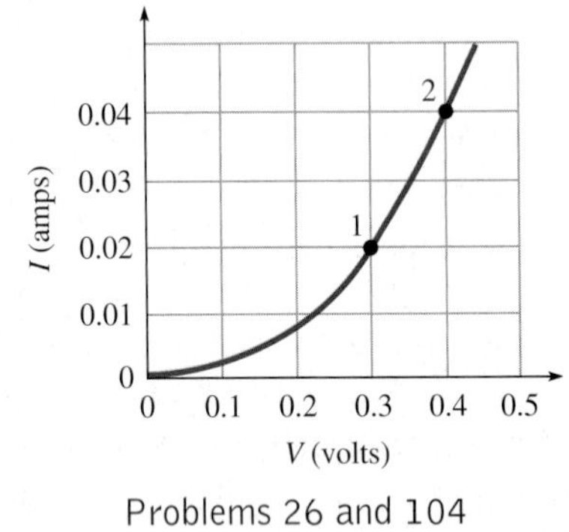

Problems 26 and 104

27. If 46 m of nichrome wire is to have a resistance of 10.0 Ω at 20°C, what diameter wire should be used?

28. The resistance of a conductor is 19.8 Ω at 15.0°C and 25.0 Ω at 85.0°C. What is the temperature coefficient of resistance of the material?

29. A common flashlight bulb is rated at 0.300 A and 2.90 V (the values of current and voltage under operating conditions). If the resistance of the bulb's tungsten filament at room temperature (20.0°C) is 1.10 Ω, estimate the temperature of the tungsten filament when the bulb is turned on.

30. Find the maximum current that a fully charged D-cell can supply—if only briefly—such that its terminal voltage is at least 1.0 V. Assume an emf of 1.5 V and an internal resistance of 0.10 Ω. (W tutorial: internal resistance of a battery)

31. A battery has a terminal voltage of 12.0 V when no current flows. Its internal resistance is 2.0 Ω. If a 1.0-Ω resistor is connected across the battery terminals, what is the terminal voltage and what is the current through the 1.0-Ω resistor?

32. (a) What are the ratios of the resistances of (a) silver and (b) aluminum wire to the resistance of copper wire (R_{Ag}/R_{Cu} and R_{Al}/R_{Cu}) for wires of the same length and the same diameter? (c) Which material is the best conductor, for wires of equal length and diameter?

33. A wire with cross-sectional area A carries a current I. Show that the electric field strength E in the wire is proportional to the current per unit area (I/A) and identify the constant of proportionality. [*Hint:* Assume a length L of wire. How is the potential difference across the wire related to the electric field in the wire? (Which is uniform?) Use $V = IR$ and the connection between resistance and resistivity.]

34. A copper wire has a resistance of 24 Ω at 20°C. An aluminum wire has three times the length and twice the radius of the copper wire. The resistivity of copper is 0.6 times that of aluminum. Both Al and Cu have temperature coefficients of resistivity of 0.004°C^{-1}. (a) What is the resistance of the aluminum wire at 20°C? (b) The graph shows a V-I plot for the copper wire. What is the resistance of the wire when operating steadily at a current of 10 A? (c) What must the temperature of the copper wire have been when operating at 10 A? Ignore changes in the wire's dimensions.

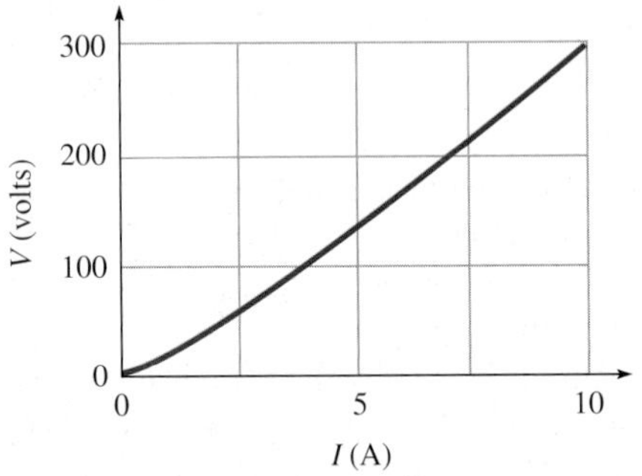

35. Refer to Problem 34. With the copper wire connected to an ideal battery, the current increases greatly when the wire is immersed in liquid nitrogen. Ignoring changes in the wire's dimensions, state whether each of the following quantities increases, decreases, or stays the same as the wire is cooled: the electric field in the wire, the resistivity, and the drift speed. Explain your answers.

18.6 Series and Parallel Circuits

36. Suppose a collection of five batteries is connected as shown. (a) What is the equivalent emf of the collection? Treat them as ideal sources of emf. (b) What is the current through the resistor if its value is 3.2 Ω?

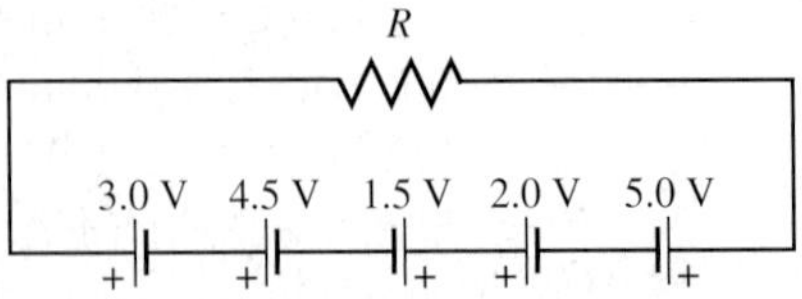

37. Suppose four batteries are connected in series as shown. (a) What is the equivalent emf of the set of four batteries? Treat them as ideal sources of emf. (b) If the current in the circuit is 0.40 A, what is the value of the resistor R?

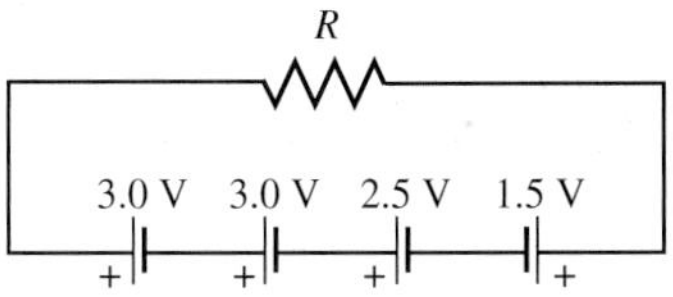

38. (a) Find the equivalent capacitance between points A and B for the three capacitors. (b) What is the charge on the 6.0-μF capacitor if a 44.0-V emf is connected to the terminals A and B for a long time?

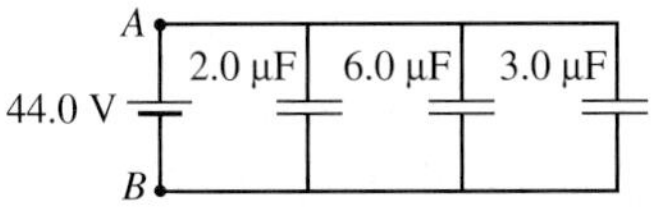

39. (a) Find the equivalent capacitance between points A and B for the five capacitors. (b) If a 16.0-V emf is connected to the terminals A and B, what is the charge on a single equivalent capacitor that replaces all five? (c) What is the charge on the 3.0-μF capacitor?

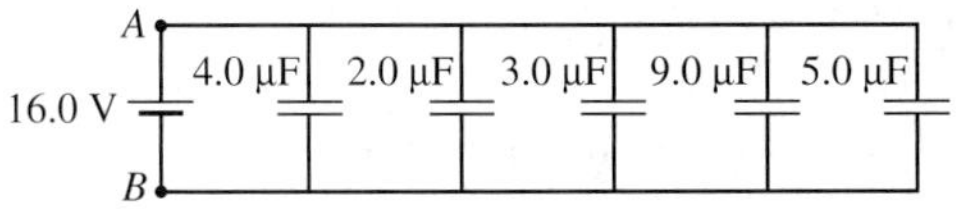

40. (a) What is the equivalent resistance between points A and B? (b) A 276-V emf is connected to the terminals A and B. What is the current in the 12-Ω resistor?

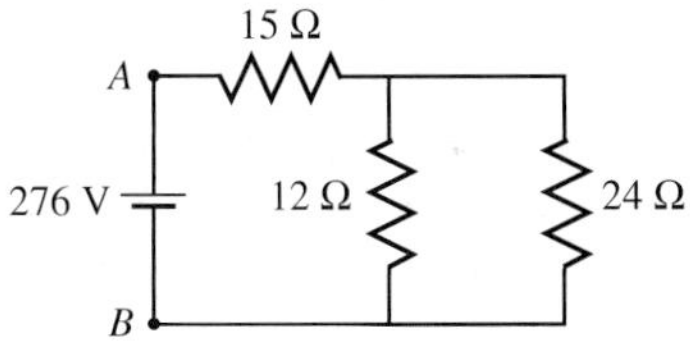

Problems 40, 72, and 73

41. (a) What is the equivalent resistance between points A and B if $R = 1.0\ \Omega$? (b) If a 20-V emf is connected to the terminals A and B, what is the current in the 2.0-Ω resistor?

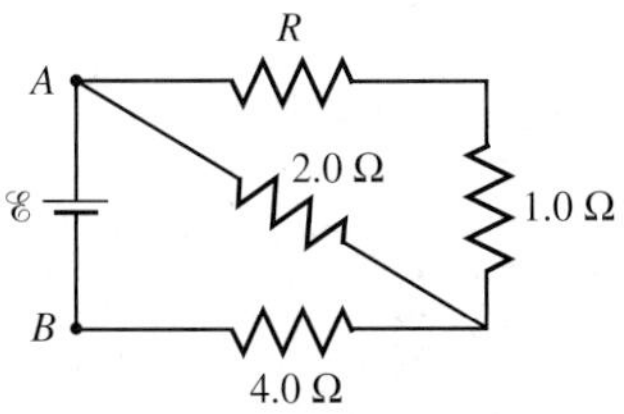

Problems 41 and 42

42. If a 93.5-V emf is connected to the terminals A and B and the current in the 4.0-Ω resistor is 17 A, what is the value of the unknown resistor R?

43. (a) What is the equivalent capacitance between points A and B if $C = 1.0\ \mu F$? (b) What is the charge on the 4.0-μF capacitor when it is fully charged?

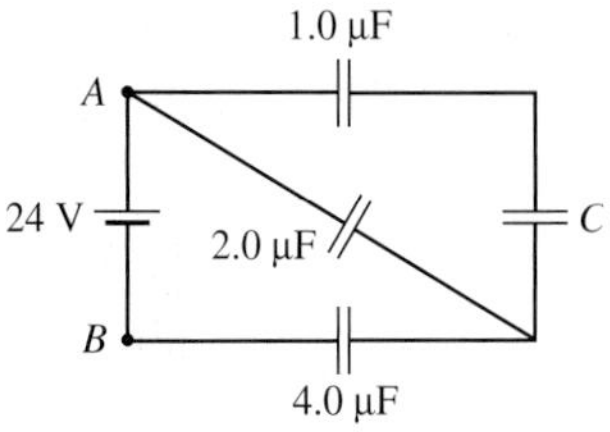

Problems 43 and 44

44. The equivalent capacitance between points A and B is 1.63 μF. (a) What is the capacitance of the unknown capacitor C? (b) What is the charge on the 4.0-μF capacitor when it is fully charged?

✦45. A 24-V emf is connected to the terminals A and B. (a) What is the current in one of the 2.0-Ω resistors? (b) What is the current in the 6.0-Ω resistor? (c) What is the current in the leftmost 4.0-Ω resistor?

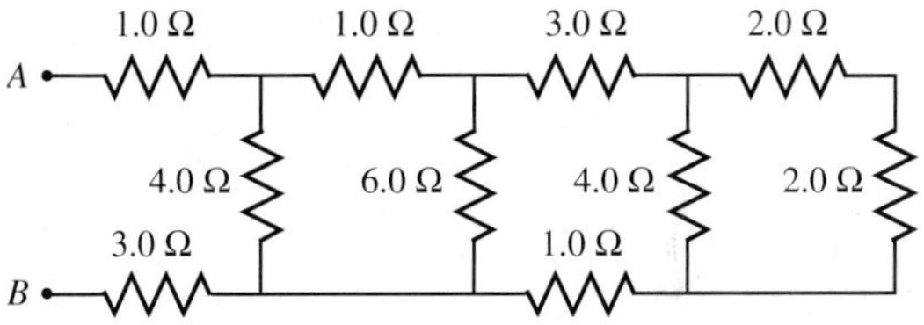

✦46. (a) Find the equivalent resistance between points A and B for the combination of resistors shown. (b) An 18-V emf is connected to the terminals A and B. What is the current through the 1.0-Ω resistor connected directly to point A? (c) What is the current in the 8.0-Ω resistor?

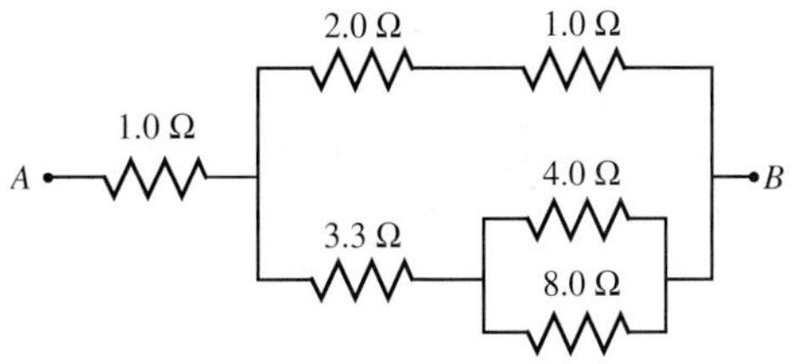

✦47. (a) What is the resistance between points A and B? Each resistor has the same resistance R. [*Hint:* Redraw the circuit.] (b) What is the resistance between points B and C? (c) If a 32-V emf is connected to terminals A and B and if each $R = 2.0\ \Omega$, what is the current in one of the resistors?

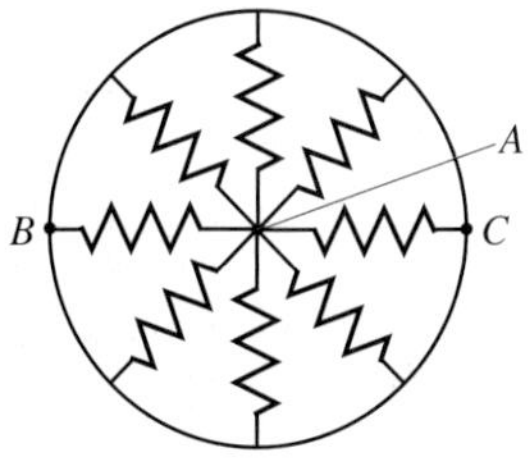

48. (a) Find the equivalent resistance between points A and B for the combination of resistors shown. (b) What is the potential difference across each of the 4.0-Ω resistors? (c) What is the current in the 3.0-Ω resistor?

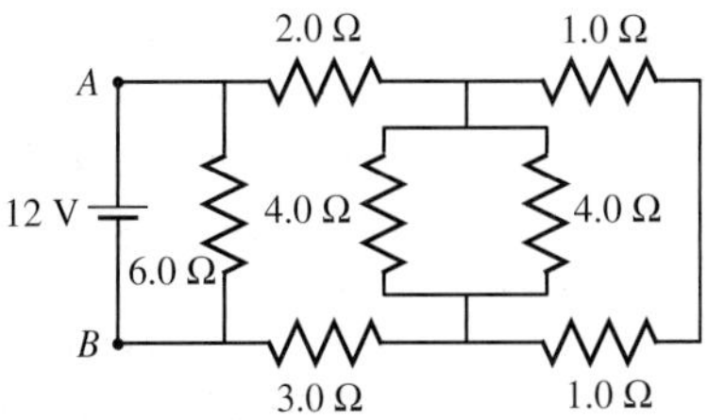

49. (a) Find the value of a single capacitor to replace the three capacitors in the diagram. (b) What is the potential difference across the 12-μF capacitor at the left side of the diagram? (c) What is the charge on the 12-μF capacitor to the far right side of the circuit?

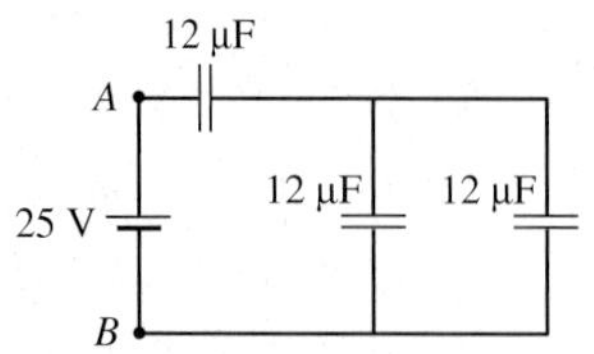

50. A 6.0-pF capacitor is needed to construct a circuit. The only capacitors available are rated as 9.0 pF. How can a combination of three 9.0-pF capacitors be assembled so that the equivalent capacitance of the combination is 6.0 pF?

51. (a) Find the equivalent resistance between terminals A and B to replace all of the resistors in the diagram. (b) What current flows through the emf? (c) What is the current through the 4.00-Ω resistor at the bottom?

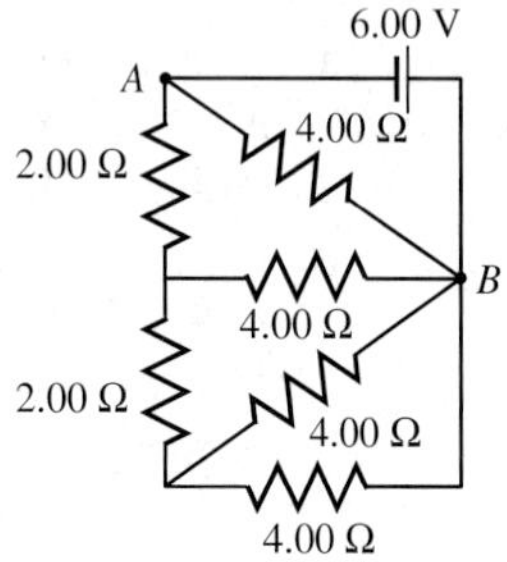

18.7 Circuit Analysis Using Kirchhoff's Rules

52. Find the current in each branch of the circuit. Specify the direction of each.

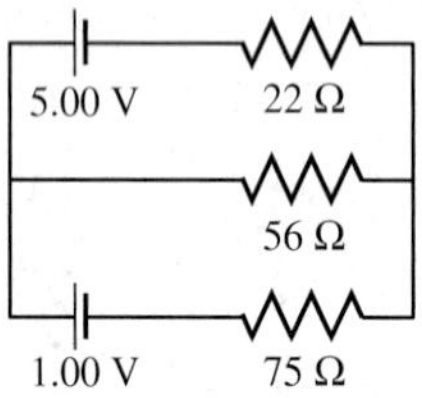

53. Find the current in each branch of the circuit. Specify the direction of each.

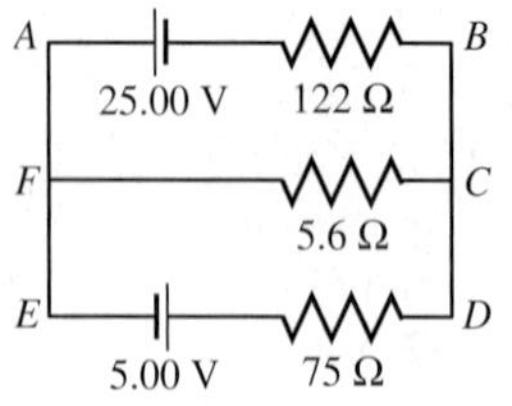

54. Find the unknown emf and the unknown currents in the circuit.

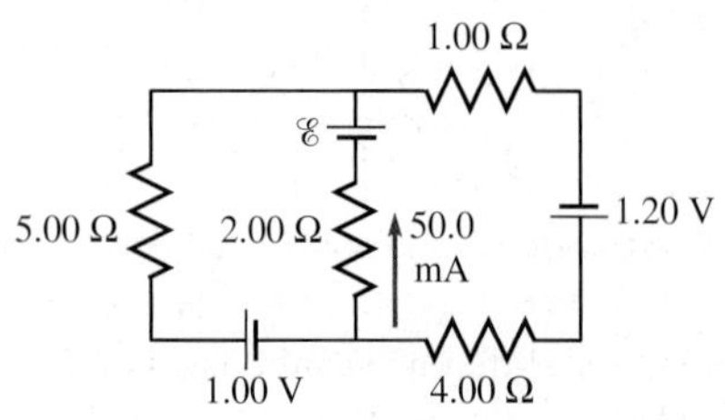

55. Find the unknown emf and the unknown resistor in the circuit.

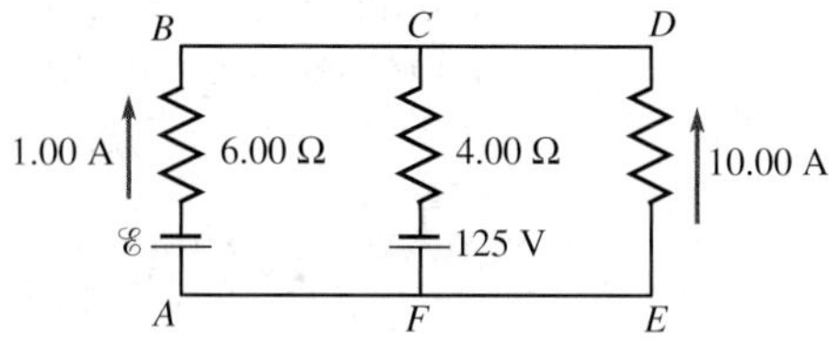

56. The figure shows a simplified circuit diagram for an automobile. The equivalent resistor R represents the total electrical load due to spark plugs, lights, radio, fans, starter, rear window defroster, and the like in parallel. If $R = 0.850\ \Omega$, find the current in each branch. What is the terminal voltage of the battery? Is the battery charging or discharging?

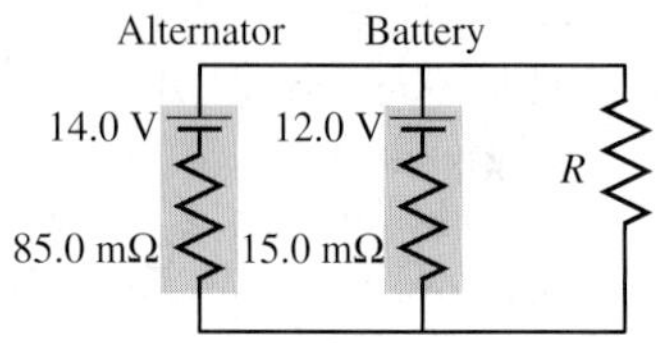

18.8 Power and Energy in Circuits

57. What is the power dissipated by the resistor in the circuit if the emf is 2.00 V?

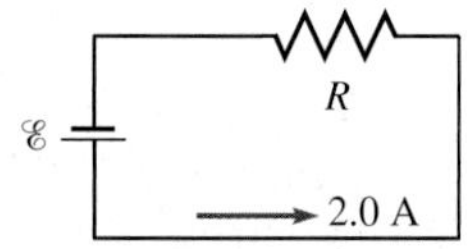

58. Refer to the figure with Problem 57. What is the power dissipated by the resistor in the circuit if $R = 5.00\ \Omega$?

59. What is the current in a 60.0-W bulb when connected to a 120-V emf?

60. What is the resistance of a 40.0-W, 120-V lightbulb?

61. If a chandelier has a label stating 120 V, 5.0 A, can its power rating be determined? If so, what is it?

62. A portable CD player does not have a power rating listed, but it has a label stating that it draws a maximum current of 250.0 mA. The player uses three 1.50-V batteries connected in series. What is the maximum power consumed?

63. How much work are the batteries in the circuit doing in every 10.0-s time interval?

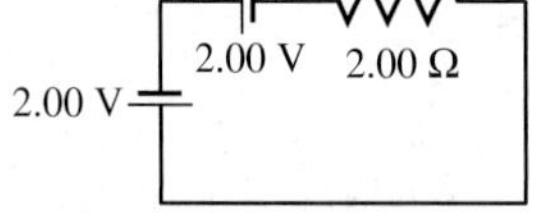

64. Show that $A^2 \times \Omega = W$ (amperes squared times ohms = watts).

65. Consider the circuit in the diagram. (a) Draw the simplest equivalent circuit and label the values of the resistor(s). (b) What current flows from the battery? (c) What is the potential difference between points A and B? (d) What current flows through each branch between points A and B? (e) Determine the power dissipated in the 50.0-Ω resistor, the 70.0-Ω resistor, and the 40.0-Ω resistor.

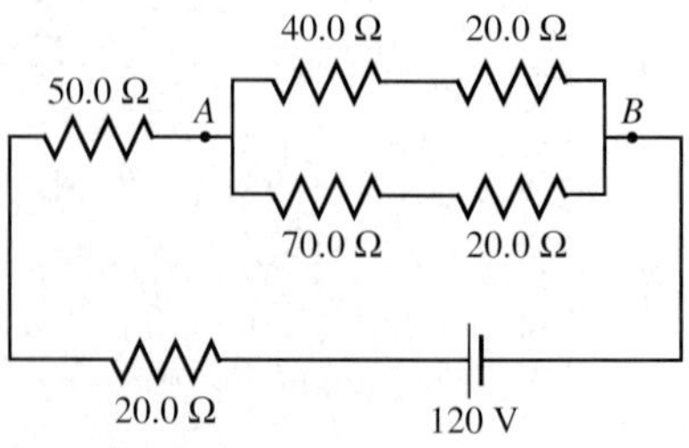

66. (a) What is the equivalent resistance of this circuit if $R_1 = 10.0\ \Omega$ and $R_2 = 15.0\ \Omega$? (b) What current flows through R_1? (c) What is the voltage drop across R_2? (d) What current flows through R_2? (e) How much power is dissipated in R_2?

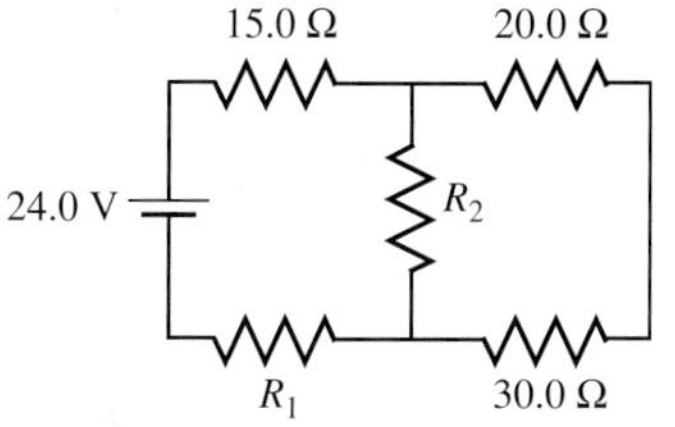

67. In her bathroom, Mindy has an overhead heater that consists of a coiled wire made of nichrome that gets hot when turned on. The wire has a length of 3.0 m when it is uncoiled. The heating element is attached to the normal 120-V wiring and when the wire is glowing red hot it has a temperature of about 420°C and dissipates 2200 W of power. Nichrome has a resistivity of $108 \times 10^{-8}\ \Omega\cdot$m at 20°C and a temperature coefficient of resistivity of $0.00040°C^{-1}$. (a) What is the resistance of the heater when it is turned on? (b) What current does the wire carry? (c) If the wire has a circular cross section, what is its diameter? Ignore the small changes in the wire's diameter and length due to changes in temperature. (d) When the heater is first turned on, it has not yet heated up, so it is operating at 20°C. What is the current through the wire when it is first turned on?

68. At what rate is electric energy converted to internal energy in the 4.00-Ω and 5.00-Ω resistors in the figure?

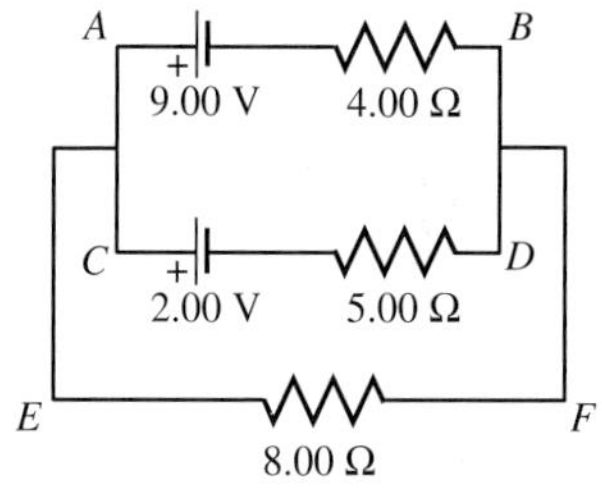

69. A battery has a 6.00-V emf and an internal resistance of 0.600 Ω. (a) What is the voltage across its terminals when the current drawn from the battery is 1.20 A? (b) What is the power supplied by the battery?

70. During a "brownout," which occurs when the power companies cannot keep up with high demand, the voltage of the household circuits drops below its normal 120 V. (a) If the voltage drops to 108 V, what would be the power consumed by a "100-W" lightbulb (that is, a lightbulb that consumes 100.0 W when connected to 120 V)? Ignore (for now) changes in the resistance of the lightbulb filament. (b) More realistically, the lightbulb filament will not be as hot as usual during the brownout. Does this make the power drop more or less than that you calculated in part (a)? Explain.

71. A source of emf $\mathscr{E}$ has internal resistance r. (a) What is the terminal voltage when the source supplies a current I? (b) The net power supplied is the terminal voltage times the current. Starting with $P = I\Delta V$, derive Eq. (18-22) for the net power supplied by the source. Interpret each of the two terms. (c) Suppose that a battery of emf $\mathscr{E}$ and internal resistance r is being recharged: another emf sends a current I through the battery in the reverse direction (from positive terminal to negative). At what rate is electric energy converted to chemical energy in the recharging battery? (d) What is the power supplied by the recharging circuit to the battery?

18.9 Measuring Currents and Voltages

72. Redraw the circuit in Problem 40 to show how an ammeter would be connected to measure (a) the current through the 15-Ω resistor and (b) the current through the 24-Ω resistor.

73. Redraw the circuit in Problem 40 to show how a voltmeter would be connected to measure (a) the potential drop across the 15-Ω resistor and (b) the potential drop across the 24-Ω resistor.

74. (a) Redraw the circuit to show how an ammeter would be connected to measure the current through the 1.40-kΩ resistor. (b) Assuming the ammeter to be ideal, what is its reading? (c) If the ammeter has a resistance of 120 Ω, what is its reading?

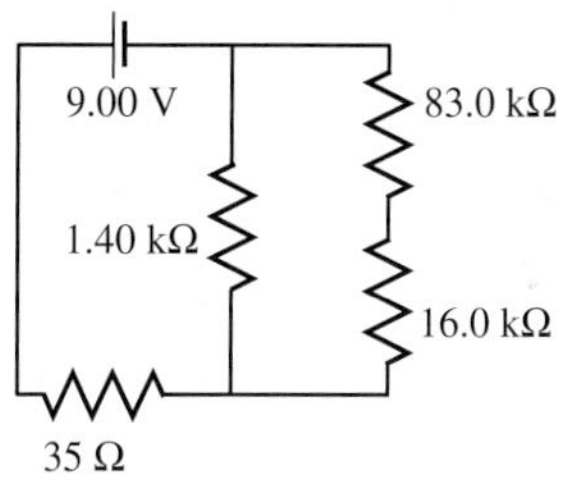

Problems 74, 75, and 120

75. (a) Redraw the circuit to show how a voltmeter would be connected to measure the voltage across the 83.0-kΩ resistor. (b) Assuming the voltmeter to be ideal, what is its reading? (c) If the voltmeter has a resistance of 1.00 MΩ, what is its reading?

76. A galvanometer has a coil resistance of 50.0 Ω. It is to be made into an ammeter with a full-scale deflection equal to 10.0 A. If the galvanometer deflects full scale for a current of 0.250 mA, what size shunt resistor should be used?

77. A galvanometer has a coil resistance of 34.0 Ω. It is to be made into a voltmeter with a full-scale deflection equal to 100.0 V. If the galvanometer deflects full scale for a current of 0.120 mA, what size resistor should be placed in series with the galvanometer?

78. A galvanometer is to be turned into a voltmeter that deflects full scale for a potential difference of 100.0 V. What size resistor should be placed in series with the

galvanometer if it has an internal resistance of 75 Ω and deflects full scale for a current of 2.0 mA?

79. Many voltmeters have a switch by which one of several series resistors can be selected. Thus, the same meter can be used with different full-scale voltages. What size series resistors should be used in the voltmeter of Problem 78 to give it full-scale voltages of (a) 50.0 V and (b) 500.0 V?

80. An ammeter with a full scale deflection for $I = 10.0$ A has an internal resistance of 24 Ω. We need to use this ammeter to measure currents up to 12.0 A. The lab instructor advises that we get a resistor and use it to protect the ammeter. (a) What size resistor do we need and how should it be connected to the ammeter, in series or in parallel? (b) How do we interpret the ammeter readings?

81. A voltmeter has a switch that enables voltages to be measured with a maximum of 25.0 V or 10.0 V. For a range of voltages to 25.0 V, the switch connects a resistor of magnitude 9850 Ω in series with the galvanometer; for a range of voltages to 10.0 V, the switch connects a resistor of magnitude 3850 Ω in series with the galvanometer. Find the coil resistance of the galvanometer and the galvanometer current that causes a full-scale deflection. [*Hint:* There are two unknowns, so you will need to solve two equations simultaneously.]

18.10 *RC* Circuits

82. In the circuit, $R = 30.0$ kΩ and $C = 0.10$ μF. The capacitor is allowed to charge fully and then the switch is changed from position *a* to position *b*. What will the voltage across the resistor be 8.4 ms later? (tutorial: capacitor discharge)

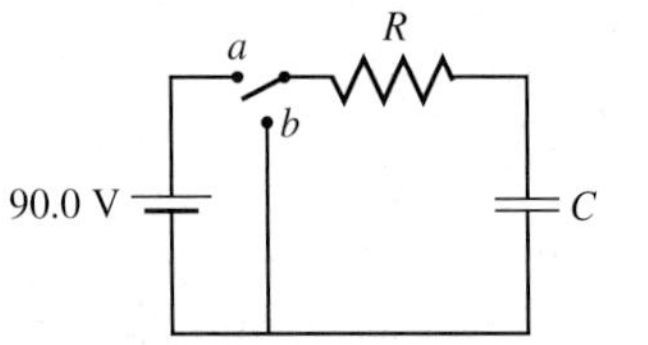

83. In the circuit shown, assume the battery emf is 20.0 V, $R = 1.00$ MΩ, and $C = 2.00$ μF. The switch is closed at $t = 0$. At what time t will the voltage across the capacitor be 15.0 V?

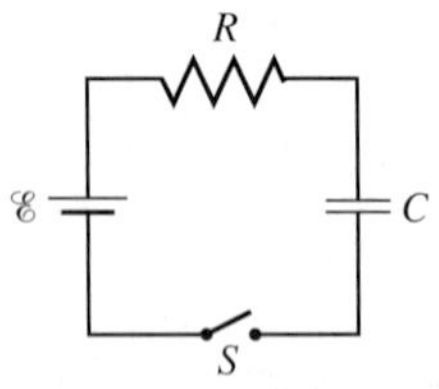

84. A charging *RC* circuit controls the intermittent windshield wipers in a car. The emf is 12.0 V. The wipers are triggered when the voltage across the 125-μF capacitor reaches 10.0 V; then the capacitor is quickly discharged (through a much smaller resistor) and the cycle repeats. What resistance should be used in the charging circuit if the wipers are to operate once every 1.80 s?

85. A capacitor is charged to an initial voltage $V_0 = 9.0$ V. The capacitor is then discharged by connecting its terminals through a resistor. The current $I(t)$ through this resistor, determined by measuring the voltage $V_R(t) = I(t)R$ with an oscilloscope, is shown in the graph. (a) Find the capacitance C, the resistance R, and the total energy dissipated in the resistor. (b) At what time is the energy in the capacitor half its initial value? (c) Graph the voltage across the capacitor, $V_C(t)$, as a function of time.

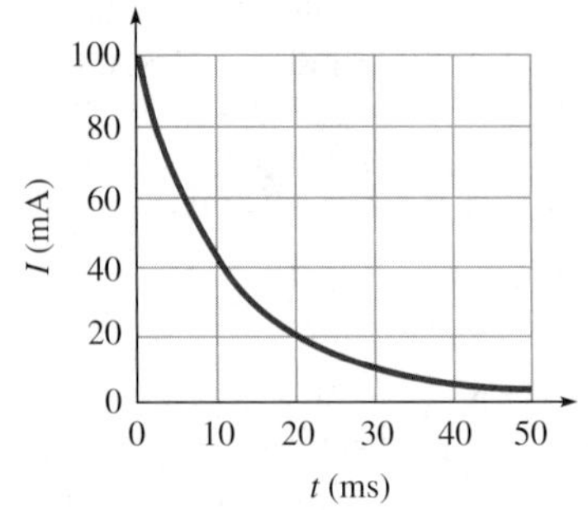

86. A defibrillator passes a brief burst of current through the heart to restore normal beating. In one such defibrillator, a 50.0-μF capacitor is charged to 6.0 kV. Paddles are used to make an electric connection to the patient's chest. A pulse of current lasting 1.0 ms partially discharges the capacitor through the patient. The electrical resistance of the patient (from paddle to paddle) is 240 Ω. (a) What is the initial energy stored in the capacitor? (b) What is the initial current through the patient? (c) How much energy is dissipated in the patient during the 1.0 ms? (d) If it takes 2.0 s to recharge the capacitor, compare the average power supplied by the power source with the average power delivered to the patient. (e) Referring to your answer to part (d), explain one reason a capacitor is used in a defibrillator.

87. Capacitors are used in many applications where one needs to supply a short burst of relatively large current. A 100.0-μF capacitor in an electronic flash lamp supplies a burst of current that dissipates 20.0 J of energy (as light and heat) in the lamp. (a) To what potential difference must the capacitor initially be charged? (b) What is its initial charge? (c) Approximately what is the resistance of the lamp if the current reaches 5.0% of its original value in 2.0 ms?

88. Consider the circuit shown with $R_1 = 25$ Ω, $R_2 = 33$ Ω, $C_1 = 12$ μF, $C_2 = 23$ μF, $C_3 = 46$ μF, and $V = 6.0$ V. (a) Draw an equivalent circuit with one resistor and one capacitor and label it with the values of the equivalent resistor and capacitor. (b) A long time after switch S is closed, what are the charge on capacitor C_1 and the current in resistor R_1? (c) What is the time constant of the circuit? (d) At what time after switch S is closed is the voltage across the combination of three capacitors 50% of its final value?

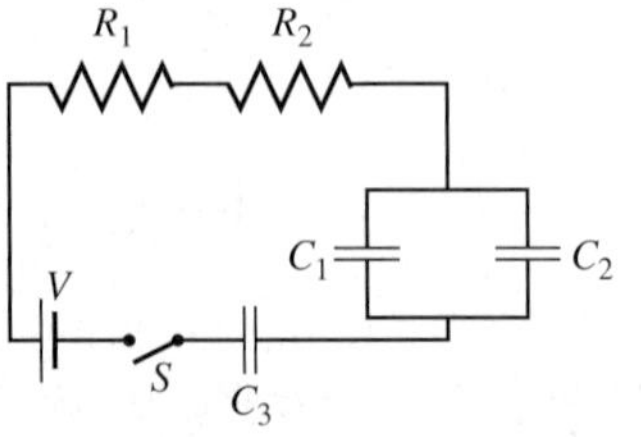

89. In the circuit, the capacitor is initially uncharged. At $t = 0$ switch S is closed. Find the currents I_1 and I_2 and voltages V_1 and V_2 (assuming $V_3 = 0$) at points 1 and 2 at the following times: (a) $t = 0$ (i.e., just after the switch is closed), (b) $t = 1.0$ ms, and (c) $t = 5.0$ ms.

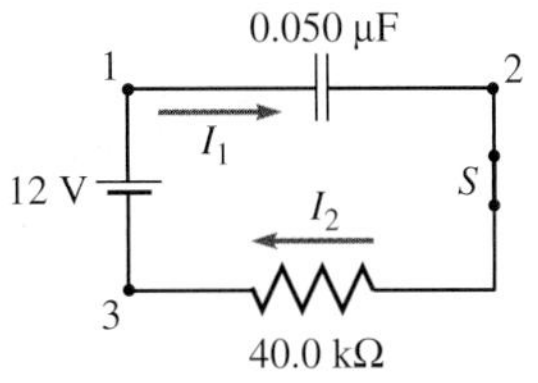

90. In the circuit, the initial energy stored in the capacitor is 25 J. At $t = 0$ the switch is closed. (a) Sketch a graph of the voltage across the resistor (V_R) as a function of t. Label the vertical axis with key numerical value(s) and units. (b) At what time is the energy stored in the capacitor 1.25 J?

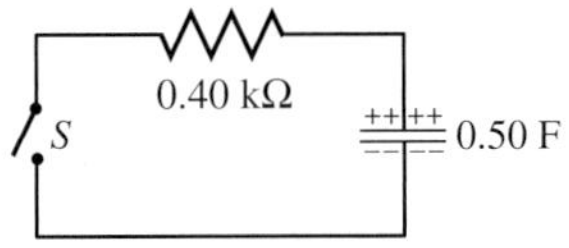

91. A 20-μF capacitor is discharged through a 5-kΩ resistor. The initial charge on the capacitor is 200 μC. (a) Sketch a graph of the current through the resistor as a function of time. Label both axes with numbers and units. (b) What is the initial power dissipated in the resistor? (c) What is the total energy dissipated?

92. (a) In a charging RC circuit, how many time constants have elapsed when the capacitor has 99.0% of its final charge? (b) How many time constants have elapsed when the capacitor has 99.90% of its final charge? (c) How many time constants have elapsed when the current has 1.0% of its initial value?

93. A capacitor is charged by a 9.0-V battery. The charging current $I(t)$ is shown. (a) What, approximately, is the total charge on the capacitor in the end? [*Hint:* During a short time interval Δt, the amount of charge that flows in the circuit is $I\Delta t$.] (b) Using your answer to (a), find the capacitance C of the capacitor. (c) Find the total resistance R in the circuit. (d) At what time is the stored energy in the capacitor half of its maximum value?

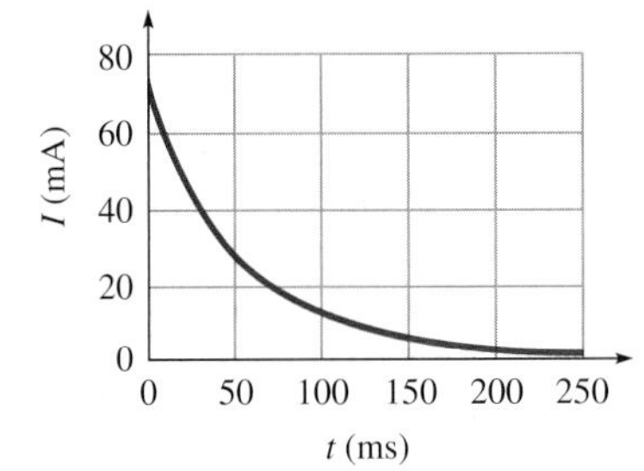

94. A charged capacitor is discharged through a resistor. The current $I(t)$ through this resistor, determined by measuring the voltage $V_R(t) = I(t)R$ with an oscilloscope, is shown in the graph. The total energy dissipated in the resistor is 2.0×10^{-4} J. (a) Find the capacitance C, the resistance R, and the initial charge on the capacitor. [*Hint:* You will need to solve three equations simultaneously for the three unknowns. You can find both the initial current and the time constant from the graph.] (b) At what time is the stored energy in the capacitor 5.0×10^{-5} J?

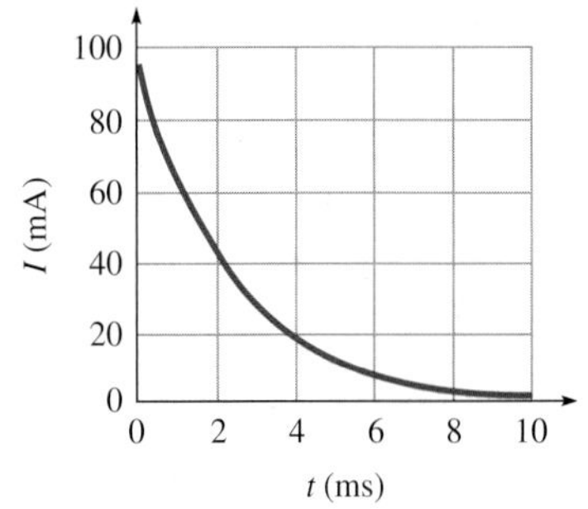

18.11 Electrical Safety

95. In the physics laboratory, Oscar measured the resistance between his hands to be 2.0 kΩ. Being curious by nature, he then took hold of two conducting wires that were connected to the terminals of an emf with a terminal voltage of 100.0 V. (a) What current passes through Oscar? (b) If one of the conducting wires is grounded and the other has an alternate path to ground through a 15-Ω resistor (so that Oscar and the resistor are in parallel), how much current would pass through Oscar if the maximum current that can be drawn from the emf is 1.00 A?

96. Chelsea inadvertently bumps into a set of batteries with an emf of 100.0 V that can supply a maximum power of 5.0 W. If the resistance between the points where she contacts the batteries is 1.0 kΩ, how much current passes through her?

97. The wiring circuit for a typical room is shown schematically. (a) Of the six locations for a circuit breaker indicated by A, B, C, D, E, and F, which one would best protect the household against a short circuit in any one of the three appliances? Explain. (b) The room circuit is supplied with 120 V. Suppose the heater draws 1500 W, the lamp draws 300 W, and the microwave draws 1200 W. The circuit breaker is rated at 20.0 A. Can all three devices be operated simultaneously without tripping the breaker? Explain.

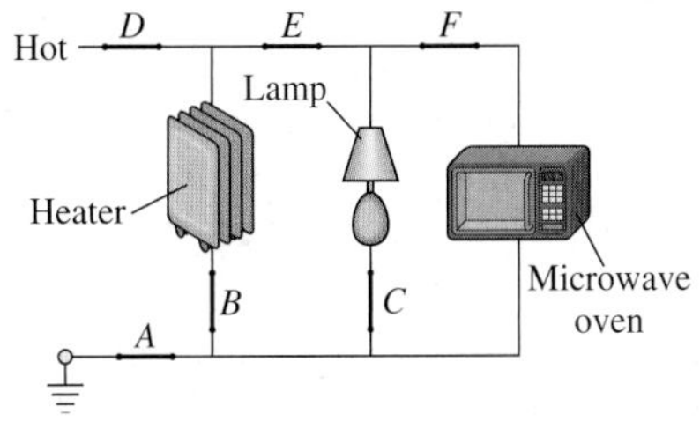

98. Several possibilities are listed for what might or might not happen if the insulation in the current-carrying wires of the figure breaks down and point b makes electric contact with point c. Discuss each possibility. (a) The

person touching the microwave oven gets a shock; (b) the cord begins to smoke; (c) a fuse blows out; (d) an electrical fire breaks out inside the kitchen wall.

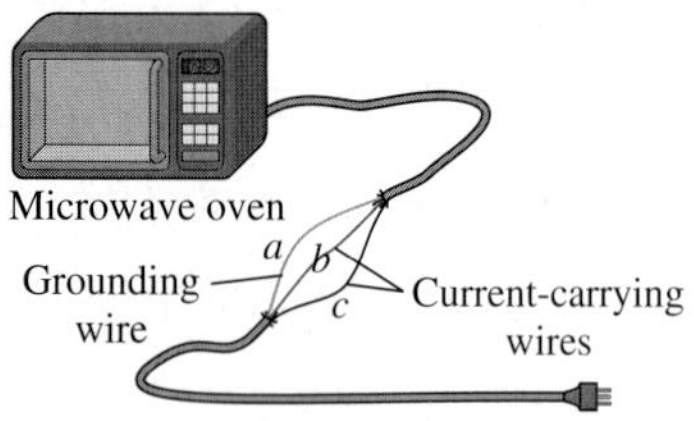

Comprehensive Problems

99. A 1.5-V flashlight battery can maintain a current of 0.30 A for 4.0 h before it is exhausted. How much chemical energy is converted to electrical energy in this process? (Assume zero internal resistance of the battery.)

100. In the diagram, the positive terminal of the 12-V battery is grounded—it is at zero potential. At what potential is point X?

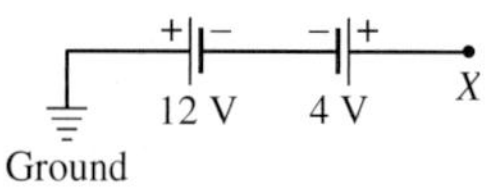

101. A_1 and A_2 represent ammeters with negligible resistance. What are the values of the currents (a) in A_1 and (b) in A_2?

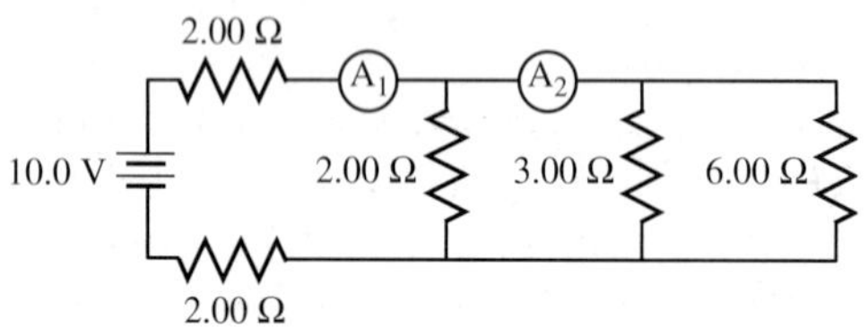

Problems 101 and 102

102. Repeat Problem 101 if each of the ammeters has resistance 0.200 Ω.

103. In a pacemaker used by a heart patient, a capacitor with a capacitance of 25 μF is charged to 1.0 V and then discharged through the heart every 0.80 s. What is the average discharge current?

104. A certain electric device has the current-voltage (I-V) graph shown with Problem 26. What is the power dissipated at points 1 and 2?

105. A 1.5-horsepower motor operates on 120 V. Ignoring I^2R losses, how much current does it draw?

106. (a) Given two identical, ideal batteries (emf = $\mathcal{E}$) and two identical lightbulbs (resistance = R assumed constant), design a circuit to make both bulbs glow as brightly as possible. (b) What is the power dissipated by each bulb? (c) Design a circuit to make both bulbs glow, but one more brightly than the other. Identify the brighter bulb.

107. Two circuits are constructed using identical, ideal batteries (emf = $\mathcal{E}$) and identical lightbulbs (resistance = R). If each bulb in circuit 1 dissipates 5.0 W of power, how much power does each bulb in circuit 2 dissipate? Ignore changes in the resistance of the bulbs due to temperature changes.

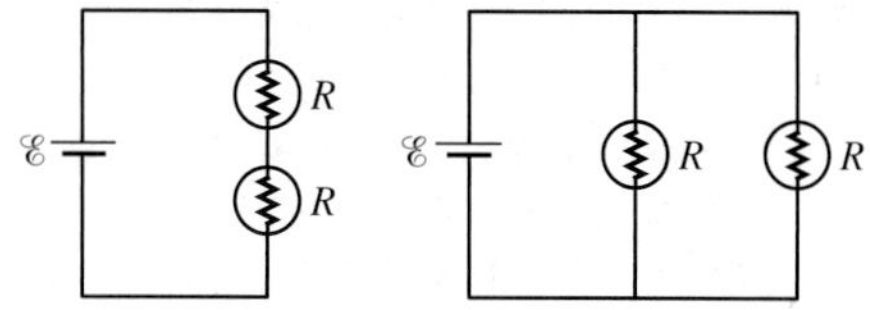

Circuit 1 Circuit 2

108. Given two identical, ideal batteries of emf $\mathcal{E}$ and two identical lightbulbs of resistance R (assumed constant), find the total power dissipated in the circuit in terms of $\mathcal{E}$ and R.

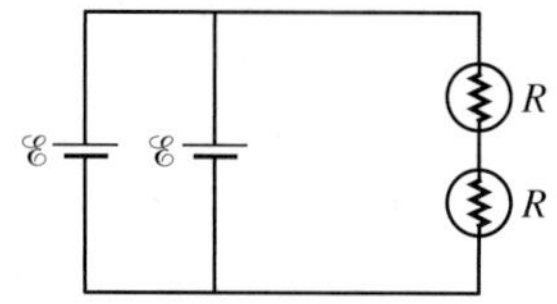

109. Consider a 60.0-W lightbulb and a 100.0-W lightbulb designed for use in a household lamp socket at 120 V. (a) What are the resistances of these two bulbs? (b) If they are wired together in a series circuit, which bulb shines brighter (dissipates more power)? Explain. (c) If they are connected in parallel in a circuit, which bulb shines brighter? Explain.

110. A 500-W electric heater unit is designed to operate with an applied potential difference of 120 V. (a) If the local power company imposes a voltage reduction to lighten its load, dropping the voltage to 110 V, by what percentage does the heat output of the heater drop? (Assume the resistance does not change.) (b) If you took the variation of resistance with temperature into account, would the actual drop in heat output be larger or smaller than calculated in part (a)?

111. The *Wheatstone bridge* is a circuit used to measure unknown resistances. The bridge in the figure is balanced—no current flows through the galvanometer. (a) What is the unknown resistance R_x? [*Hint:* What is the potential difference between points A and B?] (b) Does the resistance of the galvanometer affect the measurement? Explain.

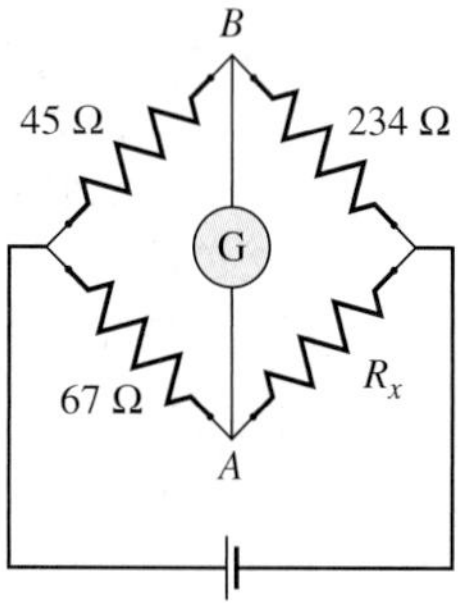

112. In the circuit shown, an emf of 150 V is connected across a resistance network. What is the current through R_2? Each of the resistors has a value of 10 Ω.

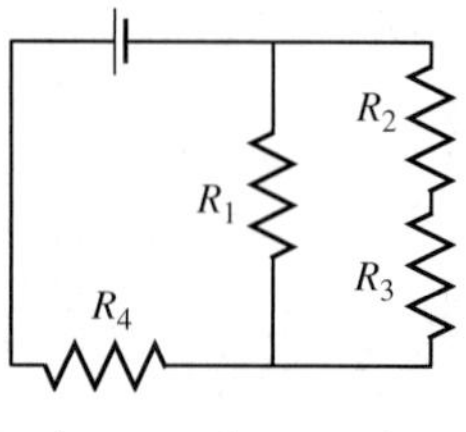

113. (a) What is the resistance of the heater element in a 1500-W hair dryer that plugs into a 120-V outlet? (b) What is the current through the hair dryer when it is turned on? (c) At a cost of \$0.10 per kW·h, how much does it cost to run the hair dryer for 5.00 min? (d) If you were to take the hair dryer to Europe where the voltage is 240 V, how much power would your hair dryer be using in the brief time before it is ruined? (e) What current would be flowing through the hair dryer during this time?

114. A string of 25 decorative lights has bulbs rated at 9.0 W and the bulbs are connected in parallel. The string is connected to a 120-V power supply. (a) What is the resistance of each of these lights? (b) What is the current through each bulb? (c) What is the total current coming from the power supply? (d) The string of bulbs has a fuse that will blow if the current is greater than 2.0 A. How many of the bulbs can you replace with 10.4-W bulbs without blowing the fuse?

115. A 2.00-μF capacitor is charged using a 5.00-V battery and a 3.00-μF capacitor is charged using a 10.0-V battery. (a) What is the total energy stored in the two capacitors? (b) The batteries are disconnected and the two capacitors are connected together (+ to + and − to −). Find the charge on each capacitor and the total energy in the two capacitors after they are connected. (c) Explain what happened to the "missing" energy. [*Hint:* The wires that connect the two have some resistance.]

116. Three identical lightbulbs are connected with wires to an ideal battery. The two terminals on each socket connect to the two terminals of its lightbulb. Wires do *not* connect with one another where they appear to cross in the picture. Ignore the change of the resistances of the filaments due to temperature changes. (a) Which of the schematic circuit diagrams correctly represent(s) the circuit? (List more than one choice if more than one diagram is correct.) (b) Which bulb(s) is/are the brightest? Which is/are the dimmest? Or are they all the same? Explain. (c) Find the current through each bulb if the filament resistances are each 24.0 Ω and the emf is 6.0 V.

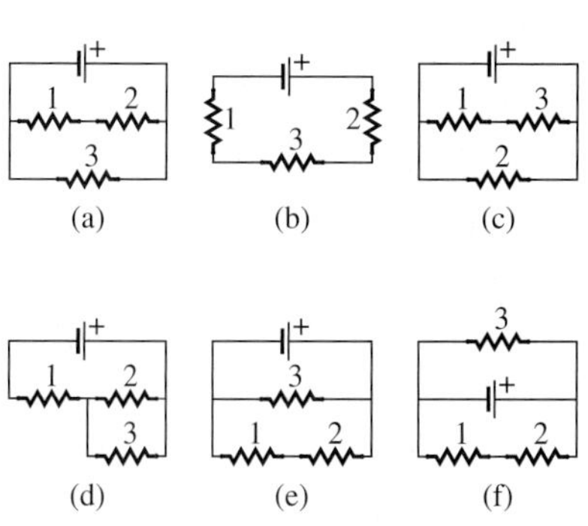

117. A portable radio requires an emf of 4.5 V. Olivia has only two nonrechargeable 1.5-V batteries, but she finds a larger 6.0-V battery. (a) How can she arrange the batteries to produce an emf of 4.5 V? Draw a circuit diagram. (b) Is it advisable to use this combination with her radio? Explain.

118. We can model some of the electrical properties of an unmyelinated axon as an electric cable covered with defective insulation so that current leaks out of the axon to the surrounding fluid. We assume the axon consists of a cylindrical membrane filled with conducting fluid. A current of ions can travel along the axon in this fluid and can also leak out through the membrane. The inner radius of the cylinder is 5.0 μm; the membrane thickness is 8.0 nm. (a) If the resistivity of the axon fluid is 2.0 Ω·m, calculate the resistance of a 1.0-cm length of axon to current flow along its length. (b) If the resistivity of the porous membrane is 2.5×10^7 Ω·m, calculate the resistance of the wall of a 1.0-cm length of axon to current flow across the membrane. (c) Find the length of axon for which the two resistances are equal. This length is a rough measure of the distance a signal can travel without amplification.

119. A piece of gold wire of length L has a resistance R_0. Suppose the wire is drawn out so that its length increases by a factor of three. What is the new resistance R in terms of the original resistance?

120. A voltmeter with a resistance of 670 kΩ is used to measure the voltage across the 83.0-kΩ resistor in the figure with Problems 74 and 75. What is the voltmeter reading?

121. A gold wire and an aluminum wire have the same dimensions and carry the same current. The electron density (in electrons/cm^3) in aluminum is three times larger than the density in gold. How do the drift speeds of the electrons in the two wires, v_{Au} and v_{Al}, compare?

122. Copper and aluminum are being considered for the cables in a high-voltage transmission line where each must carry a current of 50 A. The resistance of each cable is to be 0.15 Ω per kilometer. (a) If this line carries power from Niagara Falls to New York City (approximately 500 km), how much power is lost along the way in the cable? Compute for each choice of cable material (b) the necessary cable diameter and (c) the mass per meter of the cable. The electrical resistivities for copper and aluminum are given in Table 18.1; the mass density of copper is 8920 kg/m^3 and that of aluminum is 2702 kg/m^3.

123. The circuit is used to study the charging of a capacitor. (a) At $t = 0$, the switch is closed. What initial charging current is measured by the ammeter? (b) After the current has decayed to zero, what are the voltages at points A, B, and C?

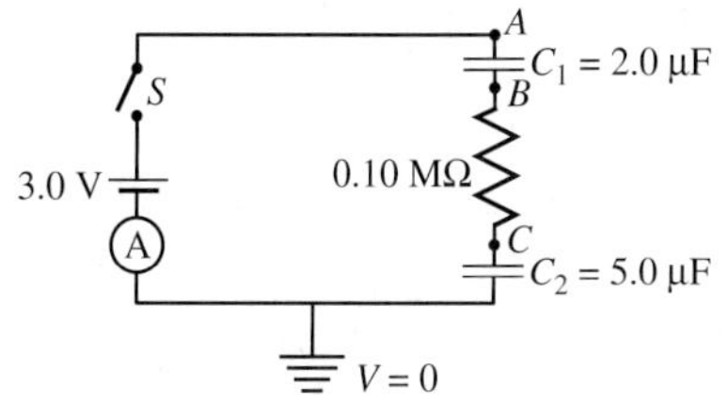

124. A parallel plate capacitor is constructed from two square conducting plates of length $L = 0.10$ m on a side. There is air between the plates, which are separated by a distance $d = 89$ μm. The capacitor is connected to a 10.0-V battery. (a) After the capacitor is fully charged, what is the charge on the upper plate? (b) The battery is disconnected from the plates and the capacitor is discharged through a resistor $R = 0.100$ MΩ. Sketch the current through the resistor as a function of time t ($t = 0$ corresponds to the time when R is connected to the capacitor). (c) How much energy is dissipated in R over the whole discharging process?

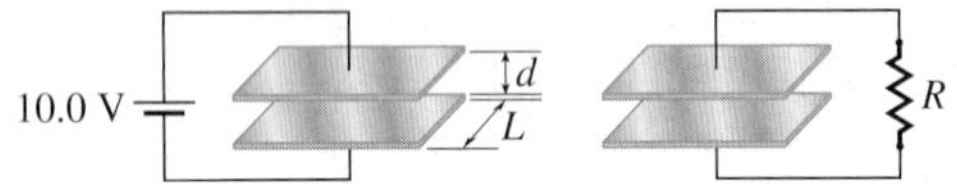

125. About 5.0×10^4 m above Earth's surface, the atmosphere is sufficiently ionized that it behaves as a conductor. The Earth and the ionosphere form a giant spherical capacitor, with the lower atmosphere acting as a leaky dielectric. (a) Find the capacitance C of the Earth-ionosphere system by treating it as a *parallel plate* capacitor. Why is it OK to do that? [*Hint:* Compare Earth's radius to the distance between the "plates."] (b) The fair-weather electric field is about 150 V/m, downward. How much energy is stored in this capacitor? (c) Due to radioactivity and cosmic rays, some air molecules are ionized even in fair weather. The resistivity of air is roughly 3.0×10^{14} Ω·m. Find the resistance of the lower atmosphere and the total current that flows between Earth's surface and the ionosphere. [*Hint:* Since we treat the system as a parallel plate capacitor, treat the atmosphere as a dielectric of *uniform thickness* between the plates.] (d) If there were no lightning, the capacitor would discharge. In this model, how much time would elapse before Earth's charge were reduced to 1% of its normal value? (Thunderstorms are the sources of emf that maintain the charge on this leaky capacitor.)

126. Near Earth's surface the air contains both negative and positive ions, due to radioactivity in the soil and cosmic rays from space. As a simplified model, assume there are 600.0 singly charged positive ions per cm^3 and 500.0 singly charged negative ions per cm^3; ignore the presence of multiply charged ions. The electric field is 100.0 V/m, directed downward. (a) In which direction do the positive ions move? The negative ions? (b) What is the direction of the current due to these ions? (c) The measured resistivity of the air in the region is 4.0×10^{13} Ω·m. Calculate the drift speed of the ions, assuming it to be the same for positive and negative ions. [*Hint:* Consider a vertical tube of air of length L and cross-sectional area A. How is the potential difference across the tube related to the electric field strength?] (d) If these conditions existed over the entire surface of the Earth, what is the total current due to the movement of ions in the air?

127. A battery with an emf of 1.0 V is connected to a 1.0-kΩ resistor and a diode (a nonohmic device) as shown in part (a) of the figure. The current that flows through the diode for a given voltage drop is shown in part (b) of the figure. (a) What is the current through the diode? (b) What is the current through the battery? (c) What is the total power dissipated in the diode and resistor? (d) Suppose the battery emf were increased so that the power dissipated in the 1.0-kΩ resistor doubled. Would you expect the power dissipated in the diode to double? If not, would it increase by a factor greater than 2 or less than 2? Explain briefly.

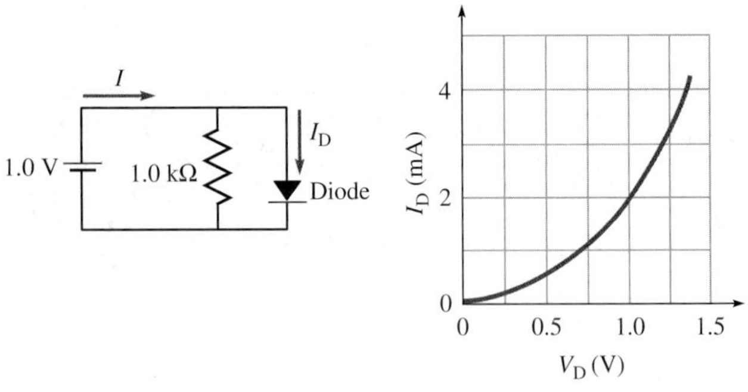

(a) (b)

128. Poiseuille's law [Eq. (9-15)] gives the volume flow rate of a viscous fluid through a pipe. (a) Show that Poiseuille's law can be written in the form $\Delta P = IR$, where $I = \Delta V/\Delta t$ represents the volume flow rate and R is a constant of proportionality called the fluid flow *resistance.* (b) Find R in terms of the viscosity of the fluid and the length and radius of the pipe. (c) If two or more pipes are connected in series so that the volume flow rate through them is the same, do the resistances of the pipes add as for electrical resistors ($R_{eq} = R_1 + R_2 + \cdots$)? Explain. (d) If two or more pipes are connected in parallel, so the pressure drop across them is the same, do the reciprocals of the resistances add as for electrical resistors ($1/R_{eq} = 1/R_1 + 1/R_2 + \cdots$)? Explain.

Answers to Practice Problems

18.1 (a) 2.00×10^{15} electrons; (b) 52 min

18.2 (a) 0.33 A; (b) 6.7 μm/s

18.3 6.9 Ω

18.4 292°C

18.5 1.495 V

18.6 1.0 Ω

18.7 $\frac{1}{3}R$ (the resistors are in parallel)

18.8 +(0.58 A)(4.0 Ω) − 1.5 V − 3.0 V + (0.72 A)(3.0 Ω) = 0.0

18.9 1.1 W; 190 J

18.10 10.0 mΩ; 10.0 mΩ

18.11 2.4 μA; 0.38 μC; 44 nA; 0.60 μC

Answers to Checkpoints

18.1 No. Equal quantities of positive and negative charge are being transported in the same direction at the same rate. There is no *net* transport of charge, so the electric current in the pipe is zero.

18.3 The thinner wire has fewer conduction electrons in a given length—the number per unit *volume* is the same, but the thinner wire has a smaller cross-sectional area. To produce the same current using fewer electrons, the electrons must move faster (on average). The thinner wire has a larger drift speed. This reasoning is confirmed by Eq. (18-3). Since I, n, and e are the same for both wires, the wire with smaller A has a larger v_D.

18.4 Resistivity is a property of the material that is independent of size or shape. Resistance depends on the size and shape.

18.6 $1/R_{eq} = 1/R + 1/R = 2/R \quad \Rightarrow \quad R_{eq} = R/2$

Review & Synthesis: Chapters 16–18

Review Exercises

1. A hollow metal sphere carries a charge of 6.0 μC. An identical sphere carries a charge of 18.0 μC. The two spheres are brought into contact with each other, then separated. How much charge is on each?

2. A hollow metal sphere carries a charge of 6.0 μC. A second hollow metal sphere with a radius that is double the size of the first carries a charge of 18.0 μC. The two spheres are brought into contact with each other, then separated. How much charge is on each?

3. Three point charges are placed on the corners of an equilateral triangle having sides of 0.150 m. What is the total electric force on the 2.50-μC charge?

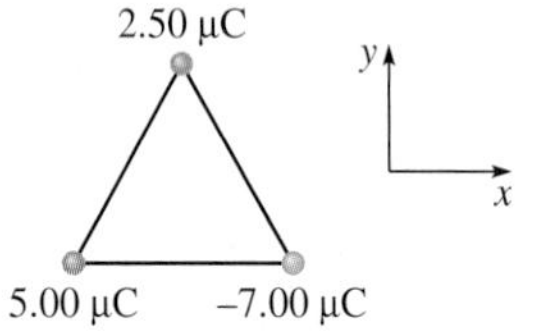

4. Two point charges are located on a coordinate system as follows: $Q_1 = -4.5$ μC at $x = 1.00$ cm and $y = 1.00$ cm and $Q_2 = 6.0$ μC at $x = 3.00$ cm and $y = 1.00$ cm. (a) What is the electric field at point P located at $x = 1.00$ cm and $y = 4.00$ cm? (b) When a 5.0-g tiny particle with a charge of −2.0 μC is placed at point P and released, what is its initial acceleration?

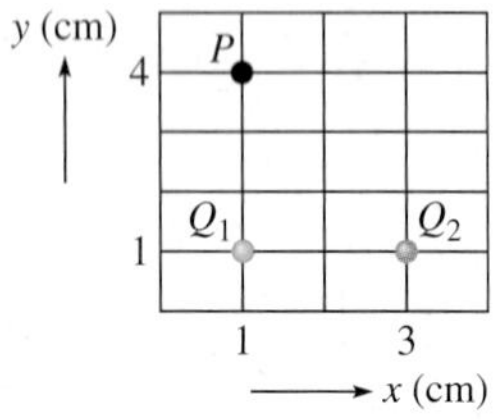

5. Object A has mass 90.0 g and hangs from an insulated thread. When object B, which has a charge of +130 nC, is held nearby, A is attracted to it. In equilibrium, A hangs at an angle $\theta = 7.20°$ with respect to the vertical and is 5.00 cm to the left of B. (a) What is the charge on A? (b) What is the tension in the thread?

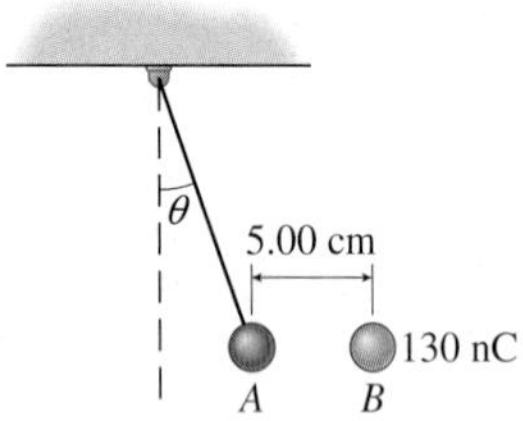

6. A lightbulb filament is made of tungsten. At room temperature of 20.0°C the filament has a resistance of 10.0 Ω. (a) What is the power dissipated in the lightbulb immediately after it is connected to a 120-V emf (when the filament is still at 20.0°C)? (b) After a brief time, the lightbulb filament has changed temperature and it glows brightly. The current is now 0.833 A. What is the resistance of the lightbulb now? (c) What is the power dissipated in the lightbulb when it is glowing brightly as in part (b)? (d) What is the temperature of the filament when it is glowing brightly? (e) Explain why lightbulbs usually burn out when they are first turned on rather than after they have been glowing for a long time.

7. Electrons in a cathode ray tube start from rest and are accelerated through a potential difference of 12.0 kV. They are moving in the $+x$-direction when they enter the space between the plates of a parallel plate capacitor. There is a potential difference of 320 V between the plates. The plates have length 8.50 cm and are separated by 1.10 cm. The electron beam is deflected in the negative y-direction by the electric field between the plates. What is the change in the y-position of the beam as it leaves the capacitor?

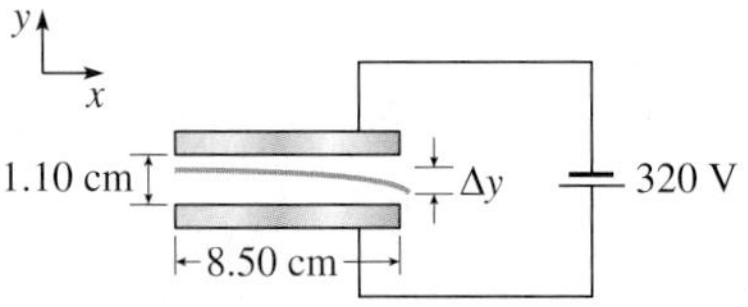

8. A 35.0-nC charge is placed at the origin and a 55.0-nC charge is placed on the $+x$-axis, 2.20 cm from the origin. (a) What is the electric potential at a point halfway between these two objects? (b) What is the electric potential at a point on the $+x$-axis 3.40 cm from the origin? (c) How much work does it take for an external agent to move a 45.0-nC charge from the point in (b) to the point in (a)?

9. In the circuit shown, $R_1 = 15.0\ \Omega$, $R_2 = R_4 = 40.0\ \Omega$, $R_3 = 20.0\ \Omega$, and $R_5 = 10.0\ \Omega$. (a) What is the equivalent resistance of this circuit? (b) What current flows through resistor R_1? (c) What is the total power dissipated by this circuit? (d) What is the potential difference across R_3? (e) What current flows through R_3? (f) What is the power dissipated in R_3?

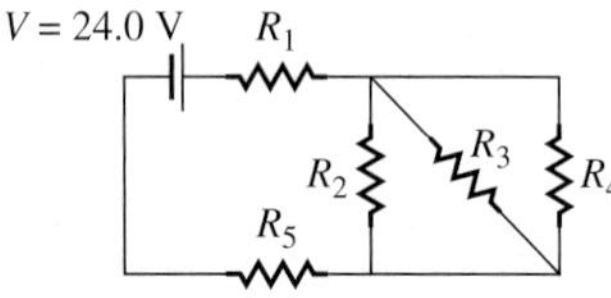

10. An electron with a velocity of 10.0 m/s in the positive y-direction enters a region where there is a uniform electric field of 200 V/m in the positive x-direction. What are the x- and y-components of the electron's displacement 2.40 μs after entering the electric-field region if no other forces act on it?

11. A proton is fired directly at a lithium nucleus. If the proton's velocity is 5.24×10^5 m/s when it is far from the nucleus, how close will the two particles get to each other before the proton stops and turns around?

12. An electron is suspended in a vacuum between two oppositely charged horizontal parallel plates. The separation between the plates is 3.00 mm. (a) What are the signs of the charge on the upper and on the lower plates? (b) What is the voltage across the plates?

13. Consider the circuit in the diagram. (a) After the switch S has been closed for a long time, what is the current through the 12-Ω resistor? (b) What is the voltage across the capacitor?

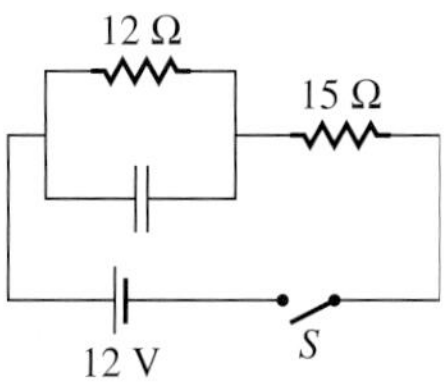

14. Consider the circuit in the diagram. Current $I_1 = 2.50$ A. Find the values of (a) I_2, (b) I_3, and (c) R_3.

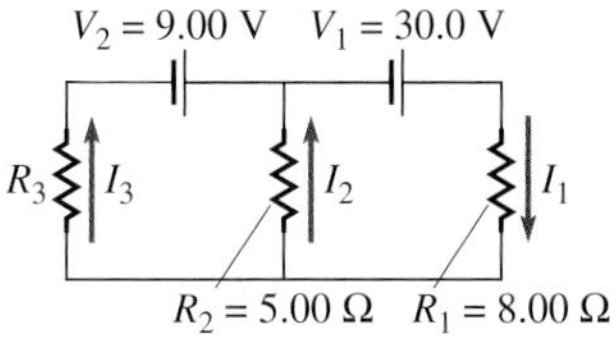

15. A large parallel plate capacitor has plate separation of 1.00 cm and plate area of 314 cm^2. The capacitor is connected across an emf of 20.0 V and has air between the plates. With the emf still connected, a slab of strontium titanate is inserted so that it completely fills the gap between the plates. Does the charge on the plates increase or decrease? By how much?

16. A *potentiometer* is a circuit to measure emfs. In the diagram with switch S_1 closed and S_2 open, there is no current through the galvanometer G for $R_1 = 20.0$ Ω with a standard cell $\mathscr{E}_s$ of 2.00 V. With switch S_2 closed and S_1 open, there is no current through the galvanometer G for $R_2 = 80.0$ Ω. (a) What is the unknown emf $\mathscr{E}_x$? (b) Explain why the potentiometer accurately measures the emf even for a source with substantial internal resistance.

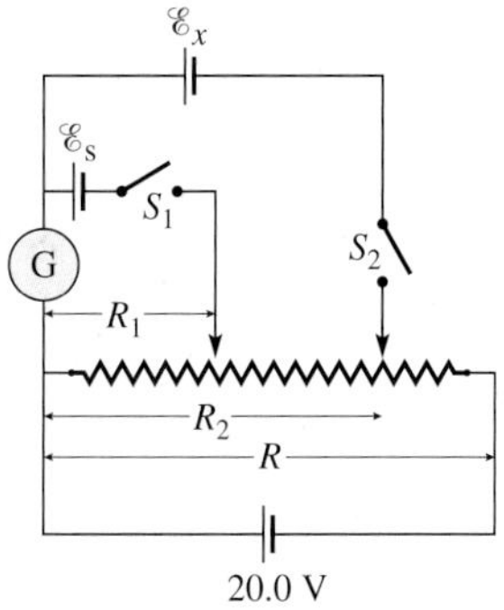

17. In the circuit, $\mathscr{E} = 45.0$ V and $R = 100.0$ Ω. If a voltage $V_x = 30.0$ V is needed for a circuit, what should resistance R_x be?

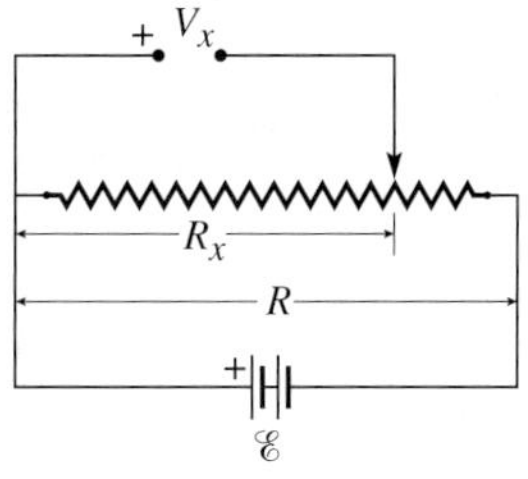

18. Two immersion heaters, A and B, are both connected to a 120-V supply. Heater A can raise the temperature of 1.0 L of water from 20.0°C to 90.0°C in 2.0 min, while heater B can raise the temperature of 5.0 L of water from 20.0°C to 90.0°C in 5.0 min. What is the ratio of the resistance of heater A to the resistance of heater B?

19. A parallel plate capacitor has 10.0-cm-diameter circular plates that are separated by 2.00 mm of dry air. (a) What is the maximum charge that can be on this capacitor? (b) A neoprene dielectric is placed between the plates, filling the entire region between the plates. What is the new maximum charge that can be placed on this capacitor?

20. What are the ratios of the resistances of (a) silver and (b) aluminum wire to the resistance of copper wire (R_{Ag}/R_{Cu} and R_{Al}/R_{Cu}) for wires of the same length and the same *mass* (not the same diameter)? (c) Which material is the best conductor, for wires of equal length and equal mass? The densities are: silver 10.1×10^3 kg/m^3; copper 8.9×10^3 kg/m^3; aluminum 2.7×10^3 kg/m^3.

21. A parallel plate capacitor used in a flash for a camera must be able to store 32 J of energy when connected to 300 V. (Most electronic flashes actually use a 1.5- to 6.0-V battery, but increase the effective voltage using a dc–dc inverter.) (a) What should be the capacitance of this capacitor? (b) If this capacitor has an area of 9.0 m^2, and a distance between the plates of 1.1×10^{-6} m, what is the dielectric constant of the material between the plates? (The large effective area can be put into a small volume by rolling the capacitor tightly in a cylinder.) (c) Assuming the capacitor completely discharges to produce a flash in 4.0×10^{-3} s, what average power is dissipated in the flashbulb during this time? (d) If the distance between the plates of the capacitor could be reduced to half its value, how much energy would the capacitor store if charged to the same voltage?

22. Consider the camera flash in Problem 21. If the flash really discharges according to Eq. (18-26), then it takes an infinite amount of time to discharge. When Problem 21 assumes that the capacitor discharges in 4.0×10^{-3} s, we mean that the capacitor has almost no charge stored on it after that amount of time. Suppose that after 4.0×10^{-3} s the capacitor has only 1.0% of the original charge still on it. (a) What is the time constant of this RC circuit? (b) What is the resistance of the flashbulb in this case? (c) What is the maximum power dissipated in the flashbulb?

23. A coffee maker can be modeled as a heating element (resistance R) connected to the outlet voltage of 120 V (assumed to be dc). The heating element boils small amounts of water at a time as it brews the coffee. When bubbles of water vapor form, they carry liquid water up through the tubing. Because of this, the coffee maker boils 5.0% of the water that passes through it; the rest is heated to 100°C but remains liquid. Starting with water at 10°C, the coffee maker can brew 1.0 L of coffee in 8.0 min. Find the resistance R.

24. Deuterium (^{2}D) is an isotope of hydrogen with a nucleus containing one proton and one neutron. In a ^{2}D–^{2}D fusion reaction, two deuterium nuclei combine to form a helium nucleus plus a neutron, releasing energy in the process.

The two ^{2}D nuclei must overcome the electrical repulsion of the positively charged nuclei ($q = +e$) to get close enough for the reaction to occur. The radius of a deuterium nucleus is about 1 fm, so the *centers* of the nuclei must get within about 2 fm of one another. To estimate the temperature that a gas of deuterium atoms must have for this fusion reaction to occur, find the temperature at which the average kinetic energy of the deuterium atoms is 5% of the required activation energy for the reaction.

25. Many home heating systems operate by pumping hot water through radiator pipes. The flow of the water to different "zones" in the house is controlled by zone valves that open in response to thermostats. The opening and closing of a zone valve is commonly performed by a wax actuator, as shown in the diagram. When the thermostat signals the valve to open, a dc voltage of 24 V is applied across a heating element (resistance $R = 200\ \Omega$) in the actuator. As the wax melts, it expands and pushes a cylindrical rod (radius 2.0 mm) out a distance 1.0 cm to open the zone switch. The actuator contains 2.0 mL of solid wax of density 0.90 g/cm^3 at room temperature (20°C). The specific heat of the wax is 0.80 J/(g·°C), its latent heat of fusion is 60 J/g, and its melting point is 90°C. When the wax melts its volume expands by 15%. How long does it take until the valve is fully open?

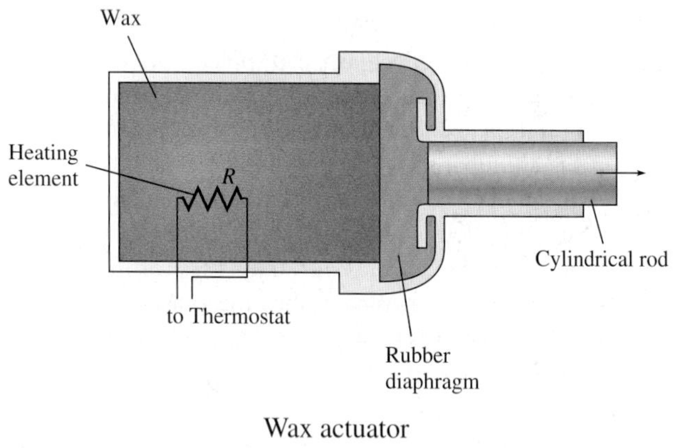

Wax actuator

Problems 26–28. *Hints:* When moving at terminal velocity, the net force on an object is zero. The viscous drag force is on a spherical object is given by Stokes's law, Eq. (9–16), where v is the speed of the object with respect to the surrounding fluid, and the buoyant force is given by Archimedes' principle (Section 9.6).

✦26. This problem illustrates the ideas behind the Millikan oil drop experiment—the first measurement of the electron charge. Millikan examined a fine spray of spherical oil droplets falling through air; the drops had picked up an electric charge as they were sprayed through an atomizer. He measured the terminal speed v_t of a drop when there was no electric field and then the electric field E that kept the drop motionless between the plates of a capacitor (plate spacing d). (a) With no electric field, the forces acting on the oil droplet were the gravitational force, the buoyant force, and viscous drag. The droplets used were so tiny (a radius of about 1 μm) that they rapidly reached terminal velocity. Find the radius R of a drop in terms of v_t, g, the densities of the oil and of air ρ_{oil} and ρ_{air}, and the viscosity of air η. (b) Find the charge q of a drop in terms of g, E, d, R, g, ρ_{oil}, and ρ_{air}. [*Hint:* The drag force is now zero because the drop is at rest.]

✦27. Ⓒ An air ionizer filters particles of dust, pollen, and other allergens from the air using electric forces. In one type of ionizer (see diagram), a stream of air is drawn in with a speed of 3.0 m/s. The air passes through a fine, highly charged wire mesh that transfers electric charge to the particles. Then the air passes through parallel "collector" plates that attract the charged particles and trap them in a filter. Consider a dust particle of radius 6.0 μm, mass 2.0×10^{-13} kg, and charge 1000e. The plates are 10 cm long and are separated by a distance of 1.0 cm. (a) *Ignoring* drag forces, what would be the minimum potential difference between the plates to ensure that the particle gets trapped by the filter? (b) At what speed would the particle be moving relative to the stream of air just before hitting the filter? (c) Calculate the viscous drag force on the particle when moving at the speed found in (b). (d) Is it realistic to ignore drag? Taking drag into consideration, should the potential difference be larger or smaller than the answer to (a)?

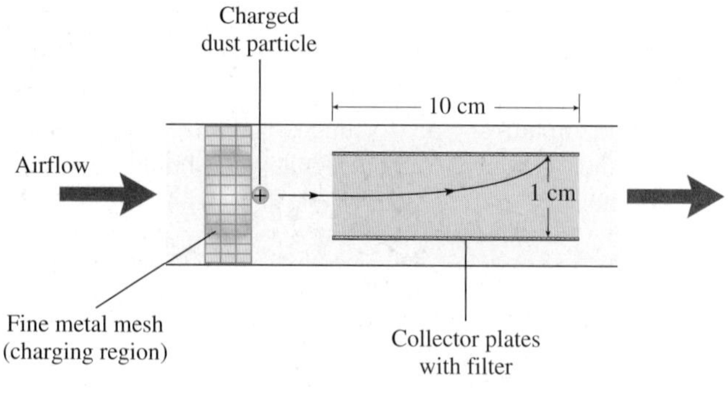

✦28. A spherical rain drop of radius 1.0 mm has a charge of +2 nC. The electric field in the vicinity is 2000 N/C downward. The terminal speed of an identical but *uncharged* drop is 6.5 m/s. The drag force is related to the drop's speed by $F_d = bv^2$ (turbulent drag rather than viscous drag). Calculate the terminal speed of the charged rain drop.

MCAT Review

The section that follows includes MCAT exam material and is reprinted with permission of the Association of American Medical Colleges (AAMC).

1. At a given temperature, the resistance of a wire to direct current depends only on the
   A. voltage applied across the wire.
   B. resistivity, length, and voltage.
   C. voltage, length, and cross-sectional area.
   D. resistivity, length, and cross-sectional area.

Refer to the two paragraphs about the holding tank for synthetic lubricating oil in the MCAT Review section for Chapters 13–15. Based on those paragraphs, answer the following two questions.

2. What electric current is required to run all of the heaters at maximum power output from a single 600-V power supply?
   A. 7.2 A
   B. 24.0 A
   C. 83.0 A
   D. 120.0 A
3. In another test, the 10 heaters are exchanged for 5 larger heaters that each use a current of 20 A from an 800-V power supply. What is the total power usage of the 5 new heaters?
   A. 16 kW
   B. 32 kW
   C. 80 kW
   D. 320 kW

Read the paragraph and then answer the following questions:

The diagram shows a small water heater that uses an electric current to supply energy to heat water. A heating element, R_L, is immersed in the water and acts as a 1.0-Ω load resistor. A dc source is mounted on the outside of the water heater and is wired in parallel with a 2.0-Ω resistor (R_S) and the load resistor. When the water is being heated, the current source supplies a steady current (I) of 0.5 A to the circuit. The water heater has a heat capacity of C and holds 1.0×10^{-3} m^3 of water. The water has a mass of 1.0 kg. The entire system is thermally isolated and designed to maintain an approximately constant temperature of 60°C. [*Note:* The specific heat of water $(c_w) = 4.2 \times 10^3$ J/(kg·°C).]

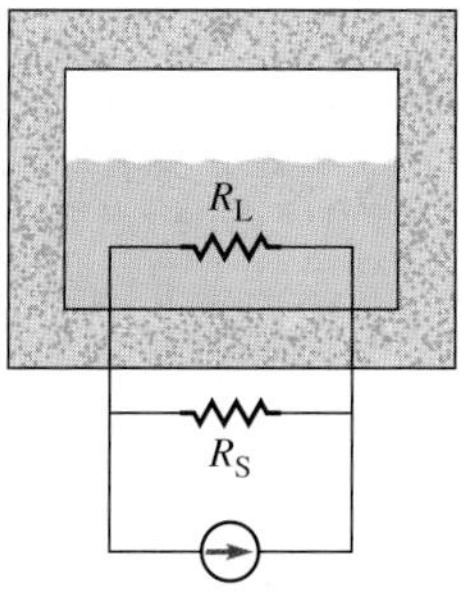

4. What is the voltage drop across R_L?
   A. 0.22 V
   B. 0.33 V
   C. 0.75 V
   D. 1.50 V
5. If the equipment outside the water heater is changed so that I is 1.2 A and R_S is 3.0 Ω, how much power will be dissipated in R_S?
   A. 0.27 W
   B. 0.40 W
   C. 1.08 W
   D. 4.32 W
6. As current flows through R_L, which of the following quantities does *not* increase?
   A. Entropy of the system
   B. Temperature of the system
   C. Total energy in the water
   D. Power dissipated in R_L
7. If the power source used for the water heater is a battery, which of the following best describes the energy transfers that take place when the current is flowing through the circuit in the water-heater system?
   A. Chemical to electrical to heat
   B. Chemical to heat to electrical
   C. Electrical to chemical to heat
   D. Electrical to heat to chemical
8. If the resistance of R_L increased as a function of time, which of the following quantities would also increase with time?
   A. Power dissipated in R_L
   B. Current through R_L
   C. Current through R_S
   D. Resistance of R_S
9. If a different current source caused R_L to dissipate power into the water at a rate of 1.0 W, how long would it take to increase the temperature of the water by 1.0°C? [*Note:* Assume that the heat used to heat the heating element and insulation is negligible.]
   A. 70 s
   B. 420 s
   C. 700 s
   D. 4200 s

Read the passage and then answer the following questions:

Electric power is generally transmitted to consumers by overhead wires. To reduce power loss due to heat, utility companies strive to reduce the magnitudes of both the current (I) through the wires and the resistance (R) of the wires.

A reduction in R requires the use of highly conductive materials and large wires. The size of wires is limited by the cost of materials and weight. The table lists the resistances and masses of 1000-m sections of copper wires of different diameters at two different temperatures.

Diameter (m)	**Resistance per 10^3 m at 25°C (Ω)**	**Resistance per 10^3 m at 65°C (Ω)**	**Mass per 10^3 m (kg)**
6.6×10^{-2}	7.2×10^{-3}	8.2×10^{-3}	2.4×10^{4}
2.9×10^{-2}	3.5×10^{-2}	4.1×10^{-2}	4.6×10^{3}
2.1×10^{-2}	7.1×10^{-2}	8.2×10^{-2}	2.3×10^{3}
9.5×10^{-3}	3.4×10^{-1}	3.8×10^{-1}	4.9×10^{2}

Safety and technical equipment considerations limit voltage. Because electricity is transmitted at high-voltage levels for long-distance transmission, transformers are needed to lower the voltage to safer levels before entering residences.

10. If a residence uses 1.2×10^4 W at 120 V, how much current is required?
    A. 10 A
    B. 12 A
    C. 100 A
    D. 120 A

11. Based on the table, if the temperature changes from 25°C to 65°C in a 10^5-m section of 9.5×10^{-3}-m diameter wire, approximately how much will the wire's resistance change?
    A. 0.04 Ω
    B. 0.4 Ω
    C. 4.0 Ω
    D. 40 Ω

12. How much power is lost to heat in a transmission line with a resistance of 3 Ω that carries 2 A?
    A. 1.5 W
    B. 6 W
    C. 12 W
    D. 18 W

13. In order to supply 10 residences with 10^4 W of power each over a grid that loses 5×10^3 W of power to heat, how much power is needed?
    A. 1.5×10^4 W
    B. 5.25×10^4 W
    C. 1.05×10^5 W
    D. 1.5×10^5 W

CHAPTER

19

Magnetic Forces and Fields

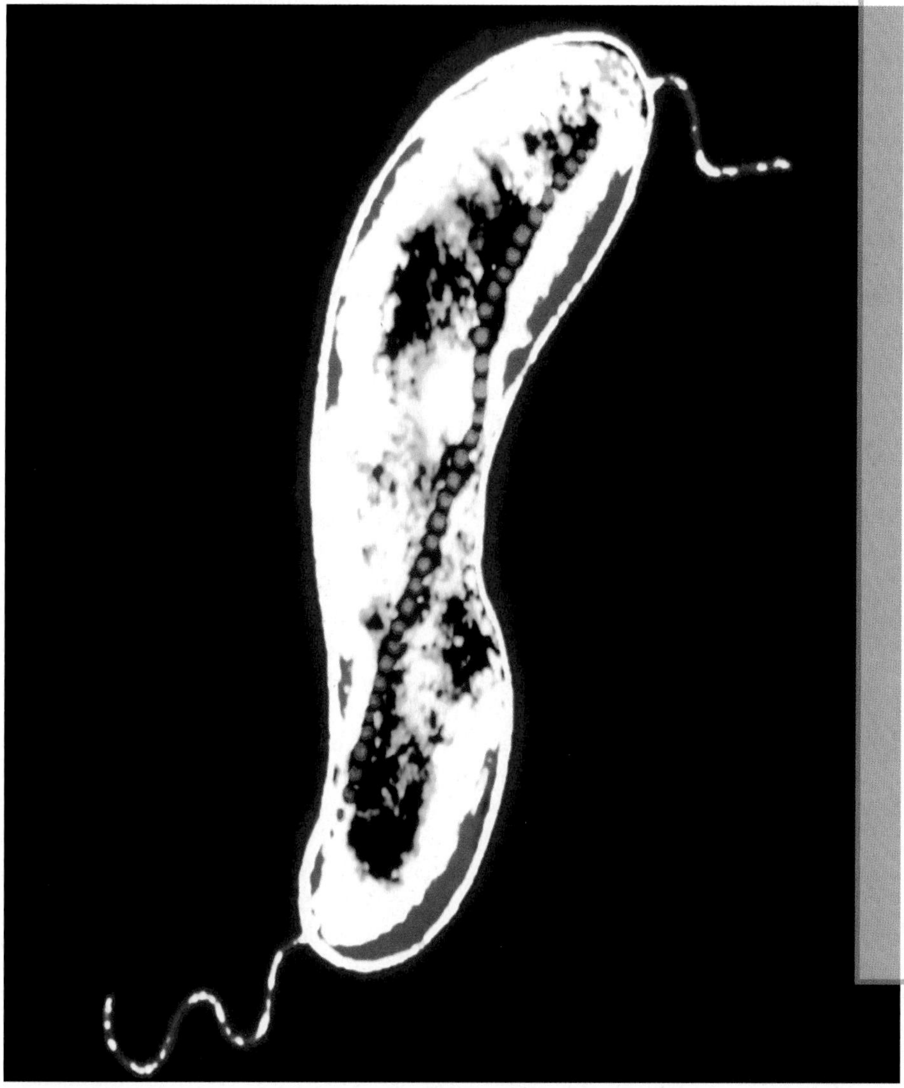

Electron micrograph of a magnetotactic bacterium

Some bacteria live in the mud at the bottom of the sea. As long as they are in the mud, all is well. Suppose that the mud gets stirred up, perhaps by a crustacean walking by. Now things are not so rosy. The bacteria cannot survive for long in the water, so it is imperative that they swim back down to the mud as soon as possible. The problem is that knowing which direction is down is not so easy. The mass density of the bacteria is almost identical to that of water, so the buoyant force prevents them from "feeling" the downward pull of gravity. Nevertheless, the bacteria are somehow able to swim in the correct direction to get back to the mud. How do they do it? (See p. 697 for the answer.)

Concepts & Skills to Review

- sketching and interpreting electric field lines (Section 16.4)
- uniform circular motion; radial acceleration (Section 5.2)
- torque; lever arm (Section 8.2)
- relation between current and drift velocity (Section 18.3)

Working model of a spoon-shaped compass from the Han Dynasty (202 B.C.E. to 220 C.E.). The spoon, made of lodestone (magnetite ore) rests on a bronze plate called a "heaven-plate" or diviner's board. The earliest Chinese compasses were used for prognostication; only much later were they used as navigation aids. The model was constructed by Susan Silverman.

19.1 MAGNETIC FIELDS

Permanent Magnets

Permanent magnets have been known at least since the time of the ancient Greeks, about 2500 years ago. A naturally occurring iron ore called lodestone (now called magnetite) was mined in various places, including the region of modern-day Turkey called Magnesia. Some of the chunks of lodestone were permanent magnets; they exerted magnetic forces on each other and on iron and could be used to turn a piece of iron into a permanent magnet. In China, the magnetic compass was used as a navigational aid at least a thousand years ago—possibly much earlier. Not until 1820 was a connection between electricity and magnetism established, when Danish scientist Hans Christian Oersted (1777–1851) discovered that a compass needle is deflected by a nearby electric current.

Figure 19.1a shows a plate of glass lying on top of a bar magnet. Iron filings have been sprinkled on the glass and then the glass has been tapped to shake the filings a bit and allow them to move around. The filings have lined up with the **magnetic field** (symbol: $\vec{\mathbf{B}}$) due to the bar magnet. Figure 19.1b shows a sketch of the magnetic field lines representing this magnetic field. As is true for electric field lines, the magnetic field lines represent both the magnitude and direction of the magnetic field vector. The magnetic field vector at any point is tangent to the field line and the magnitude of the field is proportional to the number of lines per unit area perpendicular to the lines.

Symbol for magnetic field: $\vec{\mathbf{B}}$

Figure 19.1b may strike you as being similar to a sketch of the electric field lines for an electric dipole (see Fig. 16.29). The similarity is not a coincidence; the bar magnet is one instance of a **magnetic dipole**. By *dipole* we mean *two opposite poles.* In an electric dipole, the electric poles are positive and negative electric charges. A magnetic dipole consists of two opposite magnetic poles. The end of the bar magnet where the field lines emerge is called the **north pole** and the end where the lines go back in is called the **south pole**. If two magnets are near one another, opposite poles (the north pole of one magnet and the south pole of the other) exert attractive forces on one another; like poles (two north poles or two south poles) repel one another.

CONNECTION:

Electric dipole: one positive charge and one negative charge. Magnetic dipole: one north pole and one south pole.

Application: magnetic compass

The names *north pole* and *south pole* are derived from magnetic compasses. A compass is simply a small bar magnet that is free to rotate. Any magnetic dipole, including a compass needle, feels a torque that tends to line it up with an external magnetic

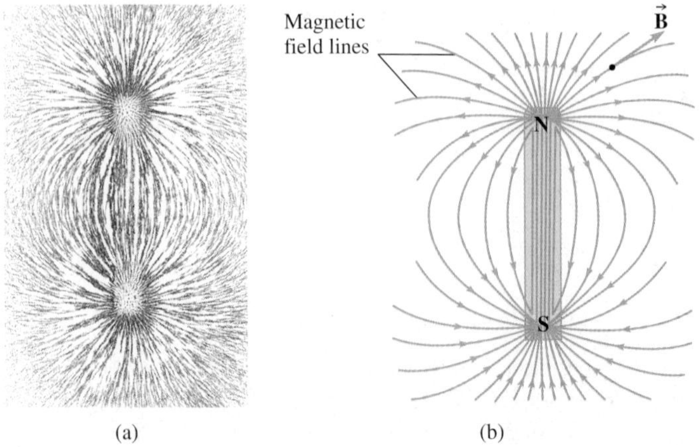

Figure 19.1 (a) Photo of a bar magnet. Nearby iron filings line up with the magnetic field. (b) Sketch of the magnetic field lines due to the bar magnet. The magnetic field vectors are tangent to the field lines.

field (Fig. 19.2). The north pole of the compass needle is the end that points in the direction of the magnetic field. In a compass, the bar magnet needle is mounted to minimize frictional and other torques so it can swing freely in response to a magnetic field.

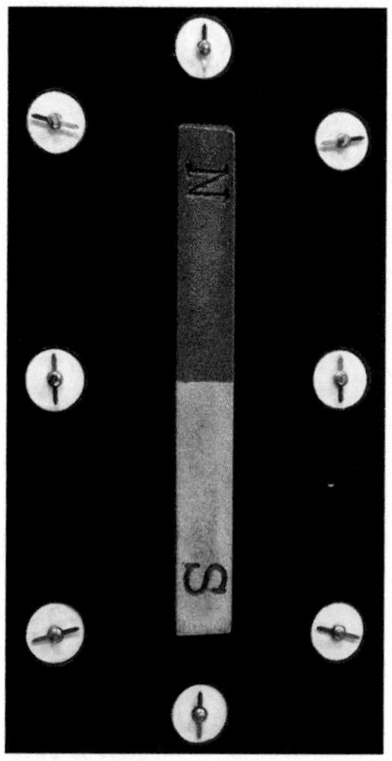

Figure 19.2 Each compass needle is aligned with the magnetic field due to the bar magnet. The "north" (red) end of each needle points in the direction of the magnetic field.

PHYSICS AT HOME

Obtain a flexible magnetic card. With scissors, cut a thin strip (about 2 mm wide) from one edge. Rub the back of the strip across the back of the remaining piece. Try both orientations (both parallel and perpendicular to the side from which the strip was cut). Repeat with a strip cut from an adjacent side of the card. Estimate the orientation and the size of the magnetized strips. (See Fig. 19.3b.)

Permanent magnets come in many shapes other than the bar magnet. Figure 19.3 shows some others, with the magnetic field lines sketched. Notice in Fig. 19.3a that if the pole faces are parallel and close together, the magnetic field between them is nearly uniform. A magnet need not have only two poles; it must have *at least* one north pole and *at least* one south pole. Some magnets are designed to have a large number of north and south poles. The flexible magnetic card (Fig. 19.3b), commonly found on refrigerator doors, is designed to have many poles, both north and south, on one side and no poles on the other. The magnetic field is strong near the side with the poles and weak near the other side; the card sticks to an iron surface (such as a refrigerator door) on one side but not on the other.

No Magnetic Monopoles Coulomb's law for *electric* forces gives the force acting between two point charges—two electric *monopoles.* However, as far as we know, there are no *magnetic* monopoles—that is, there is no such thing as an isolated north pole or an isolated south pole. If you take a bar magnet and cut it in half, you do not obtain one piece with a north pole and another piece with a south pole. Both pieces are magnetic dipoles (Fig. 19.4). There have been theoretical predictions of the existence of magnetic monopoles, but years of experiments have yet to turn up a single one. If magnetic monopoles do exist in our universe, they must be extremely rare.

Magnetic Field Lines

Figure 19.1 shows that magnetic field lines do not begin on north poles and end on south poles: *magnetic field lines are always closed loops.* If there are no magnetic monopoles, there is no place for the field lines to begin or end, so they *must* be closed loops. Contrast Fig. 19.1b with Fig. 16.29—the field lines for an electric dipole. The field line patterns are similar *away* from the dipole, but nearby and between the poles they are quite different. The electric field lines are not closed loops; they start on the positive charge and end on the negative charge.

Magnetic field lines are always closed loops.

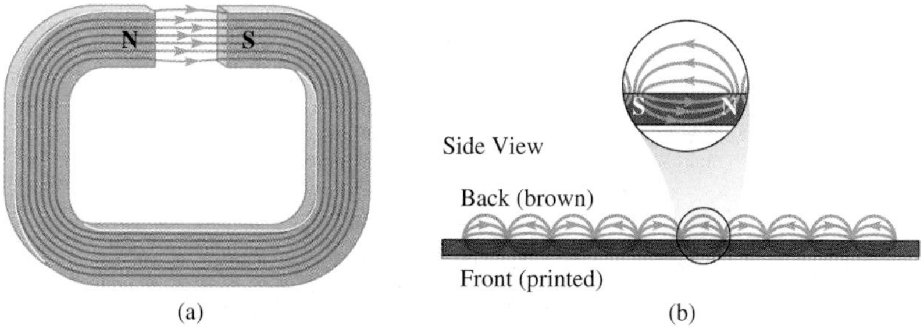

Figure 19.3 Two permanent magnets with their magnetic field lines. In (a), the magnetic field between the pole faces is nearly uniform. (b) A refrigerator magnet (shown here in a side view) has many poles.

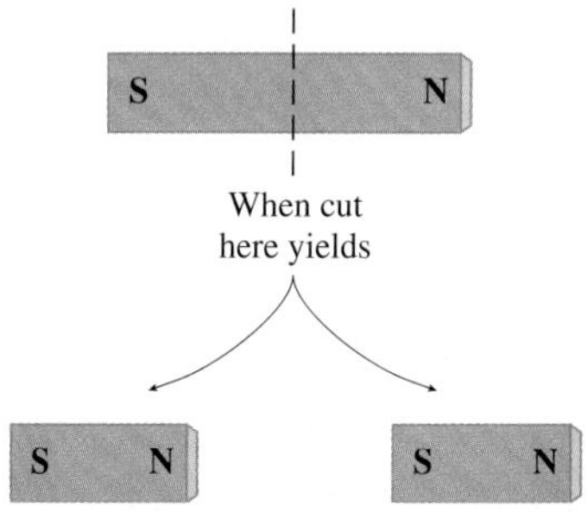

Figure 19.4 Sketch of a bar magnet that is subsequently cut in half. Each piece has both a north and a south pole.

Despite these differences between electric and magnetic field lines, the *interpretation* of magnetic field lines is exactly the same as for electric field lines:

CONNECTION:

Magnetic field lines help us visualize the magnitude and direction of the magnetic field vectors, just as electric field lines do for the magnitude and direction of $\vec{\mathbf{E}}$.

Interpretation of Magnetic Field Lines

- The direction of the magnetic field vector at any point is *tangent to the field line* passing through that point and is in the direction indicated by arrows on the field line (as in Fig. 19.1b).
- The magnetic field is strong where field lines are close together and weak where they are far apart. More specifically, if you imagine a small surface perpendicular to the field lines, the magnitude of the magnetic field is proportional to the number of lines that cross the surface, divided by the area.

The Earth's Magnetic Field

Figure 19.5 shows field lines for Earth's magnetic field. Near Earth's surface, the magnetic field is approximately that of a dipole, as if a bar magnet were buried at the center of the Earth. Farther away from Earth's surface, the dipole field is distorted by the solar wind—charged particles streaming from the Sun toward Earth. As discussed in Section 19.8, moving charged particles create their own magnetic fields, so the solar wind has a magnetic field associated with it.

In most places on the surface, Earth's magnetic field is not horizontal; it has a significant vertical component. The vertical component can be measured directly using a *dip meter*, which is just a compass mounted so that it can rotate in a vertical plane. In the northern hemisphere, the vertical component is downward, while in the southern hemisphere it is upward. In other words, magnetic field lines emerge from Earth's surface in the southern hemisphere and reenter in the northern hemisphere. A magnetic dipole that is free to rotate aligns itself with the magnetic field such that the north end of the dipole points in the direction of the field. Figure 19.2 shows a bar magnet with several compasses in the vicinity. Each compass needle points in the direction of the local magnetic field, which in this case is due to the magnet. A compass is normally used to detect Earth's magnetic field. In a horizontally mounted compass, the needle is free to rotate only in a horizontal plane, so its north end points in the direction of the *horizontal component* of Earth's field.

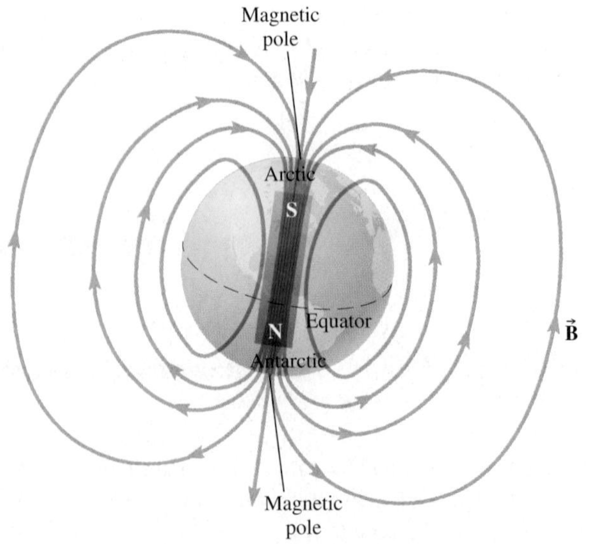

Figure 19.5 Earth's magnetic field. The diagram shows the magnetic field lines in one plane. In general, the magnetic field at the surface has both horizontal and vertical components. The magnetic poles are the points where the magnetic field at the surface is purely vertical. The magnetic poles do not coincide with the geographic poles, which are the points at which the axis of rotation intersects the surface. Near the surface, the field is approximately that of a dipole, like that of the fictitious bar magnet shown. Note that the south pole of this bar magnet points toward the Arctic and the north pole points toward the Antarctic.

Note the orientation of the fictitious bar magnet in Fig. 19.5: the south pole of the magnet faces roughly toward geographic *north* and the north pole of the magnet faces roughly toward geographic *south*. The field lines emerge from Earth's surface in the southern hemisphere and return in the northern hemisphere.

Origin of Earth's Magnetic Field The origin of the Earth's magnetic field is still under investigation. According to a leading theory, the field is created by electric currents in the molten iron and nickel of Earth's outer core, more than 3000 km below the surface. Earth's magnetic field is slowly changing. In 1948, Canadian scientists discovered that the location of Earth's magnetic pole in the Arctic was about 250 km away from where it was found in 1831 by a British explorer. The magnetic poles move about 40 km per year. The magnetic poles have undergone a complete reversal in polarity (north becomes south and south becomes north) roughly 100 times in the past 5 million years. The most recent Geological Survey of Canada, completed in May 2001, located the north magnetic pole—the point on Earth's surface where the magnetic field points straight down—at 81°N latitude and 111°W longitude, about 1600 km south of the geographic north pole (the point where Earth's rotation axis intersects the surface, at 90°N latitude).

Application: Magnetotactic Bacteria

In the electron micrograph of the bacterium shown with the chapter opener, a line of crystals (stained orange) stands out. They are crystals of magnetite, the same iron oxide (Fe_3O_4) that was known to the ancient Greeks. The crystals are tiny permanent magnets that function essentially as compass needles. When the bacteria get stirred up into the water, their compass needles automatically rotate to line up with the magnetic field. As the bacteria swim along, they follow a magnetic field line. In the northern hemisphere, the north end of the "compass needle" faces forward. The bacteria swim in the direction of the magnetic field, which has a downward component, so they return to their home in the mud. Bacteria in the southern hemisphere have the south pole forward; they must swim opposite to the magnetic field since the field has an *upward* component. If some of these *magnetotactic* (*-tactic* = feeling or sensing) bacteria are brought from the southern hemisphere to the northern, or vice versa, they swim up instead of down!

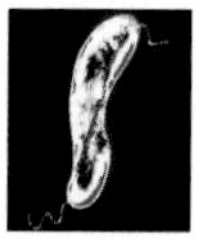

How do the bacteria swim in the correct direction?

There is evidence of magnetic navigation in several species of bacteria and also in some higher organisms. Experiments with homing pigeons, robins, and bees have shown that these organisms have some magnetic sense. On sunny days, they primarily use the Sun's location for navigation, but on overcast days they use Earth's magnetic field. Permanently magnetized crystals, similar to those found in the mud bacteria, have been found in the brains of these organisms, but the mechanism by which they can sense Earth's field and use it to navigate is not understood. Some experiments have shown that even humans may have some sense of Earth's magnetic field, which is not out of the realm of possibility since tiny magnetite crystals have been found in the brain.

19.2 MAGNETIC FORCE ON A POINT CHARGE

Before we go into more detail on the magnetic forces and torques on a magnetic dipole, we need to start with the simpler case of the magnetic force on a moving point charge. Recall that in Chapter 16 we defined the electric field as the electric force per unit charge. The electric force is either in the same direction as $\vec{\mathbf{E}}$ or in the opposite direction, depending on the sign of the point charge.

The magnetic force on a point charge is more complicated—it is *not* the charge times the magnetic field. The magnetic force *depends on the point charge's velocity* as well as on the magnetic field. If the point charge is at rest, there is no magnetic force. The magnitude and direction of the magnetic force depend on the direction and speed of

The magnetic force is velocity-dependent.

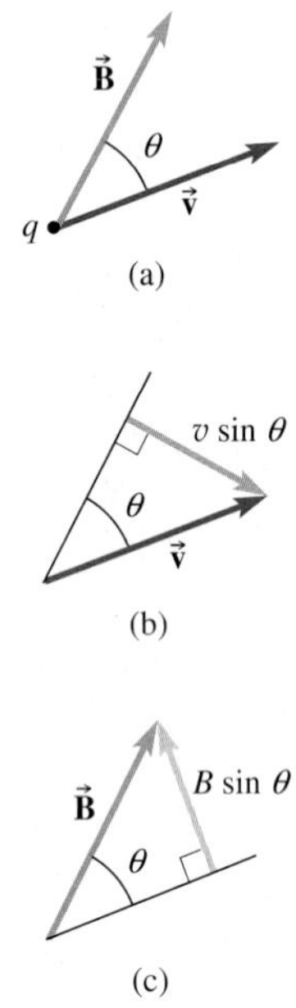

Figure 19.6 A positive charge moving in a magnetic field. (a) The particle's velocity vector $\vec{v}$ and the magnetic field vector $\vec{B}$ are drawn starting at the same point. θ is the angle between them. (b) The component of $\vec{v}$ perpendicular to $\vec{B}$ is $v \sin \theta$. (c) The component of $\vec{B}$ perpendicular to $\vec{v}$ is $B \sin \theta$.

the charge's motion. We have learned about other velocity-dependent forces, such as the drag force on an object moving through a fluid. Like drag forces, the magnetic force increases in magnitude with increasing velocity. However, the direction of the drag force is always opposite to the object's velocity, while the direction of the magnetic force on a charged particle is *perpendicular* to the velocity of the particle.

Imagine that a positive point charge q moves at velocity $\vec{v}$ at a point where the magnetic field is $\vec{B}$. The angle between $\vec{v}$ and $\vec{B}$ is θ (Fig. 19.6a). The magnitude of the magnetic force acting on the point charge is the product of

- The magnitude of the charge $|q|$,
- The magnitude of the field B, and
- The component of the velocity perpendicular to the field (Fig. 19.6b).

Magnitude of the magnetic force on a moving point charge:

$$F_B = |q|v_\perp B = |q|(v \sin \theta)B$$
$$(\text{since } v_\perp = v \sin \theta) \qquad (19\text{-}1a)$$

Note that if the point charge is at rest ($v = 0$) or if its motion is along the same line as the magnetic field ($v_\perp = 0$), then the magnetic force is zero.

In some cases it is convenient to look at the factor $\sin \theta$ from a different point of view. If we associate the factor $\sin \theta$ with the magnetic field instead of with the velocity, then $B \sin \theta$ is the component of the magnetic field perpendicular to the velocity of the charged particle (Fig. 19.6c):

$$F_B = |q|v(B \sin \theta) = |q|vB_\perp \qquad (19\text{-}1b)$$

SI Unit of Magnetic Field From Eq. (19-1), the SI unit of magnetic field is

$$\frac{\text{force}}{\text{charge} \times \text{velocity}} = \frac{\text{N}}{\text{C·m/s}} = \frac{\text{N}}{\text{A·m}}$$

This combination of units is given the name tesla (symbol T) after Nikola Tesla (1856–1943), an American engineer who was born in Croatia.

$$1\text{ T} = 1\frac{\text{N}}{\text{A·m}} \qquad (19\text{-}2)$$

CHECKPOINT 19.2

An electron is moving with speed v in a uniform downward magnetic field $\vec{B}$. (a) In what direction(s) can it be moving if the magnetic force on it is zero? (b) In what direction(s) can it be moving if the magnetic force on it has the largest possible magnitude?

CONNECTION:

The cross product of two vectors is a vector quantity. The cross product is a different mathematical operation than the dot product of two vectors, which is a *scalar* (see Section. 6.2). The cross product has its maximum magnitude when the two vectors are perpendicular; the dot product is maximum when the two vectors are parallel.

Cross Product of Two Vectors

The direction and magnitude of the magnetic force depend on the vectors $\vec{v}$ and $\vec{B}$ in a special way that occurs often in physics and mathematics. The magnetic force can be written in terms of the **cross product** (or *vector product*) of $\vec{v}$ and $\vec{B}$. The cross product of two vectors $\vec{a}$ and $\vec{b}$ is written $\vec{a} \times \vec{b}$. The magnitude of the cross product is the magnitude of one vector times the perpendicular component of the other; it doesn't matter which is which.

The cross product of two vectors is perpendicular to both vectors.

$$|\vec{a} \times \vec{b}| = |\vec{b} \times \vec{a}| = a_\perp b = ab_\perp = ab \sin \theta \qquad (19\text{-}3)$$

However, the order of the vectors *does* matter in determining the *direction* of the result. Switching the order reverses the direction of the product:

$$\vec{\mathbf{b}} \times \vec{\mathbf{a}} = -\vec{\mathbf{a}} \times \vec{\mathbf{b}} \qquad (19\text{-}4)$$

The cross product of two vectors $\vec{\mathbf{a}}$ and $\vec{\mathbf{b}}$ is a vector that is perpendicular to both $\vec{\mathbf{a}}$ and $\vec{\mathbf{b}}$. Note that $\vec{\mathbf{a}}$ and $\vec{\mathbf{b}}$ do not have to be perpendicular to one another. For any two vectors that are neither in the same direction nor in opposite directions, there are *two* directions perpendicular to both vectors. To choose between the two, we use a **right-hand rule**.

Using Right-Hand Rule 1 to Find the Direction of a Cross Product $\vec{a} \times \vec{b}$

1. Draw the vectors $\vec{\mathbf{a}}$ and $\vec{\mathbf{b}}$ starting from the same origin (Fig. 19.7a).
2. The cross product is in one of the two directions that are perpendicular to both $\vec{\mathbf{a}}$ and $\vec{\mathbf{b}}$. Determine these two directions.
3. Choose one of these two perpendicular directions to test. Place your right hand in a "karate chop" position with your palm at the origin, your fingertips pointing in the direction of $\vec{\mathbf{a}}$, and your thumb in the direction you are testing (Fig. 19.7b).
4. Keeping the thumb and palm stationary, curl your fingers inward toward your palm until your fingertips point in the direction of $\vec{\mathbf{b}}$ (Fig. 19.7c). If you can do it, sweeping your fingers through an angle less than 180°, then your thumb points in the direction of the cross product $\vec{\mathbf{a}} \times \vec{\mathbf{b}}$. If you can't do it because your fingers would have to sweep through an angle greater than 180°, then your thumb points in the direction *opposite* to $\vec{\mathbf{a}} \times \vec{\mathbf{b}}$.

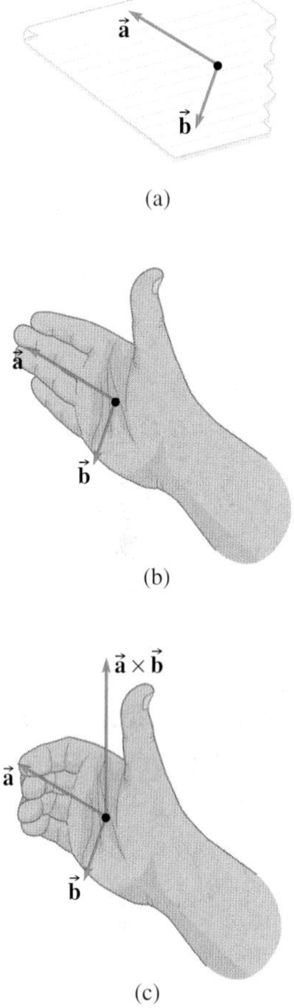

Figure 19.7 Using the right-hand rule to find the direction of a cross product. (a) First we draw the two vectors, $\vec{\mathbf{a}}$ and $\vec{\mathbf{b}}$, starting at the same point. (b) Initial hand position to test whether $\vec{\mathbf{a}} \times \vec{\mathbf{b}}$ is up. The thumb points up and the fingers point along $\vec{\mathbf{a}}$. (c) The fingers are curled in through an angle < 180° until they point along $\vec{\mathbf{b}}$. Therefore, $\vec{\mathbf{a}} \times \vec{\mathbf{b}}$ is up.

Vector symbols: • or ⊙ = out of the page; × or ⊗ = into the page

Since magnetism is inherently three-dimensional, we often need to draw vectors that are perpendicular to the page. The symbol • (or ⊙) represents a vector arrow pointing out of the page; think of the tip of an arrow coming toward you. The symbol × (or ⊗) represents a vector pointing into the page; it suggests the tail feathers of an arrow moving away from you.

Direction of the Magnetic Force

The magnetic force on a charged particle can be written as the charge times the cross product of $\vec{\mathbf{v}}$ and $\vec{\mathbf{B}}$:

Magnetic force on a moving point charge:

$$\vec{\mathbf{F}}_B = q\vec{\mathbf{v}} \times \vec{\mathbf{B}} \qquad (19\text{-}5)$$

$$\text{Magnitude: } F_B = qvB \sin\theta$$

Direction: perpendicular to both $\vec{\mathbf{v}}$ and $\vec{\mathbf{B}}$; use the right-hand rule to find $\vec{\mathbf{v}} \times \vec{\mathbf{B}}$, then reverse it if q is negative.

The direction of the magnetic force is not along the same line as the field (as is the case for the electric field); instead it is *perpendicular*. The force is also perpendicular to the charged particle's velocity. Therefore, if $\vec{\mathbf{v}}$ and $\vec{\mathbf{B}}$ lie in a plane, the magnetic force is always perpendicular to that plane; magnetism is inherently three-dimensional. A negatively charged particle feels a magnetic force in the direction *opposite* to $\vec{\mathbf{v}} \times \vec{\mathbf{B}}$; multiplying a *negative* scalar (q) by $\vec{\mathbf{v}} \times \vec{\mathbf{B}}$ reverses the direction of the magnetic force.

The magnetic force on a point charge is *perpendicular* to the magnetic field and perpendicular to the velocity.

Problem-Solving Technique: Finding the Magnetic Force on a Point Charge

1. The magnetic force is zero if (a) the particle is not moving ($\vec{\mathbf{v}} = 0$), (b) its velocity has no component perpendicular to the magnetic field ($v_\perp = 0$), or (c) the magnetic field is zero.
2. Otherwise, determine the angle θ between the velocity and magnetic field vectors when the two are drawn starting at the same point.
3. Find the magnitude of the force from $F_B = |q|vB \sin \theta$ [Eq. (19-1)], using the *magnitude* of the charge (since magnitudes of vectors are nonnegative).
4. Determine the direction of $\vec{\mathbf{v}} \times \vec{\mathbf{B}}$ using the right-hand rule. The magnetic force is in the direction of $\vec{\mathbf{v}} \times \vec{\mathbf{B}}$ if the charge is positive. If the charge is negative, the force is in the direction *opposite* to $\vec{\mathbf{v}} \times \vec{\mathbf{B}}$.

Work Done by the Magnetic Field Because the magnetic force on a point charge is always perpendicular to the velocity, the magnetic force does no work. If no other forces act on the point charge, then its kinetic energy does not change. The magnetic force, acting alone, changes the *direction* of the velocity *but not the speed* (the magnitude of the velocity).

Conceptual Example 19.1

Deflection of Cosmic Rays

Cosmic rays are charged particles moving toward Earth at high speeds. The origin of the particles is not fully understood, but explosions of supernovae may produce a significant fraction of them. About seven eighths of the particles are protons that move toward Earth with an average speed of about two thirds the speed of light. Suppose that a proton is moving straight down, directly toward the equator. (a) What is the direction of the magnetic force on the proton due to Earth's magnetic field? (b) Explain how Earth's magnetic field shields us from bombardment by cosmic rays. (c) Where on Earth's surface is this shielding least effective?

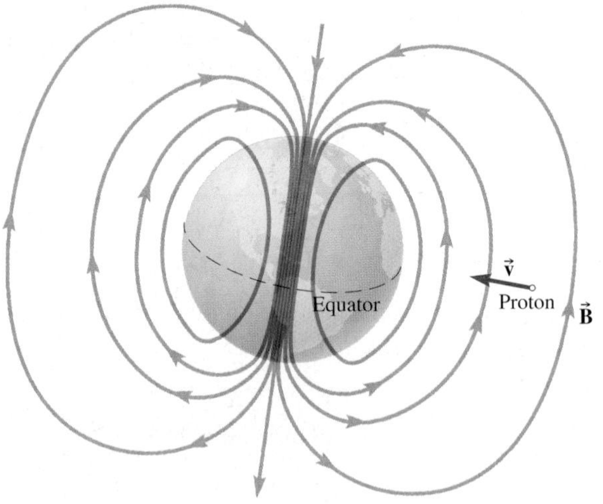

Figure 19.8

A sketch of Earth, its magnetic field lines, and the velocity vector $\vec{\mathbf{v}}$ of the proton.

Strategy and Solution (a) First we sketch Earth's magnetic field lines and the velocity vector for the proton (Fig. 19.8). The field lines run from southern hemisphere to northern; high above the equator, the field is approximately horizontal (due north). To find the direction of the magnetic force, first we determine the two directions that are perpendicular to both $\vec{\mathbf{v}}$ and $\vec{\mathbf{B}}$; then we use the right-hand rule to determine which is the direction of $\vec{\mathbf{v}} \times \vec{\mathbf{B}}$. Figure 19.9 is a sketch of $\vec{\mathbf{v}}$ and $\vec{\mathbf{B}}$ in the xy-plane. The x-axis points away from the equator (up) and the y-axis points north. The two directions that are perpendicular to both vectors are perpendicular to the xy-plane: into the page and out of the page. Using the right-hand rule, if the thumb points out of the page, the fingers of the right hand would have to curl from $\vec{\mathbf{v}}$ to $\vec{\mathbf{B}}$ through an angle of 270°. Therefore, $\vec{\mathbf{v}} \times \vec{\mathbf{B}}$ is into the page (Fig. 19.10). Since $\vec{\mathbf{F}}_B = q\vec{\mathbf{v}} \times \vec{\mathbf{B}}$ and q is positive, the magnetic force is into the page or east.

(b) Without Earth's magnetic field, the proton would move straight down toward Earth's surface. The magnetic field deflects the particle sideways and keeps it from reaching the surface. Many fewer cosmic ray particles reach the surface than would do so if there were no magnetic field.

continued on next page

Conceptual Example 19.1 continued

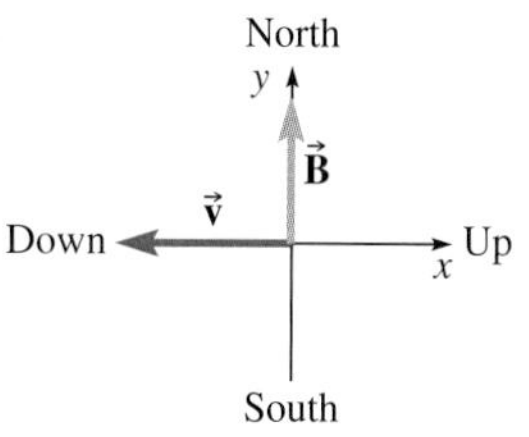

Figure 19.9
The vectors $\vec{v}$ and $\vec{B}$. The y-axis points north; the x-axis points away from the equator.

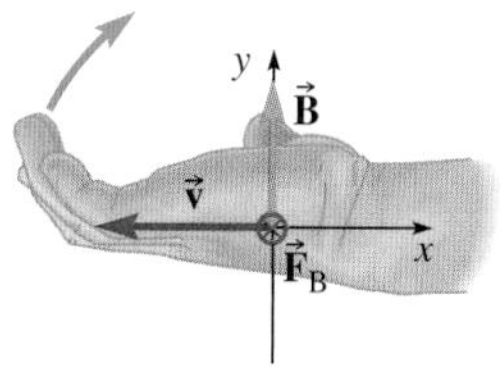

Figure 19.10
The right-hand rule shows that $\vec{v} \times \vec{B}$ is into the page. With the thumb pointing into the page, the fingers sweep from $\vec{v}$ to $\vec{B}$ through an angle of 90°.

(c) Near the poles, the component of $\vec{v}$ perpendicular to the field ($v_\perp$) is a small fraction of v. Since the magnetic force is proportional to $v_\perp$, the deflecting force is much less effective near the poles.

Discussion When finding the direction of the magnetic force (or any cross product), a good sketch is essential. Since all three dimensions come into play, we must choose the two axes that lie in the plane of the sketch. In this example, both vectors $\vec{v}$ and $\vec{B}$ lie in the plane of the sketch, so we know that $\vec{F}_B$ is perpendicular to that plane.

Practice Problem 19.1 Acceleration of Cosmic Ray Particle

If $v = 6.0 \times 10^7$ m/s and $B = 6.0$ μT, what is the magnitude of the magnetic force on the proton and the magnitude of the proton's acceleration?

Example 19.2

Magnetic Force on an Ion in the Air

At a certain place, Earth's magnetic field has magnitude 0.50 mT. The field direction is 70.0° below the horizontal; its horizontal component points due north. (a) Find the magnetic force on an oxygen ion (O_2^-) moving due east at 250 m/s. (b) Compare the magnitude of the magnetic force with the ion's weight, 5.2×10^{-25} N, and to the electric force on it due to Earth's fair-weather electric field (150 N/C downward).

Strategy Since there are two equivalent ways to find the magnitude of the magnetic force [Eq. (19-1)], we choose whichever seems most convenient. To find the direction of the force, first we determine the two directions that are perpendicular to both $\vec{v}$ and $\vec{B}$; then we use the right-hand rule to determine which one is the direction of $\vec{v} \times \vec{B}$. Since we are finding the force on a negatively charged particle, the direction of the magnetic force is *opposite* to the direction of $\vec{v} \times \vec{B}$. Note that the magnitude of the field is specified in *milli*teslas (1 mT = 10^{-3} T).

Solution (a) The ion is moving east; the field has northward and downward components, but no east/west component. Therefore, $\vec{v}$ and $\vec{B}$ are perpendicular; $\theta = 90°$ and $\sin \theta = 1$. The magnitude of the magnetic force is then

$$F = |q|vB = (1.6 \times 10^{-19}\ \text{C}) \times 250\ \text{m/s} \times (5.0 \times 10^{-4}\ \text{T})$$
$$= 2.0 \times 10^{-20}\ \text{N}$$

Since $\vec{v}$ is east and the force must be perpendicular to $\vec{v}$, the force must lie in a plane perpendicular to the east/west axis. We draw the velocity and magnetic field vectors in this plane, using axes that run north/south and up/down (Fig. 19.11a, where east is out of the page). Since north is to the right in this sketch, the viewer looks westward; west is into the page and east is out of the page. The force $\vec{F}$ must lie in this plane and be perpendicular to $\vec{B}$. There are two possible directions, shown with a dashed line in Fig. 19.11a. Now we try these two directions with the right-hand rule; the correct direction for $\vec{v} \times \vec{B}$ is shown in Fig. 19.11b. Since the ion is negatively charged, the magnetic force is in the direction opposite to $\vec{v} \times \vec{B}$; it is 20.0° below the horizontal, with its horizontal component pointing south.

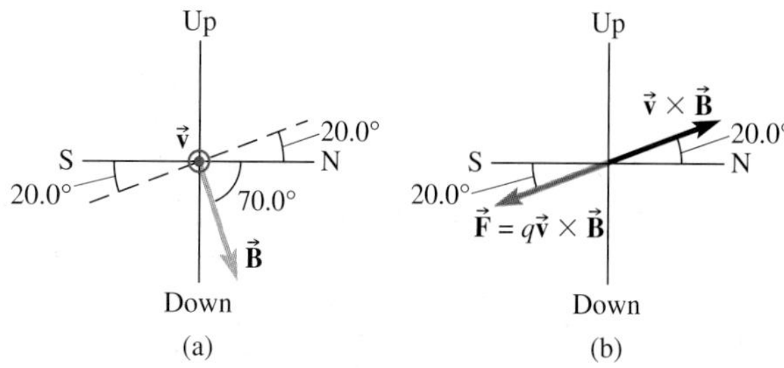

Figure 19.11
(a) The vectors $\vec{v}$ and $\vec{B}$, with $\vec{v}$ out of the page. West is into the page and east is out of the page. Since $\vec{F}$ is perpendicular to both $\vec{v}$ and $\vec{B}$, it must lie along the dashed line. (b) The direction for $\vec{v} \times \vec{B}$ given by the right-hand rule. Since the ion is negatively charged, the magnetic force direction is *opposite* $\vec{v} \times \vec{B}$.

continued on next page

Example 19.2 continued

(b) The electric force has magnitude

$$F_E = |q|E = (1.6 \times 10^{-19}\text{ C}) \times 150\text{ N/C} = 2.4 \times 10^{-17}\text{ N}$$

The magnetic force on the ion is much stronger than the gravitational force and much weaker than the electric force.

Discussion Again, a key to solving this sort of problem is drawing a convenient set of axes. If one of the two vectors $\vec{\mathbf{v}}$ and $\vec{\mathbf{B}}$ lies along a reference direction—a point of the compass, up or down, or along one of the *xyz*-axes—and the other does not, a good choice is to sketch axes in a plane *perpendicular* to that reference direction. In this case, $\vec{\mathbf{v}}$ is in a reference direction (east) but $\vec{\mathbf{B}}$ is not, so we sketch axes in a plane perpendicular to east.

Practice Problem 19.2 Magnetic Force on an Electron

Find the magnetic force on an *electron* moving straight up at 3.0×10^6 m/s in the same magnetic field. [*Hint:* The angle between $\vec{\mathbf{v}}$ and $\vec{\mathbf{B}}$ is *not* 90°.]

Example 19.3

Electron in a Magnetic Field

An electron moves with speed 2.0×10^6 m/s in a uniform magnetic field of 1.4 T directed due north. At one instant, the electron experiences an upward magnetic force of 1.6×10^{-13} N. In what direction is the electron moving at that instant? [*Hint:* If there is more than one possible answer, find all the possibilities.]

Strategy This example is more complicated than Examples 19.1 and 19.2. We need to apply the magnetic force law again, but this time we must deduce the direction of the velocity from the directions of the force and field.

Solution The magnetic force is always perpendicular to both the magnetic field and the particle's velocity. The force is upward, therefore the velocity must lie in a horizontal plane.

Figure 19.12 shows the magnetic field pointing north and a variety of possibilities for the velocity (all in the horizontal plane). The direction of the magnetic force is up, so the direction of $\vec{\mathbf{v}} \times \vec{\mathbf{B}}$ must be down since the charge is negative. Pointing the thumb of the right hand downward, the fingers curl in the clockwise sense. Since we curl from $\vec{\mathbf{v}}$ to $\vec{\mathbf{B}}$, the velocity must be somewhere in the left half of the plane; in other words, it must have a west component in addition to a north or south component.

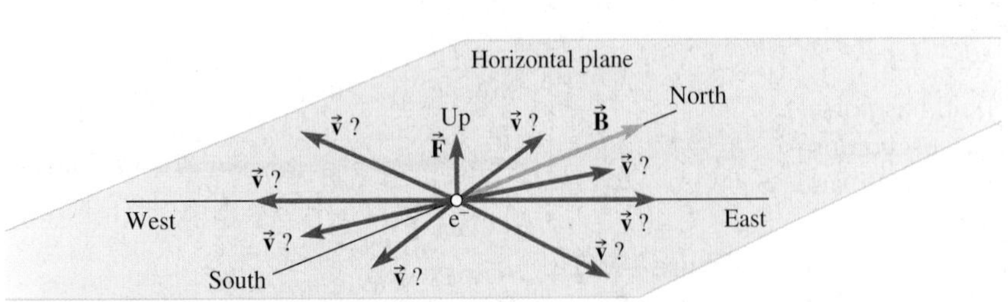

Figure 19.12
The velocity must be perpendicular to the force and thus in the plane shown. Various possibilities for the direction of $\vec{\mathbf{v}}$ are considered. Only those in the west half of the plane give the correct direction for $\vec{\mathbf{v}} \times \vec{\mathbf{B}}$.

The westward component is the component of $\vec{\mathbf{v}}$ that is perpendicular to the field. Using the magnitude of the force, we can find the perpendicular component of the velocity:

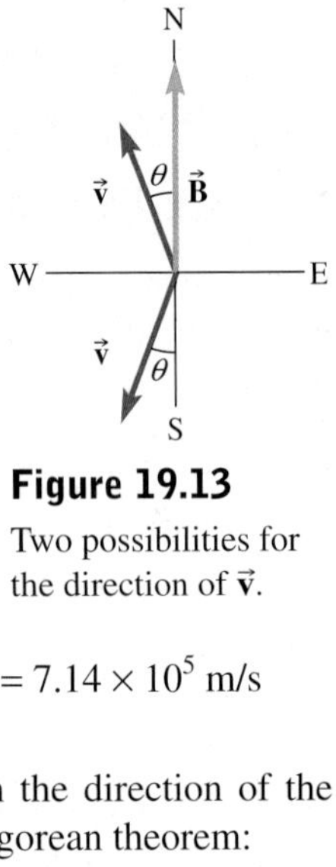

Figure 19.13
Two possibilities for the direction of $\vec{\mathbf{v}}$.

$$F_B = |q|v_\perp B$$

$$v_\perp = \frac{F_B}{|q|B} = \frac{1.6 \times 10^{-13}\text{ N}}{1.6 \times 10^{-19}\text{ C} \times 1.4\text{ T}} = 7.14 \times 10^5\text{ m/s}$$

The velocity also has a component in the direction of the field that can be found using the Pythagorean theorem:

$$v^2 = v_\perp^2 + v_\parallel^2$$

$$v_\parallel = \pm\sqrt{v^2 - v_\perp^2} = \pm 1.87 \times 10^6\text{ m/s}$$

The ± sign would seem to imply that $v_\parallel$ could either be a north or a south component. The two possibilities are shown in Fig. 19.13. Use of the right-hand rule confirms that *either* gives $\vec{\mathbf{v}} \times \vec{\mathbf{B}}$ in the correct direction.

Now we need to find the direction of $\vec{\mathbf{v}}$ given its components. From Fig. 19.13,

$$\sin\theta = \frac{v_\perp}{v} = \frac{7.14 \times 10^5\text{ m/s}}{2.0 \times 10^6\text{ m/s}} = 0.357$$

$$\theta = 21°\text{ W of N or } 159°\text{ W of N}$$

continued on next page

Example 19.3 continued

Since 159° W of N is the same as 21° W of S, the direction of the velocity is either 21° W of N or 21° W of S.

Discussion We *cannot* assume that $\vec{\mathbf{v}}$ is perpendicular to $\vec{\mathbf{B}}$. The magnetic force is always perpendicular to both $\vec{\mathbf{v}}$ and $\vec{\mathbf{B}}$, but there can be any angle between $\vec{\mathbf{v}}$ and $\vec{\mathbf{B}}$.

Practice Problem 19.3 Velocity Component Parallel to the Field

Suppose the electron moves with the same speed in the same magnetic field. If the magnetic force on the electron has magnitude 2.0×10^{-13} N, what is the component of the electron's velocity parallel to the magnetic field?

19.3 CHARGED PARTICLE MOVING PERPENDICULARLY TO A UNIFORM MAGNETIC FIELD

Using the magnetic force law and Newton's second law of motion, we can deduce the trajectory of a charged particle moving in a uniform magnetic field with no other forces acting. In this section, we discuss a case of particular interest: when the particle is initially moving perpendicularly to the magnetic field.

Figure 19.14a shows the magnetic force on a positively charged particle moving perpendicularly to a magnetic field. Since $v_\perp = v$, the magnitude of the force is

$$F = |q|vB \qquad (19\text{-}6)$$

Since the force is perpendicular to the velocity, the particle changes direction but not speed. The force is also perpendicular to the field, so there is no acceleration component in the direction of $\vec{\mathbf{B}}$. Thus, the particle's velocity remains perpendicular to $\vec{\mathbf{B}}$. As the velocity changes direction, the magnetic force changes direction to stay perpendicular to both $\vec{\mathbf{v}}$ and $\vec{\mathbf{B}}$. The magnetic force acts as a steering force, curving the particle around in a trajectory of radius r at constant speed. The particle undergoes uniform circular motion, so its acceleration is directed radially inward and has magnitude v^2/r [Eq. (5-12)]. From Newton's second law,

CONNECTION:

The expression for the radially inward acceleration of a particle in uniform circular motion, $a_r = v^2/r$, is the same one used for other kinds of circular motion.

$$a_r = \frac{v^2}{r} = \frac{\Sigma F}{m} = \frac{|q|vB}{m} \qquad (19\text{-}7)$$

where m is the mass of the particle. Since the radius of the trajectory is constant—r depends only on q, v, B, and m, which are all constant—the particle moves in a circle at constant speed (Fig. 19.14b). Negative charges move in the opposite sense from positive charges in the same field (Fig. 19.14c).

Magnetic fields can cause charges to move along circular paths.

Application: Bubble Chamber

The circular motion of charged particles in uniform magnetic fields has many applications. The *bubble chamber*, invented by American physicist Donald Glaser (1926–), is a particle detector that was used in high-energy physics experiments from the 1950s into the 1970s. The chamber is filled with liquid hydrogen and is immersed in a magnetic field. When a charged particle moves through the liquid, it leaves a trail of bubbles.

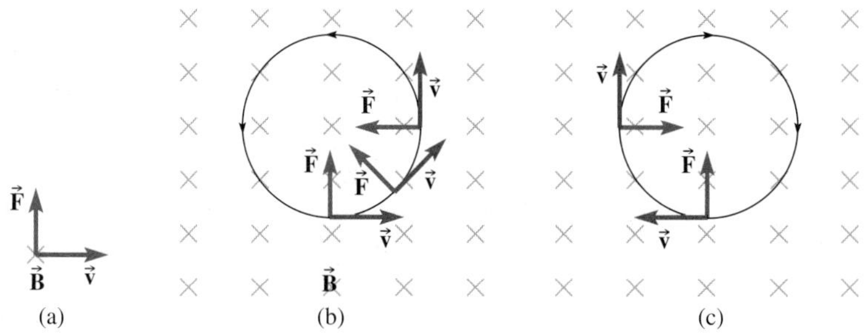

Figure 19.14 (a) Force on a positive charge moving to the right in a magnetic field that is into the page. (b) As the velocity changes direction, the magnetic force changes direction to stay perpendicular to both $\vec{\mathbf{v}}$ and $\vec{\mathbf{B}}$. The force is constant in magnitude, so the particle moves along the arc of a circle. (c) Motion of a negative charge in the same magnetic field.

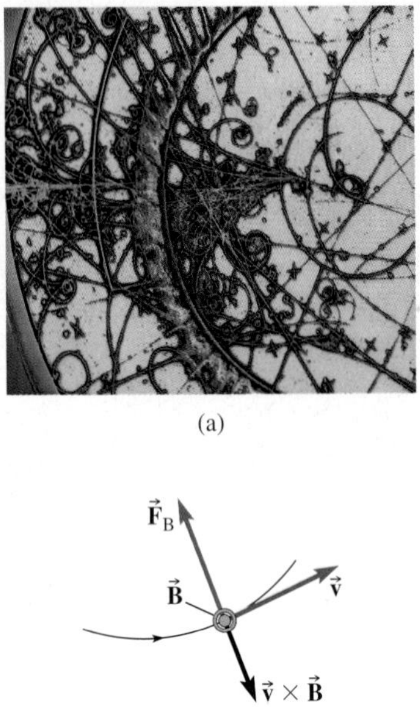

Figure 19.15 (a) Artistically enhanced tracks left by charged particles moving through the BEBC (Big European Bubble Chamber). The tracks are curved due to the presence of a magnetic field. The direction of curvature reveals the sign of the charge. (b) Analysis of the magnetic force on one particular particle. This particle must have a negative charge since the force is opposite in direction to $\vec{v} \times \vec{B}$.

Figure 19.15a shows tracks made by particles in a bubble chamber. The magnetic field is out of the page. The magnetic force on any particle points toward the center of curvature of the particle's trajectory. Figure 19.15b shows the directions of $\vec{v}$ and $\vec{B}$ for one particle. Using the right-hand rule, $\vec{v} \times \vec{B}$ is in the direction shown in Fig. 19.15b. Since $\vec{v} \times \vec{B}$ points away from the center of curvature, which is the direction of $\vec{F}$, the particle must have a negative charge. The magnetic force law lets us determine the sign of the charge on the particle.

Application: Mass Spectrometer

The basic purpose of a *mass spectrometer* is to separate ions (charged atoms or molecules) by mass and measure the mass of each type of ion. Although originally devised to measure the masses of the products of nuclear reactions, mass spectrometers are now used by researchers in many different scientific fields and in medicine to identify what atoms or molecules are present in a sample and in what concentrations. Even ions present in minute concentrations can be isolated, making the mass spectrometer an essential tool in toxicology and in monitoring the environment for trace pollutants. Mass spectrometers are used in food production, petrochemical production, the electronics industry, and in the international monitoring of nuclear facilities. They are also an important tool for investigations of crime scenes, as several popular TV shows demonstrate weekly.

Today, many different types of mass spectrometer are in use. The oldest type, now called a magnetic-sector mass spectrometer, is based on the circular motion of a charged particle in a magnetic field. The atoms or molecules are first ionized so that they have a known electric charge. They are then accelerated by an electric field that can be varied to adjust their speeds. The particles then enter a region of uniform magnetic field $\vec{B}$ oriented perpendicular to their velocities $\vec{v}$ so that they move in circular arcs. From the charge, speed, magnetic field strength, and radius of the circular arc, we can determine the mass of the particle.

In some magnetic-sector spectrometers, the ions start at rest or at low speed and are accelerated through a fixed potential difference. If the ions all have the same charge, then they all have the *same kinetic energy* when they enter the magnetic field but, if they have different masses, their speeds are not all the same. Another possibility is to use a *velocity selector* (Section 19.5) to make sure that all ions, regardless of mass or charge, have the same *speed* when they enter the magnetic field. In the spectrometer of Example 19.4, ions of different masses travel in circular paths of different radii (Fig. 19.16a). In other spectrometers, only ions that travel along a path of *fixed radius* reach the detector; either the speed of the ions or the magnetic field is varied to select which ions move with the correct radius (Fig. 19.16b).

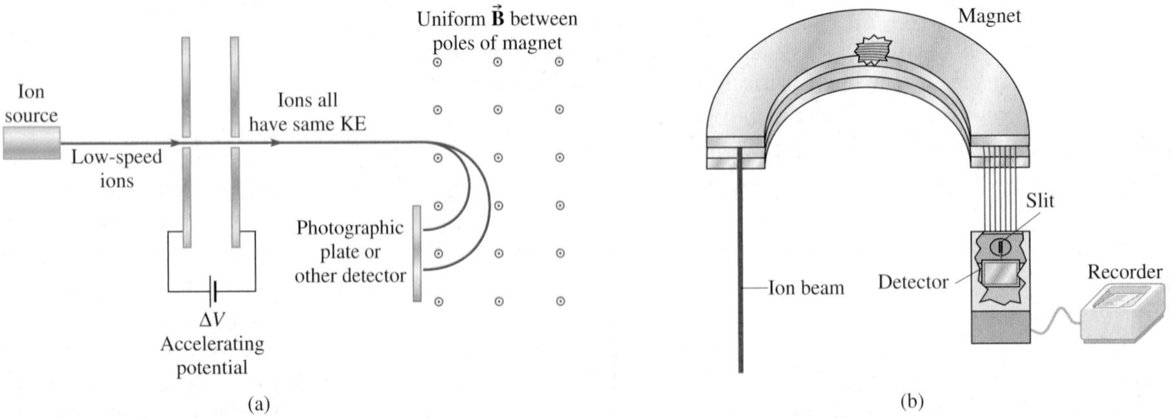

Figure 19.16 (a) A simplified diagram of a magnetic-sector mass spectrometer that accelerates ions through a fixed potential difference so that they all enter the magnetic field with the same kinetic energy. (b) A mass spectrometer in which ions travel around a path of fixed radius.

Example 19.4

Separation of Lithium Ions in a Mass Spectrometer

In a mass spectrometer, a beam of $^6Li^+$ and $^7Li^+$ ions passes through a velocity selector so that the ions all have the same velocity. The beam then enters a region of uniform magnetic field. If the radius of the orbit of the $^6Li^+$ ions is 8.4 cm, what is the radius of the orbit of the $^7Li^+$ ions?

Strategy Much of the information in this problem is implicit. The charge of the $^6Li^+$ ions is the same as the charge of the $^7Li^+$ ions. The ions enter the magnetic field with the same speed. We do not know the magnitudes of the charge, velocity, or magnetic field, but they are the same for the two types of ion. With so many common quantities, a good strategy is to try to find the *ratio* between the radii for the two types of ion so that the common quantities cancel out.

Solution From Appendix B we find the masses of $^6Li^+$ and $^7Li^+$:

$$m_6 = 6.015 \text{ u}$$

$$m_7 = 7.016 \text{ u}$$

where $1 \text{ u} = 1.66 \times 10^{-27}$ kg. We now apply Newton's second law to an ion moving in a circle. The acceleration is that of uniform circular motion:

$$a_\perp = \frac{v^2}{r} = \frac{F}{m} = \frac{|q|vB}{m} \qquad (1)$$

Since the charge q, the speed v, and the field B are the same for both types of ion, the radius must be directly proportional to the mass.

$$r \propto m$$

$$\frac{r_7}{r_6} = \frac{m_7}{m_6} = \frac{7.016 \text{ u}}{6.015 \text{ u}} = 1.166$$

$$r_7 = 8.4 \text{ cm} \times 1.166 = 9.8 \text{ cm}$$

Discussion To solve this sort of problem, there aren't any new formulas to learn. We apply Newton's second law with the net force given by the magnetic force law ($\vec{\mathbf{F}}_B = q\vec{\mathbf{v}} \times \vec{\mathbf{B}}$) and the magnitude of the radial acceleration being what it always is for uniform circular motion (v^2/r).

If the direct proportion between r and m is not apparent, we could proceed by solving (1) for the radius:

$$r = \frac{mv^2}{|q|vB}$$

Now, if we set up a ratio between r_7 and r_6, all quantities except the masses cancel, yielding

$$\frac{r_7}{r_6} = \frac{m_7}{m_6}$$

Practice Problem 19.4 Ion Speed

The magnetic field strength used in the mass spectrometer of Example 19.4 is 0.50 T. At what speed do the Li^+ ions move through the magnetic field? (Each ion has charge $q = +e$ and moves perpendicular to the field.)

Application: Cyclotrons

Another device that was originally used in experimental physics but is now used frequently in the life sciences and medicine is the *cyclotron*, invented in 1929 by American physicist Ernest O. Lawrence (1901–1958). Figure 19.17 shows a schematic diagram of a proton cyclotron. The two hollow metal shells are called *dees* after their shape (like the letter "D"). An alternating potential difference is maintained between the dees to accelerate the protons. When the protons are inside one of the dees, there is no electric field acting on them; inside the conductor they are all at the same potential. However, the uniform magnetic field causes the protons to travel in a circular arc at constant speed. The potential difference alternates so that, whenever a proton reaches the gap between the dees, the dee toward which it moves is at a lower potential. Thus, the electric field between the dees gives the proton a little kick every time it crosses the gap. As the proton speed increases, the radius of its path increases. When protons reach the maximum radius of the dees, they are taken out of the cyclotron and the high-energy proton beam is used to bombard some target.

As the protons increase their speed and kinetic energy, the time it takes them to move around one complete circle stays constant (see Problem 33). As the speed

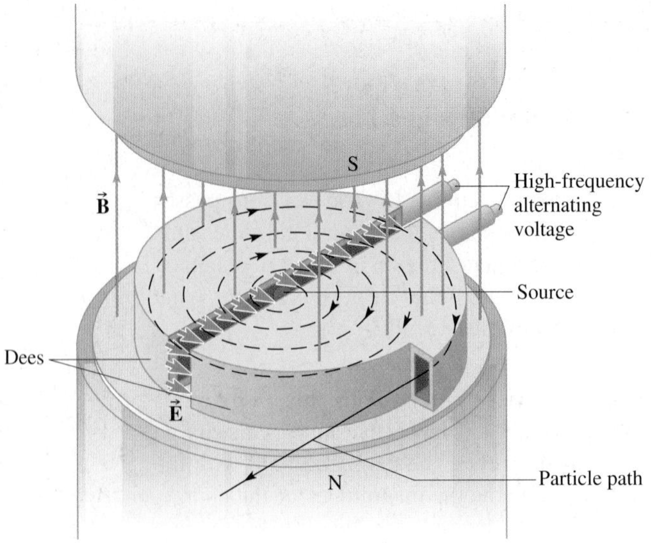

Figure 19.17 Schematic view of a cyclotron.

increases, so does the distance they travel (the circumference of the circular path), and therefore the time for one revolution stays the same. The potential difference between the dees can then be made to alternate at this same frequency (the *cyclotron frequency*) so that the protons gain rather than lose kinetic energy each time they cross the gap.

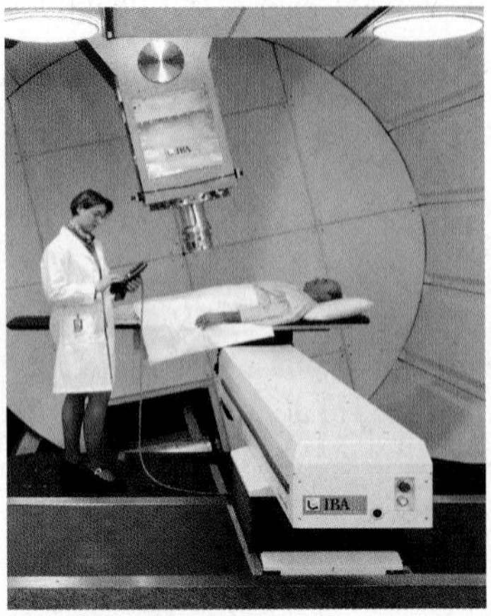

Figure 19.18 A patient is prepared for surgery at the Northeast Proton Therapy Center of Massachusetts General Hospital. The protons are accelerated by a cyclotron (not shown).

Medical Uses of Cyclotrons In hospitals, cyclotrons produce some of the radioisotopes used in nuclear medicine. While nuclear reactors also produce medical radioisotopes, cyclotrons offer certain advantages. For one thing, a cyclotron is much easier to operate and is much smaller—typically 1 m or less in radius. A cyclotron can be located in or adjacent to a hospital so that short-lived radioisotopes can be produced as they are needed. It would be difficult to try to produce short-lived isotopes in a nuclear reactor and transport them to the hospital fast enough for them to be useful. Cyclotrons also tend to produce different kinds of isotopes than do nuclear reactors.

Another medical use of the cyclotron is *proton beam radiosurgery*, in which the cyclotron's proton beam is used as a surgical tool (Fig. 19.18). Proton beam radiosurgery offers advantages over surgical and other radiological methods in the treatment of unusually shaped brain tumors. For one thing, doses to the surrounding tissue are much lower than with other forms of radiosurgery.

Example 19.5

Maximum Kinetic Energy in a Proton Cyclotron

A proton cyclotron uses a magnet that produces a 0.60-T field between its poles. The radius of the dees is 24 cm. What is the maximum possible kinetic energy of the protons accelerated by this cyclotron?

Strategy As a proton's kinetic energy increases, so does the radius of its path in the dees. The maximum kinetic energy is therefore determined by the maximum radius.

Solution While in the dees, the only force acting on the proton is magnetic. First we apply Newton's second law to a circular path.

$$F = |q|vB = \frac{mv^2}{r}$$

We can solve for v:

$$v = \frac{|q|Br}{m}$$

continued on next page

Example 19.5 continued

From v, we calculate the kinetic energy:

$$K = \frac{1}{2}mv^2 = \frac{1}{2}m\left(\frac{|q|Br}{m}\right)^2$$

For a proton, $q = +e$. The magnetic field strength is $B = 0.60$ T. For the maximum kinetic energy, we set the radius to its maximum value $r = 0.24$ m.

$$K = \frac{(qBr)^2}{2m} = \frac{(1.6 \times 10^{-19}\ \text{C} \times 0.60\ \text{T} \times 0.24\ \text{m})^2}{2 \times 1.67 \times 10^{-27}\ \text{kg}}$$

$$= 1.6 \times 10^{-13}\ \text{J}$$

Discussion Just as in Example 19.4 (the mass spectrometer), this cyclotron problem is solved using Newton's second law. Once again the net force on the moving charge is given by the magnetic force law and the radial acceleration has magnitude v^2/r for motion at constant speed along the arc of a circle.

Practice Problem 19.5 Increasing Kinetic Energy in a Proton Cyclotron

Using the same magnetic field, what would the radius of the dees have to be to accelerate the protons to a kinetic energy of 1.6×10^{-12} J (ten times the previous value)?

19.4 MOTION OF A CHARGED PARTICLE IN A UNIFORM MAGNETIC FIELD: GENERAL

What is the trajectory of a charged particle moving in a uniform magnetic field with no other forces acting? In Section 19.3, we saw that the trajectory is a circle *if* the velocity is perpendicular to the magnetic field. If $\vec{\mathbf{v}}$ has no perpendicular component, the magnetic force is zero and the particle moves at constant velocity.

In general, the velocity may have components both perpendicular to and parallel to the magnetic field. The component parallel to the field is constant, since the magnetic force is always perpendicular to the field. The particle therefore moves along a *helical* path (Text website interactive: magnetic fields). The helix is formed by circular motion of the charge in a plane perpendicular to the field superimposed onto motion of the charge at constant speed along a field line (Fig. 19.19a).

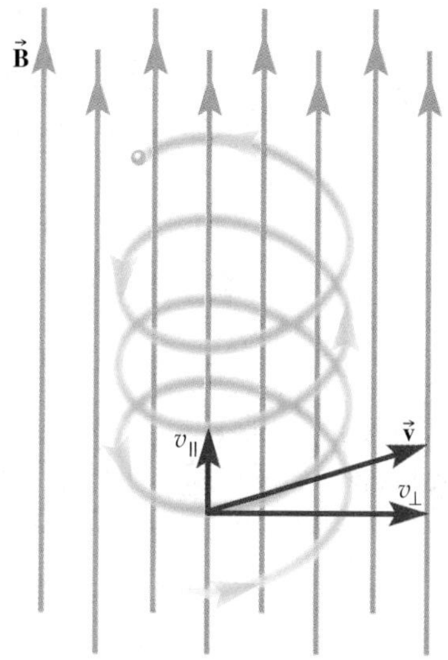

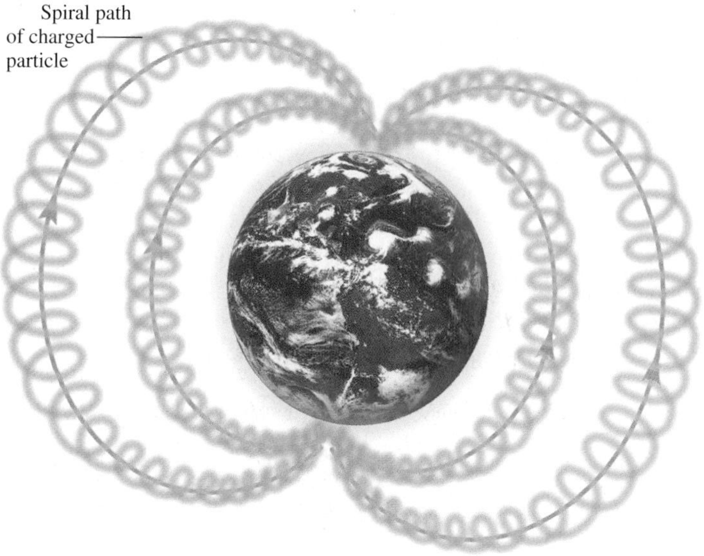

Figure 19.19 (a) Helical motion of a charged particle in a uniform magnetic field. (b) Charged particles spiral back and forth along field lines high above the atmosphere.

CHECKPOINT 19.4

A particle's helical motion is shown in Fig. 19.19a. Is the particle positively or negatively charged? Explain.

Even in nonuniform fields, charged particles tend to spiral around magnetic field lines. Above Earth's surface, charged particles from cosmic rays and the solar wind (charged particles streaming toward Earth from the Sun) are trapped by Earth's magnetic field. The particles spiral back and forth along magnetic field lines (Fig. 19.19b). Near the poles, the field lines are closer together, so the field is stronger. As the field strength increases, the radius of a spiraling particle's path gets smaller and smaller. As a result, there is a concentration of these particles near the poles. The particles collide with and ionize air molecules. When the ions recombine with electrons to form neutral atoms, visible light is emitted—the *aurora borealis* in the northern hemisphere and the *aurora australis* in the southern hemisphere. Aurorae also occur on Jupiter and Saturn, which have much stronger magnetic fields than does Earth.

Application: aurorae on Earth, Jupiter, and Saturn

19.5 A CHARGED PARTICLE IN CROSSED $\vec{\mathbf{E}}$ AND $\vec{\mathbf{B}}$ FIELDS

If a charged particle moves in a region of space where both electric and magnetic fields are present, then the electromagnetic force on the particle is the vector sum of the electric and magnetic forces:

$$\vec{\mathbf{F}} = \vec{\mathbf{F}}_E + \vec{\mathbf{F}}_B \quad (19\text{-}8)$$

A particularly important and useful case is when the electric and magnetic fields are perpendicular to one another and the velocity of a charged particle is perpendicular to both fields. Since the magnetic force is always perpendicular to both $\vec{\mathbf{v}}$ and $\vec{\mathbf{B}}$, it must be either in the same direction as the electric force or in the opposite direction. If the magnitudes of the two forces are the same and the directions are opposite, then there is zero net force on the charged particle (Fig. 19.20). For any particular combination of electric and magnetic fields, this balance of forces occurs only for one particular particle speed, since the magnetic force is velocity-dependent, but the electric force is not. The velocity that gives zero net force can be found from

The magnetic force is velocity-dependent, but the electric force is not.

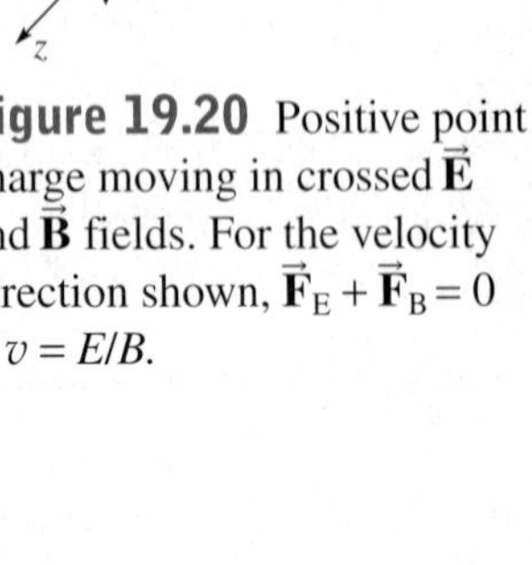

Figure 19.20 Positive point charge moving in crossed $\vec{\mathbf{E}}$ and $\vec{\mathbf{B}}$ fields. For the velocity direction shown, $\vec{\mathbf{F}}_E + \vec{\mathbf{F}}_B = 0$ if $v = E/B$.

$$\vec{\mathbf{F}} = \vec{\mathbf{F}}_E + \vec{\mathbf{F}}_B = 0$$

$$q\vec{\mathbf{E}} + q\vec{\mathbf{v}} \times \vec{\mathbf{B}} = 0$$

Dividing out the common factor of q,

$$\vec{\mathbf{E}} + \vec{\mathbf{v}} \times \vec{\mathbf{B}} = 0 \quad (19\text{-}9)$$

There is zero net force on the particle only if

$$v = \frac{E}{B} \quad (19\text{-}10)$$

and if the direction of $\vec{\mathbf{v}}$ is correct. Since $\vec{\mathbf{E}} = -\vec{\mathbf{v}} \times \vec{\mathbf{B}}$, it can be shown (see Conceptual Question 7) that the correct direction of $\vec{\mathbf{v}}$ is the direction of $\vec{\mathbf{E}} \times \vec{\mathbf{B}}$.

CHECKPOINT 19.5

An electron moves straight up in a region where the electric field is east and the magnetic field is north. (a) What is the direction of the electric force on the electron? (b) What is the direction of the magnetic force on the electron?

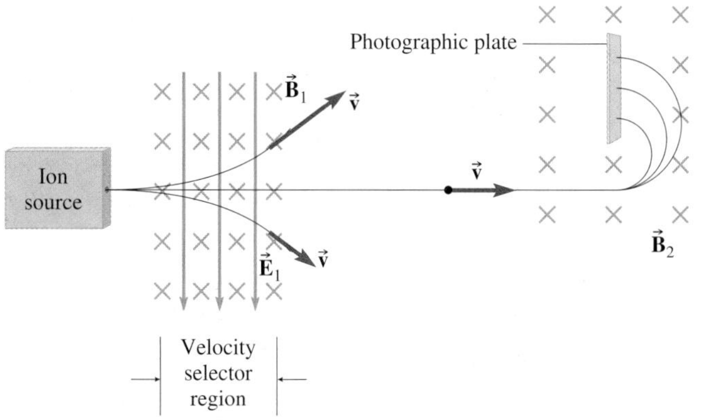

Figure 19.21 A mass spectrometer that uses a velocity selector to ensure that all ions enter the second magnetic field with the same speed. Both magnetic fields are into the page.

Application: Velocity Selector

A **velocity selector** uses crossed electric and magnetic fields to select a single velocity out of a beam of charged particles. Suppose a beam of ions is produced in the first stage of a mass spectrometer. The beam may contain ions moving at a range of different speeds. If the second stage of the mass spectrometer is a velocity selector (Fig. 19.21), only ions moving at a single speed $v = E/B$ pass through the velocity selector and into the third stage. The speed can be selected by adjusting the magnitudes of the electric and magnetic fields. For particles moving *faster* than the selected speed, the magnetic force is stronger than the electric force; fast particles curve out of the beam in the direction of the magnetic force. For particles moving *slower* than the selected speed, the magnetic force is weaker than the electric force; slow particles curve out of the beam in the direction of the electric force. The velocity selector ensures that only ions with speeds very near $v = E/B$ enter the magnetic sector of the mass spectrometer.

Example 19.6

Velocity Selector

A velocity selector is to be constructed to select ions moving to the right at 6.0 km/s. The electric field is 300.0 V/m into the page. What should be the magnitude and direction of the magnetic field?

Strategy First, in a velocity selector, $\vec{\mathbf{E}}$, $\vec{\mathbf{B}}$, and $\vec{\mathbf{v}}$ are mutually perpendicular. That allows only two possibilities for the direction of $\vec{\mathbf{B}}$. Setting the magnetic force equal and opposite to the electric force determines which of the two directions is correct and gives the magnitude of $\vec{\mathbf{B}}$. The magnitude of the magnetic field is chosen so that the electric and magnetic forces on a particle moving at the given speed are exactly opposite.

Solution Since $\vec{\mathbf{v}}$ is to the right and $\vec{\mathbf{E}}$ is into the page, the magnetic field must either be up or down. The sign of the ions' charge is irrelevant—changing the charge from positive to negative would change the directions of *both* forces, leaving them still opposite to each other. For simplicity, then, we assume the charge to be positive.

The direction of the electric force on a positive charge is the same as the direction of the field, which here is into the page. Then we need a magnetic force that is out of the page. Using the right-hand rule to evaluate both possibilities for $\vec{\mathbf{B}}$ (up and down), we find that $\vec{\mathbf{v}} \times \vec{\mathbf{B}}$ is out of the page if $\vec{\mathbf{B}}$ is up (Fig. 19.22).

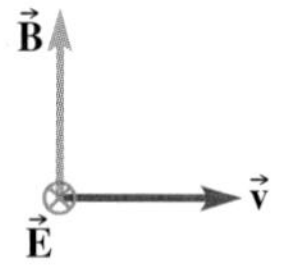

Figure 19.22 Directions of $\vec{\mathbf{E}}$, $\vec{\mathbf{v}}$, and $\vec{\mathbf{B}}$.

The magnitudes of the forces must also be equal:

$$qE = qvB$$

$$B = \frac{E}{v} = \frac{300.0 \text{ V/m}}{6000 \text{ m/s}} = 0.050 \text{ T}$$

Discussion Let's check the units; is a tesla really equal to (V/m)/(m/s)? From $\vec{\mathbf{F}} = q\vec{\mathbf{v}} \times \vec{\mathbf{B}}$, we can reconstruct the tesla:

$$[B] = \text{T} = \left[\frac{F}{qv}\right] = \frac{\text{N}}{\text{C·m/s}}$$

continued on next page

Example 19.6 continued

Recall that two equivalent units for electric field are N/C = V/m. By substitution,

$$\mathrm{T} = \frac{\mathrm{V}}{\mathrm{m^2/s}} = \frac{\mathrm{V/m}}{\mathrm{m/s}}$$

so the units check out.

Another check: for a velocity selector the correct direction of $\vec{\mathbf{v}}$ is the direction of $\vec{\mathbf{E}} \times \vec{\mathbf{B}}$. The velocity is to the right. Using the right-hand rule, $\vec{\mathbf{E}} \times \vec{\mathbf{B}}$ is to the right if $\vec{\mathbf{B}}$ is up.

Practice Problem 19.6 Deflection of a Particle Moving Too Fast

If a particle enters this velocity selector with a speed greater than 6.0 km/s, in what direction is it deflected out of the beam?

The velocity selector can be used to determine the charge-to-mass ratio q/m of a charged particle. First, the particle is accelerated from rest through a potential difference ΔV, converting electric potential energy into kinetic energy. The change in its electric potential energy is $\Delta U = q\,\Delta V$, so the charge acquires a kinetic energy

$$K = \tfrac{1}{2}mv^2 = -q\,\Delta V$$

(K is positive regardless of the sign of q: a positive charge is accelerated by decreasing its potential, while a negative charge is accelerated by increasing its potential.) Now a velocity selector is used to determine the speed $v = E/B$, by adjusting the electric and magnetic fields until the particles pass straight through. The charge-to-mass ratio q/m can now be determined (see Problem 43). In 1897, British physicist Joseph John Thomson (1856–1940) used this technique to show that "cathode rays" are charged particles. In a vacuum tube, he maintained two electrodes at a potential difference of a few thousand volts (Fig. 19.23) so that cathode rays were emitted by the negative electrode (the cathode). By measuring the charge-to-mass ratio, Thomson established that cathode rays are streams of negatively charged particles that all have the same charge-to-mass ratio—particles we now call *electrons*.

Application: Electromagnetic Blood Flowmeter

The principle of the velocity selector finds another application in the electromagnetic flowmeters used to measure the speed of blood flow through a major artery during cardiovascular surgery. Blood contains ions; the motion of the ions can be affected by a magnetic field. In an electromagnetic flowmeter, a magnetic field is applied perpendicular to

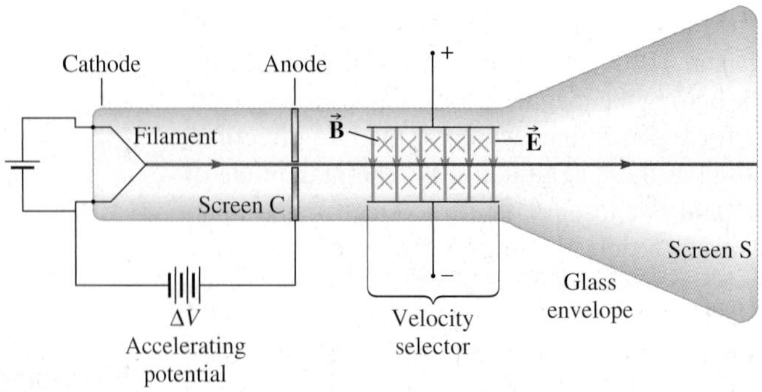

Figure 19.23 Modern apparatus, similar in principle to the one used by Thomson, to find the charge-to-mass ratio of the electron. Electrons emitted from the cathode are accelerated toward the anode by the electric field between the two. Some of the electrons pass through the anode and then enter a velocity selector. The deflection of the electrons is viewed on the screen. The electric and magnetic fields in the velocity selector are adjusted until the electrons are not deflected.

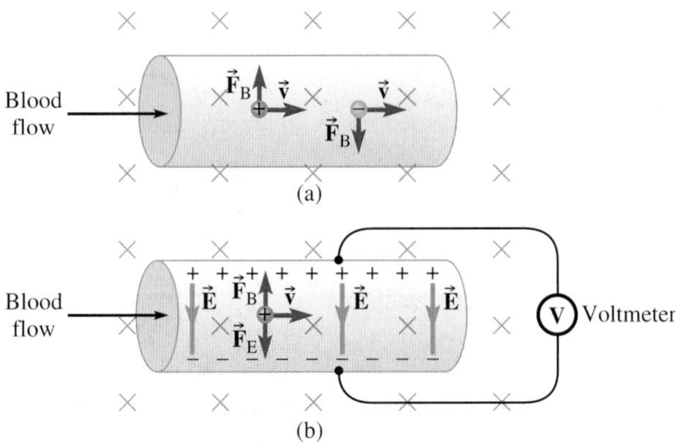

Figure 19.24 Principles behind the electromagnetic blood flowmeter. (a) When a magnetic field is applied perpendicular to the direction of blood flow, positive and negative ions are deflected toward opposite sides of the artery. (b) As the ions are deflected, an electric field develops across the artery. In equilibrium, the electric force on an ion due to this field is equal and opposite to the magnetic force; the ions move straight down the artery with an average velocity of magnitude $v = E/B$.

the flow direction. The magnetic force on positive ions is toward one side of the artery, while the magnetic force on negative ions is toward the opposite side (Fig. 19.24a). This separation of charge, with positive charge on one side and negative charge on the other, produces an electric field across the artery (Fig. 19.24b). As the electric field builds up, it exerts a force on moving ions in a direction opposite to that of the magnetic field. In equilibrium, the two forces are equal in magnitude:

$$F_E = F_B$$

$$qE = qvB$$

$$E = vB$$

where v is the average speed of an ion, equal to the average speed of the blood flow. Thus, the flowmeter is just like a velocity selector, except that the ion speed determines the electric field instead of the other way around.

A voltmeter is attached to opposite sides of the artery to measure the potential difference. From the potential difference, we can calculate the electric field; from the electric field and magnetic field magnitudes, we can determine the speed of blood flow. A great advantage of the electromagnetic flowmeter is that it does not involve inserting anything into the artery.

Application: The Hall Effect

The **Hall effect** (named after Edwin Herbert Hall, 1855–1938) is similar in principle to the electromagnetic flowmeter, but pertains to the moving charges in a current-carrying wire or other solid, not to moving ions in blood. A magnetic field perpendicular to the wire causes the moving charges to be deflected to one side. This charge separation causes an electric field across the wire. The potential difference (or **Hall voltage**) *across* the wire is measured and used to calculate the electric field (or **Hall field**) across the wire. The drift velocity of the charges is then given by $v_D = E/B$. The Hall effect enables the measurement of the drift velocity and the determination of the sign of the charges. (The carriers in metals are generally electrons, but semiconductors may have positive or negative carriers or both.)

The Hall effect is also the principle behind the **Hall probe**, a common device used to measure magnetic fields. As shown in Example 19.7, the Hall voltage across a conducting strip is proportional to the magnetic field strength. A circuit causes a fixed current flow through the strip. The probe is then calibrated by measuring the Hall voltage caused by magnetic fields of known strength. Once calibrated, measurement of the Hall voltage enables a quick and accurate determination of magnetic field strengths.

Example 19.7

Hall Effect

A flat slab of semiconductor has thickness $t = 0.50$ mm, width $w = 1.0$ cm, and length $L = 30.0$ cm. A current $I = 2.0$ A flows along its length to the right (Fig. 19.25). A magnetic field $B = 0.25$ T is directed into the page, perpendicular to the flat surface of the slab. Assume that the carriers are electrons. There are 7.0×10^{24} mobile electrons per m^3. (a) What is the magnitude of the Hall voltage across the slab? (b) Which edge (top or bottom) is at the higher potential?

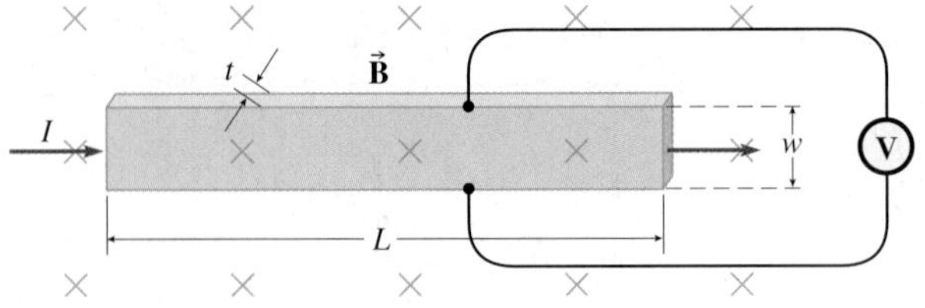

Figure 19.25
Measuring the Hall voltage.

Strategy We need to find the drift velocity of the electrons from the relation between current and drift velocity. Since the Hall field is uniform, the Hall voltage is the Hall field times the width of the slab.

Given: current $I = 2.0$ A, magnetic field $B = 0.25$ T, thickness $t = 0.50 \times 10^{-3}$ m, width $w = 0.010$ m, $n = 7.0 \times 10^{24}$ electrons/m^3

Solution (a) The drift velocity is related to the current:

$$I = neAv_D \qquad (18\text{-}3)$$

The area is the width times the thickness of the slab:

$$A = wt$$

Solving for the drift velocity,

$$v_D = \frac{I}{newt}$$

We find the Hall field by setting the magnitude of the magnetic force equal to the magnitude of the electric force caused by the Hall field across the slab:

$$F_E = eE_H = F_B = ev_D B$$

$$E_H = v_D B$$

The Hall voltage is

$$V_H = E_H w = Bv_D w$$

Substituting the expression for drift velocity,

$$V_H = \frac{BIw}{newt} = \frac{BI}{net}$$

$$= \frac{0.25\ \text{T} \times 2.0\ \text{A}}{7.0 \times 10^{24}\ \text{m}^{-3} \times 1.6 \times 10^{-19}\ \text{C} \times 0.50 \times 10^{-3}\ \text{m}}$$

$$= 0.89\ \text{mV}$$

(b) Since the current flows to the right, the electrons actually move to the left. Figure 19.26a shows that the magnetic force on an electron moving to the left is upward. The magnetic force deflects electrons toward the top of the slab, leaving the bottom with a positive charge. An upward electric field is set up across the slab (Fig. 19.26b). Therefore, the bottom edge is at the higher potential.

Discussion The width of the slab w does not appear in the final expression for the Hall voltage $V_H = BI/(net)$. Is it possible that the Hall voltage is independent of the width? If the slab were twice as wide, for instance, the same current means half the drift velocity v_D since the number of carriers per unit volume n and their charge magnitude e cannot change. With the carriers moving half as fast on average, the average magnetic force is half. Then in equilibrium, the electric force is half, which means the field is half. An electric field half as strong times a width twice as wide gives the same Hall voltage.

Practice Problem 19.7 Holes as Carriers

If the carriers had been particles with charge $+e$ instead of electrons, with everything else the same, would the Hall voltage have been any different? Explain.

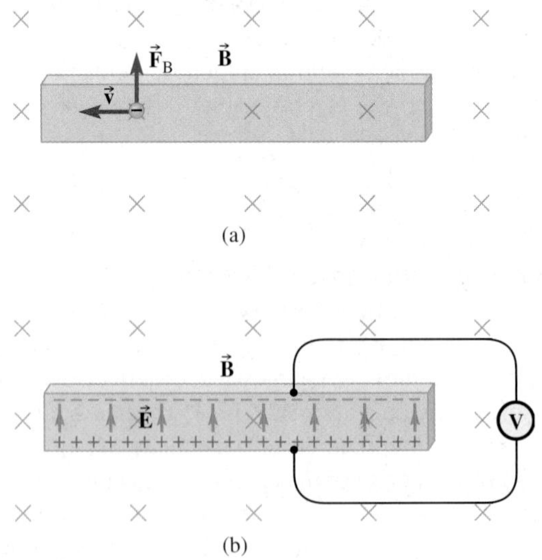

Figure 19.26
(a) Magnetic force on an electron moving to the left. (b) With electrons deflected toward the top of the slab, the top is negatively charged and the bottom is positively charged. The Hall field in this case is directed upward, from the positive charges to the negative charges.

19.6 MAGNETIC FORCE ON A CURRENT-CARRYING WIRE

A wire carrying electric current has many moving charges in it. For a current-carrying wire in a magnetic field, the magnetic forces on the individual moving charges add up to produce a net magnetic force on the wire. Although the average force on one of the charges may be small, there are so many charges that the net magnetic force on the wire can be appreciable.

Say a straight wire segment of length L in a uniform magnetic field $\vec{\mathbf{B}}$ carries a current I. The mobile carriers have charge q. The magnetic force on any one charge is

$$\vec{\mathbf{F}} = q\vec{\mathbf{v}} \times \vec{\mathbf{B}}$$

where $\vec{\mathbf{v}}$ is the instantaneous velocity of that charge. The net magnetic force on the wire is the vector sum of these forces. The sum isn't easy to carry out, since we don't know the instantaneous velocity of each of the charges. The charges move about in random directions at high speeds; their velocities suffer large changes when they collide with other particles. Instead of summing the instantaneous magnetic force on each charge, we can instead multiply the *average* magnetic force on each charge by the number of charges. Since each charge has the same average velocity—the drift velocity—each experiences the same average magnetic force $\vec{\mathbf{F}}_{av}$.

$$\vec{\mathbf{F}}_{av} = q\vec{\mathbf{v}}_D \times \vec{\mathbf{B}}$$

Then, if N is the total number of carriers in the wire, the total magnetic force on the wire is

$$\vec{\mathbf{F}} = Nq\vec{\mathbf{v}}_D \times \vec{\mathbf{B}} \qquad (19\text{-}11)$$

Equation (19-11) can be rewritten in a more convenient way. Instead of having to figure out the number of carriers and the drift velocity, it is more convenient to have an expression that gives the magnetic force in terms of the current I. The current I is related to the drift velocity:

$$I = nqAv_D \qquad (18\text{-}3)$$

Here n is the number of carriers *per unit volume*. If the length of the wire is L and the cross-sectional area is A, then

$$N = \text{number per unit volume} \times \text{volume} = nLA$$

By substitution, the magnetic force on the wire can be written

$$\vec{\mathbf{F}} = Nq\vec{\mathbf{v}}_D \times \vec{\mathbf{B}} = nqAL\vec{\mathbf{v}}_D \times \vec{\mathbf{B}}$$

Almost there! Since current is not a vector, we cannot substitute $\vec{\mathbf{I}} = nqA\vec{\mathbf{v}}_D$. Therefore, we define a *length vector* $\vec{\mathbf{L}}$ to be a vector in the direction of the current with magnitude equal to the length of the wire (Fig. 19.27). Then $nqAL\vec{\mathbf{v}}_D = I\vec{\mathbf{L}}$ and

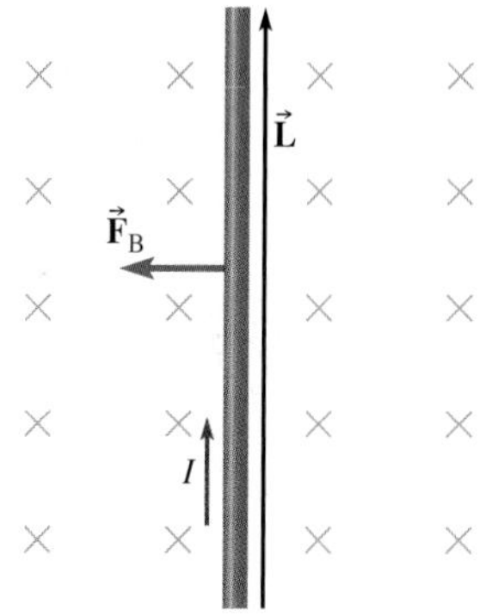

Figure 19.27 A current-carrying wire in an externally applied magnetic field experiences a magnetic force.

Magnetic force on a straight segment of current-carrying wire:

$$\vec{\mathbf{F}} = I\vec{\mathbf{L}} \times \vec{\mathbf{B}} \qquad (19\text{-}12a)$$

CONNECTION:

The magnetic force on a current-carrying wire is the sum of the magnetic forces on the charge carriers in the wire.

The current I times the cross product $\vec{\mathbf{L}} \times \vec{\mathbf{B}}$ gives the magnitude and direction of the force. The magnitude of the force is

$$F = IL_\perp B = ILB_\perp = ILB \sin \theta \qquad (19\text{-}12b)$$

The direction of the force is perpendicular to both $\vec{\mathbf{L}}$ and $\vec{\mathbf{B}}$. The same right-hand rule used for any cross product is used to choose between the two possibilities.

Problem-Solving Technique: Finding the Magnetic Force on a Straight Segment of Current-Carrying Wire

1. The magnetic force is zero if (a) the current in the wire is zero, (b) the wire is parallel to the magnetic field, or (c) the magnetic field is zero.
2. Otherwise, determine the angle θ between $\vec{\mathbf{L}}$ and $\vec{\mathbf{B}}$ when the two are drawn starting at the same point.
3. Find the magnitude of the force from Eq. (19-12b).
4. Determine the direction of $\vec{\mathbf{L}} \times \vec{\mathbf{B}}$ using the right-hand rule.

CHECKPOINT 19.6

Suppose the magnetic field in Fig. 19.27 were to the right (in the plane of the page) instead of into the page. What would be the direction of the magnetic force on the wire?

Example 19.8

Magnetic Force on a Power Line

A 125-m-long power line is horizontal and carries a current of 2500 A toward the south. The Earth's magnetic field at that location is 0.52 mT toward the north and inclined 62° below the horizontal (Fig. 19.28). What is the magnetic force on the power line? (Ignore any drooping of the wire; assume it's straight.)

Strategy We are given all the quantities necessary to calculate the force:

$I = 2500$ A;

$\vec{\mathbf{L}}$ has magnitude 125 m and direction south;

$\vec{\mathbf{B}}$ has magnitude 0.52 mT. It has a downward component and a northward component.

We find the cross product $\vec{\mathbf{L}} \times \vec{\mathbf{B}}$ and then multiply by I.

Solution The magnitude of the force is given by

$$F = IL_{\perp}B = ILB_{\perp}$$

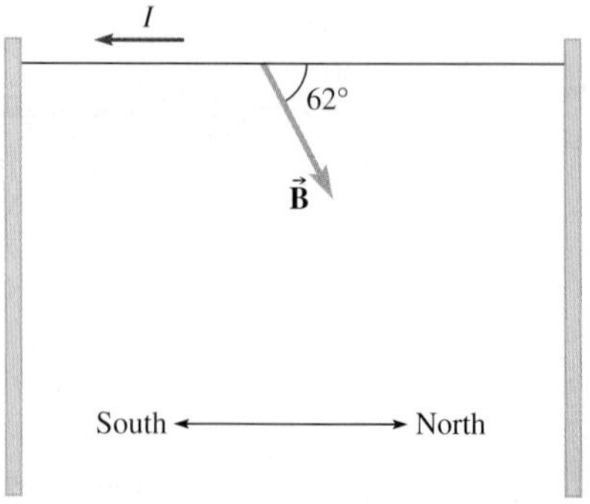

Figure 19.28
The wire and the magnetic field vector.

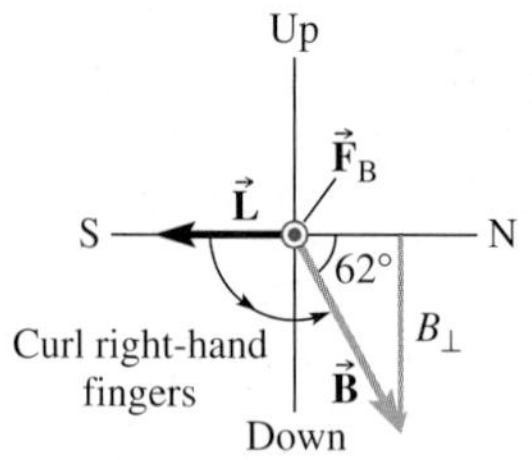

Figure 19.29
The vectors $\vec{\mathbf{L}}$ and $\vec{\mathbf{B}}$ sketched in a vertical plane. The cross product of the two must then be perpendicular to this plane—either east (out of the page) or west (into the page). The right-hand rule enables us to choose between the two possibilities.

The second form is more convenient here, since $\vec{\mathbf{L}}$ is southward. The perpendicular component of $\vec{\mathbf{B}}$ is the vertical component, which is $B \sin 62°$ (see Fig. 19.29). Then

$$F = ILB \sin 62° = 2500 \text{ A} \times 125 \text{ m} \times 5.2 \times 10^{-4} \text{ T} \times \sin 62°$$
$$= 140 \text{ N}$$

Figure 19.29 shows the vectors $\vec{\mathbf{L}}$ and $\vec{\mathbf{B}}$ sketched in the north/south–up/down plane. Since north is to the right, this is a view looking toward the west. The cross product $\vec{\mathbf{L}} \times \vec{\mathbf{B}}$ is out of the page by the right-hand rule. Therefore, the direction of the force is east.

Discussion The hardest thing in this sort of problem is choosing a plane in which to sketch the vectors. Here we chose a plane in which we could draw both $\vec{\mathbf{L}}$ and $\vec{\mathbf{B}}$; then the cross product has to be perpendicular to this plane.

Practice Problem 19.8 Magnetic Force on a Current-Carrying Wire

A vertical wire carries 10.0 A of current upward. What is the direction of the magnetic force on the wire if the magnetic field is the same as in Example 19.8?

19.7 TORQUE ON A CURRENT LOOP

Consider a rectangular loop of wire carrying current I in a uniform magnetic field $\vec{\mathbf{B}}$. In Fig. 19.30a, the field is parallel to sides 1 and 3 of the loop. There is no magnetic force on sides 1 and 3 since $\vec{\mathbf{L}} \times \vec{\mathbf{B}} = 0$ for each. The forces on sides 2 and 4 are equal in magnitude and opposite in direction. There is no net magnetic force on the loop, but the lines of action of the two forces are offset by a distance b, so there is a nonzero net torque. The torque tends to make the loop rotate about a central axis in the direction indicated in Fig. 19.30a. The magnitude of the magnetic force on sides 2 and 4 is

$$F = ILB = IaB$$

The lever arm for each of the two forces is $\frac{1}{2}b$, so the torque due to each is

$$\text{magnitude of force} \times \text{lever arm} = F \times \tfrac{1}{2}b = \tfrac{1}{2}IabB$$

Then the total torque on the loop is $\tau = IabB$. The area of the rectangular loop is $A = ab$, so

$$\tau = IAB$$

If, instead of a single turn, there are N turns forming a coil, then the magnetic torque on the coil is

$$\tau = NIAB \qquad \text{(19-13a)}$$

Equation (19-13a) holds for a planar loop or coil of *any* shape (see Problem 58).

What if the field is not parallel to the plane of the coil? In Fig. 19.30b, the same loop has been rotated about the axis shown. The angle θ is the angle between the magnetic field and a line *perpendicular* to the current loop. Which perpendicular direction is determined by a right-hand rule: curl the fingers of your right hand in toward your palm, following the current in the loop, and your thumb indicates the direction of $\theta = 0$ (Fig. 19.30c). Before, when the field was in the plane of the loop, θ was 90°. For $\theta \neq 90°$, the magnetic forces on sides 1 and 3 are no longer zero, but they are equal and opposite and act along the same line of action, so they contribute neither to the net force nor to the net torque. The magnetic forces on sides 2 and 4 are the same as before, but now the lever arms are smaller by a factor of $\sin\theta$: instead of $\frac{1}{2}b$, the lever arms are now $\frac{1}{2}b\sin\theta$. Therefore,

Torque on a current loop:

$$\tau = NIAB\sin\theta \qquad \text{(19-13b)}$$

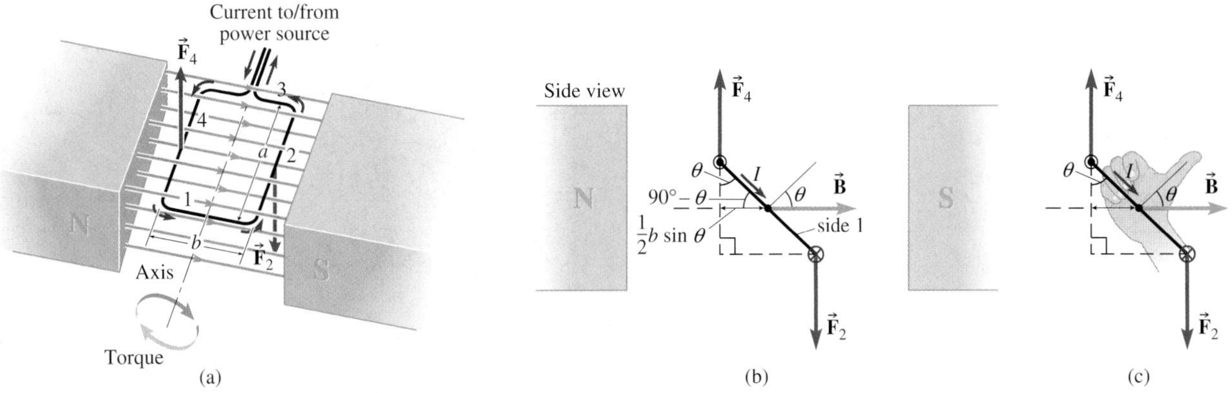

Figure 19.30 (a) A rectangular coil of wire in a uniform magnetic field. The current in the coil (counterclockwise as viewed from the top) causes a magnetic torque, which is clockwise as viewed from the front. (b) Side view of the same coil after it has been rotated in the field. The current in side 4 comes out of the page, along side 1 (diagonally down the page), and back into the page in side 2. The lever arms of the forces on sides 2 and 4 are now smaller: $\frac{1}{2}b\sin\theta$ instead of $\frac{1}{2}b$. The torque is then smaller by the same factor ($\sin\theta$). (c) Using the right-hand rule to choose the perpendicular direction from which θ is measured.

The torque has maximum magnitude if the field is in the plane of the coil ($\theta = 90°$ or $270°$). If $\theta = 0°$ or $180°$, the field is perpendicular to the plane of the loop and the torque is zero. There are *two* positions of rotational equilibrium, but they are not equivalent. The position at $\theta = 180°$ is an *unstable* equilibrium, because at angles *near* $180°$ the torque tends to rotate the coil *away* from $180°$. The position at $\theta = 0°$ is a *stable* equilibrium; the torque for angles *near* $0°$ makes the coil rotate back toward $\theta = 0°$ and thus tends to restore the equilibrium.

CHECKPOINT 19.7

Suppose the coil of wire in Fig. 19.30 is in a vertical plane with wire 2 on top and wire 4 on the bottom. The current still flows around the coil in the direction indicated in the figure. (a) What are the directions of the magnetic forces on the two wires? (b) Explain why the torque about the axis of rotation is zero. (c) Is the coil in stable or unstable equilibrium? (d) What is the angle θ as defined in Fig. 19.30?

The torque on a current loop in a uniform magnetic field is analogous to the torque on an electric dipole in a uniform electric field (see Problem 56). This similarity is our first hint that

> A current loop is a magnetic dipole.

The direction perpendicular to the loop chosen by the right-hand rule is the direction of the **magnetic dipole moment vector**. The dipole moment vector points from the dipole's south pole toward its north pole. (By comparison, the *electric* dipole moment vector points from the electric dipole's negative charge toward its positive charge.)

Application: Electric Motor

In a simple dc motor, a coil of wire is free to rotate between the poles of a permanent magnet (Fig. 19.31). When current flows through the loop, the magnetic field exerts a torque on the loop. If the direction of the current in the coil doesn't change, then the coil just oscillates about the stable equilibrium orientation ($\theta = 0°$). To make a motor we need the coil to keep turning in the same direction. The trick used to make a dc motor is to automatically reverse the direction of the current as soon as the coil passes $\theta = 0°$. In

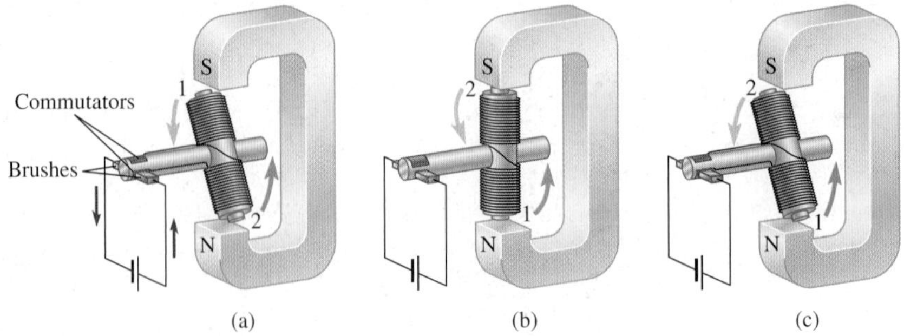

Figure 19.31 Simple dc motor. The two sides of the rotating coil are labeled 1 and 2. In position (a) the coil rotates away from unstable equilibrium. In position (b) brushes pass over the split in the commutator, interrupting the flow of current. If the current in the coil still flowed in the same direction as in (a), this would be stable equilibrium. When the coil turns a little farther, in position (c), the brushes reverse the direction of the current. Now the coil is pushed away from *unstable* equilibrium rather than pulled back toward stable equilibrium.

effect, just as the coil goes through the stable equilibrium orientation, we reverse the current to make the coil's orientation an *unstable* equilibrium. Then, instead of pulling the coil backward toward the (stable) equilibrium, the torque keeps turning the coil in the same direction by pushing it *away from* (unstable) equilibrium.

To reverse the current, the source of current is connected to the coil of the motor by means of two brushes. The brushes make electric contact with the *commutator*, which rotates with the coil. The commutator is a split ring with each side connected to one end of the coil. Every time the brushes pass over the split (Fig. 19.31b), the current to the coil is reversed.

Application: Galvanometer

The magnetic torque on a current loop is also the principle behind the operation of a galvanometer—a sensitive device used to measure current. A rectangular coil of wire is placed between the poles of a magnet (Fig. 19.32). The shape of the magnet's pole faces keeps the field perpendicular to the wires and constant in magnitude regardless of the angle of the coil, so the torque does not depend on the angle of the coil. A hairspring provides a restoring torque that is proportional to the angular displacement of the coil. When a current passes through the coil, the magnetic torque is proportional to the current. The coil rotates until the restoring torque due to the spring is equal in magnitude to the magnetic torque. Thus, the angular displacement of the coil is proportional to the current in the coil.

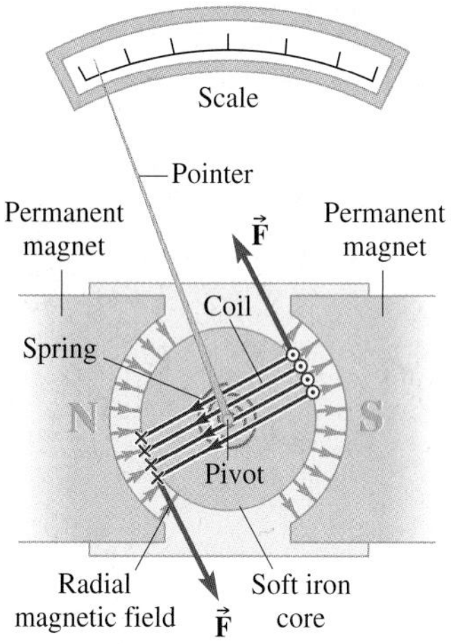

Figure 19.32 A galvanometer.

Conceptual Example 19.9

Force and Torque on a Galvanometer Coil

Show that (a) there is zero net magnetic force on the pivoted coil in the galvanometer of Fig. 19.32; (b) there is a net torque; and (c) the torque is in the correct direction to swing the pointer in the plane of the page. (d) Determine which direction the current in the coil must flow to swing the pointer to the right. Assume that the magnetic field is radial and has uniform *magnitude* in the space between the magnet pole faces and the iron core and that the field is zero in the vicinity of the two sides of the coil that cross over the iron core.

Strategy Since we do not know the direction of the current, we pick one arbitrarily; in part (d) we will find out whether the choice was correct. Only the two sides of the coil near the magnet pole faces experience magnetic forces, since the other two sides are in zero field.

Solution We choose the current in the side near the north pole to flow into the page. The current must then flow out of the page in the side of the coil near the south pole. In Fig. 19.32, the current directions are marked with symbols ⊙ and ×, which also represent the directions of the $\vec{\mathbf{L}}$ vectors used to find the magnetic force. The magnetic field vectors are also shown. Note that, since the direction of the field is radial, the two magnetic vectors are the same (same direction *and* magnitude). The direction of the magnetic force on either side is given by

$$\vec{\mathbf{F}} = NI\vec{\mathbf{L}} \times \vec{\mathbf{B}}$$

where N stands for the number of turns of wire in the coil. The force vectors are shown on Fig. 19.32.

(a) Since the $\vec{\mathbf{B}}$ vectors are the same and the $\vec{\mathbf{L}}$ vectors are equal and opposite (same length but opposite direction), the forces are equal and opposite. Then the net magnetic force on the coil is zero. (b) The net torque is not zero because the lines of action of the forces are separated. (c) The forces make the pointer rotate counterclockwise in the plane of the page. (d) Since the meter shows positive current by rotating clockwise, we have chosen the wrong direction for the current. The leads of the galvanometer should be attached so that positive current makes the current in the coil flow in the direction opposite to the one we chose initially.

Discussion The galvanometer works because the torque is proportional to the current but independent of the orientation of the coil. In Eq. (19-13b), θ is the angle between the magnetic field and a line perpendicular to the coil. In the galvanometer, the magnetic field acting on the coil is always in the plane of the coil; in essence θ is a constant 90° even while the coil swings about the pivot.

Practice Problem 19.9 Torque on a Coil

Starting with the magnetic forces on the sides of the coil, show that the torque on the coil is $\tau = NIAB$, where A is the area of the coil.

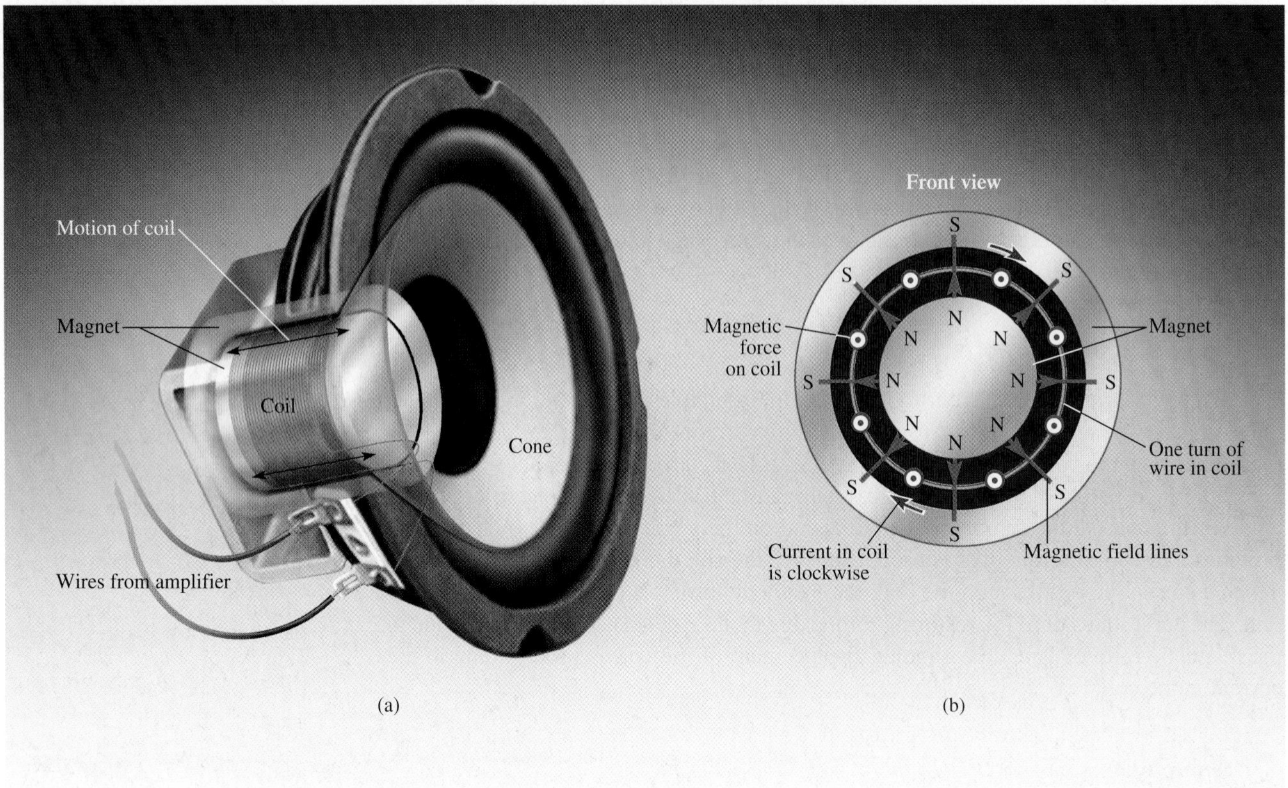

Figure 19.33 (a) Simplified sketch of a loudspeaker. A varying current from the amplifier flows through a coil. The magnetic force on the coil makes it and the attached cone move in and out. The motion of the cone displaces air in the vicinity and creates a sound wave. (b) A front view of the coil. The coil is sandwiched between cylindrical shaped poles of a magnet. The magnetic field is directed radially outward. (Compare with Fig. 19.32 to see how the radial magnetic fields and the coil orientations differ.) Applying $\vec{\mathbf{F}} = I\vec{\mathbf{L}} \times \vec{\mathbf{B}}$ to any short length of the coil shows that, for the clockwise current shown here, the magnetic force is out of the page. (In the galvanometer, the net magnetic force on the coil is zero, but there is a nonzero net magnetic torque.)

Application: Audio Speakers

In contrast to a coil in a uniform field, a coil of wire in a *radial* magnetic field may experience a nonzero net magnetic *force.* A coil in a radial field is the principle behind the operation of many audio speakers (Fig. 19.33a). An electric current passes through a coil of wire. The coil sits between the poles of a magnet shaped so that the magnetic field is radial (Fig. 19.33b). Even though the coil is not a straight wire, the field direction is such that the force on every part of the coil is in the same direction. Since the field is everywhere perpendicular to the wire, the magnetic force is $F = ILB$ where L is the *total* length of the wire in the coil. A springlike mechanism exerts a linear restoring force on the coil so that when a magnetic force acts, the displacement of the coil is proportional to the magnetic force, which in turn is proportional to the current in the coil. Thus, the motion of the coil—and the motion of the attached cone—mirrors the current sent through the speaker by the amplifier.

19.8 MAGNETIC FIELD DUE TO AN ELECTRIC CURRENT

So far we have explored the magnetic forces acting on charged particles and current-carrying wires. We have not yet looked at *sources* of magnetic fields other than permanent magnets. It turns out that *any moving charged particle* creates a magnetic field. There is a certain symmetry about the situation:

- Moving charges experience magnetic forces and moving charges create magnetic fields;
- Charges at rest feel no magnetic forces and create no magnetic fields;
- Charges feel electric forces and create electric fields, whether moving or not.

Today we know that electricity and magnetism are closely intertwined. It may be surprising to learn that they were not known to be related until the nineteenth century. Hans Christian Oersted discovered in 1820 by happy accident that electric currents flowing in wires made nearby compass needles swing around. Oersted's discovery was the first evidence of a connection between electricity and magnetism.

The magnetic field due to a single moving charged particle is negligibly small in most situations. However, when an electric current flows in a wire, there are enormous numbers of moving charges. The magnetic field due to the wire is the sum of the magnetic fields due to each charge; the principle of superposition applies to magnetic fields just as it does to electric fields.

Magnetic Field due to a Long Straight Wire

Let us first consider the magnetic field due to a long, straight wire carrying a current I. What is the magnetic field at a distance r from the wire and far from its ends? Figure 19.34a is a photo of such a wire, passing through a glass plate on which iron filings have been sprinkled. The iron bits line up with the magnetic field due to the current in the wire. The photo suggests that the magnetic field lines are circles centered on the wire. Circular field lines are indeed the only possibility, given the symmetry of the situation. If the lines were any other shape, they would be farther from the wire in some directions than in others.

The iron filings do not tell us the direction of the field. By using compasses instead of iron filings (Fig. 19.34b), the direction of the field is revealed—it is the direction indicated by the north end of each compass. The field lines due to the wire are shown in Fig. 19.34c, where the current in the wire flows upward. A right-hand rule relates the current direction in the wire to the direction of the field around the wire:

Using Right-Hand Rule 2 to Find the Direction of the Magnetic Field due to a Long Straight Wire

1. Point the thumb of the right hand in the direction of the current in the wire.
2. Curl the fingers inward toward the palm; the direction that the fingers curl is the direction of the magnetic field lines around the wire (Fig. 19.34c).
3. As always, the magnetic field at any point is tangent to a field line through that point. For a long straight wire, the magnetic field is tangent to a circular field line and, therefore, perpendicular to a radial line from the wire.

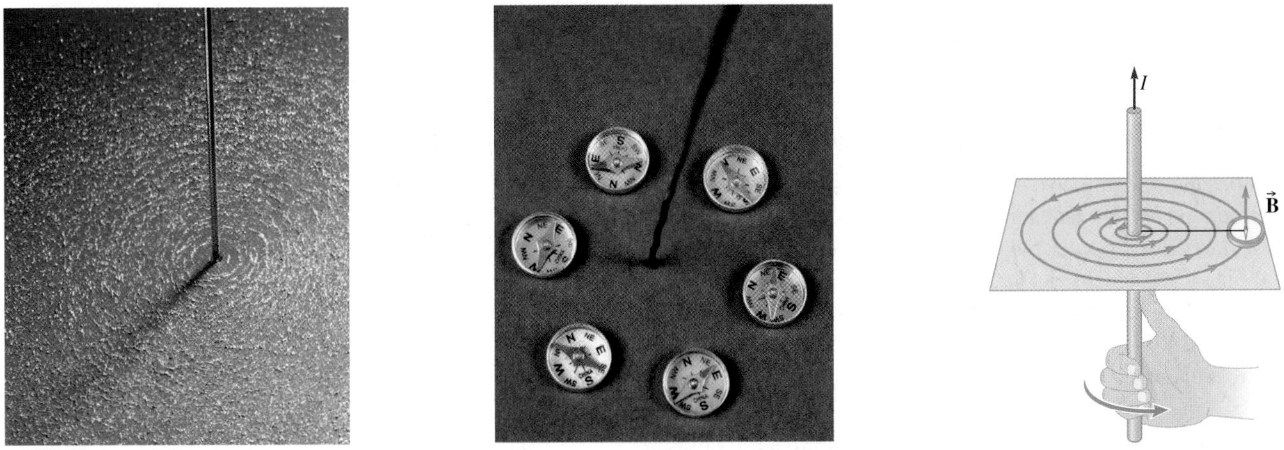

(a) (b) (c)

Figure 19.34 Magnetic field due to a long straight wire. (a) Photo of a long wire, with iron filings lining up with the magnetic field. (b) Compasses show the direction of the field. (c) Sketch illustrating how to use the right-hand rule to determine the direction of the field lines. At any point, the magnetic field is tangent to one of the circular field lines and, therefore, perpendicular to a radial line from the wire.

CHECKPOINT 19.8

What is the direction of the magnetic field at a point directly behind the wire in Fig. 19.34c?

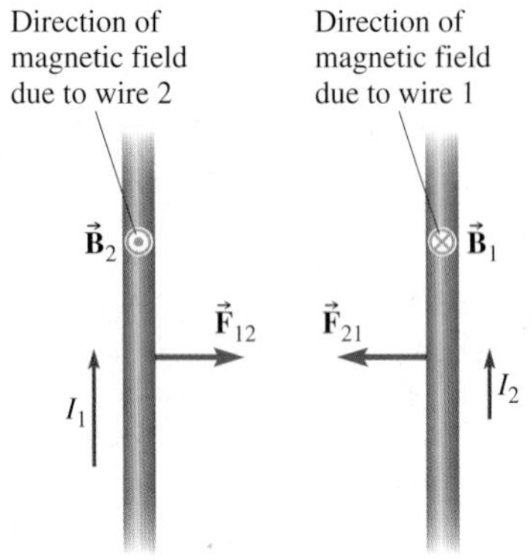

Figure 19.35 Two parallel wires exert magnetic forces on one another. The force on wire 1 due to wire 2's magnetic field is $\vec{\mathbf{F}}_{12} = I_1\vec{\mathbf{L}}_1 \times \vec{\mathbf{B}}_2$. Even if the currents are unequal, $\vec{\mathbf{F}}_{21} = -\vec{\mathbf{F}}_{12}$ (Newton's third law).

The magnitude of the magnetic field at a distance r from the wire can be found using Ampère's law (Section 19.9; see Example 19.11):

Magnetic field due to a long straight wire:

$$B = \frac{\mu_0 I}{2\pi r} \qquad (19\text{-}14)$$

where I is the current in the wire and μ_0 is a universal constant known as the **permeability of free space**. The permeability plays a role in magnetism similar to the role of the permittivity (ϵ_0) in electricity. In SI units, the value of μ_0 is

$$\mu_0 = 4\pi \times 10^{-7}\ \frac{\text{T}\cdot\text{m}}{\text{A}} \quad \text{(exact, by definition)} \qquad (19\text{-}15)$$

Two parallel current-carrying wires that are close together exert magnetic forces on one another. The magnetic field of wire 1 causes a magnetic force on wire 2; the magnetic field of wire 2 causes a magnetic force on wire 1 (Fig. 19.35). From Newton's third law, we expect the forces on the wires to be equal and opposite. If the currents flow in the same direction, the force is attractive; if they flow in opposite directions, the force is repulsive (see Problem 72). Note that for current-carrying wires, "likes" (currents in the same direction) *attract* one another and "unlikes" (currents in opposite directions) *repel* one another.

The constant μ_0 can be assigned an exact value because the magnetic forces on two parallel wires are used to *define* the ampere, which is an SI base unit. One ampere is the current in each of two long parallel wires 1 m apart such that each exerts a magnetic force on the other of exactly 2×10^{-7} N per meter of length. The ampere, not the coulomb, is chosen to be an SI base unit because it can be defined in terms of forces and lengths that can be measured accurately. The coulomb is then defined as 1 ampere-second.

Example 19.10

Magnetic Field due to Household Wiring

In household wiring, two long parallel wires are separated and surrounded by an insulator. The wires are a distance d apart and carry currents of magnitude I in opposite directions. (a) Find the magnetic field at a distance $r >> d$ from the center of the wires (point P in Fig. 19.36). (b) Find the numerical value of B if $I = 5$ A, $d = 5$ mm, and $r = 1$ m and compare with Earth's magnetic field strength at the surface ($\sim 5 \times 10^{-4}$ T).

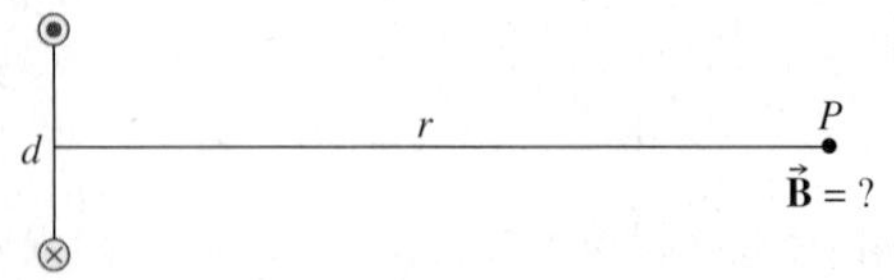

Figure 19.36

The two wires are perpendicular to the plane of the page. They are marked to show that the current in the upper wire flows out of the page and the current in the lower wire flows into the page.

Strategy The magnetic field is the vector sum of the fields due to each of the wires. The fields due to the wires at P are equal in magnitude (since the currents and distances are the same), but the directions are not the same. Equation (19-14) gives the magnitude of the field due to either wire. Since the field lines due to a single long wire are circular, the direction of the field is tangent to a circle that passes through P and

continued on next page

Example 19.10 continued

whose center is on the wire. The right-hand rule determines which of the two tangent directions is correct.

Solution (a) Since $r >> d$, the distance from either wire to P is approximately r (see Fig. 19.37). Then the magnitude of the field at P due to either wire is

$$B \approx \frac{\mu_0 I}{2\pi r}$$

In Fig. 19.37, we draw radial lines from each wire to point P. The direction of the magnetic field due to a long wire is tangent to a circle and therefore perpendicular to a radius. Using the right-hand rule, the field directions are as shown in Fig. 19.37. The y-components of the two field vectors add to zero; the x-components are the same:

$$B_x = \frac{\mu_0 I}{2\pi r} \sin\theta$$

Since $r >> d$,

$$\sin\theta = \frac{\text{opposite}}{\text{hypotenuse}} \approx \frac{\frac{1}{2}d}{r}$$

The total magnetic field due to the two wires is in the $+x$-direction and has magnitude

$$B_{\text{net}} = 2B_x = \frac{\mu_0 I d}{2\pi r^2}$$

(b) By substitution,

$$B = \frac{\mu_0}{2\pi} \times \frac{Id}{r^2} = 2 \times 10^{-7}\ \frac{\text{T}\cdot\text{m}}{\text{A}} \times \frac{5\ \text{A} \times 0.005\ \text{m}}{(1\ \text{m})^2} = 5 \times 10^{-9}\ \text{T}$$

The field due to the wires is 10^{-5} times Earth's field.

Discussion The field strength at P due to both wires is a factor of d/r times the field strength due to either wire alone.

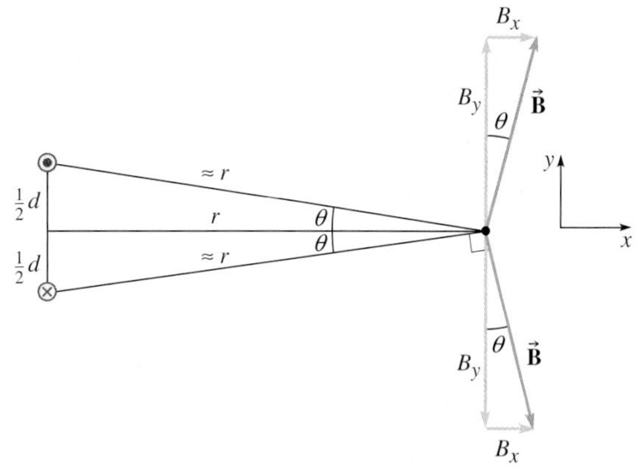

Figure 19.37

Field vectors due to each wire.

Since $d/r = 0.005$, the field strength due to both is only 0.5% of the field strength due to either one. The field due to the two wires decreases with distance proportional to $1/r^2$. It falls off much faster with distance than does the field due to a single wire, which is proportional to $1/r$. With equal currents flowing in opposite directions, we have a net current of zero. The only reason the field isn't zero is the small distance between the two wires.

Since the current in household wiring actually alternates at 60 Hz, so does the field. If 5 A is the maximum current, then 5×10^{-9} T is the maximum field strength.

Practice Problem 19.10 Field Midway Between Two Wires

Find the magnetic field at a point halfway between the two wires in terms of I and d.

Magnetic Field due to a Circular Current Loop

In Section 19.7, we saw the first clue that a loop of wire that carries current around in a complete circuit is a magnetic dipole. A second clue comes from the magnetic field produced by a circular loop of current. As for a straight wire, the magnetic field lines circulate around the wire, but for a circular current loop, the field lines are not circular. The field lines are more concentrated inside the current loop and less concentrated outside (Fig. 19.38a). The field lines emerge from one side of the current loop (the north pole) and reenter the other side (the south pole). Thus, the field due to a current loop is similar to the field of a short bar magnet.

The direction of the field lines is given by right-hand rule 3.

Using Right-Hand Rule 3 to Find the Direction of the Magnetic Field due to a Circular Loop of Current

Curl the fingers of your right hand inward toward the palm, following the current around the loop (Fig. 19.38b). Your thumb points in the direction of the magnetic field in the *interior* of the loop.

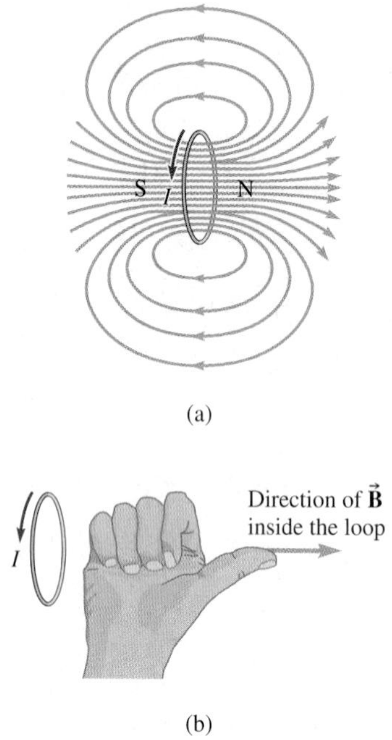

Figure 19.38 (a) Magnetic field lines due to a circular current loop. (b) Using right-hand rule 3 to determine the direction of the field inside the loop.

The magnitude of the magnetic field *at the center* of a circular loop (or coil) is given by

$$B = \frac{\mu_0 NI}{2r} \quad (19\text{-}16)$$

where N is the number of turns, I is the current, and r is the radius.

The magnetic fields due to coils of current-carrying wire are used in televisions and computer monitors to deflect the electron beam so that it lands on the screen in the desired spot.

Magnetic Field due to a Solenoid

An important source of magnetic field is that due to a **solenoid** because the field inside a solenoid is nearly uniform. In magnetic resonance imaging (MRI), the patient is immersed in a strong magnetic field inside a solenoid.

To construct a solenoid with a circular cross section, wire is tightly wrapped in a cylindrical shape, forming a helix (Fig. 19.39a). We can think of the field as the superposition of the fields due to a large number of circular loops. If the loops are sufficiently close together, then the field lines go straight through one loop to the next, all the way down the solenoid. Having a large number of loops, one next to the other, straightens out the field lines. Figure 19.39b shows the magnetic field lines due to a solenoid. Inside the solenoid and away from the ends, the field is nearly uniform and parallel to the solenoid's axis as long as the solenoid is long compared to its radius. To find in which direction the field points along the axis, use right-hand rule 3 exactly as for the circular loop of current.

If a long solenoid has N turns of wire and length L, then the magnetic field strength inside is given by (see Problem 82):

Magnetic field strength inside an ideal solenoid:

$$B = \frac{\mu_0 NI}{L} = \mu_0 nI \quad (19\text{-}17)$$

In Eq. (19-17), I is the current in the wire and $n = N/L$ is the number of turns per unit length. Note that the field does *not* depend on the radius of the solenoid. The magnetic field near the ends is weaker and starts to bend outward; the field outside the solenoid is quite small—look how spread out the field lines are outside. A solenoid is one way to produce a nearly uniform magnetic field.

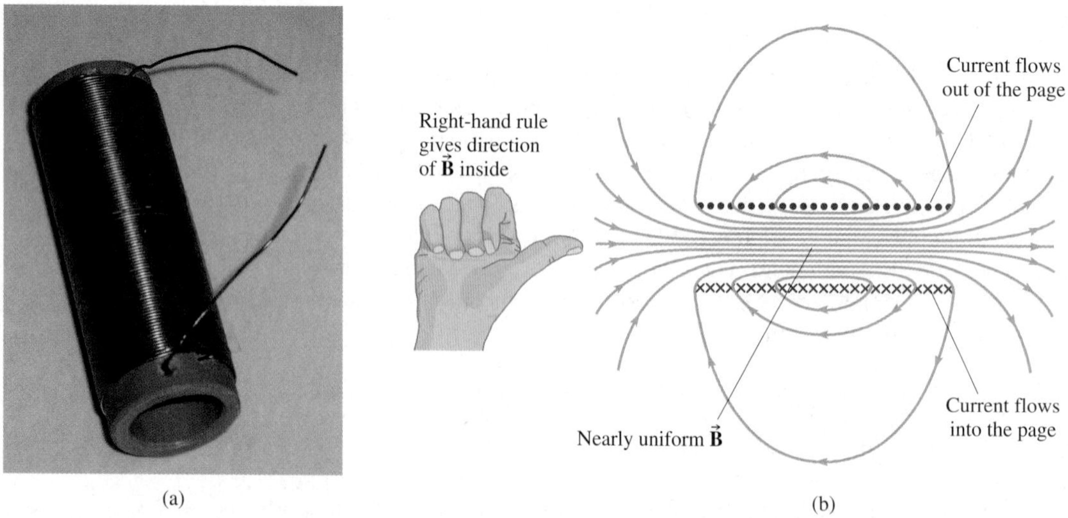

Figure 19.39 (a) A solenoid. (b) Magnetic field lines due to a solenoid. Each dot represents the wire crossing the plane of the page with current out of the page; each cross represents the wire crossing the plane of the page with current into the page.

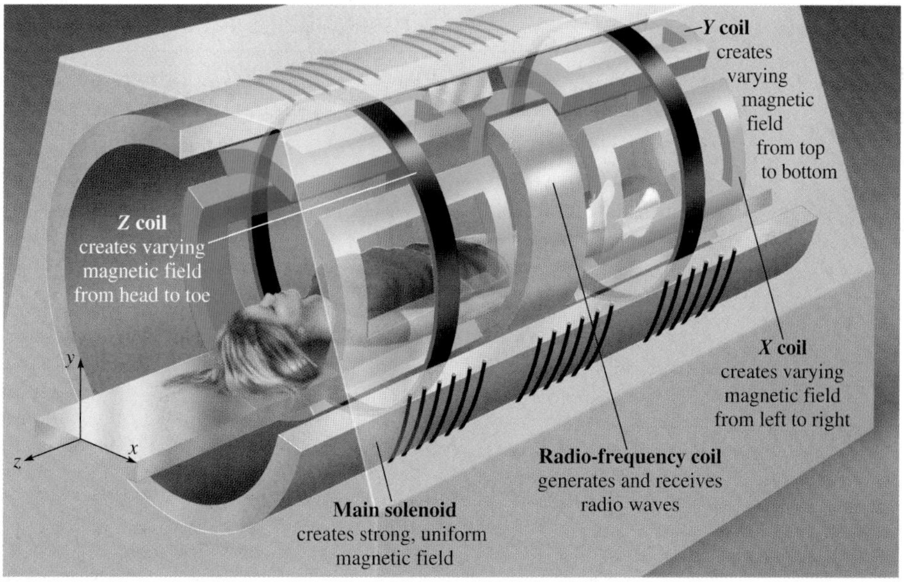

Figure 19.40 MRI apparatus.

The similarity in the magnetic field lines due to a solenoid compared with those due to a bar magnet (Fig. 19.1b) suggested to André-Marie Ampère that the magnetic field of a permanent magnet might also be due to electric currents. The nature of these currents is explored in Section 19.10.

Application: Magnetic Resonance Imaging

In magnetic resonance imaging (Fig. 19.40), the main solenoid is usually made with superconducting wire, which must be kept at low temperature (see Section 18.4). The main solenoid produces a strong, uniform magnetic field (typically 0.5 T–2 T). The nuclei of hydrogen atoms (protons) in the body act like tiny permanent magnets; a magnetic torque tends to make them line up with the magnetic field. A radio-frequency coil emits pulses of radio waves (rapidly varying electric and magnetic fields). If the radio wave has just the right frequency (the resonant frequency), the protons can absorb energy from the wave, which disturbs their magnetic alignment. When the protons flip back into alignment with the field, they emit radio wave signals of their own that can be detected by the radio-frequency coil.

The resonant frequency of the pulse that makes the protons flip depends on the total magnetic field due to the MRI machine and due to the neighboring atoms. Protons in different chemical environments have slightly different resonant frequencies. In order to image a slice of the body, three other coils create small (15 to 30-mT) magnetic fields that vary in the x-, y-, and z-directions. The magnetic fields of these coils are adjusted so that the protons are in resonance with the radio-frequency signal only in a single slice, a few millimeters thick, in any desired direction through the body.

19.9 AMPÈRE'S LAW

Ampère's law plays a role in magnetism similar to that of Gauss's law in electricity (Sec. 16.7). Both relate the field to the source of the field. For the electric field, the source is charge. Gauss's law relates the net charge inside a closed surface to the flux of the electric field through that surface. The source of magnetic fields is current. Ampère's law must take a different form from Gauss's law: since magnetic field lines are always closed loops, the magnetic flux through a *closed surface* is always zero. (This fact is called *Gauss's law for magnetism* and is itself a fundamental law of electromagnetism.)

Table 19.1 Comparison of Gauss's and Ampère's Laws

Gauss's Law	Ampère's Law
Electric field	Magnetic field (static only)
Applies to any closed *surface*	Applies to any closed *path*
Relates the electric field on the surface to the net *electric charge* inside the surface	Relates the magnetic field on the path to the net *current* cutting through interior of the path
Component of the electric field *perpendicular* to the surface ($E_\perp$)	Component of the magnetic field *parallel* to the path ($B_\parallel$)
Flux = perpendicular field component × *area* of surface	Circulation = parallel field component × *length* of path
$\sum E_\perp \Delta A$	$\sum B_\parallel \Delta l$
Flux = $1/\epsilon_0$ × net charge	Circulation = μ_0 × net current
$\sum E_\perp \Delta A = \frac{1}{\epsilon_0} q$	$\sum B_\parallel \Delta l = \mu_0 I$

Instead of a closed surface, Ampère's law concerns any closed *path* or *loop.* For Gauss's law we would find the flux: the perpendicular component of the electric field times the surface area. If $E_\perp$ is not the same everywhere, then we break the surface into pieces and sum up $E_\perp \Delta A$. For Ampère's law, we multiply the component of the magnetic field *parallel* to the path (or the tangential component at points along a closed curve) times the *length* of the path. Just as for flux, if the magnetic field component is not constant then we take parts of the path (each of length Δl) and sum up the product. This quantity is called the **circulation**.

$$\text{circulation} = \sum B_\parallel \Delta l \tag{19-18}$$

Ampère's law relates the circulation of the field to the *net* current I that crosses the interior of the path.

Ampère's Law

$$\sum B_\parallel \Delta l = \mu_0 I \tag{19-19}$$

There is a symmetry between Gauss's law and Ampère's law (Table 19.1).

Example 19.11

Magnetic Field due to a Long Straight Wire

Use Ampère's law to show that the magnetic field due to a long straight wire is $B = \mu_0 I/(2\pi r)$.

Strategy As with Gauss's law, the key is to exploit the symmetry of the situation. The field lines have to be circles around the wire, assuming the ends are far away. Choose a closed path around a circular field line (Fig. 19.41). The field is everywhere tangent to the field line and therefore tangent to the path; there is no perpendicular component. The field must also have the same magnitude at a uniform distance r from the wire.

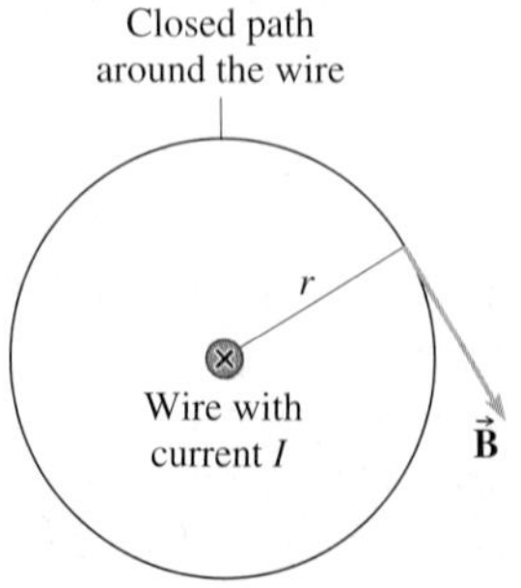

Figure 19.41
Applying Ampère's law to a long straight wire. A closed path is chosen to follow a circular magnetic field line; the magnetic field is then calculated from Ampère's law.

continued on next page

Example 19.11 continued

Solution Since the field has no component perpendicular to the path, $B_{\parallel} = B$. Going around the circular path, B is constant, so

$$\text{circulation} = B \times 2\pi r = \mu_0 I$$

where I is the current in the wire. Solving for B,

$$B = \frac{\mu_0 I}{2\pi r}$$

Discussion Ampère's law shows why the magnetic field of a long wire varies inversely as the distance from the wire. A circle of any radius r around the wire has a length that is proportional to r, while the current that cuts through the interior of the circle is always the same (I). So the field must be proportional to $1/r$.

Practice Problem 19.11 Circulation due to Three Wires

What is the circulation of the magnetic field for the path in Fig. 19.42?

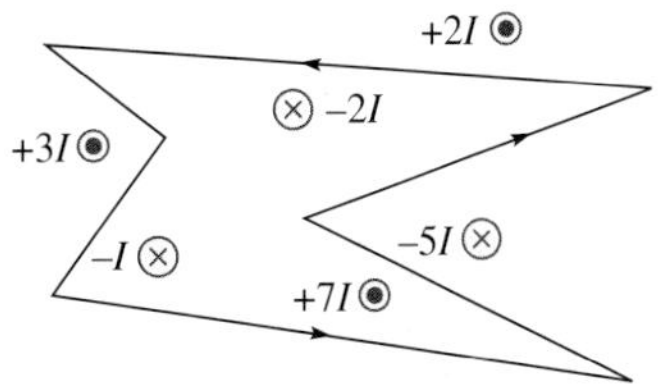

Figure 19.42
Six wires perpendicular to the page carry currents as indicated. A path is chosen to enclose three of the wires.

19.10 MAGNETIC MATERIALS

All materials are magnetic in the sense that they have magnetic properties. The magnetic properties of most substances are quite unremarkable, though. If a bar magnet is held near a piece of wood or aluminum or plastic, there is no noticeable interaction between the two. In common parlance, these substances might be called nonmagnetic. In reality, all substances experience *some* force when near a bar magnet. For most substances, the magnetic force is so weak that it is not noticed.

Substances that experience a noticeable force due to a nearby magnet are called **ferromagnetic** (*ferro-* in Latin refers to iron). Iron is a well-known ferromagnet; others include nickel, cobalt, and chromium dioxide (used to make chrome audiotapes). Ferromagnetic materials experience a magnetic force toward a region of stronger magnetic field. Refrigerator magnets stick because the refrigerator door is made of a ferromagnetic metal. When a permanent magnet is brought near, there is an attractive force on the door, and from Newton's third law there must also be an attractive force on the magnet. The surfaces of the magnet and the door are pulled together by magnetic forces. As a result, each exerts a surface contact force on the other; the component of the contact force parallel to the contact surface—the frictional force—holds the magnet up.

The so-called nonmagnetic substances can be divided into two groups. **Paramagnetic substances** are like the ferromagnets in that they are attracted toward regions of stronger magnetic field, though the force is *much* weaker. **Diamagnetic** substances are weakly repelled by a region of stronger field. All substances, including liquids and gases, are ferromagnetic, paramagnetic, or diamagnetic.

Any substance, whether ferromagnetic, diamagnetic, or paramagnetic, contains a large number of tiny magnets: the electrons. The electrons are like little magnets in two ways. First, an electron's orbital motion around the nucleus makes it a tiny current loop and thus is a magnetic dipole. Second, an electron has an *intrinsic* magnetic dipole moment *independent of its motion.* The intrinsic magnetism of the electron is one of its fundamental properties, just like its electric charge and mass. (Other particles, such as protons and neutrons, also have intrinsic magnetic dipole moments.) The net magnetic dipole moment of an atom or molecule is the vector sum of the dipole moments of its constituent particles.

In most materials—paramagnets and diamagnets—the atomic dipoles are randomly oriented. Even when the material is immersed in a strong external magnetic field, the dipoles only have a slight tendency to line up with it. The torque that tends to make dipoles line up with the external magnetic field is overwhelmed by the thermal tendency

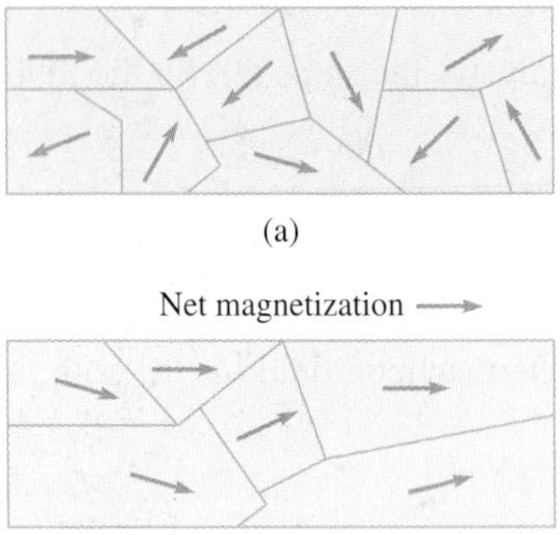

Figure 19.43 Domains within a ferromagnetic material are indicated by arrows indicating the direction of each domain's magnetic field. In (a), the domains are randomly oriented; the material is unmagnetized. In (b), the material is magnetized; the domains show a high degree of alignment to the right.

for the dipoles to be *randomly* aligned, so there is only a slight degree of large-scale alignment. The magnetic field inside the material is nearly the same as the applied field; the dipoles have little effect in paramagnets and diamagnets.

Ferromagnetic materials have much stronger magnetic properties because there is an interaction—the explanation of which requires quantum physics—that keeps the magnetic dipoles aligned, even in the *absence* of an external magnetic field. A ferromagnetic material is divided up into regions called **domains** in which the atomic or molecular dipoles line up with each other. Even though each atom is a weak magnet by itself, when all of them have their dipoles aligned in the same direction within a domain, the domain can have a significant dipole moment.

The moments of different domains are not necessarily aligned with each other, however. Some may point one way and some another (Fig. 19.43a). When the net dipole moment of all the domains is zero, the material is unmagnetized. If the material is placed in an external magnetic field, two things happen. Atomic dipoles at domain boundaries can "defect" from one domain to an adjacent one by flipping their dipole moments. Thus, domains with their dipole moments aligned or nearly aligned with the external field grow in size and the others shrink. The other thing that happens is that domains can change their direction of orientation, with all the atomic dipoles flipping to a new direction. When the net dipole moment of all the domains is nonzero, the material is magnetized (Fig. 19.43b).

PHYSICS AT HOME

If a paper clip is placed in contact with a magnet, the paper clip becomes magnetized and can attract other paper clips. This phenomenon is easily observed in paper-clip containers with magnets that hold the paper clips upright for ease in pulling one out. The magnetized paper clips often drag out other paper clips as well (Fig. 19.44). Try it.

Figure 19.44 Each magnetized paper clip is capable of magnetizing another paper clip.

Once a ferromagnet is magnetized, it does not necessarily lose its magnetization when the external field is removed. It takes some energy to align the domains with the field; there is a kind of internal friction that must be overcome. If there is a lot of this internal friction, then the domains stay aligned even after the external field is removed. The material is then a permanent magnet. If there is relatively little of this internal friction, then there is little energy required to reorient the domains. This kind of ferromagnet does not make a good permanent magnet; when the external field is removed, it retains only a small fraction of its maximum magnetization.

At high temperature, the interaction that keeps the dipoles aligned within a domain is no longer able to do so. Without the alignment of dipoles, there are no longer any domains; the material becomes paramagnetic. The temperature at which this occurs for a particular ferromagnetic material is called the *Curie temperature* of that material [after Pierre Curie (1859–1906), the French physicist famous for studies of radioactive materials done with his wife, Marie Curie]. For iron, the Curie temperature is about 770°C.

Application: Electromagnets

An *electromagnet* is made by inserting a *soft iron* core into the interior of a solenoid. Soft iron does not retain a significant permanent magnetization when the solenoid's field is turned off—soft iron does not make a good permanent magnet. When current flows in the solenoid, magnetic dipoles in the iron tend to line up with the field due to the solenoid. The net effect is that the field inside the iron is intensified by a factor known as the **relative permeability** κ_B. The relative permeability is analogous in magnetism to the dielectric constant in electricity. However, the dielectric constant is the factor by which the electric field is *weakened*, while the relative permeability is the factor by which the magnetic field is *strengthened*. The relative permeability of a ferromagnet can be in the hundreds or even thousands—the intensification of the magnetic field is significant. Not only that, but in an electromagnet the strength and even direction of the magnetic field can be changed by changing the current in the solenoid. Figure 19.45 shows the field lines in an electromagnet. Notice that the iron core channels the field lines; the windings of the solenoid need not be at the business end of the electromagnet.

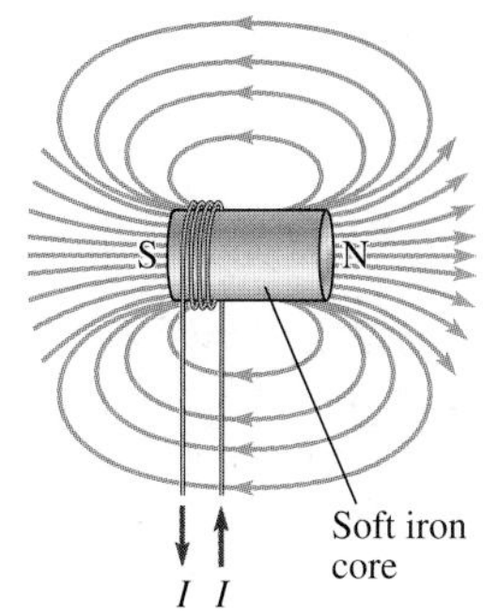

Figure 19.45 An electromagnet with field lines sketched.

Application: Magnetic Storage

The basic principles of magnetic storage are the same, whether applied to computer hard disks, removable disks, or magnetic tape used to store audio, video, or computer data. To record or write, an electromagnet called a *head* is used to magnetize ferromagnetic particles in a coating on the disk or tape surface (Fig. 19.46). The ferromagnetic particles retain their magnetization even after the head has moved away, so the data persists until it is erased or written over. Data can be erased if a tape or disk is brought close to a strong magnet.

Figure 19.46 A computer hard drive. Each platter has a magnetizable coating on each side. The spindle motor turns the platters at several thousand rpm. There is one read-write head on each surface of each platter.

Master the Concepts

- Magnetic field lines are interpreted just like electric field lines. The magnetic field at any point is tangent to the field line; the magnitude of the field is proportional to the number of lines per unit area perpendicular to the lines.
- Magnetic field lines are always closed loops because there are no magnetic monopoles.
- The smallest unit of magnetism is the magnetic dipole. Field lines emerge from the north pole and reenter at the south pole. A magnet can have more than two poles, but it must have at least one north pole and at least one south pole.

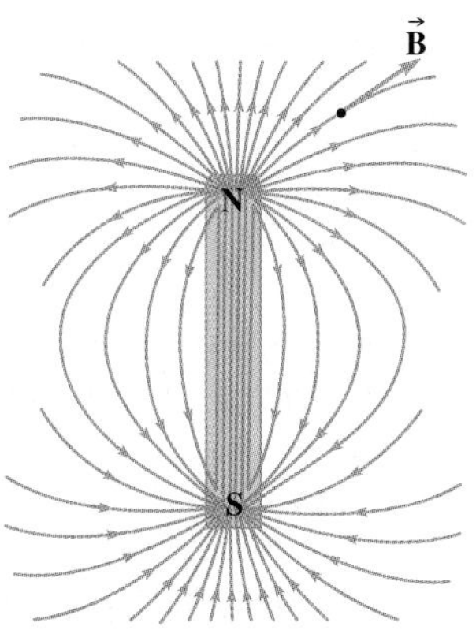

- The magnitude of the cross product of two vectors is the magnitude of one vector times the perpendicular component of the other:

$$|\vec{\mathbf{a}} \times \vec{\mathbf{b}}| = |\vec{\mathbf{b}} \times \vec{\mathbf{a}}| = a_\perp b = ab_\perp = ab \sin\theta \qquad (19\text{-}3)$$

- The direction of the cross product is the direction perpendicular to both vectors that is chosen using right-hand rule 1.

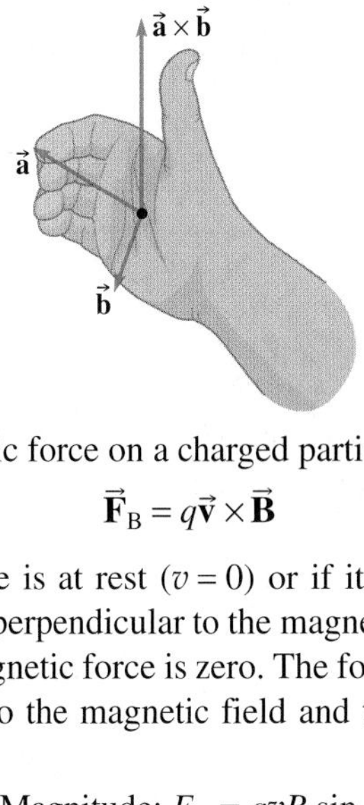

- The magnetic force on a charged particle is

$$\vec{\mathbf{F}}_B = q\vec{\mathbf{v}} \times \vec{\mathbf{B}} \qquad (19\text{-}5)$$

If the charge is at rest ($v = 0$) or if its velocity has no component perpendicular to the magnetic field ($v_\perp = 0$), then the magnetic force is zero. The force is always perpendicular to the magnetic field and to the velocity of the particle.

Magnitude: $F_B = qvB \sin\theta$

Direction: use the right-hand rule to find $\vec{\mathbf{v}} \times \vec{\mathbf{B}}$, then reverse it if q is negative.

continued on next page

Master the Concepts continued

- The SI unit of magnetic field is the tesla:

$$1\ \text{T} = 1\,\frac{\text{N}}{\text{A·m}} \qquad (19\text{-}2)$$

- If a charged particle moves at right angles to a uniform magnetic field, then its trajectory is a circle. If the velocity has a component parallel to the field as well as a component perpendicular to the field, then its trajectory is a helix.

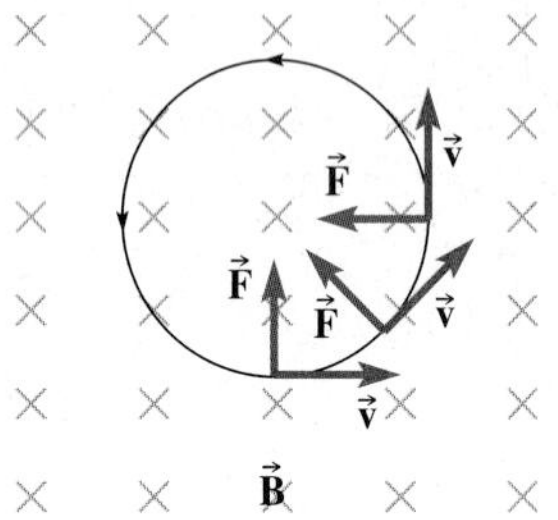

- The magnetic force on a straight wire carrying current I is

$$\vec{\mathbf{F}} = I\vec{\mathbf{L}} \times \vec{\mathbf{B}} \qquad (19\text{-}12a)$$

where $\vec{\mathbf{L}}$ is a vector whose magnitude is the length of the wire and whose direction is along the wire in the direction of the current.

- The magnetic torque on a planar current loop is

$$\tau = NIAB \sin\theta \qquad (19\text{-}13b)$$

where θ is the angle between the magnetic field and the dipole moment vector of the loop. The direction of the dipole moment is perpendicular to the loop as chosen using right-hand rule 1 (take the cross product of $\vec{\mathbf{L}}$ for any side with $\vec{\mathbf{L}}$ for the *next* side, going around in the same direction as the current).

- The magnetic field at a distance r from a long straight wire has magnitude

$$B = \frac{\mu_0 I}{2\pi r} \qquad (19\text{-}14)$$

The field lines are circles around the wire with the direction given by right-hand rule 2.

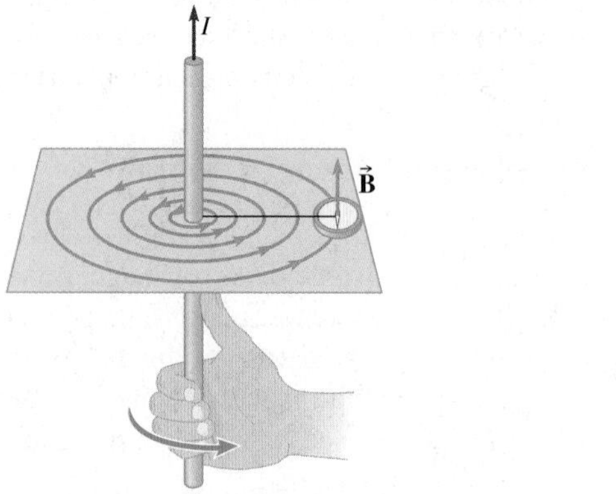

- The permeability of free space is

$$\mu_0 = 4\pi \times 10^{-7}\,\frac{\text{T·m}}{\text{A}} \qquad (19\text{-}15)$$

- The magnetic field inside a long tightly wound solenoid is uniform:

$$B = \frac{\mu_0 NI}{L} = \mu_0 nI \qquad (19\text{-}17)$$

Its direction is along the axis of the solenoid, as given by right-hand rule 3.

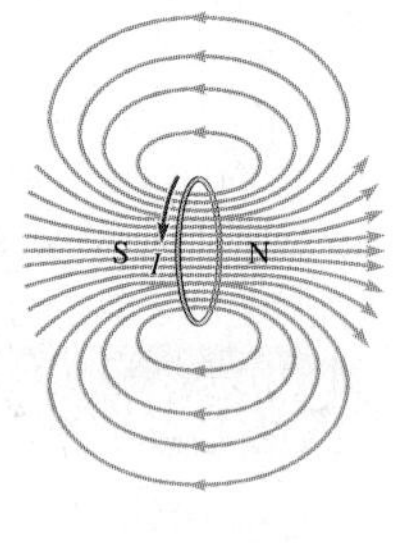

- Ampère's law relates the circulation of the magnetic field around a closed path to the *net* current I that crosses the interior of the path.

$$\Sigma B_{\parallel}\,\Delta l = \mu_0 I \qquad (19\text{-}19)$$

- The magnetic properties of ferromagnetic materials are due to an interaction that keeps the magnetic dipoles aligned within regions called domains, even in the *absence* of an external magnetic field.

Conceptual Questions

1. The electric field is defined as the electric force per unit charge. Explain why the magnetic field *cannot* be defined as the magnetic force per unit charge.
2. A charged particle moves through a region of space at constant velocity. Ignore gravity. In the region, is it possible that there is (a) a magnetic field but no electric field? (b) an electric field but no magnetic field? (c) a magnetic field and an electric field? For each possibility, what must be true about the direction(s) of the field(s)?
3. Suppose that a horizontal electron beam is deflected to the right by a uniform magnetic field. What is the direction of the magnetic field? If there is more than one possibility, what can you say about the direction of the field?
4. A circular metal loop carries a current I as shown. The points are all in the plane of the page and the loop is perpendicular to the page. Sketch the loop, and draw vector arrows at the points A, B, C, D, and E to show the direction of the magnetic field at those points.

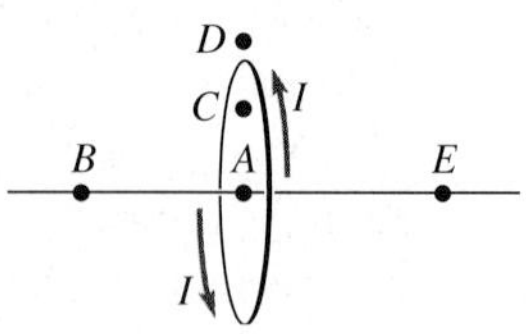

5. In a TV or computer monitor, a constant electric field accelerates the electrons to high speed; then a magnetic field is used to deflect the electrons to the side. Why can't a constant magnetic field be used to speed up the electrons?
6. A uniform magnetic field directed upward exists in some region of space. In what direction(s) could an electron be moving if its trajectory is (a) a straight line? (b) a circle? Assume that the electron is subject only to magnetic forces.

7. In a velocity selector, the electric and magnetic forces cancel if $\vec{\mathbf{E}} + \vec{\mathbf{v}} \times \vec{\mathbf{B}} = 0$. Show that $\vec{\mathbf{v}}$ must be in the same direction as $\vec{\mathbf{E}} \times \vec{\mathbf{B}}$. [*Hint:* Since $\vec{\mathbf{v}}$ is perpendicular to both $\vec{\mathbf{E}}$ and $\vec{\mathbf{B}}$ in a velocity selector, there are only two possibilities for the direction of $\vec{\mathbf{v}}$: the direction of $\vec{\mathbf{E}} \times \vec{\mathbf{B}}$ or the direction of $-\vec{\mathbf{E}} \times \vec{\mathbf{B}}$.]
8. Two ions with the same velocity and mass but different charges enter the magnetic field of a mass spectrometer. One is singly charged ($q = +e$) and the other is doubly charged ($q = +2e$). Is the radius of their circular paths the same? If not, which is larger? By what factor?
9. The mayor of a city proposes a new law to require that magnetic fields generated by the power lines running through the city be zero outside of the electric company's right of way. What would you say at a public discussion of the proposed law?
10. A horizontal wire that runs east-west carries a steady current to the east. A C-shaped magnet (see Fig. 19.3a) is placed so that the wire runs between the poles, with the north pole above the wire and the south pole below. What is the direction of the magnetic force on the wire between the poles?
11. The magnetic field due to a long straight wire carrying steady current is measured at two points, P and Q. Where is the wire and in what direction does the current flow?

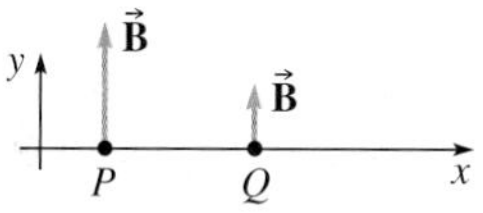

12. A circular loop of current carries a steady current. (a) Sketch the magnetic field lines in a plane perpendicular to the plane of the loop. (b) Which side of the loop is the north pole of the magnetic dipole and which is the south pole?
13. Computer speakers that are intended to be placed near a computer monitor are magnetically shielded—either they don't use magnets or they are designed so that their magnets produce only a small magnetic field nearby. Why is the shielding important? What might happen if an ordinary speaker (*not* intended for use near a monitor) is placed next to a computer monitor?
14. One iron nail does not necessarily attract another iron nail, although both are attracted by a magnet. Explain.
15. Two wires at right angles in a plane carry equal currents. At what points in the plane is the magnetic field zero?

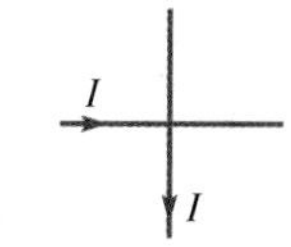

16. If a magnet is held near the screen of a TV, computer monitor, or oscilloscope, the picture is distorted. [*Don't try this*—see part (b).] (a) Why is the picture distorted? (b) With a color TV or monitor, the distortion remains even after the magnet is removed. Explain. (A color CRT has a metal mask just behind the screen with holes to line up the electrons from different guns with the red, green, and blue phosphors. Of what kind of metal is the mask made?)
17. A metal bar is shown at two different times. The arrows represent the alignment of the dipoles within each magnetic domain. (a) What happened between t_1 and t_2 to cause the change? (b) Is the metal a paramagnet, diamagnet, or ferromagnet? Explain.

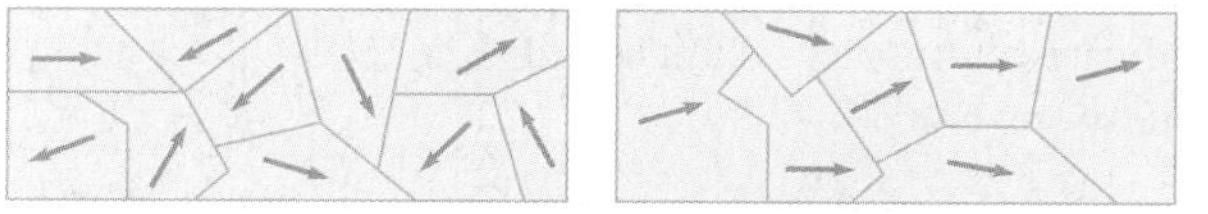

Time t_1 Time $t_2 > t_1$

18. Explain why a constant magnetic field does no work on a point charge moving through the field. Since the field does no work, what can we say about the speed of a point charge acted on only by a magnetic field?
19. Refer to the bubble chamber tracks in Fig. 19.15a. Suppose that particle 2 moves in a smaller circle than particle 1. Can we conclude that $|q_2| > |q_1|$? Explain.
20. The trajectory of a charged particle in a uniform magnetic field is a helix if $\vec{\mathbf{v}}$ has components both parallel to and perpendicular to the field. Explain how the two other cases (circular motion for $v_\parallel = 0$ and straight line motion for $v_\perp = 0$) can each be considered to be special cases of helical motion.

Multiple-Choice Questions

Multiple-Choice Questions 1–4. In the figure, four point charges move in the directions indicated in the vicinity of a bar magnet. The magnet, charge positions, and velocity vectors all lie in the plane of this page. Answer choices:

(a) ↑ (b) ↓ (c) ← (d) →
(e) × (into page) (f) • (out of page) (g) the force is zero

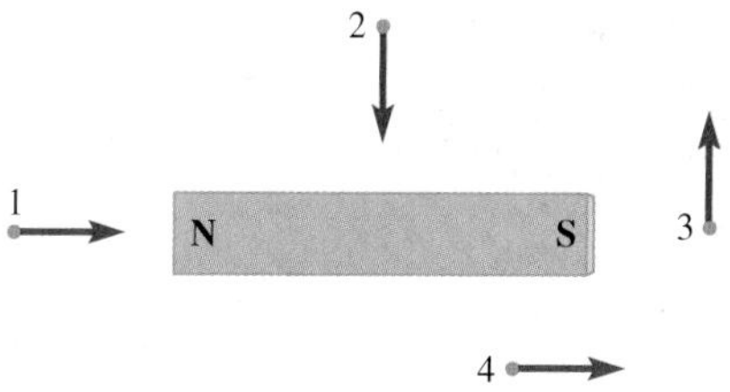

Multiple-Choice Questions 1–4

1. What is the direction of the magnetic force on charge 1 if $q_1 < 0$?
2. What is the direction of the magnetic force on charge 2 if $q_2 > 0$?
3. What is the direction of the magnetic force on charge 3 if $q_3 < 0$?
4. What is the direction of the magnetic force on charge 4 if $q_4 < 0$?
5. The magnetic force on a point charge in a magnetic field $\vec{\mathbf{B}}$ is largest, for a given speed, when it
   (a) moves in the direction of the magnetic field.
   (b) moves in the direction opposite to the magnetic field.
   (c) moves perpendicular to the magnetic field.
   (d) has velocity components both parallel to and perpendicular to the field.

Multiple-Choice Questions 6–9. A wire carries current as shown in the figure. Charged particles 1, 2, 3, and 4 move in the directions indicated. Answer choices for Questions 6–8:

(a) ↑ (b) ↓
(c) ← (d) →
(e) × (into page)
(f) ⊙ (out of page)
(g) the force is zero

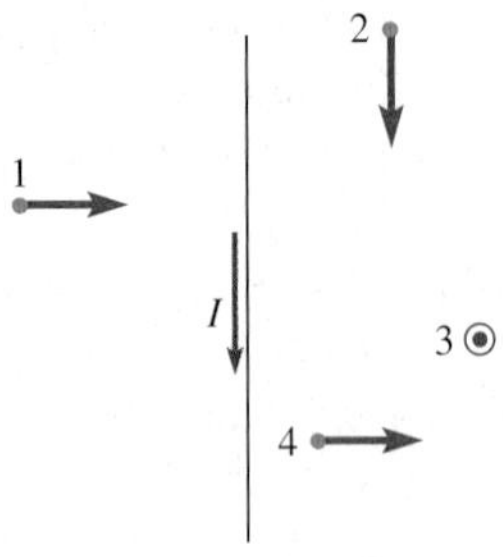

Multiple-Choice Questions 6–9

6. What is the direction of the magnetic force on charge 1 if $q_1 < 0$?
7. What is the direction of the magnetic force on charge 2 if $q_2 > 0$?
8. What is the direction of the magnetic force on charge 3 if $q_3 < 0$?
9. If the magnetic forces on charges 1 and 4 are equal and their velocities are equal,
   (a) the charges have the same sign and $|q_1| > |q_4|$.
   (b) the charges have opposite signs and $|q_1| > |q_4|$.
   (c) the charges have the same sign and $|q_1| < |q_4|$.
   (d) the charges have opposite signs and $|q_1| < |q_4|$.
   (e) $q_1 = q_4$.
   (f) $q_1 = -q_4$.
10. The magnetic field lines *inside* a bar magnet run in what direction?
    (a) from north pole to south pole
    (b) from south pole to north pole
    (c) from side to side
    (d) None of the above—there are no magnetic field lines *inside* a bar magnet.
11. The magnetic forces that two parallel wires with unequal currents flowing in opposite directions exert on each other are
    (a) attractive and unequal in magnitude.
    (b) repulsive and unequal in magnitude.
    (c) attractive and equal in magnitude.
    (d) repulsive and equal in magnitude.
    (e) both zero.
    (f) in the same direction and unequal in magnitude.
    (g) in the same direction and equal in magnitude.
12. What is the direction of the magnetic field at point *P* in the figure? (*P* is on the axis of the coil.)

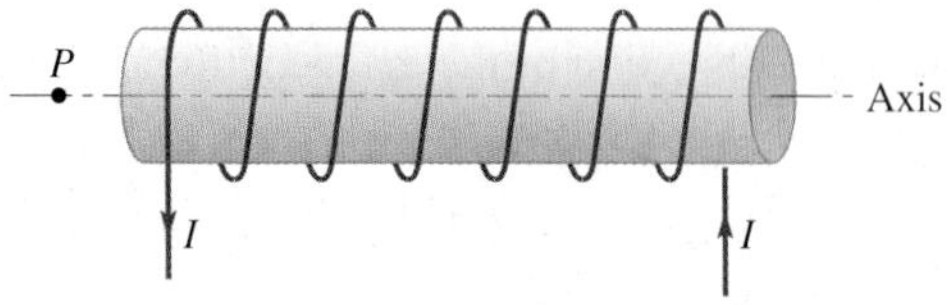

(a) ↑ (b) ↓ (c) ← (d) →
(e) × (into page) (f) • (out of page)

Problems

- Combination conceptual/quantitative problem
- Biological or medical application
- Challenging problem
- Blue # Detailed solution in the Student Solutions Manual
- 1 2 Problems paired by concept
- Text website interactive or tutorial

19.1 Magnetic Fields

1. At which point in the diagram is the magnetic field strength (a) the smallest and (b) the largest? Explain.

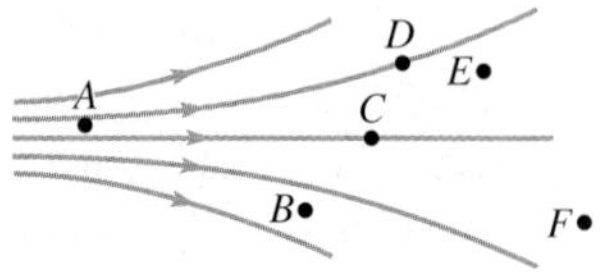

Problems 1 and 2

2. Draw vector arrows to indicate the direction and relative magnitude of the magnetic field at each of the points *A–F*.
3. Two identical bar magnets lie next to one another on a table. Sketch the magnetic field lines if the north poles are at the same end.

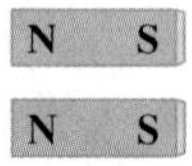

4. Two identical bar magnets lie next to one another on a table. Sketch the magnetic field lines if the north poles are at opposite ends.

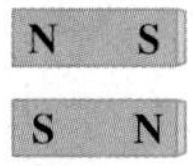

5. Two identical bar magnets lie on a table along a straight line with their north poles facing each other. Sketch the magnetic field lines.

6. Two identical bar magnets lie on a table along a straight line with opposite poles facing each other. Sketch the magnetic field lines.

7. The magnetic forces on a magnetic dipole result in a torque that tends to make the dipole line up with the magnetic field. In this problem we show that the *electric forces* on an *electric dipole* result in a torque that tends to make the electric dipole line up with the electric field. (a) For each orientation of the dipole shown in the diagram, sketch the electric forces and determine the direction of the torque—clockwise or counterclockwise—about an axis perpendicular to the page through the center of the dipole. (b) The torque always tends to make the dipole rotate toward what orientation?

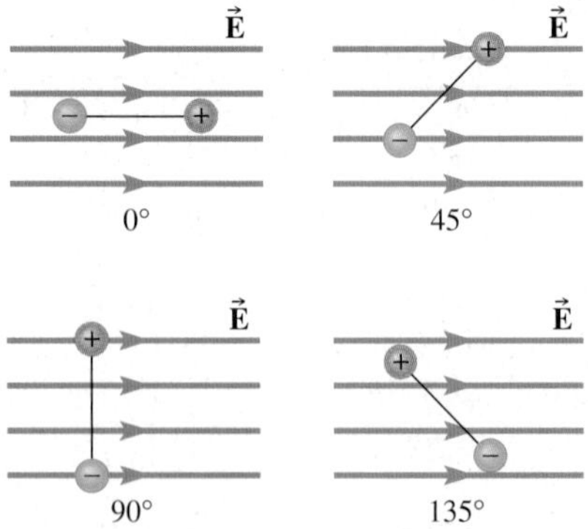

19.2 Magnetic Force on a Point Charge

8. Find the magnetic force exerted on an electron moving vertically upward at a speed of 2.0×10^7 m/s by a horizontal magnetic field of 0.50 T directed north. (tutorial: magnetic deflection of electron)

9. Find the magnetic force exerted on a proton moving east at a speed of 6.0×10^6 m/s by a horizontal magnetic field of 2.50 T directed north.

10. A uniform magnetic field points north; its magnitude is 1.5 T. A proton with kinetic energy 8.0×10^{-13} J is moving vertically downward in this field. What is the magnetic force acting on it?

11. A uniform magnetic field points vertically upward; its magnitude is 0.800 T. An electron with kinetic energy 7.2×10^{-18} J is moving horizontally eastward in this field. What is the magnetic force acting on it?

Problems 12–14. Several electrons move at speed 8.0×10^5 m/s in a uniform magnetic field with magnitude $B = 0.40$ T directed downward.

12. Find the magnetic force on the electron at point *a*.

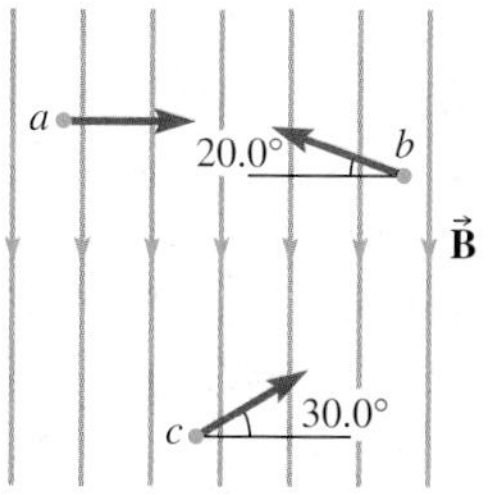

Problems 12–14

13. Find the magnetic force on the electron at point *b*.

14. Find the magnetic force on the electron at point *c*.

15. Electrons in a television's CRT are accelerated from rest by an electric field through a potential difference of 2.5 kV. In contrast to an oscilloscope, where the electron beam is deflected by an electric field, the beam is deflected by a magnetic field. (a) What is the speed of the electrons? (b) The beam is deflected by a perpendicular magnetic field of magnitude 0.80 T. What is the magnitude of the acceleration of the electrons while in the field? (c) What is the speed of the electrons after they travel 4.0 mm through the magnetic field? (d) What strength electric field would give the electrons the same magnitude acceleration as in (b)? (e) Why do we have to use an electric field in the first place to get the electrons up to speed? Why not use the large acceleration due to a magnetic field for that purpose?

16. A magnet produces a 0.30-T field between its poles, directed to the east. A dust particle with charge $q = -8.0 \times 10^{-18}$ C is moving straight down at 0.30 cm/s in this field. What is the magnitude and direction of the magnetic force on the dust particle?

17. At a certain point on Earth's surface in the southern hemisphere, the magnetic field has a magnitude of 5.0×10^{-5} T and points upward and toward the north at an angle of 55° above the horizontal. A cosmic ray muon with the same charge as an electron and a mass of 1.9×10^{-28} kg is moving directly down toward Earth's surface with a speed of 4.5×10^7 m/s. What is the magnitude and direction of the force on the muon?

18. An electron beam in vacuum moving at 1.8×10^7 m/s passes between the poles of an electromagnet. The diameter of the magnet pole faces is 2.4 cm and the field between them is 0.20×10^{-2} T. How far and in what direction is the beam deflected when it hits the screen, which is 25 cm past the magnet? [*Hint:* The electron velocity changes relatively little, so approximate the magnetic force as a constant force acting during a 2.4-cm displacement to the right.]

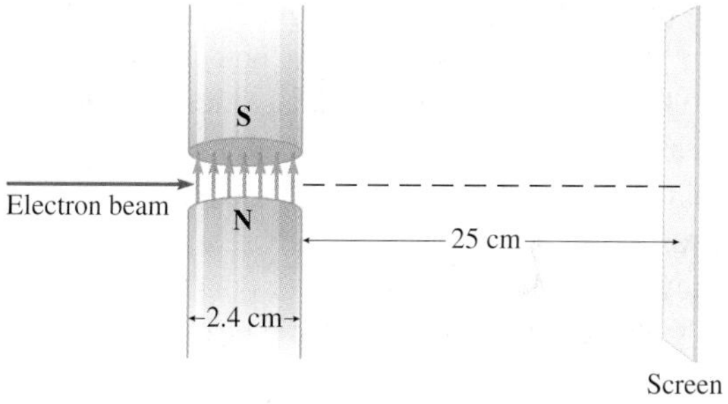

19. A positron ($q = +e$) moves at 5.0×10^7 m/s in a magnetic field of magnitude 0.47 T. The magnetic force on the positron has magnitude 2.3×10^{-12} N. (a) What is the component of the positron's velocity perpendicular to the magnetic field? (b) What is the component of the positron's velocity parallel to the magnetic field? (c) What is the angle between the velocity and the field?

20. An electron moves with speed 2.0×10^5 m/s in a 1.2-T uniform magnetic field. At one instant, the electron is moving due west and experiences an upward magnetic force of 3.2×10^{-14} N. What is the direction of the magnetic field? Be specific: give the angle(s) with respect to N, S, E, W, up, down. (If there is more than one possible answer, find all the possibilities.)

21. An electron moves with speed 2.0×10^5 m/s in a uniform magnetic field of 1.4 T, pointing south. At one instant, the electron experiences an upward magnetic force of 1.6×10^{-14} N. In what direction is the electron moving at that instant? Be specific: give the angle(s) with respect to N, S, E, W, up, down. (If there is more than one possible answer, find all the possibilities.)

19.3 Charged Particle Moving Perpendicularly to a Uniform Magnetic Field

22. The magnetic field in a cyclotron is 0.50 T. Find the magnitude of the magnetic force on a proton with speed 1.0×10^7 m/s moving in a plane perpendicular to the field.

23. An electron moves at speed 8.0×10^5 m/s in a plane perpendicular to a cyclotron's magnetic field. The magnitude

of the magnetic force on the electron is 1.0×10^{-13} N. What is the magnitude of the magnetic field?

24. When two particles travel through a region of uniform magnetic field pointing out of the plane of the paper, they follow the trajectories shown. What are the signs of the charges of each particle?

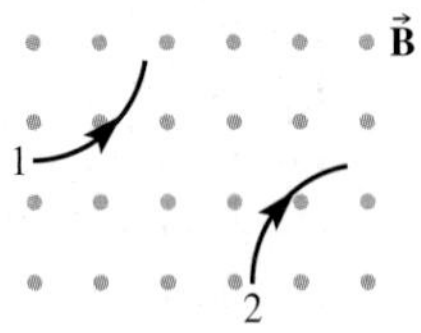

25. The magnetic field in a cyclotron is 0.360 T. The dees have radius 82.0 cm. What maximum speed can a proton achieve in this cyclotron?

26. The magnetic field in a cyclotron is 0.50 T. What must be the minimum radius of the dees if the maximum proton speed desired is 1.0×10^7 m/s?

27. A singly charged ion of unknown mass moves in a circle of radius 12.5 cm in a magnetic field of 1.2 T. The ion was accelerated through a potential difference of 7.0 kV before it entered the magnetic field. What is the mass of the ion?

Problems 28–32. The conversion between atomic mass units and kilograms is

$$1 \text{ u} = 1.66 \times 10^{-27} \text{ kg}$$

28. Natural carbon consists of two different isotopes (excluding ^{14}C, which is present in only trace amounts). The isotopes have different masses, which is due to different numbers of neutrons in the nucleus; however, the number of protons is the same, and subsequently the chemical properties are the same. The most abundant isotope has an atomic mass of 12.00 u. When natural carbon is placed in a mass spectrometer, two lines are formed on the photographic plate. The lines show that the more abundant isotope moved in a circle of radius 15.0 cm, while the rarer isotope moved in a circle of radius 15.6 cm. What is the atomic mass of the rarer isotope? (The ions have the same charge and are accelerated through the same potential difference before entering the magnetic field.)

29. After being accelerated through a potential difference of 5.0 kV, a singly charged carbon ion ($^{12}C^+$) moves in a circle of radius 21 cm in the magnetic field of a mass spectrometer. What is the magnitude of the field?

✦30. A sample containing carbon (atomic mass 12 u), oxygen (16 u), and an unknown element is placed in a mass spectrometer. The ions all have the same charge and are accelerated through the same potential difference before entering the magnetic field. The carbon and oxygen lines are separated by 2.250 cm on the photographic plate, and the unknown element makes a line between them that is 1.160 cm from the carbon line. (a) What is the mass of the unknown element? (b) Identify the element.

31. A sample containing sulfur (atomic mass 32 u), manganese (55 u), and an unknown element is placed in a mass spectrometer. The ions have the same charge and are accelerated through the same potential difference before entering the magnetic field. The sulfur and manganese lines are separated by 3.20 cm, and the unknown element makes a line between them that is 1.07 cm from the sulfur line. (a) What is the mass of the unknown element? (b) Identify the element.

✦32. In one type of mass spectrometer, ions having the *same velocity* move through a uniform magnetic field. The spectrometer is being used to distinguish $^{12}C^+$ and $^{14}C^+$ ions that have the same charge. The $^{12}C^+$ ions move in a circle of diameter 25 cm. (a) What is the diameter of the orbit of $^{14}C^+$ ions? (b) What is the ratio of the frequencies of revolution for the two types of ion?

✦33. Prove that the time for one revolution of a charged particle moving perpendicular to a uniform magnetic field is independent of its speed. (This is the principle on which the cyclotron operates.) In doing so, write an expression that gives the period T (the time for one revolution) in terms of the mass of the particle, the charge of the particle, and the magnetic field strength.

19.5 A Charged Particle in Crossed $\vec{E}$ and $\vec{B}$ Fields

34. Crossed electric and magnetic fields are established over a certain region. The magnetic field is 0.635 T vertically downward. The electric field is 2.68×10^6 V/m horizontally east. An electron, traveling horizontally northward, experiences zero net force from these fields and so continues moving in a straight line. What is the electron's speed?

◉35. A current $I = 40.0$ A flows through a strip of metal. An electromagnet is switched on so that there is a uniform magnetic field of magnitude 0.30 T directed into the page. (a) How would you hook up a voltmeter to measure the Hall voltage? Show how the voltmeter is connected on a sketch of the strip. (b) Assuming the carriers are electrons, which lead of your voltmeter is at the higher potential? Mark it with a "+" sign in your sketch. Explain briefly.

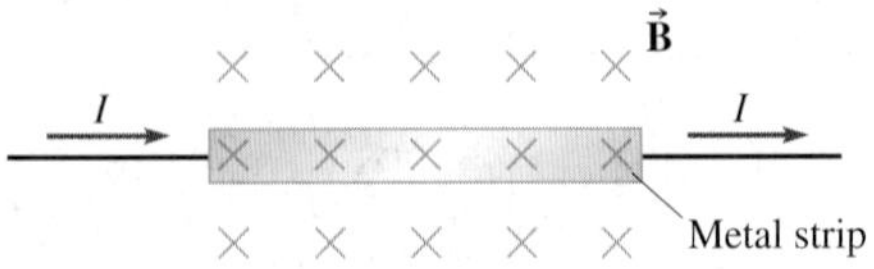

Problems 35–39

36. In Problem 35, if the width of the strip is 3.5 cm, the magnetic field is 0.43 T, and the Hall voltage is measured to be 7.2 μV, what is the drift velocity of the carriers in the strip?

37. In Problem 35, the width of the strip is 3.5 cm, the magnetic field is 0.43 T, the Hall voltage is measured to be 7.2 μV, the thickness of the strip is 0.24 mm, and the current in the wire is 54 A. What is the density of carriers (number of carriers per unit volume) in the strip?

38. The strip in the diagram is used as a Hall probe to measure magnetic fields. (a) What happens if the strip is not perpendicular to the field? Does the Hall probe still read the correct field strength? Explain. (b) What happens if the field is in the plane of the strip?

39. A strip of copper 2.0 cm wide carries a current I = 30.0 A to the right. The strip is in a magnetic field B = 5.0 T into the page. (a) What is the direction of the average magnetic force on the conduction electrons? (b) The Hall voltage is 20.0 μV. What is the drift velocity?

40. A proton is initially at rest and moves through three different regions as shown in the figure. In region 1, the proton accelerates across a potential difference of 3330 V. In region 2, there is a magnetic field of 1.20 T pointing out of the page and an electric field pointing perpendicular to the magnetic field and perpendicular to the proton's velocity. Finally, in region 3, there is no electric field, but just a 1.20-T magnetic field pointing out of the page. (a) What is the speed of the proton as it leaves region 1 and enters region 2? (b) If the proton travels in a straight line through region 2, what is the magnitude and direction of the electric field? (c) In region 3, will the proton follow path 1 or 2? (d) What will be the radius of the circular path the proton travels in region 3?

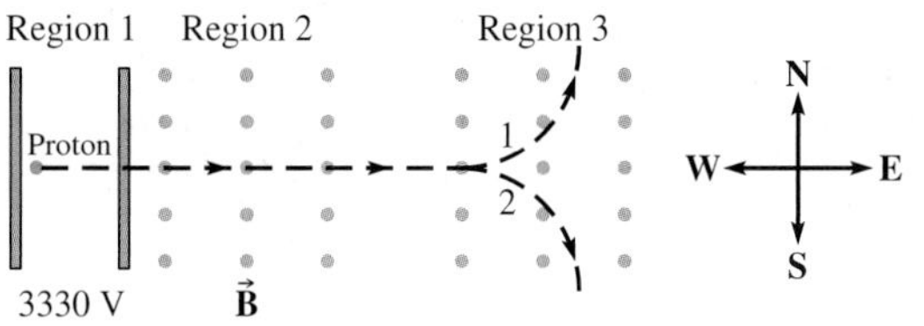

41. An electromagnetic flowmeter is used to measure blood flow rates during surgery. Blood containing Na^+ ions flows due south through an artery with a diameter of 0.40 cm. The artery is in a downward magnetic field of 0.25 T and develops a Hall voltage of 0.35 mV across its diameter. (a) What is the blood speed (in m/s)? (b) What is the flow rate (in m^3/s)? (c) The leads of a voltmeter are attached to diametrically opposed points on the artery to measure the Hall voltage. Which of the two leads is at the higher potential?

42. An electromagnetic flowmeter is used to measure blood flow rates during surgery. Blood containing ions (primarily Na^+) flows through an artery with a diameter of 0.50 cm. The artery is in a magnetic field of 0.35 T and develops a Hall voltage of 0.60 mV across its diameter. (a) What is the blood speed (in m/s)? (b) What is the flow rate (in m^3/s)? (c) If the magnetic field points west and the blood flow is north, is the top or bottom of the artery at the higher potential?

43. A charged particle is accelerated from rest through a potential difference ΔV. The particle then passes straight through a velocity selector (field magnitudes E and B). Derive an expression for the charge-to-mass ratio (q/m) of the particle in terms of ΔV, E, and B.

19.6 Magnetic Force on a Current-Carrying Wire

44. A straight wire segment of length 0.60 m carries a current of 18.0 A and is immersed in a uniform external magnetic field of magnitude 0.20 T. (a) What is the magnitude of the maximum possible magnetic force on the wire segment? (b) Explain why the given information enables you to calculate only the *maximum possible* force.

45. A straight wire segment of length 25 cm carries a current of 33.0 A and is immersed in a uniform external magnetic field. The magnetic force on the wire segment has magnitude 4.12 N. (a) What is the minimum possible magnitude of the magnetic field? (b) Explain why the given information enables you to calculate only the *minimum possible* field strength.

46. Parallel conducting tracks, separated by 2.0 cm, run north and south. There is a uniform magnetic field of 1.2 T pointing upward (out of the page). A 0.040-kg cylindrical metal rod is placed across the tracks and a battery is connected between the tracks, with its positive terminal connected to the east track. If the current through the rod is 3.0 A, find the magnitude and direction of the magnetic force on the rod.

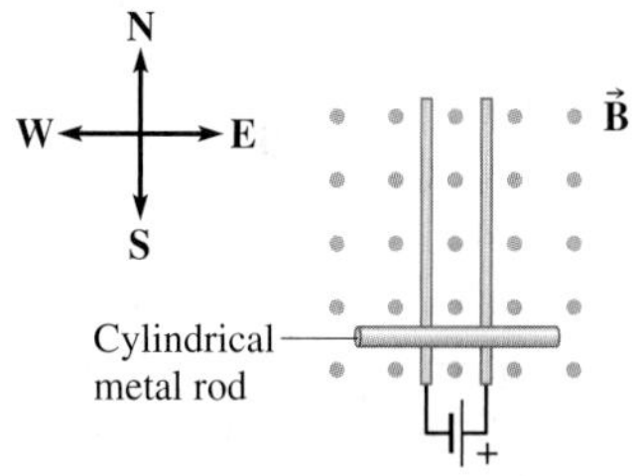

47. An electromagnetic rail gun can fire a projectile using a magnetic field and an electric current. Consider two conducting rails that are 0.500 m apart with a 50.0-g conducting rod connecting the two rails as in the figure with Problem 46. A magnetic field of magnitude 0.750 T is directed perpendicular to the plane of the rails and rod. A current of 2.00 A passes through the rod. (a) What direction is the force on the rod? (b) If there is no friction between the rails and the rod, how fast is the rod moving after it has traveled 8.00 m down the rails?

48. A straight, stiff wire of length 1.00 m and mass 25 g is suspended in a magnetic field B = 0.75 T. The wire is connected to an emf. How much current must flow in the wire and in what direction so that the wire is suspended and the tension in the supporting wires is zero?

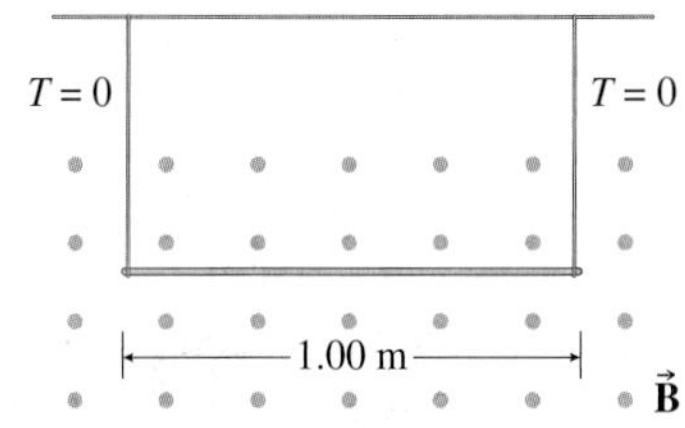

49. A 20.0 cm × 30.0 cm rectangular loop of wire carries 1.0 A of current clockwise around the loop. (a) Find the magnetic force on each side of the loop if the magnetic field is 2.5 T out of the page. (b) What is the net magnetic force on the loop?

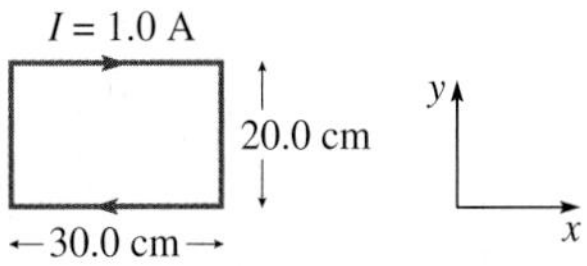

Problems 49, 50, and 107

50. Repeat Problem 49 if the magnetic field is 2.5 T to the left (in the $-x$-direction).

✦51. A straight wire is aligned east-west in a region where Earth's magnetic field has magnitude 0.48 mT and direction 72° below the horizontal, with the horizontal component directed due north. The wire carries a current I toward the west. The magnetic force on the wire per unit length of wire has magnitude 0.020 N/m. (a) What is the direction of the magnetic force on the wire? (b) What is the current I?

✦52. A straight wire is aligned north-south in a region where Earth's magnetic field $\vec{\mathbf{B}}$ is directed 58.0° above the horizontal, with the horizontal component directed due north. The wire carries a current of 8.00 A toward the south. The magnetic force on the wire per unit length of wire has magnitude 2.80×10^{-3} N/m. (a) What is the direction of the magnetic force on the wire? (b) What is the magnitude of $\vec{\mathbf{B}}$?

19.7 Torque on a Current Loop

53. In an electric motor, a circular coil with 100 turns of radius 2.0 cm can rotate between the poles of a magnet. When the current through the coil is 75 mA, the maximum torque that the motor can deliver is 0.0020 N·m. (a) What is the strength of the magnetic field? (b) Is the torque on the coil clockwise or counterclockwise as viewed from the front at the instant shown in the figure?

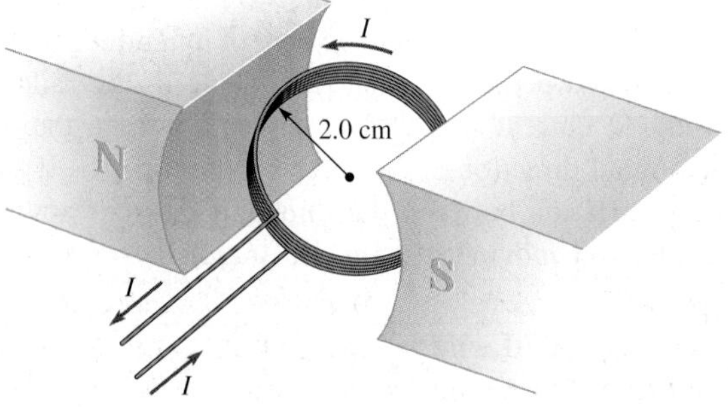

54. In an electric motor, a coil with 100 turns of radius 2.0 cm can rotate between the poles of a magnet. The magnetic field strength is 0.20 T. When the current through the coil is 50.0 mA, what is the maximum torque that the motor can deliver?

55. A square loop of wire of side 3.0 cm carries 3.0 A of current. A uniform magnetic field of magnitude 0.67 T makes an angle of 37° with the plane of the loop.

(a) What is the magnitude of the torque on the loop?
(b) What is the net magnetic force on the loop?

56. The torque on a loop of wire (a magnetic dipole) in a uniform magnetic field is $\tau = NIAB \sin \theta$, where θ is the angle between $\vec{\mathbf{B}}$ and a line perpendicular to the loop of wire. Suppose an *electric* dipole, consisting of two charges $\pm q$ a fixed distance d apart is in a uniform electric field $\vec{\mathbf{E}}$. (a) Show that the net electric force on the dipole is zero. (b) Let θ be the angle between $\vec{\mathbf{E}}$ and a line running from the negative to the positive charge. Show that the torque on the electric dipole is $\tau = qdE \sin \theta$ for all angles $-180° \le \theta \le 180°$. (Thus, for both electric and magnetic dipoles, the torque is the product of the *dipole moment* times the field strength times $\sin \theta$. The quantity qd is the electric dipole moment; the quantity NIA is the magnetic dipole moment.)

✦57. A certain fixed length L of wire carries a current I. (a) Show that if the wire is formed into a square coil, then the maximum torque in a given magnetic field B is developed when the coil has just one turn. (b) Show that the magnitude of this torque is $\tau = \frac{1}{16}L^2IB$.

✦58. Use the following method to show that the torque on an irregularly shaped planar loop is given by Eq. (19-13a). The irregular loop of current in part (a) of the figure carries current I. There is a perpendicular magnetic field B. To find the torque on the irregular loop, sum up the torques on each of the smaller loops shown in part (b) of the figure. The pairs of imaginary currents flowing across carry equal currents in opposite directions, so the magnetic forces on them would be equal and opposite; they would therefore contribute nothing to the net torque. Now generalize this argument to a loop of *any* shape. [*Hint:* Think of a curved loop as a series of tiny, straight, perpendicular segments.]

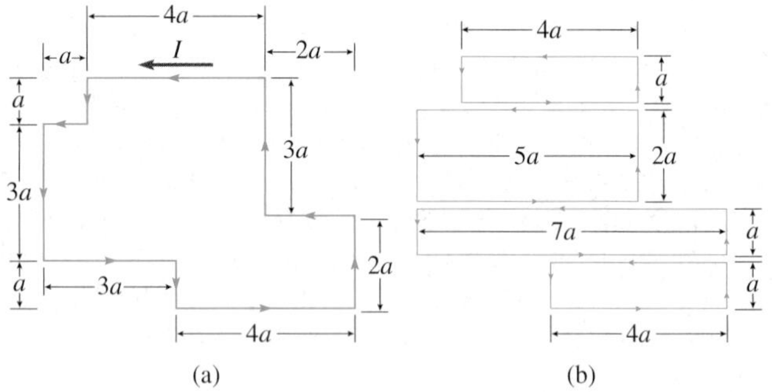

✦59. A square loop of wire with side 0.60 m carries a current of 9.0 A as shown in the figure. When there is no applied magnetic field, the plane of the loop is horizontal and the nonconducting, nonmagnetic spring ($k = 550$ N/m) is unstretched. A horizontal magnetic field of magnitude 1.3 T is now applied. At what

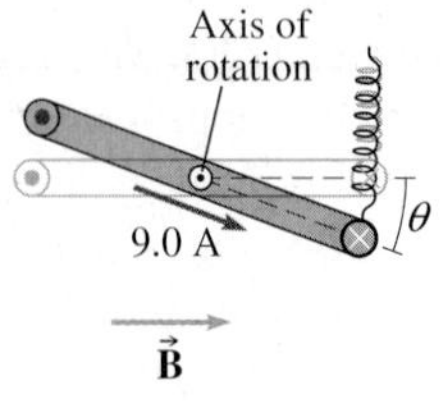

angle θ is the wire loop's new equilibrium position? Assume the spring remains vertical because θ is small. [*Hint:* Set the sum of the torques from the spring and the magnetic field equal to 0.]

19.8 Magnetic Field due to an Electric Current

60. Imagine a long straight wire perpendicular to the page and carrying a current I into the page. Sketch some $\vec{\mathbf{B}}$ field lines with arrowheads to indicate directions.

61. Two wires each carry 10.0 A of current (in *opposite directions*) and are 3.0 mm apart. Calculate the magnetic field 25 cm away at point P, in the plane of the wires.

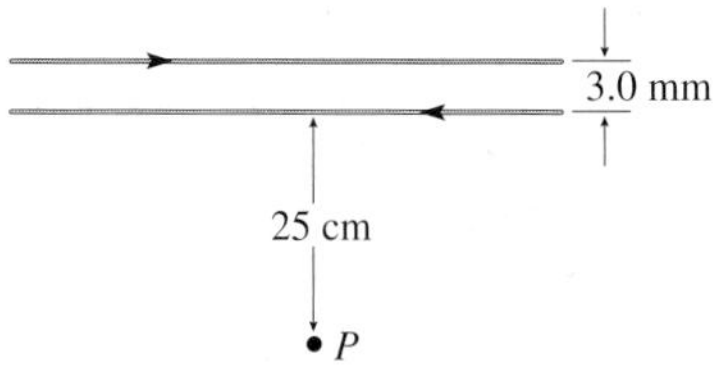

Problems 61–62

62. What is the magnetic field at point P if the currents instead both run to the left in Problem 61?

63. Point P is midway between two long, straight, parallel wires that run north-south in a horizontal plane. The distance between the wires is 1.0 cm. Each wire carries a current of 1.0 A toward the north. Find the magnitude and direction of the magnetic field at point P.

64. Repeat Problem 63 if the current in the wire on the east side runs toward the south instead.

65. A long straight wire carries a current of 50.0 A. An electron, traveling at 1.0×10^7 m/s, is 5.0 cm from the wire. What force (magnitude and direction) acts on the electron if the electron's velocity is directed toward the wire?

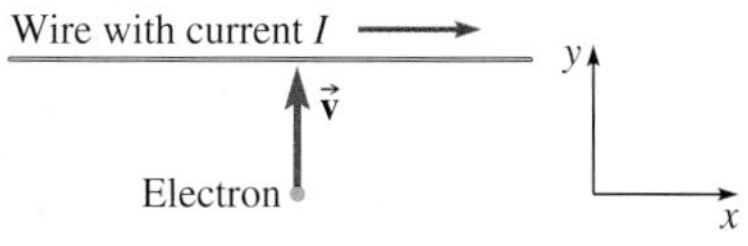

Problems 65–91

66. A long straight wire carries a current of 3.2 A in the positive x-direction. An electron, traveling at 6.8×10^6 m/s in the positive x-direction, is 4.6 cm from the wire. What force acts on the electron?

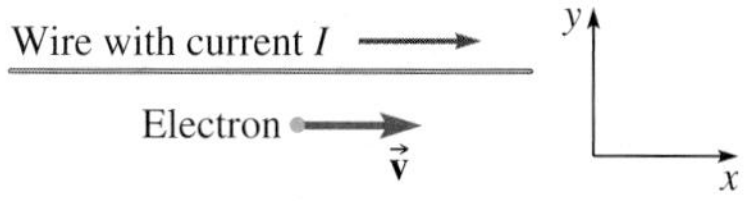

67. Two long straight wires carry the same amount of current in the directions indicated. The wires cross each other in the plane of the paper. Rank points A, B, C, and D in order of decreasing field strength.

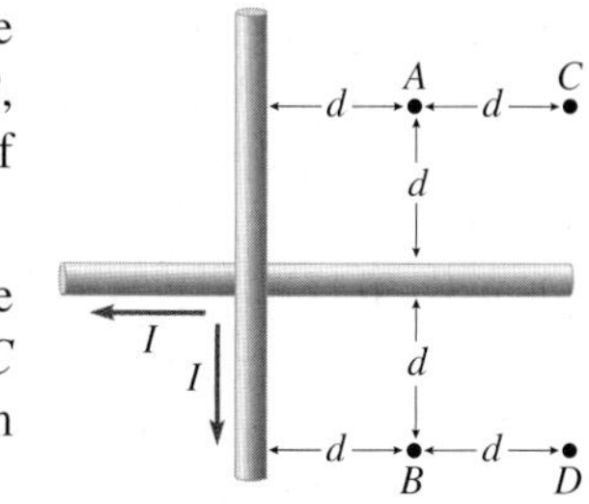

Problems 67 and 68

68. In Problem 67, find the magnetic field at points C and D when $d = 3.3$ cm and $I = 6.50$ A.

69. In Problem 67, find the magnetic field at points A and B when $d = 6.75$ cm and $I = 57.0$ mA.

70. A solenoid of length 0.256 m and radius 2.0 cm has 244 turns of wire. What is the magnitude of the magnetic field well inside the solenoid when there is a current of 4.5 A in the wire?

71. Two long straight parallel wires separated by 8.0 cm carry currents of equal magnitude but heading in opposite directions. The wires are shown perpendicular to the plane of this page. Point P is 2.0 cm from wire 1, and the magnetic field at point P is 1.0×10^{-2} T directed in the $-y$-direction. Calculate the current in wire 1 and its direction.

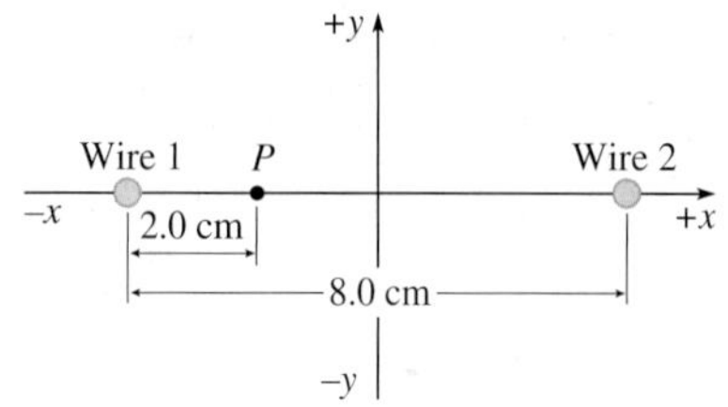

72. Two parallel wires in a horizontal plane carry currents I_1 and I_2 to the right. The wires each have length L and are separated by a distance d. (a) What are the magnitude and direction of the field due to wire 1 at the location of wire 2? (b) What are the magnitude and direction of the magnetic force on wire 2 due to this field? (c) What are the magnitude and direction of the field due to wire 2 at the location of wire 1? (d) What are the magnitude and direction of the magnetic force on wire 1 due to this field? (e) Do parallel currents in the same direction attract or repel? (f) What about parallel currents in opposite directions?

73. Two concentric circular wire loops in the same plane each carry a current. The larger loop has a current of 8.46 A circulating clockwise and has a radius of 6.20 cm. The smaller loop has a radius of 4.42 cm. What is the current in the smaller loop if the total magnetic field at the center of the system is zero? [See Eq. (19-16).]

74. A solenoid has 4850 turns *per meter* and radius 3.3 cm. The magnetic field inside has magnitude 0.24 T. What is the current in the solenoid?

Problems 75–77. Four long parallel wires pass through the corners of a square with side 0.10 m. All four wires carry the same magnitude of current $I = 10.0$ A in the directions indicated.

75. Find the magnetic field at the center of the square.

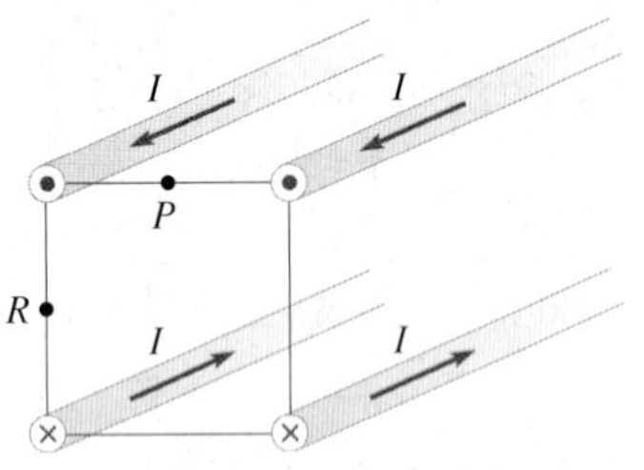

Problems 75–77

76. Find the magnetic field at point P, the midpoint of the top side of the square.

77. Find the magnetic field at point R, the midpoint of the left side of the square.

78. Four long straight wires, each with current I overlap to form a square with side $2r$. (a) Find the magnetic field at the center of the square. (b) Compare your answer with the magnetic field at the center of a circular loop of radius r carrying current I [see Eq. (19-16)].

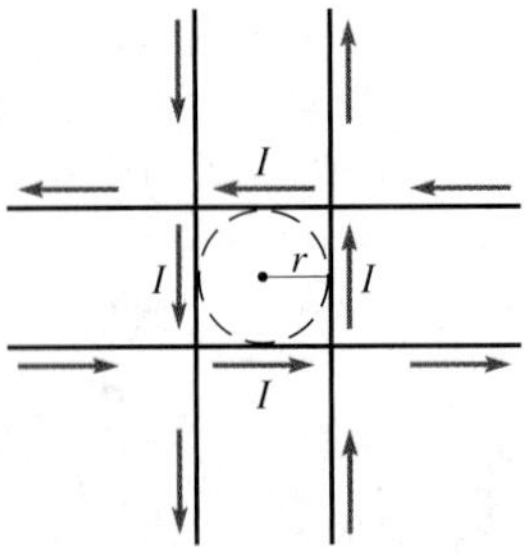

✦79. Two parallel long straight wires are suspended by strings of length $L = 1.2$ m. Each wire has mass per unit length 0.050 kg/m. When the wires each carry 50.0 A of current, the wires swing apart. (a) How far apart are the wires in equilibrium? (Assume that this distance is small compared with L.) [*Hint:* Use a small angle approximation.] (b) Are the wires carrying current in the same or opposite directions?

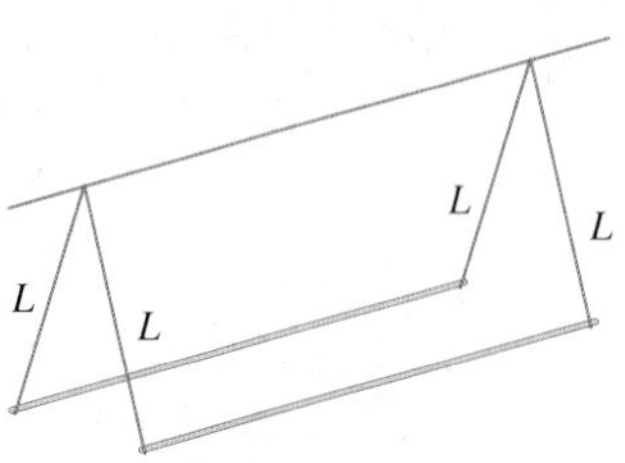

19.9 Ampère's Law

80. A number of wires carry currents into or out of the page as indicated in the figure. (a) Using loop 1 for Ampère's law, what is the net current through the interior of the loop? (b) Repeat for loop 2.

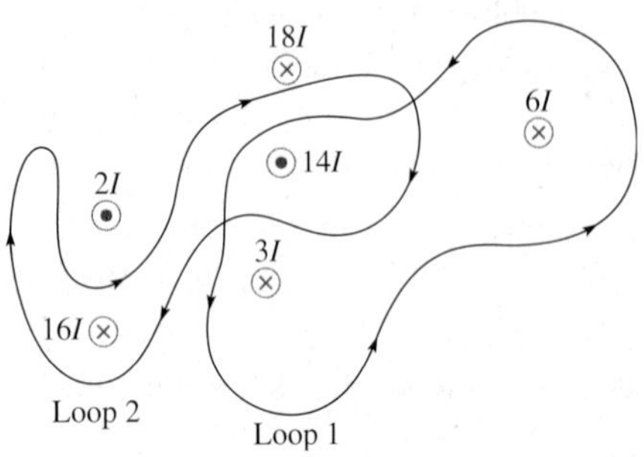

✦81. An infinitely long, thick cylindrical shell of inner radius a and outer radius b carries a current I uniformly distributed across a cross section of the shell. (a) On a sketch of a cross section of the shell, draw some magnetic field lines. The current flows out of the page. Consider all regions ($r \le a$, $a \le r \le b$, $b \le r$). (b) Find the magnetic field for $r > b$.

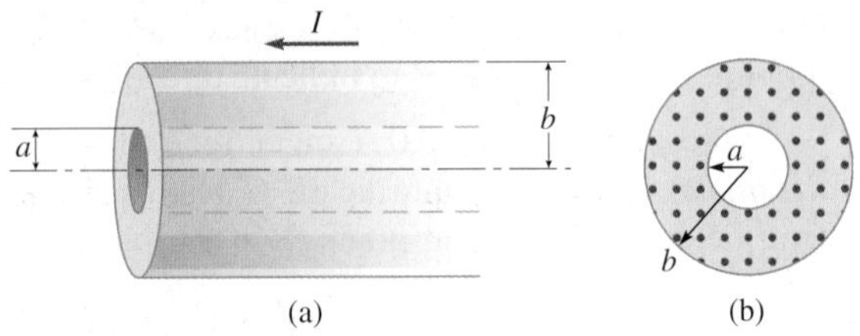

✦82. In this problem, use Ampère's law to show that the magnetic field inside a long solenoid is $B = \mu_0 nI$. Assume that the field inside the solenoid is uniform and parallel to the axis and that the field outside is zero. Choose a rectangular path for Ampère's law. (a) Write down $B_{\parallel}\Delta l$ for each of the four sides of the path, in terms of B, a, (the short side) and b (the long side). (b) Sum these to form the circulation. (c) Now, to find the current cutting through the path: each loop carries the same current I, and some number N of loops cut through the path, so the total current is NI. Rewrite N in terms of the number of turns per unit length (n) and the physical dimensions of the path. (d) Solve for B.

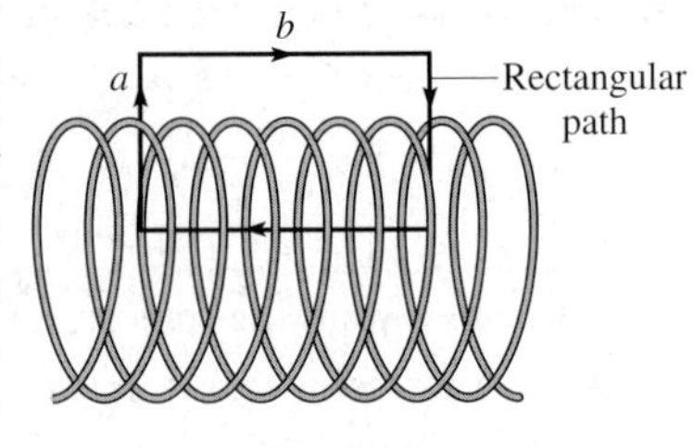

✦83. A toroid is like a solenoid that has been bent around in a circle until its ends meet. The field lines are circular, as shown in the figure. What is the magnitude of the magnetic field inside a toroid of N turns carrying current I? Apply Ampère's law, following a field line at a distance r from the center of the toroid. Work in terms of the total number of turns N, rather than the number of turns per unit length (why?). Is the field uniform, as it is for a long solenoid? Explain.

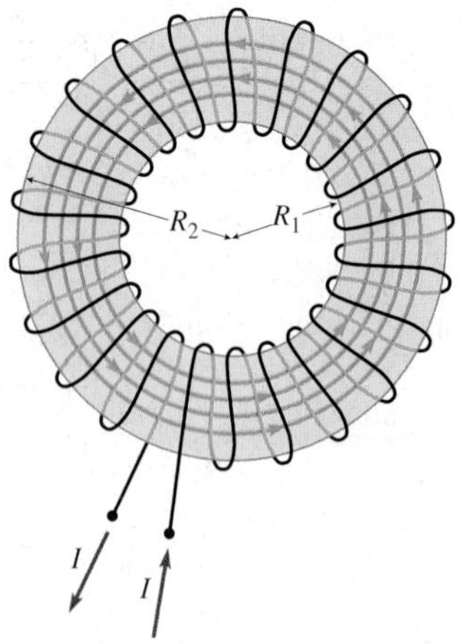

19.10 Magnetic Materials

84. A bar magnet is broken into two parts. If care is taken not to disturb the magnetic domains, what are the polarities of the new ends c and d, respectively?

85. The intrinsic magnetic dipole moment of the electron has magnitude 9.3×10^{-24} A·m^2. In other words, the electron acts as though it were a tiny current loop with $NIA = 9.3 \times 10^{-24}$ A·m^2. What is the maximum torque on an electron due to its intrinsic dipole moment in a 1.0-T magnetic field?

86. An electromagnet is made by inserting a soft iron core into a solenoid. The solenoid has 1800 turns, radius 2.0 cm, and length 15 cm. When 2.0 A of current flows through the solenoid, the magnetic field inside the iron core has magnitude 0.42 T. What is the relative permeability κ_B of the iron core? (See Section 19.10 for the definition of κ_B.)

87. The figure shows *hysteresis curves* for three different materials. A hysteresis curve is a plot of the magnetic field strength inside the material (B) as a function of the externally applied field (B_0). (a) Which material would make the best permanent magnet? Explain. (b) Which would make the best core for an electromagnet? Explain.

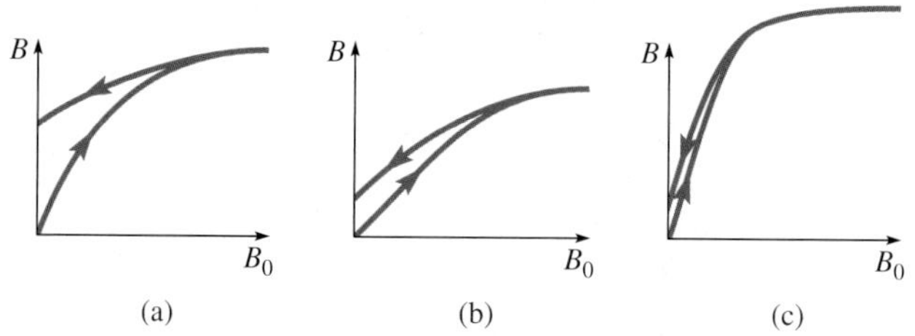

88. In a simple model, the electron in a hydrogen atom orbits the proton at a radius of 53 pm and at a constant speed of 2.2×10^6 m/s. The orbital motion of the electron gives it an orbital magnetic dipole moment. (a) What is the current I in this current loop? [*Hint:* How long does it take the electron to make one revolution?] (b) What is the orbital dipole moment IA? (c) Compare the orbital dipole moment with the intrinsic magnetic dipole moment of the electron (9.3×10^{-24} A·m^2).

Comprehensive Problems

89. A compass is placed directly on top of a wire (needle not shown). The current in the wire flows to the right. Which way does the north end of the needle point? Explain. (Ignore Earth's magnetic field.)

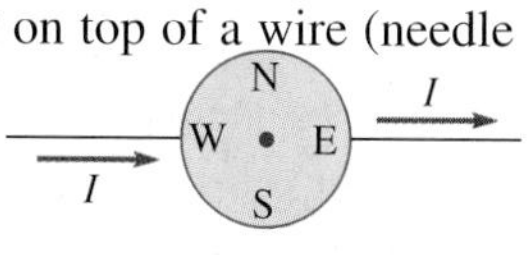

90. You want to build a cyclotron to accelerate protons to a speed of 3.0×10^7 m/s. The largest magnetic field strength you can attain is 1.5 T. What must be the minimum radius of the dees in your cyclotron? Show how your answer comes from Newton's second law.

91. A long straight wire carries a 4.70-A current in the positive x-direction. At a particular instant, an electron moving at 1.00×10^7 m/s in the positive y-direction is 0.120 m from the wire. Determine the magnetic force on the electron at this instant. See the figure with Problem 65.

92. A uniform magnetic field of 0.50 T is directed to the north. At some instant, a particle with charge +0.020 μC is moving with velocity 2.0 m/s in a direction 30° north of east. (a) What is the magnitude of the magnetic force on the charged particle? (b) What is the direction of the magnetic force?

93. Two identical long straight conducting wires with a mass per unit length of 25.0 g/m are resting parallel to each other on a table. The wires are separated by 2.5 mm and are carrying currents in opposite directions. (a) If the coefficient of static friction between the wires and the table is 0.035, what minimum current is necessary to make the wires start to move? (b) Do the wires move closer together or farther apart?

94. Two long insulated wires lie in the same horizontal plane. A current of 20.0 A flows toward the north in wire A and a current of 10.0 A flows toward the east in wire B. What is the magnitude and direction of the magnetic field at a point that is 5.00 cm above the point where the wires cross?

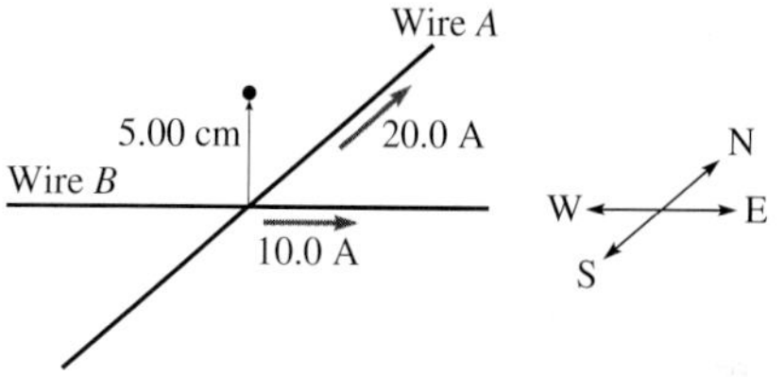

95. (a) A proton moves with uniform circular motion in a magnetic field of magnitude 0.80 T. At what frequency f does it circulate? (b) Repeat for an electron.

96. The concentration of free electrons in silver is 5.85×10^{28} per m^3. A strip of silver of thickness 0.050 mm and width 20.0 mm is placed in a magnetic field of 0.80 T. A current of 10.0 A is sent down the strip. (a) What is the drift velocity of the electrons? (b) What is the Hall voltage measured by the meter? (c) Which side of the voltmeter is at the higher potential?

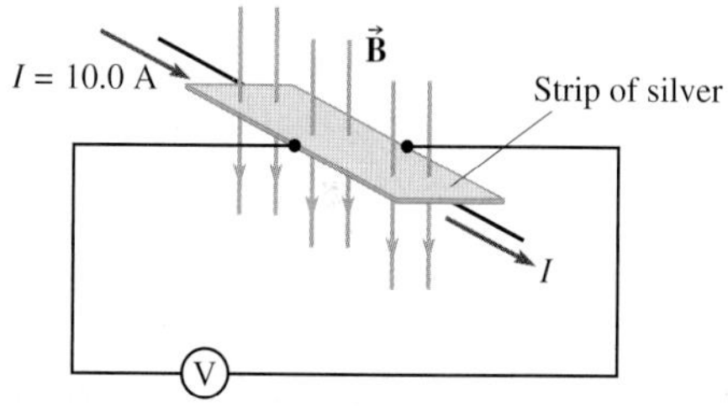

97. An electromagnetic flowmeter is to be used to measure blood speed. A magnetic field of 0.115 T is applied across an artery of inner diameter 3.80 mm. The Hall voltage is measured to be 88.0 μV. What is the average speed of the blood flowing in the artery?

98. Sketch the magnetic field as it would appear inside the coil of wire to an observer, looking into the coil from the position shown.

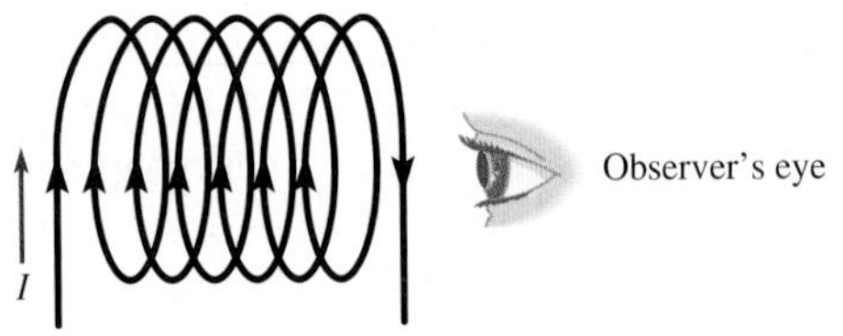

99. Two conducting wires perpendicular to the page are shown in cross section as gray dots in the figure. They each carry 10.0 A out of the page. What is the magnetic field at point P?

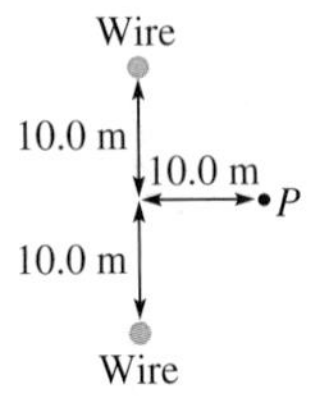

100. A strip of copper carries current in the $+x$-direction. There is an external magnetic field directed out of the page. What is the direction of the Hall electric field?

101. A bar magnet is held near the electron beam in an oscilloscope. The beam passes directly below the south pole of the magnet. In what direction will the beam move on the screen? (*Don't try this* with a color TV tube. There is a metal mask just behind the screen that separates the pixels for red, green, and blue. If you succeed in magnetizing the mask, the picture will be permanently distorted.)

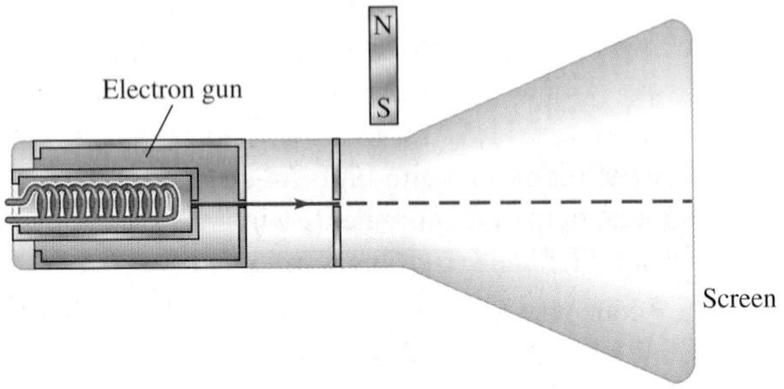

102. The strength of Earth's magnetic field, as measured on the surface, is approximately 6.0×10^{-5} T at the poles and 3.0×10^{-5} T at the equator. Suppose an alien from outer space were at the North Pole with a single loop of wire of the same circumference as his space helmet. The diameter of his helmet is 20.0 cm. The space invader wishes to cancel Earth's magnetic field at his location. (a) What is the current required to produce a magnetic field (due to the current alone) at the center of his loop of the same size as that of Earth's field at the North Pole? (b) In what direction does the current circulate in the loop, CW or CCW, as viewed from above, if it is to cancel Earth's field?

103. A tangent galvanometer is an instrument, developed in the nineteenth century, designed to measure current based on the deflection of a compass needle. A coil of wire in a vertical plane is aligned in the magnetic north-south direction. A compass is placed in a horizontal plane at the center of the coil. When no current flows, the compass needle points directly toward the north side of the coil. When a current is sent through the coil, the compass needle rotates through an angle θ. Derive an equation for θ in terms of the number of coil turns N, the coil radius r, the coil current I, and the horizontal component of Earth's field B_H. [*Hint:* The name of the instrument is a clue to the result.]

104. In the mass spectrometer of the diagram, neon ions ($q = +e$) come from the ion source and are accelerated through a potential difference V. The ions then pass through an aperture in a metal plate into a uniform magnetic field where they travel in semicircular paths until exiting into the detector. Neon ions having a mass of 20.0 u leave the field at a distance of 50.0 cm from the aperture. At what distance from the aperture do neon ions having a mass of 22.0 u leave the field? ($1 \text{ u} = 1.66 \times 10^{-27}$ kg.)

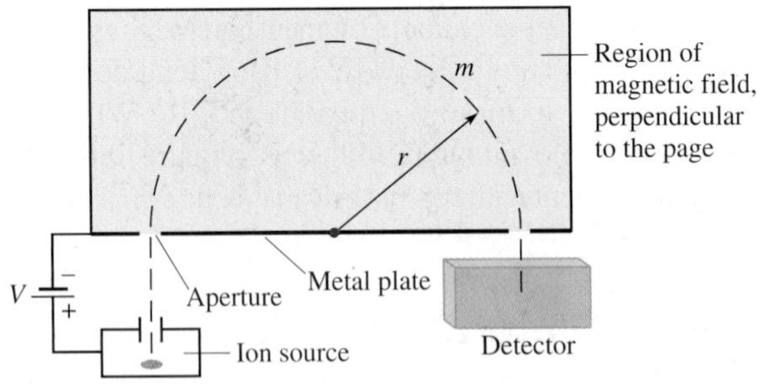

105. A rectangular loop of wire, carrying current $I_1 = 2.0$ mA, is next to a very long wire carrying a current $I_2 = 8.0$ A. (a) What is the direction of the magnetic force on each of the four sides of the rectangle due to the long wire's magnetic field? (b) Calculate the net magnetic force on the rectangular loop due to the long wire's magnetic

field. [*Hint:* The long wire does *not* produce a uniform magnetic field.]

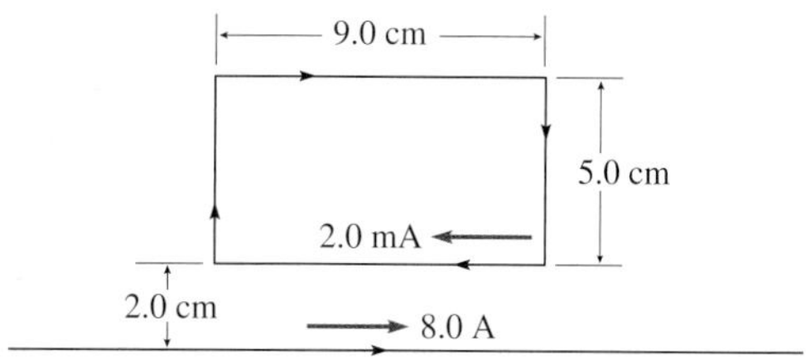

106. In a carbon-dating experiment, a particular type of mass spectrometer is used to separate ^{14}C from ^{12}C. Carbon ions from a sample are first accelerated through a potential difference ΔV_1 between the charged accelerating plates. Then the ions enter a region of uniform vertical magnetic field $B = 0.200$ T. The ions pass between deflection plates spaced 1.00 cm apart. By adjusting the potential difference ΔV_2 between these plates, only one of the two isotopes (^{12}C or ^{14}C) is allowed to pass through to the next stage of the mass spectrometer. The distance from the entrance to the ion detector is a fixed 0.200 m. By suitably adjusting ΔV_1 and ΔV_2, the detector counts only one type of ion, so the relative abundances can be determined. (a) Are the ions positively or negatively charged? (b) Which of the accelerating plates (east or west) is positively charged? (c) Which of the deflection plates (north or south) is positively charged? (d) Find the correct values of ΔV_1 and ΔV_2 in order to count $^{12}C^+$ ions (mass 1.993×10^{-26} kg). (e) Find the correct values of ΔV_1 and ΔV_2 in order to count $^{14}C^+$ ions (mass 2.325×10^{-26} kg).

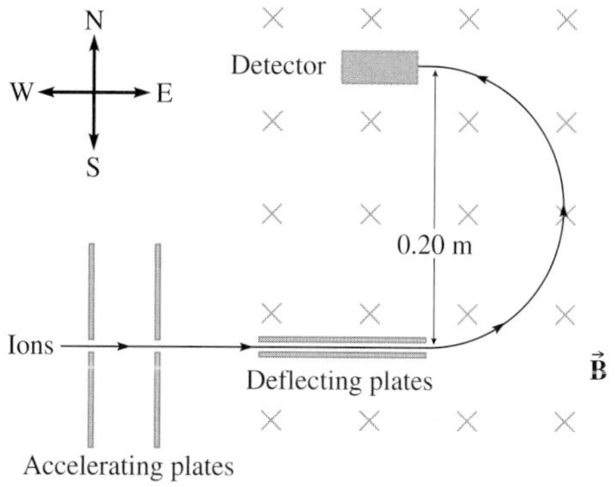

107. Repeat Problem 49 if the magnetic field is 2.5 T in the plane of the loop, 60.0° below the $+x$-axis.

108. A *current balance* is a device to measure magnetic forces. It is constructed from two parallel coils, each with an average radius of 12.5 cm. The lower coil rests on a balance; it has 20 turns and carries a constant current of 4.0 A. The upper coil, suspended 0.314 cm above the lower coil, has 50 turns and a current that can be varied. The reading of the balance changes as the magnetic force on the lower coil changes. What current is needed in the upper coil to exert a force of 1.0 N on the bottom coil? [*Hint:* Since the distance between the coils is small relative to the radius of the coils, approximate the setup as two long parallel straight wires.]

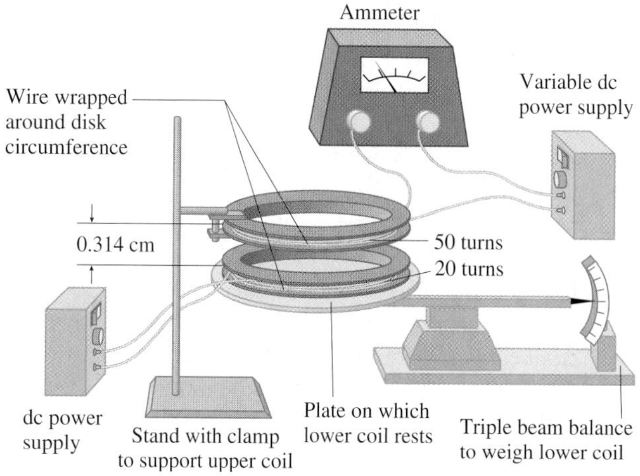

109. In a certain region of space, there is a uniform electric field $\vec{\mathbf{E}} = 3.0 \times 10^4$ V/m directed due east and a uniform magnetic field $\vec{\mathbf{B}} = 0.080$ T *also directed due east.* What is the electromagnetic force on an electron moving *due south* at 5.0×10^6 m/s?

110. An early cyclotron at Cornell University was used from the 1930s to the 1950s to accelerate protons, which would then bombard various nuclei. The cyclotron used a large electromagnet with an iron yoke to produce a uniform magnetic field of 1.3 T over a region in the shape of a flat cylinder. Two hollow copper dees of inside radius 16 cm were located in a vacuum chamber in this region. (a) What is the frequency of oscillation necessary for the alternating voltage difference between the dees? (b) What is the kinetic energy of a proton by the time it reaches the outside of the dees? (c) What would be the equivalent voltage necessary to accelerate protons to this energy from rest in one step (say between parallel plates)? (d) If the potential difference between the dees has a magnitude of 10.0 kV each time the protons cross the gap, what is the minimum number of revolutions each proton has to make in the cyclotron?

111. In a certain region of space, there is a uniform electric field $\vec{\mathbf{E}} = 2.0 \times 10^4$ V/m to the east and a uniform magnetic field $\vec{\mathbf{B}} = 0.0050$ T to the west. (a) What is the electromagnetic force on an electron moving north at 1.0×10^7 m/s? (b) With the electric and magnetic fields as specified, is there some velocity such that the net electromagnetic force on the electron would be zero? If so, give the magnitude and direction of that velocity. If not, explain briefly why not.

112. An electron moves in a circle of radius R in a uniform magnetic field $\vec{\mathbf{B}}$. The field is into the page. (a) Does the electron move clockwise or counterclockwise? (b) How much time does the electron take to make one complete revolution? *Derive* an expression for the time, starting with the magnetic force on the electron. Your answer may include R, B, and any fundamental constants.

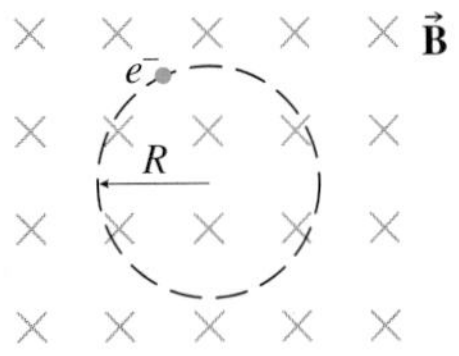

113. A proton moves in a helical path at speed $v = 4.0 \times 10^7$ m/s high above the atmosphere, where Earth's magnetic field has magnitude $B = 1.0 \times 10^{-6}$ T. The proton's velocity makes an angle of 25° with the magnetic field. (a) Find the radius of the helix. [*Hint:* Use the perpendicular component of the velocity.] (b) Find the *pitch* of the helix—the distance between adjacent "coils." [*Hint:* Find the time for one revolution; then find how far the proton moves along a field line during that time interval.]

Answers to Practice Problems

19.1 5.8×10^{-17} N; 3.4×10^{10} m/s^2

19.2 magnitude = 8.2×10^{-17} N, direction = east

19.3 $\pm 1.8 \times 10^6$ m/s

19.4 6.7×10^5 m/s

19.5 76 cm

19.6 out of the page (if the speed is too great, the magnetic force is larger than the electric force)

19.7 same magnitude Hall voltage, but opposite polarity: the top edge would be at the higher potential

19.8 west

19.9 (proof)

19.10 $\vec{\mathbf{B}} = \dfrac{2\mu_0 I}{\pi d}$ in the $+x$-direction

19.11 $+4\mu_0 I$

Answers to Checkpoints

19.2 (a) The magnetic force is zero if the velocity $\vec{\mathbf{v}}$ is along the same line as the magnetic field $\vec{\mathbf{B}}$. Therefore, the magnetic force on the electron is zero if it is moving straight down or straight up. (b) For a given v and $\vec{\mathbf{B}}$, the magnetic force is largest when $\vec{\mathbf{v}}$ is perpendicular to $\vec{\mathbf{B}}$. Therefore, the magnetic force on the electron is largest if it is moving in any horizontal direction.

19.4 At the point where the velocity vector is shown in Fig. 19.19a, $\vec{\mathbf{v}} \times \vec{\mathbf{B}}$ is out of the page. The magnetic force $\vec{\mathbf{F}} = q\vec{\mathbf{v}} \times \vec{\mathbf{B}}$ on the particle must be *into* the page, toward the central axis of the helix. The particle is negatively charged.

19.5 (a) $\vec{\mathbf{F}}_E = q\vec{\mathbf{E}}$, $\vec{\mathbf{E}}$ points east, and q is negative, so $\vec{\mathbf{F}}_E$ points west. (b) From the right-hand rule, $\vec{\mathbf{v}} \times \vec{\mathbf{B}}$ points west. $\vec{\mathbf{F}}_B = q\vec{\mathbf{v}} \times \vec{\mathbf{B}}$ and q is negative, so $\vec{\mathbf{F}}_B$ points east.

19.6 The magnetic force is in the direction of $\vec{\mathbf{L}} \times \vec{\mathbf{B}}$. $\vec{\mathbf{L}}$ is the same as before, but now $\vec{\mathbf{B}}$ is to the right. The two directions perpendicular to both $\vec{\mathbf{L}}$ and $\vec{\mathbf{B}}$ are into the page and out of the page. Using the right-hand rule, the direction of the magnetic force is into the page.

19.7 (a) $\vec{\mathbf{L}}_2$, $\vec{\mathbf{L}}_4$, and $\vec{\mathbf{B}}$ are all in the same directions as in Fig. 19.30, so the directions of $\vec{\mathbf{F}}_2$ and $\vec{\mathbf{F}}_4$ are the same: down and up, respectively. (b) The torque due to each of these forces is zero because the lever arm is zero. That is, the forces act along the line from the axis of rotation to the point of application of the force. (c) The equilibrium is unstable. Imagine the coil rotated slightly away from equilibrium. The forces on wires 2 and 4 make the coil rotate away from equilibrium, not toward equilibrium. (d) $\theta = 180°$.

19.8 To the left.

CHAPTER

20

Electromagnetic Induction

In many electric stoves, an electric current passes through the coiled heating elements. As electric energy is dissipated in the elements, they get hot. A different kind of electric stove—the induction stove—has an advantage over stoves with resistance heating elements. If you touch the heating element of a traditional stove, you will burn your hand; a potholder carelessly left in contact with a hot coil may catch on fire. With an induction stove, the surface does not feel hot to the touch and the potholder does not catch fire. (*Caution*: Most stoves with flat stovetops are *not* induction stoves, so do not try this at home unless you are *certain.*) How can heat get to the food in a pot or pan if the stove surface is not hot? (See p. 759 for the answer.)

Concepts & Skills to Review

- emf (Section 18.2)
- microscopic view of current in a metal (Section 18.3)
- magnetic fields and forces (Sections 19.1, 19.2, 19.8)
- electric potential (Section 17.2)
- angular velocity (Section 5.1)
- angular frequency (Section 10.6)
- right-hand rule 1 to find the direction of a cross product (Section 19.2)
- right-hand rule 2 to determine the direction of magnetic fields caused by currents (Section 19.8)
- exponential function; time constant (Appendix A.3; Section 18.10)

20.1 MOTIONAL EMF

The only sources of electric energy (emf) we've discussed so far are batteries. The amount of electric energy that can be supplied by a battery before it needs to be recharged or replaced is limited. Most of the world's electric energy is produced by generators. In this section we study **motional emf**—the emf induced when a conductor is moved in a magnetic field. Motional emf is the principle behind the electric generator.

Imagine a metal rod of length L in a uniform magnetic field $\vec{\mathbf{B}}$. When the rod is at rest, the conduction electrons move in random directions at high speeds, but their average velocity is zero. Since their average velocity is zero, the average magnetic force on the electrons is zero; therefore, the total magnetic force on the rod is zero. The magnetic field affects the motion of individual electrons, but the rod as a whole feels no net magnetic force.

Now imagine a vertical rod that is moving instead of being at rest. Figure 20.1a shows a uniform magnetic field into the page, the velocity $\vec{\mathbf{v}}$ of the rod is to the right, and the rod is vertical—the field, velocity, and axis of the rod are mutually perpendicular. Now the electrons have a nonzero average velocity: it is $\vec{\mathbf{v}}$, since the electrons are being carried to the right along with the rod. Then the average magnetic force on each conduction electron is

$$\vec{\mathbf{F}}_B = -e\vec{\mathbf{v}} \times \vec{\mathbf{B}}$$

By right-hand rule 1 (Sec. 19.2), the direction of this force is down (toward the lower end of the rod). The magnetic force causes electrons to accumulate at the lower end, giving it a negative charge and leaving positive charge at the upper end (Fig. 20.1b). This separation of charge by the magnetic field is similar to the Hall effect, but here the charges are moving due to the motion of the rod itself rather than due to a current flowing in a stationary rod.

As charge accumulates at the ends, an electric field develops in the rod, with field lines running from the positive to the negative charge. Eventually an equilibrium is reached: the electric field builds up until it causes a force equal and opposite to the magnetic force on electrons in the middle of the rod (Fig. 20.1c). Then there is no further accumulation of charge at the ends. Thus, in equilibrium,

$$\vec{\mathbf{F}}_E = q\vec{\mathbf{E}} = -\vec{\mathbf{F}}_B = -(q\vec{\mathbf{v}} \times \vec{\mathbf{B}})$$

or

$$\vec{\mathbf{E}} = -\vec{\mathbf{v}} \times \vec{\mathbf{B}}$$

just as for the Hall effect. Since $\vec{\mathbf{v}}$ and $\vec{\mathbf{B}}$ are perpendicular, $E = vB$. The potential difference between the ends is

$$\Delta V = EL = vBL \quad \text{(20-1a)}$$

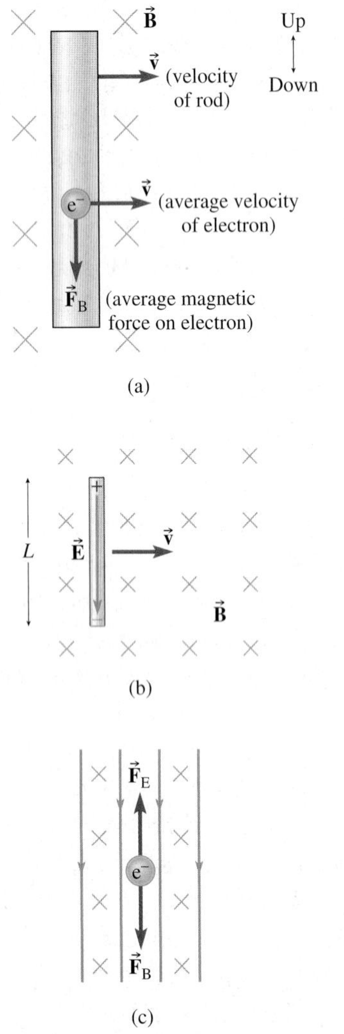

Figure 20.1 (a) An electron in a metal rod that is moving to the right with velocity $\vec{\mathbf{v}}$. The magnetic field is into the page. The average magnetic force on the electron is $\vec{\mathbf{F}}_B = -e\vec{\mathbf{v}} \times \vec{\mathbf{B}}$. (b) The magnetic force pushes electrons toward the bottom of the rod, leaving the top end positively charged. This separation of charge gives rise to an electric field in the rod. (c) In equilibrium, the sum of the electric and magnetic forces on the electron is zero.

In this case, the direction of $\vec{\mathbf{E}}$ is parallel to the rod. If it were not, then the potential difference between the ends is found using only the *component* of $\vec{\mathbf{E}}$ parallel (||) to the rod:

$$\Delta V = E_{\parallel}L \qquad (20\text{-}1b)$$

CHECKPOINT 20.1

If the rod in Fig. 20.1 were moving out of the page instead of to the right, what would be the induced emf?

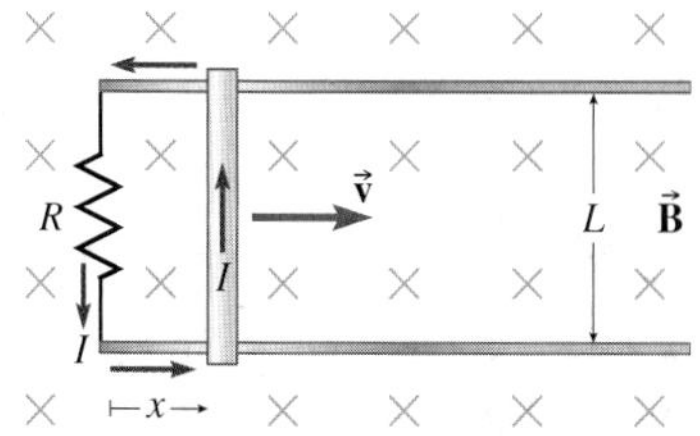

Figure 20.2 When the rod is connected to a circuit with resistance R, current flows around the circuit.

As long as the rod keeps moving at constant speed, the separation of charge is maintained. The moving rod acts like a battery that is not connected to a circuit; positive charge accumulates at one terminal and negative charge at the other, maintaining a constant potential difference. Now the important question: if we connect this rod to a circuit, does it act like a battery and cause current to flow?

Figure 20.2 shows the rod connected to a circuit. The rod slides on metal rails so that the circuit stays complete even as the rod continues to move. We assume the resistance R is large relative to the resistances of the rod and rails—in other words, the internal resistance of our source of emf (the moving rod) is negligibly small. The resistor R sees a potential difference ΔV across it, so current flows. The current tends to deplete the accumulated charge at the ends of the rod, but the magnetic force pumps more charge to maintain a constant potential difference. So the moving rod *does* act like a battery with an emf given by

Motional emf:

$$\mathscr{E} = vBL \qquad (20\text{-}2a)$$

More generally, if $\vec{\mathbf{E}}$ is not parallel to the rod, then

$$\mathscr{E} = (\vec{\mathbf{v}} \times \vec{\mathbf{B}})_{\parallel}L \qquad (20\text{-}2b)$$

A sliding rod would be a clumsy way to make a generator. No matter how long the rails are, the rod will eventually reach the end. In Section 20.2, we see that the principle of the motional emf can be applied to a *rotating coil* of wire instead of a sliding rod.

Where does the electric energy come from? The rod is acting like a battery, supplying electric energy that is dissipated in the resistor. How can energy be conserved? The key is to recognize that as soon as current flows through the rod, a magnetic force acts on the rod in the direction opposite to the velocity (Fig. 20.3). Left on its own, the rod would slow down as its kinetic energy gets transformed into electric energy. To maintain a constant emf, the rod must maintain a constant velocity, which can only happen if some other force pulls the rod. The work done by the force pulling the rod is the source of the electric energy (Problem 3).

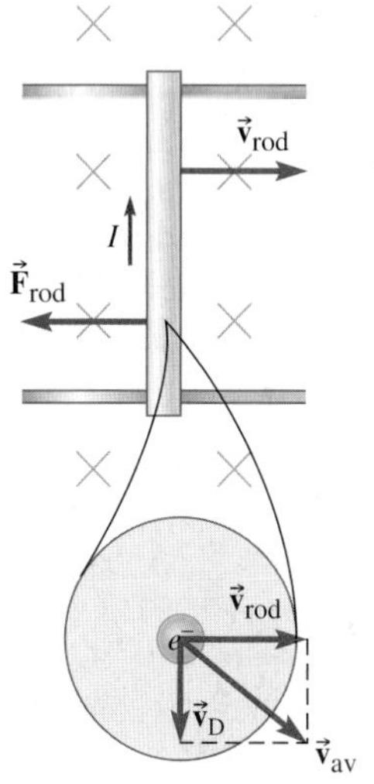

Figure 20.3 The magnetic force on the rod is $\vec{\mathbf{F}}_{\text{rod}} = I\vec{\mathbf{L}} \times \vec{\mathbf{B}}$ and is directed to the left, opposite the velocity of the rod ($\vec{\mathbf{v}}_{\text{rod}}$). The average velocity of an electron in the rod is $\vec{\mathbf{v}}_{\text{av}} = \vec{\mathbf{v}}_{\text{rod}} + \vec{\mathbf{v}}_{\text{D}}$; the electrons drift downward relative to the rod as the rod carries them to the right. The average magnetic force on an electron has two perpendicular components. One is $-e\vec{\mathbf{v}}_{\text{rod}} \times \vec{\mathbf{B}}$, which is directed downward and causes the electron to drift relative to the rod. The other is $-e\vec{\mathbf{v}}_{\text{D}} \times \vec{\mathbf{B}}$, which pulls the electron to the left side of the rod and, because each electron in turn pulls on the rest of the rod, contributes to the leftward magnetic force on the rod.

Example 20.1

Loop Moving Through a Magnetic Field

A square metal loop made of four rods of length L moves at constant velocity $\vec{\mathbf{v}}$ (Fig. 20.4). The magnetic field in the central region has magnitude B; elsewhere the magnetic field is zero. The loop has resistance R. At each position 1–5, state the direction (CW or CCW) and the magnitude of the current in the loop.

Strategy If current flows in the loop, it is due to the motional emf that pumps charge around. The vertical sides (a, c) have motional emfs as they move through the magnetic field, just as in Fig. 20.2. We need to look at the horizontal sides (b, d) to see whether they also give rise to motional emfs. Once we figure out the emf in each side, then we can determine whether they cooperate with each other—pumping charge around in the same direction—or tend to cancel each other.

Solution The vertical sides (a, c) have motional emfs as they move through the region of magnetic field. The emf acts to pump current upward (toward the top end). The magnitude of the emf is

$$\mathscr{E} = vBL$$

For the horizontal sides (b, d), the average magnetic force on a current-carrying electron is $\vec{\mathbf{F}}_{av} = -e\vec{\mathbf{v}} \times \vec{\mathbf{B}}$. Since the velocity is to the right and the field is into the page, the right-hand rule shows that the direction of the force is down, just as in sides a and c. However, now the magnetic force does not move charge along the length of the rod; the magnetic force instead moves charge across the diameter of the rod. An electric field then develops *across* the rod. In equilibrium, the magnetic and electric forces cancel, exactly as in the Hall effect. The magnetic force does not push charge along the length of the rod, so there is no motional emf in sides b and d.

In positions 1 and 5, the loop is completely out of the region of magnetic field. There is no motional emf in any of the sides; no current flows.

In position 2, there is a motional emf in side c only; side a is still outside the region of $\vec{\mathbf{B}}$ field. The emf makes current flow upward in side c, and therefore counterclockwise in the loop. The magnitude of the current is

$$I = \frac{\mathscr{E}}{R} = \frac{vBL}{R}$$

In position 3, there are motional emfs in both sides a and c. Since the emfs in both sides push current toward the top of the loop, the net emf around the loop is zero—as if two identical batteries were connected as in Fig. 20.5. No current flows around the loop.

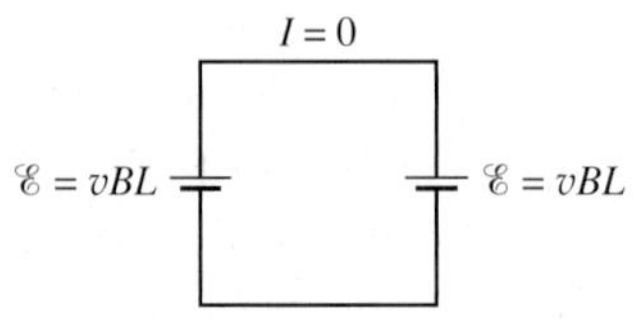

Figure 20.5
At position 3, the emfs induced in sides a and c can be represented with battery symbols in a circuit diagram.

In position 4, there is a motional emf only in side a, since side c has left the region of the $\vec{\mathbf{B}}$ field. The emf makes current flow upward in side a, and therefore *clockwise* in the loop. The magnitude of the current is again

$$I = \frac{\mathscr{E}}{R} = \frac{vBL}{R}$$

Discussion Figure 20.5 illustrates a useful technique: it often helps to draw battery symbols to represent the directions of the induced emfs.

Note that if the loop were *at rest* instead of moving to the right at constant velocity, there would be no motional emf at any of the positions 1–5. The motional emf does not arise simply because one of the vertical sides of the loop is immersed in magnetic field while the other is not; it arises because one side *moves through* a magnetic field while the other does not.

Conceptual Practice Problem 20.1
Loop of Different Metal

Suppose a loop made of a different metal but with identical size, shape, and velocity moved through the same magnetic field. Of these quantities, which would be different: the magnitudes of the emfs, the directions of the emfs, the magnitudes of the currents, or the directions of the currents?

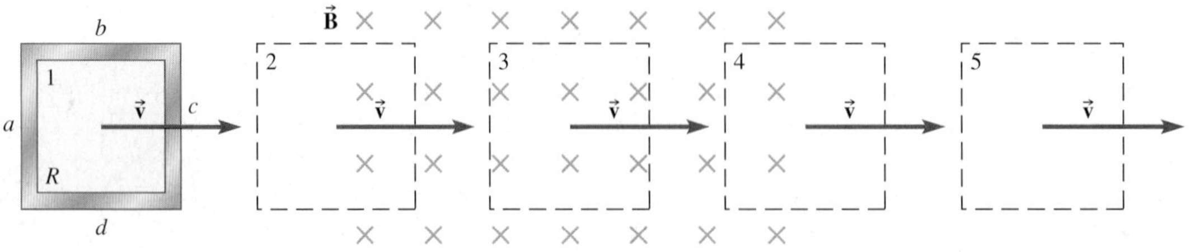

Figure 20.4
Loop moving into, through, and then out of a region of uniform magnetic field $\vec{\mathbf{B}}$ perpendicular to the loop.

20.2 ELECTRIC GENERATORS

Application: electric generators

For practical reasons, electric generators use coils of wire that rotate in a magnetic field rather than rods that slide on rails. The rotating coil is called an *armature*. A simple ac electric generator is shown in Fig. 20.6. The rectangular coil is mounted on a shaft that is turned by some external power source such as the turbine of a steam engine.

Let us begin with a single turn of wire—a rectangular loop—that rotates at a constant angular speed ω. The loop rotates in the space between the poles of a permanent magnet or an electromagnet that produces a nearly uniform field of magnitude B. Sides 2 and 4 are each of length L and are a distance r from the axis of rotation; the length of sides 1 and 3 is therefore $2r$ each.

None of the four sides of the loop moves perpendicularly to the magnetic field at all times, so we must generalize the results of Section 20.1. In Problem 9, you can verify that there is zero induced emf in sides 1 and 3, so we concentrate on sides 2 and 4. Since these two sides do not, in general, move perpendicularly to $\vec{\mathbf{B}}$, the magnitude of the average magnetic force on the electrons is reduced by a factor of sin θ, where θ is the angle between the velocity of the wire and the magnetic field (Fig. 20.7):

Generators at Little Goose Dam in the state of Washington.

$$F_{av} = evB \sin \theta$$

The induced emf is then reduced by the same factor:

$$\mathscr{E} = vBL \sin \theta$$

Note that the induced emf is proportional to the component of the velocity perpendicular to $\vec{\mathbf{B}}$ ($v_\perp = v \sin \theta$). For a visual image, think of the induced emf as proportional to the *rate* at which the wire *cuts through magnetic field lines*. The component of the velocity *parallel* to $\vec{\mathbf{B}}$ moves the wire along the magnetic field lines, so it does not contribute to the rate at which the wire cuts through the field lines.

The loop turns at constant angular speed ω, so the speed of sides 2 and 4 is

$$v = \omega r$$

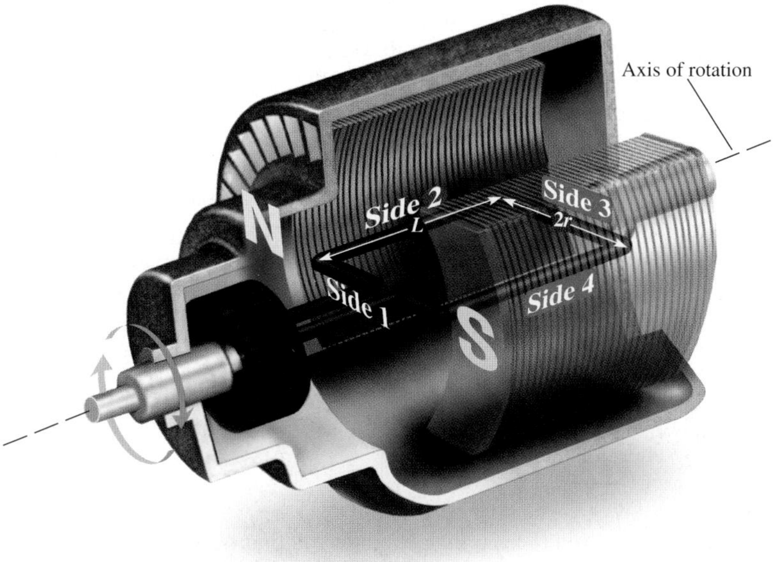

Figure 20.6 A simple ac generator, in which a rectangular loop or coil of wire rotates at constant angular speed between the poles of a permanent magnet or electromagnet. Emfs are induced in sides 2 and 4 of the loop due to their motion through the magnetic field as the loop rotates. (Sides 1 and 3 have zero induced emf.) A magnetic torque opposes the rotation of the coil, so an external torque must be applied to keep the loop rotating at constant angular velocity.

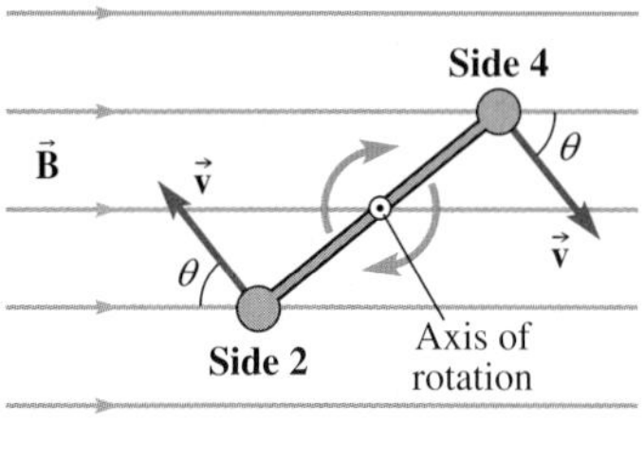

Figure 20.7 Side view of the rectangular loop, looking straight down the axis of rotation. The velocity vectors of sides 2 and 4 make an angle θ with the magnetic field.

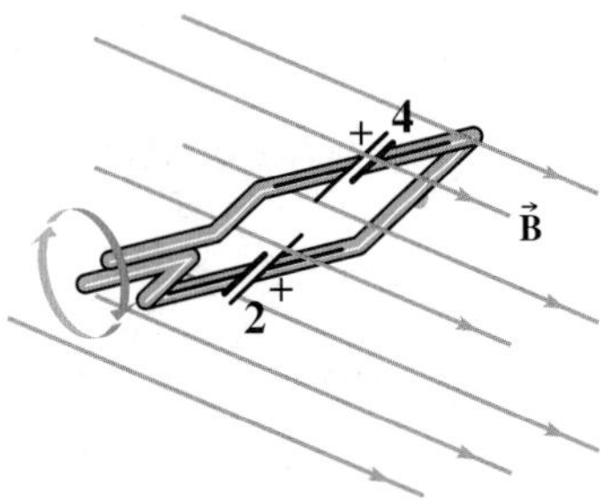

Figure 20.8 Battery symbols indicate the direction of the emfs in the rotating loop of wire.

The angle θ changes at a constant rate ω. For simplicity, we choose $\theta = 0$ at $t = 0$, so that $\theta = \omega t$ and the emf $\mathscr{E}$ as a function of time t in each of sides 2 and 4 is

$$\mathscr{E}(t) = vBL \sin\theta = (\omega r)BL \sin \omega t$$

Sides 2 and 4 move in opposite directions, so current flows in opposite directions; in side 2, current flows into the page (as viewed in Fig. 20.7), while in side 4 it flows out of the page. *Both* sides tend to send current counterclockwise around the loop as viewed in Fig. 20.8. Therefore, the *total* emf in the loop is the *sum* of the two:

$$\mathscr{E}(t) = 2\omega rBL \sin \omega t$$

The rectangular loop has sides L and $2r$, so the area of the loop is $A = 2rL$. Therefore, the total emf $\mathscr{E}$ as a function of time t is

$$\mathscr{E}(t) = \omega BA \sin \omega t \qquad (20\text{-}3a)$$

When written in terms of the area of the loop, Eq. (20-3a) is true for a planar loop of *any* shape. If the coil consists of N turns of wire (N identical loops), the emf is N times as great:

Emf produced by an ac generator:

$$\mathscr{E}(t) = \omega NBA \sin \omega t \qquad (20\text{-}3b)$$

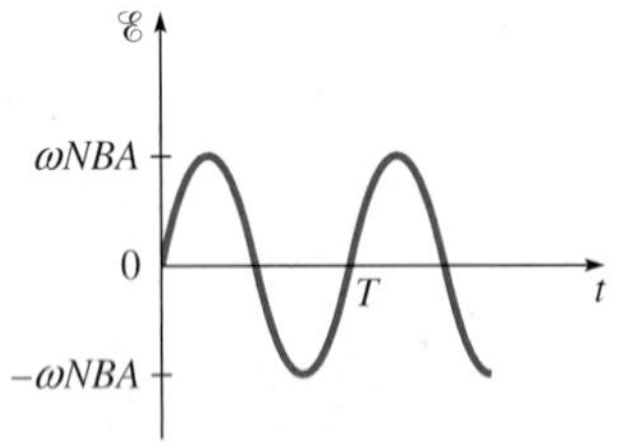

Figure 20.9 Generator-produced emf is a sinusoidal function of time.

The emf produced by a generator is not constant; it is a sinusoidal function of time (see Fig. 20.9 and interactive: induction). The maximum emf ($= \omega NBA$) is called the **amplitude** of the emf (just as in simple harmonic motion, where the maximum displacement is called the amplitude). Sinusoidal emfs are used in ac (alternating current) circuits. Household electric outlets in the United States and Canada provide an emf with an amplitude of approximately 170 V and a frequency $f = \omega/(2\pi) = 60$ Hz. In much of the rest of the world, the amplitude is about 310–340 V and the frequency is 50 Hz.

CHECKPOINT 20.2

A generator produces an emf of amplitude 18 V when rotating with a frequency of 12 Hz. How will the frequency and amplitude of the emf change if the frequency of rotation drops to 10 Hz?

The energy supplied by a generator does not come for free; work must be done to turn the generator shaft. As current flows in the coil, the magnetic force on sides 2 and 4 cause a torque in the direction opposing the coil's rotation (Problem 69). To keep the coil rotating at constant angular speed, an equal and oppositely directed torque must be applied to the shaft. In an ideal generator, this external torque would do work at exactly the same rate as electric energy is generated. In reality, some energy is dissipated by friction and by the electrical resistance of the coil, among other things. Then the external torque does more work than the amount of electric energy generated. Since the rate at which electric energy is generated is

$$P = \mathscr{E}I$$

the external torque required to keep the generator rotating depends not only on the emf but also on the current it supplies. The current supplied depends on the *load*—the external circuit through which the current must flow.

In most power stations that supply our electricity, the work to turn the generator shaft is supplied by a steam engine. The steam engine is powered by burning coal, natural gas, or oil, or by a nuclear reactor. In a hydroelectric power plant, the gravitational potential energy of water is the energy source used to turn the generator shaft.

Application: regenerative braking

In electric and hybrid gas-electric cars, the drive train of the vehicle is connected to an electric generator when brakes are applied, which charges the batteries. Thus, instead of the kinetic energy of the vehicle being completely dissipated, much of it is stored in the batteries. This energy is used to propel the car after braking is finished.

Application: The DC Generator

Note that the induced emf produced in an ac generator reverses direction twice per period. Mathematically, the sine functions in Eqs. (20-3) are positive half the time and negative half the time. When the generator is connected to a load, the current also reverses direction twice per period—which is why we call it alternating current.

What if the load requires a direct current (dc) instead? Then we need a dc generator, one in which the emf does *not* reverse direction. One way to make a dc generator is to equip the ac generator with a split-ring commutator and brushes, exactly as for the dc motor (Section 19.7). Just as the emf is about to change direction, the connections to the rotating loop are switched as the brushes pass over the gap in the split ring. The commutator effectively reverses the connections to the outside load so that the emf and current supplied maintain the same direction. The emf and current are *not* constant, though. The emf is described by

$$\mathscr{E}(t) = \omega NBA\,|\sin \omega t| \qquad (20\text{-}3c)$$

which is graphed in Fig. 20.10.

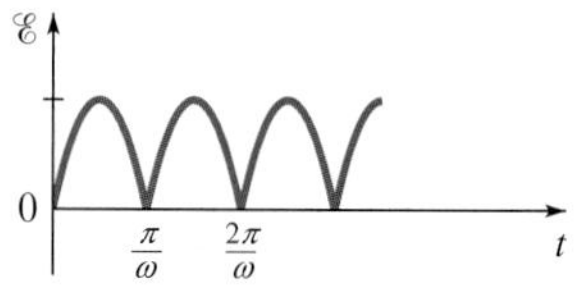

Figure 20.10 The emf in a dc generator as a function of time.

A simple dc *motor* can be used as a dc *generator*, and vice versa. When configured as a motor, an external source of electric energy such as a battery causes current to flow through the loop. The magnetic torque makes the motor rotate. In other words, the current is the input and the torque is the output. When configured as a generator, an external torque makes the loop rotate, the magnetic field induces an emf in the loop, and the emf makes current flow. Now the torque is the input and the current is the output. The conversion between mechanical energy and electric energy can proceed in either direction.

CONNECTION:

A dc generator is a dc motor with its input and output reversed.

More sophisticated dc generators have many coils distributed evenly around the axis of rotation. The emf *in each coil* still varies sinusoidally, but each coil reaches its peak emf at a different time. As the commutator rotates, the brushes connect selectively to the coil that is nearest its peak emf. The output emf has only small fluctuations, which can be smoothed out by a circuit called a voltage regulator if necessary.

Example 20.2

A Bicycle Generator

A simple dc generator in contact with a bicycle's tire can be used to generate power for the headlight. The generator has 150 turns of wire in a circular coil of radius 1.8 cm. The magnetic field strength in the region of the coil is 0.20 T. When the generator supplies an emf of amplitude 4.2 V to the lightbulb, the lightbulb consumes an average power of 6.0 W and a maximum instantaneous power of 12.0 W. (a) What is the rotational speed in rpm of the armature of the generator? (b) What is the average torque and maximum instantaneous torque that must be applied by the bicycle tire to the generator, assuming the generator to be ideal? (c) The radius of the tire is 32 cm and the radius of the shaft of the generator where it contacts the tire is 1.0 cm. At what linear speed must the bicycle move to supply an emf of amplitude 4.2 V?

Strategy The amplitude is the maximum value of the time-dependent emf [Eq. (20-3c)]. To find the torques, two methods are possible. One is to find the current in the coil, then the torque on the coil due to the magnetic field. To keep the armature moving at a constant angular velocity, an equal magnitude but oppositely directed torque must be applied to it. Another method is to analyze the energy transfers. The external torque applied to the armature must do work at the same rate that electric energy is dissipated in the lightbulb. The second approach is easiest, especially since the problem states the power in the lightbulb. To find the linear speed of the bicycle, we set the tangential speeds of the tire and shaft equal (the shaft is "rolling" on the tire).

continued on next page

Example 20.2 continued

Solution (a) The emf as a function of time is

$$\mathscr{E}(t) = \omega NBA\,|\sin \omega t| \qquad (20\text{-}3c)$$

The emf has its maximum value when $\sin \omega t = \pm 1$. Thus, the amplitude of the emf is

$$\mathscr{E}_m = \omega NBA$$

where $N = 150$, $A = \pi r^2$, and $B = 0.20$ T. Solving for the angular frequency,

$$\omega = \frac{\mathscr{E}_m}{NAB} = \frac{4.2 \text{ V}}{150 \times \pi \times (0.018 \text{ m})^2 \times 0.20 \text{ T}} = 137.5 \text{ rad/s}$$

A check of the units verifies that $1 \text{ V/(T·m}^2) = 1 \text{ s}^{-1}$. The question asks for the number of rpm, so we convert the angular frequency to rev/min:

$$\omega = 137.5 \frac{\text{rad}}{\text{s}} \times \frac{1 \text{ rev}}{2\pi \text{ rad}} \times \frac{60 \text{ s}}{1 \text{ min}} = 1300 \text{ rpm}$$

(b) Assuming the generator to be ideal, the torque applied to the crank must do work at the same rate that electric energy is generated:

$$P = \frac{W}{\Delta t}$$

Since for a small angular displacement $\Delta\theta$ the work done is $W = \tau\Delta\theta$,

$$P = \tau\frac{\Delta\theta}{\Delta t} = \tau\omega$$

The average torque is then

$$\tau_{av} = \frac{P_{av}}{\omega} = \frac{6.0 \text{ W}}{137.5 \text{ rad/s}} = 0.044 \text{ N·m}$$

and the maximum torque is

$$\tau_m = \frac{P_m}{\omega} = \frac{12.0 \text{ W}}{137.5 \text{ rad/s}} = 0.087 \text{ N·m}$$

(c) The tangential speed of the generator shaft is

$$v_{tan} = \omega r = 137.5 \text{ rad/s} \times 0.010 \text{ m} = 1.4 \text{ m/s}$$

The tangential speed of the tire where it touches the generator shaft is the same, since the shaft rolls without slipping on the tire. Since the generator is almost at the outside edge of the tire, the tangential speed at the outer radius of the tire is approximately the same. Assuming that the bicycle rolls without slipping on the road, its linear speed is approximately 1.4 m/s.

Discussion To check the result, we can find the maximum current in the coil and use it to find the maximum torque. The maximum current occurs when the power dissipated is maximum:

$$P_m = \mathscr{E}_m I_m$$

$$I_m = \frac{12.0 \text{ W}}{4.2 \text{ V}} = 2.86 \text{ A}$$

The magnetic torque on a current loop is

$$\tau = NIAB \sin \theta$$

where $\theta = \omega t$ is the angle between the magnetic field and the *normal* to the loop. At the position where the emf is maximum, $|\sin \theta| = 1$. Then

$$\tau_m = NI_m AB = 150 \times 2.86 \text{ A} \times \pi \times (0.018 \text{ m})^2 \times 0.20 \text{ T}$$
$$= 0.087 \text{ N·m}$$

Practice Problem 20.2 Riding More Slowly

What would the maximum power be if the bicycle moves half as fast? Assume that the resistance of the lightbulb does not change. Remember that the angular velocity affects the emf, which in turn affects the current. How does the power in the lightbulb depend on the bicycle's speed?

20.3 FARADAY'S LAW

In 1820, Hans Christian Oersted accidentally discovered that an electric current produces a magnetic field (Section 19.1). Soon after hearing the news of that discovery, the English scientist Michael Faraday (1791–1867) started experimenting with magnets and electric circuits in an attempt to do the reverse—use a magnetic field to produce an electric current. Faraday's brilliant experiments led to the development of the electric motor, the generator, and the transformer.

A Changing $\vec{\mathbf{B}}$ Field Can Cause an Induced Emf In 1831, Faraday discovered two ways to produce an induced emf. One is to move a conductor in a magnetic field (motional emf). The other does *not* involve movement of the conductor. Instead, Faraday found that a changing magnetic field induces an emf in a conductor even if the conductor is stationary. The induced emf due to a changing $\vec{\mathbf{B}}$ field cannot be understood in terms of the magnetic force on the conduction electrons: if the conductor is stationary, the average velocity of the electrons is zero, and the average magnetic force is zero.

Consider a circular loop of wire between the poles of an electromagnet (Fig. 20.11). The loop is perpendicular to the magnetic field; field lines cross the interior of the loop. Since the strength of the magnetic field is related to the spacing of the field lines, if the strength of the field varies (by changing the current in the electromagnet), the number of field lines passing through the conducting loop changes. Faraday found that the emf induced in the loop is proportional to the *rate of change* of the number of field lines that cut through the interior of the loop.

We can formulate *Faraday's law* mathematically so that numbers of field lines are not involved. The magnitude of the magnetic field is proportional to the number of field lines *per unit cross-sectional area*:

$$B \propto \frac{\text{number of lines}}{\text{area}}$$

If a flat, open surface of area A is perpendicular to a uniform magnetic field of magnitude B, then the number of field lines that cross the surface is proportional to BA, since

$$\text{number of lines} = \frac{\text{number of lines}}{\text{area}} \times \text{area} \propto BA \qquad (20\text{-}4)$$

Equation (20-4) is correct only if the surface is perpendicular to the field. In general, the number of field lines crossing a surface is proportional to the *perpendicular component* of the field times the area:

$$\text{number of lines} \propto B_\perp A = BA \cos\theta$$

where θ is the angle between the magnetic field and the direction normal to the surface. The component of the magnetic field parallel to the surface $B_\parallel$ doesn't contribute to the number of lines crossing the surface; only $B_\perp$ does (see Fig. 20.12a). Equivalently, Fig. 20.12b shows that the number of lines crossing the surface area A is the same as the number crossing a surface of area $A \cos\theta$, which is perpendicular to the field.

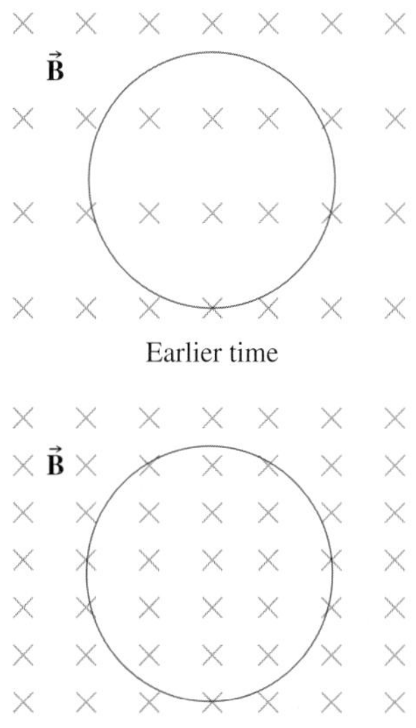

Figure 20.11 Circular loop in a magnetic field of increasing magnitude.

The word *normal* means *perpendicular* in geometry. *The normal to the loop* means *the direction perpendicular to the plane of the loop.*

Magnetic Flux The mathematical quantity that is proportional to the number of field lines cutting through a surface is called the **magnetic flux**. The symbol Φ (Greek capital phi) is used for flux; in Φ_B the subscript B indicates *magnetic* flux.

CONNECTION:

Magnetic flux is analogous to electric flux (Section 16.7).

Magnetic flux through a flat surface of area A:

$$\Phi_B = B_\perp A = BA_\perp = BA\cos\theta \qquad (20\text{-}5)$$

(θ is the angle between $\vec{\mathbf{B}}$ and the *normal* to the surface)

The SI unit of magnetic flux is the weber (1 Wb = 1 T·m^2).

Faraday's Law

Faraday's law says that the magnitude of the induced emf around a loop is equal to the rate of change of the magnetic flux through the loop.

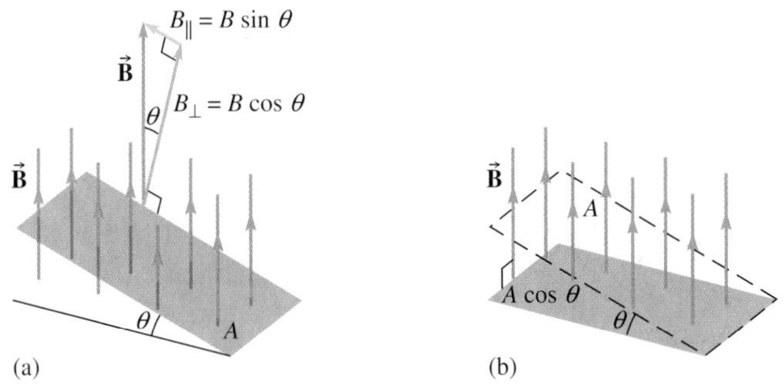

Figure 20.12 (a) The component of $\vec{\mathbf{B}}$ perpendicular to the surface of area A is $B\cos\theta$. (b) The projection of the area A onto a plane perpendicular to $\vec{\mathbf{B}}$ is $A\cos\theta$, showing that the magnetic flux is $BA\cos\theta$.

Faraday's Law

$$\mathscr{E} = -\frac{\Delta\Phi_B}{\Delta t} \quad (20\text{-}6a)$$

Problem 22 asks you to verify the units of Eq. (20-6a). Faraday's law, if it is to give the *instantaneous* emf, must be taken in the limit of a very small time interval Δt. However, Faraday's law can be applied just as well to longer time intervals; then $\Delta\Phi_B/\Delta t$ represents the *average* rate of change of the flux, and $\mathscr{E}$ represents the *average* emf during that time interval.

The negative sign in Eq. (20-6a) concerns the sense of the induced emf around the loop (clockwise or counterclockwise). The interpretation of the sign depends on a formal definition of the emf direction that we do not use. Instead, in Section 20.4, we introduce *Lenz's law*, which gives the direction of the induced emf.

If, instead of a single loop of wire, we have a coil of N turns, then Eq. (20-6a) gives the emf induced in each turn; the total emf in the coil is then N times as great:

$$\mathscr{E} = -N\frac{\Delta\Phi_B}{\Delta t} \quad (20\text{-}6b)$$

The quantity $N\Phi_B$ is called the total **flux linkage** through the coil.

Example 20.3

Induced Emf due to Changing Magnetic Field

A 40.0-turn coil of wire of radius 3.0 cm is placed between the poles of an electromagnet. The field increases from 0 to 0.75 T at a constant rate in a time interval of 225 s. What is the magnitude of the induced emf in the coil if (a) the field is perpendicular to the plane of the coil? (b) the field makes an angle of 30.0° with the plane of the coil?

Strategy First we write an expression for the flux through the coil in terms of the field. The only thing changing is the strength of the field, so the rate of flux change is proportional to the rate of change of the field. Faraday's law gives the induced emf.

Solution (a) The magnetic field is perpendicular to the coil, so the flux through one turn is

$$\Phi_B = BA$$

where B is the field strength and A is the area of the loop. Since the field increases at a constant rate, so does the flux. The rate of change of flux is then equal to the change in flux divided by the time interval. The flux changes at a constant rate, so the emf induced in the loop is constant.

By Faraday's law,

$$\mathscr{E} = -N\frac{\Delta\Phi_B}{\Delta t} = -N\frac{B_f A - 0}{\Delta t}$$

$$|\mathscr{E}| = 40.0 \times \frac{0.75\ \text{T} \times \pi \times (0.030\ \text{m})^2}{225\ \text{s}} = 3.77 \times 10^{-4}\ \text{V}$$

$$= 0.38\ \text{mV}$$

(b) In Eq. (20-5), θ is the angle between $\vec{\mathbf{B}}$ and the direction *normal* to the coil. If the field makes an angle of 30.0° with the plane of the coil, then it makes an angle

$$\theta = 90.0° - 30.0° = 60.0°$$

with the normal to the coil. The magnetic flux through one turn is

$$\Phi_B = BA\cos\theta$$

The induced emf is therefore,

$$|\mathscr{E}| = N\frac{\Delta\Phi_B}{\Delta t} = N\frac{B_f A\cos\theta - 0}{\Delta t} = 3.77 \times 10^{-4}\ \text{V} \times \cos 60.0°$$

$$= 0.19\ \text{mV}$$

Discussion If the rate of change of the field were not constant, then 0.38 mV would be the *average* emf during that time interval. The instantaneous emf would be sometimes higher and sometimes lower.

Practice Problem 20.3 Using the Perpendicular Component of $\vec{B}$

Draw a sketch that shows the coil, the direction normal to the coil, and the magnetic field lines. Find the component of $\vec{\mathbf{B}}$ in the normal direction. Now use $\Phi_B = B_\perp A$ to verify the answer to part (b).

Faraday's Law and Motional Emfs

CONNECTION:

Faraday's law gives the induced emf due to a changing magnetic flux, including the motional emfs of Sections 20.1 and 20.2.

Earlier in this section, we wrote Faraday's law to give the magnitude of the induced emf due to a changing magnetic field. But that's only part of the story. Faraday's law gives the induced emf due to a changing magnetic flux, *no matter what the reason for the flux change.* The flux change can occur for reasons other than a changing magnetic field. A conducting loop might be moving through regions where the field is not constant, or it can be rotating, or changing size or shape. In all of these cases, Faraday's law as already stated gives the correct emf, regardless of why the flux is changing (Text website tutorial: magnetic flux). Recall that flux can be written

$$\Phi_B = BA\cos\theta \qquad (20\text{-}5)$$

Then the flux changes if the magnetic field strength (B) changes, or if the area of the loop (A) changes, or if the angle between the field and the normal changes.

Faraday's law says that, no matter what the reason for the change in flux, the induced emf is

$$\mathscr{E} = -N\frac{\Delta\Phi_B}{\Delta t} \qquad (20\text{-}6b)$$

For example, the moving rod of Fig. 20.2 is one side of a conducting loop. The magnetic flux through the loop is increasing as the rod slides to the right because the loop's *area* is increasing. Faraday's law gives the same induced emf in the loop that we found in Eq. (20-2a)—see Problem 20.

The mobile charges in a moving conductor are pumped around due to the magnetic force on the charges. Since the conductor as a whole is moving, the mobile charges have a nonzero average velocity and therefore a nonzero average magnetic force. In the case of a changing magnetic field and a stationary conductor, the mobile charges aren't set into motion by the magnetic force—they have zero average velocity before current starts to flow. Exactly what *does* make current flow is considered in Section 20.8.

Sinusoidal Emfs

Emfs that are sinusoidal (sine or cosine) functions of time, such as in Example 20.2, are common in ac generators, motors, and circuits. A sinusoidal emf is generated whenever the flux is a sinusoidal function of time. It can be shown (see Fig. 20.13 and Problem 26) that:

$$\text{If } \Phi(t) = \Phi_0 \sin\omega t, \text{ then } \frac{\Delta\Phi}{\Delta t} = \omega\Phi_0\cos\omega t \quad (\text{for small } \Delta t); \qquad (20\text{-}7a)$$

$$\text{if } \Phi(t) = \Phi_0 \cos\omega t, \text{ then } \frac{\Delta\Phi}{\Delta t} = -\omega\Phi_0\sin\omega t \quad (\text{for small } \Delta t). \qquad (20\text{-}7b)$$

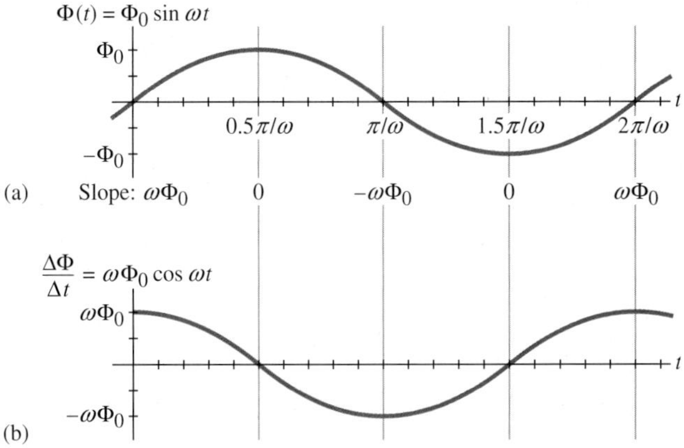

Figure 20.13 (a) A graph of a sinusoidal emf $\Phi(t) = \Phi_0 \sin\omega t$ as a function of time. (b) A graph of the slope $\Delta\Phi/\Delta t$, which represents the rate of change of $\Phi(t)$.

Example 20.4

Applying Faraday's Law to a Generator

The magnetic field between the poles of an electromagnet has constant magnitude B. A circular coil of wire immersed in this magnetic field has N turns and area A. An externally applied torque causes the coil to rotate with constant angular velocity ω about an axis perpendicular to the field (as in Fig. 20.6). Use Faraday's law to find the emf induced in the coil.

Strategy The magnetic field does not vary, but the orientation of the coil does. The number of field lines crossing through the coil depends on the angle that the field makes with the normal (the direction perpendicular to the coil). The changing magnetic flux induces an emf in the coil, according to Faraday's law.

Solution Let us choose $t = 0$ to be an instant when the field is perpendicular to the coil. At this instant, $\vec{\mathbf{B}}$ is parallel to the normal, so $\theta = 0$. At a later time $t > 0$, the coil has rotated through an angle $\Delta\theta = \omega t$. Thus, the angle that the field makes with the normal as a function of t is

$$\theta = \omega t$$

The flux through the coil is

$$\Phi = BA \cos \theta = BA \cos \omega t$$

To find the instantaneous emf, we need to know the instantaneous rate of change of the flux. Using Eq. (20-7b), where $\Phi_0 = BA$,

$$\frac{\Delta\Phi}{\Delta t} = -\omega BA \sin \omega t$$

From Faraday's law,

$$\mathscr{E} = -N\frac{\Delta\Phi}{\Delta t} = \omega NBA \sin \omega t$$

which is what we found in Section 20.2 [Eq. (20-3b)].

Discussion Equation (20-3b) was obtained using the magnetic force on the electrons in a rectangular loop to find the motional emfs in each side. It would be difficult to do the same for a *circular* loop or coil. Faraday's law is easier to use and shows clearly that the induced emf doesn't depend on the particular shape of the loop or coil, as long as it is flat. Only the area and number of turns are relevant.

Practice Problem 20.4 Rotating Coil Generator

In a rotating coil generator, the magnetic field between the poles of an electromagnet has magnitude 0.40 T. A circular coil between the poles has 120 turns and radius 4.0 cm. The coil rotates with *frequency* 5.0 Hz. Find the *maximum* emf induced in the coil.

Technology Based on Electromagnetic Induction

An enormous amount of our technology depends on electromagnetic induction. There are so many applications of Faraday's law that it's hard to even begin a list. Certainly first on the list has to be the electric generator. Almost all of the electricity we use is produced by generators—either moving coil or moving field—that operate according to Faraday's law. Our entire system for distributing electricity is based on *transformers*, devices that use magnetic induction to change ac voltages (Section 20.6). Transformers raise voltages for transmission over long distances across power lines; transformers then reduce the voltages for safe use in homes and businesses. So our entire system for generating *and* distributing electricity depends on Faraday's law of induction.

Application: ground fault interrupter

Application: moving coil microphone

A *ground fault interrupter* (GFI) is a device commonly used in ac electric outlets in bathrooms and other places where the risk of electric shock is great. In Fig. 20.14, the two wires that supply the outlet normally carry equal currents in opposite directions at all times. These ac currents reverse direction 120 times per second. If a person with wet hands accidentally comes into contact with part of the circuit, a current may flow to ground through the person instead of through the return wiring. Then the currents in the two wires are unequal. The magnetic field lines due to the unequal currents are channeled by a ferromagnetic ring through a coil. The flux through the coil reverses direction 120 times per second, so there is an induced emf in the coil, which trips a circuit breaker that disconnects the circuit from the power lines. GFIs are sensitive and fast, so they are a significant safety improvement over a simple circuit breaker.

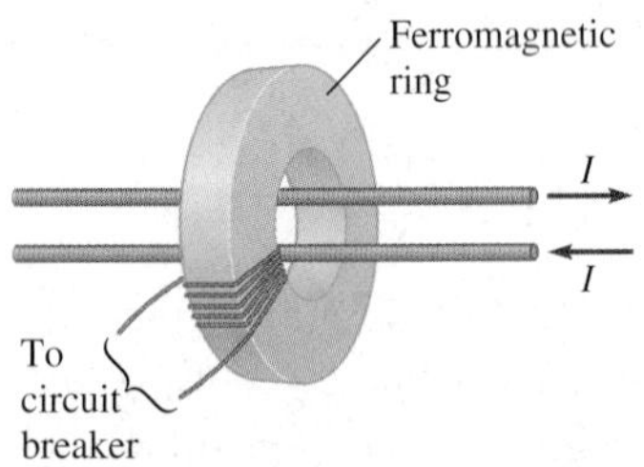

Figure 20.14 A ground fault interrupter.

Figure 20.15 is a simplified sketch of a moving coil microphone. The coil of wire is attached to a diaphragm that moves back and forth in response to sound waves in the air. The magnet is fixed in place. An induced emf appears in the coil due to the changing magnetic flux. In another common type of microphone, the magnet is attached to the diaphragm and the coil is fixed in place.

Faraday's law provides another way to detect currents that flow in the human body. In addition to measuring potential differences between points on the skin, we can measure the magnetic fields generated by these currents. Since the currents are small, the magnetic fields are weak, so sensitive detectors called SQUIDs (superconducting quantum interference devices) are used. When the currents change, changes in the magnetic field induce emfs in the SQUIDs. In a magnetoencephalogram, the induced emfs are measured at many points just outside the cranium (Fig. 20.16); then a computer calculates the location, magnitude, and direction of the currents in the brain that produce the field. Similarly, a magnetocardiogram detects the electric currents in the heart and surrounding nerves.

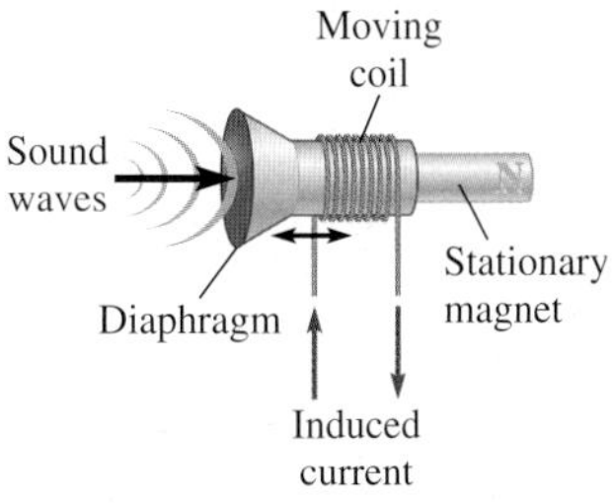

Figure 20.15 A moving coil microphone.

Application: magnetoencephalography

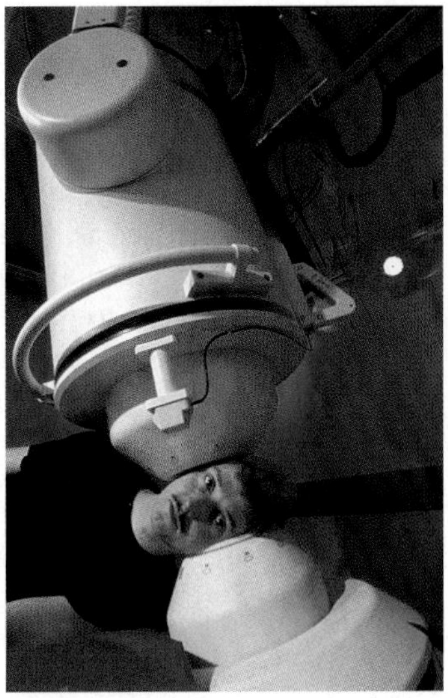

Figure 20.16 In magnetoencephalography, brain function can be observed in real time through noninvasive means. The two white cryostats seen here contain sensitive magnetic field detectors cooled by liquid helium.

20.4 LENZ'S LAW

The directions of the induced emfs and currents caused by a changing magnetic flux can be determined using **Lenz's law**, named for the Baltic German physicist Heinrich Friedrich Emil Lenz (1804–1865):

Lenz's Law

The direction of the induced current in a loop always opposes the *change* in magnetic flux that induces the current.

Note that induced emfs and currents do not necessarily oppose the magnetic field or the magnetic flux; they oppose the *change* in the magnetic flux.

One way to apply Lenz's law is to look at the direction of the magnetic field produced by the induced current. The induced current around a loop produces its own magnetic field. This field may be weak compared with the external magnetic field. It cannot prevent the magnetic flux through the loop from changing, but its direction is always such that it "*tries*" to prevent the flux from changing. The magnetic field direction is related to the direction of the current by right-hand rule 2 (Section 19.8).

CHECKPOINT 20.4

In Fig. 20.11, the magnetic field is increasing in strength. (a) In what direction does induced current flow in the circular loop of wire? (b) In what direction would current flow if the field were decreasing in strength instead?

CONNECTION:

Lenz's law is really an expression of energy conservation. (See Conceptual Example 20.5.)

Conceptual Example 20.5

Faraday's and Lenz's Laws for the Moving Loop

Verify the emfs and currents calculated in Example 20.1 using Faraday's and Lenz's laws—that is, find the directions and magnitudes of the emfs and currents by looking at the changing magnetic flux through the loop.

Strategy To apply Faraday's law, look for the reason why the flux is changing. In Example 20.1, a loop moves to the right at constant velocity into, through, and then out of a region of magnetic field. The magnitude and

continued on next page

Conceptual Example 20.5 continued

direction of the magnetic field within the region are not changing, nor is the area of the loop. What does change is the *portion* of that area that is immersed in the region of magnetic field.

Solution At positions 1, 3, and 5, the flux is *not* changing even though the loop is moving. In each case, a small displacement of the loop causes no flux change. The flux is zero at positions 1 and 5, and nonzero but constant at position 3. For these three positions, the induced emf is zero and so is the current.

If the loop were *at rest* at position 2, the magnetic flux would be constant. However, since the loop is moving into the region of field, the area of the loop through which magnetic field lines cross is increasing. Thus, the flux is increasing. According to Lenz's law, the direction of the induced current opposes the change in flux. Since the field is into the page, and the flux is increasing, the induced current flows in the direction that produces a magnetic field *out of* the page. By the right-hand rule, the current is counterclockwise.

At position 2, a length x of the loop is in the region of magnetic field. The area of the loop that is immersed in the field is Lx. The flux is then

$$\Phi_B = BA = BLx$$

Only x is changing. The rate of change of flux is

$$\frac{\Delta\Phi_B}{\Delta t} = BL\frac{\Delta x}{\Delta t} = BLv$$

Therefore,

$$|\mathscr{E}| = BLv$$

and

$$I = \frac{|\mathscr{E}|}{R} = \frac{BLv}{R}$$

At position 4, the flux is decreasing as the loop leaves the region of magnetic field. Once again, let a length x of the loop be immersed in the field. Just as at position 2,

$$\Phi_B = BLx$$

$$|\mathscr{E}| = \left|\frac{\Delta\Phi_B}{\Delta t}\right| = BL\left|\frac{\Delta x}{\Delta t}\right| = BLv$$

and

$$I = \frac{|\mathscr{E}|}{R} = \frac{BLv}{R}$$

This time the flux is *decreasing*. To oppose a *decrease*, the induced current makes a magnetic field in the *same* direction as the external field—into the page. Then the current must be clockwise.

The magnitudes and directions of the emfs and currents are exactly as found in Example 20.1.

Discussion Another way to use Lenz's law to find the direction of the current is by looking at the magnetic force on the loop. The changing flux is due to the motion of the loop to the right. In order to oppose the change in flux, current flows in the loop in whatever direction gives a magnetic force to the *left*, to try to bring the loop to rest and stop the flux from changing. At position 2, the magnetic forces on sides b and d are equal and opposite; there is no magnetic force on side a since $B = 0$ there. Then there must be a magnetic force on side c to the left. From $\vec{\mathbf{F}} = I\vec{\mathbf{L}} \times \vec{\mathbf{B}}$, the current in side c is up and thus flows counterclockwise in the loop. Similarly, at position 4, the current in side a is upward to give a magnetic force to the left.

The connection between Lenz's law and energy conservation is more apparent when looking at the force on the loop. When current flows in the loop, electric energy is dissipated at a rate $P = I^2R$. Where does this energy come from? If there is no external force pulling the loop to the right, the magnetic force slows down the loop; the dissipated energy comes from the kinetic energy of the loop. To keep the loop moving to the right at constant velocity while current is flowing, an external force must pull it to the right. The work done by the external force replenishes the loop's kinetic energy.

Practice Problem 20.5 The Magnetic Force on the Loop

(a) Find the magnetic force on the loop at positions 2 and 4 in terms of B, L, v, and R. (b) Verify that the rate at which an external force does work ($P = Fv$) to keep the loop moving at constant velocity is equal to the rate at which energy is dissipated in the loop ($P = I^2R$).

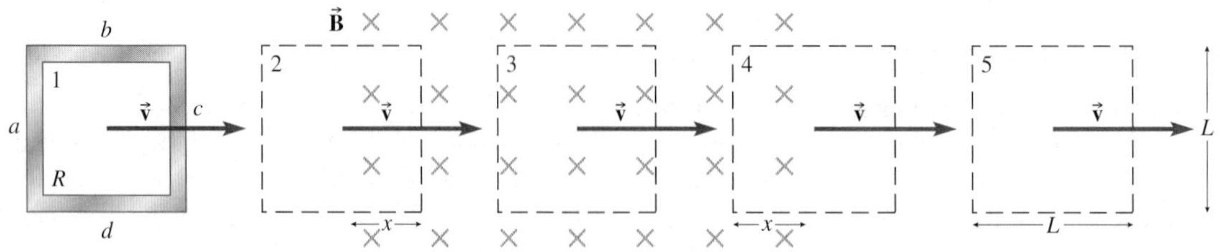

Conceptual Example 20.6

Lenz's Law for a Conducting Loop in a Changing Magnetic Field

A circular loop of wire moves toward a bar magnet at constant velocity (Fig. 20.17). The loop passes around the magnet and continues away from it on the other side. Use Lenz's law to find the direction of the current in the loop at positions 1 and 2.

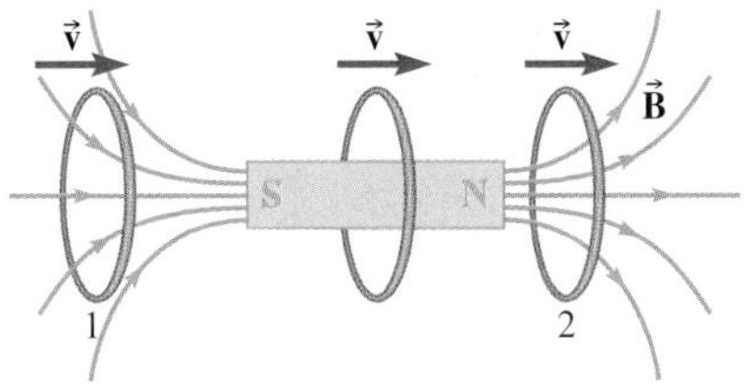

Figure 20.17 Conducting loop passing over a bar magnet.

Strategy The magnetic flux through the loop is changing because the loop moves from weaker to stronger field (at position 1), and vice versa (at position 2). We can specify current directions as counterclockwise or clockwise as viewed from the left (with the loop moving away).

Solution At position 1, the magnetic field lines enter the magnet at the south pole, so the field lines cross the loop from left to right (Fig. 20.18a). Since the loop is moving closer to the magnet, the field is getting stronger; the number of field lines crossing the loop increases (Fig. 20.18b). The flux is therefore increasing. To oppose the increase, the current makes a magnetic field to the left (Fig. 20.18c). The right-hand rule gives the current direction to be counterclockwise as viewed from the left.

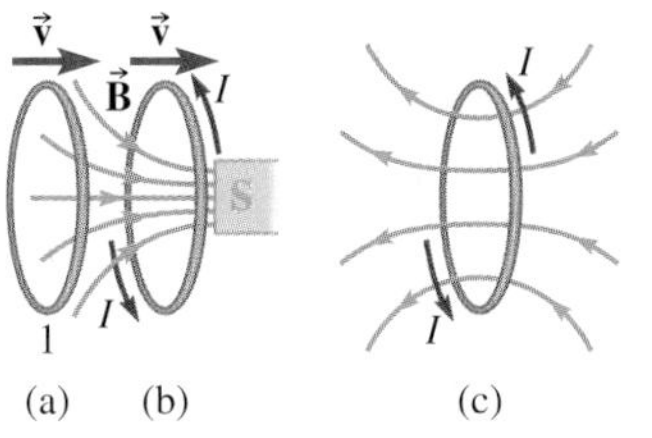

Figure 20.18 Loop moving toward magnet from position (a) to (b); (c) current induced in loop to produce a $\vec{\mathbf{B}}$ field opposing the increasing strength of the nearing bar magnet.

At position 2, the field lines still cross the loop from left to right (Fig. 20.19a), but now the field is getting weaker

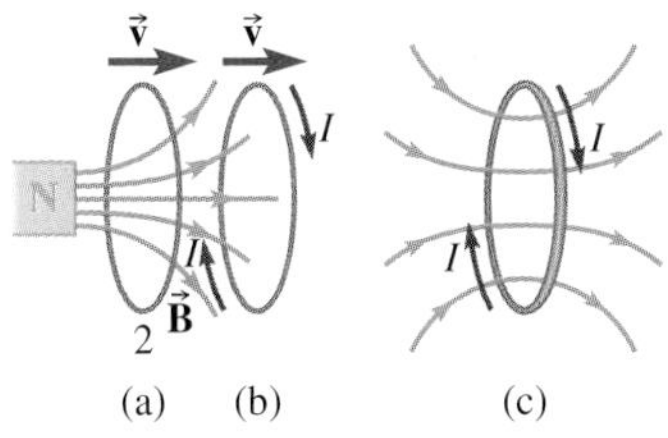

Figure 20.19 Loop moving away from magnet from position (a) to (b); (c) current induced in loop to produce a $\vec{\mathbf{B}}$ field opposing the decreasing strength of the retreating bar magnet.

(Fig. 20.19b). The current must flow in the opposite direction—clockwise as viewed from the left (Fig. 20.19c).

Discussion There's almost always more than one way to apply Lenz's law. An alternative way to think about the situation is to remember the current loop is a magnetic dipole and we can think of it as a little bar magnet. At position 1, the current loop is repelled by the (real) bar magnet. The flux change is due to the motion of the loop toward the magnet; to oppose the change there should be a force pushing away. Then the poles of the current loop must be as in Fig. 20.20a; like poles repel. Point the thumb of the right hand in the direction of the north pole, and curl the fingers to find the current direction.

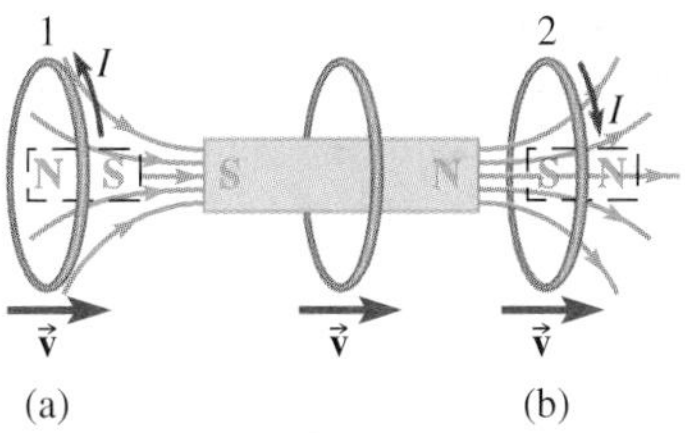

Figure 20.20 Current loops can be represented by small bar magnets.

The same procedure can be used at position 2. Now the flux change is due to the loop moving away from the magnet, so to oppose the change in flux there must be a force attracting the loop toward the magnet (Fig. 20.20b).

Conceptual Practice Problem 20.6 Direction of Induced Emf in Coil

(a) In Fig. 20.21, just after the switch is closed, what is the direction of the magnetic field in the iron core? (b) In what direction does current flow through the resistor connected to coil 2? (c) If the switch remains closed, does current continue to flow in coil 2? Why or why not? (d) Make a drawing in which coils 1 and 2, just after the switch is closed, are replaced by equivalent little bar magnets.

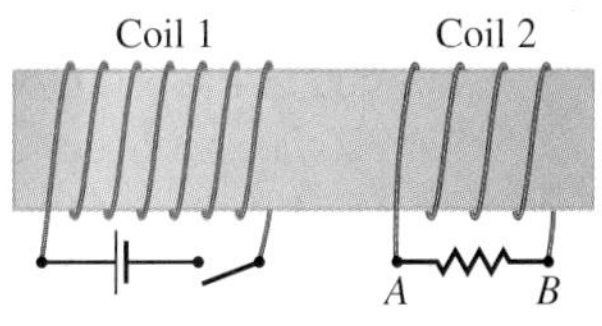

Figure 20.21 Two coils wrapped about a common soft iron core.

20.5 BACK EMF IN A MOTOR

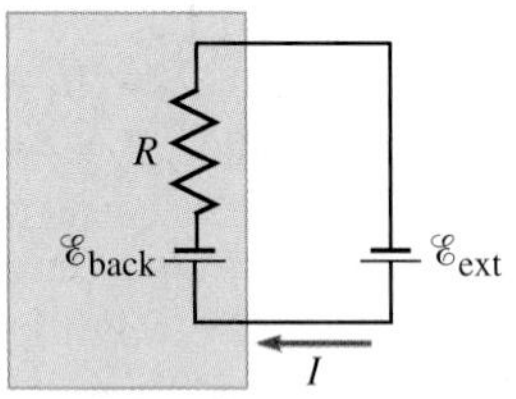

Figure 20.22 An external emf ($\mathscr{E}_{ext}$) is connected to a dc motor. The back emf ($\mathscr{E}_{back}$) is due to the changing flux through the windings. As the motor's rotational speed increases, the back emf increases and the current decreases.

The current through a load that is connected to an emf is sometimes called the current *drawn* by the load.

If a generator and a motor are essentially the same device, is there an induced emf in the coil (or windings) of a motor? There must be, according to Faraday's law, since the magnetic flux through the coil changes as the coil rotates. By Lenz's law, this induced emf—called a **back emf**—opposes the flow of current in the coil, since it is the current that makes the coil rotate and thus causes the flux change. The magnitude of the back emf depends on the rate of change of the flux, so the back emf increases as the rotational speed of the coil increases.

Figure 20.22 shows a simple circuit model of the back emf in a dc motor. We assume that this motor has many coils (also called windings) at all different angles so that the torques, emfs, and currents are all constant. When the external emf is first applied, there is no back emf because the windings are not rotating. Then the current has a maximum value $I = \mathscr{E}_{ext}/R$. The faster the motor turns, the greater the back emf, and the smaller the current: $I = (\mathscr{E}_{ext} - \mathscr{E}_{back})/R$.

You may have noticed that when a large motor—as in a refrigerator or washing machine—first starts up, the room lights dim a bit. The motor draws a large current when it starts up because there is no back emf. The voltage drop across the wiring in the walls is proportional to the current flowing in them, so the voltage across lightbulbs and other loads on the circuit is reduced, causing a momentary "brownout." As the motor comes up to speed, the current drawn is much smaller, so the brownout ends.

If a motor is overloaded, so that it turns slowly or not at all, the current through the windings is large. Motors are designed to withstand such a large current only momentarily, as they start up; if the current is sustained at too high a level the motor "burns out"—the windings heat up enough to do damage to the motor.

20.6 TRANSFORMERS

Making the Connection: transformers

Circuit symbol for a transformer

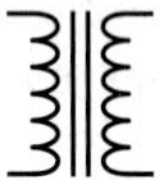

In the late nineteenth century, there were ferocious battles over what form of current should be used to supply electric power to homes and businesses. Thomas Edison was a proponent of direct current, while George Westinghouse, who owned the patents for the ac motor and generator invented by Nikola Tesla, was in favor of alternating current. Westinghouse won mainly because ac permits the use of transformers to change voltages and to transmit over long distances with less power loss than dc, as we see in this section.

Figure 20.23 shows two simple transformers. In each, two separate strands of insulated wire are wound around an iron core. The magnetic field lines are guided through the iron, so the two coils enclose the same magnetic field lines. An alternating voltage is applied to the *primary* coil; the ac current in the primary causes a changing magnetic flux through the *secondary* coil.

If the primary coil has N_1 turns, an emf $\mathscr{E}_1$ is induced in the primary coil according to Faraday's law:

$$\mathscr{E}_1 = -N_1 \frac{\Delta\Phi_B}{\Delta t} \tag{20-8a}$$

Here $\Delta\Phi_B/\Delta t$ is the rate of change of the flux through *each turn* of the primary. Ignoring resistance in the coil and other energy losses, the induced emf is equal to the ac voltage applied to the primary.

If the secondary coil has N_2 turns, then the emf induced in the secondary coil is

$$\mathscr{E}_2 = -N_2 \frac{\Delta\Phi_B}{\Delta t} \tag{20-8b}$$

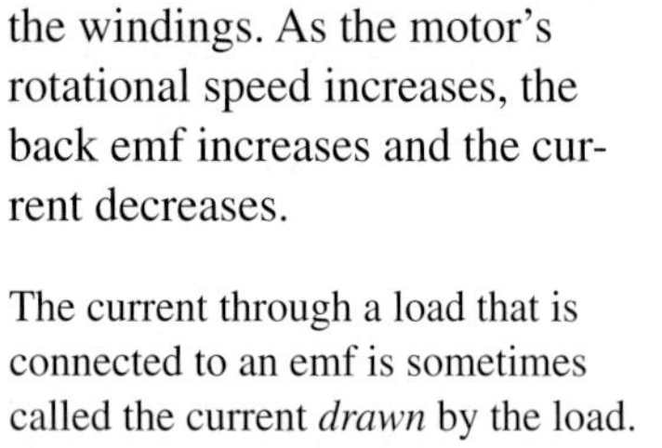

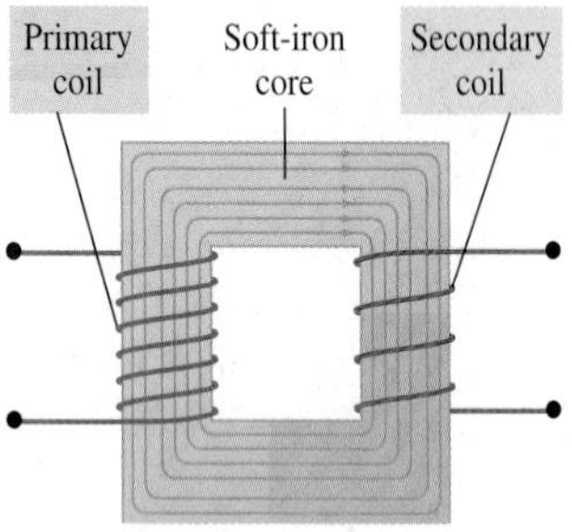

Figure 20.23 Two simple transformers. Each consists of two coils wound on a common iron core so that nearly all the magnetic field lines produced by the primary coil pass through each turn of the secondary.

At any instant, the flux through each turn of the secondary is equal to the flux through each turn of the primary, so $\Delta\Phi_B/\Delta t$ is the same quantity in Eqs. (20-8a) and (20-8b). Eliminating $\Delta\Phi_B/\Delta t$ from the two equations, we find the ratio of the two emfs to be

$$\frac{\mathscr{E}_2}{\mathscr{E}_1} = \frac{N_2}{N_1} \quad (20\text{-}9)$$

The output—the emf in the secondary—is N_2/N_1 times the input emf applied to the primary. The ratio N_2/N_1 is called the **turns ratio**. A transformer is often called a *step-up* or a *step-down* transformer, depending on whether the secondary emf is larger or smaller than the emf applied to the primary. The same transformer may often be used as a step-up or step-down transformer depending on which coil is used as the primary.

Current Ratio In an *ideal transformer*, power losses in the transformer itself are negligible. Most transformers are very efficient, so ignoring power loss is usually reasonable. Then the rate at which energy is supplied to the primary is equal to the rate at which energy is supplied by the secondary ($P_1 = P_2$). Since power equals voltage times current, the ratio of the currents is the inverse of the ratio of the emfs:

$$\frac{I_2}{I_1} = \frac{\mathscr{E}_1}{\mathscr{E}_2} = \frac{N_1}{N_2} \quad (20\text{-}10)$$

Example 20.7

A CD Player's Transformer

A transformer inside the power supply for a portable CD player has 500 turns in the primary coil. It supplies an emf of amplitude 6.8 V when plugged into the usual sinusoidal household emf of amplitude 170 V. (a) How many turns does the secondary coil have? (b) If the current drawn by the CD player has amplitude 1.50 A, what is the amplitude of the current in the primary?

Strategy The ratio of the emfs is the same as the turns ratio. We know the two emfs and the number of turns in the primary, so we can find the number of turns in the secondary. To find the current in the primary, we assume an ideal transformer. Then the currents in the two are inversely proportional to the emfs.

Solution (a) The turns ratio is equal to the emf ratio:

$$\frac{\mathscr{E}_2}{\mathscr{E}_1} = \frac{N_2}{N_1}$$

Solving for N_2 yields

$$N_2 = \frac{\mathscr{E}_2}{\mathscr{E}_1}N_1 = \frac{6.8\text{ V}}{170\text{ V}} \times 500 = 20 \text{ turns}$$

(b) The currents are inversely proportional to the emfs:

$$\frac{I_1}{I_2} = \frac{\mathscr{E}_2}{\mathscr{E}_1} = \frac{N_2}{N_1}$$

$$I_1 = \frac{\mathscr{E}_2}{\mathscr{E}_1}I_2 = \frac{6.8\text{ V}}{170\text{ V}} \times 1.50\text{ A} = 0.060\text{ A}$$

Discussion The most likely error would be to get the turns ratio upside down. Here we need a step-down transformer, so N_2 must be smaller than N_1. If the same transformer were hooked up backward, interchanging the primary and the secondary, then it would act as a step-up transformer. Instead of supplying 6.8 V to the CD player, it would supply

$$170\text{ V} \times \frac{500}{20} = 4250\text{ V}$$

We can check that the power input and the power output are equal:

$$P_1 = \mathscr{E}_1 I_1 = 170\text{ V} \times 0.060\text{ A} = 10.2\text{ W}$$

$$P_2 = \mathscr{E}_2 I_2 = 6.8\text{ V} \times 1.50\text{ A} = 10.2\text{ W}$$

(Since emfs and currents are sinusoidal, the instantaneous power is not constant. By multiplying the amplitudes of the current and emf, we calculate the *maximum* power.)

Practice Problem 20.7 An Ideal Transformer

An ideal transformer has five turns in the primary and two turns in the secondary. If the average power input to the primary is 10.0 W, what is the average power output of the secondary?

Figure 20.24 Voltages are transformed in several stages. This step-up transformer raises the voltage from a generating station to 345 kV for transmission over long distances. Voltages are transformed back down in several stages. The last transformer in the series reduces the 3.4 kV on the local power lines to the 170 V used in the house.

Application: The Distribution of Electricity

Why is it so important to be able to transform voltages? The main reason is to minimize energy dissipation in power lines. Suppose that a power plant supplies a power P to a distant city. Since the power supplied is $P_S = I_S V_S$, where I_S and V_S are the current and voltage supplied to the load (the city), the plant can either supply a higher voltage and a smaller current, or a lower voltage and a larger current. If the power lines have total resistance R, the rate of energy dissipation in the power lines is I_S^2R. Thus, to minimize energy dissipation in the power lines, we want as small a current as possible flowing through them, which means the potential differences must be large—hundreds of kilovolts in some cases. Transformers are used to raise the output emf of a generator to high voltages (Fig. 20.24). It would be unsafe to have such high voltages on household wiring, so the voltages are transformed back down before reaching the house.

20.7 EDDY CURRENTS

Whenever a conductor is subjected to a changing magnetic flux, the induced emf causes currents to flow. In a solid conductor, induced currents flow simultaneously along many different paths. These **eddy currents** are so named due to their resemblance to swirling eddies of current in air or in the rapids of a river. Though the pattern of current flow is complicated, we can still use Lenz's law to get a general idea of the direction of the current flow (clockwise or counterclockwise). We can also determine the qualitative effects of eddy current flow using energy conservation. Since they flow in a resistive medium, the eddy currents dissipate electric energy.

Conceptual Example 20.8

Eddy-Current Damping

A balance must have some damping mechanism. Without one, the balance arm would tend to oscillate for a long time before it settles down; determining the mass of an object would be a long, tedious process. A typical device used to damp out the oscillations is shown in Fig. 20.25.

A metal plate attached to the balance arm passes between the poles of a permanent magnet. (a) Explain the damping effect in terms of energy conservation. (b) Does the damping force depend on the speed of the plate?

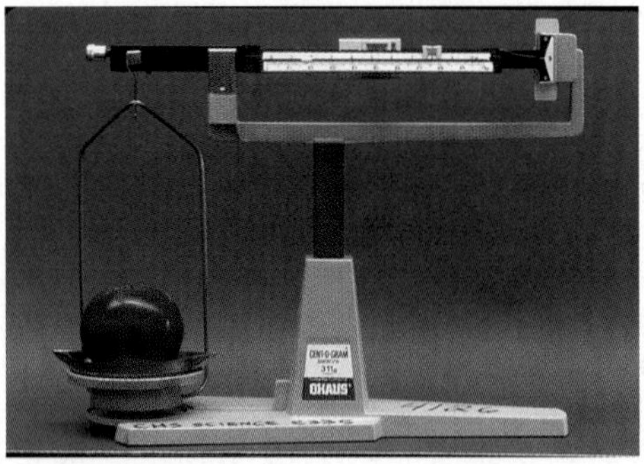

Figure 20.25
A balance. The damping mechanism is at the far right; as the balance arm oscillates, the metal plate moves between the poles of a magnet.

Strategy As portions of the metal plate move into or out of the magnetic field, the changing magnetic flux induces emfs. These induced emfs cause the flow of eddy currents. Lenz's law determines the direction of the eddy currents.

Solution (a) As the plate moves between the magnet poles, parts of it move into the magnetic field while other parts move out of the field. Due to the changing magnetic flux, induced emfs cause eddy currents to flow. The eddy currents dissipate energy; the energy must come from the kinetic energy of the balance arm, pan, and object on the pan. As the currents flow, the kinetic energy of the balance decreases and it comes to rest much sooner than it would otherwise.

(b) If the plate is moving faster, the flux is changing faster. Faraday's law says that the induced emfs are proportional to the rate of change of the flux. Larger induced emfs cause larger currents to flow. The damping force is the magnetic force acting on the eddy currents. Therefore, the damping force is larger.

continued on next page

Conceptual Example 20.8 continued

Discussion Another way to approach part (a) is to use Lenz's law. The magnetic force acting on the eddy currents must oppose the flux change, so it must oppose the motion of the plate through the magnet. Slowing down the plate lessens the rate of flux change, while speeding up the plate would increase the rate of flux change—and increase the balance's kinetic energy, violating energy conservation.

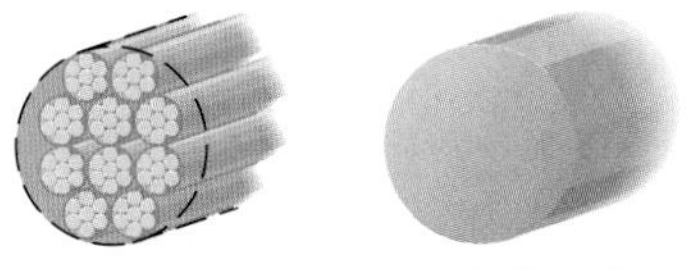

(a) Bundles of iron wires (b) Solid soft iron core

Figure 20.26 Transformer cores.

Conceptual Practice Problem 20.8 Choosing a Core for a Transformer

In some transformers, the core around which wire is wrapped consists of parallel, insulated iron wires instead of solid iron (Fig. 20.26). Explain the advantage of using the insulated wires instead of the solid core. [*Hint:* Think about eddy currents. Why are eddy currents a disadvantage here?]

PHYSICS AT HOME

If either the magnets or the metal plate are removed from a balance, it takes much longer for the oscillations of the balance arm to die out. If your instructor consents, test this on a laboratory balance. Usually a few screws need to be removed.

Application: Eddy-Current Braking

The phenomenon described in Example 20.8 is called *eddy-current braking.* The eddy-current brake is ideal for a sensitive instrument such as a balance. The damping mechanism never wears out or needs adjustment and we are guaranteed that it exerts no force when the balance arm is not moving. Eddy-current brakes are also used with modern rail vehicles such as the maglev monorail, tramways, locomotives, passenger coaches, freight cars, and the latest high-speed maglev trains.

The damping force due to eddy currents automatically acts opposite to the motion; its magnitude is also larger when the speed is larger. The damping force is much like the viscous force on an object moving through a fluid (see Problem 39).

Application: The Induction Stove

How does an induction stove work?

The induction stove discussed in the opening of this chapter operates via eddy currents. Under the cooking surface is an electromagnet that generates an oscillating magnetic field. When a metal pan is put on the stove, the emf causes currents to flow, and the energy dissipated by these currents is what heats the pan (Fig. 20.27). The pan must be made of metal; if a pan made of Pyrex glass is used, no currents flow and no heating occurs. For the same reason, there is no risk of starting a fire if a pot holder or sheet of paper is accidentally put on the induction stove. The cooking surface itself is a nonconductor; its temperature only rises to the extent that heat is conducted to it from the pan. The cooking surface therefore gets no hotter than the bottom of the pan.

Figure 20.27 The eddy currents induced in a metal pan on an induction stove.

20.8 INDUCED ELECTRIC FIELDS

When a conductor moves in a magnetic field, a motional emf arises due to the magnetic force on the mobile charges. Since the charges move along with the conductor, they have a nonzero average velocity. The magnetic force on these charges pushes them around the circuit if a complete circuit exists.

Table 20.1 Comparison of Conservative and Nonconservative $\vec{E}$ Fields

	Conservative $\vec{E}$ Fields	Nonconservative (Induced) $\vec{E}$ Fields
Source	Charges	Changing $\vec{B}$ fields
Field lines	Start on positive charges and end on negative charges	Closed loops
Can be described by an electric potential?	Yes	No
Work done over a closed path	Always zero	Can be nonzero

What causes the induced emf in a stationary conductor in a changing magnetic field? Now the conductor is at rest and the mobile charges have an average velocity of zero. The average magnetic force on them is then zero, so it cannot be the magnetic force that pushes the charges around the circuit. An **induced electric field**, created by the changing magnetic field, acts on the mobile charge in the conductor, pushing it around the circuit. The same force law ($\vec{F} = q\vec{E}$) applies to induced electric fields as to any other electric field.

The induced emf around a loop is the work done per unit charge on a charged particle that moves around the loop. Thus, an induced electric field does nonzero work on a charge that moves around a closed path, starting and ending at the same point. In other words, the induced electric field is nonconservative. The work done by the induced $\vec{E}$ field *cannot* be described as the charge times the potential difference. The concept of potential depends on the electric field doing zero work on a charge moving around a closed path—only then can the potential have a unique value at each point in space. Table 20.1 summarizes the differences between conservative and nonconservative $\vec{E}$ fields.

Electromagnetic Fields

How can Faraday's law give the induced emf regardless of why the flux is changing—whether because of a changing magnetic field or because of a conductor moving in a magnetic field? A conductor that is moving in one frame of reference is at rest in another frame of reference (see Section 3.6). As we will see in Chapter 26, Einstein's theory of special relativity says that either reference frame is equally valid. In one frame, the induced emf is due to the motion of the conductor; in the other, the induced emf is due to a changing magnetic field.

CONNECTION:

Relativity unifies the electric and magnetic fields.

The electric and magnetic fields are not really separate entities. They are intimately connected. Though it is advantageous in many circumstances to think of them as distinct fields, a more accurate view is to think of them as two aspects of the **electromagnetic field**. To use a loose analogy: a vector has different x- and y-components in different coordinate systems, but these components represent the same vector quantity. In the same way, the electromagnetic field has electric and magnetic parts (analogous to vector components) that depend on the frame of reference. A purely electric field in one frame of reference has both electric and magnetic "components" in another reference frame.

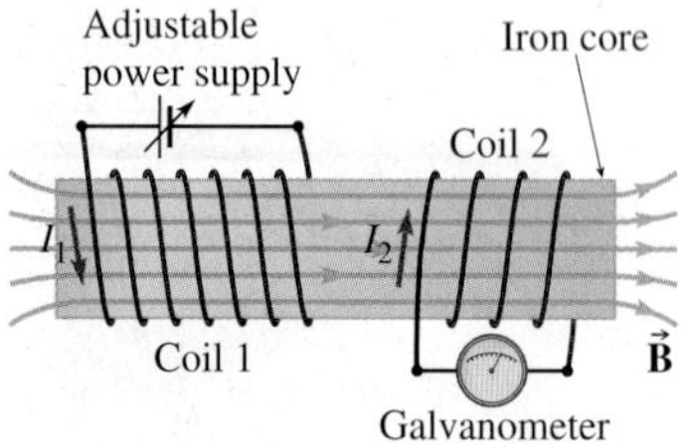

Figure 20.28 An induced emf appears in coil 2 due to the changing current in coil 1.

You may notice a missing symmetry. If a changing $\vec{B}$ field is always accompanied by an induced $\vec{E}$ field, what about the other way around? Does a changing electric field make an induced magnetic field? The answer to this important question—central to our understanding of light as an electromagnetic wave—is yes (Chapter 22).

20.9 INDUCTANCE

Mutual Inductance

Figure 20.28 shows two coils of wire. A power supply with variable emf causes current I_1 to flow in coil 1; the current produces magnetic field lines as shown. Some of these field lines cross through the turns of coil 2. If we adjust the power supply so that I_1 changes, the flux through coil 2 changes and an induced emf appears in coil 2. **Mutual inductance**—when a changing current in one device causes an induced emf in another device—can occur between two circuit elements in the same circuit as well as between circuit elements in two different circuits. In either case, a changing current through one element induces an emf in the other. The effect is truly mutual: a changing current in coil 2 induces an emf in coil 1 as well.

At any point, the magnetic field due to coil 1 is proportional to I_1. For instance, if we double I_1, the magnetic field everywhere would be twice as large. The total flux linkage through coil 2 is proportional to the magnetic field, and therefore to the current I_1:

$$N_2\Phi_{21} \propto I_1$$

where the subscripts remind us that Φ_{21} stands for the total flux through coil 2 due to the field produced by coil 1. The constant of proportionality is called the *mutual inductance* (M):

$$N_2\Phi_{21} = MI_1 \tag{20-11}$$

The mutual inductance depends on the shape and size of the two circuit elements, their separation, and their relative orientation. It is exceedingly difficult to calculate mutual inductances from the geometry of the two elements. In every case, the mutual inductance M turns out to be the same regardless of whether we consider the flux linkage through coil 2 due to the current in coil 1 or vice versa:

$$M = \frac{N_2\Phi_{21}}{I_1} = \frac{N_1\Phi_{12}}{I_2} \tag{20-12}$$

From Faraday's law, the induced emf in coil 2 is

$$\mathscr{E}_{21} = -N_2\frac{\Delta\Phi_{21}}{\Delta t} = -M\frac{\Delta I_1}{\Delta t} \tag{20-13}$$

Similarly, the induced emf in coil 1 is

$$\mathscr{E}_{12} = -N_1\frac{\Delta\Phi_{12}}{\Delta t} = -M\frac{\Delta I_2}{\Delta t}$$

Recall that, to give the *instantaneous* emf, Faraday's law must be applied to a very short time interval. If Δt represents a longer time interval, then Faraday's law gives the *average* emf over that time interval.

From Eq. (20-13), we can find the SI units of M:

$$[M] = \frac{[\mathscr{E}]}{[\Delta I/\Delta t]} = \frac{\text{V}}{\text{A/s}} = \frac{\text{V}\cdot\text{s}}{\text{A}}$$

This combination of units is given the name henry (symbol: H) after Joseph Henry (1797–1878), the American scientist who was the first to wrap insulated wires around an iron core to make an electromagnet. Henry actually discovered induced emfs before Faraday, but Faraday published first.

Example 20.9

Mutual Inductance

A circular loop of wire is placed near a solenoid (Fig. 20.29). When the current in the solenoid is 550 mA, the flux through the circular loop is 2.7×10^{-5} Wb. When the current in the solenoid changes at 6.0 A/s, the induced current in the circular loop is 0.36 mA. What is the resistance of the circular loop?

Strategy The mutual inductance is the proportionality constant between the current in the solenoid and the flux through the loop. M is also the proportionality constant between the rate of change of current in the solenoid and the emf in the loop. From the induced emf and the induced current, we can find the resistance of the loop.

Solution The mutual inductance is

$$M = \frac{N_1\Phi_{1s}}{I_s} = \frac{1 \times 2.7 \times 10^{-5}\ \text{Wb}}{0.550\ \text{A}} = 4.91 \times 10^{-5}\ \text{H}$$

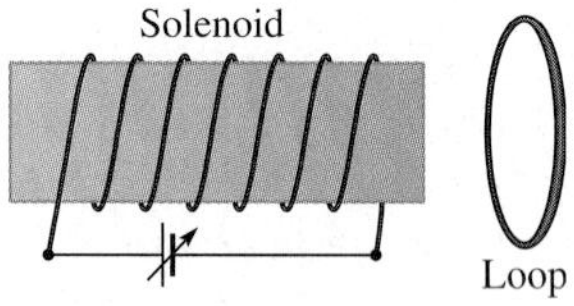

Figure 20.29 A changing current in the solenoid induces an emf in the loop; a changing current in the loop also induces an emf in the solenoid.

The subscript "l" stands for "loop" and "s" stands for "solenoid." Then when the current in the solenoid changes at the rate $\Delta I_s/\Delta t = 6.0$ A/s, the induced emf in the loop is

$$|\mathscr{E}_{1s}| = M\left|\frac{\Delta I_s}{\Delta t}\right| = 4.91 \times 10^{-5}\ \frac{\text{V·s}}{\text{A}} \times 6.0\ \frac{\text{A}}{\text{s}} = 2.95 \times 10^{-4}\ \text{V}$$

The resistance of the loop is

$$R = \frac{|\mathscr{E}_{1s}|}{I} = \frac{0.295\ \text{mV}}{0.36\ \text{mA}} = 0.82\ \Omega$$

Discussion The mutual inductance determines the induced emf in the loop for a given rate of change of current in the solenoid. How much current flows in the loop in response to the induced emf depends on the loop's electrical resistance. A loop with a much higher resistance would have the *same* emf, but a much smaller current would flow.

Practice Problem 20.9 Flux Through the Solenoid

If a 1.5-V power supply is connected to the loop, what would be the total magnetic flux through the solenoid due to the loop's magnetic field?

Self-Inductance

The circuit symbol for an inductor is 

Henry was the first to suggest that a changing current in a coil induces an emf in the *same* coil as well as in other coils—an effect called **self-inductance**. When the current through the coil is changing, the changing magnetic flux inside the coil produces an induced electric field that gives rise to an induced emf. When a coil, solenoid, toroid, or other circuit element is used in a circuit primarily for its self-inductance effects, it is often referred to as an **inductor**. *Self-inductance* is often shortened to *inductance.* A few inductors are shown in Fig. 20.30.

To calculate the self-inductance of coil 1 in Fig. 20.28, we follow the same steps that we used to find mutual inductance. If a current I_1 flows in coil 1, then the total flux through coil 1 is proportional to I_1:

$$N_1\Phi_{11} \propto I_1$$

The subscripts remind us that Φ stands for the total flux through coil 1 due to the current in coil 1. When the context is clearly one of self-inductance, we write simply $N\Phi \propto I$. The self-inductance L of the coil is defined as the constant of proportionality between self-flux and current:

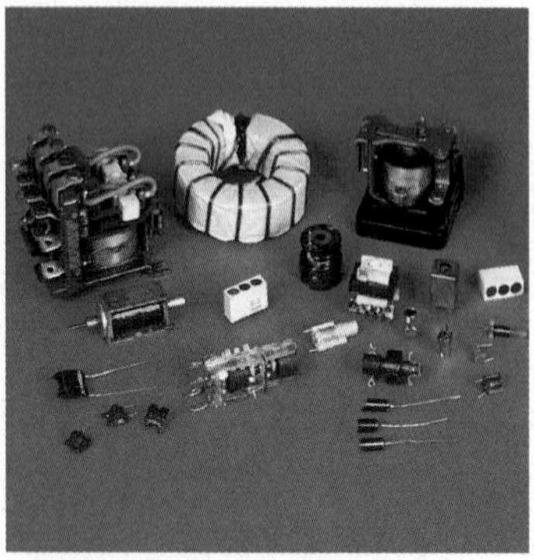

Figure 20.30 Inductors come in many sizes and shapes.

Definition of self-inductance:

$$N\Phi = LI \qquad (20\text{-}14)$$

The most common form of inductor is the solenoid. In Problem 49, the self-inductance L of a long air core solenoid of n turns per unit length, length ℓ, and radius r is found to be

$$L = \mu_0 n^2 \pi r^2 \ell \qquad (20\text{-}15a)$$

For a solenoid with $N = n\ell$ turns,

$$L = \frac{\mu_0 N^2 \pi r^2}{\ell} \qquad (20\text{-}15b)$$

According to Faraday's law, the induced emf in coil 1 is then

$$\mathscr{E} = -N\frac{\Delta\Phi}{\Delta t} = -L\frac{\Delta I}{\Delta t} \qquad (20\text{-}16)$$

The SI unit of self-inductance is the same as that of mutual inductance: the henry.

Inductors in Circuits The behavior of an inductor in a circuit can be summarized as current *stabilizer*. The inductor "likes" the current to be constant—it "tries" to maintain the status quo. If the current is constant, there is no induced emf; to the extent that we can ignore the resistance of its windings, the inductor acts like a short circuit. When the current is changing, the induced emf is proportional to the rate of change. According to Lenz's law, the direction of the emf opposes the change that produces it. If the current is increasing, the direction of the emf in the inductor pushes back as if to make it harder for the current to increase (Fig. 20.31a). If the current is decreasing, the direction of the emf in the inductor is forward, as if to help the current keep flowing (Fig. 20.31b).

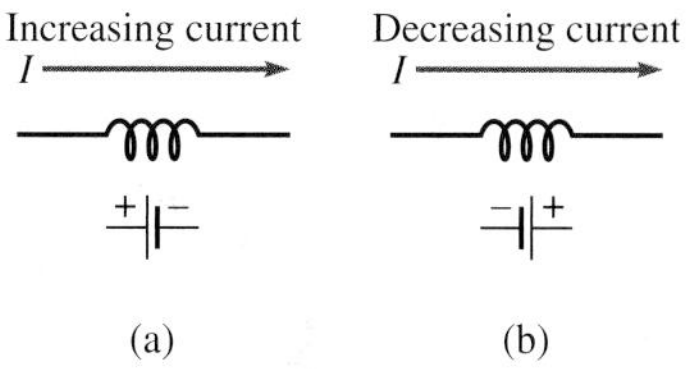

Figure 20.31 The current through both these inductors flows to the right. In (a), the current is increasing; the induced emf in the inductor "tries" to prevent the increase. In (b), the current is decreasing; the induced emf in the inductor "tries" to prevent the decrease.

Inductors Store Energy An inductor stores energy in a magnetic field, just as a capacitor stores energy in an electric field. Suppose the current in an inductor increases at a constant rate from 0 to I in a time T. We let lowercase i stand for the instantaneous current at some time t between 0 and T, and let uppercase I stand for the *final* current. The instantaneous rate at which energy accumulates in the inductor is

$$P = \mathscr{E}i$$

Since current increases at a constant rate, the magnetic flux increases at a constant rate, so the induced emf is constant. Also, since the current increases at a constant rate, the average current is $I_{av} = I/2$. Then the *average* rate at which energy accumulates is

$$P_{av} = \mathscr{E}I_{av} = \tfrac{1}{2}\mathscr{E}I$$

Using Eq. (20-16) for the emf, the average power is

$$P_{av} = \tfrac{1}{2}L\frac{\Delta i}{\Delta t}I$$

and the total energy stored in the inductor is

$$U = P_{av}T = \tfrac{1}{2}\left(L\frac{\Delta i}{\Delta t}\right)IT$$

Since the current changes at a constant rate, $\Delta i/\Delta t = I/T$. The total energy stored in the inductor is

Magnetic energy stored in an inductor:

$$U = \tfrac{1}{2}LI^2 \qquad (20\text{-}17)$$

Although to simplify the calculation we assumed that the current was increased from zero at a constant rate, Eq. (20-17) for the energy stored in an inductor depends only on the current I and not on how the current reached that value (see Problem 54).

CONNECTION:

Compare the energy stored in an inductor and the energy stored in a capacitor: $U_C = \frac{1}{2}C^{-1}Q^2$ [Eq. (17-18c)] The energy in the magnetic field of an inductor is proportional to the square of the current, just as the energy in the electric field of a capacitor is proportional to the square of the charge.

Magnetic Energy Density We can use the inductor to find the magnetic energy density in a magnetic field. Consider a solenoid so long that we can ignore the magnetic energy stored in the field outside it. The inductance is

$$L = \mu_0 n^2 \pi r^2 \ell$$

where n is the number of turns per unit length, ℓ is the length of the solenoid, and r is its radius. The energy stored in the inductor when a current I flows is

$$U = \tfrac{1}{2}LI^2 = \tfrac{1}{2}\mu_0 n^2 \pi r^2 \ell I^2$$

The volume of space inside the solenoid is the length times the cross-sectional area:

$$\text{volume} = \pi r^2 \ell$$

Then the magnetic energy density—energy per unit volume—is

$$u_B = \frac{U}{\pi r^2 \ell} = \frac{1}{2}\mu_0 n^2 I^2$$

To express the energy density in terms of the magnetic field strength, recall that $B = \mu_0 nI$ [Eq. (19-17)] inside a long solenoid. Therefore,

Magnetic energy density:

$$u_B = \frac{1}{2\mu_0}B^2 \qquad (20\text{-}18)$$

Equation (20-18) is valid for more than the interior of an air core solenoid; it gives the energy density for *any* magnetic field except for the field inside a ferromagnet. Both the magnetic energy density and the electric energy density are proportional to the square of the field strength: recall that the electric energy density is

$$u_E = \tfrac{1}{2}\kappa\epsilon_0 E^2 \qquad (17\text{-}19)$$

Example 20.10

Energy Stored in a Solenoid

An ideal air-core solenoid has radius 2.0 cm, length 12 cm, and 9000.0 turns. The solenoid carries a current of 2.0 A. (a) Find the magnetic field inside the solenoid. (b) How much energy is stored in the solenoid?

Strategy Since the solenoid is ideal, we ignore the nonuniformity in the magnetic field near the ends. We consider the magnetic field to be uniform in the entire volume inside. There are two ways to find the energy. We can either find the self-inductance and then use Eq. (20-17) for the energy stored in an inductor, or we can find the energy density [Eq. (20-18)] and multiply by the volume.

Given: $N = 9000.0$, $r = 0.020$ m, $\ell = 0.12$ m

Find: B, U_B

Solution (a) The magnetic field inside an ideal solenoid is

$$B = \mu_0 nI = \frac{\mu_0 NI}{\ell}$$

$$= \frac{4\pi \times 10^{-7}\ \text{H/m} \times 9000.0 \times 2.0\ \text{A}}{0.12\ \text{m}} = 0.19\ \text{T}$$

(b) The magnetic energy density is

$$u_B = \frac{1}{2\mu_0}B^2 \qquad (20\text{-}18)$$

The total energy stored in the solenoid is

$$U_B = \frac{1}{2\mu_0}B^2 \times \pi r^2 \ell$$

since $\pi r^2 \ell$ is the volume of the interior of the solenoid. Substituting the expression for B yields

$$U_B = \frac{1}{2\mu_0}\left(\frac{\mu_0 NI}{\ell}\right)^2 \times \pi r^2 \ell = \mu_0 N^2 I^2 \pi r^2/(2\ell)$$

Now we can substitute numerical values.

$$U_B = \frac{4\pi \times 10^{-7}\ \text{H/m} \times 9000.0^2 \times (2.0\ \text{A})^2 \times \pi(0.020\ \text{m})^2}{2 \times 0.12\ \text{m}}$$

$$= 2.1\ \text{J}$$

continued on next page

Example 20.10 continued

Discussion Let's use the alternative method as a check. For a solenoid with N turns,

$$L = \frac{\mu_0 N^2 \pi r^2}{\ell} \quad (20\text{-}15b)$$

The magnetic energy stored in an inductor is

$$U_B = \tfrac{1}{2} L I^2 \quad (20\text{-}17)$$

By substitution,

$$U_B = \frac{1}{2}\frac{\mu_0 N^2 \pi r^2}{\ell} I^2$$

which agrees with the expression found previously.

We should also verify the units. Since $1\ \text{H} = \frac{1\ \text{V}}{\text{A/s}}$,

$$\frac{\text{H/m} \times \text{A}^2 \times \text{m}^2}{\text{m}} = \text{H} \times \text{A}^2 = \frac{\text{V} \times \text{A}^2}{\text{A/s}} = \text{V} \times \text{A} \times \text{s} = \text{V} \times \text{C} = \text{J}$$

Practice Problem 20.10 Power in an Inductor

Suppose the current in the inductor of Example 20.10 increases from 0 to 2.0 A during a time interval of 4.0 s. Calculate the average rate at which energy is stored in the inductor during this time interval. [*Hint:* Use one method to calculate the answer and another as a check.]

20.10 *LR* CIRCUITS

To get an idea of how inductors behave in circuits, let's first study them in dc circuits—that is, in circuits with batteries or other constant-voltage power supplies. Consider the ***LR* circuit** in Fig. 20.32. The inductor is assumed to be ideal: its windings have negligible resistance. At $t = 0$, the switch S is closed. What is the subsequent current in the circuit?

The current through the inductor just before the switch is closed is zero. As the switch is closed, the current is initially zero. An instantaneous change in current through an inductor would mean an instantaneous change in its stored energy, since $U \propto I^2$. An instantaneous change in energy means that energy is supplied in zero time. Since nothing can supply infinite power,

> Current through an inductor must always change *continuously*, never instantaneously.

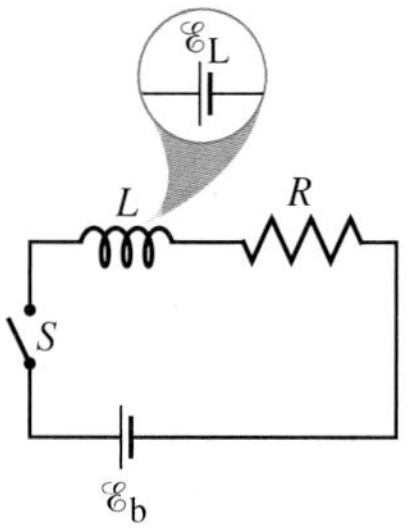

Figure 20.32 A dc circuit with an inductor L, a resistor R, and a switch S. When the current is changing, an emf is induced in the inductor (represented by a battery symbol above the inductor).

The initial current is zero, so there is no voltage drop across the resistor. The magnitude of the induced emf in the inductor ($\mathscr{E}_L$) is *initially* equal to the battery's emf ($\mathscr{E}_b$). Therefore, the current is rising at an initial rate given by

$$\frac{\Delta I}{\Delta t} = \frac{\mathscr{E}_b}{L}$$

As current builds up, the voltage drop across the resistor increases. Then the induced emf in the inductor ($\mathscr{E}_L$) gets smaller (Fig. 20.33) so that

$$(\mathscr{E}_b - \mathscr{E}_L) - IR = 0 \quad (20\text{-}19a)$$

or

$$\mathscr{E}_b = \mathscr{E}_L + IR \quad (20\text{-}19b)$$

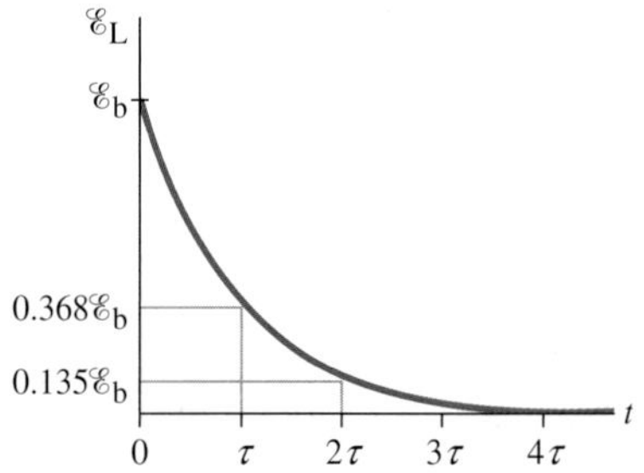

Figure 20.33 The voltage drop across the inductor as the current builds up.

Since the voltage across an *ideal* inductor is the induced emf, we can substitute $\mathscr{E}_L = L(\Delta I/\Delta t)$: [The minus sign has already been written explicitly in Eq. (20-19); $\mathscr{E}_L$ here stands for the *magnitude* of the emf.]

$$\mathscr{E}_b = L\frac{\Delta I}{\Delta t} + IR \quad (20\text{-}20)$$

The battery emf is constant. Thus, as the current increases, the voltage drop across the resistor gets larger and the induced emf in the inductor gets smaller. Therefore, the *rate* at which the current increases gets smaller (Fig. 20.34).

After a very long time, the current reaches a stable value. Since the current is no longer changing, there is no voltage drop across the inductor, so $\mathcal{E}_b = I_f R$ or

$$I_f = \frac{\mathcal{E}_b}{R}$$

The current as a function of time $I(t)$ is:

$$I(t) = I_f(1 - e^{-t/\tau}) \quad (20\text{-}21)$$

CONNECTION:

$I(t)$ for this *LR* circuit has the same form as $q(t)$ for the charging *RC* circuit.

The time constant τ for this circuit must be some combination of L, R, and $\mathcal{E}$. Dimensional analysis (Problem 63) shows that τ must be some dimensionless constant times L/R. It can be shown with calculus that the dimensionless constant is 1:

Time constant, *LR* circuit:

$$\tau = \frac{L}{R} \quad (20\text{-}22)$$

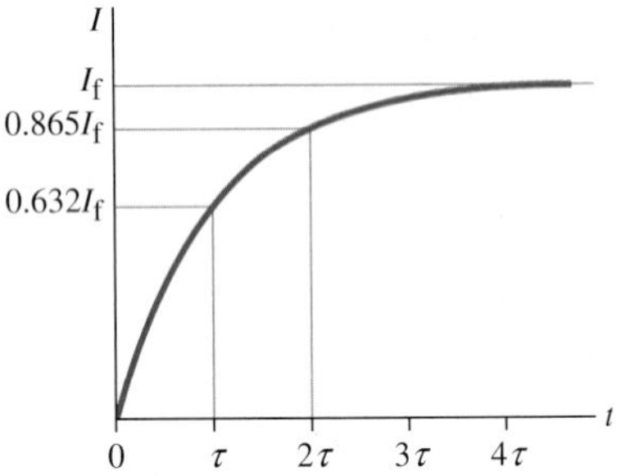

Figure 20.34 The current in the circuit as a function of time.

The induced emf as a function of time is

$$\mathcal{E}_L(t) = \mathcal{E}_b - IR = \mathcal{E}_b - \frac{\mathcal{E}_b}{R}(1 - e^{-t/\tau})R = \mathcal{E}_b e^{-t/\tau} \quad (20\text{-}23)$$

The *LR* circuit in which the current is initially zero is analogous to the charging *RC* circuit. In both cases, the device starts with no stored energy and gains energy after the switch is closed. In charging a capacitor, the *charge* eventually reaches a nonzero equilibrium value, while for the inductor the *current* reaches a nonzero equilibrium value.

CHECKPOINT 20.10

In Fig. 20.32, $\mathcal{E}_b = 1.50$ V, $L = 3.00$ mH, and $R = 12.0\ \Omega$. (a) At what rate is the current through the inductor changing just after the switch is closed? (b) When does the induced emf in the inductor fall to $e^{-1} \approx 0.368$ times its initial value?

What about an *LR* circuit analogous to the discharging *RC* circuit? That is, once a steady current is flowing through an inductor, and energy is stored in the inductor, how can we stop the current and reclaim the stored energy? Simply opening the switch in Fig. 20.32 would *not* be a good way to do it. The attempt to suddenly stop the current would induce a *huge* emf in the inductor. Most likely, sparks would complete the circuit across the open switch, allowing the current to die out more gradually. (Sparking generally isn't good for the health of the switch.)

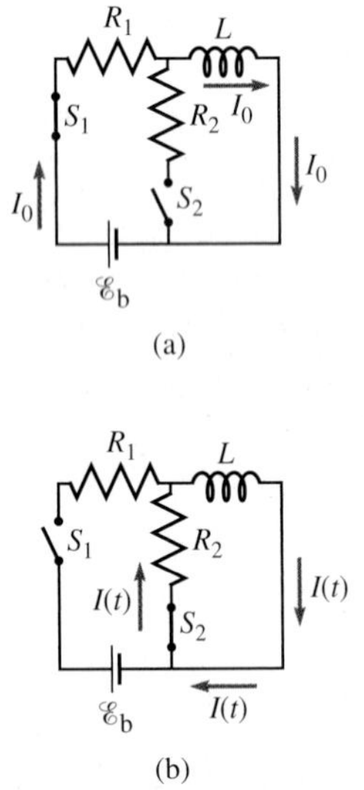

Figure 20.35 A circuit that allows the current in the inductor circuit to be safely stopped. (a) Initially switch 1 is closed and switch 2 is open. (b) At $t = 0$, switch 2 is closed and then switch 1 immediately opened.

A better way to stop the current is shown in Fig. 20.35. Initially switch S_1 is closed and a current $I_0 = \mathcal{E}_b/R_1$ is flowing through the inductor (Fig. 20.35a). Switch S_2 is closed and then S_1 is immediately opened at $t = 0$. Since the current through the inductor can only change continuously, the current flows as shown in Fig. 20.35b. At $t = 0$, the current is $I_0 = \mathcal{E}_b/R_1$. The current gradually dies out as the energy stored in the inductor is dissipated in resistor R_2. The current as a function of time is a decaying exponential:

$$I(t) = I_0 e^{-t/\tau} \quad (20\text{-}24)$$

CONNECTION:

$I(t)$ in this *LR* circuit is analogous to $q(t)$ for a discharging *RC* circuit.

where

$$\tau = \frac{L}{R_2}$$

The voltages across the inductor and resistor can be found from the loop rule and Ohm's law.

CONNECTION

This summary shows that *RC* and *LR* circuits are closely analogous.

	Capacitor	Inductor
Voltage is proportional to	Charge	Rate of change of current
Can change discontinuously	Current	Voltage
Cannot change discontinuously	Voltage	Current
Energy stored (U) is proportional to	V^2	I^2
When $V = 0$ and $I \neq 0$	$U = 0$	$U =$ maximum
When $I = 0$ and $V \neq 0$	$U =$ maximum	$U = 0$
Energy stored (U) is proportional to	E^2	B^2
Time constant =	RC	L/R
"Charging" circuit	$I(t) \propto e^{-t/\tau}$	$I(t) \propto (1 - e^{-t/\tau})$
	$V_C(t) \propto (1 - e^{-t/\tau})$	$V_L(t) = \mathscr{E}_L(t) \propto e^{-t/\tau}$
"Discharging" circuit	$I(t) \propto e^{-t/\tau}$	$I(t) \propto e^{-t/\tau}$
	$V_C(t) \propto e^{-t/\tau}$	$V_L(t) = \mathscr{E}_L(t) \propto e^{-t/\tau}$

Example 20.11

Switching on a Large Electromagnet

A large electromagnet has an inductance $L = 15$ H. The resistance of the windings is $R = 8.2\ \Omega$. Treat the electromagnet as an ideal inductor in series with a resistor (as in Fig. 20.32). When a switch is closed, a 24-V dc power supply is connected to the electromagnet. (a) What is the ultimate current through the windings of the electromagnet? (b) How long after closing the switch does it take for the current to reach 99.0% of its final value?

Strategy When the current reaches its final value, there is no induced emf. The ideal inductor in Fig. 20.32 therefore has no potential difference across it. Then the entire voltage of the power source is across the resistor. The current follows an exponential curve as it builds to its final value. When it is at 99.0% of its final value, it has 1.0% left to go.

Solution (a) After the switch has been closed for many time constants, the current reaches a steady value. When the current is no longer changing, there is no induced emf. Therefore, the entire 24 V of the power supply is dropped across the resistor:

$$\mathscr{E}_b = \mathscr{E}_L + IR$$

$$\text{when } \mathscr{E}_L = 0, \quad I_f = \frac{\mathscr{E}_b}{R} = \frac{24\text{ V}}{8.2\ \Omega} = 2.9\text{ A}$$

(b) The factor $e^{-t/\tau}$ represents the fraction of the current yet to build up. When the current reaches 99.0% of its final value,

$$1 - e^{-t/\tau} = 0.990 \quad \text{or} \quad e^{-t/\tau} = 0.010$$

There is 1.0% yet to go. To solve for t, first take the natural logarithm (ln) of both sides to get t out of the exponent:

$$\ln(e^{-t/\tau}) = -t/\tau = \ln 0.010 = -4.61$$

Now solve for t:

$$t = -\tau \ln 0.010 = -\frac{L}{R}\ln 0.010 = -\frac{15\text{ H}}{8.2\ \Omega} \times (-4.61) = 8.4\text{ s}$$

It takes 8.4 s for the current to build up to 99.0% of its final value.

Discussion A slightly different approach is to write the current as a function of time:

$$I(t) = \frac{\mathscr{E}_b}{R}(1 - e^{-t/\tau}) = I_f(1 - e^{-t/\tau})$$

We are looking for the time t at which $I = 99.0\%$ of 2.9 A or $I/I_f = 0.990$. Then

$$0.990 = 1 - e^{-t/\tau} \quad \text{or} \quad e^{-t/\tau} = 0.010$$

as before.

continued on next page

Example 20.11 continued

Practice Problem 20.11 Switching Off the Electromagnet

When the electromagnet is to be turned off, it is connected to a 50.0-Ω resistor, as in Fig. 20.36, to allow the current to decrease gradually. How long after the switch is opened does it take for the current to decrease to 0.1 A?

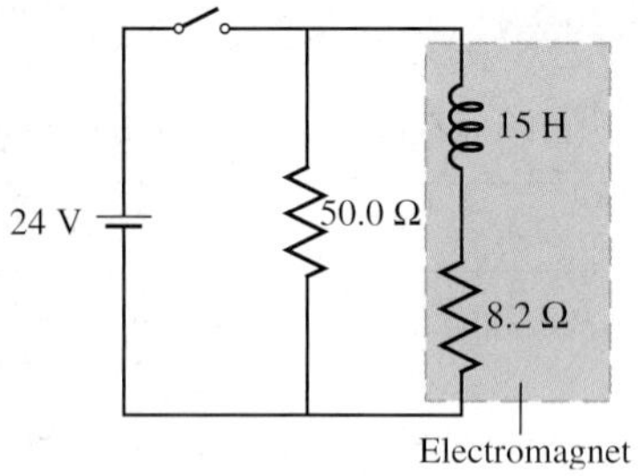

Figure 20.36
Practice Problem 20.11.

Master the Concepts

- A conductor moving through a magnetic field develops a motional emf given by

$$\mathscr{E} = vBL \quad (20\text{-}2a)$$

if both $\vec{\mathbf{v}}$ and $\vec{\mathbf{B}}$ are perpendicular to the rod.

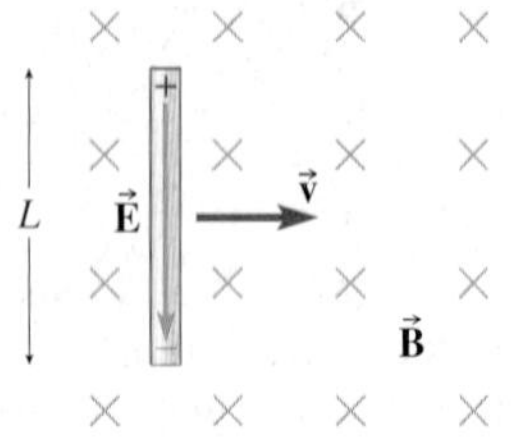

- The emf due to an ac generator with one planar coil of wire turning in a uniform magnetic field is sinusoidal and has amplitude ωNBA:

$$\mathscr{E}(t) = \omega NBA \sin \omega t \quad (20\text{-}3b)$$

Here ω is the angular speed of the coil, A is its area, and N is the number of turns.

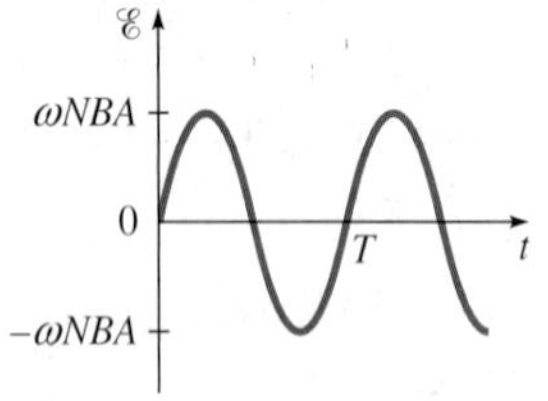

- Magnetic flux through a planar surface:

$$\Phi_B = B_\perp A = BA_\perp = BA \cos \theta \quad (20\text{-}5)$$

(θ is the angle between $\vec{\mathbf{B}}$ and the *normal.*)

The magnetic flux is proportional to the number of magnetic field lines that cut through a surface. The SI unit of magnetic flux is the weber (1 Wb = 1 T·m^2).

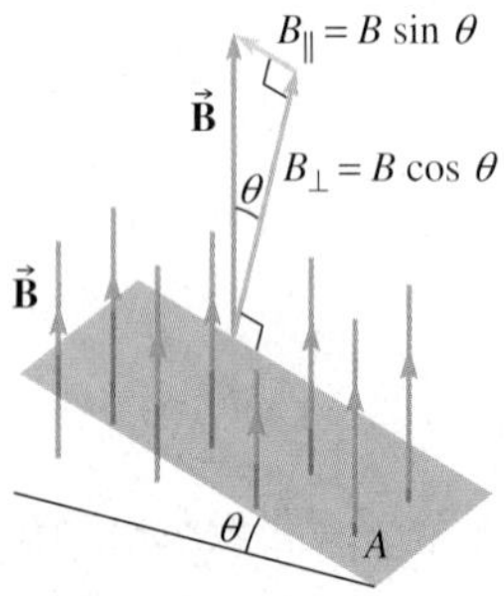

- Faraday's law gives the induced emf whenever there is a changing magnetic flux, regardless of the reason the flux is changing:

$$\mathscr{E} = -N\frac{\Delta\Phi_B}{\Delta t} \quad (20\text{-}6b)$$

- Lenz's law: the direction of an induced emf or an induced current *opposes* the *change* that caused it.
- The back emf in a motor increases as the rotational speed increases.
- For an ideal transformer,

$$\frac{\mathscr{E}_2}{\mathscr{E}_1} = \frac{N_2}{N_1} = \frac{I_1}{I_2} \quad (20\text{-}9, 10)$$

The ratio N_2/N_1 is called the turns ratio. There is no energy loss in an ideal transformer, so the power input is equal to the power output.

- Whenever a solid conductor is subjected to a changing magnetic flux, the induced emf causes eddy currents to flow simultaneously along many different paths. Eddy currents dissipate energy.
- A changing magnetic field gives rise to an induced electric field. The induced emf is the circulation of the induced electric field.

continued on next page

Master the Concepts continued

- A changing current in one circuit element induces an emf in another circuit element. The mutual inductance is the constant of proportionality between the rate of change of the current and the induced emf.

$$M = \frac{N_2\Phi_{21}}{I_1} = \frac{N_1\Phi_{12}}{I_2} \qquad (20\text{-}12)$$

$$\mathscr{E}_{21} = -N_2\frac{\Delta\Phi_{21}}{\Delta t} = -M\frac{\Delta I_1}{\Delta t} \qquad (20\text{-}13)$$

- Self-inductance is when a changing current induces an emf in the same device:

$$N\Phi = LI \qquad (20\text{-}14)$$

$$\mathscr{E} = -L\frac{\Delta I}{\Delta t} \qquad (20\text{-}16)$$

- The energy stored in an inductor is

$$U = \tfrac{1}{2}LI^2 \qquad (20\text{-}17)$$

- The energy density (energy per unit volume) in a magnetic field is

$$u_B = \frac{1}{2\mu_0}B^2 \qquad (20\text{-}18)$$

- Current through an inductor must always change *continuously*, never instantaneously. In an *LR* circuit, the time constant is

$$\tau = \frac{L}{R} \qquad (20\text{-}22)$$

The current in an *LR* circuit is

$$\text{If } I_0 = 0, \quad I(t) = I_f(1 - e^{-t/\tau}) \qquad (20\text{-}21)$$

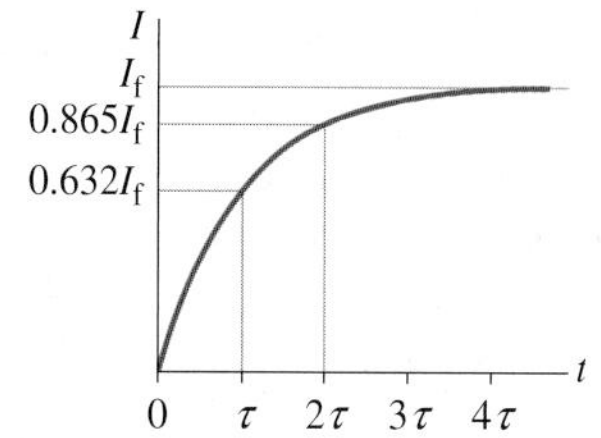

$$\text{If } I_f = 0, \quad I(t) = I_0e^{-t/\tau} \qquad (20\text{-}24)$$

Conceptual Questions

1. A vertical magnetic field is perpendicular to the horizontal plane of a wire loop. When the loop is rotated about a horizontal axis in the plane, the current induced in the loop reverses direction twice per rotation. Explain why there are *two* reversals for *one* rotation.
2. A transformer is essentially a mutual inductance device. Two coils are wound around an iron core; an alternating current in one coil induces an emf in the second. The core is normally made of either laminated iron—thin sheets of iron with an insulating material between them—or of a bundle of parallel insulated iron wires. Why not just make it of solid iron?
3. A certain amount of energy must be supplied to increase the current through an inductor from 0 mA to 10 mA. Does it take the same amount of energy, more, or less to increase the current from 10 mA to 20 mA?
4. The primary coil of a transformer is connected to a dc battery. Is there an emf induced in the secondary coil? If so, why do we not use transformers with dc sources?
5. A metal plate is attached to the end of a rod and positioned so that it can swing into and out of a perpendicular magnetic field pointing out of the plane of the paper as shown. In position 1, the plate is just swinging into the field; in position 2, the plate is swinging out of the field. Does an induced eddy current circulate clockwise or counterclockwise in the metal plate when it is in (a) position 1 and (b) position 2? (c) Will the induced eddy currents act as a braking force to stop the pendulum motion? Explain.

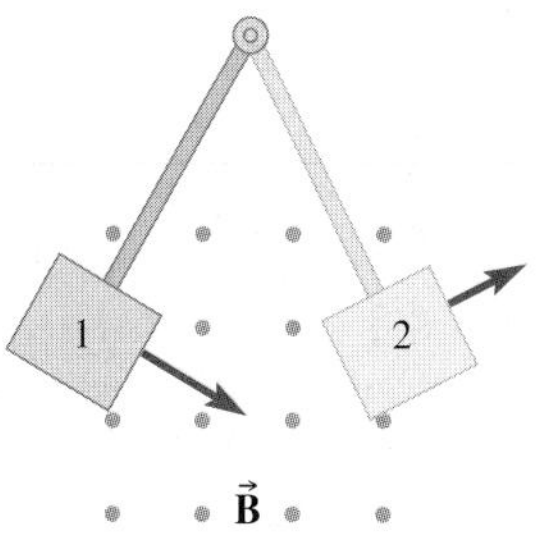

6. Magnetic induction is the principle behind the operation of mechanical speedometers used in automobiles and bicycles. In the drawing, a simplified version of the speedometer, a metal disk is free to spin about the vertical axis passing through its center. Suspended above the disk is a horseshoe magnet. (a) If the horseshoe magnet is connected to the drive shaft of the vehicle so that it rotates about a vertical axis, what happens to the disk? [*Hint:* Think about eddy currents and Lenz's law.] (b) Instead of being free to rotate, the disk is restrained by a hairspring. The hairspring exerts a restoring torque on the disk proportional to its angular displacement from equilibrium. When the horseshoe magnet rotates, what happens to the disk? A pointer attached to the disk indicates the speed of the vehicle. How does the angular *position* of the pointer depend on the angular *speed* of the magnet?

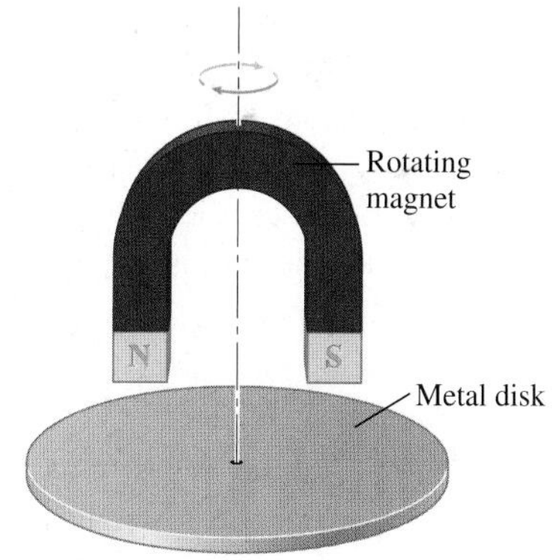

7. Wires that carry telephone signals are twisted. The twisting reduces the noise on the line from nearby electric

devices that produce changing currents. How does the twisting reduce noise pickup?

8. If the magnetic fields produced by the x-, y-, and z-coils in an MRI (see Section 19.8) are changed too rapidly, the patient may experience twitching or tingling sensations. What do you think might be the cause of these sensations? Why does the much stronger static field not cause twitching or tingling?
9. The magnetic flux through a flat surface is known. The area of the surface is also known. Is that information enough to calculate the average magnetic field on the surface? Explain.
10. Would a ground fault interrupter work if the circuit used dc current instead of ac? Explain.
11. In the study of thermodynamics, we thought of a refrigerator as a reversed heat engine. (a) Explain how a generator is a reversed electric motor. (b) What kind of device is a reversed loudspeaker?
12. Two identical circular coils of wire are separated by a fixed center-to-center distance. Describe the orientation of the coils that would (a) maximize or (b) minimize their mutual inductance.
13. (a) Explain why a transformer works for ac but not for dc. (b) Explain why a transformer designed to be connected to an emf of amplitude 170 V would be damaged if connected to a dc emf of 170 V.
14. Credit cards have a magnetic strip that encodes information about the credit card account. Why do devices that read the magnetic strip often include the instruction to swipe the card rapidly? Why can't the magnetic strip be read if the card is swiped too slowly?
15. Think of an example that illustrates why an "anti-Lenz" law would violate the conservation of energy. (The "anti-Lenz" law is: The direction of induced emfs and currents always *reinforces* the *change* that produces them.)
16. A 2-m-long copper pipe is held vertically. When a marble is dropped down the pipe, it falls through in about 0.7 s. A magnet of similar size and shape dropped down the pipe takes *much* longer. Why?
17. An electric mixer is being used to mix up some cake batter. What happens to the motor if the batter is too thick, so the beaters are turning slowly?
18. A circular loop of wire can be used as an antenna to sense the changing magnetic fields in an electromagnetic wave (such as a radio transmission). What is the advantage of using a coil with many turns rather than a single loop?
19. Some low-cost tape recorders do not have a separate microphone. Instead, the speaker is used as a microphone when recording. Explain how this works.
20. High-voltage power lines run along the edge of a farmer's field. Describe how the farmer might be able to steal electric power without making any electrical connection to the power line. (Yes, it works. Yes, it has been done. Yes, it is illegal.)

Multiple-Choice Questions

1. An electric current is induced in a conducting loop by all but one of these processes. Which one does *not* produce an induced current?
   (a) rotating the loop so that it cuts across magnetic field lines
   (b) placing the loop so that its area is perpendicular to a changing magnetic field
   (c) moving the loop parallel to uniform magnetic field lines
   (d) expanding the area of the loop while it is perpendicular to a uniform magnetic field
2. A split-ring commutator is used in a dc generator to
   (a) rotate a loop so that it cuts through magnetic flux.
   (b) reverse the connections to an armature so that the current periodically reverses direction.
   (c) reverse the connections to an armature so that the current does not reverse direction.
   (d) prevent a coil from rotating when the magnetic field is changing.
3. Suppose the switch in Fig. 20.21 has been closed for a long time but is suddenly opened at $t = t_0$. Which of these graphs best represents the current in coil 2 as a function of time? I_2 is positive if it flows from A to B through the resistor.

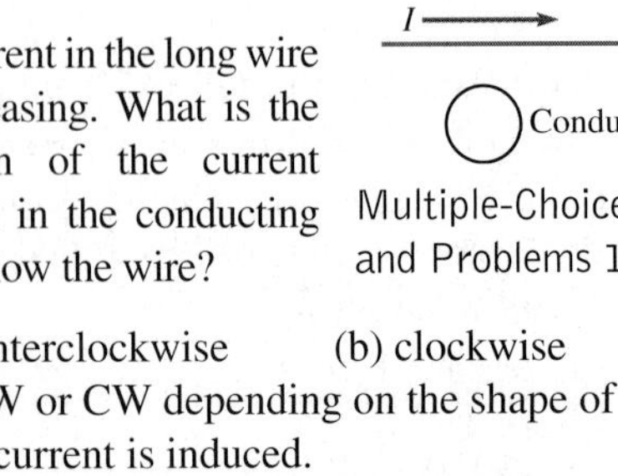

4. The current in the long wire is decreasing. What is the direction of the current induced in the conducting loop below the wire?

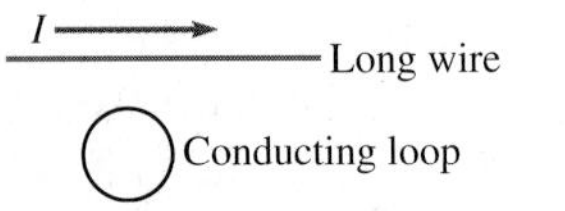

Multiple-Choice Question 4 and Problems 15 and 16

   (a) counterclockwise (b) clockwise
   (c) CCW or CW depending on the shape of the loop
   (d) No current is induced.
5. In a bicycle speedometer, a bar magnet is attached to the spokes of the wheel and a coil is attached to the frame so that the north pole of the magnet moves past it once for every revolution of the wheel. As the magnet moves past the coil, a pulse of current is induced in the coil. A computer then measures the time between pulses and computes the bicycle's speed. The figure shows the

magnet about to move past the coil. Which of the graphs shows the resulting current pulse? Take current counterclockwise in part (a) of the figure to be positive.

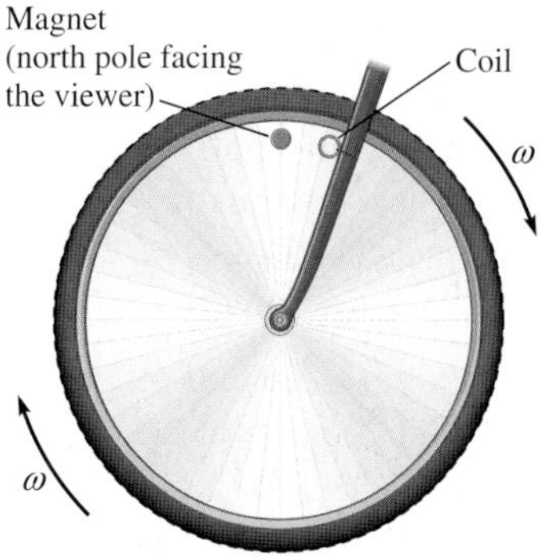

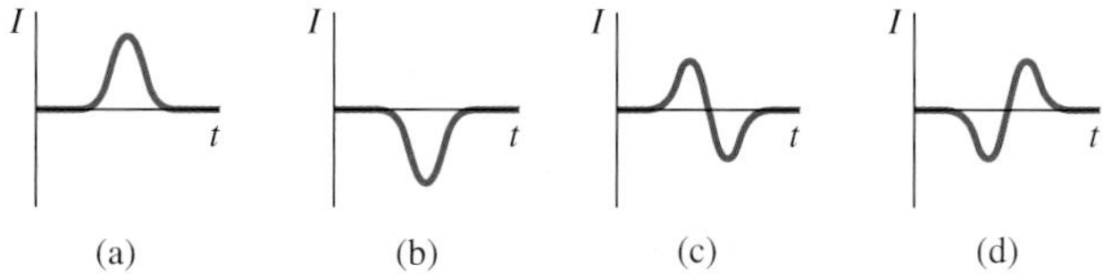

6. For each of the experiments (1, 2, 3, 4) shown, in what direction does current flow *through the resistor*? Note that the wires are not always wrapped around the plastic tube in the same way.

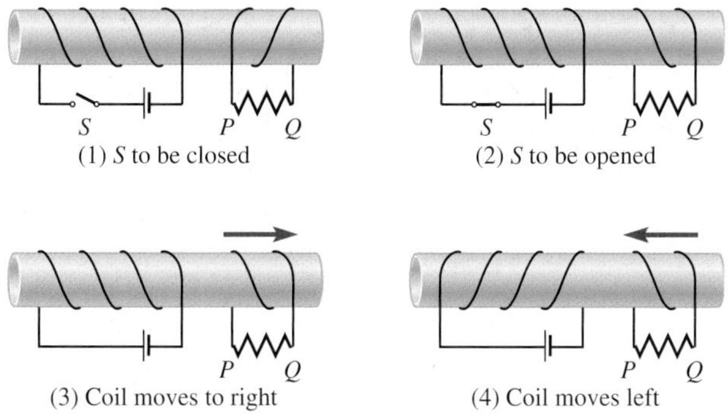

	(1)	(2)	(3)	(4)
(a)	*P* to *Q*	*P* to *Q*	*P* to *Q*	*P* to *Q*
(b)	*P* to *Q*	*Q* to *P*	*P* to *Q*	*Q* to *P*
(c)	*Q* to *P*	*P* to *Q*	*Q* to *P*	*P* to *Q*
(d)	*Q* to *P*	*P* to *Q*	*P* to *Q*	*Q* to *P*
(e)	*Q* to *P*	*Q* to *P*	*Q* to *P*	*Q* to *P*
(f)	*Q* to *P*	*Q* to *P*	*P* to *Q*	*P* to *Q*

7. The figure shows a region of uniform magnetic field out of the page. Outside the region, the magnetic field is zero. Some rectangular wire loops move as indicated. Which of the loops would feel a magnetic force directed to the right?

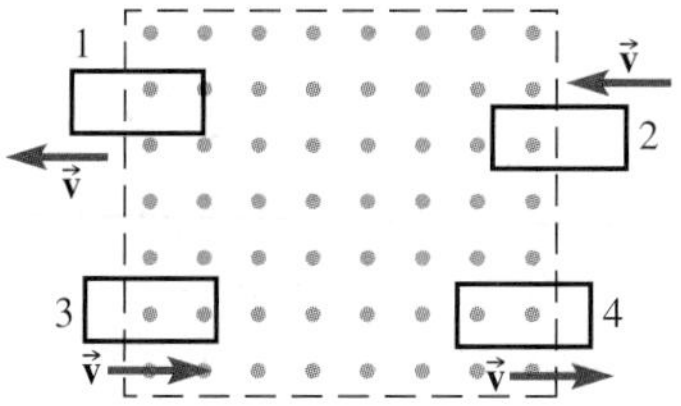

(a) 1 (b) 2 (c) 3 (d) 4
(e) 1 and 2 (f) 2 and 4 (g) 3 and 4
(h) none of them

8. In a moving coil microphone, the induced emf in the coil at any instant depends mainly on
(a) the displacement of the coil.
(b) the velocity of the coil.
(c) the acceleration of the coil.

9. A moving magnet microphone is similar to a moving coil microphone (Fig. 20.15) except that the coil is stationary and the magnet is attached to the diaphragm, which moves in response to sound waves in the air. If, in response to a sound wave, the magnet moves according to $x(t) = A \sin \omega t$, the induced emf in the coil would be (approximately) proportional to which of these?
(a) $\sin \omega t$ (b) $\cos \omega t$ (c) $\sin 2\omega t$ (d) $\cos 2\omega t$

10. An airplane is flying due east. Earth's magnetic field has a downward vertical component and a horizontal component due north. Which point on the plane's exterior accumulates positive charge due to the motional emf?
(a) the nose (the point farthest east)
(b) the tail (the point farthest west)
(c) the tip of the left wing (the point farthest north)
(d) the tip of the right wing (the point farthest south)

Problems

- Combination conceptual/quantitative problem
- Biological or medical application
- Challenging problem
- Blue # Detailed solution in the Student Solutions Manual
- 1 2 Problems paired by concept
- Text website interactive or tutorial

20.1 Motional Emf; 20.2 Electric Generators

1. In Fig. 20.2, a metal rod of length L moves to the right at speed v. (a) What is the current in the rod, in terms of v, B, L, and R? (b) In what direction does the current flow? (c) What is the direction of the magnetic force on the rod? (d) What is the magnitude of the magnetic force on the rod (in terms of v, B, L, and R)?

2. Suppose that the current were to flow in the direction *opposite* to that found in Problem 1. (a) In what direction would the magnetic force on the rod be? (b) In the absence of an external force, what would happen to the rod's kinetic energy? (c) Why is this not possible? Returning to the correct direction of the current, sketch a rough graph of the kinetic energy of the rod as a function of time.

3. To maintain a constant emf, the moving rod of Fig. 20.2 must maintain a constant velocity. In order to maintain a constant velocity, some external force must pull it to the right. (a) What is the magnitude of the external force required, in terms of v, B, L, and R? (See Problem 1.) (b) At what rate does this force do work on

the rod? (c) What is the power dissipated in the resistor? (d) Overall, is energy conserved? Explain.

4. In Fig. 20.2, what would the magnitude (in terms of v, L, R, and B) and direction (CW or CCW) of the current be if the direction of the magnetic field were: (a) into the page; (b) to the right (in the plane of the page); (c) up (in the plane of the page); (d) such that it has components both out of the page and to the right, with a 20.0° angle between the field and the plane of the page?

5. A 15.0-g conducting rod of length 1.30 m is free to slide downward between two vertical rails without friction. The rails are connected to an 8.00-Ω resistor, and the entire apparatus is placed in a 0.450-T uniform magnetic field. Ignore the resistance of the rod and rails. (a) What is the terminal velocity of the rod? (b) At this terminal velocity, compare the magnitude of the change in gravitational potential energy per second with the power dissipated in the resistor.

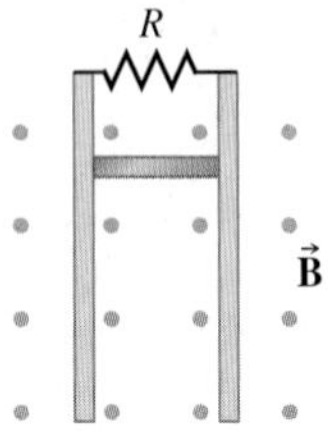

6. When the armature of an ac generator rotates at 15.0 rad/s, the amplitude of the induced emf is 27.0 V. What is the amplitude of the induced emf when the armature rotates at 10.0 rad/s? (tutorial: generator)

7. The armature of an ac generator is a circular coil with 50 turns and radius 3.0 cm. When the armature rotates at 350 rpm, the amplitude of the emf in the coil is 17.0 V. What is the strength of the magnetic field (assumed to be uniform)?

8. The armature of an ac generator is a rectangular coil 2.0 cm by 6.0 cm with 80 turns. It is immersed in a uniform magnetic field of magnitude 0.45 T. If the amplitude of the emf in the coil is 17.0 V, at what angular speed is the armature rotating?

9. In Fig. 20.6, side 3 of the rectangular coil in the electric generator rotates about the axis at constant angular speed ω. The figure with this problem shows side 3 by itself. (a) First consider the right half of side 3. Although the speed of the wire differs depending on the distance from the axis, the direction is the same for the entire right half. Use the magnetic force law to find the direction of the force on electrons in the right half of the wire. (b) Does the magnetic force tend to push electrons along the wire, either toward or away from the axis? (c) Is there an induced emf along the *length* of this half of the wire? (d) Generalize your answers to the left side of wire 3 and the two sides of wire 1. What is the net emf due to these two sides of the coil?

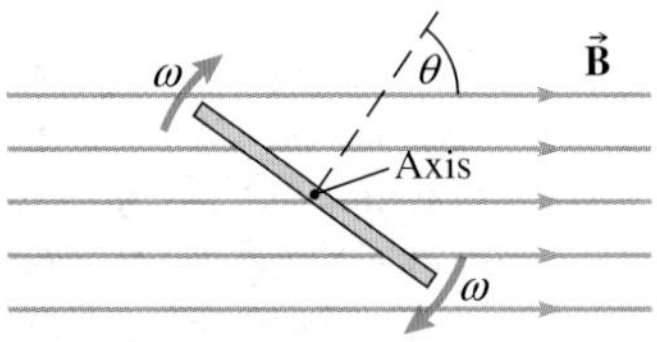

10. A solid copper disk of radius R rotates at angular velocity ω in a perpendicular magnetic field B. The figure shows the disk rotating clockwise and the magnetic field into the page. (a) Is the charge that accumulates on the edge of the disk positive or negative? Explain. (b) What is the potential difference between the center of the disk and the edge? [*Hint:* Think of the disk as a large number of thin wedge-shaped rods. The center of such a rod is at rest, and the outer edge moves at speed $v = \omega R$. The rod moves through a perpendicular magnetic field at an *average* speed of $\frac{1}{2}\omega R$.]

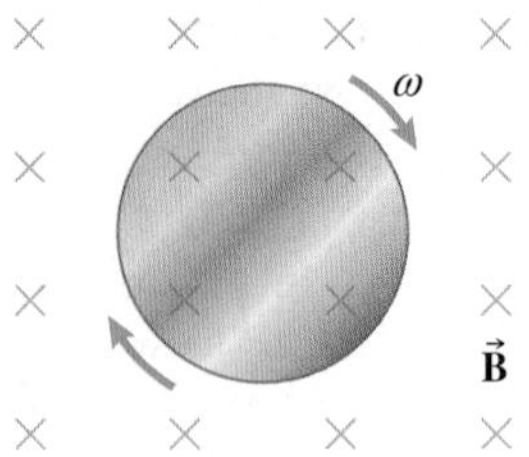

11. A square loop of wire of side 2.3 cm and electrical resistance 79 Ω is near a long straight wire that carries a current of 6.8 A in the direction indicated. The long wire and loop both lie in the plane of the page. The left side of the loop is 9.0 cm from the wire. (a) If the loop is at rest, what is the induced emf in the loop? What are the magnitude and direction of the induced current in the loop? What are the magnitude and direction of the magnetic force on the loop? (b) Repeat if the loop is moving to the right at a constant speed of 45 cm/s. (c) In (b), find the electric power dissipated in the loop and show that it is equal to the rate at which an external force, pulling the loop to keep its speed constant, does work.

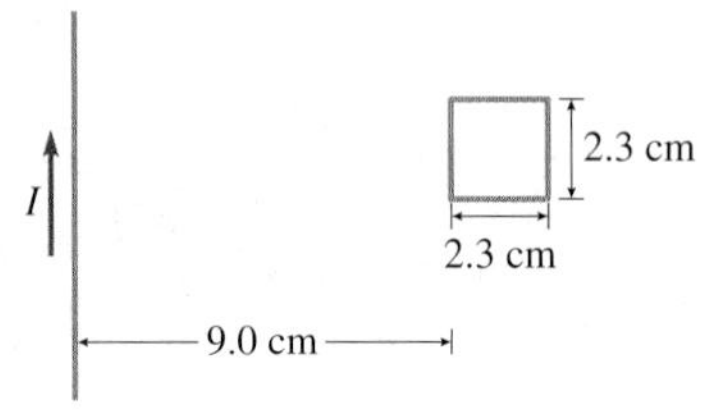

12. A solid metal cylinder of mass m rolls down parallel metal rails spaced a distance L apart with a constant acceleration of magnitude a_0 [part (a) of figure]. The rails are inclined at an angle θ to the horizontal. Now the rails are connected electrically at the top and immersed in a magnetic field of magnitude B that is perpendicular to the plane of the rails [part (b) of figure]. (a) As it rolls down the rails, in what direction does current flow in the cylinder? (b) What direction is the magnetic force on the cylinder? (c) Instead of rolling at constant acceleration, the cylinder now approaches a terminal speed v_t. What is v_t in terms of L, m, R, a_0, θ, and B? R is the total electrical resistance of the circuit consisting of the cylinder, rails, and wire; assume R is constant (that is, the resistances of the rails themselves are negligible).

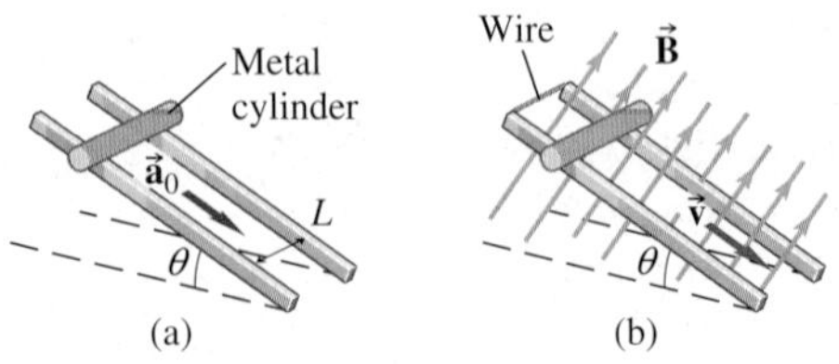

20.3 Faraday's Law; 20.4 Lenz's Law

13. A horizontal desk surface measures 1.3 m × 1.0 m. If Earth's magnetic field has magnitude 0.44 mT and is directed 65° below the horizontal, what is the magnetic flux through the desk surface?

14. A square loop of wire, 0.75 m on each side, has one edge along the positive z-axis and is tilted toward the yz-plane at an angle of 30.0° with respect to the horizontal (xz-plane). There is a uniform magnetic field of 0.32 T pointing in the positive x-axis direction. (a) What is the flux through the loop? (b) If the angle increases to 60°, what is the new flux through the loop? (c) While the angle is being increased, which direction will current flow through the top side of the loop?

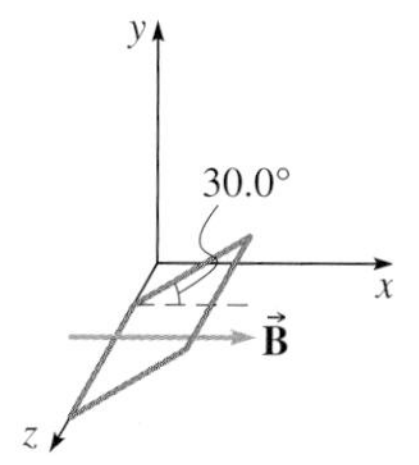

15. A long straight wire carrying a steady current is in the plane of a circular loop of wire. See the figure with Multiple-Choice Question 4. (a) If the loop is moved closer to the wire, what direction does the induced current in the loop flow? (b) At one instant, the induced emf in the loop is 3.5 mV. What is the rate of change of the magnetic flux through the loop at that instant in webers per second?

16. A long straight wire carrying a current I is in the plane of a circular loop of wire. See the figure with Multiple-Choice Question 4. The current I is decreasing. Both the loop and the wire are held in place by external forces. The loop has resistance 24 Ω. (a) In what direction does the induced current in the loop flow? (b) In what direction is the external force holding the loop in place? (c) At one instant, the induced current in the loop is 84 mA. What is the rate of change of the magnetic flux through the loop at that instant in webers per second?

Problems 17–20. Two wire loops are side by side, as shown. The current I_1 in loop 1 is supplied by an external source (not shown) and is clockwise as viewed from the right.

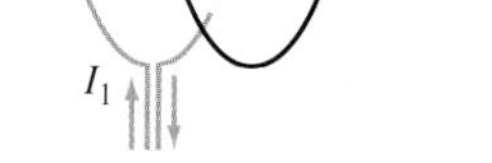

Problems 17–20 and 41–43

17. While I_1 is increasing, does current flow in loop 2? If so, does it flow clockwise or counterclockwise as viewed from the right? Explain. (tutorial: coupled loops)

18. While I_1 is increasing, what is the direction of the magnetic force exerted on loop 2, if any? Explain.

19. While I_1 is constant, does current flow in loop 2? If so, does it flow clockwise or counterclockwise as viewed from the right? Explain.

20. Refer to Fig. 20.2. The rod has length L and its position is x at some instant, as shown in the figure. Express your answers in terms of x, L, v, B (the magnetic field strength), and R, as needed. (a) What is the area enclosed by the conducting loop at this instant? (b) What is the magnetic flux through the loop at this instant? (c) The rod moves to the right at speed v. At what rate is the flux changing? (d) According to Faraday's law, what is the induced emf in the loop? Compare your answer with Eq. (20-2a). (e) What is the induced current I? (f) Explain why the induced current flows counterclockwise around the loop.

21. A circular conducting coil with radius 3.40 cm is placed in a uniform magnetic field of 0.880 T with the plane of the coil perpendicular to the magnetic field. The coil is rotated 180° about the axis in 0.222 s. (a) What is the average induced emf in the coil during this rotation? (b) If the coil is made of copper with a diameter of 0.900 mm, what is the average current that flows through the coil during the rotation?

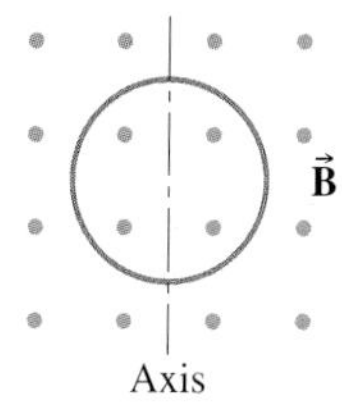

22. Verify that, in SI units, $\Delta\Phi_B/\Delta t$ can be measured in volts—in other words, that 1 Wb/s = 1 V.

23. The component of the external magnetic field along the central axis of a 50-turn coil of radius 5.0 cm increases from 0 to 1.8 T in 3.6 s. (a) If the resistance of the coil is 2.8 Ω, what is the magnitude of the induced current in the coil? (b) What is the direction of the current if the axial component of the field points away from the viewer?

24. In the figure, switch S is initially open. It is closed, and then opened again a few seconds later. (a) In what direction does current flow through the ammeter when switch S is closed? (b) In what direction does current flow when switch S is then opened? (c) Sketch a qualitative graph of the current through the ammeter as a function of time. Take the current to be positive to the right.

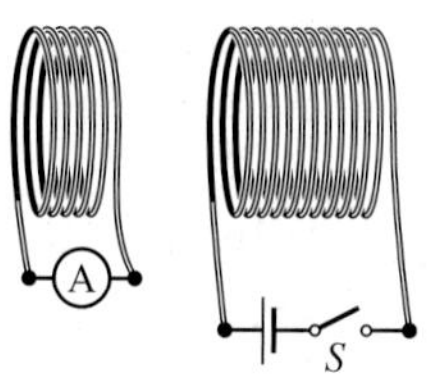

25. Another example of motional emf is a rod attached at one end and rotating in a plane perpendicular to a uniform magnetic field. We can analyze this motional emf using Faraday's law. (a) Consider the area that the rod sweeps out in each revolution and find the magnitude of the emf in terms of the angular frequency ω, the length of the rod R, and the strength of the uniform magnetic field B. (b) Write the emf magnitude in terms of the speed v of the tip of the rod and compare this with motional emf magnitude of a rod moving at constant velocity perpendicular to a uniform magnetic field.

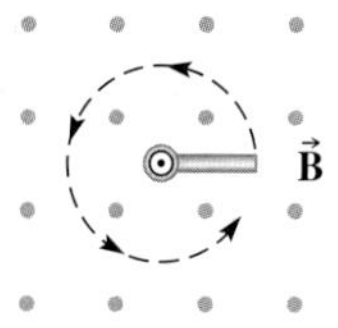

26. (a) For a particle moving in simple harmonic motion, the position can be written $x(t) = x_m \cos \omega t$. What is the velocity $v_x(t)$ as a function of time for this particle? (b) Using the small-angle approximation for the sine function, find the slope of the graph of $\Phi(t) = \Phi_0 \sin \omega t$ at $t = 0$. Does your result agree with the value of $\Delta\Phi/\Delta t = \omega\Phi_0 \cos \omega t$ at $t = 0$?

27. Two loops of wire are next to one another in the same plane. (a) If the switch S is closed, does current flow in loop 2? If so, in what direction? (b) Does the current in loop 2 flow for only a brief moment, or does it continue? (c) Is there a magnetic force on loop 2? If so, in what direction? (d) Is there a magnetic force on loop 1? If so, in what direction?

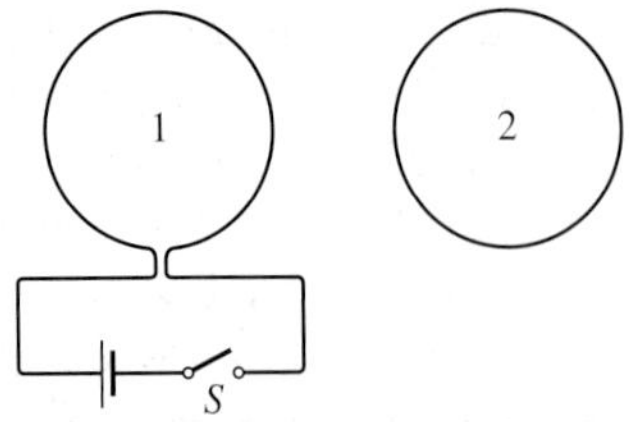

20.5 Back Emf in a Motor

28. A dc motor has coils with a resistance of 16 Ω and is connected to an emf of 120.0 V. When the motor operates at full speed, the back emf is 72 V. (a) What is the current in the motor when it first starts up? (b) What is the current when the motor is at full speed? (c) If the current is 4.0 A with the motor operating at less than full speed, what is the back emf at that time?

29. Tim is using a cordless electric weed trimmer with a dc motor to cut the long weeds in his back yard. The trimmer generates a back emf of 18.00 V when it is connected to an emf of 24.0 V dc. The total electrical resistance of the electric motor is 8.00 Ω. (a) How much current flows through the motor when it is running smoothly? (b) Suddenly the string of the trimmer gets wrapped around a pole in the ground and the motor quits spinning. What is the current through the motor when there is no back emf? What should Tim do?

30. A dc motor is connected to a constant emf of 12.0 V. The resistance of its windings is 2.0 Ω. At normal operating speed, the motor delivers 6.0 W of mechanical power. (a) What is the initial current drawn by the motor when it is first started up? (b) What current does it draw at normal operating speed? (c) What is the back emf induced in the windings at normal speed?

20.6 Transformers

31. A step-down transformer has 4000 turns on the primary and 200 turns on the secondary. If the primary voltage amplitude is 2.2 kV, what is the secondary voltage amplitude?

32. A step-down transformer has a turns ratio of 1/100. An ac voltage of amplitude 170 V is applied to the primary. If the primary current amplitude is 1.0 mA, what is the secondary current amplitude?

33. A doorbell uses a transformer to deliver an amplitude of 8.5 V when it is connected to a 170-V amplitude line. If there are 50 turns on the secondary, (a) what is the turns ratio? (b) How many turns does the primary have?

34. The primary coil of a transformer has 250 turns; the secondary coil has 1000 turns. An alternating current is sent through the primary coil. The emf in the primary is of amplitude 16 V. What is the emf amplitude in the secondary? (tutorial: transformer)

35. When the emf for the primary of a transformer is of amplitude 5.00 V, the secondary emf is 10.0 V in amplitude. What is the transformer turns ratio (N_2/N_1)?

36. A transformer with a primary coil of 1000 turns is used to step up the standard 170-V amplitude line voltage to a 220-V amplitude. How many turns are required in the secondary coil?

37. A transformer with 1800 turns on the primary and 300 turns on the secondary is used in an electric slot car racing set to reduce the input voltage amplitude of 170 V from the wall output. The current in the secondary coil is of amplitude 3.2 A. What is the voltage amplitude across the secondary coil and the current amplitude in the primary coil?

38. A transformer for an answering machine takes an ac voltage of amplitude 170 V as its input and supplies a 7.8-V amplitude to the answering machine. The primary has 300 turns. (a) How many turns does the secondary have? (b) When idle, the answering machine uses a maximum power of 5.0 W. What is the amplitude of the current drawn from the 170-V line?

20.7 Eddy Currents

39. A 2-m-long copper pipe is held vertically. When a marble is dropped down the pipe, it falls through in about 0.7 s. A magnet of similar size and shape takes *much* longer to fall through the pipe. (a) As the magnet is falling through the pipe with its north pole below its south pole, what direction do currents flow around the pipe above the magnet? Below the magnet (CW or CCW as viewed from the top)? (b) Sketch a graph of the speed of the magnet as a function of time. [*Hint:* What would the graph look like for a marble falling through honey?]

40. In Problem 39, the pipe is suspended from a spring scale. The weight of the pipe is 12.0 N; the weight of the marble and magnet are each 0.3 N. Sketch graphs to show the reading of the spring scale as a function of time for the fall of the marble and again for the fall of the magnet. Label the vertical axis with numerical values.

20.9 Inductance

Problems 41–43. Two wire loops are side by side, as shown in the figure with Problems 17–20. The current I_1 in loop 1 is supplied by an external source (not shown) and is clockwise as viewed from the right.

41. When the current in loop 1 is $I_1 = 0.75$ A, the magnetic flux through loop 2 is 2.35×10^{-8} T·m^2. What is the mutual inductance of the two loops?

42. When the current in loop 1 decreases at a steady rate of 28 A/s, the induced emf in loop 2 is 1.40 mV. (a) What

is the direction of the current in loop 2 as viewed from the right? (b) What is the mutual inductance of the two loops?

43. When the current in loop 1 increases at a steady rate of 9.0 A/s, the current in loop 2 is 0.185 mA. The resistance of loop 2 is 0.66 Ω. (a) What is the direction of the current in loop 2 as viewed from the right? (b) What is the mutual inductance of the two loops?

44. Two solenoids, of N_1 and N_2 turns respectively, are wound on the same form. They have the same length L and radius r. (a) What is the mutual inductance of these two solenoids? (b) If an ac current

$$I_1(t) = I_m \sin \omega t$$

flows in solenoid 1 (N_1 turns), write an expression for the total flux through solenoid 2. (c) What is the maximum induced emf in solenoid 2? [*Hint:* Refer to Eq. (20-7).]

45. A solenoid is made of 300.0 turns of wire, wrapped around a hollow cylinder of radius 1.2 cm and length 6.0 cm. What is the self-inductance of the solenoid?

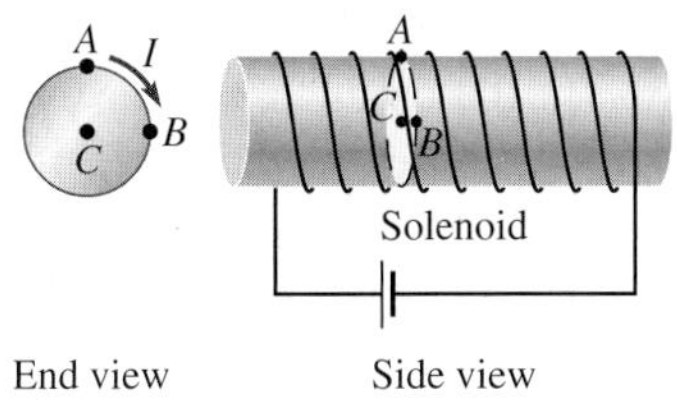

46. A solenoid of length 2.8 cm and diameter 0.75 cm is wound with 160 turns per cm. When the current through the solenoid is 0.20 A, what is the magnetic flux through *one* of the windings of the solenoid?

47. If the current in the solenoid in Problem 46 is decreasing at a rate of 35.0 A/s, what is the induced emf (a) in one of the windings? (b) in the entire solenoid?

48. An ideal solenoid has length ℓ. If the windings are compressed so that the length of the solenoid is reduced to 0.50ℓ, what happens to the inductance of the solenoid?

49. In this problem, you derive the expression for the self-inductance of a long solenoid [Eq. (20-15a)]. The solenoid has n turns per unit length, length ℓ, and radius r. Assume that the current flowing in the solenoid is I. (a) Write an expression for the magnetic field inside the solenoid in terms of n, ℓ, r, I, and universal constants. (b) Assume that all of the field lines cut through each turn of the solenoid. In other words, assume the field is uniform right out to the ends of the solenoid—a good approximation if the solenoid is tightly wound and sufficiently long. Write an expression for the magnetic flux through one turn. (c) What is the total flux linkage through all turns of the solenoid? (d) Use the definition of self-inductance [Eq. (20-14)] to find the self-inductance of the solenoid.

50. Compare the electric energy that can be stored in a capacitor to the magnetic energy that can be stored in an inductor of the same size (that is, the same volume). For the capacitor, assume that air is between the plates; the maximum electric field is then the breakdown strength of air, about 3 MV/m. The maximum magnetic field attainable in an ordinary solenoid with an air core is on the order of 10 T.

51. The current in a 0.080-H solenoid increases from 20.0 mA to 160.0 mA in 7.0 s. Find the average emf in the solenoid during that time interval.

52. Calculate the equivalent inductance L_{eq} of two ideal inductors, L_1 and L_2, connected in series in a circuit. Assume that their mutual inductance is negligible. [*Hint:* Imagine replacing the two inductors with a single equivalent inductor L_{eq}. How is the emf in the series equivalent related to the emfs in the two inductors? What about the currents?]

53. Calculate the equivalent inductance L_{eq} of two ideal inductors, L_1 and L_2, connected in parallel in a circuit. Assume that their mutual inductance is negligible. [*Hint:* Imagine replacing the two inductors with a single equivalent inductor L_{eq}. How is the emf in the parallel equivalent related to the emfs in the two inductors? What about the currents?]

54. In Section 20.9, in order to find the energy stored in an inductor, we assumed that the current was increased from zero at a constant rate. In this problem, you will prove that the energy stored in an inductor is $U_L = \frac{1}{2}LI^2$—that is, it only depends on the current I and not on the previous time dependence of the current. (a) If the current in the inductor increases from i to $i + \Delta i$ in a very short time Δt, show that the energy added to the inductor is $\Delta U = Li\Delta i$. [*Hint:* Start with $\Delta U = P\Delta t$.] (b) Show that, on a graph of Li versus i, for any small current interval Δi, the energy added to the inductor can be interpreted as the area under the graph for that interval. (c) Now show that the energy stored in the inductor when a current I flows is $U = \frac{1}{2}LI^2$.

20.10 *LR* Circuits

55. A 5.0-mH inductor and a 10.0-Ω resistor are connected in series with a 6.0-V dc battery. (a) What is the voltage across the resistor immediately after the switch is closed? (b) What is the voltage across the resistor after the switch has been closed for a long time? (c) What is the current in the inductor after the switch has been closed for a long time?

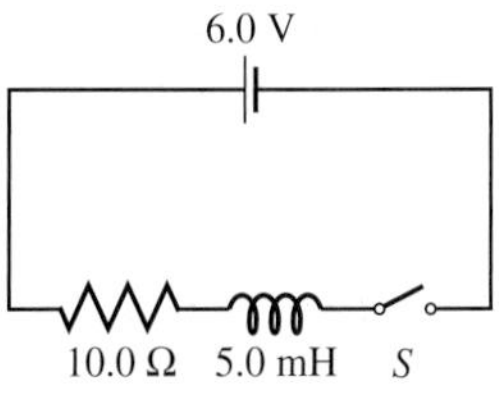

56. In a circuit, a parallel combination of a 10.0-Ω resistor and a 7.0-mH inductor is connected in

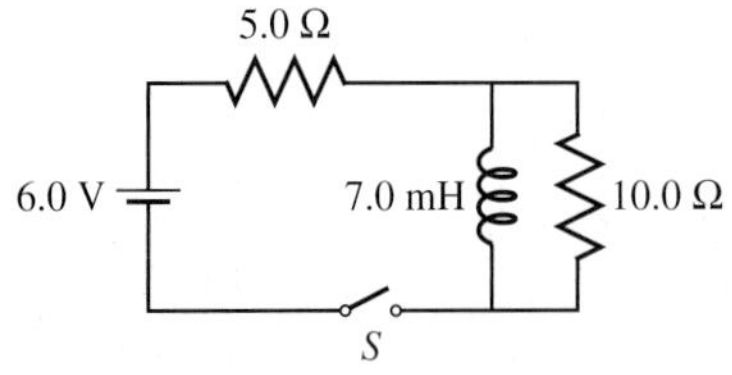

Problems 56, 57, and 65

series with a 5.0-Ω resistor, a 6.0-V dc battery, and a switch. (a) What are the voltages across the 5.0-Ω resistor and the 10.0-Ω resistor, respectively, immediately after the switch is closed? (b) What are the voltages across the 5.0-Ω resistor and the 10.0-Ω resistor, respectively, after the switch has been closed for a long time? (c) What is the current in the 7.0-mH inductor after the switch has been closed for a long time?

57. Refer to Problem 56. After the switch has been closed for a very long time, it is opened. What are the voltages across (a) the 5.0-Ω resistor and (b) the 10.0-Ω resistor immediately after the switch is opened?

58. No currents flow in the circuit before the switch is closed. Consider all circuit elements to be ideal. (a) At the instant the switch is closed, what are the values of the currents I_1 and I_2, the potential differences across the resistors, the power supplied by the battery, and the induced emf in the inductor? (b) After the switch has been closed for a long time, what are the values of the currents I_1 and I_2, the potential differences across the resistors, the power supplied by the battery, and the induced emf in the inductor?

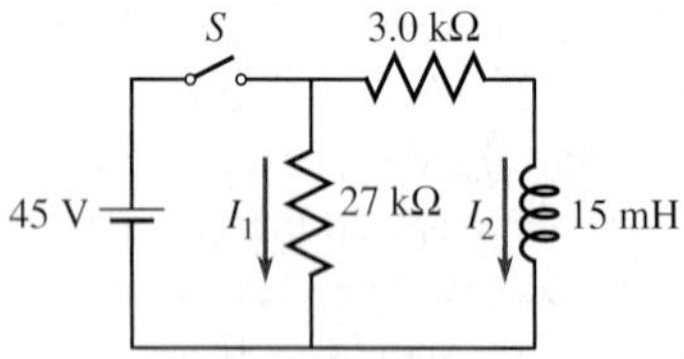

Problems 58 and 65

59. A coil of wire is connected to an ideal 6.00-V battery at $t = 0$. At $t = 10.0$ ms, the current in the coil is 204 mA. One minute later, the current is 273 mA. Find the resistance and inductance of the coil. [*Hint:* Sketch $I(t)$.]

60. A 0.67-mH inductor and a 130-Ω resistor are placed in series with a 24-V battery. (a) How long will it take for the current to reach 67% of its maximum value? (b) What is the maximum energy stored in the inductor? (c) How long will it take for the energy stored in the inductor to reach 67% of its maximum value? Comment on how this compares to the answer in part (a).

61. The windings of an electromagnet have inductance $L = 8.0$ H and resistance $R = 2.0$ Ω. A 100.0-V dc power supply is connected to the windings by closing switch S_2. (a) What is the current in the windings? (b) The electromagnet is to be shut off. Before disconnecting the power supply by opening switch S_2, a shunt resistor with resistance 20.0 Ω is connected in parallel across the windings. Why is the shunt resistor needed? Why must it be connected *before* the power supply is disconnected? (c) What is the maximum power dissipated in the shunt resistor? The shunt resistor must be chosen so that it can handle at least this much power without damage. (d) When the power supply is disconnected by opening switch S_2, how long does it take for the current in the windings to drop to 0.10 A? (e) Would a larger shunt resistor dissipate the energy stored in the electromagnet faster? Explain.

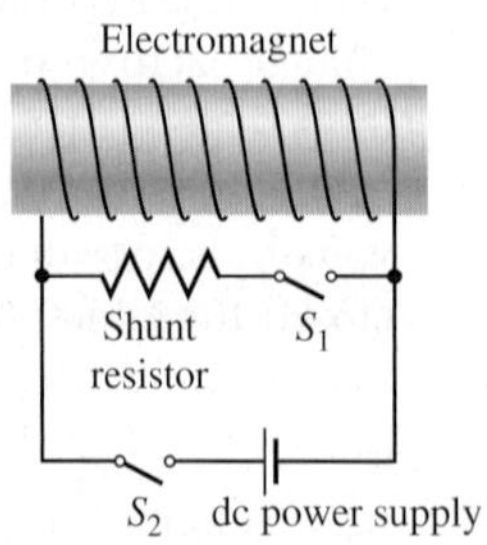

62. A coil has an inductance of 0.15 H and a resistance of 33 Ω. The coil is connected to a 6.0-V ideal battery. When the current reaches half its maximum value: (a) At what *rate* is magnetic energy being stored in the inductor? (b) At what rate is energy being dissipated? (c) What is the total power that the battery supplies?

63. The time constant τ for an *LR* circuit must be some combination of *L*, *R*, and $\mathscr{E}$. (a) Write the units of each of these three quantities in terms of V, A, and s. (b) Show that the only combination that has units of seconds is *L*/*R*.

64. In the circuit, switch *S* is opened at $t = 0$ after having been closed for a long time. (a) How much energy is stored in the inductor at $t = 0$? (b) What is the instantaneous rate of change of the inductor's energy at $t = 0$? (c) What is the *average* rate of change of the inductor's energy between $t = 0.0$ and $t = 1.0$ s? (d) How long does it take for the current in the inductor to reach 0.0010 times its initial value?

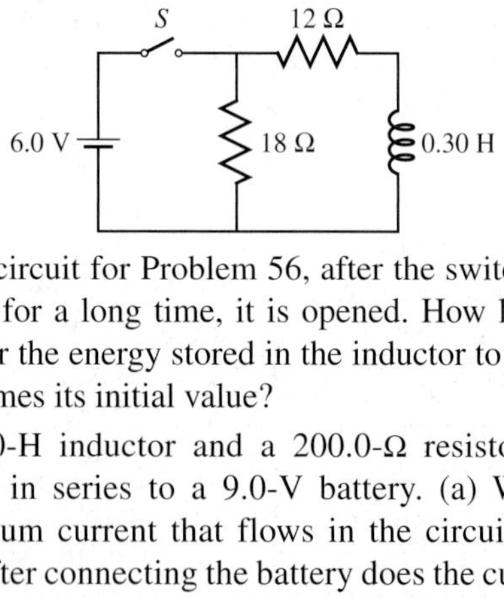

65. In the circuit for Problem 56, after the switch has been closed for a long time, it is opened. How long does it take for the energy stored in the inductor to decrease to 0.10 times its initial value?

66. A 0.30-H inductor and a 200.0-Ω resistor are connected in series to a 9.0-V battery. (a) What is the maximum current that flows in the circuit? (b) How long after connecting the battery does the current reach half its maximum value? (c) When the current is half its maximum value, find the energy stored in the inductor, the rate at which energy is being stored in the inductor, and the rate at which energy is dissipated in the resistor. (d) Redo parts (a) and (b) if, instead of being negligibly small, the internal resistances of the inductor and battery are 75 Ω and 20.0 Ω, respectively.

67. A coil has an inductance of 0.15 H and a resistance of 33 Ω. The coil is connected to a 6.0-V battery. After a long time elapses, the current in the coil is no longer changing. (a) What is the current in the coil? (b) What is the energy stored in the coil? (c) What is the rate of energy dissipation in the coil? (d) What is the induced emf in the coil?

Comprehensive Problems

68. Switch S_2 has been closed for a long time. (a) If switch S_1 is closed, will a current flow in the left-hand coil? If so, what direction will it flow across the ammeter? (b) After some time, switch S_1 is opened again while switch S_2 remains closed. Will a current flow in the left coil? If so, what direction will it flow across the ammeter?

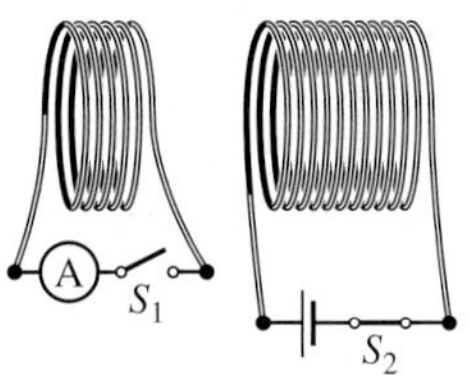

69. In the ac generator of Fig. 20.6, the emf produced is $\mathscr{E}(t) = \omega BA \sin \omega t$. If the generator is connected to a load of resistance R, then the current that flows is

$$I(t) = \frac{\omega BA}{R} \sin \omega t$$

(a) Find the magnetic forces on sides 2 and 4 at the instant shown in Fig. 20.7. (Remember that $\theta = \omega t$.) (b) Why do the magnetic forces on sides 1 and 3 not cause a torque about the axis of rotation? (c) From the magnetic forces found in (a), calculate the torque on the loop about its axis of rotation at the instant shown in Fig. 20.7. (d) In the absence of other torques, would the magnetic torque make the loop increase or decrease its angular velocity? Explain.

70. A circular metal ring is suspended above a solenoid. The magnetic field due to the solenoid is shown. The current in the solenoid is increasing. (a) What is the direction of the current in the ring? (b) The flux through the ring is proportional to the current in the solenoid. When the current in the solenoid is 12.0 A, the magnetic flux through the ring is 0.40 Wb. When the current increases at a rate of 240 A/s, what is the induced emf in the ring? (c) Is there a net magnetic force on the ring? If so, in what direction? (d) If the ring is cooled by immersing it in liquid nitrogen, what happens to its electrical resistance, the induced current, and the magnetic force? The change in size of the ring is negligible. (With a sufficiently strong magnetic field, the ring can be made to shoot up high into the air.)

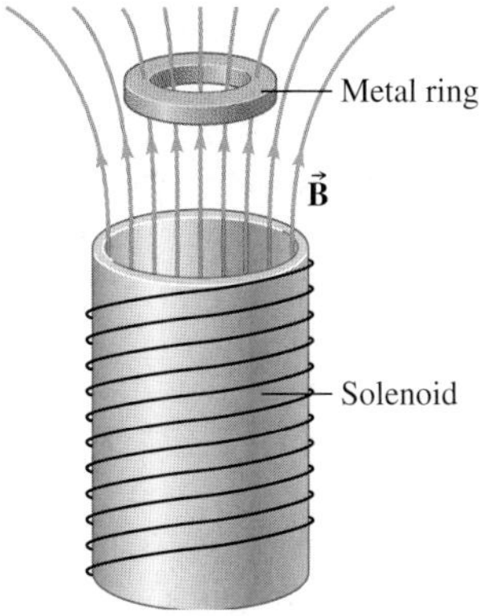

71. The strings of an electric guitar are made of ferromagnetic metal. The pickup consists of two components. A magnet causes the part of the string near it to be magnetized. The vibrations of the string near the pickup coil produce an induced emf in the coil. The electrical signal in the coil is then amplified and used to drive the speakers. In the figure, the string is moving away from the coil. What is the direction of the induced current in the coil?

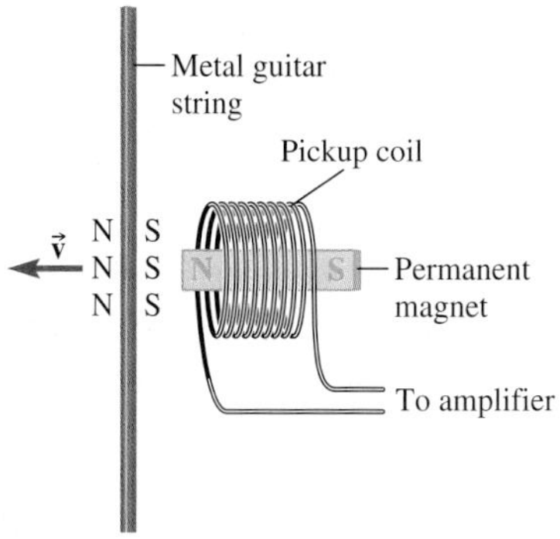

72. A toroid has a square cross section of side a. The toroid has N turns and radius R. The toroid is narrow ($a \ll R$) so that the magnetic field inside the toroid can be considered to be uniform in magnitude. What is the self-inductance of the toroid?

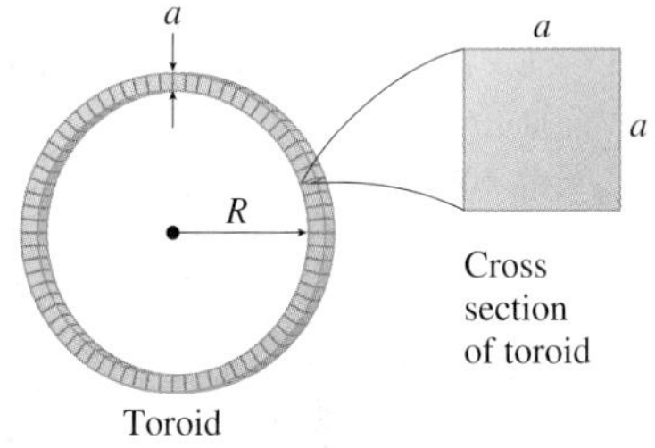

73. An ideal toroid has N turns and self-inductance L. A single turn of wire is wrapped around the toroid [see part (a) of the figure]. (a) What is the mutual inductance between the toroid and the single turn of wire? (b) What would the mutual inductance be if the turn of wire had an area twice as large as the cross-sectional area of the toroid [see part (b) of the figure]?

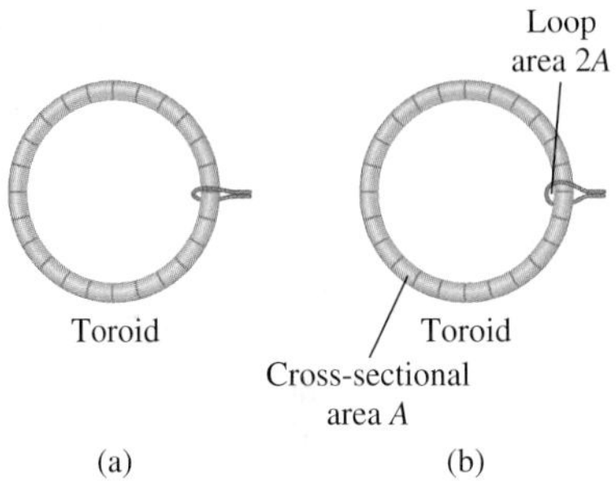

74. Suppose you wanted to use Earth's magnetic field to make an ac generator at a location where the magnitude of the field is 0.50 mT. Your coil has 1000.0 turns and a radius of 5.0 cm. At what angular velocity would you have to rotate it in order to generate an emf of amplitude 1.0 V?

75. A uniform magnetic field of magnitude 0.29 T makes an angle of 13° with the plane of a circular loop of wire. The loop has radius 1.85 cm. What is the magnetic flux through the loop?

76. A solenoid is 8.5 cm long, 1.6 cm in diameter, and has 350 turns. When the current through the solenoid is 65 mA, what is the magnetic flux through one turn of the solenoid?

77. How much energy due to Earth's magnetic field is present in 1.0 m^3 of space near Earth's surface, at a place where the field has magnitude 0.45 mT?

78. The largest constant magnetic field achieved in the laboratory is about 40 T. (a) What is the magnetic energy density due to this field? (b) What magnitude electric field would have an equal energy density?

79. A TV tube requires a 20.0-kV-amplitude power supply. (a) What is the turns ratio of the transformer that raises the 170-V-amplitude household voltage to 20.0 kV? (b) If the tube draws 82 W of power, find the currents in the primary and secondary windings. Assume an ideal transformer.

80. The alternator in an automobile generates an emf of amplitude 12.6 V when the engine idles at 1200 rpm. What is the amplitude of the emf when the car is being driven on the highway with the engine at 2800 rpm?

81. The outside of an ideal solenoid (N_1 turns, length L, radius r) is wound with a coil of wire with N_2 turns. (a) What is the mutual inductance? (b) If the current in the solenoid is changing at a rate $\Delta I_1/\Delta t$, what is the magnitude of the induced emf in the coil?

82. A standard ammeter must be inserted in series into the circuit (Section 18.9). An *induction ammeter* has the great advantage of being able to measure currents without making any electrical connection to the circuit. An iron ring is hinged so that it can be snapped around a wire. A coil is wrapped around the iron ring; the ammeter uses the induced emf in the coil to determine the current flowing in the wire. (a) Does the induction ammeter work equally well for ac and dc currents? Explain. (b) Can the induction ammeter be placed around both wires connected to an appliance to measure the current drawn by the appliance? Explain.

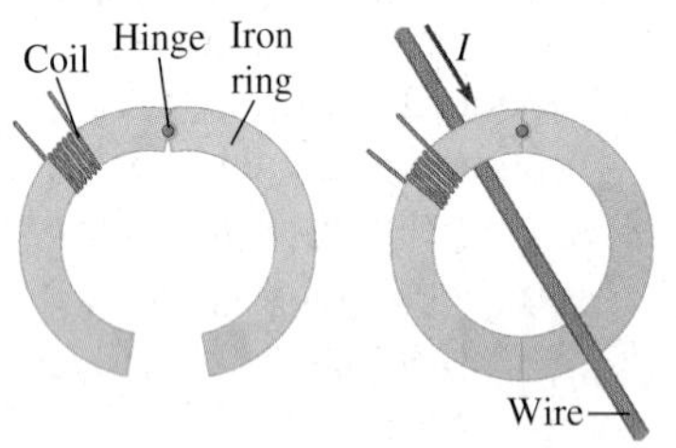

83. A *flip coil* is a device used to measure a magnetic field. A coil of radius r, N turns, and electrical resistance R is initially perpendicular to a magnetic field of magnitude B. The coil is connected to a special kind of galvanometer that measures the total charge Q that flows through it. To measure the field, the flip coil is rapidly flipped upside down. (a) What is the change in magnetic flux through the coil in one flip? (b) If the time interval during which the coil is flipped is Δt, what is the average induced emf in the coil? (c) What is the average current that flows through the galvanometer? (d) What is the total charge Q in terms of r, N, R, and B?

84. A 100-turn coil with a radius of 10.0 cm is mounted so the coil's axis can be oriented in any horizontal direction. Initially the axis is oriented so the magnetic flux from Earth's field is maximized. If the coil's axis is rotated through 90.0° in 0.080 s, an emf of 0.687 mV is induced in the coil. (a) What is the magnitude of the horizontal component of Earth's magnetic field at this location? (b) If the coil is moved to another place on Earth and the measurement is repeated, will the result be the same?

85. A bar magnet is initially far from a circular loop of wire. The magnet is moved at constant speed along the axis of the loop. It moves toward the loop, proceeds to pass through it, and then continues until it is far away on the right side of the loop. Sketch a qualitative graph of the current in the loop as a function of the position of the bar magnet. Take the current to be positive when it is counterclockwise as viewed from the left.

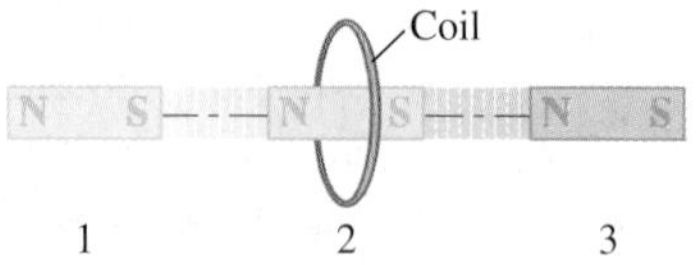

86. A bar magnet approaches a coil [part (a) of figure]. (a) In which direction does current flow through the galvanometer as the magnet approaches? (b) In part (b) of the figure, the magnet is initially at rest inside the coil. It is then pulled out from the left side. In which direction does current flow through the galvanometer as the magnet is pulled away? (c) In both situations, how does the magnitude of the current depend on the number of turns in the coil? (The resistance of the coil is negligible compared to the resistance of the galvanometer.) (d) How does the current depend on the speed of the magnet? (e) How does the magnitude of the current change if two such magnets were used, held together with the north poles together and the south poles together? (f) How does the magnitude of the current change if two such magnets were used, held together with the *opposite* poles together? (g) Would the experiment give similar results if the magnet remains stationary and the coil moves instead? Explain.

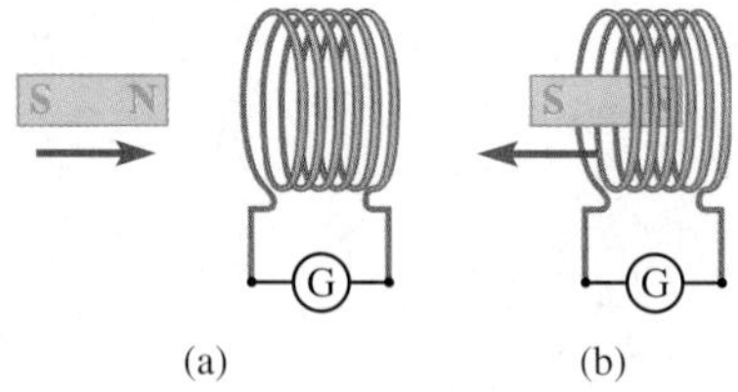

87. An ideal inductor of inductance L is connected to an ac power supply, which provides an emf $\mathscr{E}(t) = \mathscr{E}_m \sin \omega t$. (a) Write an expression for the current in the inductor as a function of time. [*Hint:* See Eq. (20-7).] (b) What is the ratio of the maximum emf to the maximum current? This ratio is called the *reactance*. (c) Do the maximum emf and maximum current occur at the same time? If not, how much time separates them?

88. An airplane is flying due north at 180 m/s. Earth's magnetic field has a northward component of 0.30 mT and an upward component of 0.38 mT. (a) If the wingspan (distance between the wingtips) is 46 m, what is the motional emf between the wingtips? (b) Which wingtip is positively charged?

89. Repeat Problem 88 if the plane flies 30.0° west of south at 180 m/s instead.

90. The magnetic field between the poles of an electromagnet is 2.6 T. A coil of wire is placed in this region so that the field is parallel to the axis of the coil. The coil has electrical resistance 25 Ω, radius 1.8 cm, and length 12.0 cm. When the current supply to the electromagnet is shut off, the total charge that flows through the coil is 9.0 mC. How many turns are there in the coil?

91. An ideal solenoid (N_1 turns, length L_1, radius r_1) is placed inside another ideal solenoid (N_2 turns, length $L_2 > L_1$, radius $r_2 > r_1$) such that the axes of the two coincide. (a) What is the mutual inductance? (b) If the current in the outer solenoid is changing at a rate $\Delta I_2/\Delta t$, what is the magnitude of the induced emf in the inner solenoid?

Answers to Practice Problems

20.1 only the magnitudes of the currents

20.2 3.0 W. The power is proportional to the bicycle's speed *squared.*

20.3 $B_\perp = B\cos 60.0°$

20.4 7.6 V

20.5 (a) $F = B^2L^2v/R$ to the left at position 2 and position 4; (b) $P = B^2L^2v^2/R$

20.6 (a) to the left; (b) from A to B through the resistor; (c) no; current only flows in coil 2 while the flux is *changing*. When the magnetic field due to coil 1 is constant, no current flows in coil 2. (d)

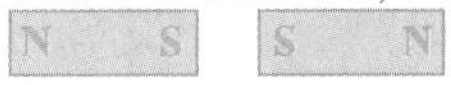

Coil 1 Coil 2

20.7 10.0 W

20.8 In a solid core, eddy currents would flow around the axis of the core. The insulation between wires prevents these eddy currents from flowing. Since energy is dissipated by eddy currents, their existence reduces the efficiency of the transformer.

20.9 9.0×10^{-5} Wb

20.10 0.53 W

20.11 0.9 s

Answers to Checkpoints

20.1 The average velocity of the electrons in the rod is out of the page and the magnetic field is into the page, so the average magnetic force on the electrons is zero. Therefore, the induced emf is zero.

20.2 Both the amplitude and frequency of the emf will change. The frequency is reduced from 12 to 10 Hz. The amplitude of the emf is proportional to the frequency, so the new amplitude is 18 V × (10/12) = 15 V.

20.4 (a) The flux through the loop due to the external magnetic field is increasing. From Lenz's law, the induced current opposes the change in flux. Therefore, the induced current creates its own magnetic field out of the page. From the right-hand rule, the induced current is counterclockwise. (b) Now the flux is decreasing. To oppose this change, the induced current produces a magnetic field into the page. The current is clockwise.

20.10 (a) The induced emf in the inductor is initially equal to the battery emf: $\mathscr{E}_b = \mathscr{E}_L = L(\Delta I/\Delta t)$. Then $\Delta I/\Delta t = \mathscr{E}_b/L =$ 500 A/s. (b) The induced emf in the inductor falls to e^{-1} times its initial value at $t = \tau = L/R = 0.250$ ms.

CHAPTER

21 Alternating Current

Look closely at the overhead power lines that supply electricity to a house. Why are there three cables—aren't two sufficient to make a complete circuit? Do the three cables correspond to the three prongs of an electric outlet? (See p. 784 for the answer.)

Concepts & Skills to Review

- ac generators; sinusoidal emfs (Sections 20.2 and 20.3)
- resistance; Ohm's law; power (Sections 18.4 and 18.8)
- emf and current (Sections 18.1 and 18.2)
- period, frequency, angular frequency (Section 10.6)
- capacitance and inductance (Sections 17.5 and 20.9)
- vector addition (Sections 3.1 and 3.2; Appendix A.8)
- graphical analysis of SHM (Section 10.7)
- resonance (Section 10.10)

21.1 SINUSOIDAL CURRENTS AND VOLTAGES: RESISTORS IN AC CIRCUITS

In an alternating current (ac) circuit, currents and emfs periodically change direction. An ac power supply periodically reverses the polarity of its emf. The sinusoidally varying emf due to an ac generator (also called an ac source) can be written (Fig. 21.1a)

$$\mathscr{E}(t) = \mathscr{E}_{\text{m}} \sin \omega t$$

The emf varies continuously between $+\mathscr{E}$_m and $-\mathscr{E}_{\text{m}}$; $\mathscr{E}_{\text{m}}$ is called the **amplitude** (or **peak** value) of the emf. In a circuit with a sinusoidal emf connected to a resistor (Fig. 21.1b), the potential difference across the resistor is equal to $\mathscr{E}(t)$, by Kirchhoff's loop rule. Then the current $i(t)$ varies sinusoidally with amplitude $I = \mathscr{E}_{\text{m}}/R$:

$$i(t) = \frac{\mathscr{E}(t)}{R} = \frac{\mathscr{E}_{\text{m}}}{R} \sin \omega t = I \sin \omega t$$

It is important to distinguish the time-dependent quantities from their amplitudes. Note that lowercase i stands for the instantaneous current, but capital I stands for the amplitude of the current. We use this convention for all time-dependent quantities in this chapter except for emf: $\mathscr{E}$ is the instantaneous emf and $\mathscr{E}_{\text{m}}$ ("m" for *maximum*) is the amplitude of the emf.

As simple as it may appear, the circuit of Fig. 21.1 has many applications. Electric heating elements found in toasters, hair dryers, electric baseboard heaters, electric stoves, and electric ovens are just resistors connected to an ac source. So is an incandescent lightbulb: the filament is a resistor whose temperature rises due to energy dissipation until it is hot enough to radiate a significant amount of visible light.

The time T for one complete cycle is the period. The frequency f is the inverse of the period:

$$\text{cycles per second} = \frac{1}{\text{seconds per cycle}}$$

$$f = \frac{1}{T}$$

CONNECTION: In Section 20.2 we learned how a generator produces a sinusoidal emf.

Circuit symbol for an ac generator (source of sinusoidal emf): ⏦

Application: Resistance heating

CONNECTION: The definitions of period, frequency, and angular frequency used in ac circuits are the same as for simple harmonic motion.

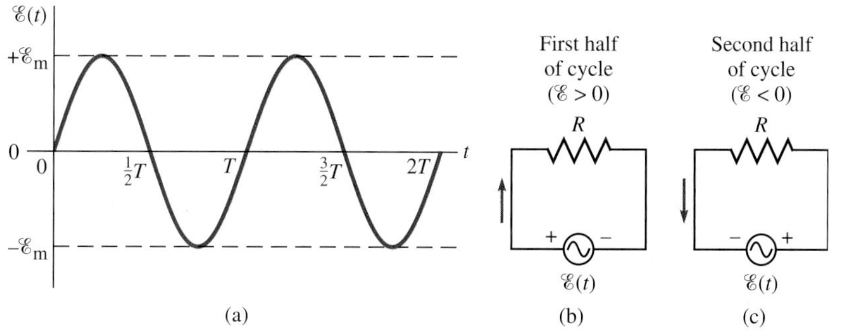

Figure 21.1 (a) A sinusoidal emf as a function of time. (b) The emf connected to a resistor, indicating the direction of the current and the polarity of the emf during the first half of the cycle ($0 < t < \frac{1}{2}T$). (c) The same circuit, indicating the direction of current and the polarity of the emf during the second half of the cycle ($\frac{1}{2}T < t < T$).

Since there are 2π radians in one complete cycle, the angular frequency in radians is

$$\omega = 2\pi f$$

In SI units the period is measured in seconds, the frequency is measured in hertz (Hz), and the angular frequency is measured in rad/s. The usual voltage at a wall outlet in a home in the United States has an amplitude of about 170 V and a frequency of 60 Hz.

Power Dissipated in a Resistor

Remember that *power dissipated* means *the rate at which energy is dissipated.*

The instantaneous power dissipated by a resistor in an ac circuit is

$$p(t) = i(t)v(t) = I \sin \omega t \times V \sin \omega t = IV \sin^2 \omega t \qquad (21\text{-}1)$$

where $i(t)$ and $v(t)$ represent the current through and potential difference across the resistor, respectively. Since $v = ir$, the power can also be written as

$$p = I^2R \sin^2 \omega t = \frac{V^2}{R} \sin^2 \omega t$$

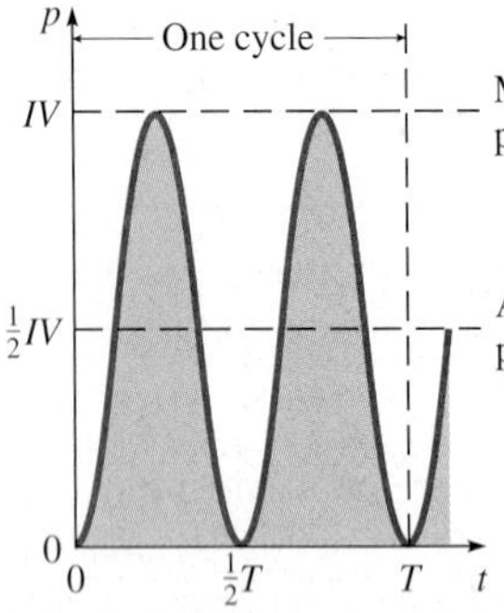

Figure 21.2 Power p dissipated by a resistor in an ac circuit as a function of time during one cycle. The area under the graph of $p(t)$ represents the energy dissipated. The *average* power is $IV/2$.

Figure 21.2 shows the instantaneous power delivered to a resistor in an ac circuit; it varies from 0 to a maximum of IV. Since the sine function *squared* is always nonnegative, the power is always nonnegative. The direction of *energy* flow is always the same—energy is dissipated in the resistor—no matter what the direction of the *current.*

The maximum power is given by the product of the peak current and the peak voltage (IV). We are usually more concerned with average power than with instantaneous power, since the instantaneous power varies rapidly. In a toaster or lightbulb, the fluctuations in instantaneous power are so fast that we usually don't notice them. The average power is IV times the average value of $\sin^2 \omega t$, which is 1/2 (see Problem 11).

Average power dissipated by a resistor:

$$P_{av} = \tfrac{1}{2}IV = \tfrac{1}{2}I^2R \qquad (21\text{-}2)$$

RMS Values

What dc current I_{dc} would dissipate energy at the same average rate as an ac current of amplitude I? Clearly $I_{dc} < I$ since I is the maximum current. To find I_{dc}, we set the average powers (through the same resistance) equal:

$$P_{av} = I_{dc}^2 R = \tfrac{1}{2}I^2R$$

Solving for I_{dc} yields

$$I_{dc} = \sqrt{\tfrac{1}{2}I^2}$$

This effective dc current is called the **root mean square (rms)** current because it is the square *root* of the *mean* (average) of the *square* of the ac current $i^2(t) = I^2 \sin^2 \omega t$. Using angle brackets to represent averaging:

$$\langle i^2 \rangle = \langle I^2 \sin^2 \omega t = I^2 \rangle \times \tfrac{1}{2}$$

$$I_{rms} = \sqrt{\langle i^2 \rangle} = \frac{1}{\sqrt{2}} I$$

CONNECTION:

Rms speed of a gas molecule (Section 13.6) is defined the same way: $v_{rms} = \sqrt{\langle v^2 \rangle}$

Thus, the rms current is equal to the peak current divided by $\sqrt{2}$. Similarly, the rms values of *sinusoidal* emfs and potential differences are also equal to the peak values divided by $\sqrt{2}$.

$$\text{rms} = \frac{1}{\sqrt{2}} \times \text{amplitude} \qquad (21\text{-}3)$$

Rms values have the advantage that they can be treated like dc values for finding the average power dissipated in a resistor:

$$P_{av} = I_{rms} V_{rms} = I_{rms}^2 R = \frac{V_{rms}^2}{R} \qquad (21\text{-}4)$$

Meters designed to measure ac voltages and currents are usually calibrated to read rms values instead of peak values. In the United States, most electric outlets supply an ac voltage of approximately 120 V rms; the peak voltage is 120 V × $\sqrt{2}$ = 170 V. Electric devices are usually labeled with rms values.

CHECKPOINT 21.1

A hair dryer is labeled "120 V, 10 A," where both quantities are rms values. What is the average power dissipated?

Example 21.1

Resistance of a 100-W Lightbulb

A 100-W lightbulb is designed to be connected to an ac voltage of 120 V (rms). (a) What is the resistance of the lightbulb filament at normal operating temperature? (b) Find the rms and peak currents through the filament. (c) When the cold filament is initially connected to the circuit by flipping a switch, is the average power larger or smaller than 100 W?

Strategy The *average* power dissipated by the filament is 100 W. Since the rms voltage across the bulb is 120 V, if we connected the bulb to a *dc* power supply of 120 V, it would dissipate a constant 100 W.

Solution (a) Average power and rms voltage are related by

$$P_{av} = \frac{V_{rms}^2}{R} \qquad (21\text{-}4)$$

We solve for R:

$$R = \frac{V_{rms}^2}{P_{av}} = \frac{(120\text{ V})^2}{100\text{ W}} = 144\ \Omega$$

(b) Average power is rms voltage times rms current:

$$P_{av} = I_{rms} V_{rms}$$

We can solve for the rms current:

$$I_{rms} = \frac{P_{av}}{V_{rms}} = \frac{100\text{ W}}{120\text{ V}} = 0.833\text{ A}$$

The amplitude of the current is a factor of $\sqrt{2}$ larger.

$$I = \sqrt{2}\, I_{rms} = 1.18\text{ A}$$

(c) For metals, resistance increases with increasing temperature. When the filament is cold, its resistance is smaller. Since it is connected to the same voltage, the current is larger and the average power dissipated is larger.

Discussion Check: The power dissipated can also be found from peak values:

$$P_{av} = \tfrac{1}{2}IV = \tfrac{1}{2}(1.18\text{ A} \times 170\text{ V}) = 100\text{ W}$$

Another check: the amplitudes should be related by Ohm's law.

$$V = IR = 1.18\text{ A} \times 144\ \Omega = 170\text{ V}$$

Practice Problem 21.1 European Wall Outlet

The rms voltage at a wall outlet in Europe is 220 V. Suppose a space heater draws an rms current of 12.0 A. What are the amplitudes of the voltage and current? What are the peak power and the average power dissipated in the heating element? What is the resistance of the heating element?

21.2 ELECTRICITY IN THE HOME

In a North American home, most electric outlets supply an rms voltage of 110–120 V at a frequency of 60 Hz. However, some appliances with heavy demands—such as electric heaters, water heaters, stoves, and large air conditioners—are supplied with 220–240 V rms. At twice the voltage amplitude, they only need to draw half as much current for the same power to be delivered, reducing energy dissipation in the wiring (and the need for extra thick wires).

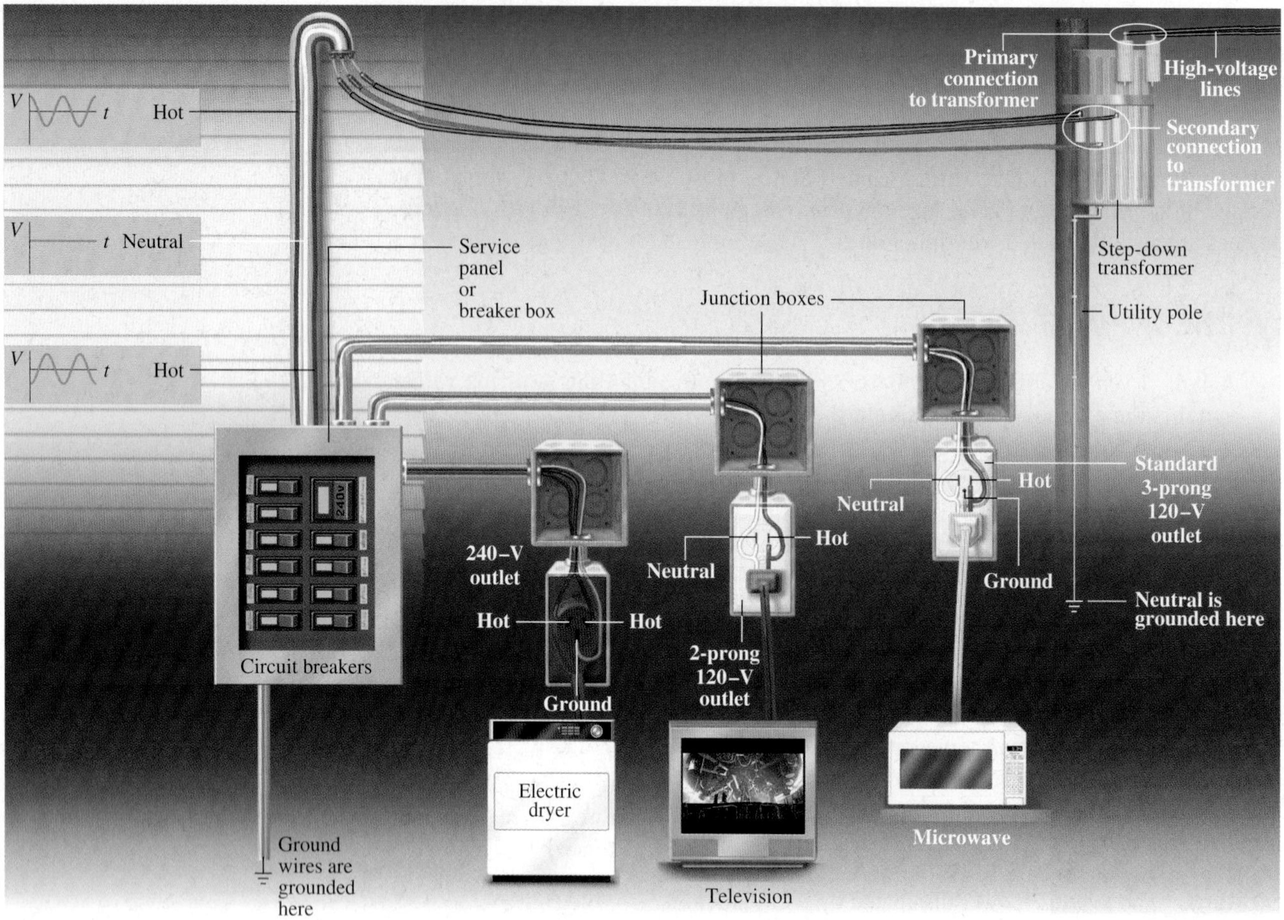

Figure 21.3 Electric wiring in a North American home.

What are the three cables that supply electricity to a house?

Local power lines are at voltages of several kilovolts. Step-down transformers reduce the voltage to 120/240 V rms. You can see these transformers wherever the power lines run on poles above the ground; they are the metal cans mounted to some of the poles (Fig. 21.3). The transformer has a center tap—a connection to the middle of the secondary coil; the voltage across the entire secondary coil is 240 V rms, but the voltage between the center tap and either end is only 120 V rms. The center tap is grounded at the transformer and runs to a building by a cable that is often uninsulated. There it is connected to the *neutral* wire (which usually has white insulation) in every 120-V circuit in the building.

The other two connections from the transformer run to the building by insulated cables and are called *hot.* The hot wires in an outlet box usually have either black or red insulation. Relative to the neutral wire, each of the hot wires is at 120 V rms, but the two are 180° out of phase with one another. Half of the 120-V circuits in the building are connected to one of the hot cables and half to the other. Appliances needing to be supplied with 240 V are connected to both hot cables; they have no connection to the neutral cable.

Older 120-V outlets have only two prongs: hot and neutral. The slot for the neutral prong is slightly larger than the hot; a *polarized* plug can only be connected one way, preventing the hot and neutral connections from being interchanged. This safety feature is now superseded in devices that use the third prong on modern outlets. The third prong is connected directly to ground through its own set of wires (usually uninsulated or with green insulation)—it is not connected to the neutral wires. The metal case of most electric appliances is connected to ground as a safety measure. If something goes wrong with the wiring inside the appliance so that the case becomes electrically connected to the hot wire, the third prong provides a low-resistance path for the current to flow to ground; the large current trips a circuit breaker or fuse. Without the ground connection, the case of the appliance would be at 120 V rms with respect to ground; someone touching the case could get a shock by providing a conducting path to ground.

21.3 CAPACITORS IN AC CIRCUITS

Figure 21.4a shows a capacitor connected to an ac source. The ac source pumps charge as needed to keep the voltage across the capacitor equal to the voltage of the source. Since the charge on the capacitor is proportional to the voltage v,

$$q(t) = Cv(t)$$

The current is proportional to the *rate of change* of the voltage $\Delta v/\Delta t$:

$$i(t) = \frac{\Delta q}{\Delta t} = C\frac{\Delta v}{\Delta t} \tag{21-5}$$

The time interval Δt must be small for i to represent the instantaneous current.

Figure 21.4b shows the voltage $v(t)$ and current $i(t)$ as functions of time for the capacitor. Note some important points:

- The current is maximum when the voltage is zero.
- The voltage is maximum when the current is zero.
- The capacitor repeatedly charges and discharges.

The voltage and the current are both sinusoidal functions of time with the same frequency, but they are out of phase: the current starts at its maximum positive value but the voltage reaches its maximum positive value one quarter cycle later. The voltage stays a quarter cycle behind the current at all times. The period T is the time for one complete cycle of a sinusoidal function; one cycle corresponds to 360° since

$$\omega T = 2\pi \text{ rad} = 360°$$

For one quarter cycle, $\frac{1}{4}\omega T = \pi/2$ rad $= 90°$. Thus, we say that the voltage and current are one quarter cycle out of phase or 90° out of phase. The current *leads* the capacitor voltage by a phase constant of 90°; equivalently, the voltage *lags* the current by the same phase angle.

Current leads voltage by 90° in a capacitor in an ac circuit.

If the voltage across the capacitor is given by

$$v(t) = V \sin \omega t$$

then the current varies in time as

$$i(t) = I \sin(\omega t + \pi/2)$$

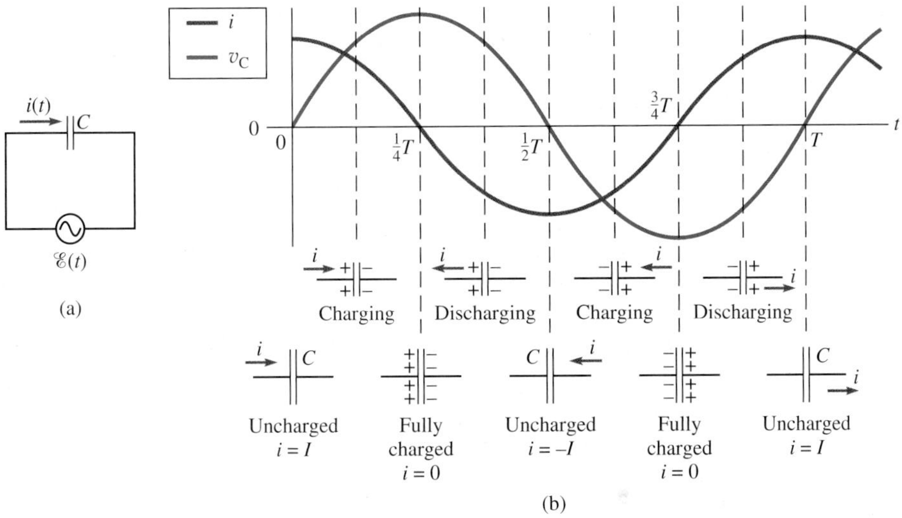

Figure 21.4 (a) An ac generator connected to a capacitor. (b) One complete cycle of the current and voltage for a capacitor connected to an ac source as a function of time. Signs are chosen so that positive current (to the right) gives the capacitor a positive charge (left plate positive).

We add the $\pi/2$ radians to the argument of the sine function to give the current a head start of $\pi/2$ rad. (We use radians rather than degrees since angular frequency ω is generally expressed in rad/s.)

In the general expression

$$i = I \sin(\omega t + \phi)$$

the angle ϕ is called the **phase constant**, which, for the case of the current in the capacitive circuit, is $\phi = \pi/2$. A sine function shifted $\pi/2$ radians ahead is a cosine function, as can be seen in Fig. 21.4; that is,

$$\sin(\omega t + \pi/2) = \cos \omega t$$

so

$$i(t) = I \cos \omega t$$

Reactance

The amplitude of the current I is proportional to the voltage amplitude V. A larger voltage means that more charge needs to be pumped onto the capacitor; to pump more charge in the same amount of time requires a larger current. We write the proportionality as

Reactance: ratio of voltage amplitude to current amplitude for a capacitor or inductor

$$V_C = IX_C \qquad (21\text{-}6)$$

where the quantity X_C is called the **reactance** of the capacitor. Compare Eq. (21-6) to Ohm's law for a resistor ($v = iR$); reactance must have the same SI unit as resistance (ohms). We have written Eq. (21-6) in terms of the amplitudes (V, I), but it applies equally well if *both* V and I are rms values (since both are smaller by the same factor, $\sqrt{2}$).

CONNECTION:

Reactance is a generalization of the definition of resistance (ratio of voltage to current).

By analogy with Ohm's law, we can think of the reactance as the "effective resistance" of the capacitor. The reactance determines how much current flows; the capacitor reacts in a way to impede the flow of current. A larger reactance means a smaller current, just as a larger resistance means a smaller current.

There are, however, important differences between reactance and resistance. A resistor dissipates energy, but an ideal capacitor does *not*; the average power dissipated by an ideal capacitor is zero, not $I_{rms}^2 X_C$. Note also that Eq. (21-6) relates only the *amplitudes* of the current and voltage. Since the current and voltage in a capacitor are 90° out of phase, it does *not* apply to the instantaneous values:

$$v(t) \neq i(t)X_C$$

For a resistor, on the other hand, the current and voltage are *in phase* (phase difference of zero); it *is* true for a resistor that $v(t) = i(t)R$.

Another difference is that reactance depends on frequency. Recall from Chapter 20 that

$$\text{If } \Phi(t) = \Phi_0 \sin \omega t, \text{ then } \frac{\Delta\Phi}{\Delta t} = \omega\Phi_0 \cos \omega t \quad (\text{for small } \Delta t); \qquad (20\text{-}7a)$$

$$\text{if } \Phi(t) = \Phi_0 \cos \omega t, \text{ then } \frac{\Delta\Phi}{\Delta t} = -\omega\Phi_0 \sin \omega t \quad (\text{for small } \Delta t). \qquad (20\text{-}7b)$$

CONNECTION:

These are general mathematical relationships giving the rates of change of sinusoidal functions. For example, if the position of a particle is $x(t) = A \sin \omega t$ then its velocity, the rate of change of position, is $v_x(t) = \Delta x/\Delta t = \omega A \cos \omega t$ [Eq. (10-25b)].

For a capacitor in an ac circuit, if the charge as a function of time is

$$q(t) = Q \sin \omega t$$

then the current (the rate of change of the charge on the capacitor) must be

$$i(t) = \frac{\Delta q}{\Delta t} = \omega Q \cos \omega t$$

Therefore, the peak current is

$$I = \omega Q$$

Since $Q = CV$, we can find the reactance:

$$X_C = \frac{V}{I} = \frac{V}{\omega Q} = \frac{V}{\omega CV}$$

$$X_C = \frac{1}{\omega C} \quad (21\text{-}7)$$

Reactance of a capacitor

The reactance is inversely proportional to the capacitance and to the angular frequency. To understand why, let us focus on the first quarter of a cycle ($0 \le t \le T/4$) in Fig. 21.4b. During this quarter cycle, a total charge $Q = CV$ flows onto the capacitor plates since the capacitor goes from being uncharged to fully charged. For a larger value of C, a proportionately larger charge must be put on the capacitor to reach a potential difference of V; to put more charge on in the same amount of time ($T/4$), the current must be larger. Thus, when the capacitance is larger, the reactance must be lower because more current flows for a given ac voltage amplitude.

The reactance is also inversely proportional to the frequency. For a higher frequency, the time available to charge the capacitor ($T/4$) is shorter. For a given voltage amplitude, a larger current must flow to achieve the same maximum voltage in a shorter amount of time. Thus, the reactance is smaller for a higher frequency.

At very high frequencies, the reactance approaches zero. The capacitor no longer impedes the flow of current; ac current flows in the circuit as if there were a conducting wire short-circuiting the capacitor. For the other limiting case, very low frequencies, the reactance approaches infinity. At a very low frequency, the applied voltage changes slowly; the current stops as soon as the capacitor is charged to a voltage equal to the applied voltage.

Example 21.2

Capacitive Reactance for Two Frequencies

(a) Find the capacitive reactance and the rms current for a 4.00-μF capacitor when it is connected to an ac source of 12.0 V rms at 60.0 Hz. (b) Find the reactance and current when the frequency is changed to 15.0 Hz while the rms voltage remains at 12.0 V.

Strategy The reactance is the proportionality constant between the rms values of the voltage across and current through the capacitor. The capacitive reactance is given by Eq. (21-7). Frequencies in Hz are given; we need *angular* frequencies to calculate the reactance.

Solution (a) Angular frequency is

$$\omega = 2\pi f$$

Then the reactance is

$$X_C = \frac{1}{2\pi f C}$$

$$= \frac{1}{2\pi \times 60.0\ \text{Hz} \times 4.00 \times 10^{-6}\ \text{F}} = 663\ \Omega$$

The rms current is

$$I_{rms} = \frac{V_{rms}}{X_C} = \frac{12.0\ \text{V}}{663\ \Omega} = 18.1\ \text{mA}$$

(b) We could redo the calculation in the same way. An alternative is to note that the frequency is multiplied by a factor $\frac{15}{60} = \frac{1}{4}$. Since reactance is *inversely* proportional to frequency,

$$X_C = 4 \times 663\ \Omega = 2650\ \Omega$$

A larger reactance means a smaller current:

$$I_{rms} = \frac{1}{4} \times \frac{12.0\ \text{V}}{663\ \Omega} = 4.52\ \text{mA}$$

Discussion When the frequency is increased, the reactance decreases and the current increases. As we see in Section 21.7, capacitors can be used in circuits to filter out low frequencies because at lower frequency, less current flows. When a PA system makes a humming sound (60-Hz hum), a capacitor can be inserted between the amplifier and the speaker to block much of the 60-Hz noise while letting the higher frequencies pass through.

Practice Problem 21.2 Capacitive Reactance and rms Current for a New Frequency

Find the capacitive reactance and the rms current for a 4.00-μF capacitor when it is connected to an ac source of 220.0 V rms and 4.00 Hz.

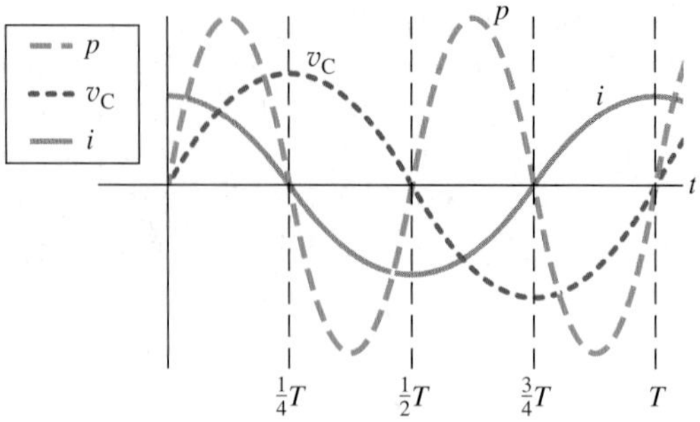

Figure 21.5 Current, voltage, and power for a capacitor in an ac circuit.

Power

Figure 21.5 shows a graph of the instantaneous power $p(t) = v(t)i(t)$ for a capacitor superimposed on graphs of the current and voltage. The 90° ($\pi/2$ rad) phase difference between current and voltage has implications for the power in the circuit. During the first quarter cycle ($0 \le t \le T/4$), both the voltage and the current are positive. The power is positive: the generator is delivering energy to the capacitor to charge it. During the second quarter cycle ($T/4 \le t \le T/2$), the current is negative while the voltage remains positive. The power is negative; as the capacitor discharges, energy is returned to the generator from the capacitor.

The average power is zero for an ideal capacitor in an ac circuit.

The power continues to alternate between positive and negative as the capacitor stores and then returns electric energy. The average power is zero since all the energy stored is given back and none of it is dissipated.

21.4 INDUCTORS IN AC CIRCUITS

An inductor in an ac circuit develops an induced emf that opposes changes in the current, according to Faraday's law [Eq. (20-6)]. We use the same sign convention as for the capacitor: the current i through the inductor in Fig. 21.6a is positive when it flows to the right and the voltage across the inductor v_L is positive if the left side is at a higher potential than the right side. If current flows in the positive direction and is *increasing*, the induced emf *opposes the increase* (Fig. 21.6b) and v_L is positive. If current flows in the positive direction and is *decreasing*, the induced emf *opposes the decrease* (Fig. 21.6c) and v_L is negative. Since in the first case $\Delta i/\Delta t$ is positive and in the second case $\Delta i/\Delta t$ is negative, the voltage has the correct sign if we write

$$v_L = L\frac{\Delta i}{\Delta t} \qquad (21\text{-}8)$$

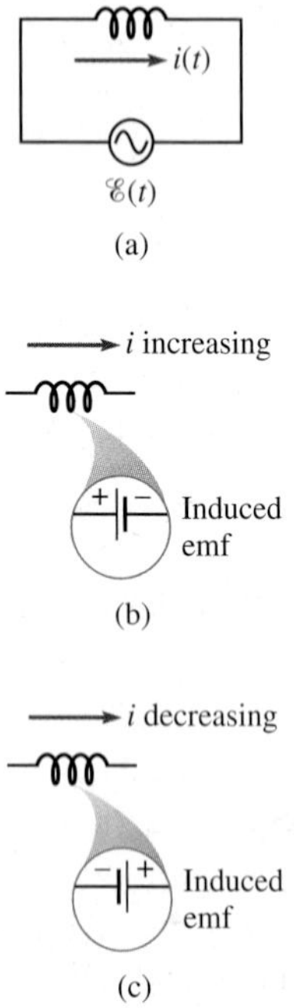

Figure 21.6 (a) An inductor connected to an ac source. (b) and (c) The potential difference across the inductor for current flowing to the right depends on whether the current is increasing or decreasing.

In Problem 29 you can verify that Eq. (21-8) also gives the correct sign when current flows to the left.

The voltage amplitude across the inductor is proportional to the amplitude of the current. The constant of proportionality is called the **reactance** of the inductor (X_L):

$$V_L = IX_L \qquad (21\text{-}9)$$

As for the capacitive reactance, the inductive reactance X_L has units of ohms. As in Eq. (21-6), V and I in Eq. (21-9) can be *either* amplitudes *or* rms values, but be careful not to mix amplitude and rms in the same equation.

In Problem 31 you can show, using reasoning similar to that used for the capacitor, that the reactance of an inductor is

Reactance of an inductor

$$X_L = \omega L \qquad (21\text{-}10)$$

Note that the inductive reactance is directly proportional to the inductance L and to the angular frequency ω, in contrast to the capacitive reactance, which is *inversely* proportional to the angular frequency and to the capacitance. The induced emf in the inductor always acts to oppose changes in the current. At higher frequency, the more rapid changes in current are opposed by a greater induced emf in the inductor. Thus, the ratio of the amplitude of the induced emf to the amplitude of the current—the reactance—is greater at higher frequency.

CHECKPOINT 21.4

Suppose an inductor and a capacitor have equal reactance at some angular frequency ω_0. (a) Which has the larger reactance for $\omega > \omega_0$? (b) Which has the larger reactance for $\omega < \omega_0$?

Figure 21.7 shows the potential difference across the inductor and the current through the inductor as functions of time. We assume an ideal inductor—one with no resistance in its windings. Since $v_L = L\,\Delta i/\Delta t$, the graph of $v_L(t)$ is proportional to the *slope* of the graph of $i(t)$ at any time t. The voltage and current are out of phase by a quarter cycle, but this time the current *lags* the voltage by 90° ($\pi/2$ rad); current reaches its maximum a quarter cycle *after* the voltage reaches a maximum. A mnemonic device for remembering what leads and what lags is that the letter *c* (for *current*) appears in the second half of the word *induc̲tor* (current *lags* inductor voltage) and at the *beginning* of the word *c̲apacitor* (current *leads* capacitor voltage).

Current lags voltage across an inductor in an ac circuit.

In Fig. 21.7, the voltage across the inductor can be written

$$v_L(t) = V \sin \omega t$$

The current is

$$i(t) = -I \cos \omega t = I \sin (\omega t - \pi/2)$$

where we have used the trigonometric identity $-\cos \omega t = \sin (\omega t - \pi/2)$. We see explicitly that the current lags behind the voltage from the phase constant $\phi = -\pi/2$.

Power

As for the capacitor, the 90° phase difference between current and voltage means that the average power is zero. No energy is dissipated in an *ideal* inductor (one with no resistance). The generator alternately sends energy to the inductor and receives energy back from the inductor.

The average power is zero for an *ideal* inductor in an ac circuit.

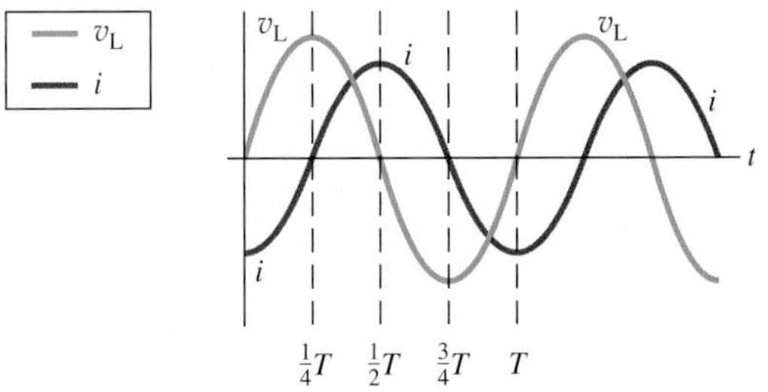

Figure 21.7 Current and potential difference across an inductor in an ac circuit. Note that when the current is maximum or minimum, its instantaneous rate of change—represented by its slope—is zero, so $v_L = 0$. On the other hand, when the current is zero, it is changing the fastest, so v_L has its maximum magnitude.

Example 21.3

Inductor in a Radio Tuning Circuit

A 0.56-μH inductor is used as part of the tuning circuit in a radio. Assume the inductor is ideal. (a) Find the reactance of the inductor at a frequency of 90.9 MHz. (b) Find the amplitude of the current through the inductor if the voltage amplitude is 0.27 V. (c) Find the capacitance of a capacitor that has the same reactance at 90.9 MHz.

Strategy The reactance of an inductor is the product of angular frequency and inductance. The reactance in ohms is the ratio of the voltage amplitude to the amplitude of the current. For the capacitor, the reactance is $1/(\omega C)$.

Solution (a) The reactance of the inductor is

$$X_L = \omega L = 2\pi f L$$
$$= 2\pi \times 90.9\text{ MHz} \times 0.56\ \mu\text{H} = 320\ \Omega$$

(b) The amplitude of the current is

$$I = \frac{V}{X_L}$$
$$= \frac{0.27\text{ V}}{320\ \Omega} = 0.84\text{ mA}$$

(c) We set the two reactances equal ($X_L = X_C$) and solve for C:

$$\omega L = \frac{1}{\omega C}$$

$$C = \frac{1}{\omega^2 L} = \frac{1}{4\pi^2 \times (90.9 \times 10^6\text{ Hz})^2 \times 0.56 \times 10^{-6}\text{ H}}$$
$$= 5.5\text{ pF}$$

Discussion We can check by calculating the reactance of the capacitor:

$$X_C = \frac{1}{\omega C} = \frac{1}{2\pi \times 90.9 \times 10^6\text{ Hz} \times 5.5 \times 10^{-12}\text{ F}} = 320\ \Omega$$

In Section 21.6 we study tuning circuits in more detail.

Practice Problem 21.3 Reactance and rms Current

Find the inductive reactance and the rms current for a 3.00-mH inductor when it is connected to an ac source of 10.0 mV (rms) at a frequency of 60.0 kHz.

21.5 *RLC* SERIES CIRCUITS

Figure 21.8a shows an *RLC* series circuit. Kirchhoff's junction rule tells us that the instantaneous current through each element is the same, since there are no junctions. The loop rule requires the sum of the instantaneous voltage drops across the three elements to equal the applied ac voltage:

$$\mathscr{E}(t) = v_L(t) + v_R(t) + v_C(t) \qquad (21\text{-}11)$$

The three voltages are sinusoidal functions of time with the same frequency but different phase constants.

Suppose that we choose to write the current with a phase constant of zero. The voltage across the resistor is in phase with the current, so it also has a phase constant of zero (see Fig. 21.8b). The voltage across the inductor leads the current by 90°, so it has a

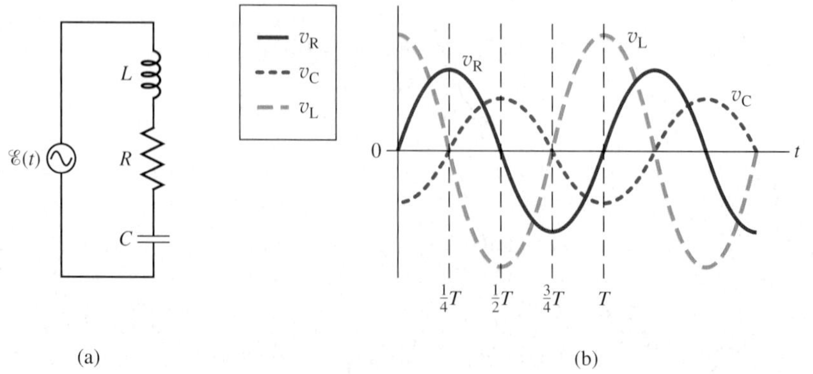

Figure 21.8 (a) An *RLC* series circuit. (b) The voltages across the circuit elements as functions of time. The current is in phase with v_R, leads v_C by 90°, and lags v_L by 90°.

phase constant of $+\pi/2$. The voltage across the capacitor lags the current by 90°, so it has a phase constant of $-\pi/2$.

$$\mathscr{E}(t) = \mathscr{E}_m \sin(\omega t + \phi) = V_L \sin\left(\omega t + \frac{\pi}{2}\right) + V_R \sin \omega t + V_C \sin\left(\omega t - \frac{\pi}{2}\right) \quad (21\text{-}12)$$

Phasor Diagrams We could simplify this sum using trigonometric identities, but there is an easier method. We can represent each sinusoidal voltage by a vector-like object called a **phasor**. The magnitude of the phasor represents the amplitude of the voltage; the angle of the phasor represents the phase constant of the voltage. We can then add phasors the same way we add vectors. (See Problem 43 for insight into why the phasor method works.) Although we draw them like vectors and *add like vectors*, they are not vectors in the usual sense. A phasor is not a quantity with a direction in space, like real vectors such as acceleration, momentum, or magnetic field.

Figure 21.9a shows three phasors representing the voltages $v_L(t)$, $v_R(t)$, and $v_C(t)$. An angle counterclockwise from the +*x*-axis represents a positive phase constant. First we add the phasors representing $v_L(t)$ and $v_C(t)$, which are in opposite directions. Then we add the sum of these two to the phasor that represents $v_R(t)$ (Fig. 21.9b). The vector sum represents $\mathscr{E}(t)$. The amplitude of $\mathscr{E}(t)$ is the length of the sum; from the Pythagorean theorem,

$$\mathscr{E}_m = \sqrt{V_R^2 + (V_L - V_C)^2} \quad (21\text{-}13)$$

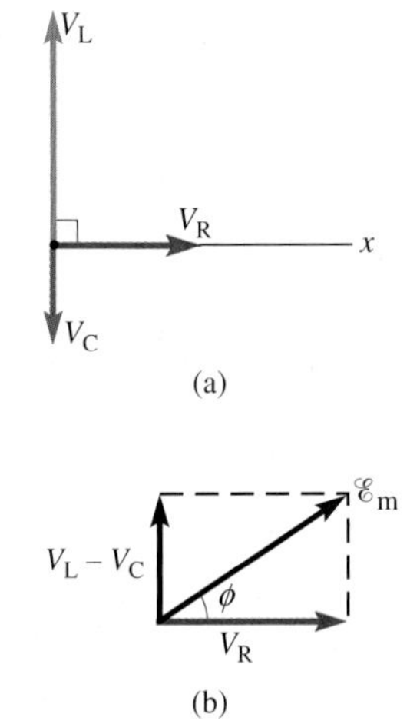

Figure 21.9 (a) Phasor representation of the voltages. (b) The phase angle ϕ between the source emf and the voltage across the resistor (which is in phase with the current).

CHECKPOINT 21.5

In a series *RLC* circuit, the voltage amplitudes across the capacitor and inductor are 90 mV and 50 mV, respectively. The applied emf has amplitude $\mathscr{E}_m$ = 50 mV. What is the voltage amplitude across the resistor?

Impedance Each of the voltage amplitudes on the right side of Eq. (21-13) can be rewritten as the amplitude of the current times a reactance or resistance:

$$\mathscr{E}_m = \sqrt{(IR)^2 + (IX_L - IX_C)^2}$$

Factoring out the current yields

$$\mathscr{E}_m = I\sqrt{R^2 + (X_L - X_C)^2}$$

Thus, the amplitude of the ac source voltage is proportional to the amplitude of the current. The constant of proportionality is called the **impedance** *Z* of the circuit.

$$\mathscr{E}_m = IZ \quad (21\text{-}14a)$$

$$Z = \sqrt{R^2 + (X_L - X_C)^2} \quad (21\text{-}14b)$$

Impedance is measured in ohms.

From Fig. 21.9b, the source voltage $\mathscr{E}(t)$ leads $v_R(t)$—and the current $i(t)$—by a phase angle ϕ where

$$\tan \phi = \frac{V_L - V_C}{V_R} = \frac{IX_L - IX_C}{IR} = \frac{X_L - X_C}{R} \quad (21\text{-}15)$$

We assumed $X_L > X_C$ in Figs. 21.8 and 21.9. If $X_L < X_C$, the phase angle ϕ is negative, which means that the source voltage *lags* the current. Figure 21.9b also implies that

$$\cos \phi = \frac{V_R}{\mathscr{E}_m} = \frac{IR}{IZ} = \frac{R}{Z} \quad (21\text{-}16)$$

If one or two of the elements *R*, *L*, and *C* are not present in a circuit, the foregoing analysis is still valid. Since there is no potential difference across a missing element, we can set the resistance or reactance of the missing element(s) to zero. For instance, since an inductor is made by coiling a long length of wire, it usually has an appreciable resistance. We can model a real inductor as an ideal inductor in series with a resistor. The impedance of the inductor is found by setting $X_C = 0$ in Eq. (21-14b).

Example 21.4

An *RLC* Series Circuit

In an *RLC* circuit, the following three elements are connected in series: a resistor of 40.0 Ω, a 22.0-mH inductor, and a 0.400-μF capacitor. The ac source has a peak voltage of 0.100 V and an angular frequency of 1.00×10^4 rad/s. (a) Find the amplitude of the current. (b) Find the phase angle between the current and the ac source. Which leads? (c) Find the peak voltages across each of the circuit elements.

Strategy The impedance is the ratio of the source voltage amplitude to the amplitude of the current. By finding the reactances of the inductor and capacitor, we can find the impedance and then solve for the amplitude of the current. The reactances also enable us to calculate the phase constant ϕ. If ϕ is positive, the source voltage leads the current; if ϕ is negative, the source voltage lags the current. The peak voltage across any element is equal to the peak current times the reactance or resistance of that element.

Solution (a) The inductive reactance is

$$X_L = \omega L = 1.00 \times 10^4 \text{ rad/s} \times 22.0 \times 10^{-3} \text{ H} = 220\ \Omega$$

The capacitive reactance is

$$X_C = \frac{1}{\omega C} = \frac{1}{1.00 \times 10^4 \text{ rad/s} \times 0.400 \times 10^{-6} \text{ F}} = 250\ \Omega$$

Then the impedance of the circuit is

$$Z = \sqrt{R^2 + X^2} = \sqrt{(40.0\ \Omega)^2 + (-30\ \Omega)^2} = 50\ \Omega$$

For a source voltage amplitude $V = 0.100$ V, the amplitude of the current is

$$I = \frac{V}{Z} = \frac{0.100 \text{ V}}{50\ \Omega} = 2.0 \text{ mA}$$

(b) The phase angle ϕ is

$$\phi = \tan^{-1} \frac{X_L - X_C}{R} = \tan^{-1} \frac{-30\ \Omega}{40.0\ \Omega} = -0.64 \text{ rad} = -37°$$

Since $X_L < X_C$, the phase angle ϕ is negative, which means that the source voltage *lags* the current.

(c) The voltage amplitude across the inductor is

$$V_L = IX_L = 2.0 \text{ mA} \times 220\ \Omega = 440 \text{ mV}$$

For the capacitor and resistor,

$$V_C = IX_C = 2.0 \text{ mA} \times 250\ \Omega = 500 \text{ mV}$$

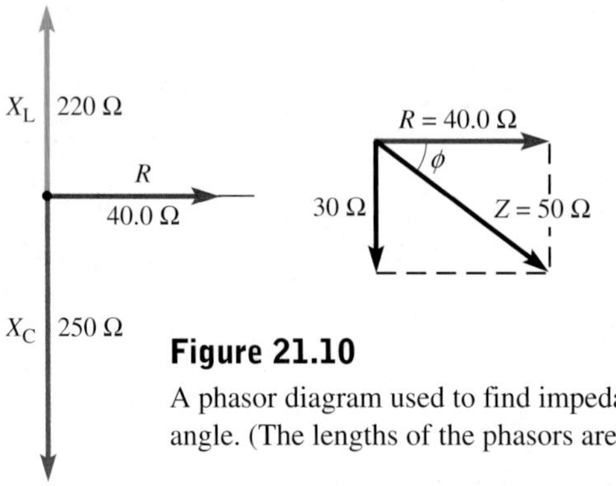

Figure 21.10

A phasor diagram used to find impedance and phase angle. (The lengths of the phasors are not to scale.)

and

$$V_R = IR = 2.0 \text{ mA} \times 40.0\ \Omega = 80 \text{ mV}$$

Discussion Since the voltage phasors in Fig. 21.9 are each proportional to I, we can divide each by I to form a phasor diagram where the phasors represent reactances or resistances (Fig. 21.10). Such a phasor diagram can be used to find the impedance of the circuit and the phase constant, instead of using Eqs. (21-14b) and (21-15).

Note that the sum of the voltage amplitudes across the three circuit elements is not the same as the source voltage amplitude:

$$100 \text{ mV} \neq 440 \text{ mV} + 80 \text{ mV} + 500 \text{ mV}$$

The voltage amplitudes across the inductor and capacitor are each *larger* than the source voltage amplitude. The voltage amplitudes are *maximum* values; since the voltages are not in phase with each other, they do not attain their maximum values at the same instant of time. What is true is that the sum of the *instantaneous* potential differences across the three elements at any given time is equal to the instantaneous source voltage at the same time [Eq. (21-12)].

Practice Problem 21.4 Instantaneous Voltages

If the current in this same circuit is written as $i(t) = I \sin \omega t$, what would be the corresponding expressions for $v_C(t)$, $v_L(t)$, $v_R(t)$, and $\mathcal{E}(t)$? (The main task is to get the phase constants correct.) Using these expressions, show that at $t = 80.0$ μs, $v_C(t) + v_L(t) + v_R(t) = \mathcal{E}(t)$. (The loop rule is true at *any* time t; we just verify it at one particular time.)

Power Factor

No power is dissipated in an ideal capacitor or an ideal inductor; the power is dissipated only in the resistance of the circuit (including the resistances of the wires of the circuit and the windings of the inductor):

$$P_{av} = I_{rms} V_{R,rms} \quad (21\text{-}4)$$

We want to rewrite the average power in terms of the rms source voltage.

$$\frac{V_{R,rms}}{\mathscr{E}_{rms}} = \frac{I_{rms}R}{I_{rms}Z} = \frac{R}{Z}$$

From Eq. (21-16), $R/Z = \cos\phi$. Therefore,

$$V_{R,rms} = \mathscr{E}_{rms} \cos\phi$$

and

$$P_{av} = I_{rms}\mathscr{E}_{rms} \cos\phi \quad (21\text{-}17)$$

The factor $\cos\phi$ in Eq. (21-17) is called the **power factor**. When there is only resistance and no reactance in the circuit, $\phi = 0$ and $\cos\phi = 1$; then $P_{av} = I_{rms}\mathscr{E}_{rms}$. When there is only capacitance or inductance in the circuit, $\phi = \pm 90°$ and $\cos\phi = 0$, so that $P_{av} = 0$. Many electric devices contain appreciable inductance or capacitance; the load they present to the source voltage is not purely a resistance. In particular, any device with a transformer has some inductance due to the windings. The label on an electric device sometimes includes a quantity with units of V·A and a smaller quantity with units of W. The former is the product $I_{rms}\mathscr{E}_{rms}$; the latter is the average power consumed.

PHYSICS AT HOME

Find an electric device that has a label with two numerical ratings, one in V·A and one in W. The windings of a transformer have significant inductance, so try something with an external transformer (inside the power supply) or an internal transformer (such as a desktop computer). The windings of motors also have inductance, so something with a motor is also a good choice. Calculate the power factor for the device. Now find a device that has little reactance relative to its resistance, such as a heater or a lightbulb. Why is there no numerical rating in V·A?

Example 21.5

Laptop Power Supply

A power supply for a laptop computer is labeled as follows: "45 W AC Adapter. AC input: 1.0 A max, 120 V, 60.0 Hz." A simplified circuit model for the power supply is a resistor R and an ideal inductor L in series with an ideal ac emf. The inductor represents primarily the inductance of the windings of the transformer; the resistor represents primarily the load presented by the laptop computer. Find the values of L and R when the power supply draws the maximum rms current of 1.0 A.

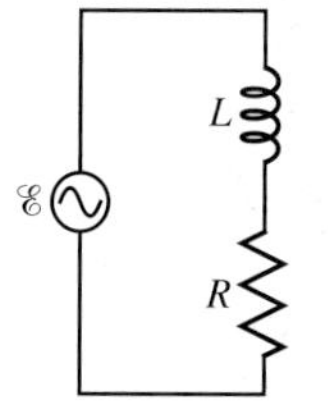

Figure 21.11 A circuit diagram for the power supply.

Strategy First we sketch the circuit (Fig. 21.11). The next step is to identify the quantities given in the problem, taking care to distinguish rms quantities from amplitudes and average power from $I_{rms}\mathscr{E}_{rms}$. Since power is dissipated in the resistor but not in the inductor, we can find the resistance from the average power. Then we can use the power factor to find L. We assume no capacitance in the circuit, which means we can set $X_C = 0$.

Solution The problem tells us that the maximum rms current is $I_{rms} = 1.0$ A. The rms source voltage is $\mathscr{E}_{rms} = 120$ V. The frequency is $f = 60.0$ Hz. The average power is 45 W when the power supply draws 1.0 A rms; the average power is smaller when the current drawn is smaller. Then

$$\mathscr{E}_{rms} I_{rms} = 120\text{ V} \times 1.0\text{ A} = 120\text{ V·A}$$

continued on next page

Example 21.5 continued

Note that the average power is less than $I_{rms}\mathscr{E}_{rms}$; it can never be greater than $I_{rms}\mathscr{E}_{rms}$ since $\cos\phi \le 1$.

Since power is dissipated only in the resistor,

$$P_{av} = I_{rms}^2 R$$

The resistance is therefore

$$R = \frac{P_{av}}{I_{rms}^2} = \frac{45\ \text{W}}{(1.0\ \text{A})^2} = 45\ \Omega$$

The ratio of the average power to $I_{rms}\mathscr{E}_{rms}$ gives the power factor:

$$\frac{\mathscr{E}_{rms} I_{rms} \cos\phi}{\mathscr{E}_{rms} I_{rms}} = \cos\phi = \frac{45\ \text{W}}{120\ \text{V·A}} = 0.375$$

The phase angle is $\phi = \cos^{-1} 0.375 = 68.0°$. From the phasor diagram of Fig. 21.12,

$$\tan\phi = \frac{V_L}{V_R} = \frac{IX_L}{IR} = \frac{X_L}{R}$$

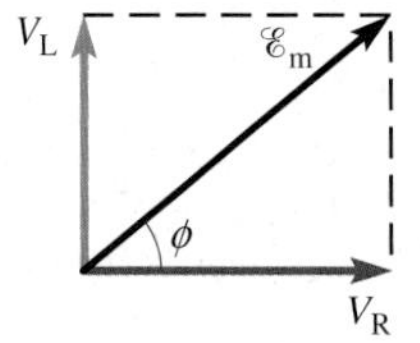

Figure 21.12
Phasor addition of the voltages across the inductor and resistor.

Solving for X_L,

$$X_L = R\tan\phi = (45\ \Omega)\tan 68.0° = 111.4\ \Omega = \omega L$$

Now we can solve for L:

$$L = \frac{X_L}{\omega} = \frac{111.4\ \Omega}{2\pi \times 60.0\ \text{Hz}} = 0.30\ \text{H}$$

Discussion Check: $\cos\phi$ should be equal to R/Z.

$$\frac{R}{Z} = \frac{R}{\sqrt{R^2 + X_L^2}} = \frac{45\ \Omega}{\sqrt{(45\ \Omega)^2 + (111.4\ \Omega)^2}} = 0.375$$

which agrees with $\cos\phi = 0.375$.

Practice Problem 21.5 A More Typical Current Draw

The adapter rarely draws the maximum rms current of 1.0 A. Suppose that, more typically, the adapter draws an rms current of 0.25 A. What is the average power? Use the same simplified circuit model with the same value of L but a *different* value of R. [*Hint:* Begin by finding the impedance $Z = \sqrt{R^2 + X_L^2}$.]

21.6 RESONANCE IN AN *RLC* CIRCUIT

Suppose an *RLC* circuit is connected to an ac source with a fixed amplitude but variable frequency. The impedance depends on frequency, so the amplitude of the current depends on frequency. Figure 21.13 shows three graphs (called **resonance curves**) of the amplitude of the current $I = \mathscr{E}_m/Z$ as a function of angular frequency for a circuit with $L = 1.0$ H, $C = 1.0$ μF, and $\mathscr{E}_m = 100$ V. Three different resistors were used: 200 Ω, 500 Ω, and 1000 Ω.

The shape of these graphs is determined by the frequency dependence of the inductive and capacitive reactances (Fig. 21.14). At low frequencies, the reactance of the capacitor $X_C = 1/(\omega C)$ is much greater than either R or X_L, so $Z \approx X_C$. At high frequencies, the reactance of the inductor $X_L = \omega L$ is much greater than either R or X_C, so $Z \approx X_L$. At extreme frequencies, either high or low, the impedance is larger and the amplitude of the current is therefore small.

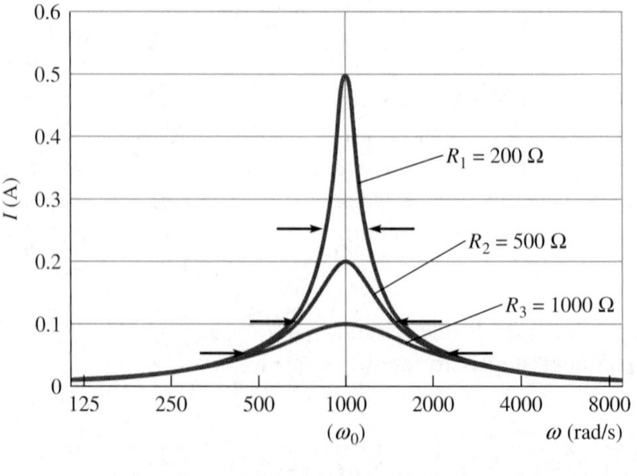

Figure 21.13 The amplitude of the current I as a function of angular frequency ω for three different resistances in a series *RLC* circuit. The widths of each peak at half-maximum current are indicated. The horizontal scale is logarithmic.

The impedance of the circuit is

$$Z = \sqrt{R^2 + (X_L - X_C)^2} \quad \text{(21-14b)}$$

Since R is constant, the minimum impedance $Z = R$ occurs at an angular frequency ω_0—called the **resonant** angular frequency—for which the reactances of the inductor and capacitor are equal so that $X_L - X_C = 0$.

$$X_L = X_C$$

$$\omega_0 L = \frac{1}{\omega_0 C}$$

Solving for ω_0,

$$\omega_0 = \frac{1}{\sqrt{LC}} \quad \text{(21-18)}$$

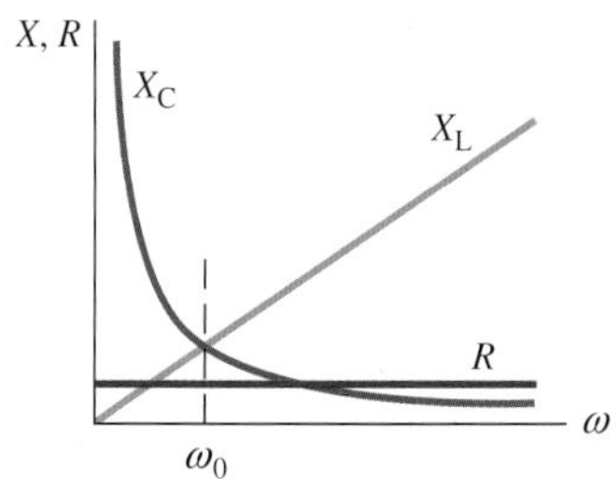

Figure 21.14 Frequency dependence of the inductive and capacitive reactances and of the resistance as a function of frequency.

Resonant angular frequency of *RLC* circuit

Note that the resonant frequency of a circuit depends only on the values of the inductance and the capacitance, not on the resistance. In Fig. 21.13, the maximum current occurs at the resonant frequency for any value of R. However, the value of the maximum current depends on R since $Z = R$ at resonance. The resonance peak is higher for a smaller resistance. If we measure the width of a resonance peak where the amplitude of the current has half its maximum value, we see that the resonance peaks get narrower with decreasing resistance.

Resonance in an *RLC* circuit is analogous to resonance in mechanical oscillations (see Section 10.10 and Table 21.1). Just as a mass-spring system has a single resonant frequency, determined by the spring constant and the mass, the *RLC* circuit has a single resonant frequency, determined by the capacitance and the inductance. When either system is driven externally—by a sinusoidal applied force for the mass-spring or by a sinusoidal applied emf for the circuit—the amplitude of the system's response is greatest when driven at the resonant frequency. In both systems, energy is being converted back and forth between two forms. For the mass-spring, the two forms are kinetic and elastic potential energy; for the *RLC* circuit, the two forms are electric energy stored in the capacitor and magnetic energy stored in the inductor. The resistor in the *RLC* circuit fills the role of friction in a mass-spring system: dissipating energy.

CONNECTION:

Resonance in *RLC* circuits and in mechanical systems

Application: Tuning Circuits A sharp resonance peak enables the tuning circuit in a TV or radio to select one out of many different frequencies being broadcast. With one type of tuner, common in old radios, the tuning knob adjusts the capacitance by rotating one set of parallel plates relative to a fixed set so that the area of overlap is varied (Fig. 21.15). By changing the capacitance, the resonant frequency can be varied. The tuning circuit is driven by a mixture of many different frequencies coming from the antenna, but only frequencies very near the resonance frequency produce a significant response in the tuning circuit.

Table 21.1 Analogy Between *RLC* Oscillations and Mechanical Oscillations

RLC	Mechanical
$q, i, \Delta i/\Delta t$	x, v_x, a_x
$\frac{1}{C}, R, L$	k, b, m
$\frac{1}{C}q + Ri + L\,\Delta i/\Delta t = 0$	$kx + bv_x + ma_x = 0$
$\frac{1}{2}\left(\frac{1}{C}\right)q^2$	$\frac{1}{2}kx^2$
$\frac{1}{2}Li^2$	$\frac{1}{2}mv_x^2$
Ri^2	bv_x^2
$\omega_0 = \sqrt{\frac{1/C}{L}}$	$\omega_0 = \sqrt{\frac{k}{m}}$

Figure 21.15 The variable capacitor inside an old radio. The radio is tuned to a particular resonant frequency by adjusting the capacitance. This is done by rotating the knob which changes the overlap of the two sets of plates.

Example 21.6

A Tuner for a Radio

A radio tuner has a 400.0-Ω resistor, a 0.50-mH inductor, and a variable capacitor connected in series. Suppose the capacitor is adjusted to 72.0 pF. (a) Find the resonant frequency for the circuit. (b) Find the reactances of the inductor and capacitor at the resonant frequency. (c) The applied emf at the resonant frequency coming in from the antenna is 20.0 mV (rms). Find the rms current in the tuning circuit. (d) Find the rms voltages across each of the circuit elements.

Strategy The resonant frequency can be found from the values of the capacitance and the inductance. The reactances at the resonant frequency must be equal. To find the current in the circuit, we note that the impedance is equal to the resistance since the circuit is in resonance. The rms current is the ratio of the rms voltage to the impedance. The rms voltage across a circuit element is the rms current times the element's reactance or resistance.

Solution (a) The resonant angular frequency is given by

$$\omega_0 = \frac{1}{\sqrt{LC}} = \frac{1}{\sqrt{0.50 \times 10^{-3}\ \text{H} \times 72.0 \times 10^{-12}\ \text{F}}} = 5.27 \times 10^6\ \text{rad/s}$$

The resonant frequency in Hz is

$$f_0 = \frac{\omega_0}{2\pi} = 840\ \text{kHz}$$

(b) The reactances are

$$X_L = \omega L = 5.27 \times 10^6\ \text{rad/s} \times 0.50 \times 10^{-3}\ \text{H} = 2.6\ \text{k}\Omega$$

and

$$X_C = \frac{1}{\omega C} = \frac{1}{5.27 \times 10^6\ \text{rad/s} \times 72.0 \times 10^{-12}\ \text{F}} = 2.6\ \text{k}\Omega$$

They are equal, as expected.

(c) At the resonant frequency, the impedance is equal to the resistance.

$$Z = R = 400.0\ \Omega$$

The rms current is

$$I_{rms} = \frac{\mathscr{E}_{rms}}{Z} = \frac{20.0\ \text{mV}}{400.0\ \Omega} = 0.0500\ \text{mA}$$

(d) The rms voltages are

$$V_{L\text{-rms}} = I_{rms} X_L = 0.0500\ \text{mA} \times 2.6 \times 10^3\ \Omega = 130\ \text{mV}$$

$$V_{C\text{-rms}} = I_{rms} X_C = 0.0500\ \text{mA} \times 2.6 \times 10^3\ \Omega = 130\ \text{mV}$$

$$V_{R\text{-rms}} = I_{rms} R = 0.0500\ \text{mA} \times 400.0\ \Omega = 20.0\ \text{mV}$$

Discussion The resonant frequency of 840 kHz is a reasonable result since it lies in the AM radio band (530–1700 kHz).

The rms voltages across the inductor and across the capacitor are equal at resonance, but the instantaneous voltages are opposite in phase (a phase difference of π rad or 180°), so the sum of the potential difference across the two is always zero. In a phasor diagram, the phasors for v_L and v_C are opposite in direction and equal in length, so they add to zero. Then the voltage across the resistor is equal to the applied emf in both amplitude and phase.

Practice Problem 21.6 Tuning the Radio to a Different Station

Find the capacitance required to tune to a station broadcasting at 1420 kHz.

21.7 CONVERTING AC TO DC; FILTERS

Diodes

The circuit symbol for a diode is ⊳|. The arrow indicates the direction of current flow.

Application: rectifiers

A *diode* is a circuit component that allows current to flow much more easily in one direction than in the other. An *ideal* diode has zero resistance for current in one direction, so that the current flows without any voltage drop across the diode, and infinite resistance for current in the other direction, so that no current flows. The circuit symbol for a diode has an arrowhead to indicate the direction of allowed current.

The circuit in Fig. 21.16a is called a *half-wave rectifier.* If the input is a sinusoidal emf, the output (the voltage across the resistor) is as shown in Fig. 21.16b. The output signal can be smoothed out by a capacitor (Fig. 21.16c). The capacitor charges up when

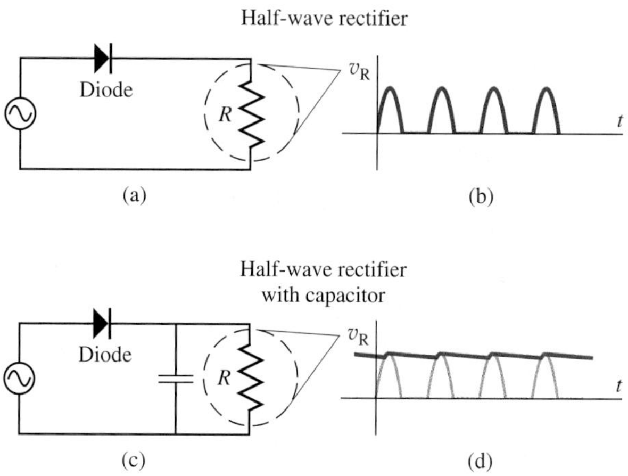

Figure 21.16 (a) A half-wave rectifier. (b) The voltage across the resistor. When the input voltage is negative, the output voltage v_R is zero, so the negative half of the "wave" has been cut off. (c) A capacitor inserted to smooth the output voltage. (d) The dark graph line shows the voltage across the resistor, assuming the RC time constant is much larger than the period of the sinusoidal input voltage. The light graph line shows what the output would have been without the capacitor.

current flows through the diode; when the source voltage starts to drop and then changes polarity, the capacitor discharges through the resistor. (The capacitor cannot discharge through the diode because that would send current the wrong way through the diode.) The discharge keeps the voltage v_R up. By making the RC time constant ($\tau = RC$) long enough, the discharge through the resistor can be made to continue until the source voltage turns positive again (Fig. 21.16d).

Circuits involving more than one diode can be arranged to make a *full-wave rectifier.* The output of a full-wave rectifier (without a capacitor to smooth it) is shown in Fig. 21.17a. Circuits like these are found inside the ac adapter used with devices such as portable CD players, radios, and laptop computers (Fig. 21.17b). Many other devices have circuits to do ac-to-dc conversion inside of them.

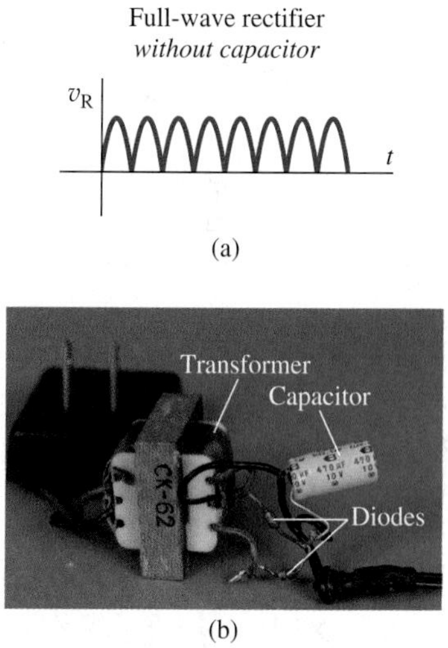

Figure 21.17 (a) Output of a full-wave rectifier. (b) This ac adapter from a portable CD player contains a transformer (labeled "CK-62") to reduce the amplitude of the ac source voltage. The two red diodes serve as a full-wave rectifier circuit and the capacitor (labeled "470 μF") smooths out the ripples. The output is a nearly constant dc voltage.

Filters

The capacitor in Fig. 21.16c serves as a *filter.* Figure 21.18 shows two *RC filters* commonly used in circuits. Figure 21.18a is a *low-pass filter.* For a high-frequency ac signal, the capacitor serves as a low reactance path to ground ($X_C \ll R$); the voltage across the resistor is much larger than the voltage across the capacitor, so the voltage across the output terminals is a small fraction of the input voltage. For a low-frequency signal, $X_C \gg R$, so the output voltage is nearly as great as the input voltage. For a signal consisting of a mixture of frequencies, the high frequencies are "filtered out" while the low frequencies "pass through."

The *high-pass filter* of Fig. 21.18b does just the opposite. Suppose a circuit connected to the input terminals supplies a mixture of a dc potential difference plus ac voltages at a range of frequencies. The reactance of the capacitor is large at low frequencies, so most of the voltage drop for low frequencies occurs across the capacitor; most of the high-frequency voltage drop occurs across the resistor and thus across the output terminals.

Combinations of capacitors and inductors are also used as filters. For both RC and LC filters, there is a gradual transition between frequencies that are blocked and frequencies

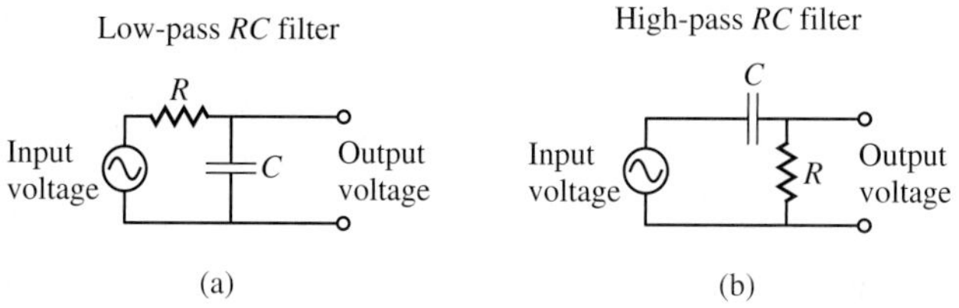

Figure 21.18 Two RC filters: (a) low-pass and (b) high-pass.

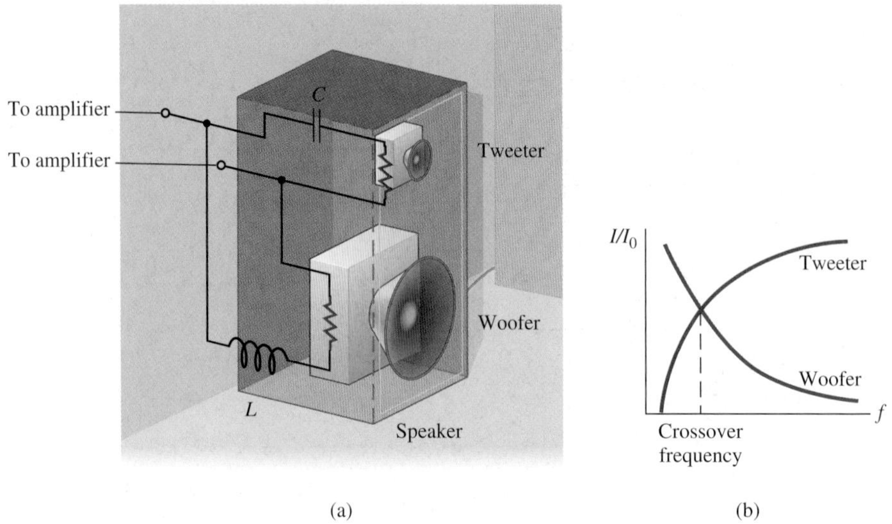

Figure 21.19 (a) Two speaker drivers are connected to an amplifier by a crossover network. (b) The amplitude of the current I going to each of the drivers (expressed as a fraction of the input amplitude I_0), graphed as a function of frequency.

that pass through. The frequency range where the transition occurs can be selected by choosing the values of R and C (or L and C).

Application: Crossover Networks A speaker used with an audio system, often has two vibrating cones (the *drivers*) that produce the sounds. A *crossover network* (Fig. 21.19) separates the signal from the amplifier, sending the low frequencies to the woofer and the high frequencies to the tweeter.

Master the Concepts

- In the equation

$$v = V \sin(\omega t + \phi)$$

the lowercase letter (v) represents the instantaneous voltage while the uppercase letter (V) represents the *amplitude* (peak value) of the voltage. The quantity ϕ is called the *phase constant.*
- The *rms value* of a sinusoidal quantity is $1/\sqrt{2}$ times the amplitude.
- *Reactances* (X_C, X_L) and *impedance* (Z) are generalizations of the concept of resistance and are measured in ohms. The amplitude of the voltage across a circuit element or combination of elements is equal to the amplitude of the current through the element(s) times the reactance or impedance of the element(s). Except for a resistor, there is a phase difference between the voltage and current:

	Amplitude	Phase
Resistor	$V_R = IR$	v_R, i are in phase
Capacitor	$V_C = IX_C$ $X_C = 1/(\omega C)$	i leads v_C by 90°
Inductor	$V_L = IX_L$ $X_L = \omega L$	v_L leads i by 90°
RLC series circuit	$\mathscr{E}_m = IZ$ $Z = \sqrt{R^2 + (X_L - X_C)^2}$	$\mathscr{E}$ leads/lags i by $\phi = \tan^{-1} \dfrac{X_L - X_C}{R}$

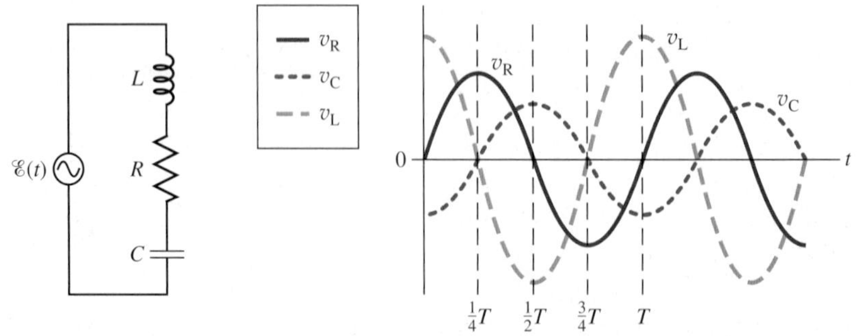

continued on next page

Master the Concepts continued

- The average power dissipated in a resistor is

$$P_{av} = I_{rms} V_{rms} = I_{rms}^2 R = \frac{V_{rms}^2}{R} \quad (21\text{-}4)$$

The average power dissipated in an ideal capacitor or ideal inductor is zero.

- The average power dissipated in a series *RLC* circuit can be written

$$P_{av} = I_{rms} \mathscr{E}_{rms} \cos \phi \quad (21\text{-}17)$$

where ϕ is the phase difference between $i(t)$ and $\mathscr{E}(t)$. The *power factor* $\cos \phi$ is equal to R/Z.

- To add sinusoidal voltages, we can represent each voltage by a vector-like object called a *phasor*. The magnitude of the phasor represents the amplitude of the voltage; the angle of the phasor represents the phase constant of the voltage. We can then add phasors the same way we add vectors.

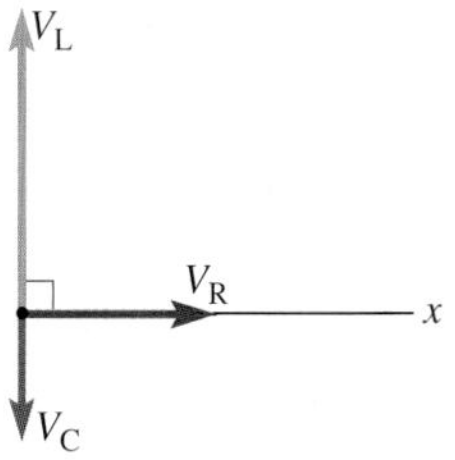

- The angular frequency at which *resonance* occurs in a series *RLC* circuit is

$$\omega_0 = \frac{1}{\sqrt{LC}} \quad (21\text{-}18)$$

At resonance, the current amplitude has its maximum value, the capacitive reactance is equal to the inductive reactance, and the impedance is equal to the resistance. If the resistance in the circuit is small, the resonance curve (the graph of current amplitude as a function of frequency) has a sharp peak. By adjusting the resonant frequency, such a circuit can be used to select a narrow range of frequencies from a signal consisting of a broad range of frequencies, as in radio or TV broadcasting.

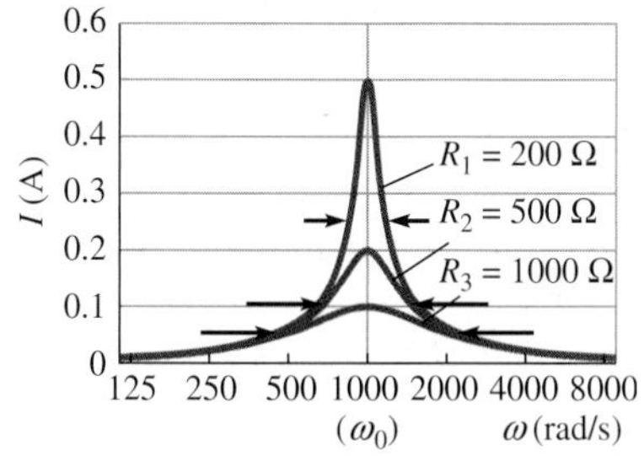

- An *ideal* diode has zero resistance for current in one direction, so that the current flows without any voltage drop across the diode, and infinite resistance for current in the other direction, so that no current flows. Diodes can be used to convert ac to dc.
- Capacitors and inductors can be used to make filters to selectively remove unwanted high or low frequencies from an electrical signal.

Conceptual Questions

1. Explain why there is a phase difference between the current in an ac circuit and the potential difference across a capacitor in the same circuit.
2. Electric power is distributed long distances over transmission lines by using high ac voltages and therefore small ac currents. What is the advantage of using high voltages instead of safer low voltages?
3. Explain the differences between average current, rms current, and peak current in an ac circuit.
4. The United States and Canada use 120 V rms as the standard household voltage, while most of the rest of the world uses 240 V rms for the household standard. What are the advantages and disadvantages of the two systems?
5. Some electric appliances are able to operate equally well with either dc or ac voltage sources, but other appliances require one type of source or the other and cannot run on both. Explain and give a few examples of each type of appliance.
6. For an ideal inductor in an ac circuit, explain why the voltage across the inductor must be zero when the current is maximum.
7. For a capacitor in an ac circuit, explain why the current must be zero when the voltage across the capacitor is maximum.
8. An electric heater is plugged into an ac outlet. Since the ac current changes polarity, there is no net movement of electrons through the heating element; the electrons just tend to oscillate back and forth. How, then, does the heating element heat up? Don't we need to send electrons *through* the element? Explain.
9. An electric appliance is rated 120 V, 5 A, 500 W. The first two are rms values; the third is the average power consumption. Why is the power not 600 W (= 120 V × 5 A)?
10. How does adjusting the tuning knob on a radio tune in different stations?
11. A circuit has a resistor and an unknown component in series with a 12-V (rms) sinusoidal ac source. The current in the circuit decreases by 20% when the frequency decreases from 240 Hz to 160 Hz. What is the second component in the circuit? Explain your reasoning.
12. What happens if a 40-W lightbulb, designed to be connected to an ac voltage with amplitude 170 V and frequency 60 Hz, is instead connected to a 170-V dc power supply? Explain. What dc voltage would make the lightbulb burn with the same brightness as the 170 V peak 60-Hz ac?
13. How can the lights in a home be dimmed using a coil of wire and a soft iron core?
14. Explain what is meant by a *phase difference.* Sketch graphs of $i(t)$ and $v_C(t)$, given that the current leads the voltage by $\pi/2$ radians.

15. What does it mean if the power factor is 1? What does it mean if it is zero?
16. A circuit has a resistor and an unknown component in series with a 12-V (rms) sinusoidal ac source. The current in the circuit decreases by 25% when the frequency increases from 150 Hz to 250 Hz. What is the second component in the circuit? Explain your reasoning.
17. Suppose you buy a 120-W lightbulb in Europe (where the rms voltage is 240 V). What happens if you bring it back to the United States (where the rms voltage is 120 V) and plug it in?
18. Let's examine the crossover network of Fig. 21.19 in the limiting cases of very low and very high frequencies. (a) How do the reactances of the capacitor and inductor compare for very low frequencies? (b) How do the rms currents through the tweeter and woofer compare for very low frequencies? (c) Answer these two questions in the case of very high frequencies. (d) With what should a frequency be compared to determine if it is "very low" or "very high"? (tutorial: reactance)

Multiple-Choice Questions

1. Graphs (1, 2) could represent:

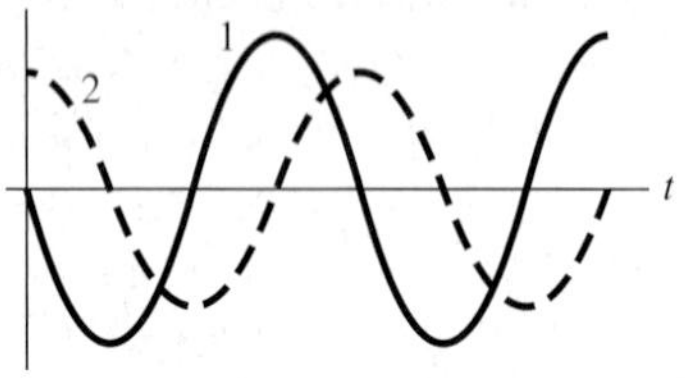

(a) the (1-voltage, 2-current) for a capacitor in an ac circuit.
(b) the (1-current, 2-voltage) for a capacitor in an ac circuit.
(c) the (1-voltage, 2-current) for a resistor in an ac circuit.
(d) the (1-current, 2-voltage) for a resistor in an ac circuit.
(e) the (1-voltage, 2-current) for an inductor in an ac circuit.
(f) the (1-current, 2-voltage) for an inductor in an ac circuit.
(g) either (a) or (e).
(h) either (a) or (f).
(i) either (b) or (e).
(j) either (b) or (f).

2. For a capacitor in an ac circuit, how much energy is stored in the capacitor at the instant when current is zero?
(a) zero
(b) maximum
(c) half of the maximum amount
(d) $1/\sqrt{2}\times$ the maximum amount
(e) impossible to answer without being given the phase angle

3. For an ideal inductor in an ac circuit, how much energy is stored in the inductor at the instant when current is zero?
(a) zero
(b) maximum
(c) half of the maximum amount
(d) $1/\sqrt{2}\times$ the maximum amount
(e) impossible to tell without being given the phase angle

4. For an ideal inductor in an ac circuit, the current through the inductor
(a) is in phase with the induced emf.
(b) leads the induced emf by 90°.
(c) leads the induced emf by an angle less than 90°.
(d) lags the induced emf by 90°.
(e) lags the induced emf by an angle less than 90°.

5. A capacitor is connected to the terminals of a variable frequency oscillator. The peak voltage of the source is kept fixed while the frequency is increased. Which statement is true?
(a) The rms current through the capacitor increases.
(b) The rms current through the capacitor decreases.
(c) The phase relation between the current and source voltage changes.
(d) The current stops flowing when the frequency change is large enough.

6. A voltage of $v(t) = (120\text{ V})\sin[(302\text{ rad/s})t]$ is produced by an ac generator. What is the rms voltage and the frequency of the source?
(a) 170 V and 213 Hz
(b) 20 V and 427 Hz
(c) 60 V and 150 Hz
(d) 85 V and 48 Hz

7. An ac source is connected to a series combination of a resistor, capacitor, and an inductor. Which statement is correct?
(a) The current in the capacitor leads the current in the inductor by 180°.
(b) The current in the inductor leads the current in the capacitor by 180°.
(c) The current in the capacitor and the current in the resistor are in phase.
(d) The voltage across the capacitor and the voltage across the resistor are in phase.

8. A series *RLC* circuit is connected to an ac generator. When the generator frequency varies (but the peak emf is constant), the average power is:
(a) a minimum when $|X_L - X_C| = R$.
(b) a minimum when $X_C = X_L$.
(c) equal to $I_{rms}^2 R$ only at the resonant frequency.
(d) equal to $I_{rms}^2 R$ at all frequencies.

Questions 9 and 10. The graphs show the peak current as a function of frequency for various circuit elements placed in the diagrammed circuit. The amplitude of the generator emf is constant, independent of the frequency.

9. Which graph is correct if the circuit element is a capacitor?

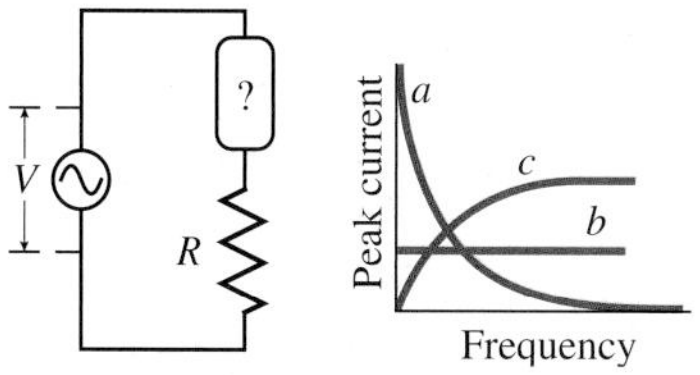

Multiple-Choice Questions 9 and 10

10. Which graph is correct if the circuit element is a resistor?

Problems

- Combination conceptual/quantitative problem
- Biological or medical application
- Challenging problem
- Blue # Detailed solution in the Student Solutions Manual
- 1 2 Problems paired by concept
- Text website interactive or tutorial

21.1 Sinusoidal Currents and Voltages: Resistors in ac Circuits; 21.2 Electricity in the Home

1. A lightbulb is connected to a 120-V (rms), 60-Hz source. How many times per second does the current reverse direction?
2. A European outlet supplies 220 V (rms) at 50 Hz. How many times per second is the magnitude of the voltage equal to 220 V?
3. A 1500-W heater runs on 120 V rms. What is the peak current through the heater? (tutorial: power in ac circuits)
4. A circuit breaker trips when the rms current exceeds 20.0 A. How many 100.0-W lightbulbs can run on this circuit without tripping the breaker? (The voltage is 120 V rms.)
5. A 1500-W electric hair dryer is designed to work in the United States, where the ac voltage is 120 V rms. What power is dissipated in the hair dryer when it is plugged into a 240-V rms socket in Europe? What may happen to the hair dryer in this case?
6. A 4.0-kW heater is designed to be connected to a 120-V rms source. What is the power dissipated by the heater if it is instead connected to a 120-V dc source?
7. (a) What rms current is drawn by a 4200-W electric room heater when running on 120 V rms? (b) What is the power dissipation by the heater if the voltage drops to 105 V rms during a brownout? Assume the resistance stays the same.
8. A television set draws an rms current of 2.50 A from a 60-Hz power line. Find (a) the average current, (b) the average of the square of the current, and (c) the amplitude of the current.
9. The instantaneous sinusoidal emf from an ac generator with an rms emf of 4.0 V oscillates between what values?
10. A hair dryer has a power rating of 1200 W at 120 V rms. Assume the hair dryer circuit contains only resistance. (a) What is the resistance of the heating element? (b) What is the rms current drawn by the hair dryer? (c) What is the maximum instantaneous power that the resistance must withstand?
11. Show that over one complete cycle, the average value of a sine function squared is $\frac{1}{2}$. [*Hint:* Use the following trigonometric identities: $\sin^2 a + \cos^2 a = 1$; $\cos 2a = \cos^2 a - \sin^2 a$.]
12. The diagram shows a simplified household circuit. Resistor $R_1 = 240.0\ \Omega$ represents a lightbulb; resistor $R_2 = 12.0\ \Omega$ represents a hair dryer. The resistors $r = 0.50\ \Omega$ (each) represent the resistance of the wiring in the walls. Assume that the generator supplies a constant 120.0 V rms. (a) If the lightbulb is on and the hair dryer is off, find the rms voltage across the lightbulb and the power dissipated by the lightbulb. (b) If both the lightbulb and the hair dryer are on, find the rms voltage across the lightbulb, the power dissipated by the lightbulb, and the rms voltage between point A and ground. (c) Explain why lights sometimes dim when an appliance is turned on. (d) Explain why the neutral and ground wires in a junction box are not at the same potential even though they are both grounded.

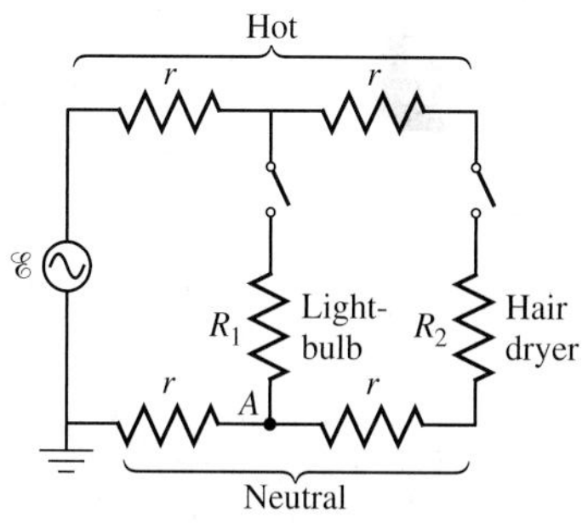

21.3 Capacitors in ac Circuits

13. A variable capacitor with negligible resistance is connected to an ac voltage source. How does the current in the circuit change if the capacitance is increased by a factor of 3.0 and the driving frequency is increased by a factor of 2.0?
14. At what frequency is the reactance of a 6.0-μF capacitor equal to 1.0 kΩ?
15. A 0.400-μF capacitor is connected across the terminals of a variable frequency oscillator. (a) What is the frequency when the reactance is 6.63 kΩ? (b) Find the reactance for half of that same frequency.

16. A 0.250-μF capacitor is connected to a 220-V rms ac source at 50.0 Hz. (a) Find the reactance of the capacitor. (b) What is the rms current through the capacitor?

17. A capacitor is connected across the terminals of a 115-V rms, 60.0-Hz generator. For what capacitance is the rms current 2.3 mA?

18. Show, from $X_C = 1/(\omega C)$, that the units of capacitive reactance are ohms.

19. A parallel plate capacitor has two plates, each of area $3.0 \times 10^{-4}\ m^2$, separated by 3.5×10^{-4} m. The space between the plates is filled with a dielectric. When the capacitor is connected to a source of 120 V rms at 8.0 kHz, an rms current of 1.5×10^{-4} A is measured. (a) What is the capacitive reactance? (b) What is the dielectric constant of the material between the plates of the capacitor?

20. A capacitor (capacitance = C) is connected to an ac power supply with peak voltage V and angular frequency ω. (a) During a quarter cycle when the capacitor goes from being uncharged to fully charged, what is the *average* current (in terms of C, V, and ω)? [*Hint:* $i_{av} = \Delta Q/\Delta t$.] (b) What is the rms current? (c) Explain why the average and rms currents are not the same.

21. Three capacitors (2.0 μF, 3.0 μF, 6.0 μF) are connected in series to an ac voltage source with amplitude 12.0 V and frequency 6.3 kHz. (a) What are the peak voltages across each capacitor? (b) What is the peak current that flows in the circuit?

22. A capacitor and a resistor are connected in parallel across an ac source. The reactance of the capacitor is equal to the resistance of the resistor. Assuming that $i_C(t) = I \sin \omega t$, sketch graphs of $i_C(t)$ and $i_R(t)$ on the same axes.

21.4 Inductors in ac Circuits

23. A variable inductor with negligible resistance is connected to an ac voltage source. How does the current in the inductor change if the inductance is increased by a factor of 3.0 and the driving frequency is increased by a factor of 2.0?

24. At what frequency is the reactance of a 20.0-mH inductor equal to 18.8 Ω?

25. What is the reactance of an air core solenoid of length 8.0 cm, radius 1.0 cm, and 240 turns at a frequency of 15.0 kHz?

26. A solenoid with a radius of 8.0×10^{-3} m and 200 turns/cm is used as an inductor in a circuit. When the solenoid is connected to a source of 15 V rms at 22 kHz, an rms current of 3.5×10^{-2} A is measured. Assume the resistance of the solenoid is negligible. (a) What is the inductive reactance? (b) What is the length of the solenoid?

27. A 4.00-mH inductor is connected to an ac voltage source of 151.0 V rms. If the rms current in the circuit is 0.820 A, what is the frequency of the source?

28. Two ideal inductors (0.10 H, 0.50 H) are connected in series to an ac voltage source with amplitude 5.0 V and frequency 126 Hz. (a) What are the peak voltages across each inductor? (b) What is the peak current that flows in the circuit?

29. Suppose that current flows to the *left* through the inductor in Fig. 21.6a so that i is negative. (a) If the current is increasing in magnitude, what is the sign of $\Delta i/\Delta t$? (b) In what direction is the induced emf that opposes the increase in current? (c) Show that Eq. (21-8) gives the correct sign for v_L. [*Hint:* v_L is positive if the left side of the inductor is at a higher potential than the right side.] (d) Repeat these three questions if the current flows to the left through the inductor and is *decreasing* in magnitude.

30. Suppose that an ideal capacitor and an ideal inductor are connected in series in an ac circuit. (a) What is the phase difference between $v_C(t)$ and $v_L(t)$? [*Hint:* Since they are in series, the same current $i(t)$ flows through both.] (b) If the rms voltages across the capacitor and inductor are 5.0 V and 1.0 V, respectively, what would an ac voltmeter (which reads rms voltages) connected across the series combination read?

31. The voltage across an inductor and the current through the inductor are related by $v_L = L\,\Delta i/\Delta t$. Suppose that $i(t) = I \sin \omega t$. (a) Write an expression for $v_L(t)$. [*Hint:* Use one of the relationships of Eq. (20-7).] (b) From your expression for $v_L(t)$, show that the reactance of the inductor is $X_L = \omega L$. (c) Sketch graphs of $i(t)$ and $v_L(t)$ on the same axes. What is the phase difference? Which one leads?

32. Make a figure analogous to Fig. 21.4 for an ideal *inductor* in an ac circuit. Start by assuming that the voltage across an ideal inductor is $v_L(t) = V_L \sin \omega t$. Make a graph showing one cycle of $v_L(t)$ and $i(t)$ on the same axes. Then, at each of the times $t = 0, \frac{1}{8}T, \frac{2}{8}T, \ldots, T$, indicate the direction of the current (or that it is zero), whether the current is increasing, decreasing, or (instantaneously) not changing, and the direction of the induced emf in the inductor (or that it is zero).

21.5 *RLC* Series Circuits

33. A 25.0-mH inductor, with internal resistance of 25.0 Ω, is connected to a 110-V rms source. If the average power dissipated in the circuit is 50.0 W, what is the frequency? (Model the inductor as an ideal inductor in series with a resistor.)

34. An inductor has an impedance of 30.0 Ω and a resistance of 20.0 Ω at a frequency of 50.0 Hz. What is the inductance? (Model the inductor as an ideal inductor in series with a resistor.)

35. A 6.20-mH inductor is one of the elements in a simple *RLC* series circuit. When this circuit is connected to a 1.60-kHz sinusoidal source with an rms voltage of

960.0 V, an rms current of 2.50 A lags behind the voltage by 52.0°. (a) What is the impedance of this circuit? (b) What is the resistance of this circuit? (c) What is the average power dissipated in this circuit?

36. A series combination of a resistor and a capacitor are connected to a 110-V rms, 60.0-Hz ac source. If the capacitance is 0.80 μF and the rms current in the circuit is 28.4 mA, what is the resistance?

37. A 300.0-Ω resistor and a 2.5-μF capacitor are connected in series across the terminals of a sinusoidal emf with a frequency of 159 Hz. The inductance of the circuit is negligible. What is the impedance of the circuit?

38. A series *RLC* circuit has a 0.20-mF capacitor, a 13-mH inductor, and a 10.0-Ω resistor, and is connected to an ac source with amplitude 9.0 V and frequency 60 Hz. (a) Calculate the voltage amplitudes V_L, V_C, V_R, and the phase angle. (b) Draw the phasor diagram for the voltages of this circuit.

39. (a) Find the power factor for the *RLC* series circuit of Example 21.4. (b) What is the average power delivered to each element (*R*, *L*, *C*)?

40. A computer draws an rms current of 2.80 A at an rms voltage of 120 V. The average power consumption is 240 W. (a) What is the power factor? (b) What is the phase difference between the voltage and current?

41. An *RLC* series circuit is connected to an ac power supply with a 12-V amplitude and a frequency of 2.5 kHz. If $R = 220\ \Omega$, $C = 8.0\ \mu F$, and $L = 0.15$ mH, what is the average power dissipated?

42. An ac circuit has a single resistor, capacitor, and inductor in series. The circuit uses 100 W of power and draws a maximum rms current of 2.0 A when operating at 60 Hz and 120 V rms. The capacitive reactance is 0.50 times the inductive reactance. (a) Find the phase angle. (b) Find the values of the resistor, the inductor, and the capacitor.

43. Suppose that two sinusoidal voltages at the same frequency are added:

$$V_1 \sin \omega t + V_2 \sin (\omega t + \phi_2) = V \sin (\omega t + \phi)$$

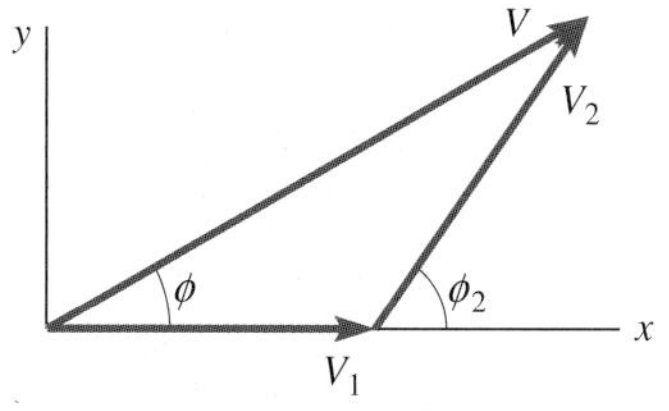

A phasor representation is shown in the diagram. (a) Substitute $t = 0$ into the equation. Interpret the result by referring to the phasor diagram. (b) Substitute $t = \pi/(2\omega)$ and simplify using the trigonometric identity $\sin(\theta + \pi/2) = \cos\theta$. Interpret the result by referring to the phasor diagram.

44. An ac circuit contains a 12.5-Ω resistor, a 5.00-μF capacitor, and a 3.60-mH inductor connected in series to an ac generator with an output voltage of 50.0 V (peak) and frequency of 1.59 kHz. Find the impedance, the power factor, and the phase difference between the source voltage and current for this circuit.

45. A 0.48-μF capacitor is connected in series to a 5.00-kΩ resistor and an ac source of voltage amplitude 2.0 V. (a) At $f = 120$ Hz, what are the voltage amplitudes across the capacitor and across the resistor? (b) Do the voltage amplitudes add to give the amplitude of the source voltage (i.e., does $V_R + V_C = 2.0$ V)? Explain. (c) Draw a phasor diagram to show the addition of the voltages.

46. A series combination of a 22.0-mH inductor and a 145.0-Ω resistor are connected across the output terminals of an ac generator with peak voltage 1.20 kV. (a) At $f = 1250$ Hz, what are the voltage amplitudes across the inductor and across the resistor? (b) Do the voltage amplitudes add to give the source voltage (i.e., does $V_R + V_L = 1.20$ kV)? Explain. (c) Draw a phasor diagram to show the addition of the voltages.

47. A 3.3-kΩ resistor is in series with a 2.0-μF capacitor in an ac circuit. The rms voltages across the two are the same. (a) What is the frequency? (b) Would each of the rms voltages be half of the rms voltage of the source? If not, what fraction of the source voltage are they? (In other words, $V_R/\mathscr{E}_m = V_C/\mathscr{E}_m = ?$) [*Hint:* Draw a phasor diagram.] (c) What is the phase angle between the source voltage and the current? Which leads? (d) What is the impedance of the circuit?

48. A 150-Ω resistor is in series with a 0.75-H inductor in an ac circuit. The rms voltages across the two are the same. (a) What is the frequency? (b) Would each of the rms voltages be half of the rms voltage of the source? If not, what fraction of the source voltage are they? (In other words, $V_R/\mathscr{E}_m = V_L/\mathscr{E}_m = ?$) (c) What is the phase angle between the source voltage and the current? Which leads? (d) What is the impedance of the circuit?

49. A series circuit with a resistor and a capacitor has a time constant of 0.25 ms. The circuit has an impedance of 350 Ω at a frequency of 1250 Hz. What are the capacitance and the resistance?

50. (a) What is the reactance of a 10.0-mH inductor at the frequency $f = 250.0$ Hz? (b) What is the impedance of a series combination of the 10.0-mH inductor and a 10.0-Ω resistor at 250.0 Hz? (c) What is the maximum current through the same circuit when the ac voltage source has a peak value of 1.00 V? (d) By what angle does the current lag the voltage in the circuit?

21.6 Resonance in an *RLC* Circuit

51. The FM radio band is broadcast between 88 MHz and 108 MHz. What range of capacitors must be used to tune in these signals if an inductor of 3.00 μH is used?

52. An *RLC* series circuit is built with a variable capacitor. How does the resonant frequency of the circuit change when the area of the capacitor is increased by a factor of 2?

53. A series RLC circuit has $R = 500.0\ \Omega$, $L = 35.0$ mH, and $C = 87.0$ pF. What is the impedance of the circuit at resonance? Explain.

54. In an RLC series circuit, these three elements are connected in series: a resistor of 60.0 Ω, a 40.0-mH inductor, and a 0.0500-F capacitor. The series elements are connected across the terminals of an ac oscillator with an rms voltage of 10.0 V. Find the resonant frequency for the circuit.

55. An RLC series circuit is driven by a sinusoidal emf at the circuit's resonant frequency. (a) What is the phase difference between the voltages across the capacitor and inductor? [*Hint:* Since they are in series, the same current $i(t)$ flows through them.] (b) At resonance, the rms current in the circuit is 120 mA. The resistance in the circuit is 20 Ω. What is the rms value of the applied emf? (c) If the frequency of the emf is changed without changing its rms value, what happens to the rms current? (tutorial: resonance)

56. An RLC series circuit has a resistance of $R = 325\ \Omega$, an inductance $L = 0.300$ mH, and a capacitance $C = 33.0$ nF. (a) What is the resonant frequency? (b) If the capacitor breaks down for peak voltages in excess of 7.0×10^2 V, what is the maximum source voltage amplitude when the circuit is operated at the resonant frequency?

57. An RLC series circuit has $L = 0.300$ H and $C = 6.00\ \mu$F. The source has a peak voltage of 440 V. (a) What is the angular resonant frequency? (b) When the source is set at the resonant frequency, the peak current in the circuit is 0.560 A. What is the resistance in the circuit? (c) What are the peak voltages across the resistor, the inductor, and the capacitor at the resonant frequency?

58. Finola has a circuit with a 4.00-kΩ resistor, a 0.750-H inductor, and a capacitor of unknown value connected in series to a 440.0-Hz ac source. With an oscilloscope, she measures the phase angle to be 25.0°. (a) What is the value of the unknown capacitor? (b) Finola has several capacitors on hand and would like to use one to tune the circuit to maximum power. Should she connect a second capacitor in parallel across the first capacitor or in series in the circuit? Explain. (c) What value capacitor does she need for maximum power?

59. Repeat Problem 38 for an operating frequency of 98.7 Hz. (a) What is the phase angle for this circuit? (b) Draw the phasor diagram. (c) What is the resonant frequency for this circuit?

21.7 Converting ac to dc; Filters

60. An RC filter is shown. The filter resistance R is variable between 180 Ω and 2200 Ω and the filter capacitance is $C = 0.086\ \mu$F. At what frequency is the output amplitude equal to $1/\sqrt{2}$ times the input amplitude if $R =$ (a) 180 Ω? (b) 2200 Ω? (c) Is this a low-pass or high-pass filter? Explain.

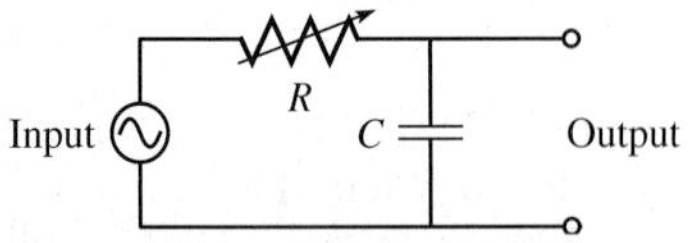

61. In the crossover network of the figure, the crossover frequency is found to be 252 Hz. The capacitance is $C = 560\ \mu$F. Assume the inductor to be ideal. (a) What is the impedance of the tweeter branch (the capacitor in series with the 8.0-Ω resistance of the tweeter) at the crossover frequency? (b) What is the impedance of the woofer branch at the crossover frequency? [*Hint:* The current amplitudes in the two branches are the same.] (c) Find L. (d) Derive an equation for the crossover frequency f_{co} in terms of L and C.

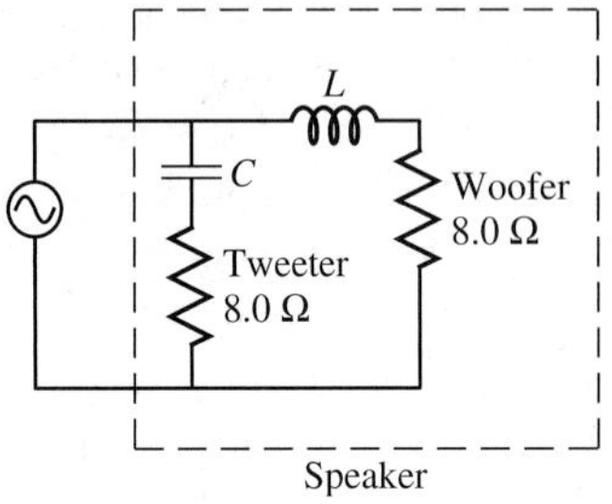

Problems 61 and 62

62. In the crossover network of Problem 61, the inductance L is 1.20 mH. The capacitor is variable; its capacitance can be adjusted to set the crossover point according to the frequency response of the woofer and tweeter. What should the capacitance be set to for a crossover point of 180 Hz? [*Hint:* At the crossover point, the currents are equal in amplitude.]

63. The circuit shown has a source voltage of 440 V rms, resistance $R = 250\ \Omega$, inductance $L = 0.800$ H, and capacitance $C = 2.22\ \mu$F. (a) Find the angular frequency ω_0 for resonance in this circuit. (b) Draw a phasor diagram for the circuit at resonance. (c) Find these rms voltages measured between various points in the circuit: V_{ab}, V_{bc}, V_{cd}, V_{bd}, and V_{ad}. (d) The resistor is replaced with one of $R = 125\ \Omega$. Now what is the angular frequency for resonance? (e) What is the rms current in the circuit operated at resonance with the new resistor?

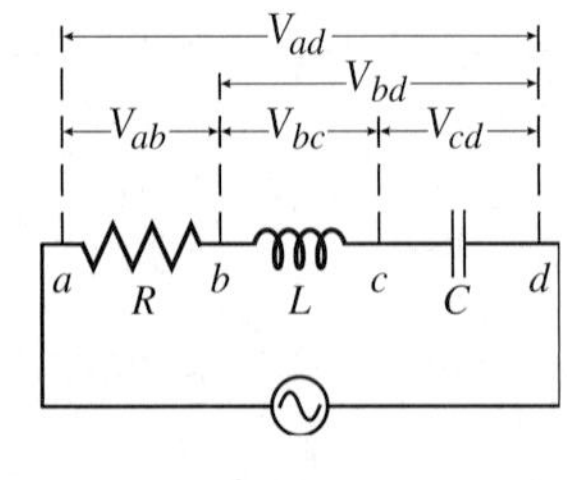

Comprehensive Problems

64. For a particular RLC series circuit, the capacitive reactance is 12.0 Ω, the inductive reactance is 23.0 Ω, and the maximum voltage across the 25.0-Ω resistor is 8.00 V. (a) What is the impedance of the circuit? (b) What is the maximum voltage across this circuit?

65. The phasor diagram for a particular *RLC* series circuit is shown in the figure. If the circuit has a resistance of 100 Ω and is driven at a frequency of 60 Hz, find (a) the current amplitude, (b) the capacitance, and (c) the inductance.

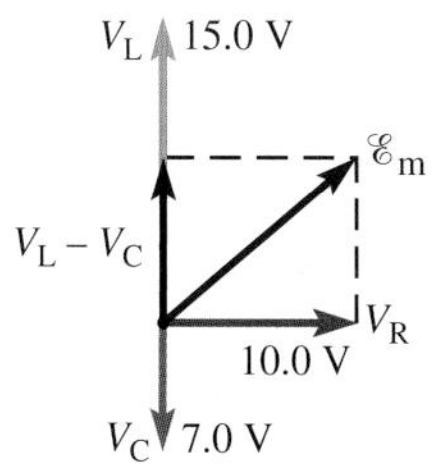

66. A portable heater is connected to a 60-Hz ac outlet. How many times per second is the instantaneous power a maximum?

67. What is the rms voltage of the oscilloscope trace of the figure, assuming that the signal is sinusoidal? The central horizontal line represents zero volts. The oscilloscope voltage knob has been clicked into its calibrated position.

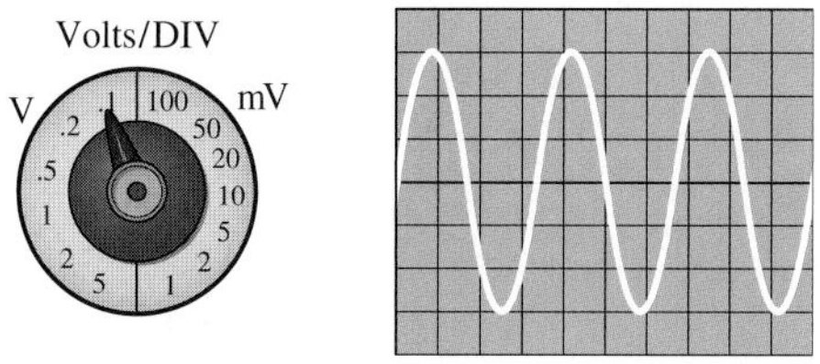

68. A 22-kV power line that is 10.0 km long supplies the electric energy to a small town at an average rate of 6.0 MW. (a) If a pair of aluminum cables of diameter 9.2 cm are used, what is the average power dissipated in the transmission line? (b) Why is aluminum used rather than a better conductor such as copper or silver?

69. An x-ray machine uses 240 kV rms at 60.0 mA rms when it is operating. If the power source is a 420-V rms line, (a) what must be the turns ratio of the transformer? (b) What is the rms current in the primary? (c) What is the average power used by the x-ray tube?

70. A coil with an internal resistance of 120 Ω and inductance of 12.0 H is connected to a 60.0-Hz, 110-V rms line. (a) What is the impedance of the coil? (b) Calculate the current in the coil.

71. The field coils used in an ac motor are designed to have a resistance of 0.45 Ω and an impedance of 35.0 Ω. What inductance is required if the frequency of the ac source is (a) 60.0 Hz? and (b) 0.20 kHz?

72. A capacitor is rated at 0.025 μF. How much rms current flows when the capacitor is connected to a 110-V rms, 60.0-Hz line?

73. A capacitor to be used in a radio is to have a reactance of 6.20 Ω at a frequency of 520 Hz. What is the capacitance?

74. An alternator supplies a peak current of 4.68 A to a coil with a negligibly small internal resistance. The voltage of the alternator is 420-V peak at 60.0 Hz. When a capacitor of 38.0 μF is placed in series with the coil, the power factor is found to be 1.00. Find (a) the inductive reactance of the coil and (b) the inductance of the coil.

75. At what frequency does the maximum current flow through a series *RLC* circuit containing a resistance of 4.50 Ω, an inductance of 440 mH, and a capacitance of 520 pF?

76. What is the rms current flowing in a 4.50-kW motor connected to a 220-V rms line when (a) the power factor is 1.00 and (b) when it is 0.80?

77. A variable capacitor is connected in series to an inductor with negligible internal resistance and of inductance 2.4×10^{-4} H. The combination is used as a tuner for a radio. If the lowest frequency to be tuned in is 0.52 MHz, what is the maximum capacitance required?

78. A large coil used as an electromagnet has a resistance of $R = 450$ Ω and an inductance of $L = 2.47$ H. The coil is connected to an ac source with a voltage amplitude of 2.0 kV and a frequency of 9.55 Hz. (a) What is the power factor? (b) What is the impedance of the circuit? (c) What is the peak current in the circuit? (d) What is the average power delivered to the electromagnet by the source?

79. An ac series circuit containing a capacitor, inductor, and resistance is found to have a current of amplitude 0.50 A for a source voltage of amplitude 10.0 V at an angular frequency of 200.0 rad/s. The total resistance in the circuit is 15.0 Ω. (a) What are the power factor and the phase angle for the circuit? (b) Can you determine whether the current leads or lags the source voltage? Explain.

80. A generator supplies an average power of 12 MW through a transmission line that has a resistance of 10.0 Ω. What is the power loss in the transmission line if the rms line voltage $\mathscr{E}_{rms}$ is (a) 15 kV and (b) 110 kV? What percentage of the total power supplied by the generator is lost in the transmission line in each case?

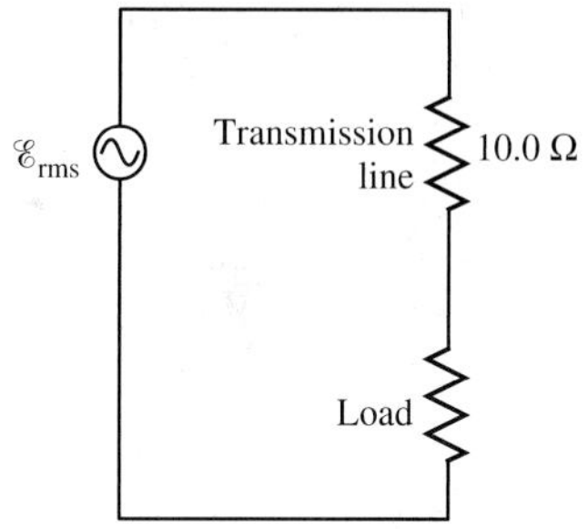

Problems 80 and 81

81. (a) Calculate the rms current drawn by the load in the figure with Problem 80 if $\mathscr{E}_{rms} = 250$ kV and the average power supplied by the generator is 12 MW. (b) Suppose that the average power supplied by the generator is still 12 MW, but the load is not purely resistive; rather, the load has a power factor of 0.86. What is the rms current drawn? (c) Why would the power company want to charge more in the second case, even though the average power is the same?

82. Transformers are often rated in terms of kilovolt-amps. A pole on a residential street has a transformer rated at 35 kV·A to serve four homes on the street. (a) If each home has a fuse that limits the incoming current to 60 A rms at 220 V rms, find the maximum load in kV·A on the transformer. (b) Is the rating of the transformer adequate? (c) Explain why the transformer rating is given in kV·A rather than in kW.

83. A variable inductor can be placed in series with a lightbulb to act as a dimmer. (a) What inductance would reduce the current through a 100-W lightbulb to 75% of its maximum value? Assume a 120-V rms, 60-Hz source. (b) Could a variable resistor be used in place of the variable inductor to reduce the current? Why is the inductor a much better choice for a dimmer?

84. A certain circuit has a 25-Ω resistor and one other component in series with a 12-V (rms) sinusoidal ac source. The rms current in the circuit is 0.317 A when the frequency is 150 Hz and increases by 25.0% when the frequency increases to 250 Hz. (a) What is the second component in the circuit? (b) What is the current at 250 Hz? (c) What is the numerical value of the second component?

85. A 40.0-mH inductor, with internal resistance of 30.0 Ω, is connected to an ac source

$$\mathscr{E}(t) = (286\text{ V}) \sin [(390\text{ rad/s})t]$$

(a) What is the impedance of the inductor in the circuit? (b) What are the peak and rms voltages across the inductor (including the internal resistance)? (c) What is the peak current in the circuit? (d) What is the average power dissipated in the circuit? (e) Write an expression for the current through the inductor as a function of time.

86. In an *RLC* circuit, these three elements are connected in series: a resistor of 20.0 Ω, a 35.0-mH inductor, and a 50.0-μF capacitor. The ac source of the circuit has an rms voltage of 100.0 V and an angular frequency of 1.0×10^3 rad/s. Find (a) the reactances of the capacitor and inductor, (b) the impedance, (c) the rms current, (d) the current amplitude, (e) the phase angle, and (f) the rms voltages across each of the circuit elements. (g) Does the current lead or lag the voltage? (h) Draw a phasor diagram.

87. (a) What is the reactance of a 5.00-μF capacitor at the frequencies $f = 12.0$ Hz and 1.50 kHz? (b) What is the impedance of a series combination of the 5.00-μF capacitor and a 2.00-kΩ resistor at the same two frequencies? (c) What is the maximum current through the circuit of part (b) when the ac source has a peak voltage of 2.00 V? (d) For each of the two frequencies, does the current lead or lag the voltage? By what angle?

88. An *RLC* series circuit is connected to a 240-V rms power supply at a frequency of 2.50 kHz. The elements in the circuit have the following values: $R = 12.0$ Ω, $C = 0.26$ μF, and $L = 15.2$ mH. (a) What is the impedance of the circuit? (b) What is the rms current? (c) What is the phase angle? (d) Does the current lead or lag the voltage? (e) What are the rms voltages across each circuit element?

Answers to Practice Problems

21.1 $V = 310$ V; $I = 17.0$ A; $P_{max} = 5300$ W; $P_{av} = 2600$ W; $R = 18$ Ω

21.2 9950 Ω; 22.1 mA

21.3 1.13 kΩ; 8.84 μA

21.4 $v_C(t) = (500\text{ mV}) \sin (\omega t - \pi/2)$,
$v_L(t) = (440\text{ mV}) \sin (\omega t + \pi/2)$, $v_R(t) = (80\text{ mV}) \sin \omega t$,
and $\mathscr{E}(t) = (100\text{ mV}) \sin (\omega t - 0.64)$.
At $t = 80.0$ μs, $\omega t = 0.800$ rad.
$v_C(t) = (500\text{ mV}) \sin (-0.771\text{ rad}) = -350$ mV,
$v_L(t) = (440\text{ mV}) \sin (2.371\text{ rad}) = +310$ mV,
$v_R(t) = (80\text{ mV}) \sin (0.80\text{ rad}) = +57$ mV,
and $\mathscr{E}(t) = (100\text{ mV}) \sin (0.16\text{ rad}) = +16$ mV.
$v_C + v_L + v_R = +17$ mV (discrepancy comes from roundoff error)

21.5 29 W

21.6 25 pF

Answers to Checkpoints

21.1 The average power is the product of the rms voltage and current: $P_{av} = I_{rms}V_{rms} = 10\text{ A} \times 120\text{ V} = 1200$ W.

21.4 The inductive reactance X_L increases with increasing frequency. The capacitive reactance X_C decreases with increasing frequency. (a) For $\omega > \omega_0$, $X_L > X_C$. (b) For $\omega < \omega_0$, $X_C > X_L$.

21.5 $\mathscr{E}_m = \sqrt{V_R^2 + (V_L - V_C)^2}$, so $V_R = \sqrt{\mathscr{E}_m^2 - (V_L - V_C)^2} = 30$ mV.

Review & Synthesis: Chapters 19–21

Review Exercises

1. A solenoid with 8500 turns per meter has radius 65 cm. The current in the solenoid is 25.0 A. A circular loop of wire with 100 turns and radius 8.00 cm is put inside the solenoid. The current in the circular loop is 2.20 A. What is the maximum possible magnetic torque on the loop? What orientation does the loop have if the magnetic torque has its maximum value?

2. Two long, straight wires, each with a current of 5.0 A, are placed on two corners of an equilateral triangle with sides of length 3.2 cm as shown. One of the wires has a current into the page and one has a current out of the page. (a) What is the magnetic field at the third corner of the triangle? (b) A proton has a velocity of 1.8×10^7 m/s out of the page when it crosses the plane of the page at the third corner of the triangle. What is the magnetic force on the proton at that point due to the two wires?

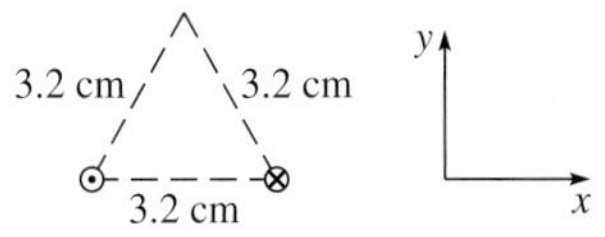

3. Two long, straight wires, each with a current of 12.0 A, are placed on two corners of an equilateral triangle with sides of length 2.50 cm as shown. Both of the wires have a current into the page. (a) What is the magnetic field at the third corner of the triangle? (b) Another wire is placed at the third corner, parallel to the other two wires. Which direction should current flow in the third wire so that the force on it is in the $+y$-direction. (c) If the third wire has a linear mass density of 0.150 g/m, what current should it have so that the magnetic force on the wire is equal in magnitude to the gravitational force, and the third wire can "hover" above the other two?

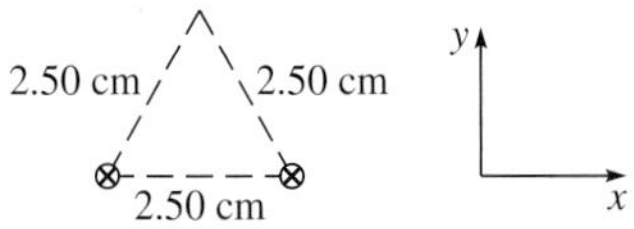

4. A loop of wire is connected to a battery and a variable resistor as shown. Two other loops of wire, B and C, are placed inside the large loop and outside the large loop, respectively. As the resistance in the variable resistor is increased, are there currents induced in the loops B and C? If so, do the currents circulate CW or CCW?

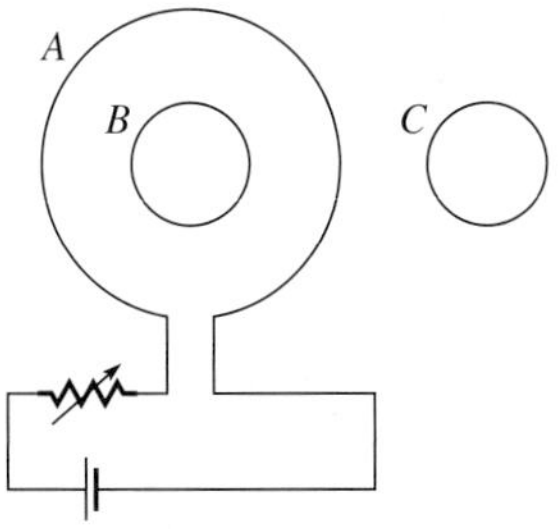

5. A cosmic ray muon with the same charge as an electron and a mass of 1.9×10^{-28} kg is moving toward the ground at an angle of 25° from the vertical with a speed of 7.0×10^7 m/s. As it crosses point P, the muon is at a horizontal distance of 85.0 cm from a high-voltage power line. At that moment, the power line has a current of 16.0 A. What is the magnitude and direction of the force on the muon at the point P in the diagram?

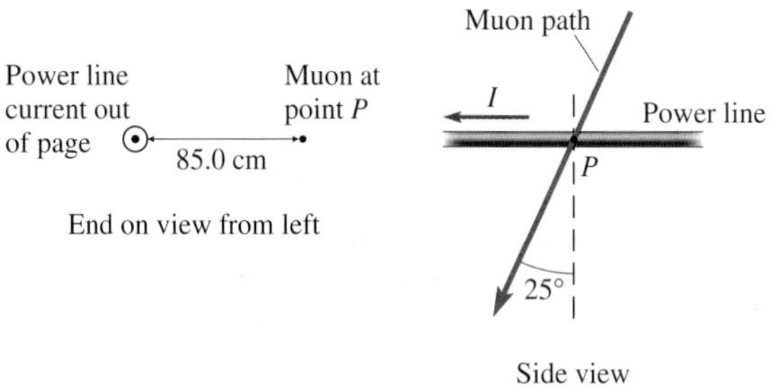

6. A variable capacitor is connected to an ac source. The rms current in the circuit is I_i. If the frequency of the source is reduced by a factor of 2.0 while the overlapping area of the capacitor plates is increased by a factor of 3.0, what will be the new rms current in the circuit? The resistance in the circuit is negligible.

7. Power lines carry electricity to your house at high voltage. This problem investigates the reason for that. Suppose a power plant produces 800 kW of power and wants to send that power for many miles over a copper wire with a total resistance of 12 Ω. (a) If the power is sent at a voltage of 120 V rms as used in houses in the United States, how much current flows through the copper wires? [*Hint:* The 12-Ω resistance of the wires is in series with the load in the house, and the 120-V rms voltage is connected across the series combination.] (b) What is the power dissipated due to the resistance of the copper wires? (c) If transformers are used so that the power is sent across the copper wires at 48 kV rms, how much current flows through the wires? (d) What is the power dissipated due to the resistance of the wires at this current? What percent of the total power output of the plant is this? (e) Although a series of transformers step the voltage down to the 120 V used for household voltage, assume you are using a single transformer to do the job. If the single transformer has 10,000 primary turns, how many secondary turns should it have?

8. A square loop of wire is made up of 50 turns of wire, 45 cm on each side. The loop is immersed in a 1.4-T magnetic field perpendicular to the plane of the loop. The loop of wire has little resistance but it is connected to two resistors in parallel as shown. (a) When the loop of wire is rotated by 180°, how much charge flows through the circuit? (b) How much charge goes through the 5.0-Ω resistor?

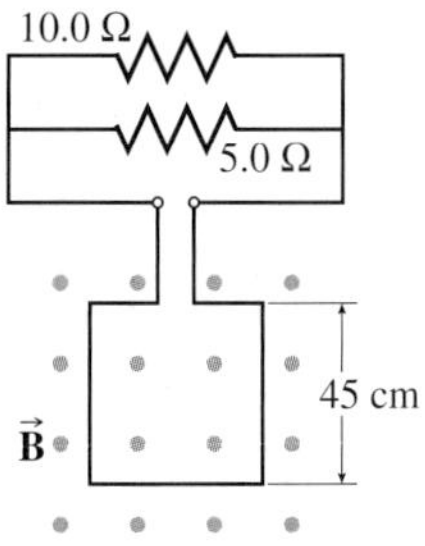

9. A circular loop of wire is placed near a long current carrying wire. What happens while the loop is moved in each of the three directions? Does current flow? If so, is it CW or CCW? In what direction does a magnetic force act on the loop, if any?

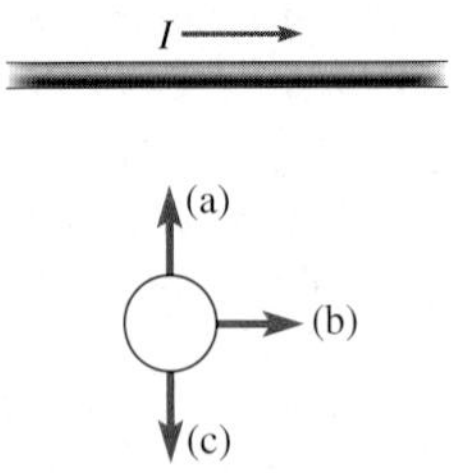

10. You are working as an electrical engineer designing transformers for transmitting power from a generating station producing 2.5×10^6 W to a city 120 km away. The power will be carried on two transmission lines to complete a circuit, each line constructed out of copper with a radius of 5.0 cm. (a) What is the total resistance of the transmission lines? (b) If the power is transmitted at 1200 V rms, find the average power dissipated in the wires. [*Hint:* See Section 20.2.] (c) The rms voltage is increased from 1200 V by a factor of 150 using a transformer with a primary coil of 1000 turns. How many turns are in the secondary coil? (d) What is the new rms current in the transmission lines after the voltage is stepped up with the transformer? (e) How much average power is dissipated in the transmission lines when using the transformer?

11. An electromagnetic rail gun can fire a projectile using a magnetic field and an electric current. Consider two conducting rails that are 0.500 m apart with a 50.0-g conducting projectile that slides along the two rails. A magnetic field of 0.750 T is directed perpendicular to the plane of the rails and points upward. A constant current of 2.00 A passes through the projectile. (a) What direction is the force on the projectile? (b) If the coefficient of kinetic friction between the rails and the projectile is 0.350, how fast is the projectile moving after it has traveled 8.00 m down the rails? (c) As the projectile slides down the rails, does the applied emf have to increase, decrease, or stay the same to maintain a constant current?

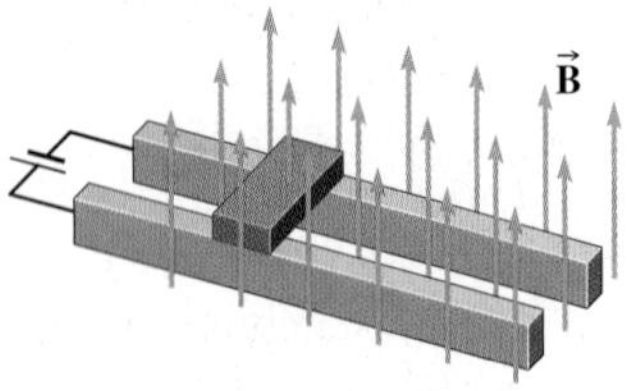

12. An air-filled parallel plate capacitor is used in a simple series *RLC* circuit along with a 0.650-H inductor. At a frequency of 220 Hz, the power output is found to be less than the maximum possible power output. After the space between the plates is filled with a dielectric with $\kappa = 5.50$, the circuit dissipates the maximum possible power. (a) What is the capacitance of the air-filled capacitor? (b) What was the resonant frequency of this circuit *before* inserting the dielectric?

13. (a) When the resistance of an *RLC* series circuit that is at resonance is doubled, what happens to the power dissipated? (b) Now consider an *RLC* series circuit that is not at resonance. For this circuit, the initial resistance and impedance are related by $R = X_C = X_L/2$. Determine how the power output changes when the resistance doubles for this circuit.

14. An *RLC* circuit has a resistance of 10.0 Ω, an inductance of 15.0 mH, and a capacitance of 350 μF. By what factor does the impedance of this circuit change when the frequency at which it is driven changes from 60 Hz to 120 Hz? Does the impedance increase or decrease?

15. An *RLC* circuit has a resistance of 255 Ω, an inductance of 146 mH, and a capacitance of 877 nF. (a) What is the resonant frequency of this circuit? (b) If this circuit is connected to a sinusoidal generator with a frequency 0.50 times the resonant frequency and a maximum voltage of 480 V, which will lead, the current or the voltage? (c) What is the phase angle of this circuit? (d) What is the rms current in this circuit? (e) How much average power is dissipated in this circuit? (f) What is the maximum voltage across each circuit element?

16. A variable inductor is connected to a voltage source whose frequency can vary. The rms current is I_i. If the inductance is increased by a factor of 3.0 and the frequency is reduced by a factor of 2.0, what will be the new rms current in the circuit? The resistance in the circuit is negligible.

17. Kieran measures the magnetic field of an electron beam. The beam strength is such that 1.40×10^{11} electrons pass a point every 1.30 μs. What magnetic field strength does Kieran measure at a distance of 2.00 cm from the beam center?

Problems 18–22. A mass spectrometer (see the figure) is designed to measure the mass *m* of the $^{238}U^+$ ion. A source of $^{238}U^+$ ions (not shown) sends ions into the device with negligibly small initial kinetic energies. The ions pass between parallel accelerating plates and then through a velocity selector designed to allow only ions moving at speed v to pass straight through. The ions that emerge from the velocity selector move in a semicircle of diameter *D* in a uniform magnetic field of magnitude *B*, which is the same as the magnetic field in the velocity selector. (Express your answers in terms of quantities given in the problems and universal constants as necessary.)

18. The accelerating plates have area *A* and are a distance *d* apart. (a) What should the charges on the plates be so the ions emerge at speed v, ignoring their initial kinetic energies? Indicate which plate is positive and which negative. (b) Sketch the electric field lines between the plates.

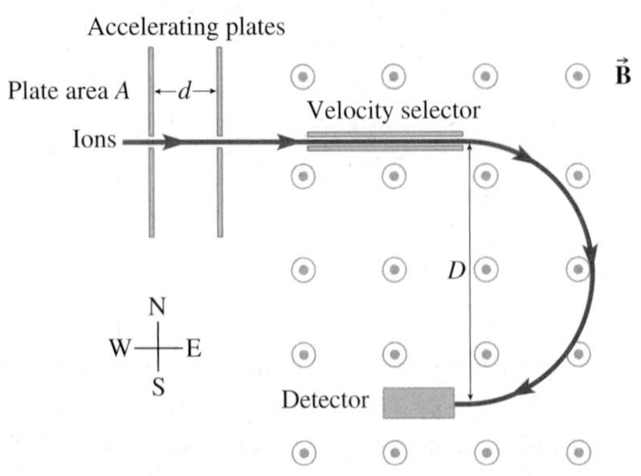

Problems 18–22

19. The uniform magnetic field in the velocity selector is directed out of the page and has magnitude B. (a) What should the magnitude and direction of the electric field in the selector be to allow ions with speed v to pass straight through? (b) Sketch the trajectory inside the velocity selector for ions that enter with speeds slightly less than v.

20. Find the mass of the $^{238}U^+$ ions in terms of v, B, D, and universal constants.

21. Suppose some $^{235}U^+$ ions are present in the beam. They have the same charge as the $^{238}U^+$ ions but a smaller mass (approximately $0.98737m$). (a) With what speed do the $^{235}U^+$ ions emerge from the accelerating plates, assuming $^{238}U^+$ ions emerge with speed v? (b) Sketch the trajectory of $^{235}U^+$ ions inside the velocity selector. (c) Now the velocity selector is removed. $^{238}U^+$ ions move in a circular path of diameter D in the uniform magnetic field. What is the diameter of the path of the $^{235}U^+$ ions?

22. Suppose some $^{238}U^{2+}$ ions are present in the beam. They have the same mass m as the $^{238}U^+$ ions but twice the charge ($+2e$). (a) With what speed do the $^{238}U^{2+}$ ions emerge from the accelerating plates, assuming $^{238}U^+$ ions emerge with speed v? (b) Sketch the trajectory of $^{238}U^{2+}$ ions inside the velocity selector. (c) Now the velocity selector is removed. $^{238}U^+$ ions move in a circular path of diameter D in the uniform magnetic field. What is the diameter of the path of the $^{238}U^{2+}$ ions?

23. A hydroelectric power plant is situated at the base of a large dam. Water flows into the intake near the bottom of the reservoir at a depth of 100 m. The water flows through 10 turbine generators and exits the power plant 120 m below the top of the reservoir at a speed of about 10 m/s (at atmospheric pressure). The average volume flow rate of water through each generator is 100 m^3/s. Each generator produces a peak voltage of 10 kV and operates with an energy efficiency of 80%. Estimate the maximum possible peak current that a single generator can supply.

24. A Faraday flashlight uses electromagnetic induction to produce energy when shaken. A magnet in the handle is free to slide back and forth through a loop with 50000 turns. The energy is stored in a 1.0-F capacitor, which can then be used to power a 0.50-W LED bulb. Suppose the area of the loop is 3.0 cm^2 and that the magnetic field in the loop when the magnet is at its farthest position is 1.0 T. A voltage rectifier is used to convert the induced ac emf to dc (so the capacitor is charging when the magnet moves either way). The resistance in the circuit is 500 Ω. Estimate how many shakes (one motion of the magnet back and forth) it takes to produce enough energy for 5.0 min of operation.

25. Consider an induction stove utilizing a primary heating coil located beneath the stove top. The coil is a solenoid with diameter 5.0 cm, length 1.0 cm, and 18 turns. The circuit elements in the stove supply the coil with a peak ac voltage of 340 V at a frequency of 50 kHz. The resistance of the coil is 1.0 Ω. (a) What average power is dissipated in the coil when the stove is turned on but with nothing on the stove top? (b) What average power must the stove deliver to 1.0 L of water initially at 20°C to bring it to boiling temperature in 5.0 min?

26. A toy race track has a 1.0-m-long straight section connected to a vertical circular loop-the-loop section of radius 15 cm. The straight section has a uniform magnetic field directed upward of magnitude 0.10 T and contains two metal strips with negligible resistance. The toy car has mass 40 g and contains a 2.0-cm-long perpendicular rod with resistance 100 mΩ that connects the strips when placed on the track. With a toy car placed on the starting line, a dc voltage is applied to the strips and the cars accelerate along the straight portion of the track. The operation is similar to that of a rail gun (see Problem 11). Ignoring friction, what minimum voltage must be applied to the strips for a toy car to make it around the loop-the-loop section without losing contact with the track?

MCAT Review

The section that follows includes MCAT exam material and is reprinted with permission of the Association of American Medical Colleges (AAMC).

Read the paragraph and then answer the following questions.

An electromagnetic railgun is a device that can fire projectiles using electromagnetic energy instead of chemical energy. A schematic of a typical railgun is shown here.

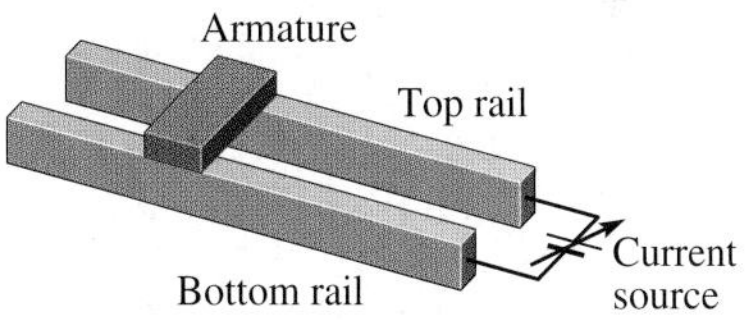

Schematic of a railgun

The operation of the railgun is simple. Current flows from the current source into the top rail, through a movable, conducting armature into the bottom rail, then back to the current source. The current in the two rails produces a magnetic field directly proportional to the amount of current. This field produces a force on the charges moving through the movable armature. The force pushes the armature and the projectile along the rails.

The force is proportional to the square of the current running through the railgun. For a given current, the force and the magnetic field will be constant along the entire length of the railgun. The detectors placed outside the railgun give off a signal when the projectile passes them. This information can be used to determine the exit speed v_i and kinetic energy of the projectile. The projectile mass, rail current, and exit speed for four different trials are listed in the table.

Projectile Mass (kg)	Rail Current (A)	Exit Speed (km/s)
0.01	10.0	2.0
0.01	15.0	3.0
0.02	10.0	1.4
0.04	10.0	1.0

1. Which of the following diagrams best represents the magnetic field created by the rail currents in the region between the rails?

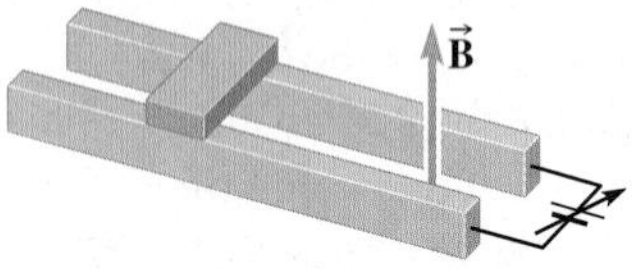

A.

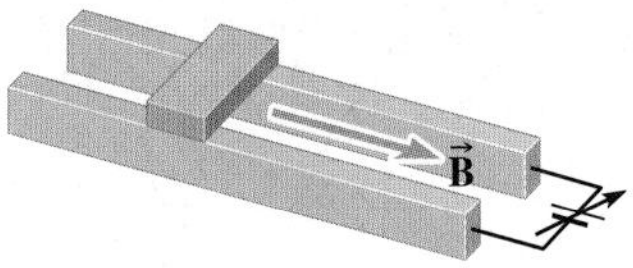

B.

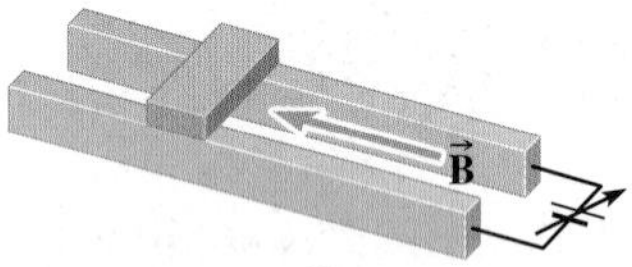

C.

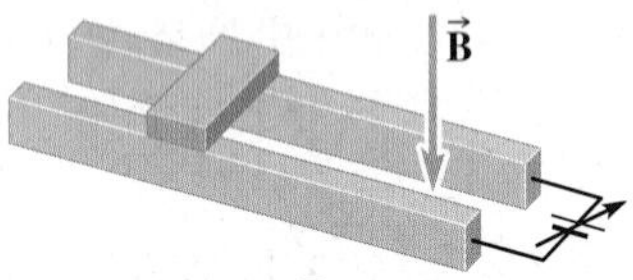

D.

2. For a given mass, if the current were decreased by a factor of 2, the new exit speed v would be equal to
   A. $2v_i$
   B. $\sqrt{2}v_i$
   C. $v_i/\sqrt{2}$
   D. $v_i/2$
3. Lengthening the rails would increase the exit speed because of
   A. an increased rail resistance.
   B. a stronger magnetic field between the rails.
   C. a larger force on the armature.
   D. a longer distance over which the force is present.
4. What change made to the railgun would reduce power consumption without lowering the exit speeds?
   A. Lowering the rail current
   B. Lowering the rail resistivity
   C. Lowering the rail cross-sectional area
   D. Reducing the magnetic field strength
5. If a projectile with a mass of 0.10 kg accelerates from a resting position to a speed of 10.0 m/s in 2.0 s, what will be the average power supplied by the railgun to the projectile?
   A. 0.5 W
   B. 2.5 W
   C. 5.0 W
   D. 10.0 W
6. A projectile with a mass of 0.08 kg that is propelled by a rail current of 20.0 A will have approximately what exit speed?
   A. 0.7 km/s
   B. 1.0 km/s
   C. 1.4 km/s
   D. 2.0 km/s

Questions 7 and 8. Refer to the three paragraphs about power being transmitted to consumers by utility companies in the MCAT Review section for Chapters 16–18. Based on those paragraphs, answer the following two questions.

7. When delivering a constant amount of power, why does the power lost to heat decrease as the transmission-line voltage increases?
   A. Increasing the voltage decreases the required current.
   B. Increasing the voltage increases the required current.
   C. Increasing the voltage decreases the required resistance.
   D. Increasing the voltage increases the required resistance.
8. Which of the following figures best illustrates the direction of the magnetic field ($\vec{B}$) associated with a section of wire carrying a current?

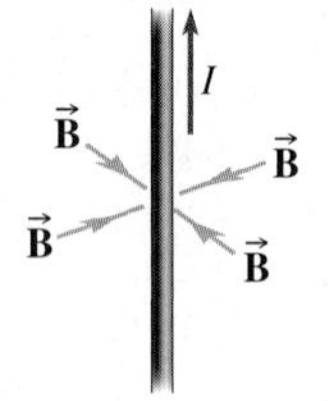

A.

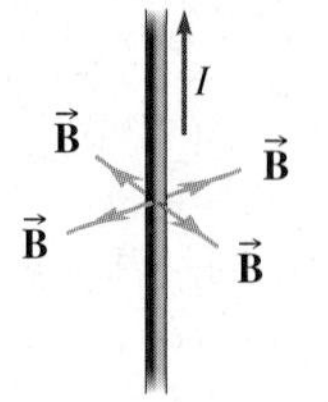

B.

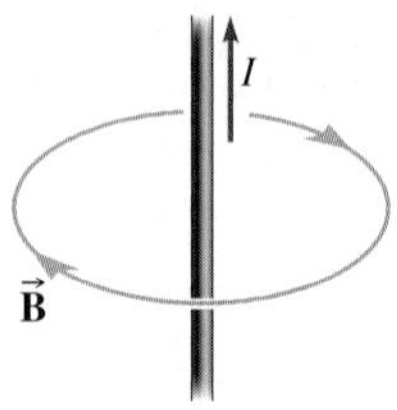

C.

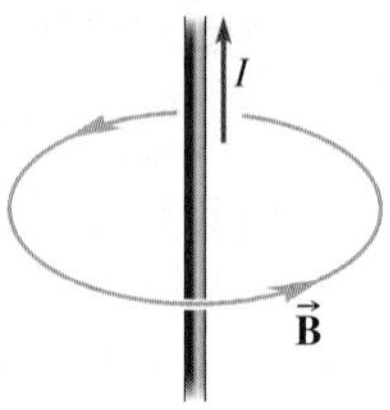

D.

CHAPTER

22

Electromagnetic Waves

Bees use the position of the Sun in the sky to navigate and find their way back to their hives. This is remarkable in itself—since the Sun moves across the sky during the day, the bees navigate with respect to a moving reference point rather than a fixed reference point. Even if the bees are kept in the dark for part of the day, they still navigate with reference to the Sun; they compensate for the motion of the Sun during the time they were in the dark. They must have some sort of internal clock that enables them to keep track of the Sun's motion.

What do they do when the Sun's position is obscured by clouds? Experiments have shown that the bees can still navigate as long as there is a patch of blue sky. How is this possible? (See p. 837 for the answer.)

Concepts & Skills to Review

- simple harmonic motion (Section 10.5)
- energy transport by waves; transverse waves; amplitude, frequency, wavelength, wavenumber, and angular frequency; equations for waves (Sections 11.1–11.5)
- Ampère's and Faraday's laws (Sections 19.9 and 20.3)
- dipoles (Sections 16.4 and 19.1)
- rms values (Section 21.1)
- thermal radiation (Section 14.8)
- Doppler effect (Section 12.8)
- relative velocity (Section 3.6)

22.1 MAXWELL'S EQUATIONS AND ELECTROMAGNETIC WAVES

Accelerating Charges Produce Electromagnetic Waves

EM waves are produced only by *accelerating* charges.

In our study of electromagnetism so far, we have considered the electric and magnetic fields due to charges whose accelerations are small. A point charge at rest gives rise to an electric field only. A charge moving at constant velocity gives rise to both electric and magnetic fields. Charges at rest or moving at constant velocity do not generate **electromagnetic waves**—waves that consist of oscillating electric and magnetic fields. Electromagnetic (EM) waves are produced only by charges that *accelerate.* EM waves, also called **electromagnetic radiation**, consist of oscillating electric and magnetic fields that travel away from the accelerating charges.

To create an EM wave that lasts longer than a pulse, the charges must continue to accelerate. Let's consider two point charges $\pm q$ that move in simple harmonic motion along the same line with the same amplitude and frequency but half a cycle out of phase. What do the electric and magnetic fields due to this oscillating electric dipole look like? The fields don't just look like oscillating versions of the fields of static electric and magnetic dipoles. The charges emit EM radiation because the oscillating fields affect each other. The magnetic field is not constant, since the motion of the charges is changing. According to Faraday's law of induction, a changing magnetic field induces an electric field. The electric field of the oscillating dipole at any instant is therefore different from the electric field of a static dipole. Faraday's law liberates the electric field lines: they do not have to start and end on the source charges. Instead, they can be closed loops far from the oscillating dipole.

According to Ampère's law, as we have stated it, the magnetic field lines must enclose the current that is their source. Scottish physicist James Clerk Maxwell (1831–1879) was puzzled by a lack of symmetry in the laws of electromagnetism. If a changing magnetic field gives rise to an electric field, might not a changing electric field give rise to a magnetic field? The answer turns out to be yes (see text website for more information). A changing electric field does give rise to a magnetic field. The magnetic field lines need not enclose a current; they can circulate around electric field lines, which extend far from the oscillating dipole.

Figure 22.1 shows the electric and magnetic field lines due to an oscillating dipole. With changing electric fields as a source of magnetic fields, the field lines (both electric and magnetic) can break free of the dipole, form closed loops, and travel away from the dipole as an electromagnetic wave. The electric and magnetic fields sustain one another as the wave travels outward. Although the fields do diminish in strength, they do so much less rapidly than if the field lines were tied to the dipole. Since changing electric fields are a source of magnetic fields, a wave consisting of just an oscillating electric field without an oscillating magnetic field is impossible. Since changing magnetic fields are a source of electric fields, a wave consisting of just an oscillating magnetic field without an oscillating electric field is also impossible.

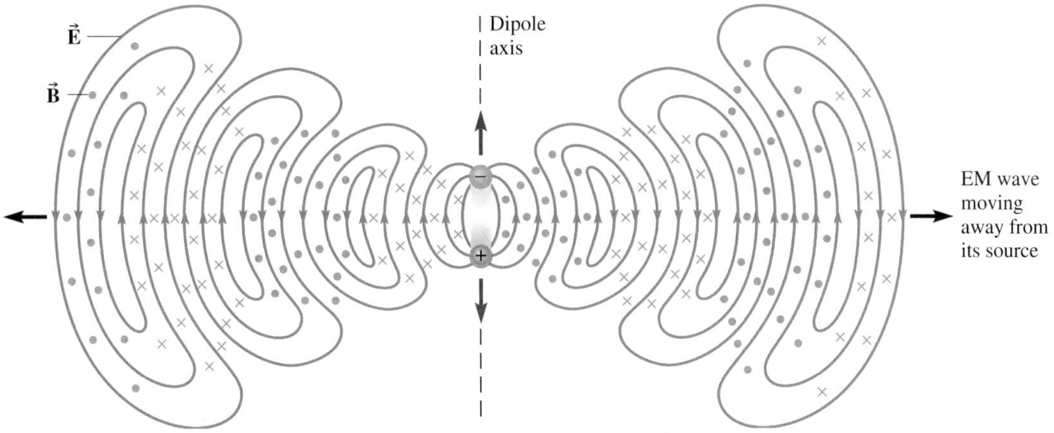

Figure 22.1 Electric and magnetic field lines due to an oscillating dipole. The green lines are electric field lines in the plane of the page. The orange dots and crosses are magnetic field lines crossing the plane of the page. The field lines break free of the dipole and travel away from it as an electromagnetic wave. Far from the dipole, the fields are strongest in directions perpendicular to the dipole axis and weakest in directions along the axis.

There are no electric waves or magnetic waves; there are only electromagnetic waves.

Maxwell's Equations

Maxwell modified Ampère's law and then used it with the three other basic laws of electromagnetism to show that electromagnetic waves exist and to derive their properties. In honor of this achievement, the four laws are collectively called **Maxwell's equations**. They are

CONNECTION:

Maxwell's equations: A collection of the four basic laws of electromagnetism.

1. **Gauss's law** [Eqs. (16-8) and (16-9)]: If an electric field line is not a closed loop, it can only start and stop on electric charges. Electric charges produce electric fields.
2. **Gauss's law for magnetism**: Magnetic field lines are always closed loops since there are no magnetic charges (*monopoles*). The magnetic flux *through a closed surface* (or the *net* number of field lines leaving the surface) is zero.

$$\Phi_B = \sum B_\perp A = 0 \qquad (22\text{-}1)$$

3. **Faraday's law** [Eq. (20-6)]: Changing magnetic fields are another source of electric fields.
4. **The Ampère-Maxwell law** says that changing electric fields as well as currents are sources of magnetic fields. Magnetic field lines are still always closed loops, but the loops do not have to surround currents; they can surround changing electric fields as well.

$$\sum B_\parallel \Delta l = \mu_0 \left(I + \epsilon_0 \frac{\Delta \Phi_E}{\Delta t} \right) \qquad (22\text{-}2)$$

22.2 ANTENNAS

Electric Dipole Antenna as Transmitter The **electric dipole antenna** consists of two metal rods lined up as if they were a single long rod (Fig. 22.2). The rods are fed from the center with an oscillating current. For half of a cycle, the current flows upward; the

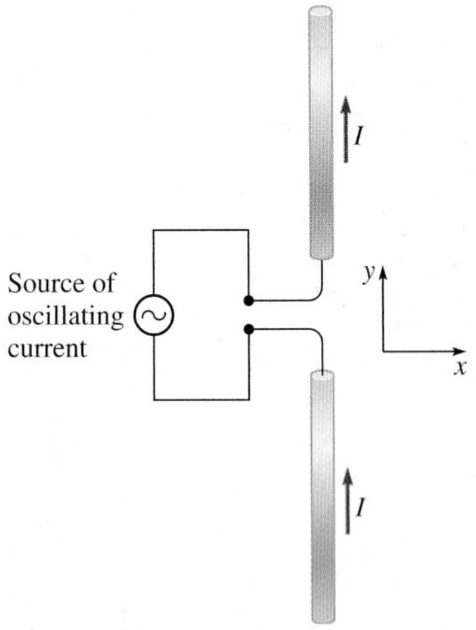

Figure 22.2 Current in an electric dipole antenna.

top of the antenna acquires a positive charge and the bottom acquires an equal negative charge. When the current reverses direction, these accumulated charges diminish and then reverse direction so that the top of the antenna becomes negatively charged and the bottom becomes positively charged. The result of feeding an alternating current to the antenna is an oscillating electric dipole.

The field lines for the EM wave emitted by an electric dipole antenna are similar to the field lines for an oscillating electric dipole (see Fig. 22.1). From the field lines, some of the properties of EM waves can be observed:

- For equal distances from the antenna, the amplitudes of the fields are smallest along the antenna's axis (in the $\pm y$-direction in Fig. 22.2) and largest in directions perpendicular to the antenna (in any direction perpendicular to the y-axis).
- In directions perpendicular to the antenna, the electric field is parallel to the antenna's axis. In other directions, $\vec{\mathbf{E}}$ is *not* parallel to the antenna's axis, but is perpendicular to the *direction of propagation* of the wave—that is, perpendicular to the direction that energy travels from the antenna to the observation point.
- The magnetic field is perpendicular to both the electric field and to the direction of propagation.

An electric dipole antenna used as a receiver should be aligned with the electric field of the wave.

Electric Dipole Antenna as Receiver An electric dipole antenna can be used as a receiver or detector of EM waves as well. In Fig. 22.3a, an EM wave travels past an electric dipole antenna. The electric field of the wave acts on free electrons in the antenna, causing an oscillating current. This current can then be amplified and the signal processed to decode the radio or TV transmission. The antenna is most effective if it is aligned with the electric field of the wave. If it is not, then only the component of $\vec{\mathbf{E}}$ parallel to the antenna acts to cause the oscillating current. The emf and the oscillating current are reduced by a factor of cos θ, where θ is the angle between $\vec{\mathbf{E}}$ and the antenna (Fig. 22.3b). If the antenna is perpendicular to the $\vec{\mathbf{E}}$ field, no oscillating current results.

CHECKPOINT 22.2

What happens if an electric dipole antenna (being used as a receiver) is oriented perpendicular to the $\vec{\mathbf{E}}$ field of the wave?

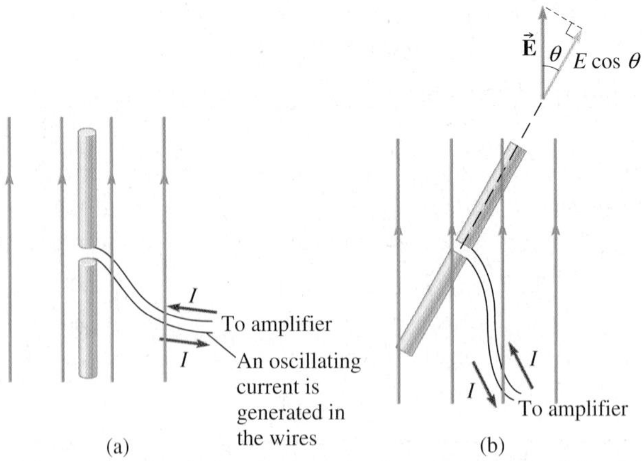

Figure 22.3 (a) The $\vec{\mathbf{E}}$ field of an EM wave makes an oscillating current flow in an electric dipole antenna. (The magnetic field lines are omitted for clarity.) (b) The current in the antenna is smaller when it is not aligned with the electric field. Only the component of $\vec{\mathbf{E}}$ parallel to the antenna accelerates electrons along the antenna's length.

Example 22.1

Electric Dipole Antenna

An electric dipole antenna at the origin has length 84 cm. It is used as a receiver for an EM wave traveling in the $+z$-direction. The electric field of the wave is always in the $\pm y$-direction and varies sinusoidally with time. The electric field in the vicinity of the antenna is

$$E_y(t) = E_m \cos \omega t;\ E_x = E_z = 0$$

where the amplitude—the maximum magnitude—of the electric field is $E_m = 3.2$ V/m. (a) How should the antenna be oriented for best reception? (b) What is the emf in the antenna if it is oriented properly?

Strategy For maximum amplitude, the antenna must be oriented so that the full electric field can drive current along the length of the antenna. The emf is defined as the work done by the electric field per unit charge.

Solution (a) We want the electric field of the wave to push free electrons along the antenna's length with a force directed along the length of the antenna. The electric field is always in the $\pm y$-direction, so the antenna should be oriented along the y-axis.

(b) The work done by the electric field E as it moves a charge q along the length of the antenna is

$$W = F_y \Delta y = qEL$$

The emf is the work per unit charge:

$$\mathscr{E} = \frac{W}{q} = EL$$

The emf varies with time because the electric field oscillates. The emf as a function of time is

$$\mathscr{E}(t) = EL = E_m L \cos \omega t$$

Therefore, it is a sinusoidally varying emf with the same frequency as the wave. The amplitude of the emf is

$$\mathscr{E}_m = E_m L = 3.2 \text{ V/m} \times 0.84 \text{ m} = 2.7 \text{ V}$$

Discussion The oscillating electric field has the same amplitude and phase at every point on the antenna. As a result, the emf is proportional to the length of the antenna. If the antenna is so long that the phase of the electric field varies with position along the antenna, then the emf is no longer proportional to the length of the antenna and may even start to decrease with additional length.

Practice Problem 22.1 Location of Transmitting Antenna

(a) If the wave in Example 22.1 is transmitted from a distant electric dipole antenna, where is the transmitting antenna located relative to the receiving antenna? (Answer in terms of xyz-coordinates.) (b) Write an equation for the electric field components as a function of position and time.

Magnetic Dipole Antenna Another kind of antenna is the **magnetic dipole antenna**. Recall that a loop of current is a magnetic dipole. (The right-hand rule establishes the direction of the north pole of the dipole: if the fingers of the right hand are curled around the loop in the direction of the current, the thumb points "north.") To make an oscillating magnetic dipole, we feed an alternating current into a loop or coil of wire. When the current reverses directions, the north and south poles of the magnetic dipole are interchanged.

If we consider the antenna axis to be the direction perpendicular to the coil, then the three observations made for the electric dipole antenna still hold, if we just substitute *magnetic* for *electric* and vice versa.

The magnetic dipole antenna works as a receiver as well (Fig. 22.4). The oscillating magnetic field of the wave causes a changing magnetic flux through the antenna. According to Faraday's law, an induced emf is present that makes an alternating current flow in the antenna. To maximize the rate of change of flux, the magnetic field should be perpendicular to the plane of the antenna.

A magnetic dipole antenna used as a receiver should be aligned so the magnetic field of the wave is perpendicular to the plane of the antenna.

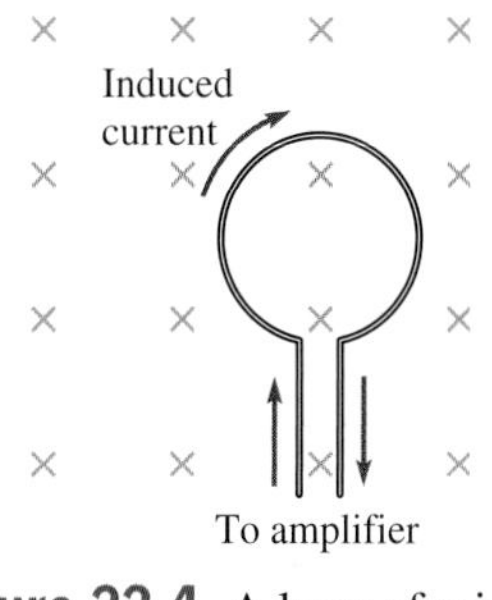

Figure 22.4 A loop of wire serves as a magnetic dipole antenna. As the magnetic field of the wave changes, the magnetic flux through the loop changes, causing an induced current in the loop. (The electric field lines are omitted for clarity.)

Antenna Limitations Antennas can generate only EM waves with long wavelengths and low frequencies. It isn't practical to use an antenna to generate EM waves with short wavelengths and high frequencies such as visible light; the frequency at which the current would have to alternate to generate such waves is far too high to be achieved in an antenna, while the antenna itself cannot be made short enough. (To be most effective, the length of an antenna should not be larger than half the wavelength.)

Problem-Solving Strategy: Antennas

- Electric dipole antenna (rod): antenna axis is along the rod.
- Magnetic dipole antenna (loop): antenna axis is perpendicular to the loop.
- Used as a transmitter, a dipole antenna radiates most strongly in directions perpendicular to its axis. In these directions, the wave's electric field is parallel to the antenna axis if transmitted by an electric dipole antenna and the wave's magnetic field is parallel to the antenna axis if transmitted by a magnetic dipole antenna.
- An antenna does not radiate in the two directions along its axis.
- For maximum sensitivity when used as a receiver, the axis of an electric dipole antenna should be aligned with the electric field of the wave and the *axis* of a magnetic dipole antenna should be aligned with the magnetic field of the wave.

22.3 THE ELECTROMAGNETIC SPECTRUM

EM waves can exist at every frequency, without restriction. The properties of EM waves and their interactions with matter depend on the frequency of the wave. The **electromagnetic spectrum**—the range of frequencies (and wavelengths)—is traditionally divided into six or seven named regions (Fig. 22.5). The names persist partly for historical reasons—the regions were discovered at different times—and partly because the EM radiation of different regions interacts with matter in different ways. The boundaries between the regions are fuzzy and somewhat arbitrary. Throughout this section, the wavelengths given are those *in vacuum*; EM waves in vacuum or in air travel at a speed of 3.00×10^8 m/s.

Visible Light

Visible light is the part of the spectrum that can be detected by the human eye. This seems like a pretty cut-and-dried definition, but actually the sensitivity of the eye falls

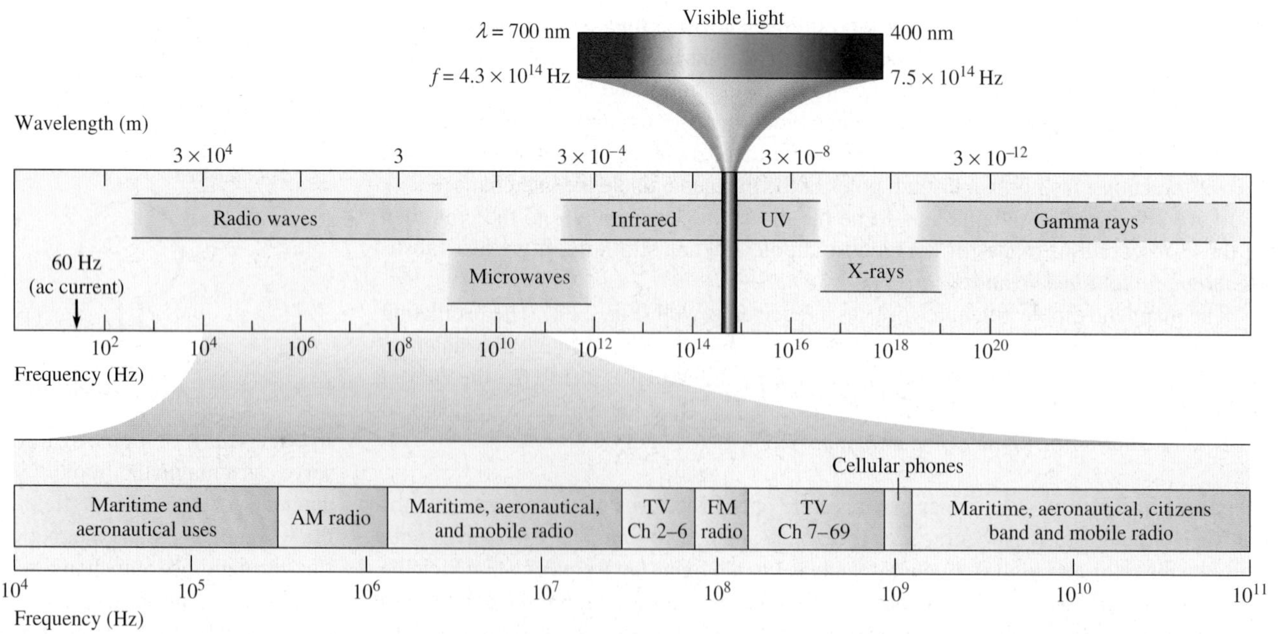

Figure 22.5 Regions of the EM spectrum. Note that the wavelength and frequency scales are logarithmic.

off gradually at both ends of the visible spectrum. Just as the range of frequencies of sound that can be heard varies from person to person, so does the range of frequencies of light that can be seen. For an average range we take frequencies of 430 THz (1 THz = 10^{12} Hz) to 750 THz, corresponding to wavelengths in vacuum of 700–400 nm. Light containing a mixture of all the wavelengths in the visible range appears white. White light can be separated by a prism into the colors red (700–620 nm), orange (620–600 nm), yellow (600–580 nm), green (580–490 nm), blue (490–450 nm), and violet (450–400 nm). Red has the lowest frequency (longest wavelength) and violet has the highest frequency (shortest wavelength).

It is not a coincidence that the human eye evolved to be most sensitive to the range of EM waves that are most intense in sunlight (Fig. 22.6). However, other animals have visible ranges that differ from that of humans; the range is often well suited to the particular needs of the animal.

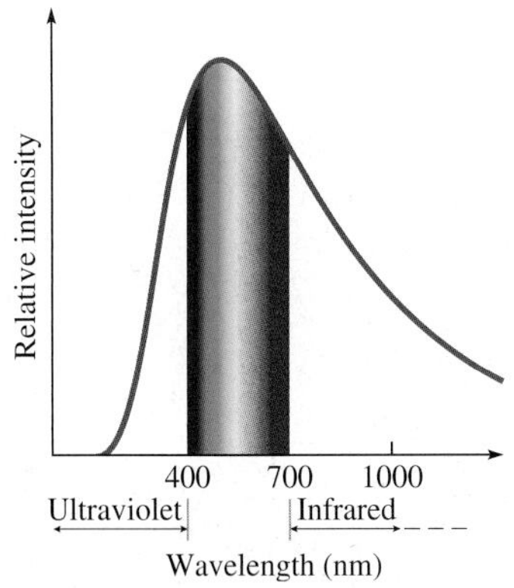

Figure 22.6 Graph of relative intensity (average power per unit area) of sunlight incident on Earth's atmosphere as a function of wavelength.

Lightbulbs, fire, the Sun, and fireflies are some *sources* of visible light. Most of the things we see are *not* sources of light; we see them by the light they *reflect.* When light strikes an object, some may be absorbed, some may be transmitted through the object, and some may be reflected. The relative amounts of absorption, transmission, and reflection usually differ for different wavelengths. A lemon appears yellow because it reflects much of the incident yellow light and absorbs most of the other spectral colors.

The wavelengths of visible light are small on an everyday scale but large relative to atoms. The diameter of an average-sized atom—and the distance between atoms in solids and liquids—is about 0.2 nm. Thus, the wavelengths of visible light are 2000–4000 times larger than the size of an atom.

Infrared

After visible light, the first parts of the EM spectrum to be discovered were those on either side of the visible: infrared and ultraviolet (discovered in 1800 and 1801, respectively). The prefix *infra-* means *below*; **infrared** radiation (IR) is lower in frequency than visible light. IR extends from the low-frequency (red) edge of the visible to a frequency of about 300 GHz ($\lambda = 1$ mm). The astronomer William Herschel (1738–1822) discovered IR in 1800 while studying the temperature rise caused by the light emerging from a prism. He discovered that the thermometer reading was highest for levels just *outside* the illuminated region, adjacent to the red end of the spectrum. Since the radiation was not *visible*, Herschel deduced that there must be some invisible radiation beyond the red.

The thermal radiation given off by objects near room temperature is primarily infrared (Fig. 22.7), with the peak of the radiated IR at a wavelength of about 0.01 mm = 10 μm. At higher temperatures, the power radiated increases as the wavelength of peak radiation decreases. A roaring wood stove with a surface temperature of 500°F has an absolute temperature about 1.8 times room temperature (530 K); it radiates about 11 times more power than when at room temperature since $P \propto T^4$ [Stefan's law, Eq. (14-16)]. Nevertheless, the peak is still in the infrared. The wavelength of peak radiation is about 5.5 μm = 5500 nm since $\lambda_{max} \propto 1/T$ [Wien's law, Eq. (14-17)]. If the stove gets even hotter, its radiation is still mostly IR but glows red as it starts to radiate significantly in the red part of the visible spectrum. (Call the fire department!) Even the filament of a lightbulb ($T \approx 3000$ K) radiates much more IR than it does visible. The *peak* of the Sun's thermal radiation is in the visible; nevertheless about half the energy reaching us from the Sun is IR.

CONNECTION:

Thermal radiation was discussed as a type of heat flow in Section 14.8.

Ultraviolet

The prefix *ultra-* means *above*; **ultraviolet** (UV) radiation is higher in frequency than visible light. UV ranges in wavelength from the shortest visible wavelength (about 380 nm) down to about 10 nm. There is plenty of UV in the Sun's radiation; its effects on human skin include tanning, sunburn, formation of vitamin D, and melanoma. Water vapor transmits much of the Sun's UV, so tanning and sunburn can occur on overcast

Application: thermograms

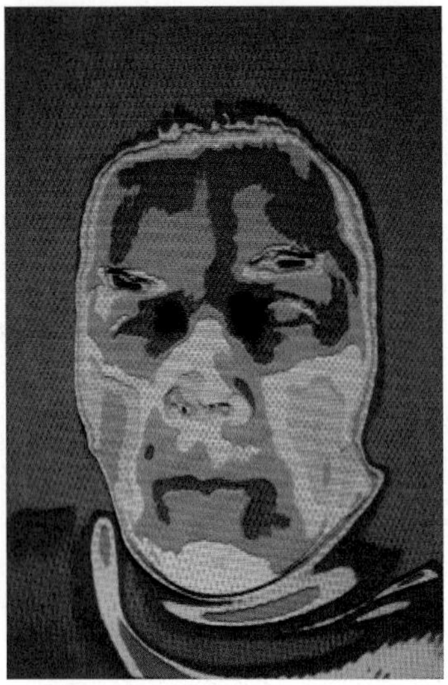

(a)

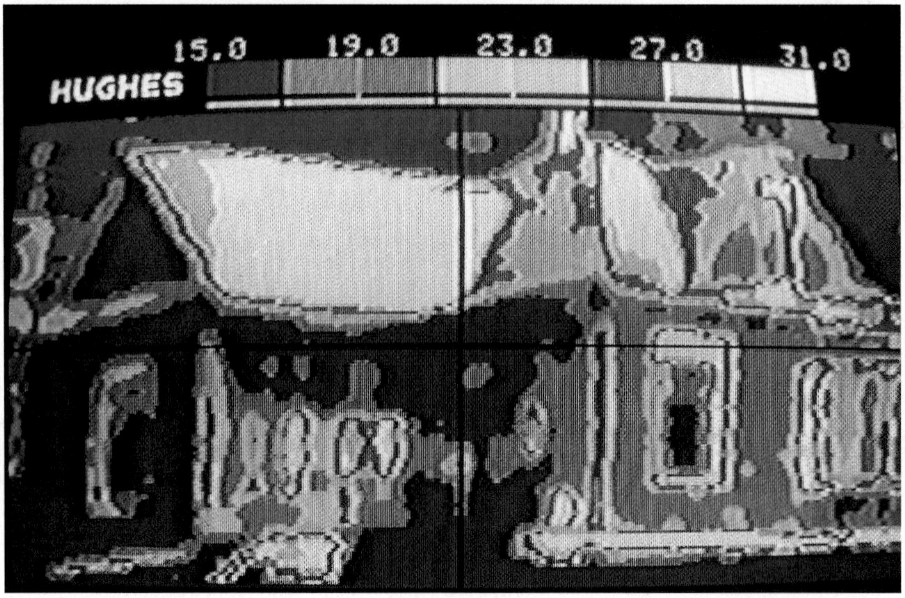

(b)

Figure 22.7 (a) False-color thermogram of a man's head. The red areas show regions of pain from a headache; these areas are warmer, so they give off more infrared radiation. (b) False-color thermogram of a house in winter, showing that most of the heat escapes through the roof. The scale shows that the blue areas are the coolest, while the pink areas are the warmest. Note that some heat escapes around the window frame, while the window itself is cool due to double-pane glass.

Application: fluorescence

days. On the other hand, ordinary glass absorbs most UV. Black lights emit UV; certain *fluorescent* materials—such as the coating on the inside of the glass tube in a fluorescent light—can absorb UV and then emit visible light (Fig. 22.8).

Radio Waves

After IR and UV were identified, most of the nineteenth century passed before any of the outlying regions of the EM spectrum were discovered. The lowest frequencies (up to about 1 GHz) and longest wavelengths (down to about 0.3 m) are called **radio waves**. AM and FM radio, VHF and UHF TV broadcasts, and ham radio operators occupy assigned frequency bands within the radio wave part of the spectrum.

(a) (b)

Figure 22.8 (a) The large star coral (*Montastraea cavernosa*) is dull brown when illuminated by white light. (b) When illuminated with an ultraviolet source, the coral absorbs UV and emits visible light that appears bright yellow. A small sponge (bottom right corner) looks bright red in white light due to selective reflection. It appears black when illuminated with UV because it does not fluoresce.

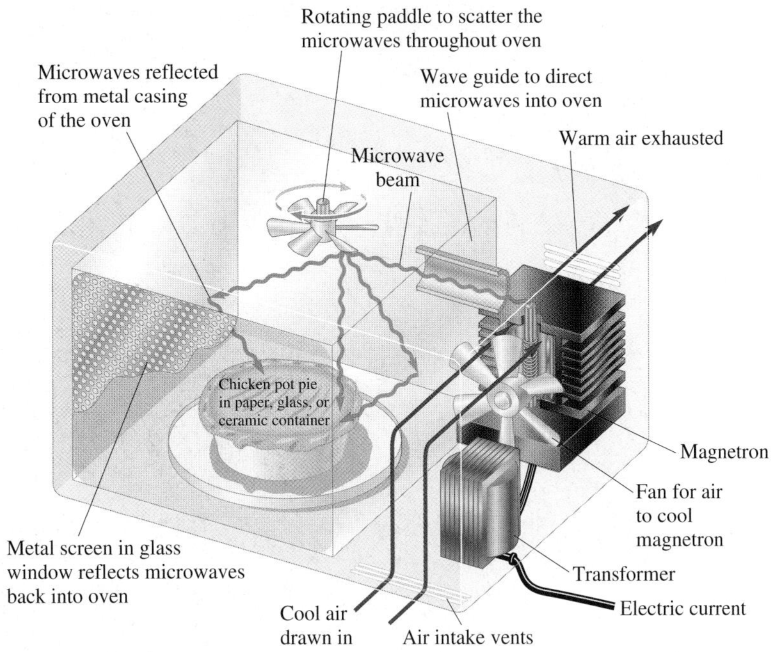

Figure 22.9 A microwave oven. The microwaves are produced in a *magnetron*, a resonant cavity that produces the oscillating currents that give rise to microwaves at the desired frequency. Since metals reflect microwaves well, a metal waveguide directs the microwaves toward the rotating metal stirrer, which reflects the microwaves in many different directions to distribute them throughout the oven. (This reflective property is one reason why metal containers and aluminum foil should generally not be used in a microwave oven; no microwaves could reach the food inside the container or foil.) The oven cavity is enclosed by metal to reflect microwaves back in and minimize the amount leaking out of the oven. The sheet of metal in the door has small holes so we can see inside, but since the holes are much smaller than the wavelength of the microwaves, the sheet still reflects microwaves.

Microwaves

Microwaves are the part of the EM spectrum lying between radio waves and IR, with vacuum wavelengths roughly from 1 mm to 30 cm. Microwaves were first generated and detected in the laboratory in 1888 by Heinrich Hertz (1857–1894). Microwaves are used in communications (cell phones and satellite TV) and in radar. After the development of radar in World War II, the search for peacetime uses of microwaves resulted in the development of the microwave oven.

Application: microwave ovens

A microwave oven (Fig 22.9) immerses food in microwaves with a wavelength in vacuum of about 12 cm. Water is a good absorber of microwaves because the water molecule is polar. An electric dipole in an electric field feels a torque that tends to align the dipole with the field, since the positive and negative charges are pulled in opposite directions. As a result of the rapidly oscillating electric field of the microwaves ($f = 2.5$ GHz), the water molecules rotate back and forth; the energy of this rotation then spreads throughout the food.

Application: cosmic microwave background radiation

In the early 1960s, Arno Penzias (b. 1933) and Robert Wilson (b. 1936) were having trouble with their radio telescope; they were plagued by noise in the microwave part of the spectrum. Subsequent investigation led them to discover that the entire universe is bathed in microwaves that correspond to blackbody radiation at a temperature of 2.7 K (peak wavelength about 1 mm). This *cosmic microwave background radiation* is left over from the origin of the universe—a huge explosion called the *Big Bang*.

X-Rays and Gamma Rays

Higher in frequency and shorter in wavelength than UV are **x-rays** and **gamma rays**, which were discovered in 1895 and 1900, respectively. The two names are still used,

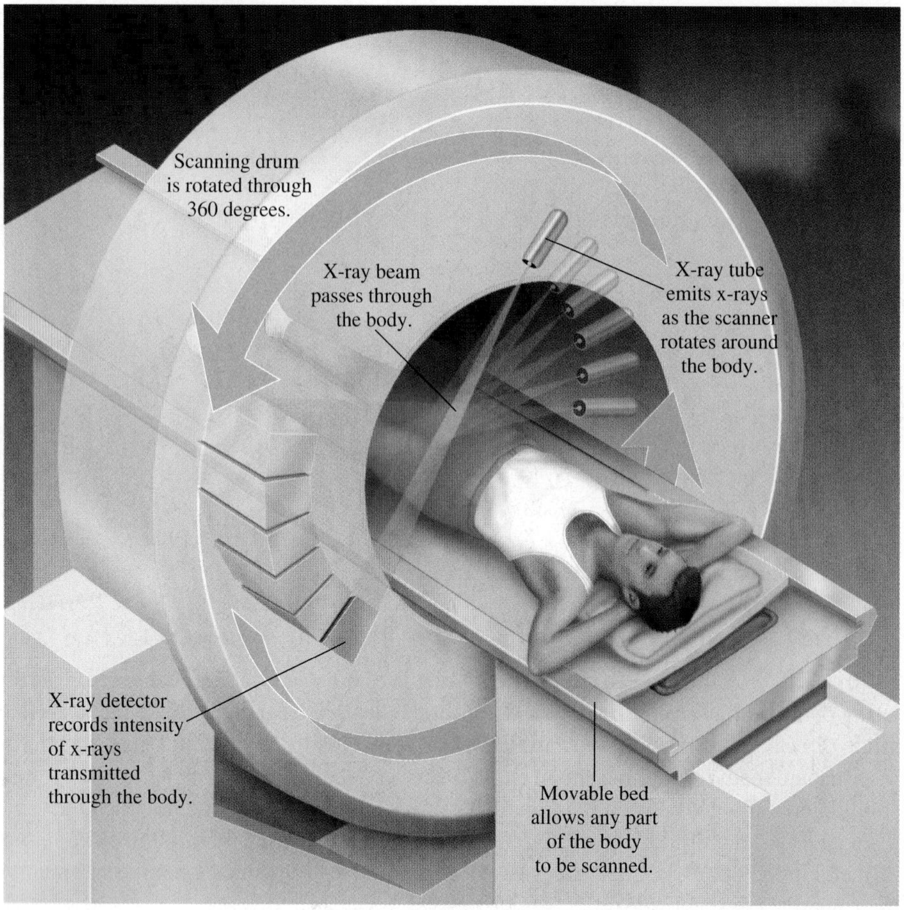

Figure 22.10 Apparatus used for a CAT scan.

based on the source of the waves, mostly for historical reasons. There is considerable overlap in the frequencies of the EM waves generated by these two methods, so today the distinction is somewhat arbitrary.

X-rays were unexpectedly discovered by Wilhelm Konrad Röntgen (1845–1923) when he accelerated electrons to high energies and smashed them into a target. The large deceleration of the electrons as they come to rest in the target produces the x-rays. Röntgen received the first Nobel Prize in physics for the discovery of x-rays.

Application: x-rays in medicine and dentistry, CAT scans

Most diagnostic x-rays used in medicine and dentistry have wavelengths between 10 and 60 pm (1 pm = 10^{-12} m). In a conventional x-ray, film records the amount of x-ray radiation that passes through the tissue. Computer-assisted tomography (CAT or CT scan) allows a cross-sectional image of the body. An x-ray source is rotated around the body in a plane and a computer measures the x-ray transmission at many different angles. Using this information, the computer constructs an image of that slice of the body (Fig. 22.10).

Gamma rays were first observed in the decay of radioactive nuclei on Earth. Pulsars, neutron stars, black holes, and explosions of supernovae are sources of gamma rays that travel toward Earth, but—fortunately for us—gamma rays are absorbed by the atmosphere. Only when detectors were placed high in the atmosphere and above it by using balloons and satellites did the science of gamma-ray astronomy develop. In the late 1960s, scientists first observed bursts of gamma rays from deep space that last for times ranging from a fraction of a second to a few minutes; these bursts occur about once a day. A gamma-ray burst can emit more energy in 10 s than the Sun will emit in its entire lifetime. The source of the gamma-ray bursts is still under investigation.

22.4 SPEED OF EM WAVES IN VACUUM AND IN MATTER

Light travels so fast that it is not obvious that it takes any time at all to go from one place to another. Since high-precision electronic instruments were not available, early measurements of the speed of light had to be cleverly designed. In 1849, French scientist Armand Hippolyte Louis Fizeau (1819–1896) measured the speed of visible light to be approximately 3×10^8 m/s (Fig. 22.11).

Speed of Light in Vacuum

In Chapters 11 and 12 we saw that the speed of a mechanical wave depends on properties of the wave medium. Sound travels faster through steel than it does through water and faster through water than through air. In every case, the wave speed depended on two characteristics of the wave medium: one that characterizes the restoring force and another that characterizes the inertia.

Unlike mechanical waves, electromagnetic waves can travel through vacuum; they do not require a material medium. Light reaches Earth from galaxies billions of light-years away, traveling the vast distances between galaxies without problem; but a sound wave can't even travel a few meters between two astronauts on a space walk, since there is no air or other medium to sustain a sound wave's pressure variations. What, then, determines the speed of light in vacuum?

Looking back at the laws that describe electric and magnetic fields, we find two universal constants. One of them is the permittivity of free space ϵ_0, found in Coulomb's law and Gauss's law; it is associated with the electric field. The second is the permeability of free space μ_0, found in Ampère's law; it is associated with the magnetic field. Since these are the only two quantities that can determine the speed of light in vacuum, there must be a combination of them that has the dimensions of speed.

> **CONNECTION:**
>
> The speed of a mechanical wave depends on properties of the medium (e.g., tension and linear mass density for a transverse wave on a string). The speed of EM waves *through a transparent material* such as glass depends on the electric and magnetic properties of that material. The speed of EM waves in vacuum is a universal constant related to the constants ϵ_0 and μ_0.

The values of these constants in SI units are

$$\epsilon_0 = 8.85 \times 10^{-12}\ \frac{\text{C}^2}{\text{N}\cdot\text{m}^2} \qquad \text{and} \qquad \mu_0 = 4\pi \times 10^{-7}\ \frac{\text{T}\cdot\text{m}}{\text{A}}$$

The tesla can be written in terms of other SI units. Using $\vec{\mathbf{F}} = q\vec{\mathbf{v}} \times \vec{\mathbf{B}}$ as a guide,

$$1\ \text{T} = 1\ \frac{\text{N}}{\text{C}\cdot\text{m/s}}$$

The only combination of these constants that has the dimensions of a velocity is

$$\frac{1}{\sqrt{\epsilon_0\mu_0}} = \left(8.85 \times 10^{-12}\ \frac{\cancel{\text{C}^2}}{\cancel{\text{N}}\cdot\text{m}^2} \times 4\pi \times 10^{-7}\ \frac{\cancel{\text{N}}\cdot\cancel{\text{m}}}{\cancel{\text{C}}\cdot(\cancel{\text{m}}/\text{s})\cdot(\cancel{\text{C}}/\text{s})}\right)^{-1/2} = 3.00 \times 10^8\ \text{m/s}$$

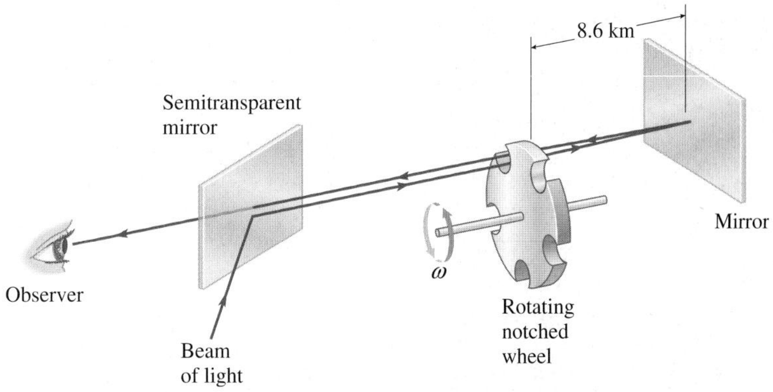

Figure 22.11 Fizeau's apparatus to measure the speed of light. The notched wheel rotates at an angular speed ω that can be varied. At certain values of ω, the beam of light passes through one of the notches in the wheel, travels a long distance to a mirror, reflects, and passes back through another notch to the observer. At other values of ω, the reflected beam is interrupted by the rotating wheel. The speed of light can be calculated from the measured angular speeds at which the observer sees the reflected beam.

The dimensional analysis done here leaves the possibility of a multiplying factor such as $\frac{1}{2}$ or $\sqrt{\pi}$. In the midnineteenth century, Maxwell proved mathematically that an electromagnetic wave—a wave consisting of oscillating electric and magnetic fields propagating through space—could exist in vacuum. Starting from Maxwell's equations (Section 22.1), he derived the *wave equation*, an equation of a special mathematical form that describes wave propagation for *any* kind of wave. In the place of the wave speed appeared $(\epsilon_0\mu_0)^{-1/2}$. Using the values of ϵ_0 and μ_0 that had been measured in 1856, Maxwell showed that electromagnetic waves in vacuum travel at 3.00×10^8 m/s—very close to what Fizeau measured. Maxwell's derivation was the first evidence that light is an electromagnetic wave.

The speed of electromagnetic waves in vacuum is represented by the symbol c (for the Latin *celeritas*, "speed").

Speed of electromagnetic waves in vacuum:

$$c = \frac{1}{\sqrt{\epsilon_0\mu_0}} = 3.00 \times 10^8 \text{ m/s} \qquad (22\text{-}3)$$

While c is usually called *the speed of light*, it is the speed of *any* electromagnetic wave in vacuum, regardless of frequency or wavelength, not just the speed for frequencies visible to humans.

Example 22.2

Light Travel Time from a "Nearby" Supernova

A supernova is an exploding star; a supernova is billions of times brighter than an ordinary star. Most supernovae occur in distant galaxies and cannot be observed with the naked eye. The last two supernovae visible to the naked eye occurred in 1604 and 1987. Supernova SN1987a (Fig. 22.12) occurred 1.6×10^{21} m from Earth. *When* did the explosion occur?

Strategy The light from the supernova travels at speed c. The time that it takes light to travel a distance 1.6×10^{21} m tells us how long ago the explosion occurred.

Figure 22.12
Photo of the sky after light from Supernova SN1987a reached Earth.

Solution The time for light to travel a distance d at speed c is

$$\Delta t = \frac{d}{c} = \frac{1.6 \times 10^{21} \text{ m}}{3.00 \times 10^8 \text{ m/s}} = 5.33 \times 10^{12} \text{ s}$$

To get a better idea how long that is, we convert seconds to years:

$$5.33 \times 10^{12} \text{ s} \times \frac{1 \text{ yr}}{3.156 \times 10^7 \text{ s}} = 170\,000 \text{ yr}$$

Discussion When we look at the stars, the light we see was radiated by the stars long ago. By looking at distant galaxies, astronomers get a glimpse of the universe in the past. Beyond the Sun, the closest star to Earth is about 4 ly (light-years) away, which means that it takes light 4 yr to reach us from that star. The most distant galaxies observed are at a distance of over 10^{10} ly; looking at them, we see over 10 billion yr into the past.

Practice Problem 22.2 A Light-Year

A light-year is the distance traveled by light (in vacuum) in one Earth year. Find the conversion factor from light-years to meters.

Speed of Light in Matter

When an EM wave travels through a material medium, it travels at a speed v that is less than c. For example, visible light travels through glass at speeds between about 1.6×10^8 m/s and 2.0×10^8 m/s, depending on the type of glass and the frequency of the light. Instead of specifying the speed, it is common to specify the **index of refraction** n:

The speed of an EM wave through matter is less than c.

Index of refraction:

$$n = \frac{c}{v} \tag{22-4}$$

Refraction refers to the bending of a wave as it passes from one medium to another; we study refraction in detail in Section 23.3. Since the index of refraction is a ratio of two speeds, it is a dimensionless number. For glass in which light travels at 2.0×10^8 m/s, the index of refraction is

$$n = \frac{3.0 \times 10^8 \text{ m/s}}{2.0 \times 10^8 \text{ m/s}} = 1.5$$

The speed of light in air (at 1 atm) is only slightly less than c; the index of refraction of air is 1.0003. Most of the time this 0.03% difference is not important, so we can use c as the speed of light in air. The speed of visible light in an optically transparent medium is less than c, so the index of refraction is greater than 1.

When an EM wave passes from one medium to another, the frequency and wavelength cannot *both* remain unchanged since the wave speed changes and $v = f\lambda$. As is the case with mechanical waves, it is the wavelength that changes; the frequency remains the same. The incoming wave (with frequency f) causes charges in the atoms at the boundary to oscillate with the same frequency f, just as for the charges in an antenna. The oscillating charges at the boundary radiate an EM wave at that same frequency into the second medium. Therefore, the electric and magnetic fields in the second medium *must* oscillate at the same frequency as the fields in the first medium. In just the same way, if a transverse wave of frequency f traveling down a string reaches a point at which an abrupt change in wave speed occurs, the incident wave makes that point oscillate up and down at the same frequency f as any other point on the string. The oscillation of that point sends a wave of the same frequency to the other side of the string. Since the wave speed has changed but the frequency is the same, the wavelength has changed as well.

A wave passing from one medium into another changes wavelength but retains the same frequency.

We sometimes need to find the wavelength λ of an EM wave in a medium of index n, given its wavelength λ_0 in vacuum. Since the frequencies are equal,

$$f = \frac{c}{\lambda_0} = \frac{v}{\lambda}$$

Solving for λ gives

$$\lambda = \frac{v}{c}\lambda_0 = \lambda_0 \tag{22-5}$$

Since $n > 1$, the wavelength is shorter than the wavelength in vacuum. The wave travels more slowly in the medium than in vacuum; since the wavelength is the distance traveled by the wave in one period $T = 1/f$, the wavelength in the medium is shorter.

If blue light of wavelength $\lambda_0 = 480$ nm enters glass that has an index of refraction of 1.5, it is still visible light, even though its wavelength in glass is 320 nm; it has not been transformed into UV radiation. When light of a given frequency enters the eye, it has the same frequency in the fluid in the eye regardless of how many material media it has passed through, since the frequency remains the same at each boundary.

CHECKPOINT 22.4

A light wave travels from water ($n = 4/3$) into air. Its wavelength in water is 480 nm. What is its wavelength in air?

Example 22.3

Wavelength Change from Glass to Water

The index of refraction of glass is 1.50 and that of water is 1.33. If light of wavelength 285 nm *in glass* passes into water, what is the wavelength in the water?

Strategy The key is to remember that the frequency is the same as the wave passes from one medium to another.

Solution Frequency, wavelength, and speed are related by

$$v = \lambda f$$

Solving for frequency, $f = v/\lambda$. Since the frequencies are equal,

$$\frac{v_w}{\lambda_w} = \frac{v_g}{\lambda_g}$$

The speed of light in a material is $v = c/n$. Solving for λ_w and substituting $v = c/n$ gives

$$\lambda_w = v_w \frac{\lambda_g}{v_g} = \frac{c}{n_w} \times \frac{n_g \lambda_g}{c} = \frac{1.50 \times 285 \text{ nm}}{1.33} = 321 \text{ nm}$$

Discussion Water has a smaller index of refraction, so the speed of light in water is greater than in glass. Since wavelength is the distance traveled in one period, the wavelength in water is longer (321 nm > 285 nm).

Practice Problem 22.3 Wavelength Change from Air to Water

The speed of visible light in water is 2.25×10^8 m/s. When light of wavelength 592 nm in air passes into water, what is its wavelength in water?

Figure 22.13 A prism separates a beam of white light (coming in from the left) into the colors of the spectrum.

Dispersion

Although EM waves of every frequency travel through vacuum at the same speed c, the speed of EM waves in a material medium *does* depend on frequency. Therefore, the index of refraction is not a constant for a given material; it is a function of frequency. Variation of the speed of a wave with frequency is called **dispersion**. Dispersion causes white light to separate into colors when it passes through a glass prism (Fig. 22.13). The dispersion of the light into different colors arises because each color travels at a slightly different speed in the same medium.

A **nondispersive** medium is one for which the variation in the index of refraction is negligibly small for the range of frequencies of interest. No medium (apart from vacuum) is truly nondispersive, but many can be treated as nondispersive for a restricted range of frequencies. For most optically transparent materials, the index of refraction increases with increasing frequency; blue light travels more slowly through glass than does red light. In other parts of the EM spectrum, or even for visible light in unusual materials, n can decrease with increasing frequency instead.

22.5 CHARACTERISTICS OF TRAVELING ELECTROMAGNETIC WAVES IN VACUUM

The various characteristics of traveling EM waves in vacuum (Fig. 22.14) can be derived from Maxwell's equations (Section 22.1). Such a derivation requires higher level mathematics, so we state the characteristics without proof.

CONNECTION:

The wavelength, wavenumber, frequency, angular frequency, and period of an EM wave are defined exactly as for mechanical waves.

- EM waves in vacuum travel at speed $c = 3.00 \times 10^8$ m/s, independent of frequency. The speed is also independent of amplitude.
- The electric and magnetic fields oscillate at the *same frequency*. Thus, a single frequency f and a single wavelength $\lambda = c/f$ pertain to both the electric and magnetic fields of the wave.
- The electric and magnetic fields oscillate *in phase* with one another. That is, at a given instant, the electric and magnetic fields are at their maximum magnitudes at

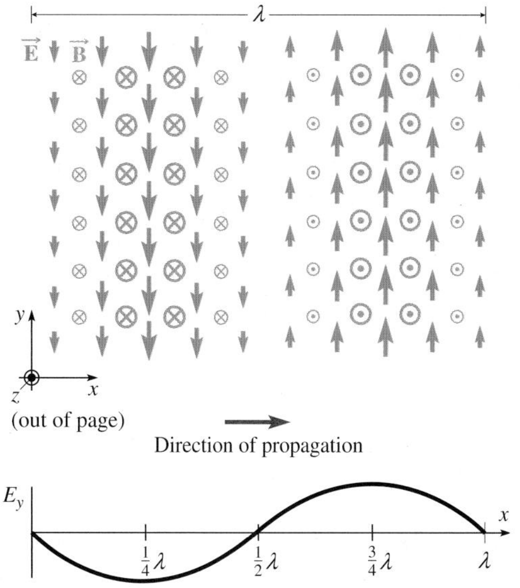

Figure 22.14 One wavelength of an EM wave traveling in the $+x$-direction (to the right). The electric field is represented by green vector arrows sketched at a few points, pointing in the $-y$-direction for $0 < x < \frac{1}{2}\lambda$ and in the $+y$-direction for $\frac{1}{2}\lambda < x < \lambda$. The magnetic field is perpendicular to the plane of the page and is represented by orange vector symbols. The magnetic field is in the $-z$-direction for $0 < x < \frac{1}{2}\lambda$ and in the $+z$-direction for $\frac{1}{2}\lambda < x < \lambda$. The magnitude of $\vec{\mathbf{E}}$ is represented by the length of the green arrows. The magnitude of $\vec{\mathbf{B}}$ is represented by the size of the orange vector symbols. The graph shows the y-component of $\vec{\mathbf{E}}$ as a function of x at some instant. A graph of the z-component of $\vec{\mathbf{B}}$ at the same instant would look the same because the electric and magnetic fields are *in phase*.

a common set of points. Similarly, the fields are both zero at a common set of points at any instant.

- The amplitudes of the electric and magnetic fields are proportional to one another. The ratio is c:

$$E_{\rm m} = cB_{\rm m} \qquad (22\text{-}6)$$

- Since the fields are in phase and the amplitudes are proportional, the instantaneous magnitudes of the fields are proportional at any point:

$$|\vec{\mathbf{E}}(x, y, z, t)| = c|\vec{\mathbf{B}}(x, y, z, t)| \qquad (22\text{-}7)$$

- The EM wave is *transverse*; that is, the electric and magnetic fields are each perpendicular to the direction of propagation of the wave.
- The fields are also perpendicular to *one another*. Therefore, $\vec{\mathbf{E}}$, $\vec{\mathbf{B}}$, and the velocity of propagation are three mutually perpendicular vectors.
- At any point, $\vec{\mathbf{E}} \times \vec{\mathbf{B}}$ is always in the direction of propagation (Fig. 22.15).
- The electric energy density is equal to the magnetic energy density at any point. The wave carries exactly half its energy in the electric field and half in the magnetic field.

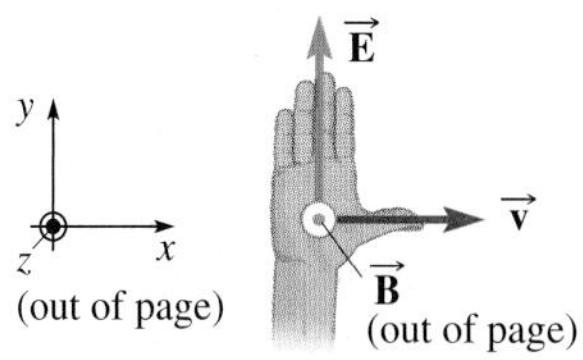

Figure 22.15 Using the right-hand rule to check the directions of the fields in Fig. 22.14. At $x > \frac{1}{2}\lambda$, $\vec{\mathbf{E}}$ is in the $+y$-direction and $\vec{\mathbf{B}}$ is in the $+z$-direction. The cross product $\vec{\mathbf{E}} \times \vec{\mathbf{B}}$ is in the direction of propagation ($+x$).

CHECKPOINT 22.5

An EM wave travels in the $+x$-direction. The wave's electric field at a point P and at time t has magnitude 0.009 V/m and is in the $-y$-direction. What is the magnetic field at P at the same instant?

Example 22.4

Traveling EM Wave

The x-, y-, and z-components of the electric field of an EM wave in vacuum are

$$E_y(x, y, z, t) = -60.0\,\frac{\text{V}}{\text{m}} \times \cos\,[(4.0\ \text{m}^{-1})x + \omega t],\ E_x = E_z = 0$$

(a) In what direction does the wave travel? (b) Find the value of ω. (c) Write an expression for the components of the magnetic field of the wave.

Strategy Parts (a) and (b) require some general knowledge about waves, but nothing specific to EM waves. Turning back to Chapter 11 may help refresh your memory. Part (c) involves the relationship between the electric and magnetic fields, which *is* particular to EM waves. The instantaneous magnitude of the magnetic field is given by $B(x, y, z, t) = E(x, y, z, t)/c$. We must also determine the direction of the magnetic field: $\vec{\mathbf{E}}$, $\vec{\mathbf{B}}$, and the velocity of propagation are three mutually perpendicular vectors and $\vec{\mathbf{E}} \times \vec{\mathbf{B}}$ must be in the direction of propagation.

Solution (a) Since the electric field depends on the value of x but not on the values of y or z, the wave moves parallel to the x-axis. Imagine riding along a crest of the wave—a point where

$$\cos\,[(4.0\ \text{m}^{-1})x + \omega t] = 1$$

Then

$$(4.0\ \text{m}^{-1})x + \omega t = 2\pi n$$

where n is some integer. A short time later, t is a little bigger, so x must be a little smaller so that $(4.0\ \text{m}^{-1})x + \omega t$ is still equal to $2\pi n$. Since the x-coordinate of a crest gets smaller as time passes, the wave is moving in the $-x$-direction.

(b) The constant multiplying x, $4.0\ \text{m}^{-1}$, is the *wavenumber*, a quantity related to the wavelength. Since the wave repeats in a distance λ and the cosine function repeats every 2π radians, $k(x + \lambda)$ must be 2π radians greater than kx:

$$k(x + \lambda) = kx + 2\pi$$

or

$$k = \frac{2\pi}{\lambda}$$

Therefore, the wavenumber is $k = 4.0\ \text{m}^{-1}$. The speed of the wave is c. Since any periodic wave travels a distance λ in a time T,

$$T = \frac{\lambda}{c}$$

$$\omega = \frac{2\pi}{T} = \frac{2\pi c}{\lambda} = kc = 4.0\ \text{m}^{-1} \times 3.00 \times 10^8\ \text{m/s}$$
$$= 1.2 \times 10^9\ \text{rad/s}$$

(c) Since the wave moves in the $-x$-direction and the electric field is in the $\pm y$-direction, the magnetic field must be in the $\pm z$-direction to make three perpendicular directions. Since the magnetic field is in phase with the electric field, with the same wavelength and frequency, it must take the form

$$B_z(x, y, z, t) = \pm B_\text{m} \cos\,[(4.0\ \text{m}^{-1})x + (1.2 \times 10^9\ \text{s}^{-1})t],$$
$$B_x = B_y = 0$$

The amplitudes are proportional:

$$B_\text{m} = \frac{E_\text{m}}{c} = \frac{60.0\ \text{V/m}}{3.00 \times 10^8\ \text{m/s}} = 2.00 \times 10^{-7}\ \text{T}$$

The last step is to decide which sign is correct. At $x = t = 0$, the electric field is in the $-y$-direction. $\vec{\mathbf{E}} \times \vec{\mathbf{B}}$ must be in the $-x$-direction (the direction of propagation). Then

$$(-y\text{-direction}) \times (\text{direction of } \vec{\mathbf{B}}) = (-x\text{-direction})$$

Trying both possibilities with the right-hand rule (Fig. 22.16), we find that $\vec{\mathbf{B}}$ is in the $+z$-direction at $x = t = 0$. Then the magnetic field is written

$$B_z(x, y, z, t) = (2.00 \times 10^{-7}\ \text{T}) \cos\,[(4.0\ \text{m}^{-1})x + (1.2 \times 10^9\ \text{s}^{-1})t],$$
$$B_x = B_y = 0$$

Discussion When $\cos\,[(4.0\ \text{m}^{-1})x + (1.2 \times 10^9\ \text{s}^{-1})t]$ is negative, then $\vec{\mathbf{E}}$ is in the $+y$-direction and $\vec{\mathbf{B}}$ is in the $-z$-direction. Since both fields reverse direction, it is still true that $\vec{\mathbf{E}} \times \vec{\mathbf{B}}$ is in the direction of propagation.

Practice Problem 22.4 Another Traveling Wave

The x-, y-, and z-components of the electric field of an EM wave in vacuum are

$$E_x(x, y, z, t) = 32\,\frac{\text{V}}{\text{m}} \times \cos\,[ky - (6.0 \times 10^{11}\ \text{s}^{-1})t],$$
$$E_y = E_z = 0$$

where k is positive. (a) In what direction does the wave travel? (b) Find the value of k. (c) Write an expression for the components of the magnetic field of the wave.

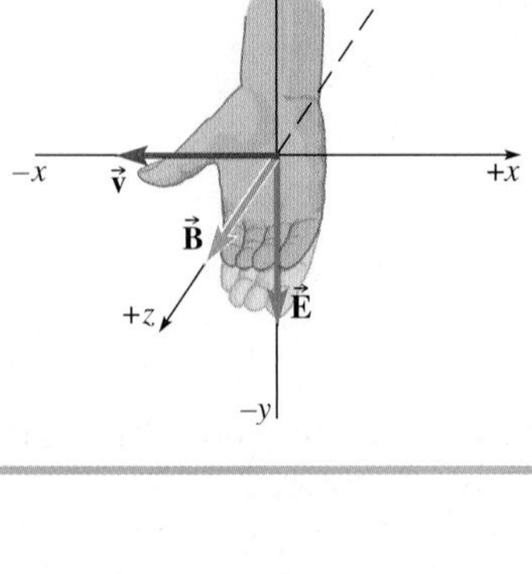

Figure 22.16
Using the right-hand rule to find the direction of $\vec{\mathbf{B}}$.

22.6 ENERGY TRANSPORT BY EM WAVES

Electromagnetic waves carry energy, as do all waves. Life on Earth exists only because the energy of EM radiation from the Sun can be harnessed by green plants, which through photosynthesis convert some of the energy in light to chemical energy. Photosynthesis sustains not only the plants themselves, but also animals that eat plants and fungi that derive their energy from decaying plants and animals—the entire food chain can be traced back to the Sun as energy source. Only a few exceptions exist, such as the bacteria that live in geothermal vents on the ocean floor. The heat flow from the interior of the Earth does not originate with the Sun; it comes from radioactive decay.

Most industrial sources of energy are derived from electromagnetic energy from the Sun. Fossil fuels—petroleum, coal, and natural gas—come from the remains of plants and animals. Solar cells convert the incident sunlight's energy directly into electricity (Fig. 22.17); the Sun is also used to heat water and homes directly. Hydroelectric power plants rely on the Sun to evaporate water, in a sense pumping it back uphill so that it can once again flow down rivers and turn turbines. Wind can be harnessed to generate electricity, but the winds are driven by uneven heating of Earth's surface by the Sun. The only energy sources we have that do not come from the Sun's EM radiation are nuclear fission and geothermal energy.

Figure 22.17 A solar panel farm in the Sierra Nevada Mountains.

Energy Density

The energy in light is stored in the oscillating electric and magnetic fields in the wave. For an EM wave in vacuum, the energy densities (SI unit: J/m^3) are

$$u_E = \frac{1}{2}\epsilon_0 E^2 \qquad (17\text{-}19)$$

and

$$u_B = \frac{1}{2\mu_0}B^2 \qquad (20\text{-}18)$$

CONNECTION:

The expressions for electric and magnetic energy densities in an EM wave are the same as introduced in Chapters 17 and 20.

It can be proved (Problem 41) that the two energy densities are equal for a traveling EM wave in vacuum, using the relationship between the magnitudes of the fields [Eq. (22-7)]. Thus, for the total energy density, we can write

$$u = u_E + u_B = \epsilon_0 E^2 = \frac{1}{\mu_0}B^2 \qquad (22\text{-}8)$$

Since the fields vary from point to point and also change with time, so do the energy densities. Since the fields oscillate rapidly, in most cases we are concerned with the *average* energy densities—the average of the squares of the fields. Recall that an rms (root mean square) value is defined as the square root of the average of the square (Section 21.1):

$$E_{rms} = \sqrt{\langle E^2 \rangle} \quad \text{and} \quad B_{rms} = \sqrt{\langle B^2 \rangle} \qquad (22\text{-}9)$$

The angle brackets around a quantity denote the average value of that quantity. Squaring both sides, we have

$$E_{rms}^2 = \langle E^2 \rangle \quad \text{and} \quad B_{rms}^2 = \langle B^2 \rangle$$

Then the average energy density can be written in terms of the rms values of the fields:

$$\langle u \rangle = \epsilon_0 \langle E^2 \rangle = \epsilon/0 E_{rms}^2 \qquad (22\text{-}10)$$

$$\langle u \rangle = \frac{1}{\mu_0}\langle B^2 \rangle = \frac{1}{\mu_0}B_{rms}^2 \qquad (22\text{-}11)$$

If the electric and magnetic fields are *sinusoidal* functions of time, the rms values are $1/\sqrt{2}$ times the amplitudes (see Section 21.1).

Normal incidence: the direction of propagation of the light is *perpendicular* to the surface.

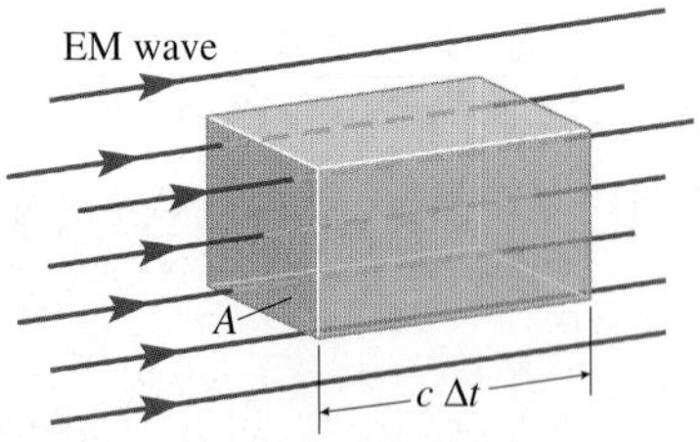

Figure 22.18 Geometry for finding the relationship between energy density and intensity.

Intensity

The energy density tells us how much energy is stored in the wave per unit volume; this energy is being carried with the wave at speed c. Suppose light falls at normal incidence on a surface (e.g., a photographic film or a leaf) and we want to know how much energy hits the surface. For one thing, the energy arriving at the surface depends on how long it is exposed—the reason exposure time is a critical parameter in photography. Also important is the surface area; a large leaf receives more energy than a small one, everything else being equal. Thus, the most useful quantity to know is how much energy arrives at a surface per unit time per unit area—or the average power per unit area. If light hits a surface of area A *at normal incidence*, the **intensity** (I) is

$$I = \frac{\langle P \rangle}{A} \tag{22-12}$$

The SI units of I are

$$\frac{\text{energy}}{\text{time·area}} = \frac{\text{J}}{\text{s·m}^2} = \frac{\text{W}}{\text{m}^2}$$

CONNECTION:

Intensity of an EM wave is defined exactly as for mechanical waves (Section 11.1)—average power per cross-sectional area.

The intensity depends on how much energy is in the wave (measured by u) and the speed at which the energy moves (which is c). If a surface of area A is illuminated by light at normal incidence, how much energy falls on it in a time Δt? The wave moves a distance $c\Delta t$ in that time, so all the energy in a volume $Ac\ \Delta t$ hits the surface during that time (Fig. 22.18). (We are not concerned with what happens to the energy—whether it is absorbed, reflected, or transmitted.) The intensity is then

$$I = \frac{\langle u \rangle V}{A\ \Delta t} = \frac{\langle u \rangle A c \Delta t}{A\ \Delta t} = \langle u \rangle c \tag{22-13}$$

Intensity is proportional to amplitude *squared.*

From Eq. (22-13), the intensity I is proportional to average energy density $\langle u \rangle$, which is proportional to the squares of the rms electric and magnetic fields [Eqs. (22-10) and (22-11)]. If the fields are sinusoidal functions of time, the rms values are $1/\sqrt{2}$ times the amplitudes [Eq. (21-3)]. Therefore, *the intensity is proportional to the squares of the electric and magnetic field amplitudes.*

Example 22.5

EM Fields of a Lightbulb

At a distance of 4.00 m from a 100.0-W lightbulb, what are the intensity and the rms values of the electric and magnetic fields? Assume that all of the electric power goes into EM radiation (mostly in the infrared) and that the radiation is isotropic (equal in all directions).

Strategy Since the radiation is isotropic, the intensity depends only on the distance from the lightbulb. Imagine a sphere surrounding the lightbulb at a distance of 4.00 m. Radiant energy must pass through the surface of the sphere at a rate of 100.0 W. We can figure out the intensity (average power per unit area) and from it the rms values of the fields.

Solution All of the energy radiated by the lightbulb crosses the surface of a sphere of radius 4.00 m. Therefore, the intensity at that distance is the power radiated divided by the surface area of the sphere:

$$I = \frac{\langle P \rangle}{A} = \frac{\langle P \rangle}{4\pi r^2} = \frac{100.0\ \text{W}}{4\pi \times 16.0\ \text{m}^2} = 0.497\ \text{W/m}^2$$

To solve for E_{rms}, we relate the intensity to the average energy density and then the energy density to the field:

$$\langle u \rangle = \frac{I}{c} = \epsilon_0 E_{\text{rms}}^2$$

$$E_{\text{rms}} = \sqrt{\frac{I}{\epsilon_0 c}} = \sqrt{\frac{0.497\ \text{W/m}^2}{8.85 \times 10^{-12}\ \frac{\text{C}^2}{\text{N·m}^2} \times 3.00 \times 10^8\ \text{m/s}}}$$

$$= 13.7\ \text{V/m}$$

continued on next page

Example 22.5 continued

Similarly, for B_{rms},

$$B_{rms} = \sqrt{\frac{\mu_0 I}{c}} = \sqrt{\frac{4\pi \times 10^{-7} \frac{T\cdot m}{A} \times 0.497 \text{ W/m}^2}{3.00 \times 10^8 \text{ m/s}}}$$

$$= 4.56 \times 10^{-8} \text{ T}$$

Discussion A good check would be to calculate the ratio of the two rms fields:

$$\frac{E_{rms}}{B_{rms}} = \frac{13.7 \text{ V/m}}{4.56 \times 10^{-8} \text{ T}} = 3.00 \times 10^8 \text{ m/s} = c$$

as expected.

Practice Problem 22.5 Field of Lightbulb at Greater Distance

What are the rms fields 8.00 m away from the lightbulb? [*Hint:* Look for a shortcut rather than redoing the whole calculation.]

Power and Angle of Incidence If a surface is illuminated by light of intensity I, but the surface is not perpendicular to the incident light, the rate at which energy hits the surface is less than IA. As Fig. 22.19 shows, a perpendicular surface of area $A \cos \theta$ casts a shadow over the surface of area A and thus intercepts all the energy. The **angle of incidence** θ is measured between the direction of the incident light and the normal (a direction *perpendicular* to the surface). Thus, a surface that is not perpendicular to the incident wave receives energy at a rate

$$\langle P \rangle = IA \cos \theta \quad (22\text{-}14)$$

If Eq. (22-14) reminds you of flux, then congratulations on your alertness! The intensity is often called the *flux density.* Electric and magnetic fields are sometimes called *electric flux density* and *magnetic flux density.* However, the flux involved with intensity is not the same as the electric or magnetic fluxes that we defined in Eqs. (16-8) and (20-5). The intensity is the *power* flux density.

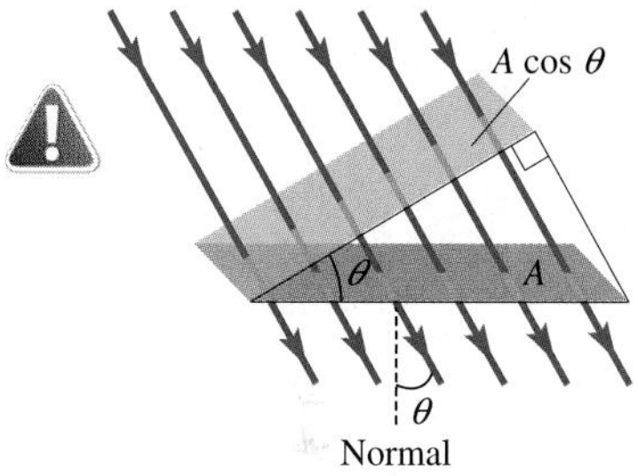

Figure 22.19 The surface of area $A \cos \theta$, which is perpendicular to the incoming wave, intercepts the same light energy as would a surface of area A for which the incoming wave is incident at an angle θ from the normal.

Example 22.6

Power per Unit Area from the Sun on the Summer Solstice

The intensity of sunlight reaching Earth's surface on a clear day is about 1.0 kW/m^2. At a latitude of 40.0° north, find the average power per unit area reaching Earth at noon on the summer solstice (Fig. 22.20a). (The difference is due to the 23.5° inclination of Earth's rotation axis. In summer, the axis is inclined toward the Sun, while in winter it is inclined away from the Sun.)

Strategy Because Earth's surface is not perpendicular to the Sun's rays, the power per unit area falling on Earth is less than 1.0 kW/m^2. We must find the angle that the Sun's rays make with the *normal* to the surface.

Solution A radius going from Earth's center to the surface is normal to the surface at that point, assuming Earth to be a sphere. We need to find the angle between the normal and an incoming ray. At a latitude of 40.0°, the angle between the radius and Earth's axis of rotation is 90.0° − 40.0° = 50.0° (Fig. 22.20a). From the figure, θ + 50.0° + 23.5° = 90.0° and therefore θ = 16.5°. The average power per unit area is then

$$\frac{\langle P \rangle}{A} = I \cos \theta = 1.0 \times 10^3 \text{ W/m}^2 \times \cos 16.5° = 960 \text{ W/m}^2$$

Discussion In Practice Problem 22.6, you will find that the power per unit area at the winter solstice is less than half that at the summer solstice. The intensity of sunlight hasn't changed; what changes is how the energy is spread out on the surface. Fewer of the Sun's rays hit a given surface area when the surface is tilted more.

Earth is actually a bit *closer* to the Sun in the northern hemisphere's winter than in summer. The angle at which the Sun's radiation hits the surface and the number of hours of daylight are much more important in determining the incident power than is the small difference in distance from the Sun.

continued on next page

Example 22.6 continued

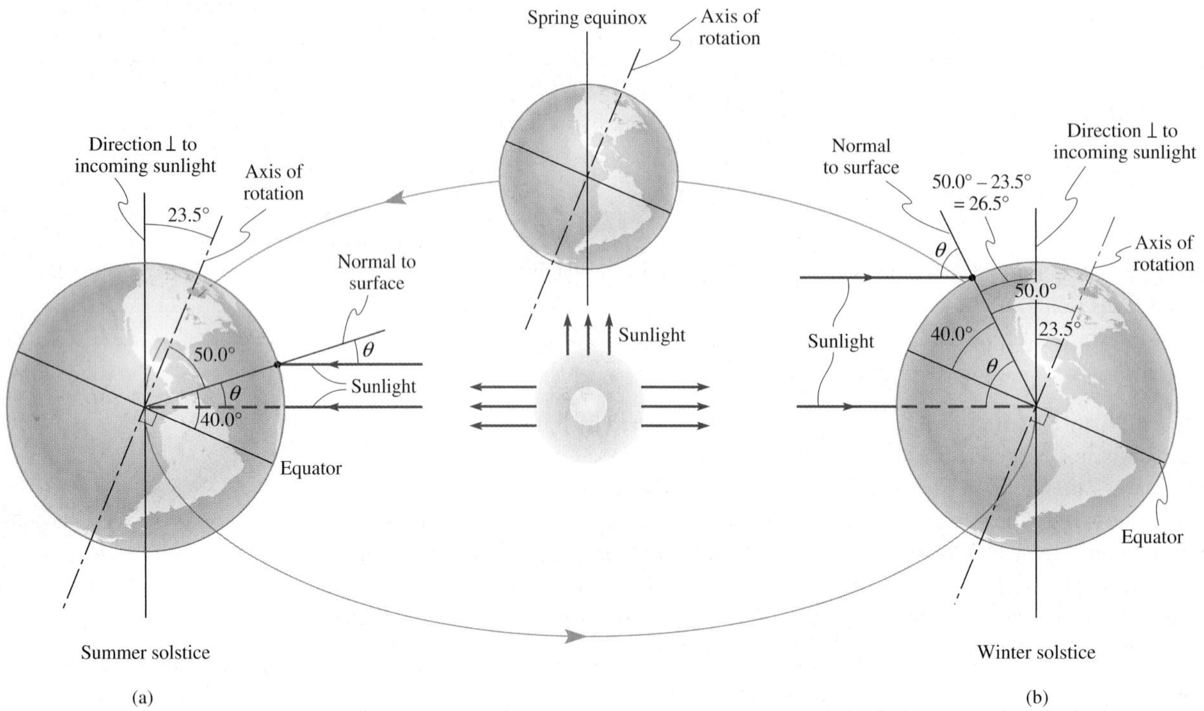

Figure 22.20

(a) At noon on the summer solstice in the northern hemisphere, the rotation axis is inclined 23.5° toward the Sun. At a latitude of 40.0° north, the incoming sunlight is nearly normal to the surface of the Earth. (b) At noon on the winter solstice in the northern hemisphere, the rotation axis is inclined 23.5° *away from* the Sun. At a latitude of 40.0° north, the incoming sunlight makes a large angle with the normal to the surface. (Diagram is *not* to scale.)

Practice Problem 22.6 Average Power on the Winter Solstice

What is the average power per unit area at a latitude of 40.0° north at noon on the winter solstice (Fig. 22.20b)?

22.7 POLARIZATION

Linear Polarization

Imagine a transverse wave traveling along the z-axis. Since this discussion applies to any transverse wave, let us use a transverse wave on a string as an example. In what directions can the string be displaced to produce transverse waves on this string? The displacement could be in the $\pm x$-direction, as in Fig. 22.21a. Or it could be in the $\pm y$-direction, as in Fig. 22.21b. Or it could be in any direction in the xy-plane. In Fig. 22.21c, the displacement of any point on the string from its equilibrium position is parallel to a line that makes an angle θ with the x-axis. These three waves are said to be **linearly polarized**. For the wave in Fig. 22.21a, we would say that the wave is polarized in the $\pm x$-direction (or, for short, in the x-direction).

Linearly polarized waves are also called **plane-polarized**; the two terms are synonymous, despite what you might guess. Each wave in Fig. 22.21 is characterized by a single plane, called the **plane of vibration**, in which the entire string vibrates. For example, the plane of vibration for Fig. 22.21a is the xz-plane. Both the direction of propagation of the wave and the direction of motion of every point of the string lie in the plane of vibration.

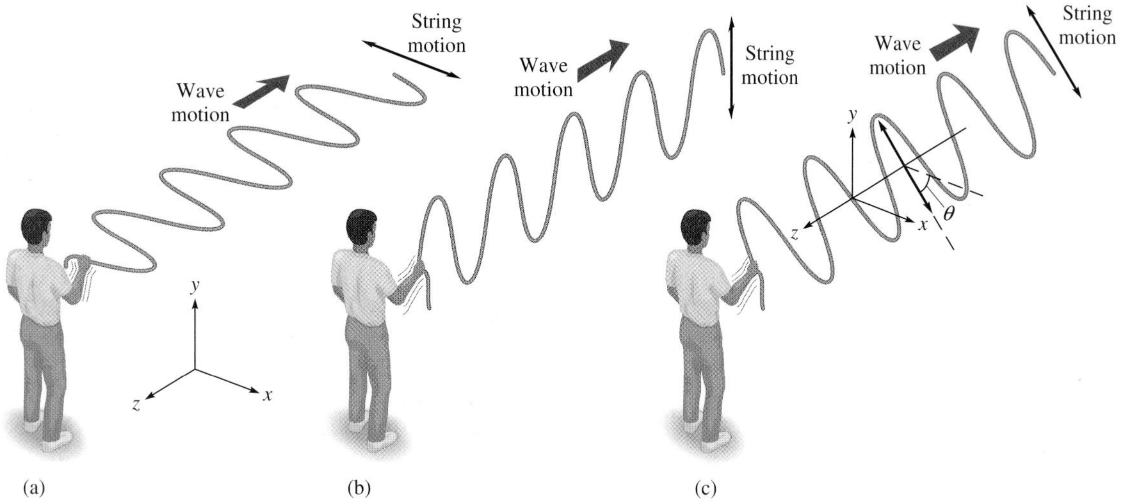

Figure 22.21 Transverse waves on a string with three different linear (plane) polarizations.

Polarization of an EM wave: direction of its electric field

Any transverse wave can be linearly polarized in any direction perpendicular to the direction of propagation. EM waves are no exception. But there are two fields in an EM wave, which are perpendicular to one another. Knowing the direction of one of the fields is sufficient, since $\vec{\mathbf{E}} \times \vec{\mathbf{B}}$ must point in the direction of propagation. By convention, the direction of polarization of EM waves is taken to be the *electric* field direction.

Both electric and magnetic dipole antennas emit radio waves that are linearly polarized. If an FM radio broadcast is transmitted using a horizontal electric dipole antenna, the radio waves at any receiver are linearly polarized. The direction of polarization varies from place to place. If you are due west of the transmitter, the waves that reach you are polarized along the north-south direction, since they must be in the horizontal plane and perpendicular to the direction of propagation (which is west in this case). For best reception, an electric dipole antenna should be aligned with the direction of polarization of the radio waves, since it is the electric field that drives current in the antenna.

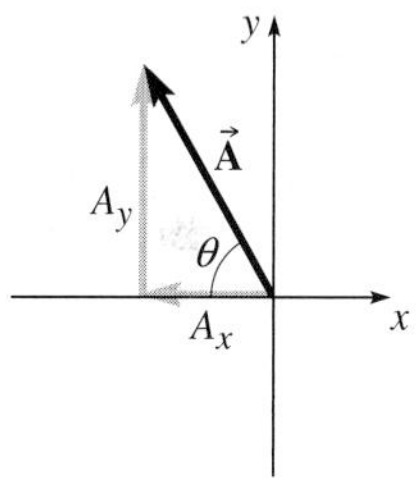

Figure 22.22 Any linearly polarized wave can be thought of as a superposition of two perpendicular polarizations, since displacements—as well as electric and magnetic fields—are vectors.

In Section 22.3, we said that if an electric dipole antenna is not lined up with the electric field of the wave, then the emf is reduced by a factor of cos θ, where θ is the angle between $\vec{\mathbf{E}}$ and the antenna. Think about this in terms of polarization. Any linearly polarized wave can be thought of as the superposition of two perpendicular linearly polarized waves along any axes we choose. Displacements are vectors and vectors can always be written as the sum of perpendicular components; therefore, the transverse wave on the string in Fig. 22.21c can be thought of as the superposition of two waves, one polarized in the x-direction and the other in the y-direction. If the amplitude of the wave is A, the amplitude of the "x-component wave" is $A \cos \theta$ and the amplitude of the "y-component wave" is $A \sin \theta$ (see Fig. 22.22).

The same is true for EM waves, since the electric and magnetic fields are vectors. Any linearly polarized EM wave can be regarded as the sum of two waves polarized along perpendicular axes. If an electric dipole antenna makes an angle θ with the electric field of a wave, only the component of $\vec{\mathbf{E}}$ along the antenna makes electrons move back and forth along the antenna. If we think of the wave as two perpendicular polarizations, the antenna responds to the polarization parallel to it while the perpendicular polarization has no effect.

Random Polarization

The light coming from an incandescent lightbulb is **unpolarized** or **randomly polarized**. The direction of the electric field changes rapidly and in a random way. Antennas emit linearly polarized waves because the motion of the electrons up and down the

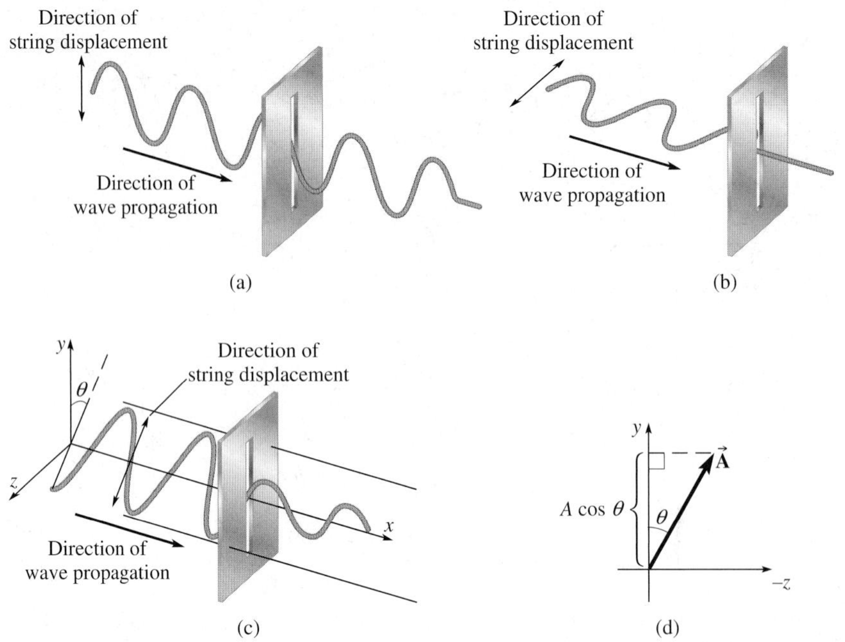

Figure 22.23 For transverse waves *on a string*, a vertical slot allows vertically polarized waves to pass through (a), but not horizontally polarized waves (b). If the incident wave is polarized at an angle θ to the vertical (c), a vertically polarized wave of amplitude $A \cos \theta$ is transmitted (d).

antenna is orderly and always along the same line. Thermal radiation (which is mostly IR, but also includes visible light) from a lightbulb is caused by the thermal vibrations of huge numbers of atoms. The atoms are essentially independent of each other; nothing makes them vibrate in step or in the same direction. The wave is therefore made up of the superposition of a huge number of waves whose electric fields are in random, uncorrelated directions. Thermal radiation is always unpolarized, whether it comes from a lightbulb, from a wood stove (mostly IR), or from the Sun.

Polarizers

Devices called *polarizers* transmit linearly polarized waves in a fixed direction regardless of the polarization state of the incident waves. A polarizer for transverse waves on a string is shown in Fig. 22.23. The vertical slot enables the string to slide vertically without friction, but prevents any horizontal motion. When a vertically polarized wave is sent down the string toward the slot, it passes through (Fig. 22.23a). A horizontally polarized wave does not pass through (Fig. 22.23b); it is reflected since the slot acts like a fixed end for horizontal motion. The direction of the slot is called the *transmission axis* since the polarizer transmits waves polarized in that direction.

What if a linearly polarized wave is sent toward the polarizer, as in Fig. 22.23c, where the incident wave is polarized at an angle θ to the transmission axis? The incident wave can be decomposed into components parallel and perpendicular to the transmission axis; the parallel wave passes through. If the incident wave has amplitude A, then the transmitted wave has amplitude $A \cos \theta$ (Fig. 22.23d).

A polarizer for microwaves consists of many parallel strips of metal (Fig. 22.24). The spacing of the strips must be significantly less than the wavelength of the microwaves. The strips act as little antennas. The parallel component of the electric field of the incident wave makes currents flow up and down the metal strips. These currents dissipate energy, so some of the wave is absorbed. The antennas also produce a wave of their own; it is out of phase with the incident wave, so it cancels the parallel-component of $\vec{\mathbf{E}}$ in the forward-going wave and sends a reflected wave back. Between absorption and reflection, none of the electric field parallel to the metal strips gets through the polarizer. The microwaves that are transmitted are linearly polarized *perpendicular to the strips.* Although the microwave polarizer *looks* similar to the polarizer for waves on a string, the electric field does not pass through the "slots" between the metal strips! The transmission axis of the polarizer is *perpendicular* to the strips.

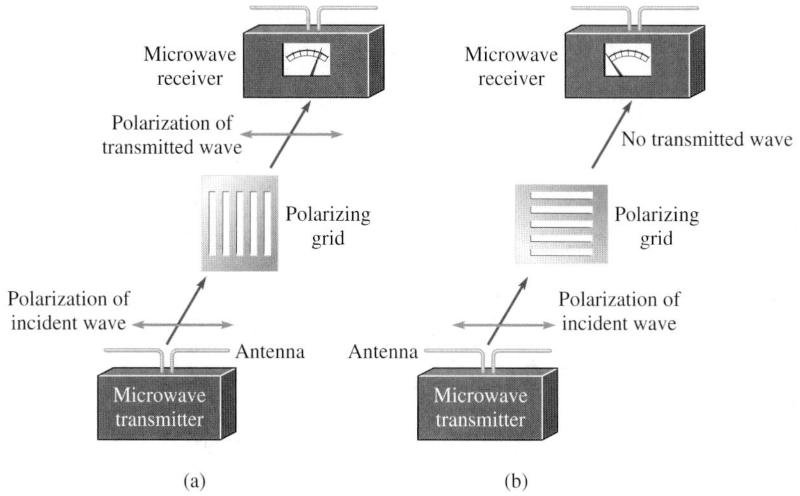

Figure 22.24 A polarizing grid for microwaves. (a) A horizontally polarized microwave beam passes through the polarizer if its strips are oriented vertically, but (b) is blocked by a polarizer with horizontal strips.

Sheet polarizers for visible light operate on a principle similar to that of the wire grid polarizer. A sheet polarizer contains many long hydrocarbon chains with iodine atoms attached. In production, the sheet is stretched so that these long molecules are all aligned in the same direction. The iodine atoms allow electrons to move easily along the chain, so the aligned polymers behave as parallel conducting wires, and their spacing is close enough that it does to visible light what a wire grid polarizer does to microwaves. The sheet polarizer has a transmission axis perpendicular to the aligned polymers.

Ideal Polarizers

If randomly polarized light is incident on an ideal polarizer, the transmitted intensity is half the incident intensity, regardless of the orientation of the transmission axis (Fig. 22.25a). The randomly polarized wave can be thought of as two perpendicular polarized waves that are *uncorrelated*—the relative phase of the two varies rapidly with time. Half of the energy of the wave is associated with each of the two perpendicular polarizations.

$$I = \tfrac{1}{2}I_0 \quad \text{(incident wave unpolarized, ideal polarizer)} \qquad (22\text{-}15)$$

If, instead, the incident wave is linearly polarized, then the component of $\vec{\mathbf{E}}$ parallel to the transmission axis gets through (Fig. 22.25b). If θ is the angle between the incident polarization and the transmission axis, then

$$E = E_0 \cos\theta \quad \text{(incident wave polarized, ideal polarizer)} \qquad (22\text{-}16a)$$

Since intensity is proportional to the square of the amplitude, the transmitted intensity is

$$I = I_0 \cos^2\theta \quad \text{(incident wave polarized, ideal polarizer)} \qquad (22\text{-}16b)$$

Equation (22-16b) is called **Malus's law** after its discoverer Étienne-Louis Malus (1775–1812), an engineer and one of Napoleon's captains.

When applying Malus's law, be sure to use the correct angle. In Eqs. (22-16), θ is the angle between the *polarization direction of the incident light* and the transmission axis of the polarizer.

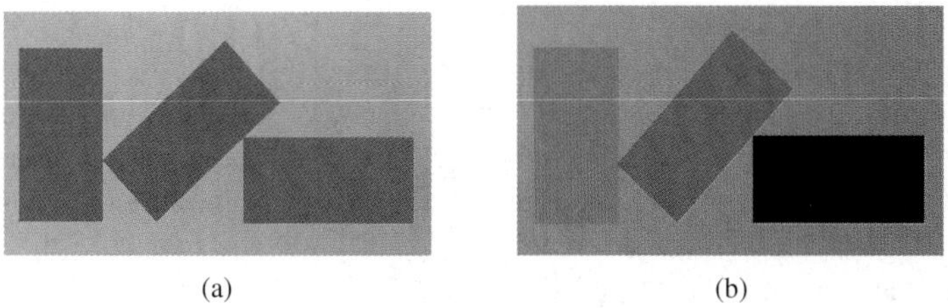

Figure 22.25 (a) Unpolarized light is incident on three polarizers oriented in different directions. The transmitted intensity is the same for all three. (b) Linearly polarized light is incident on the same three polarizers. Note that the transmitted intensity for $\theta = 0$ is slightly less than the incident intensity—these are not *ideal* polarizers.

CHECKPOINT 22.7

Light with intensity I_0 is incident on an ideal polarizing sheet. The transmitted intensity is $\frac{1}{2}I_0$. How can you determine whether the incident light is randomly polarized or linearly polarized? If it is linearly polarized, what is the direction of its polarization?

Example 22.7

Unpolarized Light Incident on Two Polarizers

Randomly polarized light of intensity I_0 is incident on two sheet polarizers (Fig. 22.26). The transmission axis of the first polarizer is vertical; that of the second makes a 30.0° angle with the vertical. What is the intensity and polarization state of the light after passing through the two?

Strategy We treat each polarizer separately. First we find the intensity of light transmitted by the first polarizer. The light transmitted by a polarizer is always linearly polarized parallel to the transmission axis of the polarizer, since only the component of $\vec{\mathbf{E}}$ parallel to the transmission axis gets through. Then we know the intensity and polarization state of the light that is incident on the second polarizer.

Solution When randomly polarized light passes through a polarizer, the transmitted intensity is half the incident intensity [Eq. (22-15)] since the wave has equal amounts of energy associated with its two perpendicular (but uncorrelated) components.

$$I_1 = \tfrac{1}{2}I_0$$

The light is now linearly polarized parallel to the transmission axis of the first polarizer, which is vertical.

The component of the electric field parallel to the transmission axis of the second polarizer passes through. The amplitude is thus reduced by a factor cos 30.0° and, since intensity is proportional to amplitude squared, the intensity is reduced by a factor $\cos^2 30.0°$ (Malus's law). The intensity transmitted through the second polarizer is

$$I_2 = I_1 \cos^2 30.0° = \tfrac{1}{2}I_0 \cos^2 30.0° = 0.375I_0$$

The light is now linearly polarized 30.0° from the vertical.

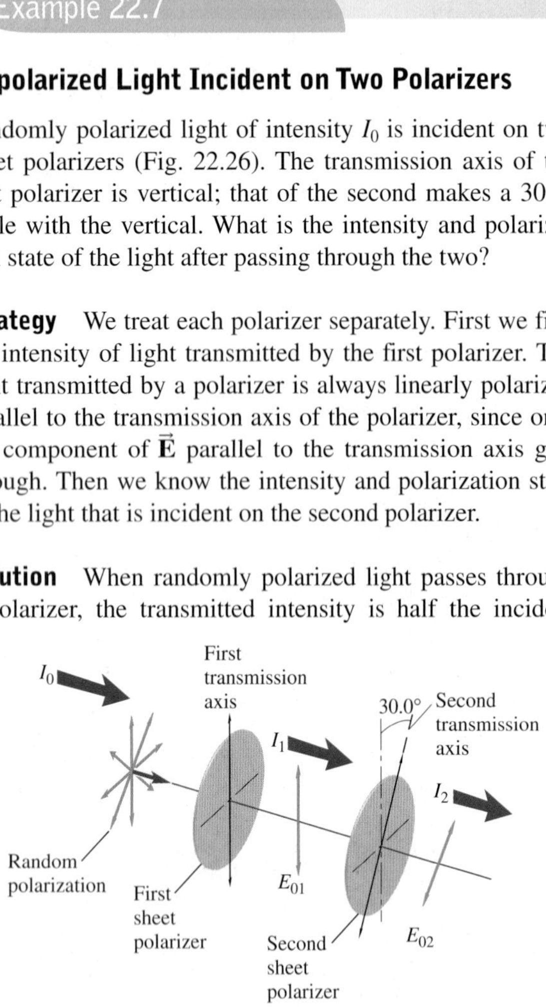

Figure 22.26

The circular disks are polarizing sheets with their transmission axes marked.

Discussion For problems involving two or more polarizers in series, treat each polarizer in turn. Use the intensity and polarization state of the light that emerges from one polarizer as the incident intensity and polarization for the next polarizer.

Practice Problem 22.7 Minimum and Maximum Intensities

If randomly polarized light of intensity I_0 is incident on two polarizers, what are the maximum and minimum possible intensities of the transmitted light as the angle between the two transmission axes is varied?

Application: Liquid Crystal Displays

Liquid crystal displays (LCDs) are commonly found in flat-panel computer screens, calculators, digital watches, and digital meters. In each segment of the display, a liquid crystal layer is sandwiched between two finely grooved surfaces with their grooves perpendicular (Fig. 22.27a). As a result the molecules twist 90° between the two surfaces. When a voltage is applied across the liquid crystal layer, the molecules line up in the direction of the electric field (Fig. 22.27b).

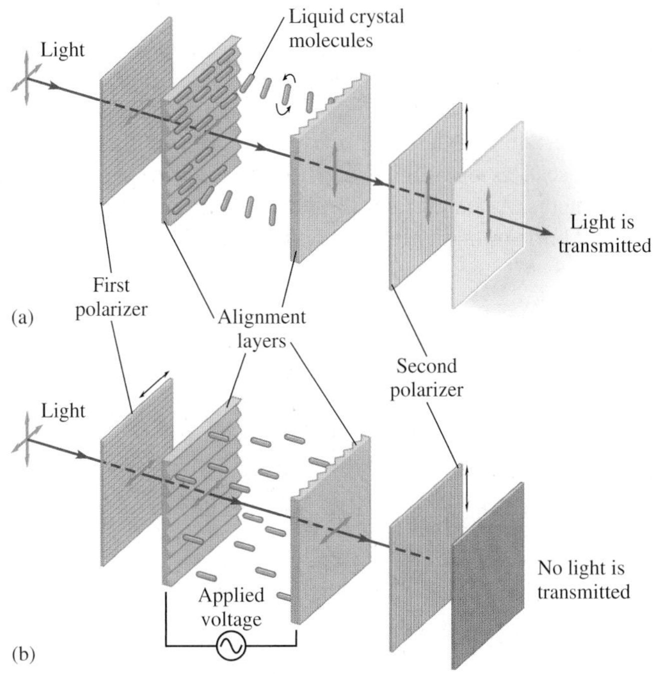

Figure 22.27 (a) When no voltage is applied to the liquid crystal, it rotates the polarization of the light so it can pass through the second polarizing sheet. (b) When a voltage is applied to the liquid crystal, no light is transmitted through the second polarizing sheet.

Unpolarized light from a small fluorescent bulb is polarized by one polarizing sheet. The light then passes through the liquid crystal and then through a second polarizing sheet with its transmission axis perpendicular to the first. When no voltage is applied, the liquid crystal rotates the polarization of the light by 90° and the light can pass through the second polarizer (Fig. 22.27a). When a voltage is applied, the liquid crystal transmits light without changing its polarization; the second polarizer blocks transmission of the light (Fig. 22.27b). When you look at an LCD display, you see the light transmitted by the second sheet. If a segment has a voltage applied to it, no light is transmitted; we see a black segment. If a segment of liquid crystal does not have an applied voltage, it transmits light and we see the same gray color as the background.

Polarization by Scattering

Although the radiation emitted by the Sun is unpolarized, much of the sunlight that we see is **partially polarized**. Partially polarized light is a mixture of unpolarized and linearly polarized light. A sheet polarizer can be used to distinguish linearly polarized, partially polarized, and unpolarized light. The polarizer is rotated and the transmitted intensity at different angles is noted. If the incident light is unpolarized, the intensity stays constant as the polarizer is rotated. If the incident light is linearly polarized, the intensity is zero in one orientation and maximum at a perpendicular orientation. If partially polarized light is analyzed in this way, the transmitted intensity varies as the polarizer is rotated, but it is not zero for *any* orientation; it is maximum in one orientation and minimum (but nonzero) in a perpendicular orientation. A polarizer used to analyze the polarization state of light is often called an *analyzer.*

Natural, unpolarized light becomes partially polarized when it is scattered or reflected. So, unless you look straight at the Sun (which can cause severe eye damage—do not try it!), the sunlight that reaches you has been scattered or reflected and thus is partially polarized. Common polarized sunglasses consist of a sheet polarizer, oriented to absorb the preferential direction of polarization of light reflected from horizontal surfaces, such as a road or the water on a lake, and to reduce the glare of scattered light in the air. Polarized sunglasses are often used in boating and aviation because they preferentially cut down on glare rather than indiscriminately reducing the intensity for all polarization states (Fig. 22.28).

(a)

(b)

Figure 22.28 Photo of a lake in Yosemite National Park taken without (a) and with (b) a polarizing filter in front of the camera's lens. The filter reduces the amount of reflected glare from the surface of the lake.

Application: polarized sunglasses

How do scattering and reflection make the light partially polarized? Let's look at scattering; polarization by reflection is discussed in Section 23.5. The blue sky we see on sunny days is sunlight that is scattered by molecules in the air. On the Moon, there is no blue sky because there is no atmosphere. Even during the day, the sky is as black as at night, although the Sun and the Earth may be brightly shining above (Fig. 22.29). Earth's atmosphere scatters blue light, with its shorter wavelengths, more than light with longer wavelengths. At sunrise and sunset, we see the light left over after much of the blue is scattered out—primarily red and orange.

Application: colors of the sky during the day and at sunset

Figure 22.29 An astronaut walks away from the lunar module *Intrepid* while a brilliant Sun shines above the Apollo 12 base. Notice that the sky is dark even though the Sun is above the horizon; the Moon lacks an atmosphere to scatter sunlight and form a blue sky.

PHYSICS AT HOME

Take a pair of inexpensive polarized sunglasses outside on a sunny day and analyze the polarization of the sky in various directions (but do not look directly at the Sun, even through sunglasses!). Get a second pair of sunglasses so you can put two polarizers in series. Rotate the one closest to you while holding the other in the same orientation. When is the transmitted intensity maximum? When is it minimum?

The same scattering process that makes the sky blue and the sunset red also polarizes the scattered light. Figure 22.30 shows unpolarized sunlight being scattered by a molecule in the atmosphere. In this case, the incident light is horizontal, as would occur shortly before sunset. In response to the electric field of the wave, charges in the molecule oscillate—the molecule becomes an oscillating dipole. Since the incoming wave is unpolarized, the dipole does not oscillate along a single axis, but does so in random directions perpendicular to the incident wave. As an oscillating dipole, the molecule radiates EM waves. An oscillating dipole radiates most strongly in directions perpendicular to its axis; *it does not radiate at all in directions parallel to its axis.*

North-south oscillation of the molecular dipole radiates in the three directions A, B, and C equally, since those directions are all perpendicular to the north-south axis of the dipole. Vertical oscillation of the molecular dipole radiates most strongly in a horizontal plane (including A). Vertical oscillation radiates more weakly in direction B and not at all in direction C. Therefore, in direction C, the light is linearly polarized in the north-south direction. Generalizing this observation, light scattered through 90° is polarized in a direction that is perpendicular both to the direction of the incident light and to the direction of the scattered light.

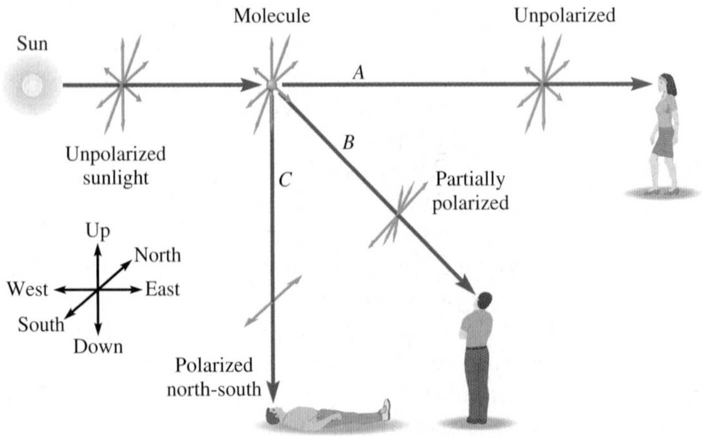

Figure 22.30 Unpolarized sunlight is scattered by the atmosphere. (In this illustration, it is early evening, so the incident light from the Sun comes in horizontally from west to east.) A person looking straight up at the sky sees light that is scattered through 90°. This light (C) is polarized north-south, which is perpendicular both to the direction of propagation of incident light (east) and to the direction of propagation of scattered light (down).

Conceptual Example 22.8

Light Polarized by Scattering

At noon, if you look at the sky just above the horizon toward the east, in what direction is the light polarized?

Strategy At noon, sunlight travels straight down (approximately). Some of the light is scattered by the atmosphere through roughly 90° and then travels westward toward the observer. We consider the unpolarized light from the Sun to be a random mixture of two perpendicular polarizations. Looking at each polarization by itself, we determine how effectively a molecule can scatter the light downward. A sketch of the situation is crucial.

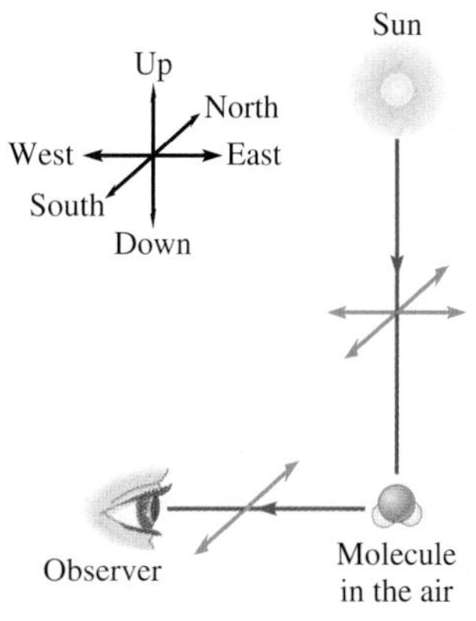

Figure 22.31 Light traveling downward from the Sun is an uncorrelated mixture of both east-west and north-south polarizations. The two polarizations are represented by double-headed arrows. The light scattered westward is polarized along the north-south direction.

Solution and Discussion Figure 22.31 shows light traveling downward from the Sun as a mixture of north-south and east-west polarizations. Now we treat the two polarizations one at a time.

The north-south electric fields cause charges in the molecule to oscillate along a north-south axis. An oscillating dipole radiates most strongly in all directions perpendicular to the dipole axis, including in the westward direction of the scattered light we want to analyze.

The east-west electric fields produce an oscillating dipole with an east-west axis. An oscillating dipole radiates only weakly in directions nearly parallel to its axis. Therefore, the light scattered westward is polarized in the north-south direction.

Conceptual Practice Problem 22.8 Looking North

Just before sunset, if you look north at the sky just over the horizon, in what direction is the light partially polarized?

Application: Bees Can Detect the Polarization of Light

How can bees navigate on cloudy days?

A bee has a compound eye consisting of thousands of transparent fibers called the ommatidia. Each ommatidium has one end on the hemispherical surface of the compound eye (Fig. 22.32) and is sensitive to light coming from the direction along which the fiber is aligned.

Each ommatidium is made up of nine cells. One of these cells is sensitive to the polarization of the incident light. The bee can therefore detect the polarization state of light coming from various directions. When the Sun is not visible, the bee can infer the position of the Sun from the polarization of scattered light, as was established by a series of ingenious experiments by Karl von Frisch and others in the 1960s. Using polarizing sheets, von Frisch and his colleagues could change the apparent polarization state of the scattered sunlight and watch the effects on the flight of the bees.

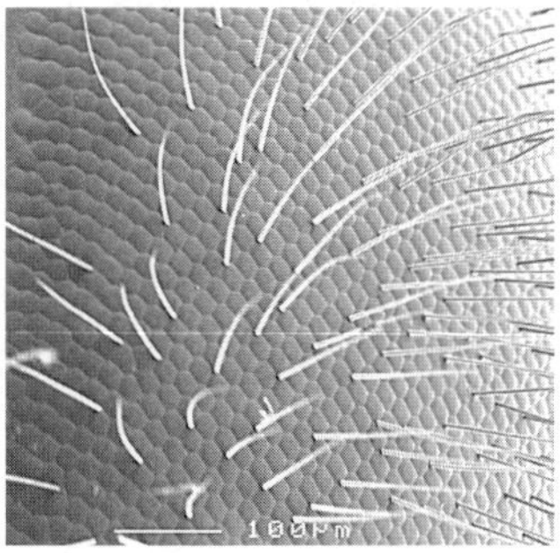

Figure 22.32 Electron micrograph of the compound eye of a bee. The "bumps" are the outside surfaces of the ommatidia.

CONNECTION:

Doppler effect: The observed frequency of a wave is affected by the motion of the source or observer (Section 12.8). With sound, the motion of source and observer are measured with respect to the wave medium. For EM waves in vacuum, the Doppler shift depends only on the relative motion of source and observer.

22.8 THE DOPPLER EFFECT FOR EM WAVES

The Doppler effect exists for all kinds of waves, including EM waves. However, the Doppler formula [Eq. (12-14)] derived for sound cannot be correct for EM waves. Those equations involve the velocity of the source and the observer *relative to the medium through which the sound travels.* For sound waves in air, v_s and v_o are measured *relative to the air.* Since EM waves do not require a medium, the Doppler shift for light involves only the *relative* velocity of the observer and the source.

Using Einstein's relativity, the Doppler shift formula for light can be derived:

$$f_o = f_s \sqrt{\frac{1 + v_{rel}/c}{1 - v_{rel}/c}} \qquad (22\text{-}17)$$

In Eq. (22-17), v_{rel} is positive if the source and observer are approaching (getting closer together) and negative if receding (getting farther apart). If the relative speed of source and observer is much less than c, a simpler expression can be found using the binomial approximations found in Appendix A.5:

$$\left(1 + \frac{v_{rel}}{c}\right)^{1/2} \approx 1 + \frac{v_{rel}}{2c} \qquad \text{and} \qquad \left(1 - \frac{v_{rel}}{c}\right)^{-1/2} \approx 1 + \frac{v_{rel}}{2c}$$

Substituting these approximations into Eq. (22-17),

$$f_o \approx f_s \left(1 + \frac{v_{rel}}{2c}\right)^2$$

$$f_o \approx f_s \left(1 + \frac{v_{rel}}{c}\right) \qquad (22\text{-}18)$$

where in the last step we used the binomial approximation once more.

Example 22.9

A Speeder Caught by Radar

A police car is moving at 38.0 m/s (85.0 mi/h) to catch up with a speeder directly ahead. The speed limit is 29.1 m/s (65.0 mi/h). A police car radar "clocks" the speed of the other car by emitting microwaves with frequency 3.0×10^{10} Hz and observing the frequency of the reflected wave. The reflected wave, when combined with the outgoing wave, produces beats at a rate of 1400 s^{-1}. How fast is the speeder going? [*Hint:* First find the frequency "observed" by the speeder. The electrons in the metal car body oscillate and emit the reflected wave with this same frequency. For the reflected wave, the speeder is the source and the police car is the observer.]

Strategy There are *two* Doppler shifts, since the EM wave is reflected off the car. We can first think of the car as the observer, receiving a Doppler-shifted radar wave from the police car (Fig. 22.33a). Then the car "rebroadcasts" this wave back to the police car (Fig. 22.33b). This time the speeder's car is the source and the police car is the observer. The relative speed of the two cars is *much* less than the speed of light, so we use the approximate formula [Eq. (22-18)].

There are three different frequencies in the problem. Let's call the frequency emitted by the police car $f_1 = 3.0 \times 10^{10}$ Hz, the frequency received by the speeder f_2, and the frequency of the reflected wave as observed by the police car f_3. The police car is catching up to the speeder, so the source and observer are approaching; therefore, v_{rel} is positive and the Doppler shift is toward higher frequencies.

Solution The beat frequency is

$$f_{beat} = f_3 - f_1 \qquad (12\text{-}11)$$

The frequency observed by the speeder is

$$f_2 = f_1\left(1 + \frac{v_{rel}}{c}\right)$$

continued on next page

Example 22.9 continued

Now the speeder's car emits a microwave of frequency f_2. The frequency observed by the police car is

$$f_3 = f_2\left(1 + \frac{v_{\text{rel}}}{c}\right) = f_1\left(1 + \frac{v_{\text{rel}}}{c}\right)^2$$

We need to solve for v_{rel}. We can avoid solving a quadratic equation by using the binomial approximation:

$$f_3 = f_1\left(1 + \frac{v_{\text{rel}}}{c}\right)^2 \approx f_1\left(1 + 2\frac{v_{\text{rel}}}{c}\right)$$

Solving for v_{rel},

$$v_{\text{rel}} = \frac{1}{2}c\left(\frac{f_3}{f_1} - 1\right) = \frac{1}{2}c\left(\frac{f_3 - f_1}{f_1}\right) = \frac{1}{2}c\left(\frac{f_{\text{beat}}}{f_1}\right)$$

$$= \frac{1}{2} \times 3.00 \times 10^8 \text{ m/s} \times \frac{1400 \text{ Hz}}{3.0 \times 10^{10} \text{ Hz}} = 7.0 \text{ m/s}$$

Since the two are approaching, the speeder is moving at less than 38.0 m/s. Relative to the road, the speeder is moving at

$$38.0 \text{ m/s} - 7.0 \text{ m/s} = 31.0 \text{ m/s} \quad (= 69.3 \text{ mi/h})$$

Perhaps the police officer will be kind enough to give only a warning this time.

Discussion Using the approximate form for the Doppler shift greatly simplifies the algebra. Using the exact form would be much more difficult and in the end would give the same answer. The speeds involved are so much less than c that the error is truly negligible.

Practice Problem 22.9 Reflection from Stationary Objects

Suppose the police car is moving at 23 m/s. What beat frequency results when the radar is reflected from stationary objects?

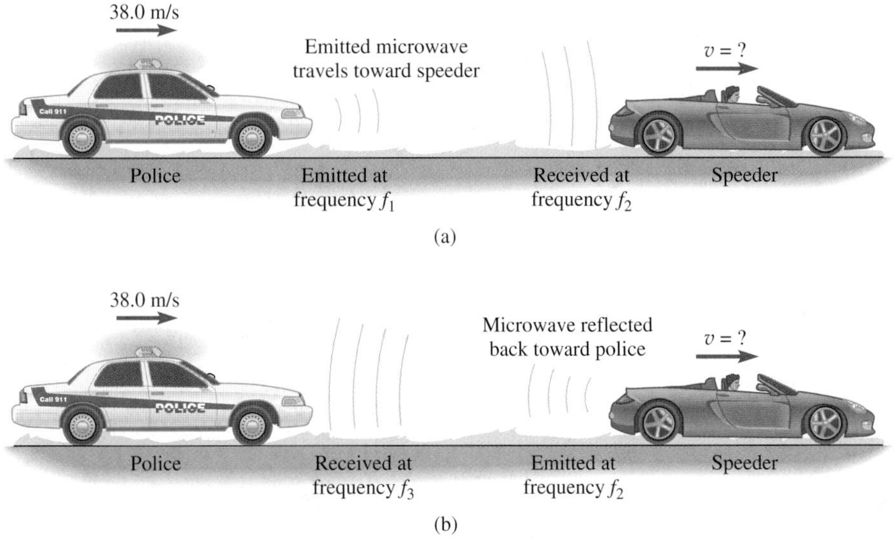

Figure 22.33
(a) The police car emits microwaves at frequency f_1. The speeder receives them at a Doppler-shifted frequency f_2. (b) The wave is reflected at frequency f_2; the police car receives the reflected wave at frequency f_3.

Applications: Doppler radar and evidence for the expansion of the universe

Radar used by meteorologists can provide information about the position of storm systems. Now they use *Doppler radar*, which also provides information about the velocity of storm systems. Another important application of the Doppler shift of visible light is the evidence it gives for the expansion of the universe. Light reaching Earth from distant stars is *red-shifted.* That is, the spectrum of visible light is shifted downward in frequency toward the red. According to *Hubble's law* (named for American astronomer Edwin Hubble, 1889–1953), the speed at which a galaxy moves away from ours is proportional to how far from us the galaxy is. Thus, the Doppler shift can be used to determine a star or galaxy's distance from Earth.

Looking out at the universe, the red shift tells us that other galaxies are moving away from ours in all directions; the farther away the galaxy, the faster it is receding from us and the greater the Doppler shift of the light that reaches Earth. This doesn't mean that Earth is at the center of the universe; in an expanding universe, observers on a planet *anywhere* in the universe would see distant galaxies moving away from it in all directions. Ever since the Big Bang, the universe has been expanding. Whether it continues to expand forever, or whether the expansion will stop and the universe collapse into another big bang, is a central question studied by cosmologists and astrophysicists.

Master the Concepts

- EM waves consist of oscillating electric and magnetic fields that propagate away from their source. EM waves always have both electric and magnetic fields.
- The Ampère-Maxwell law is Ampère's law modified by Maxwell so that a changing electric field generates a magnetic field.
- The Ampère-Maxwell law, along with Gauss's law, Gauss's law for magnetism, and Faraday's law, are called Maxwell's equations. They describe completely the electric and magnetic fields. Maxwell's equations say that $\vec{\mathbf{E}}$- and $\vec{\mathbf{B}}$-field lines do not have to be tied to matter. Instead, they can break free and electromagnetic waves can travel far from their sources.
- Radiation from a dipole antenna is weakest along the antenna's axis and strongest in directions perpendicular to the axis. Electric dipole antennas and magnetic dipole antennas can be used either as sources of EM waves or as receivers of EM waves.
- The electromagnetic spectrum—the range of frequencies and wavelengths of EM waves—is traditionally divided into named regions. From lowest to highest frequency, they are: radio waves, microwaves, infrared, visible, ultraviolet, x-rays, and gamma rays.
- EM waves of any frequency travel through vacuum at a speed

$$c = \frac{1}{\sqrt{\epsilon_0 \mu_0}} = 3.00 \times 10^8 \text{ m/s} \qquad (22\text{-}3)$$

- EM waves can travel through matter, but they do so at speeds less than c. The index of refraction for a material is defined as

$$n = \frac{c}{v} \qquad (22\text{-}4)$$

where v is the speed of EM waves through the material.

- The speed of EM waves (and therefore also the index of refraction) in *matter* depends on the frequency of the wave.

- When an EM wave passes from one medium to another, the wavelength changes; the frequency remains the same. The wave in the second medium is created by the oscillating charges at the boundary, so the fields in the second medium must oscillate at the same frequency as the fields in the first.
- Properties of EM waves in vacuum:

The electric and magnetic fields oscillate at the *same frequency* and are *in phase.*

$$|\vec{\mathbf{E}}(x, y, z, t)| = c\,|\vec{\mathbf{B}}(x, y, z, t)| \qquad (22\text{-}7)$$

$\vec{\mathbf{E}}$, $\vec{\mathbf{B}}$, and the direction of propagation are three mutually perpendicular directions.

$\vec{\mathbf{E}} \times \vec{\mathbf{B}}$ is always in the direction of propagation.

The electric energy density is equal to the magnetic energy density.

- Energy density (SI unit: J/m^3) of an EM wave in vacuum:

$$\langle u \rangle = \epsilon_0 \langle E^2 \rangle = \epsilon_0 E_{\text{rms}}^2 = \frac{1}{\mu_0}\langle B^2 \rangle = \frac{1}{\mu_0} B_{\text{rms}}^2 \qquad (22\text{-}10, 11)$$

- The intensity (SI unit: W/m^2) is

$$I = \langle u \rangle c \qquad (22\text{-}13)$$

Intensity is proportional to the squares of the electric and magnetic field amplitudes.

- The average power incident on a surface of area A is

$$\langle P \rangle = IA \cos \theta \qquad (22\text{-}14)$$

where θ is 0° for normal incidence and 90° for grazing incidence.

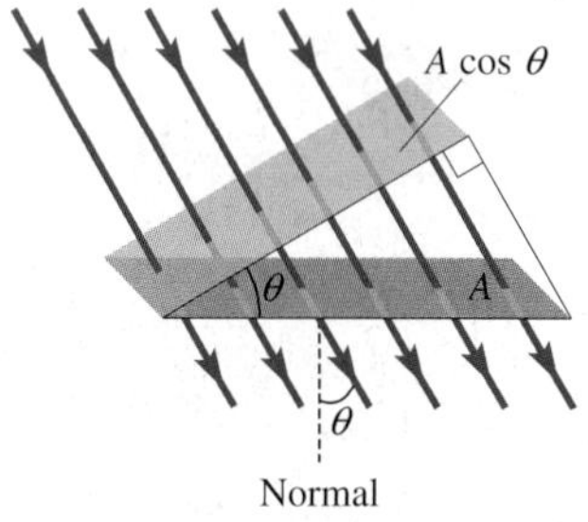

- The polarization of an EM wave is the direction of its electric field.
- If unpolarized waves pass through a polarizer, the transmitted intensity is half the incident intensity:

$$I = \tfrac{1}{2} I_0 \qquad (22\text{-}15)$$

- If a linearly polarized wave is incident on a polarizer, the component of $\vec{\mathbf{E}}$ parallel to the transmission axis gets through. If θ is the angle between the incident polarization and the transmission axis, then

$$E = E_0 \cos \theta \qquad (22\text{-}16a)$$

Since intensity is proportional to the square of the amplitude, the transmitted intensity is

$$I = I_0 \cos^2 \theta \qquad (22\text{-}16b)$$

continued on next page

Master the Concepts continued

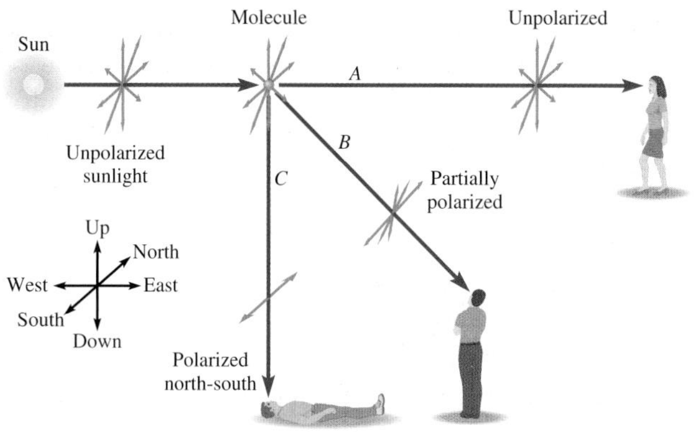

- Unpolarized light can be partially polarized due to scattering or reflection.
- The Doppler effect for EM waves:

$$f_o = f_s \sqrt{\frac{1 + v_{rel}/c}{1 - v_{rel}/c}} \qquad (22\text{-}17)$$

where v_{rel} is positive if the source and observer are approaching, and negative if receding. If the relative speed of source and observer is much less than c,

$$f_o \approx f_s\left(1 + \frac{v_{rel}}{c}\right) \qquad (22\text{-}18)$$

Conceptual Questions

1. In Section 22.3, we stated that an electric dipole antenna should be aligned with the electric field of an EM wave for best reception. If a magnetic dipole antenna is used instead, should its axis be aligned with the magnetic field of the wave? Explain.
2. A magnetic dipole antenna has its axis aligned with the vertical. The antenna sends out radio waves. If you are due south of the antenna, what is the polarization state of the radio waves that reach you?
3. Linearly polarized light of intensity I_0 shines through two polarizing sheets. The second of the sheets has its transmission axis perpendicular to the polarization of the light before it passes through the first sheet. Must the intensity transmitted through the second sheet be zero, or is it possible that some light gets through? Explain.
4. Using Faraday's law, explain why it is impossible to have a magnetic wave without any electric component.
5. According to Maxwell, why is it impossible to have an electric wave without any magnetic component?
6. Zach insists that the seasons are caused by the elliptical shape of Earth's orbit. He says that it is summer when Earth is closest to the Sun and winter when it is farthest away from the Sun. What evidence can you think of to show that the seasons are *not* due to the change in distance between Earth and the Sun?
7. Why are days longer in summer than in winter?
8. Describe the polarization of radio waves transmitted from a horizontal electric dipole antenna that travel parallel to Earth's surface.
9. The figure shows a magnetic dipole antenna transmitting an electromagnetic wave. At a point P far from the antenna, what are the directions of the electric and magnetic fields of the wave?

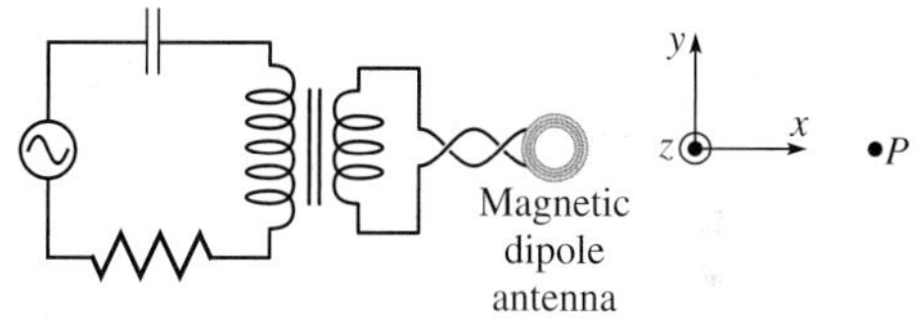

Conceptual Question 9 and Problem 33

10. In everyday experience, visible light seems to travel in straight lines while radio waves do not. Explain.
11. A light wave passes through a hazy region in the sky. If the electric field vector of the emerging wave is one quarter that of the incident wave, what is the ratio of the transmitted intensity to the incident intensity?
12. Can sound waves be polarized? Explain.
13. Until the Supreme Court ruled it to be unconstitutional, drug enforcement officers examined buildings at night with a camera sensitive to infrared. How did this help them identify marijuana growers?
14. The amplitudes of an EM wave are related by $E_m = cB_m$. Since $c = 3.00 \times 10^8$ m/s, a classmate says that the electric field in an EM wave is much larger than the magnetic field. How would you reply?
15. Why is it warmer in summer than in winter?
16. Why is the antenna on a cell phone shorter than the radio antenna on a car?

Multiple-Choice Questions

1. The radio station that broadcasts your favorite music is located exactly north of your home; it uses a horizontal electric dipole antenna directed north-south. In order to receive this broadcast, you need to
   (a) orient the receiving antenna horizontally, north-south.
   (b) orient the receiving antenna horizontally, east-west.
   (c) use a vertical receiving antenna.
   (d) move to a town farther to the east or to the west.
   (e) use a magnetic dipole antenna instead of an electric dipole antenna.

2. Which of these statements correctly describes the orientation of the electric field ($\vec{\mathbf{E}}$), the magnetic field ($\vec{\mathbf{B}}$), and the velocity of propagation ($\vec{\mathbf{v}}$) of an electromagnetic wave?
   (a) $\vec{\mathbf{E}}$ is perpendicular to $\vec{\mathbf{B}}$; $\vec{\mathbf{v}}$ may have any orientation relative to $\vec{\mathbf{E}}$.
   (b) $\vec{\mathbf{E}}$ is perpendicular to $\vec{\mathbf{B}}$; $\vec{\mathbf{v}}$ may have any orientation perpendicular to $\vec{\mathbf{E}}$.
   (c) $\vec{\mathbf{E}}$ is perpendicular to $\vec{\mathbf{B}}$; $\vec{\mathbf{B}}$ is parallel to $\vec{\mathbf{v}}$.
   (d) $\vec{\mathbf{E}}$ is perpendicular to $\vec{\mathbf{B}}$; $\vec{\mathbf{E}}$ is parallel to $\vec{\mathbf{v}}$.
   (e) $\vec{\mathbf{E}}$ is parallel to $\vec{\mathbf{B}}$; $\vec{\mathbf{v}}$ is perpendicular to both $\vec{\mathbf{E}}$ and $\vec{\mathbf{B}}$.
   (f) Each of the three vectors is perpendicular to the other two.
3. An electromagnetic wave is created by
   (a) all electric charges.
   (b) an accelerating electric charge.
   (c) an electric charge moving at constant velocity.
   (d) a stationary electric charge.
   (e) a stationary bar magnet.
   (f) a moving electric charge, whether accelerating or not.
4. The speed of an electromagnetic wave in vacuum depends on
   (a) the amplitude of the electric field but not on the amplitude of the magnetic field.
   (b) the amplitude of the magnetic field but not on the amplitude of the electric field.
   (c) the amplitude of both fields.
   (d) the angle between the electric and magnetic fields.
   (e) the frequency and wavelength.
   (f) none of the above.
5. If the wavelength of an electromagnetic wave is about the diameter of an apple, what type of radiation is it?
   (a) X-ray (b) UV (c) Infrared
   (d) Microwave (e) Visible light (f) Radio wave
6. The Sun is directly overhead and you are facing toward the north. Light coming to your eyes from the sky just above the horizon is
   (a) partially polarized north-south.
   (b) partially polarized east-west.
   (c) partially polarized up-down.
   (d) randomly polarized.
   (e) linearly polarized up-down.
7. A dipole radio transmitter has its rod-shaped antenna oriented vertically. At a point due south of the transmitter, the radio waves have their magnetic field
   (a) oriented north-south. (b) oriented east-west.
   (c) oriented vertically.
   (d) oriented in any horizontal direction.
8. A vertical electric dipole antenna
   (a) radiates uniformly in all directions.
   (b) radiates uniformly in all horizontal directions, but more strongly in the vertical direction.
   (c) radiates most strongly and uniformly in the horizontal directions.
   (d) does not radiate in the horizontal directions.
9. A beam of light is linearly polarized. You wish to rotate its direction of polarization by 90° using one or more *ideal* polarizing sheets. To get maximum transmitted intensity, you should use how many sheets?
   (a) 1
   (b) 2
   (c) 3
   (d) As many as possible
   (e) There is no way to rotate the direction of polarization 90° using polarizing sheets.
10. Light passes from one medium (in which the speed of light is v_1) into another (in which the speed of light is v_2). If $v_1 < v_2$, as the light crosses the boundary,
   (a) both f and λ decrease.
   (b) neither f nor λ change.
   (c) f increases, λ decreases.
   (d) f does not change, λ increases.
   (e) both f and λ increase.
   (f) f does not change, λ decreases.
   (g) f decreases, λ increases.

Problems

Ⓒ Combination conceptual/quantitative problem
Biological or medical application
✦ Challenging problem
Blue # Detailed solution in the Student Solutions Manual
1 2 Problems paired by concept
Text website interactive or tutorial

22.1 Maxwell's Equations and Electromagnetic Waves

Problems 1–4. Apply the Ampère-Maxwell law to one of the circular paths in Fig. 22.3 to find the magnitude of the magnetic field at the locations specified.

✦ 1. Find B outside the wire at a distance $r \geq R$ from the central axis. [*Hint:* The electric field inside the wire is constant, so there is no changing electric flux.]

✦ 2. Find B outside the gap in the wire at a distance $r \geq R$ from the central axis. [*Hint:* What is the rate of change of electric flux through the circle in terms of the current I?]

✦ 3. Find B inside the gap in the wire at a distance $r \leq R$ from the central axis. [*Hint:* Only the rate of change of electric flux $\Delta\Phi_E/\Delta t$ *through the interior of the circular path* goes into the Ampère-Maxwell law.]

✦ 4. Find B inside the wire at a distance $r \leq R$ from the central axis. [*Hint:* Only the current through the interior of the circular path goes into the Ampère-Maxwell law.]

22.2 Antennas

Problems 5–7. An electric dipole antenna used to transmit radio waves is oriented vertically.

5. At a point due south of the transmitter, what is the direction of the wave's magnetic field?
6. At a point due north of the transmitter, how should a second electric dipole antenna be oriented to serve as a receiver?
7. At a point due north of the transmitter, how should a *magnetic* dipole antenna be oriented to serve as a receiver?

Problems 8–9. An electric dipole antenna used to transmit radio waves is oriented horizontally north-south.

8. At a point due east of the transmitter, what is the direction of the wave's electric field?
9. At a point due east of the transmitter, how should a *magnetic* dipole antenna be oriented to serve as a receiver?
10. Using Faraday's law, show that if a magnetic dipole antenna's axis makes an angle θ with the magnetic field of an EM wave, the induced emf in the antenna is reduced from its maximum possible value by a factor of $\cos\theta$. [*Hint:* Assume that, at any instant, the magnetic field everywhere inside the loop is uniform.]
11. A magnetic dipole antenna is used to detect an electromagnetic wave. The antenna is a coil of 50 turns with radius 5.0 cm. The EM wave has frequency 870 kHz, electric field amplitude 0.50 V/m, and magnetic field amplitude 1.7×10^{-9} T. (a) For best results, should the axis of the coil be aligned with the electric field of the wave, or with the magnetic field, or with the direction of propagation of the wave? (b) Assuming it is aligned correctly, what is the amplitude of the induced emf in the coil? (Since the wavelength of this wave is *much* larger than 5.0 cm, it can be assumed that at any instant the fields are uniform within the coil.) (c) What is the amplitude of the emf induced in an electric dipole antenna of length 5.0 cm aligned with the electric field of the wave?

22.3 The Electromagnetic Spectrum; 22.4 Speed of EM Waves in Vacuum and in Matter

12. What is the wavelength of the radio waves broadcast by an FM radio station with a frequency of 90.9 MHz?
13. What is the frequency of the microwaves in a microwave oven? The wavelength is 12 cm.
14. What is the speed of light in a diamond that has an index of refraction of 2.4168?
15. The speed of light in topaz is 1.85×10^8 m/s. What is the index of refraction of topaz?
16. How long does it take sunlight to travel from the Sun to Earth?
17. How long does it take light to travel from this text to your eyes? Assume a distance of 50.0 cm.
18. The index of refraction of water is 1.33. (a) What is the speed of light in water? (b) What is the wavelength in water of a light wave with a vacuum wavelength of 515 nm?
19. Light of wavelength 692 nm in air passes into window glass with an index of refraction of 1.52. (a) What is the wavelength of the light inside the glass? (b) What is the frequency of the light inside the glass?
20. How far does a beam of light travel in 1 ns?
21. The currents in household wiring and power lines alternate at a frequency of 60.0 Hz. (a) What is the wavelength of the EM waves emitted by the wiring? (b) Compare this wavelength with Earth's radius. (c) In what part of the EM spectrum are these waves?
22. In order to study the structure of a crystalline solid, you want to illuminate it with EM radiation whose wavelength is the same as the spacing of the atoms in the crystal (0.20 nm). (a) What is the frequency of the EM radiation? (b) In what part of the EM spectrum (radio, visible, etc.) does it lie?
23. In musical acoustics, a frequency ratio of 2:1 is called an octave. Humans with extremely good hearing can hear sounds ranging from 20 Hz to 20 kHz, which is approximately 10 octaves (since $2^{10} = 1024 \approx 1000$). (a) Approximately how many octaves of visible light are humans able to perceive? (b) Approximately how many octaves wide is the microwave region?
24. You and a friend are sitting in the outfield bleachers of a major league baseball park, 140 m from home plate on a day when the temperature is 20°C. Your friend is listening to the radio commentary with headphones while watching. The broadcast network has a microphone located 17 m from home plate to pick up the sound as the bat hits the ball. This sound is transferred as an EM wave a distance of 75,000 km by satellite from the ball park to the radio. (a) When the batter hits a hard line drive, who will hear the "crack" of the bat first, you or your friend, and what is the shortest time interval between the bat hitting the ball and one of you hearing the sound? (b) How much later does the other person hear the sound?
25. In the United States, the ac household current oscillates at a frequency of 60 Hz. In the time it takes for the current to make one oscillation, how far has the electromagnetic wave traveled from the current carrying wire? This distance is the wavelength of a 60-Hz EM wave. Compare this length with the distance from Boston to Los Angeles (4200 km).
26. By expressing ϵ_0 and μ_0 in base SI units (kg, m, s, A), prove that the *only* combination of the two with dimensions of speed is $(\epsilon_0\mu_0)^{-1/2}$.

22.5 Characteristics of Traveling Electromagnetic Waves in Vacuum

27. The electric field in a microwave traveling through air has amplitude 0.60 mV/m and frequency 30 GHz. Find the amplitude and frequency of the magnetic field.

28. The magnetic field in a microwave traveling through vacuum has amplitude 4.00×10^{-11} T and frequency 120 GHz. Find the amplitude and frequency of the electric field.

29. The magnetic field in a radio wave traveling through air has amplitude 2.5×10^{-11} T and frequency 3.0 MHz. (a) Find the amplitude and frequency of the electric field. (b) The wave is traveling in the $-y$-direction. At $y = 0$ and $t = 0$, the magnetic field is 1.5×10^{-11} T in the $+z$-direction. What are the magnitude and direction of the electric field at $y = 0$ and $t = 0$?

30. The electric field in a radio wave traveling through vacuum has amplitude 2.5×10^{-4} V/m and frequency 1.47 MHz. (a) Find the amplitude and frequency of the magnetic field. (b) The wave is traveling in the $+x$-direction. At $x = 0$ and $t = 0$, the electric field is 1.5×10^{-4} V/m in the $-y$-direction. What are the magnitude and direction of the magnetic field at $x = 0$ and $t = 0$?

◆31. The magnetic field of an EM wave is given by $B_y = B_m \sin(kz + \omega t)$, $B_x = 0$, and $B_z = 0$. (a) In what direction is this wave traveling? (b) Write expressions for the components of the electric field of this wave.

◆32. The electric field of an EM wave is given by $E_z = E_m \sin(ky - \omega t + \pi/6)$, $E_x = 0$, and $E_z = 0$. (a) In what direction is this wave traveling? (b) Write expressions for the components of the magnetic field of this wave.

◆33. An EM wave is generated by a magnetic dipole antenna as shown in the figure with Conceptual Question 9. The current in the antenna is produced by an LC resonant circuit. The wave is detected at a distant point P. Using the coordinate system in the figure, write equations for the x-, y-, and z-components of the EM fields at a distant point P. (If there is more than one possibility, just give one consistent set of answers.) Define all quantities in your equations in terms of L, C, E_m (the electric field amplitude at point P), and universal constants.

22.6 Energy Transport by EM Waves

34. The intensity of the sunlight that reaches Earth's upper atmosphere is approximately 1400 W/m^2. (a) What is the average energy density? (b) Find the rms values of the electric and magnetic fields.

35. The cylindrical beam of a 10.0-mW laser is 0.85 cm in diameter. What is the rms value of the electric field?

36. In astronomy it is common to expose a photographic plate to a particular portion of the night sky for quite some time in order to gather plenty of light. Before leaving a plate exposed to the night sky, Matt decides to test his technique by exposing two photographic plates in his lab to light coming through several pinholes. The source of light is 1.8 m from one photographic plate and the exposure time is 1.0 h. For how long should Matt expose a second plate located 4.7 m from the source if the second plate is to have equal exposure (that is, the same energy collected)?

37. A 1.0-m^2 solar panel on a satellite that keeps the panel oriented perpendicular to radiation arriving from the Sun absorbs 1.4 kJ of energy every second. The satellite is located at 1.00 AU from the Sun. (The Earth-Sun distance is defined to be 1.00 AU.) How long would it take an identical panel that is also oriented perpendicular to the incoming radiation to absorb the same amount of energy, if it were on an interplanetary exploration vehicle 1.55 AU from the Sun? (tutorial: sunlight on Io)

38. Fernando detects the electric field from an isotropic source that is 22 km away by tuning in an electric field with an rms amplitude of 55 mV/m. What is the average power of the source?

39. A certain star is 14 million light-years from Earth. The intensity of the light that reaches Earth from the star is 4×10^{-21} W/m^2. At what rate does the star radiate EM energy?

40. The intensity of the sunlight that reaches Earth's upper atmosphere is approximately 1400 W/m^2. (a) What is the total average power output of the Sun, assuming it to be an isotropic source? (b) What is the intensity of sunlight incident on Mercury, which is 5.8×10^{10} m from the Sun?

41. Prove that, in an EM wave traveling in vacuum, the electric and magnetic energy densities are equal; that is, prove that

$$\frac{1}{2}\epsilon_0 E^2 = \frac{1}{2\mu_0} B^2$$

at any point and at any instant of time.

42. Verify that the equation $I = \langle u \rangle c$ is dimensionally consistent (i.e., check the units).

⊙43. The solar panels on the roof of a house measure 4.0 m by 6.0 m. Assume they convert 12% of the incident EM wave's energy to electric energy. (a) What average power do the panels supply when the incident intensity is 1.0 kW/m^2 and the panels are perpendicular to the incident light? (b) What average power do the panels supply when the incident intensity is 0.80 kW/m^2 and the light is incident at an angle of 60.0° from the normal? (tutorial: solar collector) (c) Take the average daytime power requirement of a house to be about 2 kW. How do your answers to (a) and (b) compare? What are the implications for the use of solar panels?

44. The radio telescope in Arecibo, Puerto Rico, has a diameter of 305 m. It can detect radio waves from space with intensities as small as 10^{-26} W/m^2. (a) What is the average power incident on the telescope due to a wave at normal incidence with intensity 1.0×10^{-26} W/m^2? (b) What is the average power incident on Earth's surface? (c) What are the rms electric and magnetic fields?

22.7 Polarization

45. Unpolarized light passes through two polarizers in turn with polarization axes at 45° to one another.

What is the fraction of the incident light intensity that is transmitted?

46. Light polarized in the x-direction shines through two polarizing sheets. The first sheet's transmission axis makes an angle θ with the x-axis and the transmission axis of the second is parallel to the y-axis. (a) If the incident light has intensity I_0, what is the intensity of the light transmitted through the second sheet? (b) At what angle θ is the transmitted intensity maximum?

47. Unpolarized light is incident on a system of three polarizers. The second polarizer is oriented at an angle of 30.0° with respect to the first and the third is oriented at an angle of 45.0° with respect to the first. If the light that emerges from the system has an intensity of 23.0 W/m^2, what is the intensity of the incident light?

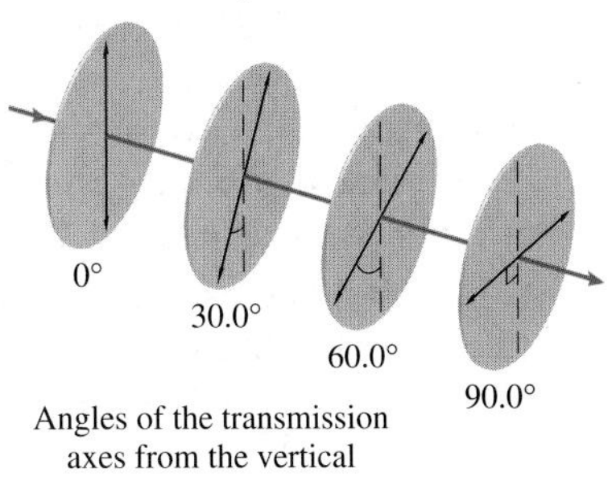

48. Unpolarized light is incident on four polarizing sheets with their transmission axes oriented as shown in the figure. What percentage of the initial light intensity is transmitted through this set of polarizers?

49. A polarized beam of light has intensity I_0. We want to rotate the direction of polarization by 90.0° using polarizing sheets. (a) Explain why we must use at least two sheets. (b) What is the transmitted intensity if we use two sheets, each of which rotates the direction of polarization by 45.0°? (c) What is the transmitted intensity if we use four sheets, each of which rotates the direction of polarization by 22.5°?

50. Vertically polarized microwaves traveling into the page are directed at each of three metal plates (a, b, c) that have parallel slots cut in them. (a) Which plate transmits microwaves best? (b) Which plate reflects microwaves best? (c) If the intensity *transmitted through* the *best* transmitter is I_1, what is the intensity transmitted through the second-best transmitter?

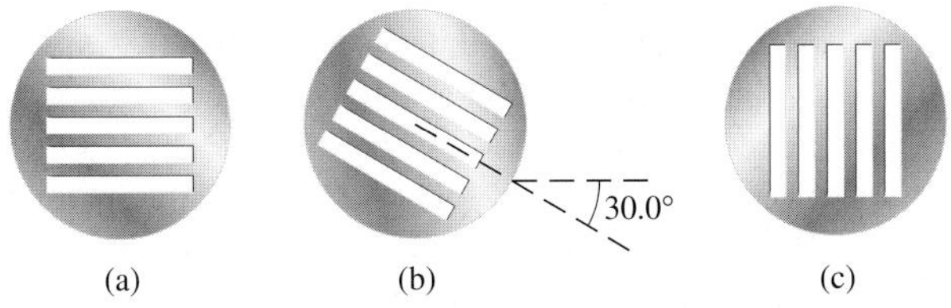

51. Two sheets of polarizing material are placed with their transmission axes at right angles to one another. A third polarizing sheet is placed between them with its transmission axis at 45° to the axes of the other two. (a) If unpolarized light of intensity I_0 is incident on the system, what is the intensity of the transmitted light? (b) What is the intensity of the transmitted light when the middle sheet is removed?

52. Vertically polarized light with intensity I_0 is normally incident on an ideal polarizer. As the polarizer is rotated about a horizontal axis, the intensity I of light transmitted through the polarizer varies with the orientation of the polarizer (θ), where $\theta = 0$ corresponds to a vertical transmission axis. Sketch a graph of I as a function of θ for one *complete* rotation of the polarizer ($0 \le \theta \le 360°$). (tutorial: polarized light)

53. Just after sunrise, you look north at the sky just above the horizon. Is the light you see polarized? If so, in what direction?

54. Just after sunrise, you look straight up at the sky. Is the light you see polarized? If so, in what direction?

22.8 The Doppler Effect for EM Waves

55. If the speeder in Example 22.9 were going *faster* than the police car, how fast would it have to go so that the reflected microwaves produce the same number of beats per second?

56. Light of wavelength 659.6 nm is emitted by a star. The wavelength of this light as measured on Earth is 661.1 nm. How fast is the star moving with respect to Earth? Is it moving toward Earth or away from it?

57. A star is moving away from Earth at a speed of 2.4×10^8 m/s. Light of wavelength 480 nm is emitted by the star. What is the wavelength as measured by an Earth observer?

58. A police car's radar gun emits microwaves with a frequency of $f_1 = 7.50$ GHz. The beam reflects from a speeding car, which is moving toward the police car at 48.0 m/s with respect to the police car. The speeder's radar detector detects the microwave at a frequency f_2. (a) Which is larger, f_1 or f_2? (b) Find the frequency difference $f_2 - f_1$.

59. A police car's radar gun emits microwaves with a frequency of $f_1 = 36.0$ GHz. The beam reflects from a speeding car, which is moving away at 43.0 m/s with respect to the police car. The frequency of the *reflected* microwave as observed by the police is f_2. (a) Which is larger, f_1 or f_2? (b) Find the frequency difference $f_2 - f_1$. [*Hint:* There are two Doppler shifts. First think of the police as source and the speeder as observer. The speeding car "retransmits" a reflected wave at the same frequency at which it observes the incident wave.]

60. What must be the relative speed between source and receiver if the wavelength of an EM wave as measured by the receiver is twice the wavelength as measured by the source? Are source and observer moving closer together or farther apart?

61. How fast would you have to drive in order to see a red light as green? Take $\lambda = 630$ nm for red and $\lambda = 530$ nm for green.

Comprehensive Problems

62. Calculate the frequency of an EM wave with a wavelength the size of (a) the thickness of a piece of paper (60 μm), (b) a 91-m-long soccer field, (c) the diameter of Earth, (d) the distance from Earth to the Sun.

63. The intensity of solar radiation that falls on a detector on Earth is 1.00 kW/m^2. The detector is a square that measures 5.00 m on a side and the normal to its surface makes an angle of 30.0° with respect to the Sun's radiation. How long will it take for the detector to measure 420 kJ of energy?

64. Astronauts on the Moon communicated with mission control in Houston via EM waves. There was a noticeable time delay in the conversation due to the round-trip transit time for the EM waves between the Moon and the Earth. How long was the time delay?

65. The antenna on a cordless phone radiates microwaves at a frequency of 2.0 GHz. What is the maximum length of the antenna if it is not to exceed half of a wavelength?

66. Two identical television signals are sent between two cities that are 400.0 km apart. One signal is sent through the air, and the other signal is sent through a fiber optic network. The signals are sent at the same time but the one traveling through air arrives 7.7×10^{-4} s before the one traveling through the glass fiber. What is the index of refraction of the glass fiber?

✦67. An AM radio station broadcasts at 570 kHz. (a) What is the wavelength of the radio wave in air? (b) If a radio is tuned to this station and the inductance in the tuning circuit is 0.20 mH, what is the capacitance in the tuning circuit? (c) In the vicinity of the radio, the amplitude of the electric field is 0.80 V/m. The radio uses a coil antenna of radius 1.6 cm with 50 turns. What is the maximum emf induced in the antenna, assuming it is oriented for best reception? Assume that the fields are sinusoidal functions of time.

68. A 60.0-mW pulsed laser produces a pulse of EM radiation with wavelength 1060 nm (in air) that lasts for 20.0 ps (picoseconds). (a) In what part of the EM spectrum is this pulse? (b) How long (in centimeters) is a single pulse in air? (c) How long is it in water ($n = 1.33$)? (d) How many wavelengths fit in one pulse? (e) What is the total electromagnetic energy in one pulse?

69. The range of wavelengths allotted to the radio broadcast band is from about 190 m to 550 m. If each station needs exclusive use of a frequency band 10 kHz wide, how many stations can operate in the broadcast band?

70. Polarized light of intensity I_0 is incident on a pair of polarizing sheets. Let θ_1 and θ_2 be the angles between the direction of polarization of the incident light and the transmission axes of the first and second sheets, respectively. Show that the intensity of the transmitted light is $I = I_0 \cos^2 \theta_1 \cos^2 (\theta_1 - \theta_2)$.

71. An unpolarized beam of light (intensity I_0) is moving in the x-direction. The light passes through three ideal polarizers whose transmission axes are (in order) at angles 0.0°, 45.0°, and 30.0° counterclockwise from the y-axis in the yz-plane. (a) What is the intensity and polarization of the light that is transmitted by the last polarizer? (b) If the polarizer in the middle is removed, what is the intensity and polarization of the light transmitted by the last polarizer?

72. What are the three lowest angular speeds for which the wheel in Fizeau's apparatus (see Fig. 22.13) allows the reflected light to pass through to the observer? Assume the distance between the notched wheel and the mirror is 8.6 km and that there are 5 notches in the wheel.

73. A microwave oven can heat 350 g of water from 25.0°C to 100.0°C in 2.00 min. (a) At what rate is energy absorbed by the water? (b) Microwaves pass through a waveguide of cross-sectional area 88.0 cm^2. What is the average intensity of the microwaves in the waveguide? (c) What are the rms electric and magnetic fields inside the waveguide?

74. A sinusoidal EM wave has an electric field amplitude $E_m = 32.0$ mV/m. What are the intensity and average energy density? [*Hint:* Recall the relationship between amplitude and rms value for a quantity that varies sinusoidally.]

75. Energy carried by an EM wave coming through the air can be used to light a bulb that is not connected to a battery or plugged into an electric outlet. Suppose a receiving antenna is attached to a bulb and the bulb is found to dissipate a maximum power of 1.05 W when the antenna is aligned with the electric field coming from a distant source. The wavelength of the source is large compared to the antenna length. When the antenna is rotated so it makes an angle of 20.0° with the incoming electric field, what is the power dissipated by the bulb?

76. A 10-W laser emits a beam of light 4.0 mm in diameter. The laser is aimed at the Moon. By the time it reaches the Moon, the beam has spread out to a diameter of 85 km. Ignoring absorption by the atmosphere, what is the intensity of the light (a) just outside the laser and (b) where it hits the surface of the Moon?

77. To measure the speed of light, Galileo and a colleague stood on different mountains with covered lanterns. Galileo uncovered his lantern and his friend, seeing the light, uncovered his own lantern in turn. Galileo measured the elapsed time from uncovering his lantern to seeing the light signal response. The elapsed time should be the time for the light to make the round trip plus the reaction time for his colleague to respond. To determine reaction time, Galileo repeated the experiment while he and his friend were close to one another. He found the same time whether his colleague was nearby or far away and concluded that light traveled almost instantaneously. Suppose the reaction time of Galileo's colleague was

0.25 s and for Galileo to observe a difference, the complete round trip would have to take 0.35 s. How far apart would the two mountains have to be for Galileo to observe a finite speed of light? Is this feasible?

78. Suppose some astronauts have landed on Mars. (a) When Mars and Earth are on the same side of the Sun and as close as they can be to one another, how long does it take for radio transmissions to travel one way between the two planets? (b) Suppose the astronauts ask a question of mission control personnel on Earth. What is the shortest possible time they have to wait for a response? The average distance from Mars to the Sun is 2.28×10^{11} m.

Answers to Practice Problems

22.1 (a) EM waves from the transmitting antenna travel outward in all directions. Since the wave travels from the transmitter to the receiver in the $+z$-direction (the direction of propagation), the direction from the receiver to the transmitter is the $-z$-direction. (b) $E_y(t) = E_m \cos(kz - \omega t)$, where $k = 2\pi/\lambda$ is the wavenumber; $E_x = E_z = 0$.

22.2 1 ly $= 9.5 \times 10^{15}$ m

22.3 444 nm

22.4 (a) $+y$-direction; (b) $2.0 \times 10^3\ \text{m}^{-1}$; (c) $B_z(x, y, z, t) = (-1.1 \times 10^{-7}\ \text{T}) \cos [(2.0 \times 10^3\ \text{m}^{-1})y - (6.0 \times 10^{11}\ \text{s}^{-1})t]$, $B_x = B_y = 0$

22.5 The rms fields are proportional to $\sqrt{I}$ and I is proportional to $1/r^2$, so the rms fields are proportional to $1/r$. $E_{rms} = 6.84$ V/m; $B_{rms} = 2.28 \times 10^{-8}$ T

22.6 450 W/m^2

22.7 minimum zero (when transmission axes are perpendicular); maximum is $\frac{1}{2}I_0$ (when transmission axes are parallel)

22.8 vertically

22.9 4.6 kHz

Answers to Checkpoints

22.2 The component of the electric field parallel to the antenna is zero. As a result, the wave does not cause an oscillating current to flow along the antenna.

22.4 The frequency of the wave does not change. With $n_{air} \approx 1$, $\lambda_{air} \approx \lambda_0$ (the vacuum wavelength). $\lambda_0 = n_{water}\,\lambda_{water} =$ 640 nm. [More generally, if neither medium is air, set the frequencies equal: $f = v_1/\lambda_1 = v_2/\lambda_2$. Then $\lambda_2 = \lambda_1(v_2/v_1) = \lambda_1(n_1/n_2)$.]

22.5 The magnitude of the magnetic field is $B = E/c = 3 \times 10^{-11}$ T. The direction of $\vec{\mathbf{B}}$ must be perpendicular to both the direction of propagation ($+x$) and the electric field ($-y$), so it's either in the $+z$- or $-z$-direction. From the right-hand rule, the direction of $\vec{\mathbf{B}}$ is in the $-z$-direction.

22.7 Rotate the polarizing sheet. If the incident light is randomly polarized, the transmitted intensity does not change. If the incident light is linearly polarized, the transmitted intensity does change as you rotate the polarizer. To get transmitted intensity of $\frac{1}{2}I_0$, the incident polarization must be at a 45° angle to the transmission axis of the polarizer ($\cos^2 45° = \frac{1}{2}$).

CHAPTER

23 Reflection and Refraction of Light

Alexander Graham Bell (1847–1922) is famous today for the invention of the telephone in the 1870s. However, Bell believed his most important invention was the *Photophone.* Instead of sending electrical signals over metal wires, the Photophone sent light signals through the air, relying on focused beams of sunlight and reflections from mirrors. What prevented Bell's Photophone from becoming as commonplace as the telephone many years ago? (See p. 863 for the answer.)

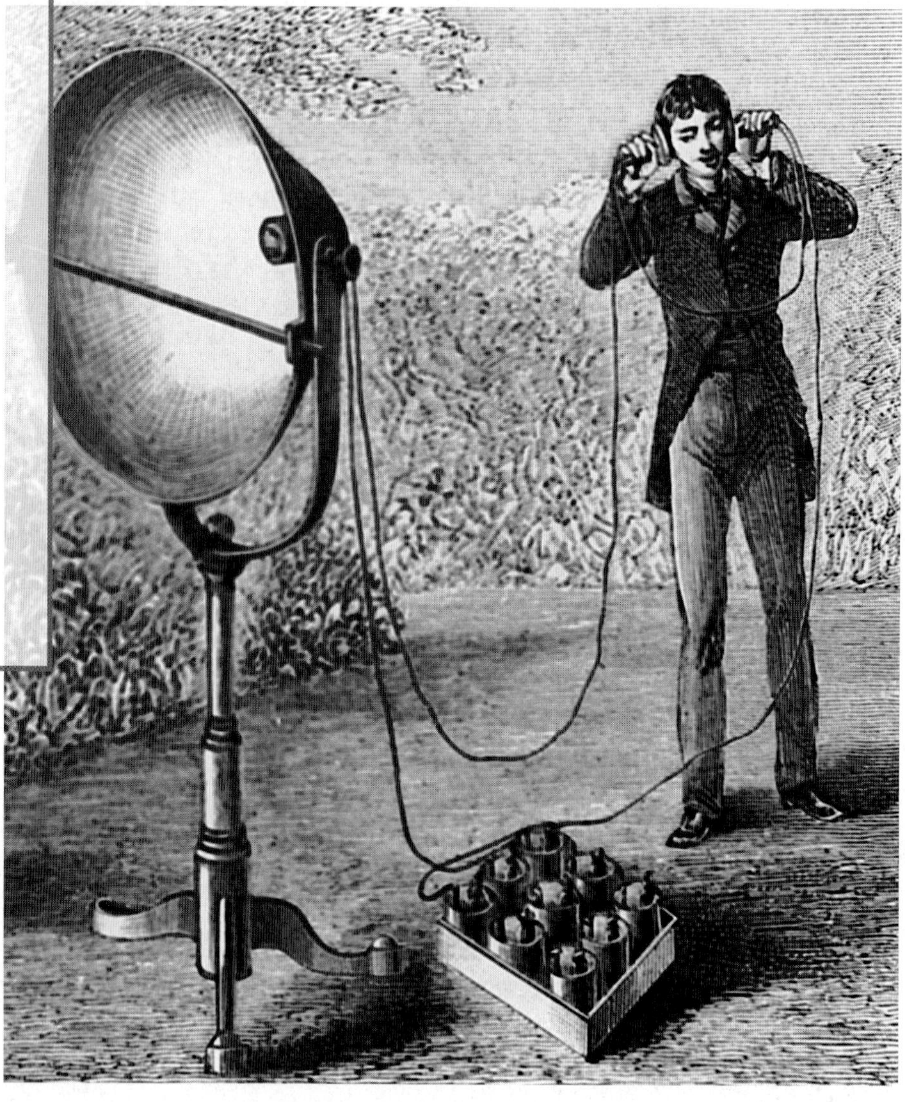

Concepts & Skills to Review

- phase (Section 11.5)
- reflection and refraction (Section 11.8)
- index of refraction; dispersion (Section 22.4)
- polarization by scattering and by reflection (Section 22.7)

23.1 WAVEFRONTS, RAYS, AND HUYGENS'S PRINCIPLE

Sources of Light

When we speak of *light*, we mean electromagnetic radiation that we can see with the unaided eye. Light is produced in many different ways. The filament of an incandescent lightbulb emits light due to its high surface temperature; at $T \approx 3000$ K, a significant fraction of the thermal radiation occurs in the visible range. The light emitted by a firefly is the result of a chemical reaction, not of a high surface temperature (Fig. 23.1). A fluorescent substance—such as the one painted on the inside of a fluorescent lightbulb—emits visible light after absorbing ultraviolet radiation.

Figure 23.1 The light flash of a firefly is caused by a chemical reaction between oxygen and the substance luciferin. The reaction is catalyzed by the enzyme luciferase.

Most objects we see are not sources of light; we see them by the light they reflect or transmit. Some fraction of the light incident on an object is absorbed, some fraction is transmitted through the object, and the rest is reflected. The nature of the material and its surface determine the relative amounts of absorption, transmission, and reflection at a given wavelength. Grass appears green because it reflects wavelengths that the brain interprets as green. Terra-cotta roof tiles reflect wavelengths that the brain interprets as red-orange (Fig. 23.2).

Application: colors from reflection and absorption of light

Wavefronts and Rays

Since EM waves share many properties in common with all waves, we can use other waves (e.g., water waves) to aid visualization. A pebble dropped into a pond starts a disturbance that propagates radially outward in all directions on the surface of the water (Fig. 23.3). A **wavefront** is a set of points of equal phase. Each of the circular wave crests in Fig. 23.3 can be considered a wavefront. A water wave with straight line wavefronts can be created by repeatedly dipping a long bar into water.

A **ray** points in the direction of propagation of a wave and is perpendicular to the wavefronts. For a circular wave, the rays are radii pointing outward from the point of

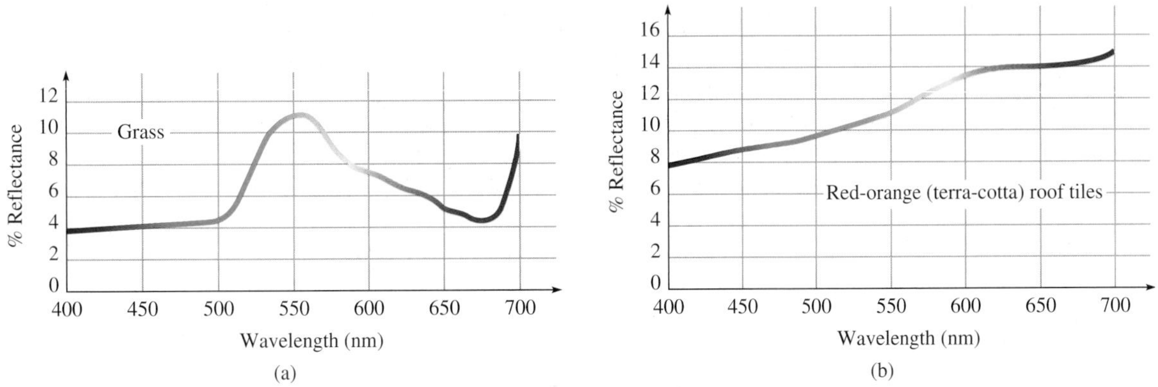

Figure 23.2 Reflectance—percentage of incident light that is reflected—as a function of wavelength for (a) grass and (b) some terra-cotta roof tiles.

Source: Reproduced from the ASTER Spectral Library through the courtesy of the Jet Propulsion Laboratory, California Institute of Technology, Pasadena, California.

Figure 23.3 Concentric circular ripples travel on the surface of a pond outward from the point where a fish broke the water surface to catch a bug. Each of the circular wave crests is a wavefront. Rays are directed radially outward from the center and are perpendicular to the wavefronts.

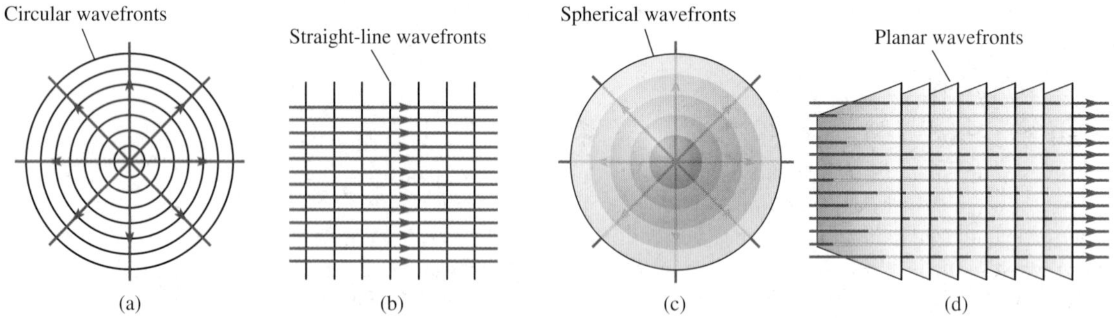

Figure 23.4 Wavefronts and rays for waves with (a) circular wavefronts, (b) straight-line wavefronts, (c) spherical wavefronts, and (d) planar wavefronts.

origin of the wave (Fig. 23.4a); for a linear wave, the rays are a set of lines parallel to each other, perpendicular to the wavefronts (Fig. 23.4b).

Whereas a surface water wave can have wavefronts that are circles or lines, a wave traveling in three dimensions, such as light, has wavefronts that are spheres, planes, or other *surfaces.* If a small source emits light equally in all directions, the wavefronts are spherical and the rays point radially outward (Fig. 23.4c). Far away from such a point source, the rays are nearly parallel to each other and the wavefronts nearly planar, so the wave can be represented as a plane wave (Fig. 23.4d). The Sun can be considered a point source when viewed from across the galaxy; even on Earth we can treat the sunlight falling on a small lens as a collection of nearly parallel rays.

Huygens's Principle

Long before the development of electromagnetic theory, the Dutch scientist Christiaan Huygens (1629–1695) developed a geometric method for visualizing the behavior of light when it travels through a medium, passes from one medium to another, or is reflected.

Huygens's Principle

At some time t, consider every point on a wavefront as a source of a new spherical wave. These *wavelets* move outward at the same speed as the original wave. At a later time $t + \Delta t$, each wavelet has a radius $v\Delta t$, where v is the speed of propagation of the wave. The wavefront at $t + \Delta t$ is a surface tangent to the wavelets. (In situations where no reflection occurs, we ignore the backward-moving wavefront.)

Geometric Optics

Geometric optics is an *approximation* to the behavior of light that applies only when interference and diffraction (Section 11.9) are negligible. In order for diffraction to be negligible, the sizes of objects and apertures must be *large* relative to the wavelength of the light. In the realm of geometric optics, the propagation of light can be analyzed using rays alone. In a homogeneous material, the rays are straight lines. At a boundary between two different materials, both reflection and transmission may occur. Huygens's principle enables us to derive the laws that determine the directions of the reflected and transmitted rays.

Conceptual Example 23.1

Wavefronts from a Plane Wave

Apply Huygens's principle to a plane wave. In other words, draw the wavelets from points on a planar wavefront and use them to sketch the wavefront at a later time.

Strategy Since we are limited to a two-dimensional sketch, we draw a wavefront of a plane wave as a straight line. We choose a few points on the wavefront as sources of wavelets. Since there is no backward-moving wave, the wavelets are hemispheres; we draw them as semicircles. Then we draw a line tangent to the wavelets to represent the surface tangent to the wavefronts; this surface is the new wavefront.

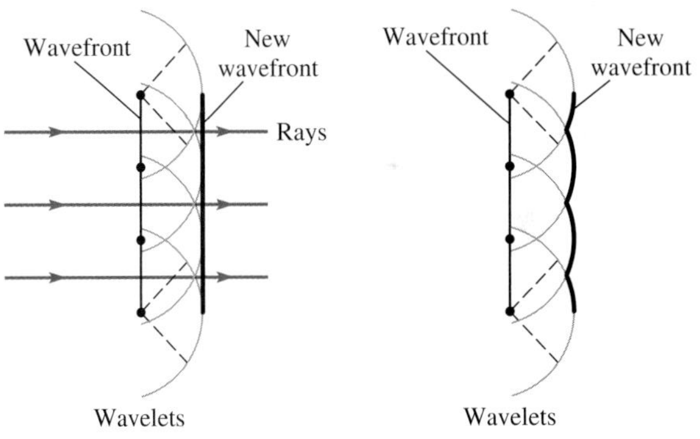

Figure 23.5
(a) Application of Huygens's principle to a plane wave. (b) This construction is not complete because it does not show wavelets coming from *every point* on the wavefront.

Solution and Discussion In Fig. 23.5a, we first draw a wavefront and four points. We imagine each point as a source of wavelets, so we draw four semicircles of equal radius, one centered on each of the four points. Finally, we draw a line tangent to the four semicircles; this line represents the wavefront at a later time.

Why draw a straight line instead of a wavy line that follows the semicircles along their edges as in Fig. 23.5b? Remember that Huygens's principle says that *every* point on the wavefront is a source of wavelets. We only draw wavelets from a few points, but we must remember that wavelets come from every point on the wavefront. Imagine drawing in more and more wavelets; the surface tangent to them would get less and less wavy, ultimately becoming a plane.

At the edges, the new wavefront is curved. This distortion of the wavefront at the edges is an example of diffraction. If a plane wavefront is large, then the wavefront at a later time is a plane with only a bit of curvature at the edges; for many purposes, the diffraction at the edges is negligible.

Conceptual Practice Problem 23.1
A Spherical Wave

Repeat Example 23.1 for the spherical light wave due to a point source.

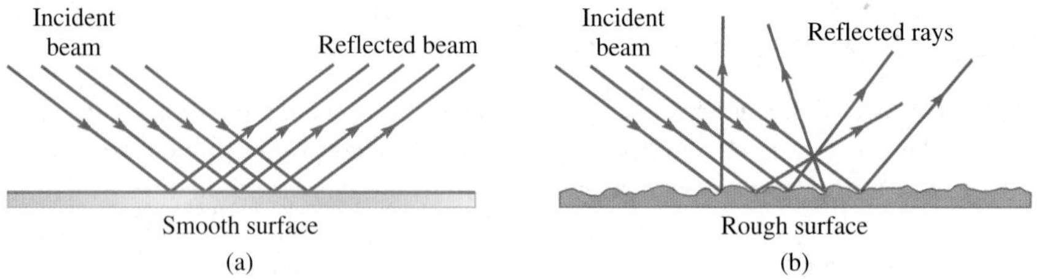

Figure 23.6 (a) A beam of light reflecting from a mirror illustrates specular reflection. (b) Diffuse reflection occurs when the same laser reflects from a rough surface.

23.2 THE REFLECTION OF LIGHT

Specular and Diffuse Reflection

Reflection from a smooth surface is called *specular reflection*; rays incident at a given angle all reflect at the same angle (Fig. 23.6a). Reflection from a rough, irregular surface is called *diffuse reflection* (Fig. 23.6b). Diffuse reflection is more common in everyday life and enables us to see our surroundings. Specular reflection is more important in optical instruments.

The roughness of a surface is a matter of degree; what appears smooth to the unaided eye can be quite rough on the atomic scale. Thus, there is not a sharp distinction between diffuse and specular reflection. If the sizes of the pits and holes in the rough surface of Fig. 23.6b were small compared with the wavelengths of visible light, the reflection would be specular. When the sizes of the pits are much larger than the wavelengths of visible light, the reflection is diffuse. A polished glass surface looks smooth to visible light, because the wavelengths of visible light are thousands of times larger than the spacing between atoms in the glass. The same surface looks rough to x-rays with wavelengths smaller than the atomic spacing. The metal mesh in the door of a microwave oven reflects microwaves well because the size of the holes is small compared to the 12-cm wavelength of the microwaves.

The Laws of Reflection

Huygens's principle illustrates how specular reflection occurs. In Fig. 23.7, plane wavefronts travel toward a polished metal surface. Every point on an incident wavefront serves as a source of secondary wavelets. Points on an incident wavefront just make the wavefront advance toward the surface. When a point on an incident wavefront contacts the metal, the wavelet propagates *away* from the surface—forming the reflected wavefront—since light cannot penetrate the metal. Wavelets emitted from these points all travel at the same speed, but they are emitted at different times. At any given instant, a wavelet's radius is proportional to the time interval since it was emitted.

Although Huygens's principle is a geometric construction, the construction is validated by modern wave theory. We now know that the reflected wave is generated by charges at the surface that oscillate in response to the incoming electromagnetic wave; the oscillating charges emit EM waves, which add up to form the reflected wave.

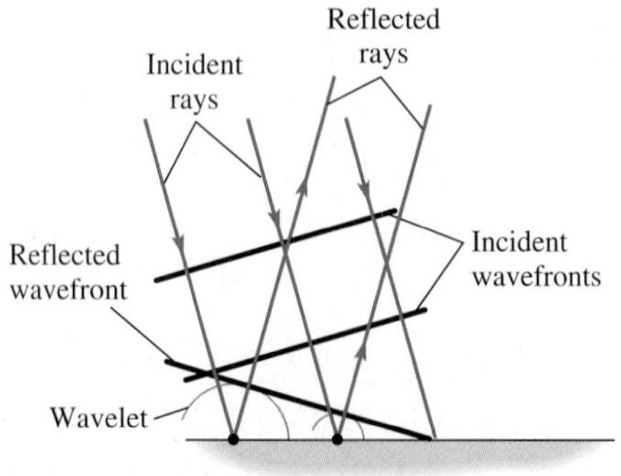

Figure 23.7 A plane wave strikes a metal surface. The wavelets emitted by each point on an incident wavefront when it reaches the surface form the reflected wave.

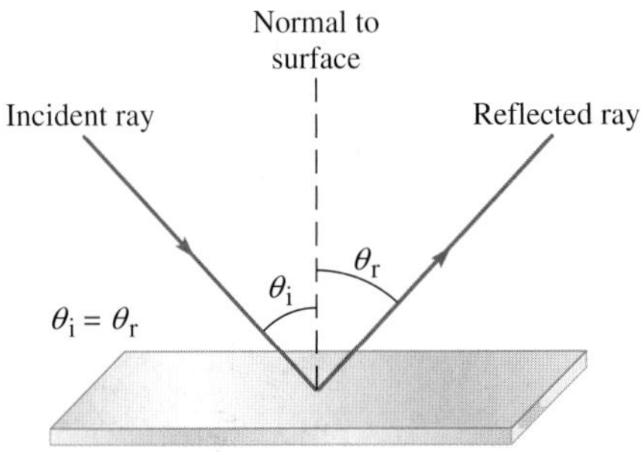

Figure 23.8 The angles of incidence and of reflection are measured between the ray and the *normal* to the surface (not between the ray and the surface). The incident ray, the reflected ray, and the normal all lie in the same plane.

The laws of reflection summarize the relationship between the directions of the incident and reflected rays. The laws are formulated in terms of the angles between a ray and a *normal*—a line *perpendicular* to the surface where the ray touches the surface. The **angle of incidence** (θ_i) is the angle between an incident ray and the normal (Fig. 23.8); the **angle of reflection** (θ_r) is the angle between the reflected ray and the normal. In Problem 9 you can go on to prove that

$$\theta_i = \theta_r \qquad (23\text{-}1)$$

The other law of reflection says that the incident ray, the reflected ray, and the normal all lie in the same plane (the **plane of incidence**).

Laws of Reflection

1. The angle of incidence equals the angle of reflection.
2. The reflected ray lies in the same plane as the incident ray and the normal to the surface at the point of incidence. The two rays are on opposite sides of the normal.

For diffuse reflection from rough surfaces, the angles of reflection for the incoming rays are still equal to their respective angles of incidence. However, the normals to the rough surface are at random angles with respect to each other, so the reflected rays travel in many directions (Fig. 23.6b).

Reflection and Transmission

So far we have considered only specular reflection from a totally reflecting surface such as polished metal. When light reaches a boundary between two *transparent* media, such as from air to glass, some of the light is reflected and some is transmitted into the new medium. The reflected light still follows the same laws of reflection (as long as the surface is smooth so that the reflection is specular). For *normal* incidence on an air-glass surface, only 4% of the incident intensity is reflected; 96% is transmitted.

23.3 THE REFRACTION OF LIGHT: SNELL'S LAW

In Section 22.4, we showed that when light passes from one transparent medium to another, the wavelength changes (unless the speeds of light in the two media are the same) while the frequency stays the same. In addition, Huygens's principle helps us understand why *light rays change direction* as they cross the boundary between the two media—a phenomenon known as **refraction**.

Refraction: the changing of direction of a light ray as it passes from one medium into another.

We can use Huygens's principle to understand how refraction occurs. Figure 23.9a shows a plane wave incident on a planar boundary between air and glass. In the air, a series of planar wavefronts moves toward the glass. The distance between the wavefronts is equal to one wavelength. Once the wavefront reaches the glass boundary and enters the new material, the wave slows down—light moves more slowly through glass

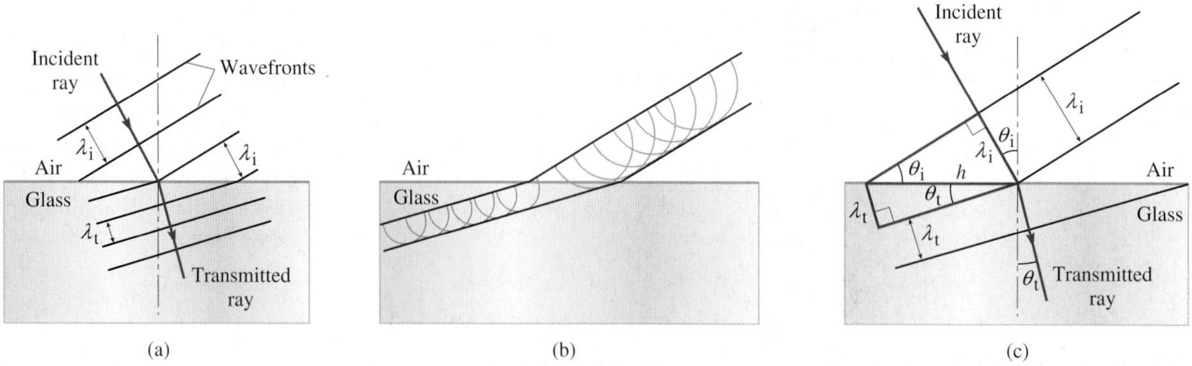

Figure 23.9 (a) Wavefronts and rays at a glass-air boundary. The reflected wavefronts are omitted. Note that the wavefronts are closer together in glass because the wavelength is smaller. (b) Huygens's construction for a wavefront partly in air and partly in glass. (c) Geometry for finding the angle of the transmitted ray.

than through air. Since the wavefront approaches the boundary at an angle to the normal, the portion of the wavefront that is still in air continues at the same merry pace while the part that has entered the glass moves more slowly. Figure 23.9b shows a Huygens's construction for a wavefront that is partly in glass. The wavelets have smaller radii in glass since the speed of light is smaller in glass than in air.

Figure 23.9c shows two right triangles that are used to relate the angle of incidence θ_i to the angle of the transmitted ray (or angle of refraction) θ_t. The two triangles share the same hypotenuse (h). Using some trigonometry, we find that

$$\sin\theta_i = \frac{\lambda_i}{h} \quad \text{and} \quad \sin\theta_t = \frac{\lambda_t}{h}$$

Eliminating h yields

$$\frac{\sin\theta_i}{\sin\theta_t} = \frac{\lambda_i}{\lambda_t} \tag{23-2}$$

It is more convenient to rewrite this relationship in terms of the indices of refraction. Recall that when light passes from one transparent medium to another, the *frequency f does not change* (see Section 22.4). Since $v = f\lambda$, λ is directly proportional to v. By definition [$n = c/v$, Eq. (22-4)], the index of refraction n is *inversely* proportional to v. Therefore, λ is inversely proportional to n:

$$\frac{\lambda_i}{\lambda_t} = \frac{v_i/f}{v_t/f} = \frac{v_i}{v_t} = \frac{c/n_i}{c/n_t} = \frac{n_t}{n_i} \tag{23-3}$$

By replacing λ_i/λ_t with n_t/n_i in Eq. (23-2) and cross multiplying, we obtain

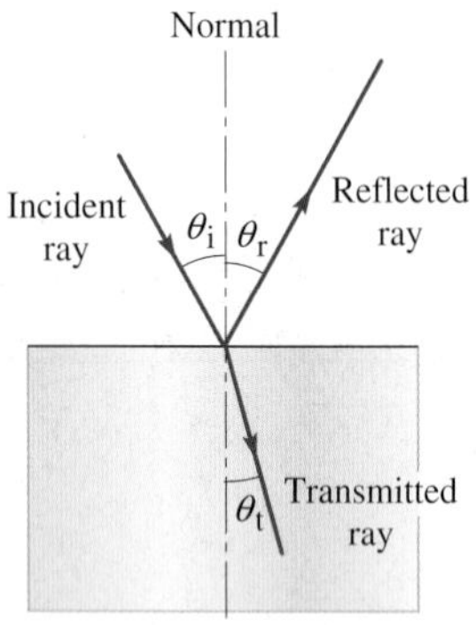

Figure 23.10 The incident ray, the reflected ray, the transmitted ray, and the normal all lie in the same plane. All angles are measured with respect to the normal. Notice that the reflected and transmitted rays are always on the opposite side of the normal from the incident ray.

Snell's Law

$$n_i \sin\theta_i = n_t \sin\theta_t \tag{23-4}$$

This law of refraction was discovered experimentally by Dutch professor Willebrord Snell (1580–1626). To determine the direction of the transmitted ray *uniquely*, two additional statements are needed:

Laws of Refraction

1. $n_i \sin\theta_i = n_t \sin\theta_t$, where the angles are measured from the normal.
2. The incident ray, the transmitted ray, and the normal all lie in the same plane—the plane of incidence.
3. The incident and transmitted rays are on *opposite sides* of the normal (Fig. 23.10).

Table 23.1 Indices of Refraction for $\lambda = 589.3$ nm in Vacuum (at 20°C unless otherwise noted)

Material	Index	Material	Index
Solids		**Liquids**	
Ice (at 0°C)	1.309	Water	1.333
Fluorite	1.434	Acetone	1.36
Fused quartz	1.458	Ethyl alcohol	1.361
Polystyrene	1.49	Carbon tetrachloride	1.461
Lucite	1.5	Glycerine	1.473
Plexiglas	1.51	Sugar solution (80%)	1.49
Crown glass	1.517	Benzene	1.501
Plate glass	1.523	Carbon disulfide	1.628
Sodium chloride	1.544	Methylene iodide	1.74
Light flint glass	1.58		
Dense flint glass	1.655	**Gases at 0°C, 1 atm**	
Sapphire	1.77		
Zircon	1.923	Helium	1.000 036
Diamond	2.419	Ethyl ether	1.000 152
Titanium dioxide	2.9	Water vapor	1.000 250
Gallium phosphide	3.5	Dry air	1.000 293
		Carbon dioxide	1.000 449

Mathematically, Snell's law treats the two media as interchangeable, so the path of a light ray transmitted from one medium to another is correct if the direction of the ray is reversed.

The index of refraction of a material depends on the temperature of the material and on the frequency of the light. Table 23.1 lists indices of refraction for several materials for yellow light with a *wavelength in vacuum* of 589.3 nm. (It is customary to specify the vacuum wavelength instead of the frequency.) In many circumstances the slight variation of n over the visible range of wavelengths can be ignored.

CHECKPOINT 23.3

A glass ($n = 1.5$) fish tank is filled with water ($n = 1.33$). When a light ray in the glass is transmitted into the water, does it refract toward the normal or away from the normal? Explain. (Assume the light ray is not normal to the glass surface.)

PHYSICS AT HOME

Fill a clear drinking glass with water and then put a pencil in the glass. Look at the pencil from many different angles. Why does the pencil look as if it is bent?

PHYSICS AT HOME

Place a coin at the far edge of the bottom of an empty mug. Sit in a position so that you are just unable to see the coin. Then, without moving your head, utter the magic word *REFRACTION* as you pour water carefully into the mug on the near side; pour slowly so that the coin does not move. The coin becomes visible when the mug is filled with water (Fig. 23.11).

Figure 23.11 (a) The coin at the bottom of the mug is not visible. (b) After the mug is filled with water, the coin is visible. (c) Refraction at the water-air boundary bends light rays from the coin so they enter the eye.

Example 23.2

Ray Traveling Through a Window Pane

A beam of light strikes one face of a window pane with an angle of incidence of 30.0°. The index of refraction of the glass is 1.52. The beam travels through the glass and emerges from a parallel face on the opposite side. Ignore reflections. (a) Find the angle of refraction for the ray inside the glass. (b) Show that the rays in air on either side of the glass (the incident and emerging rays) are parallel to each other.

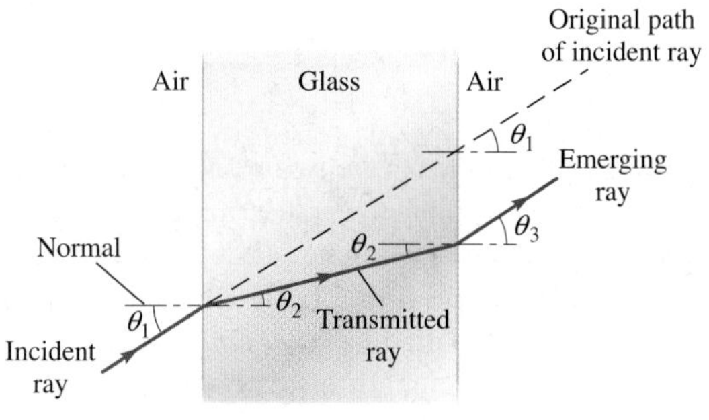

Figure 23.12
A ray of light travels through a window pane.

Strategy First we draw a ray diagram. We are only concerned with the rays transmitted at each boundary, so we omit reflected rays from the diagram. At each boundary we draw a normal, label the angles of incidence and refraction, and apply Snell's law. When the ray passes from air ($n = 1.00$) to glass ($n = 1.52$), it bends *closer to* the normal: since $n_1 \sin\theta_1 = n_2 \sin\theta_2$, a larger n means a smaller θ. Likewise, when the ray passes from glass to air, it bends *away from* the normal.

Solution (a) Figure 23.12 is a ray diagram. At the first air-glass boundary, Snell's law yields

$$n_1 \sin\theta_1 = n_2 \sin\theta_2$$

$$\sin\theta_2 = \frac{n_1}{n_2}\sin\theta_1 = \frac{1.00}{1.52}\sin 30.0° = 0.3289$$

The angle of refraction is

$$\theta_2 = \sin^{-1} 0.3289 = 19.2°$$

(b) At the second boundary, from glass to air, we apply Snell's law again. Since the surfaces are parallel, the two normals are parallel. The angle of refraction at the first boundary and the angle of incidence at the second are alternate interior angles, so the angle of incidence at the second boundary must be θ_2.

$$n_2 \sin\theta_2 = n_3 \sin\theta_3$$

We do not need to solve for θ_3 numerically. From the first boundary we know that $n_1 \sin\theta_1 = n_2 \sin\theta_2$; therefore, $n_1 \sin\theta_1 = n_3 \sin\theta_3$. Since $n_1 = n_3$, $\theta_3 = \theta_1$. The two rays—emerging and incident—are parallel to one another.

Discussion Note that the emerging ray is parallel to the incident ray, but it is *displaced* (see the dashed line in Fig. 23.12). If the two glass surfaces were not parallel, then the two normals would not be parallel. Then the angle of incidence at the second boundary would not be equal to the angle of refraction at the first; the emerging ray would *not* be parallel to the incident ray.

Practice Problem 23.2 Fish Eye View

A fish is at rest beneath the still surface of a pond. If the Sun is 33° above the horizon, at what angle above the horizontal does the fish see the Sun? [*Hint:* Draw a diagram that includes the normal to the surface; be careful to correctly identify the angles of incidence and refraction.]

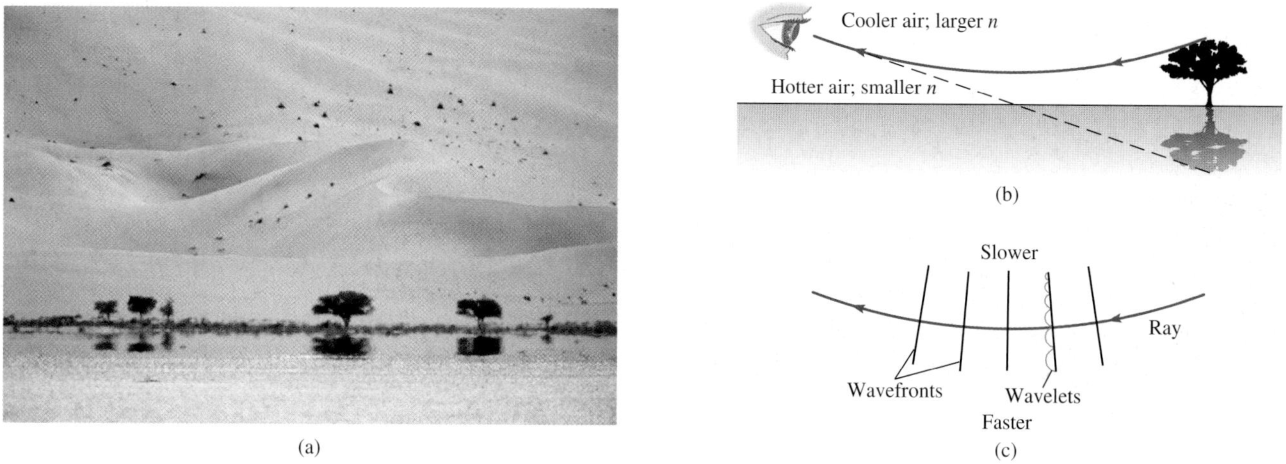

Figure 23.13 (a) Mirage seen in the desert in Namibia. Note that the images are upside down. (b) A ray from the Sun bends upward into the eye of the observer. (c) The bottom of the wavefront moves faster than the top.

Application: Mirages

Refraction of light in the air causes the *mirages* seen in the desert or on a hot road in summer (Fig. 23.13a). The hot ground warms the air near it, so light rays from the sky travel through warmer and warmer air as they approach the ground. Since the speed of light in air increases with increasing temperature, light travels faster in the hot air near the ground than in the cooler air above. The temperature change is gradual, so there is no *abrupt* change in the index of refraction; instead of being bent abruptly, rays gradually curve upward (Fig. 23.13b).

The wavelets from points on a wavefront travel at different speeds; the radius of a wavelet closer to the ground is larger than that of a wavelet higher up (Fig. 23.13c). The brain interprets the rays coming upward into the eye as coming from the ground even though they really come from the sky. What may look like a body of water on the ground is actually an image of the blue sky overhead.

A *superior mirage* occurs when the layer of air near Earth's surface is *colder* than the air above, due to a snowy field or to the ocean. A ship located just *beyond* the horizon can sometimes be seen because light rays from the ship are gradually bent downward (Fig. 23.14). Ships and lighthouses seem to float in the sky or appear much taller than they are. Refraction also allows the Sun to be seen before it actually rises above the horizon and after it is already below the horizon at sunrise and sunset.

Dispersion in a Prism

When natural white light enters a triangular prism, the light emerging from the far side of the prism is separated into a continuous spectrum of colors from red to violet (Fig. 23.15). The separation occurs because the prism is **dispersive**—that is, the speed of light in the prism depends on the frequency of the light (see Section 22.4).

Dispersion: the speed of light in matter depends on frequency.

Natural white light is a mixture of light at all the frequencies in the visible range. At the front surface of the prism, each light ray of a particular frequency refracts at an angle determined by the index of refraction of the prism at that frequency. The index of refraction increases with increasing frequency, so it is smallest for red and increases gradually until it is largest for violet. As a result, violet bends the most and red the least. Refraction occurs again as light leaves the prism. The geometry of the prism is such that the different colors are spread apart farther at the back surface.

(a)

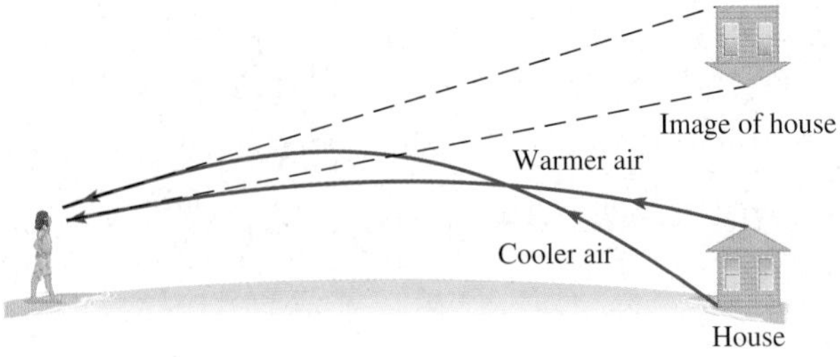

(b)

Figure 23.14 (a) Superior mirage of a house seen in Finland's southwestern archipelago; (b) a sketch of the light rays that form a superior mirage of a house.

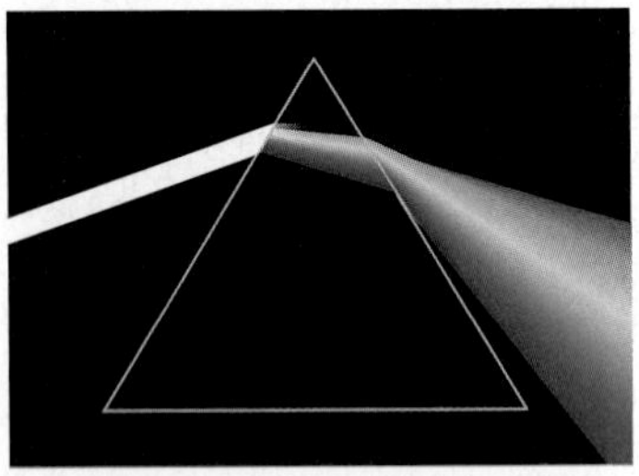

Figure 23.15 Dispersion of white light by a prism. (See also the photo in Fig. 22.13.)

Application: Rainbows

Rainbows are formed by the dispersion of light in water. A ray of sunlight that enters a raindrop is separated into the colors of the spectrum. At each air-water boundary there may be both reflection and refraction. The rays that contribute to a *primary rainbow*—the brightest and often the only one seen—pass into the raindrop, reflect off the back of the raindrop, and then are transmitted back into the air (Fig. 23.16a). Refraction occurs both where the ray enters the drop (air-water) and again when it leaves (water-air), just as for a prism. Since the index of refraction varies with frequency, sunlight is separated into the spectral colors. For relatively large water droplets, as occur in a gentle summer shower, the rays emerge with an angular separation between red and violet of about 2° (Fig. 23.16b).

A person looking into the sky away from the Sun sees red light coming from raindrops higher in the sky and violet light coming from lower droplets (Fig. 23.16c). The rainbow is a circular arc that subtends an angle of 42° for red and 40° for violet, with the other colors falling in between.

In good conditions, a double rainbow can be seen. The secondary rainbow has a larger radius, is less intense, and has its colors reversed (Fig. 23.16d). It arises from rays that undergo *two reflections* inside the raindrop before emerging. The angles subtended by a secondary rainbow are 50.5° for red and 54° for violet.

23.4 TOTAL INTERNAL REFLECTION

According to Snell's law, if a ray is transmitted from a slower medium into a faster medium (from a higher index of refraction to a lower one), the refracted ray bends *away* from the normal (Fig. 23.17, ray *b*). That is, the angle of refraction is greater than the angle of incidence. As the angle of incidence is increased, the angle of refraction eventually reaches 90° (Fig. 23.17, ray *c*). At 90°, the refracted ray is parallel to the surface. It isn't transmitted into the faster medium; it just moves along the surface. The angle of incidence for which the angle of refraction is 90° is called the **critical angle** θ_c for the boundary between the two media. From Snell's law,

$$n_i \sin \theta_c = n_t \sin 90°$$

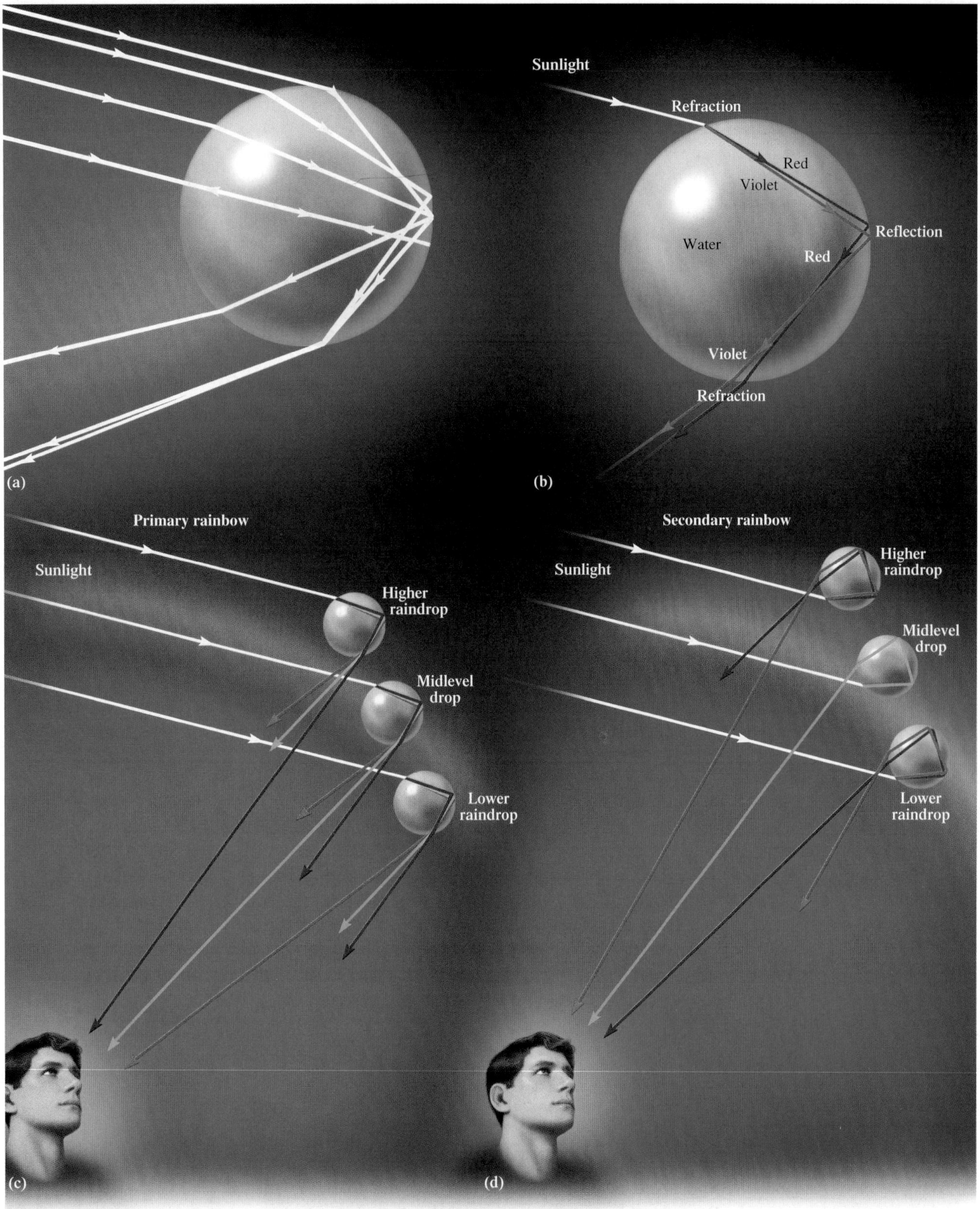

Figure 23.16 (a) Rays of sunlight that are incident on the upper half of a raindrop and reflect once inside the raindrop. Although the incident rays are parallel, the emerging rays are not. The pair of rays along the bottom edge shows where the emerging light has the highest intensity. Only the rays of maximum intensity are shown in parts (b) through (d). (b) Because the index of refraction of water depends on frequency, the angle at which the light leaves the drop depends on frequency. At each boundary, both reflection and transmission occur. Reflected or transmitted rays that do not contribute to the primary rainbow are omitted. (c) Light from many different raindrops contributes to the appearance of a rainbow. Angles are exaggerated for clarity. (d) Light rays that reflect twice inside the raindrop form the secondary rainbow. Note that the order of the colors is reversed: now violet is highest and red is lowest.

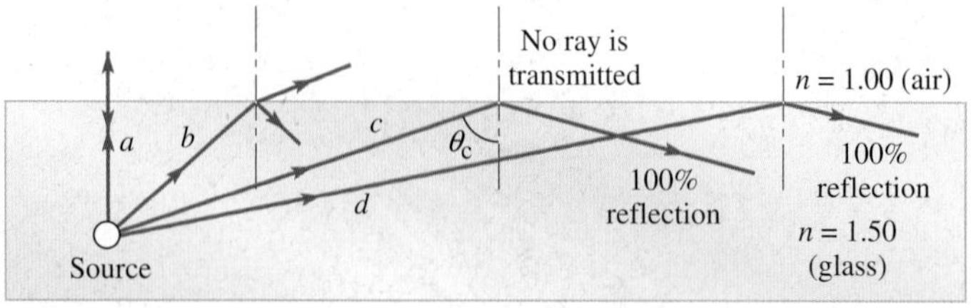

Figure 23.17 Partial reflection and total internal reflection at the upper surface of a rectangular glass block. The angles of incidence of rays *a* and *b* are less than the critical angle, ray *c* is incident at the critical angle θ_c, and ray *d* is incident at an angle greater than θ_c. (Angles exaggerated for clarity.)

Critical angle:

$$\theta_c = \sin^{-1}\frac{n_t}{n_i} \tag{23-5a}$$

where the subscripts "i" and "t" refer to the media in which the incident and transmitted rays travel. Since we are discussing an incident ray in a slower medium, $n_i > n_t$.

The **critical angle** is the minimum angle of incidence for which no light is transmitted past the boundary.

For an angle of incidence greater than θ_c, the refracted ray can't bend away from the normal *more* than 90°; to do so would be reflection rather than refraction, and a different law governs the angle of reflection. Mathematically, there is no angle whose sine is greater than 1 (= sin 90°), so it is impossible to satisfy Snell's law if $n_i \sin\theta_i > n_t$ (which is equivalent to saying $\theta_i > \theta_c$). If the angle of incidence is greater than θ_c, there cannot be a transmitted ray; if there is no ray transmitted into the faster medium, all the light must be reflected from the boundary (Fig. 23.17, ray *d*). This is called **total internal reflection**.

$$\text{no transmitted ray for } \theta_i \geq \theta_c \tag{23-5b}$$

Total reflection cannot occur when a ray in a faster medium hits a boundary with a slower medium. In that case the refracted ray bends *toward* the normal, so the angle of refraction is always less than the angle of incidence. Even at the largest possible angle of incidence, 90°, the angle of refraction is less than 90°. Total internal reflection can only occur when the incident ray is in the slower medium.

Example 23.3

Total Internal Reflection in a Triangular Glass Prism

A beam of light is incident on the triangular glass prism in air. What is the largest angle of incidence θ_i below the normal (as shown in Fig. 23.18) so that the beam undergoes total reflection from the back of the prism (the hypotenuse)? The prism has an index of refraction $n = 1.50$.

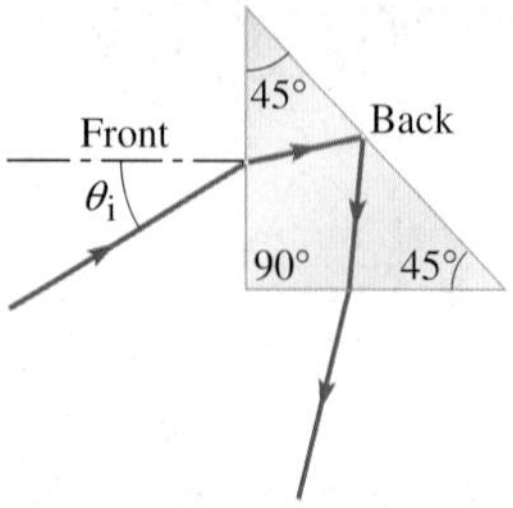

Figure 23.18 Example 23.3.

Strategy In this problem it is easiest to work backward. Total internal reflection occurs if the angle of incidence at the back of the prism is greater than or equal to the critical angle. We start by finding the critical angle and then work backward using geometry and Snell's law to find the corresponding angle of incidence at the front of the prism.

continued on next page

Example 23.3 continued

Solution To find the critical angle from Snell's law, we set the angle of refraction equal to 90°.

$$n_i \sin \theta_c = n_a \sin 90°$$

The incident ray is in the internal medium (glass). Therefore, $n_i = 1.50$ and $n_a = 1.00$.

$$\sin \theta_c = \frac{n_a}{n_i} \sin 90° = \frac{1.00}{1.50} \times 1.00 = 0.667$$

$$\theta_c = \sin^{-1} 0.667 = 41.8°$$

In Fig. 23.19, we draw an enlarged ray diagram and label the angle of incidence at the back of the prism as θ_c. The angles of incidence and refraction at the front are labeled θ_i and θ_t; they are related through Snell's law:

$$1.00 \sin \theta_i = 1.50 \sin \theta_t$$

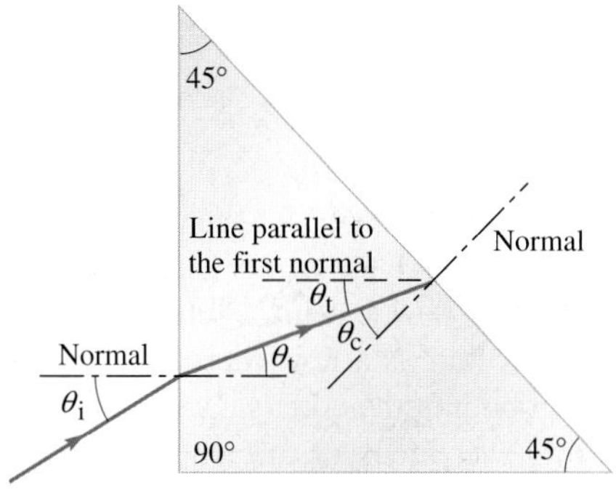

Figure 23.19
Ray diagram to show the three angles θ_i, θ_t, and θ_c.

What remains is to find the relationship between θ_t and θ_c. By drawing a line at the second boundary that is parallel to the normal at the first boundary, we can use alternate interior angles to label θ_t (see Fig. 23.19). The angle between the two normals is 45.0°, so

$$\theta_t = 45.0° - \theta_c = 45.0° - 41.8° = 3.2°$$

Then

$$\sin \theta_i = 1.50 \sin \theta_t = 1.50 \times 0.05582 = 0.0837$$

$$\theta_i = \sin^{-1} 0.0837 = 4.8°$$

Discussion For a beam incident below the normal at angles from 0 to 4.8°, total internal reflection occurs at the back. If a beam is incident at an angle greater than 4.8°, then the angle of incidence at the back is less than the critical angle, so transmission into the air occurs there. Conceptual Practice Problem 23.3 considers what happens to a beam incident above the normal.

If we had mixed up the two indices of refraction, we would have wound up trying to take the inverse sine of 1.5. That would be a clue that we made a mistake.

Conceptual Practice Problem 23.3 Ray Incident from Above the Normal

Draw a ray diagram for a beam of light incident on the prism of Fig. 23.18 from *above* the normal. Show that at *any* angle of incidence, the beam undergoes total internal reflection at the back of the prism.

Total Internal Reflection in Prisms

Application: periscope

Optical instruments such as periscopes, single-lens reflex (SLR) cameras, binoculars, and telescopes often use prisms to reflect a beam of light. Figure 23.20a shows a simple periscope. Light is reflected through a 90° angle by each of two prisms; the net result is a displacement of the beam. A similar scheme is used in binoculars (Fig. 23.20b). In an SLR camera, one of the prisms is replaced by a movable mirror. When the mirror is in place, the light through the camera lens is diverted up to the viewfinder, so you can see exactly what will appear on film. Depressing the shutter moves the mirror out of the way so the light falls onto the film instead. In binoculars and telescopes, *erecting prisms* are often used to turn an upside down image right side up.

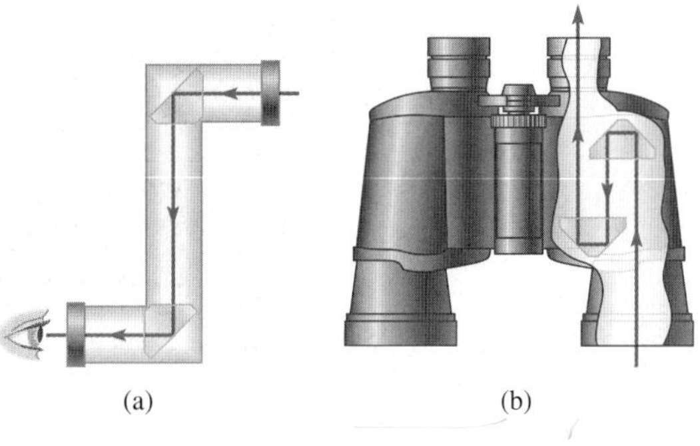

Figure 23.20 (a) A periscope uses two reflecting prisms to shift the beam of light. (b) In binoculars, the light undergoes total internal reflection twice in each prism.

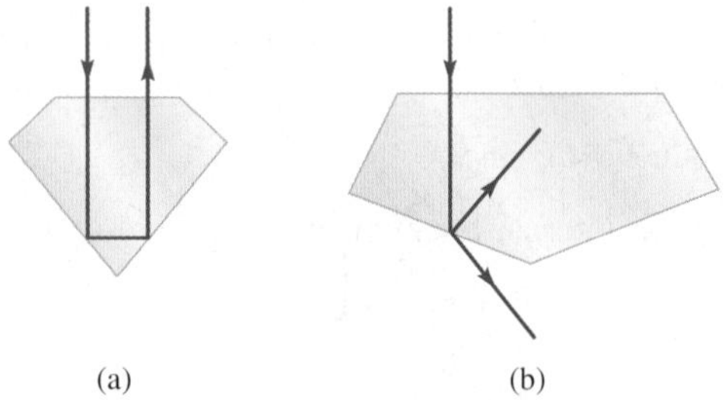

Figure 23.21 (a) This ray undergoes total internal reflection twice before re-emerging from a front face of the diamond. (b) Due to a poor cut, a similar ray in this diamond would be incident on one of the back faces at less than the critical angle. The ray is mostly transmitted out the back of the diamond.

An advantage of using prisms instead of mirrors in these applications is that 100% of the light is reflected. A typical mirror reflects only about 90%—remember that the oscillating electrons that produce the reflected wave are moving in a metal with some electrical resistance, so energy dissipation occurs.

The brilliant sparkle of a diamond is due to total internal reflection. The cuts are made so that most of the light incident on the front faces is totally reflected several times inside the diamond and then re-emerges toward the viewer. A poorly cut diamond allows too much light to emerge away from the viewer (Fig. 23.21).

Application: Fiber Optics

Total internal reflection is the principle behind fiber optics, a technology that has revolutionized both communications and medicine. At the center of an optical fiber is a transparent cylindrical core made of glass or plastic with a relatively high index of refraction (Fig. 23.22). The core may be as thin as a few micrometers in diameter—quite a bit thinner than a human hair. Surrounding the core is a coating called the cladding, which is also transparent but has a lower index of refraction than the core. The "mismatch" in the indices of refraction is maximized so that the critical angle at the core-cladding boundary is as small as possible.

Light signals travel nearly parallel to the axis of the core. It is impossible to have light rays enter the fiber *perfectly* parallel to the axis of the fiber, so the rays eventually hit the cladding *at a large angle of incidence.* As long as the angle of incidence is greater than the critical angle, the ray is totally reflected back into the core; no light leaks out into the cladding. A ray may typically reflect from the cladding thousands of times per meter of fiber, but since the ray is totally reflected each time, the signal can travel long distances—kilometers in some cases—before any appreciable signal loss occurs.

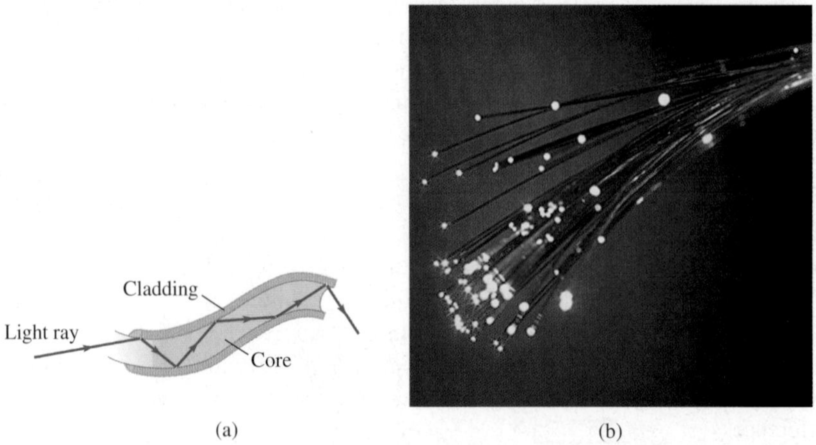

Figure 23.22 (a) An optical fiber. (b) A bundle of optical fibers.

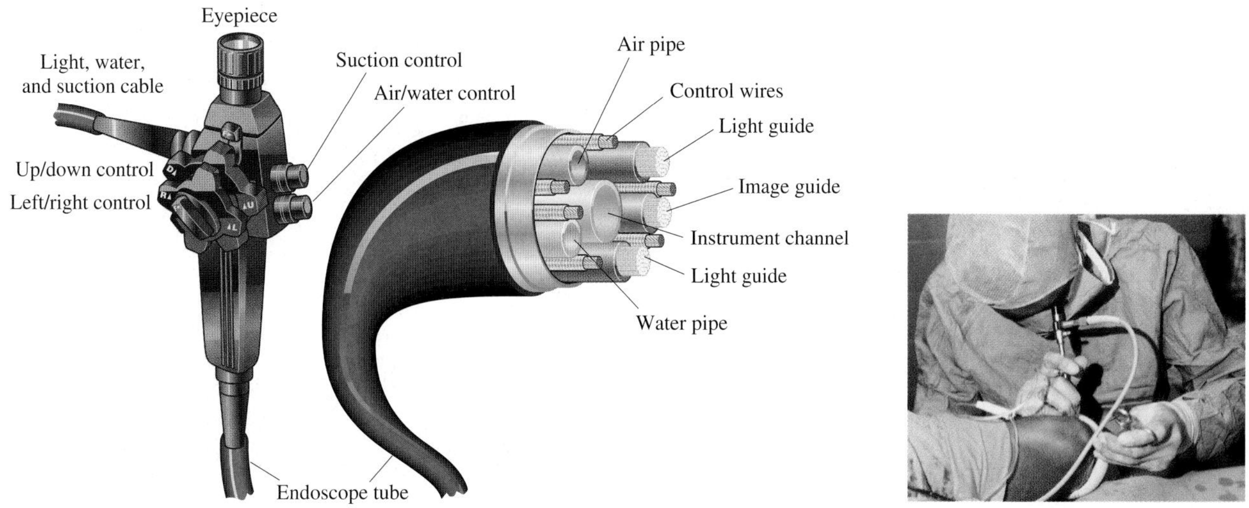

(a) (b)

Figure 23.23 (a) An endoscope. (b) Arthroscopic knee surgery. An arthroscope is similar to an endoscope, but is used in the diagnosis and treatment of injuries to the joints.

The fibers are flexible so they can be bent as necessary. The smaller the critical angle, the more tightly a fiber can be bent. If the fiber is kinked (bent too tightly), rays strike the boundary at less than the critical angle, resulting in dramatic signal loss as light passes into the cladding.

Optical fiber is far superior to copper wire in its capacity to carry information. A single optical fiber can carry tens of thousands of phone conversations, but a pair of copper wires can only carry about 20 at the most. Electrical signals in copper wires also lose strength much more rapidly (due in part to the electrical resistance of the wires) and are susceptible to electrical interference. Over 80% of the long-distance phone calls in the world are carried by fiber optics; computer networks and video increasingly use fiber optics as well.

In medicine, bundles of optical fibers are at the heart of the endoscope (Fig. 23.23), which is fed through the nose, mouth, or rectum, or through a small incision, into the body. One bundle of fibers carries light into a body cavity or an organ and illuminates it; another bundle transmits an image back to the doctor for viewing.

Application: endoscope

The endoscope is not limited to diagnosis; it can be fitted with instruments enabling a physician to take tissue samples, perform surgery, cauterize blood vessels, or suction out debris. Surgery performed using an endoscope uses much smaller incisions than traditional surgery; as a result, recovery is much faster. A gallbladder operation that used to require an extended hospital stay can now be done on an outpatient basis in many cases.

Bell's Photophone

Almost a century before the invention of fiber optics, Bell's Photophone used light to carry a telephone signal. The Photophone projected the voice toward a mirror, which vibrated in response. A focused beam of sunlight reflecting from the mirror captured the vibrations. Other mirrors were used to reflect the signal as necessary until it was transformed back into sound at the receiving end. The light traveled in straight line paths through air between the mirrors.

What kept Bell's Photophone from becoming commonplace?

Bell's Photophone worked only intermittently. Many things could interfere with a transmission, including cloudy weather. With nothing to keep the beam from spreading out, it worked only over short distances. Not until the invention of fiber optics in the 1970s could light signals travel reliably over long distances without significant loss or interference.

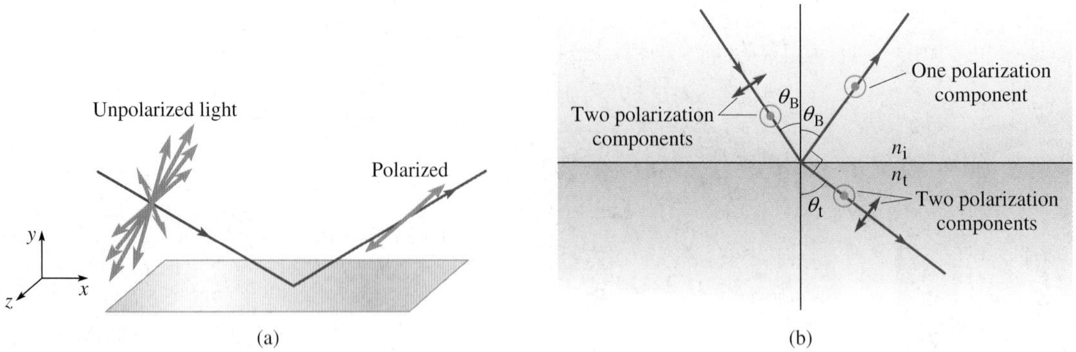

Figure 23.24 (a) Unpolarized light is partially or totally polarized by reflection. (b) When light is incident at Brewster's angle, the reflected and transmitted rays are perpendicular and the reflected light is totally polarized perpendicular to the plane of the page. [The polarization directions are shown in different colors in (b) merely to help distinguish them; these colors have nothing to do with the color of the light.]

23.5 POLARIZATION BY REFLECTION

Brewster's angle: the angle of incidence for which the reflected light is totally polarized.

In Section 22.7 we mentioned that unpolarized light is partially or totally polarized by reflection (Fig. 23.24a). Using Snell's law, we can find the angle of incidence for which the reflected light is totally polarized. This angle of incidence is called **Brewster's angle** θ_B, after David Brewster (1781–1868).

The reflected light is totally polarized *when the reflected and transmitted rays are perpendicular to each other* (Fig. 23.24b). These rays are perpendicular if $\theta_B + \theta_t = 90°$. Since the two angles are complementary, $\sin\theta_t = \cos\theta_B$. Then

$$n_i \sin\theta_B = n_t \sin\theta_t = n_t \cos\theta_B$$

$$\frac{\sin\theta_B}{\cos\theta_B} = \frac{n_t}{n_i} = \tan\theta_B$$

Brewster's angle:

$$\theta_B = \tan^{-1}\frac{n_t}{n_i} \qquad (23\text{-}6)$$

The value of Brewster's angle depends on the indices of refraction of the two media. Unlike the critical angle for total internal reflection, Brewster's angle exists regardless of which index of refraction is larger.

Why Is the Reflected Light Totally Polarized When the Reflected and Transmitted Rays Are Perpendicular? In Fig. 23.24, the unpolarized incident light is represented as a mixture of two perpendicular polarization components: one perpendicular to the plane of incidence and one in the plane of incidence. Note that the polarization components in the plane of incidence, represented by red and blue arrows, are not in the same direction; polarization components must be perpendicular to the ray since light is a transverse wave.

The same oscillating charges at the surface of the second medium radiate both the reflected light and the transmitted light. The oscillations are along the blue and green directions. The blue direction of oscillation contributes nothing to the reflected ray because an oscillating charge does not radiate along its axis of oscillation. Thus, when the light is incident at Brewster's angle, *the reflected light is totally polarized perpendicular to the plane of incidence.* At other angles of incidence, the reflected light is *partially* polarized perpendicular to the plane of incidence. If light is incident at Brewster's angle and is polarized in the plane of incidence (that is, it has no polarization component perpendicular to the plane of incidence), no light is reflected.

CHECKPOINT 23.5

Polarized sunglasses are useful for cutting out reflected glare due to reflection from horizontal surfaces. In which direction should the transmission axis of the sunglasses be oriented: vertically or horizontally? Explain.

23.6 THE FORMATION OF IMAGES THROUGH REFLECTION OR REFRACTION

When you look into a mirror, you see an image of yourself. What do we mean by an *image*? It *appears* as if your identical twin were standing behind the mirror. If you were looking at an actual twin, each point on your twin would reflect light in many different directions. Some of that light enters your eye. In essence, what your eye does is take the rays diverging from a given point and trace them backward to figure out where they come from. Your brain interprets light reflected from the mirror in the same way: all the light rays from any point on you (the object whose image is being formed) must reflect from the mirror *as if they came from a single point behind the mirror.*

Ideally, in the formation of an image, there is a one-to-one correspondence of points on the object and points on the image. If rays from one point on the object seem to come from many different points, the overlap of light from different points would look blurred. (A real lens or mirror may deviate somewhat from ideal behavior, causing some degree of blurring in the image.)

There are two kinds of images. For the plane mirror, the light rays *seem* to come from a point behind the mirror, but we know there aren't actually any light rays back there. In a **virtual image**, we trace light rays from a point on the object back to a point from which they *appear* to diverge, even though the rays do not actually come from that point. In a **real image**, the rays actually *do* pass through the image point. A camera lens forms a real image of the object being photographed on the film. The light rays have to actually be there to expose the film! The rays from a point on the object must all reach the same point on the film or else the picture will come out blurry. If the film and the back of the camera were not there to interrupt the light rays, they would diverge from the image point (Fig. 23.25). An image must be real if it is projected onto a surface such as film, a viewing screen, or a detector.

An image is *real* if light rays from a point on the object converge to the corresponding point on the image. An image is *virtual* if light rays from a point on the object are directed *as if* they diverged from a point on the image, even though the rays do not actually pass through the image point.

Projecting a real image onto a screen is only one way to view it. Real images can also be viewed directly (as virtual images are viewed) by looking into the lens or mirror. However, to view a real image, the viewer must be located *beyond the image* so that the rays from a point on the object all diverge from a point on the image. In Fig. 23.25, if the film is removed, the image can be viewed by looking into the lens from points beyond the image (that is, to the right of where the film is placed).

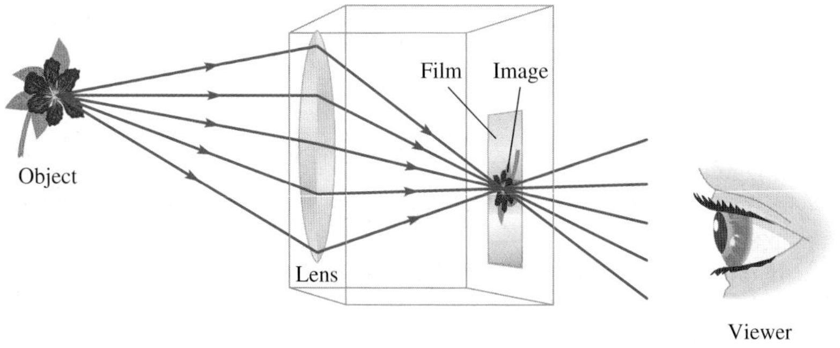

Figure 23.25 Formation of a real image by a camera lens. If the film and the back of the camera were not there, the rays would continue on, diverging from the image point. A viewer could then see the real image directly.

CHECKPOINT 23.6

In Figure 23.26, is the image of the fish real or virtual? Explain.

Finding an Image Using a Ray Diagram

- Draw two (or more) rays coming from a single off-axis point on the object toward whatever forms the image (usually a lens or mirror). (Only two rays are necessary since they *all* map to the same image point.)
- Trace the rays, applying the laws of reflection and refraction as needed, until they reach the observer.
- Extrapolate the rays backward along straight line paths until they intersect at the image point.
- If light rays actually go through the image point, the image is real. If they do not, the image is virtual.
- To find the image of an extended object, find the images of two or more points on the object.

Example 23.4

A Kingfisher Looking for Prey

A small fish is at a depth d below the surface of a still pond. What is the *apparent* depth of the fish as viewed by a belted kingfisher—a bird that dives underwater to catch fish? Assume the kingfisher is directly above the fish. Use $n = \frac{4}{3}$ for water.

Strategy The apparent depth is the depth of the *image* of the fish. Light rays coming from the fish toward the surface are refracted as they pass into the air. We choose a point on the fish and trace the rays from that point into the air; then we trace the refracted rays backward along straight lines until they meet at the image point. The kingfisher directly above sees not only a ray coming straight up ($\theta_i = 0$); it also sees rays at small but nonzero angles of incidence. We may be able to use small-angle approximations for these angles. However, for clarity in the ray diagram, we exaggerate the angles of incidence.

Solution In Fig. 23.26a we sketch a fish under water at a depth d. From a point on the fish, rays diverge toward the surface. At the surface they are bent away from the normal (since air has a lower index of refraction). The image point is found by tracing the refracted rays straight backward (dashed lines) to where they meet. We label the image depth d'. From the ray diagram, we see that $d' < d$; the apparent depth is less than the actual depth.

Only two rays need be used to locate the image. To simplify the math, one of them can be the ray normal to the surface. The other ray is incident on the water surface at angle θ_i. This ray leaves the water surface at angle θ_t, where

$$n_w \sin \theta_i = n_a \sin \theta_t \qquad (1)$$

To find d', we use two right triangles (Fig. 23.26b) that share the same side s—the distance between the points at which the two chosen rays intersect the water surface. The angles θ_i and θ_t are known since they are alternate interior angles with the angles at the surface. From these triangles,

$$\tan \theta_i = \frac{s}{d} \quad \text{and} \quad \tan \theta_t = \frac{s}{d'}$$

For small angles, we can set $\tan \theta \approx \sin \theta$. Then Eq. (1) becomes

$$n_w \frac{s}{d} = n_a \frac{s}{d'}$$

After eliminating s, we solve for the ratio d'/d:

$$\frac{\text{apparent depth}}{\text{actual depth}} = \frac{d'}{d} = \frac{n_a}{n_w} = \frac{3}{4}$$

The apparent depth of the fish is $\frac{3}{4}$ of the actual depth.

Discussion The result is valid only for small angles of incidence—that is, for a viewer directly above the fish.

continued on next page

Example 23.4 continued

The apparent depth depends on the angle at which the fish is viewed.

The image of the fish is virtual. The light rays seen by the kingfisher *seem* to come from the location of the image, but they do not.

Practice Problem 23.4 Evading the Predator

Suppose the fish looks upward and sees the kingfisher. If the kingfisher is a height h above the surface of the pond, what is its apparent height h' as viewed by the fish?

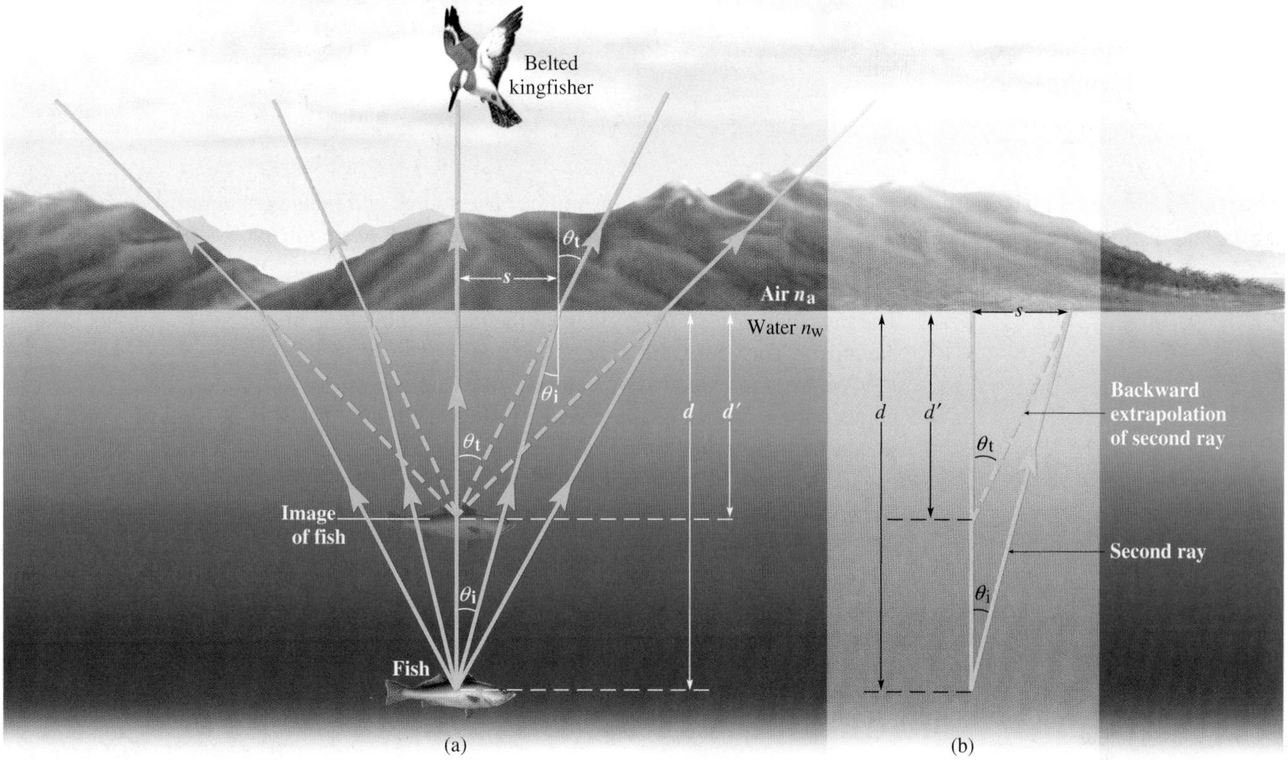

Figure 23.26 (a) Formation of the image of the fish. (b) Two right triangles that share side s enable us to solve for the image depth d' in terms of d.

23.7 PLANE MIRRORS

A shiny metal surface is a good reflector of light. An ordinary mirror is *back-silvered*; that is, a thin layer of shiny metal is applied to the *back* of a flat piece of glass. A back-silvered mirror actually produces two reflections: a faint one, seldom even noticed, from the front surface of the glass and a strong one from the metal. *Front-silvered* mirrors are used in precision work, since they produce only one reflection; they are not practical for everyday use because the metal coating is easily scratched. If we ignore the faint reflection from the glass, then back-silvered mirrors are treated the same as front-silvered mirrors.

Light reflected from a mirror follows the laws of reflection discussed in Section 23.2. Figure 23.27a shows a point source of light located in front of a plane mirror; an observer looks into the mirror. If the reflected rays are extrapolated backward through the mirror, they all intersect at one point, which is the image of the point source. Using any two rays and some geometry, you can show (Problem 45) that

For a plane mirror, a point source and its image are at the same distance from the mirror (on opposite sides); both lie on the same normal line.

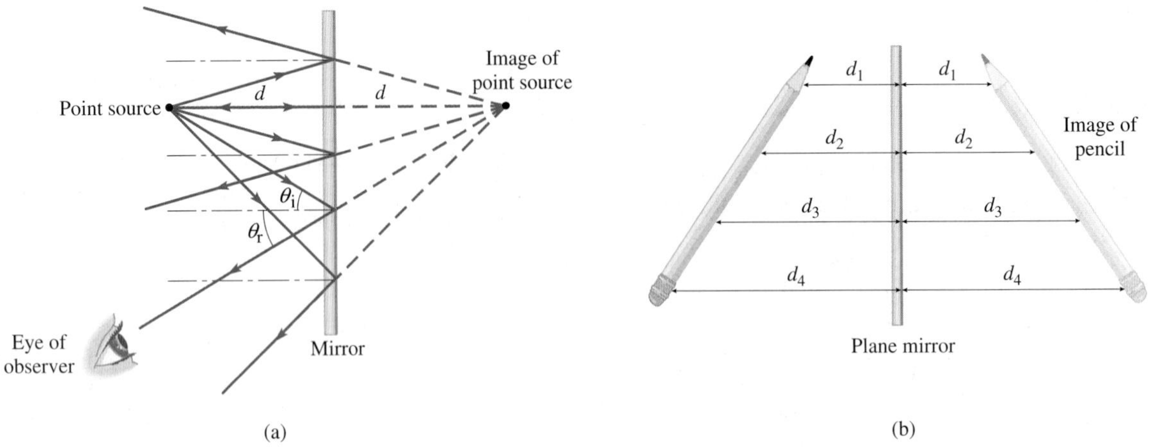

Figure 23.27 (a) A plane mirror forms a virtual image of a point source. The source and image are equidistant from the mirror and lie on the same normal line. (b) Sketching the image of a pencil formed by a plane mirror.

The rays only *appear* to originate at the image behind the mirror; no rays travel through the mirror. Therefore, the image is *virtual.*

We treat an extended object in front of a plane mirror as a set of point sources (the points on the surface of the object). In Fig. 23.27b, a pencil is in front of a mirror. To sketch the image, we first construct normals to the mirror from several points on the pencil. Then each image point is placed a distance behind the mirror equal to the distance from the mirror to the corresponding point on the object.

Conceptual Example 23.5

Mirror Length for a Full-Length Image

Grant is carrying his niece Dana on his shoulders (Fig. 23.28). What is the minimum vertical length of a plane mirror in which Grant can see a full image (from his toes to the top of Dana's head)? How should this minimum-length mirror be placed on the wall?

Strategy Ray diagrams are *essential* in geometric optics. A ray diagram is most helpful if we carefully decide which rays are most important to the solution. Here, we want to make sure Grant can see the images of two particular points: his toes and the top of Dana's head. If he can see those two points, he can see everything between them. In order for Grant to see the image of a point, a ray of light from that point must reflect from the mirror and enter Grant's eye.

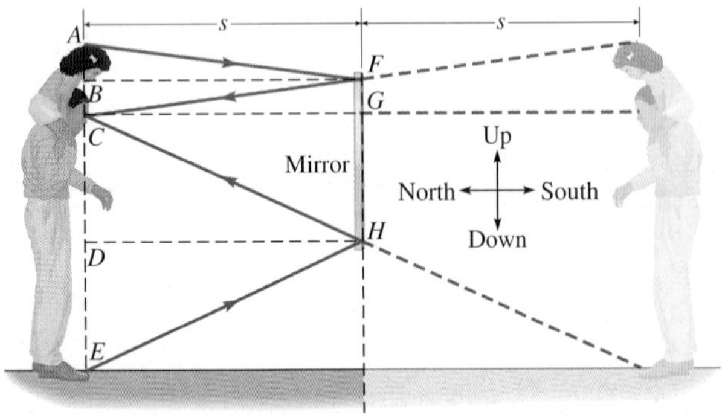

Figure 23.28
Conceptual Example 23.5.

Solution and Discussion After drawing Grant, Dana, and the mirror (Fig. 23.28), we want to draw a ray from Grant's toes that strikes the mirror and is reflected to his eye. The line *DH* is a normal to the mirror surface. Since the angle of incidence is equal to the angle of reflection, the triangles *CHD* and *EHD* are congruent and $CD = DE = GH$. Therefore,

$$GH = \tfrac{1}{2}CE$$

Similarly, we draw a ray from the top of Dana's head to the mirror that is reflected into Grant's eye and find that

$$FG = \tfrac{1}{2}AC$$

The length of the mirror is

$$FH = FG + GH = \tfrac{1}{2}(AC + CE) = \tfrac{1}{2}AE$$

Therefore, the length of the mirror must be *one half* the distance from Grant's toes to Dana's head.

continued on next page

Conceptual Example 23.5 continued

The minimum-length mirror only allows a full-length view if it is hung properly. The top of the mirror (F) must be a distance AB below the top of Dana's head. A full-length mirror is *not* necessary to get a full-length view. Extending the mirror all the way to the floor is of no use; the bottom of the mirror only needs to be halfway between the floor and the eyes of the shortest person who uses the mirror. Note that the distance s between Grant and the mirror has no effect on the result. That is, you need the same height mirror whether you're up close to it or farther back.

Practice Problem 23.5 Two Sisters with One Mirror

Sarah's eyes are 1.72 m above the floor when she is wearing her dress shoes, and the top of her head is 1.84 m above the floor. Sarah has a mirror that is 0.92 m in length, hung on the wall so she can just see a full-length image of herself. Suppose Sarah's sister Michaela is 1.62 m tall and her eyes are 1.52 m above the floor. If Michaela uses Sarah's mirror without moving it, can she see a full-length image of herself? Draw a ray diagram to illustrate.

PHYSICS AT HOME

You can easily demonstrate *multiple* images using two plane mirrors. Set up two plane mirrors at a 90° angle on a table and place an object with lettering on it between them. You should see three images. The image straight back is due to rays that reflect twice—once from each mirror. Draw a ray diagram to illustrate the formation of each image. In which of the images is the lettering reversed? (See Conceptual Question 4 for some insight into the apparent left-right reversal.)

One way to think about multiple images is that each mirror forms an image of the other mirror; then the image mirrors produce images of images. To explore further, gradually reduce the angle between the mirrors (Fig. 23.29).

Figure 23.29 Two plane mirrors at an angle of 72° form four images.

23.8 SPHERICAL MIRRORS

Convex Spherical Mirror

In a spherical mirror, the reflecting surface is a section of a sphere. A **convex mirror** curves *away from* the viewer; its *center of curvature* is *behind* the mirror (Fig. 23.30). An extended radius drawn from the center of curvature through the **vertex**—the center of the surface of the mirror—is the **principal axis** of the mirror.

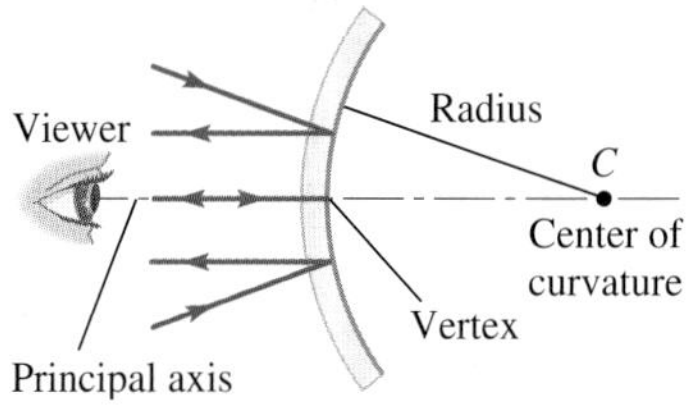

Figure 23.30 A convex mirror's center of curvature is behind the mirror.

In Fig. 23.31a, a ray parallel to the principal axis is incident on the surface of a convex mirror at point A, which is close to the vertex V. (In the diagram, the distance between points A and V is exaggerated for clarity.) A radial line from the center of

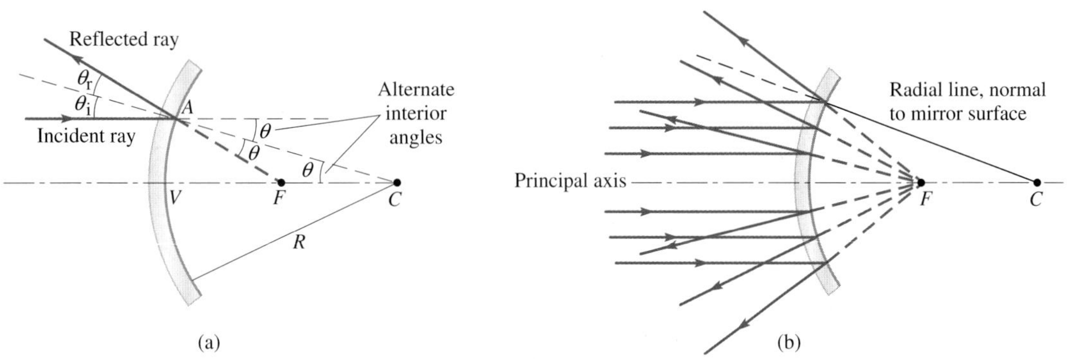

Figure 23.31 (a) Location of the focal point (F) of a convex mirror. (b) Parallel rays reflected from a convex mirror *appear to be* coming from the focal point.

curvature through point A is normal to the mirror. The angle of incidence is equal to the angle of reflection: $\theta_i = \theta_r = \theta$.

By alternate interior angles, we know that

$$\angle ACF = \theta$$

Triangle AFC is isosceles since it has two equal angles; therefore,

$$\overline{AF} = \overline{FC}$$

Since the incident ray is close to the principal axis, θ is small. As a result,

$$\overline{AF} + \overline{FC} \approx R \quad \text{and} \quad \overline{VF} \approx \overline{AF} \approx \tfrac{1}{2}R$$

The notation $\overline{AF}$ means the length of the line segment from A to F.

where $\overline{AC} = \overline{VC} = R$ is the radius of curvature of the mirror. Note that this derivation is true for *any* angle θ *as long as it is sufficiently small.* Thus, all rays parallel to the axis *that are incident near the vertex* are reflected by the convex mirror so that they *appear* to originate from point F, which is called the **focal point** of the mirror (Fig. 23.31b). A convex mirror is also called a **diverging mirror** since the reflection of a set of parallel rays is a set of diverging rays.

> The focal point of a convex mirror is on the principal axis a distance $\frac{1}{2}R$ behind the mirror.

To find the image of an object placed in front of the mirror, we draw a few rays. Figure 23.32 shows an object in front of a convex mirror. Four rays are drawn from the point at the top of the object to the mirror surface. One ray (shown in green) is parallel to the principal axis; it is reflected as if it were coming from the focal point. Another ray (red) is headed along a radius toward the center of curvature C; it reflects back on itself since the angle of incidence is zero. A third ray (blue) heads directly toward the focal point F. Since a ray parallel to the axis is reflected as if it came from F, a ray going toward F is reflected parallel to the axis. Why? Because the law of reflection is reversible; we can reverse the direction of a ray and the law of reflection still holds. A fourth ray (brown), incident on the mirror at its vertex, reflects making an equal angle with the axis (since the axis is normal to the mirror).

These four reflected rays—as well as other reflected rays from the top of the object—meet at one point when extended behind the mirror. That is the location of the top of the image. The bottom of the image lies on the principal axis because the bottom of the object is on the principal axis; rays along the principal axis are radial rays, so they reflect back on themselves at the surface of the mirror. From the ray diagram, we can conclude that the image is upright, virtual, smaller than the object, and closer to the mirror than the object. Note that the image is *not* at the focal point; the rays coming from a point on the object are *not* all parallel to the principal axis. If the object were far from the mirror, then the rays from any point would be nearly parallel to each other. Rays from a point on the principal axis would meet at the focal point; rays from a point not

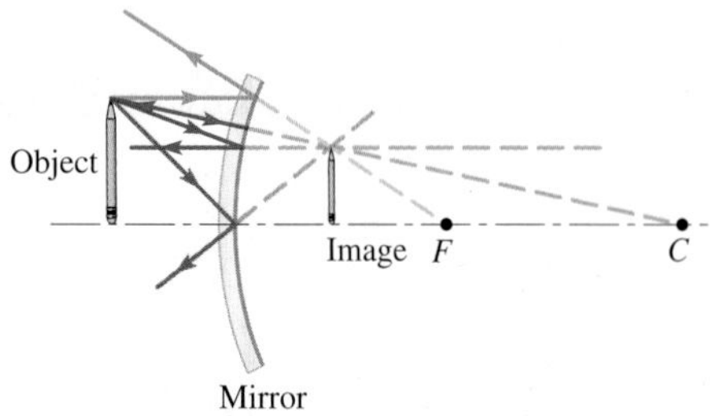

Principal rays for convex mirrors

1. A ray parallel to the principal axis is reflected as if it came from a focal point.
2. A ray along a radius is reflected back on itself.
3. A ray directed toward the focal point is reflected parallel to the principal axis.
4. A ray incident on the vertex of the mirror reflects at an equal angle to the axis.

Figure 23.32 Using the principal rays to locate the image formed by a convex mirror. The rays are shown in different colors merely to help distinguish them; the actual color of the light along each ray is the same—whatever the color of the top of the object is.

Figure 23.33 The same scene viewed in plane (left) and convex (right) mirrors. The convex mirror provides a larger field of view.

on the axis would meet at a point in the **focal plane**—the plane perpendicular to the axis passing through the focal point.

The four rays we chose to draw are called the **principal rays** only because they are easier to draw than other rays. Principal rays are easier to draw, but they are not more important than other rays in forming an image. Any two of them can be drawn to locate an image, but it is wise to draw a third as a check.

A convex mirror enables one to see a larger area than the same size plane mirror (Fig. 23.33). The outward curvature of the convex mirror enables the viewer to see light rays coming from larger angles. Convex mirrors are sometimes used in stores to help clerks watch for shoplifting. The passenger's side mirror in most cars is convex to enable the driver to see farther out to the side.

Concave Spherical Mirror

A **concave mirror** curves *toward* the viewer; its center of curvature is *in front* of the mirror. A concave mirror is also called a **converging mirror** since it makes parallel rays converge to a point (Fig. 23.34). In Problem 55 you can show that rays parallel to the mirror's principal axis pass through the focal point F at a distance $R/2$ from the vertex, assuming the angles of incidence are small.

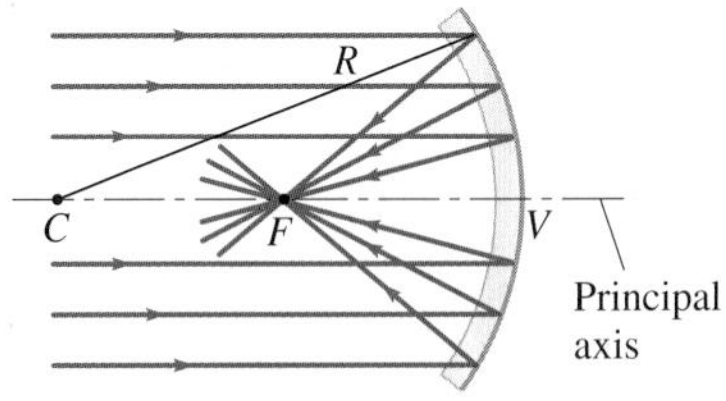

Figure 23.34 Reflection of rays parallel to the principal axis of a concave mirror. Point C is the mirror's center of curvature and F is the focal point. Both points are in *front* of the mirror, in contrast to the convex mirror.

The location of the image of an object placed in front of a concave mirror can be found by drawing two or more rays. As for the convex mirror, there are four principal rays—rays that are easiest to draw. The principal rays are similar to those for a convex mirror, the difference being that the focal point is in *front* of a concave mirror.

Figure 23.35 illustrates the use of principal rays to find an image. In this case, the object is between the focal point and the center of curvature. The image is real because

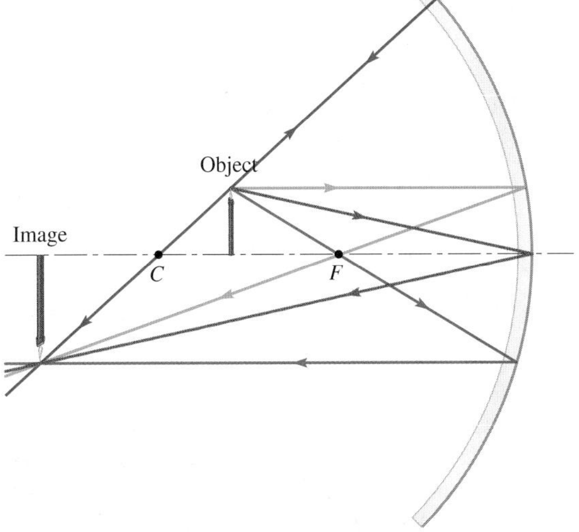

Principal rays for concave mirrors

1. A ray parallel to the principal axis is reflected through the focal point.
2. A ray along a radius is reflected back on itself.
3. A ray along the direction from the focal point to the mirror is reflected parallel to the principal axis.
4. A ray incident on the vertex of the mirror reflects at an equal angle to the axis.

Figure 23.35 An object between the focal point and the center of curvature of a concave mirror forms a real image that is inverted and larger than the object. (The angles and the curvature of the mirror are exaggerated for clarity.)

(a)

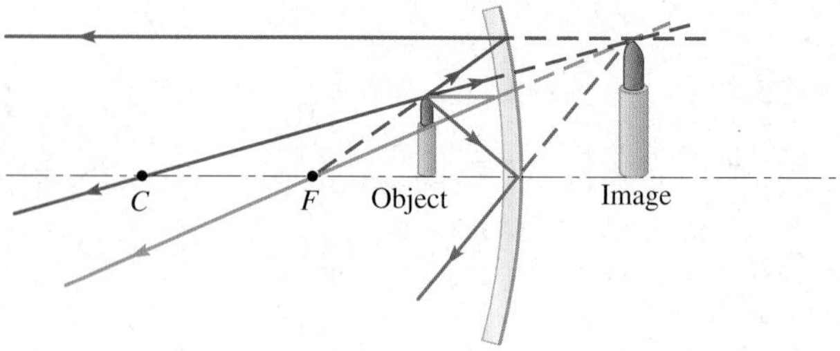

(b)

Figure 23.36 (a) Putting on makeup is made easier because the image is enlarged. (b) Formation of an image when the object is between the focal point and the concave mirror's surface.

it is in front of the mirror; the principal rays actually do pass through the image point. Depending on the location of the object, a concave mirror can form either real or virtual images. The images can be larger or smaller than the object.

Applications: cosmetic mirrors and automobile headlights

Mirrors designed for shaving or for applying cosmetics are often concave in order to form a magnified image (Fig. 23.36a). Dentists use concave mirrors for the same reason. Whenever an object is within the focal point of a concave mirror, the image is virtual, upright, and larger than the object (Fig. 23.36b).

In automobile headlights, the lightbulb filament is placed at the focal point of a concave mirror. Light rays coming from the filament are reflected out in a beam of parallel rays. (Sometimes the shape of the mirror is parabolic rather than spherical; a parabolic mirror reflects *all* the rays from the focal point into a parallel beam, not just those close to the principal axis.)

Example 23.6

Scale Diagram for a Concave Mirror

Make a scale diagram showing a 1.5-cm-tall object located 10.0 cm in front of a concave mirror with a radius of curvature of 8.0 cm. Locate the image graphically and estimate its position and height.

Strategy For a scale diagram, we should use a piece of graph paper and choose a scale that fits on the paper—although sometimes it is helpful to make a rough sketch first to get some idea of where the image is. Drawing two principal rays enables us to find the image. Using the third principal ray is a good check. Since the mirror is concave, the center of curvature and the focal point are both in front of the mirror.

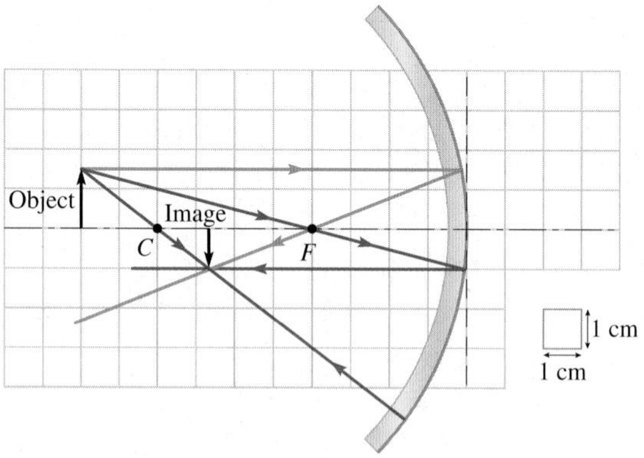

Figure 23.37
Example 23.6.

Solution To start, we draw the mirror and the principal axis; then we mark the focal point and center of curvature at the correct distances from the vertex (Fig. 23.37). The green ray goes from the top of the object to the mirror parallel to the principal axis. It is reflected by the mirror so that it passes through the focal point. The blue ray travels from the tip of the object through the focal point *F*. This ray is reflected from the mirror along a line parallel to the principal axis. The intersection of the two rays determines the location of the tip of the image. By measuring the image on the graph paper, we find that the image is 6.7 cm from the mirror and is 1.0 cm high.

continued on next page

Example 23.6 continued

Discussion As a check, the red ray travels through the center of curvature along a radius. Assuming the mirror extends far enough to reflect this ray, it strikes the mirror perpendicular to the surface since it is on a radial line. The reflected ray travels back along the same radial line and intersects the other two rays at the tip of the image, verifying our result.

Practice Problem 23.6 Another Graphical Solution

Draw a scale diagram to locate the image of an object 1.5 cm tall and 6.0 cm in front of the same mirror. Estimate the position and height of the image. Is it real or virtual? [*Hint:* Draw a rough sketch first.]

Transverse Magnification

The image formed by a mirror or a lens is, in general, not the same size as the object. It may also be inverted (upside down). The **transverse magnification** m (also called the *lateral* or *linear* magnification) is a ratio that gives both the relative size of the image—in any direction perpendicular to the principal axis—and its orientation. The magnitude of m is the ratio of the image size to the object size:

$$|m| = \frac{\text{image size}}{\text{object size}} \quad (23\text{-}7)$$

If $|m| < 1$, the image is smaller than the object. The sign of m is determined by the orientation of the image. For an inverted (upside down) image, $m < 0$; for an upright (right side up) image, $m > 0$.

Let h be the height of the object (really the *displacement* of the top of the object from the axis) and h' be the height of the image. If the image is inverted, h' and h have opposite signs. Then the definition of the transverse magnification is

$$m = \frac{h'}{h} \quad (23\text{-}8)$$

Using Fig. 23.38, we can find a relationship between the magnification and p and q, the **object distance** and the **image distance**. Note that p and q are measured along the principal axis to the vertex of the mirror. The two right triangles ΔPAV and ΔQBV are similar, so

$$\frac{h}{p} = \frac{-h'}{q}$$

Why the negative sign? In this case, if h is positive, then h' is negative, since the image is on the opposite side of the axis from the object. The magnification is then

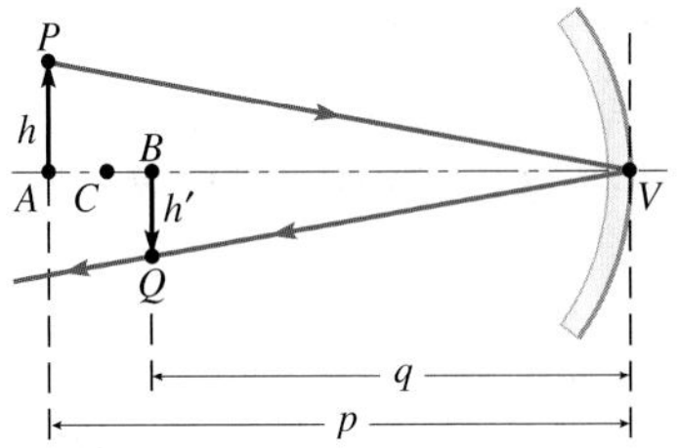

Figure 23.38 Right triangles ΔPAV and ΔQBV are similar because the angle of incidence for the ray equals the angle of reflection.

Magnification equation:

$$m = \frac{h'}{h} = -\frac{q}{p} \quad (23\text{-}9)$$

Although in Fig 23.38 the object is beyond the center of curvature, Eq. (23-9) is true regardless of where the object is placed. It applies to any spherical mirror, concave or convex (see Problem 53), as well as to plane mirrors.

CHECKPOINT 23.8

A plane mirror forms an image of an object in front of it. Is the image real or virtual? What is the transverse magnification (m)?

Table 23.2 Sign Conventions for Mirrors

Quantity	When Positive (+)	When Negative (−)
Object distance p	Real object*	Virtual object*
Image distance q	Real image	Virtual image
Focal length f	Converging mirror (concave): $f=\frac{1}{2}R$	Diverging mirror (convex): $f=-\frac{1}{2}R$
Magnification m	Upright image	Inverted image

*In Chapter 23, we consider only real objects. Chapter 24 discusses multiple-lens systems, in which *virtual* objects are possible.

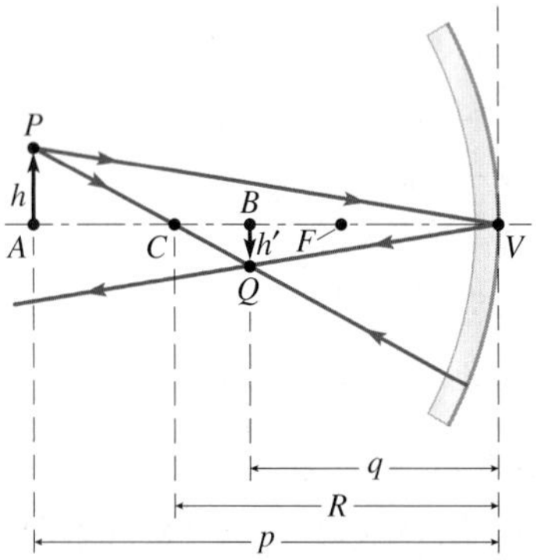

Figure 23.39 Similar triangles ΔPAC and ΔQBC used to derive the lens equation.

The Mirror Equation

From Fig. 23.39, we can derive an equation relating the object distance p, the image distance q, and the **focal length** $f=\frac{1}{2}R$ (the distance from the focal point to the mirror). Note that p, q, and f are all measured along the principal axis to the vertex V of the mirror. Triangles ΔPAC and ΔQBC are similar. Note that $\overline{AC}=p-R$ and $\overline{BC}=R-q$, where R is the radius of curvature. Then

$$\frac{\overline{PA}}{\overline{AC}}=\frac{\overline{QB}}{\overline{BC}} \quad \text{or} \quad \frac{h}{p-R}=\frac{-h'}{R-q}$$

Rearranging yields

$$\frac{h'}{h}=-\frac{R-q}{p-R}$$

Since h'/h is the magnification,

$$\frac{h'}{h}=-\frac{q}{p}=-\frac{R-q}{p-R} \qquad (23\text{-}9)$$

Substituting $f=R/2$, cross multiplying, and dividing by p, q, and f (Problem 56), we obtain the **mirror equation**.

Mirror equation:

$$\frac{1}{p}+\frac{1}{q}=\frac{1}{f} \qquad (23\text{-}10)$$

We derived the magnification and mirror equations for a concave mirror forming a real image, but the equations apply as well to convex mirrors and to virtual images if we use the sign conventions for q and f listed in Table 23.2. Note that q is negative when the image is behind the mirror and f is negative when the focal point is behind the mirror.

The magnification equation and the sign convention for q imply that *real images of real objects are always inverted* (if both p and q are positive, m is negative); *virtual images of real objects are always upright* (if p is positive and q is negative, m is positive). The same rule can be established by drawing ray diagrams. A real image is always in front of the mirror (where the light rays are); a virtual image is behind the mirror.

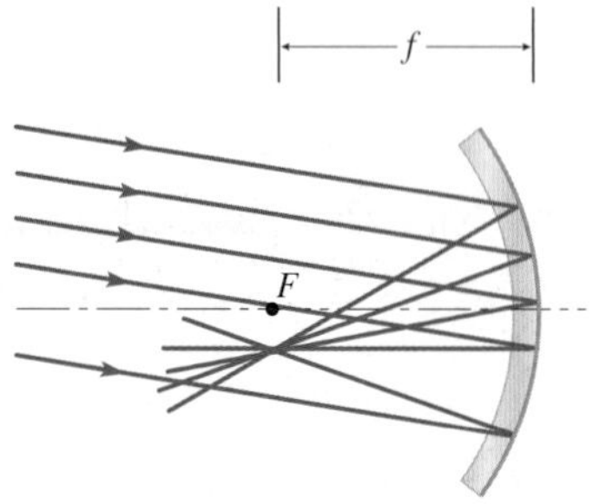

Figure 23.40 A faraway object above the principal axis forms an image at $q=f$.

If an object is far from the mirror ($p=\infty$), the mirror equation gives $q=f$. Rays coming from a faraway object are nearly parallel to each other. After reflecting from the mirror, the rays converge at the focal point for a concave mirror or appear to diverge from the focal point for a convex mirror. If the faraway object is not on the principal axis, the image is formed above or below the focal point (Fig. 23.40).

Example 23.7

Passenger's Side Mirror

An object is located 30.0 cm from a passenger's side mirror. The image formed is upright and one third the size of the object. (a) Is the image real or virtual? (b) What is the focal length of the mirror? (c) Is the mirror concave or convex?

Strategy The magnitude of the magnification is the ratio of the image size to the object size, so $|m| = \frac{1}{3}$. The sign of the magnification is positive for an upright image and negative for an inverted image. Therefore, we know that $m = +\frac{1}{3}$. The object distance is $p = 30.0$ cm. The magnification is also related to the object and image distances, so we can find q. The sign of q indicates whether the image is real or virtual. Then the mirror equation can be used to find the focal length of the mirror. The sign of the focal length tells us whether the mirror is concave or convex.

Solution (a) The magnification is related to the image and object distances:

$$m = -\frac{q}{p} \qquad (23\text{-}9)$$

Solving for the image distance,

$$q = -mp = -\tfrac{1}{3} \times 30.0\text{ cm} = -10.0\text{ cm}$$

Since q is negative, the image is virtual.

(b) Now we can use the mirror equation to find the focal length:

$$\frac{1}{f} = \frac{1}{p} + \frac{1}{q} = \frac{q+p}{pq}$$

$$f = \frac{pq}{q+p}$$

$$= \frac{30.0\text{ cm} \times (-10.0\text{ cm})}{-10.0\text{ cm} + 30.0\text{ cm}}$$

$$= -15.0\text{ cm}$$

(c) Since the focal length is negative, the mirror is convex.

Discussion As expected, the passenger's side mirror is convex. With all the distances known, we can sketch a ray diagram (Fig. 23.41) to check the result.

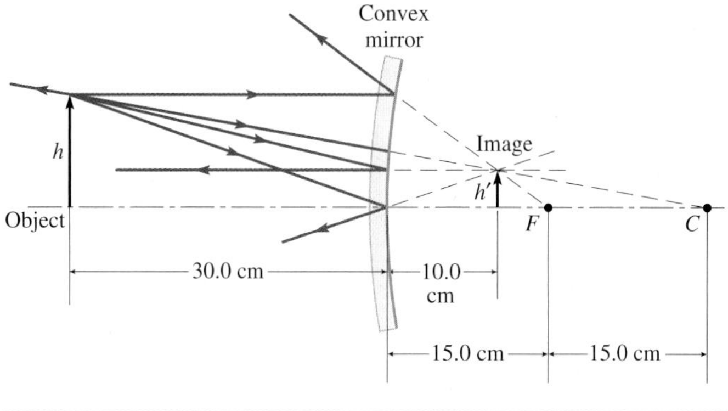

Figure 23.41
Ray diagram for convex mirror (Example 23.7).

Practice Problem 23.7 A Spherical Mirror of Unknown Type

An object is in front of a spherical mirror; the image of the object is upright and twice the size of the object and it appears to be 12.0 cm behind the mirror. What is the object distance, what is the focal length of the mirror, and what type of mirror is it (convex or concave)?

PHYSICS AT HOME

Look at each side of a *shiny* metal spoon. (Stainless steel gets dull with use; the newer the spoon the better. A polished silver spoon would be ideal.) One side acts as a convex mirror; the other acts as a concave mirror. For each, notice whether your image is upright or inverted and enlarged or diminished in size. Next, decide whether each image is real or virtual. Which side gives you a larger field of view (in other words, enables you to see a bigger part of the room)? Try holding the spoon at different distances to see what changes. (Keep in mind that the focal length of the spoon is small. If you hold the spoon less than a focal length from your eye, you won't be able to see clearly—your eye cannot focus at such a small distance. Thus, it is not possible to get close enough to the concave side to see a virtual image.)

23.9 THIN LENSES

Whereas mirrors form images through reflection, lenses form images through refraction. In a spherical lens, each of the two surfaces is a section of a sphere. The **principal axis** of a lens passes through the centers of curvature of the lens surfaces. The **optical center** of a lens is a point on the principal axis through which rays pass without changing direction.

We can understand the behavior of a lens by regarding it as an assembly of prisms (Fig. 23.42). The angle of deviation of the ray—the angle that the ray emerging from the prism makes with the incident ray—is proportional to the angle between the two faces of the prism (see Fig. 23.43 and Problem 19). The two faces of a lens are parallel where they intersect the principal axis. A ray striking the lens at the center emerges in the same direction as the incident ray since the refraction of an entering ray is undone as the ray emerges. However, the ray is *displaced*; it is not along the same line as the incident ray. As long as we consider only *thin lenses*—lenses whose thickness is small compared with the focal length—the displacement is negligible; the ray passes straight through the lens.

The curved surfaces of a lens mean that the angle β between the two faces gradually increases as we move away from the center. Thus, the angle of deviation of a ray increases as the point where it strikes the lens moves away from the center. We restrict our consideration to **paraxial rays**: rays that strike the lens close to the principal axis (so that β is small) and do so at a small angle of incidence. For paraxial rays and thin lenses, a ray incident on the lens at a distance d from the center has an angle of deviation δ that is proportional to d (Fig. 23.44; see Problem 99).

Lenses are classified as **diverging** or **converging**, depending on what happens to the rays as they pass through the lens. A diverging lens bends light rays outward, away from the principal axis. A converging lens bends light rays inward, toward the principal axis (Fig. 23.45a). If the incident rays are already diverging, a converging lens may not be able to make them converge; it may only make them diverge less (Fig. 23.45b). Lenses take many possible shapes (Fig. 23.46); the two surfaces may have different radii of curvature. Note that converging lenses are thickest at the center and diverging lenses are thinnest at the center, assuming the index of refraction of the lens is greater than that of the surrounding medium.

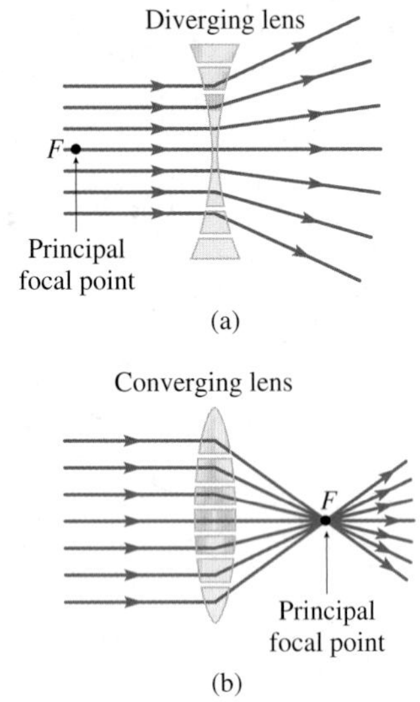

Figure 23.42 (a) and (b) Lenses made by combining prism sections.

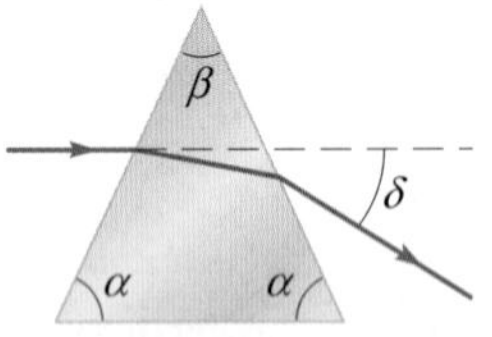

Figure 23.43 The angle of deviation δ increases as the angle β between the two faces increases. For small β, δ is proportional to β.

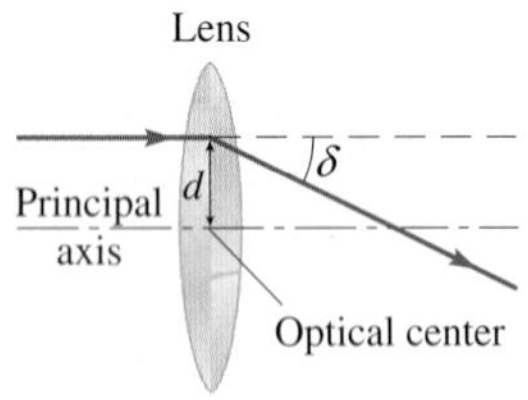

Figure 23.44 The angle of deviation of a paraxial ray striking the lens a distance d from the principal axis is proportional to d. To simplify ray diagrams, we draw rays as if they bend at a vertical line through the optical center rather than bending at each of the two lens surfaces.

Focal Points and Principal Rays

Any lens has two focal points. The distance between each focal point and the optical center is the magnitude of the **focal length** of the lens. The focal length of a lens with spherical surfaces depends on four quantities: the radii of curvature of the two surfaces and the indices of refraction of the lens and of the surrounding medium (usually, but not necessarily, air). For a diverging lens, incident rays parallel to the principal axis are refracted by the lens so that they appear to diverge from the **principal focal point**,

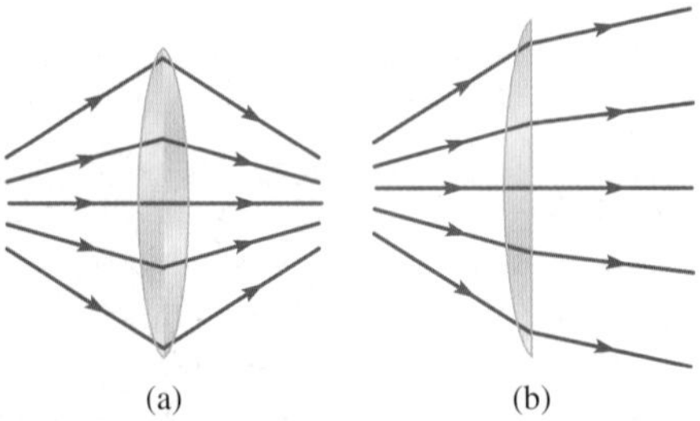

Figure 23.45 (a) When diverging rays strike a converging lens, the lens bends them inward. (b) If they are diverging too rapidly, the lens may not be able to bend them enough to make them converge. In that case, the rays diverge less rapidly after they leave the lens.

Table 23.3 Principal Rays and Principal Focal Points for Thin Lenses

Principal Ray/Focal Point	Converging Lens	Diverging Lens
Ray 1. An incident ray parallel to the principal axis	Passes through the principal focal point	Appears to come from the principal focal point
Ray 2. A ray incident at the optical center	Passes straight through the lens	Passes straight through the lens
Ray 3. A ray that *emerges* parallel to the principal axis	Appears to come from the secondary focal point	Appears to have been heading for the secondary focal point
Location of the principal focal point	Past the lens	Before the lens

which is *before* the lens (see Fig. 23.42a). For a converging lens, incident rays parallel to the axis are refracted by the lens so they converge to the principal focal point *past* the lens (Fig. 23.42b).

Two rays suffice to locate the image formed by a thin lens, but a third ray is useful as a check. The three rays that are generally the easiest to draw are called the **principal rays** (Table 23.3). The third principal ray makes use of the **secondary focal point**, which is on the opposite side of the lens from the principal focal point. The behavior of ray 3 can be understood by reversing the direction of all the rays, which also interchanges the two focal points. Figure 23.47 illustrates how to draw the principal rays.

CHECKPOINT 23.9

Is the image formed by a converging lens always real, always virtual, or can it be either real or virtual? Explain. [*Hint*: Refer to Fig. 23.45.]

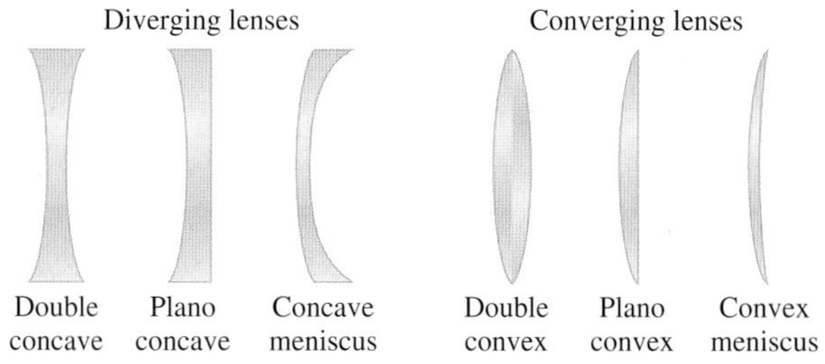

Figure 23.46 Shapes of some diverging and converging lenses. Diverging lenses are thinnest at the center; converging lenses are thickest at the center.

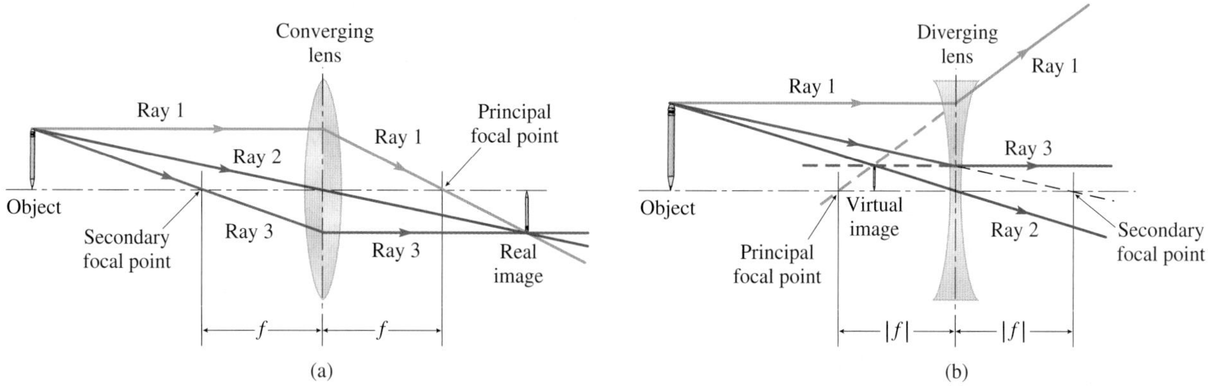

Figure 23.47 (a) The three principal rays for a converging lens forming a real image. (b) The three principal rays for a diverging lens forming a virtual image.

Conceptual Example 23.8

Orientation of Virtual Images

A lens forms an image of an object placed before the lens. Using a ray diagram, show that if the image is virtual, then it must also be upright, regardless of whether the lens is converging or diverging.

Strategy The principal rays are usually the easiest to draw. Principal rays 1 and 3 behave differently for converging and diverging lenses. They also deal with focal points, whereas the problem implies that the location of the object with respect to the focal points is irrelevant (except that we know a virtual image is formed). Ray 2 passes undeviated through the center of the lens. It behaves the same way for both types of lens and does not depend on the location of the focal points.

Solution and Discussion Figure 23.48 shows an object in front of a lens (which could be either converging or diverging). Principal ray 2 from the top of the object passes straight through the center of the lens. We extrapolate the refracted ray backward and sketch a few possibilities for the location of the image—with only one ray we do not know the actual location. We do know that a point on a virtual image is located not where the rays emerging from the lens meet, but rather where the *backward extrapolation of those rays* meet. In other words, the position of a virtual image is always before the lens (on the same side as the *incident* rays). Therefore, the image is on the same side of the lens as the object. From Fig. 23.48, we see that, just as for mirrors, the virtual image is upright—the image of the point at the top of the object is always above the principal axis.

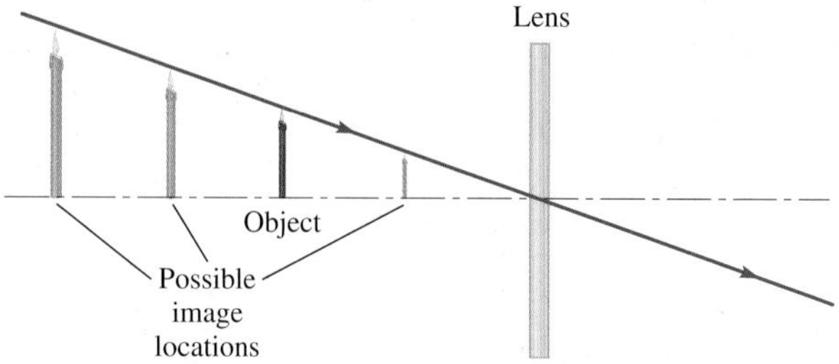

Figure 23.48
The principal ray passing undeviated through the center of a lens shows that virtual images of real objects are upright.

Conceptual Practice Problem 23.8 Orientation of Real Images

A converging lens forms a real image of an object placed before the lens. Using a ray diagram, show that the image is inverted.

The Magnification and Thin Lens Equations

We can derive the thin lens equation and the magnification equation from the geometry of Fig. 23.49. From the similar right triangles ΔEGC and ΔDBC, we write

$$\tan \alpha = \frac{h}{p} = \frac{-h'}{q}$$

As in the derivation of the mirror equation, h' is a signed quantity. For an inverted image, h' is negative; $-h'$ is the (positive) length of side BD. The magnification is given by

Magnification equation:

$$m = \frac{h'}{h} = -\frac{q}{p} \qquad (23\text{-}9)$$

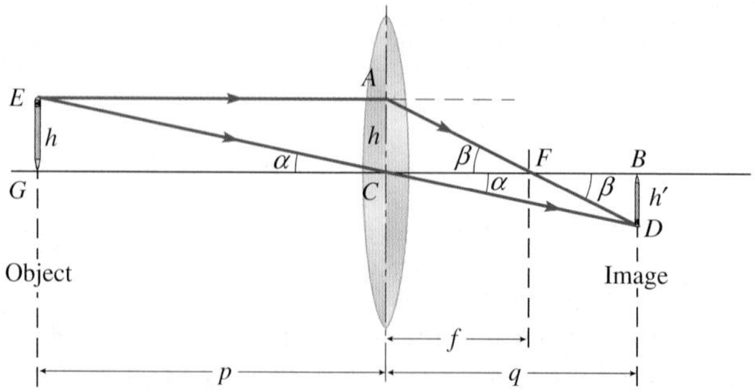

Figure 23.49 Ray diagram showing two of the three principal rays used in the derivation of the thin lens equation and the magnification.

Table 23.4 Sign Conventions for Mirrors and Lenses

Quantity	When Positive (+)	When Negative (−)
Object distance p	Real object*	Virtual object*
Image distance q	Real image	Virtual image
Focal length f	Converging lens or mirror	Diverging lens or mirror
Magnification m	Upright image	Inverted image

*In Chapter 23, we consider only real objects. Chapter 24 discusses multiple-lens systems, in which *virtual* objects are possible.

From two other similar right triangles ΔACF and ΔDBF,

$$\tan\beta = \frac{h}{f} = \frac{-h'}{q-f}$$

or

$$\frac{q-f}{f} = \frac{-h'}{h} = \frac{q}{p}$$

After dividing through by q and rearranging, we obtain the **thin lens equation**.

Thin lens equation:

$$\frac{1}{p} + \frac{1}{q} = \frac{1}{f} \quad (23\text{-}10)$$

CONNECTION:

The magnification and thin lens equations have exactly the same form as the corresponding equations derived for mirrors. The derivations used a converging lens and a real image, but they apply to all cases—either kind of lens and either kind of image—as long as we use the same sign conventions for q and f as for spherical mirrors (Table 23.4).

Example 23.9

Zoom Lens

A wild daisy 1.2 cm in diameter is 90.0 cm from a camera's zoom lens. The focal length of the lens has magnitude 150.0 mm. (a) Find the distance between the lens and the film. (b) How large is the image of the daisy?

Strategy The problem can be solved using the lens and magnification equations. The lens must be *converging* to form a real image on the film, so $f = +150.0$ mm. The image must be formed on the film, so the distance from the lens to the film is q. After finishing the algebraic solution, we sketch a ray diagram as a check.

Given: $h = 1.2$ cm; $p = 90.0$ cm; $f = +15.00$ cm

Find: q, h'

Solution (a) Since p and f are known, we find q from the thin lens equation

$$\frac{1}{p} + \frac{1}{q} = \frac{1}{f}$$

Let us solve for q.

$$q = \left(\frac{1}{f} - \frac{1}{p}^{-1}\right)$$

Substituting numerical values,

$$q = \left(\frac{1}{15.00\text{ cm}} - \frac{1}{90.0\text{ cm}}\right)^{-1} = +18.0\text{ cm}$$

The film is 18.0 cm from the lens.

(b) From the magnification equation,

$$m = \frac{h'}{h} = -\frac{q}{p} = -\frac{18.0\text{ cm}}{90.0\text{ cm}} = -0.200$$

$$h' = mh = -0.200 \times 1.2\text{ cm}$$

$$= -0.24\text{ cm}$$

The image of the daisy is 0.24 cm in diameter.

Discussion Figure 23.50 shows a ray diagram using the three principal rays that confirms the algebraic solution.

Practice Problem 23.9 Finding the Focal Length of a Lens

A 3.00-cm-tall object is placed 60.0 cm in front of a lens. The virtual image formed is 0.50 cm tall. What is the focal length of the lens? Is it converging or diverging?

continued on next page

Example 23.9 continued

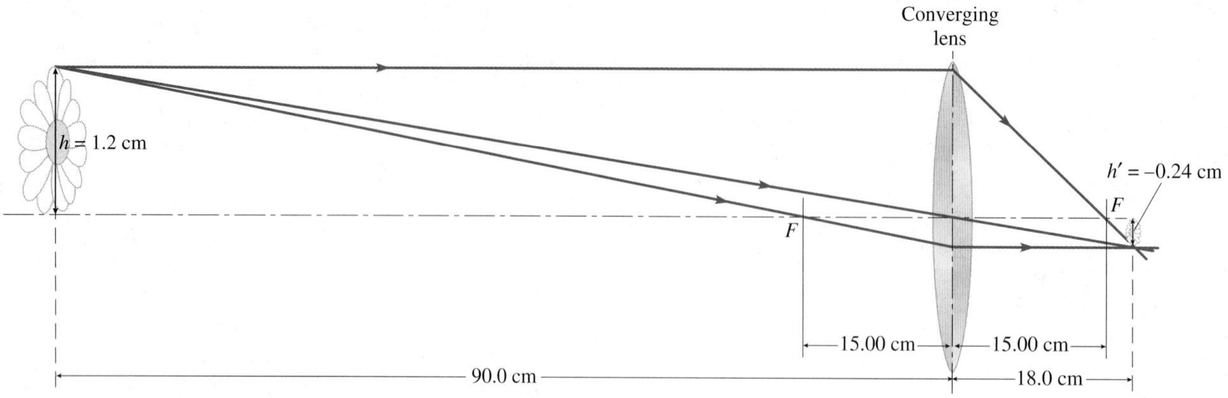

Figure 23.50
Ray diagram for Example 23.9.

PHYSICS AT HOME

If you or a friend are farsighted and have eyeglasses, put the glasses on a table so that you can look through the lenses. Increase your distance from the lenses until you see a clear image of distant objects. Why is the image inverted? Is the image real or virtual? The eyeglasses certainly form an upright image when they are worn as intended. Are the lenses converging or diverging?

Objects and Images at Infinity

Suppose an object is a large distance from a lens ("at infinity"). Substituting $p = \infty$ in the lens equation yields $q = f$. The rays from a faraway object are nearly parallel to each other when they strike the lens, so the image is formed in the principal **focal plane** (the plane perpendicular to the axis passing through the principal focal point). Similarly, if an object is placed in the principal focal plane of a converging lens, then $p = f$ and $q = \infty$. The image is at infinity—that is, the rays emerging from the lens are parallel, so they appear to be coming from an object at infinity.

Master the Concepts

- A wavefront is a set of points of equal phase. A ray points in the direction of propagation of a wave and is perpendicular to the wavefronts. Huygens's principle is a geometric construction used to analyze the propagation of a wave. Every point on a wavefront is considered a source of spherical wavelets. A surface tangent to the wavelets at a later time is the wavefront at that time.
- Geometric optics deals with the propagation of light when interference and diffraction are negligible. The chief tool used in geometric optics is the ray diagram. At a boundary between two different media, light can be reflected as well as transmitted. The laws of reflection and refraction give the directions of the reflected and transmitted rays. In the laws of reflection and refraction, angles are measured between rays and a normal to the boundary.
- Laws of reflection:
  1. The angle of incidence equals the angle of reflection.
  2. The reflected ray lies in the same plane as the incident ray and the normal to the surface at the point of incidence.

continued on next page

Master the Concepts continued

- Laws of refraction:
  1. $n_i \sin \theta_i = n_t \sin \theta_t$ (Snell's law).
  2. The incident ray, the transmitted ray, and the normal all lie in the same plane—the plane of incidence.
  3. The incident and transmitted rays are on *opposite sides* of the normal.

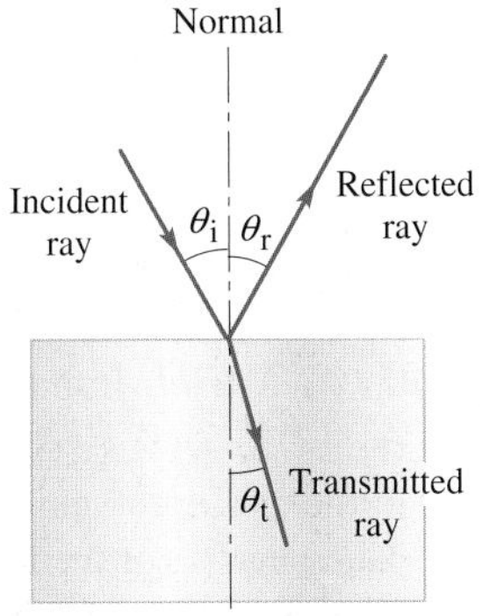

- When a ray is incident on a boundary from a material with a higher index of refraction to one with a lower index of refraction, total internal reflection occurs (there is no transmitted ray) if the angle of incidence exceeds the critical angle

$$\theta_c = \sin^{-1} \frac{n_t}{n_i} \qquad (23\text{-}5a)$$

- When a ray is incident on a boundary, the reflected ray is totally polarized perpendicular to the plane of incidence if the angle of incidence is equal to Brewster's angle

$$\theta_B = \tan^{-1} \frac{n_t}{n_i} \qquad (23\text{-}6)$$

- In the formation of an image, there is a one-to-one correspondence of points on the object and points on the image. In a virtual image, light rays *appear* to diverge from the image point, but they really don't. In a real image, the rays actually *do* pass through the image point.
- Finding an image using a ray diagram:
  1. Draw two (or more) rays coming from a single point on the object toward the lens or mirror.
  2. Trace the rays, applying the laws of reflection and refraction as needed, until they reach the observer.
  3. For a real image, the rays intersect at the image point. For a virtual image, extrapolate the rays backward along straight line paths until they intersect at the image point.
- The easiest rays to trace for a mirror or lens are called the principal rays.

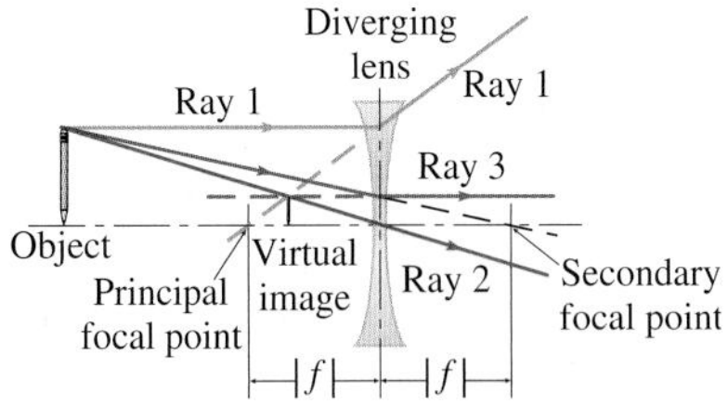

- A plane mirror forms an upright, virtual image of an object that is located at the same distance behind the mirror as the object is in front of the mirror. The object and image points are both located on the same normal line from the object to the mirror surface. The image of an extended object is the same size as the object.

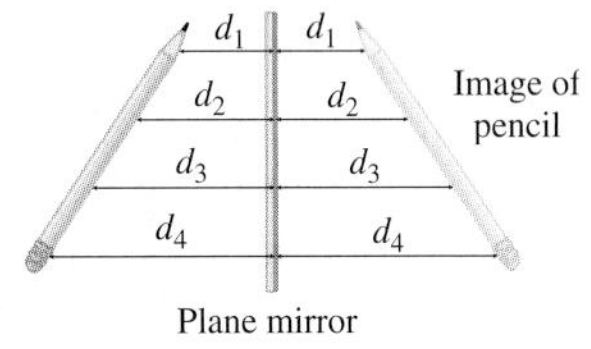

- The magnitude of the transverse magnification m is the ratio of the image size to the object size; the sign of m is determined by the orientation of the image. For an inverted (upside-down) image, $m < 0$; for an upright (right-side-up) image, $m > 0$. For either lenses or mirrors,

$$m = \frac{h'}{h} = -\frac{q}{p} \qquad (23\text{-}9)$$

- The mirror/thin lens equation relates the object and image distances to the focal length:

$$\frac{1}{p} + \frac{1}{q} = \frac{1}{f} \qquad (23\text{-}10)$$

- These sign conventions enable the magnification and mirror/thin lens equations to apply to all kinds of mirrors and lenses and both kinds of image:

Quantity	When Positive (+)	When Negative (−)
Object distance p	Real object	Virtual object
Image distance q	Real image	Virtual image
Focal length f	Converging lens or mirror	Diverging lens or mirror
Magnification m	Upright image	Inverted image

Conceptual Questions

1. Describe the difference between specular and diffuse reflection. Give some examples of each.
2. What is the difference between a virtual and a real image? Describe a method for demonstrating the presence of a real image.
3. Water droplets in air create rainbows. Describe the physical situation that causes a rainbow. Should you look toward or away from the Sun to see a rainbow? Why is the secondary rainbow fainter than the primary rainbow?
4. Why does a mirror hanging in a vertical plane seem to interchange left and right but not up and down?

[*Hint:* Refer to Fig. 23.28. Instead of calling Grant's hands left and right, call them east and west. In Grant's image, are the east and west hands reversed? Note that Grant faces south while his image faces north.]

5. A framed poster is covered with glass that has a rougher surface than regular glass. How does a rough surface reduce glare?
6. Explain how a plane mirror can be thought of as a special case of a spherical mirror. What is the focal length of a plane mirror? Does the spherical mirror equation work for plane mirrors with this choice of focal length? What is the transverse magnification for any image produced by a plane mirror?
7. A ray of light passes from air into water, striking the surface of the water with an angle of incidence of 45°. Which of these quantities change as the light enters the water: wavelength, frequency, speed of propagation, direction of propagation?
8. If the angle of incidence is greater than the angle of refraction for a light beam passing an interface, what can be said about the relative values of the indices of refraction and the speed of light in the first and second media?
9. A concave mirror has focal length f. (a) If you look into the mirror from a distance less than f, is the image you see upright or inverted? (b) If you stand at a distance greater than $2f$, is the image upright or inverted? (c) If you stand at a distance between f and $2f$, an image is formed but you cannot see it. Why not? Sketch a ray diagram and compare the locations of the object and image.
10. The focal length of a concave mirror is 4.00 m and an object is placed 3.00 m in front of the mirror. Describe the image in terms of real, virtual, upright, and inverted.
11. Why is the passenger's side mirror in many cars convex rather than plane or concave?
12. When a virtual image is formed by a mirror, is it in front of the mirror or behind it? What about a real image?
13. Light rays travel from left to right through a lens. If a virtual image is formed, on which side of the lens is it? On which side would a real image be found?
14. Why is the brilliance of an artificial diamond made of cubic zirconium ($n = 1.9$) distinctly inferior to the real thing ($n = 2.4$) even if the two are cut exactly the same way? How would an artificial diamond made of glass compare?
15. The surface of the water in a swimming pool is completely still. Describe what you would see looking straight up toward the surface from under water. [*Hint:* Sketch some rays. Consider both reflected and transmitted rays at the water surface.]
16. A ray reflects from a spherical mirror at point *P*. Explain why a radial line from the center of curvature through point *P* always bisects the angle between the incident and reflected rays.
17. Why must projectors and cameras form real images? Does the lens in the eye form real or virtual images on the retina?
18. Is it possible for a plane mirror to produce a real image of an object in front of the mirror? Explain. If it is possible, sketch a ray diagram to demonstrate. If it is not possible, sketch a ray diagram to show which way a curved mirror must curve (concave or convex) to produce a real image.
19. A slide projector forms a real image of the slide on a screen using a converging lens. If the bottom half of the lens is blocked by covering it with opaque tape, does the bottom half of the image disappear, or does the top half disappear, or is the entire image still visible on the screen? If the entire image is visible, is anything different about it? [*Hint:* It may help to sketch a ray diagram.]
20. A lens is placed at the end of a bundle of optical fibers in an endoscope. The purpose of the lens is to make the light rays parallel before they enter the fibers (in other words, to put the image at infinity). What is the advantage of using a lens with the same index of refraction as the core of the fibers?
21. A converging lens made from dense flint glass is placed into a container of transparent glycerine. Describe what happens to the focal length.
22. Suppose you are facing due north at sunrise. Sunlight is reflected by a store's display window as shown. Is the reflected light partially polarized? If so, in what direction?

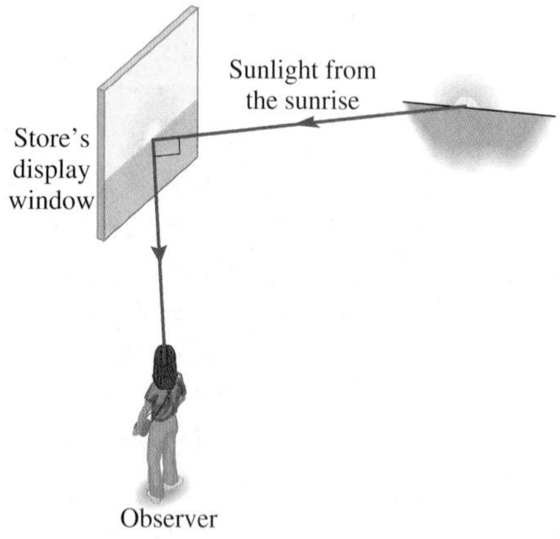

Multiple-Choice Questions

1. The image of an object in a plane mirror
   (a) is always smaller than the object.
   (b) is always the same size as the object.
   (c) is always larger than the object.
   (d) can be larger, smaller, or the same size as the object, depending on the distance between the object and the mirror.

2. Which statements are true? The rays in a plane wave are
   1. parallel to the wavefronts.
   2. perpendicular to the wavefronts.
   3. directed radially outward from a central point.
   4. parallel to each other.

   (a) 1, 2, 3, 4 (b) 1, 4 (c) 2, 3 (d) 2, 4
3. The image of a slide formed by a slide projector is correctly described by which of the listed terms?
   (a) real, upright, enlarged
   (b) real, inverted, diminished
   (c) virtual, inverted, enlarged
   (d) virtual, upright, diminished
   (e) real, upright, diminished
   (f) real, inverted, enlarged
   (g) virtual, inverted, diminished
4. During a laboratory experiment with an object placed in front of a concave mirror, the image distance q is determined for several different values of object distance p. How might the focal length f of the mirror be determined from a graph of the data?
   (a) Plot q versus p; slope $= f$
   (b) Plot q versus p; slope $= 1/f$
   (c) Plot $1/p$ versus $1/q$; y-intercept $= 1/f$
   (d) Plot q versus p; y-intercept $= 1/f$
   (e) Plot q versus p; y-intercept $= f$
   (f) Plot $1/p$ versus $1/q$; slope $= 1/f$
5. A man runs toward a plane mirror at 5 m/s and the mirror, on rollers, simultaneously approaches him at 2 m/s. The speed at which his image moves (relative to the ground) is
   (a) 14 m/s (b) 7 m/s (c) 3 m/s
   (d) 9 m/s (e) 12 m/s
6. Two converging lenses, of exactly the same size and shape, are held in sunlight, the same distance above a sheet of paper. The figure shows the paths of some rays through the two lenses. Which lens is made of material with a higher index of refraction? How do you know?
   (a) Lens 1, because its focal length is smaller
   (b) Lens 1, because its focal length is greater
   (c) Lens 2, because its focal length is smaller
   (d) Lens 2, because its focal length is greater
   (e) Impossible to answer with the information given

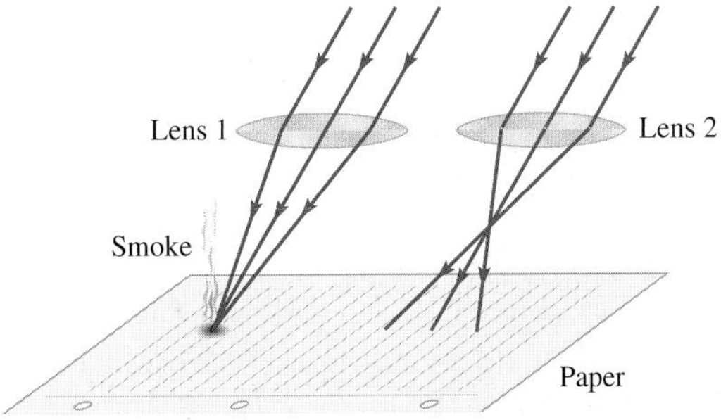

7. Which of these statements correctly describe the images formed by an object placed before a single thin lens?
   1. Real images are always enlarged.
   2. Real images are always inverted.
   3. Virtual images are always upright.
   4. Convex lenses never produce virtual images.

   (a) 1 and 3
   (b) 2 and 3
   (c) 2 and 4
   (d) 2, 3, and 4
   (e) 1, 2, and 3
   (f) 4 only
8. Light reflecting from the surfaces of lakes, roads, and automobile hoods is
   (a) partially polarized in the horizontal direction.
   (b) partially polarized in the vertical direction.
   (c) polarized only if the Sun is directly overhead.
   (d) polarized only if it is a clear day.
   (e) randomly polarized.
9. A point source of light is placed at the focal point of a converging lens; the rays of light coming out of the lens are parallel to the principal axis. Now suppose the source is moved closer to the lens but still on the axis. Which statement is true about the light rays coming out of the lens?
   (a) They diverge from each other.
   (b) They converge toward each other.
   (c) They still emerge parallel to the principal axis.
   (d) They emerge parallel to each other but not parallel to the axis.
   (e) No rays emerge because a virtual image is formed.
10. A light ray inside a glass prism is incident at Brewster's angle on a surface of the prism with air outside. Which of these is true?
    (a) There is no transmitted ray; the reflected ray is plane polarized.
    (b) The transmitted ray is plane polarized; the reflected ray is partially polarized.
    (c) There is no transmitted ray; the reflected ray is partially polarized.
    (d) The transmitted ray is partially polarized; the reflected ray is plane polarized.
    (e) The transmitted ray is plane polarized; there is no reflected ray.

Problems

- Ⓒ Combination conceptual/quantitative problem
- Biological or medical application
- ✦ Challenging problem
- Blue # Detailed solution in the Student Solutions Manual
- (1 2) Problems paired by concept
- Text website interactive or tutorial

23.1 Wavefronts, Rays, and Huygens's Principle

1. Sketch the wavefronts and rays for the light emitted by an isotropic point source (*isotropic* = same in all directions). Use Huygens's principle to illustrate the propagation of one of the wavefronts.

2. Apply Huygens's principle to a 5-cm-long planar wavefront approaching a reflecting wall at normal incidence. The wavelength is 1 cm and the wall has a wide opening (width = 4 cm). The center of the incoming wavefront approaches the center of the opening. Repeat the procedure until you have wavefronts on both sides of the wall. Without worrying about the details of edge effects, what are the general shapes of the wavefronts on each side of the reflecting wall?

3. Repeat Problem 2 for an opening of width 0.5 cm.

23.2 The Reflection of Light

4. A plane wave reflects from the surface of a sphere. Draw a ray diagram and sketch some wavefronts for the reflected wave.

5. A spherical wave (from a point source) reflects from a planar surface. Draw a ray diagram and sketch some wavefronts for the reflected wave.

6. Light rays from the Sun, which is at an angle of 35° above the western horizon, strike the still surface of a pond. (a) What is the angle of incidence of the Sun's rays on the pond? (b) What is the angle of reflection of the rays that leave the pond surface? (c) In what direction and at what angle from the pond surface are the reflected rays traveling?

7. A light ray reflects from a plane mirror as shown in the figure. What is the angle of deviation δ?

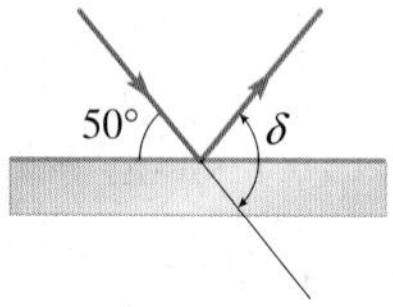

8. Two plane mirrors form a 70.0° angle as shown. For what angle θ is the final ray horizontal?

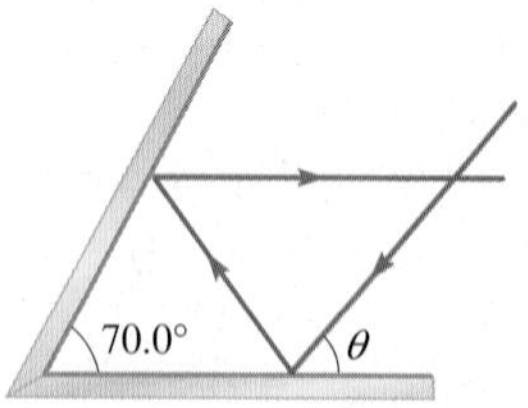

9. Choose two rays in Fig. 23.7 and use them to prove that the angle of incidence is equal to the angle of reflection. [*Hint:* Choose a wavefront at two different times, one before reflection and one after. The time for light to travel from one wavefront to the other is the same for the two rays.]

23.3 The Refraction of Light: Snell's Law

10. Sunlight strikes the surface of a lake at an angle of incidence of 30.0°. At what angle with respect to the normal would a fish see the Sun?

11. Sunlight strikes the surface of a lake. A diver sees the Sun at an angle of 42.0° with respect to the vertical. What angle do the Sun's rays in air make with the vertical?

12. A beam of light in air is incident on a stack of four flat transparent materials with indices of refraction 1.20, 1.40, 1.32, and 1.28. If the angle of incidence for the beam on the first of the four materials is 60.0°, what angle does the beam make with the normal when it emerges into the air after passing through the entire stack?

13. At a marine animal park, Alison is looking through a glass window and watching dolphins swim underwater. If the dolphin is swimming directly toward her at 15 m/s, how fast does the dolphin appear to be moving?

14. A light ray in the core ($n = 1.40$) of a cylindrical optical fiber travels at an angle $\theta_1 = 49.0°$ with respect to the *axis of the fiber.* A ray is transmitted through the cladding ($n = 1.20$) and into the air. What angle θ_2 does the exiting ray make with the outside surface of the cladding?

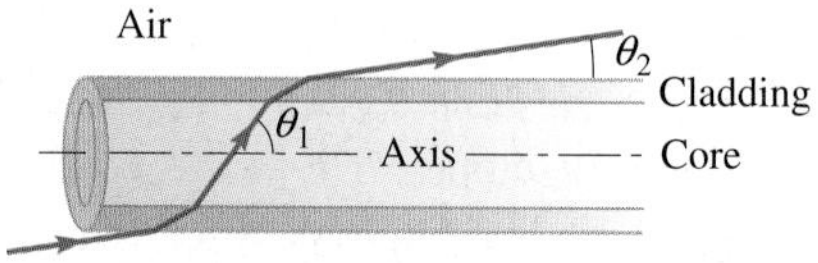

Problems 14 and 15

15. A light ray in the core ($n = 1.40$) of a cylindrical optical fiber is incident on the cladding. See the figure with Problem 14. A ray is transmitted through the cladding ($n = 1.20$) and into the air. The emerging ray makes an angle $\theta_2 = 5.00°$ with the outside surface of the cladding. What angle θ_1 did the ray in the core make with the axis?

16. A glass lens has a scratch-resistant plastic coating on it. The speed of light in the glass is $0.67c$ and the speed of light in the coating is $0.80c$. A ray of light in the coating is incident on the plastic-glass boundary at an angle of 12.0° with respect to the normal. At what angle with respect to the normal is the ray transmitted into the glass?

17. In Figure 23.12, a coin is right up against the far edge of the mug. In picture (a) the coin is just hidden from view and in picture (b) we can almost see the whole coin. If the mug is 6.5 cm in diameter and 8.9 cm tall, what is the diameter of the coin?

18. A horizontal light ray is incident on a crown glass prism as shown in the figure where $\beta = 30.0°$. Find the angle of deviation δ of the ray—the angle that the ray emerging from the prism makes with the incident ray.

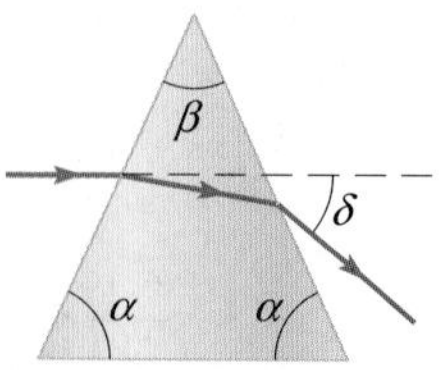

Problems 18 and 19

19. A horizontal light ray is incident on a prism as shown in the figure with Problem 18 where β is a *small angle*

(exaggerated in the figure). Find the angle of deviation δ of the ray—the angle that the ray emerging from the prism makes with the incident ray—as a function of β and n, the index of refraction of the prism and show that δ is proportional to β.

20. A diamond in air is illuminated with white light. On one particular facet, the angle of incidence is 26.00°. Inside the diamond, red light (λ = 660.0 nm in vacuum) is refracted at 10.48° with respect to the normal; blue light (λ = 470.0 nm in vacuum) is refracted at 10.33°. (a) What are the indices of refraction for red and blue light in diamond? (b) What is the ratio of the speed of red light to the speed of blue light in diamond? (c) How would a diamond look if there were no dispersion?

21. The prism in the figure is made of crown glass. Its index of refraction ranges from 1.517 for the longest visible wavelengths to 1.538 for the shortest. Find the range of refraction angles for the light transmitted into air through the right side of the prism.

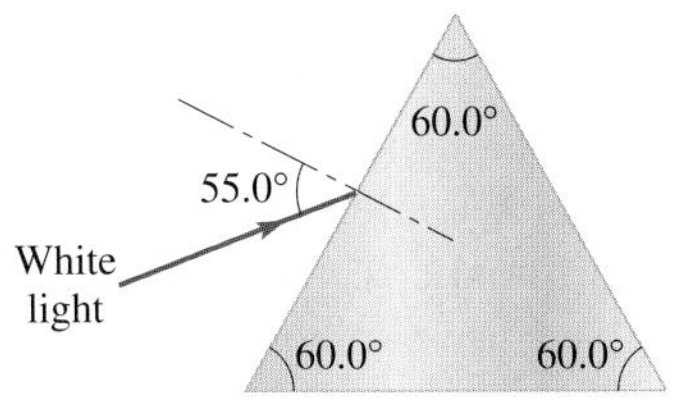

23.4 Total Internal Reflection

22. Calculate the critical angle for a sapphire surrounded by air.

23. (a) Calculate the critical angle for a diamond surrounded by air. (b) Calculate the critical angle for a diamond under water. (c) Explain why a diamond sparkles less under water than in air.

24. Is there a critical angle for a light ray coming from a medium with an index of refraction 1.2 and incident on a medium that has an index of refraction 1.4? If so, what is the critical angle that allows total internal reflection in the first medium?

25. The figure shows some light rays reflected from a small defect in the glass toward the surface of the glass. (a) If θ_c = 40.00°, what is the index of refraction of the glass? (b) Is there any point above the glass at which a viewer would not be able to see the defect? Explain.

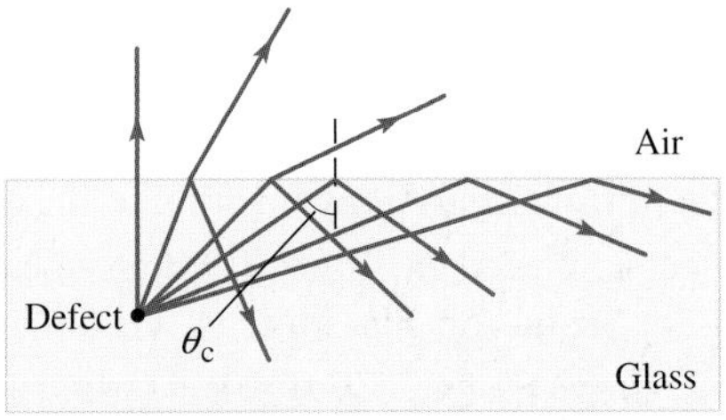

26. A 45° prism has an index of refraction of 1.6. Light is normally incident on the left side of the prism. Does light exit the back of the prism (for example, at point P)? If so, what is the angle of refraction with respect to the normal at point P? If not, what happens to the light?

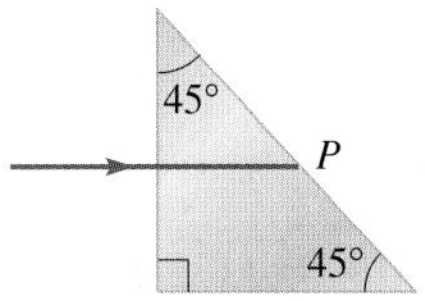

Problems 26, 27, and 91

27. Light incident on a 45.0° prism as shown in the figure undergoes total internal reflection at point P. What can you conclude about the index of refraction of the prism? (Determine either a minimum or maximum possible value.)

28. The angle of incidence θ of a ray of light in air is adjusted gradually as it enters a shallow tank made of Plexiglas and filled with carbon disulfide. Is there an angle of incidence for which light is transmitted into the carbon disulfide but not into the Plexiglas at the bottom of the tank? If so, find the angle. If not, explain why not.

29. Repeat Problem 28 for a Plexiglas tank filled with carbon tetrachloride instead of carbon disulfide.

30. What is the index of refraction of the core of an optical fiber if the cladding has n = 1.20 and the critical angle at the core-cladding boundary is 45.0°?

23.5 Polarization by Reflection

31. Some glasses used for viewing 3D movies are polarized, one lens having a vertical transmission axis and the other horizontal. While standing in line on a winter afternoon for a 3D movie and looking through his glasses at the road surface, Maurice notices that the left lens cuts down reflected glare significantly, but the right lens does not. The glare is minimized when the angle between the reflected light and the horizontal direction is 37°. (a) Which lens has the transmission axis in the vertical direction? (b) What is Brewster's angle for this case? (c)What is the index of refraction of the material reflecting the light?

32. (a) Sunlight reflected from the still surface of a lake is totally polarized when the incident light is at what angle with respect to the horizontal? (b) In what direction is the reflected light polarized? (c) Is any light incident at this angle transmitted into the water? If so, at what angle below the horizontal does the transmitted light travel?

33. (a) Sunlight reflected from the smooth ice surface of a frozen lake is totally polarized when the incident light is at what angle with respect to the horizontal? (b) In what direction is the reflected light polarized? (c) Is any light incident at this angle transmitted into the ice? If so, at what angle below the horizontal does the transmitted light travel?

34. Light travels in a medium with index n_1 toward a boundary with another material of index $n_2 < n_1$. (a) Which is larger, the critical angle or Brewster's angle? Does the answer depend on the values of n_1 and n_2 (other than

assuming $n_2 < n_1$)? (b) What can you say about the critical angle and Brewster's angle for light coming the other way (from the medium with index n_2 toward the medium with n_1)?

23.6 The Formation of Images Through Reflection or Refraction

35. A defect in a diamond appears to be 2.0 mm below the surface when viewed from directly above that surface. How far beneath the surface is the defect?

36. An insect is trapped inside a piece of amber ($n = 1.546$). Looking at the insect from directly above, it appears to be 7.00 mm below a smooth surface of the amber. How far below the surface is the insect?

♦37. A penny is at the bottom of a bowl full of water. When you look at the water surface from the side, with your eyes at the water level, the penny appears to be just barely under the surface and a horizontal distance of 3.0 cm from the edge of the bowl. If the penny is actually 8.0 cm below the water surface, what is the horizontal distance between the penny and the edge of the bowl? [*Hint:* The rays you see pass from water to air with refraction angles close to 90°.]

23.7 Plane Mirrors

38. Norah wants to buy a mirror so that she can check on her appearance from top to toe before she goes off to work. If Norah is 1.64 m tall, how tall a mirror does she need?

39. Daniel's eyes are 1.82 m from the floor when he is wearing his dress shoes, and the top of his head is 1.96 m from the floor. Daniel has a mirror that is 0.98 m in length. How high from the floor should the bottom edge of the mirror be located if Daniel is to see a full-length image of himself? Draw a ray diagram to illustrate your answer.

40. A rose in a vase is placed 0.250 m in front of a plane mirror. Nagar looks into the mirror from 2.00 m in front of it. How far away from Nagar is the image of the rose?

41. Entering a darkened room, Gustav strikes a match in an attempt to see his surroundings. At once he sees what looks like another match about 4 m away from him. As it turns out, a mirror hangs on one wall of the room. How far is Gustav from the wall with the mirror?

42. In an amusement park maze with all the walls covered with mirrors, Pilar sees Hernando's reflection from a series of three mirrors. If the reflected angle from mirror 3 is 55° for the mirror arrangement shown in the figure, what is the angle of incidence on mirror 1?

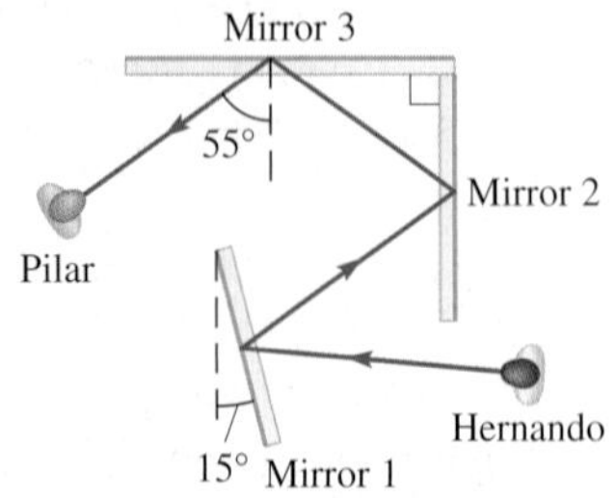

♦43. Maurizio is standing in a rectangular room with two adjacent walls and the ceiling all covered by plane mirrors. How many images of himself can Maurizio see?

44. Hannah is standing in the middle of a room with two opposite walls that are separated by 10.0 m and covered by plane mirrors. There is a candle in the room 1.50 m from one mirrored wall. Hannah is facing the opposite mirrored wall and sees many images of the candle. How far from Hannah are the closest four images of the candle that she can see?

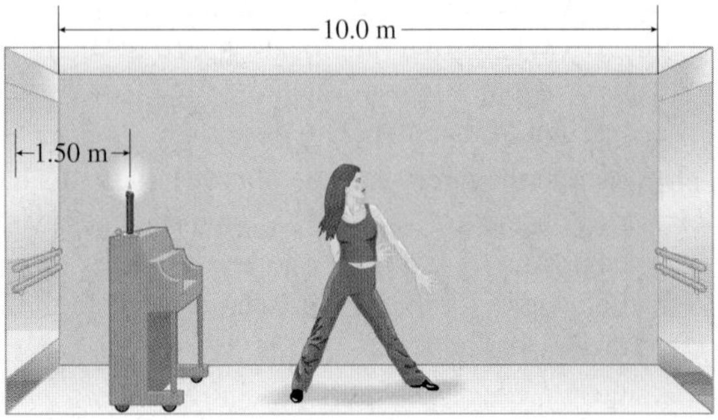

♦45. A point source of light is in front of a plane mirror. (a) Prove that all the reflected rays, when extended back behind the mirror, intersect in a single point. [*Hint:* See Fig. 23.27a and use similar triangles.] (b) Show that the image point lies on a line through the object and perpendicular to the mirror, and that the object and image distances are equal. [*Hint:* Use any pair of rays in Fig. 23.27a.]

23.8 Spherical Mirrors

46. An object 2.00 cm high is placed 12.0 cm in front of a convex mirror with radius of curvature of 8.00 cm. Where is the image formed? Draw a ray diagram to illustrate.

47. A 1.80-cm-high object is placed 20.0 cm in front of a concave mirror with a 5.00-cm focal length. What is the position of the image? Draw a ray diagram to illustrate.

48. In her job as a dental hygienist, Kathryn uses a concave mirror to see the back of her patient's teeth. When the mirror is 1.20 cm from a tooth, the image is upright and 3.00 times as large as the tooth. What are the focal length and radius of curvature of the mirror?

49. An object is placed in front of a concave mirror with a 25.0-cm radius of curvature. A real image twice the size of the object is formed. At what distance is the object from the mirror? Draw a ray diagram to illustrate.

50. An object is placed in front of a convex mirror with a 25.0-cm radius of curvature. A virtual image half the size of the object is formed. At what distance is the object from the mirror? Draw a ray diagram to illustrate.

51. The right-side rearview mirror of Mike's car says that objects in the mirror are closer than they appear. Mike

decides to do an experiment to determine the focal length of this mirror. He holds a plane mirror next to the rearview mirror and views an object that is 163 cm away from each mirror. The object appears 3.20 cm wide in the plane mirror, but only 1.80 cm wide in the rearview mirror. What is the focal length of the rearview mirror?

52. A concave mirror has a radius of curvature of 5.0 m. An object, initially 2.0 m in front of the mirror, is moved back until it is 6.0 m from the mirror. Describe how the image location changes.

53. Derive the magnification equation, $m = h'/h = -q/p$, for a *convex* mirror. Draw a ray diagram as part of the solution. [*Hint:* Draw a ray that is not one of the three principal rays, as was done in the derivation for a concave mirror.]

54. In a subway station, a convex mirror allows the attendant to view activity on the platform. A woman 1.64 m tall is standing 4.5 m from the mirror. The image formed of the woman is 0.500 m tall. (a) What is the radius of curvature of the mirror? (b) The mirror is 0.500 m in diameter. If the woman's shoes appear at the bottom of the mirror, does her head appear at the top—in other words, does the image of the woman fill the mirror from top to bottom? Explain.

55. Show that when rays parallel to the principal axis reflect from a concave mirror, the reflected rays all pass through the focal point at a distance $R/2$ from the vertex. Assume that the angles of incidence are small. [*Hint:* Follow the similar derivation for a *convex* mirror in the text.]

56. Starting with Fig. 23.39, perform all the algebraic steps to obtain the mirror equation in the form of Eq. (23-10).

23.9 Thin Lenses

57. (a) For a converging lens with a focal length of 3.50 cm, find the object distance that will result in an inverted image with an image distance of 5.00 cm. Use a ray diagram to verify your calculations. (b) Is the image real or virtual? (c) What is the magnification?

58. Sketch a ray diagram to show that when an object is placed more than twice the focal length away from a converging lens, the image formed is inverted, real, and diminished in size. (tutorial: lens)

59. Sketch a ray diagram to show that when an object is placed at twice the focal length from a converging lens, the image formed is inverted, real, and the same size as the object.

60. Sketch a ray diagram to show that when an object is placed between twice the focal length and the focal length from a converging lens, the image formed is inverted, real, and enlarged in size.

61. Sketch a ray diagram to show that when an object is a distance equal to the focal length from a converging lens, the emerging rays from the lens are parallel to each other, so the image is at infinity.

62. When an object is placed 6.0 cm in front of a converging lens, a virtual image is formed 9.0 cm from the lens. What is the focal length of the lens?

63. An object of height 3.00 cm is placed 12.0 cm from a diverging lens of focal length −12.0 cm. Draw a ray diagram to find the height and position of the image.

64. A diverging lens has a focal length of −8.00 cm. (a) What are the image distances for objects placed at these distances from the lens: 5.00 cm, 8.00 cm, 14.0 cm, 16.0 cm, 20.0 cm? In each case, describe the image as real or virtual, upright or inverted, and enlarged or diminished in size. (b) If the object is 4.00 cm high, what is the height of the image for the object distances of 5.00 cm and 20.0 cm?

65. A converging lens has a focal length of 8.00 cm. (a) What are the image distances for objects placed at these distances from the thin lens: 5.00 cm, 14.0 cm, 16.0 cm, 20.0 cm? In each case, describe the image as real or virtual, upright or inverted, and enlarged or diminished in size. (b) If the object is 4.00 cm high, what is the height of the image for the object distances of 5.00 cm and 20.0 cm?

66. Sketch a ray diagram to show that if an object is placed less than the focal length from a converging lens, the image is virtual and upright. (tutorial: magnifying glass)

67. For each of the lenses in the figure, list whether the lens is converging or diverging.

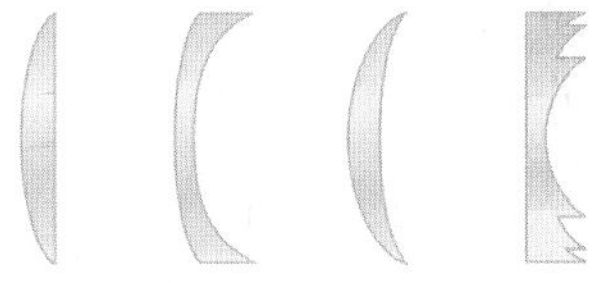

68. In order to read his book, Stephen uses a pair of reading glasses. When he holds the book at a distance of 25 cm from his eyes, the glasses form an upright image a distance of 52 cm from his eyes. (a) Is this a converging or diverging lens? (b) What is the magnification of the lens? (c) What is the focal length of the lens?

69. A standard "35-mm" slide measures 24.0 mm by 36.0 mm. Suppose a slide projector produces a 60.0-cm by 90.0-cm image of the slide on a screen. The focal length of the lens is 12.0 cm. (a) What is the distance between the slide and the screen? (b) If the screen is moved farther from the projector, should the lens be moved closer to the slide or farther away?

70. An object that is 6.00 cm tall is placed 40.0 cm in front of a diverging lens. The magnitude of the focal length of the lens is 20.0 cm. Find the image position and size. Is the image real or virtual? Upright or inverted?

Comprehensive Problems

71. Samantha puts her face 32.0 cm from a makeup mirror and notices that her image is magnified by 1.80 times. (a) What kind of mirror is this? (b) Where is her face relative to the radius of curvature or focal length? (c) What is the radius of curvature of the mirror?

72. A converging lens made of glass ($n = 1.5$) is placed under water ($n = 1.33$). Describe how the focal length of the lens under water compares to the focal length in air. Draw a diagram to illustrate your answer.

73. An object 8.0 cm high forms a virtual image 3.5 cm high located 4.0 cm behind a mirror. (a) Find the object distance. (b) Describe the mirror: is it plane, convex, or concave? (c) What are its focal length and radius of curvature?

74. A point source of light is placed 10 cm in front of a concave mirror; the reflected rays are parallel. What is the radius of curvature of the mirror?

75. A glass prism bends a ray of blue light more than a ray of red light since its index of refraction is slightly higher for blue than for red. Does a diverging glass lens have the same focal point for blue light and for red light? If not, for which color is the focal point closer to the lens?

76. A laser beam is traveling through an unknown substance. When it encounters a boundary with air, the angle of reflection is 25.0° and the angle of refraction is 37.0°. (a) What is the index of refraction of the substance? (b) What is the speed of light in the substance? (c) At what minimum angle of incidence would the light be totally internally reflected?

77. In many cars the passenger's side mirror says: "Objects in the mirror are closer than they appear." (a) Does this mirror form real or virtual images? (b) Since the image is diminished in size, is the mirror concave or convex? Why? (c) Show that the image must actually be *closer* to the mirror than is the object. (d) How then can the image seem to be farther away?

78. A scuba diver in a lake aims her underwater spotlight at the lake surface. (a) If the beam makes a 75° angle of incidence with respect to a normal to the water surface, is it reflected, transmitted, or both? Find the angles of the reflected and transmitted beams (if they exist). (b) Repeat for a 25° angle of incidence.

79. Laura is walking directly toward a plane mirror at a speed of 0.8 m/s relative to the mirror. At what speed is her image approaching the mirror?

80. Xi Yang is practicing for his driver's license test. He notices in the rearview mirror that a tree, located directly behind the automobile, is approaching his car as he is backing up. If the car is moving at 8.0 km/h in reverse, how fast *relative to the car* does the image of the tree appear to be approaching?

81. A plane mirror reflects a beam of light. Show that the rotation of the mirror by an angle α causes the beam to rotate through an angle 2α.

82. A 3.00-cm-high pin, when placed at a certain distance in front of a concave mirror, produces an upright image 9.00 cm high, 30.0 cm from the mirror. Find the position of the pin relative to the mirror and the image. Draw a ray diagram to illustrate.

83. A dentist holds a small mirror 1.9 cm from a surface of a patient's tooth. The image formed is upright and 5.0 times as large as the object. (a) Is the image real or virtual? (b) What is the focal length of the mirror? Is it concave or convex? (c) If the mirror is moved closer to the tooth, does the image get larger or smaller? (d) For what range of object distances does the mirror produce an upright image?

84. An object of height 5.00 cm is placed 20.0 cm from a converging lens of focal length 15.0 cm. Draw a ray diagram to find the height and position of the image.

85. A letter on a page of the compact edition of the *Oxford English Dictionary* is 0.60 mm tall. A magnifying glass (a single thin lens) held 4.5 cm above the page forms an image of the letter that is 2.4 cm tall. (a) Is the image real or virtual? (b) Where is the image? (c) What is the focal length of the lens? Is it converging or diverging?

86. An object is placed 10.0 cm in front of a lens. An upright, virtual image is formed 30.0 cm away from the lens. What is the focal length of the lens? Is the lens converging or diverging?

87. A manufacturer is designing a shaving mirror, which is intended to be held close to the face. If the manufacturer wants the image formed to be upright and as large as possible, what characteristics should he choose? (type of mirror? long or short focal length relative to the object distance of face to mirror?)

88. The focal length of a thin lens is –20.0 cm. A screen is placed 160 cm from the lens. What is the y-coordinate of the point where the light ray shown hits the screen? The incident ray is parallel to the central axis and is 1.0 cm from that axis.

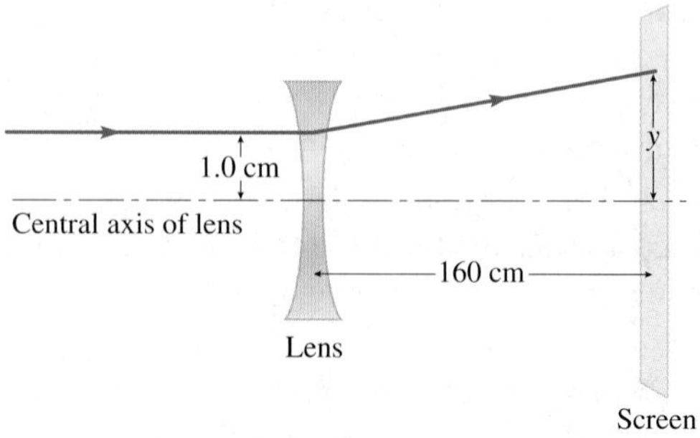

89. A ray of light is reflected from two mirrored surfaces as shown in the figure. If the initial angle of incidence is 34°, what are the values of angles α and β? (The figure is not to scale.)

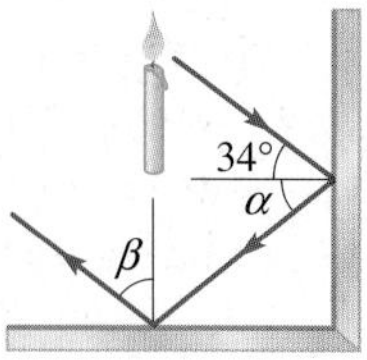

90. A beam of light consisting of a mixture of red, yellow, and blue light originates from a source submerged in some carbon disulfide. The light beam strikes an interface between the carbon disulfide and air at an angle of incidence of 37.5° as shown in the figure. The carbon disulfide has the following indices of refraction for the wavelengths present: red (656.3 nm), $n = 1.6182$; yellow (589.3 nm), 1.6276; blue (486.1 nm), 1.6523. Which color(s) is/are recorded by the detector located above the surface of the carbon disulfide?

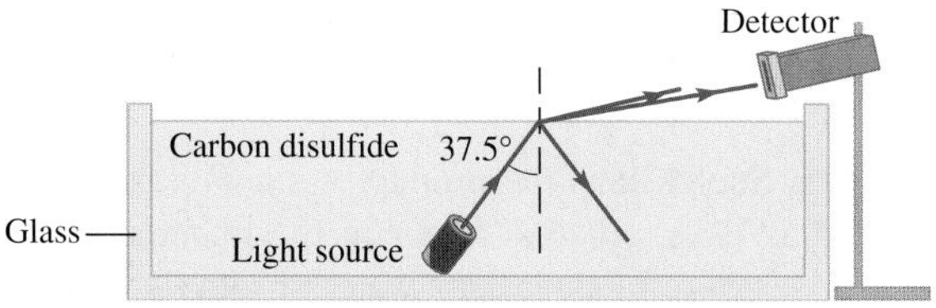

91. A ray of light is incident normally from air onto a glass ($n = 1.50$) prism as shown in the figure with Problem 26. (a) Draw all of the rays that emerge from the prism and give angles to represent their directions. (b) Repeat part (a) with the prism immersed in water ($n = 1.33$). (c) Repeat part (a) with the prism immersed in a sugar solution ($n = 1.50$).

92. A concave mirror has a radius of curvature of 14 cm. If a pointlike object is placed 9.0 cm away from the mirror on its principal axis, where is the image?

93. A glass block ($n = 1.7$) is submerged in an unknown liquid. A ray of light inside the block undergoes total internal reflection. What can you conclude concerning the index of refraction of the liquid?

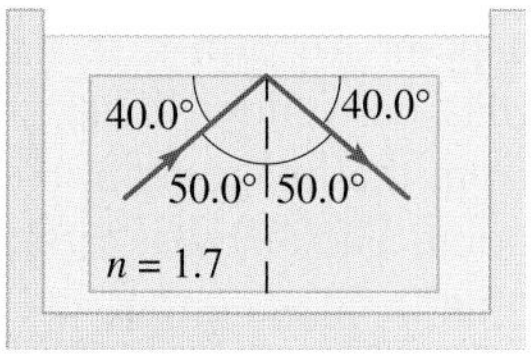

94. Ray diagrams often show objects that conveniently have one end on the principal axis. Draw a ray diagram and locate the image for the object shown in the figure that extends beyond the principal axis.

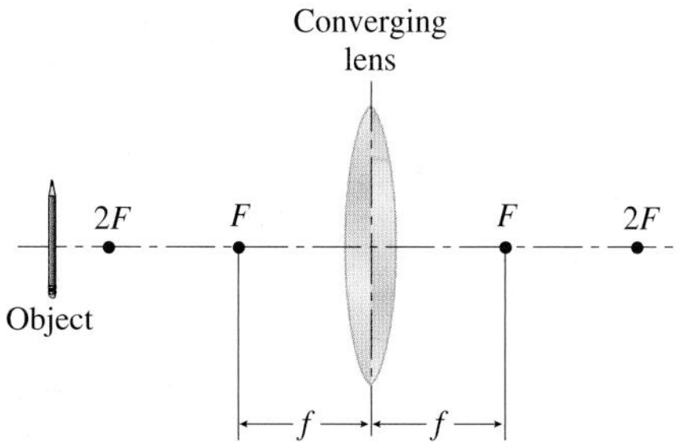

95. A 5.0-cm-tall object is placed 50.0 cm from a lens with focal length –20.0 cm. (a) How tall is the image? (b) Is the image upright or inverted?

✦96. A radar station is located at a height of 24.0 m above the shoreline. When the radar is aimed at a spot 150.0 m out to sea, it detects a whale at the bottom of the ocean. If it takes 2.10 μs for the radar to send out a beam and receive it again, how deep is the ocean where the whale is swimming?

✦97. A ray of light in air is incident at an angle of 60.0° with the surface of some benzene contained in a shallow tank made of crown glass. What is the angle of refraction of the light ray when it enters the glass at the bottom of the tank?

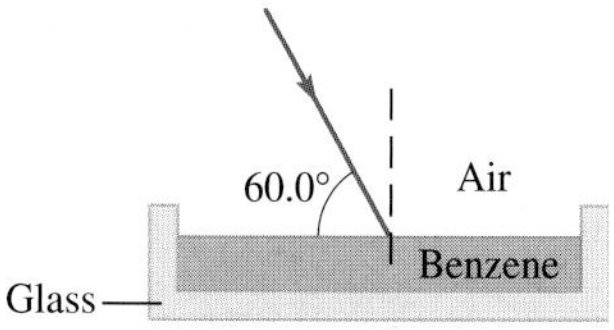

✦98. A ray of light passes from air through dense flint glass and then back into air. The angle of incidence on the first glass surface is 60.0°. The thickness of the glass is 5.00 mm; its front and back surfaces are parallel. How far is the ray displaced as a result of traveling through the glass?

✦99. Show that the deviation angle δ for a ray striking a thin converging lens at a distance d from the principal axis is given by $\delta = d/f$. Therefore, a ray is bent through an angle δ that is proportional to d and does *not* depend on the angle of the incident ray (as long as it is paraxial). [*Hint:* Look at the figure and use the small-angle approximation $\sin\theta \approx \tan\theta \approx \theta$ (in radians).]

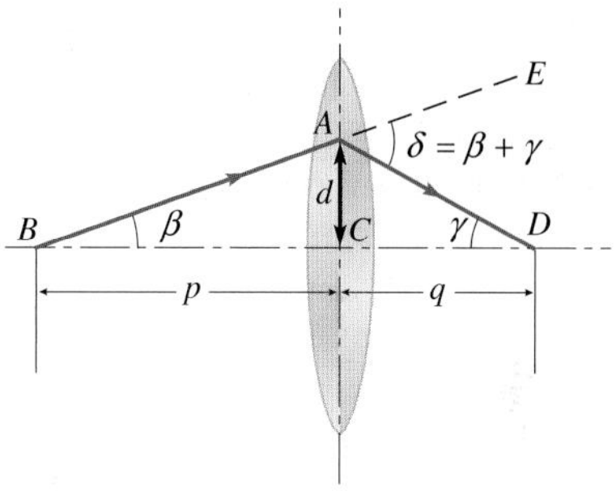

(Angles are greatly exaggerated for ease in labeling.)

✦100. The angle of deviation through a triangular prism is defined as the angle between the incident ray and the emerging ray (angle δ). It can be shown that when the angle of incidence i is equal to the angle of refraction r' for the emerging ray, the angle of deviation is at a minimum. Show that the minimum deviation angle ($\delta_{min} = D$) is related to the prism angle A and the index of refraction n, by

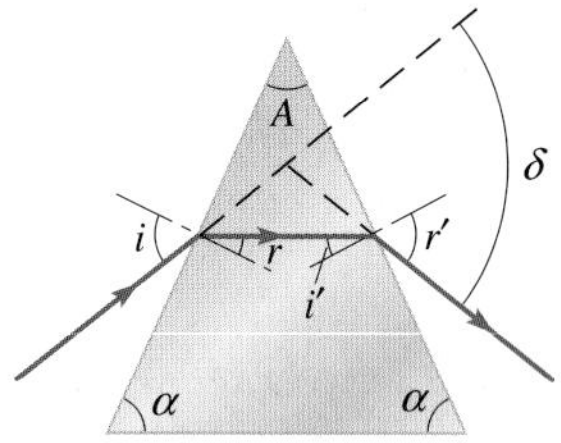

$$n = \frac{\sin\frac{1}{2}(A + D)}{\sin\frac{1}{2}A}$$

[*Hint:* For an isosceles triangular prism, the minimum angle of deviation occurs when the ray inside the prism is parallel to the base, as shown in the figure.]

✦101. The vertical displacement d of light rays parallel to the axis of a lens is measured as a function of the vertical displacement h of the incident ray from the principal axis as shown in part (a) of the figure. The data are graphed in part (b) of the figure. The distance D from the lens to the screen is 1.0 m. What is the focal length of the lens for paraxial rays?

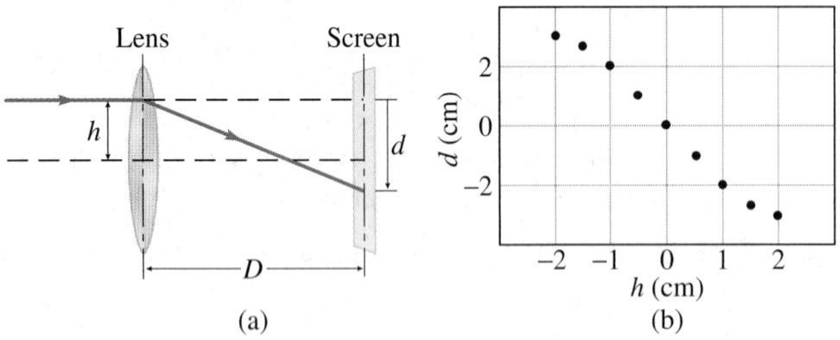

Answers to Practice Problems

23.1

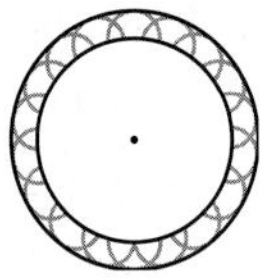

23.2 51°

23.3 If $\theta_i = 0$, then $\theta_t = 0$ and the angle of incidence at the back of the prism is 45°, which is larger than the critical angle (41.8°). If $\theta_i > 0$, then $\theta_t > 0$ and the angle of incidence at the back is greater than 45°.

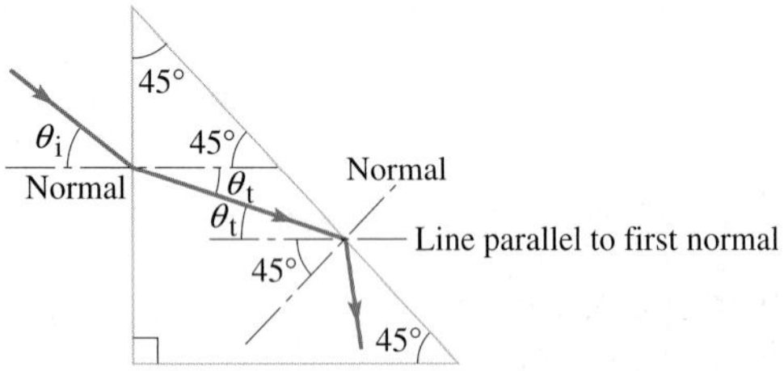

23.4 $\frac{4}{3}h$

23.5 No, she can't see her feet; the bottom of the mirror is 10 cm too high.

23.6 12 cm in front of the mirror, 3.0 cm tall, real

23.7 $p = 6.00$ cm, $f = +12$ cm, concave

23.8

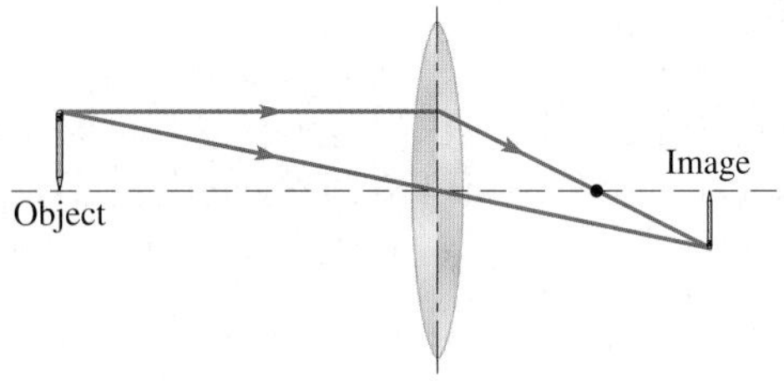

23.9 −12 cm (diverging)

Answers to Checkpoints

23.3 From Snell's law, the product $n \sin \theta$ is the same in both media. Thus, $\sin \theta$ is larger in the material with the smaller index of refraction (here, water). From 0 to 90°, $\sin \theta$ increases as θ increases, the angle that the ray makes with the normal (θ). Therefore, θ_{water} is larger than θ_{glass}. Since θ is the angle the ray makes with the normal, the ray refracts away from the normal when it enters the water.

23.5 Light reflected from a horizontal surface is partially polarized horizontally (parallel to the reflecting surface). To reduce reflected glare, the transmission axis of the polarized sunglasses should be oriented vertically.

23.6 Light rays from a point on the fish are refracted by the water-air interface. The figure shows that the rays are bent outward (away from the normal). The rays never converge to a point to form a real image. Tracing the rays backward (dashed yellow lines) shows that they *appear* to diverge from the image point but do not actually pass through that point. The image is virtual.

23.8 A plane mirror forms a virtual image that is the same size as the object (see Fig. 23.27). The magnification is $m = +1$.

23.9 The image can be either real or virtual. Figure 23.45a shows a converging lens forming a *real* image because the rays from a point on the object converge to a point on the image. Figure 23.45b shows a converging lens forming a *virtual* image. In this case, the rays from a point on the object do not converge to a point on the image. If we trace the rays coming out of the lens backward, they appear to diverge from a point on the image.

CHAPTER

24

Optical Instruments

The Hubble Space Telescope, orbiting Earth at an altitude of about 600 km, was launched in 1990 by the crew of the Space Shuttle *Discovery.* What is the advantage of having a telescope in space when there are telescopes on Earth with larger light-gathering capabilities? What justifies the cost of $2 billion to place this 12.5-ton instrument into orbit? (See p. 910 for the answer.)

Concepts & Skills to Review

- distinction between real and virtual images (Section 23.6)
- magnification (Section 23.8)
- refraction (Section 23.3)
- thin lenses (Section 23.9)
- finding images with ray diagrams (Section 23.6)
- small-angle approximations (Appendix A.7)

24.1 LENSES IN COMBINATION

Optical instruments generally involve two or more lenses in combination. Let's start this chapter by considering what happens when light rays emerging from a lens pass through another lens. We will find that the image formed by the first lens serves as the object for the second lens.

In a series of lenses, the image formed by one lens serves as the object for the next lens.

Suppose that light rays diverge from a point on the image formed by the first lens. These rays are refracted by the second lens the same way as if they were coming from a point on an object. Therefore, the location and size of the image formed by the second lens can be found by applying the lens equation, where the object distance p is the distance from the image formed by the first lens to the second lens. For lenses in combination, we apply the lens equation to each lens in turn, where the object for a given lens is the image formed by the previous lens. Remember that for any application of the lens equation, the object and image distances p and q are measured from the center of the same lens. This same procedure holds true for combinations of lenses and mirrors.

Rays from a real object are diverging as they enter a lens; rays from a virtual object are *converging* as they enter a lens.

In Chapter 23, all objects were real; p was always positive. With more than one lens, it is possible to have a **virtual object** for which p is *negative*. If one lens produces a real image that would have formed *past* the second lens—so that the rays are converging to a point past the second lens—that image becomes a virtual object for the second lens (Fig. 24.1). Before the real image could form from the first lens, the presence of the second lens intervenes; the rays striking the second lens are converging to a point rather than diverging from a point. This seemingly complicated situation is treated simply by using a negative object distance for a virtual object.

When a lens forms a real image, *its position with respect to the second lens* determines whether it is a real or a virtual object for the second lens. If the first lens would have formed a real image past the second lens, the image becomes a virtual object for the second lens. If the first lens forms a real or virtual image before the second lens, the image is a real object for the second lens.

CONNECTION:

For a system of two (or more) lenses, apply the thin lens equation to each lens in turn.

For a system of two thin lenses separated by a distance s, we can apply the thin lens equation separately to each lens:

$$\frac{1}{p_1} + \frac{1}{q_1} = \frac{1}{f_1}$$

$$\frac{1}{p_2} + \frac{1}{q_2} = \frac{1}{f_2}$$

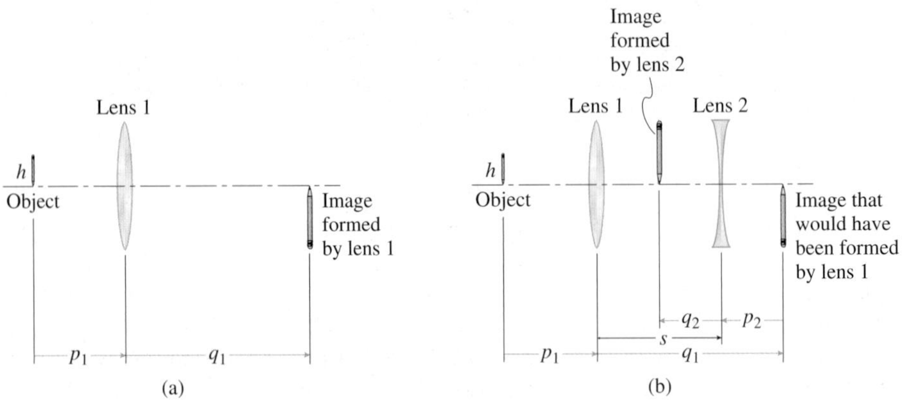

Figure 24.1 (a) Lens 1, a converging lens, forms a real image of an object. (b) Now lens 2 is placed a distance $s < q_1$ past lens 1. Lens 2 interrupts the light rays before they come together to form the real image, but we can think of the image that *would have* formed as the *virtual object* for lens 2. For a virtual object, p is negative.

The object distance p_2 for the second lens is

$$p_2 = s - q_1 \qquad (24\text{-}1)$$

Equation (24-1) gives the correct sign for p_2 in every case. If $q_1 < s$, then the image formed by the first lens is on the incident side of the second lens and, thus, is a real object for the second lens ($p_2 > 0$). If $q_1 > s$, then the second lens interrupts the light rays before they form an image. The image that would have been formed by the first lens is beyond the second lens, so the image becomes a virtual object for the second lens ($p_2 < 0$).

Ray Diagrams for Two Lenses In a ray diagram for a two-lens system, *only one of the principal rays for the first lens is a principal ray for the second lens.* Figure 24.2 shows a ray diagram for a system where lens 1 is converging and lens 2 is diverging. (The text website interactive virtual optics lab constructs ray diagrams for systems of two or more lenses and/or mirrors.)

Transverse Magnification

Suppose N lenses are used in combination. Let h be the size of the object, h_1 the size of the image formed by the first lens, and so forth. Since

$$\frac{h_N}{h} = \frac{h_1}{h} \times \frac{h_2}{h_1} \times \frac{h_3}{h_2} \times \cdots \times \frac{h_N}{h_{N-1}}$$

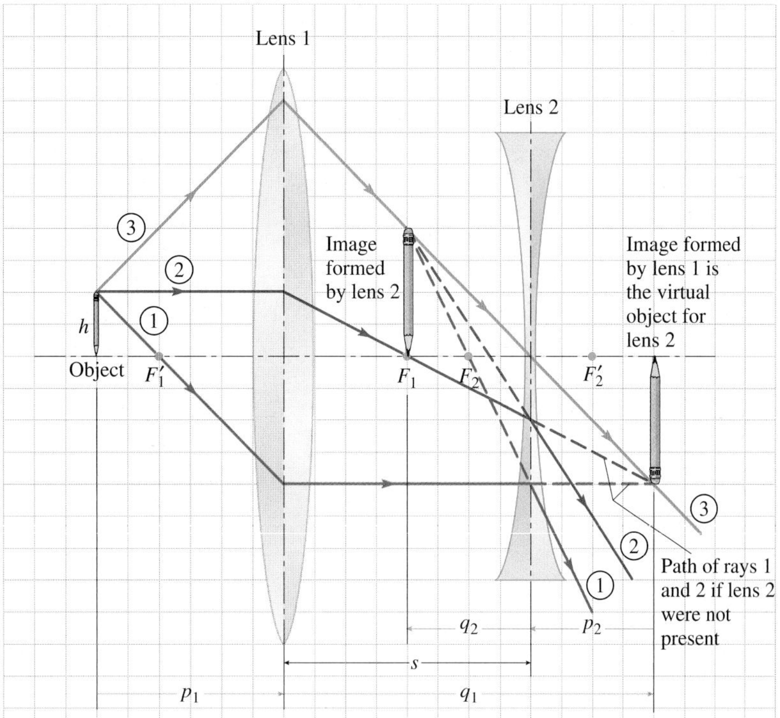

Figure 24.2 Ray diagram for a two-lens combination. Ray 1 comes from the object through focal point F'_1 and emerges from lens 1 parallel to the principal axis. Ray 1 is a principal ray for lens 2, emerging as if it came directly from F_2. In the absence of lens 2, ray 1 would have continued parallel to the axis. To locate the image formed by lens 1, we choose another principal ray (ray 2) and trace it, ignoring lens 2. These two rays locate the image formed by lens 1. Since it lies beyond lens 2, it becomes a virtual object. We do not yet know what happens to ray 2 when it strikes lens 2. To find the final image, we need another principal ray for lens 2. Ray 3 passes undeflected through the center of lens 2; we extrapolate it back through lens 1 to the object. The intersection of rays 1 and 3 locates the final image, which is virtual. Now we can finish ray 2; it must emerge from lens 2 as if coming from the image point.

the total transverse magnification due to the N lenses is the *product* (*not* the sum) of the magnifications due to the individual lenses:

Total transverse magnification:

$$m_{total} = m_1 \times m_2 \times \cdots \times m_N \quad (24\text{-}2)$$

CHECKPOINT 24.1

The grid in Fig. 24.2 represents 1 cm × 1 cm. What are the distances p_1, q_1, p_2, and q_2? What are the transverse magnifications due to lens 1 and to lens 2? What is the overall transverse magnification? Be sure to include correct algebraic signs with your answers.

Conceptual Example 24.1

Virtual Image as Object

Two lenses are used in combination. Suppose the first lens forms a virtual image. Does that image serve as a virtual object for the second lens?

Strategy The distinction between a real and virtual object depends on whether the rays incident on the second lens are converging or diverging.

Solution and Discussion If the first lens forms a virtual image, then the rays from any point on the object *diverge* as they emerge from the first lens. To find the image point, we trace those rays backward to find the point from which they seem to originate. Since the rays incident on the second lens are diverging, the image must become a *real* object for the second lens.

Another approach: the image formed by the first lens is located *before* the second lens (that is, on the same side as the incident light rays). Thus, the rays behave as if they diverge from an actual object at the same location—as a real object.

Conceptual Practice Problem 24.1 Real Image as Object

Two lenses are used in combination. Suppose the first lens forms a *real* image. Does that image serve as a real object or as a virtual object for the second lens? If either is possible, what determines whether the object is real or virtual?

Example 24.2

Two Converging Lenses

Two converging lenses, separated by a distance of 40.0 cm, are used in combination. The focal lengths are $f_1 = +10.0$ cm and $f_2 = +12.0$ cm. An object, 4.00 cm high, is placed 15.0 cm in front of the first lens. Find the intermediate and final image distances, the total transverse magnification, and the height of the final image.

Strategy We draw a diagram to help visualize what is happening and then apply the lens equation to each lens in turn. The total magnification is the product of the separate magnifications due to the two lenses.

Given: $p_1 = +15.0$ cm; $f_1 = +10.0$ cm; $f_2 = +12.0$ cm; separation $s = 40.0$ cm; $h = 4.00$ cm

To find: q_1; q_2; m; h'

Solution Figure 24.3 is a ray diagram that uses two principal rays for each lens to find the intermediate and final images. From the ray diagram, we expect that the intermediate image is real and to the left of lens 2; the final image is virtual, inverted, to the left of lens 1, and greatly enlarged.

continued on next page

Example 24.2 continued

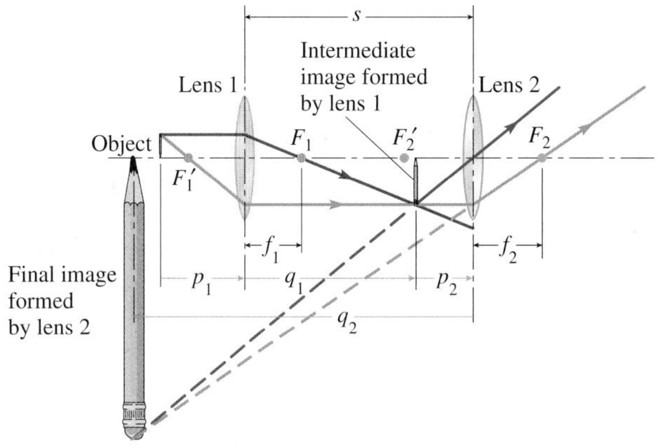

Figure 24.3

Ray diagram for Example 24.2. The intermediate real image formed by lens 1 is found using two of the principal rays, shown in red and green. The green ray is also a principal ray for lens 2. The principal ray that passes straight through the center of lens 2, shown in blue, is not actually present—lens 1 is not large enough to send a ray toward lens 2 in that direction. Nevertheless, we can still use it to locate the final image.

The thin lens equation, applied to lens 1, enables us to solve for q_1.

$$\frac{1}{p_1}+\frac{1}{q_1}=\frac{1}{f_1} \qquad (23\text{-}10)$$

Rearranging the equation and substituting values, we have

$$\frac{1}{q_1}=\frac{1}{f_1}-\frac{1}{p_1}=\frac{1}{10.0\text{ cm}}-\frac{1}{15.0\text{ cm}}=\frac{1}{30\text{ cm}}$$

Therefore, $q_1 = +30$ cm.

From Fig. 24.3, the object distance for lens 2 (p_2) is the separation of the two lenses (s) minus the image distance for the image formed by lens 1 (q_1).

$$p_2 = s - q_1 = 40.0\text{ cm} - 30\text{ cm} = 10\text{ cm}$$

The object distance is positive because the object is real: it is on the left of lens 2 and the rays from the object are diverging as they enter lens 2. We apply the thin lens equation to the second lens to find q_2.

$$\frac{1}{q_2}=\frac{1}{f_2}-\frac{1}{p_2}=\frac{1}{12.0\text{ cm}}-\frac{1}{10\text{ cm}}=-\frac{1}{60\text{ cm}}$$

$$q_2 = -60\text{ cm}$$

The image is 60 cm to the left of lens 2 or, equivalently, 20 cm to the left of lens 1. The image distance is negative, so the image is virtual.

For a single lens the magnification is

$$m=-\frac{q}{p} \qquad (23\text{-}9)$$

For a combination of two lenses the total magnification is

$$m = m_1 \times m_2 = -\frac{q_1}{p_1}\times\left(-\frac{q_2}{p_2}\right)$$
$$=\left(-\frac{30\text{ cm}}{15.0\text{ cm}}\right)\times\left(-\frac{-60\text{ cm}}{10\text{ cm}}\right)=-12$$

The final image is inverted, as indicated by the negative value of m, and its height is

$$4.00\text{ cm}\times 12 = 48\text{ cm}$$

Discussion Now we compare the numerical results with the ray diagram. As expected, the intermediate image is real and to the left of lens 2 ($q_1 = 30\text{ cm} < s = 40.0\text{ cm}$). The final image is virtual ($q_2 < 0$), inverted ($m < 0$), and enlarged ($|m| > 1$).

Practice Problem 24.2 Object Located at More than Twice the Focal Length

Repeat Example 24.2 if the same object is placed 25.0 cm before the first lens and the second lens is moved so it is only 10.0 cm from the first lens. Are you able to predict anything about the final image by sketching a ray diagram?

24.2 CAMERAS

One of the simplest optical instruments is the camera, which often has only one lens to produce an image, or even—in a pinhole camera—no lens. Figure 24.4 shows a simple 35-mm camera. The camera uses a converging lens to form a real image on the film. The image must be real in order to *expose* the film (i.e., cause a chemical reaction). Light rays from a point on an object being photographed must converge to a corresponding point on the film. A digital camera does away with film, replacing it with a CCD (charge-coupled device) array.

Cameras, slide projectors, and movie projectors form real images.

In good-quality cameras, the distance between the lens and the film can be adjusted in accordance with the lens equation so that a sharp image forms on the film. For distant objects, the lens must be one focal length from the film. For closer objects, the lens must be a little farther than that, since the image forms past the focal point. Simple fixed focus cameras have a lens that cannot be moved. Such cameras may give good results for faraway objects, but for closer objects it is more important that the lens position be adjustable.

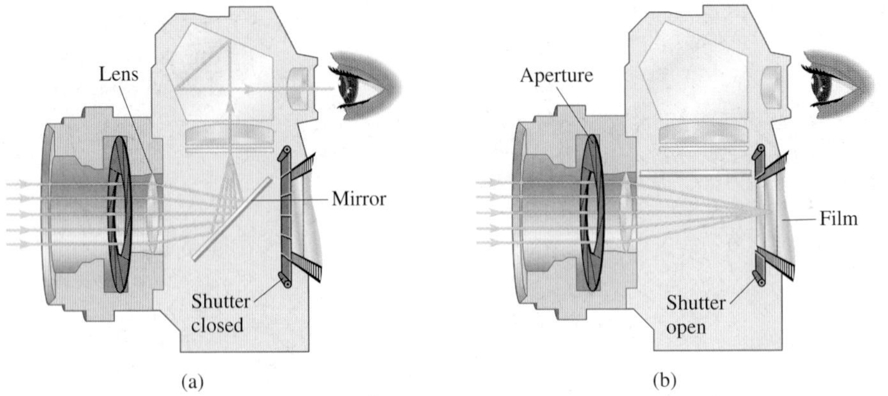

Figure 24.4 This 35-mm camera uses a single converging lens to form real images on the film. 35 mm is not the focal length of the lens; it is the width of the film. The camera is focused on objects at different distances by moving the lens closer to or farther away from the film. (a) The shutter is closed, preventing exposure of the film. (b) The mirror swings out of the way and the shutter opens for a short time to expose the film.

A slide or movie projector is the inverse of a camera. A light source is placed at the focal point of a converging lens so that nearly parallel light rays exit the lens and illuminate the slide. Another converging lens then forms an inverted, real image on a distant screen.

Example 24.3

Fixed-Focus Camera

A camera lens has a focal length of 50.0 mm. Photographs are taken of objects located at various positions, from an infinite distance away to as close as 6.00 m from the lens. (a) For an object at infinity, at what distance from the lens is the image formed? (b) For an object at a distance of 6.00 m, at what distance from the lens is the image formed?

Strategy We apply the thin lens equation for the two object distances and find the two image distances.

Solution (a) The thin lens equation is

$$\frac{1}{p}+\frac{1}{q}=\frac{1}{f}$$

For an object at infinity, $1/p = 1/\infty = 0$. Then

$$0+\frac{1}{q}=\frac{1}{f}$$

Therefore, $q = f$. The image distance is equal to the focal length; the image is 50.0 mm from the lens.

(b) This time $p = 6.00$ m from the camera:

$$\frac{1}{6.00\text{ m}}+\frac{1}{q}=\frac{1}{50.0\times 10^{-3}\text{ m}}$$

Solving for q yields

$$\frac{1}{q}=\frac{1}{50.0\times 10^{-3}\text{ m}}-\frac{1}{6.00\text{ m}}$$

or

$$q = 50.4\text{ mm}$$

Discussion The images are formed within 0.4 mm of each other, so the camera can form reasonably well focused images for objects from 6 m to infinity with a fixed distance between lens and film.

Practice Problem 24.3 Close-Up Photograph

Suppose the same lens is used with an adjustable camera to take a photograph of an object at a distance of 1.50 m. To what distance from the film should the lens be moved?

Regulating Exposure

A diaphragm made of overlapping metal blades acts like the iris of the eye; it regulates the size of the *aperture*—the opening through which light is allowed into the camera (see Fig. 24.4). The *shutter* is the mechanism that regulates the *exposure time*—the time interval during which light is allowed through the aperture. The aperture size and exposure time are selected so that the correct amount of light energy reaches the film. If they are chosen incorrectly, the film is over- or underexposed.

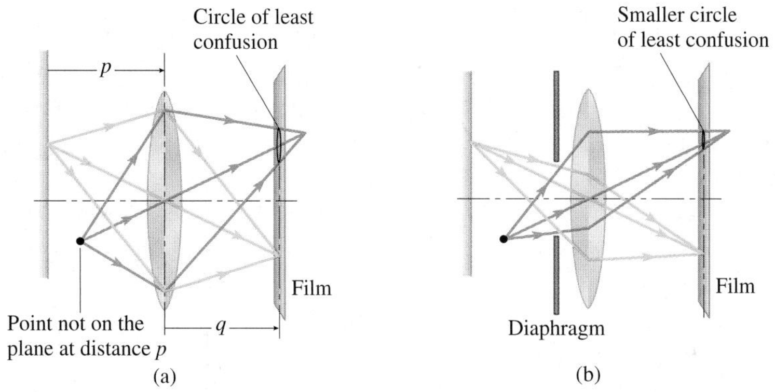

Figure 24.5 (a) The circle of least confusion for a point not on the plane in focus. (b) Reduction of the aperture size reduces the circle of least confusion and thereby increases the depth of field.

Depth of Field

Once a lens is focused by adjusting its distance q from the film, only objects in a plane at a particular distance p from the lens form sharp images on the film. Rays from a point on an object not in this plane expose a *circle* on the film (the *circle of least confusion*) instead of a single point (Fig. 24.5a). For some range of distances from the plane, the circle of least confusion is small enough to form an acceptably clear image on the film. This range of distances is called the *depth of field.*

A diaphragm can be placed before the lens to reduce the aperture size, reducing the size of the circle of least confusion (Fig. 24.5b). Thus, reducing the aperture size causes an increase in the depth of field. The trade-off is that, with a smaller aperture, a longer exposure time is necessary to correctly expose the film, which can be problematic if the subject is in motion or if the camera is not held steady by a tripod. Some compromise must be made between using a small aperture—so that more of the surroundings are focused—and using a short exposure time so that motion of the subject or the camera does not blur the image.

Pinhole Camera

Even simpler than a camera with one lens is a **pinhole camera**, or *camera obscura* (latin "dark room"). To make a pinhole camera, a tiny pinhole is made in one side of a box (Fig. 24.6a). An inverted, real "image" is formed on the opposite side of the box. A photographic plate (a glass plate coated with a photosensitive emulsion) or film placed on the back wall can record the image.

Artists made use of the camera obscura by working in a chamber with a small opening that admitted light rays from a scene outside the chamber. The image could be projected onto a canvas and the artist could trace the outline of the scene on the canvas. Jan van Eyck, Titian, Caravaggio, Vermeer, and Canaletto are just a few of the artists known or believed to have used a camera obscura to achieve realistic naturalism (Fig. 24.6b). In the eighteenth and nineteenth centuries, the camera obscura was commonly used to copy paintings and prints.

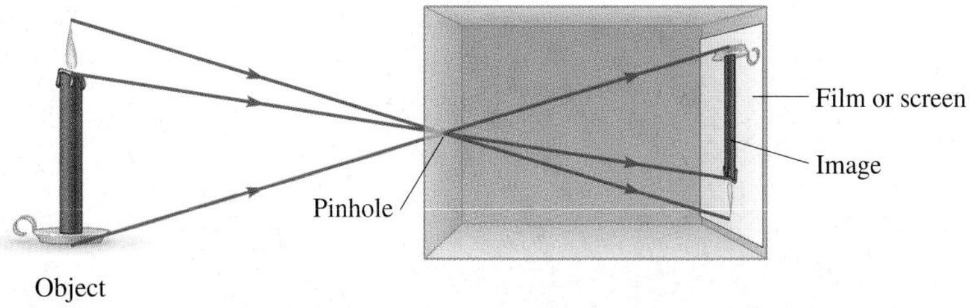

(a)

(b)

Figure 24.6 (a) A small pinhole camera. (b) *The Concert* was painted by Jan Vermeer around 1666. A camera obscura probably contributed to the accuracy of the perspective and the near-photographic detail in Vermeer's paintings.

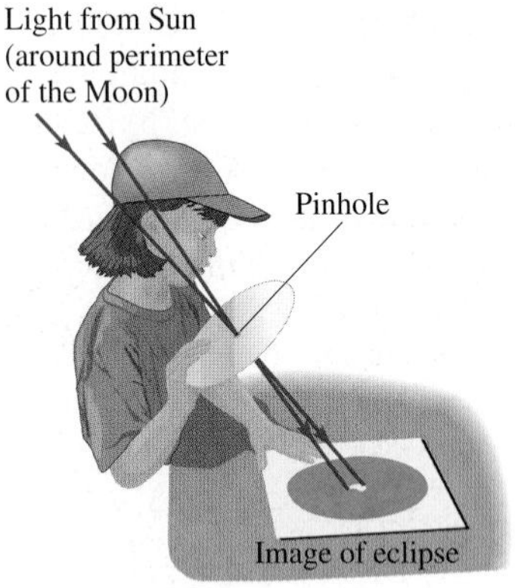

Figure 24.7 A pinhole camera arrangement for viewing an eclipse of the Sun.

PHYSICS AT HOME

A safe way to view the Sun is through a pinhole camera arrangement (Fig. 24.7). (This is a good way to view a solar eclipse.) Poke a pinhole in a piece of cardboard, a paper plate, or an aluminum pie pan. Then hold a white sheet of cardboard below the pinhole and view the image of the Sun on it. (Remember not to look directly at the Sun, even during an eclipse; severe damage to your eyes can occur.)

The pinhole camera does not form a *true* image—rays from a point on an object do not converge to a single point on the wall. The pinhole admits a narrow cone of rays diverging from each point on the object; the cone of rays makes a small circular spot on the wall. If this spot is small enough, the image appears clear to the eye. A smaller pinhole results in a dimmer, *sharper* "image" unless the hole is so small that diffraction spreads the spots out significantly.

24.3 THE EYE

The human eye is similar to a digital camera. The camera forms a real image on a CCD array; the eye forms a real image on the *retina*, a membrane with approximately 125 million photoreceptor cells (the *rods* and *cones*). The focusing mechanism is different, though. In the camera, the lens moves toward or away from the film to keep the image on the film as the object distance changes. In the eye, the lens is at a fixed distance from the retina, but it has a variable focal length; the focal length is adjusted to keep the image distance constant as the object distance varies.

Figure 24.8 shows the anatomy of the eye. It is approximately spherical, with an average diameter of 2.5 cm. A bulge in front is filled with the *aqueous fluid* (or aqueous "humor") and covered on the outside by a transparent membrane called the *cornea*. The aqueous fluid is kept at an overpressure to maintain the slight outward bulge. The curved surface of the cornea does most of the refraction of light rays entering the eye. The adjustable *lens* does the fine tuning. For most purposes, we can consider the cornea and the lens to act like a single lens, about 2.0 cm from the retina, with adjustable focal length. In order to see objects at distances of 25 cm or greater from the eye, which is

Simplified model of the human eye: a single converging lens of variable focal length at a fixed distance from the retina.

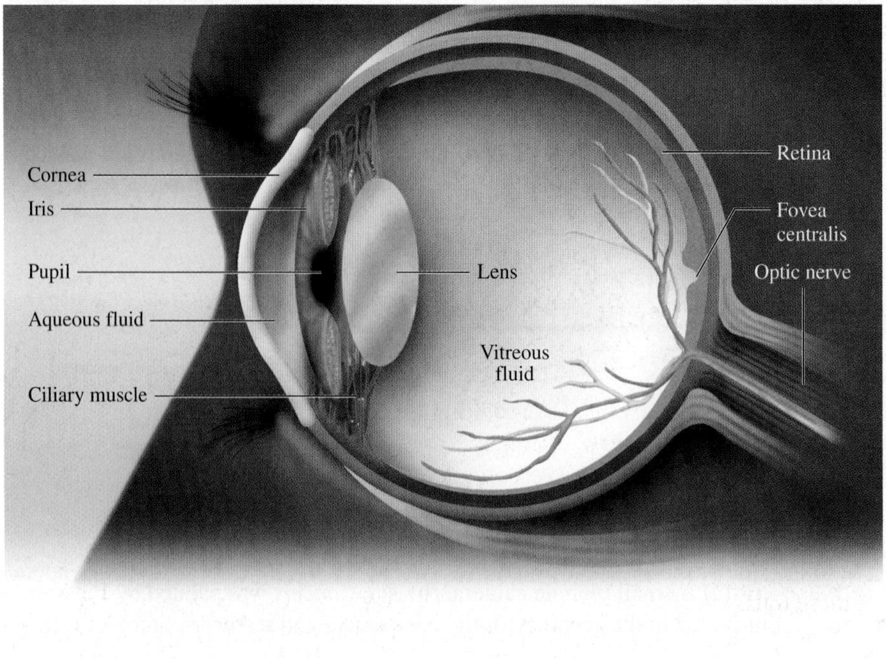

Figure 24.8 Anatomy of the human eye.

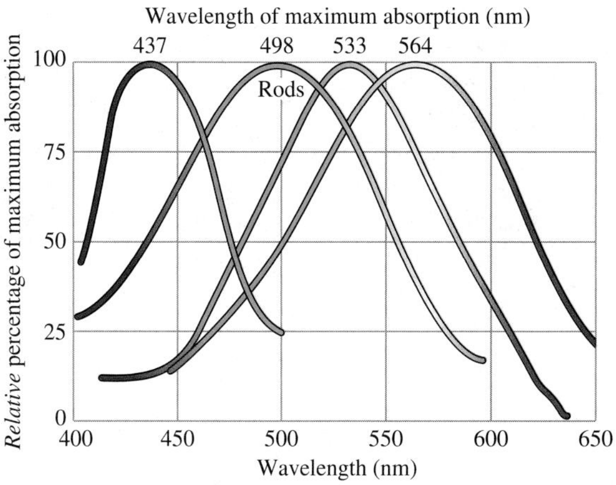

Figure 24.9 How the sensitivities of the rods and the three types of cones depend on the vacuum wavelength of the incident light. (Rods are *much* more sensitive than cones, so if the vertical scale were absolute instead of relative, the graph for the rods would be much taller than the others.)

considered normal vision, the focal length of the eye must vary between 1.85 cm and 2.00 cm if the retina is 2.00 cm from the eye (see Problem 22).

The spherical volume of the eye behind the lens is filled with a jelly-like material called the *vitreous fluid*. The indices of refraction of the aqueous fluid and the vitreous fluid are approximately the same as that of water (1.333). The index of the lens, made of a fibrous, jelly-like material, is a bit higher (1.437). The cornea has an index of refraction of 1.351.

The eye has an adjustable aperture (the *pupil*) that functions like the diaphragm in a camera to control the amount of light that enters. The size of the pupil is adjusted by the *iris*, a ring of muscular tissue (the colored portion of the eye). In bright light, the iris expands to reduce the size of the pupil and limit the amount of light entering the eye. In dim light, the iris contracts to allow more light to enter through the dilated pupil. The expansion and contraction of the iris is a *reflex* action in response to changing light conditions. In ordinary light the diameter of the pupil is about 2 mm; in dim light it is about 8 mm.

On the retina, the photoreceptor cells are densely concentrated in a small region called the *macula lutea*. The cones come in three different types that respond to different wavelengths of light (Fig. 24.9). Thus, the cones are responsible for color vision. Centered within the macula lutea is the *fovea centralis*, of diameter 0.25 mm, where the cones are tightly packed together and where the most acute vision occurs in bright light. The muscles that control eye movement ensure that the image of an object being examined is centered on the fovea centralis.

PHYSICS AT HOME

Each retina has a *blind spot* with no rods or cones, located where the optic nerve leaves the retina. The blind spot is not usually noticed because the brain fills in the missing information. To observe the blind spot, draw a cross and a dot, about 10 cm apart, on a sheet of white paper. Cover your left eye and hold the paper far from your eyes with the dot on the right. Keep your eye focused on the cross as you slowly move the paper toward your face. The dot disappears when the image falls on the blind spot. Continue to move the paper even closer to your eye; you will see the spot again when its image moves off the blind spot.

The rods are more sensitive to dim light than the cones but do not have different types sensitive to different wavelengths, so we cannot distinguish colors in very dim light. Outside the macula the photoreceptor cells are much less densely packed and they are all rods. However, the rods outside the macula are more densely packed than the rods inside the macula. If you are trying to see a dim star in the sky, it helps to look a little to the side of the star so the image of the star falls outside the macula where there are more rods.

Figure 24.10 The lens of the eye has (a) a longer focal length when viewing distant objects and (b) a shorter focal length when viewing nearby objects.

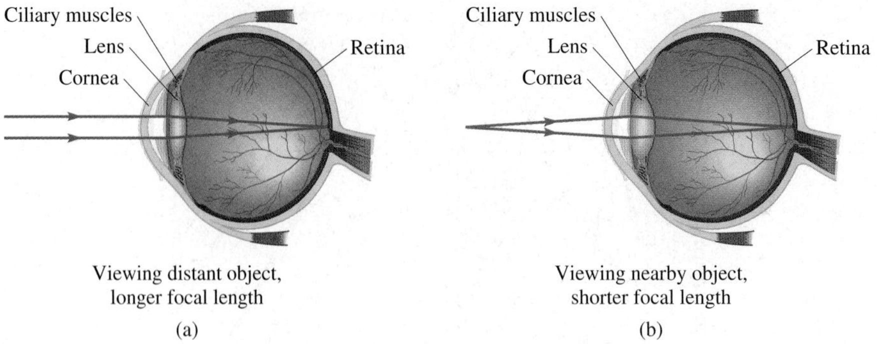

Accommodation

Variation in the focal length of the flexible lens is called **accommodation**; it is the result of an actual change in the shape of the lens of the eye through the action of the *ciliary muscles*. The adjustable shape of the lens allows for accommodation for various object distances, while still forming an image at the fixed image distance determined by the separation of lens and retina. When the object being viewed is far away, the ciliary muscles relax; the lens is relatively flat and thin, giving it a longer focal length (Fig. 24.10a). For closer objects, the ciliary muscles squeeze the lens into a thicker, more rounded shape (Fig. 24.10b), giving the lens a shorter focal length.

Accommodation enables an eye to form a sharp image on the retina of objects at a range of distances from the **near point** to the **far point**. A young adult with good vision has a near point at 25 cm or less and a far point at infinity. A child can have a near point as small as 10 cm. Corrective lenses (eyeglasses or contact lenses) or surgery can compensate for an eye with a near point greater than 25 cm or a far point less than infinity.

Optometrists write prescriptions in terms of the **refractive power** (P) of a lens rather than the focal length. (Refractive power is different from "magnifying power," which is a synonym for the angular magnification of an optical instrument.) The refractive power is simply the reciprocal of the focal length:

$$P = \frac{1}{f} \tag{24-3}$$

Refractive power is usually measured in **diopters** (symbol D). One diopter is the refractive power of a lens with focal length $f = 1$ m (1 D = 1 m^{-1}). The shorter the focal length, the more "powerful" the lens because the rays are bent more. Converging lenses have positive refractive powers and diverging lenses have negative refractive powers.

Why use refractive power instead of focal length? When two or more thin lenses with refractive powers $P_1, P_2, \ldots$ are sufficiently close together, they act as a single thin lens with refractive power

$$P = P_1 + P_2 + \cdots \tag{24-4}$$

as can be shown in Problem 8 by substituting P for $1/f$.

Application: Correcting Myopia

A myopic eye can see nearby objects clearly but not distant objects. Myopia (nearsightedness) occurs when the shape of the eyeball is elongated or when the curvature of the cornea is excessive. A myopic eye forms the image of a distant object *in front of* the retina (Fig. 24.11a). The refractive power of the lens is too large; the eye makes the rays converge too soon. A diverging corrective lens (with negative refractive power) can compensate for nearsightedness by bending the rays outward (Fig. 24.11b).

For objects at any distance from the eye, the diverging corrective lens forms a virtual image closer to the eye than is the object. For an object at infinity, the corrective lens forms an image *at the far point* of the eye (Fig. 24.11c). For less distant objects, the

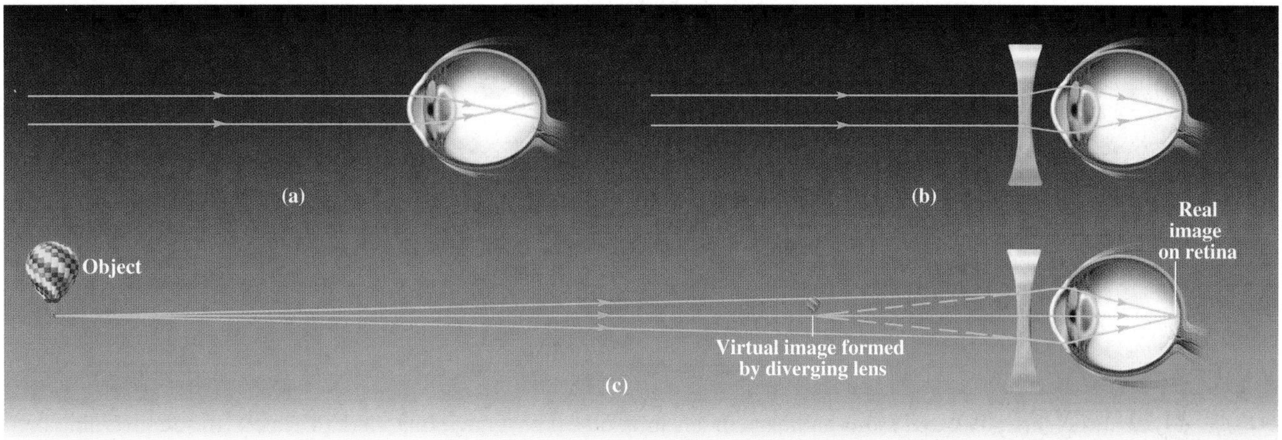

Figure 24.11 (a) In a nearsighted eye, parallel rays from a point on a distant object converge before they reach the retina. (b) A diverging lens corrects for the nearsighted eye by bending the rays outward just enough that the eye brings them back together at the retina. (c) The diverging lens forms a virtual image closer to the eye than the object; the eye can make the rays from this image converge into a real image on the retina. (Not to scale.)

virtual image is closer than the far point. The eye is able to focus rays from this image onto the retina since it is never past the far point.

Example 24.4

Correction for a Nearsighted Eye

Without her contact lenses, Dana cannot see clearly an object more than 40.0 cm away. What refractive power should her contact lenses have to give her normal vision?

Strategy The far point for Dana's eyes is 40.0 cm. For an object at infinity, the corrective lens must form a virtual image 40.0 cm from the eye. We use the lens equation with $p = \infty$ and $q = -40.0$ cm to find the focal length or refractive power of the corrective lens. The image distance is negative because the image is virtual—it is formed on the same side of the lens as the object.

Solution The thin lens equation is

$$\frac{1}{p} + \frac{1}{q} = \frac{1}{f} = P$$

Since $p = \infty$, $1/p = 0$. Then

$$0 + \frac{1}{-40.0\text{ cm}} = \frac{1}{f}$$

Solving for the focal length,

$$f = -40.0\text{ cm}$$

The refractive power of the lens in diopters is the inverse of the focal length in meters.

$$P = \frac{1}{f} = \frac{1}{-40.0 \times 10^{-2}\text{ m}} = -2.50\text{ D}$$

Discussion The focal length and refractive power are negative, as expected for a diverging lens. We might have anticipated that $f = -40.0$ cm without using the thin lens equation. Rays coming from a distant source are nearly parallel. Parallel rays incident on a diverging lens emerge such that they appear to come from the focal point before the lens. Thus, the image is at the focal point on the incident side of the lens.

Practice Problem 24.4 What Happens to the Near Point?

Suppose Dana's *near* point (without her contact lenses) is 10.0 cm. What is the closest object she can see clearly with her contact lenses on? [*Hint:* For what object distance do the contact lenses form a virtual image 10.0 cm before the lenses?]

Application: Correcting Hyperopia

A hyperopic (farsighted) eye can see distant objects clearly but not nearby objects; the near point is too large. The refractive power of the eye is too small; the cornea and lens do not refract the rays enough to make them converge on the retina (Fig. 24.12a).

Figure 24.12 (a) A farsighted eye forms an image of a nearby object past the retina. (Not to scale.) (b) A converging corrective lens forms a virtual image farther away from the eye than the object. Rays from this virtual image can be brought together by the eye to form a real image on the retina.

A converging lens can correct for hyperopia by bending the rays inward so they converge sooner (Fig. 24.12b). In order to have normal vision, the near point should be 25 cm (or less). Thus, for an object at 25 cm from the eye, the corrective lens forms a virtual image at the eye's near point.

Example 24.5

Correction for Farsighted Eye

Winifred is unable to focus on objects closer than 2.50 m from her eyes. What refractive power should her corrective lenses have?

Strategy For an object 25 cm from Winifred's eye, the corrective lens must form a virtual image at the near point of Winifred's eye (2.50 m from the eye). We use the thin lens equation with $p = 25$ cm and $q = -2.50$ m to find the focal length. As in the last example, the image distance is negative because it is a virtual image formed on the same side of the lens as the object.

Solution From the thin lens equation,

$$\frac{1}{p} + \frac{1}{q} = \frac{1}{f}$$

Substituting $p = 0.25$ m and $q = -2.50$ m,

$$\frac{1}{0.25\text{ m}} + \frac{1}{-2.50\text{ m}} = \frac{1}{f}$$

Solving for the focal length,

$$f = 0.28\text{ m}$$

The refractive power is

$$P = \frac{1}{f} = +3.6\text{ D}$$

Discussion This solution assumes that the corrective lens is very close to the eye, as for a contact lens. If Winifred wears eyeglasses that are 2.0 cm away from her eyes, then the object and image distances we should use—since they are measured from the *lens*—are $p = 23$ cm and $q = -2.48$ m. The thin lens equation then gives $P = +3.9$ D.

Practice Problem 24.5 Using Eyeglasses

A man can clearly see an object that is 2.00 m away (or more) without using his eyeglasses. If the eyeglasses have a refractive power of +1.50 D, how close can an object be to the eyeglasses and still be clearly seen by the man? Assume the eyeglasses are 2.0 cm from the eye.

Presbyopia

As a person ages, the lens of the eye becomes less flexible and the eye's ability to accommodate decreases, a phenomenon known as presbyopia. Older people have difficulty focusing on objects held close to the eyes; from the age of about 40 years many people need eyeglasses for reading. At age 60, a near point of 50 cm is typical; in some people it may be 1 m or even more. Reading glasses for a person suffering from presbyopia are similar to those used by a farsighted person.

CHECKPOINT 24.3

On a camping trip, you discover that no one has brought matches. A friend suggests using his eyeglasses to focus sunlight onto some dry grass and shredded bark to get a fire started. Could this scheme work if your friend is nearsighted? What about if he is farsighted? Explain.

24.4 ANGULAR MAGNIFICATION AND THE SIMPLE MAGNIFIER

Angular Magnification

We use magnifiers and microscopes to enlarge objects too small to see with the naked eye. But what do we mean by *enlarged* in this context? The apparent size of an object depends on the size of the image *formed on the retina* of the eye. For the unaided eye, the retinal image size is proportional to the angle subtended by the object. Figure 24.13 shows two identical objects being viewed from different distances. Imagine rays from the top and bottom of each object that are incident on the center of the lens of the eye. The angle θ is called the **angular size** of the object. The image on the retina subtends the same angle θ; the angular size of the image is the same as that of the object. Rays from the object at a greater distance subtend a smaller angle; the angular size depends on distance from the eye.

A magnifying glass, microscope, or telescope serves to make the image on the retina larger *than it would be if viewed with the unaided eye*. Since the size of the image on the retina is proportional to the angular size, we measure the usefulness of an optical instrument by its **angular magnification**—the ratio of the angular size using the instrument to the angular size with the unaided eye.

Definition of angular magnification:

$$M = \frac{\theta_{\text{aided}}}{\theta_{\text{unaided}}} \tag{24-5}$$

The *transverse* magnification (the ratio of the image size to the object size) isn't as useful here. The transverse magnification of a telescope-eye combination is minute: the Moon is *much* larger than the image of the Moon on the retina, even using a telescope. The telescope makes the image of the Moon larger than it would be in unaided viewing.

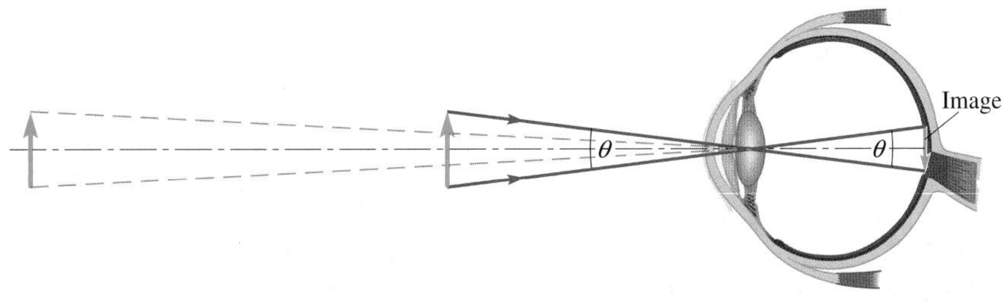

Figure 24.13 Identical objects viewed from different distances. Rays drawn from the top and bottom of the nearer object illustrate the angle θ subtended by the object.

PHYSICS AT HOME

On a clear night with the Moon visible, go outside, shut one eye, and hold a pencil at arm's length between your open eye and the Moon so it blocks your view of the Moon. Compare the angular size of the Moon with the angular width of the pencil. Estimate the distance from your eye to the pencil and the pencil's width. Use this information and the Earth-Moon distance (4×10^5 km) to estimate the diameter of the Moon. Compare your estimate to the actual diameter of the Moon (3.5×10^3 km).

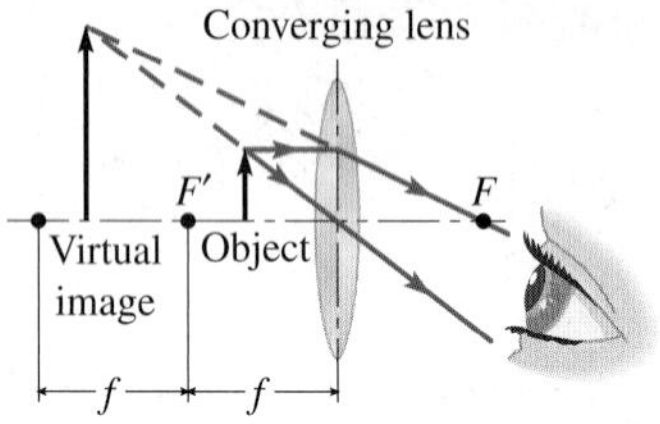

Figure 24.14 A converging lens used as a magnifying glass forms an enlarged virtual image. The object distance is less than the focal length.

Simple Magnifier

When you want to see something in greater detail, you naturally move your eye closer to the object to increase the angular size of the object. But the eye's ability to accommodate for nearby objects is limited; anything closer than the near point cannot be seen clearly. Thus, the maximum angle subtended at the unaided eye by an object occurs when the object is located at the near point.

A **simple magnifier** is a converging lens placed so that the object distance is less than the focal length. The virtual image formed is enlarged, upright, and farther away from the lens than the object (Fig. 24.14). Typically, the image is put well beyond the near point so that it is viewed by a more relaxed eye at the expense of a small reduction in angular magnification. The angle subtended by the enlarged virtual image seen by the eye is much larger than the angle subtended by the object when placed at the near point.

If a small object of height h is viewed with the unaided eye (Fig. 24.15a), the angular size when it is placed at the near point (a distance N from the eye) is

$$\theta \approx \frac{h}{N} \quad \text{(in radians)}$$

where we assume $h \ll N$ and, thus, θ is small enough that $\tan\theta \approx \theta$. If the object is now placed at the focal point of a converging lens, the image is formed at infinity and can be viewed with a relaxed eye (Fig. 24.15b). The angular size of the image is

$$\theta_\infty \approx \frac{h}{f} \quad \text{(in radians)}$$

Then the angular magnification M is

$$M = \frac{\theta_\infty}{\theta} = \frac{h/f}{h/N} = \frac{N}{f} \tag{24-6}$$

When calculating the angular magnification of an optical instrument, it is customary to assume a typical near point of $N = 25$ cm.

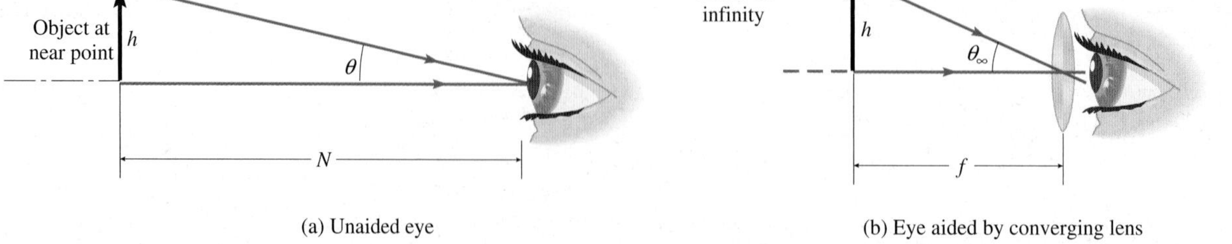

Figure 24.15 (a) The angle θ subtended at the eye by an object placed at the near point. (b) The magnifier forms a virtual image of the object at infinity. The angle θ_∞ subtended by the virtual image is larger than θ.

Equation (24-6) gives the angular magnification when the object is placed at the focal point of the magnifier. If the object is placed closer to the magnifier, the angular magnification is somewhat larger. In many cases, the small increase in angular magnification is not worth the added eyestrain of viewing an image closer to the eye (see Problem 35).

Example 24.6

A Magnifying Glass

A converging lens with a focal length of 4.00 cm is used as a simple magnifier. The lens forms a virtual image at your near point, 25.0 cm from your eye. Where should the object be placed and what is the angular magnification? Assume that the magnifier is held close to your eye.

Strategy We can use 25.0 cm as the image distance from the *lens*; if the magnifier is near the eye, distances from the lens are approximately the same as distances from the eye. We apply the thin lens equation to find the object distance with the focal length and image distance known.

Solution By rearranging the thin lens equation to solve for the object distance, we obtain

$$p = \frac{fq}{q-f}$$

We now substitute $q = -25.0$ cm (negative for a virtual image) and $f = +4.00$ cm.

$$p = \frac{4.00 \text{ cm} \times (-25.0 \text{ cm})}{-25.0 \text{ cm} - 4.00 \text{ cm}}$$

$$= 3.45 \text{ cm}$$

The object is placed 3.45 cm from the lens. The angular size (in radians) of the image formed is

$$\theta = \frac{h}{p}$$

where h is the size of the object. The object is *not* at the focal point of the lens, so the angular size is not h/f as it is in Fig. 24.15b. If the object were to be viewed without the magnifier, while placed at the near point of $N = 25.0$ cm, the angular size would be

$$\theta_0 = \frac{h}{N}$$

The angular magnification is

$$M = \frac{h/p}{h/N} = \frac{N}{p} = \frac{25.0 \text{ cm}}{3.45 \text{ cm}} = 7.25$$

Discussion If the object had been placed at the principal focal point, 4.00 cm from the lens, to form a final image at infinity, the angular magnification would have been

$$M = \frac{N}{f} = \frac{25.0 \text{ cm}}{4.00 \text{ cm}} = 6.25$$

Practice Problem 24.6 Where to Place an Object with a Magnifier

The focal length of a simple magnifier is 12.0 cm. Assume the viewer's eye is held close to the lens. (a) What is the angular magnification of an object if the magnifier forms a final image at the viewer's near point (25.0 cm)? (b) What is the angular magnification if the final image is at infinity?

24.5 COMPOUND MICROSCOPES

The simple magnifier is limited to angular magnifications of 15–20 at most. By contrast, the **compound microscope**, which uses two converging lenses, enables angular magnifications of 2000 or more. The compound microscope was probably invented in Holland around 1600.

A small object to be viewed under the microscope is placed *just beyond* the focal point of a converging lens called the **objective**. The function of the objective is to form an enlarged real image. A second converging lens, called the **ocular** or **eyepiece**, is used to view the real image formed by the objective lens (Fig. 24.16). The eyepiece acts as a simple magnifier; it forms an enlarged virtual image. The position of the final image can be anywhere between the near point of the observer and infinity. Usually it is placed at infinity, since that enables viewing with a relaxed eye and doesn't decrease the angular magnification very much. To form a final image at infinity, the image formed by the objective is located at the focal point of the eyepiece. Inside the barrel of the microscope, the positions of the two lenses are adjusted so that the image formed by the objective falls at or within the focal point of the eyepiece.

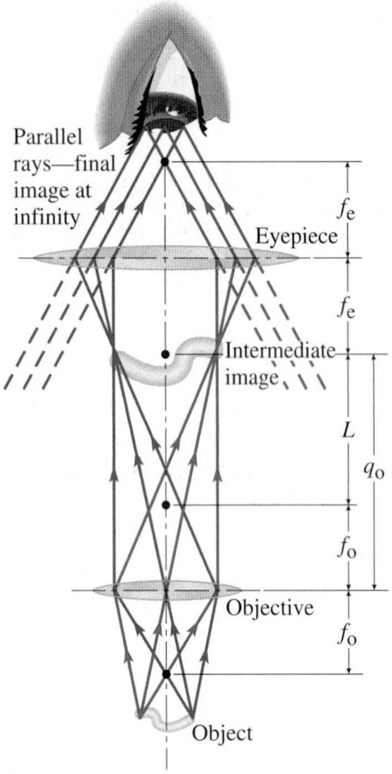

Figure 24.16 A compound microscope. To form a final image at infinity, the intermediate image must be located at the focal point of the eyepiece.

Remember that the tube length L in Eq. (24-8) is *the distance between the focal points* of the two lenses in a compound microscope.

If we used just the eyepiece as a simple magnifier to view the object, the angular magnification would be

$$M_e = \frac{N}{f_e} \quad \text{(due to eyepiece)}$$

where f_e is the focal length of the eyepiece and the virtual image is at infinity for ease of viewing. Customarily we assume $N = 25$ cm. The objective forms an image that is larger than the object; as shown in Problem 44, the transverse magnification due to the objective is

$$m_o = -\frac{L}{f_o} \quad \text{(due to objective)}$$

where L (the **tube length**) is the distance between the focal points of the two lenses. Since the image of the objective is placed at the focal point of the eyepiece, as in Fig. 24.16, the tube length is

$$L = q_o - f_o \tag{24-7}$$

Many microscopes are designed with a tube length of 16 cm.

When we view the image with the eyepiece, the eyepiece provides the same angular magnification as before (M_e), but it magnifies an image already m_o times as large as the object. The total angular magnification is the product of m_o and M_e:

Angular magnification due to a microscope:

$$M_{total} = m_o M_e = -\frac{L}{f_o} \times \frac{N}{f_e} \tag{24-8}$$

The negative sign in Eq. (24-8) means that the final image is inverted. Sometimes with microscopes and telescopes the sign is ignored and the angular magnification (also called the *magnifying power*) is reported as a positive number.

Equation (24-8) shows that, for large magnification, both focal lengths should be small. Microscopes are often made so that any one of several different objective lenses can be swung into position, depending on the magnification desired. The manufacturer usually provides the values of the magnification ($|m_o|$ and M_e) instead of the focal lengths of the lenses. For example, if an eyepiece is labeled "5 ×," then $M_e = 5$.

Example 24.7

Magnification by a Microscope

A compound microscope has an objective lens of focal length 1.40 cm and an eyepiece with a focal length of 2.20 cm. The objective and the eyepiece are separated by 19.6 cm. The final image is at infinity. (a) What is the angular magnification? (b) How far from the objective should the object be placed?

Strategy Since the final image is at infinity, Eq. (24-8) can be used to find the angular magnification M. We first find the tube length L of the microscope. From Fig. 24.16, the distance between the lenses is the sum of the two focal lengths plus the tube length. We assume the typical near point of $N = 25$ cm. To find where the object should be placed, we apply the thin lens equation to the objective. The image formed by the objective is at the focal point of the eyepiece since the final image is at infinity.

Given: $f_o = 1.40$ cm, $f_e = 2.20$ cm, lens separation = 19.6 cm
To find: (a) total angular magnification M; (b) object distance p_o

Solution (a) The tube length is

$$L = \text{distance between lenses} - f_o - f_e$$
$$= 19.6 \text{ cm} - 1.40 \text{ cm} - 2.20 \text{ cm} = 16.0 \text{ cm}$$

Then the angular magnification is

$$M = -\frac{L}{f_o} \times \frac{N}{f_e}$$
$$= -\frac{16.0 \text{ cm}}{1.40 \text{ cm}} \times \frac{25 \text{ cm}}{2.20 \text{ cm}} = -130$$

continued on next page

Example 24.7 continued

The negative magnification indicates that the final image is inverted.

(b) To have the final image at infinity, the image formed by the objective lens must be located at the focal point of the eyepiece. From Fig. 24.16, the intermediate image distance is

$$q_o = L + f_o = 16.0 \text{ cm} + 1.40 \text{ cm} = 17.4 \text{ cm}$$

Then the object distance is found using the thin lens equation:

$$\frac{1}{p_o} + \frac{1}{q_o} = \frac{1}{f_o}$$

Solving for the object distance, p_o:

$$p_o = \frac{f_o q_o}{q_o - f_o}$$

$$= \frac{1.40 \text{ cm} \times 17.4 \text{ cm}}{17.4 \text{ cm} - 1.40 \text{ cm}}$$

$$= 1.52 \text{ cm}$$

Discussion We can check the result for part (b) to see if the object is just past the focal point of the objective. The object is 1.52 cm from the objective and the focal point is 1.40 cm, so the object is just 1.2 mm past the focal point.

Practice Problem 24.7 Object Distance for Good Focus

An observer with a near point of 25 cm looks through a microscope with an objective lens of focal length $f_o = 1.20$ cm. When an object is placed 1.28 cm from the objective, the angular magnification is −198 and the final image is formed at infinity. (a) What is the tube length L for this microscope? (b) What is the focal length of the eyepiece?

The Transmission Electron Microscope

Many other kinds of microscope, both optical and nonoptical, are in use. The one most similar to the optical compound microscope is the *transmission electron microscope* (TEM). In the 1920s, the German physicist Ernst Ruska (1906–1988) found that a magnetic field due to a coil could act as a lens for electrons. An optical lens functions by changing the directions of the light rays; the magnetic coil changes the directions of the electrons' trajectories. Ruska was able to use the lens to form an image of an object irradiated with electrons. Eventually he coupled two such lenses together to form a microscope. By 1933 he had produced the first electron microscope, using an electron beam to form images of tiny objects with far greater clarity than the conventional optical microscope. Ruska's microscope is called a *transmission* microscope because the electron beam passes right through the thin slice of a sample being studied.

Resolution

A large magnification is of little use if the image is blurry. *Resolution* is the ability to form clear and distinct images of points very close to each other on an object. High resolution is a desirable quality in a microscope. The ultimate limit on the resolution of an optical instrument is limited by diffraction—the spreading out of light rays. Due to diffraction, the size of an object that can be clearly imaged by an optical instrument cannot be much smaller than the wavelength of the light used. Thus, we cannot expect to see anything smaller than about 400 nm using a compound optical microscope. An atom, roughly 0.2–0.5 nm in diameter, is much smaller than the wavelength of light, so a microscope cannot resolve details on the atomic scale. Ultraviolet microscopes can do a little better (about 100 nm) due to the shorter wavelength. Transmission electron microscopes can resolve details down to about 0.5 nm.

24.6 TELESCOPES

Refracting Telescopes

The most common type of telescope for nonscientific work is the **refracting telescope**, which has two converging lenses that function just as those in a compound microscope. The refracting telescope has an objective lens that forms a real image of the object; the

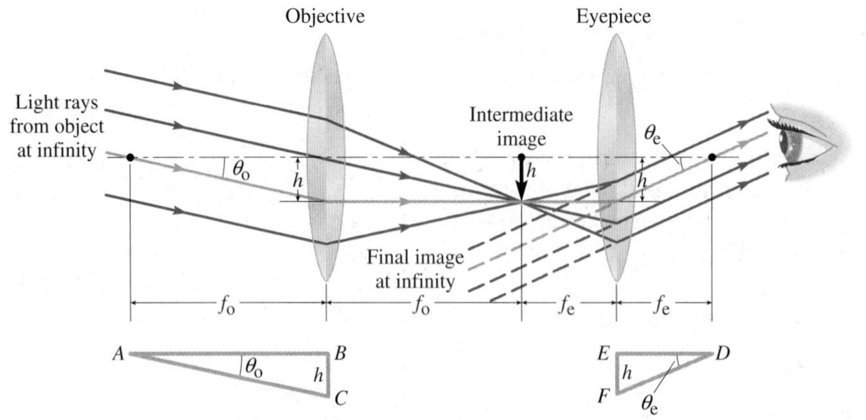

Figure 24.17 An astronomical refracting telescope. A highlighted ray passing through the secondary focal point of the objective leaves the lens parallel to the principal axis, then continues to the eyepiece and is refracted so that it goes through the principal focal point of the eyepiece. Two small right triangles are redrawn below the diagram for clarity. The hypotenuse (*AC*, *FD*) of each triangle is along the highlighted ray. The leg (*BC*, *EF*) of each triangle from the principal axis to the hypotenuse is of length *h* because the line connecting *C* to *F* is parallel to the principal axis and passes through the tip of the image.

eyepiece (ocular) is used to view this real image. The microscope is used to view *tiny* objects placed close to the objective lens; the purpose of the objective is to form an *enlarged* image. The telescope is used to view objects whose *angular* sizes are small because they are far away; the objective forms an image that is tiny compared with the object, but the image is available for closeup viewing through the eyepiece.

Astronomical Telescope In an *astronomical* refracting telescope, the object is so far away that the rays from a point on the object can be assumed to be parallel; the object distance is taken as infinity (Fig. 24.17). The objective forms a real, diminished image at its principal focus. By placing this image at the secondary focal point of the eyepiece, the final image is at infinity for ease of viewing. Thus, the principal focal point of the objective must coincide with the secondary focal point of the eyepiece, in contrast to the microscope in which the two are separated by a distance *L* (the tube length). When an astronomical telescope is connected to a camera to record the image, the camera lens replaces the eyepiece and the image formed by the objective is *not* placed at the focal point of the camera lens because the camera lens must form a real image *on the film*.

The objective is located at one end of the telescope barrel and the eyepiece is at the other end. Then the *barrel length* of the telescope is the sum of the focal lengths of the objective and the eyepiece.

$$\text{barrel length} = f_o + f_e \tag{24-9}$$

The angle that would be subtended if viewed by the unaided eye is the same as the angle subtended at the objective (θ_o). The angle subtended at the observer's eye looking through the eyepiece at the final image formed at infinity is θ_e. From the two small right triangles in Fig. 24.17 and the small angle approximation, the angular size of the object for the unaided eye is

$$\theta_o \approx \tan\theta_o = \frac{h}{AB} = \frac{h}{f_o}$$

The angular size of the final image is

$$\theta_e \approx \tan\theta_e = -\frac{h}{DE} = -\frac{h}{f_e}$$

The final image is inverted, so its angular size is negative. With a telescope, the magnification that is of interest is again the *angular* magnification: the ratio of the angle subtended at the eye by the final magnified image to the angle subtended for the unaided eye. Then the angular magnification is

Angular magnification due to an astronomical telescope:

$$M = \frac{\theta_e}{\theta_o} = -\frac{f_o}{f_e} \qquad (24\text{-}10)$$

where the negative sign indicates an inverted image. As for microscopes, the angular magnification is usually reported as a positive number. For the greatest magnification, the objective lens has as long a focal length as possible, but the eyepiece has as short a focal length as possible.

CHECKPOINT 24.6

For greatest magnifying power, the objective lens of a microscope should have a small focal length while the objective lens of a telescope should have a large focal length. Explain why.

Example 24.8

Yerkes Refracting Telescope

The Yerkes telescope in southern Wisconsin (Fig. 24.18) is the largest refracting telescope in the world. Its objective lens is 1.016 m (40 in.) in diameter and has a focal length of 19.8 m (65 ft). If the magnifying power is 508, what is the focal length of the eyepiece?

Figure 24.18

The Yerkes refracting telescope, first operated in 1897, is still the largest *refracting* telescope in the world.

Strategy The magnifying power is the magnitude of the angular magnification. For an astronomical refracting telescope, the angular magnification is negative.

Solution From Eq. (24-10), the angular magnification is

$$M = \frac{\theta_e}{\theta_o} = -\frac{f_o}{f_e}$$

Solving for f_e yields

$$f_e = -\frac{f_o}{M}$$

Now we substitute $M = -508$ and $f_o = 19.8$ m:

$$f_e = -\frac{19.8 \text{ m}}{-508} = 3.90 \text{ cm}$$

Discussion The focal length of the eyepiece is positive, which is correct. The eyepiece serves as a simple magnifier used to view the image formed by the objective. The simple magnifier is a converging lens—that is, a lens with positive focal length.

Practice Problem 24.8 Replacing the Eyepiece

If the eyepiece used with the Yerkes telescope in Example 24.8 is changed to one with focal length 2.54 cm that produces a final image at infinity, what is the new angular magnification?

Terrestrial Telescopes An inverted image is no problem when the telescope is used as an astronomical telescope. When a telescope is used to view terrestrial objects, such as a bird perched high on a tree limb or a rock singer on stage at an outdoor concert, the final image must be upright. Binoculars are essentially a pair of telescopes with reflecting prisms that invert the image so the final image is upright.

Another way to make a terrestrial telescope is to add a third lens between the objective and the eyepiece to invert the image again so that the final image is upright. The Galilean telescope, invented by Galileo in 1609, produces an upright image without using a third lens. The upright image is obtained by using a *diverging* lens as the eyepiece (see Problem 53). The eyepiece is located so that the image formed by the objective becomes a *virtual* object for the eyepiece, which then forms an upright virtual image. The barrel length for a Galilean telescope is shorter than for telescopes with only converging lenses.

Reflecting Telescopes

Reflecting telescopes use one or more mirrors in place of lenses. There are several advantages to mirrors over lenses; these advantages become overwhelming in the large telescopes that must be used to gather enough light rays to be able to see distant, faint stars. (Large telescopes also minimize the loss of resolution due to diffraction.) Since the index of refraction varies with wavelength, a lens has slightly different focal lengths for different wavelengths; thus, dispersion distorts the image. A mirror works by reflection rather than refraction, so it has the same focal length for all wavelengths. Large mirrors are much easier to build than large lenses. When making a large glass lens, the glass becomes so heavy that it deforms due to its own weight. It also suffers from stresses and strains as it cools from a molten state; such stresses reduce the optical quality of the lens. A large mirror need not be so heavy, since only the *surface* is important; it can be supported everywhere under its surface, while a lens can only be supported at the edge. Another advantage of the reflecting telescope is that the heaviest part—the large concave mirror—is located at the base of the telescope, making the instrument stable. The largest lens used with a refracting telescope—a little over 1 m (3.3 ft) in diameter—is in the Yerkes telescope. By comparison, the primary mirror in each of the twin Keck reflecting telescopes in Hawaii has a diameter ten times as large—10 m (33 ft).

Figure 24.19 shows one kind of reflecting telescope, known as the Cassegrain arrangement (after Laurent Cassegrain, 1629–1693). Parallel light rays from a distant star are reflected from a concave mirror toward its focal point *F*. Before the rays can reach the focal point, they are intercepted in their path by a smaller convex mirror. The convex mirror directs the rays through a hole in the center of the large concave mirror so that they come to a focus at a point *P*. A camera or an electronic recording instrument can be placed at point *P*, or a lens can be used to direct the rays to a viewer's eye.

Application: Hubble Space Telescope

Why put a telescope in space?

A famous telescope using the Cassegrain arrangement is the Hubble Space Telescope (HST). The HST orbits Earth at an altitude of over 600 km; its primary mirror is 2.4 m in diameter. Why put a telescope in orbit? The atmosphere limits the amount of detail that is seen by any telescope on Earth. The density of the air in the atmosphere at any location is continually fluctuating; as a result, light rays from distant stars are bent by different amounts, making it impossible to bring the rays to a sharp focus. There are systems that correct for atmospheric fluctuations, but since the HST is above the atmosphere, it avoids the whole problem.

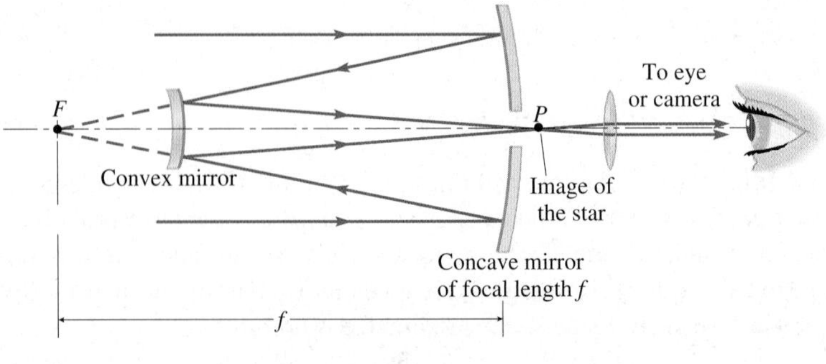

Figure 24.19 Cassegrain focus arrangement of a reflecting telescope directing rays from a distant star to an eye or a camera.

(a) (b) (c)

Figure 24.20 Three stunning images captured by the Advanced Camera for Surveys aboard the Hubble Space Telescope. (a) The Cone Nebula, a pillar of cold gas and dust. Hydrogen atoms absorb ultraviolet radiation and emit light, causing the red "halo" around the pillar. (b) Collision of two spiral galaxies known as the "Mice." A similar fate may await our galaxy a few billion years from now. (c) The center of the Omega Nebula, a region of flowing gas and newly formed stars surrounded by a cloud of hydrogen. Light emitted by excited atoms of nitrogen and sulfur produces the rose-colored region right of center. Other colors are produced by excited atoms of hydrogen and oxygen.

Accomplishments of the HST (Fig. 24.20) include clear images of quasars, the most energetic objects of the universe; the first surface map of Pluto; the discovery of intergalactic helium left over from the Big Bang (the birth of the universe); and clear evidence for the existence of black holes (objects so dense that nothing, not even light, can escape their gravitational pull). The HST has provided evidence of gravitational lensing, in which the gravity from massive galaxies bends light rays inward like a lens to form images of even more distant objects behind them.

The HST has provided a deeper look back in time than any other optical telescope, providing views of galaxies at an early stage of the universe and evidence for the age of the universe. In 2013, NASA plans to launch the James Webb Space Telescope, with a mirror 6.5 m in diameter. It will be placed 1.5 million kilometers from Earth on the side away from the Sun.

Application: Radio Telescopes

The EM radiation traveling to Earth from celestial bodies is not limited to the visible part of the spectrum. Radio telescopes detect radio waves from space. The radio telescope at Arecibo, Puerto Rico (Fig. 24.21), is the most sensitive radio telescope in the world. Arecibo takes only a few minutes to gather information from a radio source that would require several hours of observation with a smaller radio telescope.

A home satellite dish is a small version of a radio telescope. It is directed toward a satellite and forms a real image of the microwaves beamed down to Earth from the satellite. When the dish is properly aimed to receive the signal sent by the satellite from a TV station, the microwaves of that station are focused on the antenna of the receiver.

24.7 ABERRATIONS OF LENSES AND MIRRORS

Real lenses and mirrors deviate from the behavior of an ideal lens or mirror, a phenomenon known as an **aberration**. There are many kinds of aberrations; in this section we consider only two of them. In Chapter 25, we consider in more detail the limits on the resolution of optical instruments due to diffraction.

Figure 24.21 The radio telescope at Arecibo, Puerto Rico, occupies nearly twenty acres of a remote hilltop region. The bowl of the telescope, 305 m (0.19 mi) in diameter and 51 m (167 ft) deep, is made from metallic mesh panels instead of solid metal; it reflects just as well as a solid metal surface because the holes are much smaller than the wavelengths of the radio waves. A detector is suspended in midair at the focal point, 137 m above the bowl.

Spherical Aberration

The derivations of the lens and mirror equations assume that light rays are paraxial—that they are nearly parallel to the principal axis and not too far away from the axis. That assumption enabled us to use small-angle approximations. Rays for which the angle of incidence on the lens or mirror is *not* small do not all converge to the same focal point (or appear to diverge from the same focal point). Then rays from a point on the object do not correspond to a single image point; the image is blurred due to **spherical aberration**. Spherical aberration can be minimized by placing a diaphragm before the lens or mirror so that only rays traveling rather close to the principal axis can reach the lens and pass through it. The image formed is in better focus (sharper) but is less bright.

For mirrors, spherical aberration can be avoided by using a *parabolic* mirror instead of a spherical mirror. A parabolic mirror focuses parallel rays to a point even if they are not paraxial. Large astronomical reflecting telescopes use parabolic mirrors. Since light rays are reversible, if a point light source is placed at the focal point of a parabolic mirror, the reflected rays form a parallel beam. Searchlights and automobile headlights use parabolic reflectors to send out fairly parallel rays in a well-defined beam of light.

Chromatic Aberration

Another aberration of lenses (but not of mirrors) is due to dispersion. When light composed of several wavelengths passes through a lens, the various wavelengths are bent by differing amounts due to dispersion; this lens defect is called **chromatic aberration**. One way to correct for chromatic aberration is to use a combination of lenses consisting of different materials so that the aberrations from one lens cancel those from another.

PHYSICS AT HOME

Look at a TV or computer monitor with your unaided eye (or through your usual eyeglasses). Then make a fist with one hand, leaving an opening like a small tunnel to peer through with one eye. Look at the same screen through your hand; you will see a much sharper image. Slowly expand the opening made by your fist and watch the image become fuzzy. People with poor or changing eyesight sometimes squint in order to see more clearly.

Master the Concepts

- In a series of lenses, the image formed by one lens becomes the object for the next lens.
- If one lens produces a real image that would have formed *past* a second lens—so that the rays are converging to a point past the second lens—that image becomes a *virtual object* for the second lens. In the thin lens equation, p is *negative* for a virtual object.
- When the image formed by one lens serves as the object for a second lens a distance s away, the object distance p_2 for the second lens is

$$p_2 = s - q_1 \qquad (24\text{-}1)$$

- The total transverse magnification of an image formed by two or more lenses is the product of the magnifications due to the individual lenses.

$$m_{\text{total}} = m_1 \times m_2 \times \cdots \times m_N \qquad (24\text{-}2)$$

- A typical camera has a single converging lens. To focus on an object, the distance between the lens and the film (or CCD array) is adjusted so that a real image is formed on the film.
- The aperture size and the exposure time must be chosen to allow just enough light to expose the film. The *depth of field* is the range of distances from the plane of sharp focus for which the lens forms an acceptably clear image on the film. Greater depth of field is possible with a smaller aperture.
- In the human eye, the cornea and the lens refract light rays to form a real image on the photoreceptor cells in the retina. For most purposes, we can consider the cornea and the lens to act like a single lens with an adjustable focal length. The adjustable shape of the lens allows for accommodation for various object distances, while still forming an image at the fixed image distance determined by the separation of lens and retina. The nearest and farthest object distances that the eye can accommodate are called the near point and far point. A young adult with good vision has a near point at 25 cm or less and a far point at infinity.
- The refractive power of a lens is the reciprocal of the focal length:

$$P = \frac{1}{f} \qquad (24\text{-}3)$$

Refractive power is measured in diopters (1 D = 1 m^{-1}).
- A myopic (nearsighted) eye has a far point less than infinity; for objects past the far point, it forms an image before the retina. A diverging corrective lens (with negative refractive power) can compensate for nearsightedness by bending light rays outward.
- A hyperopic (farsighted) eye has too large a near point; the refractive power of the eye is too small. For objects closer than the near point, the eye forms an image past the retina. A converging lens can correct for hyperopia by bending the rays inward so they converge sooner.
- As a person ages, the lens of the eye becomes less flexible and the eye's ability to accommodate decreases, a phenomenon known as presbyopia.
- *Angular* magnification is the ratio of the angular size using the instrument to the angular size as viewed by the unaided eye.

$$M = \frac{\theta_{\text{aided}}}{\theta_{\text{unaided}}} \qquad (24\text{-}5)$$

- The simple magnifier is a converging lens placed so that the object distance is less than the focal length. The virtual image formed is enlarged and upright. If the image is formed at infinity for ease of viewing, the angular magnification M is

$$M = \frac{N}{f} \qquad (24\text{-}6)$$

where N, the near point, is usually taken to be 25 cm.
- The compound microscope consists of two converging lenses. A small object to be viewed is placed *just beyond* the focal point of the objective, which forms an enlarged real image. The eyepiece (ocular) acts as a simple magnifier to view the image formed by the objective. If the final image is at infinity, the angular magnification due to the microscope is

$$M_{\text{total}} = m_o M_e = -\frac{L}{f_o} \times \frac{N}{f_e} \qquad (24\text{-}8)$$

where N is the near point (usually 25 cm) and L (the tube length) is the distance between the focal points of the two lenses.

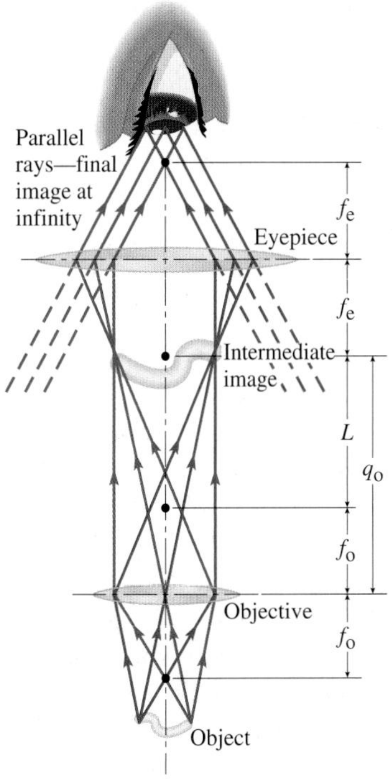

continued on next page

Master the Concepts continued

- An astronomical refracting telescope uses two converging lenses. As in the microscope, the objective forms a real image and the eyepiece functions as a magnifier for viewing the real image. The angular magnification is

$$M = -\frac{f_o}{f_e} \qquad (24\text{-}10)$$

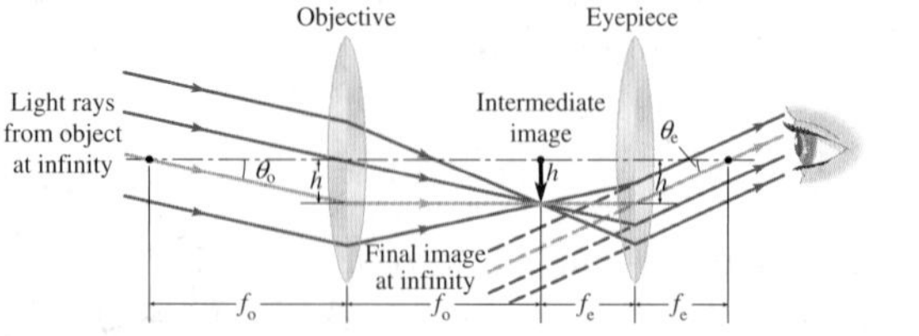

- In a reflecting telescope, a concave mirror takes the place of the objective lens.
- Spherical aberration occurs because rays that are not paraxial are brought to a focus at a different spot than are paraxial rays.
- Chromatic aberration is caused by dispersion in the lens.

Conceptual Questions

1. Why must a camera or a slide projector use a converging lens? Why must the objective of a microscope or telescope be a converging lens (or a converging mirror)? Why can the eyepiece of a telescope be either converging or diverging?
2. A magnifying glass can be held over a piece of white paper and its position adjusted until the image of an overhead light is formed on the paper. Explain.
3. If a piece of white cardboard is placed at the location of a virtual object, what (if anything) would be seen on the cardboard?
4. Why is a refracting telescope with a large angular magnification longer than one with a smaller magnification?
5. Why are astronomical observatories often located on mountaintops?
6. Why do some telescopes produce an inverted image?
7. Why is the receiving antenna of a satellite dish placed at a set distance from the dish?
8. Two magnifying glasses are labeled with their angular magnifications. Glass *A* has a magnification of "2×" ($M = 2$) and glass *B* has a magnification of 4×. Which has the longer focal length? Explain.
9. What causes chromatic aberration? What can be done to compensate for chromatic aberration?
10. For human eyes, about 70% of the refraction occurs at the cornea; less than 25% occurs at the two surfaces of the lens. Why? [*Hint:* Consider the indices of refraction.] Is the same thing true for fish eyes?
11. If rays from points on an object are converging as they enter a lens, is the object real or virtual?
12. What are some of the advantages of using mirrors rather than lenses for astronomical telescopes?
13. When snorkeling, you wear goggles in order to see clearly. Why is your vision blurry without the goggles? A nearsighted person notices that he is able to see nearby objects *more* clearly when he is underwater (without goggles or corrective lenses) than in air (without corrective lenses). Why might this be true?
14. When the muscles of the eye remain tensed for a significant period of time, eyestrain results. How much is this a concern for a person using (a) a microscope, (b) a telescope, and (c) a simple magnifier?
15. Both a microscope and a telescope can be constructed from two converging lenses. What are the differences? Why can't a telescope be used as a microscope? Why can't a microscope be used as a telescope?
16. In her bag, a photographer is carrying three exchangeable camera lenses with focal lengths of 400.0 mm, 50.0 mm, and 28.0 mm. Which lens should she use for (a) wide angle shots (a cathedral, taken from the square in front), (b) everyday use (children at play), and (c) telephoto work (lions in Africa taken from across a river)?
17. The figure shows a schematic diagram of a defective eye. What is this defect called?

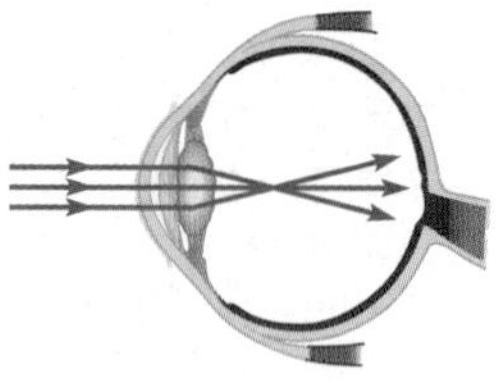

18. Draw a simple eye diagram labeling the cornea, the lens, the iris, the retina, and the aqueous and vitreous fluids.

Multiple-Choice Questions

1. The compound microscope is made from two lenses. Which statement is true concerning the operation of the compound microscope?
   (a) Both lenses form real images.
   (b) Both lenses form virtual images.
   (c) The lens closest to the object forms a virtual image; the other lens forms a real image.
   (d) The lens closest to the object forms a real image; the other lens forms a virtual image.

2. Which of these statements best explains why a telescope enables us to see details of a distant object such as the Moon or a planet more clearly?
   (a) The image formed by the telescope is larger than the object.
   (b) The image formed by the telescope subtends a larger angle at the eye than the object does.
   (c) The telescope can also collect radio waves that sharpen the visual image.
3. Siu-Ling has a far point of 25 cm. Which statement here is true?
   (a) She may have normal vision.
   (b) She is myopic and requires diverging lenses to correct her vision.
   (c) She is myopic and requires converging lenses to correct her vision.
   (d) She is hyperopic and requires diverging lenses to correct her vision.
   (e) She is hyperopic and requires converging lenses to correct her vision.
4. The figure shows a schematic diagram of a defective eye and some lenses. Which of the lenses shown can correct for this defect?

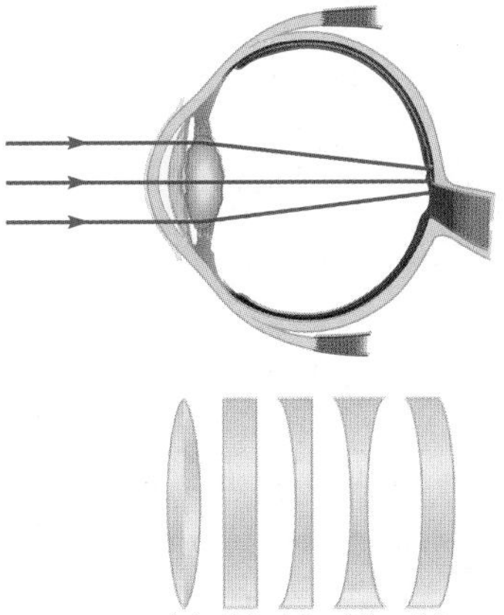

5. A nearsighted person wears corrective lenses. One of the focal points of the corrective lenses should be
   (a) at the cornea. (b) at the retina.
   (c) at infinity. (d) past the retina.
   (e) at the near point. (f) at the far point.
6. An astronomical telescope has an angular magnification of 10. The length of the barrel is 33 cm. What are the focal lengths of the objective and the eyepiece, in that order respectively, from the choices listed?
   (a) 3 cm, 30 cm (b) 30 cm, 3 cm
   (c) 20 cm, 13 cm (d) 0.3 m, 3 m
7. A compound microscope has three possible objective lenses (focal lengths f_o) and two eyepiece lenses (focal lengths f_e). For maximum angular magnification, the objective and eyepiece should be chosen such that
   (a) f_o and f_e are both the largest available.
   (b) f_o and f_e are both the smallest available.
   (c) f_o is the largest available; f_e is the smallest available.
   (d) f_e is the largest available; f_o is the smallest available.
   (e) f_e and f_o are nearly the same.
8. What causes chromatic aberration?
   (a) Light is an electromagnetic wave and has intrinsic diffraction properties.
   (b) Different wavelengths of light give different angles of refraction at the lens-air interface.
   (c) The coefficient of reflection is different for light of different wavelengths.
   (d) The outer edges of the lens produce a focus at a different point from that formed by the central portion of the lens.
   (e) The absorption of light in the glass varies with wavelength.
9. What causes spherical aberration?
   (a) Light is an electromagnetic wave and has intrinsic diffraction properties.
   (b) Different wavelengths of light give different angles of refraction at the lens-air interface.
   (c) The lens surface is not perfectly smooth.
   (d) The outer edges of the lens produce a focus at a different point from that formed by the central portion of the lens.
10. Reducing the aperture on a camera
   (a) reduces the depth of field and requires a longer exposure time.
   (b) reduces the depth of field and requires a shorter exposure time.
   (c) increases the depth of field and requires a longer exposure time.
   (d) increases the depth of field and requires a shorter exposure time.
   (e) does not change the depth of field and requires a longer exposure time.
   (f) does not change the depth of field and requires a shorter exposure time.

Problems

- Combination conceptual/quantitative problem
- Biological or medical application
- Challenging problem
- Blue # Detailed solution in the Student Solutions Manual
- 1 2 Problems paired by concept
- Text website interactive or tutorial

24.1 Lenses in Combination

1. An object is placed 12.0 cm in front of a lens of focal length 5.0 cm. Another lens of focal length 4.0 cm is placed 2.0 cm past the first lens. (a) Where is the final image? Is it real or virtual? (b) What is the overall magnification? (interactive: virtual optics lab)

2. A converging lens and a diverging lens, separated by a distance of 30.0 cm, are used in combination. The converging lens has a focal length of 15.0 cm. The diverging lens is of unknown focal length. An object is placed 20.0 cm in front of the converging lens; the final image is virtual and is formed 12.0 cm *before* the diverging lens. What is the focal length of the diverging lens?

3. Two converging lenses are placed 88.0 cm apart. An object is placed 1.100 m to the left of the first lens, which has a focal length of 25.0 cm. The final image is located 15.0 cm to the right of the second lens. (a) What is the focal length of the second lens? (b) What is the total magnification?

4. A converging lens with a focal length of 15.0 cm and a diverging lens are placed 25.0 cm apart, with the converging lens on the left. A 2.00-cm-high object is placed 22.0 cm to the left of the converging lens. The final image is 34.0 cm to the left of the converging lens. (a) What is the focal length of the diverging lens? (b) What is the height of the final image? (c) Is the final image upright or inverted?

5. An object is located 16.0 cm in front of a converging lens with focal length 12.0 cm. To the right of the converging lens, separated by a distance of 20.0 cm, is a diverging lens of focal length –10.0 cm. Find the location of the final image by ray tracing and verify using the lens equations.

6. An object is located 10.0 cm in front of a converging lens with focal length 12.0 cm. To the right of the converging lens is a second converging lens, 30.0 cm from the first lens, of focal length 10.0 cm. Find the location of the final image by ray tracing and verify by using the lens equations.

7. Verify the locations and sizes of the images formed by the two lenses in Fig. 24.1b using the lens equation and the following data: $f_1 = +4.00$ cm, $f_2 = -2.00$ cm, $s =$ 8.00 cm (where s is the distance between the lenses), $p_1 = +6.00$ cm, and $h = 2.00$ mm. (Note that the vertical scale is different from the horizontal scale.)

✦ 8. Show that if two thin lenses are close together (s, the distance between the lenses, is negligibly small), the two lenses can be replaced by a single equivalent lens with focal length f_{eq}. Find the value of f_{eq} in terms of f_1 and f_2.

24.2 Cameras

9. You would like to project an upright image at a position 32.0 cm to the right of an object. You have a converging lens with focal length 3.70 cm located 6.00 cm to the right of the object. By placing a second lens at 24.65 cm to the right of the object, you obtain an image in the proper location. (a) What is the focal length of the second lens? (b) Is this lens converging or diverging? (c) What is the total magnification? (d) If the object is 12.0 cm high, what is the image height?

10. You plan to project an inverted image 30.0 cm to the right of an object. You have a diverging lens with focal length –4.00 cm located 6.00 cm to the right of the object. Once you put a second lens at 18.0 cm to the right of the object, you obtain an image in the proper location. (a) What is the focal length of the second lens? (b) Is this lens converging or diverging? (c) What is the total magnification? (d) If the object is 12.0 cm high, what is the image height?

11. A camera uses a 200.0-mm focal length telephoto lens to take pictures from a distance of infinity to as close as 2.0 m. What are the minimum and maximum distances from the lens to the film?

12. Kim says that she was less than 10 ft away from the president when she took a picture of him with her 50-mm focal length camera lens. The picture shows the upper half of the president's body (or 3.0 ft of his total height). On the negative of the film, this part of his body is 18 mm high. How close was Kim to the president when she took the picture?

13. A statue is 6.6 m from the opening of a pinhole camera, and the screen is 2.8 m from the pinhole. (a) Is the image erect or inverted? (b) What is the magnification of the image? (c) To get a brighter image, we enlarge the pinhole to let more light through, but then the image looks blurry. Why? (d) To admit more light and still have a sharp image, we replace the pinhole with a lens. Should it be a converging or diverging lens? Why? (e) What should the focal length of the lens be?

14. Esperanza uses a 35-mm camera with a standard lens of focal length 50.0 mm to take a photo of her son Carlos, who is 1.2 m tall and standing 3.0 m away. (a) What must be the distance between the lens and the film to get a properly focused picture? (b) What is the magnification of the image? (c) What is the height of the image of Carlos on the film?

15. A person on a safari wants to take a photograph of a hippopotamus from a distance of 75.0 m. The animal is 4.00 m long and its image is to be 1.20 cm long on the film. (a) What focal length lens should be used? (b) What would be the size of the image if a lens of 50.0-mm focal length were used? (c) How close to the hippo would the person have to be to capture a 1.20-cm-long image using a 50.0-mm lens?

16. Jim plans to take a picture of McGraw Tower with a 35-mm camera that has a 50.0-mm focal length lens. A roll of 35-mm film is 35 mm wide; each frame is 24 mm by 36 mm. The tower has a height of 52 m and Jim wants a detailed close-up picture. How close to the tower should Jim be to capture the largest possible image of the entire tower on his film?

A strip of 35-mm film
(Problems 16, 18, 19, 58, and 67)

17. A photographer wishes to take a photograph of the Eiffel Tower (300 m tall) from across the Seine River, a distance of 300 m from the tower. What focal length lens should she use to get an image that is 20 mm high on the film?

18. If a slide of width 36 mm (see the figure with Problem 16) is to be projected onto a screen of 1.50 m width located 12.0 m from the projector, what focal length lens is required to fill the width of the screen?

19. A slide projector has a lens of focal length 12 cm. Each slide is 24 mm by 36 mm (see the figure with Problem 16). The projector is used in a room where the screen is 5.0 m from the projector. How large must the screen be?

20. A converging lens with focal length 3.00 cm is placed 4.00 cm to the right of an object. A diverging lens with focal length −5.00 cm is placed 17.0 cm to the right of the converging lens. (a) At what location(s), if any, can you place a screen in order to display an image? (b) Repeat part (a) for the case where the lenses are separated by 10.0 cm.

✦21. A converging lens with a focal length of 3.00 cm is placed 24.00 cm to the right of a concave mirror with a focal length of 4.00 cm. An object is placed between the mirror and the lens, 6.00 cm to the right of the mirror and 18.00 cm to the left of the lens. Name three places where you could find an image of this object. For each image tell whether it is inverted or upright and give the total magnification.

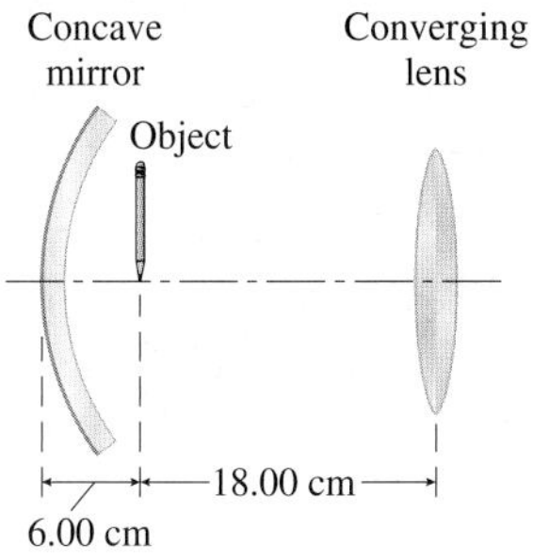

24.3 The Eye

Unless the problem states otherwise, assume that the distance from the cornea-lens system to the retina is 2.0 cm and the normal near point is 25 cm.

22. If the distance from the lens system (cornea + lens) to the retina is 2.00 cm, show that the focal length of the lens system must vary between 1.85 cm and 2.00 cm to see objects from 25.0 cm to infinity.

23. Suppose that the lens system (cornea + lens) in a particular eye has a focal length that can vary between 1.85 cm and 2.00 cm, but the distance from the lens system to the retina is only 1.90 cm. (a) Is this eye nearsighted or farsighted? Explain. (b) What range of distances can the eye see clearly without corrective lenses?

24. If Michaela needs to wear reading glasses with refractive power of +3.0 D, what is her uncorrected near point? Neglect the distance between the glasses and the eye. (tutorial: reading glasses)

25. The uncorrected far point of Colin's eye is 2.0 m. What refractive power contact lens enables him to clearly distinguish objects at large distances?

26. The distance from the lens system (cornea + lens) of a particular eye to the retina is 1.75 cm. What is the focal length of the lens system when the eye produces a clear image of an object 25.0 cm away?

✦27. A nearsighted man cannot clearly see objects more than 2.0 m away. The distance from the lens of the eye to the retina is 2.0 cm, and the eye's power of accommodation is 4.0 D (the focal length of the cornea-lens system increases by a maximum of 4.0 D over its relaxed focal length when accommodating for nearby objects). (a) As an amateur optometrist, what corrective eyeglass lenses would you prescribe to enable him to clearly see distant objects? Assume the corrective lenses are 2.0 cm from the eyes. (b) Find the nearest object he can see clearly with and without his glasses.

28. Anne is farsighted; the nearest object she can see clearly without corrective lenses is 2.0 m away. It is 1.8 cm from the lens of her eye to the retina. (a) Sketch a ray diagram to show (qualitatively) what happens when she tries to look at something closer than 2.0 m without corrective lenses. (b) What should the focal length of her contact lenses be so that she can see clearly objects as close as 20.0 cm from her eye?

24.4 Angular Magnification and the Simple Magnifier

29. Thomas wants to use his 5.5-D reading glasses as a simple magnifier. What is the angular magnification of this lens when Thomas's eye is relaxed?

30. (a) What is the focal length of a magnifying glass that gives an angular magnification of 8.0 when the image is at infinity? (b) How far must the object be from the lens? Assume the lens is held close to the eye.

31. Keesha is looking at a beetle with a magnifying glass. She wants to see an upright, enlarged image at a distance of 25 cm. The focal length of the magnifying glass is +5.0 cm. Assume that Keesha's eye is close to the magnifying glass. (a) What should be the distance between the magnifying glass and the beetle? (b) What is the angular magnification? (tutorial: magnifying glass II)

32. Callum is examining a square stamp of side 3.00 cm with a magnifying glass of refractive power +40.0 D. The magnifier forms an image of the stamp at a distance of 25.0 cm. Assume that Callum's eye is close to the magnifying glass. (a) What is the distance between the stamp and the magnifier? (b) What is the angular magnification? (c) How large is the image formed by the magnifier?

33. A simple magnifying glass can focus sunlight enough to heat up paper or dry grass and start a fire. A magnifying glass with a diameter of 4.0 cm has a focal length of 6.0 cm. (a) Using the information found on the inside back cover of the book, estimate the size of the image of the Sun when the magnifying glass focuses the image to its smallest size. (b) If the intensity of the Sun falling on

the magnifying glass is 0.85 kW/m^2, what is the intensity of the image of the Sun?

34. An insect that is 5.00 mm long is placed 10.0 cm from a converging lens with a focal length of 12.0 cm. (a) What is the position of the image? (b) What is the size of the image? (c) Is the image upright or inverted? (d) Is the image real or virtual? (e) What is the angular magnification if the lens is close to the eye?

35. A simple magnifier gives the *maximum* angular magnification when it forms a virtual image at the near point of the eye instead of at infinity. For simplicity, assume that the magnifier is right up against the eye, so that distances from the magnifier are approximately the same as distances from the eye. (a) For a magnifier with focal length f, find the object distance p such that the image is formed at the near point, a distance N from the lens. (b) Show that the angular size of this image as seen by the eye is

$$\theta = \frac{h(N+f)}{Nf}$$

where h is the height of the object. [*Hint:* Refer to Fig. 24.15.] (c) Now find the angular magnification and compare it to the angular magnification when the virtual image is at infinity.

24.5 Compound Microscopes

36. The figure shows a schematic diagram of a microscope. For the object and image locations shown, which of the points (A, B, C, or D) represents a focal point of the eyepiece? Draw a ray diagram.

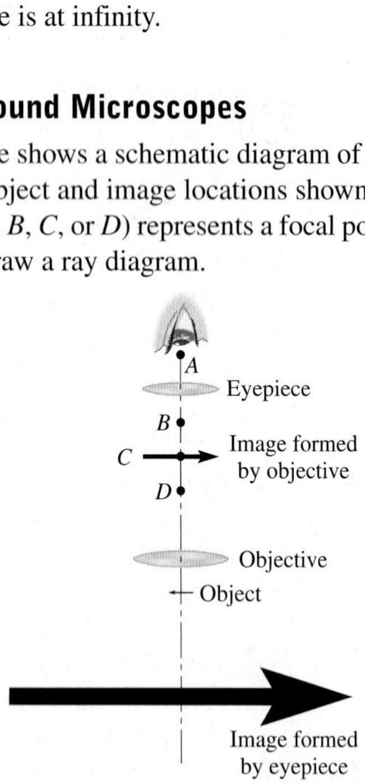

37. The eyepiece of a microscope has a focal length of 1.25 cm and the objective lens focal length is 1.44 cm. (a) If the tube length is 18.0 cm, what is the angular magnification of the microscope? (b) What objective focal length would be required to double this magnification?

38. Jordan is building a compound microscope using an eyepiece with a focal length of 7.50 cm and an objective with a focal length of 1.500 cm. He will place the specimen a distance of 1.600 cm from the objective. (a) How far apart should Jordan place the lenses? (b) What will be the angular magnification of this microscope?

39. The wing of an insect is 1.0 mm long. When viewed through a microscope, the image is 1.0 m long and is located 5.0 m away. Determine the angular magnification.

40. A microscope has an eyepiece that gives an angular magnification of 5.00 for a final image at infinity and an objective lens of focal length 15.0 mm. The tube length of the microscope is 16.0 cm. (a) What is the transverse magnification due to the objective lens alone? (b) What is the angular magnification due to the microscope? (c) How far from the objective should the object be placed?

41. Repeat Problem 40(c) using a different eyepiece that gives an angular magnification of 5.00 for a final image at the viewer's near point (25.0 cm) instead of at infinity.

42. A microscope has an objective lens of focal length 5.00 mm. The objective forms an image 16.5 cm from the lens. The focal length of the eyepiece is 2.80 cm. (a) What is the distance between the lenses? (b) What is the angular magnification? The near point is 25.0 cm. (c) How far from the objective should the object be placed?

43. Repeat Problem 42 if the eyepiece location is adjusted slightly so that the final image is at the viewer's near point (25.0 cm) instead of at infinity.

44. Use the thin-lens equation to show that the transverse magnification due to the objective of a microscope is $m_o = -L/f_o$. [*Hints:* The object is *near* the focal point of the objective; do not assume it is *at* the focal point. Eliminate p_o to find the magnification in terms of q_o and f_o. How is L related to q_o and f_o?]

24.6 Telescopes

45. (a) If you were stranded on an island with only a pair of 3.5-D reading glasses, could you make a telescope? If so, what would be the length of the telescope and what would be the best possible angular magnification? (b) Answer the same questions if you also had a pair of 1.3-D reading glasses.

46. A telescope mirror has a radius of curvature of 10.0 m. It is used to take a picture of the Moon. What is the diameter of the image of the Moon produced by this mirror? (See the inside back cover for necessary information.)

47. (a) What is the angular size of the Moon as viewed from Earth's surface? See the inside back cover for necessary information. (b) The objective and eyepiece of a refracting telescope have focal lengths 80 cm and 2.0 cm, respectively. What is the angular size of the Moon as viewed through this telescope?

48. What is the distance between the objective and eyepiece in the Yerkes telescope? (See Example 24.8.)

49. You have a set of converging lenses with focal lengths 1.00 cm, 10.0 cm, 50.0 cm, and 80.0 cm. (a) Which two lenses would you select to make a telescope with the

largest magnifying power? What is the angular magnification of the telescope when viewing a distant object? (b) Which lens is used as objective and which as eyepiece? (c) What should be the distance between the objective and eyepiece?

50. A refracting telescope is 45.0 cm long and the caption states that the telescope magnifies images by a factor of 30.0. Assuming these numbers are for viewing an object an infinite distance away with minimum eyestrain, what is the focal length of each of the two lenses?

51. The objective lens of an astronomical telescope forms an image of a distant object at the focal point of the eyepiece, which has a focal length of 5.0 cm. If the two lenses are 45.0 cm apart, what is the angular magnification?

52. A refracting telescope is used to view the Moon. The focal lengths of the objective and eyepiece are +2.40 m and +16.0 cm, respectively. (a) What should be the distance between the lenses? (b) What is the diameter of the image produced by the objective? (c) What is the angular magnification?

✦53. The eyepiece of a Galilean telescope is a diverging lens. The focal points F_o and F'_e coincide. In one such telescope, the lenses are a distance $d = 32$ cm apart and the focal length of the objective is 36 cm. A rhinoceros is viewed from a large distance. (a) What is the focal length of the eyepiece? (b) At what distance from the eyepiece is the final image? (c) Is the final image formed by the eyepiece real or virtual? Upright or inverted? (d) What is the angular magnification? [*Hint:* The angular magnification is β/α.]

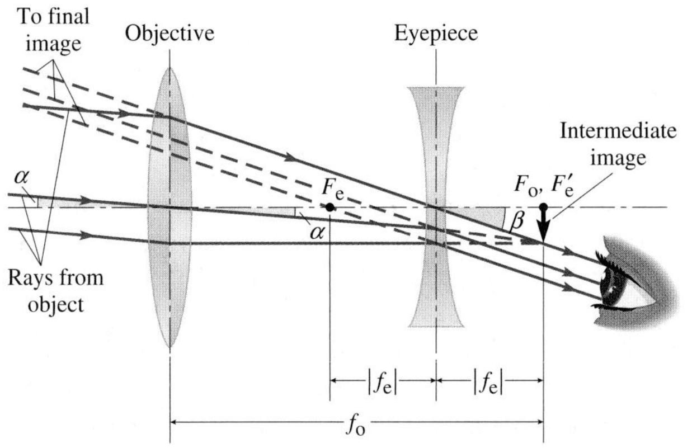

Comprehensive Problems

54. Good lenses used in cameras and other optical devices are actually compound lenses, made of five or more lenses put together to minimize distortions, including chromatic aberration. Suppose a converging lens with a focal length of 4.00 cm is placed right next to a diverging lens with focal length of −20.0 cm. An object is placed 2.50 m to the left of this combination. (a) Where will the image be located? (b) Is the image real or virtual?

55. Two converging lenses, separated by a distance of 50.0 cm, are used in combination. The first lens, located to the left, has a focal length of 15.0 cm. The second lens, located to the right, has a focal length of 12.0 cm. An object, 3.00 cm high, is placed at a distance of 20.0 cm in front of the first lens. (a) Find the intermediate and final image distances relative to the corresponding lenses. (b) What is the total magnification? (c) What is the height of the final image?

56. A camera has a telephoto lens of 240-mm focal length. The lens can be moved in and out a distance of 16 mm from the film plane by rotating the lens barrel. If the lens can focus objects at infinity, what is the closest object distance that can be focused?

◉57. You have two lenses of focal length 25.0 cm (lens 1) and 5.0 cm (lens 2). (a) To build an astronomical telescope that gives an angular magnification of 5.0, how should you use the lenses (which for objective and which for eyepiece)? Explain. (b) How far apart should they be?

58. The Ortiz family is viewing slides from their summer vacation trip to the Grand Canyon. Their slide projector has a projection lens of 10.0-cm focal length and the screen is located 2.5 m from the projector. (a) What is the distance between the slide and the projection lens? (b) What is the magnification of the image? (c) How wide is the image of a slide of width 36 mm on the screen? (See the figure with Problem 16.)

59. A slide projector, using slides of width 5.08 cm, produces an image that is 2.00 m wide on a screen 3.50 m away. What is the focal length of the projector lens?

60. An object is placed 20.0 cm from a converging lens with focal length 15.0 cm (see the figure, not drawn to scale). A concave mirror with focal length 10.0 cm is located 75.0 cm to the right of the lens. (a) Describe the final image—is it real or virtual? Upright or inverted? (b) What is the location of the final image? (c) What is the total transverse magnification?

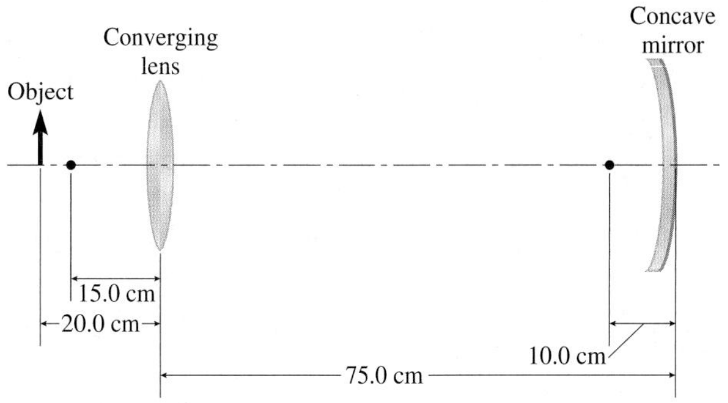

61. Two lenses, of focal lengths 3.0 cm and 30.0 cm, are used to build a small telescope. (a) Which lens should be the objective? (b) What is the angular magnification? (c) How far apart are the two lenses in the telescope?

62. (a) If Harry has a near point of 1.5 m, what focal length contact lenses does he require? (b) What is the power of these lenses in diopters?

63. An astronomical telescope provides an angular magnification of 12. The two converging lenses are 66 cm apart. Find the focal length of each of the lenses.

64. Two lenses, separated by a distance of 21.0 cm, are used in combination. The first lens has a focal length of +30.0 cm; the second has a focal length of −15.0 cm. An object, 2.0 mm long, is placed 1.8 cm before the first lens. (a) What are the intermediate and final image distances relative to the corresponding lenses? (b) What is the total magnification? (c) What is the height of the final image?

65. A camera lens has a fixed focal length of magnitude 50.0 mm. The camera is focused on a 1.0-m-tall child who is standing 3.0 m from the lens. (a) Should the image formed be real or virtual? Why? (b) Is the lens converging or diverging? Why? (c) What is the distance from the lens to the film? (d) How tall is the image on the film? (e) To focus the camera, the lens is moved away from or closer to the film. What is the total distance the lens must be able to move if the camera can take clear pictures of objects at distances anywhere from 1.00 m to infinity?

66. A camera with a 50.0-mm lens can focus on objects located from 1.5 m to an infinite distance away by adjusting the distance between the lens and the film. When the focus is changed from that for a distant mountain range to that for a flower bed at 1.5 m, how far does the lens move with respect to the film?

67. The area occupied by one frame on 35-mm film is 24 mm by 36 mm—see the figure with Problem 16. The focal length of the camera lens is 50.0 mm. A picture is taken of a person 182 cm tall. What is the minimum distance from the camera for the person to stand so that the image fits on the film? Give two answers; one for each orientation of the camera.

68. A dissecting microscope is designed to have a large distance between the object and the objective lens. Suppose the focal length of the objective of a dissecting microscope is 5.0 cm, the focal length of the eyepiece is 4.0 cm, and the distance between the lenses is 32.0 cm. (a) What is the distance between the object and the objective lens? (b) What is the angular magnification?

69. A cub scout makes a simple microscope by placing two converging lenses of +18 D at opposite ends of a 28-cm-long tube. (a) What is the tube length of the microscope? (b) What is the angular magnification? (c) How far should an object be placed from the objective lens?

70. A convex lens of power +12 D is used as a magnifier to examine a wildflower. What is the angular magnification if the final image is at (a) infinity or (b) the near point of 25 cm?

71. A refracting telescope has an objective lens with a focal length of 2.20 m and an eyepiece with a focal length of 1.5 cm. If you look through this telescope the wrong way, that is, with your eye placed at the objective lens, by what factor is the angular size of an observed object reduced?

72. Suppose the distance from the lens system of the eye (cornea + lens) to the retina is 18 mm. (a) What must the power of the lens be when looking at distant objects? (b) What must the power of the lens be when looking at an object 20.0 cm from the eye? (c) Suppose that the eye is farsighted; the person cannot see clearly objects that are closer than 1.0 m. Find the power of the contact lens you would prescribe so that objects as close as 20.0 cm can be seen clearly.

73. An object is placed 7.00 cm to the left of a converging lens of focal length 3.00 cm. A convex mirror with a radius of curvature of 4.00 cm is placed 12.00 cm to the right of the lens. (a) Where is the intermediate image formed by the lens? (b) Is the intermediate image real or virtual and is it upright or inverted with respect to the object? (c) What is the magnification of this image? (d) Where is the image formed by the mirror? (e) Is the second image real or virtual and is it upright or inverted with respect to the original object? (f) What is the magnification due to the mirror? (g) What is the total magnification?

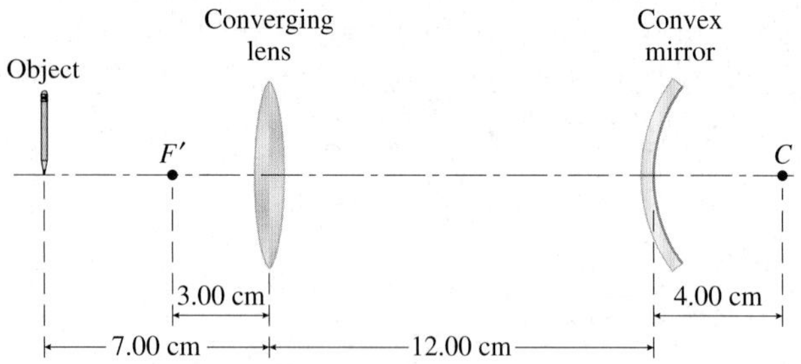

74. Refer to Problem 73. Draw ray diagrams for the two images. Mark the focal points with dots and draw three rays to determine each image. [*Hint:* The second image is very small, so start with a fairly large object.]

75. Veronique is nearsighted; she cannot see clearly anything more than 6.00 m away without her contacts. One day she doesn't wear her contacts; rather, she wears an old pair of glasses prescribed when she could see clearly up to 8.00 m away. Assume the glasses are 2.0 cm from her eyes. What is the greatest distance an object can be placed so that she can see it clearly with these glasses?

76. A man requires reading glasses with +2.0 D power to read a book held 40.0 cm away *with a relaxed eye.* Assume the glasses are 2.0 cm from his eyes. (a) What is his uncorrected far point? (b) What refractive power lenses should he use for distance vision? (c) His uncorrected near point is 1.0 m. What should the refractive powers of the two lenses in his bifocals be to give him clear vision from 25 cm to infinity?

♦77. A microscope has an eyepiece of focal length 2.00 cm and an objective of focal length 3.00 cm. The eyepiece produces a virtual image at the viewer's near point (25.0 cm from the eye). (a) How far from the eyepiece is the image formed by the objective? (b) If the lenses are 20.0 cm apart, what is the distance from the objective lens to the object being viewed? (c) What is the angular magnification?

♦78. An object is located at $x = 0$. At $x = 2.50$ cm is a converging lens with a focal length of 2.00 cm, at $x = 16.5$ cm is an unknown lens, and at $x = 19.8$ cm is another converging lens with focal length 4.00 cm. An upright image is formed at $x = 39.8$ cm. For each lens, the object distance exceeds the focal length. The magnification of the system is 6.84. (a) Is the unknown lens diverging or converging? (b) What is the focal length of the unknown lens? (c) Draw a ray diagram to confirm your answer.

Answers to Practice Problems

24.1 The object can be either real or virtual. If the real image forms before the second lens, it becomes a real object; if the second lens interrupts the light rays before they form the real image, it becomes a virtual object.

24.2 $q_1 = +16.7$ cm; $q_2 = 4.3$ cm; $m = -0.43$; $h' = 1.7$ cm

24.3 51.7 mm

24.4 13.3 cm

24.5 49.9 cm

24.6 (a) 3.08; (b) 2.08

24.7 (a) 18 cm; (b) 1.9 cm

24.8 −780

Answers to Checkpoints

24.1 $p_1 = +6$ cm (real object); $q_1 = +12$ cm (a real image would be formed if lens 2 were not there); $p_2 = -4$ cm (virtual object); $q_2 = -4$ cm (virtual image); $m_1 = -2$ (image formed by lens 1 is twice as tall as the object and inverted); $m_2 = -1$ (object and image are the same size and the image is inverted); $m = m_1 m_2 = +2$ (final image is twice as tall as the original object and is upright)

24.3 If your friend is nearsighted, the scheme won't work. Diverging lenses are used to correct for nearsightedness. A diverging lens can't be used to start a fire because it can't make the light rays converge onto a small spot. If your friend is farsighted, the scheme could work. Converging lenses are used to correct for farsightedness.

24.6 In a microscope, a small object is placed just beyond the focal point of the objective lens. The *angular size* of the object is larger when it is closer to the objective, so we want a small focal length. In a telescope, the object is far away so its angular size is fixed. An objective with a larger focal length produces a larger real image of the distant object.

CHAPTER 25

Interference and Diffraction

When we look at plants and animals, most of the colors we see—brown eyes, green leaves, yellow sunflowers—are due to the selective absorption of light by pigments. In the leaves and stems of green plants, chlorophyll is the chief pigment that absorbs light with some wavelengths and reflects light with other wavelengths that we perceive as green.

In some animals, color is produced in a different way. The shimmering, intense blue color of the wing of many species of the *Morpho* butterfly of Central and South America makes colors produced by pigments look flat. When the wing or the viewer moves, the color of the wing changes slightly, causing the shimmering quality we call *iridescence*. Iridescent colors are found in the wings or feathers of the Oregon swallowtail butterflies, ruby-throated hummingbirds, and many other species of butterflies and birds. Iridescent colors also appear in some beetles, in the scales of fish, and in the skins of snakes. How are these iridescent colors produced? (See p. 934 for the answer.)

Concepts & Skills to Review

- phase (Section 11.5)
- principle of superposition (Section 11.7)
- interference and diffraction (Section 11.9)
- wavefronts, rays, and Huygens's principle (Section 23.1)
- reflection and refraction (Sections 23.2 and 23.3)
- electromagnetic spectrum (Section 22.3)
- intensity (Section 22.6)

25.1 CONSTRUCTIVE AND DESTRUCTIVE INTERFERENCE

Chapters 23 and 24 dealt with topics in *geometric optics*: reflection, refraction, and image formation. For the most part, we were able to trace light rays propagating in straight-line paths; the rays changed direction only due to reflection or refraction at boundaries. Geometric optics is a useful approximation when objects and apertures are large relative to the wavelength of the light.

The present chapter is concerned with *physical optics*, in which the wave nature of light is made manifest. In physical optics we consider what happens when light propagates around obstacles or through apertures that are *not* large compared with the wavelength. In such situations we encounter interference and diffraction.

The distinction between interference and diffraction is not always clear-cut. Generally, **interference** refers to situations where waves from a small number of sources travel different paths and arrive at an observer with different phases. **Diffraction** is the spreading of waves after they travel around obstacles or through apertures. According to Huygens's construction, *every point* on a wavefront is a source of wavelets. The superposition of light from all these point sources must account for the phase differences due to the different paths traveled. Thus, instead of a small number of sources, in diffraction we have the superposition of waves from an *infinite* number of sources.

Any kind of wave can exhibit interference and diffraction because they are just manifestations of the principle of superposition, which says that the net wave disturbance at any point due to two or more waves is the sum of the disturbances due to each wave individually. Superposition is not a new principle for light. We used it earlier in our study of sound and other mechanical waves (see Sections 11.7 and 11.9). We also used it to find the electric and magnetic fields due to more than one source; the electric and magnetic fields are the vector sums of the fields due to each source individually (see Sections 16.4 and 19.8). Now we apply the principle of superposition to EM waves.

CONNECTION:

Interference and diffraction phenomena are manifestations of the principle of superposition.

Coherent and Incoherent Sources

Why do we not commonly see interference effects with visible light? With light from a source such as the Sun, an incandescent bulb, or a fluorescent bulb, we do not see regions of constructive and destructive interference; rather, the *intensity* at any point is the sum of the intensities due to the individual waves. Light from any one of these sources is, at the atomic level, emitted by a vast number of *independent* sources. Waves from independent sources are **incoherent**; they do not maintain a fixed phase relationship with each other. We cannot accurately predict the phase (e.g., whether the wave is at a maximum or at a zero) at one point given the phase at another point. Incoherent waves have *rapidly fluctuating* phase relationships. The result is an averaging out of interference effects, so that the total intensity (or power per unit area) is just the sum of the intensities of the individual waves.

In the superposition of *incoherent* waves, the intensity is the sum of the intensities due to each wave individually.

Only the superposition of **coherent** waves produces interference. Coherent waves must be locked in with a fixed phase relationship. *Coherent* and *incoherent* waves are idealized extremes; all real waves fall somewhere between the extremes. The light emitted by a laser can be highly coherent—two points in the beam can be coherent even if separated by as much as several kilometers. Light from a distant point source (e.g., a star other than the Sun) has some degree of coherence.

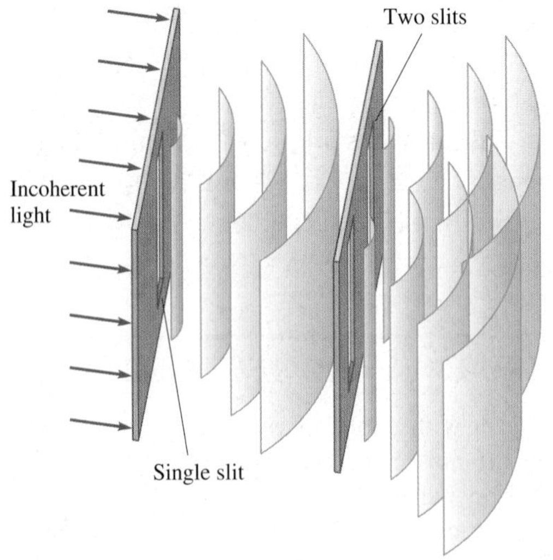

Figure 25.1 Young's technique for illuminating two slits with coherent light. The single slit on the left serves as a source of coherent light.

The British physicist Thomas Young (1773–1829) performed the first visible-light interference experiments using a clever technique to obtain two coherent light sources from a single source (Fig. 25.1). When a single narrow slit is illuminated, the light wave that passes through the slit diffracts or spreads out. The single slit acts as a single coherent source to illuminate two other slits. These two other slits then act as sources of coherent light for interference.

Interference of Two Coherent Waves

If two waves are in step with one another, with the crest of one falling at the same point as the crest of the other, they are said to be *in phase*. The phase difference between the two waves that are in phase is an integral multiple of 2π rad: 0, $\pm 2\pi$ rad, $\pm 4\pi$ rad, and so forth, which we can express as $2m\pi$ rad, where m is any integer. The superposition of two waves that are in phase has an amplitude equal to the sum of the amplitudes of the two waves. For instance, in Fig. 25.2 two sinusoidal waves are in phase. The amplitudes of the two are $2A$ and $5A$. When the two waves are added together, the resulting wave has an amplitude of $2A + 5A = 7A$. The superposition of two waves that are *in phase* is called **constructive interference**.

CONNECTION:

In the superposition of coherent waves, the intensities *cannot* be added together, whether for light or any other kind of wave (see Section 11.9).

For constructive interference, the intensity of the resulting wave (I) is *greater than* the sum of the intensities of the two waves individually ($I_1 + I_2$). Since intensity is proportional to amplitude squared (see Section 22.6), let $I = CA^2$, $I_1 = CA_1^2$, and $I_2 = CA_2^2$, where C is a constant. (For light and other EM waves, A_1, A_2, and A can represent either electric field amplitudes or magnetic field amplitudes since they are proportional to one another.) Since $A = A_1 + A_2$,

$$CA^2 = C(A_1 + A_2)^2 = CA_1^2 + CA_2^2 + 2CA_1A_2$$

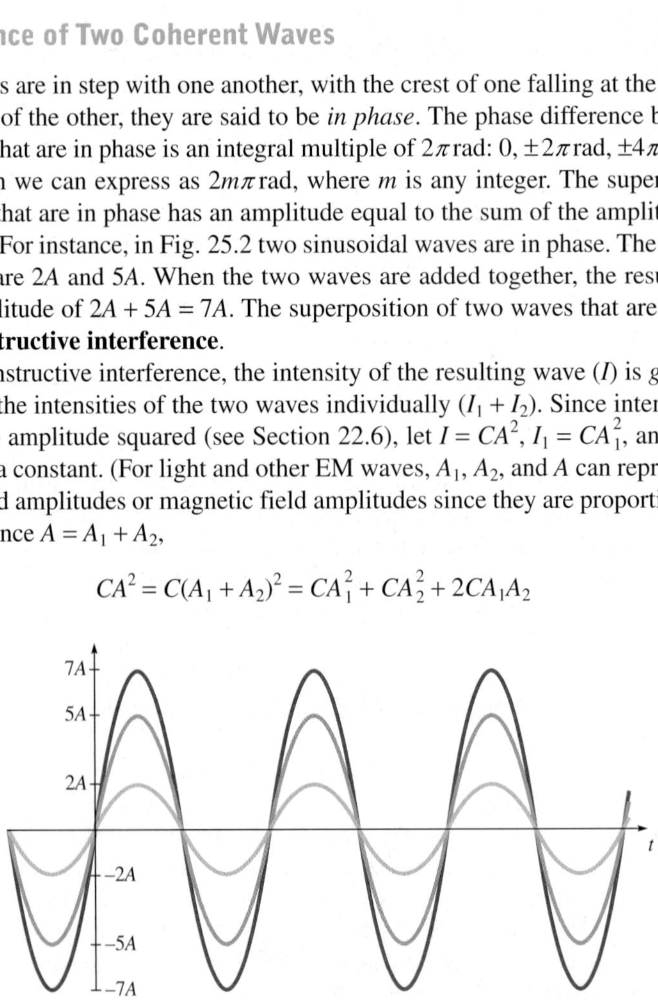

Figure 25.2 Two coherent waves (green and blue) with amplitudes $2A$ and $5A$. Since they are in phase, they interfere constructively. The superposition of the two (red) has an amplitude of $7A$. Note that shifting either of the waves a whole number of cycles to the right or left would not change the superposition of the two.

Therefore,

$$I = I_1 + I_2 + 2\sqrt{I_1 I_2}$$

Since intensity is power per unit area, where does the extra energy come from? Don't worry; energy is still conserved. If in some places $I > I_1 + I_2$, then in other places $I < I_1 + I_2$. To summarize:

Constructive interference of two waves:

Phase difference $\Delta\phi = 2m\pi$ rad $(m = 0, \pm1, \pm2, \ldots)$ (25-1)

Amplitude $A = A_1 + A_2$ (25-2)

Intensity $I = I_1 + I_2 + 2\sqrt{I_1 I_2}$ (25-3)

Two waves that are 180° out of phase are a half cycle apart; where one is at a crest the other is at a trough (Fig. 25.3). The superposition of two such waves is called **destructive interference**. The phase difference for destructive interference is π rad plus any integral multiple of 2π rad. Then $\Delta\phi = (\pi + 2m\pi)$ rad $= (m + \frac{1}{2})2\pi$ rad, where m is any integer. The destructive interference of two waves with amplitudes $2A$ and $5A$ gives a resulting amplitude of $3A$. If the two waves had the same amplitude, there would be complete cancellation—the superposition would have an amplitude of zero. To summarize:

Destructive interference of two waves:

Phase difference $\Delta\phi = (m + \frac{1}{2})2\pi$ rad $(m = 0, \pm1, \pm2, \ldots)$ (25-4)

Amplitude $A = |A_1 - A_2|$ (25-5)

Intensity $I = I_1 + I_2 - 2\sqrt{I_1 I_2}$ (25-6)

CHECKPOINT 25.1

Can the phase difference between two coherent waves be $\pi/3$ rad? If so, is the interference of the waves constructive, destructive, or something in between? Explain.

Phase Difference due to Different Paths

In interference, two or more coherent waves travel different paths to a point where we observe the superposition of the two. The paths may have different lengths, or pass through different media, or both. The difference in path lengths introduces a phase difference—it changes the phase relationship between the waves.

Suppose two waves start in phase but travel different paths in the same medium to a point where they interfere (Fig. 25.4). If the difference in path lengths Δl is an integral number of wavelengths,

$$\Delta l = m\lambda \quad (m = 0, \pm1, \pm2, \ldots) \tag{25-7}$$

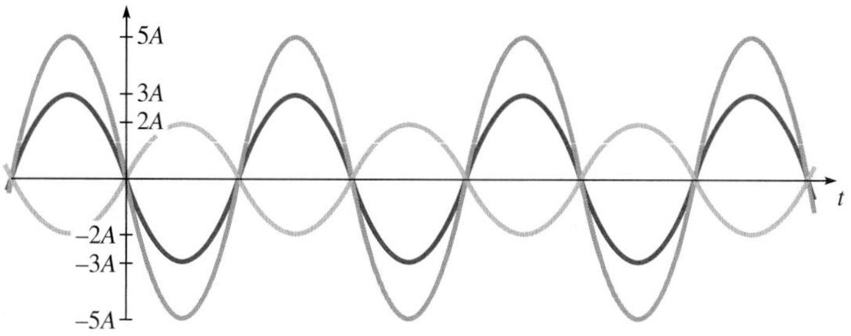

Figure 25.3 Destructive interference of two waves (green and blue) with amplitudes $2A$ and $5A$. The superposition of the two (red) has amplitude $3A$. Note that shifting either of the waves a whole number of cycles to the right or left would not change their superposition. Shifting one of the waves a *half* cycle right or left would change the superposition into *constructive* interference instead of destructive.

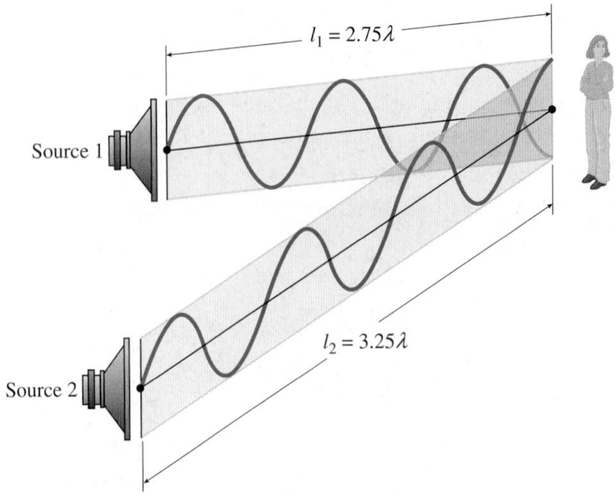

Figure 25.4 Two loudspeakers are fed the same electrical signal. The sound waves travel different distances to reach the observer. The phase difference between the two waves depends on the difference in the distances traveled. In this case, $l_2 - l_1 = 0.50\lambda$, so the waves arrive at the observer 180° out of phase. (The blue graphs represent pressure variations due to the two longitudinal sound waves.)

then one wave is simply going through a whole number of extra cycles, which leaves them in phase—they interfere constructively. Remember that one wavelength of path difference corresponds to a phase difference of 2π rad (see Section 11.9). Path lengths that are integral multiples of λ can be ignored since they do not change the relative phase between the two waves.

A path difference equal to an integral number of wavelengths does not change the superposition of two waves.

On the other hand, suppose two waves start in phase but the difference in path lengths is an odd number of *half* wavelengths:

$$\Delta l = \pm\tfrac{1}{2}\lambda, \pm\tfrac{3}{2}\lambda, \pm\tfrac{5}{2}\lambda, \ldots = (m + \tfrac{1}{2})\lambda \quad (m = 0, \pm 1, \pm 2, \ldots) \qquad (25\text{-}8)$$

One wave travels a half cycle farther than the other (plus a whole number of cycles, which can be ignored). Now the waves are 180° out of phase; they interfere destructively.

In cases where the two paths are not completely in the same medium, we have to keep track of the number of cycles in each medium separately (since the wavelength changes as a wave passes from one medium into another).

Example 25.1

Interference of Microwave Beams

A microwave transmitter (T) and receiver (R) are set up side by side (Fig. 25.5a). Two flat metal plates (M) that are good reflectors for microwaves face the transmitter and receiver, several meters away. The beam from the transmitter is broad enough to reflect from both metal plates. As the lower plate is slowly moved to the right, the microwave power measured at the receiver is observed to oscillate between minimum and maximum values (Fig. 25.5b). Approximately what is the wavelength of the microwaves?

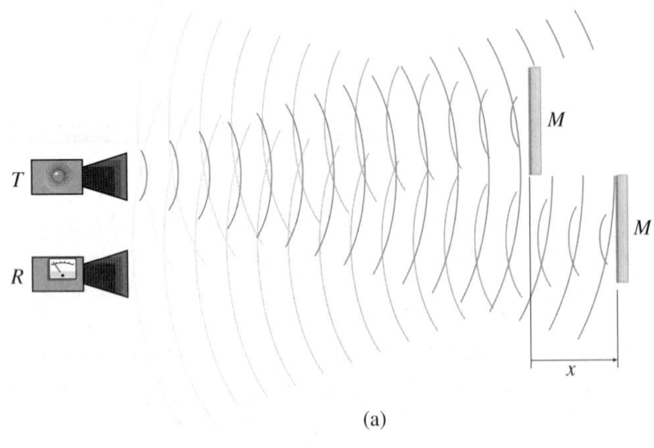

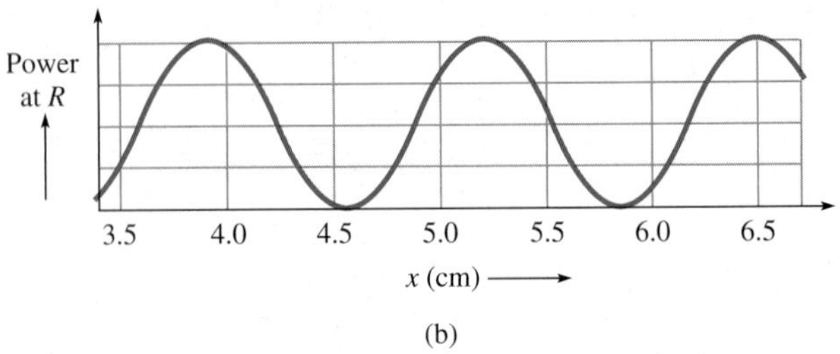

Figure 25.5
(a) Microwave transmitter and receiver and reflecting plates; (b) microwave power detected as a function of x.

continued on next page

Example 25.1 continued

Strategy Maximum power is detected when the waves reflected from the two plates interfere *constructively* at the receiver. Thus, the positions of the mirror that give maximum power must occur when the path difference is an integral number of wavelengths.

Solution When the lower plate is farther from the transmitter and receiver, the wave reflected from it travels some extra distance before reaching the receiver. If the metal plates are far enough from the transmitter and receiver, then the microwaves approach the plates and return almost along the same line. Then the extra distance traveled is approximately $2x$.

Constructive interference occurs when the path lengths differ by an integral number of wavelengths:

$$\Delta l = 2x = m\lambda \quad (m = 0, \pm 1, \pm 2, \ldots)$$

From one position of constructive interference to an adjacent one, the path length difference must change by one wavelength:

$$2\Delta x = \lambda$$

The maxima are at $x = 3.9$, 5.2, and 6.5 cm, so $\Delta x = 1.3$ cm. Then

$$\lambda = 2.6 \text{ cm}$$

Discussion Note that the distance the lower plate is moved between maxima is *half* a wavelength, since the wave makes a round trip.

Practice Problem 25.1 Path Difference for Destructive Interference

Verify that at positions where minimum power is detected, the difference in path lengths is a half-integral number of wavelengths [$\Delta l = (m + \frac{1}{2})\lambda$].

Application: How a CD Is Read

In Example 25.1, EM waves from a single source are reflected from metal surfaces at two different distances from the source; the two reflected waves interfere at the detector. A similar system is used in reading an audio CD or a CD-ROM.

To manufacture a CD, a disk of polycarbonate plastic 1.2 mm thick is impressed with a series of "*pits*" arranged in a single spiral track (Fig. 25.6). The pits are 0.5 μm wide and at least 0.83 μm long. The disk is coated with a thin layer of aluminum and then with acrylic to protect the aluminum. To read the CD, a laser beam illuminates the aluminum layer from below; the reflected beam enters a detector. The laser beam is wide enough that when it reflects from a pit, part of it also reflects off the *land* (the flat part of the aluminum layer) on either side of the track. The height h of the pits is chosen so that light reflected from the land interferes destructively with light reflected from the pit (see Problem 62). Thus, a "pit" causes a minimum intensity to be detected. On the other hand, when the laser reflects from the land between pits, the intensity at the detector is a maximum. Changes between the two intensity levels represent the binary digits (the 0's and the 1's).

25.2 THE MICHELSON INTERFEROMETER

The concept behind the Michelson interferometer (Fig. 25.7) is not complicated, yet it is an extremely precise tool. A beam of coherent light is incident on a beam splitter S (a half-silvered mirror) that reflects only half of the incident light, while transmitting the rest. Thus, a single beam of coherent light from the source is separated into two beams, which travel different paths down the *arms* of the interferometer and are reflected back by fully silvered mirrors (M_1, M_2). At the half-silvered mirror, again half of each beam is reflected and half transmitted. Light sent back toward the source leaves the interferometer. The remainder combines into a single beam and is observed on a screen. A phase difference between the two beams may arise because the arms have different lengths or because the beams travel through different media in the two arms. If the two beams arrive at the screen in phase, they interfere constructively to produce maximum intensity (a *bright fringe*) at the screen; if they arrive 180° out of phase, they interfere destructively to produce a minimum intensity (a *dark fringe*).

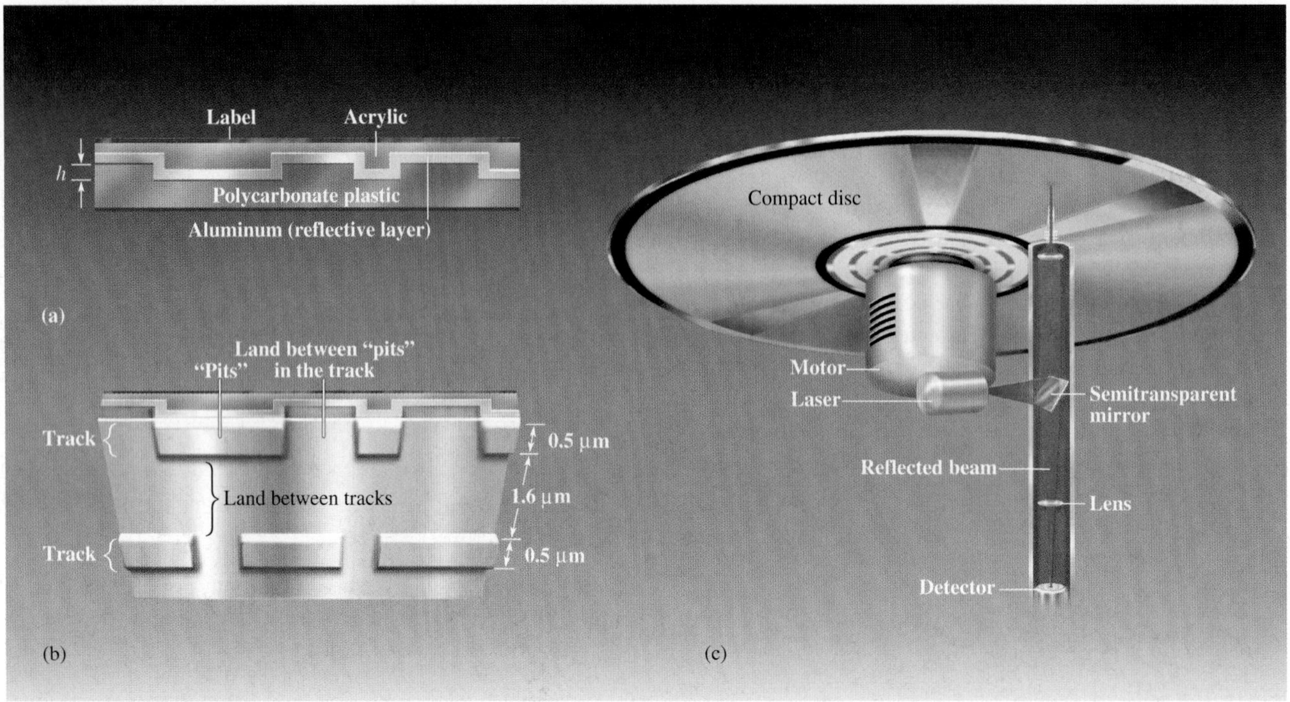

Figure 25.6 (a) Cross-sectional view of a CD. A laser beam passes through the polycarbonate plastic and reflects from the aluminum layer. (b) The "pits" are arranged in a spiral track. Surrounding the pits, the flat aluminum surface is called *land*. When the laser reflects from the bottom of a pit, it also reflects from the land on either side. (c) A motor spins the CD at between 200 and 500 rpm, keeping the track speed constant. Light from a laser is reflected by a semitransparent mirror toward the CD; light reflected by the CD is transmitted through this same mirror to the detector. The detector produces an electrical signal proportional to the variations in the intensity of reflected light.

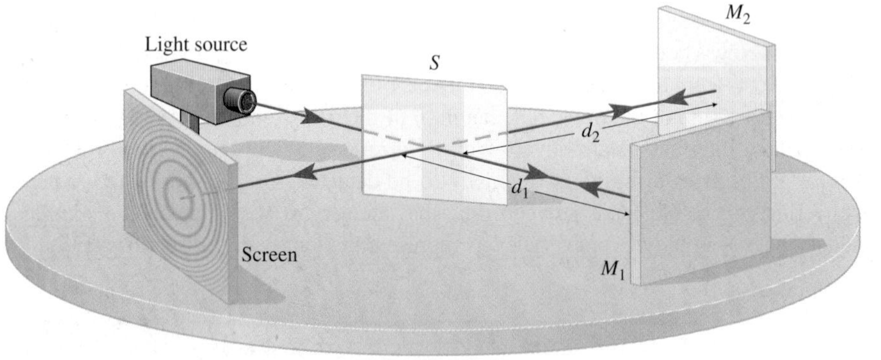

Figure 25.7 A Michelson interferometer. The American physicist Albert Michelson (1852–1931) invented the interferometer to determine whether the Earth's motion has any effect on the speed of light as measured by an observer on Earth.

Example 25.2

Measuring the Index of Refraction of Air

Suppose a transparent vessel 30.0 cm long is placed in one arm of a Michelson interferometer. The vessel initially contains air at 0°C and 1 atm. With light of vacuum wavelength 633 nm, the mirrors are arranged so that a bright spot appears at the center of the screen. As air is gradually pumped out of the vessel, the central region of the screen changes from bright to dark and back to bright 274 times—that is, 274 bright fringes are counted (not including the initial bright fringe). Calculate the index of refraction of air.

Strategy As air is pumped out, the path lengths traveled in each of the two arms do not change, but the *number of*

continued on next page

Example 25.2 continued

wavelengths traveled does change, since the index of refraction inside the vessel begins at some initial value n and decreases gradually to 1. Each new bright fringe means that the number of wavelengths traveled has changed by one more wavelength.

Solution Let the index of refraction of air at 0°C and 1 atm be n. If the *vacuum* wavelength is $\lambda_0 = 633$ nm, then the wavelength in air is $\lambda = \lambda_0/n$. Initially, the number of wavelengths traveled during a round-trip through the air in the vessel is

$$\text{initial number of wavelengths} = \frac{\text{round-trip distance}}{\text{wavelength in air}} = \frac{2d}{\lambda} = \frac{2d}{\lambda_0/n}$$

where $d = 30.0$ cm is the length of the vessel. As air is removed, the number of wavelengths decreases since, as n decreases, the wavelength gets longer. Assuming that the vessel is completely evacuated in the end (or nearly so), the final number of wavelengths is

$$\text{final number of wavelengths} = \frac{\text{round-trip distance}}{\text{wavelength in vacuum}} = \frac{2d}{\lambda_0}$$

The change in the number of wavelengths traveled, N, is equal to the number of bright fringes observed:

$$N = \frac{2d}{\lambda_0/n} - \frac{2d}{\lambda_0} = \frac{2d}{\lambda_0}(n-1)$$

Since $N = 274$, we can solve for n.

$$n = \frac{N\lambda_0}{2d} + 1 = \frac{274 \times 6.33 \times 10^{-7}\ \text{m}}{2 \times 0.300\ \text{m}} + 1 = 1.000\,289$$

Discussion The measured value for the index of refraction of air is close to that given in Table 23.1 ($n = 1.000\,293$).

Conceptual Practice Problem 25.2 A Possible Alternative Method

Instead of counting the fringes, another way to measure the index of refraction of air might be to move one of the mirrors as the air is slowly pumped out of the vessel, maintaining a bright fringe at the screen. The distance the mirror moves could be measured and used to calculate n. If the mirror moved is the one in the arm that does *not* contain the vessel, should it be moved in or out? In other words, should that arm be made longer or shorter?

Application: The Interference Microscope

An *interference microscope* enhances contrast in the image when viewing objects that are transparent or nearly so. A cell in a water solution is difficult to see with an ordinary microscope. The cell reflects only a small fraction of the light incident on it, so it transmits almost the same intensity as the water does and there is little contrast between the cell and the surrounding water. However, if the cell's index of refraction is different from that of water, light transmitted through the cell is phase-shifted compared with the light that passes through water. The interference microscope exploits this phase difference. As with the Michelson interferometer, a single beam of light is split into two and then recombined. The light in *one* arm of the interferometer passes through the sample. When the beams are recombined, interference translates the phase differences that are invisible in an ordinary microscope into intensity differences that are easily seen.

Figure 25.8 Viewing a film of soapy water by reflected light. (The background is dark so that only reflected light is shown in the photo; the camera and the light source are both on the same side of the film.) The thickness of the film gradually increases from the top of the frame toward the bottom.

25.3 THIN FILMS

The rainbow-like colors seen in soap bubbles and oil slicks are produced by interference. Suppose a wire frame is dipped into soapy water and then held vertically aloft with a thin film of soapy water clinging to the frame (Fig. 25.8). Due to gravity pulling downward, the film at the top of the wire frame is very thin—just a few molecules thick—while the film gets thicker and thicker toward the bottom. The film is illuminated with white light from behind the camera; the photo shows light *reflected* from the film. Unless otherwise stated, we will consider thin-film interference for *normal incidence* only. However, ray diagrams will show rays at *near*-normal incidence so they don't all lie on the same line in the diagram.

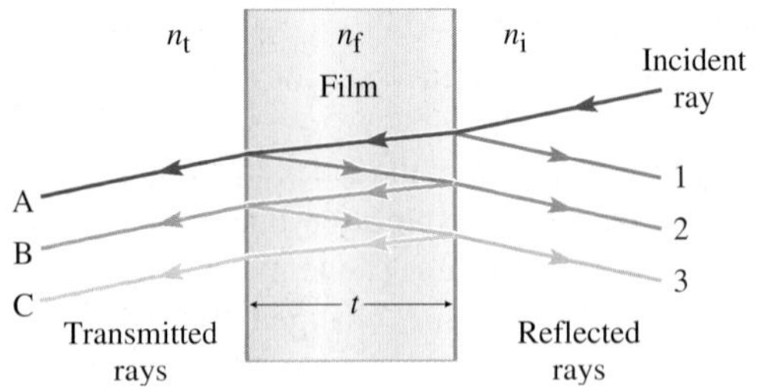

Figure 25.9 Rays reflected and transmitted by a thin film.

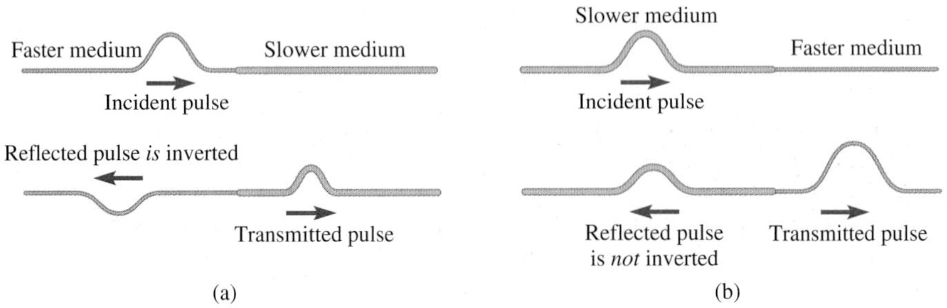

Figure 25.10 (a) A wave pulse on a string heads for a boundary with a slower medium (greater mass per unit length). The reflected pulse is inverted. (b) A pulse reflected from a *faster* medium is *not* inverted.

Figure 25.9 shows a light ray incident on a portion of a thin film. At each boundary, some light is reflected while most is transmitted. When looking at the light *reflected* from the film, we see the superposition of all the reflected rays (of which only the first three—labeled 1, 2, and 3— are shown). The interference of these rays determines what color we see. In most cases, we can consider the interference of the first two reflected rays and ignore the rest. Unless the indices of refraction on either side of a boundary are nearly the same, the amplitude of a reflected wave is a small fraction of the amplitude of the incident wave. Rays 1 and 2 each reflect only once; their amplitudes are nearly the same. Ray 3 reflects three times, so its amplitude is much smaller. Other reflected rays are even weaker.

Interference effects are much less pronounced in the transmitted light. Ray A is strong since it does not suffer a reflection. Ray B suffers two reflections, so it is much weaker than A. Ray C is even weaker since it goes through four reflections. Thus, the amplitude of the transmitted light for constructive interference is not much larger than the amplitude for destructive interference. Nevertheless, interference in the transmitted light must occur for energy to be conserved: if more of the energy of a particular wavelength is reflected, less is transmitted. In Problem 24 you can show that if a certain wavelength interferes constructively in reflected light, then it interferes destructively in transmitted light, and vice versa.

Phase Shifts due to Reflection

Whenever light hits a boundary where the wave speed suddenly changes, reflection occurs. Just as for waves on a string (Fig. 25.10), the reflected wave is inverted if it reflects off a slower medium (a medium in which the wave travels more slowly); it is *not* inverted if it reflects off a faster medium. The transmitted wave is never inverted.

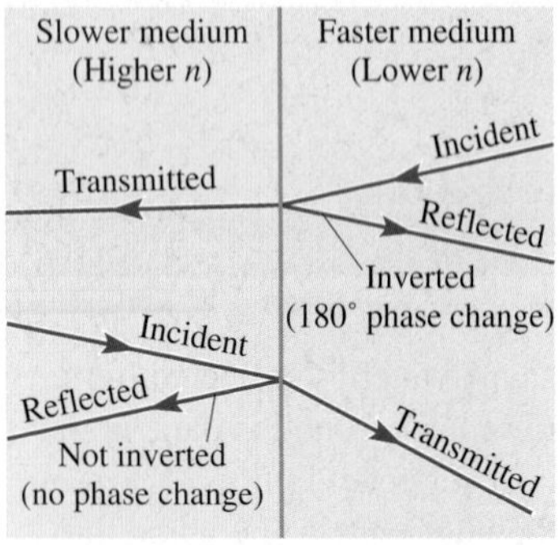

Figure 25.11 A 180° phase change due to reflection occurs when light reflects from a boundary with a slower medium.

> When light reflects at normal or near-normal incidence from a boundary with a slower medium (*higher* index of refraction), it is inverted (180° phase change); when light reflects from a faster medium (*lower* index of refraction), it is *not* inverted (no phase change). (Fig. 25.11.)

To determine whether rays 1 and 2 in Fig. 25.9 interfere constructively or destructively, we must consider both the relative phase change due to reflection and the extra path length traveled by ray 2 in the film. Depending on the indices of refraction of the three media (the film and the media on either side), it may be that *neither* of the rays is inverted on reflection, or that *both* are, or that one of the two is. If the index of refraction of the film n_f is *between* the other two indices (n_i and n_t), there is no *relative* phase difference due to reflection; either both are inverted or neither is. If the index of the film is the largest of the three or the smallest of the three, then one of the two rays is inverted; in either case there is a relative phase difference of 180°.

CONNECTION:

In Section 11.8, we saw that reflected waves are sometimes inverted, which is to say they are phase-shifted 180° with respect to the incident wave.

CHECKPOINT 25.3

In Fig. 25.9, suppose $n_i = 1.2$, $n_f = 1.6$, and $n_t = 1.4$. Which of rays 1 and 2 are phase-shifted 180° due to reflection?

Problem-Solving Strategy for Thin Films

- Sketch the first two reflected rays. Even if the problem concerns normal incidence, draw the incident ray with a *nonzero* angle of incidence to separate the various rays. Label the indices of refraction.
- Decide whether there is a relative phase difference of 180° between the rays due to reflection.
- If there is no relative phase difference due to reflection, then an extra path length of $m\lambda$ keeps the two rays in phase, resulting in constructive interference. An extra path length of $(m + \frac{1}{2})\lambda$ causes destructive interference. Remember that λ is the wavelength *in the film*, since that is the medium in which ray 2 travels the extra distance.
- If there is a 180° relative phase difference due to reflection, then an extra path length of $m\lambda$ preserves the 180° phase difference and leads to *destructive* interference. An extra path length of $(m + \frac{1}{2})\lambda$ causes *constructive* interference.
- Remember that ray 2 makes a round-trip in the film. For normal incidence, the extra path length is $2t$.

Example 25.3

Appearance of a Film of Soapy Water

A film of soapy water in air is held vertically and viewed in reflected light (as in Fig. 25.8). The film has index of refraction $n = 1.36$. (a) Explain why the film appears black at the top. (b) The light reflected perpendicular to the film at a certain point is missing the wavelengths 504 nm and 630.0 nm. No wavelengths between these two are missing. What is the thickness of the film at that point? (c) What other visible wavelengths are missing, if any?

Strategy First we sketch the first two reflected rays, labeling the indices of refraction and the thickness t of the film (Fig. 25.12). The sketch helps determine whether there is a relative phase difference of 180° due to reflection. Since the top of the film appears black, there must be destructive interference for all visible wavelengths. Farther down the film, the wavelengths missing in reflected light are those that interfere destructively; we consider phase shifts both due to

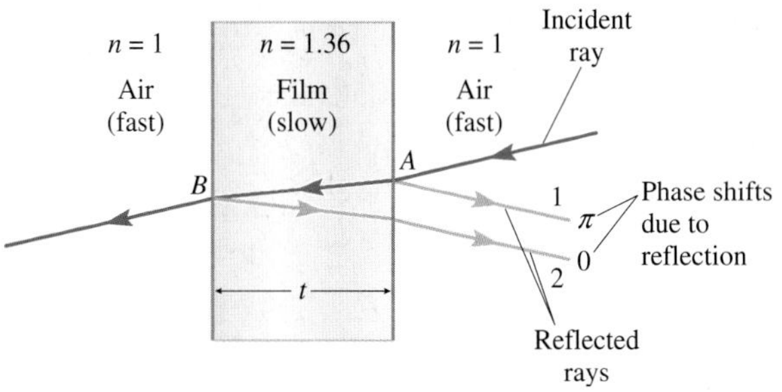

Figure 25.12
The first two rays reflected by the soap film. At A, reflected ray 1 is inverted. At B, reflected ray 2 is not inverted.

continued on next page

Example 25.3 continued

reflection and due to the extra path ray 2 travels in the film. We must remember to use the wavelength *in the film*, not the wavelength in vacuum, because ray 2 travels its extra distance within the film.

Solution (a) The speed of light in the film is slower than in air. Therefore ray 1, which reflects from a slower medium (the film), is inverted; ray 2, which reflects from a faster medium (air), is not inverted. There is a relative phase difference of 180° between the two *regardless of wavelength*. Due to gravity, the film is thinnest at the top and thickest at the bottom. Ray 2 has a phase shift compared with ray 1 due to the extra distance traveled in the film. The only way to preserve destructive interference for *all* wavelengths is if the top of the film is thin relative to the wavelengths of visible light; then the phase change of ray 2 due to the extra path traveled is negligibly small.

(b) For light reflected perpendicular to the film (normal incidence), reflected ray 2 travels an extra distance $2t$ compared with ray 1, which introduces a phase difference between them. Since there is already a relative phase difference of 180° due to reflection, the path difference $2t$ must be an *integral* number of wavelengths to preserve destructive interference:

$$2t = m\lambda = m\frac{\lambda_0}{n}$$

Suppose $\lambda_{0,m} = 630.0$ nm is the vacuum wavelength for which the path difference is $m\lambda$ for a certain value of m. Since there are no missing wavelengths between the two, $\lambda_{0,(m+1)} = 504$ nm must be the vacuum wavelength for which the path difference is $m + 1$ times the wavelength in the film. Why not $m - 1$? Because 504 nm is smaller than 630.0 nm, so a *larger* number of wavelengths fits in the path difference $2t$.

$$2nt = m\lambda_{0,m} = (m + 1)\lambda_{0,(m+1)}$$

We can solve for m:

$$m \times 630.0 \text{ nm} = (m + 1) \times 504 \text{ nm} = m \times 504 \text{ nm} + 504 \text{ nm}$$

$$m \times 126 \text{ nm} = 504 \text{ nm}$$

$$m = 4.00$$

Then the thickness is

$$t = \frac{m\lambda_0}{2n} = \frac{4.00 \times 630.0 \text{ nm}}{2 \times 1.36} = 926.47 \text{ nm} = 926 \text{ nm}$$

(c) We already know the missing wavelengths for $m = 4$ and $m = 5$. Let's check other values of m.

$$2nt = 2 \times 1.36 \times 926.47 \text{ nm} = 2520 \text{ nm}$$

For $m = 3$,

$$\lambda_0 = \frac{2nt}{m} = \frac{2520 \text{ nm}}{3} = 840 \text{ nm}$$

which is IR rather than visible. There is no need to check $m = 1$ or 2 since they give wavelengths even larger than 840 nm—wavelengths even farther from the visible range. Therefore, we try $m = 6$:

$$\lambda_0 = \frac{2nt}{m} = \frac{2520 \text{ nm}}{6} = 420 \text{ nm}$$

This wavelength is generally considered to be visible. What about $m = 7$?

$$\lambda_0 = \frac{2nt}{m} = \frac{2520 \text{ nm}}{7} = 360 \text{ nm}$$

A wavelength of 360 nm is UV. Thus, the only other missing visible wavelength is 420 nm.

Discussion As a check, we can verify directly that the three missing wavelengths in vacuum travel an integral number of wavelengths in the film:

λ_0	$\lambda = \frac{\lambda_0}{1.36}$	$m\lambda$
420 nm	308.8 nm	6×308.8 nm = 1853 nm
504 nm	370.6 nm	5×370.6 nm = 1853 nm
630 nm	463.2 nm	4×463.2 nm = 1853 nm

Since the path difference is $2t = 2 \times 926.47$ nm = 1853 nm, the extra path is an integral number of wavelengths for all three.

Practice Problem 25.3 Constructive Interference in Reflected Light

What visible wavelengths interfere *constructively* in the reflected light where $t = 926$ nm?

Thin Films of Air

A thin air gap between two solids can produce interference effects. Figure 25.13a is a photograph of two glass slides separated by an air gap. The thickness of the air gap varies because the glass surfaces are not perfectly flat. The photo shows colored fringes. Each fringe of a given color traces out a curve along which the thickness of the air film is constant. In Fig. 25.13b, an instrument pushes gently down on the top slide. The resulting distortion in the surface of the top slide causes the fringes to move.

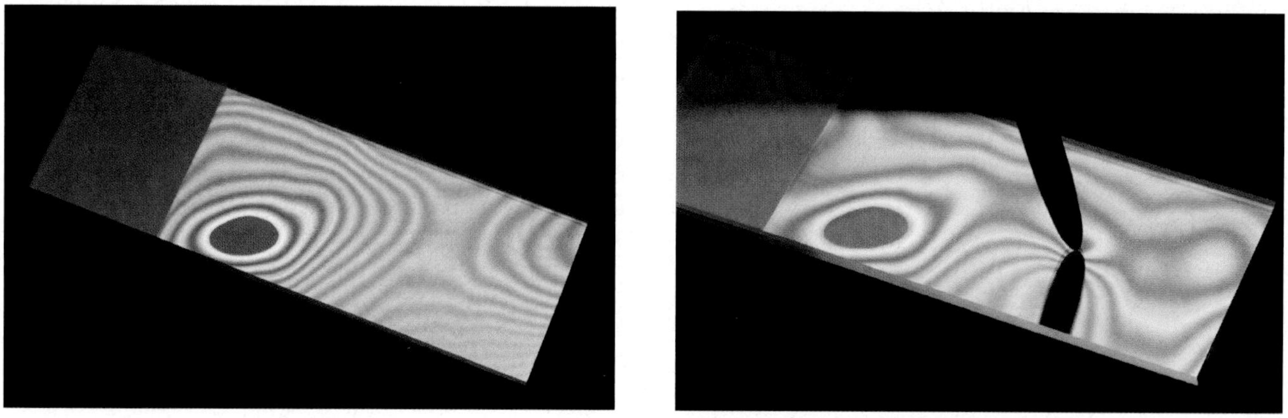

(a) (b)

Figure 25.13 (a) Two glass slides with a narrow air gap between them. When illuminated with white light, interference fringes form in the reflected light. (b) Pressing on the glass changes the thickness of the air gap and distorts the interference fringes.

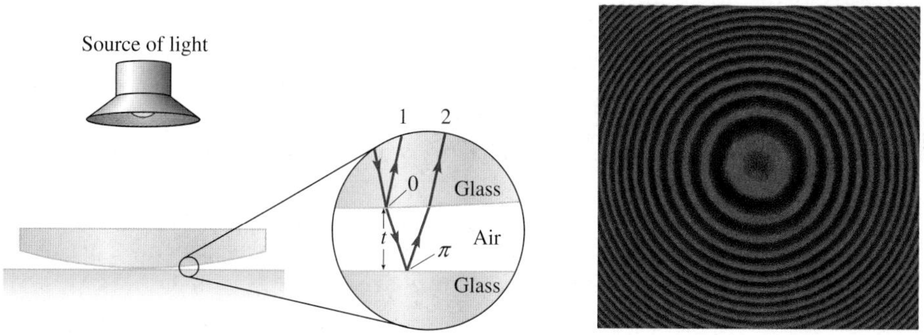

(a) (b) (c)

Figure 25.14 (a) The air gap between a convex, spherical glass surface and an optically flat glass plate. The curvature of the lens is exaggerated here. In reality, the air gap would be very thin and the glass surfaces *almost* parallel. (b) Light rays reflected from the top and bottom of the air gap. Ray 2 has a phase shift of π rad due to reflection, while ray 1 does not. Ray 2 also has a phase shift due to the extra path traveled in the air gap. For normal incidence, the extra path length is $2t$, where t is the thickness of the air gap. When viewed from above, we see the superposition of reflected rays 1 and 2. (c) A pattern of circular interference fringes, known as Newton's rings, is seen in reflected light.

If a glass lens with a convex spherical surface is placed on a flat plate of glass, the air gap between the two increases in thickness as we move out from the contact point (Fig. 25.14). Assuming a perfect spherical shape, we expect to see alternating bright and dark circular fringes in reflected light. The fringes are called Newton's rings (after Isaac Newton). Well past Newton's day, it was a puzzle that the center was a *dark* spot. Thomas Young figured out that the center is dark because of the phase shift on reflection. Young did an experiment producing Newton's rings with a lens made of crown glass ($n = 1.5$) on top of a flat plate made of flint glass ($n = 1.7$). When the gap between the two was filled with air, the center was dark in reflected light. Then he immersed the experiment in sassafras oil (which has an index of refraction between 1.5 and 1.7). Now the center spot was bright, since there was no longer a relative phase difference of 180° due to reflection.

Newton's rings can be used to check a lens to see if its surface is spherical. A perfectly spherical surface gives circular interference fringes that occur at predictable radii (see Problem 23).

Figure 25.15 The left side of this lens has an antireflective coating; the right side does not.

Application: Antireflective Coatings

A common application of thin film interference is the antireflective coatings on lenses (Fig. 25.15). The importance of these coatings increases as the number of lenses in an instrument increases—if even a small percentage of the incident light intensity is reflected at each surface, reflections at each surface of each lens can add up to a large fraction of the incident intensity being reflected and a small fraction being transmitted through the instrument.

The most common material used as an antireflective coating is magnesium fluoride (MgF_2). It has an index of refraction $n = 1.38$, between that of air ($n = 1$) and glass ($n \approx 1.5$ or 1.6). The thickness of the film is chosen so destructive interference occurs for wavelengths in the middle of the visible spectrum.

Application: Iridescent Colors in Butterfly Wings

Interference from light reflected by step structures or partially overlapping scales produces the iridescent colors seen in many butterflies, moths, birds, and fish. A stunning example is the shimmering blue of the *Morpho* butterfly. Figure 25.16a shows the *Morpho* wing as viewed under an electron microscope. The treelike structures that project up from the top surface of the wing are made of a transparent material. Light is thus reflected from a series of steps. Let us concentrate on two rays reflected from the tops of successive steps of thickness t_1 with spacing t_2 between the steps (Fig. 25.16b). Both rays are inverted on reflection, so there is no relative phase difference due to reflection. At normal incidence, the path difference is $2(t_1 + t_2)$. However, the ray passes through a thickness t_1 of the step where the index of refraction is $n = 1.5$. We cannot find the wavelength for constructive interference simply by setting the path difference equal to a whole number of wavelengths: which wavelength would we use?

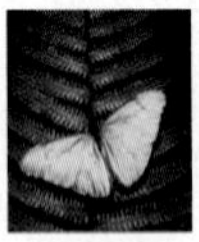

To solve this sort of problem, we think of path differences in terms of numbers of wavelengths. The number of wavelengths traveled by ray 2 in a distance $2t_1$ (round-trip) through a thickness t_1 of the wing structure is

$$\frac{2t_1}{\lambda} = \frac{2t_1}{\lambda_0/n}$$

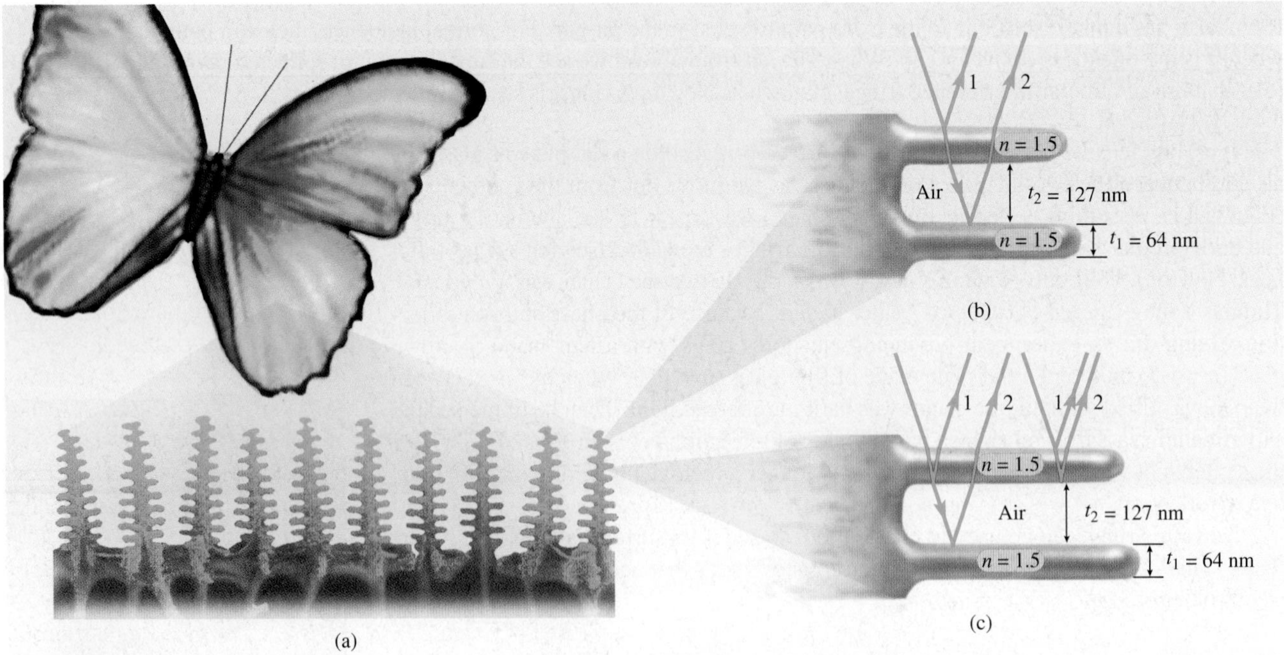

Figure 25.16 (a) *Morpho* wing as viewed under an electron microscope. (b) Light rays reflected from two successive steps interfere. Constructive interference produces the shimmering blue color of the wing. For clarity, the rays shown are not at normal incidence. (c) Two other pairs of rays that interfere.

where λ_0 is the wavelength in vacuum and $\lambda = \lambda_0/n$ is the wavelength in the medium with index of refraction n. The number of wavelengths traveled in a distance $2t_2$ in air is

$$\frac{2t_2}{\lambda} = \frac{2t_2}{\lambda_0}$$

For constructive interference, the number of extra wavelengths traveled by ray 2, relative to ray 1, must be an integer:

$$\frac{2t_1}{\lambda_0/n} + \frac{2t_2}{\lambda_0} = m$$

We can solve this equation for λ_0 to find the wavelengths that interfere constructively:

$$\lambda_0 = \frac{2}{m}(nt_1 + t_2)$$

For $m = 1$,

$$\lambda_0 = 2(1.5 \times 64 \text{ nm} + 127 \text{ nm}) = 2 \times 223 \text{ nm} = 446 \text{ nm}$$

This is the dominant wavelength in the light we see when looking at the butterfly wing at normal incidence. We only considered reflections from two adjacent steps, but if those interfere constructively, so do all the other reflections from the tops of the steps. Constructive interference at higher values of m are outside the visible spectrum (in the UV).

Since the path length traveled by ray 2 depends on the angle of incidence, the wavelength of light that interferes constructively depends on the angle of view (see Conceptual Question 16). Thus, the color of the wing changes as the viewing angle changes, which gives the wing its shimmering iridescence.

So far we have ignored reflections from the bottoms of the steps. Rays reflected from the bottoms of two successive steps interfere constructively at the same wavelength of 446 nm, since the path difference is the same. The interference of two other pairs of rays (Fig. 25.16c) gives constructive interference only in UV since the path length difference is so small.

25.4 YOUNG'S DOUBLE-SLIT EXPERIMENT

In 1801, Thomas Young performed a double-slit interference experiment that not only demonstrated the wave nature of light, but also allowed the first measurement of the wavelength of light. Figure 25.17 shows the setup for Young's experiment. Coherent light of wavelength λ illuminates a mask in which two parallel slits have been cut. Each slit has width a, which is comparable to the wavelength λ, and length $L \gg a$; the centers of the slits are separated by a distance d. When light from the slits is observed on a screen at a great distance D from the slits, what pattern do we see—how does the intensity I of light falling on the screen depend on the angle θ, which measures the direction from the slits to a point on the screen?

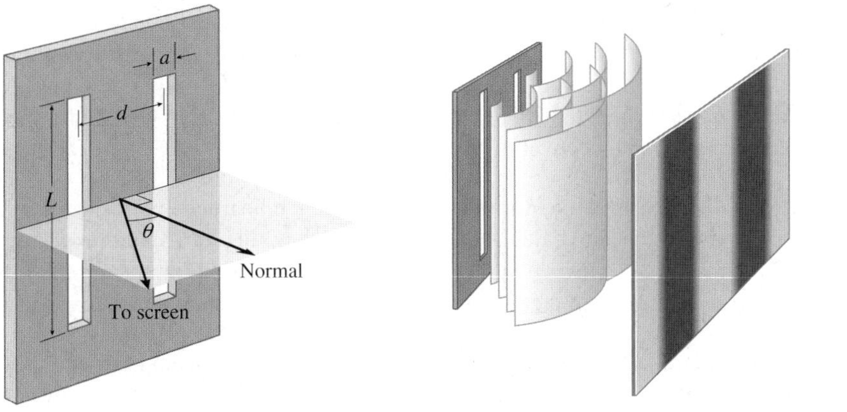

Figure 25.17 Young's double-slit interference experiment. (a) The slit geometry. The center-to-center distance between the slits is d. From the point midway between the slits, a line perpendicular to the mask extends toward the center of the interference pattern on the screen and a line making an angle θ to the normal can be used to locate a particular point to either side of the center of the interference pattern. (b) Cylindrical wavefronts emerge from the slits and interfere to form a pattern of fringes on the screen.

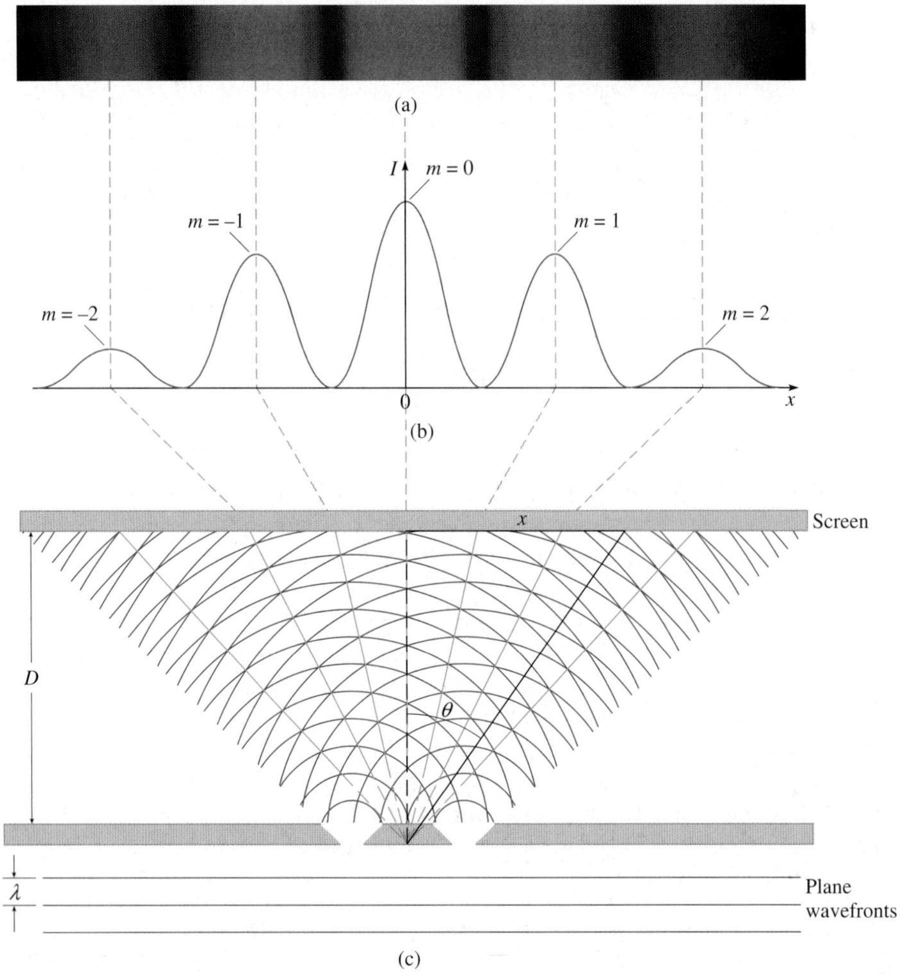

Figure 25.18 Double-slit interference pattern using red light. (a) Photo of the interference pattern on the screen. Constructive interference produces a high intensity of red light on the screen while destructive interference leaves the screen dark. (b) The intensity as a function of position x on the screen. The maxima (positions where the interference is constructive) are labeled with the associated value of m. (c) A Huygens construction for the double-slit experiment. The blue lines represent antinodes (points where the waves interfere constructively). Note the relationship between x, the position on the screen, and the angle θ: $\tan \theta = x/D$, where D is the distance from the slits to the screen.

Light from a *single* narrow slit spreads out primarily in directions perpendicular to the slit, since the wavefronts coming from it are cylindrical. Thus, the light from one narrow slit forms a band of light on the screen. The light does *not* spread out significantly in the direction *parallel* to the slit since the slit length L is *large* relative to the wavelength.

With *two* narrow slits, the two bands of light on the screen interfere with one another. The light from the slits starts out in phase, but travels different paths to reach the screen. We expect constructive interference at the center of the interference pattern ($\theta = 0$) since the waves travel the same distance and so are in phase when they reach the screen. Constructive interference also occurs wherever the path difference is an integral multiple of λ. Destructive interference occurs when the path difference is an odd number of half wavelengths. A gradual transition between constructive and destructive interference occurs since the path difference increases continuously as θ increases. This leads to the characteristic alternation of bright and dark bands (fringes) that are shown in Fig. 25.18a, a photograph of the screen from a double-slit experiment. Figure 25.18b and c are a graph of the intensity on the screen and a Huygens construction for the same interference pattern, respectively.

Locations of Maxima and Minima To find where constructive or destructive interference occurs, we need to calculate the path difference. Figure 25.19a shows two rays going from the slits to a *nearby* screen. If the screen is moved farther from the slits, the angle α gets smaller. When the screen is far away, α is small and the rays are nearly parallel. In Fig. 25.19b, the rays are drawn as parallel for a distant screen. The distances that the rays travel from points A and B to the screen are equal; the path difference is the distance from the right slit to point B:

$$\Delta l = d \sin \theta \qquad (25\text{-}9)$$

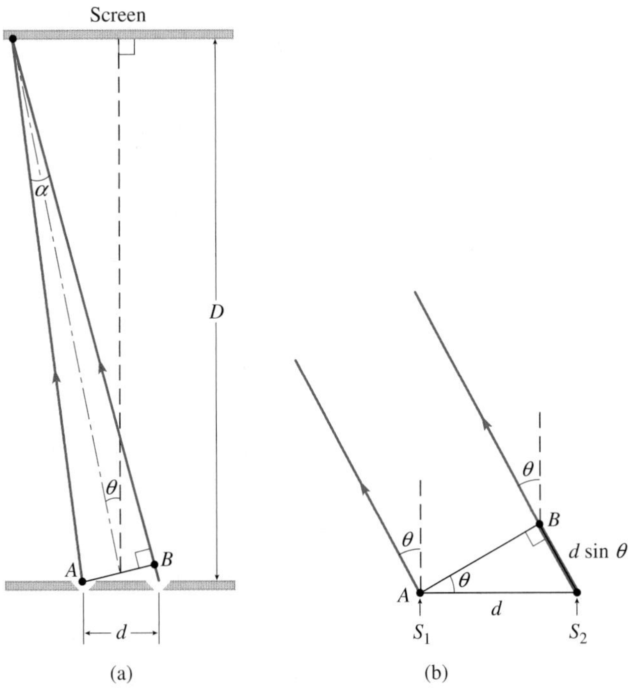

Figure 25.19 (a) Rays from two slits to a nearby screen. As the screen is moved farther away, α decreases—the rays become more nearly parallel. (b) In the limit of a distant screen, the two rays are parallel (but still meet at the same point on the screen). The difference in path lengths is $d \sin \theta$.

Maximum intensity at the screen is produced by constructive interference; for constructive interference, the path difference is an integral multiple of the wavelength:

Double-slit maxima:

$$d \sin \theta = m\lambda \quad (m = 0, \pm 1, \pm 2, \ldots) \tag{25-10}$$

The absolute value of m is often called the **order** of the maximum. Thus, the third-order maxima are those for which $d \sin \theta = \pm 3\lambda$. Minimum (zero) intensity at the screen is produced by destructive interference; for destructive interference, the path difference is an odd number of half wavelengths:

Double-slit minima:

$$d \sin \theta = (m + \tfrac{1}{2})\lambda \quad (m = 0, \pm 1, \pm 2, \ldots) \tag{25-11}$$

CONNECTION:

Antinodes are locations of maximum amplitude and nodes are locations of minimum amplitude, whether in EM waves or mechanical waves (see Sections 11.10 and 12.4).

In Fig. 25.18, the bright and dark fringes appear to be equally spaced. In Problem 28, you can show that the interference fringes *are* equally spaced near the center of the interference pattern, where θ is a small angle.

Water Wave Analogy to the Double-Slit Experiment Figure 25.20 shows the interference of *water waves* in a ripple tank. Surface waves are generated in the water by two point sources that vibrate up and down at the same frequency and in phase with one another, so they are coherent sources. The pattern of interference of the water waves far from the two sources is similar to the double-slit interference pattern for light. If d represents the distance between the sources, Eqs. (25-10) and (25-11) give the correct angles θ for constructive and destructive interference of *water* waves at a large distance. The advantage of the ripple tank is that it lets us see what the wavefronts look like. Notice the similarity between Fig. 25.20 and Fig. 25.18c.

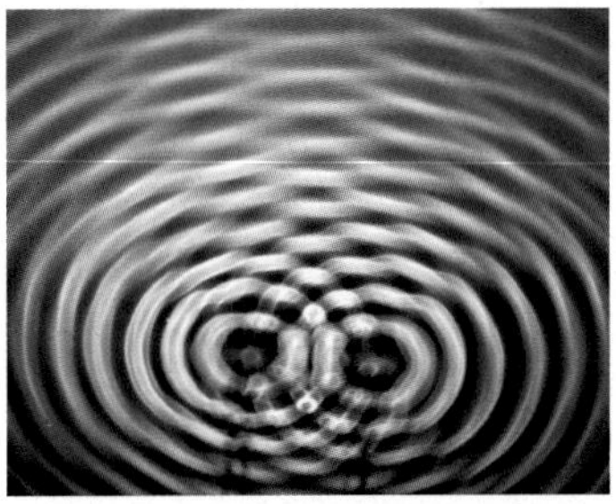

Figure 25.20 Water waves in a ripple tank exhibit two-source interference. Lines of antinodes correspond to the directions of maximum intensity in double-slit interference for light; lines of nodes correspond to the minima.

Points where the interference is constructive are called **antinodes**. Just as for standing waves, here the superposition of two coherent waves causes some points—the antinodes—to have maximum amplitudes. There are also **nodes**—points of complete destructive interference. In a one-dimensional standing wave on a string, nodes and antinodes were single points. For the two-dimensional water waves in a ripple tank, the nodes and antinodes are *curves*. For three-dimensional light waves (or three-dimensional sound waves), the nodes and antinodes are *surfaces*.

Example 25.4

Interference from Two Parallel Slits

A laser ($\lambda = 690.0$ nm) is used to illuminate two parallel slits. On a screen that is 3.30 m away from the slits, interference fringes are observed. The distance between adjacent bright fringes in the center of the pattern is 1.80 cm. What is the distance between the slits?

Strategy Bright fringes occur at angles θ given by $d \sin \theta = m\lambda$. The distance between the $m = 0$ and $m = 1$ maxima is $x = 1.80$ cm. A sketch helps us see the relationship between the angle θ and the distances given in the problem.

Solution The central bright fringe ($m = 0$) is at $\theta_0 = 0$. The next bright fringe ($m = 1$) is at an angle given by

$$d \sin \theta_1 = \lambda$$

Figure 25.21 is a sketch of the geometry of the situation. The angle between the lines going to the $m = 0$ and $m = 1$ maxima is θ_1. The distance between these two maxima on the screen is x and the distance from the slits to the screen is D. We can find θ_1 from x and D using trigonometry:

$$\tan \theta_1 = \frac{x}{D} = \frac{0.0180 \text{ m}}{3.30 \text{ m}} = 0.005455$$

$$\theta_1 = \tan^{-1} 0.005455 = 0.3125°$$

Now we substitute θ_1 into the condition for the $m = 1$ maximum.

$$d = \frac{\lambda}{\sin \theta_1} = \frac{690.0 \text{ nm}}{\sin 0.3125°} = \frac{690.0 \text{ nm}}{0.005454} = 0.127 \text{ mm}$$

Discussion We might have noticed that since $x \ll D$, θ_1 is a small angle—that's why the sine and the tangent are the same to three significant figures. Using the small angle approximation ($\sin \theta \approx \tan \theta \approx \theta$ in radians) from the start gives

$$d\theta_1 = \lambda$$

and

$$\theta_1 = \frac{x}{D}$$

so

$$d = \frac{\lambda D}{x} = \frac{690.0 \text{ nm} \times 3.30 \text{ m}}{0.0180 \text{ m}} = 0.127 \text{ mm}$$

Practice Problem 25.4 Fringe Spacing When the Wavelength Is Changed

In a particular double-slit experiment, the distance between the slits is 50 times the wavelength of the light. (a) Find the angles in radians at which the $m = 0$, 1, and 2 maxima occur. (b) Find the angles at which the first two minima occur. (c) What is the distance between two maxima at the center of the pattern on a screen 2.0 m away?

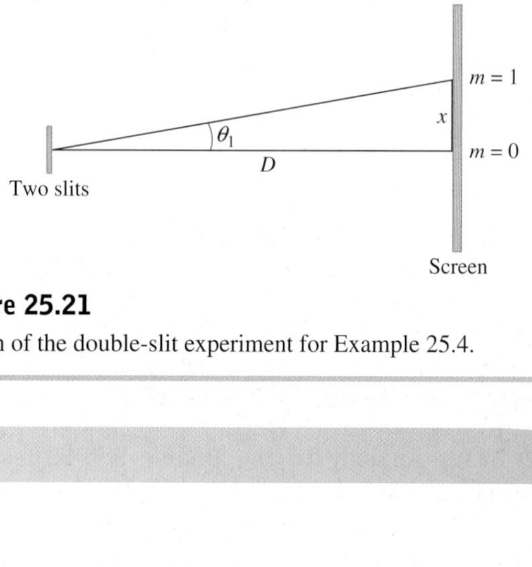

Figure 25.21
Sketch of the double-slit experiment for Example 25.4.

Conceptual Example 25.5

Changing the Slit Separation

A laser is used to illuminate two narrow parallel slits. The interference pattern is observed on a distant screen. What happens to the pattern observed if the distance between the slits is slowly decreased?

Solution and Discussion When the slits are closer together, the path difference $d \sin \theta$ for a given angle gets smaller. Larger angles are required to produce a path difference that is a given multiple of the wavelength. The interference pattern therefore spreads out, with each maximum (other than $m = 0$) and minimum moving out to larger and larger angles.

Conceptual Practice Problem 25.5 Interference Pattern for $d < \lambda$

If the distance between two slits in a double-slit experiment is less than the wavelength of light, what would you see at a distant screen?

25.5 GRATINGS

Instead of having two parallel slits, a **grating** (sometimes called a "diffraction grating") consists of a large number of parallel, narrow, evenly spaced slits. Typical gratings have hundreds or thousands of slits. The slit separation of a grating is commonly characterized by the number of slits per cm (or the number per any other unit of distance), which is the reciprocal of the slit separation d:

$$\text{slits per cm} = \frac{1}{\text{cm per slit}} = \frac{1}{d}$$

Gratings are made with up to about 50,000 slits/cm, so slit separations are as small as 200 nm. The smaller the slit separation, the more widely different wavelengths of light are separated by the grating.

Figure 25.22 shows light rays traveling from the slits of a grating to a distant screen. Suppose light from the first two slits is in phase at the screen because the path difference $d \sin \theta$ is a whole number of wavelengths $m\lambda$. Then, since the slits are evenly spaced, the light from *all* the slits arrives at the screen in phase. The path difference between any pair of slits is an integral multiple of $d \sin \theta$ and therefore an integral multiple of λ. Therefore, the angles for constructive interference for a grating are the same as for two slits with the same separation:

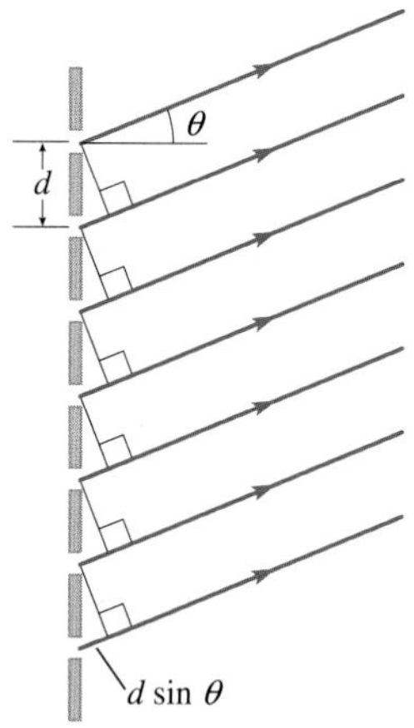

Figure 25.22 Light rays traveling from the slits of a grating to a point on a screen. Since the screen is far away, the rays are nearly parallel to each other; they all leave the grating at (nearly) the same angle θ. Since the distance between any two adjacent slits is d, the path difference between two adjacent rays is $d \sin \theta$.

Maxima for a grating:

$$d \sin \theta = m\lambda \quad (m = 0, \pm 1, \pm 2, \ldots) \qquad (25\text{-}10)$$

As for two slits, $|m|$ is called the *order* of the maximum.

For two slits, there is a gradual change in intensity from maximum to minimum and back to maximum. By contrast, for a grating with a large number of slits, the maxima are narrow and the intensity everywhere else is negligibly small. How does the presence of many slits make the maxima so narrow?

CONNECTION:

Maxima for a grating are at the same angles as maxima for two slits with the same d.

Suppose we have a grating with $N = 100$ slits, numbered 0 to 99. The first-order maximum occurs at angle θ such that the path length difference between slits 0 and 1 is $d \sin \theta = \lambda$. Now suppose we look at a *slightly* greater angle $\theta + \Delta\theta$ such that $d \sin (\theta + \Delta\theta) = 1.01\lambda$. The rays from slits 0 and 1 are almost in phase; if there were only two slits, the intensity would be almost as large as the maximum. With 100 slits, each ray is 1.01λ longer than the previous ray. If the length of ray 0 is l_0, then the length of ray 1 is $l_0 + 1.01\lambda$, the length of ray 2 is $l_0 + 2.02\lambda$, and so forth. The length of ray 50 is $l_0 + 50.50\lambda$; thus, rays 0 and 50 interfere destructively since the path difference is an odd number of half wavelengths. Likewise, slits 1 and 51 interfere destructively $(51.51\lambda - 1.01\lambda = 50.50\lambda)$; slits 2 and 52 interfere destructively; and so on. Since the light from every slit interferes destructively with the light from some other slit, the intensity at the screen is *zero*. The intensity goes from maximum at θ to zero at $\theta + \Delta\theta$.

The angle $\Delta\theta$ is called the *half-width* of the maximum since it is the angle from the center of the maximum to one edge of the maximum (rather than from one edge to the other). By generalizing the argument, we find that the widths of the maxima are inversely proportional to the number of slits ($\Delta\theta \propto 1/N$). The larger the number of slits, the narrower the maxima. Increasing N also makes the maxima *brighter*. More slits let more light pass through and bunch the light energy into narrower maxima. Since light from N slits interferes constructively, the amplitudes of the maxima are proportional to N and the intensities are proportional to N^2. The maxima for a grating are narrow and occur at different angles for different wavelengths. Therefore:

A grating separates light with a mixture of wavelengths into its constituent wavelengths.

CHECKPOINT 25.5

How are the maxima produced by a grating different from those produced by a double slit with the same spacing d?

Example 25.6

Slit Spacing for a Grating

Bright white light shines on a grating. A cylindrical strip of color film is exposed by light emerging at all angles (−90° to +90°) from the grating (Fig. 25.23a). Figure 25.23b shows the resulting photograph. Estimate the number of slits per centimeter in the grating.

Strategy The grating separates white light into the colors of the visible spectrum. Each color forms a maximum at angles given by $d \sin \theta = m\lambda$. From Fig. 25.23b, we see that more than just first-order maxima are present. If we can estimate the wavelength of the light that exposed the edge of the photo—light that left the grating at ±90°—and if we know what order maximum that is, we can find the slit separation.

Solution The central ($m = 0$) maximum appears white due to constructive interference for *all* wavelengths. On either side of the central maximum lie the first-order maxima. The first-order violet (shortest wavelength) comes first (at the smallest angle), and red is last. Next comes a gap where there are no maxima. Then the second-order maxima begin with violet. The colors do not progress through the pure spectral colors as before because the third-order maxima start to appear before the second order is finished. The third-order spectrum is not complete; the last color we see at either extreme ($\theta = \pm 90°$) is blue-green. Thus, the third-order maximum for blue-green light occurs at ±90°.

Wavelengths that appear blue-green are around 500 nm (Section 22.3). Using $\lambda = 500$ nm and $m = 3$ for the third-order maximum, we can solve for the slit separation.

$$d \sin \theta = m\lambda$$

$$d = \frac{m\lambda}{\sin \theta} = \frac{3 \times 500\ \text{nm}}{\sin 90°} = 1500\ \text{nm}$$

Then the number of slits per cm is

$$\frac{1}{d} = \frac{1}{1500 \times 10^{-9}\ \text{m}} = 670{,}000\ \text{slits/m} = 6700\ \text{slits/cm}$$

Discussion The final answer is reasonable for the number of slits per cm in a grating. We would suspect an error if it came out to be 67 million slits/cm, or 67 slits/cm.

For a maximum occurring at 90°, we *cannot* use the small angle approximation! We often look at maxima formed by gratings that occur at large angles for which the small angle approximation is not valid.

Practice Problem 25.6 Slit Spacing for a Full Third Order

How many slits per centimeter would a grating have if it just barely produced the full third-order spectrum? Would any of the fourth-order spectrum be produced by such a grating?

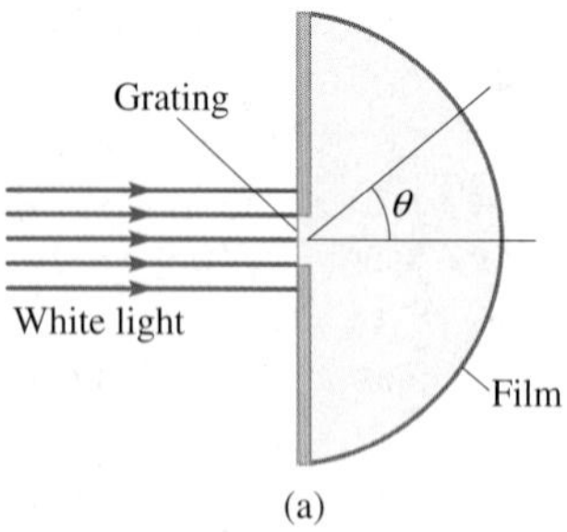

Figure 25.23
(a) White light incident on a grating. (b) The developed film.

Application: CD Tracking

Data on a CD is encoded as bumps arranged in a spiral track 0.5 μm wide (Fig. 25.24). A 1.6-μm width of flat aluminum on either side separates two adjacent tracks. One of the hardest jobs of a CD player (or CD-ROM drive) is to keep the laser beam centered on the spiral track. One method to keep the laser on track uses a grating to split the laser beam into three beams. The central ($m = 0$) maximum is centered on the track.

The first-order ($m = \pm 1$) maxima are tracking beams. They reflect from the flat aluminum surfaces (called *land*) on either side of the track onto detectors. Normally the reflected intensity is constant. If one of the tracking beams encounters the bumps in an adjacent track, the changes in reflected intensity signal that the position of the laser must be corrected.

Some fast CD-ROM drives use a grating to split a laser beam into seven beams—zeroth- through third-order maxima. Each beam is used to read a different track; thus, the drive reads seven tracks simultaneously.

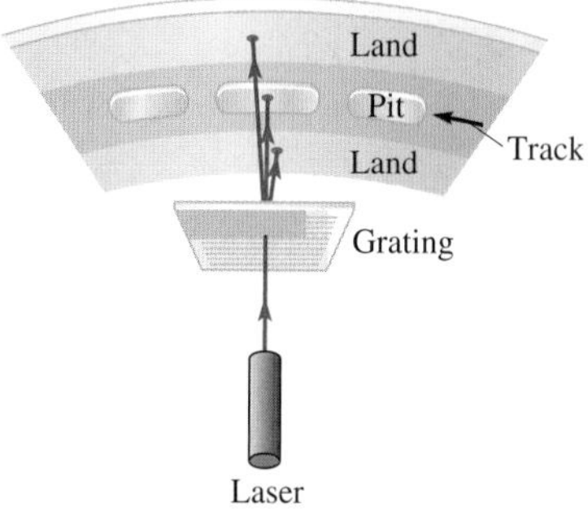

Figure 25.24 A three-beam tracking system.

Application: Spectroscopy

The **grating spectroscope** is a precision instrument to measure wavelengths of visible light (Fig. 25.25). *Spectroscope* means (roughly) *spectrum-viewing*. The angles at which maxima occur are used along with the spacing of slits in the grating to determine the wavelength(s) present in the light source. The maxima are often called *spectral lines*—they appear as thin lines because they have the shape of the entry slit of the collimator.

Although thermal radiation (e.g., sunlight and incandescent light) contains a continuous spectrum of wavelengths, other sources of light contain a discrete spectrum composed of only a few narrow bands of wavelengths. A discrete spectrum is also called a line spectrum due to its appearance as a set of lines when viewed through a spectroscope. For example, fluorescent lights and gas discharge tubes produce discrete spectra. In a gas discharge tube, a glass tube is filled with a single gas at low pressure and an electrical current is passed through the gas. The light that is emitted is a discrete spectrum that is characteristic of the gas. Some older streetlights are sodium discharge tubes; they have a characteristic yellow color.

Figure 25.26 shows the spectrum of a sodium discharge tube, which includes a pair of yellow lines. Imagine using a grating with fewer slits. The maxima would be wider; if they were too wide, the two yellow lines would overlap and appear as a single line. So a large number of slits is an advantage if we need to *resolve* (distinguish) wavelengths that are close together.

Reflection Gratings

In the **transmission gratings** we have been discussing, the light viewed is that transmitted by the transparent slits of the grating. Another common kind of grating is the **reflection grating**. Instead of slits, a reflection grating has a large number of parallel,

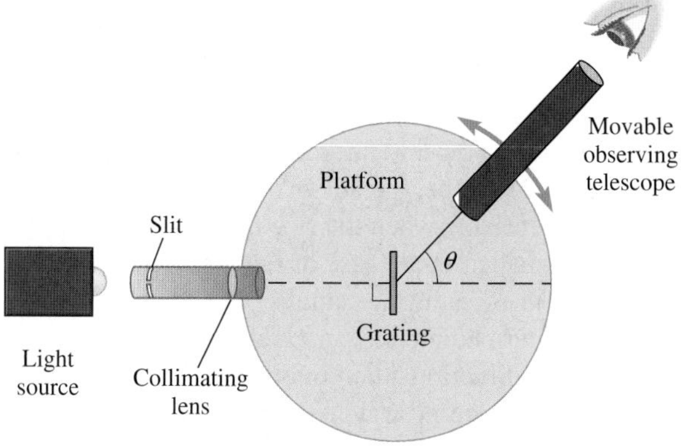

Figure 25.25 Overhead view of a grating spectroscope. Light from the source first passes through a narrow, vertical slit, which is at the focal point of the collimating lens. Thus, the rays emerging from the lens are parallel to each other. The grating rests on a platform that is adjusted so that the incident rays strike the grating at normal incidence. The telescope can be moved in a circle around the grating to observe the maxima and to measure the angle θ at which each one occurs.

Figure 25.26 Emission spectrum of sodium. The spectrum includes two yellow lines at wavelengths of 589.0 nm and 589.6 nm (the *sodium doublet*).

thin reflecting surfaces separated by absorbing surfaces. Using Huygens's principle, the analysis of the reflection grating is exactly the same as for the transmission grating, except that the direction of travel of the wavelets is reversed. Reflection gratings are used in high-resolution spectroscopy of astronomical x-ray sources. The spectra enable scientists to identify elements such as iron, oxygen, silicon, and magnesium in the corona of a star or in the remnants of a supernova.

Application: reflection gratings in astronomy

PHYSICS AT HOME

A compact disc can be used as a reflection grating, since it has a large number of equally spaced reflective tracks. Hold a CD at an angle so that the side without the label reflects light from the Sun or another light source. Tilt the CD back and forth slightly and look for the rainbow of colors that results from the interference of light reflecting from the narrowly spaced grooves of the CD. Next place the CD, label side down, on the floor directly below a ceiling light. Look down at the CD as you slowly walk away from it. The first-order maxima form a band of colors (violet to red). Once you are a meter or so away, gradually lower your head to the floor, watching the CD the whole time. You have now observed from $\theta = 0$ to $\theta = 90°$. Count how many orders of maxima you see for the different colors. Now estimate the spacing between tracks on the CD.

25.6 DIFFRACTION AND HUYGENS'S PRINCIPLE

Suppose a plane wave approaches an obstacle. Using geometric optics, we would expect the rays not blocked by the obstacle to continue straight ahead, forming a sharp, well-defined shadow on a screen beyond the obstacle. If the obstacle is large relative to the wavelength, then geometric optics gives a good *approximation* to what actually happens. If the obstacle is *not* large compared with the wavelength, we must return to Huygens's principle to show how a wave diffracts.

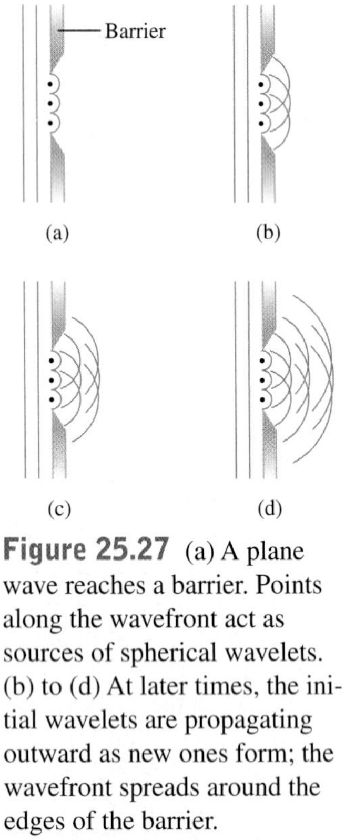

Figure 25.27 (a) A plane wave reaches a barrier. Points along the wavefront act as sources of spherical wavelets. (b) to (d) At later times, the initial wavelets are propagating outward as new ones form; the wavefront spreads around the edges of the barrier.

In Fig. 25.27a, a wavefront just reaches a barrier with an opening in it. Every point on that wavefront acts as a source of *spherical* wavelets. Points on the wavefront that are behind the barrier have their wavelets absorbed or reflected. Therefore, the propagation of the wave is determined by the wavelets generated by the unobstructed part of the wavefront. The Huygens constructions in Figs. 25.27b–d show that the wave diffracts around the edges of the barrier, something that would not be expected in geometric optics.

Figure 25.28 shows water waves in a ripple tank that pass through three openings of different widths. For the opening that is much wider than the wavelength (Fig. 25.28a), the spreading of the wavefronts is a small effect. Essentially, the part of the wavefront that is not obstructed just travels straight ahead, producing a sharp shadow. As the opening gets narrower (Fig. 25.28b), the spreading of the wavefronts becomes more pronounced. Diffraction is appreciable when the size of the opening approaches the size of the wavelength or is even smaller. In the case of Fig. 25.28c, where the opening is about the size of the wavelength, the opening acts almost like a point source of circular waves.

For the openings of intermediate size, a careful look at the waves shows that the amplitude is larger in some directions than in others (Fig. 25.28b). The source of this structure, due to the interference of wavelets from different points, is examined in Section 25.7.

Since EM waves are three-dimensional, we must be careful when interpreting two-dimensional sketches of Huygens wavelets. Figure 25.29a might represent light incident on a small circular hole or a long, thin slit. If it represents a hole, the light spreads in all directions, yielding spherically shaped wavefronts (Fig. 25.29b). If the opening represents a slit, we can think of the two directions in turn. The more narrowly restricted the wavefront, the more it spreads out. In the direction of the length of the slit, we get

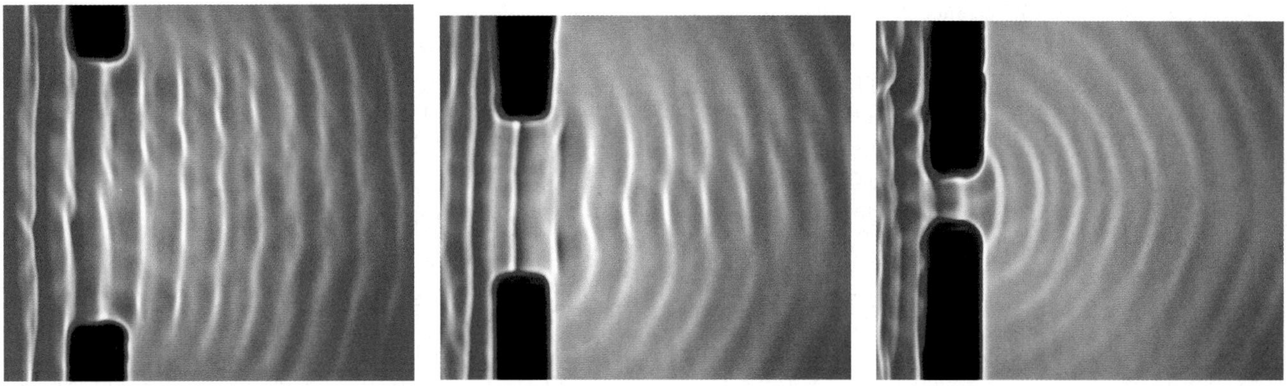

(a) (b) (c)

Figure 25.28 Water waves moving from left to right in a ripple tank pass through openings of different widths.

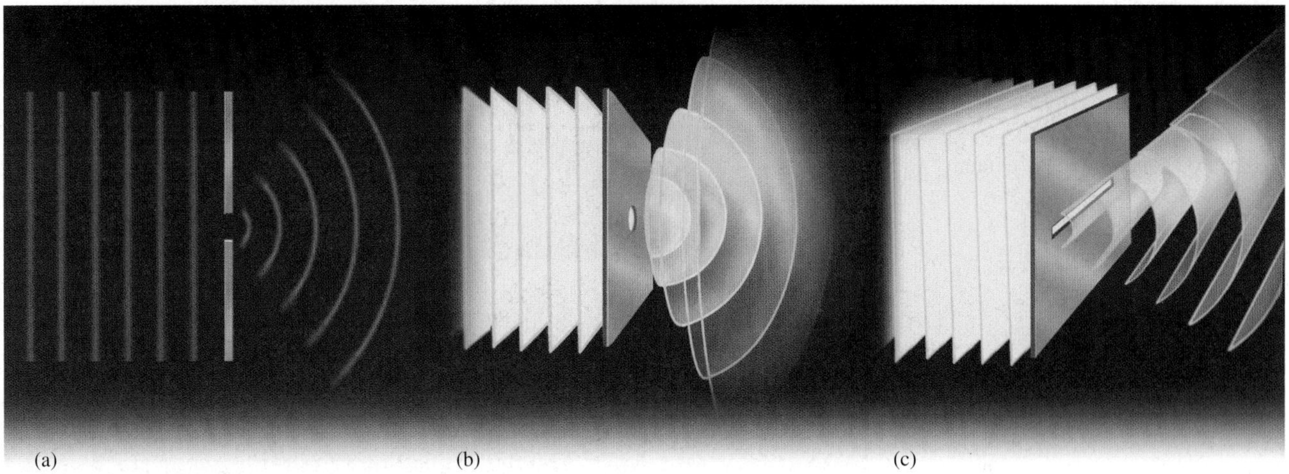

(a) (b) (c)

Figure 25.29 (a) Sketch of wavefronts that could represent either a small circular hole or a slit. (b) For a small circular hole, the emerging wavefronts are spherical. (c) For a slit, the emerging wavefronts are cylindrical.

essentially a geometric shadow. In the direction of the width, the wavefront is restricted to a short distance, so the wave spreads out in that direction. The wavefronts past the slit are cylindrically shaped (Fig. 25.29c).

Conceptual Example 25.7

Diffraction and Photolithography

The CPU (central processing unit) chip in a computer contains about 10^8 transistors, numerous other circuit elements, and the electric connections between them, all in a very small package. One process used to fabricate such a chip is called photolithography. In photolithography, a silicon wafer is coated with a photosensitive material. The chip is then exposed to ultraviolet radiation through a mask that contains the desired pattern of material to be removed. The wafer is then etched. The areas of the wafer not exposed to UV are not susceptible to etching. In areas that were exposed to UV, the photosensitive material and part of the silicon underneath are removed. Why does this process work better with UV than it would with visible light? Why are researchers trying to develop x-ray lithography to replace UV lithography?

Strategy Without knowing details of the chemical processes involved, we think about the implications of different wavelengths. X-ray wavelengths are shorter than UV wavelengths, which are in turn shorter than visible wavelengths.

continued on next page

Conceptual Example 25.7 continued

Solution and Discussion The photolithography process depends on the formation of a *sharp shadow* of the mask. To make smaller chips contain more and more circuit elements, the lines in the mask must be made as thin as possible. The danger is that if the lines are too thin, diffraction will spread out the light going through the mask. To minimize diffraction effects, the wavelength should be small compared with the openings in the mask. UV has smaller wavelengths than visible light, so the openings in the mask can be made smaller. X-ray lithography would permit even smaller openings.

Conceptual Practice Problem 25.7 Sunlight Through a Window

Sunlight streams through a rectangular window, illuminating a bright area on the floor. The edges of the illuminated area are fuzzy rather than sharp. Is the fuzziness due to diffraction? Explain. If not diffraction, what does blur the boundaries of the illuminated area?

Figure 25.30 (a) Diffraction pattern formed when an old-fashioned razor blade is illuminated with laser light. (b) Diffraction pattern formed by a small sphere. Note the bright Poisson spot at the center.

PHYSICS AT HOME

Figure 25.30a shows the shadow of an old-fashioned razor blade illuminated by coherent laser light. Bright and dark fringes are formed by the diffraction of light. One of the counterintuitive predictions of the wave theory of light is that the shadow of a circular or spherical object in coherent light should have a *bright spot at the center* due to diffraction (Fig. 25.30b). Augustin-Jean Fresnel's (1788–1827) prediction of this bright spot was considered by some eminent scientists of the nineteenth century (such as Siméon-Denis Poisson, 1781–1840) to be ridiculous—until it was shown experimentally to exist. You can see the Poisson spot yourself. Use superglue to attach a ball bearing or small opaque marble to a glass microscope slide. View the shadow of the ball bearing by holding it near your eye so that it blocks a distant source of light. The source must be distant enough to act like a point source; there should be a minimal amount of light from other sources. Try a bright streetlight at night or a single lightbulb in an otherwise dark room. Remember that it is never safe to stare directly into a bright light.

PHYSICS AT HOME

Find a finely woven piece of cloth with a regular mesh pattern, such as a piece of silk, a nylon curtain, an umbrella, or a piece of lingerie. Look through the cloth at a distant, bright light source in an otherwise darkened room—or at a streetlight outside at night. Can you explain the origin of the pattern you see? Could it be simply a geometric shadow of the threads in the cloth? Observe the pattern as you rotate the cloth. Also try stretching the cloth slightly in one direction.

25.7 DIFFRACTION BY A SINGLE SLIT

In a more detailed treatment of diffraction, we must consider the *phases* of all the Huygens wavelets and apply the principle of superposition. Interference of the wavelets causes structure in the diffracted light. In the ripple tanks of Fig. 25.28, we saw structure in the diffraction pattern. In some directions, the wave amplitude was large; in other directions it was small. Figure 25.31 shows the diffraction pattern formed by light passing through a single slit. A wide central maximum contains most of the light energy. (**Central maximum** is the usual way to refer to the entire bright band in the center of the pattern, although the actual *maximum* is just at $\theta = 0$. A more accurate name is *central bright fringe*.) The intensity is brightest right at the center and falls off gradually until the first minimum on either side, where the screen is dark (intensity is zero). Continuing away from the center, maxima and minima alternate, with the intensity changing gradually between them. The lateral maxima are quite weak compared with the central maximum and they are not as wide.

According to Huygens's principle, the diffraction of the light is explained by considering every point along the slit as a source of wavelets (Fig. 25.32a). The light intensity at any point beyond the slit is the *superposition* of these wavelets. The wavelets start out in phase, but travel different distances to reach a given point on the screen. The structure in the diffraction pattern is a result of *the interference of the wavelets*. This is

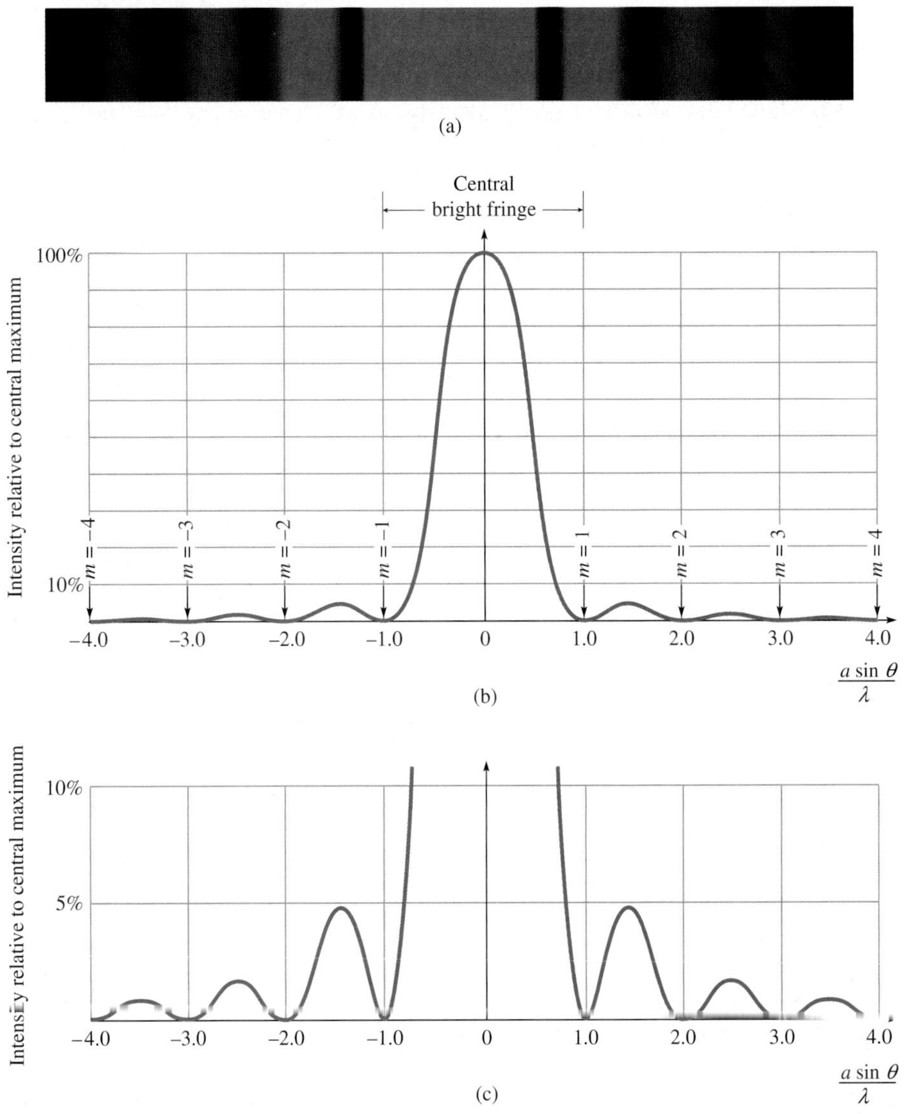

Figure 25.31 Single-slit diffraction. (a) Photo of the diffraction pattern as viewed on a screen. (b) Intensity (as a percentage of the intensity of the central maximum) as a function of the number of wavelengths difference in the path length from top and bottom rays [$(a \sin \theta)/\lambda$]. Minima occur at angles where $(a \sin \theta)/\lambda$ is an integer other than zero. (c) Close-up of the same graph. Intensities of the first three lateral maxima (as percentages) are 4.72%, 1.65%, and 0.834%. The first three lateral maxima occur when $a \sin \theta = 1.43\lambda$, 2.46λ, and 3.47λ.

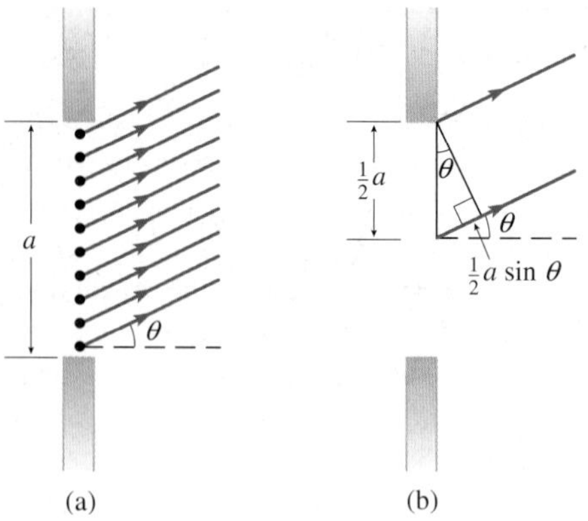

Figure 25.32 (a) Every point along a slit serves as a source of Huygens's wavelets. (b) The ray from the center of the slit travels a greater distance to reach the screen than the ray from the top of the slit; the extra distance is $\frac{1}{2}a \sin\theta$.

a much more complicated interference problem than any we have considered, because an *infinite* number of waves interfere—*every point* along the slit is a source of wavelets. Despite this complication, a clever insight—similar to one we used with the grating—lets us find out where the minima are without the need to resort to complicated math.

Finding the Minima Figure 25.32b shows two rays that represent the propagation of two wavelets: one from the top edge of the slit and one from exactly halfway down. The rays are going off at the same angle θ to reach the same point on a *distant* screen. The lower one travels an extra distance $\frac{1}{2}a \sin\theta$ to reach the screen. If this extra distance is equal to $\frac{1}{2}\lambda$, then *these two wavelets* interfere destructively. Now let's look at two other wavelets, shifted down a distance Δx so that they are still separated by half the slit width ($\frac{1}{2}a$). The path difference between these two must also be $\frac{1}{2}\lambda$, so these two interfere destructively. *All the wavelets* can be paired off; since each pair interferes destructively, no light reaches the screen at that angle. Therefore, the first diffraction minimum occurs where

$$\frac{1}{2}a \sin\theta = \frac{1}{2}\lambda$$

or

$$a \sin\theta = \lambda$$

The other minima are found in a similar way, by pairing off wavelets separated by a distance of $\frac{1}{4}a, \frac{1}{6}a, \frac{1}{8}a, \ldots, \frac{1}{2m}a$, where m is any integer *other than zero*. The diffraction minima are given by

$$\frac{1}{2m} a \sin\theta = \frac{1}{2}\lambda \quad (m = \pm1, \pm2, \pm3, \ldots)$$

Simplifying algebraically yields:

Single-slit diffraction *minima*:

$$a \sin\theta = m\lambda \quad (m = \pm1, \pm2, \pm3, \ldots) \qquad (25\text{-}12)$$

Be careful: Eq. (25-12) looks a lot like Eq. (25-10) for the interference maxima due to N slits, but it gives the locations of the diffraction *minima*. Also, $m = 0$ is excluded in Eq. (25-12); a maximum, not a minimum, occurs at $\theta = 0$.

What happens if the slit is made narrower? As a gets smaller, the angles θ for the minima get larger—the diffraction pattern spreads out. If the slit is made wider, then the diffraction pattern shrinks as the angles for the minima get smaller.

The angles at which the lateral maxima occur are much harder to find than the angles of the minima; there is no comparable simplification we can use. The central maximum is at $\theta = 0$, since the wavelets all travel the same distance to the screen and arrive in phase. The other maxima are *approximately* (not exactly) halfway between adjacent minima (see Fig. 25.31c).

Example 25.8

Single-Slit Diffraction

The diffraction pattern from a single slit of width 0.020 mm is viewed on a screen. If the screen is 1.20 m from the slit and light of wavelength 430 nm is used, what is the width of the central maximum?

Strategy The central maximum extends from the $m = -1$ minimum to the $m = +1$ minimum. Since the pattern is symmetrical, the width is twice the distance from the center to the $m = +1$ minimum. A sketch helps relate the angles and distances in the problem.

Solution The $m = 1$ minimum occurs at an angle θ satisfying

$$a \sin \theta = \lambda$$

We draw a sketch (Fig. 25.33) showing the angle θ for the $m = 1$ minimum, the distance x from the center of the diffraction pattern to the first minimum, and the distance D from the slit to the screen. The width of the central maximum is $2x$. From Fig. 25.33,

$$\tan \theta = \frac{x}{D}$$

Assuming that $x \ll D$, θ is a small angle. Therefore, $\sin \theta \approx \tan \theta$:

$$\frac{x}{D} = \frac{\lambda}{a}$$

$$x = \frac{\lambda D}{a} = \frac{430 \times 10^{-9}\ \text{m} \times 1.20\ \text{m}}{0.020 \times 10^{-3}\ \text{m}} = 0.026\ \text{m}$$

Comparing the values of x and D, our assumption that $x \ll D$ is justified. The width of the central maximum is $2x = 5.2$ cm.

Discussion The width of the central maximum depends on the angle θ for the first minimum and the distance D between the slit and the screen. The angle θ, in turn, depends on the wavelength of light and the slit width. For larger values of θ, which means either a longer wavelength or a smaller slit width, the diffraction pattern is more spread out on the screen. For a given wavelength, narrowing the slit increases the diffraction. For a given slit width, the diffraction pattern is wider for longer wavelengths so the pattern for red light ($\lambda = 690$ nm) is more spread out than that for violet light ($\lambda = 410$ nm).

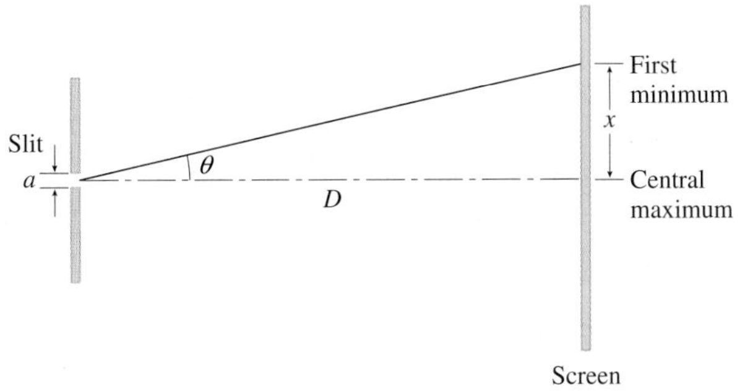

Figure 25.33

A diffraction pattern is formed on a distant screen by light of wavelength λ from a single slit of width a at a distance D from the screen.

Practice Problem 25.8 Location of First Lateral Maximum

Approximately how far from the center of the diffraction pattern is the first lateral maximum?

Intensities of the Maxima in Double-Slit Interference

In a double-slit interference experiment, the bright fringes are equally spaced but are not equal in intensity (see Fig. 25.18). Light diffracts from each slit; the light reaching the screen from either slit forms a diffraction pattern (see Fig. 25.31). The two diffraction patterns have the same amplitude at any point on the screen, but different phases. Where the interference is constructive, the amplitude is twice what it would be at that point for a single slit (and therefore four times the intensity).

Figure 25.18 shows only interference maxima within the central diffraction maximum of each slit. If the light incident on the slits is bright enough, interference maxima beyond the first diffraction minimum can be observed (Fig. 25.34).

25.8 DIFFRACTION AND THE RESOLUTION OF OPTICAL INSTRUMENTS

Cameras, telescopes, binoculars, microscopes—practically all optical instruments, including the human eye—admit light through circular apertures. Thus, the diffraction of light through a circular aperture is of great importance. If an instrument is to resolve

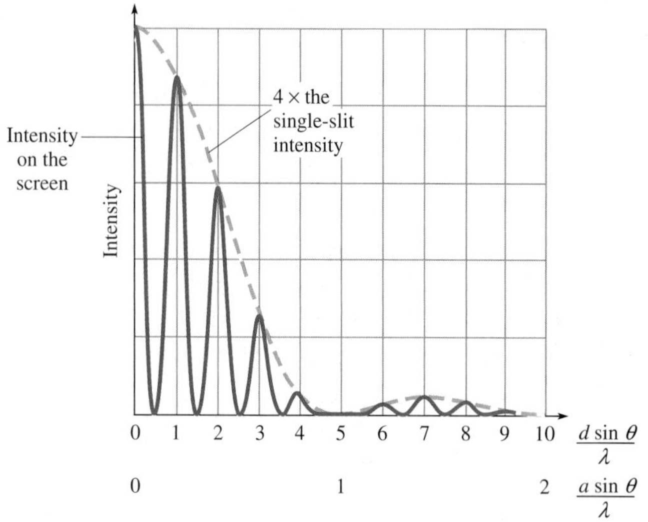

Figure 25.34 A graph of the intensity for double-slit interference where the spacing d between the two slits is five times the slit width a (that is, $d = 5a$). The first *diffraction* minimum occurs where $a \sin \theta = \lambda$; at that same angle, $5a \sin \theta = d \sin \theta = 5\lambda$. The fifth-order interference maximum is missing because it falls at the first diffraction minimum, where no light reaches the screen. The peak heights follow the intensity pattern for a single slit. At points of constructive interference, the amplitude is twice what it would be from one slit alone, so the intensity is *four* times what it would be from one slit.

Figure 25.35 Diffraction pattern from a circular aperture on a distant screen.

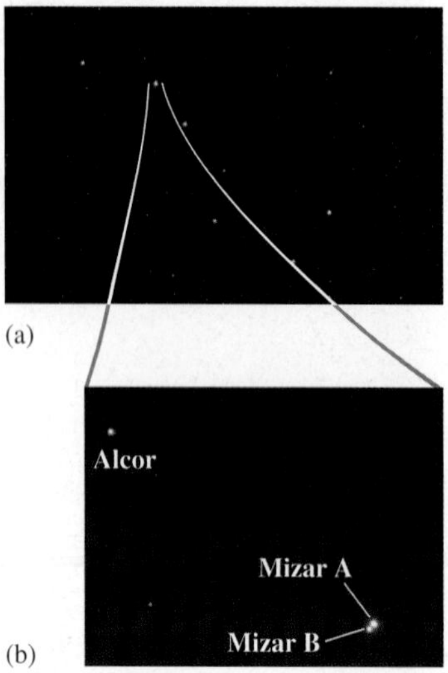

Figure 25.36 (a) The Big Dipper, a part of the constellation Ursa Major. (b) A telescope with a wide aperture reveals distinct images for Mizar A, Mizar B, and Alcor.

(distinguish) two objects as being separate entities, it must form separate images of the two. If diffraction spreads out the image of each object enough that they overlap, the instrument cannot resolve them.

When light passes through a circular aperture of diameter a, the light is restricted (the wavefronts are blocked) in *all* directions rather than being restricted primarily in a single direction (as for a slit). Thus, for a circular opening, light spreads out in all directions. The diffraction pattern due to a circular aperture (Fig. 25.35) reflects the circular symmetry of the aperture. The diffraction pattern has many similarities to that of a slit. It has a wide, bright central maximum, beyond which minima and weaker maxima alternate; but now the pattern consists of concentric circles reflecting the circular shape of the aperture.

Calculating the angles for the minima and maxima is a difficult problem. Of greatest interest to us is the location of the *first* minimum, which is given by

$$a \sin \theta \approx 1.22\lambda \qquad (25\text{-}13)$$

The reason that the first minimum is of particular interest is that it tells us the diameter of the central maximum, which contains 84% of the intensity of the diffracted light. The size of the central maximum is what limits the resolution of an optical instrument.

When we look at a distant star through a telescope, the star is far enough to be considered a point source, but since the light passes through the circular aperture of the telescope, it spreads out into a circular diffraction pattern like Fig. 25.35. What if we look at two or more stars that appear close to each other? With the unaided eye, people with good vision can see two separate stars, Mizar and Alcor, in the handle of the Big Dipper (Fig. 25.36a). With a telescope, one can see that Mizar is actually *two* stars, called Mizar A and Mizar B (Fig. 25.36b); the eye cannot resolve (separate) the images of these two stars, but a telescope with its much wider aperture can. Spectroscopic observations reveal periodic Doppler shifts in the light coming from Mizar A and Mizar B, showing that each is a *binary star*—a pair of stars so close together that they rotate about their common center of mass. The companion stars to Mizar A and Mizar B cannot be seen with even the best telescopes available. When light rays from these five stars pass through a circular aperture, diffraction spreads out the images, so that we see only three stars through a telescope or two stars when viewed directly.

Rayleigh's Criterion

Light from a single star (or other point source) forms a circular diffraction pattern after passing through a circular aperture. Two stars with a small angular separation form two

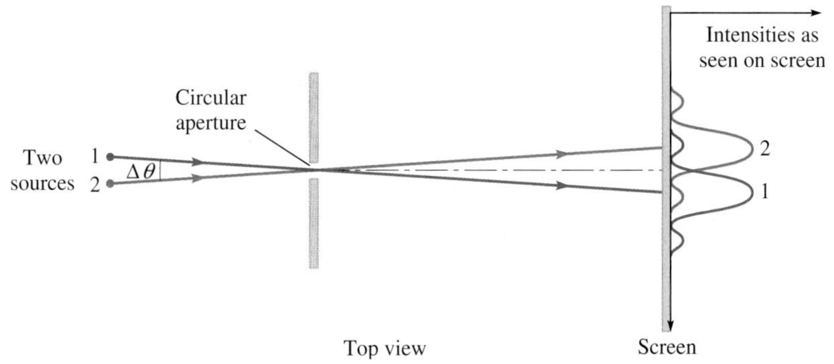

Figure 25.37 Two point sources with an angular separation $\Delta\theta$ form overlapping diffraction patterns when the light passes through a circular aperture. In this case, the images *can* be resolved according to Rayleigh's criterion.

overlapping diffraction patterns. Since the stars are incoherent sources, their diffraction patterns overlap without interfering with each other (Fig. 25.37). How far apart must the diffraction patterns be in order to resolve the stars?

A somewhat arbitrary but conventional criterion is due to Baron Rayleigh (John William Strutt, 1842–1919) who said that the images must be separated by at least half the width of each of the diffraction patterns. In other words, **Rayleigh's criterion** says that two sources can just barely be resolved if the center of one diffraction pattern falls at the first minimum of the other one. Suppose light from two sources travels through vacuum (or air) and enters a circular aperture of diameter a. If $\Delta\theta$ is the angular separation of the two sources as measured from the aperture and λ_0 is the wavelength of the light in vacuum (or air), then the sources can be resolved if

Rayleigh's criterion:

$$a \sin \Delta\theta \geq 1.22\lambda_0 \tag{25-14}$$

Example 25.9

Resolution with a Laser Printer

A laser printer puts tiny dots of ink (toner) on the page. The dots should be sufficiently close together (and therefore small enough) that we don't see individual dots; rather, we see letters or graphics. Approximately how many dots per inch (dpi) ensure that you don't see individual dots when viewing a page 0.40 m from the eye in bright light? Use a pupil diameter of 2.5 mm.

Strategy If the angular separation of the dots exceeds Rayleigh's criterion, then you are able to resolve individual dots. Therefore, the angular separation of the dots must be *smaller* than that given by Rayleigh's criterion—we do *not* want to be able to resolve individual dots.

Solution Call the distance between the centers of two adjacent dots Δx, the diameter of the pupil a, and the angular separation of the dots $\Delta\theta$ (Fig. 25.38). The page is held a distance $D = 0.40$ m from the eye. Then, since $\Delta x \ll D$, the angular separation of the dots is

$$\Delta\theta \approx \frac{\Delta x}{D}$$

In order for the dots to merge, the angular separation $\Delta\theta$ must be *smaller* than the angle given by the Rayleigh criterion for resolution. The minimum $\Delta\theta$ for resolution is given by

$$a \sin \Delta\theta \approx a\,\Delta\theta = 1.22\lambda_0$$

Since we do *not* want the dots to be resolved, we want

$$a\,\Delta\theta < 1.22\lambda_0$$

Substituting for $\Delta\theta$,

$$a\frac{\Delta x}{D} < 1.22\lambda_0$$

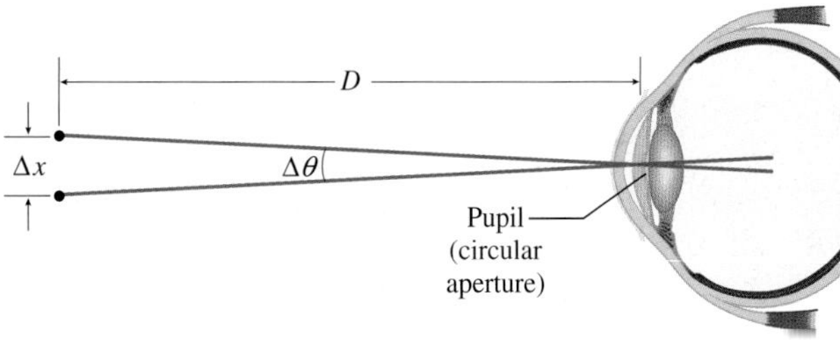

Figure 25.38
Angular separation $\Delta\theta$ of two adjacent dots.

continued on next page

Example 25.9 continued

To guarantee that Δx is small enough so that the dots blend together for *all* visible wavelengths, we take $\lambda_0 = 400.0$ nm, the smallest wavelength in the visible range. Now we solve for the distance between dots Δx:

$$\Delta x < \frac{1.22\lambda_0 D}{a} = \frac{1.22 \times 400.0 \text{ nm} \times 0.40 \text{ m}}{0.0025 \text{ m}}$$

$$= 7.81 \times 10^{-5} \text{ m} = 0.0781 \text{ mm}$$

To find the minimum number of dots per *inch*, first convert the dot separation Δx to inches.

$$\Delta x = 0.0781 \text{ mm} \div 25.4 \frac{\text{mm}}{\text{in}} = 0.00307 \text{ in}$$

$$\text{dots per inch} = \frac{1}{\text{inches per dot}}$$

$$\frac{1}{0.00307 \text{ inches/dot}} = 330 \text{ dpi}$$

Discussion Based on this estimate, we expect the printout from a 300-dpi printer to be slightly grainy, since we can just barely resolve individual dots. Output from a 600-dpi printer should look smooth.

You might wonder whether Eq. (25-14) applies to diffraction that occurs within the eye since it uses the wavelength in vacuum (λ_0). The wavelength in the vitreous fluid of the eye is $\lambda = \lambda_0/n$, where $n \approx 1.36$ is the index of refraction of the vitreous fluid. Equation (25-14) *does* apply in this situation because the factor of n in the wavelength is canceled by a factor of n due to refraction (see Problem 57).

Practice Problem 25.9 Pointillist Paintings

The Postimpressionist painter Georges Seurat perfected a technique known as *pointillism*, in which paintings are composed of closely spaced dots of different colors, each about 2 mm in diameter (Fig. 25.39). A close-up view reveals the individual dots; from farther away the dots blend together. Estimate the minimum distance away a viewer should be in order to see the dots blend into a smooth variation of colors. Assume a pupil diameter of 2.2 mm.

(a) (b)

Figure 25.39
(a) *La grève du Bas Butin à Honfleur* by Georges Seurat (1859–1891). (b) A close-up view of the same painting.

Application: Resolution of the Human Eye

In bright light, the pupil of the eye narrows to about 2 mm; diffraction caused by this small aperture limits the resolution of the human eye. In dim light, the pupil is much wider. Now the limit on the eye's resolution in dim light is not diffraction, but the spacing of the photoreceptor cells on the fovea (where they are most densely packed). For an *average* pupil diameter, the spacing of the cones is optimal (see Problem 54). If the cones were less densely packed, resolution would be lost; if they were more densely packed, there would be no gain in resolution due to diffraction.

25.9 X-RAY DIFFRACTION

Application: x-ray diffraction

The interference and diffraction examples discussed so far have dealt mostly with visible light. However, the same effects occur for wavelengths longer and shorter than those visible to our eyes. Is it possible to do an experiment that shows interference or diffraction

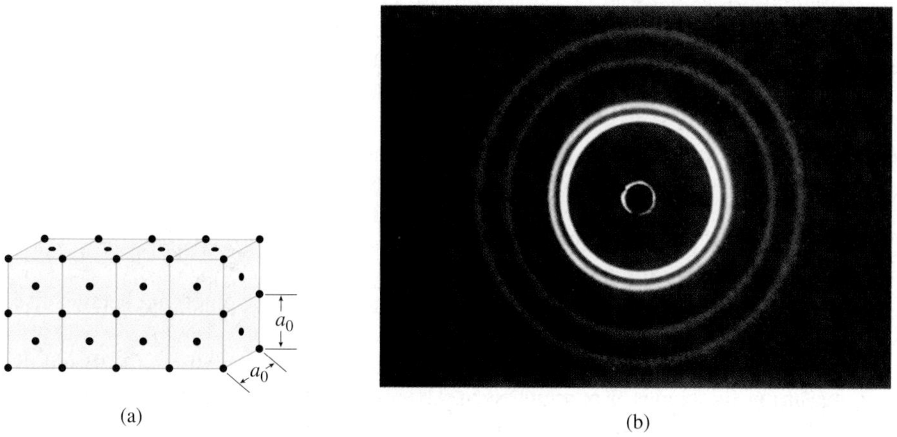

Figure 25.40 (a) Crystal structure of aluminum. The dots represent the positions of the aluminum atoms. (b) The x-ray diffraction pattern formed by polycrystalline aluminum (a large number of randomly oriented aluminum crystals). The central spot, which is formed by x-rays that are not scattered by the sample, has been mostly blocked from reaching the film. Rings form at angles for which the scattered x-rays interfere constructively.

effects with x-rays? X-ray radiation has wavelengths much shorter than those of visible light, so to do such an experiment, the size and spacing of the slits in a grating (for example) would have to be much smaller than in a visible-light grating. Typical x-ray wavelengths range from about 10 nm to about 0.01 nm. There is no way to make a parallel-slit grating small enough to work for x-rays: the diameter of an atom is typically around 0.2 nm, so the slit spacing would be about the size of a single atom.

In 1912, the German physicist Max von Laue (1879–1960) realized that the regular arrangement of atoms in a crystal makes a perfect grating for x-rays. The regular arrangement and spacing of the atoms is analogous to the regular spacing of the slits in a conventional grating, but a crystal is a *three-dimensional* grating (as opposed to the two-dimensional gratings we use for visible light).

Figure 25.40a shows the atomic structure of aluminum. When a beam of x-rays passes through the crystal, the x-rays are scattered in all directions by the atoms. The x-rays scattered in a particular direction from different atoms interfere with each other. In certain directions they interfere constructively, giving maximum intensity in those directions. Photographic film records those directions as a collection of spots for a single crystal, or as a series of rings for a sample consisting of many randomly oriented crystals (Fig. 25.40b).

Determining the directions for constructive interference is a difficult problem due to the three-dimensional structure of the grating. Australian physicist William Lawrence Bragg (1890–1971) discovered a great simplification. He showed that we can think of the x-rays *as if they reflect from planes of atoms* (Fig. 25.41a). Constructive interference occurs if the path difference between x-rays reflecting from an adjacent pair of planes is an integral multiple of the wavelength. Figure 25.41b shows that the path difference is $2d \sin \theta$, where d is the distance between the planes and θ is the angle that the incident and reflected beams make with the plane (*not* with the normal). Then, constructive interference occurs at angles given by **Bragg's law**:

X-ray diffraction maxima:

$$2d \sin \theta = m\lambda \quad (m = 1, 2, 3, \ldots) \tag{25-15}$$

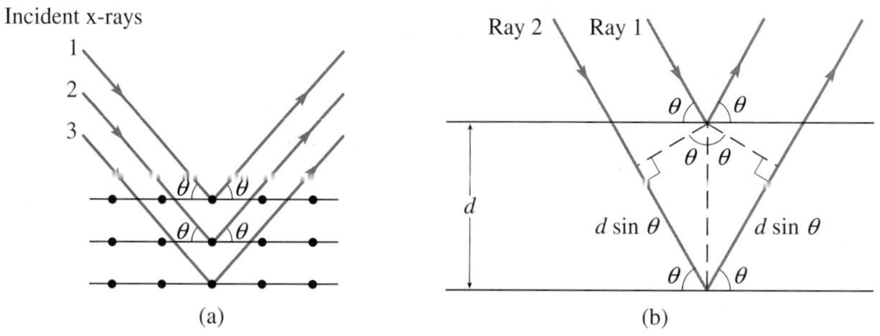

Figure 25.41 (a) Incident x-rays behave as if they reflect from parallel planes of atoms. (b) Geometry for finding the path difference for rays reflecting from two adjacent planes.

Although Bragg's law is a great simplification, x-ray diffraction is still complicated because there are many sets of parallel planes in a crystal, each with its own plane spacing. In practice, the largest plane spacings contain the largest number of scattering centers (atoms) per unit area, so they produce the strongest maxima.

Applications of X-Ray Diffraction

- Just as a grating separates white light into the colors of the spectrum, a crystal is used to extract an x-ray beam with a narrow range of wavelengths from a beam with a continuous x-ray spectrum.
- If the structure of the crystal is known, then the angle of the emerging beam is used to determine the wavelength of the x-rays.
- The x-ray diffraction pattern can be used to determine the structure of a crystal. By measuring the angles θ at which strong beams emerge from the crystal, the plane spacings d are found and from them the crystal structure.
- X-ray diffraction patterns are used to determine the molecular structures of biological molecules such as proteins. X-ray diffraction studies by Rosalind Franklin (1920–1958) were a key clue to James Watson (b. 1928) and Francis Crick (1916–2004) in their 1953 discovery of the double helix structure of DNA (Fig. 25.42). Intense beams of x-rays radiated by electrons in synchrotrons have even been used to study the structure of viruses.

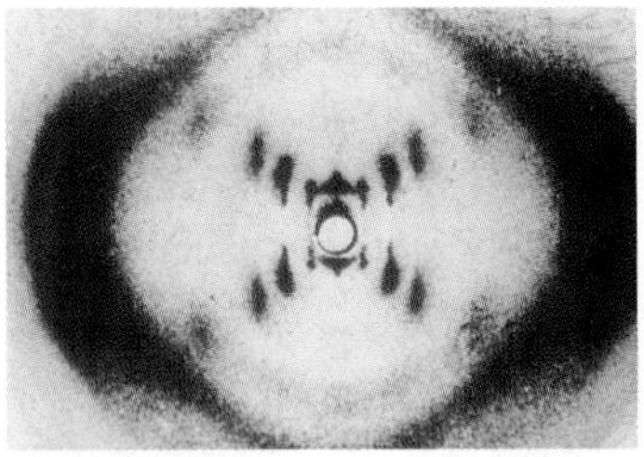

Figure 25.42 This x-ray diffraction pattern of DNA (deoxyribonucleic acid) was obtained by Rosalind Franklin in 1953. Some aspects of the structure of DNA can be deduced from the pattern of spots and bands. Franklin's data convinced James Watson and Francis Crick of DNA's helical structure, which is revealed by the cross of bands in the diffraction pattern.

25.10 HOLOGRAPHY

An ordinary photograph is a record of the intensity of light that falls on the film at each point. For incoherent light, the phases vary randomly, so it would not be useful to record phase information. A hologram is made by illuminating the subject with *coherent* light; the hologram is a record of the intensity *and the phase* of the light incident on the film. Holography was invented in 1948 by Hungarian-British physicist Dennis Gabor (1900–1979), but holograms were difficult to make until lasers became available in the 1960s.

Imagine using a laser, a beam splitter, and some mirrors to produce two coherent plane waves of light that overlap but travel in different directions (Fig. 25.43). Let the waves fall on a photographic plate. The exposure of the plate at any point depends on the intensity of the light falling on it. Since the two waves are coherent, a series of parallel fringes of constructive and destructive interference occur. The spacing between fringes depends on the angle θ_0 between the two waves; a smaller angle makes the spacing between fringes larger. In Problem 80, the spacing between fringes is found to be

$$d = \frac{\lambda}{\sin \theta_0}$$

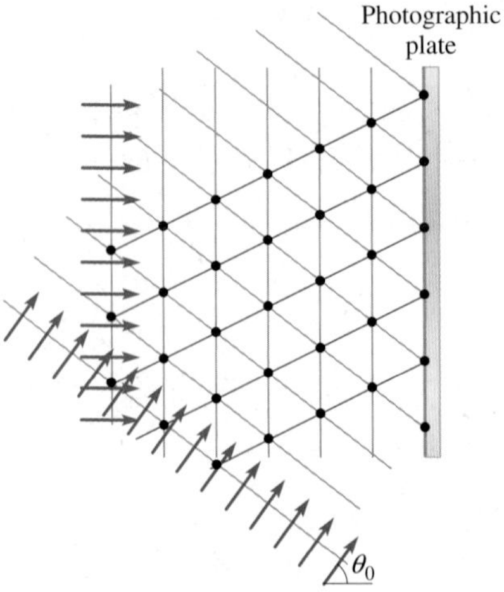

Figure 25.43 Two coherent plane waves traveling in different directions expose a photographic plate. An interference pattern is formed on the plate. The red lines indicate points of constructive interference between the two waves. Bright fringes occur where these lines intersect the photographic plate.

When the plate is developed, the equally spaced fringes make a grating. If the plate is illuminated at normal incidence with coherent light at the same wavelength λ, the central ($m = 0$) maximum is straight ahead, while the $m = 1$ maximum is at an angle given by

$$\sin\theta = \frac{\lambda}{d} = \sin\theta_0$$

Thus, the $m = 0$ and $m = 1$ maxima re-create the original two waves.

Now imagine a plane wave with a point object (Fig. 25.44). The point object scatters light, producing spherical waves just as a point source does. The interference of the original plane wave with the scattered spherical wave gives a series of circular fringes. When this plate is developed and illuminated with laser light, both the plane and spherical waves are re-created. The spherical waves appear to come from a point behind the plate, which is a virtual image of the point object. The plate is a hologram of the point object.

With a more complicated object, each point on the surface of the object is a source of spherical waves. When the hologram is illuminated with coherent light, a virtual image of the object is created. This image can be seen from different perspectives (Fig. 25.45) since the hologram *re-creates wavefronts just as if they were coming from the object.*

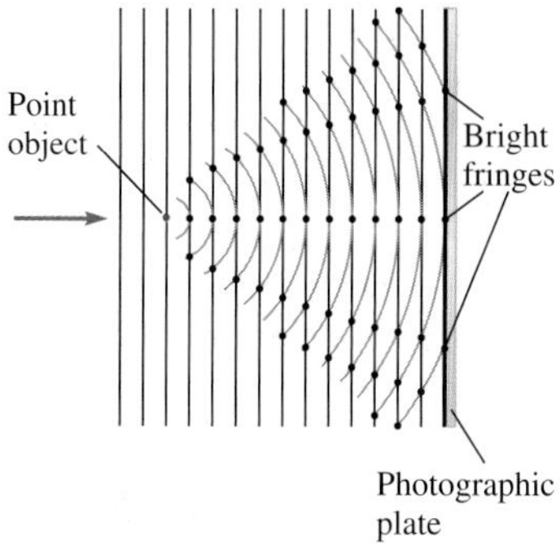

Figure 25.44 Coherent plane waves are scattered by a point object. The spherical waves scattered by the object interfere with the plane wave to form a set of circular interference fringes on a photographic plate.

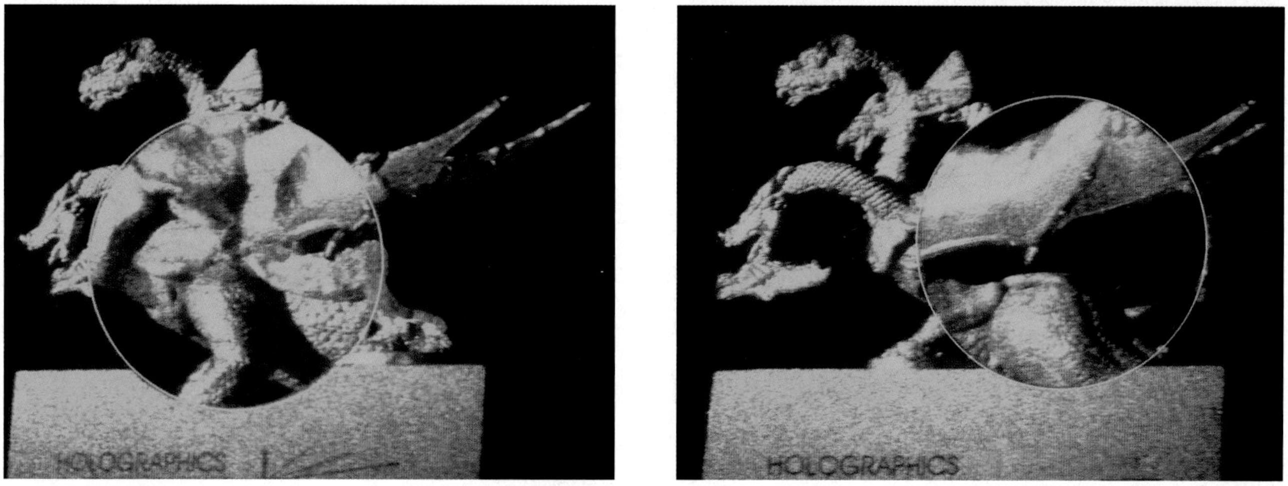

Figure 25.45 A holographic image of a dragon behind a lens. Notice that the part of the dragon that is magnified by the lens in the hologram depends on the viewing angle.

Master the Concepts

- When two coherent waves are in phase, their superposition results in constructive interference:

  Phase difference $\Delta\phi = 2m\pi$ rad ($m = 0, \pm1, \pm2, \ldots$) (25-1)

  Amplitude $A = A_1 + A_2$ (25-2)

  Intensity $I = I_1 + I_2 + 2\sqrt{I_1 I_2}$ (25-3)

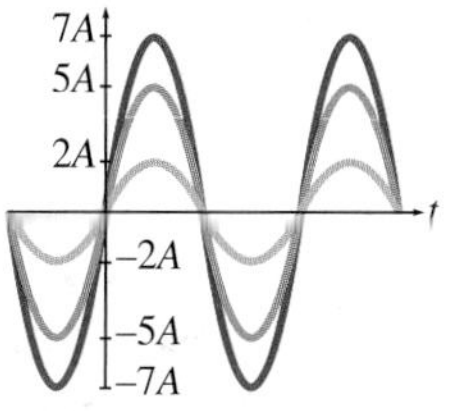

- When two coherent waves are 180° out of phase, their superposition results in destructive interference:

  Phase difference $\Delta\phi = (m + \frac{1}{2})\,2\pi$ rad ($m = 0, \pm1, \pm2, \ldots$) (25-4)

  Amplitude $A = |A_1 - A_2|$ (25-5)

  Intensity $I = I_1 + I_2 - 2\sqrt{I_1 I_2}$ (25-6)

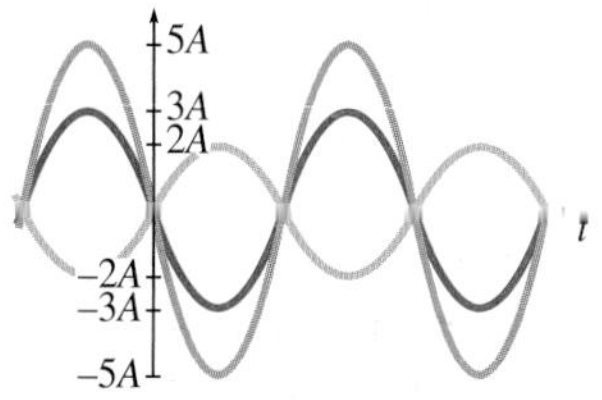

continued on next page

Master the Concepts continued

- A path length difference equal to λ causes a phase shift of 2π (360°). A path length difference of $\frac{1}{2}\lambda$ causes a phase shift of π (180°).
- When light reflects from a boundary with a slower medium (higher index of refraction), it is inverted (180° phase change); when light reflects from a faster medium (lower index of refraction), it is not inverted (no phase change).
- The angles at which the maxima and minima occur in a double-slit interference experiment are

  Maxima: $d \sin \theta = m\lambda \quad (m = 0, \pm 1, \pm 2, \ldots)$ (25-10)

  Minima: $d \sin \theta = (m + \frac{1}{2})\lambda \quad (m = 0, \pm 1, \pm 2, \ldots)$ (25-11)

  The distance between the slits is d. The absolute value of m is called the order.

- A grating with N slits produces maxima that are narrow (width $\propto 1/N$) and bright (intensity $\propto N^2$). The maxima occur at the same angles as for two slits.
- The minima in a single-slit diffraction pattern occur at angles given by

  $$a \sin \theta = m\lambda \quad (m = \pm 1, \pm 2, \pm 3, \ldots) \quad (25\text{-}12)$$

  A wide central maximum contains most of the light energy. The other maxima are approximately (not exactly) halfway between adjacent minima.

- The first minimum in the diffraction pattern due to a circular aperture is given by

  $$a \sin \theta = 1.22\lambda \quad (25\text{-}13)$$

- Rayleigh's criterion says that two sources can just barely be resolved if the center of one diffraction pattern falls at the first minimum of the other one. If $\Delta\theta$ is the angular separation of the two sources, then the sources can be resolved if

  $$a \sin \Delta\theta \geq 1.22\lambda_0 \quad (25\text{-}14)$$

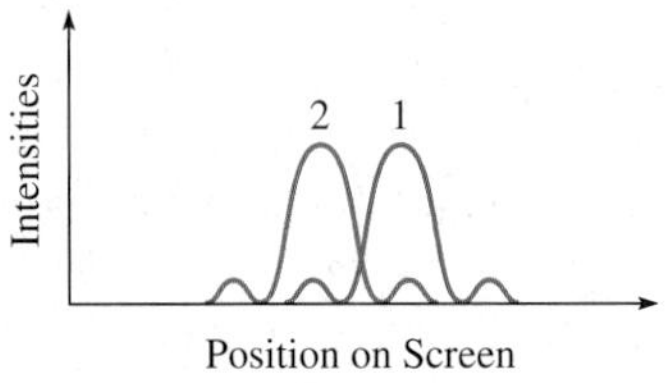

- The regular arrangement of atoms in a crystal makes a grating for x-rays. We can think of the x-rays as if they reflect from planes of atoms. Constructive interference occurs if the path difference between x-rays reflecting from an adjacent pair of planes is an integral multiple of the wavelength.
- A hologram is made by illuminating the subject with coherent light; the hologram is a record of the intensity and the phase of the light incident on the film. The hologram re-creates wavefronts just as if they were coming from the object.

Conceptual Questions

1. Explain why two waves of significantly different frequencies cannot be coherent.
2. Why do eyeglasses, cameras lenses, and binoculars with antireflective coatings often look faintly purple?
3. Telescopes used in astronomy have large lenses (or mirrors). One reason is to let a lot of light in—important for seeing faint astronomical bodies. Can you think of another reason why it is an advantage to make these telescopes so large?
4. The Hubble Space Telescope uses a mirror of radius 1.2 m. Is its resolution better when detecting visible light or UV? Explain.
5. Why can you easily hear sound around a corner due to diffraction, while you cannot see around the same corner?
6. Stereo speakers should be wired with the same polarity. If by mistake they are wired with opposite polarities, the bass (low frequencies) sound much weaker than if they are wired correctly. Why? Why is the bass (low frequencies) weakened more than the treble (high frequencies)?
7. Two antennas driven by the same electrical signal emit coherent radio waves. Is it possible for two antennas driven by *independent* signals to emit radio waves that are coherent with each other? If so, how? If not, why not?
8. A radio station wants to ensure good reception of its signal everywhere inside a city. Would it be a good idea to place several broadcasting antennas at roughly equal intervals around the perimeter of the city? Explain.
9. The size of an atom is about 0.1 nm. Can a light microscope make an image of an atom? Explain.
10. What are some of the advantages of a UV microscope over a visible light microscope? What are some of the disadvantages?
11. The *f-stop* of a camera lens is defined as the ratio of the focal length of lens to the diameter of the aperture. A large f-stop therefore means a small aperture. If diffraction is the only consideration, would you use the largest or the smallest f-stop to get the sharpest image?

12. In Section 25.3 we studied interference due to thin films. Why must the film be *thin*? Why don't we see interference effects when looking through a window or at a poster covered by a plate of glass—even if the glass is optically flat?
13. Describe what happens to a single-slit diffraction pattern as the width of the slit is slowly decreased.
14. Explain, using Huygens's principle, why the Poisson spot is expected.
15. What effect places a lower limit on the size of an object that can be clearly seen with the best optical microscope?
16. Make a sketch (similar to Fig. 25.16b) of the reflected rays from two adjacent steps of the *Morpho* butterfly wing for a large angle of incidence (around 45°). Refer to your sketch to explain why the wavelength at which constructive interference occurs depends on the viewing angle.
17. A lens ($n = 1.51$) has an antireflective coating of MgF_2 ($n = 1.38$). Which of the first two reflected rays has a phase shift of 180°? Suppose a different antireflective coating on a similar lens had $n = 1.62$. Now which of the first two reflected rays has a phase shift of 180°?
18. In the microwave experiment of Example 25.1 and in the Michelson interferometer, we ignored phase changes due to reflection from a metal surface. Microwaves and light *are* inverted when they reflect from metal. Why were we able to ignore the 180° phase shifts?
19. Why does a crystal act as a three-dimensional grating for x-rays but not for visible light?

Multiple-Choice Questions

1. If the figure shows the wavefronts for a double-slit interference experiment with light, at which of the labeled points is the intensity zero?
(a) *A* only (b) *B* only (c) C only (d) *A* and *B*
(e) *B* and *C* (f) *A* and *C* (g) *A*, *B*, and *C*

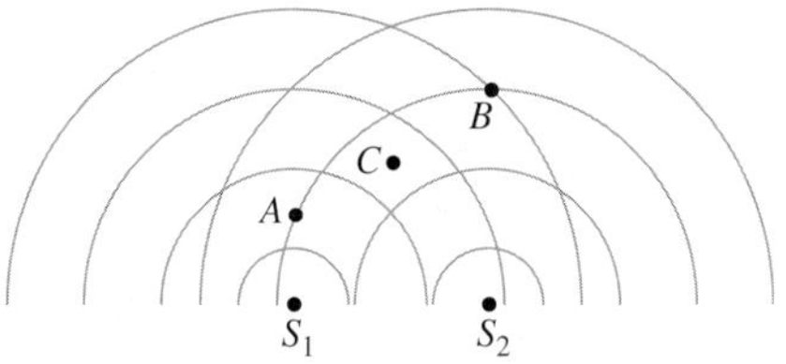

Multiple-Choice Questions 1 and 2. The wavefronts represent wave *crests* only (not crests and troughs).

2. If the figure shows the surface water waves in a ripple tank with two coherent sources, at which of the labeled points would a bit of floating cork bob up and down with the greatest amplitude? (Same answer choices as Question 1.)
3. In a double-slit experiment, light rays from the two slits that reach the second maximum on one side of the central maximum travel distances that differ by
(a) 2λ (b) λ (c) $\lambda/2$ (d) $\lambda/4$
4. A Michelson interferometer is set up for microwaves. Initially the reflectors are placed so that the detector reads a maximum. When one of the reflectors is moved 12 cm, the needle swings to minimum and back to maximum six times. What is the wavelength of the microwaves?
(a) 0.5 cm (b) 1 cm (c) 2 cm (d) 4 cm
(e) Cannot be determined from the information given.
5. In a double-slit experiment with coherent light, the intensity of the light reaching the center of the screen from one slit alone is I_0 and the intensity of the light reaching the center from the other slit alone is $9I_0$. When both slits are open, what is the intensity of the light at the interference *minima* nearest the center? The slits are very narrow.
(a) 0 (b) I_0 (c) $2I_0$
(d) $3I_0$ (e) $4I_0$ (f) $8I_0$
6. Which of these actions will improve the resolution of a microscope?
(a) increase the wavelength of the light
(b) decrease the wavelength of the light
(c) increase the diameter of the lenses
(d) decrease the diameter of the lenses
(e) both (b) and (c) (f) both (b) and (d)
(g) both (a) and (c) (h) both (a) and (d)
7. Coherent light of a single frequency passes through a double slit, with slit separation *d*, to produce a pattern of maxima and minima on a screen a distance *D* from the slits. What would cause the separation between adjacent minima on the screen to decrease? (tutorial: double slit 2)
(a) decrease the frequency of the incident light
(b) increase of the screen distance *D*
(c) decrease the separation *d* between the slits
(d) increase the index of refraction of the medium in which the setup is immersed
8. Two narrow slits, of width *a*, separated by a distance *d*, are illuminated by light with a wavelength of 660 nm. The resulting interference pattern is labeled (1) in the figure. The same light source is then used to illuminate another group of slits and produces pattern (2). The second slit arrangement is

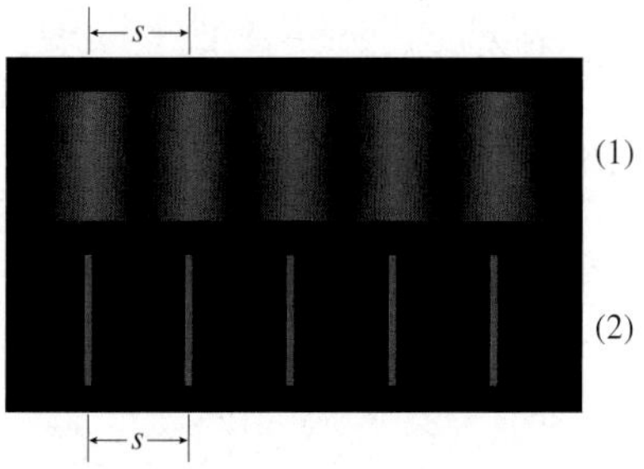

(a) many slits, spaced *d* apart.
(b) many slits, spaced 2*d* apart.
(c) two slits, each of width 2*a*, spaced *d* apart.
(d) two slits, each of width *a*/2, spaced *d* apart.

9. The figure shows the interference pattern obtained in a double-slit experiment. Which letter indicates a third-order maximum?

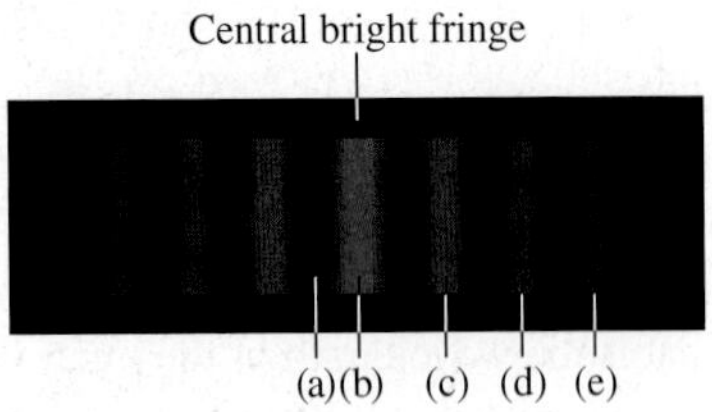

10. The intensity pattern in the diagram is due to
(a) two slits. (b) a single slit.
(c) a grating. (d) a circular aperture.

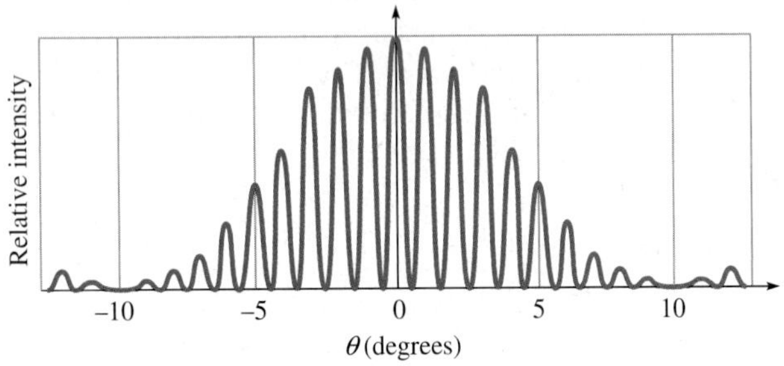

Problems

Combination conceptual/quantitative problem
Biological or medical application
Challenging problem
Blue # Detailed solution in the Student Solutions Manual
1 2 Problems paired by concept
Text website interactive or tutorial

25.1 Constructive and Destructive Interference

1. A 60-kHz radio transmitter sends an electromagnetic wave to a receiver 21 km away. The signal also travels to the receiver by another path where it reflects from a helicopter as shown. Assume that there is a 180° phase shift when the wave is reflected. (a) What is the wavelength of this EM wave? (b) Will this situation give constructive interference, destructive interference, or something in between?

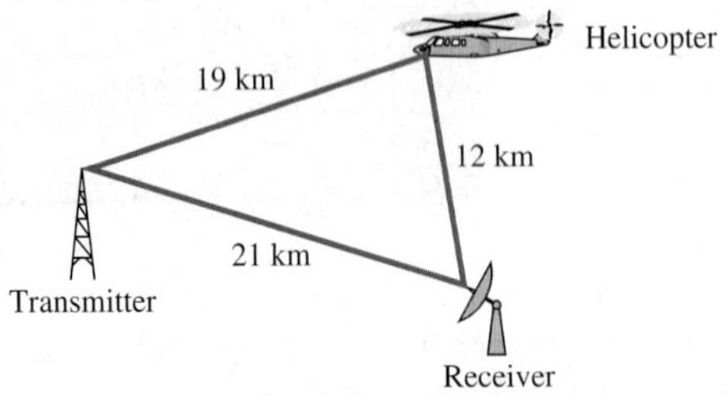

2. A steep cliff west of Lydia's home reflects a 1020-kHz radio signal from a station that is 74 km due east of her home. If there is destructive interference, what is the minimum distance of the cliff from her home? Assume there is a 180° phase shift when the wave reflects from the cliff.

3. Roger is in a ship offshore and listening to a baseball game on his radio. He notices that there is destructive interference when seaplanes from the nearby Coast Guard station are flying directly overhead at elevations of 780 m, 975 m, and 1170 m. The broadcast station is 102 km away. Assume there is a 180° phase shift when the EM waves reflect from the seaplanes. What is the frequency of the broadcast?

4. Sketch a sinusoidal wave with an amplitude of 2 cm and a wavelength of 6 cm. This wave represents the electric field portion of a visible EM wave traveling to the right with intensity I_0. (a) Sketch an identical wave beneath the first. What is the amplitude (in centimeters) of the sum of these waves? (b) What is the intensity of the new wave? (c) Sketch two more coherent waves beneath the others, one of amplitude 3 cm and one of amplitude 1 cm, so all four are in phase. What is the amplitude of the four waves added together? (d) What intensity results from adding the four waves?

5. Draw a sketch like that of Problem 4 but this time draw the third wave 180° out of phase with the others. (a) What is the amplitude of the sum of these waves? (b) What is the intensity for the four waves together? (c) Consider the case for the first three waves in phase and the fourth wave 180° out of phase. What is the amplitude for the sum of these waves? (d) What is the intensity of the wave?

6. Two incoherent EM waves of intensities $9I_0$ and $16I_0$ travel in the same direction in the same region of space. What is the intensity of EM radiation in this region?

7. When Albert turns on his small desk lamp, the light falling on his book has intensity I_0. When this is not quite enough, he turns the small lamp off and turns on a high-intensity lamp so that the light on his book has intensity $4I_0$. What is the intensity of light falling on the book when Albert turns both lamps on? If there is more than one possibility, give the range of intensity possibilities.

8. Coherent light from a laser is split into two beams with intensities I_0 and $4I_0$, respectively. What is the intensity of the light when the beams are recombined? If there is more than one possibility, give the range of possibilities. (tutorial: two waves)

9. An experiment similar to Example 25.1 is performed; the power at the receiver as a function of x is shown in the figure. (a) Approximately what is the wavelength of the microwaves? (b) What is the ratio of the *amplitudes* of the microwaves entering the detector for the two maxima shown?

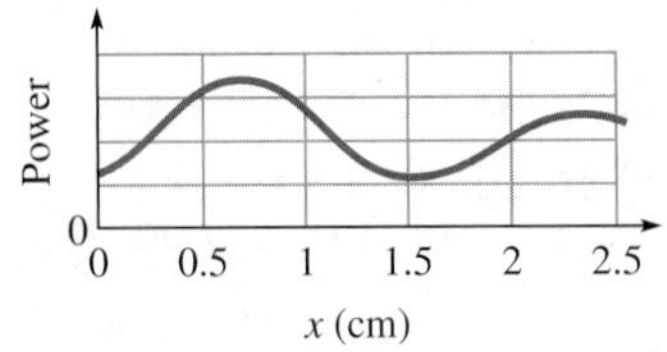

10. The feathers of the ruby-throated hummingbird have an iridescent green color due to interference. A simplified model of the step structure of the feather is shown in the figure. If the strongest reflection for normal incidence is at $\lambda = 520$ nm, what is the step height h? Assume h has the smallest possible value.

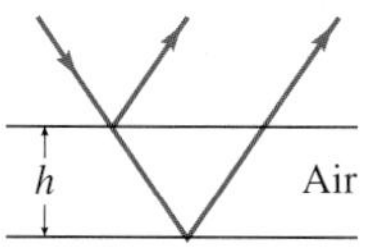

25.2 The Michelson Interferometer

11. A Michelson interferometer is adjusted so that a bright fringe appears on the screen. As one of the mirrors is moved 25.8 μm, 92 bright fringes are counted on the screen. What is the wavelength of the light used in the interferometer?

12. Suppose a transparent vessel 30.0 cm long is placed in one arm of a Michelson interferometer, as in Example 25.2. The vessel initially contains air at 0°C and 1.00 atm. With light of vacuum wavelength 633 nm, the mirrors are arranged so that a bright spot appears at the center of the screen. As air is slowly pumped out of the vessel, one of the mirrors is gradually moved to keep the center region of the screen bright. The distance the mirror moves is measured to determine the value of the index of refraction of air, n. Assume that, outside of the vessel, the light travels through vacuum. Calculate the distance that the mirror would be moved as the container is emptied of air.

13. A Michelson interferometer is set up using white light. The arms are adjusted so that a bright white spot appears on the screen (constructive interference for all wavelengths). A slab of glass ($n = 1.46$) is inserted into one of the arms. To return to the white spot, the mirror in the other arm is moved 6.73 cm. (a) Is the mirror moved in or out? Explain. (b) What is the thickness of the slab of glass?

25.3 Thin Films

14. At a science museum, Marlow looks down into a display case and sees two pieces of very flat glass lying on top of each other with light and dark regions on the glass. The exhibit states that monochromatic light with a wavelength of 550 nm is incident on the glass plates and that the plates are sitting in air. The glass has an index of refraction of 1.51. (a) What is the minimum distance between the two glass plates for one of the dark regions? (b) What is the minimum distance between the two glass plates for one of the light regions? (c) What is the next largest distance between the plates for a dark region? [*Hint:* Do not worry about the thickness of the glass plates, the *thin* film is the air between the plates.]

15. See Problem 14. This time the glass plates are immersed in clear oil with an index of refraction of 1.50. (a) What is the minimum distance between the two glass plates for one of the dark regions? (b) What is the minimum distance between the two glass plates for one of the light regions? (c) What is the next largest distance between the plates for a dark region?

16. A thin film of oil ($n = 1.50$) is spread over a puddle of water ($n = 1.33$). In a region where the film looks red from directly above ($\lambda = 630$ nm), what is the minimum possible thickness of the film? (tutorial: thin film)

17. A thin film of oil ($n = 1.50$) of thickness 0.40 μm is spread over a puddle of water ($n = 1.33$). For which wavelength *in the visible spectrum* do you expect constructive interference for reflection at normal incidence?

18. A transparent film ($n = 1.3$) is deposited on a glass lens ($n = 1.5$) to form a nonreflective coating. What is the minimum thickness that would minimize reflection of light with wavelength 500.0 nm in air?

19. A camera lens ($n = 1.50$) is coated with a thin film of magnesium fluoride ($n = 1.38$) of thickness 90.0 nm. What wavelength in the visible spectrum is most strongly transmitted through the film?

20. A soap film has an index of refraction $n = 1.50$. The film is viewed in reflected light. (a) At a spot where the film thickness is 910.0 nm, which wavelengths are missing in the reflected light? (b) Which wavelengths are strongest in reflected light?

21. A soap film has an index of refraction $n = 1.50$. The film is viewed in *transmitted* light. (a) At a spot where the film thickness is 910.0 nm, which wavelengths are weakest in the transmitted light? (b) Which wavelengths are strongest in transmitted light?

22. Two optically flat plates of glass are separated at one end by a wire of diameter 0.200 mm; at the other end they touch. Thus, the air gap between the plates has a thickness ranging from 0 to 0.200 mm. The plates are 15.0 cm long and are illuminated from above with light of wavelength 600.0 nm. How many bright fringes are seen in the reflected light? (tutorial: wedge film)

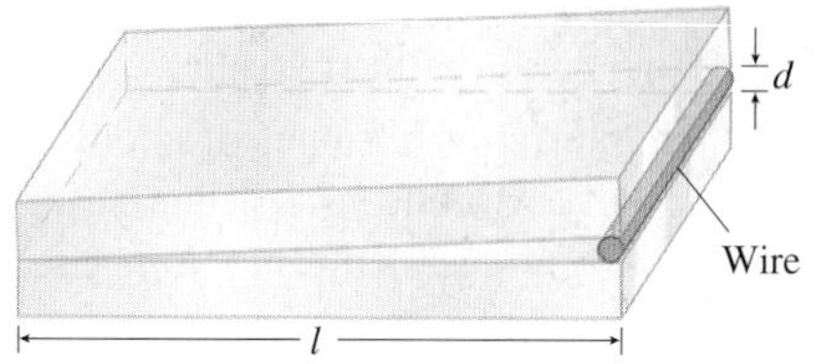

23. A lens is placed on a flat plate of glass to test whether its surface is spherical. (a) Show that the radius r_m of the mth dark ring should be

$$r_m = \sqrt{m\lambda R}$$

where R is the radius of curvature of the lens surface facing the plate and the wavelength of the light used is λ. Assume that $r_m \ll R$. [*Hint:* Start by finding the thickness t of the air gap at a radius $r = R\sin\theta \approx R\theta$. Use small-angle approximations.] (b) Are the dark fringes equally spaced? If not, do they get closer together or farther apart as you move out from the center?

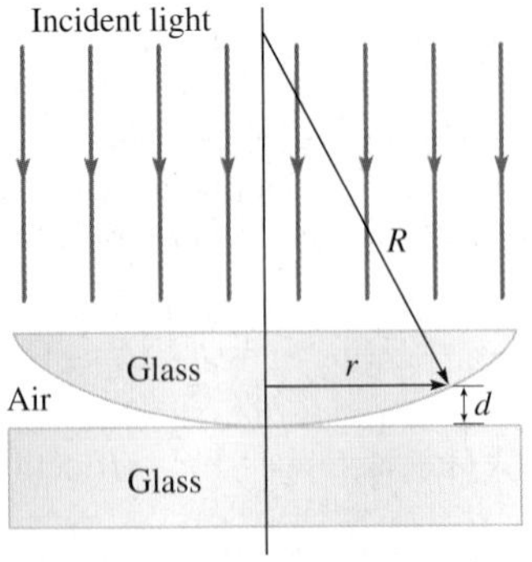

24. A thin film is viewed both in reflected and transmitted light at normal incidence. The figure shows the strongest two rays for each. Show that if rays 1 and 2 interfere constructively, then rays 3 and 4 must interfere destructively, and if rays 1 and 2 interfere destructively, then rays 3 and 4 interfere constructively. Assume that n_2 is the largest of the three indices of refraction.

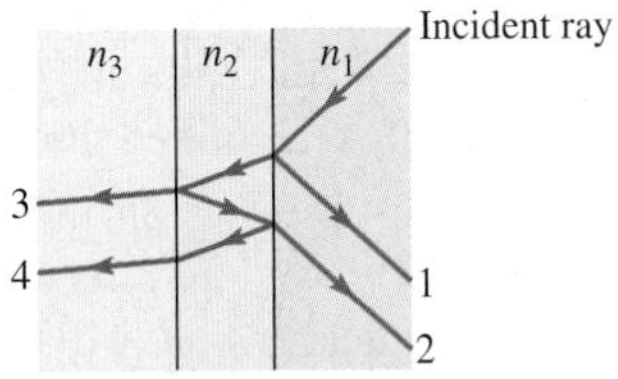

Problems 24–25

25. Repeat Problem 24 assuming that $n_1 < n_2 < n_3$.

25.4 Young's Double-Slit Experiment

26. Light of 650 nm is incident on two slits. A maximum is seen at an angle of 4.10° and a minimum of 4.78°. What is the order m of the maximum and what is the distance d between the slits?

27. You are given a slide with two slits cut into it and asked how far apart the slits are. You shine white light on the slide and notice the first-order color spectrum that is created on a screen 3.40 m away. On the screen, the red light with a wavelength of 700 nm is separated from the violet light with a wavelength of 400 nm by 7.00 mm. What is the separation of the two slits?

28. Show that the interference fringes in a double-slit experiment are equally spaced on a distant screen near the center of the interference pattern. [*Hint:* Use the small-angle approximation for θ.]

29. Use a compass to make an accurate drawing of the wavefronts in a double-slit interference experiment similar to Fig. 25.18c. Place the slits 2.0 cm apart and let the wavelength of the incident wave be 1.0 cm. Using a straightedge, draw lines of constructive interference (antinodes) and use them to find the locations of the $m = \pm 1$ maxima on a screen 12 cm from the slits. Measure the angles of the maxima with a protractor; do they agree with those given by Eq. (25-10)? Explain any discrepancy.

30. In a double-slit interference experiment, the wavelength is 475 nm, the slit separation is 0.120 mm, and the screen is 36.8 cm away from the slits. What is the linear distance between adjacent maxima on the screen? [*Hint:* Assume the small-angle approximation is justified and then check the validity of your assumption once you know the value of the separation between adjacent maxima.] (tutorial: double slit 1)

31. Light incident on a pair of slits produces an interference pattern on a screen 2.50 m from the slits. If the slit separation is 0.0150 cm and the distance between adjacent bright fringes in the pattern is 0.760 cm, what is the wavelength of the light? [*Hint:* Is the small-angle approximation justified?]

32. Ramon has a coherent light source with wavelength 547 nm. He wishes to send light through a double slit with slit separation of 1.50 mm to a screen 90.0 cm away. What is the minimum width of the screen if Ramon wants to display five interference maxima?

33. Light from a helium-neon laser (630 nm) is incident on a pair of slits. In the interference pattern on a screen 1.5 m from the slits, the bright fringes are separated by 1.35 cm. What is the slit separation? [*Hint:* Is the small angle approximation justified?]

34. Light of wavelength 589 nm incident on a pair of slits produces an interference pattern on a distant screen in which the separation between adjacent bright fringes at the center of the pattern is 0.530 cm. A second light source, when incident on the same pair of slits, produces an interference pattern on the same screen with a separation of 0.640 cm between adjacent bright fringes at the center of the pattern. What is the wavelength of the second source? [*Hint:* Is the small-angle approximation justified?]

35. A double slit is illuminated with monochromatic light of wavelength 600.0 nm. The $m = 0$ and $m = 1$ bright fringes are separated by 3.0 mm on a screen 40.0 cm away from the slits. What is the separation between the slits? [*Hint:* Is the small angle approximation justified?]

25.5 Gratings

36. A grating has exactly 8000 slits uniformly spaced over 2.54 cm and is illuminated by light from a mercury vapor discharge lamp. What is the expected angle for the third-order maximum of the green line ($\lambda = 546$ nm)?

37. A red line (wavelength 630 nm) in the third order overlaps with a blue line in the fourth order for a particular grating. What is the wavelength of the blue line?

38. Red light of 650 nm can be seen in three orders in a particular grating. About how many slits per centimeter does this grating have?

39. A grating has 5000.0 slits/cm. How many orders of violet light of wavelength 412 nm can be observed with this grating? (tutorial: grating)

40. A grating is made of exactly 8000 slits; the slit spacing is 1.50 μm. Light of wavelength 0.600 μm is incident normally on the grating. (a) How many maxima are seen in the pattern on the screen? (b) Sketch the pattern that would appear on a screen 3.0 m from the grating. Label distances from the central maximum to the other maxima.

41. A reflection grating spectrometer is used to view the spectrum of light from a helium discharge tube. The three brightest spectral lines seen are red, yellow, and blue in color. These lines appear at the positions labeled A, B, and C in the figure, though not necessarily in that order of color. In this spectrometer, the distance between the grating and slit is 30.0 cm and the slit spacing in the grating is 1870 nm. (a) Which is the red line? Which is the yellow line? Which is the blue line? (b) Calculate the wavelength (in nanometers) of spectral line C. (c) What is the highest order of spectral line C that it is possible to see using this grating?

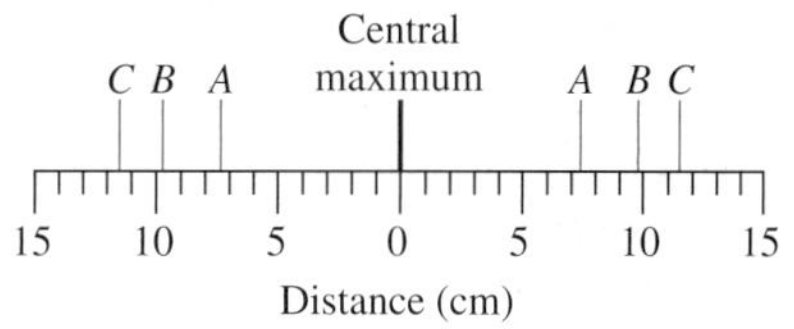

42. A spectrometer is used to analyze a light source. The screen-to-grating distance is 50.0 cm and the grating has 5000.0 slits/cm. Spectral lines are observed at the following angles: 12.98°, 19.0°, 26.7°, 40.6°, 42.4°, 63.9°, and 77.6°. (a) How many different wavelengths are present in the spectrum of this light source? Find each of the wavelengths. (b) If a different grating with 2000.0 slits/cm were used, how many spectral lines would be seen on the screen on one side of the central maximum? Explain.

43. White light containing wavelengths from 400 nm to 700 nm is shone through a grating. Assuming that at least part of the third-order spectrum is present, show that the second- and third-order spectra always overlap, regardless of the slit separation of the grating.

44. A grating 1.600 cm wide has exactly 12 000 slits. The grating is used to resolve two nearly equal wavelengths in a light source: $\lambda_a = 440.000$ nm and $\lambda_b = 440.936$ nm. (a) How many orders of the lines can be seen with the grating? (b) What is the angular separation $\theta_b - \theta_a$ between the lines in each order? (c) Which order best resolves the two lines? Explain.

45. A grating spectrometer is used to resolve wavelengths 660.0 nm and 661.4 nm in second order. (a) How many slits per centimeter must the grating have to produce both wavelengths in second order? (The answer is either a maximum or a minimum number of slits per centimeter.) (b) The minimum number of slits required to resolve two closely spaced lines is $N = \lambda/(m\Delta\lambda)$, where λ is the average of the two wavelengths, $\Delta\lambda$ is the difference between the two wavelengths, and m is the order. What minimum number of slits must this grating have to resolve the lines in second order?

25.7 Diffraction by a Single Slit

46. The central bright fringe in a single-slit diffraction pattern from light of wavelength 476 nm is 2.0 cm wide on a screen that is 1.05 m from the slit. (a) How wide is the slit? (b) How wide are the first two bright fringes on either side of the central bright fringe? (Define the width of a bright fringe as the linear distance from minimum to minimum.)

47. The first two dark fringes on one side of the central maximum in a single-slit diffraction pattern are 1.0 mm apart. The wavelength of the light is 610 nm and the screen is 1.0 m from the slit. What is the slit width?

48. Light of wavelength 630 nm is incident on a single slit with width 0.40 mm. The figure shows the pattern observed on a screen positioned 2.0 m from the slit. Determine the distance from the center of the central bright fringe to the second minimum on one side.

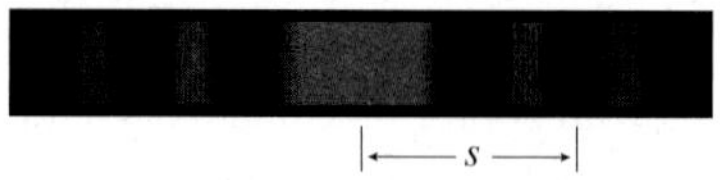

49. Light from a red laser passes through a single slit to form a diffraction pattern on a distant screen. If the width of the slit is increased by a factor of two, what happens to the width of the central maximum on the screen?

50. The diffraction pattern from a single slit is viewed on a screen. Using blue light, the width of the central maximum is 2.0 cm. (a) Would the central maximum be narrower or wider if red light is used instead? (b) If the blue light has wavelength 0.43 μm and the red light has wavelength 0.70 μm, what is the width of the central maximum when red light is used?

51. Light of wavelength 490 nm is incident on a narrow slit. The diffraction pattern is viewed on a screen 3.20 m from the slit. The distance on the screen between the central maximum and the third minimum is 2.5 cm. What is the width of the slit?

25.8 Diffraction and the Resolution of Optical Instruments

52. The Hubble Space Telescope (HST) has excellent resolving power because there is no atmospheric distortion of the light. Its 2.4-m-diameter primary mirror can collect light from distant galaxies that formed early in the history of the universe. How far apart can two galaxies be from each other if they are 10 billion light-years away from Earth and are barely resolved by the HST using visible light with a wavelength of 400 nm?

53. A beam of yellow laser light (590 nm) passes through a circular aperture of diameter 7.0 mm. What is the angular width of the central diffraction maximum formed on a screen?

54. The photosensitive cells (rods and cones) in the retina are most densely packed in the fovea—the part of the retina used to see straight ahead. In the fovea, the cells are all cones spaced about 1 μm apart. Would our vision have much better resolution if they were closer together? To answer this question, assume two light sources are just far enough apart to be resolvable according to Rayleigh's criterion. Assume an average pupil diameter of 5 mm and an eye diameter of 25 mm. Also assume that the index of refraction of the vitreous fluid in the eye is 1; in other words, treat the pupil as a circular aperture with air on both sides. What is the spacing of the cones if the centers of the diffraction maxima fall on two nonadjacent cones with a single intervening cone? (There must be an intervening dark cone in order to resolve the two sources; if two *adjacent* cones are stimulated, the brain assumes a single source.)

55. The radio telescope at Arecibo, Puerto Rico, has a reflecting spherical bowl of 305 m (1000 ft) diameter. Radio signals can be received and emitted at various frequencies with appropriate antennae at the focal point of the reflecting bowl. At a frequency of 300 MHz, what is the angle between two stars that can barely be resolved? (WW tutorial: radio telescope)

56. A pinhole camera doesn't have a lens; a small circular hole lets light into the camera, which then exposes the film. For the sharpest image, light from a distant point source makes as small a spot on the film as possible. What is the optimum size of the hole for a camera in which the film is 16.0 cm from the pinhole? A hole smaller than the optimum makes a larger spot since it diffracts the light more. A larger hole also makes a larger spot because the spot cannot be smaller than the hole itself (think in terms of geometrical optics). Let the wavelength be 560 nm.

57. To understand Rayleigh's criterion as applied to the pupil of the eye, notice that rays do *not* pass straight through the center of the lens system (cornea + lens) of the eye except at normal incidence because the indices of refraction on the two sides of the lens system are different. In a simplified model, suppose light from two point sources travels through air and passes through the pupil (diameter a). On the other side of the pupil, light travels through the vitreous fluid (index of refraction n). The figure shows two rays, one from each source, that pass through the center of the pupil. (a) What is the relationship between $\Delta\theta$, the angular separation of the two *sources*, and β, the angular separation of the two *images*? [*Hint:* Use Snell's law.] (b) The first diffraction minimum for light from source 1 occurs at angle ϕ, where $a \sin \phi = 1.22\lambda$ [Eq. (25-13)]. Here, λ is the wavelength *in the vitreous fluid*. According to Rayleigh's criterion, the sources can be resolved if the center of image 2 occurs no closer than the first diffraction minimum for image 1; that is, if $\beta \geq \phi$ or, equivalently, $\sin \beta \geq \sin \phi$. Show that this is equivalent to Eq. (25-14), where λ_0 is the wavelength *in air*.

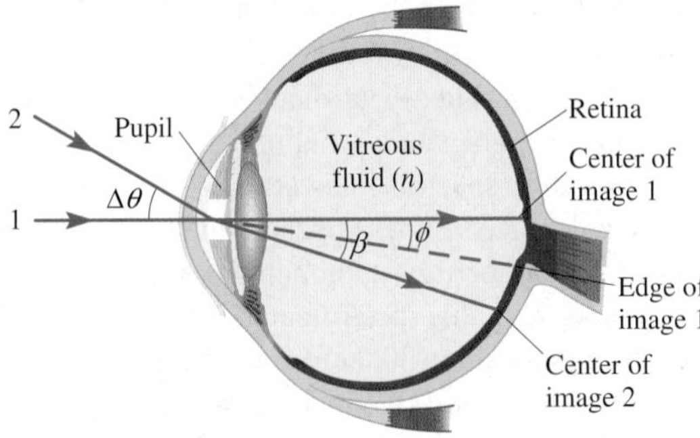

Comprehensive Problems

58. A beam of coherent light of wavelength 623 nm in air is incident on a rectangular block of glass with index of refraction 1.40. If, after emerging from the block, the wave that travels through the glass is 180° out of phase with the wave that travels through air, what are the possible lengths d of the glass in terms of a positive integer m? Ignore reflection.

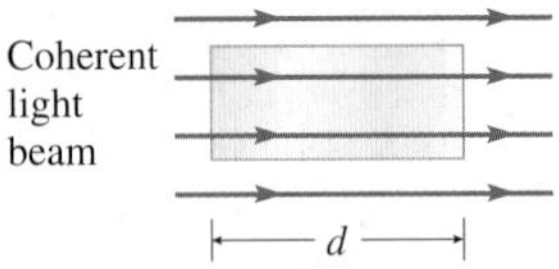

59. Light with a wavelength of 660 nm is incident on two slits and the pattern shown in the figure is viewed on a screen. Point A is directly opposite a point midway between the two slits. What is the path length difference of the light that passes through the two different slits for light that reaches the screen at points A, B, C, D, and E?

60. A thin layer of an oil ($n = 1.60$) floats on top of water ($n = 1.33$). One portion of this film appears green ($\lambda = 510$ nm) in reflected light. How thick is this portion of the film? Give the three smallest possibilities.

61. If diffraction were the only limitation, what would be the maximum distance at which the headlights of a car could be resolved (seen as two separate sources) by the naked human eye? The diameter of the pupil of the eye is about 7 mm when dark-adapted. Make reasonable estimates for the distance between the headlights and for the wavelength.

62. Find the height h of the pits on a CD (Fig. 25.6a). When the laser beam reflects partly from a pit and partly from land (the flat aluminum surface) on either side of the "pit," the two reflected beams interfere destructively; h is chosen to be the smallest possible height that causes destructive interference. The wavelength of the laser is 780 nm and the index of refraction of the polycarbonate plastic is $n = 1.55$.

63. The Very Large Array (VLA) is a set of 30 dish radio antennas located near Socorro, New Mexico. The dishes are spaced 1.0 km apart and form a Y-shaped pattern, as in the diagram. Radio pulses from a distant pulsar (a rapidly rotating neutron star) are detected by the dishes; the arrival time of each pulse is recorded using atomic clocks. If the pulsar is located 60.0° above the horizontal direction parallel to the right branch of the Y, how much time elapses between the arrival of the pulses at adjacent dishes in that branch of the VLA?

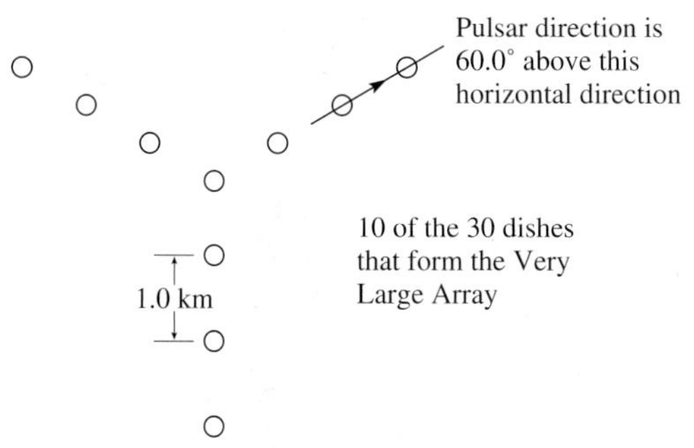

64. Two narrow slits with a center-to-center distance of 0.48 mm are illuminated with coherent light at normal incidence. The intensity of the light falling on a screen 5.0 m away is shown in the figure, where x is the distance from the central maximum on the screen. (a) What would be the intensity of the light falling on the screen if only one slit were open? (b) Find the wavelength of the light.

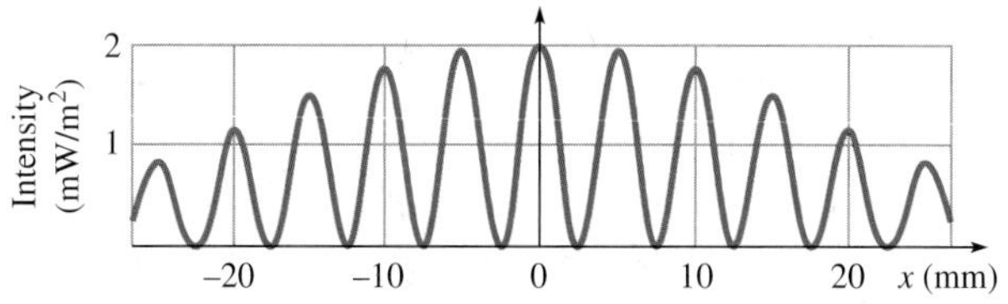

65. When a double slit is illuminated with light of wavelength 510 nm, the interference maxima on a screen 2.4 m away gradually decrease in intensity on either side of the 2.40-cm-wide central maximum and reach a minimum in a spot where the fifth-order maximum is expected. (a) What is the width of the slits? (b) How far apart are the slits?

66. Sonya is designing a diffraction experiment for her students. She has a laser that emits light of wavelength 627 nm and a grating with a distance of 2.40×10^{-3} mm between slits. She hopes to shine the light through the grating and display a total of nine interference maxima on a screen. She finds that no matter how she arranges her setup, she can see only seven maxima. Assuming that the intensity of the light is not the problem, why can't Sonya display the $m = 4$ interference maxima on either side?

67. A lens ($n = 1.52$) is coated with a magnesium fluoride film ($n = 1.38$). (a) If the coating is to cause destructive interference in reflected light for $\lambda = 560$ nm (the peak of the solar spectrum), what should its minimum thickness be? (b) At what two wavelengths closest to 560 nm does the coating cause *constructive* interference in reflected light? (c) Is any visible light reflected? Explain.

68. A thin soap film ($n = 1.35$) is suspended in air. The spectrum of light reflected from the film is missing two visible wavelengths of 500.0 nm and 600.0 nm, with no missing wavelengths between the two. (a) What is the thickness of the soap film? (b) Are there any other visible wavelengths missing from the reflected light? If so, what are they? (c) What wavelengths of light are strongest in the *transmitted* light?

69. Instead of an antireflective coating, suppose you wanted to coat a glass surface to *enhance* the reflection of visible light. Assuming that $1 < n_{coating} < n_{glass}$, what should the minimum thickness of the coating be to maximize the reflected intensity for wavelength λ?

70. A mica sheet 1.00 μm thick is suspended in air. In reflected light, there are gaps in the visible spectrum at 450, 525, and 630 nm. Calculate the index of refraction of the mica sheet.

71. In bright light, the pupils of the eyes of a cat narrow to a vertical *slit* 0.30 mm across. Suppose that a cat is looking at two mice 18 m away. What is the smallest distance between the mice for which the cat can tell that there are two mice rather than one using light of 560 nm? Assume the resolution is limited by diffraction only.

72. Parallel light of wavelength λ strikes a slit of width a at normal incidence. The light is viewed on a screen that is 1.0 m past the slits. In each case that follows, sketch the intensity on the screen as a function of x, the distance from the center of the screen, for $0 \le x \le 10$ cm. (a) $\lambda = 10a$. (b) $10\lambda = a$. (c) $30\lambda = a$.

73. About how close to each other are two objects on the Moon that can just barely be resolved by the 5.08-m-(200-in.)-diameter Mount Palomar reflecting telescope? (Use a wavelength of 520 nm.)

74. A grating in a spectrometer is illuminated with red light ($\lambda = 690$ nm) and blue light ($\lambda = 460$ nm) simultaneously. The grating has 10,000.0 slits/cm. Sketch the pattern that would be seen on a screen 2.0 m from the grating. Label distances from the central maximum. Label which lines are red and which are blue.

75. Two slits separated by 20.0 μm are illuminated by light of wavelength 0.50 μm. If the screen is 8.0 m from the slits, what is the distance between the $m = 0$ and $m = 1$ bright fringes?

76. In a double-slit experiment, what is the linear distance on the screen between adjacent maxima if the wavelength is 546 nm, the slit separation is 0.100 mm, and the slit-screen separation is 20.0 cm?

Problems 77 and 78. Two radio towers are a distance d apart as shown in the overhead view. Each antenna *by itself* would radiate equally in all directions in a horizontal plane. The radio waves have the same frequency and start out in phase. A detector is moved in a circle around the towers at a distance of 100 km.

✦77. The power radiated in a horizontal plane by both antennas together is measured by the detector and is found to vary with angle. (a) Is the power detected at $\theta = 0$ a maximum or a minimum? Explain. (b) Sketch a graph of P versus θ to show qualitatively how the power varies with angle θ (from $-180°$ to $+180°$) if $d = \lambda$. Label your graph with values of θ at which the power is maximum or minimum. (c) Make a qualitative graph of how the power varies with angle for the case $d = \lambda/2$. Label your graph with values of θ at which the power is maximum or minimum.

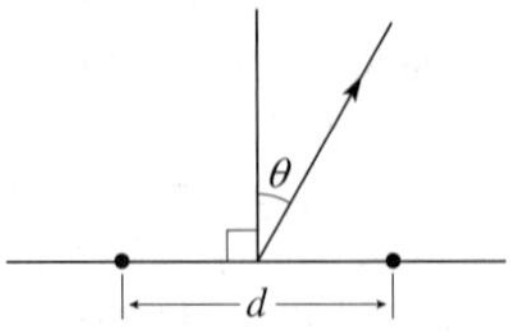

Problems 77 and 78

✦78. The waves have frequency 3.0 MHz and the distance between antennas is $d = 0.30$ km. (a) What is the difference in the path lengths traveled by the waves that arrive at the detector at $\theta = 0°$? (b) What is the difference in the path lengths traveled by the waves that arrive at the detector at $\theta = 90°$? (c) At how many angles ($0 \le \theta < 360°$) would you expect to detect a maximum intensity? Explain. (d) Find the angles (θ) of the maxima in the first quadrant ($0 \le \theta \le 90°$). (e) Which (if any) of your answers to parts (a) to (d) would change if the detector were instead only 1 km from the towers? Explain. (Don't calculate new values for the answers.)

✦79. If you shine a laser (wavelength 0.60 μm) with a small aperture at the Moon, diffraction makes the beam spread out and the spot on the Moon is large. Making the aperture smaller only makes the spot on the Moon *larger*. On the other hand, shining a wide searchlight at the Moon can't make a tiny spot—the spot on the Moon is at least as wide as the searchlight. What is the radius of the *smallest* possible spot you can make on the Moon by shining a light from Earth? Assume the light is perfectly parallel before passing through a circular aperture.

✦80. Two coherent plane waves travel at angle θ_0 toward a photographic plate. Show that the distance between fringes of constructive interference on the plate is given by $d = \lambda/\sin \theta_0$. See Fig. 25.43.

Answers to Practice Problems

25.2 The mirror should be moved in (shorter path length). Since the number of wavelengths traveled in the arm with the vessel decreases, we must decrease the number of wavelengths traveled in the other arm.

25.3 560 nm and 458 nm

25.4 (a) 0, 0.020 rad, 0.040 rad; (b) 0.010 rad, 0.030 rad; (c) 4.0 cm

25.5 The intensity is maximum at the center ($\theta = 0$) and gradually decreases to either side but never reaches zero.

25.6 4760 slits/cm; fourth-order maxima are present for wavelengths up to 525 nm.

25.7 No; the window is large compared with the wavelength of light, so we expect diffraction to be negligible. The Sun is not distant enough to treat it as a point source; rays from different points on the Sun's surface travel in slightly different directions as they pass through the window.

25.8 3.9 cm

25.9 9 m

Answers to Checkpoints

25.1 Yes, the phase difference between two coherent waves can be $\pi/3$ rad. The phase difference is not an integral multiple of π, so interference is neither constructive (maximum amplitude) or destructive (minimum amplitude). It is in between the two.

25.3 A 180° phase shift occurs when the wave reflects from a boundary with a slower medium (higher index of refraction). Ray 1 has a 180° phase shift due to reflection because $n_f > n_i$. Ray 2 has no phase shift due to reflection because $n_t < n_f$.

25.5 For the double slit, the intensity varies gradually between maxima and minima (see Fig. 25.18). The maxima due to a grating are much brighter (constructive interference due to many slits rather than just two slits). The widths of the maxima are inversely proportional to the number of slits, so for a grating with many slits, the maxima are very narrow.

Review & Synthesis: Chapters 22–25

Review Exercises

1. You are watching a baseball game on television that is being broadcast from 4500 km away. The batter hits the ball with a loud "crack" of the bat. A microphone is located 22 m from the batter, and you are 2.0 m from the television set. On a day when sound travels 343 m/s in air, what is the minimum time it takes for you to hear the crack of the bat after the batter hit the ball?

2. Sketch a sinusoidal wave with an amplitude of 3 V/m and a wavelength of 600 nm. This represents the electric field portion of a visible EM wave traveling to the right with intensity I_0. Beneath the first wave, sketch another that is identical to the first in wavelength, but with an amplitude of 2 V/m and 180° out of phase. Beneath the second wave, sketch a third wave of the same wavelength with an amplitude of 0.5 V/m and in phase with the first wave. The three waves are coherent. What is the intensity of light when the three waves are added?

3. On a cold, autumn day, Tuan is staring out of the window watching the leaves blow in the wind. One bright yellow leaf is reflecting light that has a predominant wavelength of 580 nm. (a) What is the frequency of this light? (b) If the window glass has an index of refraction of 1.50, what are the speed, wavelength, and frequency of this light as it passes through the window?

4. You are standing 1.2 m from a 1500-W heat lamp. (a) Assuming that the energy of the heat lamp is radiated uniformly in a hemispherical pattern, what is the intensity of the light on your face? (b) If you stand in front of the heat lamp for 2.0 min, how much energy is incident on your face? Assume your face has a total area of 2.8×10^{-2} m^2. (c) What are the rms electric and magnetic fields?

5. Juanita is lying in a hammock in her garden and listening to music on her portable radio tuned to WMCB (1408 kHz), a station located 98 km from her home. A plane, about to land at the airport, flies directly overhead and causes destructive interference. Juanita estimates the plane to be at least 500 m over her head. Assume there is a 180° phase shift when the radio wave reflects from the airplane. (a) What is the closest possible distance between Juanita and the plane if her estimation is correct? (b) Find two other possible elevations, the next one lower and the next one higher, for the plane in case her estimation is off.

6. Consider the three polarizing filters shown in the figure. The angles listed indicate the direction of the transmission axis of each polarizer with respect to the vertical. (a) If unpolarized light of intensity I_0 is incident from the left, what is the intensity of the light that exits the last polarizer? (b) If vertically polarized light of intensity I_0 is incident from the left, what is the intensity of the light that exits the last polarizer? (c) Can you remove one polarizer from this series of filters so that light incident from the left is not transmitted at all if unpolarized light is incident as in part (a)? If so, which polarizer should you remove? Answer the same questions for vertically polarized incident light as in part (b). (d) If you can remove one polarizer to maximize the amount of light transmitted in part (a), which one should you remove? Answer the same question for part (b).

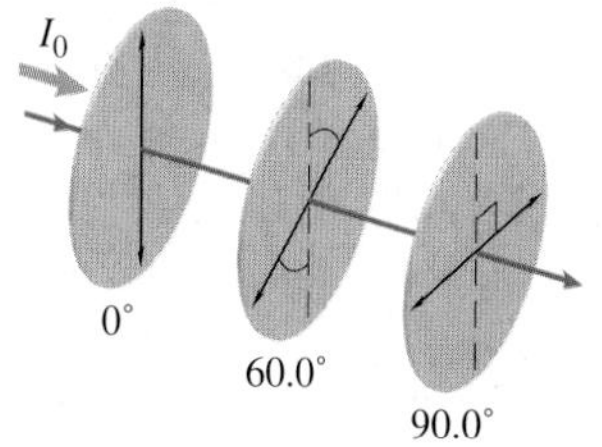

7. The projector in a movie theater has a lens with a focal length of 29.5 cm. It projects an image of the 70.0-mm-wide film onto a screen that is 38.0 m from the projector. (a) How wide is the image on the screen? (b) What kind of lens is used in the projector? (c) Is the image on the screen upright or inverted compared with the film?

8. A converging lens with a focal length of 5.500 cm is placed 8.00 cm to the left of a diverging lens with a radius of curvature of 8.40 cm. An object that is 1.0 cm tall is placed 9.000 cm to the left of the converging lens. (a) Where is the final image formed? (b) How tall is the final image? (c) Is the final image upright or inverted?

9. Why don't you see an interference pattern on your desk when you have light from two different lamps illuminating the surface?

10. A radio wave with a wavelength of 1200 m follows two paths to a receiver that is 25.0 km away. One path goes directly to the receiver and the other reflects from an airplane that is flying above the point that is exactly halfway between the transmitter and the receiver. Assume there is no phase change when the wave reflects off the airplane. If the receiver experiences destructive interference, what is the minimum possible distance that the reflected wave has traveled? For this distance, how high is the airplane?

11. A thin film of oil with index of refraction of 1.50 sits on top of a pool of water with index of refraction of 1.33. When light is incident on this film, a maximum is observed in reflected light at 480 nm and a minimum is observed in reflected light at 600 nm, with no maxima or minima for any wavelengths between these two. What is the thickness of the film?

12. A camera lens ($n = 1.50$) is coated with a thin layer of magnesium fluoride ($n = 1.38$). The purpose of the

coating is to allow all the light to be transmitted by canceling out reflected light. What is the minimum thickness of the coating necessary to cancel out reflected visible light of wavelength 550 nm?

13. A grating made of 5550 slits/cm has red light of 0.680 μm incident on it. The light shines through the grating onto a screen that is 5.50 m away. (a) What is the distance between adjacent slits on the grating? (b) How far from the central bright spot is the first-order maximum on the screen? (c) How far from the central bright spot is the second-order maximum on the screen? (d) Can you assume in this problem that $\sin\theta = \tan\theta$? Why or why not?

14. When using a certain grating, third-order violet light of wavelength 420 nm falls at the same angle as second-order light of a different wavelength. What is that wavelength?

15. (a) In double-slit interference, how does the slit separation affect the distance between adjacent interference maxima? (b) How does the distance between the slits and screen affect that separation? (c) If you are trying to resolve two closely spaced maxima, how might you design your double-slit spectrometer?

16. The radio telescope at Arecibo, Puerto Rico, has a reflecting spherical bowl of 305 m (1000 ft) diameter. Radio signals can be received and emitted at various frequencies with the appropriate antennae at the focal point. If two Moon craters 499 km apart are to be resolved, what wavelength radio waves must be used?

17. Geraldine uses a 423-nm coherent light source and a double slit with a slit separation of 20.0 μm to display three interference maxima on a screen that is 20.0 cm wide. If she wants to spread the three maxima across the full width of the screen, from a minimum on one side to a minimum on the other side, how far from the screen should she place the double slit?

18. A convex mirror produces an image located 18.4 cm behind the mirror when an object is placed 32.0 cm in front of the mirror. What is the focal length of this mirror?

19. Simon wishes to display a double-slit experiment for his class. His coherent light source has a wavelength of 510 nm and the slit separation is $d = 0.032$ mm. He must set up the light on a desk 1.5 m away from the screen that is only 10 cm wide. How many interference maxima will Simon be able to display for his students?

20. Bruce is trying to remove an eyelash from the surface of his eye. He looks in a shaving mirror to locate the eyelash, which is 0.40 cm long. If the focal length of the mirror is 18 cm and he puts his eye at a distance of 11 cm from the mirror, how long is the image of his eyelash?

21. Coherent green light with a wavelength of 520 nm and coherent violet light with a wavelength of 412 nm are incident on a double slit with slit separation of 0.020 mm. The interference pattern is displayed on a screen 72.0 cm away. (a) Find the separation between the $m = 1$ interference maxima of the two colors. (b) What is the separation between the $m = 2$ maxima for the two beams?

22. When the NASA Rover Spirit successfully landed on Mars in January of 2004, Mars was 170.2×10^6 km from Earth. Twenty-one days later, when the Rover Opportunity landed on Mars, Mars was 198.7×10^6 km from Earth. (a) How long did it take for a one-way transmission to the scientists on Earth from Spirit on its landing day? (b) How long did it take for scientists to communicate with Opportunity on its landing day?

23. A spaceship traveling 12.3 km/s relative to Earth sends out an EM pulse with a wavelength of 850.00 nm (as measured by the source). The pulse is reflected from another spaceship that is moving toward the first spaceship at a speed of 24.6 km/s relative to Earth. What will be the wavelength of the reflected pulse as measured by the first spaceship?

24. Jamila has a set of reading glasses with a refractive power of +2.00 D. (a) What is the focal length of each lens? (b) Are the lenses converging or diverging? (c) An object is placed 40.0 cm in front of one of the lenses. Where is the image formed? (d) What is the size of the image relative to the size of the object? (e) Is the image upright or inverted?

25. A beam of light in air enters a glass block at an angle of 30° to the glass surface, as shown. The glass has an index of refraction of 1.35. (a) Find the angle labeled θ_1. (b) Calculate the critical angle between the glass and air. (c) Does the light follow path *A*, path *B*, or both? Explain. (d) Find the angle(s) θ_A, if light follows path *A*, and θ_B, if light follows path *B*.

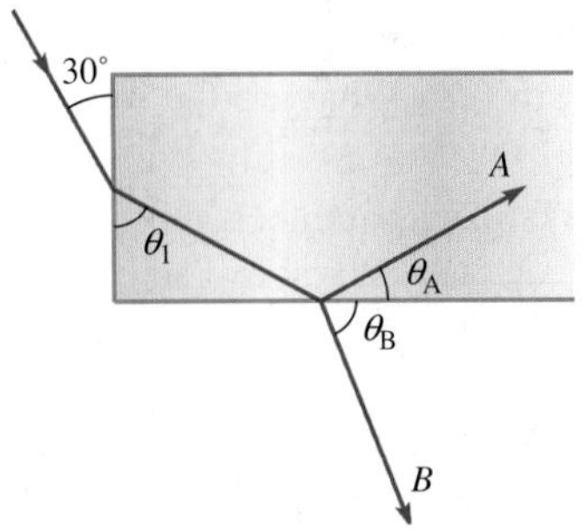

26. An object is placed between a concave mirror with a radius of curvature of 18.0 cm and a diverging lens with a focal length of magnitude 12.5 cm. The object is 15.0 cm from the mirror and 20.0 cm from the lens. Looking through the lens, you see two images. Image 1 is formed by light rays that reflect from the mirror before passing through the lens. Image 2 is formed by light rays that pass through the lens without reflecting from the mirror. Find the location of each image and determine whether it is inverted or upright and real or virtual. [*Hint:* Treat the mirror-lens combination in the same way you would treat two lenses.]

MCAT Review

The section that follows includes MCAT exam material and is reprinted with permission of the Association of American Medical Colleges (AAMC).

1. An object is placed upright on the axis of a thin convex lens at a distance of three focal lengths ($3f$) from the center of the lens. An inverted image appears at a distance of $\frac{3}{2}f$ on the other side of the lens. What is the ratio of the height of the image to the height of the object?
   A. $\frac{1}{2}$
   B. $\frac{2}{3}$
   C. $\frac{3}{2}$
   D. $\frac{2}{1}$
2. A concave spherical mirror has a radius of curvature of 50 cm. At what distance from the surface of this mirror should an object be placed to form a real, inverted image the same size as the object?
   A. 25 cm
   B. 37.5 cm
   C. 50 cm
   D. 100 cm

Read the paragraph and then answer the following questions:

The Hubble Space Telescope (HST) is the largest telescope ever placed into orbit. Its primary concave mirror has a diameter of 2.4 m and a focal length of approximately 13 m. In addition to having optical detectors, the telescope is equipped to detect ultraviolet light, which does not easily penetrate Earth's atmosphere.

3. When the HST is focused on a very distant object, the image from its primary mirror is
   A. real and upright.
   B. real and inverted.
   C. virtual and upright.
   D. virtual and inverted.
4. The magnification of a telescope is determined by dividing the focal length of the primary mirror by the focal length of the eyepiece. If an eyepiece with a focal length of 2.5×10^{-2} m could be used with the primary mirror of the HST, it would produce an image magnified by a factor of approximately
   A. 10.
   B. 96.
   C. 520.
   D. 960.
5. The image of a very distant object on the axis of a mirror such as that used in the HST will be at what location in relationship to the mirror and focal point?
   A. behind the mirror
   B. between the mirror and the focal point
   C. very close to the focal point
   D. very close to twice the distance from the mirror to the focal point
6. Which of the following is the best explanation of why ultraviolet light does not penetrate Earth's atmosphere as easily as visible light does?
   A. Ultraviolet light has a shorter wavelength and is more readily absorbed by the atmosphere.
   B. Ultraviolet light has a lower frequency and is more readily absorbed by the atmosphere.
   C. Ultraviolet light contains less energy and cannot travel as far through the atmosphere.
   D. Ultraviolet light undergoes destructive interference as it travels through the atmosphere.

CHAPTER 26

Relativity

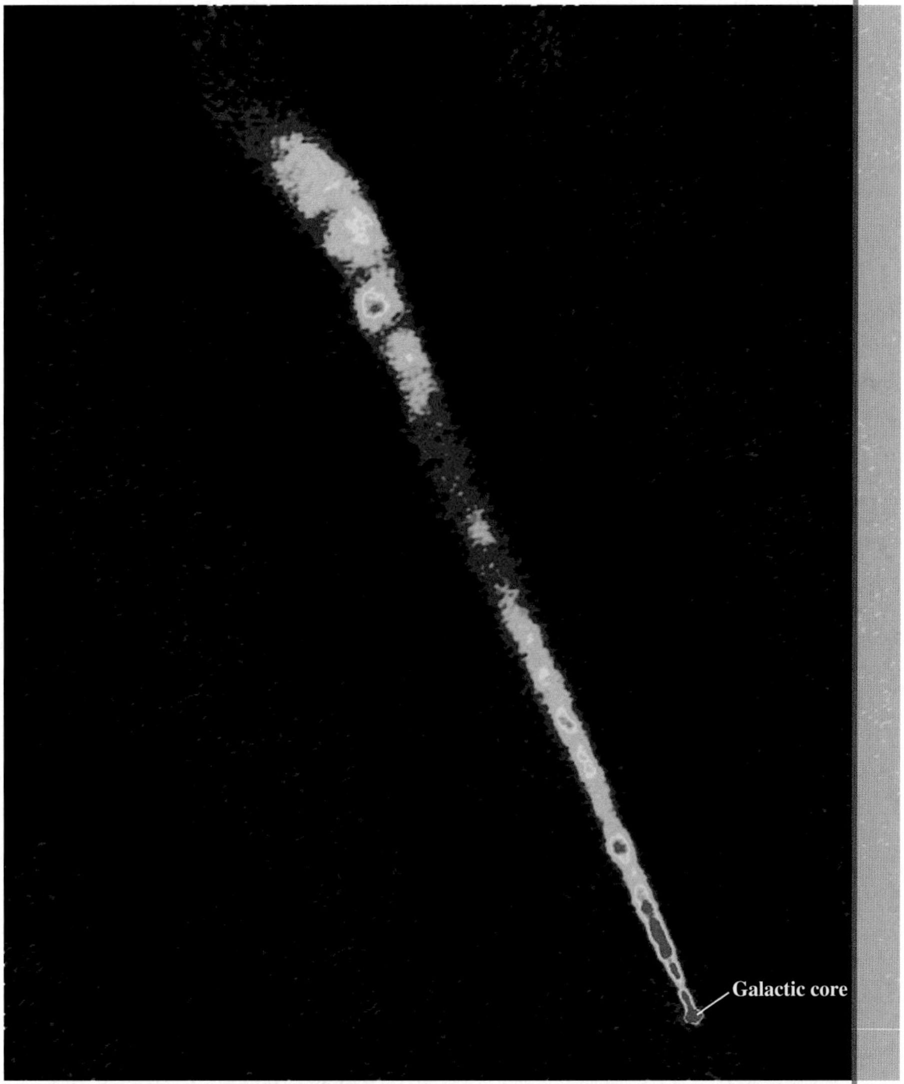

Jet emitted by the core of galaxy NGC 6251.

The centers of some galaxies are much brighter than the rest of the galaxy. These active galactic nuclei, which may be only about as big as our solar system, can give off 20 billion times as much light as the Sun. The core of the galaxy NGC 6251 emits a narrow, extremely energetic jet of charged particles in a direction roughly toward Earth. The photo shows the jet as imaged by the Very Large Array of radiotelescopes in New Mexico; the galactic core is at the lower right.

When scientists first measured the speed of the tip of this jet, they used two radiotelescope images, taken on two successive days. They measured how far the tip of the jet moved, divided by the time elapsed between the two images, and came up with a speed greater than the speed of light! Is it possible for the charged particles in the jet to move faster than light? If not, what was the scientists' mistake? (See p. 972 for the answer.)

Concepts & Skills to Review

- inertial reference frames (Section 3.6)
- relative velocity (Section 3.6)
- kinetic energy (Section 6.3)
- energy conservation (Section 6.1)
- conservation of momentum; collisions in one dimension (Sections 7.4 and 7.7)

26.1 POSTULATES OF RELATIVITY

Reference Frames

CONNECTION:

The idea of relativity is not something new introduced by Einstein. It goes all the way back to Galileo and Newton.

The idea of *relativity* is not something entirely new; it goes all the way back to Galileo. Aristotle had previously said that a body continues to move only if a force continues to propel it; take away the force and the body comes to rest. The authority of Aristotle's opinion prevailed for many centuries. Galileo turned this thinking around by saying that a body maintains a constant velocity (which can be zero or nonzero) in the absence of any external forces acting on it; this concept is the basis for the law of inertia as stated by Newton.

All motion must be measured in some particular reference frame, which we usually represent as a set of coordinate axes. Suppose two people walk hand-in-hand on a moving sidewalk in an airport. They might walk at 1.3 m/s with respect to a reference frame attached to the moving sidewalk and at 2.4 m/s with respect to a reference frame attached to the building. The two reference frames are *equally valid.*

An **inertial reference frame** is one in which no accelerations are observed in the absence of external forces. In noninertial reference frames, bodies have accelerations in the absence of applied forces because *the reference frame itself is accelerating* with respect to an inertial frame. For example, suppose two people sit across from each other on a rapidly rotating merry-go-round (Fig. 26.1). When one tosses a ball to the other, the ball is deflected sideways as viewed by observers on the merry-go-round. The sideways acceleration is not caused by any force acting on the ball; the reference frame attached to the merry-go-round is *noninertial.* The law of inertia does not hold in noninertial frames.

For many purposes, Earth's surface can be considered to be an inertial reference frame, even though strictly speaking it is not. Earth's rotation causes phenomena such as the rotary motion of hurricanes and trade winds, which, in a reference frame attached to Earth's surface, are accelerations not caused by applied forces.

Any reference frame that moves with constant velocity with respect to an inertial frame is itself inertial; if the acceleration of a body in one inertial frame is zero, its acceleration in any of the other inertial frames is also zero. In our earlier example, if a

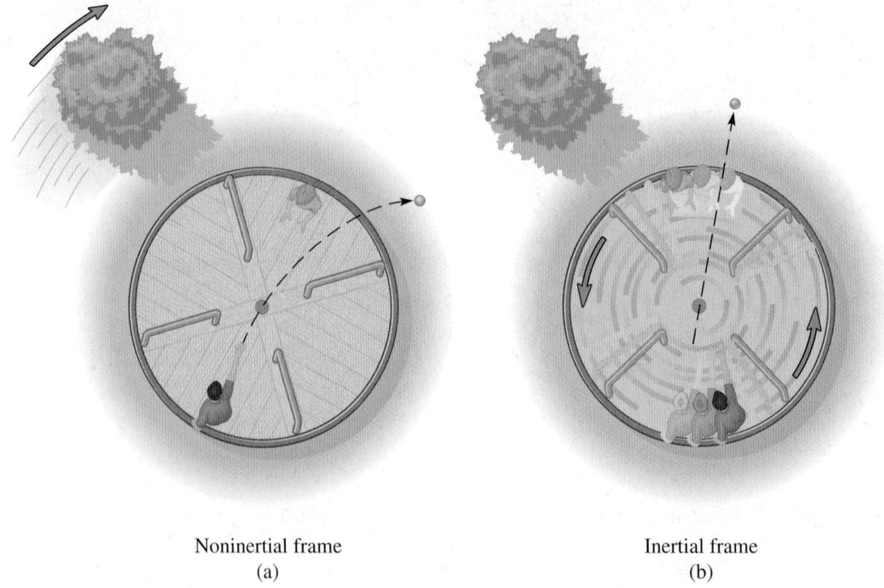

Figure 26.1 A ball tossed across a merry-go-round. (a) Trajectory of the ball as viewed in the *noninertial* frame fixed to the platform. In this frame, the platform is at rest and the tree is moving. The ball is thrown straight toward the catcher but then is deflected sideways, even though no sideways force acts on it. (b) Straight-line trajectory of the ball as viewed in the *inertial* frame fixed to the ground. In this frame, the law of inertia holds and the ball is not deflected. The catcher rotates away from the path of the ball.

reference frame fixed to the airport terminal is inertial and the moving sidewalk moves at constant velocity with respect to the terminal, then a reference frame fixed to the moving sidewalk is also inertial.

Principle of Relativity

Ever since Galileo and Newton, scientists have been careful to formulate the laws of physics so that the *same laws* hold in any inertial reference frame. Particular quantities (velocity, momentum, kinetic energy) have different values in different inertial reference frames, but the **principle of relativity** requires that the *laws* of physics (e.g., the conservation of momentum and energy) be the *same* in all inertial frames.

Principle of Relativity

The *laws* of physics are the *same* in all inertial frames.

The laws and equations in this chapter—just like those of all other chapters in this book—are only valid in inertial frames. The laws of physics must be modified if they are to apply in noninertial (accelerated) reference frames.

CHECKPOINT 26.1

You are in a special compartment on a train that admits no light, sound, or vibration. Is there any way you can tell whether the train is at rest or moving at constant nonzero velocity with respect to the ground? Explain.

Apparent Contradictions with the Principle of Relativity

In the nineteenth century, James Clerk Maxwell used the four basic laws that describe electromagnetic fields (*Maxwell's equations*, Section 22.1) to show that electromagnetic waves travel through vacuum at a speed of $c = 3.00 \times 10^8$ m/s. In fact, Maxwell's equations show that EM waves travel at the *same speed in every inertial reference frame*, regardless of the motion of the source or of the observer.

This conclusion, that the speed of light is the same in any inertial reference frame, contradicts the Galilean laws of relative velocity (Section 3.6). Suppose a car travels at velocity $\vec{\mathbf{v}}_{CG}$ with respect to the ground (Fig. 26.2). Light coming from the car's headlights travels at velocity $\vec{\mathbf{v}}_{LC}$ with respect to the car. Galilean velocity addition says that the speed of the light beam with respect to the ground is

$$\vec{\mathbf{v}}_{LG} = \vec{\mathbf{v}}_{LC} + \vec{\mathbf{v}}_{CG} \qquad (3\text{-}24)$$

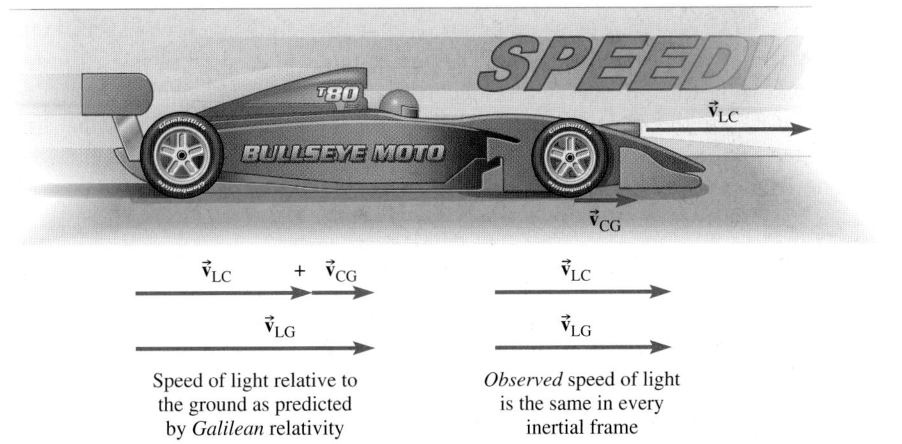

Figure 26.2 According to Galilean relativity, the speed of light would have different values in different inertial reference frames. If $\vec{\mathbf{v}}_{LC}$ is the velocity of the light beam with respect to the car and $\vec{\mathbf{v}}_{CG}$ is the velocity of the car with respect to the ground, Galilean relativity would predict the velocity of the light beam with respect to the ground to be $\vec{\mathbf{v}}_{LC} + \vec{\mathbf{v}}_{CG}$. However, the *observed* speed of light is the same in all inertial reference frames.

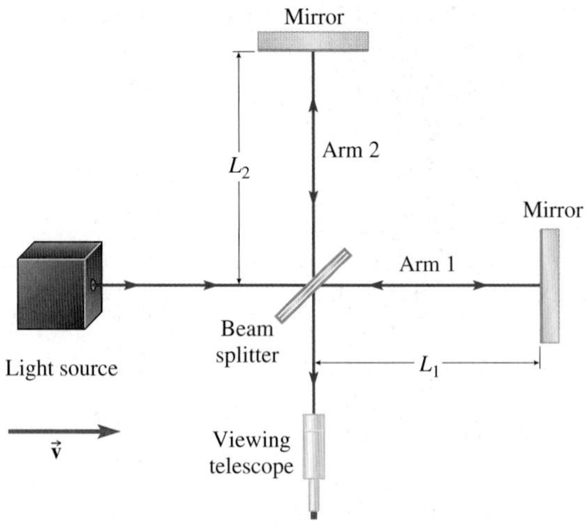

Figure 26.3 Simplified version of the Michelson-Morley experiment as seen from above. Assume the apparatus moves to the right at speed v with respect to the ether. With respect to the lab, the light beam in arm 1 moves at speed $c - v$ to the right and at speed $c + v$ to the left after reflecting from the mirror. The round-trip time in arm 1 is then $\Delta t_1 = L_1/(c - v) + L_1/(c + v) \neq (2L_1)/c$. The number of cycles of the wave in arm 1 is $\Delta t_1/T = f\,\Delta t_1$, where f is the frequency of the light. Thus, the number of cycles in arm 1 depends on the speed of the apparatus with respect to the ether. As the entire apparatus is rotated in a horizontal plane, the interference pattern viewed through the telescope should change as the difference in the number of cycles in arms 1 and 2 changes. No change in the interference pattern was observed by Michelson and Morley.

Thus, the speed of light would have two *different* values (v_{LG} and v_{LC}) in two different inertial reference frames.

A possible resolution to the contradiction would be if Maxwell's equations give the speed of light with respect to the medium in which light travels. Nineteenth-century scientists believed that light was a vibration in an invisible, elusive medium called the *ether*. If the speed of light c derived by Maxwell is the speed *with respect to the ether*, then in any inertial frame moving with respect to the ether, the speed of light should differ as predicted by Galilean relativity.

Does the speed of light as measured on Earth really depend on the motion of the Earth through the ether? In 1881, Albert Michelson designed a sensitive instrument, now called the Michelson interferometer (Section 25.2), to find out. In a later, more sensitive version of the experiment, Michelson was joined by Edward Williams Morley (1838–1923). The Michelson-Morley experiment showed no observable change in the speed of light due to the motion of Earth relative to the ether (Fig. 26.3). This led to the conclusion that there is no ether.

Einstein's Postulates

Albert Einstein (1879–1955) resolved these contradictions in his theory of special relativity (1905), now recognized as one of the cornerstones of modern physics. Einstein started with two postulates. The first is identical to Galileo's principle of relativity: the laws of physics are the same in any inertial reference frame. The second is that light travels at the same speed through vacuum in any inertial reference frame, regardless of the motion of the source or of the observer.

CONNECTION:

In essence, the second postulate extends the principle of relativity to include electromagnetism. The basic laws of electric and magnetic fields are the same in all inertial reference frames.

Einstein's Postulates of Special Relativity

(I) The laws of physics are the same in all inertial reference frames.

(II) The speed of light in vacuum is the same in all inertial reference frames, regardless of the motion of the source or of the observer.

Albert Einstein in 1910. In 1921, he was awarded the Nobel Prize in physics. Although Einstein is best known for his work on relativity, the Nobel committee cited "his discovery of the law of the photoelectric effect," which we study in Section 27.3.

The consequences Einstein derived from these two postulates deliver fatal blows to our intuitive notions of space and time. Our intuition about the physical world is based on experience, which is limited to things moving much slower than light. If moving at speeds approaching the speed of light were part of our everyday experience, then relativity would not seem strange at all. The theory of relativity has been confirmed by many experiments—which is the true test of any theory.

Einstein's theory of *special* relativity concerns inertial reference frames. In 1915, Einstein published his theory of general relativity, which concerns noninertial reference frames and the effect of gravity on intervals of space and time. In this chapter we study inertial reference frames only.

The Correspondence Principle

Galilean relativity and Newtonian physics do a great job of explaining and predicting motion at low speeds because they are excellent *approximations* when the speeds involved are much less than c. Therefore, the equations of special relativity must all reduce to their Newtonian counterparts for speeds much less than c.

The idea that a newer and more general theory must make the same predictions as an older theory, under experimental conditions that have proved the older theory successful, is called the **correspondence principle**.

CONNECTION:

Special relativity doesn't require us to throw out Newtonian physics; it is just *more general* than Newtonian physics.

26.2 SIMULTANEITY AND IDEAL OBSERVERS

The postulate that the speed of light is the same in all inertial reference frames leads to a startling conclusion: observers in different inertial reference frames disagree about whether two events are simultaneous if the events occur at different places. In Newtonian physics, time is absolute. That is, observers in different reference frames can use the same clock to measure time and they all agree on whether or not two events are simultaneous. Einstein's relativity does away with the notion of absolute time.

The idea of an event is crucial in relativity. The location of an event can be specified by three spatial coordinates (x, y, z); the time at which the event occurs is specified by t. Einstein's relativity treats space and time as four-dimensional *space-time* in which an event has four space-time coordinates (x, y, z, t).

Imagine two spaceships piloted by astronauts named Abe and Bea. Each ship has zero acceleration because external forces are negligible and they are not firing their engines. Then Abe and Bea are observers in inertial reference frames. Abe is at rest in his own reference frame and measures all velocities with respect to himself. The same can be said for Bea. They are not at rest with respect to one other, though. According to Abe, Bea moves past him at speed v; according to Bea, *Abe* moves past *her* at speed v.

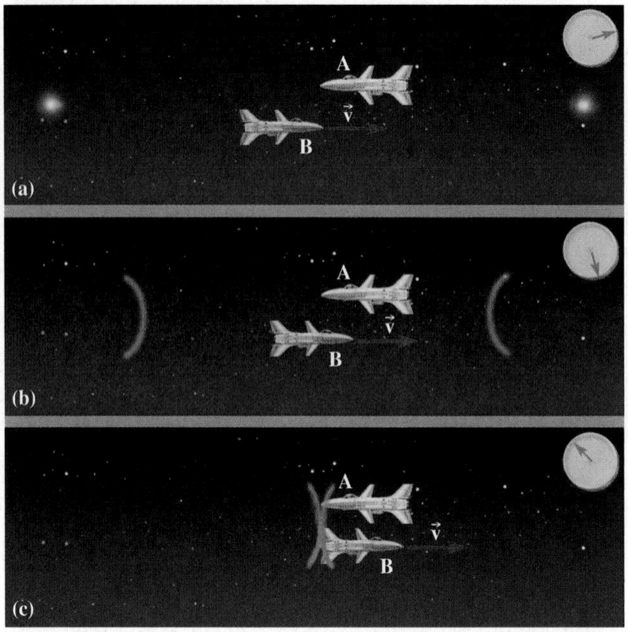

Figure 26.4 Events at three different times as viewed in Abe's reference frame. In this frame, Abe is at rest and Bea moves to the right at constant speed v. Abe's clock shows the time for each frame. (a) Two space probes flash simultaneously. The probes are at equal distances from Abe when they flash. (b) Bea travels toward the probe on the right, so she will run into that flash before the flash from the left catches up to her. (c) The two flashes reach Abe simultaneously. The pulse from the right has already passed Bea, but the pulse from the left has not yet reached her.

Abe and Bea observe two events: two space probes each emit a flash of light. Sitting in the cockpit at the nose of his ship, Abe sees the two flashes of light simultaneously. From the long measuring sticks in front of and behind his ship, which record the positions at which events occur, he finds that the flashes were emitted at *equal distances* from the nose of his ship. In Abe's frame of reference (Fig. 26.4), the flashes travel the same distance at the same speed (c) and arrive at the same time, so they must have been emitted *simultaneously*. The nose of Bea's ship happened to be alongside Abe's at the instant the flashes were emitted, but the flash from the probe on our right reaches Bea before the flash from the probe on our left. In Abe's reference frame, that happens because Bea is moving toward one probe and away from the other; the flashes do not travel equal distances to reach Bea.

In Bea's reference frame (Fig. 26.5), the right flash arrives before the left flash. Bea has measuring sticks similar to those of Abe; when she consults them, she finds that the flashes were emitted at equal distances from the nose of *her* ship. Since the flashes from each probe travel equal distances at the same speed (c), the right flash must arrive first because it was emitted first. In Bea's reference frame, *the flashes are not emitted simultaneously*. Bea's explanation for why the flashes reach Abe at the same time is that Abe is moving away from the first flash and toward the second at just the right speed. In Bea's reference frame, the flashes do not travel equal distances to reach Abe.

According to Einstein's postulates, the two reference frames are equally valid and light travels at the same speed in each. The inescapable conclusion is that the events *are* simultaneous in one frame and *are not* in the other.

Can the jet of charged particles move faster than light?

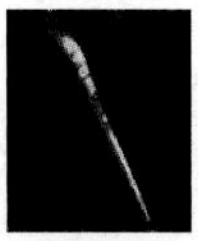

Ideal Observers

Since the high-speed jet of charged particles from the core of NGC 6251 moves toward Earth, the time it takes light to travel from the tip of the jet to Earth is continually decreasing. If we assume (incorrectly) that light arriving at Earth 1 day later was *emitted* 1 day

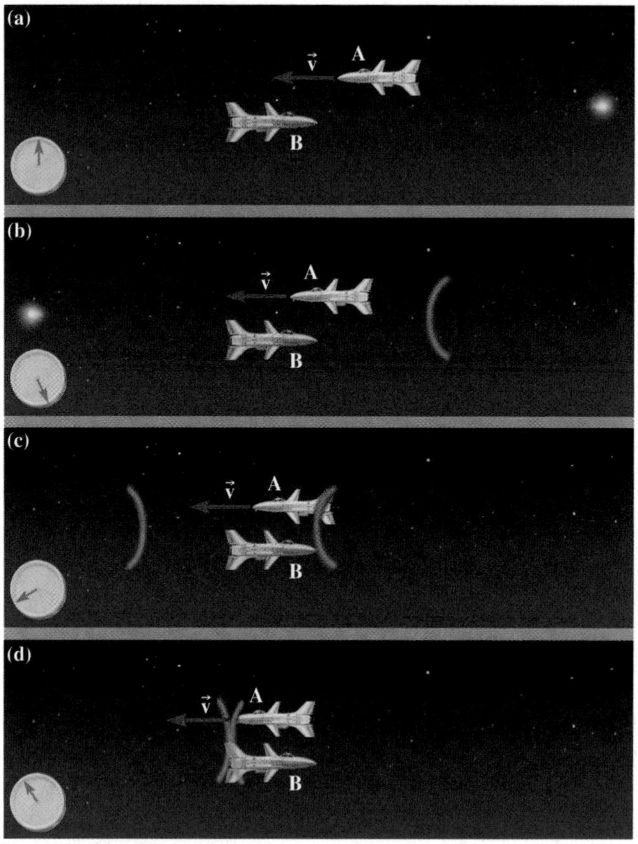

Figure 26.5 Events at four different times as viewed in Bea's reference frame. In this frame, Bea is at rest and Abe moves to the left at constant speed v. Bea's clock shows the time for each frame. (a) The space probe on the right flashes. (b) The probe on the left flashes. The two flashes take place at equal distances from Bea, but they are not simultaneous. (c) The right flash reaches Bea, but since Abe is moving to the left, it hasn't caught up with him yet. (d) The two flashes reach Abe simultaneously because he was moving away from the earlier flash and toward the later flash at just the right speed. The flash from the left still hasn't reached Bea.

later, then we calculate the apparent speed of the jet to be greater than c. The correct calculation recognizes that the light arriving 1 day later had a shorter distance to travel (Fig. 26.6), so it was emitted *more than* 1 day later. The jet is fast, but not as fast as light.

Application: observing active galactic nuclei

In Abe and Bea's disagreement about simultaneity, we were careful not to make a similar mistake. Each of them sees two flashes that travel *equal distances* to reach them. To avoid the confusion of light signals traveling different distances, we can imagine *ideal observers* who have placed sensors with synchronized clocks at rest *at every point in space* in their own reference frames. Each sensor records the time at which any event occurs at its location. Even if Abe and Bea were ideal observers, the data recorded by their sensors would still show that they reach different conclusions about the time sequence of the two flashes.

Cause and Effect

Continuing with the same reasoning, an observer moving to the left with respect to Abe would say that the *left* flash occurs first. Thus, the time order of the two events is different in different reference frames. How can there possibly be any cause-effect

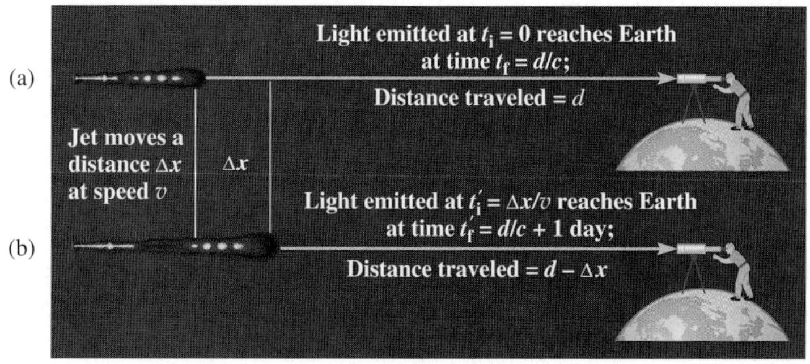

Figure 26.6 Calculating the speed of the jet. (a) Light emitted at $t_i = 0$ travels a distance d to reach Earth, arriving at $t_f = d/c$ because it travels at speed c. (b) Light emitted at t_i' travels a shorter distance $d - \Delta x$ and reaches Earth 1 day later, at $t_f' = t_f + 1$ day. The speed of the jet is $v = \Delta x/(t_i' - t_i)$, which is less than c, rather than $v_{apparent} = \Delta x/(t_f' - t_f)$.

relationships if the time order of events depends on the observer? What would it mean if, in some reference frames, the effect occurs *before* the cause?

In order for event 1 to cause event 2, some sort of signal—some information—must travel from event 1 to event 2. One conclusion of Einstein's postulates is that no signal can travel faster than c. If there is enough time, in some reference frame, for a signal at light speed to travel from event 1 to event 2, then it can be shown—through a more advanced analysis than we can do here—that a signal can travel from event 1 to event 2 in *all* inertial reference frames. The cause comes before the effect for all observers. On the other hand, if there is *not* enough time for a signal at light speed to travel from event 1 to event 2, then the two cannot have a causal relationship in *any* reference frame. For such events, some observers say that event 1 happens first, some say that event 2 happens first, and one particular observer says the events are simultaneous.

26.3 TIME DILATION

Since inertial observers in relative motion disagree about simultaneity, can two such observers agree about the time kept by clocks in relative motion? Two ideal clocks that are not moving relative to one another keep the same time by ticking simultaneously. However, if the clocks are in relative motion, the ticks are events that occur at different spatial locations, so two different inertial observers may disagree on whether the ticks are simultaneous, or about which clock ticks first.

The situation is easiest to analyze by imagining a conceptually simple kind of clock—a light clock (Fig. 26.7). A light clock is a tube of length L with mirrors at each end. A light pulse bounces back and forth between the two mirrors. One tick of the clock is one round-trip of the light pulse. The time interval between ticks for a stationary clock is $\Delta t_0 = 2L/c$.

Imagine now that Abe and Bea have two identical light clocks. Bea holds the clock vertically as she flies past Abe in her spaceship at speed $v = 0.8c$. What is the time interval between ticks of Bea's clock, as measured by Abe?

The velocity of the light pulse in Bea's clock as measured in Abe's reference frame has both x- and y-components (Fig. 26.8). The pulse must have an x-component of velocity if it is to meet up with the mirrors, which move to the right at speed v. During one tick, the light pulse moves along the *diagonal paths* shown.

Let us analyze one tick of Bea's clock as observed in Abe's reference frame. Suppose the time interval for one tick of Bea's clock as measured by Abe is Δt. Then the distance traveled by the light pulse during one tick is $c\,\Delta t$. During this same time interval, the clock moves horizontally a distance $v\,\Delta t$. By the Pythagorean theorem (see Fig. 26.8):

$$L^2 + \left(\frac{v\,\Delta t}{2}\right)^2 = \left(\frac{c\,\Delta t}{2}\right)^2$$

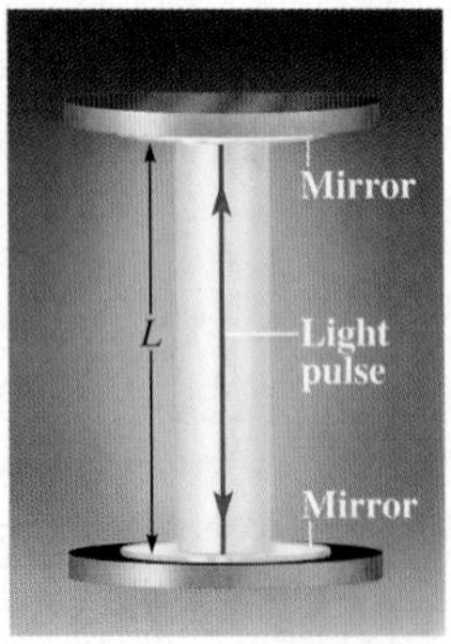

Figure 26.7 A light pulse reflects back and forth between the two parallel mirrors in a light clock. The time interval for one "tick" of the clock is the round-trip time for the light pulse, $\Delta t_0 = 2L/c$.

In Bea's reference frame, the clock is at rest. The distance traveled by the light pulse during one tick is $2L$, so the time interval for one tick as measured in Bea's reference frame is $\Delta t_0 = 2L/c$. We can therefore make the substitution

$$L = \frac{c\,\Delta t_0}{2}$$

into the Pythagorean equation:

$$\left(\frac{c\,\Delta t_0}{2}\right)^2 + \left(\frac{v\,\Delta t}{2}\right)^2 = \left(\frac{c\,\Delta t}{2}\right)^2$$

Solving for Δt yields (Problem 12)

$$\Delta t = \frac{1}{\sqrt{1 - v^2/c^2}}\,\Delta t_0 \qquad (26\text{-}1)$$

The factor multiplying Δt_0 in Eq. (26-1) occurs in many relativity equations, so we assign it a symbol (γ, the Greek letter gamma) and a name (the **Lorentz factor**, after Dutch physicist Hendrik Lorentz, 1853–1928).

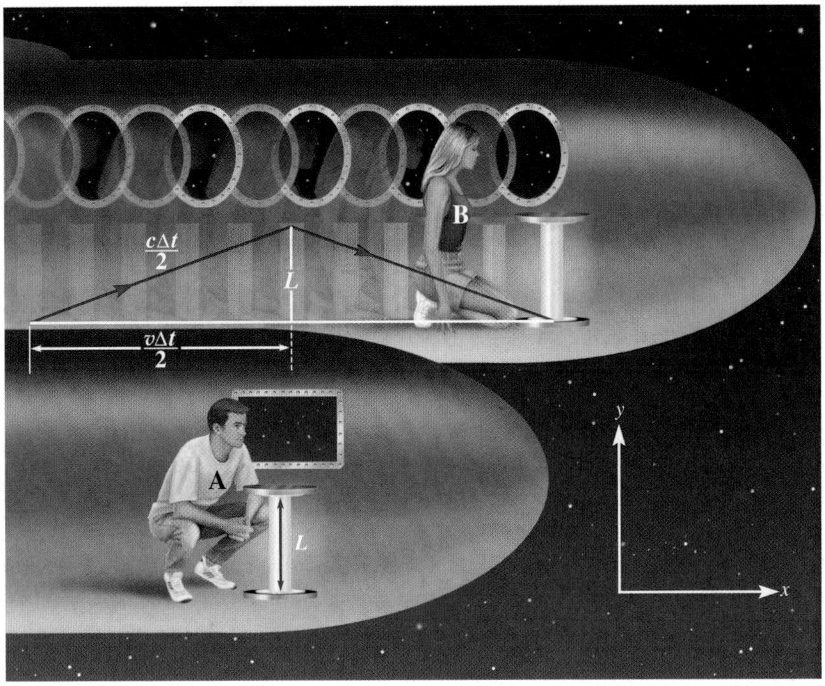

Figure 26.8 In Abe's reference frame, Bea's light clock moves to the right at speed $v = 0.8c$. The path of the light pulse in Bea's clock is along the diagonal red lines. (Not to scale.)

$$\gamma = \frac{1}{\sqrt{1 - v^2/c^2}} \tag{26-2}$$

See Fig. 26.9 for a graph of γ as a function of v/c. Using γ, Eq. (26-1) becomes

Time dilation:

$$\Delta t = \gamma\,\Delta t_0 \tag{26-3}$$

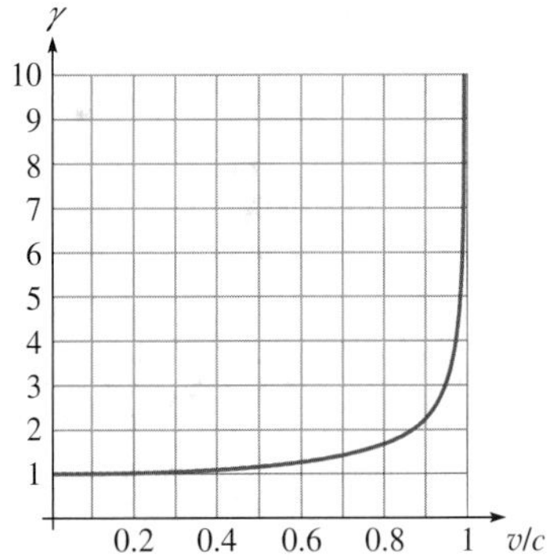

Figure 26.9 A graph of the Lorentz factor γ as a function of v/c. For low speeds, $\gamma \approx 1$. For speeds approaching that of light, γ increases without bound.

Notice that when $v \ll c$, $\gamma \approx 1$. Thus, for an object moving at nonrelativistic speeds (speeds small relative to the speed of light), $\Delta t = \gamma \Delta t_0 \approx \Delta t_0$.

Since $\gamma > 1$ for any $v \neq 0$, the time interval between ticks as measured in the reference frame in which the clock is moving, Δt, is *longer* than the time interval Δt_0 as measured in the clock's **rest frame**—the frame in which the clock is at rest. In a short phrase, *moving clocks run slow*. This effect is called **time dilation**; the time between ticks of the moving clock is dilated or expanded.

Abe's and Bea's reference frames are equally valid. Wouldn't Bea say that it is *Abe's* clock that runs slow? Yes, and they are *both* correct. Imagine that both of the clocks tick just as Abe and Bea pass each other—when they're at the same place. They agree that the clocks tick simultaneously. To see which clock runs slow, we compare the time of the *next* tick of the two clocks. Since the clocks are then at different places, the two observers disagree about the sequence of the ticks. Abe observes his clock ticking first, while Bea observes hers ticking first. They are both correct: there is no absolute or preferred reference frame from which to measure the time intervals.

The time interval Δt_0 measured in the rest frame of the clock is called the **proper** time interval; in that frame, the clock is at the same position for both ticks. When using the time dilation relation $\Delta t = \gamma\,\Delta t_0$, Δt_0 always represents the proper time interval—the time interval between two events measured in an inertial reference frame in which the events *occur in the same place*. The proper time interval is always shorter than the time interval Δt measured in any other inertial frame. The time dilation equation does *not* apply to a time interval between events that occur at different locations for *all* inertial observers.

Although we have analyzed time dilation using light clocks, *any* clock must show the same effect; otherwise there would be a preferred reference frame—the one in which light clocks behave the same as other clocks. Furthermore, a *clock* can be anything that measures a time interval. Biological processes such as the beating of a heart or the aging process are subject to time dilation. The nature of space and time, not the workings of a particular kind of device, is responsible for time dilation.

Time dilation may seem strange, but it has been verified in many experiments. One straightforward one was done in 1971 by J. C. Hafele and R. E. Keating using extremely precise cesium beam atomic clocks. The clocks were loaded onto airplanes and flown around for nearly 2 d. When the clocks were compared with reference clocks at the U.S. Naval Observatory, the clocks that had been in the air were behind those on the ground by an amount consistent with relativity.

Problem-Solving Strategy: Time Dilation

- Identify the two events that mark the beginning and end of the time interval in question. The "clock" is whatever measures this time interval.
- Identify the reference frame in which the clock is at rest. In that frame, the clock measures the proper time interval Δt_0.
- In any other reference frame, the time interval is *longer* by a factor of $\gamma = (1 - v^2/c^2)^{-1/2}$, where v is the speed of one frame relative to the other.

Speed and Distance Units Commonly Used in Relativity

- Speeds are usually written as a fraction times the speed of light (e.g., $0.13c$).
- Distances are often measured in **light-years** (symbol ly). A light-year is the distance that light travels in 1 yr. Calculations involving light-years are simplified by writing the speed of light as 1 ly/yr.

Example 26.1

Slowing the Aging Process

A 20.0-yr-old astronaut named Ashlin leaves Earth in a spacecraft moving at $0.80c$. How old is Ashlin when he returns from a trip to a star 30.0 light-years from Earth, assuming that he moves at $0.80c$ relative to Earth during the entire trip?

Strategy and Solution According to an earthbound observer, the trip takes (60.0 ly) ÷ (0.80 ly/yr) = 75 yr to complete. Since the astronaut is moving at high speed relative to Earth, all clocks on board—including biological processes such as aging—run slow as observed by Earth observers. Therefore, when the astronaut returns he is *less than* 95 yr old. Maybe he'll have time for another trip!

The two events that measure the time interval are the departure and return of the astronaut. Let the "clock" be the astronaut's aging process. This "clock" measures a 75-yr time interval *according to Earth observers*, for whom the clock is *moving*. The proper time interval is that measured by the astronaut himself. Thus, $\Delta t = 75$ yr and we want to find Δt_0. The Lorentz factor is calculated using the relative speed of the two reference frames, which is $0.80c$.

$$\gamma = (1 - 0.80^2)^{-1/2} = \tfrac{5}{3}$$

From the time dilation relation $\Delta t = \gamma \Delta t_0$,

$$\Delta t_0 = \frac{1}{\gamma}\Delta t = \frac{3}{5} \times 75 \text{ yr} = 45 \text{ yr}$$

If the astronaut ages 45 yr during the trip, he is 65 yr old on his return.

Discussion If the astronaut could travel 60.0 ly in 45 yr, then he would be traveling faster than light; his speed would be (60.0 ly)/(45 yr) = $1.3c$. As we see in Section 26.4, the

continued on next page

Example 26.1 continued

astronaut has traveled *less than 60.0 ly* in his reference frame. Just as time intervals are different in different reference frames, so are distances.

Suppose that Ashlin has a twin brother, Earnest, who stays behind on Earth. When Ashlin returns, he is 65 yr old, but Earnest is 95 yr old. Why can't Ashlin say that, in *his* reference frame, *Earnest* is the one who was moving at $0.80c$, and therefore *Earnest's* biological clock should run slow so that Earnest is *younger* rather than older? This question is sometimes called the *twin paradox*.

We analyzed the situation in the reference frame of Earnest, which is assumed to be inertial. The analysis from Ashlin's point of view is much more difficult because Ashlin is not traveling at constant velocity with respect to Earnest for the entire trip—if he were, he could never return to Earth. Nevertheless, analysis of the trip from Ashlin's perspective confirms that when he returns to Earth he is younger than Earnest.

Practice Problem 26.1 Journey to Newly Formed Stars near Earth

In 1998, using the Keck II telescope, scientists discovered some previously undetected, young stars that are only 150 ly from Earth. Suppose a space probe is flown to one of these stars at a speed of $0.98c$. The battery that powers the communications systems can run for 40 yr. Will the battery still be good when the space probe reaches the star?

26.4 LENGTH CONTRACTION

Suppose Abe has two identical metersticks, which he has verified to have exactly the same length. He gives one to Bea, the astronaut. As Bea flies past Abe with speed $v = 0.6c$, holding her meterstick in the direction of motion, they compare the lengths of the two metersticks (Fig. 26.10). Are they still equal?

No; Abe finds that Bea's meterstick is *less than* 1 m in length. To measure the length of Bea's moving meterstick, Abe might start a timer when the front end of her meterstick passes a reference point and stop the timer when the other end passes. The length L that Abe measures for Bea's stick is the measured time interval Δt_0 multiplied by the speed at which she is moving:

$$L = v\,\Delta t_0 \qquad \text{(Abe; moving stick)}$$

Since Abe measures the time interval between two events that occur at the same place—at his reference point—Δt_0 is the *proper* time interval between the events.

Bea can measure the length of her own stick (L_0) the same way—by recording the time interval Δt between when Abe's reference point passes the two ends of her meterstick.

$$L_0 = v\,\Delta t \qquad \text{(Bea; stick at rest)}$$

The time interval Bea measures is dilated; it is longer than the proper time interval by a factor of γ:

$$\Delta t = \gamma\,\Delta t_0$$

Therefore, the length of Bea's meterstick as measured by Abe (L) is *shorter* than the length measured by Bea ($L_0 = 1$ m):

$$\frac{L}{L_0} = \frac{\Delta t_0}{\Delta t} = \frac{1}{\gamma}$$

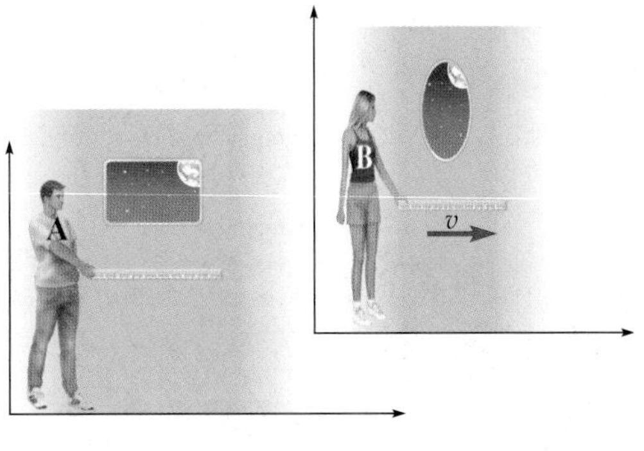

Figure 26.10 In Abe's reference frame, Bea moves to the right at speed $0.6c$. Abe measures everything on board Bea's ship—including the meterstick and even Bea herself—as shortened along the direction of motion.

Length contraction:

$$L = \frac{L_0}{\gamma} \qquad (26\text{-}4)$$

In the length contraction relation $L = L_0/\gamma$, L_0 represents the **proper** length or rest length—the length of an object in its rest frame. L is the length measured by an observer for whom the object is moving.

Meanwhile, Bea can also measure Abe's meterstick. Bea would say that *Abe's* meterstick is the one that is shorter. How can they both be right? Which meterstick *really* is shorter?

To resolve the issue once and for all, they might want to hold the metersticks together, but they cannot: the metersticks are in relative motion. To compare the lengths, they could wait until the left ends of the two metersticks coincide (Fig. 26.11). They must compare the positions of the right ends of the metersticks *at the same instant.* Since Abe and Bea disagree about simultaneity, they disagree about which meterstick is shorter, but they are *both* correct. Just as an observer always finds that a moving clock runs slow compared with a stationary clock, an observer always finds that a moving object is contracted (shortened) *along the direction of its motion.* Lengths perpendicular to the direction of motion are not contracted.

Problem-Solving Strategy: Length Contraction

- Identify the object whose length is to be measured in two different frames. The length is contracted only in the direction of the object's motion. If the length in question is a distance rather than the length of an actual object, it often helps to imagine the presence of a long measuring stick.
- Identify the reference frame in which the object is at rest. The length in that frame is the proper length L_0.
- In any other reference frame, the length L is *contracted*: $L = L_0/\gamma$. $\gamma = (1 - v^2/c^2)^{-1/2}$, where v is the speed of one frame relative to the other.

CHECKPOINT 26.4

A sprinter crosses the start line (event 1) and runs at constant velocity until she crosses the finish line (event 2). In what reference frame would an observer measure the proper time interval between these two events? In what reference frame would an observer measure the proper length of the track from start line to finish line?

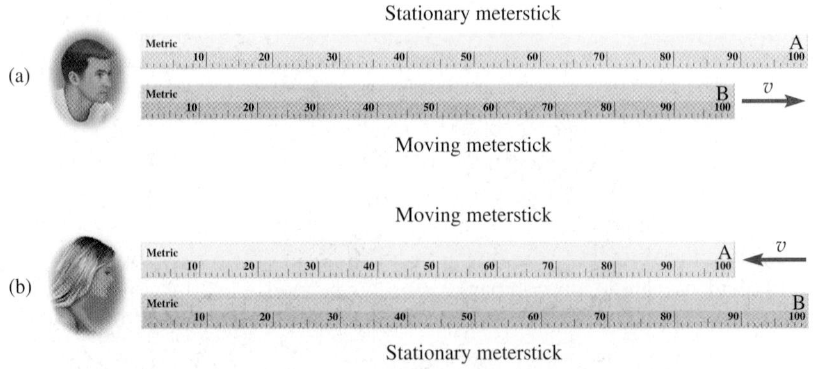

Figure 26.11 Comparing two metersticks by lining up the left ends. (a) As seen in Abe's rest frame. (b) As seen in Bea's rest frame.

Example 26.2

Muon Survival

Cosmic rays are energetic particles—mostly protons—that enter Earth's upper atmosphere from space. The particles collide with atoms or molecules in the upper atmosphere and produce showers of particles. One of the particles produced is the muon, which is something like a heavy electron. The muon is unstable. Half of the muons present at any particular instant of time still exist 1.5 μs later; the other half decay into an electron plus two other particles. In a shower of muons streaming toward Earth's surface, some decay before reaching the ground. If 1 million muons are moving toward the ground at speed $0.995c$ at an altitude of 4500 m above sea level, how many survive to reach sea level? (tutorial: muon lifetime)

Strategy Imagine a measuring stick extending from the upper atmosphere to sea level (Fig. 26.12). In the reference frame of the Earth, the measuring stick is at rest; its proper length is $L_0 = 4500$ m. In the reference frame of the muons, the muons are at rest and the measuring stick moves past them at speed $0.995c$. In the muon frame, the measuring stick is contracted. In the muon frame, sea level is *not* 4500 m away when the upper end of the measuring stick passes by; the distance is shorter due to length contraction. Once we find the contracted distance L, the time the measuring stick takes to move past them at speed v is $\Delta t = L/v$. From the elapsed time, we can determine how many muons decay and how many are left.

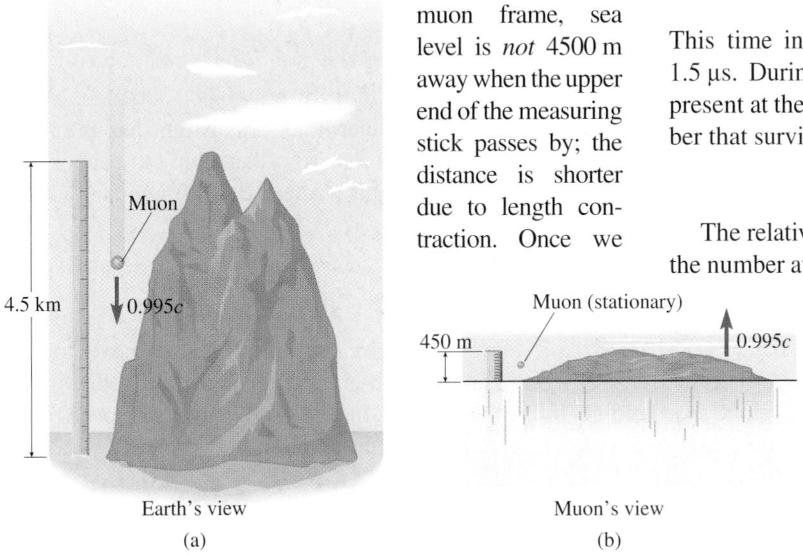

Figure 26.12
The muons' trip as viewed by (a) an Earth observer and (b) the muons.

Solution The contracted distance is $L = L_0/\gamma$, where the Lorentz factor is

$$\gamma = (1 - 0.995^2)^{-1/2} = 10$$

Therefore, the contracted distance is $L = \frac{1}{10} \times 4500 \text{ m} = 450$ m. The elapsed time is

$$\Delta t = L/v = (450 \text{ m})/(0.995 \times 3 \times 10^8 \text{ m/s}) = 1.5 \text{ μs}$$

During 1.5 μs, half of the muons decay, so 500 000 muons reach sea level.

Discussion If there were no length contraction, the elapsed time would be

$$\Delta t = \frac{4500 \text{ m}}{0.995 \times 3 \times 10^8 \text{ m/s}} = 15 \text{ μs}$$

This time interval is equal to ten successive intervals of 1.5 μs. During each of those intervals, half of the muons present at the start of the interval decay. Therefore, the number that survive to reach sea level would be only

$$1\,000\,000 \times \left(\tfrac{1}{2}\right)^{10} \approx 980 \text{ muons}$$

The relative number of muons at sea level compared with the number at higher elevations has been studied experimentally; the results are consistent with relativity.

Practice Problem 26.2
Rocket Velocity

An astronaut in a rocket passes a meterstick moving parallel to its long dimension. The astronaut measures the meterstick to be 0.80 m long. How fast is the rocket moving with respect to the meterstick?

26.5 VELOCITIES IN DIFFERENT REFERENCE FRAMES

Figure 26.13 shows Abe and Bea in their spaceships; in Abe's reference frame, Bea moves at velocity v_{BA}. Bea launches a space probe, which, in *her* reference frame, moves at velocity v_{PB}. What is the velocity of the probe in Abe's reference frame (v_{PA})? (Since we consider only velocities along a straight line—in this case, along a horizontal line—we write the velocities as *components* along that line.)

If v_{PB} and v_{BA} are small compared to c, time dilation and length contraction are negligible; then v_{PA} is given by the Galilean velocity addition formula (Eq. 3-24):

$$v_{PA} = v_{PB} + v_{BA}$$

CONNECTION:

Velocities are relative even in classical physics (see Section 3.6). The Galilean velocity addition formula may seem correct intuitively, but it is only *approximately* correct when the speeds are much less than c.

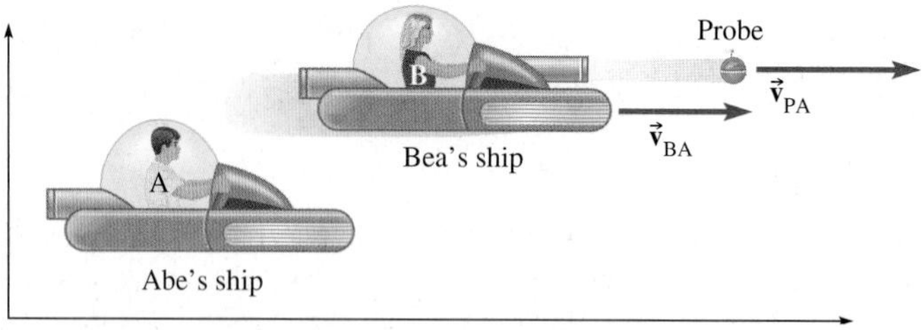

Figure 26.13 As viewed in Abe's reference frame, Bea's ship has velocity $\vec{v}_{BA}$ and the space probe has velocity $\vec{v}_{PA}$. How can we find $\vec{v}_{PA}$ given $\vec{v}_{BA}$ and $\vec{v}_{PB}$, the velocity of the probe with respect to *Bea*?

However, Eq. (3-24) cannot be correct in general because the probe cannot move faster than light in *any* inertial reference frame. (If $v_{PB} = +0.6c$ and $v_{BA} = +0.7c$, the Galilean formula gives $v_{PA} = 1.3c$.) The relativistic equation that replaces Eq. (3-24) takes time dilation and length contraction into account and predicts $|v_{PA}| < c$ for *any* values of v_{PB} and v_{BA} whose magnitudes are less than c.

The relativistic velocity transformation formula is

$$v_{PA} = \frac{v_{PB} + v_{BA}}{1 + v_{PB}v_{BA}/c^2} \qquad (26\text{-}5)$$

The denominator in Eq. (26-5) can be thought of as a correction factor to account for both time dilation and length contraction. When v_{PB} and v_{BA} are small compared to c, the denominator is approximately 1; then Eq. (26-5) reduces to the Galilean approximation [Eq. (3-24)]. For example, if $v_{PB} = v_{BA} = +3$ km/s (fast by ordinary standards, but small compared to the speed of light), the denominator is

$$1 + \frac{v_{PB}v_{BA}}{c^2} = 1 + \frac{(3 \times 10^3 \text{ m/s})^2}{(3 \times 10^8 \text{ m/s})^2} = 1 + 10^{-10}$$

In this case the Galilean velocity addition formula is off by only 0.000 000 01%.

Next, we verify that Einstein's second postulate holds—that light has the same speed in any inertial reference frame. Suppose that instead of launching a space probe, Bea turns on her headlights. The velocity of the light beam in Bea's frame is $v_{LB} = +c$. The speed of the light beam in Abe's frame is

$$v_{LA} = \frac{v_{LB} + v_{BA}}{1 + v_{LB}v_{BA}/c^2} = \frac{c + v_{BA}}{1 + cv_{BA}/c^2} = \frac{c(1 + v_{BA}/c)}{1 + v_{BA}/c} = c$$

Thus, even if the jet of charged particles from the active nucleus of a galaxy moves toward Earth at a speed close to that of light in an Earth observer's reference frame, the light it emits travels at the speed of light in the Earth frame. In the calculation of the speed of the jet outlined in Section 26.2, it is correct to use c for the speed of light emitted by the jet.

Problem-Solving Strategy: Relative Velocity

- Sketch the situation as seen in two different reference frames. Label the velocities with subscripts to help keep them straight. The subscripts in v_{BA} mean *the velocity of B as measured in A's reference frame*.
- The velocity transformation formula [Eq. (26-5)] is written in terms of the *components* of the three velocities along a straight line. The components are positive for one direction (your choice) and negative for the other.
- If A moves to the right in B's frame, then B moves to the left in A's frame:

$$v_{BA} = -v_{AB}$$

- To get Eq. (26-5) right, make sure that the inner subscripts on the right side are the same and "cancel" to leave the left-side subscripts *in order*:

$$v_{LA} = \frac{v_{L\cancel{B}} + v_{\cancel{B}A}}{1 + v_{L\cancel{B}}v_{\cancel{B}A}/c^2}$$

Example 26.3

Observation in Space

Two spaceships travel at high speed in the same direction along the same straight line. As measured by an observer on a nearby planet, ship 1 is behind ship 2 and moves at speed $0.90c$; ship 2 moves at speed $0.70c$. According to an observer aboard ship 1, how fast and in what direction is ship 2 moving?

Strategy The two reference frames of interest are that of the planet and that of ship 1. Then a sketch of the planet and the two ships as seen in each of the reference frames helps us assign subscripts to the velocities. After choosing a positive direction, we carefully assign the correct algebraic signs to each velocity. Then we are ready to apply the velocity transformation formula.

Solution First we draw Fig. 26.14a showing the two ships moving to the right in the reference frame of the planet. Let the right be the positive direction. In this frame, the velocities of the ships are $v_{1P} = +0.90c$ and $v_{2P} = +0.70c$. We want to find the velocity of ship 2 as seen by observers on ship 1 (v_{21}), so we draw Fig. 26.14b in the reference frame of ship 1. Since ship 1 moves to the right in the planet frame, the planet moves to the left in ship 1's frame: $v_{P1} = -v_{1P} = -0.90c$.

Now we apply Eq. (26-5). Since we want to calculate v_{21}, it goes on the left side of the equation. The two velocities on the right side are v_{2P} and v_{P1} so that the P's "cancel" to leave v_{21}.

$$v_{21} = \frac{v_{2P} + v_{P1}}{1 + v_{2P}v_{P1}/c^2}$$

Substituting $v_{2P} = +0.70c$ and $v_{P1} = -0.90c$ yields

$$v_{21} = \frac{0.70c + (-0.90c)}{1 + [0.70c \times (-0.90c)]/c^2} = \frac{-0.20c}{1 - 0.63} = -0.54c$$

So according to observers on spaceship 1, spaceship 2 is moving to the left at speed $0.54c$.

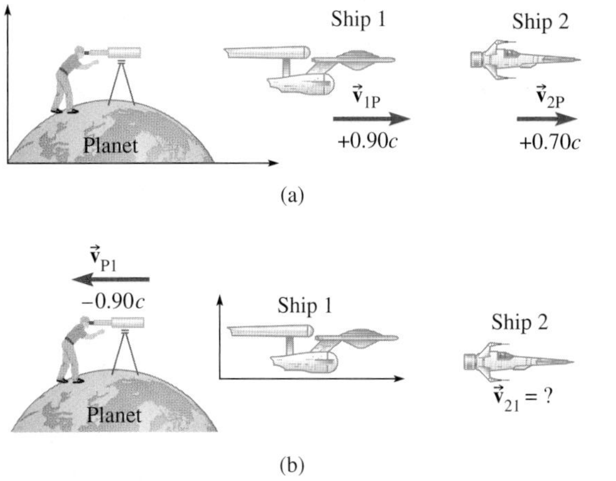

Figure 26.14

(a) An observer on the planet measures the velocities of the two spaceships. (b) The same events seen by an observer on spaceship 1.

Discussion Note how important it is to get the signs correct. For instance, if we had made an error by writing $v_{P1} = +0.90c$, then we would have calculated $v_{21} = +0.98c$. This answer has spaceship 2 moving to the *right* relative to spaceship 1, which doesn't make sense. In the planet frame, ship 1 is catching up to ship 2, so in ship 1's frame, ship 2 must move toward ship 1. In one dimension, the velocity is always in the same *direction* that you would expect in Galilean velocity addition; only the speed is different.

Practice Problem 26.3 Relative Velocity of Approaching Rocket

According to an observer on a space station, two rocket ships are moving toward one another in opposite directions along the x-axis, ship A with velocity $0.40c$ to the right and ship B with velocity $0.80c$ to the left. With what speed does an observer on ship B observe ship A to be moving?

26.6 RELATIVISTIC MOMENTUM

When a particle's speed is not small relative to the speed of light, the nonrelativistic expressions for momentum and kinetic energy are not valid. If we try to use them for particles moving at high speeds, it appears that the momentum and energy conservation laws are violated. We must redefine momentum and kinetic energy so that the conservation laws hold for *any* speed. The nonrelativistic expressions $\vec{\mathbf{p}} = m\vec{\mathbf{v}}$ and $K = \frac{1}{2}mv^2$ are *good approximations* as long as $v \ll c$. The relativistic expressions are more general; they give the correct momentum and kinetic energy for *any* speed. Thus, the relativistic expressions must give the same momentum and kinetic energy as the nonrelativistic expressions when $v \ll c$.

CONNECTION:

Relativity maintains conservation of momentum and energy as fundamental principles of physics, but requires modified definitions of momentum and kinetic energy. The classical definitions $\vec{\mathbf{p}} = m\vec{\mathbf{v}}$ and $K = \frac{1}{2}mv^2$ are excellent *approximations* when $v \ll c$.

The relativistically correct expression for the momentum of a particle with mass m and speed v is

Momentum:

$$\vec{\mathbf{p}} = \gamma m\vec{\mathbf{v}} \qquad (26\text{-}6)$$

where γ is calculated using the particle's speed v.

For speeds small relative to c, $\gamma \approx 1$ and $\vec{\mathbf{p}} \approx m\vec{\mathbf{v}}$. For example, consider an airplane traveling at 300 m/s (670 mi/h), which is just under the speed of sound in air. Compared with the speed of light, 300 m/s is quite slow; it's just one one-millionth of the speed of light. When $v = 300$ m/s, the Lorentz factor is $\gamma = 1.000\,000\,000\,000\,5$. In this case, using the nonrelativistic expression $\vec{\mathbf{p}} = m\vec{\mathbf{v}}$ to find the momentum of the plane is a *very* good approximation! But for a proton ejected from the Sun at nine-tenths the speed of light, $\gamma = 2.3$. The proton's momentum is more than twice as large as we would expect from the nonrelativistic expression $\vec{\mathbf{p}} = m\vec{\mathbf{v}}$. Therefore, $\vec{\mathbf{p}} = m\vec{\mathbf{v}}$ should not be used for a proton traveling at $0.9c$.

What is the cutoff speed below which the nonrelativistic formula can be used? There's no clear boundary. As a rule of thumb, the nonrelativistic formula is less than 1% off as long as $\gamma < 1.01$. Setting $\gamma = 1.01$ and solving for v, we get $v = 0.14c \approx 4 \times 10^7$ m/s. As long as the speed of the particle is less than about $\frac{1}{7}$ the speed of light, the nonrelativistic momentum expression is correct to within 1%.

The relativistic expression for momentum has some dramatic consequences. Consider the momentum of a particle as v gets close to the speed of light (Fig. 26.15). As v approaches the speed of light, the momentum increases *without bound.* The momentum can get as large as you want *without the speed ever reaching the speed of light.* Or, in other words, it is impossible to accelerate something to the speed of light. You can give something as much momentum as you like, but you can never get the speed up to c.

With relativistic momentum, it is still true that the impulse delivered equals the change in momentum ($\Sigma\vec{\mathbf{F}}\,\Delta t = \Delta\vec{\mathbf{p}}$), but $\Sigma\vec{\mathbf{F}} = m\vec{\mathbf{a}}$ is *not* true: the acceleration due to a constant net force gets smaller and smaller as the particle's speed approaches c. The longer the force is applied, the larger the momentum, but the speed never reaches the speed of light. This fact is verified in the daily operation of particle accelerators, which are used in high-energy physics research. Particles such as electrons and protons are "accelerated" to larger and larger momenta (and kinetic energies) as their speeds get closer and closer to—but never exceed—the speed of light.

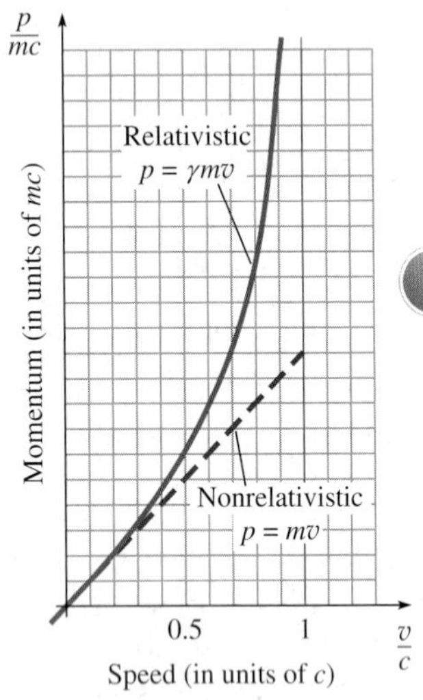

Figure 26.15 A graph of momentum versus v showing the nonrelativistic and relativistic expressions. At low speeds the two expressions are in agreement.

Example 26.4

Collision in the Upper Atmosphere

Cosmic rays collide with atoms or molecules in the upper atmosphere (Fig. 26.16). If a proton moving at $0.70c$ makes a head-on collision with a nitrogen atom, initially at rest, and the proton recoils at $0.63c$, what is the speed of the nitrogen atom after the collision? (The mass of a nitrogen atom is about 14 times the mass of a proton.)

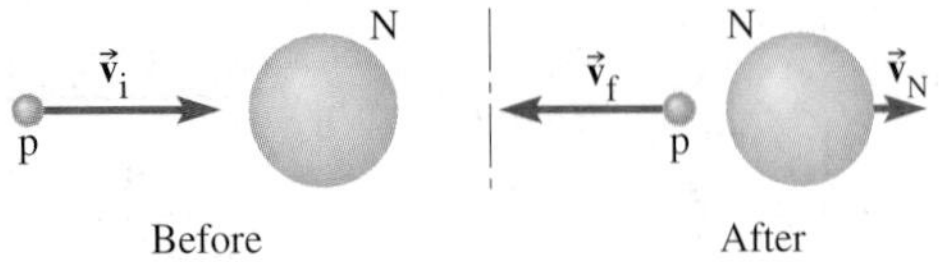

Figure 26.16

A proton moving with speed v_i collides head-on with a nitrogen atom at rest in the upper atmosphere. The nitrogen atom moves with speed v_N after the collision, while the proton rebounds with speed v_f.

Strategy We apply the principle of momentum conservation to solve this collision problem. The only change from the way we have analyzed collisions previously is that we must use the relativistic momentum expression for the proton. It remains to be seen whether the nitrogen atom is moving at relativistic speed after the collision. If it is not, we can simplify the calculations by using the nonrelativistic momentum expression. It is perfectly fine to "mix" the two; they aren't different kinds of momentum. The nonrelativistic

continued on next page

Example 26.4 continued

expression is just an approximation—when it is a good approximation, we use it.

Solution Choose the direction of the proton's initial velocity as the $+x$-direction. The initial momentum of the proton is

$$p_{ix} = \gamma m_p v_{ix}$$

where $v_{ix} = +0.70c$ and

$$\gamma = (1 - 0.70^2)^{-1/2} = 1.4003$$

Therefore, the x-component of the initial momentum is

$$p_{ix} = 1.4003m_p \times 0.70c = +0.9802m_p c$$

After the collision, the proton's momentum is

$$p_{fx} = \gamma m_p v_{fx}$$

where $v_f = -0.63c$ since it moves in the $-x$-direction and

$$\gamma = (1 - 0.63^2)^{-1/2} = 1.288$$

The final momentum of the proton is

$$p_{fx} = -0.8114m_p c$$

The change in the proton's x-component of momentum is

$$\Delta p_x = -0.8114m_p c - 0.9802m_p c = -1.7916m_p c$$

To conserve momentum, the nitrogen atom's final momentum is $P_x = +1.7916m_p c$. To find the velocity of the atom, we set $P_x = \gamma M v_{Nx}$.

Since the mass of a nitrogen atom is about 14 times that of a proton,

$$1.7916m_p c = \gamma \times 14m_p \times v_{Nx}$$

Canceling m_p from both sides and simplifying, we have

$$0.1280c = \gamma v_{Nx} = [1 - (v_{Nx}/c)^2]^{-1/2} \times v_{Nx}$$

This equation can be solved for v_{Nx} with some very messy algebra, but it's better to realize that an approximation is appropriate. Since γ is never less than 1, v_{Nx} cannot be greater than $0.1280c$. Therefore, v_{Nx} is small enough to use the nonrelativistic momentum expression $P_x = Mv_{Nx}$—or, in other words, to set $\gamma = 1$. Then $v_{Nx} = 0.1280c$. Rounding to two significant figures, the speed of the nitrogen atom is $0.13c$.

Discussion Using the nonrelativistic momentum *throughout*, for the proton as well as the atom, would have given:

$$0.70m_p c = -0.63m_p c + 14m_p v_{Nx}$$

$$v_{Nx} = \frac{1.33c}{14} = 0.095c$$

which is 26% smaller than the correct value. On the other hand, you can verify (by doing a lot of algebraic manipulation) that using the relativistic expression for the nitrogen without approximating would have given $0.13c$—the same answer (to within two significant figures). All that extra algebra would have yielded nothing. It pays to decide whether it is necessary to use relativistic expressions or whether the nonrelativistic ones are perfectly adequate.

Practice Problem 26.4 A Change in Momentum

A chunk of space debris with mass 1.0 kg is moving with a speed of $0.707c$. A constant force of magnitude 1.0×10^8 N, in the direction opposite to the chunk's motion, acts on it. How long must this force act to bring the space debris to rest? [*Hint:* The impulse delivered is equal to the change in momentum.]

26.7 MASS AND ENERGY

A particle at rest has no *kinetic* energy, but that doesn't mean it has no energy. Relativity tells us that mass* is a measure of rest energy. The **rest energy** E_0 of a particle is its energy as measured in its rest frame. Thus, rest energy does *not* include kinetic energy. The relationship between rest energy and mass is

Rest energy:

$$E_0 = mc^2 \qquad (26\text{-}7)$$

The interpretation of mass as a measure of rest energy is confirmed by observations of radioactive decay, in which particles at rest decay into products of smaller total mass; the products carry off kinetic energy equal to the decrease in total mass times c^2.

A kilogram of coal has a rest energy of $(1 \text{ kg}) \times (3 \times 10^8 \text{ m/s})^2 = 9 \times 10^{16}$ J. If the entire rest energy of the coal were converted into electric energy, it would be enough to

*In this book, *mass* is used exclusively to mean *invariant mass* or *rest mass.*

supply the electricity needs of a typical American household for *millions of years*. When coal is burned, only a tiny fraction (about one part in a billion) of the coal's rest energy is released. The change in mass—the difference between the mass of the coal and the total mass of all the products—is immeasurably small. In chemical reactions it *seems* as if mass is conserved.

On the other hand, in nuclear reactions and radioactive decays, a much larger fraction of the mass of a nucleus is transformed into the kinetic energy of the reaction products. The total mass of the **daughter** particles (particles present after the reaction) is not the same as the total mass of the **parent** particles (particles present before the reaction). Mass is *not* conserved, but total energy (the sum of rest energy and kinetic energy) is conserved. If there is a decrease in mass, then energy is released by the reaction. Total energy is still conserved; it has just been changed from one form to another—from rest energy to kinetic energy or radiation (or both). If there is an increase in rest energy (i.e., if the daughter particles have more total mass than the parent particles), the reaction does not occur spontaneously. The reaction can occur only if the energy deficit is supplied by the initial kinetic energies of the parent particles.

The Electron-Volt

A unit of energy commonly used in atomic and nuclear physics is the electron-volt (symbol eV). One electron-volt is equal to the kinetic energy that a particle with charge $\pm e$ (e.g., an electron or a proton) gains when it is accelerated through a potential difference of magnitude 1 V. Since 1 V = 1 J/C and $e = 1.60 \times 10^{-19}$ C, the conversion between electron-volts and joules is:

$$1 \text{ eV} = e \times 1 \text{ V} = 1.60 \times 10^{-19} \text{ C} \times 1 \text{ J/C} = 1.60 \times 10^{-19} \text{ J}$$

For larger amounts of energy, keV (pronounced *kay-ee-vee*) represents kilo-electron-volts (10^3 eV) and MeV (pronounced *em-ee-vee*) represents mega-electron-volts (10^6 eV).

To facilitate calculations in electron-volts, momentum can be expressed in units of eV/c and mass can be expressed in eV/c^2. Instead of multiplying or dividing by the numerical value of c, factors of c get carried in the units. For example, an electron's rest energy is 511 keV. Using $E_0 = mc^2$, the electron's mass is

$$m = E_0/c^2 = 511 \text{ keV}/c^2$$

The momentum of an electron moving at speed $0.80c$ is

$$p = \gamma m v = 1.667 \times 511 \text{ keV}/c^2 \times 0.80c = 680 \text{ keV}/c$$

Example 26.5

Energy Released in Radioactive Decay

Carbon dating is based on the radioactive decay of a carbon-14 nucleus (a nucleus with 6 protons and 8 neutrons) into a nitrogen-14 nucleus (with 7 protons and 7 neutrons). In the process, an electron (e^-) and a particle called an antineutrino ($\overline{\nu}$) are created. The reaction is written as

$$^{14}\text{C} \rightarrow {}^{14}\text{N} + e^- + \overline{\nu}$$

Find the energy released by this reaction. The masses of the nuclei are 13.999 950 u for ^{14}C and 13.999 234 u for ^{14}N. [The *atomic mass unit* (u) is commonly used in atomic and nuclear physics; 1 u = 1.66×10^{-27} kg.] The antineutrino's mass is negligibly small.

Strategy We compare the total masses of the particles before and after the decay. A decrease in total mass means that rest energy has been transformed into other forms: the kinetic energies of the nitrogen atom and electron and the energy of the antineutrino. The energy released by the reaction is equal to the change in the amount of rest energy.

continued on next page

Example 26.5 continued

Solution Before decay, the total mass is 13.999 950 u. After decay, the total mass is

$$13.999\ 234\ \text{u} + \frac{9.11 \times 10^{-31}\ \text{kg}}{1.66 \times 10^{-27}\ \text{kg/u}} = 13.999\ 783\ \text{u}$$

The change in mass is

$$\Delta m = -0.000\ 167\ \text{u}$$

Let Q be the quantity of energy released. Since total energy is conserved, the rest energy before decay is equal to the rest energy after decay plus the energy released:

$$m_i c^2 = m_f c^2 + Q$$

$$\begin{aligned} Q &= m_i c^2 - m_f c^2 \\ &= -\Delta m \times c^2 \\ &= 0.000\ 167\ \text{u} \times 1.66 \times 10^{-27}\ \text{kg/u} \times (3.00 \times 10^8\ \text{m/s})^2 \\ &= 2.495 \times 10^{-14}\ \text{kg·m}^2/\text{s}^2 = 2.50 \times 10^{-14}\ \text{J} \end{aligned}$$

In units of keV,

$$Q = \frac{2.495 \times 10^{-14}\ \text{J}}{1.60 \times 10^{-19}\ \text{J/eV}} \times 10^{-3}\ \frac{\text{keV}}{\text{eV}} = 156\ \text{keV}$$

Discussion The fraction of the original rest energy of the ^{14}C nucleus that is released is (0.000 167 u)/(13.999 950 u) = 0.0012%. This may not seem like a huge fraction, but it is about 10^4 times larger than the fractional decrease in mass that occurs when carbon is burned. In nuclear fusion, the fractional mass change approaches 1%.

Practice Problem 26.5 How Fast Is the Sun Losing Mass?

The Sun radiates energy at a rate of 4×10^{26} J/s. At what rate is the mass of the Sun decreasing?

Invariance

So far we have seen two quantities that are *invariant*—that have the same value as measured in all inertial frames. One is the speed of light; the other is mass. Distances and time intervals are not the same in different frames of reference, so they are not invariant.

Let us emphasize the difference between a conserved quantity and an invariant quantity. A conserved quantity maintains the same value *in a given reference frame*; the value may differ from one frame to another, but in any given frame its value is constant. An invariant is a quantity that, at a given moment in history, has the same numerical value in *all inertial frames*. Thus, momentum is conserved but is not invariant. Mass is invariant but is not conserved; as in Example 26.5, the total mass can change in a radioactive decay or other nuclear reaction. The total energy *is* conserved in such a reaction; but total energy is not invariant, since particles have different kinetic energies in different frames.

CHECKPOINT 26.7

An *invariant* is a quantity that has the same value in all inertial reference frames. (a) According to Galilean relativity, which of these quantities are invariant: position, displacement, length, time interval, velocity, acceleration, force, momentum, mass, kinetic energy, the speed of light in vacuum? (b) Which of them are invariant according to Einstein's special relativity?

26.8 RELATIVISTIC KINETIC ENERGY

A more general, relativistic expression for momentum is required in order to preserve the principle of momentum conservation for particles moving at relativistic speeds. We need to do the same for kinetic energy.

With the relation between force and momentum $(\Sigma\vec{\mathbf{F}} = \Delta\vec{\mathbf{p}}/\Delta t)$ and the concept of work as the product of force and distance, we can deduce a formula for the kinetic energy of a particle. As in the nonrelativistic case, the kinetic energy of a body is equal to the work done to accelerate it from rest to its present velocity. The result is

Kinetic energy:

$$K = (\gamma - 1)mc^2 \tag{26-8}$$

where γ is calculated using the particle's speed v.

Kinetic energy is energy of motion—the *additional* energy that a moving object has, compared with the energy of the same body when at rest. Einstein proposed identifying the kinetic energy expression above as the difference of two terms. The first term in Eq. (26-8), γmc^2, is the **total energy** E of the particle, which includes both kinetic energy and rest energy. The second term, mc^2, is the rest energy E_0—the energy of the particle when at rest. Therefore, we can rearrange Eq. (26-8) as

$$E = K + mc^2 = K + E_0 = \gamma mc^2 \tag{26-9}$$

With total energy and kinetic energy defined in this way, we find that if any reaction conserves total energy in one inertial reference frame, the total energy is automatically conserved in all other inertial frames. In other words, energy conservation is restored to the status of a universal law of physics.

No object with mass can travel at the speed of light.

Recall that as v approaches c, γ increases without bound. Then from Eq. (26-9), we can conclude that no object with mass can travel at the speed of light since it would need to have an *infinite total energy* to do so.

At first it may look as if kinetic energy, $K = (\gamma - 1)mc^2$, doesn't depend on speed, but remember that γ is a function of speed. As the speed of a particle increases, γ increases, and therefore so does the kinetic energy. There is no limit to the kinetic energy of a particle. As is true with momentum, the kinetic energy gets large without bound as the speed gets closer to c.

It's not at all obvious that the relativistic expression for K approaches the nonrelativistic expression $\frac{1}{2}mv^2$ for objects moving much slower than the speed of light, but it does. To show that, we make use of the binomial approximation $(1 - x)^n \approx 1 - nx$ for $x \ll 1$ (see Appendix A.5). If we let $x = v^2/c^2$ and $n = -\frac{1}{2}$ in the binomial approximation, γ becomes

$$\gamma = \left(1 - \frac{v^2}{c^2}\right)^{-1/2} \approx 1 + \frac{1}{2}\left(\frac{v^2}{c^2}\right)$$

The kinetic energy is then

$$K = (\gamma - 1)mc^2 \approx \left[1 + \frac{1}{2}\left(\frac{v^2}{c^2}\right) - 1\right]mc^2 = \frac{1}{2}\left(\frac{v^2}{c^2}\right)mc^2 = \frac{1}{2}mv^2$$

The relativistic expression for K is valid for both relativistic and nonrelativistic motion. The nonrelativistic expression $\frac{1}{2}mv^2$ is an approximation that is only valid for speeds much less than c. If $K \ll mc^2$, then γ is very close to 1; the particle is not moving at a relativistic speed, so nonrelativistic approximations can be used.

Example 26.6

An Energetic Electron

The radioactive decay of carbon-14 into nitrogen-14 ($^{14}C \rightarrow {}^{14}N + e^- + \bar{\nu}$) releases 156 keV of energy (Fig. 26.17). If all of the energy released appears as the kinetic energy of the electron, how fast is the electron moving?

Strategy We are given the kinetic energy of the electron and need to find its speed. The electron-volt (eV) and its multiples, keV or MeV, are commonly used energy units for atomic, nuclear, and high-energy particle physics. How do

continued on next page

Example 26.6 continued

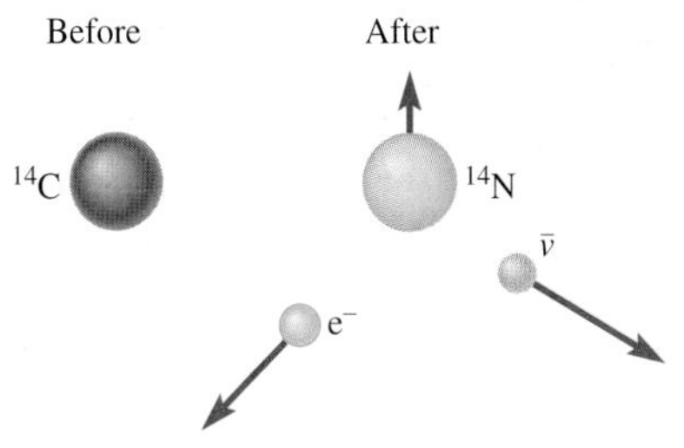

Figure 26.17
Radioactive decay of carbon-14.

we know whether it is appropriate to use the nonrelativistic expression for kinetic energy? We compare the kinetic energy of the electron (156 keV) to its rest energy (mc^2).

Solution The rest energy of the electron is

$$E_0 = mc^2 = 9.109 \times 10^{-31}\ \text{kg} \times (2.998 \times 10^8\ \text{m/s})^2$$
$$= 8.187 \times 10^{-14}\ \text{J}$$

Since we know K in keV, let us convert E_0 to keV.

$$E_0 = 8.187 \times 10^{-14}\ \text{J} \times \frac{1\ \text{eV}}{1.602 \times 10^{-19}\ \text{J}} \times \frac{1\ \text{keV}}{1000\ \text{eV}} = 511\ \text{keV}$$

Thus, K is of the same order of magnitude as E_0. Since the electron is moving at a relativistic speed, the relativistic equations must be used.

The Lorentz factor is

$$\gamma = 1 + \frac{K}{mc^2} = 1 + \frac{156\ \text{keV}}{511\ \text{keV}} = 1.3053$$

From γ, we determine the speed. First we square γ:

$$\gamma^2 = \frac{1}{1 - v^2/c^2}$$

Now we solve for v/c to find the speed of the electron as a fraction of c.

$$1 - \frac{v^2}{c^2} = \frac{1}{\gamma^2}$$

$$v/c = \sqrt{1 - 1/\gamma^2} = 0.6427$$

$$v = 0.6427c$$

Discussion The Lorentz factor is *not* very close to 1, which is another indication that we cannot use $K = \frac{1}{2}mv^2$ for this electron. Doing so would give

$$v = (2K/m)^{1/2} = 2.342 \times 10^8\ \text{m/s} = 0.781c$$

A result this close to c should cause concern about using a nonrelativistic approximation.

The speed $0.6427c$ is an upper limit on the kinetic energy of the electron. The electron cannot carry off all of the energy released in the reaction; the kinetic energy must be divided among the three particles in such a way as to conserve momentum.

Practice Problem 26.6 Accelerating a Proton

How much work must be done to accelerate a proton from rest to $0.75c$? Express the answer in MeV.

Momentum-Energy Relationships

In Newtonian physics, the relation between kinetic energy and momentum is $K = p^2/(2m)$ (Problem 52), but that relation no longer holds for particles moving at relativistic speeds. From the relativistic definitions of $\vec{\mathbf{p}}$, E, and K, you can derive these useful relations (try it—see Problems 56 and 57):

$$E^2 = E_0^2 + (pc)^2 \qquad (26\text{-}10)$$

$$(pc)^2 = K^2 + 2KE_0 \qquad (26\text{-}11)$$

In particle physics, momentum is usually written in units of eV/c (or multiples such as MeV/c) to avoid lots of unit conversions. Masses are commonly written in units of eV/c^2, keV/c^2, MeV/c^2, or GeV/c^2.

Equations (26-10) and (26-11) are valuable for calculating total energy or kinetic energy from momentum or vice versa without going through the intermediate step of calculating the speed of the particle. Since $E = \gamma mc^2$ and $\vec{\mathbf{p}} = \gamma m\vec{\mathbf{v}}$, another useful relationship is

$$\frac{\vec{\mathbf{v}}}{c} = \frac{\vec{\mathbf{p}}c}{E} \qquad (26\text{-}12)$$

Equation (26-12) makes it easier to calculate the velocity or momentum or total energy when any two of these three quantities are known. It also shows that pc can never exceed the total energy, but approaches E as $v \to c$.

Example 26.7

Speed and Momentum of an Electron

An electron has a kinetic energy of 1.0 MeV. Find the electron's speed and its momentum.

Strategy Use energy-momentum relations and momentum units of MeV/c to simplify the calculation.

Solution In Example 26.6, we found that the rest energy of an electron is $E_0 = 0.511$ MeV. Since the kinetic energy is almost twice the rest energy, we definitely must do relativistic calculations. The total energy of the electron is $E = K + E_0 = 1.511$ MeV. We can immediately find the momentum:

$$E^2 = E_0^2 + (pc)^2$$

$$(pc)^2 = E^2 - E_0^2$$

$$pc = \sqrt{(1.511\text{ MeV})^2 - (0.511\text{ MeV})^2} = 1.422\text{ MeV}$$

Dividing both sides of this equation by c gives the momentum in units of MeV/c:

$$p = 1.4\text{ MeV}/c$$

Now that we know the momentum, we can find the speed:

$$\frac{v}{c} = \frac{pc}{E} = \frac{1.422}{1.511} = 0.9411$$

$$v = 0.94c$$

Discussion A good check is to use the speed to calculate the kinetic energy. First we find the Lorentz factor:

$$\gamma = \sqrt{\frac{1}{1 - 0.9411^2}} = 2.957$$

Then the kinetic energy is

$$K = (\gamma - 1)mc^2 = 1.957 \times 0.511\text{ MeV} = 1.0\text{ MeV}$$

The momentum in SI units can be obtained as

$$p = 1.422\,\frac{\text{MeV}}{c} \times \frac{10^6\text{ eV}}{\text{MeV}} \times \frac{1.60 \times 10^{-19}\text{ J}}{\text{eV}} \times \frac{c}{3.00 \times 10^8\text{ m/s}}$$

$$= 7.6 \times 10^{-22}\text{ kg·m/s}$$

Practice Problem 26.7 Protons and Antiprotons at Fermilab

The Tevatron at the Fermi National Accelerator Laboratory accelerates protons and antiprotons to kinetic energies of 0.980 TeV (tera-electron-volts). Antiprotons have the same mass as protons (938.3 MeV/c^2) but charge $-e$ instead of $+e$. (a) What is the magnitude of the momentum of the protons and antiprotons in units of TeV/c? (b) At what speed are the protons and antiprotons moving relative to the lab? [*Hint:* Since $\gamma \gg 1$, use the binomial approximation: $v/c = \sqrt{1 - 1/\gamma^2} \approx 1 - 1/(2\gamma^2)$.]

Deciding When to Use Relativistic Equations

There are several ways of deciding when relativistic calculations are called for, depending on the information given in a particular problem. Section 26.6 suggested that for $v = 0.14c$, the nonrelativistic expression for momentum differs from the relativistic by about 1%. We may not need that degree of accuracy. Even for $v = 0.2c$, the differences between the nonrelativistic and relativistic expressions for momentum and kinetic energy are only about 2% and 3%, respectively. For speeds higher than about $0.3c$, γ rises rapidly and the differences between nonrelativistic and relativistic physics become appreciable.

Comparing a particle's kinetic energy with its rest energy is another way to decide whether to use relativistic or nonrelativistic expressions. If $K \ll mc^2$, then γ is very close to 1 and the particle's speed is nonrelativistic.

A particle is *nonrelativistic* if any of the following equivalent conditions are true:

$$v \ll c \tag{26-13}$$

$$\gamma - 1 \ll 1 \tag{26-14}$$

$$K \ll mc^2 \tag{26-15}$$

$$p \ll mc \tag{26-16}$$

Master the Concepts

- The two postulates of relativity are
  - (I) The laws of physics are the same in all inertial frames.
  - (II) The speed of light in vacuum is the same in all inertial frames.
- The speed of light in vacuum in any inertial reference frame is

$$c = 3.00 \times 10^8 \text{ m/s}$$

- Observers in different reference frames disagree about the time order of two events (including whether the events are simultaneous) if there is *not* enough time for a signal at light speed to travel from one event to the other.
- The Lorentz factor occurs in many relativity equations.

$$\gamma = \frac{1}{\sqrt{1 - v^2/c^2}} \qquad (26\text{-}2)$$

When γ is used in expressions for time dilation or length contraction, v in Eq. (26-2) stands for the *relative speed of the two reference frames*. When γ is used in expressions for the momentum, kinetic energy, or total energy of a particle, v in Eq. (26-2) stands for the *particle's speed.*

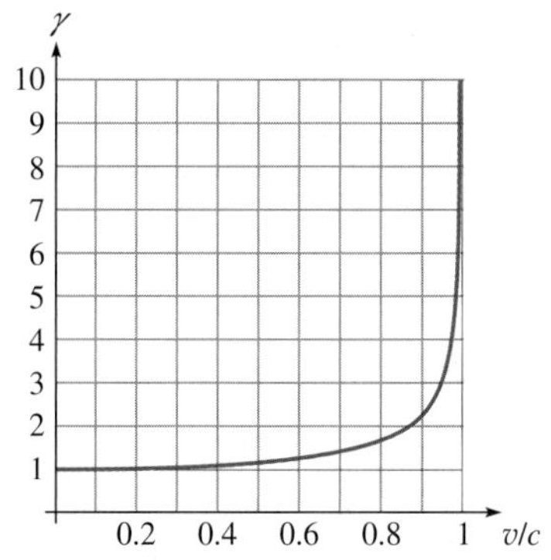

- In time dilation problems, identify the two events that mark the beginning and end of the time interval in question. The "clock" is whatever measures this time interval. Identify the reference frame in which the clock is at rest. In that frame, the clock measures the proper time interval Δt_0. In any other reference frame, the time interval is *longer*:

$$\Delta t = \gamma\, \Delta t_0 \qquad (26\text{-}3)$$

- In length contraction problems, identify the object whose length is to be measured in two different frames. The length is contracted only in the direction of the object's motion. If the length in question is a distance rather than the length of an actual object, it often helps to imagine the presence of a long measuring stick. Identify the reference frame in which the object is at rest. The length in that frame is the proper length L_0. In any other reference frame, the length L is *contracted*:

$$L = \frac{L_0}{\gamma} \qquad (26\text{-}4)$$

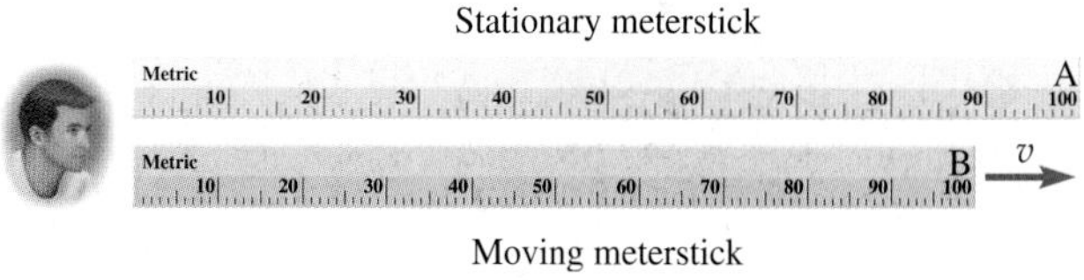

- Velocities in different reference frames are related by

$$v_{PA} = \frac{v_{PB} + v_{BA}}{1 + v_{PB}v_{BA}/c^2} \qquad (26\text{-}5)$$

The subscripts in v_{BA} mean *the velocity of B as measured in A's reference frame*. Equation (26-5) is written in terms of the *components* of the three velocities along a straight line. The components are positive for one direction (your choice) and negative for the other. If A moves to the right in B's frame, then B moves to the left in A's frame: $v_{BA} = -v_{AB}$.

- The relativistic expression for momentum is

$$\vec{\mathbf{p}} = \gamma m \vec{\mathbf{v}} \qquad (26\text{-}6)$$

With relativistic momentum, it is still true that the impulse delivered equals the change in momentum ($\Sigma\vec{\mathbf{F}}\,\Delta t = \Delta\vec{\mathbf{p}}$), but $\Sigma\vec{\mathbf{F}} = m\vec{\mathbf{a}}$ is *not* true: the acceleration due to a constant net force gets smaller and smaller as the particle's speed approaches c. Thus, it is impossible to accelerate something to the speed of light.

- The rest energy E_0 of a particle is its energy as measured in its rest frame. The relationship between rest energy and mass is:

$$E_0 = mc^2 \qquad (26\text{-}7)$$

Kinetic energy is

$$K = (\gamma - 1)mc^2 \qquad (26\text{-}8)$$

Total energy is rest energy plus kinetic energy:

$$E = \gamma mc^2 = K + E_0 \qquad (26\text{-}9)$$

- Useful relations between momentum and energy:

$$E^2 = E_0^2 + (pc)^2 \qquad (26\text{-}10)$$

$$(pc)^2 = K^2 + 2KE_0 \qquad (26\text{-}11)$$

$$\frac{\vec{\mathbf{v}}}{c} = \frac{\vec{\mathbf{p}}c}{E} \qquad (26\text{-}12)$$

Conceptual Questions

1. A friend argues with you that relativity is absurd: "It's obvious that moving clocks don't run slow and that moving objects aren't shorter than when they're at rest." How would you reply?
2. An electron is moving at nearly light speed. A constant force of magnitude F is acting on the electron in the direction of its motion. Is the acceleration of the electron less than, equal to, or greater in magnitude than F/m? Explain.
3. As you talk on a cell phone, does the mass of the phone's battery change at all? If so, does it increase or decrease?
4. A particle with nonzero mass m can never move faster than the speed of light. Is there also a maximum momentum that the particle can have? A maximum kinetic energy? Explain.
5. An astronaut in top physical condition has an average resting pulse on Earth of about 52 beats per minute. Suppose the astronaut is in a spaceship traveling at $0.87c$ ($\gamma = 2$) with respect to Earth when he takes his own resting pulse. Does he measure about 52 beats per minute, about 26 beats per minute, or about 104 beats per minute? Explain.
6. A constant force is applied to a particle initially at rest. Sketch qualitative graphs of the particle's speed, momentum, and acceleration as functions of time. Assume that the force acts long enough so the particle achieves relativistic speeds.
7. Harry and Sally are on opposite sides of the room at a wedding reception. They simultaneously (in the frame of the room) take flash pictures of the bride and groom cutting the cake in the center of the room. What would an observer moving at constant velocity from Harry to Sally say about the time order of the two flashes?
8. In an Earth laboratory, an astronaut measures the length of a rod to be 1.00 m. The astronaut takes the rod aboard a spaceship and flies away from Earth at speed $0.5c$. Is the length of the rod measured by an observer on Earth greater than, less than, or equal to 1.00 m as measured by the astronaut in the spaceship? Explain. Does the answer depend on the orientation of the rod?
9. In Section 26.2, suppose that another astronaut, Celia, moves in a spaceship to the *left* with respect to Abe (see Fig. 26.4). What would Celia conclude about the time order of the two flashes?
10. Explain why it is impossible for a particle with mass to move faster than the speed of light.
11. Does a stretched spring have the same mass as when it is relaxed? Explain.
12. A quasar is a bright center in a far distant galaxy where some energetic action is taking place (probably due to energy being released as matter falls into a black hole at the center of the galaxy). Through her telescope Mavis observes a quasar 12×10^9 ly (light-years) away. She wishes she could travel to the quasar to observe it more closely. If she were able to travel that far in her lifetime, would she be able to observe the activity she sees through her telescope?

Multiple-Choice Questions

1. Which of these statements are *postulates* of Einstein's special relativity?
   (1) The speed of light is the same in all inertial reference frames.
   (2) Moving clocks run slow.
   (3) Moving objects are contracted along the direction of motion.
   (4) The laws of nature are the same in all inertial reference frames.
   (5) $E_0 = mc^2$
   (a) 1 only (b) 2 and 3 only (c) all 5
   (d) 4 only (e) 1 and 4 only (f) 4 and 5 only
2. Which of these statements correctly defines an inertial frame?
   (a) An inertial frame is a frame in which there are no forces.
   (b) An inertial frame is one in which Newton's second and third laws hold, but not his first.
   (c) An inertial frame is a frame of reference in which Newtonian mechanics holds true, but relativistic mechanics does not.
   (d) An inertial frame is a frame where there are no accelerations without applied forces.
   (e) An inertial frame is a frame of reference in which relativistic mechanics holds true, but Newtonian mechanics does not.
3. An astronaut in a rocket moving with a speed $v = 0.6c$ relative to Earth performs a collision experiment with two small steel balls and concludes that both momentum and energy are conserved in his reference frame. What would an Earth observer conclude?
   (a) Momentum and energy are conserved.
   (b) Momentum is conserved, but energy is not.
   (c) Energy is conserved, but momentum is not.
   (d) The collision never takes place because the two balls are never at the same place at the same time.
   (e) Neither energy nor momentum is conserved.
4. A spaceship moves away from Earth at constant velocity $0.60c$, according to Earth observers. In the reference frame of the spaceship,
   (a) Earth moves away from the spaceship at $0.60c$.
   (b) Earth moves away from the spaceship at a speed less than $0.60c$.
   (c) Earth moves away from the spaceship at a speed greater than $0.60c$.

(d) The speed of Earth cannot be accurately measured because the reference frame is moving.
(e) The speed of Earth is not constant.

5. A clock ticks once each second and is 10 cm long when at rest. If the clock is moving at $0.80c$ parallel to its length with respect to an observer, the observer measures the time between ticks to be _________ and the length of the clock to be _________.
(a) more than 1 s; more than 10 cm
(b) less than 1 s; more than 10 cm
(c) more than 1 s; less than 10 cm
(d) less than 1 s; less than 10 cm
(e) equal to 1 s; equal to 10 cm

6. Which best describes the *proper time interval* between two events?
(a) the time interval measured in a reference frame in which the two events occur at the same place
(b) the time interval measured in a reference frame in which the two events are simultaneous
(c) the time interval measured in a reference frame in which the two events occur a maximum distance away from each other
(d) the longest time interval measured by any inertial observer

7. Before takeoff, an astronaut measures the length of the Space Shuttle orbiter to be 37.24 m long using a steel rule. Once aboard the shuttle, with it traveling at $0.10c$, he measures the length again using the same steel rule and finds a value of
(a) 37.05 m.
(b) 37.24 m.
(c) 37.43 m.
(d) Either 37.05 m or 37.24 m, depending on whether the ship's length is parallel or perpendicular to the direction of motion.

8. An observer sees an asteroid with a radioactive element moving by at a speed of $0.20c$ and notes that the half-life of the radioactivity is T. Another observer is moving with the asteroid and measures the half-life to be
(a) less than T.
(b) equal to T.
(c) greater than T.
(d) either (a) or (c) depending on whether the asteroid is approaching or receding from the first observer.

9. Twin sisters become astronauts. One sister goes on a space mission lasting several decades while the other remains behind on Earth. Which of the following statements concerning their relative ages is true?
(a) The sister who was on the mission in space is older than her twin once they reunite on Earth.
(b) The sister who remained on Earth is older than her traveling twin once they are reunited on Earth.
(c) The sisters are the same age when the traveling twin returns to Earth because each sister was traveling at the same speed relative to the other as measured in each other's reference frames.
(d) This is a paradox so there is no possibility of comparing their ages.

Problems

- Combination conceptual/quantitative problem
- Biological or medical application
- Challenging problem
- Blue # Detailed solution in the Student Solutions Manual
- 1 2 Problems paired by concept
- Text website interactive or tutorial

26.1 Postulates of Relativity

1. An engineer in a train moving toward the station with a velocity $v = 0.60c$ lights a signal flare as he reaches a marker 1.0 km from the station (according to a scale laid out on the ground). By how much time, on the stationmaster's clock, does the arrival of the optical signal precede the arrival of the train?

2. The light-second is a unit of distance; 1 light-second is the distance that light travels in 1 second. (a) Find the conversion between light-seconds and meters: 1 light-second = ? m. (b) What is the speed of light in units of light-seconds per second?

3. A spaceship traveling at speed $0.13c$ away from Earth sends a radio transmission to Earth. (a) According to Galilean relativity, at what speed would the transmission travel relative to Earth? (b) Using Einstein's postulates, at what speed does the transmission travel relative to Earth?

4. Event A happens at the spacetime coordinates $(x, y, z, t) = (2\text{ m}, 3\text{ m}, 0, 0.1\text{ s})$ and event B happens at the spacetime coordinates $(x, y, z, t) = (0.4 \times 10^8\text{ m}, 3\text{ m}, 0, 0.2\text{ s})$. (a) Is it possible that event A caused event B? (b) If event B occurred at $(0.2 \times 10^8\text{ m}, 3\text{ m}, 0, 0.2\text{ s})$ instead, would it then be possible that event A caused event B? [*Hint:* How fast would a signal need to travel to get from event A to the location of B before event B occurred?]

26.3 Time Dilation

5. An astronaut wears a new Rolex watch on a journey at a speed of 2.0×10^8 m/s with respect to Earth. According to mission control in Houston, the trip lasts 12.0 h. How long is the trip as measured on the Rolex?

6. An unstable particle called the *pion* has a mean lifetime of 25 ns in its own rest frame. A beam of pions travels through the laboratory at a speed of $0.60c$. (a) What is the mean lifetime of the pions as measured in the laboratory frame? (b) How far does a pion travel (as measured by laboratory observers) during this time?

7. Suppose your handheld calculator will show six places beyond the decimal point. How fast (in meters per second) would an object have to be traveling so that the decimal places in the value of γ can actually be seen on your calculator display? That is, how fast should an object travel so that $\gamma = 1.000001$? [*Hint:* Use the binomial approximation.]

8. A spaceship is traveling away from Earth at 0.87c. The astronauts report home by radio every 12 h (by their own clocks). At what interval are the reports *sent* to Earth, according to Earth clocks?

9. A spaceship travels at constant velocity from Earth to a point 710 ly away as measured in Earth's rest frame. The ship's speed relative to Earth is 0.9999c. A passenger is 20 yr old when departing from Earth. (a) How old is the passenger when the ship reaches its destination, as measured by the ship's clock? (b) If the spaceship sends a radio signal back to Earth as soon as it reaches its destination, in what year, by Earth's calendar, does the signal reach Earth? The spaceship left Earth in the year 2000.

10. A clock moves at a constant velocity of 8.0 km/s with respect to Earth. If the clock ticks at intervals of one second in its rest frame, how much more than a second elapses between ticks of the clock as measured by an observer at rest on Earth? [*Hint:* Use the binomial approximation $(1 - v^2/c^2)^{-1/2} \approx 1 + v^2/(2c^2)$.]

11. A plane trip lasts 8.0 h; the average speed of the plane during the flight relative to Earth is 220 m/s. What is the time difference between an atomic clock on board the plane and one on the ground, assuming they were synchronized before the flight? (Ignore general relativistic complications due to gravity and the acceleration of the plane.)

12. Fill in the missing algebraic steps in the derivation of the time dilation equation [Eq. (26-1)].

26.4 Length Contraction

13. A spaceship travels toward Earth at a speed of 0.97c. The occupants of the ship are standing with their torsos parallel to the direction of travel. According to Earth observers, they are about 0.50 m tall and 0.50 m wide. What are the occupants' (a) height and (b) width according to others on the spaceship?

14. While the spaceship in Problem 13 continues to travel in the same direction, one of the occupants lies on his side, so that now his torso is perpendicular to the direction of travel and his width is parallel to the travel direction. What are the (a) height and (b) width of this occupant according to an Earth observer?

15. A cosmic ray particle travels directly over a football field, from one goal line to the other, at a speed of 0.50c. (a) If the length of the field between goal lines in the Earth frame is 91.5 m (100 yd), what length is measured in the rest frame of the particle? (b) How long does it take the particle to go from one goal line to the other according to Earth observers? (c) How long does it take in the rest frame of the particle?

16. A laboratory measurement of the coordinates of the ends of a moving meterstick, taken at the same time in the laboratory, gives the result that one end of the stick is 0.992 m due north of the other end. If the stick is moving due north, what is its speed with respect to the lab? (tutorial: Lorentz contraction)

17. Two spaceships are moving directly toward one another with a relative velocity of 0.90c. If an astronaut measures the length of his own spaceship to be 30.0 m, how long is the spaceship as measured by an astronaut in the other ship?

18. A spaceship is moving at a constant velocity of 0.70c relative to an Earth observer. The Earth observer measures the length of the spaceship to be 40.0 m. How long is the spaceship as measured by its pilot?

19. A spaceship moves at a constant velocity of 0.40c relative to an Earth observer. The pilot of the spaceship is holding a rod, which he measures to be 1.0 m long. (a) The rod is held perpendicular to the direction of motion of the spaceship. How long is the rod according to the Earth observer? (b) After the pilot rotates the rod and holds it parallel to the direction of motion of the spaceship, how long is it according to the Earth observer?

20. A rectangular plate of glass, measured at rest, has sides 30.0 cm and 60.0 cm. (a) As measured in a reference frame moving parallel to the 60.0-cm edge at speed 0.25c with respect to the glass, what are the lengths of the sides? (b) How fast would a reference frame have to move in the same direction so that the plate of glass viewed in that frame is square?

21. A futuristic train moving in a straight line with a uniform speed of 0.80c passes a series of communications towers. The spacing between the towers, according to an observer on the ground, is 3.0 km. A passenger on the train uses an accurate stopwatch to see how often a tower passes him. (a) What is the time interval the passenger measures between the passing of one tower and the next? (b) What is the time interval an observer on the ground measures for the train to pass from one tower to the next?

22. An astronaut in a rocket moving at 0.50c toward the Sun finds himself halfway between Earth and the Sun. According to the astronaut, how far is he from Earth? In the frame of the Sun, the distance from Earth to the Sun is 1.50×10^{11} m.

23. The mean (average) lifetime of a muon in its rest frame is 2.2 μs. A beam of muons is moving through the lab with speed 0.994c. How far on average does a muon travel through the lab before it decays?

24. The Tevatron is a particle accelerator at Fermilab that accelerates protons and antiprotons to high energies in

an underground ring. Scientists observe the results of collisions between the particles. The protons are accelerated until they have speeds only 100 m/s slower than the speed of light. The circumference of the ring is 6.3 km. What is the circumference according to an observer moving with the protons? [*Hint:* Let $v = c - u$ where v is the proton speed and $u = 100$ m/s.]

26.5 Velocities in Different Reference Frames

25. Kurt is measuring the speed of light in an evacuated chamber aboard a spaceship traveling with a constant velocity of $0.60c$ with respect to Earth. The light is moving in the direction of motion of the spaceship. Siu-Ling is on Earth watching the experiment. With what speed does the light in the vacuum chamber travel, according to Siu-Ling's observations?
26. The rogue starship *Galaxa* is being chased by the battlecruiser *Millenia.* The *Millenia* is catching up to the *Galaxa* at a rate of $0.55c$ when the captain of the *Millenia* decides it is time to fire a missile. First the captain shines a laser range finder to determine the distance to the *Galaxa* and then he fires a missile that is moving at a speed of $0.45c$ with respect to the *Millenia*. What speed does the *Galaxa* measure for (a) the laser beam and (b) the missile as they both approach the starship?
27. Particle A is moving with a constant velocity $v_{AE} = +0.90c$ relative to an Earth observer. Particle B moves in the opposite direction with a constant velocity $v_{BE} = -0.90c$ relative to the same Earth observer. What is the velocity of particle B as seen by particle A?
28. A man on the Moon observes two spaceships coming toward him from opposite directions at speeds of $0.60c$ and $0.80c$. What is the relative speed of the two ships as measured by a passenger on either one of the spaceships?
29. Rocket ship *Able* travels at $0.400c$ relative to an Earth observer. According to the same observer, rocket ship *Able* overtakes a slower moving rocket ship *Baker* that moves in the same direction. The captain of *Baker* sees *Able* pass her ship at $0.114c$. Determine the speed of *Baker* relative to the Earth observer.
30. Relative to the laboratory, a proton moves to the right with a speed of $\frac{4}{5}c$, while relative to the proton, an electron moves to the left with a speed of $\frac{5}{7}c$. What is the speed of the electron relative to the lab?
31. As observed from Earth, rocket *Alpha* moves with speed $0.90c$ and rocket *Bravo* travels with a speed of $0.60c$. They are moving along the same line toward a head-on collision. What is the speed of rocket *Alpha* as measured from rocket *Bravo*? (tutorial: adding velocities)
32. Electron A is moving west with speed $\frac{3}{5}c$ relative to the lab. Electron B is also moving west with speed $\frac{4}{5}c$ relative to the lab. What is the speed of electron B in a frame of reference in which electron A is at rest?

26.6 Relativistic Momentum

33. A proton moves at $0.90c$. What is its momentum in SI units?
34. An electron has momentum of magnitude 2.4×10^{-22} kg·m/s. What is the electron's speed?
35. The fastest vehicle flown by humans on a regular basis is the Space Shuttle which has a mass of about 1×10^5 kg and travels at a speed of about 8×10^3 m/s. Find the momentum of the Space Shuttle using relativistic formulae and then again by using nonrelativistic formulae. By what percent do the relativistic and nonrelativistic momenta of the Space Shuttle differ? [*Hint:* You might want to use one of the approximations in Appendix A.5.]
36. A body has a mass of 12.6 kg and a speed of $0.87c$. (a) What is the magnitude of its momentum? (b) If a constant force of 424.6 N acts in the direction opposite to the body's motion, how long must the force act to bring the body to rest?
37. A constant force, acting for 3.6×10^4 s (10 h), brings a spaceship of mass 2200 kg from rest to speed $0.70c$. (a) What is the magnitude of the force? (b) What is the *initial* acceleration of the spaceship? Comment on the magnitude of the answer.

26.7 Mass and Energy

38. How much energy is released by a nuclear reactor if the total mass of the fuel decreases by 1.0 g?
39. Two lumps of putty are moving in opposite directions, each one having a speed of 30.0 m/s. They collide and stick together. After the collision the combined lumps are at rest. If the mass of each lump was 1.00 kg before the collision, and no energy is lost to the environment, what is the change in mass of the system due to the collision? (tutorial: colliding bullets)
40. The solar energy arriving at the outer edge of Earth's atmosphere from the Sun has intensity 1.4 kW/m^2. (a) How much mass does the Sun lose per day? (b) What percent of the Sun's mass is this?
41. A white dwarf is a star that has exhausted its nuclear fuel and lost its outer mass so that it consists only of its dense, hot inner core. It will cool unless it gains mass from some nearby star. It can form a binary system with such a star and gradually gain mass up to the limit of 1.4 times the mass of the Sun. If the white dwarf were to start to exceed the limit, it would explode into a supernova. How much energy is released by the explosion of a white dwarf at its limiting mass if 80.0% of its mass is converted to energy?
42. A lambda hyperon Λ^0 (mass = 1115 MeV/c^2) at rest decays into a neutron n (mass = 940 MeV/c^2) and a pion (mass = 135 MeV/c^2):

$$\Lambda^0 \rightarrow \text{n} + \pi^0$$

What is the total kinetic energy of the neutron and pion?

43. Radon decays as follows: $^{222}Rn \rightarrow {}^{218}Po + \alpha$. The mass of the radon-222 nucleus is 221.970 39 u, the mass of the polonium-218 nucleus is 217.962 89 u, and the mass of the alpha particle is 4.001 51 u. How much energy is released in the decay? ($1\ u = 931.494\ MeV/c^2$.)

26.8 Relativistic Kinetic Energy

44. The energy to accelerate a starship comes from combining matter and antimatter. When this is done the total rest energy of the matter and antimatter is converted to other forms of energy. Suppose a starship with a mass of 2.0×10^5 kg accelerates to $0.3500c$ from rest. How much matter and antimatter must be converted to kinetic energy for this to occur?

45. A laboratory observer measures an electron's energy to be 1.02×10^{-13} J. What is the electron's speed?

46. A muon with rest energy 106 MeV is created at an altitude of 4500 m and travels downward toward Earth's surface. An observer on Earth measures its speed as $0.980c$. (a) What is the muon's total energy in the Earth observer's frame of reference? (b) What is the muon's total energy in the muon's frame of reference?

47. An object of mass 0.12 kg is moving at 1.80×10^8 m/s. What is its kinetic energy in joules?

48. When an electron travels at $0.60c$, what is its total energy in mega-electron-volts?

49. An observer in the laboratory finds that an electron's total energy is $5.0mc^2$. What is the magnitude of the electron's momentum (as a multiple of mc), as observed in the laboratory?

50. The rest energy of an electron is 0.511 MeV. What momentum (in MeV/c) must an electron have in order that its total energy be 3.00 times its rest energy?

51. An electron has a total energy of 6.5 MeV. What is its momentum (in MeV/c)?

52. For a *nonrelativistic* particle of mass m, show that $K = p^2/(2m)$. [*Hint:* Start with the nonrelativistic expressions for kinetic energy K and momentum p.]

53. Find the conversion between the momentum unit MeV/c and the SI unit of momentum.

54. Find the conversion between the mass unit MeV/c^2 and the SI unit of mass.

55. In a beam of electrons used in a diffraction experiment, each electron is accelerated to a kinetic energy of 150 keV. (a) Are the electrons relativistic? Explain. (b) How fast are the electrons moving?

56. Derive the energy-momentum relation

$$E^2 = E_0^2 + (pc)^2 \qquad (26\text{-}10)$$

Start by squaring the definition of total energy ($E = K + E_0$) and then use the relativistic expressions for momentum and kinetic energy [Eqs. (26-6) and (26-8)].

57. Starting with the energy-momentum relation $E^2 = E_0^2 + (pc)^2$ and the definition of total energy, show that $(pc)^2 = K^2 + 2KE_0$ [Eq. (26-11)].

58. Show that Eq. (26-11) reduces to the nonrelativistic relationship between momentum and kinetic energy, $K \approx p^2/(2m)$, for $K \ll E_0$.

59. Show that each of these statements implies that $v \ll c$, which means that v can be considered a nonrelativistic speed: (a) $\gamma - 1 \ll 1$ [Eq. (26-14)]; (b) $K \ll mc^2$ [Eq. (26-15)]; (c) $p \ll mc$ [Eq. (26-16)]; (d) $K \approx p^2/(2m)$.

Comprehensive Problems

60. Octavio, traveling at a speed of $0.60c$, passes Tracy and her barn. Tracy, who is at rest with respect to her barn, says that the barn is 16 m long in the direction in which Octavio is traveling, 4.5 m high, and 12 m deep. (a) What does Tracy say is the volume of her barn? (b) What volume does Octavio measure?

61. A spaceship resting on Earth has a length of 35.2 m. As it departs on a trip to another planet, it has a length of 30.5 m as measured by the Earthbound observers. The Earthbound observers also notice that one of the astronauts on the spaceship exercises for 22.2 min. How long would the astronaut herself say that she exercises?

62. At the 10.0-km-long Stanford Linear Accelerator, electrons with rest energy of 0.511 MeV have been accelerated to a total energy of 46 GeV. How long is the accelerator as measured in the reference frame of the electrons?

63. Consider the following decay process: $\pi^+ \rightarrow \mu^+ + \nu$. The mass of the pion ($\pi^+$) is 139.6 MeV/$c^2$, the mass of the muon ($\mu^+$) is 105.7 MeV/$c^2$, and the mass of the neutrino (ν) is negligible. If the pion is initially at rest, what is the total kinetic energy of the decay products?

64. A neutron (mass 1.008 66 u) disintegrates into a proton (mass 1.007 28 u), an electron (mass 0.000 55 u), and an antineutrino (mass 0). What is the sum of the kinetic energies of the particles produced, if the neutron is at rest? ($1\ u = 931.5\ MeV/c^2$.)

65. A starship takes 3.0 days to travel between two distant space stations according to its own clocks. Instruments on one of the space stations indicate that the trip took 4.0 days. How fast did the starship travel, relative to that space station?

66. Two spaceships are observed from Earth to be approaching each other along a straight line. Ship A moves at $0.40c$ relative to the Earth observer, and ship B moves at $0.50c$ relative to the same observer. What speed does the captain of ship A report for the speed of ship B?

67. A neutron, with rest energy 939.6 MeV, has momentum 935 MeV/c downward. What is its total energy?

68. Suppose that as you travel away from Earth in a spaceship, you observe another ship pass you heading in the same direction and measure its speed to be $0.50c$. As you look back at Earth, you measure Earth's speed relative to you to be $0.90c$. What is the speed of the ship that passed you according to Earth observers?

69. (a) If you measure the ship that passes you in Problem 68 to be 24 m long, how long will the observers on Earth measure that ship to be? (b) If there is a rod on your spaceship that you measure to be 24 m long, how long will the observers on Earth measure your rod to be? (c) How long do the observers on the passing ship measure your rod to be?

70. Verify that the collision between the proton and the nitrogen nucleus in Example 26.4 is elastic.

71. Muons are created by cosmic-ray collisions at an elevation h (as measured in Earth's frame of reference) above Earth's surface and travel downward with a constant speed of $0.990c$. During a time interval of 1.5 μs in the rest frame of the muons, half of the muons present at the beginning of the interval decay. If one fourth of the original muons reach Earth before decaying, about how big is the height h?

72. Refer to Example 26.1. Ashlin travels at speed $0.800c$ to a star 30.0 ly from Earth. (a) Find the distance between Earth and the star in the astronaut's frame of reference. (b) How long (as measured by the astronaut) does it take to travel this distance at a speed of $0.800c$? Compare your answer to the result of Example 26.1 and explain any discrepancy.

73. An extremely relativistic particle is one whose kinetic energy is much larger than its rest energy. Show that for an extremely relativistic particle $E \approx pc$.

74. A spaceship is moving away from Earth with a constant velocity of $0.80c$ with respect to Earth. The spaceship and an Earth station synchronize their clocks, setting both to zero, at an instant when the ship is near Earth. By prearrangement, when the clock on Earth reaches a reading of 1.0×10^4 s, the Earth station sends out a light signal to the spaceship. (a) In the frame of reference of the Earth station, how far must the signal travel to reach the spaceship? (b) According to an Earth observer, what is the reading of the clock on Earth when the signal is received?

75. Refer to Example 26.2. One million muons are moving toward the ground at speed $0.9950c$ from an altitude of 4500 m. *In the frame of reference of an observer on the ground*, what are (a) the distance traveled by the muons; (b) the time of flight of the muons; (c) the time interval during which half of the muons decay; and (d) the number of muons that survive to reach sea level. [*Hint:* The answers to (a) to (c) are *not* the same as the corresponding quantities in the muons' reference frame. Is the answer to (d) the same?]

76. A particle decays in flight into two pions, each having a rest energy of 140.0 MeV. The pions travel at right angles to each other with equal speeds of $0.900c$. Find (a) the momentum magnitude of the original particle, (b) its kinetic energy, and (c) its mass in units of MeV/c^2.

77. A charged particle is observed to have a total energy of 0.638 MeV when it is moving at $0.600c$. If this particle enters a linear accelerator and its speed is increased to $0.980c$, what is the new value of the particle's total energy?

78. A cosmic-ray proton entering the atmosphere from space has a kinetic energy of 2.0×10^{20} eV. (a) What is its kinetic energy in joules? (b) If all of the kinetic energy of the proton could be harnessed to lift an object of mass 1.0 kg near Earth's surface, how far could the object be lifted? (c) What is the speed of the proton? *Hint:* v is so close to c that most calculators do not keep enough significant figures to do the required calculation, so use the binomial approximation:

$$\text{if } \gamma \gg 1, \quad \sqrt{1 - \frac{1}{\gamma^2}} \approx 1 - \frac{1}{2\gamma^2}$$

79. A spaceship is traveling away from Earth at $0.87c$. The astronauts report home by radio every 12 h (by their own clocks). (a) At what interval are the reports *sent* to Earth, according to Earth clocks? (b) At what interval are the reports *received* by Earth observers, according to their own clocks?

80. *Derivation of the Doppler formula for light.* A source and receiver of EM waves move relative to one another at velocity v. Let v be positive if the receiver and source are moving apart from one another. The source emits an EM wave at frequency f_s (in the source frame). The time between wavefronts as measured by the source is $T_s = 1/f_s$. (a) In the receiver's frame, how much time elapses between the *emission* of wavefronts by the source? Call this T_r'. (b) T_r' is *not* the time that the receiver measures between the *arrival* of successive wavefronts because the wavefronts travel different distances. Say that, according to the receiver, one wavefront is emitted at $t = 0$ and the next at $t = T_o'$. When the first wavefront is emitted, the distance between source and receiver is d_r. When the second wavefront is emitted, the distance between source and receiver is $d_r + vT_r'$. Each wavefront travels at speed c. Calculate the time T_r between the arrival of these two wavefronts as measured by the receiver. (c) The frequency detected by the receiver is $f_r = 1/T_r$. Show that f_r is given by

$$f_r = f_s \sqrt{\frac{1 - v/c}{1 + v/c}}$$

81. You are trying to communicate with a rocket ship that is traveling at 1.2×10^8 m/s away from Earth on its way into space. If you send a message at a frequency of 55 kHz, to what frequency should the astronauts tune to

receive your message? [*Hint:* Use the result derived in Problem 80.]

✦82. An astronaut has spent a long time in the Space Shuttle traveling at 7.860 km/s. When he returns to Earth, he is 1.0 s younger than his twin brother. How long was he on the shuttle? [*Hint:* Use an approximation from Appendix A.5 and beware of calculator round off errors.]

✦83. Two atomic clocks are synchronized. One is put aboard a spaceship that leaves Earth at $t = 0$ at a speed of $0.750c$. (a) When the spaceship has been traveling for 48.0 h (according to the atomic clock on board), it sends a radio signal back to Earth. When would the signal be received on Earth, according to the atomic clock on Earth? (b) When the Earth clock says that the spaceship has been gone for 48.0 h, it sends a radio signal to the spaceship. At what time (according to the spaceship's clock) does the spaceship receive the signal?

✦84. A spaceship passes over an observation station on Earth. Just as the nose of the ship passes the station, a light in the nose of the ship flashes. As the tail of the ship passes the station, a light flashes in the ship's tail. According to an Earth observer, 50.0 ns elapses between the two events. In the astronaut's reference frame, the length of the ship is 12.0 m. (a) How fast is the ship traveling according to an Earth observer? (b) What is the elapsed time between light flashes in the astronaut's frame of reference?

✦85. A lambda hyperon Λ^0 (mass = 1115.7 MeV/c^2) at rest in the lab frame decays into a neutron n (mass = 939.6 MeV/c^2) and a pion (mass = 135.0 MeV/c^2):

$$\Lambda^0 \rightarrow \text{n} + \pi^0$$

What are the kinetic energies (in the lab frame) of the neutron and pion after the decay? [*Hint:* Use Eq. (26-11) to find momentum.]

✦86. Radon decays as $^{222}\text{Rn} \rightarrow {}^{218}\text{Po} + \alpha$. The mass of the radon-222 nucleus is 221.970 39 u, the mass of the polonium-218 nucleus is 217.962 89 u, and the mass of the alpha particle is 4.001 51 u. (1 u = 931.5 MeV/c^2.) If the radon nucleus is initially at rest in the lab frame, at what speeds (in the lab frame) do the (a) polonium-218 nucleus and (b) alpha particle move?

Answers to Practice Problems

26.1 Yes. The trip takes 30 yr as measured in the rest frame of the battery.

26.2 $0.60c$

26.3 $0.909c$

26.4 3.0 s

26.5 4×10^9 kg/s

26.6 480 MeV

26.7 (a) 0.981 TeV/c; (b) $0.999\,999\,54c$

Answers to Checkpoints

26.1 An observer on the train cannot tell whether he is at rest or moving at constant velocity with respect to the ground. The laws of physics would be the same in either case, so no experiment can be devised to distinguish the two.

26.4 The proper time interval is measured in a reference frame in which the two events occur at the same place. Therefore, an observer in the sprinter's reference frame would measure the proper time interval. The proper length is measured in a reference frame in which the object is at rest. An observer at rest with respect to the track would measure the proper length.

26.7 (a) Length, time interval, acceleration, force, and mass; (b) mass and the speed of light in vacuum.

CHAPTER

27

Early Quantum Physics and the Photon

The police respond to a call and find a hallway that looks clean in normal light. A detective sprays a colorless liquid on the walls and floor and sees spots on the floor that give off a blue glow. The detective tells a uniformed officer to cordon off the hallway as a possible crime scene. What causes the blue glow? Why does the detective suspect that a crime took place in the hallway? (See p. 1019 for the answer.)

Concepts & Skills to Review

- heat transfer by radiation (Section 14.8)
- the spectroscope (Section 25.5)
- relativistic momentum and kinetic energy (Sections 26.6 and 26.8)
- rest energy (Section 26.7)

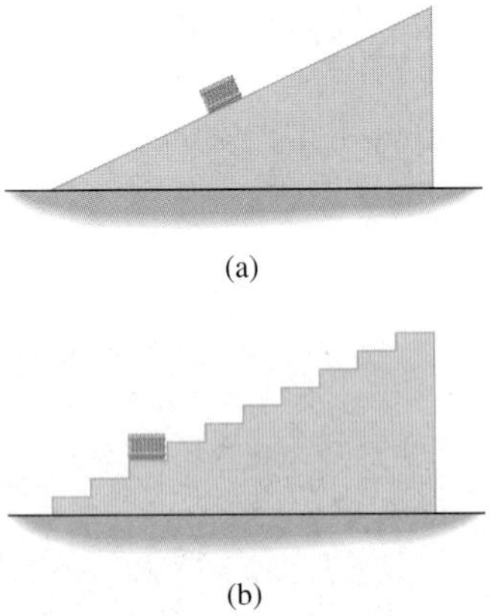

Figure 27.1 (a) A crate resting on a ramp; gravitational potential energy is continuous. (b) A crate resting on a staircase. Gravitational potential energy is quantized; it can have only one of a set of discrete values.

27.1 QUANTIZATION

As the nineteenth century ended, much progress had been made in physics—so much that some physicists feared that everything had been discovered. Newton had laid the foundations of mechanics in his *Principia*, the laws of thermodynamics were well established, and Maxwell had explained electromagnetism. Nevertheless, as scientists developed new experimental techniques and new kinds of equipment, questions arose that could not be explained by the set of physical laws that had seemed nearly complete until then—the laws now known as *classical physics*. The new laws that were developed in the first decades of the twentieth century were the foundation of what we now call *quantum physics*.

In classical physics, most quantities are continuous: they can take any value in a continuous range. As an analogy, Fig. 27.1a shows a crate resting on a ramp. The gravitational potential energy of the crate is continuous—it can have *any* value between the minimum and maximum. By contrast, a crate resting on a staircase can only have certain allowed values (Fig. 27.1b). A quantity is **quantized** when its possible values are limited to a discrete set. A salient feature of quantum physics is the quantization of quantities that were thought to be continuous in classical physics.

The staircase is an imperfect analogy of quantization. While a crate is being moved from one step to another, the gravitational potential energy passes through all the intermediate values. By contrast, something that is truly quantized does *not* pass through intermediate values; it changes suddenly from one value to another.

Standing waves provide an example of quantization in classical physics. The frequency of a standing wave on a string fixed at both ends is quantized (Fig. 27.2). The allowed frequencies are integral multiples of the fundamental frequency ($f_n = nf_1$).

This chapter considers several experiments whose results are difficult or impossible to explain with the laws of classical physics, but relatively easy to explain once electromagnetic waves are assumed to be quantized.

CONNECTION:

The quest to understand thermal EM radiation (Section 14.8) was a major step in the development of quantum physics.

27.2 BLACKBODY RADIATION

A major problem vexing late nineteenth-century physics was blackbody radiation (see Section 14.8). An ideal blackbody absorbs all the radiant energy that falls on it; the radiation emitted by a blackbody is a continuous spectrum characteristic only of the temperature of the body. Figure 27.3 shows experimental blackbody radiation curves—graphs

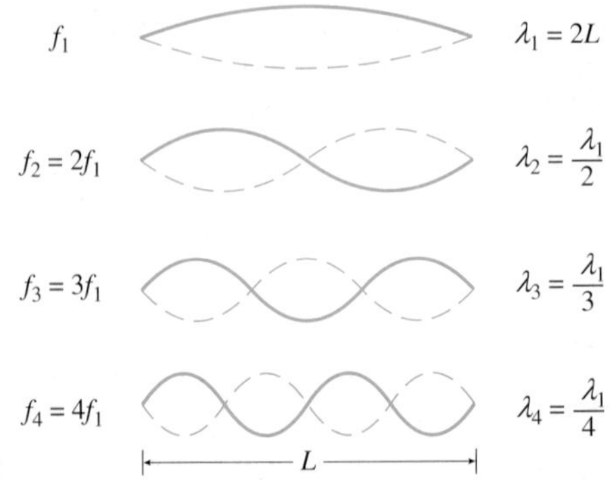

Figure 27.2 Quantization in classical physics: standing wave patterns for a string fixed at both ends. The frequencies and wavelengths are quantized.

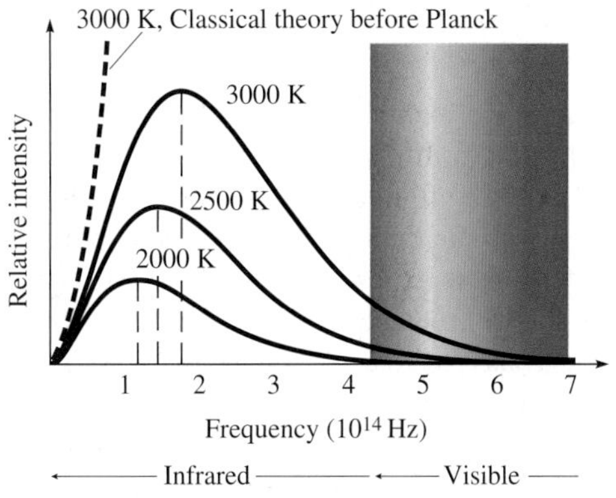

Figure 27.3 Three blackbody curves showing the relative intensity of blackbody radiation as a function of frequency for three different temperatures: 2000 K, 2500 K, and 3000 K. Also, the blackbody curve as predicted by classical theory before Max Planck's proposal.

of the relative intensity of the EM radiation as a function of the frequency—at three temperatures. As the temperature increases, the peak of the radiation curve shifts to higher frequencies. At 2000 K, almost all of the power is radiated in the infrared. At 2500 K, the object is red hot—it radiates significantly in the red and orange parts of the visible spectrum. An object at 3000 K, such as the filament of an incandescent lightbulb, radiates light that we perceive as white, but most of the radiation is still infrared. The total area under the curve, for any absolute temperature T, represents the total radiated power per unit surface area; the total power is proportional to T^4.

However, classical theory predicted that the blackbody radiation curve should continue to increase with increasing frequency (into the ultraviolet and beyond), instead of peaking and then decreasing to zero (Fig. 27.3). As a result, classical theory predicted that a blackbody should radiate an *infinite amount of energy*, an impossibility dubbed the *ultraviolet catastrophe*.

In 1900, the German physicist Max Planck (1858–1947) found a mathematical expression that fit the experimental radiation curves. He then sought a physical model to be the basis for his mathematical expression. He proposed something revolutionary: that the energy emitted and absorbed by oscillating charges must occur only in discrete amounts called **quanta** (singular, **quantum**). He associated a fundamental amount of energy E_0 with each oscillator; the oscillator could emit E_0, or $2E_0$, or any integral multiple of E_0, but nothing in between. As an analogy, imagine that the economy of the oscillator is limited to $10 bills; an oscillator can have in its bank $10, $20, $30, but no intermediate amounts such as $15 or $4. When it spends its capital, it can only give away multiples of $10.

Planck found that the theoretical expression based on quantization matched the experimental radiation curves if E_0 is directly proportional to the frequency f of the oscillator:

$$E_0 = hf \qquad (27\text{-}1)$$

where the constant of proportionality has the unique value $h = 6.626 \times 10^{-34}$ J·s.

Planck's assumption of quantization was a bold break with the fundamental ideas of classical physics. No one knew it at the time, but Planck had launched a half century of exciting developments in physics. He chose the value of h so that his theory would match the experimental data; now h is called **Planck's constant** and is included among the fundamental physical constants such as the speed of light c and the elementary charge e.

CHECKPOINT 27.2

An incandescent lightbulb is connected to a dimmer switch. When the bulb operates at full power, it appears white, but as it is dimmed it looks more and more red. Explain.

27.3 THE PHOTOELECTRIC EFFECT

Photoelectric effect: EM radiation supplies the energy to remove electrons from a metal.

In 1886 and 1887, Heinrich Hertz did experiments that confirmed Maxwell's classical theory of electromagnetic waves. In the course of these experiments, Hertz discovered the effect that Einstein later used to introduce the *quantum* theory of EM waves. Hertz produced sparks between two metal knobs by applying a large potential difference. He noticed that when the knobs were exposed to ultraviolet light, the sparks became stronger. He had discovered the **photoelectric effect** in which EM radiation incident on a metal surface causes electrons to be ejected from a metal.

Later experiments by another German physicist, Philipp von Lenard (1862–1947), found results that were puzzling in the framework of classical physics and were first explained by Einstein in 1905. Figure 27.4 shows an apparatus similar to one invented by Lenard to study the photoelectric effect. EM radiation (visible light or UV) falls on the metal plate; some of the emitted electrons make their way to the collecting wire, which completes the circuit.

An applied potential difference holds the collecting wire at a lower potential than the plate so that electrons lose kinetic energy as they move from the plate to the wire. The larger the potential difference, the smaller the number of electrons with enough kinetic energy to complete the circuit. The **stopping potential** V_s is the magnitude of the potential difference that stops even the most energetic electrons. Therefore, the maximum kinetic energy of the electrons is equal to the increase in potential energy for an electron moving through a potential difference $-V_s$:

$$K_{max} = q\,\Delta V = (-e) \times (-V_s) = eV_s \quad (27\text{-}2)$$

Experimental Results

The photoelectric effect itself seems reasonable according to classical physics: the EM wave supplies the energy needed by the electrons to break free from the metal. However, several *details* of the photoelectric effect were puzzling.

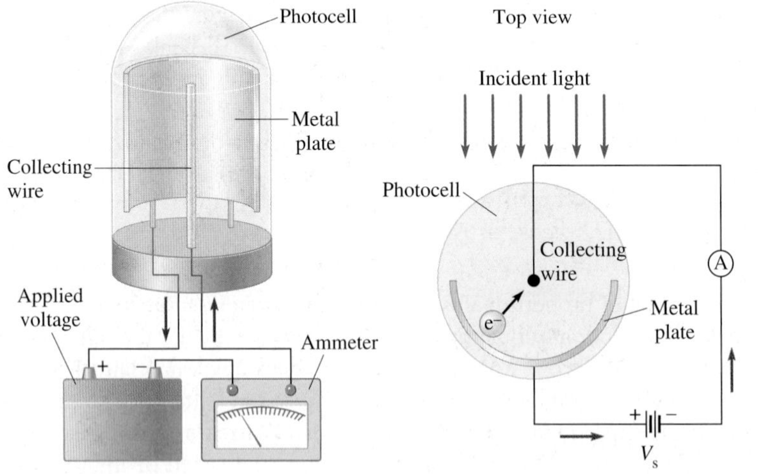

Figure 27.4 Apparatus used to study the photoelectric effect. A photocell is made by enclosing a metal plate and a collecting wire in an evacuated glass tube. EM radiation (visible light or UV) falls on the metal plate; some of the emitted electrons make their way to the collecting wire, which completes the circuit. An ammeter measures the current in the circuit and thus the number of electrons per second that move from the plate to the collecting wire. An applied potential difference holds the collecting wire at a lower potential than the plate so that electrons lose kinetic energy as they move from the plate to the wire.

1. *Brighter* light causes an increase in current (more electrons ejected) but does *not* give the individual electrons higher kinetic energies. In other words, the maximum kinetic energy of the electrons is independent of the intensity of the light. Classically, more intense light has larger amplitude EM fields and thus delivers more energy. That should not only enable more electrons to escape from the metal; it should also give the electrons emitted more kinetic energy.
2. The maximum kinetic energy of the emitted electrons *does* depend on the *frequency* of the incident radiation (Fig. 27.5). Thus, if the incident light is very dim (low intensity) but high in frequency, electrons with large kinetic energies are released. Classically, there is no explanation for a frequency dependence.
3. For a given metal, there is a **threshold frequency** f_0. If the frequency of the incident light is below the threshold, *no electrons are emitted*—no matter what the intensity of the incident light. Again, classical physics has no explanation for the frequency dependence.
4. When EM radiation falls on the metal, electrons are emitted virtually instantaneously; the time delay observed experimentally is about 10^{-9} s, regardless of the light intensity. If the EM radiation behaves as a classical wave, its energy is evenly distributed across the wavefronts. If the intensity of the light is low, it should take some time for enough energy to accumulate on a particular spot to liberate an electron. Experiments have used intensities so low that, classically, there ought to be a time delay of hours before the first electrons escape the metal. Instead, electrons are detected almost immediately!

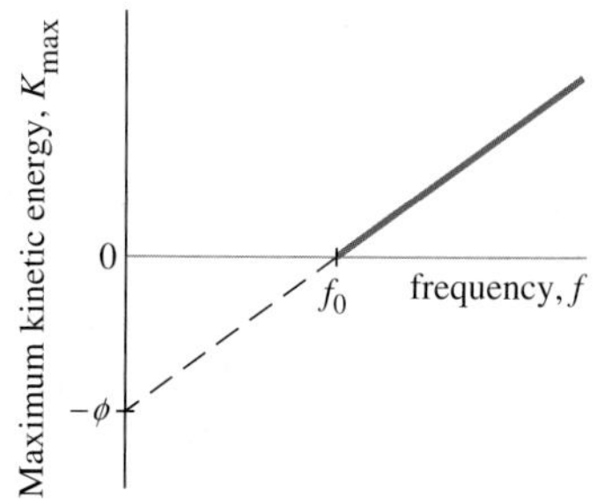

Figure 27.5 Maximum kinetic energy of the electrons ejected from a metal as a function of the frequency f of the incident light.

The Photon

Planck's explanation of blackbody radiation said that the possible energies of the oscillating charges in matter are quantized; the energy of an oscillator at frequency f can only have the values $E = nhf$, where n is an integer and

$$h = 6.626 \times 10^{-34}\ \text{J·s} \qquad (27\text{-}3)$$

In 1905, the same year that he published his special theory of relativity, Einstein explained the photoelectric effect and correctly predicted the results of some experiments that had not yet been performed. Einstein said that *EM radiation itself* is quantized. The quantum of EM radiation—that is, the smallest indivisible unit—is now called the **photon**. The energy of a photon of EM radiation with frequency f is

$$E = hf \qquad (27\text{-}4)$$

Energy of a photon

According to Einstein, the reason a blackbody can only emit or absorb energy in integral multiples of hf is because the EM radiation emitted or absorbed by a blackbody is itself quantized. A blackbody can emit or absorb only a whole number of photons.

The key to understanding the photoelectric effect is that the electron has to absorb a whole photon (Fig. 27.6); it cannot absorb a fraction of a photon's energy. The energy of a photon is proportional to frequency; thus, the photon theory explains the frequency dependence in the photoelectric effect that had mystified scientists.

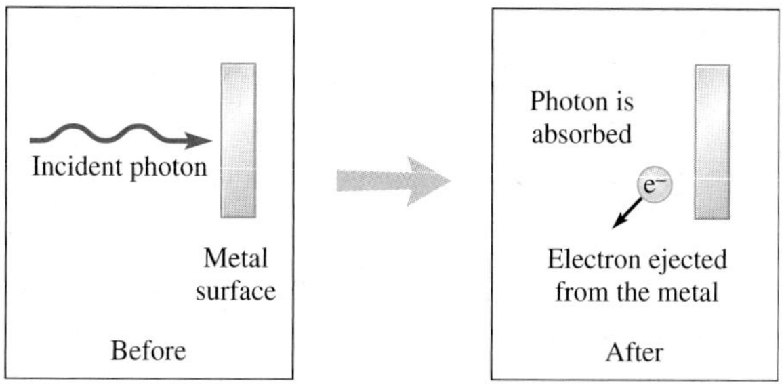

Figure 27.6 In the photoelectric effect, a photon is absorbed. If the energy of the photon is sufficient, an electron can be ejected from the metal.

Example 27.1

Energies of Visible and X-Ray Photons

Find the energy of a photon of visible red light of wavelength 670 nm and compare it with the energy of an x-ray photon with frequency 1.0×10^{19} Hz.

Strategy The product of Planck's constant with each frequency gives the corresponding photon energy. For the 670-nm photon, the frequency and wavelength are related by $c = f\lambda$.

Solution The frequency of the red light is

$$f = \frac{c}{\lambda}$$

To find the energy we multiply the frequency by Planck's constant.

$$E = hf = \frac{hc}{\lambda}$$

$$= \frac{6.626 \times 10^{-34}\ \text{J·s} \times 3.00 \times 10^{8}\ \text{m/s}}{670 \times 10^{-9}\ \text{m}} = 3.0 \times 10^{-19}\ \text{J}$$

For the x-ray photon,

$$E = hf = 6.626 \times 10^{-34}\ \text{J·s} \times 1.0 \times 10^{19}\ \text{Hz} = 6.6 \times 10^{-15}\ \text{J}$$

The energy of the x-ray photon is more than 20 000 times the energy of a photon of red light.

Discussion $E = 3.0 \times 10^{-19}$ J is the smallest amount of energy for red light of wavelength 670 nm that can be absorbed or emitted in any process. Similarly, 6.6×10^{-15} J is the energy of one quantum—one photon—of x-rays at the given frequency. The much larger energy of an x-ray photon is the reason that x-rays can be far more damaging to the human body and that exposure to x-rays should be minimized (Fig. 27.7).

Practice Problem 27.1 Energy of a Photon of Blue Light

Find the energy of one photon of visible blue light of frequency 6.3×10^{14} Hz.

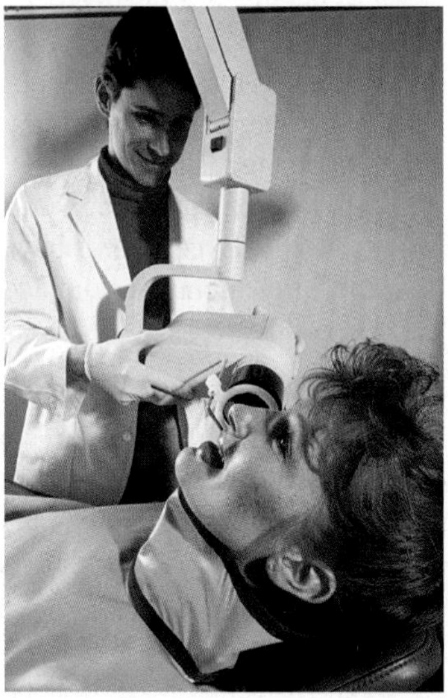

Figure 27.7
The body of a person having an x-ray film taken for dental purposes is protected by a lead apron. Lead is a good absorber of x-rays, so the apron minimizes the exposure of the rest of the body to x-rays.

Example 27.2

Photons Emitted by a Laser

A laser produces a beam of light 2.0 mm in diameter. The wavelength is 532 nm and the output power is 20.0 mW. How many photons does the laser emit per second?

Strategy The photons all have the same energy since the beam has a single wavelength. The output power is the energy output per unit time. Then the energy output per second is the energy of each photon times the number of photons emitted per second.

Solution

energy per second = energy per photon × photons per second

Since $\lambda f = c$, the energy of a photon of wavelength λ is

$$E = hf = h \times \frac{c}{\lambda} = \frac{hc}{\lambda}$$

The energy of each photon emitted by the laser is

$$E = \frac{6.626 \times 10^{-34}\ \text{J·s} \times 3.00 \times 10^{8}\ \text{m/s}}{532 \times 10^{-9}\ \text{m}} = 3.736 \times 10^{-19}\ \text{J}$$

The number of photons emitted per second is

$$\text{photons per second} = \frac{\text{energy per second}}{\text{energy per photon}}$$

$$= \frac{0.0200\ \text{J/s}}{3.736 \times 10^{-19}\ \text{J/photon}} = 5.35 \times 10^{16}\ \text{photons/s}$$

Discussion Notice that the diameter of the beam is irrelevant to the solution. If the power output were the same but the diameter of the beam were larger, the same number of photons per second would be emitted; they would just be

continued on next page

Example 27.2 continued

spread across a wider beam. If the problem had stated the *intensity* (power per unit area) of the beam rather than the total power output, the diameter of the beam would have been relevant.

The quantization of light is not noticed in many situations due to the extremely large numbers of photons. An ordinary 100-W incandescent lightbulb or a 23-W compact fluorescent bulb both emit about 10 W of power as visible light. Thus, the number of photons per second in the visible range emitted by an ordinary lightbulb is around 3×10^{19}.

Practice Problem 27.2 Radio Wave Photons

A radio station broadcasts at 90.9 MHz. The power output of the transmitter is 50.0 kW. How many radio wave photons per second are emitted by the transmitter?

The Electron-Volt

The energies of the photons in Examples 27.1 and 27.2 are small compared with energies of macroscopic bodies, so it is often convenient to express them in electron-volts (symbol eV) rather than in joules. One electron-volt is equal to the kinetic energy that a particle with charge $\pm e$ (e.g., an electron or a proton) gains when it is accelerated through a potential difference of magnitude 1 V. Since 1 V = 1 J/C and $e = 1.60 \times 10^{-19}$ C, the conversion between electron-volts and joules is

$$1 \text{ eV} = e \times 1 \text{ V} = 1.60 \times 10^{-19} \text{ C} \times 1 \text{ J/C} = 1.60 \times 10^{-19} \text{ J} \qquad (27\text{-}5)$$

For larger amounts of energy, keV represents kilo-electron-volts (10^3 eV) and MeV represents mega-electron-volts (10^6 eV). The photon of red light in Example 27.1 has energy 1.9 eV; the x-ray photon has energy 41 keV.

When finding the energy of a photon given its wavelength (or vice versa) using $E = hc/\lambda$, the energy of a photon is often expressed in electron-volts (eV) and wavelengths are often stated in nanometers (nm). For this reason, it is useful to express the constant hc in units of eV·nm:

$$h = \frac{6.626 \times 10^{-34} \text{ J·s}}{1.602 \times 10^{-19} \text{ J/eV}} = 4.136 \times 10^{-15} \text{ eV·s}$$

$$c = 2.998 \times 10^{8} \text{ m/s} \times 10^{9} \text{ nm/m} = 2.998 \times 10^{17} \text{ nm/s}$$

$$hc = 4.136 \times 10^{-15} \text{ eV·s} \times 2.998 \times 10^{17} \text{ nm/s} = 1240 \text{ eV·nm} \qquad (27\text{-}6)$$

The Photon Theory Explains the Photoelectric Effect

The amount of energy that must be supplied to break the bond between a metal and one of its electrons is called the **work function** (ϕ). Each metal has its own characteristic work function. According to Einstein, if the photon energy (hf) is at least equal to the work function, then absorption of a photon can eject an electron. If the photon energy is greater than the work function, some or all of the extra energy can appear as the ejected electron's kinetic energy. The maximum kinetic energy of an electron is the difference between the photon energy and the work function.

Einstein's photoelectric equation:

$$K_{\max} = hf - \phi \qquad (27\text{-}7)$$

CONNECTION:

The photoelectric equation is an expression of energy conservation.

According to Eq. (27-7), a graph of $K_{\max}$ versus f is a straight line with a slope of h and a "y"-intercept of $-\phi$; this equation fits the experimental relationship (see Fig. 27.5). The intercept on the frequency axis is the threshold frequency f_0. Setting

$$K_{\max} = hf_0 - \phi = 0$$

predicts a **threshold frequency** of

$$f_0 = \frac{\phi}{h} \tag{27-8}$$

The four puzzling results of photoelectric effect experiments are easily explained using the photon concept:

1. Light of greater intensity (but constant frequency) delivers more photons per unit time to the metal surface, so the *number* of electrons ejected per second increases as the intensity of the light increases. However, the energy of each photon remains the same. The maximum kinetic energy of the emitted electrons does not depend on the number of photons striking the metal per second because each emitted electron is the result of the absorption of *one photon.*
2. Higher-frequency light has larger energy photons. As the frequency of the light increases, the photons have more excess energy that can potentially become the electron's kinetic energy. Thus, K_{max} increases with increasing frequency.
3. Below the threshold frequency, a photon does not have enough energy to free an electron from the metal, so no electrons are emitted.
4. At low intensities, the number of photons per second is small, but the energy is still delivered in discrete packets. Just after the light is turned on, some photons hit the surface; some of them are absorbed and eject electrons from the metal. There is no time delay because the electrons cannot gradually accumulate energy; they either absorb a photon or they don't.

CHECKPOINT 27.3

In the photoelectric effect, why are no electrons emitted from the metal when the incident light is below the threshold frequency?

Example 27.3

A Photoelectric Effect Experiment

Cesium has a work function of 1.8 eV. When cesium is illuminated with light of a certain wavelength, the electrons ejected from the surface have kinetic energies ranging from 0 to 2.2 eV. What is the wavelength of the light?

Strategy The work function and the maximum kinetic energy (2.2 eV) are given. To eject an electron, the photon must supply 1.8 eV of energy. Some or all of the remainder of the photon's energy $(hf - \phi)$ gives the electron its kinetic energy. The maximum kinetic energy occurs when all of the remainder goes to the electron's kinetic energy.

Solution The energy of a photon is hf. The maximum kinetic energy of the photoelectrons is

$$K_{max} = hf - \phi = 2.2 \text{ eV}$$

The problem asks for the wavelength, so we substitute $f = c/\lambda$ and solve for λ:

$$K_{max} = \frac{hc}{\lambda} - \phi$$

$$\lambda = \frac{hc}{K_{max} + \phi}$$

Substituting $hc = 1240$ eV·nm yields

$$\lambda = \frac{1240 \text{ eV·nm}}{2.2 \text{ eV} + 1.8 \text{ eV}} = 310 \text{ nm}$$

Discussion The photon has energy 2.2 eV + 1.8 eV = 4.0 eV. 1.8 eV raises the potential energy of the electron so that it is free to leave the metal. The remaining 2.2 eV does not necessarily all become the electron's kinetic energy; some of it can be absorbed by the metal. Therefore, 2.2 eV is the *maximum* kinetic energy of a photoelectron. Since the wavelength is less than 400 nm, the photon is in the ultraviolet part of the spectrum.

Practice Problem 27.3 Wavelength of Incident Light

A metal with a work function of 2.40 eV is illuminated with monochromatic light. If the stopping potential that prevents electrons from being ejected is 0.82 V, what is the wavelength of the light? [*Hint:* What is the maximum kinetic energy of the electrons ejected from the surface in electron-volts?]

Applications of the Photoelectric Effect

Although our principal interest in the photoelectric effect is how clearly it illustrates the concept of the photon, many practical applications also exist. Most of them exploit the dependence of the photoelectric current on the intensity of the light. A movie sound track is a strip along one side of the film that encodes sound as variations in the transparency of the film (Fig. 27.8). Light shines through the sound track and then onto a photocell. The photocell generates a current proportional to the intensity of light incident on it. This current is amplified and fed to the speakers.

Devices such as garage door openers, burglar alarms, and smoke detectors often use a light beam and a photocell as a switch. When the light beam is interrupted, the current through the photocell stops. A child walking underneath a garage door that is being closed interrupts a light beam; when the current stops, a switch stops the motion of the door. In some smoke alarms, particles of smoke in the air reduce the intensity of the light that reaches a photocell; when the current drops below a certain level, the alarm is activated.

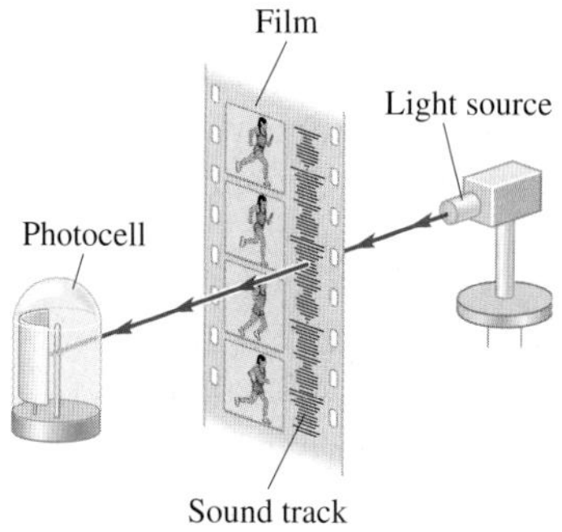

Figure 27.8 An optical sound track.

27.4 X-RAY PRODUCTION

Another confirmation of the quantization of EM radiation is found in the production of x-rays. Figure 27.9a shows an x-ray tube; it looks something like a photocell operated in reverse. In the photoelectric effect, EM radiation incident on a target causes the emission of electrons; in an x-ray tube, electrons incident on a target cause the emission of EM radiation. Electrons move through a large potential difference V to give them large kinetic energies $K = eV$. In the target, they are deflected as they pass by atomic nuclei (Fig. 27.9b). Sometimes an x-ray photon is emitted; the energy of the photon comes from the electron's kinetic energy, so the electron slows down. This process for creating x-rays is called *bremsstrahlung*, from the German for *braking radiation*, since the x-rays are emitted as electrons slow down.

Charges radiate only when they are accelerating, not when they move at constant velocity.

Cutoff Frequency The x-rays produced in this way do not all have the same frequency; there is a continuous spectrum of frequencies up to a maximum, called the **cutoff frequency** (Fig. 27.10). Typically an electron emits many photons as it slows down; each of the photons takes away *part of* the electron's kinetic energy. The maximum frequency occurs when all of the electron's kinetic energy is carried away by *a single photon*:

$$hf_{\max} = K \tag{27-9}$$

CONNECTION:

Eq. (27-9) is once again a consequence of energy conservation.

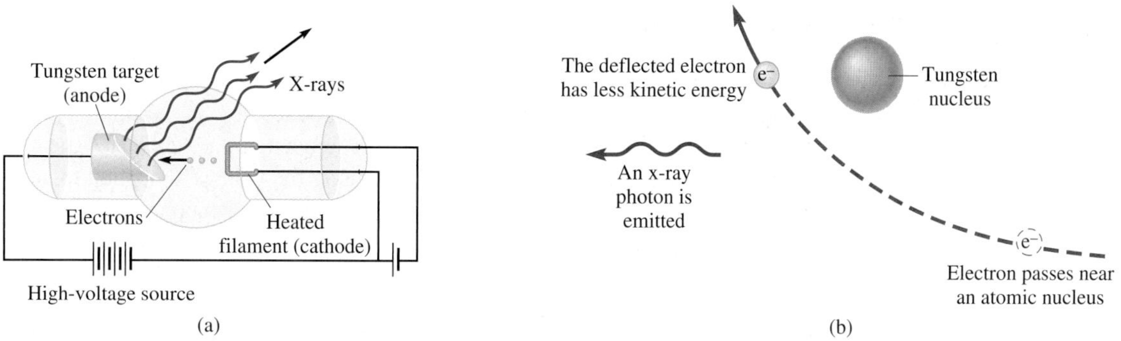

Figure 27.9 (a) An x-ray tube. An electric current heats the filament to "boil off" electrons. The electrons are accelerated through a large potential difference between the filament and the target. When electrons strike the target, x-rays are emitted as the electrons lose kinetic energy. (b) An electron is deflected by an atomic nucleus. An x-ray photon is emitted, carrying away some of the electron's kinetic energy.

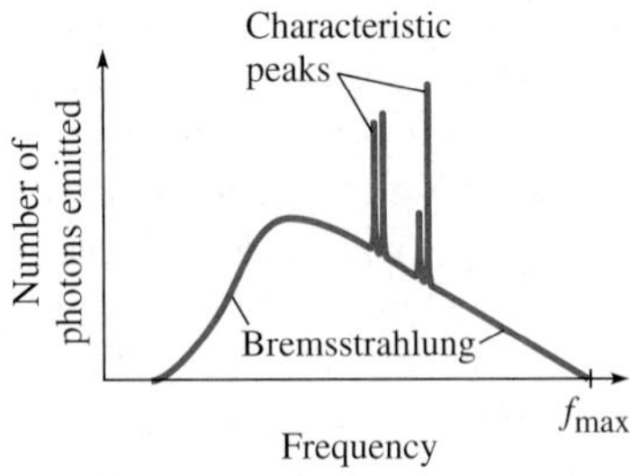

Figure 27.10 Spectrum of x-rays produced by an x-ray tube. The continuous spectrum is due to bremsstrahlung. The cutoff frequency f_{max} depends only on the voltage applied to the x-ray tube. The frequencies of the characteristic peaks depend only on the material used as the target in the tube.

Example 27.4

Diagnostic X-Rays in Medicine

A potential difference of 87.0 kV is applied between the filament and target in the x-ray tube used at the local clinic to look for broken bones. What are the shortest wavelength x-rays produced by this tube?

Strategy The shortest wavelength corresponds to the highest frequency. The highest frequency is produced when all of the electron's kinetic energy is given up in the emission of a single photon.

The accelerating potential of 87.0 kV supplies the electrons with kinetic energy before they hit the target. We do not need to use the numerical value of e to find the kinetic energy. An electron traveling through 1 V of potential difference gains an energy of 1 eV, so an electron traveling through a potential difference of 87.0 kV gains 87.0 keV of kinetic energy. The constants h and c can be looked up individually, but it is easier to use the combination $hc = 1240$ eV·nm.

Solution The maximum frequency occurs when the energy of the photon is equal to the electrons' kinetic energy:

$$hf_{max} = K = 87.0 \text{ keV}$$

$$f_{max} = \frac{K}{h}$$

The minimum wavelength is

$$\lambda_{min} = \frac{c}{f_{max}}$$

Substituting for f_{max} yields

$$\lambda_{min} = \frac{hc}{K} = \frac{1240 \text{ eV·nm}}{87.0 \times 10^3 \text{ eV}} = 0.0143 \text{ nm} = 14.3 \text{ pm}$$

Discussion Notice how much simpler the calculation is made by using the electron-volt for energy. The electron-volt saves physicists from having to constantly multiply and divide by the numerical value of the elementary charge e.

Practice Problem 27.4 Potential Difference Across an X-Ray Tube

If the shortest wavelength detected for x-rays from an x-ray tube is 0.124 nm, what is the potential difference applied to the tube?

Characteristic X-Rays

Notice that an x-ray spectrum (see Fig. 27.10) includes some sharp, intense peaks superimposed on the continuous spectrum of x-rays produced by bremsstrahlung. These peaks are called **characteristic x-rays** because their frequencies are characteristic of the material used as the target in the x-ray tube. Changing the voltage V applied to an x-ray tube changes the cutoff frequency f_{max} but does *not* change the frequencies of the characteristic peaks. The process that gives rise to the characteristic x-rays is described in Section 27.7.

27.5 COMPTON SCATTERING

In 1922, American physicist Arthur Holly Compton (1892–1962) noticed that when x-rays of a single wavelength impinged on matter, some of the radiation was scattered in various directions. Further study showed that some of the scattered radiation had longer

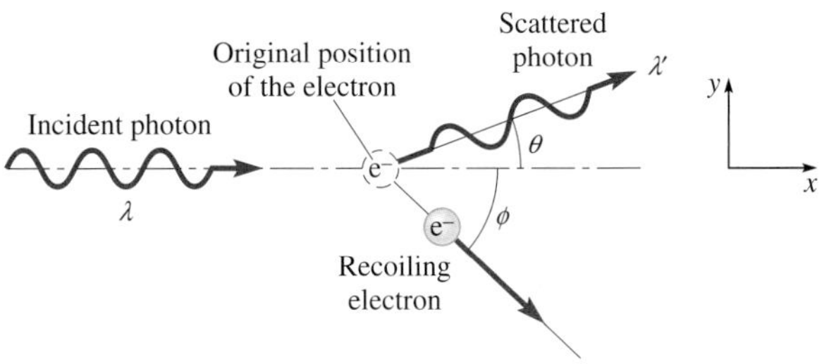

Figure 27.11 In Compton scattering, momentum and energy are transferred to the electron. Since momentum and energy are both conserved, the scattered photon has less energy—and therefore a longer wavelength—than the incident photon. The interaction can be analyzed as an elastic collision.

wavelengths than the incident radiation. The increase in the wavelength depended only on the angle between the incident radiation and the scattered radiation. According to classical theory, the incident radiation should induce vibration of the electrons in the target material *at the same frequency* as the incident wave. A scattered wave results when some of the incident energy is absorbed and reemitted in a different direction. Thus, according to classical EM theory, the scattered radiation should have the same frequency and wavelength as the incident radiation.

In the photon picture, **Compton scattering** is viewed as a collision between a photon and an electron (Fig. 27.11). The scattered photon must have less energy than the incident photon since some energy is given to the recoiling electron. Thus, conservation of energy requires that

$$E = K_e + E'$$

or

$$\frac{hc}{\lambda} = K_e + \frac{hc}{\lambda'} \tag{27-10}$$

where E is the energy of the incident photon, K_e is the kinetic energy given to the recoiling electron, and E' is the energy of the scattered photon. Since the scattered photon has less energy, its wavelength is longer. Although the scattered photon has less energy, it does *not* move any more slowly than the incident photons. All photons move at the same speed c.

Energy conservation alone does not explain why the wavelength of the photons scattered in a particular direction (at angle θ with respect to the incident photons) is *always the same* for a given incident wavelength λ. If energy conservation were the only restriction, photons of *any* energy $E' < E$ could be scattered at *any* angle θ. Just as in other collisions, we must consider conservation of *momentum.*

According to classical electromagnetic theory, EM waves carry momentum of magnitude E/c, where E is the energy of the wave and c is the speed of light. In the photon picture, each photon carries a little bit of that momentum in proportion to the amount of energy it carries. The **momentum of a photon** is

$$p = \frac{\text{photon energy}}{c} = \frac{hf}{c} = \frac{h}{\lambda} \tag{27-11}$$

The direction of the photon's momentum is in its direction of propagation.

In most cases the initial energy and momentum of the electron are negligible compared with the energy and momentum imparted by the collision. The energy of an x-ray photon is large relative to the work function of the target material, so we can ignore the work function and treat the electron as free. Compton's explanation ignores the initial energy and momentum of the electron and the work function; the scattering process is viewed as a collision between a photon and a *free* electron *initially at rest.*

Conservation of momentum requires:

$$\vec{\mathbf{p}} = \vec{\mathbf{p}}_e + \vec{\mathbf{p}}'$$

Using the incident photon's direction as the x-axis, we can separate this into two component equations:

$$\frac{h}{\lambda} = p_e \cos \phi + \frac{h}{\lambda'} \cos \theta \quad (x\text{-component}) \qquad (27\text{-}12)$$

and

$$0 = -p_e \sin \phi + \frac{h}{\lambda'} \sin \theta \quad (y\text{-component}) \qquad (27\text{-}13)$$

From the equations for conservation of energy and momentum [Eqs. (27-10), (27-12), and (27-13)], Compton derived this relationship:

Compton shift: difference in wavelength between the scattered and incident photons

Compton shift:

$$\lambda' - \lambda = \frac{h}{m_e c}(1 - \cos \theta) \qquad (27\text{-}14)$$

CONNECTION:

Calculation of the shift in the photon wavelength requires putting together just two principles: conservation of energy and conservation of momentum.

In Eq. (27-14), the incident photon has wavelength λ, the scattered photon has wavelength λ', m_e is the mass of the electron, and θ is called the scattering angle. Equation (27-14) correctly predicts the wavelength shifts observed in experiment.

In many cases, the electron's recoil speed is fast enough that we *cannot* use the nonrelativistic equations $K_e = \frac{1}{2}mv^2$ and $p_e = mv$ for the kinetic energy and momentum of the electron. Compton used the relativistic equations for the momentum [Eq. (26-6)] and kinetic energy [Eq. (26-8)] of the electron in his derivation, so Eq. (27-14) is valid for any recoil speed.

The quantity $h/(m_e c)$ is known as the **Compton wavelength** because it has the *dimensions* of a wavelength.

$$\frac{h}{m_e c} = \frac{6.626 \times 10^{-34}\ \text{J·s}}{9.109 \times 10^{-31}\ \text{kg} \times 2.998 \times 10^8\ \text{m/s}}$$

$$= 2.426 \times 10^{-3}\ \text{nm} = 2.426\ \text{pm} \qquad (27\text{-}15)$$

Since $\cos \theta$ can vary between +1 and −1, the quantity $(1 - \cos \theta)$ varies from 0 to 2 and the wavelength change varies from 0 to twice the Compton wavelength (4.853 pm). The Compton shift is difficult to observe if the wavelength of the incident photon is large compared to 4.853 pm.

CHECKPOINT 27.5

Why does a photon that has been scattered from an electron, initially at rest, have a longer wavelength than the incident photon?

Example 27.5

Energy of a Recoiling Electron

An x-ray photon of wavelength 10.0 pm is scattered through 110.0° by an electron. What is the kinetic energy of the recoiling electron?

Strategy Since we know the scattering angle, we can find the Compton shift [Eq. (27-14)]. The Compton shift and the wavelength of the incident photon enable us to find the wavelength of the scattered photon; from the wavelength we can find the energy of the scattered photon. By energy conservation, the kinetic energy of the electron plus the energy of the scattered photon is equal to the energy of the incident photon.

Solution The Compton shift formula is

$$\Delta\lambda = \lambda' - \lambda = \frac{h}{m_e c}(1 - \cos \theta)$$

continued on next page

Example 27.5 continued

where $h/(m_e c) = 2.426$ pm. With $\lambda = 10.0$ pm and $\theta = 110.0°$,

$$\Delta\lambda = \lambda' - \lambda = 2.426 \text{ pm} \times (1 - \cos 110.0°) = 3.256 \text{ pm}$$

Then the scattered photon has wavelength

$$\lambda' = \lambda + \Delta\lambda = 10.0 \text{ pm} + 3.256 \text{ pm} = 13.26 \text{ pm}$$

The kinetic energy of the electron is

$$K_e = E - E' = \frac{hc}{\lambda} - \frac{hc}{\lambda'}$$

$$= 1240 \text{ eV·nm} \times \left(\frac{1}{0.0100 \text{ nm}} - \frac{1}{0.01326 \text{ nm}}\right)$$

$$= 30.5 \text{ keV}$$

Discussion Avoid the common algebraic mistake of substituting $hc/\Delta\lambda$ for $hc/\lambda - hc/\lambda'$. That error would have given an answer of 380 keV for the kinetic energy of the electron—wrong by more than a factor of 12.

Practice Problem 27.5 Change in Wavelength

In a Compton scattering experiment, x-rays scattered through an angle of 37.0° with respect to the incident x-rays have a wavelength of 4.20 pm. What is the wavelength of the incident x-rays?

27.6 SPECTROSCOPY AND EARLY MODELS OF THE ATOM

Line Spectra

In 1853, the Swedish spectroscopist Anders Jonas Ångström (1814–1874) used spectroscopy to study the light emitted by various low-pressure gases in a *discharge tube* (Fig. 27.12a). The gas is kept at low pressure so that the atoms are far apart from one another; thus, the light is emitted by a collection of essentially independent atoms. Electrons are injected into the gas, either by heating the electrodes or by applying a large potential difference between them. The electrons collide with gas atoms in the tube. Electric current flows between the electrodes; electrons move in one direction and positive gas ions in the other. As long as the current is maintained, the tube emits light. A neon sign (Fig. 27.12b) is a familiar example of a discharge tube. A fluorescent lamp is a mercury discharge tube with a coating on the inside of the glass.

Applications: neon signs and fluorescent lights

In the spectroscopic analysis of the light emitted by a discharge tube, the light first passes through a thin slit. Then it passes through either a prism or a grating so that light of different wavelengths emerges at different angles. Although the light emitted by a hot solid object forms a *continuous* spectrum, Ångström discovered that the light from a gas discharge tube forms a *discrete* spectrum (Fig. 27.13a–d). A discrete spectrum is also called a *line* spectrum because each discrete wavelength forms an image of the slit; the spectrum appears as a set of narrow, parallel lines of different colors with dark space between the lines.

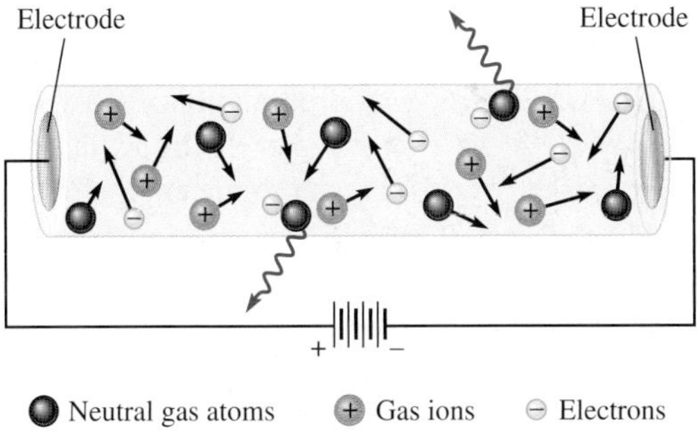

(a)

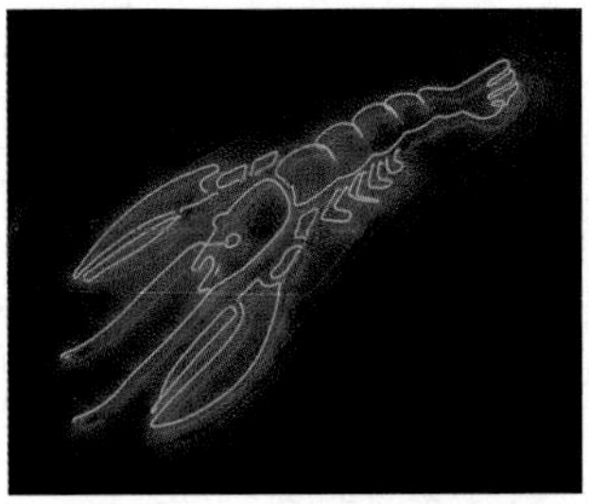

(b)

Figure 27.12 (a) A gas discharge tube. (b) A neon sign is a discharge tube containing neon gas.

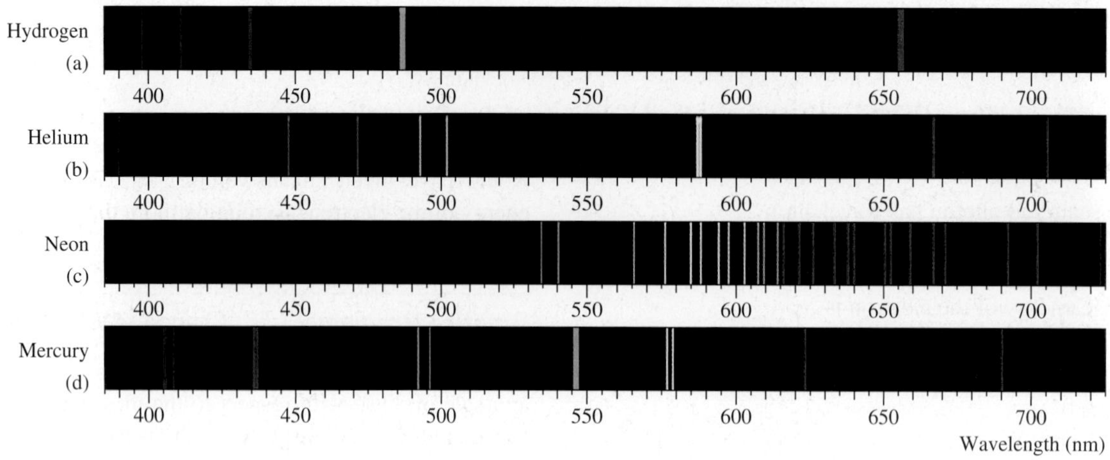

Figure 27.13 (a-d) Emission spectra for atomic hydrogen, helium, neon, and mercury. The intensities of the brightest spectral lines have been reduced to enhance the visibility of the weaker spectral lines.

In addition to examining the light emitted by a gas, scientists also studied the light absorbed by a gas. A beam of white light is sent through the gas and the transmitted light is analyzed with a spectrometer (Fig. 27.14). The resulting absorption spectrum is the continuous spectrum expected for white light except for some dark lines. Most wavelengths are transmitted through the gas, but the dark lines show that a few discrete wavelengths are absorbed. The wavelengths absorbed are a subset of the wavelengths emitted by the same gas when in a discharge tube.

Application: spectroscopic analysis of the elements

Each element has its own characteristic emission spectrum. For instance, the characteristic red color of a neon sign is caused by the emission spectrum of neon. Scientists soon began to use spectroscopy to identify the elements present in substances. Many previously unknown elements were discovered through spectroscopy. Cesium was named for its bright blue spectral lines (in Latin, *cesius* = "sky blue"); rubidium was named for its prominent red lines (in Latin, *rubidius* = "dark red"). Turning their spectroscopes toward the Sun and stars, scientists identified elements such as helium, which had not yet been discovered on Earth. (The Greek word for the Sun is *helios.*)

The spectra of most elements show no obvious pattern, but hydrogen—the simplest atom—does show a striking pattern. Figure 27.15 shows an emission spectrum for hydrogen that includes lines in the ultraviolet, visible, and near infrared. In 1885, the Swiss mathematician Johann Jakob Balmer (1825–1898) found a simple formula for the four wavelengths of the hydrogen emission lines in the *visible* range:

$$\frac{1}{\lambda} = R\left(\frac{1}{n_f^2} - \frac{1}{n_i^2}\right) \tag{27-16}$$

where $n_f = 2$ and $n_i = 3, 4, 5,$ or 6. The experimentally measured quantity $R = 1.097 \times 10^7\ \text{m}^{-1}$ is called the Rydberg constant after Swedish spectroscopist Johannes Rydberg (1854–1919).

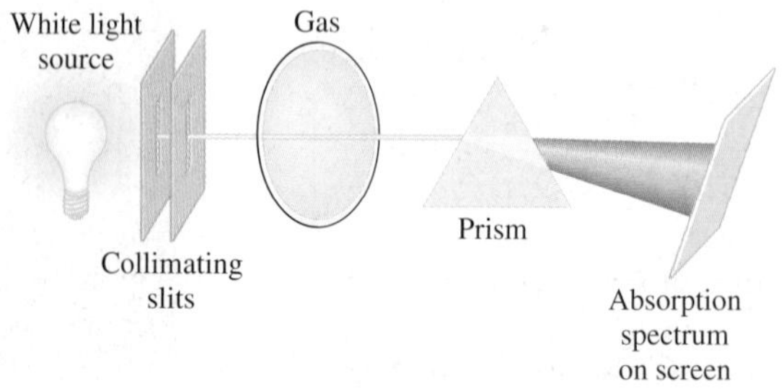

Figure 27.14 Setup for obtaining an absorption spectrum. When white light passes through a gas, some wavelengths are absorbed. The missing wavelengths cause dark lines to appear in the otherwise continuous spectrum on the screen.

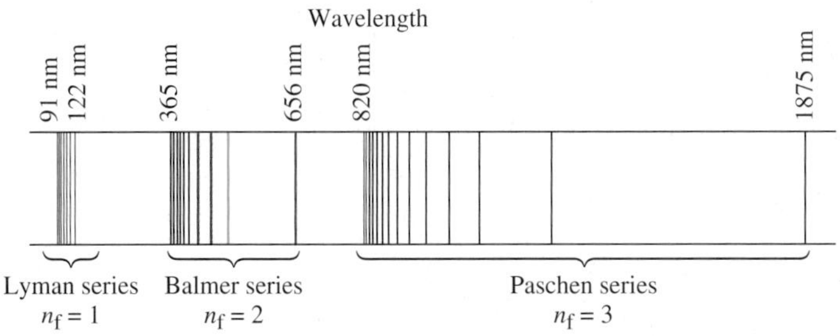

Figure 27.15 Emission spectrum of atomic hydrogen. Four lines in the Balmer series are in the visible part of the spectrum. The rest of the Balmer series and the entire Lyman series are in the ultraviolet. The Paschen series and other series with higher values of n_f are in the infrared. In each series, the wavelength that corresponds to $n_i = \infty$ is called the *series limit.*

Subsequently, it was found that Eq. (27-16) gives the wavelengths of *all* of the hydrogen lines, not just the four visible lines. Each value of n_f (1, 2, 3, 4, . . .) gives rise to a series of lines; each line in a series has a unique value of $n_i > n_f$.

The experimental observation that individual atoms in a gas absorb and emit EM radiation only at discrete wavelengths was impossible to explain using early models of atomic structure.

Discovery of the Atomic Nucleus

At the beginning of the twentieth century, the most common model of the atom was the *plum pudding model.* The positive charge and most of the mass of the atom were thought to be spread evenly throughout the volume of the atom, with negatively charged electrons sprinkled here and there like plums in a pudding (Fig. 27.16). J. J. Thomson, who discovered the electron, accepted the uniform distribution of positive charge but said that the electrons in the atom were moving.

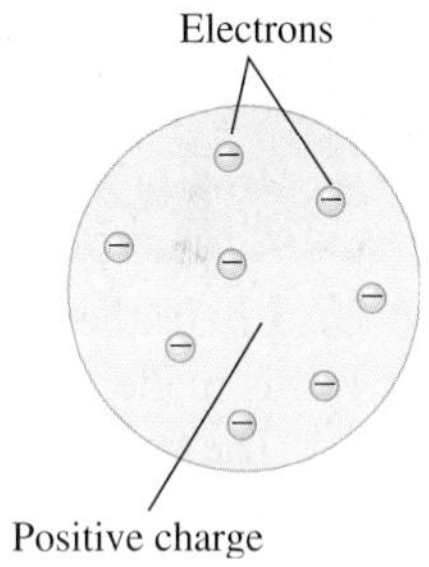

Figure 27.16 Thomson's plum pudding model of the atom, in which the positive charge and most of the mass of the atom are spread out. The discovery of the atomic nucleus showed this model to be incorrect.

Rutherford Experiment In 1911, the New Zealander Ernest Rutherford (1871–1937) designed an experiment in which a thin gold foil was bombarded with alpha particles. (*Alpha particles* are emitted in the decay of some radioactive elements. They have charge $+2e$ and approximately four times the mass of a hydrogen atom. We now know that an alpha particle consists of 2 protons and 2 neutrons.) After striking the foil, the alpha particles were detected by observing the flashes of light produced when they hit a fluorescent screen (Fig. 27.17a).

In the plum pudding model of the atom, positive charge and mass are distributed evenly, with no points at which mass or charge are concentrated. Based on this model, Rutherford expected the alpha particles to pass through the atoms of the foil barely deflected at all. He was surprised to find alpha particles that were deflected through large angles—sometimes more than 90°, so they bounced back from the foil instead of passing through it. Rutherford expressed his surprise by saying, "it was almost as incredible as if you fired a fifteen-inch [artillery] shell at a piece of tissue paper and it came back and hit you."

The alpha particles deflected through large angles must have collided with something massive; the massive object must be tiny since most of the alpha particles are deflected through much smaller angles. Based on the results of scattering experiments, Rutherford proposed a new model of the atom in which a central dense nucleus with a radius of about 10^{-15} m contains all of the positive charge and most of the mass of the atom (Fig. 27.17b). The positively charged nucleus repels the positively charged alpha particles that come near it. The radius of the nucleus is only one hundred-thousandth (10^{-5}) times the radius of the atom; thus, most of the alpha particles pass right through the gold foil without significant deflection. The few alpha particles that pass near to the nucleus feel a large repulsive force and are deflected through large angles.

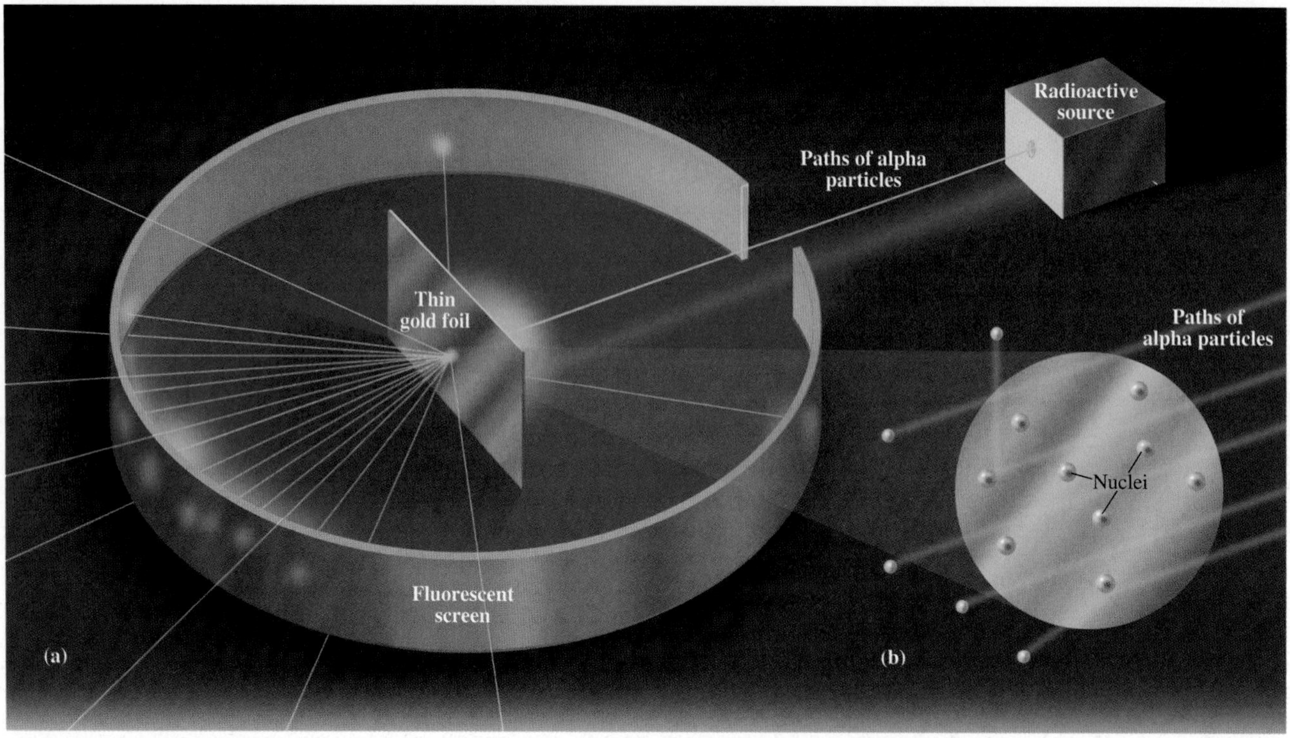

Figure 27.17 Rutherford scattering experiment. (a) Alpha particles from a radioactive source are aimed at a thin gold foil. The foil is made as thin as possible to minimize the chance that an alpha particle might be scattered by more than one nucleus. The scattered particles are detected by light emitted when they hit a fluorescent screen. (b) Alpha particles that get close to a gold nucleus are deflected through large angles; alpha particles that do not pass near a nucleus are barely deflected at all.

After the discovery of the nucleus, the planetary model of the atom replaced the plum pudding model: electrons were thought to revolve around the nucleus like a small solar system, with the electric force on the electrons due to the nucleus playing the role that gravity plays in the solar system.

Serious Questions Unanswered by the Planetary Model of the Atom Two serious questions bothered scientists. First, in classical electromagnetic theory, an accelerating electric charge gives off EM radiation. An electron orbiting the nucleus has an acceleration—the direction of its velocity is always changing—so it ought to be continually radiating. As the radiation carries off energy, the electron's energy should decrease, causing the electron to spiral into the nucleus. Thus, atoms ought to radiate for a short while—only about 0.01 μs—until they collapse; atoms could not be stable according to classical electromagnetism. The second question: when atoms do radiate, as in a discharge tube, why only at certain wavelengths? In other words, why are emission spectra from atoms seen as line spectra rather than as continuous spectra?

27.7 THE BOHR MODEL OF THE HYDROGEN ATOM; ATOMIC ENERGY LEVELS

In 1913, the Danish physicist Niels Bohr (1885–1962) published the first atomic model that addressed these questions. Bohr's model is of the hydrogen atom—the simplest atom, with one electron and a single proton as the nucleus.

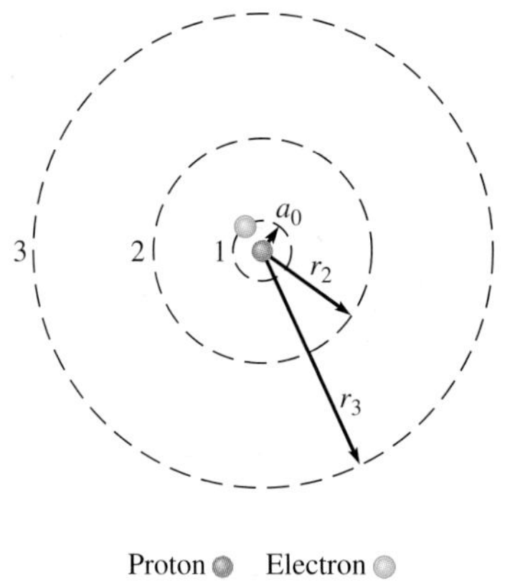

Figure 27.18 In the Bohr model of the hydrogen atom, the electron orbits the nucleus in a circle. The radius of the orbit must be one of a discrete (quantized) set of radii.

Assumptions of the Bohr Model

1. *The electron can exist without radiating energy only in certain circular orbits* (Fig. 27.18). Bohr asserts that, since the accelerating electron does not radiate, some aspects of classical electromagnetic theory *do not apply* to an electron orbiting the nucleus at certain discrete radii. The electron is allowed to be in only one of a discrete set of orbits called **stationary states**. (The *electron* is not stationary; it orbits the nucleus. The *state* of the electron is stationary because the electron orbits at a fixed radius without radiating.) Each stationary state has a definite energy associated with it; the set of energies of the states are called **energy levels**. Thus, Bohr extends quantum theory to the *structure of the atom* itself: both the radii and energies of the orbits are quantized.
2. *The laws of Newtonian mechanics apply to the motion of the electron in any of the stationary states.* The force on the electron due to the nucleus is given by Coulomb's law. Newton's second law ($\Sigma\vec{\mathbf{F}} = m\vec{\mathbf{a}}$) relates the Coulomb force to the radial acceleration of the electron in its circular orbit. The energy of the orbit is the electron's kinetic energy plus the electric potential energy of the interaction between the electron and the nucleus.
3. *The electron can make a transition between stationary states through the emission or absorption of a single photon* (Fig. 27.19). The energy of the photon is equal to the difference between the energies of the two stationary states:

$$|\Delta E| = hf \qquad (27\text{-}17)$$

Since the electron energy levels have only certain discrete values, emission and absorption spectra are made up of photons of discrete energies—they are line spectra. Bohr made no attempt to explain *how* an electron "jumps" from one orbit to another.

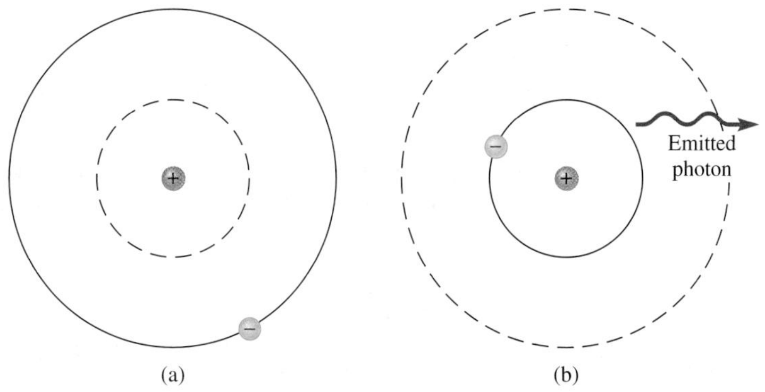

Figure 27.19 (a) A hydrogen atom in an allowed orbit. (b) The atom emits a photon and the electron drops down into a different allowed orbit with lower energy. Absorption of a photon is just the reverse: the photon "donates" its energy to the atom, moving the electron to a higher energy level.

4. *The stationary states are those circular orbits in which the electron's angular momentum is quantized in integral multiples of $h/(2\pi)$.*

$$L_n = n\frac{h}{2\pi} = n\hbar \quad (n = 1, 2, 3, \ldots) \qquad \text{(27-18)}$$

$\hbar = h/(2\pi)$

The combination of constants $h/(2\pi)$ is commonly abbreviated as $\hbar$ ("h-bar"). Bohr chose these values of angular momentum because they gave agreement with the experimental data on the hydrogen emission spectrum.

Radii of Bohr Orbits The Bohr orbit with the smallest radius, is known as the **Bohr radius**:

$$a_0 = \frac{\hbar^2}{m_e k e^2} = 52.9 \text{ pm} = 0.0529 \text{ nm} \qquad \text{(27-19)}$$

See for more details about the Bohr model.

The allowed orbital radii of the electron are

$$r_n = n^2 a_0 \quad (n = 1, 2, 3, \ldots) \qquad \text{(27-20)}$$

Energy Levels of the Hydrogen Atom

The energy of a stationary state is the sum of the electron's kinetic energy and the electric potential energy when the electron and nucleus are separated by a distance r:

$$E = K + U = \frac{1}{2}m_e v^2 - \frac{ke^2}{r} \qquad \text{(27-21)}$$

The potential energy U is negative because we assume that the potential energy is zero at infinite separation; the potential energy decreases as the distance between the oppositely charged electron and proton decreases.

The energy E is negative because the energy of the atom with the electron bound to the nucleus is less than the energy of the ionized atom. In the ionized atom, the electron is at rest infinitely far from the nucleus, so $E = 0$ (both the kinetic and potential energies are zero). An electron in one of the bound states must be supplied with energy for it to escape from the nucleus and cause the atom to become ionized.

For the state $n = 1$, called the **ground state**, the orbit has the smallest possible radius and the lowest possible energy. The ground state energy is

$$E_1 = -\frac{m_e k^2 e^4}{2\hbar^2} = -2.18 \times 10^{-18} \text{ J} = -13.6 \text{ eV} \qquad \text{(27-22)}$$

The energies of the **excited states** ($n > 1$) are given by

$$E_n = \frac{E_1}{n^2}, \quad n = 1, 2, 3, \ldots \qquad \text{(27-23)}$$

Figure 27.20 is an energy level diagram for hydrogen. Each horizontal line represents an energy level. The vertical arrows show transitions between levels, accompanied by either the emission or absorption of a photon of the appropriate energy. The energy of the photon emitted when the electron goes from initial state n_i to a lower-energy final state n_f is

$$E = \frac{hc}{\lambda} = E_i - E_f = E_1\left(\frac{1}{n_i^2} - \frac{1}{n_f^2}\right) \qquad \text{(27-24)}$$

If we take the general form of the Balmer formula [Eq. (27-16)] and multiply both sides by hc, we get

$$\frac{hc}{\lambda} = hcR\left(\frac{1}{n_f^2} - \frac{1}{n_i^2}\right) = -hcR\left(\frac{1}{n_i^2} - \frac{1}{n_f^2}\right)$$

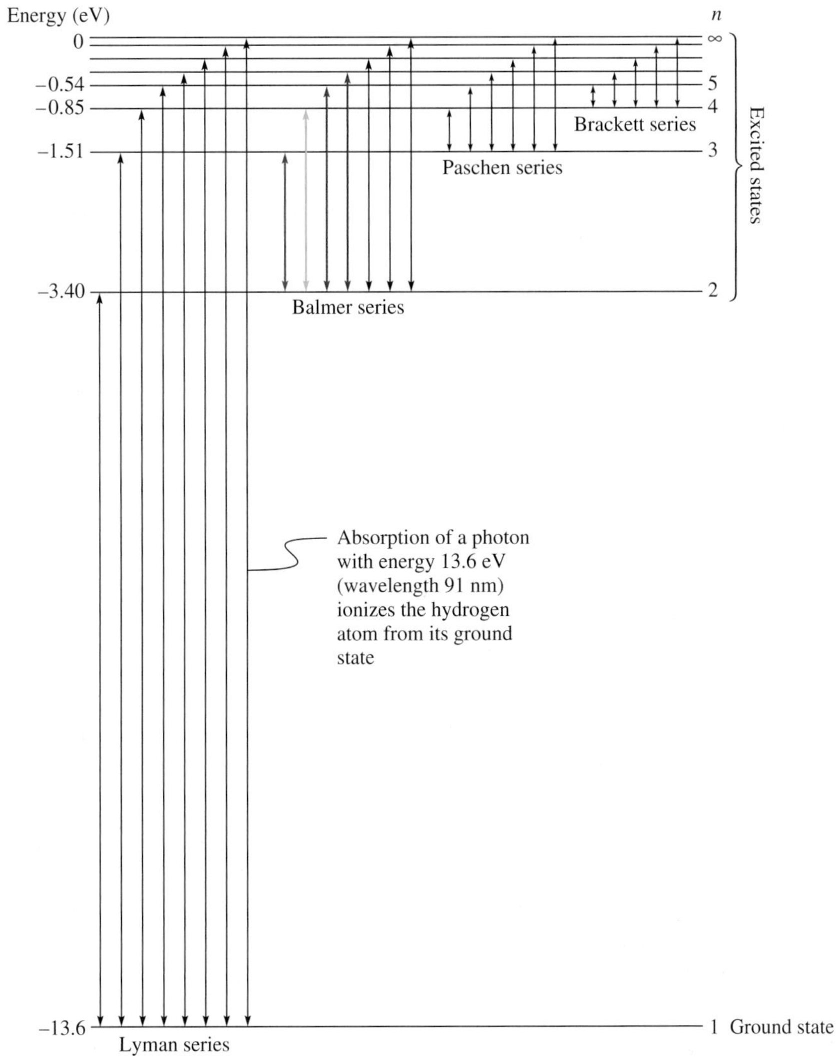

Figure 27.20 Energy level diagram for the hydrogen atom. Energy $E = 0$ at level $n = \infty$ corresponds to the ionized atom (electron and proton separated). Arrows represent transitions between energy levels. The length of an arrow represents the energy of the photon emitted or absorbed. Compare with Fig. 27.15.

where R is the Rydberg constant. Thus, Bohr's theory is in perfect agreement with the spectroscopic data as long as $E_1 = -hcR$. When Bohr did the calculation, he found the two in agreement to within 1%.

CHECKPOINT 27.7

What is the energy of the photon emitted when a hydrogen atom makes a transition from the $n = 5$ state to the $n = 2$ state? (Refer to Fig. 27.20.)

Example 27.6

Identifying Initial and Final States

One wavelength in the infrared part of the hydrogen emission spectrum has wavelength 1.28 μm. What are the initial and final states of the transition that results in this wavelength being emitted?

Strategy The energy of the 1.28-μm photon must be the difference in two energy levels. Rather than trying to solve an equation with two unknowns (the initial and final values of n), we can use the energy level diagram to narrow down the choices first.

continued on next page

Example 27.6 continued

Solution The energy of the photon emitted is

$$E = \frac{hc}{\lambda} = \frac{1240\ \text{eV}\cdot\text{nm}}{1280\ \text{nm}} = 0.969\ \text{eV}$$

Looking at the energy level diagram (see Fig. 27.20), the photon must be in the Paschen series. The smallest photon energy in the Balmer series is

$$(-1.51\ \text{eV}) - (-3.40\ \text{eV}) = 1.89\ \text{eV}$$

The photons in the Lyman series have even larger energies. The *largest* energy photon in the Brackett series has energy 0.85 eV. Only the Paschen series can include a photon around 1 eV. Therefore, the final state is $n = 3$ and the final energy is $E_3 = -1.51$ eV. Now we solve for the initial state n.

$$\text{energy of photon} = E_i - E_f$$

$$0.969\ \text{eV} = \frac{-13.6\ \text{eV}}{n^2} - (-1.51\ \text{eV})$$

$$n = \sqrt{\frac{13.6\ \text{eV}}{1.51\ \text{eV} - 0.969\ \text{eV}}} = 5$$

The 1.28-μm photon is emitted when the electron goes from $n = 5$ to $n = 3$.

Discussion For a photon in the hydrogen spectrum, identifying the lower of the two energy levels is simplified by noting that the various series do not overlap. All of the photons in the Lyman series (lower energy level $n = 1$) have larger energies than any of the photons in the Balmer series (lower energy level $n = 2$); all of the photons in the Balmer series have larger energies than any in the Paschen series; and so on.

Practice Problem 27.6 Fifth Balmer Line

The first four Balmer lines are easily visible. What is the wavelength of the fifth Balmer line?

Example 27.7

Thermal Excitation

Absorbing or emitting a photon is not the *only* way an atom can make a transition between energy levels. One of the other ways is called thermal excitation. If their kinetic energies are sufficiently large, two atoms can undergo an inelastic collision in which one of them makes a transition into an excited state, leaving the atoms with less total translational kinetic energy after the collision than before. (a) What is the average translational kinetic energy of an atom in a gas at room temperature (300 K)? (b) Explain why, in atomic hydrogen gas at room temperature, almost all of the atoms are in the ground state.

Strategy In Section 13.6, we found the average translational kinetic energy of an ideal gas to be $\langle K_{tr} \rangle = \frac{3}{2}kT$ [Eq. (13-20)]. To facilitate comparison with the energy levels in hydrogen, we convert the average kinetic energy to eV. The key is to see whether the translational kinetic energies of the hydrogen atoms are large enough that an inelastic collision can excite one of the atoms.

Solution and Discussion (a) At $T = 300$ K,

$$\langle K_{tr} \rangle = \frac{3}{2}kT \qquad (13\text{-}20)$$

$$= \frac{3}{2} \times 1.38 \times 10^{-23}\ \text{J/K} \times 300\ \text{K} \times \frac{1\ \text{eV}}{1.60 \times 10^{-19}\ \text{J}}$$

$$= 0.04\ \text{eV}$$

(b) Suppose two atoms, both in the ground state, collide. To excite one of them into $n = 2$ (the transition from the ground state that requires the smallest energy) requires

$$\Delta E = E_2 - E_1 = (-3.40\ \text{eV}) - (-13.6\ \text{eV}) = 10.2\ \text{eV}$$

This is 260 times the average kinetic energy. At any given instant, some atoms have more than the average and some have less; very few have much more than average. The number of atoms with kinetic energies *hundreds of times* the average kinetic energy is extremely small. (See the Maxwell-Boltzmann distribution curves in Fig. 13.13.) The tiny number of atoms excited in this way quickly decay back to the ground state by emitting a photon. Thus, at any given instant, a negligibly small fraction of the atoms are excited; for all practical purposes they are all in the ground state.

Conceptual Practice Problem 27.7
Absorption Spectrum of Hydrogen Atoms

At room temperature, the *absorption* spectrum of atomic hydrogen has no black lines in the visible part of the spectrum. At high temperatures, the absorption spectrum of atomic hydrogen has four dark lines in the visible—at the same wavelengths as the four visible lines in the Balmer series of the emission spectrum. Explain.

Successes of the Bohr Model

The Bohr model has been replaced by the quantum mechanical version of the atom (Chapter 28). Despite serious deficiencies, the Bohr model was an important step in the development of quantum physics. Some important ideas that carry over from the Bohr atom to the quantum mechanical atom include:

- The electron can be in one of a discrete set of stationary states with quantized energy levels.
- The atom can make a transition between energy levels by emitting or absorbing a photon.
- Angular momentum is quantized.
- Stationary states can be described by quantum numbers (n is now called the *principal quantum number*).
- The electron makes a discontinuous transition ("quantum jump") between energy levels.

Bohr's model gives the correct numerical values—even if for the wrong reasons—of the energy levels in the hydrogen atom. It also correctly predicts the size of the H atom: the Bohr radius a_0 is now understood as the *most likely* distance between the electron and the nucleus when the H atom is in the ground state.

Problems with the Bohr Model

- The whole idea of the electron orbiting the nucleus—indeed, of the electron having any kind of trajectory—is incorrect. Newtonian mechanics does *not* apply to the motion of the electron. The electron must be described quantum mechanically; quantum mechanics gives the *probabilities* of the electron being located at various distances from the nucleus.
- Angular momentum is quantized, but *not* in integral multiples of $\hbar$.
- The Bohr model gives no way to calculate the probabilities of an electron absorbing or emitting a photon.
- The Bohr model cannot be extended to atoms with more than one electron.

Applications of the Bohr Model to Other One-Electron Atoms

The Bohr model can be applied to ions that have a *single electron* such as ionized helium (He^+) and doubly ionized lithium (Li^{2+}). Instead of having nuclear charge $+e$, these ions have a nuclear charge of $+Ze$, where Z is the atomic number (the number of protons in the nucleus). Every time e^2 appears in equations for the hydrogen atom, one factor of e came from the electron's charge and one from the charge of the nucleus. For a nucleus with charge Ze, we replace each factor of e^2 with Ze^2. Then the orbital radii are smaller by a factor of Z:

$$r_n = \frac{n^2}{Z} a_0 \quad (n = 1, 2, 3, \ldots) \tag{27-25}$$

and the energy levels are larger by a factor of Z^2:

$$E_n = -\frac{Z^2}{n^2} \times 13.6 \text{ eV} \quad (n = 1, 2, 3, \ldots) \tag{27-26}$$

Example 27.8

He^+ Energy Levels

Calculate the first five energy levels for singly ionized helium. Draw an energy level diagram for singly ionized helium and compare it with that for hydrogen.

Strategy Helium has an atomic number $Z = 2$. We use the ground state energy for hydrogen along with Z and the various values of n to find the energy levels.

continued on next page

Example 27.8 continued

Solution and Discussion The ground state energy for *hydrogen* is −13.6 eV. A one-electron atom in which the nucleus has charge $+Ze$ has energy levels

$$E_n = -\frac{Z^2}{n^2} \times 13.6 \text{ eV} \quad (n = 1, 2, 3, \ldots)$$

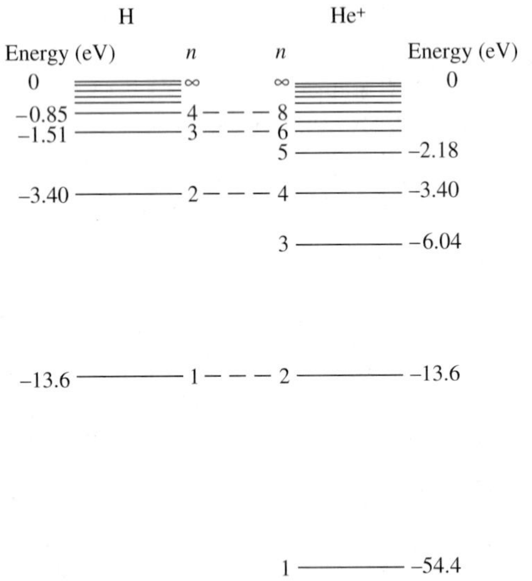

Figure 27.21

Energy levels of H and He^+.

For He^+, $Z = 2$:

$$E_n = -\frac{4}{n^2} \times 13.6 \text{ eV} = -\frac{1}{n^2} \times 54.4 \text{ eV} \quad (n = 1, 2, 3, \ldots)$$

The first six energy levels for He^+ are

$$E_1 = -54.4 \text{ eV};\ E_2 = -13.6 \text{ eV};\ E_3 = -6.04 \text{ eV};$$
$$E_4 = -3.40 \text{ eV};\ E_5 = -2.18 \text{ eV}$$

Now we draw an energy level diagram (not to scale) for He^+ next to one for hydrogen (Fig. 27.21). Due to the factor of Z^2, each energy level in He^+ is four times the energy level for the same value of n in hydrogen. The first excited state ($n = 2$) for He^+ has the same energy as the ground state of hydrogen; the third excited state ($n = 4$) for He^+ has the same energy as the first excited state ($n = 2$) of hydrogen. In general, the energy of state $2n$ in He^+ is the same as the energy of state n in hydrogen.

Conceptual Practice Problem 27.8 Ionization Energy

The *ionization energy* is the energy that must be supplied to an atom in its ground state to separate the electron from the nucleus. (a) What is the ionization energy for H? (b) What is the ionization energy for He^+? (c) Give a qualitative explanation for why He^+ has a larger ionization energy than H.

Applications: Fluorescence, Phosphorescence, and Chemiluminescence

Suppose atomic hydrogen gas is illuminated with ultraviolet radiation of wavelength 103 nm. Some of the atoms absorb a photon and are excited into the $n = 3$ level. When one of the excited atoms decays back to the ground state, it does not necessarily emit a 103-nm photon. It can decay first to $n = 2$ (by emitting a 656-nm photon) and then to $n = 1$ (by emitting a 122-nm photon). *The presence of intermediate energy levels enables the atom to absorb a photon of one wavelength and emit photons of longer wavelengths.*

Fluorescent materials absorb ultraviolet radiation and decay in a series of steps; at least one of the steps involves the emission of a photon of visible light. In a molecule or solid, not all of the transitions involve the emission of a photon. Some of the transitions increase the vibrational or rotational energy of the molecule of the solid; this energy is ultimately dissipated into the surroundings.

Applications: fluorescent lamps; fluorescent dyes in laundry detergent

A fluorescent lamp is a mercury discharge tube whose interior is coated with a mixture of fluorescent materials called phosphors. The phosphors absorb ultraviolet radiation emitted by the mercury atoms and in turn emit visible light. A "black light"—a source of ultraviolet radiation—makes fluorescent dyes glow brightly in the dark. Fluorescent dyes are also added to laundry detergents. The dyes absorb UV and emit blue light to "make whites whiter" (Fig. 27.22) by compensating for the yellowing of a fabric as it ages.

Phosphorescence is similar to fluorescence but involves a time delay. Most excited states of atoms and molecules decay quickly (typically within a few nanoseconds), but certain *metastable* excited states last for several seconds or even longer before a transition

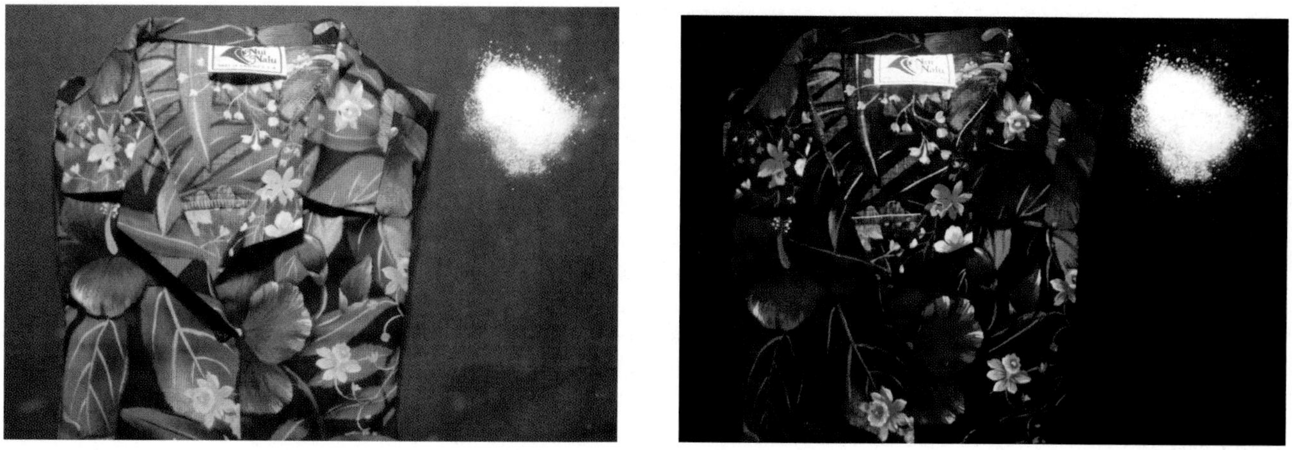

(a) (b)

Figure 27.22 A freshly laundered blouse and the laundry detergent used viewed in (a) natural light and (b) ultraviolet light.

occurs. Watch dials, wall switch plates, and toys that glow in the dark absorb photons when illuminated and get stuck in a metastable state so that the emission of light occurs much later.

In Rutherford's scattering experiment, alpha particles were detected by a phosphor screen. The phosphors were excited by a collision with an alpha particle rather than by absorbing a photon. The phosphor dots on a CRT television screen are excited by a beam of high-speed electrons; the decay back to the ground state involves emitting a visible photon. The screen uses three different phosphors to produce blue, green, and red.

PHYSICS AT HOME

Watch a CRT television for a while in a dark room. Turn off the TV but continue to look at the screen. What do you see? Does the screen go dark immediately?

The blue glow from the floor described at the beginning of this chapter is caused by *chemiluminescence.* The colorless liquid solution contains luminol (3-aminophthalhydrazide) and hydrogen peroxide. Traces of hemoglobin (which is found in blood) catalyze an oxidation reaction between the luminol and the hydrogen peroxide. The reaction leaves one of the products in an excited state, which then decays to the ground state by emitting a photon. The luminol test is effective even on clothing or surfaces that have been carefully washed. Thus, the blue glow reveals the location of possible bloodstains. Fireflies light up due to a similar chemical reaction. The reaction is controlled by enzymes (biological catalysts), allowing the firefly to turn the light on and off.

What causes the blue glow at the crime scene?

Energies of Characteristic X-Rays

The energies of the characteristic x-ray peaks superimposed on the continuous spectrum of bremsstrahlung (see Fig. 27.10) are determined by the energy levels of atoms in the target. When an incident electron strikes the target in an x-ray tube, it can supply the energy to free one of the tightly bound inner electrons from the atom. Then an electron in one of the higher energy levels will drop into the vacant energy level, emitting an x-ray photon whose energy is the difference in the two energy levels.

27.8 PAIR ANNIHILATION AND PAIR PRODUCTION

The Positron

In 1929, the British physicist Paul Dirac (1902–1984) made a theoretical prediction of the existence of a particle with the same mass as the electron but opposite charge ($q = +e$). Experiments later verified the existence of this particle, now called the *positron*. Some radioactive elements emit a positron spontaneously as they decay.

The positron was the first *antiparticle* discovered. Each of the particles that make up ordinary matter (electron, proton, and neutron) has an antiparticle (positron, antiproton, and antineutron). Cosmologists struggle with the question of why there is apparently more matter than antimatter in the universe. We introduce the positron here because two processes, the production of and the annihilation of an electron-positron pair, provide some of the clearest and most direct evidence for the photon model of EM radiation.

Pair Production

An energetic photon can *create* a positron and an electron where no such particles existed before. The photon is totally absorbed in the process. Energy must be conserved in any process, so in order for **pair production** to occur,

$$E_{\text{photon}} = E_{\text{electron}} + E_{\text{positron}}$$

The total energy of a particle with mass is the sum of its kinetic energy and its rest energy (the energy of the particle when at rest). A particle of mass m has *rest energy*

$$E_0 = mc^2 \qquad (26\text{-}7)$$

(see Section 26.7). Thus, a photon must have an energy of at least $2m_ec^2$ in order to create an electron-positron pair. If the photon's energy is greater than $2m_ec^2$, the excess energy appears as kinetic energy of the electron and positron. A photon is massless and thus, has no rest energy; the total energy of a photon is $E = hf = hc/\lambda$.

Momentum must also be conserved. For the photon, $p = E/c$. For an electron or a positron,

$$p = \frac{1}{c}\sqrt{E^2 - (m_ec^2)^2} < \frac{E}{c}$$

An electron or positron with total energy E has a momentum less than E/c—that is, less than the momentum of a photon with the same energy. Even if the electron and positron move in the same direction, their total momentum cannot be as large as the momentum of the photon. Therefore, it is impossible for both the pair's total momentum and total energy to be equal to the photon's momentum and total energy. Another particle must take part in the reaction: pair production can only occur when the photon passes near a massive particle such as an atomic nucleus (Fig. 27.23). The recoil of the massive particle satisfies momentum conservation without carrying off a significant amount of energy, so our assumption that all of the energy of the photon goes into the electron-positron pair is a good approximation.

Figure 27.23 Pair production. (a) A photon passes near an atomic nucleus. (b) The photon vanishes by creating an electron-positron pair. The nucleus recoils with an insignificant kinetic energy but, due to its large mass, with a significant momentum.

Figure 27.24 Pair annihilation. (a) An electron and positron vanish, creating (b) a pair of photons.

Pair Annihilation

Since ordinary matter contains plenty of electrons, sooner or later a positron gets near an electron. For a short while, the pair forms something like an atom; then—poof!—both particles disappear by creating *two* photons (Fig. 27.24). Pair annihilation cannot create just one photon; two photons are required to conserve both energy and momentum. The total energy of the two photons must be equal to the total energy of the electron-positron pair. Ordinarily the kinetic energies of the electron and positron are negligible compared with their rest energies, so for simplicity we assume they are at rest; then their total energy is just their rest energy, $2m_ec^2$, and their total momentum is zero. Annihilation of the pair then produces two photons, each with energy $E = hf = m_ec^2 = 511$ keV, traveling in opposite directions. Annihilation is the ultimate fate of positrons; the characteristic 511-keV photons are the sign that pair annihilation has taken place.

Besides confirming the photon model of EM radiation, pair annihilation and pair production clearly illustrate Einstein's ideas about mass and rest energy.

Example 27.9

Threshold Wavelength for Pair Production

Find the threshold wavelength for a photon to produce an electron-positron pair.

Strategy The photon must have at least enough energy to create the electron and positron, each of which has a rest energy of $m_ec^2 = 511$ keV. From the minimum photon energy, we find the threshold wavelength—the *maximum* wavelength, since larger wavelengths correspond to smaller photon energies.

Solution The minimum photon energy to create an electron-positron pair is

$$E = 2m_ec^2 = 1.022 \text{ MeV}$$

Now we find the wavelength of a photon with this energy.

$$E = hf = \frac{hc}{\lambda}$$

Solving for the wavelength,

$$\lambda = \frac{hc}{E} = \frac{1240 \text{ eV}\cdot\text{nm}}{1.022 \times 10^6 \text{ eV}} = 0.00121 \text{ nm} = 1.21 \text{ pm}$$

Discussion Quick check: a visible photon has a wavelength of about 500 nm and an energy of about 2 eV. Here, the photon energy is half a million times that of a visible photon, so the wavelength is about 500 nm/500 000 = 0.001 nm = 1 pm.

Practice Problem 27.9 Muon-Antimuon Pair Production

What is the longest wavelength that a photon can have if it is to supply enough energy to create a muon and an antimuon? The masses of the muon and the antimuon are 106 MeV/c^2 (their rest energies are 106 MeV). The pair production occurs as a result of the interaction between a photon and a nucleus.

Application: Positron Emission Tomography

Positron emission tomography (PET) is a medical imaging technique based on pair annihilation that is used to diagnose diseases of the brain and heart as well as certain types of cancer. A tracer is first injected into the body. The tracer is a compound—commonly glucose, water, or ammonia—that incorporates radioactive atoms. When one of the radioactive atoms emits a positron in the body, the positron annihilates with an electron, producing

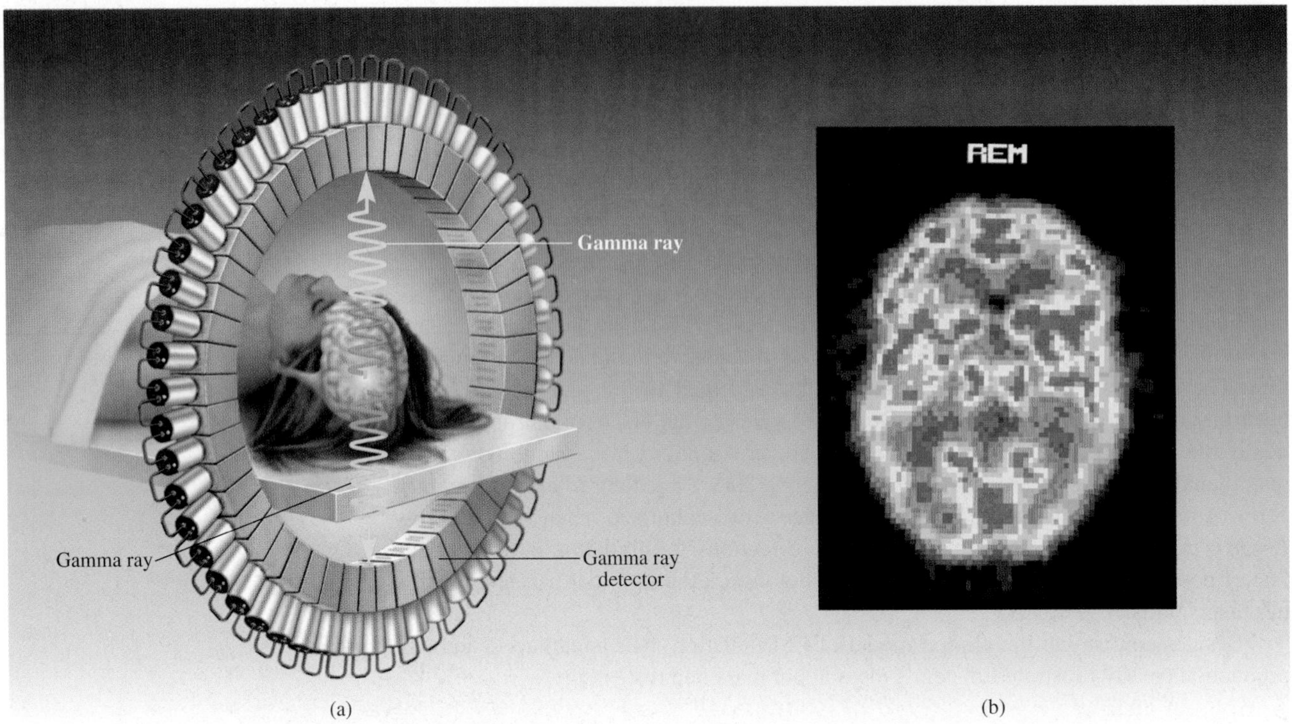

Figure 27.25 (a) A PET scan detects the gamma rays emitted when a positron and an electron annihilate within the body. (b) A PET scan of the brain. Color is used to distinguish regions with differing levels of positron emission.

two 511-keV gamma rays traveling in opposite directions. The two photons are detected by a ring of detectors around the body (Fig. 27.25a); then the atom that emitted the positron lies along the line between the two detectors. A computer analyzes the directions of many gamma rays and locates the regions of highest concentration of the tracer. Then the computer constructs an image of that slice of the body (Fig. 27.25b).

Other imaging techniques such as x-ray films, CAT scans, and MRIs show the structure of body tissues, but PET scans show the biochemical activity of an organ or tissue. For example, a PET scan of the heart can differentiate normal heart tissue from nonfunctioning heart tissue, which helps the cardiologist determine whether the patient can benefit from bypass surgery or from angioplasty.

Because rapidly growing cancer cells gobble up a glucose tracer faster than healthy cells, PET scans can accurately distinguish malignant from benign tumors. They help oncologists to determine the best treatment for a patient with cancer as well as to monitor the efficacy of a course of treatment. A brain tumor can be precisely located without cutting into the patient's skull for a biopsy. PET is used to evaluate diseases of the brain such as Alzheimer's, Huntington's, and Parkinson's diseases; epilepsy; and stroke.

Master the Concepts

- A quantity is quantized when its possible values are limited to a discrete set.
- Max Planck found an equation to match experimental results for blackbody radiation. The equation led him to postulate that the energy of an oscillator must be quantized in integral multiples of hf, where f is the frequency of the oscillator. Planck's constant is now recognized as one of the fundamental constants in physics:

$$h = 6.626 \times 10^{-34}\ \text{J·s} \qquad (27\text{-}3)$$

- In the photoelectric effect, EM radiation incident on a metal surface causes electrons to be ejected from the

continued on next page

Master the Concepts continued

metal. To explain the photoelectric effect, Einstein said that *EM radiation itself* is quantized. The quantum of EM radiation—that is, the smallest indivisible unit—is now called the photon. The energy of a photon with frequency f is

$$E = hf \quad (27\text{-}4)$$

The maximum kinetic energy of an electron is the difference between the photon energy and the work function ϕ, which is the amount of energy that must be supplied to break the bond between an electron and the metal.

$$K_{max} = hf - \phi \quad (27\text{-}7)$$

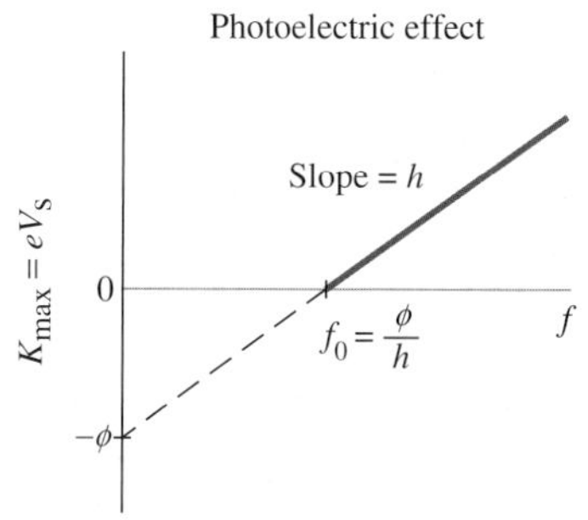

- One electron-volt is equal to the kinetic energy that a particle with charge $\pm e$ (such as an electron or a proton) gains when it is accelerated through a potential difference of magnitude 1 V.

$$1 \text{ eV} = 1.60 \times 10^{-19} \text{ J} \quad (27\text{-}5)$$

- In an x-ray tube, electrons are accelerated to kinetic energy K and then strike a target. The maximum frequency of the x-ray radiation emitted occurs when all of the electron's kinetic energy is carried away by *a single photon*:

$$hf_{max} = K \quad (27\text{-}9)$$

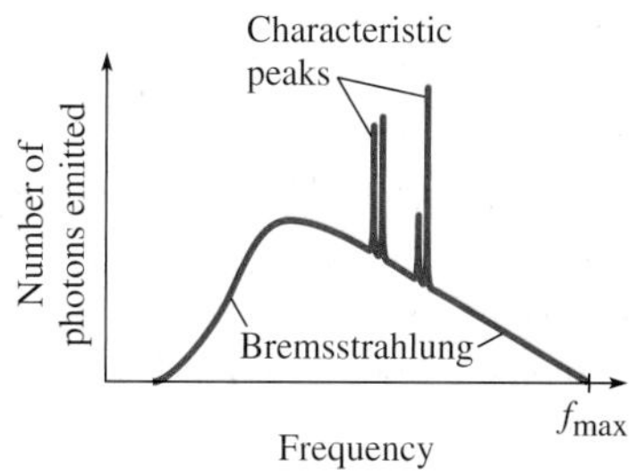

- In Compton scattering, x-rays scattered from a target have longer wavelengths than the incident x-rays; the wavelength shift depends on the scattering angle θ:

$$\lambda' - \lambda = \frac{h}{m_e c}(1 - \cos\theta) \quad (27\text{-}14)$$

Compton scattering can be viewed as a collision between a photon and a free electron at rest. The momentum and kinetic energy of the incident photon must equal the total momentum and kinetic energy of the scattered photon and recoiling electron.

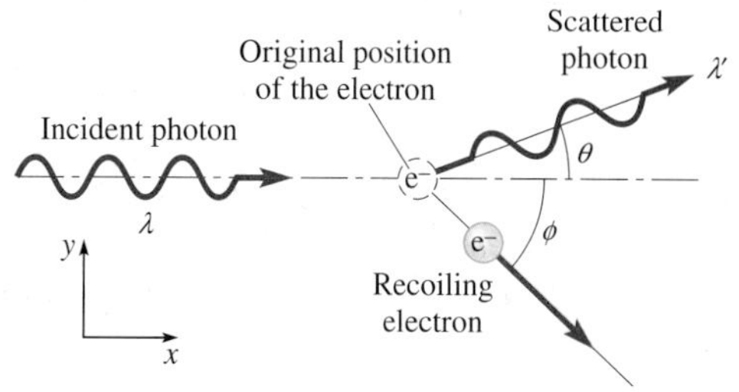

- Emission and absorption of EM radiation by individual atoms form *line spectra* at discrete wavelengths. Each element has its own characteristic spectrum determined by its discrete (quantized) set of energy levels. The energy of the photon emitted or absorbed when an atom makes a transition between energy levels is equal to the difference between the atomic energy levels:

$$|\Delta E| = hf \quad (27\text{-}17)$$

- The energy levels of a hydrogen atom are

$$E_n = \frac{E_1}{n^2} \quad (27\text{-}23)$$

where the ground state energy (lowest energy level) is

$$E_1 = -13.6 \text{ eV} \quad (27\text{-}22)$$

- The Bohr model of the hydrogen atom assumed that the electron moves in a circular orbit around the nucleus. The radii and energies of these orbits are quantized. Although calculations using the Bohr model give the correct energy levels [Eqs. (27-22) and (27-23)], it has serious deficiencies and has been replaced by the quantum mechanical description of the hydrogen atom (Chapter 28).
- Fluorescent materials absorb ultraviolet radiation and decay in a series of steps; one or more of the steps involve the emission of a photon of visible light.
- In pair production, an energetic photon passing by a massive particle creates an electron-positron pair. In pair annihilation, an electron-positron pair is annihilated and two photons are created.

Conceptual Questions

1. Describe the photoelectric effect and four aspects of the experimental results that were puzzling to nineteenth-century physicists. How does the photon model of light explain the experimental results in each case?
2. Use the photon model to explain why ultraviolet radiation is harmful to your skin while visible light is not.
3. An experiment shines visible light on a target and measures the wavelengths of light scattered at different angles. Would the experiment show that the scattered photons are Compton-shifted? Explain.
4. Some stars are reddish in color, others bluish, and others yellowish-white (like the Sun). How is the color related to the surface temperature of the star? What color are the hottest stars? What color are the coolest?
5. How does the observation of the sharp lines seen in the hydrogen emission spectrum verify the notion that all electrons have the same charge?
6. In the photoelectric effect, what is the relationship between the maximum kinetic energy of ejected electrons and the frequency of the light incident on the surface?
7. Describe the process by which a continuous spectrum of x-rays is produced. Does the spectrum have a maximum wavelength or a minimum wavelength? Explain.
8. A darkroom used for developing black-and-white film can be dimly lit by a red lightbulb without ruining the film. Why is a red lightbulb used rather than white or blue or some other color?
9. List the assumptions of the Bohr theory of the hydrogen atom.
10. If green light causes the ejection of electrons from a metal in a photoelectric effect experiment and yellow light does not, what would you expect to happen if red light were used to illuminate the same metal? Do you expect more intense yellow light to eject electrons? What about very faint violet light?
11. In both Compton scattering and the photoelectric effect, an electron gains energy from an incident photon. What is the essential difference between the two processes?
12. Why is the Compton shift more noticeable for an incident x-ray photon than for a photon of visible light?
13. What process becomes especially important for photons with energies in excess of 1.02 MeV?
14. Explain how Rutherford's experiment, in which alpha particles are incident on a thin gold foil, refutes the plum pudding model of the atom.
15. In a photoelectric effect experiment, how is the stopping potential determined? What does the stopping potential tell us about the electrons emitted from the metal surface?
16. A fluorescent substance absorbs EM radiation of one wavelength and then emits EM radiation of a different wavelength. Which wavelength is longer? Explain.
17. Explain why every line in the absorption spectrum of hydrogen is present in the emission spectrum, but not every line in the emission spectrum is present in the absorption spectrum. [*Hint:* The excited states are very short-lived.]
18. A solar cell is used to generate electricity when sunlight falls on it. How would you expect the current produced by a solar cell to depend on the intensity of the incident light? How would you expect the current to depend on the wavelength of the incident light?
19. The photoresponse of the retina of the human eye at low light levels depends on individual photosensitive molecules in rod cells being excited by the incident light. When excited, these molecules change shape, leading to other changes in the cell that trigger a nerve impulse to the brain. How does the photon model of light do a better job than the wave model in explaining how these changes can happen even at low light levels?
20. Explain why the annihilation of an electron and a positron creates a *pair* of photons rather than a single photon.
21. In a photoelectric effect experiment, two different metals (1 and 2) are subjected to EM radiation. Metal 1 produces photoelectrons for both red and blue light; metal 2 produces photoelectrons for blue light but not for red. Which metal produces photoelectrons for ultraviolet radiation? Which *might* produce photoelectrons for infrared radiation? Which has the larger work function?
22. When a plot is made of x-ray intensity versus wavelength for a particular x-ray tube, two sharp peaks are superimposed on the continuous x-ray spectrum. These sharp peaks are called "characteristic" x-rays. Explain the origin of this name. In other words, of what are these x-rays characteristic?

Multiple-Choice Questions

1. How many emission lines are possible for atomic hydrogen gas with atoms excited to the $n = 4$ state?
   (a) 1 (b) 2 (c) 4 (d) 5 (e) 6
2. An electron, passing close to a target nucleus, slows and radiates away some of its energy. What is this process called?
   (a) Compton effect (b) photoelectric effect
   (c) bremsstrahlung (d) blackbody radiation
   (e) stimulated emission
3. In the Compton effect a photon of wavelength λ and frequency f is scattered from an electron, initially at rest. In this process,
   (a) the electron gains energy from the photon so that the scattered photon's wavelength is less than λ.
   (b) the electron gives energy to the scattered photon so that the photon's frequency is greater than f.
   (c) momentum is not conserved, but energy is conserved.
   (d) the photon loses energy so that the scattered photon has a frequency less than f.

4. The number of electrons per second ejected from a metal in the photoelectric effect
   (a) is proportional to the intensity of the incident light.
   (b) is proportional to the frequency of the incident light.
   (c) is proportional to the wavelength of the incident light.
   (d) is proportional to the threshold frequency of the metal.
5. Two lasers emit equal numbers of photons per second. If the first laser emits blue light and the second emits red light, the power radiated by the first is
   (a) greater than that emitted by the second.
   (b) less than that emitted by the second.
   (c) equal to that emitted by the second.
   (d) impossible to determine without knowing the time interval during which emission occurs.
6. If a photoelectric material has a work function ϕ, the threshold wavelength for the material is given by

   (a) $\frac{\phi}{hc}$ (b) hf (c) $\frac{hc}{\phi}$ (d) $\frac{\phi}{e}$ (e) $\frac{\phi}{hf}$
7. In analyzing data from a spectroscopic experiment, the inverse of each experimentally determined wavelength of the Balmer series is plotted versus $1/(n_i^2)$, where n_i is the initial energy level from which a transition to the $n = 2$ level takes place. The slope of the line is
   (a) the shortest wavelength of the Balmer series.
   (b) $-h$, where h is Planck's constant.
   (c) one divided by the longest wavelength in the Balmer series.
   (d) $-hc$, where h is Planck's constant.
   (e) $-R$, where R is the Rydberg constant.
8. Electrons are accelerated through a potential difference V and then strike a dense target. In the x-rays that are produced, there is
   (a) a maximum wavelength.
   (b) a minimum wavelength.
   (c) a single wavelength.
   (d) neither a maximum nor a minimum wavelength.
   (e) both a maximum and a minimum wavelength.
9. In a photoelectric effect experiment, light of a single wavelength is incident on the metal surface. As the intensity of the incident light is increased,
   (a) the stopping potential increases.
   (b) the stopping potential decreases.
   (c) the work function increases.
   (d) the work function decreases.
   (e) none of the above.
10. In a photoelectric effect experiment, the stopping potential is determined by
    (a) the work function of the metal.
    (b) the wavelength of the incident light.
    (c) the intensity of the incident light.
    (d) all three (a), (b), and (c). (e) (b) and (c).
    (f) (a) and (b). (g) (a) and (c).

Problems

Combination conceptual/quantitative problem
Biological or medical application
Challenging problem
Blue # Detailed solution in the Student Solutions Manual
1 2 Problems paired by concept
Text website interactive or tutorial

27.3 The Photoelectric Effect

1. A 200-W infrared laser emits photons with a wavelength of 2.0×10^{-6} m, and a 200-W ultraviolet light emits photons with a wavelength of 7.0×10^{-8} m. (a) Which has greater energy, a single infrared photon or a single ultraviolet photon? (b) What is the energy of a single infrared photon and the energy of a single ultraviolet photon? (c) How many photons of each kind are emitted per second?
2. What is the energy of a photon of light of wavelength 0.70 μm?
3. Find the (a) wavelength and (b) frequency of a 3.1-eV photon.
4. The photoelectric threshold frequency of silver is 1.04×10^{15} Hz. What is the minimum energy required to remove an electron from silver?
5. A rubidium surface has a work function of 2.16 eV. (a) What is the maximum kinetic energy of ejected electrons if the incident radiation is of wavelength 413 nm? (b) What is the threshold wavelength for this surface?
6. A clean iron surface is illuminated by ultraviolet light. No photoelectrons are ejected until the wavelength of the incident UV light falls below 288 nm. (a) What is the work function (in electron-volts) of the metal? (b) What is the maximum kinetic energy for electrons ejected by incident light of wavelength 140 nm?
7. The minimum energy required to remove an electron from a metal is 2.60 eV. What is the longest wavelength photon that can eject an electron from this metal?
8. Photons of wavelength 350 nm are incident on a metal plate in a photocell and electrons are ejected. A stopping potential of 1.10 V is able to just prevent any of the ejected electrons from reaching the opposite electrode. What is the maximum wavelength of photons that will eject electrons from this metal?
9. Ultraviolet light of wavelength 220 nm illuminates a tungsten surface and electrons are ejected. A stopping potential of 1.1 V is able to just prevent any of the ejected electrons from reaching the opposite electrode. What is the work function for tungsten?

10. Photons with a wavelength of 400 nm are incident on an unknown metal, and electrons are ejected from the metal. However, when photons with a wavelength of 700 nm are incident on the metal, no electrons are ejected. (a) Could this metal be cesium with a work function of 1.8 eV? (b) Could this metal be tungsten with a work function of 4.6 eV? (c) Calculate the maximum kinetic energy of the ejected electrons for each possible metal when 200-nm photons are incident on it.

11. Two different monochromatic light sources, one yellow (580 nm) and one violet (425 nm), are used in a photoelectric effect experiment. The metal surface has a photoelectric threshold frequency of 6.20×10^{14} Hz. (a) Are both sources able to eject photoelectrons from the metal? Explain. (b) How much energy is required to eject an electron from the metal? (Use $h = 4.136 \times 10^{-15}$ eV·s.)

12. (a) Light of wavelength 300 nm is incident on a metal that has a work function of 1.4 eV. What is the maximum speed of the emitted electrons? (b) Repeat part (a) for light of wavelength 800 nm incident on a metal that has a work function of 1.6 eV. (c) How would your answers to parts (a) and (b) vary if the light intensity were doubled?

13. Calculate the value of (a) Planck's constant and (b) the work function of the metal from the data obtained by Robert A. Millikan in 1916. Millikan was attempting to disprove Einstein's photoelectric equation; instead he found that his data supported Einstein's prediction.

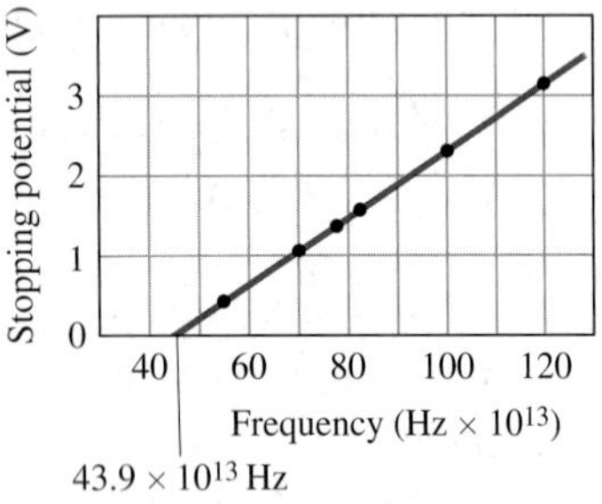

14. A 640-nm laser emits a 1-s pulse in a beam with a diameter of 1.5 mm. The rms electric field of the pulse is 120 V/m. How many photons are emitted per second? [*Hint:* Review Section 22.6.]

27.4 X-Ray Production

15. If the shortest wavelength produced by an x-ray tube is 0.46 nm, what is the voltage applied to the tube?

16. What is the minimum potential difference applied to an x-ray tube if x-rays of wavelength 0.250 nm are produced?

17. What is the cutoff frequency for an x-ray tube operating at 46 kV?

18. The potential difference in an x-ray tube is 40.0 kV. What is the minimum wavelength of the continuous x-ray spectrum emitted from the tube?

19. In a color TV tube, electrons are accelerated through a potential difference of 20.0 kV. Some of the electrons strike the metal mask (instead of the phosphor dots behind holes in the mask), causing x-rays to be emitted. What is the smallest wavelength of the x-rays emitted?

20. Show that the cutoff frequency for an x-ray tube is proportional to the potential difference through which the electrons are accelerated.

27.5 Compton Scattering

21. X-rays of wavelength 10.0 pm are incident on a target. Find the wavelengths of the x-rays scattered at (a) 45.0° and (b) 90.0°. (tutorial: Compton scattering)

22. An x-ray photon of wavelength 0.150 nm collides with an electron initially at rest. The scattered photon moves off at an angle of 80.0° from the direction of the incident photon. Find (a) the Compton shift in wavelength and (b) the wavelength of the scattered photon.

23. An incident beam of photons is scattered through 100.0°; the wavelength of the scattered photons is 124.65 pm. What is the wavelength of the incident photons?

24. A photon of wavelength 0.14800 nm, traveling due east, is scattered by an electron initially at rest. The wavelength of the scattered photon is 0.14900 nm and it moves at an angle θ north of east. (a) Find θ. (b) What is the south component of the electron's momentum?

25. What is the velocity of the scattered electron in Problem 24?

26. An x-ray photon of initial frequency 3.0×10^{19} Hz collides with a free electron at rest; the scattered photon moves off at 90°. What is the frequency of the scattered photon?

27. A photon is incident on an electron at rest. The scattered photon has a wavelength of 2.81 pm and moves at an angle of 29.5° with respect to the direction of the incident photon. (a) What is the wavelength of the incident photon? (b) What is the final kinetic energy of the electron?

28. A photon of energy 240.0 keV is scattered by a free electron. If the recoil electron has a kinetic energy of 190.0 keV, what is the wavelength of the scattered photon?

29. An incident photon of wavelength 0.0100 nm is Compton scattered; the scattered photon has a wavelength of 0.0124 nm. What is the change in kinetic energy of the electron that scattered the photon?

30. A Compton scattering experiment is performed using an aluminum target. The incident photons have wavelength λ. The scattered photons have wavelengths λ' and energies E that depend on the scattering angle θ. (a) At what angle θ are scattered photons with the smallest energy detected? (b) At this same scattering angle θ, what is the ratio λ'/λ for $\lambda = 10.0$ pm?

27.6 Spectroscopy and Early Models of the Atom; 27.7 The Bohr Model of the Hydrogen Atom; Atomic Energy Levels

31. What is the orbital radius of the electron in the $n = 3$ state of hydrogen?

32. Find the energy for a hydrogen atom in the stationary state $n = 4$.

33. (a) What is the difference in radius between the $n = 1$ state and the $n = 2$ state for hydrogen? (b) What is the difference in radius between the $n = 100$ state and the $n = 101$ state for hydrogen? How do the neighboring orbital separations compare for large and small n values?

34. Find the Bohr radius of doubly ionized lithium (Li^{2+}).

35. Find the energy in electron-volts required to remove the remaining electron from a doubly ionized lithium (Li^{2+}) atom.

36. How much energy must be supplied to a hydrogen atom to cause a transition from the ground state to the $n = 4$ state?

37. A hydrogen atom in its ground state absorbs a photon of energy 12.1 eV. To what energy level is the atom excited?

38. Use the Bohr theory to find the energy necessary to remove the electron from a hydrogen atom initially in its ground state.

39. How much energy is required to ionize a hydrogen atom initially in the $n = 2$ state?

40. What is the smallest energy photon that can be absorbed by a hydrogen atom in its ground state?

41. Find the wavelength of the radiation emitted when a hydrogen atom makes a transition from the $n = 6$ to the $n = 3$ state. (WW tutorial: hydrogen atom)

42. One line in the helium spectrum is bright yellow and has the wavelength 587.6 nm. What is the difference in energy (in electron-volts) between two helium levels that produce this line?

43. A hydrogen atom has an electron in the $n = 5$ level. (a) If the electron returns to the ground state by emitting radiation, what is the *minimum* number of photons that can be emitted? (b) What is the *maximum* number that might be emitted?

44. The *Paschen series* in the hydrogen emission spectrum is formed by electron transitions from $n_i > 3$ to $n_f = 3$. (a) What is the longest wavelength in the Paschen series? (b) What is the wavelength of the series limit (the lower bound of the wavelengths in the series)? (c) In what part or parts of the EM spectrum is the Paschen series found (IR, visible, UV, etc.)?

45. A fluorescent solid absorbs a photon of ultraviolet light of wavelength 320 nm. If the solid dissipates 0.500 eV of the energy and emits the rest in a single photon, what is the wavelength of the emitted light?

46. (a) Find the energies of the first four levels of doubly ionized lithium (Li^{2+}), starting with $n = 1$. (b) What are the energies of the photons emitted or absorbed when the electron makes a transition between these levels? (c) Are any of the photons in the visible part of the EM spectrum?

47. A photon with a wavelength in the visible region (between 400 and 700 nm) causes a transition from the n to the $(n + 1)$ state in doubly ionized lithium (Li^{2+}). What is the lowest value of n for which this could occur?

48. The Bohr theory of the hydrogen atom ignores gravitational forces between the electron and the proton. Make a calculation to justify this omission. [*Hint:* Find the ratio of the gravitational and electrostatic forces acting on the electron due to the proton.]

49. By directly substituting the values of the fundamental constants, show that the Bohr radius $a_0 = \hbar^2/(m_e ke^2)$ has the numerical value 5.29×10^{-11} m.

50. By directly substituting the values of the fundamental constants, show that the ground state energy for hydrogen in the Bohr model $E_1 = -m_e k^2 e^4/(2\hbar^2)$ has the numerical value −13.6 eV.

51. Calculate, according to the Bohr model, the speed of the electron in the ground state of the hydrogen atom.

52. A particle collides with a hydrogen atom in the $n = 2$ state, transferring 15.0 eV of energy to the atom. As a result, the electron breaks away from the hydrogen nucleus. What is the kinetic energy of the electron when it is far from the nucleus?

27.8 Pair Annihilation and Pair Production

53. What is the maximum wavelength of a photon that can create an electron-positron pair?

54. An electron-positron pair is created in a particle detector. If the tracks of the particles indicate that each one has a kinetic energy of 0.22 MeV, what is the energy of the photon that created the two particles?

55. A photon passes near a nucleus and creates an electron and a positron, each with a total energy of 8.0 MeV. What was the wavelength of the photon?

56. A positron and an electron that were at rest suddenly vanish and two photons of identical frequency appear. What is the wavelength of each of these photons?

57. A muon and an antimuon, each with a mass that is 207 times greater than an electron, were at rest when they annihilated and produced two photons of equal energy. What is the wavelength of each of the photons?

58. In gamma-ray astronomy, the existence of positrons (e^+) can be inferred by a characteristic gamma ray that is emitted when a positron and an electron (e^-) annihilate. For simplicity, assume that the electron and positron are at rest with respect to an Earth observer when they annihilate and that nothing else is in the vicinity. (a) Consider the reactions $e^- + e^+ \rightarrow \gamma$, where the annihilation

of the two particles at rest produces one photon, and $e^- + e^+ \rightarrow 2\gamma$, where the annihilation produces two photons. Explain why the first reaction does not occur, but the second does. (b) Suppose the reaction $e^- + e^+ \rightarrow 2\gamma$ occurs and one of the photons travels toward Earth. What is the energy of the photon?

Comprehensive Problems

59. A surgeon is attempting to correct a detached retina by using a pulsed laser. (a) If the pulses last for 20.0 ms and if the output power of the laser is 0.500 W, how much energy is in each pulse? (b) If the wavelength of the laser light is 643 nm, how many photons are present in each pulse?

60. These data are obtained for photoelectric stopping potentials using light of four different wavelengths. (a) Plot a graph of the stopping potential versus the reciprocal of the wavelength. (b) Read the values of the work function and threshold wavelength for the metal used directly from the graph. (c) What is the slope of the graph? Compare the slope with the expected value (calculated from fundamental constants).

Color	Wavelength (nm)	Stopping Potential (V)
Yellow	578	0.40
Green	546	0.60
Blue	436	1.10
Ultraviolet	366	1.60

61. An FM radio station broadcasts at a frequency of 89.3 MHz. The power radiated from the antenna is 50.0 kW. (a) What is the energy in electron-volts of each photon radiated by the antenna? (b) How many photons per second does the antenna emit?

62. An owl has good night vision because its eyes can detect a light intensity as faint as 5.0×10^{-13} W/m^2. What is the minimum number of photons per second that an owl eye can detect if its pupil has a diameter of 8.5 mm and the light has a wavelength of 510 nm?

63. What is the shortest wavelength x-ray produced by a 0.20-MV x-ray machine?

64. How much energy is required to remove an electron from a hydrogen atom in the $n = 4$ state?

65. The output power of a laser pointer is about 1 mW. (a) What are the energy and momentum of one laser photon if the laser wavelength is 670 nm? (b) How many photons per second are emitted by the laser? (c) What is the average force on the laser due to the momentum carried away by these photons?

66. In a photoelectric experiment using sodium, when incident light of wavelength 570 nm and intensity 1.0 W/m^2 is used, the measured stopping potential is 0.28 V. (a) What would the stopping potential be for incident light of wavelength 400.0 nm and intensity 1.0 W/m^2? (b) What would the stopping potential be for incident light of wavelength 570 nm and intensity 2.0 W/m^2? (c) What is the work function of sodium?

67. A 100-W lightbulb radiates *visible* light at a rate of about 10 W; the rest of the EM radiation is mostly infrared. Assume that the lightbulb radiates uniformly in all directions. Under ideal conditions, the eye can see the lightbulb if at least 20 visible photons per second enter a dark-adapted eye with a pupil diameter of 7 mm. (a) Estimate how far from the source the lightbulb can be seen under these rather extreme conditions. Assume an average wavelength of 600 nm. (b) Why do we not normally see lightbulbs at anywhere near this distance?

68. A thin aluminum target is illuminated with photons of wavelength λ. A detector is placed at 90.0° to the direction of the incident photons. The scattered photons detected are found to have half the energy of the incident photons. (a) Find λ. (b) What is the wavelength of backscattered photons (detector at 180°)? (c) What (if anything) would change if a copper target were used instead of an aluminum one?

69. What potential difference must be applied to an x-ray tube to produce x-rays with a minimum wavelength of 45.0 pm?

70. What happens to the energies of the characteristic x-rays when the potential difference accelerating the electrons in an x-ray tube is doubled?

71. Nuclei in a radium-226 radioactive source emit photons whose energy is 186 keV. These photons are scattered by the electrons in a metal target; a detector measures the energy of the scattered photons as a function of the angle θ through which they are scattered. Find the energy of the γ-rays scattered through $\theta = 90.0°$ and 180.0°.

72. What is the ground state energy, according to Bohr theory, for (a) He^+, (b) Li^{2+}, (c) deuterium (an isotope of hydrogen whose nucleus contains a neutron as well as a proton)?

73. Follow the steps outlined in this problem to estimate the time lag (predicted classically but *not* observed experimentally) in the photoelectric effect. Let the intensity of the incident radiation be 0.01 W/m^2. (a) If the area of the atom is (0.1 nm)^2, find the energy per second falling on the atom. (b) If the work function is 2.0 eV, how long would it take (classically) for enough energy to fall on this area to liberate one photoelectron? (c) Explain briefly, using the photon model, why this time lag is not observed.

74. The photoelectric effect is studied using a tungsten target. The work function of tungsten is 4.5 eV. The incident photons have energy 4.8 eV. (a) What is the threshold frequency? (b) What is the stopping potential? (c) Explain why, in classical physics, no threshold frequency is expected.

75. An x-ray photon with wavelength 6.00 pm collides with a free electron initially at rest. What is the maximum possible kinetic energy acquired by the electron?

76. A photoelectric effect experiment is performed with tungsten. The work function for tungsten is 4.5 eV. (a) If ultraviolet light of wavelength 0.20 μm is incident on the tungsten, calculate the stopping potential. (b) If the stopping potential is turned off (i.e., the cathode and anode are at the same voltage), the 0.20-μm incident light produces a photocurrent of 3.7 μA. What is the photocurrent if the incident light has wavelength 400 nm and the same intensity as before?

77. In a CRT television, electrons of kinetic energy 2.0 keV strike the screen. No EM radiation is emitted below a certain wavelength. Calculate this wavelength.

78. The *Lyman series* in the hydrogen emission spectrum is formed by electron transitions from an excited state to the ground state. Calculate the longest three wavelengths in the Lyman series.

79. Consider the emission spectrum of singly ionized helium (He^+). Find the longest three wavelengths for the series in which the electron makes a transition from a higher excited state to the first excited state (*not* the ground state).

✦80. During a Compton scattering experiment, an electron that was initially at rest recoils in the direction of motion of the incident x-ray photon. If the recoil electron has a kinetic energy of 0.20 keV, what is the wavelength of the incident x-ray? What is the wavelength of the scattered x-ray?

✦81. Compare the orbital radii of the He^+ and H atoms for levels of *equal energy* (not the same value of *n*). Can you draw a general conclusion from your results?

✦82. A hydrogen atom in its ground state is immersed in a continuous spectrum of ultraviolet light with wavelengths ranging from 96 nm to 110 nm. After absorbing a photon, the atom emits one or more photons to return to the ground state. (a) What wavelength(s) can be absorbed by the H atom? (b) For each of the possibilities in (a), if the atom is at rest before absorbing the UV photon, what is its recoil speed after absorption (but before emitting any photons)? (c) For each of the possibilities in (a), how many different ways are there for the atom to return to the ground state?

✦83. A hydrogen atom in its ground state absorbs a 97-nm ultraviolet photon. It then emits one or more photons to return to the ground state. (a) If the atom is at rest before absorbing the UV photon, what is its recoil speed after absorption? (b) There are several different possible ways for the atom to return to the ground state. How many? (c) For each of the possibilities in (b), give the wavelength of each photon emitted, and classify it as visible, UV, IR, x-ray, etc.

✦84. Photons of energy $E = 4.000$ keV undergo Compton scattering. What is the largest possible change in photon energy, measured as a fraction of the incident photon's energy $(E - E')/E$?

✦85. Suppose that you have a glass tube filled with atomic hydrogen gas (H, not H_2). Assume that the atoms start out in their ground states. You illuminate the gas with monochromatic light of various wavelengths, ranging through the entire IR, visible, and UV parts of the spectrum. At some wavelengths, visible light is emitted from the H atoms. (a) If there are two and only two visible wavelengths in the emitted light, what is the wavelength of the incident radiation? (b) What is the largest wavelength of incident radiation that causes the H atoms to emit visible light? What wavelength(s) is/are emitted for incident radiation at that wavelength? (c) For what wavelengths of incident light are hydrogen ions (H^+) formed?

Answers to Practice Problems

27.1 4.2×10^{-19} J

27.2 8.30×10^{29} photons per second

27.3 385 nm ($K_{max} = 0.82$ eV)

27.4 10.0 kV

27.5 3.71 pm

27.6 397 nm—difficult to see for most people

27.7 At room temperature, the atoms are almost all in the ground state. The absorption spectrum shows only transitions that start from the ground state—the Lyman series; all of them are in the ultraviolet. At high temperatures, some of the atoms are excited into the $n = 2$ energy level by collisions. These atoms can absorb photons in the Balmer series, causing a transition from $n = 2$ to a higher energy level.

27.8 (a) 13.6 eV; (b) 54.4 eV; (c) In He^+, the electron is more tightly bound since the nucleus has twice the charge.

27.9 5.85 fm

Answers to Checkpoints

27.2 At full power, the filament is hot enough to emit EM radiation across the entire visible spectrum (plus even more infrared radiation), so the light looks white. At lower power, the filament temperature is lower. As a result, the peak of the emitted EM radiation is shifted toward lower frequencies, which increases the relative amount of red light in the mixture relative to other colors.

27.3 The energy of a photon is proportional to its frequency. At the threshold frequency, a photon has just enough energy to liberate an electron from the metal. Below the threshold frequency, a photon has insufficient energy to liberate an electron.

27.5 The collision conserves both momentum and energy. The electron initially has zero kinetic energy. The electron recoils, moving off with some kinetic energy. Therefore, the scattered photon must have less energy (a longer wavelength) than the incident photon.

27.7 $E = E_i - E_f = (-0.54 \text{ eV}) - (-3.40 \text{ eV}) = 2.86$ eV

CHAPTER

28 Quantum Physics

Biologists and medical researchers commonly use electron microscopes instead of light microscopes when very fine detail is desired. What enables an electron microscope to achieve a greater resolution than a light microscope? Are there any limits to the resolution of an electron microscope? (See p. 1036 for the answer.)

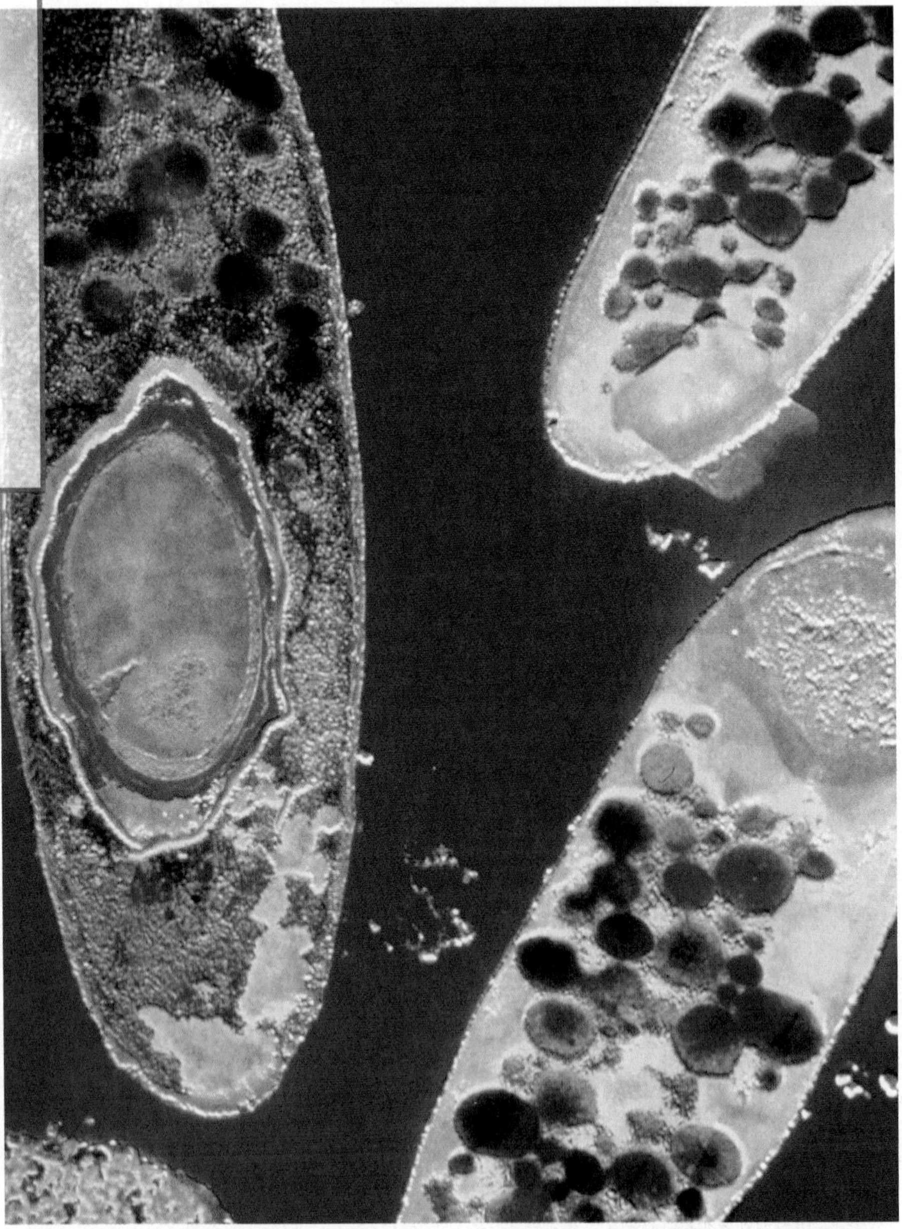

A colorized transmission electron micrograph of *Clostridium butyricum* bacteria, magnified approximately 50 000 times.

Concepts & Skills to Review

- quantization (Section 27.1)
- the photon (Section 27.3)
- double-slit interference experiment (Section 25.4)
- diffraction and the resolution of optical instruments (Section 25.8)
- intensity of an EM wave (Section 22.6)
- x-ray diffraction (Section 25.9)
- atomic energy levels and the Bohr model (Section 27.7)
- calculating wavelengths and frequencies of standing waves (Section 11.10)

28.1 THE WAVE-PARTICLE DUALITY

Classical physics maintains a sharp distinction between particles and waves; quantum physics blurs the distinction. Interference and diffraction experiments (Chapter 25) demonstrate that light propagates as a wave. On the other hand, in the photoelectric effect, Compton effect, and pair production and annihilation (Chapter 27), EM radiation interacts with matter as if it is composed of *particles* called photons. In quantum physics the two descriptions, particle and wave, are complementary. In some circumstances, light behaves more like a wave and less like a particle; in other circumstances, more like a particle and less like a wave.

Double-Slit Interference Experiment

Imagine a double-slit interference experiment in which the screen is replaced by a set of photomultipliers—devices that can count individual photons. Each photomultiplier records the number of photons that arrive during a set time interval. Since the intensity is proportional to the number of photons counted, a graph of the number of photons as a function of position along the "screen" looks just like a graph of the intensity pattern that would be recorded by photographic film. The photomultiplier records alternating maxima and minima with smooth transitions between them (Fig. 28.1).

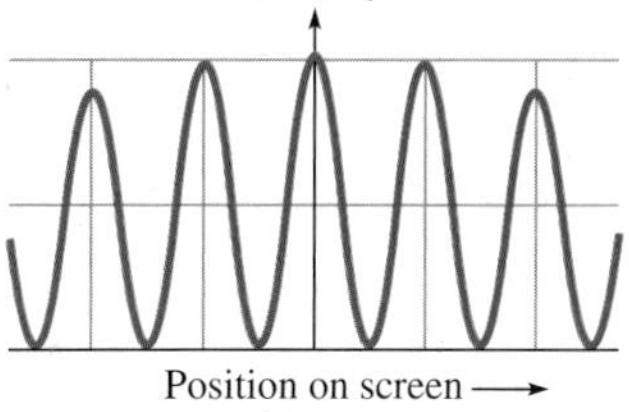

Figure 28.1 Double-slit interference pattern: the intensity as a function of position on the screen. Compare with Fig. 25.18.

Now suppose the intensity of the incident light is reduced until only *one photon at a time* leaves the source. In the wave picture, the interference pattern arises from the superposition of EM waves from each of the slits. When only one photon at a time leaves the source, will there be an interference pattern? Common sense suggests that each photon reaching the detector must have gone through either one slit or the other, but not both.

What are the results of this experiment? At first, photons seem to appear at random places (Fig. 28.2a); there is no way to predict where the next photon will be detected. As the experiment continues, the photons are clearly more numerous in some places than in others (Fig. 28.2b). We still cannot predict where the next photon will land, but the *probability* of detecting a photon is higher in some places than in others. If the experiment is allowed to run for a long time, the photons form distinct interference fringes (Fig. 28.2c). After a very long time, the intensity pattern is exactly that of Fig. 28.1—the double-slit interference pattern—*even though only one photon at a time passes through the slits.* Nevertheless, even after a clear interference pattern forms, we still cannot predict where the *next* photon will be detected.

CONNECTION:

For large numbers of photons, quantum physics predicts the same double-slit interference pattern as classical wave theory.

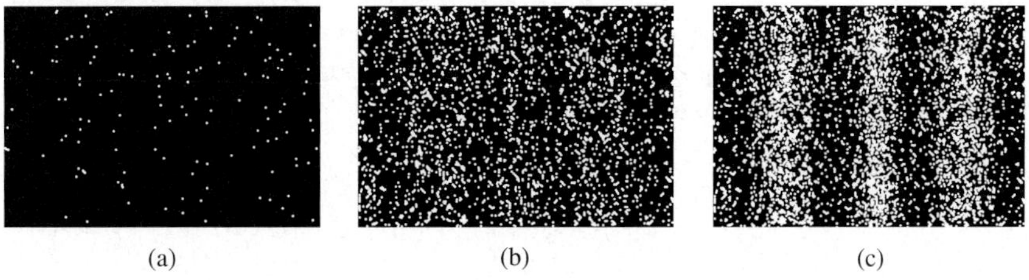

Figure 28.2 A double-slit experiment in which only one photon at a time passes through the slits. The experiment replicates the usual double-slit interference pattern once a large number of photons are recorded.

If this wave-particle duality seems strange, rest assured that even the greatest physicists have felt the same way. Niels Bohr said: "Anyone who has not been shocked by quantum mechanics has not understood it." Common sense is formed from observations in which quantum effects are not noticeable. While studying quantum mechanics, don't be discouraged when it seems confusing; quantum mechanics never seems obvious to anyone, but that's partly what makes it fascinating. The U.S. physicist Richard P. Feynman (1918–1988) put it this way: "I am going to tell you what nature behaves like. If you will simply admit that maybe she does behave like this, you will find her a delightful, entrancing thing."

Probability

In the double-slit experiment, we can never predict where any one photon will end up, but we can calculate the *probability* that it will fall in a given location. Two photons that are initially *identical* can end up at different places on the screen. The intensity pattern calculated by treating light as a wave is a statistical average that assumes a large number of photons.

The intensity of an EM wave is the energy flow per unit time per unit cross-sectional area:

$$I = \frac{\text{energy}}{\text{time} \cdot \text{area}}$$

In the wave picture, the intensity is proportional to the square of the electric field amplitude:

$$I \propto E^2$$

In the photon picture, each photon carries a definite quantity of energy, so

$$I = \frac{\text{number of photons}}{\text{time} \cdot \text{area}} \times \text{energy of one photon}$$

The number of photons that cross a given area is proportional to the *probability* that a photon crosses the area:

$$I \propto \frac{\text{number of photons}}{\text{time} \cdot \text{area}} \propto \frac{\text{probability of finding a photon}}{\text{time} \cdot \text{area}}$$

Probability ∝ square of the wave function

Therefore, the probability of finding a photon is proportional to the square of the electric field amplitude. The electric field as a function of position and time can be regarded as the *wave function*—the mathematical function that describes the wave—so the probability of finding a photon in some region of space is proportional to the square of the wave function in that region.

28.2 MATTER WAVES

In 1923, French physicist Louis de Broglie (1892–1987; his name is pronounced roughly *lwee duh broy*) suggested that this wave-particle duality may pertain to particles such as electrons and protons as well as to light. If light, which was so successfully established as a wave by Maxwell, could also have particle properties, why couldn't an electron have wave properties? But what would the wavelength of an electron be? De Broglie proposed that the relationship between the momentum and wavelength of any particle is the same as that for a photon [Eq. (27-11)]. It was not long before overwhelming experimental evidence confirmed de Broglie's hypothesis of the wave nature of electrons and other particles. The wavelength of the matter wave describing the behavior of a particle is now called its **de Broglie wavelength.**

CONNECTION:

The relationship between λ and p is the same for photons, electrons, neutrons, or any other particle.

de Broglie wavelength:

$$\lambda = \frac{h}{p} \qquad (28\text{-}1)$$

Electron Diffraction

How can wave characteristics of particles such as electrons be observed? The hallmarks of a wave are interference and diffraction. In 1925, the American physicists Clinton Davisson (1881–1958) and Lester H. Germer (1896–1971) directed a low-energy electron beam toward a crystalline nickel target and observed the number of electrons scattered as a function of the scattering angle ϕ (Fig. 28.3). The maximum number of electrons was detected at a scattering angle $\phi = 50°$. What could make the number of scattered electrons maximum at one particular angle? Could the maximum be due to interference or diffraction? If so, then electrons must have wavelike properties.

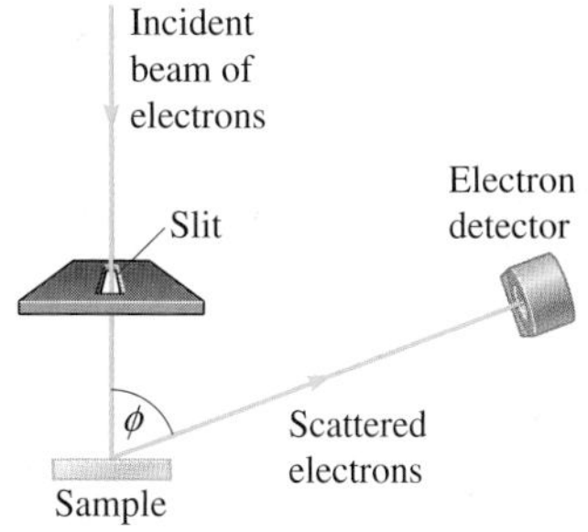

Figure 28.3 The Davisson-Germer experimental setup.

Later analysis showed that the maximum occurred at the angle predicted by Bragg's law for x-ray diffraction [Eq. (25-15)] if the wavelength of the electrons is given by de Broglie's relation (see Problem 73). The scattered electrons interfere just as scattered x-rays interfere, giving a maximum intensity at angles where the path difference is an integral number of wavelengths.

Davisson and Germer observed a broad maximum. The low-energy electrons they used did not penetrate very far into the crystal, so the electrons were scattered from a relatively small number of planes. Just as a large number of slits in a grating makes the maxima narrow, if the electrons scatter from all of the planes in the crystal, the electron diffraction maxima become sharp. In 1927, the British physicist George Paget Thomson (1892–1975) performed an electron diffraction experiment using higher-energy electrons. Instead of a single crystal, his sample was polycrystalline—many small crystals with random orientations. In x-ray diffraction, a polycrystalline sample produces maxima in a series of bright concentric rings due to constructive interference. Thomson saw a ring pattern for electron diffraction that was identical to an x-ray diffraction pattern when the x-rays had the same wavelength as the electrons (Fig. 28.4). These experiments showed that de Broglie's hypothesis was correct; electrons with a wavelength $\lambda = h/p$ diffract just as do x-rays of the same wavelength.

An interesting historical aside: J. J. Thomson is credited with the discovery of the electron in the late 1890s due to his measurement of the electron's charge-to-mass ratio. His son, G. P. Thomson, performed groundbreaking experiments in electron diffraction. The experiments of the father showed that electrons are *particles*; those of his son demonstrated the *wave* nature of electrons.

CHECKPOINT 28.2

When an electron is accelerated to a higher speed, what happens to its de Broglie wavelength?

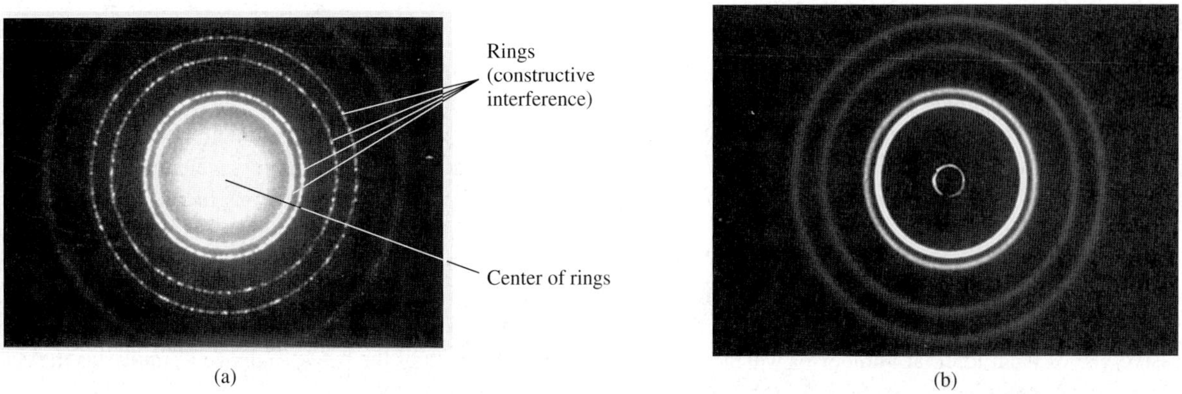

Figure 28.4 (a) Electron diffraction pattern from a polycrystalline aluminum sample. Before reaching the sample, the electrons move toward the center of the pattern on the screen. Each ring is formed by constructive interference of electrons scattered at a particular angle. (b) X-ray diffraction pattern from the same sample. Since the electron energy in (a) was chosen so that the de Broglie wavelength of the electrons was the same as the wavelength of the x-rays in (b), the bright fringes occur at the same angles.

Example 28.1

Electron Diffraction Experiment

An electron diffraction experiment is performed using electrons that have been accelerated through a potential difference of 8.0 kV. (a) Find the de Broglie wavelength of the electrons. (b) Find the wavelength and energy of x-ray photons that would give the same diffraction pattern on the same sample.

Strategy The relationship between wavelength and momentum is the same for both electrons and photons, but the relationship between wavelength and energy is *not* the same. The Bragg condition [Eq. (25-15)] for diffraction maxima in x-ray diffraction requires the path difference between x-rays reflecting off adjacent planes to be an integral multiple of the wavelength. The conditions for interference and diffraction maxima and minima always relate path differences to wavelengths. So to give the same diffraction pattern, the x-rays must have the same wavelength as the electrons. We expect the energy of the x-ray photons to be different from the kinetic energy of the electrons—the relationship between momentum and energy is not the same for a photon as for a particle with mass.

Solution (a) If electrons are accelerated through a potential difference of magnitude 8.0 kV, they have a kinetic energy of 8.0 keV. We need the kinetic energy in SI units to find the momentum in SI units:

$$K = 8000 \text{ eV} \times 1.6 \times 10^{-19} \text{ J/eV} = 1.28 \times 10^{-15} \text{ J}$$

The electron's kinetic energy (8.0 keV) is small compared with its rest energy (511 keV), so the electron is nonrelativistic—we can use $p = mv$ and $K = \frac{1}{2}mv^2$. Solving for p in terms of K by eliminating the speed v, the momentum is

$$p = \sqrt{2mK} = \sqrt{2 \times 9.11 \times 10^{-31} \text{ kg} \times 1.28 \times 10^{-15} \text{ J}}$$
$$= 4.83 \times 10^{-23} \text{ kg·m/s}$$

The wavelength is then

$$\lambda = \frac{h}{p} = \frac{6.626 \times 10^{-34} \text{ J·s}}{4.83 \times 10^{-23} \text{ kg·m/s}} = 1.372 \times 10^{-11} \text{ m} = 13.7 \text{ pm}$$

(b) The x-rays would need to have the same wavelength, 13.7 pm. The energy of a photon with this wavelength is

$$E = hf = \frac{hc}{\lambda} = \frac{1.24 \text{ keV·nm}}{0.01372 \text{ nm}} = 90.4 \text{ keV}$$

Discussion An alternative solution to part (a) does not require conversion to SI units. Multiplying both sides of $p = \sqrt{2mK}$ by c yields $pc = \sqrt{2mc^2K}$. For an electron, $mc^2 = 511$ keV. Then

$$\lambda = \frac{h}{p} = \frac{hc}{pc} = \frac{hc}{\sqrt{2mc^2K}} = \frac{1.24 \text{ keV} \cdot \text{nm}}{\sqrt{2 \times 511 \text{ keV} \times 8.0 \text{ keV}}}$$
$$= 0.0137 \text{ nm} = 13.7 \text{ pm}$$

Practice Problem 28.1 A Neutron's de Broglie Wavelength

Find the kinetic energy of a neutron with the same de Broglie wavelength as a 22 keV photon.

Conceptual Example 28.2

Size of Diffraction Pattern and Electron Energy

An electron diffraction experiment is performed on a polycrystalline aluminum sample. The electrons produce a ring pattern as in Fig. 28.4a. If the accelerating potential of the electrons is increased, what happens to the radius of the rings? See Fig. 28.5, which shows the formation of one of the rings.

Strategy A ring is formed by constructive interference; the path difference between electrons reflecting from two successive planes is an integral number of wavelengths. As the accelerating potential is increased, the wavelength changes. Then we determine how ϕ must change to keep the extra path length equal to a fixed number of wavelengths.

Solution and Discussion A larger accelerating potential gives the electrons a larger kinetic energy and a larger momentum. With a larger momentum, the de Broglie wavelength is smaller. For a smaller wavelength, it takes a smaller path difference to produce constructive interference since the path difference must remain equal to a fixed integer times a smaller wavelength. From Fig. 28.5b, a smaller path difference is produced by a smaller ϕ; from Fig. 28.5a, a smaller ϕ makes the radius of the ring smaller. Thus, the radius of each of the bright rings gets smaller as the electron energy is increased.

continued on next page

Conceptual Example 28.2 continued

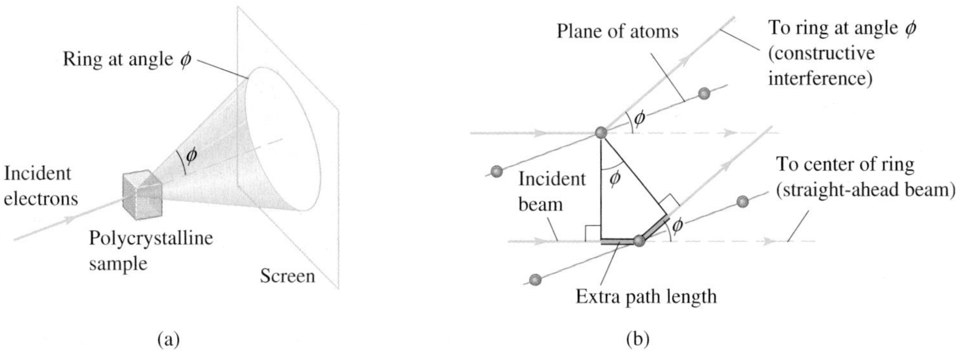

Figure 28.5
(a) One ring in the diffraction pattern is formed by electrons scattered at an angle ϕ from the incident beam. (b) Rays of electrons reflected from two successive planes of atoms, showing the path length difference.

Conceptual Practice Problem 28.2 Double-Slit Pattern

In a double-slit experiment using a beam of monoenergetic electrons (electrons that all have the same kinetic energy) instead of light, the same interference pattern is obtained as for light. The interference maxima are found at angles satisfying $d \sin \theta = m\lambda$ [Eq. (25-10)], where d is the slit separation and λ is the de Broglie wavelength of the electron beam. What happens to the interference pattern as the accelerating potential is increased?

Later, *neutron* diffraction experiments were performed on crystals; again, the results confirmed de Broglie's hypothesis that $\lambda = h/p$. Today, x-ray, electron, and neutron diffraction are commonly used tools for probing microscopic structures. There are some differences among them. Electrons do not penetrate as well as x-rays, so electrons are better for studying microscopic structures of surfaces. X-rays primarily interact with the atomic electrons. If a sample is made primarily of lighter elements, which have few electrons, x-ray diffraction studies are not as effective. In these cases, neutron diffraction is often used. Neutrons interact with the nuclei in the sample; since they are electrically neutral they hardly interact with electrons at all. Neutron diffraction is especially useful in determining the position of hydrogen atoms within the structure of a protein or other biological macromolecule.

In recent years, interference and diffraction experiments have been performed using beams of atoms or molecules. Even beams of "buckyballs," molecules composed of 60 tightly bound carbon atoms and shaped like a soccer ball, have been shown to interfere according to quantum theory.

Matter Waves and Probability

Consider a double-slit interference experiment using an *electron beam* rather than light. The interference pattern emerges even if we send only one electron at a time toward the slits. Each electron hits the screen as a localized particle and makes a small spot, just as a photon does. After many electrons have hit the screen, the interference pattern becomes evident—just as for photons (see Fig. 28.2). The interference of the matter waves emerging from the two slits determines the probability that an electron lands at a particular spot on the screen. Where the matter wave interferes constructively, the probability is high; where it interferes destructively, the probability is low.

CONNECTION:

A double-slit experiment with *electrons* produces an interference pattern like the one for light.

The interference pattern is evidence that the electron wave propagates through *both slits.* Suppose we add a detector to record which slit each electron passes through. Such a detector always finds that an electron goes through *one slit or the other* but never both. However, when this detector is in place, the interference pattern *disappears*!

28.3 ELECTRON MICROSCOPES

The resolution of a conventional light microscope is limited by diffraction (see Section 25.8). Under ideal conditions, the smallest distance on the object that can be resolved (distinguished in the image formed by the microscope) is roughly half the wavelength of the light. Using 400 nm as the shortest wavelength in the visible part of the spectrum, a light microscope can resolve distances of about 200 nm. That's a large distance on the scale of atoms and molecules; the distance between atoms in a solid is typically only about 0.2 nm.

To get better resolution, one possibility is to use an ultraviolet microscope. These microscopes use wavelengths down to about 200 nm. For wavelengths shorter than that, making effective lenses becomes too difficult.

Are there limits to the resolution of an electron microscope?

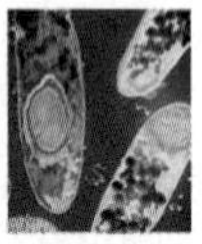

A beam of electrons can *easily* be made to have a wavelength around 0.2 nm or smaller. To make electrons with a wavelength of 0.2 nm, we would need to accelerate them through a potential difference of only 37.4 V. Typically the electrons used in an electron microscope are more energetic than that, and so have shorter wavelengths. However, the resolution of an electron microscope is also limited by lens aberrations—imperfections in the electromagnetic "lenses" used to focus the electron beam and form the image.

The workings of an electron microscope can be explained *without* talking explicitly about the wave nature of the electrons. We described light microscopes using geometric optics by tracing light rays. Similarly, we can follow the trajectories of electrons as they are bent by magnetic lenses and scattered by the sample being studied. The advantage of the electron microscope over the light microscope is the smaller wavelength of the electrons, which extends "geometric electron optics" to much smaller objects. A disadvantage is that the electron microscope requires a vacuum.

Transmission Electron Microscope Electron microscopes come in several forms. The one closest to the familiar light microscope is called the transmission electron microscope, or TEM (Fig. 28.6a,b). When a beam of parallel electrons passes through the sample, electrons that are scattered by a point within the sample are focused back to a point on a screen by magnetic lenses, forming a real image of the sample on the screen. The electrons must pass through the sample without being slowed down appreciably, so the TEM works only for thin samples—about 100 nm of thickness maximum. The TEM can resolve details as small as 0.2 nm—about 500 times better than an ultraviolet microscope with wavelength 200 nm.

Scanning Electron Microscope Another kind of electron microscope, the scanning electron microscope (SEM), uses a magnetic lens to focus a beam of electrons onto one point on the sample at a time (Fig. 28.6c,d). These primary electrons knock *secondary* electrons out of the sample; an electron collector detects the number of secondary electrons produced. The primary electron beam is swept across the sample by a beam deflector. The number of secondary electrons emitted at each spot on the sample is measured and fed to a computer that constructs an image of the specimen. The resolution of the SEM is not as good as the TEM—about 10 nm at best. But the SEM doesn't require thin samples and, since it is sensitive to the surface contour of the specimen, it is much better at imaging three-dimensional structures.

Other Electron Microscopes The scanning transmission electron microscope (STEM) scans the sample point by point, like the SEM, but it detects the electrons transmitted through the sample. Another kind of electron microscope, the scanning tunneling microscope (STM), is discussed in Section 28.10.

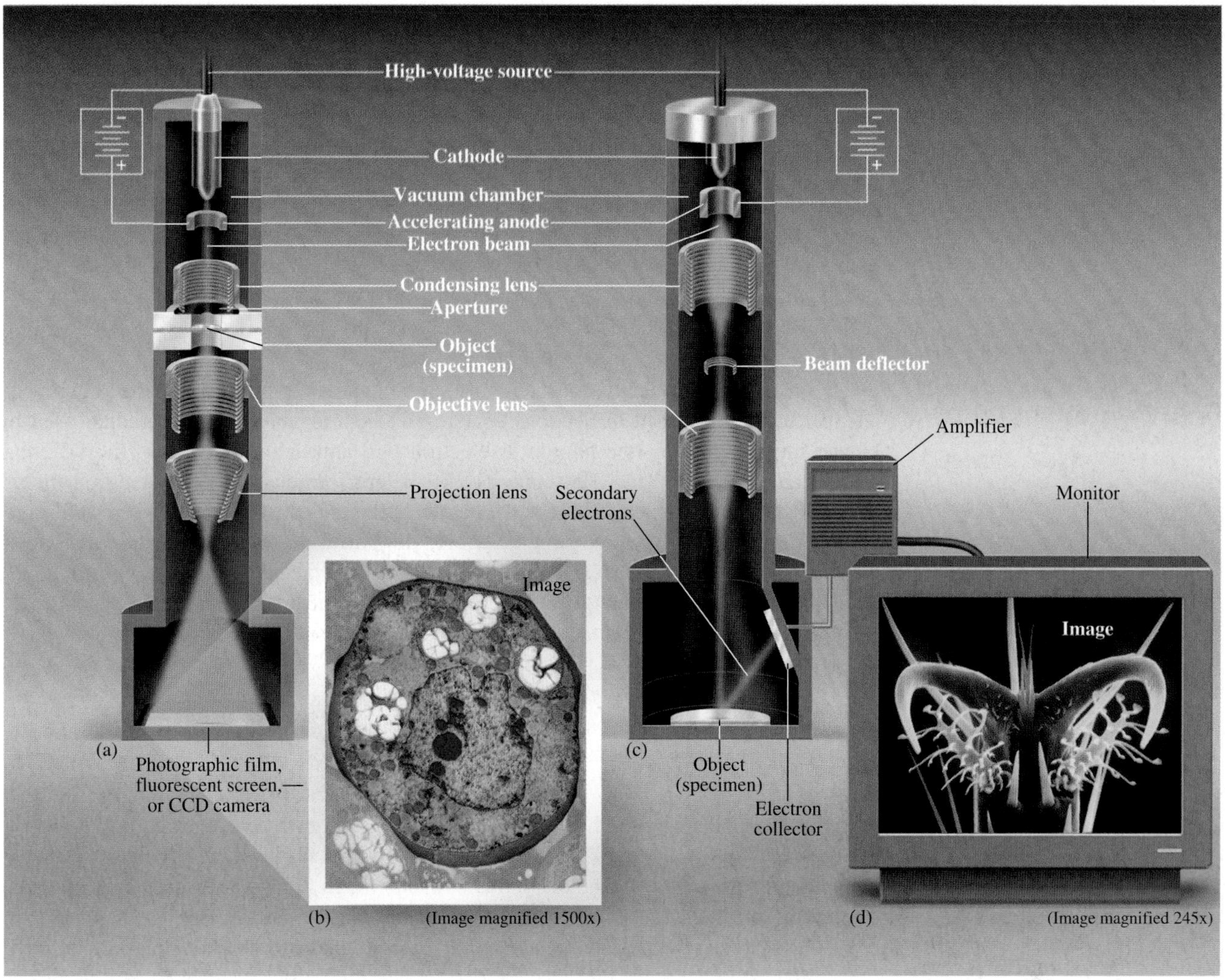

Figure 28.6 Two types of electron microscope. In both types, electrons emitted from a heated filament are accelerated by the electric field between the cathode and the anode. (a) In a TEM, a condensing lens forms a parallel beam and an aperture restricts its diameter. After the beam passes through the specimen, the objective lens forms a real image. One or more projection lenses magnify the image and project it onto film, a fluorescent screen, or a CCD (charge-coupled device) camera (similar to a video camera). (b) Colorized transmission electron micrograph of a parenchyma cell from the voodoo lily plant (*Sauromatum guttatum*). The image shows the nucleus (light green), DNA (blue), mitochondria (red), cell wall (dark green), and starch grains (light yellow). (c) In an SEM, the condensing lens forms a narrow beam. The beam deflector is a series of coils that sweep the beam across the sample. The objective lens focuses the electron beam into a small spot on the specimen. Secondary electrons knocked out of the specimen at that spot are detected by the electron collector and the electrical signal is fed to a monitor or computer. (d) Colorized scanning electron micrograph of the fruit fly claw and pulvillar pad.

28.4 THE UNCERTAINTY PRINCIPLE

In the latter part of the nineteenth century, Newtonian mechanics, Maxwell's equations of electromagnetism, and thermodynamics were thought to be so highly developed—and so well confirmed by experiment—that some scientists thought no new basic laws were left to discover. Some people even became complete determinists. Their rationale was that the state of the universe at one instant of time (the position and velocity of every particle) determined the state at all later times. In principle, the future position and velocity of every particle could be calculated using Newton's laws.

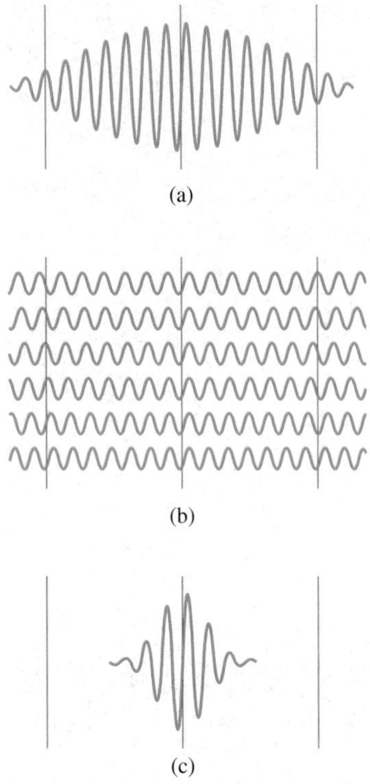

Figure 28.7 (a) A wave packet representing a localized particle. The uncertainty in the particle's position is the width of the wave packet. (b) These six waves have slightly different wavelengths; they are all in phase at the center. Moving away from the center, phase differences accumulate due to the differing wavelengths. The sum of these six waves is the wave packet in (a). (Adding just these six actually produces a recurring packet like a beat pattern. To get a true localized wave packet that does not repeat, we need to add an infinite number of waves over a small range of wavelengths.) (c) A larger range of wavelengths—around the same *average* wavelength—is needed to form a narrower wave packet like this one. A particle with a smaller uncertainty in its position is represented as a wave packet with a larger range of wavelengths and therefore a larger uncertainty in its momentum.

Quantum mechanics does not allow complete determinism. In the double-slit experiments described in Sections 28.1 and 28.2, it is impossible to predict, even in principle, where any one photon or electron will appear on the screen. In 1927, the German physicist Werner Heisenberg (1901–1976) formulated the **Heisenberg uncertainty principle**, which describes the nature of this indeterminacy. Suppose we design an experiment to determine simultaneously the position and momentum of a particle. The uncertainty principle says there are limits to how precisely they can be simultaneously measured, even in an ideal experiment. If Δx is the uncertainty in the x-coordinate of position and Δp_x is the uncertainty in the x-component of the momentum, then

Position-momentum uncertainty principle:

$$\Delta x\,\Delta p_x \geq \frac{1}{2}\hbar \qquad (28\text{-}2)$$

Rigorous application of the uncertainty principle requires precise definitions of the uncertainty in x and in p_x. Those definitions are beyond the level of this text. Instead, we apply the uncertainty principle only to make rough, order-of-magnitude estimates, which means we can get by with rough estimates of the uncertainties.

Why should the precise determination of position and momentum be incompatible? It is a result of the wave-particle duality. In quantum physics a localized particle is represented as a *wave packet*—a wave with a finite extent in space (Fig. 28.7a). The momentum of a particle is related to the wavelength. To make a localized wave packet, we need to add waves *with different wavelengths* (Fig. 28.7b). These waves cancel each other everywhere except in the wave packet. The shorter the length of the wave packet, the larger the range of wavelengths that must go into the mix (Fig. 28.7c). Equivalently, the smaller the uncertainty in the particle's position, the larger the uncertainty in the momentum. The superposition of waves with a smaller range of wavelengths produces a longer wave packet—since the wavelengths are close together, they stay in phase with each other over a longer distance. Therefore, the smaller the uncertainty in momentum, the larger is the uncertainty in the position.

In Newtonian mechanics, the forces acting on a particle determine the object's motion. There is no fundamental limit to how precisely a particle's trajectory can be calculated or measured. By contrast, the uncertainty principle places a fundamental limit on the precision with which the position and momentum can simultaneously be known. The more precisely we know the position of a particle at time t, the less precisely its momentum at the same instant can be known. Uncertainty in the momentum at time t means that we cannot predict precisely where the particle will be at time $t + \Delta t$. Thus, it is not possible, even in principle, to track the motion of a particle as a function of time.

CHECKPOINT 28.4

Why is the Bohr model of the hydrogen atom incompatible with the uncertainty principle?

Example 28.3

Uncertainty in a Single-Slit Experiment

An electron diffraction experiment is performed using a single horizontal slit of width a (Fig. 28.8). Let the center of the slit be at $y = 0$. The y-coordinates of the electrons that pass through the slit are between $y = -a/2$ and $y = +a/2$. Thus, y is within $\pm a/2$ of the average position ($y = 0$), so an estimate of the uncertainty in the y-coordinate (Δy) is $a/2$. (a) What is the y-component of the momentum of an electron that leaves the slit at angle θ? Write the answer in terms of p and θ. (b) Most of the electrons fall within the central diffraction maximum. Use this fact to estimate the uncertainty Δp_y of the electrons as they pass through the slit. (c) Find the product $\Delta y\, \Delta p_y$. How does it compare with the limiting value given by the uncertainty principle?

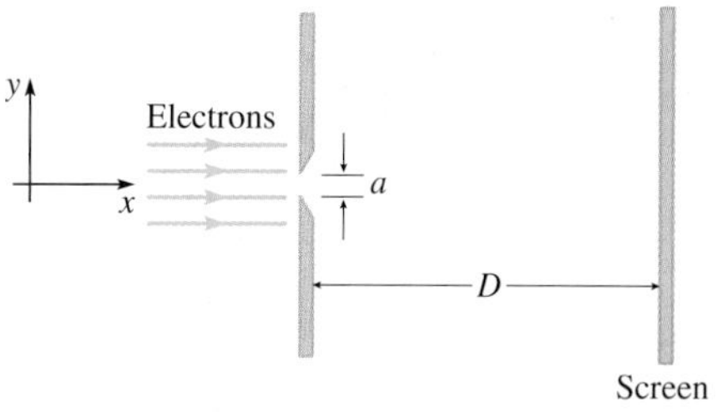

Figure 28.8
A single-slit electron diffraction experiment.

Strategy For a wide slit ($a \gg \lambda$), we expect little diffraction; a large uncertainty in y allows for a small uncertainty in p_y, and the electrons travel straight ahead to form a geometric shadow. For a narrow slit, the electrons form a diffraction pattern on the screen. The electrons spread out into the diffraction pattern because their y-components of momentum vary as the electrons pass through the slit. The wider the diffraction pattern, the greater is Δp_y as they pass through the slit.

Solution (a) Figure 28.9 shows the momentum vector of an electron moving toward the screen at angle θ. The y-component is

$$p_y = p \sin \theta$$

(b) The angle of the first diffraction minimum is

$$\sin \theta = \frac{\lambda}{a} \qquad (25\text{-}12)$$

Thus, the range of momentum y-components for electrons that land in the central maximum is

$$-\frac{p\lambda}{a} < p_y < \frac{p\lambda}{a}$$

The uncertainty in the y-component of momentum is approximately

$$\Delta p_y = \frac{p\lambda}{a}$$

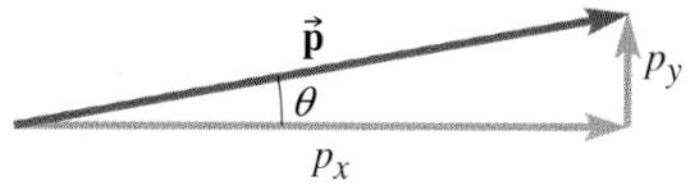

Figure 28.9
An electron heading off at an angle θ has a momentum vector $\vec{\mathbf{p}}$ as shown. Components of $\vec{\mathbf{p}}$ are found using a right triangle.

(c) The product of the uncertainties is

$$\Delta y\, \Delta p_y = \frac{a}{2} \times \frac{p\lambda}{a} = \frac{p\lambda}{2}$$

Since $\lambda = h/p$,

$$\Delta y\, \Delta p_y = \frac{ph}{2p} = \frac{1}{2}h$$

This estimate of $\Delta y\, \Delta p_y$ is a factor of 2π larger than the minimum value required by the uncertainty principle ($\Delta y\, \Delta p_y \geq \frac{1}{2}\hbar$).

Discussion This rough calculation shows that the product $\Delta y\, \Delta p_y$ is on the order of Planck's constant h, *regardless of the width of the slit or the wavelength of the electrons.* In accordance with the uncertainty principle, the two uncertainties are inversely related. A wide slit (Δy large) produces little diffraction (Δp_y small); a narrow slit (Δy small) produces a large diffraction pattern (Δp_y large).

Practice Problem 28.3 Confined Electron

An electron is confined to a "quantum wire" of length 150 nm. What is the minimum uncertainty in the electron's momentum component along the length of the wire? What is the minimum uncertainty in the electron's velocity component along the length of the wire?

Energy-Time Uncertainty Principle

Another uncertainty principle has to do with energy. If a system (e.g., an atom) is in a quantum state for a time interval Δt, then the uncertainty in the energy of that state is related to the lifetime of that state (Δt) by

$$\Delta E\, \Delta t \geq \frac{1}{2}\hbar \qquad (28\text{-}3)$$

28.5 WAVE FUNCTIONS FOR A CONFINED PARTICLE

An unconfined particle can have *any* momentum and energy. In an electron diffraction experiment or electron microscope, there is no theoretical restriction on the de Broglie wavelength of the electrons used. By contrast, electrons in atoms have only certain discrete or quantized energy levels available to them. The difference is due to the *confinement* of the electron. *A confined particle has quantized energy levels.*

A good analogy is that of a transverse wave on a string. Any wavelength is possible for a traveling wave on a long string. However, for a standing wave, in which the wave is confined to a length L of the string, only certain wavelengths are possible (see Section 11.10). If the string is fixed at both ends, the allowed wavelengths are

$$\lambda_n = \frac{2L}{n} \quad (n = 1, 2, 3, \ldots) \qquad (11\text{-}12)$$

For the longest wavelength ($\lambda = 2L$), the string vibrates at its lowest possible frequency (the fundamental). The standing wave is a classical example of quantization.

The same thing is true for particles such as electrons. If they are not confined, there is no restriction on their de Broglie wavelengths or energies. When they are confined, then only certain allowed values of the wavelength and energy are possible.

The wave function $y(x, t)$ for a string wave is the displacement y as a function of position along the string (x) and time (t). For the quantum mechanical wave function of a particle in one dimension we write $\psi(x, t)$, where ψ is the Greek letter psi. The interpretation of the wave function for a transverse wave on a string is easy: it tells how far a certain point on the string is displaced from its equilibrium position. For now we defer the question of what ψ stands for.

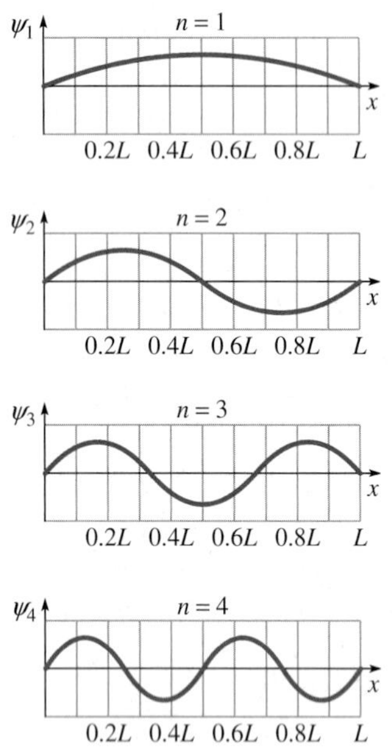

Figure 28.10 Wave functions for a particle in a box (n = 1, 2, 3, and 4).

CONNECTION:

Wavelengths for the particle in a box are the same as for waves on a string fixed at both ends (see Section 11.10).

Particle in a Box

The simplest model of a confined particle is a particle that can only move in one dimension and is confined by absolutely impenetrable "walls" to a length L. The particle is free—that is, its potential energy does not change—in the region between $x = 0$ and $x = L$, but it cannot leave that region, no matter how much energy it has. This model is called the **particle in a box** (but remember that the "box" is one-dimensional).

The wave function of the particle confined in this way is exactly analogous to a transverse wave on a string fixed at both ends, so we obtain the same result for the possible wavelengths:

$$\lambda_n = \frac{2L}{n} \quad (n = 1, 2, 3, \ldots) \qquad (28\text{-}4)$$

The de Broglie wavelength of the particle is related to its momentum:

$$p_n = \frac{h}{\lambda_n} = \frac{nh}{2L} \qquad (28\text{-}5)$$

Figure 28.10 shows the wave functions for the ground state (the quantum state of lowest energy) and the first three excited states—that is, for n = 1, 2, 3, and 4.

What is the energy of the confined particle? The energy is the sum of the potential and kinetic energies. The potential energy is the same everywhere inside the box; for simplicity we choose $U = 0$ inside the box. The kinetic energy can be found from the momentum:

$$K = \frac{1}{2}mv^2 = \frac{(mv)^2}{2m} = \frac{p^2}{2m} \qquad (28\text{-}6)$$

$$E = K + U = \frac{p^2}{2m} + 0 = \frac{n^2h^2}{8mL^2} \qquad (28\text{-}7)$$

Just as the string wave has a fundamental mode with the lowest possible frequency, the confined particle has a *minimum possible energy* in its ground state ($n = 1$). The energy of the ground state is

$$E_1 = \frac{h^2}{8mL^2} \quad (28\text{-}8)$$

The existence of a nonzero minimum energy has important ramifications. A confined particle *cannot* have zero kinetic energy. A particle confined to a smaller box has a *larger* ground-state energy. This conclusion is supported by the uncertainty principle: a smaller box means a smaller uncertainty in position ($\Delta x \approx L/2$) and therefore a greater uncertainty in momentum. Although the *magnitude* of the momentum is well defined for the particle in a box ($p = h/\lambda$), the momentum *x-component* can be either $+p$ or $-p$. Thus, $\Delta p_x \approx h/(2L)$ for the ground state. The product of the uncertainties is

$$\Delta x\, \Delta p_x \approx \frac{1}{2}L \times \frac{h}{2L} = \frac{1}{4}h = \frac{\pi}{2}\hbar$$

If the uncertainty principle is used to estimate the ground-state energy for the particle in a box, using *estimates* for the two uncertainties, the result is only off by a factor of π.

The energies of the excited states are

$$E_n = n^2 E_1 \quad (28\text{-}9)$$

Just as for the H atom, the particle in a box can make a transition from an excited state n to a lower energy state m by radiating a photon with energy

$$E = E_n - E_m \quad (28\text{-}10)$$

Note that the energy levels get farther apart as n increases. In contrast, the energy levels of the H atom get closer together as n increases. Why the difference? The particle in a box is confined to the same length L, no matter how much energy it has. The potential energy that confines the electron in the H atom changes gradually (Fig. 28.11). For electrons with higher energies, the box is longer.

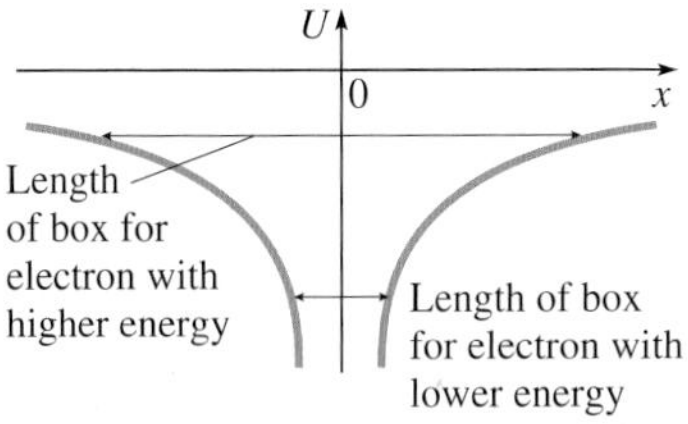

Figure 28.11 Potential energy of the electron in a hydrogen atom as a function of x; we assume for simplicity that the electron is confined in a one-dimensional box. The nucleus is at $x = 0$.

Finite Box

A slightly more realistic model of a particle confined in one dimension is the *particle in a finite box.* In this model, the "walls" are not impenetrable; as shown in Fig. 28.12, the potential energy outside the box ($U = U_0$) is higher than that inside the box ($U = 0$). For a particle in a finite box, the energies are still quantized for *bound states* ($E < U_0$), but the number of bound states is finite. If the particle has an energy E greater than U_0, then it is no longer confined to the box. For these states, since the particle is not confined to the box, a continuum of wavelengths and energies is possible.

In a *finite* box, the wave functions for bound states do not have to be zero at the walls and everywhere outside; instead, they extend past the walls a bit, decaying exponentially as the distance from the wall increases (Fig. 28.13). According to classical physics, a particle with $E < U_0$ can *never* be in the region outside the box since that would make the kinetic energy negative. Many experiments have verified that the wave function of a confined particle *does* extend outside the box, in accordance with the predictions of quantum mechanics.

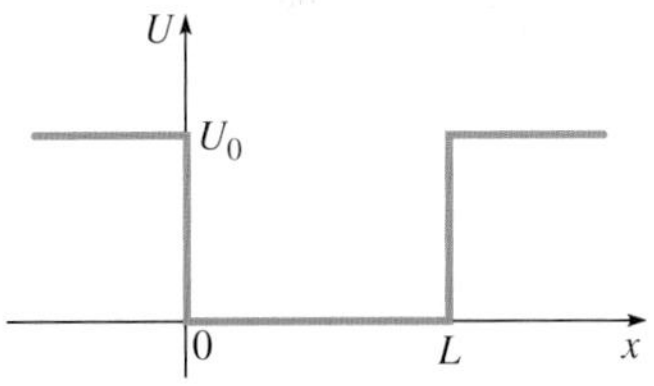

Figure 28.12 Potential energy for a particle in a finite box.

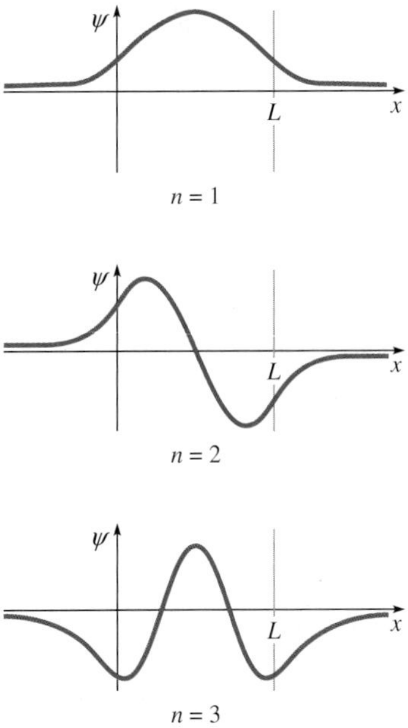

Figure 28.13 Wave functions for a particle in a finite box ($n = 1$, 2, and 3).

Application: Quantum Corral

In 1993, researchers at IBM fabricated a quantum corral (Fig. 28.14)—a two-dimensional finite box for electrons. The corral is a circular ring of radius 7.13 nm composed of 48 iron atoms on the surface of a copper crystal. The ring of iron atoms traps some electrons inside. The ripples show the circular standing wave formed by the electrons' wave functions. A scanning tunneling microscope (Section 28.10) was used first to move the iron atoms into place one at a time and then to form an image of the corral.

Figure 28.14 Scanning tunneling microscope picture of a quantum corral. Clearly visible in this false-color image is an electron standing wave. The electrons are confined by 48 iron atoms that form the "fence" of the corral on a copper surface. The radius of the corral is 7.13 nm.

Interpretation of the Wave Function

In 1925, Austrian physicist Erwin Schrödinger (1887–1961) obtained de Broglie's thesis concerning the wavelike nature of particles. Within a few weeks, Schrödinger formulated a fundamental equation of quantum mechanics. Quantum-mechanical wave functions are solutions of the Schrödinger equation.

> The probability of finding a particle in a certain location is proportional to the *square* of the magnitude of the wave function: $P \propto |\psi|^2$.

The statistical interpretation of the wave function is due to the German physicist Max Born (1882–1970). The probability that a photon in an interference experiment hits the screen at a certain location is proportional to the intensity of light at that location. The intensity, in turn, is proportional to the square of the electric field amplitude. Classically, the field serves as the wave function for an EM wave, so the probability of finding a photon at a certain location is proportional to the square of the magnitude of the wave function.

To be more precise, we can't ever expect to find a particle exactly at a single mathematical point; rather, we want to know the probability of finding a particle in a small region of space. In one dimension, $|\psi(x)|^2 \Delta x$ is the probability of finding the particle between x and $x + \Delta x$.

Quantum physics is probabilistic in a way that classical physics is not. A particle's future is *not* completely determined by its present. Two particles, even if identical and in identical environments, may not behave in the same way. Two hydrogen atoms in the same excited state may not return to the ground state in the same way or at the same time. One may stay in the excited state longer than the other, and they may take different intermediate steps. The best we can do is to find the probability per unit time that photons of various energies are radiated.

Probability is central in nuclear physics (Chapter 29). A collection of identical radioactive nuclei, for instance, decay at different times and possibly by different processes. We can predict and measure the half-life—the time interval during which half of the nuclei decay—but there is no way to know *which* nuclei will decay *when*, or by which process.

Figure 28.15 Electron cloud representation of the ground state of the hydrogen atom. The cloud represents the probability density—the electron is more likely to be found where the cloud is darker. The cloud is centered on the nucleus (not shown).

28.6 THE HYDROGEN ATOM: WAVE FUNCTIONS AND QUANTUM NUMBERS

The quantum picture of the hydrogen atom is quite different from Bohr's model. The electron doesn't orbit the proton in a circular orbit—or any other kind of orbit. The best we can expect is to calculate the *probability* of finding the electron in a given place.

You may have seen the electron depicted as an electron cloud similar to Fig. 28.15. The electron cloud is one way to represent the electron's probability distribution.

But the electron is *not* spread out into a fuzzy cloud; any measurement to locate the electron would find a point particle. (If the electron is not a point particle, experiments have shown that its size is less than 10^{-17} m, which is $\frac{1}{100}$ the size of the proton and less than 10^{-7} times the size of an atom.) Although the electron does not follow an orbit, it does have kinetic energy and can have angular momentum associated with its motion.

Since an electron bound to a nucleus is confined in space to the region surrounding the nucleus, its energies are quantized. A confined particle in a stationary state of definite energy is a standing wave. The wave function for the electron is a three-dimensional standing wave.

The potential energy for an electron at a distance r from the proton is

$$U = -\frac{ke^2}{r}$$

Recall that $a_0 = \hbar^2/(m_e k e^2) =$ 52.9 pm is the Bohr radius for the hydrogen atom and $k = 1/(4\pi\epsilon_0) = 8.99 \times 10^9\ \text{N·m}^2/\text{C}^2$ is the Coulomb constant.

The electron in the ground state has energy $E_1 = -13.6$ eV, just as in the Bohr model. As you can show in Problem 47, the electric potential energy is equal to E_1 at a distance $2a_0$ from the nucleus (Fig. 28.16). Since $E = K + U$, the kinetic energy at $r = 2a_0$ is zero. According to classical physics, the electron could never be found at distances $r > 2a_0$; but the wave function of the electron extends into the region $r > 2a_0$, just as the wave function extends past the walls of a finite box.

Since the potential energy is not constant, the wave function does not have a single, constant wavelength. The wave function $\psi(r)$ for the ground state ($n = 1$) is shown in Fig. 28.17a. Although the wave function has its maximum value at $r = 0$, the distance from the nucleus at which the electron is most likely to be found is not 0 but a_0 (see Fig. 28.17b,c).

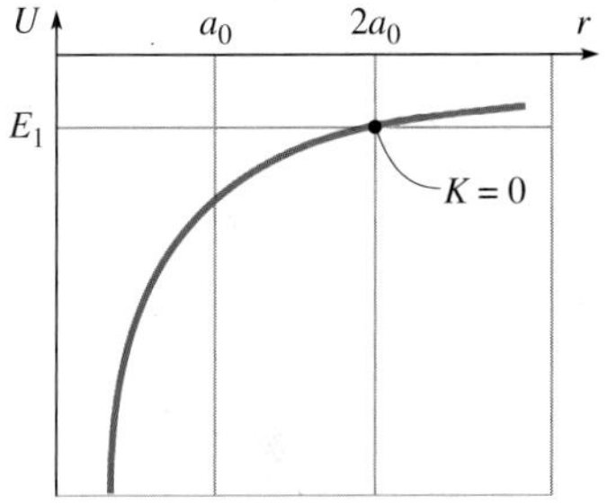

Figure 28.16 The graph shows the electric potential energy of the electron as a function of distance r from the nucleus ($U = -ke^2/r$). E_1 is the ground-state energy. Since $E = K + U$, the kinetic energy at any distance r is the difference between the horizontal line representing E_1 and the curve representing $U(r)$.

Quantum Numbers

It turns out that the quantum state of the electron is not determined by n alone. Specifying the quantum state requires four **quantum numbers**. The integer n is called the **principal quantum number**. The energy levels are the same as the Bohr energies:

$$E_n = \frac{E_1}{n^2}, \quad E_1 = -\frac{m_e k^2 e^4}{2\hbar^2} = -13.6\ \text{eV} \tag{28-11}$$

For a given principal quantum number n, the electron can have n different quantized magnitudes of orbital angular momentum $\vec{\mathbf{L}}$.

$$L = \sqrt{\ell(\ell+1)}\,\hbar, \quad \ell = 0, 1, 2, \ldots, n-1 \tag{28-12}$$

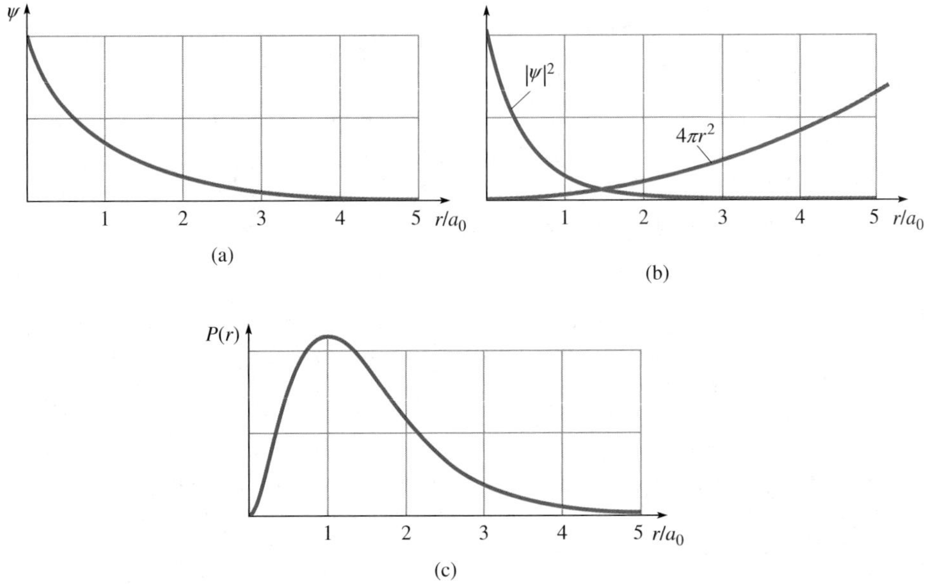

Figure 28.17 (a) Ground-state wave function of the electron in the H atom. (b) Graphs of $|\psi|^2$ and $4\pi r^2$, the two competing factors that determine the probability of finding the electron at a given *distance* from the proton. $|\psi|^2$ is the probability *per unit volume.* The volume of space at distances between r and $r + \Delta r$ from the nucleus is the area of a thin spherical shell ($4\pi r^2$) times its thickness Δr. (c) Graph of $4\pi r^2|\psi|^2$, which is proportional to the probability of finding the electron at distances between r and $r + \Delta r$ from the nucleus. The probability is maximum at $r = a_0$.

For a given n, the **orbital angular momentum quantum number** ℓ can be any integer from 0 to $n-1$. In the ground state ($n = 1$), $\ell = 0$ is the only possible value; the angular momentum in the ground state must be $L = 0$. For higher n, there are states both with nonzero and zero L. Note that L is called the *orbital* angular momentum because it is associated with the *motion* of the electron, but remember that the electron does not follow a well-defined orbit.

The orbital angular momentum quantum number ℓ determines only the magnitude L of the orbital angular momentum; what about the direction? The direction also turns out to be quantized. For a given n and ℓ, the component of $\vec{\mathbf{L}}$ along some direction that we'll call the z-axis can have one of $2\ell + 1$ quantized values:

$$L_z = m_\ell \hbar, \quad m_\ell = -\ell, -\ell+1, \ldots, -1, 0, +1, \ldots, \ell-1, \ell \qquad (28\text{-}13)$$

The **orbital magnetic quantum number** m_ℓ can be any integer from $-\ell$ to $+\ell$.

Figure 28.18 shows the probability density $|\psi|^2$ for several quantum states of the H atom. Notice that the states with zero orbital angular momentum ($\ell = 0$) are spherically symmetrical, while $\ell \neq 0$ states are not.

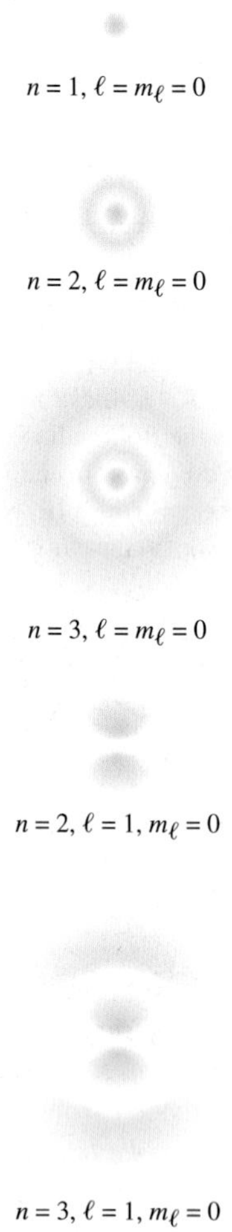

Figure 28.18 Electron cloud representations of the probability density $|\psi|^2$ for a few of the quantum states of the hydrogen atom. The sketches show the probability densities in a single plane. For an idea of what the electron clouds look like in three dimensions, imagine rotating each of the sketches about a vertical axis.

In addition to the angular momentum associated with its motion, an electron has an intrinsic angular momentum $\vec{\mathbf{S}}$ whose magnitude is $S = (\sqrt{3}/2)\hbar$. Originally, it was thought that the electron was spinning about an axis—we still call S the *spin angular momentum*—but it can't be. The electron is, as far as we know, a point particle; to generate this angular momentum by spinning, the electron would have to be large and would have to violate relativity. The spin angular momentum is an intrinsic property of the electron, like its charge or mass.

Electrons always have the same *magnitude* spin angular momentum, but the z-component of $\vec{\mathbf{S}}$ has two possible values:

$$S_z = m_s \hbar, \quad m_s = \pm\tfrac{1}{2} \qquad (28\text{-}14)$$

The two values of the **spin magnetic quantum number** m_s are often referred to as *spin up* and *spin down.* (The quantum numbers m_ℓ and m_s are called *magnetic* because the energy of a state depends on their values when the atom is in an external magnetic field.)

The state of the electron in a hydrogen atom is completely determined by the values of the four quantum numbers n, ℓ, m_ℓ, and m_s.

CHECKPOINT 28.6

List the quantum numbers for all possible electron states in the hydrogen atom with principal quantum number $n = 2$.

28.7 THE EXCLUSION PRINCIPLE; ELECTRON CONFIGURATIONS FOR ATOMS OTHER THAN HYDROGEN

According to the **Pauli exclusion principle**—named after the Austrian-Swiss physicist Wolfgang Pauli (1900–1958)—no two electrons in an atom can be in the same quantum state. The quantum state of an electron in any atom is specified by the same four quantum numbers used for hydrogen: n, ℓ, m_ℓ, and m_s (Table 28.1). However, the electron energy levels are not the same as those of hydrogen. In atoms with more than one electron, interactions between electrons must be taken into account. In addition, the nuclear charge varies from one element to another. Thus, the same set of four quantum numbers do not correspond to the same energy level from one species of atom to another.

Table 28.1 Quantum Numbers for Electron States in an Atom

Symbol	Quantum Number	Possible Values
n	principal	$1, 2, 3, \ldots$
ℓ	orbital angular momentum	$0, 1, 2, 3, \ldots, n-1$
m_ℓ	orbital magnetic	$-\ell, -\ell+1, \ldots, -1, 0, 1, \ldots, \ell-1, \ell$
m_s	spin magnetic	$-\frac{1}{2}, +\frac{1}{2}$

Shells and Subshells The set of electron states with the same value of n is called a **shell**. Each shell is composed of one or more **subshells**. A subshell is a unique combination of n and ℓ. Subshells are often represented by the numerical value of n followed by a lowercase letter representing the value of ℓ. The letters s, p, d, f, g, and h stand for $\ell = 0$, 1, 2, 3, 4, and 5, respectively (Table 28.2). For example, $3p$ is the subshell with $n = 3$ and $\ell = 1$. The letters s, p, and d came from the appearance of the associated spectral lines long before the advent of quantum theory. The dominant or **p**rincipal spectral lines came from the $\ell = 1$ subshell; the spectral lines from the $\ell = 0$ subshell were especially **s**harp in appearance; and those from the $\ell = 2$ subshell looked more **d**iffuse than the others.

Since the orbital angular momentum quantum number ℓ can be any integer from 0 to $n - 1$, with n possible values, there are n subshells in a given shell. Thus, there are three subshells in the $n = 3$ shell: $3s$, $3p$, and $3d$. A superscript following the subshell label indicates how many electrons are present in that subshell. This compact notation represents the configuration of electrons in an atom. For example, the ground state of the nitrogen atom is $1s^2 2s^2 2p^3$; it has two electrons in the $1s$ subshell, two in the $2s$ subshell, and three in the $2p$ subshell.

Application: electron orbitals, electronic configuration of atoms

Orbitals Each subshell, in turn, consists of one or more **orbitals**, which are specified by n, ℓ, and m_ℓ. Since m_ℓ can be any integer from $-\ell$ to $+\ell$, there are $2\ell + 1$ orbitals in a subshell. Therefore, s subshells have only one orbital, p subshells have three orbitals, d subshells have five orbitals, and so forth. Each orbital can accommodate two electrons: one spin up $\left(m_s = +\frac{1}{2}\right)$ and one spin down $\left(m_s = -\frac{1}{2}\right)$. It can be shown (Problem 79) that

the number of electron states in a subshell is $4\ell + 2$, and the number of states in a shell is $2n^2$ (28-15)

Ground-State Configuration The ground-state (lowest energy) electronic configuration of an atom is found by filling up electron states, starting with the lowest energy, until all the electrons have been placed. According to the exclusion principle, there can only be one electron in each state. Generally, the subshells in order of increasing energy are

$$1s, 2s, 2p, 3s, 3p, 4s, 3d, 4p, 5s, 4d, 5p, 6s, 4f, 5d, 6p, 7s \quad (28\text{-}16)$$

However, there are some exceptions. The energies of the subshells are not the same in different atoms; different nuclear charges and the interaction of the electrons make the energy levels differ from one atom to another. So, for example, the ground state of chromium (Cr, atomic number 24) is $1s^2 2s^2 2p^6 3s^2 3p^6 4s^1 3d^5$ instead of $1s^2 2s^2 2p^6 3s^2 3p^6 4s^2 3d^4$. Similarly, the ground state of copper (Cu, atomic number 29) is $1s^2 2s^2 2p^6 3s^2 3p^6 4s^1 3d^{10}$ instead of $1s^2 2s^2 2p^6 3s^2 3p^6 4s^2 3d^9$. There are only 8 elements among the first 56 that are exceptions to the subshell order in Eq. (28-16):

Cr, Cu, Nb, Mo, Ru, Rh, Pd, Ag

Table 28.2 Electron Subshells Summarized

$\ell =$	0	1	2	3	4	5
Spectroscopic notation	s	p	d	f	g	h
Number of states in subshell	2	6	10	14	18	22

Many more exceptions are found in the electron configurations of elements with atomic numbers greater than 56.

Example 28.4

Electron Configuration of Arsenic

What is the ground-state electron configuration of arsenic (atomic number 33)?

Strategy Arsenic has atomic number 33, so there are 33 electrons in the neutral atom. Arsenic is not one of the above-mentioned exceptions for atomic numbers ≤ 56, so subshells are filled in the order listed in Eq. (28-16) until the total number of electrons reaches 33. A subshell can hold up to $4\ell + 2$ electrons. Each s ($\ell = 0$) subshell holds a maximum of $4 \times 0 + 2 = 2$ electrons, each p ($\ell = 1$) subshell holds $4 \times 1 + 2 = 6$, and each d ($\ell = 2$) subshell holds $4 \times 2 + 2 = 10$.

Solution We fill up subshells and keep track of the total number of electrons: $1s^2 2s^2 2p^6 3s^2 3p^6 4s^2 3d^{10}$ has $2 + 2 + 6 + 2 + 6 + 2 + 10 = 30$ electrons. Then the remaining 3 go into the subshell with the next highest energy—$4p$. The ground-state configuration of arsenic is therefore

$$1s^2 2s^2 2p^6 3s^2 3p^6 4s^2 3d^{10} 4p^3$$

Discussion To double-check an electron configuration for an element that is not one of the exceptions:

- Add up the total number of electrons.
- Check that the subshells go in the order of Eq. (28-16).
- Make sure that all subshells except the last are full (s^2, p^6, d^{10}).

If the configuration passes those three tests, it is correct.

Practice Problem 28.4 Electron Configuration of Phosphorus

What is the electron configuration of phosphorus (atomic number 15)?

Filling the Orbitals If a subshell is not full, how are the electrons distributed among that subshell's orbitals? Recall that a subshell contains $2\ell + 1$ orbitals and each orbital contains two electron states. As a rule, electrons do not double up in an orbital until each orbital has one electron in it. The two electrons in an orbital have the same spatial distribution—the same electron cloud. Thus, the two electrons in a single orbital are closer together, on average, than are two electrons in different orbitals. Due to the electrical repulsion, the energy is lower if the electrons are in different orbitals, since they are farther apart. For example, the three $4p$ electrons in arsenic (Example 28.4) are in different orbitals in the ground state: one has $m_\ell = 0$, one has $m_\ell = +1$, and one has $m_\ell = -1$.

Application: Understanding the Periodic Table

The elements in the periodic table (see inside back cover) are arranged in order of increasing atomic number Z. The nucleus of an element has charge $+Ze$ and the neutral atom has Z electrons. Furthermore, the elements are arranged in columns according to the configuration of their electrons (Table 28.3). Elements with similar electronic configurations tend to have similar chemical properties.

Although the energy level of a subshell differs from one atom to another, Fig. 28.19 gives a *general* idea of the energies of the various atomic subshells. Note the larger than

Table 28.3 The Periodic Table Organizes the Elements According to Electronic Configuration

1A	2A	3B–8B, 1B, 2B	3A	4A	5A	6A	7A	8A
Alkali Metals	Alkaline Earths	Transition Elements, Lanthanides, and Actinides					Halogens	Noble Gases
s^1	s^2	d^ns^2, d^ns^1, or $f^md^ns^2$	s^2p^1	s^2p^2	s^2p^3	s^2p^4	s^2p^5	s^2p^6 (except He)

The periodic table of the elements is arranged in columns by electronic configuration. Elements with similar electronic configurations tend to have similar chemical properties. The table lists only the subshells beyond the configuration of the previously occurring noble gas.

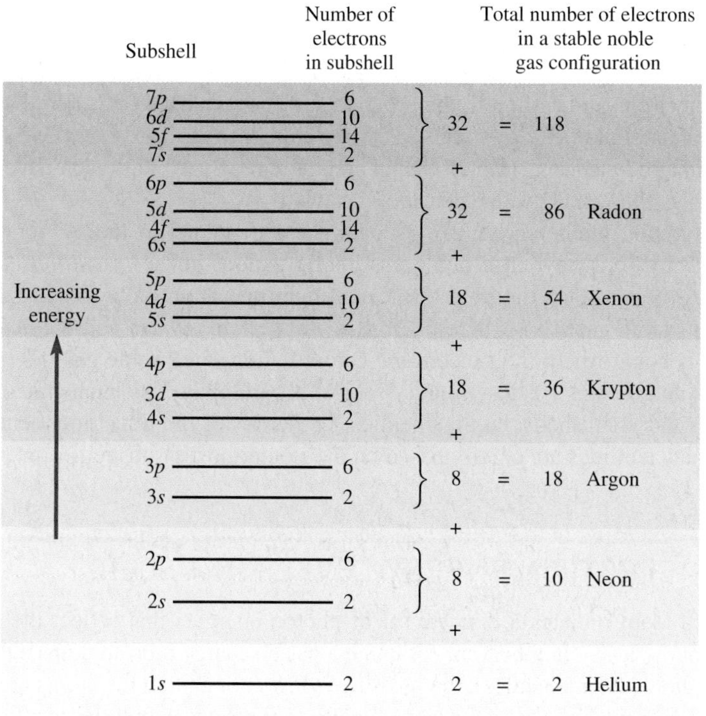

Figure 28.19 Energy level diagram for atomic subshells. The energies of the subshells differ from one atom to another, but this diagram gives a general idea of the relative spacing of the energies. The subshells are filled from the bottom (lowest energy) up.

usual spacing between each *s*-subshell and the subshell below it. The *s*-subshell is the lowest energy subshell in a given shell. When starting a new shell (with a higher value of *n*), the electrons are farther from the nucleus and more weakly bound. The most stable electronic configurations—those that are difficult to ionize and are chemically nonreactive—are those that have all the subshells below an *s*-subshell full. Elements with this stable configuration are called the *noble gases* (Group 8A). Helium has configuration $1s^2$—the only subshell below 2*s* is full. The rest of the noble gases have a full *p*-subshell as their highest energy subshell: neon (all subshells below 3*s* full), argon (full below 4*s*), krypton (full below 5*s*), xenon (full below 6*s*), and radon (full below 7*s*).

The energy required to excite a helium atom into its first excited state ($1s^12s^1$) is quite large—about 20 eV—due to the large energy gap between the 1*s* and 2*s* subshells. The energy required to excite a lithium atom into its first excited state is much smaller (about 2 eV). Lithium and the other *alkali metals* (Group 1A) have one electron beyond a noble gas configuration. As a shorthand, we often write the spectroscopic notation only for the electrons in an atom that are beyond the configuration of a noble gas, since only those electrons participate in chemical reactions; thus, lithium's configuration is [He]$2s^1$, sodium's is [Ne]$3s^1$, and so on. The single electron in the *s*-subshell is quite weakly bound, so it can easily be removed from the atom, making the alkali metals highly reactive. They can easily give up their "extra" electron to achieve a noble gas configuration as an ion with charge $+e$ (alkali metals have valence +1).

Valence: the number of electrons that an atom will gain, lose, or share in chemical reactions

The alkali metals form ionic bonds with the highly reactive *halogens* (Group 7A), which are one electron shy of a noble gas configuration. For instance, chlorine (Cl, [Ne]$3s^2 3p^5$) needs to gain only one electron to have the electron configuration of the noble gas argon (Ar, [Ne]$3s^2 3p^6$). Thus, the halogens have valence −1. Sodium can give its weakly bound electron to chlorine, leaving both ions (Na^+ and Cl^-) in stable noble gas configurations. The electrostatic attraction between the two forms an ionic bond: NaCl.

The *alkaline earths* (Group 2A) all have a full *s*-subshell (s^2) beyond a noble gas configuration. They are not as reactive as the alkali metals, since the full *s*-subshell lends some stability, but they can give up both *s* electrons to achieve a noble gas configuration, so alkaline earths usually act with valence +2.

Toward the middle of the periodic table, the chemical properties of elements are more subtle. Covalent bonds tend to form when two or more elements have unpaired electrons in orbitals that they can share. Carbon is particularly interesting. Its ground state is $1s^2 2s^2 2p^2$. The two $2p$ electrons are in different orbitals; then there are two unpaired electrons and carbon in the ground state has a valence of 2. However, it takes only a small amount of energy to raise a carbon atom into the state $1s^2 2s^1 2p^3$. Now there are four unpaired electrons (the $2s$ orbital and the three $2p$ orbitals each have one electron). Thus, carbon can have a valence of 4 as well.

In the groups numbered 1A, 2A, . . . , 7A, the numeral before the "A" represents the number of electrons beyond a noble gas configuration. In the *transition elements*, a *d*-subshell is being filled; their electronic configurations are usually [noble gas]$d^n s^2$ but sometimes [noble gas]$d^n s^1$, where $0 \le n \le 10$. In the *lanthanides* and the *actinides*, an *f*-subshell is being filled; their electronic configurations are [noble gas] $f^m d^n s^2$, where $0 \le m \le 14$ and $0 \le n \le 10$. The *d*- and *f*-subshells participate less in chemical reactions than the *s*- and *p*-subshells, so the chemical properties of the transition elements, lanthanides, and actinides are chiefly based on their outermost *s*-subshell.

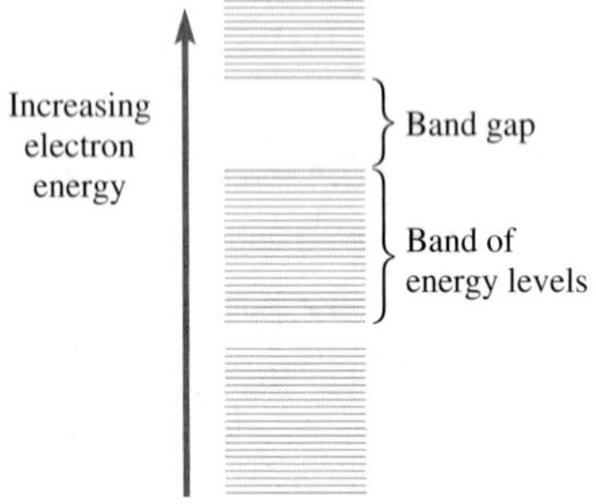

Figure 28.20 Electron energies in a solid form bands of closely spaced energy levels. Band gaps are ranges of energy in which there are no electron energy levels.

See [icon] for more information on energy levels in solids.

28.8 ELECTRON ENERGY LEVELS IN A SOLID

An isolated atom radiates a discrete set of photon energies that reflect the quantized electron energy levels in the atom. Although a gas discharge tube contains a large number of gas atoms (or molecules), the pressure is low enough that the atoms are, on average, quite far apart. As long as the wave functions of electrons in different atoms do not overlap appreciably, each atom radiates photons of the same energies as would a single isolated atom.

On the other hand, solids radiate a continuous spectrum rather than a line spectrum. What has happened to the quantization of electron energies? The energy levels are still quantized, but they are so close together that in many circumstances we can think of them as continuous **bands** of energy levels. A **band gap** is a range of energies in which no electron energy levels exist (Fig. 28.20).

Constructing the electronic ground state of a solid is similar to constructing the ground state of an atom: the electron states are filled up in order of increasing energy starting from the lowest energy states, according to the exclusion principle. A solid at room temperature isn't in the ground state, but its electron configuration is not very different from the ground state; the extra thermal energy available promotes a small fraction of the electrons (still a large number, though) into higher energy levels, leaving some lower energy states vacant. The energy range of electron states that are thermally excited is small—of the order of $k_B T$, where k_B is the Boltzmann constant ($k_B = 1.38 \times 10^{-23}$ J/K $= 8.62 \times 10^{-5}$ eV/K).

Conductors, Semiconductors, and Insulators

The ground-state electron configuration (i.e., the configuration at absolute zero) of a solid determines its electrical conductivity. If the highest-energy electron state filled at

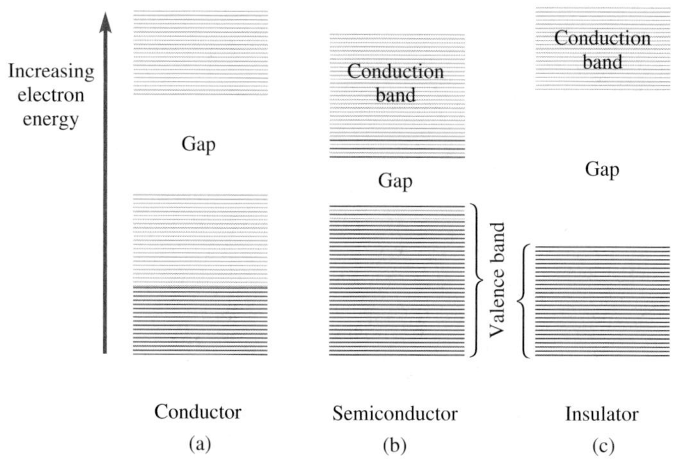

Figure 28.21 Electron energy bands in (a) a conductor, (b) a semiconductor, and (c) an insulator. Horizontal lines indicate electron energy levels; the darker lines are those levels that are occupied by electrons. In a semiconductor at room temperature (b), the valence band is mostly full but a relatively small number of electrons are thermally excited into the conduction band, leaving some vacancies near the top of the valence band.

$T = 0$ is in the middle of a band, so that this band is only partially filled, the solid is a conductor (Fig. 28.21). In order for current to flow, an electric field (perhaps due to a battery connected to the conductor) must be able to change the momentum and energy of the conduction electrons; this can only happen if there are vacant electron states nearby into which the conduction electrons can make transitions. Since the band is only partly full, there are plenty of available electron states at energies only slightly higher than the highest occupied.

On the other hand, if the ground-state configuration fills up the electron states right to the top of a band, then the solid is a semiconductor or an insulator. The difference between the two depends on how the size of the band gap E_g above the completely occupied band (the *valence band*) compares to the available thermal energy ($\approx k_B T$) and thus depends on the temperature of the solid.

Most materials considered semiconductors at room temperature have band gaps between about 0.1 eV and 2.2 eV. The technologically most important semiconductor, silicon, has a gap of 1.1 eV, which is about 40 times the available thermal energy at room temperature ($\approx$ 0.025 eV). The number of electrons excited to higher energy states is much smaller than in a conductor, since there are no available energy levels nearby in the same band. The only electrons that can carry current are those promoted to the mostly empty band above the gap (the *conduction band*).

Because a relatively small number of electrons move into the conduction band, an equal number of vacant electron states exist near the top of the valence band. Electrons in nearby states can easily "fall" into these **holes**, filling one vacancy and creating another. The holes act like particles of charge $+e$ that, in response to an external electric field, move in a direction opposite that of the conduction electrons. The electric current in a semiconductor has two components: the electron current and the hole current.

28.9 LASERS

A laser produces an intense, parallel beam of coherent, monochromatic light. The word **laser** is an acronym for *light amplification by stimulated emission of radiation.*

Stimulated Emission

When a photon has energy $\Delta E = E' - E$, where E' is a vacant energy level in an atom and E is a lower energy level that is filled, the photon can be absorbed, kicking the electron up into the higher energy level (Fig. 28.22a). If the higher energy level is filled and the lower one is vacant, the electron can drop into the lower energy level by spontaneously emitting a photon of energy ΔE (Fig. 28.22b).

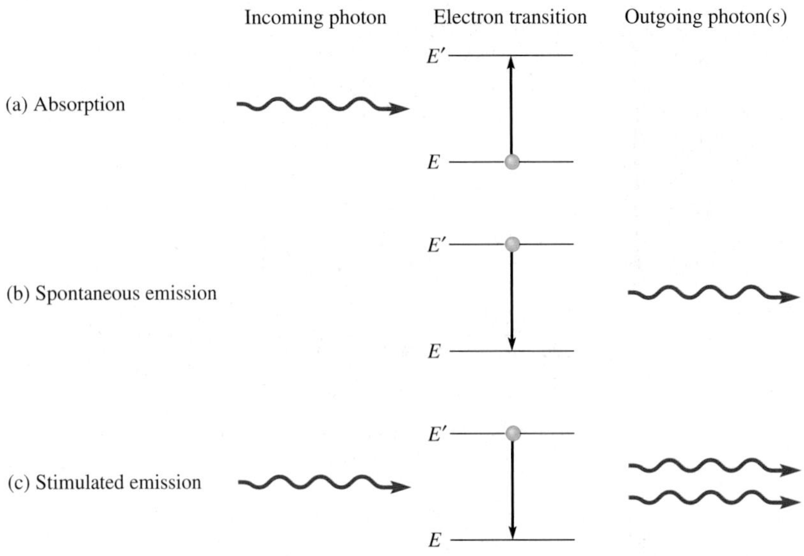

Figure 28.22 Absorption, spontaneous emission, and stimulated emission of a photon by an atom. All of the photons have energy $E' - E$ (the difference between the two energy levels). For a photon to be emitted, either spontaneously or when stimulated by an incoming photon, the electron must initially be in the higher energy level E'. In stimulated emission, an incident photon of energy $E' - E$ stimulates the atom to emit a photon. The two photons are identical in energy, phase, and direction.

In addition to absorption and spontaneous emission, a third interaction between an atom and a photon was first proposed by Einstein in 1917. Called **stimulated emission** (Fig. 28.22c), this process is a kind of resonance. If the electron is in the higher energy level and the lower level is vacant, an incident photon of energy ΔE can stimulate the emission of a photon as the electron drops to the lower energy level. The photon emitted by the atom is *identical* to the incident photon that stimulates the emission: they have the same energy and wavelength, move in the same direction, and are in phase with one another.

If a cascade of stimulated emissions occur, the number of identical photons increases—the *light amplification* in the acronym *laser*. The beam is *coherent* because the photons are all in phase; the beam is *monochromatic* because the photons all have the same wavelength; and the beam is *parallel* because the photons all move in the same direction.

Metastable States

How can a cascade of stimulated emissions occur when most of the atoms are in their ground states, with the electrons populating the lowest energy levels? An atom in an excited state returns to the ground state quickly by spontaneous emission of a photon. In such circumstances, the probability that a photon of energy ΔE stimulates emission is extremely small, because very few of the atoms are in the excited state; the photon is overwhelmingly more likely to be absorbed by an atom in the ground state.

To produce a cascade of identical photons, stimulated emission must be *more likely than absorption*: more of the atoms must be in the higher-energy state than are in the lower-energy state. Since this is the reverse of the usual case, it is called a **population inversion**. A population inversion is difficult to achieve if the higher-energy state is short lived—that is, if the atom quickly emits a photon. However, some excited states—called **metastable states**—last for a relatively long time before spontaneous emission occurs. If atoms can be *pumped* up into a metastable state fast enough, a population inversion can occur.

The Ruby Laser

One way to achieve a population inversion is called **optical pumping**. Incident light of the correct wavelength is absorbed, causing the atoms to make transitions into a short-lived excited state, from which the atoms spontaneously decay to the metastable

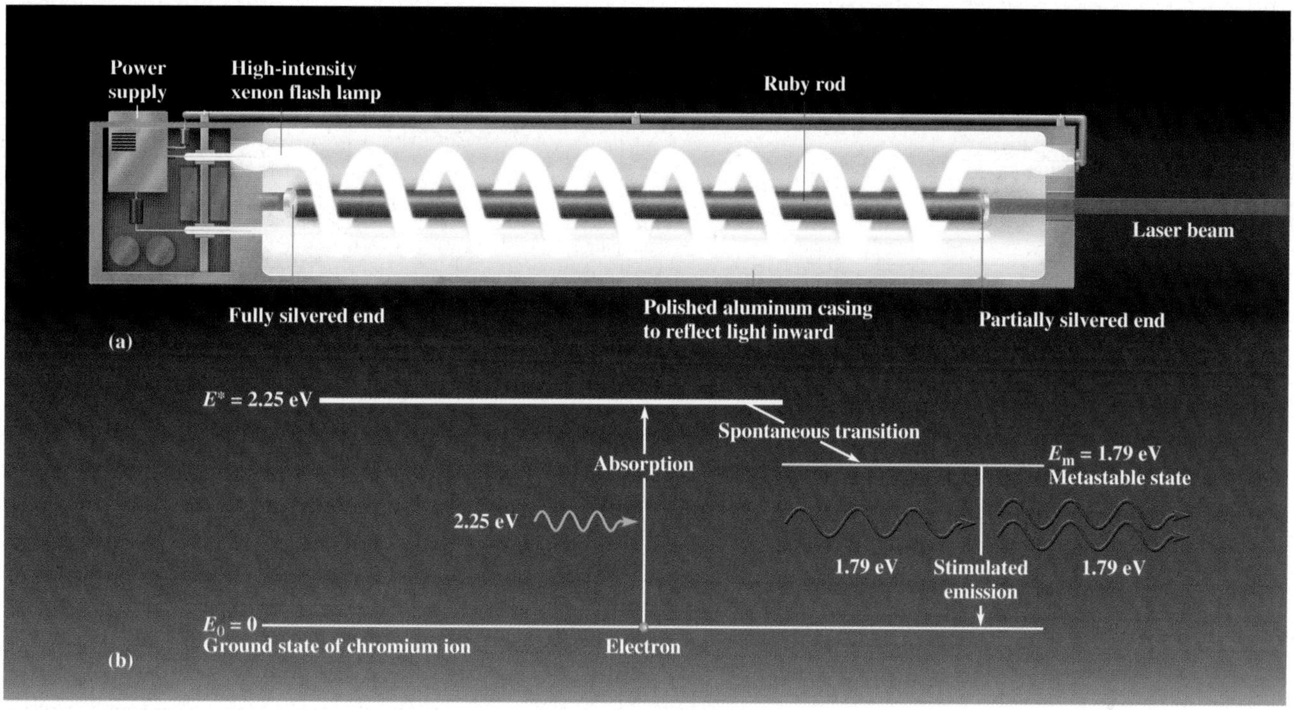

Figure 28.23 (a) A ruby laser. (b) Energy level diagram for a ruby laser. Optical pumping occurs when incident 2.25-eV photons are absorbed by the chromium ion, leaving it in one of the excited states E^*. The ion can then decay to the metastable state E_m. While the ion is in the metastable state, a 1.79-eV photon passing by can stimulate emission of an identical 1.79-eV photon.

state. The ruby laser (Fig. 28.23a), developed in 1960, uses optical pumping. Ruby is an aluminum oxide crystal (sapphire) in which some of the aluminum atoms are replaced by chromium atoms. The energy levels of the chromium ion Cr^{3+} are shown in Fig. 28.23b. The state labeled E_m is a metastable state of energy 1.79 eV above the ground state E_0. At an energy of about 2.25 eV above the ground state, a band of closely spaced energy levels E^* exists. If an atom excited to one of the E^* energy levels quickly decays to the metastable E_m state, the atom remains in the metastable state for a relatively long time.

To make a laser, a ruby rod has its ends polished and silvered to become mirrors. One end is partially transparent. A high-intensity flash lamp coils around the rod. The lamp produces a series of rapid, high-intensity bursts of light. Absorption of light with wavelength 550 nm (photon energy 2.25 eV) pumps Cr^{3+} ions to the E^* states, from which some spontaneously decay to the metastable state E_m. (Others spontaneously decay right back to the ground state.) Strong optical pumping results in a population inversion in which the number of ions in the metastable state exceeds the number in the ground state. Eventually a few ions decay from the metastable state to the ground state by spontaneously emitting photons of wavelength 694 nm (energy 1.79 eV, in the red part of the spectrum). These photons then cause stimulated emission by other chromium ions in the metastable state. Only photons emitted parallel to the axis of the rod are reflected back and forth many times by the mirrors at the ends to continue stimulating emissions. Some of these photons escape through the end of the rod that is partially silvered to form a narrow, intense, coherent beam of light.

Other Lasers

Similar to the ruby laser, the Nd:YAG laser consists of an optically pumped rod. Nd:YAG is yttrium aluminum garnet (YAG), a colorless crystal once used to make imitation

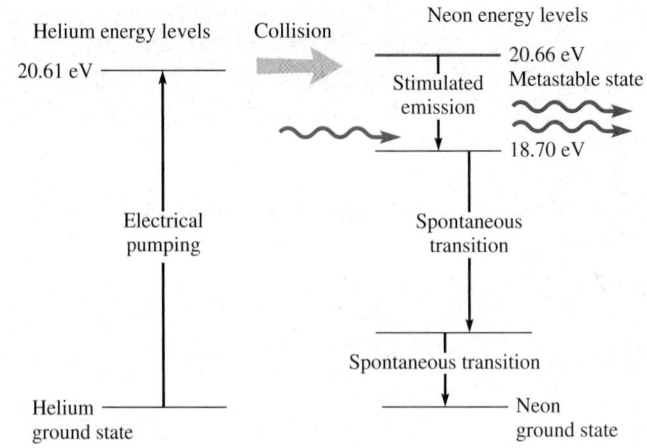

Figure 28.24 Simplified energy level diagram for the He-Ne laser.

diamonds, into which some neodymium atoms (Nd) have been introduced as impurities. The Nd ions have a metastable state suitable for lasing. Unlike ruby, which can only operate as a pulsed laser, Nd:YAG can operate either as a continuous beam or as a pulsed beam (see Conceptual Question 11). The Nd:YAG laser can produce a high-power beam at wavelength 1064 nm (in the infrared); it is commonly used in industry and in medicine.

Helium-neon (He-Ne) lasers are commonly used in school laboratories and in older barcode readers. A gas discharge tube contains a low pressure mixture of helium and neon. The He-Ne laser is *electrically* pumped: the electrical discharge excites helium atoms into a metastable state with energy 20.61 eV above the ground state (Fig. 28.24). Neon has a metastable state 20.66 eV above its ground state—0.05 eV higher than the energy of the metastable state of helium. An excited helium atom can make an inelastic collision with a neon atom in the ground state, leaving the neon atom in its metastable state and returning the helium atom to its ground state; the extra 0.05 eV of energy comes from the kinetic energies of the atoms. Stimulated emission leaves the atom in an excited state of energy 18.70 eV; spontaneous transitions quickly take it back to the ground state.

The carbon dioxide laser, which produces an infrared beam (10.6-μm wavelength), is similar in operation to the He-Ne laser. A gas discharge tube contains a low-pressure mixture of CO_2 and N_2. The N_2 molecule is excited by the electrical discharge; the CO_2 molecule is excited into a metastable state by colliding with an excited N_2 molecule. The most powerful continuous wave lasers in common use are carbon dioxide lasers; the power of a single beam can exceed 10 kW. An almost perfectly parallel beam can be focused onto a very small spot, allowing CO_2 lasers to cut, drill, weld, and machine the hardest metal with ease. CO_2 lasers are also widely used in medicine.

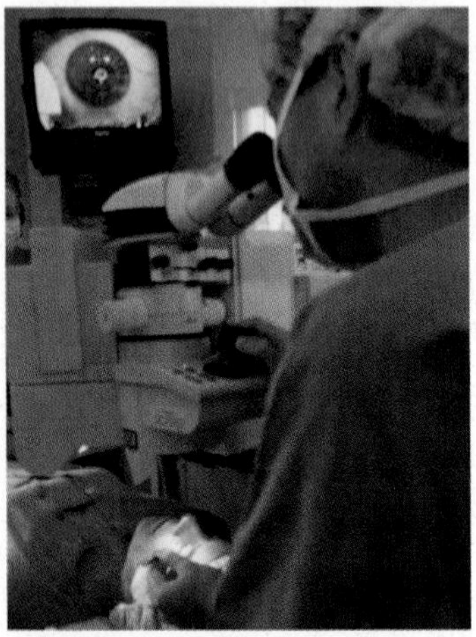

Figure 28.25 A patient undergoes laser optical surgery to correct her vision. The procedure is called LASIK (laser-assisted in situ keratomileusis).

Semiconductor lasers are small, inexpensive, efficient, and reliable. They are used in CD and DVD players, barcode readers, laser printers, and laser pointers. A semiconductor laser is electrically pumped: an electric current pumps electrons from the valence band to the conduction band. A photon is emitted when an electron jumps back from the conduction band to the valence band. Thus, the wavelength of the laser light depends on the band gap in the semiconductor.

Application: Lasers in Medicine

Lasers are widely used in surgery to destroy tumors, to cauterize blood vessels, and to pulverize kidney stones and gallstones. A detached retina can be "welded" back into place by a laser beam shone through the pupil of the eye. Laser surgery is used to reshape the cornea of the eye to correct nearsightedness (Fig. 28.25). The laser beam can be guided by an optical fiber (Section 23.4) in an endoscope to the site of a tumor;

an optical fiber can also guide a laser beam into an artery to remove plaque from the artery walls. In photodynamic cancer therapy, a photosensitizing drug is injected into the bloodstream; the drug accumulates selectively in cancerous tissues. Laser light of the correct frequency is delivered to the tumor site by an endoscope. The light causes a chemical reaction that activates the drug; it becomes toxic, destroying tumor cells and the blood vessels that supply oxygen to the tumor.

Conceptual Example 28.5

Photocoagulation

An argon ion laser is used to repair vascular abnormalities and fissures in the retina of the eye in a process known as photocoagulation. Laser light *absorbed* by the tissue raises its temperature until proteins become coagulated, forming the scar tissue that repairs the split. The principal wavelengths emitted by the argon laser are 514 nm and 488 nm. (a) What are the photon energies for these wavelengths? (b) What are the colors associated with these two wavelengths? (See Fig. 22.5.) Are both wavelengths effective on blood vessels?

Strategy The energy of a photon is related to its wavelength by $E = hc/\lambda$. Section 22.3 lists the colors of the visible spectrum and the associated wavelengths. A wavelength is effective if it is strongly absorbed.

Solution and Discussion (a) The photon energies are

$$E = \frac{hc}{\lambda} = \frac{1240\ \text{eV}\cdot\text{nm}}{514\ \text{nm}} = 2.41\ \text{eV}$$

and

$$E = \frac{hc}{\lambda} = \frac{1240\ \text{eV}\cdot\text{nm}}{488\ \text{nm}} = 2.54\ \text{eV}$$

(b) The color associated with 514 nm is green and with 488 nm is blue. Both wavelengths are effective on blood vessels because red blood vessels reflect red and absorb radiation of other colors.

Conceptual Practice Problem 28.5 Ruby Laser and Blood

Would red light from a ruby laser be effective on blood and thus useful in the treatment of vascular abnormalities?

28.10 TUNNELING

The wave function of a particle in a finite box extends into regions where, according to classical physics, the particle can never go because it has insufficient energy (see Section 28.5). In these *classically forbidden regions*, the wave function decays exponentially. If the classically forbidden region is of finite length, a curious but significant phenomenon called **tunneling** can occur.

Figure 28.26a shows a situation in which tunneling is possible. A particle is initially confined to a one-dimensional box. On the right side, the barrier is of finite thickness a. According to classical physics, if $E < U_0$, the particle can *never* get out of the box; it doesn't have enough energy.

However, if $U_1 < E < U_0$, the classical prediction is wrong; instead, there is a nonzero probability of finding the particle *outside the box* at a later time. The wave function of the particle decays exponentially only from $x = 0$ to $x = a$; for $x > a$ it becomes sinusoidal again, although with a reduced amplitude due to the exponential decay in the barrier (Fig. 28.26b). The amplitude of the wave function for $x > a$ determines the probability per second that the particle is found outside the box.

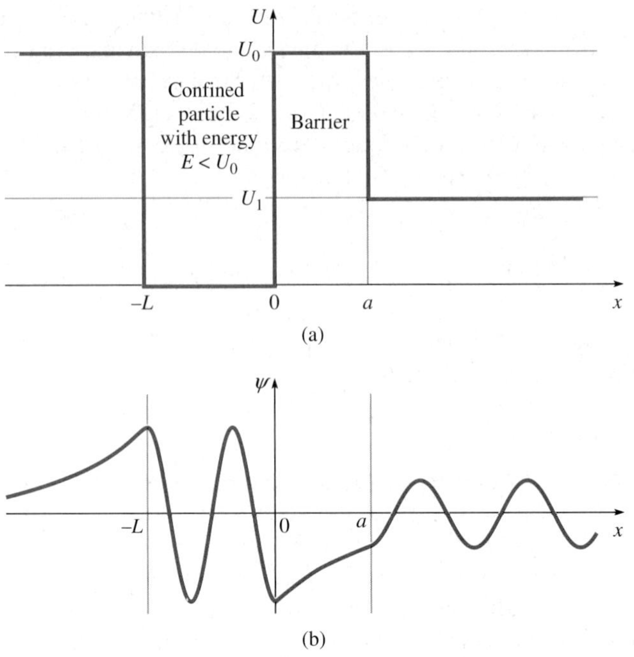

Figure 28.26 (a) A particle of energy $E < U_0$ is confined to a finite box of length L. The potential energy inside the box is taken to be zero; the potential energy on either side of the box is U_0. To the right of the barrier, the potential energy is U_1. For $U_1 < E < U_0$, the particle can tunnel out of the box. (b) Sketch of the wave function for a particle that can tunnel out of the box.

Since the wave function decays exponentially in the barrier, the tunneling probability decreases dramatically as the barrier thickness increases. For a relatively wide barrier, the tunneling probability decreases exponentially with barrier thickness:

$$P \propto e^{-2\kappa a} \qquad (28\text{-}17)$$

In Eq. (28-17), P is the probability per unit time that tunneling occurs, a is the barrier thickness, and κ is a measure of the barrier height:

$$\kappa = \sqrt{\frac{2m}{\hbar^2}(U_0 - E)} \qquad (28\text{-}18)$$

Equation (28-17) is an approximation valid when $e^{-2\kappa a} \ll 1$. The tunneling probability's dependence on barrier thickness is more complicated for an extremely thin barrier.

It is also possible for a particle to tunnel *into* a box. A particle initially to the right of the barrier in Fig. 28.26 can later be found inside the box (on the left side of the barrier).

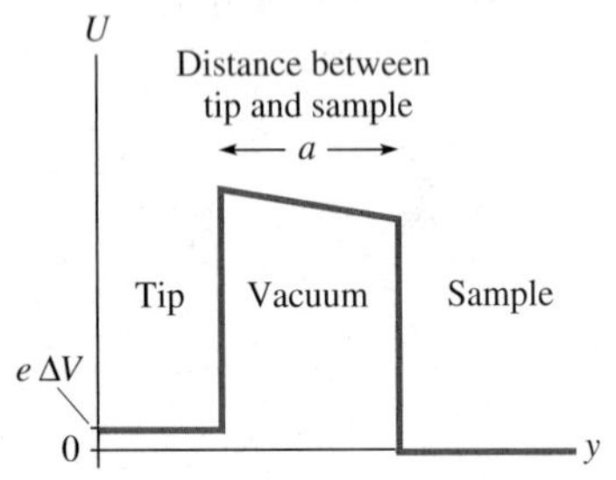

Figure 28.27 Simplified model of the potential energy of an electron that tunnels from the tip of an STM to a sample a distance a away. An applied potential difference ΔV causes a potential energy difference of magnitude $e\,\Delta V$ between tip and sample. Normally, an electron must be supplied with an energy equal to the work function of the metal—a few electron-volts—to break free of the metal. Here, because the tip and sample are only a few nanometers apart, an electron can tunnel through the barrier presented by the work functions of the metals.

Application: The Scanning Tunneling Microscope

The scanning tunneling microscope (STM) exploits the exponential dependence of tunneling probability on barrier thickness to produce highly magnified images of surfaces. In an STM, a very fine metal tip is placed very close to a surface of interest. The tip must be much finer than an ordinary needle—it ideally should have a single atom at the tip. The distance between the tip and the sample is typically only a few nanometers. The apparatus must be isolated from vibrations, which under ordinary circumstances have amplitudes of 1000 nm or more. The sample and tip are in an evacuated chamber.

A small potential difference $\Delta V \approx 10$ mV is applied between the tip and the sample. Electrons now tunnel between the tip and the sample. The barrier they tunnel through is due to the work functions of the tip and the sample (Fig. 28.27); an electron bound to the metal has a lower energy than one that is outside of the metal.

As the tip is scanned over the surface, its distance from the sample is adjusted to keep the tunneling current constant (Fig. 28.28). Since the current depends exponentially

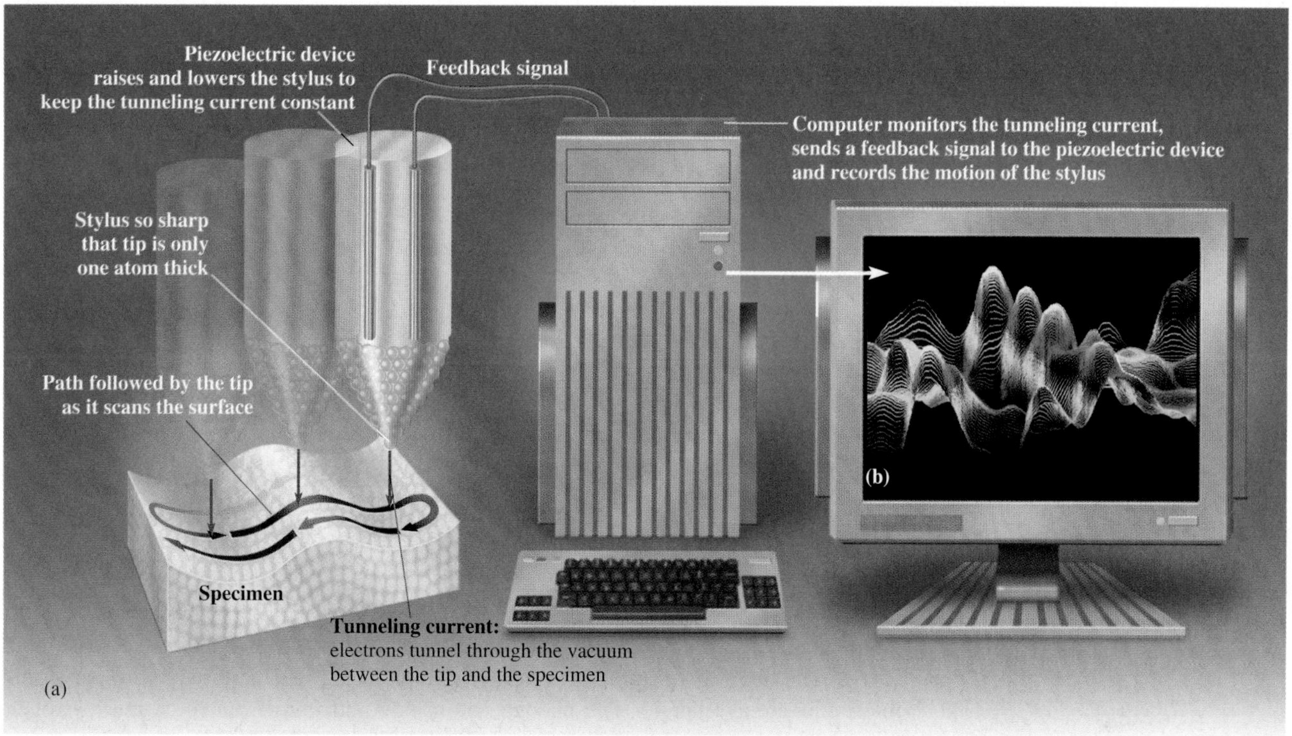

Figure 28.28 (a) Schematic of an STM. (b) Scanning tunneling micrograph of a section of a DNA molecule. The average distance between the coils of the helix (seen as yellow peaks) is 3.5 nm.

on the distance a, the tip is moved to keep a constant. Thus, the movements of the tip accurately reflect the surface beneath. An STM is easily able to image individual atoms on a surface (see Fig. 28.14).

Example 28.6

Change in Tunneling Current

Suppose that an STM scans a surface at a distance of $a = 1.000$ nm. Take the height of the potential energy barrier to be $U_0 - E = 2.00$ eV. If the distance between the surface and the STM tip decreases by 1.0% (= 0.010 nm, which is about one fifth the radius of the smallest atom), estimate the percentage change in the tunneling current.

Strategy The tunneling current is proportional to the number of electrons that tunnel per second, which is in turn proportional to the tunneling probability per second [P in Eq. (28-17)]. Thus, the ratio of the probabilities per second is equal to the ratio of the tunneling currents.

Solution The tunneling probability per unit time is

$$P \propto e^{-2\kappa a} \qquad (28\text{-}17)$$

where

$$\kappa = \sqrt{\frac{2m}{\hbar^2}(U_0 - E)} \qquad (28\text{-}18)$$

$$= \sqrt{\frac{2 \times 9.109 \times 10^{-31}\ \text{kg}}{[6.626 \times 10^{-34}\ \text{J·s}/(2\pi)]^2} \times (2.00\ \text{eV} \times 1.602 \times 10^{-19}\ \text{J/eV})}$$

$$= 7.245 \times 10^{9}\ \text{m}^{-1}$$

Since the tip moves 0.010 nm closer to the surface, the distance changes from $a = 1.000$ nm to $a' = 0.990$ nm. The ratio of the tunneling probabilities is

$$\frac{P_{a'}}{P_a} = \frac{e^{-2\kappa a'}}{e^{-2\kappa a}} = e^{-2\kappa(a' - a)}$$

The quantity in the exponent is

$$2\kappa(a' - a) = 2 \times 7.245 \times 10^{9}\ \text{m}^{-1} \times (-0.010 \times 10^{-9}\ \text{m})$$
$$= -0.1449$$

continued on next page

Example 28.6 continued

The ratio of the probabilities per unit time is

$$\frac{P_{a'}}{P_a} = e^{0.1449} = 1.16$$

Then the ratio of the currents is also 1.16. A 1.0% decrease in the distance between tip and sample causes a 16% increase in the tunneling current.

Discussion A decrease in distance means an increase in tunneling current, as expected. The large change in current for a small change in distance is due to the exponential fall-off of the wave function in the forbidden region; it makes the STM a very sensitive instrument.

Let us check the units in the calculation for κ:

$$\sqrt{\frac{\text{kg}}{\text{J}^2\cdot\text{s}^2}\times \text{J}} = \sqrt{\frac{\text{kg}}{\text{s}^2}\times\frac{1}{\text{J}}} = \sqrt{\frac{\text{kg}}{\text{s}^2}\times\frac{\text{s}^2}{\text{kg}\cdot\text{m}^2}} = \text{m}^{-1}$$

Practice Problem 28.6 Change in Tunneling Current When Tip Moves Away

Estimate the percentage change in tunneling current if the tip moves *away* by 1.00% (from 1.0000 nm to 1.0100 nm).

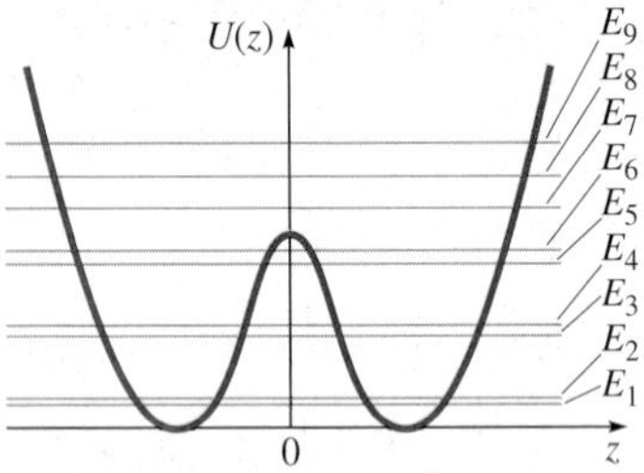

Figure 28.29 Potential energy of the nitrogen atom in the NH_3 molecule as a function of its position along the z-axis, which is perpendicular to the plane of the three H atoms. For the lowest six vibrational energy levels, the nitrogen atom tunnels from one side to the other.

Application: An Atomic Clock Based on Tunneling

Tunneling in the ammonia molecule (NH_3) was exploited to make the first atomic clocks. The three-dimensional structure of the molecule has the three hydrogen atoms in an equilateral triangle. The nitrogen atom is equidistant from the three hydrogen atoms. The nitrogen atom has two possible equilibrium positions: it can be on either side of the plane of the H atoms.

The potential energy of the nitrogen atom is shown in Fig. 28.29. The equilibrium positions are the two minima in $U(z)$. The barrier between the two is due to the Coulomb repulsion between the atoms. In the ground state of the NH_3 molecule, the N atom does not have enough energy to move back and forth along the z-axis between the two equilibrium positions. However, it *does* oscillate back and forth between the two positions: the N atom *tunnels* back and forth through the potential energy barrier. The tunneling probability determines the frequency of oscillation, which is 2.4×10^{10} Hz. Since the oscillation depends on tunneling, this frequency is much lower than a typical molecular vibration frequency, making it easier to use as a time standard for the first atomic clocks.

Master the Concepts

- In quantum physics the two descriptions, particle and wave, are complementary. The wavelength of a particle is called its de Broglie wavelength:

$$\lambda = \frac{h}{p} \quad (28\text{-}1)$$

- The uncertainty principle sets limits on how precisely we can simultaneously determine the position and momentum of a particle:

$$\Delta x\, \Delta p_x \geq \frac{1}{2}\hbar \quad (28\text{-}2)$$

- If a system is in a quantum state for a time interval Δt, then the uncertainty in the energy of that state is related to the lifetime of that state by the energy-time uncertainty principle:

$$\Delta E\, \Delta t \geq \frac{1}{2}\hbar \quad (28\text{-}3)$$

- Confined particles have wave functions that are standing waves. Confinement leads to the quantization of de Broglie wavelengths and energies.
- A particle in a one-dimensional box has wavelengths analogous to those of a standing wave on a string:

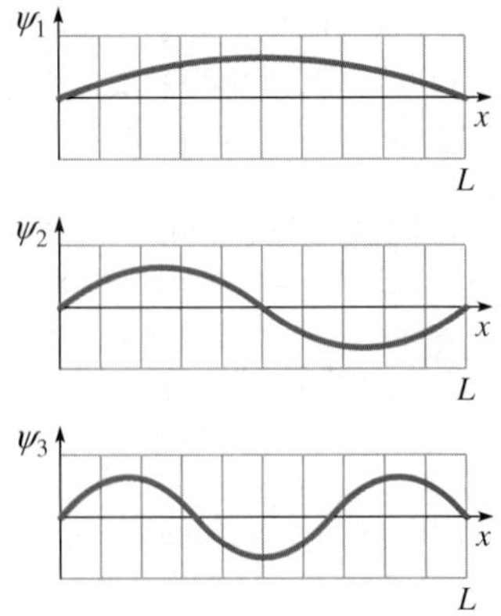

$$\lambda_n = \frac{2L}{n} \quad (n = 1, 2, 3, \ldots) \quad (28\text{-}4)$$

continued on next page

Master the Concepts continued

- The square of the magnitude of the wave function is proportional to the probability of locating the particle in a given region of space.
- The quantum state of the electron in an atom can be described by four quantum numbers:

  principal quantum number $n = 1, 2, 3, \ldots$
  orbital angular momentum quantum number $\ell = 0, 1, 2, 3, \ldots, n-1$
  magnetic quantum number $m_\ell = -\ell, -\ell+1, \ldots, -1, 0, 1, \ldots, \ell-1, \ell$
  spin magnetic quantum number $m_s = -\frac{1}{2}, +\frac{1}{2}$
- According to the exclusion principle, no two electrons in an atom can be in the same quantum state.
- The set of electron states with the same value of n is called a shell. A subshell is a unique combination of n and ℓ. Spectroscopic notation for a subshell is the numerical value of n followed by a letter representing the value of ℓ.
- In a solid, the electron states form bands of closely spaced energy levels. Band gaps are ranges of energy in which there are no electron energy levels. Conductors, semiconductors, and insulators are distinguished by their band structure.
- If an electron is in a higher energy level and a lower level is vacant, an incident photon of energy ΔE can stimulate the emission of a photon. The photon emitted by the atom is *identical* to the incident photon.
- Lasers are based on stimulated emission. In order for stimulated emission to occur more often than absorption, a population inversion must exist (the state of higher energy must be more populated than the state of lower energy).
- The wave function of a confined particle extends into regions where, according to classical physics, the particle can never go because it has insufficient energy. If the classically forbidden region is of finite length, tunneling can occur.

Conceptual Questions

1. An electron diffraction experiment gives the same pattern as an x-ray diffraction experiment with the same sample. How do we know the wavelengths of the electrons and x-rays are the same? Would they give the same pattern if their *energies* were the same?
2. In the Bohr model, the electron in the ground state of the hydrogen atom is in a circular orbit of radius 0.0529 nm. Explain how the Bohr model is incompatible with the uncertainty principle. How does the quantum mechanical picture of the H atom differ from the Bohr model? In what ways are the two similar?
3. It is sometimes said that, at absolute zero, all molecular motion, vibration, and rotation would cease. Do you agree? Explain.
4. The uncertainty principle does not allow us to think of the electron in an atom as following a well-defined trajectory. Why, then, are we able to define trajectories for golf balls, comets, and the like? [*Hint:* How are the uncertainties in momentum and velocity related?]
5. We often refer to the state of the hydrogen atom as "the $n = 3$ state," for example. Under what circumstances do we only need to specify one of the four quantum numbers? Under what circumstances would we have to be more specific?
6. Why does a particle confined to a finite box have only a finite number of bound states?
7. How should we interpret electron cloud representations of electron states in atoms?
8. Describe some differences between the beam of light from a flashlight and from a laser.
9. In an optically pumped laser, the light that causes optical pumping is always shorter in wavelength than the laser beam. Explain.
10. Explain why a population inversion is necessary in a laser.
11. The Nd:YAG laser operates in a four-state cycle as shown in the figure, and the ruby laser operates in a three-state cycle (compare with Fig. 28.24b). In which laser is it easier to maintain a population inversion? Why? Explain why the Nd:YAG laser can produce a continuous beam, but the ruby laser can produce only brief pulses of laser light.

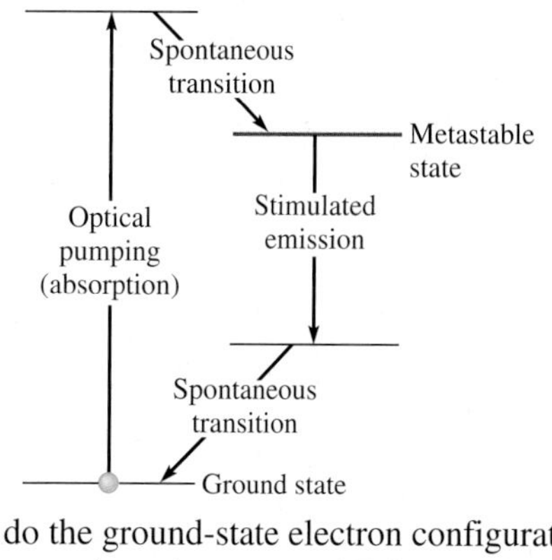

12. What do the ground-state electron configurations of the noble gases have in common? Why are the noble gases chemically nonreactive?
13. Central to the operation of a photocopy machine (see Section 16.2) is a drum coated with a photoconductor—

a semiconductor that is a good insulator in the dark but allows charge to flow freely when illuminated with light. How does light allow charge to flow freely through the semiconducting material? How large should the band gap be for a good photoconductor? If the drum gets hot, is the contrast between light and dark areas on the image improved or degraded?

14. Why does a confined particle have quantized energy levels?
15. How can we demonstrate the existence of matter waves?
16. When a particle's kinetic energy increases, what happens to its de Broglie wavelength?
17. Explain why the electrical resistivity of a semiconductor decreases with increasing temperature.
18. When aluminum is exposed to oxygen, a *very thin* layer of aluminum oxide forms on the outside. Aluminum oxide is a good insulator. Nevertheless, if two aluminum wires are twisted together, electric current can flow from one to the other, even if the oxide layer has not been cleaned off. How is this possible?

Multiple-Choice Questions

1. What happens to the energy level spacings for a particle in a box when the box is made longer?
   (a) The spacings decrease.
   (b) The spacings increase.
   (c) The spacings stay the same.
   (d) Insufficient information to answer this question.
2. Which of the lettered transitions on the energy level diagram would be the best candidate for light amplification by stimulated emission?

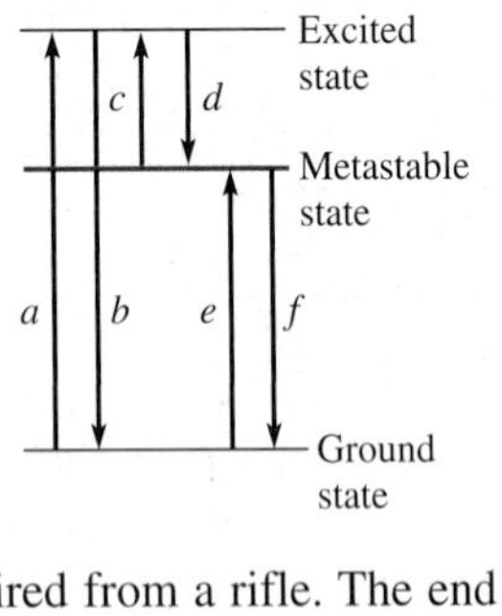

3. A bullet is fired from a rifle. The end of the barrel is a circular aperture. Is diffraction a measurable effect?
   (a) No, because only charged particles have de Broglie wavelengths.
   (b) No, because a circular aperture never causes diffraction.
   (c) No, because the de Broglie wavelength of the bullet is too large.
   (d) No, because the de Broglie wavelength of the bullet is too small.
   (e) Yes.
4. An electron and a neutron have the same de Broglie wavelength. Which is true?
   (a) The electron has more kinetic energy and a higher speed.
   (b) The electron has less kinetic energy but a higher speed.
   (c) The electron has less kinetic energy and a lower speed.
   (d) The electron and neutron have the same kinetic energy but the electron has the higher speed.
   (e) The neutron has more kinetic energy but the two have the same speed.
5. Which one of these statements is true?
   (a) The atomic spacing in crystals is too fine to produce observable diffraction effects with matter waves.
   (b) Only charged particles have matter waves associated with them.
   (c) Identical diffraction patterns are obtained when either electrons or neutrons of the same kinetic energy are incident on a single crystal.
   (d) Electrons, neutrons, and x-rays of appropriate energies can all produce similar diffraction patterns when incident on single crystals.
   (e) Wave phenomena are not observed for macroscopic objects such as baseballs because the de Broglie wavelength associated with such macroscopic objects is too long.
6. A particle is incident from the left on a slit of width w and the particle passes through the slit opening. The uncertainty principle restricts which of these quantities?

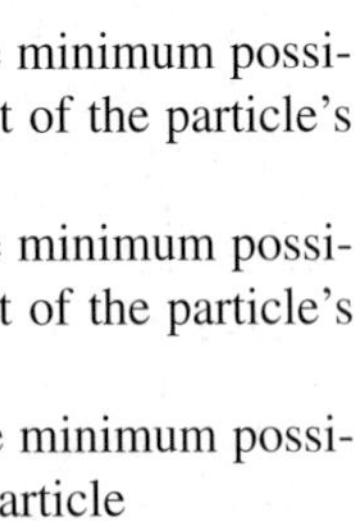

   (a) the product of the width w and the minimum possible uncertainty in the y-component of the particle's momentum
   (b) the product of the width w and the minimum possible uncertainty in the x-component of the particle's momentum
   (c) the product of the width w and the minimum possible uncertainty in the z-component of the particle's momentum
   (d) the product of the width w and the minimum possible de Broglie wavelength of the particle
   (e) the maximum possible width w
7. The exclusion principle:
   (a) Implies that in an atom no two electrons can have identical sets of quantum numbers.
   (b) Says that no two electrons in an atom can have the same orbit.
   (c) Excludes electrons from atomic nuclei.
   (d) Excludes protons from electron orbits.
8. Which one of these statements is true?
   (a) The principal quantum number of the electron in a hydrogen atom does not affect its energy.

(b) The principal quantum number of an electron in its ground state is zero.

(c) The orbital angular momentum quantum number of an electron state is always less than the principal quantum number of that state.

(d) The electron spin quantum number can take on any one of four different values.

9. If a particle is confined to a three-dimensional, cubical region of length L on each side:

(a) The particle has an uncertainty in each component of momentum of about $h/(\pi L)$.

(b) The particle cannot have a wavelength less than $2L$.

(c) The components of the particle's momentum in the y- and z-directions can be determined exactly as long as there is a finite uncertainty in the x-component of momentum.

(d) The particle's kinetic energy has an upper limit but no lower limit.

10. Which of these statements about electron energy levels in hydrogen atoms is true?

(a) An electron in the hydrogen atom is best represented as a traveling wave.

(b) An electron with positive total energy is a bound electron.

(c) An electron in a stable energy level radiates electromagnetic waves because the electron is accelerating as it moves around the nucleus.

(d) The orbital angular momentum of an electron in the ground state is zero.

(e) An electron in state n can make transitions only to the state $n + 1$ or $n - 1$.

Problems

- Ⓒ Combination conceptual/quantitative problem
- Biological or medical application
- ✦ Challenging problem
- Blue # Detailed solution in the Student Solutions Manual
- (1 2) Problems paired by concept
- Text website interactive or tutorial

28.2 Matter Waves

1. What is the de Broglie wavelength of a basketball of mass 0.50 kg when it is moving at 10 m/s? Why don't we see diffraction effects when a basketball passes through the circular aperture of the hoop?

2. A fly with a mass of 1.0×10^{-4} kg crawls across a table at a speed of 2 mm/s. Compute the de Broglie wavelength of the fly and compare it with the size of a proton (about 1 fm, 1 fm $= 10^{-15}$ m).

3. An 81-kg student who has just studied matter waves is concerned that he may be diffracted as he walks through a doorway that is 81 cm across and 12 cm thick. (a) If the wavelength of the student must be about the same size as the doorway to exhibit diffraction, what is the fastest the student can walk through the doorway to exhibit diffraction? (b) At this speed, how long would it take the student to walk through the doorway?

4. What is the magnitude of the momentum of an electron with a de Broglie wavelength of 0.40 nm?

5. What is the de Broglie wavelength of an electron moving at speed $\frac{3}{5}c$?

6. The distance between atoms in a crystal of NaCl is 0.28 nm. The crystal is being studied in a neutron diffraction experiment. At what speed must the neutrons be moving so that their de Broglie wavelength is 0.28 nm?

7. An x-ray diffraction experiment using 16-keV x-rays is repeated using electrons instead of x-rays. What should the kinetic energy of the electrons be in order to produce the same diffraction pattern as the x-rays (using the same crystal)?

8. What are the de Broglie wavelengths of electrons with the following values of kinetic energy? (a) 1.0 eV; (b) 1.0 keV. (tutorial: de Broglie wavelength)

9. What is the ratio of the wavelength of a 0.100-keV photon to the wavelength of a 0.100-keV electron?

✦10. Neutron diffraction by a crystal can be used to make a velocity selector for neutrons. Suppose the spacing between the relevant planes in the crystal is $d = 0.20$ nm. A beam of neutrons is incident at an angle $\theta = 10.0°$ with respect to the planes. The incident neutrons have speeds ranging from 0 to 2.0×10^4 m/s. (a) What wavelength(s) are strongly reflected from these planes? [*Hint:* Bragg's law, Eq. (25-15), applies to neutron diffraction as well as to x-ray diffraction.] (b) For each of the wavelength(s), at what angle with respect to the incident beam do those neutrons emerge from the crystal?

11. A nickel crystal is used as a diffraction grating for x-rays. Then the same crystal is used to diffract electrons. If the two diffraction patterns are identical, and the energy of each x-ray photon is $E = 20.0$ keV, what is the kinetic energy of each electron?

28.3 Electron Microscopes

12. If diffraction were the only limitation on resolution, what would be the smallest structure that could be resolved in an electron microscope using 10-keV electrons?

13. To resolve details of an object, you must use a wavelength that is about the same size, or smaller, than the details you want to observe. Suppose you want to study a molecule that is about 1.000×10^{-10} m in length. (a) What minimum photon energy is required to study this molecule? (b) What is the minimum kinetic energy of electrons that could study this? (c) Through what potential difference should the electrons be accelerated to reach this energy?

14. A scanning electron microscope is used to look at cell structure with 10-nm resolution. A beam of electrons from a hot filament is accelerated with a voltage of 12 kV and then focused to a small spot on the specimen. (a) What is the wavelength in nanometers of the beam of incoming electrons? (b) If the size of the focal spot were determined only by diffraction, and if the diameter of the electron lens is one fifth the distance from the lens to the specimen, what would be the minimum separation resolvable on the specimen? (In practice, the resolution is limited much more by aberrations in the magnetic lenses and other factors.)

15. An image of a biological sample is to have a resolution of 5 nm. (a) What is the kinetic energy of a beam of electrons with a de Broglie wavelength of 5.0 nm? (b) Through what potential difference should the electrons be accelerated to have this wavelength? (c) Why not just use a light microscope with a wavelength of 5 nm to image the sample?

16. The phenomenon of Brownian motion is the random motion of microscopically small particles as they are buffeted by the still smaller molecules of a fluid in which they are suspended. For a particle of mass 1.0×10^{-16} kg, the fluctuations in velocity are of the order of 0.010 m/s. For comparison, how large is the change in this particle's velocity when the particle absorbs a photon of light with a wavelength of 660 nm, such as might be used in observing its motion under a microscope?

28.4 The Uncertainty Principle

17. If the momentum of the basketball in Problem 1 has a fractional uncertainty of $\Delta p/p = 10^{-6}$, what is the uncertainty in its position?

18. An electron passes through a slit of width 1.0×10^{-8} m. What is the uncertainty in the electron's momentum component in the direction perpendicular to the slit but in the plane containing the slit?

19. At a baseball game, a radar gun measures the speed of a 144-g baseball to be 137.32 ± 0.10 km/h. (a) What is the minimum uncertainty of the position of the baseball? (b) If the speed of a proton is measured to the same precision, what is the minimum uncertainty in its position?

20. A hydrogen atom has a radius of about 0.05 nm. (a) Estimate the uncertainty in any component of the momentum of an electron confined to a region of this size. (b) From your answer to (a), estimate the electron's kinetic energy. (c) Does the estimate have the correct order of magnitude? (The ground-state kinetic energy predicted by the Bohr model is 13.6 eV.)

21. A bullet with mass 10.000 g has a speed of 300.00 m/s; the speed is accurate to within 0.04%. (a) Estimate the minimum uncertainty in the position of the bullet, according to the uncertainty principle. (b) An electron has a speed of 300.00 m/s, accurate to 0.04%. Estimate the minimum uncertainty in the position of the electron. (c) What can you conclude from these results?

22. A radar pulse has an average wavelength of 1.0 cm and lasts for 0.10 μs. (a) What is the average energy of the photons? (b) Approximately what is the least possible uncertainty in the energy of the photons?

23. A beam of electrons passes through a single slit 40.0 nm wide. The width of the central fringe of a diffraction pattern formed on a screen 1.0 m away is 6.2 cm. What is the kinetic energy of the electrons passing through the slit?

24. Electrons are accelerated through a potential difference of 38.0 V. The beam of electrons then passes through a single slit. The width of the central fringe of a diffraction pattern formed on a screen 1.00 m away is 1.13 mm. What is the width of the slit?

25. The omega particle (Ω) decays on average about 0.1 ns after it is created. Its rest energy is 1672 MeV. Estimate the fractional uncertainty in the Ω's rest energy ($\Delta E_0/E_0$). [*Hint:* Use the energy-time uncertainty principle, Eq. (28-3).]

26. Nuclei have energy levels just as atoms do. An excited nucleus can make a transition to a lower energy level by emitting a gamma-ray photon. The lifetime of a typical nuclear excited state is about 1 ps. What is the uncertainty in the energy of the gamma-rays emitted by a typical nuclear excited state? [*Hint:* Use the energy-time uncertainty principle, Eq. (28-3).]

28.5 Wave Functions for a Confined Particle

27. What is the minimum kinetic energy of an electron confined to a region the size of an atomic nucleus (1.0 fm)?

28. An electron is confined to a box of length 1.0 nm. What is the magnitude of its momentum in the $n = 4$ state?

29. A marble of mass 10 g is confined to a box 10 cm long and moves at a speed of 2 cm/s. (a) What is the marble's quantum number n? (b) Why can we not observe the quantization of the marble's energy? [*Hint:* Calculate the energy difference between states n and $n + 1$. How much does the marble's speed change?]

30. Suppose the electron in a hydrogen atom is modeled as an electron in a one-dimensional box of length equal to the Bohr diameter, $2a_0$. What would be the ground-state energy of this "atom"? How does this compare with the actual ground-state energy?

31. The particle in a box model is often used to make rough estimates of energy level spacings. For a metal wire 10 cm long, treat a conduction electron as a particle confined to a one-dimensional box of length 10 cm. (a) Sketch the wave function ψ as a function of position for the electron in this box for the ground state and each of the first three excited states. (b) Estimate the spacing

between energy levels of the conduction electrons by finding the energy *spacing* between the ground state and the first excited state.

32. The particle in a box model is often used to make rough estimates of ground-state energies. Suppose that you have a *neutron* confined to a one-dimensional box of length equal to a nuclear diameter (say 10^{-14} m). What is the ground-state energy of the confined neutron?

33. An electron confined to a one-dimensional box has a ground-state energy of 40.0 eV. (a) If the electron makes a transition from its first excited state to the ground state, what is the wavelength of the emitted photon? (b) If the box were somehow made twice as long, how would the photon's energy change for the same transition (first excited state to ground state)?

34. An electron is confined to a one-dimensional box. When the electron makes a transition from its first excited state to the ground state, it emits a photon of energy 1.2 eV. (a) What is the ground-state energy (in electron-volts) of the electron? (b) List all energies (in electron-volts) of photons that could be emitted when the electron starts in its second excited state and makes transitions downward to the ground state either directly or through intervening states. Show all these transitions on an energy level diagram. (c) What is the length of the box (in nanometers)?

28.6 The Hydrogen Atom: Wave Functions and Quantum Numbers; 28.7 The Exclusion Principle; Electron Configurations for Atoms Other than Hydrogen

35. What is the ground state electron configuration of a K^+ ion?

36. How many electron states of the H atom have the quantum numbers $n = 3$ and $\ell = 1$? Identify each state by listing its quantum numbers.

37. What are the possible values of L_z (the component of angular momentum along the z-axis) for the electron in the second excited state ($n = 3$) of the hydrogen atom?

38. What is the largest number of electrons with the same pair of values for n and ℓ that an atom can have? (WW tutorial: neon atom)

39. List the number of electron states in each of the subshells in the $n = 7$ shell. What is the total number of electron states in this shell?

40. What is the ground-state electron configuration of nickel (Ni, atomic number 28)?

41. What is the ground-state electron configuration of bromine (Br, atomic number 35)?

42. What is the maximum possible value of the angular momentum for an outer electron in the ground state of a bromine atom?

43. (a) What are the electron configurations of the ground states of lithium ($Z = 3$), sodium ($Z = 11$), and potassium ($Z = 19$)? (b) Why are these elements placed in the same column of the periodic table?

44. (a) What are the electron configurations of the ground states of fluorine ($Z = 9$) and chlorine ($Z = 17$)? (b) Why are these elements placed in the same column of the periodic table?

45. What is the electronic configuration of the ground state of the carbon atom? Write it in the following ways: (a) using spectroscopic notation ($1s^2$ . . .); (b) listing the four quantum numbers for each of the electrons. Note that there may be more than one possibility in (b).

46. (a) Find the magnitude of the angular momentum $\vec{\mathbf{L}}$ for an electron with $n = 2$ and $\ell = 1$ in terms of $\hbar$. (b) What are the allowed values for L_z? (c) What are the angles between the positive z-axis and $\vec{\mathbf{L}}$ so that the quantized components, L_z, have allowed values?

47. (a) Show that the ground-state energy of the hydrogen atom can be written $E_1 = -ke^2/(2a_0)$, where a_0 is the Bohr radius. (b) Explain why, according to classical physics, an electron with energy E_1 could never be found at a distance greater than $2a_0$ from the nucleus.

28.8 Electron Energy Levels in a Solid

48. A light-emitting diode (LED) has the property that electrons can be excited into the conduction band by the electrical energy from a battery; a photon is emitted when the electron drops back to the valence band. (a) If the band gap for this diode is 2.36 eV, what is the wavelength of the light emitted by the LED? (b) What color is the light emitted? (c) What is the minimum battery voltage required in the electrical circuit containing the diode?

49. A photoconductor (see Conceptual Question 13) allows charge to flow freely when photons of wavelength 640 nm or less are incident on it. What is the band gap for this photoconductor?

28.9 Lasers

50. What is the wavelength of the light usually emitted by a helium-neon laser? (See Fig. 28.24.)

51. Many lasers, including the helium-neon, can produce beams at more than one wavelength. Photons can stimulate emission and cause transitions between the 20.66-eV metastable state and several different states of lower energy. One such state is 18.38 eV above the ground state. What is the wavelength for this transition? If only these photons leave the laser to form the beam, what color is the beam?

52. In a ruby laser, laser light of wavelength 694.3 nm is emitted. The ruby crystal is 6.00 cm long, and the index of refraction of ruby is 1.75. Think of the light in the

ruby crystal as a standing wave along the length of the crystal. How many wavelengths fit in the crystal? (Standing waves in the crystal help to reduce the range of wavelengths in the beam.)

53. The beam emerging from a ruby laser passes through a circular aperture 5.0 mm in diameter. (a) If the spread of the beam is limited only by diffraction, what is the angular spread of the beam? (b) If the beam is aimed at the Moon, how large a spot would be illuminated on the Moon's surface?

28.10 Tunneling

✦54. A proton and a deuteron (which has the same charge as the proton but 2.0 times the mass) are incident on a barrier of thickness 10.0 fm and "height" 10.0 MeV. Each particle has a kinetic energy of 3.0 MeV. (a) Which particle has the higher probability of tunneling through the barrier? (b) Find the ratio of the tunneling probabilities.

✦55. Refer to Example 28.6. Estimate the percentage change in the tunneling current if the distance between the sample surface and the STM tip increases 2.0%.

Comprehensive Problems

56. Mitch drops a 2.0-g coin into a 3.0-m-deep wishing well. What is the de Broglie wavelength of the coin just before it hits the bottom of the well?

57. A magnesium ion Mg^{2+} is accelerated through a potential difference of 22 kV. What is the de Broglie wavelength of this ion?

58. The energy-time uncertainty principle allows for the creation of virtual particles, that appear from a vacuum for a very brief period of time Δt, then disappear again. This can happen as long as $\Delta E\,\Delta t = \hbar/2$, where ΔE is the rest energy of the particle. (a) How long could an electron created from the vacuum exist according to the uncertainty principle? (b) How long could a shot put with a mass of 7 kg created from the vacuum exist according to the uncertainty principle?

59. An electron moving in the positive x-direction passes through a slit of width $\Delta y = 85$ nm. What is the minimum uncertainty in the electron's velocity in the y-direction?

60. In Fig. 28.4b, the x-rays had a frequency of 1.0×10^{19} Hz. Through what potential difference were the electrons in Fig. 28.4a accelerated?

61. The neutrons produced in fission reactors have a wide range of kinetic energies. After the neutrons make several collisions with atoms, they give up their excess kinetic energy and are left with the same *average* kinetic energy as the atoms, which is $\frac{3}{2}k_BT$. If the temperature of the reactor core is $T = 400.0$ K, find (a) the average kinetic energy of the thermal neutrons, and (b) the de Broglie wavelength of a neutron with this kinetic energy.

62. A double-slit interference experiment is performed with 2.0-eV photons. The same pair of slits is then used for an experiment with electrons. What is the kinetic energy of the electrons if the interference pattern is the same as for the photons (i.e., the spacing between maxima is the same)?

63. An electron is confined in a one-dimensional box of length L. Another electron is confined in a box of length $2L$. Both are in the ground state. What is the ratio of their energies E_{2L}/E_L?

◉64. Before the discovery of the neutron, one theory of the nucleus proposed that the nucleus contains protons and electrons. For example, the helium-4 nucleus would contain 4 protons and 2 electrons instead of—as we now know to be true—2 protons and 2 neutrons. (a) *Assuming that the electron moves at nonrelativistic speeds*, find the ground-state energy in mega-electron-volts of an electron confined to a one-dimensional box of length 5.0 fm (the approximate diameter of the ^{4}He nucleus). (The electron actually does move at relativistic speeds. See Problem 80.) (b) What can you conclude about the electron-proton model of the nucleus? The binding energy of the ^{4}He nucleus—the energy that would have to be supplied to break the nucleus into its constituent particles—is about 28 MeV. (c) Repeat (a) for a neutron confined to the nucleus (instead of an electron). Compare your result with (a) and comment on the viability of the proton-neutron theory relative to the electron-proton theory.

65. What is the ground-state electron configuration of tellurium (Te, atomic number 52)?

66. A free neutron (that is, a neutron on its own rather than in a nucleus) is not a stable particle. Its average lifetime is 15 min, after which it decays into a proton, an electron, and an antineutrino. Use the energy-time uncertainty principle [Eq. (28-3)] and the relationship between mass and rest energy to estimate the inherent uncertainty in the mass of a free neutron. Compare with the average neutron mass of 1.67×10^{-27} kg. (Although the uncertainty in the neutron's mass is far too small to be measured, unstable particles with extremely short lifetimes have marked variation in their measured masses.)

67. A beam of electrons is accelerated across a potential of 15 kV before passing through two slits. The electrons form a interference pattern on a screen 2.5 m in front of the slits. The first-order maximum is 8.3 mm from the central maximum. What is the distance between the slits?

◉68. A bullet leaves the barrel of a rifle with a speed of 300.0 m/s. The mass of the bullet is 10.0 g. (a) What is the de Broglie wavelength of the bullet? (b) Compare λ

with the diameter of a proton (about 1 fm). (c) Is it possible to observe wave properties of the bullet, such as diffraction? Explain.

69. A beam of neutrons is used to study molecular structure through a series of diffraction experiments. A beam of neutrons with a wide range of de Broglie wavelengths comes from the core of a nuclear reactor. In a time-of-flight technique, used to select neutrons with a small range of de Broglie wavelengths, a pulse of neutrons is allowed to escape from the reactor by opening a shutter very briefly. At a distance of 16.4 m downstream, a second shutter is opened very briefly 13.0 ms after the first shutter. (a) What is the speed of the neutrons selected? (b) What is the de Broglie wavelength of the neutrons? (c) If each shutter is open for 0.45 ms, estimate the *range* of de Broglie wavelengths selected.

70. The particle in a box model is often used to make rough estimates of energy level spacings. Suppose that you have a proton confined to a one-dimensional box of length equal to a nuclear diameter (about 10^{-14} m). (a) What is the energy difference between the first excited state and the ground state of this proton in the box? (b) If this energy is emitted as a photon as the excited proton falls back to the ground state, what is the wavelength and frequency of the electromagnetic wave emitted? In what part of the spectrum does it lie? (c) Sketch the wave function ψ as a function of position for the proton in this box for the ground state and each of the first three excited states.

71. An electron is confined to a one-dimensional box of length L. When the electron makes a transition from its first excited state to the ground state, it emits a photon of energy 0.20 eV. (a) What is the ground-state energy (in electron-volts) of the electron in this box? (b) What are the energies (in electron-volts) of the photons that can be emitted when the electron starts in its third excited state and makes transitions downwards to the ground state (either directly or through intervening states)? (c) Sketch the wave function of the electron in the third excited state. (d) If the box were somehow made longer, how would the electron's new energy level spacings compare with its old ones? (Would they be greater, smaller, or the same? Or is more information needed to answer this question? Explain.)

72. An electron in an atom has an angular momentum quantum number of 2. (a) What is the magnitude of the angular momentum of this electron in terms of $\hbar$? (b) What are the possible values for the z-components of this electron's angular momentum? (c) Draw a diagram showing possible orientations of the angular momentum vector $\vec{\mathbf{L}}$ relative to the z-axis. Indicate the angles with respect to the z-axis.

73. In the Davisson-Germer experiment (Section 28.2), the electrons were accelerated through a 54.0-V potential difference before striking the target. (a) Find the de Broglie wavelength of the electrons. (b) Bragg plane spacings for nickel were known at the time; they had been determined through x-ray diffraction studies. The largest plane spacing (which gives the largest intensity diffraction maxima) in nickel is 0.091 nm. Using Bragg's law [Eq. (25-15)], find the Bragg angle for the first-order maximum using the de Broglie wavelength of the electrons. (c) Does this agree with the observed maximum at a scattering angle of 130°? [*Hint:* The scattering angle and the Bragg angle are not the same. Make a sketch to show the relationship between the two angles.]

74. A beam of neutrons has the same de Broglie wavelength as a beam of photons. Is it possible that the energy of each photon is equal to the kinetic energy of each neutron? If so, at what de Broglie wavelength(s) does this occur? [*Hint:* For the neutron, use the relativistic energy-momentum relation $E^2 = E_0^2 + (pc)^2$.]

75. (a) Make a qualitative sketch of the wave function for the $n = 5$ state of an electron in a *finite* box [$U(x) = 0$ for $0 < x < L$ and $U(x) = U_0 > 0$ elsewhere]. (b) If $L = 1.0$ nm and $U_0 = 1.0$ keV, *estimate* the number of bound states that exist.

76. An electron is confined to a one-dimensional box of length L. (a) Sketch the wave function for the third excited state. (b) What is the energy of the third excited state? (c) The potential energy can't really be infinite outside of the box. Suppose that $U(x) = +U_0$ outside the box, where U_0 is large but finite. Sketch the wave function for the third excited state of the electron in the finite box. (d) Is the energy of the third excited state for the finite box less than, greater than, or equal to the value calculated in part (b)? *Explain your reasoning.* [*Hint:* Compare the wavelengths inside the box.] (e) Give a rough estimate of the number of bound states for the electron in the *finite* box in terms of L and U_0.

77. An electron in a one-dimensional box has ground-state energy 0.010 eV. (a) What is the length of the box? (b) Sketch the wave functions for the lowest three energy states of the electron. (c) What is the wavelength of the electron in its second excited state ($n = 3$)? (d) The electron is in its ground state when it absorbs a photon of wavelength 15.5 μm. Find the wavelengths of the photon(s) that could be emitted by the electron subsequently.

78. A particle is confined to a *finite* box of length L. In the nth state, the wave function has $n - 1$ nodes. The wave function must make a smooth transition from sinusoidal inside the box to a decaying exponential outside—there can't be a kink at the wall. (a) Make some sketches to show that the wavelength λ_n inside the box must fall in the range $2L/n < \lambda_n < 2L/(n - 1)$. (b) Show that the energy levels E_n in the finite box satisfy $(n - 1)^2E_1 < E_n < n^2E_1$, where $E_1 = h^2/(8mL^2)$ is the ground-state energy for an *infinite* box of length L.

79. (a) Show that the number of electron states in a subshell is $4\ell + 2$. [*Hint:* First, how many states are in each orbital? Second, how many orbitals are in each subshell? See tutorial: quantum numbers.] (b) By summing the number of states in each of the subshells, show that the number of states in a shell is $2n^2$. [*Hint:* The sum of the first n odd integers, from 1 to $2n - 1$, is n^2. That comes from regrouping the sum in pairs, starting by adding the largest to the smallest:

$$1 + 3 + 5 + \cdots + (2n - 5) + (2n - 3) + (2n - 1)$$
$$= [1 + (2n - 1)] + [3 + (2n - 3)] + [5 + (2n - 5)] + \cdots$$
$$= 2n + 2n + 2n + \cdots = 2n \times \frac{n}{2} = n^2$$

80. Repeat Problem 64(a), this time assuming the electron is ultra-relativistic ($E \approx pc$). Is the assumption justified? (tutorial: electron in box)

Answers to Practice Problems

28.1 0.26 eV

28.2 Increasing energy ⇒ decreasing wavelength; decreasing the wavelength decreases θ for a given fringe, so the spacing between fringes decreases (the pattern contracts).

28.3 $\Delta p_x \approx 10^{-27}$ kg·m/s; $\Delta v_x = \Delta p_x/m \approx 1$ km/s

28.4 $1s^2 2s^2 2p^6 3s^2 3p^3$

28.5 A ruby laser would be ineffective. Blood appears red because it *reflects* red light; the red light emitted by a ruby laser would be largely reflected rather than absorbed.

28.6 −13.5% (a decrease)

Answers to Checkpoints

28.2 At higher speed, the electron's momentum is larger and its de Broglie wavelength is smaller ($\lambda = h/p$).

28.4 In the Bohr model, the electron moves around the nucleus in a well-defined trajectory (a circular orbit). Such a trajectory violates the uncertainty principle for reasons explained in the preceding text paragraph.

28.6 For $n = 2$, $\ell = 0$ or 1. For $\ell = 0$, $m_\ell = 0$. For $\ell = 1$, $m_\ell = -1$, 0, or 1. For any m_ℓ, $m_s = +\frac{1}{2}$ or $-\frac{1}{2}$. There are eight electron states: $(n, \ell, m_\ell, m_s) = \left(2, 0, 0, +\frac{1}{2}\right)$, $\left(2, 0, 0, -\frac{1}{2}\right)$, $\left(2, 1, -1, +\frac{1}{2}\right)$, $\left(2, 1, -1, -\frac{1}{2}\right)$, $\left(2, 1, 0, +\frac{1}{2}\right)$, $\left(2, 1, 0, -\frac{1}{2}\right)$, $\left(2, 1, 1, +\frac{1}{2}\right)$, and $\left(2, 1, 1, -\frac{1}{2}\right)$.

CHAPTER

29

Nuclear Physics

After more than 300 yr, Rembrandt's 1653 painting *Aristotle with a Bust of Homer* needed to be cleaned. Aristotle's black apron showed signs of damage; it was unclear whether any of the original paint had survived underneath the apron. Conservators at the Metropolitan Museum of Art (New York) needed to know as much as possible about the damaged area before undertaking the painting's restoration and cleaning. Art historians wanted to know whether Rembrandt altered the composition as he worked on the painting. To help provide such information, the painting was taken to a nuclear reactor at the Brookhaven National Laboratory. How can a nuclear reactor help conservators and art historians learn about a painting? (See p. 1091 for the answer.)

Concepts & Skills to Review

- Rutherford scattering experiment; discovery of the nucleus (Section 27.6)
- fundamental forces (Section 4.12)
- mass and rest energy (Section 26.7)
- exclusion principle (Section 28.7)
- exponential functions (Appendix A.3, Section 18.10)
- tunneling (Section 28.10)

29.1 NUCLEAR STRUCTURE

In an atom, electrons are bound electrically to a positively charged nucleus. In Chapters 27 and 28, we generally treated the nucleus as a point charge so massive that it is not affected by electric forces on it due to the electrons. In reality, the atomic nucleus is several thousand times more massive than the electrons in an atom and occupies only a tiny fraction of the atom's volume (about 1 part in 10^{12} or less). The finite mass and volume of the nucleus have subtle effects on the electronic configuration and thus on the chemical properties of atoms. However, the nucleus has a complex structure of its own that manifests itself in radioactive decay and nuclear reactions.

The nucleus is a bound collection of protons and neutrons. Together, protons and neutrons are referred to as **nucleons** (particles found inside the nucleus). The **atomic number** Z is the number of protons in the nucleus. Each proton has a charge of $+e$ and the neutron is uncharged, so the electric charge of a nucleus is $+Ze$. The number of electrons in a neutral atom is also equal to Z. The number of protons determines to which element, or chemical species, an atom belongs.

Once it was thought that all atoms of a given element were identical. However, we now know that there exist different **isotopes** of a given element. The isotopes of an element all have the same number of protons in the nucleus, but they have different masses because the number of neutrons (N) differs. The total number of nucleons therefore also differs from one isotope to another. The **nucleon number** A is the total number of protons and neutrons:

$$A = Z + N \tag{29-1}$$

Any particular species of nucleus, called a **nuclide**, is characterized by the values of A and Z. The nucleon number A is also called the **mass number**. Since almost all of the mass of an atom is found in the nucleus, and since protons and neutrons have *approximately* the same mass, the mass of an atom is roughly proportional to the number of nucleons.

Since their masses differ, the isotopes of an element can be separated using a mass spectrometer (Section 19.3). Sometimes the differing masses of isotopes have an effect on chemical reaction rates, but on the whole, the chemical properties of different isotopes are virtually identical. On the other hand, different nuclides have *very* different nuclear properties. The number of neutrons present affects how strongly the nucleus is held together, so that some are stable and others are unstable (**radioactive**). Nuclear energy levels, radioactive half-lives, and radioactive decay modes are all particular to a specific nuclide; they are very different for two isotopes of the same element.

Several notations are used to distinguish nuclides. The chemical symbol O stands for the element oxygen. To specify a particular isotope of oxygen, the mass number must also be specified. Oxygen-18, O-18, O^{18}, and ^{18}O all stand for the isotope of oxygen with $A = 18$. Sometimes it is helpful to include the atomic number as well, even though it is redundant; oxygen by definition has 8 protons. When including the atomic number, the preferred form is $^{18}_{8}O$, although $_8O^{18}$ is found in some older texts.

CHECKPOINT 29.1

In the nuclide $^{23}_{11}Na$, how many protons are in the nucleus? How many neutrons? What is the mass number?

Example 29.1

Finding the Number of Neutrons

How many neutrons are present in an ^{18}O nucleus?

Strategy The superscript gives the number of nucleons (A). We consult the periodic table to find the atomic number (Z) for oxygen. The number of neutrons is $N = A - Z$.

Solution An ^{18}O nucleus has 18 nucleons. Oxygen has atomic number 8, so there are 8 protons in the nucleus. That leaves $18 - 8 = 10$ neutrons.

Discussion Different isotopes of oxygen have different numbers of neutrons but the same number of protons.

Practice Problem 29.1 Identifying the Element

Write the symbol (in the form $^{A}_{Z}X$) for the nuclide with 44 protons and 60 neutrons and identify the element.

Atomic Mass Units It is usually more convenient to write the mass of a nucleus in **atomic mass units** instead of kilograms. The modern symbol for the atomic mass unit is "u"; in older literature it is often written "amu." The atomic mass unit is defined as exactly $\frac{1}{12}$ the mass of a neutral ^{12}C atom. The conversion factor between u and kg is

$$1 \text{ u} = 1.660\,539 \times 10^{-27} \text{ kg} \quad (29\text{-}2)$$

Nucleons have masses of *approximately* 1 u, but the electron is much less massive (Table 29.1). Therefore, the mass of a nucleus (or an atom) is *approximately A* atomic mass units—which is why A is called the mass number.

The atomic mass of an element given in the periodic table is an average over the isotopes of that element in their natural relative abundances on Earth. In nuclear physics we must consult a table of nuclides (Appendix B) for the mass of a specific nuclide.

Table 29.1 Masses and Charges of the Proton, Neutron, and Electron

Particle	Mass (u)	Charge
Proton	1.007 276 5	$+e$
Neutron	1.008 664 9	0
Electron	0.000 548 6	$-e$

Example 29.2

Estimating Mass

Estimate the mass in kilograms of 1 mol of ^{14}C.

Strategy We can estimate 1 u of mass for each nucleon and ignore the relatively small mass of the electrons. One mole contains Avogadro's number of atoms. Then we convert atomic mass units to kilograms.

Solution A ^{14}C nucleus has 14 nucleons, so the mass of the ^{14}C atom is roughly 14 u. One mole contains Avogadro's number of atoms; therefore the mass of 1 mol is roughly

$$M = N_A m = 6.02 \times 10^{23} \times 14 \text{ u} = 8.4 \times 10^{24} \text{ u}$$

Now we convert to kilograms:

$$8.4 \times 10^{24} \text{ u} \times 1.66 \times 10^{-27} \text{ kg/u} = 0.014 \text{ kg}$$

Discussion Note that the mass of 1 mol of an isotope with mass number 14 is approximately 14 g. The atomic mass unit is defined so that the mass of one atom in atomic mass units is numerically equal to the mass of 1 mol of atoms in grams.

The mass of a nucleus is not exactly equal to A atomic mass units for two reasons. The masses of the proton and neutron are not exactly 1 u. Even if they were, as we see in Section 29.2, the mass of a nucleus is *less than* the total mass of its individual protons and neutrons.

Practice Problem 29.2 Estimating the Mass of a Nucleus in u

Approximately what is the mass in atomic mass units of an oxygen nucleus that has nine neutrons?

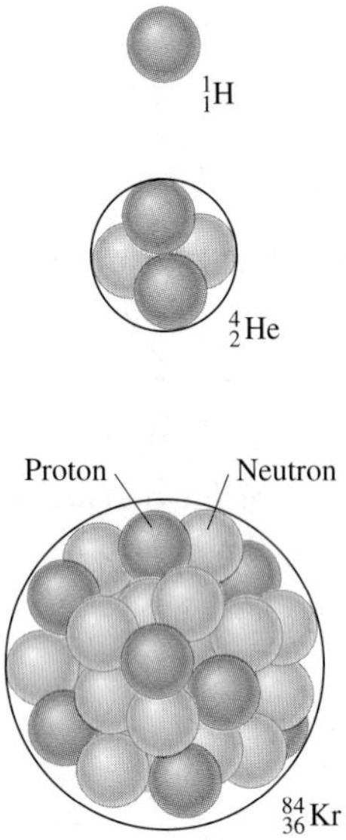

Figure 29.1 Simplified model of the nucleus as a set of hard spheres (representing the nucleons) packed together into a sphere.

Sizes of Nuclei

How do we know the *sizes* of nuclei? The first experimental evidence came from the Rutherford scattering of alpha particles from gold nuclei (Section 27.6). From analysis of the number of alpha particles observed at different scattering angles, we can estimate the size of the gold nucleus. Similar experiments were performed on other nuclei. More recently, electron diffraction has been used to probe the structure of the nucleus. Using electrons of very short wavelength, we can determine not only the size of the nucleus but learn about its internal structure as well.

These and other experiments show that the mass density of all nuclei is approximately the same—the volume of a nucleus is proportional to its mass. Imagine a nucleus to be like a spherical container full of marbles (Fig. 29.1); each marble represents a nucleon. The nucleons are tightly packed together, as if touching each other. Both the mass and volume of the nucleus are proportional to the number of nucleons, so the mass per unit volume (density ρ) is approximately independent of the number of nucleons. If m is the mass of a nucleus, V is its volume, and A is its mass number, then

$$m \propto A \quad \text{and} \quad V \propto A$$

$$\Rightarrow \rho = \frac{m}{V} \text{ is independent of } A$$

Most nuclei are approximately spherical in shape, so

$$V = \frac{4}{3}\pi r^3 \propto A$$

$$\Rightarrow r^3 \propto A \quad \text{or} \quad r \propto A^{1/3} \tag{29-3}$$

The radius of a nucleus is proportional to the cube root of its mass number. Experiment shows that the constant of proportionality is approximately 1.2×10^{-15} m:

Radius of a nucleus:

$$r = r_0 A^{1/3} \tag{29-4}$$

$$r_0 = 1.2 \times 10^{-15}\ \text{m} = 1.2\ \text{fm} \tag{29-5}$$

The SI prefix "f-" stands for *femto*; the fm is properly called a *femtometer* but is also called a *fermi*, after the Italian physicist Enrico Fermi (1901–1954). The nuclear radius ranges from 1.2 fm (for $A = 1$) to 7.7 fm (for $A \approx 260$).

Although nuclei all have about the same mass density, *atoms* do not. More massive atoms are generally denser than lighter atoms. The increase in volume of an atom does not keep pace with the increase in mass. Although larger atoms have more electrons, these electrons are on average more tightly bound, due to the increased charge of the nucleus. Thus, some solids and liquids (in which the atoms are tightly packed) are denser than others.

Example 29.3

Radius and Volume of Barium Nucleus

What are the radius and volume of the barium-138 nucleus?

Strategy To find the radius of a nucleus, all we need to know is the mass number A, which in this case is 138. To find the volume, we assume that the nucleus is approximately spherical.

Solution To find the radius we apply Eq. (29-4), substituting $A = 138$:

$$r = r_0 A^{1/3}$$

$$r = 1.2\ \text{fm} \times 138^{1/3} = 6.2\ \text{fm}$$

continued on next page

Example 29.3 continued

The approximate volume of the nucleus is

$$V = \tfrac{4}{3}\pi r^3$$

Cubing Eq. (29-4) yields

$$r^3 = r_0^3 A$$

Therefore, the volume of a nucleus is approximately

$$V = \tfrac{4}{3}\pi r_0^3 A$$

Now we substitute numerical values.

$$V = \tfrac{4}{3}\pi \times (1.2 \times 10^{-15}\ \text{m})^3 \times 138 = 1.0 \times 10^{-42}\ \text{m}^3$$

Discussion The radius (6.2 fm) is within the expected range of 1.2 fm to 7.7 fm. The equation $V = \tfrac{4}{3}\pi r_0^3 A$ says the volume of a nucleus is proportional to the number of nucleons (A), as expected; each nucleon occupies a volume of $\tfrac{4}{3}\pi r_0^3$.

Practice Problem 29.3 Volume of a Radium Nucleus

What is the volume of a radium-226 nucleus?

29.2 BINDING ENERGY

The Strong Force

What holds the nucleons together in a nucleus? Gravity is far too weak to do it; electric forces push protons *away* from each other. The nucleons are held together by the **strong force**, one of the four fundamental forces (along with gravity, electromagnetism, and the weak force) discussed in Section 4.12. The strong force makes little distinction between protons and neutrons.

Unlike gravity and the electromagnetic forces, the strong force is extremely short range. The ranges of the gravitational and electromagnetic forces are infinite, with the magnitude of the force between point objects falling off with distance as $1/r^2$. By contrast, the strong force between two nucleons is significant only at distances of about 3.0 fm or less. Because the strong force is so short range, a nucleon is attracted only to its *nearest neighbors* in the nucleus. On the other hand, since electrical repulsion is long range, every proton in the nucleus repels *all* the other protons. These two competing forces determine which nuclei are stable.

Binding Energy and Mass Defect

The **binding energy** E_B of a nucleus is the energy that must be supplied to separate a nucleus, a system of bound protons and neutrons, into individual, free protons and neutrons. Since the nucleus is a bound system, its total energy is *less* than the energy of Z protons and N neutrons that are far apart and at rest.

CONNECTION:

The concept of binding energy is a way to look at how the nucleus is held together in terms of energy instead of forces.

Binding energy:

$$E_B = (\text{total energy of } Z \text{ protons and } N \text{ neutrons}) - (\text{total energy of nucleus}) \quad (29\text{-}6)$$

The concept of binding energy applies to systems other than nuclei. The total energy of a proton and an electron far from each other is 13.6 eV higher than the energy when the two are bound together in a hydrogen atom (in its ground state). Thus, the binding energy of the hydrogen *atom* is 13.6 eV. In atoms with more than one electron, the binding energy is not the same as the ionization energy. The ionization energy is the energy required to remove *one* electron; the binding energy is the energy required to remove *all of the electrons*.

The mass of a particle is a measure of its *rest energy*—its total energy in a reference frame in which it is at rest (see Section 26.7):

$$E_0 = mc^2 \quad (26\text{-}7)$$

Since the rest energy of a nucleus is *less than* the total rest energy of Z protons and N neutrons, the mass of the nucleus is less than the total mass of the protons and neutrons. The difference, called the **mass defect** Δm, comes about because we would have to *add* energy to a nucleus to break it up into Z individual protons and N individual neutrons. The mass defect is related to the binding energy via Eq. (26-7).

Mass defect and binding energy:

$$\Delta m = (\text{mass of } Z \text{ protons and } N \text{ neutrons}) - (\text{mass of nucleus}) \quad (29\text{-}7)$$

$$E_B = (\Delta m)c^2 \quad (29\text{-}8)$$

The energy unit most commonly used in nuclear physics is the MeV (mega-electron-volt). When using MeV for energy and atomic mass units for mass in Eq. (29-8), it is convenient to know the value of c^2 in units of MeV/u. It can be shown (Problem 18) that

$$c^2 = 931.494 \text{ MeV/u} \quad (29\text{-}9)$$

Mass tables such as Appendix B give the masses of *neutral atoms*, which include the masses of the electrons as well as the mass of the nucleus. To find the mass of a nucleus with atomic number Z, subtract the mass of Z electrons from the mass of the neutral atom. The binding energy of the electrons to the nucleus is much smaller than the rest energy of the electrons and can be ignored.

Example 29.4

Binding Energy of a Nitrogen-14 Nucleus

Find the binding energy of the ^{14}N nucleus.

Strategy From Appendix B, the mass of the ^{14}N *atom* is 14.003 074 0 u. The mass of the N atom includes the mass of 7 electrons. Subtracting $7m_e$ from the mass of the atom gives the mass of the nucleus. Then we can find the mass defect and the binding energy.

Solution

$$\text{mass of } ^{14}\text{N nucleus} = 14.003\,074\,0 \text{ u} - 7m_e$$
$$= 14.003\,074\,0 \text{ u} - 7 \times 0.000\,548\,6 \text{ u}$$
$$= 13.999\,233\,8 \text{ u}$$

The ^{14}N nucleus has 7 protons and 7 neutrons. The mass defect is

$$\Delta m = (\text{mass of 7 protons and 7 neutrons}) - (\text{mass of nucleus})$$
$$= 7 \times 1.007\,276\,5 \text{ u} + 7 \times 1.008\,664\,9 \text{ u} - 13.999\,233\,8 \text{ u}$$
$$= 0.112\,356\,0 \text{ u}$$

The binding energy is therefore,

$$E_B = (\Delta m)c^2 = 0.112\,356\,0 \text{ u} \times 931.494 \text{ MeV/u}$$
$$= 104.659 \text{ MeV}$$

Discussion Since the binding energy of the electrons in an atom is so small, we can assume that the mass of an atom is equal to the mass of its nucleus plus the mass of the electrons. As a shortcut, we can calculate the mass defect using the mass of the nitrogen *atom* instead of the nitrogen nucleus and the mass of the *hydrogen atom* instead of the proton. Since each term contains the extra mass of 7 electrons, the masses of the electrons subtract out:

$$\Delta m = (\text{mass of 7 } ^1\text{H atoms and 7 neutrons}) - (\text{mass of } ^{14}\text{N atom})$$
$$= 7 \times 1.007\,825\,0 \text{ u} + 7 \times 1.008\,664\,9 \text{ u} - 14.003\,074\,0 \text{ u}$$
$$= 0.112\,355\,3 \text{ u}$$

Practice Problem 29.4 Binding Energy of Nitrogen-15

Calculate the binding energy of the ^{15}N nucleus. The mass of the ^{15}N *nucleus* is 14.996 269 u. [*Hint:* This time you have been given the mass of the *nucleus*, not the mass of the atom.]

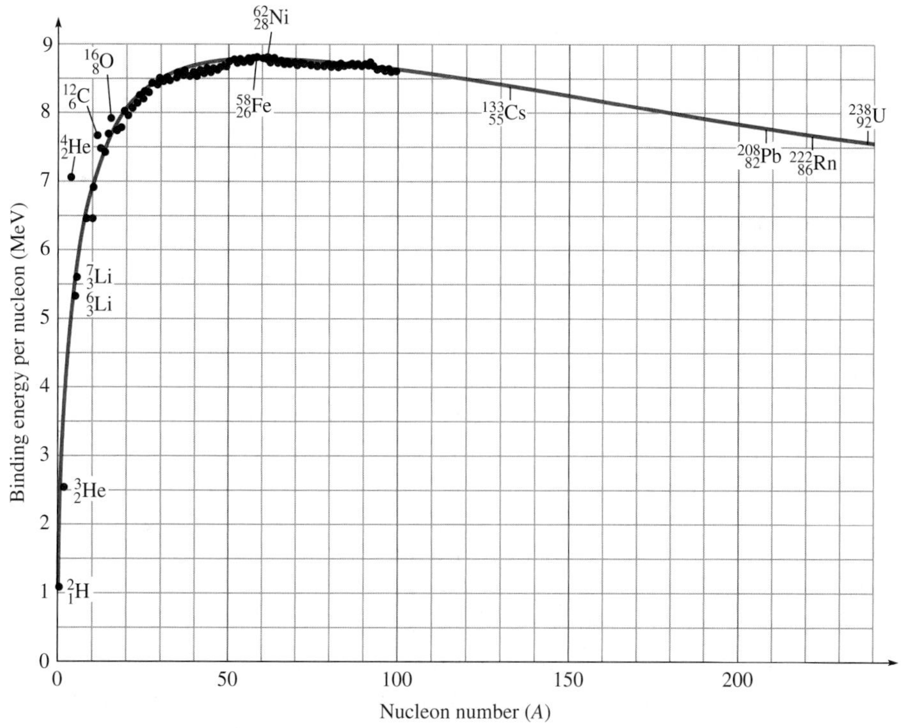

Figure 29.2 Binding energy per nucleon (E_B/A) for the most stable nuclide with nucleon number A. Individual data points are shown for $A < 100$; a smooth curve showing the general trend is shown in red. (Data points are omitted for $A \geq 100$ since they differ little from the values given by the red curve.) $^{62}_{28}$Ni has the largest binding energy per nucleon of all the nuclides (8.795 MeV), followed by $^{58}_{26}$Fe and $^{56}_{26}$Fe (8.792 MeV and 8.790 MeV, respectively). Data points for $^{4}_{2}$He, $^{12}_{6}$C, and $^{16}_{8}$O lie significantly above the red curve—these nuclides are particularly stable compared with nuclides with similar values of A.

Binding Energy Curve

Figure 29.2 shows a graph of the binding energy *per nucleon* as a function of mass number. Recall that the strong force binds nucleons only to their nearest neighbors. In small nuclides there are not enough nucleons for all to fully bind since the average number of nearest neighbors is small. Increasing the number of nucleons leads to a larger binding energy per nucleon, up to a point, because the average number of nearest neighbors is increasing. Thus, we see a steep increase in the binding energy per nucleon as A increases.

Once nuclei reach a certain size, all nucleons except those on the surface have as many nearest neighbors as possible. Adding more nucleons doesn't increase the average binding energy per nucleon due to the strong force very much, but the Coulomb repulsion keeps adding up since it is long range. Thus, above $A \approx 60$, adding more nucleons *decreases* the average binding energy per nucleon. The decrease is relatively gentle, compared to the steep increase for small A, since the Coulomb repulsion is weak compared to the strong force.

The binding energy per nucleon is within the range 7–9 MeV for all but the smallest nuclides. For example, in Example 29.4 we found that the binding energy of ^{14}N is 104.659 MeV. The binding energy *per nucleon* for ^{14}N is

$$\frac{104.659 \text{ MeV}}{14 \text{ nucleons}} = 7.47564 \text{ MeV/nucleon}$$

The most tightly bound nuclides are around $A \approx 60$, where the binding energy is about 8.8 MeV/nucleon.

Nuclear Energy Levels

Neutrons and protons obey the Pauli exclusion principle: no two identical nucleons in the same nucleus can be in the same quantum state. As for atomic energy levels, a group of closely spaced nuclear energy levels is called a *shell*. We can describe the quantum state of the nucleus by specifying how the proton and neutron states are occupied, much as we did for electron states in the atom (Fig. 29.3). Two protons can occupy each proton energy level (one spin up, one spin down) and two neutrons can occupy each neutron energy level. The energy levels for the proton and neutron are similar; as far as the nuclear

CONNECTION:

The Pauli exclusion principle applies to electrons in an atom (Section 28.7) and to nucleons in the nucleus.

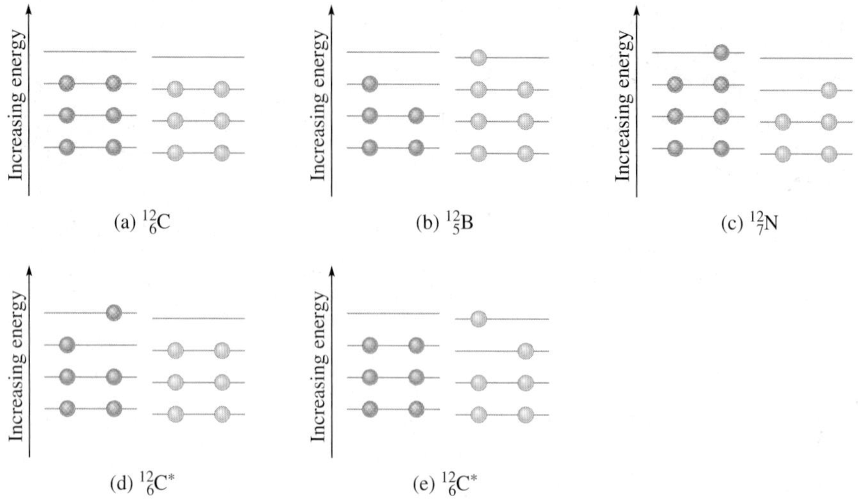

Figure 29.3 Qualitative energy level diagrams for some nuclides with $A = 12$. Red spheres represent protons and gray spheres represent neutrons. Compare the *atomic* energy level diagram in Figure 28.19. $^{12}_{6}C$ is stable, while $^{12}_{5}B$ and $^{12}_{7}N$ are unstable. The asterisks in (d) and (e) indicate that $^{12}_{6}C^*$ is in an excited state. $^{12}_{6}C^*$ can return to the ground state ($^{12}_{6}C$) by emitting a photon whose energy equals the difference in the energy levels. $^{12}_{5}B$ and $^{12}_{7}N$ can emit an electron or positron, respectively, to change into $^{12}_{6}C$ (see Beta decay, Section 29.3).

force is concerned, protons and neutrons are pretty much the same. The main difference is that the protons are affected by the Coulomb repulsion in addition to the strong force.

In Problem 20, an order of magnitude calculation shows that the energy level spacings in nuclei are expected to be in the MeV range. The structure of the nucleus is complex; energy level spacings range from tens of keV to several MeV. A nucleus in an excited state can return to the ground state by emitting one or more *gamma-ray* photons. [The distinction between gamma rays and x-rays is based more on the source than the energy. A photon emitted by an excited nucleus or in pair annihilation (Section 27.8) is called a gamma ray; a high-energy photon emitted by an excited *atom*, by an electron slowing down on striking a target (Section 27.4), or by a circulating charged particle in a synchrotron is usually called an x-ray.] Just as the energy levels of atoms can be deduced by measuring the wavelengths of photons radiated by excited atoms, measurement of the gamma-ray energies emitted by excited nuclei enables us to deduce the nuclear energy levels. Each nuclide has its own characteristic gamma-ray spectrum, which can be used to identify it. A gamma-ray spectrum usually identifies the energy of the photons, in contrast to a visible spectrum where the wavelength is usually specified. In both cases, the quantity used is the one that is easier to measure.

Energy level diagrams help explain why, in stable light nuclides, the number of neutrons and protons tends to be approximately equal. Figure 29.3 shows energy level diagrams for three different nuclides, each of which has 12 nucleons. The energy levels are *not* quantitatively correct, but serve to illustrate the general idea. A maximum of two protons can be in any proton energy level and a maximum of two neutrons can be in any neutron energy level. The proton and neutron energy levels are similar; the proton levels are slightly higher in energy than the neutron levels to account for the Coulomb repulsion between the protons. The energy is lower with 6 protons and 6 neutrons than is possible with 5 of one and 7 of the other.

The story is more complicated for heavier nuclides. The Coulomb repulsion between protons favors more neutrons ($N > Z$) since the neutrons are immune to the Coulomb repulsion. For larger nuclides, the Coulomb repulsion becomes more and more important since it is long range: each proton repels *every other proton* in the nucleus. The proton energy levels get higher and higher with respect to the neutron energy levels as the electric potential energy of all those repelling protons adds up. Thus, large nuclides tend to have an excess of neutrons ($N > Z$). On the other hand, there is a limit to the neutron excess: neutrons are slightly more massive than protons, so if there is too much of a neutron excess, the mass (and therefore the energy) of the nucleus is higher than it would be if one or more neutrons were changed into protons.

Figure 29.4 shows the number of protons (Z) and number of neutrons (N) for the stable nuclides (represented as points in green). For the smallest nuclides, $N \approx Z$. As the

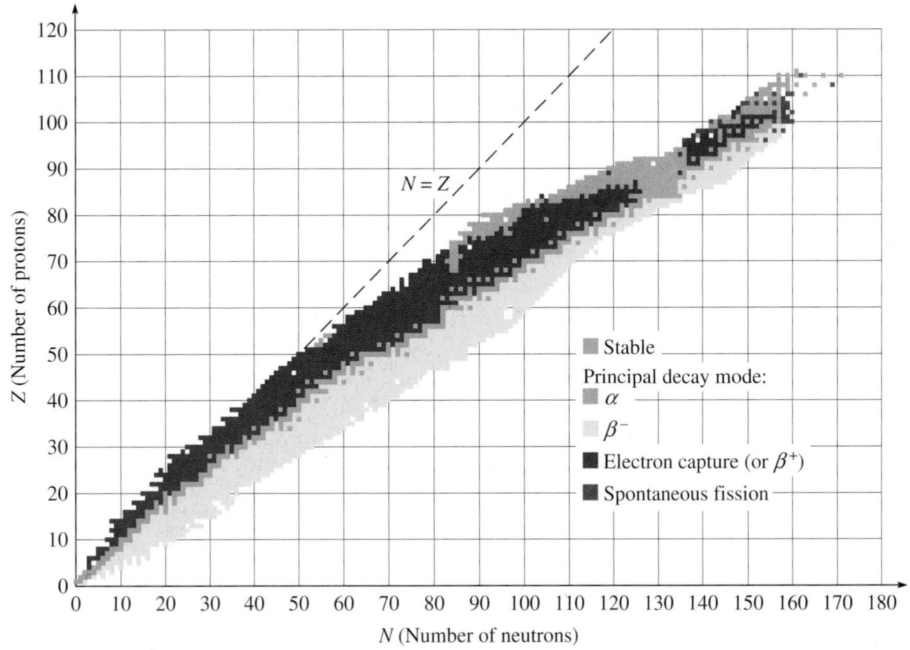

Figure 29.4 Chart of the most common nuclides. Stable nuclides are shown as green points. Note the general trend of increasing N/Z ratio for stable nuclides.

total number of nucleons ($A = Z + N$) increases, the number of neutrons increases faster than the number of protons. The largest stable nuclides have about 1.5 times as many neutrons as protons.

29.3 RADIOACTIVITY

The French physicist Henri Becquerel (1852–1908) discovered radioactivity in 1896 when, quite by accident, he found that a uranium salt spontaneously emitted radiation in the absence of an external source of energy, such as sunlight. The radiation exposed a photographic plate even though the plate was wrapped in black paper to keep light out.

Nuclides can be divided into two broad categories. Some are stable; others are unstable, or **radioactive**. An unstable nuclide **decays**—takes part in a spontaneous nuclear reaction—by emitting radiation. (The radiation may include but is not limited to *electromagnetic* radiation.) Depending on the kind of radiation emitted, the reaction may change the nucleus into a different nuclide, with a different charge or nucleon number or both.

Scientists studying radioactivity soon identified three different kinds of radiation emitted by radioactive nuclei; they were named alpha (α) rays, beta (β) rays, and gamma (γ) rays after the first three letters of the Greek alphabet. The initial distinction between the three was their differing abilities to penetrate matter (Fig. 29.5). Alpha rays are the least penetrating; they can only make it through a few centimeters of air and are completely blocked by human skin, thin paper, and other solids. Beta rays can travel farther in air—about a meter typically—and can penetrate a hand or a thin metal foil. Gamma

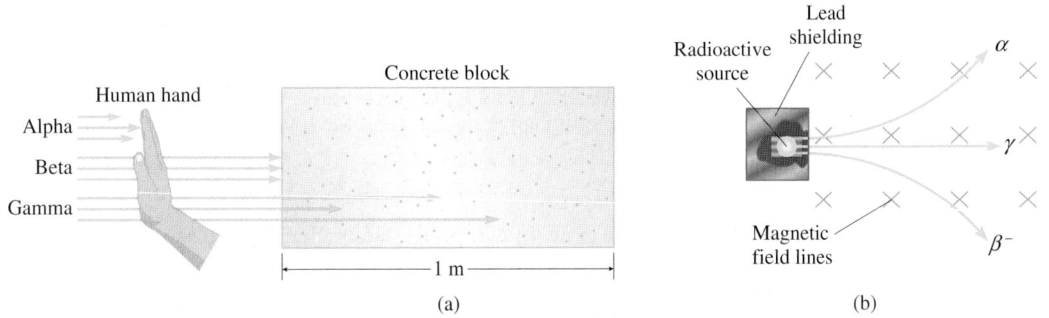

Figure 29.5 Alpha, beta, and gamma rays differ (a) in their abilities to penetrate matter, as well as (b) in their electric charges.

Table 29.2 Particles Commonly Involved in Radioactive Decay and Other Nuclear Reactions

Particle Name	Symbols	Charge (in units of e)	Nucleon Number
Electron	e^-, β^-, ${}^{0}_{-1}e$	−1	0
Positron	e^+, β^+, ${}^{0}_{1}e$	+1	0
Proton	p, ${}^{1}_{1}p$, ${}^{1}_{1}H$	+1	1
Neutron	n, ${}^{1}_{0}n$	0	1
Alpha particle	α, ${}^{4}_{2}\alpha$, ${}^{4}_{2}He$	+2	4
Photon	γ, ${}^{0}_{0}\gamma$	0	0
Neutrino	ν, ${}^{0}_{0}\nu$	0	0
Antineutrino	$\bar{\nu}$, ${}^{0}_{0}\nu$	0	0

rays are much more penetrating than either α or β rays. Later, the electric charge and mass were determined and used to distinguish the three types of radiation—and ultimately to identify them.

Of the approximately 1500 known nuclides, only about 20% are stable. All of the largest nuclides (those with $Z > 83$) are radioactive. As far as we know, stable nuclei last forever without decaying spontaneously. Each radioactive nuclide decays with an average lifetime characteristic of that nuclide. The lifetimes cover an enormous range, from about 10^{-22} s (roughly the time it takes light to travel a distance equal to the diameter of a nucleus) to 10^{+28} s (10^{10} times the age of the universe).

Conservation Laws in Radioactive Decay

In a nuclear reaction, whether spontaneous or not, the total electric charge is conserved. Another conservation law says that the total number of nucleons must stay the same. We *balance* a nuclear reaction by applying these two conservation laws. It is helpful to write symbols for electrons, positrons, and neutrons as if they were nuclei, with a superscript for the number of nucleons and a subscript for the electric charge in units of e (Table 29.2). Then the reaction is balanced with regard to nucleon number if the sum of the superscripts is the same on both sides; it is balanced with regard to charge if the sum of the subscripts is the same on both sides.

Another conservation law is important in radioactive decay: all nuclear reactions also conserve energy. How can a nucleus with little or no kinetic energy decay, leaving products with significant kinetic energies? Where did this energy come from? In a spontaneous nuclear reaction, some of the rest energy of the radioactive nucleus is converted into kinetic energy of the products. The amount of rest energy that is converted into other forms of energy is called the **disintegration energy**. In order for kinetic energy to increase, there must be a corresponding decrease in rest energy. The total mass of the products must be less than the mass of the original radioactive nucleus in order for that nucleus to decay spontaneously. In other words, the products must be more tightly bound than is the original nucleus; the disintegration energy is the difference between the binding energy of the radioactive nucleus and the total binding energy of the products.

Disintegration energy = binding energy of radioactive nucleus − total binding energy of products

Alpha Decay

Alpha "rays" are now known to be ${}^{4}He$ nuclei. The helium nucleus is a group of two protons and two neutrons and it is very tightly bound. The mass of an alpha particle is 4.001 506 u and its charge is $+2e$.

In alpha decay, the original (*parent*) nuclide is converted to a "*daughter*" by the emission of an α particle. Balancing the reaction shows that the daughter nuclide has a nucleon number reduced by four and a charge reduced by two. Using P for the parent nuclide and D for the daughter nuclide, the spontaneous reaction in which an α particle is emitted is

Alpha decay:

$$^{A}_{Z}P \rightarrow {}^{A-4}_{Z-2}D + {}^{4}_{2}\alpha \qquad (29\text{-}10)$$

Emission of an α particle is the most common type of radioactive decay for large nuclides ($Z > 83$). Since no nuclide with $Z > 83$ is stable, emitting an α particle moves toward stability most directly by decreasing both Z and N by 2. Emission of an α particle increases the ratio of neutrons to protons. For example, $^{238}_{92}\text{U}$ has a neutron-to-proton ratio of $(238 - 92)/92 = 1.587$. By emitting an α particle, $^{238}_{92}\text{U}$ becomes $^{234}_{90}\text{Th}$ with a higher neutron-to-proton ratio: $(234 - 90)/90 = 1.600$. Thus, large nuclides with a smaller neutron-to-proton ratio are more likely to α decay than are similar nuclides with a greater neutron-to-proton ratio.

Example 29.5

An Alpha Decay

Polonium-210 decays via α decay. Identify the daughter nuclide.

Strategy First we look up the atomic number of polonium in the periodic table (inside back cover). Next we write the nuclear reaction with an unknown nuclide and an α particle as the products. Balancing the reaction gives us the values of Z and A of the daughter nucleus.

Solution Polonium is atomic number 84. Then the reaction is

$$^{210}_{84}\text{Po} \rightarrow {}^{A}_{Z}(?) + {}^{4}_{2}\alpha$$

where A and Z are the nucleon number and atomic number of the daughter nucleus. To conserve charge,

$$84 = Z + 2$$

Thus, $Z = 82$. To conserve nucleon number,

$$210 = A + 4$$

and $A = 206$. Looking up atomic number 82 in the periodic table reveals that the element is lead. Thus, the daughter nucleus is lead-206 ($^{206}_{82}\text{Pb}$).

Discussion Writing out the reaction makes it easy to check that the total number of nucleons and the total electric charge are both conserved by the reaction:

$$^{210}_{84}\text{Po} \rightarrow {}^{206}_{82}\text{Pb} + {}^{4}_{2}\alpha$$

Practice Problem 29.5 Finding the Parent Given the Daughter

Radon-222, the radioactive gas that poses a significant health risk in some areas, is itself produced by the α decay of another nuclide. Identify its parent nuclide.

Energy in Alpha Decay In α decay, the disintegration energy released is shared between the daughter nucleus and the α particle. Momentum conservation determines exactly how the energy is shared. Therefore, the α particles released in a particular radioactive decay have a characteristic energy (assuming that the initial kinetic energy of the parent is insignificant and can be taken to be zero).

Example 29.6

Alpha Decay of Uranium-238

The ^{238}U nuclide can decay by emitting an α particle:

$$^{238}U \rightarrow {}^{234}Th + \alpha$$

The *atomic* masses of ^{238}U, ^{234}Th, and $^{4}_{2}He$ are 238.050 782 6 u, 234.043 595 4 u, and 4.002 603 2 u, respectively. (a) Find the disintegration energy. (b) Find the kinetic energy of the α particle, assuming the parent ^{238}U nucleus was initially at rest.

Strategy The calculations can be performed using *atomic* masses. The mass of the $^{238}_{92}U$ atom includes 92 electrons; the combined masses of the $^{234}_{90}Th$ and $^{4}_{2}He$ atoms also include 90 + 2 = 92 electrons.

We expect *most* of the kinetic energy to go to the α particle, since its mass is much smaller than that of the thorium nucleus. Momentum conservation determines exactly how the kinetic energy splits between the two particles.

Solution (a) The total mass of the products is

$$234.043\,595\,4\text{ u} + 4.002\,603\,2\text{ u} = 238.046\,198\,6\text{ u}$$

which is less than the mass of the parent nucleus. The change in mass is

$$\Delta m = 238.046\,198\,6\text{ u} - 238.050\,782\,6\text{ u} = -0.004\,584\,0\text{ u}$$

Δm stands for the *change* in mass: final mass minus initial mass. (When we write the mass defect of a nucleus as Δm, we imagine a reaction that separates the nucleus into its constituent protons and neutrons.) The decrease in mass for this reaction means that the rest energy decreases. According to Einstein's mass-energy relation, the change in rest energy is

$$E = (\Delta m)c^2 = -0.004\,584\,0\text{ u} \times 931.494\text{ MeV/u}$$
$$= -4.2700\text{ MeV}$$

By conservation of energy, the kinetic energy of the products is 4.2700 MeV more than the kinetic energy of the parent. The disintegration energy is 4.2700 MeV.

(b) Assuming for the moment that the daughter nucleus and the α particle can be treated nonrelativistically, their kinetic energies are related to their momenta by

$$K = \frac{p^2}{2m}$$

Momentum conservation says that their momenta must be equal in magnitude and opposite in direction. Therefore, the ratio of the kinetic energies is

$$\frac{K_\alpha}{K_{Th}} = \frac{p^2/(2m_\alpha)}{p^2/(2m_{Th})} = \frac{m_{Th}}{m_\alpha} = \frac{234.043\,595\,5}{4.002\,603\,2} = 58.4728$$

The two kinetic energies must add up to 4.2699 MeV.

$$K_\alpha + K_{Th} = 4.2699\text{ MeV}$$

Substituting for K_{Th} from the kinetic energy ratio,

$$K_\alpha + \frac{K_\alpha}{58.4728} = 4.2699\text{ MeV}$$

Solving yields $K_\alpha = 4.1981$ MeV.

Discussion The change in mass is *negative*: the total mass after the decay is less than the mass before. Some of the mass (or, more accurately, rest energy) of the U nucleus is converted into the kinetic energy of the products. The disintegration energy is positive because it is the quantity of energy *released.*

Since the α particle's kinetic energy is much smaller than its rest energy (about $4\text{ u} \times 931.494\text{ MeV/u} \approx 3700$ MeV), the nonrelativistic expression for kinetic energy was appropriate. A relativistic calculation shows that our answer is correct to three significant figures.

Practice Problem 29.6 Alpha Energy in the Decay of Polonium-210

Find the kinetic energy of the α particle emitted by the decay of ^{210}Po:

$$^{210}_{84}Po \rightarrow {}^{206}_{82}Pb + \alpha$$

Beta Decay

The positron is the antiparticle of the electron (see Section 27.8); it has the same mass as the electron but a positive charge of $+e$.

Beta particles are electrons or positrons (sometimes still called beta-minus [β^-] and beta-plus [β^+] particles). In β^- decay, an electron is emitted and a neutron in the nucleus is converted into a proton. Thus, the mass number does not change, but the charge of the nucleus increases by one:

Beta-minus decay:

$$^{A}_{Z}P \rightarrow {}^{A}_{Z+1}D + {}^{0}_{-1}e + {}^{0}_{0}\bar{\nu} \quad (29\text{-}11)$$

The symbol $\overline{\nu}$ represents an **antineutrino**, an uncharged particle with negligible mass. In β^+ decay, a positron is emitted and a proton in the nucleus is converted into a neutron. This time the charge of the nucleus decreases by one:

Beta-plus decay:

$$^{A}_{Z}P \rightarrow {}^{A}_{Z-1}D + {}^{0}_{+1}\text{e} + {}^{0}_{0}\nu \qquad (29\text{-}12)$$

The symbol ${}^{0}_{+1}\text{e}$ represents the emitted positron and ν is a **neutrino** with no charge and negligible mass. Before long, the positron will run into one electron and the pair will be annihilated producing a pair of photons (Section 27.8).

Unlike α decay, β decay of a radionuclide does not change the number of nucleons. In essence, β decay changes a neutron into a proton or vice versa. Since the mass of the neutron is greater than the combined mass of a proton plus an electron, free neutrons decay spontaneously by β^- emission. The half-life of this process is 10.2 min. A free proton cannot spontaneously decay into a neutron plus a positron; that would violate energy conservation. But within a nucleus, a proton can change into a neutron by emitting a positron; the energy required to make this happen comes from the change in the binding energy of the nucleus. Thus, the basic β decay reactions that take place inside the nucleus are

$$\beta^-: \quad {}^{1}_{0}\text{n} \rightarrow {}^{1}_{1}\text{p} + {}^{0}_{-1}\text{e} + {}^{0}_{0}\overline{\nu}$$

$$\beta^+: \quad {}^{1}_{1}\text{p} \rightarrow {}^{1}_{0}\text{n} + {}^{0}_{+1}\text{e} + {}^{0}_{0}\nu$$

Beta decay does not change the mass number, but it does change the ratio of neutrons to protons. A nuclide that has too many neutrons to be stable is likely to decay via β^-. By emitting an electron, a neutron is changed into a proton inside the nucleus. A nuclide that has too few neutrons is likely to decay by β^+, emitting a positron and turning a proton into a neutron. In either case, total electric charge is conserved.

Prediction and Discovery of the Neutrino Beta decay was a puzzle at first because a *continuous spectrum* of electron (or positron) energies was observed. In α decay, the definite kinetic energy of the α particles emitted in a given decay reaction is understood to come from conservation of both energy and linear momentum. For the same reasons, scientists thought that β particles emitted in a given decay reaction should also be monoenergetic. However, when the kinetic energies were measured, the emitted β particles had a continuous range of kinetic energies up to a maximum value (Fig. 29.6). The *maximum* kinetic energy was consistent with what scientists thought the β particle's kinetic energy should have been.

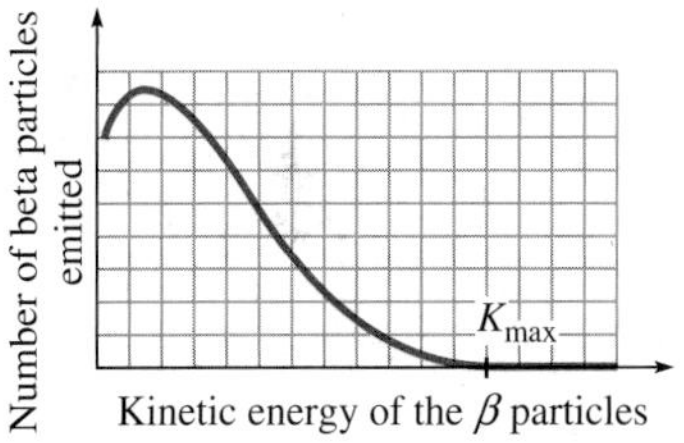

Figure 29.6 Typical continuous energy spectrum of electrons emitted in β decay from a particular nuclide.

Why did many of the β particles have lower energies than expected? Had scientists found an exception to one of the conservation laws (energy or momentum)? Although some quite respectable scientists—including Niels Bohr—started to think that energy conservation had been violated, Wolfgang Pauli finally suggested another explanation, which turned out to be correct. Pauli speculated that not one, but two particles were being emitted, the β particle and another, as yet undetected, particle. If a nucleus emits two particles instead of one, then they can conserve both energy and momentum while splitting up the kinetic energy in every possible way. Two momentum vectors that add to zero must be equal in magnitude and opposite in direction, but *three* momentum vectors can add to zero in an infinite number of ways and still share the same total kinetic energy.

Enrico Fermi named this hypothetical particle the neutrino. The symbol for the neutrino is the Greek letter "nu" (ν). An antineutrino is written with a bar over it ($\overline{\nu}$). For reasons that we study in Chapter 30, an antineutrino ($\overline{\nu}$) is emitted in β^- decay, while a neutrino (ν) is emitted in β^+ decay. Neutrinos are famously hard to detect because they do not interact via the electromagnetic or strong interactions. It took 25 years after Pauli predicted their existence before one was actually observed. A neutrino can pass through the Earth with only about a 1 in 10^{12} chance of interacting. Enormous numbers of neutrinos, streaming toward us from the Sun, pass through your body every second but cause no ill effects.

Example 29.7

Beta Decay of Nitrogen-13

The isotope of nitrogen with mass number 13 (^{13}N) is unstable to beta decay. (a) ^{14}N and ^{15}N are the stable isotopes of nitrogen. Do you expect ^{13}N to decay via β^- or β^+? Explain. (b) Write the decay reaction. (c) Calculate the maximum kinetic energy of the emitted beta particle.

Strategy The key in deciding between β^- and β^+ is whether the nucleus has too many or too few neutrons to be stable.

Solution (a) The stable isotopes of nitrogen have more neutrons than ^{13}N, so ^{13}N has too few neutrons to be stable. The beta decay should convert a proton into a neutron to increase the neutron-to-proton ratio. That means the charge of the nucleus decreases by e, so a positron (charge $= +e$) must be created to conserve charge. We expect the isotope ^{13}N to undergo β^+ decay.

(b) Since a positron is emitted, it must be accompanied by a neutrino (not an antineutrino). Z decreases by 1, from 7 (for nitrogen) to 6 (which is carbon). A is unchanged. The reaction is

$$^{13}_{7}N \rightarrow {}^{13}_{6}C + {}^{0}_{+1}e + {}^{0}_{0}\nu$$

Both charge and nucleon number are conserved: $13 = 13 + 0$ and $7 = 6 + 1$.

(c) From Appendix B, the atomic masses of $^{13}_{7}N$ and $^{13}_{6}C$ are 13.005 738 6 u and 13.003 354 8 u. To get the masses of the nuclei, we subtract Zm_e from each. The mass of the positron is the same as that of the electron: $m_e = 0.000\,548\,6$ u. The neutrino mass is negligibly small. If M_N and M_C represent atomic masses, then

$$\Delta m = [(M_C - 6m_e) + m_e] - (M_N - 7m_e)$$
$$= M_C - M_N + 2m_e$$
$$= 13.003\,354\,8 \text{ u} - 13.005\,738\,6 \text{ u} + 2 \times 0.000\,548\,6 \text{ u}$$
$$= -0.001\,286\,6 \text{ u}$$

The mass decreases, as it must for a spontaneous decay. The disintegration energy is

$$E = |\Delta m|c^2 = 0.001\,286\,6 \text{ u} \times 931.494 \text{ MeV/u} = 1.1985 \text{ MeV}$$

This *is* the maximum kinetic energy of the positron, since it can get virtually all of the energy and leave the neutrino and daughter nucleus with a negligibly small amount.

Discussion It is *usually* possible to determine whether a radioactive nuclide decays via β^+ or β^-, but there are exceptions. For example, $^{40}_{19}K$ can decay by *either* β^+ or β^-. The only way to be sure is to compare the masses of the products with the mass of the radionuclide to see whether the spontaneous decay is energetically possible.

Note that in β^+ decay the electron masses (which are included in the atomic masses) do not automatically "cancel out" as they did for α decay.

Practice Problem 29.7 Maximum Electron Energy in the Decay of Potassium-40

Find the maximum energy of the electron emitted in the β^- decay of $^{40}_{19}K$.

Electron Capture

Any nuclide that can decay via β^+ can also decay by **electron capture**. Both processes convert a proton into a neutron. In electron capture, instead of emitting a positron, the nucleus absorbs one of the atom's electrons. The basic reaction is

$$^{0}_{-1}e + {}^{1}_{1}p \rightarrow {}^{1}_{0}n + {}^{0}_{0}\nu \qquad (29\text{-}13)$$

When a nucleus captures an electron, the only reaction products are the daughter nucleus and the neutrino. With only two particles, conservation of momentum and energy determine what fraction of the energy released is taken by each particle. The neutrino, with its tiny mass, takes almost all of the kinetic energy, leaving the daughter to recoil with only a few electron-volts of kinetic energy. Some nuclides can decay by electron capture but *not* by β^+ because the difference in mass between the parent and daughter is less than the mass of a positron.

CONNECTION:

A γ ray is a photon.

Gamma Decay

Gamma rays are high-energy photons. Emission of a gamma ray does not change the nucleus into a different nuclide, since neither the charge nor the number of nucleons is changed. A γ-ray photon is emitted when a nucleus in an excited state makes a transition to a lower energy state, just as photons are emitted when electrons in atoms make transitions between energy levels.

Figure 29.7 shows some of the energy levels of the thallium-208 ($^{208}_{81}$Tl) nucleus. The nucleus in an excited state can radiate a photon to jump to a state of lower energy. For instance, the third arrow from the right, from 492 keV to 40 keV, shows a transition that results in the emission of a 452-keV photon.

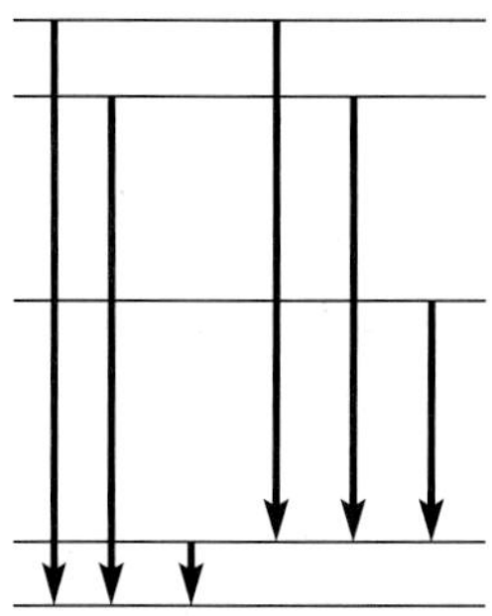

Figure 29.7 An energy level diagram for $^{208}_{81}$Tl. Downward arrows show the allowed transitions for γ decay.

To emphasize that a nucleus is in an excited state, we put an asterisk as a superscript after the symbol: $^{208}_{81}\text{Tl}^*$. The gamma decay of an excited Tl-208 nucleus by emitting one photon is written as

$$^{208}_{81}\text{Tl}^* \rightarrow \, ^{208}_{81}\text{Tl} + \gamma$$

Alpha and beta decay do not always leave the daughter nucleus in its ground state. Sometimes the daughter nucleus is left in an excited state that then emits one or more γ-ray photons until it reaches the ground state. In α decay, therefore, there may be different possible kinetic energies of the α particles emitted, depending on which excited state of the daughter nucleus is produced by the decay. For example, $^{212}_{83}$Bi can α decay to form any of the five energy states of $^{208}_{81}$Tl shown in Fig. 29.7 (the ground state and four excited states). The α-particle spectrum in the decay of $^{212}_{83}$Bi is still discrete, but there are five discrete values instead of one (see Problem 80). In β decay, if the daughter nucleus can be left in an excited state, then the amount of kinetic energy shared by the electron (or positron), the antineutrino (or neutrino), and the daughter nucleus is smaller. The spectrum of electron (or positron) kinetic energies is still continuous.

29.4 RADIOACTIVE DECAY RATES AND HALF-LIVES

What determines when an unstable nucleus decays? Radioactive decay is a quantum-mechanical process that can only be described in terms of *probability.* Given a collection of identical nuclides, they do not all decay at the same time, and there is no way to predict which one decays when. The decay probability for one nucleus is independent of its past history and of the other nuclei. Each radioactive nuclide has a certain decay probability per unit time, written λ (no relation to wavelength). The decay probability per unit time is also called the **decay constant**. Since probability is a pure number, the decay constant has SI units s^{-1} (probability *per second*).

$$\text{decay constant } \lambda = \frac{\text{probability of decay}}{\text{unit time}} \qquad (29\text{-}14)$$

The probability that a nucleus decays *during a short time interval* Δt is $\lambda \, \Delta t$.

In a collection of a large number N of identical radioactive nuclei, each one has the same decay probability per unit time. The nuclei are independent—the decay of one has nothing to do with the decay of another. Since the decays are independent, the average number that decay during a short time interval Δt is just N times the probability that any one decays:

$$\Delta N = -N\lambda \, \Delta t \qquad (29\text{-}15)$$

The negative sign is necessary because as nuclei decay, the number of nuclei that *remain* is *decreasing*, so the change in N is *negative.* Equation (29-15) gives the *average* number that are expected to decay during Δt. Since radioactive decay is a statistical process, we may not observe exactly that number of decays. If N is sufficiently large, then we expect Eq. (29-15) to be very close to what we observe; for small N, however, deviations from the expected number can be significant. Statistical fluctuations in the number of decays $|\Delta N|$ actually measured are of order $\sqrt{|\Delta N|}$; that is, if the average number of decays expected is 10 000, the actual number of decays that occur in a particular experiment can vary by about $\sqrt{10\,000} = 100$ above or below the average number.

Equation (29-15) is only valid for a *short* time interval $\Delta t \ll 1/\lambda$ because it assumes that the number of nuclei is a constant N. If the time interval is long enough that N changes significantly, then what N should we use? The initial number? The final number? Some sort of average? As long as $\Delta t \ll 1/\lambda$, then we are sure that $|\Delta N| \ll N$, which means that N has not changed significantly.

Activity The number of radioactive decays from a sample per unit time is called the decay rate or **activity** (symbol R). The SI unit of activity is the becquerel (Bq), named for Henri Becquerel. These three ways of writing the SI unit of activity are equivalent:

$$1 \text{ Bq} = 1 \frac{\text{decay}}{\text{s}} = 1 \text{ s}^{-1} \quad (29\text{-}16)$$

Another unit of activity in common use is the curie (Ci) named for the Polish-French physicist Marie Sklodowska Curie (1867–1934) who discovered polonium and radium:

$$1 \text{ Ci} = 3.7 \times 10^{10} \text{ Bq} \quad (29\text{-}17)$$

If the number of decays during a short interval Δt is $|\Delta N|$, then the activity is

$$R = \frac{\text{number of decays}}{\text{unit time}} = \frac{-\Delta N}{\Delta t} = \lambda N \quad (29\text{-}18)$$

CONNECTION:

Whenever the rate of change of a quantity is a negative constant times the quantity, the quantity is an *exponential* function of time.

In Eq. (29-18), the rate of change of N ($\Delta N/\Delta t$) is a negative constant ($-\lambda$) times N. The number of remaining nuclei N in radioactive decay (the number that have *not* decayed) is

$$N(t) = N_0 e^{-t/\tau} \quad (29\text{-}19)$$

A graph of N versus t is shown in Fig. 29.8. For radioactive decay, the time constant is

$$\tau = \frac{1}{\lambda} \quad (29\text{-}20)$$

and N_0 is the number of nuclei at $t = 0$. The time constant is also called the **mean lifetime** since it is the *average* time that a nucleus survives before decaying. However, it would be a misconception to think that nuclei "get old." A uranium-238 nucleus that has been sitting in rock for millions of years has the *same* probability per second of decay as one that has just been created seconds ago in a nuclear reaction; no more, no less. Equations such as (29-18) and (29-19) tell us *how many* nuclei are expected to decay, but not *which ones.*

Since the decay rate is proportional to the number of nuclei, the rate also decays exponentially:

$$R(t) = R_0 e^{-t/\tau} \quad (29\text{-}21)$$

As for exponential decay in any other context, the time constant τ is the time for the quantity to decrease to $1/e \approx 36.8\%$ of its initial value. During a time interval τ, 63.2% of the nuclei decay, leaving 36.8%. After a time interval of 2τ, $1/e^2 \approx 13.5\%$ of the nuclei still have not decayed, while $1 - 1/e^2 \approx 86.5\%$ have decayed.

Half-life Radioactive decay is often described in terms of the **half-life** $T_{1/2}$ instead of the time constant τ. The half-life is the time during which half of the nuclei decay. After

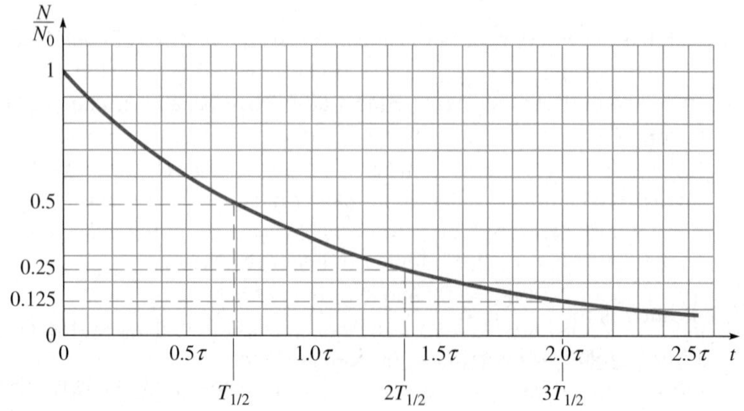

Figure 29.8 Fraction of radioactive nuclei remaining (N/N_0) as a function of time.

two half-lives, one quarter of the nuclei remain; after m half-lives, $\left(\frac{1}{2}\right)^m$ remain. It can be shown (Problem 43) that

$$T_{1/2} = \tau \ln 2 \approx 0.693\tau \quad (29\text{-}22)$$

where ln 2 is the natural (base-e) logarithm of 2. Then

$$N(t) = N_0(2^{-t/T_{1/2}}) = N_0\left(\tfrac{1}{2}\right)^{t/T_{1/2}} \quad (29\text{-}23)$$

CHECKPOINT 29.4

Manganese-54 has a half-life of 312.0 d. What fraction of nuclei in a sample of Mn-54 decay during a period of 936.0 d?

Example 29.8

Radioactive Decay of Nitrogen-13

The half-life of ^{13}N is 9.965 min. (a) If a sample contains 3.20×10^{12} ^{13}N atoms at $t = 0$, how many ^{13}N nuclei are present 40.0 min later? (b) What is the ^{13}N activity at $t = 0$ and at $t = 40.0$ min? Express the activities in Bq. (c) What is the probability that any one ^{13}N nucleus decays during a 1-s time interval?

Strategy (a, b) The number of nuclei at $t = 0$ is $N_0 = 3.20 \times 10^{12}$ and the half-life is $T_{1/2} = 9.965$ min. The problem asks for N at $t = 40.0$ min and for R at both $t = 0$ and at $t = 40.0$ min. Since the time interval of 40.0 min is approximately four times the half-life, we can first estimate the solution: both N and R are multiplied by $\frac{1}{2}$ during each half-life.

(c) The probability of decay during a 1-s interval is λ *only if* 1 s can be considered a short time interval. Since the half-life is 9.965 min = 597.9 s, 1 s is a tiny fraction of the half-life and therefore *can* be considered a short time interval.

Solution (a) Half of the nuclei are left after one half-life, $\frac{1}{2} \times \frac{1}{2} = \left(\frac{1}{2}\right)^2$ after two half-lives, and $\left(\frac{1}{2}\right)^4$ after four half-lives. Therefore, the number remaining after four half-lives is

$$N = \left(\tfrac{1}{2}\right)^4 \times 3.20 \times 10^{12} = 2.00 \times 10^{11}$$

Using Eq. (29-23) gives the precise result:

$$N(t) = N_0\left(\tfrac{1}{2}\right)^{t/T_{1/2}} = N_0\left(\tfrac{1}{2}\right)^{40.0/9.965} = 1.98 \times 10^{11}$$

(b) The activity and number of nuclei are related by Eq. (29-18):

$$R = \lambda N = \frac{N}{\tau}$$

The time constant τ is related to the half-life by Eq. (29-22):

$$\tau = \frac{T_{1/2}}{\ln 2} = \frac{9.965 \text{ min} \times 60 \text{ s/min}}{0.693\,15} = 862.6 \text{ s}$$

Next we substitute the number of nuclei N at $t = 0$ and at $t = 40.0$ min to determine the rate of decay at those two times. The time constant does not change.

At $t = 0$,

$$R_0 = \frac{N_0}{\tau} = \frac{3.20 \times 10^{12}}{862.6 \text{ s}} = 3.71 \times 10^9 \text{ Bq}$$

At $t = 40.0$ min,

$$R = \frac{N}{\tau} = \frac{1.98 \times 10^{11}}{862.6 \text{ s}} = 2.30 \times 10^8 \text{ Bq}$$

(c) The probability per second is

$$\lambda = \frac{1}{\tau} = 1.1593 \times 10^{-3} \text{ s}^{-1}$$

A nucleus has a 0.001 159 3 probability of decaying during a 1-s interval.

Discussion As a check, R after four half-lives should be $\frac{1}{16}$ of R_0:

$$\tfrac{1}{16} \times 3.71 \times 10^9 \text{ Bq} = 2.32 \times 10^8 \text{ Bq}$$

Since 40.0 min is slightly more than four half-lives, the activity at $t = 40.0$ min is slightly less than 2.32×10^8 Bq.

The probability of decay in one second would *not* be equal to λ if the half-life were not large compared to 1 s. For a longer time interval, we find the decay probability as follows:

$$\text{probability of decay} = \frac{\text{number expected to decay}}{\text{original number}}$$

$$= \frac{|\Delta N|}{N_0} = \frac{N_0 - N}{N_0} = 1 - e^{-t/\tau}$$

Practice Problem 29.8 Number Remaining After One Half of a Half-Life

How many ^{13}N atoms are present at $t = 5.0$ min?

PHYSICS AT HOME

You can simulate radioactive decay by tossing pennies. Assemble a collection of as many pennies (or other coins) as you can; around 100 would be ideal. The pennies serve as models of radioactive nuclei. Record the original number of pennies, then toss them. Let each penny that comes up "tails" represent a nucleus that has decayed. Remove all the tails and count them. Then toss the remaining pennies. Continue until only 1–3 are left. Make a table like this:

t (Number of Tosses)	$\lvert\Delta N\rvert$ (Number of Tails)	N (Number Remaining)
0	—	100
1	54	46 (= 100 – 54)
2	22	24 (= 46 – 22)
3	. . .	. . .

One toss serves as a unit of time. The half-life of the pennies is expected to be 1.0 toss. Plot your data points N as a function of t on graph paper. Also plot the function $N(t) = N_0 2^{-t/T_{1/2}}$ as a smooth curve. (Use semilog graph paper if you have some and know how to use it.) Are the deviations of your data points from the curve reasonable, given that the expected size of statistical fluctuations is approximately $\sqrt{\lvert\Delta N\rvert}$? The fluctuations are relatively large here because of the small number of "nuclei."

Application: Radiocarbon Dating

The immensely useful radiocarbon dating technique (or carbon dating as it is frequently called) is based on the radioactive decay of a rare isotope of carbon. Almost all of the naturally occurring carbon on Earth is one of the two stable isotopes—98.9% is ^{12}C and 1.1% is ^{13}C. However, there is also a trace amount of ^{14}C—about one in every 10^{12} carbon atoms. The carbon-14 isotope, ^{14}C, has a relatively short half-life of 5730 yr. Since the Earth is about 4.5×10^9 yr old, we would expect to find no carbon-14 at all if it were not continually being replenished.

The production of carbon-14 occurs because the Earth's atmosphere is bombarded by cosmic rays. Cosmic rays are extremely high-energy charged particles—mostly protons—from space. When one of these particles hits an atom in Earth's upper atmosphere, a shower of secondary particles is created, which includes a large number of neutrons. Typically about 1 million neutrons are produced by each cosmic ray particle. Some of these neutrons then react with ^{14}N nuclei in the atmosphere to form ^{14}C:

$$\mathrm{n} + {}^{14}\mathrm{N} \rightarrow {}^{14}\mathrm{C} + \mathrm{p} \qquad (29\text{-}24)$$

The ^{14}C forms CO_2 molecules and drifts down to the surface, where it is absorbed from the air by plants and incorporated into carbonate minerals. Animals take in the ^{14}C by eating plants and other animals. The ^{14}C in an organism or mineral decays via beta decay:

$$^{14}\mathrm{C} \rightarrow {}^{14}\mathrm{N} + \mathrm{e}^- + \bar{\nu} \qquad (29\text{-}25)$$

Balance between the rate at which ^{14}C is continually being created by cosmic rays and the rate at which the ^{14}C decays results in an equilibrium ratio of ^{14}C to ^{12}C atoms in the atmosphere equal to 1.3×10^{-12}. While an organism is alive, carbon is exchanged with the environment, so the organism maintains the same relative abundance of ^{14}C as the environment. The carbon-14 activity in the atmosphere or in a living organism is 0.25 Bq per gram of carbon (see Problem 41). When an organism dies, or when ^{14}C is incorporated into a mineral, carbon exchange with the environment stops. As the ^{14}C present in the organism decays, the ratio of ^{14}C to ^{12}C decreases. The ratio of ^{14}C to ^{12}C in a sample can be measured and used to determine the age of the sample. One way to do this is to measure the carbon-14 activity per gram of carbon.

Example 29.9

Dating a Charcoal Sample

A piece of charcoal (essentially 100% carbon) from an archaeological site in Egypt is subjected to radiocarbon dating. The sample has a mass of 3.82 g and a ^{14}C activity of 0.64 Bq. What is the age of the charcoal sample?

Strategy While a tree is alive, it maintains the same relative abundance of ^{14}C as the environment. After a tree is cut down to make charcoal, the relative abundance of ^{14}C decreases since ^{14}C is no longer being replaced from the environment. As the number of ^{14}C nuclei decreases, so does the ^{14}C activity. The activity decreases exponentially from its initial value with a *half-life* of 5730 yr. We assume the relative abundance in the environment in ancient Egypt was similar to today, so the initial activity is 0.25 Bq per gram of carbon.

Solution The activity of ^{14}C decreases exponentially:

$$R = R_0\, e^{-t/\tau}$$

The initial activity is

$$R_0 = 0.25 \text{ Bq/g} \times 3.82 \text{ g} = 0.955 \text{ Bq}$$

The present activity is $R = 0.64$ Bq. Now we solve for t from the values of R and R_0.

$$\frac{R}{R_0} = e^{-t/\tau}$$

Taking the natural logarithm of each side gets t out of the exponent:

$$\ln \frac{R}{R_0} = \ln e^{-t/\tau} = -\frac{t}{\tau}$$

$$t = -\tau \ln \frac{R}{R_0} = -\frac{T_{1/2}}{\ln 2} \ln \frac{R}{R_0}$$

$$= -\frac{5730 \text{ yr}}{\ln 2} \times \ln \frac{0.64 \text{ Bq}}{0.955 \text{ Bq}} = 3300 \text{ yr}$$

The charcoal is 3300 yr old.

Discussion As a check, we can test to see whether

$$R_0(2^{-t/T_{1/2}}) = R$$

$$R_0(2^{-t/T_{1/2}}) = 0.955 \text{ Bq} \times 2^{-3300 \text{ yr}/5730 \text{ yr}} = 0.955 \text{ Bq} \times 0.671$$

$$= 0.64 \text{ Bq} = R$$

Practice Problem 29.9 The Age of Ötzi

In 1991, a hiker found the frozen, naturally mummified remains of a man protruding from a glacier in the Italian Alps. The man was called Ötzi by researchers and became popularly known as the Iceman. The ^{14}C activity of the Iceman's remains was measured to be 0.131 Bq per gram of carbon. How long ago did the Iceman die?

Example 29.10

Yearly Decrease in Carbon-14 Activity of a Nonliving Sample

By what percentage does the ^{14}C activity of a nonliving sample decrease in one year?

Strategy We are given neither the activity at the beginning nor at the end of the one year period, but we only want to find the change *expressed as a percentage of the initial activity.* The percentage change is a way to express the fractional change (the change in activity as a fraction of the initial activity). Let the initial activity be R_0 and the activity one year later be R. The quantity to be determined is

$$\frac{\Delta R}{R_0} = \frac{R - R_0}{R}$$

expressed as a percentage.

Solution The activities R_0 and R are related by

$$R(t) = R_0(2^{-t/T_{1/2}})$$

We choose this form rather than the exponential form $R = R_0 e^{-t/\tau}$ because we are given the half-life rather than the time constant. We don't know R_0 or R, but we can find the ratio of the two.

$$\frac{R}{R_0} = 2^{-t/T_{1/2}} = 2^{-1/5730} = 0.999\ 879$$

Now we find the fractional change during 1 yr.

$$\frac{\Delta R}{R_0} = \frac{R - R_0}{R_0} = \frac{R}{R_0} - 1 = 0.999\ 879 - 1 = -0.000\ 121$$

The carbon-14 activity decreases 0.012% in a year.

Discussion The tiny change in activity illustrates one reason why we do not expect carbon-14 dating to give dates precise to a specific year.

Practice Problem 29.10 Dating Precision

If the ^{14}C activity of a shard of pottery can be determined to a precision of ±0.1%, to what precision can we expect to date the shard (assuming no other sources of imprecision)? [*Hint:* In what time interval does the activity change by 0.1%?]

Carbon dating can be used for specimens up to about 60 000 yr old, which is about 10 half-lives of ^{14}C. The older a specimen is, the smaller its ^{14}C activity; for very old samples, it is difficult to measure the ^{14}C activity accurately. The half-life also imposes constraints on the precision with which a sample can be dated. Even if all our assumptions were true, we cannot expect carbon dating to give ages down to the exact year. One year is only a small fraction of the half-life, so the activity changes very little during a year's time (as shown in Example 29.10).

A major assumption of the simplest kind of carbon dating presented here is that the equilibrium ratio of ^{14}C to ^{12}C in Earth's atmosphere has been the same for the past 60 000 yr. Is that a good assumption? How can we test it? One way to test it for relatively short times is by taking core samples from very old trees—or from the remains of ancient trees—and measuring ^{14}C activities from various times. The tree rings give an independent way to determine the age of different parts of the sample.

At present, scientists believe that the relative abundance of ^{14}C in the atmosphere hasn't changed much in the past 1000 yr (until the beginning of the twentieth century) although it has varied considerably during the past 60 000 yr, reaching peaks as much as 40% higher than at present. Fortunately, radiocarbon dating can be adjusted for the changes in the relative abundance of ^{14}C in the atmosphere. Tree rings allow such adjustment going back about 11 000 yr. In Japan's Lake Suigetsu, layers of dead algae sink to the bottom annually and are covered by a layer of clay sediment before the next algae layer. The alternating layers of light-colored algae and dark clay can be read like tree rings, allowing radiocarbon data to be adjusted for the varying abundance of ^{14}C in the atmosphere going back about 43 000 yr.

The relative abundance of ^{14}C in the atmosphere began changing rapidly in the twentieth century due to human activity. An enormous increase in the burning of fossil fuels introduced large quantities of old carbon—that is, carbon with a low abundance of ^{14}C—into the atmosphere. Beginning about 1940, open-air nuclear testing, nuclear bombs, and nuclear reactors have increased the relative abundance of ^{14}C in the atmosphere. In the distant future it will be difficult to use radiocarbon dating for artifacts from the twentieth century.

Other Isotopes Used in Radioactive Dating

Application: radioactive dating of geologic formations

Besides ^{14}C, other radioactive nuclides are also used for radioactive dating. Isotopes commonly used to date geologic formations (with their half-lives in billions of years) include uranium-235 (0.7), potassium-40 (1.248), uranium-238 (4.5), thorium-232 (14), and rubidium-87 (49). One direct way to calculate Earth's age is based on the abundances of various lead isotopes in terrestrial samples and in meteorites. Pb-206 and Pb-207 are the final products of long chains of radioactive decays that begin with U-238 and U-235, respectively.

Lead-210, with a half-life of only 22.20 yr, is used for geologic dating over the last 100 to 150 yr. It forms in rocks containing uranium-238 as a decay product of radon gas. After forming from radon in the atmosphere, the lead isotope falls to Earth, where it collects on the surface and is stored in the soil, or in the sediment of lakes and oceans, or in glacial ice. The age of a sediment layer can be determined by measuring the amount of lead-210 present.

Quantum-Mechanical Tunneling Explains Radioactive Half-Lives for Alpha Decay

An early triumph of quantum mechanics was its explanation of the correlation between the half-life of a particular alpha decay and the kinetic energy of the alpha particle. The kinetic energies vary over a narrow range (4–9 MeV) but the half-lives range from 10^{-5} s to 10^{25} s (10^{17} yr). Despite this discrepancy in ranges, the two quantities are closely correlated (Fig. 29.9a); higher alpha particle energies consistently go with shorter half-lives.

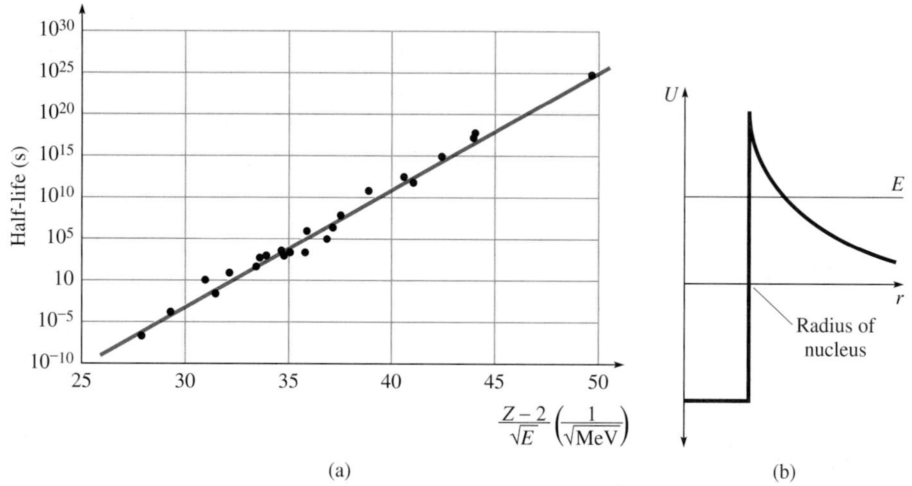

Figure 29.9 (a) Correlation between half-life and the energy (E) of the α particle. Z is the atomic number of the parent nuclide. Note that the vertical scale is logarithmic and that increasing numbers on the horizontal axis represent *decreasing* energies. (b) Simplified model of the potential energy U of an α particle as a function of its distance r from the center of the nucleus. The energy of the α particle is E.

The correlation arises because the alpha particle must tunnel (see Section 28.10) out of the nucleus. Think of an alpha particle in a nucleus as facing the simplified potential energy graph of Fig. 29.9b. Inside the nucleus, the potential energy of the alpha particle is roughly constant. Beyond the edge of the nucleus, where the strong attractive force no longer pulls the alpha particle toward the nucleus, the alpha particle feels only a Coulomb repulsion from the nucleus [which has charge $+(Z-2)e$ since it has lost two protons]. The potential energy barrier is higher than the energy E of the alpha particle. Since the barrier tapers off gradually, decreasing with distance as $1/r$, lower energy alpha particles are not only farther below the top of the barrier; they face a much *wider* barrier as well. Higher energy alpha particles have much higher tunneling probabilities and therefore much shorter half-lives.

29.5 BIOLOGICAL EFFECTS OF RADIATION

We are all continually exposed to radiation. The biological effects of radiation depend on what kind of radiation it is, how much of it is absorbed by the body, and the duration of the exposure. *Ionizing* radiation is radiation with enough energy to ionize an atom or molecule—typically between 1 eV and a few tens of eV. An α particle, β particle, or γ ray with a typical energy of about 1 MeV can potentially ionize tens of thousands of molecules. Molecules in living cells that are ionized due to radiation become chemically active and can interfere with the normal operation and reproduction of the cell.

The **absorbed dose** of ionizing radiation is the amount of radiation energy absorbed per unit mass of tissue. The SI unit of absorbed dose is the gray (Gy):

$$1 \text{ Gy} = 1 \text{ J/kg} \tag{29-26}$$

Another common unit for absorbed dose is the rad:

$$1 \text{ rad} = 0.01 \text{ Gy} \tag{29-27}$$

The name "rad" stands for radiation absorbed dose.

Different kinds of radiation cause different amounts of biological damage, even if the absorbed dose is the same. The health effects also depend on what kind of tissue is exposed. To account for these factors, a quantity called the quality factor (QF)—sometimes called the relative biological effectiveness (RBE)—is assigned to each type of radiation. The QF is a relative measure of the biological damage caused by different kinds of radiation compared to 200-keV x-rays (which are assigned QF = 1). The QF varies depending on the kind of radiation (alpha, beta, gamma), the energy of the radiation, the kind of tissue exposed, and the biological effect under consideration. Table 29.3 gives some typical QF values.

Table 29.3 Typical Quality Factors (QFs) for Various Forms of Radiation

Gamma rays	0.5–1
Beta particles	1
Protons, neutrons	2–10
Alpha particles	10–20

To measure the biological damage caused by exposure to radiation, we calculate the **biologically equivalent dose**. The SI unit for biologically equivalent dose is the sievert (Sv).

$$\text{equivalent dose (in sieverts)} = \text{absorbed dose (in grays)} \times \text{QF} \quad (29\text{-}28a)$$

$$\text{equivalent dose (in rem)} = \text{absorbed dose (in rad)} \times \text{QF} \quad (29\text{-}28b)$$

Example 29.11

Biologically Equivalent Dose in a Brain Scan

A 60.0-kg patient about to have a brain scan is injected with 20.0 mCi of the radionuclide technetium-99, $^{99}\text{Tc}^{\text{m}}$. (The superscript "m" stands for metastable. The metastable state $^{99}\text{Tc}^{\text{m}}$ decays to the ground state with a half-life of 6.0 h.) The $^{99}\text{Tc}^{\text{m}}$ nucleus decays by emitting a 143-keV photon. Assuming that half of these photons escape the body without interacting, what biologically equivalent dose does the patient receive? The QF for these photons is 0.97. Assume that all of the $^{99}\text{Tc}^{\text{m}}$ decays while in the body.

Strategy The activity (20.0 mCi) together with the half-life (6.0 h) enable us to calculate the number of $^{99}\text{Tc}^{\text{m}}$ nuclei. Then we can determine how many photons are absorbed in the body; multiplying the number of photons absorbed by the energy of each photon (143 keV) gives the total radiation energy absorbed. The absorbed dose is the radiation energy absorbed per unit mass of tissue. The biologically equivalent dose is the absorbed dose times the quality factor.

Solution The activity of the injected material in becquerels (Bq) is

$$R_0 = 20.0 \times 10^{-3}\ \text{Ci} \times 3.7 \times 10^{10}\ \text{Bq/Ci} = 7.4 \times 10^8\ \text{Bq}$$

The activity is related to the number of nuclei N by

$$R_0 = \lambda N_0 = \frac{N_0}{\tau}$$

Then the number of nuclei injected is

$$N_0 = \tau R_0 = \frac{T_{1/2}}{\ln 2} R_0 = \frac{6.0\ \text{h} \times 3600\ \text{s/h}}{\ln 2} \times 7.4 \times 10^8\ \text{s}^{-1}$$

$$= 2.306 \times 10^{13}$$

Each of these nuclei emits a photon and half of the photons are absorbed by the body. The energy of each photon is 143 keV. Therefore, the total energy absorbed in joules is

$$E = \frac{1}{2} \times (2.306 \times 10^{13}\ \text{photons}) \times 1.43 \times 10^5\ \frac{\text{eV}}{\text{photon}} \times (1.60 \times 10^{-19}\ \frac{\text{J}}{\text{eV}})$$

$$= 0.264\ \text{J}$$

The absorbed dose is

$$\frac{0.264\ \text{J}}{60.0\ \text{kg}} = 0.0044\ \text{Gy}$$

The biologically equivalent dose is the absorbed dose times the quality factor:

$$0.0044\ \text{Gy} \times 0.97 = 0.0043\ \text{Sv}$$

Discussion A quantity of radioactive material is often specified by its activity ("20.0 mCi of $^{99}\text{Tc}^{\text{m}}$") rather than by mass, number of moles, or number of nuclei. As already shown, the number of radioactive nuclei can be calculated from the activity and the half-life.

Practice Problem 29.11 Determining Mass from Activity

What is the mass of 5.0 mCi of $^{60}_{27}\text{Co}$?

Average Radiation Doses due to Natural Sources The average radiation dose received by a person in one year due to natural sources is about 0.003 Sv. Of this, an average of 0.002 Sv is due to inhaled radon-222 gas and its decay products. Radon-222 is constantly produced by the α decay of radium-226 present in soil and rocks. Radon gas usually enters houses through cracks in the foundation. When ^{222}Rn nuclei are inhaled, they can give a significant dose of α-particle radiation to the lungs. The amount of radon gas that enters a building varies greatly from one place to another. In some localities, radon is not much of a problem. In other places, with large amounts of radium in the soil and geological formations that make it easy for radon gas to find its way into a basement, it is a major cause of lung cancer (second only to smoking). Fortunately, an inexpensive test can be used to determine the concentration of radon gas in the air. Where radon is a problem, sealing cracks in the basement and adding ventilation are often all that is needed.

Of the average annual dose, about 0.0007 Sv is due to radioactive nuclides that enter the body in food and water (such as ^{14}C and ^{40}K) or are present in the soil and in building materials (e.g., polonium, radium, thorium, and uranium). Another 0.0003 Sv is due to cosmic rays. The cosmic ray dose is significantly higher for people living at high altitudes or who spend a lot of time in airplanes. In a commercial jet at 35 000 ft, the dose received is about 7×10^{-6} Sv/h, so 40 h of flight doubles the average person's cosmic ray dose; 90 h of flight is equivalent to the average annual dose due to medical and dental exposure.

Average Doses due to Human Activity Human activities have added an average annual dose of 0.0006 Sv to the 0.003 Sv from natural sources. Most of this additional radiation dose comes from medical and dental diagnosis and treatment (principally diagnostic x-rays). The average annual dose due to fallout from testing of nuclear weapons and due to nuclear reactors is about 10^{-5} Sv, but is much higher in some places (for instance in Ukraine, due to the Chernobyl disaster).

Short- and Long-Term Effects of Radiation A single large dose of radiation causes radiation sickness. Symptoms can include nausea, diarrhea, vomiting, and hair loss. Radiation sickness can be fatal if the dose is large enough. A single dose of about 4–5 Sv is fatal about half of the time. Long-term effects of much smaller doses of radiation include increased risk of cancer and genetic mutations. In the United States, the Nuclear Regulatory Commission limits occupational radiation exposure for adults who work with radioactive material to less than 0.050 Sv/yr above background levels (0.001 Sv/yr for the general public).

Penetration of Radiation

Different kinds of radiation have different abilities to penetrate biological tissue (or other materials). The range of an alpha particle in human tissue is about 0.03 mm to 0.3 mm, depending on the energy of the particle. Alpha particles are stopped by a few centimeters of air or by an aluminum foil only 0.02 mm thick. Alpha particles are potentially the most damaging form of radiation, since each can ionize large numbers of molecules. On the other hand, they cannot penetrate the skin, so α-emitters outside the body are not so dangerous. Radon gas is dangerous because the α decay occurs *inside* the body, exposing the lung tissue directly. Similarly, if alpha emitters are present in food, they can deliver a significant dose of radiation to the digestive tract, and those with longer half-lives may then be incorporated into other body tissue (e.g., radioactive iodine collects in the thyroid and radioactive iron collects in the blood).

Beta-minus particles (electrons) are more penetrating than αs. Their range in human tissue can be as much as a few centimeters (again, depending on energy). They can penetrate several meters of air; it takes an aluminum plate about 1 cm thick to stop them. High-speed electrons not only ionize molecules, but also emit x-rays through bremsstrahlung (Section 27.4); the x-rays are much more penetrating than are the electrons themselves. β^+ particles (positrons) have a very limited range—they quickly annihilate with an electron, producing two photons.

Beta emitters are more dangerous when found inside the body, though the difference is not as striking as for alpha-emitters. Atmospheric tests of nuclear weapons in the 1950s produced many dangerous radioactive nuclides. One of them, radioactive strontium-90, is produced by the fission of ^{235}U. Strontium is chemically similar to calcium. Both are alkali metals; Sr is directly below Ca in the periodic table. The strontium-90 produced by atmospheric tests entered the human food supply and was incorporated into the bones and teeth of growing children. Strontium-90 undergoes β decay with a half-life of 28 yr, but since calcium (and strontium) stays in the body for a long time, the presence of this radionuclide in the bones ends up delivering a significant radiation dose and probably increases the incidence of leukemia and other cancers. Fortunately, atmospheric tests are now banned internationally, and the incidence of strontium-90 and other artificially produced radionuclides is smaller than it once was.

Both alphas and electrons have a fairly definite range for a given material and energy. They lose their energy through a large number of collisions with molecules. By contrast, a γ-ray photon can lose a large proportion or even all of its energy in a *single* interaction (via the photoelectric effect, Compton scattering, or pair production).

The probability of one of these interactions occurring can be calculated using quantum mechanics. For photons of a certain energy, we can only predict the *average* distance traveled in a given material. For example, half of 5-MeV photons can penetrate 23 cm or more into the body. Half of 5-MeV photons can penetrate 1.5 cm or more in lead. The penetrating ability of photons is measured as a *half-value layer*, which is the thickness of material that half of the photons can penetrate.

Medical Applications of Radiation

Radioactive Tracers in Medical Diagnosis There are many medical applications of radioactive materials and of radiation. **Radioactive tracers** are important diagnostic tools. One example was mentioned in Example 29.11. The metastable state of technetium-99 is the product of the β decay of molybdenum-99. Most nuclear excited states decay to the ground state in very short times, ranging from about 10^{-15} s to 10^{-8} s. The metastable state of technetium-99 has an unusually long half-life of 6.0 h, perfect for use as a radioactive tracer. If the half-life were much shorter, much of the $^{99}Tc^{m}$ would decay before it reached the tumor cells. If the half-life were much longer, then the activity would be small and only a small fraction of the γ rays could be detected within a reasonable length of time.

The blood-brain barrier prevents the technetium (which is injected as technetium oxide and attaches to red blood cells) from diffusing into normal brain cells, but the abnormal cells in a tumor do not have such a barrier. Therefore, the tumor can be located and imaged by observation of the γ rays emitted from the brain.

One way to do the imaging is to use an **Anger camera** (pronounced ahn-zhay; Fig. 29.10). A lead collimating plate has parallel holes drilled in it. The lead absorbs γ rays, so only photons emitted parallel to one of the holes can get through the plate. Behind the plate is a scintillation crystal; when a gamma photon hits this crystal, a pulse of light is produced. Photomultiplier tubes, one for each hole in the collimator, detect these light pulses. By moving the Anger camera around at different angles, we can "triangulate" back and figure out where the tumor is.

Similarly, TlCl (thallium chloride) tends to collect at the site of a blood clot. Thallium-201 has a half-life of 73 h. When thallium-201 undergoes β decay in the body, γ rays are also emitted as the daughter nucleus drops down into its ground state. An Anger camera can then be used to locate the clot.

Radioactive tracers are used in research as well as in clinical diagnosis. For example, radioactive iron-59 was used to determine that iron, unlike most other elements, is not constantly being eliminated from the body and then replaced. Rather, once an iron atom is incorporated into a hemoglobin molecule, it stays there for the

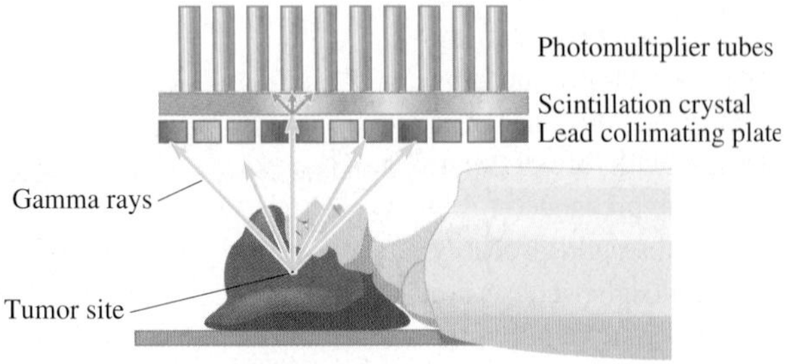

(a)

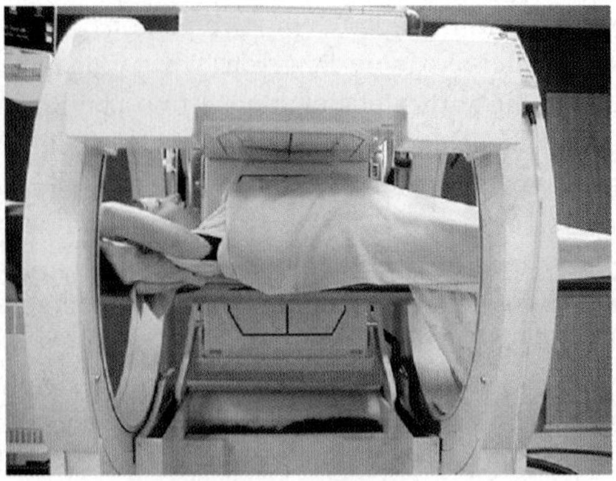
(b)

Figure 29.10 (a) Simplified diagram of an Anger camera. A radioactive tracer has accumulated at the tumor site and emits γ rays. A γ-ray photon that passes through a hole in the collimating plate is detected by the apparatus. (b) This photo shows an Anger camera with two detector heads, one above the patient's chest and the other to her left. The lead plates, scintillation crystals, and photomultiplier tubes are hidden behind the grids marking the detector heads.

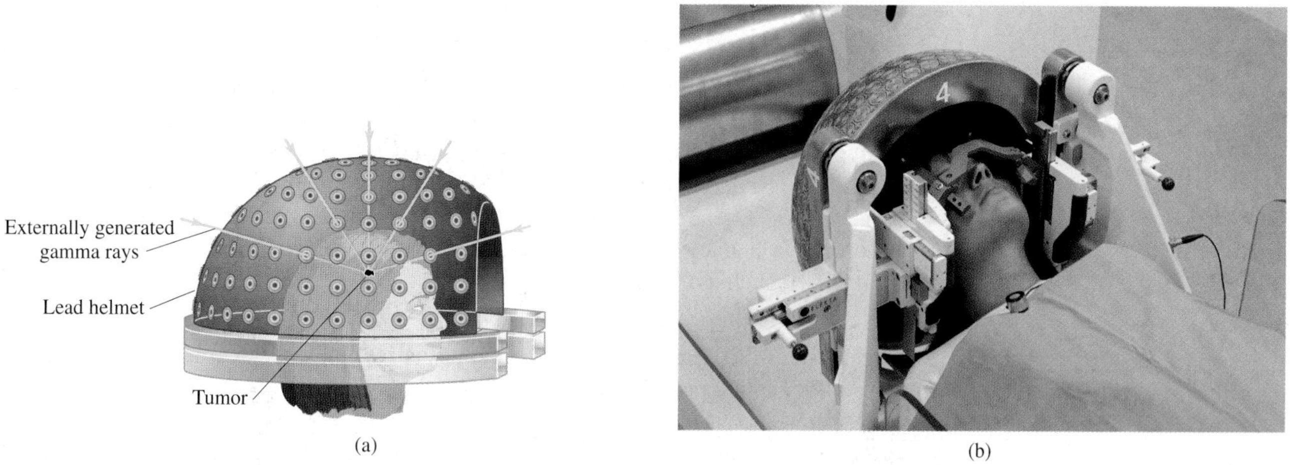

Figure 29.11 (a) Diagram of the lead "helmet" used in gamma knife radiosurgery. (b) The patient is carefully positioned in the helmet to ensure that the γ rays converge at the desired point in the brain. A lead apron protects the body from exposure to radiation.

entire life of the red blood cell. Even when a red blood cell dies, the iron is recycled for use in another cell.

Positron Emission Tomography (PET) In positron emission tomography (PET), positron-emitters (radioisotopes whose decay mode is β^+) are injected into the body. The tracer most commonly used in PET is the sugar fluorodeoxyglucose. The fluorine nuclide in the molecule is a positron-emitter (^{18}F). A positron emitted in the body quickly annihilates with an electron to produce two γ rays traveling in opposite directions. The photons are detected by a ring of detectors around the body (see Fig. 27.25). Among the uses of PET are detecting tumors and metastatic cancer, assessing coronary artery disease, locating heart damage caused by a heart attack, and diagnosing central nervous system disorders.

Radiation Therapy **Radiation therapy** is used in cancer treatment. Cancer cells are more vulnerable to the destructive effects of radiation, in part because they are rapidly dividing. Thus, the idea of radiation therapy is to supply enough radiation to destroy the malignant cells without causing too much damage to normal cells. The radiation can be administered internally or externally. Internally applied radiation treatment uses radionuclides that are either injected into the tumor, or which collect at the tumor site (much as tracers do). In a promising new technique for targeting cancer cells with radiation, a single radioactive atom is enclosed in a microscopic cage made of carbon and nitrogen atoms. Attached to the cage is a protein that locks onto a specific protein on the surface of a cancerous cell, after which the cage moves inside the cell. The α particles emitted in a series of radioactive decays then kill the cell.

Externally applied radiation can be x-rays produced by bremsstrahlung or by other means. Cobalt-60 emits gamma rays that are also used for radiation therapy. The cobalt-60 is kept in a lead box with a small hole so that the γ rays can be limited to the site of the tumor.

Gamma Knife Radiosurgery An advanced form of cobalt-60 therapy is called **gamma knife radiosurgery**. In this technique, a spherical lead "helmet" with hundreds of holes (Fig. 29.11) enables the γ rays to converge at a small region in the brain. In this way, the radiation dose to the tumor, where all the γ rays converge, can be much larger than the dose to the surrounding tissue.

Particle Accelerators in Hospitals Some hospitals have cyclotrons (see Section 19.3) or other particle accelerators on site. Their purpose is twofold. The accelerator can be used to manufacture radionuclides that have short half-lives. Radionuclides with longer half-lives can be manufactured offsite at either an accelerator or at a nuclear reactor. Second, beams of accelerated charged particles are used in radiation therapy.

29.6 INDUCED NUCLEAR REACTIONS

In radioactivity, an unstable nucleus decays in a *spontaneous* nuclear reaction, releasing energy in the process. An *induced* nuclear reaction is one that does not occur spontaneously; it is caused by a collision between a nucleus and something else. The other reactant can be another nucleus, a proton, a neutron, an α particle, or a photon.

We have already seen an example of an induced nuclear reaction; carbon-14 is formed in a nuclear reaction induced when an energetic neutron collides with a nitrogen-14 nucleus:

$$\text{n} + {}^{14}\text{N} \rightarrow {}^{14}\text{C} + \text{p} \qquad (29\text{-}24)$$

Equation (29-24) is an example of **neutron activation**, in which a stable nucleus is transformed into a radioactive one by absorbing a neutron.

A spontaneous nuclear reaction always releases energy, so that the total mass of the products is always less than the total mass of the reactants. By contrast, an induced reaction can convert some of the kinetic energy of the reactants into rest energy. Thus, the total mass of the products can be greater than, less than, or equal to the total mass of the reactants. A nucleus involved in such a reaction does not have to be radioactive; a stable nucleus can participate in a reaction when struck by some other particle. The first such reaction ever observed, by Rutherford in 1919, was

$$\alpha + {}^{14}\text{N} \rightarrow {}^{17}\text{O} + \text{p} \qquad (29\text{-}29)$$

A reaction takes place when the target nucleus absorbs the incident particle, forming an intermediate compound nucleus. In the reaction of Eq. (29-29), the compound nucleus is ${}^{18}\text{F}$:

$${}^{4}_{2}\text{He} + {}^{14}_{7}\text{N} \rightarrow {}^{18}_{9}\text{F} \rightarrow {}^{17}_{8}\text{O} + {}^{1}_{1}\text{H}$$

CHECKPOINT 29.6

What is the intermediate compound nucleus formed in the induced reaction of Eq. (29-24)?

Example 29.12

A Neutron Activation

Consider the reaction

$$\text{n} + {}^{24}\text{Mg} \rightarrow \text{p} + ?$$

(a) Determine the product nucleus and the intermediate compound nucleus. (b) Is this reaction exoergic or endoergic? That is, does it release energy, or does it require the input of energy to occur? Calculate either the energy released (if exoergic) or the energy absorbed (if endoergic).

Strategy The product nucleus and compound nucleus can be identified by balancing the reaction: the total charge and total number of nucleons must remain the same. The energetics are determined by whether the total mass of the products is greater or less than the total mass of the reactants.

Solution (a) Magnesium is atomic number 12. The reaction, written out more fully, is

$${}^{1}_{0}\text{n} + {}^{24}_{12}\text{Mg} \rightarrow {}^{25}_{12}(?) \rightarrow {}^{24}_{11}(?) + {}^{1}_{1}\text{p}$$

where we have made sure that the total electric charge and the total number of nucleons remain unchanged. From the periodic table, atomic number 11 is Na and we already know that atomic number 12 is Mg. Therefore, the product nucleus is ${}^{24}_{11}\text{Na}$ and the intermediate nucleus is ${}^{25}_{12}\text{Mg}$.

(b) We compare the total mass of the reactants with the total mass of the products. From Appendix B, the atomic masses are

$$\text{mass of } {}^{24}\text{Mg} = 23.985\,041\,9 \text{ u}$$
$$\text{mass of } {}^{24}\text{Na} = 23.990\,963\,3 \text{ u}$$
$$\text{mass of } {}^{1}\text{H} = 1.007\,825\,0 \text{ u}$$
$$\text{mass of n} = 1.008\,664\,9 \text{ u}$$

Using atomic masses is fine, since both sides include the extra mass of the same number of electrons (12). Then the total mass of the reactants is

$$1.008\,664\,9 \text{ u} + 23.985\,041\,9 \text{ u} = 24.993\,706\,8 \text{ u}$$

continued on next page

Example 29.12 continued

and the total mass of the products is

$$1.007\,825\,0 \text{ u} + 23.990\,963\,3 \text{ u} = 24.998\,788\,3 \text{ u}$$

Thus, the total mass *increases* when this reaction takes place:

$$\Delta m = 24.998\,788\,3 \text{ u} - 24.993\,706\,8 \text{ u} = +0.005\,081\,5 \text{ u}$$

Since the mass of the products is greater than the mass of the reactants, the reaction is endoergic; there is less kinetic energy after the reaction than there was before. The energy absorbed is

$$E = (\Delta m)c^2 = 0.005\,081\,5 \text{ u} \times 931.494 \text{ MeV/u}$$
$$= 4.7334 \text{ MeV}$$

Discussion We expect this reaction to be *possible* only if the total kinetic energy of the reactants is at least 4.7334 MeV more than the total kinetic energy of the products. It is not necessarily the *most likely* outcome. Other reactions compete, such as the emission of one or more photons:

$${}^{1}_{0}\text{n} + {}^{24}_{12}\text{Mg} \rightarrow {}^{25}_{12}\text{Mg}^* \rightarrow {}^{25}_{12}\text{Mg} + \gamma$$

In other cases, the competing reaction might include α decay, β decay, or fission.

Practice Problem 29.12 The Reaction That Produces Carbon-14

Determine whether the reaction $\text{n} + {}^{14}\text{N} \rightarrow {}^{14}\text{C} + \text{p}$ is exoergic or endoergic. How much energy is released or absorbed?

Application: Neutron Activation Analysis

Neutron activation analysis (NAA) is a technique used to study precious works of art, rare archaeological specimens, geological objects, and the like. It is used to determine which elements are present in the sample being studied—even those present in only trace amounts. The great advantage of NAA over other methods of analysis is that it is minimally invasive. An entire painting can be analyzed without the need to scrape off some of the paint, as would have to be done to use a mass spectrometer. Art historians know that different paint pigments were used in different historical periods. Determination of the pigments used can help establish the date of a painting; it can also detect forgeries, repairs, and canvasses that have been painted over.

How can a nuclear reactor help conservators and art historians learn about a painting?

The elements present in a sample are identified by the characteristic γ-ray energies emitted by the activated nuclei when they decay. By taking γ-ray spectra at different times, the half-lives can also be used for identification purposes. Quantitative analysis of the γ-ray spectrum yields the concentrations of various elements in the samples being studied. Neutron activation analysis of this type has been used to study lunar samples from the Apollo missions, bullets and gunshot residue swabs used as forensic evidence in criminal investigations, oceanographic fossils and sediments, textiles, and artifacts from archaeological excavations, just to name a few examples.

NAA enables the art historian to determine which pigments have been used on which parts of the painting, even in the underlying layers, without damaging the painting. In *Aristotle with a Bust of Homer*, NAA helped reveal the extent of the damage to the apron and to the hat. Art historians also drew some conclusions about how Rembrandt's composition was altered as he worked, such as changes in Aristotle's costume, changes in the positions of the arms and shoulders, a change in the position of the medal, and a change in the height of the bust of Homer. Historians knew that the canvas had lost 14 inches of its original height; the early position of the bust helped them conclude that most of the missing canvas was at the bottom.

29.7 FISSION

As shown in Fig. 29.2, very large nuclei have a smaller binding energy per nucleon than do nuclides of intermediate mass. The binding energy of large nuclides is reduced by the long-range Coulomb repulsion of the protons. Each proton in the nucleus repels every other proton. The strong force, which holds the nucleons together in a nucleus, is short range. Each nucleon is bound only to its nearest neighbors. In large nuclides the average number of nearest neighbors is approximately constant, so the strong force

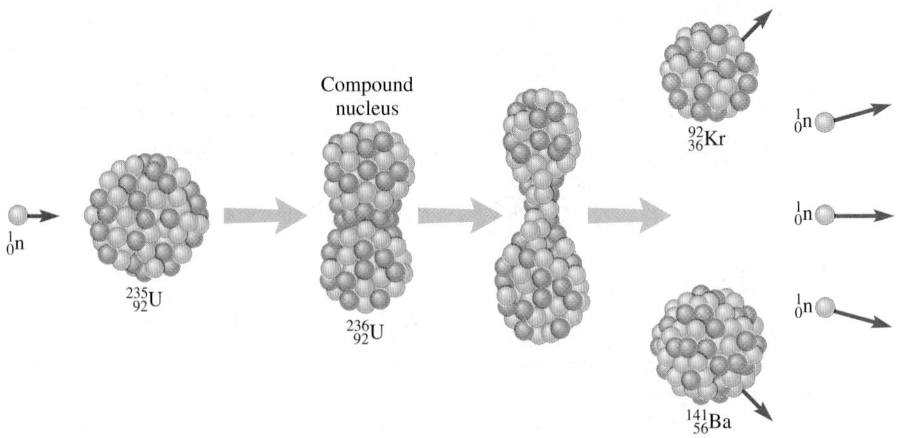

Figure 29.12 Fission of ^{235}U induced by the capture of a slow neutron. In addition to the two daughter nuclei, some neutrons are released.

does not increase the binding energy per nucleon to compensate for the Coulomb repulsion, which decreases the binding energy per nucleon.

A large nucleus can therefore release energy by splitting into two smaller, more tightly bound nuclei in the process called **fission**. The term is borrowed from biology; a cell fissions when it splits into two. Nuclear fission was discovered in 1938 by German and Austrian scientists Otto Hahn, Fritz Strassman, Lise Meitner, and Otto Frisch.

Some very large nuclei can fission spontaneously. Radioactive uranium-238, for instance, can break apart into two fission products, though it is much more likely to decay by emitting an α particle. Fission can also be induced by an incident neutron, proton, deuteron (a ^{2}H nucleus), alpha particle, or photon. Fission due to the capture of a slow neutron allows the possibility of a chain reaction. Uranium-235 is the only naturally occurring nuclide that can be induced to fission by slow neutrons.

Suppose that a slow neutron is captured by a ^{235}U nucleus. The compound nucleus formed, ^{236}U, is in an excited state since the neutron gives up energy when it becomes bound to the nucleus (Problem 53). The excited nucleus is elongated in shape (Fig. 29.12). The attractive force between nucleons tends to pull the nucleus back into a sphere, while the Coulomb repulsion between protons tends to push the ends apart. If the excitation energy is sufficient, a neck forms and the nucleus splits into two parts. The Coulomb repulsion then pushes the two fragments apart so they do not recombine into a single nucleus.

Figure 29.13 shows the potential energy of a nucleus as it elongates and splits into two. To form an elongated shape, the potential energy of the nucleus must increase about 6 MeV. In the absence of an incident particle to supply this energy, *spontaneous* fission can occur only by quantum-mechanical tunneling through the 6-MeV energy

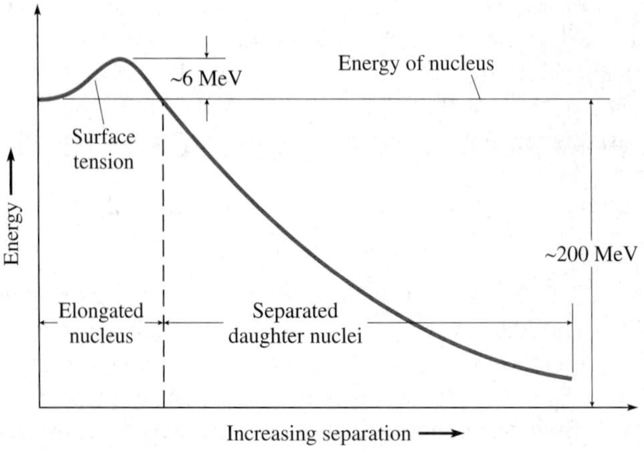

Figure 29.13 Potential energy as a function of separation of the two daughter nuclei in spontaneous fission.

barrier (Section 28.10). The tunneling probability is much lower than the probability of α decay. If ^{238}U decayed only by spontaneous fission, its half-life would be about 10^{16} yr (instead of 4×10^9 yr).

Many different fission reactions are possible for a given parent nuclide. Here are two examples of the induced fission of ^{235}U after it captures a slow neutron:

$$^1_0\text{n} + ^{235}_{92}\text{U} \rightarrow ^{236}_{92}\text{U}^* \rightarrow ^{141}_{56}\text{Ba} + ^{92}_{36}\text{Kr} + 3^1_0\text{n} \quad (29\text{-}30)$$

$$^1_0\text{n} + ^{235}_{92}\text{U} \rightarrow ^{236}_{92}\text{U}^* \rightarrow ^{139}_{54}\text{Xe} + ^{95}_{38}\text{Sr} + 2^1_0\text{n} \quad (29\text{-}31)$$

Notice that the masses of the two daughter nuclei differ significantly in these two examples. The ratio of the masses of the two ^{235}U fission fragments varies from 1 (equal masses) to a little over 2 (one slightly more than twice as massive as the other). The most likely split is a mass ratio of approximately 1.4–1.5 (Fig. 29.14).

Besides the daughter nuclei, neutrons are released in a fission reaction. Large nuclei are more neutron-rich than are smaller nuclei; a few excess neutrons are released when a large nucleus fissions. As many as five neutrons can be released in the fission of ^{235}U; the average number released in a large number of fission reactions is about 2.5. The fission fragments themselves are often still too neutron-rich. The unstable fragments undergo β decay one or more times, stopping when a stable nuclide is formed. In a fission chain reaction (Fig. 29.15), hundreds of different radioactive nuclides—most of which do not occur naturally—are produced.

Example 29.13 shows that the energy released in a fission reaction is enormous—typically around 200 MeV for the split of a single nucleus. To get a *macroscopically* significant amount of energy from fission, a large number of nuclei must split. A neutron can induce fission in ^{235}U, and each fission produces an average of 2.5 neutrons, each of which can go on to induce other nuclei to fission in a **chain reaction**. An uncontrolled chain reaction is the basis of the fission bomb. To make constructive use of the energy released by fission, the chain reaction must be controlled.

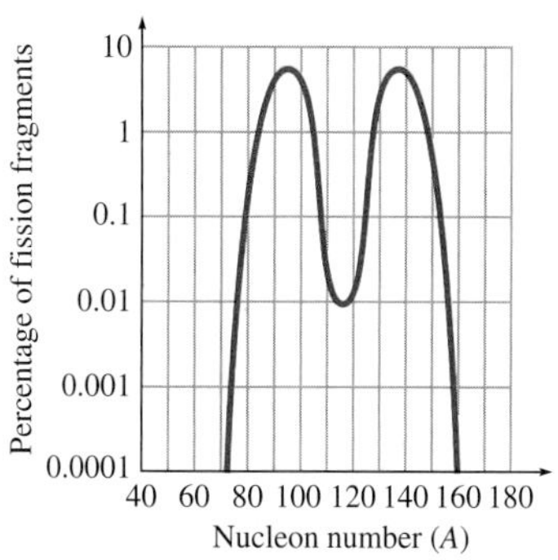

Figure 29.14 Mass distribution of fission fragments from ^{235}U. Note that the vertical scale is logarithmic.

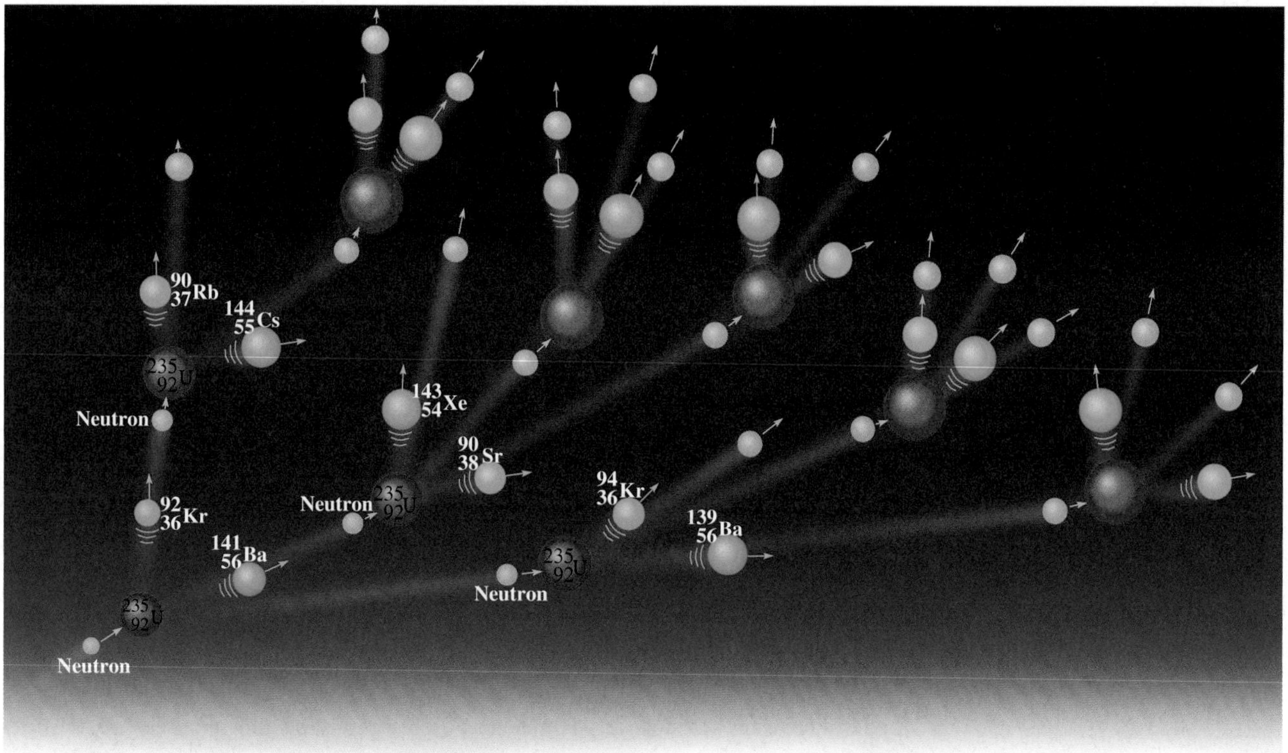

Figure 29.15 A fission chain reaction. Neutrons released when fission occurs can go on to induce fission in other nuclei.

Example 29.13

Energy Produced in a Fission Reaction

Estimate the energy released in the fission reaction of Eq. (29-30). Use Fig. 29.2 to estimate the binding energies per nucleon for $^{235}_{92}U$, $^{141}_{56}Ba$, and $^{92}_{36}Kr$.

Strategy The energy released is equal to the increase in binding energy. The binding energies are estimated by reading the binding energy per nucleon from Fig. 29.2 and multiplying by the number of nucleons.

Solution From Fig. 29.2, the binding energies per nucleon of $^{235}_{92}U$, $^{141}_{56}Ba$, $^{92}_{36}Kr$ are approximately 7.6 MeV, 8.25 MeV, and 8.75 MeV, respectively. We find the total binding energies by multiplying by the number of nucleons. Binding energy:

$$\text{for } ^{235}_{92}\text{U} \approx 235 \times 7.6 \text{ MeV} = 1786 \text{ MeV}$$

$$\text{for } ^{141}_{56}\text{Ba} \approx 141 \times 8.25 \text{ MeV} = 1163 \text{ MeV}$$

$$\text{for } ^{92}_{36}\text{Kr} \approx 92 \times 8.75 \text{ MeV} = 805 \text{ MeV}$$

The increase in binding energy is

$$1163 \text{ MeV} + 805 \text{ MeV} - 1786 \text{ MeV} = 182 \text{ MeV}$$

The energy released by the fission reaction is about 180 MeV.

Discussion The energy released doesn't vary much from one fission reaction to another. A nuclide with $A \approx 240$ has a binding energy of about 7.6 MeV/nucleon. The fission products have an average binding energy of about 8.5 MeV/nucleon. Thus, we expect the energy released to be a little less than 1 MeV per nucleon.

To refine this estimate, we can do a precise calculation of the energy released in the reaction using the masses of the parent and daughter nuclides (Problem 55).

Conceptual Practice Problem 29.13 Can Smaller Nuclides Fission?

Suppose that a $^{54}_{24}Cr$ nucleus captures a slow neutron:

$$^{1}_{0}\text{n} + ^{54}_{24}\text{Cr} \rightarrow ^{55}_{24}\text{Cr}^*$$

Explain why fission does not occur. What might happen instead?

Application: Fission Reactors

Most modern fission reactors (Fig. 29.16) use *enriched uranium* as fuel. Only ^{235}U sustains the chain reaction; ^{238}U can capture neutrons without splitting. Naturally occurring uranium is 99.3% ^{238}U and only 0.7% ^{235}U; with so much ^{238}U absorbing neutrons, it would be difficult to maintain a chain reaction. In enriched uranium, the ^{235}U content is increased to a few percent. The neutrons produced in a fission reaction have large energies. These fast neutrons are equally likely to be captured by ^{238}U nuclei or ^{235}U nuclei. But if the neutrons are slowed down, then they are much more likely to be captured by ^{235}U and induce fission. For this reason, a substance called a *moderator* is included in the fuel core. Moderators include hydrogen (in water or zirconium hydride), deuterium (^2H, in molecules of heavy water), beryllium, or carbon (as graphite). The moderator's function is to slow down the neutrons by colliding with them without capturing too many. Light nuclei are most effective at slowing down the neutrons, since the fractional loss in kinetic energy decreases with increasing target mass.

To control the chain reaction, a substance that readily absorbs neutrons, such as cadmium or boron, is formed into *control rods*. The control rods are lowered into the fuel core to absorb more neutrons, or retracted to absorb fewer neutrons. In normal operation, the reactor is *critical*: on average one neutron from each fission goes on to initiate another fission. A critical reactor produces a steady power output. If the reactor is *subcritical*, on average less than one of the neutrons produced by a fission reaction goes on to cause another fission. As fewer and fewer fission reactions occur, the chain reaction dies out. A reactor is shut down by lowering the control rods to make the reactor subcritical. If the reactor is *supercritical*, then on average more than one neutron from each fission causes another fission. Thus, the number of fission reactions per second is increasing in a supercritical reactor. A reactor must be allowed to be supercritical for a *brief* time while it is starting up.

Fission reactors have purposes other than power generation. They also provide the neutron sources for neutron activation analysis and neutron diffraction experiments. The neutrons from a reactor are also used to produce artificial radioisotopes for medical use.

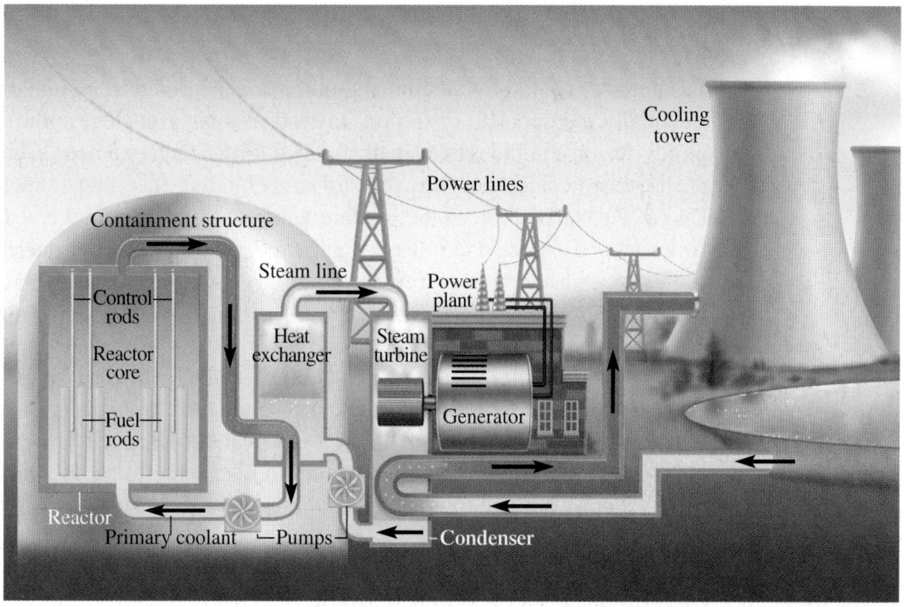

Figure 29.16 A pressurized water fission reactor. Water under high pressure is the primary coolant that carries heat from the core into a heat exchanger in a closed loop. (In some other reactors, liquid sodium is used as the primary coolant.) The heat exchanger extracts heat from the primary coolant to make steam; the steam drives a turbine connected to an electric generator. In essence, the fission reactions in the core act like a furnace to supply the heat needed to run a heat engine. As in all heat engines, waste heat must be exhausted to the environment. In this case, cool water is taken in from a nearby body of water. This water takes heat from the steam engine and then evaporates in the cooling tower so the waste heat goes into the air.

A by-product of the fission reactions in a *breeder reactor* is to produce more fissionable material (plutonium-239) from uranium-238 in its fuel core than is consumed. The ^{239}Pu can be left in the core, to fission and generate power, or it can be extracted and used to make bombs. Thus, breeder reactors are a concern for those working to halt the proliferation of nuclear weapons.

Problems with Fission Reactors

Although fission reactors do not contribute to global warming because they do not produce greenhouse gases, they must be carefully designed to prevent the release of harmful radioactive materials into the environment. By far the worst accident at a nuclear reactor occurred in 1986 at the Chernobyl reactor in Ukraine, then part of the Soviet Union (Fig. 29.17). Poor reactor design and a series of mistakes by the operators of the reactor resulted in two explosions, releasing radioactive fission products into the atmosphere and allowing the graphite moderator in the core to burst into flame. The graphite fire continued for 9 days. The estimated amount of radiation that escaped is of the order of 10^{19} Bq; winds dispersed radioactive material over Ukraine, Belarus, Russia, Poland, Scandinavia, and eastern Europe.

Figure 29.17 Aerial view of the exploded fourth reactor of the Chernobyl nuclear power plant.

Another major problem is the safe transport and storage of radioactive waste. Spent fuel rods, which are removed from the reactor core when the fissionable material is depleted, contain highly radioactive fission products that must be stored for thousands of years. Other parts of the reactor become radioactive by being exposed to neutrons. After about 30 yr of operation, the structural materials of the reactor have been weakened by radiation and the reactor must be decommissioned.

In 1988, the U.S. government chose Yucca Mountain in Nevada as the nation's permanent repository for approximately 77 000 tons of high-level radioactive waste from fission reactors. Subsequent studies showed that the desert site might not be as geologically stable as was originally thought. The Nevada state government and some other groups oppose the planned facility; as this book goes to press, the legal battles continue and high-level nuclear waste continues to be stored on site at the reactors.

29.8 FUSION

The energy radiated by the Sun and other stars is produced by nuclear **fusion**. Fusion is essentially the opposite of fission. Instead of a large nucleus splitting into two smaller pieces, fusion combines two small nuclei to form a larger nucleus. Both fission and fusion release energy, since they move toward larger binding energies per nucleon (see Fig. 29.2). Due to the steep slope of the binding energy per nucleon curve at low mass numbers, fusion can be expected to release significantly more energy per nucleon than fission.

Here is an example of a fusion reaction:

$$^{2}\text{H} + {}^{3}\text{H} \rightarrow {}^{4}\text{He} + \text{n} \qquad (29\text{-}32)$$

Two hydrogen nuclei fuse to form a helium nucleus. The reaction releases 17.6 MeV of energy, which is 3.52 MeV per nucleon—much more than the 0.75–1 MeV per nucleon typical of fission reactions. Although this reaction produces a tremendous amount of energy, it cannot occur at room temperature. The deuterium (^{2}H) and tritium (^{3}H) nuclei must get close enough to react. At room temperature, the two positively charged nuclei have kinetic energies much too small to overcome their mutual Coulomb repulsion. However, in the Sun's interior the temperature is about 2×10^7 K and the average kinetic energy of the nuclei is $\frac{3}{2}k_BT \approx 2.52$ keV. This *average* kinetic energy is still far too small to allow a fusion reaction (see Example 29.14), but some of the more energetic nuclei do have enough kinetic energy to overcome the Coulomb repulsion. Fusion reactions are also called thermonuclear reactions because they depend on the large kinetic energies available at high temperatures.

Proton-Proton Cycle Two fusion cycles to explain energy production in stars were proposed by the German-U.S. physicist Hans Bethe (1906–2005). One cycle is called the **proton-proton cycle**:

$$\text{p} + \text{p} \rightarrow {}^{2}\text{H} + \text{e}^{+} + \nu \qquad (29\text{-}33\text{a})$$

$$\text{p} + {}^{2}\text{H} \rightarrow {}^{3}\text{He} \qquad (29\text{-}33\text{b})$$

$$^{3}\text{He} + {}^{3}\text{He} \rightarrow {}^{4}\text{He} + 2\text{p} \qquad (29\text{-}33\text{c})$$

The net effect of the proton-proton cycle is to fuse four protons into a ^{4}He nucleus. (The first two reactions must each take place twice to form the two ^{3}He nuclei needed for the third reaction.) The three steps can be summarized as

$$4\text{p} \rightarrow {}^{4}\text{He} + 2\text{e}^{+} + 2\nu$$

Each positron annihilates with an electron, so the overall reaction due to the proton-proton cycle is

$$4\text{p} + 2\text{e}^{-} \rightarrow {}^{4}\text{He} + 2\nu \qquad (29\text{-}34)$$

Carbon Cycle Another cycle of fusion reactions that occurs in some stars is the **carbon cycle**:

$$\text{p} + {}^{12}\text{C} \rightarrow {}^{13}\text{N} \qquad (29\text{-}35\text{a})$$

$$^{13}\text{N} \rightarrow {}^{13}\text{C} + \text{e}^{+} + \nu \qquad (29\text{-}35\text{b})$$

$$\text{p} + {}^{13}\text{C} \rightarrow {}^{14}\text{N} \qquad (29\text{-}35\text{c})$$

$$\text{p} + {}^{14}\text{N} \rightarrow {}^{15}\text{O} \qquad (29\text{-}35\text{d})$$

$$^{15}\text{O} \rightarrow {}^{15}\text{N} + \text{e}^{+} + \nu \qquad (29\text{-}35\text{e})$$

$$\text{p} + {}^{15}\text{N} \rightarrow {}^{12}\text{C} + {}^{4}\text{He} \qquad (29\text{-}35\text{f})$$

Here the carbon-12 nucleus acts as a catalyst; it is present in the beginning and at the end. After the annihilation of the two positrons, the net effect is the same as the proton-proton cycle:

$$4p + 2e^- \rightarrow {}^4\text{He} + 2\nu$$

The *total* energy released by the carbon cycle is the same as the total energy released by the proton-proton cycle. By "total energy released" we mean the total energy of all the photons [not shown in Eqs. (29-33) through (29-35)] and neutrinos produced plus the kinetic energy of the ^{4}He nucleus minus the initial kinetic energies of the protons and electrons.

Example 29.14

First Step of the Carbon Cycle

(a) Calculate the energy released in the first step of the carbon cycle. (b) Estimate the minimum kinetic energy of the proton and ^{12}C nucleus required for this reaction to take place.

Strategy To calculate the energy released, we must determine the mass difference between reactants and products. For the minimum initial kinetic energy, we know that the two positively charged particles repel one another. We can find the distance between the two when they just "touch," and find the electric potential energy in that position.

Solution (a) The reaction in question is

$$p + {}^{12}\text{C} \rightarrow {}^{13}\text{N}$$

We use atomic masses in the calculations since the extra mass of seven electrons is equally present in the atomic masses of the reactants and products. The initial mass is then

$$1.007\,825\,0\text{ u} + 12.000\,000\,0\text{ u} = 13.007\,825\,0\text{ u}$$

The mass change is

$$\Delta m = 13.005\,738\,6\text{ u} - 13.007\,825\,0\text{ u} = -0.002\,086\,4\text{ u}$$

The energy released is

$$E = 0.002\,086\,4\text{ u} \times 931.494\text{ MeV/u} = 1.9435\text{ MeV}$$

(b) From Eq. (29-4), the radii of the proton and ^{12}C nucleus are 1.2 fm and

$$1.2\text{ fm} \times 12^{1/3} = 2.75\text{ fm}$$

For an *estimate* of the electric potential energy when the proton and ^{12}C nucleus are just "touching," we find the electric potential energy of two *point charges*, $+e$ and $+6e$, at a separation of 3.95 fm.

$$U_E = \frac{6ke^2}{r} = \frac{6 \times (9 \times 10^9\text{ N·m}^2/\text{C}^2) \times (1.60 \times 10^{-19}\text{ C})^2}{3.95 \times 10^{-15}\text{ m}}$$

$$= 3.50 \times 10^{-13}\text{ J} = 2\text{ MeV}$$

The minimum total kinetic energy of the proton and ^{12}C nucleus that allows the reaction to take place is about 2 MeV.

Discussion The energy released, 1.9435 MeV, includes both the increase in kinetic energy and the energy of the photon.

Practice Problem 29.14 Second Step of the Carbon Cycle

Calculate the energy released in the second step in the carbon cycle.

Stars act as factories to form heavier nuclides from lighter nuclides. In a star like our Sun, most of the fusion reactions produce helium from hydrogen. At higher core temperatures, heavier nuclides can take part in fusion reactions. Nuclides all the way up to the peak of the binding energy curve (see Fig. 29.2), around $A = 60$, are formed by fusion reactions in the interiors of stars. Once a star core is rich in iron and nickel, elements near the top of the binding energy curve, fusion reactions die out. Heavier nuclides are *less* tightly bound than iron and nickel, so fusion reactions no longer release energy. Eventually the large star implodes under its own gravity; the implosion provides the energy for the fusion of heavier nuclides. Ultimately the star may explode, an event called a supernova. Additional fusion and neutron capture reactions occur in the shock waves caused by the explosion, forming the heaviest nuclides. The nuclides formed in the supernova, plus the ones that had already been formed in the star's core, are dispersed by the explosion into space. The atoms that make up all of us and our surroundings were distributed into space by one or more supernovae before making their way to

Application: nuclear fusion in stars

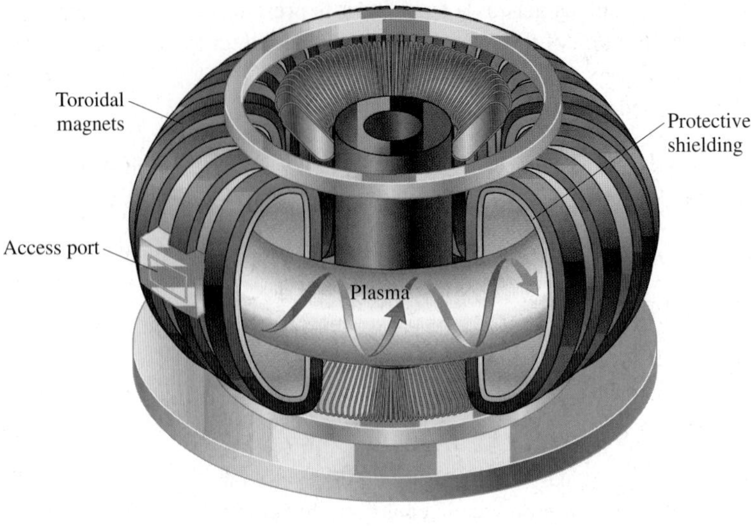

Figure 29.18 The tokamak is one of the most promising methods for containing a controlled fusion reaction. At such high temperatures, the atoms in the fuel break apart to form a *plasma*—a mixture of electrons and positively charged nuclei. Magnetic fields confine the charged nuclei to the interior of a doughnut-shaped, evacuated chamber. The nuclei spiral around magnetic field lines and are confined without colliding with the walls of the vacuum chamber.

Earth. Other than hydrogen (and a small fraction of some of the other light elements), all of the elements found on Earth were either made in the core of a star or in a supernova (or are radioactive decay products of these elements).

Application: Fusion Reactors

In a thermonuclear bomb (or hydrogen bomb), a fission bomb creates the high temperatures that enable an uncontrolled fusion reaction to take place. For decades, researchers have been trying to make possible a sustained, *controlled* fusion reaction. Fusion as an energy source would have several advantages over fission. The fuel for fusion is more easily obtained than the fuel for fission. The most promising reactions for controlled fusion are deuteron-deuteron fusion ($^2H + {}^2H$) or deuteron-triton fusion [$^2H + {}^3H$, as in Eq. (29-32)]. Deuterium is readily available in seawater; about 0.015% of the water molecules in the ocean contain a deuterium atom. Tritium's natural abundance is very small, but is not difficult to produce.

One of the biggest problems associated with *fission* reactors is the radioactive waste that must be safely stored for thousands of years. A fusion reactor would produce less radioactive waste and the waste would not have to be stored for as long.

However, a sustained, controlled fusion reaction has not yet been achieved. The main problem is containing the fuel at the extremely high temperatures (estimated to be about 10^8 K, which is higher than the temperature of the Sun's interior) needed for fusion to take place while maintaining a high density of nuclei so that they collide into each other. An ordinary container cannot be used; the nuclei would lose too much kinetic energy when they collide with the walls of such a container and the container would be vaporized by the high temperatures. Two principal confinement schemes are being tried. One is *magnetic confinement* (Fig. 29.18). The other is *inertial confinement*, in which a small fuel pellet is heated rapidly by intense laser beams from all sides, causing the fuel pellet to implode and the fusion reactions to take place before the pellet is vaporized.

Master the Concepts

- A particular nuclide is characterized by its atomic number Z (the number of protons) and its nucleon number A (the total number of protons and neutrons). The isotopes of an element have the same atomic number but different numbers of neutrons.
- The mass density of all nuclei is approximately the same. The radius of a nucleus is

$$r = r_0 A^{1/3} \qquad (29\text{-}4)$$

where

$$r_0 = 1.2 \times 10^{-15}\ \text{m} = 1.2\ \text{fm} \qquad (29\text{-}5)$$

continued on next page

Master the Concepts continued

- The binding energy E_B of a nucleus is the energy that must be supplied to separate a nucleus into individual protons and neutrons. Since the nucleus is a bound system, its total energy is *less* than the energy of Z protons and N neutrons that are far apart and at rest.

$$E_B = (\Delta m)c^2 \quad (29\text{-}8)$$

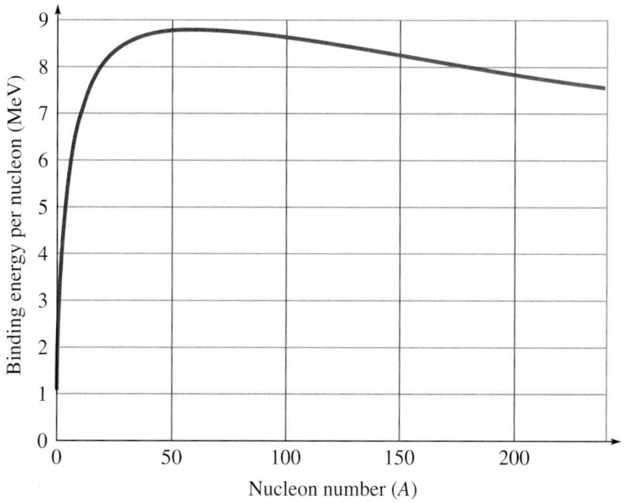

- In any nuclear reaction, the total electric charge and the total number of nucleons are conserved.
- An unstable or radioactive nuclide decays by emitting radiation.

Alpha decay $\quad {}^{A}_{Z}P \rightarrow {}^{A-4}_{Z-2}D + {}^{4}_{2}\alpha \quad$ (29-10)

Beta-minus decay $\quad {}^{A}_{Z}P \rightarrow {}^{A}_{Z+1}D + {}^{0}_{-1}\text{e} + {}^{0}_{0}\overline{\nu} \quad$ (29-11)

Beta-plus decay $\quad {}^{A}_{Z}P \rightarrow {}^{A}_{Z-1}D + {}^{0}_{+1}\text{e} + {}^{0}_{0}\nu \quad$ (29-12)

Gamma decay $\quad P^* \rightarrow P + \gamma$

- Each radioactive nuclide has a characteristic decay probability per unit time λ. The activity R of a sample with N nuclei is

$$R = \frac{\text{number of decays}}{\text{unit time}} = \frac{-\Delta N}{\Delta t} = \lambda N \quad (29\text{-}18)$$

Activity is commonly measured in becquerels (1 Bq = 1 decay per second) or curies (1 Ci = 3.7×10^{10} Bq).

- The number of remaining nuclei N in radioactive decay (the number that have *not* decayed) is an exponential function:

$$N(t) = N_0\, e^{-t/\tau} \quad (29\text{-}19)$$

where the time constant is $\tau = 1/\lambda$. The half-life is the time during which half of the nuclei decay:

$$T_{1/2} = \tau \ln 2 \approx 0.693\tau \quad (29\text{-}22)$$

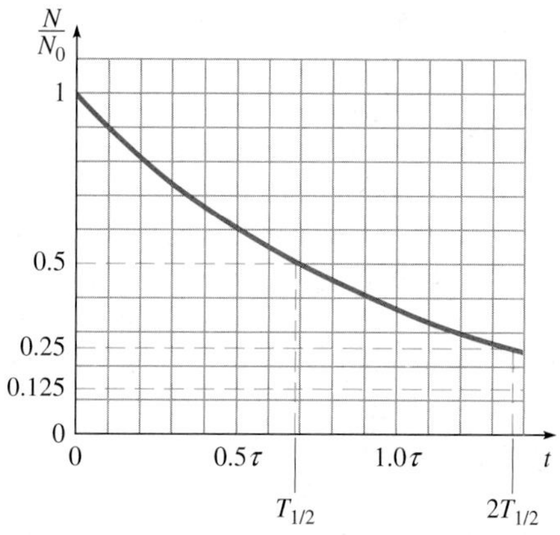

- The absorbed dose is the amount of radiation energy absorbed per unit mass of tissue, measured in grays (1 Gy = 1 J/kg) or rads (1 rad = 0.01 Gy).
- The quality factor (QF) is a relative measure of the biological damage caused by different kinds of radiation. The biologically equivalent dose is

$$\text{biologically equivalent dose} = \text{absorbed dose} \times \text{QF} \quad (29\text{-}28)$$

- A large nucleus can release energy by splitting into two smaller, more tightly bound nuclei in the process called fission. The energy released in a fission reaction is enormous—typically around 200 MeV for the split of a single nucleus.
- Nuclear fusion combines two small nuclei to form a larger nucleus. Fusion typically releases significantly more energy per nucleon than fission.

Conceptual Questions

1. How could Henri Becquerel and other scientists determine that there were three *different* kinds of radiation *before* having determined the electric charges or masses of the α, β, and γ rays?
2. What technique could Becquerel and others have used to determine that α rays are positively charged, β rays negatively charged, and γ rays uncharged? Explain how they could find that α rays have a charge-to-mass ratio half that of the H^+ ion, and β rays have the same charge-to-mass ratio as "cathode rays" (electrons). (See Chapter 19 for some ideas.)
3. Why is a slow neutron more likely to induce a nuclear reaction (as in neutron activation and induced fission) than a proton with the same kinetic energy?
4. Explain why neutron-activated nuclides tend to decay by β^- rather than β^+.
5. Why can we ignore the binding energies of the atomic *electrons* in calculations such as Example 29.4? Isn't there a mass defect due to the binding energy of the electrons?
6. Why would we expect atmospheric testing of nuclear weapons to increase the relative abundance of carbon-14 in the atmosphere? Why would we expect the widespread burning of fossil fuels to *decrease* the relative abundance of carbon-14 in the atmosphere?

7. Isolated atoms (or atoms in a dilute gas) radiate photons at discrete energies characteristic of that atom. In dense matter, the spectrum radiated is quasi-continuous. Why doesn't the same thing happen with nuclear spectra: why do the γ rays have the same characteristic energies even when emitted from a solid?
8. Section 29.8 states that the total energy released by the proton-proton cycle is the same as that released by the carbon cycle. Why must the total energy released be the same?
9. Iodine is eliminated from the body through *biological* processes with an effective half-life of about 140 days. The radioactive half-life of iodine-131 is 8 days. Suppose some radioactive ^{131}I nuclei are present in the body. Assuming that no new ^{131}I nuclei are introduced into the body, how much time must pass until only half as much ^{131}I is left in the body: less than 8 days, between 8 and 140 days, or more than 140 days? Explain your reasoning.
10. Why does a fission reaction tend to release one or more neutrons? Why is the release of neutrons necessary in order to sustain a chain reaction?
11. Radioactive α-emitters are relatively harmless outside the body, but can be dangerous if ingested or inhaled. Explain.
12. Fission reactors and cyclotrons tend to produce different kinds of isotopes. A reactor produces isotopes primarily through neutron activation; thus, the isotopes tend to be neutron-rich (high neutron-to-proton ratio). A cyclotron can only accelerate charged particles such as protons or deuterons. When stable nuclei are bombarded with protons or deuterons, the resulting radioisotopes are neutron-deficient (low neutron-to-proton ratio). (a) Explain why a cyclotron cannot accelerate neutrons. (b) Suppose a hospital needs a supply of radioisotopes to use in positron-emission tomography (PET). Would the radioisotopes more likely come from a reactor or a cyclotron? Explain.
13. Why would a fusion reactor produce less radioactive waste than a fission reactor? [*Hint:* Compare the products of a fission reaction with those from a fusion reaction.]
14. Radon-222 is created in a series of radioactive decays starting with $^{238}_{92}$U and ending with $^{206}_{82}$Pb. The half-life of ^{222}Rn is 3.8 days. (a) If the half-life is so short, why hasn't all the ^{222}Rn gas decayed by now? (b) If the half-life of ^{222}Rn were much shorter, say a few seconds, would it be more dangerous to us or less dangerous? What if the half-life were much longer, say thousands of years?

Multiple-Choice Questions

1. Radioactive $^{215}_{83}$Bi decays into $^{215}_{84}$Po. Which of these particles is released in the decay?
(a) a proton (b) an electron (c) a positron
(d) an α particle (e) a neutron (f) none of the above
2. For all stable nuclei
(a) there are equal numbers of protons and neutrons.
(b) there are more protons than neutrons.
(c) there are more neutrons than protons.
(d) none of the above have to be true.
3. For all stable nuclei
(a) the mass of the nucleus is less than $Zm_p + (A - Z)m_n$.
(b) the mass of the nucleus is greater than $Zm_p + (A - Z)m_n$.
(c) the mass of the nucleus is equal to $Zm_p + (A - Z)m_n$.
(d) none of the above have to be true.
4. Of the *hypothetical* nuclear reactions listed here, which would violate conservation of charge?
(a) $^{10}_{5}\text{B} + ^{4}_{2}\text{He} \rightarrow ^{13}_{7}\text{N} + ^{1}_{1}\text{H}$ (b) $^{10}_{5}\text{B} + ^{1}_{0}\text{n} \rightarrow ^{11}_{5}\text{B} + \beta^- + \overline{\nu}$
(c) $^{23}_{11}\text{Na} + ^{1}_{1}\text{H} \rightarrow ^{20}_{10}\text{Ne} + ^{4}_{2}\text{He}$
(d) $^{14}_{7}\text{N} + ^{1}_{1}\text{H} \rightarrow ^{13}_{6}\text{C} + \beta^+ + \nu$
(e) none of them (f) all of them
(g) all but (c) (h) (a) and (d)
5. Of the *hypothetical* nuclear reactions listed in Multiple-Choice Question 4, which would violate conservation of nucleon number?
6. In a fusion reaction, two deuterons produce a helium-3 nucleus. What is the other product of the reaction?
(a) an electron (b) a proton (c) a neutron
(d) an α particle (e) a positron (f) a neutrino
7. The activity of a radioactive sample (with a single radioactive nuclide) decreases to one eighth its initial value in a time interval of 96 days. What is the half-life of the radioactive nuclide present?
(a) 6 days (b) 8 days (c) 12 days
(d) 16 days (e) 24 days (f) 32 days
8. Solid lead has more than four times the mass density of solid aluminum. What is the main reason that lead is so much more dense?
(a) The Pb atom is smaller than the Al atom.
(b) The Pb nucleus is smaller than the Al nucleus.
(c) The Pb nucleus is more massive than the Al nucleus.
(d) The Pb nucleus is more dense than the Al nucleus.
(e) The Pb atom has many more electrons than the Al atom.
9. Which of these are appropriate units for the decay constant λ of a radioactive nuclide?
(a) s (b) Ci (c) rd
(d) s^{-1} (e) rem (f) MeV
10. Which of the units listed in Multiple-Choice Question 9 are appropriate for the biologically equivalent dose that results when a person is exposed to radiation?

Problems

Combination conceptual/quantitative problem
Biological or medical application
Challenging problem
Blue # Detailed solution in the Student Solutions Manual
1 2 Problems paired by concept
Text website interactive or tutorial

29.1 Nuclear Structure

1. Estimate the number of nucleons found in the body of a 75-kg person.
2. Calculate the mass density of nuclear matter.
3. A neutron star is a star that has collapsed into a collection of tightly packed neutrons. Thus, it is something like a giant nucleus; but since it is electrically neutral, there is no Coulomb repulsion to break it up. The force holding it together is gravity. Suppose the Sun were to collapse into a neutron star. What would its radius be? Assume that the density is about the same as for a nucleus. Express your answer in kilometers.
4. Write the symbol (in the form ${}^{A}_{Z}X$) for the nuclide with 38 protons and 50 neutrons and identify the element.
5. Write the symbol (in the form ${}^{A}_{Z}X$) for the isotope of potassium with 21 neutrons.
6. How many neutrons are found in a ${}^{35}Cl$ nucleus?
7. How many protons are found in a ${}^{136}Xe$ nucleus?
8. Write the symbol (in the form ${}^{A}_{Z}X$) for the nuclide that has 78 neutrons and 53 protons.
9. Find the radius and volume of the ${}^{107}_{43}Tc$ nucleus.

29.2 Binding Energy

10. What is the binding energy of an α particle (a ${}^{4}He$ nucleus)? The mass of an α particle is 4.001 51 u.
11. Find the binding energy of a deuteron (a ${}^{2}H$ nucleus). The mass of a deuteron (*not* the deuterium atom) is 2.013553 u.
12. What is the average binding energy per nucleon for ${}^{40}_{18}Ar$?
13. (a) Find the binding energy of the ${}^{16}O$ nucleus. (b) What is the average binding energy per nucleon? Check your answer using Fig. 29.2.
14. Calculate the binding energy per nucleon of the ${}^{31}_{15}P$ nucleus.
15. What is the mass defect of the ${}^{14}N$ nucleus?
16. What is the mass of an ${}^{16}O$ atom in units of MeV/c^2? (1 MeV/c^2 is the mass of a particle with rest energy 1 MeV.)
17. (a) What is the mass defect of the ${}^{1}H$ *atom* due to the binding energy of the *electron* (in the ground state)? (b) Should we worry about this mass defect when we calculate the mass of the ${}^{1}H$ nucleus by subtracting the mass of one electron from the mass of the ${}^{1}H$ atom?
18. Show that $c^2 = 931.494$ MeV/u. [*Hint:* Start with the conversion factors to SI units for MeV and atomic mass units.]
19. Using a mass spectrometer, the mass of the ${}^{238}_{92}U^{+}$ *ion* is found to be 238.050 24 u. (a) Use this result to calculate the mass of the ${}^{238}_{92}U$ *nucleus.* (b) Now find the binding energy of the ${}^{238}_{92}U$ nucleus.
20. To make an order-of-magnitude estimate of the energy level spacings in the nucleus, assume that a nucleon is confined to a one-dimensional box of width 10 fm (a typical nuclear diameter). Calculate the energy of the ground state.

29.3 Radioactivity

21. Identify the daughter nuclide when ${}^{40}_{19}K$ decays via β^- decay.
22. Thorium-232 (${}^{232}_{90}Th$) decays via α decay. Write out the reaction and identify the daughter nuclide.
23. Write out the reaction and identify the daughter nuclide when ${}^{22}_{11}Na$ decays by electron capture.
24. Write out the reaction and identify the daughter nuclide when ${}^{22}_{11}Na$ decays by emitting a positron.
25. Radium-226 decays as ${}^{226}_{88}Ra \rightarrow {}^{222}_{86}Rn + {}^{4}_{2}He$. If the ${}^{226}_{88}Ra$ nucleus is at rest before the decay and the ${}^{222}_{86}Rn$ nucleus is in its ground state, *estimate* the kinetic energy of the α particle. (Assume that the ${}^{222}_{86}Rn$ nucleus takes away an insignificant fraction of the kinetic energy.)
26. Which decay mode would you expect for radioactive ${}^{31}_{14}Si$: α, β^-, or β^+? Explain. [*Hint:* Look at the neutron-to-proton ratio.]
27. Calculate the maximum kinetic energy of the β particle when ${}^{40}_{19}K$ decays via β^- decay.
28. Calculate the energy of the antineutrino when ${}^{90}_{38}Sr$ decays via β^- decay if the β particle has a kinetic energy of 435 keV.
29. Show that the spontaneous α decay of ${}^{19}O$ is not possible.
30. Calculate the kinetic energy of the α particle in Problem 25. This time, do *not* assume that the ${}^{222}_{86}Rn$ nucleus is at rest after the reaction. Start by figuring out the ratio of the kinetic energies of the α particle and the ${}^{222}_{86}Rn$ nucleus.
31. An isotope of sodium, ${}^{22}_{11}Na$, decays by β^+ emission. *Estimate* the maximum possible kinetic energy of the positron by assuming that the kinetic energy of the daughter nucleus and the total energy of the neutrino emitted are both zero. [*Hint:* Remember to keep track of the electron masses.]
32. The nucleus in a ${}^{12}_{7}N$ atom captures one of the atom's electrons, changing the nucleus to ${}^{12}_{6}C$ and emitting a neutrino. What is the total energy of the emitted neutrino? [*Hint:* You can use the classical expression for the kinetic energy of the ${}^{12}_{6}C$ atom and the extremely relativistic expression for the kinetic energy of the neutrino.]

29.4 Radioactive Decay Rates and Half-Lives

33. A certain radioactive nuclide has a half-life of 200.0 s. A sample containing just this one radioactive nuclide has an initial activity of 80,000.0 s^{-1}. (a) What is the activity 600.0 s later? (b) How many nuclei were there initially? (c) What is the probability per second that any one of the nuclei decays?

34. The half-life of I-131 is 8.0 days. A sample containing I-131 has an activity of 6.4×10^8 Bq. How many days later will the sample have an activity of 2.5×10^6 Bq? (tutorial: radioactive decay)

35. Some bones discovered in a crypt in Guatemala are carbon-dated. The ^{14}C activity of the bones is measured to be 0.242 Bq per gram of carbon. Approximately how old are the bones? (tutorial: radio carbon dating)

36. Carbon-14 dating is used to date a bone found at an archaeological excavation. If the ratio of C-14 to C-12 atoms is 3.25×10^{-13}, how old is the bone? [*Hint:* Note that this ratio is one fourth the ratio of 1.3×10^{-12} that is found in a living sample.]

37. A sample of radioactive $^{214}_{83}Bi$, which has a half-life of 19.9 min, has an activity of 0.058 Ci. What is its activity 1.0 h later?

38. The activity of a sample containing radioactive ^{108}Ag is 6.4×10^4 Bq. Exactly 12 min later, the activity is 2.0×10^3 Bq. Calculate the half-life of ^{108}Ag.

39. Calculate the activity of 1.0 g of radium-226 in Ci.

40. What is the activity in becquerels of 1.0 kg of ^{238}U?

41. In this problem, you will verify the statement (in Section 29.4) that the ^{14}C activity in a living sample is 0.25 Bq per gram of carbon. (a) What is the decay constant λ for ^{14}C? (b) How many ^{14}C atoms are in 1.00 g of carbon? One mole of carbon atoms has a mass of 12.011 g, and the relative abundance of ^{14}C is 1.3×10^{-12}. (c) Using your results from parts (a) and (b), calculate the ^{14}C activity per gram of carbon in a living sample.

42. A radioactive sample has equal numbers of ^{15}O and ^{19}O nuclei. Use the half-lives found in Appendix B to determine how long it will take before there are twice as many ^{15}O nuclei as ^{19}O. What percent of the ^{19}O nuclei have decayed during this time?

43. Show mathematically that $2^{-t/T_{1/2}} = \left(\frac{1}{2}\right)^{t/T_{1/2}} = e^{-t/\tau}$ if and only if $T_{1/2} = \tau \ln 2$. [*Hint:* Take the natural logarithm of each side.]

44. The Physics at Home in Section 29.4 suggests tossing coins as a model of radioactive decay. An improved version is to toss a large number of dice instead of coins: each die that comes up a "one" represents a nucleus that has decayed. Suppose that N dice are tossed. (a) What is the average number of dice you expect to decay on one toss? (b) What is the average number of dice you expect to remain undecayed after three tosses? (c) What is the average number of dice you expect to remain undecayed after four tosses? (d) What is the half-life in numbers of tosses?

29.5 Biological Effects of Radiation

45. An α particle produced in radioactive α decay has a kinetic energy of typically about 6 MeV. When an α particle passes through matter (e.g., biological tissue), it makes ionizing collisions with molecules, giving up some of its kinetic energy to supply the binding energy of the electron that is removed. If a typical ionization energy for a molecule in the body is around 20 eV, roughly how many molecules can the alpha particle ionize before coming to rest?

46. If meat is irradiated with 2000.0 Gy of x-rays, most of the bacteria are killed and the shelf life of the meat is greatly increased. (a) How many 100.0-keV photons must be absorbed by a 0.30-kg steak so that the absorbed dose is 2000.0 Gy? (b) Assuming steak has the same specific heat as water, what temperature increase is caused by a 2000.0-Gy absorbed dose?

47. Some types of cancer can be effectively treated by bombarding the cancer cells with high energy protons. Suppose 1.16×10^{17} protons, each with an energy of 950 keV, are incident on a tumor of mass 3.82 mg. If the quality factor for these protons is 3.0, what is the biologically equivalent dose?

48. Make an order-of-magnitude estimate of the amount of radon-222 gas, measured in curies, found in the lungs of an average person. Assume that 0.1 rem/yr is due to the alpha particles emitted by radon-222. The half-life is 3.8 days. You will need to calculate the energy of the alpha particles emitted.

29.6 Induced Nuclear Reactions

49. A certain nuclide absorbs a neutron. It then emits an electron, and then breaks up into two α particles. (a) Identify the original nuclide and the two intermediate nuclides (after absorbing the neutron and after emitting the electron). (b) Would any (anti)neutrino(s) be emitted? Explain.

50. A neutron-activated sample emits gamma rays at energies that are consistent with the decay of mercury-198 nuclei from an excited state to the ground state. If the reaction that takes place is $n + (?) \rightarrow {}^{198}Hg^* + e^- + \bar{\nu}$, what is the nuclide "(?)" that was present in the sample before neutron activation? (tutorial: neutron detector)

51. Irène and Jean Frédéric Joliot-Curie, in an experiment that led to the 1935 Nobel Prize in chemistry, bombarded aluminum $^{27}_{13}Al$ with α particles to form a highly unstable isotope of phosphorus, $^{31}_{15}P$. The phosphorus immediately decayed into another isotope of phosphorus, $^{30}_{15}P$, plus another product. Write out these reactions, identifying the other product.

52. The reactions listed in Problem 51 did not stop there. To the surprise of the Curies, the phosphorus decay continued after the α bombardment ended with the phosphorus $^{30}_{15}P$ emitting a β^+ to form yet another product. Write out this reaction, identifying the other product.

29.7 Fission

53. A ^{235}U nucleus captures a low-energy neutron to form the compound nucleus $^{236}U^*$. Find the excitation energy of the compound nucleus. Ignore the small initial kinetic energy of the captured neutron.

54. *Estimate* the energy released in the fission reaction of Eq. (29-31). Look up the binding energy per nucleon of the nuclides in Fig. 29.2.

55. Calculate the energy released in the fission reaction of Eq. (29-30). The atomic masses of $^{141}_{56}Ba$ and $^{92}_{36}Kr$ are 140.914 u and 91.926 u, respectively.

56. One possible fission reaction for ^{235}U is $^{235}U + n \rightarrow$ $^{141}Cs + ^{93}Rb + ?n$, where "?n" represents one or more neutrons. (a) How many neutrons? (b) From the graph in Fig. 29.2, you can read the approximate binding energies per nucleon for the three nuclides involved. Use that information to *estimate* the total energy released by this fission reaction. (c) Do a precise calculation of the energy released. (d) What fraction of the rest energy of the ^{235}U nucleus is released by this reaction?

29.8 Fusion

57. Consider the fusion reaction of a proton and a deuteron: $^1_1H + ^2_1H \rightarrow X$. (a) Identify the reaction product X. (b) The binding energy of the deuteron is about 1.1 MeV *per nucleon* and the binding energy of "X" is about 2.6 MeV *per nucleon.* Approximately how much energy (in MeV) is released in this fusion reaction? (c) Why is this reaction unlikely to occur in a room temperature setting?

58. What is the total energy released by the proton-proton cycle [Eq. (29-34)]? (The total energy released is the total energy of the neutrinos and γ rays plus the kinetic energy of the ^4He nucleus minus the initial kinetic energies of the protons and electrons.)

59. Estimate the minimum total kinetic energy of the ^2H and ^3H nuclei necessary to allow the fusion reaction of Eq. (29-32) to take place.

60. Compare the amount of energy released when 1.0 kg of the uranium isotope ^{235}U undergoes the fission reaction of Eq. (29-30) with the energy released when 1.0 kg of hydrogen undergoes the fusion reaction of Eq. (29-32).

Comprehensive Problems

61. Which of these unidentified nuclides are isotopes of each other? $^{175}_{71}(?)$, $^{71}_{32}(?)$, $^{175}_{74}(?)$, $^{167}_{71}(?)$, $^{71}_{30}(?)$, and $^{180}_{74}(?)$.

62. What is the average binding energy per nucleon for $^{23}_{11}Na$?

63. The carbon isotope ^{15}C decays much faster than ^{14}C. (a) Using Appendix B, write a nuclear reaction showing the decay of ^{15}C. (b) How much energy is released when ^{15}C decays?

64. A radioactive sample of radon has an activity of 2050 Bq. How many kilograms of radon are present?

65. Figure 29.7 is an energy level diagram for ^{208}Tl. What are the energies of the photons emitted for the six transitions shown?

66. Approximately what is the total energy of the neutrino emitted when $^{22}_{11}Na$ decays by electron capture?

67. $^{106}_{52}Te$ is radioactive; it α decays to $^{102}_{50}Sn$. $^{102}_{50}Sn$ is itself radioactive and has a half-life of 4.5 s. At $t = 0$, a sample contains 4.00 mol of $^{106}_{52}Te$ and 1.50 mol of $^{102}_{50}Sn$. At $t =$ 25 μs, the sample contains 3.00 mol of $^{106}_{52}Te$ and 2.50 mol of $^{102}_{50}Sn$. How much $^{102}_{50}Sn$ will there be at $t = 50$ μs?

68. In 1988 the shroud of Turin, a piece of cloth that some people believe is the burial cloth of Jesus, was dated using ^{14}C. The measured ^{14}C activity of the cloth was about 0.23 Bq/g. According to this activity, when was the cloth in the shroud made?

69. Radon gas (Rn) is produced by the α decay of radium $^{226}_{88}Ra$ (a) How many neutrons and how many protons are present in the nucleus of the isotope of Rn produced by this decay? (b) In the air in an average size room in a student basement apartment in Ithaca, NY, there are about 10^7 Rn nuclei. The Rn nucleus itself is radioactive; it too decays by emitting an α particle. The half-life of Rn is 3.8 days. How many α particles per second are emitted by decaying Rn nuclei in the room?

70. (a) What fraction of the ^{238}U atoms present at the formation of the Earth still exist? Take the age of the Earth to be 4.5×10^9 yr. (b) Answer the same question for ^{235}U. Could this explain why there are more than 100 times as many ^{238}U atoms as ^{235}U atoms in the Earth today?

71. The radioactive decay of ^{238}U produces α particles with a kinetic energy of 4.17 MeV. (a) At what speed do these α particles move? (b) Put yourself in the place of Rutherford and Geiger. You know that α particles are positively charged (from the way they are deflected in a magnetic field). You want to measure the speed of the α particles using a velocity selector. If your magnet produces a magnetic field of 0.30 T, what strength electric field would allow the α particles to pass through undeflected? (c) Now that you know the speed of the α particles, you measure the radius of their trajectory in the same magnetic field (without the electric field) to determine their charge-to-mass ratio. Using the charge and mass of the α particle, what would the radius be in a 0.30-T field? (d) Why can you determine only the charge-to-mass ratio (q/m) by this experiment, but not the individual values of q and m?

72. Once Rutherford and Geiger determined the charge-to-mass ratio of the α particle (Problem 71), they performed another experiment to determine its charge. An α source was placed in an evacuated chamber with a fluorescent screen. Through a glass window in the chamber, they could see a flash on the screen every time an α particle hit it. They used a magnetic field to deflect β particles away from the screen so they were sure that every flash represented an alpha particle. (a) Why is the deflection of a β particle in a magnetic field much larger than the deflection of an α particle moving at the same speed? (b) By counting the flashes, they could determine the number of αs per second striking the screen (R). Then they replaced the screen with a metal plate

connected to an electroscope and measured the charge Q accumulated in a time Δt. What is the α-particle charge in terms of R, Q, and Δt?

73. A water sample is found to have 0.016% deuterium content (that is, 0.016% of the hydrogen nuclei in the water are ^{2}H). If the fusion reaction (^{2}H + ^{2}H) yields 3.65 MeV of energy on average, how much energy could you get from 1.00 L of the water? (There are two reactions with approximately equal probabilities; one yields 4.03 MeV and the other 3.27 MeV.) Assume that you are able to extract and fuse 87.0% of the deuterium in the water. Give your answer in kilowatt hours.

74. (a) Find the approximate number of water molecules in 1.00 L of water. (b) What fraction of the liter's volume is occupied by water *nuclei*?

75. Radioactive iodine, ^{131}I, with a half-life of 8.0252 d, is used in some forms of medical diagnostics. (a) If the initial activity of a sample is 64.5 mCi, what is the mass of ^{131}I in the sample? (b) What will the activity be 4.5 d later?

76. An α particle with a kinetic energy of 1.0 MeV is headed straight toward a gold nucleus. (a) Find the distance of closest approach between the centers of the α particle and gold nucleus. (Assume the gold nucleus remains stationary. Since its mass is much larger than that of the α particle, this assumption is a fairly good approximation.) (b) Will the two get close enough to "touch"? (c) What is the minimum initial kinetic energy of an α particle that will make contact with the gold nucleus?

77. A space rock contains 3.00 g of $^{147}_{62}$Sm and 0.150 g of $^{143}_{60}$Nd. $^{147}_{62}$Sm α decays to $^{143}_{60}$Nd with a half-life of 1.06×10^{11} yr. If the rock originally contained no $^{143}_{60}$Nd, how old is it?

78. In naturally occurring potassium, 0.0117% of the nuclei are radioactive ^{40}K. (a) What mass of ^{40}K is found in a broccoli stalk containing 300 mg of potassium? (b) What is the activity of this broccoli stalk due to ^{40}K?

79. The power supply for a pacemaker is a small amount of radioactive ^{238}Pu. This nuclide decays by α decay with a half-life of 86 yr. The pacemaker is typically replaced every 10.0 yr. (a) By what percentage does the activity of the ^{238}Pu source decrease in 10 yr? (b) The energy of the α particles emitted is 5.6 MeV. Assume an efficiency of 100%—all of the α-particle energy is used to run the pacemaker. If the pacemaker starts with 1.0 mg of ^{238}Pu, what is the power output initially and after 10.0 yr?

80. $^{212}_{83}$Bi can α decay to the ground state of $^{208}_{81}$Tl, or to any of the four excited states of $^{208}_{81}$Tl shown in Fig. 29.7. The maximum kinetic energy of the α particles emitted by $^{212}_{83}$Bi is 6.090 MeV. What other α-particle kinetic energies are possible? [*Hint:* Estimate the atomic mass of $^{208}_{81}$Tl.]

81. Suppose that a radioactive sample contains equal numbers of two radioactive nuclides A and B at $t = 0$. A has a half-life of 3.0 h, while B has a half-life of 12.0 h. Find the ratio of the decay rates or activities R_A/R_B at (a) $t = 0$, (b) $t = 12.0$ h, and (c) $t = 24.0$ h.

82. The first nuclear reaction ever observed (in 1919 by Ernest Rutherford) was $\alpha + {}^{14}_{7}\text{N} \rightarrow \text{p} + \text{X}$. (a) Show that the reaction product "X" must be $^{17}_{8}$O. (b) For this reaction to take place, the α particle must come in contact with the nitrogen nucleus. Calculate the distance d between their centers when they just make contact. (c) If the α particle and the nitrogen nucleus are initially far apart, what is the minimum value of their kinetic energy necessary to bring the two into contact? Express your answer in terms of the elementary charge e, the contact distance d, and whatever else you need. (d) Is the total kinetic energy of the reaction products more or less than the initial kinetic energy in part (c)? Why? Calculate this kinetic energy difference.

83. The last step in the carbon cycle that takes place inside stars is $\text{p} + {}^{15}\text{N} \rightarrow {}^{12}\text{C} + (?)$. This step releases 5.00 MeV of energy. (a) Show that the reaction product "(?)" must be an α particle. (b) Calculate the atomic mass of helium-4 from the information given. (c) In order for this reaction to occur, the proton must come into contact with the nitrogen nucleus. Calculate the distance d between their centers when they just "touch." (d) If the proton and nitrogen nucleus are initially far apart, what is the minimum value of their total kinetic energy necessary to bring the two into contact?

Answers to Practice Problems

29.1 $^{104}_{44}$Ru (ruthenium) **29.2** 17 u
29.3 1.6×10^{-42} m^3 **29.4** 115.492 MeV
29.5 $^{226}_{88}$Ra (radium-226) **29.6** 5.3044 MeV
29.7 1.3111 MeV **29.8** 2.26×10^{12}
29.9 5300 yr ago **29.10** ±8 yr **29.11** 4.4 μg
29.12 exoergic; 0.6259 MeV released
29.13 From Fig. 29.2, nuclides around $A \approx 60$ are the most tightly bound; they have the highest binding energies per nucleon. Fission cannot occur because the total mass of the daughter nuclides and any neutrons released would be *greater* than the mass of the $^{55}_{24}$Cr* compound nucleus. More likely, $^{55}_{24}$Cr* would emit an electron and one or more γ rays, leaving a stable $^{55}_{25}$Mn nucleus as the final product.
29.14 1.1985 MeV

Answers to Checkpoints

29.1 $^{23}_{11}$Na has 11 protons and 23 − 11 = 12 neutrons. The mass number is 23.

29.4 After 3 half-lives have passed, $(1/2)^3 = 1/8$ of the Mn-54 nuclei remain. Therefore, during 3 half-lives, 7/8 of them decay.

29.6 Balancing the charge and nucleon numbers reveals that the intermediate nucleus is

$$^{15}_{7}\text{N}\ ({}^{1}_{0}\text{n} + {}^{14}_{7}\text{N} \rightarrow {}^{15}_{7}\text{N} \rightarrow {}^{1}_{1}\text{H} + {}^{14}_{6}\text{C})$$

CHAPTER

30

Particle Physics

The Large Hadron Collider (LHC) tunnel is about 100 m below ground and is 27 km in circumference; it straddles the border of France and Switzerland near Geneva. The LHC will have four detectors. ATLAS (A toroidal LHC apparatus), about the size of a five-story building, and CMS, which weighs almost 14 000 tons, are general-purpose detectors. ALICE will concentrate on the products of heavy ion collisions. LHCb will focus on proton-proton collisions that produce a special type of particle called the B meson, among other things.

The Large Hadron Collider (LHC) at the European Organization for Nuclear Research (CERN) near Geneva, Switzerland, was designed to achieve collisions between protons with kinetic energies up to 7 TeV (= 7000 GeV) so that collisions at an energy of 14 TeV can be observed. What are the goals of studying particle collisions with higher and higher energies? (See p. 1114 for the answer.)

Concepts & Skills to Review

- antiparticles (Section 27.8)
- fundamental interactions; unification (Section 4.12)
- mass and rest energy (Section 26.7)

30.1 FUNDAMENTAL PARTICLES

One of the overarching goals of physics is to find the fundamental building blocks of the universe and to understand their interactions. In the fifth century B.C.E., the Greek philosopher Democritus speculated that all matter was composed of fundamental units so tiny they could not be seen; the English word *atom* comes from the Greek word *atomos*, meaning indivisible. However, atoms are *not* indivisible: they consist of one or more electrons bound to a nucleus. The nucleus, in turn, is a bound collection of protons and neutrons. Are electrons, protons, and neutrons the fundamental building blocks of matter?

Quarks

We now know that protons and neutrons have internal structure and thus are *not* fundamental particles. Each proton or neutron contains three **quarks**. Quarks are fundamental particles whose existence was proposed independently in 1963 by Murray Gell-Mann (b. 1929) and George Zweig (b. 1937). Gell-Mann took the name *quark* from a line in *Finnegan's Wake* by James Joyce: "Three quarks for Muster Mark." Although three quarks were originally proposed, subsequent experiments have shown that there are six altogether (Table 30.1). The quark masses are expressed in GeV/c^2, a mass unit commonly used in high-energy physics. Since $c^2 = 0.931\ 494$ GeV/u [see Eq. (29-9)], $1\ \text{u} = 0.931\ 494\ \text{GeV}/c^2$.

For each of the six quarks, there is a corresponding *antiquark* with the same mass and opposite electric charge. In Section 27.8, we saw that the electron and its antiparticle, the positron, can *annihilate*, producing two photons to carry away the energy and momentum. Electron-positron pairs can also be *created.* Similarly, other particle-antiparticle pairs can be created or annihilated. Annihilation does not always produce a pair of photons; it can, for instance, produce a different particle-antiparticle pair. The antiquarks are written with a bar over the symbol; for example, the antiparticle of the u quark is written $\bar{\text{u}}$ (read *u-bar*).

Quarks were first detected in a scattering experiment similar to the way the nucleus was discovered in Rutherford's experiment (Section 27.6). In 1968–1969, scientists at the Stanford Linear Accelerator Center (SLAC) studied the effects of scattering high-energy electrons from protons and neutrons. The experiment showed that the electrons scattered from pointlike objects inside each proton or neutron.

Although many experiments have looked for them, an *isolated* quark has never been observed. We now think that it is impossible, even in principle, to observe an isolated quark because of the unusual properties of the interaction between quarks—the **strong interaction** (Section 30.2). A bound quark-antiquark pair is called a **meson**; a bound triplet of quarks or antiquarks is called a **baryon** (Fig. 30.1). Collectively, the

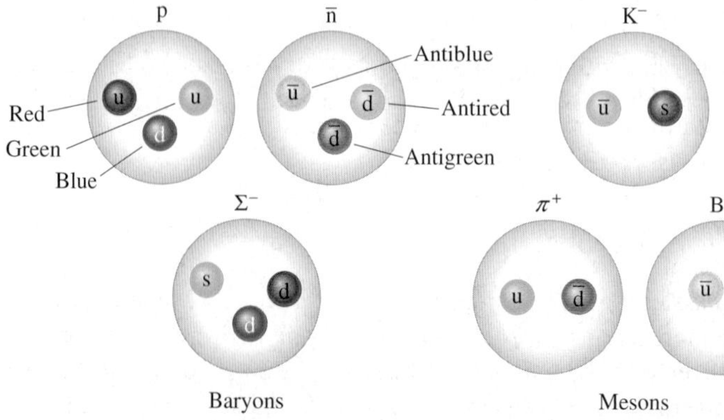

Figure 30.1 The quark content of a few hadrons. (The significance of the colors is discussed in Section 30.2.)

Table 30.1 The Six Quarks

Name	Symbol	Antiparticle	Charge	Mass (GeV/c^2)	Generation
Up	u	$\bar{u}$	$\pm\frac{2}{3}e$	0.0015–0.003	I
Down	d	$\bar{d}$	$\mp\frac{1}{3}e$	0.003–0.007	
Strange	s	$\bar{s}$	$\mp\frac{1}{3}e$	0.07–0.12	II
Charm	c	$\bar{c}$	$\pm\frac{2}{3}e$	1.16–1.34	
Bottom	b	$\bar{b}$	$\mp\frac{1}{3}e$	4.1–4.4	III
Top	t	$\bar{t}$	$\pm\frac{2}{3}e$	169–175	

The upper symbol in ± or ∓ is for the particle and the lower is for the antiparticle.
Precise values of the quark masses are not known. The table lists ranges of the masses that are consistent with experiment.

mesons and baryons are called **hadrons**. The proton is a baryon containing two up quarks and one down quark (uud); the neutron is a baryon containing one up quark and two down quarks (udd).

Hundreds of hadrons have been observed and to date every one is consistent with the quark model. Other than the proton and neutron, all of them have short half-lives—less than 0.1 μs for the longest-lived ones. A neutron *inside a nucleus* can be stable, but an *isolated* neutron decays with a mean lifetime of 14.8 min into a proton, an electron, and an antineutrino ($n \rightarrow p + e^- + \bar{\nu}_e$). The proton appears to be stable; experiments have shown that if it is unstable, its half-life is at least 10^{29} yr—roughly 10^{19} times the age of the universe.

Table 30.1 organizes the quarks into three generations (indicated by roman numerals). Each generation has two quarks, one with charge $+\frac{2}{3}e$ and one with charge $-\frac{1}{3}e$. The quark charges are fractions of the elementary charge e, but they always occur in mesons or hadrons such that the smallest *observable* unit of charge is e. The quarks in each successive generation are more massive than those in the previous generation. Ordinary matter contains only quarks from the first generation (u and d).

Super-Kamiokande, the world's largest underground neutrino observatory, is located 1 km under Mt. Ikenoyama, Japan. This photo shows a crew cleaning the 50-cm diameter faces of some of the 11 200 photomultiplier tubes that line the walls of the cylindrical inner detector. When in operation, this inner detector is filled with 32 000 tons of ultrapure water. When charged particles move through the water at speeds greater than the speed of light in water, they emit blue light that is detected by the photomultiplier tubes. In 1998, the Super-Kamiokande collaboration announced conclusive experimental evidence for nonzero neutrino masses.

Leptons

Although the proton and neutron are composed of quarks, no experiment has suggested that the electron has any internal structure. The electron belongs to another group of fundamental particles called the **leptons** (Table 30.2).

The six leptons (and their antiparticles) are grouped into three generations like the quarks. Each generation has one particle with charge $-e$ and an uncharged neutrino. The masses again increase from one generation to the next. As is true for the quarks, ordinary matter contains only first-generation leptons. Electrons are a basic building block of atoms. The positron (e^+) is the antiparticle of the electron and is emitted in the β^+ decay of a radioactive nucleus, along with an electron neutrino ν_e (Section 29.3). In β^- decay, an electron antineutrino ($\bar{\nu}_e$) is emitted in addition to the electron. Electron neutrinos and antineutrinos are also released in nuclear fusion (Section 29.8); Earth is bathed in a steady stream of billions of neutrinos per square centimeter of cross-sectional area per second from the fusion reactions taking place in the Sun's interior.

Neutrinos are difficult to observe because they can pass through matter with only a small probability of interacting with anything. For a long time the neutrinos were thought to be massless, but experiments have recently shown that neutrinos do have mass. There are more neutrinos in the universe than all of the other leptons and quarks combined. However, even with their large numbers, neutrinos do not make a significant contribution to the mass of the universe because their masses are so small.

Table 30.2 The Six Leptons

Name	Symbol	Antiparticle	Charge	Mass (GeV/c^2)	Generation
Electron	e^-	e^+	$\mp e$	0.000 511	I
Electron neutrino	ν_e	$\bar{\nu}_e$	0	< 0.000 000 002	
Muon	μ^-	μ^+	$\mp e$	0.106	II
Muon neutrino	ν_μ	$\bar{\nu}_\mu$	0	< 0.000 19	
Tau	τ^-	τ^+	$\mp e$	1.777	III
Tau neutrino	ν_τ	$\bar{\nu}_\tau$	0	< 0.0182	

The table gives the largest values of the neutrino masses that are consistent with experiments to date. The upper symbol in $\mp$ is for the particle and the lower is for the antiparticle. Note that the antiparticles of the negatively charged leptons are written with plus signs to indicate their positive charges but without a bar over them.

Muons were the first second-generation particles to be observed. Cosmic rays—streams of energetic particles, mostly protons, traveling from outer space—continually bombard Earth's upper atmosphere. The cosmic-ray particles usually have energies in the GeV range, but some have been observed with energies over 10^{11} GeV—much higher than the energies that can be achieved by particle accelerators. When cosmic ray particles collide with atoms high in Earth's atmosphere, the resulting shower of secondary particles—including electrons, muons, and gamma rays—can be detected at Earth's surface. Muons rain down on us at the rate of about 1/cm^2/min.

Neither the muon nor the tau is stable; they are considered to be fundamental, or *elementary*, particles because they do not appear to have any substructure. A neutrino can transform from one type of neutrino to another. This effect, called **neutrino oscillation**, explains why the number of electron neutrinos reaching Earth from the Sun is smaller than had been predicted—some of the electron neutrinos are transformed into muon or tau neutrinos before they reach Earth.

CHECKPOINT 30.1

Which quarks and leptons are found in an atom?

30.2 FUNDAMENTAL INTERACTIONS

Quarks and leptons are not the whole story; what about the interactions between them? In Section 4.12 we described the four fundamental interactions in the universe: strong, electromagnetic, weak, and gravitational. The interactions are sometimes called *forces* but in a sense much broader than in Newtonian physics (where force is the rate of change of momentum). The fundamental "forces" do much more than push or pull; they include every change that occurs between particles: annihilation and creation of particle-antiparticle pairs, decay of unstable particles, binding of quarks into hadrons, and all kinds of reactions.

Each interaction can be understood as the exchange of a particle called a **mediator**, or **exchange, particle** (Table 30.3). The exchange particle is emitted by one particle and absorbed by another; it can transfer momentum and energy from one particle to another. The photon mediates the electromagnetic interaction. The weak interaction is mediated by one of three particles (W^+, W^-, and Z^0) whose existence was predicted in the 1960s by Steven Weinberg (b. 1933), Sheldon Glashow (b. 1932), and Abdus Salam (1926–1996). A team of scientists led by Carlo Rubbia (b. 1934) first observed the three particles in 1982–1983. The strong interaction is mediated by *gluons*; gravity is believed

Table 30.3 The Four Fundamental Interactions and Their Exchange Particles

Interaction	Relative Strength	Range (m)	Affects Which Fundamental Particles?	Exchange Particles	Masses of Exchange Particles (GeV/c^2)
Strong	1	10^{-15}	Quarks	Gluons (g)	0
Electromagnetic	10^{-2}	∞	Electrically charged	Photon (γ)	0
Weak	10^{-6}	10^{-17}	Quarks and leptons	W^+, W^-, Z^0	80.4, 80.4, 91.2
Gravitational	10^{-43}	∞	All	Graviton*	0

*Predicted but not observed (to date).
The relative strengths are for a pair of u quarks a distance of 0.03 fm apart.

to be mediated by a particle called the *graviton*, which has not yet been observed. Like the photon, the gluons and graviton have no electric charge and are massless. Like the quarks and leptons, the exchange particles are considered to be fundamental; they apparently have no substructure.

PHYSICS AT HOME

To get a feel for a particle that mediates a force, collect a friend and a heavy medicine ball. Then toss the heavy ball back and forth. The ball is the mediating or exchange particle that carries momentum and energy from one of you to the other. This is a repulsive force. It is even more dramatic if you are both standing on skateboards or wearing ice skates. Unfortunately, this analogy can only illustrate a repulsive force.

The Strong Interaction

The strong interaction holds quarks together to form hadrons. Quarks interact via the strong interaction but leptons do not. Quarks carry *strong charge* (or *color charge*) that determines their strong interactions, just as a particle's electric charge determines its electromagnetic interactions. Electric charge comes in only one kind (positive) and its opposite (negative), but strong charge comes in *three* kinds (called red, blue, and green), each of which has an opposite (called antired, antiblue, and antigreen). Color charge has nothing to do with the colors of light that we perceive visually. Rather, they are based on an analogy: just as the red, blue, and green dots produced by different phosphors on a TV screen combine to make white, a red quark, blue quark, and green quark combine to form a colorless (white) combination.

A baryon always contains one quark of each color, an antibaryon always contains one antiquark of each anticolor, and a meson always contains a quark of one color and an antiquark of the corresponding anticolor, such as a red quark and an antired antiquark (see Fig. 30.1). In each case, the strong force holds quarks together in *colorless* combinations, similar to the way the electromagnetic force holds negative and positive charges together to form a neutral atom with zero net electric charge. (Figure 30.1 depicts each hadron as a colorless combination of quarks and antiquarks.)

Although it is possible to pull an electron from an atom, leaving an ion with a net electrical charge, the theory of quark confinement says that the strong force does not allow a quark to be pulled out of a colorless group—which is why isolated quarks are not observed. Just as two ions exert a much greater electromagnetic force on one another than do two neutral atoms, pulling a quark out of a colorless group would leave two groups of quarks with colors other than white; the force between the two groups would be extremely strong and, unlike the electromagnetic force, the strong force grows *stronger* with increasing distance within its short range.

Gluons, the mediators of the strong force, are the "glue" that holds the quarks together. Quarks continually emit and absorb gluons; the gluons carry strong charges so that emission or absorption of a gluon changes the color of the quark. Gluons themselves can emit and absorb gluons. If a quark is pulled out of a colorless combination, the gluons exchanged between the quarks have to travel a greater distance; this greater distance allows more time for the gluons to emit additional gluons, so the force gets *stronger* as the distance between the quarks increases. If there is enough energy to pull the quarks apart, some of the energy is used to create a quark-antiquark pair. The newly created quark goes off with one group and the newly created antiquark goes with the other group in such a way that both groups are colorless. Thus, even in high-energy collisions where hadrons are created and decay into other particles, quarks always end up in colorless combinations.

When we say that a proton contains three quarks (uud), we really mean that its *net* quantum numbers match that picture. Quarks are surrounded by clouds of gluons being emitted and absorbed; from these gluons quark-antiquark pairs are continually created and annihilated, all within the volume of the proton. The energy of the clouds of gluons and the quark-antiquark pairs contribute to the rest energy of the proton (0.938 GeV), which is much larger than the sum of the rest energies of two up quarks and one down quark (less than 0.02 GeV). The same fundamental interaction that holds three quarks together to form a nucleon also binds nucleons together to form a nucleus. However, the force between quarks is much stronger than the force between the colorless nucleons, just as the electromagnetic force between two ions is much stronger than the electromagnetic force between two neutral atoms.

CHECKPOINT 30.2

Why are there no observed particles composed of two quarks (qq) or four quarks (qqqq)? [*Hint:* Consider the color charges of the quarks.]

The Weak Interaction

The weak interaction proceeds by the exchange of one of three particles (W^+, W^-, Z^0), two of which are electrically charged. All three of these particles have mass, which effectively limits the range of the weak interaction. Although leptons do not take part in the strong interaction because they have no color charge, both leptons and quarks have weak charge and thus can take part in weak interactions.

CONNECTION:

The weak interaction can change quark flavors and can change neutrino flavors (Section 30.1).

The weak interaction allows one quark *flavor* (u, d, s, c, b, t) to change into another. Since isolated quarks cannot be observed, the transformation of one quark flavor into another occurs within a hadron.

For example, the β^- decay of a radioactive nucleus was described as the transformation of a neutron into a proton within the nucleus (see Section 29.3):

$$\mathrm{n} \rightarrow \mathrm{p} + \mathrm{e}^- + \bar{\nu}_\mathrm{e}$$

Since a neutron is udd and a proton is uud, at a more fundamental level the d quark within the neutron is transformed into a u quark by emitting a W^-:

$$\mathrm{d} \rightarrow \mathrm{u} + \mathrm{W}^-$$

The W^- then quickly decays into an electron and an electron antineutrino.

The Standard Model

The successful quantum mechanical description of the strong, weak, and electromagnetic interactions and the three generations of quarks and leptons is called the **standard model** of particle physics. The standard model, equipped with experimentally measured quantities (e.g., the particle masses and force charges), correctly predicts the results of decades of experiments in particle physics to a precision unparalleled in any other theory.

Although it is a remarkable achievement, the standard model is at best incomplete; it generates as many questions as it does answers. We introduce some of these questions in the remainder of the chapter.

30.3 UNIFICATION

Newton's law of gravity is an early example of unification. Before Newton, scientists did not understand that the same force that makes an apple fall to the ground from a tree also keeps the planets in orbit around the Sun. In the nineteenth century, Maxwell showed that the electric and magnetic forces are aspects of the same fundamental electromagnetic interaction.

CONNECTION:

One of the main goals of physics is to understand how the world works at its most basic and fundamental level. Part of that goal is to describe the immense variety of forces in the universe in terms of the fewest number of *fundamental* interactions.

A more recent success of unification is the electroweak theory. In ordinary matter, the electromagnetic and weak interactions have entirely different ranges, strengths, and effects. Glashow, Salam, and Weinberg showed that at energies of about 1 TeV or higher, the differences between the two fade until they are indistinguishable; they merge into a single **electroweak interaction**.

The ultimate goal is to describe all forces in terms of a single interaction. Many physicists believe that there was only one fundamental interaction immediately after the **Big Bang**—the explosion that gave birth to the universe (Fig. 30.2). As the universe cooled and expanded, first gravity split off; then the strong force split, leaving three fundamental interactions (gravity, strong, and electroweak). Finally, the electroweak split into the weak and electromagnetic interactions. The splitting apart of the interactions all took place in about the first 10^{-11} s after the Big Bang. Higher energy accelerators may tell us whether the electroweak and strong interactions are unified into a single interaction.

Gravity remains a stumbling block, even though it has been a goal of physics since Einstein to unify gravity with other forces. The standard model does not include gravity; attempts to develop a quantum theory of gravity have led to some of the most exciting ideas in contemporary physics, such as string theory and the possible existence of more than four spacetime dimensions. General relativity, Einstein's theory of gravity, works well for large-scale phenomena—astronomically large distances and long times—but not for the microcosmos of quantum mechanics. Physicists working on the small scale can use quantum mechanics without worrying about general relativity, since gravitational effects are negligible. Thus, the standard model has successfully explained experiments in particle physics even though it omits gravity.

Supersymmetry

In the quest to unify the strong and electroweak interactions some physicists have developed theories of *supersymmetry.* In the standard model, the fundamental particles are divided into two main groups. *Fermions* are the quarks and leptons that make up matter, while *bosons* are the exchange particles that mediate the forces acting on matter. Supersymmetry is based on treating bosons and fermions on equal footing; it predicts equal numbers of fermions and bosons for each type of fundamental particle. For example, supersymmetry predicts a bosonic partner to the electron (the *selectron*) and a fermionic partner to the photon (the *photino*). One of the main goals of the LHC is to seek experimental confirmation of the existence of some of these supersymmetrical particles.

Higher Dimensions

Theorists have found that including extra dimensions in their models of the universe may enable them to reconcile gravity with quantum mechanics and to unify gravity with the other fundamental forces. Higher dimensions also provide a theoretical basis for quantities such as the masses of the quarks and leptons that, within the standard model,

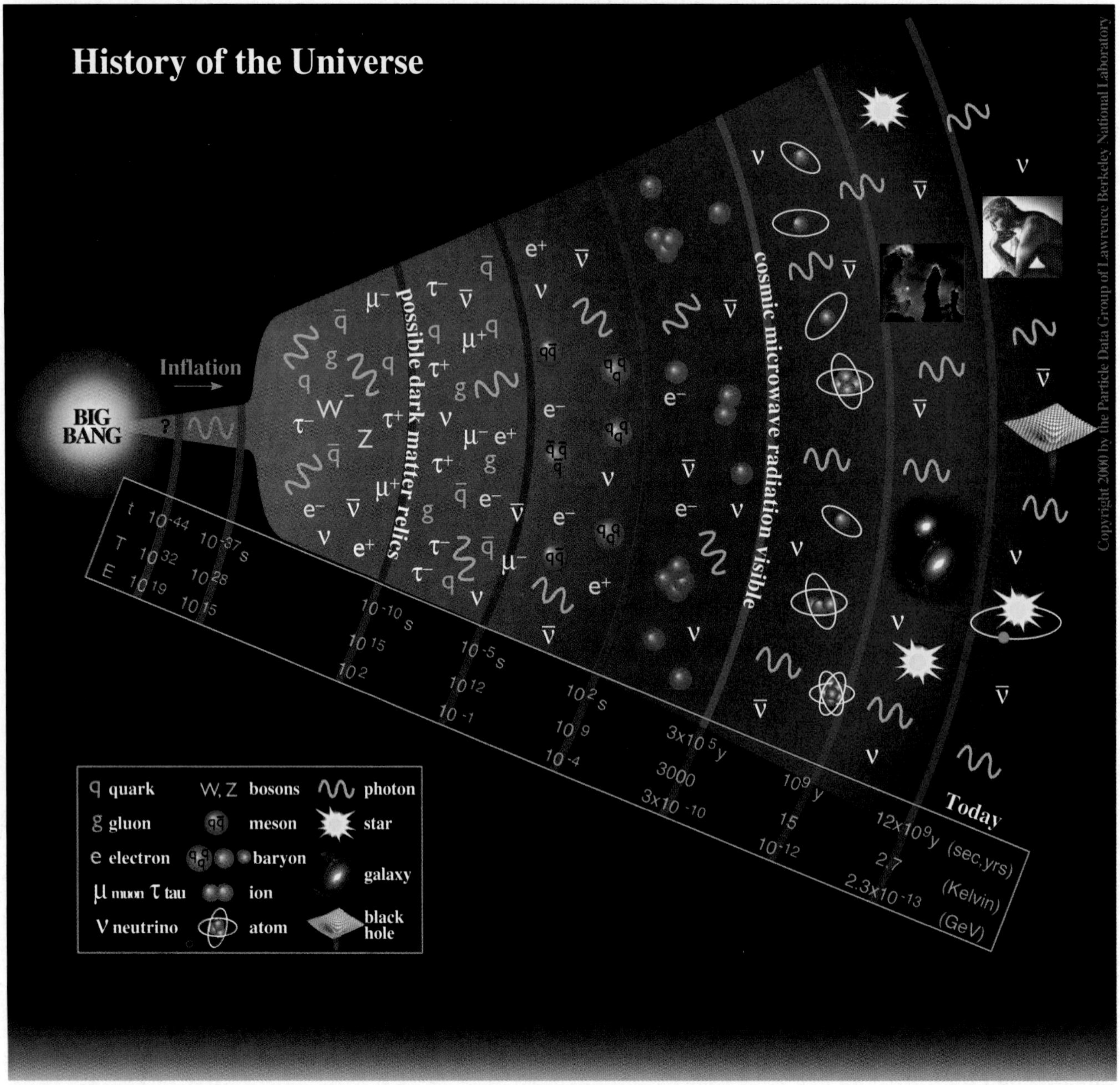

Figure 30.2 The History of the Universe

- For the first 10^{-43} s after the Big Bang, the smooth fabric of spacetime does not yet exist; the universe is a frothing, bubbling "quantum foam." There is just one fundamental interaction: a single superforce.
- At about 10^{-43} s, gravity splits off from the other interactions, leaving two fundamental interactions (gravity and strong-electroweak); spacetime comes into existence. Quarks, leptons, and exchange particles exist. Particle-antiparticle pairs are created and annihilated. Radiation dominates the universe: the total energy of the photons is much greater than the total energy of matter.
- At 10^{-34} s, the strong force splits off from the electroweak.
- At 10^{-11} s, the weak and electromagnetic interactions split, so there are now four fundamental interactions.
- At 10^{-5} s, quark confinement begins as hadrons are formed. As the universe continues to cool, the heavy hadrons annihilate or decay, leaving light hadrons (e.g., protons, neutrons, and pions), leptons, and photons.
- Nuclei begin to form at 10 s, but there are few if any atoms due to the large numbers of photons with more than enough energy to ionize an atom.
- At 3×10^5 yr, the temperature of the universe has cooled to about 3000 K. The energies of the photons in equilibrium at this temperature are mostly too small to ionize atoms, so atoms begin to form and the universe is for the first time transparent to photons. The cosmic microwave background radiation that we observe today is left over from this era, but the photon energies are much smaller now since the universe has cooled to only 2.7 K.

can only be measured experimentally. String theory and brane-world theory represent radical changes in our ideas about space and time and what a particle is.

According to *string theory* and *M-theory*, the various leptons and quarks are not fundamental entities; they are different vibrational patterns of a one-dimensional entity called a *string* that exists in a universe with 10 or 11 dimensions. The extra 6 or 7 dimensions are so small that we cannot observe them directly. As a visual aid, imagine the surface of a thin wire: it is a two-dimensional surface, but one of the dimensions is very small. In a similar way, the extra dimensions proposed by string theory are "curled up" over a length scale of about 10^{-35} m. To probe distances so small requires accelerator energies of about 10^{16} TeV—not possible in the foreseeable future. Experiments must look for indirect tests of string theory.

Other theories propose that the particles we can observe live in a four-dimensional membrane (the familiar three space dimensions and one time dimension) within a six- or seven-dimensional universe. In *brane-world theory*, the additional dimensions don't have to be as small as in string theory. They could be as large as a fraction of a millimeter, while the brane in which we're trapped extends only $\frac{1}{1000}$ the radius of a proton into the additional dimensions. If there are two extra dimensions, the force of gravity is predicted to be proportional to $1/r^4$ at distances less than a millimeter; scientists are now trying to measure the strength of gravity over small distances to see if this prediction is correct. Even if there were seven extra dimensions, as in the leading variant of string theory, it would cause a measurable effect on the strength of gravity over distances about the size of a nucleus.

How can we know if there really are higher dimensions in the universe? Even the most elegant theory in physics would be abandoned if it cannot describe—or predict—the results of experiments. In brane-world theories, if a particle is given enough energy, it may be able to pop out of the membrane and move through the extra dimensions. Another possibility is that a particle moving into the higher dimensions will look like a new particle from our point of view; the discovery of new particles predicted by theory would be evidence for the extra dimensions.

30.4 PARTICLE ACCELERATORS

A *synchrotron* is a ring-shaped particle accelerator with many separate radio frequency (RF) cavities. Each time a bunch of particles passes through an RF cavity, an electric field gives the particles a little boost in kinetic energy. The machine is ring-shaped so that the particles can pass through the RF cavities many times. The particles also pass between the poles of strong magnets placed all around the ring to bend the paths of the particles approximately into a circle. Because charged particles radiate when they are accelerated, the particles lose energy each time a magnet changes their direction of motion. This lost kinetic energy must also be replenished by the RF cavities. Once the particles reach the desired energy, they continue to circulate in a *storage ring* until they are made to collide inside a detector. A storage ring is similar to a synchrotron; it has magnets to bend the paths of the particles and accelerating tubes to replenish the lost kinetic energy. In some cases, the same machine acts both as synchrotron and storage ring.

In a *linear accelerator*, the charged particles move in a straight line rather than around a circular ring. Therefore, much less energy is lost due to radiation because there is no radial acceleration. On the other hand, the same RF cavities cannot be used to repeatedly accelerate the charged particles, as can be done in a synchrotron. Small linear accelerators are used to feed particles into a synchrotron. SLAC is currently the only large linear accelerator in operation. The next large accelerator to be built after the LHC is likely to be the proposed International Linear Collider.

After increasing the kinetic energies of charged particles, particle accelerators then slam them together. The resulting cascade of decays proceeds until particles are formed that live long enough to be detected. By creating the high energies that existed in the early moments of the universe, unstable particles that once existed are created in the

Why study higher and higher energy collision?

laboratory. Going to higher energies allows us to probe matter on shorter and shorter length scales; recall that a particle's de Broglie wavelength is inversely proportional to its momentum [$\lambda = h/p$, Eq. (28-1)]. Higher collision energies also allow the creation of particles with larger mass.

The Large Hadron Collider occupies a circular tunnel of circumference 27 km that formerly housed the Large Electron Positron Collider. The LHC was designed to achieve proton-proton collisions with energies up to 14 TeV. Five thousand superconducting magnets located along the circumference of the ring will steer and focus the two particle beams, traveling in opposite directions, through the 27-km-long, high-vacuum tube. Large detectors will be located at four crossing points of the beams. In addition to proton-proton collisions, the LHC will also collide beams of heavy ions, such as lead, sulfur, calcium, and oxygen. A head-on collision between two lead ions results in a total collision energy of 1148 TeV. Scientists expect these collisions to produce a mixture of quarks and gluons similar to one that existed one-millionth of a second after the Big Bang, before the quarks and gluons coalesced into hadrons.

30.5 TWENTY-FIRST-CENTURY PARTICLE PHYSICS

Particle physics is at the brink of a revolution, according to many physicists. The standard model is extremely successful so far, but it is incomplete. It cannot unify gravity with the other forces. It cannot predict the masses of the quarks and leptons, and it cannot predict what happens at the energies available at the LHC. (See text website for updates on the LHC and on recent research since the printing of this text.)

Some of the many questions that particle physicists are trying to answer include:

Is there a pattern to the masses of quarks and leptons? Scientists searched for the top quark believing that the third generation should include two quarks like the other generations, but there was no way to predict the mass of the top quark. Why is there such a large range of masses of fundamental particles? The top quark has a mass about 175 times that of the proton. Can something with so much mass really be fundamental?

Are there exactly three generations of quarks and leptons, or are there more yet to be discovered? Why are there three generations (or however many there turn out to be)? Are there fundamental particles or forces we don't know about?

Will the Higgs particle be observed? The Higgs particle and the force associated with it resists the motion of particles and effectively gives them mass. Without the Higgs particle, quarks and leptons would be massless and the weak force would be much stronger because the W and Z would be massless (like the photon).

Are quarks and leptons truly fundamental?

Is the proton truly stable?

Why is there a vast range of strengths of forces? Why is gravity so weak?

What makes up the dark matter in the universe? In the last few years, we have learned that the universe is made of only about 4% ordinary matter (that makes up stars and planets), 23% *dark matter*, and 73% *dark energy*. There is more gravitational attraction between nearby galaxies and between inner and outer parts of individual galaxies than can be accounted for by the mass of the ordinary matter making up the galaxies. What is the nature of the invisible (hence, *dark*) matter that supplies this extra gravitational force? One candidate for dark matter is the lightest supersymmetrical particle, the neutralino. The search for supersymmetrical particles at the LHC may help resolve the puzzle of dark matter.

What is the nature of the dark energy that constitutes about 73% of the universe? Recently, scientists have discovered that the expansion of the universe is *speeding up* rather than slowing down. The accelerating expansion of the universe is attributed to the presence of *dark energy*, but we know very little about what it is.

What happened to the antimatter? If there is a symmetry between matter and antimatter, why do we observe almost no antimatter in the universe? If the Big Bang created equal amounts of matter and antimatter, what happened to the antimatter? To help answer this question, experiments planned for the LHC will look for differences in the behavior of particles and antiparticles.

Can gravity be unified with the other fundamental interactions?

Why is the universe four-dimensional (three spatial dimensions and one time dimension)? Or does it actually have more than four dimensions; if so, why does it *appear* to have four?

With all these open questions, particle physics in the twenty-first century promises to be even more exciting than it was in the twentieth.

Master the Concepts

- Protons and neutrons are not fundamental particles; they contain quarks.
- According to the standard model, the fundamental particles are the six quarks (up, down, strange, charm, bottom, and top), the six leptons (electron, muon, tau, and the three kinds of neutrinos), the antiparticles of the quarks and leptons, and the exchange particles for the strong, weak, and electromagnetic interactions.
- Isolated quarks are not observed; quarks are always confined by the strong force to colorless groups. Color charge plays a role in the strong interaction similar to, but more complicated than, that of electric charge in the electromagnetic interaction.

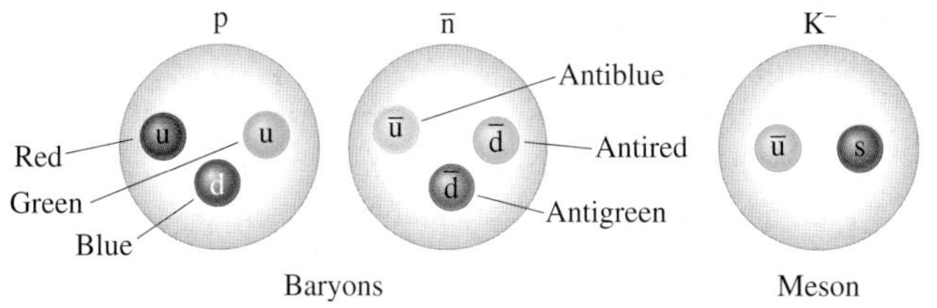

- Only the first generation of quarks and leptons (up, down, electron, and electron neutrino) are found in ordinary matter.
- Just after the Big Bang, there was only a single interaction. First gravity split off, then the strong interaction; finally the weak and electromagnetic interactions split, giving the four fundamental interactions we now recognize.
- New particle accelerators at higher energies will put the standard model, as well as theories competing to be its successor, to the test.

Conceptual Questions

1. What *fundamental* particles make up an atom?
2. What tool enabled scientists to create hundreds of different hadrons in the latter half of the twentieth century?
3. Why is the number of electron neutrinos reaching Earth from the Sun smaller than had originally been predicted?
4. How many different hadrons are stable (as far as we know)?
5. What particles are in the lepton family?
6. Why is the muon sometimes called a "heavy electron"?
7. Is *e* the smallest *fundamental* unit of charge? The smallest *observable* unit of charge? [*Hint:* Try to come up with a meson or baryon with a charge that is not an integral multiple of *e*.] Explain.
8. Describe the use of "color" as a quantum number for the quarks.
9. Why do we not notice the effects of billions of neutrinos passing through each square centimeter of our body surface every second?
10. In a synchrotron, charged particles are accelerated as they travel around in circles; in a linear accelerator they move in a straight line. What are some of the advantages and disadvantages of each?
11. In a fixed-target experiment, high-energy charged particles from an accelerator are smashed into a stationary target. By contrast, in a colliding beam experiment, two beams of particles are accelerated to high energies; particles moving in opposite directions suffer head-on collisions when the two beams are steered together. Describe one advantage of each type of experiment over the other. [*Hint:* For an advantage of the colliding beam experiment, consider not only the total kinetic energy of the particles involved in the collision, but also how much of that energy is available to create new particles. Remember that momentum must be conserved in the collision.]
12. Why can a neutron within a nucleus be stable, while an isolated neutron is unstable? What determines whether a neutron within a nucleus is stable? [*Hint:* Consider conservation of energy.]

Multiple-Choice Questions

1. A baryon can be composed of
(a) any odd number of quarks.
(b) three quarks with three different colors.
(c) three quarks of matching color.
(d) a colorless quark-antiquark pair.
2. Mesons are composed of
(a) any odd number of quarks.
(b) three quarks with three different colors.
(c) three quarks of matching color.
(d) a colorless quark-antiquark pair.
3. Quark flavors include
(a) up, down. (b) red, green. (c) muon, pion.
(d) cyan, magenta. (e) lepton, gluon.
4. Hadrons that contain one or more strange quarks are called strange particles. The particles were originally called *strange*—before quark theory had been formulated—due to their anomalously long lifetimes of 10^{-10} to 10^{-7} s (compared with about 10^{-23} to 10^{-20} s for the other hadrons known at the time). When a strange hadron decays into particles that are not strange, the decay is a manifestation of the
(a) strong interaction. (b) weak interaction.
(c) electromagnetic interaction.
(d) gravitational interaction.
5. The weak interaction is mediated by
(a) leptons. (b) photons. (c) gluons.
(d) W^+, W^-, Z^0. (e) mesons.
6. The exchange particle that mediates the electromagnetic interaction is the
(a) graviton. (b) photon. (c) gluon.
(d) hadron. (e) neutrino.
7. The exchange particle that mediates the strong interaction is the
(a) graviton. (b) photon. (c) gluon.
(d) hadron. (e) neutrino.
8. The exchange particle that mediates the gravitational interaction is called the
(a) graviton. (b) photon. (c) gluon.
(d) hadron. (e) neutrino.
9. Which of the following particles interact via the strong interaction?
(a) quarks (b) gravitons (c) electrons
(d) leptons (e) neutrinos
10. The strong force is __________, and over that range it __________ as the distance between quarks increases.
(a) short-range; becomes weaker
(b) short-range; becomes stronger
(c) short-range; does not vary
(d) long-range; becomes weaker
(e) long-range; becomes stronger

Comprehensive Problems

- Combination conceptual/quantitative problem
- Biological or medical application
- Challenging problem
- Blue # Detailed solution in the Student Solutions Manual
- 1 2 Problems paired by concept
- Text website interactive or tutorial

Note: a particle is extremely relativistic when its rest energy is negligible compared to its kinetic energy. Then

$$E = K + E_0 \gg E_0 \quad and \quad E = \sqrt{(pc)^2 + E_0^2} \approx pc$$

1. A pion (mass 0.140 GeV/c^2) at rest decays by the weak interaction into a muon of mass 0.106 GeV/c^2 and a muon antineutrino: $\pi^- \rightarrow \mu^- + \bar{\nu}_\mu$. Ignoring the rest energy of the antineutrino, what is the total kinetic energy of the muon and the antineutrino?
2. Two factors that can determine the distance over which a force can act are the mass of the exchange particle that carries the force and the Heisenberg uncertainty principle [Eq. (28-3)]. Assume that the uncertainty in the energy of an exchange particle is given by its rest energy and that the particle travels at nearly the speed of light. What is the range of the weak force carried by the Z particle that has a mass of 92 GeV/c^2? Compare this with the range of the weak force given in Table 30.3.
3. What is the quark content of an antiproton? [*Hint:* Replace each of the three quarks that compose a proton with its corresponding antiquark.]
4. Show that the charge of the neutron and the charge of the proton can be derived from their constituent quark content.
5. Which fundamental force is responsible for each of the decays shown here? [*Hint:* In each case, one of the decay products reveals the interaction force.] (a) $\pi^+ \rightarrow \mu^+ + \nu_\mu$, (b) $\pi^0 \rightarrow \gamma + \gamma$, (c) $n \rightarrow p^+ + e^- + \bar{\nu}_e$.
6. Three types of sigma baryons can be created in accelerator collisions. Their quark contents are given by uus, uds, and dds, respectively. What are the electric charges of each of these sigma particles, respectively?
7. The energy at which the fundamental forces are expected to unify is about 10^{19} GeV. Find the mass (in kilograms) of a particle with rest energy 10^{19} GeV.
8. What is the de Broglie wavelength of a proton with kinetic energy 1.0 TeV?
9. What is the de Broglie wavelength of an electron with kinetic energy 7.0 TeV?
10. In the Cornell Electron Storage Ring, electrons and positrons circulate in opposite directions with kinetic energies of 6.0 GeV each. When an electron collides with a positron and the two annihilate, one possible (though unlikely) outcome is the production of one or more

proton-antiproton pairs. What is the *maximum possible* number of proton-antiproton *pairs* that could be formed?

11. The K^0 meson can decay to two pions: $K^0 \rightarrow \pi^+ + \pi^-$. The rest energies of the particles are: $K^0 = 497.7$ MeV, $\pi^+ = \pi^- = 139.6$ MeV. If the K^0 is at rest before it decays, what are the kinetic energies of the π^+ and the π^- after the decay?

12. A proton in Fermilab's Tevatron is accelerated through a potential difference of 2.5 MV during each revolution around the ring of radius 1.0 km. In order to reach an energy of 1 TeV, how many revolutions must the proton make? How far has it traveled?

13. Estimate the magnetic field strength required at the LHC to make 7.0-TeV protons travel in a circle of circumference 27 km. Start by deriving an expression, using Newton's second law, for the field strength B in terms of the particle's momentum p, its charge q, and the radius r. Even though derived using classical physics, the expression is relativistically correct. (The estimate will come out much lower than the actual value of 8.33 T. In the LHC, the protons do not travel in a constant magnetic field; they move in straight-line segments between magnets.)

14. A muon decay is described by $\mu^- \rightarrow e^- + \nu_\mu + \bar{\nu}_e$. What is the maximum kinetic energy of the electron, if the muon was at rest? Assume that the electron is extremely relativistic and ignore the small masses of the neutrinos.

15. A neutral pion (mass 0.135 GeV/c^2) decays via the electromagnetic interaction into two photons: $\pi^0 \rightarrow \gamma + \gamma$. What is the energy of each photon, assuming the pion was at rest?

16. A proton of mass 0.938 GeV/c^2 and an antiproton, at rest relative to an observer, annihilate each other as described by $p + \bar{p} \rightarrow \pi^- + \pi^+$. What are the kinetic energies of the two pions, each of which has mass 0.14 GeV/c^2?

17. In an accelerator, two protons with equal kinetic energies collide head-on. The following reaction takes place: $p + p \rightarrow p + p + p + \bar{p}$. What is the minimum possible kinetic energy of each of the incident proton beams?

Problems 18–21. Determine the quark content of these particles:

18. A meson with charge +*e* composed of up and/or strange quarks and/or antiquarks.

19. A baryon with charge 0 composed of up and/or strange quarks and/or antiquarks.

20. An antibaryon with charge +*e* composed of up and/or strange quarks and/or antiquarks.

21. A meson with charge −*e* composed of up and/or down quarks and/or antiquarks.

22. According to Figure 30.2, higher energies correspond with times that are closer to the origin of the universe, so particle accelerators at higher energies probe conditions that existed shortly after the Big Bang. At Fermilab's Tevatron, protons and antiprotons are accelerated to kinetic energies of approximately 1 TeV. Estimate the time after the Big Bang that corresponds to proton-antiproton collisions in the Tevatron.

23. A charged pion can decay either into a muon or an electron. The two decay modes of a π^- are: $\pi^- \rightarrow \mu^- + \bar{\nu}_\mu$ and $\pi^- \rightarrow e^- + \bar{\nu}_e$. Write the two decay modes for the π^+. [*Hint:* π^+ is the antiparticle of π^-. Replace each particle in the decay reaction with its corresponding antiparticle.]

24. When a proton and an antiproton annihilate, the annihilation products are usually pions. (a) Suppose three pions are produced. What combination(s) of π^+, π^-, and π^0 are possible? (b) Suppose five pions are produced. What combination(s) of π^+, π^-, and π^0 are possible? (c) What is the maximum number of pions that could be produced if the kinetic energies of the proton and antiproton are negligibly small? The mass of a charged pion is 0.140 GeV/c^2 and the mass of a neutral pion is 0.135 GeV/c^2.

25. In Problem 1, what is the kinetic energy of the muon? [*Hint:* The muon is nonrelativistic, so its kinetic energy-momentum relationship is $K = p^2/(2m)$. The antineutrino is extremely relativistic.]

26. At the Stanford Linear Accelerator, electrons and positrons collide together at very high energies to create other elementary particles. Suppose an electron and a positron, each with rest energies of 0.511 MeV, collide to create a proton (rest energy 938 MeV), an electrically neutral kaon (498 MeV), and a negatively charged sigma baryon (1197 MeV). The reaction can be written as:

$$e^+ + e^- \rightarrow p^+ + K^0 + \Sigma^-$$

(a) What is the minimum kinetic energy the electron and positron must have to make this reaction go? Assume they each have the same energy. (b) The sigma can decay in the reaction $\Sigma^- \rightarrow n + \pi^-$ with rest energies of 940 MeV (neutron) and 140 MeV (pion). What is the kinetic energy of each decay particle if the sigma decays at rest?

27. A sigma baryon at rest decays into a lambda baryon and a photon: $\Sigma^0 \rightarrow \Lambda^0 + \gamma$. The rest energies of the baryons are given by $\Sigma^0 = 1192$ MeV and $\Lambda^0 = 1116$ MeV. What is the photon wavelength? [*Hint:* Use relativistic formulas and be sure momentum is conserved as well as energy.]

Answers to Checkpoints

30.1 Two quarks (u and d) and one lepton (the electron).

30.2 Bound systems of quarks exist only in colorless combinations. No observed particles contain two quarks or four quarks because such a combination cannot be colorless.

Review & Synthesis: Chapters 26–30

Review Exercises

1. A starship is traveling at a speed of $0.78c$ toward Earth when it experiences a major malfunction and the crew is forced to evacuate. An escape pod that is 12.0 m long with respect to its passengers is ejected from the starship and sent toward Earth at a speed of $0.63c$ with respect to the starship. How long does the escape pod appear to the people on Earth?
2. An electron is accelerated through a potential difference of 25.00 MV. (a) What would you calculate for the speed of the electron if relativistic equations were not used? (b) What is the actual speed of the electron in this case?
3. According to the special theory of relativity, no object that has mass can travel faster than the speed of light. Yoo Jin says she knows something that moves faster than the speed of light. She tells you to consider a rotating beacon on Earth with a powerful laser that can send a beam to the Moon. (a) If the beacon rotates with a period of 6.00 s, how fast will light from the laser travel across the Moon's surface? (b) How do you explain to Yoo Jin that this does not violate the results of the theory of special relativity?
4. Harvey claims that he annihilated a 1.00-lb bag of chocolate-chip cookies after playing basketball for 3 h. (a) If Harvey had truly annihilated the mass in the cookies, how much energy would be produced? (b) How many killowatt-hours of electric energy is this?
5. A laboratory observer measures an electron's kinetic energy to be 1.02×10^{-13} J. What is the electron's speed?
6. When photons with a wavelength of 120.0 nm are incident on a metal, electrons are ejected that can be stopped with a stopping potential of 6.00 V. (a) What stopping potential is needed when the photons have a wavelength of 240.0 nm? (b) What happens when the photons have a wavelength of 360 nm?
7. (a) Light of wavelength 300 nm is incident on a metal that has a work function of 1.4 eV. What is the maximum speed of the emitted electrons? (b) If light of wavelength 800 nm is incident on a metal that has a work function of 1.6 eV, are any electrons ejected? (c) How would your answers to parts (a) and (b) change if the light intensity were doubled?
8. A 220-W laser fires a 0.250-ms pulse of light with a wavelength of 680 nm. (a) What is the energy of each photon in the laser beam? (b) How many photons are in this pulse?
9. Electrons are accelerated through a potential difference of 8.95 kV and pass through a single slit of width 6.6×10^{-10} m. How wide is the central maximum on a screen that is 2.50 m from the slit?
10. A particle traveling at a speed of 6.50×10^{6} m/s has the uncertainty in its position given by its de Broglie wavelength. What is the minimum uncertainty in the speed of the particle?
11. Strontium-90 $\left({}^{90}_{38}\text{Sr}\right)$ is a radioactive element that is produced in nuclear fission. It decays by β^- decay to yttrium (Y) with a half-life of 28.8 yr. (a) Write down the decay scheme for ${}^{90}_{38}\text{Sr}$. (b) What is the initial activity of 2.0 kg of ${}^{90}_{38}\text{Sr}$? (c) What will be the activity in 1000 yr?
12. How much energy is released in the fusion reaction ${}^{2}\text{H} + {}^{3}\text{H} \rightarrow {}^{4}\text{He} + \text{n}$?
13. A lambda particle (Λ) decays at rest to a proton and pion through the reaction $\Lambda \rightarrow \text{p} + \pi^-$. The rest energies of the particles are Λ: 1116 MeV, p: 938 MeV, π^-: 139.6 MeV. Use conservation of energy and momentum to determine the kinetic energies of the proton and pion.
14. (a) A particle is made up of the quarks $s\bar{u}$. Is this a meson or a baryon? What is the charge of this particle? (b) A particle is made up of the quarks udc. Is this a meson or a baryon? What is the charge of this particle? (c) The particle in (b) can decay to $\Lambda + e^+ + \nu_e$. Through what fundamental force did this decay occur?
15. ◆ UV light with a wavelength of 180 nm is incident on a metal and electrons are ejected. Instead of determining the maximum kinetic energy of the electrons with a stopping potential, the maximum kinetic energy is determined by injecting the electrons into a uniform magnetic field that is perpendicular to the velocity of the electrons. For a certain metal, the electrons with maximum kinetic energy follow a path with a radius of 6.7 cm in a magnetic field of 7.5×10^{-5} T. (a) What is the work function for this metal? (b) Do electrons with maximum kinetic energy follow a path with the maximum or minimum radius?
16. The isotope ${}^{12}_{7}\text{N}$ undergoes radioactive decay to form ${}^{12}_{6}\text{C}$. What charged particle is emitted in the decay process, and what is its charge?
17. A sample of gold, ${}^{198}_{79}\text{Au}$, decays radioactively with an initial rate of 1.00×10^{10} Bq into ${}^{198}_{80}\text{Hg}$. The half-life is 2.70 days. (a) What is the decay rate after 8.10 days? (b) What particle or particles are emitted during this decay process?
18. A charged particle has a total energy of 0.638 MeV when it is moving at $0.600c$. If this particle then enters a linear accelerator and its speed is increased to $0.980c$, what will be the new value of the particle's total energy?
19. Suppose that as we travel away from Earth, the spaceship *Eagle* overtakes and passes us, headed in the same direction we are heading. We measure its speed as $0.50c$. We look back at Earth and note that Earth is receding from us at $0.90c$. If people on Earth measure the speed of the Eagle, what value do they find?
20. An electron in a hydrogen atom has quantum numbers: $n = 8$; $m_\ell = 4$. What are the possible values for the orbital angular momentum quantum number ℓ of the electron?
21. A microscope using photons locates an electron in an atom to within a distance of 0.01 nm. What is the order of

magnitude of the minimum uncertainty in the momentum of the electron located in this way?

22. What is the de Broglie wavelength of an electron with a speed of $0.60c$?

23. If an atom had only four distinct energy levels between which electrons could make transitions, how many spectral lines of different wavelengths could the atom emit?

24. Let the kinetic energy of the electron in the hydrogen atom in its ground state ($n = 1$) be given by K. What multiple of K represents the electron's kinetic energy in the first excited state ($n = 2$)?

25. You are given two x-ray tubes, A and B. In tube A, electrons are accelerated through a potential difference of 10 kV. In tube B, the electrons are accelerated through 40 kV. What is the ratio of the minimum wavelength of x-rays in tube A to the minimum wavelength in tube B?

✦26. The figure shows the lowest six energy levels of the outer electron in sodium. In the ground state, the electron is in the "3*s*" level. (a) What is the ionization energy of sodium? (b) What is the wavelength of the radiation emitted in the transition from the 3*d* to the 3*p* level? (c) What is the transition that gives rise to the characteristic yellow light of sodium at 589 nm?

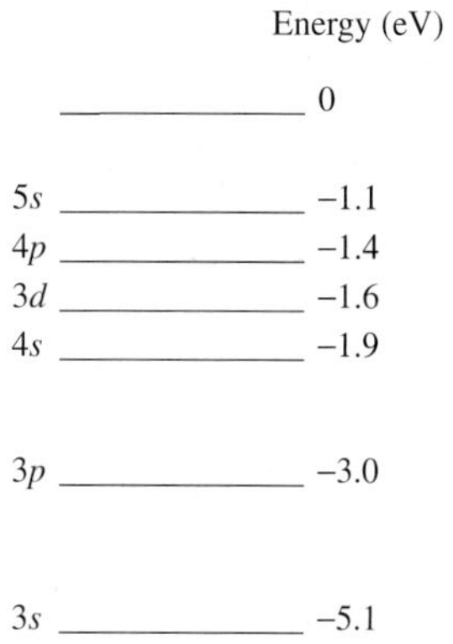

✦27. A $^{208}_{81}Tl^*$ nucleus (mass 208.0 u) emits a 452-keV photon to jump to a state of lower energy. Assuming the nucleus is initially at rest, calculate the kinetic energy of the nucleus after the photon has been emitted. [*Hint:* Assume the nucleus can be treated nonrelativistically.]

28. A spaceship of mass m is traveling away from Earth at speed v. Its momentum has magnitude $2.5\,mv$. (a) Find v. (b) An astronaut on the spaceship has a watch that ticks once every second. How often does the watch tick as measured by an Earth observer?

29. A 65-kg patient undergoes a diagnostic chest x-ray and receives a biologically equivalent dose of 20 mrad distributed over one-third of the patient's body mass. If the x-rays have a quality factor of 0.90, how much energy is absorbed by the patient's body?

30. In the LHC, protons are accelerated to a total energy of 7 TeV. (a) What is the speed of these protons? (b) The LHC tunnel is 27 km in circumference. As measured by an Earth observer, how long does it take the protons to go around the tunnel once? (c) In the reference frame of the protons, how long does it take? (d) What is the de Broglie wavelength of these protons in Earth's reference frame? (e) How fast would a mosquito of mass 1.0 mg be moving if its kinetic energy is 7 TeV?

MCAT Review

The section that follows includes MCAT exam material and is reprinted with permission of the Association of American Medical Colleges (AAMC).

Read the paragraph and then answer the following questions:

The following is a discussion of three common types of radioactive decay.

Alpha Decay

Some heavy nuclei decay by spontaneously emitting alpha (α) particles, which consist of two neutrons and two protons and are indistinguishable from ^4He nuclei. One example is the α decay of ^{238}Pu

$$^{238}_{94}Pu \rightarrow {}^{234}_{92}U + {}^{4}_{2}\alpha$$

The α particle has a mass of 4 u (6.6×10^{-27} kg) and carries a positive charge equal to twice the charge on the proton.

Beta Decay

A beta (β^-) particle is an energetic electron. A typical β^- decay is illustrated by the ^{36}Cl decay

$$^{36}_{17}Cl \rightarrow {}^{36}_{18}Ar + \beta^- + \bar{\nu}$$

where $\bar{\nu}$ is an *antineutrino*, an additional particle that is created in the decay.

The electrons emitted in β^- decay carry a wide range of kinetic energies. The maximum energy carried is called the *endpoint energy* and is determined by the relative energy states of the nuclei involved.

The net effect of β^- decay in a nucleus is the replacement of a neutron with a proton.

Gamma Decay

Gamma (γ) rays are very high-energy electromagnetic waves that have no mass and no charge. They are emitted by excited nuclei during transitions from higher to lower energy levels within the nucleus.

[*Note:* 1 u (atomic mass unit) = 1.66×10^{-27} kg; the proton, neutron, and electron masses are 1.0073 u; 1.0087 u; and 9.11×10^{-31} kg, respectively; the energy equivalent of 1 u is 931 MeV (10^6 electron volts); and 1 eV = 1.6×10^{-19} J.]

1. If an atom of radium initially at rest emits a 4.8-MeV α particle, which of the following is the α particle's approximate speed?

   A. 1.5×10^6 m/s B. 2.3×10^6 m/s
   C. 1.5×10^7 m/s D. 3.0×10^7 m/s

2. In a *radioactive series*, a nucleus decays through several steps. One thorium series starts with the ^{232}Th nucleus, which then successively emits the following particles: an α, two β^-, four α, a β^-, an α, and finally a β^- to reach a stable nucleus. The final product of this series is which of the following?

   A. $^{208}_{82}Pb$ B. $^{208}_{88}Ra$
   C. $^{220}_{82}Pb$ D. $^{220}_{88}Ra$

3. The energy required to break the nucleus into its constituent parts is called its *binding energy*, which is the energy equivalent of the difference between the sum of the masses of the constituent parts of the nucleus and the mass of the nucleus itself. Given that the mass of ^{7}Li is 7.014 u, which of the following is the approximate binding energy of this isotope?
   A. 0.038 MeV B. 0.043 MeV
   C. 35.0 MeV D. 40.0 MeV
4. Which of the following best describes the subsequent motion of α, β^-, and γ rays on entering a uniform magnetic field region if each ray's initial velocity vector points perpendicularly to the magnetic field direction?
   A. All three rays are not bent.
   B. All three rays are bent in the same direction.
   C. The γ rays are not bent; the α and β^- rays are bent in the same direction.
   D. The γ rays are not bent; the α and β^- rays are bent in opposite directions.
5. When a nucleus emits a 2.5-MeV γ ray, the nuclear mass decreases by:
   A. 2.8×10^{-28} kg. B. 1.2×10^{-28} kg.
   C. 4.5×10^{-30} kg. D. 8.6×10^{-31} kg.
6. A sample of $^{24}_{11}$Na has a half-life of 15 h. If the sample's activity (disintegrations per unit time) is 100 mCi after 24 h, the sample's original activity must have been closest to:
   A. 200 mCi. B. 300 mCi.
   C. 600 mCi. D. 1000 mCi.
7. The mass equivalent of the energy required to break the nucleus of the $^{20}_{10}$Ne atom into its constituent parts is 0.173 u. Which of the following is the atomic mass of the atom?
   A. 19.987 u B. 20.002 u
   C. 20.219 u D. 20.333 u
8. A radioactive series begins with $^{238}_{92}$U and has an intermediate product of $^{234}_{92}$U. Which of the following decay sequences will produce $^{234}_{92}$U from $^{238}_{92}$U?
   A. $\beta^-, \beta^-, \beta^-, \beta^-$ B. $\alpha, \beta^-, \beta^-, \beta^-$
   C. $\alpha, \alpha, \beta^-, \beta^-$ D. α, β^-, β^-
9. If a radioactive isotope has a half-life of 8 months, what fraction of a sample of the isotope will still remain after 2 years?
   A. $\frac{1}{32}$ B. $\frac{1}{16}$ C. $\frac{1}{8}$ D. $\frac{1}{4}$

Read the paragraph and then answer the following questions:

Thallium-201 stress imaging is a noninvasive clinical procedure used for the diagnostic assessment of the extent and severity of cardiovascular disease. Physicians use the results of the imaging as a tool to select the most appropriate therapy. Thallium-201 is a radioisotope that undergoes electron capture

$$^{201}_{81}\text{Tl} + e^- \rightarrow {}^{201}_{80}\text{Hg}^* + \nu$$

(In *electron capture*, an orbital electron—most often a 1*s* or a 2*s*—is captured by the nucleus and a *neutrino* ν is created.) The resulting Hg nucleus often undergoes γ decay.

The half-life of $^{201}_{81}$Tl is 73 h, and its energy spectrum consists of about 88% x-rays in the energy range 69–83 keV and about 12% γ rays with energies of 135 and 167 keV.

In a typical imaging procedure, the patient walks on a treadmill until a high level of stress is sustained. The level of stress required for imaging depends on physiological parameters (e.g., pulse rate, blood pressure) and patient responses (chest pains, muscle fatigue). The average time required for a patient to reach a useful level of stress is typically 20–30 min, although increasing the speed and the inclination angle θ_{in} of the treadmill can substantially reduce this time. Five minutes after establishing cardiac stress, a thallium-labeled pharmaceutical is introduced into the bloodstream of the patient. This radiopharmaceutical is distributed within the heart via the circulatory system. Photons emitted from the decaying radioisotope are recorded by a photon detector positioned close to the patient's chest.

(*Note:* Planck's constant is 4.15×10^{-15} eV·s, and the speed of light is 3.0×10^8 m/s.)

10. During the γ decay of $^{201}_{80}$Hg, which of the following happens to its atomic number *Z* and mass number *A*?
    A. *Z* stays constant, and *A* increases.
    B. Both *Z* and *A* stay constant.
    C. *Z* increases, and *A* stays constant.
    D. Both *Z* and *A* decrease.
11. Which of the following is the correct composition of a $^{201}_{81}$Tl atom?
    A. 201 protons, 201 neutrons, and 201 electrons
    B. 120 protons, 81 neutrons, and 120 electrons
    C. 81 protons, 201 neutrons, and 81 electrons
    D. 81 protons, 120 neutrons, and 81 electrons
12. How does the activity (disintegrations per unit time) of a sample of $^{201}_{81}$Tl change as it decays? The activity:
    A. increases exponentially with time.
    B. decreases linearly with time.
    C. decreases exponentially with time.
    D. remains constant with time.
13. Which of the following is the correct expression for the wavelength of the 135-keV γ ray emitted from $^{201}_{81}$Tl?
    A. $\dfrac{(4.15 \times 10^{-15})(3.0 \times 10^8)}{1.35 \times 10^5}$ m
    B. $\dfrac{(4.15 \times 10^{-15})(3.0 \times 10^8)}{1.35 \times 10^3}$ m
    C. $\dfrac{4.15 \times 10^{-15}}{(3.0 \times 10^8)(1.35 \times 10^3)}$ m
    D. $(4.15 \times 10^{-15})(3.0 \times 10^8)(1.35 \times 10^5)$ m

Appendix **A**

Mathematics Review

A.1 ALGEBRA

There are two basic kinds of algebraic manipulations.

- The same operation can always be performed on both sides of an equation.
- Substitution is always permissible (if $a = b$, then any occurrence of a in any equation can be replaced with b).

Products distribute over sums

$$a(b + c) = ab + ac \tag{A-1}$$

The reverse—replacing $ab + ac$ with $a(b + c)$—is called *factoring*. Since dividing by c is the same as multiplying by $1/c$,

$$\frac{a + b}{c} = \frac{a}{c} + \frac{b}{c} \tag{A-2}$$

Equation (A-2) is the basis of the procedure for adding fractions. To add fractions, they must be expressed with a *common denominator.*

$$\frac{a}{b} + \frac{c}{d} = \frac{a}{b} \times \frac{d}{d} + \frac{c}{d} \times \frac{b}{b} = \frac{ad}{bd} + \frac{bc}{bd}$$

Now applying Eq. (A-2), we end up with

$$\frac{a}{b} + \frac{c}{d} = \frac{ad + bc}{bd} \tag{A-3}$$

To divide fractions, remember that dividing by c/d is the same as multiplying by d/c:

$$\frac{a}{b} \div \frac{c}{d} = \frac{\frac{a}{b}}{\frac{c}{d}} = \frac{a}{b} \times \frac{d}{c} = \frac{ad}{bc}$$

A product in a square root can be separated:

$$\sqrt{ab} = \sqrt{a} \times \sqrt{b} \tag{A-4}$$

Pitfalls to Avoid

These are some of the most common *incorrect* algebraic substitutions. Don't fall into any of these traps!

$$\sqrt{a + b} \neq \sqrt{a} + \sqrt{b}$$

$$\frac{a}{b + c} \neq \frac{a}{b} + \frac{a}{c}$$

$$\frac{a}{b} + \frac{c}{d} \neq \frac{a + c}{b + d}$$

$$(a + b)^2 \neq a^2 + b^2$$

In the last one, the cross term is missing: $(a + b)^2 = a^2 + 2ab + b^2$.

Graphs of Linear Functions

If the graph of y as a function of x is a straight line, then y is a *linear function* of x. The relationship can be written in the standard form

$$y = mx + b \tag{A-5}$$

where m is the *slope* and b is the *y-intercept.* The slope measures how steep the line is. It tells how much y changes for a given change in x:

$$m = \frac{\Delta y}{\Delta x} = \frac{y_2 - y_1}{x_2 - x_1} \tag{A-6}$$

The y-intercept is the value of y when $x = 0$. On the graph, the line crosses the y-axis at $y = b$.

Example A.1

What is the equation of the line graphed in Fig. A.1?

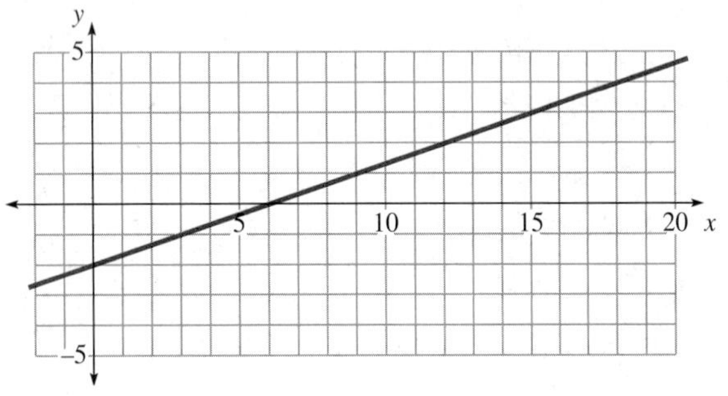

Figure A.1

Solution The y-intercept is -2. To find the slope, we choose two points on the line and then divide the "rise" (Δy) by the "run" (Δx). Using the points $(0, -2)$ and $(18, 4)$,

$$m = \frac{\text{rise}}{\text{run}} = \frac{y_2 - y_1}{x_2 - x_1} = \frac{4 - (-2)}{18 - 0} = \frac{1}{3}$$

Then $y = mx + b = \frac{1}{3}x - 2$.

A.2 SOLVING EQUATIONS

Solving an equation means using algebraic operations to isolate one variable. Many students tend to substitute numerical values into an equation as soon as possible. In many cases, that's a mistake. Although at first it may seem easier to manipulate numerical quantities than to manipulate algebraic symbols, there are several advantages to working with symbols:

- Symbolic algebra is much easier to follow than a series of numerical calculations. Plugging in numbers tends to obscure the logic behind your solution. If you need to trace back through your work (to find an error or review for an exam), it'll be much clearer if you have worked through the problem symbolically. It will also help your instructor when grading your homework papers or exams. When your work is clear, your instructor is better able to help you understand your mistakes. You may also get more partial credit on exams!
- Symbolic algebra lets you draw conclusions about how one quantity depends on another. For instance, working symbolically you might see that the horizontal range of a projectile is proportional to the *square* of the initial speed. If you had substituted the numerical value of the initial speed, you wouldn't notice that. In particular, when an algebraic symbol cancels out of the equation, you know that the answer doesn't depend on that quantity.
- On the most practical level, it's easy to make arithmetic or calculation errors. The later on in your solution that numbers are substituted, the fewer number of steps you have to check for such errors.

When solving equations that contain square roots, be careful not to assume that a square root is positive. The equation $x^2 = a$ has *two* solutions for x, $x = \pm\sqrt{a}$. (The symbol $\pm$ means *either* + *or* –.)

Solving Quadratic Equations

An equation is quadratic in x if it contains terms with no powers of x other than a squared term (x^2), a linear term (x^1), and a constant (x^0). Any quadratic equation can be put into the standard form:

$$ax^2 + bx + c = 0 \quad \text{(A-7)}$$

The quadratic formula gives the solutions to any quadratic equation written in standard form:

$$x = \frac{-b \pm \sqrt{b^2 - 4ac}}{2a} \quad \text{(A-8)}$$

The symbol "±" (read "plus or minus") indicates that in general there are two solutions to a quadratic equation; that is, two values of x will satisfy the equation. One solution is found by taking the + sign and the other by taking the – sign in the quadratic formula. If $b^2 - 4ac = 0$, then there is only one solution (or, technically, the two solutions happen to be the same). If $b^2 - 4ac < 0$, then there is no solution to the equation (for x among the real numbers).

The quadratic formula still works if $b = 0$ or $c = 0$, although in such cases the equation can easily be solved without recourse to the quadratic formula.

Example A.2

Solve the equation $5x(3 - x) = 6$.

Solution First put the equation in standard quadratic form:

$$15x - 5x^2 = 6$$

$$-5x^2 + 15x - 6 = 0$$

We identify $a = -5$, $b = 15$, $c = -6$. Then

$$x = \frac{-b \pm \sqrt{b^2 - 4ac}}{2a}$$

$$= \frac{-15 \pm \sqrt{15^2 - 4 \times (-5) \times (-6)}}{-10}$$

$$\approx \frac{-15 \pm 10.25}{-10} = 0.475 \text{ or } 2.525$$

Solving Simultaneous Equations

Simultaneous equations are a set of N equations with N unknown quantities. We wish to solve these equations *simultaneously* to find the values of all of the unknowns. We *must* have at least as many equations as unknowns. It pays to keep track of the number of unknown quantities and the number of equations in solving more challenging problems. If there are more unknowns than equations, then look for some other relationship between the quantities—perhaps some information given in the problem that has not been used.

One way to solve simultaneous equations is by *successive substitution*. Solve one of the equations for one unknown (in terms of the other unknowns). Substitute this expression into each of the other equations. That leaves $N - 1$ equations and $N - 1$ unknowns. Repeat until there is only one equation left with one unknown. Find the value of that unknown quantity, and then work backward to find all the others.

Example A.3

Solve the equations $2x - 4y = 3$ and $x + 3y = -5$ for x and y.

Solution First solve the second equation for x in terms of y:

$$x = -5 - 3y$$

Substitute $-5 - 3y$ for x in the first equation:

$$2 \times (-5 - 3y) - 4y = 3$$

This can be solved for y:

$$-10 - 10y = 3$$

$$-10y = 13$$

$$y = \frac{13}{-10} = -1.3$$

Now that y is known, use it to find x:

$$x = -5 - 3y = -5 - 3 \times (-1.3) = -1.1$$

It's a good idea to check the results by substituting into the original equations.

A.3 EXPONENTS AND LOGARITHMS

These identities show how to manipulate exponents.

$$a^{-x} = \frac{1}{a^x} \qquad \text{(A-9)}$$

$$(a^x) \times (a^y) = a^{x+y} \qquad \text{(A-10)}$$

$$\frac{a^x}{a^y} = (a^x) \times (a^{-y}) = a^{x-y} \qquad \text{(A-11)}$$

$$(a^x) \times (b^x) = (ab)^x \qquad \text{(A-12)}$$

$$(a^x)^y = a^{xy} \qquad \text{(A-13)}$$

$$a^{1/n} = \sqrt[n]{a} \qquad \text{(A-14)}$$

$$a^0 = 1 \quad (\text{for any } a \neq 0) \qquad \text{(A-15)}$$

$$0^a = 0 \quad (\text{for any } a \neq 0) \qquad \text{(A-16)}$$

A common mistake to avoid: $(a^x) \times (a^y) \neq a^{xy}$ [see Eq. (A-10)].

Logarithms

Taking a logarithm is the inverse of exponentiation:

$$x = \log_b y \quad \text{means that} \quad y = b^x \qquad \text{(A-17)}$$

Thus, one undoes the other:

$$\log_b b^x = x \qquad \text{(A-18)}$$

$$b^{\log_b x} = x \qquad \text{(A-19)}$$

The two commonly used bases b are 10 (the *common* logarithm) and $e = 2.71828\ldots$ (the *natural* logarithm). The common logarithm is written "$\log_{10}$," or sometimes just "log" if base 10 is understood. The natural logarithm is usually written "ln" rather than "$\log_e$."

These identities are true for any base logarithm.

$$\log xy = \log x + \log y \qquad \text{(A-20)}$$

$$\log \frac{x}{y} = \log x - \log y \qquad \text{(A-21)}$$

$$\log x^a = a \log x \qquad \text{(A-22)}$$

Here are some common mistakes to avoid:

$$\log (x + y) \neq \log x + \log y$$

$$\log(x + y) \neq \log x \times \log y$$
$$\log xy \neq \log x \times \log y$$
$$\log x^a \neq (\log x)^a$$

Semilog Graphs

A semilog graph uses a logarithmic scale on the vertical axis and a linear scale on the horizontal axis. Semilog graphs are useful when the data plotted is thought to be an exponential function. If

$$y = y_0 e^{ax}$$

then

$$\ln y = ax + \ln y_0$$

so a graph of ln y versus x will be a straight line with slope a and y-intercept ln y_0.

Rather than calculating ln y for each data point and plotting on regular graph paper, it is convenient to use special semilog paper. The vertical axis is marked so that the values of y can be plotted directly, but the markings are spaced proportional to the log of y. (If you are using a plotting calculator or a computer to make the graph, log scale should be chosen for one axis from the menu of options.) The slope a on a semilog graph is *not* $\Delta y/\Delta x$ since the logarithm is actually being plotted. The correct way to find the slope is

$$a = \frac{\Delta(\ln y)}{\Delta x} = \frac{\ln y_2 - \ln y_1}{x_2 - x_1}$$

Note that there cannot be a zero on a logarithmic scale.

The two graphs of Figs. A.2 and A.3 are linear and semilog plots, respectively, of the function $y = 3e^{-2x}$.

Log-Log Graphs

A log-log graph uses logarithmic scales for both axes. Log-log graphs are useful when the data plotted is thought to be a power function

$$y = Ax^n$$

For such a function,

$$\log y = n \log x + \log A$$

so a graph of log y versus log x will be a straight line with slope n and y-intercept log A. The slope (n) on a log-log graph is found as

$$n = \frac{\Delta(\log y)}{\Delta(\log x)} = \frac{\log y_2 - \log y_1}{\log x_2 - \log x_1}$$

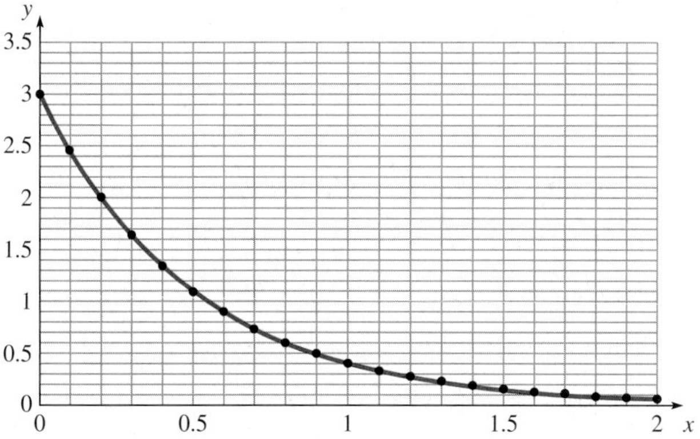

Figure A.2 Graph of the exponential function $y = 3e^{-2x}$ on linear graph paper.

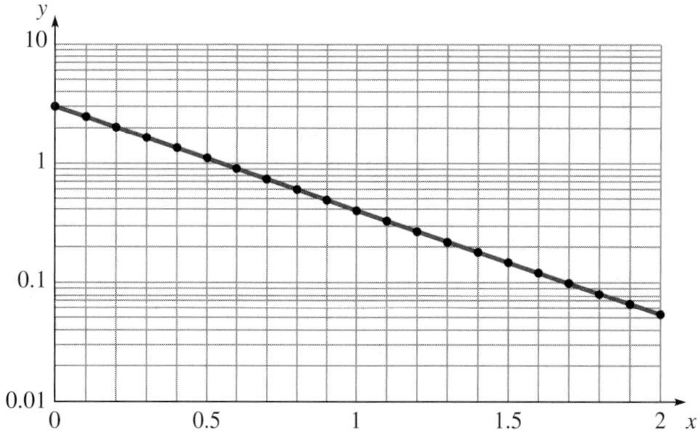

Figure A.3 Graph of the exponential function $y = 3e^{-2x}$ on semilog graph paper.

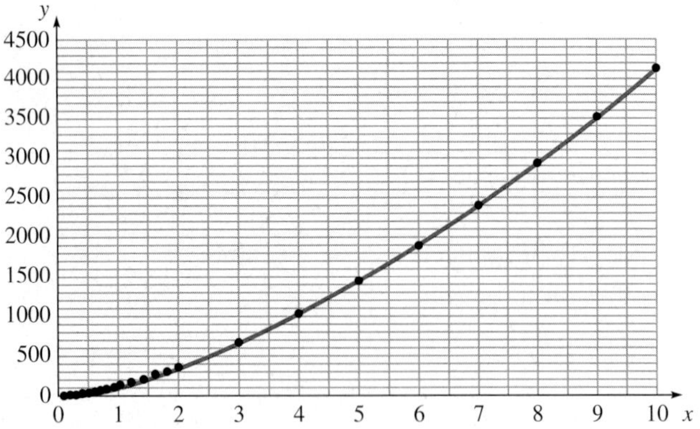

Figure A.4 Graph of the power function $y = 130x^{3/2}$ on linear graph paper.

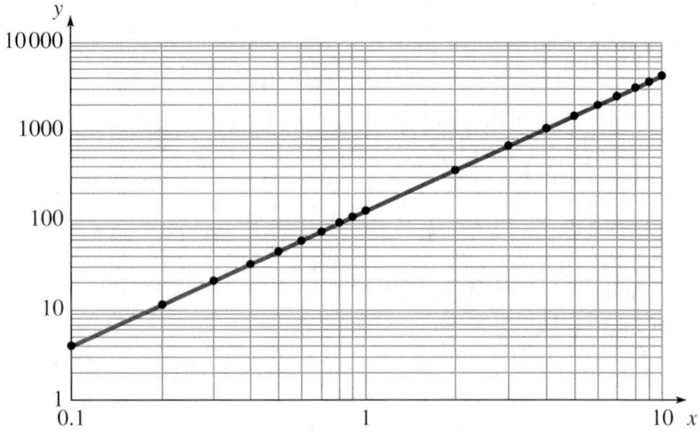

Figure A.5 Graph of the power function $y = 130x^{3/2}$ on log-log graph paper.

The graphs of Figs. A.4 and A.5 are linear and log-log plots, respectively, of the function $y = 130x^{3/2}$.

A.4 PROPORTIONS AND RATIOS

The notation

$$y \propto x$$

means that y is directly proportional to x. A proportionality can be written as an equation

$$y = kx$$

if the constant of proportionality k is written explicitly. Be careful: an equation can *look like* a proportionality without being one. For example, $V = IR$ means that $V \propto I$ *if and only if* R is constant. If R depends on I, then V is not proportional to I.

The notation

$$y \propto \frac{1}{x}$$

means that y is inversely proportional to x. The notation

$$y \propto x^n$$

means that y is proportional to the nth power of x.

Writing out proportions as ratios usually simplifies solutions when some common items in an equation are unknown but we do know the values of all but one of the proportional quantities (tutorial: ball toss). For example if $y \propto x^n$, we can write

$$\frac{y_1}{y_2} = \left(\frac{x_1}{x_2}\right)^n$$

Percentages

Percentages require careful attention. Look at these four examples:

"B is 30% of A" means $B = 0.30A$

"B is 30% larger than A" means $B = (1 + 0.30)A = 1.30A$

"B is 30% smaller than A" means $B = (1 - 0.30)A = 0.70A$

"A increases by 30%" means $\Delta A = +0.30A$

Example A.4

If $P \propto T^4$, and T increases by 10.0%, by what percentage does P increase?

Solution

$$\Delta T = +0.100T_i$$

$$T_f = T_i + \Delta T = 1.100T_i$$

$$\frac{P_f}{P_i} = \left(\frac{T_f}{T_i}\right)^4 = 1.100^4 \approx 1.464$$

Therefore, P increases by about 46.4%.

A.5 APPROXIMATIONS

Binomial Approximations

A binomial is the sum of two terms. The general rule for the nth power of an algebraic sum is given by the binomial expansion:

$$(a+b)^n = a^n + na^{n-1}b + \frac{n(n-1)}{1\times 2}a^{n-2}b^2 + \frac{n(n-1)(n-2)}{1\times 2\times 3}a^{n-3}b^3 + \cdots$$

The binomial approximations are used when a binomial in which one term is much smaller than the other is raised to a power n. Only the first two terms of the binomial expansion are of significant value; the other terms are dropped. A common case for physics problems is that in which $a = 1$, or can be made equal to one by factoring. The basic approximation forms are then given by

$$(1+x)^n \approx 1 + nx \quad \text{when } |x| \ll 1 \qquad \text{(A-23)}$$

$$(1-x)^n \approx 1 - nx \quad \text{when } |x| \ll 1 \qquad \text{(A-24)}$$

The power n can be any real number, including negative as well as positive numbers. It does not have to be an integer. (W tutorial: small percentage changes) An *estimate* of the error—the difference between the approximation and the exact expression—is given by

$$\text{error} \approx \tfrac{1}{2}n(n-1)x^2 \qquad \text{(A-25)}$$

Of course, the larger term in a binomial is not necessarily 1, but the larger term can be factored out and then Eq. (A-23) or Eq. (A-24) applied. For instance, if $A \gg b$, then

$$(A+b)^n = \left[A\times\left(1+\frac{b}{A}\right)\right]^n = A^n\left(1+\frac{b}{A}\right)^n$$

Another common expansion is

$$e^x = 1 + x + \frac{x^2}{2!} + \frac{x^3}{3!} + \cdots$$

where, for any integer n, $n! = n\times(n-1)\times(n-2)\times\cdots\times[n-(n-1)]$; for example, $3! = 3\times 2\times 1 = 6$.

Small-Angle Approximations

Approximations for small angles appear in Section A.7 on trigonometry.

Table A.1 Properties of Common Geometric Shapes

Geometric Shape	Properties	Geometric Shape	Properties
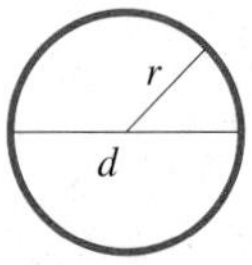 Circle	Diameter $d = 2r$ Circumference $C = \pi d = 2\pi r$ Area $A = \pi r^2$	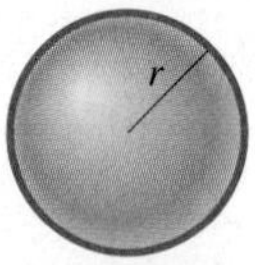 Sphere	Surface area $A = 4\pi r^2$ Volume $V = \frac{4}{3}\pi r^3$
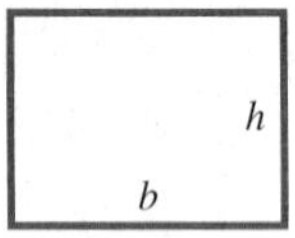 Rectangle	Perimeter $P = 2b + 2h$ Area $A = bh$	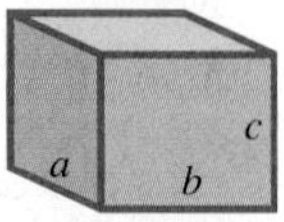 Parallelepiped	Surface area $A = 2(ab + bc + ac)$ Volume $V = abc$
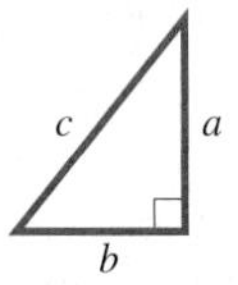 Right triangle	Perimeter $P = a + b + c$ Area $A = \frac{1}{2}\text{base} \times \text{height} = \frac{1}{2}ba$ Pythagorean theorem $c^2 = a^2 + b^2$ Hypotenuse $c = \sqrt{a^2 + b^2}$	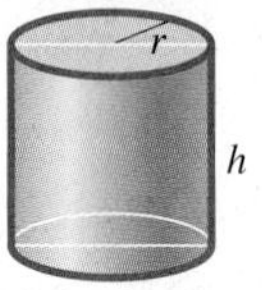 Right circular cylinder	Surface area $A = 2\pi r^2 + 2\pi rh$ $= 2\pi r(r + h)$ Volume $V = \pi r^2 h$
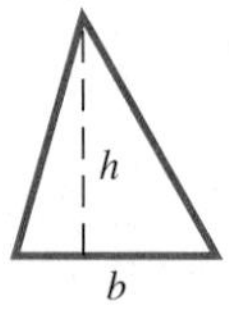 Triangle	Area $A = \frac{1}{2}bh$		

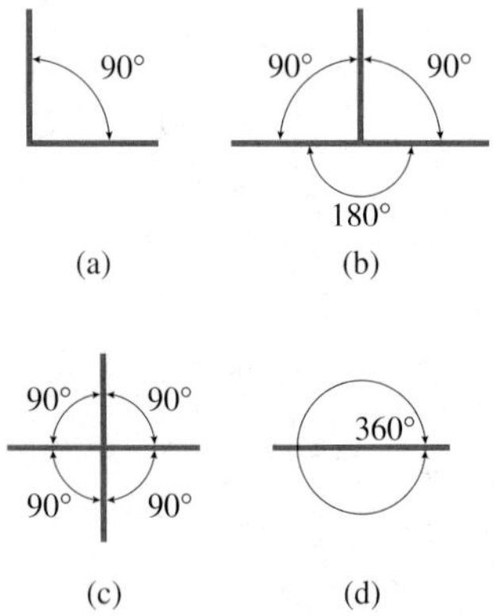

Figure A.6 (a) A right angle; (b) two adjacent right angles, or a straight line; (c) four adjacent right angles; (d) a full circle.

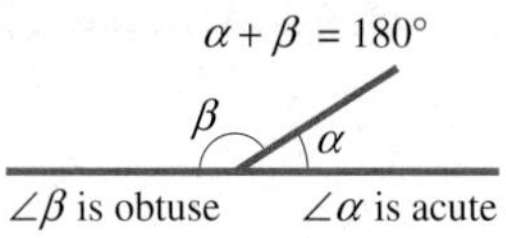

Figure A.7 Acute and obtuse angles.

A.6 GEOMETRY

Geometric Shapes

Table A.1 lists the geometric shapes that commonly appear in physics problems. It is often necessary to determine the area or volume of one of these simple shapes to complete the solution of a problem. The formulae for the properties associated with each geometric form are listed in the column to the right.

Angular Measure

An angle between two straight lines that meet at a single point is specified in degrees. If the two lines are perpendicular to each other, as shown in Fig. A.6a, the angular separation is said to be a right angle or 90°. In Fig. A.6b two such 90° angles are placed side by side; they add to 180°, so a straight line represents an angular separation of 180°. When four right angles are grouped as shown in Fig. A.6c, the angles add to 360° and a full circle contains 360° as shown in Fig. A.6d. An angle that is less than 90° is called an **acute** angle; one greater than 90° is called an **obtuse** angle.

When two lines meet, as shown in Fig. A.7, there are two possible angles that might be specified; one is the acute angle α and the other is the obtuse angle β. The symbol used to indicate an angle is $\angle$. When two angles placed adjacent to each other form a straight line, they are called supplementary angles; angles α and β in Fig. A.7 are supplementary angles. When two angles placed adjacent to each other form a right angle, they are called complementary angles.

Various triangles are shown in Fig. A.8. The sum of the interior angles of any triangle is 180°. An isosceles triangle has two sides of equal length; the angles opposite to the equal sides are equal angles. An equilateral triangle has all three sides of equal length; it is also equiangular. Right triangles have one right angle, 90°, and the sum of the other two angles is 90°, so those angles are acute angles. Commonly used right triangles have sides in the ratio of 3:4:5 and 5:12:13.

Triangles are similar when all three angles of one are equal to the three angles of the other. If two angles of one triangle are equal to two angles of the other, the third angles are necessarily equal and the triangles are similar. The ratio of corresponding sides of similar triangles are equal, as shown in Fig. A.9. Similar triangles of the same size are called congruent triangles.

Figure A.10 shows other useful relations among angles between intersecting lines. When two angles add to 180°, as $\angle\alpha + \angle\beta$ in one of the small figures, the angles are supplementary. Another small figure shows two angles, $\angle\alpha + \angle\beta$ adding to 90°, so in that case the angles are complementary.

For many physics problems it is convenient to use angles measured in radians rather than in degrees; the abbreviation for radians is rad. The arc length s measured along a circle is proportional to the angle between the two radii that define the arc, as shown in Fig. A.11. One radian is defined as the angle subtended when the arc length is equal to the length of the radius.

For θ measured in radians,

$$s = r\theta$$

When the angular displacement is all the way around the circle, 360°, the arc length is equal to the circumference of the circle:

$$s = 2\pi r = r\theta$$

The equivalent to 360° measured in radians is thus $\theta = 2\pi$ radians and the equivalence between radians and degrees is

$$1 \text{ rad} = \frac{360°}{2\pi} \approx 57.3° \qquad \text{or} \qquad 1° = \frac{2\pi}{360°} \approx 0.01745 \text{ rad}$$

Note that the radian has no physical dimensions; it is a ratio of two lengths so it is a pure number. We use the term *rad* to remind us of the angular units being used.

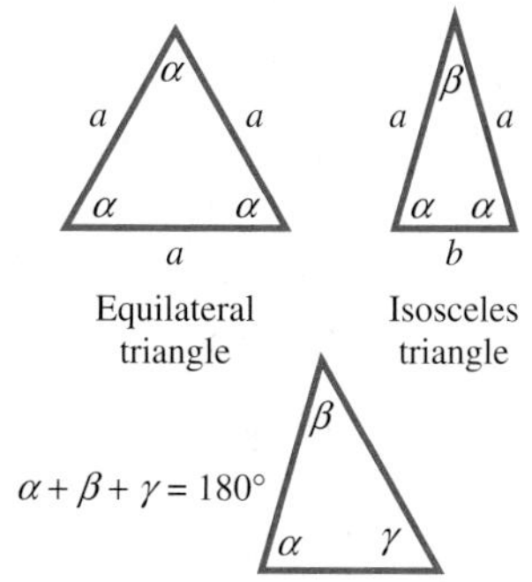

Figure A.8 Triangles.

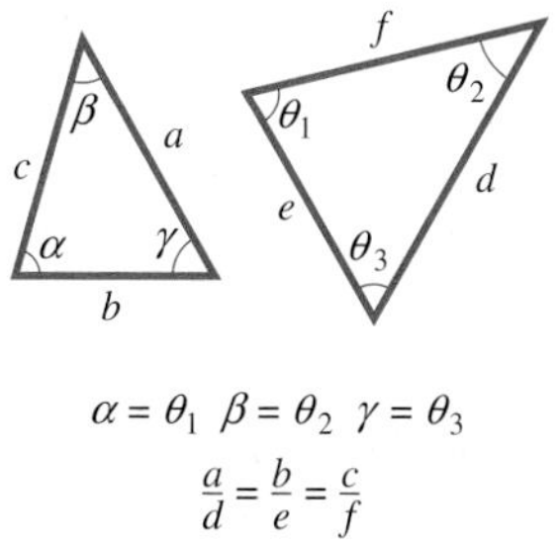

Figure A.9 Similar triangles.

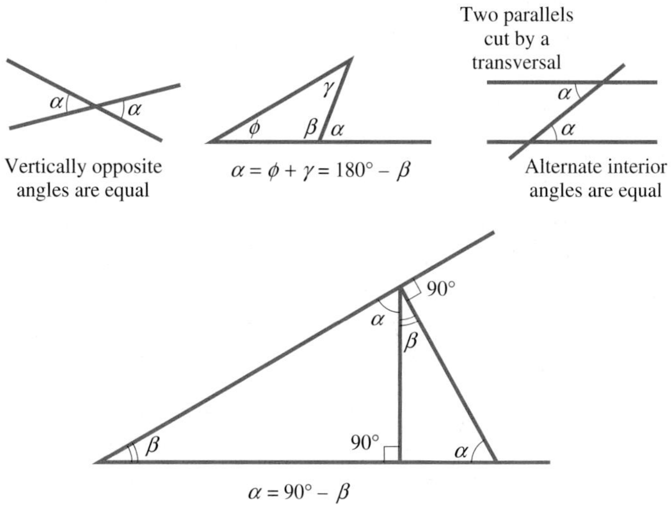

Figure A.10 Angles formed by intersecting lines.

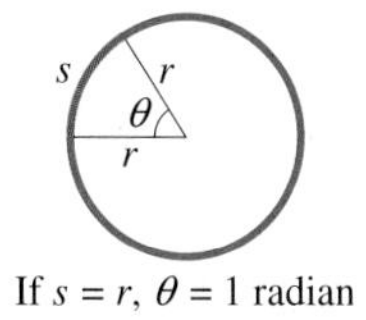

Figure A.11 Radian measure.

A.7 TRIGONOMETRY

The basic trigonometric functions used in physics are shown in Fig. A.12. Note that in determining the function values, the units of length cancel, so the sine, cosine, and tangent functions are dimensionless.

The side opposite and the side adjacent to either of the acute angles in the right triangle are of lesser length than the hypotenuse, according to the Pythagorean theorem. Therefore, the absolute values of the sine and cosine cannot exceed 1. The absolute value of the tangent can exceed 1.

Figure A.13 shows the signs (positive or negative) associated with the trigonometric functions for an angle θ located in each of the four quadrants. The hypotenuse r is positive, so the sign for the sine or cosine is determined by the signs of x or y as measured along the positive or negative x- and y-axes. The sign of the tangent then depends on the signs of the sine and cosine. The angle θ is measured in a counterclockwise direction starting from the positive x-axis, which represents 0°. Angles measured from the x-axis going in a clockwise direction (below the x-axis) are negative angles; an angle of −60°, which is located in the fourth quadrant, is the same as an angle of +300°. Figure A.14 shows graphs of $y = \sin\theta$ and $y = \cos\theta$ as functions of θ in radians. Also graphed

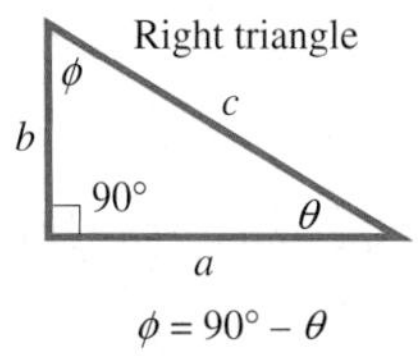

$$\sin\theta = \frac{\text{side opposite } \angle\theta}{\text{hypotenuse}} = \frac{b}{c} = \cos\phi$$

$$\cos\theta = \frac{\text{side adjacent } \angle\theta}{\text{hypotenuse}} = \frac{a}{c} = \sin\phi$$

$$\tan\theta = \frac{\text{side opposite } \angle\theta}{\text{side adjacent } \angle\theta} = \frac{b}{a} = \frac{\sin\theta}{\cos\theta} = \frac{1}{\tan\phi}$$

Figure A.12 Trigonometric functions used in physics problems; angles θ and ϕ are complementary angles.

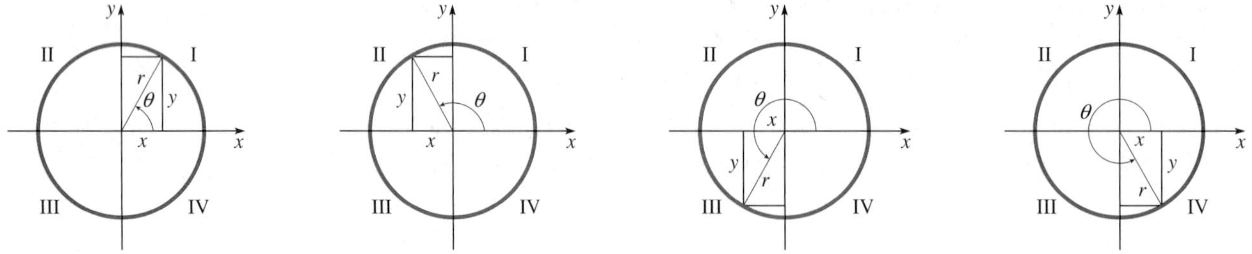

Quadrant I: $0 < \theta < 90°$
$\sin\theta = y/r$ is positive
$\cos\theta = x/r$ is positive
$\tan\theta = y/x$ is positive

Quadrant II: $90° < \theta < 180°$
$\sin\theta = y/r$ is positive
$\cos\theta = x/r$ is negative
$\tan\theta = y/x$ is negative

Quadrant III: $180° < \theta < 270°$
$\sin\theta = y/r$ is negative
$\cos\theta = x/r$ is negative
$\tan\theta = y/x$ is positive

Quadrant IV: $270° < \theta < 360°$
$\sin\theta = y/r$ is negative
$\cos\theta = x/r$ is positive
$\tan\theta = y/x$ is negative

Figure A.13 Signs of trigonometric functions in various quadrants.

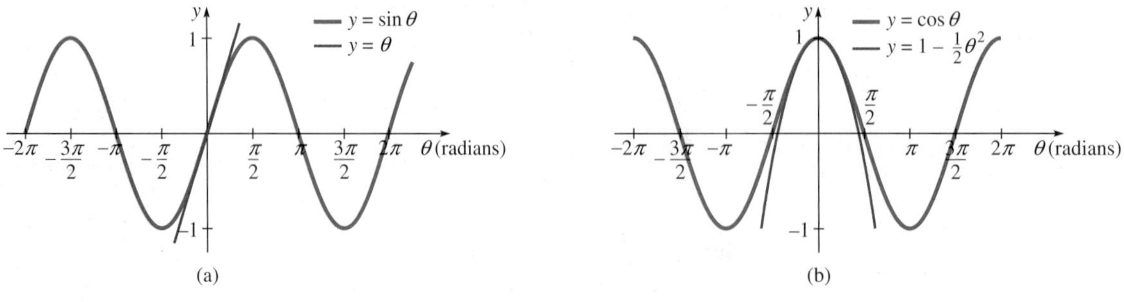

Figure A.14 (a) Graphs of $y = \sin\theta$ and $y = \theta$. Note that $\sin\theta \approx \theta$ for small θ. (b) Graphs of $y = \cos\theta$ and $y = 1 - \frac{1}{2}\theta^2$. Note that $\cos\theta \approx 1 - \frac{1}{2}\theta^2$ for small θ.

Table A.2 Useful Trigonometric Identities

$$\sin^2\theta + \cos^2\theta = 1$$

$$\sin(-\theta) = -\sin\theta$$

$$\cos(-\theta) = \cos\theta$$

$$\tan(-\theta) = -\tan\theta$$

$$\sin(180° \pm \theta) = \mp\sin\theta$$

$$\cos(180° \pm \theta) = -\cos\theta$$

$$\tan(180° \pm \theta) = \pm\tan\theta$$

$$\sin(90° \pm \beta) = \cos\beta$$

$$\cos(90° \pm \beta) = \mp\sin\beta$$

$$\sin 2\theta = 2\sin\theta\cos\theta$$

$$\cos 2\theta = \cos^2\theta - \sin^2\theta = 2\cos^2\theta - 1 = 1 - 2\sin^2\theta$$

$$\tan 2\theta = \frac{2\tan\theta}{1 - \tan^2\theta}$$

$$\sin(\alpha \pm \beta) = \sin\alpha\cos\beta \pm \cos\alpha\sin\beta$$

$$\cos(\alpha \pm \beta) = \cos\alpha\cos\beta \mp \sin\alpha\sin\beta$$

$$\tan(\alpha \pm \beta) = \frac{\tan\alpha \pm \tan\beta}{1 \mp \tan\alpha\tan\beta}$$

$$\sin\alpha + \sin\beta = 2\sin\left[\tfrac{1}{2}(\alpha + \beta)\right]\cos\left[\tfrac{1}{2}(\alpha - \beta)\right]$$

$$\sin\alpha - \sin\beta = 2\cos\left[\tfrac{1}{2}(\alpha + \beta)\right]\sin\left[\tfrac{1}{2}(\alpha - \beta)\right]$$

$$\cos\alpha + \cos\beta = 2\cos\left[\tfrac{1}{2}(\alpha + \beta)\right]\cos\left[\tfrac{1}{2}(\alpha - \beta)\right]$$

$$\cos\alpha - \cos\beta = -2\sin\left[\tfrac{1}{2}(\alpha + \beta)\right]\sin\left[\tfrac{1}{2}(\alpha - \beta)\right]$$

Table A.3 Inverse Trigonometric Functions

Function	Principal Value Range	(Quadrants)	To Find Value in a Different Quadrant
$\sin^{-1}$	$-\frac{\pi}{2}$ to $\frac{\pi}{2}$	(I and IV)	Subtract principal value from π
$\cos^{-1}$	0 to π	(I and II)	Subtract principal value from 2π
$\tan^{-1}$	$-\frac{\pi}{2}$ to $\frac{\pi}{2}$	(I and IV)	Add principal value to π

are two functions that are useful approximations for the sine and cosine functions when $|\theta|$ is sufficiently small (see the section titled Small-Angle Approximations, p. A-12).

Table A.2 lists some of the most useful trigonometric identities.

Inverse Trigonometric Functions The inverse trigonometric functions can be written in either of two ways. To use the inverse cosine as an example: $\cos^{-1} x$ or arccos x. Both of these expressions mean *an angle whose cosine is equal to x*. A calculator returns only the *principal value* of an inverse trigonometric function (Table A.3), which may or may not be the correct solution in a given problem.

Law of Sines and Law of Cosines These two laws apply to any triangle labeled as shown in Fig. A.15:

Law of sines: $$\frac{\sin\alpha}{a} = \frac{\sin\beta}{b} = \frac{\sin\gamma}{c}$$

Law of cosines: $c^2 = a^2 + b^2 - 2ab\cos\gamma$ (where γ is the interior angle formed by the intersection of sides a and b)

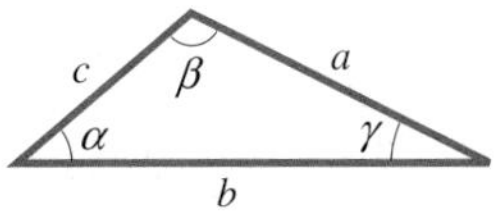

Figure A.15 A general triangle.

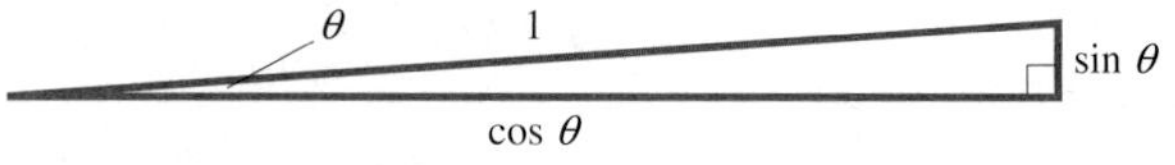

Figure A.16 Illustration of the small angle approximations $\sin\theta \approx \theta$ and $\cos\theta \approx 1 - \frac{1}{2}\theta^2$ (for θ in radians) using a right triangle with $\theta \ll 1$ rad.

Small-Angle Approximations

These approximations are written for θ *in radians* and are valid when $\theta \ll 1$ rad.

$$\sin\theta \approx \theta \tag{A-26}$$

$$\cos\theta \approx 1 - \tfrac{1}{2}\theta^2 \tag{A-27}$$

$$\tan\theta \approx \theta \tag{A-28}$$

The sizes of the errors involved in using these approximations are roughly $\frac{1}{6}\theta^3$, $\frac{1}{24}\theta^4$, and $\frac{2}{3}\theta^3$, respectively. In *some* circumstances it may be all right to ignore the $\frac{1}{2}\theta^2$ term and write

$$\cos\theta \approx 1 \tag{A-29}$$

The origin of these approximations can be illustrated using a right triangle of hypotenuse 1 with one very small angle θ (Fig. A.16). If θ is very small, then the adjacent side ($\cos\theta$) will be nearly the same length as the hypotenuse (1). Then we can think of those two sides as radii of a circle that subtend an angle θ. The relationship between the arc length s and the angle subtended is

$$s = \theta r$$

Since $\sin\theta \approx s$ and $r = 1$, we have $\sin\theta \approx \theta$. To find an approximate form for $\cos\theta$ (but one more accurate than $\cos\theta \approx 1$), we can use the Pythagorean theorem:

$$\sin^2\theta + \cos^2\theta = 1$$

$$\cos\theta = \sqrt{1 - \sin^2\theta} \approx \sqrt{1 - \theta^2}$$

Now, using a binomial approximation,

$$\cos\theta \approx (1 - \theta^2)^{1/2} \approx 1 - \tfrac{1}{2}\theta^2$$

A.8 VECTORS

The distinction between vectors and scalars is discussed in Section 3.1. Scalars have magnitude while vectors have magnitude and direction. A vector is represented graphically by an arrow of length proportional to the magnitude of the vector and aligned in a direction that corresponds to the vector direction.

In print, the symbol for a vector quantity is sometimes written in bold font, or in roman font with an arrow over it, or in bold font with an arrow over it (as done in this book). When writing by hand, a vector is designated by drawing an arrow over the symbol: $\vec{A}$. When we write just plain A, that stands for the *magnitude* of the vector. We also use absolute value bars to stand for the magnitude of a vector, so $A = |\vec{\mathbf{A}}|$.

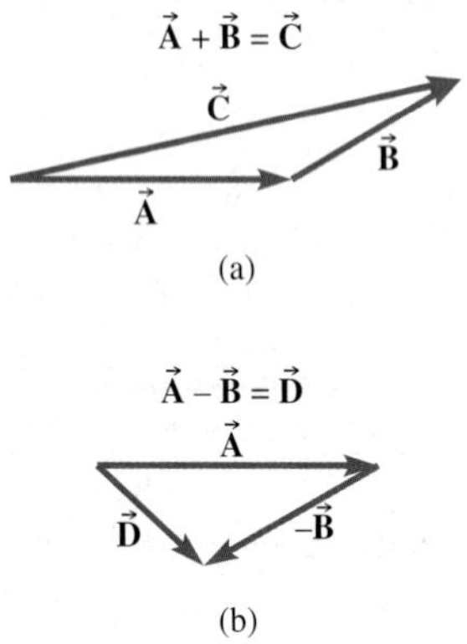

Figure A.17 Graphical (a) addition and (b) subtraction of two vectors.

Addition and Subtraction of Vectors

When vectors are added or subtracted, the magnitudes and directions must be taken into account. Details on vector addition and subtraction are found in Sections 3.1 and 3.2. Here we provide a brief summary.

The graphical method for adding vectors involves placing the vectors tip to tail and then drawing from the tail of the first to the tip of the second, as shown in Fig. A.17.

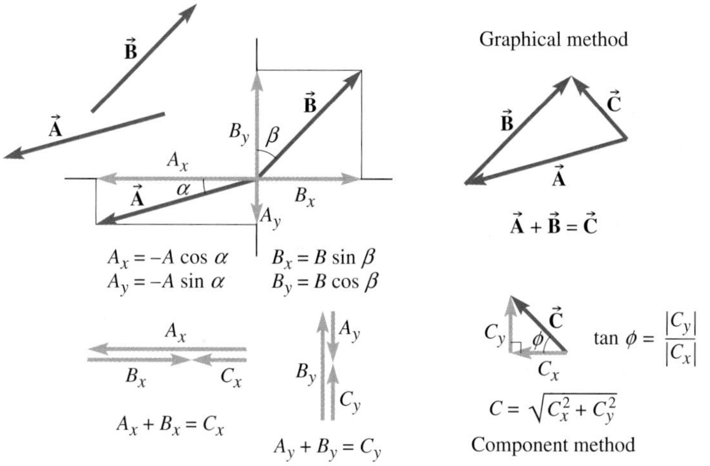

Figure A.18 Adding two arbitrary vectors by two different methods.

Figure A.19 Multiplication of a vector by a scalar.

To subtract a vector, add its opposite. In Fig. A.17, $-\vec{\mathbf{B}}$ has the same magnitude as $\vec{\mathbf{B}}$ but is opposite in direction. Then $\vec{\mathbf{A}} + \vec{\mathbf{B}} = \vec{\mathbf{A}} - (-\vec{\mathbf{B}})$.

Figure A.18 shows both the graphical and component methods of vector addition.

Product of a Vector and a Scalar

When a vector is multiplied by a scalar, the magnitude of the vector is multiplied by the absolute value of the scalar, as shown in Fig. A.19. The direction of the vector does not change unless the scalar factor is negative, in which case the direction is reversed.

Scalar Product of Two Vectors

One type of product of two vectors is the *scalar product* (also called the *dot product*). The notation for it is

$$C = \vec{\mathbf{A}} \cdot \vec{\mathbf{B}}$$

As its name implies, the scalar product of two vectors is a scalar quantity; it can be positive, negative, or zero but has no direction.

The scalar product depends on the magnitudes of the two vectors and on the angle θ between them. To find the angle, draw the two vectors starting *at the same point* (Fig. A.20). Then the scalar product is defined by

$$\vec{\mathbf{A}} \cdot \vec{\mathbf{B}} = AB \cos \theta$$

Reversing the order of the two vectors does not change the scalar product: $\vec{\mathbf{B}} \cdot \vec{\mathbf{A}} = \vec{\mathbf{A}} \cdot \vec{\mathbf{B}}$. The scalar product can be written in terms of the components of the two vectors:

$$\vec{\mathbf{A}} \cdot \vec{\mathbf{B}} = A_x B_x + A_y B_y + A_z B_z$$

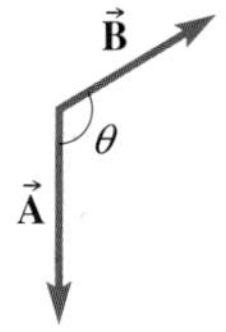

Figure A.20 Two vectors are drawn starting at the same point. The angle θ between the vectors is used to find the scalar product and the cross product of the vectors.

Cross Product of Two Vectors

Another type of product of two vectors is the *cross product* (also called the *vector product*), which is introduced in Chapter 19. It is denoted by

$$\vec{\mathbf{A}} \times \vec{\mathbf{B}} = \vec{\mathbf{C}}$$

The cross product is a *vector* quantity; it has magnitude and direction. $\vec{\mathbf{A}} \times \vec{\mathbf{B}}$ is read as "$\vec{\mathbf{A}}$ cross $\vec{\mathbf{B}}$."

For two vectors, $\vec{\mathbf{A}}$ and $\vec{\mathbf{B}}$, separated by an angle θ (with θ chosen to be the *smaller* angle between the two as in Fig. A.20), the magnitude of the cross product $\vec{\mathbf{C}}$ is

$$|\vec{\mathbf{C}}| = |\vec{\mathbf{A}} \times \vec{\mathbf{B}}| = AB \sin \theta$$

The direction of the cross product $\vec{\mathbf{C}}$ is one of the two directions perpendicular to both $\vec{\mathbf{A}}$ and $\vec{\mathbf{B}}$. To choose the correct direction, use the right-hand rule explained in Section 19.2.

The cross product depends on the order of the multiplication.

$$\vec{\mathbf{A}} \times \vec{\mathbf{B}} = -\vec{\mathbf{B}} \times \vec{\mathbf{A}}$$

The magnitude is $AB \sin \theta$ in both cases, but the direction of one cross product is opposite to the direction of the other.

A.9 SELECTED MATHEMATICAL SYMBOLS

$\times$ or $\cdot$	multiplication
Δ	change in, small increment, or uncertainty in
$\approx$	is approximately equal to
$\neq$	is not equal to
$\leq$	is less than or equal to
$\geq$	is greater than or equal to
$\ll$	is much less than
$\gg$	is much greater than
$\propto$	is proportional to
$\lvert Q \rvert$	absolute value of Q
$\lvert\vec{\mathbf{a}}\rvert$	magnitude of vector $\vec{\mathbf{a}}$
$\perp$	perpendicular
$\parallel$	parallel
∞	infinity
$'$	prime (used to distinguish different values of the same variable)
Q_{av}, $\overline{Q}$, or $\langle Q \rangle$	average of Q
Σ	sum
Π	product
$\log_b x$	the logarithm (base b) of x
$\ln x$	the natural (base e) logarithm of x
$\pm$	plus or minus
$\mp$	minus or plus
$\ldots$	ellipsis (indicates continuation of a series or list)
$\angle$	angle
$\Rightarrow$	implies
$\therefore$	therefore

Table of Selected Nuclides

Atomic Number Z	Element	Symbol	Mass Number *A*	Mass of *neutral atom* (u)	Percentage Abundance (or Decay Mode)	Half-life (if Unstable)
0	(Neutron)	n	1	1.008 664 9	β^-	10.24 min
1	Hydrogen	H	1	1.007 825 0	99.985	
	(Deuterium)	(D)	2	2.014 101 8	0.015	
	(Tritium)	(T)	3	3.016 049 3	β^-	12.32 yr
2	Helium	He	3	3.016 029 3	0.000 137	
			4	4.002 603 2	99.999 863	
3	Lithium	Li	6	6.015 122 3	7.6	
			7	7.016 004 0	92.4	
4	Beryllium	Be	7	7.016 929 2	EC	53.22 d
			8	8.005 305 1	2α	6.8×10^{-17}s
			9	9.012 182 1	100	
5	Boron	B	10	10.012 937 0	19.8	
			11	11.009 305 5	80.2	
6	Carbon	C	11	11.011 433 8	EC	20.334 min
			12	12.000 000 0	98.89	
			13	13.003 354 8	1.11	
			14	14.003 242 0	β^-	5730 yr
			15	15.010 599 3	β^-	2.449 s
7	Nitrogen	N	12	12.018 613 2	EC	11.00 ms
			13	13.005 738 6	EC	9.965 min
			14	14.003 074 0	99.634	
			15	15.000 108 9	0.366	
8	Oxygen	O	15	15.003 065 4	EC	122.24 s
			16	15.994 914 6	99.762	
			17	16.999 131 5	0.038	
			18	17.999 160 4	0.200	
			19	19.003 579 3	β^-	26.88 s
9	Fluorine	F	19	18.998 403 2	100	
10	Neon	Ne	20	19.992 440 2	90.48	
			22	21.991 385 5	9.25	
11	Sodium	Na	22	21.994 436 8	EC	2.6019 yr
			23	22.989 769 7	100	
			24	23.990 963 3	β^-	14.9590 h
12	Magnesium	Mg	24	23.985 041 9	78.99	
13	Aluminum	Al	27	26.981 538 4	100	
14	Silicon	Si	28	27.976 926 5	92.230	
15	Phosphorus	P	31	30.973 761 5	100	
			32	31.973 907 2	β^-	14.262 d
16	Sulfur	S	32	31.972 070 7	95.02	
17	Chlorine	Cl	35	34.968 852 7	75.77	
18	Argon	Ar	40	39.962 383 1	99.6003	
19	Potassium	K	39	38.963 706 9	93.2581	
			40	39.963 998 7	0.0117; β^-	1.248×10^9 yr
20	Calcium	Ca	40	39.962 591 2	96.94	
24	Chromium	Cr	52	51.940 511 9	83.789	
25	Manganese	Mn	54	53.940 363 2	EC	312.0 d
			55	54.938 049 6	100	
26	Iron	Fe	56	55.934 942 1	91.754	

(continued)

Atomic Number Z	Element	Symbol	Mass Number *A*	Mass of *neutral atom* (u)	Percentage Abundance (or Decay Mode)	Half-life (if Unstable)
27	Cobalt	Co	59	58.933 200 2	100	
			60	59.933 822 2	β^-	5.271 yr
28	Nickel	Ni	58	57.935 347 9	68.077	
			60	59.930 790 6	26.223	
29	Copper	Cu	63	62.929 601 1	69.17	
30	Zinc	Zn	64	63.929 146 6	48.63	
36	Krypton	Kr	84	83.911 507 3	57.0	
			86	85.910 610 3	17.3	
			92	91.926 153 1	β^-	1.840 s
37	Rubidium	Rb	85	84.911 789 3	72.17	
			93	92.922 033 8	β^-	5.84 s
38	Strontium	Sr	88	87.905 614 3	82.58	
			90	89.907 737 6	β^-	28.90 yr
39	Yttrium	Y	89	88.905 847 9	100	
			90	89.907 151 4	β^-	64.00 h
47	Silver	Ag	107	106.905 093 6	51.839	
50	Tin	Sn	120	119.902 196 6	32.58	
53	Iodine	I	131	130.906 124 2	β^-	8.0252 d
55	Cesium	Cs	133	132.905 447 3	100	
			141	140.920 044 1	β^-	24.84 s
56	Barium	Ba	138	137.905 241 3	71.698	
			141	140.914 406 9	β^-	18.27 min
60	Neodymium	Nd	143	142.909 810 3	12.2	
62	Samarium	Sm	147	146.914 893 3	14.99; α	1.06×10^{11} yr
79	Gold	Au	197	196.966 552 3	100	
82	Lead	Pb	204	203.973 029 3	1.4	$\geq 1.4 \times 10^{17}$ yr
			206	205.974 449 3	24.1	
			207	206.975 881 3	22.1	
			208	207.976 636 3	52.4	
			210	209.984 173 3	β^-	22.20 yr
			211	210.988 731 3	β^-	36.1 min
			212	211.991 887 5	β^-	10.64 h
			214	213.999 798 1	β^-	26.8 min
83	Bismuth	Bi	209	208.980 383 3	100	
			211	210.987 258 6	α	2.14 min
			214	213.998 699 1	β^-	19.9 min
84	Polonium	Po	210	209.982 857 3	α	138.376 d
			214	213.995 186 3	α	164.3 μs
			218	218.008 965 8	α	3.10 min
86	Radon	Rn	222	222.017 570 5	α	3.8235 d
88	Radium	Ra	226	226.025 402 6	α	1600 yr
			228	228.031 064 1	β^-	5.75 yr
90	Thorium	Th	228	228.028 731 3	α	1.91 yr
			232	232.038 050 4	100; α	1.405×10^{10} yr
			234	234.043 595 4	β^-	24.10 d
92	Uranium	U	235	235.043 923 1	0.7204; α	7.038×10^{8} yr
			236	236.045 561 9	α	2.342×10^{7} yr
			238	238.050 782 6	99.2742; α	4.468×10^{9} yr
			239	239.054 287 8	β^-	23.45 min
93	Neptunium	Np	237	237.048 167 3	α	2.144×10^{6} yr
94	Plutonium	Pu	239	239.052 156 5	α	24 110 yr
			242	242.058 736 8	α	3.75×10^{5} yr
			244	244.064 198 5	α	8.00×10^{7} yr

EC = electron capture.

Answers to Selected Questions and Problems

CHAPTER 1

Multiple-Choice Questions

1. (b) **2.** (a) **3.** (b) **4.** (c) **5.** (d) **6.** (b) **7.** (d) **8.** (b) **9.** (d) **10.** (c)

Problems

1. 2.5 m **3.** 7.7% **5.** $6/s$ **7.** 10^{-8} **9.** 11.8 yr **11.** 36.0% **13.** (a) 1.29×10^8 kg (b) 1.3×10^8 m/s **15.** (a) 3.63×10^7 g (b) 1.273×10^2 m **17.** 1.7×10^{-10} m^3 **19.** 459 m/s **21.** 2.8×10^{-7} inches **23.** (a) 4.863×10^2 m; 10^2 (b) 1.834×10^3 m; 10^3 **25.** (a) 8.6 m/s (b) 19 mi/h **27.** 0.12 or 12% **29.** 13.6 g/cm^3 **31.** 1.7×10^{-10} km^3 **33.** (a) 2.7×10^{-3} ft/s (b) 1.9×10^{-3} mi/h **35.** kg·m^2·s^{-2}

37. $[T]^2 = \dfrac{[L]^3}{\dfrac{[L]^3}{[M][T]^2} \times [M]} = \dfrac{[L]^3}{[M]} \times \dfrac{[M][T]^2}{[L]^3} = [T]^2$ **39.** (a) $[L^3]$ (b) volume **41.** 30–40 cm **43.** (a) 10 kg (b) 10 m **45.** Answers may vary. **47.** 100 m

49.

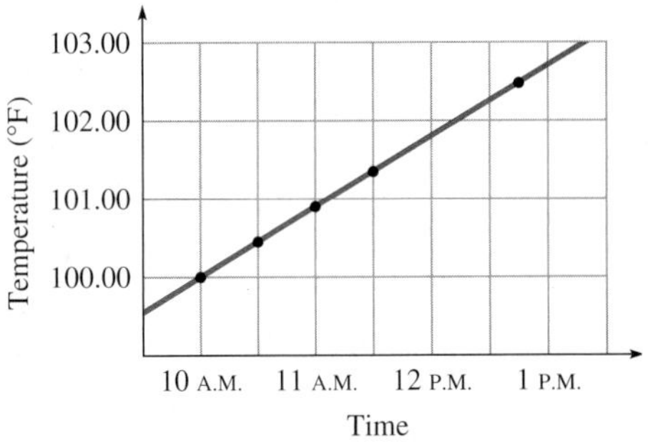

(a) 101.8°F (b) 0.9°F/h (c) No; the patient would die before 12 hours passed and the temperature reached 113°F.

51. 104.5°F **53.** (a) a (b) $+v_0$

55. (a)

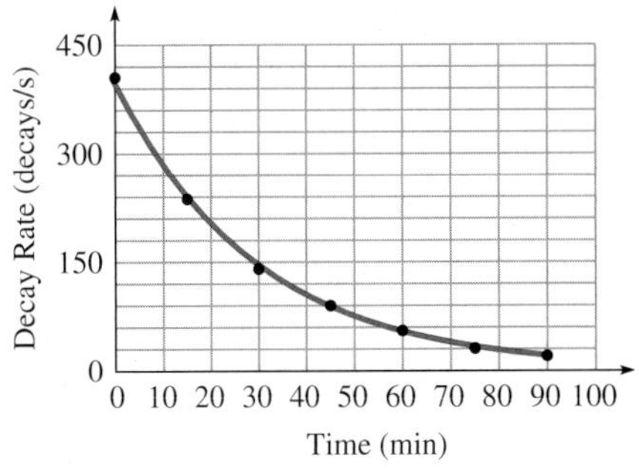

(b)

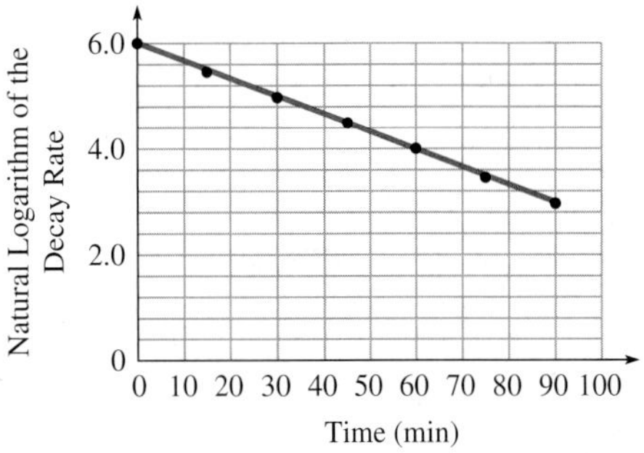

The presentation is useful because the graph is linear.

57. (a) 186.303 (b) 186.297 (c) 0.56 (d) 62 000 (e) Case (a): 0.0016%; Case (b): 0.0016%; For case (c), ignoring 0.0030 causes you to multiply by zero and get a zero result. For case (d), ignoring 0.0030 causes you to divide by zero. (f) You can neglect small values when they are added to or subtracted from sufficiently large values. The term "sufficiently large" is determined by the number of significant figures required.

59. 4.0 **61.** 434 m/s **63.** (a) 3; 5.74×10^{-3} kg (b) 1; 2 m (c) 3; 4.50×10^{-3} m (d) 3; 4.50×10^1 kg (e) 4; 1.009×10^5 s (f) 4; 9.500×10^3 mL **65.** (a) 6 Mm (b) 2 m (c) 1 μm (d) 3 nm (e) 0.3 nm **67.** (a) 3.3×10^{-8} m (b) 3.3×10^{-2} μm (c) 1.3×10^{-6} in **69.** 2.2×10^2 m^3 **71.** (a) $a = K\frac{v^2}{r}$, where K is a dimensionless constant. (b) 21.0% **73.** 2.24 mi/h = 1 m/s; for a quick, approximate conversion, multiply by 2.

75. 10^{11} gal **77.** $\dfrac{\text{kg} \cdot \text{m}}{\text{s}^2}$ **79.** \$59 000 000 000

81. (a) 2.4×10^5 km/h (b) 10 min **83.** (a) $\sqrt{\dfrac{hG}{c^5}}$ (b) 1.3×10^{-43} s **85.** 0.46 s^{-1}

87. (a)

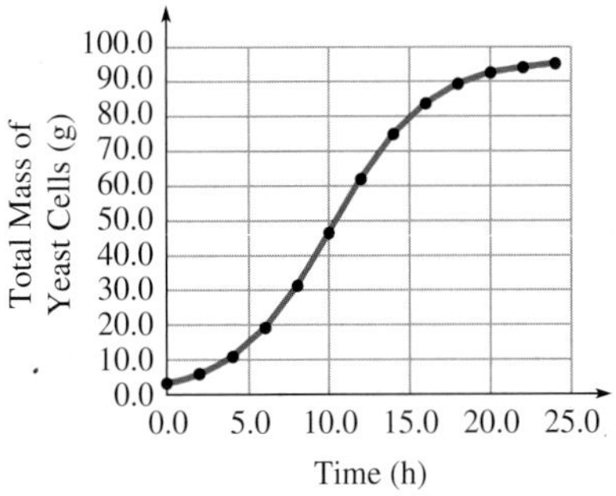

(b) about 100 g

(c)

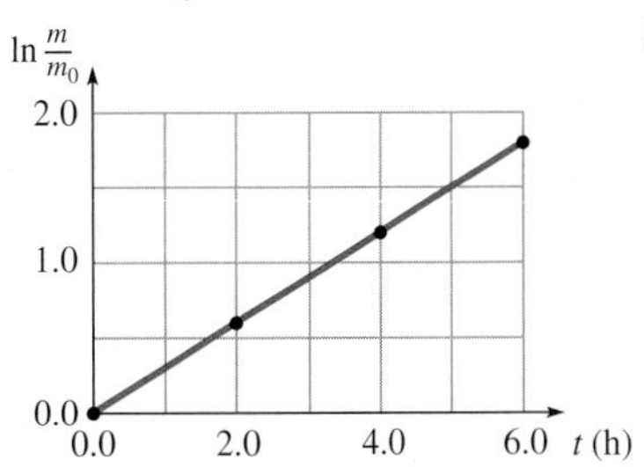

; 0.30 s^{-1}

CHAPTER 2

Multiple-Choice Questions

1. (c) **2.** (d) **3.** (a) **4.** (b) **5.** (c) **6.** (a) **7.** (b) **8.** (a) **9.** (a) **10.** (c) **11.** (a) **12.** (a) **13.** (d) **14.** (c) **15.** (d) **16.** (a) **17.** **(a)** (b) **(b)** (a) **(c)** (d) **(d)** (c)

Problems

1. 16 cm, east **3.** (a) −80 m, or 80 m west (b) −20 m, or 20 m west (c) +80 m, or 80 m east (d) 240 m **5.** (a) 8 km, north of its position at 3 P.M. (b) 116 km, south of its starting point (c) 104 km, north of its position at 4 P.M. **7.** 14.3 m/s, east **9.** 53.1 mi/h due west **11.** 160 m **13.** (a) DE (b) 4 s and 5 s (c) 20 m **15.** 27 m/s west **17.** 16.5 m **19.** 1.0 m/s **21.** (a) 170 cm to the left (b) 28 cm/s (c) 9.4 cm/s to the left **23.** 1.05 m/s to the north **25.** 7.0 m/s^2 in the direction opposite the car's velocity **27.** 28 m/s^2 toward the paddle **29.** 2.5 m/s^2 **31.** (a) −10 m/s^2 (b) 0 (c) 5.0 m

33. (a)

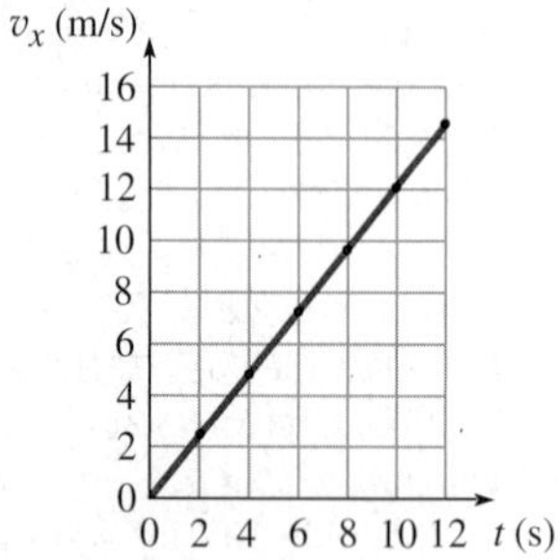

(b) 86.4 m (c) 14.4 m/s

(d)

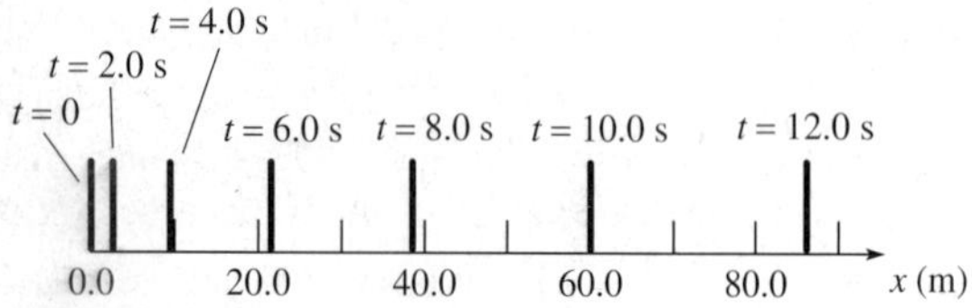

35. (a)

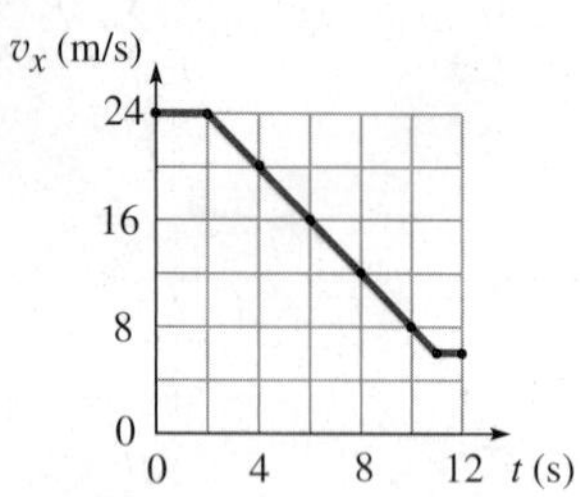

(b) 2.00 m/s^2 north (c) 135 m

37. (a) 4.0 m/s (b) 5.0 m/s **39.** No; it takes 236 m for the train to stop. **41.** 80 m

43. (a)

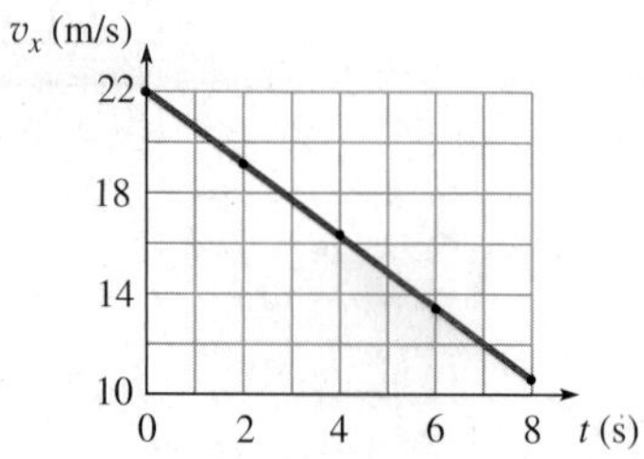

(b) 11 m/s (c) 130 m

(d)

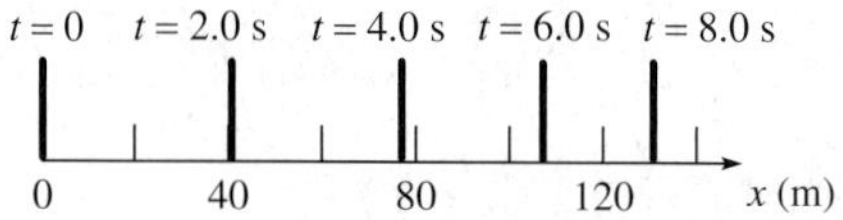

45. 85.0 m/s down **47.** 5.0 m/s **49.** 1.22 s **51.** (a) 44 m (b) 7.0 m/s (c) 29 m/s (d) 17.1 m below the top of the tower **53.** (a) 120 m/s^2 toward Lois (b) 170 m/s **55.** 46 m **57.** (a) 224 m (b) 0.99 m/s^2 **59.** (a) 330 m/s up (b) 16 m/s^2 up **61.** $2v$ **63.** (a) 420 m/s^2 opposite the direction of motion (b) 4200 m/s^2 opposite the direction of motion **65.** 59 mi north; 96 mi/h north; 11 mi/h^2 south **67.** (a) higher

(b)

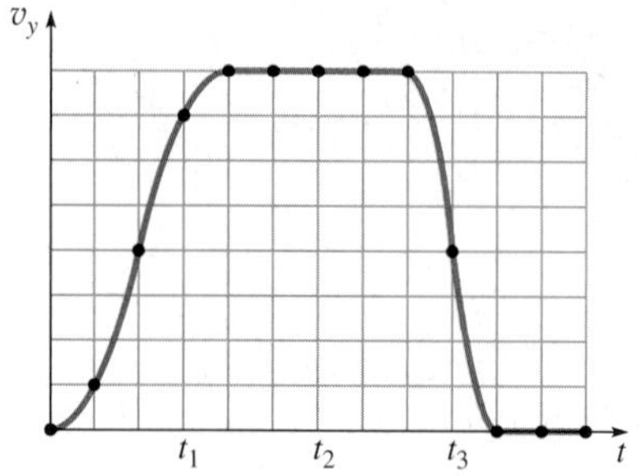

(c)

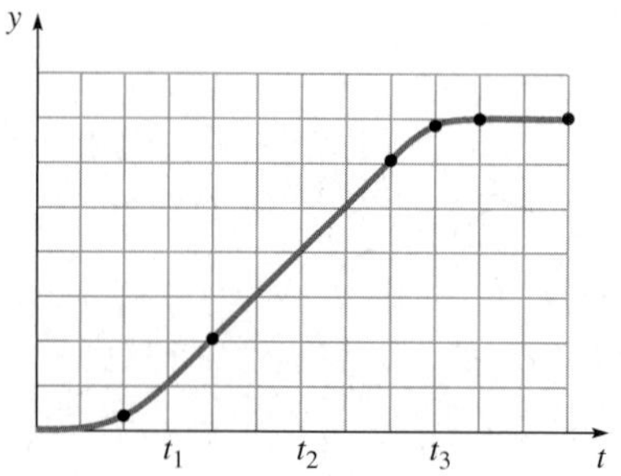

69. (a) 25.0 km (b) 152 s (c) 76.0 km (d) 1220 m/s downward **71.** 3.0 cm/s^2 in the direction of the velocity **73.** (a) 1.0 mm/s (b) 20 ms (c) 100 m/s

CHAPTER 3

Multiple-Choice Questions

1. (d) **2.** (b) **3.** (b) **4.** (d) **5.** (a) **6.** (c) **7.** (e) **8.** (c) **9.** (a) **10.** (d) **11.** (d) **12.** (a) **13.** (b) **14.** (c)

Problems

1. (a) 10.00 km west (b) 4.88 km east (c) 4.88 km west **3.** (a) same direction (b) perpendicular (c) opposite directions; 1.0

5.

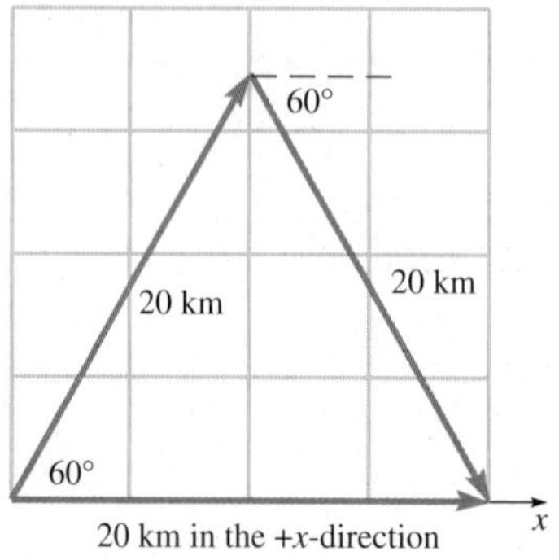

7. (a)

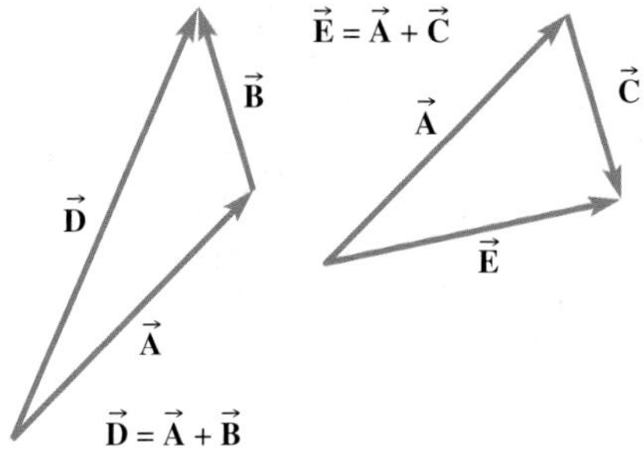

(b)

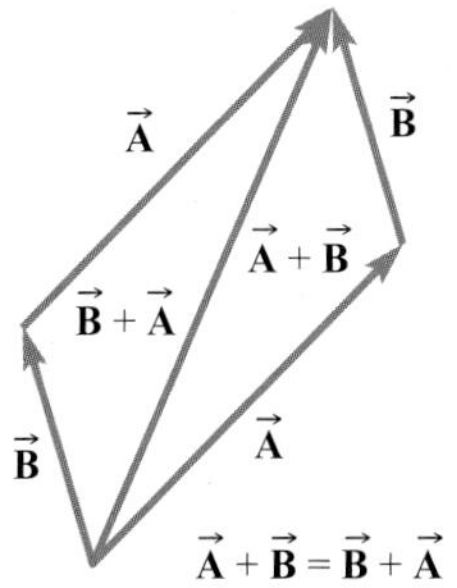

9. (a) 45 km (b) 16 km **13.** x-comp $= -17.3$ m; y-comp $= 10.0$ m **15.** (a) 2.0 units at 30° CCW from the $+y$-axis (b) 2.0 units at 30° CW from the $+y$-axis (c) x-comp $= -1.0$ unit; y-comp $= -\sqrt{3.0}$ units **17.** 1.4 cm **19.** $A_x = 6.6$ m, $A_y = 2.4$ m; $B_x = 6.6$ m/s, $B_y = -2.4$ m/s; $C_x = -2.4$ m, $C_y = 6.6$ m; $D_x = -2.4$ m/s, $D_y = -6.6$ m/s **21.** (a) $\vec{A} = 9.4$ m/s, 32° CCW from the $+y$-axis (b) $\vec{B} = 130$ m, 27° CW from the $+x$-axis (c) $\vec{C} = 16.3$ m/s, 33° CCW from the $-x$-axis (d) $\vec{D} = 2.3$ m/s^2, 1.6° CCW from the $+x$-axis **23.** (a) $B_x = 6.9$, $B_y = -1.7$ (b) 6.9 at 15° CW from the $-y$-axis (c) 9.8 at 31° CCW from the $-y$-axis (d) 10 at 30° CCW from the $-x$-axis (e) x-component: -8.7, y-component: -5.0 **25.** 4.92 mi at 24.0° north of east **27.** 29 nautical miles at 17° south of east **29.** (a) 5.03 m/s (b) 0.996 m/s at 12.4° west of north

31. (a)

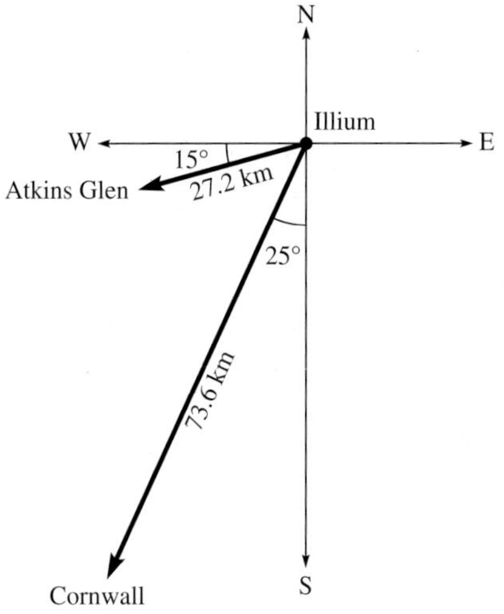

(b) 59.9 km at 85° north of east (c) 80 km/h at 85° north of east **33.** 26 km/h at 31° north of east **35.** (a) 102 km/h (b) 90.8 km/h at 16.6° south of west **37.** (a) 76.2 km (b) 102 km/h at 22.0° north of west (c) 32.6 km/h at 22.0° south of east (d) 0 (e) 63.3 km/h **39.** 13 m/s^2 up **41.** (a) 9.82 m/s (b) 13.9 m/s southeast (c) 8.68 m/s^2 southeast **43.** (a) 180 km/h at 24° south of east (b) 280 km/h^2 at 24° south of east **45.** 0.8 s **47.** It is on the ground after 1.32 s, so the horizontal distance along the ground is 26.3 m. **49.** (a) 5.9 m (b) 17.0 m/s **51.** (a) 202 m (b) 51.1° below the horizontal

53. (a) 37 m (b) 170 m (c) 32 m/s; −27 m/s
55. (a)

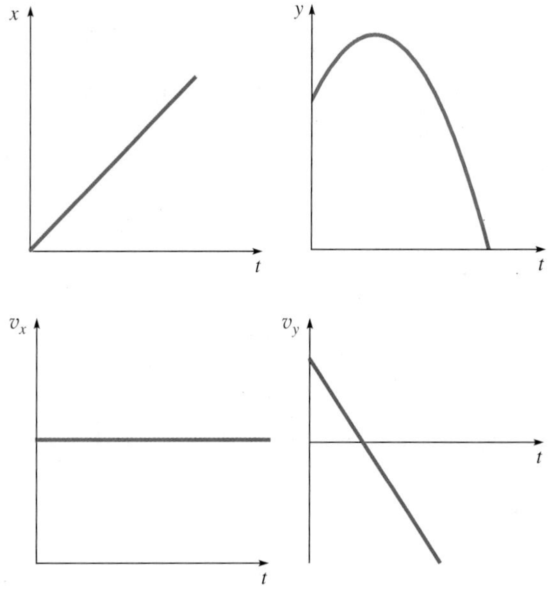

(b) 27.6 m/s at 25.0° above the horizontal (c) 37.5 m (d) 44.4 m above the ground **57.** (a) $\dfrac{2v_i \sin\theta}{g}$ (c) $\dfrac{v_i^2}{g}$ **59.** (a) $v_x = v_i \cos\theta$ and $v_y = 0$ (b) $v_i \sin\theta/g$ **61.** 15.8 m **63.** 11 m/s; 2.5 m **65.** 130 km/h north **67.** 63 km/h at 40° south of west **69.** 50 m/s east **71.** (a) 9.6° north of west (b) 38 m/s **73.** $v_x = 50$ km/h east; $v_y = 40$ km/h south **75.** (a) 30.0° north of west (b) 9.1 min **77.** (a) 1.00 m/s (b) 1.12 m/s **79.** (a) 1.1 m (b) down **81.** (a) 873 km (b) 9.90° south of east (c) 2.250 h (d) 2.18 h **83.** (a) 160 km/h at 20° north of east (b) 150 km/h at 21° north of east (c) 10 km/h west **85.** 68 m/s **87.** 12 m east and 40 m north **89.** (a) 28.6 cm (b) smaller (c) larger (d) $H = 21.3$ cm; $R = 85.1$ cm **91.** (a) 33.1 h (b) 34.1 h (c) 33.6 h **93.** 63° below the horizontal **95.** (a) 15 km/h due west (b) 5.8° west of north **97.** step 4

CHAPTER 4

Multiple-Choice Questions

1. (b) **2.** (b) **3.** (a) **4.** (d) **5.** (b) **6.** (c) **7.** (b) **8.** (c) **9.** (b) **10.** (d) **11.** (e) **12.** (e) **13.** (a) **14.** (b)

Problems

1. the weight of the person **3.** 778 N **5.** 20 N in the positive x-direction **7.** 2 N to the east **9.** 120 N north **11.** 13° from the vertical
13

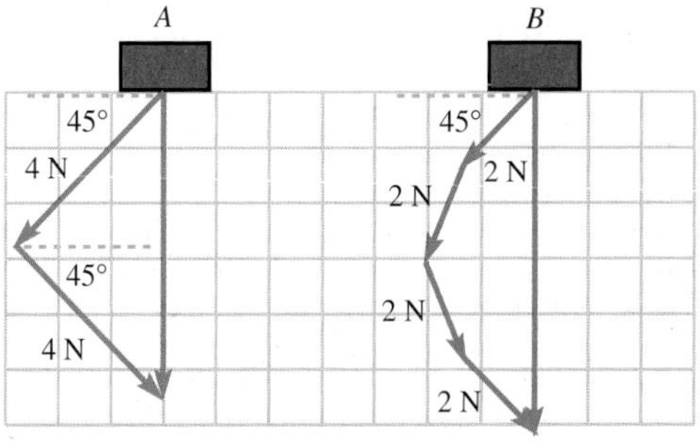

The net force magnitude on object B is greater than that on object A because two of the forces acting on B are directed at an angle greater than 45° with respect to the horizontal and contribute more to the downward directed net force.
15. 2 kN east **17.** 0.30 N **19.** (a) 1.3 kg (b) 0.90 m/s at 18° south of west **21.** 4.0 kg **23.** (a) 3.5 m/s^2 up (b) 15 m/s up **25.** 20 N **27.** 2.40 m/s^2 forward **29.** 4.22 km
31. (a) (b) (c) (d)

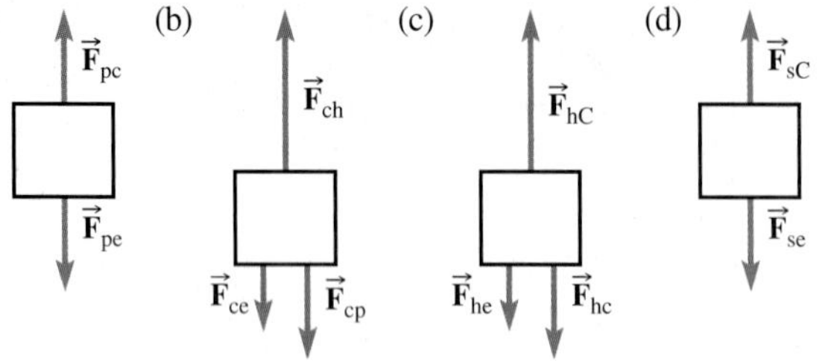

33. (a) 50.0 N upward (total for both feet) (b) 650.0 N upward (c) s = woman and chair system; e = Earth; f = floor

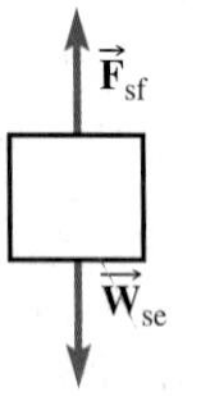

35. One force acting on the fish is an upward force on the fish by the line; its interaction partner is a downward force on the line by the fish. A second force acting on the fish is the downward gravitational force on the fish; its interaction partner is the upward gravitational force on the Earth by the fish. **37.** (a) 543 N (b) contact force of Margie's feet (c) 588 N (d) contact force on the Earth due to the scale
39. (a) 670 N (b) 2.5 N (c) a stick of butter **41.** 82 kg **43.** 2 nN **45.** 9.3×10^3 km **47.** (a) 1432 km (b) 4664 km **49.** 2639 km **51.** 1.5×10^{-9} N **53.** (a) and (b) **57.** (a) 160 N up the slope (b) 0.19 **59.** 61 N up the ramp **61.** b = book; t = table; e = Earth; h = hand
(a)

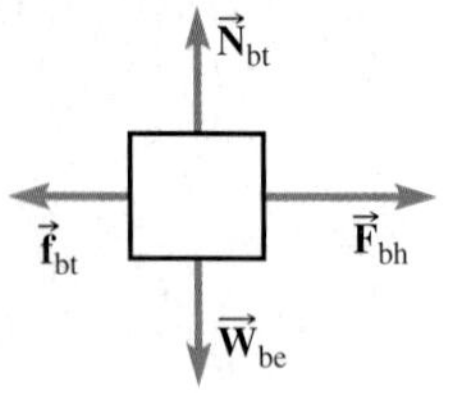

(b)

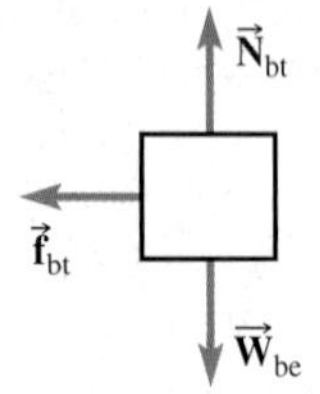

(c)

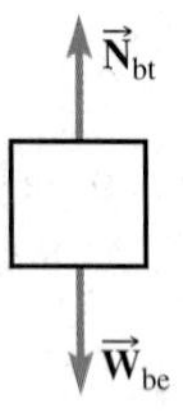

(d) (a) and (b) (e) 2.0 N opposite the direction of motion (f) The FBD would look just like the diagram for part (c). The book would not slow down because there is no net force on the book. **63.** (a) 0.41 (b) 40 N **65.** 0.4 **67.** 17°

69.

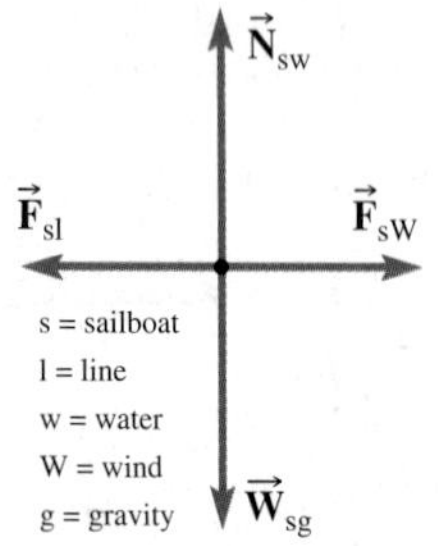

1) the force of gravity 2) the vertical force of the water opposing gravity 3) the force of the wind 4) the force of the line tied to the mooring **71.** 1.38 kN **73.** lower cord: 8.3 N; upper cord: 12.4 N **75.** Scale A reads 120 N. Scale B reads 240 N.

77. (a) $\sqrt{2}\,Mg$ (b) 45° **79.** $T_{15} = 30$ N; $T_{25} = 18$ N **81.** 2.0 N toward the back of the mouth **83.** $\dfrac{m_1}{m_1 + m_2}$ **85.** (a) 20 N (b) 12 N **87.** (a) $\vec{a}_1 = 3.9$ m/s^2 to the right; $\vec{a}_2 = 3.9$ m/s^2 downward (b) 4.7 m/s to the right (c) 2.8 m to the right (d) block 1: 0.31 m to the right; block 2: 0.31 m down **89.** (a) m_1: 2.5 m/s^2 up; m_2: 2.5 m/s^2 down (b) 37 N **91.** 0.81 s
93. 0.365°;

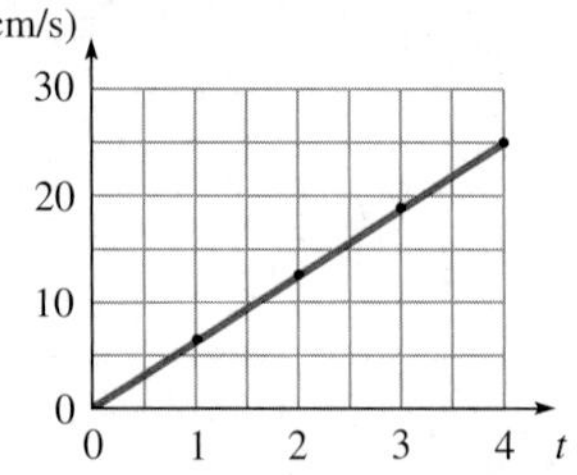

95. (a) $\mu_s > 0.48$ (b) 0.60 (c) 0.48 **97.** 642 N **99.** 1 kN upward **101.** (a) 1.4 m/s^2 downward (b) no **103.** 620 N **105.** the weak force **107.** electromagnetic and gravitational forces **109.** the strong force **111.** (a) 23 m/s (b) 0.19 **113.** (a) zero (b) 2.6×10^4 N
115.

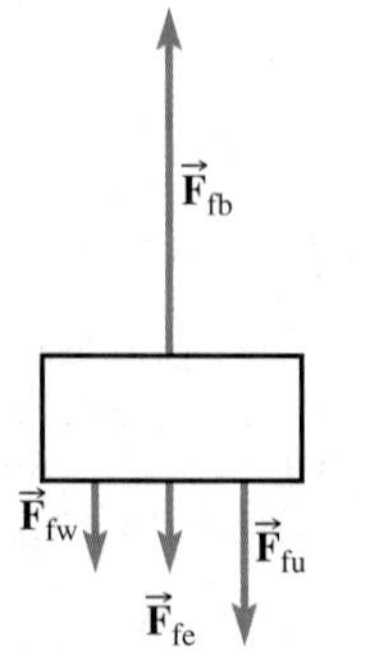

117. (a)

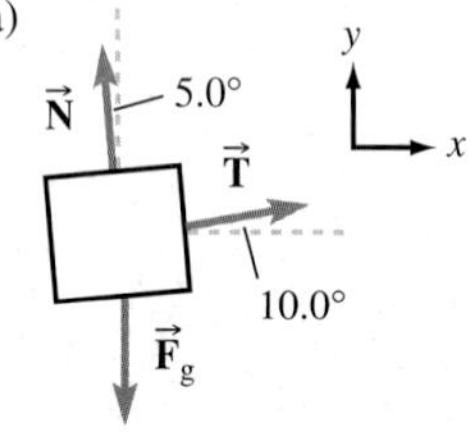

(b) 860 N **119.** 0.027 N/kg **121.** (a) 80 N (b) 160 N **123.** (a) 1360 N (b) 21.5° **125.** $F/3$ to the right
127. (a) 22° with respect to the horizontal (b) 0.9 m/s^2 down the incline **129.** (a) no (b) yes (c) 0.6 m/s^2 (d) The force of friction will be less on the Moon; more. **131.** 120 N **133.** 90.0% of the Earth-Moon distance **135.** (a) 1.5 N upward (b) 2.4 N upward (c) 0.85 N downward (d) 0.85 N downward; 9.80 m/s^2 downward **137.** (a) 110.0 N (b) $T_A = 115.0$ N $= T_C$ and $T_B = 110.0$ N $= T_D$. **139.** (a) 1) the gravitational forces between the magnet and the Earth 2) The contact forces, normal and frictional, between the magnet and the photo 3) The magnetic forces between the magnet and the refrigerator
(b)

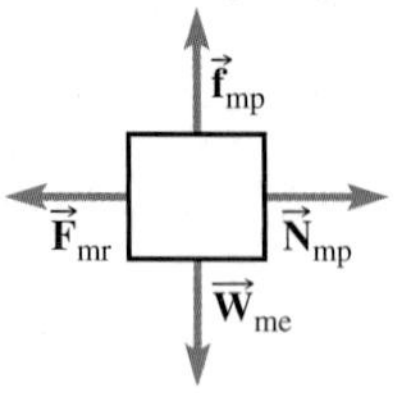

(c) The long-range forces are gravity and magnetism. The contact forces are friction and the normal force. (d) $W_{me} = 0.14$ N, $F_{mr} = 2.10$ N, $f_{mp} = W_{me} = 0.14$ N, and $N_{mp} = F_{mr} = 2.10$ N
141. (a) A = 137 N; B = 39 N (b) A = 147 N; B = 39 N
143. (a) $mg \tan\theta$ (b) $mg \tan\theta$ (c) $mg \tan\theta + \dfrac{ma}{\cos\theta}$
145. (a) $1.10mg$ (b) $1.10mg$ **147.** 1810 N; 5 times the force with which Yoojin pulls; the oak tree supplies additional force.
149. (a) 2.60×10^8 m from Earth (b) away from **151.** 15 m/s^2

153. (a) $a = \frac{m_2 - \mu_k m_1}{m_1 + m_2} g$; $T = (1 + \mu_k)\frac{m_1 m_2}{m_1 + m_2} g$
(b) For $m_1 \ll m_2$: $a \approx g$ and $T \approx (1 + \mu_k)\, m_1 g \ll m_2 g$, so the tension is negligible compared to the weight of m_2; it's essentially in free fall. For $m_1 \gg m_2$: $a = 0$ and $T = m_2 g$. For $m_1 = m_2$: $a = \frac{1}{2}(1 - \mu_k)g$ and $T = \frac{1}{2}(1 + \mu_k)mg$. (c) $a = 0$ only for $m_2 = 0$; thus, there is no value at which the two masses slide with constant velocity. For $m_2 = 0$, there is no tension in the cord.

CHAPTER 5

Multiple-Choice Questions

1. (b) **2.** (a) **3.** (f) **4.** (b) **5.** (a) **6.** (b) **7.** (b) **8.** (a) **9.** (e) **10.** (c) **11.** (b) **12.** (b)

Problems

1. 17 m **3.** 0.105 rad/s **5.** 26 rad/s **7.** (a) 3.49 rad/s (b) 0.45 m/s **9.** 3800 ft **11.** (a) 31 m/s (b) 31 rad/s **13.** 3.37 cm/s^2 **15.** 5.74 m/s **17.** (a) $\frac{mv^2}{L}$ (b) $T = m\sqrt{g^2 + \left(\frac{v^2}{L\cos\theta}\right)^2}$ **19.** (a) $\sqrt{\mu_s g R}$ (b) The static frictional force is not large enough to keep the car in a circular path; the car skids toward the outside of the curve. **21.** 7.9 m/s **23.** 59° **25.** (a) 2300 N (b) 19 m/s **27.** $\tan^{-1}\frac{v^2}{rg}$ **29.** 2.99×10^4 m/s **31.** 130 h **33.** $r_{Io} = 420\ 000$ km; $r_{Europa} = 670\ 000$ km **35.** 2.04×10^7 m **37.** 16 h **39.** (a) 13 N (b) The bob has an upward acceleration, so the net F_y must be upward and greater than the weight of the bob. **41.** 23.2 m/s **43.** 4.0 rad/s^2 **47.** $a_t = 2.54$ m/s^2; $a_r = 2.45$ m/s^2; 11.9 N **49.** (a) 1.7 rad/s^2 (b) 0.56 rev **51.** (a) 17.7 m/s (b) 6.28 m/s^2 (c) 6.59 m/s^2 at an angle of 17.7° east of south **53.** (a) 1.3×10^6 s (b) 5.0×10^{10} rev **55.** $16g$ **57.** 7.0 rad/s **59.** (a) 0.034 m/s^2 (b) less (c) 0.34% smaller (d) at the poles **61.** (a) 518.5 N (b) 521.5 N (c) 45 m **63.** 0.40ω **65.** 150 m/s **67.** (a) 3.00 m/s east (b) 3.00 m/s west **69.** 2.9 rotations for A; 5.7 rotations for B **71.** (a) 38 m/s (b) You would need 135 km of tape to record one hour. **73.** (a) $8.0\pi^2$ m/s^2 = 79 m/s^2 (b) $4.0\pi^2$ N = 39 N **75.** smallest; 4.1 s **77.** 110 μm/s **79.** 8 cm **81.** 120 km/h **83.** (a) $90g$ (b) 7.9×10^{-11} N (c) 4.4×10^{-18} N (d) $5.0 \times 10^5\ g$ **85.** 1.4 rev/s **87.** 42200 km

REVIEW & SYNTHESIS: CHAPTERS 1–5

Review Exercises

1. N/m = kg/s^2 **3.** (a) 220 markers (b) 221 markers **5.** (a) 1.74 m/s (b) 0.332 m/s in his original direction of motion **7.** (a) 3300 ft or 1000 m **9.** The cart will go off the road toward south. **11.** (a) 19 m (b) 3.6 m/s **13.** 1.7 m/s **15.** (a) 15.1 N (b) 34.3 N **17.** Stefan's plan is superior and thus more likely to work. **19.** 29 km/h at 83° north of west **21.** (a) The rocks will have the same speed when they hit the ground. (b) 19.8 m/s **23.** 2.40 s **25.** (a) 3.6×10^7 m (b) 55 N **27.** 11.5 m/s **29.** 2.02 s; 1.65 m to the left of B's initial position **31.** 0.98 m/s directed downward **33.** (a) $R = \frac{2v_i^2 \sin\theta\cos\theta}{g}$ (b) 221 m (c) 4 m **35.** (a) 283 m (b) 84.9 m **37.** (a) 216 billion solar masses (b) 0.46

MCAT Review

1. D **2.** C **3.** D **4.** C **5.** C **6.** D **7.** A

CHAPTER 6

Multiple-Choice Questions

1. (c) **2.** (b) **3.** (b) **4.** (a) **5.** (c) **6.** (c) **7.** (c) **8.** (c) **9.** (b) **10.** (b) **11.** (f)

Problems

1. 75 J **3.** No work is done. **5.** 210 kJ **7.** (a) 0 (b) 8.8 J **9.** 1.3 m **11.** 15.6 J **13.** (a) 0.70 J (b) 0.37 m/s **15.** 0 **17.** −4.17 kJ **19.** 5.8 MJ (meteoroid); 0.46 MJ (car); the meteoroid has more than 12 times the kinetic energy of the car. **21.** (a) 0 (b) 3.4 kJ (c) dissipated as heat **23.** (a) 0 (b) −2.9 J **25.** (a) 2 (b) 1.88 kJ (c) 1.88 kJ (d) 8.00 m **27.** (a) 14.3 m/s (b) Yes; the cart will reach position 4. **29.** 8.42 m/s **31.** −52 kJ **33.** (a) $\sqrt{v^2 + 2gh}$ (b) The final speed is independent of the angle. **35.** (a) 0.286 N/cm (b) 11.0 cm **37.** 2.37 km/s **39.** 13.0 km/s **41.** 2 **43.** 10.0 km/s **45.** 1.6 km/s **47.** 8 J **49.** 5.2 J **51.** (a) 4.9 cm (b) 1.4 N/cm (c) 88 mJ **53.** (a) 1.9 N/cm (b) 0.49 J (c) 2.4 kg **55.** zero **57.** 0.35 m **59.** 13 m **61.** 8.7 cm **63.** (a) 2.2 m/s (b) 0.21 m (c) 0.50 m **67.** 22 W **69.** (a) 20 N (b) 6.7 m/s **71.** 60 kW **73.** 6.2 g; the other 90% of the energy is dissipated as heat. **75.** 930 kW **77.** 4.8 m/s **79.** 16 m/s **83.** (a) $k/2$ (b) $2k$ **85.** 27 N **87.** 0.33 m **89.** 1.6 m/s **91.** (a) 10 kW (b) 5.8° **93.** (a) 2.62 kW (b) 7.85 kW **95.** (a) 2200 kcal/day (b) more than 0.51 lb **97.** (b) 4.9 m/s (c) 1.24 m **99.** $\frac{2}{3}R$ **101.** (a) $k = k_1 + k_2$ (b) 0.16 J **103.** 1.3 cm; 32 J **105.** (a) 26 cm (b) 34 cm **109.** No; because the kinetic energy cannot be negative as would be the case in the region 3 cm < x < 8 cm. The particle must remain in the region x < 3 cm.

CHAPTER 7

Multiple-Choice Questions

1. (c) **2.** (d) **3.** (c) **4.** (b) **5.** (d) **6.** (b) **7.** (f) **8.** (d) **9.** (a) **10.** (e) **11.** (d) **12.** (b)

Problems

1. 0 **5.** 3 kg·m/s north **7.** 20 kg·m/s in the −x-direction **9.** 1.0×10^2 kg·m/s downward **11.** 320 s **13.** 6.0×10^3 N opposite the car's direction of motion **15.** (a) 750 kg·m/s upward (b) 990 N·s downward (c) 2500 N downward **17.** (a) 11 m/s

(b) 1300 N **19.** (a) 33 m/s (b) 0.94 N down **21.** 2.6×10^5 m/s **23.** 0.30 m/s **25.** 0.10 m/s **27.** (8.0 cm, 20 cm) **29.** 4.0 cm in the positive x-direction **31.** (0.900 m, −2.15 m) **33.** 21 cm **35.** (6 m/s, −4 m/s) **37.** (a) (−0.13 m/s, −4.1 m/s) (b) The center of mass of the system remains at the origin after the explosion. **39.** 270 m/s to the right
41. (a) 0.20 m/s (b) 0.25 m/s **43.** 5.0 m/s **45.** 2.0 kg·m/s to the right **47.** 43 m/s **49.** 3.0 m/s **51.** 0.20 kg
53. 0.49 m **55.** 5.0 m/s **57.** 170 m/s
59. (a) $\Delta p_{1x} = -1.00 m_1 v_i$; $\Delta p_{1y} = 0.751 m_1 v_i$ (b) $\Delta p_{2x} = m_1 v_i$; $\Delta p_{2y} = -0.751 m_1 v_i$; the momentum changes for each mass are equal and opposite. **61.** $1.73 v_{1f}$ **63.** 8.7 kg·m/s **65.** 6.0 m/s at 21° south of east **67.** 1.7 m/s at 30° below the x-axis
69. 20 m/s at 18° west of north **71.** 10.2 m/s
73. 5.0×10^9 kg·m/s **75.** 34 N **77.** (2.0, 0.75, 0.25) in
79. Inexperienced: 5000 N; experienced: 500 N **80.** 37 m/s in the $+x$-direction **81.** (a) 5.3 kg·m/s opposite the ball's direction of motion (b) 5.3 kg·m/s opposite the ball's direction of motion (c) 1.8 kN opposite the ball's direction of motion
83. (a) 148.6° CCW from the electron's direction
(b) 9.60×10^{-19} kg·m/s in the direction found in (a)
85. The lighter car was speeding. **87.** 10^{-18} N **89.** $\frac{1}{9}h$
91. 10 m/s **93.** (a) $\frac{111}{2}$ (b) 1 (c) $\frac{111}{2}$

CHAPTER 8

Multiple-Choice Questions

1. (b) **2.** (d) **3.** (a) **4.** (c) **5.** (e) **6.** (b) **7.** (a) **8.** (f) **9.** (e) **10.** (c)

Problems

3. (a) reduced by a factor of 8 (b) reduced by a factor of 32
5. (a) 1.5 kg·m² (b) 0.75 kg·m² (c) 1.5 kg·m²
7. $\frac{2}{5}\frac{R_E^2}{R_o^2}$, where R_E is the Earth's radius and R_o is Earth's orbital radius about the Sun. **9.** (a) no (b) 0.017 **11.** 4.0 N·m
13. 780 N·m **15.** 25 N·m **19.** 57.4 N·m **21.** 1.2 cm toward the doorknob as measured from the center of the door
23. 150 J **25.** (a) 5.5 kJ (b) 29 N·m **27.** 98 N **29.** 17.0°
31. The center of mass = 0.8542 m < 0.8600 m; the system balances. **33.** 180 N toward the wall **35.** $T = 350$ N; $F_x = 290$ N; $F_y = -2$ N; the magnitude of F_y is small. **37.** 22.3° **39.** palms: 390 N; feet: 270 N **41.** tendon, 2100 N upward and tibia, 2800 N downward **43.** 130 N **45.** 3.0 kN; about 5.5 times larger
49. 0.0012 N·m **51.** 4.3 N·m **53.** 0.09 N·m
55. (a) 0.11 rad/s² (b) 0.44 rad/s **59.** 2.9 m/s **61.** 1.79 m
63. 2.75 s **65.** (a) $\frac{5}{2}r$ (b) $\frac{27}{10}r$ **67.** h will decrease. The smaller the rotational inertia, the less gravitational energy will go into rotational energy, and the more will go into translational energy. Problem 66 had a minimum of $h = 3r$. With a solid sphere, the minimum is $h = 2.7r$, which is a little less than $3r$.
69. 0.0864 kg·m²/s **71.** 1.4×10^7 kg·m²/s **73.** 1.60 s
75. 15.6 rad/s **77.** 0.125 rad/s **79.** (a) 3.0 (b) 1.6
81. 2.10×10^6 N·m **83.** 1.14 **85.** 98 N·m **87.** 5.4 rad/s
89. The objects reach the bottom in the following order from first to last: cube, solid sphere, solid cylinder, hollow sphere, and hollow cylinder. **91.** 0.792 m **93.** (a) 16 kg·m² (b) 8.0×10^7 J
(c) 320 (d) 120 km **97.** 110 N **99.** (a) 1.35×10^{-5} kg·m² (b) 524 N **101.** 0.19 kg·m²/s **103.** (a) 9.4×10^{-4} kg·m²/s (b) 1.2×10^{-6} kg·m²/s **105.** 230 N **107.** (a) 735.0 N (b) 0.88 m (c) $0.55h$ **109.** 23 N **111.** 1.3 rev/s
113. (a) 9.6 m/s (b) 3.1 m/s (c) 21 m/s

REVIEW & SYNTHESIS: CHAPTERS 6–8

Review Exercises

1. (a) 0.20 m (b) 250 N/m **3.** $2mg$ **5.** (a) 940 J (b) 0.734
7. 2.3 m **9.** 30 m/s **11.** (a) 0.502 kg·m² (b) 17 N·m
13. 1.53 m/s **15.** 0.73 m **17.** 10.3 J **19.** $h_A = 0.57$ m; $h_B = 2.3$ m **21.** 1.27 m **23.** 2.0 m/s **25.** 2.06 m/s at 41.6° south of east
27. (a) $\dfrac{\omega_i}{1 + \dfrac{mr^2}{MR^2}}$
(b) The total angular momentum does not change, since no external torques act on the system. (c) Yes; the kinetic energy changes.
29. (a) The Vulcan ship will have the greater kinetic energy. The ships will have the same momentum. (b) The ships will have the same kinetic energy. The Romulan ship will have the greater momentum. (c) In part (a), the momenta are the same, 9.5×10^8 kg·m/s, but the kinetic energies differ: Vulcan at 6.9×10^{12} J and Romulan at 3.5×10^{12} J. In part (b), the kinetic energies are the same, 9.5×10^8 J, but the momenta differ: Vulcan at 1.1×10^7 kg·m/s and Romulan at 1.6×10^7 kg·m/s.
31. (a) $U = -550$ J; $K = 450$ J (b) $E = -100$ J; $U = -100$ J; $K = 0$ (c) 200 J (d) The particle has a kinetic energy of 450 J at $t = 0$, and we are told the motion is to the left. The particle will continue moving left but the kinetic energy will decrease by 450/4.5 J for every cm of travel until it reaches $x = 1$ cm. At this point $K = 0$, and the particle has stopped instantaneously. It will next move to the right with an increasing K until it reaches $x = 5.5$ cm. At this point $K = 450$ J, and this kinetic energy will be maintained as it continues moving right until it reaches $x = 11$ cm. At this point, its kinetic energy will decrease by 450/2.5 J for every cm of travel until it reaches $x = 13.5$ cm. At this point $K = 0$, and the particle has again stopped instantaneously. It will then turn around again. **33.** (a) 19.4 m/s (b) no **35.** 13 m/s

MCAT Review

1. D **2.** D **3.** B **4.** B **5.** D **6.** C **7.** B **8.** B **9.** D **10.** A **11.** B **12.** D **13.** B **14.** C **15.** A **16.** D **17.** C

CHAPTER 9

Multiple-Choice Questions

1. (b) **2.** (b) **3.** (d) **4.** (a) **5.** (a) **6.** (b) **7.** (a) **8.** (a) **9.** (d) **10.** (c)

Problems

1. 49 atm **3.** 22 kPa **5.** The baby applies 2.0 times as much pressure as the adult. **7.** 4.0 kN southward **9.** 1.0 m

11. (a) 30 N (b) 5.8 N·m **13.** 2.0 atm **15.** $\frac{V_{Pt}}{V_{Al}} = 0.126$ **17.** 1.0 MN **19.** (a) 343 kPa (b) 410 Pa **21.** (a) 2.2×10^5 Pa (b) 1700 torr (c) 2.2 atm **23.** 114.0 cm Hg **25.** (a) 5.6 cm (b) 0.37 cm **27.** 250 kg/m^3 **29.** (a) 91.7% (b) 0.917 **31.** (a) 8.8 N upward (b) 9.6 N upward **33.** 100% **35.** 0.78 **37.** yes **39.** (a) 9.8 m/s^2 upward (b) 3.3 m/s^2 upward (c) 68.6 m/s^2 upward **41.** 50 m/s **43.** (a) 39.1 cm/s (b) 78.5 cm^3/s (c) 78.5 g/s **45.** 1.12×10^5 Pa **47.** 1.9×10^5 N **49.** 310 kPa **51.** 8.6 m **55.** (a) 6850 Pa (b) 0.685 N **57.** 0.040 m^3/s **59.** (a) 50 Pa (b) 1100 Pa (c) approximately 13 kPa **61.** 0.4 Pa·s **63.** (a) 1.3×10^{-10} N (b) 2.6×10^{-14} W **65.** Since m/v_t is constant, the drag force is primarily viscous. **67.** 3.0 mm/s **69.** 5 Pa **71.** (a) $\gamma L \Delta s$ (b) $\Delta E = \gamma \Delta A$ **73.** (a) 1.54 N (b) 1.54×10^4 N (c) For a given depth the pressure is the same everywhere, so the very tall, narrow column of water is as effective as having a whole barrel of water filled to the same height and pushing upward on the barrel top. **75.** (a) 7.43% (b) 1060 kg **77.** (a) 5.94 m/s (b) As long as we can assume Bernoulli's equation applies, it doesn't matter what fluid is in the vat. (c) The speed would be reduced by a factor of 0.40. **79.** 230 kg **81.** 23.0 m **83.** 110 m **85.** 1.1 cm **87.** 27 kPa **89.** (a) 2.2 m/s up (b) 21 kPa/s **91.** (a) 1.4 N (b) 0.43 N upward (c) 6.8 m/s^2 downward **93.** d is not a linear function of ρ: $d = \frac{m}{\pi \rho r^2}$, **95.** (a) 26 m/s (b) 2.6 m/s **97.** 0.83 g/cm^3 **99.** (a) 5.2 kPa = 0.051 atm (b) 11.8 Pa/m (c) 8.61 km (d) A decreasing air density means that the atmosphere extends to a higher altitude. **101.** (a) 220% (b) 0.68

CHAPTER 10

Multiple-Choice Questions

1. (c) **2.** (b) **3.** (b) **4.** (a) **5.** (a) **6.** (a) **7.** (c) **8.** (b) **9.** (c) **10.** (a) **11.** (f) **12.** (e) **13.** (e) **14.** (f) **15.** (f) **16.** (e) **17.** (c) **18.** (c) **19.** (j) **20.** (k)

Problems

1. 0.097 mm **3.** 2.2 cm **5.** 0.80 mm **7.** 5.0 mm **9.** tension: 1.5×10^{10} N/m^2; compression: 9.0×10^9 N/m^2 **11.** 8.7×10^{-5} m **13.** 630 N **15.** human: 3 cm^2; horse: 7.1 cm^2 **17.** (a) 2.8×10^7 Pa (b) 4.7×10^{-4} (c) 9.3×10^{-4} m (d) 5.0×10^5 N **19.** volume: 7.7×10^{-4}; radius: 2.6×10^{-4} **21.** The volume of the steel sphere would decrease by 57×10^{-6} cm^3. **23.** 7.5×10^5 N **25.** 0.30 N **27.** 7.9 m/s^2 **29.** 3.10 m/s; 8560 m/s^2 **33.** (a) high frequency (b) 1.3×10^{-6} m/s; 1.6×10^{-4} m/s^2 (c) 0.0013 m/s; 160 m/s^2 **35.** 5.0 rad/s **37.** 2.5 Hz **39.** (a) 1.7×10^{-4} m (b) 0.13 m/s (c) 510 N **41.** (a) 1.4 kN (b) 0.13 J **43.** (a) 0.39 m (b) 2.0 m/s **45.** 0.70 s **47.** 0.250 Hz **49.** (a) a vertical straight line of length 24 cm (b) a positive cosine plot of amplitude 12 cm

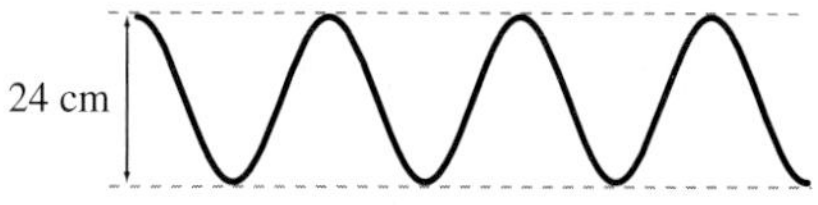

51. (a) $\frac{2}{\pi}\omega A$ (b) ωA (c) $\frac{2}{\pi}$

(d)

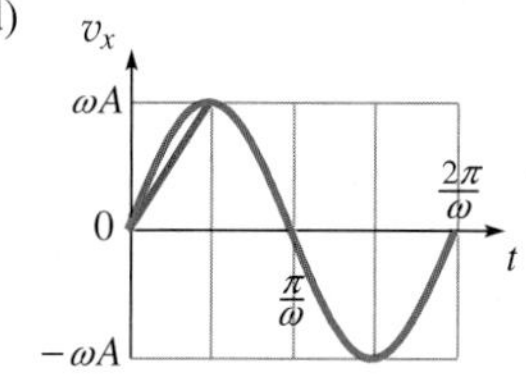

If the acceleration were constant so that the speed varied linearly, the average speed would be 1/2 of the maximum velocity. Since the actual speed is always larger than what it would be for constant acceleration, the average speed must be larger.

53. (a)

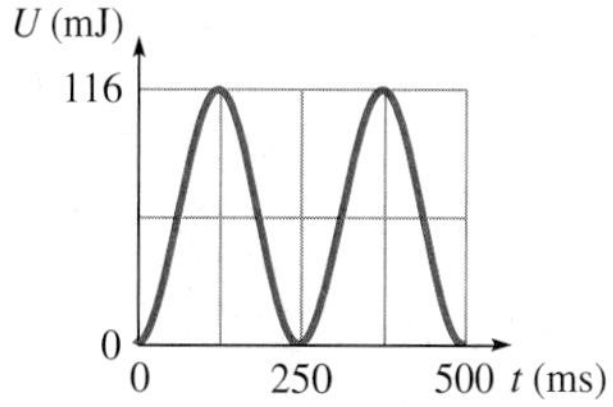

(b)

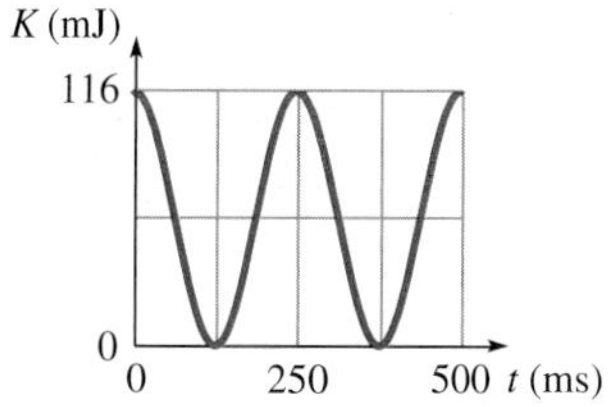

(c)

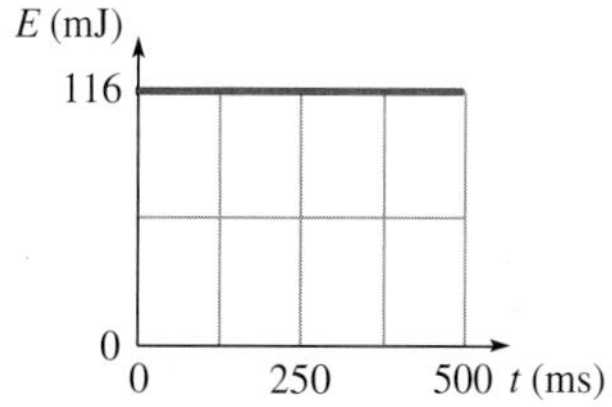

(d) U, K, and E would gradually be reduced to zero.

55. 4.0 s **57.** 1.5 s **59.** (a) $v_x = \omega A \cos \omega t$

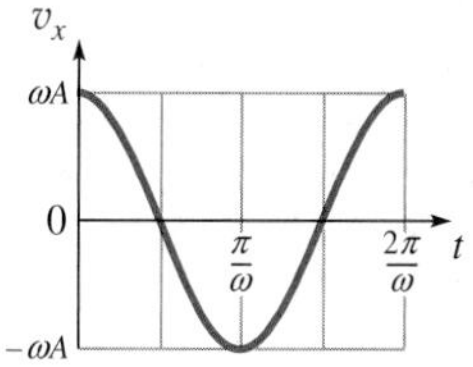

(b) $\frac{1}{2}m\omega^2A^2$

61. 1.11 **63.** 1st method: 3.14 cm/s; 2nd method: 3.14 cm/s **65.** 11 mJ **67.** (a) 6.1 mJ (b) 1.1% **69.** −9.75% **71.** 2.5 s **73.** (a) more (b) 56 N **75.** (a) The frequency and period don't vary with amplitude, they only vary with m and k. Since these two values remain constant, so do the frequency and period. (b) The total energy for an amplitude of $2D$ is four times that for an amplitude of D. (c) The frequency and period are still the same. (d) The energy is greater when given an initial push, since it has an amplitude > $2D$. The increase in energy is $\frac{1}{2}mv_i^2$.

77.

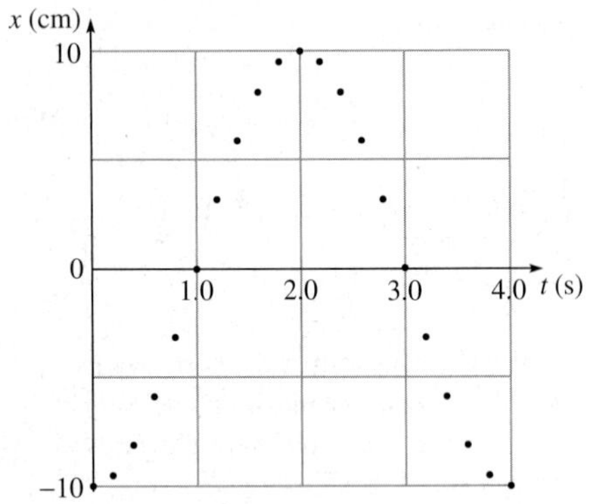

The distance between adjacent dots should be the least at the endpoints and greatest at the center, so its speed is lowest at the endpoints and fastest at its equilibrium position. **81.** $y = (1.6\text{ cm})\cos[(25\text{ rad/s})t]$ **83.** (a) 0.395 m (b) 1.11 m/s (c) 0.960 m/s **85.** 8.0×10^8 Pa; it is just under the elastic limit. **87.** 0.63 Hz **89.** (a) ρgh (b) 7.6 km (c) no **91.** (a) 42.2° (b) 48 g (c) 9.1 cm **93.** (a) $\sqrt{2gL}$ (b) $\frac{\pi}{2}\sqrt{gL}$; larger

95. (a) $2\pi\sqrt{\dfrac{L\left(\frac{m_1}{3}+m_2\right)}{g\left(\frac{m_1}{2}+m_2\right)}} = 2\pi\sqrt{\dfrac{2L(m_1+3m_2)}{3g(m_1+2m_2)}}$

(b) For $m_1 >> m_2$, $T = 2\pi\sqrt{\dfrac{2L}{3g}}$, and for $m_1 << m_2$, $T = 2\pi\sqrt{\dfrac{L}{g}}$.

CHAPTER 11

Multiple-Choice Questions

1. (b) **2.** (c) **3.** (d) **4.** (f) **5.** (a) **6.** (b) **7.** (d) **8.** (a) **9.** (b) **10.** (d)

Problems

1. 52 W/m^2 **3.** 170 mW/m^2 **5.** 4.0×10^{26} W **7.** (a) 6.0 m (b) 1.7 s **9.** 168 m/s **11.** 16 ms **13.** 0.375 m **15.** (a) 340 Hz (b) 3.0×10^8 Hz **17.** 0.33 Hz **19.** 0.83 cm/s **21.** (a) 4.0 mm (b) 1.0 m (c) 0.010 s (d) 100 m/s (e) in the $+x$-direction (to the right) **23.** $y(x, t) = (0.120\text{ m})\sin[(134\text{ s}^{-1})t + (20.9\text{ m}^{-1})x]$ **25.** (a) 2.6 cm (b) 14 m (c) 20 m/s (d) 1.4 Hz (e) 0.70 s **27.** $v_m = 0.063$ m/s; $a_m = 0.79$ m/s^2

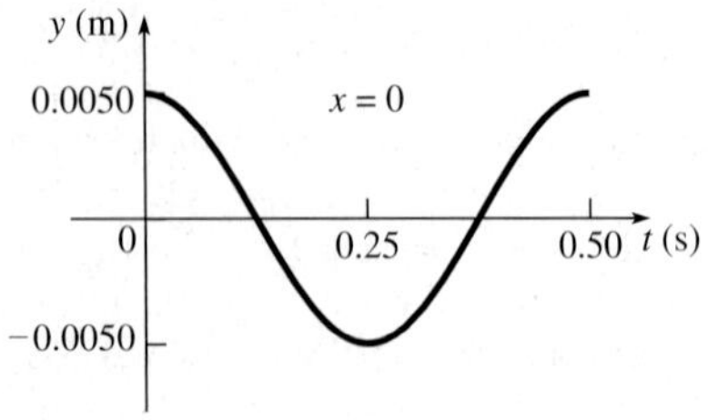

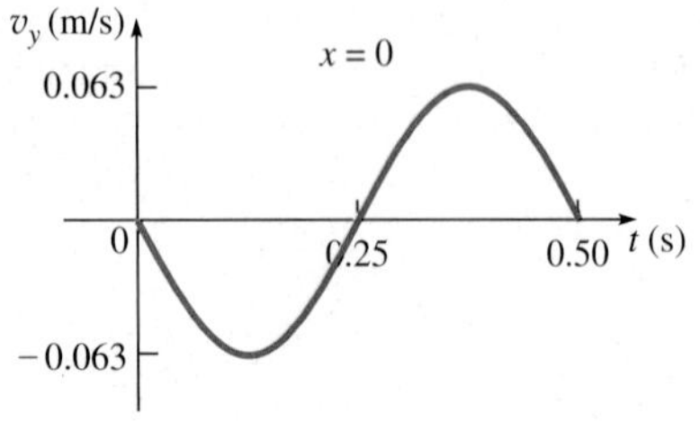

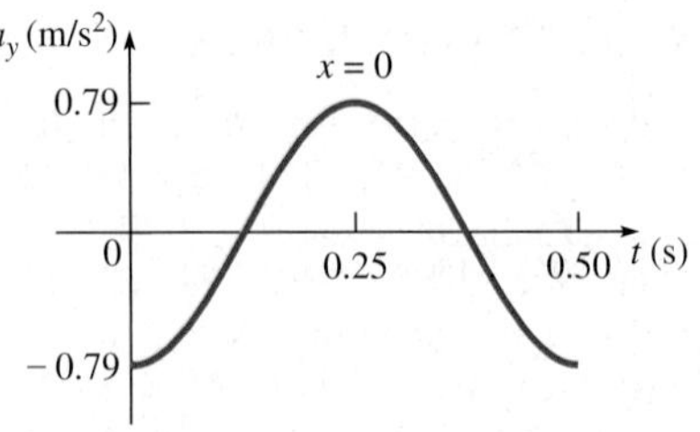

29. (a)

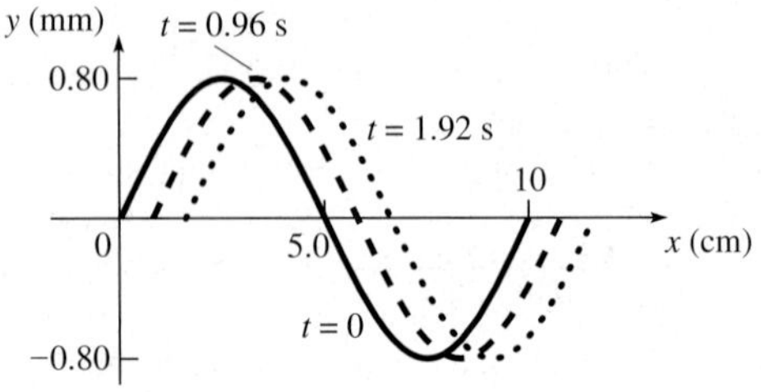

(b)

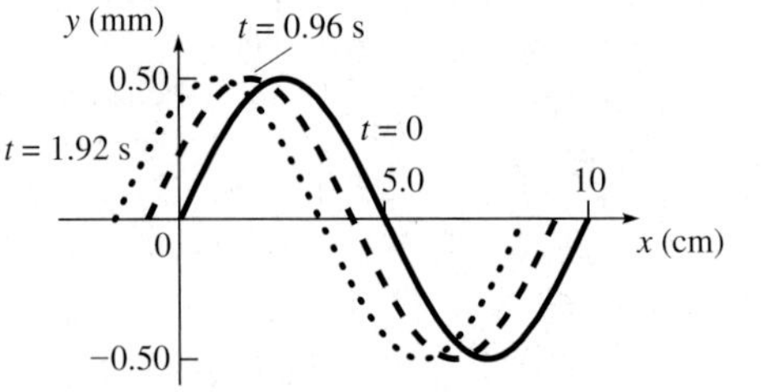

(c) $y(x, t) = (0.80\text{ mm})\sin(kx - \omega t)$ represents a wave traveling in the $+x$-direction.
$y(x, t) = (0.50\text{ mm})\sin(kx + \omega t)$ represents a wave traveling in the $-x$-direction.

31.

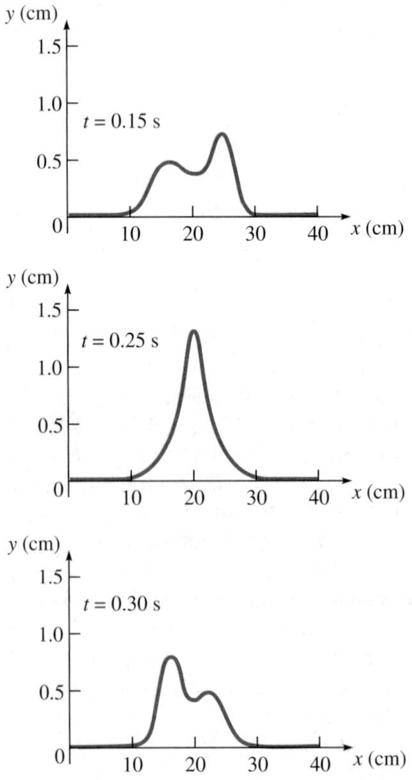

33. (a) 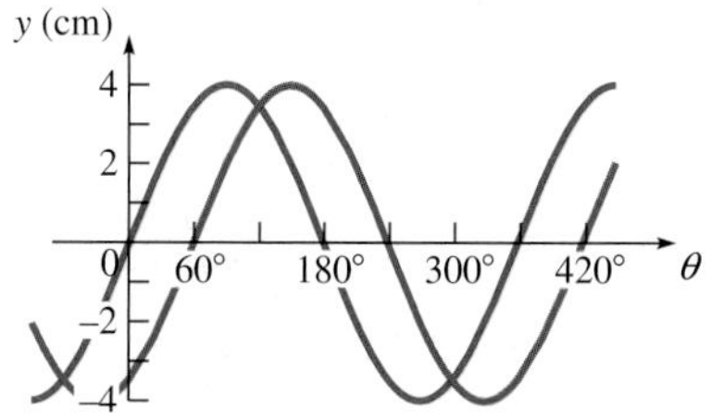

; 6.9 cm

(b) 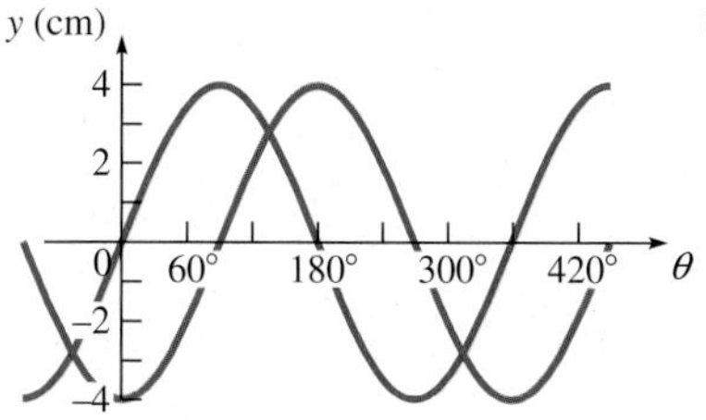

; 5.7 cm

35. 96.0° **37.** 1.7 s **39.** (a) 0°; 8.0 cm (b) 180°; 2.0 cm (c) 4:1 **41.** 79 mW/m^2 **43.** (a) 0.25 W/m^2 (b) 0.010 W/m^2 (c) 0.130 W/m^2 **45.** 7.8% **47.** 0.016 m **49.** (a) 33 Hz (b) 300 N **51.** 4.5 × 10^{-4} kg/m **53.** (a) 260 Hz (b) 2.8 g **55.** 0.050 kg **57.** 190 m **59.** 3.3 m **61.** 80 km **63.** 3.64 cm, 7.07 cm, 10.32 cm **65.** 470 Hz **67.** (a) Hooke's law: $T = k(x - x_0) \approx kx$ for $x >> x_0$. (b) 4.00 s **69.** $v \propto \sqrt{\dfrac{\gamma}{\lambda\rho}}$; dispersive **71.** (a) upward (b) downward (c) A **73.** 12
77. 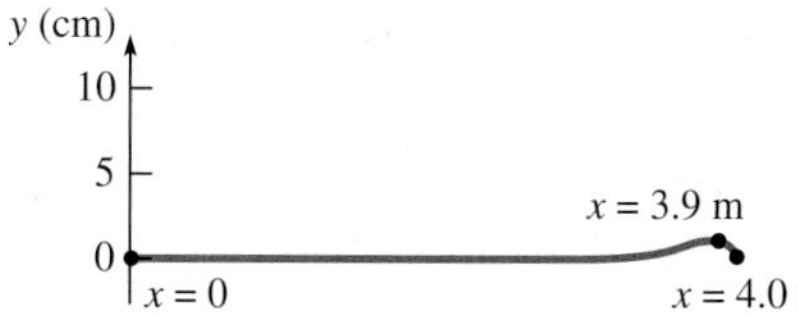

CHAPTER 12

Multiple-Choice Questions

1. (c) **2.** (a) **3.** (b) **4.** (c) **5.** (b) **6.** (c) **7.** (b) **8.** (c) **9.** (b) **10.** (d)

Problems

1. 3.4 mm **3.** 173 ms **5.** 4.7 s ≈ 5 s **7.** 1.4 km/s **11.** 1.1 μJ **13.** 95 dB; this is not much different than with only one machine running. **15.** (a) 28.7 N/m^2 (b) 1.58 mN **19.** 8.58 mm **21.** (a) 65.6 cm (b) 252.4 Hz **23.** 43.3 cm **25.** 34°C **27.** 3/4 **29.** (a) There is a displacement node (pressure antinode) at the center of the rod and displacement antinodes (pressure nodes) at the ends. (b) 5100 m/s (c) 13.1 cm (d) The ends move in opposite directions and, thus, they are out of phase. **31.** (a) 290.0 Hz (b) 1.4% **33.** (a) 85.6 N (b) 432 m/s (c) 335 Hz (d) 0.256 m **35.** 580 Hz **37.** 6.35 Hz **39.** (a) 1.5 kHz (b) 500 Hz **41.** (a) 3.0 kHz (b) 330 Hz (c) 1.0 kHz **45.** (a) 670 m (b) 2.8 s **47.** 403 m **49.** 83.6 kHz **51.** 640 Hz **53.** (a) 319 Hz (b) 319 Hz; 1.1 m **55.** 17.9 Hz; 53.6 Hz; 89.3 Hz; 125 Hz **57.** (b) First object: 110%; second object: 46% **59.** 2.3 kHz **61.** 0.0955 s **63.** (a) 5.05 m (b) 16.35 Hz **65.** 196 Hz **67.** 0.019 **69.** 29.0 dB

REVIEW & SYNTHESIS: CHAPTERS 9–12

Review Exercises

1. (a) Aluminum, since it is less dense it occupies more volume. (b) Wood, since it displaces more water than the steel. (c) Lead: 0.87 N; aluminum: 3.6 N; steel: 1.2 N; wood: 9.8 N **3.** 0.116 m/s **5.** 0.88 m/s **7.** (a) Eq. I; 1.50 cm/s (b) Eq. II; 2.09 cm (c) Eq. II; 13.5 cm/s (d) Eq. II **9.** (a) 58 N (b) 49 cm **11.** 21.4 cm **13.** 1500 Hz; 22.9 cm **15.** about 1 min **17.** 346 Hz **19.** (a) 41.7 cm/s; 118 kPa (b) 5.98 cm **21.** (a) 1.28 m (b) 141 m/s (c) 4.48 × 10^{-4} kg/m (d) 1.60 m/s (e) 110.0 Hz (f) 3.12 m **23.** (a) 5.13 × 10^{-2} N (b) 2.69 s **25.** (a) 6.17 × 10^{-4} m (b) 8.61 J (c) 0.536 s

MCAT Review

1. A **2.** A **3.** D **4.** C **5.** C **6.** B **7.** D **8.** B **9.** B **10.** A **11.** B **12.** D **13.** C **14.** C

CHAPTER 13

Multiple-Choice Questions

1. (e) **2.** (d) **3.** (b) **4.** (c) **5.** (b) **6.** (d) **7.** (a) **8.** (c) **9.** (c) **10.** (e)

Problems

1. (a) 29°C (b) 302 K **3.** (a) −40 (b) 575 **5.** $T_J = (0.750°J/°C)T_C + 85.5°J$ **7.** 2.0 mm **9.** (a) 3.6 mm (b) 10.8 mm **11.** 3.8 × 10^{-4} mm^2 **13.** (b) 2.4 × 10^{-3} **15.** 1.67 mL **17.** 75°C **19.** 1.3 m **21.** 150°C **23.** 24.98 cm **29.** 7.31 × 10^{-26} kg **31.** 1.7 × 10^{27} **33.** 2.650 × 10^{25} atoms **35.** 8.9985 mol **37.** 2.5 × 10^{19} molecules **39.** 10^{18} atoms **41.** 400°C **45.** 135 kPa **47.** (a) 1.3 kg/m^3 (b) 1.2 kg/m^3 **49.** 1.3 × 10^3 m^3 **51.** 1.50 **53.** 1.3 × 10^{26} **55.** 2.1 mm **57.** (a) 28 min (b) 11 min **59.** (b) 3410 × 10^{-6} K^{-1} **61.** 152 J **65.** 3.4 kJ **67.** $\frac{1}{\sqrt{2}}$ **69.** (a) 493 m/s (b) 461 m/s (c) 393 m/s **71.** yes **73.** 2220 K **77.** 0.14°C **79.** (a) 100 nm (b) 200 nm (c) 8 μm **81.** 2.5 × 10^4 s **83.** 140 atm **85.** 165°C **87.** HNO_3 **89.** (a) 6.42 × 10^{-21} J (b) 0.25% **91.** (a) The number of moles decreases by 25%. (b) −48°C **93.** average: 78.1; rms: 78.6; 83 **95.** (a) 0.400 mm Hg/°C (b) 3.21 × 10^{-3} mol **97.** 4 nm **99.** (a) 5.2 × 10^{24} m^{-3} (b) 1.9% **101.** 630°C **103.** 1.9 × 10^{14} molecules **105.** 25 m/s **107.** 3.05 mm **109.** 7.4 × 10^3 N/m

CHAPTER 14

Multiple-Choice Questions

1. (a) **2.** (b) **3.** (d) **4.** (d) **5.** (c) **6.** (b) **7.** (c) **8.** (d) **9.** (d) **10.** (b) **11.** (c) **12.** (c)

Problems

1. (a) 34 J (b) Yes; the increase in internal energy causes a slight temperature increase. **3.** 4.90 kJ **5.** (a) 250 J (b) all three **7.** 5.4 J **9.** 2.78×10^{-4} kW·h **11.** 6.40×10^{-4} kJ/K **13.** 0.50 MJ **15.** 700 m **17.** (a) 2430 kJ/K (b) 3500 kJ/K **19.** 742 kJ **21.** 0.13 kJ/(kg·K) **23.** 0.090 J **25.** 57 kJ **27.** 58°C **29.** (a) B to C, solid to liquid; D to E, liquid to gas (b) B (c) D **31.** 330 J/g **33.** 461 g **35.** 157 g **37.** 242 g; 35% **39.** 46.3 g **41.** 2 g **43.** 36 g **45.** 22.8 kJ/kg **47.** (a) 2.0 cm (b) 29 m **49.** (a) 0.12 K/W (b) 2.5×10^{-4} K/W (c) 5.0×10^{-5} K/W **51.** 6.67 W/m^2 **53.** −37°C **55.** (a) 300 W (b) 4500 W **57.** (a) 0.32 W (b) 800 K/m (c) 0.16 W (d) 0.64 W (e) 64°C **59.** 1.76 μm **61.** 150 W **63.** 390 W **65.** 2800 K **67.** 2.24 kW **69.** Coffeepot: 4.5 W; teapot: 24 W **71.** (a) 39°C (b) 182 W/m^2 **73.** 320 s **75.** (a) 180°C (b) 20.9°C **77.** (a) 9.9 kJ (b) 360 g **79.** 0.0065°C **81.** 4.0 times higher **83.** 0.792 kJ/(kg·K) **85.** 10.4 W **87.** 5400 kcal/h **89.** 4.0 g **91.** (a) 190 W (b) 31°C (c) Wearing clothing slows heat loss by radiation because air layers trapped between clothing layers act as insulation. **93.** 140 m **95.** 35°C **97.** (a) 7.00 times higher (b) 35.7°C; the dog is a much better regulator of temperature and, as a result, has more endurance. **99.** 0.84 kJ/(kg·K) **101.** 15.2 kJ/mol

CHAPTER 15

Multiple-Choice Questions

1. (b) **2.** (d) **3.** (c) **4.** (c) **5.** (d) **6.** (c) **7.** (a) **8.** (c) **9.** (d) **10.** (d) **11.** (e) **12.** (b) **13.** (d)

Problems

1. 2.9 J **3.** 100 J of heat flows out of the system. **5.** 202.6 J **7.** (a) 98.0 kPa; 1180 K (b) −200 J (c) 66 J (d) $\Delta U = 0$ because $\Delta T = 0$ in a cycle. **9.** (a) 436 J (b) 1.23 L (c) 125 J (d) 312 J **11.** (a) −1372 J (b) ΔU = 1216 J; Q = 2588 J **13.** (a) 182 kJ (b) 182 kJ **15.** 240 MJ **17.** (a) 210 J (b) 790 J **19.** (a) 1.2×10^{17} J (b) 1.4×10^{13} kg **21.** 0.182 **23.** 3.0 kJ **25.** 171 K **27.** 25.0 kJ **29.** 14 W **31.** The coal-fired plant and the nuclear plant exhaust 0.43 MJ and 0.60 MJ of heat, respectively. **33.** (a) 0.3436 (b) 275.7 kJ **35.** 4.2% **37.** 0.0174 **39.** 110 kJ **41.** 4.5 GW **43.** +250 W **47.** (b), (a), (c), (d) **49.** +6.05 kJ/K **51.** (a) 3.4×10^{-3} J/K (b) -2.8×10^{-3} J/K (c) 6×10^{-4} J/K **53.** 0.102 J/(K·s) **55.** (a) 97 W (b) 0.33 W/K **57.** The engine will not work. **59.** 15 kJ **61.** (a) 304 kJ (b) 2350 K (c) 13.0 mol **63.** (a) 15.9°C (b) −0.03 J/K (c) The entropy of the universe never decreases. **65.** 24°C **67.** 0.401 or 40.1% **69.** 350 J/K **71.** (a) 0.90 J/K (b) −2.7 J/K **73.** 15 min **75.** (a) 6.2 mJ (b) 22 mJ (c) 1.2 mK **77.** (a) 0.051 (b) 31 m^3 (c) yes

REVIEW & SYNTHESIS: CHAPTERS 13–15

Review Exercises

1. 108 kJ **3.** 28.4°C **5.** (a) 74 g (b) 11°C **5.** 467 mol **7.** The ice will melt completely; 32°C **9.** (a) 4140 K (b) 1.09×10^{26} W (c) 1.01×10^{-9} W/m^2 **11.** (a) 8.87 kPa; 1200 K (b) 23 kJ (c) 20.0 kJ (d) 0 **13.** 2.44 kJ/K **15.** 10.9°C **17.** reduced to 75% of the original **19.** 12 kJ **21.** (a) The boiling temperature of water varies with pressure. If the pressure is high, the water molecules are pushed close together, making it harder for them to form a gas. (Gas molecules are farther apart from each other than are liquid molecules.) A higher pressure raises the temperature at which the coolant fluid will boil. (b) If you were to remove the cap on your radiator without first bringing the radiator pressure down to atmospheric pressure, the fluid would suddenly boil, sending out a jet of hot steam that could burn you. **23.** (a) if they have the same mass (b) Since they are at the same temperature, there is no net energy transfer between the two blocks. (c) The blocks need not touch each other in order to be in thermal contact. They can be in thermal contact due to convection and radiation.

25. (a)

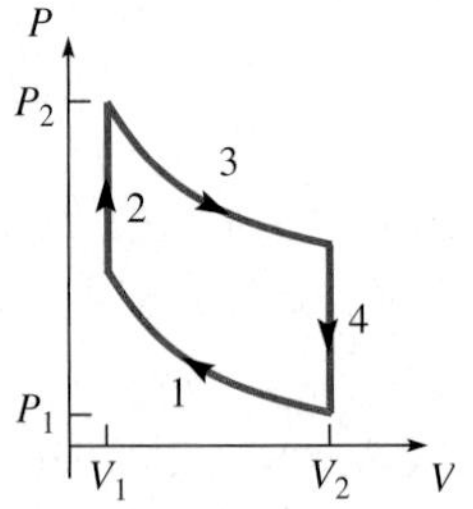

(b)

Process	W (kJ)	ΔU (kJ)	Q (kJ)
Step 1	11.2	0	−11.2
Step 2	0	27.4	27.4
Step 3	−34.1	0	34.1
Step 4	0	−27.4	−27.4
Total	−22.8	0	22.8

(c) 0.371 or 37.1% (d) 0.670 or 67.0% **27.** 132°C **29.** (a) 11 200 m/s (b) 1850 m/s (c) 461 m/s (d) The atoms in the high end of the distribution are much faster than the average. Some of the hydrogen atoms have speeds greater than the escape speed, thus they can escape. This is not the case for oxygen, which is much more massive and, thus, much slower.

MCAT Review

1. C **2.** B **3.** C **4.** B **5.** A **6.** D **7.** A **8.** C **9.** B **10.** A **11.** A

CHAPTER 16

Multiple-Choice Questions

1. (j) **2.** (c) **3.** (e) **4.** (a) **5.** (c) **6.** (b) **7.** (d) **8.** (c) **9.** (b) **10.** (b)

Problems

1. 9.6×10^5 C **3.** (a) added (b) 3.7×10^9 **5.** (a) negative charge (b) an equal magnitude of positive charge **7.** $Q/4$; 0 **9.** 30 km **11.** 2.268×10^{39} **13.** (a) 6.0×10^{-5} N toward the −3.0-nC charge (b) 6.0×10^{-5} N toward the 2.0-nC charge **15.** $kq^2/(2d^2)$ to the left **17.** 2.8×10^{-12} N toward the Cl^- ion **19.** 1.2 N at 28° below the negative x-axis **21.** 6.21 μC and 1.29 μC **23.** 6.8 mN **25.** 0.72 N to the east **27.** 3.2×10^{12} m/s² up **29.** 1.5×10^8 N/C directed toward the −15-μC charge
31.

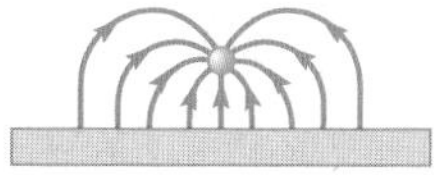

33. $7k|q|/(4d^2)$ to the left **35.** $0 < x < 3d$
37.

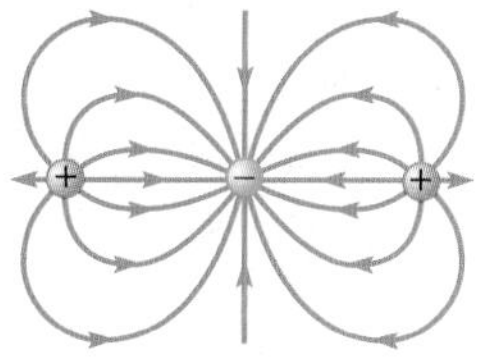

39. 1.12×10^6 N/C up **41.** 9.78×10^5 N/C at 14.6° CCW from a vertical axis through the left side of the square **43.** 1.61 N in the $+x$-direction **45.** $-0.43q$ **47.** 400 N/C **49.** (a) 8.010×10^{-17} N down (b) 2.40×10^{-19} J **51.** (a) The gravitational force is about 1/3 of the electrical force, so the gravitational force can't be neglected. (b) 1.78 m **53.** 1.3×10^5 N/C **55.** (a) vertically downward (b) 2600 N/C (c) 4.3×10^{-17} m **57.** (a) −6 μC (b) 12 μC **59.** (a) 6.8×10^6 N/C (b) 0 (c) 2.3×10^6 N/C **61.** (a) -6.8×10^5 C; −1.3 nC/m² (b) 1×10^{-12} C/m³ **63.** $0.866EA$ **65.** (a) $4\pi r^2 E$

67. (a) $E(r \geq R) = \dfrac{kq}{r^2}$ (b) $E(r \leq R) = \dfrac{kq}{R^3} r$
(c)

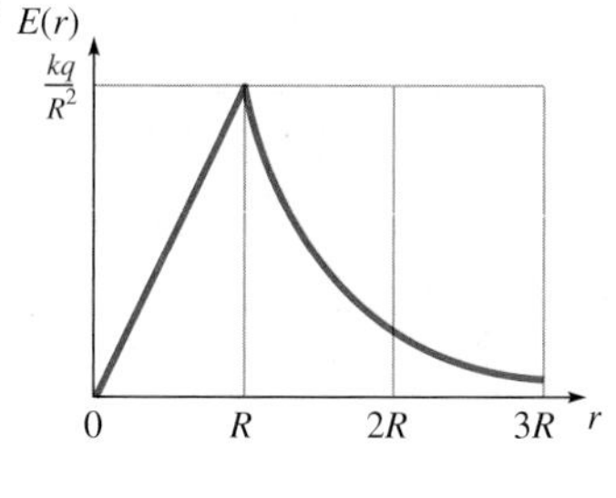

69. (a)

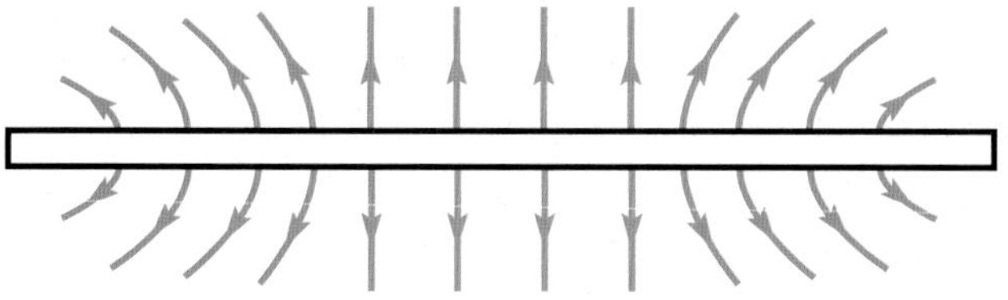

(b)

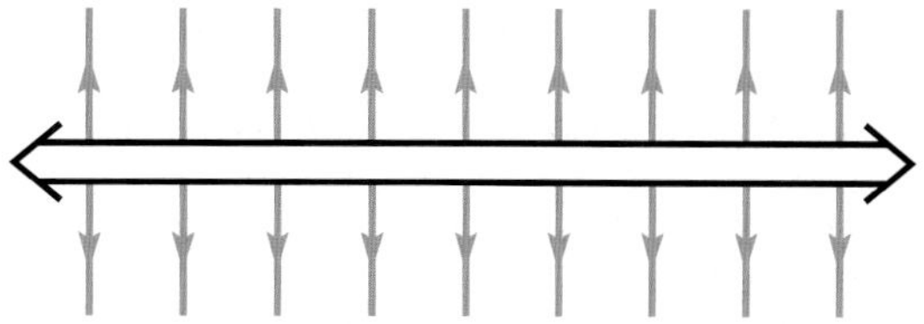

(c) The field strength is independent of the distance from the sheet. (d) yes **71.** (a)

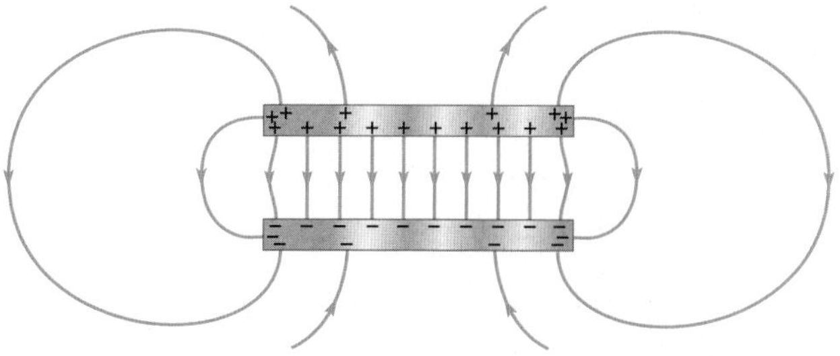

The fields due to each plate have the same direction (adding fields) between the plates and opposite directions (canceling fields) outside. Thus, the field is much stronger between the plates. (c) The result agrees. **73.** kq/r^2 ($r > R$) directed radially away from the center of the shell **75.** 1.3 m **77.** 3.20×10^{-14} N upward **79.** 2.2×10^6 m/s **81.** −1.5 nC **83.** 0.80 **85.** $E_x = 2.89 \times 10^5$ N/C; $E_y = 2.77 \times 10^6$ N/C **87.** (a) 8.4×10^7 m/s (b) 6.6 ns
89.

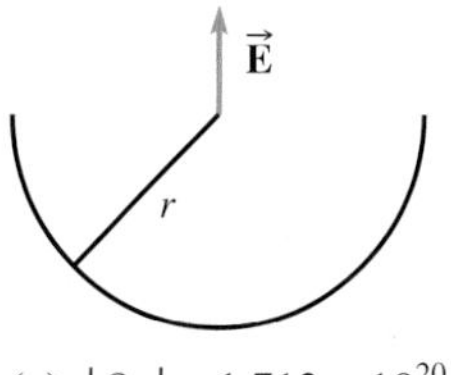

91. (a) $|Q_S| = 1.712 \times 10^{20}$ C and $|Q_E| = 5.148 \times 10^{14}$ C (b) No, the force would be repulsive. **93.** (a) $\dfrac{kqd}{x^3}$ (b) negative y-direction for all x **95.** (a) 0 (b) $-qEd \sin\theta$
(c)

θ (°)	Torque (N·m)
0	0
36.9	−0.0025
90.0	−0.0042

CHAPTER 17

Multiple-Choice Questions

1. (a) **2.** (f) **3.** (c) **4.** (e) **5.** (d) **6.** (b) **7.** (f) **8.** (e) **9.** (b) **10.** (b) **11.** (b) **12.** (d)

Problems

1. −18 mJ **3.** 8.4 J **5.** −3.0 J **7.** −17.5 μJ **9.** −11.2 μJ **11.** −2.70 μJ **13.** 4.49 μJ **15.** 75 nJ **17.** $\vec{E} = 0$; $V = 2.3 \times 10^7$ V **19.** (a) −1.5 kV (b) −900 V (c) 600 V; increase (d) -6.0×10^{-7} J; decrease (e) 6.0×10^{-7} J **21.** (a) positive (b) 10.0 cm
23. (a)

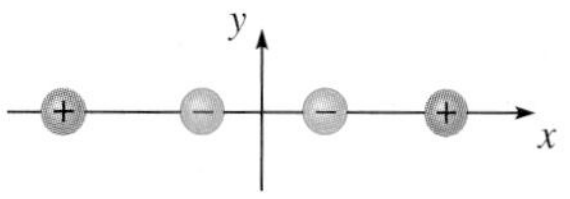

(b) 36 kV **25.** 9.0 V **27.** (a) $V_a = 300$ V; $V_b = 0$ (b) 0 **29.** (a) $V_b = -899$ V; $V_c = 0$ (b) 1.80 μJ **31.** (a) 1.0 μN to the right (b) 0.25 μJ (c) 60 V **33.** 1.0 cm **35.** cylinders;

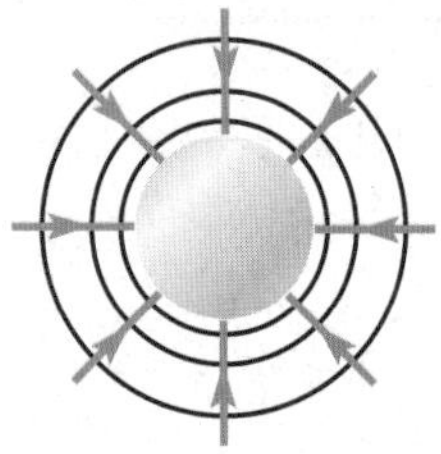

37. (a)

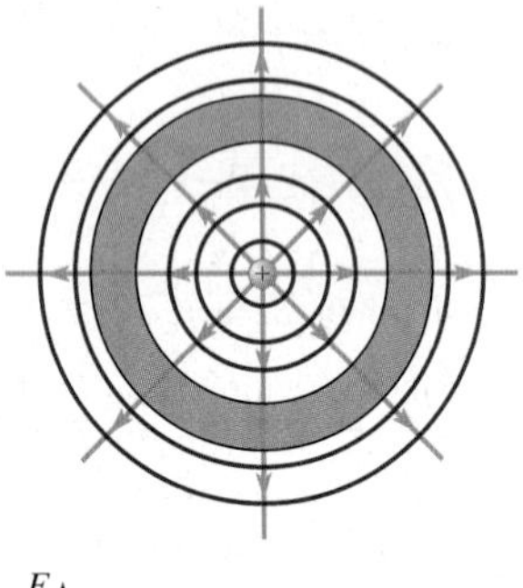

(b)

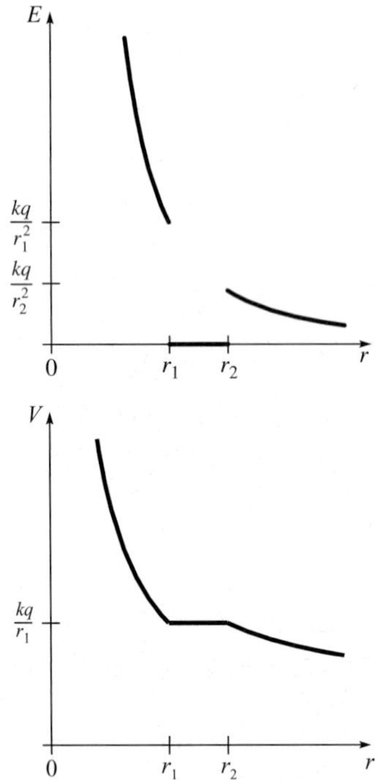

39. 1.6×10^{-19} C = e **41.** 150 V **43.** 4.6×10^7 m/s **45.** (a) upward (b) $\frac{v_y md}{e\,\Delta V}$ (c) decreases **47.** 2.8×10^{-16} J **49.** 2.56×10^{-17} J **51.** 18 μC **53.** 612 μC **55.** (a) stays the same (b) increases **57.** (a) decreases (b) stays the same (c) stays the same **59.** (a) 0.347 pF (b) 0.463 pF **61.** 8.0 pF **63.** 4.51×10^6 m/s **65.** (a) 3.3×10^3 V/m (b) 6.0×10^2 V/m **67.** (a) 1.1×10^5 V/m toward the hind legs (b) Cow A **69.** 0.30 mm **71.** 89 nF **73.** (a) 7.1 μF (b) 1.1×10^4 V **75.** The energy increases by 50%. **77.** (a) 0.18 μF (b) 8.9×10^8 J **79.** (a) 18 nC (b) 1.3 μJ **81.** (a) 630 V (b) 0.063 C **83.** 0.27 mJ **85.** (a) 0.14 C (b) 0.30 MW **87.** (a) 10.0 GJ (b) 443 kg (c) 0.694 month **89.** (a) $U_a = -6.3$ μJ; $U_b = U_c = 0$ (b) −6.3 μJ **91.** 450 kV

93. (a)

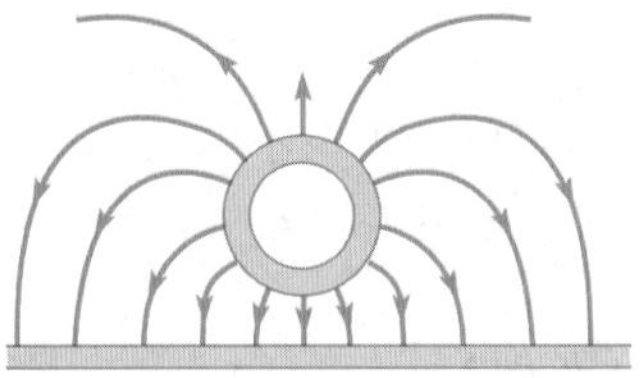

(b)

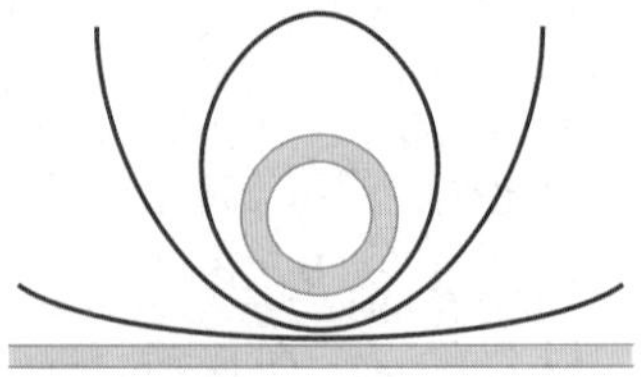

95. 9×10^6 V/m **97.** 51 **99.** 8.29×10^6 **101.** 3.44 mK **103.** 5×10^{-14} F **105.** (a) 2.7 kV (b) 6.8 μJ **107.** 1.44×10^{-20} J **109.** The energy is reduced by 20.0%. **111.** (a) 3.2×10^{-7} F/m (b) the outside of the membrane; 8.8×10^{-4} C/m^2 **113.** (a) 1.25 μm (b) 1600 m^2 **115.** 8.0 V/m

CHAPTER 18

Multiple-Choice Questions

1. (a) **2.** (d) **3.** (f) **4.** (d) **5.** (c) **6.** (b) **7.** (b) **8.** (d) **9.** (d) **10.** (b)

Problems

1. 4.3×10^4 C **3.** (a) from the anode to the filament (b) 0.96 μA **5.** 2.0×10^{15} electrons/s **7.** 22.1 mA **9.** 810 J **11.** (a) 264 C (b) 3.17 kJ **13.** $v_1 = 4v_2$ **15.** 17.8 min **17.** 81 μm **19.** 0.11 mm/s **21.** 1.3 A **23.** 0.794 **25.** (a) 50 V (b) to avoid becoming part of the circuit **27.** 2.5 mm **29.** 1750°C **31.** 4.0 V; 4.0 A **33.** $E = \rho\frac{I}{A}$, where ρ is the resistivity. **35.** The electric field stays the same, the resistivity decreases, and the drift speed increases. **37.** (a) 7.0 V (b) 18 Ω **39.** (a) 23.0 μF (b) 368 μC (c) 48 μC **41.** (a) 5.0 Ω (b) 2.0 A **43.** (a) 1.5 μF (b) 37 μC **45.** (a) 0.50 A (b) 1.0 A (c) 2.0 A **47.** (a) $R/8$ (b) 0 (c) 16 A **49.** (a) 8.0 μF (b) 17 V (c) 1.0×10^{-4} C **51.** (a) 2.00 Ω (b) 3.00 A (c) 0.375 A

53.

Branch	I(A)	Direction
AB	0.20	right to left
FC	0.12	left to right
ED	0.076	left to right

55. 75 V; 8.1 Ω **57.** 4.0 W **59.** 0.50 A **61.** yes; 600 W **63.** 80.0 J **65.** (a)

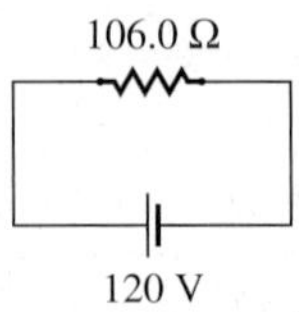

(b) 1.1 A (c) 41 V (d) upper branch: 0.68 A; lower branch: 0.45 A (e) $P_{50} = 64$ W; $P_{70} = 14$ W; $P_{40} = 18$ W

67. (a) 6.5 Ω (b) 18 A (c) 0.86 mm (d) 21 A **69.** (a) 5.28 V (b) 6.34 W **71.** (a) $\mathscr{E} - Ir$ (b) $P = \mathscr{E}I - I^2r$; $\mathscr{E}I$ is the rate that an ideal emf supplies power, and I^2r is the rate of energy dissipation by the internal resistance. (c) $\mathscr{E}I$ (d) $\mathscr{E}I + I^2r$

73. (a)

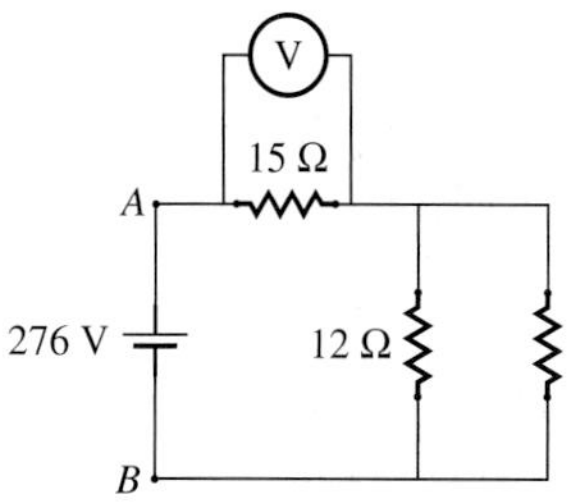

(b)

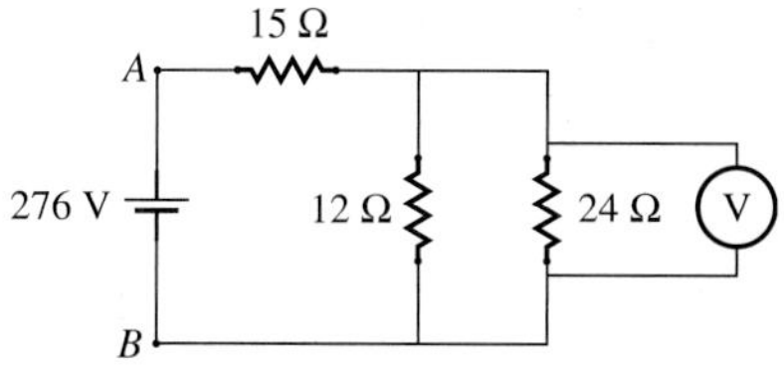

75. (a)

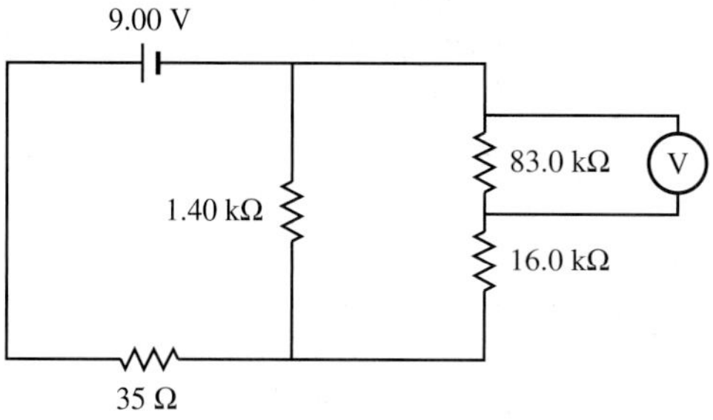

(b) 7.36 V (c) 7.86 mV **77.** 833 kΩ **79.** (a) 25 kΩ (b) 250 kΩ **81.** 150 Ω; 2.50 mA **83.** 2.77 s **85.** (a) 140 μF; 90 Ω; 5.8 mJ (b) 4.4 ms

(c)

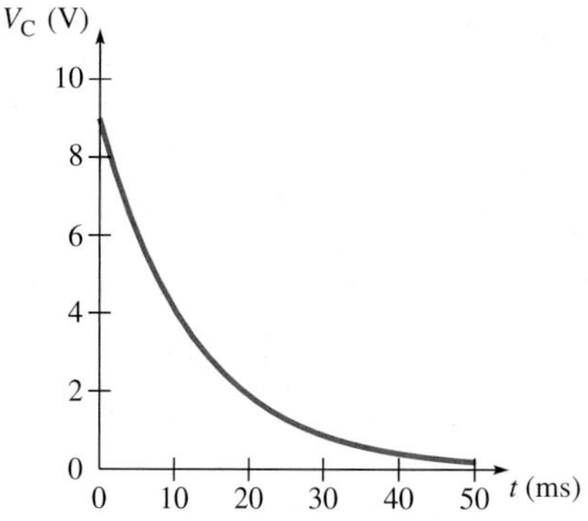

87. (a) 632 V (b) 63.2 mC (c) 6.7 Ω **89.** (a) $I_1 = I_2 = 0.30$ mA; $V_1 = V_2 = 12$ V (b) $I_1 = I_2 = 0.18$ mA; $V_1 = 12$ V; $V_2 = 7.3$ V (c) $I_1 = I_2 = 25$ μA; $V_1 = 12$ V; $V_2 = 0.99$ V

91. (a)

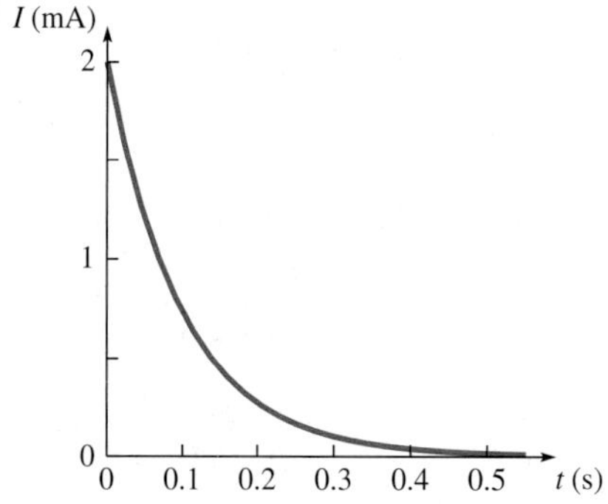

(b) 20 mW (c) 1 mJ **93.** (a) 4.2 mC (b) 470 μF (c) 130 Ω (d) 74 ms **95.** (a) 50 mA (b) 7.4 mA **97.** (a) *D* (b) The devices cannot all be operated at the same time since the total current would be 25 A, which is greater than the rated 20.0 A. **99.** 6.5 kJ **101.** (a) 2.00 A (b) 1.00 A **103.** 31 μA **105.** 9.3 A **107.** 20 W **109.** (a) $R_{60} = 240\ \Omega$; $R_{100} = 140\ \Omega$ (b) 60.0-W bulb (c) 100.0-W bulb **111.** (a) 350 Ω (b) no **113.** (a) 9.6 Ω (b) 13 A (c) 1.3 cents (d) 6.0 kW (e) 25 A **115.** (a) 175 μJ (b) $Q_{2f} = 16.0$ μC; $Q_{3f} = 24.0$ μC; 160 μJ (c) heat loss

117. (a)

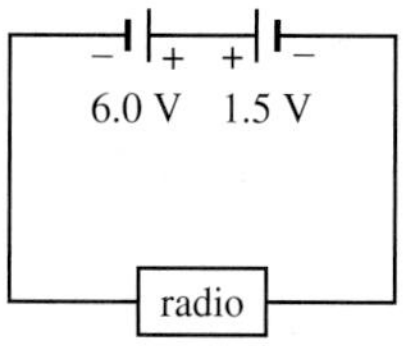

(b) The 1.5-V battery is not meant to be recharged. **119.** $9R_0$ **121.** $v_{Au} = 3v_{Al}$ **123.** (a) 30 μA (b) *A*: 3.0 V; *B* and *C*: 0.86 V **125.** (a) 0.090 F (b) 2.5 TJ (c) 29 kΩ; 260 A (d) 200 min **127.** (a) 2 mA (b) 3 mA (c) 3 mW (d) increase by a factor greater than two

REVIEW & SYNTHESIS: CHAPTERS 16–18

Review Exercises

1. 12.0 μC **3.** 6.24 N at 16.1° below the +*x*-axis **5.** (a) −238 nC (b) 0.889 N **7.** 4.4 mm **9.** (a) 35.0 Ω (b) 0.686 A (c) 16.5 W (d) 6.9 V (e) 0.34 A (f) 2.4 W **11.** 3.01×10^{-12} m **13.** (a) 0.44 A (b) 5.3 V **15.** The charge on the plates increases by a factor of 310. **17.** 66.7 Ω **19.** (a) 200 nC (b) 5.6 μC **21.** (a) 710 μF (b) 9.8 (c) 8.0 kW (d) 64 J **23.** 14 Ω **25.** 51 s **27.** (a) 220 V (b) 0.60 m/s (c) 1.2 nN (d) It is not realistic to ignore drag, since $F_E \ll F_D$. The potential difference should be larger.

MCAT Review

1. D **2.** C **3.** C **4.** B **5.** A **6.** D **7.** A **8.** C **9.** D **10.** C **11.** C **12.** C **13.** C

CHAPTER 19

Multiple-Choice Questions

1. (g) **2.** (f) **3.** (e) **4.** (e) **5.** (c) **6.** (b) **7.** (c) **8.** (g) **9.** (b) **10.** (b) **11.** (d) **12.** (d)

Problems

1. (a) *F* (b) *A*; highest density of field lines at point *A* and lowest density at point *F*

3.

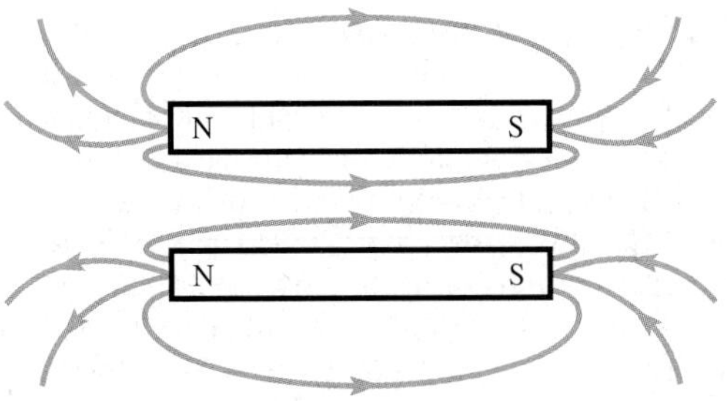

5.

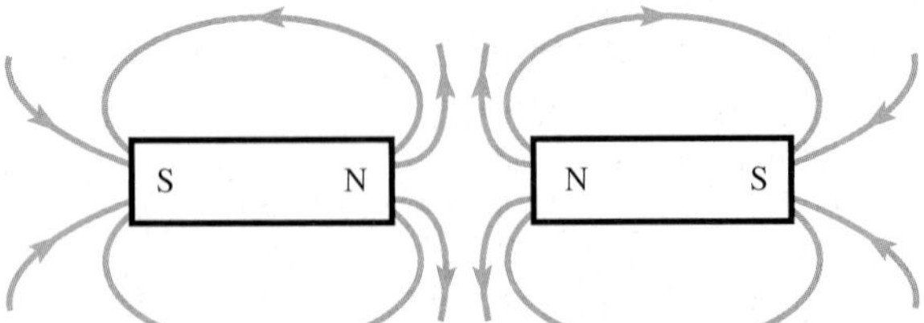

7. (a)

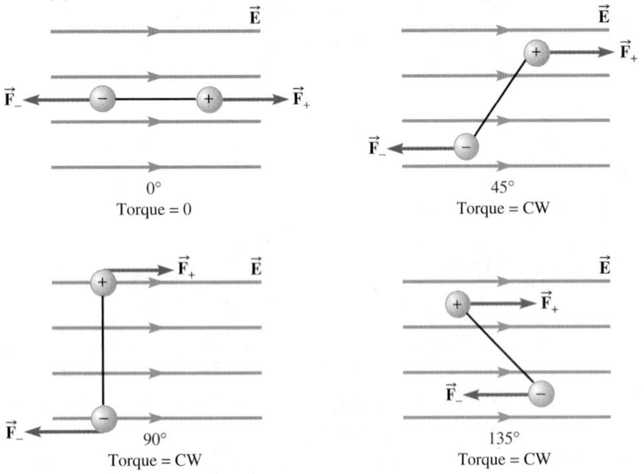

(b) parallel to the electric field lines **9.** 2.4×10^{-12} N up **11.** 5.1×10^{-13} N north **13.** 4.8×10^{-14} N into the page **15.** (a) 3.0×10^7 m/s (b) 4.2×10^{18} m/s^2 (c) 3.0×10^7 m/s (d) 2.4×10^7 V/m (e) Since the force due to the magnetic field is always perpendicular to the velocity of the electrons, it does not increase the electrons' speed but only changes their direction. **17.** 2.1×10^{-16} N to the west **19.** (a) 3.1×10^7 m/s (b) 4.0×10^7 m/s (c) 38° **21.** There are two possibilities: 21°E of N and 21°E of S. **23.** 0.78 T **25.** 2.83×10^7 m/s **27.** 2.6×10^{-25} kg **29.** 0.17 T **31.** (a) 39 u (b) potassium

33. $\dfrac{2\pi m}{qB}$ **35.** (a), (b)

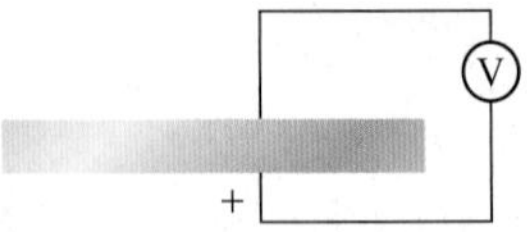

37. 8.4×10^{28} m^{-3} **39.** (a) upward (b) 0.20 mm/s **41.** (a) 0.35 m/s (b) 4.4×10^{-6} m^3/s (c) the east lead **43.** $\dfrac{E^2}{2B^2\,\Delta V}$ **45.** (a) 0.50 T (b) We do not know the directions of the current and the field; therefore, we set $\sin\theta = 1$ and get the minimum field strength. **47.** (a) north (b) 15.5 m/s **49.** (a) $\vec{\mathbf{F}}_{\rm top} = 0.75$ N in the $-y$-direction; $\vec{\mathbf{F}}_{\rm bottom} = 0.75$ N in the $+y$-direction; $\vec{\mathbf{F}}_{\rm left} = 0.50$ N in the $+x$-direction; $\vec{\mathbf{F}}_{\rm right} = 0.50$ N in the $-x$-direction (b) 0 **51.** (a) 18° below the horizontal with the horizontal component due south (b) 42 A **53.** (a) 0.21 T (b) clockwise **55.** (a) 0.0014 N·m (b) 0 **59.** 4.9°

61. 9×10^{-8} T out of the page **63.** 0 **65.** 3.2×10^{-16} N parallel to the current **67.** *B*, *D*, *C*, *A* **69.** $\vec{\mathbf{B}}_A = 0$; $\vec{\mathbf{B}}_B = 3.38 \times 10^{-7}$ T out of the page **71.** 750 A into the page **73.** 6.03 A, CCW **75.** 80 μT to the right **77.** 0.11 mT to the right **79.** (a) 4.9 cm (b) opposite **81.** (a)

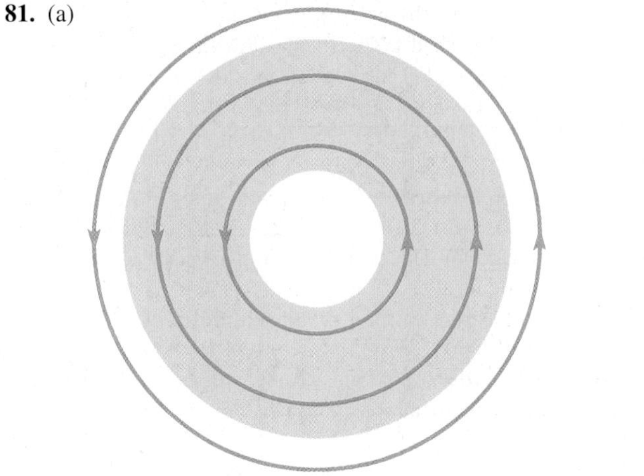

(b) $\dfrac{\mu_0 I}{2\pi r}$ CCW as seen from above **83.** n depends upon r; $B = \dfrac{\mu_0 NI}{2\pi r}$; the field is not uniform since $B \propto \dfrac{1}{r}$. **85.** 9.3×10^{-24} N·m **87.** (a) graph a (b) graph c **89.** south **91.** 1.25×10^{-17} N in the $+x$-direction **93.** (a) 10 A (b) farther apart **95.** (a) 1.2×10^7 Hz (b) 2.2×10^{10} Hz **97.** 20.1 cm/s **99.** 2.00×10^{-7} T up **101.** into the page **103.** $\tan^{-1}\dfrac{\mu_0 NI}{2rB_{\rm H}}$ **105.** (a)

Side	Current direction	Field direction	Force direction
top	right	out of the page	attracted to long wire
bottom	left	out of the page	repelled by long wire
left	up	out of the page	right
right	down	out of the page	left

(b) 1.0×10^{-8} N away from the long wire **107.** (a) $\vec{\mathbf{F}}_{\rm top} = 0.65$ N into the page; $\vec{\mathbf{F}}_{\rm bottom} = 0.65$ N out of the page; $\vec{\mathbf{F}}_{\rm right} = 0.25$ N out of the page; $\vec{\mathbf{F}}_{\rm left} = 0.25$ N into the page (b) 0 **109.** 6.4×10^{-14} N at 86° below west **111.** (a) 8.6×10^{-15} N at 68° below west (b) No; since $\vec{\mathbf{F}}_{\rm E}$ and $\vec{\mathbf{F}}_{\rm B}$ are perpendicular **113.** (a) 180 km (b) 2.4×10^6 m

CHAPTER 20

Multiple-Choice Questions

1. (c) **2.** (c) **3.** (d) **4.** (b) **5.** (d) **6.** (e) **7.** (e) **8.** (b) **9.** (b) **10.** (c)

Problems

1. (a) $\dfrac{vBL}{R}$ (b) CCW (c) left (d) $\dfrac{vB^2L^2}{R}$ **3.** (a) $\dfrac{vB^2L^2}{R}$ (b) $\dfrac{v^2B^2L^2}{R}$ (c) $\dfrac{v^2B^2L^2}{R}$ (d) Energy is conserved since the rate at which the external force does work is equal to the

power dissipated in the resistor. **5.** (a) 3.44 m/s (b) The magnitude of the change in gravitational potential energy per second and the power dissipated in the resistor are the same, 0.505 W. **7.** 3.3 T
9. (a) into the page (b) no (c) no (d) $\vec{\mathbf{F}}$ is out of the page; no, electrons are pushed perpendicular to the length of the wire; there is no induced emf. The situation for side 1 is identical to that of side 3. **11.** (a) no magnetic force; $\mathscr{E} = 0$; $I = 0$
(b) 32 nV; 400 pA; 2.9×10^{-17} N to the left (c) 1.3×10^{-17} W
13. 5.2×10^{-4} Wb **15.** (a) CCW (b) 3.5×10^{-3} W/s
17. yes; counterclockwise as viewed from the right **19.** no
21. (a) 28.8 mV (b) 5.13 A **23.** (a) 0.070 A (b) CCW
25. (a) $B\omega R^2/2$ (b) The result is half of the value of the motional emf of a rod moving at constant speed v; which is reasonable, since different points on the rotating rod have different speeds ranging from 0 to v. **27.** (a) CW (b) for a brief moment
(c) to the right (d) to the left **29.** (a) 0.750 A (b) 3.00 A; Tim should shut the trimmer off because the wires in the motor were not meant to sustain this much current. The wires will burn up if this current flows through them for very long. **31.** 110 V
33. (a) 1/20 (b) 1000 **35.** 2.00 **37.** 28 V; 0.53 A
39. (a) CCW; CW
(b)

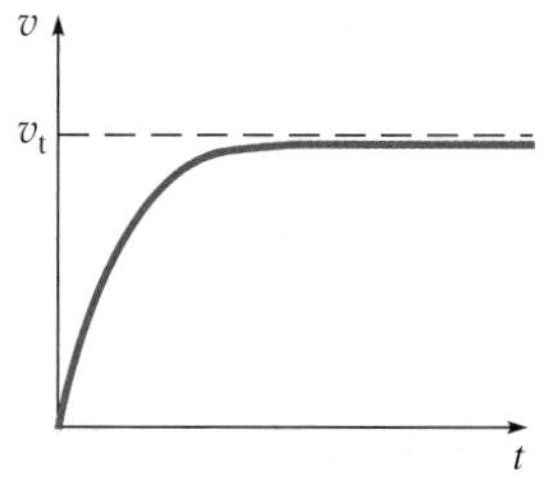

41. 3.1×10^{-8} **43.** (a) CCW (b) 1.4×10^{-5} **45.** 8.5×10^{-4} H
47. (a) 3.1×10^{-5} V (b) 14 mV **51.** 1.6 mV
53. $L_{eq} = \dfrac{L_1 L_2}{L_1 + L_2}$

55. (a) 0 (b) 6.0 V (c) 0.60 A **57.** (a) 0 (b) 12 V
59. 22.0 Ω; 0.160 H **61.** (a) 50 A (b) The attempt to suddenly stop the current would induce a huge emf in the windings of the electromagnet, possibly damaging it. It is likely that sparks would complete the circuit across the open switch. The shunt resistor is used to reduce the rate of change of the current, and allow the electromagnet to be safely shut off. (c) 50 kW (d) 2.3 s
(e) $\tau \propto \dfrac{1}{R_L + R_{shunt}}$ **63.** (a) V·s/A; V/A; V **65.** 580 ns
67. (a) 180 mA (b) 2.5 mJ (c) 1.1 W (d) 0
69. (a) $\vec{\mathbf{F}}_2 = \dfrac{\omega B^2 AL}{R}\sin\omega t$ down and $\vec{\mathbf{F}}_4 = \dfrac{\omega B^2 AL}{R}\sin\omega t$ up
(b) The magnetic forces on sides 1 and 3 are always parallel to the axis of rotation. (c) $\dfrac{2\omega r B^2 AL}{R}\sin^2\omega t$ CCW
(d) The torque is counterclockwise, and the angular velocity is clockwise, so the magnetic torque would tend to decrease the angular velocity. **71.** CCW as viewed from the left **73.** (a) L (b) L
75. 7.0×10^{-5} Wb **77.** 81 mJ **79.** (a) 120 (b) $I_1 = 0.48$ A; $I_2 = 4.1$ mA **81.** (a) $\dfrac{\mu_0 N_1 N_2 \pi r^2}{L}$ (b) $\dfrac{\mu_0 N_1 N_2 \pi r^2}{L}\dfrac{\Delta I_1}{\Delta t}$
83. (a) $2\pi r^2 NB$ (b) $\dfrac{2\pi r^2 NB}{\Delta t}$ (c) $\dfrac{2\pi r^2 NB}{R\,\Delta t}$ (d) $\dfrac{2\pi r^2 NB}{R}$
85.

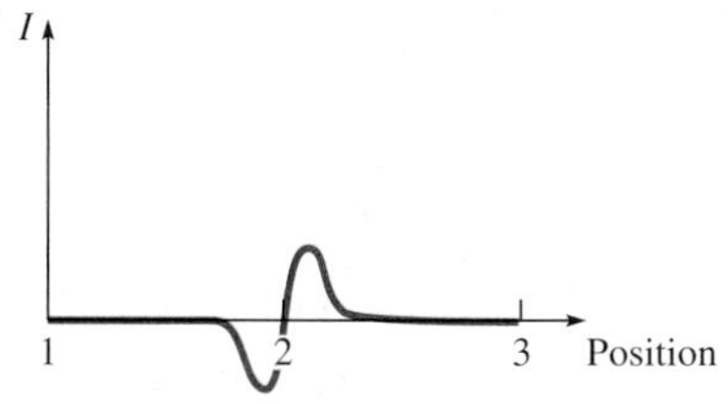

87. (a) $I(t) = \dfrac{\mathscr{E}_m}{\omega L}\cos\omega t$ (b) ωL (c) $\dfrac{\pi}{2\omega}$ **89.** (a) 3.1 V (b) the northernmost wingtip **91.** (a) $\dfrac{\mu_0 N_1 N_2 \pi r_1^2}{L_2}$
(b) $\dfrac{\mu_0 N_1 N_2 \pi r_1^2}{L_2}\dfrac{\Delta I_2}{\Delta t}$

CHAPTER 21

Multiple-Choice Questions

1. (i) **2.** (b) **3.** (a) **4.** (d) **5.** (a) **6.** (d) **7.** (c) **8.** (c)
9. (c) **10.** (b)

Problems

1. 120 times per second **3.** 18 A **5.** 6000 W; the heating element of the hair dryer will burn out because it is not designed for this amount of power. **7.** (a) 35 A (b) 3.2 kW **9.** −5.7 V and 5.7 V **13.** The current is increased by a factor of 6.0.
15. (a) 60.0 Hz (b) 13.3 kΩ **17.** 53 nF **19.** (a) 8.0×10^5 Ω
(b) 3.3 **21.** (a)

C (μF)	V (V)
2.0	6.0
3.0	4.0
6.0	2.0

(b) 0.48 A

23. The current is reduced by a factor of 1/6.0. **25.** 27 Ω
27. 7.33 kHz **29.** (a) negative (b)

(d) positive;

31. (a) $\omega LI\cos\omega t$
(c) $v_L(t)$ leads $i(t)$ by $\mu/2$.

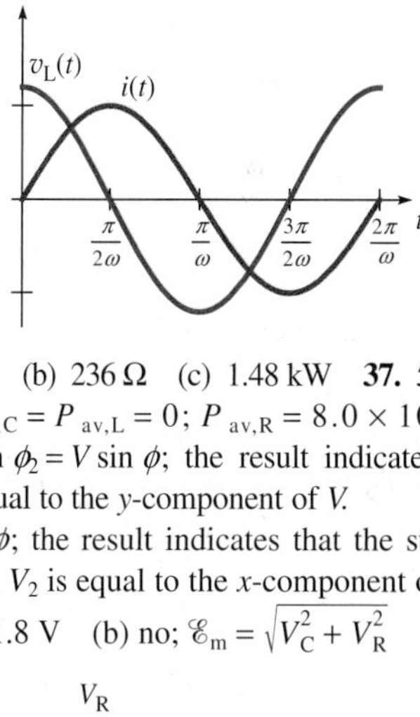

33. 470 Hz **35.** (a) 384 Ω (b) 236 Ω (c) 1.48 kW **37.** 500 Ω
39. (a) 0.800 (b) $P_{av,C} = P_{av,L} = 0$; $P_{av,R} = 8.0 \times 10^{-5}$ W
41. 0.33 *W* **43.** (a) $V_2 \sin\phi_2 = V\sin\phi$; the result indicates that the y-component of V_2 is equal to the y-component of V.
(b) $V_1 + V_2\cos\phi_2 = V\cos\phi$; the result indicates that the sum of the x-components of V_1 and V_2 is equal to the x-component of V.
45. (a) $V_C = 0.97$ V; $V_R = 1.8$ V (b) no; $\mathscr{E}_m = \sqrt{V_C^2 + V_R^2}$
(c)

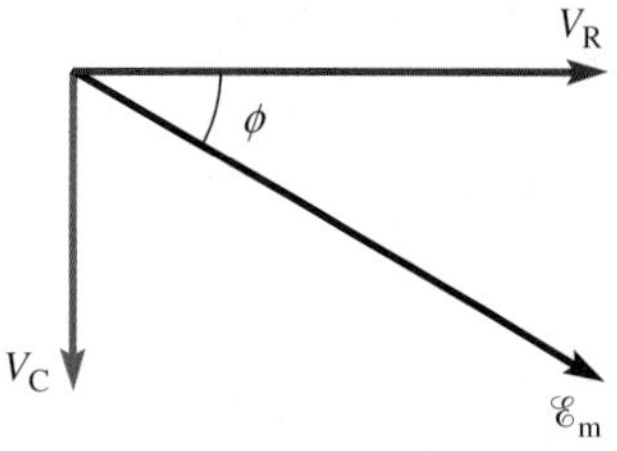

47. (a) 24 Hz (b) $\dfrac{V_R}{\mathscr{E}_m} = \dfrac{V_C}{\mathscr{E}_m} = \dfrac{1}{\sqrt{2}}$ (c) I leads $\mathscr{E}$ by $\dfrac{\pi}{4}$ rad $= 45°$.
(d) 4.7 kΩ **49.** 8.0×10^{-7} F; 310 Ω **51.** 0.724 pF to 1.1 pF **53.** 500.0 Ω **55.** (a) 0° (b) 2.4 V (c) I_{rms} decreases

57. (a) 745 rad/s (b) 790 Ω (c) $V_R = \mathscr{E}_m = 440$ V; $V_C = V_L =$ 125 V **59.** (a) 0° (b)

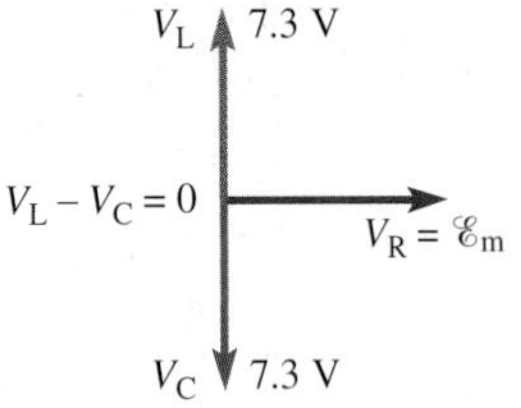

(c) 98.7 Hz

61. (a) 8.1 Ω (b) 8.1 Ω (c) 7×10^{-4} H (d) $f_{co} = \dfrac{1}{2\pi\sqrt{LC}}$
63. (a) 750 rad/s
(b)

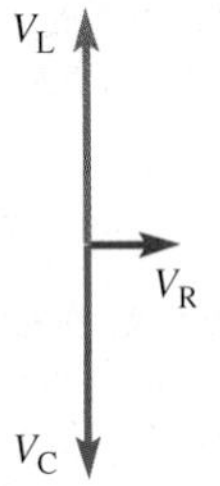

(c) $V_{ab} = 440$ V; $V_{bc} = 1.1$ kV; $V_{cd} = 1.1$ kV; $V_{bd} = 0$; $V_{ad} = 440$ V (d) 750 rad/s (e) 3.5 A
65. (a) 0.10 A (b) 38 μF (c) 0.40 H
67. 0.2 V **69.** (a) 570 (b) 34 A (c) 14 kW
71. (a) 92.8 mH (b) 28 mH **73.** 49 μF
75. 11 kHz **77.** 390 pF **79.** (a) 0.75; ±41° (b) No; V_L and V_C cannot be determined.
81. (a) 48 A (b) 56 A (c) The power lost in transmission is greater.
83. (a) 0.3 H (b) Yes; an inductor reduces the output with little energy loss and, therefore, it is a much better choice for a dimmer.
85. (a) 33.8 Ω (b) $\mathscr{E}_m = 286$ V; $V_{rms} = 202$ V (c) 8.46 A (d) 1.07 kW (e) $i(t) = (8.46\text{ A}) \sin [(390\text{ rad/s})t - 0.480\text{ rad}]$
87. (a) $X_1 = 2.65$ kΩ; $X_2 = 21.2$ Ω (b) $Z_1 = 3.32$ kΩ; $Z_2 = 2.00$ kΩ (c) 1.00 mA (d) $\phi_{12.0} = 53.0°$; $\phi_{1.50} = 0.608°$

REVIEW & SYNTHESIS: CHAPTERS 19–21

Review Exercises

1. 1.2 N·m; when the plane of the loop is parallel to the axis of the solenoid. **3.** (a) 1.66×10^{-4} T along the +x-axis (b) out of the page (c) 8.84 A **5.** 1.8×10^{-17} N out of the plane of the paper in the side view (or to the right in the end on view)
7. (a) 6.7×10^3 A (b) 530 MW; all of the power is dissipated, since 530 MW is greater than the total power output of the power plant. (c) 17 A (d) 3.3 kW; 0.42% (e) 25 **9.** (a) A counterclockwise current is induced because the flux through the loop is increasing as it nears the wire; the net force on the loop is away from the wire. (b) No current is induced because there is no change in flux through the loop; there is no magnetic force acting on the loop. (c) A clockwise current is induced because the flux through the loop is decreasing as it moves away from the wire; the net force on the loop is toward the wire. **11.** (a) to the right (b) 13.6 m/s (c) The applied emf has to increase because of the increased resistance in the longer rail lengths in the circuit and because of the increasingly large induced emf as the rod moves faster ($\mathscr{E} = vBL$). **13.** (a) The power is cut in half. (b) The power is 4/5 of its original value. **15.** (a) 445 Hz (b) current (c) −67° (d) 0.51 A (e) 67 W (f) $V_R = 180$ V; $V_L = 150$ V; $V_C = 590$ V **17.** 1.73×10^{-7} T
19. (a) vB north (b)

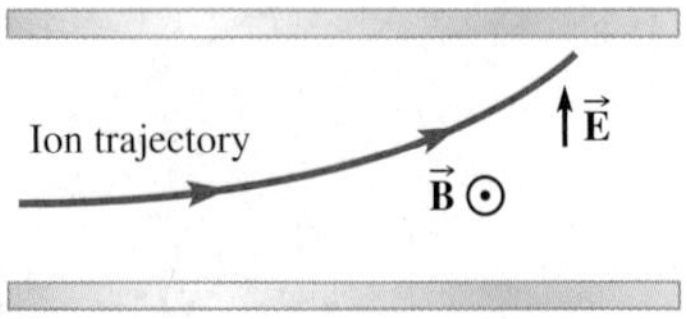

21. (a) $1.00638v$ (b)

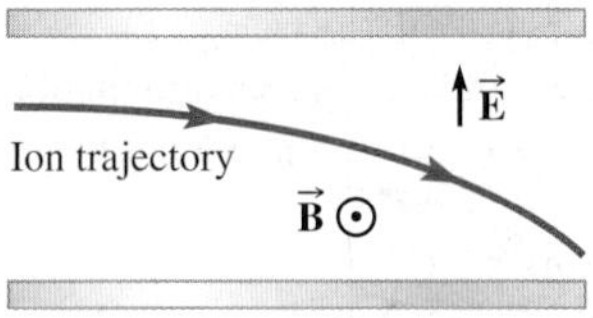

(c) $0.99366D$
23. 920 A
25. (a) 91 W (b) 1.1 kW

MCAT Review

1. A **2.** D **3.** D **4.** B **5.** B **6.** C **7.** A **8.** D

CHAPTER 22

Multiple-Choice Questions

1. (d) **2.** (f) **3.** (b) **4.** (f) **5.** (d) **6.** (b) **7.** (b) **8.** (c) **9.** (d) **10.** (d)

Problems

1. $\dfrac{\mu_0 I}{2\pi r}$ **3.** $\dfrac{\mu_0 I r}{2\pi R^2}$ **5.** east-west **7.** In the vertical plane defined by the vertical electric dipole antenna and the direction of wave propagation **9.** The axis of the magnetic dipole antenna should be oriented up-down. **11.** (a) magnetic field (b) 3.6 mV (c) 25 mV **13.** 2.5 GHz **15.** 1.62 **17.** 1.67 ns **19.** (a) 455 nm (b) 4.34×10^{14} Hz **21.** (a) 5.00×10^6 m (b) The radius of the Earth is 6.4×10^6 m, which is close in value to the wavelength. (c) radio waves **23.** (a) about one octave (b) approximately 8 octaves **25.** 5000 km; this means that in one oscillation, 1/60th of a second, the EM wave created from household current has traveled the entire length of the U.S. **27.** 2.0×10^{-12} T; 30 GHz **29.** (a) 7.5 mV/m; 3.0 MHz (b) 4.5 mV/m; in the +x-direction **31.** (a) −z-direction (b) $E_x = -cB_m \sin(kz + \omega t)$, $E_y = E_z = 0$ **33.** $E_y = E_m \cos(kx - \omega t)$, $E_x = E_z = 0$ and $B_z = \dfrac{E_m}{c}\cos(kx - \omega t)$, $B_x = B_y = 0$ where $\omega = \dfrac{1}{\sqrt{LC}}$ and $k = \dfrac{1}{c\sqrt{LC}}$ **35.** 260 V/m **37.** 2.4 s **39.** 9×10^{26} W **43.** (a) 2.9 kW (b) 1.2 kW (c) The excess energy produced at midday should be stored for use at night; in general, a supplemental energy source is necessary. **45.** 0.25 **47.** 65.7 W/m^2 **49.** (a) $I_1 = I_0 \cos^2 90.0° = 0$ (b) $0.250I_0$ (c) $0.531I_0$ **51.** (a) $0.125I_0$ (b) 0 **53.** yes; up-down **55.** 45.0 m/s **57.** 1440 nm **59.** (a) f_1 (b) −5.2 kHz **61.** 5×10^7 m/s **63.** 19 s **65.** 7.5 cm **67.** (a) 530 m (b) 390 pF (c) 380 μV **69.** 100 **71.** (a) $0.233I_0$ (b) $0.375I_0$ **73.** (a) 0.92 kW (b) 1.0×10^5 W/m^2 (c) 2.1×10^{-5} T **75.** 0.927 W **77.** 15 000 km; this scenario is certainly not feasible since it is almost halfway around the entire Earth.

CHAPTER 23

Multiple-Choice Questions

1. (b) **2.** (d) **3.** (f) **4.** (c) **5.** (d) **6.** (c) **7.** (b) **8.** (d) **9.** (a) **10.** (d)

Problems

1.

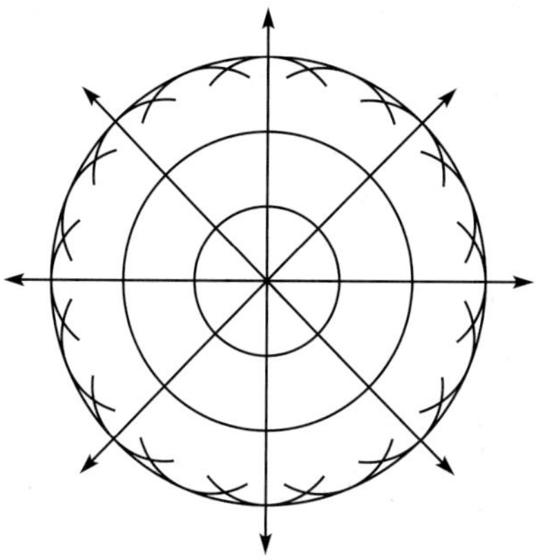

3.

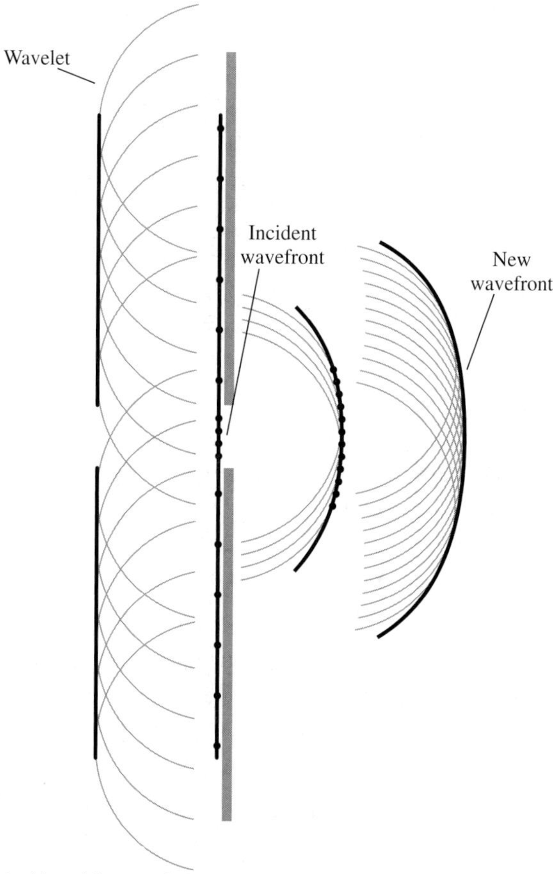

incident side: two planar waves;
transmitted side: one hemispherical wave

5.

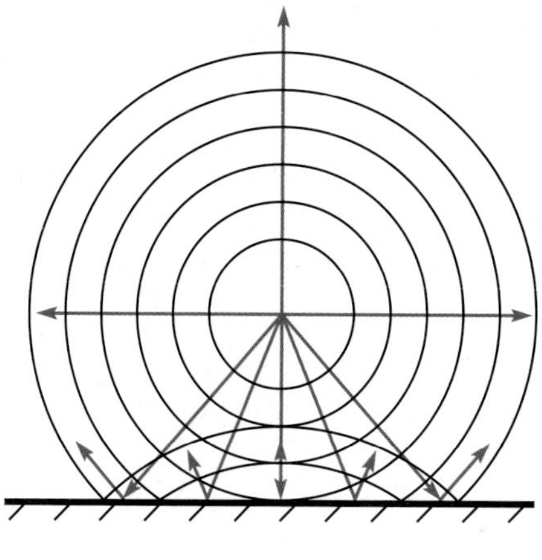

7. 100° **11.** 63.1° **13.** 11 m/s **15.** 44.6° **17.** 2.1 cm **19.** $\delta = \beta(n-1)$ **21.** $44.1° \leq \theta \leq 45.9°$ **23.** (a) 24.42° (b) 33.44° (c) Under water, the larger critical angle means that fewer light rays are totally reflected at the bottom surfaces of the diamond. Thus, less light is reflected back toward the viewer. **25.** (a) 1.556 (b) No; for $0 \leq \theta_i \leq \theta_c$, $0 \leq \theta_t \leq 90°$. **27.** The minimum index of refraction is 1.41. **29.** no **31.** (a) the left lens (b) 53° (c) 1.3 **33.** (a) 37.38° (b) perpendicular to the plane of incidence (c) 52.62° **35.** 4.8 mm **37.** 12.1 cm **39.** 0.91 m;

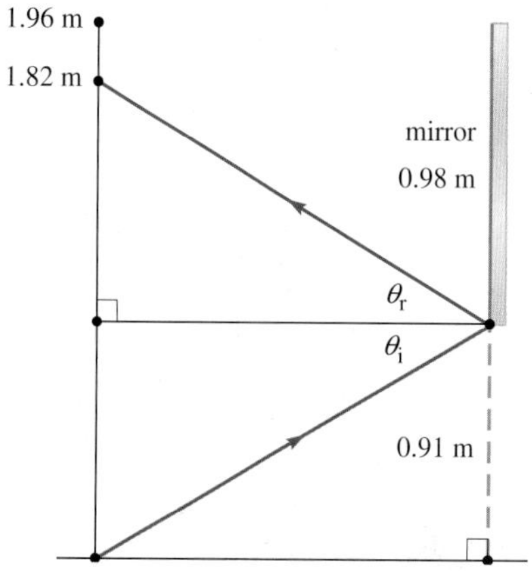

41. 2 m **43.** He sees 7 images total. **47.** 6.67 cm in front of the mirror

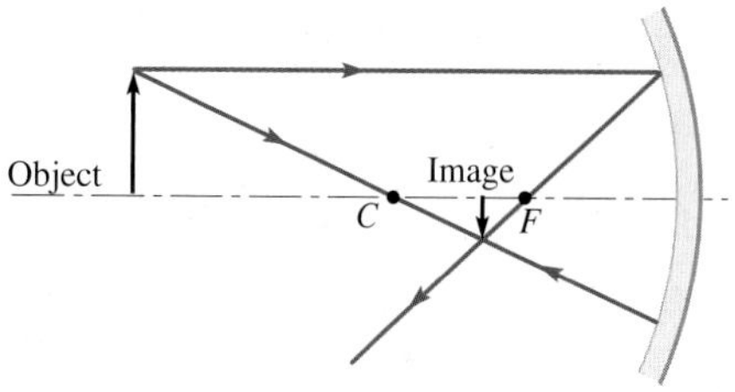

49. 18.8 cm in front of the mirror;

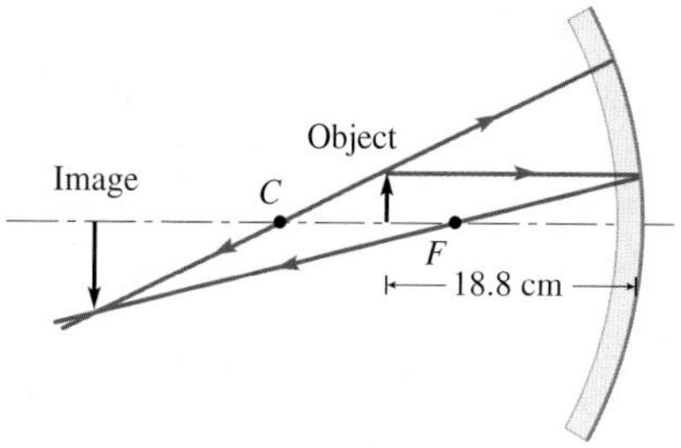

51. −210 cm **57.** (a) 11.7 cm;

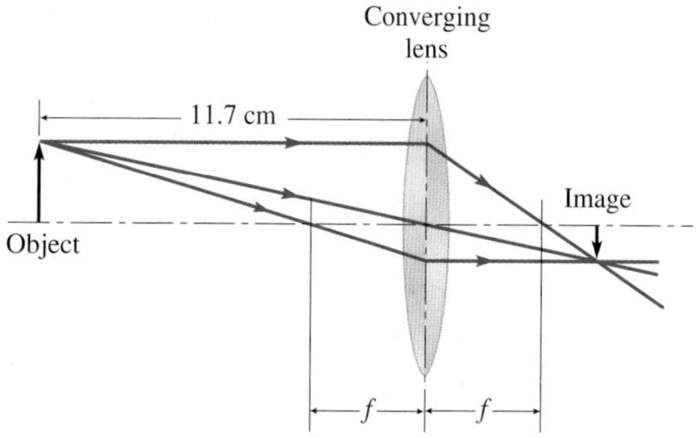

(b) real (c) −0.429

59.

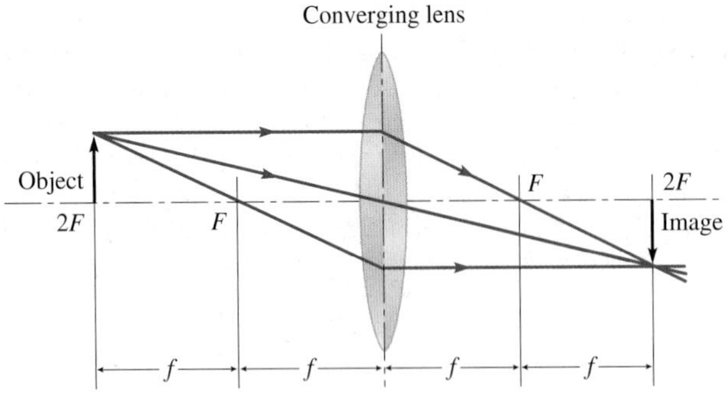

61.

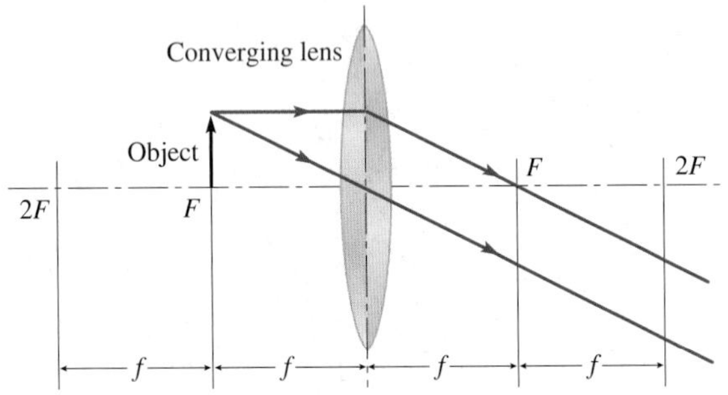

63.

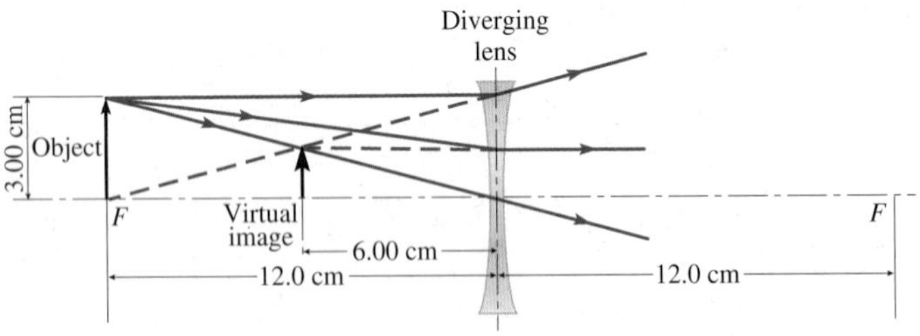

The image is located 6.00 cm from the lens on the same side as the object and has a height of 1.50 cm.

65. (a)

p (cm)	*q* (cm)	*m*	Real or virtual	Orientation	Relative size
5.00	−13.3	2.67	virtual	upright	enlarged
14.0	18.7	−1.33	real	inverted	enlarged
16.0	16.0	−1.00	real	inverted	same
20.0	13.3	−0.667	real	inverted	diminished

(b) $h' = 10.7$ cm; $h' = -2.67$ cm **67.** (a) converging (b) diverging (c) converging (d) diverging **69.** (a) 3.24 m (b) closer **71.** (a) concave (b) inside the focal length (c) 144 cm **73.** (a) 9.1 cm (b) convex (c) $f = -7.1$ cm; $R = 14$ cm **75.** Red light and blue light will have different focal points. The focal point for blue light will be closer to the lens. **77.** (a) virtual (b) convex (d) The image seems to be farther away than the object because it appears smaller than the object. **79.** 0.8 m/s **83.** (a) virtual (b) 2.4 cm; concave (c) smaller (d) $p < f$ **85.** (a) virtual (b) 180 cm behind the lens (c) 4.6 cm; converging **87.** The mirror should be concave; the focal length must be long compared to the object distance. **89.** $\alpha = 34°$; $\beta = 56°$

91. (a)

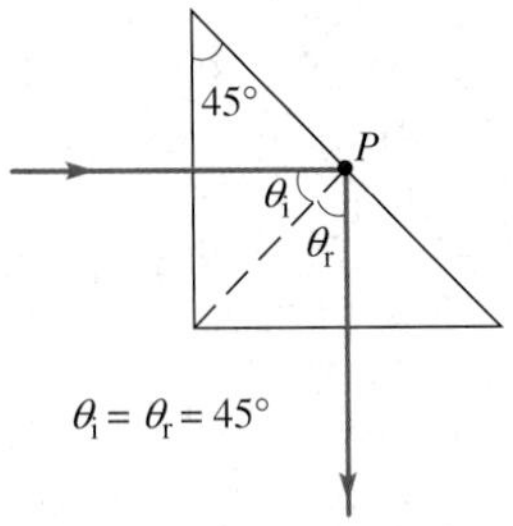

(b)

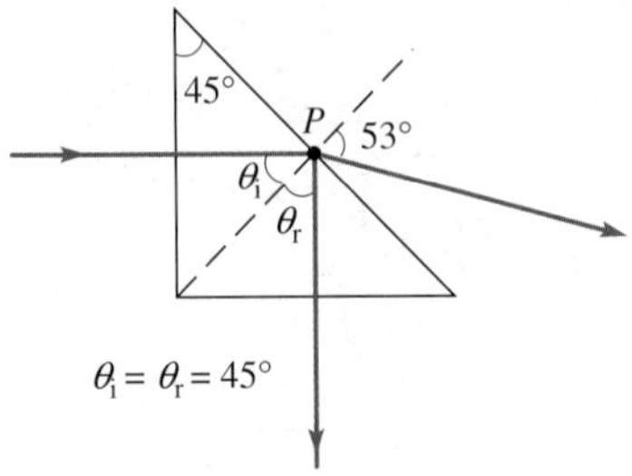

(c)

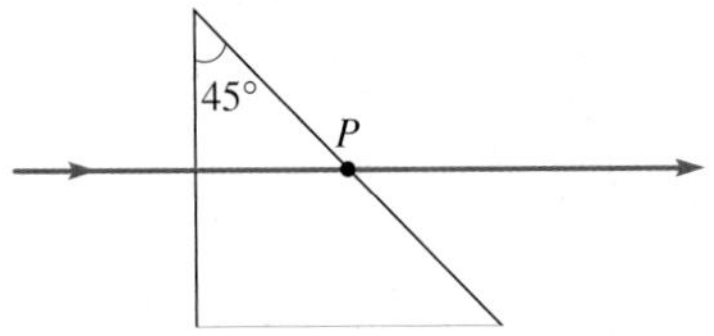

93. $n_{\text{liquid}} < 1.3$ **95.** (a) 1.4 cm (b) upright **97.** 19.2° **101.** 50 cm

CHAPTER 24

Multiple-Choice Questions

1. (d) **2.** (b) **3.** (b) **4.** (a) **5.** (f) **6.** (b) **7.** (b) **8.** (b) **9.** (d) **10.** (c)

Problems

1. (a) 2.5 cm past the 4.0-cm lens; real (b) −0.27 **3.** (a) 11.8 cm (b) 0.0793 **5.** 15.6 cm to the left of the diverging lens **7.** $q_1 = 12.0$ cm; $q_2 = -4.0$ cm; $h_1' = -4.00$ mm; $h_2' = 4.0$ mm **9.** (a) 4.05 cm (b) converging (c) 1.31 (d) 15.8 cm **11.** minimum: 20.00 cm; maximum: 22.2 cm **13.** (a) inverted (b) −0.42 (c) The cone of light rays, diverging from each point on the object, is larger, so the eye can detect that the rays do not converge to a single point. (d) converging (e) 2.0 m **15.** (a) 224 mm (b) 2.67 mm (c) 16.7 m **17.** 20 mm **19.** 98 cm by 150 cm **21.** An inverted image with magnification −2.00 is formed 12.0 cm to the right of the mirror. An inverted image with magnification −0.200 is formed 3.60 cm to the right of the lens. An upright image (with respect to the original object) with magnification 0.667 (with respect to the original object) is formed 4.00 cm to the right of the lens. **23.** (a) farsighted (b) 70 cm to infinity **25.** −0.50 D **27.** (a) −0.51 D (b) Without his glasses, his near point is 22 cm. With his glasses, his near point is 25 cm. **29.** 1.4 **31.** (a) 4.2 cm (b) 6.0 **33.** (a) 0.56 mm in diameter (b) 4.4×10^3 kW/m^2 **35.** (a) $\frac{Nf}{N+f}$ (c) $M = \frac{N}{f} + 1 = M_\infty + 1$ **37.** (a) −250 (b) 0.720 cm **39.** 50 **41.** 1.63 cm

43. (a) 19.0 cm (b) −318 (c) 5.16 mm **45.** (a) There would be no magnification, so you can't really make a telescope. (b) Using a lens from each pair of glasses, the telescope would be 1.05 m long and have an angular magnification of −2.7. **47.** (a) 0.00904 rad (b) 0.36 rad **49.** (a) 1.00 cm, 80.0 cm; −80.0 (b) $f_e = 1.00$ cm, $f_o = 80.0$ cm (c) 81.0 cm **51.** −8.0 **53.** (a) −4 cm (b) infinity (c) virtual; upright (d) 9 **55.** (a) $q_1 = 60.0$ cm; $q_2 = 5.45$ cm (b) −1.64 (c) 4.91 cm **57.** (a) Lens 1 is the objective and lens 2 is the eyepiece. (b) 30.0 cm **59.** 8.67 cm **61.** (a) the lens with the 30.0-cm focal length (b) −10 (c) 33.0 cm **63.** $f_e = 5.1$ cm; $f_o = 61$ cm **65.** (a) real (b) converging (c) 51 mm (d) 17 mm (e) 2.6 mm **67.** 3.8 m and 2.6 m for 24 mm and 36 mm, respectively **69.** (a) 17 cm (b) −14 (c) 7.4 cm **71.** −0.0068 **73.** (a) 5.25 cm to the right of the lens (b) real and inverted (c) −0.750 (d) 1.54 cm to the right of the mirror (e) virtual and inverted with respect to the original object (f) 0.229 (g) −0.171 **75.** 23.9 m **77.** (a) 1.85 cm (b) 3.59 cm (c) −68.2

CHAPTER 25

Multiple-Choice Questions

1. (a) **2.** (e) **3.** (a) **4.** (d) **5.** (e) **6.** (e) **7.** (d) **8.** (a) **9.** (e) **10.** (a)

Problems

1. (a) 5.0 km (b) Destructive interference occurs, since the path difference is 10 km and there is a $\lambda/2$ phase shift.
3. 1530 kHz **5.**

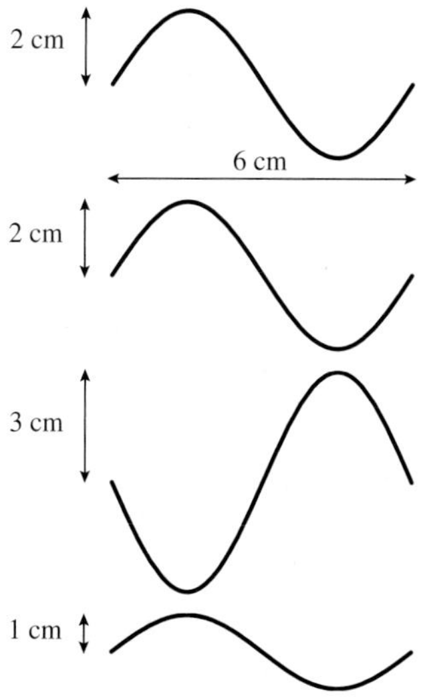

(a) 2 cm (b) I_0 (c) 6 cm (d) $9I_0$ **7.** $5I_0$ **9.** (a) 3.2 cm (b) 1.1 **11.** 560 nm **13.** (a) out (b) 15 cm **15.** (a) touching; zero (b) 92 nm (c) 180 nm **17.** 480 nm **19.** 497 nm **21.** (a) 607 nm, 496 nm, and 420 nm (b) 683 nm, 546 nm, and 455 nm **23.** (b) closer together **27.** 0.15 mm **29.** maxima: ± 6.9 cm from the central maximum; ± 30°; yes, they agree. **31.** 456 nm **33.** 7.0×10^{-5} m **35.** 8.0×10^{-5} m **37.** 470 nm **39.** four **41.** (a) *A* is the blue line, *B* is the yellow line, and *C* is the red line. (b) 669 nm (c) 2 **45.** (a) 7560 slits/cm (b) 240 **47.** 0.61 mm **49.** The new width is half the old width. **51.** 0.19 mm **53.** 0.012° **55.** 4.0 mrad **57.** (a) $\sin \Delta\theta = n \sin \beta$ **59.** 0, 330 nm, 660 nm, 990 nm, and 1.3 μm, respectively **61.** 20 km **63.** 1.7 μs **65.** (a) 0.10 mm (b) 0.51 mm **67.** (a) 100 nm (b) 280 nm; 140 nm (c) Yes; although perfectly constructive interference does not occur for any visible wavelength, some visible light is reflected at all visible wavelengths except 560 nm (the only wavelength with perfectly destructive interference). **69.** $t = \dfrac{\lambda}{2n_{\text{coating}}}$ **71.** 3.4 cm **73.** 48 m **75.** 20 cm **77.** (a) maximum
(b)

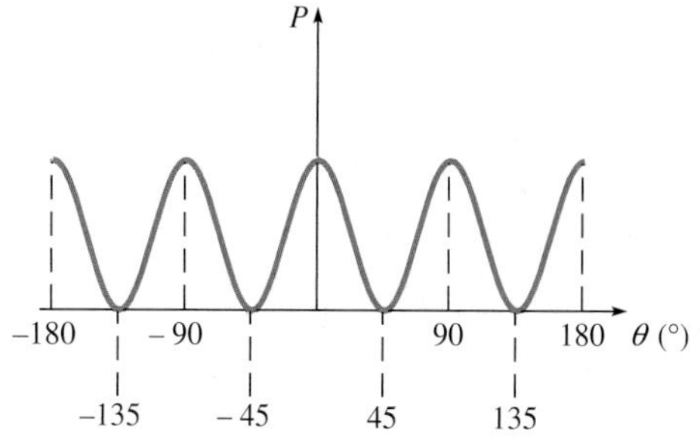

(c)

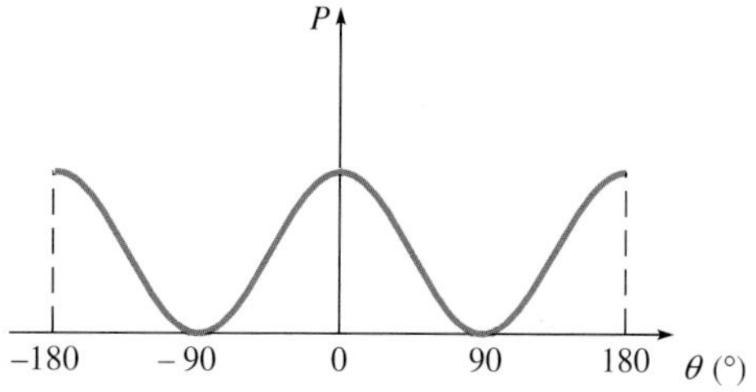

79. 12 m

REVIEW & SYNTHESIS: CHAPTERS 22–25

Review Exercises

1. 85 ms **3.** (a) 5.2×10^{14} Hz (b) 2.00×10^8 m/s; 390 nm; 5.2×10^{14} Hz **5.** (a) The plane must be at least $3\lambda = 638.8$ m over Juanita's head. (b) 425.9 m and 851.7 m **7.** (a) 8.95 m (b) converging lens (c) inverted **9.** The lamps are not emitting coherent light. **11.** 400 nm **13.** (a) 1.80×10^{-6} m (b) 2.24 m (c) 6.33 m (d) The assumption that $\sin \theta = \tan \theta$ is not valid because the angles are not small. **15.** (a) The distance between adjacent maxima is inversely proportional to the slit separation. (b) The distance between adjacent maxima increases linearly as the distance between the slit and the screen increases. (c) You would want a large distance to the screen or a small slit separation or both. **17.** 3.15 m **19.** five **21.** (a) 3.9 mm (b) 7.8 mm **23.** 849.9 nm **25.** (a) 50.1° (b) 47.8° (c) Path *A* (d) 39.9°

MCAT Review

1. A **2.** C **3.** B **4.** C **5.** C **6.** A

CHAPTER 26

Multiple-Choice Questions

1. (e) **2.** (d) **3.** (a) **4.** (a) **5.** (c) **6.** (a) **7.** (b) **8.** (a) **9.** (b)

Problems

1. 2.2 μs **3.** (a) 0.87c (b) c **5.** 8.9 h **7.** 0.001c **9.** (a) 30 years old (b) 3420 **11.** 7.7 ns **13.** (a) 2 m (b) 0.50 m **15.** (a) 79 m (b) 610 ns (c) 530 ns **17.** 13 m **19.** (a) 1.0 m (b) 0.92 m **21.** (a) 7.5 μs (b) 13 μs **23.** 6.0 km **25.** 3.00×10^8 m/s **27.** −0.994c **29.** 0.300c **31.** 0.974c **33.** 1.0×10^{-18} kg·m/s **35.** 800000000.3 kg·m/s; 800000000 kg·m/s; 4×10^{-8} % **37.** (a) 1.8×10^7 N (b) 8200 m/s^2; this is much larger than any human could survive. **39.** increased by 1.00×10^{-14} kg **41.** 2.0×10^{47} J **43.** 5.58 MeV **45.** 0.595c **47.** 2.7×10^{15} J **49.** 4.9mc **51.** 6.5 MeV/c **53.** 1 MeV/c $= 5.344 \times 10^{-22}$ kg·m/s **55.** (a) The electrons are relativistic. (b) 0.63c **61.** 19.2 min **63.** 33.9 MeV **65.** 0.66c **67.** 1.326 GeV **69.** (a) 7.2 m (b) 10 m (c) 21 m **71.** 6.3 km **75.** (a) 4500 m (b) 15 μs (c) 15 μs (d) 500000 **77.** 2.6 MeV **79.** (a) every 24 h (b) 46 h **81.** 36 kHz **83.** (a) 127 h (b) 127 h **85.** neutron: 5.7 MeV; pion: 35.4 MeV

CHAPTER 27

Multiple-Choice Questions

1. (e) **2.** (c) **3.** (d) **4.** (a) **5.** (a) **6.** (c) **7.** (e) **8.** (b) **9.** (e) **10.** (f)

Problems

1. (a) ultraviolet (b) infrared: 9.9×10^{-20} J; ultraviolet: 2.8×10^{-18} J (c) infrared: 2.0×10^{21} photons/s; ultraviolet: 7.0×10^{19} photons/s **3.** (a) 400 nm (b) 7.5×10^{14} Hz **5.** (a) 0.84 eV (b) 574 nm **7.** 477 nm **9.** 4.5 eV **11.** (a) No; violet light (b) 2.56 eV **13.** (a) 6.66×10^{-34} J·s (b) 1.82 eV **15.** 2.7 kV **17.** 1.1×10^{19} Hz **19.** 62.0 pm **21.** (a) 10.7 pm (b) 12.4 pm **23.** 121.80 pm **25.** 4.45×10^6 m/s at 62.6° south of east **27.** (a) 2.50×10^{-12} m (b) 55.6 keV **29.** 2.4×10^4 eV **31.** 0.476 nm **33.** (a) 1.59×10^{-10} m (b) 1.06×10^{-8} m; the orbital separations are much larger for larger n values. **35.** 122 eV **37.** $n = 3$ **39.** 3.40 eV **41.** 1.09 μm **43.** (a) one (b) four **45.** 370 nm **47.** $n = 4$ **51.** 2.19×10^6 m/s **53.** 1.21 pm **55.** 7.75×10^{-14} m **57.** 1.17×10^{-14} m **59.** (a) 10.0 mJ (b) 3.23×10^{16} **61.** (a) 3.69×10^{-7} eV (b) 8.45×10^{29} photons/s **63.** 6.2 pm **65.** (a) 1.9 eV; 9.9×10^{-28} kg·m/s (b) 3×10^{15} photons/s (c) 3×10^{-12} N **67.** (a) 2000 km (b) Light is scattered by the atmosphere and other light sources overwhelm the light from a single bulb. **69.** 27.6 kV **71.** $E_{90} = 136$ keV; $E_{180} = 108$ keV **73.** (a) 1×10^{-22} J/s (b) 3.2×10^3 s (c) An electron is ejected immediately when a single photon of sufficient energy strikes it. **75.** 92.4 keV **77.** 0.62 nm **79.** 164 nm, 122 nm, and 109 nm **81.** $r_H/r_{He} = 1/2$; for levels of equal energy, the ratio of orbital radii appears to equal the ratio of atomic numbers. **83.** (a) 4.1 m/s (b) 4 ways emitting 6 different photons (c)

Transition	λ (nm)	Class
$4 \to 3$	1875	IR
$4 \to 2$	486	visible
$4 \to 1$	97	UV
$3 \to 2$	656	visible
$3 \to 1$	103	UV
$2 \to 1$	122	UV

85. (a) 97.3 nm (b) 102.6 nm; 656.3 nm (c) $\lambda \le 91.2$ nm

CHAPTER 28

Multiple-Choice Questions

1. (a) **2.** (f) **3.** (d) **4.** (a) **5.** (d) **6.** (a) **7.** (a) **8.** (c) **9.** (a) **10.** (d)

Problems

1. 1.3×10^{-34} m; the wavelength is much smaller than the diameter of the hoop—a factor of 10^{-34} smaller! **3.** (a) 1.0×10^{-35} m/s (b) 3.8×10^{26} yr **5.** 3.23 pm **7.** 250 eV **9.** 101 **11.** 391 eV **13.** (a) 12.4 keV (b) 150 eV (c) 150 V **15.** (a) 0.060 eV (b) 0.060 V (c) 5 nm is an x-ray wavelength. **17.** 1×10^{-29} m **19.** (a) 1.3×10^{-32} m (b) 1.1×10^{-6} m **21.** (a) 4×10^{-32} m (b) 0.5 mm (c) The uncertainty principle can be neglected in the macroscopic world, but not on the atomic scale. **23.** 0.98 eV **25.** 2×10^{-15} **27.** 380 GeV **29.** (a) 6×10^{28} (b) The energy difference between levels is too small to observe. **31.** (a)

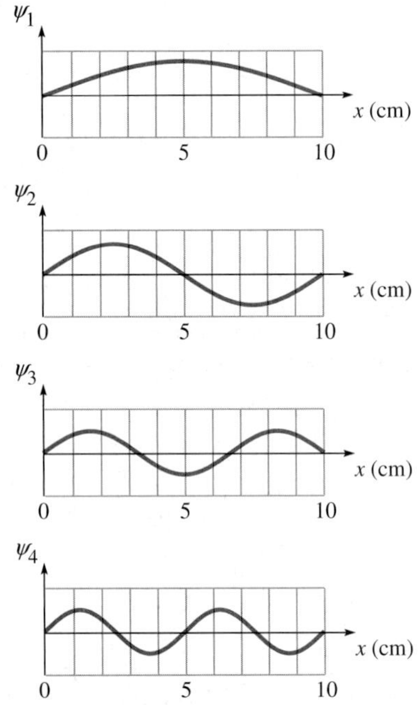

(b) 1.1×10^{-16} eV **33.** (a) 10.3 nm (b) one fourth as much **35.** $1s^2 2s^2 2p^6 3s^2 3p^6$ **37.** $-2\hbar$, $-\hbar$, 0, $\hbar$, and $2\hbar$ **39.** For $\ell = 0$, 1, 2, 3, 4, 5, and 6, there are 2, 6, 10, 14, 18, 22, and 26 electron states, respectively; the total is 98. **41.** $1s^2 2s^2 2p^6 3s^2 3p^6 4s^2 3d^{10} 4p^5$ **43.** (a) Li: $1s^2 2s^1$; Na: $1s^2 2s^2 2p^6 3s^1$; K: $1s^2 2s^2 2p^6 3s^2 3p^6 4s^1$ (b) s^1 subshell **45.** (a) $1s^2 2s^2 2p^2$ (b)

n	ℓ	m_ℓ	m_s
1	0	0	$\pm\frac{1}{2}$
2	0	0	$\pm\frac{1}{2}$
2	1	−1	$\pm\frac{1}{2}$
2	1	0	$\pm\frac{1}{2}$
2	1	1	$\pm\frac{1}{2}$

47. (b) The kinetic energy of an electron at $r = 2a_0$ is zero, so there is no available energy to move the electron past this point. **49.** 1.9 eV **51.** 544 nm; green **53.** (a) 0.0097°

(b) 130 km in diameter **55.** −25% **57.** 2.8×10^{-14} m **59.** 680 m/s **61.** (a) 8.286×10^{-21} J (b) 125.8 pm **63.** $\frac{1}{4}$ **65.** $1s^2 2s^2 2p^6 3s^2 3p^6 4s^2 3d^{10} 4p^6 5s^2 4d^{10} 5p^4$ **67.** 3.0 nm **69.** (a) 1.26 km/s (b) 314 pm (c) 303–324 pm **71.** (a) 0.067 eV (b) 0.20 eV, 0.33 eV, 0.47 eV, 0.53 eV, 0.80 eV, and 1.0 eV (c)

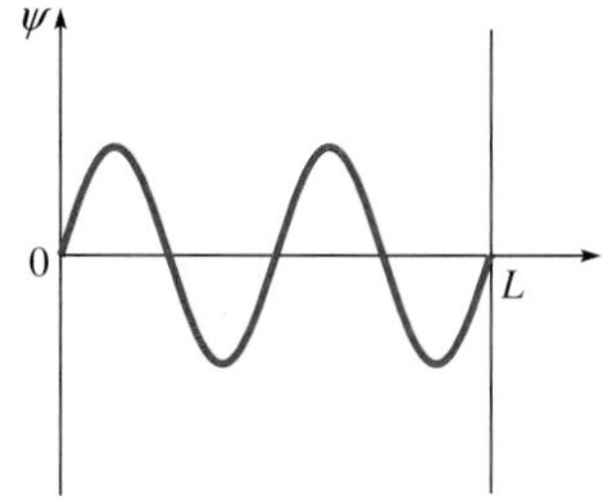

(d) smaller since $E \propto 1/L^2$ **73.** (a) 167 pm (b) 66.5° (c) yes **75.** (a)

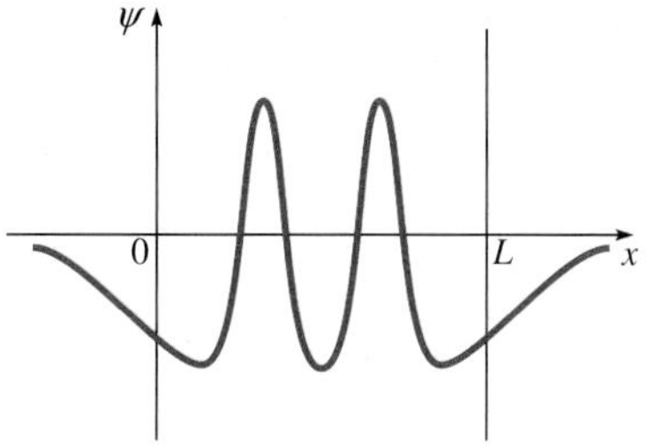

(b) 51 **77.** (a) 6.1 nm
(b)

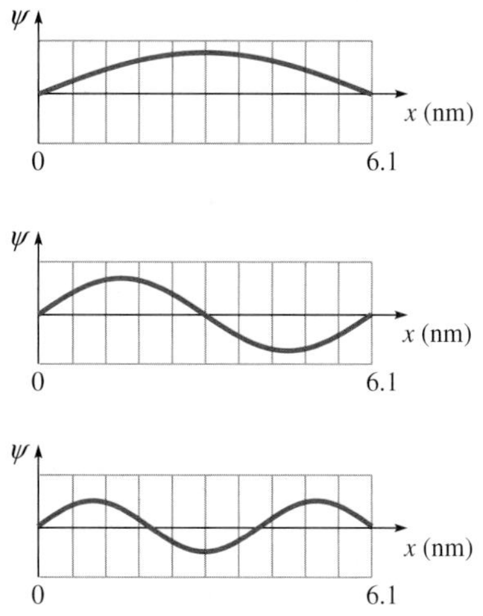

(c) 4.1 nm
(d) $\lambda_{31} = 15.5$ μm; $\lambda_{32} = 25$ μm; $\lambda_{21} = 41$ μm

CHAPTER 29

Multiple-Choice Questions

1. (b) **2.** (d) **3.** (a) **4.** (g) **5.** (d) **6.** (c) **7.** (f) **8.** (c) **9.** (d) **10.** (e)

Problems

1. 4.5×10^{28} **3.** 13 km **5.** $^{40}_{19}$K **7.** 54 **9.** 5.7 fm; 7.7×10^{-43} m^3 **11.** 2.225 MeV **13.** (a) 127.619 MeV (b) 7.976 19 MeV/nucleon **15.** 0.112 355 3 u **17.** (a) 1.46×10^{-8} u (b) no **19.** (a) 238.000 32 u (b) 1.801 69 GeV **21.** $^{40}_{20}$Ca **23.** $^{22}_{11}\text{Na} + {}^{0}_{-1}\text{e} \rightarrow {}^{22}_{10}\text{Ne} + {}^{0}_{0}\nu$; $^{22}_{10}$Ne **25.** 4.8707 MeV **27.** 1.3111 MeV **31.** 1.820 MeV **33.** (a) 10 000 s^{-1} (b) 2.308×10^7 (c) 3.466×10^{-3} s^{-1} **35.** 270 yr **37.** 0.0072 Ci **39.** 0.99 Ci **41.** (a) 3.83×10^{-12} s^{-1} (b) 6.5×10^{10} atoms (c) 0.25 Bq/g **45.** 3×10^5 molecules **47.** 1.4×10^{10} Sv **49.** (a) $^7_3$Li; $^8_3$Li; $^8_4$Be (b) Yes; the emission of an electron (beta-minus decay) is accompanied by the emission of one antineutrino. **51.** $^{27}_{13}\text{Al} + {}^4_2\text{He} \rightarrow {}^{31}_{15}\text{P} \rightarrow {}^{30}_{15}\text{P} + {}^1_0\text{n}$ **53.** 6.5448 MeV **55.** 173 MeV **57.** (a) $^3_2$He (b) 5.6 MeV (c) At room temperature, the kinetic energies of the proton and the deuteron are much too small to overcome their Coulomb repulsion. **59.** 0.44 MeV **61.** $^{175}_{71}$(Lu) and $^{167}_{71}$(Lu); $^{175}_{74}$(W) and $^{180}_{74}$(W) **63.** (a) $^{15}_6\text{C} \rightarrow {}^{15}_7\text{N} + {}^0_{-1}\text{e} + {}^0_0\bar{\nu}$ (b) 9.771 74 MeV **65.** From left to right, the energies are: 492 keV, 472 keV, 40 keV, 452 keV, 432 keV, 287 keV **67.** 3.25 mol **69.** (a) 136 neutrons; 86 protons (b) 20 alpha particles per second **71.** (a) 1.42×10^7 m/s (b) 4.3×10^6 V/m (c) 98 cm (d) Both m and q affect the radius of the trajectory. **73.** 760 kW·h **75.** (a) 5.2×10^{-7} g (b) 44 mCi **77.** 7.67×10^9 yr **79.** (a) 8% (b) $P_0 = 0.58$ mW; $P_{10.0} = 0.53$ mW **81.** (a) 4.0 (b) 0.50 (c) 0.063 **83.** (b) 4.0026 u (c) 4.2 fm (d) 2.4 MeV

CHAPTER 30

Multiple-Choice Questions

1. (b) **2.** (d) **3.** (a) **4.** (b) **5.** (d) **6.** (b) **7.** (c) **8.** (a) **9.** (a) **10.** (b)

Problems

1. 34 MeV **3.** $\bar{u}\bar{u}\bar{d}$ **5.** (a) weak (b) electromagnetic (c) weak **7.** 20 ng **9.** 1.8×10^{-19} m **11.** 109.3 MeV **13.** 5.4 T **15.** 67.5 MeV **17.** 938 MeV **19.** uss **21.** $\bar{u}$d **23.** $\pi^+ \rightarrow \mu^+ + \nu_\mu$ and $\pi^+ \rightarrow e^+ + \nu_e$ **25.** 4.2 MeV **27.** 1.7×10^{-14} m

REVIEW & SYNTHESIS: CHAPTERS 26–30

Review Exercises

1. 3.91 m **3.** (a) 4.03×10^8 m/s (b) The disturbance that moves across the Moon's surface is not an object that has mass, so there is no violation. **5.** $0.895c = 2.68 \times 10^8$ m/s **7.** (a) 9.8×10^5 m/s (b) No electrons are ejected. (c) Doubling the intensity has no effect on the electron speed, nor does it cause electrons to be ejected if none were ejected prior to the doubling of the intensity. **9.** 9.8 cm **11.** (a) $^{90}_{38}\text{Sr} \rightarrow {}^{90}_{39}\text{Y} + {}^0_{-1}\text{e} + {}^0_0\bar{\nu}$ (b) 1.0×10^{16} Bq (c) 3.6×10^5 Bq **13.** proton: 5.5 MeV; pion: 33 MeV **15.** (a) 4.7 eV (b) maximum **17.** (a) 1.25×10^9 Bq (b) an electron and an antineutrino **19.** $0.966c$ **21.** 10^{-4} eV·s/m **23.** six **25.** 4:1 **27.** 0.527 eV **29.** 4.8×10^{-3} J

MCAT Review

1. C **2.** A **3.** D **4.** D **5.** C **6.** B **7.** A **8.** D **9.** C **10.** B **11.** D **12.** C **13.** A

Credits

PHOTOGRAPHS

About the Authors

Photo courtesy of Phil Krasicky, Department of Physics, Cornell University.

Chapter 1

Opener: NASA/JPL/Cornell; p. 2: © Royalty-Free/CORBIS; 1.1a: © Science VU/Visuals Unlimited; 1.1b: Photo by Jennifer Merlis; 1.1c: © Vol. 86/PhotoDisc RF; 1.1d: © Vol. 56/ CORBIS RF; 1.1e: © Hans Gelderblom/ Getty Images; 1.1f: © Jonathan Blair/ CORBIS; 1.1g: © Vol. 56/CORBIS RF; p. 12: © Vol. 86/PhotoDisc RF; 1.3: © Dennis Kunkel Microscopy.

Chapter 2

Opener: © J & A SCOTT/NHPA/ Photoshot.

Chapter 3

Opener: © Peter Johnson/CORBIS; p. 58: © Stone/Getty Images; p. 65: © Eye Wire/RF; 3.19: © Loren M. Winters/Visuals Unlimited.

Chapter 4

Opener: Courtesy Durango Soaring Club, Inc.; 4.2: © Frank Seguin/ CORBIS; p. 94: © Vol. 125 PhotoDisc/ Getty Images RF; p. 98: © Felicia Martinez/PhotoEdit, Inc.; p. 140: © Vol. 85/CORBIS RF.

Chapter 5

Opener: © AFP/Getty Images; p. 150: © Vol. 29/PhotoDisc RF; 5.14: © Chris Sattlberger/Getty Images; p. 169: © CORBIS RF; p. 170: © Angelo Hornak/CORBIS; 5.23: Supplied by Globe Photos; p. 175: © Richard T. Nowitz; p. 176: © Matthew Stockman/Allsport/Getty Images.

Chapter 6

Opener: © Mark Newman/PictureQuest/ Jupiter Images, Inc.; 6.1: © Owaki-Kulla/ CORBIS; 6.2: © AP/Wide World Photo; 6.24: USA Archery; 6.32a-c: © Dwight Kuhn; p. 215: © Brand X Pictures/Getty RF; p. 216: Courtesy Pileco, Inc.; p. 217: U.S. Navy Photo; p. 221: © Free Agents Limited/CORBIS.

Chapter 7

Opener: © 911 Pictures; 7.3: © Scott Leva—Precision Stunts; 7.8: © Vol. 56/ CORBIS RF; 7.18: © Loren M. Winters/ Visuals Unlimited; p. 251: © Stone/ Getty Images.

Chapter 8

Opener: © Warren Morgan/CORBIS; 8.3: © AP/Wide World Photo; 8.20: Courtesy The Frank Lloyd Wright Foundation; 8.27: © Mike Powell/ Allsport/Getty Images; 8.33: © Michael Newman/PhotoEdit, Inc.; p. 285: © Kevin Fleming/CORBIS; 8.37a,b: © Photography by Leah; p. 298: © Doug Pensinger/Allsport/Getty Images; p. 305(left): © Tony Duffy/Allsport/Getty Images; p. 305(right): © Jonathan Daniel/Allsport/Getty Images; p. 306: © Gilbert Lundt, Jean-Yves Ruszniewski/TempSport/CORBIS.

Chapter 9

Opener: © CORBIS RF; 9.14: James Gathany/CDC; p. 329: © CORBIS RF; 9.17: © PhotoDisc/Getty Images RF; 9.18: Ames Research Center; 9.21: Photo courtesy of Kohler Co.; 9.29: © Gary S. Settles/Photo Researchers, Inc.; 9.30: © Dennis Drenner/Visuals Unlimited.

Chapter 10

Opener: © Hubert Stadler/CORBIS; 10.1: © Amoz Echerson/Visuals Unlimited; 10.5: © SMARTEC SA, Switzerland; p. 362: Greater Houston Convention and Visitors Bureau; p. 363: © John Springer Collection/ CORBIS; 10.9a: © Doug Pensinger/ Getty Images; 10.9b: © SIU/Visuals Unlimited; 10.16: © Vol. 54 PhotoDisc/Getty Images RF; 10.28a,b: © Bettmann/CORBIS; p. 383(top): © Ray Malace Photo; p. 383(bottom): © Adam Eastland/Alamy.

Chapter 11

Opener: © AP/Wide World Photo; p. 393: © Vol. 44/PhotoDisc/Getty Images RF; 11.6a,b: © Loren M. Winters/Visuals Unlimited; 11.19: © Larry Lo Wing-Kei (Hong Kong). Licensed under Creative Commons Attribution-Share Alike 2.5.

Chapter 12

Opener: © Alan Giambattista; p. 425: © Brian E. Small/www.briansmallphoto.com; p. 429: U.S. Navy photo by Ensign John Gay; 12.4: © billkelleyphoto.com/Bill Kelley; 12.6: © Grant Heilman Photography/Joshua Slaymaker; 12.17: © Loren M. Winters/Visuals Unlimited; p. 444: © Stephen Dalton/Photo Researchers, Inc.; 12.19: © AP/Wide World Photo.

Chapter 13

Opener: © Stone/Getty Images; 13.5: © John Sohlden/Visuals Unlimited; p. 470: © Geoff Tompkinson/Photo Researchers, Inc.; 13.15a: © Joe McDonald/CORBIS; 13.15b: © Craig Lovell/CORBIS; 13.15c: © Ken Kaminesky/CORBIS.

Chapter 14

Opener: © AP/Wide World Photo; 14.6: Photo by author R.C. Richardson; 14.8: Staffan Widstrand/CORBIS; 14.18: © 2008 David Montrose, M.D.—Custom Medical Stock Photo, All Rights Reserved.

Chapter 15

Opener: © David Davis/Index Stock Imagery/PictureQuest/Jupiter Images, Inc.

Chapter 16

Opener: Image courtesy Massashi Kawasaki, University of Virginia; 16.2: © The McGraw-Hill Companies, Inc./ Joe Franek, photographer; 16.3: © Tony Freeman/PhotoEdit, Inc.; 16.38: © Paul McMahon.

Chapter 17

Opener: © Lester Lefkowitz/CORBIS; 17.13: © Roger Ressmeyer/CORBIS; 17.15: © Melanie Brown/PhotoEdit, Inc.; 17.22: © Loren M. Winters/Visuals

Unlimited; 17.24: © Tom Pantages; 17.30: © Dan Robinson, WVlightning.com; 17.34: © Adam Hart-Davis/Photo Researchers, Inc.

Chapter 18

Opener: © Chuck Swartzell/Visuals Unlimited; 18.4: © Tom McHugh/Photo Researchers, Inc.; 18.6: © Richard Megna/Fundamental Photographs, NYC; 18.7a: © The McGraw-Hill Companies, Inc./Joe Franek, photographer; 18.11: Photo courtesy of IBM Corporation; 18.12: © Tony Freeman/PhotoEdit, Inc.; 18.31: © Oleksiy Maksymenko/Alamy RF; 18.40: © Tom Pantages.

Chapter 19

Opener: Image courtesy Hojatollah Vali, McGill University; p. 694: Photo by Stan Sherer; 19.1a: © Phil Degginger/Color-Pic, Inc.; 19.2: © The McGraw-Hill Companies, Inc./Joe Franek, photographer; 19.15a: © CERN Geneva; 19.18: © IBA Proton Therapy Cyclotron; 19.34a: © Richard Megna/Fundamental Photographs, NYC; 19.34b: © Loren M. Winters/Visuals Unlimited; 19.39a: © The McGraw-Hill Companies, Inc./ Joe Franek, photographer; 19.44: © Tom Pantages; p. 738: © Richard Paselk/ Photos by Richard A. Paselk.

Chapter 20

Opener: © The Picture Source/Terry Oakley; p. 745: U.S. Army Corps of Engineers; 20.16: © 2008 SPL—Custom Medical Stock Photo, All Rights Reserved; 20.24: © Mark Antman/The Image Works; 20.25: © Tom Pantages; 20.30: © The Image Works Archives.

Chapter 21

Opener: Alan Giambattista; 21.15: © Loren M. Winters/Visuals Unlimited; 21.17b: © The Image Works Archives.

Chapter 22

Opener: © George D. Lepp/CORBIS; 22.7a: © Dan McCoy/DoctorStock.com; 22.7b: © Richard Lowenberg/Science Source/Photo Researchers, Inc.; 22.8a,b: © Charles Mazel; 22.12: © Anglo-Australian Observatory/Photograph by David Malin; 22.13: © D. Parker/Photo Researchers, Inc.; 22.17: © Inga Spence/ Visuals Unlimited; 22.28a,b: © Timothy Edberg; 22.29: © CORBIS; 22.32: Image courtesy Lewis Ling, Carleton University; p. 840: © D. Parker/Photo Researchers, Inc.

Chapter 23

Opener: © Bettmann/CORBIS; 23.1: © 1998 Jeff J. Daly/Fundamental Photographs; 23.3: © Thinkstock/Getty RF; 23.11a,b: © The Picture Source/Terry Oakley; 23.13a: © Peter Turner/Getty Images; 23.14a: © Pekka Parviainen/ Polar Image; 23.22b: © Yoav Levy/ Phototake; 23.23b: © Collection CNRI/ Phototake; 23.29: © Tom Pantages; 23.33: © Bob Crook/Photographers Direct; 23.36a: © Felicia Martinez/PhotoEdit, Inc.; p. 875: © RDF/Visuals Unlimited.

Chapter 24

Opener: NASA; 24.6b: © Bettmann/ CORBIS; 24.18: © Roger Ressmeyer/ CORBIS; 24.20a-c: NASA, H. Ford (JHU), G. Illingworth (USCS/LO), M. Clampin (STScI), G. Hartig (STScI), the ACS Science Team, and ESA; 24.21: Courtesy of NAIC, Cornell University.

Chapter 25

Opener: © Kevin Schafer/Getty Images; 25.8, 25.13a,b: © The Picture Source/Terry Oakley; 25.14c: © Tom Pantages; 25.15, 25.18a: © The Picture Source/Terry Oakley; 25.20: © 1986 Richard Megna—Fundamental Photographs; 25.23b: Alan Giambattista; 25.26: © Tom Pantages; 25.28a-c: © The Picture Source/Terry Oakley; 25.30a: © 1987 Ken Kay/ Fundamental Photographs, NYC; 25.30b: © The Picture Source/Terry Oakley; 25.31a, 25.35: © Tom Pantages; 25.36a: © Noel Carboni; 25.36b: © David Nash; 25.39a,b: © Erich Lessing/Art Resource, NY; 25.40b: © Education Development Center, Inc.; 25.42: © Science Source/ Photo Researchers, Inc.; 25.45(left),(right): © Holographics North, Inc.; p. 954(top): © The Picture Source/Terry Oakley; p. 954(bottom): © Tom Pantages.

Chapter 26

Opener: Map of the core and jet of the radio galaxy NGC6251 at 1.4GHz made by P.N. Werner, M. Birkinshaw & D.M. Worrall using the NRAO Very Large Array; p. 971: © Hulton-Deutsch Collection/CORBIS.

Chapter 27

Opener: © Bluestar®; 27.7: © Richard T. Nowitz/CORBIS; 27.12b: © W. Cody/ CORBIS; 27.22a,b: © Charles Mazel; 27.25b: © Hank Morgan/Photo Researchers, Inc.

Chapter 28

Opener: © Collection CNRI/Phototake; 28.4a,b: © Education Development Center, Inc.; 28.6b,d: © Dennis Kunkel/ Phototake; 28.14: Courtesy: IBM Research, Almaden Research Center. Unauthorized use not permitted; 28.25: © Yoav Levy/Phototake; 28.28b: © Lawrence Berkeley Lab/Photo Researchers, Inc.; p. 1057: Courtesy: IBM Research, Almaden Research Center. Unauthorized use not permitted.

Chapter 29

Opener: © Geoffrey Clements/CORBIS; 29.10b: Vesper V. Grantham & Lindsay Huckabaa, University of Oklahoma Health Sciences Center; 29.11b: Image courtesy Elekta Instruments, Inc.; 29.17: © Vladimir Repik/Reuters/CORBIS.

Chapter 30

Opener: © CERN Geneva; p. 1107: Kamioka Observatory Institute for Cosmic Ray Research, University of Tokyo.

TEXT

Some examples and problems adapted from Alan H. Cromer, *Physics for the Life Sciences*. Copyright ©1977 by the McGraw-Hill Companies, Inc., NewYork. All rights reserved. Reprinted by permission of Alan H. Cromer.

Some examples and problems adapted from Alan H. Cromer, Study Guide for *Physics for the Life Sciences*. Copyright ©1977 by the McGraw-Hill Companies, Inc., New York. All rights reserved. Reprinted by permission of Alan H. Cromer.

Some examples and problems adapted from Alan H. Cromer, *Physics for the Life Sciences*, 2d. ed. Copyright © 1994 by the McGraw-Hill Companies, Inc. Primis Custom Publishing, Dubuque, Iowa. All rights reserved. Reprinted by permission of Alan H. Cromer.

Chapter 28, p.1016: Quote by Richard Feynman from Richard Phillips Feynman, *The Character of Physical Law*; introduction by James Gleick. Copyright ©1994 by Modern Library, New York. All rights reserved. Reprinted by permission of MIT Press.

Index

Page numbers followed by *f* and *t* refer to figures and tables respectively.

A

B

C

D

E

F

G

H

I

J

K

L

M

N

O

P

S

T

U

V

W

X

Y

Z

Metals (main-group)
Metals (transition)
Metals (inner transition)
Metalloids
Nonmetals

Key: 1 — Atomic number; H — Symbol; 1.00794 — Atomic mass

Numbers in parentheses ( ) are for longest lived isotopes of elements without stable isotopes.

MAIN-GROUP ELEMENTS: 1A (1), 2A (2), 3A (13)–8A (18)

TRANSITION ELEMENTS: 3B (3)–2B (12)

Period	1A (1)	2A (2)	3B (3)	4B (4)	5B (5)	6B (6)	7B (7)	8B (8)	8B (9)	8B (10)	1B (11)	2B (12)	3A (13)	4A (14)	5A (15)	6A (16)	7A (17)	8A (18)
1	1 **H** 1.00794																	2 **He** 4.00260
2	3 **Li** 6.941	4 **Be** 9.01218											5 **B** 10.811	6 **C** 12.011	7 **N** 14.0067	8 **O** 15.9994	9 **F** 18.99840	10 **Ne** 20.1797
3	11 **Na** 22.98977	12 **Mg** 24.3050											13 **Al** 26.98154	14 **Si** 28.0855	15 **P** 30.97376	16 **S** 32.065	17 **Cl** 35.453	18 **Ar** 39.948
4	19 **K** 39.0983	20 **Ca** 40.078	21 **Sc** 44.95591	22 **Ti** 47.867	23 **V** 50.9415	24 **Cr** 51.9961	25 **Mn** 54.93805	26 **Fe** 55.845	27 **Co** 58.93320	28 **Ni** 58.6934	29 **Cu** 63.546	30 **Zn** 65.409	31 **Ga** 69.723	32 **Ge** 72.64	33 **As** 74.92160	34 **Se** 78.96	35 **Br** 79.904	36 **Kr** 83.798
5	37 **Rb** 85.4678	38 **Sr** 87.62	39 **Y** 88.90585	40 **Zr** 91.224	41 **Nb** 92.90638	42 **Mo** 95.94	43 **Tc** (98)	44 **Ru** 101.07	45 **Rh** 102.90550	46 **Pd** 106.42	47 **Ag** 107.8682	48 **Cd** 112.411	49 **In** 114.818	50 **Sn** 118.710	51 **Sb** 121.760	52 **Te** 127.60	53 **I** 126.90447	54 **Xe** 131.29
6	55 **Cs** 132.90545	56 **Ba** 137.327	57 **La** 138.9055	72 **Hf** 178.49	73 **Ta** 180.9479	74 **W** 183.84	75 **Re** 186.207	76 **Os** 190.23	77 **Ir** 192.217	78 **Pt** 195.084	79 **Au** 196.96657	80 **Hg** 200.59	81 **Tl** 204.3833	82 **Pb** 207.2	83 **Bi** 208.98040	84 **Po** (209)	85 **At** (210)	86 **Rn** (222)
7	87 **Fr** (223)	88 **Ra** (226)	89 **Ac** 227.0278	104 **Rf** (261)	105 **Db** (262)	106 **Sg** (266)	107 **Bh** (264)	108 **Hs** (277)	109 **Mt** (268)	110 **Ds** (271)	111 **Rg** (272)	112 (285)		114 (289)				

Period

INNER TRANSITION ELEMENTS

6	Rare Earths (Lanthanides)	58 **Ce** 140.116	59 **Pr** 140.90765	60 **Nd** 144.24	61 **Pm** (145)	62 **Sm** 150.36	63 **Eu** 151.964	64 **Gd** 157.25	65 **Tb** 158.92535	66 **Dy** 162.500	67 **Ho** 164.93032	68 **Er** 167.259	69 **Tm** 168.93421	70 **Yb** 173.04	71 **Lu** 174.967
7	Actinides	90 **Th** 232.0381	91 **Pa** 231.036	92 **U** 238.0289	93 **Np** (237)	94 **Pu** (244)	95 **Am** (243)	96 **Cm** (247)	97 **Bk** (247)	98 **Cf** (251)	99 **Es** (252)	100 **Fm** (257)	101 **Md** (258)	102 **No** (259)	103 **Lr** (262)